D0907007

Fifth Edition

SHERRIS MEDICAL MICROBIOLOGY

EDITORS

KENNETH J. RYAN, MD

C. GEORGE RAY, MD

New York Chicago San Francisco Lisbon London Madrid Mexico City
Milan New Delhi San Juan Seoul Singapore Sydney Toronto

The McGraw-Hill Companies

Sherris Medical Microbiology, Fifth Edition

1 2 3 4 5 6 7 8 9 0 CTP/CTP 0 9

ISBN 978-0-07-160402-4
MHID 0-07-160402-2

NOTICE

Medicine is an ever-changing science. As new research and clinical experience broaden our knowledge, changes in treatment and drug therapy are required. The authors and the publisher of this work have checked with sources believed to be reliable in their efforts to provide information that is complete and generally in accord with the standards accepted at the time of publication. However, in view of the possibility of human error or changes in medical sciences, neither the authors nor the publisher nor any other party who has been involved in the preparation or publication of this work warrants that the information contained herein is in every respect accurate or complete, and they disclaim all responsibility for any errors or omissions or for the results obtained from use of the information contained in this work. Readers are encouraged to confirm the information contained herein with other sources. For example and in particular, readers are advised to check the product information sheet included in the package of each drug they plan to administer to be certain that the information contained in this work is accurate and that changes have not been made in the recommended dose or in the contraindications for administration. This recommendation is of particular importance in connection with new or infrequently used drugs.

This book was set in Minion by Glyph International.
The editors were Michael Weitz and Regina Brown.
The production supervisor was Sherri Souffrance.
Project management was provided by Rajni Pisharody, Glyph International.
The designer was Alan Barnett.
The cover designer was Thomas DePierro.
China Translation & Printing Services, Ltd. was printer and binder.

This book is printed on acid-free paper.

Cataloging-in-Publication data for this title are on file at the Library of Congress.

International Edition ISBN 978-0-07-163854-8; MHID 0-07-163854-7

EDITORS

KENNETH J. RYAN, MD
Emeritus Professor of Pathology
and Microbiology
College of Medicine
University of Arizona
Tucson, Arizona

C. GEORGE RAY, MD
Clinical Professor of Pathology
and Medicine
College of Medicine
University of Arizona
Tucson, Arizona

CONSULTING EDITOR

JOHN C. SHERRIS, MD, FRCPATH
Professor Emeritus
Department of Microbiology
School of Medicine
University of Washington
Seattle, Washington

AUTHORS

NAFEES AHMAD, PHD
Professor of Immunobiology
Department of Immunobiology
College of Medicine
University of Arizona
Tucson, Arizona

JAMES J. PLORDE, MD
Professor Emeritus
Department of Medicine and Department
of Laboratory Medicine
School of Medicine
University of Washington
Seattle, Washington

W. LAWRENCE DREW, MD, PHD
Professor of Laboratory Medicine
Professor of Medicine
School of Medicine
University of California, San Francisco
Mount Zion Medical Center
San Francisco, California

Key Features of Sherris Medical Microbiology, 5th Edition

- **66 chapters** simply and clearly describe the strains of viruses, bacteria, fungi, and parasites that can bring about infectious diseases
- **Core sections on viral, bacterial, fungal, and parasitic diseases open with new chapters** detailing basic biology, pathogenesis, and antimicrobial agents and feature a consistent presentation covering Organism, Disease, and Clinical Aspects
- **Explanations** of host-parasite relationship, dynamics of infection, and host response
- **USMLE-style questions and a clinical case** conclude each chapter on the major viral, bacterial, fungal, and parasitic diseases
- **Full-color** tables, photographs, and illustrations
- **Clinical Capsules** cover the essence of the disease(s) caused by major pathogens
- **Margin Notes** highlight key points within a paragraph to facilitate review

New full-color art illuminates important concepts

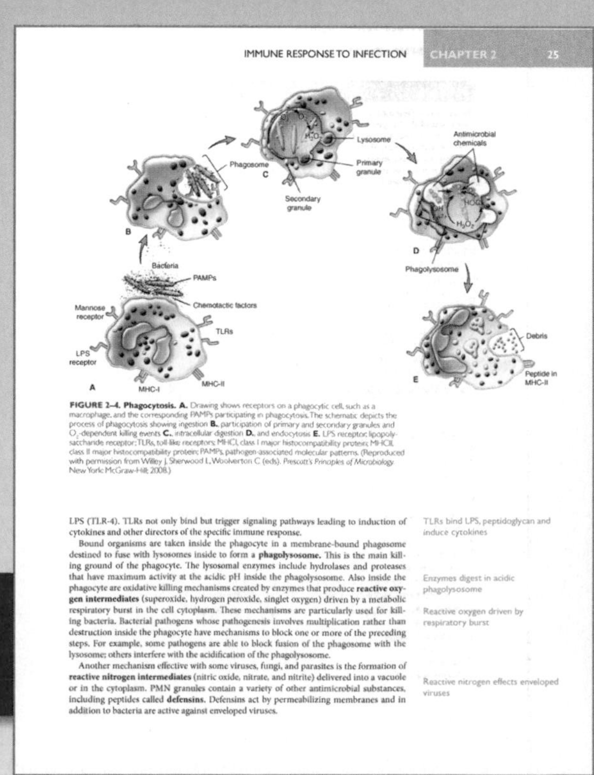
IMMUNE RESPONSE TO INFECTION CHAPTER 2 25

FIGURE 2–4. Phagocytosis. A. Drawing shows receptors on a phagocytic cell, such as a macrophage, and the corresponding PAMPs participating in phagocytosis. The schematic depicts the process of phagocytosis showing ingestion **B.**, participation of primary and secondary granules and O_2-dependent killing events **C.**, intracellular digestion **D.**, and endocytosis **E.** LPS receptor, lipopolysaccharide receptor; TLRs, toll-like receptors; MHCI, class I major histocompatibility protein; MHCII, class II major histocompatibility protein; PAMPs, pathogen-associated molecular patterns. (Reproduced with permission from Willey J, Sherwood L, Woolverton C (eds). *Prescott's Principles of Microbiology*. New York: McGraw-Hill; 2008.)

LPS (TLR-4). TLRs not only bind but trigger signaling pathways leading to induction of cytokines and other directors of the specific immune response.

TLRs bind LPS, peptidoglycan and induce cytokines

Bound organisms are taken inside the phagocyte in a membrane-bound phagosome destined to fuse with lysosomes inside to form a **phagolysosome.** This is the main killing ground of the phagocyte. The lysosomal enzymes include hydrolases and proteases that have maximum activity at the acidic pH inside the phagolysosome. Also inside the phagocyte are oxidative killing mechanisms created by enzymes that produce **reactive oxygen intermediates** (superoxide, hydrogen peroxide, singlet oxygen) driven by a metabolic respiratory burst in the cell cytoplasm. These mechanisms are particularly used for killing bacteria. Bacterial pathogens whose pathogenesis involves multiplication rather than destruction inside the phagocyte have mechanisms to block one or more of the preceding steps. For example, some pathogens are able to block fusion of the phagosome with the lysosome; others interfere with the acidification of the phagolysosome.

Enzymes digest in acidic phagolysosome

Reactive oxygen driven by respiratory burst

Another mechanism effective with some viruses, fungi, and parasites is the formation of **reactive nitrogen intermediates** (nitric oxide, nitrate, and nitrite) delivered into a vacuole or in the cytoplasm. PMN granules contain a variety of other antimicrobial substances, including peptides called **defensins.** Defensins act by permeabilizing membranes and in addition to bacteria are active against enveloped viruses.

Reactive nitrogen effects enveloped viruses

518 PART III PATHOGENIC BACTERIA

which still exists along with five other genera. Of these, *Fusobacterium, Porphyromonas,* and *Prevotella* are medically the most important. The *Bacteroides fragilis* group contains *B fragilis* and 10 similar species noted for their virulence and production of β-lactamases. (Species outside this group generally lack these features and are more similar to the other anaerobic Gram-negative bacilli.) *B fragilis* is a relatively short Gram-negative bacillus with rounded ends sometimes giving a coccobacillary appearance. The lipopolysaccharide (LPS) in its outer membrane has a much lower lipid content and thus lower toxic activity than that of most other Gram-negative bacteria. Almost all *B fragilis* strains have a polysaccharide capsule and are relatively oxygen-tolerant through production of superoxide dismutase. *Prevotella, Porphyromonas,* and *Fusobacterium* are distinguished by biochemical and other taxonomic features. *Prevotella melaninogenica* forms a black pigment in culture, and *Fusobacterium,* as its name suggests, is typically elongated and has tapered ends.

Five genera are medically important

B fragilis group produces β-lactamase and superoxide dismutase

ANAEROBIC INFECTIONS

EPIDEMIOLOGY

Despite our constant immersion in air, anaerobes are able to colonize the many oxygen-deficient or oxygen-free microenvironments of the body. Often these are created by the presence of facultative organisms whose growth reduces oxygen and decreases the local oxidation-reduction potential. Such sites include the sebaceous glands of the skin, the gingival crevices of the gums, the lymphoid tissue of the throat, and the lumina of the intestinal and urogenital tracts. Except for infections with some environmental clostridia, anaerobic infections are almost always endogenous with the infective agent(s) derived from the patient's normal flora. The specific anaerobes involved are linked to their prevalence in the flora of the relevant sites as shown in Table 29–1. In addition to the presence of clostridia in the lower intestinal tract of humans and animals, their spores are widely distributed in the environment, particularly in soil exposed to animal excreta. The spores may contaminate any wound caused by a nonsterile object (eg, splinter, nail) or exposed directly to soil.

Low redox normal flora sites are the origin of most infections

Spore-forming clostridia also come from the environment

PATHOGENESIS

The anaerobic flora normally lives in a harmless commensal relationship with the host. However, when displaced from their niche on the mucosal surface into normally sterile tissues, these organisms may cause life-threatening infections. This can occur as the result of trauma (eg, gunshot, surgery), disease (eg, diverticulosis), or isolated events (eg, aspiration). Host factors such as malignancy or impaired blood supply increase the probability that the dislodged flora will eventually produce an infection. The organisms most often involved are the anaerobes, which are both present at the mucosal site adjacent to the infection and have enhanced virulence. For example, *B fragilis* represents less than 1% of the normal colonic flora but is the organism most frequently isolated from intra-abdominal abscesses.

Anaerobes displaced from normal flora to deeper sites may cause disease

Trauma and host factors create the opportunity for infection

The relation between normal flora and site of infection may be indirect. For example, aspiration pneumonia, lung abscess, and empyema typically involve anaerobes found in the oropharyngeal flora. The brain is not a particularly anaerobic environment, but brain abscess is most often caused by these same oropharyngeal anaerobes. This presumably occurs by extension across the cribriform plate to the temporal lobe, the typical location of brain abscess. In contaminated open wounds, clostridia can come from the intestinal flora or from spores surviving in the environment.

Flora may be aspirated or displaced at a distance

Brain abscess typically involves anaerobic bacteria

Although gaining access to tissue sites provides the opportunity, additional virulence factors are needed for anaerobes to produce infection. Some anaerobic pathogens produce disease even when present as a minor part of the displaced resident flora, and other common members of the normal flora rarely cause disease. Classic virulence factors such as toxins and capsules are known only for the toxigenic clostridia and *B fragilis,* but a feature such as the ability to survive brief exposures to oxygenated environments can also be viewed as a virulence factor. Anaerobes found in human infections are far more likely to produce

Margin Notes speed your review and highlight must-know points

CHAPTER 5

Emergence and Global Spread of Infection

Epidemiology, the study of the distribution of determinants of disease and injury in human populations, is a discipline that includes both infectious and noninfectious diseases. Most epidemiologic studies of infectious diseases have concentrated on the factors that influence acquisition and spread, because this knowledge is essential for developing methods of prevention and control. Historically, epidemiologic studies and the application of the knowledge gained from them have been central to the control of the great epidemic diseases, such as cholera, plague, smallpox, yellow fever, and typhus.

An understanding of the principles of epidemiology and the spread of disease is essential to all medical personnel, whether their work is with the individual patient or with the community. Most infections must be evaluated in their epidemiologic setting. For example, what infections, especially viral, are currently prevalent in the community? Has the patient recently traveled to an area of special disease prevalence? Is there a possibility of nosocomial infection from recent hospitalization? What is the risk to the patient's family, schoolmates, and work or social contacts?

The recent recognition of emerging infectious diseases has heightened appreciation of the importance of epidemiologic information. A few examples of these newly identified infections are cryptosporidiosis, hantavirus pulmonary syndrome, and severe acute respiratory syndrome (SARS) coronavirus disease. In addition, some well-known pathogens have assumed new epidemiologic importance by virtue of acquired antimicrobial resistance (eg, penicillin-resistant pneumococci, vancomycin-resistant enterococci, and multiresistant *Mycobacterium tuberculosis*).

Factors that increase the emergence or reemergence of various pathogens include:

- Population movements and the intrusion of humans and domestic animals into new habitats, particularly tropical forests
- Deforestation, with development of new farmlands and exposure of farmers and domestic animals to new arthropods and primary pathogens
- Irrigation, especially primitive irrigation systems, which fail to control arthropods and enteric organisms
- Uncontrolled urbanization, with vector populations breeding in stagnant water
- Increased long-distance air travel, with contact or transport of arthropod vectors and primary pathogens
- Social unrest, civil wars, and major natural disasters, leading to famine and disruption of sanitation systems, immunization programs, etc.
- Global climate change
- Microbial evolution, leading to natural selection of multi-resistant agents (eg, methicillin-resistant staphylococci, new, highly virulent strains of influenza A virus). In some instances, these changes can be accelerated considerably by indiscriminate use of anti-infective agents.

89

Important new global considerations of infectious disease

INFLUENZA, PARAINFLUENZA, RESPIRATORY SYNCYTIAL VIRUS CHAPTER 9 187

REOVIRUSES

The reoviruses (respiratory enteric orphans) are naked virions that contain segmented, double-stranded RNA genomes that replicate and assemble in the cytoplasm of infected cells. They are ubiquitous and have been found in humans, simians, rodents, cattle, and a variety of other hosts. They have been studied in great detail as experimental models, revealing much basic knowledge about viral genetics and pathogenesis at the molecular level. Three serotypes are known to infect humans; however, their role and importance in human disease remain uncertain. Reoviruses causing arboviral diseases are discussed in Chapter 16.

Association with human disease is uncertain

CASE STUDY

AN INFANT WITH RESPIRATORY DISTRESS

This 9-month-old boy was born prematurely, requiring treatment in a neonatal intensive care unit for the first month of life. After discharge, he remained well until 3 days ago, when symptoms of a common cold progressed to cough, rapid and labored respiration, lethargy, and refusal to eat.

On examination, his temperature was 38.5°C, respiratory rate 60/min, and pulse 140/min. Auscultation of the chest revealed coarse crackles and occasional wheezes.

Abnormal laboratory findings included hypoxemia and hypercarbia. A chest radiograph showed hyperinflation, interstitial perihilar infiltrates, and right upper lobe atalectasis.

QUESTIONS

■ Which of these viruses is the least likely cause of this baby's illness?
A. Influenza A
B. Parainfluenza 3
C. Influenza C
D. Respiratory syncytial virus
E. Adenovirus

■ The mechanism of "antigenic drift" in influenza viruses includes all but one of the following:
A. Can involve either H or N antigens
B. Mutations caused by viral RNA polymerase
C. Can predominate under selective host population immune pressures
D. Reassortment between human and animal or avian reservoirs
E. Can involve genes encoding structural or nonstructural proteins

■ Which of the following agents can be used to prevent RSV pneumonia?
A. Amantadine
B. Vaccine to F protein
C. Oseltamivir
D. Zanamivir
E. Monoclonal antibody

ANSWERS

1(C), 2(D), 3(E)

Case Studies put the material in clinical context

CONTENTS

Preface xi

Acknowledgments xiii

PART I

The Nature of Infection 1
Kenneth J. Ryan and C. George Ray

CHAPTER 1 Infection 3

CHAPTER 2 Immune Response to Infection 19

CHAPTER 3 Sterilization, Disinfection, and Infection Control 45

CHAPTER 4 Principles of Laboratory Diagnosis of Infectious Diseases 59

CHAPTER 5 Emergence and Global Spread of Infection 89

PART II

Pathogenic Viruses 99
Nafees Ahmad, C. George Ray and W. Lawrence Drew

CHAPTER 6 The Nature of Viruses 101

CHAPTER 7 Pathogenesis of Viral Infection 137

CHAPTER 8 Antiviral Antimicrobics and Resistance 157

CHAPTER 9 Influenza, Parainfluenza, Respiratory Syncytial Virus, Adenovirus, and Other Respiratory Viruses 167

CHAPTER 10 Mumps Virus, Measles, Rubella, and Other Childhood Exanthems 189

CHAPTER 11 Poxviruses 205

CHAPTER 12 Enteroviruses 213

CHAPTER 13 Hepatitis Viruses 225

CHAPTER 14 Herpesviruses 247

CHAPTER 15 Viruses of Diarrhea 271

CHAPTER 16 Arthropod-Borne and Other Zoonotic Viruses 279

CHAPTER 17 Rabies 297

CHAPTER 18 Retroviruses: Human T-Lymphotropic Virus, Human Immunodeficiency Virus, and Acquired Immunodeficiency Syndrome 305

CHAPTER 19 Papilloma and Polyoma Viruses 327

CHAPTER 20 Persistent Viral Infections of the Central Nervous System 337

PART III

Pathogenic Bacteria 345
Kenneth J. Ryan and W. Lawrence Drew

CHAPTER 21 The Nature of Bacteria 347

CHAPTER 22 Pathogenesis of Bacterial Infection 387

CHAPTER 23 Antibacterial Agents and Resistance 403

CHAPTER 24 Staphylococci 429

CHAPTER 25 Streptococci and Enterococci 443

CHAPTER 26 *Corynebacterium, Listeria,* and *Bacillus* 471

CHAPTER 27 Mycobacteria 489

CHAPTER 28 *Actinomyces* and *Nocardia* 507

CHAPTER 29 *Clostridium, Peptostreptococcus, Bacteroides,* and Other Anaerobes 515

CHAPTER 30 *Neisseria* 535

CHAPTER 31 *Haemophilus* and *Bordetella* 551

CHAPTER 32 *Vibrio, Campylobacter,* and *Helicobacter* 565

CHAPTER 33 Enterobacteriaceae 579

CHAPTER 34 *Legionella* 609

CHAPTER 35 *Pseudomonas* and Other Opportunistic Gram-negative Bacilli 617

CHAPTER 36 Plague and Other Bacterial Zoonotic Diseases 629

CHAPTER 37 Spirochetes 643

CHAPTER 38 *Mycoplasma* and *Ureaplasma* 665

CHAPTER 39 *Chlamydia* 671

CHAPTER 40 *Rickettsia, Ehrlichia, Coxiella,* and *Bartonella* 681

PART IV

Pathogenic Fungi 691
Kenneth J. Ryan

CHAPTER 41 The Nature of Fungi 693

CHAPTER 42 Pathogenesis of Fungal Infection 703

CHAPTER 43 Antifungal Agents and Resistance 707

CHAPTER 44 Dermatophytes, *Sporothrix,* and Other Superficial and Subcutaneous Fungi 713

CHAPTER 45 *Candida, Aspergillus, Pneumocystis,* and Other Opportunistic Fungi 723

CHAPTER 46 *Cryptococcus, Histoplasma, Coccidioides,* and Other Systemic Fungal Pathogens 739

PART V

Pathogenic Parasites 757
C. George Ray and James J. Plorde

CHAPTER 47 The Nature of Parasites 759

CHAPTER 48 General Principles of Pathogenesis, Immunology, and Diagnosis of Parasitic Infection 767

CHAPTER 49 Antiparasitic Antimicrobics and Resistance 771

CHAPTER 50 Sporozoa 779

CHAPTER 51 Rhizopods 803

CHAPTER 52 Flagellates 813

CHAPTER 53 Intestinal Nematodes 835

CHAPTER 54 Tissue Nematodes 851

CHAPTER 55 Cestodes 865

CHAPTER 56 Trematodes 879

PART VI

Clinical Aspects of Infection 893
C. George Ray, Kenneth J. Ryan, and W. Lawrence Drew

CHAPTER 57 Skin and Wound Infections 895

CHAPTER 58 Bone and Joint Infections 901

CHAPTER 59 Eye, Ear, and Sinus Infections 905

CHAPTER 60 Dental and Periodontal Infections 911

CHAPTER 61 Respiratory Tract Infections 919

CHAPTER 62 Enteric Infections and Food Poisoning 929

CHAPTER 63 Urinary Tract Infections 939

CHAPTER 64 Genital Infections 945

CHAPTER 65 Central Nervous System Infections 951

CHAPTER 66. Intravascular Infections, Bacteremia, and Endotoxemia 959

Glossary 971

Index 989

PREFACE

With this 5th edition, *Sherris Medical Microbiology,* passes the quarter century mark. This longevity carries with it some need for evolution of authorship as well as text. We are pleased to welcome to the fold, Nafees Ahmad, a virologist with extensive success in teaching medical students. For those authors not continuing with this edition, we direct your attention to a special recognition on the Acknowledgments page. John Sherris, the founding editor, continues to act as an advisor to all of us.

The goal of *Sherris Medical Microbiology* remains unchanged from that of the first edition (1984). This book is intended to be the primary text for students of medicine and medical science who are encountering microbiology and infectious diseases for the first time. **Part I** opens with a chapter that explains the nature of infection and the infectious agents at the level of a general reader. The following four chapters give more detail on the immunologic, diagnostic, and epidemiologic nature of infection with minimal detail about the agents themselves. **Parts II–V** form the core of the text with chapters on the major viral, bacterial, fungal, and parasitic diseases, and each now begins with its own chapters on basic biology, pathogenesis, and antimicrobial agents. In the specific organism/disease chapters, the presentation sequence of **Organism** (structure, replication, genetics, etc.) followed by **Disease** (epidemiology, pathogenesis, immunity), and concluding with **Clinical Aspects** (manifestations, diagnosis, treatment, prevention) is maintained throughout the book. The opening of each section is marked with an icon and the **Clinical Capsule,** a snapshot of the disease, is placed at the juncture of the Organism and Disease sections. A **Clinical Case** followed by questions in USMLE format concludes each of these chapters. The 10 brief chapters of **Part VI** re-sort the material in the rest of the book into infectious syndromes. It is hoped these chapters will be of particular value when the student prepares for case discussions or sees patients.

In *Sherris Medical Microbiology,* the emphasis is on the text narrative, which is designed to be read comprehensively, not as a reference work. Considerable effort has been made to supplement this text with other learning aids such as the above-mentioned cases and questions as well as tables, photographs, and illustrations. These are now in full color, including over 300 new figures. The marginal notes, a popular feature since the first edition, are nuggets of information designed as an aid for the student during review. If a marginal note is unfamiliar, the relevant text is immediately adjacent. Many additional cases, questions, and study aids can be found in our online learning center. Visit **www.LangeTextbooks.com** to see the additional resources available.

For any book, lecture, case study, or other materials aimed at students, dealing with the onslaught of new information is a major challenge. In this edition, much new material has been included, but to keep the student from being overwhelmed, older or less important information has been deleted to keep the size of this book approximately the same as that of the 4th edition. As a rule of thumb, material on classic microbial structures, toxins, and the like in the Organism section has been trimmed unless its role is clearly explained in the Disease section. At the same time, we have tried not to eliminate detail to the point of becoming synoptic and uninteresting. Genetics is one of the greatest challenges in this regard. Without doubt this is where major progress is being made in understanding infectious diseases, but an intelligent discussion may require using the names of genes, their products, and multiple regulators to tell the complete story. Here we have tried to fully describe some of the major mechanisms and refer to them later when they reappear with other organisms. For example, *Neisseria gonorrhoeae* is used as an example of genetic mechanisms for antigenic variation in bacterial pathogenesis (Chapter 22), but how it may influence its disease, gonorrhea, is taken up in Chapter 30.

A saving grace is that our topic is important, dynamic, and fascinating—not just to us but to the public at large. Newspaper headlines now carry not only the name but the antigenic formula of *E coli* O157:H7 when it is the cause of a national outbreak of hemorrhagic colitis and kidney failure. Who could have predicted that HIV/AIDS, which occupied less than a page in the first edition of this book, would rival tuberculosis as the leading cause of premature death in the world; or that the Nobel Prize would be awarded for showing that gastritis and ulcers that were attributed to stress in the past are in fact an infection with the bacterium *Helicobacter pylori*? Just as we hit the presses, a new infectious threat has emerged in the form of Influenza virus H1N1 (another antigenic formula), or swine flu. We will keep you apprised of these developments in our online learning center, but we are confident that the basis for understanding them has already been laid out in the pages of this book. If you can't wait, read Chapter 9.

Kenneth J. Ryan
C. George Ray
Editors

ACKNOWLEDGMENTS

From its inception, *Sherris Medical Microbiology* has been a collaborative effort among microbiologists committed to making the study of infectious diseases intellectually and professionally rewarding. With this 5th edition, the editors would like to acknowledge the past contributions and continuing importance of authors who have participated over a number of editions, particularly recent editions. They are listed below. Their influence will long be felt, and in some instances their words and phrases continue to be found in the *Sherris* chapters on the topics beside their names.

AUTHORS

JAMES J. CHAMPOUX, PHD—THE NATURE OF VIRUSES, RETROVIRUSES

Professor and Chair
Department of Microbiology
School of Medicine
University of Washington
Seattle, Washington

STANLEY FALKOW, PHD—PATHOGENESIS OF INFECTION, BACTERIAL DISEASES

Professor of Microbiology and Immunology
Professor of Medicine
School of Medicine
Stanford University
Stanford, California

FREDERICK C. NEIDHARDT, PHD—THE NATURE OF BACTERIA

Frederick G. Novy Distinguished
University Professor of
Microbiology and Immunology, Emeritus
University of Michigan
Medical School
Ann Arbor, Michigan

CONTRIBUTORS

JOHN J. MARCHALONIS, PHD*—THE IMMUNE RESPONSE TO INFECTION

Professor and Head
Department of Microbiology and Immunology
College of Medicine
University of Arizona
Tucson, Arizona

MURRAY R. ROBINOVITCH, DDS, PHD—DENTAL INFECTIONS

Professor and Chairman
Department of Periodontics
School of Dentistry
University of Washington
Seattle, Washington

*Deceased

The editors and authors also wish to acknowledge the collaboration and continuous support of Michael Weitz, Regina Brown, and the McGraw-Hill staff, who guided us with remarkable speed and flexibility. We also wish to thank Diane Ray and Alexa Suslow for administrative support and manuscript review. New illustrations for this edition were prepared by Thomson Digital and its talented staff of artists under the direction of Dr. Anuradha Majumdar. Finally, we wish to acknowledge our students, past and present, who provide the stimulation for continuation of this work, and our families who provide the encouragement and support that make it possible.

PART I

The Nature of Infection

Kenneth J. Ryan and
C. George Ray

Infection CHAPTER 01

Immune Response to Infection CHAPTER 02

Sterilization, Disinfection, and Infection Control CHAPTER 03

Principles of Laboratory Diagnosis of Infectious Diseases CHAPTER 04

Emergence and Global Spread of Infection CHAPTER 05

CHAPTER 1

Infection

Humanity has but three great enemies:
fever, famine and war;
of these by far the greatest,
by far the most terrible, is fever.

—Sir William Osler, 1896*

When Sir William Osler, the great physician/humanist wrote these words, fever (infection) was indeed the scourge of the world. Tuberculosis and other forms of pulmonary infection were the leading causes of premature death among the well to do and the less fortunate. The terror was due to the fact that although some of the causes of infection were being discovered, little could be done to prevent or alter the course of disease. In the 20th century, advances in public sanitation and the development of vaccines and antimicrobials changed this (**Figure 1–1**), but only for the nations that could afford these interventions. As the 21st century begins, the world is divided into countries in which heart attacks, cancer, and stroke have surpassed infection as a cause of death and those in which infection is still the leading cause of death.

A new uneasiness that is part evolutionary, part discovery, and part diabolic has taken hold. Infectious agents once conquered have demonstrated resistance to established therapy, such as multiresistant *Mycobacterium tuberculosis,* and new diseases, such as acquired immunodeficiency syndrome (AIDS), have emerged. The spectrum of infection has widened, with discoveries that organisms once thought to be harmless can cause disease under certain circumstances. Who could have guessed that *Helicobacter pylori,* not even mentioned in the first edition of this book, would be the major cause of gastric and duodenal ulcers and an officially declared carcinogen? Finally, bioterrorist forces have unearthed two previously controlled infectious diseases, anthrax and smallpox, and threatened their distribution as agents of biological warfare. For students of medicine, understanding the fundamental basis of infectious diseases has more relevance than ever.

BACKGROUND

The science of medical microbiology dates back to the pioneering studies of Pasteur and Koch, who isolated specific agents and proved that they could cause disease by introducing the

*Osler W. *JAMA* 1896; 26:999.

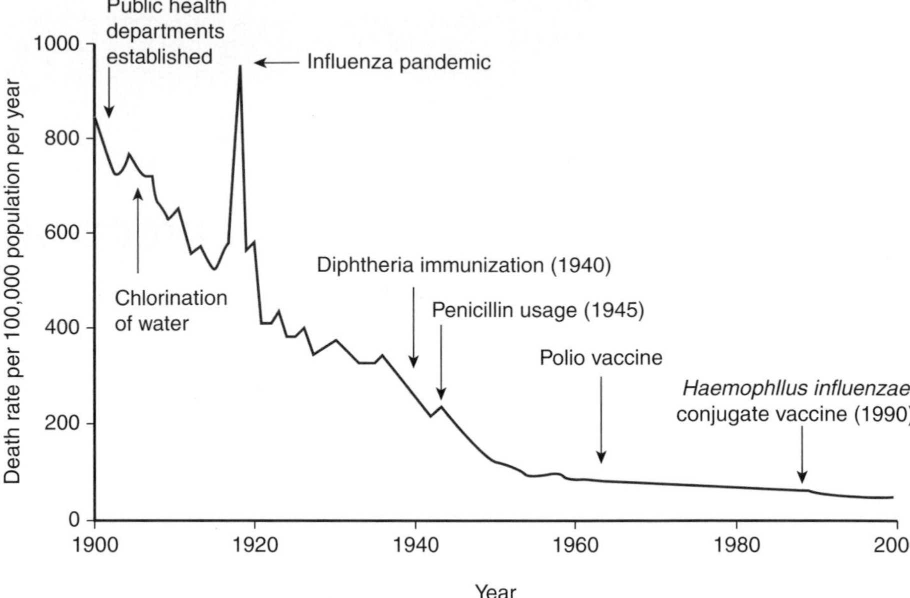

FIGURE 1–1. Death rates for infectious disease in the United States in the 20th century. Note the steady decline in death rates related to the introduction of public health, immunization, and antimicrobial interventions.

experimental method. The methods they developed lead to the first golden age of microbiology (1875-1910), when many bacterial diseases and the organisms responsible for them were defined. These efforts, combined with work begun by Semmelweis and Lister, which showed how these diseases spread, led to the great advances in public health that initiated the decline in disease and death. In the first half of the 20th century, scientists studied the structure, physiology, and genetics of microbes in detail and began to answer questions relating to the links between specific microbial properties and disease. By the end of the 20th century, the sciences of molecular biology, genetics, genomics, and proteomics extended these insights to the molecular level. Genetic advances have reached the point at which it is possible to know not only the genes involved but to understand how they are regulated. The discoveries of penicillin by Fleming in 1929 and of sulfonamides by Domagk in 1935 opened the way to great developments in chemotherapy. These gradually extended from bacterial diseases to fungal, parasitic, and finally viral infections. Almost as quickly, virtually all categories of infectious agents developed resistance to all categories of antimicrobics to counter these chemotherapeutic agents.

INFECTIOUS AGENTS: THE MICROBIAL WORLD

Microbes are small

Microbiology is a science defined by smallness. Its creation was made possible by the invention of the microscope (Gr. *micro,* small + *skop,* to look, see), which allowed visualization of structures too small to see with the naked eye. This definition of microbiology as the study of microscopic living forms still holds if one can accept that some organisms can live only in other cells (eg, all viruses and some bacteria) and that others have macroscopic forms (eg, fungal molds, parasitic worms). The relative sizes of some microorganisms are shown in **Figure 1–2**.

Most play benign roles in the environment

Microorganisms are responsible for much of the breakdown and natural recycling of organic material in the environment. Some synthesize nitrogen-containing compounds that contribute to the nutrition of living things that lack this ability; others (oceanic algae) contribute to the atmosphere by producing oxygen through photosynthesis. Because microorganisms have an astounding range of metabolic and energy-yielding abilities, some can exist under conditions that are lethal to other life forms. For example, some bacteria can oxidize inorganic compounds such as sulfur and ammonium ions to generate energy, and some can survive and multiply in hot springs at temperatures higher than 75°C.

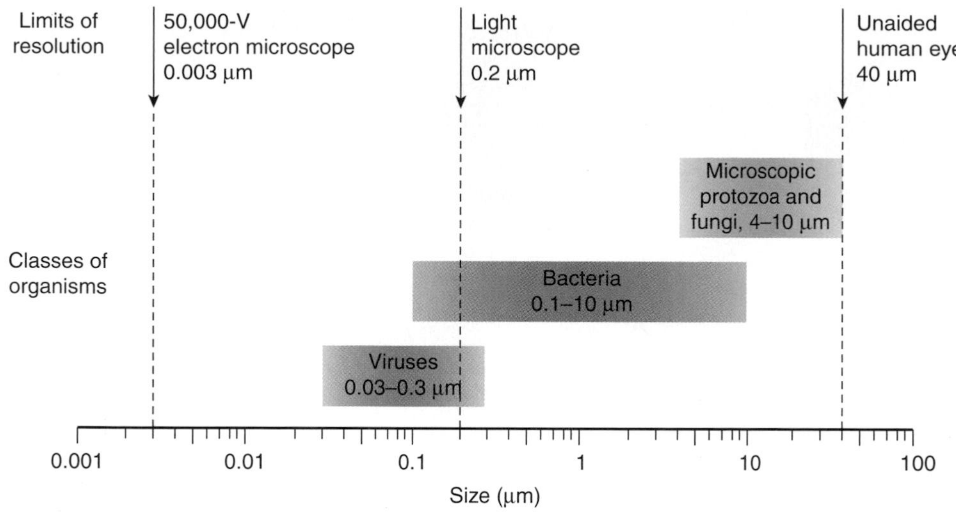

FIGURE 1–2. Relative size of microorganisms.

Some microbial species have adapted to a symbiotic relationship with higher forms of life. For example, bacteria that can fix atmospheric nitrogen colonize root systems of legumes and of a few trees such as alders and provide the plants with their nitrogen requirements. When these plants die or are plowed under, the fertility of the soil is enhanced by nitrogenous compounds originally derived from the metabolism of the bacteria. Ruminants can use grasses as their prime source of nutrition, because the abundant flora of anaerobic bacteria in the rumen break down cellulose and other plant compounds to usable carbohydrates and amino acids and synthesize essential nutrients including some amino acids and vitamins. These few examples illustrate the protean nature of microbial life and their essential place in our ecosystem.

Products of microbes contribute to the atmosphere

The major classes of microorganisms in terms of ascending size and complexity are viruses, bacteria, fungi, and parasites. Parasites exist as single or multicellular structures with the same eukaryotic cell plan of our own cells. Fungi are also eukaryotic, but have a rigid external wall that makes them seem more like plants than animals. Bacteria also have a cell wall, but their cell plan is prokaryotic and lacks the organelles of eukaryotic cells. Viruses have a genome and some structural elements but must take over the machinery of another living cell (eukaryotic or prokaryotic) to replicate. The four classes of infectious agents are summarized in **Table 1–1,** and generic examples of each are shown in **Figure 1–3**.

Increasing complexity: viruses → bacteria → fungi → parasites

VIRUSES

Viruses are strict intracellular parasites of other living cells, not only of mammalian and plant cells, but also of simple unicellular organisms, including bacteria (the bacteriophages).

TABLE 1–1 Features of Infectious Agentss

	VIRUSES	BACTERIA	FUNGI	PARASITES
Size (µM)	< 1	2-8	4+	2+
Cell wall	No	Yes	Yes	No/yes[a]
Cell plan	None	Prokaryotic	Eukaryotic	Eukaryotic
Free living	No	Yes[b]	Yes	Yes
Intracellular	Yes	No/yes[b]	No	No/yes

[a]Parasitic cysts have cell walls.
[b]A few bacteria grow only within cells.
[c]The life cycle of some parasites includes intracellular multiplication.

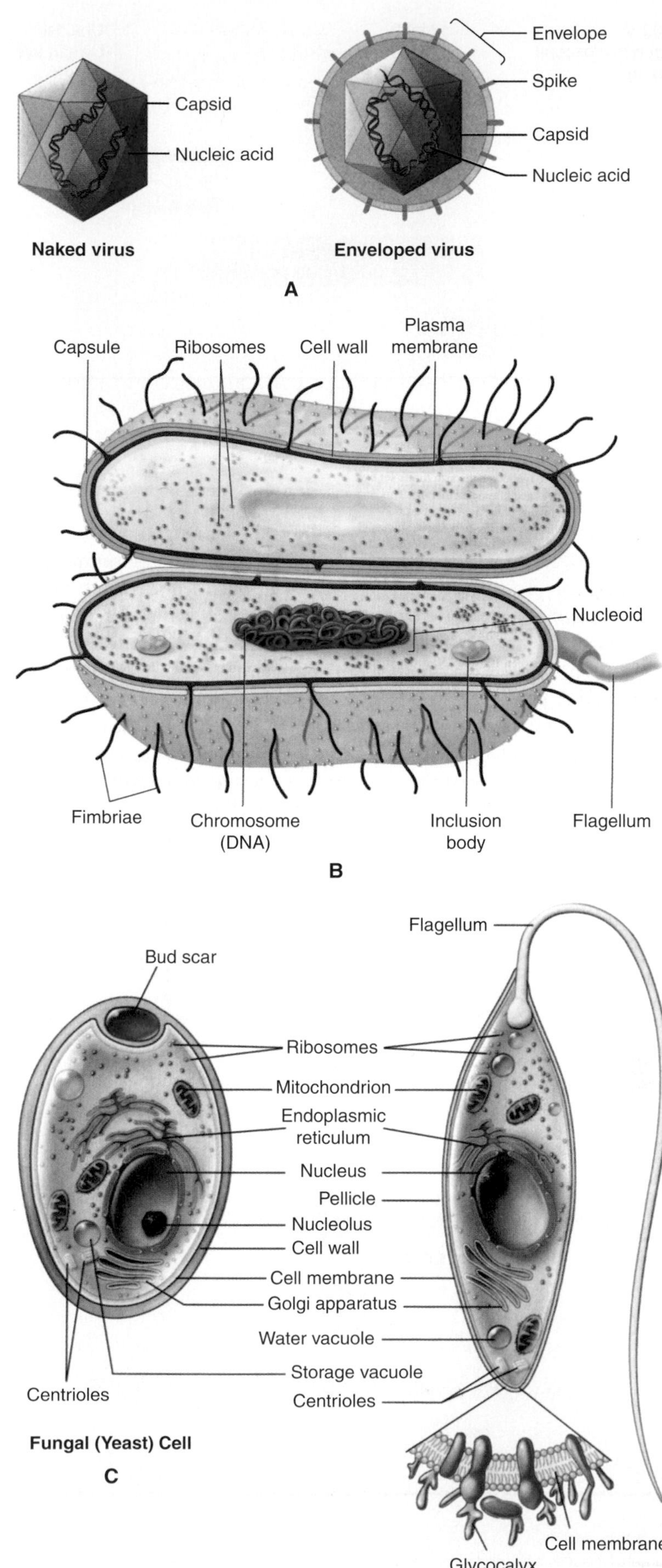

FIGURE 1–3. Infectious agents. **A.** Virus. **B.** Bacterium. **C.** Fungus. **D.** Parasite. (Reproduced with permission from Willey J, Sherwood L, Woolverton C (eds). *Prescott's Principles of Microbiology.* New York: McGraw-Hill; 2008.)

Viruses are simple forms of replicating, biologically active particles that carry genetic information in either DNA or RNA molecules, but never in both. Most mature viruses have a protein coat over their nucleic acid and sometimes a lipid surface membrane derived from the cell they infect. Because viruses lack the protein-synthesizing enzymes and structural apparatus necessary for their own replication, they bear essentially no resemblance to a true eukaryotic or prokaryotic cell.

Viruses contain little more than DNA or RNA

Viruses replicate by using their own genes to direct the metabolic activities of the cell they infect to bring about the synthesis and reassembly of their component parts. A cell infected with a single viral particle may thus yield many thousands of viral particles, which can be assembled almost simultaneously under the direction of the viral nucleic acid. With many viruses, cell death and infection of other cells by the newly formed viruses results. Sometimes, viral reproduction and cell reproduction proceed simultaneously without cell death, although cell physiology may be affected. The close association of the virus with the cell sometimes results in the integration of viral nucleic acid into the functional nucleic acid of the cell, producing a latent infection that can be transmitted intact to the progeny of the cell.

Replication is by control of the host cell metabolic machinery

Some integrate into the genome

BACTERIA

Bacteria are the smallest (0.1 to 10 μm) living cells. They have a cytoplasmic membrane surrounded by a cell wall; a unique interwoven polymer called peptidoglycan makes the wall rigid. The simple prokaryotic cell plan includes no mitochondria, lysosomes, endoplasmic reticulum, or other organelles (**Table 1–2**). In fact, most bacteria are about the size of mitochondria. Their cytoplasm contains only ribosomes and a single, double-stranded DNA chromosome. Bacteria have no nucleus, but all the chemical elements of nucleic acid and protein synthesis are present. Although their nutritional requirements vary greatly, most bacteria are free-living if given an appropriate energy source. Tiny metabolic factories, they divide by binary fission and can be grown in artificial culture, often in less than 1 day. Archaebacteria differ radically from other bacteria in structure and metabolic processes; they live in environments humans consider hostile (eg, hot springs, high salt areas) but are not associated with disease.

Smallest living cells

Prokaryotic cell plan lacks nucleus and organelles

FUNGI

Fungi exist in either yeast or mold forms. The smallest of yeasts are similar in size to bacteria, but most are larger (2 to 12 μm) and multiply by budding. Molds form tubular

TABLE 1–2 Distinctive Features of Prokaryotic and Eukaryotic Cells

CELL COMPONENT	PROKARYOTES	EUKARYOTES
Nucleus	No membrane, single circular chromosome	Membrane bounded, a number of individual chromosomes
Extrachromosomal DNA	Often present in form of plasmid(s)	In organelles
Organelles in cytoplasm	None	Mitochondria (and chloroplasts in photosynthetic organisms)
Cytoplasmic membrane	Contains enzymes of respiration; active secretion of enzymes; site of phospholipid and DNA synthesis	Semipermeable layer not possessing functions of prokaryotic membrane
Cell wall	Rigid layer of peptidoglycan (absent in *Mycoplasma*)	No peptidoglycan (in some cases cellulose present)
Sterols	Absent (except in *Mycoplasma*)	Usually present
Ribosomes	70 S in cytoplasm	80 S in cytoplasmic reticulum

Yeasts and molds are surrounded by cell wall

extensions called hyphae, which, when linked together in a branched network, form the fuzzy structure seen on neglected bread. Fungi are eukaryotic, and both yeasts and molds have a rigid external cell wall composed of their own unique polymers, called glucan, mannan, and chitin. Their genome may exist in a diploid or haploid state and replicate by meiosis or simple mitosis. Most fungi are free-living and widely distributed in nature. Generally, fungi grow more slowly than bacteria, although their growth rates sometimes overlap.

PARASITES

Range from tiny amoebas to meter-long worms

Parasites are the most diverse of all microorganisms. They range from unicellular amoebas of 10 to 12 µm to multicellular tapeworms 1 meter long. The individual cell plan is eukaryotic, but organisms such as worms are highly differentiated and have their own organ systems. Most worms have a microscopic egg or larval stage, and part of their life cycle may involve multiple vertebrate and invertebrate hosts. Most parasites are free-living, but some depend on combinations of animal, arthropod, or crustacean hosts for their survival.

NORMAL MICROBIAL FLORA

Before moving on to discuss how, when, and where the previously mentioned agents cause human disease, we should note that the presence of microbes on or in humans is not by itself abnormal. In fact from shortly after birth on it is universal; that is, normal. The term **normal flora** is used to describe microorganisms that are frequently found in various body sites in normal, healthy individuals. The constituents and numbers of the flora vary in different areas of the body and sometimes at different ages and physiologic states. They comprise microorganisms whose morphologic, physiologic, and genetic properties allow them to colonize and multiply under the conditions that exist in particular sites, to coexist with other colonizing organisms, and to inhibit competing intruders. Thus, each accessible area of the body presents a particular ecologic niche, colonization of which requires a particular set of properties of the invading microbe.

Organisms of the normal flora may have a symbiotic relationship that benefits the host or may simply live as commensals with a neutral relationship to the host. A parasitic relationship that injures the host would not be considered "normal," but in most instances not enough is known about the organism–host interactions to make such distinctions. Like houseguests, the members of the normal flora may stay for highly variable periods. **Residents** are strains that have an established niche at one of the many body sites, which they occupy indefinitely. **Transients** are acquired from the environment and establish themselves briefly but tend to be excluded by competition from residents or by the host's innate or immune defense mechanisms. The term **carrier state** is used when potentially pathogenic organisms are involved, although its implication of risk is not always justified. For example, *Streptococcus pneumoniae,* a cause of pneumonia, and *Neisseria meningitidis,* a cause of meningitis, may be isolated from the throat of 5% to 40% of healthy people. Whether these bacteria represent transient flora, resident flora, or carrier state is largely semantic. The possibility that their presence could be the prelude to disease is impossible to determine in advance.

Flora may stay for short or extended periods

If pathogens are involved, the relationship is called the carrier state

It is important for students of medical microbiology and infectious disease to understand the role of the normal flora because of its significance both as a defense mechanism against infection and as a source of potentially pathogenic organisms. It is also important for physicians to know the sites and composition of flora to avoid interpretive confusion between normal flora species and pathogens when interpreting laboratory culture results. The following excerpt indicates that the English poet W.H. Auden understood the need for balance between the microbial flora and its host. He was influenced by an article in *Scientific American* about the flora of the skin.

Balance is the desired state

On this day tradition allots
to taking stock of our lives,
my greetings to all of you, Yeasts,
Bacteria, Viruses,
Aerobics and Anaerobics:
A Very Happy New Year
to all for whom my ectoderm
is as middle earth to me.

For creatures your size I offer
a free choice of habitat,
so settle yourselves in the zone
that suits you best, in the pools
of my pores or the tropical
forests of arm-pit and crotch,
in the deserts of my fore-arms,
or the cool woods of my scalp.

Build colonies: I will supply
adequate warmth and moisture,
the sebum and lipids you need,
on condition you never
do me annoy with your presence,
but behave as good guests should,
not rioting into acne
or athlete's-foot or a boil.

—W.H. Auden,
Epistle to a Godson

ORIGIN AND NATURE

The healthy fetus is sterile until the birth membranes rupture. During and after birth, the infant is exposed to the flora of the mother's genital tract and to other organisms in the environment. During the infant's first few days of life, the flora reflects chance exposure to organisms that can colonize particular sites in the absence of competitors. Subsequently, as the infant is exposed to a broader range of organisms, those best adapted to colonize particular sites become predominant. Thereafter, the flora generally resembles that of other individuals in the same age group and cultural milieu.

Initial flora is acquired during and immediately after birth

Local physiologic and ecologic conditions determine the nature of the flora. These conditions are sometimes highly complex, differing from site to site, and sometimes vary with age. Conditions include the amounts and types of nutrients available, pH, oxidation–reduction potentials, and resistance to local antibacterial substances such as bile and lysozyme. Many bacteria have adhesin-mediated affinity for receptors on specific types of epithelial cells; this facilitates colonization and multiplication and prevents removal by the flushing effects of surface fluids and peristalsis. Various microbial interactions also determine their relative prevalence in the flora. These interactions include competition for nutrients, inhibition by the metabolic products of other organisms.

Physiologic conditions such as local pH influence colonization

Adherence factors counteract mechanical flushing

Ability to compete for nutrients is an advantage

NORMAL FLORA AT DIFFERENT SITES

The total normal flora of the body probably contains more than 1000 distinct species of microorganisms. The major members known to be important in preventing or causing disease, as well as those that may be confused with etiologic agents of local infections, are summarized in **Table 1–3** and are described in greater detail in subsequent chapters.

Blood, Body Fluids, and Tissues

In health, the blood, body fluids, and tissues are sterile. Occasional organisms may be displaced across epithelial barriers as a result of trauma or during childbirth; they may be briefly recoverable from the bloodstream before they are filtered out in the pulmonary capillaries or removed by cells of the reticuloendothelial system. Such transient bacteremia may be the source of infection when structures such as damaged heart valves and foreign bodies (prostheses) are in the bloodstream.

Tissues and body fluids such as blood are sterile in health

Transient bacteremia can result from trauma

Skin

The skin plays host to an abundant flora that varies somewhat according to the number and activity of sebaceous and sweat glands. The flora is most abundant on moist skin areas

TABLE 1–3 Predominant and Potentially Pathogenic Flora of Various Body Sites

	FLORA	
BODY SITE	**POTENTIAL PATHOGENS (CARRIER)**	**LOW VIRULENCE (RESIDENT)**
Blood	None	None[a]
Tissues	None	None
Skin	*Staphylococcus aureus*	*Propionibacterium, Corynebacterium* (diphtheroids), coagulase-negative staphylococci
Mouth	*Candida albicans*	*Neisseria* spp., viridans streptococci, *Moraxella (Branhamella), Peptostreptococcus*
Nasopharynx	*Streptococcus pneumoniae, Neisseria meningitidis, Haemophilus influenzae*, group A streptococci, *Staphylococcus aureus* (anterior nares)	*Neisseria* spp., viridans streptococci, *Moraxella, Peptostreptococcus*
Stomach	None	Streptococci, *Peptostreptococcus*, others from mouth
Small intestine	None	Scanty, variable
Colon	*Bacteroides fragilis, Escherichia coli, Pseudomonas, Candida, Clostridium (C. perfringens, C. difficile)*	*Eubacterium, Lactobacillus, Bacteroides, Fusobacterium*, Enterobacteriaceae, *Enterococcus, Clostridium*
Vagina		
Prepubertal and postmenopausal	*C. albicans*	Diphtheroids, staphylococci, Enterobacteriaceae
Childbearing	Group B streptococci, *C. albicans*	*Lactobacillus*, streptococci

[a]Organisms such as viridans streptococci may be transiently present after disruption of a mucosal site.

Propionibacteria and staphylococci are dominant bacteria

Skin flora is not easily removed

Conjunctiva resembles skin

(axillae, perineum, and between toes). Staphylococci and members of the *Propionibacterium* genus occur all over the skin, and facultative diphtheroids (corynebacteria) are found in moist areas. Propionibacteria are slim, anaerobic, or microaerophilic Gram-positive rods that grow in subsurface sebum and break down skin lipids to fatty acids. Thus, they are most numerous in the ducts of hair follicles and of the sebaceous glands that drain into them. Even with antiseptic scrubbing, it is difficult to eliminate bacteria from skin sites, particularly those bearing pilosebaceous units. Organisms of the skin flora are resistant to the bactericidal effects of skin lipids and fatty acids, which inhibit or kill many extraneous bacteria. The conjunctivae have a very scanty flora derived from the skin flora. The low bacterial count is maintained by the high lysozyme content of lachrymal secretions and by the flushing effect of tears.

Intestinal Tract

Oropharynx has streptococci and *Neisseria*

The **mouth** and **pharynx** contain large numbers of facultative and strict anaerobes. Different species of streptococci predominate on the buccal and tongue mucosa because of different specific adherence characteristics. Gram-negative diplococci of the genera *Neisseria* and *Moraxella (Branhamella)* make up the balance of the most commonly isolated facultative organisms. Strict anaerobes and microaerophilic organisms of the oral cavity have their niches in the depths of the gingival crevices surrounding the teeth and in sites such as tonsillar crypts, where anaerobic conditions can develop readily.

The total number of organisms in the oral cavity is very high, and it varies from site to site. Saliva usually contains a mixed flora of about 10^8 organisms per milliliter, derived

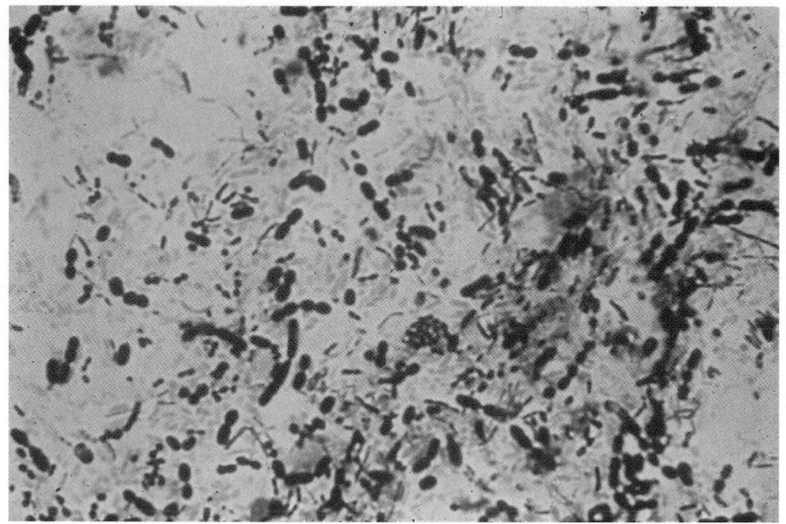

FIGURE 1–4. Stool flora. Gram smear of feces, showing great diversity of microorganisms. (Reprinted with permission of Schering Corporation, Kenilworth, New Jersey, the copyright owner. All rights reserved.)

Stomach and small bowel have few residents

Small intestinal flora is scanty but increases toward lower ileum

mostly from the various epithelial colonization sites. The stomach contains few, if any, resident organisms in health because of the lethal action of gastric hydrochloric acid and peptic enzymes on bacteria. The small intestine has a scanty resident flora, except in the lower ileum, where it begins to resemble that of the colon.

Adult colonic flora is abundant and predominantly anaerobic

Diet affects species composition

The colon carries the most prolific flora in the body (**Figure 1–4**). In the adult, feces are 25% or more bacteria by weight (about 10^{10} organisms per gram). More than 90% are anaerobes, predominantly members of the genera *Bacteroides, Fusobacterium, Eubacterium,* and *Clostridium*. The remainder of the flora is composed of facultative organisms such as *Escherichia coli,* enterococci, yeasts, and numerous other species. There are considerable differences in adult flora depending on the diet of the host. Those whose diets include substantial amounts of meat have more *Bacteroides* and other anaerobic Gram-negative rods in their stools than those on a predominantly vegetable or fish diet.

Respiratory Tract

S. aureus is carried in anterior nares

The external 1 cm of the anterior nares is lined with squamous epithelium. The nares have a flora similar to that of the skin except that it is the primary site of carriage of a pathogen, *Staphylococcus aureus*. About 25% to 30% of healthy people carry this organism as either resident or transient flora at any given time. The nasopharynx has a flora similar to that of the mouth; however, it is often the site of carriage of potentially pathogenic organisms such as pneumococci, meningococci, and *Haemophilus* species.

Lower tract is protected by mucociliary action

The respiratory tract below the level of the larynx is protected in health by the action of the epithelial cilia and by the movement of the mucociliary blanket; thus, only transient inhaled organisms are encountered in the trachea and larger bronchi. The accessory sinuses are normally sterile and are protected in a similar fashion, as is the middle ear by the epithelium of the eustachian tubes.

Genitourinary Tract

Bladder and upper urinary tract are sterile

Hormonal changes affect the vaginal flora

The urinary tract is sterile in health above the distal 1 cm of the urethra, which has a scanty flora derived from the perineum. Thus, in health the urine in the bladder, ureters, and renal pelvis is sterile. The vagina has a flora that varies according to hormonal influences at different ages. Before puberty and after menopause, it is mixed, nonspecific, and relatively scanty, and it contains organisms derived from the flora of the skin and colon. During the childbearing years, it is composed predominantly of anaerobic and microaerophilic members of the genus *Lactobacillus,* with smaller numbers of anaerobic Gram-negative rods, Gram-positive cocci, and yeasts that can survive under the acidic conditions produced by the lactobacilli. These conditions develop because glycogen is deposited in vaginal epithelial cells under the influence of estrogenic hormones and metabolized to lactic acid by lactobacilli. This process results in a vaginal pH of 4 to 5, which is optimal for growth and survival of the lactobacilli, but inhibits many other organisms.

Use of epithelial glycogen by lactobacilli produces low pH

ROLES IN HEALTH AND DISEASE

Opportunistic Infection

Flora that reach sterile sites may cause disease

Compromised defense systems increase the opportunity for invasion

Mouth flora plays a major role in dental caries

Many species among the normal flora are opportunists in that they can cause infection when they reach protected areas of the body in sufficient numbers. For example, certain strains of *E coli* can reach the urinary bladder by ascending the urethra and cause acute urinary tract infection. Perforation of the colon from a ruptured diverticulum or a penetrating abdominal wound releases feces into the peritoneal cavity; this contamination may be followed by peritonitis or intra-abdominal abscesses caused by the more opportunistic members of the flora. Reduced innate defenses or immunologic responses can result in local invasion and disease by normal floral organisms. Caries and periodontal disease are caused by organisms that are members of the normal oral flora (see Chapter 60).

Exclusionary Effect

Competing with pathogens has a protective effect

Antibiotic therapy may provide a competitive advantage for pathogens

Balancing the prospect of opportunistic infection is the tendency of the normal flora to produce conditions that compete with extraneous pathogens and thus reduce their ability to establish a niche in the host. The flora in the colon of the breastfed infant produces an environment inimical to colonization by enteric pathogens, as does a vaginal flora dominated by lactobacilli. The benefit of this exclusionary effect has been demonstrated by what happens when it is removed. Antibiotic therapy, particularly with broad-spectrum agents, may so alter the normal flora of the gastrointestinal tract that antibiotic-resistant organisms multiply in the ecologic vacuum. Under these conditions, the spore-forming *Clostridium difficile* has a selective advantage that allows it to survive, proliferate, and produce a toxic colitis.

Priming of Immune System

Sterile animals have little immunity to microbial infection

Low exposure correlates with asthma risk

Organisms of the normal flora play an important role in the development of immunologic competence. Animals delivered and raised under completely aseptic conditions ("sterile" or gnotobiotic animals) have a poorly developed reticuloendothelial system, low serum levels of immunoglobulins, and none of the antibodies to normal floral antigens that often confer a degree of protection against pathogens. There is evidence of immunologic differences between children who are raised under usual conditions and those whose exposure to diverse flora is minimized. Some studies have found a higher incidence of asthma in the more isolated children.

PROMOTING "GOOD" FLORA

Intestinal lactobacilli may protect against diarrheal agents

The field of probiotics promotes colonization with "good" flora such as with lactobacilli in the intestinal tract. Elie Metchnikoff originally suggested that the longevity of Bulgarian peasants was attributable to their consumption of large amounts of yogurt; the live lactobacilli in the yogurt presumably replaced the colonic flora to the general benefit of their health. This notion persists today in the alleged benefit of natural (unpasteurized) yogurt, which contains live lactobacilli. Although we now know that lactobacillary replacement of the flora of the adult colon does not take place so easily, there have been some successes with capsules containing lyophilized bacteria. In some studies, administration of preparations containing a particular strain of *Lactobacillus* (*L rhamnosus* strain GG, LGG) has reduced the duration of rotavirus diarrhea in children and prevented relapses of antibiotic-associated diarrhea caused by *C difficile*.

INFECTIOUS DISEASE

Of the thousands of species of viruses, bacteria, fungi, and parasites, only a tiny portion is involved in disease of any kind. These are called **pathogens**. There are plant pathogens, animal pathogens, and fish pathogens, as well as the subject of this book, human pathogens.

Pathogens are rare

Virulence varies greatly

Among pathogens, there are degrees of potency called **virulence**, which sometimes makes the dividing line between benign and virulent microorganisms difficult to draw. Other pathogens are virtually always associated with disease of varying severity. *Yersinia pestis,* the cause of plague, causes fulminant disease and death in 50% to 75% of persons who come in contact with it. It is highly virulent. Understanding the basis of these differences in virulence is a fundamental goal of this book. The better that students of medicine understand how a pathogen causes disease, the better they will be prepared to intervene and help their patients.

For any pathogen the basic aspects of how it interacts with the host to produce disease can be expressed in terms of its epidemiology, pathogenesis, and immunity. Usually our knowledge of one or more of these topics is incomplete. It is the task of the physician to relate these topics to the clinical aspects of disease and be prepared for new developments which clarify, or in some cases, alter them. We do not know everything, and not all of what we believe we know is correct.

EPIDEMIOLOGY

Epidemiology is the "who, what, when, and where" of infectious diseases. The power of the science of epidemiology was first demonstrated by Semmelweis, who by careful data analysis alone determined how streptococcal puerperal fever is transmitted. He even devised a means to prevent transmission (ie, handwashing) decades before the organism itself was discovered. Since then, each organism has built its own profile of vital statistics. Some agents are transmitted by air, others by food, and others by insects; some spread by the person-to-person route. **Figure 1–5** presents some of the variables in this regard. Some agents occur worldwide, and others only in certain geographic locations or ecologic circumstances. Knowing how an organism gains access to its victim and spreads is crucial to understanding

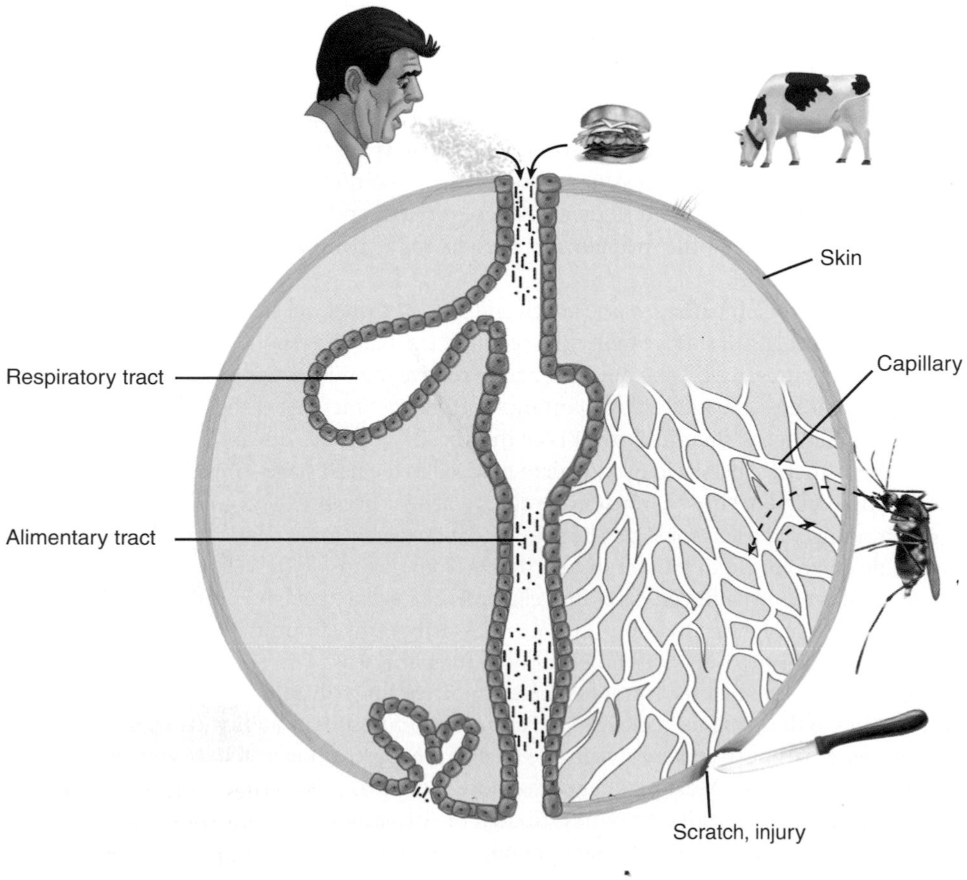

FIGURE 1–5. Infection overview. The sources and potential sites of infection are shown. Infection may be endogenous from the internal normal flora or exogenous from the sources shown around the outside.

Each agent has its own mode of spread

the disease. It is also essential in discovering the emergence of "new" diseases, whether they are truly new (AIDS) or just recently discovered (Legionnaires disease). Solving mysterious outbreaks or recognizing new epidemiologic patterns have usually pointed the way to the isolation of new agents.

Poor socioeconomic conditions foster infection

Modern society may facilitate spread

Epidemic spread and disease are facilitated by malnutrition, poor socioeconomic conditions, natural disasters, and hygienic inadequacy. In previous centuries, epidemics, sometimes caused by the introduction of new organisms of unusual virulence, often resulted in high morbidity and mortality rates. The possibility of recurrence of old pandemic infections remains, and, as with AIDS, we are currently witnessing a new and extended pandemic infection. Modern times and technology have introduced new wrinkles to epidemiologic spread. Intercontinental air travel has allowed diseases to leap continents even when they have very short incubation periods (cholera). The efficiency of the food industry has sometimes backfired when the distributed products are contaminated with infectious agents. The well-publicized outbreaks of hamburger-associated *Escherichia coli* O157:H7 infection constitute an example. The nature of massive meat-packing facilities allowed organisms from infected cattle on isolated farms to be mixed with other meat and distributed rapidly and widely. By the time outbreaks were recognized, cases of disease were widespread, and tons of meat had to be recalled. In simpler times, local outbreaks from the same source would have been detected and contained more quickly.

Anthrax and smallpox are new bioterrorism threats

Of course, the most ominous and uncertain epidemiologic threat of these times is not amplification of natural transmission but the specter of unnatural, deliberate spread. Anthrax is a disease uncommonly transmitted by direct contact of animals or animal products with humans. Under natural conditions, it produces a nasty but usually not life-threatening ulcer. The inhalation of human-produced aerosols of anthrax spores could produce a lethal pneumonia on a massive scale. Smallpox is the only disease officially eradicated from the world. It took place so long ago that most of the population has never been exposed or immunized and are thus vulnerable to its reintroduction. We do not know whether infectious bioterrorism will work on the scale contemplated by its perpetrators, but in the case of anthrax we do know that sophisticated systems have been designed to attempt it. We hope never to learn whether bioterrorism will work on a large scale.

PATHOGENESIS

Pathogenicity is multifactorial

Once a potential pathogen reaches its host, features of the organism determine whether or not disease ensues. The primary reason why pathogens are so few in relation to the microbial world is that being a successful pathogen is very complicated. Multiple features, called virulence factors, are required to persist, cause disease, and escape to repeat the cycle. The variations are many, but the mechanisms used by many pathogens are now being dissected at the molecular level.

Pathogens have molecules that bind to host cells

Invasion requires adaptation to new environments

The first step for any pathogen is to attach and persist at whatever site it gains access. This usually involves specialized surface molecules or structures that correspond to receptors on human cells. Because human cells were not designed to receive the microorganisms, the pathogens are usually exploiting some molecule important for essential functions of the cell. For some toxin-producing pathogens, this attachment is all they need to produce disease. For most pathogens, it just allows them to persist long enough to proceed to the next stage—invasion into or beyond the mucosal cells. For viruses, invasion of cells is essential, because they cannot replicate on their own. Invading pathogens must also be able to adapt to a new milieu. For example, the nutrients and ionic environment of the cell surface differs from that inside the cell or in the submucosa. Some of the steps in pathogenesis at the cellular level are illustrated in **Figure 1–6**.

Persistence and even invasion do not necessarily translate immediately to disease. The invading organisms must disrupt function in some way. For some, the inflammatory response they stimulate is enough. For example, a lung alveolus filled with neutrophils responding to the presence of *Streptococcus pneumoniae* loses its ability to exchange oxygen. The longer a pathogen can survive in the face of the host response, the greater the compromise in host function. Most pathogens do more than this. Destruction of host cells through the production of digestive enzymes, toxins, or intracellular multiplication is among the more common mechanisms. Other pathogens operate by altering the function of a cell

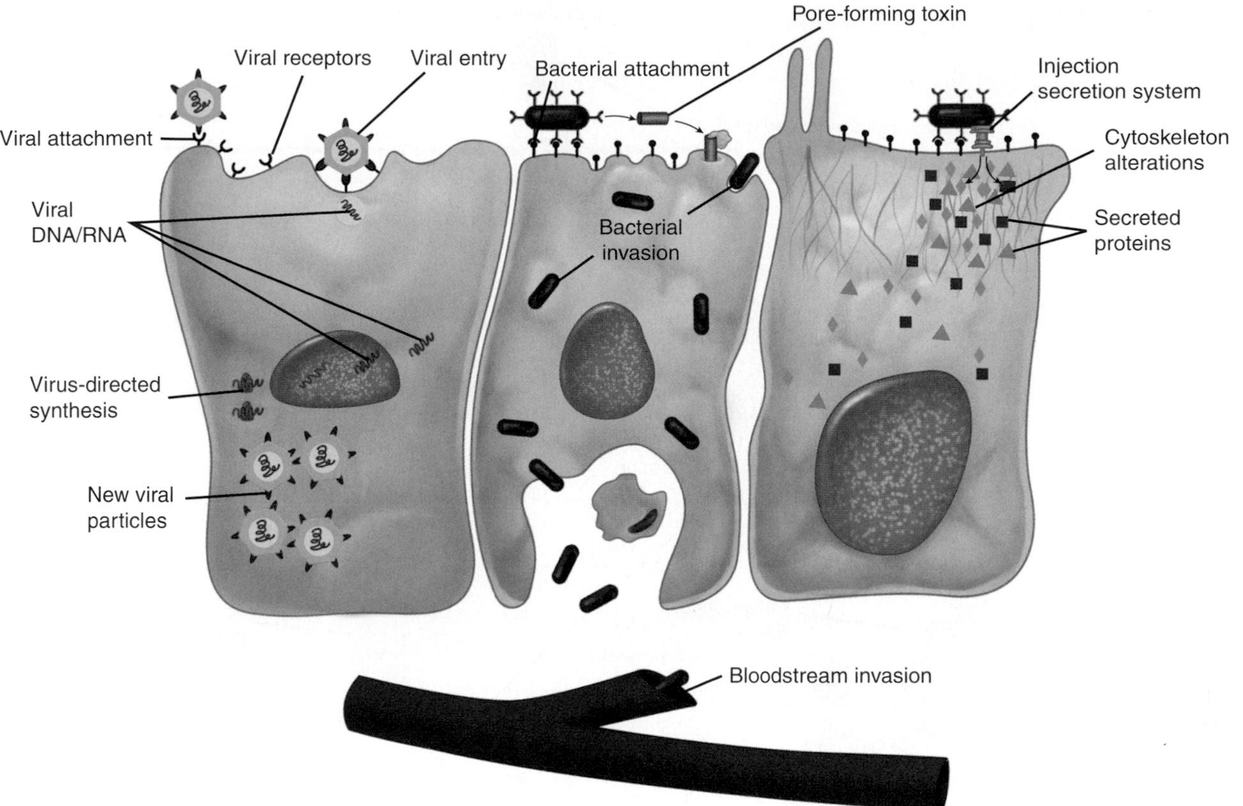

FIGURE 1–6. Infection cellular view. *Left.* A virus is attaching to the cell surface but can replicate only within the cell. *Middle.* A bacterial cell attaches to the surface, invades, and spreads through the cell to the bloodstream. *Right.* A bacterial cell attaches and injects proteins into the cell. The cell is disrupted while the organism remains on the surface.

without injury. Some of these actions are understood at a molecular level. Diphtheria is caused by a bacterial toxin that blocks protein synthesis inside the host cell. Details of the molecular mechanism for this action are illustrated in **Figure 1–7**. Some viruses cause the insertion of molecules in the host cell membrane, which cause other host cells to attack it. The variations are diverse and fascinating.

Inflammation alone can result in injury

Cells may be destroyed or their function altered

IMMUNITY

Although the science of immunology is beyond the scope of this book, understanding the immune response to infection (see Chapter 2) is an important part of appreciating pathogenic mechanisms. In fact, one of the most important virulence attributes any pathogen can have is an ability to evade the immune response. Some pathogens attack the immune effector cells, and others undergo changes that confound the immune response. The old observation that there seems to be no immunity to gonorrhea turns out to be an example of the latter mechanism. *Neisseria gonorrhoeae,* the causative agent of gonorrhea, undergoes antigenic variation of important surface structures so rapidly that antibodies directed against the bacteria become irrelevant.

Evading the immune response is a major feature of virulence

For each pathogen, the primary interest is whether there is natural immunity and, if so, whether it is based on cell-mediated (T_H1, CMI) or humoral (T_H2, antibody) mechanisms. Humoral and CMI responses are broadly stimulated with most infections, but the specific response to a particular molecular structure is usually dominant in mediating immunity to reinfection. For example, the repeated nature of strep throat (group A streptococcus) in childhood is not due to antigenic variation as described for gonorrhea. The antigen against which protective antibodies are directed (M protein) is stable but naturally exists in over 80 types.

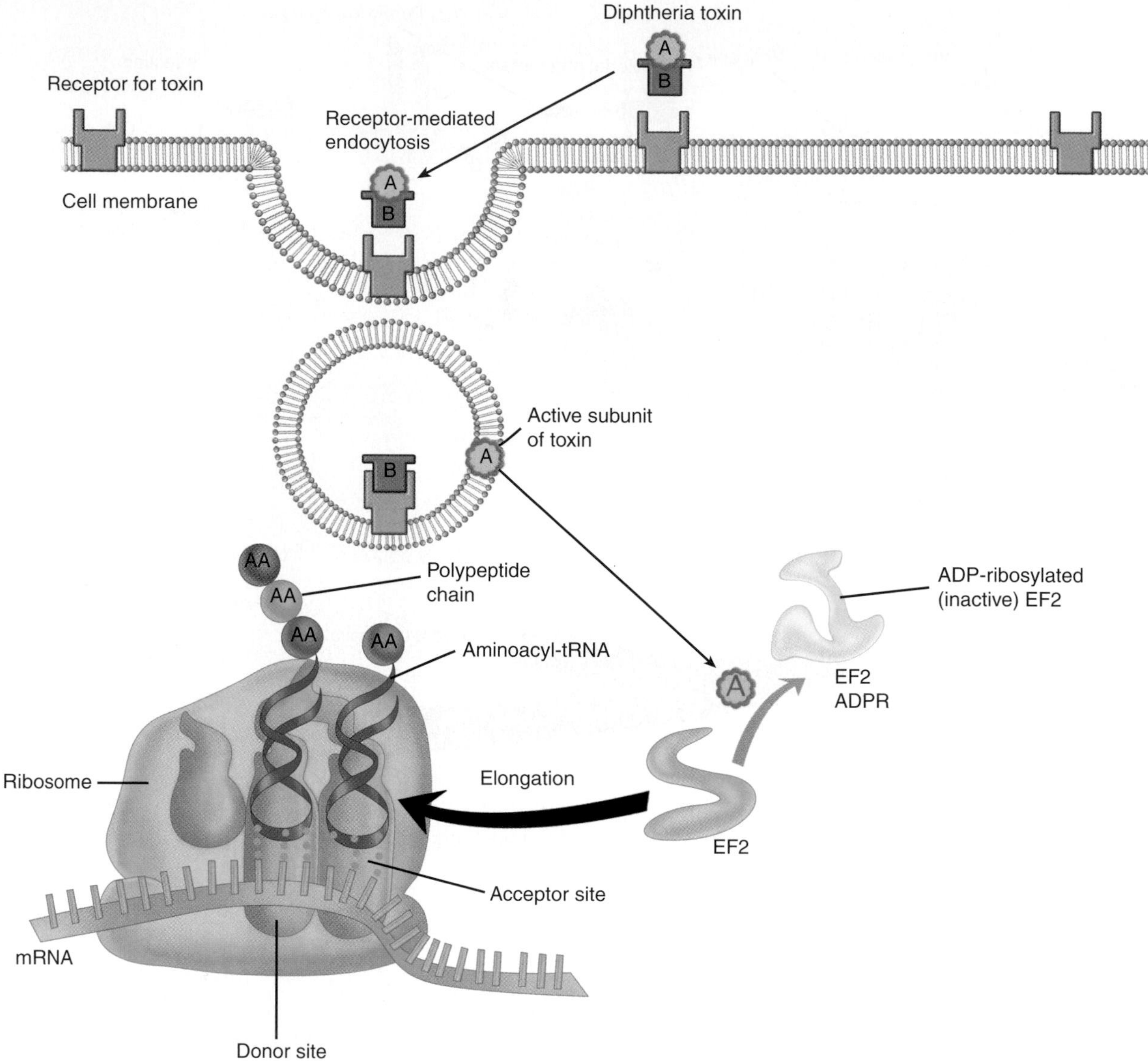

FIGURE 1–7. Action of diphtheria toxin, molecular view. The toxin-binding (B) portion attaches to the cell membrane, and the complete molecule enters the cell. In the cell, the A subunit dissociates and catalyzes a reaction that ADP-ribosylates (ADPR) and thus inactivates elongation factor 2 (EF-2). This factor is essential for ribosomal reactions at the acceptor and donor sites, which transfer triplet code from messenger RNA (mRNA) to amino acid sequences via transfer RNA (tRNA). Inactivation of EF-2 stops building of the polypeptide chain.

Antibody or cell-mediated mechanisms may be protective

Each type requires its own specific antibody. Knowing the molecule against which the protective immune response is directed is particularly important for devising preventive vaccines.

CLINICAL ASPECTS OF INFECTIOUS DISEASE

Manifestations

Fever, pain, and swelling are the universal signs of infection. Beyond this, the particular organs involved and the speed of the process dominate the signs and symptoms of disease. Cough, diarrhea, and mental confusion represent disruption of three different body systems. On the basis of clinical experience, physicians have become familiar with the range of behavior of the major pathogens. However, signs and symptoms overlap considerably. Skilled physicians use this knowledge to begin a deductive process leading to a list of suspected pathogens and a strategy to make a specific diagnosis and provide patient care. Through the probability assessment, an understanding of how the diseases work is a distinct advantage in making the correct decisions.

Body system(s) involved dictate clinical findings

Diagnosis

A major difference between infectious and other diseases is that the probabilities just described can be specifically resolved, often overnight. Most microorganisms can be isolated from the patient, grown in artificial culture, and identified. Others can be seen microscopically or detected by measuring the host-specific immune response. Preferred modalities for diagnosis of each agent have been developed and are available in clinic, hospital, and public health laboratories all over the world. Empiric diagnosis made on the basis of clinical findings can be confirmed and the treatment plan modified accordingly. The new molecular methods, which detect molecular structures or genes of the agent, are not yet practical for most infectious diseases.

Disease-causing microbes can be grown and identified

Treatment

Over the last 70 years, therapeutic tools of remarkable potency and specificity have become available for the treatment of bacterial infections. These include all the antibiotics and an array of synthetic chemicals that kill or inhibit the infecting organism, but have minimal or acceptable toxicity for the host. Antibacterial agents exploit the structural and metabolic differences between bacterial and eukaryotic cells to provide the selectivity necessary for good antimicrobial therapy. Penicillin, for example, interferes with the synthesis of the bacterial cell wall, a structure that has no analog in human cells. There are fewer antifungal and antiprotozoal agents because the eukaryotic cells of the host and those of the parasite have close metabolic and structural similarities. Nevertheless, hosts and parasites do have some significant differences, and effective therapeutic agents have been discovered or developed to exploit them.

Antibiotics are directed at structures of bacteria not present in host

Specific therapeutic attack on viral disease has posed more complex problems, because of the intimate involvement of viral replication with the metabolic and replicative activities of the cell. Thus, most substances that inhibit viral replication have unacceptable toxicity to host cells. However, recent advances in molecular virology have identified specific viral targets that can be attacked. Scientists have developed some successful antiviral agents, including agents that interfere with the liberation of viral nucleic acid from its protective protein coat or with the processes of viral nucleic acid synthesis and replication. The successful development of new agents for human immunodeficiency virus has involved targeting enzymes coded by the virus genome.

Antivirals target unique virus-coded enzymes

The success of the "antibiotic era" has been clouded by the development of resistance by the organisms. The mechanisms involved are varied but most often involve a mutational alteration in the enzyme, ribosome site, or other target against which the antimicrobial is directed. In some instances, organisms acquire new enzymes or block entry of the antimicrobic to the cell. Many bacteria produce enzymes that directly inactivate antibiotics. To make the situation worse, the genes involved are readily spread by promiscuous genetic mechanisms. New agents that are initially effective against resistant strains have been developed, but resistance by new mechanisms usually follows. The battle is by no means lost but has become a never-ending policing action.

Resistance complicates therapy

Mechanisms include mutation and inactivation

Prevention

The outcome of the scientific study of any disease is its prevention. In the case of infectious diseases, this has involved public health measures and immunization. The public health measures depend on knowledge of transmission mechanisms and on interfering with them. Water disinfection, food preparation, insect control, handwashing, and a myriad of other measures prevent humans from coming in contact with infections agents. Immunization relies on knowledge of immune mechanisms and designing vaccines that stimulate protective immunity.

Public health and immunization are primary preventive measures

Immunization follows two major strategies—live vaccines and inactivated vaccines. The former uses live but attenuated organisms that have been modified so they do not produce disease, but still stimulate a protective immune response. Such vaccines have been effective but carry the risk that the vaccine strain itself may cause disease. This event has been observed with the live oral polio vaccine. Although this rarely occurs, it has caused a shift

Attenuated strains stimulate immunity

Live vaccines can cause disease

back to the original Salk inactivated vaccine. This issue has reemerged with a debate over strategies for the use of smallpox immunization to protect against bioterrorism. This vaccine uses vaccinia virus, a cousin of smallpox, and its potential to produce disease on its own has been recognized since its original use by Jenner in 1798. Serious disease would be expected primarily in immunocompromised individuals (eg, from cancer chemotherapy or AIDS), who represent a significantly larger part of the population than when smallpox immunization was stopped in the 1970s. Could immunization cause more disease than it prevents? The question is difficult to answer.

Purified components are safe vaccines

The safest immunization strategy is the use of organisms that have been killed or, better yet, killed and purified to contain only the immunizing component. This approach requires much better knowledge of pathogenesis and immune mechanisms. Vaccines for meningitis use only the polysaccharide capsule of the bacterium, and vaccines for diphtheria and tetanus use only a formalin-inactivated protein toxin. Pertussis (whooping cough) immunization has undergone a transition in this regard. The original killed whole-cell vaccine was effective but caused a significant incidence of side effects. A purified vaccine containing pertussis toxin and a few surface components has reduced side effects while retaining efficacy.

Vaccines can be genetically engineered

The newest approaches for vaccines require neither live organisms nor killed, purified ones. As the entire genomes of more and more pathogens are being reported, an entirely genetic strategy is emerging. Armed with knowledge of molecular pathogenesis and immunity and the tools of genomics and proteomics, scientists can now synthesize an immunogenic protein without ever growing the organism itself. Such an idea would have astonished even the great microbiologists of the last two centuries.

SUMMARY

Infectious diseases remain as important and fascinating as ever. Where else do we find the emergence of new diseases, together with improved understanding of the old ones? At a time when the revolution in molecular biology and genetics has brought us to the threshold of new and novel means of infection control, the perpetrators of bioterrorism threaten us with diseases we have already conquered. Meeting this challenge requires a secure knowledge of the pathogenic organisms and how they produce disease, as well as an understanding of the clinical aspects of these diseases. In the collective judgment of the authors, this book presents the principles and facts required for students of medicine to understand the most important infectious diseases.

CHAPTER 2

Immune Response to Infection

Within a very short period immunity has been placed
in possession not only of a host of medical ideas
of the highest importance, but also of effective means
of combating a whole series of maladies
of the most formidable nature in man
and domestic animals.

—Elie Metchnikoff, 1905

The "maladies" Metchnikoff and the other pioneers of immunology were fighting were infections, and for decades their field was defined in terms of the immune response to infection. We now understand that the immune system is as much a part of everyday human biologic function as the cardiovascular or renal systems. In its adaptive and disordered states, infectious diseases play only a part, along with cancer and autoimmune diseases, which have little or no known connection to infection. Students of medicine take up immunology as a separate unit with its own text covering the field broadly. This chapter is not intended to fulfill that function or to be a shortened but comprehensive version of those sources. It is included as an overview of aspects related to infection for other students and as an internal reference for topics that reappear in later pages of this book. These include some of the greatest successes of medical science. The early and continuing development of vaccines that prevent and potentially eliminate diseases is but one example. Also, knowledge of the immune response to infection is integral to understanding the pathogenesis of infectious diseases. It turns out that one of the main attributes of a successful pathogen is evading or confounding the immune system.

The immune response to infection is presented as two major components—innate immunity and adaptive immunity. The primary effectors of both are cells that are part of the white blood cell series derived from hematopoietic stem cells in the bone marrow (**Figure 2–1**). Innate immunity includes the role of physical, cellular, and chemical systems that are in place and that respond to all aspects of foreignness. These include mucosal barriers, phagocytic cells, and the action of circulating glycoproteins such as complement. The adaptive side is sometimes called specific immunity because it has the ability to develop new responses that are highly specific to molecular components of infectious agents called **antigens**. These encounters trigger the development of new cellular responses and production of circulating antibody, which have a component of memory if the invader returns. Artificially creating this memory is, of course, the goal of vaccines.

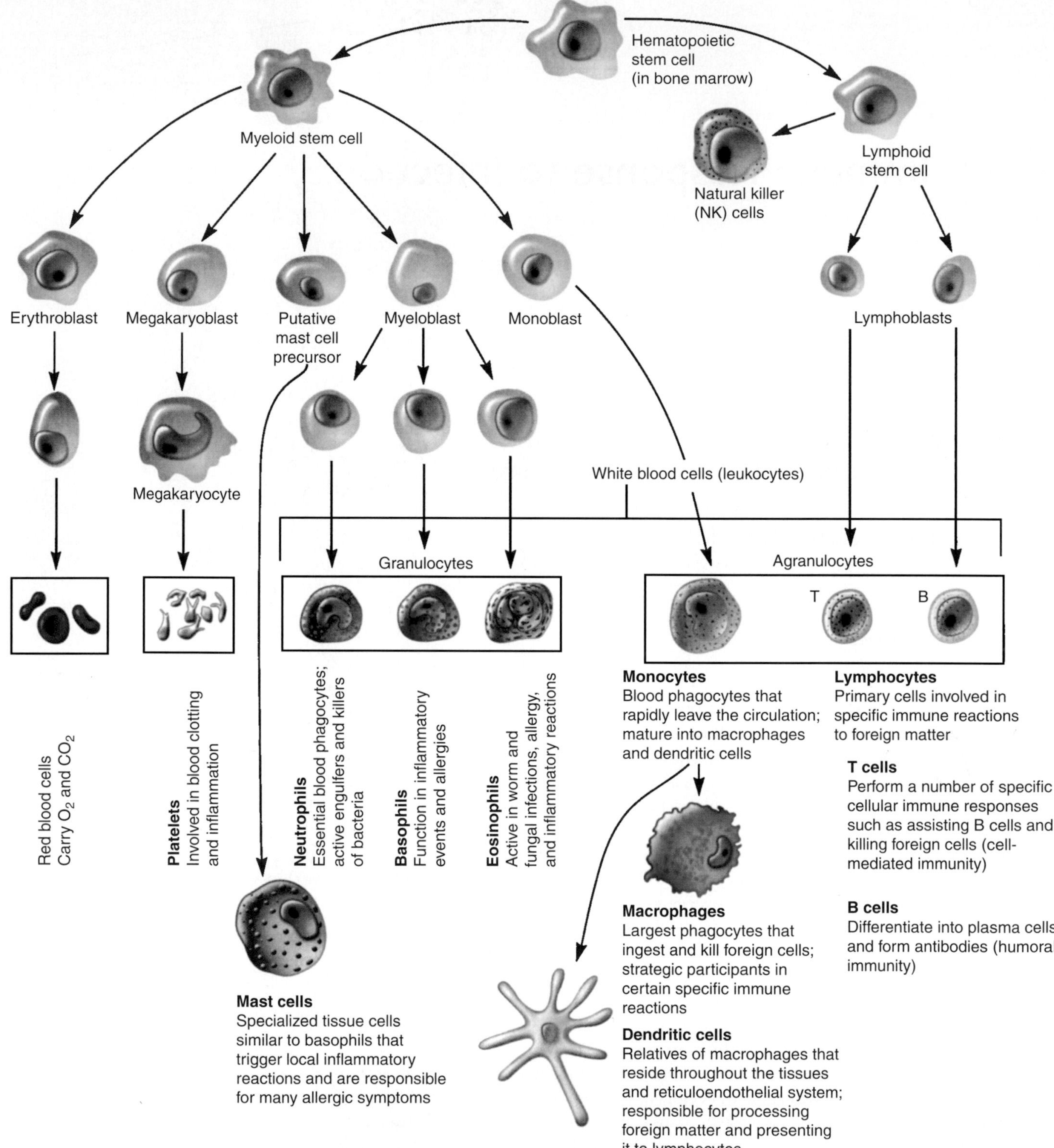

FIGURE 2–1. **Human blood cells.** Stem cells in the bone marrow divide to form two blood cell lineages: (1) the lymphoid stem cell gives rise to B cells that become antibody-secreting plasma cells, T cells that become activated T cells, and natural killer cells. (2) The common myeloid progenitor cell gives rise to granulocytes and monocytes that give rise to macrophages and dendritic cells. (Reproduced with permission from Willey J, Sherwood L, Woolverton C (eds). *Prescott's Principles of Microbiology.* New York: McGraw-Hill; 2008.)

INNATE (NONSPECIFIC) IMMUNITY

Innate immunity acts through a series of specific and nonspecific mechanisms all working to create a series of hurdles for the pathogen to navigate (**Table 2–1**). The first are mechanical barriers such as the tough multilayered skin or the softer but fused mucosal layers of internal surfaces. As discussed in Chapter 1, normal flora on these surfaces presents formidable competitors for space and nutrients. Turbulent movement of the mucosal surfaces and enzymes or acid secreted on their surface make it difficult for an organism to persist. Organisms that are able to pass the mucosa encounter a population of cells with the ability to engulf and destroy them. In addition, body fluids contain chemical agents such as complement, which can directly injure the microbe. The entire process has cross-links to the adaptive immune system. The endpoint of phagocytosis and digestion in a macrophage is the presentation of the antigen on its surface; the first step in specific immune recognition.

Skin, mucosa are barriers

Cells engulf, digest, and present antigens from microbes

TABLE 2–1 Features of Innate Immunity in Infection

	LOCATION	ACTIVITY AGAINST PATHOGENS
Cells		
Macrophage	Circulation, tissues	Phagocytosis, digestion
Dendritic cell	Tissues	Phagocytosis, digestion
Polymorphonuclear neutrophil (PMN)	Circulation, tissues (by migration)	Phagocytosis, digestion
M cell	Mucus membranes	Endocytosis and delivery to phagocytes
Surface Receptors		
Lectin	Phagocyte	Recognize carbohydrates
Arginine-glycine-arginine (RGD)	Phagocyte	Recognize arginine-glycine-aspartic acid sequence
Pathogen-associated molecular pattern (PAMP)	Phagocyte	Recognize molecular patterns unique to pathogens
Toll-like (TLR)	Phagocyte	Specialized PAMP, recognizes bacterial LPS (TLR-4), peptidoglycan[a] (TLR-2)
Inflammation		
Selectins	Endothelium	Attract and attach PMNs
Integrins	PMNs	Attach to selectins
Kallikrein	Extracellular fluid	Release bradykinin, prostaglandins
Chemical Mediators		
Cathelicidin	PMNs, macrophages, epithelial cells	Ionic membrane pores
Defensins	PMN granules	Ionic membrane pores
Complement (alternative)	Serum, extracellular fluid	Membrane pores, phagocyte receptors
Complement (lectin)	Serum, extracellular fluid	Phagocyte receptors

LPS, lipopolysaccharide of Gram-negative bacterial outer membrane.
[a]Cell wall component of Gram-positive and Gram-negative bacteria.

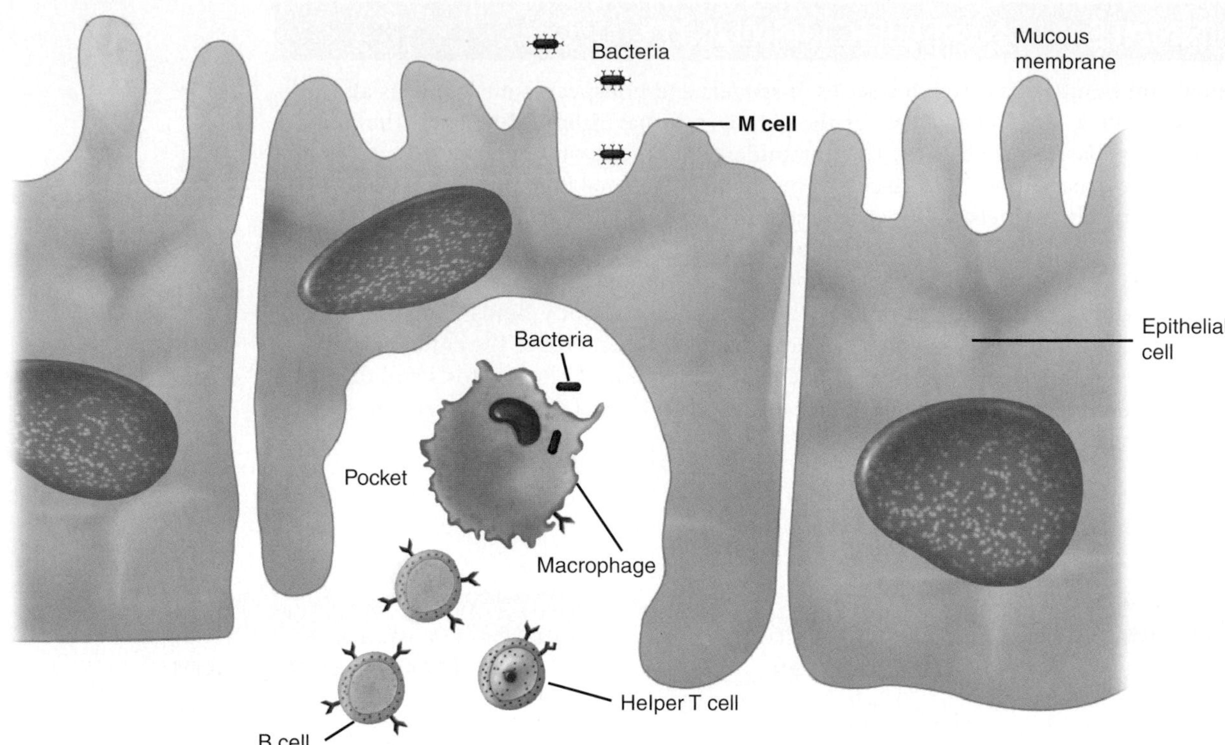

FIGURE 2–2. M cell. An M cell is shown between two epithelial cells in a mucous membrane. It has endocytosed a pathogen and released it into a pocket containing macrophages and other immune cells.

PHYSICAL BARRIERS

Lysozyme digests bacterial walls

Cilia move particles away from the alveoli

The thick layers of the skin containing insoluble keratins present the most formidable barrier to infection. The mucosal membranes of the alimentary and urogenital tract are not as tough but often are bathed in secretions inhospitable to invaders. Lysozyme is an enzyme that digests peptidoglycan, a unique structural component of the bacterial cell wall. Lysozyme is secreted onto many surfaces and is particularly concentrated in conjunctival tears. The acid pH of the vagina and particularly the stomach makes colonization difficult for most organisms. Only small particles (5-10 μm) can be inhaled deep into the lung alveoli because the lining of the respiratory includes cilia that trap and move them toward the pharynx.

M cells deliver to macrophages and lymphocytes

The skin and mucosal surfaces of the intestinal and respiratory tract also contain concentrations of lymphoid tissue within or just below their surfaces, which provide a next-level defense for invaders surviving the above-described defenses. These lymphoid collections are designed to entrap and deliver invaders to some of the phagocytes described in the following text. For example, in the intestine, M cells (**Figure 2–2**) that lack the villous brush border of their neighbors endocytose bacteria and then release them into a pocket containing macrophages and lymphocytic components (T and B cells) of the adaptive immune system. The enteric pathogen *Shigella* exploits this receptiveness of the M cell to attack the adjacent enterocytes from the side.

IMMUNORESPONSIVE CELLS AND ORGANS

Not all the cells shown in Figure 2–1 are involved in the immune system; of those that are, not all respond to infection. What the immunoresponsive cells have in common is derivation from hematopoietic stem cells in the bone marrow, which create the myeloid and lymphoid series followed by further differentiation into their mature cell types. Of the types shown, the erythroblast and megakaryocte do not participate in immune reactions. In the myeloid series, basophils and mast cells are primarily involved in allergic reactions rather than infection. The immunoresponsive cells are found throughout the body in the circulation or at fixed locations in tissues. They are concentrated in the lymph nodes and spleen and form a unified filtration network designed as a sentinel warning system. In the lymphoid series, cells destined to become T cells mature in the thymus (the source of their name). Thus, the thymus, spleen, and lymph nodes might be thought of as the organs of the immune system. These are collectively referred to as the lymphoid tissues.

Stem cells differentiate to myeloid and lymphoid series

Thymus, spleen, and lymph nodes are immune organs

Cells Responding to Infection

Monocytes

Monocyte is a general morphologic term for cells that include or quickly (hours) differentiate into macrophages or dendritic cells. These are the cells of the immune system that both phagocytose invaders and process them for presentation to the adaptive immune system. **Macrophages** are found in the circulation and tissues, where they are sometimes given regional names such as alveolar macrophage. They possess surface receptors such as mannose and fructose, which nonspecifically recognize components commonly found on pathogens and more specialized receptors able to recognize unique components of microbes such as the lipopolysaccharide (LPS) of Gram-negative bacteria. They also have receptors that recognize antibody and complement.

Macrophages in circulation or tissues

Surface receptors recognize pathogens

Dendritic cells have a distinctive star-like morphology and are present in the skin and in the mucous membranes of the respiratory and intestinal tracts. Like macrophages, they phagocytose and present foreign antigens. Surface recognition includes a process called **pathogen-associated molecular patterns (PAMPs)** in which selective molecular patterns unique to pathogens are recognized and bound. After binding and phagocytosis, dendritic cells migrate to lymphoid tissues where specific immune responses are triggered.

Star-like tissue phagocytes

Migrate to lymphoid tissues

Granulocytes

Of the cells in the granulocyte series, the most active is the **polymorphonuclear neutrophil** or **PMN.** These cells have a distinctive multilobed nucleus and cytoplasmic granules that contain lytic enzymes and antimicrobial substances including peroxidase, lysozyme, defensins, collagenase, and cathelicidins. PMNs have surface receptors for antibody and complement and are active phagocytes. In addition to the digestive enzymes, PMNs have other oxygen-dependent and oxygen-independent pathways for killing microorganisms. Unlike macrophages, they only circulate and are not present in tissues except by migration as part of an acute inflammatory response.

PMNs have digestive and killing pathways

In circulation unless they migrate in inflammation

Eosinophils are nonphagocytic cells that participate in allergic reactions along with **basophils** and **mast cells**. Eosinophils are also involved in the defense against infectious parasites by releasing peptides and oxygen intermediates into the extracellular fluid. It is felt that these products damage membranes of the parasite.

Eosinophils damage parasites

Lymphocytes

Lymphocytes are the primary effector cells of the adaptive immune system. They are produced from a lymphocyte stem cell in the bone marrow and leave in a static state marked to become T cells, B cells, or null cells after further differentiation (**Figure 2–3**). This requires activation mediated by surface binding, which then stimulates further replication and differentiation.

T, B, and null cells initially static

B cells mature in the bone marrow and then circulate in the blood to lymphoid organs. At these sites they may become activated to a form called a plasma cell, which produces antibodies. **T cells** mature in the thymus and then circulate awaiting activation. Their activation results in production of cytokines, which are effector molecules for multiple immunocytes and somatic cells. Some of the uncommitted null cells become **natural killer (NK) cells,** which have the capacity to directly kill cells infected with viruses.

B cells make antibody

T cells secrete cytokines

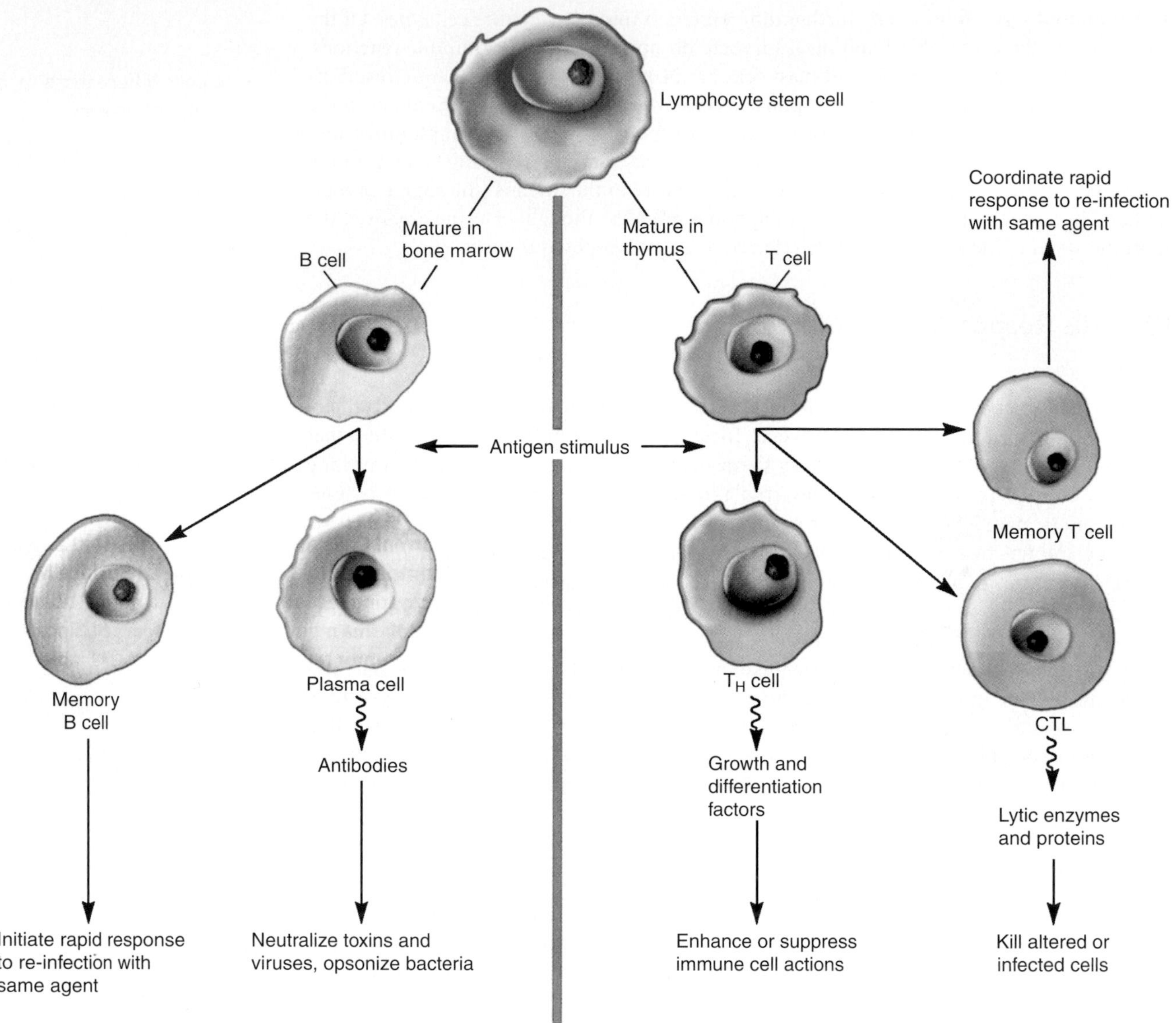

FIGURE 2–3. B and T lymphocytes. B cells and T cells arise from the same cell lineage but diverge into two functional types. Immature B cells and T cells are indistinguishable by morphology. (Reproduced with permission from Willey J, Sherwood L, Woolverton C (eds). *Prescott's Principles of Microbiology.* New York: McGraw-Hill; 2008.)

Phagocytosis

Opsonization not required

Carbohydrate and peptide sequence recognized

Phagocytosis is one of the most important defenses against microbial invaders (**Figure 2–4**). The major cells involved are PMNs, macrophages, and dendritic cells. For all, the process begins with surface pathogen recognition mechanisms, which may be either dependent on opsonization of the organism with complement or antibody or independent of opsonization. At this point only the opsonin-independent mechanisms are considered. These use the non-specific mechanisms already described and hydrophobic interactions between bacteria and the phagocyte surface. More powerful mechanisms include **lectins,** which bind carbohydrate moieties and protein-protein interactions based on a specific peptide sequence (arginine-glycine-arginine or RGD). These **RGD receptors** are present on virtually all phagocytes.

Another mechanism is use of the PAMPs already mentioned. Phagocytes have evolved a distinct class called **Toll-like receptors (TLRs),** of which at least 10 sets are known. These include sets that recognizes a molecular pattern in bacterial peptidoglycan (TLR-2) and

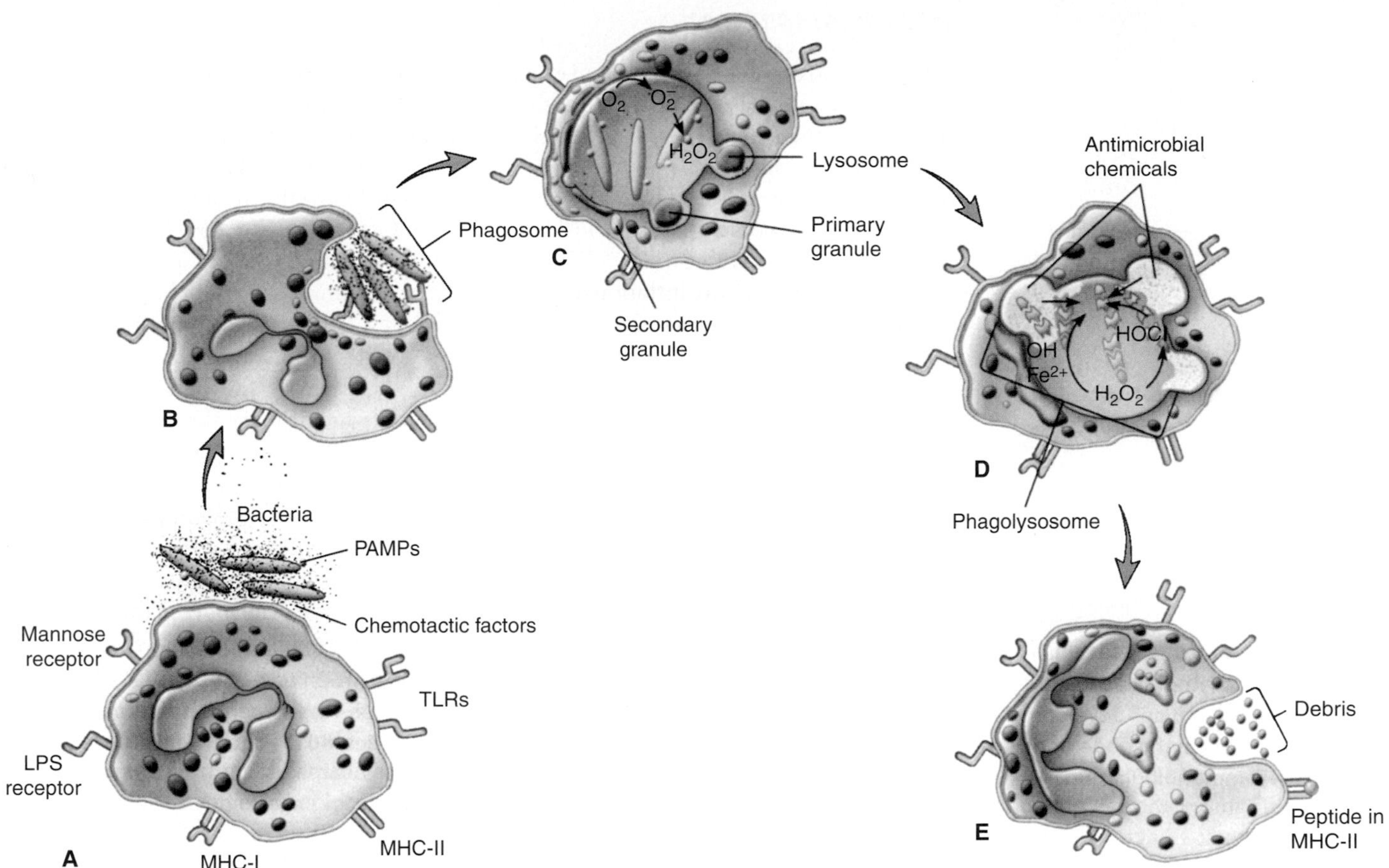

FIGURE 2–4. Phagocytosis. A. Drawing shows receptors on a phagocytic cell, such as a macrophage, and the corresponding PAMPs participating in phagocytosis. The schematic depicts the process of phagocytosis showing ingestion **B.**, participation of primary and secondary granules and O_2-dependent killing events **C.**, intracellular digestion **D.**, and endocytosis **E.** LPS receptor, lipopolysaccharide receptor; TLRs, toll-like receptors; MHCI, class I major histocompatibility protein; MHCII, class II major histocompatibility protein; PAMPs, pathogen-associated molecular patterns. (Reproduced with permission from Willey J, Sherwood L, Woolverton C (eds). *Prescott's Principles of Microbiology.* New York: McGraw-Hill; 2008.)

LPS (TLR-4). TLRs not only bind but trigger signaling pathways leading to induction of cytokines and other directors of the specific immune response.

TLRs bind LPS, peptidoglycan and induce cytokines

Bound organisms are taken inside the phagocyte in a membrane-bound phagosome destined to fuse with lysosomes inside to form a **phagolysosome.** This is the main killing ground of the phagocyte. The lysosomal enzymes include hydrolases and proteases that have maximum activity at the acidic pH inside the phagolysosome. Also inside the phagocyte are oxidative killing mechanisms created by enzymes that produce **reactive oxygen intermediates** (superoxide, hydrogen peroxide, singlet oxygen) driven by a metabolic respiratory burst in the cell cytoplasm. These mechanisms are particularly used for killing bacteria. Bacterial pathogens whose pathogenesis involves multiplication rather than destruction inside the phagocyte have mechanisms to block one or more of the preceding steps. For example, some pathogens are able to block fusion of the phagosome with the lysosome; others interfere with the acidification of the phagolysosome.

Enzymes digest in acidic phagolysosome

Reactive oxygen driven by respiratory burst

Another mechanism effective with some viruses, fungi, and parasites is the formation of **reactive nitrogen intermediates** (nitric oxide, nitrate, and nitrite) delivered into a vacuole or in the cytoplasm. PMN granules contain a variety of other antimicrobial substances, including peptides called **defensins.** Defensins act by permeabilizing membranes and in addition to bacteria are active against enveloped viruses.

Reactive nitrogen effects enveloped viruses

INFLAMMATION

Inflammation encompasses a series of events in which the above cells are deployed in response to an injury—such as a new microbial invader. At the first insult, chemical signals mobilize cells, fluids, and other mediators to the site to contain, combat, and heal. In acute inflammation, the first events may be noticed in minutes and the entire process resolved over a matter of days to a couple of weeks. Chronic inflammation may follow the incomplete resolution of an acute process or arise as a slow insidious process of its own. The natural history of infections such as tuberculosis, which follow this pattern, run for months, years, even decades.

Acute = hours to days

Chronic = weeks to months

The first event in **acute inflammation** is the release of chemical signals (chemokines) that act on adhesion molecules (selectins) in local capillaries. This slows the movement of passing PMNs and activates adhesive integrins on their surface. This leads to tight adhesion to the endothelium followed by squeezing past the endothelial wall to the tissues below. There, chemotactic factors released by the bacteria lead them to the primary site. Increasing acidity of local fluids releases enzymes (kallikrein, bradykinin) that open junctions in capillary walls and allow increased flow of fluids and more leukocytes. Histamine (from mast cells) arachidonic acid, and prostaglandin release complete the picture of swelling and pain.

PMNs migrate from capillaries

Enzymes and chemical mediators facilitate swelling

Chronic inflammation bridges the innate and adaptive immune responses. An acute phase, if present, is usually not noticed, and the cellular infiltrate is composed of lymphocytes and macrophages with relatively few PMNs. It is generally associated with slower-growing pathogens such as mycobacteria, fungi, and parasites in which cell-mediated immunity (T_H1) is the primary adaptive defense. Many of these pathogens have mechanisms that allow them to multiply in nonactivated macrophages. If the macrophages are effectively activated by T cells, the multiplication ceases and the inflammation and injury are minimal. If not, multiplication and chronic inflammation continue sometimes in the form of a **granuloma,** which is an indication of a destructive hypersensitivity component to the inflammation.

Lymphocytes and macrophages predominate

Granulomas indicate failure to resolve by adaptive cellular mechanisms

CHEMICAL MEDIATORS

Chemical mediators of innate immunity that have direct antimicrobial activity include cationic proteins and complement. The cationic proteins (cathelicidin, defensins) act on bacterial plasma membranes by the formation of ionic pores, which alter membrane permeability. The complement system is a series of glycoproteins, which can directly insert in bacterial membranes or act as receptors for antibody. Cytokines are proteins or glycoproteins released by one cell population that act as signaling molecules for another. They are generally thought of in the context of the adaptive immune system, but they can be stimulated directly by microorganisms.

Peptides alter membrane permeability

The Complement System

The complement system consists of over 30 distinct components and several other precursors. All are in the plasma of healthy individuals in inactive forms that must be enzymatically cleaved to become active. When this happens, a cascade of reactions is triggered, which activates the various components in a fixed sequence (**Figure 2–5**). The difference between the pathways is in the mechanisms for their initiation. Once started, any pathway can produce the same effects on pathogens, which include enhancing phagocytosis, activation of leukocytes, and lysis of bacterial cell walls. An important step in the process is coating of the organism with serum components, a process called **opsonization**. The coatings may be mannose-binding proteins, complement components, or antibody. There is no immunologic specificity in complement activation or in its effects.

Multiple components activated in cascade fashion when triggered

Pathways differ in initiation mechanism

Opsonization is serum coating of pathogens

Alternative Pathway

The alternative pathway is activated by bacterial cell wall components with repetitive surface structures such as LPS. The multiple components come together in the formation of the **membrane attack complex,** which inserts directly into bacterial membranes (**Figure 2–6**), particularly the outer membrane of Gram-negative bacteria. This not only injures the organism but enhances phagocytosis because the other end of the molecule has receptors for

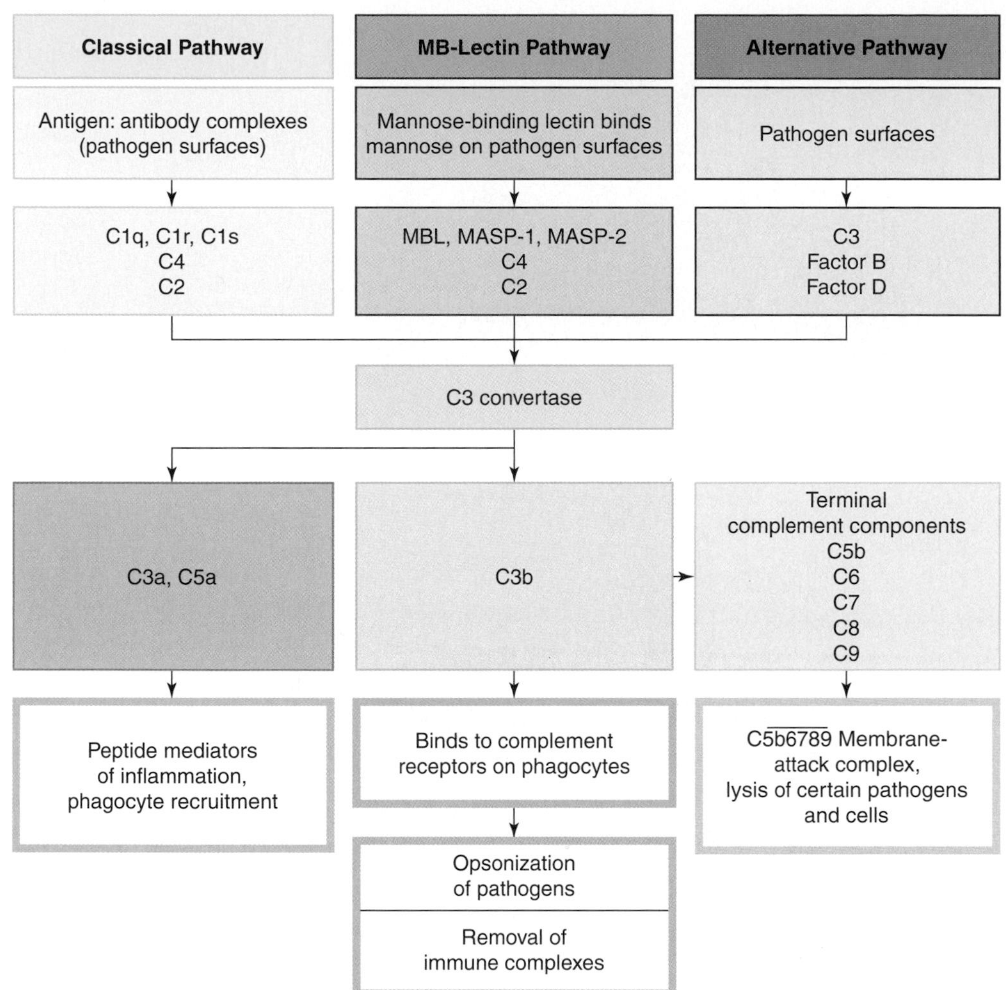

FIGURE 2–5. Components and action of complement. Complement activation involves a series of enzymatic reactions that culminate in the formation of C3 convertase, which cleaves complement component C3 into C3b and C3a. The production of the C3 convertase is where the three pathways converge. C3a is a peptide mediator of local inflammation. C3b binds covalently to the bacterial cell membrane and opsonizes the bacteria, enabling phagocytes to internalize them. C5a and C5b are generated by the cleavage of C5 by a C5 convertase. C5a is also a powerful peptide mediator of inflammation. C5b promotes the terminal components complement to assemble into a membrane attack complex. (Reproduced with permission from Willey J, Sherwood L, Woolverton C (eds). *Prescott's Principles of Microbiology.* New York: McGraw-Hill; 2008.)

phagocytes. Gram-positive bacteria are less affected because they have no exposed membrane (see Chapter 21). These actions are particularly important for the effectiveness of innate immunity in the early stages of acute infections before the adaptive immune system has time to act. The key complement component for alternate pathway activity is C3b. C3b activation and degradation are regulated by a number of serum factors (factors B, D, and H), which can modulate its activity. A major mechanism for pathogens to block alternate pathway attack is by binding factor H to their surface. This is accomplished by bacterial capsules and surface proteins. This concentration of factor H causes local degradation of C3b (see Chapter 22, Figure 22–4).

Activation is by pathogen surfaces

Membrane attack complex inserts and provides phagocyte receptors

Factor H binding accelerates C3b degradation on capsules

Lectin Pathway

Another means of activating the complement system is based on the carbohydrate building of lectins. In this case, the lectins bind to mannose, a common surface component of bacteria, fungi, and some virus envelopes. This binding opsonizes the pathogen and enhances

Lectins bind mannose on pathogens

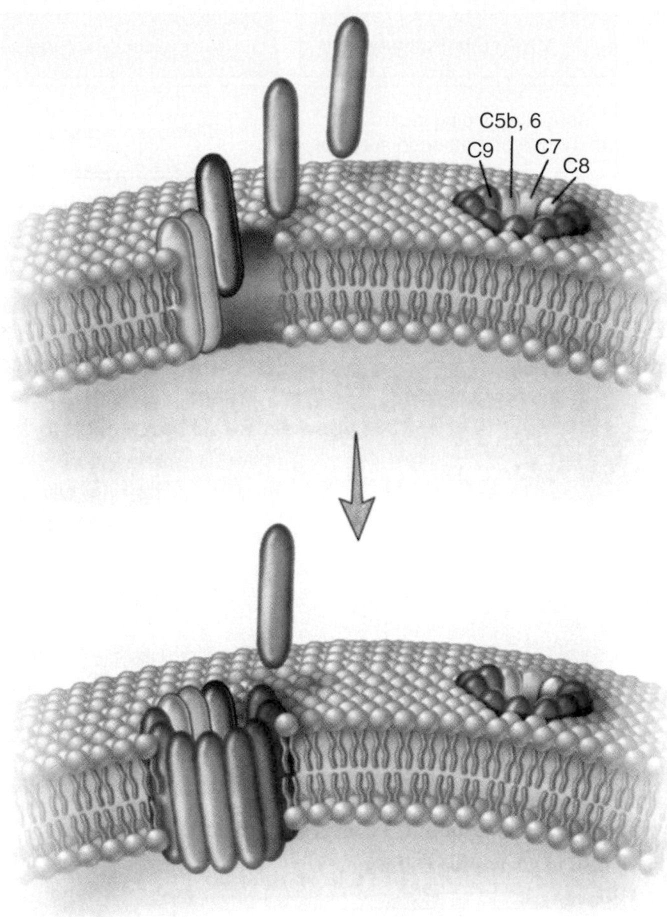

FIGURE 2–6. Complement membrane attack complex. The membrane attack complex (MAC) is a tubular structure that forms a transmembrane pore in the target cell's plasma membrane. The subunit architecture of the MAC shows that the transmembrane channel is formed by multiple polymerized molecules. (Reproduced with permission from Willey J, Sherwood L, Woolverton C (eds). *Prescott's Principles of Microbiology.* New York: McGraw-Hill; 2008.)

phagocytosis. Thus, as in the alternative pathway the activation comes from pathogen surfaces and proceeds through the same C3 convertase (Figure 2–5).

Classic Pathway

Antigen—-antibody reaction exposes complement binding sites

C3b has receptors for phagocytes

The classic complement pathway is initiated by the binding of antibodies formed during the adaptive immune response (see following text) with their specific antigens on the surface of a pathogen. This binding is highly specific but amounts to another case of opsonization activating the complement cascade. In this case, specific sites on the Fc portion of immunoglobulin molecules bind and activate the C1 component of complement to start the process. The pathway and sequence of individual complements are characteristic of the classic pathway, but it still reaches C3b, the common point for microbial directed action. As with the alternative pathway, this creates the membrane attack complex, the mediators of inflammation, and receptors for phagocytes on C3b.

Cytokines

ILs, IFNs, TNF, chemokines are all cytokines

Cytokine is a broad term referring to molecules released from one cell population destined to have an effect on another cell population (**Table 2–2**). As these proteins and glycoproteins have been discovered, they have been named and classified in relation to biologic effects observed initially only to discover that they have multiple other actions. For infectious diseases, the operative subcategories are **chemokines,** which are cytokines chemotactic for inflammatory cell migration, and **interleukins (IL-1, 2, 3,** and so on), which regulate growth and differentiation between monocytes and lymphocytes. **Tumor necrosis factor (TNF)**, so named for its cytotoxic effect on tumor cells also can induce apoptosis (programmed cell death) in phagocytes, a useful feature for pathogens they have taken in. **Interferons (INF-α,β,γ)** were originally named for their interference with viral replication (**Figure 2–7**), but are now known

TABLE 2–2 Some Cytokines Acting in Infection

	CELL SOURCE	FUNCTIONS
Interleukins (IL)		
IL-1	Macrophages, endothelium, fibroblasts, epithelial	Differentiation and function of inflammatory and immune effectors
IL-2	T cells (T_H1)	T-cell proliferation, cytolytic activity of natural killer (NK) cells
IL-4	T cells (T_H2), macrophages, B cells	Differentiation of naïve T cells to helper T cells, proliferation of B cells
IL-8	Macrophages, endothelial, T cells, keratinocytes, polymorphonuclear neutrophils (PMNs)	Chemoattractant for PMNs and T cells, PMN degranulation, migration of PMNs
IL-10	T cells (T_H2), B cells, macrophages, keratinocytes	Reduces production of IFN-γ, IL-1, TNF-α, with IL-2 proliferation of CD8+ cytotoxic T cells
Interferons (IFN)		
IFN-α/β	T cells, B cells, macrophages, fibroblasts	Antiviral activity, stimulates macrophages, MHC (major compatibility complex) class I expression
IFN-γ	T cells (T_H1, CD8+), NK cells	Activation of T cells, macrophages, PMNs, NK cells, antiviral, MHC class I and II expression
Tumor Necrosis Factor (TNF)		
TNF-α	T cells, macrophages, NK cells	Expression of multiple cytokines, (growth and transcription factors), stimulates inflammatory response, cytotoxic for tumor cells
TNF-β	T cells, B cells	Same as TNF-α

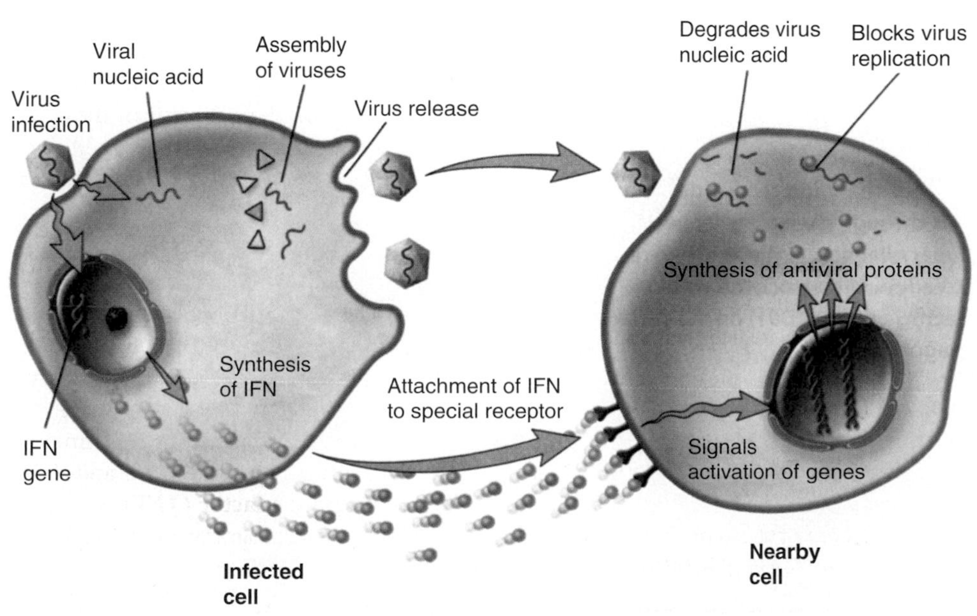

FIGURE 2–7. Antiviral action of interferon. Interferon (IFN) synthesis and release are often induced by a virus infection. IFN binds to a ganglioside receptor on the plasma membrane of a second cell and triggers the production of enzymes that render the cell resistant to virus infection. The two most important such enzymes are oligo (A) synthetase and a special protein kinase. When an IFN-stimulated cell is infected, viral protein synthesis is inhibited by an active endoribonuclease that degrades viral RNA. An active protein kinase phosphorylates and inactivates the initiation factor elf-2 required for viral protein. (Reproduced with permission from Willey J, Sherwood L, Woolverton C (eds). *Prescott's Principles of Microbiology.* New York: McGraw-Hill; 2008.)

to be central to activation of T cells and macrophages. Unless commanded to understand specific situations, cytokine is used to represent all these mediators in these pages.

THE ADAPTIVE (SPECIFIC) IMMUNE SYSTEM

The adaptive immune system differs from the innate immune response in its discrimination between self and nonself and in the magnitude and diversity of highly specific immune responses possible (**Table 2–3**). It also has a **memory** function, which is able to mount an accelerated response if an invader returns. The adaptive system operates in two broad arms—**humoral immunity** and **cell-mediated immunity**. Humoral immunity comes from bone marrow–derived **B cells** and acts through the ability of the antibodies it produces to bind foreign molecules called antigens. Cell-mediated (cellular) immunity is mediated through **T cells** that mature in the thymus and respond to antigens by directly attacking infected cells or by secreting cytokines to activate other cells. As shown in **Figure 2–8**, the B-cell and T-cell systems are interactive.

TABLE 2–3 Cells Involved in the Adaptive Immune System

CELL	FUNCTION	SPECIFIC RECEPTORS FOR ANTIGEN	CHARACTERISTIC CELL SURFACE MARKER	SPECIAL CHARACTERISTICS
B cells	Production of antibody	Surface immunoglobulin (Ig M_m)	Fc and complement C3d receptors; MHC (major histocompatibility complex) class II	Differentiate into plasma cells (major antibody producers)
	Present antigen to T cells			
Helper T cells (T_H)	Stimulate B cells by providing specific and nonspecific (cytokine) signals for activation and differentiation	α/β T-cell receptor (TCR)	CD4+	Presented by MHC class II
	Activate macrophages by cytokines			Can be classified into two types: T_H1 activates macrophages, makes interferon γ, T_H2 activates B cells, makes IL-4
Cytotoxic T cell	Lyse antigen-expressing cells such as virally infected cells or allografts	α/β TCR	CD8+	Presented by MHC class I
Natural killer (NK) cells	Spontaneous lysis of tumor and infected cells	Inhibitory; activating	Fc receptor for IgG	Recognize MHC class I
NK T cells	Amplify both cell-mediated and humoral immunity	α/β TCR	CD4+	Express a restricted subset of Vα
Macrophages (monocytes)	Phagocytosis, secretion of cytokines to activate T cells (eg, IL-1) or other accessory cells such as polymorphonuclear neutrophils (PMNs)[c]	None but can be "armed" by antibodies binding to Fc receptors	Macrophage surface antigens	Express surface receptors for the activated third component of complement (C3), kill ingested bacteria by oxidative bursts
Polymorphonuclear leukocytes (neutrophils, eosinophils)	Phagocytosis killing	None but can be "armed" by antibodies		Protective in parasitic infections, but adverse side effects such as granuloma formation can occur

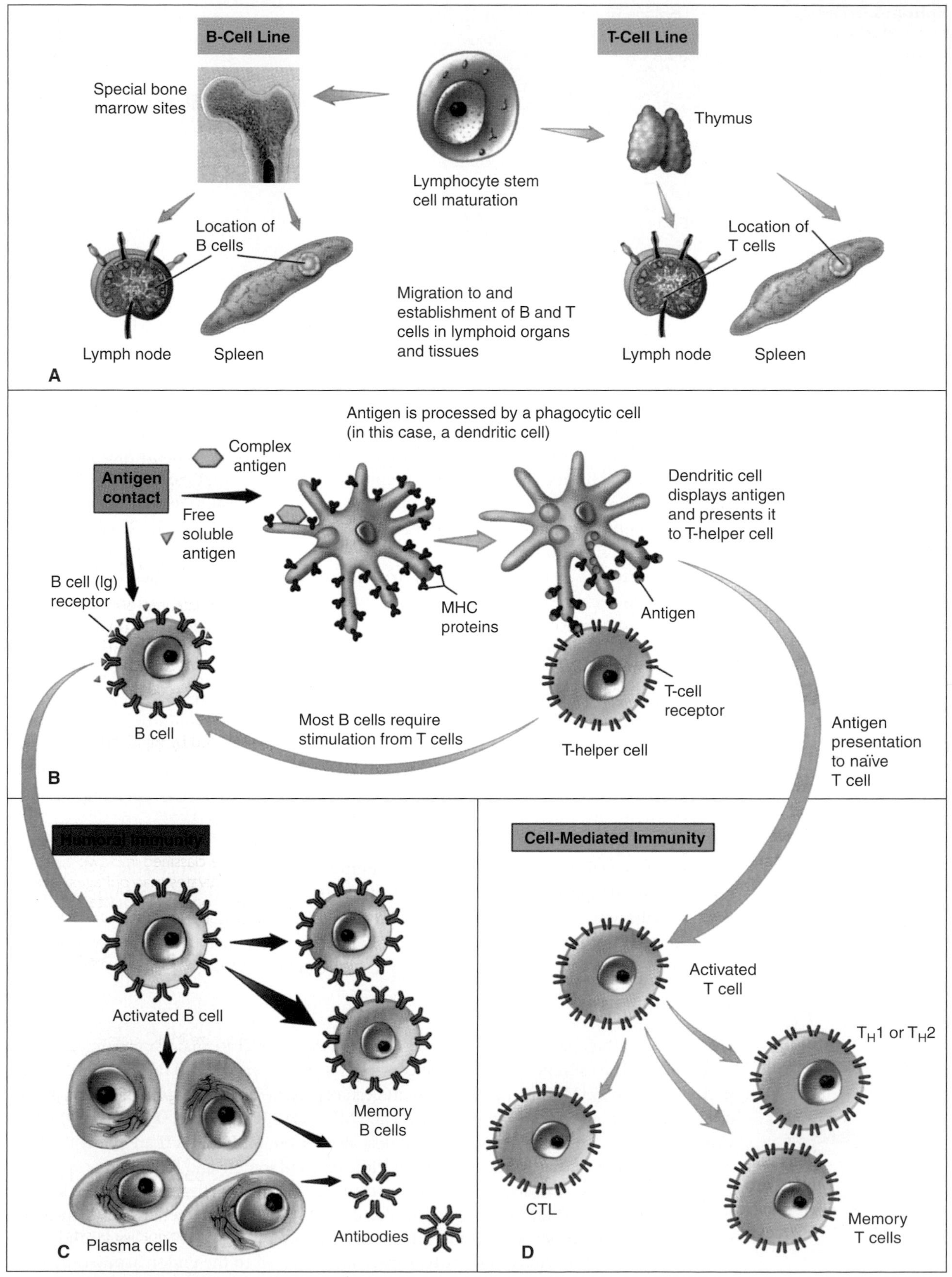

FIGURE 2–8. Acquired immune system development. A. Lymphocyte stem cells develop into B- and T-cell precursors that migrate to the bone marrow or thymus, respectively. Mature B and T cells seed secondary lymphoid tissues. **B.** Lymphocyte receptor binding of antigen activates B and T cells to become effector cells. **C.** B lymphocytes develop into memory cells and antibody-secreting plasma cells. **D.** T cells develop into memory cells, helper T cells, and cytotoxic T cells. (Reproduced with permission from Willey J, Sherwood L, Woolverton C (eds). *Prescott's Principles of Microbiology.* New York: McGraw-Hill; 2008.)

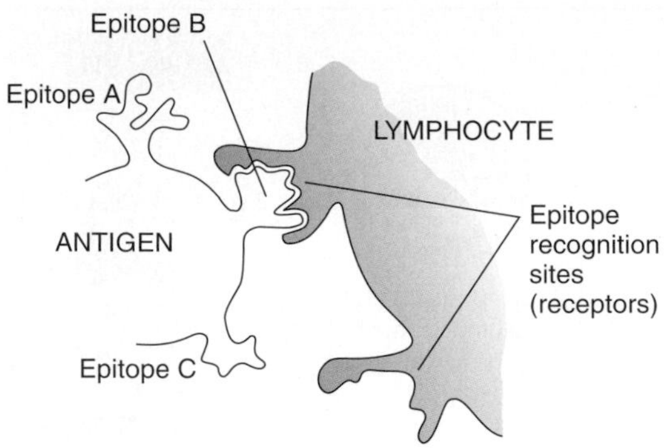

FIGURE 2–9. Epitopes. Schematic of epitope recognition by an immunoresponsive lymphocyte. Epitope B on the antigen binds to a complementary recognition site on the surface of the immunoresponsive cell. Antigens may have many different epitopes, but an immunoresponsive lymphocyte has receptors of only one specificity. In most cases, epitopes are recognized on the surface of macrophages that have processed the antigen. The receptor for antigens on B cells is the combining site of the surface immunoglobulin.

Antigens and Epitopes

Antigens stimulate immune response

Epitopes fit to the combining site of T-cell receptors and antibodies

An antigen is any substance (usually foreign) with the ability to stimulate an immune response when presented in an effective fashion. They are usually large structurally complex proteins, polysaccharides, or glycolipids. Each antigen can contain many subregions that are the actual antigenic determinants, or epitopes. These epitopes can consist of separate peptides, carbohydrates, or lipids of the correct size and three-dimensional configuration to fit the combining site of an antibody molecule or a T-cell receptor (TCR) (**Figure 2–9**). Approximately six amino acids or monosaccharide units provide a correctly sized epitope. Antigens presented by infectious agents typically contain multiple epitopes, including copies of the same epitope. Other small organic molecules that would not ordinarily stimulate an immune response may do so if bound to a larger carrier like a protein. These are called **haptens,** and the specificity of the immune response may be generated for both the hapten and its larger carrier.

B cells multiply and produce antibody

A foreign antigen entering a human host may by chance encounter a B cell whose surface antibody is able to bind it. This interaction stimulates the B cell to multiply, differentiate, and produce more surface and soluble antibody of the same specificity. Eventually, the process leads to production of enough antibody to bind more of the antigen. This mechanism is most likely to operate with antigens such as polysaccharides that have repeating subunits, thus improving the possibility that exposed epitopes are recognized.

Protein antigens must be processed first

Large, complex antigens such as proteins and viruses must be processed before their epitopes can be effectively recognized by the immune system. This processing takes place in macrophages or specialized epithelial cells found in the skin and lymphoid organs, where they are adjacent to other immunoresponsive cells. The ingested antigen is degraded to peptides of 10 to 20 amino acids that are presented by major histocompatibility molecules on the host cell surface to be recognized by T cells (Figure 2-11).

Recognition of Foreignness

MHC gene complex codes surface molecules

MHC II on macrophages, dendritic cells

Distinguishing between self and non self is obviously essential to maintaining integrity and homeostasis. The collection of genes that control these functions is called the **major histocompatibility complex (MHC),** and it codes for molecules present on the surface of almost all human cells. Of interest in infection are MHC class I and MHC class II molecules (**Figure 2–10**). **MHC class I** molecules are in the membrane of almost all cells, but **MHC class II** are present only on certain leukocytes such as macrophages, dendritic cells, and some T and B cells.

MHC I presents cytoplasmic peptides to CD8+

MHC II presents foreign peptides to CD4+

Both MHC class I and class II participate in antigen processing but by distinctly different pathways (**Figure 2–11**). MHC class I molecules bind to products generated in the cytoplasm by a natural process or a viral infection. Viral proteins are digested to peptides in a cytoplasmic structure called the **proteasome** and delivered to the endoplasmic reticulum. Here they find the binding site of the class I molecule and are transported to the surface for presentation of the peptide. MHC class II molecules bind to fragments that originally come from outside the cell, but have been taken into the endocytotic vacuole of a phagocyte. After digestion in the phagolysosome, peptide fragments are combined with class II molecules

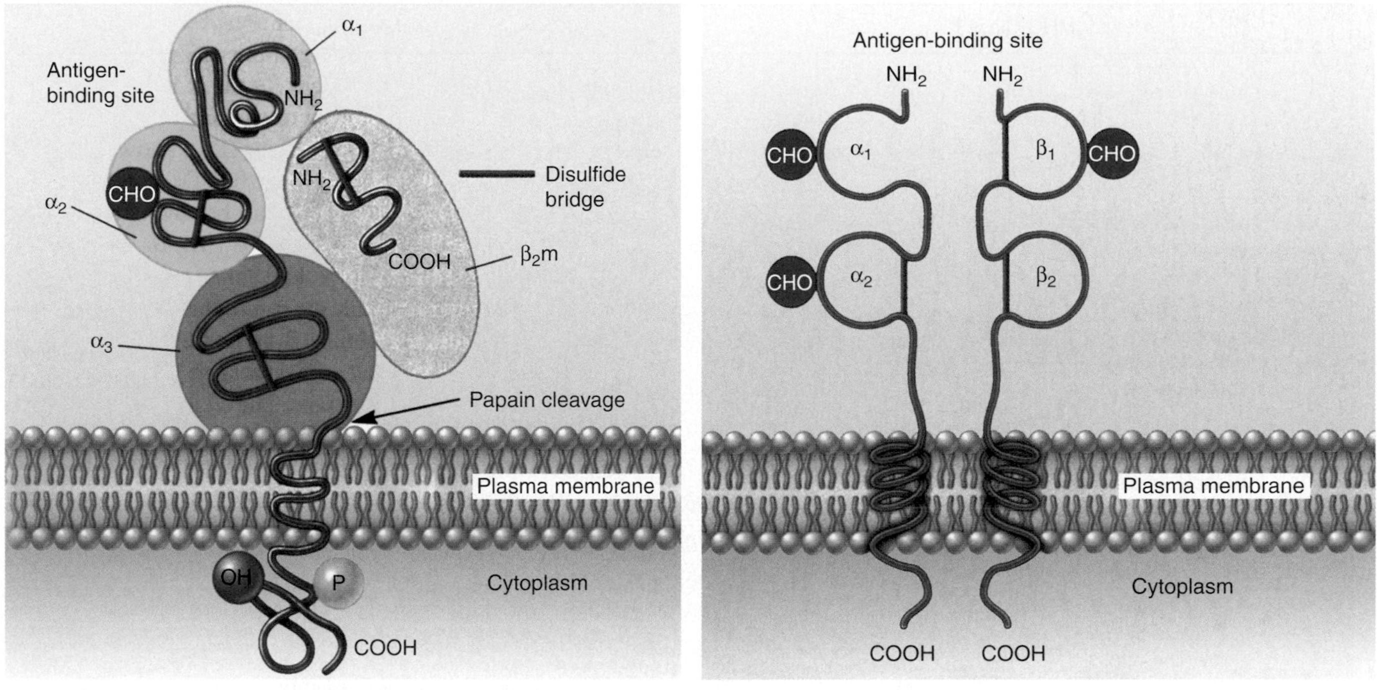

FIGURE 2–10. MHC class I and class II molecules. A. The class I molecule is a heterodimer composed of the alpha protein, which is divided into three domains: α_1, α_2, and α_3, and the protein β_2 microglobulin. **B.** The class II molecule is a heterodimer composed of two distinct proteins called alpha and beta. Each is divided into two domains α_1, α_2, and β_1, β_2, respectively. (Reproduced with permission from Willey J, Sherwood L, Woolverton C (eds). *Prescott's Principles of Microbiology.* New York: McGraw-Hill; 2008.)

and move to the surface for presentation. The presented MHC class I peptides are recognized by CD8+ T cells and the MHC class II by CD4+ T cells.

THE T-CELL RESPONSE

T cells originate in the bone marrow and migrate to the thymus for differentiation. Those that recognize self are destroyed. Those that survive are mature but still to be activated. T cells have specific **TCRs** on their surface with binding sites extending to the outside (**Figure 2–12**). The two major types of T cells are helper T (CD4+) and cytotoxic (CD8+) T cells. The major roles of T cells in the immune response are:

1. Recognition of peptide epitopes presented by MHC molecules on cell surfaces. This is followed by activation and clonal expansion of T cells in the case of epitopes associated with class II MHC molecules.
2. Production of cytokines that act as intercellular signals and mediate the activation and modulation of various aspects of the immune response and of nonspecific host defenses.
3. Direct killing of foreign cells, of host cells bearing foreign surface antigens along with class I MHC molecules (eg, some virally infected cells), and of some immunologically recognized tumor cells.

CD4+ Helper T Lymphocytes

Helper T cells (T_H cells) are stimulated by antigen in the context of MHC class II presentation and are further marked by the presence of the CD4 cell surface antigen. If T cells are of the proper MHC background to recognize the antigen specifically, T-cell activation occurs.

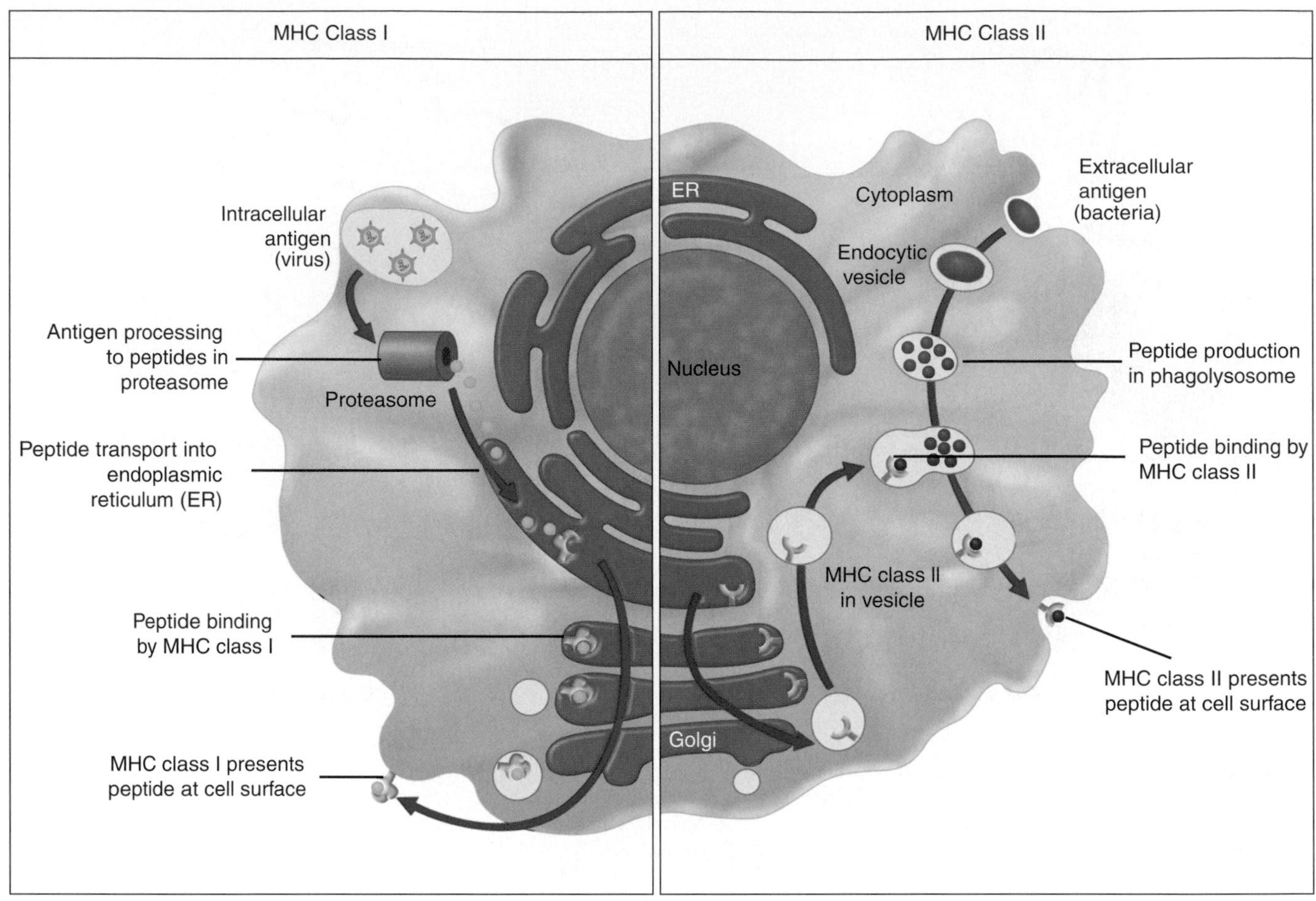

FIGURE 2–11. Antigen processing and presentation. A. Antigens originating in the cytoplasm are digested by the proteasome to peptides. The peptides are bound to the MHC class I molecules in the endoplasmic reticulum (ER) and transported to the surface for presentation. **B.** Antigens originating outside the cell are endocytosed and digested in the phagolysosome. The digested peptides are bound to MHC class II molecules in the ER and transported to the surface for presentation. MHC, major histocompatibility complex.

T_H cells are stimulated by MHC II-presented antigen

T_H1 to cell-mediated reactions

T_H2 to antibody production

The antigen–MHC complex presented to a specific T cell by the macrophage is the specific signal that induces the T cell to become activated and divide. At this point, the helper T cells follow either the **T_H1 pathway** toward cell-mediated immunity or the **T_H2 pathway** toward antibody production and humoral immunity. Before this differentiation, the helper cells are sometimes referred to as T_H0. The T_H1 and T_H2 responses are characterized by their own set of cytokines and biologic actions. In both pathways, this clonal expansion includes **memory cells** along with the T cells committed to effector functions. These pathways are illustrated in **Figure 2–13**.

CD8+ Cytotoxic T Lymphocytes

CD8+ lymphocytes react with MHC I

Eliminate virally infected cells

CD8+ cytotoxic T lymphocytes (CTLs) are a second class of effector T cells. They are lethal to cells expressing the epitope against which they are directed when the epitope is presented by class I MHC molecules. They too have specific epitope recognition sites, but they are characterized by the CD8 cell surface marker; thus, they are referred to as CD8+ cytotoxic T cells. These cells recognize the association of antigenic epitopes with class I MHC molecules on a wide variety of cells of the body. In the case of virally infected cells, cytotoxic CD8+ cells prevent viral production and release by eliminating the host cell before viral synthesis or assembly is complete (**Figure 2–14**). The destruction of the virally infected cell is accomplished through a complement-like action mediated by perforins, which also facilitates entry into the cell of enzymes (granzymes) that activate apoptosis.

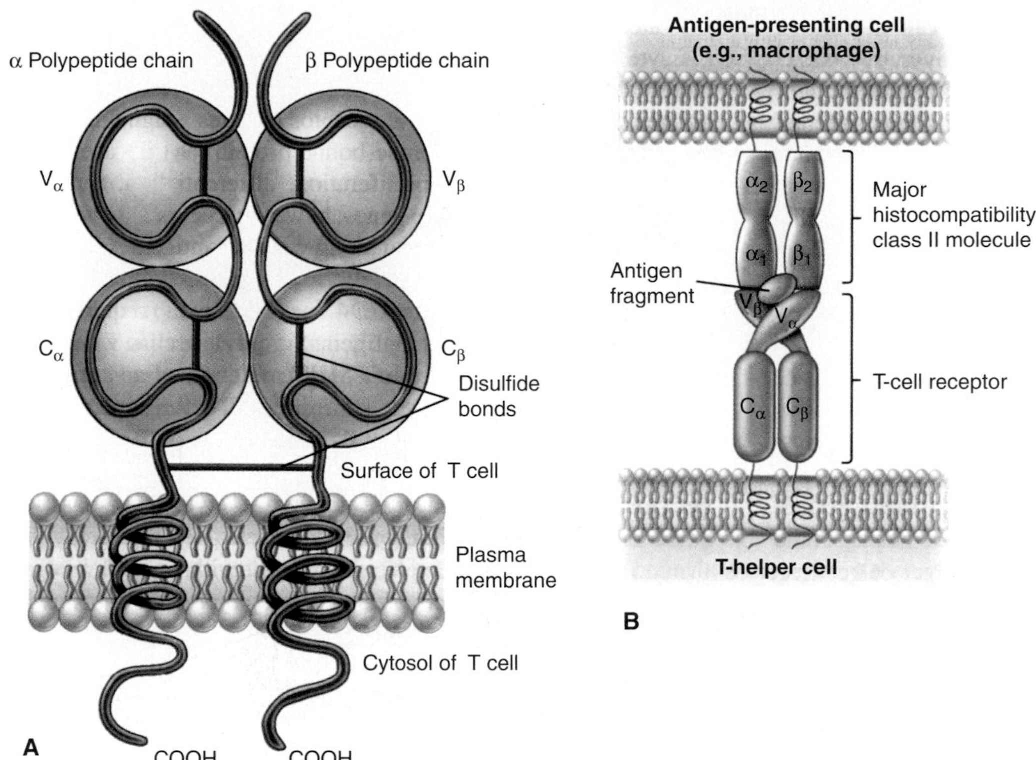

FIGURE 2–12. T-cell receptor and helper T activation. A. Structure of the T-cell antigen receptor. **B.** An antigen-presenting cell begins the activation process by displaying a peptide antigen fragment in its MHC class II molecule. A helper T cell is activated after the variable region of its receptor (Vα, Vβ) reacts with the fragment. (Reproduced with permission from Willey J, Sherwood L, Woolverton C (eds). *Prescott's Principles of Microbiology.* New York: McGraw-Hill; 2008.)

Superantigens

A group of antigens have been termed superantigens because they stimulate a much larger number of T cells than would be predicted based on the specificity of combining site diversity. This causes a massive cytokine release. The action of superantigens is based on their ability to bind directly to MHC proteins and to particular Vβ regions of the TCR without involving the antigen combining site. Individual superantigens recognize exposed portions defined by framework residues that are common to the structure of one or more Vβ regions. Any T cells bearing those Vβ sites may be directly stimulated. A variety of microbial products have been identified as superantigens. Superantigens are discussed further in Chapter 22 (see Figure 22–6) and in Chapters 24 and 25 describing their role in **toxic shock syndromes** caused by *Staphylococcus aureus* and group A streptococci.

Superantigens bind directly to MHC proteins and TCR Vβ region

Higher proportion of T cells are stimulated

Cell-Mediated Immunity

In the control of infection, cell-mediated immunity is most important in the response to obligate or facultative intracellular pathogens. These include some slow-growing bacteria, such as the mycobacteria and fungi against which antibody responses appear to be ineffective. The mechanisms are complex and involve a number of cytokines with amplifying feedback mechanisms for their production. After the initial processing of antigen to stimulate activation of the antigen-recognizing CD4+ T cell, cytokine feedback from the CD4+ T cells to macrophages further increases their clonal expansion (including memory cells) and activates CD8+ (cytotoxic) T lymphocytes. Other cytokines from CD4+ T cells attract macrophages to the site of infection, hold them there, and activate them to greatly enhance

Of primary importance with intracellular pathogens

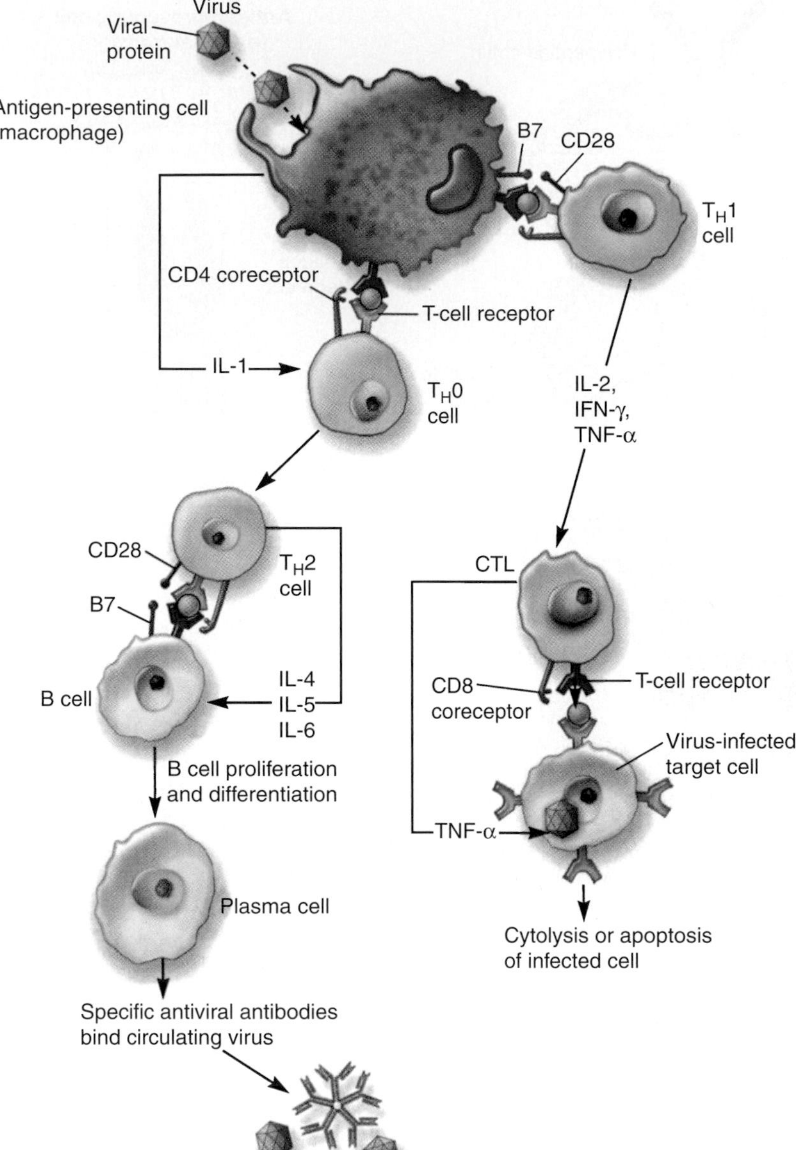

FIGURE 2–13. T-cell responses. A virus is phagocytosed by a macrophage, and a small antigen fragment (peptide) is presented to naïve T_H cells in association with class II MHC molecules. Once activated, the T_H0 cell may differentiate into a T_H2 cell that secretes the cytokines that cause B-cell proliferation and subsequent secretion of specific antiviral antibodies. Alternately, they differentiate into T_H1 cells that secrete cytokines, which regulate the proliferation of cytotoxic T cells (CTLs). Once a CTL proliferates and differentiates into an activated effector cell, it attacks and causes lysis or apoptosis of a virus-infected cell. (Reproduced with permission from Willey J, Sherwood L, Woolverton C (eds). *Prescott's Principles of Microbiology.* New York: McGraw-Hill; 2008.)

Helper and cytotoxic T lymphocytes interact

Macrophages are mobilized and enhanced

microbiocidal activity. The sum of the individual and collaborative activities of T cells, macrophages, and their products is a progressive mobilization of a range of host defenses to the site of infection and greatly enhanced macrophage activity. In the case of tuberculosis, IFN-γ inhibits the replication of the mycobacteria inside macrophages. In viral infections, CD8+ cytotoxic lymphocytes destroy their cellular habitat leaving already assembled virions accessible to circulating antibody.

B CELLS AND ANTIBODY RESPONSES

B cells carry epitope recognition sites on their surface

B lymphocytes are the cells responsible for antibody responses. They develop from precursor cells in the bone marrow before migrating to other lymphoid tissues. Each mature cell of this series carries a specific epitope recognition site on its surface. This B-cell receptor is actually a monomer of one form of antibody (IgM) oriented with its binding sites facing outward. Upon binding antigen, the receptor-antigen complex is internalized for initiation

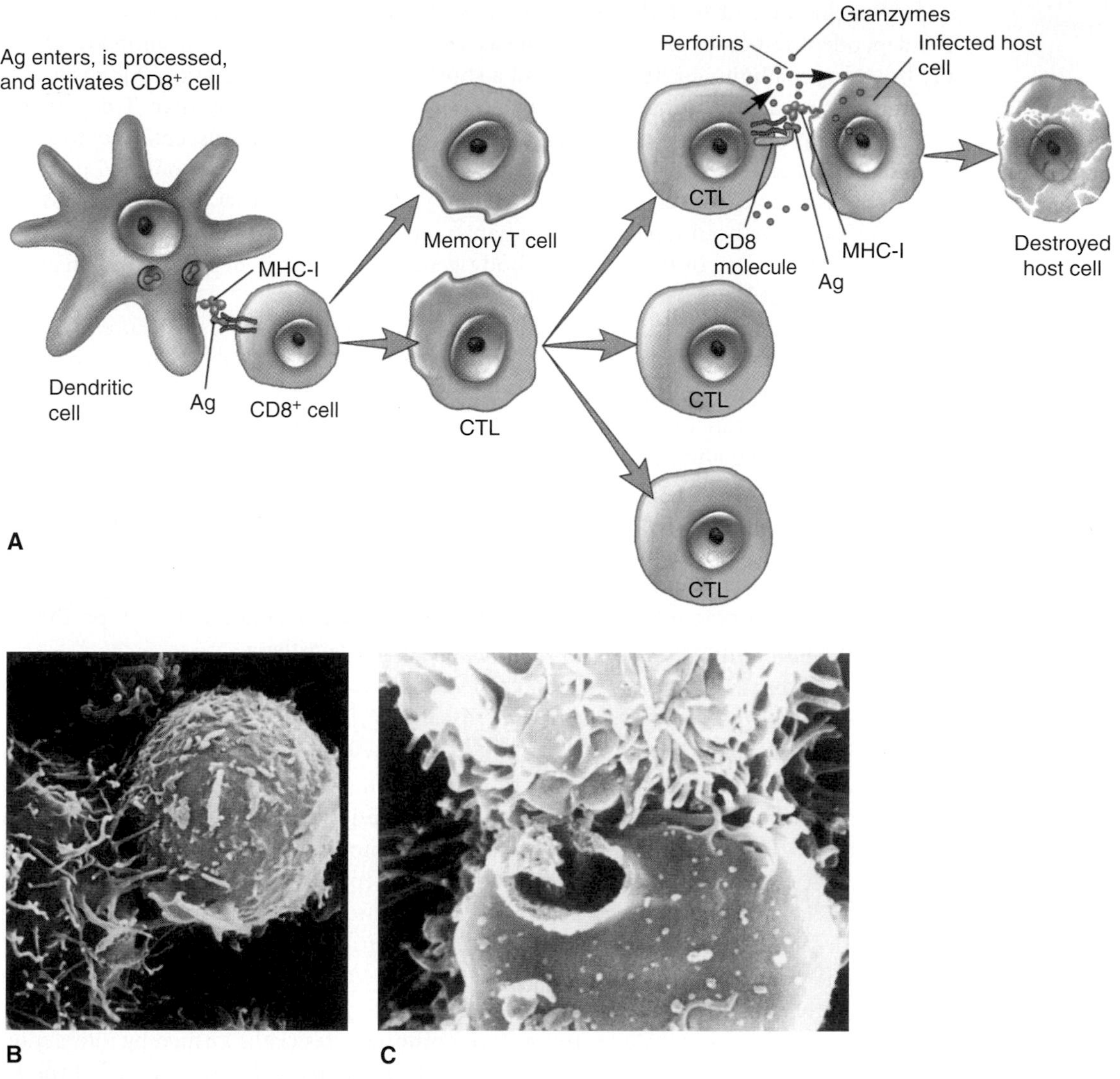

FIGURE 2–14. Cytotoxic T-cell (CTL) destruction virus-infected cells. A. Naïve CD8+ T cells are activated when they are exposed to antigen within a class I MHC molecule on an antigen-presenting cell. Antigen activation leads to development of effector CTL and memory cells. Effector CTLs and their memory cells subsequently react with antigen expressed in class I MHC molecules of any host cell to destroy it. T-cell cytotoxicity often involves the perforin pathway and leads to apoptosis or cytolysis. MHC, major histocompatibility complex. **B.** CTL (left) contacting target cell (right). **C.** Perforins form pore in target cell membrane. (Reproduced with permission from Willey J, Sherwood L, Woolverton C (eds). *Prescott's Principles of Microbiology.* New York: McGraw-Hill; 2008.)

of antibody production by the stimulated B cell. In this process, the B lymphocytes multiply, differentiating into either **memory** or **plasma cells**. Plasma cells are end cells adapted for secretion of large amounts of antibodies. In addition to their essential role in antibody production, B cells can also present antigen to T cells.

Stimulated cells differentiate to form memory, plasma cells

There are two broad types of antigen triggering, T-dependent and T-independent. **T-dependent** reactions are those that are use collaboration between helper T cells and B cells to initiate the process of antibody production in the manner shown in Figure 2–13. This is the mechanism evoked by proteins and haptens bound to proteins. The response is strong and includes memory cells so it can be boosted in the case of immunization.

T-dependent has memory

T-independent responses are those that do not require help by T cells to stimulate B-cell antibody production. It is evoked by large molecules with many repeating units such as polysaccharides. At first glance, this independence may seem to be an advantage, but T-independent responses are not the same as T-dependent responses. The antibody generally has a lower affinity for its antigen and a shorter duration in circulation. Memory cells are not produced, and T-independent responses mature more slowly than T-dependent responses. This delay in maturation may contribute to the increased susceptibility to some bacterial infections in early life. It certainly contributed to the failure of purified polysaccharide vaccines to effectively immunize children younger than 2 years. For use in children, these vaccines have been replaced with a hapten approach in which the polysaccharide is conjugated to protein. In this form antibody generated by the T-dependent mechanism (protein carrier) still has specificity for the polysaccharide epitopes.

T-cell independent responses are weaker and lack memory

Poor response under 2 years of age

After challenge with foreign antigen, there is a lag period of 4 to 6 days before antibody can be detected in serum. This period reflects the events involved in the recognition of the antigen, its processing, and the specific activation of the cells of the immune system. The first event is the clearance of antigen from the circulation by what is essentially a metabolic process in which the antigen is recognized in a nonspecific sense and ingested. The vast preponderance of antigen ends up in circulating phagocytes or in stationary macrophages. The macrophages process the antigen so that immunogenic moieties can be presented to T cells, which then cause the B cells to produce immunoglobulins. The antibody-forming system is a learning system that responds to challenge by foreign molecules by producing large amounts of specific antibody. In addition, the affinity of its binding to the specifically recognized antigen often increases with time or secondary challenge.

Antigen processing causes delay in antibody response

Learning system increases affinity with time or secondary challenge

Antibody Structure

Antibodies belong to the **immunoglobulin** family of proteins, which appear in quantity in serum and on the surfaces of B cells. Of the five known structural types, three (IgG, IgM, IgA) are involved in the defense against infection. The basic structure of an immunoglobulin is illustrated in **Figure 2–15**, which depicts an **IgG** molecule. Immunoglobulins have a basic tetrameric structure consisting of two light polypeptide chains and two heavy chains usually associated as light/heavy pairs by disulfide bonds. The two light/heavy pairs are covalently associated by disulfide bonds to form the tetramer. There are two types of light chains, κ and λ, which are the products of distinct genetic loci. The class or isotype of the immunoglobulin is defined by the type of heavy chain expressed.

Immunoglobulins have tetrameric structure combining light chains and heavy chains

Isotypes are defined by type of heavy chain

The Y-shaped structure includes two **antigen binding sites (Fab)** formed by interaction of the **variable domains** of the heavy chain and the light chain. The stalk is called the **Fc fragment.** Antibodies carry out two broad sets of functions: the recognition function is the

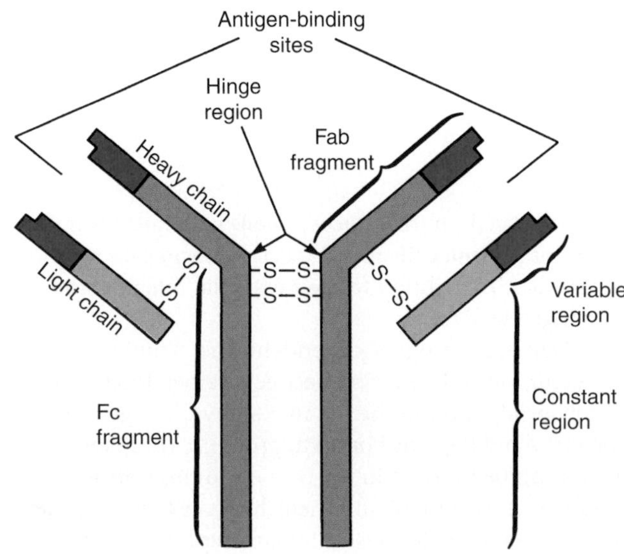

FIGURE 2–15. Immunoglobulin G structure. The IgG molecule consists of two identical light chains and two identical heavy chains held together by disulfide bonds. (Reproduced with permission from Willey J, Sherwood L, Woolverton C (eds). *Prescott's Principles of Microbiology.* New York: McGraw-Hill; 2008.)

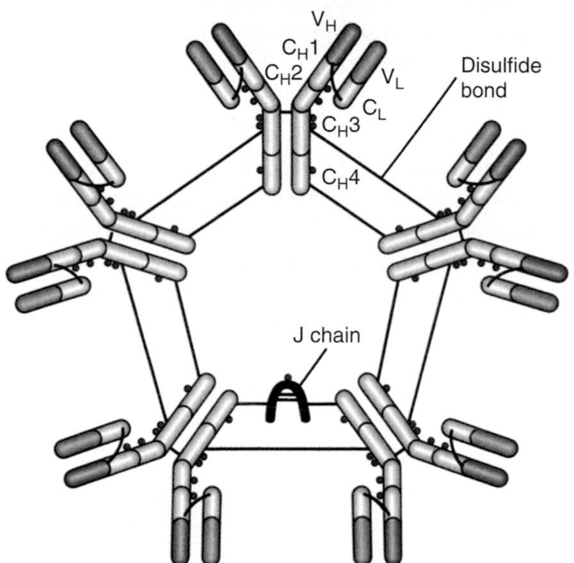

FIGURE 2–16. Immunoglobulin M structure. The pentameric structure has disulfide bonds linking peptide chains shown in black; carbohydrate side chains are in red. The J chain links the molecule together. (Reproduced with permission from Willey J, Sherwood L, Woolverton C (eds). *Prescott's Principles of Microbiology.* New York: McGraw-Hill; 2008.)

property of the Fab sites for antigen, and the effector functions are mediated by the constant regions of the heavy chains. Variations in the hypervariable region of the Fab-combining site due to mutations are called **idiotypes.** Antibodies combine with foreign antigens, but the actual destruction or removal of antigen requires the interaction of portions of the Fc fragment with other molecules such as complement components and phagocytes which have **Fc receptors.**

Fab sites bind antigen

Fc fragment recognized by complement, phagocytes

Combining site is idiotype

Figure 2–16 shows a schematic representation of a serum **IgM** immunoglobulin. This molecule consists of five subunits of the typical IgG molecule. The molecule occurs as a cyclic pentamer, and a J (joining) chain links the intact structure. When IgM is present on the surface of B cells where it serves as a primary receptor for antigen, it is present as a monomer. Other immunoglobulins showing a difference in arrangement from the typical IgG model are the **IgA** immunoglobulins. In serum, these immunoglobulins can occur as a monomer, but they can also occur in dimers in which the joining chain is required to stabilize the dimer. IgA molecules in the gut occur as dimers in which both the J chain and an additional polypeptide, termed the **secretory component,** are present in the complex.

Fab is antigen-binding region

IgM has five subunits

IgA is a monomer or dimer

Functional Properties of Immunoglobulins

Immunoglobulin G (IgG)

Immunoglobulin G is the most abundant immunoglobulin in health and provides the most extensive and long-lived antibody response to the various microbial and other antigens that are encountered throughout life. Although at least four subclasses of IgG have been characterized, they are grouped together for the purpose of this chapter. The IgG molecule is bivalent with two identical and specific combining sites. The Fc region does not vary with differences in specificity of combining sites of different antibody molecules. The Fc fragment binding sites for phagocytic cells are made available when the variable region of the antibody molecule has reacted with specific antigen, leaving the Fc facing outward.

Bivalent molecule with specific combining site and constant region

Constant region binds phagocytes

Immunoglobulin G antibody is characteristically formed in large amounts during the secondary response to an antigenic stimulus and usually follows production of IgM (see section that follows) in the course of a viral or bacterial infection. Memory cells are programmed for rapid IgG response when another antigenic stimulus of the same type occurs later. Immunoglobulin G antibodies are the most significant antibody class for neutralizing bacterial exotoxins and viruses often by blocking their attachment to cell receptors.

Antibodies produced during secondary response neutralize toxins and viruses

Accelerated IgG responses from memory cell expansion frequently confer lifelong immunity when directed against microbial antigens that are determinants of virulence. IgG is the only immunoglobulin class able to cross the placental barrier and thus provides passive immune protection to the newborn in the form of maternal antibody.

Binding may block attachment receptor

Immunoglobulin M (IgM)

Monomers of IgM constitute the specific epitope recognition sites on B cells that ultimately give rise to plasma cells producing one or another of the different immunoglobulin classes of antibody. Because of its many specific combining sites, IgM is particularly effective in agglutinating particles carrying epitopes against which it is directed. It also contains many sites for binding the first component of complement. These sites become available once the IgM molecule has reacted with antigen. IgM is particularly active in bringing about complement-mediated cytolytic damage to foreign antigen-bearing cells. It is less effective as an opsonizing antibody because its Fc portion is not available to phagocytes.

Effective agglutinating antibody

Binds complement at multiple sites

Immunoglobulin A (IgA)

Immunoglobulin A has a special role as a major determinant of so-called local immunity in protecting epithelial surfaces from colonization and infection. Certain B cells in lymphoid tissues adjacent to or draining surface epithelia of the intestines, respiratory tract, and genitourinary tract are encoded for specific IgA production. After antigenic stimulus, the clone expands locally, and some of the IgA-producing cells also migrate to other viscera and secretory glands. At the epithelia, two IgA molecules combine with another protein, termed the **secretory piece,** which is present on the surface of local epithelial cells. The complex, then termed **secretory IgA** (sIgA), passes through the cells into the mucous layer on the epithelial surface or into glandular secretions, where it exerts its protective effect. The secretory piece not only mediates secretion, but also protects the molecule against proteolysis by enzymes such as those present in the intestinal tract.

sIgA is produced at mucosal surfaces

Secretory piece combines molecules and resists proteolysis

The major role of sIgA is to prevent attachment of antigen-carrying particles to receptors on mucous membrane epithelia. Thus, in the case of bacteria and viruses, it reacts with surface antigens that mediate adhesion and colonization and prevents the establishment of local infection or invasion of the subepithelial tissues. sIgA can agglutinate particles, but has no Fc domain for activating the classic complement pathway; however, it can activate the alternative pathway. Reaction of IgA with antigen within the mucous membrane initiates an inflammatory reaction that helps mobilize other immunoglobulin and cellular defenses to the site of invasion. IgA response to an antigen is shorter lived than the IgG response.

Interferes with attachment of microbes to mucosal surfaces

Antibody Production

The major events characterizing the time course of antibody production are illustrated in **Figure 2–17** and summarized as follows: Initial contact with a new antigen evokes the **primary response,** which is characterized by a lag phase of approximately 1 week between the challenge and the detection of circulating antibodies. In general, the length of the lag phase depends on the immunogenicity of the stimulating antigen and the sensitivity of the detection system for the antibodies produced. Once antibody is detected in serum, the levels rise exponentially to attain a maximal steady state in about 3 weeks. These levels then decline gradually with time if no further antigenic stimulation is given. The first antibodies synthesized in the primary immune response are IgM and then in the latter phase IgG antibodies arise and eventually predominate. This transition is termed the **IgM/IgG switch**.

After a lag phase, the primary response lasts for weeks and then declines

IgM response switches to IgG

After a subsequent exposure or booster injection of the same antigen, a different sequence called the **secondary response** or **anamnestic response** ensues. This response involves memory. In the secondary response, the lag time between the immunization and the appearance of antibody is shortened, the rate of exponential increase to the maximum steady-state level is more rapid, and the steady-state level itself is higher, representing a larger amount of antibody. Another key factor of the secondary response is that the antibodies formed are predominantly of the IgG class. In addition to higher levels, the secondary IgG antibodies have a higher affinity for their antigen. Figure 2–17 shows the participation of memory T cells created during the primary response in these reactions.

Secondary response is primarily IgG

Affinity for antigen is greater

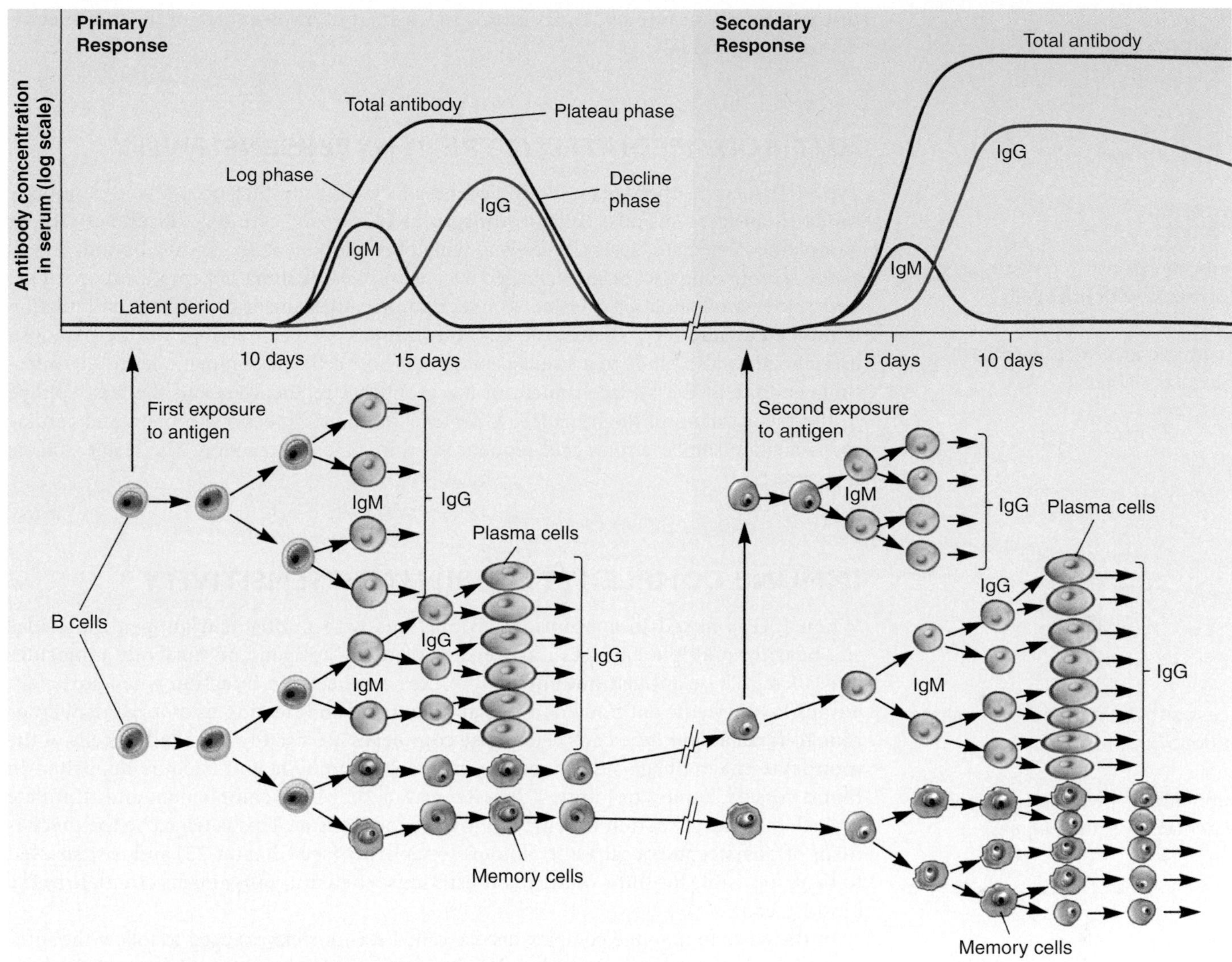

FIGURE 2–17. Antibody production and kinetics. The four phases of a primary antibody response correlate to the clonal expansion of the activated B cell, differentiation into plasma cells, and secretion of the antibody protein. The secondary response is much more rapid, and total antibody production is nearly 1000 times greater than that of the primary response. (Reproduced with permission from Willey J, Sherwood L, Woolverton C (eds). *Prescott's Principles of Microbiology.* New York: McGraw-Hill; 2008.)

ADVERSE EFFECTS OF IMMUNOLOGIC REACTIONS

The immune system is no different from any other human system. In balance we do not even know it is there, but in an exaggerated state we call **hypersensitivity,** it can cause injury and even chronic disease. Hypersensitivity reactions have been placed into four classes based on their mechanism of immunologic injury. Type I or allergic reactions relate to the action of IgE and the release of powerful mediators like histamine from mast cells. Type II or cytotoxic reactions are created when IgG or IgM antibodies are misdirected to host cells. Type III or immune complex reactions are created when an excess of antigen-antibody complexes are deposited and followed by complement-mediated inflammation. Type IV reactions are cell-mediated and often called delayed-type hypersensitivity because of the time delay in invoking the T_H1 response. The hypersensitivity diseases include allergy, anaphylaxis, asthma, transfusion reactions, rheumatoid arthritis, and type 1 diabetes. Infectious

Mechanisms I-IV involve antibody and cell-mediated injury

Allergy, asthma, and diabetes are due to hypersensitivity

Infection is a small part

diseases are a relatively small part of this spectrum, but involve three of the four mechanisms (II, III, and IV).

ANTIBODY-MEDIATED (TYPE II) HYPERSENSITIVITY

Antibody against microbe epitope also reacts with host cells

Rheumatic fever is caused by molecular mimicry

Type II hypersensitivity is antibody-dependent cytotoxicity that occurs when antibody binds to antigens on host cells, leading to phagocytosis, cytotoxic T-cell activity, or complement-mediated lysis. The cells to which the antibody is specifically bound, as well as the surrounding tissues, are damaged because of the inflammatory amplification. In the best understood situations related to infection, the mechanism of antibody stimulation is **molecular mimicry.** That is, the antibody stimulated by an epitope on the pathogen unfortunately also binds to a similar epitope on host cells. In rheumatic fever, the infectious epitope is in a surface protein of the group A streptococcus and the host epitope in the myocardium of the heart (see Chapter 25). The streptococcal protein and cardiac myosin share similar amino acid sequences, so it is a cross-reaction. The result is acute myocarditis.

IMMUNE COMPLEX (TYPE III) HYPERSENSITIVITY

Excess antigen-antibody complexes are deposited in tissues

Complement-mediated inflammation causes injury

When IgG is mixed in appropriate proportions with multivalent antigen molecules (ie, bearing multiple epitopes), aggregates of many antigen and antibody molecules may form. These antigen-antibody complexes can occur in infection when sufficient amounts of specific antibody and free antigen from an infecting microorganism combine to form an immune complex. These complexes are usually removed by cells of the monocyte-macrophage system, but in excess can circulate and become deposited in blood vessels, kidney, or joints. When deposited, they bind complement and stimulate an inflammatory reaction that may injure the local tissue. This is felt to be the mechanism of poststreptococcal acute glomerulonephritis (see Chapter 25) and is suspected to be responsible for some of the manifestations when microorganisms circulate in the bloodstream.

Serum sickness is reaction to animal immunoglobulin

In the past, an immune complex disease called **serum sickness** used to follow the infusion of antibodies (antisera) produced in horses to combat infection. Human antibody to the foreign horse immunoglobulin was formed. These diseases (diphtheria, tetanus) are now prevented by vaccines that stimulate antibody against the same epitopes in humans. When passive immunization is used, human sources of antibody are now available.

DELAYED-TYPE (TYPE IV) HYPERSENSITIVITY

DTH requires time for T_H1 response to develop

Inflammation causes continuing local injury

Type IV delayed-type hypersensitivity (DTH) is a cell-mediated immune reaction. The delay is the time required after initiation of a T_H1 response for antigen to be processed, cytokines produced, and T cells to migrate and accumulate at the antigen site. At the site, cytotoxic T cells, macrophages, and other inflammatory mediators directed at cells containing the antigen also produce injury in the surrounding tissue. The purest form of DTH is the intradermal skin test for tuberculosis. In persons already sensitized to the antigens of *Mycobacterium tuberculosis,* it takes 1 to 2 days for induration to be produced at the site of inoculation of a standardized antigen called tuberculin. This is a useful diagnostic test, but in infectious disease DTH is also the hypersensitivity mechanism that causes the most injury. This occurs in diseases in which immunity is cell-mediated with little or no effective antibody component. If T_H1 responses are successful in containing the infection at an early stage, there is little destruction. If they are not successful enough to contain growth of the pathogen, increasing amounts of antigen stimulate continuing DTH-mediated destructive inflammation. This is the primary mechanism of injury in tuberculosis, fungal infections, and many parasitic diseases.

FAVORABLE USE OF THE IMMUNE RESPONSE

NATURAL IMMUNITY TO INFECTION

The majority of encounters with microorganisms including pathogens end favorably for the host. The heightened immunologic responses following infection usually provide immunity, often for life. This is called natural immunity. In some instances, the gauntlet is long because a pathogen of the same name may exist in multiple antigenic types. Because of the specificity of the adaptive immune response, immunity must be developed individually to each antigenic type. Development of natural immunity need not require a clinical infection. There is ample evidence from population studies that individuals with no history or recollection of infection have evidence of immunity in the form of specific antibody. From the time of birth forward, we have many encounters with infectious agents, most of which lead to immunity without disease.

Natural infection often confers lifelong immunity

Clinical disease is not required

PASSIVE IMMUNITY

Passive immunity is the transfer of antibodies from one person to another. Because the antibody was not made by the recipient, this antibody is transient and lasts only a few weeks or months. This is a natural process in the case of IgG transferred transplacentally from mother to fetus. The protection provided by this antibody is limited to the immunologic experience of the mother but covers a particularly vulnerable time in life, lasting as long as 6 months after birth. Passive immunity can also be provided as a therapeutic product in which specific antibodies are infused. Such antisera are available for only a limited number of diseases such as rabies, botulism, and tetanus.

Transplacental IgG protects the fetus

VACCINES

Vaccines artificially stimulate immunity through exposure to an antigenic substance. The early vaccines such as Jenner's for smallpox and Pasteur's for anthrax (in animals) were live attenuated strains with the ability to produce a true if mild infection. We later learned how to kill the agent in a way that retained its antigenicity. These killed vaccines are practical if the number of antigens present is limited as with a virus (polio) or bacterial toxin (diphtheria), but usually too crude if whole bacteria are used. Progress with killed bacterial vaccines required knowledge of just which antigenic component provides protective immunity. This allowed inactivation followed by purification of the selected component. This approach with bacterial polysaccharide capsules has produced a dramatic reduction (more than 95%) in childhood meningitis. Genomic approaches are now aimed at producing a protective antigen without growth of the organism itself. For each of the 50 chapters in this book devoted to specific infectious agents, vaccines and the immunologic mechanisms involved are carefully examined.

Live vaccines use attenuated strains

Killed vaccines may require purification

CHAPTER 3

Sterilization, Disinfection, and Infection Control

From the time of debates about the germ theory of disease, killing microbes before they reach patients has been a major strategy for preventing infection. In fact, Ignaz Semmelweis successfully applied disinfection principles decades before bacteria were first isolated. This chapter discusses the most important methods used for this purpose in modern medical practice. Understanding how these methods work has become of increasing importance in an environment that includes immunocompromised patients, transplantation, indwelling devices, and acquired immunodeficiency syndrome (AIDS).

DEFINITIONS

Death/killing as it relates to microbial organisms is defined in terms of how we detect them in culture. Operationally, it is a loss of ability to multiply under any known conditions. This is complicated by the fact that organisms that appear to be irreversibly inactivated may sometimes recover when appropriately treated. For example, ultraviolet (UV) irradiation of bacteria can result in the formation of thymine dimers in the DNA with loss of ability to replicate. A period of exposure to visible light may then activate an enzyme that breaks the dimers and restores viability by a process known as photoreactivation. Mechanisms also exist for repair of the damage without light. Such considerations are of great significance in the preparation of safe vaccines from inactivated virulent organisms.

Absence of growth does not necessarily indicate sterility

Sterilization is complete killing, or removal, of all living organisms from a particular location or material. It can be accomplished by incineration, nondestructive heat treatment, certain gases, exposure to ionizing radiation, some liquid chemicals, and filtration.:

Sterilization is killing of all living forms

Pasteurization is the use of heat at a temperature sufficient to inactivate important pathogenic organisms in liquids such as water or milk, but at a temperature lower than that needed to ensure sterilization. For example, heating milk at a temperature of 74°C for 3 to 5 seconds or 62°C for 30 minutes kills the vegetative forms of most pathogenic bacteria that may be present without altering its quality. Obviously, spores are not killed at these temperatures.

Pasteurization uses heat to kill vegetative forms of bacteria

Disinfection is the destruction of pathogenic microorganisms by processes that fail to meet the criteria for sterilization. Pasteurization is a form of disinfection, but the term is most commonly applied to the use of liquid chemical agents known as disinfectants, which usually have some degree of selectivity. Bacterial spores, organisms with waxy coats (eg, mycobacteria), and some viruses may show considerable resistance to the common disinfectants. **Antiseptics** are disinfectant agents that can be used on body surfaces such as the skin or vaginal tract to reduce the numbers of normal flora and pathogenic contaminants. They have lower toxicity than disinfectants used environmentally, but are usually less active

Disinfection uses chemical agents to kill pathogens with varying efficiency in killing vegetative organisms. **Sanitization** is a less precise term with a meaning somewhere between disinfection and cleanliness. It is used primarily in housekeeping and food preparation contexts.

Spores are particularly resistant

Asepsis applies sterilization and disinfection to create a protective environment

Asepsis describes processes designed to prevent microorganisms from reaching a protected environment. It is applied in many procedures used in the operating room, in the preparation of therapeutic agents, and in technical manipulations in the microbiology laboratory. An essential component of aseptic techniques is the sterilization of all materials and equipment used.

MICROBIAL KILLING

Bacterial killing follows exponential kinetics

Killing of bacteria by heat, radiation, or chemicals is usually exponential with time; that is, a fixed proportion of survivors are killed during each time increment. Thus, if 90% of a population of bacteria are killed during each 5 minutes of exposure to a weak solution of a disinfectant, a starting population of 10^6/mL is reduced to 10^5/mL after 5 minutes, to 10^3/mL after 15 minutes, and theoretically to 1 organism (10^0)/mL after 30 minutes. Exponential killing corresponds to a first-order reaction or a "single-hit" hypothesis in which the lethal change involves a single target in the organism, and the probability of this change is constant with time. Thus, plots of the logarithm of the number of survivors against time are linear (**Figure 3–1A**); however, the slope of the curve varies with the effectiveness of the killing process, which is influenced by the nature of the organism, lethal agent, concentration (in the case of disinfectants), and temperature. In general, the rate of killing increases exponentially with arithmetic increases in temperature or in concentrations of disinfectant.

Achieving sterility is a matter of probability

An important consequence of exponential killing with most sterilization processes is that sterility is not an absolute term, but must be expressed as a probability. Thus, to continue the example given previously, the chance of a single survivor in 1 mL is theoretically 10^{-1} after 35 minutes. If a chance of 10^{-9} were the maximum acceptable risk for a single surviving organism in a 1-mL sample (eg, of a therapeutic agent), the procedure would require continuation for a total of 75 minutes.

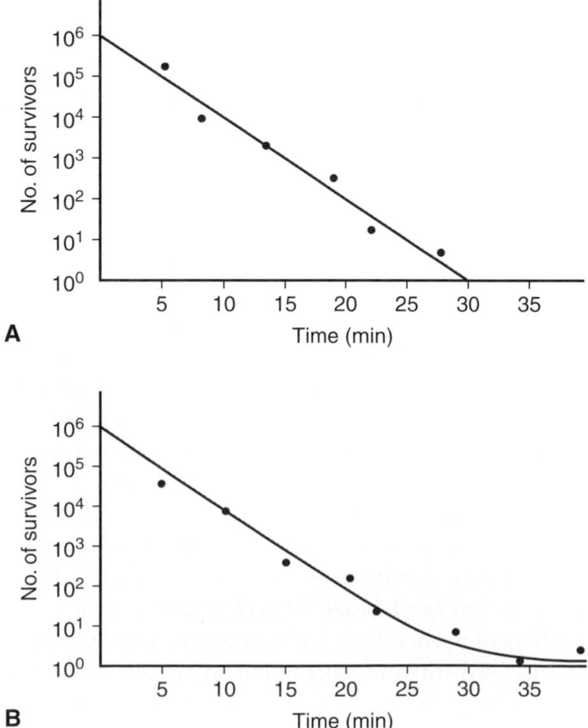

FIGURE 3–1. Kinetics of bacterial killing. A. Exponential killing is shown as a function of population size and time. **B.** Deviation from linearity, as with a mixed population, extends the time.

A simple single-hit curve often does not express the kinetics of killing adequately. In the case of some bacterial endospores, a brief period (activation) may elapse before exponential killing by heat begins. If multiple targets are involved, the experimental curve deviates from linearity. More significant is the fact that microbial populations may include a small proportion of more resistant mutants or of organisms in a physiologic state that confers greater resistance to inactivation. In these cases, the later stages of the curve are flattened (**Figure 3–1B**), and extrapolations from the exponential phase of killing may seriously underestimate the time needed for a high probability of achieving complete sterility. In practice, materials that come into contact with tissues are sterilized under conditions that allow a very wide margin of safety, and the effectiveness of inactivation of organisms in vaccines is tested directly with large volumes and multiple samples before a product is made available for use.

Heterogeneous microbial subpopulations may extend the killing kinetics

STERILIZATION

The availability of reliable methods of sterilization has made possible the major developments in surgery and intrusive medical techniques that have helped to revolutionize medicine over the last century. Furthermore, sterilization procedures form the basis of many food preservation procedures, particularly in the canning industry. The various modes of sterilization described in the text are summarized in **Table 3–1**.

TABLE 3–1 Methods of Disinfection and Sterilization

METHOD	ACTIVITY LEVEL	SPECTRUM	USES/COMMENTS
Heat			
Autoclave	Sterilizing	All	General
Boiling	High	Most pathogens, some spores	General
Pasteurization	Intermediate	Vegetative bacteria	Beverages, plastic hospital equipment
Ethylene oxide gas	Sterilizing	All	Potentially explosive; aeration required
Radiation			
Ultraviolet	Sterilizing	All	Poor penetration
Ionizing	Sterilizing	All	General, food
Chemicals			
Alcohol	Intermediate	Vegetative bacteria, fungi, some viruses	
Hydrogen peroxide	High	Viruses, vegetative bacteria, fungi	Contact lenses; inactivated by organic matter
Chlorine	High	Viruses, vegetative bacteria, fungi	Water; inactivated by organic matter
Iodophors	Intermediate	Viruses, vegetative bacteria,[a] fungi	Skin disinfection; inactivated by organic matter
Phenolics	Intermediate	Some viruses, vegetative bacteria, fungi	Handwashing
Glutaraldehyde	High	All	Endoscopes, other equipment
Quaternary ammonium compounds	Low	Most bacteria and fungi, lipophilic viruses	General cleaning; inactivated by organic matter

[a]Variable results with *Mycobacterium tuberculosis*.

Heat

Incineration is rapid and effective

Dry heat requires 160°C for 2 hours to kill

The simplest method of sterilization is to expose the surface to be sterilized to a naked flame, as is done with the wire loop used in microbiology laboratories. It can be used equally effectively for emergency sterilization of a knife blade or a needle. Of course, disposable material is rapidly and effectively decontaminated by incineration. Carbonization of organic material and destruction of microorganisms, including spores, occur after exposure to dry heat of 160°C for 2 hours in a sterilizing oven. This method is applicable to metals, glassware, and some heat-resistant oils and waxes that are immiscible in water and, therefore, cannot be sterilized in the autoclave. A major use of the dry heat sterilizing oven is in preparation of laboratory glassware.

Moisture allows for rapid denaturation of protein

Boiling water fails to kill bacterial spores

Moist heat in the form of water or steam is far more rapid and effective in sterilization than dry heat because reactive water molecules denature protein irreversibly by disrupting hydrogen bonds between peptide groups at relatively low temperatures. Most vegetative bacteria of importance in human disease are killed within a few minutes at 70°C or less, although many bacterial spores (see Chapter 21) can resist boiling for prolonged periods. For applications requiring sterility, the use of boiling water has been replaced by the autoclave, which, when properly used, ensures sterility by killing all forms of microorganisms.

Autoclave creates increased temperature of steam under pressure

In effect, the **autoclave** is a sophisticated pressure cooker (**Figure 3–2**). In its simplest form, it consists of a chamber in which the air can be replaced with pure saturated steam under pressure. Air is removed either by evacuation of the chamber before filling it with steam or by displacement through a valve at the bottom of the autoclave, which remains open until all air has drained out. The latter, which is termed a **downward displacement autoclave,** capitalizes on the heaviness of air compared with saturated steam. When the air

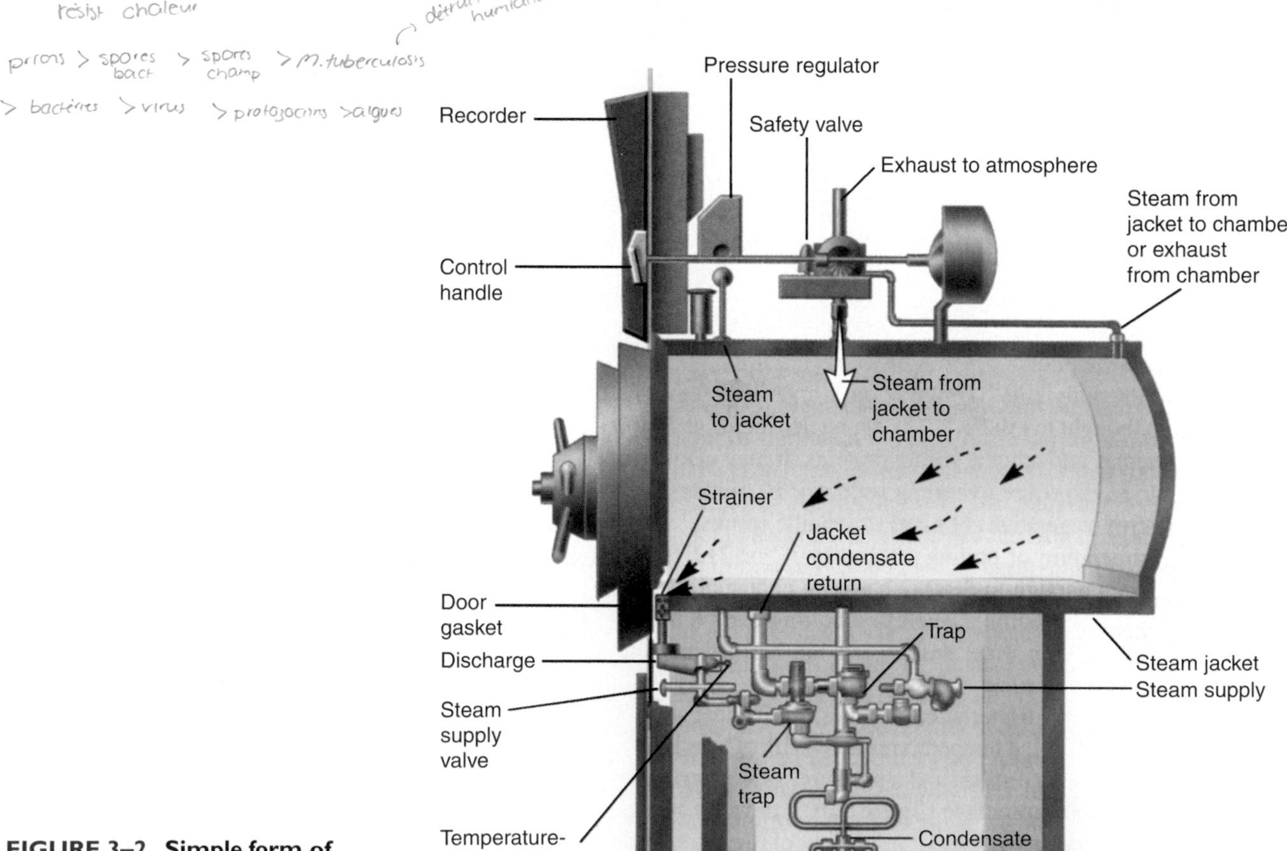

FIGURE 3–2. Simple form of downward displacement autoclave. (Reproduced with permission from Willey J, Sherwood L, Woolverton C (eds). *Prescott's Principles of Microbiology.* New York: McGraw-Hill; 2008.)

has been removed, the temperature in the chamber is proportional to the pressure of the steam; autoclaves are usually operated at 121°C, which is achieved with a pressure of 15 pounds per square inch. Under these conditions, spores directly exposed are killed in less than 5 minutes, although the normal sterilization time is 10 to 15 minutes to account for variation in the ability of steam to penetrate different materials and to allow a wide margin of safety. The velocity of killing increases logarithmically with arithmetic increases in temperature, so a steam temperature of 121°C is vastly more effective than 100°C. For example, the spores of *Clostridium botulinum,* the cause of botulism, may survive 5 hours of boiling, but can be killed in 4 minutes at 121°C in the autoclave.

Steam displaces air from the autoclave

Killing rate increases logarithmically with arithmetic increase in temperature

The use of saturated steam in the autoclave has other advantages. Latent heat equivalent to 539 cal/g of condensed steam is immediately liberated on condensation on the cooler surfaces of the load to be sterilized. The temperature of the load is thus raised very rapidly to that of the steam. Condensation also permits rapid steam penetration of porous materials such as surgical drapes by producing a relative negative pressure at the surface, which allows more steam to enter immediately. Autoclaves can thus be used for sterilizing any materials that are not damaged by heat and moisture, such as heat-stable liquids, swabs, most instruments, culture media, and rubber gloves.

Condensation and latent heat increase effectiveness of autoclave

It is essential that those who use autoclaves understand the principles involved. The effectiveness of autoclaves depends on the absence of air, pure saturated steam, and access of steam to the material to be sterilized. Pressure per se plays no role in sterilization other than to ensure the increased temperature of the steam. Failure can result from attempting to sterilize the interior of materials that are impermeable to steam or the contents of sealed containers. Under these conditions, a dry heat temperature of 121°C is obtained, which may be insufficient to kill even vegetative organisms. Large volumes of liquids require longer sterilization times than normal loads, because their temperature must reach 121°C before timing begins. When sealed containers of liquids are sterilized, it is essential for the autoclave to cool without being opened or evacuated; otherwise, the containers may explode as the external pressure falls in relation to that within.

Access of pure saturated steam is required for sterilization

Impermeable or large volume materials present special problems

"Flash" autoclaves, which are widely used in operating rooms, often use saturated steam at a temperature of 134°C for 3 minutes. Air and steam are removed mechanically before and after the sterilization cycle so that metal instruments may be available rapidly. Quality control of autoclaves depends primarily on ensuring that the appropriate temperature for the pressure used is achieved and that packing and timing are correct. Biologic and chemical indicators of the correct conditions are available and are inserted from time to time in the loads.

Flash autoclaves use 134°C for 3 minutes

Gas

A number of articles, particularly certain plastics and lensed instruments that are damaged or destroyed by autoclaving, can be sterilized with gases. **Ethylene oxide** is an inflammable and potentially explosive gas. It is an alkylating agent that inactivates microorganisms by replacing labile hydrogen atoms on hydroxyl, carboxy, or sulfhydryl groups, particularly of guanine and adenine in DNA. Ethylene oxide sterilizers resemble autoclaves and expose the load to 10% ethylene oxide in carbon dioxide at 50° to 60°C under controlled conditions of humidity. Exposure times are usually about 4 to 6 hours and must be followed by a prolonged period of aeration to allow the gas to diffuse out of substances that have absorbed it. Aeration is essential, because absorbed gas can cause damage to tissues or skin. Ethylene oxide is a mutagen, and special precautions are now taken to ensure that it is properly vented outside of working spaces. Used under properly controlled conditions, ethylene oxide is an effective sterilizing agent for heat-labile devices such as artificial heart valves that cannot be treated at the temperature of the autoclave. Other alkylating agents such as **formaldehyde** vapor can be used without pressure to decontaminate larger areas such as rooms, and oxidizing agents (hydrogen peroxide, ozone) have selective use.

Ethylene oxide sterilization is used for heat-labile materials

Aeration needed after ethylene oxide sterilization

Formaldehyde and oxidizing agents are useful in sterilization

Ultraviolet Light and Ionizing Radiation

Ultraviolet (UV) light in the wavelength range of 240 to 280 nm is absorbed by nucleic acids and causes genetic damage, including the formation of the thymine dimers discussed previously. The practical value of UV sterilization is limited by its poor ability to penetrate.

UV light causes direct damage to DNA

Use of UV light is limited by penetration and safety

Apart from the experimental use of UV light as a mutagen, its main application has been in irradiation of air in the vicinity of critical hospital sites and as an aid in the decontamination of laboratory facilities used for handling particularly hazardous organisms. In these situations, single exposed organisms are rapidly inactivated. It must be remembered that UV light can cause skin and eye damage, and workers exposed to it must be appropriately protected.

Ionizing radiation damages DNA

Use for surgical supplies, food

Killed organisms may remain morphologically intact and stainable

Ionizing radiation carries far greater energy than UV light. It, too, causes direct damage to DNA and produces toxic free radicals and hydrogen peroxide from water within the microbial cells. Cathode rays and gamma rays from cobalt-60 are widely used in industrial processes, including the sterilization of many disposable surgical supplies such as gloves, plastic syringes, specimen containers, some foodstuffs, and the like, because they can be packaged before exposure to the penetrating radiation. Ionizing irradiation does not always result in the physical disintegration of killed microbes. As a result, plasticware sterilized in this way may carry significant numbers of dead but stainable bacteria. Recent food-borne outbreaks (*Escherichia coli*) and bioterrorism (anthrax) have increased the use of ionizing radiation.

DISINFECTION

Physical Methods

Filtration

Membrane filters remove bacteria by mechanical and electrostatic mechanisms

Both live and dead microorganisms can be removed from liquids by positive- or negative-pressure filtration. Membrane filters, usually composed of cellulose esters (eg, cellulose acetate), are available commercially with pore sizes of 0.005 to 1 μm. For removal of bacteria, a pore size of 0.2 μm is effective because filters act not only mechanically but by electrostatic adsorption of particles to their surface. Filtration is used for disinfection of large volumes of fluid, especially fluid containing heat-labile components such as serum. For microorganisms larger than the pore size, filtration "sterilizes" these liquids. It is not considered effective for removing viruses.

Pasteurization

Kills vegetative bacteria but not spores

Used for foods and fragile medical equipment

Pasteurization involves exposure of liquids to temperatures in the range 55° to 75°C to remove all vegetative bacteria of significance in human disease. Spores are unaffected by the pasteurization process. Pasteurization is used commercially to render milk safe and to extend its storage quality. With the outbreaks of infection due to contamination with enterohemorrhagic *E coli* (see Chapter 33), this has been extended (reluctantly) to fruit drinks. To the dismay of some of his compatriots, Pasteur proposed application of the process to winemaking to prevent microbial spoilage and vinegarization. Pasteurization in water at 70°C for 30 minutes has been effective and inexpensive when used to render inhalation therapy equipment free of organisms that may otherwise multiply in mucus and humidifying water.

Microwaves

Microwaves kill by generating heat

The use of microwaves in the form of microwave ovens or specially designed units is another method of disinfection. These systems are not under pressure, but they but can achieve temperatures near boiling if moisture is present. In some situations, they are being used as a practical alternative to incineration for disinfection of hospital waste. These procedures cannot be considered sterilization only because heat-resistant spores may survive the process.

Chemical Methods

Most agents are general protoplasmic poisons

Given access and sufficient time, chemical disinfectants cause the death of pathogenic vegetative bacteria. Most of these substances are general protoplasmic poisons and are not used in the treatment of infections other than very superficial lesions, having been replaced by antimicrobics. Some disinfectants such as the quaternary ammonium compounds, alcohol, and the iodophors reduce the superficial flora and can eliminate contaminating pathogenic bacteria from the skin surface. Other agents such as the phenolics are valuable only for treating inanimate surfaces or for rendering contaminated materials safe. All are bound and inactivated to varying degrees by protein and dirt, and they lose considerable activity when applied to other than clean surfaces. Their activity increases exponentially with increases in temperature, but the relation between increases in concentration and killing effectiveness is complex and varies for each compound. Optimal in-use concentrations have been established for all available disinfectants. The major groups of compounds currently used are briefly discussed next.

Disinfectants are variably inactivated by organic material

Chemical disinfectants are classified on the basis of their ability to sterilize. High-level disinfectants kill all agents, except the most resistant of bacterial spores. Intermediate-level disinfectants kill all agents, but not spores. Low-level disinfectants are active against most vegetative bacteria and lipid-enveloped viruses.

Activity against spores and viruses varies

Alcohol

The alcohols are protein denaturants that rapidly kill vegetative bacteria when applied as aqueous solutions in the range of 70% to 95% alcohol. They are inactive against bacterial spores and many viruses. Solutions of 100% alcohol dehydrate organisms rapidly but fail to kill, because the lethal process requires water molecules. Ethanol (70-90%) and isopropyl alcohol (90-95%) are widely used as skin decontaminants before simple invasive procedures such as venipuncture. Their effect is not instantaneous, and the traditional alcohol wipe, particularly when followed by a vein-probing finger, is more symbolic than effective because insufficient time is given for significant killing. Isopropyl alcohol has largely replaced ethanol in hospital use because it is somewhat more active and is not subject to diversion to house/staff parties.

Alcohols require water for maximum effectiveness

Action of alcohol is slow

Halogens

Iodine is an effective disinfectant that acts by iodinating or oxidizing essential components of the microbial cell. Its original use was as a tincture of 2% iodine in 50% alcohol, which kills more rapidly and effectively than alcohol alone. This preparation has the disadvantage of sometimes causing hypersensitivity reactions and of staining materials with which it comes in contact. Tincture of iodine has now been largely replaced by preparations in which iodine is combined with carriers (povidone) or nonionic detergents. These agents, termed **iodophors**, gradually release small amounts of iodine. They cause less skin staining and dehydration than tinctures and are widely used in preparation of skin before surgery. Although iodophors are less allergenic than inorganic iodine preparations, they should not be used on patients with a history of iodine sensitivity.

Tincture of iodine in alcohol is effective

Iodophors combine iodine with detergents

Chlorine is a highly effective oxidizing agent, which accounts for its lethality to microbes. It exists as hypochlorous acid in aqueous solutions that dissociate to yield free chlorine over a wide pH range, particularly under slightly acidic conditions. In concentrations of less than one part per million, chlorine is lethal within seconds to most vegetative bacteria, and it inactivates most viruses; this efficacy accounts for its use in rendering supplies of drinking water safe and in chlorination of water in swimming pools. Chlorine reacts rapidly with protein and many other organic compounds, and its activity is lost quickly in the presence of organic material. This property, combined with its toxicity, renders it ineffective on body surfaces; however, it is the agent of choice for decontaminating surfaces and glassware that have been contaminated with viruses or spores of pathogenic bacteria. For these purposes, it is usually applied as a 5% solution called **hypochlorite.**

Chlorine oxidative action is rapid

Activity is reduced by organic matter

The use of chlorination to disinfect water supplies has proved insufficient in some hospitals because of the relative resistance of *Legionella pneumophila* to the usual concentrations of chlorine. Some institutions have been forced to augment chlorination with systems that add copper and silver ions to the water.

Legionella may resist chlorine

Hydrogen Peroxide

Hydrogen peroxide is a powerful oxidizing agent that attacks membrane lipids and other cell components. Although it acts rapidly against many bacteria and viruses, it kills bacteria that produce catalase and spores less rapidly. Hydrogen peroxide has been useful in disinfecting items such as contact lenses, which are not susceptible to its corrosive effect.

Hydrogen peroxide oxidizes cell components

Surface-Active Compounds

Surfactants are compounds with hydrophobic and hydrophilic groups that attach to and solubilize various compounds or alter their properties. Anionic detergents such as soaps are highly effective cleansers, but have little direct antibacterial effect, probably because their charge is similar to that of most microorganisms. Cationic detergents, particularly the **quaternary ammonium compounds** ("quats") such as benzalkonium chloride, are highly bactericidal in the absence of contaminating organic matter. Their hydrophobic and lipophilic groups react with the lipid of the cell membrane of the bacteria, alter the membrane's surface properties and its permeability, and lead to loss of essential cell components and death. These compounds have little toxicity to skin and mucous membranes and thus have been used widely for their antibacterial effects in a concentration of 0.1%. They are inactive against spores and most viruses. Quats in much higher concentrations than those used in medicine (eg, 5-10%) can be used for sanitizing surfaces.

Hydrophobic and hydrophilic groups of surfactants act on lipids of bacterial membranes

Little activity against viruses

The greatest care is needed in the use of quats because they adsorb to most surfaces with which they come in contact, such as cotton, cork, and even dust. As a result, their concentration may be lowered to a point at which certain bacteria, particularly *Pseudomonas aeruginosa,* can grow in the quat solutions and then cause serious infections. Many instances have been recorded of severe infections resulting from contamination of ophthalmic preparations or of solutions used for treating skin before transcutaneous procedures. It should also be remembered that cationic detergents are totally neutralized by anionic compounds. Thus, the antibacterial effect of quaternary ammonium compounds is inactivated by soap. Because of these problems, quats have been replaced by other antiseptics and disinfectants for most purposes.

"Quats" adsorbed to surfaces may become contaminated with bacteria

Cationic detergents are neutralized by soaps

Phenolics

Phenol is a potent protein denaturant and bactericidal agent. Substitutions in the ring structure of phenol have substantially improved activity and have provided a range of phenols and cresols that are the most effective environmental decontaminants available for use in hospital hygiene. Concern about their release into the environment in hospital waste and sewage has created some pressure to limit their use. This is another of the classic environmental dilemmas of our society: a compound that reduces the risk of disease for one group may raise it for another. Phenolics are less "quenched" by protein than are most other disinfectants, have a detergent-like effect on the cell membrane, and are often formulated with soaps to increase their cleansing property. They are too toxic to skin and tissues to be used as antiseptics, although brief exposures can be tolerated. They are the active ingredient in many mouthwash and sore throat preparations.

Relatively stable to protein

Environmental contamination with phenols and cresols limits use

Two diphenyl compounds, hexachlorophene and chlorhexidine, have been extensively used as skin disinfectants. **Hexachlorophene** is primarily bacteriostatic. Incorporated into a soap, it builds up on the surface of skin epithelial cells over 1 to 2 days of use to produce a steady inhibitory effect on skin flora and Gram-positive contaminants, as long as its use is continued. It was a major factor in controlling outbreaks of severe staphylococcal infections in nurseries during the 1950s and 1960s, but cutaneous absorption was found to produce neurotoxic effects in some premature infants. When it was applied in excessive concentrations, similar problems occurred in older children. It is now a prescription drug.

Skin binding of hexachlorophene enhances effectiveness for staphylococci

Absortion through skin limits use

Chlorhexidine has replaced hexachlorophene as a routine hand and skin disinfectant and for other topical applications. It has greater bactericidal activity than hexachlorophene without its toxicity but shares with hexachlorophene the ability to bind to the skin and produce a persistent antibacterial effect. It acts by altering membrane permeability of both Gram-positive and Gram-negative bacteria. It is cationic and thus its action is neutralized by soaps and anionic detergents.

Chlorhexidine also binds to skin but is less toxic

Glutaraldehyde and Formaldehyde

Glutaraldehyde and formaldehyde are alkylating agents highly lethal to essentially all microorganisms (**Figure 3–3**). Formaldehyde gas is irritative, allergenic, and unpleasant, properties that limit its use as a solution or gas. Glutaraldehyde is an effective high-level disinfecting agent for apparatus that cannot be heat treated, such as some lensed instruments and equipment for respiratory therapy. Formaldehyde vapor, an effective environmental decontaminant under conditions of high humidity, is sometimes used to decontaminate laboratory rooms that have been accidentally and extensively contaminated with pathogenic bacteria, including those such as the anthrax bacillus that form resistant spores. Such rooms are sealed for processing and thoroughly aired before reoccupancy.

Glutaraldehyde is useful for decontamination of equipment

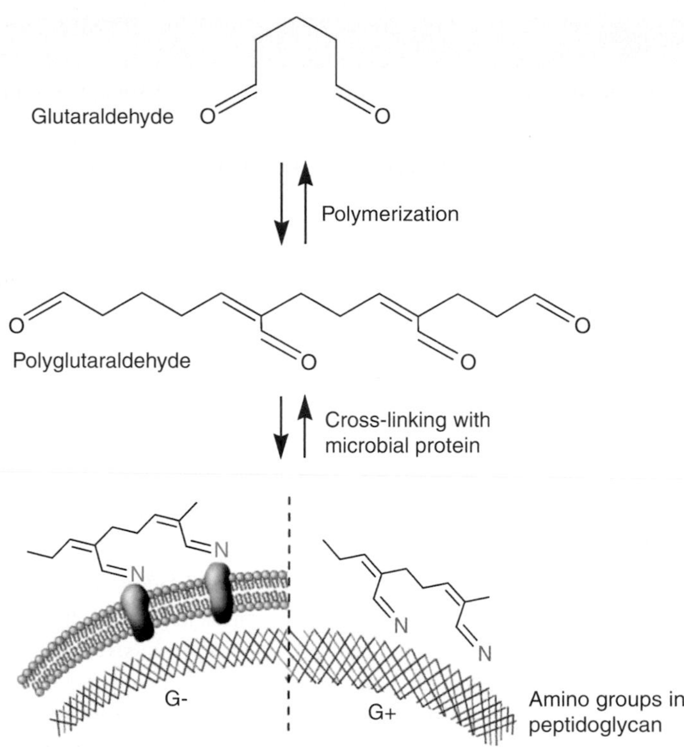

FIGURE 3–3. Action of glutaraldehyde. Glutaraldehyde polymerizes and then interacts with amino acids in proteins (*left*) or in bacterial peptidoglycan (*right*). As a result they are alkylated and inactivated. (Reproduced with permission from Willey J, Sherwood L, Woolverton C (eds). *Prescott's Principles of Microbiology.* New York: McGraw-Hill; 2008.)

INFECTION CONTROL AND NOSOCOMIAL INFECTIONS

Some risk of infection exists in all health care settings. Hospitalized patients are particularly vulnerable, and the hospital environment is complex. Infection control is the proper matching of the principles and procedures described here to general and specialized situations together with aseptic practices to reduce these risks. "Nosocomial" is a medical term for "hospital-associated." Nosocomial infections are complications that arise during hospitalizations. The morbidity, mortality, and costs associated with these infections are preventable to a substantial degree. The purpose of hospital infection control is prevention of nosocomial infections by application of epidemiologic concepts and methods.

History: Semmelweis and Childbed Fever

The shining example of the fundamental importance of epidemiology in detection and control of nosocomial infections is the work of Ignaz Semmelweis, which preceded the microbiologic discoveries of Pasteur and Koch by a decade. Semmelweis was assistant obstetrician at the Vienna General Hospital, where more than 7000 infants were delivered each year. Childbed fever (puerperal endometritis), which we now know is caused primarily by group A streptococci, was a major problem accounting for 600 to 800 maternal deaths per year. By careful review of hospital statistics between 1846 and 1849, Semmelweis clearly showed that the death rate in one of the two divisions of the hospital was 10 times that in the other. Division I, which had the high mortality rate, was the teaching unit in which all deliveries were by obstetricians and students. In division II, all deliveries were by midwives. No similar epidemic existed elsewhere in the city of Vienna, and the mortality rate was very low in mothers delivering at home.

Childbed fever was associated with obstetricians on teaching unit

Midwife and home births had lower rates

Semmelweis postulated that the key difference between divisions I and II was participation of the physicians and students in autopsies. One or more cadavers were dissected daily, some from cases of childbed fever and other infections. Handwashing was perfunctory, and Semmelweis believed this allowed the transmission of "invisible cadaver particles" by direct contact between the mother and the physician's hands during examinations and delivery. In 1847, as a countermeasure, he required handwashing with a chlorine solution until the hands were slippery and the odor of the cadaver was gone. The results were dramatic. The full effect of the chlorine handwashing can be seen by comparing mortality rates in the two

Transmission from cadavers was suspected

TABLE 3–2 Childbed Fever at the Vienna General Hospital

	DIVISION I (TEACHING UNIT)			DIVISION II (MIDWIFE UNIT)		
YEAR	BIRTHS	MATERNAL DEATHS	PERCENTAGE	BIRTHS	MATERNAL DEATHS	PERCENTAGE
1846[a]	4010	459	11.4	3754	105	2.7
1848[b]	3556	45	1.3	3219	43	1.3

[a]No handwashing.
[b]First full year of chlorine handwashing

Disinfectant handwashing reduced the infection rates

divisions for 1846 and 1848 (**Table 3–2**). The mortality rate in division I was reduced to that of division II, and both were lower than 2%.

Unfortunately, because of his personality and failure to publish his work until 1860, Semmelweis's contribution was not generally appreciated in his lifetime. As his frustration mounted over lack of acceptance of his ideas, he became abusive and irrational, eventually alienating even his early supporters. Some believe that he also suffered from Alzheimer's disease. He died in an insane asylum in 1865, unaware that his concept of spread via direct contact would later be recognized as the most important mechanism of nosocomial infection and that handwashing would remain the most important means of infection control in hospitals.

NOSOCOMIAL INFECTIONS AND THEIR SOURCES

Community-acquired infections are acquired before admission

Nosocomial infections are acquired in hospital

Infections occurring during any hospitalization are either community acquired or nosocomial. Community-acquired infections are defined as those present or incubating at the time of hospital admission. All others are considered nosocomial. For example, a hospital case of chickenpox could be community acquired if it erupted on the fifth hospital day (incubating) or nosocomial if hospitalization was beyond the limits of the known incubation period (20 days). Infections appearing shortly after discharge (2 weeks) are considered nosocomial, although some could have been acquired at home. Infectious hazards are inherent to the hospital environment; it is there that the most seriously infected and most susceptible patients are housed and often cared for by the same staff.

Endogenous infections are part of hospital risk

The infectious agents responsible for nosocomial infections arise from various sources, including patients' own normal flora. In addition to any immunocompromising disease or therapy, the hospital may impose additional risks by treatments that breach the normal defense barriers. Surgery, urinary or intravenous catheters, and invasive diagnostic procedures all may provide normal flora with access to usually sterile sites. Infections in which the source of organisms is the hospital rather than the patient include those derived from hospital personnel, the environment, and medical equipment.

Hospital Personnel

Cross-infection is usually by direct contact

Infected medical attendants are particularly dangerous

Physicians, nurses, students, therapists, and any others who come in contact with the patient may transmit infection. Transmission from one patient to another is called **cross-infection.** The vehicle of transmission is most often the inadequately washed hands of a medical attendant. Another source is the infected medical attendant. Many hospital outbreaks have been traced to hospital personnel, particularly physicians, who continue to care for patients despite an overt infection. Transmission is usually by direct contact, although airborne transmission is also possible. A third source is the person who is not ill but is carrying a virulent strain. For *Staphylococcus aureus* and group A streptococci, nasal carriage is most important, but sites such as the perineum and anus have also been involved in outbreaks. An occult carrier is less often the source of nosocomial infection than a physician covering up a boil or a nurse minimizing "the flu." The carrier is difficult to detect unless the epidemic strain has distinctive characteristics or the epidemiologic circumstances point to a single person.

Infection from carriers can transmit to patients

Environment

The hospital air, walls, floors, linens, and the like are not sterile and thus could serve as a source of organisms causing nosocomial infections, but the importance of this route has generally been exaggerated. With the exception of the immediate vicinity of an infected individual or a carrier, transmission through the air or on fomites is much less important than that caused by personnel or equipment. Notable exceptions are when the environment becomes contaminated with *Mycobacterium tuberculosis* from a patient or *L egionella pneumophila* in the water supply. These events are most likely to result in disease when the organisms are numerous or the patient is particularly vulnerable (eg, after heart surgery or bone marrow transplantation).

Environmental contamination is relatively unimportant

M tuberculosis and **Legionella** are risks

Medical Devices

Much of the success of modern medicine is related to medical devices that support or monitor basic body functions. By their very nature, devices such as catheters and respirators carry a risk of nosocomial infection because they bypass normal defense barriers, providing microorganisms access to normally sterile fluids and tissues. Most of the recognized causes are bacterial or fungal. The risk of infection is related to the degree of debilitation of the patient and various factors concerning the design and management of the device. Any device that crosses the skin or a mucosal barrier allows flora in the patient or environment to gain access to deeper sites around the outside surface. Possible access inside the device (eg, in the lumen) adds another and sometimes greater risk. In some devices, such as urinary catheters, contamination is avoidable; in others, such as respirators, complete sterility is either impossible or impractical to achieve.

Equipment that crosses epithelial barriers provides microbial access

The risk of contamination leading to infection is increased if organisms that gain access can multiply within the system. The availability of water, nutrients, and a suitable temperature largely determine which organism will survive and multiply. Many of the Gram-negative rods such as *Pseudomonas, Acinetobacter,* and members of Enterobacteriaceae can multiply in an environment containing water and little else. Gram-positive bacteria generally require more physiologic conditions.

Conditions for bacterial growth increase risk

Even with proper growth conditions, many hours are required before contaminating organisms become numerous. Detailed studies of catheters and similar devices show that the risk of infection begins to increase after 24 to 48 hours and is cumulative even if the device is changed or disinfected at intervals. It is thus important to discontinue transcutaneous procedures as soon as medically indicated. The medical devices most frequently associated with nosocomial infections are listed in the following text. The infectious risk of others can be estimated from the principles discussed previously. New devices are constantly being introduced into medical care, occasionally without adequate consideration of their potential to cause nosocomial infection.

Transcutaneous and indwelling devices should be changed as soon as possible

Urinary Catheters

Urinary tract infection (UTI) accounts for 40% to 50% of all nosocomial infections, and at least 80% of these are associated with catheterization. The infectious risk of a single urinary catheterization has been estimated at 1%, and indwelling catheters carry a risk that may be as high as 10%. The major preventive measure is maintenance of a completely closed system through the use of valves and aspiration ports designed to prevent bacterial access to the inside of the catheter or collecting bag. Unfortunately, breaks in closed systems eventually occur when the system is in place for more than 30 days. The urine itself serves as an excellent culture medium once bacteria gain access.

Closed urinary drainage systems are still violated

Vascular Catheters

Needles and plastic catheters placed in veins (or, less often, in arteries) for fluid administration, monitoring vital functions, or diagnostic procedures are a leading cause of nosocomial bacteremia. These sites should always be suspected as a source of organisms whenever blood cultures are positive with no apparent primary site for the bacteremia. Contamination at the insertion site is generally staphylococcal, with continued growth in the catheter tip. Organisms may gain access somewhere in the lines, valves, bags, or bottles of intravenous solutions proximal to the insertion site. The latter circumstance usually involves Gram-negative rods. Preventive measures include aseptic insertion technique and appropriate care of the lines, including changes at regular intervals.

Skin is primary source for intravenous contamination

Respirators

Machines that assist or control respiration by pumping air directly into the trachea have a great potential for causing nosocomial pneumonia if the aerosol they deliver becomes contaminated. Bacterial growth is significant only in the parts of the system that contain water; in systems using nebulizers, bacteria can be suspended in water droplets small enough to reach the alveoli. The organisms involved include *Pseudomonas,* Enterobacteriaceae, and a wide variety of environmental bacteria such as *Acinetobacter.* The primary control measure is periodic changing and disinfection of the tubing, reservoirs, and nebulizer jets.

Changing controls nebulizer contamination

Blood and Blood Products

Infections related to contact with blood and blood products are generally a risk for health care workers rather than patients. Manipulations ranging from phlebotomy and hemodialysis to surgery carry varying risk of blood containing an infectious agent reaching mucous membranes or skin of the health care worker. The major agents transmitted in this manner are hepatitis B, hepatitis C, and human immunodeficiency virus (HIV). Control requires meticulous attention to procedures that prevent direct contact with blood, such as the use of gloves, eyewear, and gowns. Cuts and needle sticks among health care workers carry a risk approaching 2%. Identification of hepatitis virus and HIV carriers is a part of a protective process that must be balanced by patient privacy considerations. Health care facilities all have established policies concerning serologic surveillance of patients and the procedures to follow (eg, testing, prophylaxis) when blood-related accidents occur. Similarly, products for transfusion undergo extensive screening to protect the recipient.

Risk of hepatitis B, hepatitis C, and HIV is related to blood manipulation

Screen is determined by institutional policy

INFECTION CONTROL

Infection control is the sum of all the means used to prevent nosocomial infections. Historically, such methods have been developed as an integral part of the study of infectious diseases, often serving as key elements in the proof of infectious etiology. Semmelweis's hand-washing is the first example. Later in the 19th century, Joseph Lister achieved a dramatic reduction in surgical wound infections by infusion of a phenolic antiseptic into wounds. This local destruction of organisms was known as **antisepsis,** and it sometimes included liberal applications of disinfectants, including sprays to the environment. As it became recognized that contamination of wounds was not inevitable, the emphasis gradually shifted to preventing contact between microorganisms and susceptible sites, a concept called **asepsis.** Asepsis, which combines containment with the methods of sterilization and disinfection previously discussed, is the central concept of infection control. The measures taken to achieve asepsis vary, depending on whether the circumstances and environment are most similar to the operating room, hospital ward, or outpatient clinic.

Antisepsis attacks contaminating organisms

Asepsis prevents contamination

Asepsis

Operating Room

The surgical suite and operating room represent the most controlled and rigid application of aseptic principles. The procedure begins with the use of an antiseptic scrub of the skin over the operative site and the hands and forearms of all who will have contact with the patient. The use of sterile drapes, gowns, and instruments serves to prevent spread through direct contact, and caps and face masks reduce airborne spread from personnel to the wound. As all students learn the first time they scrub, even the manner of dressing and moving in the operating room are rigidly specified, and those involved assume a strict aseptic attitude as well as their masks and gowns. The level of bacteria in the air is generally related more to the number of persons and amount of movement in the operating room than to incoming air. The net effect of these procedures is to draw a sterile curtain around the operative site, thus minimizing contact with microorganisms. Surgical asepsis is also used in other areas where invasive special procedures such as cardiac catheterization are performed.

Sterile drapes and instruments prevent contact of organisms with wound

Airborne bacteria are associated with personnel in operating room

Hospital Ward

Although theoretically desirable, strict aseptic procedures as used in the operating room are impractical in the ward setting. Asepsis is practiced by the use of sterile needles, medications, dressings, and other items that could serve as transmission vehicles if contaminated. A "no touch" technique for examining wounds and changing dressings eliminates direct contact with any nonsterile item. Invasive procedures such as catheter insertion and lumbar punctures are performed under aseptic precautions similar to those used in the operating room. In all circumstances, handwashing between patient contacts is the single most important aseptic precaution.

Handwashing is the most important measure

Outpatient Clinic

The general aseptic practices used on the hospital ward are also appropriate to the outpatient situation as preventive measures. The potential for cross-infection in the clinic or waiting room is obvious but has been little studied regarding preventive measures. Patients who may be infected should be segregated whenever possible using techniques similar to those of hospital ward isolation. The examining room may be used in a manner analogous to the private rooms on a hospital ward. Although this approach is difficult because of patient turnover, it should be attempted for infections that would require strict or respiratory isolation in the hospital.

Waiting areas present a risk

Isolation Procedures

Patients with infections pose special problems because they may transmit their infections to other patients either directly or by contact with a staff member. This additional risk is managed by the techniques of isolation, which place barriers between the infected patient and others on the ward. Because not every infected patient presents with suspect signs and/or symptoms, some precaution should be taken with all patients. In the system recommended by the Centers for Disease Control and Prevention, these are called **standard precautions** and include the use of gowns and gloves when in contact with patient blood or secretions. These are particularly directed at protecting health care workers from HIV and hepatitis infection. For those with suspected or proven infection, additional precautions are taken, the nature of which is determined by the known mode of transmission of the organism. These **transmission-based precautions** are divided into those directed at airborne, droplet, and contact routes. The **airborne** transmission precautions are for infections known to be transmitted by extremely small (less than 5 μm) particles suspended in the air. This requires that the room air circulation be maintained with negative pressure relative to the surrounding area and be exhausted to the outside. Those entering the room must wear surgical masks, and in the case of tuberculosis, specially designed respirators. **Droplet** precautions are for infections in which the organisms are suspended in larger droplets, which may be airborne, but generally do not travel more than 3 feet from the patient who generates them. These can be contained by the use of gowns, gloves, and masks when working close to the patient. **Contact** precautions are used for infections that require direct contact with organisms on or that pass in secretions of the patient. Diarrheal infections are of special concern because of the extent to which they contaminate the environment. Details of the precautions and examples of the typical infectious agents are summarized in **Table 3–3**.

Standard precautions protect health care workers from HIV

Transmission precautions block airborne, droplet, and contact routes

Organization

Modern hospitals are required to have formal infection control programs that include an infection control committee, epidemiology service, and educational activities. The infection control committee is composed of representatives of various medical, administrative, nursing, housekeeping, and support services. The committee establishes the institution's

TABLE 3–3 Precautions for Prevention of Nosocomial Infections

PRECAUTION	ROOM	HANDWASHING[a]	GLOVES	GOWNS	MASK[b]	TYPICAL DISEASES
Standard		After removing gloves, between patients	Blood, fluid contact, touching skin	Blood, fluid contact, during procedures	During procedures	All
Transmission-based						
Airborne	Private, negative pressure[c]	After removing gloves, between patients	Room entry	Room entry	Room entry or respirator[d]	Measles, chickenpox, tuberculosis[d]
Droplet	Private[e]	After removing gloves, between patients	Blood, fluid contact	Blood, fluid contact	Within 3 feet of patient	Meningitis, pertussis, plague, influenza
Contact	Private[e]	After removing gloves, between patients	Room entry	Patient contact	—	Infectious diarrhea[f], *S aureus* wounds

[a]Using a disinfectant soap.
[b]Standard surgical mask, goggles.
[c]Room pressure must be negative in relation to surrounding area and the circulation exhausted outside the building.
[d]For patients with diagnosed or suspect tuberculosis, a specially filtered respirator/mask must be worn.
[e]Door may be left open and patients with the same organism may share a room.
[f]Particularly *Clostridium difficile, Escherichia coli* O:157, *Shigella,* and incontinent patient. shedding rotavirus or hepatitis A.

Infection control programs determine and enforce policy

infection control procedures and regularly reviews information on the status of nosocomial infections in the hospital. When epidemiologic circumstances warrant it, the committee must be empowered to take drastic action such as closing a hospital unit or suspending a physician's privileges.

Epidemiologic surveillance and outbreak investigation are required

The epidemiology service is the working arm of the infection control committee. Its functions are performed by one or more epidemiologists who usually have a nursing background. This work requires familiarity with clinical microbiology, epidemiology, infectious disease, and hospital procedures, as well as immense tact. The main activities are surveillance and outbreak investigation. Surveillance is the collection of data documenting the frequency and nature of nosocomial infections in the hospital to detect deviations from the institutional or national norms. Although routine microbiologic sampling of the hospital environment is of no value, programs to sample some of the medical devices known to be nosocomial hazards can be useful. On-the-spot investigation of potential outbreaks allows early implementation of preventive measures. This activity is probably the single most important function of the epidemiology service. Suspicion of an increased number of infections leads to an investigation to verify the facts, establish basic epidemiologic associations, and relate them to preventive measures. The primary concern is cross-infection, in which a virulent organism is being transmitted from patient to patient. :

Prevention

The prevention of nosocomial infections is contingent on basic and applied knowledge drawn from all parts of this book. Applied with common sense, these principles can both prevent disease and reduce the costs of medical care.

CHAPTER 4

Principles of Laboratory Diagnosis of Infectious Diseases

The diagnosis of a microbial infection begins with an assessment of clinical and epidemiologic features, leading to the formulation of a diagnostic hypothesis. Anatomic localization of the infection with the aid of physical and radiologic findings (eg, right lower lobe pneumonia, subphrenic abscess) is usually included. This clinical diagnosis suggests a number of possible etiologic agents based on knowledge of infectious syndromes and their courses (see Chapters 57 through 66). The specific cause is then established by the application of methods described in this chapter. A combination of science and art on the part of both the clinician and laboratory worker is required: The clinician must select the appropriate tests and specimens to be processed and, where appropriate, suggest the suspected etiologic agents to the laboratory. The laboratory worker must use the methods that will demonstrate the probable agents and be prepared to explore other possibilities suggested by the clinical situation or by the findings of the laboratory examinations. The best results are obtained when communication between the clinic and laboratory is maximal.

The general approaches to laboratory diagnosis vary with different microorganisms and infectious diseases. However, the types of methods are usually some combination of direct microscopic examinations, culture, antigen detection, and antibody detection (serology). Nucleic acid amplification assays that allow direct detection of genomic components of pathogens are now numerous, but few are practical for routine use. This chapter considers the principles of infectious disease laboratory diagnosis. Details with regard to particular agents are discussed in their chapters and with regard to clinical situations in Chapters 57 through 66. All diagnostic approaches begin with some kind of specimen collected from the patient.

Microscopic, culture, antigen, and antibody detection are classic methods

Genomic approaches are being developed

THE SPECIMEN

The primary connection between the clinical encounter and the diagnostic laboratory is the specimen submitted for processing. If it is not appropriately chosen and/or collected, no degree of laboratory skill can rectify the error. Failure at the level of specimen collection is the most common reason for failing to establish an etiologic diagnosis, or worse, for suggesting a wrong diagnosis. In the case of bacterial infections, the primary problem lies in distinguishing resident or contaminating normal floral organisms from those causing the infection. The three specimen categories illustrated in **Figure 4–1A-C** are discussed in the text that follows.

Quality of the specimen is crucial

Direct Tissue or Fluid Samples

Direct specimens (Figure 4–1A) are collected from normally sterile tissues (lung, liver) and body fluids (cerebrospinal fluid, blood). The methods range from needle aspiration

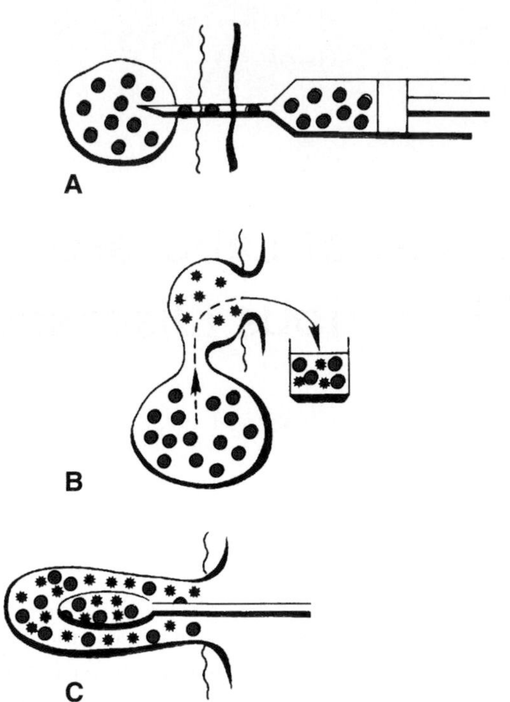

FIGURE 4–1. Specimens for the diagnosis of infection. A. Direct specimen. The pathogen is localized in an otherwise sterile site, and a barrier such as the skin must be passed to sample it. This may be done surgically or by needle aspiration as shown. The specimen collected contains only the pathogen. Examples are deep abscess and cerebrospinal fluid. **B.** Indirect sample. The pathogen is localized as in A, but must pass through a site containing normal flora in order to be collected. The specimen contains the pathogen, but is contaminated with the nonpathogenic flora. The degree of contamination is often related to the skill with which the normal floral site was "bypassed" in specimen collection. Examples are expectorated sputum and voided urine. **C.** Sample from site with normal flora. The pathogen and nonpathogenic flora are mixed at the site of infection. Both are collected and the non-pathogen is either inhibited by the use of selective culture methods or discounted in interpretation of culture results. Examples are throat and stool.

Direct samples give highest quality and risk

of an abscess to surgical biopsy. In general, such collections require the direct involvement of a physician and may carry some risk for the patient. The results are always useful because positive findings are diagnostic and negative findings can exclude infection at the suspected site.

Indirect Samples

Bypassing the normal flora requires extra effort

Results require interpretive evaluation of contamination

Indirect samples (Figure 4–1B) are specimens of inflammatory exudates (expectorated sputum, voided urine) that have passed through sites known to be colonized with normal flora. The site of origin is usually sterile in healthy persons; however, some assessment of the probability of contamination with normal flora during collection is necessary in interpretation of the results. This assessment requires knowledge of the potential contaminating flora as well as the probable pathogens to be sought. Indirect samples are usually more convenient for both physician and patient, but carry a higher risk of misinterpretation. For some specimens, such as expectorated sputum, guidelines to assess specimen quality have been developed by correlation of clinical and microbiologic findings (see Chapter 61).

::: Normal flora, p.9

Samples from Normal Flora Sites

Strict pathogens can be specifically sought

Frequently, the primary site of infection is in an area known to be colonized with many organisms (pharynx and large intestine) (Figure 4–1C). This is primarily an issue with bacterial diagnosis because they dominate the makeup of the normal flora. In such instances, examinations are selectively made for organisms known to cause infection that are not normally found at the infected site. For example, the enteric pathogens *Salmonella, Shigella,* and *Campylobacter* may be selectively sought in a stool specimen or only β-hemolytic streptococci in a throat culture. In these instances, selective media that inhibit growth of the other bacteria are used, or others growing are simply ignored.

Lack of normal viral flora simplifies interpretation

The selection of specimens for viral diagnosis is easier because there is usually little normal viral flora to confuse interpretation. This allows selection guided by knowledge of which sites are most likely to yield the suspected etiologic agent. For example, enteroviruses are the most common viruses involved in acute infection of the central nervous system.

Specimens that might be expected to yield these agents on culture include throat, stool, and cerebrospinal fluid.

Specimen Collection and Transport

The **sterile swab** is the most convenient and most commonly used tool for specimen collection; however, it provides the poorest conditions for survival and can only absorb a small volume of inflammatory exudate. The worst possible specimen is a dried-out swab; the best is a collection of 5 to 10 mL or more of the infected fluid or tissue. The volume is important because infecting organisms that are present in small numbers may not be detected in a small sample.

Swabs limit volume and survival

Specimens should be transported to the laboratory as soon after collection as possible because some microorganisms survive only briefly outside the body. For example, unless special **transport media** are used, isolation rates of the organism that causes gonorrhea (*Neisseria gonorrhoeae*) are decreased when processing is delayed beyond a few minutes. Likewise, many respiratory viruses survive poorly outside the body. On the other hand, some bacteria survive well and may even multiply after the specimen is collected. The growth of enteric Gram-negative rods in specimens awaiting culture may in fact compromise specimen interpretation and interfere with the isolation of more fastidious organisms. Significant changes are associated with delays of more than 3 to 4 hours.

Viability may be lost if specimen is delayed

Various transport media have been developed to minimize the effects of the delay between specimen collection and laboratory processing. In general, they are buffered fluid or semisolid media containing minimal nutrients and are designed to prevent drying, maintain a neutral pH, and minimize growth of bacterial contaminants. Other features may be required to meet special requirements, such as an oxygen-free atmosphere for obligate anaerobes.

Transport media stabilize conditions and prevent drying

DIRECT EXAMINATION

Of the infectious agents discussed in this book, only some of the parasites are large enough to be seen with the naked eye. Bacteria and fungi can be seen clearly with the light microscope when appropriate methods are used. Individual viruses can be seen only with the electron microscope, although aggregates of viral particles in cells (viral inclusions) may be seen by light microscopy. Various stains are used to visualize and differentiate microorganisms in smears and histologic sections.

All but some parasites require microscopy for visualization

Light Microscopy

Direct examination of stained or unstained preparations by **light (bright-field) microscopy** (**Figure 4–2A**) is particularly useful for detection of bacteria, fungi, and parasites. Even the smallest bacteria (1–2 μm wide) can be visualized, although all require staining and some require special lighting techniques. As the resolution limit of the light microscope is near 0.2 μm, the optics must be ideal if small organisms are to be seen clearly by direct microscopy. These conditions may be achieved with a 100× oil immersion objective, a 5 to 10× eyepiece, and optimal lighting.

Bacteria are visible if optics are maximized

Bacteria may be stained by a variety of dyes, including methylene blue, crystal violet, carbol-fuchsin (red), and safranin (red). The two most important methods, the Gram and acid-fast techniques, use staining, decolorization, and counterstaining in a manner that helps to classify as well as stain the organism.

Bacteria must be stained

The Gram Stain

The differential staining procedure described in 1884 by the Danish physician Hans Christian Gram has proved one of the most useful in microbiology and medicine. The procedure (**Figure 4–3A**) involves the application of a solution of iodine in potassium iodide to cells previously stained with an acridine dye such as crystal violet. This treatment produces a mordanting action in which purple insoluble complexes are formed with ribonuclear protein in the cell. The difference between Gram-positive and Gram-negative bacteria is in the

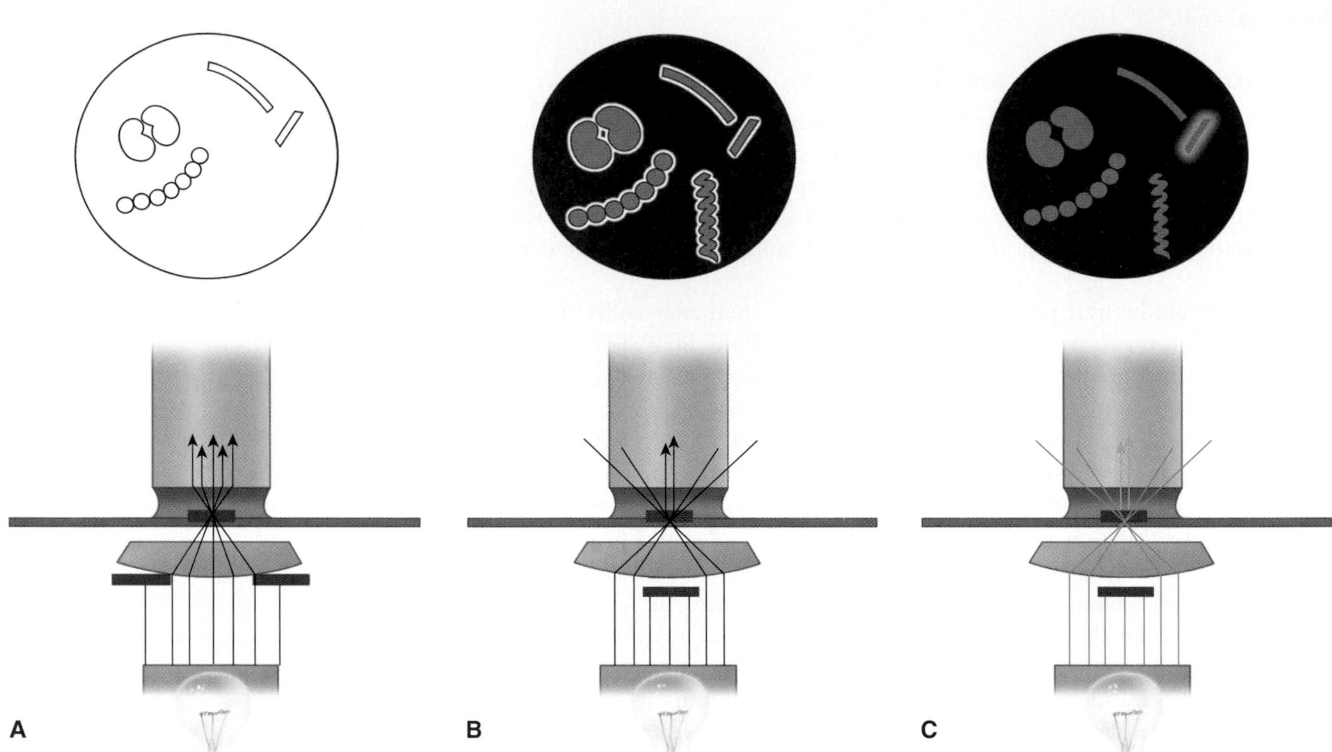

FIGURE 4–2. Bright-field, dark-field, and fluorescence microscopy. A. Bright-field illumination properly aligned. The purpose is to focus light directly on the preparation for optimal visualization against a bright background. **B.** In dark-field illumination, a black background is created by blocking the central light. Peripheral light is focused so that it will be collected by the objective only when it is reflected from the surfaces of particles (eg, bacteria). The microscopic field shows bright halos around some bacteria and reveals a spirochete too thin to be seen with bright-field illumination. **C.** Fluorescence microscopy is similar to dark-field microscopy, except that the light source is ultraviolet and the organisms are stained with fluorescent compounds. The incident light generates light of a different wavelength, which is seen as a halo (colored in this illustration) around only the organism tagged with fluorescent compounds. For the most common fluorescent compound, the light is green.

permeability of the cell wall to these complexes on treatment with mixtures of acetone and alcohol solvents. This extracts the purple iodine–dye complexes from Gram-negative cells, whereas Gram-positive bacteria retain them. An intact cell wall is necessary for a positive reaction, and Gram-positive bacteria may fail to retain the stain if the organisms are old, dead, or damaged by antimicrobial agents. No similar conditions cause a Gram-negative organism to appear Gram positive. The stain is completed by the addition of red counterstain such as safranin, which is taken up by bacteria that have been decolorized. Thus, cells stained purple are Gram positive, and those stained red are Gram negative. As indicated in Chapter 21, Gram positivity and negativity correspond to major structural differences in the cell wall.

Gram-positive bacteria retain purple iodine–dye complexes

Gram-negative bacteria do not retain complexes when decolorized

In many bacterial infections, the etiologic agents are readily seen on stained Gram smears of pus or fluids. The purple or red bacteria are seen against a Gram-negative (red) background of leukocytes, exudate, and debris (**Figures 4–3A** and **4-4C**). This information, combined with the clinical findings, may guide the management of infection before culture results are available. Interpretation requires considerable experience and knowledge of probable causes, of their morphology and Gram reaction, and of any organisms normally present in health at the infected site.

Properly decolorized background should be red

Gram reaction plus morphology guide clinical decisions

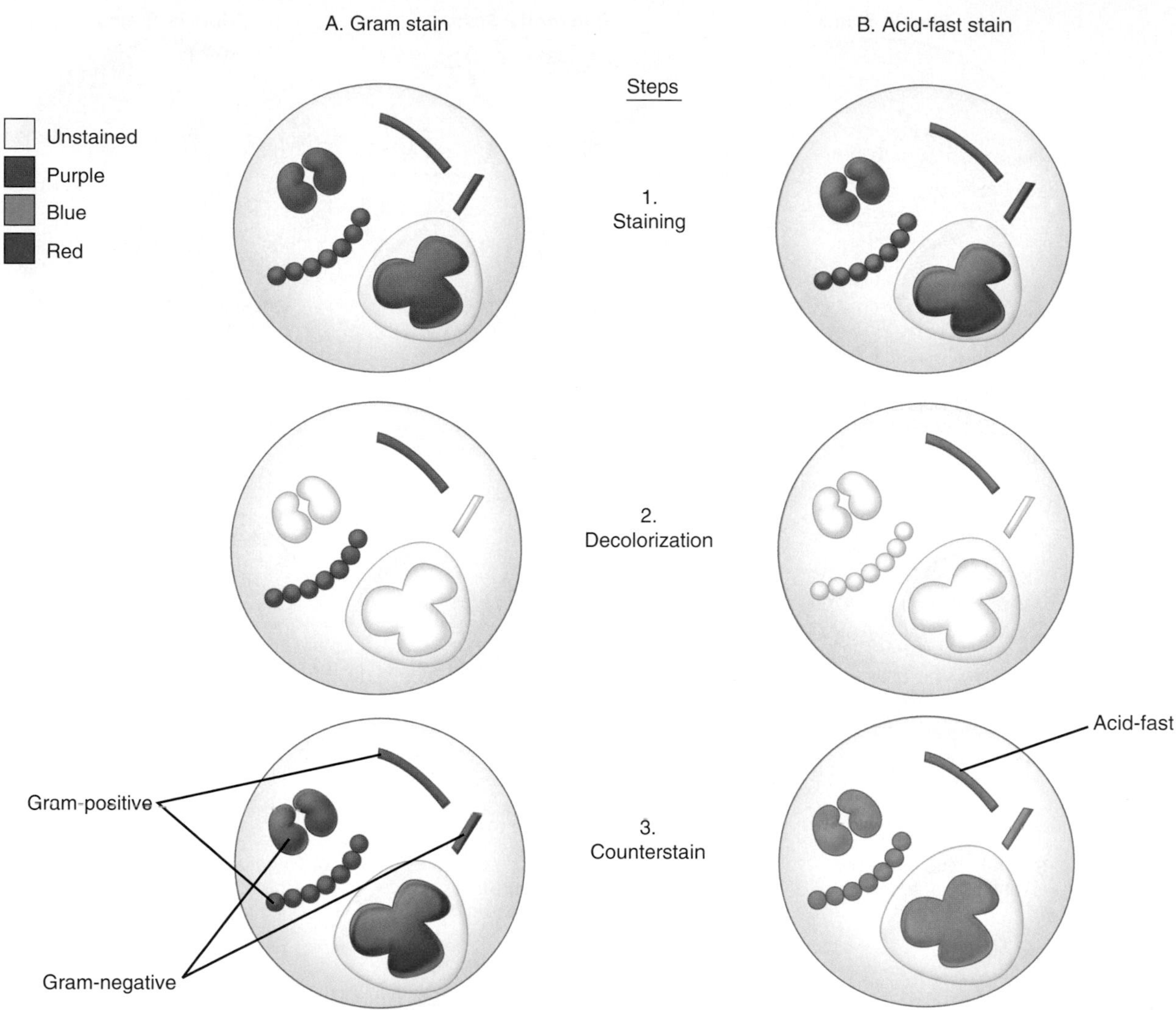

FIGURE 4–3. Gram and acid-fast stains. Four bacteria and a polymorphonuclear neutrophil are shown at each stage. All are initially stained purple by the crystal violet and iodine of the Gram stain (A1) and red by the carbol fuchsin of the acid-fast stain (B1). After decolorization, Gram-positive and acid-fast organisms retain their original stain. Others are unstained (A2, B2). The safranin of the Gram counterstain stains the Gram-negative bacteria and makes the background red (A3), and the methylene blue leaves a blue background for the contrasting red acid-fast bacillus (B3).

The Acid-Fast Stain

Acid fastness is a property of the mycobacteria (eg, *Mycobacterium tuberculosis*) and related organisms. Acid-fast organisms generally stain very poorly with dyes, including those used in the Gram stain. However, they can be stained by prolonged application of more concentrated dyes, by penetrating agents, or by heat treatment. Their unique feature is that once stained, acid-fast bacteria resist decolorization by concentrations of mineral acids and ethanol that remove the same dyes from other bacteria. This combination of weak initial staining and strong retention once stained is related to the high lipid content of the mycobacterial cell wall. Acid-fast stains are completed with a counterstain to provide a contrasting background for viewing the stained bacteria (Figure 4–3B).

Acid-fast bacteria take stains poorly

Once stained, they retain it strongly

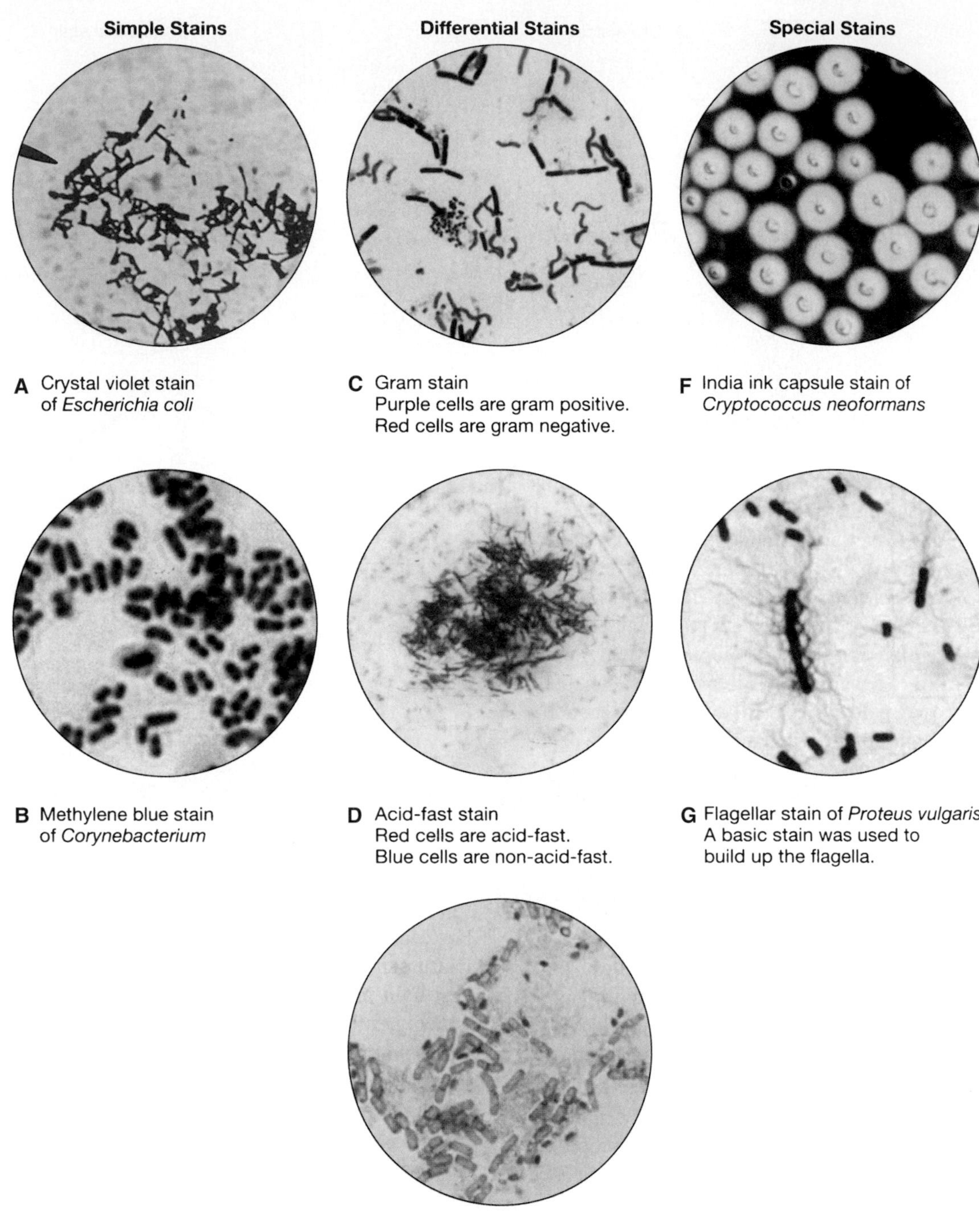

Simple Stains
Differential Stains
Special Stains
A Crystal violet stain of *Escherichia coli*
C Gram stain
Purple cells are gram positive.
Red cells are gram negative.
F India ink capsule stain of *Cryptococcus neoformans*
B Methylene blue stain of *Corynebacterium*
D Acid-fast stain
Red cells are acid-fast.
Blue cells are non-acid-fast.
G Flagellar stain of *Proteus vulgaris*.
A basic stain was used to build up the flagella.
E Endospore stain, showing endospores (red) and vegetative cells (blue)

FIGURE 4–4. Types of microbiologic stains. (Reproduced with permission from Willey J, Sherwood L, Woolverton C (eds). *Prescott's Principles of Microbiology.* New York: McGraw-Hill; 2008.)

There are multiple variants of the acid-fast stain

In the acid-fast procedure, the slide is flooded with carbol-fuchsin (red) and decolorized with hydrochloric acid in alcohol. When counterstained with methylene blue, acid-fast organisms appear red against a blue background (Figure 4–4B). A variant is the **fluorochrome stain,** which uses a fluorescent dye, (auramine, or an auramine–rhodamine mixture) followed by decolorization with acid–alcohol. Acid-fast organisms retain the fluorescent stain, which allows their visualization by fluorescence microscopy.

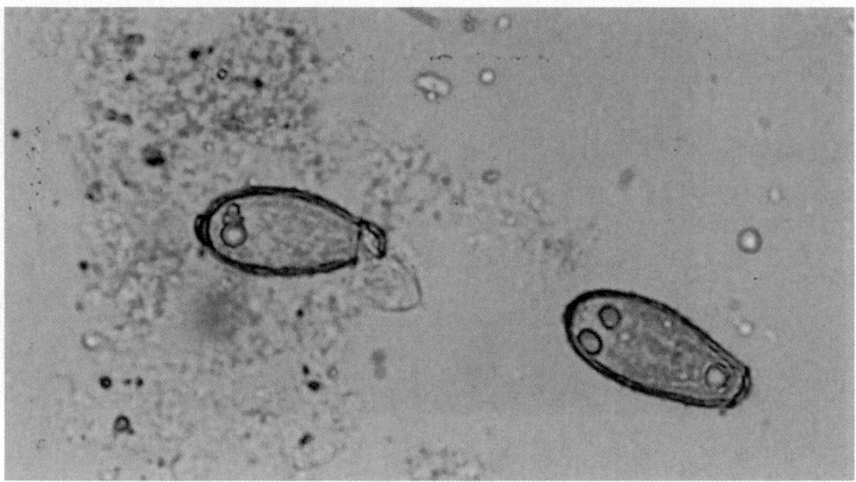

FIGURE 4–5. Iodine-stained parasite eggs. Two eggs of the intestinal fluke *Clonorchis sinensis* are present in this stool specimen. (Reproduced with permission from Connor DH, Chandler FW, Schwartz DQA, Manz HJ, Lack EE (eds). *Pathology of Infectious Diseases,* vol. 1. Stamford, CT: Appleton & Lange; 1997.)

Fungal and Parasitic Stains

The smallest fungi are the size of large bacteria and all parasitic forms are larger. This allows detection in simple wet mount preparations often without staining. Fungi in sputum or body fluids can be seen by mixing the specimen with a potassium hydroxide solution (to dissolve debris) and viewing with a medium power lens. The use of simple stains or the fluorescent calcofluor white improves the sensitivity of detection. Another technique is to mix the specimen with India ink, which outlines the fugal cells (Figure 4–4F). Detection of the cysts and eggs of parasites requires a concentration procedure if the specimen is stool, but once done they can be visualized with a simple iodine stain (**Figure 4–5**).

Fungi and parasites visible with simple stains

Dark-field and Fluorescence Microscopy

Some bacteria, such as *Treponema pallidum*, the cause of syphilis, are too thin to be visualized with the usual bright-field illumination. They can be seen by use of the dark-field technique. With this method, a condenser focuses light diagonally on the specimen in such a way that only light reflected from particulate matter such as bacteria reaches the eyepiece (Figure 4–2B). The angles of incident and reflected light are such that the organisms are surrounded by a bright halo against a black background. This type of illumination is also used in other microscopic techniques, in which a high light contrast is desired, and for observation of fluorescence. Fluorescent compounds, when excited by incident light of one wavelength, emit light of a longer wavelength and thus a different color. When the fluorescent compound is conjugated with an antibody as a probe for detection of a specific antigen, the technique is called **immunofluorescence,** or fluorescent antibody microscopy (**Figure 4–6**). The appearance is the same as in dark-field microscopy except that the halo is the emitted color of the fluorescent compound (Figures 4–2C and 4–6 C, D). For improved safety, most modern fluorescence microscopy systems direct the incident light through the objective from above (epifluorescence).

Dark-field creates a halo around organisms too thin to see by bright-field

Fluorescent stains convert dark-field to fluorescence microscopy

Electron Microscopy

Electron microscopy demonstrates structures by transmission of an electron beam and has 10 to 1000 times the resolving power of light microscopic methods. For practical reasons its diagnostic application is limited to virology, where, because of the resolution possible at high magnification, it offers results not possible by any other method. Using negative staining techniques, direct examination of fluids, and tissues from affected body sites enables visualization of viral particles. In some instances, electron microscopy has been the primary means of discovery of viruses that do not grow in the usual cell culture systems.

Viruses are visible only by electron microscopy

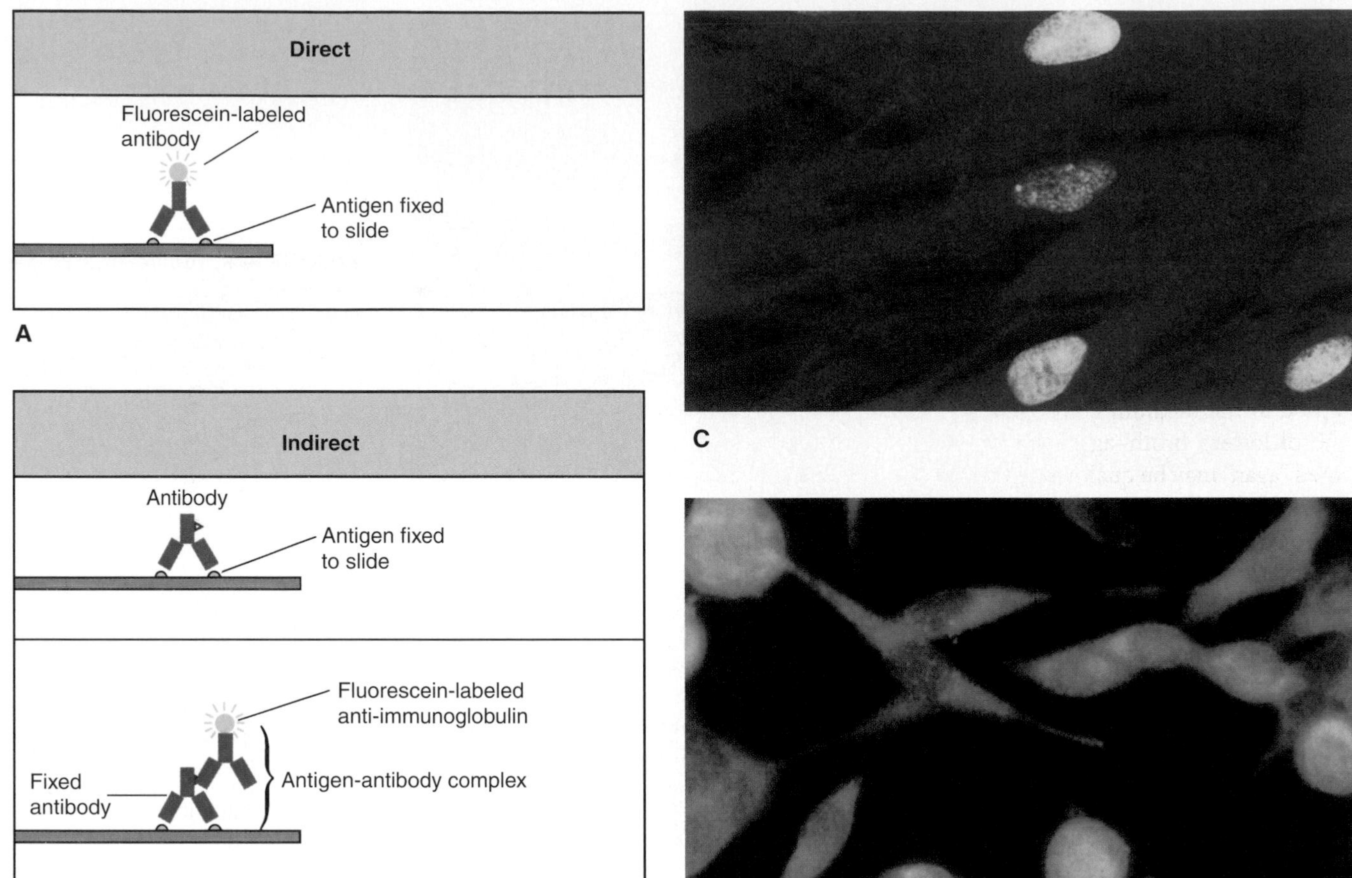

FIGURE 4–6. Direct and indirect immunofluorescence. A. In the direct fluorescent antibody (DFA), the specimen containing antigen is fixed to a slide. Fluorescently labeled antibodies that recognize the antigen are then added, and the specimen is examined with a fluorescence microscope for yellow-green fluorescence. **B.** The indirect fluorescent antibody technique (IFA) detects antigen on a slide as it reacts with an unlabeled antibody directed against it. The original antigen–antibody complex is detected with a second labeled antibody that recognizes any antibody. **C.** Three infected nuclei in a cytomegalovirus positive tissue culture. **D.** Several infected cells in a herpes simplex virus positive tissue culture. (Reproduced with permission from Willey J, Sherwood L, Woolverton C (eds). *Prescott's Principles of Microbiology.* New York: McGraw-Hill; 2008.)

CULTURE

Growth and identification of the infecting agent in vitro is usually the most sensitive and specific means of diagnosis and is thus the method most commonly used. Theoretically, the presence of a single live organism in the specimen can yield a positive result. Most bacteria and fungi can be grown in a variety of artificial media, but strict intracellular microorganisms (eg, *Chlamydia*, *Rickettsia*, and viruses) can be isolated only in cultures of living eukaryotic cells. The culture of some parasites is possible but used only in highly specialized laboratories.

Isolation and Identification of Bacteria and Fungi

Almost all medically important bacteria can be cultivated outside the host in artificial culture media. A single bacterium placed in the proper culture conditions multiplies to quantities sufficient to be seen by the naked eye. Bacteriologic media are soup-like recipes prepared from digests of animal or vegetable protein supplemented with nutrients such as glucose, yeast extract, serum, or blood to meet the metabolic requirements of the organism. Their chemical composition is complex, and their success depends on matching the nutritional requirements of most heterotrophic living things. The same approaches and some of the same culture media used for bacteria are also used for fungi.

Bacteria grow in soup-like media

Growth in media prepared in the fluid state (broths) is apparent when bacterial numbers are sufficient to produce turbidity or macroscopic clumps. Turbidity results from reflection of transmitted light by the bacteria; depending on the size of the organism, more than 10^6 bacteria per milliliter of broth are required. The addition of a gelling agent to a broth medium allows its preparation in solid form as plates in Petri dishes. The universal gelling agent for diagnostic bacteriology is **agar,** a polysaccharide extracted from seaweed. Agar has the convenient property of becoming liquid at about 95°C but not returning to the solid gel state until cooled to less than 50°C. This allows the addition of a heat-labile substance such as blood to the medium before it sets. At temperatures used in the diagnostic laboratory (37°C or lower), broth–agar exists as a smooth, solid, nutrient gel. This medium, usually termed "agar," may be qualified with a description of any supplement (eg, blood agar).

Large numbers of bacteria in broth produce turbidity

Agar is a convenient gelling agent for solid media

A useful feature of agar plates is that the bacteria can be separated by spreading a small sample of the specimen over the surface. Bacterial cells that are well separated from others grow as isolated colonies, often reaching 2 to 3 mm in diameter after overnight incubation. This allows isolation of bacteria in pure culture because the colony is assumed to arise from a single organism (**Figure 4–7**). Colonies vary greatly in size, shape, texture, color, and other features called **colonial morphology**. Colonies from different species or genera often differ substantially, whereas those derived from the same strain are usually consistent. Differences in colonial morphology are very useful for separating bacteria in mixtures and as clues to their identity. Some examples of colonial morphology are shown in **Figure 4–8**.

Bacteria may be separated in isolated colonies on agar plates

Colonies may have consistent and characteristic features

New methods that do not depend on visual changes in the growth medium or colony formation are also used to detect bacterial growth in culture. These techniques include optical, chemical, and electrical changes in the medium, produced by the growing numbers of bacterial cells or their metabolic products. Many of these methods are more sensitive than classic techniques and thus can detect growth hours, or even days, earlier than traditional methods. Some have also been engineered for instrumentation and automation. For

Optical, chemical, and electrical methods can detect growth

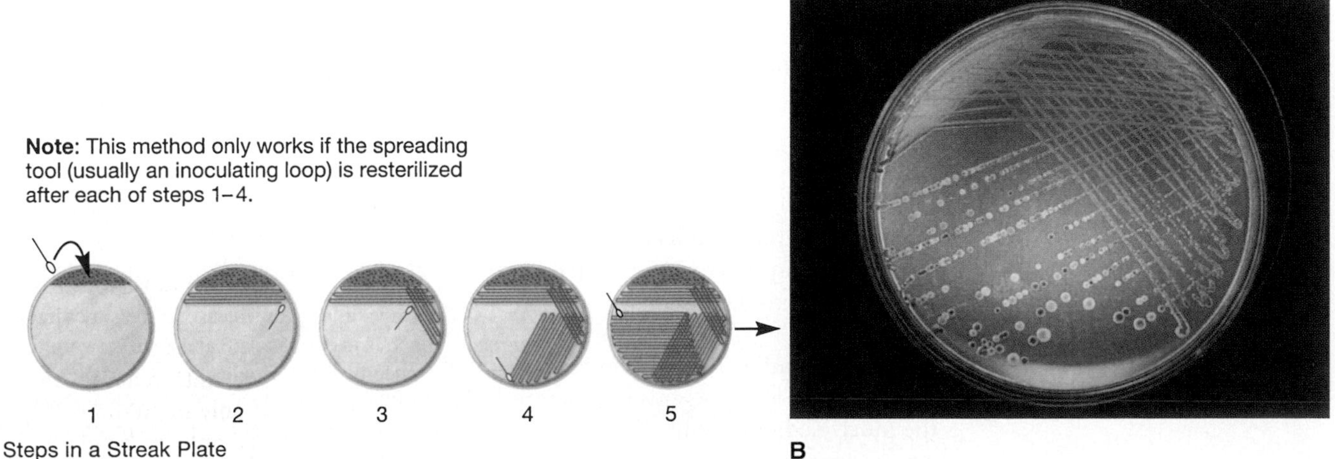

FIGURE 4–7. Bacteriologic plate streaking. Plate streaking is essentially a dilution procedure. **A.** (1)The specimen is placed on the plate with a swab, loop, or pipette and evenly spread over approximately part of plate surface with a sterilized bacteriologic loop.(2-5) The loop is flamed to remove residual bacteria, and a series of overlapping the streaks are made flaming the loop between each one. **B.** After overnight incubation, heavy growth is seen in the primary areas followed by isolated colonies. More than one organism is present because both a red and a clear colony are seen. (Reproduced with permission from Willey J, Sherwood L, Woolverton C (eds). *Prescott's Principles of Microbiology.* New York: McGraw-Hill; 2008.)

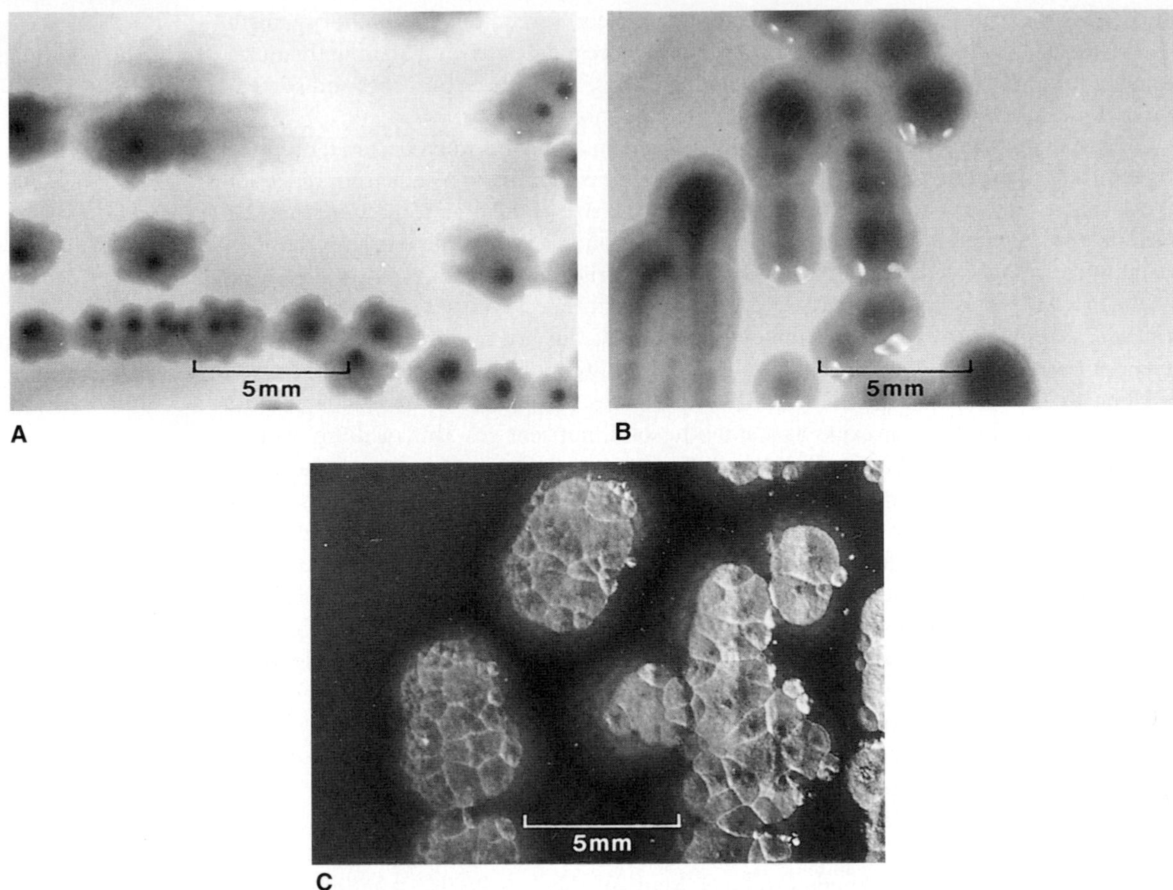

FIGURE 4–8. Bacterial colonial morphology. The colonies formed on agar plates by three different Gram-negative bacilli are shown at the same magnification. Each is typical for its species, but variations are common. **A.** *Escherichia coli* colonies are flat with an irregular scalloped edge. **B.** *Klebsiella pneumoniae* colonies with a smooth entire edge and a raised glistening surface. **C.** *Pseudomonas aeruginosa* colonies with an irregular reflective surface, suggesting hammered metal.

example, one fully automated system that detects bacterial metabolism fluorometrically can complete a bacterial identification and antimicrobial susceptibility test in 2 to 4 hours.

Culture Media

Over the last 100 years, countless media have been developed by microbiologists to aid in the isolation and identification of medically important bacteria and fungi. Only a few have found their way into routine use in clinical laboratories. These may be classified as nutrient, selective, or indicator media.

Nutrient Media The nutrient component of a medium is designed to satisfy the growth requirements of the organism to permit isolation and propagation. For medical purposes, the ideal medium would allow rapid growth of all agents. No such medium exists; however, several suffice for good growth of most medically important bacteria and fungi. These media are prepared with enzymatic or acid digests of animal or plant products such as muscle, milk, or beans. The digest reduces the native protein to a mixture of polypeptides and amino acids that also includes trace metals, coenzymes, and various undefined growth factors. For example, one common broth contains a pancreatic digest of casein (milk curd) and a papaic digest of soybean meal. To this nutrient base, salts, vitamins, or body fluids such as serum may be added to provide pathogens with the conditions needed for optimum growth.

Media are prepared from animal or plant products

Selective Media Selective media are used when specific pathogenic organisms are sought in sites with an extensive normal flora (eg, *N gonorrhoeae* in specimens from the uterine cervix or rectum). In these cases, other bacteria may overgrow the suspected etiologic species in simple nutrient media, either because the pathogen grows more slowly or because it is present in much smaller numbers. Selective media usually contain dyes, other chemical additives, or antimicrobics at concentrations designed to inhibit contaminating flora but not the suspected pathogen.

Unwanted organisms are inhibited with chemicals or antimicrobics

Indicator Media Indicator media contain substances designed to demonstrate biochemical or other features characteristic of specific pathogens or organism groups. The addition to the medium of one or more carbohydrates and a **pH indicator** is frequently used. A color change in a colony indicates the presence of acid products and thus of fermentation or oxidation of the carbohydrate by the organism. The addition of red blood cells (RBCs) to plates allows the **hemolysis** produced by some organisms to be used as a differential feature. In practice, nutrient, selective, and indicator properties are often combined to various degrees in the same medium. It is possible to include an indicator system in a highly nutrient medium and also make it selective by adding appropriate antimicrobics. Some examples of culture media commonly used in diagnostic microbiology are listed in **Appendix 4–1**, and more details of their constitution and application are provided in **Appendix 4–2**.

Metabolic properties of bacteria are demonstrated by substrate and indicator systems

Atmospheric Conditions

Aerobic Once inoculated, cultures of most aerobic bacteria are placed in an incubator with temperature maintained at 35° to 37°C. Slightly higher or lower temperatures are used occasionally to selectively favor a certain organism or organism group. Most bacteria that are not obligate anaerobes grow in air; however, CO_2 is required by some and enhances the growth of others. Incubators that maintain a 2% to 5% concentration of CO_2 in air are frequently used for primary isolation, because this level is not harmful to any bacteria and improves isolation of some. A simpler method is the candle jar, in which a lighted candle is allowed to burn to extinction in a sealed jar containing plates. This method adds 1% to 2% CO_2 to the atmosphere.

Incubation temperature and atmosphere vary with organism

Anaerobic Strictly anaerobic bacteria do not grow under the conditions just described, and many die when exposed to atmospheric oxygen or high oxidation–reduction potentials. Most medically important anaerobes grow in the depths of liquid or semisolid media containing any of a variety of **reducing agents,** such as cysteine, thioglycollate, ascorbic acid, or even iron filings. An anaerobic environment for incubation of plates can be achieved by replacing air with a gas mixture containing hydrogen, CO_2, and nitrogen and allowing the hydrogen to react with residual oxygen on a catalyst to form water. A convenient commercial system accomplishes this chemically in a packet to which water is added before the jar is sealed. Specimens suspected to contain significant anaerobes should be processed under conditions designed to minimize exposure to atmospheric oxygen at all stages.

Anaerobes require reducing conditions and protection from oxygen

Clinical Microbiology Systems

Routine laboratory systems for processing specimens from various sites are needed because no single medium or atmosphere is ideal for all bacteria. Combinations of broth and solid-plated media and aerobic, CO_2, and anaerobic incubation must be matched to the organisms expected at any particular site or clinical circumstance. Examples of such routines are shown in **Table 4–1**. In general, it is not practical to routinely include specialized media for isolation of rare organisms such as *Corynebacterium diphtheriae.* For detection of these and other uncommon organisms, the laboratory must be specifically informed of their possible presence by the physician. Appropriate media and special procedures can then be included.

Routine systems are designed to detect the most common organisms

Identification

Once growth is detected in any medium, the process of identification begins. Identification involves methods for obtaining pure cultures from single colonies, followed by tests designed to characterize and identify the isolate. The exact tests and their sequences vary

TABLE 4–1 Routine Use of Gram Smear and Isolation Systems for Selected Clinical Specimens[a]

	SPECIMEN							
MEDIUM (INCUBATION)	**BLOOD**	**CEREBROSPINAL FLUID**	**WOUND, PUS**	**GENITAL, CERVIX**	**THROAT**	**SPUTUM**	**URINE**	**STOOL**
Gram smear		×	×	×		×	×	
Soybean-casein digest broth (CO_2)	×	×	×					
Selenite F broth (air)								×
Blood agar (CO_2)		×	×	×		×	×	
Blood agar (anaerobic)			×		×[b]			
MacConkey agar (air)			×	×		×	×	×
Chocolate agar (CO_2)		×	×	×		×		
Martin-Lewis agar (CO_2)				×				
Hektoen agar (air)								×
Campylobacter agar (CO_2, 42°C)[c]								×

[a]The added sensitivty of a nutrient broth is used only when contamination by normal flora is unlikely. Exact media and isolation systems may vary between laboratories.
[b]Anaerobic incubation used to enhance hemolysis by β-hemolytic streptococci.
[c]Incubation in a reduced oxygen atmosphere.

Extent of identification is linked to medical relevance

with different groups of organisms, and the taxonomic level (genus, species, subspecies, and so on) of identification needed varies according to the medical usefulness of the information. In some cases, only a general description or the exclusion of particular organisms is important. For example, a report of "mixed oral flora" in a sputum specimen or "no *N gonorrhoeae*" in a cervical specimen may provide all the information needed.

Features Used to Classify Bacteria and Fungi

Growth under various conditions has differential value

Cultural Characteristics Cultural characteristics include the demonstration of properties such as unique nutritional requirements, pigment production, and the ability to grow in the presence of certain substances (sodium chloride, bile) or on certain media (MacConkey, nutrient agar). Demonstration of the ability to grow at a particular temperature or to cause hemolysis on blood agar plates is also used. For fungi, growth as a yeast colony or a mold is the primary separator. For molds, the morphology of the mold structures (hyphae, conidia, and so on) are the primary means of identification.

Biochemical reactions analyzed by tables and computers give identification probability

Biochemical Characteristics The ability to attack various substrates or to produce particular metabolic products has broad application to the identification of bacteria and yeast. The most common properties examined are listed in **Appendix 4–3**. Biochemical and cultural tests for bacterial identification are analyzed by reference to tables that show the reaction patterns characteristic of individual species. In fact, advances in computer analysis have now been applied to identification of many bacterial and fungal groups. These systems use the same biochemical principles along with computerized databases to determine the most probable identification from the observed test pattern.

Toxin Production and Pathogenicity Direct evidence of virulence in laboratory animals is rarely needed to confirm a clinical diagnosis. In some diseases caused by production of a specific toxin, the toxin may be detected in vitro through cell cultures or immunologic methods. Neutralization of the toxic effect with specific antitoxin is the usual approach to identify the toxin.

Detection of specific toxin may define disease

Antigenic Structure Viruses, bacteria, fungi, and parasites possess many antigens, such as capsular polysaccharides, surface proteins, and cell wall components. Serology involves the use of antibodies of known specificity to detect antigens present on whole organisms or free in extracts (soluble antigens). The methods used for demonstrating antigen–antibody reactions are discussed in a later section ::: antigen, p.32; antibody p.36.

Antigenic structures of organism demonstrated with antisera

Genomic Structure Nucleic acid sequence relatedness as determined by homology and direct sequence comparisons have become a primary determinant of taxonomic decisions. They are discussed later in the section on nucleic acid methods.

Isolation and Identification of Viruses

Cell and Organ Culture

Living cell cultures that can support their replication are the primary means of isolating pathogenic viruses. The cells are derived from a tissue source by outgrowth of cells from a tissue fragment (explant) or by dispersal with proteolytic agents such as trypsin. They are allowed to grow in nutrient media on a glass or plastic surface until a confluent layer one cell thick (monolayer) is achieved. In some circumstances, a tissue fragment with a specialized function (eg, fetal trachea with ciliated epithelial cells) is cultivated in vitro and used for viral detection. This procedure is known as organ culture.

Cell cultures derived from human or animal tissues are used to isolate viruses

Three basic types of cell culture monolayers are used in diagnostic virology. The **primary cell culture,** in which all cells have a normal chromosome count (diploid), is derived from the initial growth of cells from a tissue source. Redispersal and regrowth produce a **secondary cell culture,** which usually retains characteristics similar to those of the primary culture (diploid chromosome count and virus susceptibility). Monkey and human embryonic kidney cell cultures are examples of commonly used primary and secondary cell cultures.

Monkey kidney is used in primary and secondary culture

Further dispersal and regrowth of secondary cell cultures usually lead to one of two outcomes: the cells eventually die, or they undergo spontaneous transformation, in which the growth characteristics change, the chromosome count varies (haploid or heteroploid), and the susceptibility to virus infection differs from that of the original. These cell cultures have characteristics of "immortality"; that is, they can be redispersed and regrown many times (serial cell culture passage). They can also be derived from cancerous tissue cells or produced by exposure to mutagenic agents in vitro. Such cultures are commonly called **cell lines.** A common cell line in diagnostic use is the *Hep*-2, derived from a human epithelial carcinoma. A third type of culture is often termed a **cell strain.** This culture consists of diploid cells, commonly fibroblastic, that can be redispersed and regrown a finite number of times; usually 30 to 40 cell culture passages can be made before the strain dies out or spontaneously transforms. Human embryonic tonsil and lung fibroblasts are common cell strains in routine diagnostic use.

Primary cultures either die out or transform

Cell strains regrow a limited number of times

Cells from cancerous tissue may grow continuously

Detection of Viral Growth

Viral growth in susceptible cell cultures can be detected in several ways. The most common effect is seen with lytic or cytopathic viruses; as they replicate in cells, they produce alterations in cellular morphology (or cell death), which can be observed directly by light microscopy under low magnification (30× or 100×). This **cytopathic effect (CPE)** varies with different viruses in different cell cultures. For example, enteroviruses often produce cell rounding, pleomorphism, and eventual cell death in various culture systems, whereas measles and respiratory syncytial viruses cause fusion of cells to produce multinucleated giant cells (syncytia). The microscopic appearance of some normal cell cultures and the CPE produced in in them by different viruses are illustrated in **Figure 4–9**.

Viral CPE is due to morphologic changes or cell death

CPE is characteristic for some viruses

Other viruses may be detected in cell culture by their ability to produce **hemagglutinins.** These hemagglutinins may be present on the infected cell membranes, as well as in the

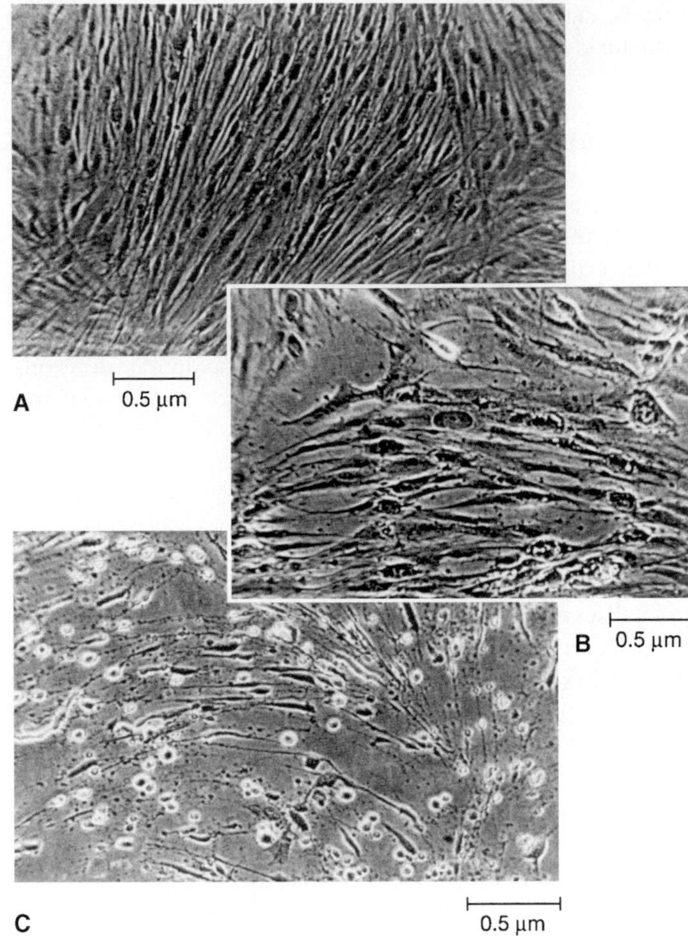

FIGURE 4–9. Viral cytopathic effect (CPE). A. Normal human diploid fibroblast cell monolayer. **B.** CPE caused by infection with adenovirus. **C.** CPE caused by infection with herpes simplex virus. (Reproduced with permission from Willey J, Sherwood L, Woolverton C (eds). *Prescott's Principles of Microbiology.* New York: McGraw-Hill; 2008.)

culture media, as a result of release of free, hemagglutinating virions from the cells. Addition of erythrocytes to the infected cell culture results in their adherence to the cell surfaces, a phenomenon known as **hemadsorption.** Another method of viral detection in cell culture is by **interference.** In this situation, the virus that infects the susceptible cell culture produces no CPE or hemagglutinin, but can be detected by "challenging" the cell culture with a different virus that normally produces a characteristic CPE. The second, or challenge, virus fails to infect the cell culture because of interference by the first virus, which is thus detected. This method is obviously cumbersome, but has been applied to the detection of rubella virus in certain cell cultures.

Hemadsorption or interference marks cells that may not show CPE

For some agents, such as Epstein-Barr virus (EBV) or human immunodeficiency virus (HIV), even more novel approaches may be applied. Both EBV and HIV can replicate in vitro in suspension cultures of normal human lymphocytes such as those derived from neonatal cord blood. Their presence may be determined in several ways; for example, EBV-infected B lymphocytes and HIV-infected T lymphocytes express virus-specified antigens and viral DNA or RNA, which can be detected with immunologic or genomic probes. In addition, HIV reverse transcriptase can be detected in cell culture by specific assay methods. Immunologic and nucleic acid probes (see text that follows) can also be used to detect virus in clinical specimens or in situations in which only incomplete, noninfective virus replication has occurred in vivo or in vitro. An example is the use of in situ cytohybridization, whereby specific labeled nucleic acid probes are used to detect and localize papillomavirus genomes in tissues where neither infectious virus nor its antigens can be detected.

EBV and HIV antigens are expressed on lymphocytes

Immunologic or genomic probes detect incomplete viruses

In Vivo Isolation Methods

In vivo methods for isolation are also sometimes necessary. The embryonated hen's egg is still often used for the initial isolation and propagation of influenza A virus.

Virus-containing material is inoculated on the appropriate egg membrane, and the egg is incubated to permit viral replication and recognition. Animal inoculation is now only occasionally used for detecting some viruses. The usual animal host for viral isolation is the mouse; suckling mice in the first 48 hours of life are especially susceptible to many viruses. Evidence of viral replication is based on the development of illness, manifested by such signs as paralysis, convulsions, poor feeding, or death. The nature of the infecting virus can be further elucidated by histologic and immunofluorescent examination of tissues or by detection of specific antibody responses. Many arboviruses and rabies virus can be detected in this system.

Embryonated eggs and animals are used for isolation of some viruses

Viral isolation from a suspect case involves a number of steps. First, the viruses that are believed to be most likely involved in the illness are considered, and appropriate specimens are collected. Centrifugation or filtration and addition of antimicrobics are frequently required with respiratory or fecal specimens to remove organic matter, cellular debris, bacteria, and fungi, which can interfere with viral isolation. The specimens are then inoculated into the appropriate cell culture systems. The time between inoculation and initial detection of viral effects varies; however, for most viruses positive cultures are usually apparent within 5 days of collection. With proper collection methods and application of the diagnostic tools discussed later, many infections can even be detected within hours. On the other hand, some viruses may require culture for a month or more before they can be detected.

Specimen preparation is required

Time to detection varies from days to weeks

Viral Identification

On isolation, a virus can usually be tentatively identified to the family or genus level by its cultural characteristics (eg, the type of CPE produced). Confirmation and further identification may require enhancement of viral growth to produce adequate quantities for testing. This result may be achieved by inoculation of the original isolate into fresh culture systems (viral passage) to amplify replication of the virus, as well as improve its adaptation to growth in the in vitro system.

Nature of CPE and cell cultures affected may suggest virus

Neutralization and Serologic Detection Of the several ways of identifying the isolate, the most common is to neutralize its infectivity by mixing it with specific antibody to known viruses before inoculation into cultures. The inhibition of the expected viral effects on the cell culture such as CPE or hemagglutination is then evidence of that virus. As in bacteriology, demonstration of specific viral antigens is a useful way to identify many agents. Immunofluorescence and enzyme immunoassay (EIA) are the most common methods.

Neutralization of biologic effect with specific antisera confirms identification

Cytology and Histology In some instances, viruses produce specific cytologic changes in infected host tissues that aid in diagnosis. Examples include specific intranuclear inclusions seen in neuronal infections due to herpes simplex (Cowdry type A bodies) and due to intracytoplasmic inclusions in rabies (Negri bodies), and cell fusion, which results in multinucleated epithelial giant cells (eg, measles and varicella-zoster). Although such findings are useful when seen, their overall diagnostic sensitivity and specificity are usually considerably less than those of the other methods discussed.

Inclusions and giant cells suggest viruses

Electron Microscopy When virions are present in sufficient numbers, they may be further characterized by specific agglutination of viral particles on mixture with type-specific antiserum. This technique, immune electron microscopy, can be used to identify viral antigens specifically or to detect antibody in serum using viral particles of known antigenicity.

Immune electron microscopy shows agglutinated viral particles

Some viruses (eg, human rotaviruses and hepatitis A and B viruses) grow poorly or not at all in the laboratory culture systems currently available. However, they can be efficiently detected by immunologic or molecular methods (to be described later in this chapter).

Not all viruses grow in culture

IMMUNOLOGIC SYSTEMS

Diagnostic microbiology makes great use of the specificity of the binding between antigen and antibody. Antisera of known specificity are used to detect their homologous antigen

Antisera detect viral antigens

Viral antigens detect immune response

in cultures, or more recently, directly in body fluids. Conversely, known antigen preparations are used to detect circulating antibodies as evidence of a current or previous infection with that agent. Many methods are in use to demonstrate the antigen–antibody binding. The greatly improved specificity of **monoclonal antibodies** has had a major impact on the quality of methods where they have been applied. Before discussing their application to diagnosis, the principles involved in the most important methods are discussed.

Methods for Detecting an Antigen–Antibody Reaction

Precipitation

Both the speed and the sensitivity of immunodiffusion are improved by CIE

When antigen and antibody combine in the proper proportions, a visible precipitate is formed. Optimum antigen–antibody ratios can be produced by allowing one to diffuse into the other, most commonly through an agar matrix **(immunodiffusion).** In the immunodiffusion procedure, wells are cut in the agar and filled with antigen and antibody. One or more precipitin lines may be formed between the antigen and antibody wells; depending on the number of different antigen–antibody reactions occurring. **Counterimmunoelectrophoresis (CIE)** is immunodiffusion carried out in an electrophoretic field. The net effect is that antigen and antibody are rapidly brought together in the space between the wells to form a precipitin line.

RBCs and latex particles coated with antigen or antibody enhance demonstration

Simple mixing on slide causes agglutination

The amount of antigen or antibody necessary to produce a visible immunologic reaction can be reduced if either is on the surface of a relatively large particle. This condition can be produced by fixing soluble antigens or antibody onto the surface of RBCs or microscopic latex particles suspended in a test tube or microtiter plate well (**Figure 4–10**). Whole bacteria are large enough to serve as the particle if the antigen is present on the microbial surface. The relative proportions of antigen and antibody thus become less critical, and antigen–antibody reactions are detectable by agglutination when immune serum and particulate antigen, or particle-associated antibody and soluble antigen, are mixed on a slide. The process is termed slide agglutination, hemagglutination, or latex agglutination depending on the nature of the sensitized particle.

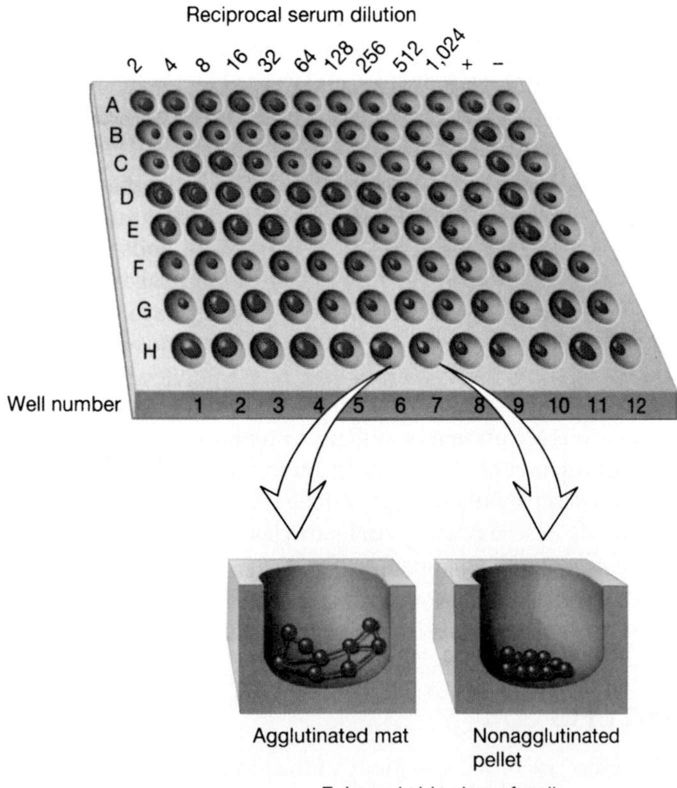

FIGURE 4–10. Agglutination. A microtiter plate demonstrating hemagglutinating antibody. Test sera (antibody) are placed in the wells (1-10) at the dilutions shown across the top. Positive controls (row 11) and negative controls (row 12) are included. Red blood cells are added to each well. If sufficient antibody is present to agglutinate the cells they sink as a mat to the bottom of the well. If insufficient antibody is present they sink to the bottom as a pellet. The endpoints for A-H can be read from left to right for each specimen. (Reproduced with permission from Willey J, Sherwood L, Woolverton C (eds). *Prescott's Principles of Microbiology.* New York: McGraw-Hill; 2008.)

Neutralization

Neutralization takes some observable function of the agent, such as cytopathic effect of viruses or the action of a bacterial toxin, and neutralizes it. This is usually done by first reacting the agent with antibody and then placing the antigen–antibody mixture into the test system. The steps involved are illustrated in **Figure 4–11**. In viral neutralization, a single antibody molecule can bind to surface components of the extracellular virus and interfere

Property of the agent is neutralized by antibody

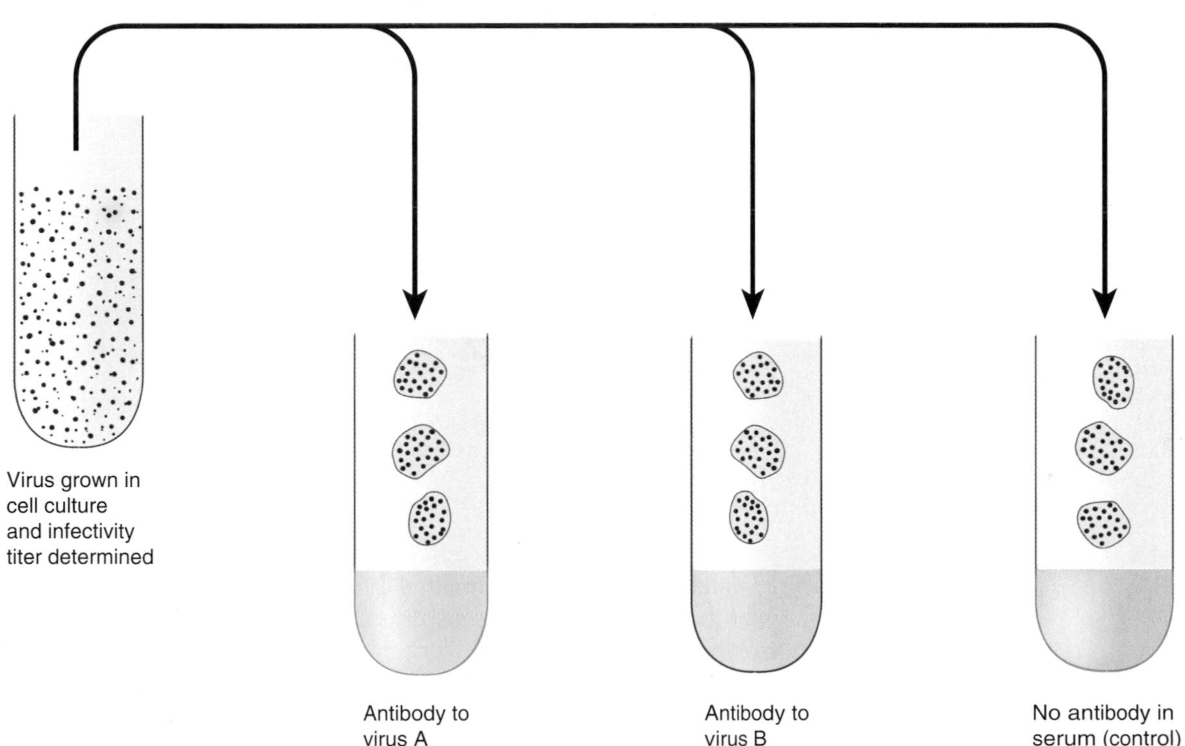
Virus grown in
cell culture
and infectivity
titer determined
Antibody to
virus A
Antibody to
virus B
No antibody in
serum (control)
Virus aliquot added to tubes containing antibody to known
viruses and to control, incubated for 1 h, then each
inoculated into cell culture tubes, incubated, and observed daily

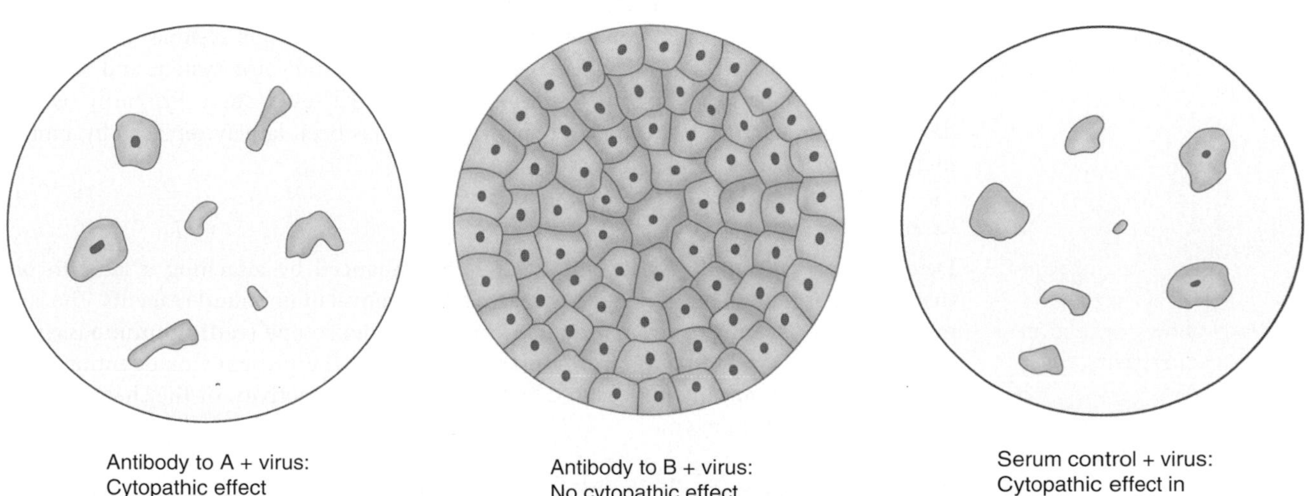
Antibody to A + virus:
Cytopathic effect
(not virus A)
Antibody to B + virus:
No cytopathic effect
(confirms virus B)
Serum control + virus:
Cytopathic effect in
control

FIGURE 4–11. Identification of a virus isolate (cytopathic virus) as "virus B."

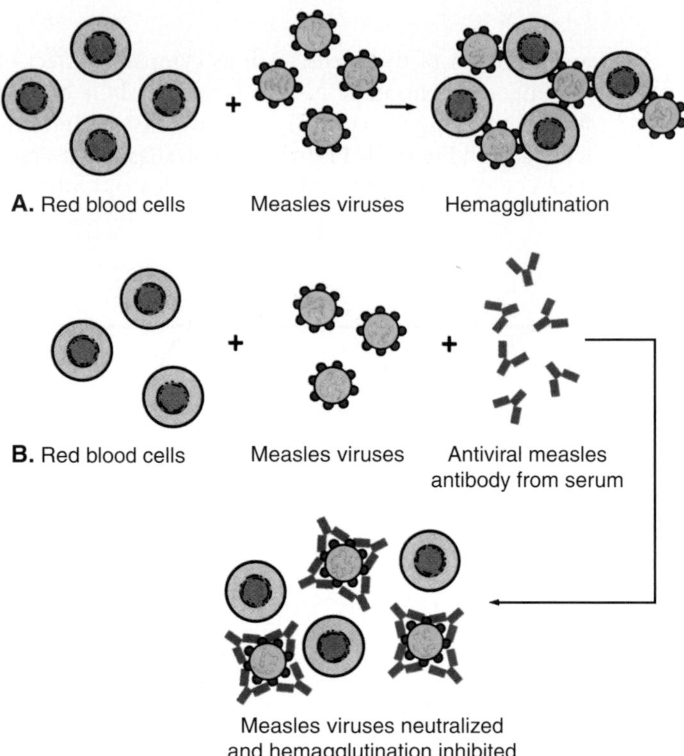

FIGURE 4–12. Viral hemagglutination. A. Certain viruses can bind to red blood cells causing cross-linking called hemagglutination. **B.** If serum containing antibody to the virus is included it will neutralize the viral effect. As shown, this is a positive test for measles antibody. (Reproduced with permission from Willey J, Sherwood L, Woolverton C (eds). *Prescott's Principles of Microbiology.* New York: McGraw-Hill; 2008.)

with one of the initial events of the viral multiplication cycle (adsorption, penetration, or uncoating). Some bacterial and viral agents directly bind to RBCs (hemagglutination). Neutralization of this reaction by antibody blocking of the receptor is called hemagglutination inhibition (**Figure 4–12**).

Complement Fixation

Action of complement on RBCs is used as indicator system

Complement fixation assays depend on two properties of complement. The first is fixation (inactivation) of complement on formation of antigen–antibody complexes. The second is the ability of bound complement to cause hemolysis of sheep (RBCs coated with anti-sheep RBC antibody ([sensitized RBCs]). Complement fixation assays are performed in two stages: The test system reacts the antigen and antibody in the presence of complement; the indicator system, which contains the sensitized RBCs, detects residual complement. Hemolysis indicates that complement was present in the indicator system and therefore that antigen–antibody complexes were not formed in the test system. Primarily used to detect and quantitate antibody, complement fixation has been largely replaced by simpler methods that can be readily automated.

Labeling Methods

Labeling antibody allows detection of fluorescence, radioactivity, or enzyme

Detection of antigen–antibody binding may be enhanced by attaching a label to one (usually the antibody) and detecting the label after removal of unbound reagents. The label may be a fluorescent dye (immunofluorescence), a radioisotope (**radioimmunoassay,** or **RIA**), or an enzyme (**enzyme immunoassay,** or **EIA**). The presence or quantitation of antigen–antibody binding is measured by fluorescence, radioactivity, or the chemical reaction catalyzed by the enzyme.

Immunofluorescence The most common labeling method in diagnostic microbiology is immunofluorescence (Figure 4–6), in which antibody labeled with a fluorescent dye, usually **fluorescein isothiocyanate (FITC)**, is applied to a slide of material that may contain the antigen sought. Under fluorescence microscopy, binding of the labeled antibody can be detected as a bright green halo surrounding bacterium, or in the case of viruses, as a fluorescent clump in or on an infected cell. The method is called "direct" if the FITC is conjugated directly to the antibody with the desired specificity. In "indirect" immunofluorescence, the specific antibody is not labeled, but its binding to an antigen is detected in an additional step using an FITC-labeled anti-immunoglobulin antibody that binds to the specific antibody. Choice between the two approaches involves purely technical considerations.

Light halo enhances microscopic visualization

Indirect methods use a second antibody

Radioimmunoassay (RIA) and Enzyme Immunoassay (EIA) The labels used in RIA and EIA are more suitable for liquid phase assays and are used particularly in virology. They are also used in direct and indirect methods and many other ingenious variations such as the "sandwich" methods, so called because the antigen of interest is "trapped" between two antibodies (**Figure 4–13**). These extremely sensitive techniques are discussed further with regard to antibody detection.

Liquid phase RIA and EIA methods have many variants

Serologic Classification

For most important antigens of diagnostic significance, antisera are commercially available. The most common test methods for bacteria are agglutination and immunofluorescence; and for viruses, neutralization. In most cases, these methods subclassify organisms

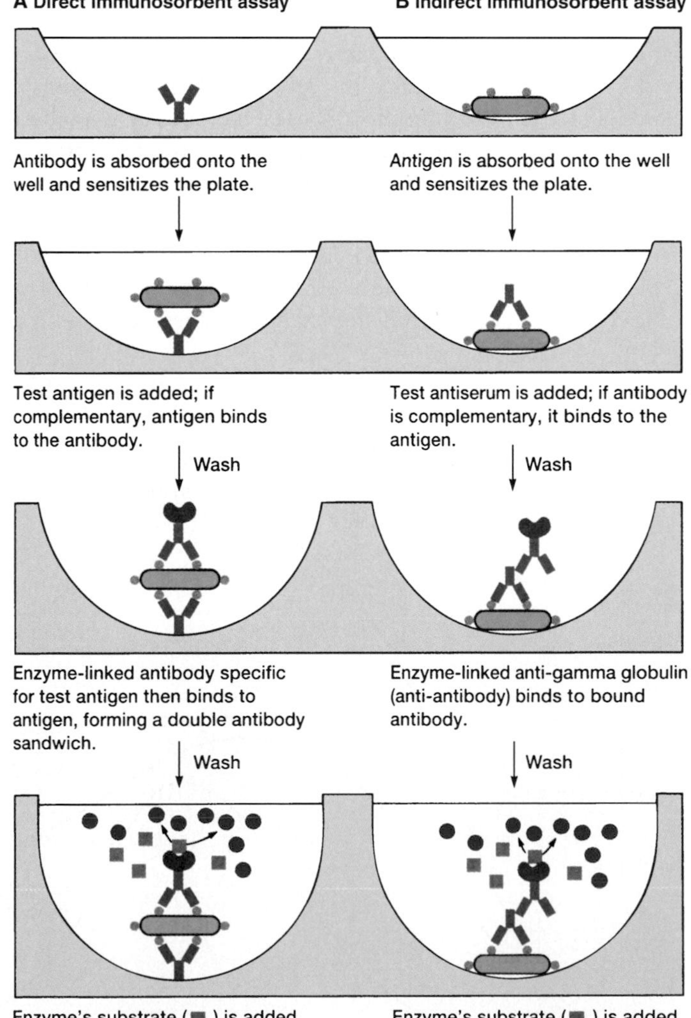
A Direct immunosorbent assay
B Indirect immunosorbent assay
Antibody is absorbed onto the well and sensitizes the plate.
Antigen is absorbed onto the well and sensitizes the plate.
Test antigen is added; if complementary, antigen binds to the antibody.
Test antiserum is added; if antibody is complementary, it binds to the antigen.
Wash
Wash
Enzyme-linked antibody specific for test antigen then binds to antigen, forming a double antibody sandwich.
Enzyme-linked anti-gamma globulin (anti-antibody) binds to bound antibody.
Wash
Wash
Enzyme's substrate (■) is added, and reaction produces a visible color change (●) that is measured spectrophotometrically.
Enzyme's substrate (■) is added, and reaction produces a visible color change (●) that is measured spectrophotometrically.

FIGURE 4–13. The ELISA or EIA test. A. The direct or double antibody sandwich method for the detection of antigens. **B.** The indirect assay for detecting antibodies. (Reproduced with permission from Willey J, Sherwood L, Woolverton C (eds). *Prescott's Principles of Microbiology.* New York: McGraw-Hill; 2008.)

Antigenic systems classify below the species level

below the species level and thus are primarily of value for epidemiologic and research purposes. The terms "serotype" and "serogroup" are used together with numbers, letters, or Roman numerals with no apparent logic other than historical precedent. For a few genera, the most fundamental taxonomic differentiation is serologic. This is the case with the streptococci, in which an existing classification based on biochemical and cultural characteristics was superseded because a serologic classification scheme developed by Rebecca Lancefield correlated better with disease.

Serologic classification is primarily of epidemiologic value

Proof of etiologic relationship depends on antigen detection

Before these techniques can be applied to the diagnosis of specific infectious diseases, considerable study of the causative agent(s) is required. Antigen–antibody systems may vary in complexity from a single epitope to scores of epitopes on several macromolecular antigens, whose chemical nature may or may not be known. The cause of the original 1976 outbreak of Legionnaires disease (caused by *Legionella pneumophila*) was proved through the development of immune reagents that detected the bacteria in tissue and antibodies directed against the bacteria in the serum of patients. Now, more than 25 years later, there are more than a dozen serotypes and many additional species, each requiring specific immunologic reagents for antigen or antibody detection for diagnosis.

Antibody Detection (Serology)

Antibodies are formed in response to infection

Antibodies may indicate current, recent, or past infection

During infection—viral, bacterial, fungal, or parasitic—the host usually responds with the formation of antibodies, which can be detected by modification of any of the methods used for antigen detection. The formation of antibodies and their time course depend on the antigenic stimulation provided by the infection. The precise patterns vary depending on the antigens used, the classes of antibody detected, and the method. An example of temporal patterns of development and increase and decline in specific antiviral antibodies measured by different tests is illustrated in **Figure 4–14**. These responses can be used to detect evidence of recent or past infection. The test methods do not inherently indicate immunoglobulin class, but can be modified to do so, usually by pretreatment of the serum to remove IgG to differentiate the IgM and IgG responses. Several basic principles must be emphasized:

Paired specimens are compared

1. In an acute infection, the antibodies usually appear early in the illness, and then rise sharply over the next 10 to 21 days. Thus, a serum sample collected shortly after the onset of illness (acute serum) and another collected 2 to 3 weeks later (convalescent serum) can be compared quantitatively for changes in specific antibody content.

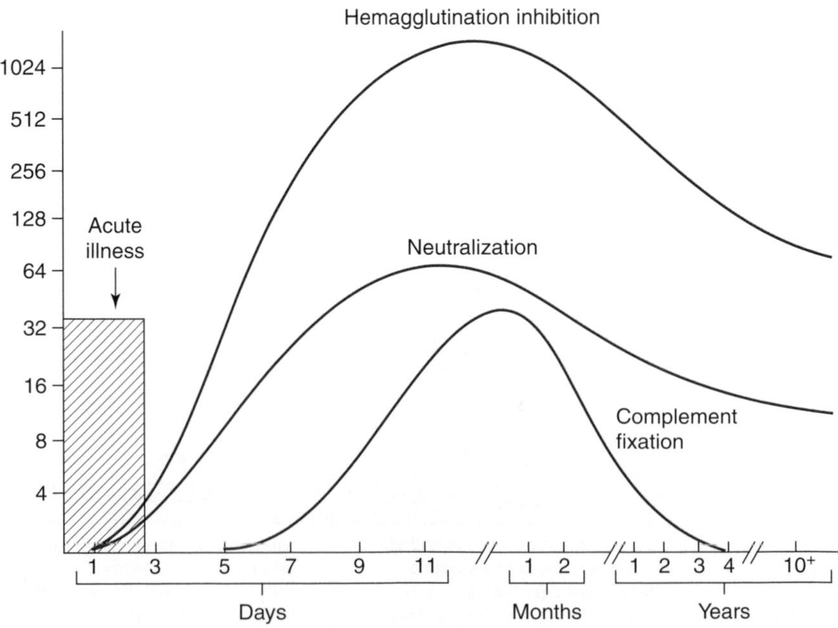

FIGURE 4–14. Examples of patterns of antibody responses to an acute infection, measured by three different methods.

2. Antibodies can be quantitated by several means. The most common method is to dilute the serum serially in appropriate media and determine the maximal dilution that will still yield detectable antibody in the test system (eg, serum dilutions of 1:4, 1:8, and 1:16). The highest dilution that retains specific activity is called the antibody titer.

Titer is the highest serum dilution demonstrating activity

3. The interpretation of significant antibody responses (evidence of specific, recent infection) is most reliable when definite evidence of seroconversion is demonstrated; that is, detectable specific antibody is absent from the acute serum (or preillness serum, if available) but present in the convalescent serum. Alternatively, a fourfold or greater increase in antibody titer supports a diagnosis of recent infection; for example, an acute serum titer of 1:4 or less and a convalescent serum titer of 1:16 or greater would be considered significant.

Seroconversion or fourfold rise in titer most conclusive

4. In instances in which the average antibody titers of a population to a specific agent are known, a single convalescent antibody titer significantly greater than the expected mean may be used as supportive or presumptive evidence of recent infection. However, this finding is considerably less valuable than those obtained by comparing responses of acute and convalescent serum samples. An alternative and somewhat more complex method of serodiagnosis is to determine which major immunoglobulin subclass constitutes the major proportion of the specific antibodies. In primary infections, the IgM-specific response is often dominant during the first days or weeks after onset, but is replaced progressively by IgG-specific antibodies; thus, by 1 to 6 months after infection, the predominant antibodies belong to the IgG subclass. Consequently, serum containing a high titer of antibodies of the IgM subclass would suggest a recent, primary infection.

Single titers may be useful in some circumstances

IgM responses indicate acute infection

The immunologic methods used to identify bacterial or viral antigens are applied to serologic diagnosis by simply reversing the detection system: that is, using a known antigen to detect the presence of an antibody. The methods of serologic diagnosis to be used are selected on the basis of their convenience and applicability to the antigen in question. As shown in Figure 4–14, the temporal relationships of antibody response to infection vary according to the method used. Of the methods for measuring antigen–antibody interaction discussed previously, those now used most frequently for serologic diagnosis are agglutination, RIA, and EIA.

Experience with systems and temporal relationships aids interpretation

Western Blot

The Western blot immunoassay is another technique that is now commonly used to detect and confirm the specificity of antibodies to a variety of epitopes. Its greatest use has been in the diagnosis of HIV infections (see Chapter 18), in which virions are electrophoresed in a polyacrylamide gel to separate the protein and glycoprotein components and then transferred onto nitrocellulose. This is then incubated with patient serum, and antibody to the different viral components is detected by using an antihuman globulin IgG antibody conjugated with an enzyme label.

Western blot confirms specificity of antibodies for protein components of the agent (eg, HIV)

Antigen Detection

Theoretically, any of the methods described for detecting antigen–antibody interactions can be applied directly to clinical specimens. The most common of these is immunofluorescence, in which antigen is detected on the surface of the organism or in cells present in the infected secretion. The greatest success with this approach has been in respiratory infections in which a nasopharyngeal, throat washing, sputum, or bronchoalveolar lavage specimen may contain bacteria or viral aggregates in sufficient amount to be seen microscopically. Although the fluorescent tag makes it easier to find organisms, these methods are generally not as sensitive as culture. With some genera and species, the immunofluorescent detection of antigens in clinical material provides the most rapid means of diagnosis, as with *Legionella* and respiratory syncytial virus.

Immunofluorescence detects agents in respiratory secretions

Another approach to detecting antigens is to detect free antigen released by the organism into body fluids. This offers the possibility of bypassing direct examination, culture, and identification tests to achieve a diagnosis. Success requires a highly specific antibody, a sensitive detection method, and the presence of the homologous antigen in an accessible body fluid. The latter is an important limitation, because not all organisms release free antigen in the course of infection. At present, diagnosis by antigen detection is limited to some bacteria and fungi with polysaccharide capsules (eg, *Haemophilus influenzae*), to *Chlamydia*, and to certain viruses. The techniques of agglutination with antibody bound to latex particles, CIE, RIA, and EIA are used to detect free antigen in serum, urine, cerebrospinal fluid, and joint fluid. Live organisms are not required for antigen detection, and these tests may still be positive when the causative organism has been eliminated by antimicrobial therapy. The procedures can yield results within 1 or 2 hours, sometimes within a few minutes. This feature is attractive for office practice because it allows diagnostic decisions to be made during the patient's visit. A number of commercial products detect group A streptococci in sore throats with over 90% sensitivity; however, because these tests are less sensitive than culture, negative results must still be confirmed by culture.

Soluble antigens may be detected in body fluids

Rapid detection can replace culture

NUCLEIC ACID ANALYSIS

As with the human genome, the genome sequence of the major human pathogens has or soon will be determined. These data are placed in widely available computer databases and have already been used for applications ranging from taxonomy to detection of antimicrobial resistance genes. Some of the methods and applications relevant to the study of infectious diseases are briefly summarized below. The student is referred to textbooks of molecular biology for more complete coverage.

Methods of Nucleic Acid Analysis

DNA Hybridization and Probes

If the DNA double helix is opened, leaving single-stranded (denatured) DNA, the nucleotide bases are exposed and thus available to interact with other single-stranded nucleic acid molecules. If complementary sequences of a second DNA molecule are brought into physical contact with the first, they hybridize to it, forming a new double-stranded molecule in that area. A probe is a cloned DNA fragment that has been labeled so that it can be detected if it hybridizes to complementary sequences in such a test system (**Figure 4–15**). The probe may be derived from the gene for a known protein of the pathogen or be empirically derived just for diagnostic purposes. The methods that allow the hybridization to take place include those that immobilize the single-stranded target DNA on a membrane, as in the figure or liquid phase assays, which can be rapid and automated. A variant in which the DNA is separated by agarose gel electrophoresis before binding to the membrane is called **Southern hybridization.**

DNA hybridization methods allow DNA from different sources to combine

Agarose Gel Electrophoresis

Nucleic acids may be separated in an electrophoretic field in an **agarose** (highly purified agar) gel. The speed of migration depends on size, with the smaller molecules moving faster and appearing at the bottom (end) of the gel. This method is able to separate DNA fragments in the range of 0.1 to 50 kilobases, which is far below the size of bacterial genomes but includes some naturally occurring genetic elements such as bacterial plasmids. This analysis can be refined by the use of restriction endonucleases, which are enzymes derived from bacteria that recognize specific nucleotide sequences in DNA molecules and digest (cut) them at all sites where the sequence appears. Thus, plasmids of the same size may be differentiated by the size of fragments generated by endonuclease digestion of DNA as shown in **Figure 4–16A-C**. Agarose gel electrophoresis, endonuclease digestion, and the specificity of a probe may all be combined in searches for the source specific genes as shown in **Figure 4–17A-E**.

Agarose gel electrophoresis separates DNA fragments or plasmids based on size

Restriction endonuclease digestion analysis

Probes detect sequences in fragments

Nucleic Acid Amplification

Nucleic acid amplification (NAA) methods such as the polymerase chain reaction (PCR) allow the detection and selective replication of a targeted portion of the genome. The basic PCR technique uses synthetic oligonucleotide primers and special DNA polymerases in a way that allows repeated cycles of synthesis of only a segment of a targeted DNA molecule that may be as large as an entire genome. The specificity is provided by the sequence of approximately 20 nucleotides in each primer pair, which are crafted to flank the desired segment of the genome. The DNA polymerases used are ones that operate at unusually high temperatures.

NAA replicates a genome segment

PCR uses temperature to manipulate primers and polymerases

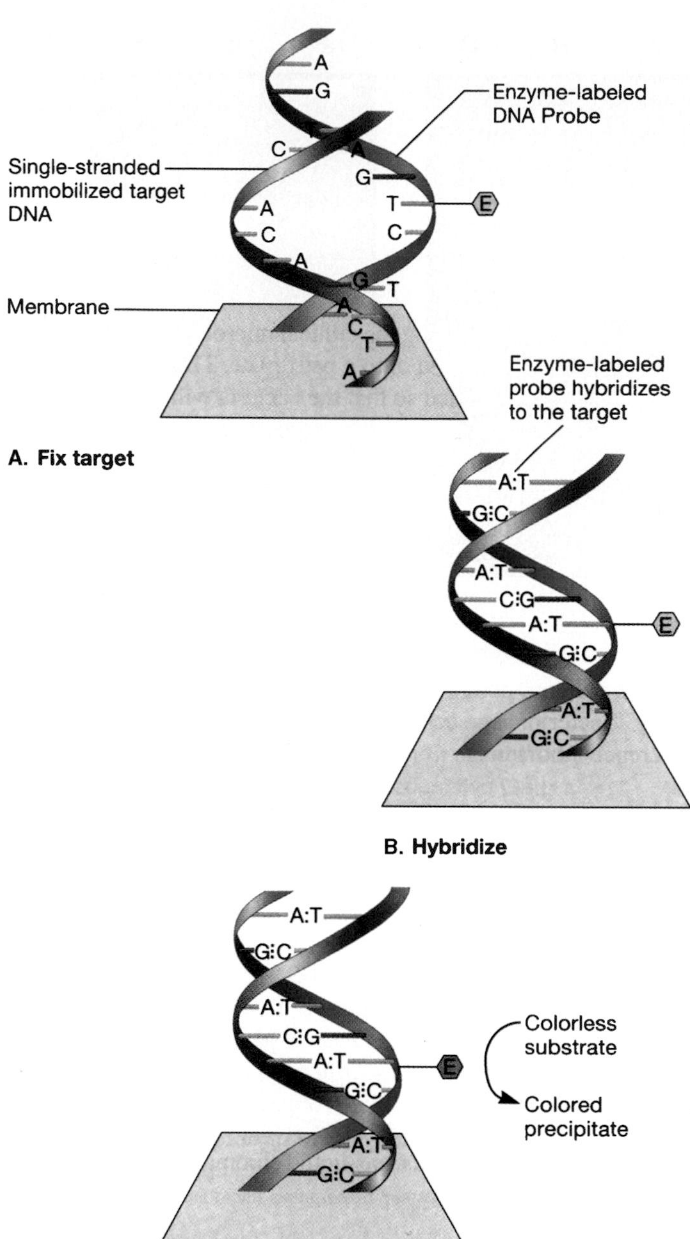

FIGURE 4–15. DNA probe hybridization. **A.** A single-stranded (denatured) target nucleic acid is bound to a membrane. A DNA probe with attached enzyme (E) is also employed. **B.** If the probe finds complementary sequences it hybridizes to the target DNA forming a double-stranded hybrid. **C.** A colorless substrate is added, which in the presence of the enzyme is converted to a colored substrate. Measuring the color development quantitates the amount of probe bound to the original target. (Reproduced with permission from Willey J, Sherwood L, Woolverton C (eds). *Prescott's Principles of Microbiology.* New York: McGraw-Hill; 2008.)

This allows the use of temperature to control shifts between separation of the complementary DNA strands (so primers can bind) and replication of the DNA sequence that lies between the two primers. Because each strand generates a new fragment, the increase is exponential. In a machine called a thermocycler, the targeted DNA can be amplified 1 million to 1 billion times in 20 to 30 cycles (**Figure 4–18A-D**). Other NAA methods use the same principles.

Application of Nucleic Acid Methods to Infectious Diseases

Bacterial and Viral Genomes

Number and size of plasmids differentiate strains

Endonuclease digestion of plasmids refines their comparison

The only intact genetic elements of infectious agents that are small enough to be directly detected and sized by agarose gel electrophoresis are bacterial plasmids. Not all bacterial species typically harbor plasmids, but those that do may carry one or a number of plasmids ranging in size from less than 1 to over 50 kilobases. This diversity makes the presence or absence, number, and sizes of plasmids of considerable value in differentiating strains for epidemiologic purposes. Because plasmids are not stable components of the bacterial genome, plasmid

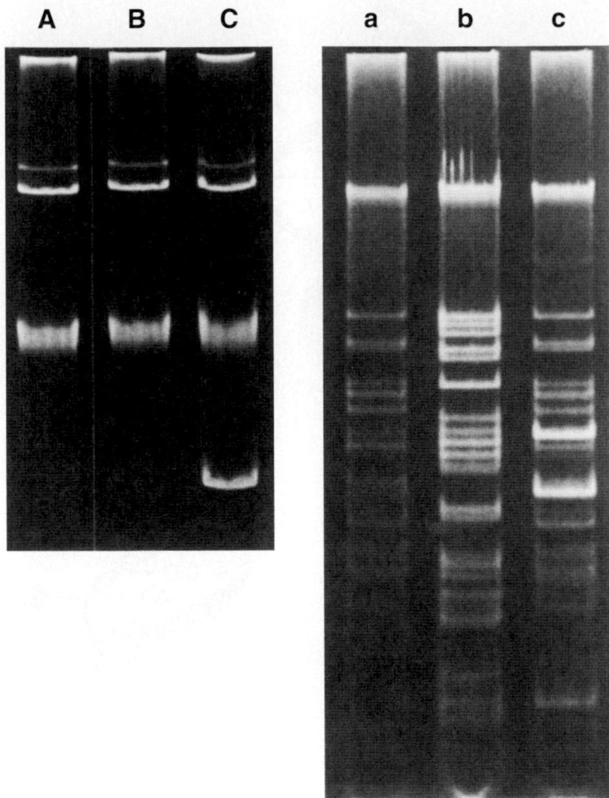

FIGURE 4-16. Plasmid fingerprinting. Agarose gel electrophoresis of plasmid DNA. **A, B, C.** Lanes show whole plasmids of various sizes indicated by extent of migration in the gel. **a, b, c.** The same plasmids after digestion with a restriction endonuclease, which produces many fragments depending on the frequency of the sequence it cuts in the DNA. The whole plasmids in A and B are the same size, but their endonuclease digests (a, b) reveal that they are different. (Reproduced with permission from Willey J, Sherwood L, Woolverton C (eds). *Prescott's Principles of Microbiology.* New York: McGraw-Hill; 2008.)

analysis also has the element of a timely "snapshot" of the circumstances of a disease outbreak. The specificity of these results can be improved by digesting the plasmids with restriction endonucleases before electrophoresis. Two plasmids of the same size from different strains may not be the same, but if an identical pattern of fragments is generated from the digestion, they almost certainly are. These principles are illustrated in Figures 4–16 and 4–17.

Bacterial chromosomes must be digested before electrophoresis

Because of their larger size, the chromosomes of bacteria must be digested with endonucleases to resolve them on gels. For viruses, the outcome is much like that with plasmids, depending on the genomic size and the endonuclease used. Digested bacterial chromosomes can be compared in this manner, but the number of fragments is very large and the patterns complex. The combined use of endonucleases, which make infrequent cuts, and electrophoretic methods able to resolve large fragments can produce a comparison comparable to that possible with plasmids. This approach is also used for analysis of the multiple chromosomes of fungi and parasites.

DNA Probes

Probes may be cloned or synthesized from known sequences

Probes may be recovered from NAA procedures or more commonly synthesized as a single chain of nucleotides (oligonucleotide probe) from known sequence data. They may contain a gene of known function or simply sequences empirically found to be useful for the application in question. When labeled with a radioisotope or other marker and used in hybridization reactions, they can detect the homologous sequences in unknown specimens (Figure 4–15) or to further refine gel electrophoresis findings (Figure 4–17).

Probes can detect DNA of pathogen directly in clinical specimens

The diagnostic use of DNA probes is to detect or identify microorganisms by hybridization of the probe to homologous sequences in DNA extracted from the entire organism. A number of probes have been developed that can quickly and reliably identify organisms already isolated in culture. The application of probes for detection of infectious agents directly in clinical specimens such as blood, urine, and sputum is more difficult because only a small number of organisms may be present. This problem of sensitively can be overcome by combining probes with NAA methods (see below). This approach offers the potential for rapid diagnosis and the detection of characteristics not possible by routine methods.

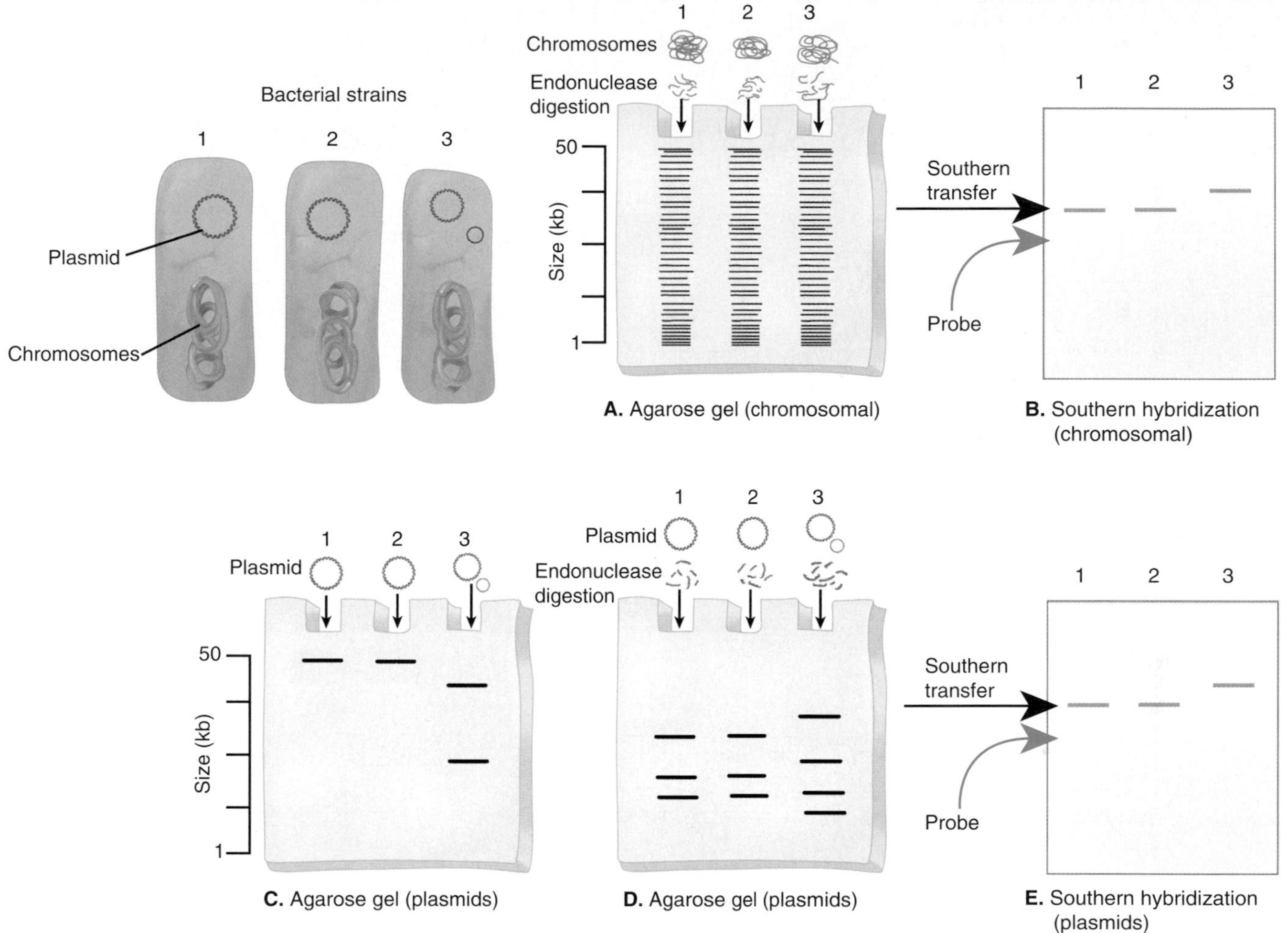

FIGURE 4–17. Molecular diagnostic methods. Three bacterial strains of the same species are shown each with chromosome and plasmid(s). **A.** The chromosomal DNA of each strain is isolated, digested with a restriction endonuclease, and separated by agarose gel electrophoresis. An almost continuous range of fragment sizes is generated for each strain, making them difficult to distinguish. **B.** The restriction fragments in A are transferred to a membrane (Southern transfer) and hybridized with a probe. The probe binds to a single fragment from each strain, but the larger size of the fragment from strain 3 indicates variation in restriction sites and thus a genomic difference between it and strains 1 and 2. **C.** Plasmids from each strain are isolated and separated in the same manner as A. The results show a plasmid of the same size from 1 and 2. Strain 3 has two plasmids, each of a different size than strains 1 and 2. **D.** The same plasmids are restriction digested before electrophoresis. The plasmids from strains 1 and 2 show three fragments of identical size, proving they are identical. The plasmids of strain 3 appear unrelated. **E.** The fragments in D are transferred and reacted with a probe. The positive result with the largest of the strain 1 and 2 fragments confirms their relatedness. The positive hybridization with one of the strain 3 fragments suggests that it contains at least some DNA that is homologous to the plasmid from strains 1 and 2.

For example, a bacterial toxin gene probe can demonstrate both the presence of the related organism and its toxigenicity without the need for culture.

Applications of Polymerase Chain Reaction (PCR)

The amplification power of the PCR offers a solution for the sensitivity problems inherent in the direct application of probes in clinical specimens. The nucleic acid segment amplified by PCR can be detected by direct hybridization with the probe (Figure 4–18D2) or for greater specificity after electrophoresis and Southern transfer (Figure 4–18D3,4). This approach has been successful for a wide range of infectious agents and awaits only further resolution of practical problems for wider use.

PCR combined with probes gives the greatest sensitivity

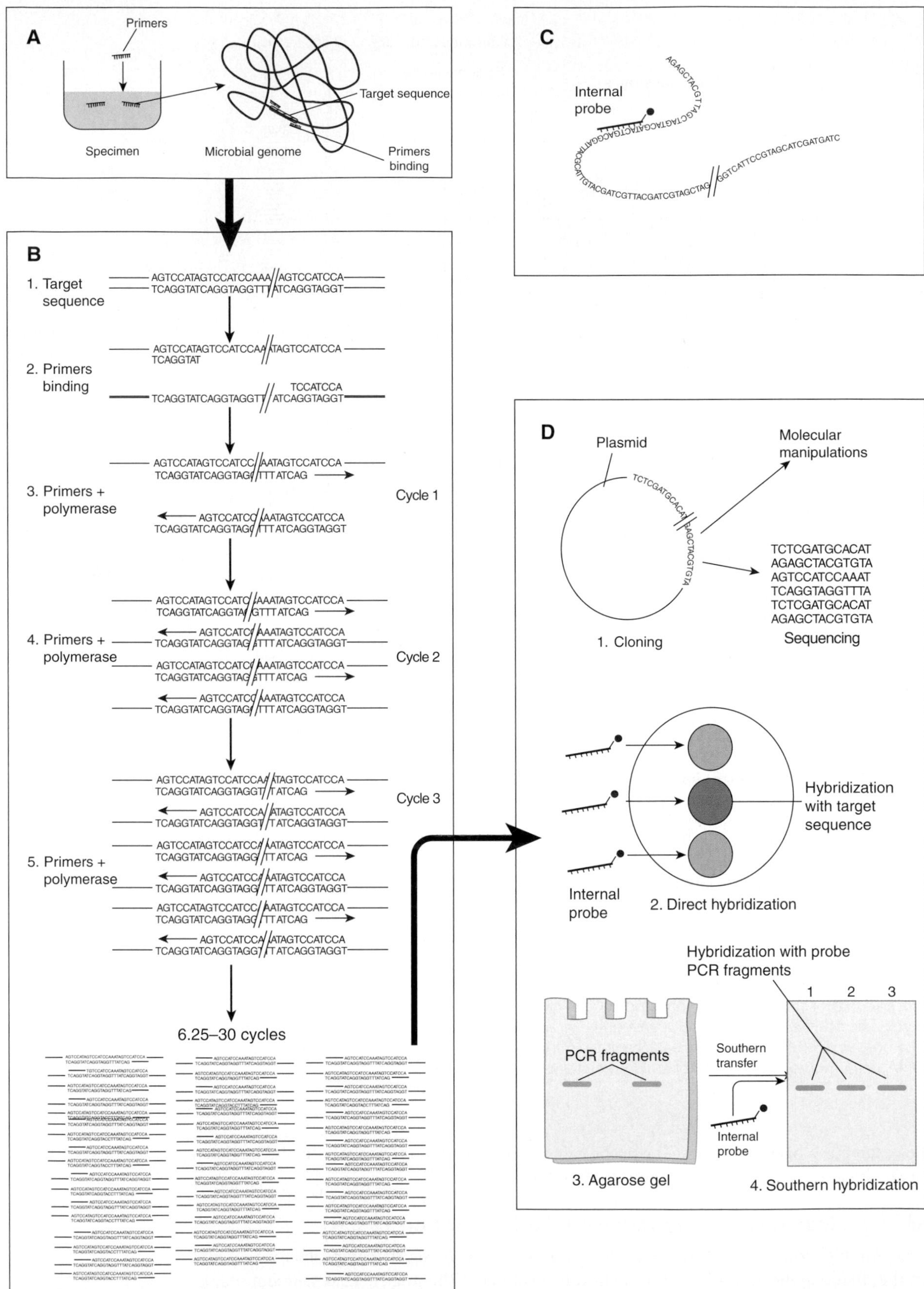

FIGURE 4–18. Diagnostic applications of the polymerase chain reaction (PCR). A. A clinical specimen (eg, pus, tissue) contains DNA from many sources as well as the chromosome of the organism of interest. If the DNA strands are separated (denatured), the PCR (*continued on next page*)

Another creative use of PCR has been in the study of infectious agents seen in tissue but not grown in culture. PCR primers derived from sequences known to be highly conserved among bacteria, such as ribosomal RNA, have been applied to tissue specimens. The amplification produces enough DNA to clone and sequence. This sequence can then be compared with sequences published for other organisms using computers. Thus, taxonomic relationships can be inferred for an organism that has never been isolated in culture.

PCR from tissue allows study of organisms that cannot be cultured

Ribotyping

Ribotyping also makes use of the conserved nature of bacterial ribosomal RNA and of the ability of RNA to hybridize to DNA under certain conditions. Labeled ribosomal RNA of one organism can be hybridized with restriction endonuclease–digested chromosomal DNA of another. In this case, ribosomal RNA is being used as a massive probe of restriction fragments separated by electrophoresis. Hybridization to multiple fragments is common, but if the organisms are genetically different, the restriction fragments, which contain the ribosomal RNA sequences, will vary in size. The pattern of bands produced by epidemiologically related strains can then be compared side by side.

Ribotyping refines comparison of chromosomal endonuclease digestion patterns

SUMMARY

The application of some combination of the principles described in this chapter is appropriate to the diagnosis of any infectious disease. The usefulness of any individual method differs among infectious agents as a result of biologic variation and uneven study. In general, for agents that can be grown in vitro, culture remains the “gold standard” as both the most sensitive and specific method. Molecular methods have the potential to replace culture and have in some areas. Aside from cost, their broader application in infectious disease diagnosis must deal with their highly specific nature. Depending on the clinical situation, the specimen introduced at the beginning of this chapter could be directed at a very narrow or very broad question. If the question is only the diagnosis of a short list of diseases (AIDS, gonorrhea, tuberculosis, malaria), a DNA probe approach can be rapid, sensitive, and practical. Very often, however, the question is “almost anything” or at least a wide range of possibilities. In this instance, culture is difficult to replace because it offers sure detection of the common along with a reasonable chance of catching the uncommon and even the rare infection.

---

primers can bind to their target sequences in the specimen itself. **B.** Amplification of the target sequence by PCR. (1) The target sequence is shown in its native state. (2) The DNA is denatured, allowing the primers to bind where they find the homologous sequence. (3) In the presence of the special DNA polymerase, new DNA is synthesized from both strands in the region between the primers. (4-6) Additional cycles are added by temperature control of the polymerase with each new sequence acting as the template for another. The DNA doubles with each cycle. After 25 to 30 cycles enough DNA is present to analyze diagnostically. **C.** Internal probe. The amplified target sequence is shown. A probe can be designed to bind to a sequence located between (internal to) the primers. **D.** Analysis of PCR amplified DNA. (1) The amplified sequence can be cloned into a plasmid vector. In this form, a variety of molecular manipulations or sequencing may be carried out. (2) Direct hybridizations usually make use of an internal probe. The example shows three specimens, each of which went through steps A and B. After amplification, each was bound to a separate spot on a filter (dot blot). The filter is then reacted with the internal probe to detect the PCR-amplified DNA. The result shows that only the middle specimen contained the target sequence. (3) The amplified DNA may be detected directly by agarose gel electrophoresis. The example shows detection of amplified fragments in two of three lanes on the gel. (4) The sensitivity of detection may be increased by use of the internal probe after Southern transfer. The example shows detection of a third fragment of the same size that was not seen on the original gel because the amount of DNA was too small.

APPENDIX 4–1 Some Media Used for Isolation of Bacterial Pathogens

MEDIUM	USES
General-Purpose Media	
Nutrient broths (eg, Soybean–Casein Digest Broth)	Most bacteria, particularly when used for blood culture
Thioglycolate broth	Anaerobes, facultative bacteria
Blood agar	Most bacteria (demonstrates hemolysis) and fungi
Chocolate agar	Most bacteria, including fastidious species (eg, *Haemophilus*) and fungi
Selective Media	
MacConkey agar	Nonfastidious Gram-negative rods
Hektoen-enteric agar	*Salmonella* and *Shigella*
Selenite F broth	*Salmonella* enrichment
Sabouraud's agar	Isolation of fungi, particularly dermatophytes
Special-Purpose Media	
Löwenstein–Jensen medium, Middlebrook agar	*Mycobacterium tuberculosis* and other mycobacteria (selective)
Martin–Lewis medium	*Neisseria gonorrhoeae* and *N* meningitidis (selective)
Fletcher medium (semisolid)	Leptospira (nonselective)
Tinsdale agar	*Corynebacterium diphtheriae* (selective)
Charcoal agar	*Bordetella pertussis* (selective)
Buffered charcoal–yeast extract agar	*Legionella* species (nonselective)
Campylobacter blood agar	*Campylobacter jejuni* (selective)
Thiosulfate-citrate-bile-sucrose agar (TCBS)	*Vibrio cholerae* and *V parahaemolyticus* (selective)

APPENDIX 4–2 Characteristics of Commonly Used Bacteriologic Media

1. **Nutrient broths.** Some form of nutrient broth is used for culture of all direct tissue or fluid samples from sites that are normally sterile to obtain the maximum culture sensitivity. Selective or indicator agents are omitted to prevent inhibition of more fastidious organisms.
2. **Blood agar.** The addition of defibrinated blood to a nutrient agar base enhances the growth of some bacteria, such as streptococci. This often yields distinctive colonies and provides an indicator system for hemolysis. Two major types of hemolysis are seen: β-hemolysis, a complete clearing of red cells from a zone surrounding the colony; and α-hemolysis, which is incomplete (ie, intact red cells are still present in the hemolytic zone), but shows a green color caused by hemoglobin breakdown products. The net effect is a hazy green zone extending 1 to 2 mm beyond the colony. A third type, α′-hemolysis, produces a hazy, incomplete hemolytic zone similar to that caused by -hemolysis, but without the green coloration.
3. **Chocolate agar.** If blood is added to molten nutrient agar at about 80°C and maintained at this temperature, the red cells are gently lysed, hemoglobin products are released, and the medium turns a chocolate brown color. The nutrients released permit the growth of some fastidious organisms such as *Haemophilus influenzae,* which fail to grow on blood or nutrient agars. This quality is particularly pronounced when the medium is further enriched with vitamin supplements. Given the same incubation conditions, any organism that grows on blood agar also grows on chocolate agar.
4. **Martin–Lewis medium.** A variant of chocolate agar, Martin–Lewis medium is a solid medium selective for the pathogenic *Neisseria* (*N gonorrhoeae* and *N meningitidis*). Growth of most other bacteria and fungi in the genital or respiratory flora is inhibited by the addition of antimicrobics. One formulation includes vancomycin, colistin, trimethoprim, and anisomycin.
5. **MacConkey agar.** This agar is both a selective and an indicator medium for Gram-negative rods, particularly members of the family Enterobacteriaceae and the genus *Pseudomonas.* In addition to a peptone base, the medium contains bile salts, crystal violet, lactose, and neutral red as a pH indicator. The bile salts and crystal violet inhibit Gram-positive bacteria and the more fastidious Gram-negative organisms, such as *Neisseria* and *Pasteurella*. Gram-negative rods that grow and ferment lactose produce a red (acid) colony, often with a distinctive colonial morphology.
6. **Hektoen enteric agar.** The Hektoen medium is one of many highly selective media developed for the isolation of *Salmonella* and *Shigella* species from stool specimens. It has both selective and indicator properties. The medium contains a mixture of bile, thiosulfate, and citrate salts that inhibits not only Gram-positive bacteria, but members of Enterobacteriaceae other than *Salmonella* and *Shigella* that appear among the normal flora of the colon. The inhibition is not absolute; recovery of *Escherichia coli* is reduced 1000- to 10,000-fold relative to that on nonselective media, but there is little effect on growth of *Salmonella* and *Shigella.* Carbohydrates and a pH indicator are also included to help to differentiate colonies of *Salmonella* and *Shigella* from those of other enteric Gram-negative rods.
7. **Anaerobic media.** In addition to meeting atmospheric requirements, isolation of some strictly anaerobic bacteria on blood agar is enhanced by reducing agents such as L-cysteine and by vitamin enrichment. Sodium thioglycolate, another reducing agent, is often used in broth media. Plate media are made selective for anaerobes by the addition of aminoglycoside antibiotics, which are active against many aerobic and facultative organisms but not against anaerobic bacteria. The use of selective media is particularly important with anaerobes because they grow slowly and are commonly mixed with facultative bacteria in infections.
8. **Highly selective media.** Media specific to the isolation of almost every important pathogen have been developed. Many allow only a single species to grow from specimens with a rich normal flora (eg, stool). The most common of these media are are listed in Appendix 4–1; they are discussed in greater detail in following chapters.

APPENDIX 4–3 Common Biochemical Tests for Microbial Identification

1. **Carbohydrate breakdown.** The ability to produce acidic metabolic products, fermentatively or oxidatively, from a range of carbohydrates (eg, glucose, sucrose, and lactose) has been applied to the identification of most groups of bacteria. Such tests are crude and imperfect in defining mechanisms, but have proved useful for taxonomic purposes. More recently, gas chromatographic identification of specific short-chain fatty acids produced by fermentation of glucose has proved useful in classifying many anaerobic bacteria.
2. **Catalase production.** The enzyme catalase catalyzes the conversion of hydrogen peroxide to water and oxygen. When a colony is placed in hydrogen peroxide, liberation of oxygen as gas bubbles can be seen. The test is particularly useful in differentiation of staphylococci (positive) from streptococci (negative), but also has taxonomic application to Gram-negative bacteria.
3. **Citrate utilization.** An agar medium that contains sodium citrate as the sole carbon source may be used to determine ability to use citrate. Bacteria that grow on this medium are termed **citrate-positive.**
4. **Coagulase.** The enzyme coagulase acts with a plasma factor to convert fibrinogen to a fibrin clot. It is used to differentiate *Staphylococcus aureus* from other, less pathogenic staphylococci.
5. **Decarboxylases and deaminases.** The decarboxylation or deamination of the amino acids lysine, ornithine, and arginine is detected by the effect of the amino products on the pH of the reaction mixture or by the formation of colored products. These tests are used primarily with Gram-negative rods.
6. **Hydrogen sulfide.** The ability of some bacteria to produce H_2S from amino acids or other sulfur-containing compounds is helpful in taxonomic classification. The black color of the sulfide salts formed with heavy metals such as iron is the usual means of detection.
7. **Indole.** The indole reaction tests the ability of the organism to produce indole, a benzopyrrole, from tryptophan. Indole is detected by the formation of a red dye after addition of a benzaldehyde reagent. A spot test can be done in seconds using isolated colonies.
8. **Nitrate reduction.** Bacteria may reduce nitrates by several mechanisms. This ability is demonstrated by detection of the nitrites and/or nitrogen gas formed in the process.
9. **O-Nitrophenyl-β-D-galactoside (ONPG) breakdown.** The ONPG test is related to lactose fermentation. Organisms that possess the β-galactoside necessary for lactose fermentation but lack a permease necessary for lactose to enter the cell are ONPG-positive and lactose-negative.
10. **Oxidase production.** The oxidase tests detect the *c* component of the cytochrome–oxidase complex. The reagents used change from clear to colored when converted from the reduced to the oxidized state. The oxidase reaction is commonly demonstrated in a spot test, which can be done quickly from isolated colonies.
11. **Proteinase production.** Proteolytic activity is detected by growing the organism in the presence of substrates such as gelatin or coagulated egg.
12. **Urease production.** Urease hydrolyzes urea to yield two molecules of ammonia and one of CO_2. This reaction can be detected by the increase in medium pH caused by ammonia production. Urease-positive species vary in the amount of enzyme produced; bacteria can thus be designated as positive, weakly positive, or negative.
13. **Voges–Proskauer test.** The Voges–Proskauer test detects acetylmethylcarbinol (acetoin), an intermediate product in the butene glycol pathway of glucose fermentation.

CHAPTER 5

Emergence and Global Spread of Infection

Epidemiology, the study of the distribution of determinants of disease and injury in human populations, is a discipline that includes both infectious and noninfectious diseases. Most epidemiologic studies of infectious diseases have concentrated on the factors that influence acquisition and spread, because this knowledge is essential for developing methods of prevention and control. Historically, epidemiologic studies and the application of the knowledge gained from them have been central to the control of the great epidemic diseases, such as cholera, plague, smallpox, yellow fever, and typhus.

An understanding of the principles of epidemiology and the spread of disease is essential to all medical personnel, whether their work is with the individual patient or with the community. Most infections must be evaluated in their epidemiologic setting. For example, what infections, especially viral, are currently prevalent in the community? Has the patient recently traveled to an area of special disease prevalence? Is there a possibility of nosocomial infection from recent hospitalization? What is the risk to the patient's family, schoolmates, and work or social contacts?

The recent recognition of emerging infectious diseases has heightened appreciation of the importance of epidemiologic information. A few examples of these newly identified infections are cryptosporidiosis, hantavirus pulmonary syndrome, and severe acute respiratory syndrome (SARS) coronavirus disease. In addition, some well-known pathogens have assumed new epidemiologic importance by virtue of acquired antimicrobial resistance (eg, penicillin-resistant pneumococci, vancomycin-resistant enterococci, and multiresistant *Mycobacterium tuberculosis*).

Factors that increase the emergence or reemergence of various pathogens include:

- Population movements and the intrusion of humans and domestic animals into new habitats, particularly tropical forests
- Deforestation, with development of new farmlands and exposure of farmers and domestic animals to new arthropods and primary pathogens
- Irrigation, especially primitive irrigation systems, which fail to control arthropods and enteric organisms
- Uncontrolled urbanization, with vector populations breeding in stagnant water
- Increased long-distance air travel, with contact or transport of arthropod vectors and primary pathogens
- Social unrest, civil wars, and major natural disasters, leading to famine and disruption of sanitation systems, immunization programs, etc.
- Global climate change
- Microbial evolution, leading to natural selection of multi-resistant agents (eg, methicillin-resistant staphylococci, new, highly virulent strains of influenza A virus). In some instances, these changes can be accelerated considerably by indiscriminate use of anti-infective agents.

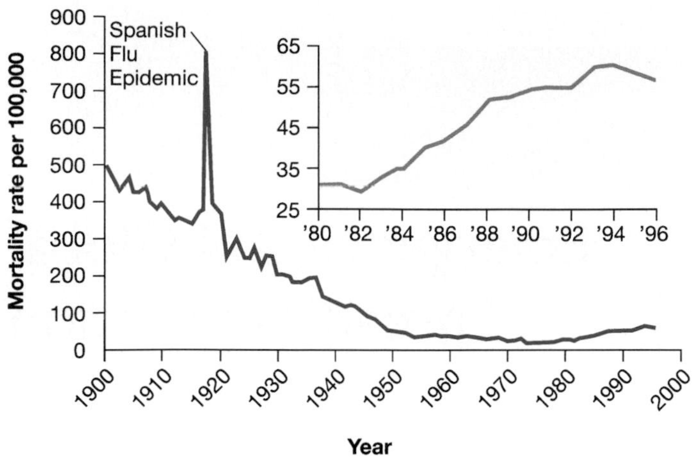

FIGURE 5–1. Infectious disease mortality rates in the United States decreased greatly during most of the 20th century. The insert is an enlargement of the right-hand portion of the graph, and the death rate has shown a rising trend since 1982. (Reproduced with permission from Willey J, Sherwood L, Woolverton C (eds). *Prescott's Principles of Microbiology.* New York: McGraw-Hill; 2008.)

There are other factors, of course, and all of these are discussed here as to their relative impacts on the specific infectious agents described in subsequent chapters.

The major general concerns for the future are that new, often unexpected, infectious diseases emerge (or in many cases simply re-emerge) for any of the reasons detailed above. Although mortality rates declined dramatically during much of the 20th century (**Figure 5–1**), an alarming upward trend has been occurring over the last 24 years. The general global nature of the problem is illustrated in **Figure 5–2**.

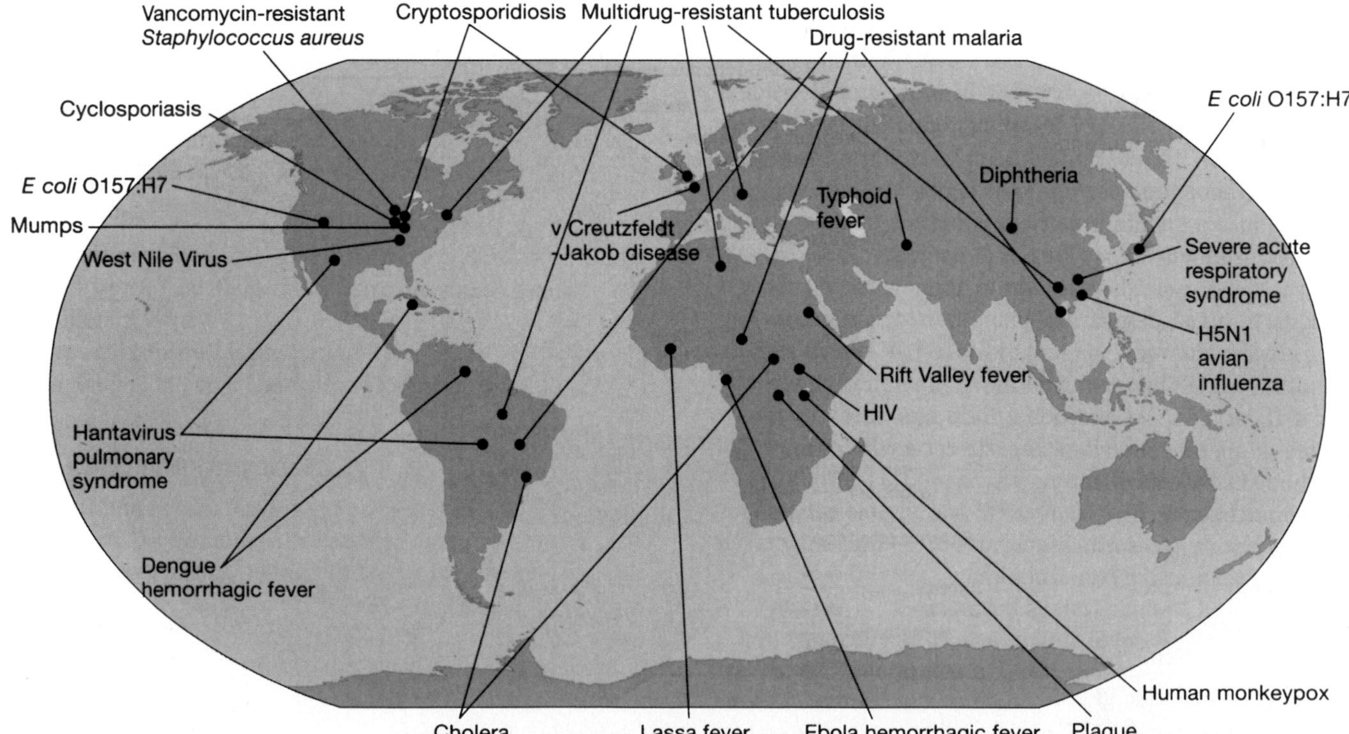

FIGURE 5–2. Examples of emerging and re-emerging infectious diseases. Although infections such as HIV are shown in a few locations here, they are indeed widespread and a threat in many regions. (Reproduced with permission from Willey J, Sherwood L, Woolverton C (eds). *Prescott's Principles of Microbiology.* New York: McGraw-Hill; 2008.)

SOURCES AND COMMUNICABILITY

Infectious diseases of humans may be caused by exclusively human pathogens such as *Shigella;* by environmental organisms such as *Legionella pneumophila;* or by organisms that have their primary reservoir in animals such as *Salmonella.*

Noncommunicable infections are those that are not transmitted from human to human and include (1) infections derived from the patient's normal flora, such as peritonitis after rupture of the appendix; (2) infections caused by the ingestion of preformed toxins, such as botulism; and (3) infections caused by certain organisms found in the environment, such as clostridial gas gangrene. Some diseases transmitted from animals to humans (zoonotic infections), such as rabies and brucellosis, are not transmitted between humans, but others such as plague may be transmitted at certain stages. Noncommunicable infections may still occur as common-source outbreaks, such as food poisoning from an enterotoxin-producing *Staphylococcus aureus*–contaminated chicken salad or multiple cases of pneumonia from extensive dissemination of *Legionella* through an air-conditioning system. Because these diseases are not transmissible to others, they do not lead to secondary spread.

Noncommunicable infections are not spread from person to person but can occur as common-source outbreaks

Communicable infections require an organism to be able to leave the body in a form that is directly infectious or to be able to become so after development in a suitable environment. The respiratory spread of the influenza virus is an example of direct communicability. In contrast, the malarial parasite requires a developmental cycle in a biting mosquito before it can infect another human. Communicable infections can be **endemic,** which implies that the disease is present at a low but fairly constant level, or **epidemic,** which involves a level of infection higher than that usually found in a community or population. In some infections, such as influenza, the infection can be endemic, persisting at a fairly low level from season to season. Introduction of a new strain, however, may result in epidemics, as illustrated in **Figure 5–3**. Communicable infections that are widespread in a region, sometimes worldwide, and have high attack rates are termed **pandemic.**

Endemic = constant presence

Epidemic = localized outbreak

Pandemic = widespread regional or global epidemic

INFECTION AND DISEASE

An important consideration in the study of the epidemiology of communicable organisms is the distinction between infection and disease. **Infection** involves multiplication of the

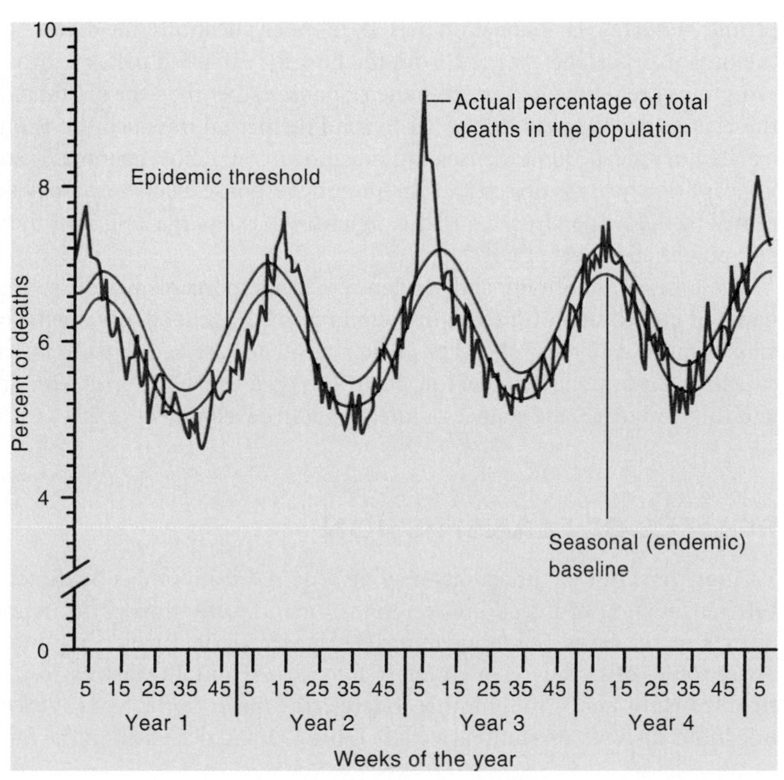

FIGURE 5–3. Endemic disease that can be epidemic. Example of yearly fluctuation of pneumonia and influenza mortality (expressed a percentage of all deaths.) (Reproduced with permission from Nester EW, Anderson DG, Roberts CE Jr, Nester MT. *Microbiology: A Human Perspective,* 6th ed. New York: McGraw-Hill; 2008.)

organism in or on the host and may not be apparent, for example, during the incubation period or latent when little or no replication is occurring (eg, with herpesviruses). **Disease** represents a clinically apparent response by, or injury to, the host as a result of infection. With many communicable microorganisms, infection is much more common than disease, and apparently healthy infected individuals play an important role in disease propagation. Inapparent infections are termed **subclinical,** and the individual is sometimes referred to as a **carrier.** The latter term is also applied to situations in which an infectious agent establishes itself as part of a patient's flora or causes low-grade chronic disease after an acute infection. For example, the clinically inapparent presence of *S aureus* in the anterior nares is termed **carriage,** as is a chronic gallbladder infection with *Salmonella* serotype Typhi that can follow an attack of typhoid fever and result in fecal excretion of the organism for years.

Infection can result in little or no illness

Carriers can be asymptomatic, but infectious to others

With some infectious diseases such as measles, infection is invariably accompanied by clinical manifestations of the disease itself. These manifestations facilitate epidemiologic control, because the existence and extent of infection in a community are readily apparent. Organisms associated with long incubation periods or high frequencies of subclinical infection, such as human immunodeficiency virus (HIV) or hepatitis B virus, may propagate and spread in a population for long periods before the extent of the problem is recognized. This makes epidemiologic control more difficult.

INCUBATION PERIOD AND COMMUNICABILITY

Incubation periods range from a few days to several months

The incubation period is the time between the exposure to the organism and the appearance of the first symptoms of the disease. Generally, organisms that multiply rapidly and produce local infections, such as gonorrhea and influenza, are associated with short incubation periods (eg, 2-4 days). Diseases such as typhoid fever, which depend on hematogenous spread and multiplication of the organism in distant target organs to produce symptoms, often have longer incubation periods (eg, 10 days to 3 weeks). Some diseases have even more prolonged incubation periods because of slow passage of the infecting organism to the target organ, as in rabies, or with slow growth of the organism, as in tuberculosis. Incubation periods for one agent may also vary widely depending on route of acquisition and infecting dose; for example, the incubation period of hepatitis B virus infection may vary from a few weeks to several months.

Transmission to others can occur before illness onset

Communicability of a disease in which the organism is shed in secretions may occur primarily during the incubation period. In other infections, the disease course is short but the organisms can be excreted from the host for extended periods. In yet other cases, the symptoms are related to host immune response rather than the organism's action, and thus the disease process may extend far beyond the period in which the etiologic agent can be isolated or spread. Some viruses can integrate into the host genome or survive by replicating very slowly in the presence of an immune response. Such dormancy or latency is exemplified by the herpesviruses, and the organism may emerge long after the original infection and potentially infect others.

The inherent infectivity and virulence of a microorganism are also important determinants of attack rates of disease in a community. In general, organisms of high infectivity spread more easily and those of greater virulence are more likely to cause disease than subclinical infection. The infecting dose of an organism also varies with different organisms and thus influences the chance of infection and development of disease.

ROUTES OF TRANSMISSION

Horizontal transmission = direct or indirect person to person

Vertical transmission = mother to fetus

Various transmissible infections may be acquired from others by direct contact, by aerosol transmission of infectious secretions, or indirectly through contaminated inanimate objects or materials. Some infections, such as malaria, involve an animate insect vector. These routes of spread are often referred to as **horizontal transmission,** in contrast to **vertical transmission**—from mother to fetus. The major horizontal routes of transmission of infectious diseases are summarized in **Table 5–1** and discussed in the following text.

TABLE 5–1 Common Routes of Transmission of Infection[a]

ROUTE OF EXIT	ROUTE OF TRANSMISSION	EXAMPLE
Respiratory	Aerosol droplet inhalation	Influenza virus; tuberculosis
	Nose or mouth → hand or object → nose	Common cold (rhinovirus)
Salivary	Direct salivary transfer (eg, kissing)	Oral-labial herpes; Epstein-Barr virus, cytomegalovirus
	Animal bite	Rabies
Gastrointestinal	Stool → hand → mouth and/or stool → object, water or food → mouth	Enterovirus; hepatitis A
	Stool → water or food → mouth	Salmonellosis; shigellosis
Skin	Skin discharge → air → respiratory tract	Varicella, smallpox, or monkeypox
	Skin to skin	Human papilloma virus (warts); syphilis
Blood	Transfusion or needle prick	Hepatitis B; cytomegalovirus infection; malaria; HIV
	Mosquito bite	Malaria; arboviruses
Genital secretions	Urethral or cervical secretions	Gonorrhea; herpes simplex; *Chlamydia*
	Semen	Cytomegalovirus
Urine	Urine → hand → catheter	Hospital-acquired urinary tract infections
Eye	Conjunctival	Adenovirus
Zoonotic	Animal bite	Rabies
	Contact with carcasses	Tularemia
	Tick bite	Rickettsia; Lyme disease

[a]The examples cited are incomplete, and in some cases more than one route of transmission exists

Respiratory Spread

Many infections are transmitted by the respiratory route, often by aerosolization of respiratory secretions with subsequent inhalation by other persons. The efficiency of this process depends in part on the extent and method of propulsion of discharges from the mouth and nose, the size of the aerosol droplets, and the resistance of the infectious agent to desiccation and inactivation by ultraviolet light. In still air, a particle 100 μm in diameter requires only seconds to fall the height of a room; a 10-m particle remains airborne for about 20 minutes, smaller particles even longer. When inhaled, particles with a diameter of 6 μm or more are usually trapped by the mucosa of the nasal turbinates, whereas particles of 0.6 to 5.0 μm attach to mucous sites at various levels along the upper and lower respiratory tract and may initiate infection. These "droplet nuclei" are most important in transmitting many respiratory pathogens (eg, *M tuberculosis*).

Respiratory secretions are often transferred on hands or inanimate objects (fomites) and may reach the respiratory tract of others in this way. For example, spread of the common cold may involve transfer of infectious secretions from nose to hand by the infected individual, with transfer to others by hand-to-hand contact and then from hand to nose by the unsuspecting victim.

Droplet nuclei are usually less than 6 μm in size

Salivary Spread

Handwashing is especially important

Some infections, such as herpes simplex and infectious mononucleosis, can be transferred directly by contact with infectious saliva through kissing. Transmission of infectious secretions by direct contact with the nasal mucosa or conjunctiva often accounts for the rapid dissemination of agents, such as respiratory syncytial virus and adenovirus. The risk of spread in these instances can be reduced by simple hygienic measures such as handwashing.

Fecal–Oral Spread

Reduced gastric hydrochloric acid can facilitate enteric infections

Fecal–oral spread involves direct or finger-to-mouth spread, the use of human feces as a fertilizer, or fecal contamination of food or water. Food handlers who are infected with an organism transmissible by this route constitute a special hazard, especially when their personal hygienic practices are inadequate. Some viruses disseminated by the fecal–oral route infect and multiply in cells of the oropharynx and then disseminate to other body sites to cause infection. However, organisms that are spread in this way commonly multiply in the intestinal tract and may cause intestinal infections. They must therefore be able to resist the acid in the stomach, the bile, and the gastric and small intestinal enzymes. Many bacteria and enveloped viruses are rapidly killed by these conditions, but members of the Enterobacteriaceae and unenveloped viral intestinal pathogens (eg, enteroviruses) are more likely to survive. Even with these organisms, the infecting dose in patients with reduced or absent gastric hydrochloric acid is often much smaller than in those with normal stomach acidity.

Skin-to-Skin Transfer

Syphilis, ringworm, and impetigo are examples of skin-to-skin transfer

Skin-to-skin transfer occurs with a variety of infections in which the skin is the portal of entry, such as the spirochete of syphilis (*Treponema pallidum*), strains of group A streptococci that cause impetigo, and the dermatophyte fungi that cause ringworm and athlete's foot. In most cases, an unapparent break in the epithelium is probably involved in infection. Other diseases may be spread through fomites such as shared towels and inadequately cleansed shower and bath floors. Skin-to-skin transfer usually occurs through abrasions of the epidermis, which may be unnoticed.

Bloodborne Transmission

Parenteral drug abuse is a major risk factor

Bloodborne transmission of infection through insect vectors requires a period of multiplication or alteration within an insect vector before the organism can infect another human host. Such is the case with the mosquito and the malarial parasite. Direct transmission from human to human through blood has become increasingly important in modern medicine because of the use of blood transfusions and blood products and the increased self-administration of illicit drugs by intravenous or subcutaneous routes using shared nonsterile equipment. Hepatitis B and C viruses, as well as HIV, were frequently transmitted in this way before the institution of blood screening tests.

Genital Transmission

Asymptomatic carriage and recurrence are common

Disease transmission through the genital tract has emerged as one of the most common infectious problems and reflects changing social and sexual mores. Spread can occur between sexual partners or from the mother to the infant at birth. A major factor in these infections has been the persistence, high rates of asymptomatic carriage, and frequency of recurrence of organisms such as *Chlamydia trachomatis*, cytomegalovirus (CMV), herpes simplex virus, and *Neisseria gonorrhoeae*.

Eye-to-Eye Transmission

Fomites, unsterile ophthalmologic instruments are associated with transmission

Infections of the conjunctiva may occur in epidemic or endemic form. Epidemics of adenovirus and *Haemophilus* conjunctivitis may occur and are highly contagious. The major endemic disease is trachoma, caused by *Chlamydia,* which remains a common cause of blindness in developing countries. These diseases may be spread by direct contact via ophthalmologic equipment or by secretions passed manually or through fomites such as towels.

Zoonotic Transmission

Zoonotic infections are spread from animals, where they have their natural reservoir, to humans. Some zoonotic infections such as rabies are directly contracted from the bite of the infected animal, whereas others are transmitted by vectors, especially arthropods (eg, ticks, mosquitoes). Many infections contracted by humans from animals are dead-ended in humans, whereas others may be transferred between humans once the disease is established in a population. Plague, for example, has a natural reservoir in rodents. Human infections contracted from the bites of rodent fleas may produce pneumonia, which may then spread to other humans by the respiratory droplet route.

Zoonotic = animals to humans

Vertical Transmission

Certain diseases can spread from mother to fetus through the placental barrier. This mode of transmission involves organisms such as rubella virus that can be present in the mother's bloodstream and may occur at different stages of pregnancy with different organisms. Another form of transmission from mother to infant occurs by contact during birth with organisms such as group B streptococci, *C trachomatis,* and *N gonorrhoeae,* which colonize the vagina. Herpes simplex virus and CMV can spread by both vertical methods as it may be present in blood or may colonize the cervix. CMV may also be transmitted by breast milk, a third mechanism of vertical transmission.

Vertical transmission can occur transplacentally, during birth, or through breast milk

EPIDEMICS

The characterization of epidemics and their recognition in a community involve several quantitative measures and some specific epidemiologic definitions. **Infectivity,** in epidemiologic terms, equates to attack rate and is measured as the frequency with which an infection is transmitted when there is contact between the agent and a susceptible individual. The **disease index** of an infection can be expressed as the number of persons who develop the disease divided by the total number infected. The **virulence** of an agent can be estimated as the number of fatal or severe cases per total number of cases. **Incidence,** the number of new cases of a disease within a specified period, is described as a rate in which the number of cases is the numerator and the number of people in the population under surveillance is the denominator. This is usually normalized to reflect a percentage of the population that is affected. **Prevalence,** which can also be described as a rate, is primarily used to indicate the total number of cases existing in a population at risk at a point in time.

Incidence and prevalence rates are usually expressed as number of cases per 100, 1000, or 100,000 population

The prerequisites for propagation of an epidemic from person to person are a sufficient degree of infectivity to allow the organism to spread, sufficient virulence for an increased incidence of disease to become apparent, and sufficient level of susceptibility in the host population to permit transmission and amplification of the infecting organism. Thus, the extent of an epidemic and its degree of severity are determined by complex interactions between parasite and host. Host factors such as age, genetic predisposition, and immune status can dramatically influence the manifestations of an infectious disease. Together with differences in infecting dose, these factors are largely responsible for the wide spectrum of disease manifestations that may be seen during an epidemic.

Interaction between host and parasite determines extent and severity of an epidemic

The effect of age can be dramatic. For example, in an epidemic of measles in an isolated population in 1846, the attack rate for all ages averaged 75%; however, mortality rate was 90 times higher in children less than 1 year of age (28%) than in those 1 to 40 years of age (0.3%). Conversely, in one outbreak of poliomyelitis, the attack rate of paralytic polio was 4% in children 0 to 4 years of age and 20% to 40% in those 5 to 50 years of age. Gender may be a factor in disease manifestations; for example, the likelihood of becoming a chronic carrier of hepatitis B is twice as high for males as for females.

Attack rates and disease severity can vary widely by age

Prior exposure of a population to an organism may alter immune status and the frequency of acquisition, severity of clinical disease, and duration of an epidemic. For example, measles is highly infectious and attacks most susceptible members of an exposed population. However, infection gives solid lifelong immunity. Thus, in unimmunized populations in which the disease is maintained in endemic form, epidemics occur at about 3-year intervals when a sufficient number of nonimmune hosts has been born to permit rapid transmission

Immune status of a population influences epidemic behavior

between them. When a sufficient immune population is reestablished, epidemic spread is blocked and the disease again becomes endemic. When immunity is short-lived or incomplete, epidemics can continue for decades if the mode of transmission is unchecked, which accounts for the present epidemic of gonorrhea.

Immunity in population influences spread

Prolonged and extensive exposure to a pathogen during previous generations selects for a higher degree of innate genetic immunity in a population. For example, extensive exposure of Western urbanized populations to tuberculosis during the 18th and 19th centuries conferred a degree of resistance greater than that among the progeny of rural or geographically isolated populations. The disease spread rapidly and in severe form, for example, when it was first encountered by Native Americans. An even more dramatic example concerns the resistance to the most serious form of malaria that is conferred on people of West African descent by the sickle celled trait (see Chapter 50). These instances are clear cases of natural selection, a process that accounts for many differences in racial immunity.

Sudden appearance of "new" agents can result in pandemic spread

Occasionally, an epidemic arises from an organism against which immunity is essentially absent in a population and that is either of enhanced virulence or appears to be of enhanced virulence because of the lack of immunity. When such an organism is highly infectious, the disease it causes may become pandemic and worldwide. A prime example of this situation is the appearance of a new major antigenic variant of influenza A virus against which there is little if any cross-immunity from recent epidemics with other strains. The 1918–1919 pandemic of influenza was responsible for more deaths than World War I (over 20 million). Subsequent but less serious pandemics have occurred at intervals because of the development of strains of influenza virus with major antigenic shifts (see Chapter 9). Another example, acquired immunodeficiency syndrome (AIDS), illustrates the same principles but also reflects changes in human ecologic and social behavior.

Social and ecological factors determine aspects of epidemic diseases

A major feature of serious epidemic diseases is their frequent association with poverty, malnutrition, disaster, and war. The association is multifactorial and includes overcrowding, contaminated food and water, an increase in arthropods that parasitize humans, and the reduced immunity that can accompany severe malnutrition or certain types of chronic stress. Overcrowding and understaffing in day-care centers or institutions for the mentally impaired can similarly be associated with epidemics of infections.

Nosocomial = hospital-acquired

In recent years, increasing attention has been given to hospital (nosocomial) epidemics of infection. Hospitals are not immune to the epidemic diseases that occur in the community; and outbreaks result from the association of infected patients or persons with those who are unusually susceptible because of chronic disease, immunosuppressive therapy, or the use of bladder, intratracheal, or intravascular catheters and tubes. Control depends on the techniques of medical personnel, hospital hygiene, and effective surveillance.

Control of Epidemics

Surveillance is the key to recognition of an epidemic

The first principle of control is recognition of the existence of an epidemic. This recognition is sometimes immediate because of the high incidence of disease, but often the evidence is obtained from ongoing surveillance activities, such as routine disease reports to health departments and records of school and work absenteeism. The causative agent must be identified, and studies to determine route of transmission (eg, food poisoning) must be initiated.

Control measures can vary widely

Measures must then be adopted to control the spread and development of further infection. These methods include (1) blocking the route of transmission, if possible (eg, improved food hygiene or arthropod control); (2) identifying, treating, and, if necessary, isolating infected individuals and carriers; (3) raising the level of immunity in the uninfected population by immunization; (4) making selective use of chemoprophylaxis for subjects or populations at particular risk of infection, as in epidemics of meningococcal infection; and (5) correcting conditions such as overcrowding or contaminated water supplies that have led to the epidemic or facilitated transfer.

GENERAL PRINCIPLES OF IMMUNIZATION

Immunization is the most effective method of providing individual and community protection against many epidemic diseases. Immunization can be active, with stimulation of the body's immune mechanisms through administration of a vaccine, or passive, through administration of plasma or globulin containing preformed antibody to the agent desired. Active immunization with living attenuated organisms generally results in a subclinical or mild illness that duplicates to a limited extent the disease to be prevented. Live vaccines generally provide both local and durable humoral immunity. Killed or subunit vaccines such as influenza vaccine and tetanus toxoid provide immunogenicity without infectivity. They generally involve a larger amount of antigen than live vaccines and must be administered parenterally with two or more spaced injections and subsequent boosters to elicit and maintain a satisfactory antibody level. Immunity usually develops more rapidly with live vaccines, but serious overt disease from the vaccine itself can occur in patients whose immune responses are suppressed. Live attenuated virus vaccines are generally contraindicated in pregnancy because of the risk of infection and damage to the developing fetus. Recent developments in molecular biology and protein chemistry have brought greater sophistication to the identification and purification of specific immunizing antigens and epitopes and to the preparation and purification of specific antibodies for passive protection. Thus, immunization is being applied to a broader range of infections.

Prophylaxis or therapy of some infections can be accomplished or aided by passive immunization. This procedure involves administration of preformed antibody obtained from humans, derived from animals actively immunized to the agent, or produced by hybridoma techniques. Animal antisera induce immune responses to their globulins that result in clearance of the passively transferred antibody within about 10 days and carry the risk of hypersensitivity reactions such as serum sickness and anaphylaxis. Human antibodies are less immunogenic and are detectable in the circulation for several weeks after administration. Two types of human antibody preparations are generally available. Immune serum globulin (gamma globulin) is the immunoglobulin G fraction of plasma from a large group of donors that contains antibody to many infectious agents. Hyperimmune globulins are purified antibody preparations from the blood of subjects with high titers of antibody to a specific disease that have resulted from natural exposure or immunization; hepatitis B immune globulin, rabies immune globulin, and human tetanus immune globulin are examples. Details of the use of these globulins can be obtained from the chapters that discuss the diseases in question. Passive antibody is most effective when given early in the incubation period.

Passive immunization has a temporary effect

CONCLUSIONS

Epidemiology is clearly the cornerstone for understanding all infectious diseases. The principles, when applied wisely, serve to understand the nature and spread of pathogens, facilitate their recognition, and suggest means of control. The latter may variously involve direct therapeutic maneuvers, prevention through selective chemoprophylaxis or immunization, implementation of environmental controls, and public education. These approaches vary among specific agents, but knowledge of their usefulness is highly important, whether dealing with a single ill patient or an entire community.

PART 

Pathogenic Viruses

Nafees Ahmad,
C. George Ray
and W. Lawrence Drew

The Nature of Viruses CHAPTER 06

Pathogenesis of Viral Infection CHAPTER 07

Antiviral Antimicrobics and Resistance CHAPTER 08

Influenza, Parainfluenza, Respiratory Syncytial Virus, Adenovirus, and Other Respiratory Viruses CHAPTER 09

Mumps Virus, Measles, Rubella, and Other Childhood Exanthems CHAPTER 10

Poxviruses CHAPTER 11

Enteroviruses CHAPTER 12

Hepatitis Viruses CHAPTER 13

Herpesviruses CHAPTER 14

Viruses of Diarrhea CHAPTER 15

Arthropod-Borne and Other Zoonotic Viruses CHAPTER 16

Rabies CHAPTER 17

Retroviruses: Human T-Lymphotropic Virus, Human Immunodeficiency Virus, and Acquired Immunodeficiency Syndrome CHAPTER 18

Papilloma and Polyoma Viruses CHAPTER 19

Persistent Viral Infections of the Central Nervous System CHAPTER 20

CHAPTER 6

The Nature of Viruses

(A virus is) "a piece of bad news wrapped in a protein coat."

—Peter Medawar

A virus is a set of genes, composed of either DNA or RNA, packaged in a protein-containing coat. Some viruses also have an outer lipid bilayer membrane external to the coat called an envelope. The resulting complete virus particle is called a **virion.** Viruses have an obligate requirement for intracellular growth and a heavy dependence on host cell structural and metabolic components. Therefore, viruses are also referred to as obligate intracellular parasites. Viruses do not have a nucleus, cytoplasm, mitochondria, or other cell organelles. Viruses that infect humans are called **human viruses** but are considered along with the general class of **animal viruses**; viruses that infect bacteria are referred to as **bacteriophages** (phages for short), and viruses that infect plants are called **plant viruses**.

A virus is an intracellular parasite composed of DNA or RNA and a protein coat and in some cases an outer lipoprotein envelope

Virus reproduction requires that a virus particle infect an appropriate host cell and program the cellular machinery to synthesize the viral components required for the assembly of new virions, generally termed **progeny virions** or **daughter viruses**. The infected host cell may produce hundreds to hundreds of thousands of new virions, usually accompanied by cell death. Tissue damage as a result of cell death accounts for the pathology of many viral diseases in humans. Many of these viruses cause **acute viral infection** followed by viral clearance. In some cases, the infected cells survive, resulting in **persistent virus production** and a **chronic infection** that can remain asymptomatic, produce a chronic disease state, or lead to relapse of an infection.

In some circumstances, a virus fails to reproduce itself and instead enters a latent state (called **lysogeny** in the case of bacteriophages), from which there is the potential for reactivation at a later time. A possible consequence of the presence of viral genome in a latent state is a new genotype for the cell. Some determinants of bacterial virulence and some malignancies of animal cells are examples of the genetic effects of latent viruses. Apparently, vertebrates have had to coexist with viruses for a long time because they have evolved the special nonspecific interferon system, which operates in conjunction with the highly specific immune system to combat virus infections.

Some viruses, instead of reproducing, enter into a latent state from which they can later be reactivated

Two classes of infectious agents exist that are structurally simpler than viruses, namely, viroids and prions. **Viroids** are infectious circular RNA molecules that lack protein shells; they are responsible for a variety of plant diseases. **Prions,** which apparently lack any genes and are composed only of protein, are agents that appear to be responsible for some transmissible and inherited spongiform encephalopathies such as scrapie in sheep; bovine spongiform encephalopathy in cattle; and kuru, Creutzfeldt-Jakob disease, and Gerstmann-Sträussler-Scheinker syndrome in humans.

Plant viroids are infectious RNA molecules

Prions are protein molecules that may cause spongiform encephalopathies

VIRUS STRUCTURE

Viruses range in size from 20 to 300 nm in diameter

Viruses are approximately 100- to 1000-fold smaller than the cells they infect. The smallest viruses, **virion size** (parvoviruses), are approximately 20 nm in diameter (1 nm = 10^{-9} m), whereas the largest human viruses (poxviruses) have a diameter of approximately 300 nm (**Figure 6–1**) and overlap the size of the smallest bacterial cells (*Chlamydia* and *Mycoplasma*). Therefore, viruses generally pass through filters designed to trap bacteria, and this property can, in principle, be used as evidence of a viral etiology.

Naked capsid viruses have a nucleic acid genome within a protein shell

Enveloped viruses have a nucleocapsid (nucleic acid-protein complex) packaged into a lipoprotein envelope

Viruses often have surface protrusions called spikes

The basic structure of all viruses places the nucleic acid genome (DNA or RNA) on the inside of a protein shell called a **capsid.** Some human viruses are further packaged into a lipid membrane, or **envelope,** which is usually acquired from the cytoplasmic membrane of the infected cell during release from the cell. Viruses that are not enveloped have a defined external capsid and are referred to as **naked capsid viruses.** The genomes of enveloped viruses form a protein complex and a structure called a **nucleocapsid,** which is often surrounded by a **matrix** protein that serves as a bridge between the nucleocapsid and the inside of the viral membrane. Protein or glycoprotein structures called **spikes,** which often protrude from the surface of virus particles, are involved in the initial contact with cells. These basic design features are illustrated schematically in **Figure 6–2** as well as in the electron micrographs in **Figures 6–3A-C** and **6–4**.

Two basic shapes: cylindrical (helical) and spherical (icosahedral)

The protein shell forming the capsid or the nucleocapsid assumes one of two basic shapes: cylindrical (helical) or spherical (icosahedral). Some of the more complex bacteriophages combine these two basic shapes. Examples of these three structural categories can be seen in the electron micrographs in Figure 6–3.

Outer shell is protective and aids in entry and packaging

The capsid or envelope of viruses functions (1) to protect the nucleic acid genome from damage during the extracellular passage of the virus from one cell to another, (2) to aid in the process of entry into the cell, and (3) in some cases to package enzymes essential for the early steps of the infection process.

Nucleic acid must be condensed during virion assembly

In general, the nucleic acid genome of a virus is hundreds of times longer than the longest dimension of the complete virion. It follows that the viral genome must be extensively condensed during the process of virion assembly. For naked capsid viruses, this condensation is achieved by the association of the viral nucleic acid with basic proteins encoded by the virus to form the **core** of the virus (Figure 6–2). For enveloped viruses, the formation of the nucleocapsid serves to condense the viral nucleic acid genome. The virion may also contain certain virus encoded essential enzymes and/or accessory/regulatory proteins.

GENOME STRUCTURE

The genomes of viruses can be either RNA or DNA, but not both

DNA or RNA genomes may be single- or double-stranded

Viral genomes can be made of either RNA or DNA and also can be either single-stranded or double-stranded. The RNA viruses can be either positive sense (indicated by a +) (polarity of mRNA) or negative sense (–) (complementary to or antisense of mRNA), double-stranded (one strand + and the second strand –) or ambisense (both + and – polarity on the same strand). Although the RNA genomes of most viruses are linear, some RNA viruses may also have segmented genomes (several segments or pieces of RNA), with each segment responsible for encoding a protein.

Genomes may be linear or circular

Some genomes are segmented

The DNA genome of viruses can be both linear and circular genomes. Most viruses contain a single copy of their genome, except that retroviruses carry two identical copies of its genome and are therefore diploid. A few viral genomes (picornaviruses, hepatitis B virus, and adenoviruses) contain covalently attached protein on the ends of the DNA or RNA chains that are remnants of the replication process. Structural diversity among the viruses is most obvious when the makeup of viral genomes is considered.

CAPSID STRUCTURE

Subunit Structure of Capsids

The capsids or nucleocapsids are virus-encoded specific proteins that protect the genome and confer shapes to viruses. The capsids of all viruses are composed of many copies of one or, at most, several different kinds of protein subunits. This fact follows from two fundamental considerations. First, all viruses code for their own capsid proteins, and even if the entire coding capacity of the genome were to be used to specify a single giant capsid protein, the protein would not be large enough to enclose the nucleic acid genome. Thus,

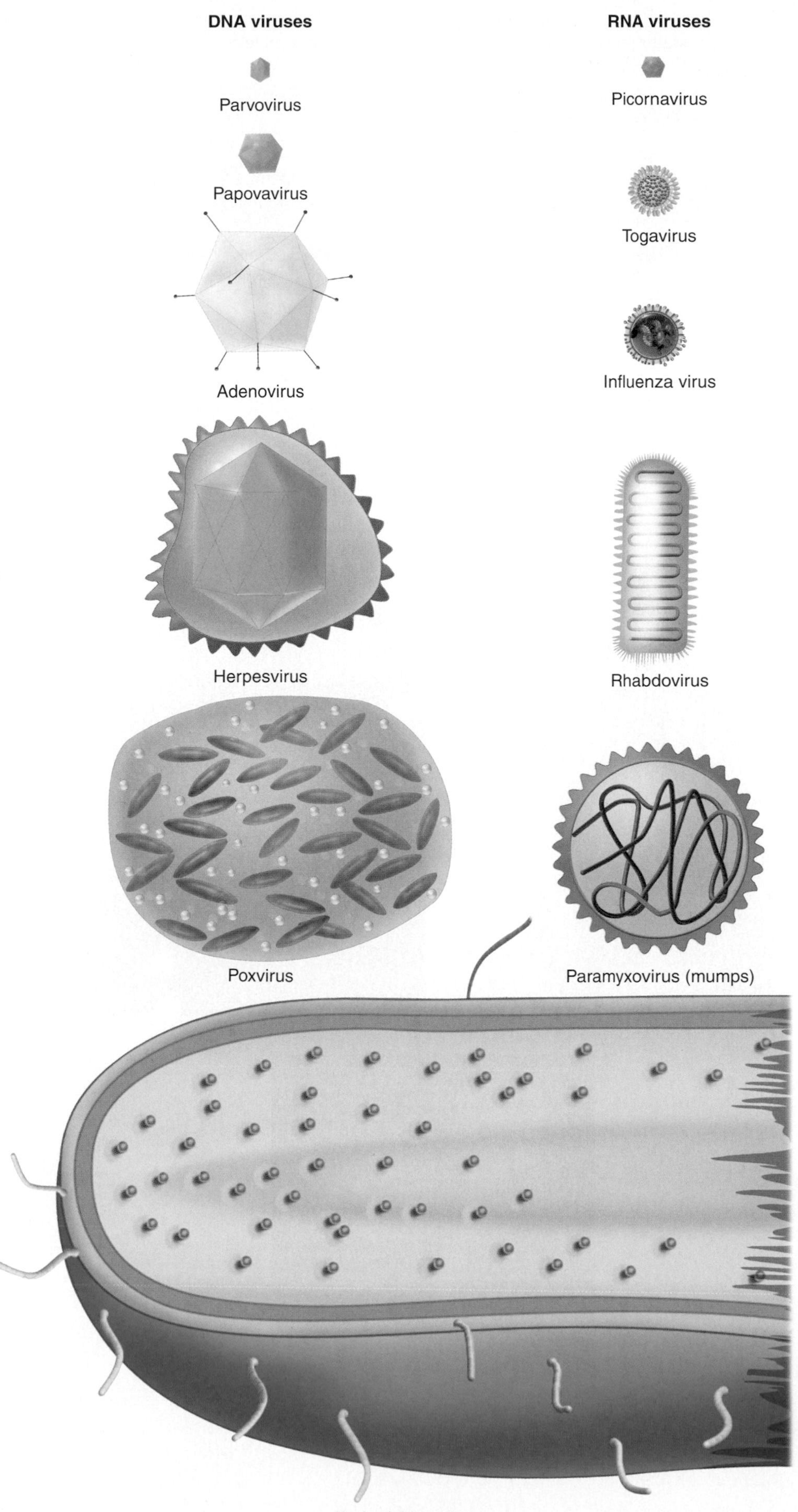

FIGURE 6–1. Size comparison of viruses with other microbes. (Reproduced with permission from Willey J, Sherwood L, Woolverton C (eds). *Prescott's Principles of Microbiology.* New York: McGraw-Hill; 2008.)

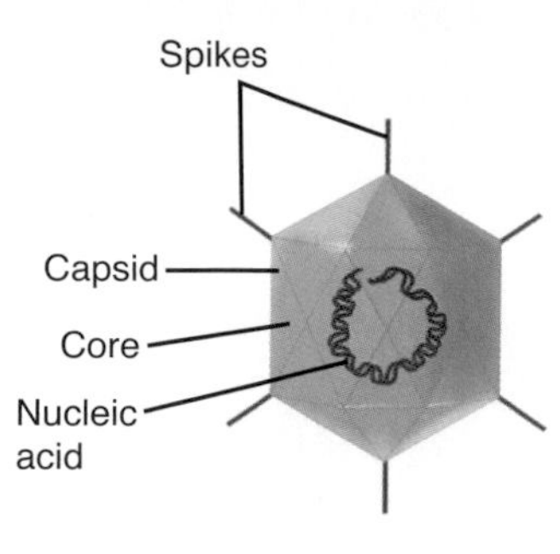

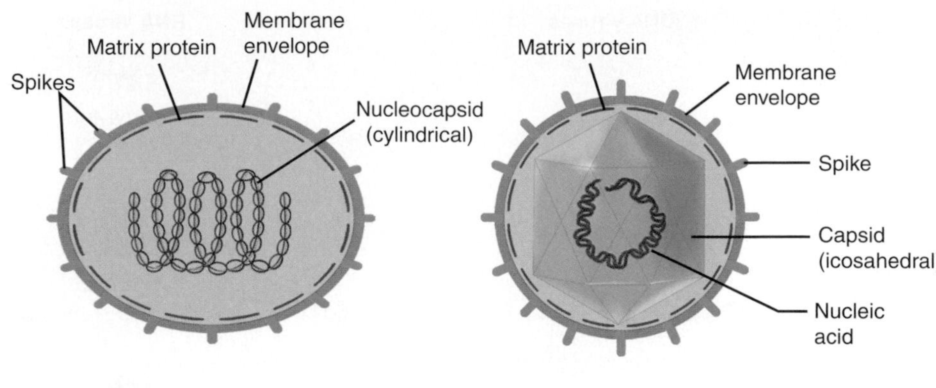

FIGURE 6–2. Schematic drawing of two basic types of virions. (Reproduced with permission from Willey J, Sherwood L, Woolverton C (eds). *Prescott's Principles of Microbiology.* New York: McGraw-Hill; 2008.)

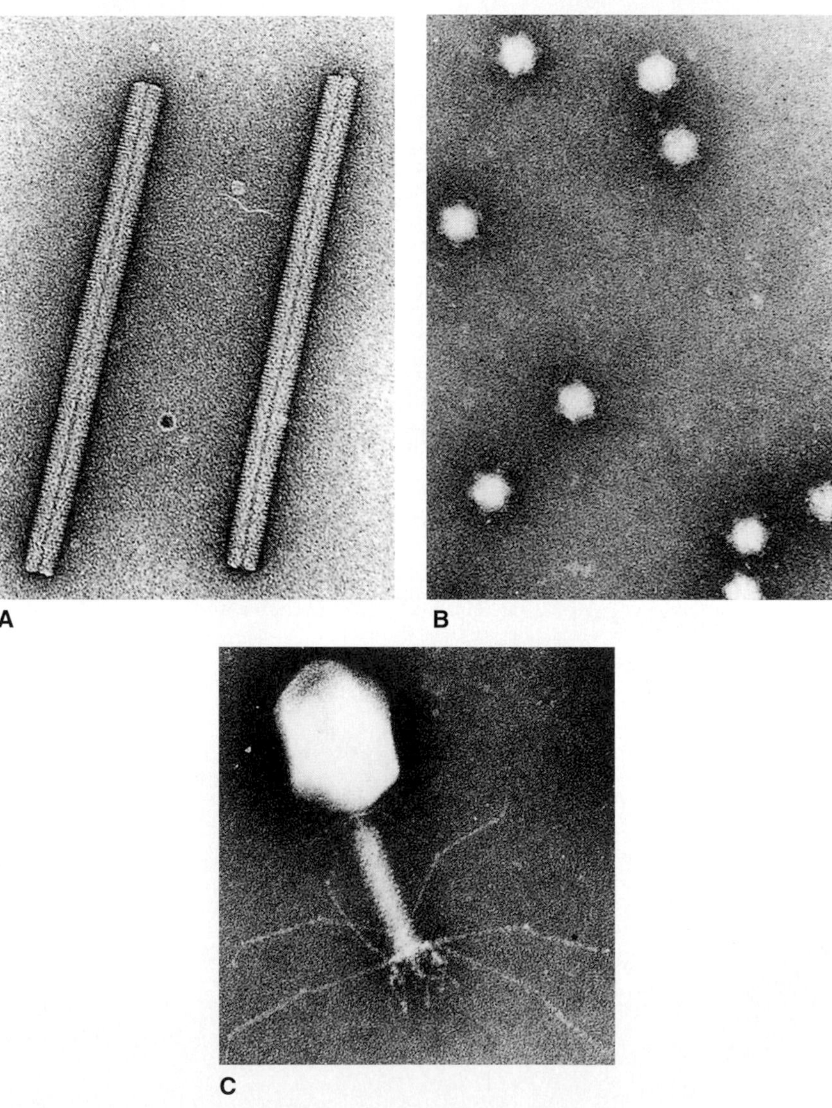

FIGURE 6–3. Three basic virus designs. A. Tobacco mosaic virus. **B.** Bacteriophage ϕX174. **C.** Bacteriophage T4. (Courtesy of Dr. Robley C. Williams.)

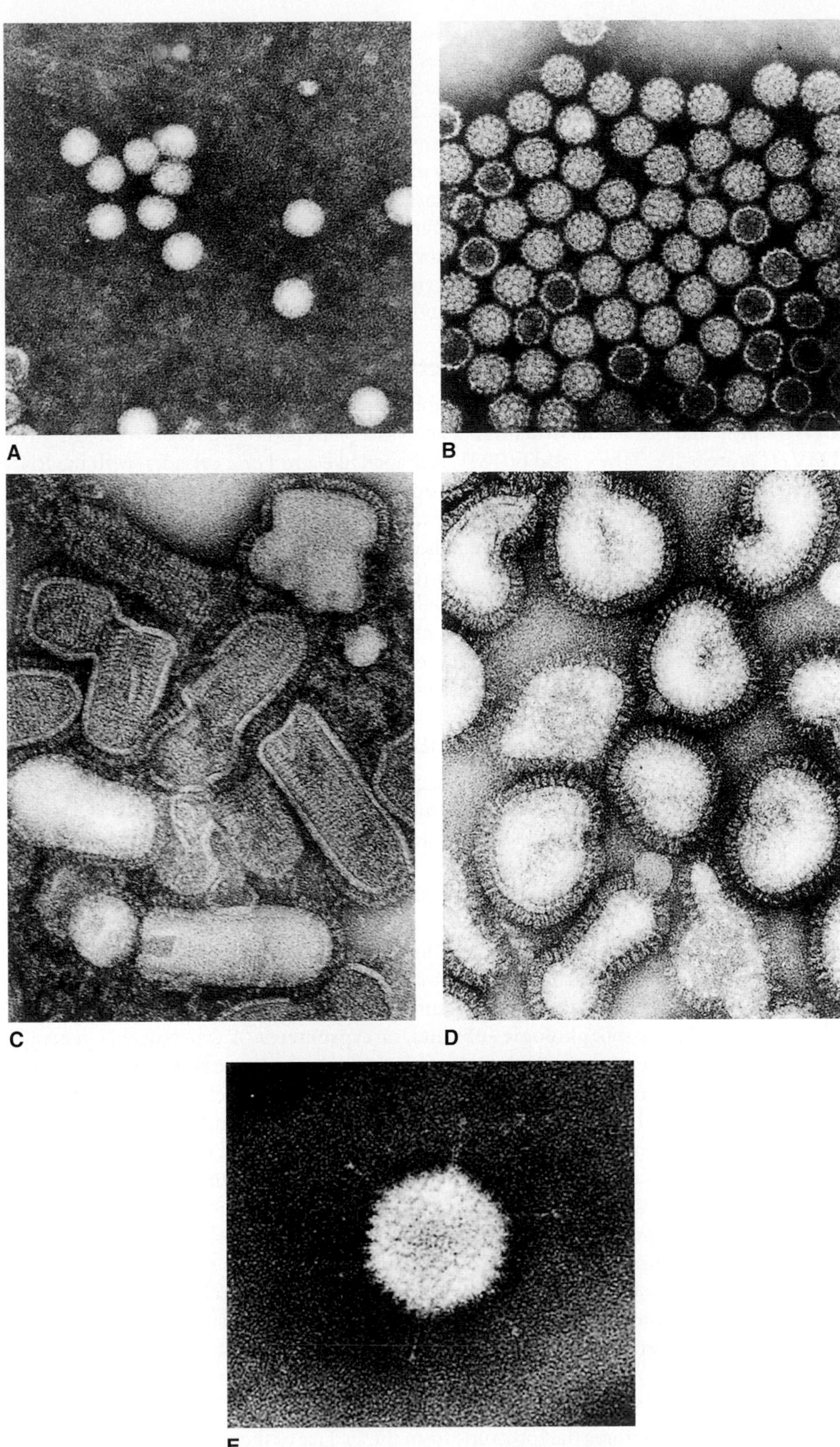

FIGURE 6–4. Representative human/animal viruses. A. Poliovirus. **B.** Simian virus 40. **C.** Vesicular stomatitis virus. **D.** Influenza virus. **E.** Adenovirus. (Courtesy of Dr. Robley C. Williams.)

Capsids and nucleocapsids are composed of multiple copies of protein molecule(s) in crystalline array

multiple protein copies are needed, and, in fact, the simplest spherical virus contains 60 identical protein subunits. Second, viruses are such highly symmetrical structures that it is not uncommon to visualize naked capsid viruses in the electron microscope as a crystalline array (eg, simian virus 40 in **Figure 6–4B**). The simplest way to construct a regular symmetrical structure out of irregular protein subunits is to follow the rules of crystallography and form an aggregate involving many identical copies of the subunits, in which each subunit bears the same relation to its neighbors as to every other subunit.

The presence of many identical protein subunits in viral capsids or the existence of many identical spikes in the membrane of enveloped viruses has important implications for adsorption, hemagglutination, and recognition of viruses by neutralizing antibodies. Two main architectures; cylindrical (helical symmetry) and spherical (icosahedral or cubic symmetry).

Cylindrical (Helical) Architecture

Cylindrical viruses have capsid protein molecules arranged in a helix

A cylindrical shape is the simplest structure for a capsid or a nucleocapsid. The first virus to be crystallized and studied in structural detail was a plant pathogen, tobacco mosaic virus (TMV) (Figure 6–3A). The capsid of TMV is shaped like a rod or a cylinder, with the RNA genome wound in a helix inside it. The capsid is composed of multiple copies of a single kind of protein subunit arranged in a close-packed helix, which places every subunit in the same microenvironment. Because of the helical arrangement of the subunits, viruses that have this type of design are often said to have helical symmetry. Although less is known about the architecture of human viruses with helical symmetry, it is likely that their structures follow the same general pattern as that of TMV. Thus, the nucleocapsids of influenza, measles, mumps, rabies, and poxviruses (**Table 6–1**) are probably constructed with a helical arrangement of protein subunits in close association with the nucleic acid genome.

Spherical (Icosahedral) Architecture

Spherical viruses exhibit icosahedral symmetry

The construction of a spherically (icosahedral) shaped virus similarly involves the packing together of many identical subunits, but in this case the subunits are placed on the surface of a geometric solid called an **icosahedron.** An icosahedron has 12 vertices, 30 sides, and 20 triangular faces (**Figure 6–5**). Because the icosahedron belongs to the symmetry group that crystallographers refer to as cubic, spherically shaped viruses are said to have cubic symmetry. (Note that the term **cubic,** as used in this context, has nothing to do with the more familiar shape called the cube.)

Capsomeres are surface structures composed of five or six protein molecules

When viewed in the electron microscope, many naked capsid viruses and some nucleocapsids appear as spherical particles with a surface topology that makes it appear that they are constructed of identical ball-shaped subunits (Figure 6–4B and E). These visible structures are referred to as **morphologic subunits,** or **capsomeres.** A capsomere is generally composed of either five or six individual protein molecules, each one referred to as a **structural subunit,** or **protomer.** In the simplest virus with cubic symmetry, five protomers are placed at each one of the 12 vertices of the icosahedron as shown in **Figure 6–5** to form a capsomere called a **pentamer.** In this case, the capsid is composed of 12 pentamers, or a total of 60 protomers. Note that in the case of helical symmetry, this arrangement places every protomer in the same microenvironment as that of every other protomer.

To accommodate the larger cavity required by viruses with large genomes, the capsids contain many more protomers. These viruses are based on a variation of the basic icosahedron in which the construction involves a mixture of pentamers and hexamers rather than only pentamers. A detailed description of this higher level of virus structure is beyond the scope of this text. Examples of icosahedral capsids are shown in **Figure 6–6.**

Special Surface Structures

Surface structures are important in adsorption and penetration

Many viruses have structures that protrude from the surface of the virion. In virtually every case, these structures are important for the two earliest steps of infection—adsorption and penetration. The most dramatic example of such a structure is the tail of some bacteriophages (Figure 6–3C), which, acts as a channel for the transfer of the genome into the cell. Other examples of surface structures include the spikes of adenovirus (Figure 6–4E) and the glycoprotein spikes found in the membrane of enveloped viruses (see influenza virus in Figure 6–4D). Even viruses without obvious surface extensions probably contain short projections, which, like the more obvious spikes, are involved in the specific binding of the virus to the cell surface.

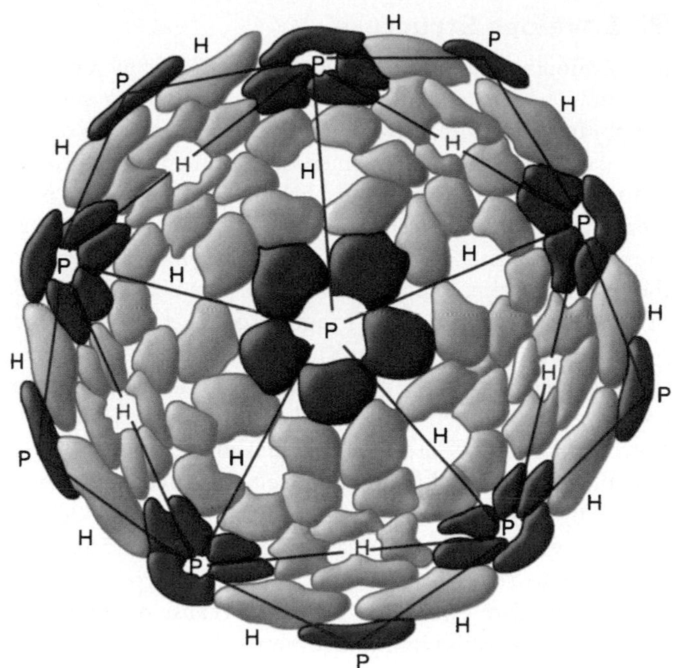

FIGURE 6–5. Diagram of an icosahedron showing 12 vertices, 20 faces, and 30 sides. The colored balls indicate the position of protomers forming a pentamer on the icosahedron. (Reproduced with permission from Willey J, Sherwood L, Woolverton C (eds). *Prescott's Principles of Microbiology.* New York: McGraw-Hill; 2008.)

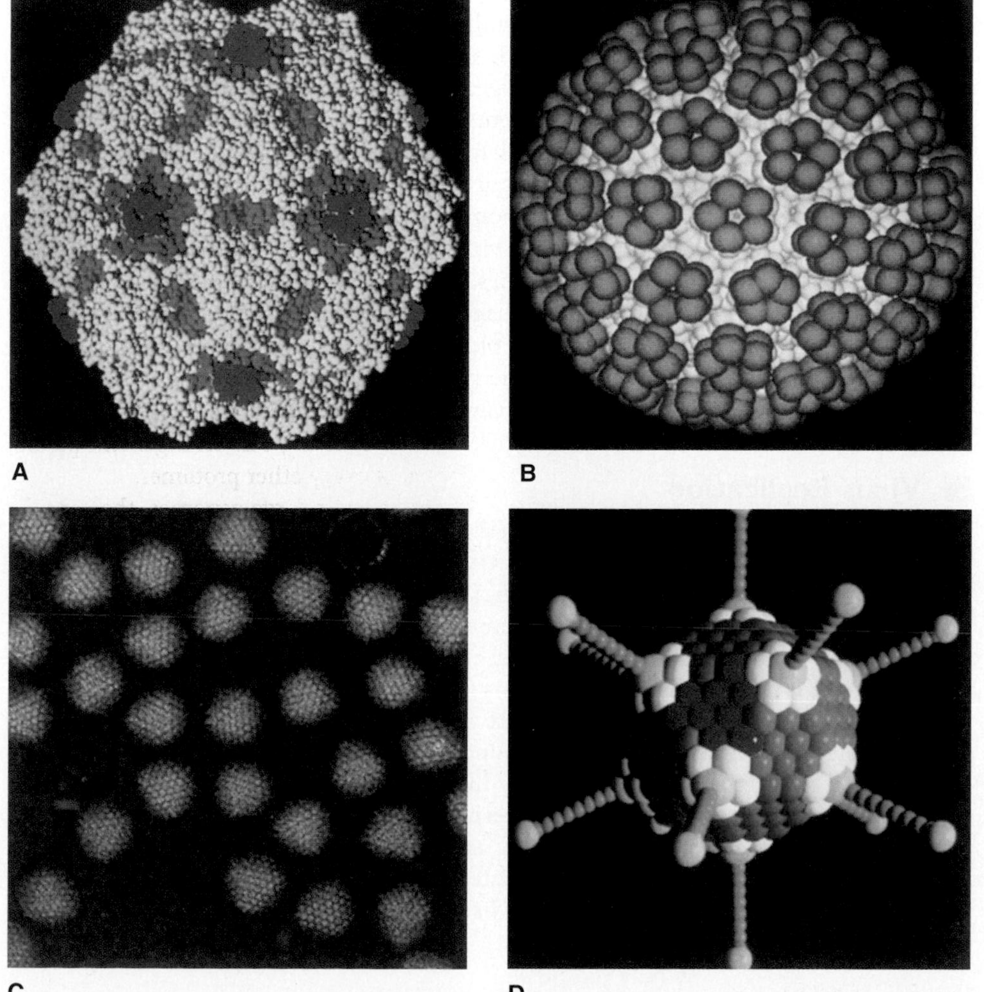

FIGURE 6–6. Examples of icosahedral capsids. (Reproduced with permission from Willey J, Sherwood L, Woolverton C (eds). *Prescott's Principles of Microbiology.* New York: McGraw-Hill; 2008.)

Envelope Structure

Viral envelopes are lipid bilayer membranes containing virus encoded glycoproteins or spikes

Many human viruses have an outer lipid bilayer membrane that is derived from cellular membranes, mainly plasma membrane, but also in some cases cytoplasmic or nuclear membranes. The viral envelope lipid layer membrane contains virus-encoded glycoproteins called "spikes" or "peplomers" or "viral envelope proteins." The envelope spikes bind to the receptor on the host cells, help the virus envelope membrane fuse with the cellular membrane of the host cells, and act as principal antigens against which the host mounts immune response for the recognition of the virus. Enveloped viruses have another protein, the matrix protein, which serves as a bridge between nucleocapsid and inner membrane of the envelope (Figure 6–2). Examples of enveloped viruses are shown in **Figure 6–7** with both helical (Figure 6–7A and B) and cubic (Figure 6–7C and D) symmetry.

Envelope glycoproteins, like surface proteins of naked capsid viruses, bind to the receptors on host cells for virus entry

Enveloped viruses are more sensitive to detergents, solvents, ethanol, ether and heat compared with nonenveloped (naked capsid) viruses whose outer coat is capsid protein. Both envelope glycoproteins, and naked capsid viruses' spikes become antigens after infection and the host mount both cell-mediated and humoral immune responses for the elimination of virus-infected cells and cell-free virus, respectively. These antigens determine the viral serotypes that are based on antigenic variation and are type-specific such as poliovirus serotypes 1, 2, and 3. Viral serotypes have cross-reactivity but often little cross-protection. Viral serotypes arise because of antigenic variations that allow viruses to escape pre existing immune response.

CLASSIFICATION OF VIRUSES

The classification of viruses has evolved at a slower pace than other microorganisms. The International Committee for Taxonomy of Viruses (ICTV) considered various properties, including virions, genome, proteins, envelope, replication, and physical and biologic properties. Based on these properties, virus families are designated with the suffix, -viridae (as in Herpesviridae), virus subfamilies with suffix -virinae (Herpesvirinae), virus genera with suffix -virus (Herpesvirus), and virus species designated by a virus type (herpes simplex virus 1). **Tables 6–1, 6–2,** and **6–3** present a classification scheme for human RNA and DNA viruses, respectively, which is based solely on their structure. The viruses are arranged in order of increasing genome size. It is important to bear in mind that phylogenetic relationships cannot be inferred from this taxonomic scheme. The tables should not be memorized, but rather used as a reference guide to virus structure. In general, viruses with similar structures exhibit similar replication strategies, as discussed later.

Representative and important bacteriophages are listed along with their properties in **Table 6–4**. In the chapters that follow, the properties of the well-studied temperate bacteriophage, λ, are described to illustrate the replicative strategies of the more medically important, but less well-studied, β phage of *Corynebacterium diphtheriae.*

Virus Replication

Virus replication cycle typically consists of six discrete phases: (1) adsorption or attachment to the host cell, (2) penetration or entry, (3) uncoating to release the genome, (4) synthetic or virion component production, (5) assembly, and (6) release from the cell. These phases are shown in a general scheme of virus replication cycle in **Figure 6–8**. This series of events, sometimes with slight variations, describes what is called the **productive** or **lytic response;** however, this is not the only possible outcome of a virus infection. Some viruses can also enter into a very different kind of relationship with the host cell in which no new virus is produced, the cell survives and divides, and the viral genetic material persists indefinitely in a latent state. This outcome of an infection is referred to as the **nonproductive response.** The nonproductive response in the case of bacteriophages is called **lysogeny** and in several human and animal viruses under some circumstances may be associated with **oncogenic transformation**. (This use of the term transformation is to be distinguished from DNA transformation of bacteria discussed in Chapter 21.)

Viral infections may be productive or nonproductive

Some human viruses can cause oncogenic transformation

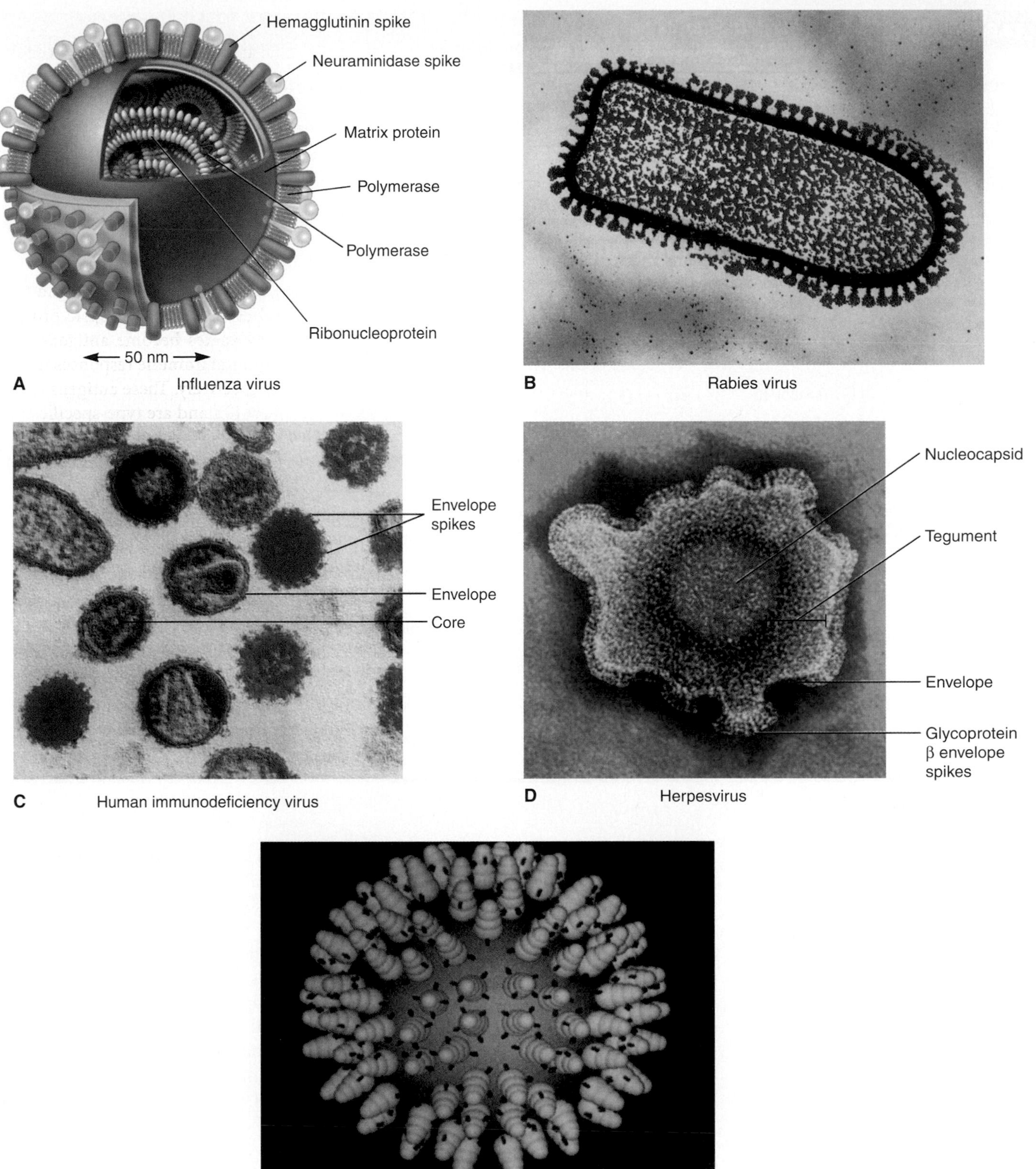

FIGURE 6–7. Examples of enveloped viruses. (Reproduced with permission from Willey J, Sherwood L, Woolverton C (eds). *Prescott's Principles of Microbiology.* New York: McGraw-Hill; 2008.)

TABLE 6–1 Classification of RNA Human Viruses

FAMILY	VIRION STRUCTURE	GENOME STRUCTURE AND MOLECULAR WEIGHT	REPRESENTATIVE MEMBERS
Picornaviruses	Cubic, naked	ss linear (+) ($2 - 3 \times 10^6$); protein attached	Human enteroviruses: poliovirus, coxsackievirus, echovirus; rhinoviruses; bovine foot-and-mouth disease virus; hepatitis A
Arenaviruses	Helical, enveloped	2 ss linear segments (+/–) (3×10^6)	Lassa virus; lymphocytic choriomeningitis virus of mice
Caliciviruses	Cubic, naked	ss linear (+) (2.6×10^6)	Vesicular exanthema virus, Norwalk-like viruses of humans
Rhabdoviruses	Helical, enveloped	ss linear (–) ($3 - 4 \times 10^6$)	Rabies virus; bovine vesicular stomatitis virus
Retroviruses	Cubic, enveloped	ss linear (+), diploid ($3 - 4 \times 10^6$)	RNA tumor viruses of mice, birds, and cats; visna virus of sheep; human immunodefi ciency viruses (acquired immunodeficiency syndrome), human T-lymphotropic viruses (adult T cell leukemia)
Togaviruses	Cubic, enveloped	ss linear (+) (4×10^6)	Alphaviruses: Western and Eastern equine encephalitis viruses, Venezuelan equine encephalitis virus, Chikungunya virus; Flaviviruses: Dengue virus, yellow fever virus, St. Louis encephalitis virus, West Nile virus, Japanese B encephalitis virus Rubiviruses: Rubella virus
Orthomyxoviruses	Helical, enveloped	8 ss linear segments (–) (5×10^6)	Type A, B, and C influenza viruses of humans, swine, horses, and avian
Coronaviruses	Helical, enveloped	ss linear (+) ($5 - 6 \times 10^6$)	Respiratory viruses of humans; calf diarrhea virus; swine enteric virus; mouse hepatitis virus
Filoviruses	Helical, enveloped	ss linear (–) (5×10^6)	Marburg and Ebola viruses
Bunyaviruses	Helical, enveloped	3 ss linear segments (+/–) (6×10^6)	Rift Valley fever virus; bunyamwera virus; hantavirus
Paramyxoviruses	Helical, enveloped	ss linear (–) ($6 - 8 \times 10^6$)	Mumps; measles; parainfluenza viruses, respiratory syncytial virus
Reoviruses	Cubic, naked	10 ds linear segments (15×10^6)	Human reoviruses; orbiviruses; Colorado tick fever virus; human rotaviruses

ss, single-stranded; ds, double-stranded.

Some viruses cause chronic or latent infection

Permissive cells allow virus replication and/or viral transformation

Nonpermissive cells do not permit virus replication, but may allow viral transformation

Some viruses can also cause a **chronic infection** where a low level of the virus is produced with little or no damage to the target tissue. Both latent infection and chronic infection are called **persistent infection**. Virus replication also depends on virus–host cell interaction such as the type of cells it infects—whether permissive or nonpermissive cells. **Permissive cells** are those that permit production of progeny virus particles and/or viral transformation. On the other hand, **nonpermissive cells** do not allow virus replication, but may allow virus transformation. Some viruses enter cells that do not support virus replication, but some early viral proteins cause cell death; this infection is termed **abortive infection**.

The outcome of an infection depends on the particular virus–host combination and on other factors such as the extracellular environment, multiplicity of infection, and physiology and developmental state of the cell. Viruses that can enter only into a

TABLE 6–2 Unclassified RNA Human Viruses

FAMILY	VIRION STRUCTURE	GENOME STRUCTURE AND MOLECULAR WEIGHT	REPRESENTATIVE MEMBERS
Hepatitis δ	Cubic, enveloped	ss circular (–) (6×10^5)	Human hepatitis δ virus
Hepatitis E virus	Cubic, naked	ss linear (+) (2.6×10^6);	Hepatitis E virus (ET non-A, non-B hepatitis)

ss, single-stranded

TABLE 6–3 Classification of DNA Human Viruses

FAMILY	VIRION STRUCTURE	GENOME STRUCTURE AND MOLECULAR WEIGHT	REPRESENTATIVE MEMBERS
Parvoviruses	Cubic, naked	ss linear ($1–2 \times 10^6$)	Human parvovirus B-19; adeno-associated viruses
Hepatitis B	Cubic, enveloped	ds circular (2×10^6), gap in one strand; protein attached	Hepatitis B virus of humans, woodchuck hepatitis virus
Papovaviruses	Cubic, naked	ds circular ($3–5 \times 10^6$)	Papillomaviruses, polyomaviruses, SV40 (monkey)
Adenoviruses	Cubic, naked	ds linear ($20–25 \times 10^6$); protein attached	Human and animal respiratory disease viruses
Herpesviruses	Cubic, enveloped	ds linear ($80–130 \times 10^6$)	Herpes simplex virus types 1 and 2; varicella-zoster virus; cytomegalovirus; Epstein-Barr virus; human herpesvirus 6, human herpesvirus 8 (Kaposi's sarcoma)
Poxviruses	Helical, enveloped	ds linear ($160–200 \times 10^6$)	Smallpox; vaccinia; monkeypox virus; cowpox virus; orf; pseudocowpox virus; yabapox virus; tanapox virus; molluscum contagiosum

ss, single-stranded; ds, double-stranded.

TABLE 6–4 Some Important Bacteriophages

BACTERIOPHAGE	HOST	GENOME STRUCTURE AND MOLECULAR WEIGHT	COMMENTS
MS2	*Escherichia coli*	ss linear RNA (1.2×10^6)	Lytic
Filamentous (M13, fd)	*Escherichia coli*	ss linear RNA (2.1×10^6)	No cell death
ϕX174	*Escherichia coli*	ss linear RNA (1.8×10^6)	Lytic
β	*Corynebacterium diphtheriae*	ds linear DNA (23×10^6)	Temperate, codes for diphtheria toxin
λ	*Escherichia coli*	ds linear DNA (31×10^6)	Temperate
T4	*Escherichia coli*	ds linear DNA (108×10^6)	Lytic

ss, single-stranded; ds, double-stranded.

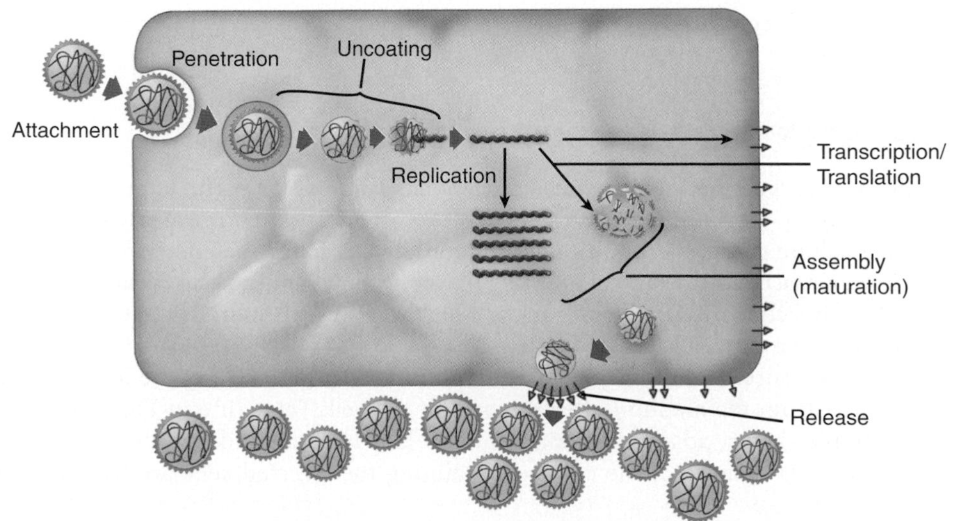

FIGURE 6–8. Virus replication cycle. A general scheme of the six discrete steps of virus replication cycle, starting from attachment to release.

productive relationship are called **lytic** or **virulent viruses.** Viruses that can establish either a productive or a nonproductive relationship with their host cells are referred to as **temperate viruses.** Some temperate viruses can be reactivated or "induced" to leave the latent state and enter into the productive response. Whether induction occurs depends on the particular virus–host combination, the physiology of the cell, and the presence of extracellular stimuli.

Temperate viruses can either replicate or enter a latent state

GROWTH AND ASSAY OF VIRUSES

Viruses are generally propagated in the laboratory by mixing the virus and susceptible cells together and incubating the infected cells until lysis occurs. After lysis, the cells and cell debris are removed by a brief centrifugation, and the resulting supernatant is called a **lysate.**

The growth of human viruses requires that the host cells be cultivated in the laboratory, mostly in human or animal cell lines (cell derived from tumors or cells transformed by viruses) and in some cases in primary cells derived from tissues. To prepare cells for growth in vitro, a tissue is removed from an animal, and the cells are disaggregated using the proteolytic enzyme trypsin. The cell suspension is seeded into a plastic Petri dish in a medium containing a complex mixture of amino acids, vitamins, minerals, and sugars. In addition to these nutritional factors, the growth of animal cells requires components present in animal serum. This method of growing cells is referred to as **tissue culture,** and the initial cell population is called a **primary culture.** The cells attach to the bottom of the plastic dish and remain attached as they divide and eventually cover the surface of the dish. When the culture becomes crowded, the cells generally cease dividing and enter a resting state. Propagation can be continued by removing the cells from the primary culture plate using trypsin and reseeding a new plate.

Viruses are cultivated in cell lines or cell cultures derived from animal tissues

Cells taken from a normal (as opposed to cancerous) tissue cannot usually be propagated in this manner indefinitely. Eventually, most of the cells die; a few may survive, and these survivors often develop into a permanent cell line. Such cell lines are very useful as host cells for isolating and assaying viruses in the laboratory, but they rarely bear much resemblance to the tissue from which they originated. When cells are taken from a tumor and cultivated in vitro, they display a very different set of growth properties, including long-term survival, reflecting their tumor phenotype.

Permanent cell lines are useful for growing viruses

When a virus is propagated in tissue culture cells, the cellular changes induced by the virus, which usually culminate in cell death, are often characteristic of a particular virus and are referred to as the **cytopathic effect** of the virus.

Cytopathic effects are characteristic for individual viruses

Viruses are quantitated by a method called the plaque assay (see Plaque Assay under Quantitation of Viruses for a detailed description of the method). Briefly, viruses are mixed with cells on a Petri plate so that each infectious particle gives rise to a zone of lysed or dead cells called a **plaque.** From the number of plaques on the plate, the titer of infectious particles in the lysate is calculated. Virus titers are expressed as the number of plaque-forming units per milliliter (pfu/mL).

Viruses are quantitated by a plaque assay

ONE-STEP GROWTH EXPERIMENT

To describe an infection in temporal and quantitative terms, it is useful to perform a one-step growth experiment (**Figure 6–9**). The objective in such an experiment is to infect every cell in a culture so that the whole population proceeds through the infection process in a synchronous fashion. The ratio of infecting plaque-forming units to cells is called the multiplicity of infection (MOI). By infecting at a high MOI (eg, 10, as in Figure 6–9), one can be certain that every cell is infected.

One-step growth experiments are useful in the study of infections

The time course and efficiency of adsorption can be followed by the loss of infectious virus from the medium after removal of the cells (blue line in Figure 6–9). In the example shown, adsorption takes about a half-hour and all but 1% of the virus is adsorbed. If samples of the culture containing the infected cells are treated so as

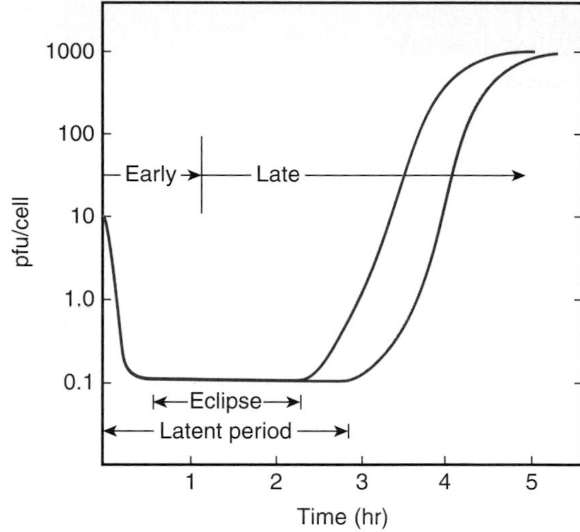

FIGURE 6–9. One-step growth experiment. pfu, plaque-forming units.

to break open the cells prior to assaying for virus (red line in Figure 6–9), it can be observed that infectious virus initially disappears, because no infectious particles are detectable above the background of unadsorbed virus. The period of infection in which no infectious viruses are found inside the cell is called the **eclipse phase** and emphasizes that the original virions lose their infectivity soon after entry. Infectivity is lost because, as is discussed later, the virus particles are dismantled as a prelude to their reproduction. Later, infectious virus particles rapidly reappear in increasing numbers and are detected inside the cell prior to their release into the environment (Figure 6–9). The length of time from the beginning of infection until progeny virions are found outside the cells is referred to as the **latent period.** Latent periods range from 20 minutes to hours for bacteriophages and from a few hours to many days for human viruses.

Shortly after infection, a virus loses its identity (eclipse phase)

Infectious virus reappears at end of eclipse phase inside the cell

The time in the infection at which genome replication begins is typically used to divide the infection operationally into early and late phases. Early viral gene expression is largely restricted to the production of the proteins required for genome replication; later, the proteins synthesized are primarily those necessary for construction of the new virus particles.

Proteins for replication are produced early and those for construction of virions are produced late

The average number of plaque-forming units released per infected cell is called the burst size for the infection. In the example shown, the burst size is about 1000. Burst sizes range from less than 10 for some relatively inefficient infections to millions for some highly virulent viruses.

VIRUS REPLICATION CYCLE

■ Adsorption or Attachment

The first step in every viral infection is the attachment or adsorption of the infecting virus particle to the surface of the host cell. A prerequisite for this interaction is a collision between the virion and the cell. Viruses do not have any capacity for locomotion, and so the collision event is simply a random process determined by diffusion. Therefore, like any bimolecular reaction, the rate of adsorption is determined by the concentrations of both the virions and the cells.

Only a small percentage of the collisions between a virus and its host cell lead to a successful infection because adsorption is a highly specific reaction that involves protein molecules on the surface of the virion called **virion attachment proteins** or **spikes** and certain molecules on the surface of the cell called **receptors.** Typically, 10^4 to 10^5 receptors are on the cell surface. Receptors for some bacteriophages are found on pili of bacteria, although

TABLE 6–5 Examples of Viral Receptors

VIRUS	RECEPTOR	CELLULAR FUNCTION
Influenza A	Sialic acid	Glycoprotein
Reoviruses	Sialic acid	Glycoprotein
	EGF receptor	Signaling
Adenoviruses	Integrins	Binding to extracellular matrix
Cytomegalovirus	Heparan sulfate	Glycoprotein
Epstein-Barr virus	CR2 (CD21)	Complement receptor
Hepatitis A virus	α_2-macroglobulin	Plasma protein (inhibitor of coagulation, fibrinolysis
Herpes simplex	Heparan sulfate	Glycoprotein
Human herpes 7	CD4	Immunoglobulin superfamily
HIV	CD4	Immunoglobulin superfamily
	CXCR4 and CCR5	Chemokine receptors
Human coronavirus	Aminopeptidase N	Protease
Human rhinoviruses	ICAM-1	Immunoglobulin superfamily
Measles	CD46	Complement regulation
Poliovirus	PVR	Immunoglobulin superfamily
Parvovirus B19	Erythrocyte P antigen	Erythroid precursors
Rabies	Acetylcholine receptor	Signaling
Rotavirus	$\alpha_2\beta_1$ and $\alpha_4\beta_1$ integrins	Cell surface receptors that interact with extracellular matrix
SV40	MHC I	Immunoglobulin superfamily
Vaccinia	EGF receptor	Signaling

EGF, endothelial growth factor; HIV, human immunodeficiency virus; ICAM, intercellular adhesion molecule; MHC, major histocompatibility complex; PVR, poliovirus receptor.

Adsorption involves attachment of viral surface proteins or spikes to the cell surface receptor proteins

most adsorb to receptors found on the bacterial cell wall. Receptors for human viruses are usually glycoproteins located in the plasma membrane of the cell. **Table 6–5** lists some of the receptors that have been identified for medically important viruses. It appears that viruses have evolved to make use of a wide variety of surface molecules as receptors, which are normally signaling devices or immune system components. Any attempts to design agents that block viral infections by binding to the receptors for a long time must consider the possibility that the loss of the normal cellular function associated with the receptors would have serious consequences for the host organism.

For some viruses, two different surface molecules, called **coreceptors,** are involved in adsorption. Although CD4 was originally thought to be the sole receptor for human immunodeficiency virus type 1 (HIV-1), the discovery of a family of coreceptors that normally function as chemokine receptors (CCR5 and CXCR4) may explain why natural resistance against the virus is found in some individuals with variant forms of these signaling molecules. Receptors for some human viruses are also found on red blood cells of certain species, such as influenza viruses, and are responsible for the phenomena of hemagglutination and hemadsorption discussed later.

Virion attachment proteins are often associated with conspicuous features on the surface of the virion. For example, the virion attachment proteins for the bacteriophages with tails are located at the very end of the tails or the tail fibers (Figure 6–3C). Likewise, the spikes found on adenoviruses (Figure 6–4E) and on virtually all the enveloped human viruses contain the virion attachment proteins.

Viral spikes and phage tails carry attachment proteins

In some cases, a region of the capsid protein serves the function of the attachment protein. For polioviruses, rhinoviruses, and probably other picornaviruses, the region on the capsid that binds to the receptor is found at the bottom of a cleft or trough that is too narrow to allow access to antibodies. This particular arrangement is clearly advantageous to the virus because it precludes the production of antibodies that might directly block receptor recognition.

The repeating subunit structure of capsids and the multiplicity of spikes on enveloped viruses are probably important in determining the strength of the binding of the virus to the cell. The binding between a single virion attachment protein and a single receptor protein is relatively weak, but the combination of many such interactions lead to a strong association between the virion and the cell. The fluid nature of the human cell membrane may facilitate the movement of receptor proteins to allow the clustering that is necessary for these multiple interactions.

Adsorption is enhanced by presence of multiple attachment and receptor proteins

A particular kind of virus is capable of infecting only a limited spectrum of cell types called its **host range.** Thus, although a few viruses can infect cells from different species, most viruses are limited to a single species. For example, dogs do not contract measles, and humans do not contract distemper. In many cases, human viruses infect only a particular subset of the cells found in their host organism. This kind of tissue tropism is clearly an important determinant of viral pathogenesis. In most cases studied, the specific host range of a virus and its associated tissue tropism are determined at the level of the binding between the cell receptors and virion attachment proteins. Thus, these two protein components must possess complementary surfaces that fit together in much the same way as a substrate fits into the active site of an enzyme. It follows that adsorption occurs only in that percentage of collisions that leads to successful binding between receptors and attachment proteins, and that the inability of a virus to infect a cell type is usually due to the absence of the appropriate receptors on the cell. The exquisite specificity of these interactions is well illustrated by the case of a particular mouse reovirus. It has been found that the tissue tropism, and therefore the resultant pathology, are altered by a point mutation that changes a single amino acid in the virion attachment protein. A few cases are known in which the host range of a virus is determined at a step after adsorption and penetration, but these are the exceptions rather than the rule.

Differences in host range and tissue tropism are due to presence or absence of receptors

Once a virus particle has penetrated to the inside of a cell, it is essentially hidden from the host immune system. Thus, if protection from a virus infection is to be accomplished at the level of antibody binding to the virions, it must occur before adsorption and prevent the virus from attaching to and penetrating the cell. It is therefore not surprising that most neutralizing antibodies, whether acquired as a result of natural infection or vaccination, are specific for virion attachment proteins.

Neutralizing antibodies are often specific for attachment proteins

PENETRATION, ENTRY, AND UNCOATING

The disappearance of infectious virus during the eclipse phase is a direct consequence of the fact that viruses are dismantled before being replicated. As discussed later in the text, the uncoating step may be simultaneous with entry or may occur in a series of steps. Ultimately, the nucleocapsid or core structure must be transported to the site or compartment in the cell where transcription and replication will occur.

Viruses are dismantled before being replicated

The Bacteriophage Strategy

The processes of penetration and uncoating are simultaneous for all bacteriophages. Thus, the viral capsids are shed at the surface, and only the nucleic acid genome enters the cell. In some cases, a small number of virion proteins may accompany the genome into the cell, but these are probably tightly associated with the nucleic acid or are essential enzymes needed to initiate the infection.

Bacteriophage capsids are shed, and only the viral genome enters the host cell

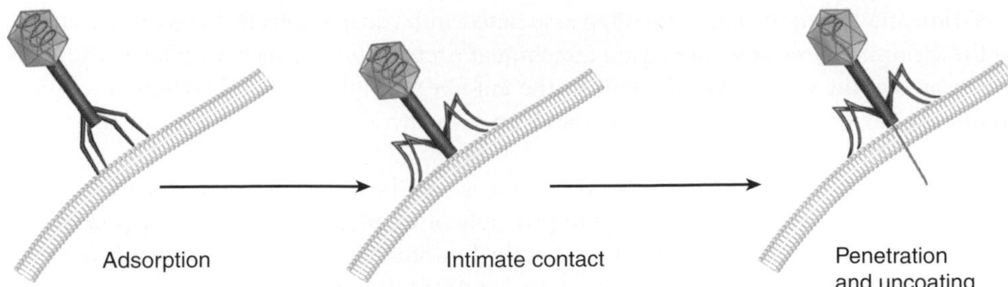

FIGURE 6–10. Bacteriophage entry.

Tailed phages attach by tail fibers and DNA is injected through the tail

Bacteriophages with tails have evolved these special appendages to facilitate the entry of the genome into the cell. The process of penetration and uncoating for bacteriophage T4 is shown schematically in **Figure 6–10**. The tail fibers extending from the end of the tail are responsible for the attachment of the virion to the cell wall, and, in the next step, the end of the tail itself makes intimate contact with the cell surface. Finally, the DNA of the virus is injected from the head directly into the cell through the hollow tail structure. The process has been likened to the action of a syringe, but the energetics and the nature of the orifice in the cell surface through which the DNA travels are poorly understood.

Enveloped Human Viruses

There are two basic mechanisms for the entry of an enveloped human virus into the cell. Both mechanisms involve fusion of the viral envelope with a cellular membrane, and the end result in both cases is the release of the free nucleocapsid into the cytoplasm. What distinguishes the two mechanisms is the nature of the cellular membrane that fuses with the viral envelope.

Some enveloped viruses enter cells by direct fusion of plasma membrane and envelope

Paramyxoviruses (eg, measles), some retroviruses (eg, HIV-1), and herpesviruses enter by a process called **direct fusion** (**Figure 6–11**). The envelopes of these viruses contain protein spikes that promote fusion of the viral membrane with the plasma membrane of the cell, releasing the nucleocapsid directly into the cytoplasm. Because the viral envelope becomes incorporated into the plasma membrane of the infected cell and still possesses its fusion proteins, infected cells have a tendency to fuse with other uninfected cells. Cell-to-cell fusion is a hallmark of infections by paramyxoviruses and HIV-1 and can be important in the pathology of diseases such as measles, respiratory syncytial virus (RSV) and acquired immunodeficiency syndrome (AIDS).

The mechanism for the entry of most of the remaining enveloped animal viruses, such as orthomyxoviruses (eg, influenza viruses), togaviruses (eg, rubella virus), rhabdoviruses (eg, rabies), and coronaviruses, is shown in **Figure 6–12**. After adsorption, the virus particles are taken up by a cellular mechanism called **receptor-mediated endocytosis,** which is normally responsible for internalizing growth factors, hormones, and some nutrients. When it involves viruses, the process is referred to as **viropexis.**

Other enveloped and naked viruses are taken in by receptor-mediated endocytosis (viropexis)

In viropexis, the adsorbed virions become surrounded by the plasma membrane in a reaction that is probably facilitated by the multiplicity of virion attachment proteins on the surface of the particle. Pinching off of the cellular membrane by fusion encloses the virion in a cytoplasmic vesicle termed the **endosomal vesicle.** The nucleocapsid is now surrounded by two membranes: the original viral envelope and the newly acquired endosomal membrane. The surface receptors are subsequently recycled back to the plasma membrane, and the endosomal vesicle is acidified by a normal cellular process. The low pH of the endosome leads to a conformational change in a viral spike protein, which results in the fusion of the two membranes and release of the nucleocapsid into the cytoplasm. In some cases, the contents of the endosomal vesicle may be transferred to a lysosome prior to the fusion step that releases the nucleocapsid.

Naked Capsid Human Viruses

Naked capsid human viruses, such as poliovirus, reovirus, and adenovirus, also appear to enter the cell by viropexis (Figure 6–12). However, in this case, the virus **cannot escape the endosomal vesicle by membrane fusion** as described earlier for some enveloped viruses. For poliovirus,

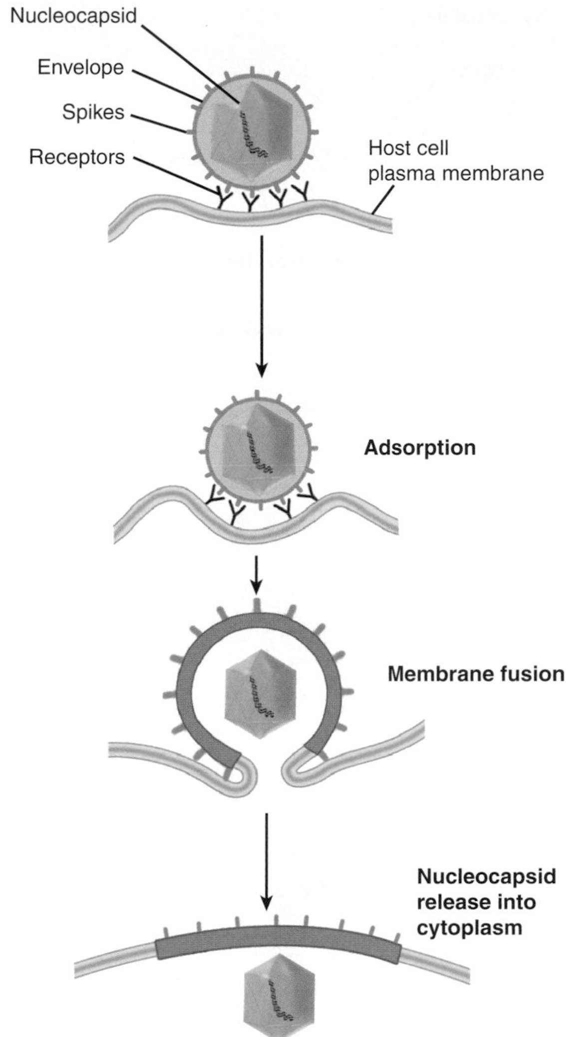

FIGURE 6–11. Entry by direct fusion.

it appears that the viral capsid proteins in the low-pH environment of the endosome expose hydrophobic domains. This process results in the binding of the virions to the membrane and release of the nucleic acid genome into the cytoplasm. In other cases, the virions may escape into the cytoplasm by simply promoting the lysis of the vesicle. This step is a potential target of antiviral chemotherapy, and some drugs have been developed that bind to the capsids of picornaviruses and prevent the release of the virus particles from the endosome.

Acidified endosome releases nucleocapsid to cytoplasm

Virions may escape endosome by dissolution of the vesicles

Reovirus is unusual in that before release into the cytoplasm, the contents of the endosome are transferred to a lysosome where the lysosomal proteases strip away part of the capsid proteins and activate virion-associated enzymes required for transcription.

SYNTHETIC OR VIRION COMPONENT PRODUCTION

Synthetic or virion production is the most important step in viral replication cycle because the virus must make mRNAs, proteins, and genomes for the assembly of progeny or daughter viruses. In the case of bacteriophages, there is evidence that the entering nucleic acid must be directed to a particular cellular locus to initiate the infection process. **Pilot proteins** have been described that accompany the phage genome into the bacterial cell and serve the function of "piloting" the nucleic acid to a particular target, such as a membrane site where transcription and replication are to occur.

For human viruses, the ultimate fate of internalized virus particles depends on the particular virus and on the cellular compartment where replication occurs. Most RNA viruses

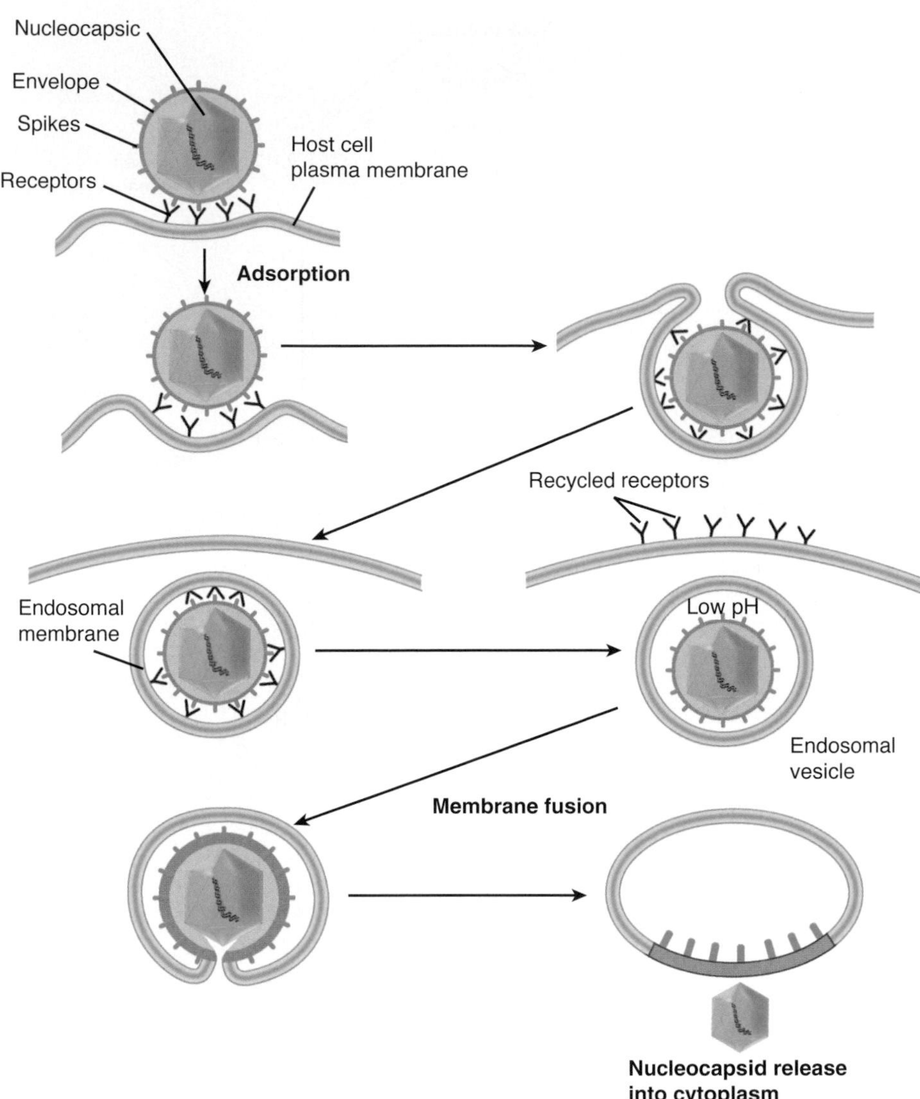

FIGURE 6–12. Viropexis.

Most RNA viruses replicate in cytoplasm except influenza viruses and retroviruses, which replicate in the nucleus

with the exception of influenza viruses and the retroviruses replicate in the cytoplasm, the immediate site of entry. Retroviruses, influenza viruses, and all the DNA viruses except the poxviruses must move from the cytoplasm to the nucleus to replicate. The larger DNA viruses, such as herpesviruses and adenoviruses, must uncoat to the level of cores before entry into the nucleus. The smaller DNA viruses, such as the parvoviruses and the papovaviruses, enter the nucleus intact through the nuclear pores and subsequently uncoat inside. The largest of the human viruses, the poxviruses, carry out their entire replicative cycle in the cytoplasm of the infected cell.

All DNA viruses replicate in the nucleu, except poxviruses, which replicate in the cytoplasm

TRANSCRIPTION

From Genome to mRNA

Virus-specified mRNAs direct synthesis of viral proteins

An essential step in every virus infection is the production of virus-specific mRNAs that program the cellular ribosomes to synthesize viral proteins. Besides the structural proteins of the virion, viruses must direct the synthesis of enzymes and other specialized proteins required for genome replication, gene expression, and virus assembly and release. The production of the first viral mRNAs at the beginning of the infection is a crucial step in the takeover of the cell by the virus.

For some viruses, the presentation of mRNA to the cellular ribosomes poses no problems. Thus, the genomes of most DNA viruses are transcribed by the host DNA-dependent RNA polymerase (RNA polymerase II) to yield the viral mRNAs. The (+)-strand RNA viruses, such as the picornaviruses, the togaviruses, and the coronaviruses, possess genomes that can be used directly as mRNAs and are translated (at least partially, as discussed later) immediately on entry into the cytoplasm of the cell.

Most DNA viruses synthesize their mRNAs by using host RNA polymerase

Positive-strand RNA virus genome serves as mRNA for early protein synthesis

However, for many viruses, the production of mRNA starting from the genome is not so straightforward. The fact that poxviruses replicate in the cytoplasm means that the cellular RNA polymerase is not available to transcribe the DNA genome. Moreover, no cellular machinery exists that can use either single-stranded or double-stranded RNA as a template to synthesize mRNA. Therefore, the poxviruses and viruses that use an RNA template to make mRNAs must provide their own transcription machinery to produce the viral mRNAs at the beginning of the infection process. This feat is accomplished by synthesizing the transcriptases in the later stages of viral development in the previous host cell and packaging the enzymes into the virions, where they remain associated with the genome as the virus enters the new cell and uncoats. In general, the presence of a transcriptase in virions is indicative that the host cell is unable to use the viral genome as mRNA or as a template to synthesize mRNA. At later times in the infection, any special enzymatic machinery required by the virus and not initially present in the cell can be supplied among the proteins translated from the first mRNA molecules.

Negative-strand RNA viruses carry virion-associated RNA-dependent RNA polymerase to produce initial mRNAs

The pathways for the synthesis of mRNA by the major virus groups are summarized in **Figure 6–13** and related to the structure of viral genomes. The polarity of mRNA is designated as (+) and the polarity of polynucleotide chains complementary to mRNA as (–). The black arrows denote synthetic steps for which host cells provide the required enzymes, whereas the colored arrows indicate synthetic steps that must be carried out by virus-encoded enzymes. Several additional points should be emphasized. The parvoviruses and some phages have single-stranded DNA genomes. Although the RNA polymerase of the cell requires double-stranded DNA as a template, these viruses need not carry special enzymes in their virions because host cell DNA polymerases can convert the genomes into double-stranded DNA. Note that the production of more mRNA by the picornaviruses and similar (+)-strand RNA viruses requires the synthesis of an intermediate (–)-strand RNA template. The enzyme required for this process is produced by translation of the genome RNA early in infection called RNA-dependent RNA polymerase.

A variety of pathways exist for synthesis of mRNA by different virus groups

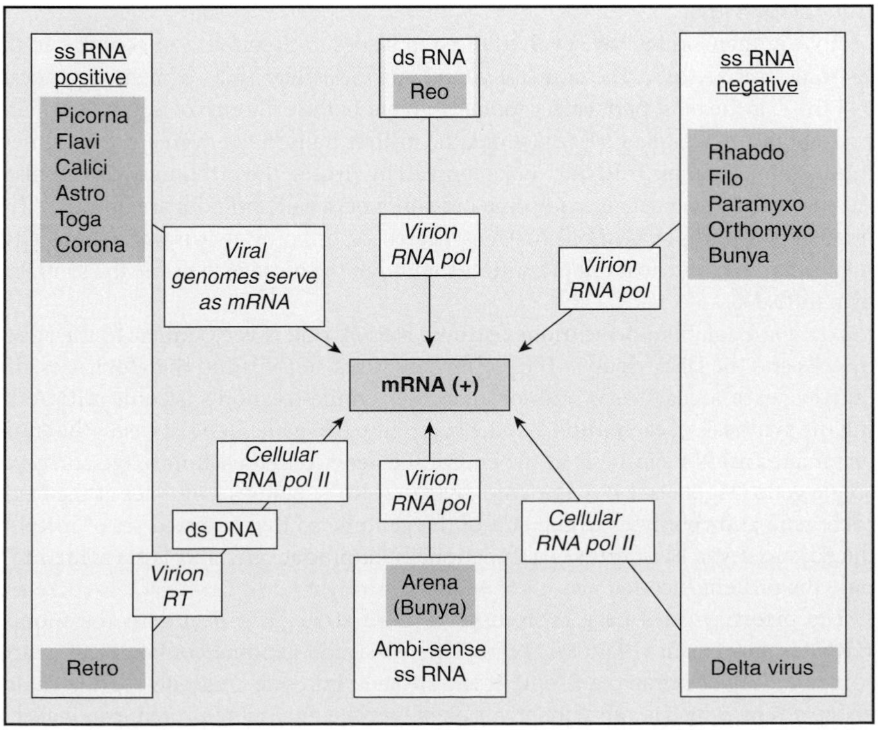

FIGURE 6–13. Pathways of mRNA synthesis for major virus groups.

The retroviruses are a special class of (+)-strand RNA viruses. Although their genomes are the same polarity as mRNA and could in principle serve as mRNAs early after infection, their replication scheme apparently precludes this. Instead, the RNA genomes of these viruses are copied into (−) DNA strands by an enzyme carried within the virion called **reverse transcriptase.** The (−) DNA strands are subsequently converted by the same enzyme to double-stranded DNA in a reaction that requires the degradation of the original genomic RNA by the RNase H activity of the reverse transcriptase. The DNA product of reverse transcription is integrated into the host cell DNA and ultimately transcribed by the host RNA polymerase to complete the replication cycle as well as produce viral mRNA. The replication of the hepatitis B DNA genome is mechanistically similar to that of a retrovirus. Thus, the viral DNA is transcribed to produce a single-stranded RNA, which in turn is reverse transcribed to produce the progeny viral DNA that is encapsidated into virions.

Retroviral RNA is copied to DNA by virion reverse transcriptase; host RNA polymerase transcribes DNA into more RNA

The Monocistronic mRNA Rule in Human Cells

The ribosome requires input of information in the form of mRNA. For a viral mRNA to be recognized by the ribosome, its production must conform to the rules of structure that govern the synthesis of the cellular mRNAs. Prokaryotic mRNA is relatively simple and can be polycistronic, which means it can contain the information for several proteins. Each cistron or coding region is translated independently beginning from its own ribosome binding site.

Prokaryotic (bacterial) mRNAs can be polycistronic

Eukaryotic mRNAs are structurally more complex, containing special 5′-cap and 3′-poly(A) attachments. In addition, their synthesis often involves removal of internal sequences by a process called **splicing.** Most important, almost all eukaryotic mRNAs are monocistronic. Accordingly, eukaryotic translation is initiated by the binding of a ribosome to the 5′-cap, followed by movement of the ribosome along the DNA until the first AUG initiation codon is encountered. The corollary to this first AUG rule is that eukaryotic ribosomes, unlike prokaryotic ribosomes, generally cannot initiate translation at internal sites on a mRNA. To conform to the monocistronic mRNA, most human viruses produce mRNAs that are translated to yield only a single polypeptide chain following initiation near the 5′ end of the mRNA.

Human virus mRNAs are almost always monocistronic

Because most DNA human viruses replicate in the nucleus, they adhere to the monocistronic mRNA rule either by having a promoter precede each gene or by programming the transcription of precursor RNAs that are processed by nuclear splicing enzymes into monocistronic mRNAs (**Figure 6–14A**). The virion transcriptase of the cytoplasmic poxviruses apparently must synthesize monocistronic mRNAs by initiation of transcription in front of each gene.

Most DNA viruses generate monocistronic mRNA through splicing

RNA human viruses have evolved three strategies to circumvent or conform to the monocistronic mRNA rule. The simplest strategy involves having a segmented genome (Figure 6–14B). For the most part, each genome segment of the orthomyxoviruses and the reoviruses corresponds to a single gene; therefore, the mRNA transcribed from a given segment constitutes a monocistronic mRNA. Unlike most RNA viruses, the orthomyxovirus virus influenza A replicates in the nucleus, and some of its monocistronic mRNAs are produced by splicing of precursor RNAs by host cell enzymes. Moreover, orthomyxoviruses use small 5′ RNA fragments derived from host cell pre-mRNAs found in the nucleus to prime the synthesis of their own mRNAs.

Some RNA viruses have segmented genomes to fulfill monocistronic mRNA rule

A second solution to the monocistronic mRNA rule is very similar to the strategy used by cells and the DNA viruses. The paramyxoviruses, togaviruses, rhabdoviruses, filoviruses, bunyaviruses, arenaviruses, and coronaviruses synthesize monocistronic mRNAs by initiating the synthesis of each mRNA at the beginning of a gene. In most cases, the transcriptase terminates mRNA synthesis at the end of the gene so that each message corresponds to a single gene (Figure 6–14C). For coronaviruses, RNA synthesis initiates at the beginning of each gene and continues to the end of the genome so that a nested set of mRNAs is produced. However, each mRNA is functionally monocistronic and is translated to produce only the protein encoded near its 5′ end.

Some viruses (mainly negative-sense RNA viruses) produce monocistronic RNAs by initiating synthesis at the start and pausing at the end of each gene

The picornaviruses have evolved yet a third strategy to deal with the monocistronic mRNA requirement (Figure 6–14D). The (+)-strand genome contains just a single ribosome binding site near the 5′ end. It is translated into one long polypeptide chain called a **polyprotein,** which is subsequently broken into the final set of protein products by a series

Positive-sense RNA viruses (picornaviruses, flaviviruses) make a polyprotein that is proteolytically cleaved later into individual proteins

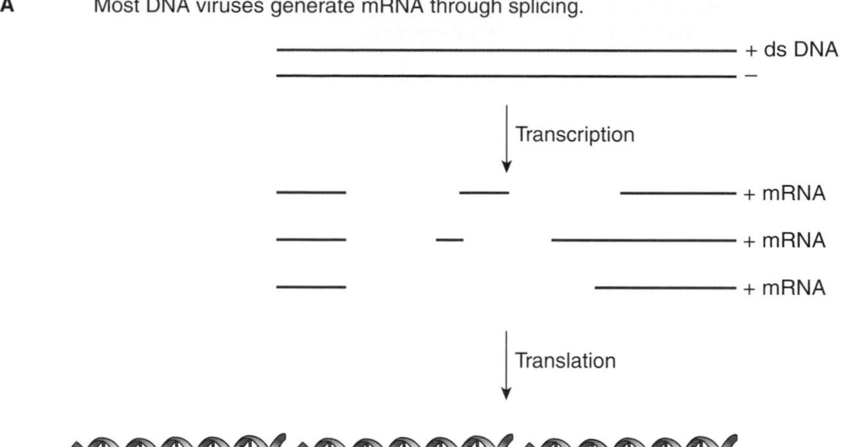

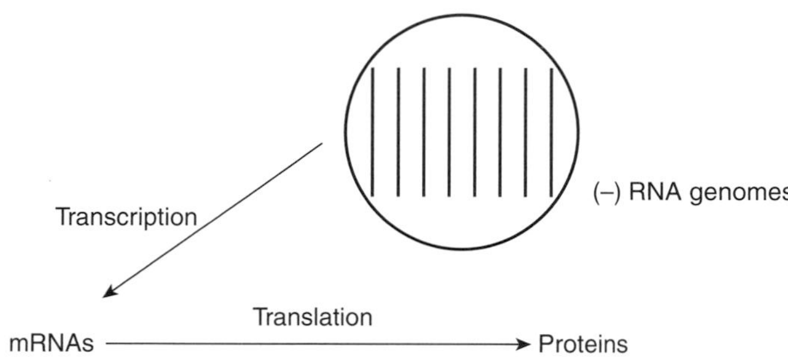

FIGURE 6–14. Monocistronic mRNA strategy for human viruses. *(continued on next page)*

of proteolytic cleavages. Most of the required protease activities reside within the polyprotein itself.

Several viruses use more than one of these strategies to conform to the monocistronic mRNA rule. For example, retroviruses, togaviruses, arenaviruses, and bunyaviruses synthesize multiple mRNAs, each one coding for a polyprotein that is subsequently cleaved into the individual protein molecules.

GENOME REPLICATION

DNA Viruses

Host cells contain the enzymes and accessory proteins that are required for the replication of DNA. In bacteria these proteins are present continuously, whereas in the eukaryotic cell they are present only during the S phase of the cell cycle, and they are restricted to the nucleus. The extent to which viruses use the cell replication machinery depends on their protein-coding potential and thus on the size of their genome.

The smallest DNA viruses depend exclusively on host DNA replication machinery

The largest DNA viruses code for enzymes important for DNA replication

The smallest of the DNA viruses, the parvoviruses, are so completely dependent on host machinery that they require the infected cells to be dividing so that a normal S phase occurs and replicates the viral DNA along with the cellular DNA. At the other end of the spectrum are the large DNA viruses, which are relatively independent of cellular functions. The largest bacteriophages such as T4 degrade the host cell chromosome early in infection and replace all the host replication machinery with virus-specified proteins. The largest human viruses, the poxviruses, are similarly independent of the host. Because they replicate in the

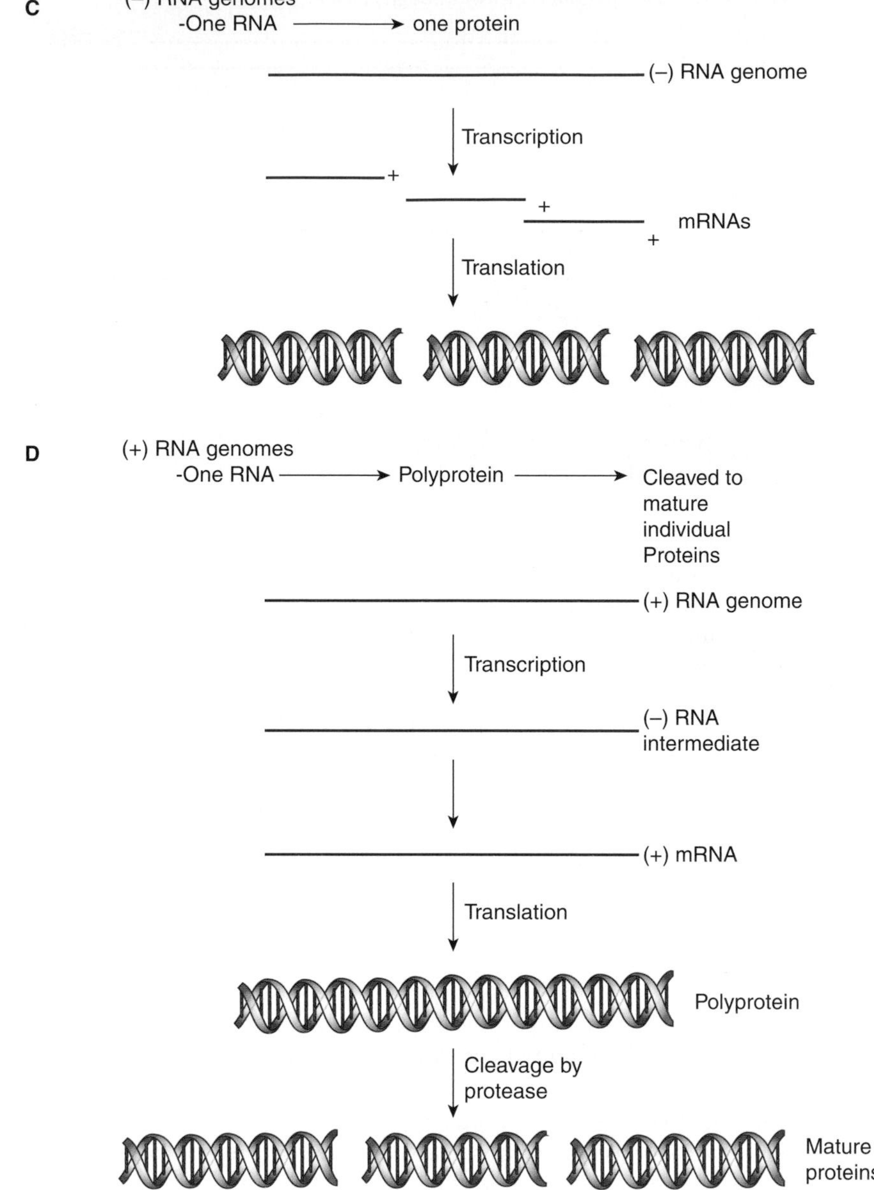

FIGURE 6–14. (*Continued*)

cytoplasm, they must code for almost all of the enzymes and other proteins required for replicating their DNA.

The remainder of the DNA viruses are only partially dependent on host machinery. For example, bacteriophages φX174 and λ code for proteins that direct the initiation of DNA synthesis to the viral origin. However, the actual synthesis of DNA occurs by the complex of cellular enzymes responsible for replication of the *Escherichia coli* DNA. Similarly, the small DNA human viruses, such as the papovaviruses, code for a protein that is involved in the initiation of synthesis at the origin, but the remainder of the replication process is carried out by host machinery. The somewhat more complex adenoviruses and herpesviruses, in addition to providing origin-specific proteins, also code for their own DNA polymerases and other accessory proteins required for DNA replication.

Herpesvirus-encoded DNA polymerase is a target of antiviral therapy (eg, acyclovir)

The fact that the herpesviruses code for their own DNA polymerase has important implications for the treatment of infections by these viruses and illustrates a central principle of antiviral chemotherapy. Certain antiviral drugs (adenine arabinoside and 5′ -iododeoxyuridine) have been found to be effective against herpesvirus infections (see Chapter 14); they are sufficiently similar to natural substrates that the virally encoded DNA polymerase mistakenly incorporates them into viral DNA, resulting in an inhibition of subsequent DNA synthesis. The host cell enzyme is more discriminating and fails to use the analogs in the synthesis of cellular DNA; thus, the drugs do not kill uninfected cells. The same principle applies to the chain-terminating drugs such as zidovudine (AZT) and dideoxyinosine (ddI) that target the HIV-1 reverse transcriptase. Similarly, the antiviral drugs acyclovir (acycloguanosine) preferentially kills herpesvirus-infected cells because the viral thymidine kinase, unlike the cellular counterpart, phosphorylate the nucleoside analog, converting it to a form that inhibits further DNA synthesis when DNA polymerases incorporate it into DNA. In principle, any viral process that is distinct from a normal cellular process is a potential target for antiviral drugs. As more becomes known about the details of viral replication, more drugs will become available that are targeted to these unique viral processes.

Viral processes that are distinct from normal cellular processes are potential targets for antiviral drugs

As noted earlier, with the exception of the poxviruses, all the DNA human viruses are at least partially dependent on host cell machinery for the replication of their genomes. However, unlike the parvoviruses, the other DNA viruses do not need to infect dividing cells for a productive infection to ensue. Instead, all these viruses code for a protein expressed early in infection that induces an unscheduled cycle of cellular DNA replication (S phase). In this way, these viruses ensure that the infected cell makes all the machinery required for the replication of their own DNA. It is noteworthy that all the DNA viruses except the parvoviruses are capable, in some circumstances, of transforming a normal cell into a cancer cell. This correlation suggests that the unlimited proliferative capacity of the cancer cells may be due to the continual synthesis of the viral protein(s) responsible for inducing the unscheduled S phase in a normal infection. The fact that these DNA viruses can induce oncogenic transformation of cell types that are nonpermissive for viral multiplication may simply be an accident related to the need to induce cellular enzymes required for DNA replication during the lytic infection.

All DNA viruses except parvoviruses can transform host cells

All DNA polymerases, including those encoded by viruses, synthesize DNA chains by the successive addition of nucleotides onto the 3′ end of the new DNA strand. Moreover, all DNA polymerases require a primer terminus containing a free 3′ -hydroxyl to initiate the synthesis of a DNA chain. In cellular replication, a temporary primer is provided in the form of a short RNA molecule. This primer (RNA) is synthesized by an RNA polymerase, and after elongation by the DNA polymerase it is removed. With circular chromosomes, such as those found in bacteria and many viruses, the unidirectional chain growth and primer requirement of the DNA polymerase pose no structural problems for replication. However, as illustrated in **Figure 6–15**, when a replication fork encounters the end of a linear DNA molecule, one of the new chains (heavy lines) cannot be completed at its 5′ end, because there exists no means of starting the DNA portion of the chain exactly at the end

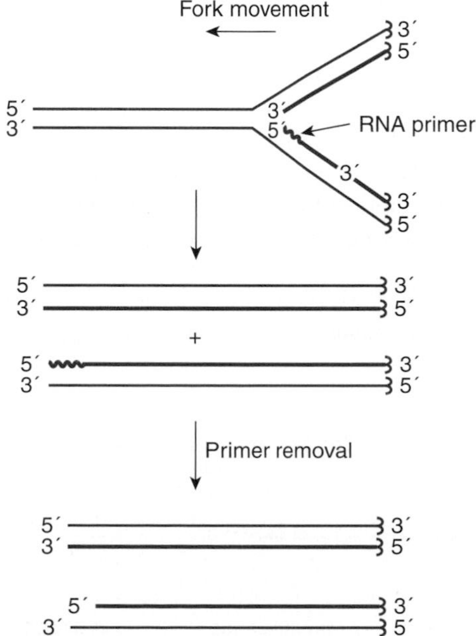

FIGURE 6–15. The end problem in DNA replication.

Replication of linear viral DNAs must solve the end problem

of the template DNA. Thus, after the RNA primer is removed, the new chain is incomplete at its 5′ end. This constraint on the completion of DNA chains on a linear template is called the **end problem** in DNA replication. Some eukaryotic cells add short repetitive sequences to chromosome ends using an enzyme called telomerase to prevent the shortening of the DNA with each successive round of replication.

Several viruses are faced with the end problem during replication of their linear genomes, but none uses the cellular telomerase to synthesize DNA ends. It is beyond the scope of this text to detail all of the strategies that viruses have evolved to deal with the end problem, but it is worth mentioning some of the structural features found in linear viral genomes whose presence is related to solutions of the end problem. These structures are diagrammed schematically in **Figure 6–16.** The linear double-stranded genome of bacteriophage λ possesses 12-bp single-stranded extensions that are complementary in sequence to each other and thus called **cohesive ends.** Very early after entry into the cell, the two ends pair up to convert the linear genome into a circular molecule to avoid the end problem in replication. The linear double-stranded adenovirus genome contains a protein molecule covalently attached to the 5′ end of both strands. These proteins provide the primers required to initiate the synthesis of the DNA chains during replication, circumventing the need for RNA primers and thus solving the end problem in replication. The single-stranded parvovirus genome contains a self-complementary sequence at the 3′ end, which causes the molecule to fold into a hairpin and make it self-priming for DNA replication. The poxviruses contain linear double-stranded genomes in which the ends are continuous. With the parvovirus and poxvirus genomes, the solutions to the end problem create additional problems that must be solved to produce replication products that are identical to the starting genomes.

RNA Viruses

Because nuclear functions are primarily designed for DNA metabolism, RNA viruses mostly replicate in the cytoplasm. Moreover, cells do not have RNA polymerases that can copy RNA templates (RNA-based RNA transcription or replication). Therefore, RNA viruses not only need to code for transcriptases or polymerases (required for transcription), as discussed earlier, but also must provide the replicases or polymerases required to duplicate the RNA genome into daughter RNA genomes. Furthermore, except in the cases of the RNA phage and the picornaviruses, in which transcription and replication are synonymous, the RNA viruses must temporally and functionally separate transcription from replication. This requirement is especially apparent for the rhabdoviruses, paramyxoviruses, togaviruses, and coronaviruses, in which a complete genome, or complementary copy of the genome, is transcribed into a set of small monocistronic mRNAs early in infection. After replication begins, these same templates are used to synthesize full-length strands for replication.

RNA viruses must encode their own polymerase or transcriptases

Two mechanisms exist to separate the process of transcription from replication.. First, in some cases, transcription is restricted to subviral particles and involves a transcriptase transported into the cell within the virion. Second, in other cases, the replication process either involves a functionally distinct RNA polymerase or depends on the presence of some other viral-specific accessory protein that directs the synthesis of full-length copies of the template rather than the shorter monocistronic mRNAs. In reoviruses, the switch from transcription to replication appears to involve the synthesis of a replicase that converts the (+)mRNAs synthesized early in infection to the double-stranded genome segments.

Transcription and replication must be separated for most RNA viruses

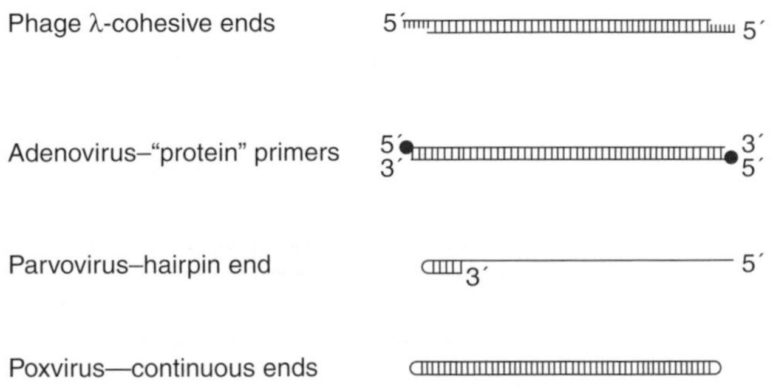

Phage λ-cohesive ends 5′ 5′

Adenovirus–"protein" primers 5′ 3′ 3′ 5′

Parvovirus–hairpin end 3′ 5′

Poxvirus—continuous ends

FIGURE 6–16. Some solutions to the end problem.

Viral RNA polymerases, like DNA polymerases, synthesize chains in only one direction; however, in general, RNA polymerases can initiate the synthesis of new chains without primers. Thus, there is no obvious end problem in RNA replication. There is one exception to this general rule. The picornaviruses contain a protein that is covalently attached to the 5′ end of the genome, called **VPg**. This protein is present on the viral RNA because it is involved in the priming of new RNA viral genomes during the infection, similar to the process described earlier for adenoviruses.

Picornaviruses use a protein to prime RNA synthesis

ASSEMBLY OF NAKED CAPSID VIRUSES AND NUCLEOCAPSIDS

The process of enclosing the viral genome in a protein capsid is called assembly or **encapsidation.** Four general principles govern the construction of capsids and nucleocapsids. First, the process generally involves self-assembly of the component parts. Second, assembly is stepwise and ordered. Third, individual protein structural subunits or protomers are usually preformed into capsomeres in preparation for the final assembly process. Fourth, assembly often initiates at a particular locus on the genome called a **packaging site.**

Capsids and nucleocapsids self-assemble from preformed capsomeres

Viruses with Helical Symmetry

The assembly of the cylindrically shaped tobacco mosaic virus (TMV) has been extensively studied and provides a model for the construction of helical capsids and nucleocapsids. For TMV, doughnut-shaped disks containing a number of individual structural subunits are preformed and added stepwise to the growing structure. Elongation occurs in both directions from a specific packaging site on the single-stranded viral RNA (**Figure 6–17**). The addition of each disk involves an interaction between the protein subunits of the disk and the genome RNA. The nature of this interaction is such that the assembly process ceases when the ends of the RNA are reached. The structural subunits as well as the RNA trace out a helical path in the final virus particle.

Tobacco mosaic virus is a model for the construction of viral components

The basic design features worked out for TMV probably apply in general to the assembly of the nucleocapsids of enveloped viruses. Thus, it is likely that the individual protein subunits are intimately associated with the RNA and that the nucleoprotein complexes are assembled by the stepwise addition of protein subunits or complexes of subunits. For influenza and other helical viruses with segmented genomes, the various genome segments are assembled into nucleocapsids independently and then brought together during virion assembly by a mechanism that is as yet poorly understood. It is notable that virtually all of the human RNA viruses with helical symmetry are enveloped.

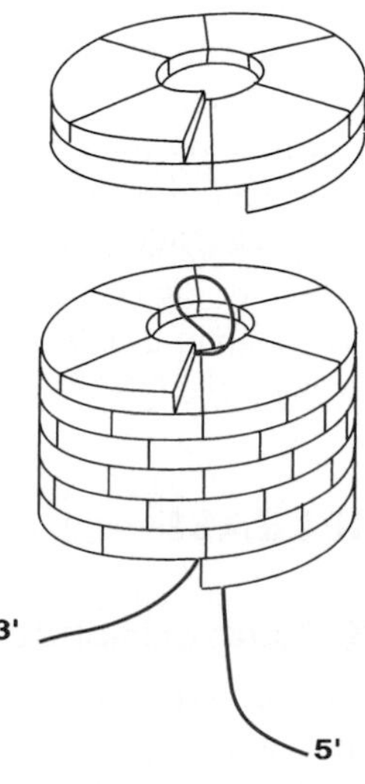

FIGURE 6–17. Tobacco mosaic virus assembly.

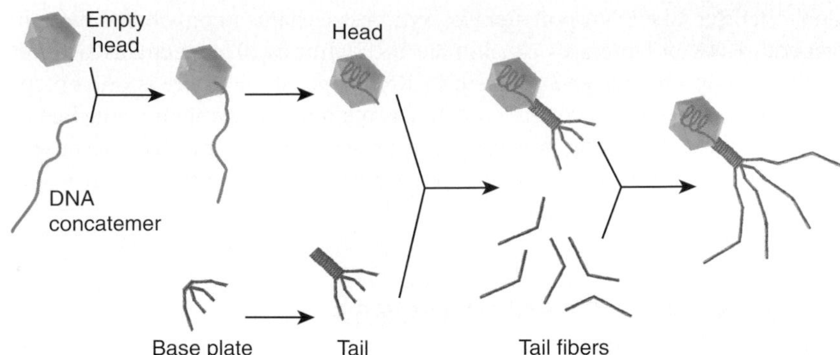

FIGURE 6–18. Assembly of bacteriophage T4.

Viruses with Icosahedral or Cubic Symmetry

Icosahedral capsids are generally preassembled, and the genomes are threaded in

For both phage and human viruses, icosahedral capsids are generally preassembled and the nucleic acid genomes, usually complexed with condensing proteins, are threaded into the empty structures. Construction of the hollow capsids appears to occur by a self-assembly process, sometimes aided by other proteins. The stepwise assembly of components involves the initial aggregation of structural subunits into pentamers and hexamers, followed by the condensation of these capsomeres to form the empty capsid. In some cases, it appears that a small complex of capsid proteins associates specifically with the viral genome and nucleates the assembly of the complete capsid around the genome.

Phage heads, tails, and tail fibers are synthesized separately and then assembled

The morphogenesis of a complex bacteriophage such as T4 involves the prefabrication of each of the major substructures by a separate pathway, followed by the ordered and sequential construction of the final particle from its component parts (**Figure 6–18**). An intermediate in the assembly of a bacteriophage head is an empty structure containing an internal protein network that is removed before insertion of the nucleic acid. The constituents of this network are often appropriately referred to as **scaffolding proteins,** which apparently provide the lattice necessary to hold the capsomeres in position during the early stages of head assembly.

Some phage DNA is replicated to produce concatemers

For many DNA bacteriophages and the herpesviruses, the products of replication are long linear DNA molecules called **concatemers,** which are made up of tandem head-to-tail repeats of genome-size units. During the threading of the DNA into the preformed capsids, these concatemers are cleaved by virus-encoded nucleases to generate genome-size pieces.

Mechanisms for cutting phage DNA during packaging involve site-specific nucleases or headful cleavage

Host DNA may be incorporated by the headful mechanism, and generalized transduction results

There are two mechanisms for determining the correct sites for nuclease cleavage during packaging of a concatemer. Bacteriophage λ and the herpesviruses typify one type of mechanism in which the enzyme that makes the cuts is a sequence-specific nuclease. The enzyme sits poised at the orifice of the capsid as the DNA is being threaded into the capsid and just before the specific cut site enters, the DNA is cleaved. For bacteriophage λ, the breaks are made in opposite strands, 12 bp apart, to generate the cohesive ends. Bacteriophages T4 and P1 are examples of bacterial viruses that illustrate the second mechanism. For these phages, the nuclease does not recognize a particular DNA sequence, but rather cuts the concatemer when the capsid is full. Because the head of the bacteriophage can accommodate slightly more than one genome equivalent of DNA and packaging can begin anywhere on the DNA, the "headful" mechanism produces genomes that are terminally redundant (the same sequence is found at both ends) and circularly permuted. The nonspecific packaging with respect to DNA sequence explains why bacteriophage P1 is capable of incorporating host DNA into phage particles, thereby promoting generalized transduction (see Chapter 21). Bacteriophage T4 does not carry out generalized transduction, because the bacterial DNA is completely degraded to nucleotides early in infection.

RELEASE

Bacteriophages

Phages encode lysozyme or peptidases that lyse bacterial cell walls

Most bacteriophages escape from the infected cell by coding for one or more enzymes synthesized late in the latent phase, which causes the lysis of the cell. The enzymes are either lysozymes or peptidases that weaken the cell wall by cleaving specific bonds in the peptidoglycan layer. The damaged cells burst as a result of osmotic pressure.

HUMAN VIRUSES

CELL DEATH

Nearly all productively infected cells die (see the text that follows for exceptions), presumably because the viral genetic program is dominant and precludes the continuation of normal cell functions required for survival. In many cases, direct viral interference with normal cellular metabolic processes leads to cell death. For example, picornaviruses shut off host protein synthesis soon after infection, and many DNA human viruses interfere with normal cell cycle controls. In many cases, the end result of such insults is a triggering of a cellular stress response called programmed cell death or **apoptosis.** Some viruses are known to code for proteins that block or delay apoptosis, probably to stave off cell death until the virus replication cycle has been completed. Ultimately, the cell lysis that accompanies cell death is responsible for the release of naked capsid viruses into the environment.

Naked capsid viruses lacking specific lysis mechanisms are released with cell death

Some viruses block or delay apoptosis to allow completion of the virus replication cycle

BUDDING

Most enveloped human viruses acquire their membrane by budding either through the plasma membrane or, in the case of herpesviruses, through the membrane of an exocytic vesicle. Thus, for these viruses, release from the cell is coupled to the final stage of virion assembly. The herpesviruses ultimately escape from the cell when the membrane of the exocytic vesicle fuses with the plasma membrane. The poxviruses appear to program the synthesis of their own outer membrane. How the poxvirus envelope is assembled on the nucleocapsid is not known.

Most enveloped viruses acquire an envelope during release by budding

Poxviruses synthesize their own envelopes

The membrane changes that accompany budding appear to be just the reverse of the entry process described before for those viruses that enter by direct fusion (compare Figure 6–11 and **Figure 6–19**). The region of the cellular membrane where budding is to occur acquires a cluster of viral glycoprotein spikes. These proteins are synthesized by the pathway that normally delivers cellular membrane proteins to the surface of the cell by way of the Golgi apparatus. At the site of the glycoprotein cluster, the inside of the membrane becomes coated with a virion structural protein called the **matrix** or **M protein.** The accumulation of the matrix protein at the proper location is probably facilitated by the presence of a binding site for the matrix protein on the cytoplasmic side of the transmembrane glycoprotein spike. The matrix protein attracts the completed nucleocapsid that triggers the envelopment process leading to the release of the completed particle to the outside (Figure 6–19).

The membrane site for budding first acquires virus-specified spikes and matrix protein

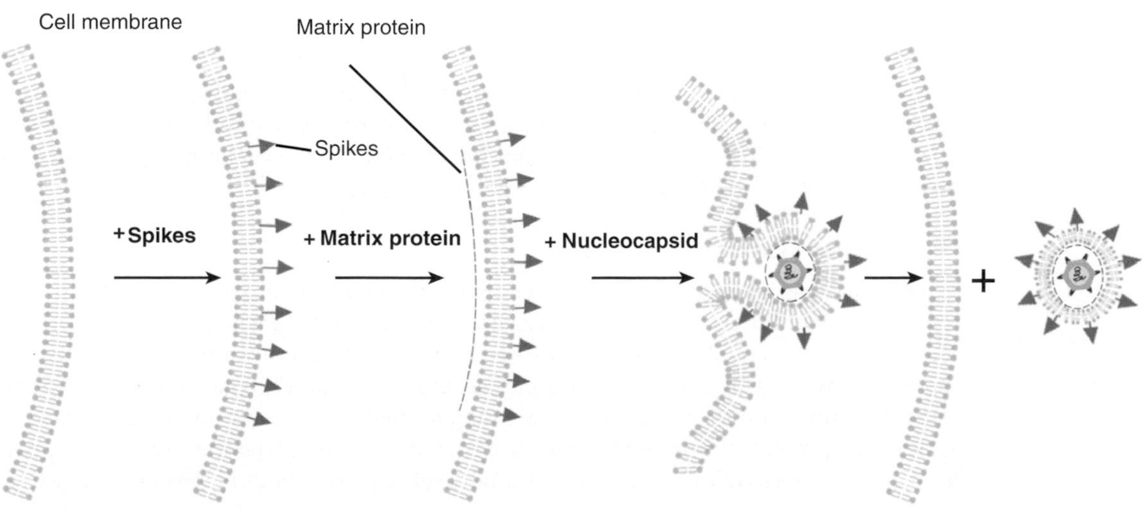

FIGURE 6–19. Viral release by budding.

For viruses that bud, it is important to note that the plasma membrane of the infected cell contains virus-specific glycoproteins that represent foreign (viral) antigens. This means that infected cells become targets for the immune system. In fact, cytotoxic T lymphocytes that recognize these antigens can be a significant factor in combating a virus infection.

The budding process rarely causes cell death

Most retroviruses (except HIV) reproduce without cell death

The process of viral budding usually does not lead directly to cell death because the plasma membrane can be repaired after budding. It is likely that cell death for most enveloped viruses, as for naked capsid viruses, is related to the loss of normal cellular functions required for survival or as a result of apoptosis. Unlike most retroviruses that do not kill the host cell, HIV-1 is cytotoxic. Although the mechanism of HIV-1 cell killing is not entirely understood, factors such as the accumulation of viral DNA in the cytoplasm, the toxic effects of certain viral proteins, alterations in plasma membrane permeability, and cell–cell fusion are believed to contribute to the cytotoxic potential of the virus.

CELL SURVIVAL

Filamentous phages assemble during extrusion without damaging cells

For retroviruses (except HIV-1 and other lentiviruses) and the filamentous bacteriophages, virus reproduction and cell survival are compatible. Retroviruses convert their RNA genome into double-stranded DNA, which integrates into a host cell chromosome and is transcribed just like any other cellular gene (see Chapter 18). Thus, the impact on cellular metabolism is minimal. Moreover, the virus buds through the plasma membrane without any permanent damage to the cell.

Because the filamentous phages are naked capsid viruses, cell survival is even more remarkable. In this case, the helical capsid is assembled onto the condensed single-stranded DNA genome as the structure is being extruded through both the membrane and the cell wall of the bacterium. How the cell escapes permanent damage in this case is unknown. As with the retroviruses, the infected cell continues to produce virus indefinitely.

QUANTITATION OF VIRUSES

Hemagglutination Assay

Virion and infected cell attachment proteins also bind red blood cells

For some human viruses, red blood cells from one or more human species contain receptors for the virion attachment proteins. Because the receptors and attachment proteins are present in multiple copies on the cells and virions, respectively, an excess of virus particles coats the cells and causes them to aggregate. This aggregation phenomenon was first discovered with influenza virus and is called **hemagglutination.** The virion attachment protein on the influenza virion is appropriately called the **hemagglutinin.** Furthermore, the presence of the hemagglutinin in the plasma membrane of the infected cell means that the cells as well as the virions bind the red blood cells. This reaction, called **hemadsorption,** is a useful indicator of infection by certain viruses.

Hemagglutination can be used to estimate the titer of virus particles in a virus-containing sample. Serially diluted samples of the virus preparation are mixed with a constant amount of red blood cells, and the mixture is allowed to settle in a test tube. Agglutinated red blood cells settle to the bottom to form a thin, dispersed layer. If there is insufficient virus to agglutinate the red blood cells, they will settle to the bottom of the tube and form a tight pellet. The difference is easily scored visually, and the endpoint of the agglutination is used as a relative measure of the virus concentration in the sample.

Plaque Assay

The plaque assay is a method for determining the titer of infectious virions in a virus preparation or lysate. The sample is diluted serially, and an aliquot of each dilution is added to a vast excess of susceptible host cells. For a human virus, the host cells are usually attached to the bottom of a plastic Petri dish; for bacterial cells, adsorption is typically carried out in a cell suspension. In both cases, the cells are then immersed in a semisolid medium such as agar, which prevents the released virions from spreading throughout the entire cell population. Thus, the virus released from the initial and subsequent rounds of infection can invade

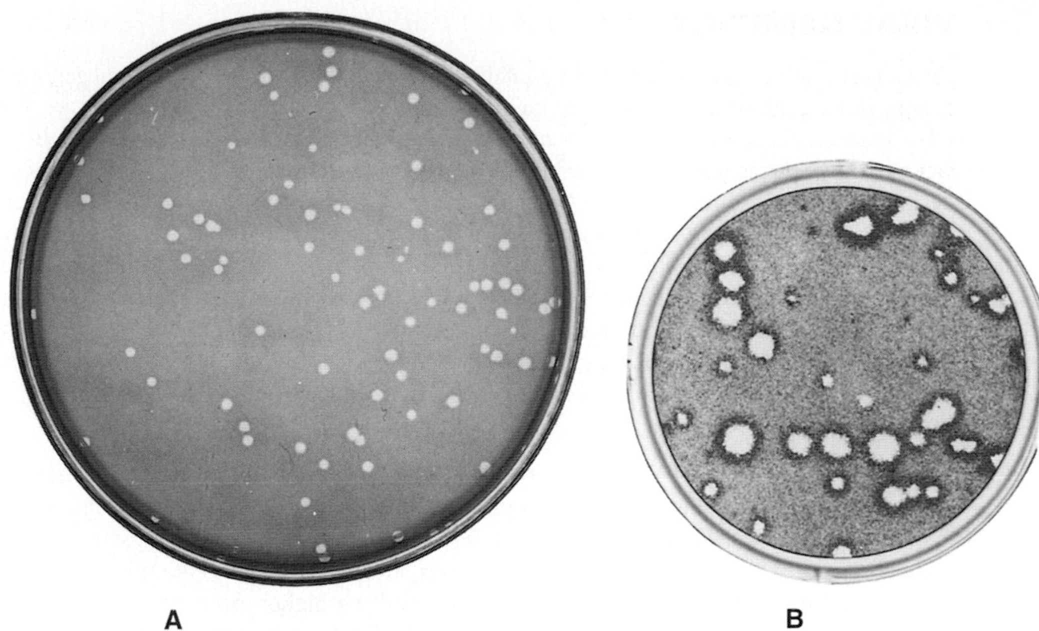

FIGURE 6–20. Plaque assays. A. Bacteriophage λ. **B.** Adenovirus.

only the cells in the immediate vicinity of the initial infected cell on the plate. The end result is an easily visible clearing of dead cells at each of the sites on the plate where one of the original infected cells was located. The clearing is called a **plaque** (**Figure 6–20**). Visualization in the case of human cells usually requires staining the cells. By counting the number of plaques and correcting for the dilution factor, the virus titer in the original sample can be calculated. The titer is usually expressed as the number of plaque-forming units per milliliter (pfu/mL).

Plaque assay: dilutions of virus are added to excess cells immobilized in agar

Replicated virus infects only neighboring cells, producing countable plaques

Immunologic Assay

Viral antigen can be quantified by using antigen-antibody specificity as measured by enzyme linked immunosorbent assay (ELISA) and immunofluoresence assay (IFA). Like other assays, in immunologic assays the antigen-antibody specificity and conditions should be worked out. For most viruses, commercial antibodies are available and can be used to detect or quantify the antigen of viruses in culture and body fluids, tissue biopsies, serum, plasma and cerebrospinal fluid (CSF). The most common example is the detection and sometimes quantification of respiratory syncytial virus by IFA in which respiratory syncytial virus antigens are be measured in nasopharyngeal and throat washing, sputum, or bronchoalveolar lavage. Also, viral antigens can be detected and quantified in blood (plasma or serum), which can then provide information on the amount of virus present in the blood. For example, HIV can be quantified by the levels of p24 (capsid) antigen in the culture fluid or blood.

Using antigen-antibody specificity, viral antigens can be quantified by ELISA

Molecular Assay

Viral genomes, both RNA and DNA, can be quantified to determine the amount of virus (viral load) in blood (serum or plasma) or any given samples. The RNA genomes of the viruses are first reversely transcribed to cDNA by reverse transcriptase enzyme and then amplified by polymerase chain reaction (PCR). However, viral DNA genomes can be directly amplified by PCR to quantify the viral genomes. Based on the number of copies of the viral genomes, the amount of viruses in any sample can be determined. This is the most sensitive and specific method to detect and quantify viral genomes. PCR is routinely used to determine viral load in HIV, hepatitis C virus, and other viral infections.

DNA and RNA genomes of viruses can be quantified by PCR

VIRAL GENETICS

Viruses generally use two mechanisms, mutation and recombination, by which viral genomes change during infection and there are medical consequences of some of these changes. For DNA bacteriophages, the ratio of infectious particles to total particles usually approaches a value of one. Such is not the case for human viruses. Typically, the majority of the particles derived from a cell infected with a human virus are noninfectious in other cells as determined by a plaque assay. Although some of this discrepancy may be attributable to inefficiencies in the assay procedures, it is clear that many defective particles are being produced. In part, this production of defective particles arises because the mutation rates for human viruses are unusually high and because many infections occur at high multiplicities, where defective genomes are complemented by nondefective viruses and therefore propagated.

Majority of the human virus particles from an infected cell are defective

Mutation

Many DNA viruses use the host DNA synthesis machinery for replicating their genomes. Therefore, they benefit from the built-in proofreading and other error-correcting mechanisms used by the cell. However, the large human viruses (adenoviruses, herpesviruses, and poxviruses) code for their own DNA polymerases, and these enzymes are not as effective at proofreading as the cellular polymerases. The resulting higher error rates in DNA replication endow the viruses with the potential for a high rate of evolution, but they are also partially responsible for the high frequency of defective viral particles.

The replication of RNA viruses is characterized by even higher error rates because viral RNA polymerases do not possess any proofreading capabilities. The result is that error rates for RNA viruses commonly approach one mistake for every 2500 to 10,000 nucleotides polymerized. Such a high misincorporation rate means that even for the smallest RNA viruses, virtually every round of replication introduces one or more nucleotide changes somewhere in the genome. If it is assumed that errors are introduced at random, most of the members of a clone (eg, in a plaque) are genetically different from all other members of the clone. The resulting mixture of different genome sequences for a particular RNA virus has been referred to as quasispecies to emphasize that the level of genetic variation is much greater than what normally exists in a species.

Because of the redundancy in the genetic code, some mutations are silent and are not reflected in changes at the protein level, but many occur in essential genes and contribute to the large number of defective particles found for RNA human viruses. The concept of genetic stability takes on a new meaning in view of these considerations, and the RNA virus population as a whole maintains some degree of homogeneity only because of the high degree of fitness exhibited by a subset of the possible genome sequences. Thus, strong selective forces continually operate on a population to eliminate most mutants that fail to compete with the few very successful members of the population. However, any time the environment changes (eg, with the appearance of neutralizing antibodies), a new subset of the population is selected and maintained as long as the selective forces remain constant.

High error rates for RNA viruses produce genetically heterogeneous populations

The high mutation rates found for RNA viruses endow them with a genetic plasticity that leads readily to the occurrence of genetic variants and permits rapid adaptation to new environmental conditions. The large number of serotypes of rhinoviruses causing the common cold, for instance, likely reflects the potential to vary by mutation. Although rapid genetic change occurs for most if not all viruses, no medically important RNA virus has exhibited this phenomenon as conspicuously as influenza virus. Point mutations accumulate in the influenza genes coding for the two envelope proteins (hemagglutinin and neuraminidase), resulting in changes in the antigenic structure of the virions. These changes lead to new variants not recognized by the immune system of previously infected individuals. This phenomenon is called **antigenic drift** (see Chapter 9). **Figure 6–21** shows the effect of mutations resulting in antigenic drift. Apparently, the domains of the two envelope proteins that are most important for immune recognition are not essential for virus entry and, as a result, can tolerate amino acid changes leading to antigenic variation. This feature may distinguish influenza from other human RNA viruses that possess the same high mutation rates, but do not exhibit such high rates of antigenic drift. Antigenic drift in epidemic influenza viruses from year to year requires continual updating of the strains used to produce immunizing vaccines.

High mutation rates permit adaptation to changed conditions

Mutations are responsible for antigenic drift in influenza viruses

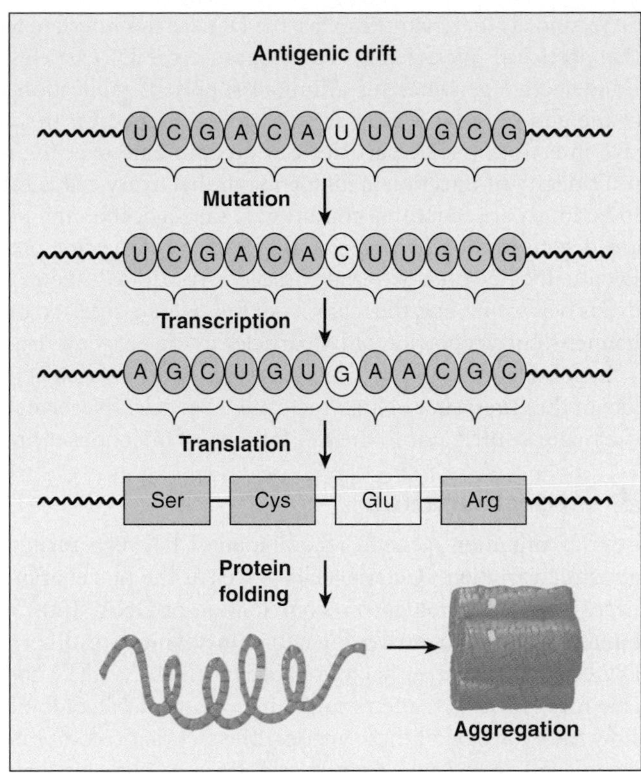

FIGURE 6-21. Point mutation resulting in antigenic drift.

The retroviruses likewise show high rates of variation because of error-prone reverse transcriptase enzyme that converts retroviral RNA into double-stranded DNA. For example, error rates for HIV-1 reverse transcriptase is approximately four to five errors per reverse transcription of the genome. Once the viral DNA has integrated into the chromosome of the host cell, the retroviral DNA is transcribed by the host RNA polymerase II, which is capable of generating errors. Accordingly, HIV-1 exhibits a high rate of mutation, and this property gives HIV-1 the ability to evolve rapidly in response to changing conditions in the infected host. Genetic variation has resulted in several clades or subtypes of HIV-1 worldwide.

High rates of mutation in retroviruses are due to error-prone reverse transcriptase

Retroviruses that exhibit high rates of antigenic variation such as HIV-1 pose particularly difficult problems for the development of effective vaccines. Attempts are being made to identify conserved, and therefore presumably essential, domains of the envelope proteins for these viruses, which might be useful in developing a genetically engineered vaccine.

HIV-1 antigenic variation makes vaccine development difficult

Von Magnus Phenomenon and Defective Interfering Particles

In early studies with influenza virus, it was noted that serial passage of virus stocks at high multiplicities of infection led to a steady decline of infectious titer with each passage. At the same time, the titer of noninfectious particles increased. As discussed later, the noninfectious genomes interfere with the replication of the infectious virus and so are called **defective interfering (DI) particles.** Later, these observations were extended to include virtually all DNA as well as RNA human viruses. The phenomenon is now named after von Magnus, who described the initial observations with influenza virus.

Defective interfering particles accumulate at high multiplicities of infection

A combination of two separate events leads to **von Magnus phenomenon**. First, deletion mutations occur at a significant frequency for all viruses. For DNA viruses, the mechanisms are not well understood, but deletions presumably occur as a result of mistakes in replication or by nonhomologous recombination. The basis for the occurrence of deletions in RNA viruses is better understood. All RNA replicases have a tendency to dissociate from the template RNA, but remain bound to the end of the growing RNA chain. By reassociating with the same or a different template at a different location, the replicase "finishes" replication, but in the process creates a shorter or longer RNA molecule. A subset of these variants possesses the proper signals for initiating RNA synthesis and continues replicating. Because the deletion variants in the population require less time to complete a replication cycle, they eventually predominate and constitute the DI particles.

Deletions result from mistakes in replication, recombination, or the dissociation–reassociation of replicases

Second, as their name implies, the DI particles interfere with the replication of nondefective particles. Interference occurs because the DI particles successfully compete with the nondefective genomes for a limited supply of replication enzymes. The virions released at the end of the infection are therefore enriched for the DI particles. With each successive infection, the DI particles can predominate over the normal particles as long as the multiplicity of infection is high enough that every cell is infected with at least one normal infectious particle. If this condition is satisfied, then the normal particle can complement any defects in the DI particles and provide all of the viral proteins required for the infection. Eventually, however, as serial passage is continued, the multiplicity of infectious particles drops below one, and the majority of the cells are infected only with DI particles. When this happens, the proportion of DI particles in the progeny virus decreases.

Defective interfering particles compete with infectious particles for replication enzymes

In good laboratory practice, virus stocks are passaged at high dilutions to avoid the problem of the emergence of high titers of DI particles. Nevertheless, the presence of DI particles is a major contributor to the low fraction of infectious virions found in all virus stocks.

Recombination

Besides mutation, genetic recombination between related viruses is a major source of genomic variation. Bacterial cells as well as the nuclei of human cells contain the enzymes necessary for homologous recombination of DNA. Thus, it is not surprising that recombinants arise from mixed infections involving two different strains of the same type of DNA virus. The larger bacteriophages such as λ and T4 code for their own recombination enzymes, a fact that attests to the importance of recombination in the life cycles and possibly the evolution of these viruses. The fact that recombination has also been observed for cytoplasmic poxviruses suggests that they too code for their own recombination enzymes.

Homologous recombination is common in DNA viruses

As far as is known, cells do not possess the machinery to recombine RNA molecules. However, recombination among at least some RNA viruses has been observed by two different mechanisms. The first, which is unique to the viruses with segmented genomes (orthomyxoviruses and reoviruses), involves reassortment of segments during a mixed infection involving two different viral strains. Recombinant progeny viruses that differ from either parent can be accounted for by the formation of new combinations of the genomic segments that are free to mix with each other at some time during the infection. Reassortment of this type occurring during infections of the same cell by human and certain animal influenza viruses is believed to account for the occasional drastic change in the antigenicity of the human influenza A virus. These dramatic changes, called **antigenic shifts** (**Figure 6–22**), produce strains to which much of the human population lacks immunity and thus can have enormous epidemiologic and clinical consequences (see Chapter 9).

Recombination for viruses with segmented RNA genomes involves reassortment of segments

Segment reassortment in mixed infections probably accounts for antigenic shifts in influenza virus

The second mechanism of RNA virus recombination is exemplified by the genetic recombination between different forms of poliovirus. Because the poliovirus RNA genome is not segmented, reassortment cannot be invoked as the basis for the observed recombinants. In this case, it appears that recombination occurs during replication by a "copy choice" type of mechanism. During RNA synthesis, the replicase dissociates from one template and resumes copying a second template at the exact place where it left off on the first. The end result is a progeny RNA genome containing information from two different input RNA molecules. Strand switching during replication, therefore, generates a recombinant virus. Although this is not frequently observed, it is likely that most of the RNA human viruses are capable of this type of recombination.

Poliovirus replicase switches templates to generate recombinants

A "copy choice" mechanism has also been invoked to explain a high rate of recombination observed with retroviruses. Early after infection, the reverse transcriptase within the virion synthesizes a DNA copy of the RNA genome by a process called reverse transcription. In the course of reverse transcription, the enzyme is required to "jump" between two sites on the RNA genome (see Chapter 18). This propensity to switch templates apparently explains how the enzyme generates recombinant viruses. Because reverse transcription takes place in subviral particles, free mixing of RNA templates brought into the cell in different virus particles is not permitted. However, retroviruses are diploid, because each particle carries two copies of the genome. This arrangement appears to be a situation ready-made for template switching during DNA synthesis and most likely accounts for retroviral recombination.

The diploid nature of retroviruses permits template switching and recombination during DNA synthesis

Occasional incorporation of host mRNA into retroviral particles may produce oncogenic variants

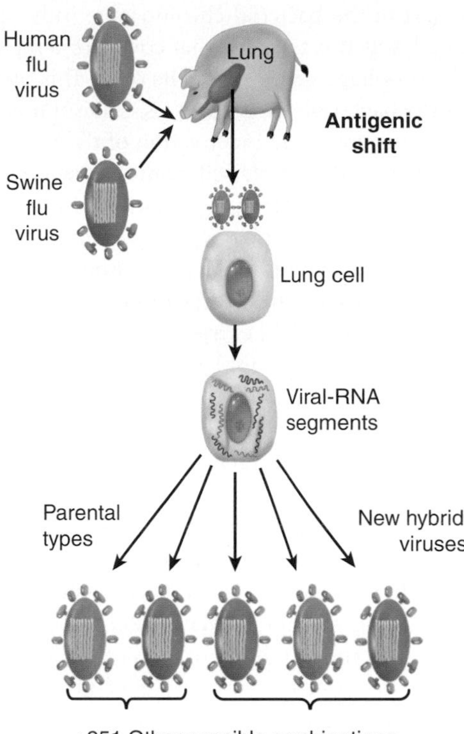

FIGURE 6-22. Reassortment of influenza virus strains (antigenic shift) resulting in new strains.

Occasionally, retroviruses package a cellular mRNA into the virion rather than a second RNA genome. This arrangement can lead to copy choice recombination between the viral genome and a cellular mRNA. The end result is sometimes the incorporation of a cellular gene into the viral genome. This mechanism is believed to account for the production of highly oncogenic retroviruses containing modified cellular genes (see below).

THE LATENT STATE

Temperate viruses can infect a cell and enter a latent state that is characterized by little or no virus production. The viral DNA genome is replicated and segregated along with the cellular DNA when the cell divides. There exist two possible states for the latent viral genome. It can exist extrachromosomally (herpesviruses) like a bacterial plasmid, or it can become integrated into the chromosome (retroviruses) like the bacterial F factor in the formation of a high-frequency recombination (HFR) strain (see Chapter 21). Because the latent genome is usually capable of reactivation and entry into the lytic cycle, it is called a **provirus** or, in the case of bacteriophages, a **prophage.** In many cases, viral latency goes undetected; however, limited expression of proviral genes can occasionally endow the cell with a new set of properties. For instance, lysogeny can lead to the production of virulence-determining toxins in some bacteria (lysogenic conversion) and latency by an human virus may produce oncogenic transformation.

The latent state involves infection of a cell with little or no virus production

Latent virus may be silent, change cell phenotype, or be induced to enter the lytic cycle

LYSOGENY

Infection of an *E coli* cell by bacteriophage λ can have two possible outcomes. A portion of the cells (as many as 90%) enters the lytic cycle and produces more phage. The remainder of the cells enter the latent state by forming stable lysogens. The proportion of the population that lyses depends on as yet undefined factors including the nutritional and physiologic state of the bacteria. In the lysogenic state, the phage DNA is physically inserted into the bacterial chromosome (see following text) and thus replicates when the bacterial DNA replicates. Lambda can thus replicate either extrachromosomally, as in the lytic cycle or as

E coli phage λ may be lytic or latent

When λ is integrated, the only active gene encodes a repressor for the other phage genes

a part of the bacterial chromosome in lysogeny. The only phage gene that remains active in a lysogen is the gene that codes for a repressor protein that turns off expression of all of the prophage genes except its own. This means that the lysogenic state can persist as long as the bacterial strain survives. Environmental insults such as exposure to ultraviolet light or mutagens cause inactivation of the repressor, resulting in induction of the lysogen. The prophage DNA is excised from the bacterial chromosome, and a lytic cycle ensues.

Inactivation of repressor causes induction and virus production

Once established, perpetuation of the lysogenic state requires a mechanism to ensure that copies of the phage genes are faithfully passed on to both daughter cells during cell division. Integration of the λ genome into the *E coli* chromosome guarantees its replication and successful segregation during cell division. In bacteriophage P1 lysogens, the viral genome exists extrachromosomally as an autonomous single-copy plasmid. Its replication is tightly coupled to chromosomal replication and the two replicated copies are precisely partitioned along with the cellular chromosomes to daughter cells during cell division.

Latent genomes can exist extrachromosomally or can be integrated

Because of its mechanistic importance and relevance to lysogenic conversion and phage transduction, λ integration and the reverse reaction called excision are described in some detail. Bacteriophage λ integrates by a site-specific, reciprocal recombination event as outlined in Figure 6–23. There exist unique sequences on both the phage and bacterial chromosomes called attachment sites where the crossover occurs. The phage attachment site is called *attP*, and the bacterial site, which is found on the *E coli* chromosome between the galactose and biotin operons, is called *attB*. The recombination reaction is catalyzed by the phage-encoded integrase protein (Int) in conjunction with two host proteins and occurs by a highly concerted reaction that requires no new DNA synthesis.

Phage λ integrates by site-specific recombination

Excision of the phage genome after induction of a lysogen is just the reverse of integration except that excision requires, in addition to the Int protein, a second phage protein called Xis. In this case, the combined activities of these two proteins catalyze site-specific recombination between the two attachment sites that flank the prophage DNA, *attL* and *attR* (**Figure 6–23**). Early after infection, when integration is to occur in the cells destined to become lysogens, synthesis of the Xis protein is blocked. Otherwise, the integrated prophage DNA would excise soon after integration, and stable lysogeny would be impossible. However, after induction of a lysogen, both the integrase and the Xis proteins are synthesized and catalyze the excision event that releases the prophage DNA from the chromosome.

Excision after λ induction involves recombination at junctions between host DNA and prophage

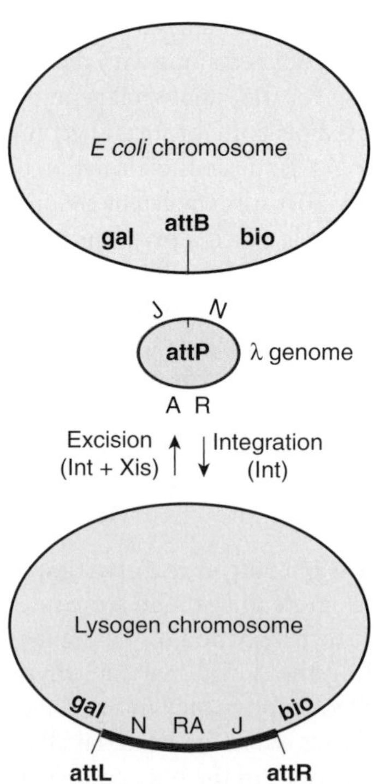

FIGURE 6–23. λ integration and excision. A, J, N, and R show the locations of some λ genes on the λ genome; *gal* and *bio* represent the *Escherichia coli* galactose and biotin operons, respectively.

At a very low frequency, excision involves sites other than the *attL* and *attR* borders of the prophage and results in the linking of bacterial genes to the phage genome. Thus, if a site to the left of the bacterial *gal* genes recombines with a site within the λ genome (to the left of the J gene, otherwise the excised genome is too large to be packaged), then the resulting phage can transduce the genes for galactose metabolism to another cell. Similarly, transducing particles can be formed that carry the genes involved in biotin biosynthesis. Because only the cellular genes adjacent to the attachment site can be acquired by an aberrant excision event, this process is called **specialized transduction** to distinguish it from generalized transduction, in which virtually any bacterial gene can be transferred by a headful packaging mechanism. ::: transduction, p.380

Specialized transduction occurs because excision occasionally includes genes adjacent to the phage genome

Occasionally, one or more phage genes, in addition to the gene coding for the repressor protein, are expressed in the lysogenic state. If the expressed protein confers a new phenotypic property on the cell, then it is said that lysogenic conversion has occurred. Diphtheria, scarlet fever, and botulism all are caused by toxins produced by bacteria that have been "converted" by a temperate bacteriophage. In each case, the gene that codes for the toxin protein resides in the phage DNA and is expressed along with the repressor gene in the lysogenic state. It remains a mystery as to how these toxin genes were acquired by the phage; it is speculated that they may have been picked up by a mechanism similar to specialized transduction.

Lysogenic conversion results from expression of a prophage gene that alters cell phenotype

Several bacterial exotoxins are encoded in temperate phages

CHAPTER 7

Pathogenesis of Viral Infection

Viral pathogenesis is the process by which viruses produce disease in the host. The factors that determine the viral transmission, multiplication, and development of disease in the host involve complex and dynamic interactions between the virus and the susceptible host. Viruses cause disease when they breach the host's primary physical and natural protective barriers; evade local, tissue, and immune defenses; spread in the body; and destroy cells either directly or via bystander immune and inflammatory responses. Viral pathogenesis can be divided into several stages, including (1) transmission and entry of the virus into the host, (2) spread in the host, (3) tropism, (4) virulence, (5) patterns of viral infection and disease, (6) host factors, (7) and host defense. The stages of a typical viral infection and its pathogenesis (eg, poliovirus pathogenesis) are shown in **Figure 7–1**.

The process by which viruses cause disease in the host is called viral pathogenesis

Complex interactions between the virus and susceptible host result in disease

An important aspect of viral pathogenesis is viral epidemiology because it allows physicians to study the distribution of determinants of disease in human populations. The factors that influence acquisition and spread of infectious disease are essential for developing methods of prevention and control. Infection in a population can be **endemic** (disease present at fairly low but constant level), **epidemic** (infection greater than usually found in the population), or **pandemic** (infections that are spread worldwide). Infection can be direct (respiratory spread of influenza virus) or indirect (involves a vector).

Epidemiology deals with distribution of determinants of disease in human populations

Several quantitative measures are expressed as infectivity, disease index, virulence, incidence, and prevalence in terms of epidemiology. **Infectivity** is the rate of attack and is measured as the frequency with which an infection is transmitted when there is contact between the virus and a susceptible host. **Disease index** is the number of persons who develop the disease divided by total number infected. **Virulence** is the number of fatal or severe cases per total number of cases. **Incidence** is the number of new cases of a disease within a specified period; it usually reflects a percentage of the population that is affected. **Prevalence** is the rate of cases existing in a population at risk during a defined period. **Propagation of epidemics** has multiple requirements (infectivity, virulence, incidence, and so on). Host factors such as age, genetic predisposition, and immune status greatly influence manifestations of disease. Age, race, and sex influence attack rates and severity. Degree and duration of immunity influence epidemic frequency. Natural selection influences the susceptibility of a population. Pandemics of infection occur when immunity is low or absent. Cross-immunity may be lost by antigenic shifts. Immunization is the most effective method of specific individual and community protection against many epidemic diseases.

TRANSMISSION AND ENTRY

Viruses are transmitted via horizontal (common route of transmission; person to person) and vertical (mother-to-child transmission) routes (**Tables 7–1** and **7–2**). Human viruses cause either systemic or localized infections by entering the host through a variety of routes,

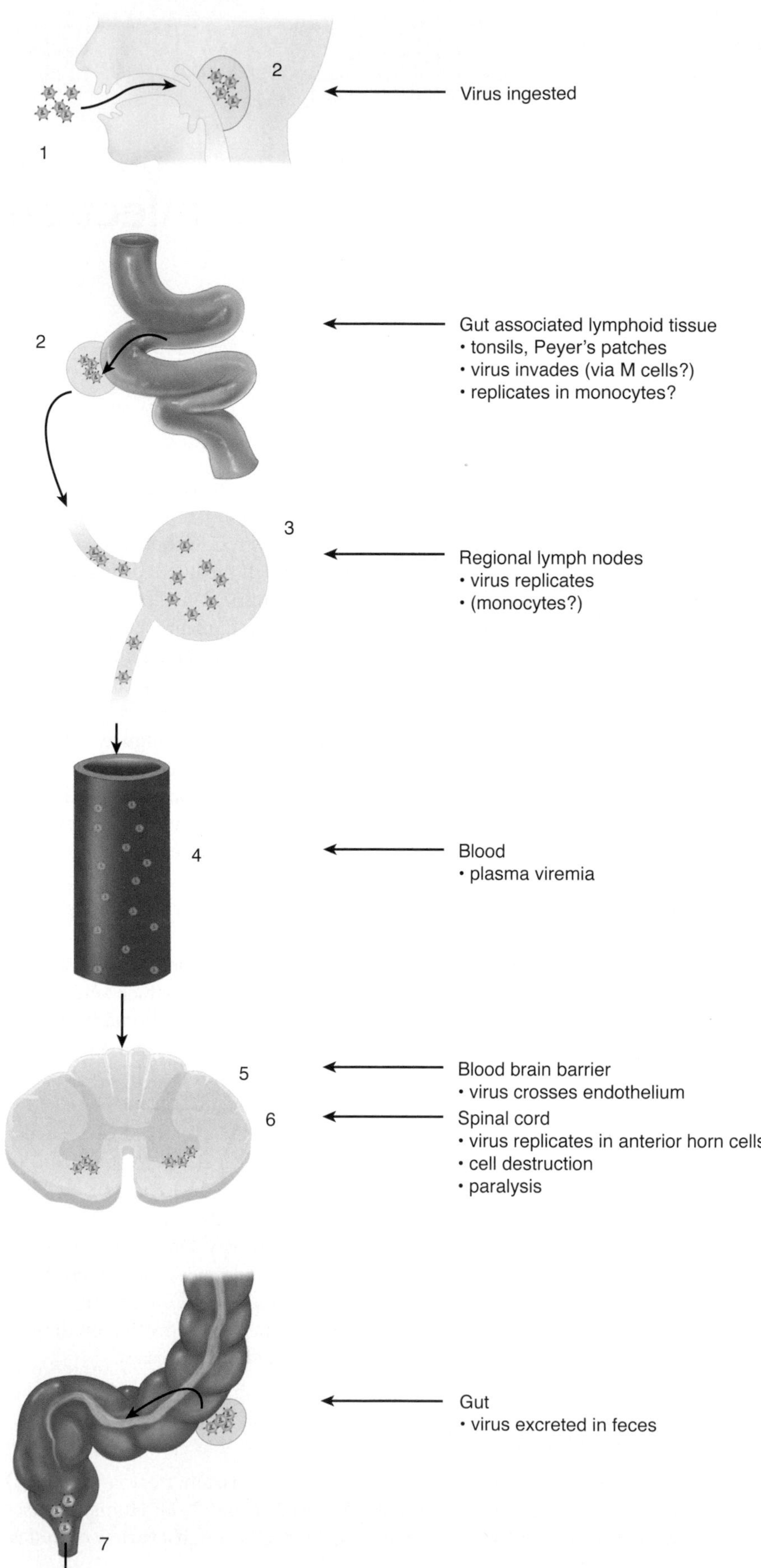

FIGURE 7–1. Stages of poliovirus pathogenesis.

TABLE 7–1 Common Routes of Transmission

ROUTE OF ENTRY	SOURCE/ MODE OF TRANSMISSION	EXAMPLES/VIRUSES
Respiratory	Aerosol droplet inhalation	Influenza virus, parainfluenza virus, respiratory syncytial virus, measles, mumps, rubella, varicella-zoster virus, hantavirus
	Nose or mouth → hand or object → nose	Common cold (rhinovirus, coronavirus, adenovirus)
Salivary	Direct salivary transfer (eg, kissing)	Herpes simplex virus (oral-labial herpes), Epstein-Barr virus (infectious mononucleosis), cytomegalovirus
Gastrointestinal	Stool → hand → mouth and/or stool → object → mouth	Enteroviruses, hepatitis A virus, poliovirus, rotavirus
Skin	Skin discharge → air → respiratory tract	Varicella-zo-ster virus, small pox virus
	Skin to skin	Human papilloma virus (warts)
	Animal bite to skin	Rabies virus
Blood	Blood products, transfusion or needle prick	Hepatitis B virus, hepatitis C virus, hepatitis D virus, human immunodeficiency virus (HIV), human T lymphotropic virus, cytomegalovirus
	Insect bite	Arboviruses, dengue virus, yellow fever virus, West Nile virus, encephalitis causing arboviruses
Genital	Genital secretions	Hepatitis B virus, HIV, herpes simplex virus, cytomegalovirus
Urine	Urine	Polyomavirus (BK virus)
Eye	Conjunctival	Adenovirus, cytomegalovirus, herpes simplex virus 1
Zoonotic	Animal bite	Rabies
	Arthropod bite	Arboviruses
	Mammals excreta	Arenavirus, hantavirus, filovirus
	Chicken, wild birds–aerosol droplets	Avian influenza virus (bird flu, H5N1)

TABLE 7–2 Vertical Transmission of Viruses

SOURCE/ MODE OF TRANSMISSION	EXAMPLES/VIRUSES
Prepartum or transplacental	Cytomegalovirus, parvovirus B19, rubella virus, human immunodeficiency virus (HIV)
Intrapartum or during delivery/birth	Hepatitis B virus, hepatitis C virus, herpes simplex virus, HIV, human papilloma virus
Postpartum or via breast feeding	Cytomegalovirus, hepatitis B virus, human T lymphotropic virus, HIV

including direct inoculation, respiratory, conjunctiva, gastrointestinal and genitourinary routes (**Figure 7–2**). In addition, viruses can enter the host through a break in the skin or via mucosal surfaces of various routes such as respiratory, gastrointestinal, and genitourinary tracts. Mother-to-child transmission (vertical transmission) can occur in utero, during delivery (via birth canal), and through breast-feeding.

Viruses are transmitted horizontally (common routes) and vertically (mother to child)

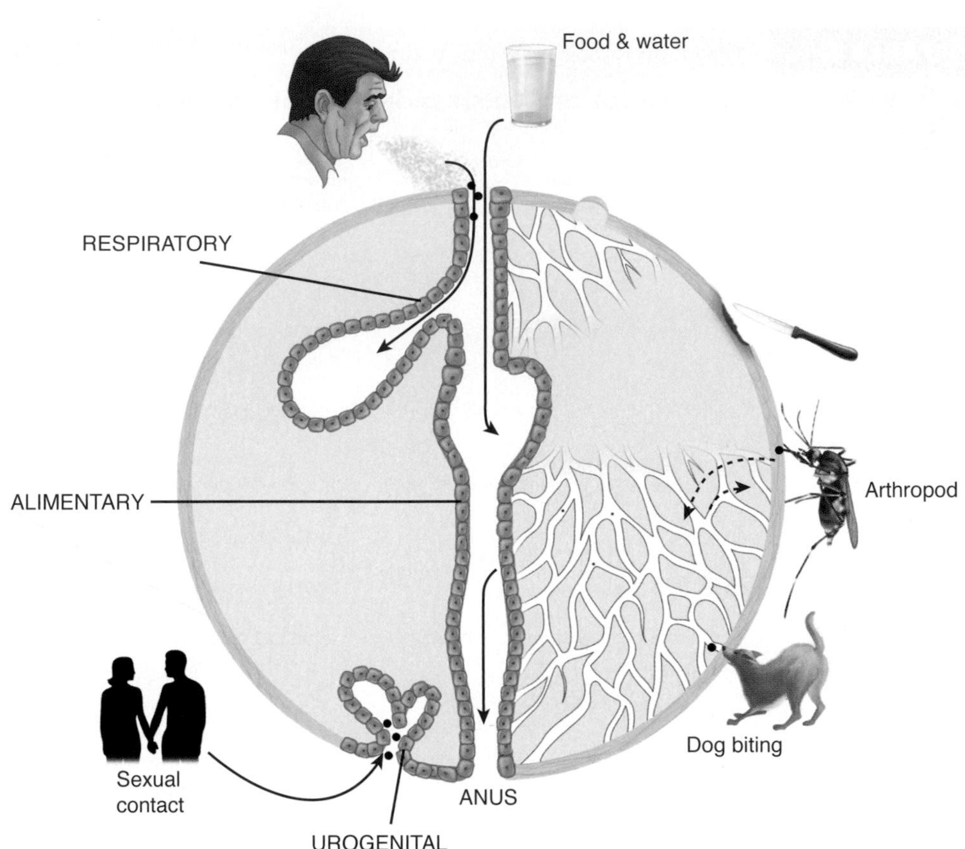

FIGURE 7–2. Routes and sites of entry of viruses into hosts.

Viruses can also be transmitted to humans by animal or insect bites

Zoonotic (animal-to-human) transmission of viral infections can occur from the bite of animals (eg, rabies) or insects (eg, dengue, yellow fever, West Nile) or from inhalation of animal excreta (eg, hantavirus, arenavirus; **Table 7–2**). In some cases, avian flu virus (bird flu) can be transmitted from birds or poultry to humans and swine flu virus can also be transmitted to humans.

After virus entry into the host, viruses have variable incubation periods. **Incubation period** is the time between exposure to the organism and appearance of the first symptoms of the disease. Some viruses have short incubation periods (influenza—2 to 4 days), whereas others have long incubation periods (eg, hepatitis B virus—weeks to several months). Incubation periods of common viral infections are shown in **Table 7–3. Communicability** of a disease is the ability of the organism to shed in secretions, which may occur early in the incubation period. Some viruses can integrate into the host genome (HIV) or survive by slow replication in the presence of an immune response (hepatitis B and C viruses [HBV, HCV] and herpes simplex virus [HSV]). This dormancy or latency is dangerous because the virus may emerge long after the original infection and potentially infect others.

SPREAD IN THE HOST

Viral infections cause either localized or systemic disease

Viral infections produce either **localized infection** at the site of entry or **disseminated infection** spread throughout the body. Localized infections include influenza, parainfluenza, common cold (rhinoviruses, coronaviruses), gastrointestinal infections (rotaviruses, Norwalk viruses), and skin infections (papilloma viruses). In localized infections, the virus spreads mainly by infecting adjacent or neighboring cells

Several viruses that cause systemic disease in the host spread from the site of entry to the target tissue, where they cause cell injury after multiplication. Viruses use two major routes to spread and cause systemic infection, namely, hematogenous (via the bloodstream) and neural (via nerves) spread. Some of the viruses that cause systemic or disseminated infection are poliovirus, flavivirus, rabies virus, hepatitis B and C viruses, HIV, and others. Poliovirus pathogenesis can be cited as an example of disseminated infection in which

TABLE 7–3 Incubation Periods of Human Pathogenic Viruses

VIRUS	INCUBATION PERIODS	DISEASE
Respiratory viruses		
Influenza virus	~2 days	Influenza (flu)
Parainfluenza virus	1-3 days	Croup
Respiratory syncytial virus (RSV)	2-4 days	Brochiolitis
Rhinovirus	2-3 days	Common cold
Coronavirus	2-10 (mean 5) days	Common cold, SARS
Adenovirus	5-7 days	Pharyngitis, febrile illness
Childhood exanthems		
Mumps virus	12-29 days (average 16-18)	Parotitis (meningitis)
Measles virus	7-18 days	Measles
Rubella virus	14-21 days (average 16)	Rubella
Parvovirus B19	4-12 days	Erythema infectiosum
Poxviruses		
Smallpox virus	12-14 days	Smallpox (variola)
Enteroviruses		
Poliovirus	4-35 days (usually 7-14)	Poliomyelitis
Coxsackievirus	2-10 days	Herpangina, pleurodynia, myocarditis
Echovirus	2-14 days	Meningitis
Enterovirus	6-12 days	Rash, febrile illness
Hepatitis viruses		
Hepatitis A virus	15-45 (mean 25) days	Hepatitis A (self-limiting)
Hepatitis B virus	30-160 (mean 60–90) days	Hepatitis B (acute, chronic)
Hepatitis C virus	15-150 (mean 50) days	Hepatitis C (chronic)
Hepatitis D virus	28-45 days	Delta hepatitis
Hepatitis E virus	21-56 (mean 40) days	Hepatitis E (self-limiting)
Herpesviruses		
Herpes simplex virus 1	7-10 days	Gingivostomatitis
Herpes simples virus 2	2-12 days	Genital herpes
Varicella-zoster virus	11-21 days	Chickenpox
Cytomegalovirus (CMV)	3-12 weeks	Heterophile-negative mononucleosis, congenital CMV
Epstein-Barr virus	30-50 days	Infectious mononucleosis
Viruses of diarrhea		
Rotavirus	1-3 days	Diarrhea
Calicivirus	0.5-2 days	Diarrhea
Astrovirus	1-2 days	Diarrhea
Adenovirus	8-10 days	Diarrhea

(Continued)

TABLE 7–3 Incubation Periods of Human Pathogenic Viruses (*Continued*)

VIRUS	INCUBATION PERIODS	DISEASE
Zoonotic viruses		
Rabies virus	10 days-1 year (average 20-90 days)	Encephalitis
Dengue virus	5-8 days	Hemorrhagic fever or febrile illness
West Nile virus	1-6 days	Muscle weakness, flaccid paralysis, encephalitis
Retroviruses		
HIV-1	1-10 years	AIDS
Human T-cell lymphotropic virus (HTLV)	15-20 years	Adult T-cell leukemia (ATL)
Papovaviruses		
Human papilloma virus	50-150 days	Warts
JC virus	long, variable	Progressive multifocal leukoencephalopathy

SARS, severe acute respiratory syndrome.

Poliovirus enters by the fecal-oral route and multiplies in the small intestine but causes the major disease in the central nervous system

poliovirus is transmitted via the fecal-oral route, and the disease (paralytic poliomyelitis) is caused in the central nervous system (CNS; Figure 7–1). Poliovirus replicates at the sites of entry in the small intestine and spreads to the regional lymph nodes where it multiplies again and enters the bloodstream, resulting in **primary viremia.** The virus is spread via the bloodstream to other organs (liver, spleen) where it multiplies and enters the bloodstream causing **secondary viremia** followed by transmission to and replication in the CNS and resulting in damage to motor neurons. The development of viremia allows the immune system to mount humoral and cell-mediated responses to control the poliovirus infection.

Some viruses are spread via nerves to the target tissue

Some of the viruses that are spread by the neural route are HSV, poliovirus, rabies virus, and certain arboviruses. HSV is transmitted in vesicle fluids, saliva, and vaginal secretion and replicated in the mucoepithelial cells, causing primary infection and then traveling via nerves in to the CNS (neurons), where they establish latent infection.

TROPISM

Tropism is the capacity of viruses to infect a specific cell type within a tissue or organ

Determined by the specific interaction between viral surface proteins and cellular receptors

Tropism is the capability of viruses to infect discrete population of cells within an organ. Cellular or tissue tropism is determined by the specific interaction of viral surface proteins (spikes) and cellular receptors on the host cells. Some of the identified cellular receptors for viruses are shown in Table 6–5. However, it should be kept in mind that the presence of a receptor for a virus is not always sufficient for viral infection in the target cells. For example, the presence of CD4 (HIV receptor) alone on cells does not allow viral entry into these cells, but it requires that target cells also express co-receptors, CXCR4 or CCR5 (chemokine receptors) for efficient viral attachment.

Some viruses such as HIV use a receptor and co-receptor

Different viruses may use the same receptor on host cells

Some other viruses such as HSV-1 and HSV-2 also use a receptor and co-receptor. Heparan sulfate serves as a receptor for HSV, whereas HveA, HveB, and HveC have been identified as co-receptors. Different viruses may use the same cellular molecule as receptors. Some examples are sialic acid residues as important components of the receptor for influenza viruses, coronaviruses, and reoviruses. Similarly, heparan sulfate is the receptor for HSV, cytomegalovirus, and adeno-associated virus (AAV).

After attachment of viral surface proteins to the cellular receptor, the viral genome-protein complex is released in the cytoplasm followed by transcription, replication, and virus assembly. Enveloped viruses use two mechanisms for entry: receptor-mediated endocytosis (viropexis) and fusion, whereas naked capsid viruses use viropexis without membrane-membrane fusion. Influenza virus is tropic to cells that express sialic acid residues containing glycoproteins where influenza virus attachment protein, hemagglutinin (HA), binds to the receptor, and the virion is internalized into an endosomal vacuole and the viral envelope membrane fuses with the vacuole membrane. For other enveloped viruses such as HIV, viral envelope gp120 binds to the cellular receptor (CD4) and co-receptor (CXCR4 or CCR5) for attachment and envelope gp41 fuses the viral envelope with the plasma membrane. Naked capsid viruses such as poliovirus and hepatitis A virus outer capsid spikes begin attachment to the cellular receptor; the virion is internalized and the viral genome is released in the cytoplasm without membrane-membrane fusion.

Enveloped viruses enter cells via receptor-mediated endocytosis (viropexis) and fusion

Naked capsid viruses enter cells via viropexis without membrane-membrane fusion

Both RNA and DNA viruses undergo genetic changes, including mutation and recombination (Chapter 6). Viral tropism can be altered in the case of some viruses because of genetic variation in the viral surface proteins. Avian influenza virus (H5N1) does not bind to the receptor of human influenza virus (H1N1), but mutation or reassortment in H5N1 may allow binding of H5N1 to H1N1 receptor (Chapter 9). Similarly, genetic changes in HIV-1 Env gp120 during infection in patients switch the co-receptor requirement from CCR5 to CXCR4. CCR5 is predominantly found on macrophages and also on T lymphocytes, whereas CXCR4 is mainly expressed on T lymphocytes (Chapter 18).

Genetic changes in viral surface proteins alter viral tropism

Although interaction of the viral surface proteins with the receptors on the host cell plays a critical role in determining the tropism, other factors such as viral gene expression, especially in case of retroviruses and papovaviruses, contribute to tropism. For example, HBV replicates more efficiently in liver cells and papilloma virus in skin cells because of regulation of individual viral promoter transcriptions.

Besides viral surface protein–cellular receptor interactions, viral gene expression also contributes to tropism

VIRULENCE AND CYTOPATHOGENICITY

The ability of a virus to cause disease in an infected host is called **virulence** or **pathogenicity.** Viral virulence is basically the degree of pathogenicity of a virus. A virus may be of high or low virulence for a particular host. Different strains of the same virus may differ in the degree of pathogenicity. The ability of a virus to cause degenerative changes in cells or cell death is called cytopathogenicity. Viral strains that kill target cells and cause disease are called virulent viruses, but other strains have mutated and lost their ability to cause cytopathic effects (CPE) and disease are termed as avirulent or **attenuated** strains. Some attenuated strains can be used as live vaccines. Examples are MMR (measles, mumps, rubella), smallpox, poliovirus (not used in the United States), and yellow fever.

Virulence is defined as the ability of a virus to cause disease in an infected host

Virulence can be measured as the degree of pathogenicity between viruses to cause disease

Three major outcomes can be attributed to a viral infection: **abortive infection**, in which no progeny virus particles are produced, but the cell may die because early viral functions can occur; **lytic infection,** in which active virus production is followed by cell death; and **persistent infection,** in which small numbers of virus particles are produced with little or no CPE. Persistent infections include **latent infection,** in which viral genetic material remains in host cell without production of virus and may be activated at a later time to produce virus and/or transform the host cell; **chronic infection,** which involves low level of virus production with little or no CPE; and **viral transformation,** in which viral infection or viral gene product induces unregulated cellular growth and cells form tumors in the host. If two closely related viruses infect a host, then infection by the first virus can inhibit the function of the second virus; this is termed **interference.**

Viruses can cause abortive, lytic, or latent infections

Virulence and cytopathogenicity depend on the nature of viruses and the characteristics of cells such as permissive and nonpermissive cells. A **permissive cell** permits production of progeny virus particles and/or viral transformation. A **nonpermissive cell** does not allow virus replication, but it may permit transformation of the cell. Replication of the virus results in alterations of cellular morphology and function as well as antigenicity of the virus. When a lytic virus infects a permissive cell, lots of daughter viruses are produced followed by lysis of the infected cells called **cytopathic effects** of the virus (Figure 4–9). The features of CPE include the following:

Features of cytopathic effects are morphologic changes of the cell organelles followed by cellular lysis

1. Nucleus: Inclusion bodies, thickening of the nucleus, swelling, nucleolar changes, margination of chromatin

2. Cytoplasm: Inclusion bodies, vacuoles
3. Membranes: Cells round up, loss of adherence, cell fusion (syncytia)
4. Cellular: Lysis (disintegration)

Molecular and genetic determinants of viral virulence are located throughout the viral genome

The molecular and genetic determinants of viral virulence are complex. Viral gene products influence pathogenesis and virulence. As previously described, viral surface proteins both in enveloped and naked capsid viruses determine tropism and spread, and alterations in these surface proteins may result in changes in tropism, spread, and virulence. However, other regions of the viral genome contribute to pathogenicity and virulence. There is no single master gene or protein that determines virulence. For example, live attenuated vaccine of poliovirus, also called oral polio vaccine (OPV), contains all three serotypes of poliovirus that are attenuated and have markedly reduced neurovirulence compared with wild-type polioviruses. The neurovirulence determinants are located in the 5′ nontranslated region of the genome involved in initiation of translation and an internal ribosomal entry site, structural capsid proteins (VP1 to VP4), and nonstructural proteins such as viral polymerase.

Some viruses such as poxviruses and herpesviruses encode virokines and viroreceptors to help cells proliferate and avoid host defenses, respectively

Some viruses encode a new class of proteins called **virokines** and **viroreceptors,** which contribute to viral virulence by mimicking cellular proteins. It is believed that some large DNA viruses such as poxviruses and herpesviruses have acquired these genes by recombination from the cells in which they replicate. Virokines are secreted from infected cells and act as cytokines, helping the cells to proliferate and increase virus production. Viroreceptors resemble cytokine receptors and attract cellular cytokines. Some viruses also encode proteins that bind antibodies or components of complement pathways to avoid lysis of virus-infected cells. For example, poxvirus member vaccinia virus (strain used in smallpox vaccine) encodes a vaccinia complement control protein (VCP) that abrogates the complement-mediated killing of virus-infected cells. Similarly, two glycoproteins of HSV act together as a receptor for the Fc domain of immunoglobulins to avoid antibody directed cell-mediated cytotoxicity (ADCC).

PATTERNS OF VIRAL INFECTION AND DISEASE

Infections are more common than disease

Infection involves multiplication of virus in the host, and disease represents clinical manifestations

Not every viral infection results in a disease. **Infection** involves multiplication of the virus in the host, whereas **disease** represents a clinically **apparent** response. Infections are much more common than disease; **unapparent** infections are termed **subclinical,** and the individual is referred to as a **carrier.** Although some primary infections are invariably accompanied by clinical manifestations of the disease (influenza, measles), other infections may propagate and spread for long periods before the extent of problem is recognized (HIV-1, hepatitis B and C viruses).

The severity of the disease depends on the role of both viral and host factors in influencing viral infection and disease progression

Relative susceptibility of a host for a viral infection in terms of severity of the disease depends on several factors such as virulence, molecular and genetic determinants of the virus, and host factors (immune status of the host, age, health, and genetic background). After viral transmission, the virus multiplies in the host; this phase is referred to as the incubation period, which varies for different viruses (Table 7–3). Initial virus replication generally results in viremia, which allows the virus to travel to the target tissues and replicate further to cause cell damage and clinical symptoms. The host immune system plays a pivotal role in determining the course of infection and progression of disease.

Viral infections could be lytic, latent, or chronic

Viral infection results in either a lytic or persistent (latent or chronic) infection. **Lytic infections** are those in which productive virus replication results in cell death because viral replication is not compatible with essential cellular functions. Several viruses interfere with the synthesis of cellular macromolecules and other factors that prevent cellular growth, maintenance, and repair leading to cell death. For example, poliovirus blocks the synthesis of cellular proteins by inhibiting the translation of cellular mRNA and competing for ribosomes. Accumulation of progeny viruses and viral proteins can destroy the structure and function and enhance the process of apoptosis, resulting in cell death. In enveloped viruses such as respiratory syncytial virus (RSV), HIV, and HSV, replication of the virus and cell surface expression of the envelope glycoproteins (spikes) cause cell-to-cell spread and formation of multinucleated giant cells (**syncytia**) causing cell death (**cytopathic effect**).

Persistent viral infections are those in which the infected cells survive the effect of viral replication. Persistent infections are of two kinds: latent (viral genome without virus production) and chronic (low level of virus production without immune clearance). Some persistent viruses also cause oncogenic transformations. All DNA viruses except parvoviruses have the potential to cause oncogenic transformation. RNA viruses, including retroviruses and HCV can cause oncogenic transformation in infected hosts. In these human oncogenic viruses, viral gene products transform the cells either by interfering with the tumor suppressor gene pathways (human papilloma virus) or increasing the expression of proto-oncogenes (human T-lymphotropic virus, HTLV).

Persistent infection could be either latent or chronic

Some persistent viruses can cause oncogenic transformation

Based on patterns and levels of detectable infectious virus in the host and the role of immune response in clearing the virus, viral infections can be divided into five categories: (1) acute infection that is cleared by the immune response; (2) acute infection that becomes latent and periodically reactivated; (3) acute infection that becomes chronic; (4) acute infection followed by persistent infection (viral set point) established by immune response and followed by virus overproduction, immune dysfunction, and opportunistic infections; and (5) slow chronic infections. These patterns are shown in **Figure 7–3 A-E.** In acute infection, the virus enters the host, then multiplies at the site of entry and in the target tissue, followed by viremia and cytopathic effects. This type of infection is a lytic infection. The immune system mounts both cellular and humoral responses and successfully eliminates the virus from the host. Examples of acute viral infections followed by clearance of the virus from the host by immune responses are hepatitis A, influenza viruses, parainfluenza viruses, rhinoviruses, and coronaviruses. After causing acute or lytic infection, some viruses are not eliminated by the immune response but persist in the host either in a noninfectious latent form or an infectious chronic form. Most of the viruses opting to persist in the host have evolved various mechanisms for persistence, including restriction of viral cytopathic effects, infection of immunologically privileged sites, maintenance of viral genomes, antigenic variation, suppression of immune components, and transformation of host cells.

Most infections have some kind of acute phase followed by either being eliminated from the host or becoming latent or chronic

Acute viral infections that are cleared by the immune system are mainly due to RNA viruses such as picornaviruses, orthomyxoviruses, and paramyxoviruses

In some viral infections, acute infection may result in either asymptomatic or symptomatic disease followed by latent infection in which the viral genome persists without any infectious virus production. This latent virus could be periodically reactivated with virus shedding at or near the primary infections with some symptomatic disease, as seen in HSV infections. In this case, productive (lytic) infection takes place in permissive cells (mucoepithelial cells), whereas latent infection occurs in nonpermissive cells (neurons).

Acute infection caused by herpes simplex virus is followed by a latent infection and periodic reactivation

In some persistent infections, acute infection causes initial disease, which is followed by a chronic infection in which a low level of infectious virus is continuously produced with little or no damage to the target tissue. Initially, the immune system controls the infection by bringing the viral load lower than seen in acute infection; however, the immune system is unable to eliminate the infection during the acute phase. During chronicity, the virus is maintained via several mechanisms, such as infection of nonpermissive cells, spread to other cell types, antigenic variation, and inability of the immune response to completely eliminate the virus. Examples of viruses that cause this type of infection are HBV and HCV.

Acute infection caused by HBV and HCV can be followed by a chronic infection accumulation of the damage over time

In other persistent infections such as HIV, the acute infection results in "acute retroviral syndrome" followed by a persistent infection in which the immune responses bring down the high viral load to a "viral set point." The viral set point is maintained because of the robust immune response against the mutating virus for a long time in most infected patients. Because of impairment of the immune system and down-regulation of immune components by HIV, the mutating and highly replicating HIV could not be contained by the immune system, which also offers an opportunity for other pathogens (opportunistic infections) to establish infection and cause full-blown AIDS.

HIV acute infection is followed by a persistent infection leading to impairment of the immune system

Some unconventional infectious agents cause slow, chronic infection without acute infection such as caused by prions. **Prions** are infectious protein molecules without any genes, causing slow, chronic infection in humans such as Creutzfeldt-Jacob disease (CJD) and bovine spongiform encephalopathy (BSE, mad cow diseases) (see Chapter 20).

Some unconventional infectious agents cause slow, chronic infection without acute symptoms

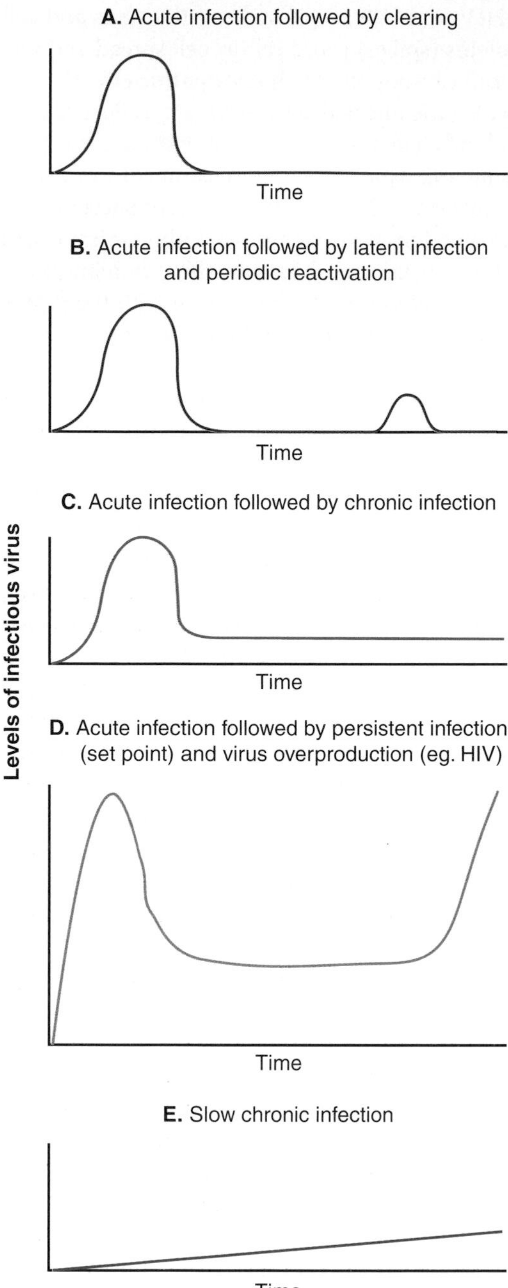

FIGURE 7–3. Viral infection patterns.

VIRAL TRANSFORMATION

Many DNA and some RNA viruses can transform normal cells into tumors

Many DNA and some RNA viruses, especially the retroviruses, can transform normal cells into abnormal cells called tumors (benign or malignant). This process is called viral transformation, and these viruses are **oncogenic viruses.** Viruses that can either cause tumors in their natural hosts or other species or can transform cells in vitro are considered to have oncogenic potential. Specifically, a tumor is an abnormal growth of cells and is classified as benign or malignant, depending on whether it remains localized or has a tendency to invade or spread by metastasis. Therefore, malignant cells have at least two defects. They fail to respond to controlling signals that normally limit the growth of nonmalignant cells, and they fail to recognize their neighbors and remain in their proper location.

When grown in tissue culture in the laboratory, these tumor cells exhibit a series of properties that correlate with the uncontrolled growth potential associated with the tumor in the organism. They have altered cell morphology and fail to grow in the organized patterns found for normal cells. They also grow to a much higher cell density than do normal cells under conditions of unlimited nutrients; therefore, they appear unable to enter the resting G0 state. Furthermore, they have lower nutritional and serum requirements than normal cells and are able to grow indefinitely in cell culture. These transformed or tumor cells often are used as cell lines for the culture or propagation of viruses in the laboratory.

Malignant cells fail to respond to signals controlling the growth and location of normal cells

In addition to the listed properties, viral transformation usually, but not always, endows the cells with the capacity to form a tumor when introduced into the appropriate animal. Although the original use of the term **transformation** referred to the changes occurring in cells grown in the laboratory, current usage often includes the initial events in the animal that lead to the development of a tumor. In recent years, it has become increasingly clear that some but not all of these viruses also cause cancers in the host species from which they were isolated.

Malignant transformation of cells in culture can be accomplished by most DNA viruses and some retroviruses

Transformation by DNA Human Viruses

The oncogenic potential of human DNA viruses is summarized in **Table 7–4**. All known DNA human viruses, except parvoviruses, are capable of causing aberrant cell proliferation under some conditions. For some viruses, transformation or tumor formation has been observed only in species other than the natural host. Apparently, infections of cells from the natural host are so cytocidal that no survivors remain to be transformed. In addition, some viruses have been implicated in human tumors without any indication that they can transform cells in culture.

Some oncogenic viruses cause tumors in species other than their natural hosts

TABLE 7–4 Oncogenicity of DNA and RNA Human Viruses

VIRUS OR VIRUS GROUP	TUMORS IN NATURAL HOST	TUMORS IN OTHER SPECIES[b]	TRANSFORM CELLS IN TISSUE CULTURE
DNA viruses			
Parvoviruses	No	No	No
Polyomaviruses (JC, BK)	No	Yes	Yes
Papilloma viruses	Yes, often benign	?	Yes
Human hepatitis B virus	Yes	?	No
Human adenoviruses	No	Yes	Yes
Human herpesviruses	Yes	Yes	Yes
Poxviruses	Occasionally, usually benign	Yes	No
RNA viruses			
Retroviruses	Yes	Yes	Yes
Human T-lymphotropic viruses I and II (HTLV I and II)			
Hepatitis C virus	Yes	Yes?	Yes

[a]"Yes" means that at least one member of the group is oncogenic.
[b]Test usually done in newborns of immunosuppressed hosts.

In nearly all cases that have been characterized, viral transformation is the result of the continual expression of one or more viral genes that are directly responsible for the loss of growth control. Two targets have been identified that appear to be critical for the transforming potential of these viruses. Adenoviruses, papilloma viruses, and simian virus 40 all code for either one or two proteins that interact with the tumor suppressor proteins known as p53 and Rb (for retinoblastoma protein) to block their normal function, which is to exert a tight control over cell cycle progression. The end result is endless cell cycling and uncontrolled growth.

Transformation by DNA viruses is analogous to lysogenic conversion

In many respects, transformation is analogous to lysogenic conversion and requires that the viral genes be incorporated into the cell as inheritable elements. Incorporation usually involves integration into the chromosome (eg, papovaviruses, adenoviruses, and retroviruses), although the DNAs of some papilloma viruses and some herpesviruses are found in transformed cells as extrachromosomal plasmids. Unlike some of the temperate bacteriophages that code for the enzymes necessary for integration, papovaviruses and adenoviruses integrate by nonhomologous recombination using enzymes present in the host cell. The recombination event is therefore nonspecific, both with respect to the viral DNA and with respect to the chromosomal locus at which insertion occurs. It follows that for transformation to be successful, the insertional recombination must not disrupt a viral gene required for transformation. In summation, two events appear to be necessary for viral transformation: a persistent association of viral genes with the cell and the expression of certain viral "transforming" proteins.

Transformation by Retroviruses

Most retroviruses produce virions without causing host cell death

A DNA copy of the retroviral genome is integrated, but not at a specific site

Two features of the replicative cycle of retroviruses are related to the oncogenic potential of this class of viruses. First, most retroviruses do not kill the host cell, but rather set up a permanent infection with continual virus production. Second, a DNA copy of the RNA genome is integrated into the host cell DNA by a virally encoded integrase (IN); however, unlike bacteriophage λ integration, a linear form of the viral DNA, rather than a circular form, is the substrate for integration. Furthermore, unlike λ, there does not appear to be a specific site in the cell DNA where integration occurs.

Retroviruses may carry transforming oncogenes

Oncogenes encode a protein that interferes with cell signaling

Retroviruses are known to transform cells by three different mechanisms. First, many animal retroviruses have acquired transforming genes called **oncogenes.** More than 30 such oncogenes have now been found since the original oncogene was identified in Rous sarcoma virus (called v-*src*, where the v stands for viral). Because normal cells possess homologs of these genes called **proto-oncogenes** (eg, c-*src*, where c stands for cellular), it is generally thought that viral oncogenes originated from host DNA. It is possible they were picked up by "copy choice" recombination involving packaged cellular mRNAs, as previously described. Because these transforming viruses carry cellular genes, they are sometimes referred to as **transducing retroviruses.** Most of the viral oncogenes have suffered one or more mutations that make them different from the cellular proto-oncogenes. These changes presumably alter the protein products so that they cause transformation. Although the mechanisms of oncogenesis are not completely understood, it appears that transformation results from inappropriate production of an abnormal protein that interferes with normal signaling processes within the cell. Uncontrolled cell proliferation is the result. Because tumor formation in vitro by retroviruses carrying an oncogene is efficient and rapid, these viruses are often referred to as **acute transforming viruses.** Although common in some animal species, this mechanism has not yet been recognized as a cause of any human cancers.

Insertional mutagenesis causes inappropriate expression of a proto-oncogene adjacent to integrated viral genome

The second mechanism is called **insertional mutagenesis** and is not dependent on continued production of a viral gene product. Instead, the presence of the viral promoter or enhancer is sufficient to cause the inappropriate expression of a cellular gene residing in the immediate vicinity of the integrated provirus. This mechanism was first recognized in avian B-cell lymphomas caused by an avian leukosis virus, a disease characterized by a very long latent period. Tumor cells from different individuals were found to have a copy of the provirus integrated at the same place in the cellular DNA. The site of the provirus insertion was found to be next to a cellular proto-oncogene called c-*myc*. The *myc* gene had previously been identified as a viral oncogene called v-*myc*. In this case, transformation occurs not because the c-*myc* gene is altered by mutation, but because the viral promoter adjacent to the gene turns on its expression continuously and the gene product is overproduced. The disease has a long latent period because, although the birds are viremic from early life, the probability of an integration occurring next to the c-*myc* gene is very low. Once such an integration event does occur, however, cell proliferation is rapid and a tumor develops. No human tumors are known for certain to result from insertional mutagenesis caused by a retrovirus; however, human cancers are known in which a chromosome translocation has placed an active cellular promoter next to a cellular proto-oncogene (Burkitt's lymphoma and chronic myelogenous leukemia).

The third mechanism was revealed by the discovery of the first human retrovirus. The virus, HTLV-I, is the causative agent of adult T-cell leukemia. HTLV-I sequences are found integrated in the DNA of the leukemic cells, and all the tumor cells from a particular individual have the proviral DNA in the same location. This observation indicates that the tumor is a clone derived from a single cell; however, the sites of integration in tumors from different individuals are different. Thus, HTLV-I does not cause malignancy by promoter insertion near a particular cellular gene. Instead, the virus has a gene called *tax* that codes for a protein that acts in trans (ie, on other genes in the same cell) to not only promote maximal transcription of the proviral DNA, but also to transcriptionally activate an array of cellular genes. The resulting cellular proteins cooperate to cause uncontrolled cell proliferation. The *tax* gene is therefore different from the oncogenes of the acute transforming retroviruses in that it is a viral gene rather than a gene derived from a cellular proto-oncogene. HTLV-I is commonly described as a **transactivating** retrovirus.

Human T-cell leukemia is caused by transactivating factor encoded in integrated HTLV-I

Transactivating factor turns on cellular genes, causing cell proliferation

Transformation by Other RNA Viruses

Hepatitis C virus (HCV) causes chronic infection in more than 80% of infected people. The chronicity in HCV infection increases the risk of cirrhosis of liver and hepatocellular carcinoma (HCC). Some studies suggest that one of the HCV nonstructural proteins (NS3) may be involved in transformation.

HOST FACTORS

Viral infection also depends on host factors. Several viral infections have repeatedly shown a variable range of outcomes from asymptomatic to symptomatic infections and even fatal disease is some cases. Furthermore, host factors probably play an important role in reversion of some of the live attenuated vaccines to a virulent state. Several of the host factors, including immune status, genetic background, age, and nutrition, play important roles in determining the outcome of viral infection. Several innate immune responses (interferons alpha and beta, natural killer cells, mucocilliary responses, and others) and adaptive immune responses (antibody and T-cell responses) influence the outcome of viral infections. Individuals with weak immune systems or those who are immunocompromised or immunosuppressed often have more severe outcomes. Details of immune responses to infection are described in Chapter 2.

Host factors such as immune status, genetic background, age, and nutrition play important roles in the outcome of viral infections

Host genetics is one of the most important factors that influence the outcome of viral infections. Several host genes, in addition to viral factors, contribute to the variable outcome of HIV infection in infected individuals; some become rapid progressors. The majority are slow progressors. Elevated levels of β-chemokines such as MIP1-α, MIP1-β, RANTES, which are natural ligands of CCR5 (HIV co-receptor), have been found to be associated with decline in the rate of HIV disease progression. These chemokines are also termed as HIV-suppressive β-chemokines. Genetic resistance to HIV-1 infection was found in individuals expressing a truncated CCR5 co-receptor, CCR5Δ32. Individuals homozygous for the Δ32 allele seem to have normal life expectancy but strongly protected (not completely) against HIV infection, whereas heterozygous Δ32 allele slows the cell-to-cell spread of HIV in infected patients. The Δ32 homozygous allele is found in 1% of Caucasians, predominantly in Northern European populations. Furthermore, long-term progressors also have a high frequency of Δ32CCR5 deletion. Although Δ32CCR5 deletion or antagonists of CCR5 provide some protection against HIV infection, it may cause a higher risk of symptomatic

Elevated levels of chemokines or Δ32CCR5 allele slow down HIV disease progression

Homozygous Δ32CCR5 allele provides strong protection against HIV infection, but increases the risk for symptomatic West Nile virus infection

West Nile virus infection and a lower likelihood of clearing hepatitis C virus (HCV). In addition, human leukocyte antigen (HLA) alleles have been associated with slow disease progression or protection against HIV infection.

Age of the host plays an important role in the severity of some viral infections

Some viruses cause severe diseases in infants; adults are more vulnerable to others

Age-related correlation between the host and several viral infections has been observed. Several viruses such as varicella-zooster virus (VZV), mumps, polio, and Epstein-Barr virus (EBV) cause less severe infection in infants, whereas others (rotaviruses, respiratory syncytial virus) result in severe disease in infants. Although the same strain of HIV infects both mothers (adults) and infants, infants develop symptomatic AIDS faster than adults. It seems that age-related increased resistance to viral infections might reflect the maturity of the immune system and other defense mechanisms.

Hormones may influence some infections

Production of hormones may also influence the outcome of some viral infections. For example, polio, hepatitis A, B, and E, and poxviruses are more severe during pregnancy, suggesting that hormones may influence viral pathogenesis. Polyomaviruses can also be reactivated during pregnancy.

Alnutrition and personal habits increase severity of some viral infections

Fever and inflammation can combat viral infections

Nutritional state and personal habits of the hosts can also have an effect on viral pathogenesis. Protein deficiency has been shown to be associated with severity of measles infection, most likely owing to weak cellular immunity. Some personal habits, such as smoking, increase the severity in influenza virus infection. In addition, host responses such as fever and inflammation have been suggested to have an important role in combating viral infections.

HOST DEFENSES

The two major types of host defenses are nonspecific (**innate**) and specific (**adaptive**) immune responses. The innate immune response includes interferons (α, β), natural killer cells, macrophages (phagocytosis), α-defensins, mucociliary clearance, apolipoprotein B RNA editing enzyme (APOBEC3G, an anti HIV enzyme) and fever, whereas the adaptive immune response involves humoral and cell-mediated immunity. Details of specific immune response to infection are described in Chapter 2.

Interferons

Interferons are chemokines produced by virally infected cells that inhibit virus production in infected and other cells

Interferons are not virus-specific

Interferons are host-encoded proteins that provide the first line of defense against viral infections. They belong to the class of molecules called **chemokines,** which are proteins or glycoproteins that are involved in cell-to-cell communication. There are three types of interferon, interferon-α (leukocyte), interferon-β (fibroblast) and interferon-γ (lymphocyte). Virus infection of all types of cells stimulates the production and secretion of either interferon-α or interferon-β, which acts on other cells to induce what is called the **antiviral state.** Unlike specific immunity, the interferons are not specific to a particular kind of virus; however, interferons usually act only on cells of the same species. Other agents such as antigens and mitogens stimulate the production of interferon-γ by lymphoid cells. In this case, interferon appears to play an important role in the immune system regardless of any role as an antiviral protein (see Chapter 2).

Interferons are produced in response to accumulation of double-stranded viral RNA during viral synthesis

The signal that leads to the production of interferon by an infected cell appears to be double-stranded RNA. This conclusion is based on the observation that treatment of cells with purified double-stranded RNA or synthetic double-stranded ribopolymers results in the secretion of interferon. Although the mechanisms are largely unclear and probably vary from one virus to another, viral infections in general lead to the accumulation of significant levels of double-stranded RNA in the cell.

Interferon is the first line of defense against viral infection by activating two pathways that degrade mRNA and inhibit protein synthesis

Changes in the synthesis of a large number of cellular proteins are characteristic of the antiviral state induced by interferon. However, the cells exhibit only minimal changes in their metabolic or growth properties. The machinery to inhibit virus production is mobilized only on infection. Interferon has multiple effects on cells, but only three systems have been extensively studied. The first system involves a protein called Mx, which is induced by interferon and specifically blocks influenza infections by interfering with viral transcription. The second system involves the up-regulation of protein kinase, which is dependent on double-stranded RNA and protein kinase and which phosphorylates and thereby inactivates one of the subunits of an initiation factor (eIF-2) necessary for protein synthesis. In some cases, viruses have evolved specific mechanisms to block the action of this protein kinase. The third system involves the induction of an enzyme called 2′, 5′-oligoadenylate synthetase, which synthesizes chains of 2′, 5′-oligo (A) up to 10 residues in length. In turn, the 2′, 5′-oligo (A) activates a constitutive ribonuclease, called RNase L, which degrades mRNA. The activities of both protein kinase and 2′, 5′-oligo (A) synthetase requires the presence of double-stranded RNA, the intracellular signal that an infection is occurring. This requirement prevents interferon from having an adverse effect on protein synthesis in uninfected cells.

In the latter two cases, viral infection of a cell that has been exposed to interferon results in a general inhibition of protein synthesis, leading to cell death and no virus production. A cell that was destined to die anyway from a viral infection is sacrificed for the benefit of the entire organism. Virus-induced interferon pathways are shown in **Figure 7–4**. In addition, interferon prepares uninfected cells to fight viral infections. Presence of interferon induces oligosynthetase and protein kinase but does not activate because there is no viral ds RNA in uninfected cells. Thus, interferon kills only infected cells but not uninfected cells.

Interferons inhibit viral protein synthesis by inducing cellular enzymes that require double-stranded RNA

Interferons inhibit protein synthesis but only in infected cells

Other Host Defenses

Natural killer (NK) cells, like interferon, are also not virus-specific but kill virus-infected cells by secreting perforins (pore forming proteins) and granzymes (serine proteases), which cause apoptosis of infected cells. NK cell-induced killing of infected cells does not require immune components such as antigen, T-cell receptor, or major histocompatibility complex. Another important cell type that limits virus infection in a nonspecific manner via phagocytosis is macrophage, especially alveolar macrophages and macrophages of the reticuloendothelial system. Furthermore, other factors show antiviral activity especially against HIV infection, including α-defensins and APOBEC3G. α-Defensins are a class of peptides known to have antiviral activity and have also been found to interfere with the interaction of HIV env gp120 with chemokine receptor CXCR4. On the other hand, APOBEC3G is an enzyme that hypermutates retroviral (HIV) DNA by deaminating cytosines in both viral DNA and mRNA, reducing viral infectivity.

Natural killer cells destroy virus-infected cells by secreting perforins and granzyme causing apoptosis

α-Defensins and APOBEC3G reduce HIV infectivity

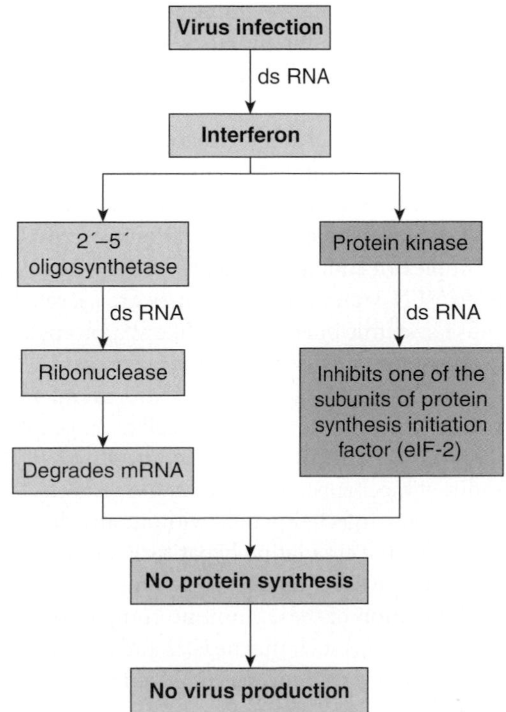

FIGURE 7–4. Virus-induced interferon pathways.

ADAPTIVE IMMUNE RESPONSES

Adaptive immunity involves elimination of the virus by neutralizing antibodies and virus-infected cells by cytotoxic T lymphocytes

Adaptive immune responses involving humoral (antibody) and cell-mediated (cytotoxic T lymphocytes) immune responses are also described in Chapter 2. These are virus-specific immune responses directed against virus proteins (antigens). **Antibody** is effective in eliminating cell-free virus and **cytotoxic T lymphocytes** (CTL) destroy virus-infected cells. The idea that adaptive or acquired immunity in patients is viral antigen–specific led the way to develop vaccines against several viral infections. Immunity could be either **active,** in that it is elicited by exposure to a pathogen or vaccine, or **passive,** in which it is transferred by immune serum. After viral infection, the first specific immune response is T-cell mediated in which **CD8 T cells** recognize viral antigen presented by class I MHC (major histocompatibility complex) and kill virus-infected cells by secreting perforins and granzyme and activating FAS proteins, causing apoptosis. It is important to differentiate that CD8 T-cell killing of virus-infected cells is viral antigen–specific, whereas NK cell killing of infected cells is nonspecific. The second important control is **neutralization** of the virus in infected hosts by antigen-antibody interactions preventing the virus to infect target cells. Antibodies are generated against all viral antigens; however, antibody against surface antigens are most effective in eliminating the virus. Antibody in conjunction with complement can also kill virus-infected cells. The evidence that viral infection elicits antibody and CTL that help the clearance of viruses in some cases (acute infection) and control or suppress the viruses in certain cases (persistent infection) has allowed researchers to develop live attenuated vaccines. Live attenuated vaccines activate both arms of the immune system, are very effective in preventing infection, and are longlasting, but they carry a very small risk of reversion. On the other hand, killed or inactivated vaccines (pathogen-killed or inactivated) and subunit vaccines (one or few proteins of the virus) predominantly activate the humoral (antibody) response, do not confer longlasting immunity, and are also needed in a larger quantity.

VIRUS-INDUCED IMMUNOPATHOLOGY

Immune responses may also destroy target cells

Antigen-antibody complex, cytotoxic T lymphocytes, complement, and cytokines mediate virus-induced immunopathology

Viral diseases are usually the result of virus-host cell interactions causing either a lytic infection and cell death or persistent infections and cell survival with some cellular dysfunction. However, sometimes both humoral and cellular immune responses against viral infections, especially those causing less cytopathic or persistent infections, mediate inflammation and disease. This could be true in viral infections in which a large number of cells are infected in an individual before the immune response is turned on and in which destruction of these infected cells by immune response may have severe or fatal pathologic outcomes. Specifically, pro-inflammatory cytokines, antigen-antibody complexes, complement activation pathways, CD4+ T-cell-induced delayed hypersensitivity, and CTL-mediated cell killing contribute to virus-induced immunopathology.

Pro-inflammatory cytokines such as IFN-γ, TNF-α, and some interleukins play roles in chronic hepatitis B, measles, and mumps are immune-mediated as opposed to direct cytopathic effects of the virus

The most important mediators of virus-induced immunopathology are the CD8+ CTLs. They release several pro-inflammatory cytokines, including interferon-gamma (IFN-γ), tumor necrosis factor alpha (TNF-α), and several interleukins (ILs), which play an important role in clinical manifestations of virus-induced immunopathology. Some selected examples of immune mediated viral diseases are shown in **Table 7–5.** After viral infections, IFN-γ and other cytokines are secreted, which stimulate multiple organ systems to cause systemic infection (flu-like symptoms), and then other immune components such as antigen-antibody complex, complement, CTL and pro-inflammatory cytokines cause cell damage. This may be the case with several viral infections of the CNS and other tissues in which "cytokine storm" causes cell damage rather than direct viral replication (**Figure 7–5**). Chronic hepatitis B virus infection provided the first clue that the disease is caused by an indirect mechanism rather than the virus itself because a low level of virus can be present in chronically infected people without any damage to the target tissue (liver) for a long time. However, the circulating hepatitis B surface antigen (HBsAg) can form immune complexes that activate the complement system, causing inflammation and tissue damage. In addition, accumulation of these immune complexes in the kidney results in renal damage. In other viral infections such the measles and mumps, many symptoms are caused by T-cell-induced inflammatory responses as opposed to the direct cytopathic effects of the virus.

TABLE 7–5 Selected Immune Mediated Viral Diseases of Humans

VIRUS	VIRAL DISEASE	IMMUNE-MEDIATED MECHANISMS
Hepatitis B virus	Hepatitis B	CD8+ T cells Antibody
Flavivirus (dengue)	Hemorrhagic fever	Immune complexes T cells
Paramyxovirus (RSV)	Bronchiolitis	CD8+ T cells Antibody
Arenavirus	Choriomenengitis	CD8+ T cells

RSV, respiratory syncytial virus.

An important example of acute antibody-mediated immunopathology is dengue hemorrhagic fever, in which a small percentage of infected patients are unable to clear the virus and develop dengue shock syndrome (DSS) with a mortality rate up to 10%. This syndrome mostly occurs in children who are either undergoing a second infection with a different serotype or in infants carrying maternal anti-dengue antibody and undergoing first infection. Also, a non-neutralizing antibody (enhancing antibody) facilitates the adsorption of flaviviruses (dengue and yellow fever viruses) into macrophages through Fc receptors followed by

Dengue hemorrhagic fever and shock syndrome is caused by antibody-mediated immunopathology

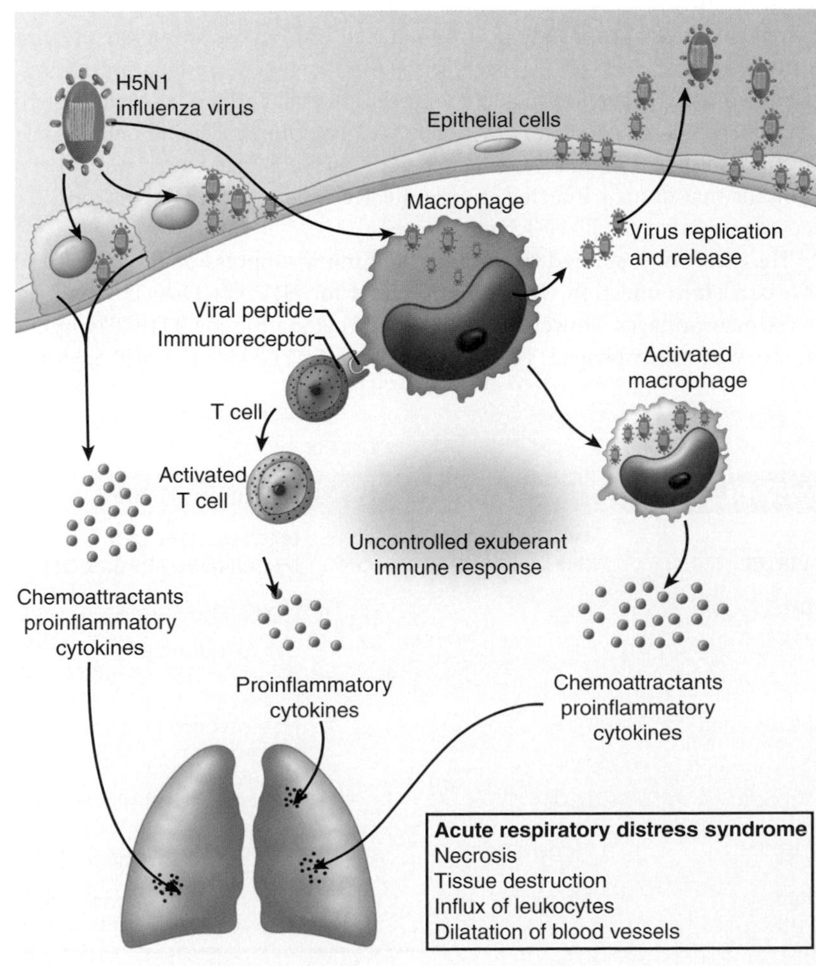

FIGURE 7–5. Cytokine storm.

replication. The infected macrophages secrete cytokines IFN-γ, TNF-α and others. In addition, dengue specific CD4+ and CD8+ T lymphocytes secrete similar types of cytokines, resulting in cytokine storm and causing hemorrhage and shock. The circulating immune complex activates the complement pathway, which also contributes to immunopathology.

Virus-initiated autoimmunity, in which a viral infection may induce an autoimmune response because the viral protein resembles a host cell protein, a phenomenon called **molecular mimicry.** Both viral epitope-specific antibody and T lymphocytes may react with cognate epitopes on the host proteins, which may elicit an autoimmune response. Viral proteins such as the polymerase of hepatitis B contain sequences similar to the encephalitogenic epitope of myelin basic protein (MBP), which is a major component of myelin sheath in the CNS. Immune responses against an epitope of hepatitis B polymerase induce an immune response against MBP, initiating an autoimmune disease process. Coxsackie virus infection has also been linked to autoimmune responses associated with type 1 diabetes as a result of molecular mimicry between a viral protein and a protein found in islet cells called glutamic acid decarboxylase (GAD).

Some autoimmune diseases are initiated by viral infections because of molecular mimicry

VIRUS-INDUCED IMMUNOSUPPRESSION

Viral infections in several instances can suppress the immune response. Immunosuppression can be achieved either by direct viral replication or by viral antigens. Some viruses specifically infect and kill immune cells. In some instances, immunosuppression is often associated with antenatal or perinatal infections. Historically, immunosuppression was first described about a century ago when patients lost their tuberculin sensitivity during and weeks after measles infection. In the last decade, immunosuppression has been the topic of discussion, concern, and treatment in the HIV/AIDS epidemic because HIV specifically infects and destroys the major type of immune cells, CD4+ T lymphocytes. **Table 7-6** shows the mechanisms of selected human viruses causing immune suppression. Several mechanisms have been proposed for virus-induced immune suppression: (1) viral replication in a major immune cells (CD4+ helper T lymphocytes) or antigen presenting cells (dendritic cells or macrophages) leading to apoptosis; (2) viral antigens stimulating pro-inflammatory cytokines causing cell death; (3) tolerance generated by clonal deletion of T lymphocytes by viral antigens, generally associated with perinatal infections; and (4) expression of viral proteins that destroy infected and uninfected cells such as HIV env gp120 depleting uninfected and infected CD4+ T lymphocytes.

Viral infections can cause suppression of the immune response

Viruses infecting either CD4+ helper T cells or antigen presenting cells cause immunosuppression

Viral gene products can cause immunosuppression by stimulating pro-inflammatory cytokines

The extensively studied virus-induced immunosuppression problem is HIV/AIDS, which is a persistent infection. The primary target for HIV is CD4+ T lymphocytes and monocytes/macrophages. However, HIV is highly cytopathic to CD4+ T lymphocytes but not to monocytes/macrophages. Therefore, depletion of CD4+ T lymphocytes in HIV-infected

TABLE 7–6 Immunosuppression by Some Human Viruses

VIRUS	DEGREE OF IMMUNOSUPPRESSION	MECHANISM OF IMMUNOSUPPRESSION
HIV	High	CD4+ T-lymphocyte depletion Env gp120–induced syncytia formation and depletion uninfected CD4+ T lymphocyte
Herpes simplex virus (HSV)	Low	HSV-encoded proteins that function as viroreceptors or virokines
Vaccinia	Low	Vaccina encodes viroreceptors and virokines
Measles	Moderate	Overproduction of cytokines
Rubella	Moderate	Immune tolerance associated with fetal infection

patients results in immunosuppression. The mechanisms of depletion of CD4+ T lymphocytes include direct killing of CD4+ T lymphocytes as a result of HIV replication and also depletion of uninfected CD4+ T lymphocytes by HIV env gp120–induced syncytia formation and apoptosis. Immunosuppression in HIV-infected patients causes opportunistic infection, whereas several other pathogens establish infection without immune challenge.

Immunosuppression in HIV-infected individuals is due to direct and indirect depletion of CD4 T lymphocytes

Measles is an acute viral infection that produces immunosuppression, which appears during the incubation period and the clinical phase of the disease. Some results of measles-induced immunosuppression include increased susceptibility to other infections, possible aggravation of chronic latent infections such as tuberculosis, and remission of autoimmune diseases. The mechanisms of measles-induced immunosuppression involve infection of several cell types and pathways. However, during measles infection the function of antigen presenting cells such as monocytes/macrophages, CD4, and CD8 T lymphocytes is compromised, which may contribute to immunosuppression. An example of immunosuppression in utero or during infancy is rubella virus infection. Fetal infections that commonly produce congenital rubella (see Chapter 10) cause greatly reduced cellular immune responses to rubella virus antigens even several years after infection. In general, several factors or determinants could be responsible for virus-induced immunosuppression, such as the strain of the virus, dose or amount of the virus entering the host, route of transmission or virus entry, age and immune status of the host, and other immunologic disorders in the host.

In measles infection, the functions of CD4 and CD8 T lymphocytes are compromised

Immunosuppression in congenital rubella is due to reduced cellular immune response during fetal infection

CONCLUSION

Over time, much has been learned about how viruses behave in their hosts and how the hosts in turn respond in ways that may be either beneficial or deleterious to their well-being. Our understanding of these processes is as yet incomplete, but the knowledge gained to date has enabled scientists to develop new strategies to deal with these issues. Two approaches that have already resulted in success are: (1) prevention, including development of effective environmental controls, and vaccines for prevention; and (2) development of specific antiviral agents that can cure, mitigate, or temporarily prevent infection. Better approaches to more advantageously manipulate specific and nonspecific host responses to such infections are expected as well. For now, all that can be stated with certainty is that exciting, meaningful progress will continue well into the future.

CHAPTER 8

Antiviral Antimicrobics and Resistance

GENERAL CONSIDERATIONS

Viruses are composed of either DNA or RNA, a protein coat (capsid), and, in many, a lipid or lipoprotein envelope. The nucleic acid codes for enzymes involved in replication and for several structural proteins. Viruses use molecules (eg, amino acids, purines, pyrimidines) supplied by the cell and cellular structures (eg, ribosomes) for synthetic functions. Thus, one of the challenges in the development of antiviral agents is identification of the steps in viral replication that are unique to the virus and not used by the normal cell. Among the unique viral events are attachment, penetration, uncoating, RNA-directed DNA synthesis (reverse transcription) or RNA-directed RNA synthesis (RNA viruses), and assembly and release of the intact virion. Each of these steps may have complex elements with the potential for inhibition. For example, assembly of some virus particles requires a unique viral enzyme, protease, and this has led to the development of protease inhibitors.

Events in the cell unique to viral replication are the target

In some cases, antiviral agents do not selectively inhibit a unique replicative event but inhibit DNA polymerase. Inhibitors of this enzyme take advantage of the fact that the virus is synthesizing nucleic acids more rapidly than the cell, so there is relatively greater inhibition of viral than cellular DNA. In many acute viral infections, especially respiratory ones, the bulk of viral replication has already occurred when symptoms are beginning to appear. Initiating antiviral therapy at this stage is unlikely to make a major impact on the illness. For these viruses, immuno- or chemoprophylaxis, rather than therapy, is a more logical approach. However, many other viral infections are characterized by ongoing viral replication and do benefit from viral inhibition, such as human immunodeficiency virus (HIV) infection and chronic hepatitis B and C.

DNA polymerase is often inhibited

The principal antiviral agents in current use are discussed according to their modes of action. Their features are summarized in **Table 8–1**.

SELECTED ANTIVIRAL AGENTS

■ Inhibitors of Attachment

Attachment to a cell receptor is a virus-specific event. Antibody can bind to extracellular virus and prevent this attachment. However, although therapy with antibody is useful in prophylaxis, it has been minimally effective in treatment.

■ Inhibitors of Cell Penetration and Uncoating

Rimantadine differs from **amantadine** by the substitution of a methyl group for a hydrogen ion. These two amines inhibit several early steps in viral replication, including viral

Amantadine and rimantadine are symmetric amines, or acyclics, which inhibit early steps in replication

TABLE 8–1 Summary of Antiviral Agents

MECHANISM OF ACTION	ANTIVIRAL AGENT	VIRAL SPECTRUM[a]
Inhibition of viral uncoating, penetration	Amantadine	Flu A
	Rimantadine	Flu A
Neuraminidase inhibition	Oseltamivir	Flu A, Flu B
	Zanamivir	Flu A, Flu B
Inhibition of viral DNA polymerase	Acyclovir	HSV, VZV
	Famciclovir	HSV, VZV
	Penciclovir	HSV
	Valacyclovir	HSV, VZV
	Ganciclovir	CMV, HSV, VZV
	Foscarnet	CMV, resistant HSV
	Cidofovir	CMV, possibly Adeno
	Trifluridine	HSV, VZV
Inhibition of viral entry	Maraviroc	HIV
Inhibition of viral fusion	Enfuvirtide	HIV
Inhibition of viral reverse transcriptase	Zidovudine	HIV
	Dideoxyinosine	HIV
	Dideoxycytidine	HIV
	Stavudine	HIV
	Lamivudine	HIV, HBV[b]
	Nevirapine	HIV
	Delavirdine	HIV
	Efavirenz	HIV
Inhibition of viral integration	Raltegravir	HIV
Inhibition of viral protease	Saquinavir	HIV
	Indinavir	HIV
	Ritonavir	HIV
	Nelfinavir	HIV
	Lopinavir	HIV
Inhibition of viral protein synthesis	Interferon α	HBV, HCV, HPV
Inhibition of viral RNA polymerase	Ribavirin	RSV, HCV,[b] Lassa fever
Antisense inhibition of viral mRNA synthesis	Fomivirsen	CMV

[a]Adeno, adenovirus; CMV, cytomegalovirus; Flu A, influenza A; Flu B, influenza B; HBV, hepatitis B virus; HCV, hepatitis C virus; HIV, human immunodeficiency virus; HPV, human papillomavirus; HSV, herpes simplex viruses; RSV, respiratory syncytial virus; VZV, varicella-zoster virus.
[b]Used in combination with interferon.

uncoating. They are extremely selective, with activity against only influenza A, where they act as M2 channel blockers. In addition, they have been used as prophylaxis. Unfortunately, since 2001 the rates of resistance to amantadine/rimantadine have increased so sharply (up to 100% for some strains) that they are no longer routinely recommended.

Effective only against influenza A viruses, but sharply rising resistance rates now preclude their routine use

Pharmacology and Toxicity

Both amantadine and rimantadine are available only as oral preparations. The pharmacokinetics of the two agents is quite different. Amantadine is excreted by the kidney without being metabolized, and its dose must be decreased in patients with impaired renal function. In contrast, rimantadine is metabolized by the liver, then excreted in the kidney, and dosage adjustment for renal failure is not necessary.

Rimantadine is metabolized by the liver

Amantadine is excreted by the kidney

Neuraminidase Inhibitors

Oseltamivir and **zanamivir** are antiviral agents that selectively inhibit the neuraminidase of influenza A and B viruses. The neuraminidase cleaves terminal sialic acid from glycoconjugates and plays a role in the release of virus from infected cells. Zanamivir was the first approved neuraminidase inhibitor. It is given by oral inhalation using a specially designed device. Oseltamivir phosphate is the oral prodrug of oseltamivir, a drug comparable to zanamivir in antineuraminidase activity.

Treatment with either oseltamivir or zanamivir reduces influenza symptoms and shortens the course of illness by 1 to 1.5 days. The activity of these compounds against both influenza A and B offers an advantage over amantadine and rimantadine, which are active only against influenza A.

Neuraminidase inhibitors are effective in treatment and prophylaxis of influenza A and B viruses

Inhibitors of Nucleic Acid Synthesis

At present, most antiviral agents are nucleoside analogs that are active against virus-specific nucleic acid polymerases or transcriptases and have much less activity against analogous host enzymes. Some of these agents serve as nucleic acid chain terminators after incorporation into nucleic acids.

Idoxuridine and Trifluorothymidine

Idoxuridine (5-iodo-2′-deoxyuridine, IUdR) is a halogenated pyrimidine that blocks nucleic acid synthesis by being incorporated into DNA in place of thymidine and producing a nonfunctional molecule (ie, by terminating synthesis of the nucleic acid chain). It is phosphorylated by thymidine kinase to the active compound. Unfortunately, it inhibits both viral and cellular DNA synthesis, and the resulting host toxicity precludes systemic administration in humans. Idoxuridine can be used topically as effective treatment of herpetic infection of the cornea (keratitis). Trifluorothymidine, a related pyrimidine analog, is effective in treating herpetic corneal infections, including those that fail to respond to IUdR. Trifluorothymidine has largely replaced IUdR.

Idoxuridine and trifluorothymidine block DNA synthesis

Acyclovir

This antiviral agent differs from the nucleoside guanosine by having an acyclic (hydroxyethoxymethyl) side chain. Acyclovir is unique in that it must be phosphorylated by thymidine kinase to be active. This phosphorylation occurs only in cells infected by certain herpesviruses. Therefore, the compound is essentially nontoxic because it is not phosphorylated or activated in uninfected host cells. Viral thymidine kinase catalyzes the phosphorylation of acyclovir to a monophosphate. From this point, host cell enzymes complete the progression to the diphosphate and finally the triphosphate.

Activity of acyclovir against herpesviruses directly correlates with the capacity of the virus to induce a thymidine kinase. Susceptible strains of herpes simplex virus types 1 and 2 (HSV-1 and HSV-2) are the most active thymidine kinase inducers and are the most readily inhibited by acyclovir. Cytomegalovirus (CMV) induces little or no thymidine kinase and is not inhibited. Varicella-zoster and Epstein-Barr viruses are between these two extremes in terms of both thymidine kinase induction and acyclovir susceptibility.

Acyclovir is effective against herpesviruses, which induce thymidine kinase

Acyclovir triphosphate inhibits viral replication by competing with guanosine triphosphate and inhibiting the function of the virally encoded DNA polymerase. The selectivity and minimal toxicity of acyclovir are aided by its 100-fold or greater affinity for viral DNA polymerase than for cellular DNA polymerase. A second mechanism of viral inhibition results from incorporation of acyclovir triphosphate into the growing viral DNA chain. This causes termination of chain growth because there is no 3′-hydroxy group on the acyclovir molecule to provide attachment sites for additional nucleotides. Resistant strains of HSV have been recovered from immunocompromised patients, including patients with acquired immunodeficiency syndrome (AIDS); in most instances, resistance results from mutations in the viral thymidine kinase gene, rendering it inactive in phosphorylation. Resistance may also result from mutations in the viral DNA polymerase. Resistant virus has rarely, if ever, been recovered from immunocompetent patients, even after years of drug exposure.

Acyclovir inhibits viral DNA polymerase and terminates viral DNA chain growth

Pharmacology and Toxicity Acyclovir is available in three forms: topical, oral, and parenteral. Topical acyclovir is rarely used. The oral form has low bioavailability (approximately 10%) but achieves concentrations in blood that inhibit HSV and, to a lesser extent, varicella-zoster virus (VZV). Intravenous acyclovir is used for serious HSV infection (eg, congenital), encephalitis, and VZV infection in immunocompromised patients. Because acyclovir is excreted by the kidney, the dosage must be reduced in patients with renal failure. Central nervous system toxicity and renal toxicity have been reported in patients treated with prolonged high intravenous doses. Acyclovir is remarkably free of bone marrow toxicity, even in patients with hematopoietic disorders.

Intravenous acyclovir used in serious HSV infections

Treatment and Prophylaxis Acyclovir is effective in the treatment of primary HSV mucocutaneous infections or for severe recurrences in immunocompromised patients. The agent is also useful in neonatal infectious herpes encephalitis, and it is also recommended for VZV infection in immunocompromised patients and for varicella in older children or adults. Acyclovir is beneficial against herpes zoster in elderly patients or any patient with eye involvement. In patients with frequent severe genital herpes, the oral form is effective in preventing recurrences. Because it does not eliminate the virus from the host, it must be taken daily to be effective. Acyclovir is minimally effective in the treatment of recurrent genital or labial herpes in otherwise healthy individuals.

Valacyclovir, Famciclovir, and Penciclovir

Valacyclovir is a prodrug of acyclovir that is better absorbed and therefore can be used in lower and less frequent dosage. Once absorbed, it becomes acyclovir. It is currently approved for use in HSV and VZV infections in immunocompetent adult patients. Dosage adjustment is necessary in patients with impaired renal function.

Famciclovir is similar to acyclovir in its structure and requirement for phosphorylation but differs slightly in its mode of action. After absorption, the agent is converted to penciclovir, the active moiety, which is also a competitive inhibitor of a guanosine triphosphate. However, it does not irreversibly terminate DNA replication. Famciclovir is currently approved for treatment of HSV and VZV infections. **Penciclovir** is approved for topical treatment of recurrent herpes labialis.

Agents that are similar to or become acyclovir after absorption are available

Ganciclovir

Ganciclovir (DHPG), a nucleoside analog of guanosine, differs from acyclovir by a single carboxyl side chain. This structural change confers approximately 50 times more activity against CMV. Acyclovir has low activity against CMV because it is not well phosphorylated in CMV-infected cells owing to the absence of the gene for thymidine kinase in CMV. However, ganciclovir is active against CMV and does not require thymidine kinase for phosphorylation. Rather, another viral-encoded phosphorylating enzyme (UL97) is present in CMV-infected cells, which is capable of phosphorylating ganciclovir and converting it to

Ganciclovir does not require viral thymidine kinase for phosphorylation

the monophosphate. Then cellular enzymes convert it to the active compound, ganciclovir triphosphate, which inhibits the viral DNA polymerase.

Oral ganciclovir is available but is inferior to the intravenous form. Oral valganciclovir, a prodrug of ganciclovir, has improved bioavailability and is equivalent to the intravenous form. Toxicity frequently limits therapy. Neutropenia, which is usually reversible, may occur early but often develops during later therapy. Discontinuation of therapy is necessary in patients whose neutrophils do not increase during dosage reduction or in response to cytokines. Thrombocytopenia (platelet count less than 20,000/mm^3) occurs in approximately 15% of patients.

Neutropenia and thrombocytopenia limit use

Clinical Use Administration of ganciclovir is indicated for the treatment of active CMV infection in immunocompromised patients, but other herpesviruses (particularly HSV-1, HSV-2, and VZV) are also susceptible. Because AIDS patients with severe CMV infection frequently have concurrent illnesses caused by other herpesviruses, treatment with ganciclovir may benefit associated HSV and VZV infections.

Resistance After several months of continuous ganciclovir therapy for treatment of CMV, between 5% and 10% of AIDS patients excrete resistant strains of CMV. In almost all isolates, a mutation is found in the phosphorylating gene, and in a lesser number a mutation may also be found in the viral DNA polymerase. Most of these strains remain sensitive to foscarnet, which may be used as alternate therapy. If only a UL97 mutation is present, the strains remain susceptible to the nucleotide analog cidofovir (see later in chapter); however, most of the strains with a ganciclovir-induced mutation in DNA polymerase are cross-resistant to cidofovir. Many clinicians tend to assume that when a patient with CMV retinitis has progression of the disease during treatment, viral resistance has developed. Progression of CMV disease during treatment is probably the result of many factors, only one of which is the susceptibility of the CMV strain to the drug. Blood and tissue concentrations of ganciclovir, penetration of ganciclovir into the retinal tissue, and the host immune response probably play important roles in determining when clinical progression of CMV disease occurs. Ganciclovir resistance has been noted in transplant recipients, especially those requiring prolonged treatment.

CMV mutant resistance increases with continuous therapy

Inhibitor of Viral RNA Synthesis: Ribavirin

Ribavirin is another analog of the nucleoside guanosine. Unlike acyclovir, which replaces the ribose moiety with a hydroxymethyl acyclic side chain, ribavirin differs from guanosine in that the base ring is incomplete and open. Like other purine nucleoside analogs, ribavirin must be phosphorylated to mono-, di-, and triphosphate forms. It is active against a broad range of viruses in vitro, but its in vivo activity is limited. The mechanism of the antiviral effect of ribavirin is not as clear as that of acyclovir. The triphosphate is an inhibitor of RNA polymerase, and it also depletes cellular stores of guanine by inhibiting inosine monophosphate dehydrogenase, an enzyme important in the synthetic pathway of guanosine. Still another mode of action is by decreasing synthesis of the mRNA 5′ cap because of interference with both guanylation and methylation of the nucleic acid base.

Ribavirin has several modes of action

Aerosol administration enables ribavirin to reach concentrations in respiratory secretions up to 10 times greater than necessary to inhibit viral replication and substantially higher than those achieved with oral administration. Problems encountered with aerosolized ribavirin include precipitation of the agent in tubing used for administration and exposure of health care personnel.

Ribavirin may be somewhat beneficial if given early by aerosol to immunocompromised patients who are infected with respiratory syncytial virus. Oral and intravenous forms have been used for patients with Lassa fever, although studies have been limited. In a recent trial of hantavirus treatment, ribavirin was ineffective. The oral form has limited activity against hepatitis C as monotherapy but provides additional benefit when combined with interferon alpha. A reversible anemia has been associated with oral administration of ribavirin.

Ribavirin is active against respiratory syncytial virus, Lassa fever virus, and hepatitis C

INHIBITORS OF HIV

Nucleoside Reverse Transcriptase Inhibitors

Azidothymidine

Azidothymidine (AZT), a nucleoside analog of thymidine, inhibits the reverse transcriptase of HIV. As with other nucleosides, AZT must be phosphorylated; host cell enzymes carry out the process. The basis for the relatively selective therapeutic effect of AZT is that HIV reverse transcriptase is more than 100 times more sensitive to AZT than is host cell DNA polymerase. Nonetheless, toxicity frequently occurs.

AZT is now used only in combination therapy

AZT was the first useful treatment for HIV infection, but now is recommended for use only in combination with other inhibitors of HIV replication (eg, lamivudine and protease inhibitors). Toxicity includes malaise, nausea, and bone marrow toxicity. All hematopoietic components may be depressed, but usually reverse with discontinuation of the drug or dose reduction. Resistance is associated with one or more mutations in the HIV reverse transcriptase gene.

Didanosine and Zalcitabine

ddI and ddC are always used in combination with other anti-HIV drugs

Didanosine (ddI, dideoxyinosine) and zalcitabine (ddC, dideoxycytidine) are nucleoside analogs that inhibit HIV replication. After intracellular phosphorylation by host enzymes to their active triphosphate form, they block viral replication by inhibiting viral reverse transcriptase, such as zidovudine. Serious adverse effects of treatment include peripheral neuropathy with either ddI or ddC, and pancreatitis with ddI; both conditions are dose-related. Dose reduction is required for impaired renal function. As with other anti-HIV drugs, these agents should be used only in combination with one or two other anti-HIV drugs to limit the development of resistance and to enhance antiviral effect.

Stavudine

D4T is a reverse transcriptase inhibitor that also terminates chain growth

Stavudine (D4T) is another nucleoside analog that inhibits HIV replication. D4T is phosphorylated by cellular enzymes to an active triphosphate form that interferes with viral reverse transcriptase. It also terminates the growth of the chain of viral nucleic acid. D4T is well absorbed and has a high bioavailability. Adverse effects include headache, nausea and vomiting, asthenia, confusion, and elevated serum transaminase and creatinine kinase. A painful sensory peripheral neuropathy that appears to be dose-related has also been noted. Dose reduction is required for impaired renal function. D4T should be used only in combination with other anti-HIV agents.

Lamivudine

3TC suppresses development of AZT resistance

Lamivudine (3TC), another reverse transcriptase inhibitor, is a comparatively safe and usually well-tolerated agent. It is used in combination with AZT or other nucleoside analogs. AZT and 3TC have a unique interaction; 3TC suppresses the development and persistence of AZT resistance mutations. When combined with interferon alpha, 3TC is also useful for treating hepatitis B.

Non-Nucleoside Reverse Transcriptase Inhibitors (NNRTIs)

NNRTIs are often active against AZT-resistant strains

Rapid development of drug resistance occurs when NNRTIs are used alone

Compounds that are not nucleoside analogs also inhibit HIV reverse transcriptase. Several compounds, such as nevirapine, delavirdine, and efavirenz, have been evaluated alone or in combination with other nucleosides. These compounds are very active against HIV-1, do not require cellular enzymes to be phosphorylated, and bind to essentially the same site on reverse transcriptase. They are active against both AZT-resistant and AZT-sensitive isolates. In addition, most of these compounds do not inhibit human DNA polymerase and are not cytotoxic at concentrations required for effective antiviral activity. Therefore, they are relatively nontoxic. Unfortunately, drug resistance readily emerges with even a single passage of virus in the presence of drug in vitro and in vivo. Thus, NNRTIs should be used only in combination regimens with other drugs active against HIV.

Protease Inhibitors

The newest agents that inhibit HIV are the protease inhibitors. These agents block the action of the viral-encoded enzyme protease, which cleaves polyproteins to produce structural proteins. Inhibition of this enzyme leads to blockage of viral assembly and release. The protease inhibitors are potent suppressors of HIV replication in vitro and in vivo, particularly when combined with other antiretroviral agents.

Protease inhibitors block viral-encoded proteases

In late 1995, **saquinavir** was the first protease inhibitor to receive approval. **Ritonavir, indinavir,** and **nelfinavir** are other potent protease inhibitors that have since been released. These drugs may cause hepatotoxicity and all agents inhibit P450, resulting in important drug interactions. Because drug resistance to all protease inhibitors develops, these agents should not be used alone without other anti-HIV drugs.

Used in combination with other anti-HIV drugs

Nucleotide Analogs: Cidofovir

In recent years, a new series of antiviral compounds, the nucleotide analogs, have been developed. The best-known example of this class of compounds is **cidofovir.** This compound mimics a monophosphorylated nucleotide by having a phosphonate group attached to the molecule. This appears to the cell as a nucleoside monophosphate, or nucleotide, and cellular enzymes then add two phosphate groups to generate the active compound. In this form, the drug inhibits both viral and cellular nucleic acid polymerases, but selectivity is provided by its higher affinity for the viral enzyme.

Cidofovir inhibits viral DNA polymerase

Nucleotide analogs do not require phosphorylation, or activation, by a viral-encoded enzyme and remain active against viruses that are resistant due to mutations in codons for these enzymes. Resistance can develop with mutations in the viral DNA polymerase, UL54. An additional feature of cidofovir is a very prolonged half-life as a result of slow clearance by the kidneys.

Cidofovir is approved for intravenous therapy of CMV retinitis, and maintenance treatment may be given as infrequently as every two weeks. It is also occasionally used to treat severe, disseminated adenovirus infections. Nephrotoxicity is a serious complication of cidofovir treatment, and patients must be monitored carefully for evidence of renal impairment.

OTHER ANTIVIRAL AGENTS

Foscarnet

Foscarnet, also known as phosphonoformate, is a pyrophosphate analog that inhibits viral DNA polymerase by blocking the pyrophosphate-binding site of the viral DNA polymerase and preventing cleavage of pyrophosphate from deoxyadenosine triphosphate. This action is relatively selective; CMV DNA polymerase is inhibited at concentrations less than 1% of that required to inhibit cellular DNA polymerase. Unlike such nucleosides as acyclovir and ganciclovir, foscarnet does not require phosphorylation to be an active inhibitor of viral DNA polymerases. This biochemical fact becomes especially important with regard to viral resistance, because the principal mode of viral resistance to nucleoside analogs is a mutation that eliminates phosphorylation of the drug in virus-infected cells. Thus, foscarnet can usually be used to treat patients with ganciclovir-resistant CMV and acyclovir-resistant HSV. Excretion is entirely renal without a hepatic component, and dosage must be decreased in patients with impaired renal function.

Foscarnet inhibits viral DNA polymerases

Effective against resistant CMV and HSV

Interferons

Interferons are host cell–encoded proteins synthesized in response to double-stranded RNA that circulate to protect uninfected cells by inhibiting viral protein synthesis. Ironically, interferons harvested in tissue culture were the first antiviral agents, but their clinical activity was disappointing. Recombinant DNA techniques now allow relatively inexpensive large-scale production of interferons by bacteria and yeasts: ::: therapeutic interferons, p. 163.

Recombinant DNA techniques allow large-scale production

Interferons inhibit viral protein synthesis

Interferon alpha is combined with ribavirin to treat chronic hepatitis C

Interferon alpha is beneficial in the treatment of chronic active hepatitis B and C infection, although its efficacy is often transient. Combinations of interferon alpha with 3TC, famciclovir, and other nucleosides are being evaluated for treatment of hepatitis B. Pegylated interferon alpha is given for 6 to 12 months to treat chronic hepatitis C disease, and combination with ribavirin usually produces improved results. Topical or intralesional interferon application is beneficial in the treatment of human papilloma virus infections. Parenteral use can cause symptomatic systemic toxicity (eg, fever, malaise), partly because of its effect on host cell protein synthesis.

Fomivirsen

Fomivirsen inhibits CMV mRNA

Fomivirsen, the first antisense compound to be approved for use in human infection, is a synthetic oligonucleotide, complementary to and presumably inhibiting a coding sequence in CMV messenger RNA (mRNA). The major immediate early transcriptional unit of CMV encodes several proteins responsible for regulation of viral gene expression. Presumably, fomivirsen inhibits production of these proteins. In this agent, oligonucleotide phosphorothioate linkages replace the usual nucleases. Fomivirsen, which exhibits greater antiviral activity than ganciclovir on a molar basis, is approved for the local (intravitreal) therapy of CMV retinitis in patients who have failed other therapies.

ANTIVIRAL RESISTANCE

Herpesviruses often develop resistance by mutations in phosphorylation

Viral genomes and their replication, as well as the mechanisms of action of the available antiviral agents, have been intensively studied. Accordingly, an understanding of resistance to antiviral drugs has evolved; investigation of resistance mechanisms has shed light on the function of specific viral genes. For example, it has become clear that a common mechanism of resistance to nucleosides (eg, acyclovir and ganciclovir) by herpesviruses consists of mutations in the viral-induced enzyme responsible for phosphorylating the nucleoside. For HSV, this is thymidine kinase; for CMV, this gene is designated UL97.

Genetic alterations (ie, mutations or deletions) are the basis for antiviral resistance. The likelihood of resistant mutants results from at least four functions:

1. **Rate of viral replication.** Herpesviruses, especially CMV and VZV do not replicate as rapidly as HIV and hepatitis B and C. Higher rates of replication are associated with higher rates of spontaneous mutations.
2. **Selective pressure of the drug.** The selective pressure increases the probability of mutations to the point that virus replication is substantially reduced.
3. **Rate of viral mutations.** In addition to viral replication, the rate of mutations differs among different viruses. In general, single-stranded RNA viruses (eg, HIV and influenza) have more rapid rates of mutation than double-stranded DNA viruses (eg, HSV).
4. **Rates of mutation in differing viral genes.** For example, within the herpesviruses, the genes for phosphorylating nucleosides (eg, UL97) are more susceptible to mutation than the viral DNA polymerase.

Resistance to antiviral agents may be detected in several ways:

Phenotypic resistance is detected by in vitro methods

- **Phenotypic.** This is the traditional method of growing virus in tissue culture in medium containing increasing concentrations of an antiviral agent. The concentration of the agent that reduces viral replication by 50% is the end point and is referred to as the inhibitory concentration (IC_{50}). The IC_{50} of resistant virus is higher than that of susceptible virus. The degree of viral replication is obtained by counting viral plaques (ie, equivalent to viral "colonies") or by measuring viral antigen or nucleic acid concentration. Unfortunately, phenotypic assays are very time-consuming, requiring days to weeks for completion. IC_{50} values increase as the percent of the viral population with the mutation increases.
- **Genotypic.** When the exact mutation or deletion responsible for antiviral resistance is known, it is possible to sequence the viral gene or detect it with restriction enzyme patterns. These tests are rapid but require knowledge of the expected mutation, and they do

not provide quantitation of the percent of the viral population harboring the mutation. If only 1% to 5% of the population has the mutation, this result may not be clinically significant when compared with a virus population that is 90% mutated.

Genotypic = molecular detection of expected mutation

- **Viral quantitation in response to treatment.** Various methods of quantitating virus (eg, culture, polymerase chain reaction, antigen assay) provide a means of assessing the decline of viral titer in response to treatment with an antiviral agent. These assays are rapid and do not require knowledge of the expected mutation. If no decline occurs despite adequate dosage and compliance, viral resistance may be responsible. Likewise, if viral titer initially decreases but subsequently recurs and/or increases, then resistance may have developed.

No reduction or increase in patient's viral burden suggests development of resistant mutants

CHAPTER 9

Influenza, Parainfluenza, Respiratory Syncytial Virus, Adenovirus, and Other Respiratory Viruses

Considering how common illness is, how tremendous the spiritual change that it brings, how astonishing, when the lights of health go down, the undiscovered countries that are then disclosed, what wastes and deserts of the soul a slight attack of influenza brings to view…

—Virginia Woolf, "On Being Ill"

Respiratory disease accounts for an estimated 75% to 80% of all acute morbidity in the US population. Most of these illnesses (approximately 80%) are viral. If episodes not requiring medical attention are included, the overall average is three to four illnesses per year per person, although incidence varies inversely with age (ie,greater among young children). Seasonality is also a feature; incidence is lowest in the summer months and highest in the winter.

Respiratory viruses represented by diverse agents

The viruses that are major causes of acute respiratory disease (ARD) include influenza viruses, parainfluenza viruses, rhinoviruses, adenoviruses, respiratory syncytial virus (RSV), human metapneumovirus (hMPV), and respiratory coronaviruses. Recently, bocaviruses have also been associated with acute respiratory illness. Reoviruses are of questionable importance but are also considered. Others, such as enterovirus and measles virus, can also cause respiratory symptoms but are discussed in other chapters.

Short incubation period

Transmission by droplet nuclei or hands

In addition to the ability to cause a variety of ARD syndromes, this somewhat heterogeneous group of viruses shares a relatively short incubation period (1-4 days) and a person-to-person mode of spread. Transmission is direct, by infective droplet nuclei, or indirect, by hand transfer of contaminated secretions to nasal or conjunctival epithelium. All these agents are associated with an increased risk of bacterial superinfection of the damaged tissue of the respiratory tract, and all have a worldwide distribution.

INFLUENZA VIRUSES

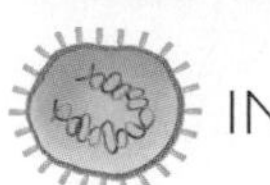

INFLUENZA VIRUS GROUP CHARACTERISTICS

Orthomyxoviruses divided into types A, B, and C

Type A has greatest virulence and epidemic spread

Influenza viruses are members of the **orthomyxovirus** group, which are enveloped, pleomorphic, single-stranded negative-sense RNA viruses. They are classified into three major serotypes, A, B, and C, based on different ribonucleoprotein antigens. Influenza A viruses are the most extensively studied, and much of the following discussion is based on knowledge of this type. They generally cause more severe disease and more extensive epidemics than the other types; naturally infect a wide variety of species, including mammals and birds; and have a great tendency to undergo significant antigenic changes (**Table 9–1**). Influenza B viruses are more antigenically stable, are only known to naturally infect humans, and usually occur in more localized outbreaks. Influenza C viruses appear to be relatively minor causes of disease, affecting humans and pigs.

Enveloped RNA virus with segmented genome

Virus-specified hemagglutinin and neuraminidase spikes

Influenza A and B viruses each consist of a nucleocapsid containing eight segments of negative-sense, **single-stranded RNA,** which is enveloped in a glycolipid membrane derived from the host cell plasma membrane. The inner side of the envelope contains a layer of virus-specified protein (M1). Two virus-specified glycoproteins, hemagglutinin (HA or H) and neuraminidase (NA or N), are embedded in the outer surface of the envelope and appear as "spikes" over the surface of the virion. **Figure 9–1** illustrates the makeup of influenza A virus. Influenza B is somewhat similar but has a unique NB protein instead of M2. Influenza C differs from the others in that it possesses only seven RNA segments and has no neuraminidase, although it does possess other receptor-destroying capability (see text that follows). In addition, the hemagglutinin of influenza C binds to a cell receptor different from that for types A and B.

Hemagglutinin acts in viral attachment

Figure 9–2 illustrates the replication cycle of influenza virus. The virus-specified glycoproteins are antigenic and have special functional importance to the virus. **Hemagglutinin** is so named because of its ability to agglutinate red blood cells from certain species (eg, chickens and guinea pigs) in vitro. Its major biologic function is to serve as a point of attachment to *N*-acetylneuraminic (sialic) acid–only containing glycoprotein or glycolipid receptor sites on human respiratory cell surfaces, which is a critical first step in initiating infection of the cell.

Neuraminidase has role in envelope fusion and viral release

Neuraminidase is an antigenic hydrolytic enzyme that acts on the hemagglutinin receptors by splitting off their terminal neuraminic (sialic) acid. The result is destruction of receptor activity, which may help in preventing superinfection of the infected cell. Neuraminidase serves several functions. It may inactivate a free mucoprotein receptor substance in respiratory secretions that could otherwise bind to viral hemagglutinin and prevent access of the virus to the cell surface. Neuraminidase is important in fusion of the viral envelope with the host cell membrane as a prerequisite to viral entry. It also aids in the release of newly formed virus particles from infected cells, thus making them available to infect other cells. Type-specific antibodies to neuraminidase appear to inhibit the spread of virus in the infected host and to limit the amount of virus released from host cells.

TABLE 9–1 Differences Among Influenza Viruses

FEATURE	INFLUENZA A	INFLUENZA B	INFLUENZA C
Gene segments	8	8	7
Unique proteins	M2	NB	HEF
Host range	Humans, swine, avians, equines, marine mammals	Humans only	Humans, swine
Disease severity	Often severe	Occasionally severe	Usually mild
Epidemic potential	Extensive; epidemics and pandemics (antigenic drift and shift)	Outbreaks; occasional epidemics (antigenic drift only)	Limited outbreaks (antigenic drift only)

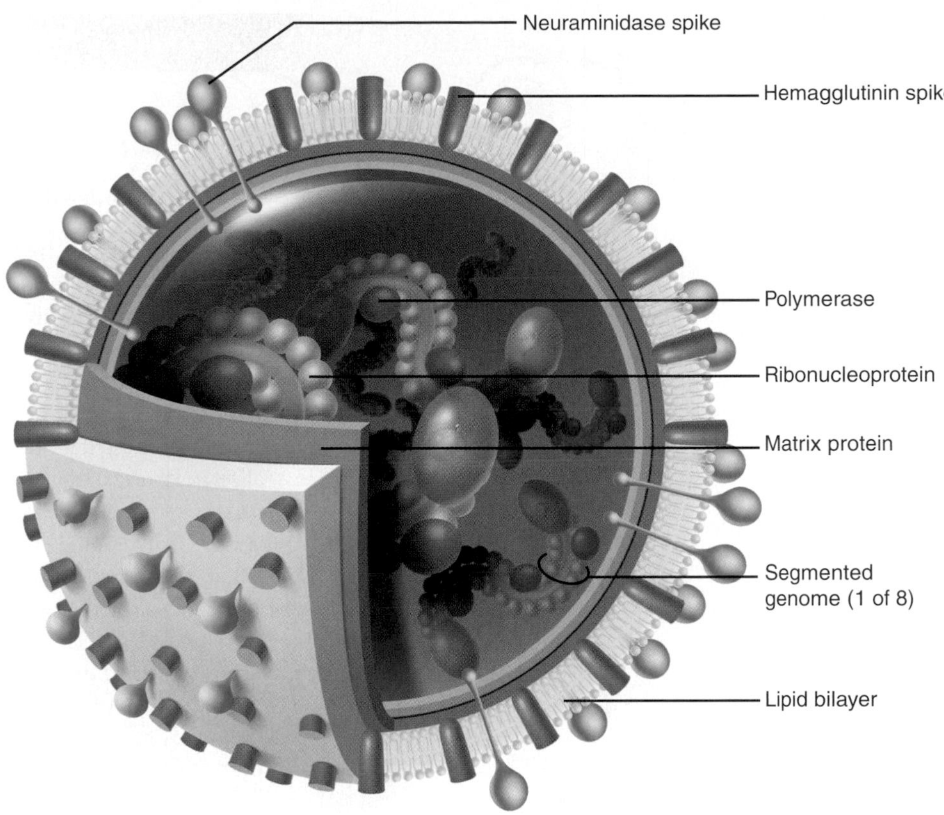

FIGURE 9–1. Diagrammatic view of influenza A virus. Three types of membrane proteins are inserted in the lipid bilayer: hemagglutinin (as trimer), neuraminidase (as tetramer), and M2 ion channel protein. The eight ribonucleoproteinz segments each contain viral RNA surrounded by nucleoprotein and associated with RNA transcriptase. (Reproduced with permission from Willey J, Sherwood L, Woolverton C (eds). *Prescott's Principles of Microbiology.* New York: McGraw-Hill; 2008.)

Nucleocapsid assembly takes place in the cell nucleus, but final virus assembly takes place at the plasma membrane. The ribonucleoproteins are enveloped by the plasma membrane, which by then contains hemagglutinin and neuraminidase. Virus "buds" are formed, and intact virions are released from the cell surface (Figure 9–2).

Nucleocapsid and virus assembly occur at different cell sites

Influenza A viruses were initially isolated in 1933 by intranasal inoculation of ferrets, which developed febrile respiratory illnesses. The viruses replicate in the amniotic sac of embryonated hen's eggs, where their presence can be detected by the hemagglutination test. Most strains can also be readily isolated in cell culture systems, such as primary monkey kidney cells. Some cause cytopathic effects in culture.

Viral isolation in eggs or cell cultures

The most efficient method of detection is demonstration of hemadsorption by adherence of erythrocytes to infected cells expressing hemagglutinin or by agglutination of erythrocytes by virus already released into the extracellular fluid. The virus can then be identified specifically by inhibition of these properties by addition of antibody directed specifically at the hemagglutinin. This method is called **hemadsorption inhibition** or **hemagglutination inhibition,** depending on whether the test is performed on infected cells or on extracellular virus, respectively. Because the hemagglutinin is antigenic, hemagglutination inhibition tests can also be used to detect antibodies in infected subjects. Research has shown that antibody directed against specific hemagglutinin is highly effective in neutralizing the infectivity of the virus.

Hemadsorption and hemagglutination inhibition used to detect presence of virus

Antihemagglutinin antibodies detectable in serum

Influenza A

Influenza A is considered in detail because of its great clinical and epidemiologic importance.

The influenza A virion contains eight segments of single-stranded RNA with defined genetic responsibilities. These functions include coding for virus-specified proteins (Figure 9–1; **Table 9–2**). A unique aspect of influenza A viruses is their ability to develop a wide variety of subtypes through the processes of mutation and whole-gene "swapping" between strains, called **reassortment.** Recombination, which occurs when new genes are assembled from sections of other genes, is thought to occur rarely, if at all. These processes result in antigenic changes called **drifts** and **shifts,** which are discussed shortly. ::: reassortment, p. 133

Influenza A genome in multiple segments

Mutability of virus produces antigenic changes

1. The endonuclease activity of the PB1 protein cleaves the cap and about 10 nucleotides from the 5′ end of host mRNA (cap snatching). The fragment is used to prime viral mRNA synthesis by the RNA-dependent RNA polymerase activity of the PB1 protein.
2. Viral mRNA is translated. Early products include more NP and PB1 proteins.
3. RNA polymerase activity of the PB1 protein synthesizes +ssRNA from genomic −ssRNA molecules.
4. RNA polymerase activity of the PB1 protein synthesizes new copies of the genome using +ssRNA made in step 3 as templates. Some of these new genome segments serve as templates for the synthesis of more viral mRNA. Later in the infection, they will become progeny genomes.
5. Viral mRNA molecules transcribed from other genome segments encode structural proteins such as hemagglutinin (HA) and neuraminidase (NA). These messages are translated by ER-associated ribosomes and delivered to the cell membrane.
6. Viral genome segments are packaged as progeny virions bud from the host cell.

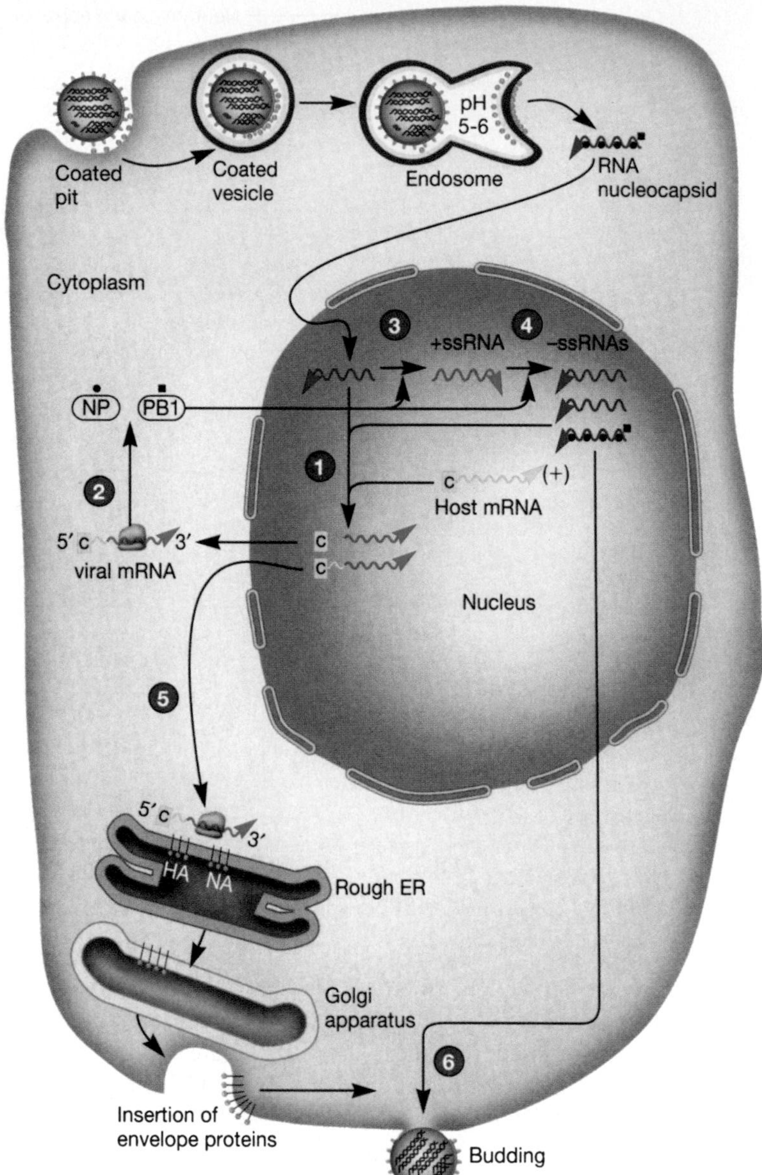

FIGURE 9–2. Diagrammatic view of influenza virus life cycle. (Reproduced with permission from Willey J, Sherwood L, Woolverton C (eds). *Prescott's Principles of Microbiology*. New York: McGraw-Hill; 2008.)

TABLE 9–2 Virus-Coded Proteins of Influenza A

RNA SEGMENT	PROTEINS	FUNCTION
1	PB2	RNA synthesis, ? virulence
2	PB1	RNA synthesis
3	PA	RNA synthesis
4	HA	Attachment
5	NP	RNA synthesis
6	NA	Virus release from infected cells
7	M1, M2	Matrix
8	NS1, NS2	Nonstructural; NS1 is interferon antagonist

The 15 recognized subtypes of hemagglutinin (H) and 9 neuraminidase (N) subtypes known to exist among influenza A viruses that circulate in birds and mammals represent a reservoir of viral genes that can undergo reassortment, or "mixing" with human strains. Although all 15 subtypes of hemagglutinins and 9 subtypes of neuraminidases have been identified in aquatic birds, pigs are infected with two major hemagglutinins (H1 and H3) and neuraminidases (N1 and N2) and horses with two H (H3 and H7) and two N (N7 and N8). Three hemagglutinins **(H1, H2, and H3)** and two neuraminidases **(N1 and N2)** appear to be of greatest importance in **human infections**. These subtypes are designated according to the H and N antigens on their surface (eg, H1N1, H3N2). There may also be more subtle, but sometimes important, antigenic differences (drifts) within each subtype. These differences are designated according to the major representative virus to which they are most closely related antigenically, using the place of initial isolation, number of the isolate, and year of detection. For example, two H3N2 strains that differ antigenically only slightly are A/Texas/1/77(H3N2) and A/Bangkok/1/79(H3N2).

Subtypes based on H and N antigens

Subtle changes known as antigenic drift

Antigenic drifts within major subtypes can involve either H or N antigens, as well as the genes encoding other structural and nonstructural proteins, and may result from as little as a single mutation in the viral RNA. These mutations are caused by viral RNA polymerase because it lacks proof reading ability. The mutant may come to predominate under selective immunologic pressures in the host population (**Figure 9–3**). Such drifts are common among influenza A viruses, occurring at least every few years and sometimes even during the course of a single epidemic. Drifts can also develop in influenza B viruses but considerably less frequently. ::: antigenic drift, p. 131

Antigenic drift every few years with type A

In contrast to the frequently occurring mutations that cause antigenic drift among influenza A strains, major changes (more than 50%) in the nucleotide sequences of the H or N genes can occur suddenly and unpredictably. These are referred to as antigenic shifts. (Figure 9–3 illustrates the difference between antigenic drifts and shifts.) They almost certainly result from reassortment that can be readily reproduced in the laboratory. Simultaneously infecting a cell with two influenza A subtypes yields progeny that contain antigens derived from either of the original viruses. For example, a cell infected simultaneously with influenza A (H3N2) and influenza A (H1N1) may produce a mixture of influenza viruses of the subtypes H3N2, H1N1, H1N2, and H3N1. When "new" epidemic strains emerge, they most likely have circulated into animal or avian reservoirs, where they have undergone genetic reassortment (and sometimes also mutation) and then are readapted and spread to human hosts when a sufficient proportion of the population has little or no immunity to the "new" subtypes. A recent example was the appearance in Hong Kong in 1997 of human cases caused by an avian influenza A (H5N1). The global spread of avian influenza (H5N1 and others) continued through 1997 and onward with several more cases every year. Studies indicated that all RNA segments were derived from an avian influenza A virus, but a single insert coding for several additional amino acids in the hemagglutinin protein facilitated cleavage by human cellular enzymes. In addition, a single amino acid substitution in the PB2 polymerase protein occurred. These two mutations together made the virus more virulent for humans; fortunately, human-to-human transmission was poor. ::: antigenic shift, p. 133

Major antigenic shifts due to reassortment

New subtype may also develop mutations

H1N1 and H5N1 target different regions of the respiratory tract

Additional molecular barriers limit human-to-human transmission of avian influenza virus (H5N1). One of the most important barriers is that avian and human influenza viruses target different regions of the human respiratory tract. Although the receptor for influenza viruses is sialic acid (SA) glycoprotein, there is a major difference in the sialic acid sugar positions with SA α 2,6 galactose for human influenza virus and SA α 2,3 galactose for avian influenza virus (H5N1). Human influenza virus receptor, SA α 2,6 galactose, is dominant on epithelial cells of nasal mucosa, paranasal sinuses, pharynx, trachea and bronchi, whereas H5N1 receptor, SA α 2,3 galactose is mainly found on nonciliated bronchiolar cells at the junction between respiratory bronchioles and alveolus. It is interesting that A/Hong Kong/213/03 (H5N1) isolated from a patient recognizes both SA α 2,6 galactose and SA α 2,3 galactose and bound extensively to both bronchial and alveolar cells.

Although receptors for H1N1 are dominant in the upper part of the respiratory tract, H5N1 receptors are found in the lower portion of the lung in humans

Major antigenic shifts, which occurred approximately every 8 to 10 years in the 20th century, often resulted in serious epidemics or pandemics among populations with little or no preexisting antibody to the new subtypes. Examples include the appearance of an H1N1 subtype in 1947, followed by an abrupt shift to an H2N2 strain in 1957, which caused the pandemic of Asian flu. A subsequent major shift in 1968 to an H3N2 subtype (the Hong Kong flu) led to another, but somewhat less severe epidemic. The Russian flu,

Major antigenic shifts correlate with epidemics

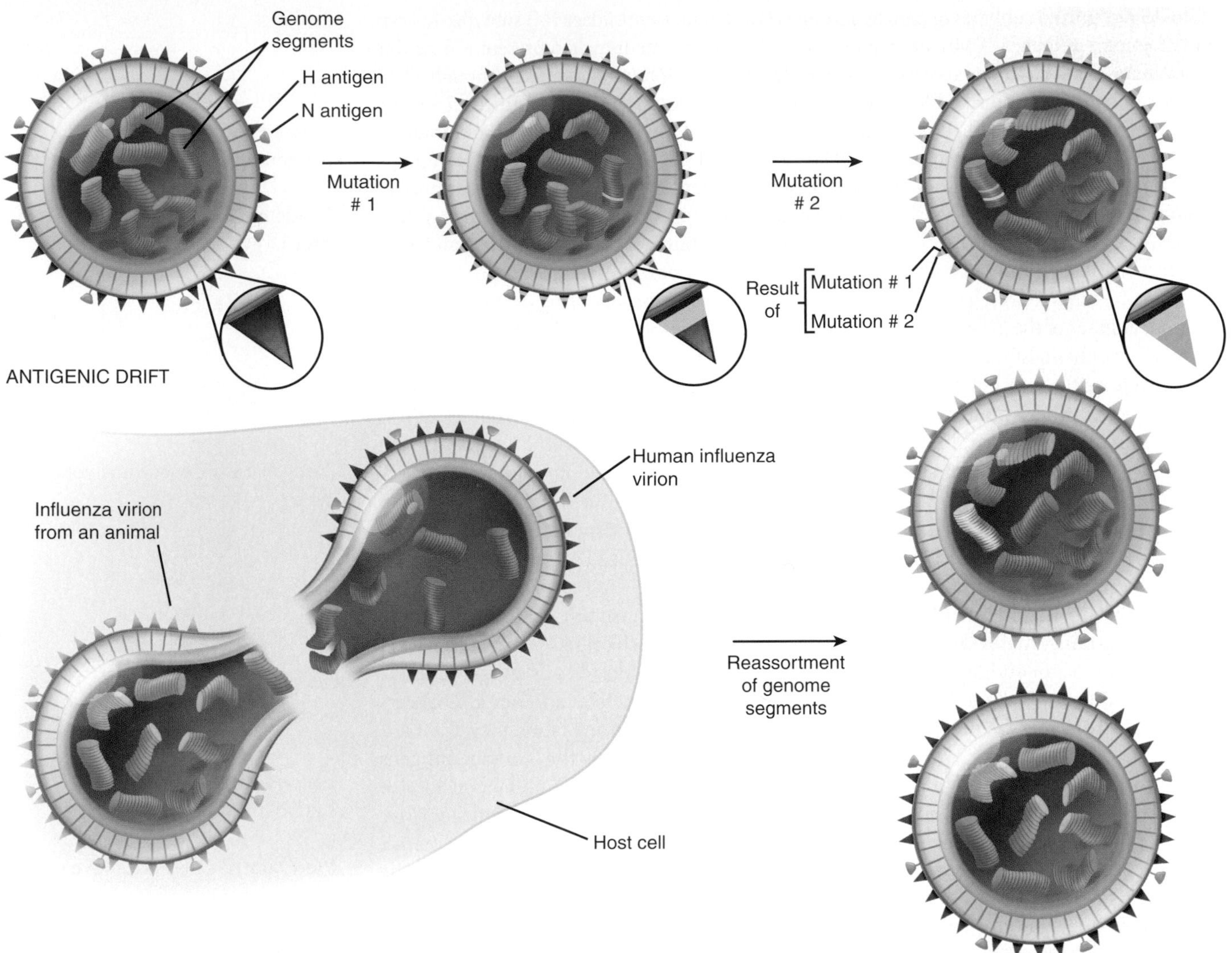

FIGURE 9–3. Influenza virus: antigenic drift and antigenic shift. With drift, repeated mutations cause a gradual change in the antigens composing hemagglutinin, so that antibody against the original virus becomes progressively less effective. With shift, there is an abrupt, major change in the hemagglutinin antigens because the virus acquires a new genome segment, which in this case codes for hemagglutinin. Changes in neuraminidase could occur by the same mechanism. (Reproduced with permission from Nester EW, Anderson DG, Roberts CE Jr, Nester MT. *Microbiology: A Human Perspective,* 6th ed. New York: McGraw-Hill; 2008.)

which appeared in late 1977, was caused by an H1N1 subtype very similar to that which dominated between 1947 and 1957 (**Table 9–3**). In April, 2009, a previously unrecognized H1N1 strain was detected in Mexico and the southwestern United States. It is a reassortant that contains genetic components from four different flu viruses—North American swine influenza, North American avian influenza, human influenza and swine influenza virus of Eurasian origin . Over the subsequent 3 months, this strain, designated swine-origin 2009 A(H1N1) rapidly spread globally.

Minor antigenic drifts allow maintenance in population

Individual variation is significant

The concepts of antigenic shift and drift in human influenza A virus infections can be approximately summarized as follows. Periodic shifts in the major antigenic components appear, usually resulting in major epidemics in populations with little or no immunologic experience with the subtype. As the population of susceptible individuals is exhausted (ie, subtype-specific immunity is acquired by increasing numbers of people), the subtype continues to circulate for a time, undergoing mutations with subtle antigenic drifts from season to season. This allows some degree of virus transmission to continue. Infectivity persists because subtype-specific immunity is not entirely protective against drifting strains;

TABLE 9–3 Major Antigenic Shifts Associated With Influenza A Pandemics, 1947-1987

YEAR	SUBTYPE	PROTOTYPE STRAIN
1947	H1N1	A/FM1/47
1957	H2N2	A/Singapore/57
1968	H3N2	A/Hong Kong/68
1977	H1N1	A/USSR/77
1987	H3N2	No pandemic occurred; various strains of H1N1 and H3N2 continue circulating worldwide through 2008; a new, pandemic swine-origin H1N1 strain emerged in 2009

for example, an individual may have antibodies reasonably protective against influenza A/Texas/77(H_3N_2), yet be susceptible in succeeding years to reinfection by influenza A/Bangkok/79(H_3N_2). Eventually, however, the overall immunity of the population becomes sufficient to minimize the epidemic potential of the major subtype and its drifting strains. Unfortunately, the battle is never entirely won; the scene is set for the sudden and usually unpredictable appearance of an entirely new subtype that may not have circulated among humans for 20 years or more.

INFLUENZA

CLINICAL CAPSULE

Influenza virus types A and B both cause more severe symptoms than does influenza virus type C. The typical illness is characterized by an abrupt onset (over several hours) of fever, diffuse muscle aches, and chills. This is followed within 12 to 36 hours by respiratory signs, such as rhinitis, cough, and respiratory distress. The acute phase usually lasts 3 to 5 days, but a complete return to normal activities may take 2 to 6 weeks. Serious complications, especially pneumonia, are common.

EPIDEMIOLOGY

Humans are the major hosts of the influenza viruses, and severe respiratory disease is the primary manifestation of infection. However, influenza A viruses closely related to those prevalent in humans circulate among many mammalian and avian species. As noted previously, some of these may undergo antigenic mutation or genetic recombination and emerge as new human epidemic strains.

Human, animal, and avian strains are similar

Characteristic influenza outbreaks have been described since the early 16th century, and outbreaks of varying severity have occurred nearly every year. Severe pandemics occurred in 1743, 1889-1890, 1918-1919 (the Spanish flu), and 1957 (the Asian flu). These episodes were associated with particularly high mortality rates; the Spanish flu was thought to have caused at least 30 million deaths, and some historians estimate the worldwide toll was closer to 100 million deaths. Usually, the elderly and persons of any age group with cardiac or pulmonary disease have the highest death rate.

Pandemic influenza may have high mortality

Direct droplet spread is the most common mode of transmission. Influenza infections in temperate climates tend to occur most frequently during midwinter months. Major epidemics of influenza A usually occur at 2- to 3-year intervals, and influenza B epidemics occur irregularly, usually every 4 to 5 years. The typical epidemic develops over a period of

Seasonality favors winter months

Epidemic intervals usually a few years

Excess mortality or increased absenteeism are indictors of epidemics

3 to 6 weeks and can involve 10% of the population. Illness rates may exceed 30% among school-aged children, residents of closed institutions, and industrial groups. One major indicator of influenza virus activity is an abrupt rise in school or industrial absenteeism. In severe influenza A epidemics, the number of deaths reported in a given area of the country often exceeds the number expected for that period. This significant increase, referred to as **excess mortality,** is another indicator of severe, widespread illness. Influenza B rarely causes such severe epidemics. In general, human influenza viruses are not stable in the environment and are sensitive to heat, acid pH, and solvents. In contrast, avian influenza viruses (H5N1 and others) retain infectivity for several weeks outside the host. The avian virus is shed in respiratory secretions and feces, and the virus survives in the feces for a long time. ::: droplet spread, p 93

PATHOGENESIS

Virus multiplies in respiratory epithelium

Synthetic blocks cause cilial damage and cell desquamation

Clearance mechanisms are compromised

Influenza viruses have a predilection for the respiratory tract, and viremia is rarely detected. They multiply in ciliated respiratory epithelial cells, leading to functional and structural ciliary abnormalities. This is accompanied by a switch-off of protein and nucleic acid synthesis in the affected cells, the release of lysosomal hydrolytic enzymes, and desquamation of both ciliated and mucus-producing epithelial cells. Thus, there is substantial interference with the mechanical clearance mechanism of the respiratory tract. The process of programmed cell death (apoptosis) results in the cleavage of complement components, leading to localized inflammation. Early in infection, the primary chemotactic stimulus is directed toward mononuclear leukocytes, which constitute the major cellular inflammatory component. The respiratory epithelium may not be restored to normal for 2 to 10 weeks after the initial insult.

Viral toxicity causes inflammation

Phagocytic host defenses compromised

The virus particles are also toxic to tissues. This toxicity can be demonstrated by inoculating high concentrations of inactivated virions into mice, which produces acute inflammatory changes in the absence of viral penetration or replication within cells. Other host cell functions are also severely impaired, particularly during the acute phase of infection. These functions include chemotactic, phagocytic, and intracellular killing functions of polymorphonuclear leukocytes and perhaps of alveolar macrophage activity.

Damage creates susceptibility to bacterial invasion

Interferon and cytotoxic T-cell responses associated with recovery

The net result of these effects is that, on entry into the respiratory tract, the viruses cause cell damage, especially in the respiratory epithelium, which elicits an acute inflammatory response and impairs mechanical and cellular host responses. This damage renders the host highly susceptible to invasive bacterial **superinfection.** In vitro studies also suggest that bacterial pathogens such as staphylococci can more readily adhere to the surfaces of influenza virus-infected cells. Recovery from infection begins with interferon production, which limits further virus replication, and with rapid generation of natural killer cells. Shortly thereafter, class I major histocompatibility complex (MHC)–restricted cytotoxic T cells appear in large numbers to participate in the lysis of virus-infected cells and thus in initial control of the infection. This is followed by the appearance of local and humoral antibody along with an evolving, more durable cellular immunity. Finally, there is repair of tissue damage. ::: cytotoxic lymphocytes, p.34

IMMUNITY

Antihemagglutinin antibody has protective effect

Antineuraminidase may limit viral spread

Although cell-mediated immune responses are undoubtedly important in influenza virus infections, humoral immunity has been investigated more extensively. Typically, patients respond to infection within a few days by producing antibodies directed toward the group ribonucleoprotein antigen, the hemagglutinin, and the neuraminidase. Peak antibody titer levels are usually reached within 2 weeks of onset and then gradually wane over the following months to varying low levels. Antibody to the ribonucleoprotein appears to confer little or no protection against reinfection. Antihemagglutinin antibody is considered the most protective; it has the ability to neutralize virus on reexposure. However, such immunity is relative, and quantitative differences in responsiveness exist among individuals. Furthermore, antigenic shifts and drifts often allow the virus to subvert the antibody response on subsequent exposures. Antibody to neuraminidase antigen is not as protective as antihemagglutinin antibody but plays a role in limiting virus spread within the host.

MANIFESTATIONS

Influenza A and B viruses tend to cause the most severe illnesses, whereas influenza C seems to occur infrequently and generally causes milder disease. The typical acute influenzal syndrome is described here.

The incubation period is brief, lasting an average of 2 days. Onset is usually abrupt, with symptoms developing over a few hours. These include fever, myalgia, headache, and occasionally shaking chills. Within 6 to 12 hours, the illness reaches its maximum severity, and a dry, nonproductive cough develops. The acute findings persist, sometimes with worsening cough, for 3 to 5 days, followed by gradual improvement. By about 1 week after onset, patients feel significantly better. However, fatigue, nonspecific weakness, and cough can remain frustrating lingering problems for an additional 2 to 6 weeks.

Short incubation period followed by acute disease with dry cough

Occasionally, patients develop a progressive infection that involves the tracheobronchial tree and lungs. In these situations, pneumonia, which can be lethal, is the result. Other unusual acute manifestations of influenza include central nervous system dysfunction, myositis, and myocarditis. In infants and children, a serious complication known as Reye's syndrome may develop 2 to 12 days after onset of the infection. It is characterized by severe fatty infiltration of the liver and by cerebral edema. This syndrome is associated not only with influenza viruses but with a wide variety of systemic viral illnesses. The risk is greatly enhanced by exposure to salicylates, such as aspirin.

Progressive respiratory infection and pneumonia may be lethal

Reye's syndrome may follow

The most common and important complication of influenza virus infection is bacterial superinfection. Such infections usually involve the lung, but bacteremia with secondary seeding of distant sites can also occur. The superinfection, which can develop at any time in the acute or convalescent phase of the disease, is often heralded by an abrupt worsening of the patient's condition after initial stabilization. The bacteria most commonly involved include *Streptococcus pneumoniae, Haemophilus influenzae,* and *Staphylococcus aureus.*

Sudden worsening suggests bacterial superinfection

In summation, there are essentially three ways in which influenza may cause death:

Underlying disease with decompensation. Individuals with limited cardiovascular or pulmonary reserves can be further compromised by any respiratory infection. Thus, the elderly and those of any age with underlying chronic cardiac or pulmonary disease are at particular risk.

Superinfection. Superinfection can lead to bacterial pneumonia and occasionally disseminated bacterial infection.

Direct rapid progression. Less commonly, progression of the viral infection can lead to overwhelming viral pneumonia with asphyxia. This phenomenon has been seen most commonly in severe pandemics; for example, the Spanish flu in 1918-1919 often produced fulminant death in healthy young soldiers.

DIAGNOSIS

During the acute phase of illness, influenza viruses can be readily isolated from respiratory tract specimens, such as nasopharyngeal and throat swabs. Most strains grow in primary monkey kidney cell cultures, and they can be detected by hemadsorption or hemagglutination. Rapid diagnosis of infection is possible by direct immunofluorescence or immunoenzymatic detection of viral antigen in epithelial cells or secretions from the respiratory tract and by polymerase chain reaction (PCR). Serologic diagnosis is of considerable help epidemiologically and is usually made by demonstrating a fourfold or greater increase in hemagglutination inhibition (HI) antibody titers in acute and convalescent specimens collected 10 to 14 days apart. For details about the HI assay, see Chapter 4. ::: polymerase chain reaction, p.86

Virus isolation detects virus

Rapid detection of antigen often used

Serodiagnosis is useful epidemiologically

TREATMENT

The two basic approaches to management of influenzal disease are symptomatic care and anticipation of potential complications, particularly bacterial superinfection. Once the

TABLE 9–4 Comparison of Antiviral Drugs for Influenza

FEATURE	AMANTADINE RIMANTADINE	ZANAMIVIR	OSELTAMIVIR
Susceptible viruses	Influenza A only	Influenza A and B	Influenza A and B
Emergent resistant strains	Yes	Yes	Yes
Administration	Oral	Inhalation	Oral

Supportive therapy indicated

diagnosis has been made, rest, adequate fluid intake, conservative use of analgesics for myalgia and headache, and antitussives for severe cough are commonly prescribed. It must be emphasized that nonprescription drugs must be used with caution. This applies particularly to drugs containing salicylates given to children, because the risk of Reye's syndrome must be considered.

Antibiotic prophylaxis does not prevent bacterial superinfection

Bacterial superinfection is often suggested by a rapid worsening of clinical symptoms after patients have initially stabilized. Antibiotic prophylaxis has not been shown to enhance or diminish the likelihood of superinfection, but can increase the risk of acquisition of more resistant bacterial flora in the respiratory tract and make the superinfection more difficult to treat. Ideally, physicians should instruct patients regarding the natural history of the influenza virus infection and be prepared to respond quickly to bacterial complications, if they occur, with specific diagnosis and therapy.

Antiviral therapy must begin early

When influenza A infection is proved or strongly suspected, 4 to 5 days of therapy with amantadine or rimantadine, two symmetric amines, may also be considered (**Table 9–4**). Such treatment has been shown to benefit some patients to a modest degree, as measured by reduction of number of days of confinement to bed, of fever, and of functional respiratory impairment. However, these effects have been observed only when the drug is administered early in the illness (within 12 to 24 hours of onset). Unfortunately, the incidence of resistance to these amines by influenza A (H3N2), the dominant circulating strains, has risen dramatically from 0.8% before 1995 to higher than 95% by 2005. Thus, neither amantadine nor rimantadine is currently recommended for routine treatment or prophylaxis. The primary mechanism of resistance is detailed below. The neuraminidase inhibitors (zanamivir or oseltamivir) have also proved beneficial, if begun early, but some strains have also developed resistance to these. They are also active against influenza B (Table 9–4).

PREVENTION

Whole virus and "split" vaccines are protective but variable and of short duration

Live attenuated influenza vaccine, FluMist, is available as nasal spray given to healthy people

Annual revaccination against most current strains is necessary

Vaccination indicated for high-risk individuals

The best available method of control of influenza is by use of **killed viral vaccines** newly formulated each year to most closely match the influenza A and B antigenic subtypes currently causing infections. These inactivated vaccines may contain whole virions or "split" subunits composed primarily of hemagglutinin antigens. They are commonly used in two doses given 1 month apart to immunize children who may not have been immunized previously. Among older children and adults, single annual doses are recommended just before influenza season. Vaccine efficacy is variable, and annual revaccination is necessary to ensure maximal protection. It is recommended that vaccination be directed primarily toward the elderly, individuals of all ages who are at high risk (eg, those with chronic lung or heart disease), and their close contacts, including medical personnel and household members. Live attenuated influenza vaccine (LAIV), which is made up of live, weakened influenza viruses (same strains used in killed vaccine) and given in the form of mist (FluMist) in the nostrils, is approved for use in healthy people 2 to 49 years of age. It is not given to pregnant women.

When antigenic drift occurs unexpectedly, vaccine efficacy in the subsequent year may fall to unacceptable levels.

A problem unique to influenza vaccinology is the inherent, often unexpected variation in antigenic drift from year to year. This often requires annual reformulation of vaccines that are hoped to provide the best protection before the onset of the next influenza season. Prediction of which strains should be used for vaccine production is based on international surveillance—always a difficult task indeed. Thus, in some years, vaccine efficacy (prevention of serologically confirmed influenza infection) has been estimated to be as high as 70% to 90%. At other times such as in the 2007-2008 season, efficacy may be only at levels estimated at 40% to 60%. The emergence of the swine-origin 2009 A(H1N1) virus has added to the dilemma. None of the previously available vaccines have been shown to confer protection; thus production of a new strain-specific vaccine is necessary.

A major factor contributing to this dilemma is related to difficulties in timely production of a vaccine. Up until very recently, all available vaccines had to be prepared in embryonated hen's eggs—a cumbersome process that required at least 22 weeks of preparation. There are now methods whereby new strains, even avian H_5N_1 viruses, can be identified quickly and mass-produced in Vero-cell cultures instead of eggs, thus reducing the production time by as much as 50%, with far higher vaccine quantities.

Switching vaccine production from hen's eggs to cell cultures could greatly improve responses to both epidemic and pandemic threats.

Both amantadine and rimantadine act by blocking the ion channel of the viral M2 protein, resulting in interference with the key role for M2 protein in early virus uncoating. Later virion assembly may also be affected. Unfortunately, virus resistance to both drugs can readily develop in vitro or in vivo. A single amino acid substitution in the transmembrane portion of the M2 protein is all that is necessary for this to occur.

Blocks virus uncoating and assembly

Resistance from single amino acid substitution in M2 protein

Zanamivir and oseltamivir, approved for use in 1999, both act by blocking the enzymatically active neuraminidase glycoprotein present on the surfaces of influenza A and B viruses, thus limiting virus release from infected cells and subsequent spread in the host. Viral resistance has now been demonstrated for some strains of influenza A (Table 9–4).

Neuraminidase inhibitors are useful for influenza A and B

Resistant mutants at low frequency, but this could change in the future

PARAINFLUENZA VIRUSES

VIROLOGY

Parainfluenza viruses belong to the paramyxovirus group. There are four serotypes of parainfluenza viruses: parainfluenza 1, 2, 3, and 4. These enveloped viruses contain linear (nonsegmented), negative-sense, single-stranded RNA genome. Like the influenza viruses, parainfluenza viruses possess a hemagglutinin and neuraminidase. The structure of paramyxovirus is shown in **Figure 9–4**. The single stranded, negative-sense linear RNA genome is bound to a nucleoprotein, and the matrix protein surrounds the nucleoprotein complex, which is packaged into a lipid bilayer envelope containing attachment protein (H and N on the same spike) and the fusion protein (F). Their mode of spread and pathogenesis

Enveloped paramyxoviruses have hemagglutinin and neuraminidase on the same spike

Four serotypes are antigenically stable

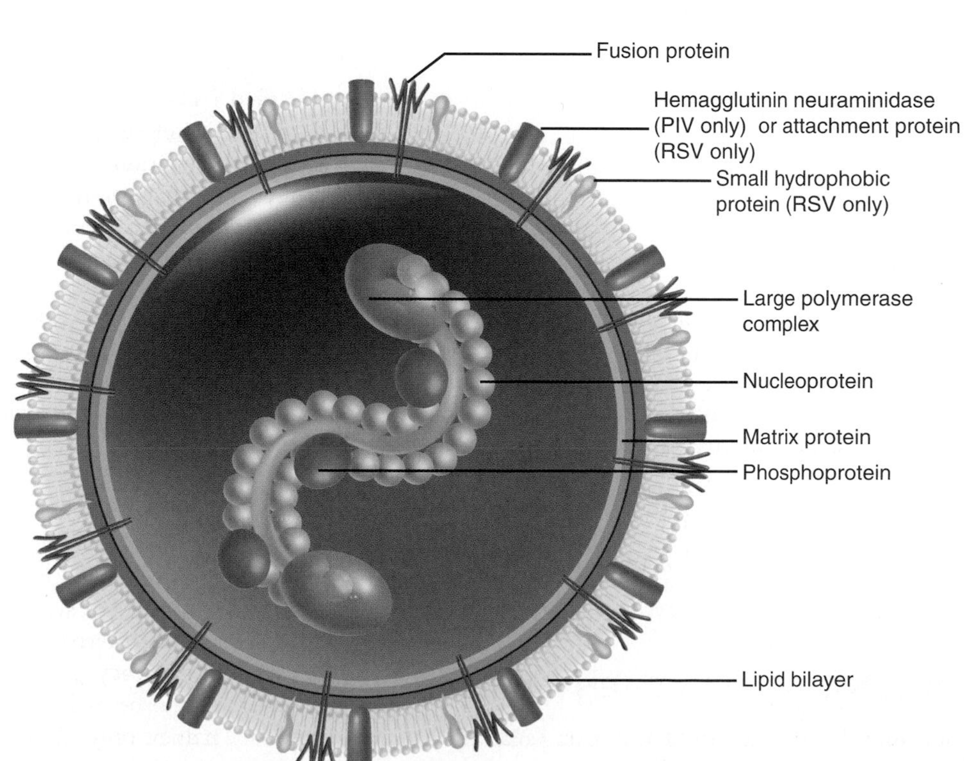

FIGURE 9–4. Diagram of a paramyxovirus PIV, parainfluenza virus; RSV, respiratory syncytial virus.

are similar to those of the influenza viruses. They differ from the influenza viruses in that RNA synthesis occurs in the cytoplasm rather than the nucleus. All events related to parainfluenza virus replication occur in the cytoplasm, like any other negative-sense RNA viruses (see Chapter 6 for details about replication of negative-sense RNA viruses). The virus buds out through plasma membranes. In addition, the antigenic makeup of the four serotypes is relatively stable, and significant antigenic shift or drift does not occur. Each serotype is considered separately.

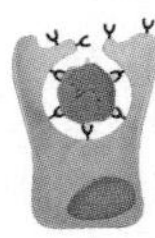

PARAINFLUENZA DISEASE

Transient Immunity

The parainfluenza viruses are important because of the serious diseases they can cause in infants and young children. Parainfluenza 1 and 3 are particularly common in this regard. Overall, the group is thought to be responsible for 15% to 20% of all nonbacterial respiratory diseases requiring hospitalization in infancy and childhood. Immunity to reinfection is transient. Although repeated infections can occur in older children and adults, they are usually milder than the illnesses of infancy and early childhood.

CLINICAL ASPECTS

MANIFESTATIONS

The onset of illness from parainfluenze virus may be abrupt, as in acute spasmodic croup, but usually begins as a mild upper respiratory infection (URI) with variable progression over 1 to 3 days to involvement of the middle or lower respiratory tract. Duration of acute illness can vary from 4 to 21 days but is usually 7 to 10 days.

Parainfluenza 1

Croup and tracheobronchitis are seen

Parainfluenza 1 is the major cause of acute croup (laryngotracheitis) in infants and young children, but also causes less severe diseases such as mild URI, pharyngitis, and tracheobronchitis in individuals of all ages. Outbreaks of infection tend to occur most frequently during the fall months.

Parainfluenza 2

Croup is primary disease

Parainfluenza 2 is of slightly less significance than parainfluenza 1 or 3. It has been associated with croup, primarily in children, with mild URI, and occasionally with acute lower respiratory disease. As with parainfluenza 1, outbreaks usually occur during the fall months.

Parainfluenza 3

Produces severe lower respiratory disease in infants, including bronchitis, pneumonia, and croup

Parainfluenza 3 is a major cause of severe lower respiratory disease in infants and young children. It often causes bronchitis, pneumonia, and croup in children younger than 1 year old. In older children and adults, it may cause URI or tracheobronchitis. Infections are common and can occur in any season; it is estimated that nearly 50% of all children have been exposed to this virus by 1 year of age.

Parainfluenza 4

Causes only URI

Parainfluenza 4 is the least common of the group. It is generally associated with mild upper respiratory illness only.

DIAGNOSIS, TREATMENT, AND PREVENTION

Specific diagnosis is based on virus isolation, usually in monkey kidney cell cultures, PCR, or serology using hemagglutination inhibition, enzyme immunoassay (EIA), or neutralization assays on paired sera to detect a rising antibody titer. Immunofluorescence or immunoenzyme assays can also be used for rapid detection of antigen in respiratory epithelial cells. Currently, there is no method of control or specific therapy for these infections.

Laboratory diagnosis by isolation or antigen detection

Croup and URI are not treatable

RESPIRATORY SYNCYTIAL VIRUS

VIROLOGY

Respiratory syncytial virus (RSV) is classified as a pneumovirus within the paramyxovirus family. Its name is derived from its ability to produce cell fusion in tissue culture (syncytium formation). Unlike influenza or parainfluenza viruses, it possesses no hemagglutinin or neuraminidase. The virion structure is similar to parainfluenza virus except that the envelope glycoproteins are an attachment (G) protein and a fusion (F) protein. The RNA genome is linear (nonsegmented), negative-sense, and single stranded and codes for at least 10 different proteins. Among these are two matrix (M) proteins in the viral envelope. One forms the inner lining of the viral envelope; the function of the other is uncertain. RSV, like other paramyxoviruses, replicates in the cytoplasm.

Pneumovirus causes syncytium formation in cell cultures

Enveloped RNA virus has a linear (unsegmented) genome

The antigens on the surface spikes of the viral envelope include the G glycoprotein, which mediates virus attachment to host cell receptors, and the fusion (F) glycoprotein, which induces fusion of the viral envelope with the host cell surface to facilitate entry. F glycoprotein is also responsible for fusion of infected cells in cell cultures, leading to the appearance of multinucleated giant cells (syncytium formation). Antibodies directed at the F glycoprotein are more efficient than anti-G glycoprotein antibodies in neutralizing the virus in vitro.

Two glycoproteins mediate attachment and syncytium formation

At least two antigenic subgroups (A and B) of RSV are known to exist. This dimorphism is due primarily to differences in the G glycoprotein. The epidemiologic and biologic significance of these variants is not yet certain; however, epidemiologic studies have suggested that group A infections tend to be more severe. RSV is the single most important etiologic agent in respiratory diseases of infancy, and it is the major cause of bronchiolitis and pneumonia among infants under 1 year of age.

RSV is the most important respiratory virus in infants

RESPIRATORY SYNCYTIAL VIRUS DISEASE

CLINICAL CAPSULE

RSV primarily infects the bronchi, bronchioles, and alveoli of the lung. The illnesses clinically categorized as croup, bronchitis, bronchiolitis, or pneumonia are extremely common in infants. The acute phase of cough, wheezing, and respiratory distress lasts 1 to 3 weeks. The severity of respiratory involvement and high prevalence during outbreaks both account for a large number of hospitalizations on pediatric units each year. Elderly or immunocompromised patients are also frequently susceptible and can be severely affected.

EPIDEMIOLOGY

Community outbreaks of RSV infection occur annually, commencing at any time from late fall to early spring. The usual outbreak lasts 8 to 12 weeks and can involve nearly 50% of all families with children. In the family setting, it appears that older siblings often introduce the virus into

High attack rate, introduced by older siblings

the home, and secondary infection rates can be almost 50%. The usual duration of virus shedding is 5 to 7 days; young infants, however, may shed virus for 9 to 20 days or longer.

Nosocomial infection reduced by careful handwashing

Spread of RSV in the hospital setting is also a major problem. Control is difficult, but includes careful attention to handwashing between contacts with patients, isolation, and exclusion of personnel and visitors who have any form of respiratory illness. Masks are not effective in controlling nosocomial spread.

PATHOGENESIS

Confined to respiratory epithelium

RSV is spread to the upper respiratory tract by contact with infective secretions. Infection appears to be confined primarily to the respiratory epithelium, with progressive involvement of the middle and lower airways. Viremia occurs rarely. The direct effect of virus on respiratory tract epithelial cells is similar to that previously described for influenza viruses, and cytotoxic T cells appear to play a similar role in early control of the acute infection.

Enhanced disease in infants may have immunologic basis

Th_2 stimulated cytokines cause injury

The apparent enhanced severity of RSV, particularly in very young infants, is not yet clearly understood but may have an immunologic basis. Factors that have been proposed to play a role include (1) qualitative or quantitative deficits in humoral or secretory antibody responses to critical virus-specified proteins, (2) formation of antigen–antibody complexes within the respiratory tract resulting in complement activation, and (3) excessive damage from inflammatory cytokines. Experimental evidence suggests that patients who respond to RSV infections with CD4+ cells that are predominantly of the T_H type 2 have more severe disease than those with predominant T_H type 1 responses. This is thought to be due to the inflammatory cytokines produced by T_H type 2 cells, including interleukin (IL)-4, IL-5, IL-6, IL-10, and IL-13.

Necrosis and inflammation plug bronchioles and alveoli

The major pathologic findings of RSV are in the bronchi, bronchioles, and alveoli. These include necrosis of epithelial cells; interstitial mononuclear cell inflammatory infiltrates, which sometimes also involve the alveoli and alveolar ducts; and plugging of smaller airways with material containing mucus, necrotic cells, and fibrin (**Figure 9–5**). Multinucleated syncytial cells with intracytoplasmic inclusions are occasionally seen in the affected tracheobronchial epithelium.

IMMUNITY

Immunity to reinfection is brief

Infection with RSV results in IgG and IgA humoral and secretory antibody responses. However, immunity to reinfection is tenuous, as demonstrated by patients who have recovered from a primary acute episode and have become reinfected with disease of similar severity in the same or succeeding year. Illness severity appears to diminish with increasing age and successive reinfection.

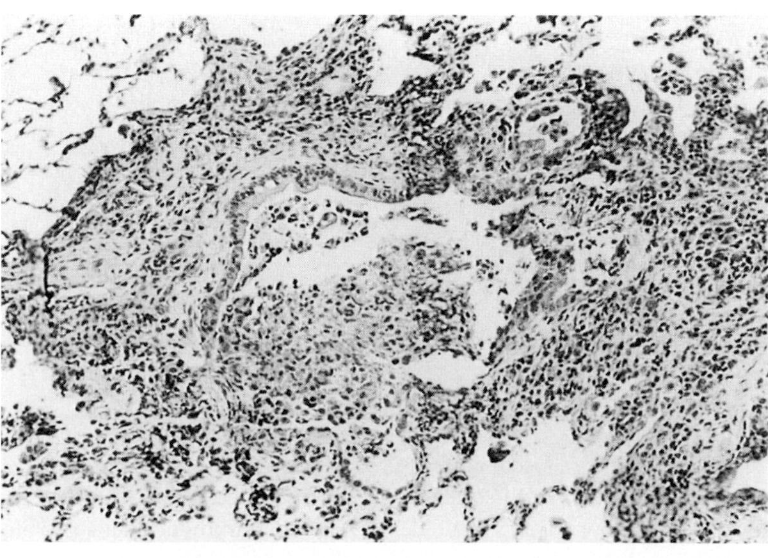

FIGURE 9–5. Photomicrograph illustrates the bronchiolar and surrounding interstitial inflammation in respiratory syncytial virus infection. (Original magnification × 100.)

CLINICAL ASPECTS

MANIFESTATIONS

The usual incubation period for RSV is 2 to 4 days, followed by the onset of rhinitis; severity of illness progresses to a peak within 1 to 3 days. In infants, this peak usually takes the form of bronchiolitis and pneumonitis, with cough, wheezing, and respiratory distress. Clinical findings include **hyperexpansion** of the lungs, **hypoxemia** (low oxygenation of blood), and **hypercapnia** (carbon dioxide retention). Interstitial infiltrates, often with areas of pulmonary collapse, may be seen on chest radiography (**Figure 9–6**). Fever is variable. The duration of acute illness is often 10 to 14 days.

RSV is the single most important agent of bronchioltis and pneumonia in infants under 1 year of age

Infant bronchiolitis and pneumonitis lasts up to 2 weeks

The fatality rate among hospitalized infected infants is estimated to be 0.5% to 1%; however, this rises to 15% or more in children receiving cancer chemotherapy, infants with congenital heart disease, and those with severe immunodeficiency. Infants with underlying chronic lung disease are also at high risk. Causes of death include respiratory failure, right-sided heart failure (cor pulmonale), and bacterial superinfection. Death has sometimes resulted from unnecessary procedures in patients in whom RSV infection was not considered. Bronchoscopy, lung biopsy, or overly aggressive therapy with corticosteroids and bronchodilators for presumed asthma all can pose a danger to such patients.

Mortality is highest with underlying disease

Older infants, children, and adults are also readily infected. The clinical illnesses in these groups are usually milder and include croup, tracheobronchitis, and URI; however, elderly persons can experience severe morbidity. RSV can also cause acute flare-ups of chronic bronchitis and trigger acute wheezing episodes in asthmatic children.

Children and adults have milder illness

Can trigger wheezing in asthmatics

DIAGNOSIS

Rapid diagnosis of RSV infection can be made by immunofluorescence or immunoenzyme detection of viral antigen, or by PCR. The virus can also be isolated from the respiratory tract by prompt inoculation of specimens into cell cultures. Syncytial cytopathic effects develop over 2 to 7 days. Serodiagnosis may also be used but requires acute and convalescent sera and is less sensitive than antigen detection methods, PCR, or culture.

Virus isolation, PCR, immunofluorescence, and immunoassay detect RSV

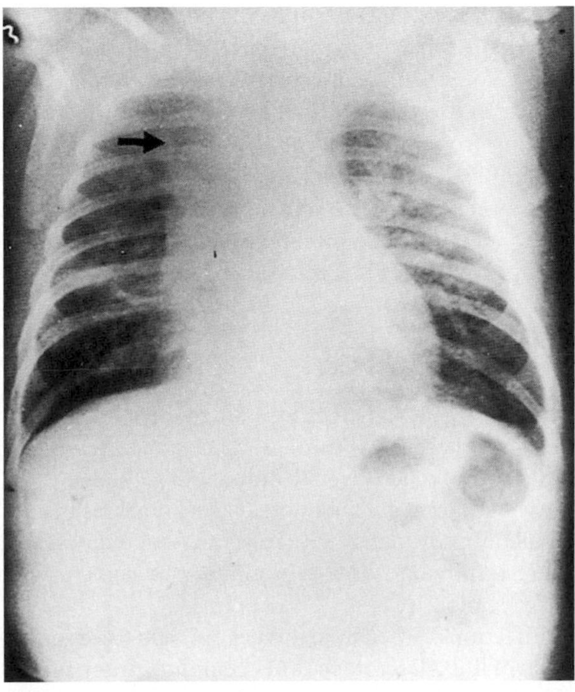

FIGURE 9–6. Chest radiograph of an infant with a severe case of respiratory syncytial virus pneumonia and bronchiolitis. Bilateral interstitial infiltrates, hyperexpansion of the lung, and right upper lobe atelectasis (*arrow*) are present.

TREATMENT AND PREVENTION

Supportive treatment is indicated

Treatment for RSV is directed primarily at the underlying pathophysiology and includes adequate oxygenation, ventilatory support when necessary, and close observation for complications such as bacterial superinfection and right-sided heart failure. Some studies suggest that ribavirin aerosol treatment may be effective in selected circumstances.

Monoclonal antibody and immune globulin used for prophylaxis

No vaccine is currently available for RSV. Attenuated live virus vaccines and immune globulin containing high antibody titers to RSV are also under active investigation; a high-titered monoclonal antibody against F protein has been used for prophylaxis in high-risk infants (those born prematurely or with chronic lung disease). This method requires monthly injections during the RSV season (usually 5 months) and is extremely expensive.

HUMAN METAPNEUMOVIRUS

Second to RSV as bronchiolitis cause

Like RSV, hMPV is a pneumovirus

Clinical and epidemiologic behaviors similar to those of RSV

Discovered in 2001, human metapneumovirus (hMPV) has subsequently been found to be a significant cause of acute respiratory disease in infants and young children. It is second only to RSV as a cause of bronchiolitis during the winter-spring seasons, and it produces illnesses that are comparable in their severity and symptoms to those of RSV. Like RSV, hMPV is a pneumovirus. Two genotypes are known to exist, but it is not known whether either produces more severe disease or protective immunity. The virus is somewhat difficult to isolate in cell cultures. The usual diagnostic methods of choice are genome amplification (PCR) or antigen detection by immunofluorescence.

ADENOVIRUSES

Multiple serotypes of naked, double-stranded DNA viruses

Potential for prolonged infection without disease

Of the almost 100 different serotypes of adenoviruses, 51 are known to affect humans and are classified into one of six subgroups (A-F) on the basis of multiple biologic properties. (Further details on the subgroups can be found in basic virology texts, but are not discussed here.) These viruses are naked and icosahedral and possess double-stranded DNA. Replication and assembly occur in the nucleus, and virions are released by cell destruction (see Chapter 6 for DNA virus replication). All adenoviruses share a common group-specific, complement-fixing antigen associated with the hexon component of the viral capsid. Adenoviruses are characterized by their ubiquity and persistence in host tissues for periods ranging from a few days to several years. Their ability to produce infection without disease is illustrated by the frequent recovery of virus from tonsils or adenoids removed from healthy children (the group name is derived from its discovery in 1953 as a latent agent in many adenoid tissue specimens) and by prolonged intermittent shedding of virus from the pharynx and intestinal tract after initial infection.

EPIDEMIOLOGY

Disease in children and military recruits is spread by respiratory or fecal-oral route

Types 1, 2 and 3 adenoviruses are highly endemic; type 5 is the next most common. Most primary infections with these viruses occur early in life and are spread by the respiratory or fecal-oral route. Overall, only about 45% of adenovirus infections result in disease. Their most significant contribution to acute illness is in children, particularly those under 2 years of age (approximately 10% of acute febrile illness). Adenoviruses are also major causes of acute respiratory disease in military recruits, usually by types 4 (prevalence over 90%), 14, 7, 3, and 21.

Infections caused by serotypes 1, 2, and 5 are generally most common during the first few years of life. All serotypes can occur during any season of the year but are encountered most frequently during late winter or early spring. Sharp outbreaks of disease caused by serotypes 3 and 7 have been traced to inadequately chlorinated swimming pools. Conjunctivitis is the illness most commonly associated with these episodes. Other outbreaks of conjunctivitis have been traced to physicians' offices and appear to have been spread by contaminated ophthalmic medications or diagnostic equipment.

Swimming pool and medication-associated conjunctivitis occur in outbreaks

PATHOGENESIS

The adenoviruses usually enter the host by inhalation of droplet nuclei or by the oral route. Direct inoculation onto nasal or conjun ctival mucosa by hands, contaminated towels, or ophthalmic medications may also occur. The virus replicates in epithelial cells, producing cell necrosis and inflammation. Viremia sometimes occurs and can result in spread to distant sites, such as the kidney, bladder, liver, lymphoid tissue (including mesenteric nodes), and occasionally the central nervous system. In the acute phase of infection, the distant sites may also show inflammation; for example, abdominal pain is occasionally seen with severe illnesses and is believed to result from mesenteric lymphadenitis caused by the viruses.

Infects by droplet, oral route, or direct inoculation

Epithelial cell replication may be followed by viremic spread and remote disease

After the acute phase of illness, the viruses may remain in tissues, particularly lymphoid structures such as tonsils, adenoids, and intestinal Peyer's patches, and may become reactivated and shed without producing illness for 6 to 18 months thereafter. This reactivation is enhanced by stressful events (stress reactivation), such as infection by other agents. Integration of adenoviral DNA into the host cell genome has been shown to occur; this latent state can persist for years in tonsillar tissue and peripheral blood lymphocytes.

Integration of adenoviral DNA produces latency

Like the viruses described previously, adenoviruses have a primary pathology involving epithelial cell necrosis with a predominantly mononuclear inflammatory response. In some instances, smudgy intranuclear inclusions may be seen in infected cells (**Figure 9–7**). A potentially important pathogenic feature of the virion is the presence of pentons, which are located at each of the 12 corners of the icosahedron. These fiber-like projections with knob-like terminal structures are believed to bind to a cellular receptor that is similar or identical to the one for group B coxsackieviruses. The pentons also appear to be responsible for a toxic effect on cells, which manifests as clumping and detachment in vitro.

Penton projections are toxic to cells

In addition, adenoviruses have developed other novel strategies to survive in the host yet produce deleterious effects. These include encoding a protein in its early E3 genomic region that binds class I MHC antigens in the endoplasmic reticulum, thus restricting their expression on the surface of infected cells and interfering with recognition and attack by cytotoxic T cells. This ability to evade immunosurveillance may be vital to establishment of latency. Another early protein (E1A) has been associated with increased susceptibility of epithelial

Proteins restrict cytotoxic T cells and enhance cytokine susceptibility

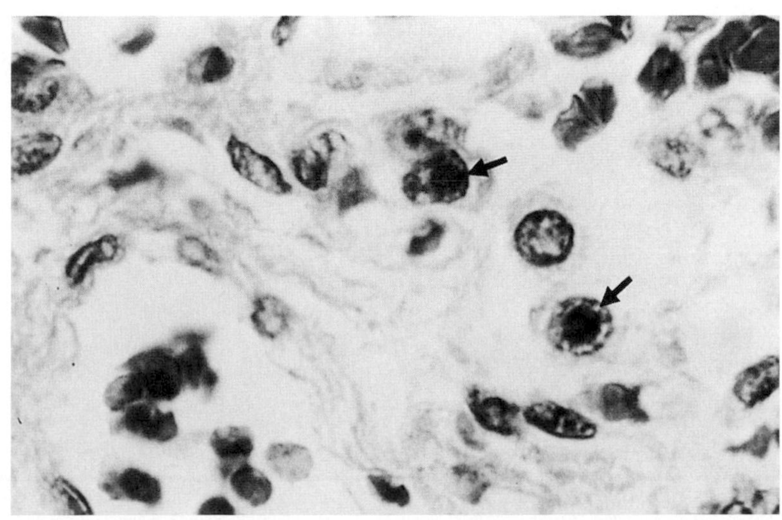

FIGURE 9–7. Lung tissue from a fatal case of adenovirus type 7 pneumonia. Large, smudgy intranuclear inclusions in alveolar epithelial cells (*arrows*), which are sometimes seen in adenovirus infections, are present. (Original magnification ×100.)

cells to destruction by tumor necrosis factor and other cytokines. Other adenoviral proteins have been described that have a variety of effects on cell function and susceptibility to cytolysis. One of these, called the **adenovirus death protein**, is considered important for efficient lysis of infected cells and release of newly formed virions.

IMMUNITY

Immunity is type-specific

Immunity to adenoviruses after infection is serotype-specific and usually longlasting. In addition to type-specific immunity, group-specific complement-fixing antibodies appear in response to infection. These antibodies are useful indicators of infection, but do not specify the infecting serotype.

CLINICAL ASPECTS

MANIFESTATIONS

Multiple upper respiratory syndromes, conjunctivitis, and pharyngitis are common

More severe disease includes hemorrhagic cystitis

The diversity of major syndromes and serotypes commonly associated with adenoviruses are summarized in **Table 9–5**. The acute respiratory syndromes vary in both clinical manifestations and severity. Symptoms include fever, rhinitis, pharyngitis, cough, and conjunctivitis. Adenoviruses are also common causes of nonstreptococcal exudative pharyngitis, particularly among children less than 3 years of age. Acute and occasionally chronic conjunctivitis and keratoconjunctivitis have been associated with several serotypes. More severe disease, such as laryngitis, croup, bronchiolitis, and pneumonia, may also occur. A syndrome of pharyngitis and conjunctivitis (pharyngoconjunctival fever) is classically associated with adenovirus infection. Adenoviruses can also cause acute hemorrhagic cystitis, in which hematuria and dysuria are prominent findings. Some serotypes are significant causes of gastroenteritis (see Chapter 15).

DIAGNOSIS

Viral isolation from oropharynx or feces may not mean disease

Many serotypes of adenoviruses, other than those associated with acute gastroenteritis, can be readily isolated in heteroploid cell cultures. There is little difficulty in relating the virus detected to the illness in question when the isolate has been obtained from a site other than the upper respiratory or gastrointestinal tract (eg, lung biopsy, conjunctival swabs, urine). However, because of the known tendency for intermittent asymptomatic shedding into the oropharynx and feces, isolates from these latter sites must be interpreted more cautiously. Serologic testing of acute and convalescent sera may be necessary to confirm the relation between the virus and the illness in question.

TABLE 9–5 Clinical Syndromes Associated With Adenovirus Infection

SYNDROME	COMMON SEROTYPES[a]
Childhood febrile illness; pharyngoconjunctival fever	1, 2, **3,** 5, 7, **7a**
Pneumonia and other acute respiratory illnesses	1, 2, **3**, 5, 7, **7a**, **7b, 14** (4 in military recruits)
Pertussis-like illness	1, 2, **3**, 5, **19**, 21
Conjunctivitis	2, 5, 7, 8, **19**, 21
Keratoconjunctivitis	**3**, 8, 9, **19**
Acute hemorrhagic cystitis	11
Acute gastroenteritis	40, 41

[a]Serotypes in **boldface** are commonly associated with outbreaks.

TREATMENT AND PREVENTION

Some in vitro and in vivo data, combined with clinical observations in patients with severe disseminated infections suggest that cidofovir might be effective for adenovirus infection. A live virus vaccine containing serotypes 4 and 7, enclosed in enteric-coated capsules and administered orally, has been used in military recruits. The viruses are released into the small intestine, where they produce an asymptomatic, nontransmissible infection. This vaccine has been found effective but is neither available nor recommended for civilian groups.

Cidofovir in severe adenovirus infections

Live vaccine used in military

RHINOVIRUSES

The rhinovirus group comprises at least 115 accepted serotypes and more that are not yet classified, all of which are members of the picornavirus family. They are small (20 to 30 nm), naked capsid virus particles containing single-stranded, positive-sense RNA genomes. They are distinguished from other picornaviruses, namely enteroviruses by their acid lability and an optimum temperature of 33°C for in vitro replication. This temperature approximates that of the nasopharynx in the human host and may be a factor in the localization of pathologic findings at that site. Rhinoviruses are most consistently isolated in cultures of human diploid fibroblasts. The receptor for most rhinoviruses (and some coxsackieviruses) is glycoprotein intercellular adhesion molecule 1 (ICAM-1), a member of the immunoglobulin supergene family. ICAM-1 is best known for its role in immunologic cell adhesion; its ligand is lymphocyte function-associated antigen-1.

Small, naked RNA viruses include multiple serotypes

Optimum growth temperature is 33°C

Virus binds to ICAM intercellular adhesion molecule

Rhinoviruses are known as the common cold viruses. They represent the major causes of mild URI syndromes in all age groups, especially older children and adults. Lower respiratory tract disease caused by rhinoviruses is uncommon. The usual incubation period is 2 to 3 days, and acute symptoms commonly last 3 to 7 days. It is interesting to note that mucosal cell damage is minimal during the illness. Data suggest that activation and an increase in kinins, particularly bradykinin, may have a major role in the pathogenesis of increased secretions, vasodilation, and sore throat. Rhinovirus infections may be seen at any time of the year. Epidemic peaks tend to occur in the early fall or spring months.

Common cold viruses cause mild URI

Minimal cell injury is produced

TREATMENT AND PREVENTION

Currently, there is no specific therapy and no methods of prevention with vaccines. Prospects for the development of an appropriate vaccine appear dim. The multiplicity of serotypes and their tendency to be type-specific in the production of antibodies seem to demand the development of a multivalent vaccine, which would be extremely difficult to accomplish. However, recent studies have suggested that a monoclonal antibody directed at the virus receptor or the use of a recombinant soluble receptor (ICAM-1) might block attachment of rhinoviruses. It remains to be seen whether these observations can be translated into effective preventive or therapeutic applications. At present, the attitude toward these viruses is best summed up by Sir Christopher Andrewes, who suggested that we should accept these infections as "one of the stimulating risks of being mortal."

Multiple serotypes make vaccine different

Pharmaceutical agents block attachment to ICAM

CORONAVIRUSES

Coronaviruses contain a single-stranded, positive-sense RNA genome, which is surrounded by an envelope that includes a lipid bilayer derived from intracellular rough endoplasmic reticulum and Golgi membranes of infected cells. Petal- or club-shaped spikes (peplomers) measuring approximately 13 nm project from the surface of the envelope, giving the appearance of a crown of thorns or a solar corona. The peplomers play an important role in inducing neutralizing and cellular immune responses. The structure of coronavirus is shown in **Figure 9–8**. Coronaviruses replicate in the cytoplasm generally like other positive-sense RNA viruses but acquire envelope from endoplasmic reticulum or Golgi apparatus. Like the rhinoviruses, coronaviruses are considered primary causes of the common cold. Based on serologic studies, it is estimated that they

Enveloped RNA viruses

Disease similar to rhinoviruses

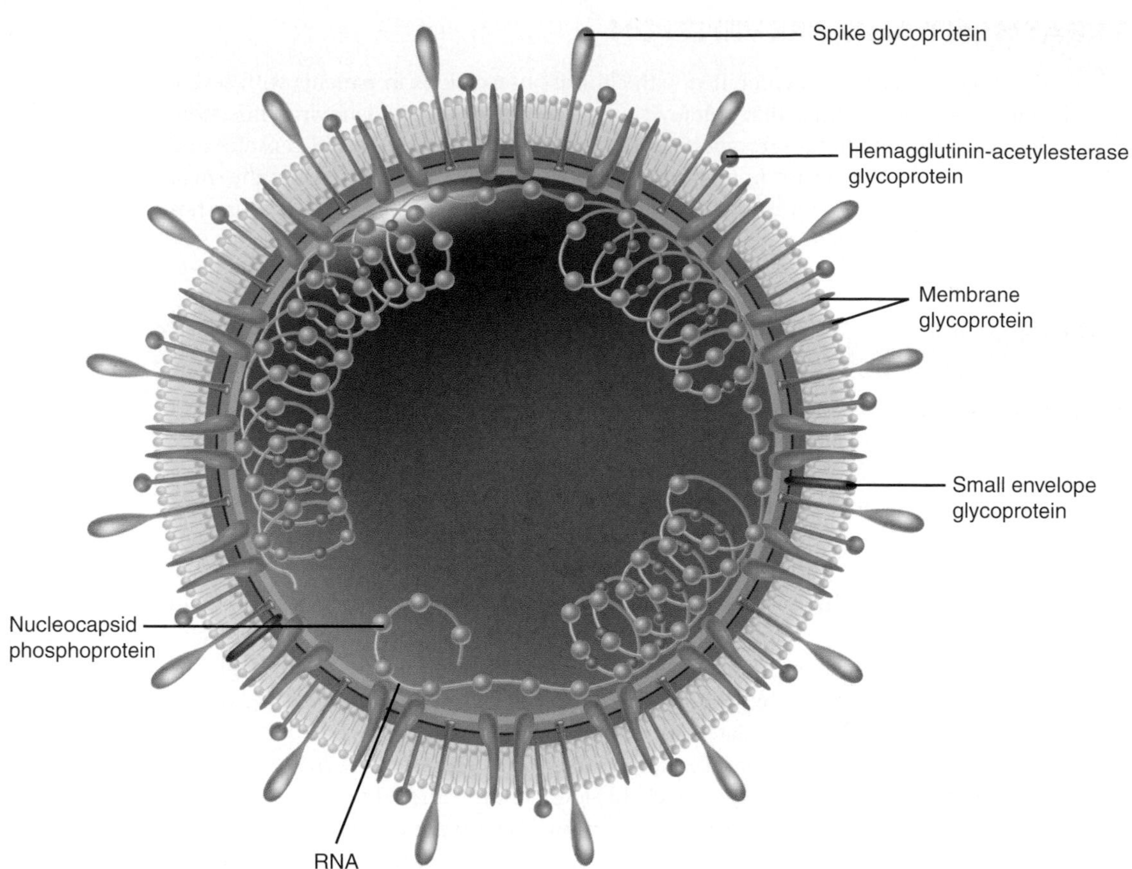

FIGURE 9–8. Virion structure of a coronavirus.

may cause up to 5% to 10% of common colds in adults and a similar proportion of lower respiratory illnesses in children.

Metalloprotease and sialic acid receptors bind some strains

The number of serotypes is unknown. Two strains (229E and OC43) have been studied to some extent; it is clear that they can cause outbreaks similar to those of the rhinoviruses and that reinfection with the same serotype can occur. The cellular receptors for these strains are a cell surface metalloprotease and a sialic acid receptor similar to that bound by influenza C virus. Recently, three other strains have also been described. They include NL63, a similar species called NH, and HKU1. All produce similar syndromes ranging from upper to lower respiratory illness.

SARS is caused by a novel, new coronavirus

Risk of transmission from an infected to an uninfected person is greatest around day 10 of illness

In late 2002, an illness called severe acute respiratory syndrome (SARS) appeared in China, spread throughout Asia, and is now found worldwide. The etiology has been identified as another previously undescribed coronavirus, with unusually high virulence for humans. The genome of SARS causing coronavirus has been sequenced, and the virus has the ability to mutate like other RNA viruses. The route of transmission is similar to that of other common cold viruses such as direct contact, via the eyes, nose, and mouth with infectious droplet. The risk of transmitting the disease to a person is greatest around day 10 of the illness, when the maximum amount of virus is shed from the respiratory tract. The older population is at a higher risk than younger people and children.

BOCAVIRUS

Human bocavirus was first discovered in 2005 by using molecular screening methods. It is a novel parvovirus with sequences similar to bovine and canine parvoviruses. Unlike another human parvovius, parvovirus B19 (see Chapter 10), it has been primarily implicated as a cause of wheezing and other respiratory illnesses in children. Diagnosis requires PCR methods. Further studies are ongoing to determine its epidemiologic behavior and relative contribution to respiratory morbidity.

REOVIRUSES

The reoviruses (respiratory enteric orphans) are naked virions that contain segmented, double-stranded RNA genomes that replicate and assemble in the cytoplasm of infected cells. They are ubiquitous and have been found in humans, simians, rodents, cattle, and a variety of other hosts. They have been studied in great detail as experimental models, revealing much basic knowledge about viral genetics and pathogenesis at the molecular level. Three serotypes are known to infect humans; however, their role and importance in human disease remain uncertain. Reoviruses causing arboviral diseases are discussed in Chapter 16.

Association with human disease is uncertain

CASE STUDY

AN INFANT WITH RESPIRATORY DISTRESS

This 9-month-old boy was born prematurely, requiring treatment in a neonatal intensive care unit for the first month of life. After discharge, he remained well until 3 days ago, when symptoms of a common cold progressed to cough, rapid and labored respiration, lethargy, and refusal to eat.

On examination, his temperature was 38.5°C, respiratory rate 60/min, and pulse 140/min. Auscultation of the chest revealed coarse crackles and occasional wheezes.

Abnormal laboratory findings included hypoxemia and hypercarbia. A chest radiograph showed hyperinflation, interstitial perihilar infiltrates, and right upper lobe atalectasis.

QUESTIONS

■ Which of these viruses is the least likely cause of this baby's illness?

A. Influenza A
B. Parainfluenza 3
C. Influenza C
D. Respiratory syncytial virus
E. Adenovirus

■ The mechanism of "antigenic drift" in influenza viruses includes all but one of the following:

A. Can involve either H or N antigens
B. Mutations caused by viral RNA polymerase
C. Can predominate under selective host population immune pressures
D. Reassortment between human and animal or avian reservoirs
E. Can involve genes encoding structural or nonstructural proteins

■ Which of the following agents can be used to prevent RSV pneumonia?

A. Amantadine
B. Vaccine to F protein
C. Oseltamivir
D. Zanamivir
E. Monoclonal antibody

ANSWERS

1(C), **2**(D), **3**(E)

CHAPTER 10

Mumps Virus, Measles, Rubella, and Other Childhood Exanthems

They wondered
If wheezeles
Could turn
Into measles,
If sneazles
Would turn
Into mumps

—A.A. Milne, *Now We Are Six*

The major viruses described in this chapter are mumps, measles, rubella, and the human parvovirus B19, which are genetically unrelated but share several common epidemiologic characteristics, including: (1) worldwide distribution, with a high incidence of infection in nonimmune individuals; (2) humans as sole reservoir of infection; and (3) person-to-person spread primarily by the respiratory (aerosol) route.

The other disease discussed in this chapter, roseola infantum, is a common illness occurring in early life. Key characteristics of these viruses are summarized in **Table 10–1**

MUMPS

VIROLOGY

Enveloped single-stranded RNA virus with hemagglutinating and neuraminidase activity

Mumps virus is a paramyxovirus, and only one major antigenic type is known. Like fellow members of its genus, it contains single-stranded, negative-sense RNA genome, and the nucleocapsid is surrounded by a lipid bilayer envelope. Two glycoproteins are on the surface of the envelope; one mediates hemagglutination and neuraminidase (HN) activity, and the other is responsible for lipid membrane fusion (F) to the host cell. Like other paramyxoviruses, mumps virus initiates

TABLE 10–1 Comparison of Mumps and Major Exanthems

FEATURE	MUMPS	MEASLES	RUBELLA	PARVOVIRUS B19	ROSEOLA
Virus type	Paramyxovirus, enveloped, single-stranded RNA	Paramyxovirus (Morbillivirus) enveloped, single-stranded RNA	Togavirus (Rubivirus) enveloped, single-stranded RNA	Parvovirus, naked capsid, single-stranded DNA	Human herpesviruses 6 or 7, enveloped, double-stranded DNA
Transmission	Respiratory	Respiratory	Respiratory	Respiratory	Oral secretions
Incubation period (days)	12-29	7-18	14-21	4-12	Unknown
Symptoms	Fever, parotitis	Fever, cough, conjunctivitis, Koplik's spots	Fever (low grade), upper respiratory symptoms	Mild fever, malaise, headache, myalgia, itching	High fever, occasional late sudden rash
Characteristic rash	None	Widespread, maculopapular	Faint, macular	Macular, reticular, often faint	Transient, faint macular
Duration of illness	7-10 days	3-5 days	1-3 days	1-2 weeks	3-5 days
Severity and/or complications	Meningitis, encephalitis, pancreatitis, orchitis, oophoritis	Bacterial superinfection, encephalitis, keratitis, reactivation of tuberculosis, subacute sclerosing panencephalitis (rare)	Overt arthritis Congenital infection	Aplastic crisis (in chronic[a] hemolytic diseases), arthritis, arthralgias	
Fetal infection	No[a]	No[a]	Yes— multiple defects	Yes-stillbirth, fetal hydrops	No[a]
Vaccine	Live attenuated	Live attenuated	Live attenuated	No	No

[a]Fetal infection may rarely occur, but with no apparent consequences.

infection by attachment of the HN spike to sialic acid on the cell surface, and F protein promotes fusion with the plasma membrane. It replicates in the cytoplasm by using its own RNA-dependent RNA polymerase, and the progeny viruses are released by budding from the cell membranes. The details about the structure and replication of paramyxoviruses are described in Chapter 9.

MUMPS INFECTION

CLINICAL CAPSULE

Before an effective vaccine against mumps was developed, the disease was a common, highly contagious childhood illness, often expressed as parotitis. It is also capable of causing aseptic meningitis, encephalitis, and (in adults) acute orchitis. In recent years, there has been a resurgence of outbreaks in the United States and elsewhere, underscoring the ongoing necessity to ensure adequate surveillance and immunization efforts.

EPIDEMOLOGY

Mumps infection is observed to occur most frequently in the 5- to 15-year age group. Infection is rarely seen in the first year of life. Although about 85% of susceptible household contacts acquire infection, approximately 30% to 40% of these contacts do not develop clinical disease. The disease is communicable from approximately 7 days before until 9 days after onset of illness; however, virus has been recovered in urine for up to 14 days after onset. The highest incidence of infection is usually during the late winter and spring months, but it can occur during any season.

High infectivity is present before and after onset of illness

PATHOGENESIS

After initial entry into the respiratory tract, the virus replicates locally in the respiratory tract epithelium and local lymph nodes. Replication is followed by viremic dissemination to target tissues such as the salivary glands and central nervous system (CNS). It is also possible that before development of immune responses, a secondary phase of viremia may result from virus replication in target tissues (eg, initial parotid involvement with later spread to other organs). Viruria is common, probably as a result of direct spread from the blood into the urine, in addition to active viral replication in the kidney. The tissue response is that of cell necrosis and inflammation, with predominantly mononuclear cell infiltration. In the salivary glands, swelling and desquamation of necrotic epithelial lining cells, accompanied by interstitial inflammation and edema, may be seen within dilated ducts.

Viremic phase follows local replication

Viruria is common

IMMUNITY

As in most viral infections, the early antibody response in mumps is predominantly with IgM, which is replaced gradually over several weeks by specific IgG antibody. The latter persists for a lifetime but can often be detected only by specific neutralization assays. Immunity is associated with the presence of neutralizing antibody. The role of cellular immune responses is not clear, but they may contribute both to the pathogenesis of the acute disease and to recovery from infection. After primary infection, immunity to reinfection is virtually always permanent.

Neutralizing antibody is protective

CLINICAL ASPECTS

MANIFESTATIONS

After an incubation period of 12 to 29 days (average 16 to 18 days), the typical case of mumps is characterized by fever and swelling with tenderness of the salivary glands, especially the parotid glands (**Figure 10–1**). Swelling may be unilateral or bilateral and persists for 7 to 10 days. Several complications can occur, usually within 1 to 3 weeks of onset of illness. All appear to be a direct result of virus spread to other sites and illustrate the extensive tissue tropism of mumps.

Incubation period is 12 to 29 days

Parotitis is unilateral or bilateral

Complications, which can occur without parotitis, include infection of the following:

1. **Meninges**: Approximately 10% of all infected patients develop meningitis. It is usually mild, but can be confused with bacterial meningitis. In about one third of these cases, associated or preceding evidence of parotitis is absent.
2. **Brain**: Encephalitis is occasionally severe.
3. **Spinal cord and peripheral nerves**: Transverse myelitis and polyneuritis are rare.
4. **Pancreas**: Pancreatitis is suggested by abdominal pain and vomiting.
5. **Testes**: Orchitis is estimated to occur in 10% to 20% of infected men. Although subsequent sterility is a concern, it appears that this outcome is rare.
6. **Ovaries**: Oophoritis is an unusual, usually benign inflammation of the ovarian glands.

Other rare and transient complications include myocarditis, nephritis, arthritis, thyroiditis, thrombocytopenic purpura, mastitis, and pneumonia. Most complications resolve without sequelae within 2 to 3 weeks. However, occasional permanent effects have been

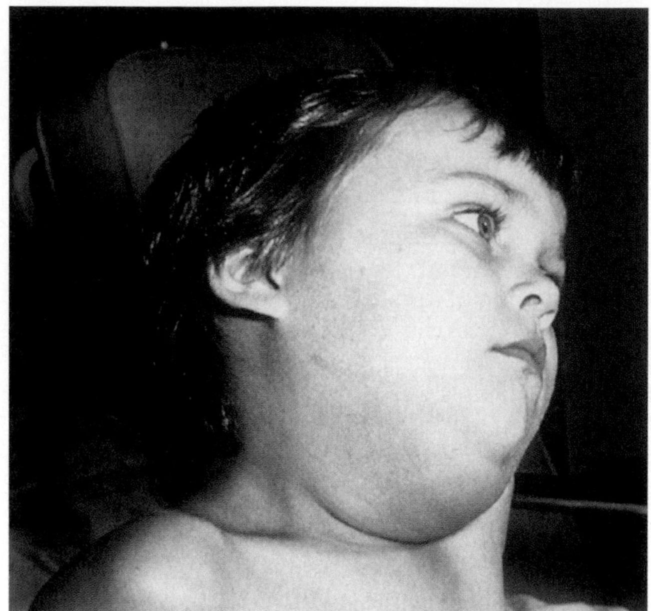

FIGURE 10–1. Mumps parotitis. The swelling just below the earlobe is due to enlargement of the parotid gland. (Reproduced with permission from Nester EW, Anderson DG, Roberts CE Jr, Nester MT. *Microbiology: A Human Perspective,* 6th ed. New York: McGraw-Hill, 2008.)

noted, particularly in severe CNS infection, in which sensorineural hearing loss and other impairment can occur.

DIAGNOSIS

Cell culture of saliva, throat, CSF, and urine; PCR

Viral antigen detected by immunofluorescence and EIA

EIA serology detects IgM and IgG

Mumps virus can be readily isolated early in the illness from the saliva, pharynx, and other affected sites, such as the cerebrospinal fluid (CSF). The urine is also an excellent source for virus isolation. Mumps virus grows well in primary monolayer cell cultures derived from monkey kidney, producing syncytial giant cells and viral hemagglutinin. Rapid diagnosis can be made by direct detection of viral antigen in pharyngeal cells or urine sediment, and by polymerase chain reaction (PCR).

The usual serologic tests are enzyme immunoassay (EIA) or enzyme-linked immunosorbent assay (ELISA) and indirect immunofluorescence to detect IgM- and IgG-specific antibody responses. Other serologic tests are also available, such as complement fixation, hemagglutination inhibition, and neutralization. Of these, the neutralization test is the most sensitive for detection of immunity to infection.:

PREVENTION

Live vaccine ideally given at 12 to 15 months of age, repeated at 4 to 6 years

No specific therapy is available for mumps. Since 1967, a live attenuated vaccine that is safe and highly effective has been available. As a result of its routine use, infections in the United States before 2005 were exceedingly rare; however, in late 2005 and into 2006, a large outbreak (over 6000 proved or probable cases) developed in Iowa and eight neighboring midwestern states. Most occurred in persons 18 to 25 years of age, many of whom had been previously vaccinated at least once. The mumps strain identified was genotype G, a common strain similar to the one that involved over 70,000 cases in the United Kingdom from 2004 to 2006. Thus, it has been re-emphasized that a two-dose vaccine regimen is essential to ensure adequate immunity. The vaccine is produced by serial propagation of virus in chick embryo cell cultures. It is commonly combined with measles, rubella, and varicella vaccines (MMRV) and given as a single injection to a child at 12 to 15 months of age. A second dose of MMRV is recommended at 4 to 6 years of age; those who have missed the second dose should receive it no later than 11 to 12 years of age. A single dose causes seroconversion in approximately 80% of recipients, and it increases only to about 90% after two doses. The vaccine must be given at least 2 to 4 weeks before exposure to be at all effective in postexposure prophylaxis.

MEASLES

VIROLOGY

The measles virus is classified in the paramyxovirus family, genus *Morbillivirus.* It contains a linear, negative-sense, single-stranded RNA genome, which encodes at least six virion structural proteins. Of these, three are in the envelope, comprising a matrix (M) protein that plays a key role in viral assembly and two types of glycoprotein projections (peplomers). One of the projections is a hemagglutinin (H), which mediates adsorption to cell surfaces; the other (F) mediates cell fusion, hemolysis, and viral entry into the cell. Unlike mumps virus, measles virus lacks neuraminidase (N) activity. The receptor for measles virus is CD46 (membrane cofactor protein), a regulator of complement activation. Replication of measles virus is similar to other paramyxoviruses. Only a single serotype restricted to human infection is recognized; however, subtle antigenic and genetic variations among wild-type measles strains do occur. These variations can be determined by sequencing analyses, enabling more precise epidemiologic tracking of outbreaks and their origins. Such ongoing molecular surveillance is also extremely important in determining whether significant antigenic drifts evolve over time.

Enveloped negative-sense, single-stranded RNA virus has hemagglutinin and fusion glycoproteins

CD46 is cell receptor

+ contagieux ! (1-2 jrs avant sympt. & 4 jrs après)

MEASLES INFECTION

CLINICAL CAPSULE

Measles infections often produce severe illness in children, associated with high fever, widespread rash, and transient immunosuppression. The virus is one the most contagious agents among humans. Serious complications include pneumonia, encephalitis, and bleeding disorders. Long-term sequelae, such as blindness, may occur, and rarely a few patients develop a slowly fatal condition called subacute sclerosing panencephalitis with onset years after the initial infection. This condition remains a major cause of mortality among malnourished children in developing countries. An effective vaccine is available.

EPIDEMIOLOGY

The highest attack rates of measles have been in children, usually sparing infants less than 6 months of age because of passively acquired antibody. However, a shift in age-specific attack rates to greater involvement of adolescents and young adults was observed in the United States in the 1980s. A marked decline in measles in the United States during the early 1990s may reflect decreased transmission as increased immunization coverage takes effect. However, in developing countries an estimated 1 million children still die from this disease each year. Furthermore, measles remains endemic in most countries in the world, including parts of Europe. In 2007-2008, large outbreaks of measles were occurring in Switzerland and Israel, resulting in imported cases leading to localized spread within the United States. Thus, continued vigilance is required for all who care for patients.

Though a childhood disease, infections in young adults is important in transmission

Dramatic decrease in United States, but importation of infections still a problem

Epidemics occur in unimmunized or partially immunized groups

Epidemics tend to occur during the winter and spring and increasingly are limited to one-dose vaccine failures or groups who do not accept immunizations. The infection rate among exposed susceptible subjects in a classroom or household setting is estimated at 85%, and more than 95% of those infected become ill. The period of communicability is estimated to be 3 to 5 days before appearance of the rash to 4 days afterward.

PATHOGENESIS

Respiratory cell multiplication disrupts cytoskeleton

Viremia disseminates to multiple sites

After implantation in the upper respiratory tract, viral replication proceeds in the respiratory mucosal epithelium. The effect within individual respiratory cells is profound. Even though measles does not directly restrict host cell metabolism, susceptible cells are damaged or destroyed by virtue of the intense viral replicative activity and the promotion of cell fusion with formation of syncytia. This results in disruption of the cellular cytoskeleton, chromosomal disorganization, and the appearance of inclusion bodies within the nucleus and cytoplasm. Replication is followed by viremic and lymphatic dissemination throughout the host to distant sites, including lymphoid tissues, bone marrow, abdominal viscera, and skin. Virus can be demonstrated in the blood during the first week after illness onset, and viruria persists for up to 4 days after the appearance of rash. Viremia also allows the infection of conjunctiva, urinary tract, small blood vessels, and the CNS.:

T and B lymphocytes are infected

Leukocyte function is impaired

Susceptibility to bacterial superinfections enhanced

During the viremic phase, measles virus infects T and B lymphocytes, circulating monocytes, and polymorphonuclear leukocytes without producing cytolysis. Profound depression of cell-mediated immunity occurs during the acute phase of illness and persists for several weeks thereafter. This is believed to be a result of virus-induced down-regulation of interleukin-12 production by monocytes and macrophages. The effect on B lymphocytes has been shown to suppress immunoglobulin synthesis; in addition, generation of natural killer cell activity appears to be impaired. There is also evidence that the capability of polymorphonuclear leukocytes to generate oxygen radicals is diminished, perhaps directly by the virus or by activated suppressor T cells. This may further explain the enhanced susceptibility to bacterial superinfections. Virion components can be detected in biopsy specimens of Koplik's spots and vascular endothelial cells in the areas of skin rash.:

Vasculitis, giant cells, and inclusions are seen

In addition to necrosis and inflammatory changes in the respiratory tract epithelium, several other features of measles virus infection are noteworthy. The skin lesions show vasculitis characterized by vascular dilation, edema, and perivascular mononuclear cell infiltrates. The lymphoid tissues show hyperplastic changes, and large multinucleated reticuloendothelial giant cells are often observed (Warthin-Finkeldey cells). Some of the giant cells contain intracytoplasmic and intranuclear inclusions. Similarly involved giant epithelial cells can be found in a variety of mucosal sites, the respiratory tract, skin, and urinary sediment.:

Encephalitis lesions are due to cytotoxic T-cell activity

In some patients with measles, an immune mediated postinfectious encephalitis occurs after the rash. The major findings in measles encephalitis include areas of edema, scattered petechial hemorrhages, perivascular mononuclear cell infiltrates, and necrosis of neurons. In most cases, perivenous demyelination in the CNS is also observed. The pathogenesis is thought to be related to infiltration by cytotoxic (CD8+) T cells, which react with myelin-forming or virus-infected brain cells.

IMMUNITY

Lifelong immunity associated with neutralizing antibody

Cell mediated immune responses to other antigens may be acutely depressed during measles infection and persist for several months. There is evidence that measles virus-specific cell-mediated immunity developing early in infection plays a role in mediating some of the features of disease, such as the rash, and is necessary to promote recovery from the illness. Antibodies to the virus appear in the first few days of illness, peak in 2 to 3 weeks, and then persist at low levels. Immunity to reinfection is lifelong and is associated with the presence of neutralizing antibody. In patients with defects in cell-mediated immunity, including those with severe protein-calorie malnutrition, infection is prolonged, tissue involvement is more severe, and complications such as progressive viral pneumonia are common. ::: cell-mediated immunity, p.35

CLINICAL ASPECTS

MANIFESTATIONS

Common synonyms for measles include **rubeola,** 5-day measles, and hard measles. The incubation period ranges from 7 to 18 days. A typical illness usually begins 9 to 11 days after exposure, with cough, coryza, conjunctivitis, and fever. One to three days after onset, pinpoint gray-white spots surrounded by erythema (grains-of-salt appearance) appear on mucous membranes. This sign, called **Koplik spots,** is usually most noticeable over the buccal mucosa opposite the molar teeth and persists for 1 to 2 days (**Figure 10–2**). Within a day of the appearance of Koplik spots, the typical measles rash begins, first on the head, then on the trunk and extremities. The rash is maculopapular and semiconfluent; it persists for 3 to 5 days before fading (**Figure 10–3**). Fever and severe systemic symptoms gradually diminish as the rash progresses to the extremities. Lymphadenopathy is also common, with particularly noticeable involvement of the cervical nodes.

Incubation period is 7 to 18 days

Koplik's spots appear on mucous membranes

Rash spreads from head to trunk and extremities

Measles can be very severe, especially in immunocompromised or malnourished patients. Death can result from overwhelming viral infection of the host, with extensive involvement of the respiratory tract and other viscera. In some developing countries, mortality rates of 15% to 25% have been recorded.

Complications

Bacterial superinfection, the most common complication, occurs in 5% to 15% of all cases. Such infections include acute otitis media, mastoiditis, sinusitis, pneumonia, and sepsis. Clinical signs of encephalitis develop in 1 of 500 to 1000 cases. This condition usually occurs 3 to 14 days after onset of illness and can be extremely severe. The mortality in measles encephalitis is approximately 15%, and permanent neurologic damage among survivors is estimated at 25%. Acute thrombocytopenic purpura may also develop during the acute phase of measles, leading to bleeding episodes. Abdominal pain and acute appendicitis can occur secondary to inflammation and swelling of lymphoid tissue.

Bacterial superinfection is common

Encephalitis can be severe

Thrombocytopenic purpura and bleeding occur in acute phase

Subacute Sclerosing Panencephalitis

Subacute sclerosing panencephalitis (SSPE) is a rare, progressive neurologic disease of children, which usually begins 2 to 10 years after a measles infection. It is characterized by insidious onset of personality change, poor school performance, progressive intellectual

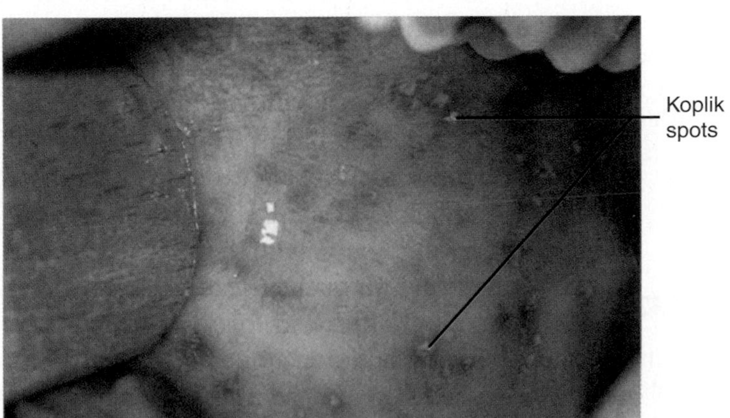

FIGURE 10–2. Oral Koplik spots on day 3 of measles. (Reproduced with permission from Nester EW, Anderson DG, Roberts CE Jr, Nester MT. *Microbiology: A Human Perspective,* 6th ed. New York: McGraw-Hill, 2008.)

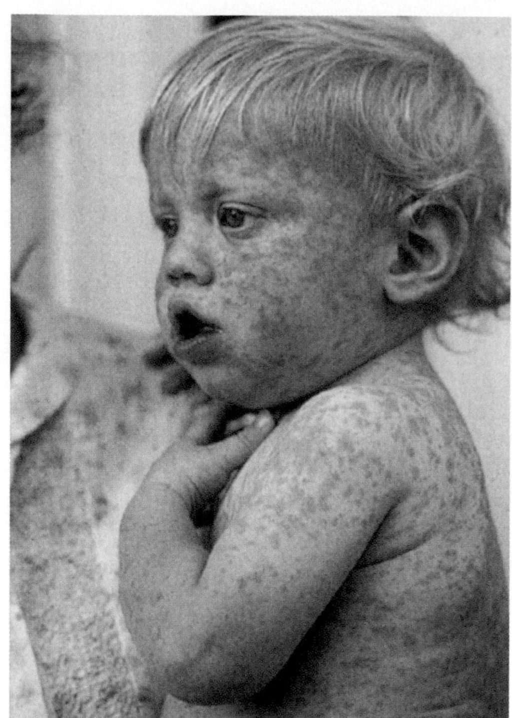

Maculopapulaire

FIGURE 10–3. Measles rash on day 4 of illness. (Reproduced with permission from Nester EW, Anderson DG, Roberts CE Jr, Nester MT. *Microbiology: A Human Perspective,* 6th ed. New York: McGraw-Hill, 2008.)

deterioration, development of myoclonic jerks (periodic muscle spasms), and motor dysfunctions such as spasticity, tremors, loss of coordination, and ocular abnormalities, including blindness. Neurologic and intellectual deterioration generally progresses over 6 to 12 months, with children eventually becoming bedridden and stuporous. Dysfunctions of the autonomic nervous system, such as difficulty with temperature regulation, may develop. Progressive inanition, superinfection, and metabolic imbalances eventually lead to death. Most of the pathologic features of the disease are localized to the CNS and retina. Both the gray matter and the white matter of the brain are involved, the most noteworthy feature being the presence of intranuclear and intracytoplasmic inclusions in oligodendroglial and neuronal cells.

Neurologic deterioration is progressive in children

Inclusions seen in neuronal cells

The disease is a result of chronic wild-type measles virus infection of the CNS. Studies have shown that patients have a variety of patterns of missing measles virus structural proteins in brain tissue. Thus, any of several defects in viral gene expression may prevent normal viral assembly, allowing persistence of defective virus at an intracellular site with failure of immune eradication. ::: persistent infection, p146

Chronic measles virus infection

Incomplete measles virus is present in brain tissue

Rarely, a similar progressive, degenerative neurologic disorder may be related to persistent rubella virus infection of the CNS. This condition is seen most often in adolescents who have had congenital rubella syndrome. Rubella virus has been isolated from brain tissue in these patients, again using cocultivation techniques.

The incidence of subacute sclerosing panencephalitis is approximately 1 per 100,000 measles cases. Its occurrence in the United States has decreased markedly over the last 25 years with the widespread use of live measles vaccine. At present, there is no accepted effective therapy for subacute sclerosing panencephalitis.

Incidence declined after introduction of measles vaccine

DIAGNOSIS

The typical measles infection can often be diagnosed on the basis of clinical findings, but laboratory confirmation is necessary. Virus isolation from the oropharynx or urine is usually most productive in the first 5 days of illness. Measles grows on a variety of cell cultures, producing multinucleated giant cells similar to those observed in infected host tissues. If rapid diagnosis is desired, measles may be identified in urinary sediment or pharyngeal cells by direct fluorescent antibody or PCR methods. Serologic diagnosis may involve hemagglutination inhibition, EIA, or indirect fluorescent antibody methods.

Rapid diagnosis is possible by immunofluorescence or PCR

TREATMENT

No specific therapy is available other than supportive measures and close observation for the development of complications such as bacterial superinfection. Intravenous ribavirin has been suggested for patients with severe measles pneumonia, but no controlled studies have been performed.

PREVENTION

Live, attenuated measles vaccine is available and highly immunogenic, most commonly administered as MMRV. To ensure effective immunization, the vaccine should be administered to infants at 12 to 15 months of age with a second dose at 4 to 6 or 11 to 12 years of age. Immunity induced by the vaccine may be life-long. Because the vaccine consists of live virus, it should not be administered to immunocompromised patients and is not recommended for pregnant women. Exceptions to these guidelines include susceptible human immunodeficiency virus-infected persons. Exposed susceptible patients who are immunologically compromised (including small infants) may be given immune serum globulin intramuscularly. This treatment can modify or prevent disease if given within 6 days of exposure, but protection is transient.:

Live, attenuated vaccine is highly immunogenic

Vaccination is contraindicated in pregnant and immunocompromised individuals

Passive protection is appropriate for immunocompromised

RUBELLA

Rubella was considered a mild, benign exanthem of childhood until 1941, when the Australian ophthalmologist Sir Norman Gregg described the profound defects that could be induced in the fetus as a result of maternal infection. Since 1962, when the virus was first isolated, knowledge regarding its extreme medical importance and biologic characteristics has increased rapidly.

VIROLOGY

Rubella virus is classified as a member of the togavirus family, *Rubivirus* genus. It is a simple, icosahedral, enveloped virus and contains a single-stranded, positive-sense RNA genome. There is a single species of capsid protein, and the lipid bilayer envelope contains two glycoproteins, E1 and E2. There is only one serotype of rubella, and no extrahuman reservoirs are known to exist. The virus can agglutinate some types of red blood cells, such as those obtained from 1-day-old chicks and trypsin-treated human type O cells.:

Enveloped togavirus contains single-stranded RNA

Rubella virus enters target cells via receptor-mediated endocytosis. Viral positive-sense RNA is translated to produce viral proteins, including RNA-dependent RNA polymerase. These proteins are required for the synthesis of replicative intermediates, genomic RNA, and subgenomic RNA. The subgenomic RNA encodes the structural proteins of the virus, including capsid and envelope proteins. Virus assembly takes place at either the Golgi complex or cytoplasmic membranes.

Genomic RNA encodes for nonstructural proteins and subgenomic RNA for structural proteins

RUBELLA INFECTION

CLINICAL CAPSULE

Infections by rubella virus are often mild, or even asymptomatic. Primary infection, when symptomatic, is often manifested as malaise, faint rash, and arthralgia .The major concerns are the profound effects of maternal infection during the first trimester of pregnancy, which can affect developing fetuses, resulting in multiple congenital malformations, such as cardiac and eye defects, deafness, and microcephaly.

EPIDEMOLOGY

Rubella virus has high infectivity but low virulence

Childbearing women are the major concern

Rubella infections are usually observed during the winter and spring months. In contrast to measles, which has a high clinical attack rate among exposed susceptible individuals, only 30% to 60% of rubella-infected susceptible persons develop clinically apparent disease. A major focus of concern is susceptible women of childbearing age, who carry a risk of exposure during pregnancy. Patients with primary acquired rubella infections are contagious from 7 days before to 7 days after the onset of rash; congenitally infected infants may spread the virus to others for 6 months or longer after birth.

PATHOGENESIS

Cellular immune responses and virus–antibody complexes mediate arthritis and rash

In acquired infection, the virus enters the host through the upper respiratory tract, replicates, and then spreads by the bloodstream to distant sites, including lymphoid tissues, skin, and organs. Viremia in these infections has been detected for as long as 8 days before to 2 days after onset of the rash, and virus shedding from the oropharynx can be detected up to 8 days after onset (**Figure 10–4**). Cellular immune responses and circulating virus-antibody immune complexes are thought to play a role in mediating the inflammatory responses to infection, such as rash and arthritis.:

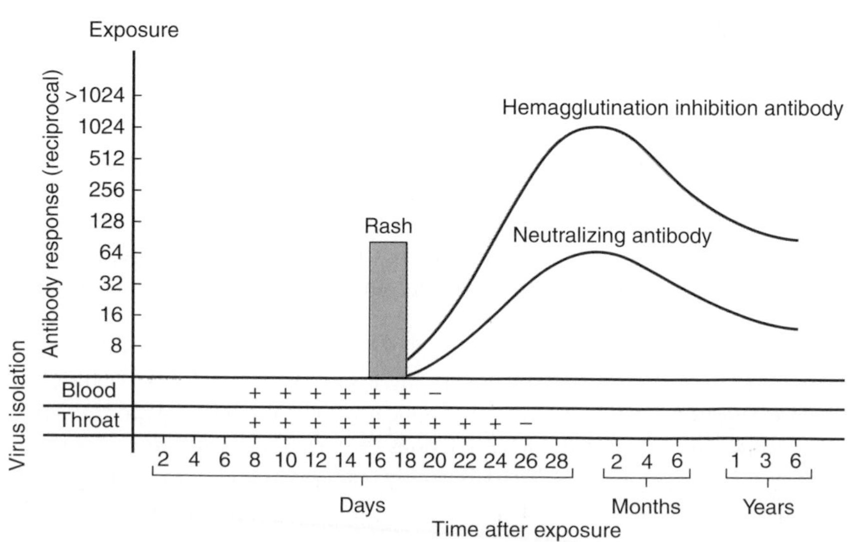

FIGURE 10–4. Antibody response and viral isolation in a typical case of acquired rubella.

Congenital infection occurs as a result of maternal viremia that leads to placental infection and then transplacental spread to the fetus. Once fetal infection occurs, it persists chronically. Such persistence is probably related to an inability to eliminate the virus by immune or interferon-mediated mechanisms. There is too little inflammatory change in the fetal tissues to explain the pathogenesis of the congenital defects. Possibilities include placental and fetal vasculitis with compromise of fetal oxygenation, chronic viral infection of cells leading to impaired mitosis, cellular necrosis, and induction of chromosomal breakage. Any or all of these factors may operate at a critical stage of organogenesis to induce permanent defects. Viral persistence with circulating virus–antibody immune complexes may evoke inflammatory changes postnatally and produce continuing tissue damage.

Transmission to fetus by viremia

Fetal infection becomes chronic

After birth, infants affected with rubella continue to excrete the virus in the throat, urine, and intestinal tract (**Figure 10–5**). Virus may be isolated from virtually all tissues in the first few weeks of life. Shedding of virus in the throat and urine, which persists for at least 6 months in most cases, has been known to continue for 30 months. Rubella virus has also been isolated from lens tissue removed 3 to 4 years later. These observations underscore the fact that such infants are important reservoirs in perpetuating virus transmission. The prolonged virus shedding is somewhat puzzling; it does not represent a typical example of immunologic tolerance. The affected infants are usually able to produce circulating IgM and IgG antibodies to the virus (Figure 10–5), although antibodies may decrease to undetectable levels after 3 to 4 years. Many infants have evidence of depressed rubella virus-specific cell-mediated immunity during the first year of life.

Infection and virus shedding continue long after birth

Virus persists despite antibody

PATHOLOGY

Because postnatally acquired disease is usually mild, little is known about the pathology of rubella. Mononuclear cell inflammatory changes can be observed in tissues, and viral antigen can be detected in the same sites (eg, skin and synovial fluid). Congenital infections are characterized primarily by the various malformations. Necrosis of tissues such as myocardium and vascular endothelium may also be seen, and quantitative studies suggest a decrease in cell quantity in affected organs. In severe cases, normal calcium deposition in the metaphyses of long bones is delayed, sometimes referred to as a "celery stalk" appearance on a radiograph.

Fetal disease includes multiple malformations

IMMUNITY

After infection with rubella, the serum antibody titer rises, reaching a peak within 2 to 3 weeks of onset (Figure 10–4). Natural infection also results in the production of specific secretory IgA antibodies in the respiratory tract. Immunity to disease is nearly always lifelong; however, re-exposure can lead to transient respiratory tract infection, with an anamnestic rise in IgG and secretory IgA antibodies, but without resultant viremia or illness.

Lasting immunity is associated with IgG and IgA

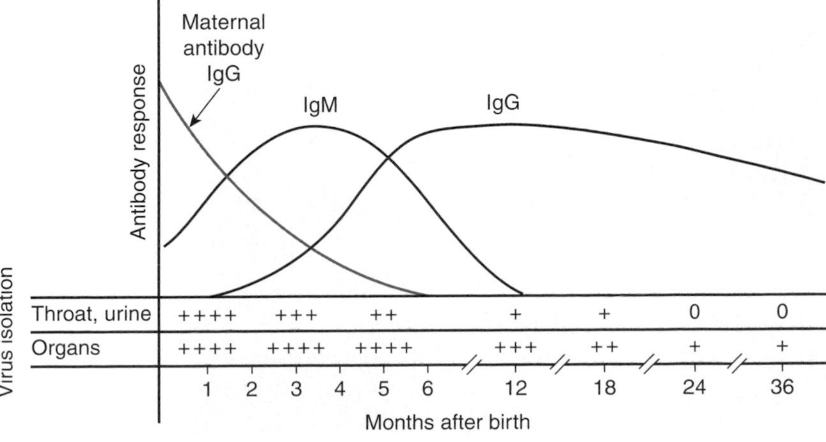

FIGURE 10–5. Persistence of rubella virus and antibody in congenitally infected infants.

CLINICAL ASPECTS

MANIFESTATIONS

Illness is mild with lymphadenopathy and macular rash

Arthralgia and arthritis is common in women

Rubella is commonly known as **German measles** or 3-day measles. The incubation period for acquired infection is 14 to 21 days (average 16 days). Illness is generally very mild, consisting primarily of low-grade fever, upper respiratory symptoms, and lymphadenopathy, which is most prominent in the posterior cervical and postauricular areas. A macular rash often follows within a day of onset and lasts 1 to 3 days. This rash, which is often quite faint, is usually most prominent over the head, neck, and trunk (**Figure 10–6**). Petechial lesions may also be seen over the soft palate during the acute phase. The most common complication is arthralgia or overt arthritis, which may affect the joints of the fingers, wrists, elbows, knees, and ankles. The joint problems, which occur most frequently in women, rarely last longer than a few days to 3 weeks. Other, rarer complications include thrombocytopenic purpura and encephalitis.

The major significance of rubella is not the acute illness but the risk of fetal damage in pregnant women, particularly when they contract either symptomatic or subclinical primary infection during the first trimester. The risk of fetal malformation and chronic fetal infection, which is estimated to be as high as 80% if infection occurs in the first 2 weeks of gestation, decreases to 6% to 10% by the 14th week. The overall risk during the first trimester is estimated at 20% to 30%.

High risk for fetal damage with infection in first trimester

Lesions of congenital rubella include multiple body systems

Clinical manifestations of congenital rubella syndrome vary, but may include any combination of the following major findings: cardiac defects, commonly patent ductus arteriosus and pulmonary valvular stenosis; eye defects such as cataracts, chorioretinitis, glaucoma, coloboma, cloudy cornea, and microphthalmia; sensorineural deafness; enlargement of liver and spleen; thrombocytopenia; and intrauterine growth restriction. Other findings include CNS defects such as microcephaly, mental retardation, and encephalitis; anemia; transient immunodeficiency; interstitial pneumonia; and intravascular coagulation; hepatitis; rash; and other congenital malformations. Late complications of congenital rubella syndrome have also been described, including an increased risk of diabetes mellitus, chronic thyroiditis, and occasionally the development of a progressive subacute panencephalitis in the second decade of life. Some congenitally infected infants may appear entirely normal at birth, and sequelae such as hearing or learning deficits may not become apparent until months later. The spectrum of defects thus varies from subtle to severe.

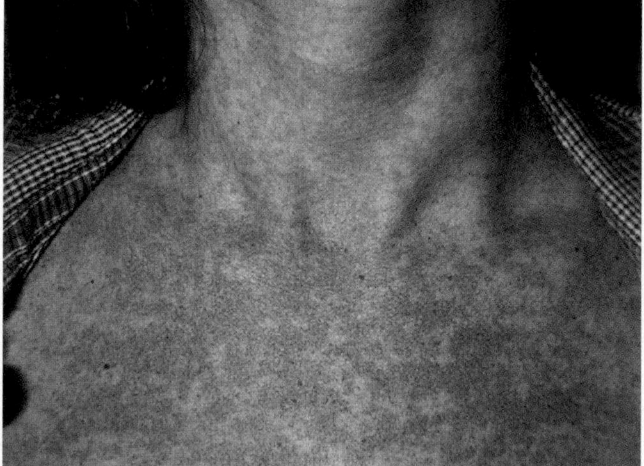

FIGURE 10–6. Rubella rash. Diffuse, macular in appearance, usually beginning on the face and spreading to the trunk. (Reproduced with permission from Nester EW, Anderson DG, Roberts CE Jr, Nester MT. *Microbiology: A Human Perspective,* 6th ed. New York: McGraw-Hill, 2008.)

DIAGNOSIS

Because of the rather nonspecific nature of the illness, a diagnosis of rubella cannot be made on clinical grounds alone. More than 30 other viral agents, which are discussed later in this chapter, can produce a similar illness. Confirmation of the diagnosis requires laboratory studies. The virus may be isolated from respiratory secretions in the acute phase (and from urine, tissues, and feces in congenitally infected infants) by inoculation into a variety of cell cultures or detected by reverse transcriptase polymerase chain reaction. Serologic diagnosis is most commonly used in acquired infections; paired acute and convalescent samples collected 10 to 21 days apart are used. Hemagglutination inhibition, indirect immunofluorescence, EIA, and other tests are available.

Acquired infections are diagnosed serologically

Determination of IgM-specific antibody is sometimes useful to ascertain whether an infection occurred in the last several months; it has also been used in the diagnosis of congenital infections. Unfortunately, there are certain pitfalls in interpreting this test. Some individuals (less than 5%) with acquired infections may have persistent elevations of IgM-specific antibodies for 200 days or more afterward, and some congenitally infected infants do not produce detectable IgM-specific antibodies.

IgM tests can help detect congenital infections

TREATMENT AND PREVENTION

Other than supportive measures, there is no specific therapy for either the acquired or the congenital rubella infection.

Since 1969, a live attenuated rubella vaccine has been available for routine immunization either alone or in combination (MMRV). As a result of the widespread use of the vaccine in the United States, the number of cases of rubella has declined dramatically. From 1990 through 1999, the median number of cases reported annually was only 232. The current vaccine virus, grown in human diploid fibroblast cell cultures (RA 27/3), has been shown to be highly effective. It causes seroconversion in approximately 95% of recipients. Routine immunization is now recommended for infants after the first year of life and for other individuals with no history of immunization and lack of immunity by serologic testing. Target groups include female adolescents and hospital personnel in high-risk settings. The vaccine is contraindicated in many immunocompromised patients and in pregnancy. To date, more than 200 instances of accidental vaccination of susceptible pregnant women have been reported, with no clinically apparent adverse effects on the fetus. However, it is strongly recommended that immunization be avoided in this setting and that nonpregnant women avoid conception for at least 3 months after receiving the vaccine.:

Live attenuated rubella vaccine is indicated for children and hospital workers

Vaccine does not produce defects in fetus

Vaccine-induced immunity may be lifelong

→ Manifest + connue = 5e maladie !

∅ Tx, mais attention ♀ enceintes !

PARVOVIRUS B19 INFECTIONS

Parvoviruses are very small (18 to 26 nm), naked icosahedral capsid virions that contain a linear single-stranded DNA genome. Diseases caused by parvoviruses have been recognized among nonhuman hosts for a number of years. Notable among these are canine parvovirus and feline panleukopenia virus, which produce particularly severe infections among puppies and kittens, respectively. These do not appear to cross species barriers. The human parvovirus B19 has been well described, but its origin is not yet known.:

Small naked, single-stranded DNA viruses

Parvovirus B19 encodes three capsid proteins (VP1, VP2, and VP3). The virus can be grown in primary cultures of human bone marrow cells, fetal liver cells, hematopoietic progenitor cells generated from peripheral blood, and a megakaryocytic leukemia cell line. The major cellular receptor for the virus is globoside (also known as blood group P antigen, which is commonly found on erythroid progenitors, erythroblasts, megakaryocytes, and endothelial cells). All represent potential targets for disease production. A primary site of replication appears to be the nucleus of an immature cell in the erythrocyte lineage that is mitotically active. Such infected cells then cease to proliferate, resulting in an impairment of normal erythrocyte development. Parvovirus enters the cells after binding to P antigen (globoside) followed by internalization, uncoating, and delivery of single-stranded DNA to

Replicates in erythroid precursor nuclei

Globoside is virus receptor

Endothelial cells and megakaryocytes can also be affected

the nucleus. The single-stranded DNA genome is converted to double-stranded DNA by host DNA polymerase, which is transcribed to produce viral mRNAs and proteins. After synthesis of single-stranded DNA genomes, progeny viruses are assembled in the nucleus and released upon cell lysis.

Aplastic crisis develops in patients with chronic hemolytic anemias

The clinical consequences of this effect on erythrocytes are generally trivial, unless patients are already compromised by a chronic hemolytic process, such as sickle cell disease or thalassemia, in which maximal erythropoiesis is continually needed to counterbalance increased destruction of circulating erythrocytes. Primary infection by parvovirus B19 in such individuals often produces an acute, severe, and sometimes fatal anemia manifested as a rapid fall in red blood cell count and hemoglobin. These patients may present initially with no clinical symptoms other than fever; this is commonly referred to as **aplastic crisis.** Immunocompromised patients such as those with acquired immunodeficiency syndrome (AIDS) sometimes have difficulty clearing the virus and develop persistent anemia with reticulocytopenia. Parvovirus B19 has also been occasionally implicated as a cause of persistent bone marrow failure and an acute hemophagocytic syndrome. In addition, it is now recognized as sometimes causing severe, protracted anemia in many settings of immunocompromise, including in AIDS patients, organ transplant recipients, and leukemic patients undergoing chemotherapy.

Erythema Infectiosum

Erythema infectiosum (also referred to as fifth disease or academy rash) is a more common disease that is clearly attributable to parvovirus B19. After an incubation period of 4 to 12 days, a mild illness appears, characterized by fever, malaise, headache, myalgia, and itching in varying degrees. A confluent, indurated rash appears on the face, giving a "slapped-cheek" appearance. The rash spreads in 1 or 2 days to other areas, particularly exposed surfaces such as the arms and legs, where it is usually macular and reticular (lace-like). During the acute phase, generalized lymphadenopathy or splenomegaly may be seen, along with a mild leukopenia and anemia.

Erythema infectiosum is usually a mild "slapped cheek" rash

Fetal infection is occasionally severe

Fetal anemia leads to hydrops fetalis

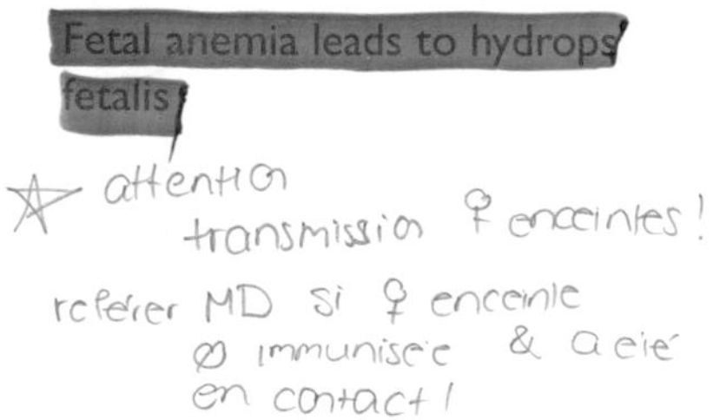

The illness of erythema infectiosum lasts 1 to 2 weeks, but rash may recur for periods of 2 to 4 weeks thereafter, exacerbated by heat, sunlight, exercise, and emotional stress. Arthralgia sometimes persists or recurs for weeks to months, particularly in adolescent or adult females. Overt arthritis or vasculitis have also been reported in some individuals. Serious complications such as hepatitis, thrombocytopenia, nephritis, or encephalitis are rare. However, like rubella, active transplacental transmission of parvovirus B19 can occur during primary infections in the first 20 weeks of pregnancy, sometimes resulting in stillbirth of fetuses that are profoundly anemic. The progress can be so severe that hypoxic damage to the heart, liver, and other tissues leads to extensive edema (hydrops fetalis). The frequency of such adverse outcomes is as yet undetermined.

It is important to be aware that erythema infectiosum is extremely variable in its clinical manifestations; even the "classic" presentation can be mimicked by other agents, such as rubella and echoviruses. Before a firm diagnosis is made on clinical grounds, especially during outbreaks, it is wise to exclude the possibility of atypical rubella infection.

Detection requires DNA probe or PCR

IgM-specific antibody supports diagnosis

Epidemiologic evidence suggests that spread of the virus is primarily by the respiratory route, and high transmission rates occur in households. Outbreaks tend to be small and localized, particularly during the spring months, with the highest rates among children and young adults. Seroepidemiologic studies have demonstrated evidence of past infection in 30% to 60% of adults. Viremia usually lasts 7 to 12 days but can persist for months in some individuals. It can be detected by specific DNA probe or polymerase chain reaction (PCR) methods. Alternatively, the presence of IgM-specific antibody late in the acute phase or during convalescence strongly supports the diagnosis.

Immunoglobulin treatment may be useful in selected cases

There is currently no definitive treatment for erythema infectiosum. Commercial immune globulins with antibodies to parvovirus B19 have been used with salutary effects and reduction of serum viral DNA in some patients with refractory infection in a setting of immunodeficiency.

Currently, a recombinant parvovirus B19 is under study, which could potentially benefit groups especially at risk because of chronic hemolytic disease, immunodeficiency, or seronegative pregnancies (to prevent hydrops fetalic), and perhaps even benefit children with acute anemia due to malaria, in whom the hematologic effects may be more profound if there is parvovirus B19 coinfection.:

Recombinant vaccine is under development

ROSEOLA INFANTUM (EXANTHEM SUBITUM)

Roseola infantum is a common illness observed in infants and children 6 months to 4 years of age. Its alternative name, exanthem subitum, means "sudden" rash. Roseola has more than one cause: the most common is human herpesvirus type 6 and, less frequently, human herpesvirus type 7 (see Chapter 14). Several other agents, including adenoviruses, coxsackieviruses, and echoviruses, have occasionally been noted to cause similar manifestations. The illness is characterized by abrupt onset of high fever, sometimes accompanied by brief, generalized convulsions and leukopenia. After 3 to 5 days, the fever diminishes rapidly, followed in a few hours by a faint, transient, macular rash.

Associated with human herpesvirus type 6 or type 7

OTHER CAUSES OF RUBELLA-LIKE RASHES

In addition to erythema infectiosum, diseases caused by numerous other agents can mimic rubella. These include at least 17 echoviruses, 9 coxsackieviruses, several adenoviral serotypes, and arboviruses such as dengue, Epstein-Barr virus, scarlet fever, and toxic drug eruptions. Because of the wide variety of diagnostic possibilities, it is not possible to diagnose or rule out rubella confidently on clinical grounds alone. Therefore, a specific diagnosis requires specific laboratory studies. Because rubella is an infection with such significant impact on the fetus, serologic study to rule out the possibility is mandatory if the diagnosis is suspected during early pregnancy, both in the woman and potentially infective contacts.

CASE STUDY

A PREVENTABLE ILLNESS?

A 12-year-old boy returned to the United States 2 days ago, after 3 weeks of travel with his family throughout southern Europe and northern Africa.

Yesterday, he developed a fever, dry cough, runny nose, and bilateral conjunctivitis. Twenty-four hours later, the fever has reached 39.1°C, and the other symptoms have worsened somewhat.

Physical examination reveals pharyngeal and conjunctival inflammation, and swollen, nontender anterior cervical lymph nodes. No rash is apparent.

He received all routinely recommended childhood immunizations by 5 years of age, but none since.

QUESTIONS

- At this stage of the boy's illness, which of the following viruses do you consider to be the most likely cause?

A. Measles
B. Mumps
C. Rubella
D. Human herpesvirus 6
E. Parvovirus B19

■ What should be your *first* course of action when evaluating the patient?

A. Obtain a complete blood count
B. Obtain viral cultures
C. IgM-specific antibody testing
D. Obtain a blood culture
E. Immediate isolation of the patient

■ The pathogenesis of infection includes a significant tropism for vascular endothelial cells in all the following viruses *except:*

A. Mumps
B. Measles
C. Rubella
D. Human herpesvirus 6
E. Parvovirus B19

ANSWERS

1(A), **2**(E), **3**(A)

CHAPTER 11

Poxviruses

You have erased from the calendar of human afflictions one of its greatest. Yours is the comfortable reflection that mankind can never forget that you have lived. Future nations will know by history only that the loathsome small-pox has existed.

—Thomas Jefferson, Letter to Edward Jenner, 1806

Poxviruses are the largest and most complex viruses infecting humans, other mammals, birds, and even insects. The agents most important in human disease are variola (smallpox), vaccinia, monkeypox, molluscum contagiosum, orf, cowpox, and pseudocowpox (**Table 11–1**).

POXVIRUSES: GROUP CHARACTERISTICS

Largest and most complex double-stranded DNA virus

All viral replication events occur in cytoplasm

Poxviruses are large, brick-shaped or ovoid, linear double-stranded DNA containing core within a double membrane and a lipoprotein envelope carrying virions measuring approximately 100 × 200 × 300 nm (**Figures 11–1 and 11–2**). The core is flanked by two lateral bodies containing several viral enzymes and proteins, including DNA-dependent RNA polymerase and transcription factors required for viral replication. The poxvirus genome encode all essential enzymes, proteins, and factors needed for viral replication in the cytoplasm of infected cells, including transcription, DNA synthesis, and virus assembly. The envelope is not acquired by budding and is also not essential for infectivity.:

Poxvirus replication is unique among DNA viruses in that the viral replication cycle takes place in the cytoplasm (**Figure 11–3**). The viral replication cycle starts with attachment, rapid adsorption to receptors followed by viral entry, and release of cores in the cytoplasm. Viral DNA-dependent RNA polymerase in the cores initiate early transcription to synthesize several proteins, including DNA and RNA polymerases, transcription factors, growth factors, and immune defense molecules. The uncoating of the cores uses viral DNA to synthesize concatemeric DNA molecules, which are eventually resolved into viral DNA genomes for progeny viruses. The late mRNAs synthesize viral structural proteins required for virus assembly and early transcription factors for packaging in the virions. Assembly of the progeny viruses begins with the formation of membrane structures followed by maturation of intracellular mature virions. The virions are further wrapped by membranes from the Golgi apparatus that are lost upon the release of extracellular enveloped virions.

TABLE 11–1 Poxviridae That Affect Humans	
GENERA	**DISEASES**
Orthopoxvirus	Variola
	Vaccinia
	Cowpox[a]
	Monkeypox[a]
Parapoxvirus	Bovine papular stomatitis[a]
	Orf[a]
	Pseudocowpox[a]
Molluscipoxvirus	Molluscum contagiosum
Yatapoxvirus	Tanapox[a]
	Yabapox[a]

[a]Viruses that have nonhuman reservoirs but can cause disease in humans (usually mild and localized).

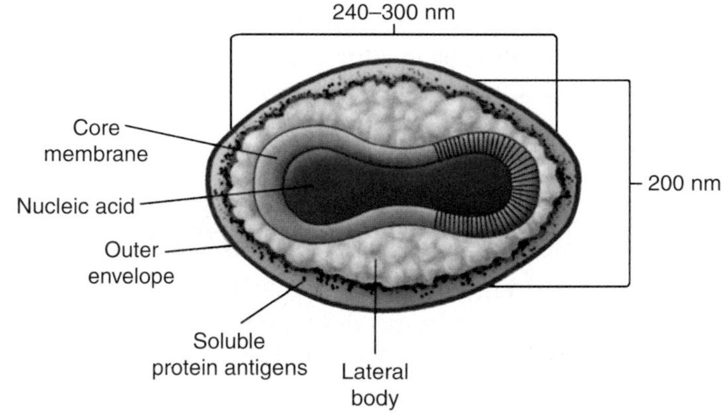

FIGURE 11–1. Schematic diagram of the structure of poxvirus virion. Viral DNA and several viral proteins within the core form the nucleosome (N). The core is covered with a 9-nm thick core membrane (CM) and assumes a dumbbell shape because of two lateral bodies (LB), which is eventually enclosed within a protein shell of 12 nm thickness (outer membrane) containing irregular surface tubules (T). The virion is enclosed in a lipid bilayer envelope containing virus-specific proteins. (Reproduced with permission from Willey J, Sherwood L, Woolverton C (eds). *Prescott's Principles of Microbiology.* New York: McGraw-Hill; 2008.)

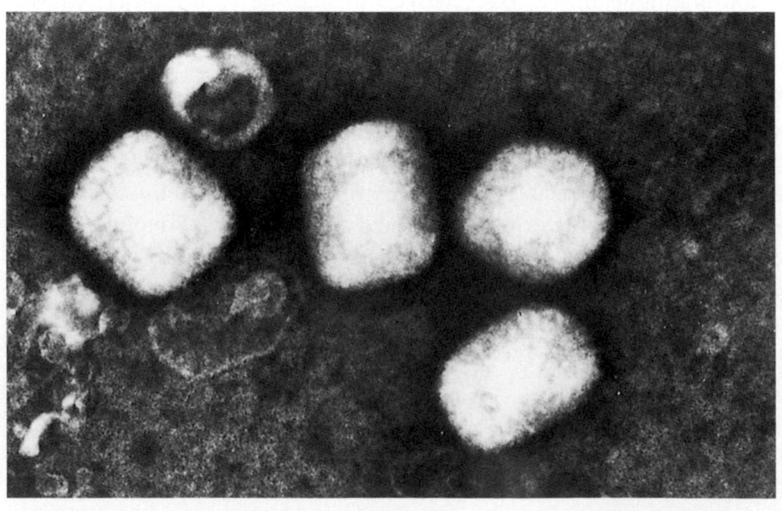

FIGURE 11–2. Electron microscopic appearance of a poxvirus (vaccinia). (Negative stain; original magnification × 60,000.) (Courtesy of Dr. Claire M. Payne.)

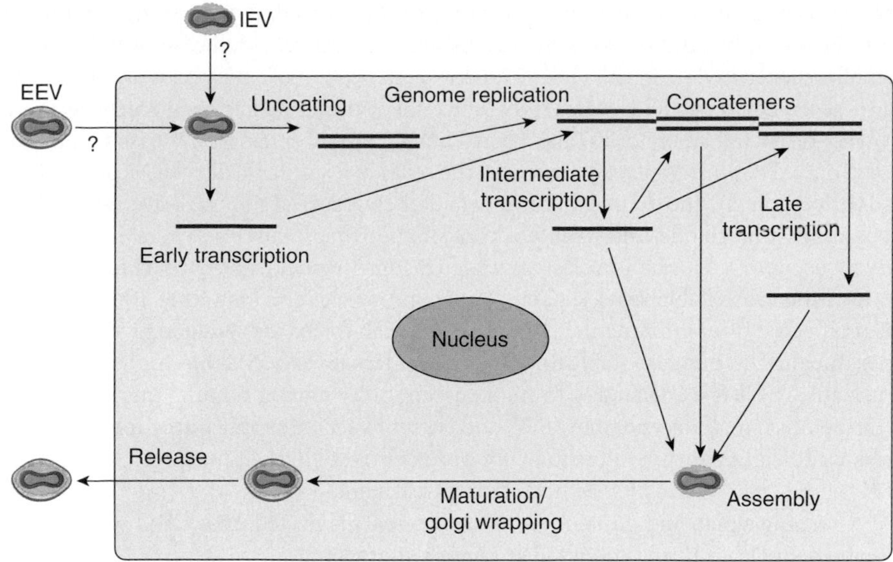

FIGURE 11–3. Replication cycle of poxviruses. All the events in sequence are listed in the diagram, including (1) attachment, (2) entry, (3) early mRNA synthesis, (4) uncoating, (5) genome replication, (6) intermediate mRNA synthesis, (7) late mRNA synthesis, (8) assembly, (9) DNA genome packaging, (10) maturation, (11) envelope wrapping from Golgi, and (12) exit or virus release. EEV (extracellular enveloped virus); IEV (intracellular enveloped mature virus.)

VARIOLA (SMALLPOX)

VIROLOGY

Two virus types are known: variola major and variola minor (alastrim). Although the viruses are indistinguishable antigenically, their fatality rates differ considerably (less than 1% for variola minor, 3% to 35% for variola major).

Variola major and variola minor are difficult to distinguish

SMALLPOX

CLINICAL CAPSULE

Smallpox is an acute infection in which the dominant feature is a uniform papulovesicular rash that evolves to pustules over 1 to 2 weeks. The potential for spread and mortality is significant, particularly in a nonimmune population. Intensive worldwide epidemiologic control measures, including vaccination, are currently thought to have achieved global eradication of the disease; nevertheless, it is imperative to continue careful surveillance in the event that the virus may unexpectedly re-emerge. Other poxviruses, such as monkeypox, are occasionally transmitted from animals to humans, and can sometimes mimic smallpox in a much milder form.

Smallpox has played a significant role in world history with respect to both the serious epidemics recorded since antiquity and the sometimes dangerous measures taken to prevent infection. Smallpox virus is highly contagious and can survive well in the extracellular environment. Acquisition of infection by infected saliva droplets or by exposure to skin lesions, contaminated articles, and fomites has been well documented.

Person-to-person communicability by respiratory droplets and fomites is high

In 1967, the World Health Organization (WHO) launched an ambitious program aimed at eradication of smallpox. This goal was considered realistic for two major reasons: (1) no extrahuman reservoir of the virus was known to exist, and (2) asymptomatic carriage apparently did not occur. The basic approach included intensive surveillance for clinical cases of smallpox, prompt quarantine of such patients and their contacts, and immunization of contacts

WHO eradication campaign based on lack of nonhuman reservoir and asymptomatic cases

with vaccinia virus (vaccination) to prevent further spread. A tremendous amount of effort was involved, but the results were astonishing: the last recorded case of naturally acquired smallpox occurred in Somalia in 1977. Global eradication of smallpox was confirmed in 1979 and accepted by WHO in May 1980. Since then, the virus has been solely secured in two WHO-restricted laboratories: one at the United States Centers for Disease Control and Prevention (CDC) in Atlanta, Georgia, and the other at a similar facility in Moscow, Russia.

Immunization and case tracing led to success in 1980

Unfortunately, the dramatic world events that occurred in 2001 have raised the chilling possibility that clandestine virus stocks of smallpox may exist elsewhere and could be effectively used for major bioterrorist attacks. Reasons for such concern include (1) smallpox is one of the most stable viruses; (2) it can remain stable for a long time, if freeze-dried; (3) it is unaffected by environmental condition; (4) scab forms are stable for 1 year at room temperature and in one case stable for 13 years in a laboratory; (5) it has high infectivity among humans; (6) it is associated with high susceptibility among populations (routine vaccination against smallpox ended in 1972, and current vaccine supplies are limited); (7) there is a risk that health care providers may not promptly recognize and respond to early cases; and (8) there is an absence of specific antiviral treatment.

Potential bioterrorist weapon

One of the most stable viruses unaffected by environmental conditions

Freeze-dried form of smallpox virus and scab form very stable for a long time

No proven antiviral treatment

A response plan and guidelines for such threats is posted on a CDC website (www.cdc.gov/nip/smallpox) and is updated at regular intervals.

Continuing surveillance also includes studies of poxviruses of animals (eg, buffalopox, monkeypox) that are antigenically somewhat similar to smallpox. Some virologists remain legitimately concerned that an animal poxvirus, such as monkeypox, could mutate to become highly virulent to humans—a further reminder that complacency could be dangerous.

Animal poxviruses could be a future threat

PATHOGENESIS

The orthopoxviruses as a group cause a dramatic effect on host cell macromolecular function, leading to a switch from cellular to viral protein synthesis, changes in cell membrane permeability, and cytolysis. Eosinophilic inclusions, called **Guarnieri bodies,** can be seen in the cytoplasm. Multiple viral proteins, such as complement regulatory protein and other factors that can interfere with induction or activities of multiple host mononuclear cell cytokines, are also synthesized. This serves to impair the host defenses that are important in early control of infection.

Profound effect on host cell protein synthesis

Viral proteins undermine host defenses

CLINICAL ASPECTS

MANIFESTATIONS AND DIAGNOSIS

The incubation period of smallpox is usually 12 to 14 days, although in occasional fulminating cases it can be as short as 4 to 5 days. The typical onset is abrupt, with fever, chills, and myalgia, followed by a rash 3 to 4 days later. The rash evolves to firm papulovesicles that become pustular over 10 to 12 days, then crust and slowly heal. Only a single crop of lesions (all in the same stage of evolution) develop; these lesions are most prominent over the head and extremities (**Figure 11–4**). Some cases are fulminant, with a hemorrhagic rash ("sledgehammer" smallpox). Death can result from the overwhelming primary viral infection or from bacterial superinfection. Diagnostic methods use vesicular scrapings and include culture, electron microscopy, gel diffusion, and polymerase chain reaction.

Single-stage rash

Vesicular scrapings used for diagnosis

PREVENTION

The first major step toward modern prevention and subsequent eradication of smallpox can be credited to Edward Jenner, who noted that milkmaids who develop mild cowpox lesions on their hands appeared immune to smallpox. In 1798, he published evidence indicating that purposeful inoculation of individuals with cowpox material could protect them against subsequent infection by smallpox. The concept of vaccination gradually evolved, with the modern use of live vaccinia virus, a poxvirus of uncertain origin to be discussed later, which produced specific immunity.

Jenner vaccinated with cowpox

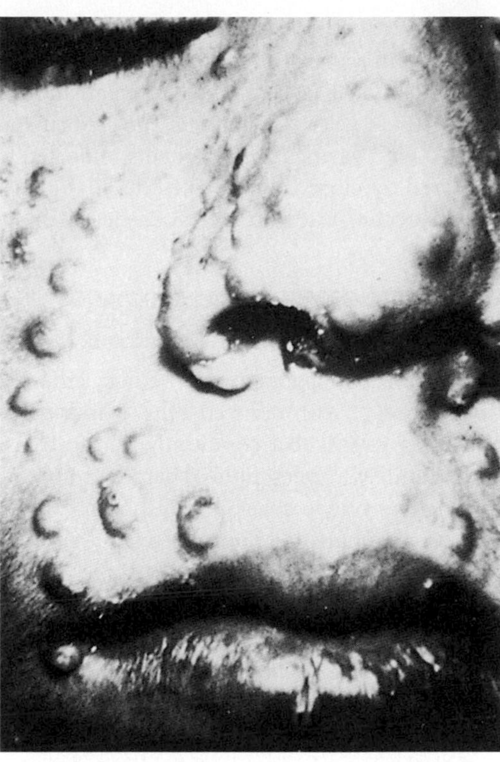

FIGURE 11–4. Closeup of facial lesions of smallpox during the first week of the illness.

VACCINIA

Vaccinia virus is serologically related to smallpox, although its exact origin is unknown. Some virologists believe it is a recombinant virus derived from smallpox and cowpox, and others suggest it originated from a poxvirus of horses. The virus is usually propagated by dermal inoculation of calves, and the resultant vesicle fluid ("lymph") is lyophilized and used as a live virus vaccine in humans. The vaccine is inoculated into the epidermis and produces a localized lesion, which indicates successful immunization. The lesion becomes vesicular, then pustular, followed by crusting and healing over 10 to 14 days. The local reaction is sometimes severe and accompanied by systemic symptoms such as fever, rash, and lymphadenopathy. Patients who are immunocompromised may experience severe reactions, such as progressive vaccinia. Vaccinia-produced immunity to smallpox wanes rapidly after 3 years, and the duration of long-term immunity beyond that time is uncertain.

Origin is unknown

Vaccination produces strong local reactions

Severe reactions seen in immunocompromised

Immunity wanes after 3 years

There has been a resurgence of scientific interest in vaccinia as a possible vector for active immunization against other diseases, such as hepatitis B, herpes simplex, and even human immunodeficiency virus. It has been shown that gene sequences coding for specific immunogenic proteins of other viruses can be inserted into the vaccinia virus genome, with subsequent expression as the virus replicates. For example, a recombinant vaccinia strain carrying the gene sequence for hepatitis B surface antigen (HbsAg) can infect cells, lead to production of HbsAg, and stimulate an antibody response to it. Theoretically, gene sequences coding for a variety of antigens could be packaged in a single viable vaccinia virus, thus allowing simultaneous active immunization against multiple agents. It has been suggested that use of other poxviruses of animal or avian origin, such as canarypox, may be even safer, yet effective vectors for use in humans. Whether such approaches become routinely applicable to clinical medicine remains to be seen.

Vaccinia of interest as mechanism for delivering the immunogenic proteins of other viruses

MONKEYPOX

The primary reservoir for monkeypox is in Central and West African rodents, not monkeys. On the other hand, two other African viruses classified in the *Yatapoxvirus* genus (tanapox and yabapox) have subhuman primates as their primary reservoirs. What all three have in common is that they can spread to humans by direct contact producing generally mild ill ness that, in more severe ceases, may be confused with smallpox.

Monkeypox, other animal poxviruses can be transmitted to humans by close animal contact

Illness can mimic smallpox

In 2003, at least 34 cases of human monkeypox occurred in the Midwestern United States. There were no fatalities. The contact sources were ill pet prairie dogs that had been housed with various exotic rodents imported from Ghana. Direct transmission to humans occurred by close contact with the ill animals. Human-to-human spread of any of these viruses occurs, but with far less efficiency than variola virus.

MOLLUSCUM CONTAGIOSUM

Transmission is direct skin-to-skin

Molluscum contagiosum is a benign, cutaneous poxvirus disease of humans, spread by direct contact with infected cells. It is usually acquired by inoculation into minute skin abrasions; events that commonly lead to transmission include "roughhousing" in shower rooms and swimming pools, sharing of towels, and sexual contact. Patients with AIDS are especially prone to develop widespread lesions.

Painless lesions express cheesy material

After an incubation period of 2 to 8 weeks, nodular, pale, firm (pearl-like) lesions usually 2 to 10 mm in diameter develop in the epidermis. These lesions are painless and umbilicated in appearance (**Figure 11–5**). A cheesy material may be expressed from the pore at the center of each lesion. Local trauma may cause spread of lesions in the involved skin area. The lesions are not associated with systemic symptoms, and they disappear in 2 to 12 months without treatment. Specific treatment, if desired, is usually by curettage or careful removal of the central core by expression with forceps.

Molluscum bodies in cytoplasm are diagnostic

Pathologic findings, which are limited to the epidermis, include hyperplasia, ballooning degeneration, and acanthosis. The diagnosis, made on clinical grounds, can be confirmed by demonstration of large, eosinophilic cytoplasmic inclusions (molluscum bodies) in the affected superficial epithelial cells (**Figure 11–6**).

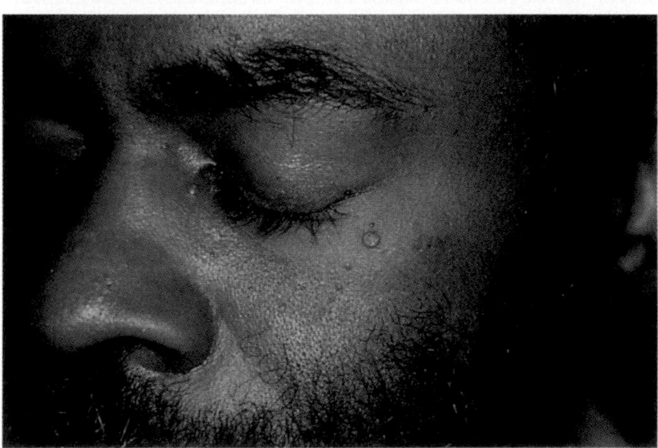

FIGURE 11–5. Several papular lesions of molluscum contagiosum on the face of a patient with AIDS. The larger lesion (near the eye) is raised, fleshy, and slightly umbilicated. (Reproduced with permission from Connor DH, Chandler FW, Schwartz DQA, Manz HJ, Lack EE (eds). *Pathology of Infectious Diseases,* vol. 1. Stamford, CT: Appleton & Lange; 1997.)

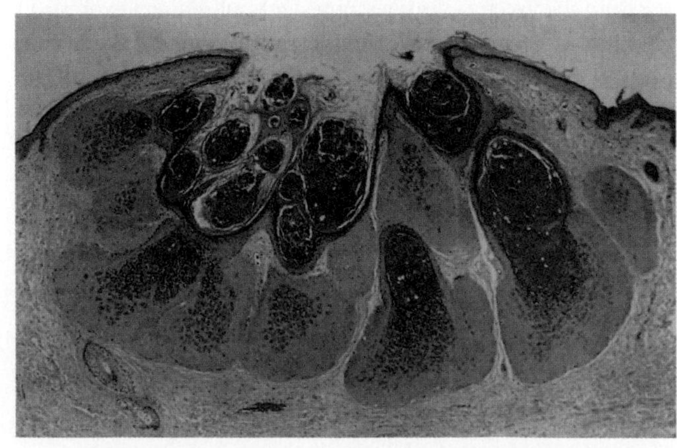

FIGURE 11–6. Molluscum contagiosum of skin. The epithelium has a craterform indentation with inverted lobules of keratinocytes containing eosinophilic inclusion. The epithelium over the edge of the lesion is raised (hematoxylin-eosin, × 40). (Reproduced with permission from Connor DH, Chandler FW, Schwartz DQA, Manz HJ, Lack EE (eds). *Pathology of Infectious Diseases,* vol. 1. Stamford, CT: Appleton & Lange; 1997.)

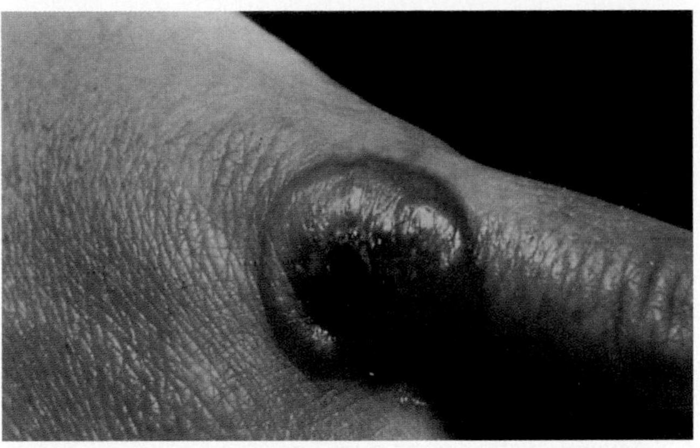

FIGURE 11–7. A boggy indurated plaque on the dorsal surface of the hand characteristic of orf, a parapoxvirus infection transmitted by sheep and goats. (Reproduced with permission from Connor DH, Chandler FW, Schwartz DQA, Manz HJ, Lack EE (eds). *Pathology of Infectious Diseases,* vol. 1. Stamford, CT: Appleton & Lange; 1997.)

ORF

Vesicular skin lesions seen in sheep- or goat-herders

Orf is an old Saxon term for a human infection caused by a parapoxvirus of sheep and goats. Synonyms for the infection in animals include contagious pustular dermatitis, ecthyma contagiosum, pustular ecthyma, and "scabby mouth." Humans usually acquire the infection by close contact with infected animals and accidental inoculation through cuts or abrasions on the hand or wrist. The typical skin lesion is solitary; it begins as a vesicle and evolves into a nodular mass that later develops central necrosis **(Figure 11–7)**. Regional lymphadenopathy sometimes develops. Dissemination is rare. The average duration of the lesion is 35 days, followed by complete resolution. The diagnosis is usually made on the basis of clinical appearance and occupational history. Serologic confirmation or electron microscopy of the lesion can be performed but is rarely necessary.

MILKER'S NODULES AND COWPOX

Localized infection acquired by direct contact with bovines

Milker's nodules (pseudocowpox) constitute a cutaneous parapoxvirus disease of cattle, distinct from cowpox, which can cause local skin infections similar to those of orf in exposed humans. Healing of the skin lesions may take 4 to 8 weeks. There is no cross-immunity to cowpox. Cowpox is now very rare in the United States. It produces a vesicular eruption on the udders of cows and similar, usually localized, vesicular skin lesions in humans who are accidentally exposed.

CASE STUDY

AN AFTERMATH OF WAR

A 22-year-old soldier has returned home after a 6-month tour of duty along the northeastern border of Afghanistan. The area consisted of scattered, small villages, where the main activities included raising goats and sheep, along with cultivation of poppies.

On arrival, the man was found to have a fever of 38.4°C, and headache. This persisted, and by the third day of illness, papulopustular skin lesions began to appear over his face and upper chest.

Laboratory studies included a mild leukocytosis (11,000/mm^3), but no other abnormalities.

QUESTIONS

1. Which of the following is the **least** likely cause of the man's condition?

A. Vaccinia
B. Variola minor
C. Cowpox
D. Monkeypox
E. Variola major

2. Which is currently the most important aspect of smallpox transmission?

A. Animal to human
B. Human to human
C. Asymptomatic human carriage
D. Evolution of mutant virus
E. Rodent contact

3. Vaccinia virus has the following attributes **except**:

A. Can cause severe localized or disseminated disease.
B. Is a live, attenuated smallpox virus.
C. Can induce immunity that lasts only a few years.
D. Has been in use for over 200 years.
E. Gene sequences coding for other viral proteins can be inserted into its genome.

ANSWERS

1(C), **2**(B), **3**(B)

CHAPTER 12

Enteroviruses

Ann Arbor. The world learned today that its hopes for finding an effective weapon against paralytic polio had been realized.

—The New York Times, April 12, 1955

Enteroviruses constitute a major subgroup of small RNA viruses (picornaviruses) that are transmitted by the fecal–oral route and readily infect the intestinal tract, where they may spread to cause paralytic disease, mild aseptic meningitis, exanthems, myocarditis, pericarditis, and nonspecific febrile illness. The enteroviruses of humans and animals are ubiquitous and have been found worldwide. Their name is derived from their ability to infect intestinal tract epithelial and lymphoid tissues and to be shed into the feces. These viruses include the polioviruses, coxsackieviruses, echoviruses, parechoviruses, and other agents that are simply designated enteroviruses. Over the years, renumbering and reclassification within the subgroups have occurred, primarily as a result of advanced sequencing analyses.

These viruses, which have many characteristics in common, are first considered as a group. Some of the special features of important serotypes are discussed in more detail later in this chapter.

ENTEROVIRUSES: GROUP CHARACTERISTICS

VIROLOGY

MORPHOLOGY AND BIOLOGICAL FEATURES

As a group, the enteroviruses are extremely small (22 to 30 nm in diameter), naked virions with icosahedral symmetry. They possess single-stranded, positive-sense RNA and a capsid formed from 60 copies of four nonglycosylated proteins (VP1, VP2, VP3, VP4). The virion structure of a picornavirus member is shown in Chapter 13 (Figure 13–1). Replication and assembly occurs exclusively in the cellular cytoplasm; one infectious cycle can occur within 6 to 7 hours. This results in cessation of host cell protein synthesis and cell lysis with release of new infectious progeny. The replication cycle is shown in **Figure 12–1**. Picornaviruses enter the host cell via receptor-mediated endocytosis (viropexis) following interaction of a viral surface protein with a specific receptor on the host cell. After removal of capsid protein, uncoating takes place, and the positive sense RNA viral genome is released in the cytoplasm, which acts as an mRNA. This genomic viral mRNA is translated into a

Small, single-stranded RNA viruses

Replication and assembly take place in cytoplasm

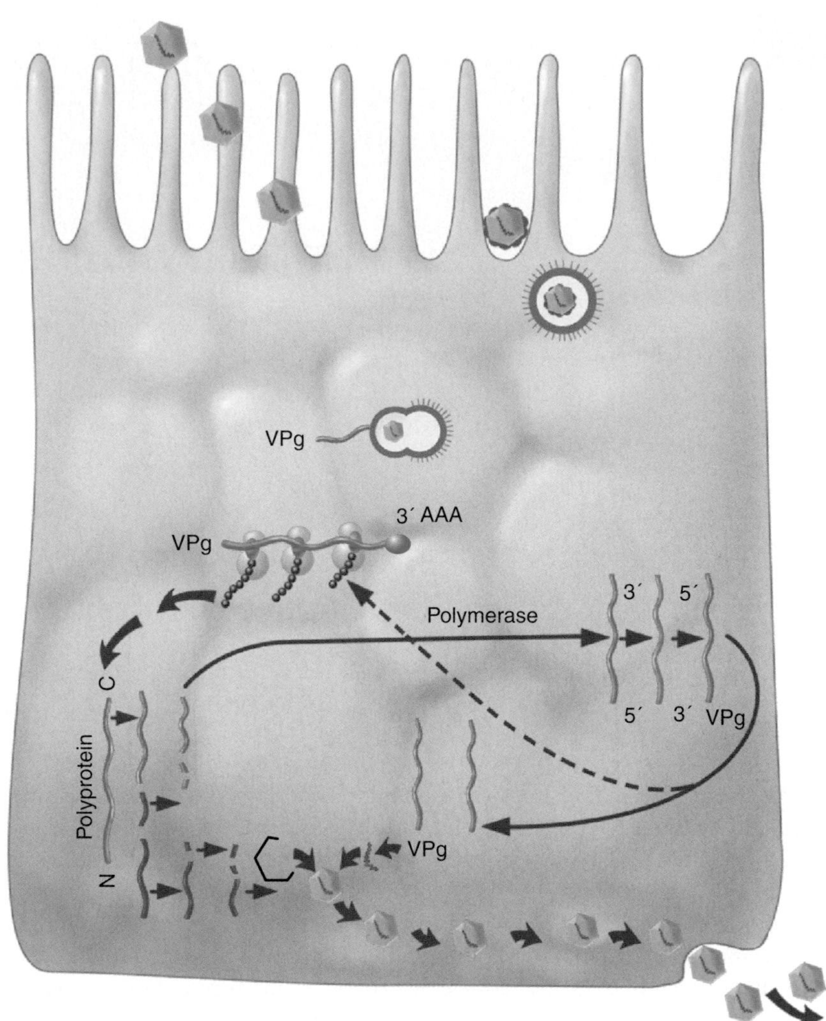

FIGURE 12–1. Replication cycle of picornavirus. VPg, viral protein genome.

polyprotein, which is processed into mature proteins, including an RNA-dependent RNA polymerase. RNA-dependent RNA polymerase directs both transcription of mRNA and synthesis of genomic RNA via negative-sense RNA intermediates. After the synthesis of viral proteins, the genomic RNA is packaged into progeny virions that are assembled in the cytoplasm and released upon cell death. ::: viropexis, p. 118

Resistant to acid, detergents, and many disinfectants

Formaldehyde and hypochlorite are active against enteroviruses

Unlike rhinoviruses, which are also members of the picornavirus family, enteroviruses are resistant to an acid pH (as low as 3.0). This feature undoubtedly helps ensure their survival during passage through the stomach to the intestines. Enteroviruses are also resistant to many common disinfectants such as 70% alcohol, substituted phenolics, ether, and various detergents that readily inactivate most enveloped viruses. Chemical agents, such as 0.3% formaldehyde or free residual chlorine at 0.3 to 0.5 ppm, are effective. However, if sufficient extraneous organic debris is present, the virus can be protected and survive long periods.

Antigenic mutations and drifts occur

Antibody to surface proteins neutralize infectivity

Some of the enterovirus serotypes share common antigens, but there are no significant serologic relationships between the currently recognized major classes listed in **Table 12–1**. Genetic variation within specific strains occurs, and mutants that exhibit antigenic drift and altered tropism for specific cell types are now recognized. Polioviruses, which have been most extensively studied as enterovirus prototypes, are known to have epitopes on three surface structural proteins (VP1, VP2, and VP3) that induce type-specific neutralizing antibodies. This appears to be generally the case for all enteroviruses; definitive identification of isolates usually requires neutralization or molecular analysis tests.

GROWTH IN THE LABORATORY

Most enteroviruses can be isolated in primate (human or simian) cell cultures and show characteristic cytopathic effects. Some strains, particularly several coxsackievirus A serotypes, are more readily detected by inoculation of newborn mice. In fact, the newborn mouse is one

TABLE 12–1 Human Enteroviruses

CLASS	NUMBER OF SEROTYPES[a]
Poliovirus	3
Coxsackievirus	
Group A	23
Group B	6
Echovirus	26
Parechovirus	4
Enterovirus	4

[a]More recently discovered enteroviruses, which have overlapping biological characteristics, are identified numerically (types 68-71). Four of the original 30 numbered echovirus serotypes have been reclassified; however, the remaining retain their original serotype number (eg, echovirus 30)

basis for originally classifying group A and B coxsackieviruses. Group A coxsackieviruses cause primarily a widespread, inflammatory, necrotic effect on skeletal muscle, leading to flaccid paralysis and death. Similar inoculation of group B coxsackieviruses causes encephalitis, resulting in spasticity and occasionally convulsions. Other enteroviruses rarely have an adverse effect on mice unless special adaptation procedures are first used. The higher-numbered enteroviruses (types 68-71), which have overlapping variable growth and host characteristics, have been classified separately.

Growth of some in primate cell cultures

Coxsackie A and B viruses have different effects on newborn mice

ENTEROVIRUS DISEASE

CLINICAL CAPSULE

Enterovirus infections can produce a great diversity of clinical disease. Some cause paralytic disease that may persist permanently (a typical feature of polioviruses), acute inflammation of the meninges with or without involvement of cerebral or spinal tissues, or sepsis-like illnesses in newborn infants. Inflammatory effects at other sites, such as the lungs, pleura, heart, and skin, have been also observed, often without concomitant or preceding central nervous system (CNS) involvement. Occasionally, infections may result in chronic, active disease processes.

EPIDEMIOLOGY

Humans are the major natural host for the polioviruses, coxsackieviruses, and echoviruses. There are enteroviruses of other animals with limited host ranges that do not appear to extend to humans. Conversely, viruses thought to be identical or related to human enteroviruses have been isolated from dogs and cats. Whether these agents cause disease in such animals is debatable, and there is no evidence of spread from animals to humans.

Animals are not involved in human disease

The enteroviruses have a worldwide distribution, and asymptomatic infection is common. The proportion of infected persons who develop illness varies from 2% to 100%, depending on the serotype or strain involved and the age of the patient. Secondary infections in households are common and range as high as 40% to 70%, depending on factors such as family size, crowding, and sanitary conditions.

Proportion of asymptomatic infections varies with strain

Dominant epidemic strains come and go

Greater prevalence during summer and fall in temperate climates

In some years, certain serotypes emerge as dominant epidemic strains; they then may wane, only to reappear in epidemic fashion years later. For example, echovirus 16 was a major cause of outbreaks in the eastern United States in 1951 and 1974. Coxsackievirus B1 was common in 1963; echovirus 9 in 1962, 1965, 1968, and 1969; and echovirus 30 in 1968 and 1969. The emergence of dominant serotypes is unpredictable from year to year. All enteroviruses show a seasonal predilection in temperate climates; epidemics are usually observed during the summer and fall months. In subtropical and tropical climates, the transmission may occur year-round.

Person-to-person fecal–oral transmission correlates with predominance in children

Direct or indirect fecal–oral transmission is considered the most common mode of spread. After infection, the virus persists in the oropharynx for 1 to 4 weeks, and it can be shed in the feces for 1 to 18 weeks. Thus, sewage-contaminated water, fecally contaminated foods, or passive transmission by insect vectors (flies, cockroaches) may occasionally be the source of infection. More commonly, however, spread is directly from person to person. This mode of transmission is suggested by the high infection rates seen among young children, whose hygienic practices tend to be less than optimal, and in crowded households. Approximately two thirds of all isolates are from children 9 years of age or younger.

Incubation periods are typically short

Incubation periods vary, but relatively short intervals (2 to 10 days) are common. Often, illness is seen concurrently in more than one family member, and the clinical features vary within the household.

PATHOGENESIS

Initial attachment binds viral surface protein to cell surface receptors

Host receptor may relate to immunoglobulin or integrin families

Initial binding of an enterovirus to the cell surface is commonly between an attachment protein in a "canyon" configuration on the virion surface and cell receptors belonging to the immunoglobulin gene superfamily. These receptors map to chromosome 19. A different receptor, belonging to the integrin group of adhesion molecules, has been identified for at least one echovirus serotype. After attachment, the virion is enveloped by the cell membrane, and its RNA is released into the cellular cytoplasm, where it binds to ribosomes and commences protein synthesis. Newly synthesized virions are released by lysis to spread to the other cells.

Initial replication in epithelial and lymphoid cells is followed by viremic spread

Injury by cell lysis localized in perivascular sites

Antibody response terminates replication

After primary replication in epithelial cells and lymphoid tissues in the upper respiratory and gastrointestinal tracts, viremic spread to other sites can occur. Potential target organs vary according to the virus strain and its tropism, but may include the CNS, heart, vascular endothelium, liver, pancreas, lungs, gonads, skeletal muscles, synovial tissues, skin, and mucous membranes. Histopathologic findings include cell necrosis and mononuclear cell inflammatory infiltrates; in the CNS, the inflammatory cells are localized most prominently in perivascular sites. The initial tissue damage is thought to result from the lytic cycle of virus replication; secondary spread to other sites may ensue. Viremia is usually undetectable by the time symptoms appear, and termination of virus replication appears to correlate with the appearance of circulating neutralizing antibody, interferon, and mononuclear cell infiltration of infected tissue. The early dominant antibody response is with immunoglobulin M (IgM), which usually wanes 6 to 12 weeks after onset to be replaced progressively by increased IgG-specific antibodies. The important role of antibodies in termination of infection, demonstrated in mouse models of group B coxsackievirus infections, is supported by the observation of persistent echovirus and poliovirus replication in patients with antibody deficiency diseases.

Although initial acute tissue damage may be caused by the lytic effects of the virus on the cell, the secondary sequelae may be immunologically mediated. Enterovirus-caused poliomyelitis, disseminated disease of the newborn, aseptic meningitis, encephalitis, and acute respiratory illnesses, thought to represent primary lytic infections, can usually be identified through routine methods of virus isolation and determination of specific antibody titer changes. On the other hand, syndromes such as myopericarditis, nephritis, and myositis have been associated with enteroviruses primarily because of serologic and epidemiologic evidence. In many of these cases, viral isolation is the exception rather than the rule. The pathogenesis of these latter infections is not clear; however, observations suggest that the acute infectious phase of the virus may be mild or subclinical and often subsides by the time clinical illness becomes evident. Illness may represent a host immunologic response to tissue injury by the virus or to viral or virus-induced antigens that persist in the affected tissues.

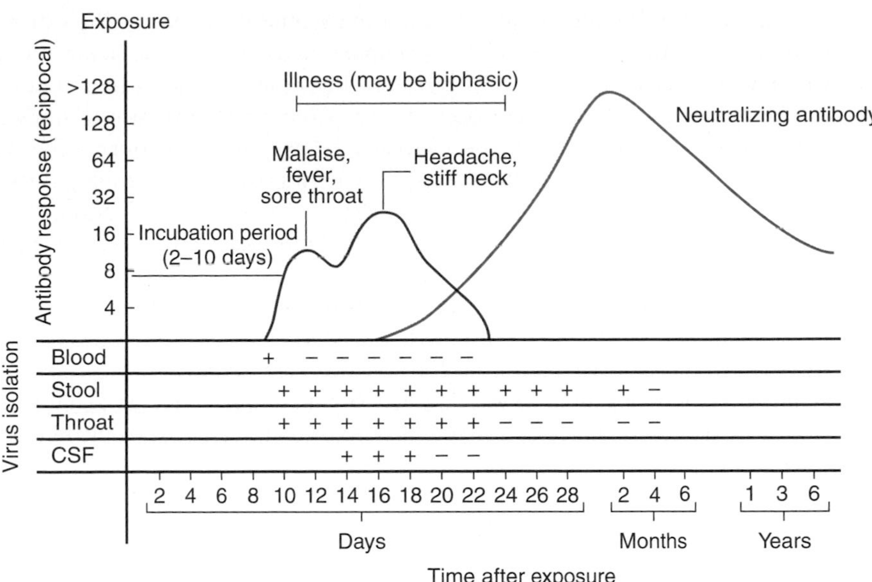

FIGURE 12–2. Antibody response and viral isolation from various sites in a typical case of enteroviral infection.

In experimental group B coxsackievirus myocarditis, mononuclear inflammatory cells (monocytes, natural killer lymphocytes) seem to play a greater role than antibody in termination of infection, and the persistence of inflammation after disappearance of detectable infectious virus or viral antigen appears to be mediated by cytotoxic T lymphocytes. Experimental findings have led to another hypothesis regarding pathogenic mechanisms, called **molecular mimicry.** This is best conceptualized as a form of virus-induced autoimmune response. It is known that small peptide sequences on viral epitopes can sometimes be shared by host tissues. Thus, an immune response produced by the virus may also generate antibodies or cytotoxic cross-reactive effector lymphocytes that recognize shared determinants located on host cells. For example, a monoclonal antibody directed against a neutralizing site of a group B coxsackievirus has also been shown to react strongly with normal myocardial cells. ::: molecular mimicry, p. 154

In addition to lytic effects of virus, there are probable immunopathologic manifestations

Disease may follow the acute infection

Coxsackie B myocarditis may involve virus-induced cross-reacting antibody

IMMUNITY

Infection by a specific serotype in an immunologically normal host is followed by a humoral antibody response, which can often be detected by neutralization methods for many years thereafter (**Figure 12–2**). There is relative immunity to reinfection by the same serotype; however, reinfection has been reported, usually resulting in subclinical infection or mild illness.

Immunity is serotype-specific

CLINICAL ASPECTS

DIAGNOSIS

Currently, the polymerase chain reaction with reverse transcription and complementary DNA amplification (RT-PCR) is being increasingly used to detect enteroviral RNA sequences in tissue and body fluids, thus greatly enhancing diagnostic sensitivity and speed. Alternatively, classicl virus isolation methods can be used. The disadvantages of the latter approach are longer time to detection (3 to 10 days versus several hours for RT-PCR) and lower sensitivity; one advantage is that virus isolates can be more readily further characterized antigenically and genetically.

RT-PCR enhances diagnostic speed, sensitivity

In acute enterovirus-caused syndromes, diagnosis is most readily established by virus detection in throat swabs, stool* or rectal swabs,* (transpose) body fluids, and occasionally tissues. Viremia may be undetectable by the time symptoms appear. When there is CNS involvement, cerebrospinal fluid (CSF) specimens taken during the acute phase of the disease may be positive in 10% to 85% of cases, depending on the stage of illness and the viral serotype involved. Direct detection of virus from affected tissues or body fluids in enclosed spaces (eg, pleural, joint, pericardial, or CSF) usually confirms the diagnosis. Detection of an enterovirus from the throat is highly suggestive of an etiologic association; the virus is usually present at this site for only 2 days to 2 weeks after infection. Detection of virus from fecal specimens only must be interpreted more cautiously; asymptomatic shedding from the bowel may persist for as long as 4 months (Figure 12–2).

Viral isolation from pharynx or closed space is significant

Prolonged shedding in stool

The diagnosis may be further supported by fourfold or greater neutralizing antibody titer changes between paired acute and convalescent serum samples. However, this method is often expensive and cumbersome, requiring careful selection of serotypes for use in antigens. Quantitative interpretations of antibody titers on single serum samples are rarely helpful, because of the wide range of titers to different serotypes that can be found among healthy individuals.

Serodiagnosis is usually impractical

TREATMENT AND PREVENTION

None of the currently available, approved antiviral agents has been shown to be effective in treatment or prophylaxis of enterovirus infections. Treatment is symptomatic and supportive. Vaccines for the prevention of poliovirus infections are discussed later in this chapter. Although proper disposal of feces and careful personal hygiene are recommended, the usual quarantine or isolation measures are relatively ineffective in controlling the spread of enteroviruses in the family or community.

Hygienic factors make prevention of spread difficult

ENTEROVIRUSES: SPECIFIC GROUPS

Polioviruses

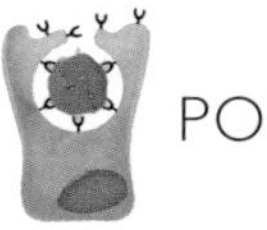

POLIO

EPIDEMIOLOGY

Worldwide, the most important enteroviruses are the three poliovirus serotypes (types 1, 2, and 3). They first emerged as important causes of disease in developed temperate zone countries during the latter part of the 19th century, and they have become increasingly important elsewhere as living conditions improve in developing countries. This somewhat paradoxical situation is related to the fact that the risk of paralytic disease resulting from infection increases with age. Improvement of sanitary conditions tends to impede spread of the viruses; thus, individuals may become infected not in early infancy but later in life, when paralysis is more likely to occur.

Risk of paralysis from infection increases with age

PATHOGENESIS

The schematic diagram of the pathogenesis of poliovirus is shown in Chapter 7 (Figure 7–1). Poliovirus is spread by fecal–oral route. Virus enters oropharynx and multiplies in mucosa, shed in oral secretions and swallowed and then multiplies in the intestine. After primary replication in epithelial cells and lymphoid tissues in the upper respiratory and gastrointestinal tracts, viremia spreads to other sites. The particular tropism of polioviruses for the

*RT-PCR is not routinely applied to these specimens.

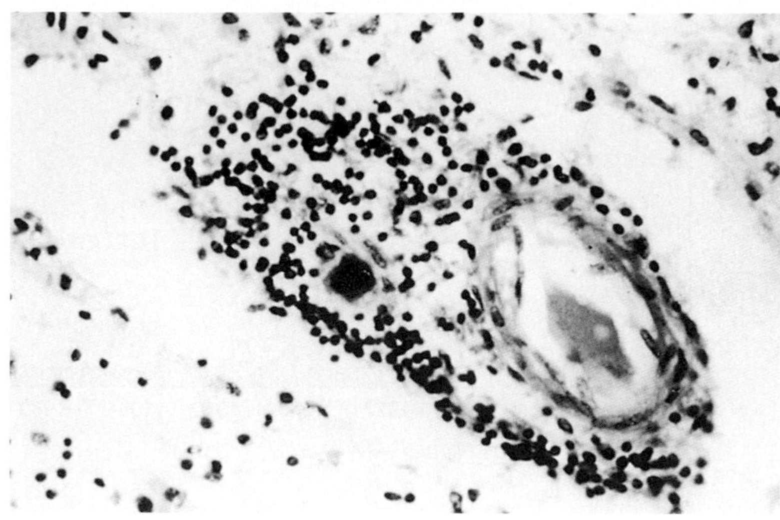

FIGURE 12–3. Section of spinal cord from a fatal case of poliomyelitis, demonstrating perivenous mononuclear cell inflammatory reaction. (Courtesy of Dr. Peter C. Johnson.)

CNS, which they usually reach by passage across the blood–CNS barrier, is perhaps favored by reflex dilatation of capillaries supplying the affected motor centers of the anterior horn of the brainstem or spinal cord. An alternate pathway is via the axons or perineural sheaths of peripheral nerves. Motor neurons are particularly vulnerable to infection and variable degrees of neuronal destruction. The histopathologic findings in the brainstem and spinal cord include necrosis of neuronal cells and perivascular "cuffing" by infiltration with mononuclear cells, primarily lymphocytes (**Figure 12–3**). ::: poliovirus pathogenesis, p. 138

CNS tropism by blood or peripheral nerves

Motor neuron cells destroyed

CLINICAL ASPECTS

MANIFESTATIONS

Most infections (perhaps 90%) are either completely subclinical or so mild that they do not come to attention. When disease does result, the incubation period ranges from 4 to 35 days, but is usually between 7 and 14 days. Three types of disease can be observed. Abortive poliomyelitis is a nonspecific febrile illness of 2 to 3 days' duration with no signs of CNS localization. Aseptic meningitis (nonparalytic poliomyelitis) is characterized by signs of meningeal irritation (stiff neck, pain, and stiffness in the back) in addition to the signs of abortive poliomyelitis; recovery is rapid and complete, usually within a few days.

Paralytic poliomyelitis occurs in less than 2% of infections. It is the major possible outcome of infection and is often preceded by a period of minor illness, sometimes with two or three intervening symptom-free days. There are signs of meningeal irritation, but the hallmark of paralytic poliomyelitis is asymmetric flaccid paralysis, with no significant sensory loss. The extent of involvement varies greatly from case to case; however, in its most serious forms, all four limbs may be completely paralyzed or the brainstem may be attacked, with paralysis of the cranial nerves and muscles of respiration (bulbar polio). The maximum extent of involvement is evident within a few days of first paralysis. Thereafter, as temporarily damaged neurons regain their function, recovery begins and may continue for as long as 6 months; paralysis persisting after this time is permanent.

Subclinical and abortive poliomyelitis common

Aseptic meningitis recovers rapidly

Paralytic poliomyelitis manifests flaccid paralysis without sensory loss

Recovery of function up to six months

PREVENTION

Two types of poliovirus vaccines are currently available: inactivated polio vaccine and live oral attenuated virus vaccine. Each contains all three viral serotypes.

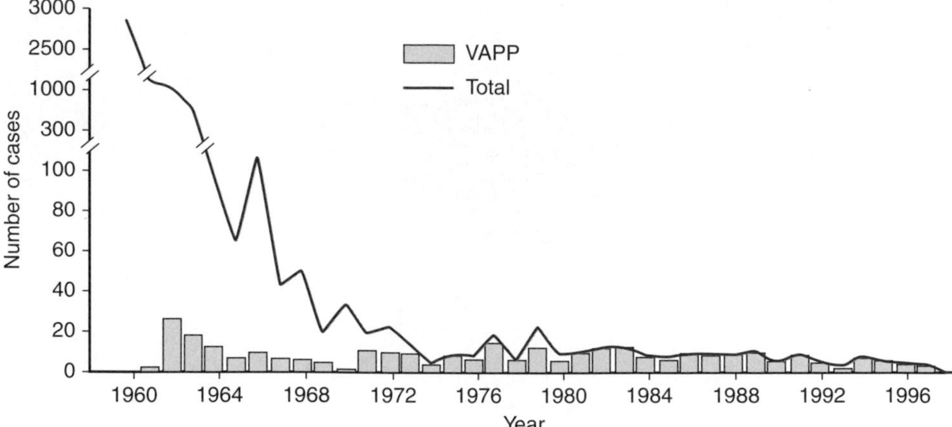

FIGURE 12–4. Total number of reported paralytic poliomyelitis cases and number of reported vaccine-associated cases (VAPP)—United States, 1950-1999. (From the Centers for Disease Control and Prevention, 2000.)

Inactivated polio vaccine (IPV) was introduced in 1955; its use was associated with a dramatic decline in paralytic cases (**Figure 12–4**). Vaccination is by subcutaneous injection. Primary vaccination with three doses of the present enhanced-potency IPV (two doses 6-8 weeks apart and the third 8-12 months later) produces antibody responses in more than 98% of recipients. The current product is considered safe, with no significant deleterious side effects. Inactivated (Salk) vaccine is used in many countries, including the United States.

Live (Sabin) vaccine is given orally (OPV)

Vaccine virus replicates and can spread

Oral polio vaccine (OPV) is composed of live, attenuated viruses that have undergone serial passage in cell cultures from humans and subhuman primates. OPV was first licensed in the United States in 1963. The vaccine is given orally as a primary series of three doses (the first two doses usually 6-8 weeks apart and the third 8-12 months later) and produces antibodies to all three serotypes in more than 95% of recipients; these antibodies persist for several years. As with IPV, recall boosters are recommended to maintain adequate antibody levels. Like wild poliovirus, OPV viruses infect and replicate in the oropharynx and intestinal tract and can be spread to other persons.

Vaccine-associated poliomyelitis is a remote risk with OPV

IPV is currently preferred in the United States

One disadvantage of OPV is the remote risk of vaccine-associated paralytic disease in some recipients or their household contacts, including immunocompromised persons. The incidence of vaccine-associated paralytic poliomyelitis is estimated at approximately 1 per 2.4 million doses distributed. Since the end of 1999, exclusive use of IPV has been recommended for all routine immunizations in the United States. OPV is recommended only in special circumstances (eg, an unvaccinated child who will be traveling in less than 4 weeks to an endemic area). It is also used for major immunization programs in countries where wild virus remains at high endemic levels.

No cases of paralytic poliomyelitis attributed to indigenously acquired wild poliovirus have occurred in the United States since 1979. Nevertheless, it must be kept in mind that importation of these strains can readily occur from endemic areas in developing nations. Once introduced into a community, the virus can spread rapidly among susceptible individuals. Thus, continuing immunization programs are of utmost importance in preventing spread of this disease. In 1988, the World Health Organization resolved to eradicate polio from the world by the year 2000. Thus far, progress toward that goal has been hampered by political strife and severe poverty in many underdeveloped nations in Africa, Asia, and the Middle East.

Coxsackieviruses and Echoviruses

EPIDEMIOLOGY

Often do not affect motor neurons

The coxsackieviruses and echoviruses are widespread throughout the world. Their epidemiology and pathogenesis are much the same as those of the polioviruses. Unlike polioviruses,

TABLE 12–2 Clinical Syndromes and Commonly Associated Enterovirus Serotypes[a]

	COXSACKIEVIRUS		
SYNDROME	**GROUP A**	**GROUP B**	**ECHOVIRUS, PARECHOVIRUS (PEV) AND ENTEROVIRUS (E)**
Aseptic meningitis, encephalitis	2, 4, 7, **9,** 10	1, **2, 3, 4, 5**	**4, 6, 9, 11, 16, 30,** E70, E71
Muscle weakness and paralysis (poliomyelitis-like disease)	7, **9**	2, 3, 4, 5	2, 4, 6, 9, 11, 18, 30, **E71**
Cerebellar ataxia	2, 4, **9**	3, 4	4, 6, 9
Exanthems and enanthems	**4, 5, 6, 9, 10, 16**	2, 3, 4, 5	**2, 4, 5, 6, 9, 11, 16, 18, 25, E71**
Pericarditis, myocarditis	4, 16	**2, 3, 4, 5**	1, 6, 8, 9, 19
Epidemic myalgia (pleurodynia), orchitis	9	1, **2, 3, 4, 5**	1, 6, 9,
Respiratory	9, 16, **21,** 24	1, 3, 4, 5	**4, 9, 11,** 20, 25
Conjunctivitis	**24**	1, 5	7, **E70**
Generalized disease (infants)	–	1, **2, 3, 4, 5**	3, 6, 9, 11, 14, 17, 19, PEV3

[a]Serotypes most commonly associated with syndrome are in **boldface.**

they have a greater tendency to affect the meninges and occasionally the cerebrum, but only a few such as enterovirus 71 affect anterior horn cells.

The consequences of infection with these agents are highly variable and related only in part to virus subgroup and serotype. Up to 60% of infections are subclinical. The main interest in these agents stems from their ability to cause more serious illness, which becomes most evident during epidemics of infection with a particular agent. Unapparent infection is common. Illness manifestations vary from mild to lethal. **Table 12–2** lists the major syndromes and serotypes commonly associated with each. However, considerable overlap occurs, and one should not be surprised if an enteroviral serotype found in connection with a specific syndrome differs from that most often encountered.

Most infections subclinical

Wide range of clinical manifestations

MANIFESTATIONS

Aseptic meningitis is the most frequently recognized clinical illness associated with enterovirus infections. This syndrome can be mild and self-limiting, lasting 5 to 14 days. However, it is sometimes accompanied by encephalitis, which can lead to permanent neurologic sequelae.

Aseptic meningitis most common syndrome

Acute inflammation of the heart muscle (myocarditis), its covering membranes (pericarditis), or both can be caused by a variety of viral agents. Group B coxsackieviruses are the most commonly implicated enteroviruses. Such infections are usually self-limiting, but may be fatal in the acute phase (arrhythmia or heart failure) or progress to chronic dilated myocardiopathy.

Myocarditis often associated with group B coxsackieviruses

The exanthems are often not associated with CNS inflammation. They can resemble rubella, roseola infantum, or adenoviral macular or maculopapular exanthems, but may also appear as vesicular or hemangioma-like lesions. One interesting syndrome is hand-foot-and-mouth disease, which usually affects children and is characterized by a vesicular eruption over the extremities and the oral cavity (**Figure 12–5**). Coxsackie virus A16 is most commonly implicated, but others, such as enterovirus 71, can cause a similar illness. When associated with enterovirus 71 infection, the illness can be especially severe, with

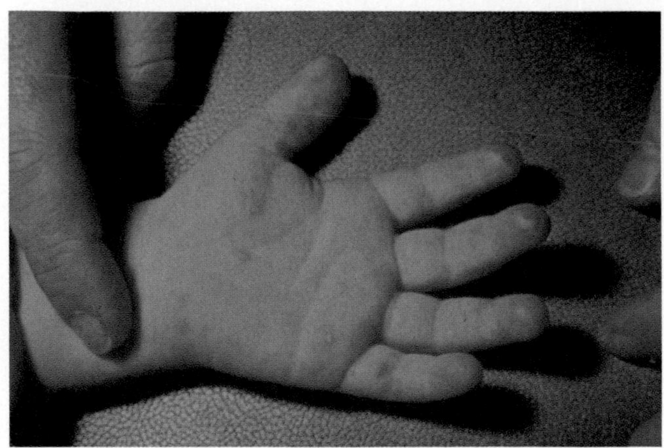

FIGURE 12–5. Vesicular lesions of hand-foot-and-mouth disease.

Exanthems can mimic other diseases

Herpangina infection of palate and tonsils

Epidemic myalgia, with pleuritic pain

encephalitis, permanent polio-like limb weakness, and often fatal cardiorespiratory failure. Herpangina is an enanthematous (mucous membrane–affecting) febrile disease in which small vesicles or white papules (lymphonodules) surrounded by a red halo are seen over the posterior palate, pharynx, and tonsillar areas (**Figure 12–6**) This mild, self-limiting (1- to 2-week) illness has usually been associated with infection by several different group A coxsackievirus serotypes.

Epidemic myalgia (pleurodynia or Bornholm disease) is characterized by fever and sudden onset of intense upper abdominal or thoracic pain. The pain may be aggravated by movement, such as breathing or coughing, and can persist as long as 14 days. Group B coxsackieviruses are often implicated.

Generalized disease of the newborn is a disseminated, often lethal, enteroviral infection characterized by pathologic changes in the heart, brain, liver, and other organs.

It is apparent from Table 12–2 that the spectrum of disease produced by these viruses is enormous and that many other illnesses may also result from infections by this subgroup. Epidemics of acute hemorrhagic keratoconjunctivitis associated with enterovirus 70 and localized outbreaks of disease resembling paralytic poliomyelitis caused by enterovirus 71 infection have been described. In addition, there is evidence that certain enteroviruses, particularly group B coxsackievirus serotypes, may sometimes participate in the pathogenesis of insulin-dependent diabetes mellitus, acute arthritis, polymyositis, and idiopathic acute nephritis. Further investigations are required to establish whether such associations are significant.

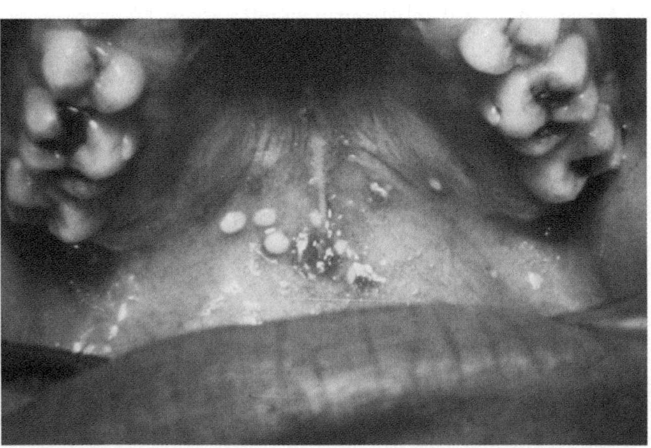

FIGURE 12–6. Herpangina. Localized lymphonodules and vesicles (mostly ruptured) in the posterior oropharynx.

CASE STUDY

A SEVERE HEADACHE

A 2-year-old girl is on a summer visit to her grandparents in the midwestern United States, when she develops irritability, vomiting, low-grade fever, and frontal headache over 2 days.

Physical examination reveals only a stiff neck, wherein the patient resists attempts to flex it.

A lumbar puncture is done to try to quickly rule out bacterial meningitis. The CSF results are 90 cells/mm^3, 70% mononuclears, glucose 60 mg/dL, and protein 45 mg/dL. Gram stain is negative for bacteria.

QUESTIONS

■ Which of the following tests would be most sensitive and specific at this stage of illness?

A. IgM-specific serology on CSF
B. Viral culture of CSF
C. RT-PCR on CSF
D. RT-PCR on rectal swab specimen
E. IgM-specific serology on serum

■ All of the below are common characteristics of enteroviruses in humans, *except*:

A. Seasonal peaks in temperate climates
B. Fecal–oral transmission
C. Resistance to 70% alcohol
D. Replication in cell cytoplasm
E. Animal reservoirs

■ Live, attenuated oral polio vaccine (OPV) and inactivated polio vaccine (IPV) are both available. In which one of the following situations is the use of OPV preferred?

A. Routine infant vaccination
B. Mass immunization programs in areas of high poliomyelitis endemicity
C. Adult immunization
D. Patients who are receiving immunosuppressive therapy
E. Family contacts of immunocompromised patients

ANSWERS

1(C), **2**(E), **3**(B)

CHAPTER 13

Hepatitis Viruses

Jaundice is the disease that your friends diagnose.

—Sir William Osler

The causes of hepatitis (inflammation of the liver) are varied and include viruses, bacteria, and protozoa, as well as drugs and toxins (eg, isoniazid, carbon tetrachloride, and ethanol). The clinical symptoms and course of acute viral hepatitis can be similar, regardless of etiology, and determination of a specific cause depends primarily on the use of laboratory tests. Hepatitis may be caused by at least five viruses whose major characteristics are summarized in **Table 13–1. Non-A, non-B hepatitis** is a term previously used to identify cases of hepatitis not due to hepatitis A or B virus. With the discovery of hepatitis viruses C, E, and G, virtually all the viral etiologies of non-A, non-B hepatitis can be specifically identified. Other viruses, such as Epstein-Barr virus and cytomegalovirus, can cause inflammation of the liver, but hepatitis is not the primary disease caused by them. Yellow fever is associated with hepatitis, but is now uncommon.

HEPATITIS A

VIROLOGY

Hepatitis A virus (HAV) belongs to the Picornaviridae family and *Hepatovirus* genus. It is an unenveloped (naked capsid), single-stranded RNA virus with a cubic (icosahedral) symmetry and a diameter of 27 nm (**Figure 13–1**). The genome of HAV is a 7.4-kb positive-sense single-stranded RNA bound to a protein called VPg, and each capsid unit comprises four proteins, VP1, 2, 3, and 4, which cover the genome and form a naked capsid icosahedral virion. VP1 is the spike of HAV that binds to the receptor on the host cells. There is only one serotype of HAV. This virus possesses several characteristics of enteroviruses; for example, it resists inactivation and is stable at –20°C with low pH. The virus has been successfully cultivated in primary marmoset liver cell cultures and in fetal rhesus monkey kidney cell cultures.

Hepatitis A virus is a picornavirus with only one serotype

HAV replicates in the cytoplasm, like other positive-sense RNA viruses (Figure 12-1). HAV interacts with the receptor (α_2-macroglobulin) on the target cells (liver cells and few other cell types) and enters via receptor-mediated endocytosis (viropexis). The positive-sense RNA is translated into a polyprotein, which is cleaved into various mature proteins, including RNA-dependent RNA polymerase. RNA-dependent RNA polymerase directs

HAV replicates in the cytoplasm

TABLE 13–1 Comparison of A, B, D (Delta), C, and E Hepatitis

FEATURE	A	B	D	C[a]	E
Virus type	Single-stranded RNA	Double-stranded DNA	Single-stranded RNA	RNA	RNA
Incubation period (days)	15-45 (mean, 25)	30-180 (mean, 60-90)	28-45	15-150 (mean 50)	21-56 (mean 40)
Onset	Usually sudden	Usually slow	Variable	Insidious	?
Age preference	Children, young adults	All ages	All ages	All ages	Young adult
Transmission					
Fecal–oral	+++	±	±	–	+++
Sexual	+	++	++	+	+?
Parenteral	–	+++	++	+++	
Chronicity (%)	None	10	50-70	85	Rare
Carrier state	None	Yes	Yes	Yes	No
Immune serum globulin protective	Yes	Yes[b]	Yes[c]	No	No
Vaccine	Yes	Yes	Yes[c]	No	No

Plus and minus signs indicate relative frequencies.
[a]Many individuals with hepatitis C virus are also infected with hepatitis G virus, which is similar to hepatitis C virus.
[b]Hyperimmune globulin is more protective.
[c]Prevention of hepatitis B prevents hepatitis D.

transcription of mRNAs to produce viral proteins as well as replication to make full-length viral genomes. The assembly of the progeny viruses takes place in the cytoplasm after the packaging of viral genomes into HAV capsid proteins. Virions are released upon cell lysis.

HEPATITIS A DISEASE

CLINICAL CAPSULE

Hepatitis A virus is the cause of what was formerly termed infectious hepatitis or short-incubation hepatitis. This virus is spread by the fecal–oral route, and outbreaks may be associated with contaminated food or water. The illness is subclinical in up to 50% of infected adults. When symptomatic, there is usually fever and jaundice. Although fatal disease may occur, self-limited illness is the rule. Chronic hepatitis A rarely if ever occurs.

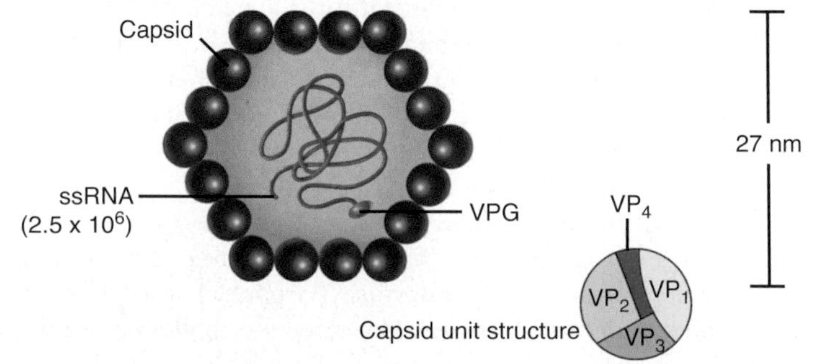

FIGURE 13–1. Diagram of the proposed structure of the hepatitis A virus. The protein capsid is made up of four viral polypeptides (VP1 to VP4). Inside the capsid is a single-stranded (ss) molecule of RNA (molecular weight 2.5×10^6), which has a genomic viral protein (VPg) on the 5 end. (Reprinted with permission of Dr. J. H. Hoofnagle and of Abbot Laboratories, Diagnostic Division, North Chicago, Illinois.)

EPIDEMIOLOGY

Humans appear to be the major natural hosts of HAV. Several other primates (including chimpanzees and marmosets) are susceptible to experimental infection, and natural infections of these animals may occur. The major mode of spread of HAV is person to person by fecal–oral exposure. Transmission through blood transfusion, though possible, is not an important means of spread, but persons with hemophilia who are given plasma products are at risk. High risk of infection is also observed in men who have sex with men, in users of illicit drugs, and in travelers from the developed world visiting developing areas. Although most cases of hepatitis A are not linked to a single contaminated source and occur sporadically, outbreaks have been described. The disease is common under conditions of crowding, and it occurs very frequently in mental hospitals, schools for the retarded, and day care centers. A chronic carrier state has not been observed with hepatitis A; perpetuation of the virus in nature presumably depends on sporadic subclinical infections and person-to-person transmission. Outbreaks of hepatitis A have been linked to the ingestion of undercooked seafood, usually shellfish from waters contaminated with human feces. Common-source outbreaks related to other foods, including vegetables as well as contaminated drinking water, have also been reported.

Fecal–oral transmission

Outbreaks linked to ingestion of uncooked seafood and contaminated food, produce, and water

No chronic carriage

Less than 50% of the general population of the United States now has serologic evidence of HAV infection, and rates have been decreasing since 1970, apparently because of better sanitation, less crowding, and the use of hepatitis A vaccine. In contrast, more than 90% of the adult population in many developing countries shows evidence of previous hepatitis A infection. The risk of clinically evident disease is much higher in infected adults than in children. Patients are most contagious in the 1 to 2 weeks before onset of clinical disease.

More than 90% of adult population is seropositive in developing countries

Subclinical infection is common in children

PATHOGENESIS

HAV is believed to replicate initially in the enteric mucosa. It can be demonstrated in feces by electron microscopy for 10 to 14 days before onset of disease. In most patients with symptoms of the disease, virus is no longer found in fecal specimens. Multiplication in the intestines is followed by a period of viremia with spread to the liver. The response to replication in the liver consists of lymphoid cell infiltration, necrosis of liver parenchymal cells, and proliferation of Kupffer cells (**Figure 13–2)**. A variable degree of biliary stasis may be present. It is also believed that cytotoxic T lymphocytes damage the hepatocytes. Except in the rare instance of acute hepatic necrosis, the infection is cleared, liver damage is reversed, and HAV does not establish a chronic infection. Initial immune response is the development of HAV-specific IgM antibody followed by appearance of IgG after a few weeks. Detectable levels of IgG antibody to HAV persist indefinitely in serum, and patients with anti–HAV antibodies are immune to reinfection. Although virus-specific IgA has been demonstrated in stool, secretory immunity has not been shown to be important for hepatitis A. The immunopathogenetic events associated with HAV infection are shown in **Figure 13–3**.

Contagion is greatest 10 to 14 days before symptoms appear

IgG-specific antibody is protective

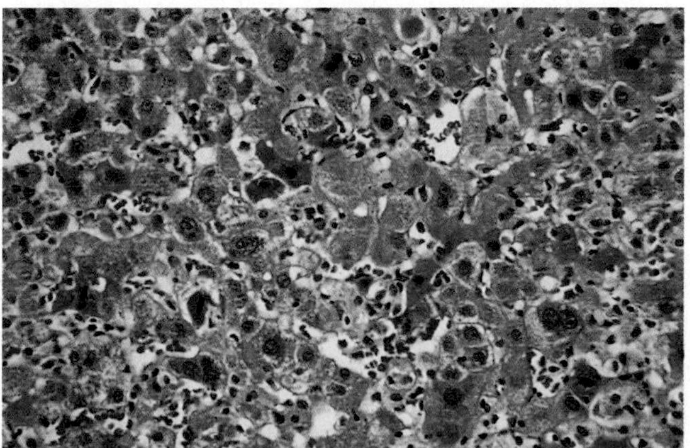

FIGURE 13–2. Acute viral hepatitis, moderately severe. There is a lobular disarray with degeneration, apoptosis, and necrosis of liver cells. Disruption of liver cell plates, hypertrophy of Kupffer cells, a predominantly lymphocytic inflammatory infiltrate, and regeneration of surviving liver cells also are seen. (Reproduced with permission from Connor DH, Chandler FW, Schwartz DQA, Manz HJ, Lack EE (eds). *Pathology of Infectious Diseases,* vol. 1. Stamford, CT: Appleton & Lange; 1997.)

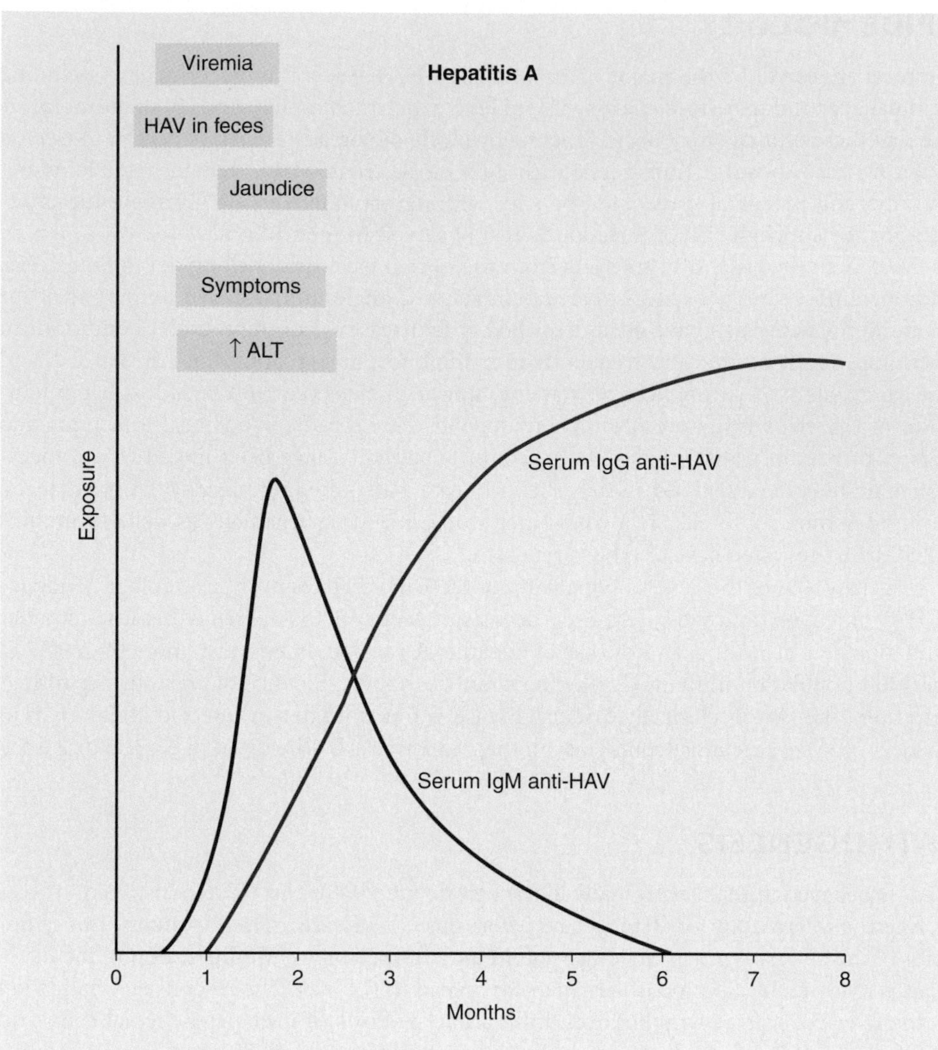

FIGURE 13–3. Sequence of appearance of viremia, virus in feces, alanine aminotransferase (ALT), symptoms, jaundice, and IgM and IgG antibodies in hepatitis A virus (HAV) infection.

CLINICAL ASPECTS

MANIFESTATIONS

Fever, anorexia, and jaundice are common

In HAV infection, an incubation period of 15-45 days (mean 25 days) is usually followed by fever; anorexia (poor appetite); nausea; pain in the right upper abdominal quadrant; and, within several days, jaundice. Dark urine and clay-colored stools may be noticed by the patient 1 to 5 days before the onset of clinical jaundice. The liver is enlarged and tender, and serum aminotransferase and bilirubin levels are elevated as a result of hepatic inflammation and damage. Recovery occurs in days to weeks.

Chronic infection does not occur

Many persons who have serologic evidence of acute HAV infection are asymptomatic or only mildly ill, without jaundice (anicteric hepatitis A). The infection-to-disease ratio is dependent on age; it may be as high as 20:1 in children and approximately 1:1 in older adults. Almost all cases (99%) of HAV are self-limiting. Chronic hepatitis such as that seen with hepatitis B is very rare. In rare cases, fulminant fatal hepatitis associated with extensive liver necrosis may occur (~0.1%).

DIAGNOSIS

Antibody to HAV can be detected during early illness, and most patients with symptoms or signs of acute HAV already have detectable antibody in serum. Early antibody responses are predominantly IgM, which can be detected for several weeks and up to several months (Figure 13–3). During convalescence, antibody of the IgG class predominates. The best method for documentation of acute HAV infection is the demonstration of high titers of virus-specific IgM antibody in serum drawn during the acute phase of illness. Because IgG antibody persists indefinitely, its demonstration in a single serum sample is not indicative of recent infection; a rise in titer between acute and convalescent sera must be documented. Immune electron microscopic identification of the virus in fecal specimens and isolation of the virus in cell cultures remain research tools.

IgM-specific antibody denotes acute infection

TREATMENT AND PREVENTION

There is no specific treatment for patients with acute hepatitis A. Supportive measures include adequate nutrition and rest. Avoidance of exposure to contaminated food or water or infected persons are important measures to reduce the risk of hepatitis A infection.

Passive Immunization

Passive (ie, antibody) prophylaxis for hepatitis A has been available for many years. Immune serum globulin (ISG), manufactured from pools of plasma from large segments of the general population, is protective if given before or during the incubation period of the disease. It has been shown to be about 80% to 90% effective in preventing clinically apparent type A hepatitis. In some cases, infection occurs but disease is ameliorated; that is, patients develop anicteric, usually asymptomatic, hepatitis A. At present, ISG should be administered to household and intimate contacts of hepatitis A patients and those known to have eaten uncooked foods prepared or handled by an infected person. When clinical symptoms have appeared, the patient is already producing antibody, and administration of ISG is not indicated.

Immune serum globulin provides temporary protection

Active Immunization

Formalin-killed HAV, which is grown in human cell culture, is used as a vaccine that induces antibody titers similar to those of wild-type virus infection, is almost 100% protective, and is now recommended for all children and for adults with a high risk of infection. Two doses are given 6 to 12 months apart to achieve long-term protection. Recent evidence suggests that immunization is as effective as ISG if given shortly after exposure. If confirmed, this would provide a preferable alternative to giving ISG.

Inactivated virus vaccine confers long-term protection

HEPATITIS B

VIROLOGY

STRUCTURE

Hepatitis B virus (HBV) is an enveloped DNA virus belonging to the family Hepadnaviridae. It is unrelated to any other human virus; however, related hepatotropic agents have been identified in woodchucks, ground squirrels, and kangaroos. A schematic of the HBV is illustrated in **Figure 13–4**. The complete virion is a 42-nm spherical particle that consists of an envelope around a 27-nm core. The core comprises a nucleocapsid that contains the DNA genome.

Smallest known human DNA virus

The viral genome consists of partially double-stranded DNA with a short, single-stranded piece. It comprises 3200 nucleotides, making it the smallest DNA virus known but capable of encoding surface (envelope) protein (HBsAg), core (nucleocapsid) protein (HBcAg),

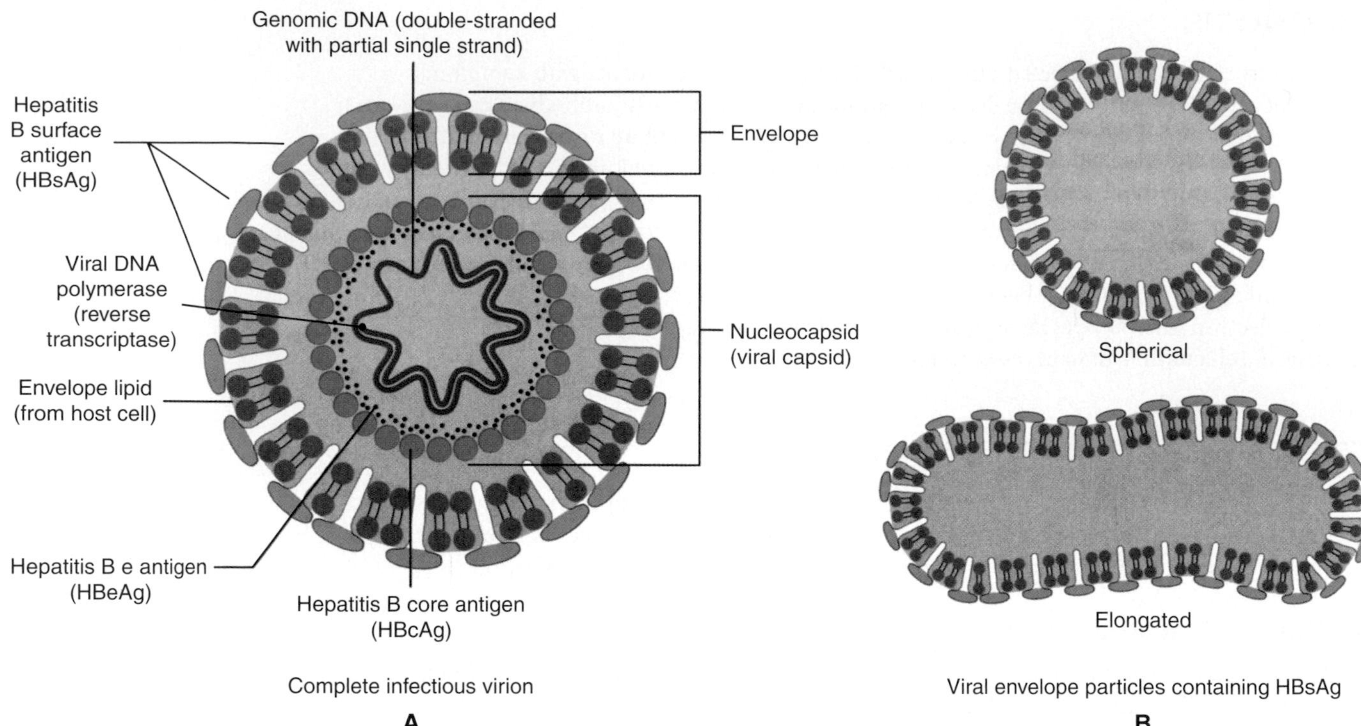

FIGURE 13–4. Schematic diagram of hepatitis B virion. A. The 42-nm particle is the "Dane particle" or the hepatitis B virus. **B.** The 22-nm particles are the filamentous and circular forms of hepatitis B surface antigen (HbsAg) or protein coat. (Reproduced with permission from Nester EW, Anderson DG, Roberts CE Jr, Nester MT. *Microbiology: A Human Perspective,* 6th ed. New York: McGraw-Hill, 2008.)

Enveloped DNA virus with viral DNA polymerase (reverse transcriptase) activity

DNA polymerase (reverse transcriptase), and HBx protein (a transcriptional activator). Closely associated with the viral DNA is a viral DNA polymerase, which has RNA-dependent DNA polymerase, DNA-dependent DNA polymerase, and RNase H activities (reverse transcriptase). Other components of the core are a hepatitis B core antigen (HBcAg) and the hepatitis B e antigen (HBeAg), which is a low-molecular-weight glycoprotein secreted from the infected cells. The virion has a lipid bilayer envelope containing the hepatitis B surface antigen (HBsAg), which is composed of one major and two other proteins. The complete virus particle is called a **Dane particle**.

HBsAg is produced in great abundance

Found in cytoplasm of infected hepatocytes

Aggregates of HBsAg are often found in great abundance in serum during infection. They may assume spherical or filamentous shapes with a mean diameter of 22 nm (Figure 13–4). Hepatitis B DNA can also be detected in serum and is an indication that infectious virions are present. In infected liver tissue, evidence of HBcAg, HBeAg, and hepatitis B DNA is found in the nuclei of infected hepatocytes, whereas HBsAg is found in cytoplasm.

There are four major serotypes of HBV (*adr, adw, ayr, ayw*) based on HBsAg antigenic epitopes. Furthermore, there are eight hepatitis B genotypes (A-H) based on nucleotide sequence variation of HBV genome, which may be associated with different clinical outcomes. These genotypes vary in geographic distribution with genotype A primarily found in North America, Northern Europe, India, and Africa, genotypes B and C in Asia, genotype D in Southern Europe, Middle East, and India, genotype E in West and South Africa, genotype F in South and Central America, genotype G in the United States and Europe, and genotype H in Central America and California.

REPLICATION CYCLE

The replication of HBV involves a reverse transcription step, and, as such, is unique among DNA viruses (**Figure 13–5**). HBV has a specific tropism for the liver. However, the receptor

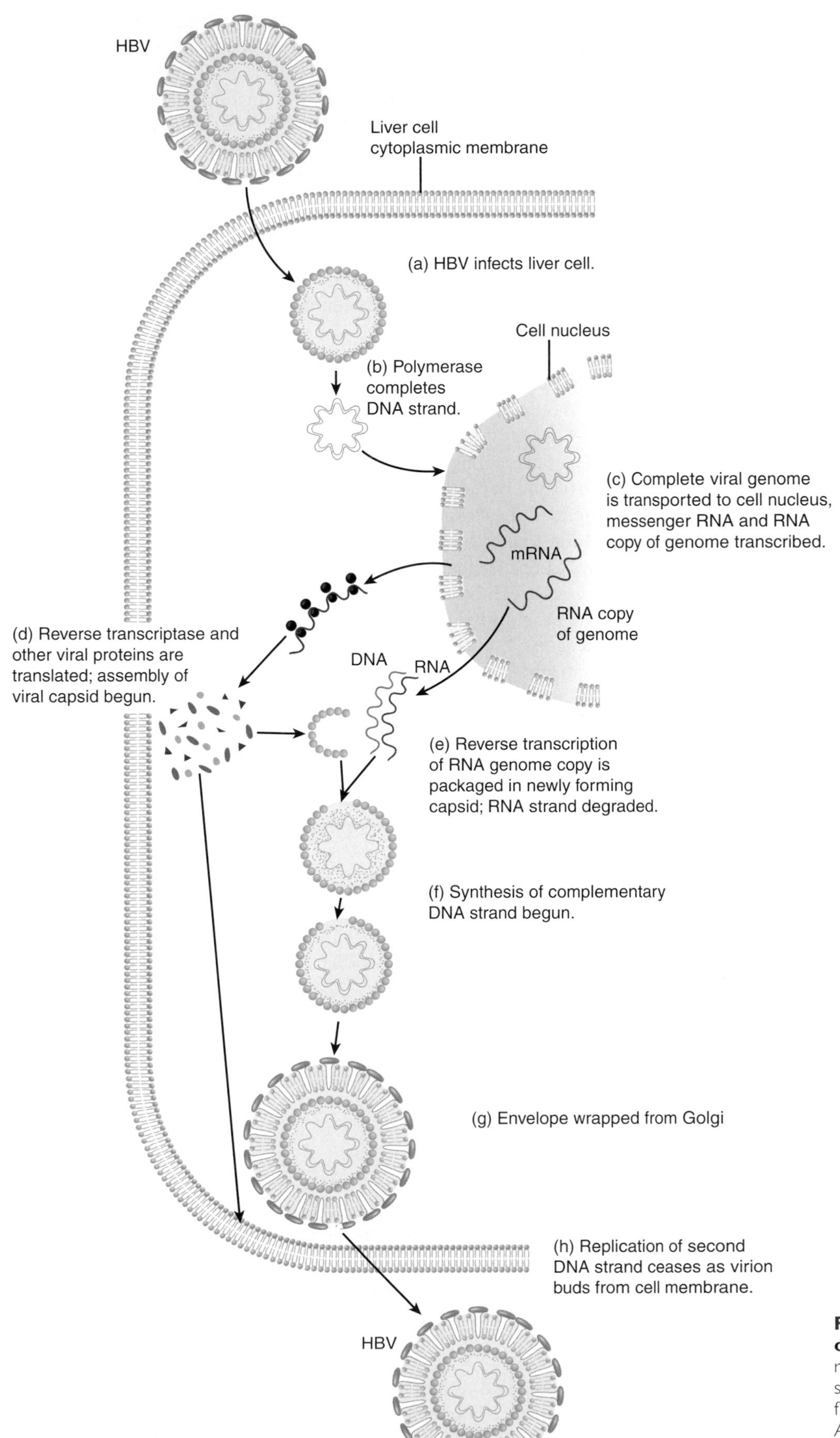
HBV
Liver cell
cytoplasmic membrane
(a) HBV infects liver cell.
Cell nucleus
(b) Polymerase
completes
DNA strand.
(c) Complete viral genome
is transported to cell nucleus,
messenger RNA and RNA
copy of genome transcribed.
mRNA
RNA copy
of genome
(d) Reverse transcriptase and
other viral proteins are
translated; assembly of
viral capsid begun.
DNA
RNA
(e) Reverse transcription
of RNA genome copy is
packaged in newly forming
capsid; RNA strand degraded.
(f) Synthesis of complementary
DNA strand begun.
(g) Envelope wrapped from Golgi
(h) Replication of second
DNA strand ceases as virion
buds from cell membrane.
HBV

FIGURE 13–5. Replication cycle of hepatitis B virus (HBV). HBV replication requires reverse transcription step, unique among DNA viruses. (Modified with permission from Nester EW, Anderson DG, Roberts CE Jr, Nester MT. *Microbiology: A Human Perspective,* 6th ed. New York: McGraw-Hill, 2008.)

for HBV and the mechanism of viral entry are not known. The attachment or adsorption of HBV to hepatocytes (liver cells) is mediated by the envelope protein (HBsAg) of the virus, probably by binding of HBsAg with polymerized human serum albumin or other serum proteins. After viral entry, the partial double-stranded DNA (incomplete) is transported to the nucleus. The double-stranded DNA is organized as two strands. One, a short strand, is associated with the viral DNA polymerase and is of positive polarity.

The complete or long strand is complementary and thus of negative polarity. The partial incomplete strand is formed into a complete double-stranded circular DNA, which is essential before transcription can take place. Host RNA polymerase directs the transcription of viral mRNAs to encode early proteins, including HBcAg, HBeAg, and viral DNA polymerase as well as full-length RNA (pre-genomic RNA). HBsAg is encoded later and associates with the membranes of endoplasmic reticulum or Golgi apparatus. HBcAg forms the core by enclosing the full-length positive-sense viral pre-genomic RNA along with viral DNA polymerase into maturing core particles late in the replication cycle. These full-length RNA strands form a template for a reverse transcription step in which negatively stranded DNA is synthesized. The RNA template strands are then degraded by ribonuclease H activity. A positive-stranded DNA is then synthesized, although this is not completed before virus maturation in which HBsAg-containing membranes of the endoplasmic reticulum or Golgi apparatus are wrapped over the nucleocapsid core, resulting in the variable-length, short, positive DNA strands found in the virions. The virions are released by exocytosis.

Unique replication using a reverse transcriptase step

HBV DNA has also been found to integrate into the host chromosomes, especially in HBV-infected patients with hepatocellular carcinoma (HCC). However, the significance of integrated HBV DNA in viral replication is not known. Despite extensive attempts, HBV has not been successfully propagated in the laboratory. Humans appear to be the major host; however, as with hepatitis A, infection of subhuman primates has been accomplished experimentally. ::: viral integration, p. 148

Humans are the major hosts

HEPATITIS B DISEASE

CLINICAL CAPSULE

Hepatitis B virus is the cause of what was formerly known as "serum hepatitis." This name was used to distinguish it from "infectious hepatitis" and reflected the association of this form of hepatitis with needle use or blood transfusion. HBV is usually an asymptomatic or limited illness with fever and jaundice for days to weeks. It becomes chronic in up to 10% of patients and may lead to cirrhosis or hepatocellular carcinoma.

EPIDEMIOLOGY

Hepatitis B infection is found worldwide, with prevalence rates varying markedly among countries, but a total of approximately 400 million persons (**Figure 13–6**). Chronic carriers constitute the main reservoir of infection: in some countries, particularly in the Far East, up to 5% to 15% of all persons carry the virus, and most are asymptomatic. About 10% of patients with human immunodeficiency virus (HIV) infection are chronic carriers of HBV.

Chronic carriers are common in the Far East

In the United States, an estimated 1.25 million people are infected with hepatitis B, and 300,000 new cases occur annually. About 300 of these patients die of acute fulminant hepatitis, and 5% to 10% of infected patients become chronic HBV carriers. As many as 4000 people die yearly of hepatitis B-related cirrhosis, and 1000 die of HCC. The virus is spread

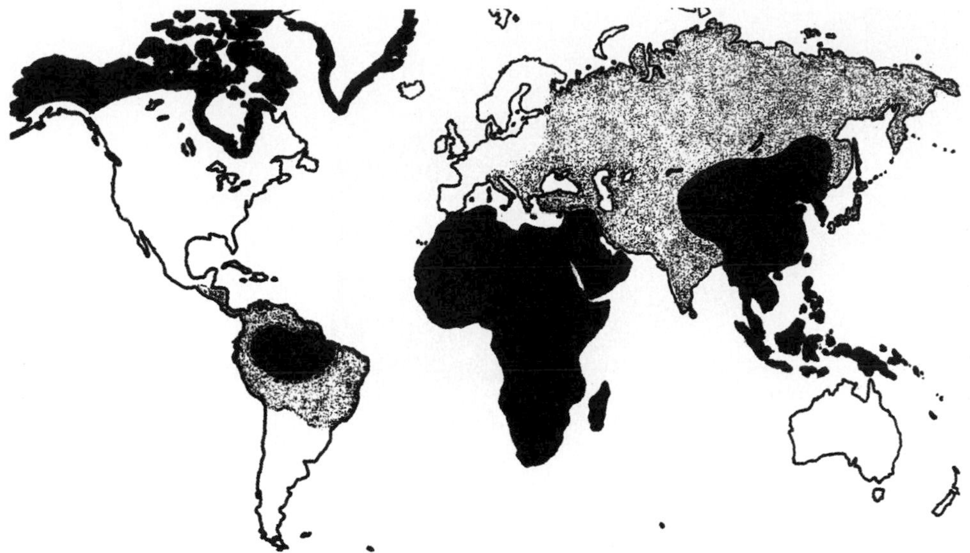

FIGURE 13–6. Worldwide distribution of hepatitis B infection. Areas with high prevalence (>8% of population) are in black, and areas with moderate prevalence (2% to 7%) are in gray. (Reproduced with permission from Connor DH, Chandler FW, Schwartz DQA, Manz HJ, Lack EE (eds). *Pathology of Infectious Diseases,* vol. 1. Stamford, CT: Appleton & Lange; 1997.)

vertically, parenterally, and by sexual contact. Approximately 50% of infections in the United States are sexually transmitted, and the prevalence of HBsAg in serum is higher in certain populations, such as among men who have sex with men, patients on hemodialysis or immunosuppressive therapy, patients with Down's syndrome, and injection drug users. Routine screening of blood donors for HbsAg and antibody to hepatitis B core antigen (anti-HBcAg) has markedly decreased the incidence of post-blood transfusion and post-plasma products hepatitis B transmission. Multiple-pool blood products still cause occasional cases. Exposure to hepatitis viruses from direct contact with blood or other body fluids, probably through needlestick injuries, has resulted in a risk of hepatitis B infection in medical personnel. Attack rates are also high in the sexual partners of infected patients.

Needlestick transmission is a risk for health care workers

Hepatitis B infection of infants does not appear to be transplacentally transmitted to the fetus in utero, but is acquired during the birth process by the swallowing of infected blood or fluids or through abrasions. The rate of virus acquisition is high (up to 90%) in infants born to mothers who have acute hepatitis B infection or are carrying HBsAg and HBeAg. Most infants do not develop clinical disease; however, infection in the neonatal period is associated with failure to produce antibody to HbsAg and cell-mediated immune responses probably as a result of an immature immune system, which allows chronic carriage to occur in nearly 90% to 100% of the infected neonates/infants.

Vertical transmission usually occurs during birth process

HCC (hepatocellular carcinoma) has been strongly associated with persistent carriage of HBV by serologic tests and by detection of viral nucleic acid sequences integrated in tumor cell genomes. In many parts of Africa and Asia, primary liver cancer accounts for 20% to 30% of all types of malignancies, but in North and South America and in Europe, it is only 1% to 2%. The estimated risk of developing the malignancy for persons with chronic HBV is increased to between 10- to more than 300-fold in different populations. The risk of HCC further increases in patients with chronic hepatitis B infection and high viral loads.

Strong association between chronic infection and hepatocellular carcinoma

PATHOGENESIS

In the past, hepatitis B was known as post-transfusion hepatitis or as hepatitis associated with the use of illicit parenteral drugs (serum hepatitis). However, over the last few years it has become clear that the major mode of acquisition is through close personal contact with body fluids of infected individuals. HBsAg has been found in most body fluids, including saliva, semen, and cervical secretions. Under experimental conditions, as little as 0.0001 mL of infectious blood has produced infection. Transmission is therefore possible by vehicles such as inadequately sterilized hypodermic needles and instruments used in tattooing and ear piercing.

Virus found in blood, saliva, and semen

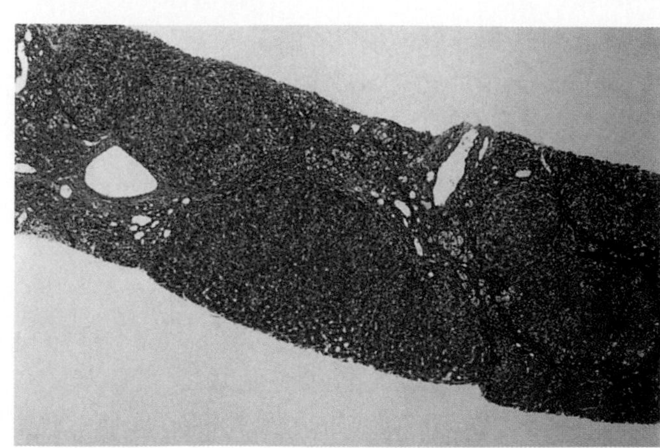

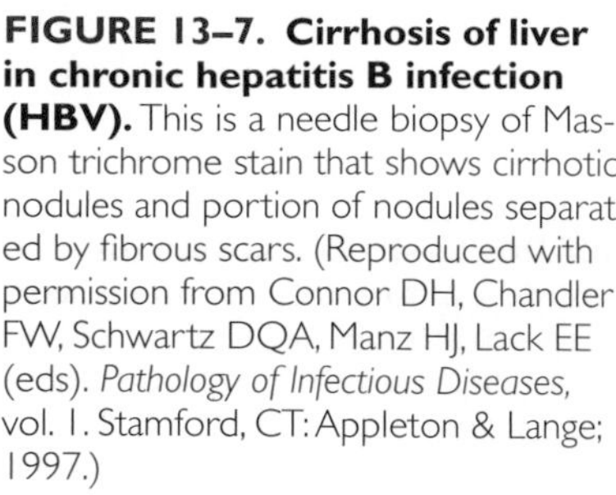

FIGURE 13–7. Cirrhosis of liver in chronic hepatitis B infection (HBV). This is a needle biopsy of Masson trichrome stain that shows cirrhotic nodules and portion of nodules separated by fibrous scars. (Reproduced with permission from Connor DH, Chandler FW, Schwartz DQA, Manz HJ, Lack EE (eds). *Pathology of Infectious Diseases,* vol. 1. Stamford, CT: Appleton & Lange; 1997.)

The factors determining the clinical manifestations of acute hepatitis B are largely unknown; however, some appear to involve immunologic responses of the host. The serum sickness–like rash and arthritis that may precede the development of symptoms and jaundice appear to be related to circulating immune complexes that activate the complement system. Antibody to HBsAg is protective and associated with resolution of the disease. Cellular immunity also may be important in the host response because patients with insufficient T-lymphocyte function have a high incidence of chronic infection with HBV. Antibody to HBcAg, which appears during infection, is present in chronic carriers with persistent hepatitis B virion production and does not appear to be protective.

Antibody to HBsAg is protective

The morphologic lesions of acute hepatitis B resemble those of other hepatitis viruses. In chronic active hepatitis B, the continued presence of inflammatory foci of infection results in necrosis of hepatocytes, collapse of the reticular framework of the liver, and progressive fibrosis. The increasing fibrosis can result in the syndrome of postnecrotic hepatic cirrhosis (**Figure 13–7**).

Chronic infection leads to progressive fibrosis and cirrhosis

Integrated hepatitis B viral DNA can be found in nearly all HCCs. The virus has not been shown to possess a transforming gene but may well activate a cellular oncogene. It is also possible that the virus does not play such a direct molecular role in oncogenicity, because the natural history of chronic hepatitis B infection involves cycles of damage or death of liver cells interspersed with periods of intense regenerative hyperplasia. This significantly increases the opportunity for spontaneous mutational changes that may activate cellular oncogenes. Whatever the mechanism, the association between chronic viral infection and HCC is clear, and liver cancer is a major cause of disease and death in countries in which chronic hepatitis B infection is common. The proven success of combined active and passive immunization in aborting hepatitis B infection in infancy and childhood makes HCC a potentially preventable disease.

Mechanism of HCC development is not clearly known

CLINICAL ASPECTS

MANIFESTATIONS

The clinical picture of hepatitis B is highly variable. The incubation period may be as brief as 30 days or as long as 180 days (mean approximately 60 to 90 days). Acute hepatitis B is usually manifested by the gradual onset of fatigue, loss of appetite, nausea and pain, and fullness in the right upper abdominal quadrant. Early in the course of disease, pain and swelling of the joints and occasional frank arthritis may occur. Some patients develop a rash. With increasing involvement of the liver, there is increasing cholestasis, and hence clay-colored stools, darkening of the urine, and jaundice. Symptoms may persist for several months before finally resolving.

Average incubation period is 10 weeks; range 30 to 180 days

In general, the symptoms associated with acute hepatitis B are more severe and more prolonged than those of hepatitis A; however, anicteric disease and asymptomatic infection occur. The infection-to-disease ratio, which varies according to patient age and method of acquisition, has been estimated to be approximately 3:1. Fulminant hepatitis, leading to extensive liver necrosis and death, develops in less than 1% of cases. One important difference between hepatitis A and hepatitis B is the development of chronic hepatitis, which occurs in approximately 10% of all patients with hepatitis B infection, with a much higher risk for newborns (~90%), children (~50%), and the immunocompromised. In immunocompetent adults, the strong cellular immune response results in acute hepatitis and only rarely (~ 1%) in chronic hepatitis. Chronic infection is associated with ongoing replication of virus in the liver and usually with the presence of HBsAg in serum. Chronic hepatitis may lead to cirrhosis, liver failure, or HCC in up to 25% of patients.

Chronic hepatitis is most common with infection in early infancy or childhood

DIAGNOSIS

The nomenclature of hepatitis B antigens and antibodies is shown in **Table 13–2** and the sequence of their appearance is shown in **Figure 13–8.** During the acute episode of disease, when there is active viral replication, large amounts of HBsAg and hepatitis B virus DNA can be detected in the serum, as can fully developed virions and high levels of DNA polymerase and HBeAg. Although HBcAg is also present, antibody against it invariably occurs and prevents its detection. Upon resolution of acute hepatitis B, HBsAg and HBeAg disappear from serum with the development of antibodies (anti-HBs and anti-HBe) against them. The development of anti-HBs is associated with elimination of infection and protection against reinfection. Anti-HBc is detected early in the course of disease and persists in serum for years. It is an excellent epidemiologic marker of infection, but is not protective. The laboratory diagnosis of acute hepatitis B is best made by demonstrating the IgM antibody to HbcAg in serum, since this antibody disappears within 6 to 12 months of the acute infection. Almost all patients who develop jaundice are anti-HBc IgM–positive at the time of clinical presentation. HBsAg may also be detected in serum. Past infection with hepatitis B is best determined by detecting IgG anti-HBc, anti-HBs, or both, whereas vaccine induces only anti-Hbs.

Appearance of anti-HBs signals elimination of infection

Acute infection associated with appearance of anti-HBc IgM

In patients with chronic hepatitis B, evidence of viral persistence can be found in serum (**Figure 13–9**). HBsAg can be detected throughout the active disease process, and anti-HBs does not develop, which probably accounts for the chronicity of the disease. However, anti-HBc *is* detected. Two types of chronic hepatitis can be distinguished. In one, HBsAg is detected, but not HBeAg; these patients usually show progressive liver dysfunction. In the other, both antigens are found; development of antibody to HBeAg is associated with

TABLE 13–2 Nomenclature for Hepatitis B Virus Antigens and Antibodies

ABBREVIATION	DESCRIPTION
HBV	Hepatitis B virus; 42-nm double-stranded DNA virus; Dane particle
HBsAg	Hepatitis B surface antigen; found on surface of virus; formed in excess and seen in serum as 22-nm spherical and tubular particles; four subdeterminants (*adw, ayw, adr,* and *ayr*) identified
HBcAg	Core antigen (nucleocapsid core); found in nucleus of infected hepatocytes by immunofluorescence
HBeAg	Glycoprotein; associated with the core antigen; used epidemiologically as marker of potential infectivity; seen only when HBsAg is also present
Anti-HBs	Antibody to HBsAg; correlated with protection against and/or resolution of disease; used as marker of past infection or vaccination
Anti-HBc	Antibody to HBcAg; seen in acute infection and chronic carriers; anti-HBc IgM used as indicator of acute infection; anti-HBc IgG used as marker of past or chronic infection; apparently not important in disease resolution; does not develop in response to vaccine
Anti-HBe	Antibody to HBeAg

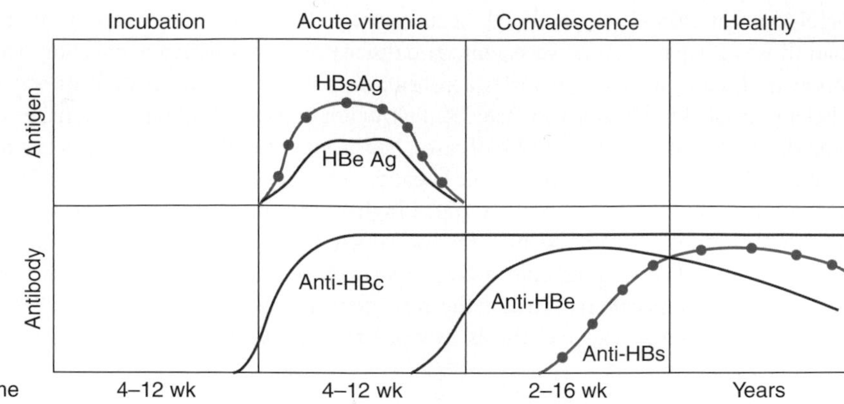

FIGURE 13–8. Sequence of appearance of viral antigens and antibodies in acute self-limiting cases of hepatitis B. Anti-HBc, antibody to hepatitis B core antigen; anti-HBe, antibody to HBeAg; anti-HBs, antibody to HBsAg; HBeAg, hepatitis B e antigen; HBsAg, hepatitis B surface antigen

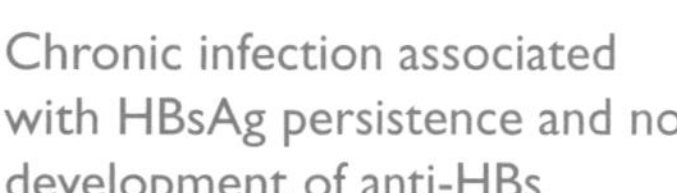
Chronic infection associated with HBsAg persistence and no development of anti-HBs

clinical improvement. Chronic infection with hepatitis B is best detected by persistence of HBsAg in blood for more than 6 to 12 months. Progression of liver disease is associated with more than 1000 IU of HBV DNA. Persons with levels lower than 1000 IU and normal liver function have a low risk of progression.

TREATMENT

No specific treatment for acute infection

Interferon and nucleoside and nucleotide analogs are of benefit

There is no specific treatment recommended for acute hepatitis B. A high-calorie diet is desirable. Treatment should be considered for patients with rapid deterioration of liver function, cirrhosis or complications such as ascites, hepatic encephalopathy, or hemorrhage as well as those who are immunosuppressed. For chronic hepatitis B diseases, pegylated or regular interferon alpha provides benefit in some patients. Lamivudine (3TC), a potent inhibitor of HIV, and other nucleoside analogs as well as certain nucleotide analogs are active against hepatitis B.

PREVENTION

Screening of blood and plasma product donors for HBsAg and anti-HBcAg has greatly reduced the incidence of hepatitis B in recipients. Similarly, screening pregnant women and treatment of exposed newborns with hepatitis B immune globulin (HBIG) and vaccine have reduced vertical transmission. Safe sexual practices and avoidance of needlestick injuries or injection drug use are approaches to diminishing the risk of hepatitis B infection. Both active prophylaxis and passive prophylaxis against hepatitis B infection can be accomplished. Most preparations of ISG contain only moderate levels of anti-HBs; however, specific HBIG with high titers of

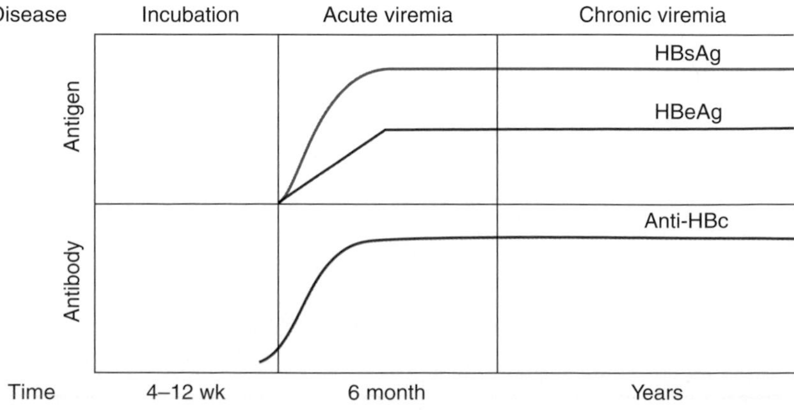

FIGURE 13–9. Sequence of appearance of viral antigens and antibodies in chronic active hepatitis B. Antibodies to HBsAg and HBeAg not detected. Anti-HBc, antibody to hepatitis B core antigen; HBeAg, hepatitis B e antigen; HBsAg, hepatitis B surface antigen.

hepatitis B antibody is now available. HBIG is prepared from sera of subjects who have high titers of antibody to HBsAg, but are free of the antigen itself. Administration of HBIG soon after exposure to the virus greatly reduces the development of symptomatic disease.

Postexposure prophylaxis with HBIG should be followed by active immunization with vaccine. Purified inactivated HBsAg vaccine from chronic carriers has been available for several years. This was developed by purification and inactivation of HBsAg from the blood of HBV-infected chronic carriers, but it is no longer in use. The current vaccine is a recombinant product derived from HBsAg expressed in yeast. Excellent protection has been shown in studies of men who have sex with men and in medical personnel. These groups and others, such as laboratory workers, injection drug users, travelers to endemic areas, persons at risk for sexually transmitted diseases, and those in contact with patients who have chronic hepatitis B, should receive hepatitis B vaccine as the preferred method of preexposure prophylaxis. Recently, immunization of newborns, all children, and adolescents has been recommended. Three doses (at 0, 1, and 6 months) are given to achieve maximum titer. Protection may not be lifelong.

Postexposure treatment with HBIG temporarily reduces risk

Recombinant (HBsAg) vaccine recommended for all children and high-risk persons

A combination of active and passive immunization is the most effective approach to prevent neonatal acquisition and chronic carriage in the neonate. Routine screening of pregnant women for the presence of HBsAg is recommended. Infants born to those who are positive should receive HBIG in the delivery room followed by three doses of hepatitis B vaccine beginning 24 hours after birth. A similar combination of passive and active immunization is used for unimmunized persons who have been exposed by needlestick or similar injuries. The procedure varies depending on the hepatitis B status of the "donor" case linked to the injury.

Combination of HBIG and vaccine significantly reduces vertical transmission

DELTA HEPATITIS (HEPATITIS D)

VIROLOGY

Delta hepatitis is caused by the hepatitis D virus (HDV). This small single-stranded circular RNA virus requires the presence HbsAg for its transmission and is thus found only in persons with acute or chronic hepatitis B infection. Strategies directed at preventing HBV are also effective in preventing HDV. Associated with the circular RNA, which forms a rod because of extensive base pairing, are proteins of 27 and 29 kDa, which constitute the delta capsid antigen. This protein–RNA complex is surrounded by HBsAg (**Figure 13–10**). Thus, although the delta virus produces its own capsid antigens, it co-opts the HBsAg in assembling its coat or envelope. Unlike other RNA viruses, HDV genome is not capable of encoding an RNA polymerase.

Hepatitis D is found only in hepatitis B-infected persons

Small single-stranded circular RNA virus

Virus uses HBsAg for transmission and assembly

The replication of HDV involves virus entry in hepatocytes (liver cells) just like HBV because HDV contains HbsAg on its surface. Since HDV lacks an RNA polymerase required for transcription and replication, it uses host cell RNA polymerase to synthesize mRNA and RNA genome in the nucleus. This is unique for an RNA virus to replicate in the nucleus without encoding its own RNA polymerase. The extensive base-pairing in some regions of the

Replication of HDV complex and unique

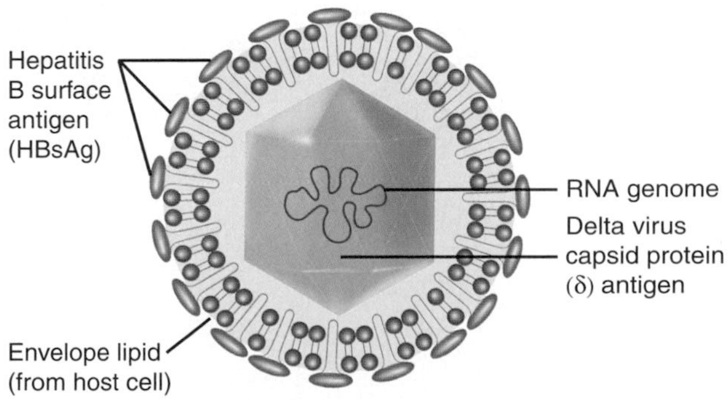

FIGURE 13–10. Schematic of delta hepatitis virus. Note outer layer derived from hepatitis B surface antigen.

Transcription and replication of HDV occurs in the nucleus using host cell RNA polymerase

Presence of HBV (HbsAg) required for HDV assembly

HDV genome allows the cellular RNA polymerase to bind the base-paired RNA sequences, as RNA polymerase binds to DNA sequences, and to transcribe HDV mRNA. The RNA genome further forms a ribozyme structure that allows self-cleaving of the RNA genome to generate mRNA. The delta capsid antigens are synthesized and associate with HDV circular RNA genomes followed by acquiring an envelope from endoplasmic reticulum or Golgi apparatus containing HBsAg. Thus, the presence of HBsAg is essential for assembly of HDV virions.

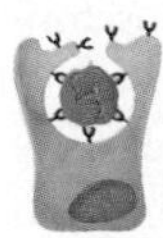

DELTA HEPATITIS DISEASE

Greatest risk is among injection drug abusers

Delta hepatitis is most prevalent in groups with a high risk for developing hepatitis B. Paradoxically, delta hepatitis is not common in East Asia, where hepatitis B is common, but it is most common in the Middle East, parts of Africa, and South America (**Figure 13–11**). Injection drug users are those at greatest risk in the western parts of the world, and up to 50% of such individuals may have IgG antibody to the delta virus antigen. Other risks include sexual transmission and dialysis. Vertical transmission can also occur.

CLINICAL ASPECTS

MANIFESTATIONS

Simultaneous hepatitis B and D infections cause more severe disease

Two major types of delta infection have been noted: simultaneous delta and hepatitis B infection or delta superinfection in those with chronic hepatitis B. Simultaneous infection with both delta and hepatitis B may result in clinical hepatitis that is indistinguishable from acute hepatitis A or B, but it may manifest as a second rise in liver enzymes (AST, ALT). Persons with chronic hepatitis B who acquire superimposed infection with hepatitis D suffer relapses of jaundice and have a high likelihood of developing chronic cirrhosis. Epidemics of delta infection have occurred in populations with a high incidence of chronic hepatitis B and have resulted in rapidly progressive liver disease, causing death in up to 20% of infected persons.

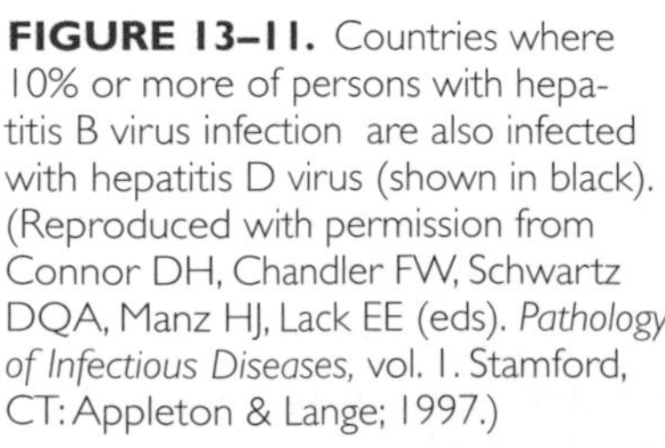

FIGURE 13–11. Countries where 10% or more of persons with hepatitis B virus infection are also infected with hepatitis D virus (shown in black). (Reproduced with permission from Connor DH, Chandler FW, Schwartz DQA, Manz HJ, Lack EE (eds). *Pathology of Infectious Diseases,* vol. 1. Stamford, CT: Appleton & Lange; 1997.)

DIAGNOSIS

Diagnosis of delta infection is made most commonly by demonstrating IgM or IgG antibodies, or both, to the delta capsid antigen in serum. IgM antibodies appear within 3 weeks of infection and persist for several weeks. IgG antibodies persist for years. In co-infection, the patient has both anti-Hbc and anti-D antibodies, whereas in superinfection, the anti-HBc is absent.

Diagnosis is by detection of antibodies to delta antigen

TREATMENT AND PREVENTION

Interferon and other anti-HBV therapies (nucleosides, nucleotides) are not active against hepatitis D. Response to treatment in patients with delta hepatitis (and hepatitis B) is less than in those with hepatitis B alone.

Because the surface of delta hepatitis is HBsAg, measures aimed at limiting the transmission of hepatitis B (eg, vaccination, blood screening) prevent the transmission of delta hepatitis. Persons infected with hepatitis B or D should not donate blood, organ, tissues, or semen. Safe sex should be practiced unless there is only a single sex partner who is already infected. Methods of reducing transmission include decreased use of contaminated needles and syringes by injection drug users and use of needle safety devices by health care workers.

Major strategies for prevention of hepatitis B also prevent hepatitis D

HEPATITIS C

VIROLOGY

Hepatitis C virus (HCV) is an RNA-enveloped virus in the Flaviviridae family (other members: yellow fever, dengue, West Nile virus) and *Hepacivirus* genus. It has a very simple positive-sense single-stranded RNA genome, consisting of just three structural (C, core; E1 and E2, envelope glycoproteins) and five nonstructural genes. The HCV virion of 50 nm in diameter contains an RNA genome of 9.5 kb, which is enclosed in an icosahedral capsid or core (C) and a lipid-bilayer envelope containing two virus specific glycoproteins E1 (gp31) and E2 (gp70) (**Figure 13–12**). The genome is encoded into a polyprotein, which is processed into individual proteins by proteases.

Enveloped RNA virus of Flaviviridae family

Positive-sense RNA genome that encodes three structural and five nonstructural proteins

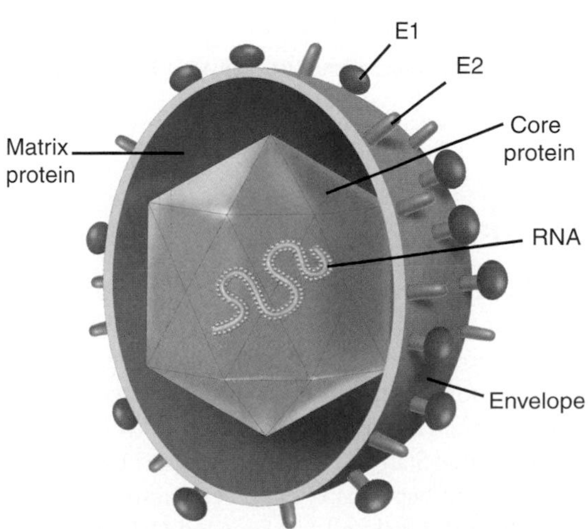

FIGURE 13–12. Structure of hepatitis C virion. Inside the core is a single stranded positive-sense RNA enclosed in a lipid bilayer membrane containing glycoproteins E1 and E2.

Highly heterogeneous virus, hypervariable regions in E2

HCV is highly heterogeneous because the genome of HCV is highly mutable, since its RNA-dependent RNA polymerase lacks proof reading ability. Mutations give rise to HCV quasispecies (variants) and antigenic variation, most noticeably in the E2 glycoprotein hypervariable regions (HVR1 and R2), which may allow the virus to escape immune response and cause chronic or persistent infection in infected persons. The hypervariable region in E2 contains the epitope for neutralization, and mutations allow the newly generated HCV variants to escape preexisting immune response.

Eleven genotypes with multiple subtypes have different geographic distribution

There are at least 11 genotypes, with multiple subtypes. The genotypes have different geographic distributions and may be associated with differing severity of disease as well as response to therapy. Genotypes 1-3 have worldwide distribution, with genotype 1a predominating in North America.

Genotypes important for therapy response

The HCV genome also encodes a nonstructural gene that is involved in sensitivity to interferon. HCV heterogeneity and generation of multiple HCV genotypes, like HIV, hinder the development of an HCV vaccine.

HCV replicates in the cytoplasm via negative-sense RNA intermediates

Like other positive-sense RNA viruses, HCV also replicates in the cytoplasm of the infected cell. Because of lack of a tissue culture system for HCV propagation, the replication cycle of HCV is not clearly understood. HCV E2 glycoprotein binds to a receptor CD81 (a member of the transmembrane 4 superfamily, tetraspanin) on the hepatocytes. It is also believed that circulating HCV particles are accompanied by low-density lipoprotein (LDL) to the liver, making a case for LDL as HCV receptor. After virus entry, uncoating takes place followed by translation of a full-length genomic positive-sense RNA into a polyprotein, which is cleaved into individual proteins. One of the proteins, RNA-dependent RNA polymerase directs transcription and replication via negative-sense RNA intermediates. Virus assembly takes place in the cytoplasm by formation of vesicles that fuses with the plasma membranes for virus release.

HEPATITIS C DISEASE

CLINICAL CAPSULE

Hepatitis C is an insidious disease in that it does not usually cause a clinically evident acute illness. Instead, its first manifestation (in 25% of those infected) may be the presence of smoldering chronic hepatitis that may ultimately lead to liver failure. Its transmission is less well understood than for hepatitis A, B, and D but causes chronicity in more than 85% of infected patients. Hepatitis C was the major cause of post-transfusion hepatitis until a serologic test for screening blood donors was developed.

EPIDEMIOLOGY

Major transmission was from blood and blood products, but is now from "needle sharing"

Like hepatitis B, hepatitis C is spread parenterally. The transmission of hepatitis C by blood was well documented. Indeed, until screening blood for transfusions was introduced, it caused most cases of post-transfusion hepatitis. Screening of donor blood for antibody has reduced post-transfusion hepatitis by 80% to 90%. Hepatitis C may be sexually transmitted but to a much lesser degree than hepatitis B. Needle sharing accounts for up to 40% of cases. In the United States, 3.5 million people (1.8%) have antibody to hepatitis C. Since the 1980s, outbreaks of hepatitis C have been associated with intravenous immune globulin (IVIG). To reduce this risk, all US-licensed IVIG products now have additional viral inactivation steps included in the manufacturing process. Furthermore, all immunoglobulin products (including intramuscular immunoglobulin products that have not been associated with hepatitis C) that lack viral inactivation steps are now excluded if hepatitis C virus is detected by polymerase chain reaction (PCR). Other individuals considered at risk for hepatitis C are health care workers becuase of needlesticks and chronic hemodialysis patients and their spouses. Vertical transmission also occurs during deliveries.

PATHOGENESIS

HCV is transmitted via blood and blood-derived products and invades and infects the peripheral blood B and T lymphocytes and monocytes and moves to the main site of infection—the liver. The rate of HCV replication in hepatocytes is very high (ie, 1×10^{12} virions per day), since 10% of the hepatic cells are infected. The high rate of viral replication results in an increased level of viral heterogeneity, which allows the virus to evade the host immune response. Although little evidence exists regarding a direct effect of HCV-induced cytopathic effects on the hepatocytes (liver cells), hepatocytes are likely killed by immune-mediated cytotoxic T cells. Several recent studies suggest that HCV replication can cause cytopathic lesions in the liver such as histologic lesions with scant inflammatory infiltrate and fulminant hepatitis C after chemotherapy in liver transplant recipients. The innate immune response results in the activation of cytokines and interferon, which initially control viral replication in some cases. However, HCV-encoded proteins help the virus to evade innate immune response, including interaction of HCV core with tumor necrosis factor (TNF) receptor, which decreases cytolytic T-cell activity and interference of HCV nonstructural proteins with interferon pathways. In addition, the natural killer (NK) cells respond to HCV infection by releasing perforins, which fragment nuclei of infected cells and induce apoptosis. HCV infection is inhibited by the release of interferon gamma, which recruits intrahepatic inflammatory cells, stimulates helper T1 (Th1) response, and induces necrosis or apoptosis of HCV-infected cells.

HCV has tropism for liver

High rate of mutations allow virus evasion of the host immune response

Hepatitis C disease mainly immune-mediated

Cytokines cause inflammation in HCV infection

Adaptive immune responses, including cell -mediated and humoral responses are elicited after expression of HCV proteins, especially the envelope glycoproteins E1 and E2. HCV antibodies appear several weeks after infection, and because of selective pressure from the host, mutations takes place in the E2/E1 proteins, allowing the virus to evade the humoral immune response and establish persistent infection. More important, HCV antibodies have been implicated in tissue damage because of immune complex formation. Examples of such tissue damage are antinuclear antibodies, autoantibodies that act against cytochrome P450 and antibodies that work against the liver and kidney.

The immune complexes are also deposited in other tissues and cause some of the other extrahepatic problems, including vasculitis, arthritis, glomerulonephritis, and others. In the absence of strong humoral immune response against HCV infection, cytotoxic T lymphocytes (CTLs) or CD8 T cells are critical to the elimination of HCV infection, and any impairment in cell-mediated immunity could be a major factor for a high level of chronicity in infected patients. The CD8 T cells eliminate HCV by apoptosis of infected hepatocytes and interferon gamma–induced inhibition of viral replication. The CTL response is less effective in chronically HCV-infected patients compared with that in acutely infected patients. Also, CD4 T cells play an important role in HCV pathogenesis by secreting several proinflammatory cytokines related to hepatocyte death. During acute infection, the rise in serum transaminases corresponds with cell damage, and the hepatic lesion is immune-mediated. The chronic infection probably progresses as a result of imbalance between Th1 and Th2 cytokines. Th1 cytokines such as interleukin 2 (IL-2) and TNF-α are associated with aggressive hepatic disease, whereas Th2 cytokines (IL-10) is related to the milder presentation. Expression of TNF-α causes hepatic injury and triggers "cytokine storm" to cause liver damage in chronically infected patients (**Figure 13–13**). ::: cytokine storm, p.153

HCV antibodies cause liver and other tissue damage owing to immune complex formation

HCV infection cause imbalance between Th1 and Th2 cytokines

Some cytokine expression triggers cytokine storm causing liver damage

In addition to immune status of the host, genetic host factors play an important role in HCV pathogenesis. One such factor is major histocompatibility complex (MHC) class II DR5 allele, which has been shown to be associated with a lower incidence of cirrhosis in HCV-infected individuals. One study identified CTLs restricted by HLA A2 in 97% of chronic hepatitis C

Host factors play important role is hepatitis C disease progression

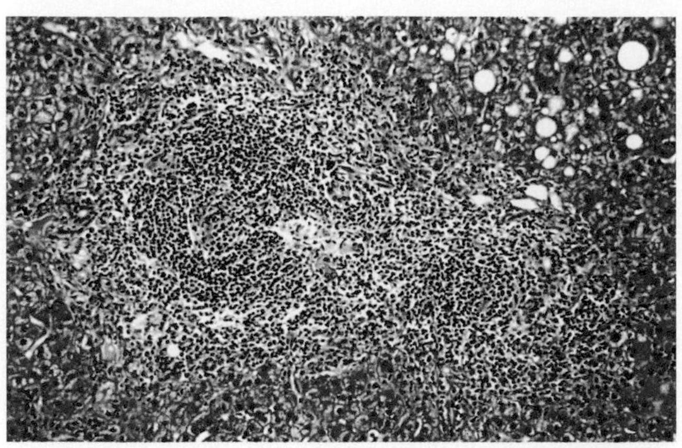

FIGURE 13–13. Inflammation in chronic hepatitis C virus (HCV) infection. Chronic inflammation of the portal area with a lymphoid aggregate in the center can be seen. At the edges of the portal area, the interface between the parenchyma and portal connective tissue, inflammation spreads outward, destroying hepatocytes and expanding the portal tract by piecemeal necrosis. (Reproduced with permission from Connor DH, Chandler FW, Schwartz DQA, Manz HJ, Lack EE (eds). *Pathology of Infectious Diseases,* vol. 1. Stamford, CT: Appleton & Lange; 1997.)

Alcohol abuse and smoking influence hepatitis severity

patients. Several extrinsic factors such as alcohol abuse and smoking are related to progression of chronic hepatitis C. The influence of age, gender, and race due to genetic factors variation has been implicated with progression of hepatitis C. Coinfection with other viruses such as HIV, HBV, HAV, and human T-lymphotropic virus influence the outcome of HCV disease.

Increased risk of HCC with chronic hepatitis C

HCV core and NS5A implicated with oncogenesis

HCV-infected patients may develop cirrhosis of liver with increased risk of HCC (hepatocellular carcinoma). It has also been suggested that alcoholism increases the rate of HCC in HCV-infected patients. It is also believed that HCC is probably caused by long-term damage followed by rapid growth rate of hepatocytes during regeneration of liver, which may be mediated by some cytokines. Recent studies suggest that various HCV protein-host–cell interactions may play a role in the development of HCC, including disturbance in the cell cycle, upregulation of oncogenes, and loss of tumor suppressor gene functions. HCV core protein has been shown to perturb and modify the growth of the cell cycle. HCV core interacts directly or indirectly with components or pathways that lead to oncogenesis such as tumor suppressor genes (*p53, p73*), protein kinase, cell cycle, and cell proliferation and differentiation. In addition, HCV nonstructural protein, NS5A, plays a role in cell transformation, differentiation, and oncogenesis.:

CLINICAL ASPECTS

MANIFESTATIONS

Acute illness usually not apparent

Chronic infection is common

The incubation period of hepatitis C averages 6 to 12 weeks. The infection is usually asymptomatic or mild and anicteric in 75%, but it results in a chronic carrier state in up to 85% of adult patients. Fulminant hepatitis due to hepatitis C is very rare in the United States. The average duration of time from infection to the development of chronic hepatitis is 10 to 18 years. Cirrhosis and HCC are late sequelae of chronic hepatitis. Chronic hepatitis tends to wax and wane, is often asymptomatic, and may be associated with either elevated or normal ALT values in serum (**Figure 13–14**). Chronic hepatitis C is the leading infectious cause of chronic liver disease and liver transplantation in the United States.

DIAGNOSIS

Antigens of hepatitis C are not detectable in blood, so diagnostic tests attempt to demonstrate antibody. Unfortunately, the antibody responses in acute disease remain negative for 1 to 3 weeks after clinical onset and may never become positive in up to 20% of patients with acute, resolving disease. Current tests measure antibodies to multiple hepatitis C antigens

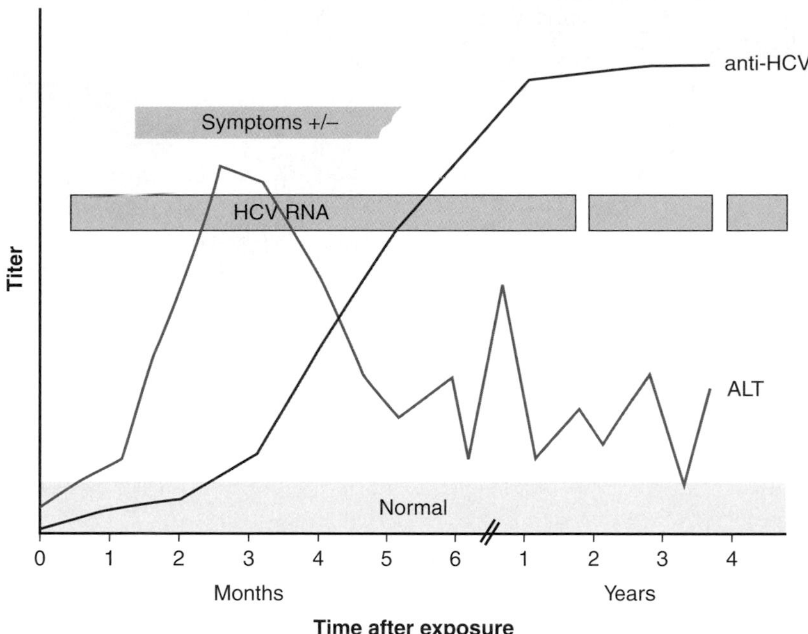

FIGURE 13–14. Sequence of appearance of viremia, alanine aminotransferase (ALT), symptoms, antibodies in acute hepatitis C virus (HCV) infection, and progression to chronic infection..

by either enzyme immunoassay or immunoblot testing. Even with these newer assays, IgG antibody to hepatitis C may not develop for up to 4 months, making the serodiagnosis of acute hepatitis C difficult. Quantitative assays of hepatitis C RNA may be used for diagnosis, predicting IFN responsiveness and monitoring therapy, but there is not a very good correlation between viral load and histology. Genotyping is important for therapy, since type 1 (the most common in the United States) requires the longest period of therapy.

Antibody responses are usually delayed

Hepatitis C RNA can be detected and quantitated by PCR

TREATMENT AND PREVENTION

Combination treatment of acute hepatitis C appears beneficial, but can be deferred for weeks to determine whether the infection resolves spontaneously. Combination therapy with IFNα and ribavirin is the current treatment of choice for patients with evidence of chronic hepatitis due to hepatitis C. Criteria for initiating treatment are controversial, but most physicians would initiate treatment in a patient with abnormal liver histology and elevated liver enzymes. Responses are better in patients with genotypes other than 1 and those with low initial titers of viral RNA. Corticosteroids are not beneficial. Avoidance of injection drug use and screening of blood products are important preventive measures. Prophylactic immune serum globulin does not protect against hepatitis C.

Combination therapy can benefit some persons with chronic infection

Immune globulin may not be protective; no vaccine exists

HEPATITIS E

Hepatitis E is the cause of another form of hepatitis that is spread by the fecal–oral route and therefore resembles hepatitis A. Hepatitis E virus is a positive sense single-stranded RNA virus that is similar to but distinct from caliciviruses. The viral particles in stool are spherical, 27 to 34 nm

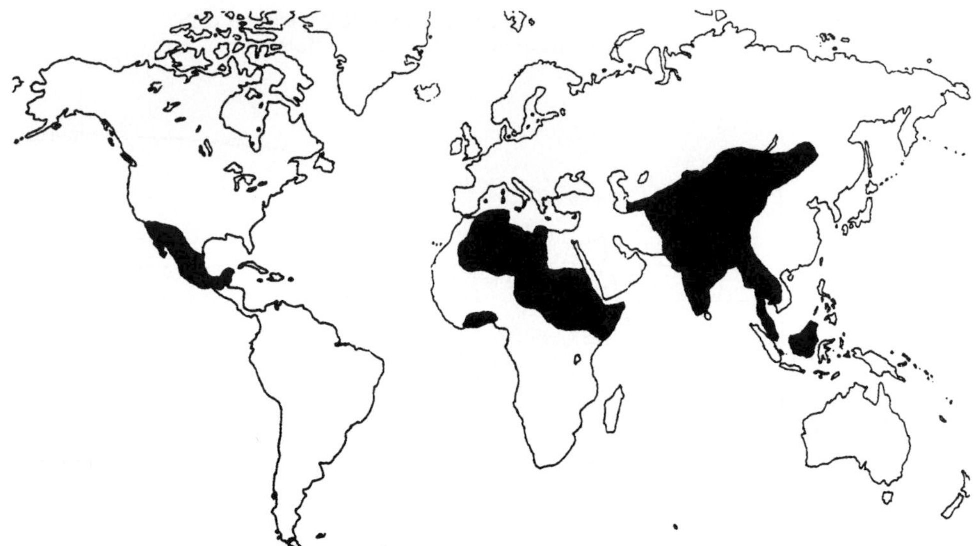

FIGURE 13–15. Distribution of hepatitis E virus infection, among countries in which outbreaks have been identified (shown in black). (Reproduced with permission from Connor DH, Chandler FW, Schwartz DQA, Manz HJ, Lack EE (eds). *Pathology of Infectious Diseases,* vol. 1. Stamford, CT: Appleton & Lange; 1997.)

Hepatitis E spreads in a similar manner to that of hepatitis A

Usually associated with contaminated drinking water

in diameter with icosahedral symmetry, and unenveloped, and they exhibit spikes on their surface. Like hepatitis A, infection with this virus is frequently subclinical. When symptomatic, it causes only acute disease that may fulminate to death, especially in pregnant women. In endemic, developing areas, hepatitis E has the highest attack rate in young adults, and infection is usually associated with contaminated drinking water. It does not appear to spread from person to person. Most cases of hepatitis E infection have been identified in developing countries with poor sanitation (eg, Asia, Africa, and the Indian subcontinent), and recurrent epidemics have been described in these areas (**Figure 13–15**). Cases have been recently recognized in developed countries such as the the United States; most have been in visitors or immigrants from endemic areas. The incubation period for hepatitis E is approximately 40 days. The diagnosis may be confirmed by demonstrating the presence of specific IgM antibody, although very few laboratories perform this test. Immune serum globulin does not appear to provide protection, and no treatment is available. Liver transplantation may be the only recourse in seriously ill patients.

HEPATITIS G

RNA virus is similar to hepatitis C

Role in human disease is currently uncertain

HGV infection may prolong survival of AIDS patients

In 1995, hepatitis G virus (HGV), or GB virus C (GBV-C), was discovered in sera from two patients. HGV and GBV-C are two isolates of the same virus. Hepatitis G is an RNA virus of 9 to 10 kb, similar to that of hepatitis C and members of the Flaviviridae family. The virion structure of hepatitis G is similar to that of HCV. HGV has been found to replicate in lymphocytes rather than in hepatocytes. An antibody assay can detect past, but not present, infection, and detection of acute infection with hepatitis G requires a PCR assay for viral RNA in serum. Up to 2% of volunteer blood donors and 35% of HIV infected patients are positive for hepatitis G RNA, which is a blood-borne virus. In addition to being closely related to hepatitis C, data suggest that 10-20% of patients infected with hepatitis C are also infected with hepatitis G. Given this association, it has been difficult to ascertain the contribution of hepatitis G to clinical disease. Patients infected with both viruses do not appear to have worse disease than those infected by HCV only.

Currently, there is no useful serologic test and no therapy is established for patients with HGV. Several studies suggest that HGV viremia prolongs the survival of HIV-positive individuals after seroconversion.

CASE STUDY

A LABORATORY DISCOVERY

A 45-year-old man has a routine physical in connection with a request for life insurance. All physical and laboratory examinations are normal except for a bilirubin of 2.6 mg/mL. The patient visited Nepal 1 year ago and acknowledged sharing intravenous drugs as a collegian. He has never had an acute hepatitis illness.

QUESTIONS

■ What was the most likely cause of the man's elevated bilirubin?

A. Hepatitis A
B. Hepatitis B
C. Hepatitis C
D. Hepatitis D
E. Hepatitis E

■ Which laboratory test would be most likely to indicate the diagnosis?

A. Specific IgM antibody assay
B. Specific IgG antibody assay
C. Quantitative viral DNA assay
D. Viral genotypic assay
E. Serum alanine aminotransferase

■ Which laboratory test is most useful for predicting response to treatment?

A. Quantitative viral load
B. Virus genotype
C. Specific IgG antibody assay
D. Western blot assay
E. Quantitative enzyme immunoassay

ANSWERS

1(C), **2**(B), **3**(B)

CHAPTER 14

Herpesviruses

The human herpesviruses belong to the Herpesviridae family, which consists of large, enveloped, double-stranded DNA viruses and which produce infections ranging from painful skin and genital ulcers to chickenpox to encephalitis to Kaposi sarcoma. There are eight members of the family that infect humans, including two herpes simplex viruses (HSV-1 and HSV-2), varicella-zoster virus (VZV), cytomegalovirus (CMV), Epstein–Barr virus (EBV), and human herpesvirus types 6, 7, and 8 (HHV-6, HHV-7, HHV-8; **Table 14–1**). In addition, a simian herpesvirus, herpes B virus, has occasionally caused human disease.

Large, enveloped double-stranded DNA viruses

HERPESVIRUSES: GROUP CHARACTERISTICS

VIROLOGY

All herpesviruses are morphologically similar, with an overall size of 180 to 200 nm. The DNA core is up to 75 nm in diameter and is surrounded by an icosahedral capsid. Over the capsid is a protein-filled region called the tegument. The outside of the viral particle is covered by a lipoprotein envelope derived from the nuclear membrane of the infected host cell. The envelope contains at least nine glycoproteins that protrude beyond it as spike-like structures. The viral genome is large, ranging from 125 to 240 kbp of DNA, which code for approximately 75 viral proteins. This large genome is necessary, because herpesviruses frequently infect nondividing cells and must therefore provide their own enzymes necessary for DNA synthesis. Example of a herpes simplex virus virion is shown in **Figure 14–1** as a representative virion structure for herpesviruses. The linear double-stranded DNA genome is surrounded by an icosahedral capsid that is enclosed by glycoproteins containing lipid bilayer membrane derived predominantly from nuclear membrane. The space between capsid and envelope is a protein-filled region called **tegument**, which contains viral proteins and enzymes that are required immediately for viral replication after infection. Despite the morphologic similarity among herpesviruses, there are substantial differences in their genomic sequences and, in turn, their structural glycoproteins and polypeptides. Antigenic analysis is an important means for differentiation among herpesviruses despite some cross-reactions (eg, between HSV and VZV).

Morphology similar among herpesviruses but genomic sequences differ

Can infect nondividing cells

Based on certain virologic similarities, the herpesviruses may be divided into three subfamilies α, β, and γ herpesviruses. HSV-1 and -2, as well as VZV, are in the α subfamily; CMV, HHV-6, and HHV-7 are in the β subfamily whereas EBV and HHV-8 are in the γ subfamily.

TABLE 14–1 Human Herpesvirusess

DESIGNATION	COMMON NAME	TRANSMISSION	PRIMARY INFECTION SITE	DISEASE	LATENT INFECTION SITE
HHV-1	Herpes simplex virus 1 (HSV-1)	Close contact	Mucoepithelial cells	Oral (fever blisters), ocular lesions; encephalitis	Nerve ganglia
HHV-2	Herpes simplex virus 2 (HSV-2)	Close contact Sexual transmission	Mucoepithelial cells	Genital, anal lesions; severe neonatal infections; meningitis	Nerve ganglia
HHV-3	Varicella-zoster virus (VZW)	Respiratory route Inhalation Close contact	Mucoepithelial cells	Chickenpox (primary infection); shingles (reactivation)	Nerve ganglia
HHV-4	Epstein-Barr virus (EBV)	Saliva Kissing	B cell	Infectious mononucleosis (primary infection); tumors, including B-cell tumors (Burkitt lymphoma, immunoblastic lymphomas of the immunosuppressed); nasopharyngeal carcinoma, some T cell tumors	B lymphocytes
HHV-5	Cytomegalovirus (CMV)	Close contact, sexual transmission Congenital Blood-to-blood Transplant	Leukocytes (T and B) Lymphocytes Monocytes	Mononucleosis; severe congenital infection; infections in immunocompromised (gastroenteritis, retinitis, pneumonia)	Monocytes, neutrophils, vascular endothelial cells
HHV-6	Human herpesvirus 6	Close contact Respiratory route	T lymphocytes	Roseola in infants (primary infection); infections in allograft recipients (pneumonia, marrow failure)	T lymphocytes monocytes, macrophages
HHV-7	Human herpesvirus 7	Saliva Close contact	T lymphocytes	Some cases of roseola (primary infection)	CD4+ T cells
HHV-8	Kaposi sarcoma-associated herpesvirus (KSHV), human herpesvirus 8	Sexual transmission	B lymphocytes Peripheral blood mononuclear cell	Tumors, including Kaposi sarcoma; some B cell lymphomas	Virus-infected tumors

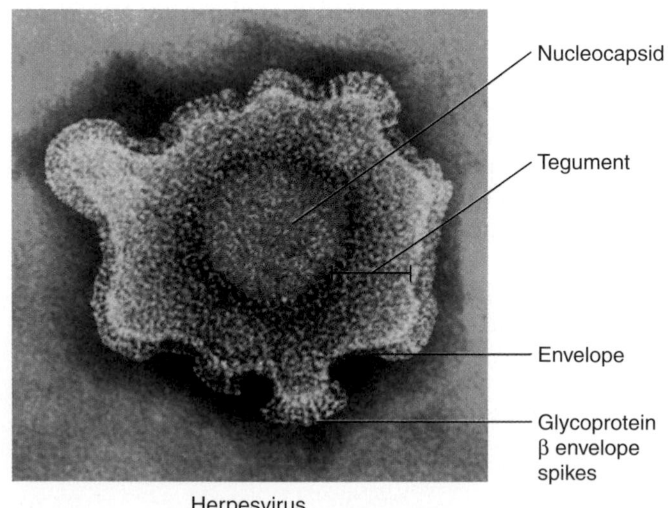

FIGURE 14–1. Virion structure of herpes simplex virus. (Reproduced with permission from Willey J, Sherwood L, Woolverton C (eds). *Prescott's Principles of Microbiology.* New York: McGraw-Hill; 2008.)

Cell tropisms for the individual viruses vary significantly. HSV has the widest range; it replicates in numerous animal and human host cells, although it affects only humans in nature. VZV infects only humans and is best grown in cells of human origin, although some laboratory-adapted strains can grow in primate cell lines. Human CMV replicates well only in human diploid fibroblast cell lines. EBV does not replicate in most commonly used cell culture systems but can be grown in continuous human or primate lymphoblastoid cell cultures HHV-6 and -7, preferentially grow in T-lymphocyte cell cultures.

Herpes simplex has widest range of cell tropism

Characteristically, all these agents produce an initial infection followed by a period of latent infection in which the genome of the virus is present in cells, but infectious virus is not recovered. During latent infection of cells, viral DNA is maintained as an episome (not integrated), with limited expression of specific virus genes required for the maintenance of latency. Reactivation of virus due to complex host–virus interactions may then result in recurrent disease. For example, immunocompromised patients, especially those with altered cellular immunity, have frequent reactivations of herpesviruses, which can lead to clinically severe disease.

Viral latency and reactivation typical for all herpesviruses

Replication

The replication of HSV is representative of all herpesviruses, as shown in **Figure 14–2**. HSV generally causes lytic infection in epithelial cells and latent infection in neuronal cells. The glycoproteins in the HSV envelope interact with cellular receptors, including heparan sulfate, to result in fusion with the cell membrane. Fusion delivers the capsid and DNA case into the cytoplasm, where it migrates to the nucleus, and the genome is circularized. Transcription of the large, complex genome is sequentially regulated in a cascade fashion. Three distinct classes of mRNAs are made: (1) immediate early (IE) mRNAs encoded by α genes (synthesized 2 to 4 hours after infection), which code for proteins initiating and regulating virus transcription; (2) early (E) mRNAs encoded by β genes, which code for further nonstructural proteins (DNA binding proteins, DNA polymerase, thymidine kinase) involved in DNA replication and minor structural proteins; and (3) late (L) mRNAs encoded by γ genes, synthesized 12 to 15 hours after infection, which code for major structural proteins (capsid and envelope glycoproteins). The early (E) proteins thymidine kinase and DNA polymerase are distinct from host cell enzymes and are therefore important targets of antiviral chemotherapy. Gene expression is coordinated (ie, synthesis of early gene products turns off IE products and initiates genome replication); some of the late structural proteins are produced independently of genome replication, whereas others are produced only after replication. The pattern of viral DNA replication is complex, resulting in the formation of high-molecular-weight DNA concatemers. Genomic concatemers are cleaved and packaged into preassembled capsids in the nucleus. The replication of viral genomes triggers late (gamma) genes transcription for the synthesis of viral structural proteins. After the synthesis of capsid proteins, these proteins are transported to the nucleus, where their assembly with DNA takes place.

Three classes of mRNAs produced

Coordinated gene expression

The envelope is acquired from the inner lamella of the nuclear membrane. Budding occurs at the inner nuclear membranes, and virions then enter the cytoplasm to be released through the endoplasmic reticulum. HSV infection appears to be a "wasteful" process: only 25% of viral DNA/protein produced is incorporated into virions. The rest accumulates in the cell, which eventually dies. Moreover, the ratio of incomplete to complete viral particles is approximately 1000 to 1. Most herpesviruses shut down host cell metabolism and ultimately cause cell death, except for CMV, which actually stimulates cellular synthesis of nucleic acids and proteins.

Most herpesviruses, except CMV, shut down host cell metabolism

HERPES SIMPLEX VIRUS

VIROLOGY

Two distinct epidemiologic and antigenic types of HSV exist (HSV-1 and HSV-2). The DNA genomes of both are linear, double-stranded molecules containing approximately 150 kbp.

1. Circularization of genome and transcription of immediate-early genes
2. α-proteins, products of immediate-early genes, stimulate transcription of early genes.
3. β-proteins, products of early genes, function in DNA replication, yielding concatemeric DNA. Late genes are transcribed.
4. γ-proteins, products of late genes, participate in virion assembly.

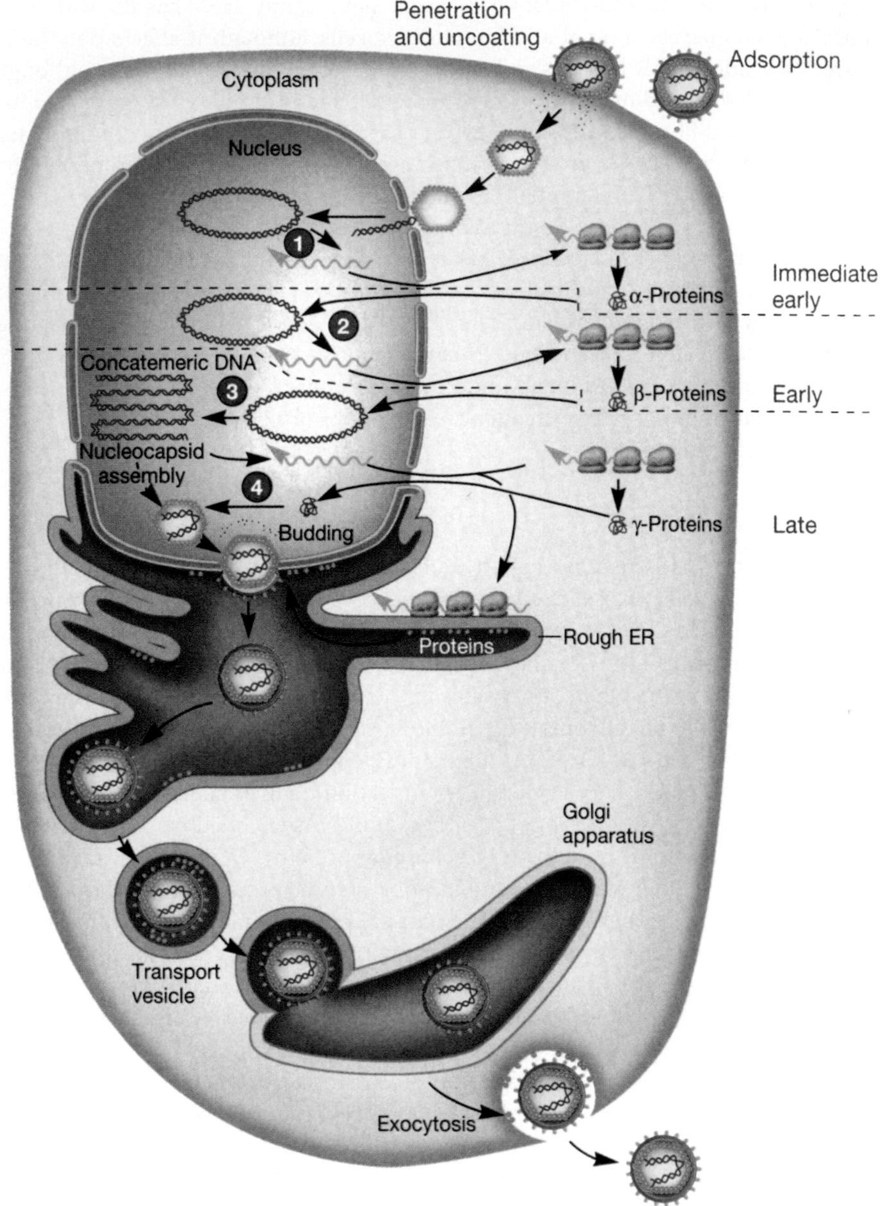

FIGURE 14–2. Replication cycle of herpes simplex virus 1. (Reproduced with permission from Willey J, Sherwood L, Woolverton C (eds). *Prescott's Principles of Microbiology.* New York: McGraw-Hill; 2008.)

HSV-1 and HSV-2 are distinct epidemiologically, antigenically, and by DNA homology

Individual strains differ by restriction endonuclease techniques

Their nucleic acids demonstrate approximately 50% base sequence homology, which is considerably greater than that shown between these viruses and other herpesviruses. HSV-1 and HSV-2 share antigens in almost all their surface glycoproteins and other structural polypeptides, but differences in glycoprotein gB enable them to be distinguished (ie, HSV-1 has gB1 and HSV-2 has gB2). Numerous strains of both HSV-1 and HSV-2 exist. In fact, by restriction endonuclease analysis of the viral genome, most strains of either HSV-1 or HSV-2 are found to differ somewhat, except in epidemiologically related cases such as mother–infant pairs and sexual partners.

HERPES SIMPLEX DISEASE

CLINICAL CAPSULE

Herpes simplex viruses are the best known of all viruses, given their frequency of infection and propensity to cause recurrent vesicles and ulcers in areas of the skin and mucous membranes. These viruses can cause progressive disease in immunocompromised persons, and encephalitis in normal hosts. Infections acquired by infants at or shortly after birth can be especially devastating. The two types differ somewhat in their predilection for causing lesions "above the waist" (HSV-1) or "below the waist" (HSV-2). As with all herpesviruses, they persist in a latent form and reactivate to cause viral excretion and/or disease.

EPIDEMIOLOGY

Herpes simplex viruses are distributed worldwide. There are no known animal vectors, and humans appear to be the only natural reservoir. Direct contact with infected secretions is the principal mode of transmission or spread. Seroepidemiologic studies indicate that the prevalence of HSV antibody varies according to the age and socioeconomic status of the population studied. In most developing countries, 90% of the population has HSV-1 antibody by the age of 30 years. In the United States, HSV-1 antibody is found in approximately 60% to 70% of adult middle-class populations; among lower socioeconomic groups, however, the percentage is higher.

No animal reservoirs

High seroprevalence among humans, which increases with age

Detection of HSV-2 antibody before puberty is unusual. The virus is associated with sexual activity, and direct sexual transmission is the major mode of spread. Approximately 15% to 30% of sexually active adults in Western industrialized countries have HSV-2 antibody. The virus can be isolated from the cervix and urethra of approximately 5% to 12% of adults attending sexually transmitted disease clinics; many of these patients are asymptomatic or have small, unnoticed lesions on penile or vulvar skin. Asymptomatic shedding accounts for transmission from a partner who has no active genital lesions and often no history of genital herpes. Genital herpes is not a reportable disease in the United States, but it is estimated that more than 1 million new cases occur per year.

Infection with HSV-2 linked to sexual activity

PATHOGENESIS

Acute Infections

Both HSV-1 and HSV-2 initially infect and replicate in the mucoepithelial cells and cause some kind of lytic or productive infection and later establish a latent infection in the neurons (sensory ganglion cells). Pathologic changes during acute infections consist of development of multinucleated giant cells (**Figure 14–3**), ballooning degeneration of epithelial cells, focal necrosis, eosinophilic intranuclear inclusion bodies, and an inflammatory response characterized by an initial polymorphonuclear neutrophil (PMN) infiltrate and a subsequent mononuclear cell infiltrate. The virus can spread intra- or interneuronally or through supporting cellular networks of an axon or nerve, resulting in latent infection of sensory and autonomic nerve ganglia. Spread of virus can occur by cell-to-cell transfer and can therefore be unaffected by circulating immune globulin.

Infection produces inflammation and giant cells

Virus can infect and spread in axons and ganglia

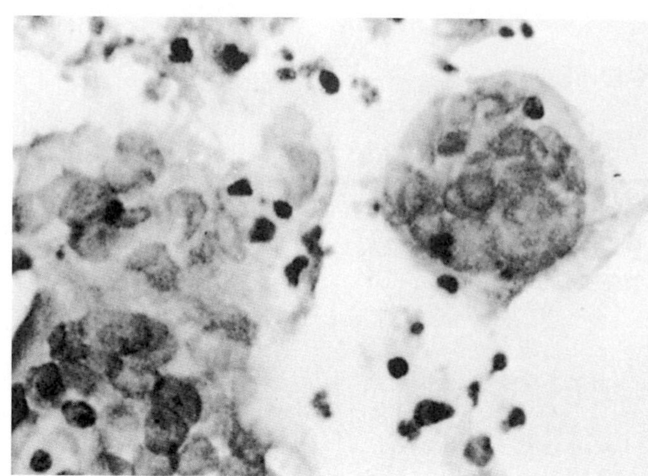

FIGURE 14–3. Multinucleated giant cells from herpes simplex virus lesion.

Latent Infection

In humans, latent infection by HSV-1 has been demonstrated by co-cultivation techniques in trigeminal, superior cervical, and vagal nerve ganglia, and occasionally in the S2-S3 dorsal sensory nerve root ganglia. Latent HSV-2 infection has been demonstrated in the sacral (S2-S3) region. Latent infection of nervous tissue by HSV does not result in the death of the cell; however, the exact mechanism of viral genome interaction with the cell is incompletely understood. Several copies of the HSV viral genomes are in each latently infected neuronal cell. They exist in a circular extrachromosomal form in the nucleus, and transcription of only a small portion of the viral genome occurs. Because latent infection does not appear to require synthesis of early or late viral polypeptides, antiviral drugs directed at the thymidine kinase enzymes or viral DNA polymerase do not eradicate the virus in its latent state.

No synthesis of early or late viral polypeptides in latent infection

Reactivation of virus from latently infected ganglionic cells with subsequent release of infectious virions appears to account for most recurrences of both genital and orolabial infections. The mechanisms by which latent infection is reactivated are unknown. Precipitating factors that are known to initiate reactivation of HSV include exposure to ultraviolet light, sunlight, fever, excitement, emotional stress, and trauma (eg, oral intubation).

Reactivation can be precipitated by sun exposure, fever, or trauma

Two theories have been proposed on how latent herpes reaches the peripheral sites, including ganglionic and skin trigger theories. In ganglionic theory, metabolic changes switch on the virus replication cycle, and the virus travels down the peripheral nerves to the skin, where it replicates in the epidermal cells and produces lesions. The skin trigger theory proposes that because of chronic multiplication of the virus in the ganglion, there is intermittent shedding of the virus through the nerve axon to the skin.

IMMUNITY

Host factors have a major effect on clinical manifestations of HSV infection. Many episodes of HSV infection are either asymptomatic or mildly symptomatic. Initial symptomatic clinical episodes of the disease are more severe than recurrent episodes, probably because of the presence of anti-HSV antibodies and immune lymphocytes in persons with recurrent infections. Prior infection with HSV-1 may protect against or shorten the duration of symptoms and lesions from subsequent infection with HSV-2 as a result of some degree of cross-protection.

Some cross-protection between HSV-1 and HSV-2

Both cellular and humoral immune responses are important in immunity to HSV. Neutralizing antibodies directed against HSV envelope glycoproteins appear to be important in preventing exogenous reinfection. Antibody-dependent cellular cytotoxicity (ADCC) may be important in limiting early spread of HSV. By the second week after infection, cytotoxic T lymphocytes can be detected, which have the ability to destroy HSV-infected cells before completion of the replication cycle. Conversely, in immunosuppressed patients, especially those with depressed cell-mediated immunity, reactivation of HSV may be associated with prolonged viral excretion and persistence of lesions.

ADCC may limit early spread of HSV; cytotoxic T lymphocytes destroy HSV-infected cells

CLINICAL ASPECTS

MANIFESTATIONS

Herpes Simplex Type 1

Infection with HSV-1 is usually, but not always, "above the waist." It consists characteristically of grouped or single vesicular lesions that become pustular and coalesce to form single or multiple ulcers. On dry surfaces, these ulcers scab before healing; on mucosal surfaces, they reepithelialize directly. HSV can be isolated from almost all ulcerative lesions, but the titer of virus decreases as the lesions evolve. Infections generally involve ectoderm (skin, mouth, conjunctiva, and nervous system).

Vesicular lesions become pustular and then ulcerate

Primary infection with HSV-1 is often asymptomatic. When symptomatic, typically in children, it appears most frequently as **gingivostomatitis**, with fever and ulcerative lesions involving the buccal mucosa, tongue, gums, and pharynx. The lesions are painful, and the acute illness usually lasts 5 to 12 days. After this initial infection, HSV may become latent within sensory nerve root ganglia of the trigeminal nerve.

Primary infections often asymptomatic

Lesions usually recur on a specific area of the lip and the immediate adjacent skin; these lesions are referred to as mucocutaneous and are commonly called "cold sores" or "fever blisters" (**Figure 14–4**). Because reactivation is usually from a single latent source, these lesions are typically unilateral. Their recurrence may be signaled by premonitory tingling or burning in the area. Systemic complaints are unusual, and the episode generally lasts approximately 7 days. It should be noted that HSV may be reactivated and excreted into the saliva with no apparent mucosal lesions present. HSV has been isolated from saliva in 5% to 8% of children and 1% to 2% of adults who were asymptomatic at the time.

Recurrent cold sores usually unilateral

Virus in saliva with asymptomatic reactivation

HSV sometimes infects the finger or nail area. This infection, termed **herpetic whitlow,** usually results from the inoculation of infected secretions through a small cut in the skin. Painful vesicular lesions of the finger develop and pustulate; they are often mistaken for bacterial infection and mistreated accordingly.

Herpetic whitlow mimics bacterial paronychia

HSV infection of the eye is one of the most common causes of corneal damage and blindness in the developed world. Infections usually involve the conjunctiva and cornea, and characteristic dendritic ulcerations are produced. With recurrence of disease, there may

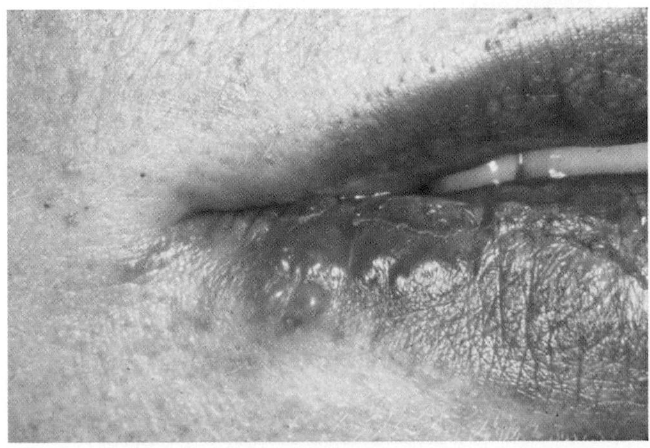

FIGURE 14–4. Coalesced, localized lesions characteristic of reactivated herpes simplex virus type 1 (HSV-1) infection.

Herpetic corneal and conjunctival infection can cause blindness

be deeper involvement with corneal scarring. Occasionally, there may be extension into deeper structures of the eye, especially when topical steroids are used.

Herpes encephalitis may be reactivation

Encephalitis may rarely result from HSV-1 infection. Most cases occur in adults with high levels of anti-HSV-1 antibody, suggesting reactivation of latent virus in the trigeminal nerve root ganglion and extension of productive (lytic) infection into the temporoparietal area of the brain. Primary HSV infection with neurotropic spread of the virus from peripheral sites up the olfactory bulb into the brain may also result in parenchymal brain infection.

Encephalitis typically localized to temporal lobe

Rapid diagnosis allows antiviral therapy

Classically, HSV encephalitis affects one temporal lobe, leading to focal neurologic signs and cerebral edema. If untreated, mortality rate is 70%. Clinically, the disease can resemble brain abscess, tumor, or intracerebral hemorrhage. Rapid diagnosis by polymerase chain reaction (PCR) of CSF has replaced brain biopsy as the diagnostic test. Intravenous acyclovir can reduce the morbidity and mortality of the disease, especially if treatment is initiated early.

Herpes Simplex Type 2

HSV-2 associated with genital infections

Genital herpes is an important sexually transmitted disease. Both HSV-1 and HSV-2 can cause genital disease, and the symptoms and signs of acute infection are similar for both viruses. Seventy percent of first episodes of genital HSV infection in the United States are caused by HSV-2, and genital HSV-2 disease is also more likely to recur than genital HSV-1 infection. Ninety percent of HSV-2 antibody-positive patients have never had a clinically evident genital HSV episode. In many instances, the first clinical episode is years after primary infection.

Primary Genital Herpes Infection

For the relatively few individuals who develop clinically evident primary genital HSV disease, the mean incubation period from sexual contact to onset of lesions is 5 days. Lesions begin as small erythematous papules, which soon form vesicles and then pustules (**Figure 14–5**). Within 3 to 5 days, the vesiculopustular lesions break to form painful coalesced ulcers that subsequently dry; some form crusts and heal without scarring. With primary disease, the

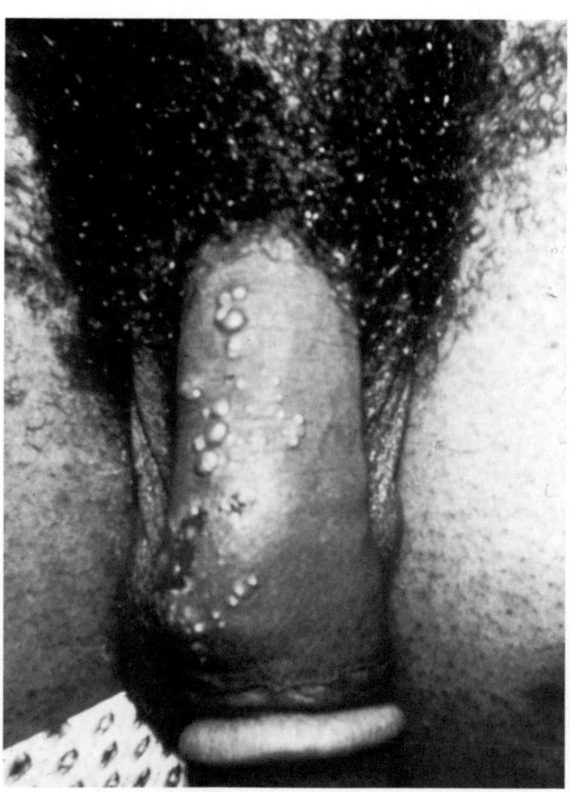

FIGURE 14–5. Multiple grouped vesicles of primary genital herpes.

genital lesions are usually multiple (mean number 20), bilateral, and extensive. The urethra and cervix are also infected frequently, with discrete or coalesced ulcers on the exocervix. Bilateral enlarged tender inguinal lymph nodes are usually present and may persist for weeks to months. About one third of patients show systemic symptoms such as fever, malaise, and myalgia, and approximately 1% develop aseptic meningitis with neck rigidity and severe headache. First episodes of disease last an average of 12 days.

Multiple painful vesiculopustular lesions

Systemic symptoms and adenopathy common

Recurrent Genital Herpes Infection

In contrast to primary infection, recurrent genital herpes is a disease of shorter duration, usually localized in the genital region and without systemic symptoms. A common symptom is prodromal paresthesias in the perineum, genitalia, or buttocks that occur 12 to 24 hours before the appearance of lesions. Recurrent genital herpes usually presents with grouped vesicular lesions in the external genital region. Local symptoms such as pain and itching are mild, lasting 4 to 5 days, and lesions usually last 2 to 5 days.

Prodromal paresthesias and shorter duration

At least 80% of patients with primary, symptomatic, genital HSV-2 infection develop recurrent episodes of genital herpes within 12 months. In patients whose lesions recur, the median number of recurrences is four or five per year. They are not evenly spaced, and some patients experience a succession of monthly attacks followed by a period of quiescence. Over time, the number of recurrences decreases by a median of one half to one recurrence per year. Most recurrences result from reactivation of virus from dorsal root ganglia. Rarely, recurrent infections may be due to reinfection with a different strain of HSV-2. Recurrent viral shedding from the genital tract may occur without clinically evident disease.

Recurrent episodes common; may involve shedding without lesions

Neonatal Herpes

Neonatal herpes usually results from transmission of virus during delivery through infected genital secretions from the mother. In utero infection, though possible, is uncommon. In most cases, severe neonatal herpes is associated with primary infection of a seronegative woman at or near the time of delivery. This results in an intense viral exposure of a seronegative infant during the birth process. The incidence of symptomatic neonatal herpes simplex infection varies greatly among populations, but it is estimated at between 1 per 6,000 and 1 per 20,000 live births in the United States. Because a normal immune response is absent in the neonate born to a mother with recent primary infection, neonatal HSV infection is an extremely severe disease with an overall mortality rate of approximately 60%, and neurologic sequelae are high in those who survive. Manifestations vary. Some infants show disseminated vesicular lesions with widespread internal organ involvement and necrosis of the liver and adrenal glands; others have involvement of the central nervous system only, with listlessness and seizures.

Usually transmitted from mother at birth

High mortality rate if disseminated

DIAGNOSIS

Herpes simplex viruses are cultured in cell lines inoculated with infected secretions or lesions. The cytopathic effects of HSV can usually be demonstrated 24 to 48 hours after inoculation of the culture. Isolates of HSV-1 and HSV-2 can be differentiated by staining virus-infected cells with type-specific monoclonal antibodies to the two types. A direct smear prepared from the base of a suspected lesion and stained by either Giemsa or Papanicolaou method may show intranuclear inclusions or multinucleated giant cells typical of herpes (Tzanck test), but this is less sensitive than viral culture and not specific. Similar changes can be seen in cells infected with VZV. Enzyme immunoassays and immunofluorescence are rapid and relatively sensitive assays for direct detection of herpes antigen in lesions. Although early versions of these noncultural tests lacked sensitivity, more recent procedures have correlations with culture that approach 90%. Serology should not be used to diagnose active HSV infections, such as those affecting the genital or central nervous systems; frequently there is no change in antibody titer when reactivation occurs. Serology can be useful in detecting those with asymptomatic HSV-2 infection. PCR on cerebrospinal fluid and blood (CSF) is the best test to diagnose HSV encephalitis.

Grow rapidly in many cell culture systems

HSV-1 and HSV-2 distinguished by type-specific monoclonal antibodies

Enzyme immunoassay, immunofluorescence, and PCR all used for rapid diagnosis

TREATMENT

Intravenous acyclovir effective in HSV encephalitis and neonatal disease

Several antiviral drugs that inhibit HSV have been developed. The most effective and commonly used is the nucleoside analog acyclovir, which is converted by a viral enzyme (thymidine kinase) to a monophosphate form and then by cellular enzymes to the triphosphate form, which is a potent inhibitor of the viral DNA polymerase. Acyclovir significantly decreases the duration of primary infection and has a lesser but definite effect on recurrent mucocutaneous HSV infections. If taken daily, it can also suppress recurrences of genital and oral–labial HSV. In its intravenous form, it is effective in reducing mortality of HSV encephalitis and neonatal herpes. Acyclovir-resistant HSV has been recovered from immunocompromised patients with persistent lesions, especially those with acquired immunodeficiency syndrome (AIDS). Foscarnet is active against acyclovir-resistant HSV.

Acyclovir or prodrugs can decrease duration of acute and recurrent disease

The US Food and Drug Administration has approved both valacyclovir and famciclovir for the treatment of recurrent genital HSV. Valacyclovir is an oral prodrug of acyclovir with better bioavailability than acyclovir (54% compared with 15% to 20%). It is rapidly converted to acyclovir and, in every characteristic except absorption, it is identical with the parent compound. Valacyclovir is not more effective than acyclovir, but can be given in lower doses and less frequently (500 mg twice daily). Famciclovir is the prodrug of another guanosine nucleoside analog, penciclovir. The bioavailability of penciclovir is also high (77%). After conversion, penciclovir must be phosphorylated, just like acyclovir. Penciclovir has a much longer tissue half-life than acyclovir and can be given as 125 mg twice daily for treatment of recurrent genital HSV. Valacyclovir and famciclovir are now also approved for chronic suppression of recurrent genital HSV.

PREVENTION

Cesarean section may be performed to avoid neonatal infection

Avoiding contact with individuals with lesions reduces the risk of spread; however, virus may be shed asymptomatically and transmitted from the saliva, urethra, and cervix by individuals with no evident lesions. Safe sexual practices should reduce transmission. Acyclovir has been shown to reduce asymptomatic shedding and transmission of genital herpes, especially from males to females. Because of the high morbidity and mortality rates of neonatal infection, special attention must be paid to preventing transmission during delivery. Where active HSV lesions are present on maternal tissues, cesarean section delivery may be used to minimize contact of the infant with infected maternal genital secretions, but cesarean delivery may not be effective if rupture of the membranes precedes delivery by more than several hours. HSV vaccines have been under study for years, and a glycoprotein D vaccine appears to be somewhat effective in preventing primary HSV-2 genital infection.

VARICELLA-ZOSTER VIRUS

VIROLOGY

Slower growth and narrower range of infected cell types

Varicella-zoster virus (VZV) has the same general structural and morphologic features of herpes simplex and other human herpesviruses, but it contains its own envelope glycoproteins and is therefore antigenically different. The genome of VZV is approximately 125 kbp, which is the smallest genome of the human herpesviruses. One of the important similarities between HSV and VZV is that VZV also encodes a thymidine kinase and is responsive to acyclovir. Cellular features of infected cells such as multinucleated giant cells and intranuclear eosinophilic inclusion bodies are similar to those of HSV. VZV is more difficult to isolate in cell culture than HSV and grows best, but slowly in human diploid fibroblast cells. The virus has a marked tendency to remain attached to the membrane of the host cell with less release of virions into fluids.

VARICELLA-ZOSTER DISEASE

CLINICAL CAPSULE

VZV causes two diseases: chickenpox (varicella) and shingles (zoster). The former usually occurs in children; the latter as reactivation of latent virus especially in the elderly. The virus remains latent in neural ganglia and reactivates as cellular immunity wanes. Almost 90% of the US population is infected with VZV by age 10 years, and the virus is spread primarily by respiratory secretions. Primary and reactivation diseases are both especially severe when they affect immunocompromised persons.

EPIDEMIOLOGY

VZV infection is ubiquitous. In temperate climates, nearly all persons contract varicella (chickenpox) before they reach adulthood, and 90% of cases occur before the age of 10 years. In contrast, the mean age at infection in tropical countries is over 20 years, and the seroprevalence at age 70 may be only 50%. The virus is highly contagious, with attack rates among susceptible contacts of 75%. Varicella occurs most frequently during the winter and spring months. The incubation period is 11 to 21 days. The major mode of transmission is respiratory, although direct contact with vesicular or pustular lesions may result in transmission. Communicability is greatest 24 to 48 hours before the onset of rash and lasts 3 to 4 days into the rash. Virus is rarely isolated from crusted lesions.

Chickenpox acquired by respiratory route, usually before adulthood

Communicability greatest before rash onset

PATHOGENESIS

Respiratory spread leads to infection of the contact patient's upper respiratory tract followed by replication in regional lymph nodes and primary viremia. The latter results in infection of the reticuloendothelial system and a subsequent secondary viremia associated with T lymphocytes. After secondary viremia, there is infection of the skin and finally a host immune response.

Secondary viremia results in skin lesions

The relation between zoster and varicella was first described by Von Bokay in 1892, when he observed several instances of varicella in households after the introduction of a case of zoster. On the basis of these epidemiologic observations, he proposed that zoster and varicella were different clinical manifestations of a single agent. The cultivation of VZV in vitro by Weller in 1954 confirmed Von Bokay's hypothesis that the viruses isolated from chickenpox and from zoster (or shingles) are identical. Latency of VZV occurs in sensory ganglia, as shown by in situ hybridization methods in dorsal root ganglia of adults many years after varicella infection. Herpes zoster (shingles) occurs when latent VZV reactivates and multiplies within a sensory ganglion and then travels back down the sensory nerve to the skin. The rash of herpes zoster is generally confined to the area of the skin (ie, dermatome) innervated by the sensory ganglion in which reactivation occurs.

Varicella virus latent in sensory ganglion cells

Reactivation produces zoster

IMMUNITY

Both humoral immunity and cell-mediated immunity are important factors in determining how frequency of reinfection and reactivation of varicella-zoster occur. Circulating

Circulating antibody prevents reinfection; cell-mediated immunity controls reactivation

antibody prevents reinfection, and cell-mediated immunity appears to control reactivation. In patients with depressed cell-mediated immune responses, especially those with bone marrow transplants, Hodgkin disease, AIDS, and lymphoproliferative disorders, reactivation can occur, and VZV infections are more frequent and more severe.

Aging associated with increasing risk of zoster

The increase in the incidence and severity of herpes zoster observed with increasing age in immunocompetent individuals is correlated with an age-related decrease in VZV-specific cellular immunity. Beginning in the fifth decade of life, there is a marked decline in cellular immunity to VZV, which can be measured by cutaneous delayed hypersensitivity as well as by a variety of in vitro assays. This occurs many years before any generalized decline in cellular immunity.

CLINICAL ASPECTS

MANIFESTATIONS

VZV produces a primary infection in normal children characterized by a generalized vesicular rash termed **chickenpox** or **varicella.** After clinical infection resolves, the virus persists for decades with no clinical manifestation. Chickenpox lesions generally appear on the back of the head and ears, then spread centrifugally to the face, neck, trunk, and proximal extremities. Involvement of mucous membranes is common, and fever may occur early in the course of disease. Lesions appear in different stages of evolution (**Figure 14–6**); this characteristic is one of the major features used to differentiate varicella from smallpox, in which lesions are concentrated on the extremities and all have a similar appearance. Varicella lesions are pruritic (itchy), and the number of lesions may vary from 10 to several hundred.

Chickenpox lesions are widespread and pruritic

Immunocompromised children may develop progressive varicella, which is associated with prolonged viremia and visceral dissemination as well as pneumonia, encephalitis, hepatitis, and nephritis. Progressive varicella has an estimated mortality rate of 20%. In thrombocytopenic patients, the lesions may be hemorrhagic. Susceptible adults have a higher risk (15 times) for VZV pneumonia during chickenpox.

Severe disease in immunocompromised patients

Reactivation of VZV is associated with the disease herpes zoster (shingles). Although zoster is seen in patients of all ages, it increases in frequency with advancing age. Clinically, pain in a sensory nerve distribution may herald the onset of the eruption, which occurs several days to 1 or 2 weeks later. The vesicular eruption is usually unilateral, involving one to three dermatomes (**Figure 14–7**). New lesions may appear over the first 5 to 7 days. Multiple attacks of VZV infection are uncommon; if recurrent attacks of a vesicular eruption occur in one area of the body, HSV infection should be considered.

Reactivation to zoster most common in elderly

Follows sensory nerve distribution

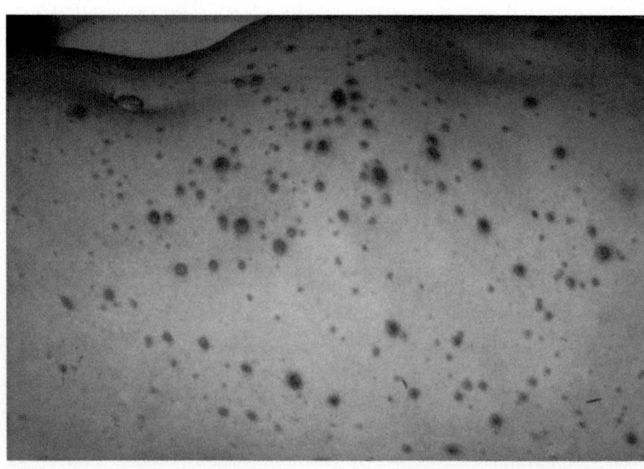

FIGURE 14–6. Primary varicella. Shows multiple stages of vesicles, papules. and crusted lesions on the abdomen.

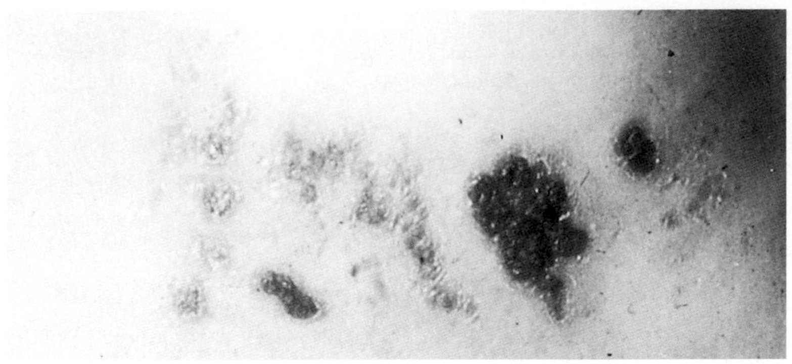

FIGURE 14–7. Herpes zoster lesion of the thorax. Note dermatomal distribution and presence of vesicles, pustules, and ulcerated and crusted lesions.

The complications of VZV infection are varied and depend on age and host immune factors. Postherpetic neuralgia is a common complication of herpes zoster in elderly adults. It is characterized by persistence of pain in the dermatome for months to years after resolution of the lesions of zoster and appears to result from damage to the involved nerve root. Immunosuppressed patients may develop localized zoster followed by dissemination of virus with visceral infection, which resembles progressive varicella. Bacterial superinfection is also possible. Maternal varicella infection during early pregnancy can result in fetal embryopathy with skin scarring, limb hypoplasia, microcephaly, cataracts, chorioretinitis, and microphthalmia. Severe varicella can also occur in seronegative neonates, with a mortality rate as high as 30%.

Postherpetic neuralgia after zoster

Dissemination with visceral infection in immunocompromised persons

DIAGNOSIS

Varicella or herpes zoster lesions can be diagnosed clinically, although they are occasionally difficult to distinguish from those caused by HSV or even vaccinia smallpox. Scrapings of lesions may reveal multinucleated giant cells characteristic of herpesviruses, but cytologic examination does not distinguish HSV lesions from those due to VZV. For rapid viral diagnosis, the best procedure is to demonstrate varicella-zoster antigen in cells from lesions by immunofluorescent antibody staining. VZV can be isolated from vesicular fluid or cells inoculated onto human diploid fibroblasts. However, the virus is difficult to grow from zoster (shingles) lesions older than 5 days, and cytopathic effects are usually not seen for 5 to 9 days. PCR of cerebrospinal fluid may be useful in the diagnosis of VZV encephalitis; culture is rarely positive.

Diagnosis usually clinical

Rapid confirmation by immunofluorescent staining

TREATMENT

Acyclovir has been shown to reduce fever and skin lesions in patients with varicella, and its use is recommended in healthy patients over 18 years of age. There are insufficient data to justify universal treatment of all healthy children and teenagers with varicella. In immunosuppressed patients, controlled trials of acyclovir have been effective in reducing dissemination, and the use of this agent is definitely indicated. In addition, controlled trials of acyclovir have demonstrated effectiveness in the treatment of herpes zoster in immunocompromised patients. Acyclovir may be used to treat herpes zoster in immunocompetent adults, but it appears to have only a modest impact on the development of postherpetic neuralgia, the most important complication of zoster. Treatment should be started within 3 days of the onset of zoster. VZV is less susceptible than HSV to acyclovir, so the dosage for treatment is substantially higher. Famciclovir or valacyclovir are more convenient and may be more effective.

Acyclovir or related prodrug therapy of immunocompromised patients

PREVENTION

High-titer immune globulin administered within 96 hours of exposure is useful in preventing infection or ameliorating disease in patients at risk for severe primary infection

TABLE 14–2 Properties of the Live Attenuated Varicella Vaccine (Oka)
• Rarely causes rash (5% in healthy children, mild)
• Two-dose schedule now recommended
• Induces cell-mediated immunity
• Lack of contact infection in most cases
• Induces long-term protective immunity
• Prevents disease when administered up to 3 days after exposure (postexposure prophylaxis)
• Incidence of herpes zoster in vaccinated children with leukemia lower than in comparable children infected naturally with wild-type virus
• >90% protection from household exposure of healthy children

Passive immunization for immunocompromised

(eg, immunosuppressed children with contact of patients with varicella or zoster). Once skin lesions have occurred, however, high-titer immune globulin has not proved useful in ameliorating disease or preventing dissemination. Immune globulin is not indicated for the treatment or prevention of reactivation (ie, zoster or shingles). In nonimmunosuppressed children, varicella is a relatively mild disease, and passive immunization is not indicated.

Live vaccine is safe and effective

Adult live vaccine now recommended for those older than 60 years

Need for isolation of cases in hospital

A live vaccine developed by a group of Japanese workers appears to be effective in both immunosuppressed and immunocompetent persons and is now recommended for routine use in healthy children older than 12 months of age (**Table 14–2**). Routine immunization at 12 to 15 months with single or combination (MMRV) vaccine, with a second dose at 4 to 6 years of age is now recommended. In immunocompromised patients who are susceptible to varicella, chickenpox can be extremely serious, even fatal. In these patients, the live vaccine appears to be protective, although it is not approved for this use in the United States. The vaccine is being used routinely in immunocompetent seronegative adults, especially those with occupational risk, such as health care workers, and it can even be helpful when given to a seronegative, immunocompetent adult shortly after exposure. A live adult V-Z vaccine is now available to stimulate waning cellular immunity and prevent or ameliorate future shingles in older individuals. It is recommended as a single dose for all adults over 60 years of age, even if they have reported shingles in the past. Chronic conditions such as renal failure, heart disease, or diabetes are not contraindications, but this vaccine should not be given to immunosuppressed patients. Varicella is a highly contagious disease and rigid isolation precautions must be instituted in all hospitalized cases.

CYTOMEGALOVIRUS

VIROLOGY

Nuclear and perinuclear cytoplasmic inclusions and cell enlargement

Human cytomegalovirus (CMV) possesses the largest genome of the herpesviruses (approximately 240 kbp) that is highly regulated by viral regulatory proteins, and its replication, though slow, is similar to that of HSV with the sequential appearance of immediate early, early, and late gene products. In addition to nuclear inclusions ("owl eye cells"), CMV produces perinuclear cytoplasmic inclusions and enlargement of the cell (cytomegaly), a property that gives the virus its name (**Figure 14–8**). Based on genomic and phenotypic heterogeneity, innumerable strains of CMV exist, and restriction endonuclease analysis of viral DNA has been useful for distinguishing strains epidemiologically. Antigenic variations have been observed, but are not of clinical importance.

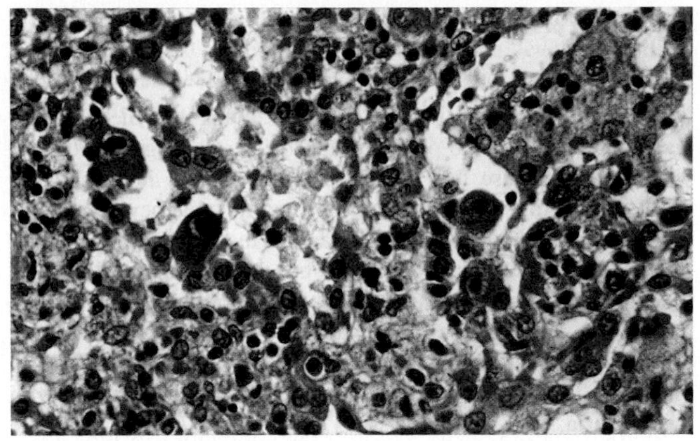

FIGURE 14–8. Cytomegalovirus-infected cells showing "owl's eye" appearance of intranuclear inclusions. (Reproduced with permission from Nester EW, Anderson DG, Roberts CE Jr, Nester MT. *Microbiology: A Human Perspective,* 6th ed. New York: McGraw-Hill, 2008.)

CYTOMEGALOVIRUS DISEASE

CLINICAL CAPSULE

CMV differs from HSV and VZV in that it doesn't cause skin disease, but CMV is similar in its ability to establish latent infection. CMV produces visceral disease, including a mononucleosis syndrome in otherwise healthy persons. Its major contribution to human misery is its high rate of congenital infection (1% of all infants; 40,000 in the United States per year). Most of those infected are asymptomatic; however, some 20% may have neurologic impairment. CMV is also an important cause of morbidity and mortality in immunocompromised patients with either primary or reactivation disease.

EPIDEMIOLOGY

CMV is ubiquitous, and in developed countries approximately 50% of adults have developed antibody. Age-specific prevalence rates show that approximately 10% to 15% of children are infected by CMV during the first 5 years of life, after which the rate of new infections levels off. The rate subsequently increases by 1% to 2% per year during adulthood, probably through close personal contact, including sexual contact with a virus-excreting person. CMV has been isolated from saliva, cervical secretions, semen, urine, and white blood cells for months to years after infection. Excretion of CMV is especially prolonged after congenital and perinatal infections, with 35% of infected infants excreting virus for as long as 5 years after birth. Transmission of infection in day care centers has been shown to occur from asymptomatic excreters to other children and, in turn, to seronegative parents. By age 18 months, up to 80% of infants in a day care center are infected and actively excreting virus in saliva and urine. Seroconversion rates in seronegative parents who have children attending day care centers are approximately 20% per year. This increases to approximately 30% if the child is shedding virus and up to 40% if the child is also less than 18 months of age. In contrast to day care centers, there is no substantial evidence of spread of CMV infection to health care workers in the hospital.

High infection rates in early childhood and early adulthood

Present in urine, saliva, semen, and cervical secretions

Latent infection, which occurs in leukocytes and their precursors, accounts for transfusion transmission, but this route is relatively infrequent—only 1% to 2% of blood units are believed to be infectious. Organ donation may also transmit latent virus, which causes primary infection in CMV-seronegative recipients and reinfection in seropositive patients.

Viral latency in leukocytes

PATHOGENESIS

CMV DNA in monocytes

As previously mentioned, CMV infects vascular endothelial cells and leukocytes and produces characteristic inclusions in the former. In vitro, CMV DNA can be demonstrated in monocytes showing no cytopathology, indicating a restricted growth potential in these cells. It is conjectured that these are the cells of latency for CMV. ::: latent infection, p 144

Immune-mediated tissue damage

CMV can cause disease by a variety of different mechanisms, including direct tissue damage and immunologic damage. Although direct infection and damage of mucosal epithelial cells in the lung is a potential mechanism for pneumonia, animal models have suggested that immunologic destruction of the lung by the host immune response to CMV infection may be the major mechanism of viral disease in this tissue. This hypothesis is supported by the observation that the degree of viral infection in lung tissue cannot account for the severity of CMV pneumonia; likewise, the disease does not respond well to antiviral therapy. Although cytolytic T-lymphocyte activity may contribute to lung pathology, cytokines released by these cells have also been implicated.

IMMUNITY

Vascular endothelial cells can be infected and support viral latency

Both humoral and cellular immune responses are important in CMV infections. In immunocompetent persons, clinical disease, if it occurs at all, results from primary infection, and reactivation with viral excretion in cervical excretions or semen is invariably subclinical. In immunocompromised patients, both primary infection and reactivation are much more likely to be symptomatic. Furthermore, CMV infection of monocytes results in dysfunction of these phagocytes in immunocompromised patients, which may increase predisposition to fungal and bacterial superinfection. When latently infected monocytes are in contact with activated T lymphocytes, the former are activated to differentiate into macrophages that produce infectious virus. These monocyte–T cell interactions may occur after transfusion or transplantation and may explain, not only transmission of CMV, but also activation of latent virus in the allograft recipient. Vascular endothelial cells may be other sites of CMV latency.

MANIFESTATIONS

Serious disease of fetus may develop with primary maternal infection

Worldwide, 1% of infants excrete CMV in urine or nasopharynx at delivery as a result of infection in utero. On physical examination, 90% of these infants appear normal or asymptomatic; however, long-term follow-up has indicated that 10% to 20% go on to develop sensory nerve hearing loss, psychomotor mental retardation, or both. Infants with symptomatic illness (about 0.1% of all births) have a variety of congenital defects or other disorders, such as hepatosplenomegaly, jaundice, anemia, thrombocytopenia, low birth weight, microcephaly, and chorioretinitis. Almost all infants with clinically evident congenital CMV infection are born of mothers who experienced primary CMV infection during pregnancy. The apparent explanation is that these babies are exposed to virus in the absence of maternal antibody. It is estimated that one third of maternal primary infections are transmitted to the fetus and that fetal damage is most likely to occur in the first trimester. Congenital infection frequently also results from reactivation in the mother with spread to the fetus, but such infection rarely leads to congenital abnormalities because the mother also transmits antibody to the fetus.

Perinatal infection asymptomatic or relatively benign

In contrast to the devastating findings with some congenital infections, neonatal infection acquired during or shortly after birth appears to be rarely associated with an adverse outcome. Most population-based studies have indicated that 10% to 15% of all mothers are excreting CMV from the cervix at delivery. Approximately one third to one half of all infants born to these mothers acquire infection. Almost all of these perinatally infected infants have no discernible illness unless the infant is premature or immunocompromised. CMV can also be efficiently transmitted from mother to child by breast milk, but these postpartum infections are also usually benign.

As with intrapartum acquisition of infection, most CMV infections during childhood and adulthood are totally asymptomatic. In healthy young adults, CMV may cause a mononucleosis-like syndrome. In immunosuppressed patients, both primary infection and reactivation may be severe. For example, in patients receiving bone marrow transplants, interstitial pneumonia caused by CMV is a leading cause of death (50% to 90% mortality rate), and in AIDS patients, CMV often disseminates to visceral organs, causing chorioretinitis, gastroenteritis, and neurologic disorders.

CMV lung, visceral, and eye infections in immunocompromised patients

DIAGNOSIS

Laboratory diagnosis of CMV infection depends on (1) detecting CMV cytopathology, antigen, or DNA in infected tissues; (2) detecting viral DNA or antigen in body fluids; (3) isolating the virus from tissue or secretions; or (4) demonstrating seroconversion. CMV can be grown readily in serially propagated diploid fibroblast cell lines. Demonstration of viral growth generally requires 1 to 14 days, depending on the concentration of virus in the specimen and whether coverslip cultures in shell vials and FA staining are used to speed detection. The presence of large inclusion-bearing cells in urine sediment may be detected in widespread CMV infection. This technique is insensitive, however, and provides positive results only when large quantities of virus are present in the urine. Culture of blood to detect viremia is now superseded by detection and quantitation of CMV antigen in peripheral blood leukocytes or detection of CMV DNA in plasma or leukocytes by PCR. These procedures are significantly more sensitive than culture.

DNA detection by PCR or antigen detection useful to find viremia

Because of the high prevalence of asymptomatic carriers and the known tendency of CMV to persist weeks or months in infected individuals, it is frequently difficult to associate a specific disease entity with the isolation of the virus from a peripheral site. Thus, the isolation of CMV from urine of immunosuppressed patients with interstitial pneumonia does not constitute evidence of CMV as the cause of that illness. CMV pneumonia or gastrointestinal disease is best diagnosed by demonstrating CMV inclusions in biopsy tissue.

Histologic detection of inclusions in lung, gastrointestinal tissues is useful

The procedures listed below are recommended to facilitate the diagnosis of CMV infection in specific clinical settings:

1. *Congenital infection*—Virus culture or viral DNA assay positive at birth or within 1 to 2 weeks (to distinguish from natally or perinatally infected infants, who will not begin to excrete virus until 3 to 4 weeks after delivery).
2. *Perinatal infection*—Culture-negative specimens at birth but positive specimens at 4 weeks or more after birth suggest natal or early postnatal acquisition. Seronegative infants may acquire CMV from exogenous sources, such as from blood transfusion.
3. *CMV mononucleosis in nonimmunocompromised patients*—Seroconversion and presence of IgM antibody specific for CMV best indicators of primary infection. Urine culture positivity supports the diagnosis of CMV infection, but may reflect remote infection because positivity may continue for months to years. A positive blood assay for CMV antigen or DNA, however, is diagnostic in this patient population.
4. *Immunocompromised patients*—Demonstration of virus by viral antigen or DNA in blood documents viremia. Demonstration of inclusions or viral antigen in diseased tissue (eg, lung, esophagus, or colon) establishes the presence of CMV infection, but does not provide proof that CMV is the cause of disease unless other pathogens are excluded. Seroconversion is diagnostic but rarely occurs, especially in AIDS patients, because more than 95% of these patients are seropositive for CMV before infection with human immunodeficiency virus (HIV). CMV-specific IgM antibody may not be present in immunocompromised transplant patients, especially during reactivation of virus. Conversely, in AIDS patients, this antibody frequently is present even when clinically important infection is absent.

TREATMENT

Ganciclovir, a nucleoside analog of guanosine structurally similar to acyclovir, has been shown to inhibit CMV replication; prevent CMV disease in AIDS patients and transplant recipients; and reduce the severity of some CMV syndromes, such as retinitis and gastrointestinal disease. Combining immune globulin with ganciclovir appears to reduce the very high mortality from CMV pneumonia in bone marrow transplant recipients more than that achieved with ganciclovir alone, Foscarnet, a second approved drug for therapy of CMV disease, is also efficacious. Its toxic effects are primarily renal, whereas ganciclovir is most apt to inhibit bone marrow function. Ganciclovir inhibits CMV DNA polymerase, like foscarnet, but the two drugs act on different sites and cross-resistance is rare. A third drug, cidofovir, a nucleotide analog, is approved for therapy of retinitis, but its use is limited because of nephrotoxicity

PREVENTION

Use of CMV-seronegative donors decreases risk

The use of blood from CMV-seronegative donors or blood that is treated to remove white cells decreases transfusion-associated CMV. Similarly, the disease can be avoided in seronegative transplant recipients by using organs from CMV-seronegative donors. Safe sexual practices including condom usage may reduce transmission. A recombinant CMV envelope glycoprotein B vaccine has been developed, which may become useful in reducing the risk of maternal and congenital infection.

EPSTEIN–BARR VIRUS

VIROLOGY

Etiologic agent of infectious mononucleosis and certain lymphomas

Cultivated only in lymphoblastoid cell lines

EBNA, VCA, and EA represent stages of viral replication

Epstein-Barr virus (EBV) is the etiologic agent of infectious mononucleosis and African Burkitt lymphoma. Its complete nucleotide sequence of 172 kbp is smaller than that of other herpesviruses, but has been thoroughly mapped. Although EBV is morphologically similar to the other herpesviruses, it can be cultured only in lymphoblastoid cell lines derived from B lymphocytes of humans and higher primates. In vivo, EBV is tropic for both human B lymphocytes and epithelial cells. The former is a nonproductive infection, whereas the latter is productive. The virus generally does not produce cytopathic effects or the characteristic intranuclear inclusions of other herpesvirus infections. After infection with EBV, lymphoblastoid cells containing viral genome can be cultivated continuously in vitro; they are thus transformed, or immortalized. Recent studies suggest that most of the viral DNA in transformed cells remains in a circular extrachromosomal, nonintegrated form as an episome, while a lesser amount is integrated into the host cell genome. Viral antigen expression has been studied by immunofluorescent staining of transformed cell lines under various conditions. One group of proteins, called EBV nuclear antigens (EBNAs), appear in the nucleus prior to virus-directed protein synthesis. Viral capsid antigen (VCA) can be detected in cell lines that produce mature virions. Other cell lines, called nonproducers, contain no mature virions, but express certain virus-associated antigens called early antigens (EAs). The latter may be seen as diffuse (D) and as restricted (R) aggregates of staining.

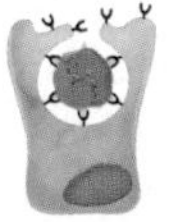

EPSTEIN-BARR VIRUS DISEASE

CLINICAL CAPSULE

Investigators discovered EBV in the course of their studies to determine the cause of Burkitt lymphoma. Serologic studies later found that the virus was the cause of infectious mononucleosis. The greatest interest in EBV hinges on its role in malignant disease, including Burkitt lymphoma, nasopharyngeal carcinoma, and lymphoproliferative disease of the immunocompromised.

EPIDEMIOLOGY

EBV can be cultured from saliva in 10% to 20% of healthy adults and is intermittently recovered from most seropositive individuals. It is of low contagiousness, and most cases of infectious mononucleosis are contracted after repeated contact between susceptible persons and those asymptomatically shedding the virus. Secondary attack rates of infectious mononucleosis are low (less than 10%) because most family or household contacts already have antibody to the agent (worldwide 90% to 95% of adults are seropositive). Infectious mononucleosis has also been transmitted by blood transfusions; most transfusion-associated mononucleosis syndromes, however, are attributable to CMV. In more highly developed countries and in individuals of higher socioeconomic status, EBV infection tends to be acquired later in life than in individuals from developing countries of lower socioeconomic status. When primary infection with EBV is delayed until the second decade of life or later, it is accompanied by symptoms of infectious mononucleosis in about 50% of cases.

Widespread asymptomatic infection; disease most common in young adults

At present, there appears to be many fewer variations of genomic strains among EBV isolates than other herpesviruses. The two strains (types A and B) both circulate widely, and both can coinfect a single individual.

PATHOGENESIS

Although EBV initially infects epithelial cells, the hallmark of EBV disease is subsequent infection of B lymphocytes and polyclonal B-lymphocyte activation with benign proliferation. The virus enters B lymphocytes by means of envelope glycoprotein binding to a surface receptor CR2 or CD21, which is the receptor for the C3d component of complement system; 18 to 24 hours later, EBV nuclear antigens are detectable within the nucleus of infected cells. Expression of the viral genome, which encodes at least two viral proteins, is associated with immortalization and proliferation of the cell. The EBV-infected B lymphocytes are polyclonally activated to produce immunoglobulin and express a lymphocyte-determined membrane antigen that is the target of host cellular immune responses to EBV-infected B lymphocytes. During the acute phase of infectious mononucleosis, up to 20% of circulating B lymphocytes demonstrate EBV antigens. After infection subsides, EBV can be isolated from only about 1% of such cells.

Infects B cells

Encodes proteins associated with immortalization of B cells

EBV has been associated with several lymphoproliferative diseases, including African Burkitt lymphoma, nasopharyngeal carcinoma, and lymphomas in immunocompromised patients. The factors that render the EBV infections oncogenic in these cases are obscure. The distribution of EBV infections in Africa has suggested an infectious cofactor, such as malaria, which may cause immunosuppression and predispose to EBV-related malignancy. In nasopharyngeal carcinoma, environmental carcinogens may create the precancerous lesion although genetic factors may also be operative. In vivo, EBV-associated lymphomas have been shown to be of both monoclonal and polyclonal origin. Translocations in B cells, most commonly involving the c-*myc* oncogene and immunoglobulin heavy or light loci, are characteristic of Burkitt lymphoma and involve specific breaks in chromosomes. These translocations lead to expression of oncogenes that may contribute to clonal activation and ultimately to malignancy. Some breakdown in immune surveillance also appears to play a role in the development of malignancy, because immunosuppressed patients are more prone to develop EBV-associated B-cell lymphomas. Recent studies suggest an association of EBV with Hodgkin's lymphoma in young adults.

Lymphomas can develop in immunocompromised patients

IMMUNITY

Virus-induced infectious mononucleosis is associated with circulating antibodies against specific viral antigens, as well as against unrelated antigens found in sheep, horse, and some beef red blood cells. The latter, referred to as heterophile antibodies, are a heterogeneous group of predominantly IgM antibodies long known to correlate with episodes of infectious mononucleosis, and are commonly used as diagnostic tests for the disease. They do not cross-react with antibodies specific for EBV, and there is not good correlation between the heterophile antibody titer and the severity of illness. Cutaneous anergy and decreased cellular immune responses to mitogens and antigens are seen early in the course of mononucleosis. The "atypical" lymphocytosis associated with infectious mononucleosis is caused by an increase in the number of circulating T cells, which appear to be activated cells developed in response to the virus-infected B lymphocytes. With recovery from illness, the atypical lymphocytosis gradually resolves, and cell-mediated immune functions return to preinfection levels, although memory T cells maintain the capacity to limit proliferation of EBV-infected B cells. In rare cases, the initial EBV-induced proliferation of B cells is not contained, and EBV lymphoproliferative disease ensues. This syndrome is most often seen in immunocompromised organ transplant recipients.

Suppressed cell-mediated immune responses in acute infection

MANIFESTATIONS

Infectious Mononucleosis

Primary infection asymptomatic or expressed as infectious mononucleosis

Although most primary EBV infections are asymptomatic, clinically apparent infectious mononucleosis is characterized by fever, malaise, pharyngitis, tender lymphadenitis, and splenomegaly. These symptoms persist for days to weeks; they slowly resolve. Complications such as laryngeal obstruction, meningitis, encephalitis, hemolytic anemia, thrombocytopenia, or splenic rupture may occur in 1% to 5% of patients.

Lymphoproliferative Syndrome

Lymphoproliferative disease occurs, especially in immunocompromised persons

Patients with primary or secondary immunodeficiency are susceptible to EBV-induced lymphoproliferative disease. For example, the incidence of these lymphomas is 1% to 2% after renal transplantations and 5% to 9% after heart–lung transplantations. The risk is greatest in patients experiencing primary EBV infection rather than reactivation. Most characteristic is persistent fever, lymphadenopathy, and hepatosplenomegaly.

Burkitt Lymphoma

Tumors may involve cofactors

Translocation may lead to clonal activation

In sub-Saharan Africa, Burkitt lymphoma is the most common malignancy in young children, with an incidence of 8 to 10 cases per 100,000 people per year. The risk is greatest in equatorial Africa, where there is a high incidence of malaria. Burkitt lymphoma is thought to result from an early EBV infection that produces a large pool of infected B lymphocytes. Malarial infection may further increase the size of this pool and provide a constant antigenic challenge. Such stimuli can lead to c-*myc* chromosomal translocations, which are pathognomonic for this lymphoma. Serologic screening for increased IgA antibody levels to both VCA and early EBV antigens can be used for early diagnostic purposes.

Nasopharyngeal Carcinoma

Endemic nasopharyngeal carcinoma in southern China; suggests environmental or genetic cofactors

Nasopharyngeal carcinoma (NPC) is endemic in southern China, where it is responsible for approximately 25% of the mortality from cancer. The high incidence of nasopharyngeal carcinoma among the southern Chinese people suggests that genetic or environmental factors in addition to EBV may also be important in the pathogenesis of the disease.

AIDS Patients

In AIDS patients, several distinct additional EBV-associated diseases may occur, including hairy leukoplakia of the tongue, interstitial lymphocytic pneumonia (especially in infants), and lymphoma.

DIAGNOSIS

Laboratory analysis of EBV infectious mononucleosis is usually documented by the demonstration of atypical lymphocytes and heterophile antibodies, or positive EBV-specific serologic findings. Hematologic examination reveals a markedly raised lymphocyte and monocyte count with more than 10% atypical lymphocytes, called Downey cells (**Figure 14–9**). Atypical lymphocytes, though not specific for EBV, are present with the onset of symptoms and disappear

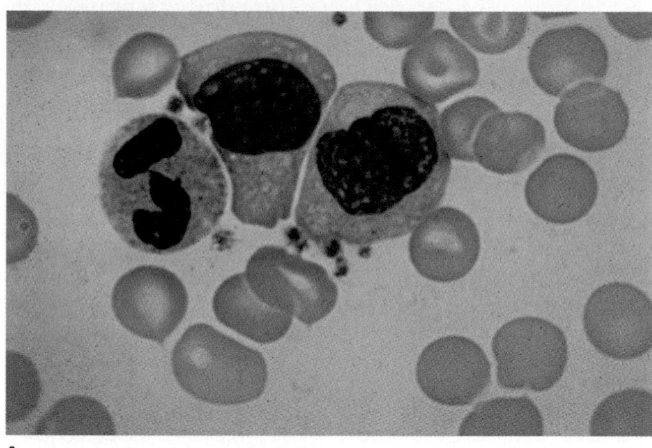
A

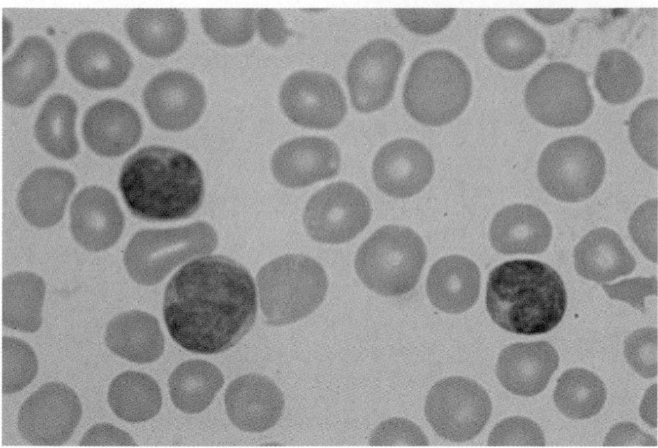
B

FIGURE 14–9. A. Atypical lymphocytes (Downey cells) in blood smear from a patient with infectious mononucleosis. Note indented cell membranes. Polymorphonuclear leukocyte is adjacent to the two affected cells. **B.** Normal lymphocytes contrast sharply with those in A.

with resolution of disease. Alterations in liver function tests may also occur, and enlargement of the liver and spleen is a common finding.

Atypical lymphocytosis common in acute infection

Though not specific for EBV, tests for heterophile antibodies are used most commonly for diagnosis of infectious mononucleosis. In commercial kits, animal erythrocytes are used in simple slide agglutination methods, which incorporate absorptions to remove cross-reacting antibodies that may develop in other illnesses, such as serum sickness. The infectious mononucleosis heterophile antibody is absorbed by sheep erythrocytes, but not by guinea pig kidney cells. Heterophile antibodies can usually be demonstrated by the end of the first week of illness, but are occasionally delayed until the third or fourth week. They may persist many months.

Approximately 5% to 15% of EBV-induced cases of infectious mononucleosis in adults and a much greater proportion in young children and infants fail to induce detectable levels of heterophile antibodies. In these cases, the EBV-specific serologic tests summarized in **Table 14–3** may be used to establish the diagnosis. The panel to be tested includes antibodies to VCA, which rise quickly and persist for life. Antibodies to EBNAs rise later in disease (after about 1 month) and also persist in low titers for life. Thus, a high titer to VCA and no titer to EBNA suggest recent EBV infection, whereas antibody titers to both antigens are indicative of past infection. The presence of IgM antibody to VCA is theoretically diagnostic of acute, primary EBV infection, but low levels may occur during reactivation of EBV, and cross-reactions with antigens of other herpesviruses occur. Persistent antibody to early antigens (anti-EA, -D, or -R) may be correlated with severe disease, nasopharyngeal carcinoma (anti-D), or African Burkitt lymphoma (anti-R), but are not useful in diagnosing infectious mononucleosis. Isolation of EBV from clinical specimens is not practical, because it requires fresh human B cells or fetal lymphocytes obtained from cord blood.

IgM antibody to VCA or high IgG antibody to VCA with negative anti-EBNA suggest, primary infection

Virus isolation is impractical for routine diagnosis

TREATMENT

Treatment of infectious mononucleosis is largely supportive. More than 95% of patients recover uneventfully. In a small percentage of patients, splenic rupture may occur; restriction of contact sports or heavy lifting during the acute illness is recommended. The DNA polymerase enzyme of EBV has been shown to be sensitive to acyclovir, and acyclovir can decrease the amount of replication of EBV in tissue culture and in vivo. Despite this antiviral activity, systemic acyclovir makes little or no impact on the clinical illness. Laryngeal obstruction should be treated with corticosteroids. Hairy leukoplakia in AIDS patients does respond to acyclovir treatment.

Treatment is supportive

TABLE 14–3 Epstein-Barr Virus–Specific Antibodies

ANTIBODY SPECIFICITY	TIME OF APPEARANCE	DURATION	COMMENTS
Viral capsid antigen (VCA)			
IgM	Early in illness	1-2 months	Indicator of primary infection
IgG	Early in illness	Lifelong	Standard Epstein-Barr virus (EBV) titer reported by most commercial and state labs; major usefulness is as marker for prior infection in epidemiologic studies; if present without EBNA (Epstein-Barr nuclear antigen) antibody, indicates current infection
EBNA IgG	3-6 weeks after onset	Lifelong	Late appearance of anti-EBNA IgG antibodies in infectious mononucleosis (IM) makes absence or seroconversion a useful marker for primary infection; persists for life
Early antigen (EA) diffuse protein (EA-D)	Peaks 3-4 weeks after onset	3-6 months	Present in IM patients; IgA antibodies useful for prediction of nasopharyngeal carcinoma in high-risk populations
EA restricted (EA-R)	Several weeks after onset	Months to years	Present in higher titer in African Burkitt lymphoma; may be useful as indicator of reactivation of EBV

PREVENTION

Immunization of humans not available

The occurrence of Burkitt lymphoma and nasopharyngeal carcinoma in restricted geographic areas offers the possibility of prevention by immunization with virus-specific antigen(s). At present, this approach is under exploration. A subunit vaccine has proved effective in preventing the development of tumors in tamarind monkeys, which are highly susceptible to the oncogenic effects of the virus under experimental conditions.

HUMAN HERPESVIRUS 6

Replicates in CD4+ T lymphocytes

In 1986, a herpesvirus, now called human herpesvirus type 6 (HHV-6), was identified in cultures of peripheral blood lymphocytes from patients with lymphoproliferative diseases. The virus, which is genetically distinct but morphologically similar to other herpesviruses, replicates in lymphoid tissue, especially CD4+ T lymphocytes, and has two distinct variants, A and B. HHV-6 is more closely related to CMV than to the other earlier known herpesviruses and is the β subfamily.

EPIDEMIOLOGY

Infection common in infancy

HHV-6 is the most rapidly spread of the herpesviruses and is shed in the throats of 10% of babies by age 5 months, 70% by 12 months, and 30% of adults. Almost all of the population has antibody to this virus by the age of 5 years.

MANIFESTATIONS

Associated with roseola in infants

HHV-6 type B is the etiologic agent of exanthem subitum (roseola), and both types A and B can cause acute febrile illnesses with or without seizures or rashes. Exanthem subitum generally occurs in infants aged 6 months to 1 year. It is characterized by fever (usually about 39°C) for 3 days, followed by a faint maculopapular rash spreading from the trunk to the extremities, which begins during defervescence.

HHV-6 also appears to reactivate in transplant recipients. It may contribute to graft rejection and clinical illnesses such as meningoencephalitis, pneumonia, and bone marrow suppression after bone marrow transplantation. The virus reactivates in other immunocompromised patients including those with AIDS, lymphoma, and leukemia, but its clinical significance is not known.

Reactivation common in immunosuppression

Initially, it was thought that HHV-6 would grow only in freshly isolated B lymphocytes, and the virus was referred to as the human B-lymphotropic virus. Now it is clear that the virus infects mainly T lymphocytes. HHV-6 establishes a latent infection in T cells, but may be activated to a productive lytic infection by mitogenic stimulation. Resting lymphocytes and lymphocytes from normal immune individuals are resistant to HHV-6 infection. In vivo, HHV-6 replication is controlled by cell-mediated factors.

Latent infection of T cells

DIAGNOSIS

Primary virus infection can be documented by seroconversion. Active virus infection can be documented by culture, antigenemia, or DNA detection in the blood (by PCR). Because asymptomatic viremic reactivation is common, it is very difficult to use these tools to identify HHV-6 as the cause of febrile or other miscellaneous syndromes.

Primary infection can be documented serologically

PCR used to detect viremic infection

TREATMENT

Definitive therapy has not been established, but HHV-6 appears to be susceptible in vitro to ganciclovir and foscarnet. It is less susceptible to acyclovir, because the virus has no thymidine kinase.

No viral thymidine kinase

HUMAN HERPESVIRUS 7

Isolation of human herpesvirus 7 (HHV-7) was first reported in 1990. The virus was isolated from activated CD4+ T lymphocytes of a healthy individual. The CD4 molecule appears to be a receptor for virus attachment. HHV-7 is distinct from all other known human herpesviruses, but is most closely related to HHV-6 and CMV and is in the β subfamily with these two viruses. Seroepidemiologic studies indicate that this virus usually does not infect children until after infancy, but that nearly 90% of children are antibody-positive by 3 years of age. As with HHV-6, this virus is frequently isolated from saliva, and close personal contact is the probable means of transmission. Also, like HHV-6, HHV-7 may be a cause of exanthem subitum. The diagnosis of acute infection can be made by the demonstration of seroconversion. No treatment has been identified.

Originally isolated from CD4+ T lymphocytes

Can cause exanthem subitum (roseola)

HUMAN HERPESVIRUS 8

Human herpesvirus 8 (Kaposi sarcoma–associated herpesvirus [KSHV]; HHV-8) was discovered in 1994 by identification of unique viral DNA sequences in Kaposi sarcoma tissue obtained from an AIDS patient, using subtractive hybridization analysis. These specific DNA sequences are found in 95% or more of Kaposi sarcoma tissues, both AIDS-related and non–AIDS-related in African cases. KSHV DNA has also been detected in cells from lymphoproliferative diseases (eg, primary effusion lymphomas, associated with AIDS and multicentric Castleman disease).

Recently, HHV-8 was isolated in culture and, when characterized, seems most closely related to EBV. Like EBV, HHV-8 preferentially infects B lymphocytes, and it is also considered to be a gamma herpesvirus. Epidemiologic and virologic studies suggest that it is a necessary but perhaps not sufficient cause of Kaposi sarcoma and that other factors (eg, immunosuppression, genetic predisposition) are cofactors in the development of this malignancy. On average, seropositivity to HHV-8 precedes the development of Kaposi sarcoma by 3 years. The virus appears to be sexually transmitted, as suggested by a higher prevalence of antibody in promiscuous gay men than in those who are not promiscuous, and by higher prevalence in gay men with HIV versus that in other HIV-positive risk groups, such as transfusion recipients and hemophiliacs. Specific and sensitive antibody assays are

Associated with Kaposi sarcoma

Infects B lymphocytes

being developed, and antibody to HHV-8 appears to be relatively rare in the general population. It is difficult to assess the impact of antivirals, because Kaposi sarcoma may improve with immune reconstitution. Interferon-alpha can be effective against Kaposi sarcoma, but this may result from immune enhancement rather than from any specific antiviral activity. Evidence of active viral replication in Kaposi sarcoma is minimal, so there may not be an appropriate target for antivirals at the time that Kaposi sarcoma becomes manifest.

CASE STUDY

A "KISSING" DISEASE

A 17-year-old girl was healthy before entering college as a freshman. Two months later, she noted an illness that progressed over a few days, beginning with fatigue and difficulty concentrating. Other symptoms followed, including fever, sore throat, headache, and "fullness" in the neck.

The physical examination revealed conjunctival and pharyngeal inflammation and enlarged, slightly tender lymph nodes in the anterior and posterior cervical triangles.

QUESTIONS

If this patient has acute, primary Epstein-Barr virus infection, which of the following would be the most sensitive and specific confirmatory test?

A. IgG–specific anti-VCA antibody and undetectable anti-EBNA antibody
B. IgG–specific anti-EBNA antibody
C. Heterophile antibodies
D. Circulating atypical lymphocytosis of 20% or greater
E. PCR of serum

The major **sites** of herpesvirus latency are listed in the **right-hand** column. Match these with the **viruses** in the **left-hand** column.

2. HSV-1 _____
3. HSV-2 _____ **A.** Nervous tissue
4. CMV _____ **B.** Monocytes
5. VZV _____ **C.** β lymphocytes
6. EBV _____

Vaccines have been demonstrated to be efficacious in preventing herpesvirus disease in which one of the following situations?

A. HSV-1 primary infection
B. Varicella-zoster reactivation
C. HSV-2 reactivation
D. CMV primary infection
E. EBV reactivation

ANSWERS

1(A), **2**(A), **3**(A), **4**(B), **5**(A), **6**(C), **7**(B)

CHAPTER 15

Viruses of Diarrhea

Acute diarrheal disease is an illness, usually of rapid evolution (within several hours), that lasts less than 3 weeks. In addition to the bacterial and protozoal agents responsible for approximately 20% to 25% of these cases, viruses are a significant cause of the balance. Rotaviruses, caliciviruses, astroviruses, and some adenoviruses are considered here.

GENERAL FEATURES

Until the 1970s, proof of viral causation of acute diarrhea was usually based on exclusion of known bacterial or protozoan pathogens and supported by feeding cell-free filtrates of diarrheal stools to volunteers in an attempt to reproduce the disease. As might be expected, the results of such experiments were variable, and the methods were impractical for routine laboratory diagnosis. One aspect of such infections that proved to be of great help was the frequent association with abundant excretion of virus particles during the acute phase of illness. Virion numbers greater than 10^8 per gram of diarrheal stool are relatively common, allowing ready visualization with an electron microscope. Direct electron microscopy and immunoelectron microscopy have been used frequently to detect and identify the presumed causative viruses; the latter method could also be used to detect humoral antibody responses to infection. More recently, polymerase chain reactions (PCR) and enzyme immunoassays (EIA) have been increasingly employed in diagnosis.

Viral diarrhea was a diagnosis of exclusion

Many viral particles seen in stool by electron microscopy

Confirmation by EIA or PCR is now possible

Detection of a specific virus in the stools of symptomatic patients is not sufficient to establish the role of the virus in causing disease. Other criteria to be fulfilled include the following:

1. Establish that the virus is detected in ill patients significantly more frequently than in asymptomatic, appropriately matched controls and that virus shedding temporally correlates with symptoms.
2. Demonstrate significant humoral or secretory antibody responses, or both, in patients shedding the virus.
3. Reproduce the disease by experimental inoculation of nonimmune human or animal hosts (usually the most difficult criterion to fulfill).
4. Exclude other known causes of diarrhea, such as bacteria, bacterial toxins, and protozoa.

Multiple criteria used for establishing etiologic relationship

Using these criteria, four groups of viruses have been clearly established as important causes of gastrointestinal disease: rotaviruses, caliciviruses, astroviruses, and some adenovirus serotypes ("enteric" adenoviruses). Other viruses have also been implicated, but many of the preceding criteria have not been fulfilled; therefore, they are currently regarded as "candidate" causes of gastrointestinal disease.

Rotaviruses, caliciviruses, astroviruses, and adenoviruses are established causes

The currently established viruses are listed in **Table 15–1,** and all have several features in common, including a tendency toward brief incubation periods; fecal–oral spread by direct or indirect routes; and production of vomiting, which generally precedes or accompanies

"Candidate" viruses meet some criteria

TABLE 15–1 Biologic and Epidemiologic Characteristics of Viruses That Cause Diarrhea

SPECIAL FEATURES	ROTAVIRUS	CALICIVIRUS	ASTROVIRUS	ADENOVIRUS
Biologic				
Nucleic acid	Double-stranded RNA	Single-stranded RNA	Single-stranded RNA	Double-stranded DNA
Diameter, shape	65-75 nm, naked, double-shelled capsid	27-38 nm, naked, round	28-38 nm, naked, star-shaped	70-90 nm, naked, icosahedral
Replication in cell culture	Usually incomplete	None	None	None or incomplete
Number of serotypes	5 important to humans	More than 4	8, perhaps more	Unknown
Pathogenic				
Site of infection	Duodenum, jejunum	Jejunum	Small intestine	Small intestine
Mechanism of immunity	Local intestinal IgA	Unknown	Unknown	Unknown
Epidemiologic				
Epidemicity	Epidemic or sporadic	Family and community outbreaks	Sporadic	Sporadic
Seasonality	Usually winter	None known	None known	None known
Ages primarily affected	Infants, children <2 years old	Older children and adults	Infants, children	Infants, children
Method of transmission	Fecal–oral	Fecal–oral; contaminated water and shellfish	Fecal–oral	Fecal–oral
Incubation period (days)	1-3	0.5-2	?1-2	8-10
Major diagnostic tests	EIA, EM	EM, IEM, PCR	EM, PCR	EIA, EM

EM, electron microscopy; EIA, enzyme immunoassay; IEM, immunoelectron microscopy; PCR, polymerase chain reaction.

Vomiting commonly follows short incubation period

the diarrhea. The last feature has influenced physicians to use the term **acute viral gastroenteritis** to describe the syndrome associated with these agents.

ROTAVIRUSES

Most common cause of winter gastroenteritis in children <2 years of age

The human intestinal rotaviruses were first found in 1973 by electron microscopic examination of duodenal biopsy specimens from infants with diarrhea (**Figure15–1**). Since then, they have been found worldwide and are believed to account for 40% to 60% of cases of acute gastroenteritis occurring during the cooler months in infants and children less than 2 years of age. Worldwide, at least 500,000 childhood deaths annually are attributed to rotavirus infections; whereas such deaths in the United States are rather infrequent, the annual morbidity rate has been nonetheless considerable in recent years (**Figure 15–2**). Recent introduction of vaccines for routine use may reduce their impact in the future. These viruses have been detected in intestinal contents and in tissues from the upper gastrointestinal tract.

VIROLOGY

The rotaviruses belong to the family Reoviridae. They are naked icosahedral capids, spherical particles 65 to 75 nm in diameter (smaller forms have also been described) with a genome containing 11 segments of double-stranded RNA, an RNA-dependent RNA polymerase, and a double-shelled outer capsid; two segments encode proteins of the outer capsid (VP4 and VP7), which are targets for neutralizing antibodies. The name is derived from the Latin *rota* ("wheel") because of the outer capsid, which resembles a wheel attached by short spokes to the inner capsid and core (**Figures 15–1, 15–3A**). The major outer capsid

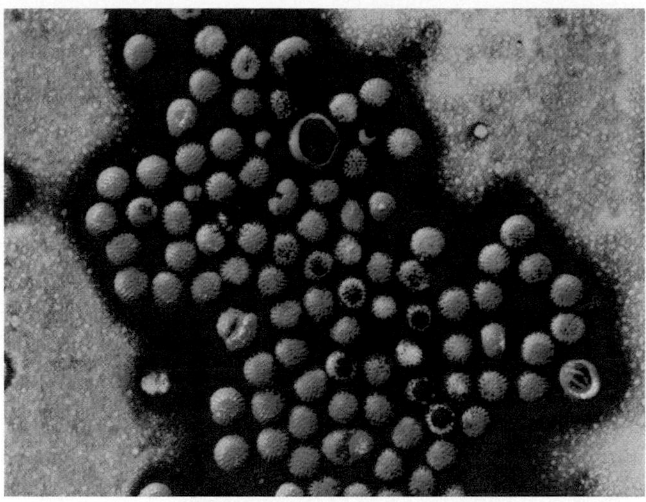

FIGURE 15–1. Rotavirus structure. (Reproduced with permission from Willey J, Sherwood L, Woolverton C (eds). *Prescott's Principles of Microbiology.* New York: McGraw-Hill; 2008.)

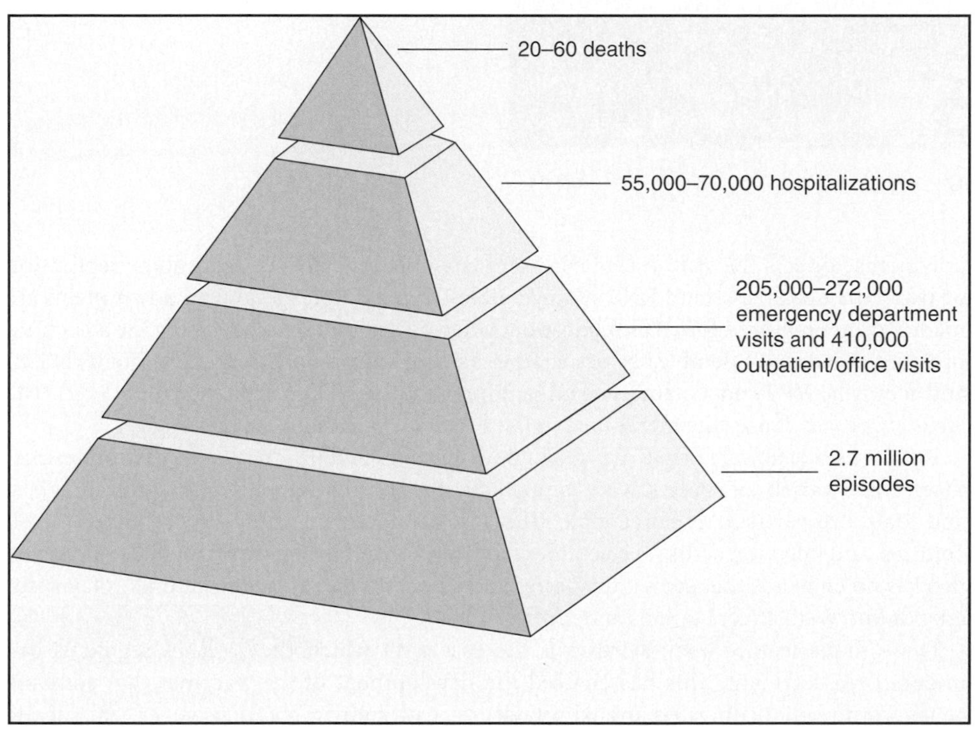

FIGURE 15–2. Estimated annual morbidity due to rotavirus infections in the United States. (Centers for Disease Control and Prevention.)

proteins are VP4 and VP7. VP4 performs several functions, including viral attachment protein, whereas VP7 is a type-specific antigen and facilitates viral attachment and entry. Five serotypes (G1, G2, G3, G4, and G9), based on VP4 and VP7 type-specific antigens on the outer capsid, are of major epidemiologic importance. The outer capsid is proteolytically cleaved in the gastrointestinal tract to generate intermediate infectious subviral particle (ISVP), which activates the virus for infection. Rotaviruses can replicate in the cytoplasm of infected cell cultures in the laboratory but are difficult to propagate because the replicative cycle is usually incomplete, and mature, infectious virions are often not produced. However, successful propagation of human strains in vitro has been achieved in some instances.

Double-stranded RNA viruses are shaped like a wheel

Antigenic types are based on capsid proteins VP4 and VP7

Rotavirus is transmitted fecal-orally, and the virus particle is partially digested in the gastrointestinal tract and activated by protease cleavage resulting in the loss of VP7 and cleavage of VP4 to generate ISVP. The VP4 binds to sialic acid containing glycoproteins on epithelial cells, and the ISVP penetrates in the target cells. The generation of ISVP is necessary for rotavirus infection because the double-shelled virus particle, after entering the cells via receptor-mediated endocytosis, is unable to establish infection owing to a dead-end pathway. After entry of the ISVP, the core containing double-stranded RNA genomes and the RNA-dependent RNA polymerase is released in the cytoplasm. Rotaviruses use negative-sense RNA strategy for transcription and replication. RNA-dependent RNA

ISVP is infectious not the whole virion

RNA-dependent RNA polymerase directs the synthesis of mRNA and genomic RNA by using negative-strand RNA of the double-stranded RNA genome

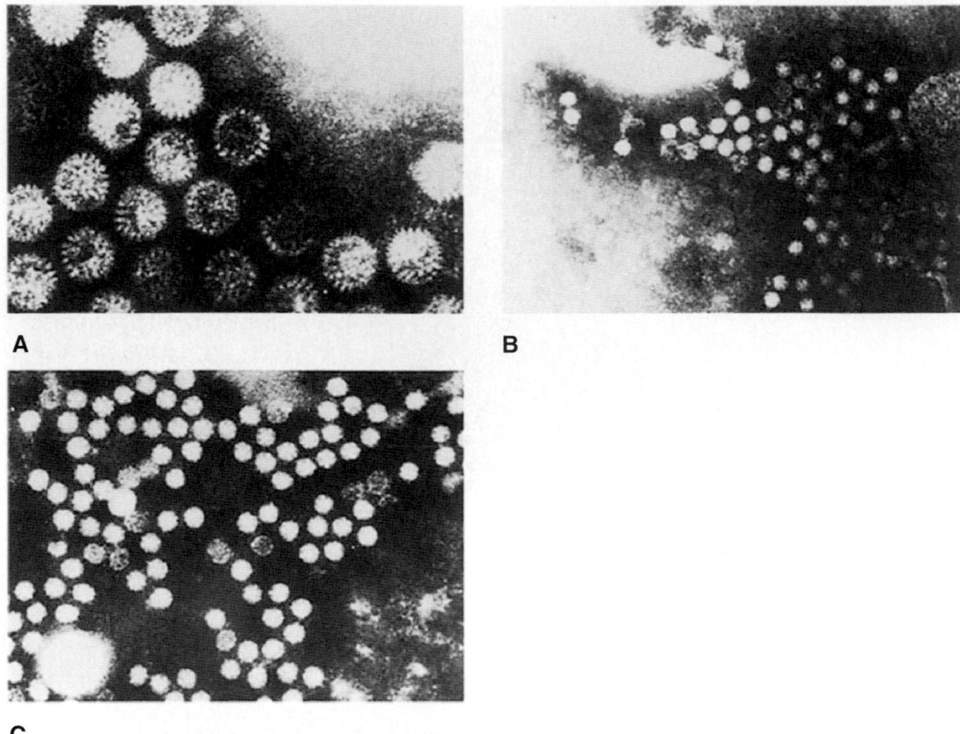

FIGURE 15–3. Viruses of diarrhea. All are photographed at the same magnification to illustrate the size and morphologic differences. **A.** Rotavirus. **B.** Calicivirus. **C.** Astrovirus. (Courtesy of Claire M. Payne.)

Virus assembly takes place at the endoplasmic reticulum

polymerase directs the synthesis of early and late mRNAs followed by genome replication by using the negative-strand RNA of the double-stranded RNA genome. Early proteins are made that are required for virus replication, whereas late proteins are mainly the structural proteins. Rotavirus assembles by associating its core with a nonstructural protein (NS28) and acquiring VP7 and a membrane on budding into the endoplasmic reticulum (ER). The virus eventually looses the membrane in the ER and is released upon cell lysis.

Animal rotaviruses produce diarrhea, but interspecies spread not demonstrated in nature

Rotaviruses of animal origin are also highly prevalent and produce acute gastrointestinal disease in a variety of species. Very young animals, such as calves, suckling mice, piglets, and foals, are particularly susceptible. The animal rotaviruses can often replicate in cell cultures, and infection across species lines has been accomplished experimentally; however, there is no evidence that such interspecies spread occurs in nature (eg, animal rotaviruses are not known to affect humans and vice versa).

Reassortment of the 11 RNA segments readily occurs

Live vaccines can incorporate genes from animal viruses

One unique feature of rotaviruses is the ease with which the 11 RNA segments can undergo reassortment. This has enabled the development of live vaccines that combine genes from readily cultivated animal rotaviruses with human rotavirus genes that encode serotype-specific capsid proteins. ::: reassortment, p. 133

HUMAN ROTAVIRUS INFECTIONS

CLINICAL CAPSULE

Worldwide, an estimated one million infants die each year as a result of rotavirus diarrhea. In the United States, the total annual deaths now are thought to be less than 100, but these viruses are still major causes of severe illness and hospitalization in early life. Vomiting, abdominal cramps, and low-grade fever, followed by watery stools that usually do not contain mucus, blood, or pus, are all characteristic of the acute phase of illness and can also be seen with infections due to caliciviruses, astroviruses, and adenoviruses.

EPIDEMIOLOGY

Outbreaks of rotavirus infection are common, particularly during the cooler months, among infants and children 1 to 24 months of age. Older children and adults can also be affected, but attack rates are usually much lower. Outbreaks among elderly, institutionalized patients have also been recognized.

Primarily infants and children in colder months

Although newborn infants can be readily infected with the virus, such infections often result in little or no clinical illness. This finding is illustrated by reported infection rates of 32% to 49% in some neonatal nurseries, but mild illness in only 8% to 28% of the infants. It is unclear whether this transient resistance to disease is a result of host maturation factors or transplacentally conferred immunity. Seroepidemiologic studies have been useful in demonstrating the ubiquity of these viruses and may help to explain the age-specific attack rates. By the age of 4 years, more than 90% of individuals have humoral antibodies, suggesting a high rate of virus infection early in life.

Most older children and adults are immune

PATHOGENESIS

Rotaviruses appear to localize primarily in the duodenum and proximal jejunum, causing destruction of villous epithelial cells with blunting (shortening) of villi and variable, usually mild, infiltrates of mononuclear and a few polymorphonuclear inflammatory cells within the villi. The gastric and colonic mucosa are unaffected; however, for unknown reasons, gastric emptying time is markedly delayed. The primary pathophysiologic effects are a decrease in absorptive surface in the small intestine and decreased production of brush border enzymes, such as the disaccharidases. The net result is a transient malabsorptive state, with defective handling of fats and sugars. It may take as long as 3 to 8 weeks to restore the normal histologic and functional integrity of the damaged mucosa. Although the specific gene product associated with virulence is not yet known, some evidence suggests that one nonstructural protein, NSP4, may behave as an enterotoxin in a manner similar to that of the heat-labile enterotoxin (LT) of *Escherichia coli* and cholera toxin. This may further explain the excess fluid and electrolyte secretion in the acute phase of illness. Viral excretion usually lasts 2 to 12 days but can be greatly prolonged in malnourished or immunodeficient patients with persistent symptoms.

Destroys villous cells of jejunum and duodenum

Absorptive surface is decreased

Enterotoxin-like effects are also present

IMMUNITY

Patients with rotavirus infection respond with production of type-specific humoral antibodies that appear to last for years, perhaps a lifetime. In addition, type-specific secretory IgA (sIgA) antibodies are produced in the intestinal tract, and their presence seems to correlate best with immunity to reinfection. Breastfeeding also seems to play a protective role against rotavirus disease in young infants. Secretory IgA antibodies to rotaviruses appear in colostrum and continue to be secreted in breast milk for several months postpartum. Human breast milk mucin glycoproteins have also been shown to bind to rotaviruses, inhibiting their replication in vitro and in vivo.

Type-specific humoral and secretory IgA antibodies are protective

IgA and mucin glycoproteins confer protective role of breastfeeding

MANIFESTATIONS

After an incubation period of 1 to 3 days, there is usually an abrupt onset of vomiting, followed within hours by frequent, copious, watery, brown stools. In severe cases, the stools may become clear; the Japanese refer to the disease as **hakuri,** the "white stool diarrhea." Fever, usually low grade, is often present. Vomiting may persist for 1 to 3 days, and diarrhea for 4 to 8 days. The major complications result from severe dehydration, occasionally associated with hypernatremia.

Severe dehydration can lead to death, particularly in very small or malnourished infants

Short incubation period, vomiting, and watery diarrhea can lead to dehydration

DIAGNOSIS

Electron microscope or EIA detect virus

Diagnosis of acute rotavirus infection is usually by detection of virus particles or antigen in the stools during the acute phase of illness. This can be accomplished by direct examination of the specimen by electron microscopy or, more conveniently, by immunologic detection of antigen with EIA methods.

TREATMENT AND PREVENTION

Live attenuated, or reassortant vaccines are available and recommended for infants

There is no specific treatment for rotavirus infection. Vigorous replacement of fluids and electrolytes is required in severe cases and can be life-saving. The rotaviruses are highly infectious and can spread quickly in family and institutional settings. Control consists of rigorous hygienic measures, including careful handwashing and adequate disposal of enteric excretions. Live attenuated or reassortant vaccines have been developed, as noted previously. In 2006, a live, oral bovine/human reassortant vaccine was licensed for routine use in the United States. To date, its efficacy after a three-dose series has been excellent, and no safety concerns have arisen. The first dose is administered at 6 to 12 weeks of age, with two subsequent doses at 4- to 10-week intervals. A second live, attenuated oral vaccine is also available in Mexico, Central and South America, and some European countries. One advantage is that it is administered as a two-dose series, rather than three.

CALICIVIRUSES

Although the caliciviruses were the first to be clearly associated with outbreaks of gastroenteritis, considerably less is known about their biology than about that of the rotaviruses. They were first associated with an outbreak in Norwalk, Ohio, in 1968, and their role was confirmed by production of disease in volunteers fed fecal filtrates. The original virus was thus called the **Norwalk agent,** and similar viruses have been given names such as Hawaii agent, Montgomery County agent, Ditchling agent, and so on.

VIROLOGY

Small, round, naked, icosahedral capsid RNA viruses are hardy

Two genera: *Norovirus* and *Sapovirus*

Several serotypes but not yet grown

The viruses are small, naked, icosahedral capsid, positive-sense RNA-containing particles 27 to 38 nm in diameter; their appearance is similar to that of the DNA-containing parvoviruses and hepatitis A virus (Figure 15–3B). They are classified as members of the Caliciviridae family. At present, two genera that cause diarrhea are recognized within this family: noroviruses (the family prototype) and sapoviruses. Norovirus particles are round, whereas other calicivirus particles are star-shaped. The viruses appear to be extremely hardy; their infectivity persists after exposure to acid, ether, and heat (60°C for 30 minutes). They have not been effectively propagated in cell or organ culture.

At least four different *Norovirus* serotypes have been demonstrated by immunoelectron microscopy with convalescent sera from affected patients. Knowledge of the antigenic characteristics and biology of these viruses has been seriously hampered by the current inability to grow them in the laboratory and by their lack of known pathogenicity for animals.

CALICIVIRUS INFECTIONS

EPIDEMIOLOGY

Sharp family and community outbreaks are common and can occur in any season. The noroviruses have been particularly a major issue in closed settings, such as cruise ships, hospitals, nursing homes, and schools. Unlike rotaviruses, caliciviruses are much more common causes of gastrointestinal illness in older children and adults. This difference in age-specific predilection is perhaps reflected in serosurveys, which have shown that the prevalence of antibodies rises slowly, reaching approximately 50% by the fifth decade of life, a striking contrast to the frequent acquisition of antibodies to rotaviruses early in life. Transmission is primarily fecal–oral; outbreaks have also been associated with consumption of contaminated water, uncooked shellfish, and other foods.

Sharp outbreaks include older children and adults

Transmission is by fecal–oral route

PATHOGENESIS

Both the pathogenesis and the pathology are similar to those described for rotaviruses, except that no enterotoxic features have yet been described for caliciviruses. The mucosal changes usually revert to normal within 2 weeks of onset of illness. Virus shedding in the feces generally lasts no more than 3 to 4 days.

Enterotoxic features are not present

IMMUNITY

Patients and experimentally infected volunteers respond to infection with the production of humoral antibodies, which persist indefinitely; their role in protection from reinfection, however, appears minimal. Reinfection and illness with the same serotype occur, and the role of local antibody has not been well defined. It is possible that nonimmune or genetic factors are essential for protection.

Reinfection can occur with same serotypes

CLINICAL ASPECTS

The incubation period is 10 to 51 hours, followed by abrupt onset of vomiting and diarrhea, a syndrome clinically indistinguishable from that caused by rotaviruses. Respiratory symptoms rarely coexist, and the duration of illness is relatively brief (usually 1 to 2 days). These viruses can be detected by electron microscopy or immunoelectron microscopy in stools during the acute phase of illness. In addition, EIA and PCR methods have been developed. As with rotavirus infection, there is no specific treatment other than fluid and electrolyte replacement. Prevention requires good hygienic measures.

Clinical picture and diagnostic tests are similar to those for rotavirus

No treatment or vaccine exists

ADENOVIRUSES, ASTROVIRUSES, AND "CANDIDATE" VIRUSES

Some adenoviruses, most of which are exceedingly difficult to cultivate in vitro (in contrast to those associated with respiratory diseases), are now recognized as significant intestinal pathogens. They may account for an estimated 5% to 15% of all viral gastroenteritis in young children. These include serotypes 40, 41, and perhaps 38.

Serotypes 40 and 41 are commonly found

Astroviruses have a shape that resembles a five- or six-pointed star (Figure 15–3C). These have been known since 1975. In recent years, astroviruses have been acknowledged as causes of often mild gastroenteritis outbreaks, primarily among toddlers, school children, and elderly nursing home residents. Seven human serotypes of astroviruses have been identified.

Illness is often, but not always, mild

Astroviruses are star-shaped 28 to 38 nm, naked capsid, positive-sense RNA viruses. The virions are spherical, and the shape and genome resembles with that of some calicivirus members. The genome of 6.8 to 7.9 nucleotides encodes a full length and a subgenomic RNA. Astroviruses are acid-stable, heat-resistant for a short period of time, and resistant to a range of detergents and lipid solvents. The replication cycle of the astroviruses is not characterized because of lack of a cell culture system.

Other agents associated with gastrointestinal diseases include coronavirus-like agents, toroviruses, and some group A coxsackieviruses (the latter primarily cause gastrointestinal symptoms in severely immunocompromised patients). This list may grow in the future; however, until more is learned about their biology, epidemiologic behavior, and impact on human health, they remain "candidate" viruses for now.

CASE STUDY

AN UNSCHEDULED TOUR STOP

This 20-year-old man was on a 3-week tour of Italy with 14 other college students. On the way to Florence, he abruptly became ill with nausea and vomiting, followed 5 hours later by abdominal cramps and watery diarrhea. No fever was noted.

QUESTIONS

■ Which of these viruses is the most likely cause of the man's illness?

A. Calicivirus
B. Rotavirus
C. Parvovirus
D. Adenovirus
E. Astrovirus

■ His illness might have been prevented by any of the following, *except*:

A. Avoidance of raw fruits
B. Live, reassortant vaccine
C. Careful handwashing
D. Avoidance of local drinking water
E. Avoidance of raw oysters

■ Infection by which of the following is localized to the duodenum and upper jejunum?

A. Rotavirus
B. Norovirus
C. Sapovirus
D. Astrovirus
E. Adenovirus

ANSWERS

1(A), **2**(B), **3**(A)

CHAPTER 16

Arthropod-Borne and Other Zoonotic Viruses

The zoonotic viruses comprise more than 400 agents, one or more of which occur in most parts of the world. Members of the group have their ultimate reservoirs in insects or lower vertebrates. They are from diverse taxonomic families of RNA viruses that primarily include the togaviruses, flaviviruses, bunyaviruses, reoviruses, arena-viruses, and filoviruses. Their major morphologic and genetic features are summarized in **Table 16–1**. The zoonotic viruses discussed here are divided into two groups. The arboviruses are transmitted to humans by infected blood-sucking insects such as mosquitoes, ticks, and *Phlebotomus* flies (sandflies). The other zoonotic RNA viruses are generally believed to be transmitted by inhalation of infected animal excretions, by the conjunctival route, or occasionally by direct contact with infected animals (nonarthropod zoon0tic viruses). Rabies virus, which is commonly transmitted by animal bites, is discussed separately in Chapter 17. Certain DNA viruses (poxviruses) are also transmissible from animals to humans, which are described in Chapter 11.

VIROLOGY

In most cases, the zoonotic viruses were first named after the place of initial isolation (eg, St. Louis encephalitis) or after the disease produced (eg, yellow fever). More recent studies have assigned the majority to families and genera on the basis of properties indicated in Table 16–1. The major characteristics of these families are summarized below.

Often named after place of initial isolation

TOGAVIRUSES

Togaviruses (*Alphavirus* genus includes arboviruses within this family) are enveloped virions containing single-stranded, positive-sense RNA measuring 70 nm in external diameter. The RNA genome is encapsidated in an icosahedral capsid that measures approximately 40 nm. The lipid bilayer envelope contains viral-encoded glycoproteins, E1 and E2. Alphaviruses have the ability to hemagglutinate via fusion of E1 glycoproten to lipids in erythrocyte membrane and E2 also participates in this process. The structure of an alphavirus virion is shown in **Figure 16–1**. Replication occurs in cells of infected arthropods and in vertebrate hosts. Virus enters via receptor-mediated endocytosis by interacting with a variety of cellular receptors, depending on the host and the cell type. It replicates like positive-sense RNA viruses by synthesizing RNA-dependent RNA polymerase and other nonstructural and structural proteins. Virions mature by budding from cellular membranes. The effect of viral replication on invertebrate and vertebrate hosts is variable, with usually a persistent infection in invertebrate hosts. The *Alphavirus* genus within these families include most arthropod-borne viruses. Viruses within this genus

Replicates in cells of arthropods and vertebrates

Enveloped RNA viruses contain glycoproteins that are hemagglutinin and lipoproteins

TABLE 16–1 Selected Arboviruses of Major Importance to Humans

GENUS AND MEMBER	MAJOR GEOGRAPHIC DISTRIBUTION	PRIMARY ARTHROPOD VECTOR	USUAL DISEASE EXPRESSION
Togaviruses			
Alphavirus			
Western equine encephalitis	North America	Mosquito	Encephalitis
Eastern equine encephalitis	North America	Mosquito	Encephalitis
Venezuelan equine encephalitis	Central and South America	Mosquito	Encephalitis
Chikungunya	Africa and Asia	Mosquito	Febrile illness
Flaviviruses			
Flavivirus			
St. Louis encephalitis	North America	Mosquito	Encephalitis
Dengue	All tropical zones	Mosquito	Febrile illness or hemorrhagic fever
Yellow fever	Africa, South America, and Caribbean	Mosquito	Hepatic necrosis, hemorrhage
West Nile fever	Africa, Eastern Europe, Middle East, Asia, North America	Mosquito	Febrile illness or encephalitis
Murray Valley encephalitis	Australia	Mosquito	Encephalitis
Russian spring–summer encephalitis	Eastern Former Soviet Union and Central Europe	Tick	Encephalitis
Powassan	Canada	Tick	Encephalitis
Japanese B encephalitis	Japan, Korea, and Philippines	Mosquito	Encephalitis
Bunyaviruses			
Bunyavirus			
California	North America	Mosquito	Encephalitis
Bunyamwera	Africa	Mosquito	Febrile illness
Rift Valley fever	Africa	Mosquito	Febrile illness
Sandfly fever	Mediterranean	*Phlebotomus*	Febrile illness
Reoviruses			
Orbivirus			
Colorado tick fever	North America	Tick	Febrile illness

are frequently serologically related to one another but not to others. Representatives are listed in Table 16–1. ::: RNA polymerase: P.119

FLAVIVIRUSES

Flaviviruses are similar to togaviruses in several respects. They are positive-sense, single-stranded RNA, icosahedral capsid, enveloped viruses. However, the virions of flaviviruses are smaller than those of togaviruses, ranging from 40 to 50 nm in diameter. The RNA genome is surrounded by multiple copies of small basic proteins; the capsid (C) protein and the matrix (M) proteins cover the core. The lipid bilayer envelope membrane contains envelope (E) protein, which is glycosylated in many flaviviruses. An example of a flavivirus virion is shown in **Figure 16–2**. The *Flavivirus* genus comprises most arboviruses within this family. *Flavivirus* members are serologically related, and there is cross-reactivity among members. The virus enters target cells via receptor-mediated endocytosis; flaviviruses can also bind to Fc receptors on macrophages, monocytes, and other cells coated with antibody.

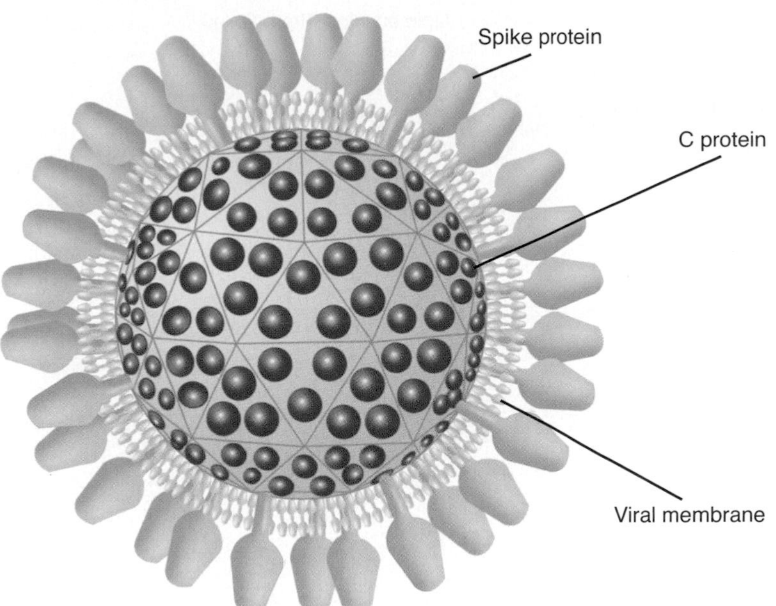

FIGURE 16–1. Virion structure of alphavirus.

The antibody, enhancing antibody, enhances viral adsorption and infectivity. The virus replicates like positive-sense RNA viruses, and the whole positive-sense RNA genome is translated into a polyprotein (like picornaviruses), which is cleaved into individual mature proteins, including a protease, an RNA-dependent RNA poymerase, a capsid and envelope proteins. Virus assembly takes place in the cytoplasm and the envelope is acquired by budding into intracellular vesicles and released upon cell lysis. Like alphaviruses, flaviviruses also cause a lytic response in vertebrate hosts and a persistent infection in invertebrate hosts.

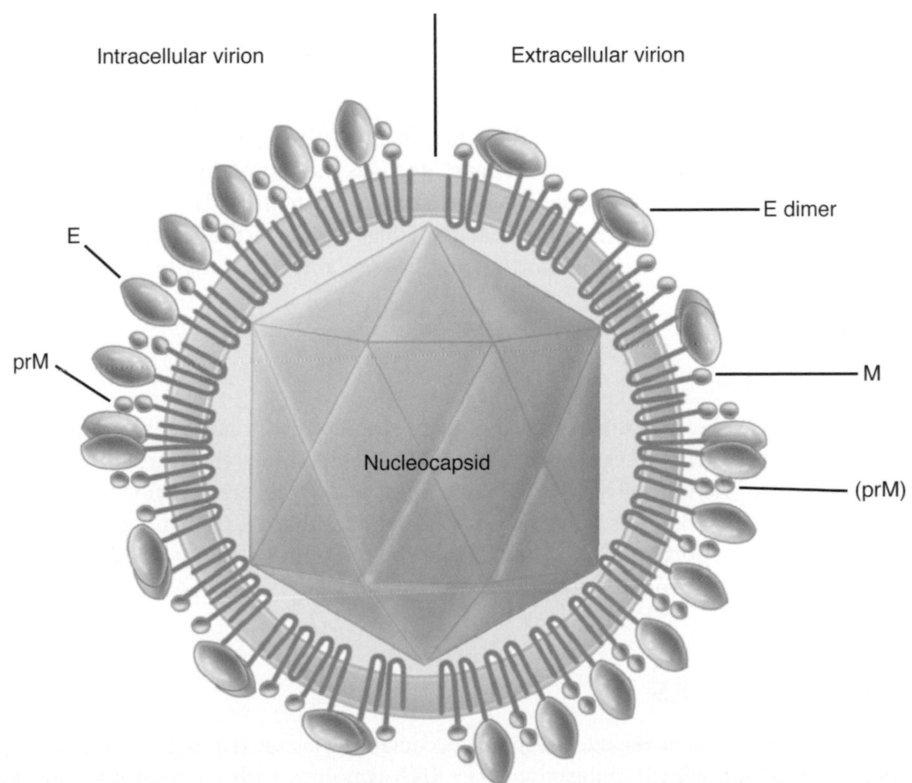

FIGURE 16–2. Virion structure of flavivirus. M, matrix; prM, precursor matrix protein.

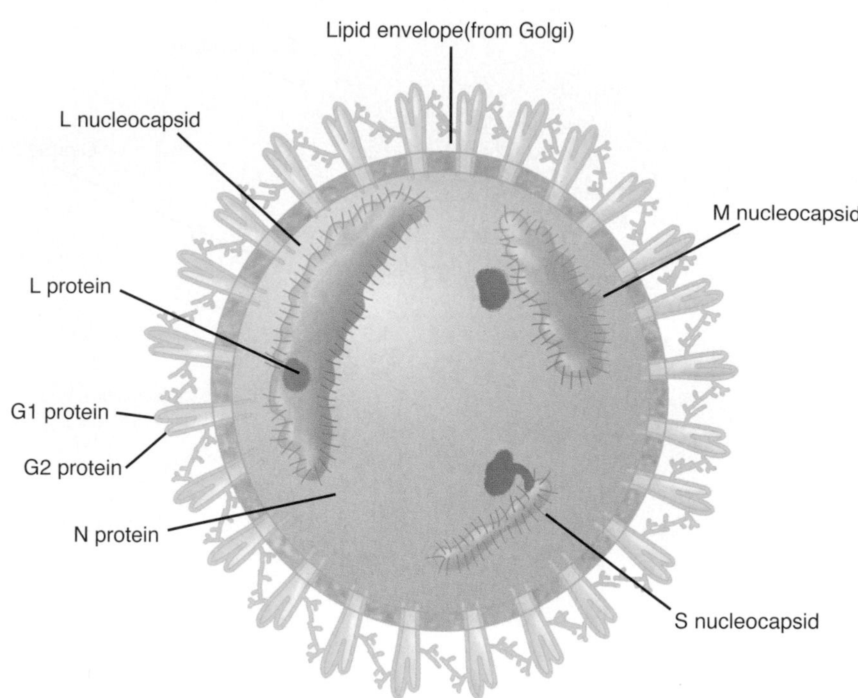

FIGURE 16–3. Bunyavirus virion structure.

BUNYAVIRUSES

Spherical, enveloped RNA viruses with three RNA-containing nucleo capsids

M segment encodes non structural proteins, and L and S segments encode polymerase and nonstructural proteins, respectively

There are four genera of Bunyaviridae family: *Bunavirus* (-) RNA, *Phlebovirus* (-) RNA, *Nairovirus* (+/-) ambisense RNA, and *Hantavirus* (-) RNA. All bunyaviruses are arboviruses, except hantavirus, which is a nonarthropod zoontic virus. Bunyaviruses are spherical, enveloped, single-stranded, negative-sense RNA viruses approximately 90 to 100 nm in external diameter. The envelope contains two glycoproteins, G1 and G2, and encloses three nucleocapsids containing RNA, namely, large (L), medium (M), and small (S) associated with an RNA-dependent RNA polymerase (L) and nonstructural proteins (N) (**Figure 16–3**). Unlike enveloped RNA viruses, bunyaviruses are devoid of a matrix protein. The viral attachment protein (G1) interacts with cellular receptors, and the virus enters the cell via receptor-mediated endocytosis. After lysis of endosomal vesicles and release of the nucleocapsids in the cytoplasm, the negative RNA strands (L, M, S) transcribes to synthesize mRNA using virion-associated RNA-dependent RNA polymerase. The M strand encodes G1 and G2 envelope, a nonstructural protein, L strand, encodes the L protein (RNA-dependent RNA polymerase), and the S strand encodes nonstructural proteins. They mature by budding into smooth-surfaced vesicles in or near the Golgi region of the infected cell. The major disease-causing bunyaviruses in North America are California virus and hantavirus.

REOVIRUSES

Unenveloped RNA viruses are prominent in North America

Reoviruses are spherical, unenveloped, double-stranded RNA viruses that measure about 80 nm in diameter with a segmented genome. The details about virus structure and replication of another member of the Reoviridae family, *Rotavirus*, are described in Chapter 15. However, the reoviruses described here are arboviruses that are transmitted through insect bites. The most important North American arbovirus of this family, which is a member of the genus *Coltivirus,* causes Colorado tick fever.

ARENAVIRUSES

The arenaviruses are enveloped, bisegmented, containing a large (L) single-stranded, negative-sense (–) and a small (S) ambisense (–/+) RNA genomes with pleomorphic morphology

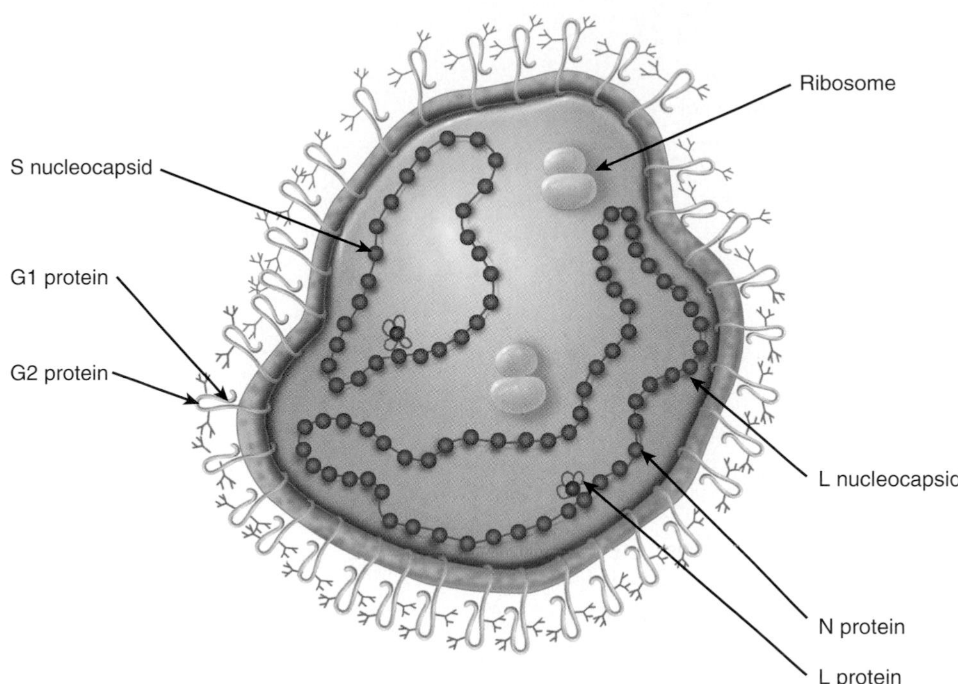

FIGURE 16–4. Virion structure of arenavirus..

ranging in size from 50 to 300 (mean 110 to 130) nm in diameter (**Figure 16–4**). There are two separate nucelocapsids, L and S, encapsidating L and S RNA segments, respectively, and the envelope contains two viral surface glycoproteins, G1 and G2. The virion contains host cell ribosomes in their interior. These ribosomes confer a granular appearance to the viruses; hence their name (from the Latin *arenosus* for "sandy"). The most significant arenavirus infections in humans are the hemorrhagic fevers, including Lassa fever. The virus of lymphocytic choriomeningitis is occasionally transmitted to humans from infected mice and other rodents.

Pleomorphic, bisegmented, enveloped RNA viruses containing host cell ribosomes

Arenaviruses replicate in the cytoplasm of the infected host's cell using the strategy of negative-sense RNA genomes. Viral attachment protein G1 interacts with a cell surface receptor (αDG), and the virions are internalized in vesicles. Viral fusion protein G2 mediates fusion, resulting in the release of nucleocapsids. Virion-associated RNA-dependent-RNA polymerase (L protein, Figure 16–4) mediates transcription, and the L RNA segment encodes the polymerase (L) protein and a Z protein, which may help the virus in assembly and release. The S RNA segment, which has ambisense (–/+) polarity, encodes nucleocapsid (N) proteins and envelope glycoproteins G1 and G2, using a negative-sense RNA strategy for transcription. The ambisense RNA strategy allows arenaviruses to regulate their gene expression, first encoding the N and later the G proteins. Like bunyaviruses, arenaviruses also lack a matrix protein, a characteristic of enveloped viruses. They mature by budding from the host cell plasma membrane. Arenaviruses cause persistent infection in rodents and are also transmitted to humans from the excreta of infected rodents.

FILOVIRUSES

Filoviruses, Marburg and Ebola viruses, are the cause of Marburg and Ebola fevers, two highly fatal hemorrhagic fevers. Although no subtypes of Marburg virus has been found, Ebola virus exits as three subtypes, including Zaire, Sudan, and Reston. Filoviruses are enveloped, single-stranded, negative-sense RNA viruses with

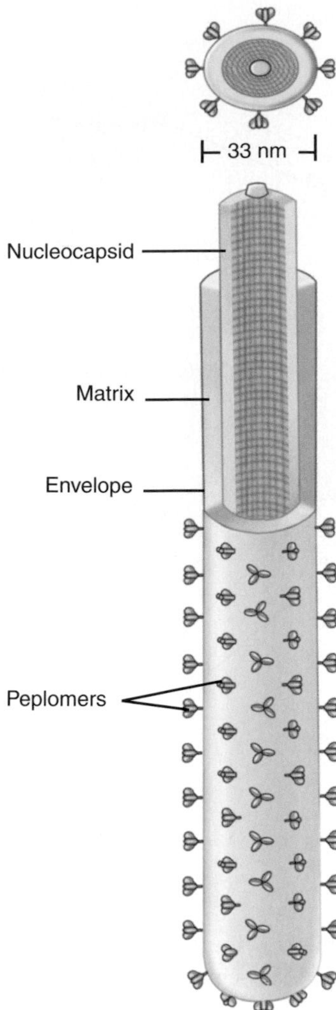

FIGURE 16–5. Morphology of filovirus virion.

Enveloped filamentous RNA viruses cause hemorrhagic fevers

filamentous and highly pleomorphic virions, averaging 80 nm in diameter and 300 to 14,000 nm in length (**Figure 16–5**). There are seven viral genes that are sequentially arranged on a 19-kb RNA genome. The nucleocapsid (NP) has a helical symmetry, and the envelope is derived from plasma membrane as result of budding. The envelope contains 10 nm peplomers or spikes, the GP glycoprotein, which mediate virus entry into susceptible cells.

Viral GP surface protein mediates virus entry into target cells. RNA-dependent RNA polymerase directs the synthesis of mRNA from a linear negative-sense RNA genome, like other negative-sense RNA viruses (rhabdoviruses, paramyxoviruses). Seven monocistronic mRNAs are generated followed by translation of viral proteins. Translation of nuclecapsid protein (NP) triggers the switch from transcription to genome replication. The NP binds to the RNA genome to form the nucleocapsid, which is enclosed in a matrix protein and buds from plasma membrane containing viral GPs.

HENIPAVIRUSES

Two zoonotic paramyxoviruses involving humans and animals appeared in Australia and Southeast Asia during the late 1990s. These are Hendra and Nipah viruses, now classifed in the Henipavirus genus of the Paramyxoviridae family.

ARBOVIRUS DISEASE

CLINICAL CAPSULE

Some arboviruses cause severe inflammation of the brain (encephalitis) with damage or destruction of neural cells that may be fatal or lead to permanent neurologic damage in survivors. Others, such as dengue viruses, can produce illnesses that range from mild flu-like symptoms to overwhelming shock with widespread hemorrhage into tissues. Still another, yellow fever virus, primarily attacks liver cells, leading to extensive destruction and sometimes fatal liver failure.

EPIDEMIOLOGY

Arboviruses of major importance in human disease are listed in Table 16–1 with summaries of their geographic distribution, the arthropod vectors that transmit them, and the usual disease syndromes that can result from infection.

With the exception of urban dengue and urban yellow fever, in which the virus may simply be transmitted between humans and mosquitoes, other arboviral diseases involve nonhuman vertebrates. These are usually small mammals, birds, or, in the case of jungle yellow fever, monkeys. Infection is transmitted within the host species by arthropods (eg, mosquitoes or ticks) that become infected. In some cases, the infection can be maintained from generation to generation in the arthropod by transovarial transmission. Infection in the arthropod usually does not appear to harm the insect; however, a period of virus multiplication (termed **extrinsic incubation period**) is required to enhance the capacity to transmit infection to vertebrates by bite.

Sometimes maintained by vertical transmission in vector

Multiplication in vector is required

The consequences of infection transmitted from the arthropod to susceptible vertebrate hosts are variable; some develop illness of varying severity with viremia, whereas others have long-term viremia without clinical disease. Vertebrate hosts are then a source of further spread of the virus by amplification, in which noninfected arthropods feeding on viremic hosts acquire the virus, thereby increasing the risk of transmission. The general features of this overall transmission cycle are illustrated in **Figure 16–6**.

Transient viremia is a feature of many of these infections in hosts other than their reservoir; those affected, including humans and higher vertebrates (eg, horses and cattle), are often referred to as blind-end hosts. In contrast, if viremia is sustained for longer periods (eg, weeks to months in a variety of togavirus, flavivirus, and bunyavirus infections of lower vertebrates), the vertebrate host becomes highly important as a reservoir for continuing transmission. Viremia may last a week or more in human dengue and yellow fever infections, and humans may then serve as a reservoir in urban disease.

Sustained viremia required for vertebrate host to be significant reservoir

Obviously, the typical arthropod vectors are rarely present during all seasons. The question then arises as to how the arboviruses survive between the time the vector disappears and the time it reappears in subsequent years. Several mechanisms can operate to sustain the virus between transmission periods (often referred to as **overwintering**): (1) sustained viremia in lower vertebrates such as small mammals, birds, and snakes, from which newly mature arthropods can be infected when taking a blood meal; (2) hibernation of infected adult arthropods that survive from one season to the next; and (3) transovarial transmission, whereby the infected female arthropod can transmit virus to its progeny.

Season-to-season survival has multiple mechanisms

The three basic specific cycles of arbovirus transmission are urban, sylvatic, and arthropod-sustained.

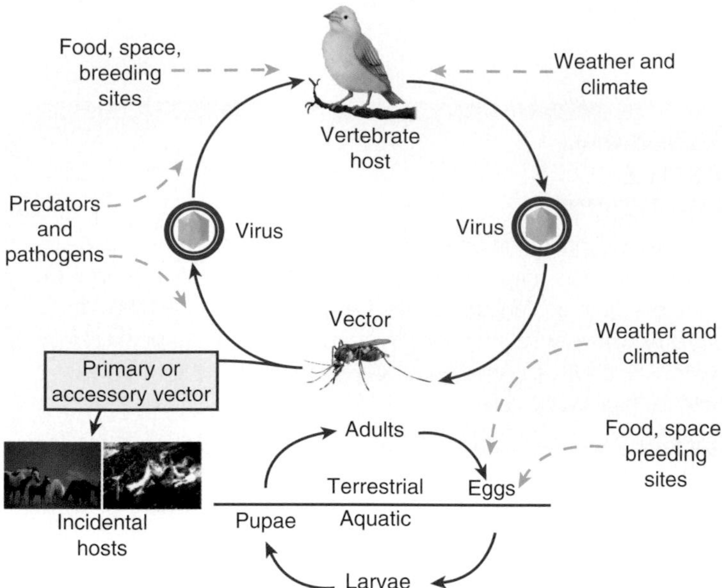

FIGURE 16–6. General features of arbovirus transmission cycles. (Reprinted from the Centers for Disease Control and Prevention.)

Urban

Urban cycle exists with dengue and yellow fever

As the term suggests, the urban cycle is favored by the presence of relatively large numbers of humans living in close proximity to arthropod (usually mosquito) species capable of virus transmission. The cycle is:

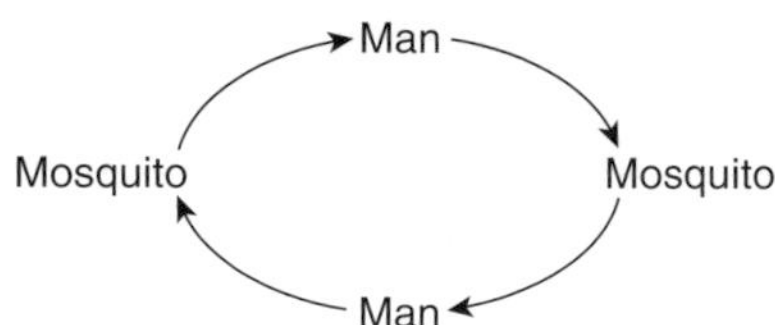

Examples of the urban cycle include urban dengue, urban yellow fever, and occasional urban outbreaks of St. Louis encephalitis.

Sylvatic

In the sylvatic cycle, a single nonhuman vertebrate reservoir may be involved.

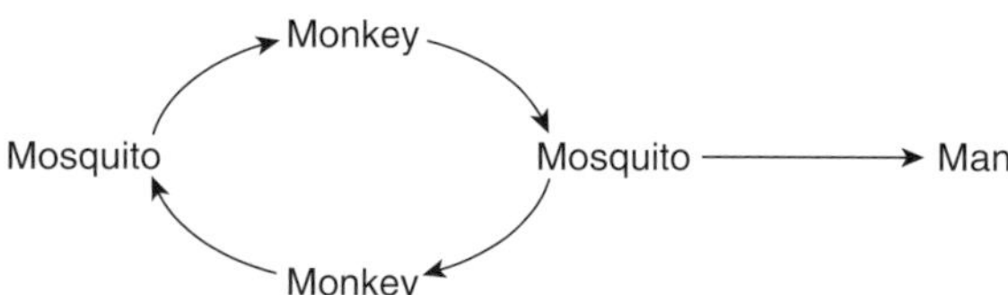

In this situation, the human, who becomes a tangential host through accidental intrusion into a zoonotic transmission cycle, is not important in maintaining the infection cycle. An example of this cycle is jungle yellow fever.

In other sylvatic cycles, multiple vertebrate reservoirs may be involved:

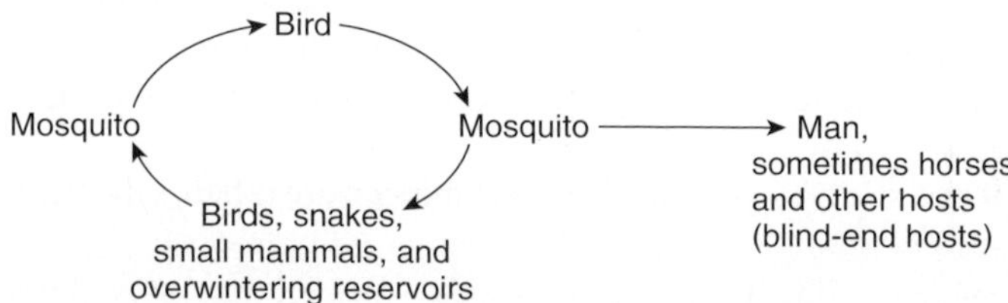

Examples include western equine encephalitis, eastern equine encephalitis, and California viruses. In some situations, such as St. Louis encephalitis and yellow fever, the urban and sylvatic cycles may operate concurrently.

Sylvatic cycle occurs with many viruses

Humans are tangential hosts

Arthropod-Sustained

Arthropods, especially ticks, may sustain the reservoir by transovarial transmission of virus to their progeny, with amplification of the cycle by spread to and from small mammals:

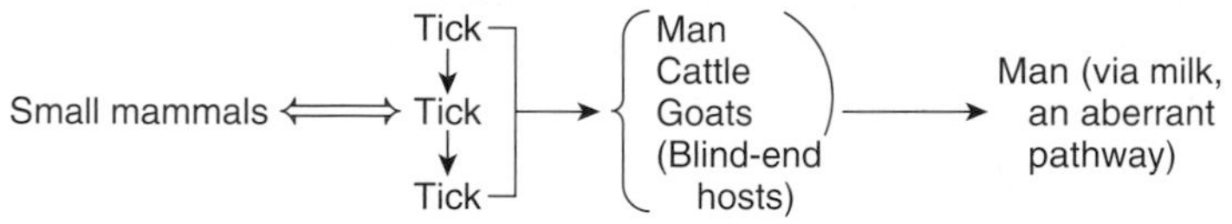

Small mammals ⟺ Tick → Tick → Tick → {Man, Cattle, Goats (Blind-end hosts)} → Man (via milk, an aberrant pathway)

Tick-borne encephalitis in Russia is transmitted by the arthropod-sustained cycle. In temperate climates such as the United States, arboviruses are major causes of disease during the summer and early fall months, the seasons of greatest activity of arthropod vectors (usually mosquitoes or ticks). When climatic conditions and ecologic circumstances (eg, swamps and ponds) are optimal for arthropod breeding and egg hatching, arbovirus amplification may begin.

Arthropod-sustained by tick transovarial transmission

Weather, swamps, and ponds alter conditions

An example of amplification is provided by western equine encephalitis. When the mosquito vectors become abundant, the level of transmission among the basic reservoir hosts (birds and small mammals) increases, and the mosquitoes also turn to other susceptible species such as the domestic fowl. These hosts experience a rapidly developing asymptomatic viremia, which permits still more arthropods to become infected on biting. At this point, spread to blind-end hosts such as humans or horses and the development of clinical disease become likely. This occurrence depends on the accessibility of the host to the infected mosquito and on mosquito feeding preferences which, for unknown reasons, vary from one season to another.

Mosquito increases create risk for blind-end human infection

PATHOGENESIS

There are three major manifestations of arbovirus diseases in humans associated with different tropisms of various viruses for human organs, although overlap can occur. In some, the central nervous system (CNS) is primarily affected, leading to aseptic meningitis or meningoencephalitis. A second syndrome involves many major organ systems, with particular damage to the liver, as in yellow fever. The third syndrome is manifested by hemorrhagic fever, in which damage is particularly severe to the small blood vessels, with skin petechiae and intestinal and other hemorrhages.

CNS, visceral, and hemorrhagic fever are major syndromes

Infection of the human by a biting, infected arthropod is followed by viremia, which is apparently amplified by extensive virus replication in the reticuloendothelial system and vascular endothelium. After replication, the virus becomes localized in various target organs, depending on its tropism, and illness results. The viruses produce cell necrosis with resultant inflammation, which leads to fever in nearly all infections. If the major viral tropism is for the CNS, virus reaching this site by crossing the blood–brain barrier or along neural pathways can cause meningeal inflammation (aseptic meningitis) or neuronal dysfunction (encephalitis). The CNS pathology consists of meningeal and perivascular mononuclear cell infiltrates, degeneration of neurons with neuronophagia, and occasionally destruction of the supporting structure of neurons.

After bite, viremia and viral tissue tropism define disease

In CNS, aseptic meningitis and encephalitis follow cell injury

In some infections, especially yellow fever, the liver is the primary target organ. Pathologic findings include hyaline necrosis of hepatocytes, which produces cytoplasmic eosinophilic masses called **Councilman bodies.** Degenerative changes in the renal tubules and myocardium may also be seen, as may microscopic hemorrhages throughout the brain. Hemorrhage is a major feature of yellow fever, largely because of the lack of liver-produced clotting factors as a result of liver necrosis.

Liver often the target, with necrosis of hepatocytes

Hemorrhagic fevers other than those related to primary hepatic destruction have a somewhat different pathogenesis, which has been studied most extensively in dengue infections. In uncomplicated dengue fever, which is associated with a rash and influenza-like symptoms, there are changes in the small dermal blood vessels. These alterations include endothelial cell swelling and perivascular edema with mononuclear cell infiltration. More severe infection, as

Dengue hemorrhagic fevers involve perivascular and endothelial injury

May progress to shock

Lymphoid hyperplasia seen

Virus–antibody complexes may trigger complement activation

in dengue hemorrhagic fever, often complicated by shock, is characterized by perivascular edema and widespread effusions into serous cavities such as the pleura and by hemorrhages. The spleen and lymph nodes show hyperplasia of lymphoid and plasma cell elements, and there is focal necrosis in the liver. The pathophysiology seems related to increased vascular permeability and disseminated intravascular coagulation, which is further complicated by liver and bone marrow dysfunction (eg, decreased platelet production and decreased production of liver-dependent clotting factors). The major vascular abnormalities may be provoked by circulating virus–antibody complexes (immune complexes), which mediate activation of complement and subsequent release of vasoactive amines. The precise reason for this phenomenon is not clear; it may be related to intrinsic virulence of the virus strains involved and to host susceptibility factors.

Cross-reacting antibodies may enhance infection

Two hypotheses are based on the existence of four distinct but antigenically related serotypes of dengue virus, any of which can generate group-specific cross-reacting antibodies that are not necessarily protective against other serotypes. One possibility is that preexisting group-specific antibody at a critical concentration serves as "enhancing" rather than neutralizing antibody. In the presence of enhancing antibody, virus–antibody complexes are more efficiently adsorbed to and engulfed by monocytes and macrophages. Subsequent replication leads to extensive spread throughout the host. Alternatively, or in concert with this, activation of previously sensitized T cells by viral antigen present on the surfaces of macrophages may result in release of cytokines, which mediate the development of shock and hemorrhage. ::: cytokines, p. 28

IMMUNITY

Neutralizing antibodies protective and last for years

Immunity is serotype-specific

The usual humoral responses (hemagglutination inhibition, IgM, neutralization,) in relation to onset of illness are illustrated in **Figure 16–7**. The rise in antibody titer generally correlates with recovery from infection. Neutralizing antibodies, which are the most serotype-specific, generally persist many years after infection. The presence of IgM-specific antibodies indicates that primary infection likely occurred within the previous 2 months. Cellular immunity and humoral immunity to reinfection are serotype-specific and appear to be permanent.

SPECIFIC ARBOVIRUSES

Western Equine Encephalitis

Human and equine illness

Encephalitis is more likely in young infants

The agent that causes western equine encephalitis is prevalent in the central valley of California, eastern Washington (Yakima valley), Colorado, and Texas. It has also been responsible for outbreaks in midwestern states (Minnesota, Wisconsin, Illinois, Missouri, and Kansas) and as far east as New Jersey. Horses and humans represent blind-end hosts; both are susceptible to infection and illness, commonly manifested as encephalitis. Although human infection in endemic areas is commonplace, overall only 1 of 1000 infections causes clinical

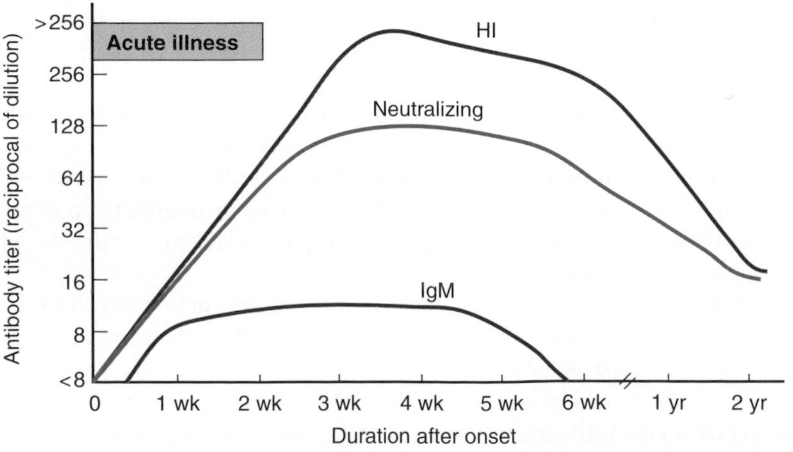

FIGURE 16–7. Typical patterns of antibody response after arbovirus infection. These patterns begin to appear about 3 days after onset and disappear after about 6 weeks. HI, hemagglutination inhibition antibodies; IgM, immunoglobulin M antibodies.

symptoms. However, in young infants, 1 of every 25 infections may produce severe illness. The attack rates are therefore far higher in young infants than in other groups. The disease spectrum may range from mild, nonspecific febrile illness to aseptic meningitis or severe, overwhelming encephalitis. Mortality rate is estimated at 5% for cases of encephalitis. It is a very serious disease in infants less than 1 year of age; as many as 60% of survivors have permanent neurologic impairment.

Eastern Equine Encephalitis

The eastern equine encephalitis virus is largely confined to the Atlantic Seaboard states from New England down the coasts of Central America and South America. The mosquito vector (principally *Culiseta melanura*) generally restricts its feeding to horses and birds, although occasional outbreaks among humans have occurred. The virus can cause severe encephalitis in horses and also in wild birds. The mortality rate for eastern equine encephalitis among humans is estimated at 50% for individuals of all ages, and the incidence of severe sequelae among survivors is high.

New England to South America

Vector feeds on horses and birds

St. Louis Encephalitis

The St. Louis encephalitis virus is a major cause of arbovirus encephalitis in the United States. Its geographic distribution and major mosquito vector (*Culex tarsalis*) are similar to those of western equine encephalitis, but St. Louis encephalitis has been much more prevalent in eastern states and in Texas, Mississippi, and Florida. It infects but causes no disease in horses. The disease spectrum in humans is similar to that of western equine encephalitis, but the major morbidity and mortality, as well as the highest attack rates, are among adults more than 40 years of age. Infants and young children are relatively spared.

Distribution and disease are similar to those of western equine encephalitis

More disease seen in adults

West Nile Virus

During the summer of 1999 in the northeastern United States, human West Nile virus infections appeared for the first time in the Western Hemisphere. A subsequent outbreak occurred again in 2000. Together, these outbreaks resulted in 78 hospitalized patients and 9 deaths, mostly among the elderly. More widespread activity was observed in 2001 (66 human cases); then in 2002, a dramatic increase in virus spread was seen across the United States, with activity in 46 states and four Canadian provinces. Now, the virus has been detected in all states in the continental United States (**Figure 16–8**). It is currently the most widespread arbovirus in North America. Before 1999, outbreaks of human West Nile virus infections were primarily confined to eastern Africa, the Middle East, eastern Europe, west Asia, and Australia.

First appeared in United States in 1999

Most important arbovirus in North America

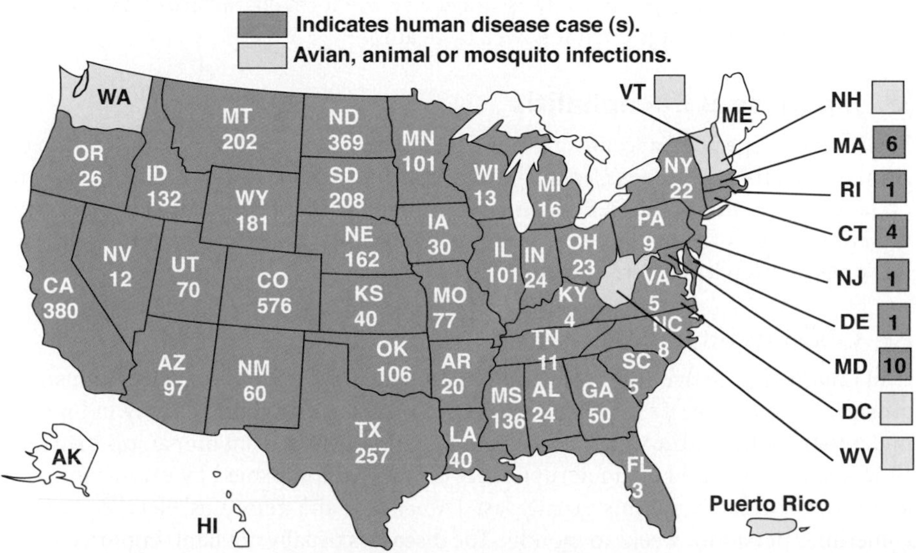

FIGURE 16–8. West Nile virus activity in the United States. (Modified from the Centers for Disease Control and Prevention.)

West Nile virus is antigenically related to St. Louis encephalitis and Japanese encephalitis. Transmission is from infected mosquitoes to birds, humans, and horses, and clinical illness leading to death can result from infections in any of these hosts. Transmission among humans via blood transfusions, breast milk, or organ transplants is also possible. Crows are particularly affected; virus has been detected in dead crows found as far south as Florida, and more recently in the midwestern United States. Clinical illness in the United States has often included muscle weakness and flaccid paralysis, suggesting an axonal polyneuropathy in addition to encephalitis. The paralysis can be asymmetric and permanent.

Dead crows often herald spread of virus in nature

Muscle weakness and flaccid paralysis can occur

California Virus

Although California virus was first isolated in that state, its major distribution in the United States has been in the Midwest; outbreaks due to the La Crosse subtype are particularly prevalent in Wisconsin, Ohio, Minnesota, Indiana, and West Virginia. In Wisconsin and Minnesota, California virus is considered an important cause of encephalitis. However, studies elsewhere in North America and throughout the world, indicate that California virus or closely related agents are present nearly everywhere. The primary mosquito vector (*Aedes triseriatus*) is commonly encountered in suburban or rural environments. The reservoir host is the chipmunk; transovarial transmission by mosquitoes to their larvae also serves to sustain the virus in nature. Unlike western equine, eastern equine, and St. Louis encephalitis viruses, the highest attack rates of California virus are seen in those aged 5 to 18 years. Infection is often characterized by abrupt onset of encephalitis, frequently with seizures.

Virus and vector common in suburban and rural areas

Chipmunk is reservoir host

Highest attack rate in those aged 5 to 18 years

Yellow Fever

Geographically, yellow fever is distributed throughout the Caribbean and Central America, the Amazon valley in South America, and a broad central zone in Africa from the Atlantic Coast to the Sudan and Ethiopia. It continues to be a potential threat to the southeastern United States because of an urban vector (*Aedes aegypti*) in that area. The clinical disease is characterized by abrupt onset of fever, chills, headache, and hemorrhage. It may progress to severe vomiting (sometimes with gastric hemorrhage), bradycardia, jaundice, and shock. If the patient recovers from the acute episode, there are no long-term sequelae.

Widespread in tropical areas

Vector persists in United States

Dengue

There are four related serotypes of dengue, any of which may exist concurrently in a given endemic area. These agents are widespread throughout the world, particularly in the Middle East, Africa, the Far East, and the Caribbean Islands, and they have invaded the United States in the past. The vector (*A aegypti*) is the same as the domestic vector of yellow fever. The known transmission cycle is human–mosquito–human, although a sylvatic cycle involving monkeys may also exist.

Vector same as yellow fever

The characteristic clinical illness usually results in fever, an erythematous rash, and severe pain in the back, head, muscles, and joints. Especially in the Far East (Philippines, Thailand, and India), dengue has periodically assumed a severe form characterized by shock, pleural effusion, and hemorrhage often followed by death.

Severe pain in back, muscles, and joints

Japanese B Encephalitis

The flavivirus species that causes Japanese B encephalitis is prevalent on the eastern coast of Asia, on its offshore islands (Japan, Taiwan, and Indonesia), and in India. Its transmission cycle resembles that of the St. Louis encephalitis and western equine encephalitis viruses. A high proportion of human infections are subclinical, especially in children; when encephalitis does develop it is severe and often fatal.

Transmission is similar to St. Louis and western equine encephalitis

Chikungunya Fever

Chikungunya (a native term for "that which bends up") is an alphavirus transmitted by mosquitoes, particularly in urban areas of Asia, Africa, and most recently in limited areas of Southern Europe. The virus may be maintained in a sylvatic subhuman primate reservoir. Illness is characterized by an abrupt onset of fever, accompanied by excruciating myalgia and polyarthritis. Symptoms usually last 1 week, but the musculoskeletal complaints can sometimes persist for weeks to months. The disease is usually not fatal. Imported cases have

Major problem in Asia and Africa

Risk to tourists traveling in endemic areas

been diagnosed in the United States, but there is no evidence that the virus has established itself in North America.

Powassan Virus

Powassan virus is the only known tick-borne *Flavivirus* species of North America. First isolated in Ontario from a fatal human case of encephalitis, it has been found in infected ticks in Ontario, British Columbia, and Colorado. Its significance to humans is not yet established; only a few patients have been described as having encephalitis proved to be caused by this agent. However, serologic evidence suggests that the virus is prevalent in many areas of North America.

Tick-borne but uncertain human importance

Colorado Tick Fever

The tick-borne *Orbivirus* species that causes Colorado tick fever has been found throughout the western United States, including Washington, Oregon, Colorado, and Idaho, and also Long Island. It is frequently found in *Dermacentor andersoni,* which are also vectors for *Rickettsia rickettsii.* The typical illness, which occurs 3 to 6 days after the tick bite, is characterized by a sudden onset with headache, muscle pains, fever, and occasionally encephalitis. Leukopenia is a consistent feature of infection. It is estimated that no more than one clinical illness occurs for every 100 infections with this agent.

Tick-borne throughout western United States

Most infections asymptomatic

CLINICAL ASPECTS

DIAGNOSIS

The arboviruses may be isolated in various culture systems including intracerebral inoculation of newborn mice, which often results in encephalitis and death. The viruses may be found in the blood (viremia) from a few days before onset of symptoms through the first 1 to 2 days of illness. Attempts at isolation from the blood are generally useful only when viremia is prolonged, as in dengue, Colorado tick fever, and some of the hemorrhagic fevers. Virus is not present in the stool and is rarely found in the throat; viral recovery from cerebrospinal fluid (CSF) is also unusual. Virus can be detected in CSF or affected tissue by reverse-transcriptase polymerase chain reaction, and sometimes by culture during the acute phase of illness. Specific diagnosis is usually accomplished by serologic techniques using acute and convalescent sera. Various tests have been used including hemagglutination inhibition, virus neutralization methods, and enzyme immunoassay. Early rapid presumptive diagnosis can sometimes be made by the detection of IgM-specific antibodies that often appear within a few days of onset (except in Colorado tick fever, where they may be delayed by 1 to 2 weeks) and may persist 1 to 2 months.

Blood is best source but must be early in disease

Multiple serologic methods used

TREATMENT AND PREVENTION

There is generally no specific treatment for arboviral infections other than supportive care; ribavirin has been used on occasion, but controlled studies have not been reported to support or refute its effectiveness. Prevention is primarily avoidance of contact with potentially infected arthropods, a task that can be extremely difficult even with the use of adequate screening and insect repellents. In some settings, vector control can be accomplished by elimination of arthropod breeding sites (stagnant pools and the like) and sometimes by attempts to eradicate the arthropods with careful use of insecticides. Such measures have been highly effective in the control of urban yellow fever, in which elimination of urban breeding sites and other measures to eradicate the principal mosquito vector species (*A. aegypti*) have been used. Viruses maintained in complex sylvatic cycles are infinitely more difficult to control without risking major environmental disruption and inestimable expense.

Treatment is supportive only

Protection from bites and vector control are primary prevention

Vaccines are available for immunization of horses against western, eastern, and Venezuelan equine encephalitis virus infections, and the latter has also been used for some laboratory personnel who work with the virus. Another arbovirus vaccine in general use for humans is a live attenuated yellow fever virus vaccine (17-D strain), which is used to protect rural populations exposed to the sylvatic cycle and international travelers to endemic areas. In fact, many countries in tropical Africa, Asia, and South America require proof of yellow fever vaccination before allowing travelers to enter. There is also a vaccine for human tick-borne encephalitis, which is endemic in areas of Western Europe; live, attenuated, and inactivated Japanese B encephalitis vaccines are widely used in endemic areas of eastern Asia and adjacent southern Pacific countries.

Yellow fever, tick-borne and Japanese B encephalitis vaccines are available

OTHER RNA VIRUSES OF ZOONOTIC ORIGIN

ARENAVIRUSES

A common feature of the arenaviruses is their zoonotic reservoir, particularly small rodents, in which they may be sustained for long periods. Primary infection (horizontal transmission) in mature rodents often results in disease and death, whereas intrauterine or perinatal infection (vertical transmission) usually leads to chronic lifelong viremia with persistent shedding of virus into the feces, urine, and respiratory secretions. Although chronically infected rodents are somewhat tolerant to the virus (ie, infection is persistent without causing illness), they produce antibodies, and evidence of deleterious effects can be found in older hosts, usually in the form of immune complex glomerulonephritis. The viruses are perpetuated by vertical transmission from infected mothers to their offspring. When environmental contact becomes close, spread from the rodent reservoir to humans (and, in some instances, subhuman primates) can occur via aerosols; through exposure to infective urine, feces, or tissues; or directly by rodent bites. This is in contrast to the arthropod spread of arboviruses. ::: persistent infection, p.146

Sustained in small rodent reservoirs

Vertical transmission in rodents

Spread to humans by aerosols and close contact

Arenaviruses Associated with Hemorrhagic Fevers

The agents of arenavirus hemorrhagic fevers are transmitted from infected rodents to humans in the manner just described, although person-to-person spread by contact with secretions and body fluids also occurs readily. The viruses in this group include the South American hemorrhagic fever agents (Junin virus, the cause of Argentinean hemorrhagic fever, and Machupo virus, the cause of Bolivian hemorrhagic fever) and Lassa virus, the cause of **Lassa fever** in West Africa.

Person-to-person spread occurs by contact with body fluids

Arenaviruses have pathogenic and pathologic features similar to those described for the arboviruses that cause hemorrhagic fevers; however, the mechanism involved in the coagulation abnormalities is not understood. All are characterized by fever, usually accompanied by hemorrhagic manifestations, shock, neurologic disturbances, and bradycardia. Lassa fever also frequently causes hepatitis, myocarditis, exudative pharyngitis, and acute deafness. The last deficit may persist after recovery. Mortality rate is estimated to be 10% to 50% for Lassa fever and 5% to 30% for the other viruses. All are considered highly dangerous in terms of infectivity. Importation of cases to nonendemic areas has occurred, with significant risk of spread to medical and laboratory personnel.

All cause fever, shock, and hemorrhage

Hepatitis and myocarditis also occur with Lassa fever

High mortality and risk of further transmission

The diagnosis of an arenavirus infection is suggested primarily by the recent travel history of the patient and the clinical syndrome. Although virus isolation and serologic diagnosis may be performed, these procedures should not be attempted in a hospital diagnostic laboratory. Any patient suspected of having such an infection should be immediately isolated and public health authorities notified. Because of the high risk of spread of infection from body fluids and excreta, even routine laboratory studies are best deferred until the diagnosis and proper disposition of specimens can be resolved. Viremia can persist 1 month, and virus shedding in the urine may continue more than 2 months after the onset of illness. Treatment is primarily supportive; however, intravenous ribavirin, if begun within 6 days of illness onset, has been shown to be helpful in Lassa fever.

Suggested by clinical findings and travel history

Diagnosis only in reference centers

Viremia may be prolonged

Lymphocytic Choriomeningitis Virus

Infection with lymphocytic choriomeningitis virus is particularly common in hamsters and mice. In the United States, most human illnesses have been traced to contact with rodent breeding colonies in research or pet supply centers and to pet hamsters in the home. The illness usually consists of fever, headache, and myalgia, although meningitis or meningoencephalitis also occurs occasionally. Such CNS infections may persist as long as 3 months. There is also evidence that transplacental infection can occur in humans, resulting in fetal death, hydrocephalus, or chorioretinitis. No person-to-person transmission of infection has been documented.

Transplacental infection in humans

The diagnosis of lymphocytic choriomeningitis is suggested by a history of rodent contact. The virus may be isolated in the early stages of disease by cell culture or intracerebral inoculation of blood or CSF into weanling mice or young guinea pigs. Serologic testing of acute and convalescent sera is usually performed by indirect immunofluorescence.

Mice and hamsters in pet stores

Meningitis may persist for months

FILOVIRUSES: MARBURG AND EBOLA VIRUSES

The association of the Marburg virus with serious disease did not become apparent until 1967, when 26 cases of hemorrhagic fever occurred among persons in Germany and Yugoslavia who were handling a group of African monkeys imported from central Uganda. The agent was later identified as Marburg virus and was apparently transmitted by the infected monkeys. In 1975, the virus was associated with a similar disease in three travelers in South Africa, and in 1980 in Kenya.

Initial cases tranmitted from monkeys

In 1976, severe outbreaks of hemorrhagic fever occurred in northern Zaire and southern Sudan, with case fatality rates from 50% to 90%. The illnesses were similar to those described for Marburg virus but were later shown to be caused by an antigenically different agent known as Ebola virus, named after a river in Zaire. More recently, another filovirus serologically related to Ebola virus was isolated from monkeys during an epizootic of simian hemorrhagic fever at a U.S. quarantine facility. The reservoir was determined to be monkeys imported from the Philippines.

Viruses differ antigenically

The reasons why these viruses can cause such fulminant, lethal hemorrhagic disease with shock in humans are not entirely clear. There is evidence that Marburg virus replicates in vascular endothelial cells, with subsequent necrosis. Other researchers have also shown that Ebola virus may exert its effects via a glycoprotein, synthesized in either a secreted or transmembrane form. The secreted glycoprotein interacts with neutrophils to inhibit early activation of the inflammatory response, whereas the transmembrane glycoprotein binds to endothelial cells. Ebola virus produces disease in humans and subhuman primates; onset is within 4 to 6 days of inoculation. The reservoir, though uncertain, is thought to be in small mammals, perhaps rodents. Serosurveys of humans residing in the areas where outbreaks have occurred suggest that human infections may be relatively common; as much as 7% of the survey group had antibodies, indicating past infection. In symptomatic infections, the mortality rate for both Marburg and Ebola viruses is extremely high (30% to 80%).

Reservoir may be small mammals

Mortality high in symptomatic infection

As with the arenavirus-associated hemorrhagic fevers, the diagnosis of infection by these agents is suggested by a similar syndrome and recent travel history. Person-to-person transmission similar to that described for Lassa fever occurs in Ebola virus infections and may be possible with Marburg virus. Diagnosis can be confirmed in a reference center by isolation of virus, as well as by serologic methods employing indirect immunofluorescence or EIA. However, as with the arenavirus-associated hemorrhagic fevers, utmost care in isolation precautions and prompt notification of public health authorities are mandatory for suspected cases before any diagnostic attempts are made. There is no specific therapy for the infections.

Diagnosis and precautions similar to arenavirus hemorrhagic fevers

HANTAVIRUSES

Hantavirus Hemorrhagic Fever

Korean hemorrhagic fever (KHF) is endemic to Korea and surrounding areas in the Far East. It is an important cause of hemorrhagic fever, often complicated by varying degrees of acute renal failure. In the 1950s, thousands of military personnel developed the disease during the Korean War. The first reported isolation of KHF was in 1978, when the antigen was detected in the lung tissues of wild rodents (*Apodemus* species) by indirect immunofluorescence using convalescent sera from affected patients. No illness was apparent in the rodents, suggesting a reservoir mechanism and mode of transmission similar to those described for the arenaviruses. Additional work indicated that the agent is a member of the family Bunyaviridae, and the generic designation of *Hantavirus* was given.

Causes of hemorrhagic fever during Korean War

Detected in lungs of wild rodents

Evidence has accumulated indicating that other agents with close antigenic similarities to KHF virus are responsible for hemorrhagic–renal syndromes occurring throughout northern Eurasia, including Russia, Eastern Europe, Finland, and Scandinavia. These syndromes

Other viruses similar to KHF throughout northern Eurasia

have been given a variety of names, including nephropathia epidemica. Methods similar to those used to diagnose KHF have detected nephropathia epidemica antigen in the lungs of small rodents (bank voles) in Finland.

Other Hantavirus Infections

Hantavirus among rodents in United States

Southwestern US outbreak related to deer mice

It has been known for some time that rodents in the United States may be infected with a hantavirus, but no associated human disease was recognized. In early 1993, an outbreak of fulminant respiratory disease with high mortality occurred in the southwestern United States. This syndrome (hantavirus pulmonary syndrome, or HPS) has been related to at least three hantaviruses, of which Sin Nombre virus is the most common (**Figure 16–9**). Infections are associated with an increased population of infected mice in and around human habitations. Since 1993, active surveillance in the United States has documented over 450 cases that have occurred in residents of 32 states, with most having been acquired in the Southwest region (**Figure 16–10**). Overall mortality rate has been 35%. The virus is

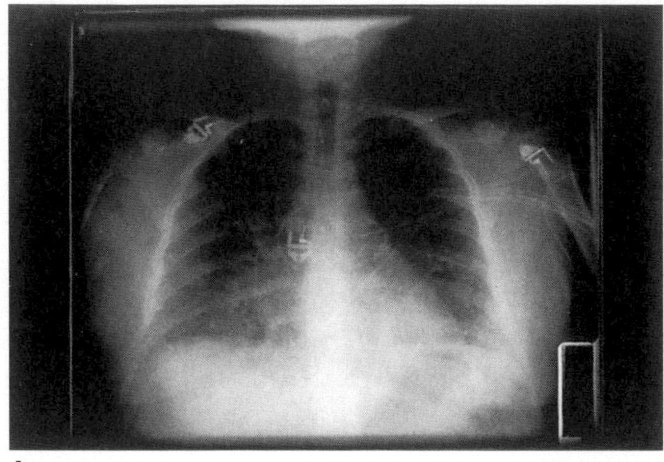
A

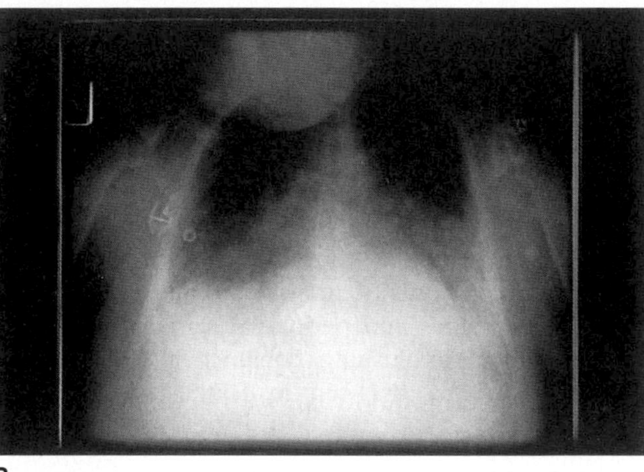
B

FIGURE 16–9. A and **B.** Serial radiographs obtained over 48 hours in a patient with hantavirus pulmonary syndrome. (Reproduced with permission from Connor DH, Chandler FW, Schwartz DQA, Manz HJ, Lack EE (eds). *Pathology of Infectious Diseases,* vol. 1. Stamford, CT: Appleton & Lange; 1997.)

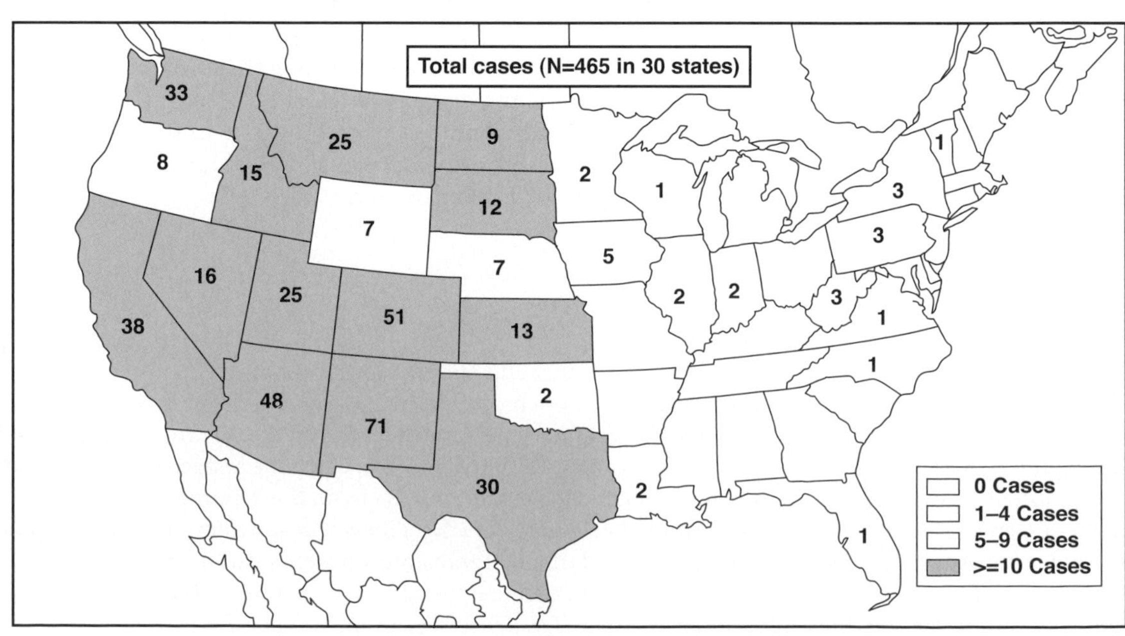

FIGURE 16–10. Hantavirus pulmonary syndrome cases in the United States. (Modified from the Centers for Disease Control and Prevention.)

believed to be transmitted to humans most often by inhalation of infectious rodent excreta, by the conjunctival route, or by direct contact with skin breaks. Human-to-human spread has not been encountered. Public health measures to inform inhabitants of routes of spread and to reduce the rodent population appear to have controlled the outbreak. Treatment has involved aggressive respiratory support. Intravenous ribavirin appears to have been of benefit in Asian hantavirus infections; however, there are no data as yet regarding its efficacy against the U.S. strains.

Humans infected by inhalation of aerosolized excreta

No human-to-human transmission

ORTHOMYXOVIRUSES

Avian and animals (pigs and horses) influenza viruses may infect humans. In the last 10 years, avian influenza viruses (bird flu), including H5N1, H7N2, H7N3, H7N7, and H9N2 have been documented to cause infections in humans. See Chapter 9 for avian influenza virus pathogenesis.

HENIPAVIRUSES

Hendra virus has been detected in Australia in two small outbreaks involving horses that also affected humans. The human cases were characterized by pneumonia and encephalitis. On the other hand, large nipah virus outbreaks have occurred in India, Bangladesh, Malaysia, and Singapore, affecting pigs, dogs, and humans. The human illnesses were similar to hendra virus, as were outcomes (more than 50% fatality rate for both). The reservoir of henipaviruses is the *Pteropus* species of fruit bats ("flying foxes") and spread to humans and animals via aerosols.

Henipaviruses spread by aerosols from bats

VESICULAR STOMATITIS VIRUS

A rhabdovirus, vesicular stomatitis virus causes outbreaks of disease in cattle, pigs, and horses that can be transmitted between animals by arthropods. Human infection is acquired by contact with infected animals, but is unusual; it consists of a self-limited febrile illness and occasional herpes-like eruptions over the lips and oral mucosa.

CASE STUDY

AN ACUTE CASE OF CONFUSION

This 70-year-old woman, who lives in a rural area in the midwestern United States, developed an illness in August that progressed over 3 days to include a moderate fever, headache, lower extremity weakness, and lethargy progressing to severe confusion.

On examination, she is unresponsive to verbal stimuli, and both pupils respond sluggishly to light. No other neurologic abnormalities are apparent. She lives with her husband on an old farm, and rodents have been frequently seen around the house and barn.

QUESTIONS

■ Which one of the following would be the most probable viral cause?

A. Western equine encephalitis
B. California (La Crosse strain) encephalitis virus
C. Colorado tick fever virus
D. West Nile encephalitis virus
E. Lymphocytic choriomeningitis virus

■ Which one of the following viruses is primarily transmitted by mosquitoes?

A. Ebola virus

B. Hantavirus

C. Yellow fever virus

D. Orbivirus

E. Henipavirus

■ Which one of the following features is least helpful in suggesting the possible cause of an arboviral illness?

A. Cerebrospinal fluid pleocytosis

B. Patient age

C. Season of occurrence

D. Knowledge of environmental reservoirs

E. Travel history

ANSWERS

1(D), **2**(C), **3**(A)

CHAPTER 17

Rabies

The dog was certainly rabid. Joseph Meister had been pulled out from under him covered with foam and blood.

—Louis Pasteur, describing the 9-year-old boy he successfully immunized against rabies in July, 1885

Rabies is an acute fatal viral illness of the central nervous system (CNS). The word rabies is derived from the Latin verb "to rage," which suggests the appearance of the rabid patient. It can affect all mammals and is transmitted between them by infected secretions, most often by bite. It was first recognized more than 3000 years ago and has been the most feared of infectious diseases. It is said that Aristotle recognized that rabies could be spread by a rabid dog.

VIROLOGY

The rabies virus is a bullet-shaped, enveloped, RNA virus, 180 by 70 nm, of the *Lyssavirus* genus within the Rhabdovirus family (**Figure 17–1**). The helical nucleocapsid (N) is composed of a single-stranded negative-sense RNA genome and an RNA-dependent RNA polymerase enclosed in a matrix (M) protein covered by a lipid bilayer envelope containing knob-like glycoprotein (G). The knob-like glycoprotein excrescences, which elicit neutralizing and hemagglutination-inhibiting antibodies, cover the surface of the virion. In the past, a single antigenically homogeneous virus was believed responsible for all rabies; however, differences in cell culture growth characteristics of isolates from different animal sources, some differences in virulence for experimental animals, and antigenic differences in surface glycoproteins have indicated strain heterogeneity among rabies virus isolates. These studies may help to explain some of the biologic differences as well as the occasional case of "vaccine failure." Other pathogens in this group include the vesicular stomatitis virus (see Chapter 16).:

Enveloped RNA virus is bullet-shaped

Knob-like envelope glycoproteins elicit neutralizing and hemagglutination antibodies

Strains from different sources are antigenically heterogeneous

Rabies virus is transmitted from the bite of an animal (usually a rabid dog or wild animal) and multiplies initially at the site of entry in muscle cells, and then the virus travels to the CNS to replicate in the brain cells. Rabies virus G protein binds to the acetylcholine or neural cell adhesion molecule (NCAM) receptor present on cell surface. The virus is internalized followed by fusion of the viral envelope with the endosomal membrane and uncoating and release of the nucleocapsid in the cytoplasm. Since rabies virus is a negative-sense RNA virus, virion-associated RNA-dependent RNA polymerase transcribes the genome to make several mRNAs in the cytoplasm. These mRNAs are translated into various proteins, including nucleocapsid, matrix, RNA polymerase, and G glycoproteins. The G glycoproteins are expressed on the infected cell surface membranes. After replication of viral RNA genomes directed by the viral RNA-dependent RNA polymerase, the progeny virions are

Negative-sense RNA virus replicates in the cytoplasm

G protein containing lipoprotein envelope acquired from plasma membrane

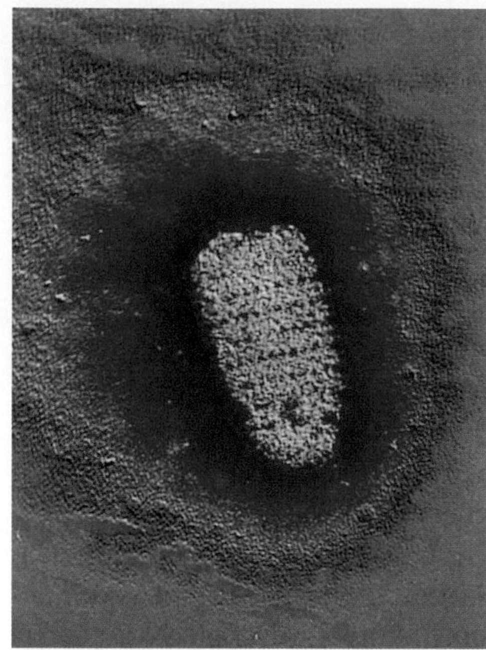

FIGURE 17–1. Electron micrograph of the rabies virus (yellow) (×36,700). Note the bullet shape. The external surface of the virus contains spike-like glycoprotein projections that bind specifically to cellular receptors. (Reproduced with permission from Willey J, Sherwood L, Woolverton C (eds). *Prescott's Principles of Microbiology.* New York: McGraw-Hill; 2008.)

assembled in the cytoplasm. The nucleocapsid protein binds the RNA genome and packages the viral RNA-dependent RNA polymerase. This nucleocapsid complex associates with the matrix protein, and the lipid bilayer envelope containing G protein is acquired as the progeny virions bud through the plasma membrane. ::: RNA polymerase, p. 119

RABIES

CLINICAL CAPSULE

Rabies involves the development of severe neurologic symptoms and signs in a patient who was previously bitten by an animal. The neurologic manifestations are very characteristic, with a relentlessly progressive excess of motor activity, agitation, hallucinations, and salivation. The patient appears to be foaming at the mouth and has severe throat contractions if swallowing is attempted. The neurologic abnormalities are explained by spread of the virus from the bite wound into the CNS and then centrifugally to the autonomic nervous system.

EPIDEMIOLOGY

Rabies exists in two epizootic forms, urban and sylvatic. The urban form is associated with unimmunized dogs or cats, and the sylvatic form occurs in wild skunks, foxes, wolves, raccoons, and bats, but not rodents or rabbits. Introduction of an infected animal into a different geographic area can lead to infection of many new members of that species (**Figure 17–2**). For example, raccoon hunters apparently are to blame for the sudden appearance of raccoon rabies in West Virginia and Virginia in 1977. Before that time, the nearest cases of raccoon rabies were found several hundred miles away in South Carolina. The hunters are believed to have imported infected raccoons from another state. Since 1977, raccoon rabies has spread from West Virginia and Virginia to 12 northeastern states.

Human infection, or the much more common infection of cattle, is incidental, is blind-ended, and does not contribute to maintenance or transmission of the disease. In the United

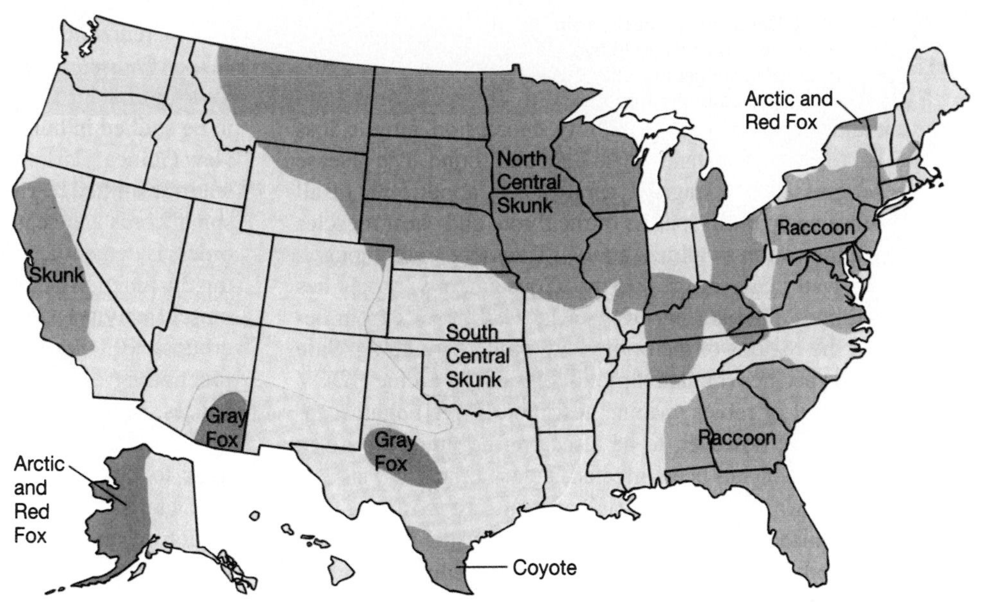

FIGURE 17–2. In the United States, rabies is found in terrestrial animals in 10 distinct geographic areas. In each area, a particular species is the reservoir, and one of five antigenic variants of the virus predominates as illustrated by the five different colors.

States, more than 75% of reported cases of rabies in animals occur among wildlife. Human exposures may be from wild animals or from unimmunized dogs or cats. In recent years, there has been a decrease of U.S. cases to less than two per year, and bat exposure has been the source in almost all cases despite a resurgence of rabies in skunks and raccoons. An occasional case has resulted from aerosol exposure (eg, bat caves and no bite). Domestic animal bites are very important sources of rabies in developing countries because of lack of enforcement of animal immunization. Infection in domestic animals usually represents a spillover from infection in wildlife reservoirs. Human infection tends to occur where animal rabies is common and where there is a large population of unimmunized domestic animals. Worldwide, the occurrence of human rabies is estimated to be about 40,000 cases per year, with the highest attack rates in Southeast Asia, the Philippines, and the Indian subcontinent. Almost all of these cases result from dog bites.

Risks to humans are from bites by infected carnivores, omnivores, and bats

Aerosol spread from exposure in bat caves

PATHOGENESIS

The sequence of events of the pathogenesis of rabies virus infection is depicted in **Figure 17–3**. The essential first event in human or animal rabies infection is the inoculation of virus through the epidermis, usually as a result of an animal bite. Inhalation of heavily contaminated material, such as bat droppings, can also cause infection. Rabies virus first replicates in striated muscle tissue at the site of inoculation. Immunization at this time is presumed to prevent migration of the virus into neural tissues. In the absence of immunity, the virus then enters the peripheral nervous system at the neuromuscular junctions and spreads to the CNS, where it replicates exclusively within the gray matter. It then passes centrifugally along autonomic nerves to reach other tissues, including the salivary glands, adrenal medulla, kidneys, and lungs. Passage into the salivary glands in animals facilitates further transmission of the disease by infected saliva. The neuropathology of rabies resembles that of other viral diseases of the CNS, with infiltration of lymphocytes and plasma cells into CNS tissue and nerve cell destruction. The pathognomonic lesion is the Negri body (**Figure 17–4**), an eosinophilic cytoplasmic inclusion distributed throughout the brain, particularly in the hippocampus, cerebral cortex, cerebellum, and dorsal spinal ganglia.

Replicates initially in muscle and then enters peripheral nervous system

Spreads to CNS gray matter

Negri bodies found in neurons

The incubation period for rabies ranges from 10 days to 1 year, depending on the amount of virus introduced, the amount of tissue involved, the host immune mechanisms, the innervation of the site, and the distance that the virus must travel from the site of inoculation to the CNS. Thus, the incubation period is generally shorter with face wounds than with leg wounds. Immunization early in the incubation period frequently aborts the infection.

Incubation period can be prolonged for months

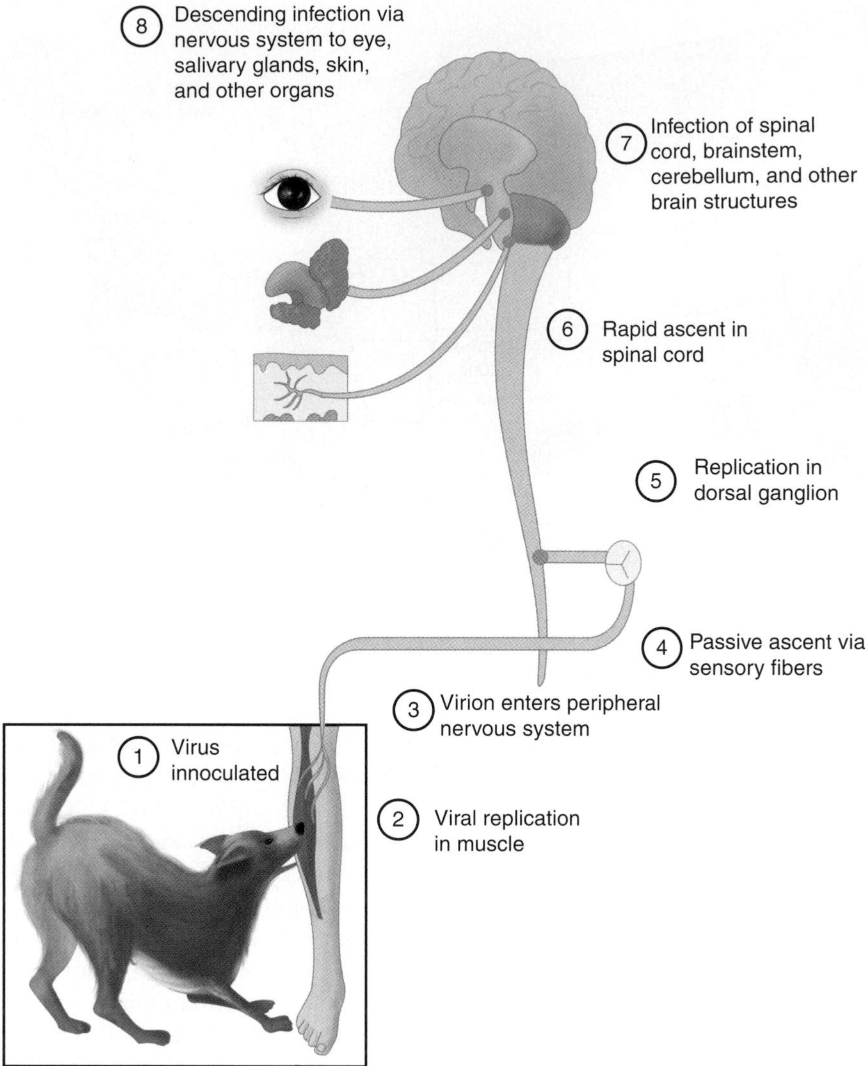

8 Descending infection via nervous system to eye, salivary glands, skin, and other organs
7 Infection of spinal cord, brainstem, cerebellum, and other brain structures
6 Rapid ascent in spinal cord
5 Replication in dorsal ganglion
4 Passive ascent via sensory fibers
3 Virion enters peripheral nervous system
1 Virus innoculated
2 Viral replication in muscle

FIGURE 17–3. Sequential steps in the pathogenesis of rabies virus infection. (Reproduced with permission from Willey J, Sherwood L, Woolverton C (eds). *Prescott's Principles of Microbiology.* New York: McGraw-Hill; 2008.)

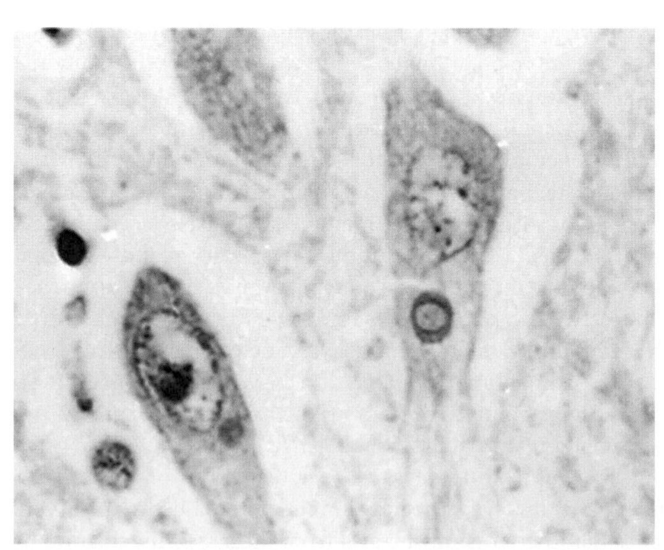

FIGURE 17–4. Negri body in cytoplasm of neuron. (Courtesy of Dr. Daniel P. Perl.)

CLINICAL ASPECTS

MANIFESTATIONS

Rabies in humans usually results from a bite by a rabid animal or contamination of a wound by its saliva. It presents as an acute, fulminant, fatal encephalitis; human survivors have been reported only occasionally. The clinical stages of rabies infection are summarized in **Table 17–1**. After an average incubation period of 20 to 90 days, the disease begins as a nonspecific illness marked by fever, headache, malaise, nausea, and vomiting. Abnormal sensations at or around the site of viral inoculation occur frequently and probably reflect local nerve involvement. The onset of encephalitis is marked by periods of excess motor activity and agitation. Hallucinations, combativeness, muscle spasms, signs of meningeal irritation, seizures, and focal paralysis occur. Periods of mental dysfunction are interspersed with completely lucid periods; however, as the disease progresses, the patient lapses into coma. Autonomic nervous system involvement often results in increased salivation. Brainstem and cranial nerve dysfunction is characteristic, with double vision, facial palsies, and difficulty in swallowing. The combination of excess salivation and difficulty in swallowing produces the fearful picture of "foaming at the mouth." Hydrophobia, the painful, violent involuntary contractions of the diaphragm and accessory respiratory, pharyngeal, and laryngeal muscles initiated by swallowing liquids including water, is seen in about 50% of cases. Involvement of the respiratory center produces respiratory paralysis, the major cause of death. Occasionally, rabies may appear as an ascending paralysis resembling Guillain–Barré syndrome. Once symptoms have developed, no drug or vaccine administration can improve survival. The median survival after onset of symptoms is 4 days, with a maximum of 20 days unless artificial supportive measures are instituted. Recovery is exceedingly rare.

Encephalitis common, sometimes with ascending paralysis

Almost uniformly fatal

TABLE 17–1 Clinical Stages of Rabies Virus Infection

STAGES OF INFECTION	TIME FRAME	SYMPTOMS	SITE OF VIRUS REPLICATION
Incubation period	10-365 days Average: 20-90 days	No symptoms	Site of bite, muscle cells
Prodrome stage	2-10 days	Nonspecific symptoms, malaise, headache, fever, nausea, vomiting, upper respiratory distress, subtle mental changes (insomnia), pain, itching, tingling at site of bite	Virus replication in the CNS
Acute neurologic stage	2-7 days	Furious or dumb presentation *Furious:* Hyperactivity, excitement, disorientation, hallucination, bizarre behavior, hydrophobia, convulsions, aggressive *Dumb (paralytic phase):* Lethargy, paralysis, (respiratory)	Virus replication in brain and transported to other sites (salivary glands and other organs)
Coma	0-14 days	Patient in coma; respiratory paralysis, cardiac arrest, drop in blood pressure, secondary infections	Virus replication in brain and transported to other organs
Death		Extremely rare survival	

DIAGNOSIS

The CSF of a rabies patient shows minimal to no abnormalities with some patients exhibiting a lymphocytic pleocytosis (5 to 30 cells/mm^3). The test of choice in a live patient is detection of rabies antigen by immunofluorescent stain of a biopsy from the nape of the neck. PCR of CSF or saliva may supplant the neck biopsy. Laboratory diagnosis of rabies in animals or deceased patients is accomplished by demonstration of virus in brain tissue. Viral antigen can be demonstrated rapidly by immunofluorescence procedures. Intracerebral inoculation of infected brain tissue or secretions into suckling mice results in death in 3 to 10 days. Histologic examination of their brain tissue shows Negri bodies in 80% of cases; electron microscopy may demonstrate both Negri bodies and rhabdovirus particles. Specific antibodies to rabies virus can be detected in serum, but generally only late in the disease.

Virus or antigen detected in brain tissue

TREATMENT

Prevention is the mainstay of controlling rabies in humans. Intensive supportive care has resulted in two or three long-term survivals; despite the best modern medical care, however, the mortality rate still exceeds 90%. In addition, because of the infrequency of the disease, many patients die without definitive diagnosis. Human hyperimmune antirabies globulin, interferon, and vaccine do not alter the disease once symptoms have developed.

No specific treatment is available

PREVENTION

In the late 1800s, Pasteur, noting the long incubation period of rabies, suggested that a vaccine to induce an immune response before the development of disease might be useful in prevention. He apparently successfully vaccinated Joseph Meister, a boy severely bitten and exposed to rabies, with multiple injections of a crude vaccine made from dried spinal cord of rabies-infected rabbits. This treatment emerged as one of the best-known and most noteworthy accomplishments in the annals of medicine. It is now believed that vaccination induces antibody that is either neutralizing or inhibits cell-to-cell spread of virus. Natural infection does not lead to an early immune response and limitation of viral migration, because the virus is replicating in muscle or neural tissue and lymphocytes do not access these sites. Cytotoxic T lymphocytes are also induced by vaccine and appear to be directed against an antigen of the virus.

Vaccine-induced antibody inhibits viral spread

Currently, the prevention of rabies is divided into preexposure and postexposure prophylaxis. Preexposure prophylaxis is recommended for individuals with high risk of contact with rabies virus, such as veterinarians, spelunkers, laboratory workers, and animal handlers. The vaccine currently used in the United States for preexposure prophylaxis uses an attenuated rabies virus grown in human diploid cell culture and inactivated with β-propiolactone. Preexposure prophylaxis consists of two subcutaneous injections of vaccine given 1 month apart, followed by a booster dose several months later.

High-risk individuals include veterinarians, spelunkers, laboratory workers

Postexposure prophylaxis requires careful evaluation and judgment. Every year more than 1 million Americans are bitten by animals, and approximately 25,000 receive postexposure rabies prophylaxis. The physician must consider (1) whether the individual came into physical contact with saliva or another substance likely to contain rabies virus; (2) whether there was significant wound or abrasion; (3) whether rabies is known or suspected in the animal species and area associated with the exposure; (4) whether the bite was provoked or unprovoked (ie, the circumstances surrounding the exposure); and (5) whether the animal is available for laboratory examination.

Any wild animal or ill, unvaccinated, or stray domestic animal involved in a possible rabies exposure, such as an unprovoked bite, should be captured and killed. The head should be sent immediately to an appropriate laboratory, usually at the state health department, for search for rabies antigen by immunofluorescence. If examination of the brain by this technique is negative for rabies virus, it can be assumed that the saliva contains no virus and that the exposed person requires no treatment. If the test is positive, the patient should be given postexposure prophylaxis. It should be noted that rodents and rabbits are not important vectors of rabies virus. There have been no rabies deaths in the United States when postexposure prophylaxis was given promptly after exposure.

Careful history and studies of biting animal are important in decision-making

Postexposure prophylaxis is based on immediate, thorough washing of the wound with soap and water; passive immunization with hyperimmune globulin, including a portion instilled around the wound site; and active immunization with antirabies vaccine on days 0, 3, 7, 14, and 28. Physicians should always seek the advice of the local health department when the question of rabies prophylaxis arises.

Rabies immune globulin plus vaccine necessary in postexposure management

CASE STUDY

THE FRIENDLY BOY AND THE UNFRIENDLY DOG

A 5-year-old boy in San Francisco reaches into a car to pet another family's dog and is bitten on the finger.

QUESTIONS

■ What is the next course of action?

A. Obtain documentation of the dog's immunization status
B. Give rabies immune globulin
C. Give rabies immune globulin plus rabies vaccine
D. Give gamma interferon
E. Examine the dog's brain for rabies antigen

■ Six weeks after the bite, the child develops fever, headache, and a seizure. He becomes combative and hallucinates. The best diagnostic test to perform on the patient to rule in rabies as a cause of his 3-day illness is:

A. Detection of serum antirabies antibody
B. Culture of CSF for virus
C. Direct fluorescent antibody (DFA) stain of a biopsy from the nape of the neck
D. Brain biopsy
E. CSF antirabies antibody

ANSWERS

1(A), **2**(C)

CHAPTER 18

Retroviruses: Human T-Lymphotropic Virus, Human Immunodeficiency Virus, and Acquired Immunodeficiency Syndrome

Retroviruses are enveloped, single-stranded, positive-sense RNA viruses. These viruses replicate via DNA intermediates because they encode an enzyme called **reverse transcriptase,** which converts the RNA genome into a double-stranded DNA copy that subsequently becomes integrated into the host chromosome. The discovery of reverse transcriptase in 1970 by two American virologists, David Baltimore and Howard Temin, earned them a Nobel Prize in Medicine. There are two major groups of retroviruses that infect humans: the **oncoretroviruses** (*onco-*, "related to a tumor") and the **lentiviruses** (*lenti-*, "slow"). There are several other groups of retroviruses that infect animals. Endogenous retrovirus sequences are found throughout the human genome. Like most enveloped viruses, all retroviruses are highly susceptible to factors that affect surface tension and are thus not transmissible through air, dust, or fomites under normal conditions, but instead require intimate contact with the infecting sources, such as bodily fluids, blood, and blood-derived products. ::: reverse transcriptase, p.120

Enveloped RNA viruses that encode reverse transcriptase enzyme, which converts retroviral RNA genome into double-stranded DNA

Members of the oncoretrovirus subgroup of retroviruses have long been associated with a variety of cancers in animals, including leukemias, lymphomas, and sarcomas, but until recent years had not been found to infect humans. The first human retrovirus, human T-cell lymphotropic virus type I (HTLV-I), was discovered in the late 1970s. It was shown to cause adult T-cell leukemia, a rare malignancy found only in Japan, Africa, and the Caribbean, although serologic evidence shows that the virus also occurs in the United States and has raised the possibility of an association with some chronic neurologic conditions. A relative of HTLV-I, HTLV-II has been associated with a few rare cases of T-cell malignancies, including hairy cell leukemia, but its precise role in these diseases remains unclear.

Oncoretroviruses cause tumors in many animals

HTLV-I and HTLV-II associated with human leukemias

The most important disease resulting from a human retrovirus infection is called **acquired immunodeficiency syndrome (AIDS),** which is caused by either of two lentiviruses known as human immunodeficiency viruses types 1 and 2 (**HIV-1** and **HIV-2**). A devastating disease worldwide, for which there is no present cure or vaccine for protection, AIDS has spurred unprecedented research efforts to determine the nature and immunopathogenic mechanisms of the viruses in the hope of finding effective drugs and vaccines. Most of our present knowledge of HIV is derived from studies on HIV-1, which is the major cause of AIDS worldwide. In 2008, two French virologists, Françoise Barré-Sinoussi and

HIV-1 and HIV-2 are lentiviruses that cause AIDS

Luc Montagnier, shared the Nobel Prize in Medicine for their work on the discovery of HIV, the virus that causes AIDS.

Oncoviruses usually not cytolytic; they transduce or activate oncogenes

Oncoretroviruses are not cytolytic and do not kill the cells they infect, but rather they transform the cells and continue to produce new virus indefinitely. This property, combined with the fact that they can transduce growth-promoting genes called **oncogenes** into a recipient cell, accounts in part for their ability to cause malignancies (see Chapter 7 and the following text). With lentivirus infections, the cell–virus relationship is different. Lentiviruses can apparently persist in infected hosts for long periods of time in a clinically latent state. Over time, the virus becomes highly cytopathic and kills infected cells and also uninfected cells, causing impairment of the host immune defenses followed by AIDS and opportunistic infections. The prototype of this type of lentivirus is HIV-1. There is another closely related sheep lentivirus called visna virus, which causes a slow degenerative neurologic disease in sheep.

Lentiviruses cause a long clinical latency period in infected patients in the presence of viremia before causing disease

HIV attacks and destroys CD4+ T lymphocytes

HIV also infects monocytes/macrophages, dendritic cells, Langerhans cells, and certain cells of the central nervous system

HIV-1 can remain clinically latent in most infected patients without causing viral latency, which means that virus is initially produced at low levels without serious disease, but when allowed to replicate in the absence of effective immune response and other factors, high levels of virus are produced causing T-lymphocyte cell death and AIDS. Although HIV-1 can infect a variety of human cell types such as T lymphocytes, monocytes/macrophages, dendritic cells, Langerhans cells, and microglia/glial cells, its most drastic effects appear to result from destruction of the CD4+ subclass of T lymphocytes, which play a central role in the capacity of the host to mount effective and protective immunologic responses to a wide range of infections.

RETROVIRUSES

STRUCTURE

Virion contains two single-stranded, positive-sense RNA molecules (diploid genome)

Three critical enzymes, reverse transcriptase, protease, and integrase, are virus-encoded

Envelope acquired during budding contains two viral glycoproteins

All retroviruses are remarkably similar in their basic composition and structure. The structure of HIV-1 is depicted in **Figure 18–1**. The virion size is about 100 nm in diameter, and because it contains two copies of the RNA genome, it is diploid. The RNA genome is coated with the nucleocapsid protein (NC), and the RNA–protein complexes are enclosed in a capsid (CA, also called p24) composed of multiple subunits. Like all enveloped viruses, the membrane is acquired during budding from the host cell, but the surface (SU, also called gp120) and transmembrane (TM, also called gp41) glycoproteins found in the envelope are virally encoded. Between the capsid and the envelope is the matrix (MA, also called p17) protein. In addition to the structural proteins shown in Figure 18–1, the virion core contains three virus-specific proteins (enzymes) that are essential for viral replication: reverse transcriptase (RT), protease (PR), and integrase (IN). The relation between the viral genes found in all retroviruses (*gag, pol,* and *env*) and the proteins they encode are presented in **Table 18–1**. Some retroviruses, including HTLV and HIV, encode additional regulatory and accessory proteins. Based on SU gp120 sequence, HIV-1 can be T-lymphotropic (X4), macrophage-tropic (R5), or both X4/R5 (dual-tropic).

RETROVIRAL REPLICATION CYCLE

Figure 18–2 depicts the life cycle of a typical retrovirus (example: HIV-1) and serves to illustrate the many unique aspects of retroviral replication that could be potential targets of therapeutic intervention.

Viral Entry

HIV-1 surface glycoprotein gp120 attaches to CD4 cell and chemokine coreceptors

Retroviral virions adsorb to cellular membrane receptors and enter the cell by direct fusion of the viral envelope with the plasma membrane of the host cell. For HIV-1, the virion attachment protein is the SU glycoprotein gp120, and the cellular receptor is the CD4 molecule with one of the chemokine receptors, CXCR4 or CCR5 acting as co-receptors. These receptors and co-receptors occur primarily in the plasma membrane of CD4+

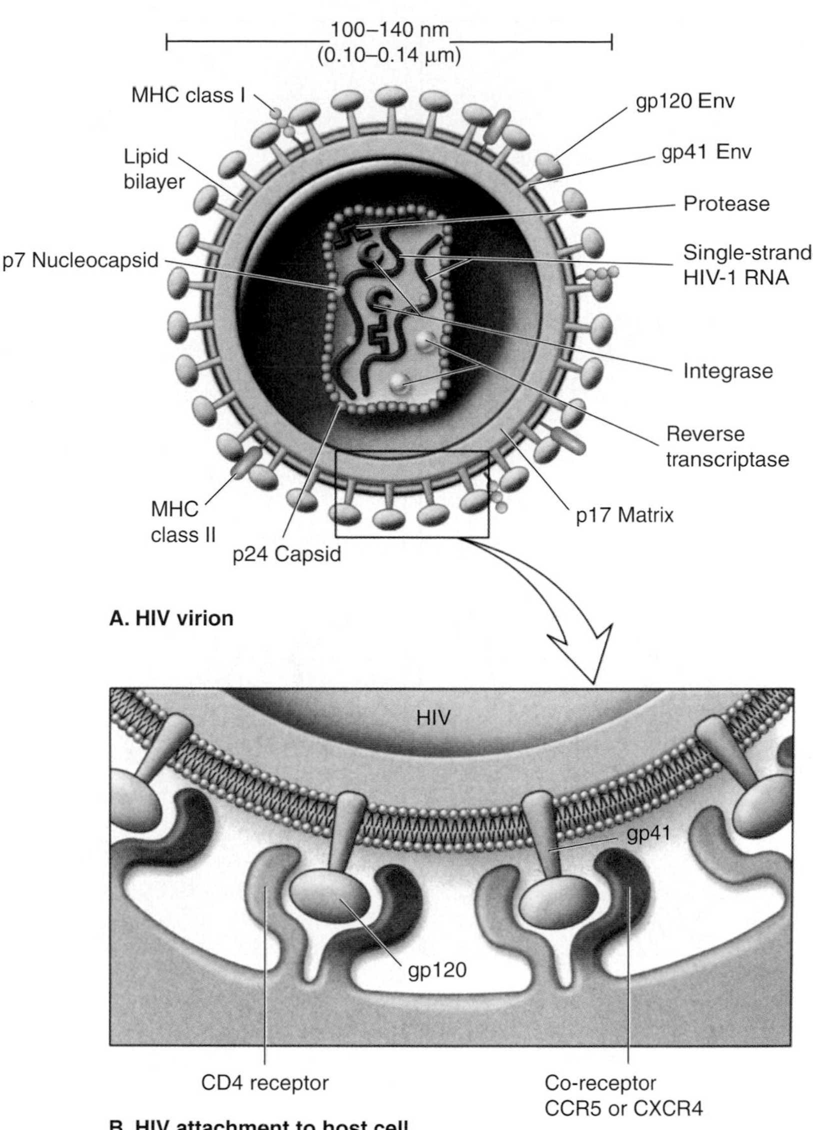

FIGURE 18–1. Structure of HIV particle. The two RNA molecules enclosed within the capsid are coated with the nucleocapsid protein. The matrix protein lies just inside the membrane envelope. The envelope contains two membrane glycoproteins, gp41 and gp120, also called transmembrane protein and surface protein, respectively. CCR5; CXCR4, chemokine receptors, acting as co-receptors.

TABLE 18–1 Major Retroviral Genes and Proteins

GENE[a]	PROTEIN PRODUCTS	FUNCTION
gag	Matrix (MA)	Structural
	Capsid (CA)	Structural
	Nucleocapsid (NC)	Structural
	Protease[b] (PR)	Protein processing
Pol	Protease[b] (PR)	Protein processing
	Reverse transcriptase (RT)	DNA synthesis
	Integrase (IN)	Integration
Env	Surface glycoprotein (SU)	Adsorption
	Transmembrane protein (TM)	Fusion of envelope with plasma membrane

[a]Each gene encodes a polyprotein that is subsequently processed by proteolysis to yield the individual proteins.
[b]The protease is encoded in either the ***gag*** gene or the ***pol*** gene, depending on the virus.

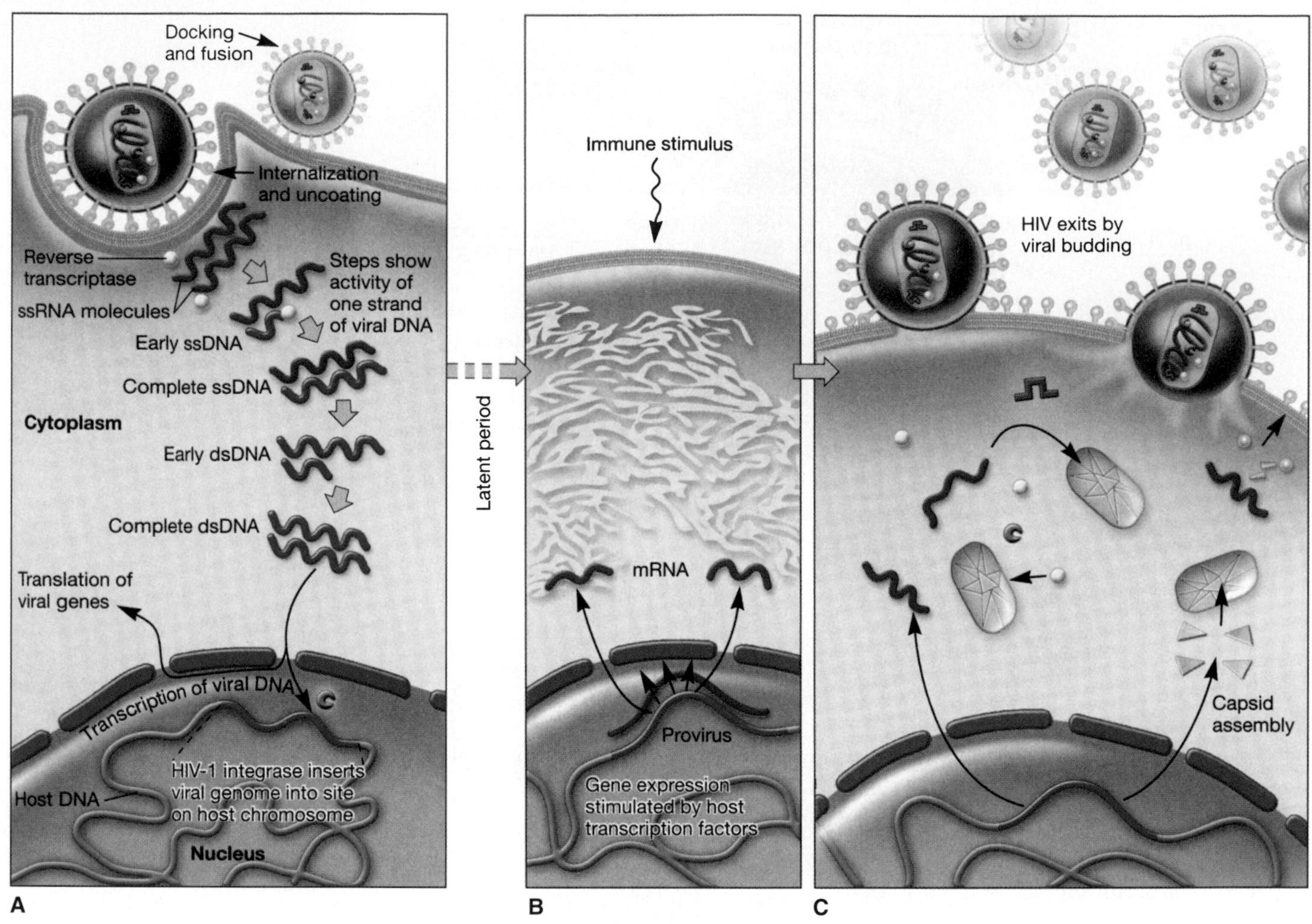

FIGURE 18–2. Retroviral (HIV-1) life cycle. A. Viral entry and post entry (reverse transcription, DNA synthesis, and integration) events; B. Viral gene expression (transcription and protein synthesis); C. Virus assembly and release.

While R5 HIV-1 binds to CD4 and CCR5, X4 HIV-1 interacts with CD4 and CXCR4

Transmembrane gp41 protein mediates fusion of viral and cell membranes

HIV-1 can infect cells expressing chemokine receptors without the CD4 molecule but with a low efficiency

Fusion provides direct cell-to-cell transmission

T lymphocytes, cells of the monocyte–macrophage lineage, and some other target cells such as Langerhans cells, dendritic cells and certain brain cells. Early in infection, the viruses are often macrophage-tropic (R5 viruses) because they preferentially use the CCR5 coreceptor. The emergence of syncytia-forming variants that use the CXCR4 co-receptor and are T-lymphotropic (X4 viruses) appears to correlate with rapid advancement to AIDS. The HIV-1 transmembrane TM protein gp41 is responsible for fusion of the viral and cell membranes, leading to entry of the virion core complex into the cytoplasm of the cell.

HIV-1 can also infect cells such as fibroblasts and certain brain cells that lack the CD4 surface molecule with a low efficiency, apparently because the chemokine receptors in combination with the fusion-inducing activity of the TM protein is sufficient in these cases to promote entry. Fusion activity may also play an important role in amplification of the effects of the virus infection, particularly during the later stages of the infection, because infected cells expressing viral glycoproteins in their membranes readily fuse with uninfected CD4+ T lymphocytes to form large syncytia. This process appears to provide a means for cell-to-cell transmission of the virus that bypasses the usual extracellular phase and may contribute to the overall depletion of CD4+ T lymphocytes in an infected person.

Viral Post-Entry Events

Among the RNA viruses, retroviral replication is unique because it involves reverse transcription. Soon after entry of the viral core into the cytoplasm of the infected cell, the RNA is copied (reverse transcribed) into a complementary DNA (cDNA) by reverse transcriptase enzyme, the virion-associated RNA-dependent DNA polymerase. The cDNA is converted

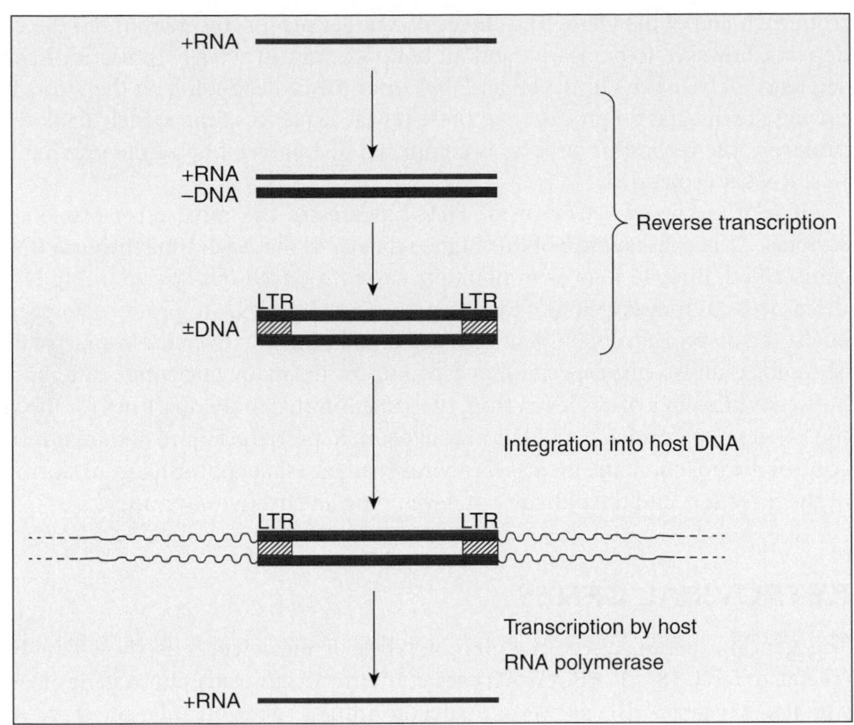

FIGURE 18–3. Retroviral RNA replication. LTR, long terminal repeat.

into double-stranded DNA by the action of the DNA-dependent DNA polymerase activity of the reverse transcriptase. The viral RNA template is removed from the RNA-DNA hybrid by RNAase H activity of the reverse transcriptase enzyme. The overall process is referred to as **reverse transcription** and results in a linear DNA molecule that circularizes and makes a preintegration complex with the help of viral and host factors. The preintegration complex enters the nucleus and integrates more-or-less at random sites into the host cell chromosome catalyzed by viral integrase. Once the viral genetic information has been converted to DNA and integrated, it essentially becomes part of the cellular genome, and the cell is permanently infected. The viral genome, called the **provirus,** is therefore replicated and faithfully inherited as long as the infected cell continues to divide.

Reverse transcriptase copies RNA to double-stranded DNA

DNA integrates into the host chromosome and replicates with the cell as a provirus

Special sequences contained within the RNA are duplicated during the reverse transcription process so that the integrated provirus contains identical long terminal repeats (LTRs) at its ends (**Figure 18–3**). The LTR sequences contain the appropriate promoter, enhancer, and other signals required for transcription of the viral genes by the host RNA polymerase II. Transcription produces a full-length RNA genome and one or more spliced mRNAs. For the oncoviruses, the predominant spliced mRNA is translated to produce the envelope glycoproteins, but in HIV-1, a series of spliced mRNAs are produced that encode, in addition to the envelope proteins, a series of viral regulatory and accessory proteins. Unlike most retroviruses, HIV-1 and the other lentiviruses apparently exert considerable control over whether the primary transcripts are allocated to full-length RNA or are spliced to produce mRNAs (see text that follows). With the exception of these regulatory and accessory proteins, all retroviral proteins are initially translated as polyproteins that are subsequently processed by proteolysis into the individual protein molecules. Although the HIV-1 envelope precursor proteins (gp160) are cleaved by cellular protease, the enzyme responsible for cleavages of Gag and Gag-Pol precursors is the virus-specific protease (PR) that is encoded by the *pol* gene of HIV-1.

Provirus includes its own promoter and signals that control transcription by host RNA polymerase

Genomic RNA and spliced mRNAs are both produced: the latter encode envelope glycoproteins and regulatory proteins

HIV-1 can control extent of genomic or spliced mRNA production

A simplified view of retroviral RNA replication is presented in **Figure 18–3**. In addition to DNA polymerase activity, the reverse transcriptase possesses an RNase H activity that is responsible for degrading the RNA portion of the DNA–RNA hybrid (+RNA/–DNA) produced in the first phase of reverse transcription. The immediate product of reverse transcription is a linear double-stranded DNA molecule that is flanked by the LTR sequences. The viral integrase (IN) catalyzes the integration of the linear DNA into host DNA. The integration process is highly specific with respect to the viral DNA, and two base pairs are generally lost

RNase H activity degrades original RNA genome

Integrase-catalyzed integration is random in host DNA

Integrated DNA is transcribed by host RNA polymerase

from each end of the DNA. The choice of a target site for integration into the cellular DNA appears, however, to be nearly random but preferably in actively transcribed genes. A short sequence of base pairs in the target DNA (four to six, depending on the virus) is duplicated during the integration process, and these repeat sequences immediately flank the integrated provirus. The replication process is completed by transcription of the proviral DNA by the host RNA polymerase II.

HIV reverse transcriptase is error-prone

Isolates from the same patient can differ in multiple properties

Of all the known retroviruses, HIV-1 possesses the most error-prone reverse transcriptase. The consequence of this high error rate is that each time the viral RNA is reverse transcribed, three to four new mutations are introduced into the resulting DNA. Because the process of transcription of the integrated proviral DNA to produce new viral genomes is also error-prone, mutant genomes accumulate rapidly over the course of an infection. The end result is a quasispecies that accounts for the many nucleotide differences observed between different isolates (even from the same infected individual) and for the variability of the SU envelope protein gp120. It may explain, in part, the failure of the immune system to control the infection, the increases in viral virulence that appear to occur during the course of the infection, and the difficulty of developing an effective vaccine.

RETROVIRAL GENES

Genome is organized into ***gag, pol,*** and ***env*** genes

The genome organization of different types of retroviruses is shown in **Figure 18–4** (see also Table 18–1). All retroviruses contain the same structural genes in the order of *gag–pol–env* genes. The *gag* (group-specific antigen) gene encodes the structural proteins (capsid, nucleocapsid, matrix) of the virus and, in some animal retroviruses, the protease. The *pol* (polymerase) gene in human retroviruses and HIV encodes the reverse transcriptase, the integrase, and the protease. The *env* (envelope) gene encodes the two membrane glycoproteins found in the viral envelope, SU gp120 and TM gp41. HIV-1 gp120 has five variable regions and several constant regions. The CD4 binding domains on gp120 are localized in the constant regions, whereas the co-receptor (CXCR4/CCR5) binding regions on gp120 are confined in the variable region 3 (V3 loop). The V3 region is also the principal neutralizing domain of the virus, and therefore contributes to antigenic variation and varying degrees of neutralization. On the other hand, gp41 is embedded in the envelope and mediates fusion of the viral envelope with the plasma membrane

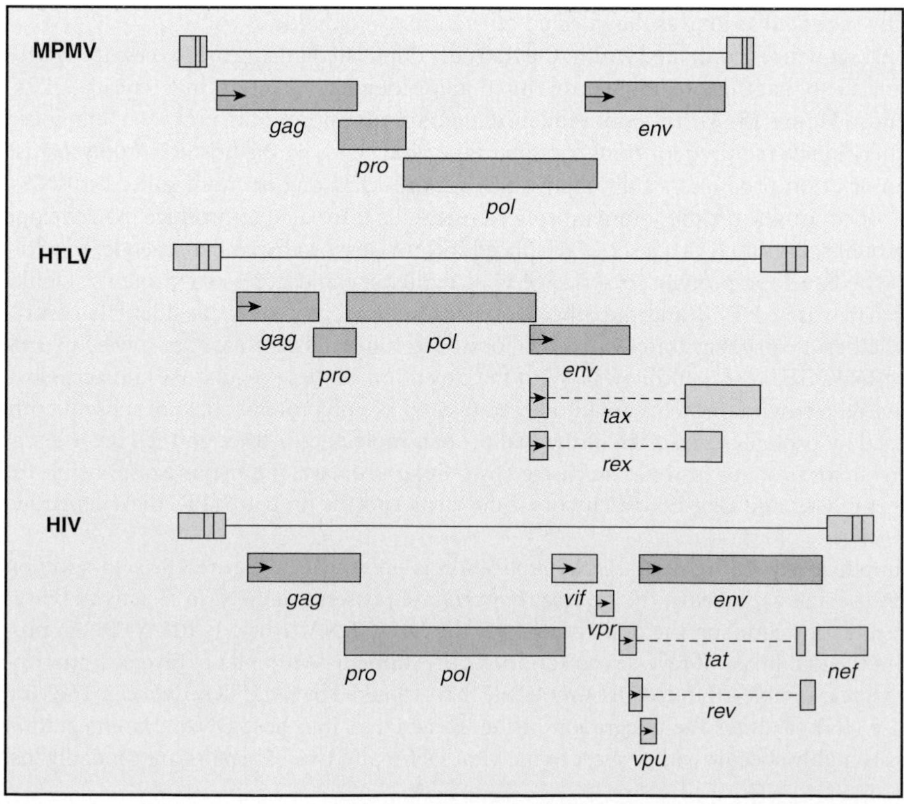

FIGURE 18-4. Structure of retroviral genes of a mouse retrovirus (MPMV), HTLV, and HIV.

TABLE 18–2 Regulatory and Accessory Proteins

GENE	PROTEIN	FUNCTION
tat	Tat	Transcriptional activator
rev	Rev	Promotes transport of unspliced mRNAs from nucleus to cytoplasm
nef	Nef	Down-regulation of cellular CD4 and MHC I proteins
vpu	Vpu	Facilitates virus assembly and release
vpr	Vpr	Facilitates nuclear entry in nondividing cells, arrests dividing cells
vif	Vif	Increases viral infectivity in certain cell types

MHC, major histocompatibility complex.

at the time of viral infection and less variable than gp120. The fusion of gp41 with the plasma membrane can be blocked by gp41 inhibitor.

A comparison of the genetic makeup of HIV-1 with that of a typical retrovirus (Figure 18–4) reveals a larger number of genes and a much more complex organization. HIV-1 contains, in addition to the *gag, pol,* and *env* genes, an array of other genes (*tat, rev, nef, vif, vpr,* and *vpu*). Expression of these genes requires mRNA splicing, and all apparently encode proteins that serve regulatory or accessory roles during the infection (see text that follows). HTLV-I encodes the regulatory proteins, Tax and Rex, which are analogous to the HIV-1 Tat and Rev proteins. The names of the genes that have been best characterized and the proteins and functions they determine are listed in **Table 18–2**.

HIV-1 has multiple regulatory genes

ROLES OF HIV-1 REGULATORY AND ACCESSORY PROTEINS

HIV-1 has the ability to produce a complex array of regulatory and accessory proteins that appear to be involved in viral replication, pathogenesis, and disease progression. These proteins also appear to interact with cellular factors to modulate the infection differently in different host cells. The roles of the two HIV-1 regulatory genes, *tat* and *rev,* and the four accessory proteins, *nef, vpu, vpr,* and *vif,* are discussed below and summarized in Table 18–2. Although the four accessory proteins are dispensable in many cell line culture systems, they appear to be important for the maximum pathogenic potential of the virus in infected individuals.

The products of the *tat* and *rev* regulatory genes are the Tat and Rev proteins, respectively. Both of these proteins are essential for viral replication. When the infected T lymphocyte is stimulated, for example, by antigen presentation, Tat and Rev play a positive role in promoting viral gene expression. In the absence of high levels of Tat, the host RNA polymerase initiates properly at the LTR promoter, but transcription is usually prematurely terminated leading to the production of short, dead-end transcripts. Tat is a transcriptional activator that acts at a sequence near the beginning of the viral mRNA, called TAR, to recruit cellular proteins to the transcribing RNA polymerase, resulting in a modification to the polymerase that prevents premature termination and allows complete transcription of the proviral genome.

Tat is a transcriptional activator that promotes synthesis of full-length viral transcripts

The Rev protein acts at the level of mRNA splicing and transport. Normally, unspliced cellular transcripts are retained in the nucleus, and only fully spliced mRNAs are transported to the cytoplasm for translation. The only viral proteins that are made from fully spliced mRNAs are Tat, Rev, and Nef, and consequently only these proteins are found early after infection, when there is no mechanism to prevent complete splicing of pre-mRNAs. To express the Vif, Vpr, and Vpu proteins, and the Env polyprotein, all of which are made from singly spliced transcripts, as well as the Gag and Pol polyproteins, which are translated from the unspliced genomic RNA, it is necessary to transport incompletely spliced RNAs to the cytoplasm. Transport of partially spliced transcripts is accomplished by Rev binding to a site on the viral RNA within the *env* gene called the Rev-responsive element (RRE). The RNA-bound Rev then interacts with normal cellular machinery responsible for protein export from the nucleus to mediate the movement of the RNA through the nuclear pore. By promoting translation of the virion structural proteins and some of the accessory proteins, Rev turns up late gene expression that leads directly to a high rate of virus production.

Rev promotes export of unspliced and partially spliced transcripts to cytoplasm

The Nef accessory protein appears to interfere with immune recognition of infected cells. Nef causes the internalization and degradation of the CD4 protein, which likely prevents superinfection and contributes to virus release by preventing the formation of complexes between the cellular receptor and newly synthesized virions. Nef also causes the down-regulation of cell surface major histocompatibility complex (MHC) I molecules, which may prevent recognition of infected cells by cytotoxic T lymphocytes. In addition, virions produced in the absence of the Nef protein are at least partially blocked at some step before integration. The combination of these and perhaps other effects allows the Nef protein to play an essential pathogenic role in an infected individual.

Nef down-regulates CD4 to avoid superinfection and also down-regulates MHC I to interfere with immune recognition

The Vpu protein of HIV-1 appears to play two separate roles during the late stages of infection. In the absence of Vpu, the Env protein forms complexes with CD4 in the endoplasmic reticulum and fails to reach the plasma membrane of the cell. One of the roles of Vpu is to target the destruction of CD4 in the endoplasmic reticulum to allow for incorporation of Env into newly synthesized virions. The second role of Vpu is to promote the release of virions from the infected cell by an unknown mechanism.

Vpu targets CD4 destruction and virion release

The Vpr protein is dispensable for HIV-1 replication in T-cell lines, but required for efficient viral replication in monocytes/macrophages. Several possible roles for Vpr in HIV-1 replication have been suggested, including modest transactivation of HIV-1 LTR, enhancement of the nuclear migration of the preintegration complex in the newly infected nondividing cells, inhibition of establishment of chronic infection, arrests of cells in the G2/M phase of the cell cycle and induce latent cells into a high level of virus production. Furthermore, successful infection of nondividing cells such as macrophages requires Vpr to allow the newly synthesized viral DNA to reach the nucleus and be integrated into the cellular DNA.

Vpr promotes transport of the preintegration complex into the nucleus of nondividing cells

Vpr arrests cells in G2/M phase of the cell cycle

HIV-2 encodes Vpx and not Vpu. Vpx has homology to Vpr and shares the functions of Vpr. The functions of HIV-2 Vpr and Vpx have been segregated, including HIV-2 Vpr maintaining the ability to induce G2 arrest, whereas Vpx retains the ability to enhance infection of nondividing cells such as macrophages.

Vif (virion infectivity factor) increases the infectivity of HIV-1 in primary T cells and certain "nonpermissive" cells in culture. In the absence of Vif, the virus fails to complete reverse transcription in these cell types. "Permissive" cell lines infected by mutants defective in the *vif* gene produce normal yields of infectious virus. One possible explanation for this observation is that "permissive" cells contain a factor that can substitute for the missing Vif protein. Thus, one role of Vif may be to extend the host range of HIV-1 to cell types that would otherwise not be infected. Vif inhibits an RNA editing enzyme, APOBEC3G (apolipoprotein B, a member of innate immune system), that causes hypermutation in HIV DNA after reverse transcription.

Vif increases efficiency of infection and yield of virus

Superimposed on this complex regulatory network is the fact that the viral promoter contains elements that are sensitive to specific cellular transcription factors. This observation may help explain why virus production in CD4+ T lymphocytes is greatly increased when the cells are activated. Clearly, the outcome of an HIV-1 infection is determined by a complex interplay between a very large number of different factors.

Activation of CD4+ T lymphocytes increases virus production

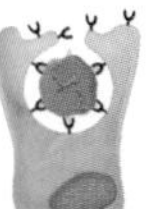

ACQUIRED IMMUNODEFICIENCY SYNDROME (AIDS)

CLINICAL CAPSULE

The primary infection in AIDS ranges from asymptomatic to an infectious mononucleosis-like illness with up to a few weeks of fever, malaise, arthralgias, and rash. A long asymptomatic period follows (usually years), after which the disease, AIDS, emerges. The progressive findings directly due to the virus are wasting, diarrhea, and neurologic degeneration. The effect of the virus on the immune system causes an extensive array of viral, bacterial, fungal, and parasitic opportunistic infections whose findings are the same or worse than those seen in patients without AIDS.

EPIDEMIOLOGY

AIDS was first recognized in the United States in 1981, when it became apparent that an unusual number of rare skin cancers (Kaposi sarcoma) and opportunistic infections were occurring among male homosexuals. These patients were found to have a marked reduction in CD4+ T lymphocytes and were subject to a wide range of opportunistic infections normally controlled by an intact immune system. The disease was found to progress relentlessly to a fatal outcome and was first identified in male homosexuals, hemophiliacs who were receiving blood-derived coagulation factors, and injection drug users.

First recognized in male homosexuals, hemophiliacs, and drug abusers

Retrospective serologic studies with material saved from patients in various studies indicate that HIV-1 infection was already occurring in Africa in the 1950s and in the United States in the 1970s. In 1985, HIV-2 was found to be endemic in parts of West Africa and to cause AIDS. To date, this virus has been relatively restricted geographically, although HIV-2 infections have occurred in the western hemisphere.

HIV-2 is endemic in West Africa

Transmission

The HIV virus is transmitted between humans in three ways: sexually, perinatally or vertically, and by exposure to contaminated blood, or blood-derived products. The virus has been demonstrated in particularly high titers in semen and cervical secretions, and the majority of cases result from sexual contact—both homosexual and heterosexual. Heterosexual contact is the major route of transmission worldwide. Infection is facilitated by breaks in epithelial surfaces, which provide direct access to the underlying tissues or bloodstream. The relative fragility of the rectal mucosa and the large numbers of sexual contacts are probable contributing factors to the predominance of the disease among promiscuous male homosexuals. HIV-1 is transmitted heterosexually to females by vaginal or cervical routes, despite natural barriers such as multicellular layers of squamous epithelial cells of vaginal mucosa and antimicrobial activity of cervicovaginal secretions.

The risk of transmission further increases with the disruption of integrity of the vaginal mucosa because of dry or traumatic sex and other infectious and inflammatory diseases. Once the virus is deposited in the vaginal mucosa, the virus can also traverse the vaginal mucous layer and probably reach the dendritic projections of Langerhans cells followed by infection of submucosal cells such as macrophages, T lymphocytes, and dendritic cells. Transmission appears to be more efficient from men to women, but the reverse is clearly documented. The probability of HIV transmission per unprotected sexual act is estimated at 0.0003 to 0.0015. The risk of perinatal transmission from an infected mother to her child has been estimated to range from 15% to 40% (average around 30%) without any antiretroviral therapy. Mother-to-child transmission can occur prepartum (via transplacental route), intrapartum (through birth canal), and postpartum (through breast milk). It is important to note that antiretroviral therapy during pregnancy can significantly reduce the risk of mother-to-child transmission of HIV-1.

Transmission is sexual and by exposure to infective fluids

Perinatal transmission can readily occur

Propagation of HIV-1 in cell culture and characterization of viral antigens allowed development of effective test procedures for detecting HIV infection. These almost eliminated the risk of transmission by blood transfusion; testing of donors and the use of recombinant or specially treated coagulation factors have now virtually eliminated these sources of infection. Until serologic tests for the infection became available in 1985, more than 10,000 cases of AIDS were probably acquired in the United States through blood transfusion, and about 80% of hemophiliacs treated with coagulation factors derived from pooled blood sources became infected. Transmission of infection by blood is now largely associated with sharing of needles and syringes by injecting drug users, and this has been an increasing source of the disease. In some areas of the world, the seroprevalence of HIV positivity among injecting drug users has been as high as 70%.

Testing of blood supply reduced risk

Intravenous drug abusers are at extremely high risk

Transmission of infection to health care workers after accidental sticks with potentially contaminated needles is very rare (considerably less than 1%), presumably because the amount of infectious virus in the blood of infected person is small and because larger volumes or repeated exposures are needed for a significant chance of infection. Nevertheless, transmission has occurred from both clinical and laboratory exposure, and extreme care in handling needles, sharps, and so on, is necessary. Transmission does not occur through day-to-day nonsexual contact with infected individuals or through insect vectors, because of the fragility of the virus and the need for direct mucosal or blood contact. It is of interest

Accidental needlesticks among health care workers mandate extreme care in prevention

Shed in breast milk, where it may infect breastfeeding infants

that the virus has been detected in saliva, tears, urine, and breast milk. With the possible exception of breast milk, these sources have not been shown to be infectious. Breast milk is considered the major route of HIV-1 vertical transmission in developing countries.

Occurrence

By the end of 2007, 33.2 million people were living with HIV globally, 2.5 million people were newly infected with HIV, and 2.1 million people died of AIDS in 2007 worldwide. Of 33.2 million people globally living with HIV/AIDS, 30.8 million were adults (15.4 million women) and 2.5 million children. Although sub-Saharan Africa has 67% of all HIV-1-infected people in the world, Southern Africa shared the disproportionate global burden of HIV infections and AIDS-related deaths in 2007 of 35% and 38%, respectively. One of the striking trends of the HIV epidemic is that 45% of infected people are between the ages of 15 to 24. At the end of 2003, approximately 1 million (1,039,000 to 1,185,000) people have been living with HIV/AIDS in the United States, including 47% blacks, 34% whites, 17% Hispanics, and about 1% Asians/Pacific Islanders and American Indians. Males accounted for 74% of the HIV-infected population, and more than one-half million people have died with HIV/AIDS. From 2002 to 2006, the HIV epidemic has stabilized in the United States. The highest prevalence rates (47%) have been in homosexual (men who have sex with men, MSM) followed by high-risk heterosexual contact (27%), intravenous drug users (22%), and those infected with both male-to-male and injection drug use (5%). In 2006, 56,300 new cases of HIV-1 infection were reported in the United States. In some areas of the United States, 40% to 60% of homosexual males attending sexually transmitted disease clinics were found to be infected.

Prevalence rates have shifted over time, with increasing cases among women and economically disadvantaged minority groups

The epidemiology of HIV infection is changing in the United States as the pandemic evolves and as the modes of transmission become more generally understood. The numbers and proportions of heterosexually transmitted, and/or drug abuse–related, cases are increasing, particularly among the poor and disadvantaged racial minorities. Antibody rates in prostitutes may be as high as 40%, depending partly on the degree of associated intravenous drug abuse. Prevalence rates in the heterosexual population, in general, are currently less than 1% but have been increasing. In 1985 in the United States, only 7% of AIDS cases were in women; by 2006, the percentage had risen to 26%. Approximately 2000 newborns per year used to be infected by HIV perinatally, but this number has significantly decreased because more pregnant women receive antiretroviral therapy during pregnancy. Black patients now account for 50% of cases, exceeding the percentages in non-Hispanic white men.

Men and women nearly equally infected in Africa and Asia

In contrast to the situation in the United States and Western Europe, heterosexual transmission is the primary route of transmission in Africa and Asia, where there is an approximately equal distribution of infection and disease between the sexes. This may be due to a high incidence in these areas of ulcerative genital lesions caused by other sexually transmitted diseases. These lesions facilitate passage of virus into the tissues of others during intercourse. In central and eastern Europe, where there is an emerging epidemic, the most common risk factor is intravenous drug use.

Increasingly widespread in Africa, South America, parts of Asia

AIDS has been reported in more than 150 countries. The disease continues to spread rapidly in Africa and South America. In sub-Saharan Africa alone, 25 million people are infected, and there are 4 million new cases per year. In some countries in Africa, 25% of the population and up to 60% of the women are HIV antibody–positive. Until recently, the Far East had few cases, but now there is epidemic spread, especially in South and Southeast Asia (India, South China, Burma, Thailand, Cambodia, Viet Nam, and Malaysia). In India and China, there are more than 2.4 million and 700,000 patients with AIDS, respectively, and the rate of new cases is increasing significantly every year. HIV-2 infection is found primarily in West Africa and is spread by heterosexual transmission. Infection by this virus has, however, been reported in Europe in homosexual men, injection drug users, transfusion recipients, and hemophiliac men. For example, in Russia, there were 940,000 AIDS cases at the end of 2007 and a prevalence rate of HIV among adults of 1.1%.

HIV Clades and Geographic Distribution

Class M most common

Clade or subtype B found in the United States

Based on genetic variation, three classes of HIV-1 have developed worldwide, including M (major), O (outlying) and N (new). However, class M accounts for more than 90% of all HIV-1 cases globally and is further classified into several subtypes or **clades**, including A to H and recombinants. In addition, the demographic distribution of individuals infected with particular clades is becoming heterogeneous with the progressing pandemic. However, several clades predominate in a given region of the world, including clade B (Americas, Europe and Australia), clade C (India and South Africa), clade E (Southeast Asia), and most major clades and recombinants (Africa). Among all clades circulating worldwide, clade C is found in more than 50% of HIV-1-infected people. The interclade variation in the envelope gene is in the range of 20% to 30%, whereas intraclade variation is 10% to 15%. There is also some argument that certain clades may have increased risk of transmission and progress to AIDS more rapidly than others. Understanding the immunopathogenesis of the emerging HIV-1 clades is key to vaccine development.

PATHOGENESIS

The pathogenesis of HIV-1 infection is very complex, but the following factors are likely to be important in the disease-causing process.

Infection

The initial target of HIV-1 is CD4 molecules and a chemokine receptor (CCR5), particularly on the surface of monocytes/macrophages and CD4+ helper T lymphocytes.. The virus can also infect other human cells expressing CD4 and CXCR4, and a wide range of CD4-negative cells, including renal and gastrointestinal epithelium and brain astrocytes. The mechanism for infection of non–CD4-bearing cells is unknown but may involve other receptors or fusion with cells already infected with HIV, probably through chemokine receptors. HIV-1 of R5 phenotype is predominantly transmitted both horizontally and vertically. The first cell type to be infected is most likely macrophages or Langerhans cells via CD4 and CCR5. The virus replicates in macrophages, and these cells could serve as a reservoir for continued expansion of the infection to other cell types, especially CD4 T lymphocytes (the major target cells) by cell-to-cell fusion, which allows the virus to spread without being exposed to neutralizing antibody. Infected macrophages may participate in breakdown of the blood–brain barrier, allowing enhanced exposure of the central nervous system (CNS). Although CNS and intestinal disturbances are a prominent part of fully developed AIDS, it is not clear whether they are a direct result of infection of these cells or mediated by cytokines from infected macrophages and T lymphocytes. Initially in infection, the predominant phenotype of HIV-1 is R5 in infected people, whereas late in infection HIV-1 becomes X4, which replicates more efficiently in CD4 T lymphocytes, causing cytopathic effects.

Major targets are CD4-bearing cells, but other cell types can also be infected

Kinetic studies of changes in viral load with antiviral therapy demonstrated that the half-life of HIV in plasma is 5 to 6 hours and an estimated 10 billion HIV particles are produced everyday in an infected individual. In other words, more than 50% of the viral load measured on any given day has been produced in the last 24 hours. Because 99% of the viral load is produced by cells that were infected within the last 48 to 72 hours, cell turnover must be equally rapid. Indeed, when similar kinetic studies are performed on changes in CD4 cell counts, it is estimated that up to 1 billion CD4 cells are produced per day in response to the infection and that the half-life of these cells is only 1.6 days.

Rapid turnover of CD4+ cells during infection

Clinical Latency

The long asymptomatic period after HIV infection (clinical latency) occurs despite active virus replication in the host. Several factors can terminate the long clinical latency period of HIV-1. Mutations occur during viral replication, which appear to enhance induction of virulent forms of the virus, with increased cytopathic capacity and altered cell tropisms. Thus, the mutated forms of HIV-1 isolated from later stages of disease infect a broader range of cell types and grow more rapidly than those isolated in the asymptomatic period. Initially, it was believed that little or no viral replication occurred during this latent period, but studies of lymph nodes of individuals with early asymptomatic disease have shown intense immunologic reactions within the lymphoid tissue at early stages of disease. This implies that the immune system is capable of controlling the virus to some degree early in the course of disease, an ability that is later lost as the disease progresses over time. **Figure 18–5** shows the temporal changes in viral load, anti-HIV immune responses, and total CD4 T-cell counts during various stages of HIV infection.

Some immune control of virus during the clinical latency period, but this is later lost

Recent studies of HIV infection have shown that the level of free virus in the plasma increases in direct relation to the stage of disease. Individuals with early-stage disease have less than 10 infectious virions per milliliter of plasma, whereas those in late-stage disease have between 100 and 1000/mL. These studies imply that either viral replication was

Level of plasma viremia directly correlates with disease progression

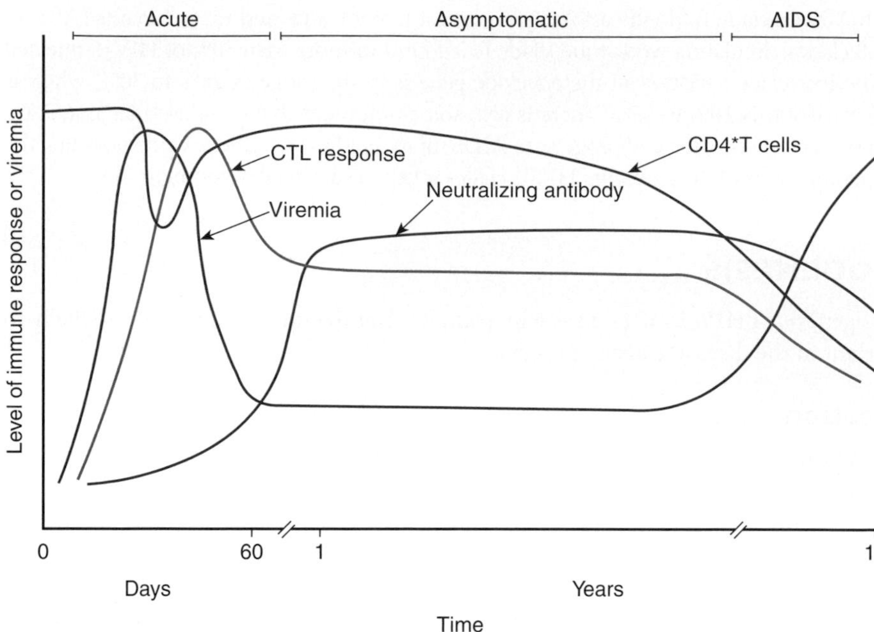

FIGURE 18–5. Temporal changes in viral load, anti-HIV immune responses, and total CD4 T-cell counts during various stages of HIV infection.

increasing during later stages of disease as a result of more virulent mutations and/or the immune system had lost its ability to clear free virus as the disease progresses.

Immune Deficiency

Immune deficiency related to reduction in numbers and normal functions of CD4+ T lymphocytes

The primary immune defect in AIDS results from the reduction in the numbers and effectiveness of CD4+ helper-inducer T lymphocytes, both in absolute numbers and relative to CD8+ suppressor T lymphocytes. This is due to direct killing of CD4+ T lymphocytes by the virus but may also involve other mechanisms. These include secondary killing of uninfected (bystander) cells during cell fusion, autoimmune processes that lead to the elimination of CD4+ T lymphocytes by opsonophagocytosis, and antibody-dependent cell-mediated cytotoxicity (ADCC) directed at gp120 expressed on the CD4+ cell surface. There are also functional defects in CD4+ T lymphocytes affecting lymphokine production and leading to inhibition of some macrophage functions.

Infected individuals are susceptible to other infections and malignancies

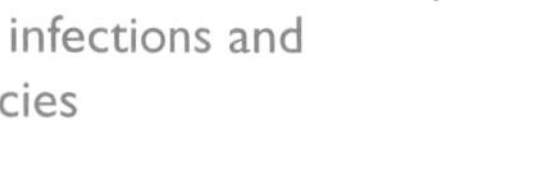

Effects on CD4+ T lymphocytes thus lead to a generalized failure of cell-mediated immune responses, but there is also an effect on antibody production due to polyclonal activation of B cells, possibly associated with other viral infections of these cells. This overwhelms the capacity of infected individuals to respond to specific antigens. The end result of these processes is a disturbance of immune balance that can give rise to malignancies as well as the susceptibility of AIDS patients to a range of opportunistic viral, fungal, and bacterial infections.

CLINICAL ASPECTS

MANIFESTATIONS

Infection is lifelong

In 1993, the Centers for Disease Control and Prevention (CDC) definition of AIDS stated that all patients who are HIV antibody–positive and have CD4+ T-lymphocyte counts lower than 200/mm^3 or less than 14% of total T lymphocytes have the disease. The initial infection with HIV is usually asymptomatic, although in some cases a mononucleosis-like illness develops 2 to 4 weeks after infection and lasts about 2 to 6 weeks. This illness may exhibit any or all of the following: fever, malaise, lymphadenopathy, hepatosplenomegaly, arthralgias, and rash. Sometimes a mild aseptic meningitis is also present. Whether or not these early manifestations of infection occur, the virus rapidly invades, persists, and integrates into the genome of some host cells, and the individual is thus infected for life.

The initial infection is followed by an asymptomatic period that, in most cases, continues for years before the disease becomes clinically apparent. During this time, virus can be isolated from blood, semen, and other bodily fluids and tissues. Approximately 50% of infected individuals develop significant disease within 10 years of infection, and the number continues to increase thereafter. It is expected that nearly all HIV-infected persons eventually develop some clinical aspects of this infection, although long-term (more than 10 years) nonprogressors are well documented. Approximately 5% of infected, untreated patients show no decrease in CD4 counts over a period of more than 10 years, but ultimately many of these individuals begin to progress. Since the late 1990s, the increases in early diagnosis, combined with more aggressive, highly active antiretroviral therapy (HAART) in the United States, has greatly reduced opportunistic infections and delayed progression to death (**Figure 18-6, 18-7**).

Progression to AIDS is highly variable among individuals

As the disease progresses in untreated patients, the number of CD4+ T lymphocytes declines. There is increasing immunodeficiency, and opportunistic infections become more frequent, severe, and difficult to treat. One of the best markers of the severity of AIDS is the absolute number of CD4+ T lymphocytes. Those individuals with overt AIDS almost always have fewer than 200 CD4+ T lymphocytes/mm^3 of blood (normal = 800 to 1200/mm^3), although opportunistic infections may occur with CD4+ T cells greater than 200/mm^3.

Individuals with overt AIDS usually have fewer than 200 CD4+ lymphocytes/mm^3

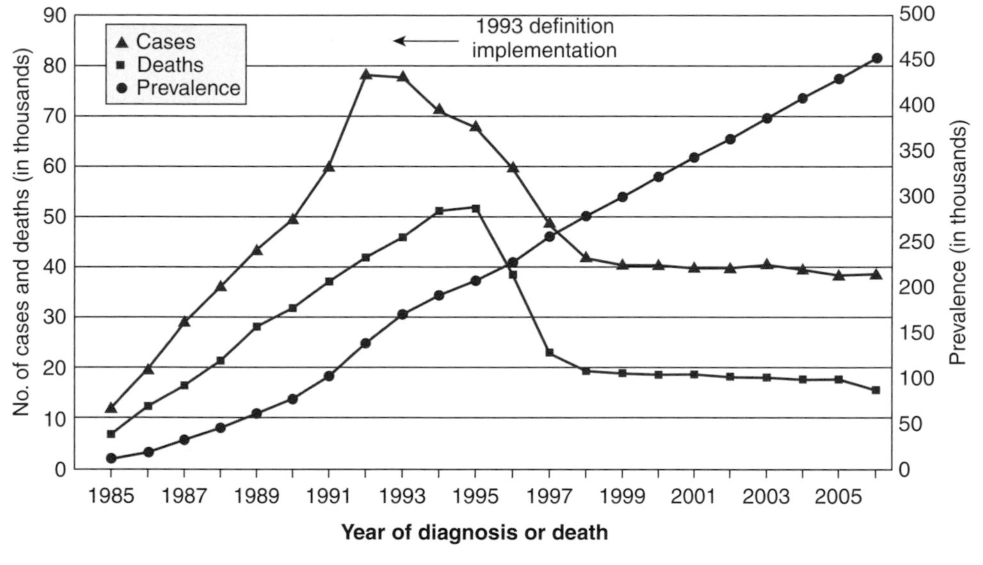

FIGURE 18–6. Estimated number of AIDS cases, deaths and persons living with AIDS, 1985-2006 (From AIDS Surveillance Trends, HIV/AIDS Statistics and Surveillance Center for Disease Control and Prevention, 2008.)

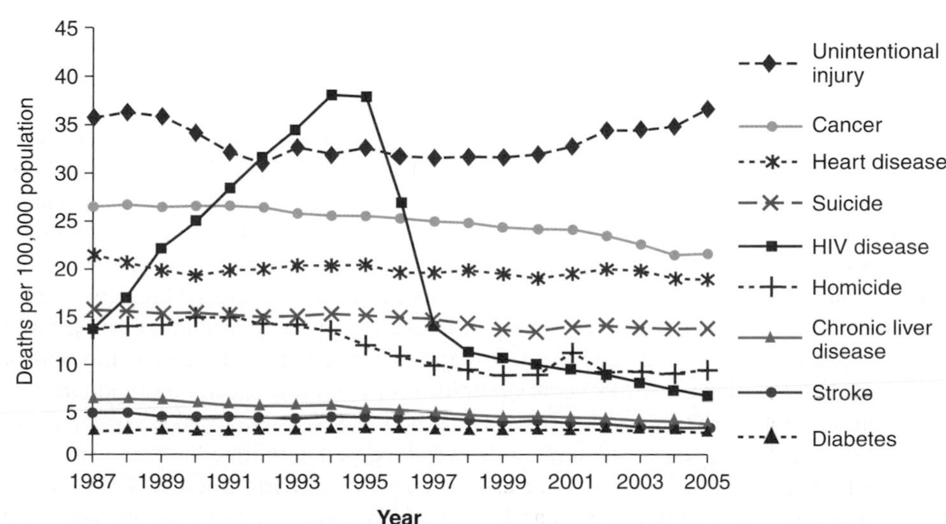

FIGURE 18–7. Trends in annual rates of death per 100,000 population from leading causes of death among persons 25 to 44 years old, United States, 1987–2005. (HIV Mortality Report, HIV/AIDS Statistics and Surveillance Center for Disease Control and Prevention, 2008.)

TABLE 18–3	Common Opportunistic Infections and Malignancies in Patients with Untreated AIDS
Protozoan	
Toxoplasmosis	
Isospora belli infection	
Cryptosporidiosis	
Fungal	
Pneumocystis jirovecii pneumonia	
Cryptococcosis	
Candidiasis	
Histoplasmosis (disseminated)	
Coccidioidomycosis (disseminated)	
Mycobacterial	
Mycobacterium tuberculosis (Disseminated tuberculosis, especially extrapulmonary)	
Mycobacterium avium-intracellulare complex infections	
Viral	
Persistent mucocutaneous herpes simplex	
Cytomegalovirus retinitis, gastrointestinal, or disseminated infection	
Varicella-zoster, persistent or disseminated	
Progressive multifocal leukoencephalopathy (JC virus)	
Opportunistic malignancies	
Kaposi sarcoma	
Lymphoma	

Patients with full-blown untreated AIDS experience a wide spectrum of infections depending on the severity of their immune defect and on the opportunistic organisms in their normal flora or those with which they come in contact (**Table 18–3**). Some clinical manifestations of AIDS may thus vary by locale. For example, disseminated histoplasmosis was a common complication in the midwest United States, as was disseminated toxoplasmosis in France. These infections are uncommon in areas where the diseases are not endemic. The diversity and anatomic sites of infection vary among patients, and any one patient may have several infections. The most common infection is pneumocystosis, and approximately 50% of AIDS patients who do not receive anti HIV therapy or prophylaxis for pneumocystosis develop *Pneumocystis jirovecii* pneumonia. In the past, about 25% of all AIDS patients developed Kaposi sarcoma, but the number of cases has been falling in the United States despite increasing numbers of cases of AIDS. The apparent explanation is that Kaposi sarcoma is due to a transmitted agent different from HIV, the Kaposi sarcoma herpesvirus (KSHV). The spread of this organism has diminished as high-risk sexual behavior has decreased, especially among homosexual men. Disease due to mycobacteria of the *Mycobacterium avium-intracellulare* complex is common, and AIDS patients are also highly susceptible to *Mycobacterium tuberculosis* infection. Oral thrush and esophagitis due to *Candida albicans* and meningitis due to *Cryptococcus* are commonly encountered fungal infections. Persistent progressive mucocutaneous herpes simplex and herpes zoster infections are common. Cytomegalovirus (CMV) chorioretinitis is one of the most common opportunistic infections and may result in unilateral or bilateral blindness. Disseminated CMV infection is also seen, and patients present with fever and visceral (eg, gastrointestinal) organ involvement.

Pneumocystosis, candidiasis, mycobacteriosis, and CMV are common

Specific opportunistic infections are associated with differing levels of CD4+ T-lymphocyte counts. For example, fungal and tuberculous pneumonia may occur with CD4+

T-lymphocyte counts of 200 to 500 cells/mm^3, whereas CMV and *M avium-intracellulare* disease are seen almost exclusively in those whose counts are lower than 50 to 100 cells/mm^3.

CMV retinitis and mycobacterial dissemination usually occur with extremely low CD4+ counts

As the duration of survival of AIDS patients became longer as a result of therapy with the earliest drugs, an increased number of patients developed neurologic manifestations of the disease and lymphoid neoplasms, especially non-Hodgkins lymphomas. HIV is a neurotropic virus and can be isolated from the cerebrospinal fluid of 50% to 70% of patients. CNS involvement may be asymptomatic, but many patients develop a subacute neurologic illness that produces clinical symptoms varying from mild cognitive dysfunction to severe dementia. Loss of complex cognitive function is usually the first sign of illness. Progression to severe memory loss, depression, seizures, and coma may ensue. Cerebral atrophy involving primarily cortical white matter can be demonstrated by computed tomography or magnetic resonance imaging. Histologically, focal vacuolation of the affected brain tissue with perivascular infiltration of macrophages is noted. Multinucleated giant cells with syncytium formation surround the perivascular infiltrates. Neurologic symptoms do not usually occur until CD4+ T-lymphocyte counts are lower than 200 cells/mm^3.

HIV is also neurotropic and can lead to dementia

The disease spectrum in Africa is similar in many respects to that in the Western world, but many more patients present with severe intractable wasting and diarrhea, known as **slim disease**. Tuberculosis is also more commonly encountered in AIDS patients in Africa, reflecting the higher incidence of the disease in the population in general. The 2-year mortality rate of persons with AIDS, once the disease has been fully established, was initially 75%, with nearly all persons eventually dying of opportunistic infections or neoplasms.

DIAGNOSIS

The diagnosis of AIDS is most commonly confirmed by demonstrating antibody to the virus or its components. Initial screening tests are performed using whole viral lysates as the target antigens in enzyme-linked immunosorbent assay (EIA) tests. These tests have a high level of sensitivity, but because false-positive results occur, all positive EIA tests must be confirmed. The confirmatory test is a Western blot analysis, which detects antibodies to specific HIV proteins. In this procedure, HIV proteins are separated by electrophoresis, transferred to nitrocellulose paper, and incubated with patient sera; antibody bound to the individual proteins is detected by enzyme-labeled anti–human globulin sera (**Figure 18–8**). Sera from infected patients

EIA screens for antibody

HIV rapid tests screen for HIV antibodies and require no instrumentation

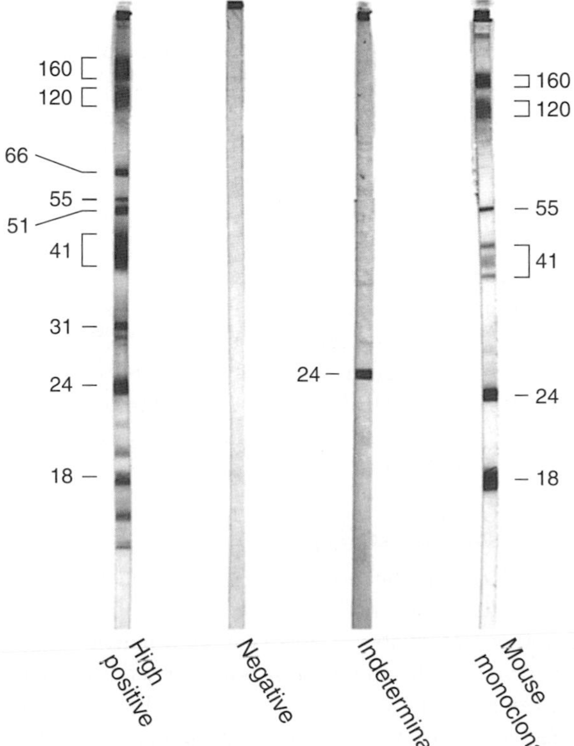

FIGURE 18–8. Western blot detection of HIV-1 antibodies. Note that the "high-positive" serum exhibits antibodies to the HIV-1 envelope glycoproteins of 160, 120, and 41 kilodaltons (kD), to the Gag (core) proteins of 24 and 17 kD, and to other HIV proteins (55 and 51 kD). The "indeterminate" serum exhibits antibody to only the Gag (core) 24-kD protein. The mouse monoclonal blot is a positive control and contains antibodies to key HIV antigens. A positive sample should exhibit antibodies to both envelope and Gag proteins or to both envelope proteins (41 and 120/160 kD).

have antibodies that react with the envelope glycoproteins, core proteins, or both. Tests made with HIV-1 detect antibody in 60% to 90% of patients infected with HIV-2. The FDA has also approved rapid HIV antibody tests that can be used in both clinical and nonclinical settings and can help to overcome some of the barriers to early diagnosis. These are screening tests that are interpreted visually and require no instrumentation. Like the EIA test, all require confirmation if reactive. In these rapid tests, HIV antigens are affixed to the test membrane and if HIV antibodies are present in the specimen being tested, they bind to the affixed antigen. The colorimetric reagent provided in the kit binds to these immunoglobulins and visually detected.

Western blot used for confirmation

The combination of EIA and Western blot tests gives a high degree of specificity to test results, but antibody is not detectable by these procedures in the first 2 to 4 weeks after infection. During this period, the individual can still transmit the infection to others by sexual contact or blood donation. Closing this detection gap is particularly important for protection of blood products for transfusion. Although the virus can be grown during this time in mixed lymphocyte cell culture, the methods are impractical and may not be positive for up to 1 month. More practical approaches include nucleic acid–based assays such as the polymerase chain reaction (PCR) for plasma HIV RNA or DNA and the branched-chain DNA (bDNA) assay. These are also useful in assessing the benefits of antiviral therapy, as well as in determining whether infants born to seropositive mothers are infected or simply demonstrating passively transmitted transplacental antibody.

Viremia precedes appearance of antibody by 2 to 4 weeks

Quantitation of plasma HIV RNA plays an especially important part in management. For example, if a patient's HIV RNA copy number rises during therapy, or fails to fall to low levels (eg, lower than 50 copies/mL), this signals that the antiviral efficacy of the drug regimen is inadequate. The most likely explanation is mutational resistance that either preexisted or developed during treatment. Other explanations to be considered include patient noncompliance and inadequate dosing.

PCR and bDNA testing used to quantitate plasma viremia and assess drug efficacy

TREATMENT

Initially, only nucleoside inhibitors of HIV reverse transcriptase were available for therapy of HIV infection, and they were used singly and found not to be beneficial to HIV-infected patients. New inhibitors targeting mainly reverse transcriptase and protease were developed. Currently, there are six classes of antiretroviral agents, including nucleoside analog reverse transcriptase inhibitors (NRTIs), non-nucleoside analog reverse transcriptase inhibitors (NNRTIs), protease inhibitors (PIs), and the gp41 fusion inhibitor (enfuvirtide). Recently, two new classes of agents, CCR5 inhibitors (maraviroc) and integrase inhibitors (raltegravir) have been approved. These inhibitors are used in a combination therapy (at least three separate inhibitors) known as highly active antiretroviral therapy (HAART). The current recommendation is to use a backbone of two NRTIs (in any number of combinations) and the third antiviral could be an NNRTI or a protease inhibitor. Other antiviral agents, especially HIV-1 entry inhibitors and those targeting HIV-1 regulatory and accessory proteins, are under development. The characteristics of current representative anti-HIV agents are further summarized in Chapter 8. Recent advances in therapy have slowed progression of the disease. Combination therapy, with the inclusion of inhibitors of HIV protease, appears to be responsible for dramatic improvement in many patients, but toxicity or the development of resistance may limit their long-term usefulness. Progression of AIDS has become much less common with the advent of HAART. However, successful suppression of HIV by HAART can reconstitute CD4 T lymphocytes numbers that cause an inflammatory response known as immune reconstitution inflammatory syndrome (IRIS). Some of the common coinfections that may be exacerbated by IRIS are tuberculous and non-tuberculous mycobacteria, CMV retinitis, cryptococcal meningitis, hepatitis B and hepatitis C.

Combinations of drugs used in treatment

Initiation of Treatment

Because viral replication proceeds at the phenomenal rate, it seems most rational to begin treatment as soon as infection is detected. However, considerations of toxicity, resistance development, quality of life, cost, and patient wishes are extremely important additional determinants. Although these issues may cause debates regarding early intervention, there is a general consensus that combination therapy has limited benefit when initiated in HIV-1 infected patients with CD4+ counts higher than 500/mm^3. In primary acute infection (acute retroviral syndrome), several studies suggest a significant short-term improvement in virologic and immunologic markers in patients who receive antiretroviral therapy. However, viral replication resumes after withdrawal of antiretroviral therapy, even after prolonged periods of virus suppression and increment of HIV-1-specific CD4 and CD8 T-cell responses. Therefore, there is no consensus at the moment regarding treatment of acute primary infection. However, the benefits of HAART in chronic HIV infection are well established. Because current therapy is unlikely to eradicate HIV-1 infection, most patients are likely to stay on therapy for life. There is broad agreement with the United States Department of Health and Human Services (DHHS) recommendation that HAART should be initiated in all patients with AIDS-defining illness or severe symptoms as well as in those patients with CD4 counts lower than 350 cells/mm^3 regardless of symptoms. In patients with CD4 counts greater than 350 cells/mm^3, treatment may be offered, but there are pros and cons. The current treatment recommendation and guidelines are available at CDC and DHHS websites.

Decision to treat aggressively is influenced by CD4+ count and AIDS-defining illness or severe symptoms

Resistance

HIV-1 error-prone reverse transcriptase enzyme and high rates of viral replication contribute to frequent mutations. As a result, resistance to an antiviral is a regular and often rapid development. Use of antiviral therapies that maximally suppress HIV viral load appear to diminish the appearance of resistant virus, especially combination therapy. The emergence of resistance occurs at a rate proportional to the frequency of preexisting variants and their relative growth benefit in the presence of antiviral. Antiviral resistance can be determined by genotypic and phenotypic assays, which are important tools in decisions related to initiation or modification of therapy.

Drug resistance is an expected development with treatment

In addition to the primary antiviral treatment of HIV, patients with CD4+ counts of less than 200/mm^3 should begin prophylactic regimens to prevent *P. carinii* pneumonia. When CD4+ counts are less than 75 to 100/mm^3, they should receive prophylaxis for mycobacterial and fungal infection.

Prophylaxis of opportunistic infections is especially important

PREVENTION

The spread of AIDS has been facilitated by changing sexual mores, injection drug use, and, in some parts of the world, disruption of family and tribal units as a consequence of industrialization and urbanization. These factors are obviously not subject to rapid change. Immediate prevention must be based on education about the means of transmission and easy access to condoms and safe needles for those large numbers of people who continue to place themselves at risk. The epidemiologic and laboratory methods used to control foci of other major epidemic diseases pose particular problems in AIDS control at present. Apart from questions of potential discrimination against infected individuals and the calamitous effects of false-positive serologic test results, the sheer magnitude and cost of case finding and contact tracing at present limit this approach.

Education is the cornerstone of prevention

Screening for asymptomatic infection in pregnancy aids effective prophylaxis

Much research is underway to develop vaccines against the virus, but the marked mutability of HIV greatly complicates this approach. Furthermore, passage of virus between fused cells and in syncytia protects it from antibody neutralization in established disease. The search continues for conserved epitopes of the surface glycopeptides that might provide possible antigenic targets. Antiviral treatment using combinations of agents may prevent infection of accidentally exposed individuals (eg, health care workers). This therapy must be initiated within hours of an accident if it is to have any chance of success. Detection and treatment of HIV-infected pregnant women is very effective in reducing perinatal infection. Cesarian section delivery, particularly that which is elective rather than emergent, is also a preventive, as is the avoidance of breastfeeding by HIV-positive mothers. Condoms, properly used, do prevent HIV transmission, bidirectionally and with efficacy rates up to 85%. Circumcision of males decreases the risk of acquisition of HIV by men but has not been clearly shown to reduce transmission to females. Screening of blood for HIV by antibody and nucleic acid testing is very effective.

Condoms, if properly used, can effectively prevent transmission

Male circumcision decreases HIV transmission in men

TRANSFORMATION BY RETROVIRUSES

Oncoretroviruses cause a variety of cancers in animals and humans, including leukemia, lymphoma, and sarcoma. Oncogenic retroviruses appear to transform cells to an oncogenic state by three distinct mechanisms, including by acquiring a cellular oncogene (acute

Typical retrovirus

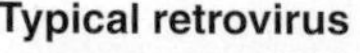

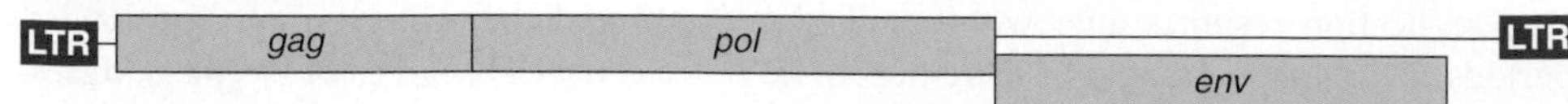

Defective acute transforming retrovirus

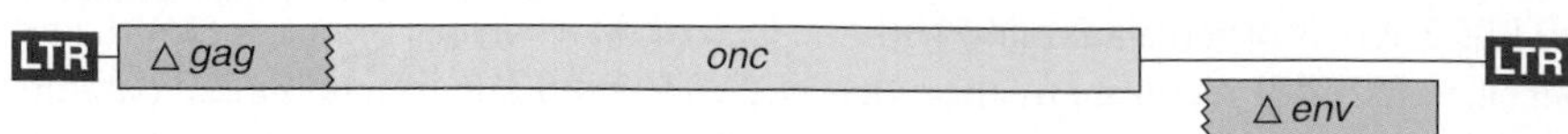

FIGURE 18–9. Comparison of a typical retrovirus with a defective acute transforming retrovirus. Onc, cellular oncogene.

Defective transforming oncogenic viruses require helper virus

Some retroviruses carry host genes rendering them oncogenic

Defective transforming oncogenic viruses require helper virus

transforming viruses), by insertional mutagenesis, and by transforming cells by continual expression of viral regulatory protein (see Chapter 7). The genomes of acute transforming oncoviruses have one feature common to nearly all of them: some viral genes are replaced by genes derived from their hosts that render them oncogenic (see later in text). In every case, the signals required for reverse transcription and for transcription of the provirus, which are located near the ends of the RNA, are retained in the infecting virus. In the example shown in **Figure 18–9**, the *pol* gene and parts of both the viral *gag* and *env* genes are deleted, but other configurations are possible. Such oncoviruses are defective and replicate only in the presence of a helper virus that can supply the missing functions. ::: viral transformation, p. 147

Noncytocidal viruses carrying cellular oncogenes can produce persistent transformation

First, the defective acute transforming viruses (Figure 18–9) have acquired a cellular gene (thereafter called an **oncogene**), which, when expressed in the infected cell, results in loss of normal growth control. On infection, the transduced oncogene is expressed from the viral LTR promoter, resulting in a rapid and acute onset of malignant disease. Persistent transformation by oncogene transduction is possible only for retroviruses that are not cytocidal. More than 30 oncogenes have been identified in a variety of animal retroviruses, but no human retroviruses are known that transform by this mechanism. ::: oncogenes, p. 149

Integration adjacent to cellular proto-oncogenes can activate them

The second mechanism is called **insertional mutagenesis.** Integration of a retrovirus in the vicinity of particular cellular genes can cause inappropriate expression of the gene, resulting in uncontrolled cell growth. These cellular genes are called **proto-oncogenes,** and insertional activation by the virus is apparently due to the close proximity of the integrated viral promoter or enhancer to the gene. Cancers that are caused by this mechanism have very long latent periods because integration is random and only rarely occurs near a cellular proto-oncogene.

HTLV-I transforms by production of Tax, which activates cellular transforming genes

The causative agent of HTLV-I exemplifies the third mechanism. In this case, the integrated provirus in the leukemic cells from any one patient is found at a unique location on a particular chromosome. Thus, the tumors are probably monoclonal. The cancer is not the result of insertional activation, however, because the chromosomal location of the provirus is never the same in any two patients. Instead, transformation results from the continual expression of the viral *tax* gene (the HTLV-I homolog of the HIV-1 *tat* gene; Table 18–2). Apparently, the Tax protein not only can transactivate viral transcription in the same manner as Tat, but Tax can also **transactivate** the expression of one or more cellular genes (possibly proto-oncogenes), resulting in malignant transformation.

HUMAN T-LYMPHOTROPIC OR T-CELL LEUKEMIA VIRUS (HTLV)

HTLV causes adult T-cell leukemia/lymphoma (ATLL), myelopathy

Human T-lymphotropic virus or human T-cell leukemia virus (HTLV) has two members, HTLV-I and HTLV-II, which cause disease in humans. HTLV-I causes two distinct diseases; adult T-cell leukemia/lymphoma (ATLL) and HTLV-associated myelopathy (a neurologic disease). HTLV-II may also cause these diseases but has been primarily linked to hairy cell leukemia.

VIROLOGY

Like other retroviruses, HTLV has the usual retroviral *gag, pol,* and *env* genes, but also encode two regulatory proteins: Tax and Rex. Tax is a transcriptional activator of HTLV

LTR and is also required for transformation. On the other hand, Rex, like HIV Rev, is a post-transcriptional activator that increases transport of structural proteins' mRNAs from nucleus to cytoplasm. In addition, other HTLV proteins are similar to HIV proteins but differ in sequence and antigenicity. The HTLV envelope glycoproteins are gp46 and gp21, whereas the capsid protein is p24. Several cellular factors interact with HTLV LTR and activate transcription. Unlike HIV, the receptors for HTLV-I and HTLV-II have not been biochemically identified. However, the receptors are found in a wide variety of human and animal cells. HTLV-I and HTLV-II use the same receptor. HTLV is able to penetrate and infect a number of cell types; however, productive infection is observed in only a few cell types such as T lymphocytes. The replication cycle of HTLV is very similar to that of HIV. Syncytia formation has been demonstrated in T lymphocytes.

Similar retroviral genes with Tax and Rex regulatory proteins

HTLV-I and II use the same unidentified receptor

Preferentially infects T lymphocytes

TRANSMISSION

Transmission of HTLV occurs via blood-to-blood, including homosexual intercourse, heterosexual intercourse, and intravenous drug use. Mother-to-child transmission of HTLV has also been documented. Unlike HIV, HTLV is not transmitted through cell free fluids, but through cell-associated fluids.

Transmission via cell-associated fluids

EPIDEMIOLOGY

HTLV is more prevalent in the Caribbean, Japan, and Hawaii. In addition, the incidence of HTLV is increasing in Western Europe and the United States among intravenous drug users. In some of these endemic areas, the rate of HTLV infection is more than 20%.

PATHOGENESIS

Adult T-cell leukemia/lymphoma (ATLL) is caused by HTLV infection of CD4 T lymphocytes leading to malignant transformation. HTLV-encoded Tax protein that binds to HTLV LTR and increases transcription of HTLV genes is also responsible for enhancing the transcription of proto-oncogenes resulting in transformation. In addition, Tax increases the production of interleukin 2 (IL-2; T-cell growth factor) and IL-2 receptor that cause uncontrolled growth of T cells resulting in transformation. The transformed cells typically do not produce HTLV progeny viruses. The other disease caused by HTLV is called HTLV-associated myelopathy (HAM), or tropical spastic paraparesis (TSP), which is a demyelinating disease of the brain and spinal cord, especially the motor neurons. It is believed that the mechanisms of HAM/TSP are immune-mediated, including an autoimmune reaction–induced damage of the neurons as well as cytotoxic T-cell–induced killing of neurons. The virus becomes latent for a long period of time (approximately 20 to 30 years) or slowly replicates to transform cells without causing cytopathic effects. In terms of immunity, antibodies are elicited against gp46 and other HTLV proteins that neutralize the slowly replicating virus and prevent cell-mediated killing of HTLV-infected cells.

HTLV Tax increases the transcription of protooncogenes

MANIFESTATIONS

HTLV-I causes ATLL, which is a highly malignant disease. There is a long period of latency (about 20 to 30 years) before onset of ATLL. Only 1% to 2.5% of infected people progress to ATLL disease, and their survival is often in months. ATLL patients present with lymphadenopathy, hepatosplenomegaly, and skin and bone lesions. The malignant T cells have a flower-shaped nucleus and are pleomorphic. Fungal and viral opportunistic infections are commonly seen in ATLL patients, especially those treated with aggressive chemotherapy. In HAM/TSP patients, gait stiffness/spasticity, lower limb weakness, and low back pain are generally seen. The flower-shaped T cells can be found in the CSF. The CSF shows lymphocytic pleocytosis, and the proteins level are elevated. In addition, hematologic malignancies, B-cell chronic lymphocytic leukemia, and immunosuppression are found in patients infected with HTLV-I. HTLV-II causes a T cell variant hairy cell leukemia, which resembles hairy cell leukemia of B cell origin.

Long latency period of 20 to 30 years

1% to 2.5% of HTLV infected patients progress to ATLL

CSF finding abnormal in HAM/TSP

DIAGNOSIS

Diagnosis by EIA or PCR

HTLV infection is detected by antibodies against HTLV by EIA; there is cross reactivity with HTLV-I and HTLV-II antigens. PCR can specifically differentiate between HTLV-I and II. ATLL is diagnosed by the presence of malignant T cells in the lesions. HAM/TSP is diagnosed by the presence of HTLV antibody in the CSF or HTLV nucleic acid in the CSF.

TREATMENT

In some patients with HAM/TSP, a combination of antiretrovirals and interferon has shown benefit, and corticosteroids may relieve symptoms. ATLL is generally treated by anticancer chemotherapy.

PREVENTION

Screening for HTLV antibodies, using condoms, and not breastfeeding babies by HTLV-infected mothers can reduce the risk of HTLV transmission. Currently, there is no vaccine to prevent HTLV infection.

CASE STUDY

A MONTH-LONG MULTISYSTEM ILLNESS

A 25-year-old man comes to a clinic accompanied by his girlfriend, complaining of increased dyspnea, fevers, and chills. He also complains of having watery diarrhea and has lost weight over the last month. His chest x-radiograph reveals a bilateral reticular infiltrate. Further lab testing reveals that he is positive for HIV antibodies; his CD4 count is 250 and viral load is more than 100,000 copies/mL. He was born in the United States and lives in Ohio.

QUESTIONS

■ Which of the following is true of HIV viral load/CD4 lymphocyte count?

A. HIV viral load is the better indicator of the risk of opportunistic infections
B. The CD4 count assesses lymphocyte quantitation and functions
C. Recovery of the CD4 count in response to antiviral therapy is a better indicator of clinical outcome than viral load results
D. Decrease in viral load in response to antiviral therapy is associated with increase in CD4 lymphocyte counts in more than 98% of patients

■ What is the most likely cause of pulmonary infection in this patient?

A. Cytomegalovirus
B. Mycobacterium tuberculosis
C. *Pneumocystis jirovecii*
D. Coccidioidomycosis
E. Herpes simplex

■ Which one of the following statements about HIV/AIDS is true?

A. Presence of HIV antibodies in this patient indicates that the infection will be cleared
B. HIV-1 arose as an endogenous virus because HIV-1 DNA is found in normal cells
C. Antibodies to HIV-1 antigens are generated in infected patients but are unable to eliminate the infection
D. If treatment reduces the plasma viral load to undetectable, the patient is cured

■ With regard to the patient's girlfriend, which of the following is true?

A. If she is negative for HIV antibody, there is no need to test her again

B. The risk of HIV transmission from male to female is remote, and she should not be concerned

C. If she is negative now and again in 6 months for HIV antibody, then she is not infected as of this current visit

D. Circumcision of her male partner would reduce the risk of HIV transmission by 50 fold.

ANSWERS

1(C), **2**(C), **3**(C), **4**(C)

CHAPTER 19

Papilloma and Polyoma Viruses

Historically, the papillomaviruses and polyomaviruses have been discussed together in microbiology textbooks, lumped under the category of *papovaviruses*. This term is now known to inadequately discern the unique characteristics that distinguish them (**Table 19–1**); thus, they are considered separately here.

PAPILLOMAVIRUSES

VIROLOGY

Papillomaviruses are small, unenveloped, double-stranded circular DNA viruses exhibiting cubic (icosahedral) symmetry of 55 nm in diameter (**Figure 19–1**). The icosahedral capsid comprises two capsid (structural) proteins, L1 (major capsid protein) and L2 (minor capsid protein). The 8-kb circular double-stranded DNA genome of human papillomavirus (HPV) encodes seven or eight early genes (E1 to E8) and two late structural capsid genes (L1 and L2). The early genes are required for regulation of viral replication and transformation. The virus does not encode any polymerases and therefore is dependent on host cell transcription and replication machinery. Based on DNA homology, there are over 100 genotypes of human papillomaviruses. Papillomaviruses cause epidermal papillomas and warts in a wide range of higher vertebrates. Different members of the group are generally species-specific. For example, bovine and human papillomaviruses infect only the hosts reflected in their names. In some cases, lesions caused by these agents can become malignant, and the role of these agents as causes of certain human cancers is increasingly recognized. Papillomaviruses have been difficult to grow in tissue culture, and most of the virologic information has been derived from molecular and gene expression studies. ::: transcription, p. 118

Naked capsid, double-stranded DNA viruses

Difficult to propagate HPV in tissue culture

The genomes of many of the papillomaviruses have now been cloned and compared by restriction endonuclease and DNA homology procedures. These studies have shown a wide genomic diversity among papillomaviruses that infect different species and also among those that infect humans. This has led to the allocation of numbers for the different genotypes.

Great genomic diversity

The replication cycle of HPV is not very well understood because of the lack of studies in a tissue culture system. However, in infected human tissue, infectious particles are found. HPV infects the basal layer of squamous epithelium and the virus is internalized and uncoated, and the viral DNA is delivered to the nucleus. Host RNA polymerase transcribes early (E) genes followed by early protein synthesis. Some of the early genes, E6 and E7, are involved in transformation that causes an increase in cell division. E6 binds to p53 (tumor suppressor) and E7 p105RB (retinoblastoma) proteins and abrogate cell cycle regulation. The dividing cells carry viral genome (as extrachromosomal DNA), allowing HPV genome

HPV infects the basal layer of the epidermis

Early replication in the basal layer of epidermis

TABLE 19–1 Characteristics of Papilloma and Polyoma Viruses

VIRUS	HUMAN SUBTYPES	TRANSMISSION	DISEASE	TREATMENT	PREVENTION
Papilloma	HPV-1–3, 10	Close contact, occupational exposure, public shower/swimming pool	Skin warts	Topical cytotoxins or surgical removal	
Papilloma	HPV-6, 11	Close contact, sexual contact	Oral, laryngeal papillomatosis Genital warts (condylomata accuminata)	Treatment of laryngeal lesions is complex, varied	Vaccine
Papilloma	HPV-16, 18 (several others are also increasingly implicated)	Sexual	Cervical neoplasia	May be removed by electrocautery	Vaccine
Polyoma	BKV	Respiratory or oral (?)	Hemorrhagic cystitis in transplant recipients; postrenal transplantation nephropathy	Cidofovir may be used, but is not proven	
Polyoma	JCV	Respiratory/oral (?)	Progressive multifocal encephalopathy (PML)	Reduce immune suppression	

BKV, BK virus; HPV, human papillomavirus; JCV, JC virus.

Vegetative DNA replication and viral assembly in terminally differentiated epithelial cells (keratinocytes)

Latent viral DNA maintained in the basal layer of epithelium

to persist in these cells. As the infected cells differentiate to early terminal stages, other viral early genes are expressed, namely, E1 and E2, which are involved in regulation of viral transcription and replication. Viral DNA synthesis occurs at two levels directed by host cell DNA polymerase: (1) in the lower portion of the epidermis to maintain a stable multicopy viral DNA for latent infection, and (2) as vegetative DNA replication, which occurs in the more differentiated epithelial cells. In some cases, papillomavirus DNA can integrate into the host chromosomes. The infected cells further differentiate to a terminal stage (keratinocytes) wherein late gene expression synthesis of late (L) capsid structural proteins and vegetative DNA synthesis take place. At this stage, there is a burst of viral DNA synthesis followed by virus assembly in the nucleus and virus release by cell lysis. ::: transformation, p.147

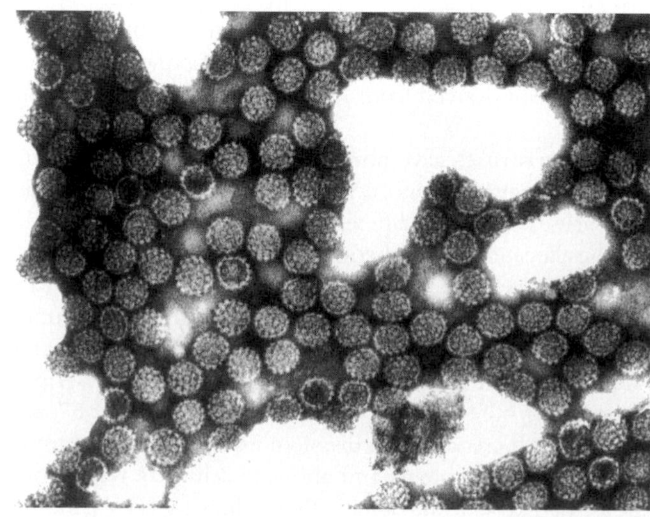

FIGURE 19–1. Electron micrograph of human papillomavirus particles isolated from a plantar wart (×300,000). (Reproduced with permission from Connor DH, Chandler FW, Schwartz DQA, Manz HJ, Lack EE (eds). *Pathology of Infectious Diseases,* vol. 1. Stamford, CT: Appleton & Lange; 1997.)

PAPILLOMAVIRUS DISEASE

CLINICAL CAPSULE

More than 100 genotypes of human papillomaviruses (HPVs) have been identified in human specimens. The genotypes are antigenically different, and groups of genotypes are associated with specific lesions. HPVs have been identified in plantar warts; in flat and papillomatous warts of other skin areas; in juvenile laryngeal papillomas; and in a variety of genital hyperplastic epithelial lesions, including cervical, vulvar, and penile warts and papillomas. In addition, they are associated with premalignant (cervical intraepithelial neoplasia) and malignant disease (cervical cancer). Lesions comparable to those occurring in the cervix are now recognized in the anus, especially among men who have sex with men and are infected by human immunodeficiency virus (HIV).

EPIDEMIOLOGY

Cutaneous nongenital warts usually occur in children and young adults; presumably immunity to the HPV genotypes causing these lesions develops and provides subsequent protection. Over 30 HPV genotypes have been identified in genital lesions of humans, and there are many apparently silent infections with these viruses. Cross-immunity does not occur, and sequential infection with multiple genotypes does take place. The genotypes causing genital lesions are different from those causing cutaneous, nongenital warts. A single sexual exposure to an infected person may transmit the infection 60% of the time; very commonly the infected person is asymptomatic. Having multiple sex partners is the major risk factor for acquiring HPV infection. From 20% to 60% of adult women in the United States are infected with one or another of the genotypes. HPV types 6 and 11 are associated most commonly with benign genital warts in males and females and with some cellular dysplasias of the cervical epithelium, but these lesions rarely become malignant. They can be perinatally transmitted and cause infantile laryngeal papillomas. HPV types 16, 18, 31, 45, and 56 may also cause lesions of the vulva, cervix, and penis. Infections with these viral types, especially type 16, may progress to malignancy. Viral genomes of at least one of these five types are found in the majority—but not all—of markedly dysplastic uterine cervical cells, in carcinoma in situ, and in cells of frankly malignant lesions.

HPV types 6 and 11 common; rarely lead to malignancy

Types 16, 18, 31, 45, and 56 associated with dysplasia and malignancy

Human papillomavirus infection is now considered to be a contributory cause of most carcinomas of the cervix. Papillomavirus infection of the anus is a clinical problem in homosexual men, especially those with acquired immunodeficiency syndrome (AIDS), and it is related to the subsequent development of anal neoplasia in these individuals.

PATHOGENESIS

Papillomaviruses have a predilection for infection at the junction of squamous and columnar epithelium (eg, in the cervix and anus). Papillomaviruses were the first DNA viruses linked to malignant changes. In the mid-1930s, Shope demonstrated that benign rabbit papillomas were due to filterable agents and could advance to become malignant squamous cell carcinomas. External cofactors, such as coal tar, could hasten this process. However, work on the biology and mechanism by which these agents foster malignant transformation has been impeded by the inability to cultivate papillomaviruses in vitro. Molecular probes to detect viral products in vivo indicate that replication and assembly of these viruses take place only in the differentiating layers of squamous epithelia, a situation that has not been reproduced in vitro.

Replication in squamous epithelium

The first evidence that HPVs could be associated with human malignant disease came from observations on epidermodysplasia verruciformis. This disease has a genetic basis that results in unusual susceptibility to HPV types 5 and 8, which produce multiple flat warts. About one third of affected patients develop squamous cell carcinoma from these lesions.

Viral genomes carry their own transforming genes, E6 and E7

E6 degrades p53 and E7 interacts with pRB to abrogate cell cycle

HPV is the major cause of cervical cancer

The mechanism of oncogenicity of HPV is less clear. Cells infected with genomes of several papillomaviruses can transform cells and produce tumors when injected into nude (T lymphocyte-deficient) mice. The viral genome exists as multiple copies of a circular episome within the nucleus of transformed cells, but is not integrated into the cellular genome. This appears also to be the case with benign human lesions. In malignant tumors, part of the viral genome is found integrated into the cellular genome, but integration is not site-specific. Both the integrated viral genome and the extrachromosomal form carry their own transforming genes. Host cells normally produce a protein that inhibits expression of papillomavirus transforming genes, but this can be inactivated by products of the virus and possibly by other infecting viruses, thus allowing malignant transformation to occur. HPV early gene products, E6 and E7, have been implicated in oncogenicity. E6 accelerates the degradation of p53, a tumor suppressor protein, and reduces its stability. E7 interacts with pRB, retinoblastoma protein, to abrogate cell cycle regulation. The inhibition of p53 and pRB functions results in cell transformation by E6 and E7, causing tumors. Another HPV gene product, E5, has been found to function in benign papillomas. HPV DNA is found in more than 95% of cervical carcinoma specimens when tested by polymerase chain reaction (PCR). The discovery that HPV causes most cervical cancers earned the 2008 Nobel Prize in Medicine for the German researcher, Harald zur Hausen.

CLINICAL ASPECTS

MANIFESTATIONS

Cutaneous warts develop at the site of inoculation within 1 to 3 months and can vary from flat to deep plantar growths (**Figure 19–2**). Although they can persist for years, they

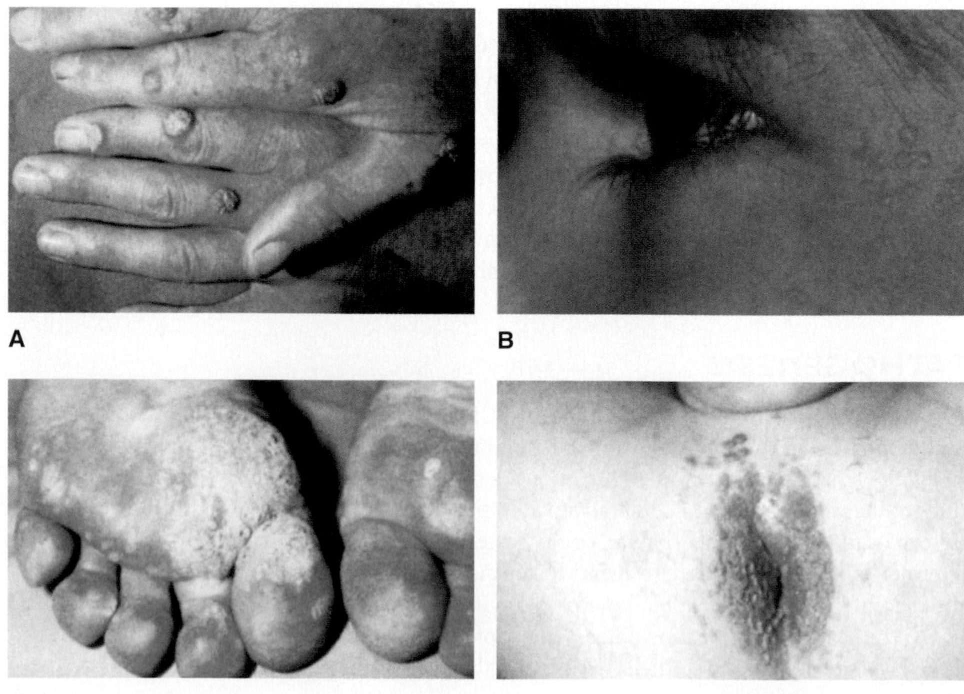

FIGURE 19–2. Warts. A. Common warts on fingers. **B.** Flat warts on the face **C.** Plantar warts on the feet **D.** Perianal condylomata acuminata. (Reproduced with permission from Willey J, Sherwood L, Woolverton C (eds). *Prescott's Principles of Microbiology.* New York: McGraw-Hill; 2008.)

ultimately spontaneously regress. Respiratory papillomatosis due most often to types 6 and 11 occurs as intraoral or laryngeal lesions. These tend to occur in infants as a result of natal exposure, or in adults. Treatment is varied and complex.

Oral or laryngeal papillomatosis in infants infected during delivery

External genital HPV infection occurs as exophytic genital warts (condylomata acuminata) caused most often by types 6 or 11 HPV (**Figures 19–2** and **19–3**). They are often found on the head or shaft of the penis, at the vaginal opening or perianal 4 to 6 weeks after exposure. Lesions may increase in size to cauliflower-like appearance during pregnancy or immunosuppression. Genital HPV infection is most often benign, and many lesions reverse spontaneously. However, they may become dysplastic and proceed through a continuum of cervical intraepithelial neoplasm to severe dysplasia and/or carcinoma (**Figure 19–4**). The most common HPV in the malignant lesions is type 16, although this genotype, as well as the others, is most apt to cause lesions that regress spontaneously.

Anal carcinoma due to HPV may be increasing

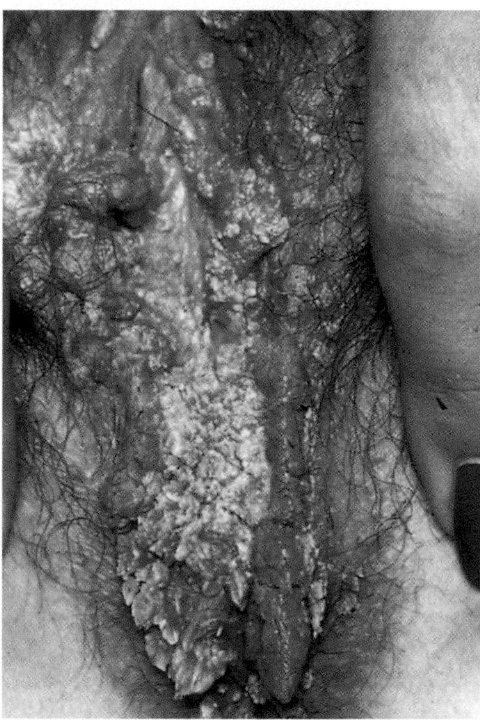

FIGURE 19–3. Extensive condylomata of vulva caused by HPV-6. (Reproduced with permission from Connor DH, Chandler FW, Schwartz DQA, Manz HJ, Lack EE (eds). *Pathology of Infectious Diseases,* vol. 1. Stamford, CT: Appleton & Lange; 1997.)

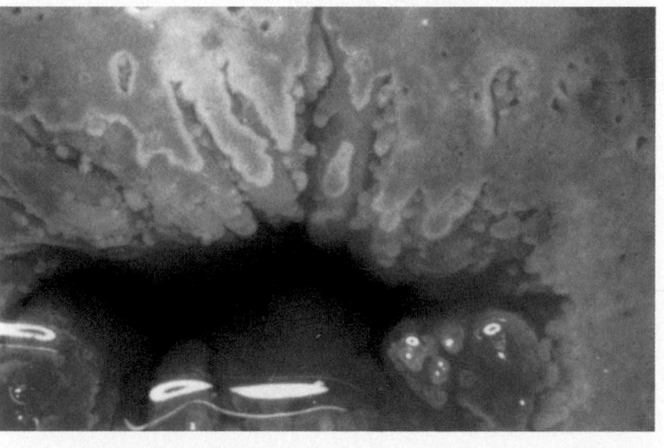

FIGURE 19–4. Colposcopic photograph of cervical transformation zone with diffusely scattered acetowhite staining, characteristic of HPV infection. (Reproduced with permission from Connor DH, Chandler FW, Schwartz DQA, Manz HJ, Lack EE (eds). *Pathology of Infectious Diseases,* vol. 1. Stamford, CT: Appleton & Lange; 1997.)

Higher-grade malignancy is most apt to occur in the cervix, but the rate of anal carcinoma related to HPV appears to be increasing, especially in AIDS patients.

DIAGNOSIS

Poikilocytosis can be seen in cytologic specimens

Molecular methods to detect specific genotypes in biopsies of cervical swabs are available

HPV does not grow in routine tissue culture, and antibody tests are rarely used, since results remain positive after the first HPV genotype infection. Papillomavirus infection leads to perinuclear cytoplasmic vacuolization and nuclear enlargement, referred to as poikilocytosis, in epithelial cells of the cervix or vagina. These changes can be seen in a routine Papanicolaou (Pap) smear **(Figure 19–5)**. The use of immunoassays to detect viral antigen and nucleic acid hybridization or PCR to detect specific viral DNA in cervical swabs or tissue is more sensitive (**Figure 19–6**) than Pap smear, but the clinical utility of detecting specific HPV types in clinical specimens remains to be determined. Detection of an abnormal cytology due to HPV should prompt colposcopy to assist in following up or treating patients with abnormal lesions.

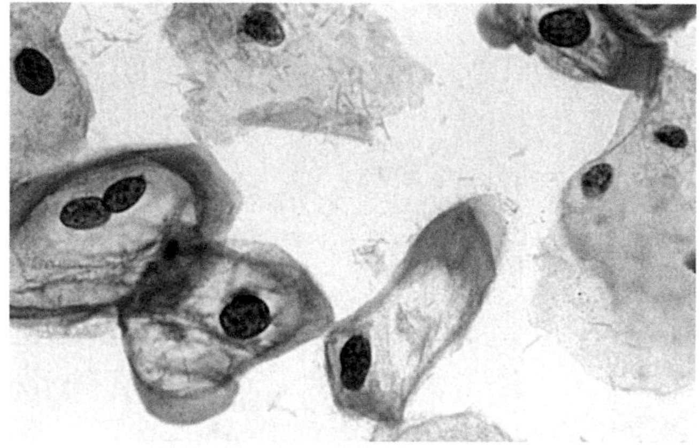

FIGURE 19–5. Abnormal Papanicolaou smear. The pink and blue objects are squamous epithelial cells; abnormalities include the doubling of the nuclei and a clear area around them. Most abnormal smears in young women are due to human papillomavirus infection, when persistent an important factor in development of cancer of the cervix. (Reproduced with permission from Nester EW, Anderson DG, Roberts CE Jr, Nester MT. *Microbiology: A Human Perspective,* 6th ed. New York: McGraw-Hill, 2008.)

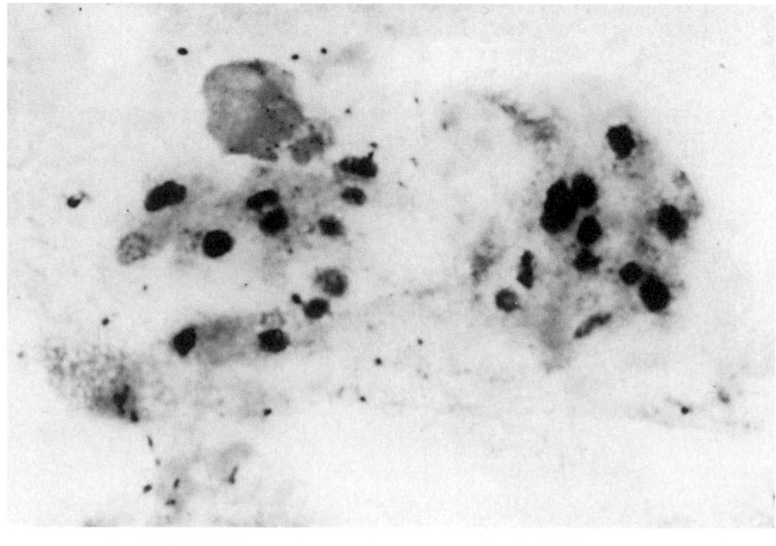

FIGURE 19–6. Human papillomavirus (HPV) type 16 DNA demonstrated in a cervical smear by in situ hybridization. The dark dots represent detection of HPV DNA sequences by the DNA probe.

TREATMENT AND PREVENTION

Current treatment of HPV is usually either cytotoxic or surgical. Among the topical cytotoxins are podophyllin, podophyllotoxin, 5-fluorouracil, and trichloroacetic acid. Warts can also be removed by laser or freezing with liquid nitrogen. Recurrences are common after cessation of treatment because of survival of virus or viral DNA in the basal layers of the epithelium. Cervical and anal lesions may be treated with electrocautery, but carcinoma may require radiation therapy or radical surgery.

Recurrences are common after topical treatment

Condom usage is encouraged to prevent transmission. Cervical Pap smears should be done regularly to detect early lesions due to HPV. A recent large prospective study of HPV-16 and 18 vaccine (an L1 polypeptide expressed in yeast) indicated prevention of infection, and a vaccine for HPV genotypes 6, 11, 16, and 18 is now approved for females, aged 9 to 26 years.

Future prospects for prevention by vaccines

POLYOMAVIRUSES

VIROLOGY

The polyomaviruses include the JC virus (JCV) and BK virus (BKV) of humans and the simian virus 40 (SV40) of monkeys. Polyomaviruses, like papillomaviruses, are members of the Papovaviridae family. They have double-stranded circular DNA of 5 kb, are naked capsid (unenveloped) viruses 45 nm in diameter, and are widely distributed among various animal species, usually without causing apparent disease. However, these viruses are able to transform cells of a variety of heterologous cell lines in culture. An electron micrograph of JCV is shown in **Figures 19–7**. Like papillomaviruses, polyomaviruses also encode early and late genes. Early genes encode the large T antigen, middle or small T antigens that are involved in mRNA transcription, DNA replication, cell growth, and transformation. Late proteins are structural capsid proteins, namely VP1, VP2, and VP3.

Naked capsid double-stranded circular DNA virus

Can transform cells in vitro

Viral replication takes place in the nucleus of infected cells. Transcription of early genes is performed by host RNA polymerase, which leads to synthesis of early proteins. The early proteins regulate viral transcription, DNA replication, cell division, and transformation. Viral DNA genomes for progeny viruses are synthesized by host cell DNA polymerase. Late mRNAs are translated into capsid proteins that are translocated in the nucleus, where assembly of progeny viruses takes place. These capsid proteins are released upon cell death.

Viral replication in the nucleus of infected cells

Host cell RNA and DNA polymerase direct virus RNA and DNA synthesis

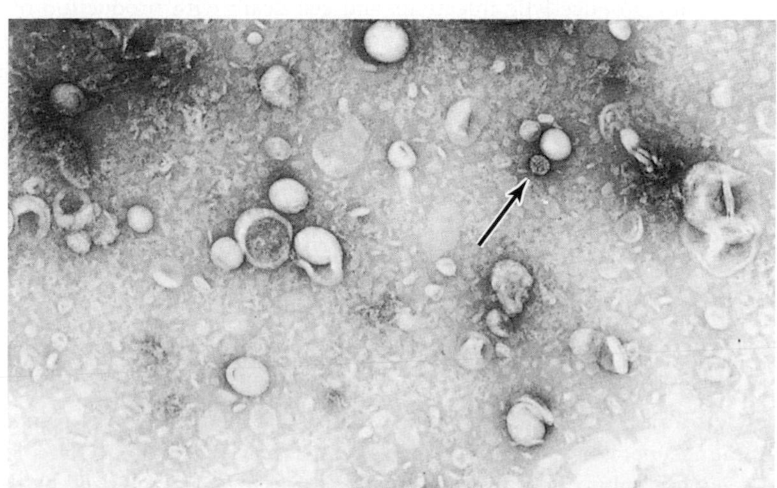

FIGURE 19–7. JC virus (*arrow*) among debris of cells from a brain biopsy of a case of progressive multifocal leukoencephalopathy. (Reprinted with permission from Palmer E, Martin ML. *An Atlas of Mammalian Viruses*. Boca Raton, FL: CRC Press; 1982. Copyright 1982 by CRC Press, Inc.)

POLYOMAVIRUS DISEASE

CLINICAL CAPSULE

Polyomaviruses are closely related to papillomaviruses, but are not known to cause clinical disease in immunocompetent patients. They can cause progressive multifocal leukoencephalopathy (PML) and hemorrhagic cystitis/nephropathy in immunocompromised patients.

EPIDEMIOLOGY

Latency is common

Disease associated with immunocompromised patients

The exact routes of polyomavirus transmission in humans are not known. However, respiratory or oral transmission (due to contaminated food or water) is suspected. Viruses are excreted in urine. Approximately 80% of adults show serologic evidence of JCV and BKV infections, all of which are usually asymptomatic. However, the viruses remain latent and may reactivate and cause disease in immunocompromised patients. BKV is estimated to cause renal disease, including graft failure in 2% to 5% of renal transplant recipients, and JCV is the cause of an uncommon neurologic disorder, progressive multifocal leukoencephalopathy (PML).

PATHOGENESIS

Do not cause malignancies in their natural hosts

Polyomaviruses can produce malignant tumors in certain experimental animals, but not in their natural hosts. For example, SV40 can produce lymphocytic leukemia and a variety of reticuloendothelial cell sarcomas in baby hamsters, but is not oncogenic in its natural monkey host. Fortunately, even though it can transform some human cells in vitro, SV40 fails to produce disease in humans, a fact that became apparent on follow-up of recipients of early batches of poliomyelitis vaccine produced in monkey kidney cell cultures that were contaminated with live SV40.

Interact with cells in a variety of ways

The reason polyomaviruses fail to produce tumors in their natural hosts is uncertain, but it may be because these viruses are usually cytocidal under these conditions. From a biologic point of view, the polyomaviruses are particularly useful models of oncogenicity because they can be readily studied in vitro and interact with cells in different ways. In some, they produce lytic infections and cell death with production of complete virions. In others, they integrate randomly into the cell genome and cause transformation by the expression of one or more of the viral genes. No human tumor has been shown to be caused by polyomaviruses.

CLINICAL ASPECTS

MANIFESTATIONS

Progressive Multifocal Leukoencephalopathy

PML is a rare, subacute, degenerative disease of the brain found primarily in adults with immunosuppressive diseases, especially AIDS and hematologic malignancies, or those receiving immunosuppressive agents. The disease is characterized by the development of impaired memory, confusion, and disorientation, followed by a multiplicity of neurologic symptoms and signs that include hemiparesis, visual disturbances, incoordination, seizures, and visual abnormalities. PML is progressive, with death usually occurring 3 to 6 months after onset of symptoms.

PML is a degenerative, progressive brain disease

In PML, cerebrospinal fluid (CSF) findings are often normal, although some patients show a slight increase in lymphocytes, and protein levels may be elevated. Pathologically, foci of demyelination are found, surrounded by giant, bizarre astrocytes containing intranuclear inclusions. The demyelination is due to viral damage to oligodendroglial cells, which synthesize and maintain myelin. Abundant JCV particles can be seen in the brain by electron microscopy (Figure 19–7) and may be concentrated within the nuclei of oligodendrocytes. JCV DNA sequences have been demonstrated by PCR in the brain of patients without PML or demyelinating lesions, suggesting that the virus may be latent in the brain before immunosuppression. There is no specific treatment for PML, although reducing the immunosuppression, if possible, may have some clinical benefit.

JCV in cell nuclei, with demyelination

No specific treatment

Urinary Tract Infection

Infection of the urinary tract with JCV and BKV can be demonstrated frequently in immunocompromised patients, but usually without symptoms or evidence of renal injury. BKV is associated with a hemorrhagic cystitis, particularly in bone marrow and renal transplant recipients. In addition, BKV is also the cause of a severe nephropathy and vasculopathy, which may lead to kidney loss in renal transplant recipients. The disease develops months after renal transplantation. Treatment consists of reducing immunosuppression, but up to 50% of patients with this syndrome may require nephrectomy. Cidofovir (a nucleotide analog) is a possible antiviral for BK disease.

BKV causes hemorrhagic cystitis and nephritis

DIAGNOSIS

Urine from patients excreting these polyomaviruses may contain "decoy" cells similar to those from patients excreting cytomegalovirus, but they can be distinguished cytologically. BKV can be isolated by routine culture in diploid fibroblast or Vero monkey kidney cells, but nephropathy is usually preceded by plasma PCR positivity, which can be monitored. At present, a kidney biopsy is required for definitive diagnosis. Viral antigens can be demonstrated in tissue by a variety of immunoassays. JCV DNA has been demonstrated in the brain of PML patients by PCR, and PCR of CSF is a diagnostic test for PML.

BKV can be isolated in cell culture

JCV and BKV can be detected by PCR

CASE STUDY

POSTCOITAL CONCERNS

A 19-year-old woman had her first and only intercourse 6 months ago and is concerned whether she has genital HPV infection. Pelvic exam reveals normal genitalia.

QUESTIONS

- What would be the best test for this purpose?

A. Serology for HPV IgG antibody
B. Serology for HPV IGM antibody
C. Cervical "Pap" smear
D. In situ hybridization for HPV DNA on cervical sample

■ Her test for HPV infection is positive. Which of the following is most appropriate?

A. Cervical "Pap" smear every other year
B. Quadrivalent HPV vaccine
C. Topical trichloroacetic acid treatment of the cervix
D. Determination whether her HPV infection is oncogenic genotype
E. Prophylactic radiation treatment of cervix

■ Her sex partner should:

A. Be counseled to practice "safe sex"
B. Have HPV in situ hybridization assay on urethral swab
C. Have HPV in situ hybridization assay on anal swab
D. Receive quadrivalent HPV vaccine

ANSWERS

1(D), **2**(B), **3**(A)

CHAPTER 20

Persistent Viral Infections of the Central Nervous System

Persistent viral infections are those in which termination of early symptoms and disease is not accompanied by elimination of the virus from the host, but persistence of viral genetic material in the host. The molecular mechanisms of persistent viral infections are not clearly understood, but three broad conditions must be satisfied for a virus to establish a persistent infection in a host:

1. Infection of the host cell by the virus should not be cytolytic. Viruses have found various cell types such as nonpermissive cells in a host to infect and remain less cytolytic to maintain persistence.

Viruses are less cytolytic to cells in which they persist

2. Viral genome must be maintained by various mechanisms. Viral genomes can be maintained in several ways, including integration and extrachromosomal episomes for DNA viruses. However, the mechanisms of viral RNA genome maintenance are not known.

DNA genomes either integrate or persist as episomes

Mechanisms of persistence of RNA genomes in cells not understood

3. Virus has to avoid detection and elimination by the host's immune system. Viruses have evolved several evasion strategies such as infection of immunologically privileged sites that are not easily accessible to the immune system (central nervous system [CNS] and other sites), antigenic variation, down-regulation of immune components, and others. Several viruses cause persistent infection of the CNS because they are not easily detected and eliminated by the host immune response.

Avoid detection and elimination by the host

Infect immune privileged sites such as CNS

Evidence has accumulated that a variety of progressive neurologic diseases in both animals and humans are caused by viral or other filterable agents that share some of the properties of viruses (**Tables 20–1, 20–2,** and **20–3**). These illnesses have been termed "slow viral diseases" because of the protracted period between infection and the onset of disease as well as the prolonged course of the illness, but a better term is "persistent viral infection."

Progressive neurologic diseases in humans and animals

Most persistent viral infections involve well-differentiated cells, such as lymphocytes and neuronal cells. They can be classified as (1) diseases associated with "conventional" viral agents that possess nucleic acid genomes and protein capsids, induce immune responses, and can be grown in cell culture systems; and (2) diseases associated with "unconventional" agents that are small, filterable infectious agents, known as "prions," which are transmissible to certain experimental animals, but do not contain nucleic acids, do not appear to be associated with immune or inflammatory responses by the host, and have not been cultivated in cell culture.

Include conventional viruses and unconventional agents

"Prions" do not produce immune or inflammatory responses

Persistence of conventional viruses can result from infection of a nonpermissive cell in the host with restrictive cytolytic effects, preservation of viral nucleic acid in infected host's cells, and mutations that interfere with or severely limit viral replication or antigenicity. ::: mutation, p.130

Persistence can be due to a variety of mechanisms

TABLE 20–1 Conventional Viruses Causing Persistent Central Nervous System Infections

DISEASE	AGENT
Subacute sclerosing panencephalitis	Measles virus
Progressive panencephalitis following congenital rubella	Rubella virus
Progressive multifocal encephalopathy	Polyoma virus (JC virus)
AIDS dementia complex	Human immunodeficiency virus
Persistent enterovirus infection of the immunodeficient	Enteroviruses

DISEASES ASSOCIATED WITH CONVENTIONAL AGENTS

The following conditions are the major persistent infections caused by conventional viral agents. They are summarized in Table 20–1.

Subacute Sclerosing Panencephalitis

Persistence of measles virus after acute childhood infection

Subacute sclerosing panencephalitis is considered in Chapter 10. It is a rare chronic measles virus infection of children that produces progressive neurologic disease characterized by an insidious onset of personality change, progressive intellectual deterioration, and both motor and autonomic nervous system dysfunctions.

Progressive Postrubella Panencephalitis

Can be a late sequela of congenital rubella infection

Even more rarely, a degenerative neurologic disorder similar to subacute sclerosing panencephalitis is associated with persistent rubella virus infection of the CNS. This condition is seen most often in adolescents who have had the congenital rubella syndrome. Rubella virus has been isolated from brain tissue in these patients using co-cultivation techniques.

Progressive Multifocal Leukoencephalopathy

Progressive neurologic disease of severely immunocompromised persons

Progressive multifocal leukoencephalopathy (PML) is a subacute, degenerative disease of the brain found primarily in adults with (1) immunosuppressive diseases, especially acquired immunodeficiency syndrome (AIDS) and hematologic malignancies; or (2) diseases requiring therapy with immunosuppressive agents. PML is due to a polyomavirus (JC virus) and is considered in Chapter 19.

Persistent Enterovirus Infection

Associated with humoral immunodeficiencies

Temporary improvement with virus type-specific hyperimmune globulin

Persons with congenital or severe acquired immunodeficiency, especially those with agammaglobulinemia, may develop a chronic CNS infection due to an echovirus or other enterovirus. Headache, confusion, lethargy, seizures, and cerebrospinal fluid (CSF) pleocytosis are common manifestations. The virus can be isolated from the CSF. Clinical improvement may be achieved by the administration of human hyperimmune globulin to the infecting virus type. Relapse, however, occurs when therapy is discontinued, indicating persistence of virus despite the therapy.

AIDS Dementia Complex

Late stages of AIDS

Human immunodeficiency virus causes a persistent infection of the CNS in many patients with symptomatic AIDS. The clinical course may vary from a mild subacute illness to severe progressive dementia (see Chapter 18).

HUMAN DISEASES CAUSED BY UNCONVENTIONAL AGENTS: SUBACUTE SPONGIFORM ENCEPHALOPATHIES

A group of progressive degenerative diseases of the CNS has been shown to be caused by infectious agents with unusual physical and chemical properties, which are now known as prions. The Nobel Prize in Medicine for 1997 was awarded to Stanley Prusiner for his work in identifying the role of prions in disease. Prions cause bovine spongiform encephalopathy in cattle, scrapie in sheep, and five fatal CNS diseases in humans (Table 20–2). Prions can be the etiologic agents of inherited, communicable, or sporadic diseases. The pathogenesis of these illnesses is not well understood, but the pathologic and clinical features are similar. Varying degrees of neuronal loss and astrocyte proliferation occur. The diseases are known as "spongiform" encephalopathies because of the vacuolar changes in the cortex and cerebellum (**Figures 20–1** and **20–2**). The incubation periods for these diseases are months to years, and their courses are protracted and inevitably fatal.

Prions affect animals and humans

Cause neuronal loss and spongiform changes in brain

A prion is a "small proteinaceous infectious particle" that is not inactivated by procedures that destroy nucleic acids (Table 20–3). They have diameters of 5 to 100 nm or less and can remain viable even in formalinized brain tissue for many years. They are resistant to ionizing radiation, boiling, and many common disinfectants. Recognizable virions have not been found in tissues by electron microscopy, and the agents have not been grown in cell culture.

Infectious agents resist inactivation

Nucleic acids absent

A prion is composed of a protein encoded by a normal cellular gene. The protein, designated PrPc, is converted from a normal form (designated as NP in **Figure 20–3**) into a disease-causing form by a change in conformation to a protein designated PrPsc (designated as PP

TABLE 20–2 Unconventional Virus (Prion) Diseases[a]

HUMANS	ANIMALS (PRIMARY HOSTS)
Creutzfeldt-Jakob disease[b]	Scrapie (sheep)
Variant Creutzfeldt-Jakob disease	Transmissible mink encephalopathy (mink)
Gerstmann-Straüssler-Scheinker syndrome	Chronic wasting disease (mule deer, elk)
Kuru	Bovine spongiform encephalopathy (cows)[b]
Fatal familial insomnia	

[a]Subacute spongiform encephalopathies.
[b]Prion agents of variant Creutzfeldt-Jakob disease and bovine spongiform encephalopathy are identical.

TABLE 20–3 Biologic and Physical Properties of Prions

- Chronic progressive pathology without remission or recovery
- No inflammatory response
- No alteration in pathogenesis by immunosuppression or immunopotentiation
- Estimated diameter of 5-100 nm
- No virion-like structures visible by electron microscopy
- Transmissible to experimental animals
- No interferon production or interference by conventional viruses
- Unusual resistance to ultraviolet irradiation, alcohol, formalin, boiling, proteases, and nucleases
- Can be inactivated by prolonged exposure to steam autoclaving or 1N or 2N NaOH

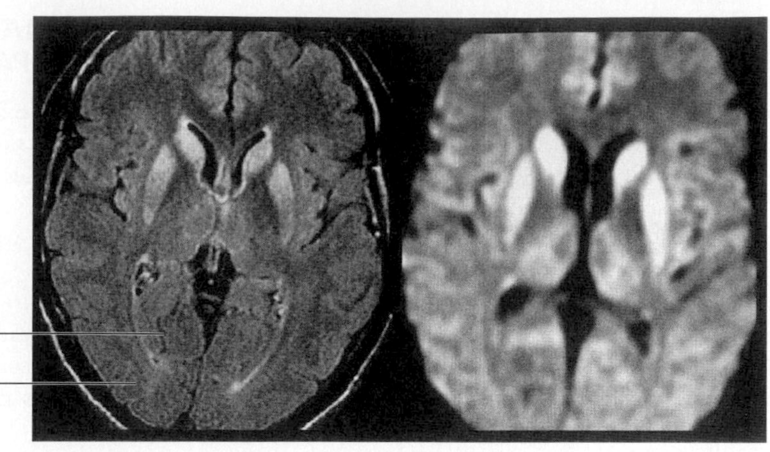

FIGURE 20–1. Appearance of brain with spongiform encephalopathy. (*Left*) Normal brain. (*Right*) Brain infected with a prion. Note the spongelike appearance. (Reproduced with permission from Nester EW, Anderson DG, Roberts CE Jr, Nester MT. *Microbiology: A Human Perspective,* 6th ed. New York: McGraw-Hill, 2008.)

PrPc is encoded by a normal cellular gene

Conformational change to PrPsc results in disease and prion proliferation

in Figure 20–3). Brain extracts from scrapie-infected animals contain PrPsc, which is not found in the brains of normal animals; PrPsc is the prion that is responsible for transmission and infection. The conformational change is also the way in which prions multiply; that is, contact with PrPsc results in a conformational change of the normal prion host cell protein (PrPc or NP) and the formation of additional abnormal or infectious prion protein, PrPsc or PP (Figure 20–3). Proliferation of PrPsc prions and the consequent pathology result from this process. During scrapie infection, prion protein may aggregate into amyloid-like birefringent rods and filamentous structures termed scrapie-associated fibrils (**Figure 20–4**), which are found in membranes of scrapie-infected brain tissues. The amino acid sequence of different prion proteins in different animal species differ from one another, and transmission across species usually does not occur. Specifically, ingestion of tissue from sheep or elk infected with abnormal prions has not been documented to lead to human disease. Tissue from infected cows did, however, transmit variant Creutzfeldt-Jacob disease (see following text).:

Kuru

Kuru was a subacute, progressive neurologic disease of the Fore people of the Eastern Highlands of New Guinea. The disease was brought to the attention of the Western world by Gadjusek and Zigas in 1957. Although the illness was localized and decreasing in incidence, its study has thrown light on the transmissibility and infectious nature of similar encephalopathies. Epidemiologic studies indicated that kuru usually afflicted adult women, or children of either sex. The disease was rarely observed outside the Fore region, and outsiders in the region did not contract the disease. The symptoms and signs were ataxia, hyperreflexia, and spasticity,

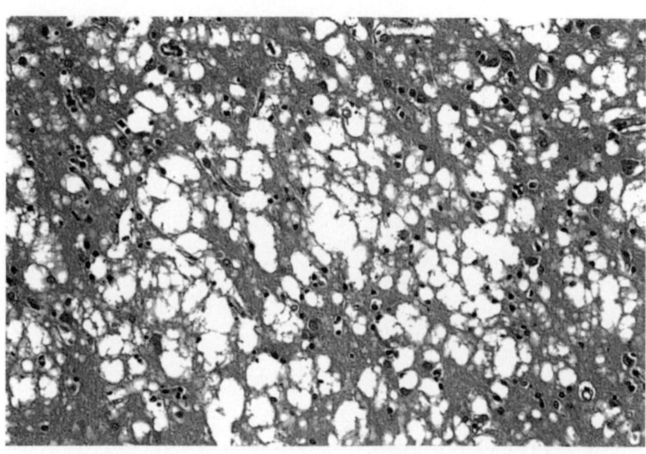

FIGURE 20–2. Spongiform changes. (Reproduced with permission from Connor DH, Chandler FW, Schwartz DQA, Manz HJ, Lack EE (eds). *Pathology of Infectious Diseases,* vol. 1. Stamford, CT: Appleton & Lange; 1997.)

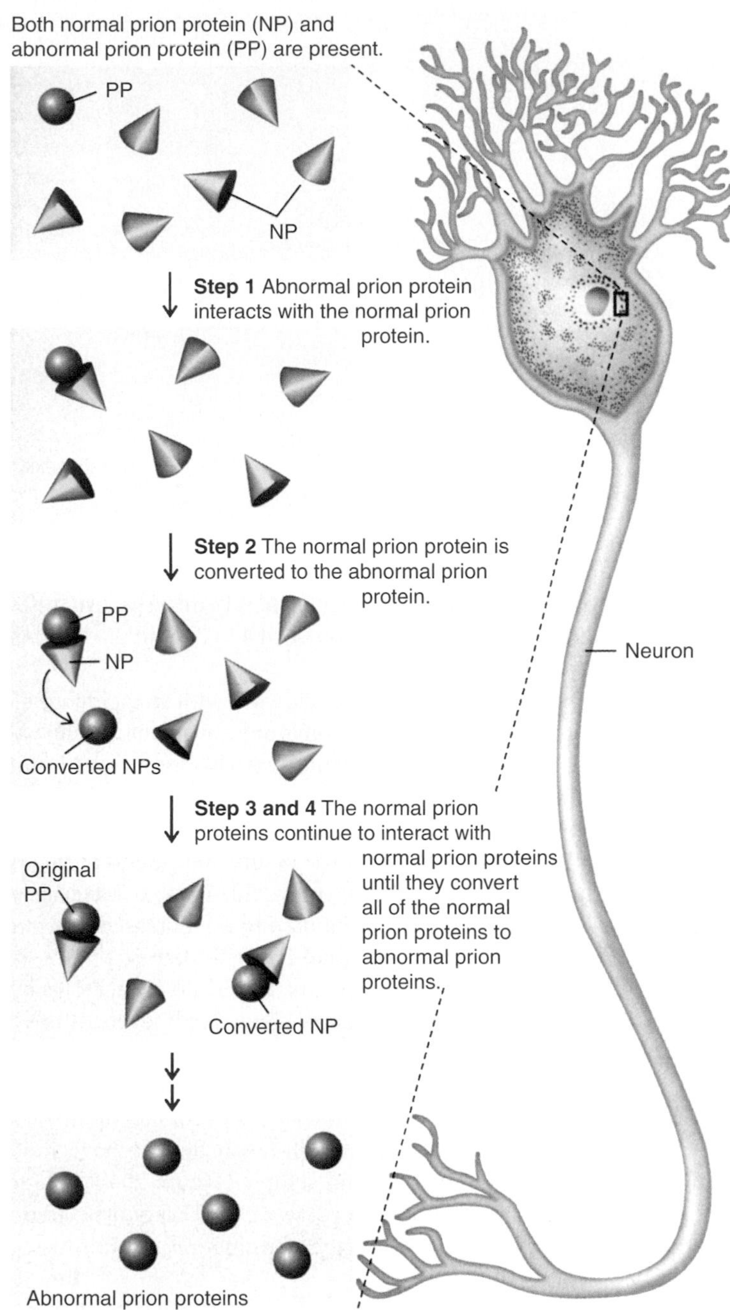
Both normal prion protein (NP) and abnormal prion protein (PP) are present.
PP
NP
Step 1 Abnormal prion protein interacts with the normal prion protein.
Step 2 The normal prion protein is converted to the abnormal prion protein.
PP
NP
Converted NPs
Neuron
Step 3 and 4 The normal prion proteins continue to interact with normal prion proteins until they convert all of the normal prion proteins to abnormal prion proteins.
Original PP
Converted NP
Abnormal prion proteins

FIGURE 20–3. Proposed mechanism of how prions are converted to abnormal proteins. The normal and abnormal prion proteins differ in their tertiary structures. (Reproduced with permission from Nester EW, Anderson DG, Roberts CE Jr, Nester MT. *Microbiology: A Human Perspective,* 6th ed. New York: McGraw-Hill, 2008.)

which led to progressive dementia, starvation, and death. Pathologic examination revealed changes only in the CNS, with diffuse neuronal degeneration and spongiform changes of the cerebral cortex and basal ganglia. No inflammatory response was apparent. Inoculation of infectious brain tissue into primates produced a disease that caused similar neurologic symptoms and pathologic manifestations after an incubation period of approximately 40 months. Epidemiologic studies indicated that transmission of the disease in humans was associated with ingestion of a soup made from the brains of dead relatives and eaten in honor of the deceased. Clinical disease developed 4 to 20 years after exposure. Since the elimination of cannibalism from the Fore culture, kuru has disappeared.

Women and children of the Fore people of New Guinea

Transmissible to primates

Associated with cannibalism

Creutzfeldt-Jakob Disease

Creutzfeldt-Jakob disease is a progressive, fatal illness of the CNS that is seen most frequently in the sixth and seventh decades of life. The initial clinical manifestations are a change in cerebral function, usually diagnosed initially as a psychiatric disorder. Forgetfulness and

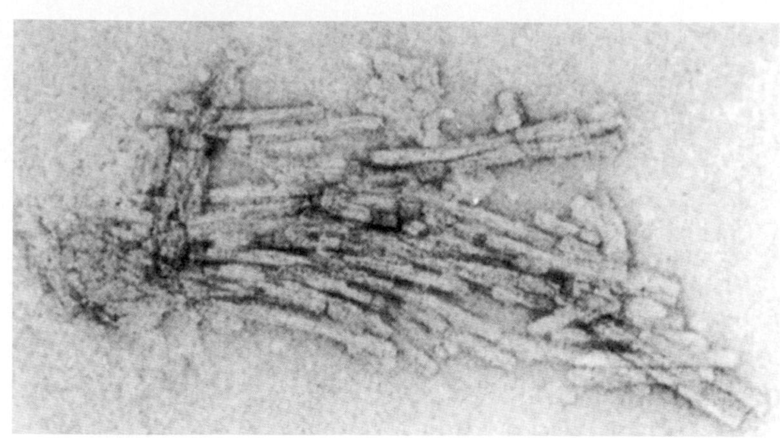

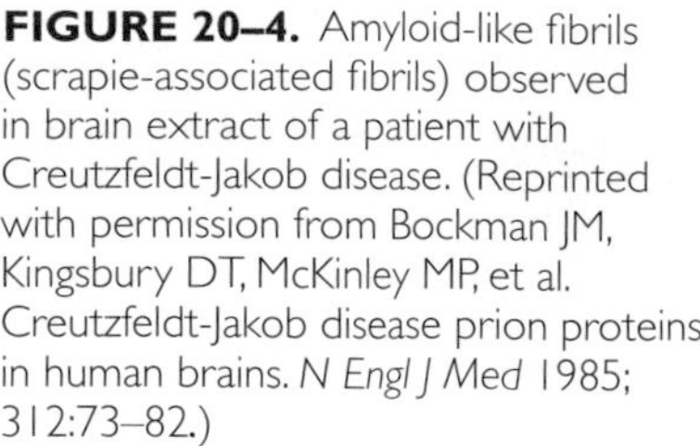

FIGURE 20–4. Amyloid-like fibrils (scrapie-associated fibrils) observed in brain extract of a patient with Creutzfeldt-Jakob disease. (Reprinted with permission from Bockman JM, Kingsbury DT, McKinley MP, et al. Creutzfeldt-Jakob disease prion proteins in human brains. *N Engl J Med* 1985; 312:73–82.)

Progressive disease, usually occurring among elderly

disorientation progress to overt dementia and the development of changes in gait, increased tone in the limbs, involuntary movement, and seizures. These manifestations resemble those of kuru. The disorder usually runs a course of 4 to 7 months, eventually leading to paralysis, wasting, pneumonia, and death.

Creutzfeldt-Jakob disease is found worldwide, with an incidence of disease of one case per million per year. The mode of acquisition is unknown, but it occurs both sporadically (85%) and in a familial pattern (15%). Infection has also been transmitted by dura mater grafts and corneal transplants, by contact with contaminated electrodes or instruments used in neurosurgical procedures, and by pituitary-derived human growth hormone. The latter was responsible for more than 100 cases. The incubation period of the disease is approximately 3 to more than 20 years. The agent of Creutzfeldt-Jakob disease has not been transmitted to animals by inoculation of body secretions, and no increased risk of disease has been noted in family members or medical personnel caring for patients.

Transmission to animals

It has been transmitted to chimpanzees, mice, and guinea pigs by inoculation of infected brain tissue, leukocytes, and certain organs. High levels of infectious agent have been found, especially in the brain, where they may reach 10^7 infectious doses per gram of brain tissue. Nonpercutaneous transmission of Creutzfeldt-Jakob disease has not been observed, and there is no evidence of transmission by direct contact or airborne spread.

Pathology identical to kuru

Scrapie-like structures seen in brain

Brains from patients with Creutzfeldt–Jakob disease have the birefringent rods and fibrillar structures noted in Kuru and scrapie (Figure 20–4). Identification of PrPsc and antibodies directed against it may become a useful diagnostic adjunct to neuropathologic examination of brain tissue. Pathologic examination of brain tissue is the only definitive diagnostic test.

Therapy

There is no effective therapy for Creutzfeldt-Jakob disease, and all cases have been fatal.

Prevention

Nosocomial infections preventable by avoidance of potentially infectious materials, careful sterilization

The small risk of nosocomial infection is related only to direct contact with infected tissue. Stereotactic neurosurgical equipment, especially that used in patients with undiagnosed dementia, should not be reused. In addition, organs from patients with undiagnosed neurologic disease should not be used for transplants. Growth hormone from human tissue has now been replaced by a recombinant genetically engineered product. Recommendations for disinfection of potentially infectious material include treatment for 1 hour with 2 N NaOH or by autoclaving at 132°C for 60 to 90 minutes. Others recommend even more extensive treatment such as combining these two procedures to ensure inactivation.

Bovine Spongiform Encephalopathy ("Mad Cow Disease") and "Variant Creutzfeldt-Jakob Disease"

Bovine spongiform encephalopathy (BSE) was identified in 1986, after it began striking cows in the United Kingdom, causing them to become uncoordinated and unusually apprehensive.

The source of the emerging epidemic was soon traced to a food supplement that included meat and bone meal from dead sheep.

To combat BSE, the British government banned the use of animal-derived feed supplements in 1988, and the epidemic among cattle, which peaked at nearly 40,000 cases in 1992, decreased to less than 4000 new cases in 1997. By February 2002, most European countries had reported cases of BSE, but new infections have ceased as a result of imposing tight controls on cattle feed. The United States had been spared, as measured by over 19,000 cattle brain examinations. The incubation period in cattle was determined to be 2 to 8 years. In addition to the incoordination and apprehension, the cows exhibited hyperesthesia, hyperreflexia, muscle fasciculations, tremors, and weight loss. Autonomic dysfunction was frequently manifested as reduced rumination, bradycardia and other cardiac arrhythmias.

Source was meat and bone meal from sheep in cattle feed

Unfortunately, the prion that causes BSE survived the heat of cooking and was transmitted to humans who inadvertently consumed infected bovine neural tissue or bone marrow (both are sometimes found in processed meats, depending on the rendering procedures used). To date, over 100 humans with "variant Creutzfeldt-Jakob disease," have died. The cases frequently present in young adults as psychiatric problems progressing to neurologic changes and dementia, with death in an average of 14 months. It appears that destruction of diseased cattle and the changes in livestock feeds have prevented further cases.

Variant Creutzfeldt-Jakob disease apparently transmitted by infected bovine tissues to humans

Gerstmann-Straüssler-Scheinker Disease

Gerstmann-Straüssler-Scheinker (GSS) disease is similar to Creutzfeldt-Jakob disease, but occurs at a younger age (fourth to fifth decade). Cerebellar ataxia and paralysis are common, but dementia is less often seen. The disease evolves over an average of 5 years. It was originally thought to be familial, but it also occurs sporadically, very rarely. GSS has been transmitted to experimental animals. The familial nature of this disease raises the question of vertical transmission versus inherited susceptibility.

Gerstmann-Straüssler-Scheinker disease similar to Creutzfeldt–Jakob disease but evolves more slowly

Fatal Familial Insomnia

This is a recently recognized familial prion disease in which a syndrome of sleeping difficulty is followed by progressive dementia. It occurs in patients aged 35 to 61, culminating in death within 13 to 25 months. The infectious agent has been transmitted to experimental animals.

Sleeping difficulties progressing to dementia

CASE STUDY

PROGRESSIVE FORGETFULNESS

During the last 3 months, a previously healthy 50-year-old man has become increasingly forgetful. Last week he was unable to find his home when returning from a walk. His walking has become unsteady, and yesterday he had a first grand mal seizure. He has not traveled outside the United States and takes no medications. Neurologic examination reveals cerebellar ataxia and spastic reflexes in his lower extremities.

QUESTIONS

- This man's most likely diagnosis is:

A. Alzheimer's disease
B. Progressive multifocal leukoencephalopathy
C. Creutzfeldt-Jacob disease
D. Mad cow disease
E. AIDS dementia

■ The most useful diagnostic test would be:

A. PCR of CSF
B. PCR of plasma
C. Brain biopsy
D. MRI of brain

■ Which of the following is true regarding therapy of this disease?

A. There is no therapy proven to be effective
B. Immunosuppressive therapy would be effective
C. Cidofovir is effective
D. Highly active antiretroviral therapy (HAART) is effective

ANSWERS

1(C), **2**(C), **3**(A)

PART 

Pathogenic Bacteria

Kenneth J. Ryan and
W. Lawrence Drew

The Nature of Bacteria CHAPTER 21

Pathogenesis of Bacterial Infection CHAPTER 22

Antibacterial Agents and Resistance CHAPTER 23

Staphylococci CHAPTER 24

Streptococci and Enterococcii CHAPTER 25

Corynebacterium, *Listeria*, and *Bacillus* CHAPTER 26

Mycobacteria CHAPTER 27

Actinomyces and *Nocardia* CHAPTER 28

Clostridium, *Peptostreptococcus*, *Bacteroides*, and Other Anaerobes CHAPTER 29

Neisseria CHAPTER 30

Haemophilus and *Bordetella* CHAPTER 31

Vibrio, *Campylobacter*, and *Helicobacter* CHAPTER 32

Enterobacteriaceae CHAPTER 33

Legionella CHAPTER 34

Pseudomonas and Other Opportunistic Gram-negative Bacilli CHAPTER 35

Plague and Other Bacterial Zoonotic Diseases CHAPTER 36

Spirochetes CHAPTER 37

Mycoplasma and *Ureaplasma* CHAPTER 38

Chlamydia CHAPTER 39

Rickettsia, *Ehrlichia*, *Coxiella*, and *Bartonella* CHAPTER 40

CHAPTER 21

The Nature of Bacteria

Bacteria are the smallest and most versatile independently living cells known. This chapter examines the structural, metabolic, and genetic features that contribute to the ubiquity and diversity of this large group of organisms. Discussion focuses particularly on the characteristics of bacteria which enable them to cause disease in humans.

BACTERIAL STRUCTURE

As discussed in Chapter 1, in the hierarchy of infectious agents, bacteria are the first and smallest organism capable of independent existence. In the wider microbial world, their prokaryotic cell plan is still considered to provide the minimum possible size for an independently reproducing organism. Individuals of different bacterial species that colonize or infect humans range from 0.1 to 10 μm (1 μm 10^{-6}m) in their largest dimension. Most spherical bacteria have diameters of 0.5 to 2 μm, and rod-shaped cells are generally 0.2 to 2 μm wide and 1 to 10 m long. As shown in Figure 1-2, bacteria overlap in at least one dimension with large viruses and some eukaryotic cells, but they are the sole possessors of the 1-μm size.

Bacteria are in the range of 1 to 10 μm

The small size and nearly colorless nature of bacteria require the use of stains for visualization with a light microscope or the use of electron microscopy. The major morphologic forms are spheres, rods, bent or curved rods, and spirals (**Figure 21–1A–E**). Spherical or oval bacteria are called **cocci** (singular: coccus) and are typically arranged in clusters or chains. Rods are called **bacilli** (singular: bacillus) and may be straight or curved. Bacilli that are small and pleomorphic to the point of resembling cocci are often called coccobacilli. Spiral-shaped bacteria may be rigid or flexible and undulating.

Bacteria exhibit sphere, rod, and spiral shapes

Whatever the overall shape of the cell, the 1-μm size could not accommodate eukaryotic mitochondria, nucleus, Golgi apparatus, lysosomes, and endoplasmic reticulum in a cell that is itself only as large as an average mitochondrion. The solution is the unique design of the **prokaryotic** bacterial cell. A generalized bacterial cell is shown in **Figure 21–2**. The major structures of the cell belong either to the multilayered **envelope** and its **appendages** or to the interior core consisting of the **nucleoid** (or nuclear body) and the **cytosol.** The cytosol is analogous to the cytoplasm of eukaryotic cells, but because there is no nucleus it is not separated from the genetic material. The general chemical nature of the bacterial design includes the familiar macromolecules of life (DNA, RNA, protein, carbohydrate, phospholipid) plus some unique to bacteria such as the peptidoglycan and lipopolysaccharide of bacterial cell walls. The smallness and simplicity of the bacterial design contribute to the ability of the cytosol to grow at least an order of magnitude faster than eukaryotic cells, a significant feature in producing disease.

Prokaryotic design includes envelope, appendages, cytosol, and nucleoid

Chemical nature is similar to eukaryotic cells plus unique components

Design facilitates rapid growth

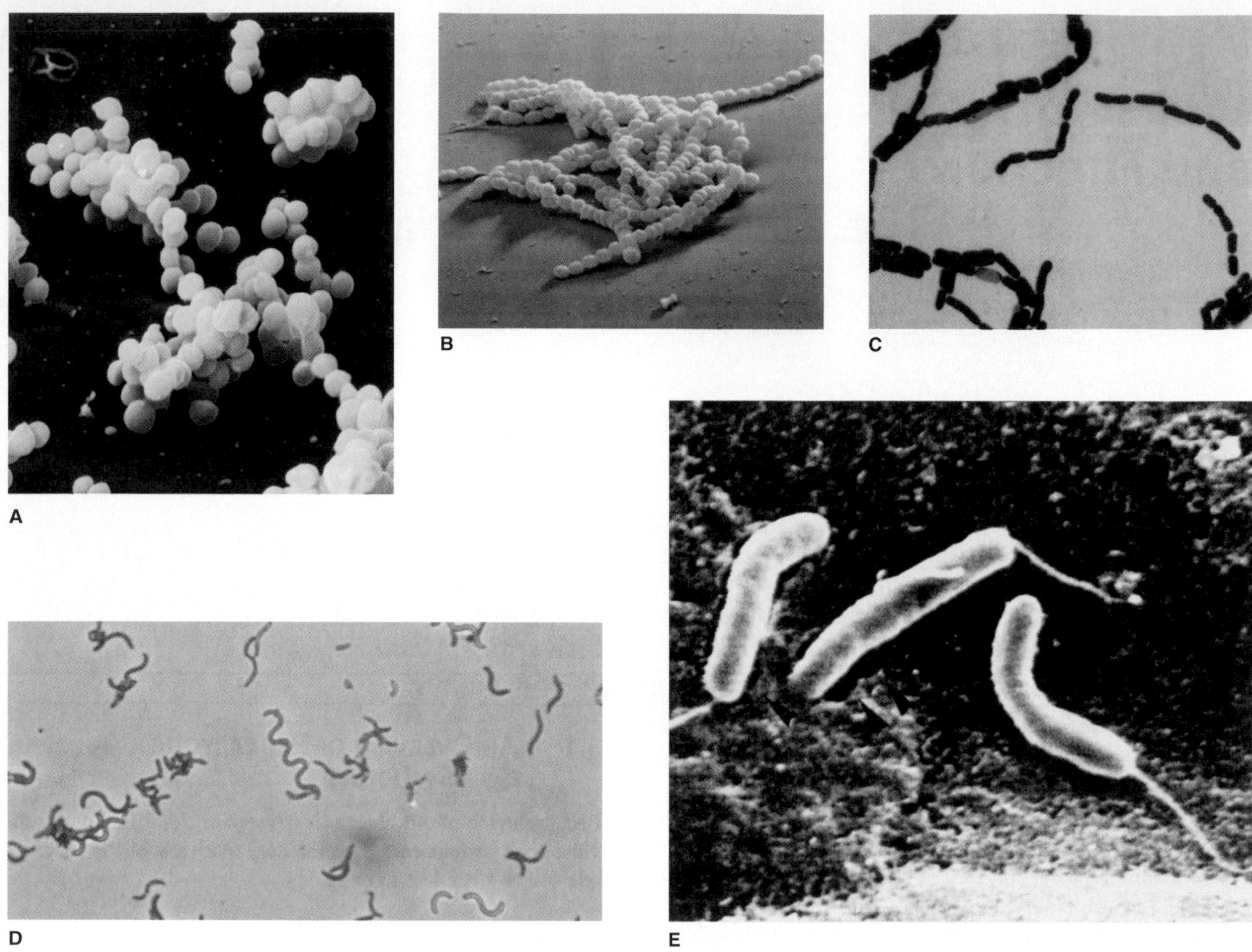

FIGURE 21–1. Shapes of bacteria. **A.** *Staphylococcus aureus*, cocci arranged in clusters; scanning electron micrograph (SEM). **B.** Group B streptococci, cocci arranged in chains; SEM. **C.** *Bacillus* species, straight rods; Gram stain. **D.** Spirochete, phase contrast, SEM. **E.** Vibrio, curved rods, SEM. (Reproduced with permission from Willey J, Sherwood L, Woolverton C (eds). *Prescott's Principles of Microbiology*. New York: McGraw-Hill; 2008.)

ENVELOPE AND APPENDAGES

Envelope and appendages carry out multiple functions

Bacteria have a very plain interior but a complex, even baroque, exterior. This can be readily understood by appreciating that the envelope not only protects the cell against chemical and biologic threats in its environment, but is also responsible for many metabolic processes that are the province of the internal organelles of eukaryotic cells. Structures in the envelope and certain appendages also mediate attachment to human cell surfaces the first step in disease. Not surprisingly, therefore, more than one fifth of the specific proteins of well-studied bacteria are located in the envelope. Some of these features are presented in **Table 21–1** in relation to the major bacterial cell wall types.

Capsule

Many bacterial cells surround themselves with one or another kind of hydrophilic gel. This layer is often thick; commonly it is thicker than the diameter of the cell. Because it is transparent and not readily stained, this layer is usually not appreciated unless made visible by its ability to exclude particulate material, such as India ink or by special capsular stains (**Figure 21–3**). If the material forms a reasonably discrete layer, it is called a **capsule;** if it is amorphous in appearance,

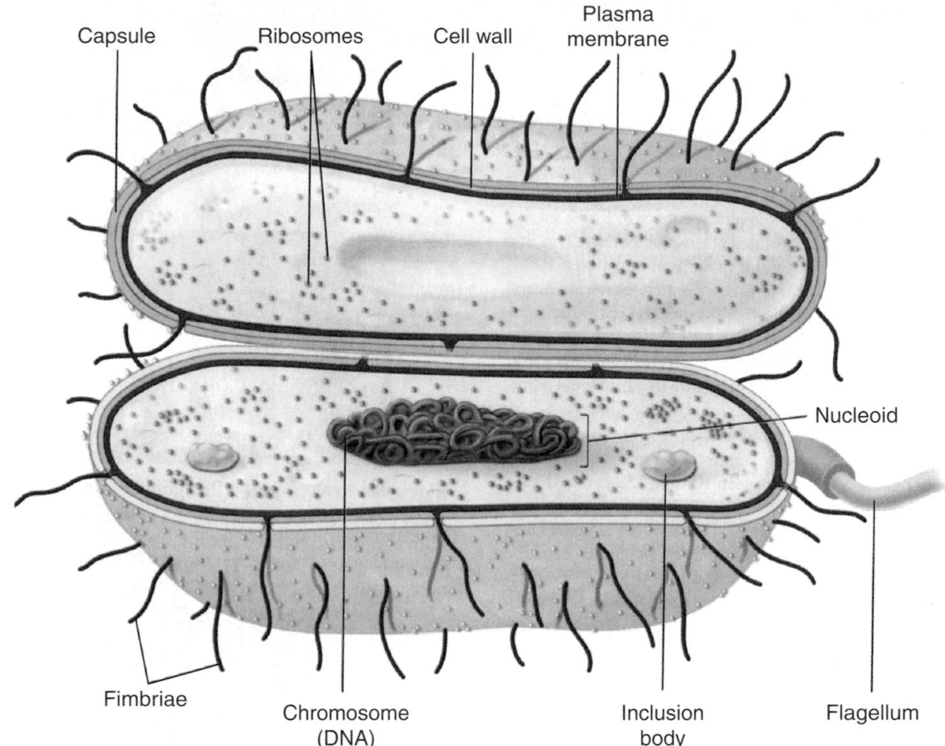

FIGURE 21–2. The prokaryotic bacterial cell. (Reproduced with permission from Willey J, Sherwood L, Woolverton C (eds). *Prescott's Principles of Microbiology.* New York: McGraw-Hill; 2008.)

it is referred to as a **slime layer.** Almost all bacterial species can make such material to some degree. Most capsules are polysaccharides made of single or multiple types of sugar residues; a few are simple polypeptides.

Hydrophilic capsules are usually polysaccharides

Capsules provide some general protection for bacteria, but their major function in pathogenic bacteria is protection from the immune system. These features are discussed in Chapter 22. Capsules do not contribute to growth and multiplication and are not essential for cell survival in artificial culture. Capsule synthesis is greatly dependent on growth conditions. For example, the capsule made by the caries-producing *Streptococcus mutans* consists of a dextran–carbohydrate polymer made in the presence of sucrose.

Capsules protect from immune system

Capsule synthesis depends on growth conditions

Cell Wall

Internal to the capsule (if one exists) but still outside the cell proper, a rigid **cell wall** surrounds all eubacterial cells except wall-less bacteria such as the mycoplasmas and *Chlamydia*. The structure and function of the bacterial wall is a hallmark of the prokaryotes; nothing like it is found elsewhere. This wall protects the cell from mechanical disruption and from bursting caused by the turgor pressure resulting from the hypertonicity of the cell interior relative to the environment. It also provides a barrier against certain toxic chemical and biologic agents. Its form is responsible for the shape of the cell. Overall, a well-constructed wall protects these minute, fragile cells from chemical and physical assault while still permitting the rapid exchange of nutrients and metabolic byproducts required for rapid growth.

Unique wall structure prevents osmotic lysis, determines shape

Bacterial evolution has led to two major solutions to cell wall structure. Although the detailed structural basis of the two is now well known, the separation derives from their reaction to a particular staining procedure devised more than a century ago. It is called the Gram stain and described in detail in Chapter 4. The staining reaction depends on the ability of cells stained with certain dyes to resist extraction of the dye with ethanol–acetone mixtures. The bacteria from which these complexes are readily extracted are called **Gram-negative,** and those that retained these complexes are termed **Gram-positive.** Thus, a positive or negative Gram stain response of a cell identifies which of the two types of wall it possesses. ::: gram stain, p. 61

Gram stain distinguishes two major envelope structures

TABLE 21–1 Components of Bacterial Cells

		CELL WALL TYPE[a]		
STRUCTURE	**COMPOSITION**	**GRAM-NEGATIVE**	**GRAM-POSITIVE**	**NONE[b]**
Envelope				
Capsule (slime layer)	Polysaccharide or polypeptide	+ or −	+ or −	−
Wall		+	+	−
Outer membrane	Proteins, phospholipids, and lipopolysaccharide	+	−	−
Peptidoglycan layer	Peptidoglycan (+ teichoate in Gram-positive)	+	+[c]	−
Periplasm	Proteins and oligosaccharides in solution	+	−	−
Cell membrane	Proteins, phospholipids	+	+	+
Appendages				
Pili (fimbriae)	Protein (pilin)	+ or −	+ or −	−
Flagella	Proteins (flagellin plus others)	+ or −	+ or −	−
Core				
Cytosol	Polyribosomes, proteins, carbohydrates (glycogen)	+	+	+
Nucleoid	DNA with associated RNA and proteins	+	+	+
Plasmids	DNA	+ or −	+ or −	+ or −
Endospore				
All cell components plus dipicolinate and special envelope components		−	+ or −	−

[a]"+" indicates that the structure is invariably present, "−" indicates it is invariably absent, and "+ or −" indicates that the structure is present is some species or strains and absent in others.
[b]*Mycoplasma* and *Ureaplasma*.
[c]In *Mycobacterium* complexed with mycolic acids and other lipids.

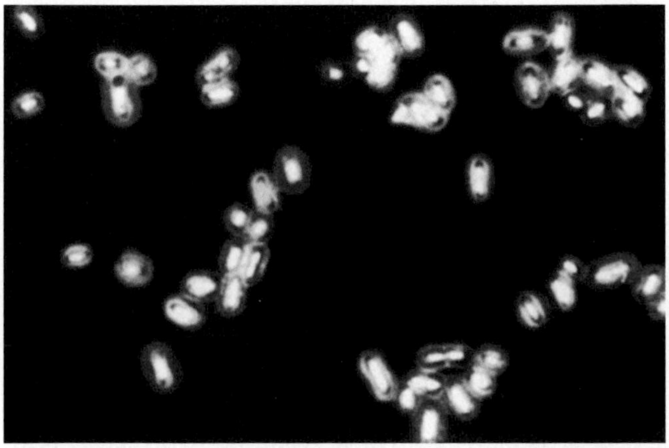

FIGURE 21–3. Bacterial capsule. This capsule surrounding cells of *Klebsiella pneumoniae* has been stained red. (Reproduced with permission from Willey J, Sherwood L, Woolverton C (eds). *Prescott's Principles of Microbiology.* New York: McGraw-Hill; 2008.)

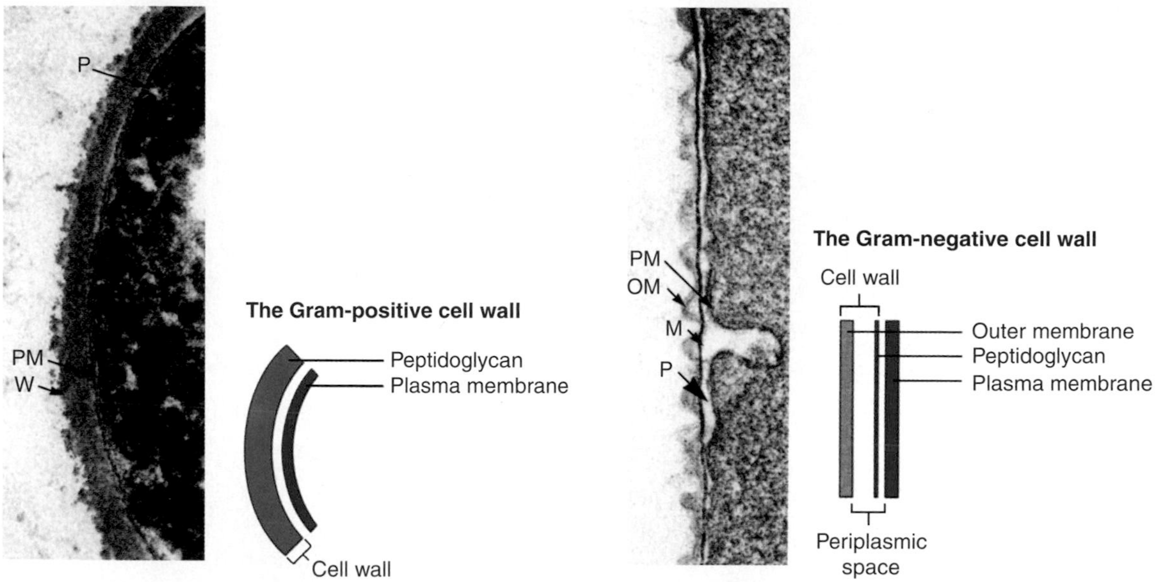

FIGURE 21–4. Gram positive and Gram negative cell walls. M—peptidoglycan or murein layer; OM, outer membrane; PM, plasma membrane; P, periplasmic space; W, Gram-positive peptidoglycan wall. (Reproduced with permission from Willey J, Sherwood L, Woolverton C (eds). *Prescott's Principles of Microbiology.* New York: McGraw-Hill; 2008.)

Virtually all bacteria with walls can now be assigned a Gram category even if they cannot be visualized with the stain itself for technical reasons. Examples include the causative agents of tuberculosis and syphilis. *Mycobacterium tuberculosis* (Gram-positive) has lipids in its cell wall that resist the uptake of most stains. *Treponema pallidum* (Gram-negative) takes stains poorly but is also too thin to be resolved in the light microscope without special illumination. In these cases, the Gram categorization is based on electron microscopy (**Figure 21–4**) and chemical analysis of the cell wall.

Poorly staining bacteria still have a Gram category

Gram-Positive Cell Wall

The Gram-positive cell wall contains two major components, peptidoglycan and teichoic acids, plus additional carbohydrates and proteins, depending on the species. A generalized scheme illustrating the arrangement of these components is shown in **Figure 21–5**. The chief component is peptidoglycan, which is found nowhere except in prokaryotes. Peptidoglycan consists of a linear glycan chain of two alternating sugars, *N*-acetylglucosamine (NAG) and *N*-acetylmuramic acid (NAM) (**Figure 21–6**). Each muramic acid residue bears a tetrapeptide of alternating L- and D-amino acids. Adjacent glycan chains are cross-linked into sheets by peptide bonds between the third amino acid of one tetrapeptide and the terminal D-alanine of another. The same cross-links between other tetrapeptides connect the sheets to form a three-dimensional, rigid matrix. The cross-links involve perhaps one third of the tetrapeptides and may be direct or may include a peptide bridge, as, for example, a pentaglycine bridge in *Staphylococcus aureus*. The cross-linking extends around the cell, producing a scaffold-like giant molecule. Peptidoglycan is much the same in all bacteria, except that there is diversity in the nature and frequency of the cross-linking bridge and in the nature of the amino acids at certain positions of the tetrapeptide.

Major components of Gram-positive walls are peptidoglycan and teichoic acid

Peptidoglycan comprises glycan chains cross-linked by peptide chains

Scaffold-like sac surrounds cell

The peptidoglycan sac derives its great mechanical strength from the fact that it is a single, covalently bonded structure. Most enzymes found in mammalian hosts and other biologic systems do not degrade peptidoglycan; one important exception is lysozyme, the hydrolase in tears and other secretions, which cleaves the β-1,4 glycosidic bond between muramic acid and glucosamine residues. The role of the peptidoglycan component of the cell wall in conferring osmotic resistance and shape on the cell is easily demonstrated by

Components of peptidoglycan provide resistance to most mammalian enzymes

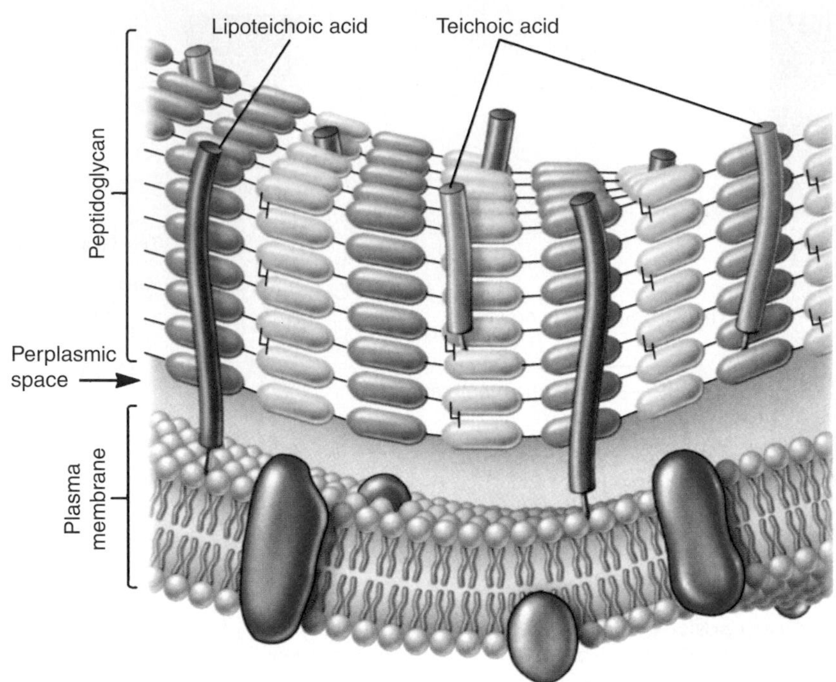

FIGURE 21–5. Gram-positive envelope. (Reproduced with permission from Willey J, Sherwood L, Woolverton C (eds). *Prescott's Principles of Microbiology*. New York: McGraw-Hill; 2008.)

Loss of cell wall leads to lysis or production of protoplasts

removing or destroying it. Treatment of a Gram-positive cell with penicillin (which blocks formation of the tetrapeptide cross-links) destroys peptidoglycan sac, and the wall is lost. Prompt lysis of the cell ensues. If the cell is protected from lysis by suspension in a medium approximately isotonic with the cell interior, such as 20% sucrose, the cell becomes round and forms a sphere called a **protoplast.** ::: Penicillin action, p. 406

A second component of the Gram-positive cell wall is a **teichoic acid**. These compounds are polymers of either glycerol phosphate or ribitol phosphate, with various sugars, amino sugars, and amino acids as substituents. The lengths of the chain and the nature and location of the substituents vary from species to species and sometimes among strains within a species. Up to 50% of the wall may be teichoic acid, some of

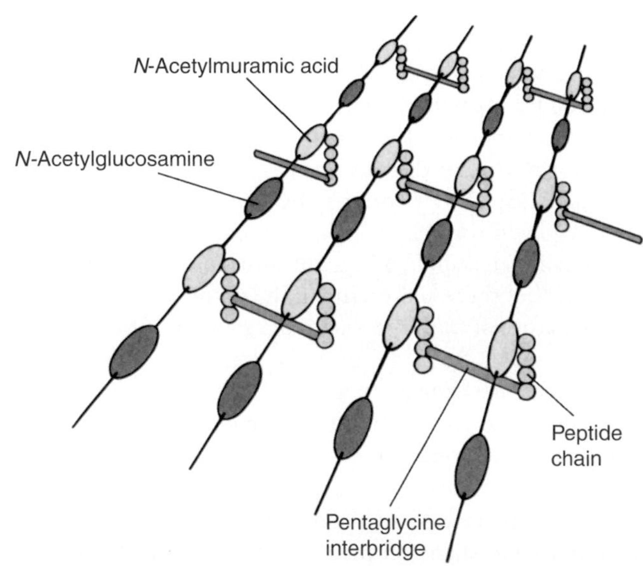

FIGURE 21–6. Peptidoglycan structure. A schematic diagram of one model of peptidoglycan. Shown are the polysaccharide chains, tetrapeptide side chains and peptide bridges. (Reproduced with permission from Willey J, Sherwood L, Woolverton C (eds). *Prescott's Principles of Microbiology*. New York: McGraw-Hill; 2008.)

which is covalently linked to occasional NAM residues of the peptidoglycan. Of the teichoic acids made of polyglycerol phosphate, much is linked not to the wall but to a glycolipid in the underlying cell membrane. This type of teichoic acid is called **lipoteichoic acid** and seems to play a role in anchoring the wall to the cell membrane and as an epithelial cell adhesin. Besides the major wall components—peptidoglycan and teichoic acids—Gram-positive walls usually have lesser amounts of other molecules characteristic of their species. Some are polysaccharides, such as the group-specific antigens of streptococci; others are proteins, such as the M protein of group A streptococci. ::: M protein, p. 446

Teichoic and lipoteichoic acids promote adhesion and anchor wall to membrane

Other cell wall components related to species

Gram-Negative Cell Wall

The second kind of cell wall found in bacteria, the Gram-negative cell wall, is depicted in **Figure 21–7**. Except for the presence of peptidoglycan, there is little chemical resemblance to cell walls of Gram-positive bacteria, and the architecture is fundamentally different. In Gram-negative cells, the amount of peptidoglycan has been greatly reduced, with some of it forming a single-layered sheet around the cell and the rest forming a gel-like substance, the **periplasmic gel,** with little cross-linking. External to this **periplasm** is an elaborate outer membrane. The proteins in solution in the periplasm consist of enzymes with hydrolytic functions, sometimes antibiotic-inactivating enzymes, and various binding proteins with roles in chemotaxis and in the active transport of solutes into the cell. Oligosaccharides secreted into the periplasm in response to external conditions serve to create an osmotic pressure buffer for the cell.

Thin peptidoglycan sac is imbedded in periplasmic gel

Periplasmic proteins have transport, chemotactic, and hydrolytic roles

The periplasm is an intermembrane structure, lying between the cell membrane and a special membrane unique to Gram-negative cells, the **outer membrane.** This has an overall structure similar to most biologic membranes with two opposing phospholipid–protein

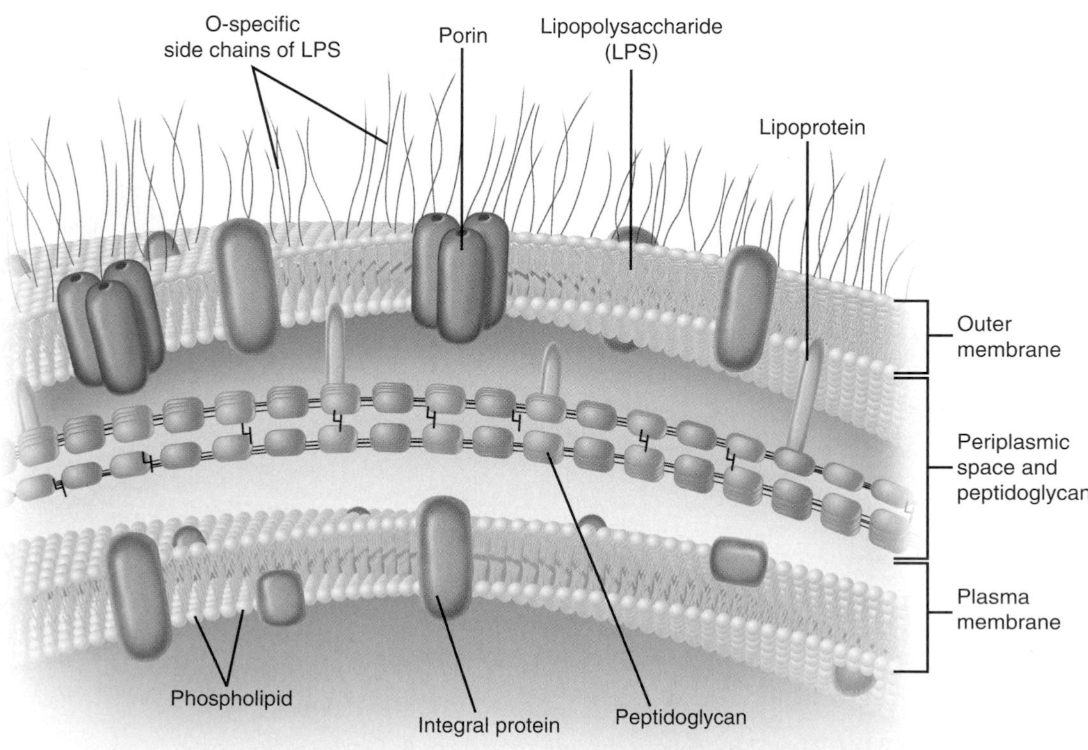

FIGURE 21–7. Gram-negative envelope. (Reproduced with permission from Willey J, Sherwood L, Woolverton C (eds). *Prescott's Principles of Microbiology.* New York: McGraw-Hill; 2008.)

Gram-negative outer membrane is phospholipoprotein bilayer, of which the outer leaflet is LPS endotoxin

leaflets. However, in terms of its chemical composition, the outer membrane is unique in all biology. Its inner leaflet consists of ordinary phospholipids, but these are replaced in the outer leaflet by a special molecule called **lipopolysaccharide** (LPS), which is extremely toxic to humans and other animals, and is called an **endotoxin.** Even in minute amounts, such as the amount released to the circulation during the course of a Gram-negative infection, this substance can produce a fever and shock syndrome called **Gram-negative shock,** or **endotoxic shock.**

Lipid A is toxic moiety of LPS; polysaccharides are antigenic determinants

LPS consists of a toxic **lipid A** (a phospholipid containing glucosamine rather than glycerol), a **core polysaccharide** (containing some unusual carbohydrate residues and fairly constant in structure among related species of bacteria), and **O antigen polysaccharide side chains** (**Figure 21–8A** and **B**). The last component constitutes the major surface antigen of Gram-negative cells.

Impermeability of outer membrane is overcome by porins

The presence of the outer membrane results in the covering of Gram-negative cells by a formidable permeability barrier. For whatever benefit is afforded by possessing a wall with an outer membrane, Gram-negative bacteria must make provision for the entry of nutrients. Special structural proteins, called **porins**, form pores through the outer membrane that make it possible for hydrophilic solute molecules to diffuse through it and into the periplasm.

In evolving a cell wall containing an outer membrane, Gram-negative bacteria have succeeded in (1) creating the periplasm, which holds digestive and protective enzymes and proteins important in transport and chemotaxis; (2) presenting an outer surface with strong negative charge, which is important in evading phagocytosis and the action of

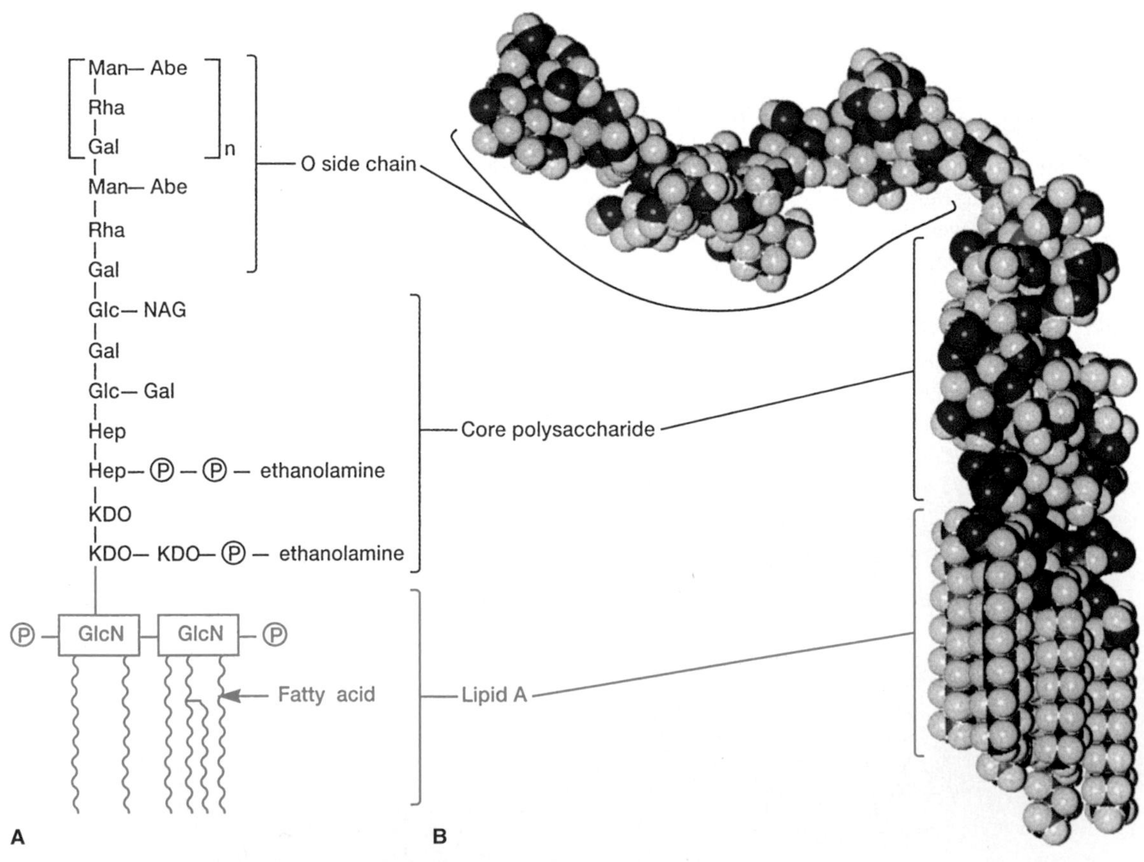

FIGURE 21–8. Lipopolysaccharide structure. A. O side chain—formed by linked sugars. Core polysaccharide—sugars linked to *N*-acetylglucoseamine (NAG) and keto-deoxycholate (KDO). Lipid A—buried in the outer membrane. **B.** Molecular model. (Reproduced with permission from Willey J, Sherwood L, Woolverton C (eds). *Prescott's Principles of Microbiology.* New York: McGraw-Hill; 2008.)

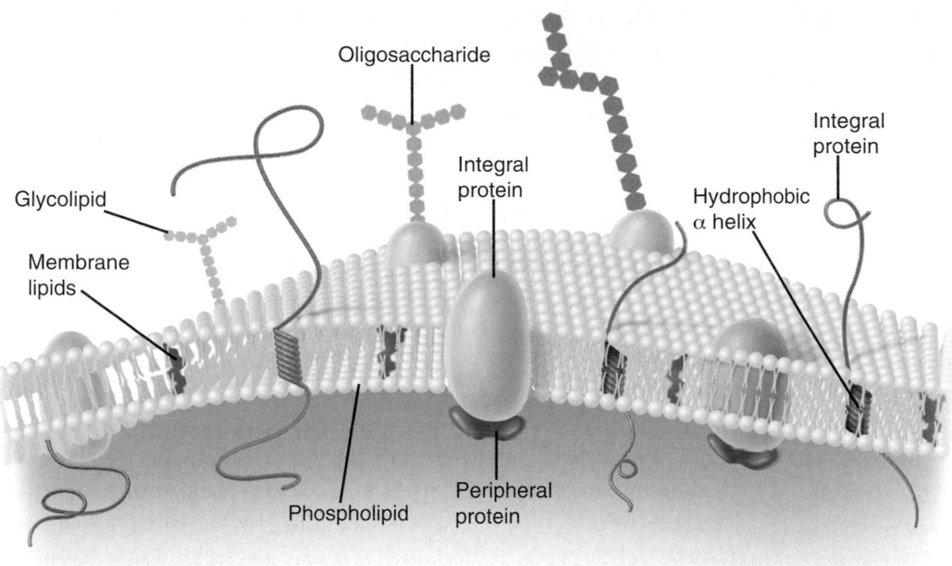

FIGURE 21–9. Bacterial plasma membrane. (Reproduced with permission from Willey J, Sherwood L, Woolverton C (eds). *Prescott's Principles of Microbiology.* New York: McGraw-Hill; 2008.)

complement; and (3) providing a permeability barrier against such dangerous molecules as host lysozyme, bile salts, digestive enzymes, and many antibiotics.

Outer membrane has many functions

Cell Membrane

Generally, the cell membrane of bacteria (**Figure 21–9**) is similar to the familiar bileaflet membrane, containing phospholipids and proteins, and which is found throughout the living world. However, there are important differences. The bacterial cell membrane is exceptionally rich in proteins and does not contain sterols (except mycoplasmas). The bacterial chromosome is attached to the cell membrane, which plays a role in segregation of daughter chromosomes at cell division, analogous to the role of the mitotic apparatus of eukaryotes. The membrane is the site of synthesis of DNA, cell wall polymers, and membrane lipids. It contains the entire electron transport system of the cell (and, hence, is functionally analogous to the mitochondria of eukaryotes). It contains receptor proteins that function in chemotaxis. Like cell membranes of eukaryotes, it is a permeability barrier and contains proteins involved in selective and active transport of solutes. It is also involved in secretion to the exterior of proteins (exoproteins), including exotoxins and hydrolytic enzymes involved in the pathogenesis of disease. The bacterial cell membrane is therefore the functional equivalent of most of the organelles of the eukaryotic cell and is vital to the growth and maintenance of the cell.

Phospholipid–protein bilayer lacking sterols

Roles in synthetic, homeostatic, secretory, and electron transport processes

Functional equivalent of many eukaryotic organelles

Flagella

Flagella are molecular organelles of motility found in many species of bacteria, both Gram-positive and Gram-negative. They may be distributed around the cell (an arrangement called peritrichous from the Greek *trichos* for "hair"), at one pole (**polar** or **monotrichous**), or at both ends of the cell (**lophotrichous**). In all cases, they are individually helical in shape and propel the cell by rotating at the point of insertion in the cell envelope. The presence or absence of flagella and their position are important taxonomic characteristics.

Flagella are rotating helical protein structures responsible for locomotion

The flagellar apparatus is complex, but consists entirely of proteins attached to the cell by a basal body consisting of several proteins organized as rings on a central rod.

Flagella have bushing rings in cell envelope

Flagellar filament is composed of the protein flagellin

Other structures include a hook that acts as a universal joint and ring-like bushings. All propel the long **filament,** which consists of polymerized molecules of a single protein species called **flagellin**. Flagellin varies in amino acid sequence from strain to strain. This makes flagella useful surface antigens for strain differentiation, particularly among the Enterobacteriaceae.

Pili

Pili are tubular hair-like projections

Pili have adherence roles and can "twitch"

Specialized pili mediate selective attachment or genetic transfer

Pili (also called fimbriae) are molecular hair-like projections found on the surface of cells of many Gram-positive and Gram-negative species. They are composed of molecules of a protein called **pilin** arranged to form a tube with a minute, hollow core. There are two general classes, common pili and sex pili (see Figure 21–33). **Common pili** cover the surface of the cell (**Figure 21–10**). They are, in many cases, adhesins**,** which are responsible for the ability of bacteria to colonize surfaces and cells. These processes are not always passive, since some pili can retract mediating movement across cell surfaces. Some pili are specialized for adherence to certain cell types such as enterocytes or uroepithelial cells. The same cell may have common and specialized pili. The **sex pilus** is involved in exchange of genetic material between some Gram-negative bacteria. There is only one per cell.

CORE

In contrast to the structural richness of the layers and appendages of the cell envelope, the interior seems relatively simple in transmission electron micrographs of thin sections of bacteria . There are two clearly visible regions, one granular (the cytosol) and one fibrous (the nucleoid). In addition, many bacteria possess plasmids that are usually circular, double-stranded DNA bodies in the cytosol that are separate from the larger nucleoid; plasmids are too small to be visible in thin sections of bacteria.

Cytosol

Cytosol contains 70 S ribosomes

Number varies with growth rate

The dense cytosol is bounded by the cell membrane. It appears granular because it is densely packed with ribosomes, which are much more abundant than in the cytoplasm of eukaryotic cells. This is a reflection of the higher growth rate of bacteria. Each ribosome is a ribonucleoprotein particle consisting of three species of rRNA (5 S, 16 S, and 23 S) and over 50 proteins. The overall subunit structure (one 50 S plus one 30 S particle) of the 70 S bacterial ribosome resembles that of eukaryotic ribosomes, but is smaller and differs sufficiently in function that a very large number of antimicrobics have the prokaryotic ribosome as their target. The number

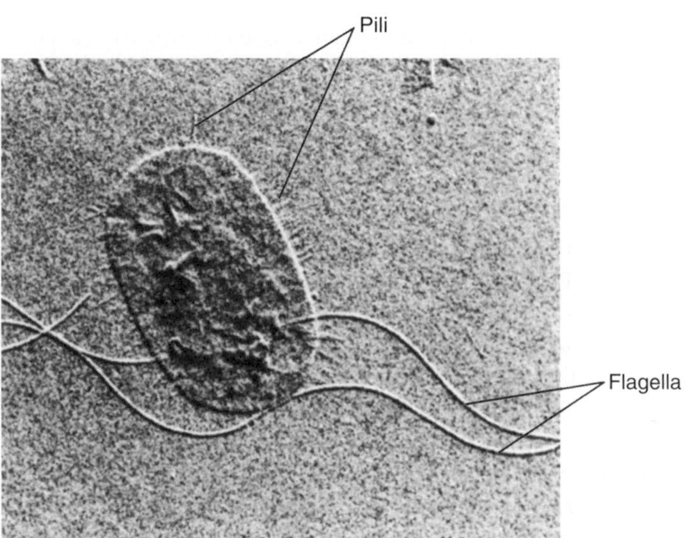

FIGURE 21–10. Flagella and pili. The long flagella and numerous shorter pili are evident in this electron micrograph of *Proteus mirabilis.* (Reproduced with permission from Willey J, Sherwood L, Woolverton C (eds). *Prescott's Principles of Microbiology.* New York: McGraw-Hill; 2008.)

of ribosomes varies directly with the growth rate of the cell. Except for the functions associated with the cell membrane, all of the metabolic reactions of the cell take place in the cytosol.

Nucleoid

The bacterial genome resides on a single chromosome (there are rare exceptions) and typically consists of about 4000 genes encoded in one, large, circular molecule of double-stranded DNA containing about 5 million nucleotide base pairs. This molecule is more than 1 mm long, and it therefore exceeds the length of the cell by about 1000 times. Tight packing displaces ribosomes and other cytosol components, creating regions that contain a chromosome, coated usually by polyamines and some specialized DNA-binding proteins. The double-helical DNA chain is twisted into supercoils and attached to the cell membrane and/or some central structure at a large number of points. This creates folds of DNA, each of which is independently coiled into a tight bundle. Each nuclear body corresponds to a DNA molecule. The number of nuclear bodies varies as a function of growth rate; resting cells have only one, and rapidly growing cells may have as many as four.

Circular chromosome of supercoiled double-stranded DNA

Attached to cell membrane and central structures

The absence of a nuclear membrane confers on the prokaryotic cell a great advantage for rapid growth in changing environments. Ribosomes can be translating mRNA molecules even as the latter are being made; no transport of mRNA from where it is made to where it functions is needed.

Plasmids

Many bacteria contain small, usually circular, covalently closed, double-stranded DNA molecules separate from the chromosome. More than one type of plasmid or several copies of a single plasmid may be present in the cell. Many plasmids carry genes coding for the production of enzymes that protect the cell from toxic substances. For example, antibiotic resistance is often plasmid-determined. Many attributes of virulence, such as production of some pili and of some exotoxins, are also determined by plasmid genes.

Plasmids are small, usually circular, double-stranded DNA molecules

SPORES

Endospores are small, dehydrated, metabolically quiescent forms that are produced by some bacteria in response to nutrient limitation or a related sign that tough times are coming. Very few species produce spores (the term is loosely used as equivalent to endospores), but they are particularly prevalent in the environment. Some spore-forming bacteria are of great importance in medicine, causing such diseases as anthrax, gas gangrene, tetanus, and botulism. All spore formers are Gram-positive rods. The bacterial endospore is not a reproductive structure. One cell forms one spore under adverse conditions (the process is called **sporulation**). The spore may persist for a long time (centuries) and then, on appropriate stimulation, give rise to a single bacterial cell (**germination**). Spores, therefore, are survival rather than reproductive devices.

Endospores are hardy, quiescent forms of some Gram-positives

Spore-forming allows survival under adverse conditions

Spores of some species can withstand extremes of pH and temperature, including boiling water, for surprising periods of time. The thermal resistance is brought about by the low water content and the presence of a large amount of a substance found only in spores, **calcium dipicolinate.** Resistance to chemicals and, to some extent, radiation is aided by extremely tough, special coats surrounding the spore. These include a **spore membrane** (equivalent to the former cell membrane); a thick **cortex** composed of a special form of peptidoglycan; a **coat** consisting of a cysteine-rich, keratin-like, insoluble structural protein; and, finally, an external lipoprotein and carbohydrate layer called an **exosporium.**

Resistance of spore is due to dehydrated state, calcium dipicolinate, and specialized coats

Sporulation is under active investigation. The molecular process by which a cell produces a highly differentiated product that is incapable of immediate growth but able to sustain growth after prolonged periods of nongrowth under extreme conditions of heat, desiccation, and starvation is of great interest. In general, the process involves the initial walling off of a nucleoid and its surrounding cytosol by invagination of the cell membrane, with later additions of special spore layers (**Figure 21–11**). Germination begins with activation by heat, acid, and reducing conditions. Initiation of germination eventually leads to outgrowth of a new vegetative cell of the same genotype as the cell that produced the spore.

Germination reproduces cell identical to that which was sporulated

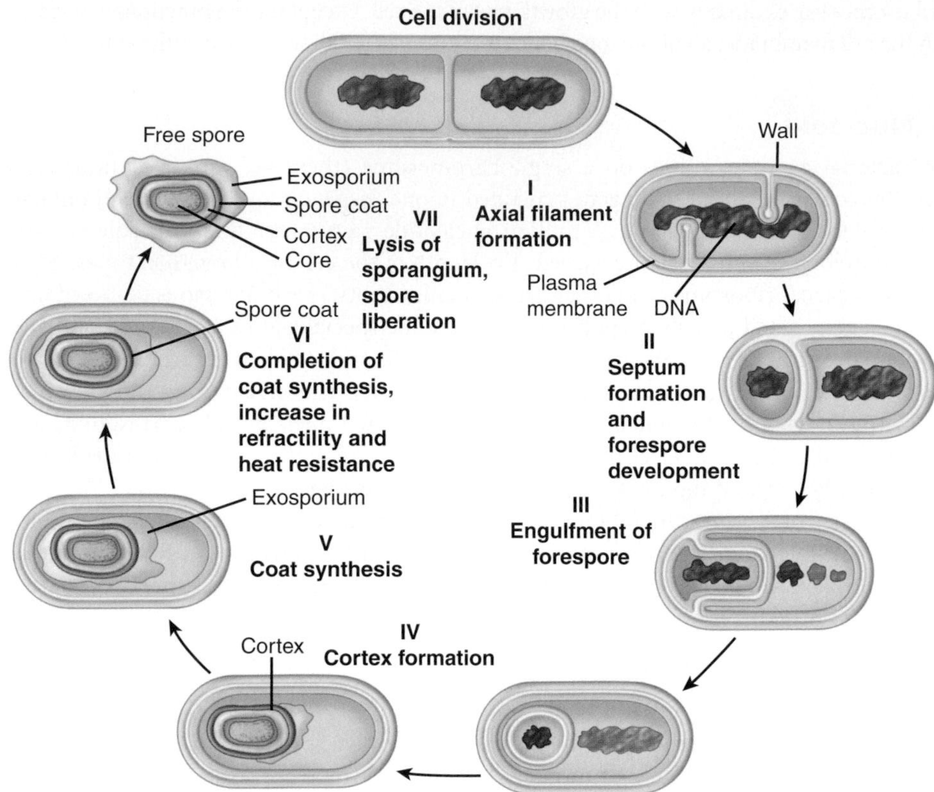

FIGURE 21–11. Stages of bacterial spore formation. (Reproduced with permission from Willey J, Sherwood L, Woolverton C (eds). *Prescott's Principles of Microbiology.* New York: McGraw-Hill; 2008.)

BACTERIAL GROWTH AND METABOLISM

Growth requires metabolism, regulation, and division by binary fission

Growth of bacteria is accomplished by an orderly progress of metabolic processes followed by cell division by binary fission. This requires metabolism, which produces cell material from the nutrient substances in the environment; regulation, which coordinates the progress of the hundreds of independent biochemical processes in an orderly way; and, finally, cell division, which produces two independent living units from one.

BACTERIAL METABOLISM

Many of the principles of metabolism are universal. This section focuses on the unique aspects of bacterial metabolism that are important in medicine. The need to compare bacterial and mammalian pathways is muted by the fact that much of what we understand about human metabolism is derived from work with *Escherichia coli.*

The broad differences between bacteria and human eukaryotic cells can be summarized as follows:

Speed. Bacteria metabolize at a rate 10 to 100 times faster.

Versatility. Bacteria use more varied compounds as energy sources and are much more diverse in their nutritional requirements.

Simplicity. The prokaryotic body plan makes it possible for bacteria to synthesize macromolecules in a streamlined way.

Uniqueness. Some biosynthetic processes, such as those producing peptidoglycan, lipopolysaccharide, and toxins, are unique to bacteria.

Bacterial metabolism is highly complex. The bacterial cell synthesizes itself and generates energy by as many as 2000 chemical reactions. These reactions can be classified according to their function in the metabolic processes of fueling, biosynthesis, polymerization, and assembly.

Fueling Reactions

Fueling reactions provide the cell with energy and with the 12 precursor metabolites used in biosynthetic reactions (**Figure 21–12**). The first step is the capture of nutrients from the environment. Other than water, oxygen, and carbon dioxide, almost no important nutrients enter the cell by **simple diffusion,** because the cell membrane is too effective a barrier. Some transport occurs by **facilitated diffusion** in which a protein carrier in the cell membrane, specific for a given compound, participates in the shuttling of molecules of that substance from one side of the membrane to the other (**Figure 21–13A** and **B**). Because no energy is involved, this process can work only with, never against, a concentration gradient of the given solute.

Substrates enter despite permeability barriers

Facilitated diffusion involves shuttling by carrier protein

Active transport mechanisms involve specific protein molecules as carriers of particular solutes, but the process is energy linked and can therefore establish a concentration gradient. That is, active transport can pump "uphill." Bacterial have multiple systems of active transport, some of which involve ATP-dependent binding proteins (**Figure 21–14**) and others that require proton pumps driven by electron transport within the energized cell membrane. Another mechanism called **group translocation** involves the chemical conversion of the solute into another molecule as it is transported.

Active transport involves binding proteins and ATP or proton gradient energy

The transport of iron and other metal ions needed in small amounts for growth is special and of particular importance in virulence. There is little free Fe^{3+} in human blood or other body fluids, because it is sequestered by iron-binding proteins (eg, **transferrin** in blood and **lactoferrin** in secretions). Bacteria must have iron to grow, and their colonization of the human host requires capture of iron. Bacteria secrete **siderophores** (iron-specific chelators) to trap Fe^{3+}; the iron-containing chelator is then transported into the bacterium by specific active transport.

Bacterial siderophores chelate iron and are actively transported into cell

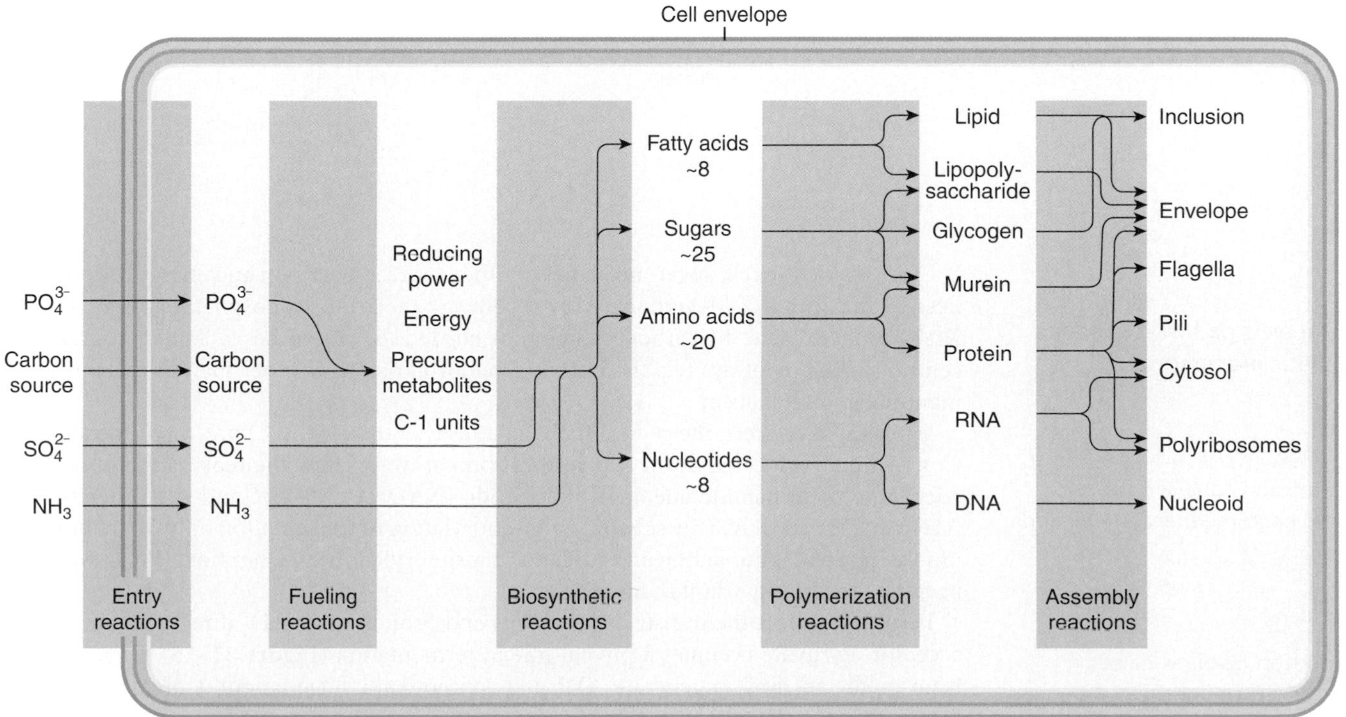

FIGURE 21–12. Bacterial metabolism. General pattern of metabolism leading to the synthesis of a bacterial cell from glucose.

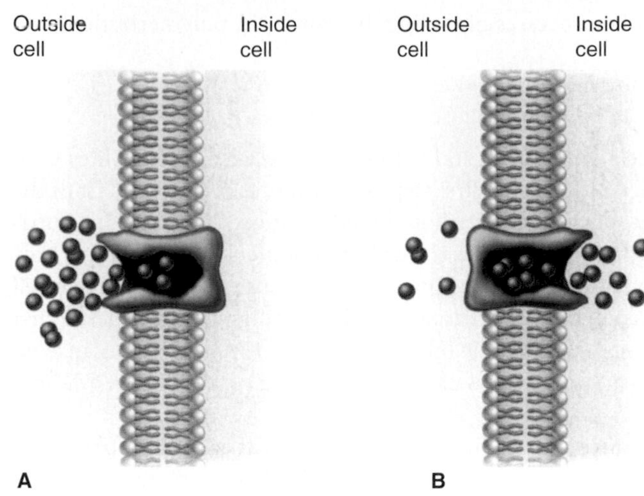

FIGURE 21–13. Facilitated diffusion. A. The membrane carrier can change conformation after binding an external molecule and subsequently release the molecule to the cell interior. **B.** It then returns to the outward oriented position and is ready to bind another solute molecule. Because there is no energy input, molecules continue to enter only as long as their concentration is greater on the outside. (Reproduced with permission from Willey J, Sherwood L, Woolverton C (eds). *Prescott's Principles of Microbiology*. New York: McGraw-Hill; 2008.)

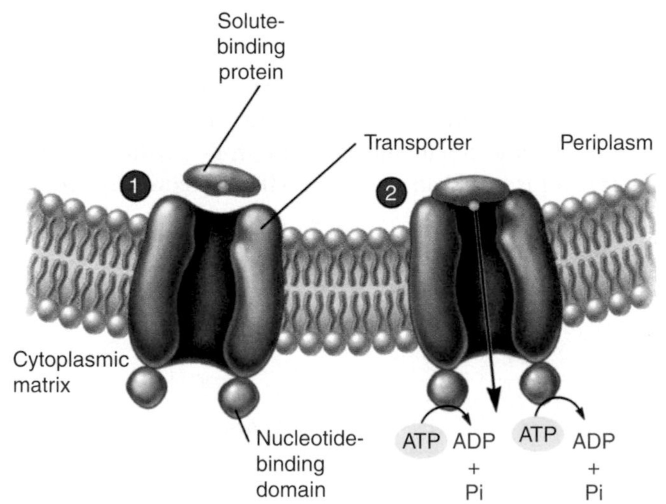

FIGURE 21–14. Active transport. 1. The solute binding protein binds the substrate to be transported and approaches the transporter complex. **2.** The solute binding which is moved across the membrane with the aid of ATP hydrolysis. (Reproduced with permission from Willey J, Sherwood L, Woolverton C (eds). *Prescott's Principles of Microbiology*. New York: McGraw-Hill; 2008.)

Central fueling pathways produce biosynthetic precursors

Once inside the cell, sugar molecules or other sources of carbon and energy are metabolized by the Embden–Meyerhof glycolytic pathway, the pentose phosphate pathway, and the Krebs cycle to yield the carbon compounds needed for biosynthesis. Some bacteria have central fueling pathways (eg, the Entner–Doudoroff pathway) other than those familiar in mammalian metabolism.

Fermentation and respiration pathways each regenerate ATP and NAD^+

Working in concert, the central fueling pathways produce the 12 precursor metabolites. Connections to **fermentation** and **respiration** pathways allow the reoxidation of reduced coenzyme nicotinamide adenine dinucleotide (NADH) to NAD^+ and the generation of ATP. Bacteria make ATP by substrate phosphorylation in fermentation or by a combination of substrate phosphorylation and oxidative phosphorylation in respiration. (Photosynthetic bacteria are not important in medicine.)

Fermentation involves direct transfer of proton and electron to organic receptor acceptor

ATP-generating efficiency is low

Fermentation is the transfer of electrons and protons via NAD^+ directly to an organic acceptor. Pyruvate occupies a pivotal role in fermentation (**Figure 21–15**). Fermentation is an inefficient way to generate ATP, and consequently huge amounts of sugar must be fermented to satisfy the growth requirements of bacteria anaerobically. Large amounts of organic acids and alcohols are produced in fermentation. Which compounds are produced depends on the particular pathway of fermentation used by a given species, and therefore the profile of fermentation products is a diagnostic aid in the clinical laboratory.

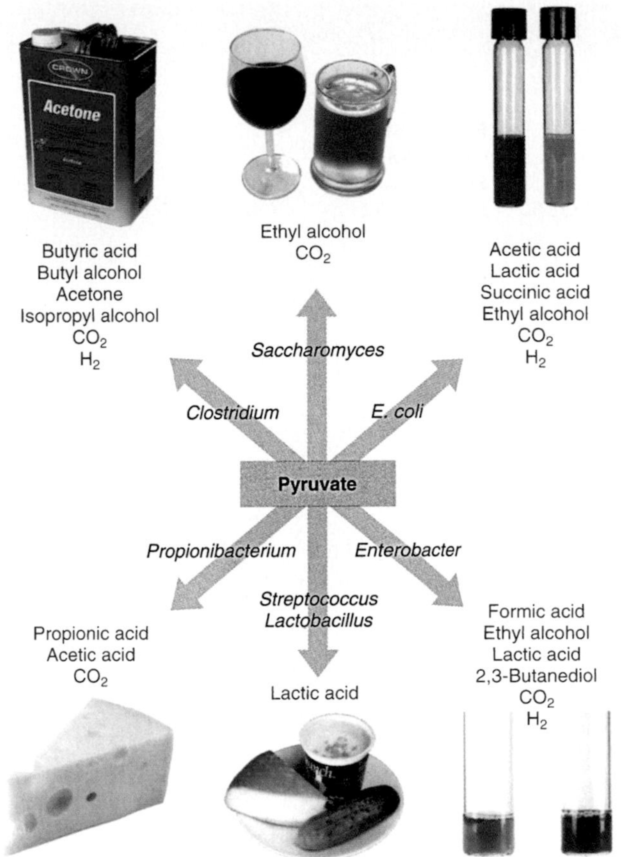

FIGURE 21–15. End products of fermentation pathways. Because a given type of organism uses a characteristic fermentation pathway, the end products can also be used as an identifying marker. (Reproduced with permission from Nester EW, Anderson DG, Roberts CE Jr, Nester MT. *Microbiology: A Human Perspective,* 6th ed. New York: McGraw-Hill, 2008.)

Respiration involves fueling pathways in which substrate oxidation is coupled to the transport of electrons through a chain of carriers to some ultimate acceptor, which is frequently, but not always, molecular oxygen. Other inorganic (eg, nitrate) as well as organic compounds (eg, succinate) can serve as the final electron acceptor, and therefore many organisms that cannot ferment can live in the absence of oxygen. Respiration is an efficient generator of ATP. Respiration in prokaryotes as in eukaryotes occurs by membrane-bound enzymes, but in prokaryotes the cell membrane rather than mitochondrial membranes provide the physical site.

Respiration uses electron chain for which oxygen is usually the terminal acceptor

Respiration is efficient energy producer

Aerobes and Anaerobes

In evolving to colonize every conceivable nook and cranny on this planet, bacteria have developed distinctive responses to oxygen. Bacteria are conveniently classified according to their fermentative and respiratory activities but much more generally by their overall response to the presence of oxygen. The response depends on their genetic ability to ferment or respire, but also on their ability to protect themselves from the deleterious effects of oxygen.

Bacteria exhibit different characteristic responses to oxygen

Oxygen, though itself only mildly toxic, gives rise to at least two extremely reactive and toxic substances, **hydrogen peroxide** (H_2O_2) and the **superoxide anion** (O^{2-}). Peroxide is produced by reactions in which electrons and protons are transferred to O_2 as final acceptor. The superoxide radical is produced as an intermediate in most reactions that reduce molecular O_2. Superoxide is partially detoxified by an enzyme, **superoxide dismutase,** found in all organisms (prokaryotes and eukaryotes) that survive the presence of oxygen. Bacteria that lack the ability to make superoxide dismutase and catalase are exquisitely sensitive to the presence of molecular oxygen and, in general, must grow anaerobically using fermentation. Bacteria that possess these protective enzymes can grow in the presence of oxygen,

Aerobic metabolism produces peroxide and toxic oxygen radicals

Superoxide dismutase and peroxidase allow growth in air; their absence requires strict anaerobiasis

TABLE 21–2 Classification of Bacteria by Response to Oxygen

TYPE OF BACTERIA	GROWTH RESPONSE: AEROBIC	GROWTH RESPONSE: ANAEROBIC	POSSESSION OF CATALASE AND SUPEROXIDE DISMUTASE	COMMENT	EXAMPLE
Aerobe	+	–	+	Requires O_2; cannot ferment	*Mycobacterium tuberculosis, Pseudomonas aeruginosa, Bacillus anthracis*
Anaerobe	–	+	–[a]	Killed by O_2; ferments in absence of O_2	*Clostridium botulinum, Bacteroides melaninogenicus*
Facultative	+	+	+	Respires with O_2; ferments in absence of O_2[c]	*Escherichia coli, Shigella dysenteriae, Staphylococcus aureus*
Microaerophilic	+[b]	+[b]	+	Grows best at low O_2 concentration; can grow without O_2	*Campylobacter jejuni*

[a]Many pathogenic anaerobes produce catalase and/or superoxide dismutase.
[b]Optimum growth at 5% to 10% O_2.
[c]Some ferment in the presence or absence of O_2.

Organisms growing in air may or may not have a respiratory pathway

but whether they use the oxygen in metabolism or not depends on their ability to respire. Whether these oxygen-resistant bacteria can grow anaerobically depends on their ability to ferment.

Aerobes require oxygen and anaerobes are killed by it

Facultative bacteria grow either way

Pathogenic anaerobes tolerate brief oxygen exposures

Various combinations of these two characteristics (oxygen resistance and the ability to use molecular oxygen as a final acceptor) are represented in different species of bacteria, resulting in the four general classes shown in **Table 21–2**. **Aerobes** require oxygen and metabolize by respiration. **Anaerobes** are inhibited or killed by oxygen and utilize fermentation exclusively. **Facultative** bacteria (the majority of pathogens) grow well under aerobic or anaerobic conditions. If oxygen is available they respire, if not they use fermentation. Some facultative bacteria ferment even if oxygen is available. **Microaerophilic** bacteria sit in the middle requiring 5% to 10% oxygen for optimal growth. There are important pathogens within each class. Although most anaerobes in the microbial world strictly follow the criteria in Table 21–2, many of the pathogenic anaerobes are in fact moderately aerotolerant and possess low levels of superoxide dismutases and peroxidases. Although they prefer anaerobic growth conditions, this allows them to survive the brief exposure to oxygen that is inherent to initiating disease. ::: Oxygen-tolerant pathogens, p. 516

Biosynthesis

Biosynthesis requires precursor metabolites, energy, amino nitrogen, sulfur, and reducing power

Nutritional requirements differ depending on synthetic ability

Biosynthetic reactions form a network of pathways that lead from precursor metabolites (provided by the fueling reactions) to the many amino acids, nucleotides, sugars, amino sugars, fatty acids, and other building blocks needed for macromolecules (Figure 21–12). In addition to the carbon precursors, large quantities of reduced nicotinamide adenine dinucleotide phosphate (NADPH), ATP, amino nitrogen, and some source of sulfur are needed for biosynthesis of these building blocks. These pathways are similar in all species of living things, but bacterial species differ greatly as to which pathways they possess. Because all cells require the same building blocks, those that cannot be produced by a given cell must be obtained preformed from the environment.

The are relatively few biosynthetic pathways that are unique to bacteria, but some form a basis for bacterial vulnerability or bacterial pathogenicity. Because bacteria must synthesize folic acid rather than use it preformed from their environment, inhibition of those

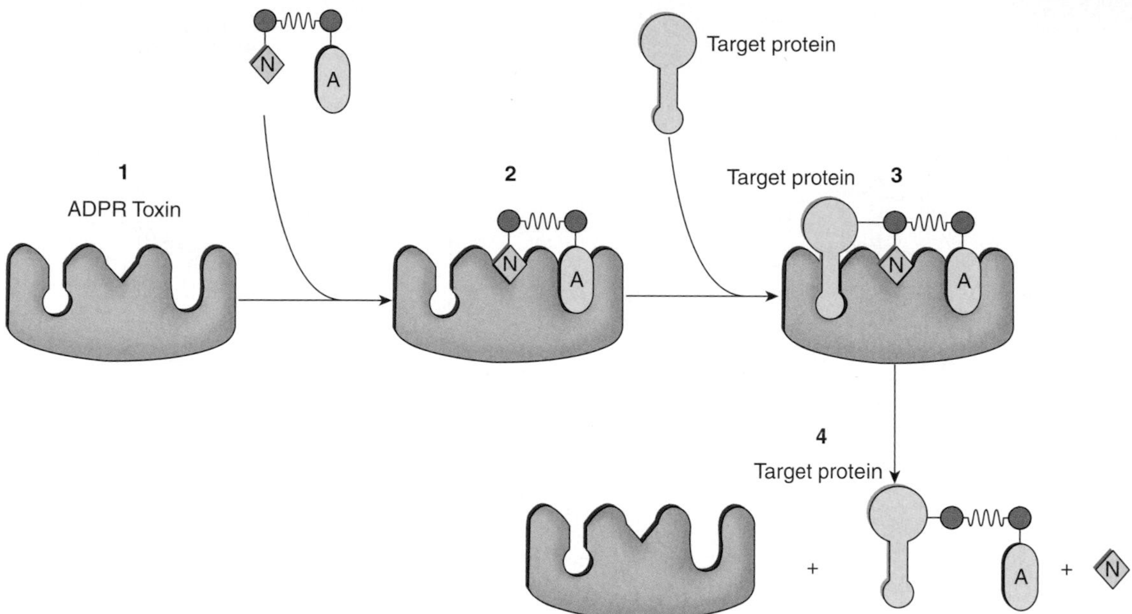

FIGURE 21–16. ADP-ribosylation (ADPR). 1. The active toxin unit binds nicotinamide adenine dinucleotide (NAD) that is present in fluids. **2.** The toxin also binds a cell protein, its target protein. **3.** An ADP-ribose group is transferred to the protein rendering it inactive. **4.** The toxin is released free to repeat the process.

pathways is the basis of the antibacterial action of sulfonamides and trimethoprim. Catalyzing ADP-ribosylation (**Figure 21–16**), a unique enzymatic reaction, is the mechanism for the action of multiple bacterial toxins including diphtheria toxin (DT) and cholera toxin (CT). To accomplish this, the active unit of the toxin binds both nicotinamide adenine dinucleotide (NAD) from body fluids and its target protein. This catalyses the transfer of an ADP-ribose group to the protein rendering it inactive. The biologic outcome of this inactivation depends on the function of the target protein. If it is crucial for a process like protein synthesis the result is cell death. If it is a regulatory protein, the process it controls may be up or down regulated.

Few pathways are unique to bacteria

ADP-ribosylation is the action of multiple toxins

Polymerization Reactions

Polymerization of DNA is called **replication.** Replication always begins at special sites on the chromosome and then precedes bidirectionally around the circular chromosome (**Figure 21–17**). Synthesis of DNA at each replication fork is termed semiconservative because each of the DNA chains serves as the template for the synthesis of its complement and, therefore, one of the two chains of the new double-stranded molecule is conserved from the original chromosome. Some chemotherapeutic agents derive their selective toxicity for bacteria from the unique features of prokaryotic DNA replication. The synthetic quinolone compounds inhibit DNA gyrase, one of the many enzymes participating in DNA replication.

Bidirectional, semiconservative replication occurs at replication forks

DNA gyrase inhibitors are selectively toxic for bacteria

Transcription is the synthesis of RNA. Transcription in bacteria differs from that in eukaryotic cells in several ways. One difference is that all forms of bacterial RNA (mRNA, tRNA, and rRNA) are synthesized by the same enzyme, **RNA polymerase**. RNA polymerase is a large, complicated molecule with a subunit (σ subunit) that locates specific DNA sequences, called promoters, which precede all transcriptional units. Remarkably, bacterial mRNA is synthesized, used, and degraded all in a matter of a few minutes. Bacterial RNA polymerase is the target of the antimicrobial **rifampin,** which blocks initiation of transcription.

A single RNA polymerase makes all forms of bacterial RNA

Rifampin inhibits RNA polymerase

Translation is the name given to protein synthesis. Bacteria activate the 20-amino-acid building blocks of protein in the course of attaching them to specific transfer RNA molecules. The aminoacyl-tRNAs are brought to the ribosomes by soluble protein factors, and there the

Amino acid residues are polymerized from specific tRNAs at the direction of mRNA

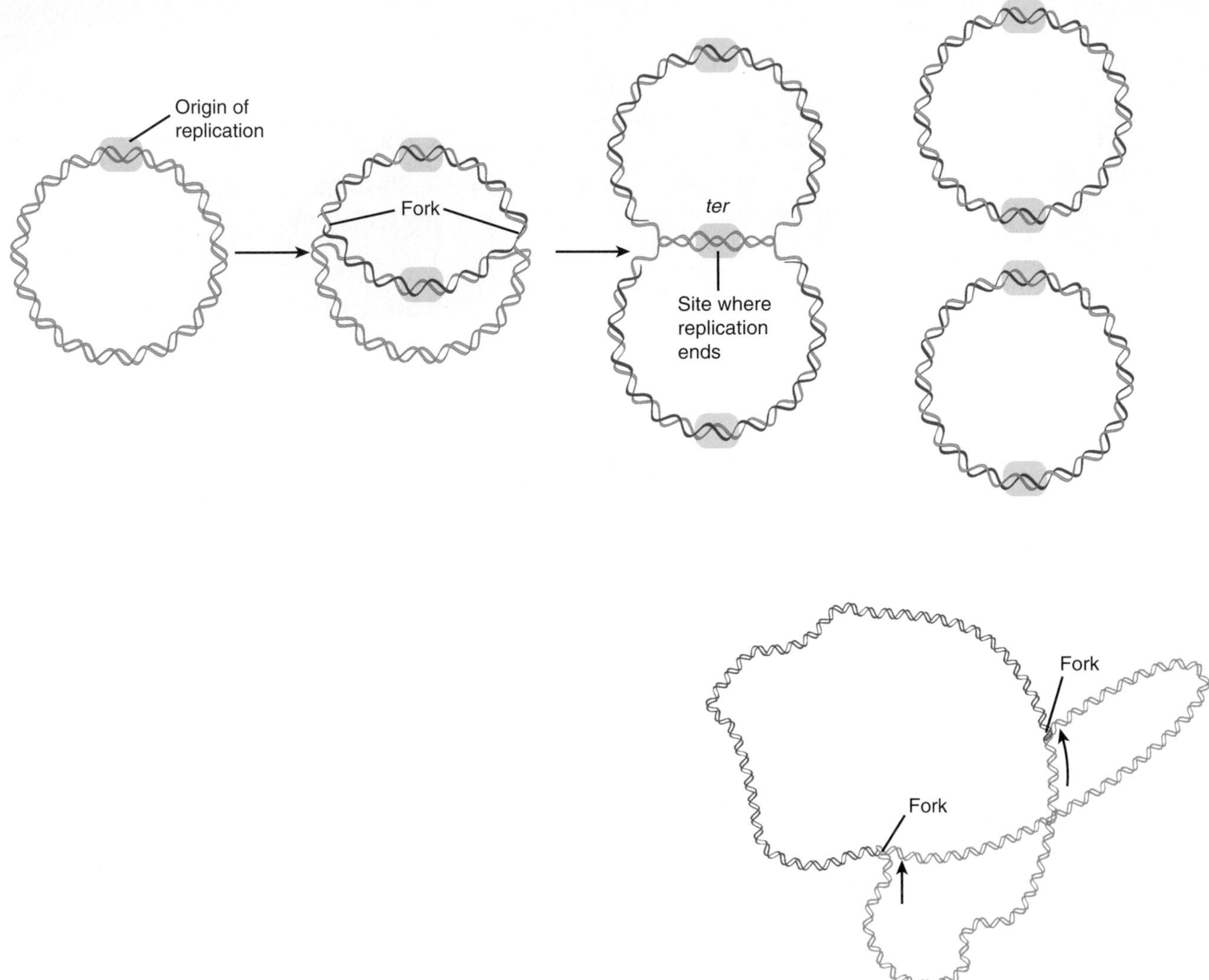

FIGURE 21–17. DNA replication in bacteria. Replication begins at the origin of replication. Two replication forks proceed in opposite directions until they meet a the replication termination site (*ter*). (Reproduced with permission from Willey J, Sherwood L, Woolverton C (eds). *Prescott's Principles of Microbiology.* New York: McGraw-Hill; 2008.)

Many antimicrobials act on bacterial translation machinery

Translation of mRNA occurs simultaneously with transcription

amino acids are polymerized into polypeptide chains according to the sequence of codons in the particular mRNA that is being translated. Having donated its amino acid, the tRNA is released from the ribosome to return for another aminoacylation cycle. Many antimicrobial agents derive their selective toxicity for bacteria from the unique features and proteins of the prokaryotic translation apparatus. In fact, protein synthesis is the target of a greater variety of antimicrobics than is any other metabolic process. Transcription and translation are illustrated in **Figure 21–18**.

Peptidoglycan Synthesis

Other polymerization reactions involve synthesis of peptidoglycan, phospholipid, LPS, and capsular polysaccharide. All of these reactions involve activated building blocks that are polymerized or assembled within or on the exterior surface of the cytoplasmic membrane. The most unique of these is **peptidoglycan,** which is completely absent from eukaryotic cells. Peptidoglycan synthesis takes place in three compartments of the cell. The steps involved are

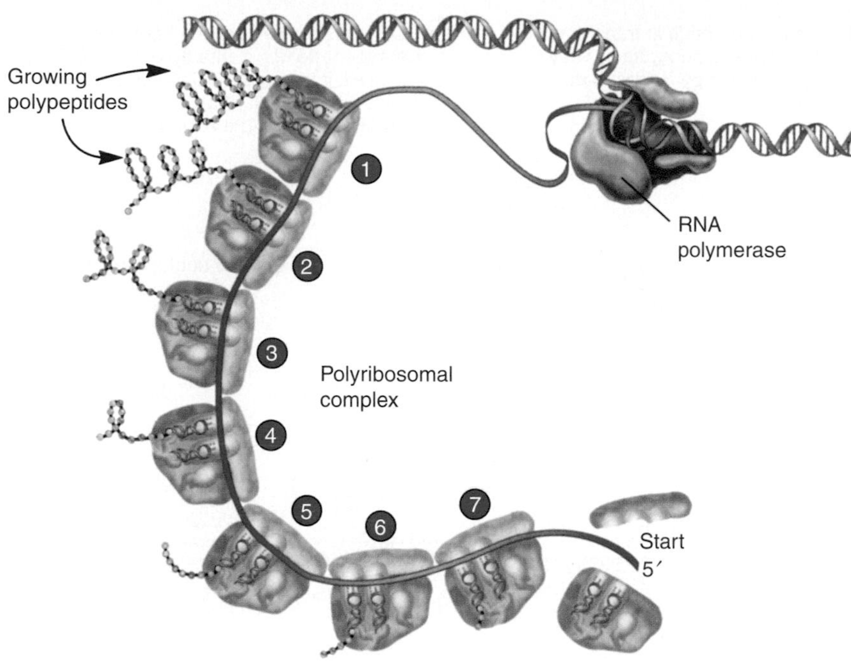

FIGURE 21–18. Coupling of transcription and translation in bacteria. As the DNA is transcribed, ribosomes bind the free 5′ end of the mRNA. Thus, translation is started before transcription is completed. Note multiple ribosomes are bound to the mRNA, forming a polyribosome. (Reproduced with permission from Willey J, Sherwood L, Woolverton C (eds). *Prescott's Principles of Microbiology.* New York: McGraw-Hill; 2008.)

summarized below and illustrated in **Figure 21–19** together with the attack points of some antimicrobials that block steps in the process.

1. **In the cytosol** a series of reactions leads to the synthesis, on a nucleotide carrier (UDP), of an *N*-acetylmuramic acid (NAM) residue bearing a pentapeptide.
2. This precursor is then attached, with the release of UMP, to a special, lipid-like carrier in the cell membrane called **bactoprenol**. Within the cell membrane *N*-acetylglucosamine (NAG) is added to the precursor, along with any amino acids that in this particular species will form the bridge between adjacent tetrapeptides
3. **Outside the cell membrane**, this disaccharide subunit is attached to the end of a growing glycan chain, and then the cross-links between chains that give the macromolecule its strength are formed by **transpeptidases** (**Figure 21–20**). These enzymes are also called **penicillin-binding proteins (PBPs)** for their property of binding this antibiotic. These transpeptidases are involved in forging, breaking, and reforging the peptide cross-links between glycan chains necessary to permit expansion of the peptidoglycan sac during cellular growth. Details of the cross-linking process vary among bacterial species.

NAM and attached peptide are synthesized in cytosol

Precursor is added to bactoprenol carrier

NAG and bridge amino acids are added

Chain cross-links are formed by transpeptidases (PBPs)

Protein Secretion

Moving macromolecules out of the cell interior and into their proper place in the wall, outer membrane, and capsule is a complex process. Moreover, many proteins are translocated through all layers of the cell envelope to the exterior environment. The later instance is of particular medical interest when the protein is an exotoxin or other protein involved in virulence. Protein secretion has become the general term to designate all these instances of translocation of proteins out of the cytosol (ie, whether the protein is to leave the cell or become part of the envelope). The process is relatively simple in Gram-positive bacteria in which proteins, after export across the cytoplasmic membrane, have only to move through the relatively porous peptidoglycan layer. In Gram-negative bacteria, the periplasmic space and the outer membrane must also be traversed.

Proteins are transported to locations in the cell structure or the exterior

In Gram-negatives, the periplasm and outer membrane are additional barriers

The simplest and most common mechanism for protein secretion called the **general secretory pathway (GSP)** is used by both Gram-positive and Gram-negative bacteria. Proteins secreted by the GSP are called preproteins because they have a signal peptide at their leading end that allows them to be guided by cytosolic chaperone proteins through the

GSP uses signal peptide and chaperone proteins

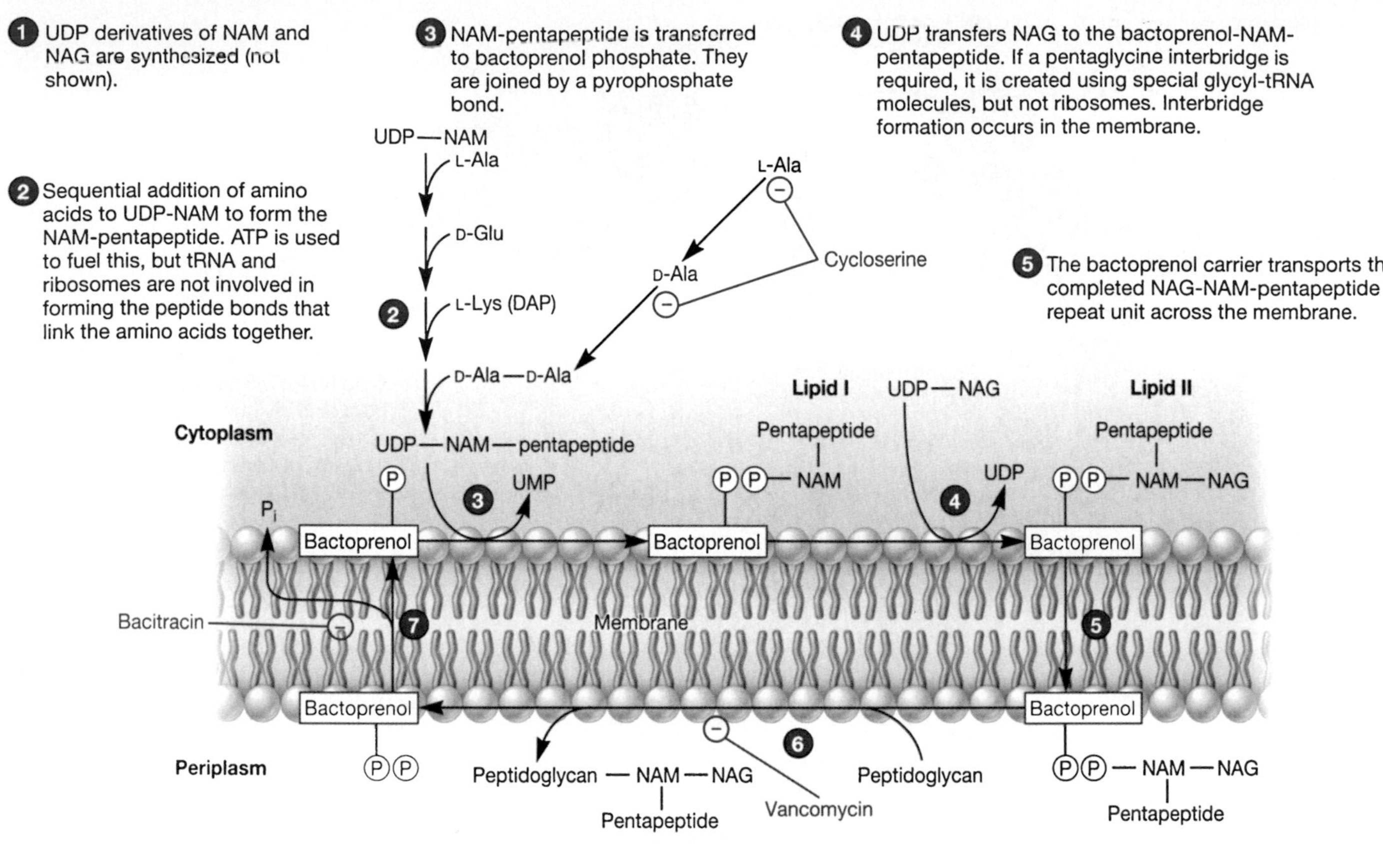

1 UDP derivatives of NAM and NAG are synthesized (not shown).
3 NAM-pentapeptide is transferred to bactoprenol phosphate. They are joined by a pyrophosphate bond.
4 UDP transfers NAG to the bactoprenol-NAM-pentapeptide. If a pentaglycine interbridge is required, it is created using special glycyl-tRNA molecules, but not ribosomes. Interbridge formation occurs in the membrane.
UDP—NAM
L-Ala
2 Sequential addition of amino acids to UDP-NAM to form the NAM-pentapeptide. ATP is used to fuel this, but tRNA and ribosomes are not involved in forming the peptide bonds that link the amino acids together.
D-Glu
L-Lys (DAP)
L-Ala
D-Ala
Cycloserine
5 The bactoprenol carrier transports the completed NAG-NAM-pentapeptide repeat unit across the membrane.
D-Ala—D-Ala
Lipid I
UDP—NAG
Lipid II
Cytoplasm
UDP—NAM—pentapeptide
Pentapeptide
NAM
Pentapeptide
NAM—NAG
UMP
UDP
P
Pi
Bactoprenol
Bactoprenol
Bactoprenol
Bacitracin
Membrane
Bactoprenol
Bactoprenol
Periplasm
Peptidoglycan — NAM—NAG
Pentapeptide
Vancomycin
Peptidoglycan
NAM—NAG
Pentapeptide
8 Peptide cross-links between peptidoglycan chains are formed by transpeptidation (not shown).
7 The bactoprenol carrier moves back across the membrane. As it does, it loses one phosphate, becoming bactoprenol phosphate. It is now ready to begin a new cycle.
6 The NAG-NAM-pentapeptide is attached to the growing end of a peptidoglycan chain, increasing the chain's length by one repeat unit.

FIGURE 21–19. Peptidoglycan synthesis. NAM is *N*-acetylmuramic acid and NAG is *N*-acetylglucosamine. The pentapeptide contains L-lysine in *Staphylococcus aureus* and diaminopimelic acid in *Escherichia coli*. Inhibition by bacitracin, cycloserine, and vancomycin are shown. Transpeptidation and the action of penicillins are shown in Figure 21–20. (Reproduced with permission from Willey J, Sherwood L, Woolverton C (eds). *Prescott's Principles of Microbiology.* New York: McGraw-Hill; 2008.)

transport machinery (**Figure 21–21**). Once through the GSP, the signal peptide is removed and the mature protein folds into its final shape.

Five systems transport across the outer membrane

Injection systems also use a syringe to penetrate host cells

In Gram-negative species, five additional pathways have been discovered that accomplish export of proteins across the outer membrane into the environment (**Figure 21–22**). Two of these (types II and V) provide a second step for proteins that have already been secreted by the GSP. The others extend across both membranes, and two of these (types III and IV) have an elaborate syringe-like apparatus that literally injects the proteins across yet a third membrane—that of a host cell. These injection secretion systems are a major mechanism for delivery of exotoxins and other proteins important in the pathogenesis of human infections. Type IV systems have the additional property of being able to inject DNA as well as proteins and are important gene transfer as discussed in the following text.

CELL DIVISION AND GROWTH

Growth of a liquid culture can be monitored by colony counts or turbidimetrically

Bacteria multiply by binary fission. In rich medium at 37°C, the entire process is completed in 20 minutes in *E coli* and many other pathogenic species. Using methods described in Chapter 4, growth of a liquid bacterial culture can be monitored by counting colonies from samples removed at timed intervals or by turbidity measured in a spectrophotometer.

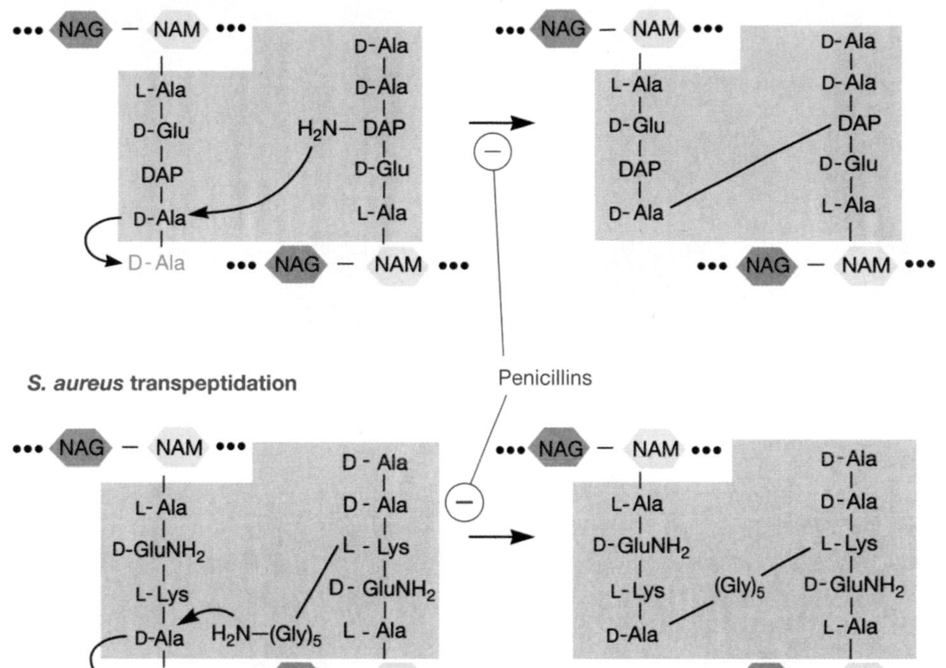

FIGURE 21–20. Transpeptidation. The transpeptidation reactions in the formation of the peptidoglycan of *Eshcerichia coli* and *Staphylococcus aureus* are shown. β-lactam antibiotics bind the transpeptidases and block cross-linking of the peptidoglycan backbone molecules. (Reproduced with permission from Willey J, Sherwood L, Woolverton C (eds). *Prescott's Principles of Microbiology.* New York: McGraw-Hill; 2008.)

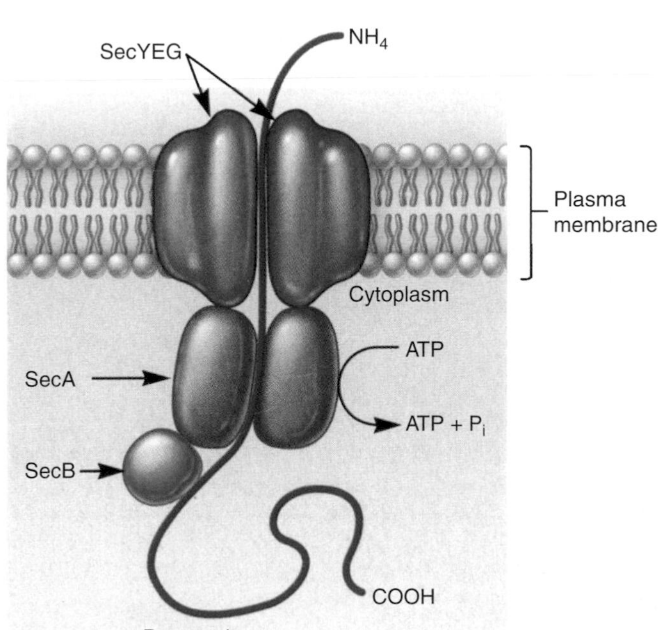

FIGURE 21–21. General secretion pathway. The amino-terminal end of the preprotein has a signal peptide that facilitates transport through the apparatus by chaperone (SecB) and proteins that form channels (SecY, SecE, SecG) or have propelling functions (SecA). The signal peptide is removed on the outside. Energy is required in the form of ATP (adenosine triphosphate) hydrolysis. (Reproduced with permission from Willey J, Sherwood L, Woolverton C (eds). *Prescott's Principles of Microbiology.* New York: McGraw-Hill; 2008.)

The growth rate of a bacterial culture depends on three factors: the species of bacterium, the chemical composition of the medium, and the temperature. The time needed for a culture to double its mass or cell number is in the range of 30 to 60 minutes for most pathogenic bacteria in rich media. Some species can double in 20 minutes (*E coli* and related organisms), and some (eg, some mycobacteria) take almost as long as mammalian cells—20 hours. There are bacteria that grow best at refrigerator temperatures (psychrophiles) and

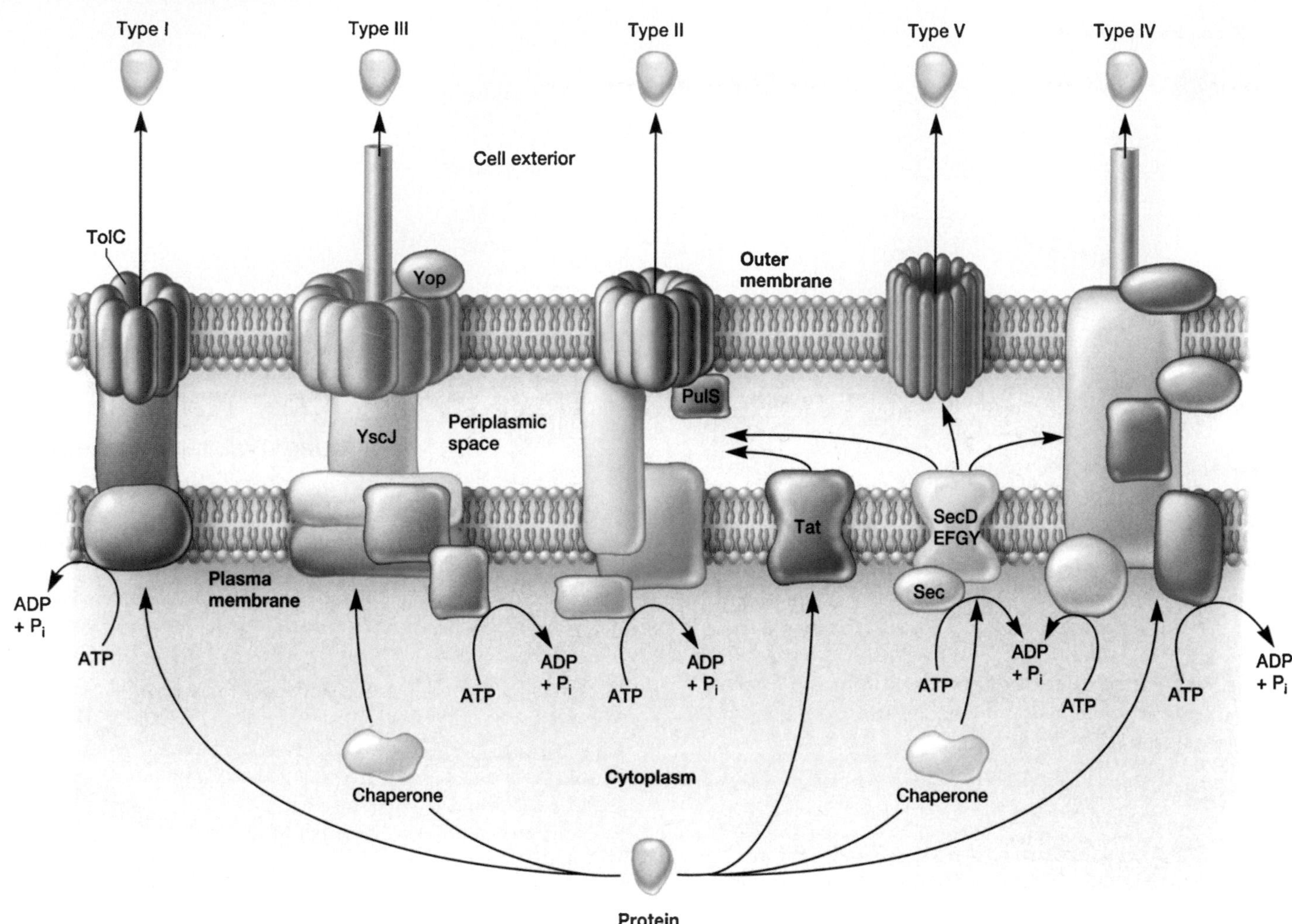

FIGURE 21–22. Gram-negative secretion systems. Type I. Proteins are exported directly across the cytoplasmic and outer membranes (OM) without use of the GSP. **Type II.** GSP or another system called Tat secrete into the periplasmic space and proteins are then transported across the OM. **Type III.** Proteins are transported across both membranes and then injected by a syringe apparatus. **Type IV.** Like type III but also injects DNA. **Type V.** Like type II except the protein is autotransported across the OM. (Reproduced with permission from Willey J, Sherwood L, Woolverton C (eds). *Prescott's Principles of Microbiology.* New York: McGraw-Hill; 2008.)

Growth rate is dependent on nutrient availability, pH, and temperature

some that grow at temperatures higher than 50°C (thermophiles). Human pathogens are in between (mesophiles). A few can grow at refrigerator temperatures and up to 42°C, but their optimums are between 30°C and 37°C.

After a lag period, liquid cultures exhibit exponential growth

Nutrient depletion or waste product accumulation terminates growth

When first inoculated, liquid cultures of bacteria characteristically exhibit a **lag period** during which growth is not detectable. This is the first phase of what is called the culture growth cycle. During this lag, the cells are actually quite active in adjusting the levels of vital cellular constituents necessary for growth in the new medium. Eventually, net growth can be detected, and after a brief period of accelerating growth, the culture enters a phase of constant, maximal growth rate, called the **exponential** or **logarithmic phase** of growth, during which the generation time is constant. During this phase, cell number, and total cell mass, and amount of any given component of the cells increase at the same exponential rate. As nutrients are depleted and waste products are accumulated, growth becomes progressively limited (**decelerating phase**) and eventually stops (**stationary phase**). The growth curve generated by this cycle is illustrated in **Figure 21–23**.

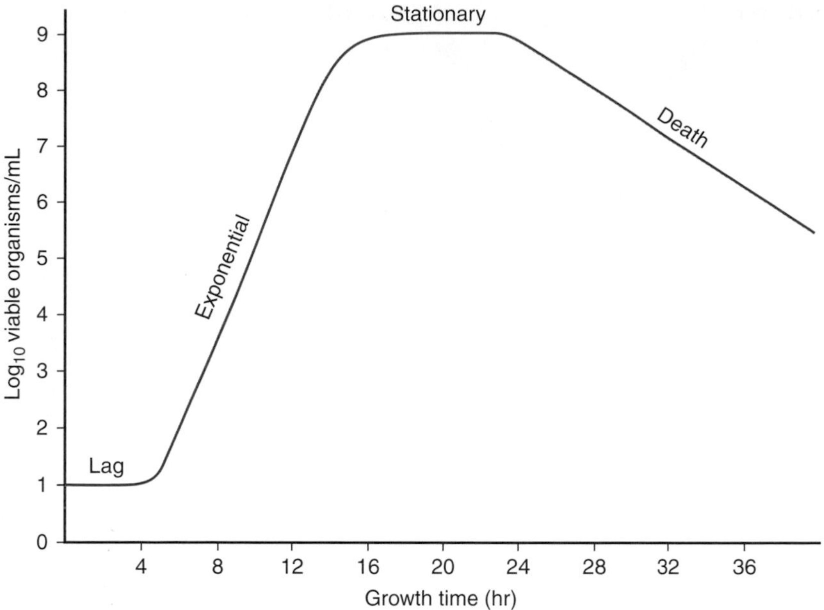

FIGURE 21–23. Growth curve. The phases of bacterial growth in liquid medium.

REGULATION AND ADAPTATION

Bacteria can do little to control their environment, so they must adjust to it in a flexible, manner. They accomplish this feat by many regulatory mechanisms, some of which operate to control enzyme activity, some to control gene expression.

Control of Enzyme Activity

By far the most prevalent means by which bacterial cells modulate the flow of material through fueling and biosynthetic pathways is by changing the activity of allosteric enzymes through the reversible binding of low-molecular-weight ligands (**Figure 21–24**). In fueling

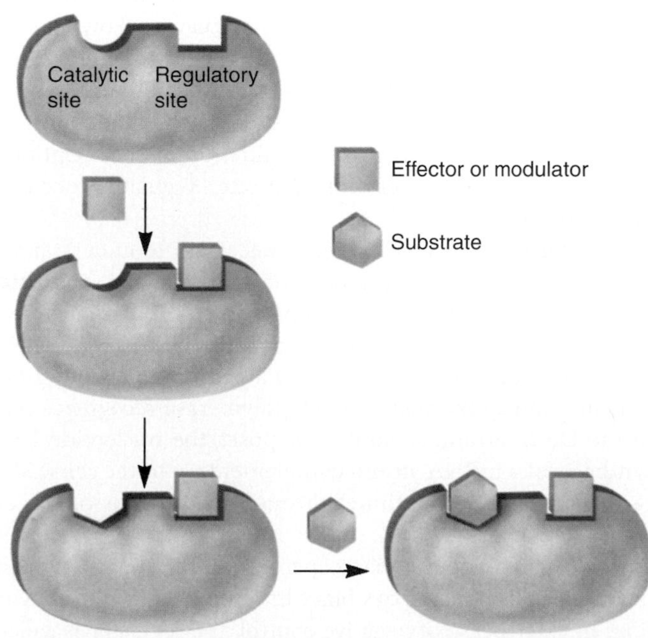

FIGURE 21–24. Allosteric regulation. In this example of the structure and function of an allosteric enzyme, the effector or modulator first binds to a separate regulatory site and causes a change in enzyme conformation that results in an alteration in the shape of the active site. The active site can now more effectively bind the substrate. This effector is a positive effector because it stimulates substrate binding and catalytic activity. (Reproduced with permission from Willey J, Sherwood L, Woolverton C (eds). *Prescott's Principles of Microbiology.* New York: McGraw-Hill; 2008.)

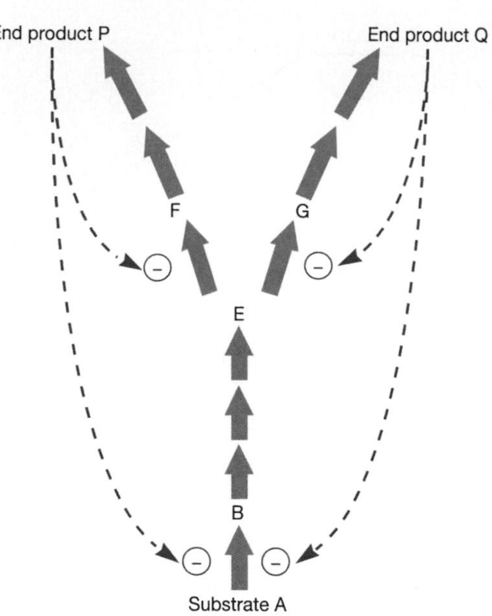

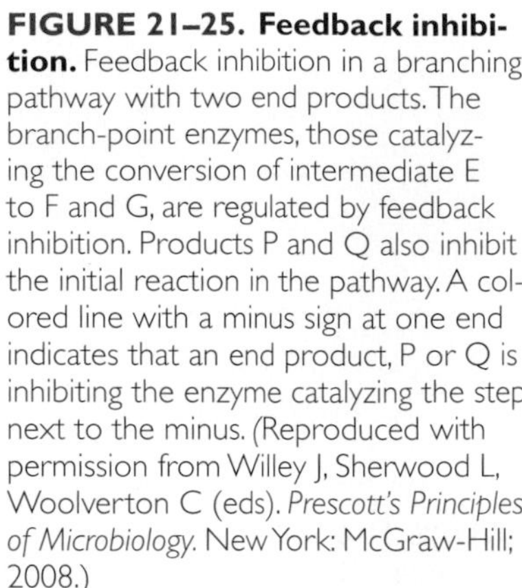
FIGURE 21–25. Feedback inhibition. Feedback inhibition in a branching pathway with two end products. The branch-point enzymes, those catalyzing the conversion of intermediate E to F and G, are regulated by feedback inhibition. Products P and Q also inhibit the initial reaction in the pathway. A colored line with a minus sign at one end indicates that an end product, P or Q is inhibiting the enzyme catalyzing the step next to the minus. (Reproduced with permission from Willey J, Sherwood L, Woolverton C (eds). *Prescott's Principles of Microbiology.* New York: McGraw-Hill; 2008.)

Most metabolic pathways are controlled by allosteric enzymes

Feedback inhibition provides economy and efficiency

pathways, it is common for AMP, ADP, and ATP to control the activity of enzymes by causing conformational changes of **allosteric enzymes**, usually located at critical branch points where pathways intersect. By this means, the flow of carbon from the major substrates through the various pathways is adjusted to be appropriate to the demands of biosynthesis. In biosynthetic pathways, it is common for the end product of the pathway to control the activity of the first enzyme in the pathway. This pattern, called **feedback inhibition** or end-product inhibition**,** ensures that each building block is made at exactly the rate it is being used for polymerization (**Figure 21–25**). It also ensures that building blocks supplied in the medium are not wastefully duplicated by synthesis.

Control of Gene Expression

Changes in transcription can rapidly change enzyme synthesis because of mRNA degradation

Most regulation operates at initiation of transcription

To a far greater extent than eukaryotic cells, bacteria regulate their metabolism by changing the amounts of different enzymes. This is accomplished chiefly by governing their rates of synthesis, that is, by controlling gene expression. This works rapidly for bacteria because of their speed of growth; shutting off the synthesis of a particular enzyme results in short order in the reduction of its cellular level owing to dilution by the growth of the cell. Most important, bacterial mRNA is degraded rapidly. The synthesis of a given enzyme can therefore be rapidly turned on and just as rapidly turned off simply by changes in the rate of transcription of its gene. Most of the regulation of gene expression occurs at or near the beginning of the process: the initiation of transcription. Once started, transcription proceeds at a more or less constant rate. Regulation occurs by a decision of whether to initiate or not.

Genes are organized as transcriptional units called operons

RNA polymerase binds to the promoter of an operon and transcribes until it meets the terminator

A closer look at transcription is necessary to understand how it is controlled. Most of the genes we know about in bacteria are organized as **multicistronic operons.** A **cistron** is a segment of DNA encoding a polypeptide. An **operon** is the unit of transcription; the cistrons that it comprises are co-transcribed as a single mRNA. The structure of a typical operon (**Figure 21–26**) consists of a **promoter** region, an **operator** region, component cistrons, and a **terminator.** RNA polymerase recognizes the promoter region and binds to the DNA. Strand separation exposes the nucleotide bases and permits initiation of synthesis of a mRNA strand complementary to the sense strand of the DNA. In a simple case, transcription continues through the cistrons of the operon until the termination signal is reached.

Near the promoter in many operons is an operator to which a specific **regulator protein** or **transcription factor** can bind. In some cases the binding of this regulator blocks initiation; in such a case of negative control, the regulator is called a **repressor.** Repressors are

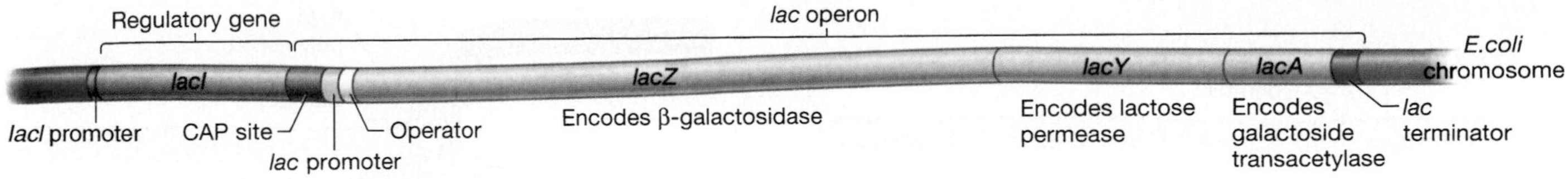

FIGURE 21–26. The *lac* operon. The *lac* operon consists of three genes: *lacZ*, *lacY*, and *lacA*, which are transcribed as a single unit from the *lac* promoter. The operon is regulated both negatively and positively. Negative control is brought about by the *lac* repressor, which is the product of the *lacI* gene. The operator is the site of *lac* repressor binding. Positive control results from the action of CAP. CAP binds the CAP site located just upstream from the *lac* promoter. CAP is, in part, responsible for a phenomenon called catabolite repression, an example of a global control network, in which numerous operons are controlled by a single protein. (Reproduced with permission from Willey J, Sherwood L, Woolverton C (eds). *Prescott's Principles of Microbiology.* New York: McGraw-Hill; 2008.)

allosteric proteins, and their binding to the operator depends on their conformation, which is determined by the binding of ligands that are called **co-repressors** if their action permits binding of the repressor and **inducers** if their action prevents binding. In some cases, the regulator protein is required for initiation of transcription, and it is then called an **activator.** The functioning of both positive and negative types of regulation on transcription initiation is illustrated in **Figure 21–27**. There are many instances known in which groups of genes that are independently controlled as members of different operons must cooperate to accomplish some response to an environmental change. When such a group of genes is subject to the control of a common regulator, the group is called a **regulon.** Some regulatory systems are able to act in multiple stages. The two-component system illustrated in **Figure 21–28** shows an environmental signal sensed in the cytoplasmic membrane leading to activation of a separate regulon. This linking of environmental sensing with regulation is taken to another level with two-compartment systems used by pathogens for deployment of virulence factors. *Bordetella pertussis* uses such a system to produce attachment proteins and toxins at just the right time during the production of whooping cough. ::: B pertussis regulation, p. 400

Activator and repressor proteins regulate transcription by binding to the operator region of operons

Regulons are groups of unlinked operons controlled by a common regulator

Two-component systems link environmental sensing with regulation

CELL SURVIVAL

■ Cell Stress Regulons

Bacterial cells have many regulons involved in survival responses during difficult circumstances. As a response to the nutritional stress of running out of glucose, the cell can redirect its pattern of gene expression to an alternative source of carbon present in the environment. When a cell suffers damage to its DNA, a set of genes involved in repair are turned on. The products of these genes repair damaged DNA and prevent cell division during the repair. In the **heat-shock response,** up to 20 genes may be transcriptionally activated on an upward shift in temperature or on imposition of several kinds of chemical stress. Fever in humans can elevate body temperature sufficiently to induce the heat-shock response, and it is suspected that this response may affect the outcome of various infections. Also, some viruses both of bacteria and of humans use the heat-shock proteins of their host cells to promote their own replication.

Bacteria have regulons that help cope with nutritional, injury, and heat stress

■ Endospores

Two of the most elaborate bacterial survival responses involve the transition of growing cells into a form that can survive long periods without growth. In a few Gram-positive bacterial species, this involves **sporulation,** the production of an endospore, as previously described. This process, extensively studied in a few species, involves cascades of RNA polymerase subunits, each sequentially activating several interrelated regulons that cooperate to produce the elaborately encased spore, which though metabolically inert and extremely resistant to environmental stress, is capable of germinating into a growing (vegetative) cell.

Sporulation involves sequential activation of interrelated regulons

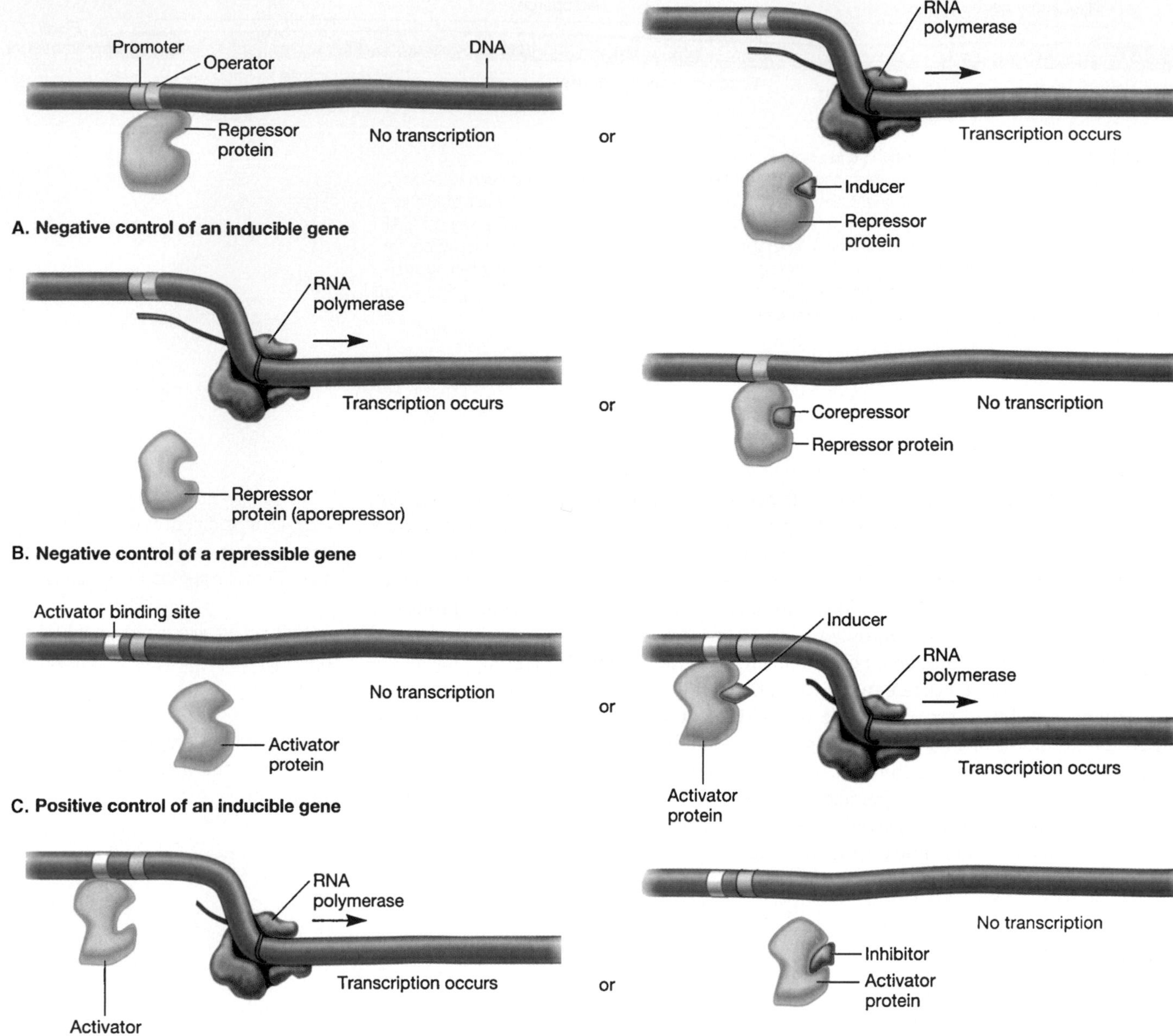

FIGURE 21–27. Bacterial regulatory proteins. Bacterial regulatory proteins have two binding sites—one for a small effector molecule and one for DNA. The binding of the effector molecule changes the regulatory protein's ability to bind DNA. **A.** In the absence of inducer, the repressor protein blocks transcription. The presence of inducer prevents the repressor from binding DNA, and transcription occurs. **B.** Without a co-repressor, the repressor is unable to bind DNA, and transcription occurs. When the co-repressor is bound to the repressor, the repressor is able to bind DNA and transcription is blocked. **C.** The activator protein is able to bind DNA and activate transcription only when it is bound to the inducer. **D.** The activator binds DNA and promotes transcription unless the inhibitor is present. When inhibitor is present, the activator undergoes a conformational change that prevents it from binding DNA; this inhibits transcription. (Reproduced with permission from Willey J, Sherwood L, Woolverton C (eds). *Prescott's Principles of Microbiology.* New York: McGraw-Hill; 2008.)

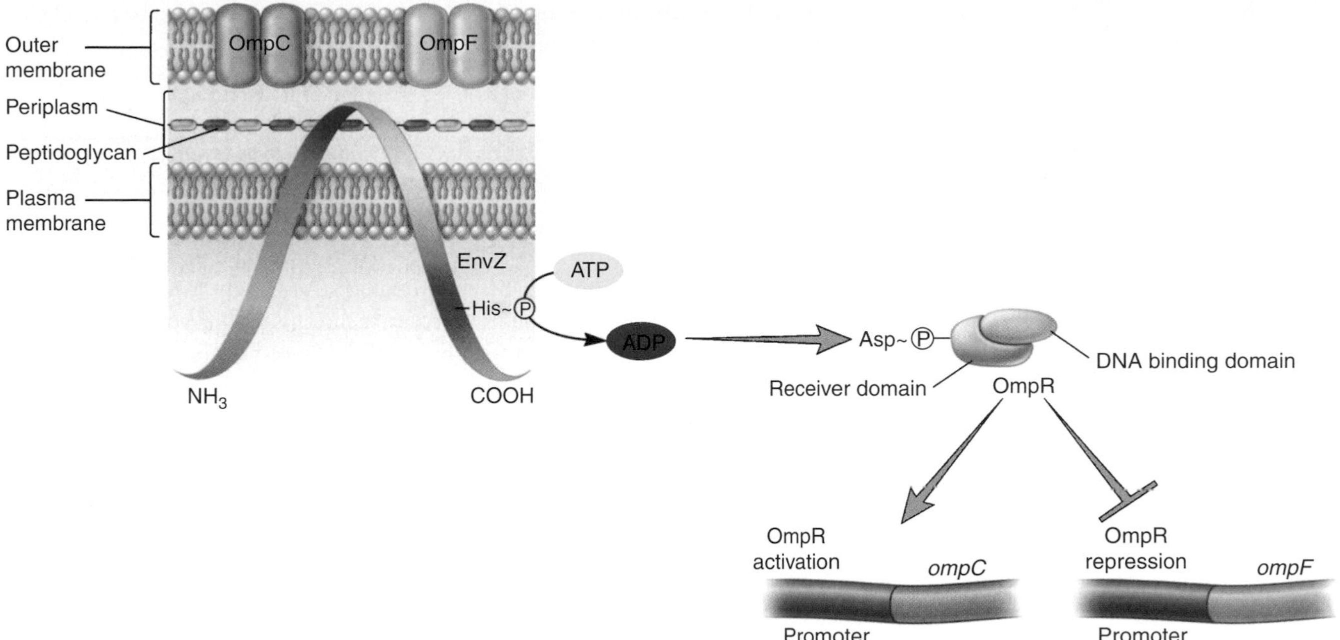

FIGURE 21–28. Two-component signal transduction system and the regulation of porin proteins. In this system, the sensor kinase protein EnvZ loops through the cytoplasmic membrane so that both its C- and N-termini are in the cytosol. When EnvZ senses an increase in osmolarity, it autophosphorylates a histidine residue at its C-terminus. EnvZ then passes the phosohoryl group to the response regulator OmpR, which accepts it on an aspartic acid residue located in its N-terminus. This activates OmpR so that it is able to bind DNA, repress *ompF* expression, and enhance that of *ompC*. (Reproduced with permission from Willey J, Sherwood L, Woolverton C (eds). *Prescott's Principles of Microbiology*. New York: McGraw-Hill; 2008.)

Stationary Phase Cells

For all other bacteria, adaptation to a nongrowing state involves formation of a differentiated cell called the stationary phase cell. The product is certainly far different morphologically from an endospore, but a tough, resistant, and metabolically quiescent cell is produced that looks distinct from its growing counterpart. Its envelope is made tougher by many modification of its structure, its chromosome is aggregated, and its metabolism is adjusted to a maintenance mode. Producing this resistance involves a process surprisingly analogous to sporulation, because, as in sporulation, cascades of signals and responses involving the sequential activation of sets of genes appear to be involved. Such states may be important in diseases such as tuberculosis, which have long latent periods after primary infection, or in cholera in which cells persist in a dormant state in the environment between epidemics. ::: Tuberculosis latency, p. 496

Formation of a stationary phase cell involves activation of many regulons in a coordinated cascade

Motility and Chemotaxis

Motility in most bacterial species is the property of swimming by means of flagellar propulsion. Chemotaxis is directed movement toward chemical attractants and away from chemical repellents. It is accomplished by a remarkable molecular sensory system that possesses many of the characteristics that would be expected of behavioral systems in higher animals, including memory and adaptation. Whether a cell is moving toward an attractant or away from a repellent, chemotaxis is achieved by **biased random walks.** These result from alterations in the frequency of productive motion called a run and tumbling in place. When a cell is, by chance, progressing toward an attractant, tumbling is suppressed and the run is long; if it is swimming away, tumbling occurs sooner and the run is brief. It is sheer chance in which direction a cell is pointed at the end of a tumble, but by regulating the frequency of tumbles in this manner, directed progress is made. Chemotaxis is both a survival device (for avoiding

Direction of flagellar rotation determines a run or a tumble

Changes in duration of runs and tumbles determine chemotactic response

Chemotaxis serves survival, growth-promoting, and pathogenic roles

toxic substances) and a growth-promoting device (for finding food). It can also be a virulence factor in facilitating colonization of the human host by bacteria.

BACTERIAL GENETICS

No feature is more central to bacterial diversity and power to produce disease than their genetic mechanisms. The news media now deliver a constant stream of reports of new antibiotic resistance and emerging pathogens. Bacteria treated successfully with an antimicrobial for decades suddenly develop resistance; diseases seemingly under control reappear; new diseases (at least new to us) emerge and spread. When traced to their origin most of these involve the speed and breadth of bacterial genetic mechanisms. Bacteria use mutation and recombination for genomic change, as do eukaryotic cells. In addition, they have powerful mechanisms for exchange of genes between cells that do not even have to be closely related. Combined with the so-called "jumping genes" (transposons), which seem to be able to go anywhere, bacteria present an astonishing array of genetic tools. The mechanisms of mutation, recombination, transformation, transduction, conjugation, and transposition form the basis of this genetic power and are discussed in the text that follows.

MUTATION

Mutations are rapidly expressed and predominate under selective conditions

The spontaneous development of mutations is a major factor in the evolution of bacteria. Mutations occur in nature at a low frequency, on the order of one mutation in every million cells for any one gene, but the large size of microbial populations ensures the presence of many mutants. Because bacteria are haploid, the consequences of a mutation, even a recessive one, are immediately evident in the mutant cell. Because the generation time of bacteria is short, it does not take many hours for a mutant cell that has arisen by chance to grow to the dominant cell type if the mutation gives it a survival advantage.

Kinds of Mutations

The several kinds of mutations all involve changes in nucleotide sequence

Mutations are heritable changes in the structure of genes. The normal, usually active, form of a gene is called the **wild-type allele;** the mutated, usually inactive, form is called the **mutant allele.** There are several kinds of mutations, based on the nature of the change in nucleotide sequence of the affected gene(s). **Replacements** involve the substitution of one base for another. **Microdeletions** and **microinsertions** involve the removal and addition, respectively, of a single nucleotide (and its complement in the opposite strand). **Insertions** involve the addition of many base pairs of nucleotides at a single site. **Deletions** remove a contiguous segment of many base pairs. **Inversions** change the direction of a segment of DNA by splicing each strand of the segment into the complementary strand. **Duplications** produce a redundant segment of DNA, usually adjacent (tandem) to the original segment.

Changes in nucleotide sequence affect the synthesis of the protein products

Frameshift mutations affect mRNA translation

By recalling the nature of genes and how their nucleotide sequence directs the synthesis of proteins, one can understand the immediate consequence of each of these biochemical changes. If a replacement mutation in a codon changes the mRNA transcript to a different amino acid, it is called a **missense mutation** (eg, an AAG [lysine] to a GAG [glutamate]). The resulting protein may be enzymatically inactive or very sensitive to environmental conditions, such as temperature. If the replacement changes a codon specifying an amino acid to one specifying none, it is called a **nonsense mutation** (eg, a UAC [tyrosine] to UAA [STOP]). Microdeletions and microinsertions cause **frameshift mutations,** changes in the reading frame by which the ribosomes translate the mRNA from the mutated gene (**Figure 21–29**). Frameshifts usually result in polymerization of a stretch of incorrect amino acids until a nonsense codon is encountered, so the product is usually a truncated polypeptide fragment with an incorrect amino acid sequence at its N-terminus. Deletion or insertion of a segment of base pairs from a gene shortens or lengthens the protein product if the number of base pairs deleted or inserted is divisible evenly by three; otherwise, it also brings about the consequence of a frame shift. Inversions of a small segment within a gene inactivate it; inverting larger segments may affect chiefly the genes at the points of inversion. Duplications, probably the most common of all

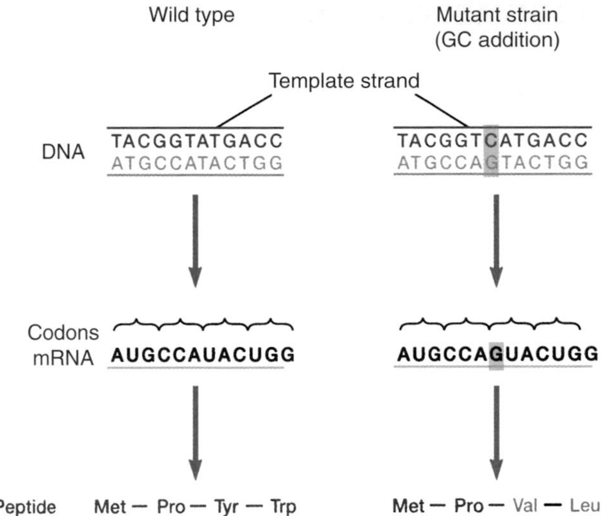

FIGURE 21–29. Frameshift mutation. A frameshift mutation resulting from the insertion of a GC base pair. The reading frameshift translates to different amino acids after the frameshift producing a different peptide.(Reproduced with permission from Willey J, Sherwood L, Woolverton C (eds). *Prescott's Principles of Microbiology*. New York: McGraw-Hill; 2008.)

mutations, serve an important role in the evolution of genes and antigenic variation. Mutations are summarized in **Table 21–3**.

Many mutations, particularly if they occur near the end of a gene, prevent the expression of all genes downstream (away from the promoter) of the mutated gene. Such **polar mutations** are thought to exert their effect on neighboring genes by the termination of transcription of downstream genes when translation of the mRNA of the mutated gene is blocked by a nonsense codon. There is a certain natural frequency of mutations brought about by errors in replication, but various environmental and biologic agents can increase the frequency greatly. Different types of mutations are increased selectively by different agents, as listed in Table 21–3. Bacteria have evolved multiple biochemical mechanisms for repairing damaged DNA.

Mutations may affect neighboring genes by termination of transcription

Mutagens increase the natural frequency of mutation

TABLE 21–3 Mutations

TYPE	CAUSATIVE AGENT	CONSEQUENCES
Replacement		
Transition: pyrimidine replaced by a pyrimidine or a purine by a purine	Base analogs, ultraviolet radiation, deaminating and alkylating agents, spontaneous	Transitions and transversions: if nonsense codon formed, truncated peptide; if missense codon formed, altered protein
Transversion: purine replaced by a pyrimidine or vice versa	Spontaneous	
Deletion		
Macrodeletion: large nucleotide segment deleted	HNO_2, radiation, bifunctional alkylating agents	Truncated peptide; other products possible, such as fusion peptides
Microdeletion: one or two nucleotides deleted	Same as macrodeletions	Frame shift, usually resulting in nonsense codon and truncated peptide
Insertion		
Macroinsertion: large nucleotide segment inserted	Transposons or insertion sequence (IS) elements	Interrupted gene yielding truncated product
Microinsertion: one or two nucleotides inserted	Acridine	Frame shift, usually resulting in nonsense codon yielding a truncated product
Inversion	IS or IS-like elements	Many possible effects

RECOMBINATION

Recombination is the process in which nucleic acid molecules from different sources are combined or rearranged to produce a new nucleotide sequence. In eukaryotes, this occurs by crossing over during meiosis. Since bacteria do not reproduce sexually or undergo meiosis, it might seem that this mechanism would be limited. In fact, it can occur any time that there is a source of recombinant DNA and strand breaks in the bacterial chromosome, which create stretches of single-stranded DNA with nucleotides exposed for potential pairing. The source of recombinant DNA may be another part of the same chromosome (endogenote) or from outside the cell (exogenote) from one of the genetic transfer mechanisms later described. If successful, a new hybrid chromosome is formed. In bacteria there are two major molecular mechanisms of recombination, homologous recombination (**Figure 21–30**) and site-specific recombination.

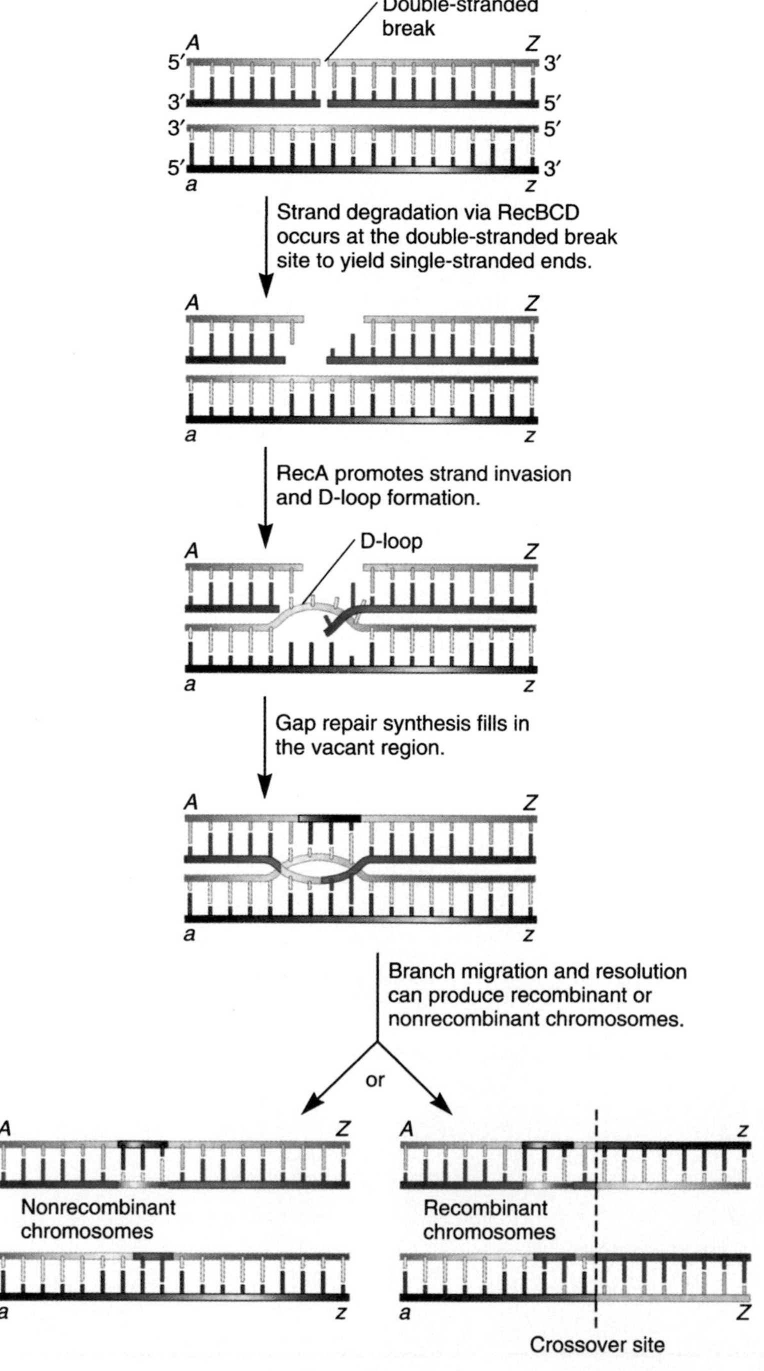

FIGURE 21–30. The double-stranded break model of homologous recombination. (Reproduced with permission from Willey J, Sherwood L, Woolverton C (eds). *Prescott's Principles of Microbiology.* New York: McGraw-Hill; 2008.)

Homologous Recombination

This term homologous recombination reflects one of the two requirements for this process: (1) the donor DNA must possess reasonably large regions of nucleotide sequence identity or similarity to segments of the host chromosome because extensive base-pairing must occur between strands of the two recombining molecules; and (2) the recipient cell must possess the genetic ability to make a set of enzymes that can bring about the covalent substitution of a segment of the donor DNA for the homologous region of the host. A protein known as RecA (recombination) controls the entire process. The same breakage and reunion process then links the second strand of each recombining DNA molecule. This crossover event repeated farther down the chromosome results in the substitution of the donor segment between the two crossovers for the homologous segment of the host.

Homologous recombination involves nucleotide similarity and specific enzymes such as RecA

Site-Specific Recombination

The second major type of recombination is site-specific recombination, which is particularly important in the integration of virus genomes into host chromosomes. Site-specific recombination relies on only limited DNA sequence similarity at the sites of crossover mediated by different sets of specialized enzymes designed to catalyze recombination of only certain DNA molecules. These recombinational events are restricted to specific sites on one or both of the recombining DNA molecules. The enzymes that bring about site-specific recombination operate not on the basis of DNA homology, but on recognition of unique DNA sequences that form the borders of the specific sites. These enzymes are commonly encoded by genes in the exogenote virus DNA. The integration of some bacteriophage genomes into the chromosome occurs only at one site on the bacterial chromosome and one site on the phage chromosome.

Site-specific recombination operates only on unique sequences

Enzymes are usually encoded by exogenote genes

Recombination and Antigenic Variation

A fascinating aspect of DNA rearrangements brought about by genetic recombination is that the expression of some chromosomal genes important in virulence can be controlled by recombinational events. In *Neisseria gonorrhoeae,* the species that causes gonorrhea, antigenic variation involves recombination between multiple genes in the same chromosome. In *Salmonella* species an **invertible element** lying between the two flagellin genes can switch between them. In one orientation, the promoter initiates transcription of one flagellar type; in the other orientation, transcription proceeds in the opposite direction to transcribe the other. A similar situation exists in *E coli,* in which an invertible segment containing a promoter shuts the transcription of adhesive (type 1) pili on and off. These kinds of antigenic variations provide a selective advantage to the bacteria by allowing invading populations to include individuals that can escape the developing immune response of the host and thus continue the infectious process. ::: antigenic variation in gonorrhea, p. 394

Antigenic variation can be brought about by a recombinational event

Invertible elements can act as a genetic switch

TRANSPOSITION

Transposition involves transposable elements that are genetic units capable of mediating their own transfer from one chromosome to another, from one location to another on the same chromosome, or between chromosome and plasmid. This transposition relies on their ability to synthesize their own site-specific recombination enzymes, called **transposases.** The major kinds of transposable elements are **insertion sequence** elements and **transposons** (**Figure 21–31**).

Genetic units move within and between chromosomes and plasmids

Insertion Sequences

Insertion sequence (IS) elements are segments of DNA that encode enzymes for site-specific recombination and have distinctive nucleotide sequences at their termini. Different IS elements have different termini, but, as illustrated in a given IS element, has the same sequence of nucleotides at each end but in an inverted order. Only genes involved in transposition (eg, one encoding a transposase) and in the regulation of its frequency are included in IS elements, and they are, therefore, the simplest transposable elements. Because IS elements contain only genes for transposition, their presence in a chromosome is not easy to detect unless they insert within a gene. Such an insertion is actually a mutation that alters or destroys the activity of the gene.

IS elements encode only proteins for their own transposition

Insertion of IS elements into a gene causes mutation

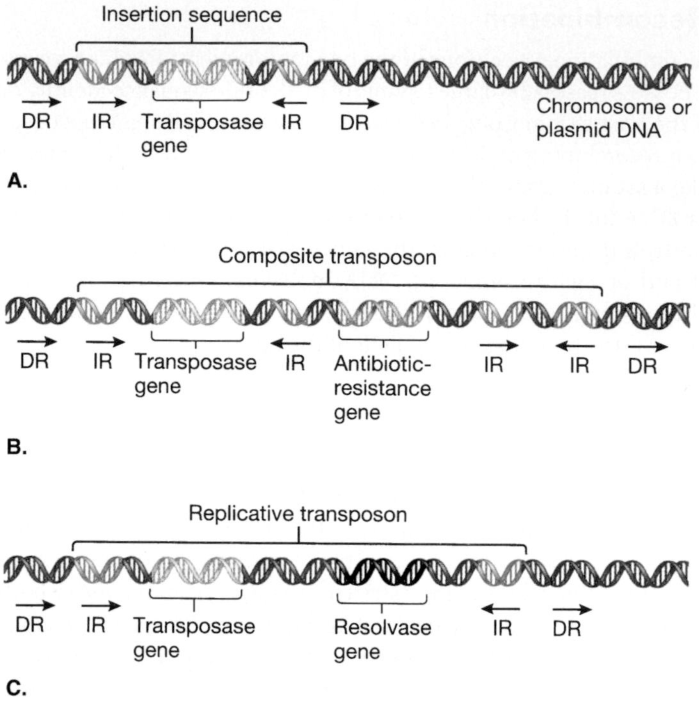

Insertion sequence
DR IR Transposase gene IR DR
Chromosome or plasmid DNA
A.

Composite transposon
DR IR Transposase gene IR Antibiotic-resistance gene IR IR DR
B.

Replicative transposon
DR IR Transposase gene Resolvase gene IR DR
C.

FIGURE 21–31. Transposable elements. All transposable elements contain common elements. These include inverted repeats (IRs) at the ends of the elements and a transposases gene. **A.** Insertion sequences consist only of the IRs on either side of the transposases gene. **B.** Composite transposons and **C.** genes. Insertion sequences and composite transposons move by simple cut-and-paste transposition. Replicative transposons move by replicative transposition. Direct repeats (DRs) in host DNA flank a transposable element. (Reproduced with permission from Willey J, Sherwood L, Woolverton C (eds). *Prescott's Principles of Microbiology.* New York: McGraw-Hill; 2008.)

Transposons

Transposons encode functions beyond those needed for their own transposition

Direct transposition moves the transposon to a new site

Replicative transposition leaves a copy behind

IS elements are components of **transposons (Tn)** that are transposable segments of DNA containing genes beyond those needed for transposition. The general structure of these composite transposons consist of a central area of genes bordered by IS elements. The genes may code for such properties as antimicrobial resistance, substrate metabolism, or other functions. Composite transposons translocate by what is called simple or **direct transposition,** in which the transposon is excised from its original location and inserted in a simple cut-and-paste manner into its new site without replication (**Figure 21–32**). Another mechanism called **replicative transposition** leaves a copy of the replicative transposon at its original site.

Transposons promote many changes in DNA

Besides the primary insertion reaction, transposable units promote other types of DNA rearrangements, including deletion of sequences adjacent to a transposon, inversion of DNA segments, and stop codons or termination sequences, which may block translation or transcription. When located in plasmids, transposable units may also participate in plasmid fusion, insertion of plasmids into the chromosome and plasmid evolution. All of these events have great significance for understanding the formation and spread of antimicrobial resistance through natural populations of pathogenic organisms.

GENETIC EXCHANGE

One-way passage of DNA from a donor to a recipient adds an exogenote to the recipient endogenote

Despite the fact that bacteria reproduce exclusively asexually, the sharing of genetic information within and between related species is common and occurs in at least three fundamentally different ways. All three processes involve a one-way transfer of DNA from a **donor cell** to a **recipient cell.** The molecule of DNA introduced into the recipient is called the **exogenote** to distinguish it from the cell's own original chromosome, called the **endogenote.**

Transformation, transduction, and conjugation are the major processes of DNA transfer

One process of DNA transfer, called **transformation,** involves the release of DNA into the environment by the lysis of some cells, followed by the direct uptake of that DNA by the recipient cells. In **transduction,** the DNA is introduced into the recipient cell by a bacteriophage that has infected the bacterial cell. The third process, called **conjugation,** involves actual contact between a donor and recipient cell during which the autonomously replicating, extrachromosomal DNA of a plasmid is transferred.

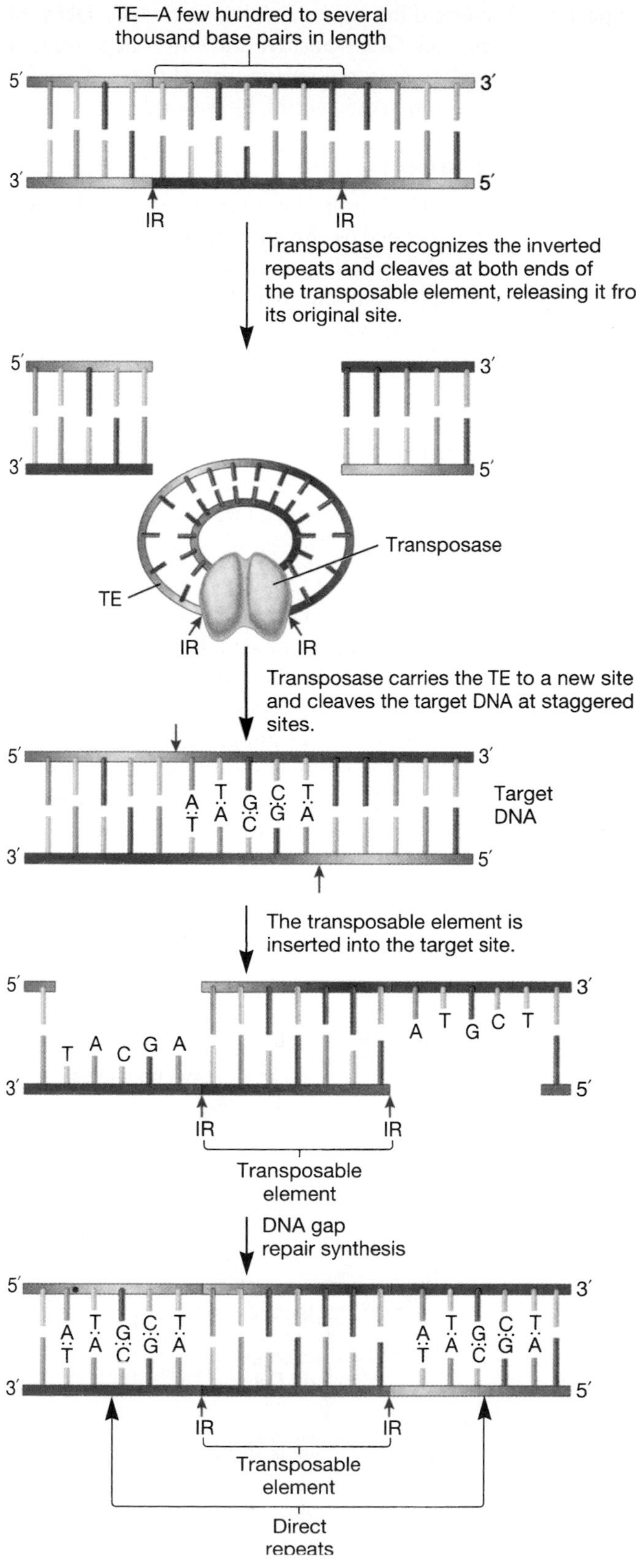

FIGURE 21–32. Simple transposition. TE, transposable element; IR, inverted repeat. (Reproduced with permission from Willey J, Sherwood L, Woolverton C (eds). *Prescott's Principles of Microbiology*. New York: McGraw-Hill; 2008.)

Transformation, transduction, and conjugation are mediated by chromosomal, viral, and plasmid genes, respectively

Species of bacteria differ in their ability to transfer DNA, but all three mechanisms are distributed among both Gram-positive and Gram-negative species; however, only transformation is governed by bacterial chromosomal genes. Transduction is totally mediated by bacteriophage genes, and conjugation, by plasmid genes.

Transformation

Competence is the ability to take DNA from the environment

Internalized DNA either recombines or is degraded

Artificial transformation involves treatment of cells

The ability to take up DNA from the environment is called **competence,** and in many species of bacteria, it is encoded by chromosomal genes that become active under certain environmental conditions. Any DNA present in the medium is bound indiscriminately. The fate of the internalized DNA fragment then depends on whether it shares homology (the same or similar base sequences) with a portion of the recipient cell's DNA. If so, recombination can occur, but heterologous DNA is degraded and causes no heritable change in the recipient (**Figure 21–33**). Other species do not naturally enter the competent state, but can be made permeable to DNA by treatment with agents that damage the cell envelope, making an **artificial transformation** possible. The experimental use of *E coli*, which has no natural competence mechanism, as a host cell in gene cloning involves treatment with salt and temperature shocks to bring about artificial transformation.

Transduction

Lytic phages produce new virions in the host bacterial cell

Transduction is the transfer of genetic information from donor to recipient cell by viruses of bacteria called **bacteriophages** or simply phages**.** The phages infect sensitive cells by adsorbing to specific receptors on the cell surface and then injecting their DNA or RNA. Phages come in two functional varieties according to what happens after injection of the viral nucleic acid. **Virulent (lytic) phages** cause lysis of the host bacterium as a culmination of the synthesis of many new virions within the infected cell. **Temperate phages** may initiate a lytic growth process of this sort or can enter a quiescent form (called a **prophage**), in which the infected host cell is permitted to proceed about its business of

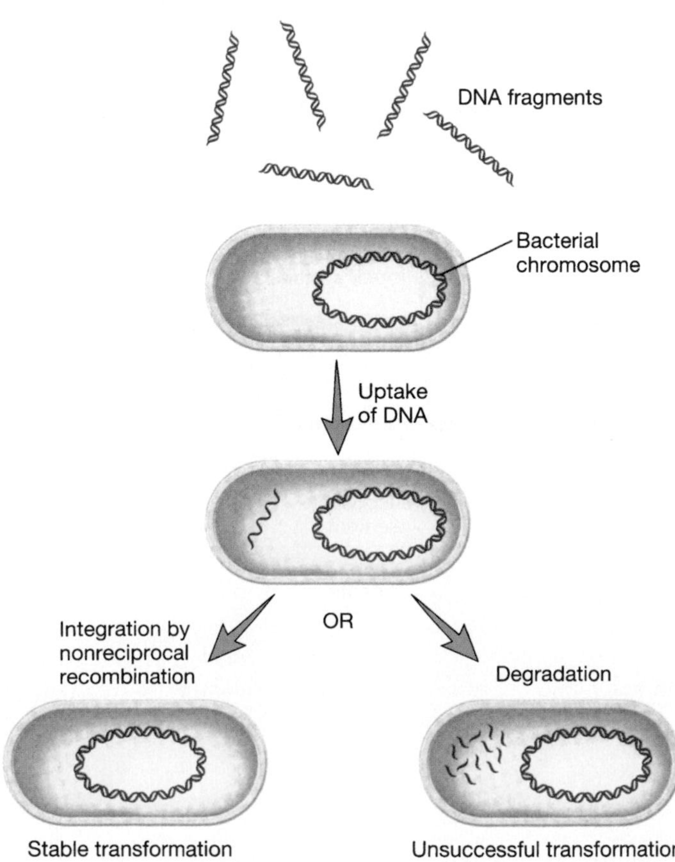

FIGURE 21–33. Bacterial transformation. The bacterial cell is transformed with DNA fragments (*purple*), which are either integrated into the chromosome (*blue*) by recombination or degraded by nucleases in the cytosol. (Reproduced with permission from Willey J, Sherwood L, Woolverton C (eds). *Prescott's Principles of Microbiology*. New York: McGraw-Hill; 2008.)

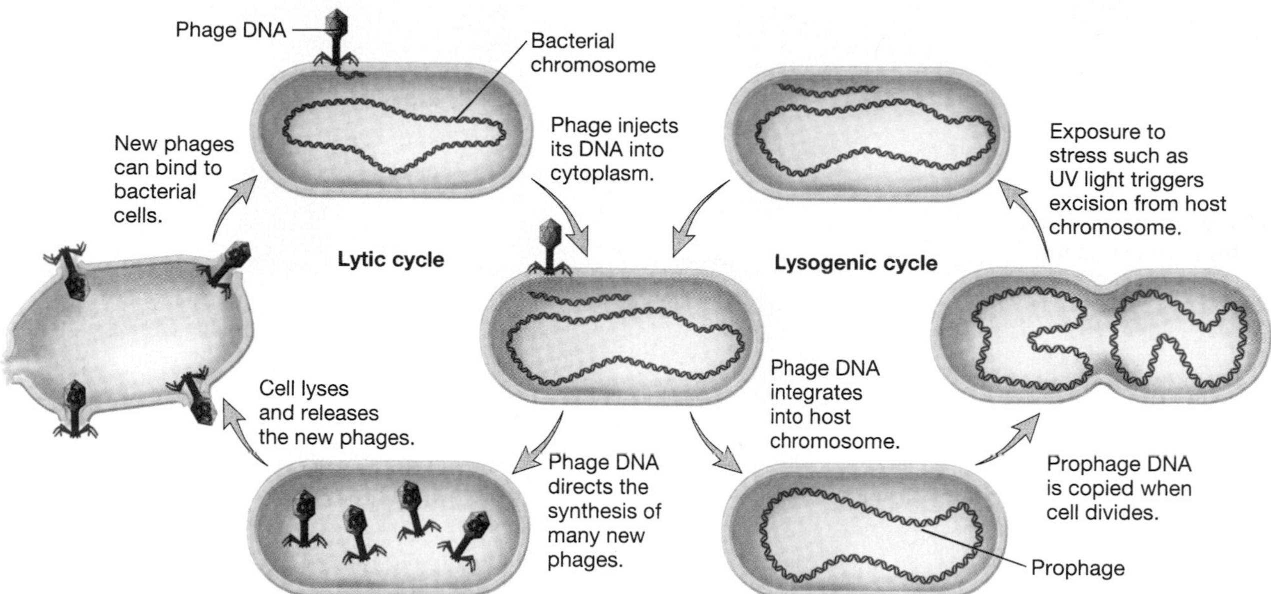

FIGURE 21–34. Transduction: lytic and lysogenic cycles of temperate phages. Temperate phages have two phases to their life cycles. The lysogenic cycle allows the genome of the virus to be replicated passively as the host cell's genome is replicated. Certain environmental factors such as UV light can cause a switch from the lysogenic cycle to the lytic cycle. In the lytic cycle, new virus particles are made and released when the host cell lyses. Virulent phages are limited to just the lytic cycle. (Reproduced with permission from Willey J, Sherwood L, Woolverton C (eds). *Prescott's Principles of Microbiology*. New York: McGraw-Hill; 2008.)

growth and division, but passes on to its descendants a prophage genome capable of being **induced** to produce phage in a process nearly identical to the growth of lytic phages. The bacterial cell that harbors a latent prophage is said to be a lysogen (capable of producing lytic phages), and its condition is referred to as **lysogeny**. Steps in this process are illustrated in **Figure 21–34**. ::: bacteriophage, p. 115

Temperate phages either lyse the bacterial host cell or lysogenize it as a prophage

For the most part, transduction is mediated by temperate phage, and the two broad types of transduction result from the different physical forms of prophage and the different means by which the transducing virion is formed. One of these is **generalized transduction,** by which any bacterial DNA picked up from the previous host cell stands an equal chance of being transduced to a recipient cell because the DNA is packaged into their capsids in a nonspecific way. As in transformation, once injected into the host cell, the exogenote DNA is lost by degradation unless it can recombine with the chromosome of the recipient cell. In the other form called **specialized transduction,** the genes that can be transduced are limited because of their placement adjacent to a special attachment site (att) in the bacterial chromosome. When these phages are induced to leave the host, errors in the excision process may cause them to carry bits of bacterial DNA, which can only integrate at locations adjacent to that same site in any new host chromosome. Because these phage genomes have a reduced chance of integration into a new chromosome, they are more restricted than those of generalized transduction.

Generalized transduction can transfer any bacterial DNA

Specialized transduction is limited to certain sites

Although both generalized transduction and specialized transduction can be regarded as the result of errors in phage production, transfer of genes between bacterial cells by phage is a reasonably common phenomenon. This includes genes for antimicrobial resistance, but transposition and conjugation are much more common mechanisms for transfer of resistance in medically important bacteria. The major impact of transduction in pathogens is the introduction and stable inheritance of virulence genes such as those coding for toxins. For example, only strains of *Corynebacterium diphtheriae,* which are lysogenic with a phage containing the diphtheria toxin gene, can cause the disease diphtheria. ::: Diphtheria toxin, p. 472

Transduction is the source of genes for bacterial toxins

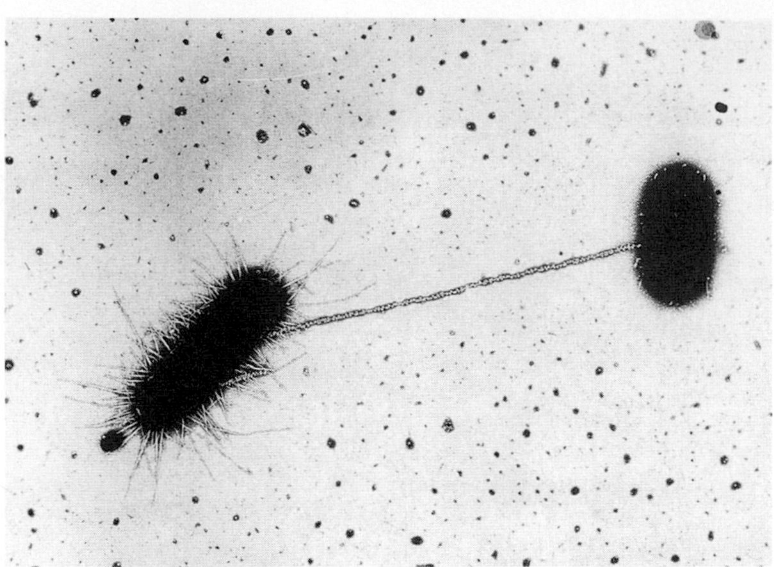

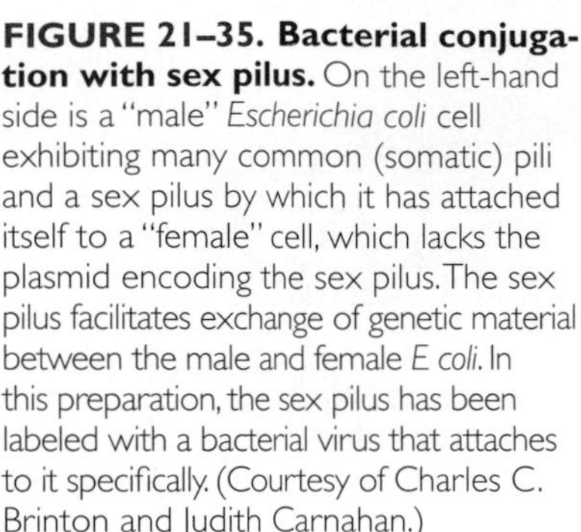

FIGURE 21–35. Bacterial conjugation with sex pilus. On the left-hand side is a "male" *Escherichia coli* cell exhibiting many common (somatic) pili and a sex pilus by which it has attached itself to a "female" cell, which lacks the plasmid encoding the sex pilus. The sex pilus facilitates exchange of genetic material between the male and female *E coli*. In this preparation, the sex pilus has been labeled with a bacterial virus that attaches to it specifically. (Courtesy of Charles C. Brinton and Judith Carnahan.)

Conjugation

One need only look at **Figure 21–35** and add the title "Sexuality in Bacteria" (as has often been done) to grasp the idea that bacteria have something special going for them in the way of gene exchange. This process called conjugation is the transfer of genetic information from donor to recipient bacterial cell in a process that requires intimate cell contact. By themselves, bacteria cannot conjugate. Only when a bacterial cell contains a self-transmissible **plasmid** (see below for definition) does DNA transfer occur. In most cases, conjugation involves transfer only of plasmid DNA; transfer of chromosomal DNA is a rarer event and is mediated by only a few plasmids. Plasmids are of enormous importance to medical microbiology. They are discussed in detail later in this chapter, but to understand conjugation we should first introduce some of their features.

Conjugation is plasmid-encoded and requires cell contact

Plasmids are autonomous extrachromosomal elements composed of circular double-stranded DNA; a few rare linear examples have been found. A single organism can harbor several distinct plasmids and single or multiple copies of each. Plasmids are found in most species of Gram-positive and Gram-negative bacteria in most environments. They replicate within the host cell (and only within the host cell) and are partitioned between the daughter cells at the time of cell division. In addition, many plasmids are able to bring about their own transfer from one cell to another by the products of a group of genes that encode the structures and enzymes required. Such plasmids are **conjugative plasmids** and those that lack these genes are **nonconjugative**.

Plasmids are small circular DNA molecules

Conjugative plasmids contain the genes for transfer

Conjugation is a highly evolved and efficient process. Suitable mixtures of donor and recipient cells can lead to nearly complete conversion of all the recipients into donor, plasmid-containing cells. Furthermore, although some conjugative plasmids can transfer themselves only between cells of the same or closely related species, others are promiscuous, promoting conjugation across a wide variety of (usually Gram-negative) species. Conjugation appears to be a carefully regulated process, normally kept in check by the production of a repressor encoded by one of the plasmid regulatory genes. It is interesting that nonconjugative plasmids that happen to inhabit a cell with a conjugative plasmid can under some circumstances be transferred owing to the conjugation apparatus of the latter; this process is called **plasmid mobilization.**

Conjugation may cross species lines

Nonconjugative plasmids transferred by plasmid mobilization

Plasmids usually include a number of genes in addition to those required for their replication and transfer to other cells. The variety of cellular properties associated with plasmids is very great and includes production of toxins, production of pili and other adhesins, resistance to antimicrobials and other toxic chemicals, production of siderophores for scavenging Fe^{3+}, and production of certain catabolic enzymes important in biodegradation of organic residues. On the other hand, plasmids can add a small metabolic burden to the cell, and in many cases, a slightly reduced growth rate results. Unless this excess baggage

Many plasmid genes promote survival and pathogenesis

provides the cell with some advantage, plasmids tend to be lost (cured is the laboratory term) during prolonged growth. Conversely, when the property conferred by the plasmid is advantageous (eg, in the presence of the antimicrobial to which the plasmid determines resistance), selective pressure favors the plasmid-carrying strain.

Without selection pressure, plasmids may be lost

Conjugation in Gram-Negative Species

After many inconclusive attempts by microbiologists to learn whether a sexual process of genetic exchange existed among bacteria, J. Lederberg and E. Tatum discovered conjugation in 1946. What they observed was a transfer of chromosomal genes between cells of two different strains of *E coli*. Their discovery stimulated an intensive analysis of the mechanism, leading to the discovery of an agent, the **F factor** (for fertility factor), which conferred on cells the ability to transfer bacterial chromosome genes to recipient cells. Now it is recognized that the F factor is a conjugative plasmid.

F factor is a conjugative plasmid that can transfer bacterial chromosome genes

Conjugative plasmids in Gram-negative bacteria contain a set of genes called *tra* (for transfer), which encode the structures and enzymes required. These include bridging structures such as a type IV secretion system (Figure 21–22) or in *E. coli* the F factor–coded **sex pilus,** shown in Figure 21–35. The sex pilus has the ability to draw the donor and recipient cell into the intimate contact needed to form a conjugal bridge through which DNA can pass. The plasmid DNA is then enzymatically cleaved, and one strand is guided through the conjugation structure into the recipient cell by the action of various *tra*-encoded proteins (**Figure 21–36**). Both the introduced strand and the strand remaining behind in

Secretion systems or sex pili form bridges between cells

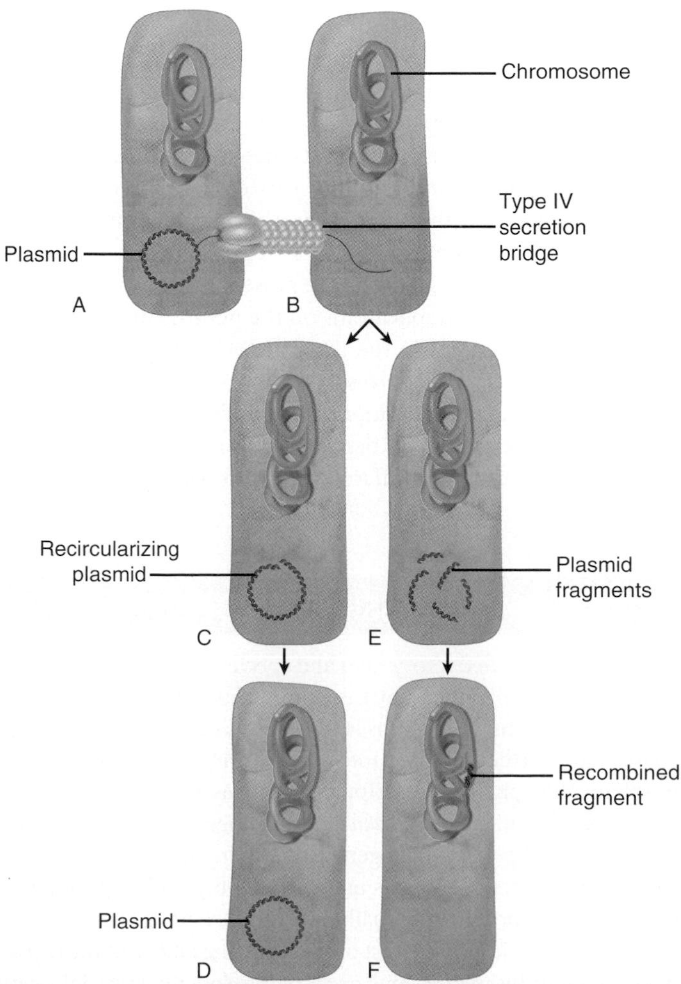

FIGURE 21–36. Conjugation. A conjugative plasmid in **A.** is donating a strand of its DNA to cell **B.** The transferred DNA either synthesizes a complementary strand and recircularizes as in **C.** and **D.** or remain in fragments as in **E.** The fragments either recombine with the recipient cell chromosome as in **F.** or are digested by nucleases in the cytosol. (Reproduced with permission from Willey J, Sherwood L, Woolverton C (eds). *Prescott's Principles of Microbiology.* New York: McGraw-Hill; 2008.)

Replication or recombination follows transfer

the donor cell direct the synthesis of their complementary strand, resulting in complete copies in both donor and recipient cells. Finally, circularization of the double-stranded molecules occurs, the conjugation bridge is broken, and both cells can now function as donor cells. An alternative outcome is recombination of fragments of the transferred plasmid with the chromosome.

Conjugation in Gram-Positive Species

Coupling results from adhesin–receptor interaction

Adhesin is produced in response to recipient pheromone

Plasmids carrying genes encoding antimicrobial resistance, common pili and other adhesins, and some exotoxins are readily transferred by conjugation among Gram-positive bacteria. However, Gram-positive species may involve chromosomal genes in the process. In *Enterococcus faecalis*, one of the most resistant Gram-positive species, donor and recipient cells do not couple by means of a secretion system or sex pilus, but rather by the clumping of cells that contain a plasmid with those that do not. This clumping is the result of interaction between a proteinaceous adhesin on the surface of the donor (plasmid-containing) cell and a receptor on the surface of the recipient (plasmid-lacking) cell. Both types of cells make the receptor, but only the plasmid-containing cell can make the adhesin, presumably because it is encoded by a plasmid gene. Note that donor cells make the adhesin only when in the vicinity of recipient cells because the recipients secrete small peptide **pheromones,** which serve to notify the donor cells of the presence of recipients. Donor cells promptly make adhesin when they sense the pheromone. As a result, clumps are formed, and plasmid DNA is transferred across conjugation bridges into the recipient cells held in the clumps.

R Plasmids

R plasmids can encode and transfer multiresistance

Plasmids that include genes conferring resistance to antimicrobial agents are of great significance in medicine. They are termed **R plasmids** or **R factors** (**resistance factors**). The genes responsible for resistance usually code for enzymes that mediate many of the resistance mechanisms discussed in Chapter 23. R plasmids of Gram-negative bacteria can be transmitted across species boundaries and, at lower frequency, even between genera. Many encode resistance to several antimicrobial agents and can thus spread multiple resistance through a diverse microbial population under selective pressure of only one of those agents to which they confer resistance. Nonpathogenic bacteria can serve as a natural reservoir of resistance determinants on plasmids that are available for spread to pathogens.

Resistance genes acquired by plasmids from transposons

Spread is facilitated by plasmid-chromosome transposon hopping

Widespread antimicrobial use selects spread R plasmids

R plasmids evolve rapidly and can easily acquire additional resistance-determining genes from fusion with other plasmids or acquisition of transposons. Most plasmids, and all R factors, contain many IS elements and transposons. In fact, almost all the resistance determinant genes on plasmids are present as transposons. As a result, these genes can be amplified by tandem duplications on the plasmid and can hop to other plasmids (or to the bacterial chromosome) in the same cell. Combined with the natural properties of many plasmids to transfer themselves by conjugation (even between dissimilar bacterial species), the rapid evolutionary development of multiple drug resistance plasmids and their spread through populations of pathogenic bacteria is a predictable result of selection as a result of the widespread use of antimicrobials in our society.

BACTERIAL CLASSIFICATION

Bacteria are classified into genera and species according to a binomial Linnean scheme similar to that used for higher organisms. For example, in the case of *Staphylococcus aureus, Staphylococcus* is the **genus** and *aureus* is the **species** designation. Some genera with common characteristics are further grouped into **families.** However, bacterial classification has posed many problems. Morphologic descriptors are not as abundant as in higher plants and animals, there is little readily interpreted fossil record to help establish phylogeny, and there is no elaborate developmental process (ontogeny) to recapitulate the evolutionary path from ancestral forms (phylogeny). These problems are minor compared with others: bacteria mutate and evolve rapidly, they reproduce asexually, and they exchange genetic material over wide boundaries. The single most important test of species—the ability of individuals within a species to reproduce sexually by mating and exchanging genetic material—cannot be applied to bacteria.

As a result, bacterial taxonomy developed pragmatically by determining multiple characteristics and weighting them according to which seemed most fundamental; for example, shape, spore formation, Gram reaction, aerobic or anaerobic growth, and temperature for growth were given special weighting in defining genera. Such properties as ability to ferment particular carbohydrates, production of specific enzymes and toxins, and antigenic composition of cell surface components were often used in defining species. As presented in Chapter 4, such properties and their weighting continue to be of central importance in identification of unknown isolates in the clinical laboratory, and the use of determinative keys is based on the concept of such weighted characteristics. These approaches are much less sound in establishing taxonomic relationships based on phylogenetic principles.

Weighted classification schemes are more valuable for identification than for taxonomy

NEW TAXONOMIC METHODS

The recognition that sound taxonomy ought to be based on the genetic similarity of organisms and to reflect their phylogenetic **relatedness** has led in recent years to the use of new methods and new principles in taxonomy. The most direct approach available in recent years involves analysis of chromosomal DNA. Analysis can be somewhat crude, such as the overall ratio of A–T to G–C base pairs; differences of greater than 10% in G–C content are taken to indicate unrelatedness, but closely similar content does not imply relatedness. Closer relationships can be assessed by determining base sequence similarity, as by DNA–DNA hybridization (see Chapter 4). However, overwhelmingly the molecular genetic technique that is introducing the greatest change in infectious disease diagnosis is the comparison of nucleotide sequences of genes highly conserved in evolution, such as 16 S ribosomal RNA genes. So successful have been the deductions of phylogenetic relatedness based on these sequences that the absence of a fossil record is now regarded as insignificant. ::: Nucleic acid analysis, p. 80

Phylogenetic relationships are assuming greater significance as the result of DNA sequence analysis

CHAPTER 22

Pathogenesis of Bacterial Infections

Pathogenicity is, in a sense, a highly skilled trade, and only a tiny minority of all the numberless tons of microbes on the earth has ever involved itself in it; most bacteria are busy with their own business, browsing and recycling the rest of life. Indeed, pathogenicity often seems to me a sort of biological accident in which signals are misdirected by the microbe or misinterpreted by the host.

—Lewis Thomas, *The Medusa and the Snail*

These words refer to all microorganisms and infectious diseases but are particularly appropriate for those caused by bacteria. In the previous chapter, we learned of their astounding diversity and adaptability made possible by simplicity, speed, and robust genetic exchange mechanisms. When antibiotics came into use in the middle of the last century, it was supposed to be the end for the bacteria. How wrong we were! Except for those prevented by immunization, the bacterial pathogens occupy as prominent position as any time since the widespread implementation of public health measures a century ago. The emergence of new pathogens and the resistance of familiar ones to the antimicrobial agents developed in the "arms race" against them are primarily responsible. *Staphylococcus aureus*, the "all-time champion" of pathogens is just as prominent and just as confounding a cause of disease today as when Sir Alexander Ogston observed it in the wounds of his surgical patients in the 1880s.

This chapter lays out the basic mechanisms that bacteria use to produce disease. The purpose is to provide a foundation for explaining how these mechanisms are used by the bacterial pathogens in Chapters 24 to 40. Before beginning, a few definitions are in order:

Pathogenicity—The ability of any bacterial species to cause disease in a susceptible human host.

Pathogen—A bacterial species able to cause such disease when presented with favorable circumstances (for the organism).

Virulence—A term which presumes pathogenicity but allows expression of degrees from low to extremely high, for example:

- **Low virulence**—*Streptococcus salivarius* is universally present in the oropharyngeal flora of humans. On its own, it seems incapable of disease production, but if during a transient bacteremia it lands on a damaged heart valve, it can stick and cause slow but steady destruction.
- **Moderate virulence**—*Escherichia coli* is universally found in the colon, but if displacement to other sites such as adjacent tissues or the urinary bladder regularly causes acute infection.

- **High virulence**—*Bordetella pertussis*, the cause of whooping cough, is not found in the normal flora, but if encountered it is highly infectious and causes disease in almost every nonimmune person it contacts.
- **Extremely high virulence**—*Yersinia pestis*, the cause of plague, is also highly infectious, but in addition leads to death in a few days in over 70% of cases.

HUMANS AND BACTERIA

Commensals coexist

Opportunistic pathogens take advantage of breaks in defense

Primary pathogens cause disease on their own

As discussed in Chapter 1 humans have a rich normal flora, and the composition of that flora is mostly bacterial. Of these bacteria, most in humans are **commensal;** that is, they eat from the same table that we do. These microbes are constant companions and often depend on humans for their existence. We also encounter transient species, which are just passing through, but some of these may be **opportunistic pathogens.** That is, they can cause disease only when one or more of the defense mechanisms designed to restrict them from the usually sterile internal tissues are breached by accident, by intent (eg, surgery), or by an underlying metabolic or an infectious disorder (eg, AIDS). Nevertheless, a small group of bacteria regularly cause infection and overt disease in seemingly healthy persons. These are the **primary pathogens** such as the typhoid bacillus, gonococcus, and the tubercle bacillus, which are never considered members of the normal flora.

Pathogens must move on to another host

Long-term survival in a primary pathogen is absolutely dependent on its ability to replicate, survive, and be transmitted to another host. To accomplish this, the primary pathogens have evolved the ability to breach human cellular and anatomic barriers that ordinarily restrict or destroy commensal and transient microorganisms. Thus, pathogens can inherently cause damage to cells to gain access by force to a new unique niche that provides them with less competition from other microorganisms, as well as a ready new source of nutrients. For microorganisms that inhabit mammals as an essential component of their survival tactic, the capacity to multiply sufficiently to be maintained or be transmitted to a new susceptible host. Thus, pathogens have not only acquired the capacity to breach cellular barriers, they also have, by necessity, learned to circumvent, exploit, subvert, and even manipulate our normal cellular mechanisms to their own selfish need to multiply at our expense.

Aerosols spread *Legionella*

Tampons enhance toxin production

E coli O157:H7 is spread by food processing

The emergence of many seemingly new bacterial diseases has as much to do with human behavior as bacterial adaptability. The Legionnaires disease outbreak of 1976 was eventually traced *Legionella pneumophila,* which is widely found in aquatic environments as an infectious agent of amoebas. However, without the aerosolization created by modern systems (cooling towers) designed to humidify large buildings, transmission to humans would not have occurred. The development of super absorbent tampons had the unintended consequence of providing conditions favorable for the production of a toxin by some strains of *S aureus*. The result was a national outbreak of toxic shock syndrome. Food poisoning by *E coli* O157:H7, *Campylobacter,* and *Salmonella* arise as much from food technology and modern food distribution networks as from any fundamental change in the virulence properties of the bacteria in question. No part of our planet is more than 3 days away by air travel, a fact known and feared by all public health officials.

ATTRIBUTES OF BACTERIAL PATHOGENICITY

Genetic manipulation can inactivate and restore virulence

Virulence genes can be studied from genome sequences

The investigation of pathogenicity is based on linking natural disease in humans with experimental infection produced by the same organism. Once pathogenicity is established, a search for bacterial virulence determinants is launched with the eventual goal of finding an immunogen for a vaccine. These approaches have been tremendously enhanced by a new genetic approach in which manipulation of genes controlling virulence properties can be isolated in an appropriate model system. This makes it possible to insert, inactivate, or restore virulence genes and their regulators as isolated variables in an experiment. With the complete sequencing of the entire genome of the major pathogens has come an understanding of common DNA sequence structures of toxins, secretion systems, and regulators so they can be sought and even studied without chemical isolation of the virulence factor itself.

The discussions that follow and in the following chapters we try to explain the conclusions of these investigations with examples of the major types of genetic control. Although much of the information is known, detailed description of virulence genes and their regulation is beyond the scope of this book.

Whether a microbe is a primary or opportunistic pathogen, it must be able to enter a host; find a unique niche; avoid, circumvent, or subvert normal host defenses; multiply; and injure the host. For long-term success as a pathogen, it must also establish itself in the host or somewhere else long enough to eventually be transmitted to a new susceptible host. This competition between the pathogen and the host can be viewed as similar to more familiar military or athletic struggles—that is, the offense against the defense. The more we learn about bacterial pathogens, the more it seems that the most successful ones not only have an excellent offense; they are also particularly able to confound the host defense.

Pathogens must establish a niche and persist long enough to produce disease

Success involves offense and confounding host defenses

ENTRY: BEATING INNATE HOST DEFENSES

Each of the portals in the body that communicates with the outside world becomes a potential site of microbial entry. Human and other animal hosts have various protective mechanisms to prevent microbial entry (**Table 22–1**). A simple, though relatively efficient, mechanical barrier to microbial invasion is provided by intact epithelium. Organisms can gain access to the underlying tissues only by breaks or by way of hair follicles, sebaceous glands, and sweat glands that traverse the stratified layers. The surface of the skin continuously desquamates and thus tends to shed contaminating organisms. The skin also inhibits the growth of most extraneous microorganisms because of low moisture, low pH, and the presence of substances with antibacterial activity. Other than those transmitted by insect bites, bacteria have no known mechanism for passing the unbroken skin. ::: innate immunity, p. 21

Microbes gain access from the environment

Skin is a major protective barrier

A viscous mucous covering secreted by goblet cells protects the epithelium lining the respiratory tract, the gastrointestinal tract, and the urogenital system. Microorganisms become trapped in the mucous layer and may be swept away before they reach the epithelial cell surface. Secretory IgA (sIgA) secreted into the mucus and other secreted antimicrobials such as lysozyme and lactoferrin aid this cleansing process. Some bacteria excrete an enzyme IgA protease, which cleaves human IgA1 in the hinge region to release the Fc portion from the Fab fragment. This enzyme may play an important role in establishing microbial species at the mucosal surface. Ciliated epithelial cells constantly move the mucus away from the lower respiratory tract. In the respiratory tract, particles larger than 5 μm are trapped in this fashion. The epithelium of the intestinal tract below the esophagus is a less efficient mechanical barrier than the skin, but there are other effective defense mechanisms. The high level of hydrochloric acid and gastric enzymes in the normal stomach kill many ingested bacteria. Other bacteria are susceptible to pancreatic digestive enzymes or to the detergent effect of bile salts. ::: Secretory IgA p. 40

Secretions coat mucosal epithelium

IgA protease aids survival

Acids and enzymes aid in cleansing

How efficiently bacterial pathogens navigate all these barriers before their initial encounter with their target cell type is in some ways summarized by their infecting dose. How many organisms must be given to a host to ensure infection in some proportion of the individuals? Estimates of the infectious doses for several pathogens are shown in **Table 22–2**. In general, pathogens that have environmental or animal reservoirs can overwhelm innate defenses with large numbers. Those that are amplified by growth in food may also deliver high numbers with or without a reservoir. Pathogens with no reservoir or amplification mechanism must be transmitted human-to-human and thus require the lowest infecting doses. Without this advantage, these pathogens would eventually die out in the population.

Infection is dose-related

ADHERENCE: THE SEARCH FOR A UNIQUE NICHE

The first major interaction between a pathogenic microorganism and its host entails attachment to a eukaryotic cell surface. In its simplest form, adherence requires the participation of two factors: an **adhesin** on the invading microbe and a **receptor** on the host cell (**Figure 22–1**). The adhesin must be exposed on the bacterial surface either alone or in association with appendages like pili. Pili seem to be "sticky" by themselves which may be enhanced

TABLE 22–1 Nonspecific Defenses Against Colonization with Pathogens

SITE	MECHANICAL BARRIER	CILIATED EPITHELIUM	COMPETITION BY NORMAL FLORA	MUCUS	sIgA	LYMPHOID FOLLICLES	LOW PH	FLUSHING EFFECTS OF CONTENTS	PERISTALSIS	SPECIAL FACTORS
Skin	+++	–	+	–	–	–	++	–	–	Fatty acids from action of normal flora on sebum
Conjunctiva	++	–	–	–	+	–	–	+++	–	Lysozyme
Oropharynx	+++	–	+++	–	+	Yes	–	++	–	
Upper respiratory tract	++	+	+++	++	++	Yes	–	++	–	Turbinate baffles
Middle ear and paranasal sinuses[a]	++	+++	–	++	?	–	–	+	–	
Lower respiratory tract[a]	++	+++	–	++	++	Yes	–	–	–	Mucociliary escalator, alveolar macrophages; cough reflex
Stomach	++	–	–	++	–	–	+++	+	+	Production of hydrochloric acid
Intestinal tract	++	–	+++	+++	+++	Yes	–	+	+++	Bile; digestive enzymes
Vagina	+++	–	+++	+	+	–	+++	–	–	Lactobacillary flora ferments
Urinary tract[a]	++	–	–	–	+	–	+	+++	–	

[a]Sterile in health

+, ++, +++, relative importance in defense at each site; – = unimportant.

TABLE 22–2 Dose of Microorganisms Required to Produce Infection in Human Volunteers

MICROBE	ROUTE	DISEASE-PRODUCING DOSE
Salmonella serotype Typhi	Oral	10^5
Shigella spp.	Oral	10–1000
Vibrio cholerae	Oral	10^8
V cholerae	Oral + HCO_3^-	10^4 [a]
Mycobacterium tuberculosis	Inhalation	1–10

[a]Lower dose reflects bicarbonate neutralizing the acid barrier of the stomach.

by specific adhesin/receptor molecular relationships. In Gram-negative bacteria, the outer membrane is a major site for adhesins. Most adhesins are proteins, but carbohydrates and teichoic acids my also be involved. The chemical nature of the receptors is less well known because of the greater difficulty in their isolation (bacteria can be grown by the gallon), but they may be thought of as general or specific. For example, two of the most common receptors, mannose and fibronectin, are widely present on human epithelial cell surfaces. Pili that bind to them can mediate attachment at many sites. Specific receptors are those unique to a particular cell type such as human enterocytes or uroepithelial cells. Where known, these receptors are usually sugar residues that are part of glycolipids or glycoproteins on the host cell surface.

Adhesin and receptor are required

Pili often bind mannose, fibronectin

Receptors may be specific to host cell type

Many bacteria have more than one mechanism of host cell attachment. In some instances, pili mediate initial attachment, which is followed by a stronger, more specific binding mediated by another protein. This may allow implementation of a second function such as cytoskeleton rearrangement or invasion. Multiple adhesins may also allow bacteria to use one set at the epithelial surface, but a different set when encountering other cell types or the immune system. The role of pili may be more than a simple adhesive one. The pili of

Many have multiple attachment mechanisms

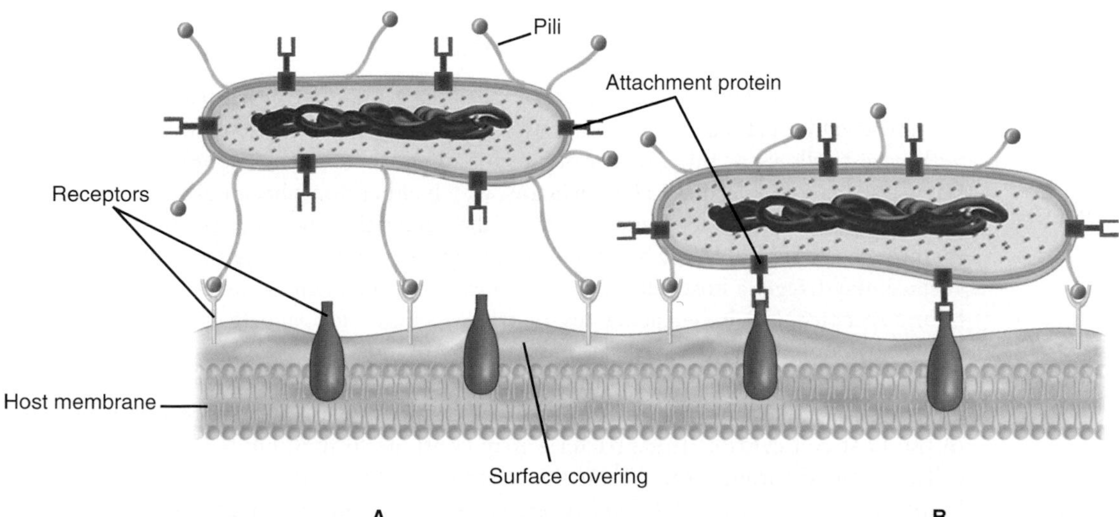

FIGURE 22–1. Bacterial attachment. A. The bacterial cell has both adhesive pili and another protein adhesin protruding from its surface. The pili are binding to a receptor present in material covering the cytoplasmic membrane. **B.** The pili have pulled the organism into closer contact allowing the second adhesin to bind its receptor, which extends from the cytoplasmic membrane through the surface coating.

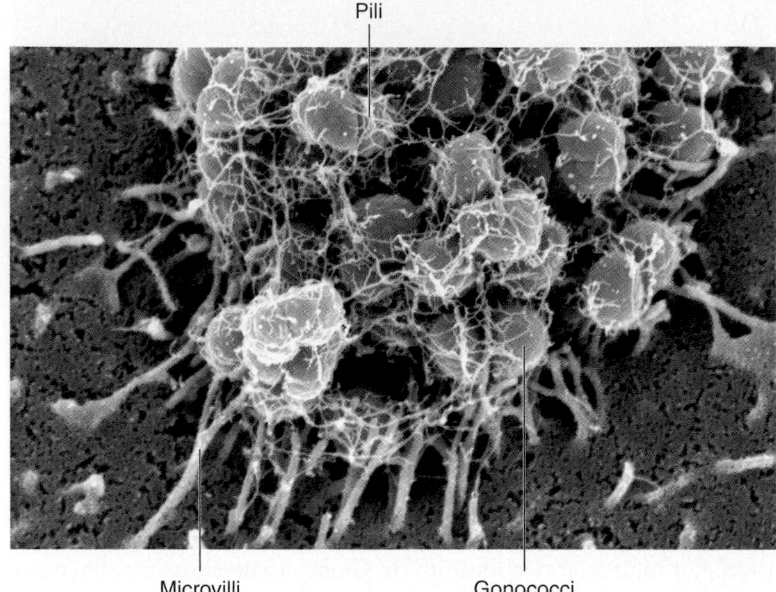

FIGURE 22–2. Pili. Pili extending from a microcolony of *Neisseria gonorrhoeae* (gonococci) are shown attaching the microvilli of an epithelial cell. The pili actively retract and mediate a movement of the colony across the cell surface called twitching motility. *(Photomicrographs kindly provided by Dustin L. Higashi and Magdalene So.)*

Neisseria gonorrhoeae, the cause of gonorrhea, mediate an active twitching motility on the cell surface with the formation of mobile microcolonies (**Figure 22–2**). ::: Pili, p.360

Strategies for Survival

Once the bacterial pathogen attaches, it must persist if it is to produce disease. Survival is less complicated if the organism can produced injury without moving from its initial niche. This is the case with some exotoxin-mediated bacterial diseases (diphtheria, whooping cough), but most pathogens must move either into the cell or beyond it. To do so requires a new set of survival strategies which include multiplying in the intracellular milieu and avoiding the attack of complement and phagocytes. The ways by which bacteria avoid, circumvent, or even subvert or manipulate such host barriers can be likened to a symphony in which each part contributes to a common theme.

INVASION: GETTING INTO CELLS

A few bacteria, like viruses, are obligate intracellular pathogens. Other bacteria are facultative intracellular pathogens and can grow as free-living cells in the environment as well as within host cells. Generally, invasive organisms adhere to host cells by one or more adhesins but use a class of molecules, called **invasins,** which either direct bacterial entry into cells or provide an intimate direct contact between the bacterial surface and the host cell plasma membrane. For example, the binding of a microbe to an integrin-like molecule on the host surface may trigger a host cell signal that causes actin filaments to link to the membrane-bound receptor, which then generates the force necessary for parasite uptake.

Invasins link to cell actin filaments

Bacteria enter cells initially within a membrane-bound, host-vesicular structure but then follow one of two distinct pathways (**Figure 22–3**). Some bacteria (*Listeria, Shigella*) enzymatically lyse the phagosome membrane and escape to the nutrient-rich safe haven of the host cell cytosol. These bacteria may continue to multiply there, infect adjacent cells, or move through the cell to the submucosa. Other invasive pathogenic species (*Salmonella typhi, Mycobacterium tuberculosis*) remain in the phagosome and replicate even in professional phagocytes. Their survival in this usually perilous location is due to thwarting normal host cell trafficking patterns and avoiding the digestive action of lysosomes. There are multiple mechanisms for the latter including preventing phagosome-lysosome fusion or, if fused, blocking acidification to the optimum pH for enzyme activity. ::: Phagosome, p. 25

Some escape the phagosome

Some multiply in phagosome by blocking lysosome function

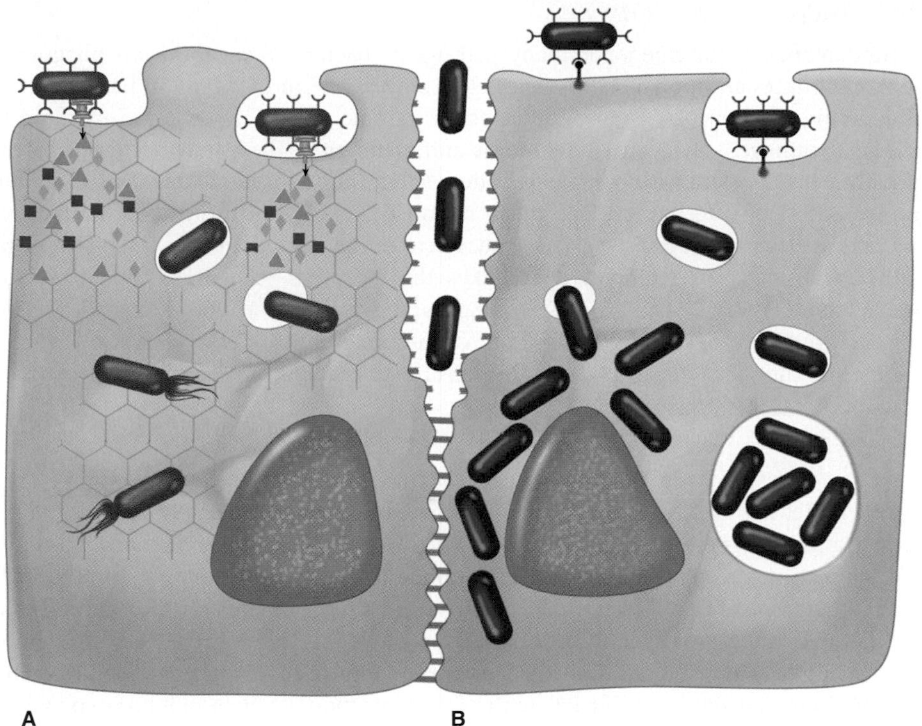

FIGURE 22–3. Bacterial invasion. A. The bacterial cell has an injection secretion system that is injecting multiple proteins into the host cell. Some of these cause cytoskeletal reorganization, which engulfs the bacteria. In the cytosol, the bacteria lyse the vacuolar membrane, escape, and move about. **B.** A bacterial surface protein binds to the cell surface and induces its own endocytosis. In the cell, some escape (as in A), and others multiply in the phagosome.

In Gram-negative bacteria with type III or IV injection secretion systems, a variation on the above scenarios is possible. The secretion systems inject many proteins, some of which disrupt cellular signaling and the cell's cytoskeleton. The cytoskeleton rearrangements may leave the bacteria tightly bound to an altered surface or trigger invasion. One pathogen even injects its own receptor, which is processed to the outer membrane where it mediates tight binding of the bacteria. ::: Secretion systems, p. 383

Injection secretion systems trigger invasion or tight binding

PERSISTING IN A NEW ENVIRONMENT

Bacteria that reach the subepithelial tissues are immediately exposed to the intercellular tissue fluids, which have defined properties that inhibit multiplication of many bacteria. For example, most tissues contain lysozyme in sufficient concentrations to disrupt the cell wall of Gram-positive bacteria. Tissue fluid itself is a suboptimal growth medium for most bacteria and is deficient in free iron. Virtually all pathogenic species come equipped with a means of extracting the essential iron that they need from the host's iron-sequestering defenses using siderophores. ::: siderophores, p. 359

Subepithelial environment is different

Iron sources are important for the pathogen

One of the most common tactics of these pathogens is to induce programmed cell death (apoptosis). This clever microbial tactic not only inactivates the killing potential of the phagocyte, but also reduces the number of defenders available to inhibit other bacterial invaders. The invading bacteria that induce apoptosis obtain the added benefit that death by apoptosis nullifies the normal cellular signaling processes of cytokine and chemokine signaling of necrotic death. Hence, the myriads of microbes that infect humans and make up their normal flora are held at bay by our innate and adaptive immune mechanisms. Pathogenic bacteria, almost by definition, can overcome these biochemical and cellular shields after they breach the mucosal barrier.

Apoptosis may be induced

Confounding the Immune System

The host immune system evolved in large part because of the selective pressure of microbial attack. To be successful, microbial pathogens must escape this system at least long enough to be transmitted to a new susceptible host or to take up residence within the host in a way that is compatible with mutual coexistence.

Antiphagocytic Activity

A fundamental requirement for many pathogenic bacteria is escape from phagocytosis by macrophages and polymorphonuclear leukocytes. The most common bacterial means of avoiding phagocytosis is an antiphagocytic capsule, which is possessed by almost all principal pathogens that cause pneumonia and meningitis. These polysaccharide capsules of pathogens interfere with complement deposition on the bacterial cell surface by binding regulators of C3b that are present in serum. When one of these, serum factor H, is concentrated on the capsular surface, it accelerates the degradation of C3b. This negates both direct complement injury and makes the receptors recognized by phagocytosis unavailable (**Figure 22–4**). This mechanism is not restricted to polysaccharide capsules. Surface proteins able to bind factor H have the same biologic effect. Antibody directed against the capsular antigen reverses this effect because C3b can then bind in association with IgG. ::: C3b & factor H, p. 27

Antigenic Variation

Surface antigens can be varied

Antigenically different subpopulations escape immune surveillance

Another method by which microorganisms avoid host immune responses is by varying surface antigens. Gonorrhea is a disease in which there appears to be no natural immunity and reinfections are common. In fact, an immune response can be mounted to the surface pili and outer membrane proteins of *N gonorrhoeae*, but the organism is continuously varying them. This happens even in the course of a single infection. The genetic mechanisms involved in this antigenic variation are illustrated in **Figure 22–5**. The effect is that when the immune system delivers specific IgG to the site of infection, it will bind it homologous antigen, but a subpopulation with an antigenically different surface can multiply and continue the infection. Therefore, the pathogen escapes immune surveillance. A number of other bacteria and parasites also undergo antigenic variation. :::Variation in gonorrhea, p. 547

Induction of Apoptosis

Apoptosis may be induced

A pathogen taken up by a phagocyte, but unable to multiply there, can still survive by killing the host cell. One of the tactics for this is to induce apoptosis. This clever microbial tactic

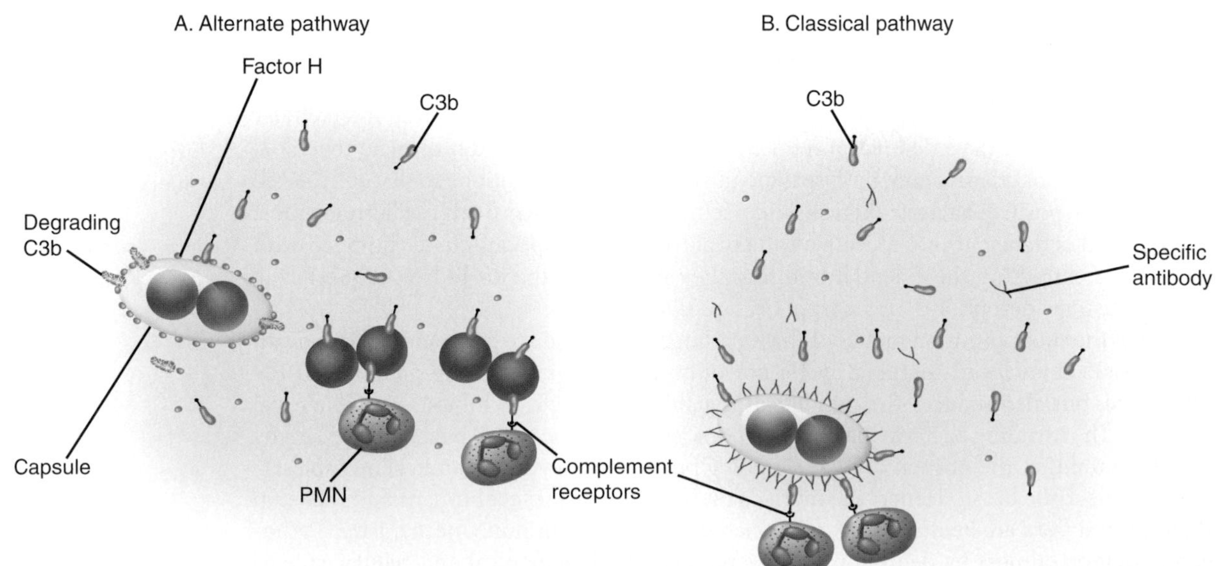

FIGURE 22–4. Bacterial resistance to opsonophagocytosis. A. Alternate pathway. In the alternate complement pathway, C3b binds to the surface of bacteria, providing a recognition site for professional phagocytes and sometimes causing direct injury. Bacteria with special surface structures such as capsules or protein are able to bind serum factor H to their surface. This interferes with complement deposition by accelerating the breakdown of C3b. **B.** Classical pathway. Specific antibody binding to an antigen on the surface provides another binding cite for C3b. Phagocyte recognition may occur even if factor H is present.

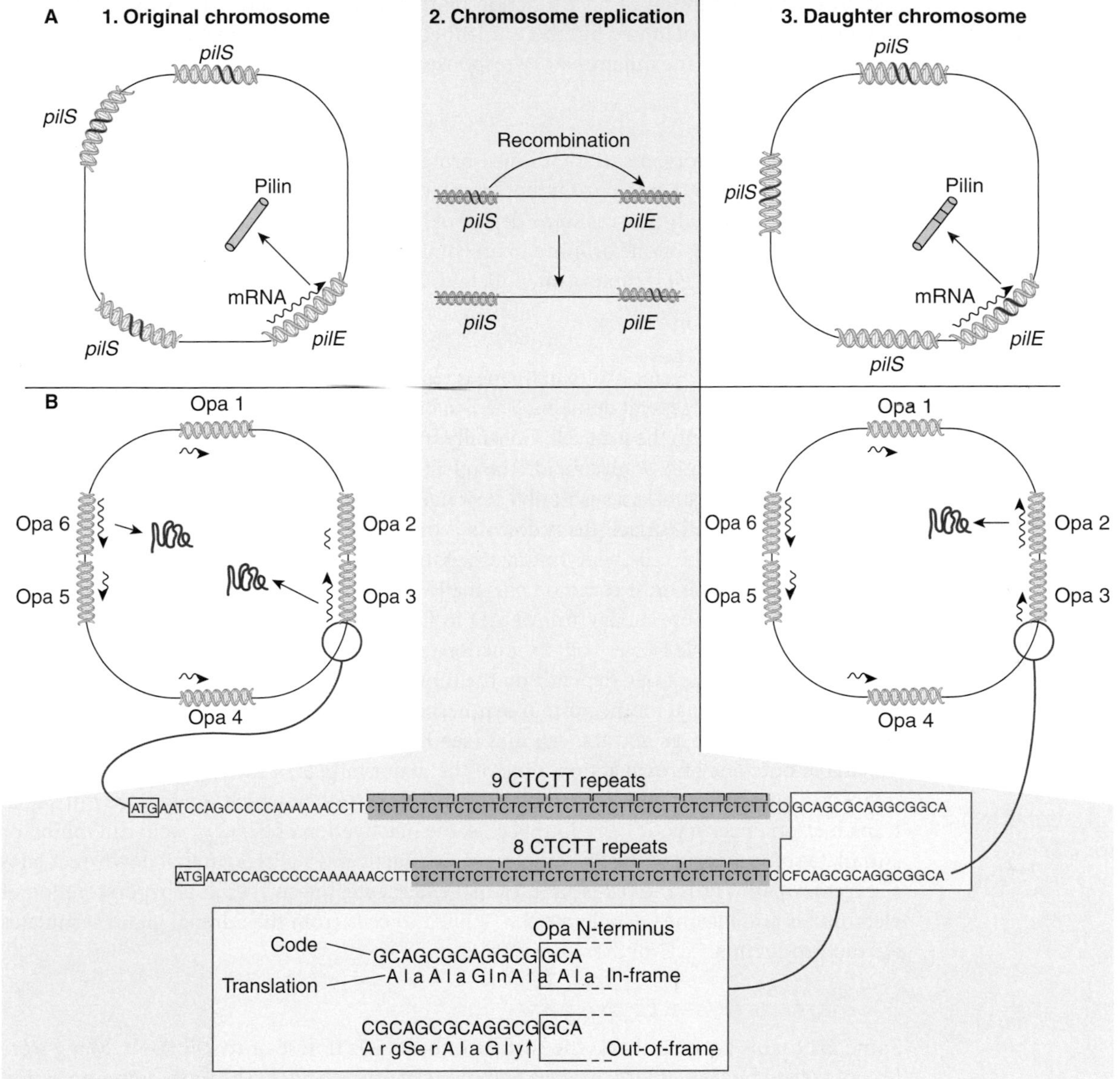

FIGURE 22–5. Antigenic variation. Mechanisms for change in the antigenic makeup of both pili and outer membrane Opa proteins of *Neisseria gonorrhoeae* are shown. **A.** The chromosome contains multiple unlinked pilin genes, which are either expressing (*pilE*) or silent (*pilS*). The expressing gene is transcribing a mature pilin protein subunit. During chromosome replication, one of the *pilS* genes recombines with one of the *pilE* genes, donating some of its DNA (red). The new daughter chromosome now produces an antigenically different pilin based on transcription of the donated (red) sequences into protein. **B.** The chromosome contains multiple Opa genes. Opa 3 and Opa 6 are "on" (producing protein), and the others are "off." During chromosome replication, replicative slippage in the leader peptide causes a five-base sequence (CTCTT) to be repeated variable numbers of times. Translation of the Opa will remain in-frame only if the number of added CTCTT nucleotides is evenly divisible by three. For the Opa gene in **B1**, the triplet code for alanine (GCA) is in-frame (9 × 5 = 45. 45 ÷ 3 = 15) but in **B3** it is out-of-frame.

not only inactivates the killing potential of the phagocyte, but also reduces the number of defenders available to inhibit other bacterial invaders. The invading bacteria that induce apoptosis obtain the added benefit that death by apoptosis nullifies the normal cellular signaling processes of cytokine and chemokine signaling of necrotic death.

INJURY

The successful pathogen must survive and multiply in the face of multiple host defenses. Although this is a formidable achievement, by itself it is not enough to cause disease. This

Secreted into their environment

requires some disruption of host function by the organism. Bacterial toxins are the most obvious mechanism of injury, but there are others and also diseases in which the only injury appears to be due to the inflammatory response to the invader.

Exotoxins

A number of microorganisms synthesize protein molecules that are toxic to their hosts and are secreted into their environment or are found associated with the microbial surface. These exotoxins usually possess some degree of host cell specificity, which is dictated by the nature of the binding of one or more toxin components to a specific host cell receptor. The distribution of host cell receptors often dictates the degree and the breadth of the toxicity.

A–B Exotoxins

B unit binds to cell receptor

A acts on target protein

The best-known pathogenic exotoxin theme is represented by the A–B exotoxins. These toxins are divisible into two general domains. The B subunit(s), is associated with the binding specificity of the molecule to the host cell. Generally speaking, the B region binds to a specific host cell surface glycoprotein or glycolipid. The other, A (active) subunit, is the catalytic domain, which enzymatically attacks a susceptible host function or structure. After attachment of the B domain to the host cell surface, the A domain is transported by direct fusion or by endocytosis into the host cell. In the cell, the A unit catalyzes the enzymatic modification of a protein called its **target protein**. The most common enzymatic reaction is called **ADP-ribosylation,** which attaches the ADP-ribose moiety from NAD to the target protein. This modifies the protein enough that it is unable to carry out its function. ::: ADP-ribosylation, p. 363

Biologic effect depends on function of target protein

Effect may be inhibitory or stimulatory

The net effect of the toxin depends on the function of the target protein and the function of the cell. If it is crucial for the protein-synthesizing apparatus of the cell (diphtheria toxin), protein synthesis ceases and the cell dies (see Figure 1–7). However, cell death is not the inevitable outcome of toxin action. One of the major targets of the ADP-ribosylating A–B toxins are guanine nucleotide-binding proteins (G proteins), which are involved in signal transduction in eukaryotic cells. In this case, the inactivation of the G protein can inhibit or stimulate some activity of the cell. Cholera toxin inactivates a G protein that down-regulates a secretory pathway. If the cell is an intestinal enterocyte, the end result is hypersecretion of electrolytes and diarrhea. Cholera toxin applied to cells from the adrenal gland stimulates steroid production. ::: Diphtheria toxin, p.16

Membrane-Active Exotoxins

Insertion in cytoplasmic membrane creates a leaking pore

Some exotoxins act directly on the surface of host cells to lyse or to kill them. Many were first observed by their ability to cause hemolysis of erythrocytes. The most common action is to create pores by directly inserting into eukaryotic membranes of a wide range of cells including phagocytes (**Figure 22–6**). These **pore-forming toxins** are produced by some of the most aggressive pathogens (*Staphylococcus aureus,* group A streptococcus, *E coli*) and cause cellular death by loss of cellular integrity and leakage through the pore. Some are called the RTX (repeats in toxin) group because of a recurrent amino acid duplication in their structure. Other membrane-active toxins have enzymatic activity such as the α-toxin of *Clostridium perfringens,* which is a lecithinase.

Hydrolytic Enzymes

Enzymatic actions cause injury, facilitate spread

Many bacteria produce one or more enzymes that are nontoxic per se, but facilitate tissue invasion or help to protect the organism against the body's defense mechanisms. For example, various bacteria produce collagenase or hyaluronidase or convert serum plasminogen to plasmin, which has fibrinolytic activity. Although the evidence is not conclusive, it is reasonable to assume that these substances facilitate spread of infection. Some bacteria also produce deoxyribonuclease, elastase, and many other biologically active enzymes, but their function in the disease process or in providing nutrients for the invaders is uncertain.

Superantigen Exotoxins

Some microbial exotoxins have a direct effect on cells of the immune system, and this interaction itself leads to disease. The most dramatic of these are the toxins causing the toxic shock syndromes of *S aureus* and group A streptococci. These syndromes are evoked when toxin

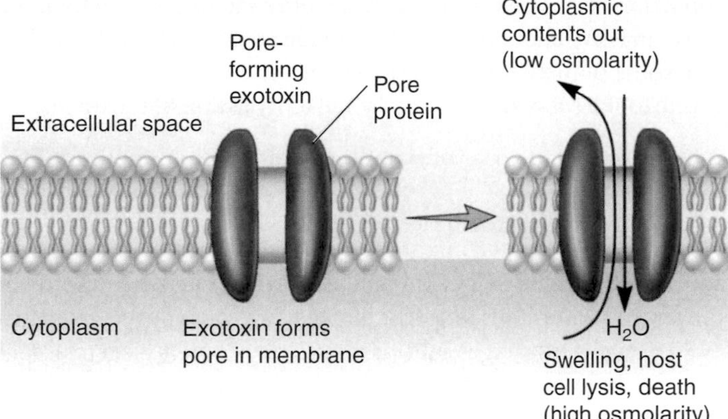

FIGURE 22–6. Pore-forming exotoxin. The pore protein has inserted itself into the host cell membrane making an open channel. Formation of multiple such pores causes cytoplasmic contents to leave the cell and water to move in. This ultimately leads to cell lysis and death. (Reproduced with permission from Willey J, Sherwood L, Woolverton C (eds). *Prescott's Principles of Microbiology.* New York: McGraw-Hill; 2008.)

is produced at an infected site and absorbed into the circulation. There these toxins are able to bind directly to class II major histocompatibility complex (MHC) molecules on antigen-presenting cells (without processing) and directly stimulate production of cytokines such as interleukin 1 (IL-1) and tumor necrosis factor TNF; (**Figure 22–7**). These molecules are called superantigens because they act as polyclonal stimulators of T cells. This means a significant

Bind directly to MHC II

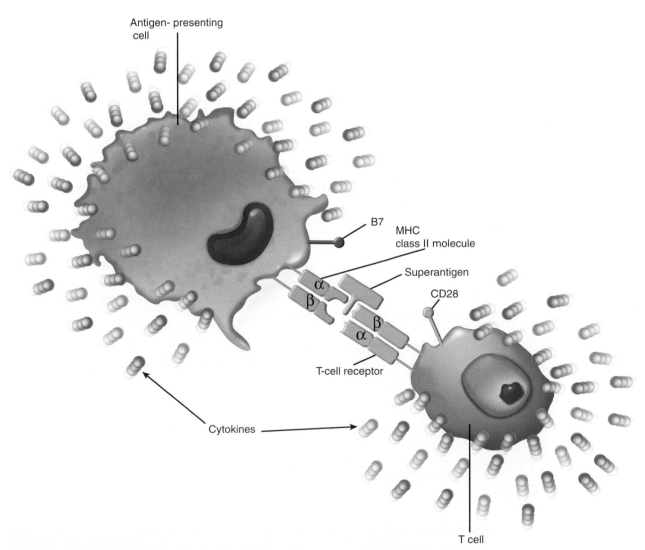

FIGURE 22–7. Superantigen exotoxin. A superantigen (*yellow*) is binding to the MHC class II molecular complex outside the groove for antigen presentation. This causes a massive secretion of cytokines.

Cytokines are released from large proportion of T cells

proportion of all T cells respond by dividing and releasing cytokines, which makes the cytokine release massive enough to cause systemic effects such as shock. When ingested preformed in food, some of these toxins cause diarrhea and vomiting. It is not known whether these effects are due to the superantigen or to some other action of the toxin.

Endotoxin

LPS in the bloodstream causes shock, DIC

Lipid A is toxic portion

In many infections caused by Gram-negative organisms, the lipopolysaccharide (LPS) endotoxin of the outer membrane is a significant component of the disease process. LPS can cause local injury, but the major effects come when Gram-negative bacteria enter the bloodstream and circulate. The lipid A portion causes fever through release of IL-1 and TNF from macrophages and dramatic physiologic effects associated with inflammation. These include hypotension, lowered polymorphonuclear leukocyte and platelet counts from increased margination of these cells to the walls of the small vessels, hemorrhage, and sometimes disseminated intravascular coagulation (DIC) from activation of clotting factors. Rapid and irreversible shock may follow passage of endotoxin into the bloodstream. ::: LPS, p. 354

Peptidoglycan fragments are not called endotoxin

The term endotoxin comes from the fact that LPS is an inherent structural component of the Gram-negative cell wall, not a secreted product of the bacteria. A comparable event with Gram-positive bacteria can occur with the release and circulation of peptidoglycan cell wall fragments. This also leads to cytokine release and systemic manifestations. Although the biology is similar, the terms endotoxin or endotoxemia are not used here because they have long been reserved for the LPS endotoxin of Gram-negative bacteria.

Damage Caused by Inflammation and Immune Responses

Many successful pathogens produce disease without using any of the known virulence factors just described. In these instances, injury can still be produced by acute or chronic inflammation or a misdirected immune response triggered by antigenic components of the pathogen.

Persistent Inflammation

PMNs cause swelling, occupy space

Prolonged DTH is destructive

The normal inflammatory response is a two-edged sword in both acute and chronic infections. Although the enzymes of polymorphonuclear neutrophils (PMNs) are killing the invader, they still cause some damage to host tissues or compromise organ function. Pulmonary alveoli filled with PMNs and macrophages are not effective at absorption of oxygen. In the closed space of the central nervous system, the swelling caused by inflammation may directly cause brain injury. In some chronic infections, the pathologic and clinical features are due largely to delayed-type hypersensitivity (DTH) reactions to the organism or its products. In tuberculosis if the host is unable to halt the growth of *M tuberculosis* by activation of T_H1 immunity, persistent growth of the pathogen will continue to stimulate DTH-mediated injury. ::: Tuberculosis DTH, p. 498

Misdirected Immune Responses

Reactions between high concentrations of antibody, soluble microbial antigens, and complement can deposit immune complexes in tissues and cause acute inflammatory reactions and immune complex disease. In poststreptococcal acute glomerulonephritis, for example, the complexes are sequestered in the glomeruli of the kidney, with serious interference in renal function from the resulting complement deposition and tissue reaction. Antibody produced against bacterial antigens can also cross-react with certain host tissues and initiate an autoimmune process. This molecular mimicry is felt to be the explanation for poststreptococcal rheumatic fever. ::: Adverse immune reactions, p. 41

GENETICS OF BACTERIAL PATHOGENICITY

All of the genetic tools described in the previous chapter are put to use in service of the complex business of being a pathogen. The multiple and sequential deployment of adherence, evasion, and injury-related virulence factors have evolved in ways that make them efficient and persistent. Part of this is the use of the plasmid and regulatory systems already described in unique ways such as pathogenicity islands. Our understanding of others remains at a more descriptive level, such as the emergence and spread of clones with enhanced virulence by unknown mechanisms.

PLASMIDS

Many of the essential determinants of pathogenicity are actually replicated as part the bacterial chromosome, but a surprising number are carried in plasmids. This often includes multiple virulence factors in the same plasmid. For example, one type of diarrhea-causing *E coli* carries the genes for pili mediating adherence to enterocytes and for the enterotoxin it delivers to those enterocytes on the same plasmid. The term **virulence plasmid** has been used for plasmids whose loss or modification causes loss of pathogenicity for the host strain. Since plasmids are inherently a less secure home for genes than the chromosome, this location must provide some efficiency for the pathogen. Perhaps the excess baggage of the plasmid is a trade-off for avoiding disruption of the organization of the bacterial chromosome. ::: Plasmid, p. 357

Genes on plasmids are multiple and related

Loss of virulence plasmids negates pathogenicity

REGULATION OF VIRULENCE GENES

In addition to the multiple steps of pathogenesis, some pathogens lurk in locations like seawater (cholera) or fleas (plague) until their opportunity to cause human disease presents itself. Some of these pathogens have evolved regulatory systems, which link sensing of environmental cues (temperature, osmolarity, iron concentration) to activation of their virulence apparatus. It appears that these signals can "tell" the pathogen whether it is in a benign environment, inside an insect vector, in body fluids, or even inside a phagocyte. The virulence factor deployment then proceeds often in a multistep manner, synthesizing the adhesin or toxin just at the time it is needed. An example of this is shown in **Figure 22–8,** which illustrates the two-compartment regulatory system used by *B pertussis* in whooping cough. This just-in time production is energy-efficient and effective in producing disease. ::: 2-compartment regulation, p. 371

Pathogens can sense their environment

Virulence factors are produced "just in time"

PATHOGENICITY ISLANDS

In recent years, large blocks of genes found on the bacterial chromosome have been given the name pathogenicity island (PAI) to describe unique regions exclusively associated with virulence (**Figure 22–9**). The "island" component of the name comes from the fact that the PAI regions themselves usually have fundamental characteristics such as guanine + cytosine content, codon usage, and tRNA genes that are different from the rest of the genome of the current host organism. This suggests that gene transfer from a foreign species sometime in the distant past is the likely origin. Many PAIs have strikingly similar homologs in bacteria that are pathogenic for plants and animals. The PAIs contain the complete package required for delivery of the pathogenic trait, even those that are the most complex involving 20 to 30 genes. In organisms that deploy type III secretion systems, the genes for the injection apparatus, the secreted proteins, and regulatory elements are all included in the PAI.

Large genomic segments transferred from an unrelated species

Genes for all components of virulence are included

CLONALITY

Bacteria cannot be a helter-skelter amalgam of genes brought about by promiscuous genetic exchange. If this were so, there would be no bacterial specialization, and all would possess a consensus chromosomal sequence. Thus, most bacteria have some degree of built-in reproductive isolation, except for members of their own or very closely related species. In this way, diversity within the species through mutation can be maximized (usually by transformation or transduction) while conserving useful gene sequences. The end result of husbanding of important genes during evolution is that at any given time in the world, many bacteria are representatives of a single or, more often, a relatively few clonal types that have

Useful genes are preserved by clonality

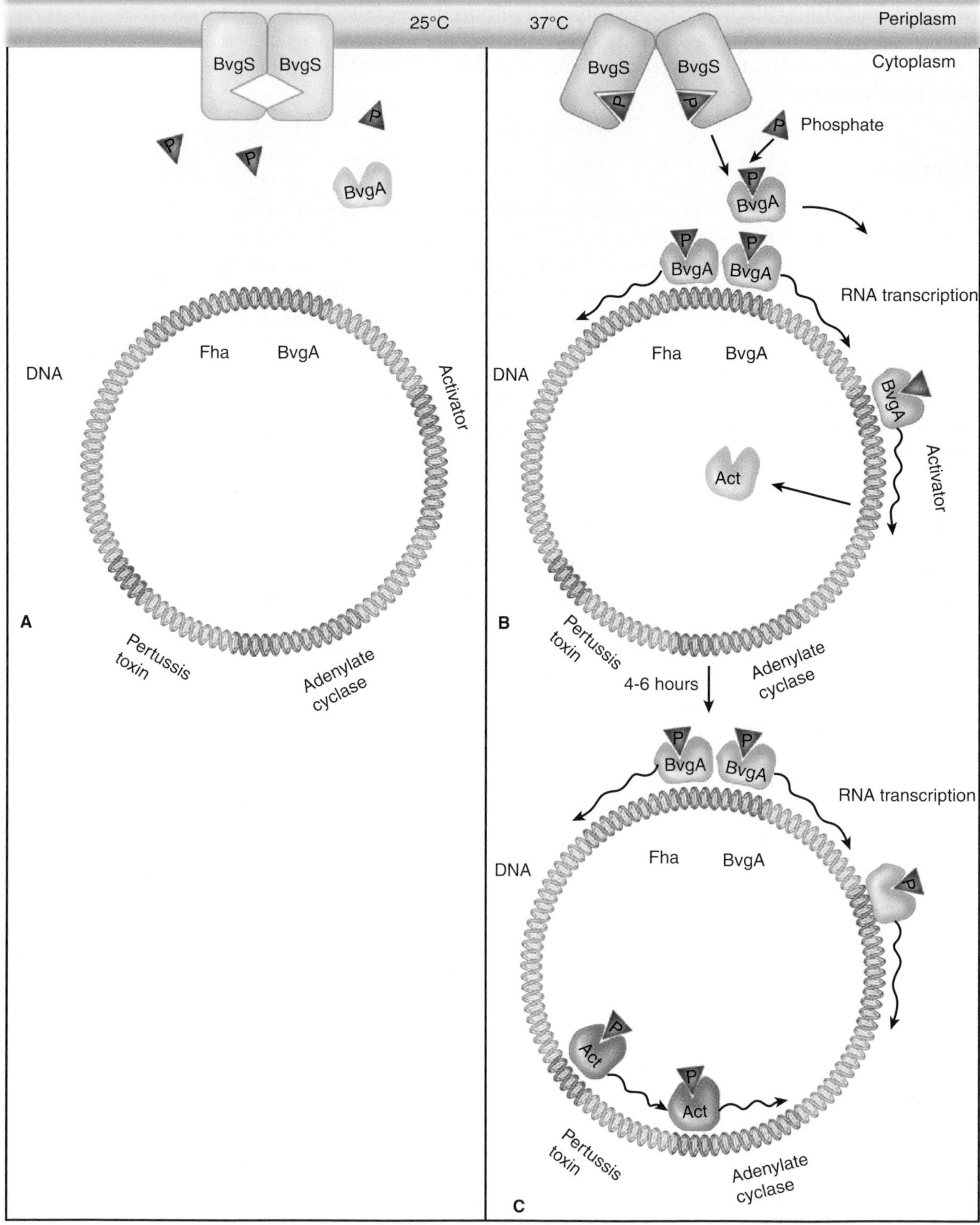

FIGURE 22–8. Regulation of *Bordetella pertussis* virulence factors. A. At 25°C, the membrane-associated regulatory protein BvgS is inactive as are the genes for virulence factors filamentous hemagglutinin (Fha), pertussis toxin, and adenylate cyclase. **B.** At 37°C, BvgS autophosphorylates and activates a cytoplasmic regulatory protein, BvgA, by phosphorylation. BvgA activates transcription of genes for production of BvgS, BvgA, Fha, and a postulated second regulator, Act. **C.** Hours later, transcription of the pertussis toxin and adenylate cyclase is activated by Act. (Adapted from Melton, AR, Weiss AA.)

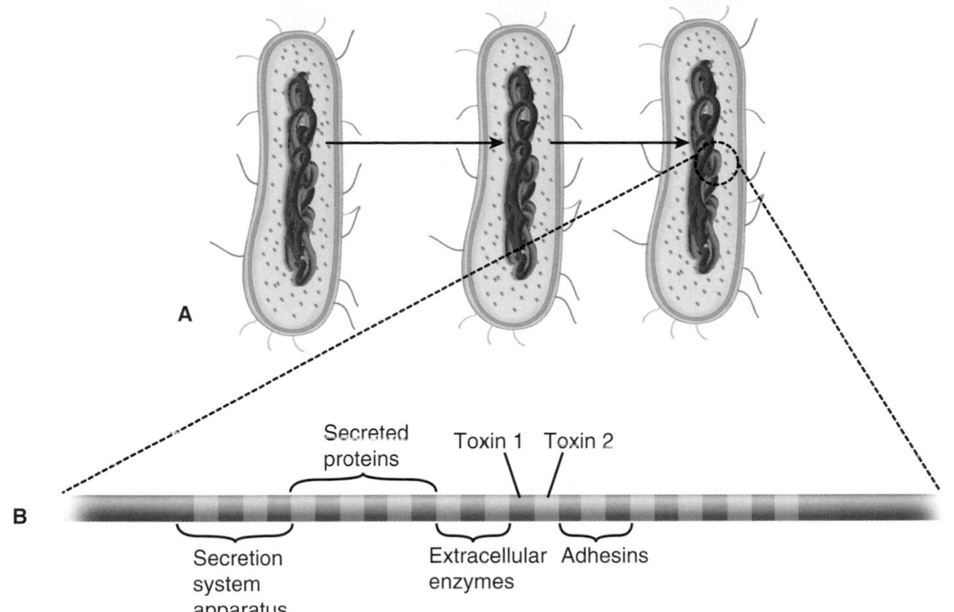

FIGURE 22–9. Pathogenicity island (PAI). A. Two bacterial strains are engaged in genetic exchange by one of the mechanisms described in Chapter 21. The recipient (right) has incorporated a large segment of the donor DNA into its chromosome. **B.** The chemical makeup of the donated segment is different from that of the host chromosome. This PAI contains genes for adhesins, toxins, and a secretion system all for production of the same disease.

become widespread for the (evolutionary) moment. The study and definition of clonality require more than the presence or absence of virulence factors. They require examining the specific alleles of various genes and the subtle differences in amino acid sequence in batteries of multiple housekeeping enzymes.

Natural populations of many pathogens are proving to have a clonal structure

In some cases single clones are responsible for geographically widespread disease

One of the discoveries to come from the application of molecular diagnostic tools to infectious diseases is the clonal nature of many infectious diseases. That is, over long periods and large geographic distances, the organisms of a given species isolated from clinical samples tend to be so similar in genetic makeup that one is forced to envision that a clone of bacteria descended from a relatively recent common ancestor is responsible for all or most of the disease incidence. The results have been striking. For example, isolates of *B pertussis* from the United States represent a single clone, whereas in Japan there is a slightly different clone. Another study has determined that only 11 multilocus genotypes (clones) of *Neisseria meningitidis* have been responsible for the major epidemics of serogroup A organisms worldwide over 60 years. When microbes establish a unique niche, they protect their selective advantage.

Although the human bacteria pathogens represent only a tiny percentage of the microbial world, they are among the most ingenious in the ways they produce disease. The independence and power of the bacterial cell are translated into some of the most feared of all diseases. The bacteriology, disease mechanisms, and clinical aspects of these diseases are explored in the following chapters.

CHAPTER 23

Antibacterial Agents and Resistance

The ability to direct therapy specifically at a disease-causing infectious agent is unique to the management of infectious diseases. Its initial success depends on exploiting differences between our own makeup and metabolism and that of the microorganism in question. The mode of action of antimicrobials on bacteria is the focus of this chapter. The continued success of antibacterial agents depends on whether the organisms to which the agent was originally directed develop resistance. Resistance to antibacterial agents is also addressed here. Specific information about pathogenic bacteria can be found in Chapters 24 through 40; a complete guide to the treatment of infectious diseases is beyond the scope of this book.

Natural materials with some activity against microbes were used in folk medicine in earlier times, such as the bark of the cinchona tree (containing quinine) in the treatment of malaria. Rational approaches to chemotherapy began with Ehrlich's development of arsenical compounds for the treatment of syphilis early in the 20th century. Many years then elapsed before the next major development, which was the discovery of the therapeutic effectiveness of a sulfonamide (prontosil rubrum) by Domagk in 1935. Penicillin, which had been discovered in 1929 by Fleming, could not be adequately purified at that time; however, this was accomplished later, and penicillin was produced in sufficient quantities so that Florey and his colleagues could demonstrate its clinical effectiveness in the early 1940s. Numerous new antimicrobial agents have been discovered or developed, and many have found their way into clinical practice.

Sulfonamides and penicillin were the first effective antibacterial agents

ANTIBACTERIAL AGENTS AND THERAPY

GENERAL CONSIDERATIONS

Clinically effective antimicrobial agents all exhibit selective toxicity toward the parasite rather than the host, a characteristic that differentiates them from most disinfectants. In most cases, selective toxicity is explained by action on microbial processes or structures that differ from those of mammalian cells. For example, some agents act on bacterial cell wall synthesis, and others on functions of the 70S bacterial ribosome but not the 80S eukaryotic ribosome. Some antimicrobials, such as penicillin, are essentially nontoxic to the host, unless hypersensitivity has developed. For others, such as the aminoglycosides, the effective therapeutic dose is relatively close to the toxic dose; as a result, control of dosage and blood level must be much more precise. ::: Disinfection, p.50

Ideally, selective toxicity is based on the ability of an antimicrobial agent to attack a target present in bacteria but not humans

Definitions

- **Antibiotic**—antimicrobials of microbial origin, most of which are produced by fungi or by bacteria of the genus *Streptomyces*.
- **Antimicrobial, antimicrobic**—any substance with sufficient antimicrobial activity that it can be used in the treatment of infectious diseases.

- **Bactericidal**—an antimicrobial that not only inhibits growth but is lethal to bacteria.
- **Bacteriostatic**—an antimicrobial that inhibits growth, but does not kill the organisms. The host defense mechanisms are ultimately responsible for eradication of infection.
- **Chemotherapeutic**—a broad term that encompasses antibiotics, antimicrobials, and drugs used in the treatment of cancer. In the context of infectious diseases, it implies that the agent is not an antibiotic.
- **Minimal inhibitory concentration (MIC)**—a laboratory term that defines the lowest concentration (μg/mL) able to inhibit growth of the microorganism.
- **Resistant**—organisms that are not inhibited by clinically achievable concentrations of an antimicrobial agent.
- **Spectrum**—an expression of the categories of microorganisms against which an antimicrobial is typically active. A narrow-spectrum agent has activity against only a few organisms. A broad-spectrum agent has activity against organisms of diverse types (eg, Gram-positive and Gram-negative bacteria).
- **Susceptible or sensitive**—terms applied to microorganisms indicating that they will be inhibited by concentrations of the antimicrobial agent that can be achieved clinically using generally recommended dosage schedules.

SOURCES OF ANTIMICROBIAL AGENTS

Antibiotics are synthesized by molds or bacteria

Production in quantity is by industrial fermentation

There are several sources of antimicrobial agents. The antibiotics are of biologic origin and probably play an important part in microbial ecology in the natural environment. Penicillin, for example, is produced by several molds of the genus *Penicillium*, and the prototype cephalosporin antibiotics were derived from other molds. The largest source of naturally occurring antibiotics is the genus *Streptomyces*, the members of which are Gram-positive, branching bacteria found in soils and freshwater sediments. Streptomycin, the tetracyclines, chloramphenicol, erythromycin, and many other antibiotics were discovered by screening large numbers of *Streptomyces* isolates from different parts of the world. Antibiotics are mass produced by techniques derived from the procedures of the fermentation industry.

Chemicals with antibacterial activity are discovered by chance or as the result of screening programs

Chemically synthesized antimicrobial agents were initially discovered among compounds synthesized for other purposes and tested for their therapeutic effectiveness in animals. The sulfonamides, for example, were discovered as a result of routine screening of aniline dyes. More recently, active compounds have been synthesized with structures tailored to be effective inhibitors or competitors of known metabolic pathways. Trimethoprim, which inhibits dihydrofolate reductase, is an excellent example.

Naturally occurring antimicrobics can be chemically modified

A third source of antimicrobials is molecular manipulation of previously discovered antibiotics or chemotherapeutics to broaden their range and degree of activity against microorganisms or to improve their pharmacologic characteristics. Examples include the development of penicillinase-resistant and broad-spectrum penicillins, as well as a large range of aminoglycosides and cephalosporins of increasing activity, spectrum, and resistance to inactivating enzymes.

SPECTRUM OF ACTION

Spectrum is the range of bacteria against which the agent is typically active

Broad-spectrum agents inhibit both Gram-positive and Gram-negative species

The **spectrum** of activity of each antimicrobial agent describes the genera and species against which it is typically active. For the most common antimicrobial agents and bacteria, these are shown in **Table 23–1**. Spectra overlap but are usually characteristic for each broad class of antimicrobial. Some antibacterial antimicrobics are known as **narrow-spectrum agents;** for example, benzyl penicillin is highly active against many Gram-positive and Gram-negative cocci but has little activity against enteric Gram-negative bacilli. Chloramphenicol, tetracycline, and the cephalosporins, on the other hand, are **broad-spectrum agents** that inhibit a wide range of Gram-positive and Gram-negative bacteria, including some obligate intracellular organisms. When resistance develops in an initially sensitive genus or species, that species is still considered within the spectrum even when the resistant subpopulation is significant. For example, the spectrum of benzyl penicillin is considered to include *Staphylococcus aureus*, although more than 80% of strains now are penicillin-resistant.

TABLE 23–1 Characteristics of Antibacterial Drugs

TARGET/REPRESENTATIVE DRUGS	CHARACTERISTICS
Cell Wall Synthesis	
β-lactams	Bactericidal against a variety of bacteria; inhibit peptidoglycan transpeptidases
Penicillins	
Natural penicillins: penicillin G, penicillin V	Active against Gram-positive bacteria and some Gram-negative cocci
Penicillinase-resistant: methicillin, dicloxacillin	Similar to the natural penicillins, but resistant to inactivation by the penicillinase of staphylococci
Broad-spectrum: ampicillin, amoxicillin	Similar to the natural penicillins, but more active against Gram-negative organisms
Extended-spectrum: ticarcillin, piperacillin	Increased activity against Gram-negative rods, including *Pseudomonas* species
Cephalosporins	Some are more effective against Gram-negative bacteria and less susceptible to destruction by β-lactamases
Cephalexin, cephradine, cefepime, cefotaxime	
Carbapenems	Resistant to inactivation by β-lactamases. Many Gram-negative bacteria are susceptible
Imipenem, meropenem	
Monobactams	Resistant to β-lactamases. Primarily active against members of the family *Enterobacteriaceae*
Aztreonam	
Non-β-lactams	
Vancomycin	Bactericidal against Gram-positive bacteria
Bacitracin	Bactericidal against Gram-positive bacteria
Protein Synthesis	
Aminoglycosides	Bactericidal against aerobic and facultative bacteria
Gentamicin, tobramycin	
Tetracyclines	Bacteriostatic against some Gram-positive and Gram-negative bacteria
Tetracycline, doxycycline	
Chloramphenicol	Bacteriostatic and broad-spectrum
Macrolides	Bacteriostatic against many Gram-positive bacteria as well as some mycobacteria
Erythromycin, clarithromycin, azithromycin	
Lincosamides	Bacteriostatic against a variety of Gram-positive and Gram-negative bacteria, including *Bacteroides fragilis*
Clindamycin	
Oxazolidinones	Bacteriostatic against a variety of Gram-positive bacteria
Linezolid	
Streptogramins	A synergistic combination of two drugs that bind to two different ribosomal sites. Individually, each drug is bacteriostatic, but together they are bactericidal. Effective against a variety of Gram-positive bacteria
Quinupristin, dalfopristin	

(Continued)

TABLE 23–1 Characteristics of Antibacterial Drugs (*Continued*)

TARGET/REPRESENTATIVE DRUGS	CHARACTERISTICS
Nucleic Acid Synthesis	
Fluoroquinolones	Bactericidal against a wide variety of Gram-positive and Gram-negative bacteria
Ciprofloxacin, ofloxacin	
Rifamycins	Bactericidal against Gram-positive and some Gram-negative bacteria. Often used to treat infections caused by *Mycobacterium tuberculosis* and as prophylaxis for close exposure to *Neisseria meningitidis* or *Haemophilus influenza* type b
Rifampin	
Folate Biosynthesis	
Sulfonamides	Bacteriostatic against a variety of Gram-positive and Gram-negative bacteria
Trimethoprim	Often used in combination with a sulfa drug for a synergistic effect
Cell Membrane Integrity	
Polymyxin B	Bactericidal against Gram-negative cells by damaging cell membranes
Daptomycin	Bactericidal against Gram-positive bacteria

SELECTED ANTIBACTERIAL AGENTS

Various aspects of the major antimicrobics are now considered in more detail, with emphasis on their modes of action and spectrum. Details on specific antimicrobial agent use, dosage, and toxicity should be sought in one of the specialized texts or handbooks written for that purpose.

Antimicrobials That Act on Cell Wall Synthesis

Cross-linking of peptidoglycan is the target of β-lactams and glycopeptides

The peptidoglycan component of the bacterial cell wall gives it its shape and rigidity. This giant molecule is formed by weaving the linear glycans *N*-acetylglucosamine and *N*-acetylmuramic acid into a basket-like structure. Mature peptidoglycan is held together by cross-linking of short peptide side chains hanging off the long glycan molecules. This cross-linking process is the target of two of the most important groups of antimicrobics, the β-lactams and the glycopeptides (vancomycin and teicoplanin) (**Figure 23–1**). Peptidoglycan is unique to bacteria so inhibition of its synthesis carries no toxic potential.

β-Lactam Antimicrobials

A β-lactam ring is part of the structure of all β-lactam antimicrobics

The β-lactam antimicrobial agents include the penicillins, cephalosporins, carbapenems, and monobactams. Their name derives from the presence of a β-lactam ring in their structure; this ring is essential for antibacterial activity. Penicillin, the first member of this class, was derived from molds of the genus *Penicillium*, and later natural β-lactams were derived from both molds and bacteria of the genus *Streptomyces*. Today it is possible to synthesize β-lactams, but most are derived from semisynthetic processes involving the chemical modification of the products of fermentation.

Interfere with peptidoglycan cross-linking by binding to transpeptidases called PBPs

The β-lactam antibacterial agents interfere with the transpeptidation reactions that seal the peptide cross-links between glycan chains. They do so by interference with the action of the transpeptidase enzymes that carry out this cross-linking. These targets of all the β-lactams are commonly called penicillin-binding proteins (PBPs), reflecting the stereochemical nature of their interference, which was first described in experiments with penicillin. Several distinct PBPs occur in any one strain, are usually species-specific, and vary in the avidity of their binding to different β-lactam antimicrobics. ::: peptidoglycan, p. 352, 364

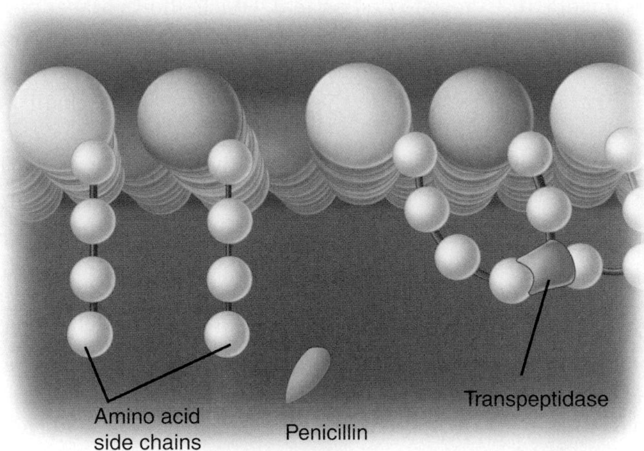

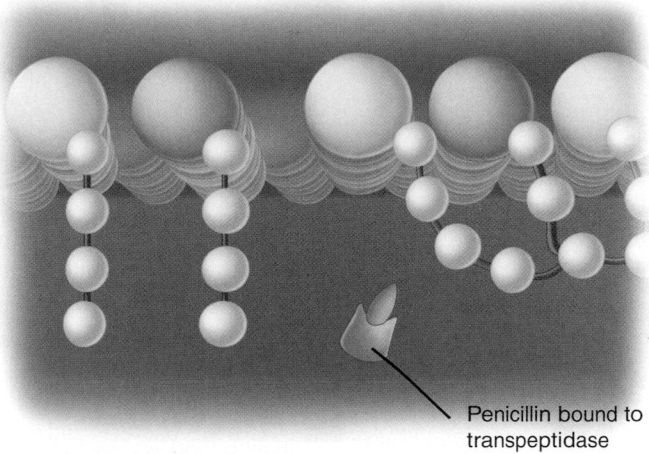

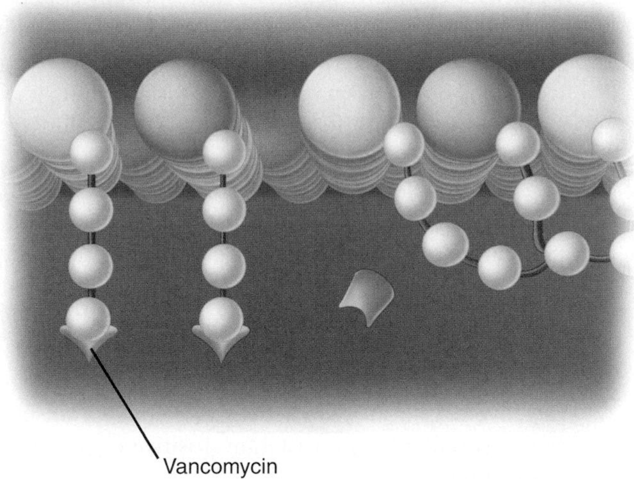

FIGURE 23–1. Action of antimicrobials on peptidoglycan synthesis. The glucan backbone and the amino acid side chains of peptidoglycan are shown. The transpeptidase enzyme catalyzes the cross-linking of the amino acid side chains. Penicillin and other β-lactams bind to the transpeptidase, preventing it from carrying out its function. Vancomycin binds directly to the amino acids, preventing the binding of transpeptidase.

The β-lactams are classified by chemical structure (**Figure 23–2**). They may have one β-lactam ring (monobactams), or a β-lactam ring fused to a five-member penem ring (penicillins, carbapenems), or a six-member cephem ring (cephalosporins). Within these major groups, differences in the side chain(s) attached to the single or double ring can have a significant effect on the pharmacologic properties and spectrum of any β-lactam.

Penicillins, cephalosporins, monobactams, and carbapenems differ in terms of the structures fused to the β-lactam ring

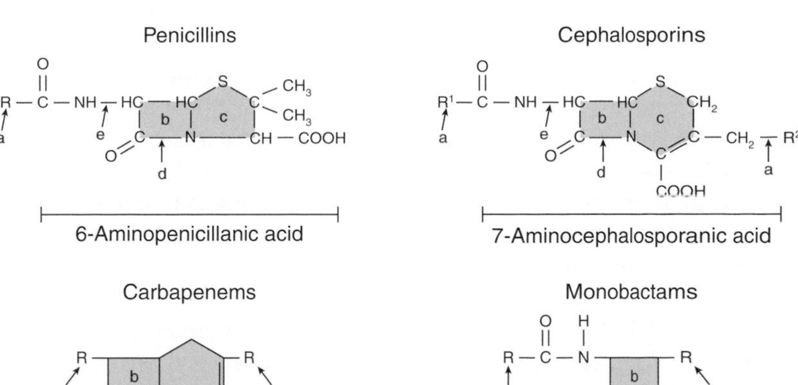

FIGURE 23–2. Structure of β-lactam antibiotics. a. Different side chains determine degree of activity, spectrum, pharmacologic properties, resistance to β-lactamases. **b.** β-lactam ring. **c.** Thiazolidine ring; c′ dihydrothiazine ring. **d.** Site of action of β-lactamases. **e.** Site of action of amidase.

The pharmacologic properties include resistance to gastric acid, which allows oral administration, and their pattern of distribution into body compartments (eg, blood, cerebrospinal fluid [CSF], joints). The features that alter the spectrum include permeability into the bacterial cell, affinity for PBPs, and vulnerability to the various bacterial mechanisms of resistance.

β-Lactam antimicrobics kill growing bacteria by lysing weakened cell walls

β-Lactam antimicrobials are usually highly bactericidal, but only to growing bacteria synthesizing new cell walls. Killing involves attenuation and disruption of the developing peptidoglycan "corset," liberation or activation of autolytic enzymes that further disrupt weakened areas of the wall, and finally osmotic lysis from passage of water through the cytoplasmic membrane to the hypertonic interior of the cell. As might be anticipated, cell wall–deficient organisms, such as *Mycoplasma*, are not susceptible to β-lactam antimicrobics.

Resistance to staphylococcal and Gram-negative β-lactamases determines spectrum

Penetration of outer membrane is often limited

Penicillins. Penicillins differ primarily in their spectrum of activity against Gram-negative bacteria and resistance to staphylococcal penicillinase. This penicillinase is one of a family of bacterial enzymes called β-lactamases, which inactivate β-lactam antimicrobials. **Penicillin G** is active primarily against Gram-positive organisms, Gram-negative cocci, and some spirochetes, including the spirochete of syphilis. They have little action against most Gram-negative bacilli because the outer membrane prevents passage of these antibiotics to their sites of action on cell wall synthesis. Penicillin G is the least toxic and least expensive of all the penicillins. Its modification as penicillin V confers acid stability, so it can be given orally.

Broad-spectrum penicillins penetrate the outer membrane of some Gram-negative bacteria

Some penicillins are inactivated by staphylococcal penicillinase

The penicillinase-resistant penicillins **(methicillin, nafcillin, oxacillin)** also have narrow spectra, but are active against penicillinase-producing *S aureus*. The broader-spectrum penicillins owe their expanded activity to the ability to traverse the outer membrane of some Gram-negative bacteria. Some, such as ampicillin, have excellent activity against a range of Gram-negative pathogens, but not *Pseudomonas aeruginosa*, an important opportunistic pathogen. Others, such as **piperacillin** and **ticarcillin,** are active against *Pseudomonas* when given in high dosage, but are less active than ampicillin against some other Gram-negative organisms. The penicillins with a Gram-negative spectrum are slightly less active than penicillin G against Gram-positive organisms and are inactivated by staphylococcal penicillinase.

Cephalosporins are penicillinase-resistant

Shifting between first- and third-generation cephalosporins gives a wider Gram-negative spectrum

Second- and third-generation cephalosporins have less activity against Gram-positive bacteria

Cephalosporins. The structure of the cephalosporins confers resistance to hydrolysis by staphylococcal penicillinase and to the β-lactamases of groups of Gram-negative bacilli, which vary with each cephalosporin. The cephalosporins are classified by generation—first, second, third, or fourth. The term "generation" relates to historical breakthroughs in expanding their spectrum through modification of the side chains. In general, a cephalosporin of a higher generation has a wider spectrum—in some instances, more quantitative activity (lower minimum inhibitory concentration; MIC) against Gram-negative bacteria. As the Gram-negative spectrum increases, these agents typically lose some of their potency (higher MIC) against Gram-positive bacteria.

First-generation cephalosporins inhibit Gram-positive bacteria and a few Enterobacteriaceae

The first-generation cephalosporins **cefazolin** and **cephalexin** have a spectrum of activity against Gram-positive organisms that resembles that of the penicillinase-resistant penicillins; in addition, they are active against some of the Enterobacteriaceae (see Table 23–1). These agents continue to have therapeutic value because of their high activity against Gram-positive organisms and because a broader spectrum may be unnecessary.

Second-generation cephalosporins are also active against anaerobes

Second-generation cephalosporins **cefoxitin** and **cefaclor** are resistant to β-lactamases of some Gram-negative organisms that inactivate first-generation compounds. Of particular importance is their expanded activity against Enterobacteriaceae species and against anaerobes such as *Bacteroides fragilis*.

Third-generation cephalosporins have increasing potency against Gram-negative organisms

Ceftriaxone and cefotaxime are preferred for meningitis

Ceftazidime is used for Pseudomonas

Third-generation cephalosporins, such as **ceftriaxone, cefotaxime,** and **ceftazidime,** have an even wider spectrum; they are active against Gram-negative organisms, often at MICs that are 10- to 100-fold lower than first-generation compounds. Of these three agents, only ceftazidime is consistently active against *P aeruginosa*. The potency, broad spectrum, and low toxicity of the third-generation cephalosporins have made them the preferred agents in life-threatening infections in which the causative organism has not yet been isolated. Selection depends on the clinical circumstances. For example, ceftriaxone or cefotaxime is preferred for childhood meningitis because it has the highest activity against the three major causes, *Neisseria meningitidis, Streptococcus pneumoniae*, and *Haemophilus influenzae*. For a febrile bone marrow transplant patient, ceftazidime might be chosen because of the prospect of *P aeruginosa* involvement.

Fourth-generation cephalosporins have enhanced ability to cross the outer membrane of Gram-negative bacteria as well as resistance to many Gram-negative β-lactamases. Compounds such as **cefepime** have activity against an even wider spectrum of Enterobacteriaceae as well as *P aeruginosa*. These cephalosporins retain the high affinity of third-generation drugs and activity against *Neisseria* and *H influenzae*.

Fourth-generation cephalosporins have enhanced ability to penetrate outer membrane

Carbapenems. The carbapenems **imipenem** and **meropenem** have the broadest spectrum of all β-lactam antibiotics. This fact appears to be due to the combination of easy penetration of Gram-negative and Gram-positive bacterial cells and high level of resistance to β-lactamases. Both agents are active against streptococci, more active than cephalosporins against staphylococci, and highly active against both β-lactamase-positive and -negative strains of gonococci and *H influenzae*. In addition, they are as active as third-generation cephalosporins against Gram-negative rods, and effective against obligate anaerobes. Imipenem is rapidly hydrolyzed by a renal tubular dehydropeptidase-1; therefore, it is administered together with an inhibitor of this enzyme (cilastatin), which greatly improves its urine levels and other pharmacokinetic characteristics. Meropenem is not significantly degraded by dehydropeptidase-1 and does not require coadministration of cilastatin; it has largely replaced imipenem in clinical use.

Carbapenems are very broad spectrum

Monobactams. Aztreonam, the first monobactam licensed in the United States, has a spectrum limited to aerobic and facultatively anaerobic Gram-negative bacteria, including Enterobacteriaceae, P *aeruginosa, Haemophilus,* and *Neisseria*. Monobactams have poor affinity for the PBPs of Gram-positive organisms and anaerobes and thus little activity against them, but they are highly resistant to hydrolysis by β-lactamases of Gram-negative bacilli. Anaerobic superinfections and major distortions of the bowel flora are less common with aztreonam therapy than with other broad-spectrum β-lactam antimicrobials, presumably because aztreonam does not produce a general suppression of gut anaerobes.

Activity is primarily against Gram-negatives

β-Lactamase Inhibitors. A number of β-lactams with little or no antimicrobial activity are capable of binding irreversibly to β-lactamase enzymes and, in the process, rendering them inactive. Three such compounds, **clavulanic acid, sulbactam,** and **tazobactam,** are referred to as suicide inhibitors, because they must first be hydrolyzed by a β-lactamase before becoming effective inactivators of the enzyme. They are highly effective against staphylococcal penicillinases and broad-spectrum β-lactamases; however, their ability to inhibit cephalosporinases is significantly less. Combinations of one of these inhibitors with an appropriate β-lactam antimicrobial agent protects the therapeutic agent from destruction by many β-lactamases and significantly enhances its spectrum. Four such combinations are now available in the United States: amoxicillin/clavulanate, ticarcillin/clavulanate, ampicillin/sulbactam, and piperacillin/tazobactam. Bacteria that produce chromosomally encoded inducible cephalosporinases are not susceptible to these combinations. Whether these combinations offer therapeutic or economic advantages compared with the β-lactamase–stable antibiotics now available remains to be determined.

β-lactamase inhibitors are β-lactams that bind β-lactamases

Other β-lactams are enhanced in the presence of β-lactamase inhibitors

Clinical Use. The β-lactam antibiotics are usually the drugs of choice for infections by susceptible organisms because of their low toxicity and bactericidal action. They have also proved to be of great value in the prophylaxis of many infections. They are excreted by the kidney and achieve very high urinary levels. Penicillins reach the CSF when the meninges are inflamed and are effective in the treatment of meningitis, but first- and second-generation cephalosporins are not. In contrast, the third-generation cephalosporins penetrate much better and have become the agents of choice in the treatment of undiagnosed meningitis and meningitis caused by most Gram-negative organisms.

Low toxicity favors use of all β-lactams

Glycopeptide Antimicrobials

Two agents, **vancomycin** and **teicoplanin,** belong to this group. Each of these antimicrobials inhibits assembly of the linear peptidoglycan molecule by binding directly to the terminal amino acids of the peptide side chains. The effect is the same as with β-lactams,

Glycopeptide antimicrobics bind directly to amino acid side chains

prevention of peptidoglycan cross-linking. Both agents are bactericidal, but are primarily active only against Gram-positive bacteria. Their main use has been against multiresistant Gram-positive infections, including those caused by strains of staphylococci that are resistant to the penicillinase-resistant penicillins and cephalosporins. Neither agent is absorbed by mouth, although both have been used orally to treat *Clostridium difficile* infections of the bowel. ::: C difficile colitis, p. 553

Inhibitors of Protein Synthesis (Figure 23–3)

Aminoglycosides

Aminoglycosides must be transported into cell by oxidative metabolism

Not active against anaerobes

All members of the aminoglycoside group of antibacterial agents have a six-member aminocyclitol ring with attached amino sugars. The individual agents differ in terms of the exact ring structure and the number and nature of the amino sugar residues. Aminoglycosides are active against a wide range of bacteria, but only against organisms that are able to transport them into the cell by a mechanism that involves oxidative phosphorylation. Thus, they have little or no activity against strict anaerobes or facultative organisms that metabolize only fermentatively (eg, streptococci). It appears highly probable that aminoglycoside activity against facultative organisms is similarly reduced in vivo when the oxidation–reduction potential is low.

Ribosome binding disrupts initiation complexes

Newer agents bind to multiple sites

Once inside bacterial cells, aminoglycosides inhibit protein synthesis by binding to the bacterial ribosomes either directly or by involving other proteins. This binding destabilizes the ribosomes, blocks initiation complexes, and thus prevents elongation of polypeptide chains. The agents may also cause distortion of the site of attachment of mRNA, mistranslation of codons, and failure to produce the correct amino acid sequence in proteins. The first aminoglycoside, streptomycin, is bound to the 30S ribosomal subunit, but the newer and more active aminoglycosides bind to multiple sites on both 30S and 50S subunits. This gives the newer agents broader spectrum and less susceptibility to resistance owing to binding site mutation.

No entry into human cells

Eukaryotic ribosomes are resistant to aminoglycosides, and the antimicrobics are not actively transported into eukaryotic cells. These properties account for their selective toxicity and also explain their ineffectiveness against intracellular bacteria such as *Rickettsia* and *Chlamydia*.

Spectrum includes P aeruginosa

Gentamicin and **tobramycin** are the major aminoglycosides; they have an extended spectrum, which includes staphylococci; Enterobacteriaceae; and of particular importance, *P aeruginosa*. **Streptomycin** and **amikacin** are now primarily used in combination with other antimicrobial agents in the therapy of tuberculosis and other mycobacterial diseases.

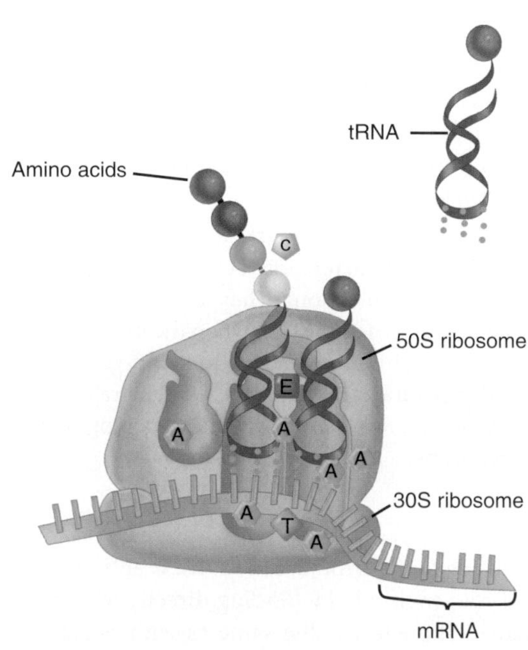

FIGURE 23–3. Action of antimicrobials on protein synthesis. Aminoglycosides (A) bind to multiple sites on both the 30S and 50S ribosomes in a manner that prevents the tRNA from forming initiation complexes. Tetracyclines (T) act in a similar manner, binding only to the 30S ribosomes. Chloramphenicol (C) blocks formation of the peptide bond between the amino acids. Erythromycin (E) and macrolides block the translocation of tRNA from the acceptor to the donor side on the ribosome.

Neomycin, the most toxic aminoglycoside, is used in topical preparations and as an oral preparation before certain types of intestinal surgery, because it is poorly absorbed.

All of the aminoglycosides are toxic to the vestibular and auditory branches of the eighth cranial nerve to varying degrees; this damage can lead to complete and irreversible loss of hearing and balance. These agents may also be toxic to the kidneys. It is often essential to monitor blood levels during therapy to ensure adequate yet nontoxic doses, especially when renal impairment diminishes excretion of the drug. For example, blood levels of gentamicin should be lower than 10 μg/mL to prevent nephrotoxicity, but many strains of *P aeruginosa* require 2 to 4 μg/mL for inhibition.

Renal and vestibular toxicity must be monitored

The clinical value of the aminoglycosides is a consequence of their rapid bactericidal effect, their broad spectrum, the slow development of resistance to the agents now most often used, and their action against *Pseudomonas* strains that resist many other antimicrobics. They cause fewer disturbances of the normal flora than most other broad-spectrum antimicrobials, probably because of their lack of activity against the predominantly anaerobic flora of the bowel and because they are only used parenterally for systemic infections. The β-lactam antibiotics often act synergistically with the aminoglycosides, most likely because their action on the cell wall facilitates aminoglycoside penetration into the bacterial cell. This effect is most pronounced with organisms such as streptococci and enterococci, which lack the metabolic pathways required to transport aminoglycosides to their interior.

Broad spectrum and slow development of resistance enhance use

Often combined with β-lactam antimicrobics

Tetracyclines

Tetracyclines are composed of four fused benzene rings. Substitutions on these rings provide differences in pharmacologic features of the major members of the group, **tetracycline** and **doxycycline.** The tetracyclines inhibit protein synthesis by binding to the 30S ribosomal subunit at a point that blocks attachment of aminoacyl-tRNA to the acceptor site on the mRNA ribosome complex. Unlike the aminoglycosides, their effect is reversible; they are bacteriostatic rather than bactericidal. A more recently developed drug in this group is **tigecycline**, which may be an especially good choice for treating polymicrobial intra-abdominal infections and other complicated deep tissue infections

Tetracyclines block tRNA attachment

Activity is bacteriostatic

The tetracyclines are broad-spectrum agents with a range of activity that encompasses most common pathogenic species, including Gram-positive and Gram-negative rods and cocci and both aerobes and anaerobes. They are active against cell wall–deficient organisms, such as *Mycoplasma* and spheroplasts, and against some obligate intracellular bacteria, including members of the genera *Rickettsia* and *Chlamydia*. Differences in spectrum of activity between members of the group are relatively minor. Acquired resistance to one generally confers resistance to all; however, tigecycline appears to overcome the two major resistance mechanisms to other tetracyclines, and thus may become an even more useful alternative.

Broad spectrum includes some intracellular bacteria

The tetracyclines are absorbed orally. In practice, they are divided into agents that generate blood levels for only a few hours and those that are longer-acting (minocycline and doxycycline), which can be administered less often. The tetracyclines are chelated by divalent cations, and their absorption and activity are reduced. Thus, they should not be taken with dairy products or many antacid preparations. Tetracyclines are excreted in the bile and urine in active form.

Orally absorbed but chelated by some foods

Tetracycline has a strong affinity for developing bone and teeth, to which it and some of the earlier derivatives give a yellowish color and enamel damage, and they are avoided in children up to 8 years of age. This does not appear to be a significant problem with doxycycline, however. Common complications of tetracycline therapy are gastrointestinal disturbance due to alteration of the normal flora, predisposing to superinfection with tetracycline-resistant organisms and vaginal or oral candidiasis (thrush) due to the opportunistic yeast *Candida albicans*.

Dental staining and enamel damage to permanent teeth limits use of older generation tetracyclines in children

Chloramphenicol

Chloramphenicol has a simple nitrobenzene ring structure that can now be mass produced by chemical synthesis. It influences protein synthesis by binding to the 50S ribosomal subunit and blocking the action of peptidyl transferase, which prevents formation of the peptide bond essential for extension of the peptide chain. Its action is reversible in most susceptible species; thus, it is bacteriostatic. It has little effect on eukaryotic ribosomes, which explains its selective toxicity.

Chloramphenicol blocks peptidyl transferase

Diffusion into body fluid compartments occurs readily

A broad-spectrum antibiotic, chloramphenicol, like tetracycline, has a wide range of activity against both aerobic and anaerobic species. Chloramphenicol is readily adsorbed from the upper gastrointestinal tract and diffuses readily into most body compartments, including the CSF. It also permeates readily into mammalian cells and is active against obligate intracellular pathogens such as *Rickettsia* and *Chlamydia.*

Marrow suppression and aplastic anemia are serious toxicities

The major drawback to this inexpensive, broad-spectrum antimicrobial with almost ideal pharmacologic features is a rare but serious toxicity. Between 1 in 50,000 and 1 in 200,000 patients treated with even low doses of chloramphenicol have an idiosyncratic reaction that results in aplastic anemia. The condition is irreversible and, before the advent of bone marrow transplantation, it was universally fatal. In high doses, chloramphenicol also causes a reversible depression of the bone marrow and in neonates, abdominal, circulatory, and respiratory dysfunction. The inability of the immature infant liver to conjugate and excrete chloramphenicol aggravates this latter condition.

Use is sharply restricted

Chloramphenicol use is now restricted to treatment of rickettsial or ehrlichial infections in which tetracyclines cannot be used because of hypersensitivity or pregnancy. Its central nervous system (CNS) penetration and activity against anaerobes continue to lend support to its use in brain abscess. In some developing countries, chloramphenicol use is more extensive because of its low cost and proven efficacy in diseases such as typhoid fever and bacterial meningitis.

Macrolides

Ribosomal binding blocks translocation

The macrolides, **erythromycin, azithromycin,** and **clarithromycin,** differ in the exact composition of a large 14- or 15-member ring structure. They affect protein synthesis at the ribosomal level by binding to the 50S subunit and blocking the translocation reaction. Their effect is primarily bacteriostatic. Macrolides, which are concentrated in phagocytes and other cells, are effective against some intracellular pathogens.

Erythromycin is active against Gram-positives and Legionella

Erythromycin, the first and still the most commonly used macrolide, has a spectrum of activity that includes most of the pathogenic Gram-positive bacteria and some Gram-negative organisms. The Gram-negative spectrum includes *Neisseria, Bordetella, Campylobacter*, and *Legionella*, but not the Enterobacteriaceae. Erythromycin and related drugs are also effective against *Chlamydia* and *Mycoplasma.*

Azithromycin and clarithromycin have enhanced Gram-negative spectrum

Bacteria that have developed resistance to erythromycin are usually resistant to the newer macrolides azithromycin and clarithromycin as well. These newer agents have the same spectrum as erythromycin, with some significant additions. Azithromycin has quantitatively greater activity (lower MICs) against most of the same Gram-negative bacteria. Clarithromycin is the most active of the three against both Gram-positive and Gram-negative pathogens. Clarithromycin is also active against mycobacteria. In addition, both azithromycin and clarithromycin have demonstrated efficacy against *Borrelia burgdorferi*, the causal agent of Lyme disease and the protozoan parasite *Toxoplasma gondii*, which causes toxoplasmosis.

Clindamycin

Spectrum is similar to macrolides with addition of anaerobes

May mitigate toxin production

Clindamycin is chemically unrelated to the macrolides, but has a similar mode of action and spectrum. It has greater activity than the macrolides against Gram-negative anaerobes, including the important *Bacteroides fragilis* group. Although clindamycin is a perfectly adequate substitute for a macrolide in many situations, its primary use is in instances in which anaerobes are or may be involved. In addition, there is experimental evidence that clindamycin may mitigate toxin production by highly virulent *Staphyloccus aureus* and *Streptococcus pyogenes* strains. For this reason, many clinicians add it to a bactericidal agent such as methicillin or vancomycin for treatment of severe deep tissue infections caused by these organisms.

Oxazolidinones

Activity against Gram-positive bacteria resistant to other agents

Linezolid is the most widely used of a new class of antibiotics that act by binding to the bacterial 50S ribosome. The exact mechanism is not known, but it does not involve peptide elongation or termination of translation. Oxazolidinones are clinically useful in pneumonia and other soft tissue infections, particularly those caused by resistant strains of staphylococci, pneumococci, and enterococci.

Streptogramins

Quinupristin and dalfopristin are used in a fixed combination known as **quinupristin-dalfopristin** in a synergistic ratio. They inhibit protein synthesis by binding to different sites on the 50S bacterial ribosome; quinupristin inhibits peptide chain elongation, and dalfopristin interferes with peptidyl transferase. Their clinical use thus far has been limited to treatment of vancomycin-resistant enterococci.

Useful against vancomycin-resistant enterococci

Inhibitors of Nucleic Acid Synthesis (Figure 23–4)

Quinolones

The quinolones have a nucleus of two fused six-member rings that when substituted with fluorine become fluoroquinolones, which are now the dominant quinolones for treatment of bacterial infections. Among the fluoroquinolones, **ciprofloxacin, norfloxacin,** and **ofloxacin,** the addition of a piperazine ring and its methylation alter the activity and pharmacologic properties of the individual compound. The primary target of all quinolones is DNA topoisomerase (gyrase), the enzyme responsible for nicking, supercoiling, and sealing bacterial DNA during replication. Bacterial topoisomerases have four subunits, one or more of which are inhibited by the particular quinolone. The enhanced activity and lower frequency of resistance seen with the newer fluoroquinolones are attributed to binding at multiple sites on the enzyme. This greatly reduces the chance a single mutation can lead to resistance, which was a problem with the first quinolone, nalidixic acid, a single-binding site agent.

Fluorinated derivatives are now dominant

Inhibition of topoisomerase blocks supercoiling

The fluoroquinolones are highly active and bactericidal against a wide range of aerobes and facultative anaerobes. However, streptococci and *Mycoplasma* are only marginally susceptible, and anaerobes are generally resistant. Ofloxacin has significant activity against *Chlamydia*, whereas ciprofloxacin is particularly useful against *P aeruginosa*.

Fluoroquinolones have a broad spectrum, including Pseudomonas

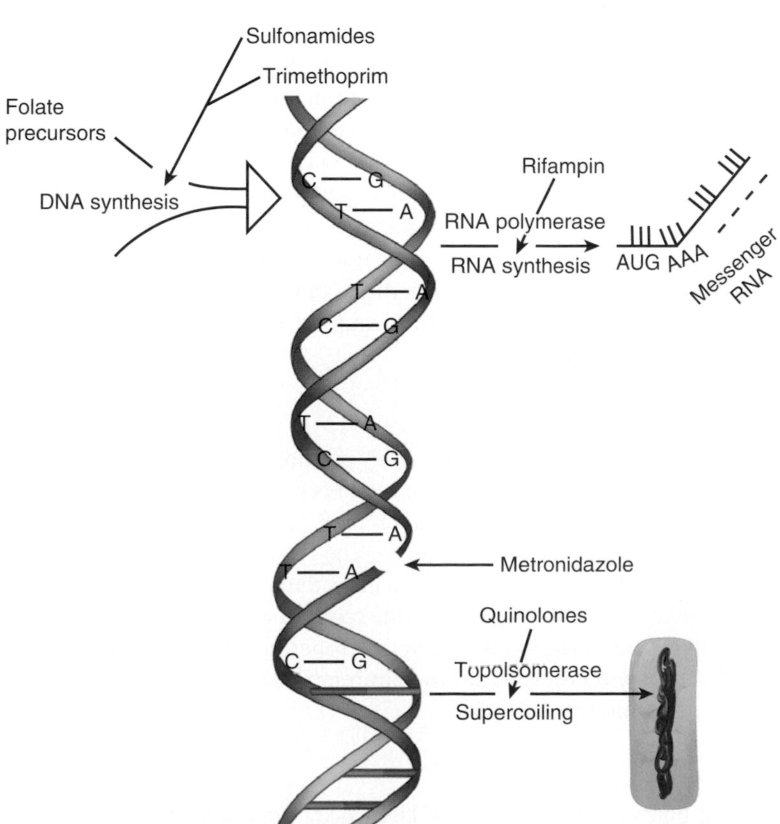

FIGURE 23–4. Antimicrobials acting on nucleic acids. Sulfonamides block the folate precursors of DNA synthesis, metronidazole inflicts breaks in the DNA itself, rifampin inhibits the synthesis of RNA from DNA by inhibiting RNA polymerase, and quinolones inhibit DNA topoisomerase and thus prevent the supercoiling required for the DNA to "fit" inside the bacterial cell.

Fluoroquinolones has several favorable pharmacologic properties in addition to their broad spectrum. These include oral administration, low protein binding, good distribution to all body compartments, penetration of phagocytes, and a prolonged serum half-life that allows once- or twice-daily dosing. Norfloxacin and ciprofloxacin are excreted by hepatic and renal routes, resulting in high drug concentrations in the bile and urine. Ofloxacin is excreted primarily by the kidney.

Well distributed after oral administration

Folate Inhibitors

Agents that interfere with synthesis of folic acid by bacteria have selective toxicity because mammalian cells are unable to accomplish this feat and use preformed folate from dietary sources. Folic acid is derived from *para*-aminobenzoic acid (PABA), glutamate, and a pteridine unit. In its reduced form, it is an essential coenzyme for the transport of one-carbon compounds in the synthesis of purines, thymidine, some amino acids, and thus indirectly of nucleic acids and proteins. The major inhibitors of the folate pathway are the sulfonamides, trimethoprim, *para*-aminosalicylic acid, and the sulfones.

Bacteria must synthesize folate that humans acquire in their diet

Sulfonamides. Sulfonamides are structural analogs of PABA and compete with it for the enzyme (dihydropteroate synthetase), which combines PABA and pteridine in the initial stage of folate synthesis. This blockage has multiple effects on the bacterial cells; the most important of these is disruption of nucleic acid synthesis. The effect is bacteriostatic, and the addition of PABA to a medium that contains sulfonamide neutralizes the inhibitory effect and allows growth to resume.

Competition with PABA disrupts nucleic acids

When introduced in the 1940s, sulfonamides had a very broad spectrum (staphylococci, streptococci, many Gram-negative bacteria), but resistance developed quickly, and this has restricted their use for systemic infections. Now their primary use is for uncomplicated urinary tract infections caused by members of the Enterobacteriaceae family, particularly *Escherichia coli*. Sulfonamides are convenient for this purpose because they are inexpensive, well absorbed by the oral route, and excreted in high levels in the urine.

Major use is urinary tract infections

Trimethoprim-Sulfamethoxazole. Trimethoprim acts on the folate synthesis pathway but at a point after sulfonamides. It competitively inhibits the activity of bacterial dihydrofolate reductase, which catalyzes the conversion of folate to its reduced active coenzyme form. When combined with sulfamethoxazole, a sulfonamide, trimethoprim leads to a two-stage blockade of the folate pathway, which often results in synergistic bacteriostatic or bactericidal effects. This quality is exploited in therapeutic preparations that combine both agents in a fixed proportion designed to yield optimum synergy.

Dihydrofolate reductase inhibition is synergistic with sulfonamides

Trimethoprim-sulfamethoxazole (TMP-SMX) has a much broader and stable spectrum than either of its components alone; this includes most of the common pathogens, whether they are Gram-positive or Gram-negative, cocci or bacilli. Anaerobes and *P. aeruginosa* are exceptions. It is also active against some uncommon agents such as *Nocardia*. TMP-SMX is widely and effectively used in the treatment of urinary tract infections, otitis media, sinusitis, prostatitis, and infectious diarrhea, and it the agent of choice for pneumonia caused by *Pneumocystis carinii*, a fungus.

Activity against common bacteria and some fungi

Metronidazole

Metronidazole is a nitroimidazole, a family of compounds with activity against bacteria, fungi, and parasites. The antibacterial action requires reduction of the nitro group under anaerobic conditions, which explains the limitation of its activity to bacteria that prefer anaerobic or at least microaerophilic growth conditions. The reduction products act on the cell at multiple points; the most lethal of these effects is induction of breaks in DNA strands.

Metronidazole is active against a wide range of anaerobes, including *B fragilis*. Clinically, it is useful for any infection in which anaerobes may be involved. Because these infections are typically polymicrobial, a second antimicrobial (eg, β-lactam) is usually added to cover aerobic and facultative bacteria.

Action requires anaerobic conditions

Rifampin

Rifampin binds to the β-subunit of DNA-dependent RNA polymerase, which prevents the initiation of RNA synthesis. This agent is active against most Gram-positive bacteria and

Blocking of RNA synthesis occurs by binding to polymerase

selected Gram-negative organisms, including *Neisseria* and *Haemophilus* but not members of Enterobacteriaceae. The most clinically useful property of rifampin is its antimycobacterial activity, which includes *Mycobacterium tuberculosis* and the other species that most commonly infect humans. Because resistance by mutation of the polymerase readily occurs, rifampin is combined with other agents in the treatment of active infections. It is used alone for chemoprophylaxis of *N meningitidis* and *H influenzae* in close contacts of infected patients.

Antimicrobials Acting on the Outer and Cytoplasmic Membranes

The polypeptide antimicrobial agents **polymyxin B** and **colistin** have a cationic detergent-like effect. They bind to the cell membranes of susceptible Gram-negative bacteria and alter their permeability, resulting in loss of essential cytoplasmic components and bacterial death. These agents react to a lesser extent with cell membranes of the host, resulting in nephrotoxicity and neurotoxicity. Their spectrum is essentially Gram-negative; they act against *P aeruginosa* and other Gram-negative rods. Although these antimicrobials were used for systemic treatment in the past, their use is now limited to topical applications. They have an advantage; resistance to them rarely develops.

Binding to cytoplasmic membrane occurs

Other Agents

Several other effective antimicrobials are in use almost exclusively for a single infectious agent or types of infections such as tuberculosis, urinary tract infections, and anaerobic infections. Where appropriate, these agents are discussed in the relevant chapter. It is beyond the scope and intent of this book to provide comprehensive coverage of all available agents.

ANTIMICROBIAL RESISTANCE

The continuing success of antimicrobial therapy depends on keeping ahead of the ability of the microorganisms to develop resistance to antimicrobial agents. At times, resistance seems to occur at a rate equal to that of the development of new antimicrobials. The mechanisms of resistance and the ways in which laboratory tests are used to guide clinicians through the uncertainties of modern treatment are the subject of this section.

SUSCEPTIBILITY AND RESISTANCE

Deciding whether any bacterium should be considered susceptible or resistant to any antimicrobial involves an integrated assessment of in vitro activity, pharmacologic characteristics, and clinical evaluation. Any agent approved for clinical use has demonstrated in vitro its potential to inhibit the growth of some target group of bacteria at concentrations that can be achieved with acceptable risks of toxicity. That is, the MIC can be comfortably exceeded by doses tolerated by the patient. Use of the antimicrobial in animal models and then human infections must have also demonstrated a therapeutic response. Because the influence of antimicrobials on the natural history of different categories of infection (eg, pneumonia, meningitis, diarrhea) varies, extensive clinical trials must include both a range of bacterial species and infected sites (eg, lung, bone, CSF). The clinical studies are important to determine whether what *should* work actually *does* work and, if so, to define the parameters of success and failure.

MICs must be below achievable blood levels

Clinical experience must validate in vitro data

Once these factors are established, the routine selection of therapy can be based on known or expected characteristics of organisms and pharmacologic features of antimicrobial agents. With regard to organisms, use of the term **susceptible** (sensitive) implies that their MIC is at a concentration attainable in the blood or other appropriate body fluid (eg, urine) using the usually recommended doses. **Resistant,** the converse of susceptible, implies that the MIC is not exceeded by normally attainable levels. As in all biological systems, the MIC of some organisms lies in between the susceptible and resistant levels. Borderline strains are called **intermediate, moderately sensitive,** or **moderately resistant,** depending on the

Susceptible bacteria are inhibited at achievable nontoxic levels; resistant strains are not

exact values and conventions of the reporting system. The antimicrobial in question may still be used to treat these organisms but at increased doses. For example, nontoxic antibiotics such as the penicillins and cephalosporins can be administered in massive doses and may thereby inhibit some pathogens that would normally be considered resistant in vitro. Furthermore, in urinary infections, urine levels of some antimicrobial agents may be very high, and organisms that are seemingly resistant in vitro may be eliminated.

Borderline isolates are called intermediate

Important pharmacologic characteristics of antimicrobial agents include dosage as well as the routes and frequency of administration. Other characteristics include whether the agents are absorbed from the upper gastrointestinal tract, whether they are excreted and concentrated in active form in the urine, whether they can pass into cells, whether and how rapidly they are metabolized, and the duration of effective antimicrobial levels in blood and tissues. Most agents are bound to some extent to serum albumin, and the protein-bound form is usually unavailable for antimicrobial action. The amount of free to bound antibiotic can be expressed as an equilibrium constant, which varies for different antibiotics. In general, high degrees of binding lead to more prolonged but lower serum levels of an active antimicrobic after a single dose.

Pharmacologic properties such as absorption, distribution, and metabolism affect the usefulness of antimicrobials

LABORATORY CONTROL OF ANTIMICROBIAL THERAPY

A unique feature of laboratory testing in microbiology is that the susceptibility of the isolate of an individual patient can be tested against a battery of potential antimicrobial agents. These tests are built around the common theme of placing the organism in the presence of varying concentrations of the antimicrobial in order to determine the MIC. The methods used are standardized, including a measured inoculum of the bacteria and the growth conditions (eg, medium, incubation, time).

Bacteria are tested against antimicrobials over a range of concentrations

In selecting therapy, the results of laboratory tests cannot be considered by themselves, but must be examined with information about the clinical pharmacology of the agent, the cause of the disease, the site of infection, and the pathology of the lesion. For example, the agent must reach the subarachnoid space and CSF in the case of meningitis. Similarly, therapy may be ineffective for an infection that has resulted in abscess formation unless the abscess is surgically drained. Previous clinical experience is also critical. In typhoid fever, for instance, ciprofloxacin is effective and aminoglycosides are not, even though the typhoid bacillus may be equally susceptible to both in vitro. This is due to the failure of aminoglycosides to achieve adequate concentrations inside the macrophages where *Salmonella typhi* multiplies.

Selection considers susceptibility, pharmacology, and clinical experience

Penetration inside cells may be important

Dilution Tests

Dilution tests determine the MIC directly by using serial dilutions of the antimicrobial agent in broth that span a clinically significant range of concentrations. The dilutions are prepared in tubes or microdilution wells, and by convention, they are doubled using a base of 1 μg/mL (0.25, 0.5, 1, 2, 4, 8, and so on). The bacterial inoculum of the patient's isolate is adjusted to a standard (10^5 to 10^6 bacteria/mL) and added to the broth. After incubation overnight (or another defined time), the tubes are examined for turbidity produced by bacterial growth. The first tube in which visible growth is absent (clear) is the MIC for that organism (**Figure 23–5**).

MIC endpoint is the lowest concentration that inhibits growth

Diffusion Tests

In diffusion testing, the inoculum is seeded onto the surface of an agar plate, and filter paper disks containing defined amounts of antimicrobials are applied. While the plates are incubating, the antimicrobic diffuses into the medium to produce a circular gradient around the disk. After incubation overnight, the size of the zone of growth inhibition around the disk (**Figure 23–6A**) can be used as an indirect measure of the MIC of the organism. It is also influenced by the growth rate of the organism, the diffusibility of the antimicrobial agent, and some technical factors. A standardized diffusion procedure accounts for these factors and includes recommendations for interpretation. The diameters of the zones of inhibition obtained with the various antibiotics are converted to "susceptible," "moderately susceptible," and "resistant" categories by referring to a table. This method is convenient and

Antimicrobial in disks produces a circular concentration gradient

Inhibition zone is a measure of the MIC

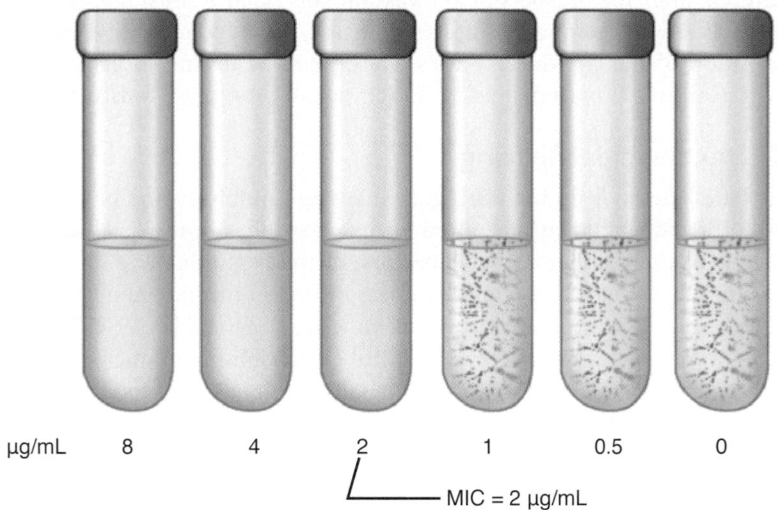

FIGURE 23–5. Broth dilution susceptibility test. The stippled tubes represent turbidity produced by bacterial growth. The MIC is 2.0 µg/mL.

flexible for rapidly growing aerobic and facultative bacteria such as the Enterobacteriaceae, *Pseudomonas,* and staphylococci. Another diffusion procedure uses gradient strips to produce elliptical zones that can be directly correlated with the MIC. This method, the E test (Figure 23–6B), can also be applied to slow-growing, fastidious, and anaerobic bacteria.

Automated Tests

Instruments are now available that carry out rapid, automated variants of the broth dilution test. In these systems, the bacteria are incubated with the antimicrobic agent in specialized modules that are read automatically every 15 to 30 minutes. The multiple readings and the increased sensitivity of determining endpoints by turbidimetric or fluorometric analysis make it possible to generate MICs in as little as 4 hours. In laboratories with sufficient volume, these methods are no more expensive than manual methods, and the rapid results have enhanced potential to influence clinical outcome, particularly when interfaced with computerized hospital information systems.

Automated methods read dilution tests in a few hours

Molecular Testing

The molecular techniques of nucleic acid hybridization, sequencing, and amplification have been applied to the detection and study of resistance. The basic strategy is to detect the resistance gene rather than measure the phenotypic expression of resistance. These

Molecular methods detect resistance genes

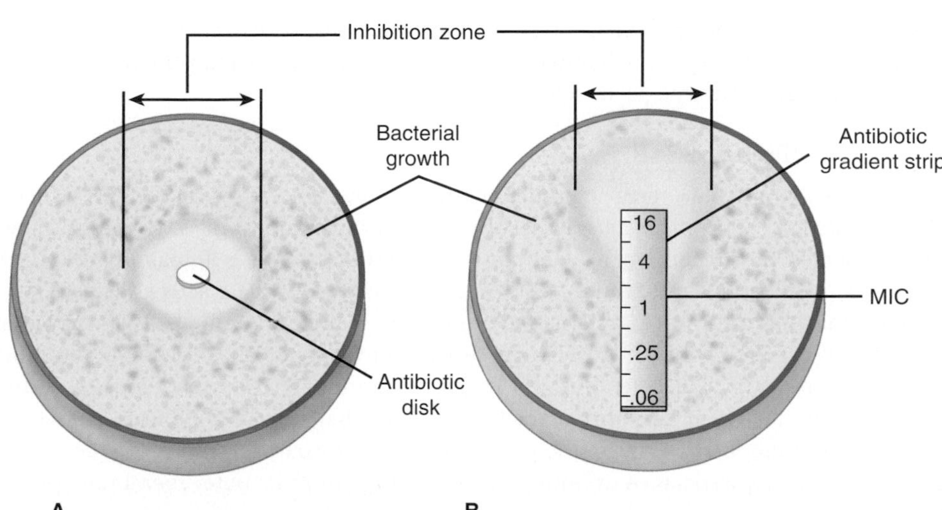

FIGURE 23–6. Diffusion tests. A. Disk diffusion. The diameter of the zone of growth inhibition around a disk of fixed antimicrobial content is inversely proportional to the minimum inhibitory concentration (MIC) for that antimicrobial, ie, the larger the zone, the lower the MIC. **B.** The E test. A strip containing a gradient of antimicrobial content creates an elliptical zone of inhibition. The conditions are empirically adjusted so that the MIC endpoint is where the growth intersects the strip.

methods offer the prospect of automation and rapid results, but as with most molecular methods, are not yet practical for routine use. Their application will also have to take consideration of the fact that the tests will be limited to known genes and that phenotypic expression is the "bottom line." ::: nucleic amplification methods, p. 80

Bactericidal Testing

Quantitation of the bactericidal effect determines the MBC

The latter methods do not distinguish between inhibitory and bactericidal activity. To do so requires quantitative subculture of the clear tubes in the broth dilution test and comparison of the number of viable bacteria at the beginning and end of the test. The least amount required to kill a predetermined portion of the inoculum (usually 99.9%) is called the **minimal bactericidal concentration (MBC).** Direct bactericidal testing is important in the initial characterization and clinical evaluation of antimicrobial agents, but is rarely needed in individual cases. Most of the antimicrobials used for acute and life-threatening infections (eg, β-lactams, aminoglycosides) act by bactericidal mechanisms.

Antimicrobial Assays

Pharmacologic monitoring is necessary in some situations

For antimicrobials with toxicity near the therapeutic range, monitoring the concentration in the serum or other appropriate body fluid is sometimes necessary. Therapeutic monitoring may also be required when the patient's pharmacologic handling of the agent is unpredictable, as in renal failure. A variety of biologic, immunoassay, and chemical procedures have been developed for this purpose.

BACTERIAL RESISTANCE TO ANTIMICROBIALS

Resistance has eroded the effectiveness of many agents

The seemingly perfect nature of antimicrobial agents, originally hailed as "wonder drugs," has been steadily eroded by the appearance of strains resistant to their action. This resistance may be inherent to the organism or may appear in a previously susceptible species by mutation or by the acquisition of new genes. The mechanisms by which bacteria develop resistance and how this resistance is spread are of great interest for the continued use of current agents and to develop strategies for the development of new antimicrobials. The following sections discuss the biochemical mechanisms of resistance, how resistance is genetically controlled, and how resistant strains survive and spread in our society. How these features relate to the antimicrobial groups is summarized in **Table 23–2** and further discussed in the chapters on specific bacteria (see Chapters 24 through 40).

Resistance and virulence are separate properties

Antimicrobial resistance has survival value for the organism, and its expression in the medical setting requires that virulence be retained despite the change that mediates resistance. There are no direct connections between resistance and virulence. Resistant bacteria have increased opportunities to produce disease, but the disease itself is the same as that produced by the bacterium's susceptible counterpart.

Mechanisms of Resistance

The major mechanisms of bacterial resistance (**Figure 23–7**) are (1) exclusion of the antimicrobial from the bacterial cell as a result of impermeability or active efflux; (2) alterations of an antimicrobial target, which render it insusceptible; and (3) inactivation of the antimicrobial agent by an enzyme produced by the microorganism.

Exclusion *(Figure 23–8)*

Cell wall and outer membrane are barriers to antimicrobials

Outer membrane protein porins restrict access to interior

An effective antimicrobial must enter the bacterial cell and achieve concentrations sufficient to act on its target. The cell wall, particularly the outer membrane, of Gram-negative bacteria presents a formidable barrier for access to the interior of the cell. Inability to traverse the outer membrane is the primary reason most β-lactams are less active against Gram-negative than Gram-positive bacteria. Outer membrane protein porin channels may allow penetration depending on the size, charge, degree of hydrophobicity, or general molecular configuration of the molecule. This is a major reason for inherent resistance to antimicrobial agents, but these transport characteristics may change even in typically susceptible species as a result of mutations in the porin proteins. For example, strains of *P aeruginosa* commonly develop resistance to imipenem owing to loss of the outer membrane protein most important in its penetration.

TABLE 23–2 Features of Bacterial Resistance to Antimicrobial Agents

	MECHANISM[a]			
ANTIMICROBIAL	**ENTRY BARRIER (EB)**	**ALTERED TARGET (AT)**	**ENZYMATIC INACTIVATION (EI)**	**EMERGING RESISTANCE[b] (ORGANISM/ ANTIMICROBIC/MECHANISM)**
β-lactams	Variable outer membrane[c] penetration	Mutant and new PBPs	β-lactamases	*Staphylococus aureus*/penicillin/EI *S aureus*/methicillin/AT *Streptococcus pneumoniae*/penicillin/AT *Haemophilus influenzae*/ampicillin/AT, EI *Neisseria gonorrhoeae*/penicillin /AT, EI *Pseudomonas aeruginosa*/ceftazidime/EB *Klebsiella, Enterobacter*/third-generation cephalosporins/EI
Glycopeptides	–	Amino acid substitution	–	*Enterococcus*/vancomycin/AT *S aureus*/vancomycin (rare)
Aminoglycosides	Oxidative transport required	Ribosomal binding site mutations	Adenylases, acetylases, phosphorylases	*Klebsiella, Enterobacter*/gentamicin/EI *P aeruginosa*/gentamicin/EB
Macrolides, clindamycin	Minimal outer membrane[c] penetration, efflux pump	Methylation of rRNA	Phosphotransferase, esterase	*Bacteroides fragilis*/clindamycin/AT *S aureus*/erythromycin/AT
Chloramphenicol	–	–	Acetyltransferase	*Salmonella*/chloramphenicol/EI
Tetracyclines	Efflux pump	New protein protects ribosome site	–	
Fluoroquinolones	Efflux pump, permeability mutation	Mutant topoisomerase	–	*Escherichia coli*/ciprofloxacin/AT *P aeruginosa*/ciprofloxacin/AT
Rifampin	–	Mutant RNA polymerase	–	*Mycobacterium tuberculosis*[d]/rifampin/AT *Neisseria meningitidis*/rifampin/AT
Folate inhibitors	–	New dihydropteroate synthetase, altered dihydrofolate reductase	–	Enterobacteriaceae/sulfonamides/AT

PBP, penicillin-binding protein.
[a]Only primary mechanisms of resistance are listed.
[b]A highly selective list of resistance emergence that has altered or threatens a major clinical use of the agent.
[c]Outer membrane of Gram-negative bacteria.
[d]See Chapter 27.

Some antimicrobials must be actively transported into the cell. For example, bacteria that lack the metabolic pathways required for transport of aminoglycosides across the cytoplasmic membrane (streptococci, enterococci, anaerobes) are therefore resistant. Conversely, other antimicrobials are actively transported *out* of the cell. A number of bacterial species have energy-dependent efflux mechanisms that literally pump antimicrobial agents that have entered the cell back out. The membrane transporter systems that drive these efflux pumps often include antimicrobials of several classes.

Active transported required for some

Efflux pumps push antimicrobials back out

Altered Target ***(Figure 23–9)***

Once in the cell, antimicrobials act by binding and inactivating their target, which is typically a crucial enzyme or ribosomal site. If the target is altered in a way that decreases its affinity for the antimicrobial, the inhibitory effect is proportionately decreased. Substitution of a single amino acid at a certain location in a protein can alter its binding to the antimicrobial without affecting its function in the bacterial cell.

Binding affinity for enzymes and ribosomes can change

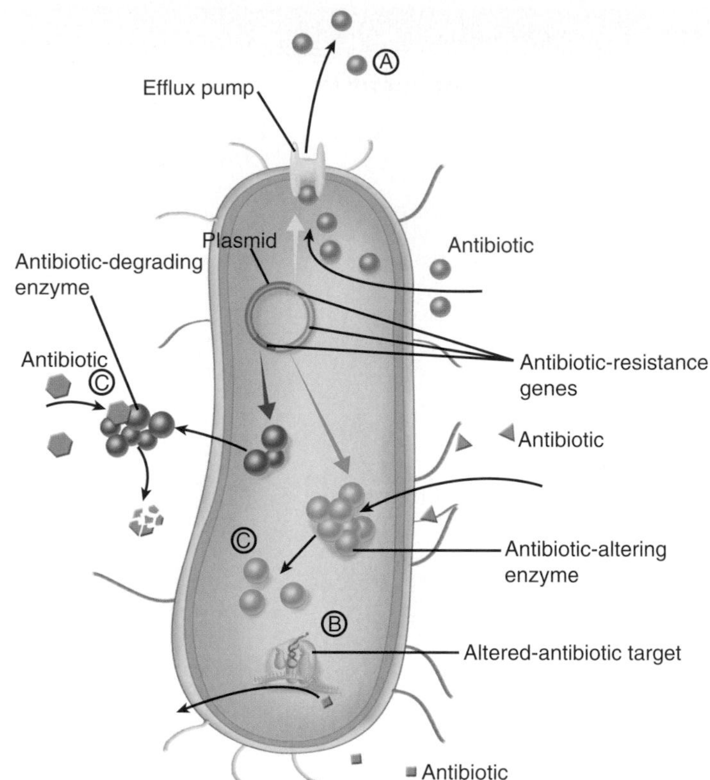

FIGURE 23–7. Antimicrobial resistance mechanisms. A. Exclusion barrier. **B.** Altered target. **C.** Enzymatic inactivation. (Modified with permission from Willey J, Sherwood L, Woolverton C (eds). *Prescott's Principles of Microbiology*. New York: McGraw-Hill; 2008.)

Multiple binding sites reduces chances for resistance

If an alteration at a single site on the target renders it insusceptible, mutation to resistance can occur in a single step, even during therapy. This occurred with the early aminoglycosides (streptomycin), which bound to a single ribosomal site, and the first quinolone (nalidixic acid), which attached to only one of four possible topoisomerase subunits. Newer agents in each class bind at multiple sites on their target, making mutation to resistance statistically much less probable.

One of the most important examples of altered target involves the β-lactam family and the peptidoglycan transpeptidase penicillin-binding proteins (PBPs) on which they act.

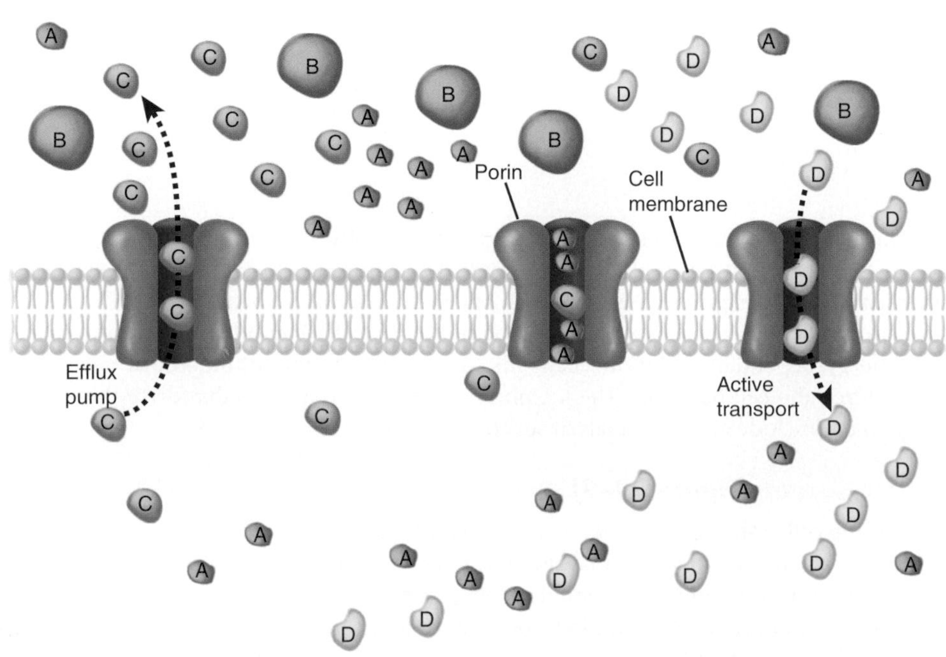

FIGURE 23–8. Exclusion barrier resistance. A, B, C, and D molecules are external to the cell wall, here shown as what could be either the outer membrane (Gram-negatives) or the cytoplasmic membrane. **A molecules** pass through and remain inside the cell, **B molecules** are unable to pass due to their size, **C molecules** pass through but are transported back out by an efflux pump, and **D molecules** must be pulled through by an active process.

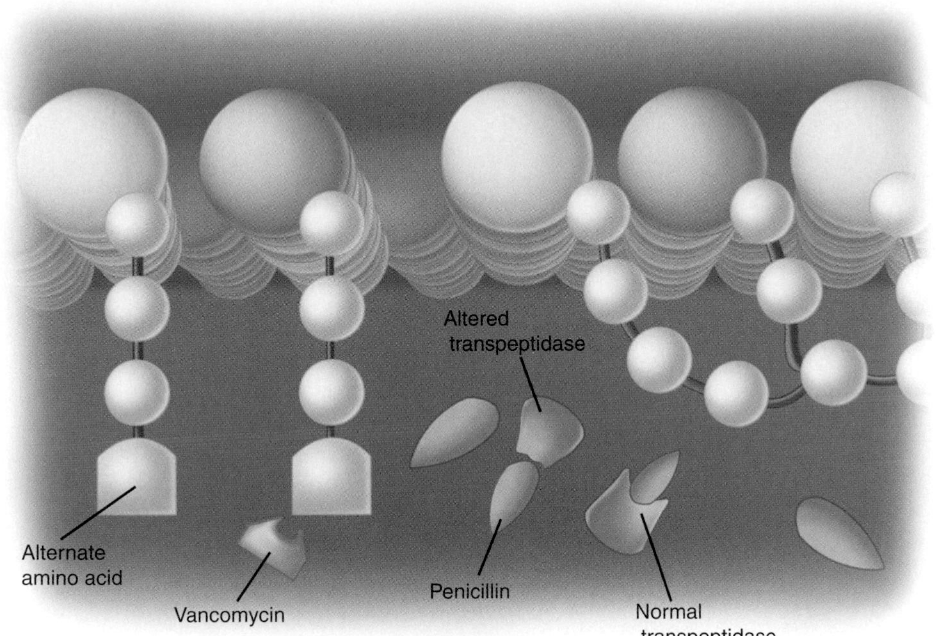

FIGURE 23–9. Altered target resistance. (Compare with Figure 23–1A and B.) A normal transpeptidase or penicillin binding protein (PBP) is inactivated by penicillin, but penicillin no longer binds to the PBP with an altered binding sites. This PBP is still able to carry out its cross-linking function so the β-lactam is no longer effective. Also shown is a terminal amino acid substitution, which will no longer bind vancomycin (Figure 23–1C).

In widely divergent Gram-positive and Gram-negative species, changes in one or more of these proteins have been correlated with decreased susceptibility to multiple β-lactams. These alterations were initially detected as changes in electrophoretic migration of one or more PBPs using radiolabeled penicillin (hence the origin of the term PBP). These changes have now been traced to point mutations, substitutions of amino acid sequences, and even synthesis of a new enzyme.

PBPs are altered transpeptidases

Because the altered binding is not absolute, decreases in susceptibility are incremental and often small. Wild-type pneumococci and gonococci are inhibited by 0.06 μg/mL of penicillin, whereas those with altered PBPs have MICs of 0.1 to 8.0 μg/mL. At the lower end, these MICs still appear to be within therapeutic range but are associated with treatment failures, even when dosage is increased. Altered PBPs generally affect all β-lactams. Although the exact MICs vary, a strain with a 10-fold decrease in susceptibility to penicillin has decreased susceptibility to cephalosporins to about the same degree.

Altered PBPs have reduced affinity for β-lactams

Penicillins and cephalosporins are affected to the same degree

PBP alterations are the prime reason for emergence of penicillin-resistant pneumococci and methicillin-resistant *Staphylococcus aureus* (MRSA). They are one of multiple mechanisms of resistance for a variety of other bacteria including enterococci, gonococci, *H influenzae,* and many other Gram-positive and Gram-negative species.

Pneumococci and MRSA have altered PBPs

Alteration of the target does not require mutation and can occur by the action of a new enzyme produced by the bacteria. Vancomycin-resistant enterococci have enzyme systems that substitute an amino acid in the terminal position of the peptidoglycan side chain (alanyl lysine for alanyl alanine). Vancomycin does not bind to the alternate amino acid, and these strains are resistant. Resistance to sulfonamides and trimethoprim occurs by acquisition of new enzymes with low affinity for these agents, but still allows bacterial cells to carry out their respective functions in the folate synthesis pathway.

New enzymes can alter bacterial targets

Clindamycin resistance involves an enzyme that methylates ribosomal RNA, preventing attachment. This modification also confers resistance to erythromycin and other macrolides, because they share binding sites. It is interesting that induction with erythromycin leads to clindamycin resistance, although the reverse is unusual.

Mutation or acquisition of a new enzyme is possible

Enzymatic Inactivation (Figure 23–10)

Enzymatic inactivation of the antimicrobial agent is the most powerful and robust of the resistance mechanisms. Literally hundreds of distinct enzymes produced by resistant bacteria

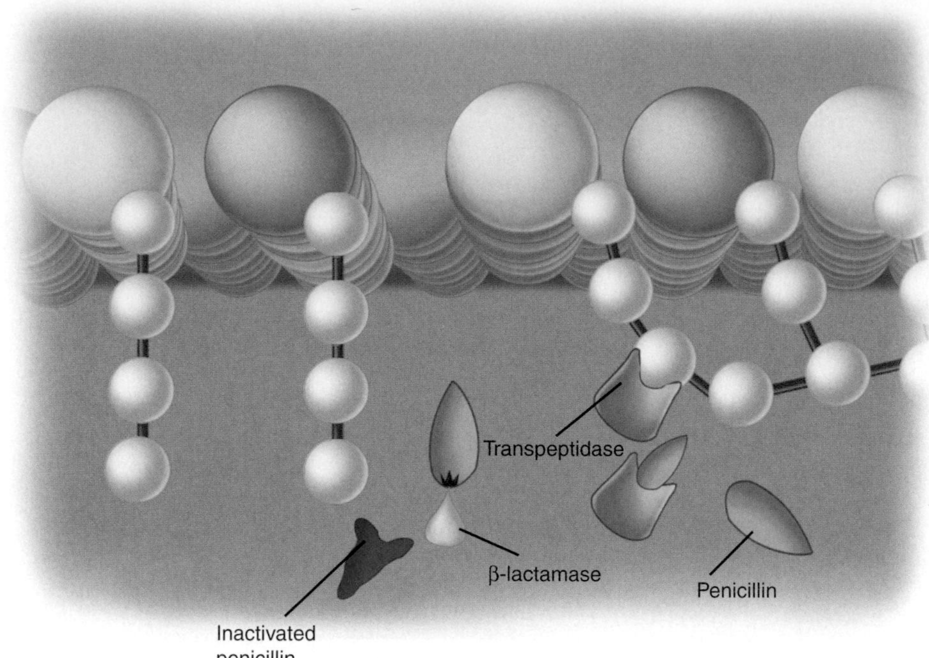

FIGURE 23–10. Enzymatic inactivation resistance. (see Figure 23–1.) The bacterium is producing a β-lactamase enzyme, which destroys penicillin by breaking open the β-lactam ring. If intact penicillin reaches a penicillin-binding protein, it can still bind and inactivate it; the more β-lactamase produced, the higher the level of resistance.

Enzymes may disrupt or chemically modify antimicrobics

may inactivate the antimicrobial in the cell, in the periplasmic space, or outside the cell. These enzymes may act on the antimicrobial molecule by disrupting its structure or by catalyzing a reaction that chemically modifies it.

Enzymes break open the β-lactam ring

β-Lactamases have variable activity against β-lactam substrates

β-Lactamases. β-Lactamase is a general term referring to any one of many bacterial enzymes able to break open the β-lactam ring and inactivate various members of the β-lactam group. The first was discovered when penicillin-resistant strains of *S aureus* emerged and were found to inactivate penicillin in vitro. The enzyme was called penicillinase, but with expansion of the β-lactam family and concomitant resistance, it has become clear that the situation is complex. Each β-lactamase is a distinct enzyme with its own physical characteristics and substrate profile. For example, the original staphylococcal penicillinase is also active against ampicillin but not against methicillin or any cephalosporin. β-Lactamases produced by *E coli* may add cephalosporinase activity, but vary in their potency against individual first-, second-, third-, and fourth-generation cephalosporins. Some β-lactamases are bound by clavulanic acid, and others are not.

May be exoenzymes or act in periplasmic space

ESBLs have broad activity against cephalosporins

To keep track of the β-lactamases they have been named (TEM-1, TEM-2, OXA, SVH etc.) and classification schemes created based on details of their action, substrate profiles, and inducibility (ie, whether enzymes are inducible or produced constitutively). A consideration of β-lactamase classification is beyond the scope of this book, but some discussion of the major types is useful. Gram-positive β-lactamases are exoenzymes with little activity against cephalosporins or the antistaphylococcal penicillins (methicillin, oxacillin). They are bound by β-lactamase inhibitors such as clavulanic acid. Gram-negative enzymes act in the periplasmic space and may have penicillinase and/or cephalosporinase activity. They may or may not be inhibited by clavulanic acid. Many of the Gram-negative β-lactamases are constitutively produced at very low levels, but can be induced to high levels by exposure to a β-lactam agent. A newer class, called extended-spectrum β-lactamases (ESBLs) because their range includes multiple cephalosporins, is particularly worrisome. The laboratory detection of ESBLs is complex because they are inducible enzymes, and the conditions for induction may not be met in the usual susceptibility tests. ::: Enzyme regulation, p. 369

Bacteria that produce β-lactamases typically demonstrate high-level resistance with MICs far outside the therapeutic range. Even weak β-lactamase producers are considered resistant because the outcome of susceptibility tests (and presumably infected sites) is strongly influenced by the number of bacteria present. Large bacterial populations act like an enzyme preparation inactivating the antimicrobial before it ever reaches an organism. Rapid direct tests for β-lactamase can provide this information in a few minutes.

Weak β-lactamase producers are still considered resistant

Modifying Enzymes. The most common cause of acquired bacterial resistance to aminoglycosides is through production of one or more of over 50 enzymes that acetylate, adenylate, or phosphorylate hydroxyl or amino groups on the aminoglycoside molecule. The modifications take place in the cytosol or in close association with the cytoplasmic membrane. The resistance conveyed by these actions is usually high level; the chemically modified aminoglycoside no longer binds to the ribosome. As with the β-lactamases, the aminoglycoside-modifying enzymes represent a large and diverse group of bacterial proteins, each with its characteristic properties and substrate profile. Inactivating enzymes have been described for a number of other antimicrobials. Most of these act by chemically modifying the antimicrobial molecule in a manner similar to the aminoglycoside-modifying enzymes. The most clinically significant enzymes convey resistance to erythromycin (esterase, phosphotransferase) and chloramphenicol (acetyltransferase).

Chemically modified aminoglycosides do not bind to ribosomes

Genetics of Resistance

Intrinsic Resistance

For any antimicrobial, there are bacterial species that are typically within its spectrum and those that are not (**Appendix 23–1**). The resistance of the latter group is referred to as **intrinsic** or **chromosomal** to reflect its inherent nature. The resistant species have features such as permeability barriers, a lack of susceptibility of the cell wall, or ribosomal targets that make them inherently insusceptible. Some species constitutively produce low levels of inactivating enzymes, particularly the β-lactamases of Gram-negative bacteria. The chromosomal genes encoding these β-lactamases may be under repressor control and subject to induction by certain β-lactam antimicrobials. This leads to increased production of β-lactamase, which usually results in resistance not only to the inducer but other β-lactams to which the organism would otherwise be susceptible. Many ESBLs operate in this manner.

Permeability barriers or enzyme production may be intrinsic

Inducible enzymes may have broad spectrum

Acquired Resistance

When an initially susceptible species develops resistance, such acquired resistance can be mutational or derived from another organism by the acquisition of new genes using one of the mechanisms of genetic exchange illustrated in **Figure 23–11**. Of the four major mechanisms of genetic exchange among bacteria (transformation, transduction, conjugation, transposition), conjugation and transposition are the most important clinically and often work in tandem.

Conjugation and transposition are most important

Mutational Resistance

Acquired resistance may occur when there is a crucial mutation in the target of the antimicrobic or in proteins related to access to the target (ie, permeability). Mutations in regulatory proteins can also lead to resistance. Mutations take place at a regular but low frequency and are expressed only when they are not associated with other effects that are disadvantageous to the bacterial cell. Mutational resistance can emerge in a single step or evolve slowly requiring multiple mutations before clinically significant resistance is achieved. Single-step mutational resistance is most likely when the antimicrobial agent binds to a single site on its target. Resistance can also emerge rapidly when it is related to gene regulation, such as mutational derepression of a chromosomally encoded cephalosporinase. A slow, progressive resistance evolving over years, even decades, is typical for β-lactam resistance related to altered PBPs. ::: Mutation, p. 374

Mutations in structural or regulatory genes can confer resistance

Mutations are usually low-frequency

Plasmids and Conjugation

The transfer of plasmids by conjugation was the first discovered mechanism for acquisition of new resistance genes, and it continues to be the most important. Resistance genes on

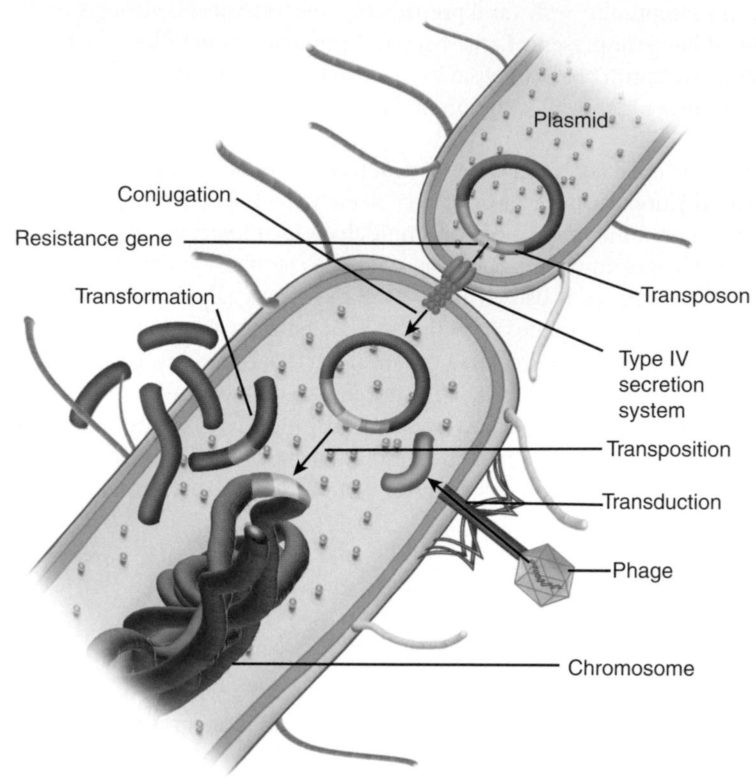

FIGURE 23–11. Genetic mechanisms of acquired resistance. Bacteria are shown exchanging genetic information by transformation, transduction, conjugation, and transposition. Conjugation and transposition are the most common in human infections and are often combined. (Modified with permission from Willey J, Sherwood L, Woolverton C (eds). *Prescott's Principles of Microbiology.* New York: McGraw-Hill; 2008.)

Plasmid conjugation allows multidrug resistance

Species may carry multiple or no plasmids

plasmids (R plasmids) can determine resistance to one antimicrobial or to several that act by different mechanisms. After conjugation, the resistance genes may remain on a recircularized plasmid or less often become integrated into the chromosome by recombination. Of course, resistance is not the only concern of plasmids. A single cell may contain more than one distinct plasmid and/or multiple copies of the same plasmid. Although most resistance mechanisms have been linked to plasmids in one species or another, plasmid distribution among the bacterial pathogens is by no means uniform. The compatibility systems that maintain plasmids from one bacteria cell generation to the next are complex. Some species of bacteria are more likely than others to contain plasmids. For example, *Neisseria gonorrhoeae* typically has multiple plasmids, whereas closely related *N meningitidis* rarely has any. ::: Conjugation, p. 382

Conjugation genes and host range enhance plasmid spread

Plasmids are most likely to be transferred to another strain if they are conjugative; that is, if the resistance plasmid also contains the genes mediating conjugation. Another factor in the spread of plasmids is their host range. Some plasmids can be transferred only to closely related strains; others can be transferred to a broad range of species in and beyond their own genus. A conjugative plasmid with a broad host range has great potential to spread any resistance genes it carries.

Transposons and Transposition

Transposons resistance genes move between chromosomes and plasmids

Transposition and conjugation combine for resistance spread

Transposons containing resistance genes can move from plasmid to plasmid or between plasmid and chromosome. Most of the resistance genes carried on plasmids are transposon insertions, which can be carried along with the rest of the plasmid genome to another strain by conjugation. Once there, the transposon is free to remain in the original plasmid, insert in a new plasmid, insert in the chromosome, or any combination of these (Figure 23–11). Theoretically, plasmids can accomplish the same events by recombination, but the nature of the transposition process is such that it is much more likely to result in the transfer of an intact gene. Transposons also have a variable host range, which in general is even broader than plasmids. Together, conjugation and transposition provide extremely efficient means for spreading resistance genes. ::: Transposition, p. 377

Other Genetic Mechanisms

Although the transfer of resistance genes by transduction has been demonstrated in the laboratory, its association to clinically significant resistance has been uncommon. Transduction of imipenem resistance by wild-type bacteriophages carried by *P aeruginosa* to other strains of the same bacteria is a recent example. Because of the high specificity of bacteriophages, transduction is typically limited to bacteria of the same species. Transformation is the most common way genes are manipulated in the laboratory, but detecting its occurrence in the clinical situation is particularly difficult. Plasmids are readily isolated and characterized, and transposons have flanking insertion sequences to flag their presence, but there is little to mark the uptake of naked DNA. Molecular epidemiologic studies suggest that the spread of PBP mutations in *S pneumoniae* is due to transformation, and there may be many more examples awaiting discovery. ::: Transformation, transduction, p. 380

Transduction is limited by specificity of bacteriophages

Importance of transformation is unknown

Epidemiology of Resistance

It seems that sooner or later microorganisms develop resistance to any antimicrobial agent to which they are exposed. Since the start of the antibiotic era, each new antimicrobic has tended to go through a remarkably similar sequence. When an agent is first introduced, its spectrum of activity seems almost completely predictable; some species are naturally resistant, and others are susceptible, with few exceptions. With clinical use, resistant strains of previously susceptible species begin to appear and become increasingly common.

Clinical use is followed by resistance

In some situations, resistance develops rapidly; in other cases it takes years, even decades. For example, when penicillin was first introduced in 1944, all strains of *S aureus* appeared to be fully susceptible, but by 1950, less than one third of isolates remained susceptible. We now know strains containing the penicillinase plasmid existed long before and were selected out when penicillin use became widespread. On the other hand, the discovery of *H influenzae* (meningitis) and *N gonorrhoeae* (gonorrhea) strains resistant to ampicillin and penicillin did not occur until those antibiotics had been used heavily for a decade or more. In these instances, resistance genes apparently not present in the species initially were acquired from other bacterial species either directly or through recombination of plasmids. There are small enclaves that have escaped resistance. After more than half a century the causes of syphilis (*Treponema pallidum*) and strep throat (group A streptococcus) have thus far retained their susceptibility to penicillin. ::: Recombination, p. 376

Preexisting resistance is selected by antimicrobial use

Resistance acquired after long delays

Whatever the genetic mechanism, the persistence and spread of resistance require an environment in which the resistant strain has a selective advantage. The primary human factors that favor this selection are the overuse of antimicrobial agents in medicine and the inclusion of antimicrobials in animal feeds. Any use of antimicrobial agents by physicians that is beyond that required for the infection at hand has the potential for an unintended consequence of selecting for resistance. This includes prescribing antibacterial agents for viral colds or using a broad-spectrum agent when penicillin would do. Exceeding guidelines for prophylactic use of antimicrobials (see text that follows) also contributes. In truth, as with any intervention in medicine the use of antimicrobial agents carries benefits and risks for the patient. The difference with antimicrobials is that the risk of resistance is for the population at large, not just the individual patient.

Antimicrobial use creates selection for resistance

Overuse increases resistance risk for patients and population at large

The use of antimicrobials added to animal feeds for their growth-promoting effects is a major source of resistant strains. Cattle or poultry that consume feed supplemented with antimicrobials rapidly develop a resistant enteric flora that spreads throughout the herd. Resistant strains can then appear in the flora of those living in proximity or handling animal products. Links from farm to human disease have been established in multiple outbreaks. As a consequence, many countries have banned the addition to animal feeds of antimicrobial agents which are used in humans.

Antimicrobials in animal feeds increase the resistant population

Outbreaks have been traced from patients back to farms

SELECTION OF ANTIBACTERIAL AGENTS

Empiric Therapy

Probable etiology and susceptibility statistics guide initial selection

Empiric therapy is that based solely on clinical findings. The first decisions are based on the physician's assessment of the probable microbial etiology of the patient's infection. The variables involved are the subject of much of this book and include the site of infection (eg, throat, lung, urine) and epidemiologic factors such as age, season, geography, and predisposing conditions. A mental list of probable etiologies must then be matched with their probable antimicrobial susceptibilities as shown in Table 23–1 and Appendix 23–1. Specific local "batting averages" for each antimicrobial against the common bacterial pathogens are available from hospital laboratories and infection control committees.

Narrow versus broad spectrum is influenced by clinical severity

This process may be as simple as selecting penicillin to treat a patient with suspected group A streptococcal pharyngitis or as complex as resorting to a cocktail of broad-spectrum antibacterial, antifungal, and antiviral agents to treat a febrile patient who has had a bone marrow transplantation. In general, the risks of broad-spectrum treatment (superinfection, overgrowth) become more tolerable as the severity of the infection increases. When the risk of not "covering" an improbable pathogen is death, it is more difficult to be selective. Empiric therapy is expected to be converted to specific therapy within a few days, although in some instances this is not possible. In otitis media, there is no safe way to culture the middle ear, so empiric therapy must be continued and the outcome evaluated on clinical grounds.

Specific Therapy

Isolation of the causative agent allows specificity

Susceptibility tests provide final guide

Combinations may be synergistic

Specific therapy is that directed only at the known pathogen. This is unique to infectious diseases and made possible by isolation and susceptibility testing of the patient's isolate in the laboratory. This is widely available for almost all bacterial infections. As the results of Gram smears, cultures, and susceptibility tests are reported, unnecessary antimicrobials can be discontinued and the spectrum of therapy narrowed. For example, a patient with suspect staphylococcal or streptococcal infection might be empirically started on a cephalosporin which covers both possibilities. Once the situation is resolved, a more specific antimicrobial is substituted for the broader-spectrum treatment. Usually, this is the single best agent, but sometimes combinations of antimicrobials that have different modes of action are used for enhanced effect. The major indications for combinations are reducing the probability of emergence of resistance, which is important in chronic infections like tuberculosis, and taking advantage of known synergy between two antimicrobials. Synergy is when the activity of a combination is far greater than would be expected from the individual MICs of the two antimicrobials.

Prophylaxis

High-risk exposures merit prophylaxis

Some surgical procedures benefit

GBS reduced in neonates

The use of antimicrobials to prevent infection is a tempting but potentially hazardous endeavor. The risk for the individual patient is infection with a different, more resistant organism. The risks for the population are in increasing the pressure for the selection and spread of resistance. After many years of experience, the indications for antimicrobial prophylaxis have now been narrowed to a limited number of situations in which antimicrobial agents have been shown to decrease transmission during a period of high risk. For example, persons known to have been exposed to highly infectious and virulent pathogens such as *Bacillus anthracis* (anthrax) or *Yersinia pestis* (plague) can abort an infection during the incubation period by the administration of ciprofloxacin. Prophylaxis can also reduce the risk of endogenous infection associated with certain surgical and dental procedures if given during the procedure (a few hours at most). The practice of administering prophylactic penicillin during labor to mothers with demonstrated vaginal group B streptococcal (GBS) colonization lead to a dramatic decrease in the leading cause of sepsis and meningitis in neonates. Chemoprophylaxis of tuberculin skin test converters is a regular part of tuberculosis prevention.

APPENDIX 23–1 Usual Susceptibility Patterns of Common Bacteria to Some Commonly Used Bacteriostatic and Bactericidal Antimicrobial Agents

(Symbols used below: ○ = 100% susceptible; ◔ = partly resistant (e.g., 25% resistant); ◕ = mostly resistant; ● = 100% resistant; ◍ = intermediate susceptibility. A number or letter before a symbol is the mark printed above it.)

Antimicrobic	Bactericidal	Bacteriostatic	*Staphylococcus aureus*	Enterococci	Other Streptococci	*Neisseria*	*Haemophilus*	*Legionella*	*Mycoplasma*	*Escherichia coli*	*Proteus mirabilis*	Other *Proteus* spp	*Klebsiella*	*Enterobacter*	*Serratia*	*Pseudomonas aeruginosa*	*Bacteroides fragilis*	Other Gram-negative Anaerobes	*Clostridium*	*Rickettsia*	*Chlamydia*	
Benzyl penicillin	+		1 ◕	C ◍	1 ◔	1 ◔	◍	●	●	●	◍	●	●	●	●	●	●	1 ◔	1 ○	●	●	Narrow-spectrum agents
Penicillinase-resistant penicillins	+		1 ◔	●	2 ◔	●	●	●	●	●	●	●	●	●	●	●	●	●	●	●	●	
Erythromycin	±	+	2 ◔	2 ◔	2 ◔	◔		1 ○	1 ○	●	●	●	●	●	●	●	–	–	–	–	2 ○	
Clindamycin	±	+	2 ◔	–	◔	●	●	–	–	●	●	●	●	●	●	●	◔	◔	◔	–	–	
Vancomycin	+		2 ◔	1 ◔	2 ○	●	●	●	●	●	●	●	●	●	●	●	–	–	1 ○	–	–	
Ampicillin	+		2 ◕	1 ◔	2 ◔	1 ◔	1 ◔	●	●	1 ◔	1 ◔	●	●	●	●	●	●	–	1 ○	●	●	Broad-spectrum agents
Piperacillin	+		–	○	○	–	◔	●	●	–	1 ○	1 ◔	1 ◔	1 ◔	1 ◔	1 ◔	2 ◔	1 ○	–	●	●	
Cephalothin	+		2 ◔	●	2 ◔	–	●	●	●	◔	○	●	◔	●	●	●	●	◔	◔	●	●	
Cefotetan	+		–	●	1 ○	1 ○	–	●	●	1 ○	1 ○	1 ○	1 ○	●	◔	●	2 ◔	○	–	●	●	
Ceftazidime	+		–	–	–	–	○	●	●	1 ○	1 ○	1 ○	1 ○	2 ◔	2 ◔	◔	◔	–	–	●	●	
Imipenem	+		2 ○	2 ○	2 ○	1 ○	1 ○	–	●	1 ○	1 ○	1 ○	1 ○	1 ○	1 ○	1 ◔	1 ○	1 ○	1 ○	–	–	
Aztreonam	+		●	●	●	1 ○	1 ○	–	●	1 ◔	1 ◔	1 ○	1 ○	◔	1 ◔	1 ◔	●	●	●	–	–	
Gentamicin	+		◔	C ●	●	–	–	●	–	1 ◔	1 ○	1 ○	1 ◔	1 ◔	1 ◔	1 ◔	●	●	●	–	–	
Amikacin	+		◔	C ●	●	–	–	●	–	1 ◔	1 ○	1 ○	1 ◔	1 ◔	1 ◔	1 ◔	●	●	●	–	–	
Tetracycline		+	◔	●	◕	◔	○	2 ○	1 ○	◔	●	●	◔	◔	◕	●	◕	◔	◔	1 ○	1 ○	
Chloramphenicol		+	◔	◔	◔	◔	◔	2 ◕	–	◔	◔	◔	◔	◔	◕	●	1 ◔	2 ○	–	1 ○	–	
Ciprofloxacin	+		◔	◍	◍	2 ○	–	–	●	1 ◔	1 ○	1 ○	1 ○	1 ○	1 ○	2 ◔	●	●	–	–	◍	
Sulfamethoxazole + trimethoprim	±	+	–	–	–	–	1 ◔	–	–	1 ◔	◔	◔	◔	◔	◔	●	◔	–	–	–	3 –	

Proportions of susceptible and resistant strains: ○, 100% susceptible ◔, 25% resistant ●, 100% resistant ◍, intermediate susceptibility.

Abbreviations: – = no present indication for therapy or insufficient data 1 = antimicrobic of choice for susceptible strains 2 = second-line agent
3 = *c. trachomatis*-sensitive, *c. psittaci*-resistant C = useful in combinations of a ß-lactam and an aminoglycoside

CHAPTER 24

Staphylococci

Thou art a boil,
A plague sore, an embossed carbuncle
In my corrupted blood.

—Shakespeare: *King Lear*

Members of the genus *Staphylococcus* (staphylococci) are Gram-positive cocci that tend to be arranged in grape-like clusters (**Figure 24–1**). Worldwide, *Staphylococcus aureus* is one of the most common causes of acute purulent infections. Other species are common in the skin flora, but produce lower grade disease, typically in association with some mechanical abridgment of the host such as an indwelling catheter.

STAPHYLOCCOCI: GROUP CHARACTERISTICS

Although staphylococci have a marked tendency to form clusters (from the Greek *staphyle*, bunch of grapes), some single cells, pairs, and short chains are also seen. Staphylococci have a typical Gram-positive cell wall structure. Like all medically important cocci, they are nonflagellate, nonmotile, and non–spore-forming. Staphylococci grow best aerobically but are facultatively anaerobic. In contrast to streptococci, staphylococci produce catalase. More than one dozen species of staphylococci colonize humans; of these, *S aureus* is by far the most virulent. The ability of *S aureus* to form coagulase separates it from other, less virulent species (**Table 24–1**). It is common to lump the other species together as coagulase-negative staphylococci (CoNS).

Staphylococci form clusters and are catalase-positive

Coagulase distinguishes S aureus from other species

Staphylococcus aureus

BACTERIOLOGY

MORPHOLOGY AND STRUCTURE

In growing cultures, the cells of *S aureus* are uniformly Gram-positive and regular in size, fitting together in clusters with the precision of pool balls. In older cultures, in resolving lesions, and in the presence of some antibiotics, the cells often become more variable in size, and many lose their Gram positivity.

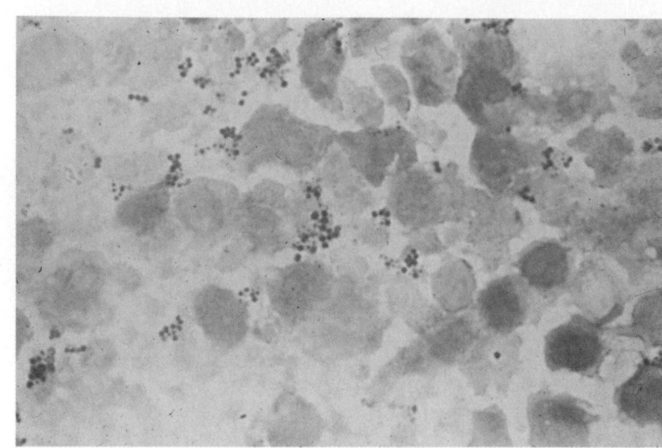

FIGURE 24–1. *Staphylococcus aureus.* Gram stain showing the Gram-positive cocci in clusters resembling bunches of grapes. (Reprinted with permission of Schering Corporation, Kenilworth, NJ, the copyright owner. All rights reserved.)

Surface proteins bind fibrinogen, fibronectin

Protein A binds Fc portion of IgG

The cell wall of *S aureus* consists of a typical Gram-positive peptidoglycan interspersed with considerable amounts of teichoic acid. The peptidoglycan of the cell wall is commonly overlain with polysaccharide and surface proteins. Although polysaccharide capsules have been identified, their significance in human infections is unknown, and they will not be discussed further. Surface proteins such as clumping factor (Clf), which binds to fibrinogen, and fibronectin-binding proteins (FnBP) likely play a role in the early stages of infection. Another protein, protein A, is unique in that it binds the Fc portion of IgG molecules, leaving the antigen-reacting Fab portion directed externally (turned around). This phenomenon has been exploited in test systems for detecting free antigens. Protein A probably contributes to the virulence of *S aureus* as an antiphagocytic factor. ::: Gram-positive cell wall, p. 351

CHARACTERISTICS FOR IDENTIFICATION AND SUBTYPING

Colonies are white or golden and hemolytic

After overnight incubation on blood agar, *S aureus* produces white colonies that tend to turn a buff-golden color with time, which is the basis of the species epithet *aureus* (golden). Most, but not all, strains show a rim of clear β-hemolysis surrounding the colony.

Coagulase produces a fibrin clot

The most important test used to distinguish *S aureus* from other staphylococci is the production of **coagulase,** which activates prothrombin without the usual proteolytic cleavages to form a fibrin clot. It is demonstrated by incubating staphylococci in plasma; this produces a fibrin clot within hours. A dense emulsion of *S aureus* cells in water also clumps

TABLE 24–1 Features of Human Staphylococci

SPECIES	COAGULASE	α-TOXIN	SAgs	HABITAT	BIOFILM	BOILS	UTI[a]	DEEP INFECTIONS
Staphylococcus aureus	+	+	+	Anterior nares, perineum	+	+	–	Pneumonia, osteomyelitis, abscesses, TSS
S epidermidis	–	–	–	Anterior nares, skin	+	–	–	Device colonization
S saprophyticus	–	–	–	Gastrointestinal tract	–	–	+	None
Others	–	–		Variable	Variable	–	–	Device colonization

N:SAgs, superantigen exotoxin production; TTS, Toxic shock syndrome; UTI, urinary tract infection.
[a]Significant cause of UTI

immediately on mixing with plasma as a result of direct binding of fibrinogen by clumping factor on the cell surface. This is the basis of a quick laboratory test called the slide clumping test, which has a high correlation with coagulase (95%). Commercial agglutination tests that correlate well with the coagulase test are also used. *S aureus* isolates can be subdivided by a variety of typing systems and by their pattern of lysis by bacteriophages (phage typing). In epidemiologic investigations, molecular methods such as pulsed field gel electrophoresis are now used to "fingerprint" the spread of virulent clones.

Slide clumping factor correlates with coagulase

Molecular methods create tracking for epidemiologic investigations

TOXINS AND BIOLOGICALLY ACTIVE EXTRACELLULAR ENZYMES

α-Toxin

S aureus produces a number of named cytolytic toxins (α, β, δ, γ), of which α-toxin is the most important. α-Toxin is a protein secreted by almost all strains of *S aureus,* but not by coagulase-negative staphylococci. It is a pore-forming cytotoxin that lyses the cytoplasmic membranes of a wide variety of host cell types by direct insertion into the lipid bilayer to form transmembrane pores (**Figure 24–2**). The resultant egress of vital molecules leads to cell death. This action is similar to other biologically active cytolysins such as streptolysin O, complement, and the effector proteins of cytotoxic T lymphocytes. ::: Pore-forming toxin, p. 396

α-Toxin inserts in lipid bilayer to form transmembrane pores

Exfoliatin

Exfoliatin produced by *S aureus* binds to a specific cell membrane ganglioside found only in the stratum granulosum of the keratinized epidermis of young children and rare adults. There it causes intercellular splitting of the epidermis between the stratum spinosum and stratum granulosum, presumably by disruption of intercellular junctions. Two variants of exfoliatin are antigenic in humans, and circulating antibody confers immunity to their effects.

Exfoliatin splits intraepidermal junctions

Staphylococcal Superantigen Toxins

The superantigens (SAgs) are a family of secreted proteins that are able to stimulate systemic effects as a result of absorption from the gastrointestinal tract after ingestion or at a site where they are produced in vivo by multiplying bacteria. There are now more than 15 described staphylococcal superantigen toxins (StaphSAgs), the most important of which in human disease are antigenic variants of the long-known staphylococcal enterotoxins and the more recently discovered toxic shock syndrome toxin (TSST-1). An individual strain may produce one or more toxins, but less than 10% of *S aureus* strains produce any StaphSAg. As superantigens they are strongly mitogenic for T cells and do not require proteolytic processing before binding with class II major histocompatibility complex (MHC) molecules on antigen-presenting cells. This process not only bypasses the specificity of antigen processing but results in massive cytokine release. The StaphSAg toxins share physiochemical and biologic activity similarities with each other and StrepSAgs produced by group A streptococci. ::: Superantigen toxin, p. 396

StaphSAgs bind MHC II without processing

Superantigens cause massive cytokine release

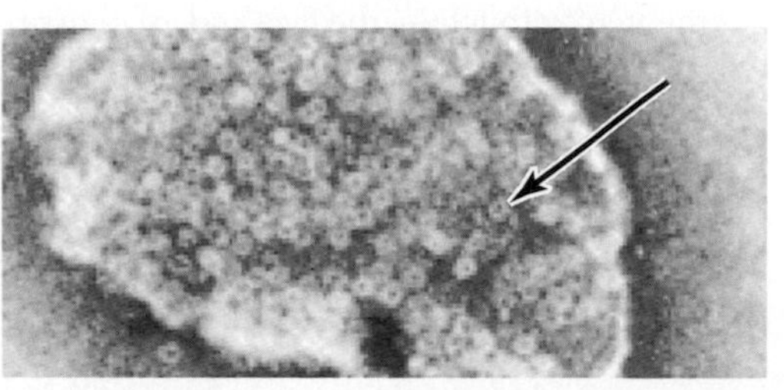

FIGURE 24–2. *Staphylococcus aureus* α-toxin. A fragment of a rabbit erythrocyte lysed with α-toxin is shown. Note the ring-shaped pores in the membrane created by insertion of the toxin. (From Bhadki S, Tranum-Jensen J. Alpha toxin of *Staphylococcus aureus*. Microbiol Rev 1991; 55:733–751, with permission.)

Staphylococcal Enterotoxins

Once formed, enterotoxins are stable to boiling and digestive enzymes

Vomiting is stimulated by brain stem mechanism

The ability of *S aureus* enterotoxins to stimulate gastrointestinal symptoms (primarily vomiting) in humans and animals has long been known. There are several antigenically distinct proteins in this class (eg, SEA, SEB, etc.); once formed, these toxins are quite stable, retaining activity even after boiling or exposure to gastric and jejunal enzymes. In addition to their superantigen actions, they appear to act by stimulating reflexes in the abdominal viscera, which are transmitted to medullary emetic centers in the brain stem via the vagus nerve.

STAPHYLOCOCCAL DISEASE

CLINICAL CAPSULE

Infections produced by *S aureus* are typified by acute, aggressive, locally destructive purulent lesions. The most familiar of these is the common boil, a painful lump in the skin that has a necrotic center and fibrous reactive shell. Infections in organs other than the skin such as the lung, kidney, or bone are also focal and destructive, but have greater potential for extension within the organ and beyond to the blood and other organs. Such infections typically produce high fever and systemic toxicity and may be fatal in only a few days. A subgroup (<10%) of *S aureus* infections has manifestations produced by secreted toxins in addition to those caused by the primary infection. Symptoms include diarrhea, rash, skin desquamation, and multiorgan effects as in staphylococcal toxic shock syndrome (TSS). Ingestion of preformed staphylococcal enterotoxin causes a form of food poisoning in which vomiting begins in only a few hours.

In many ways, *S aureus* is the "all-time champion" of microbial pathogens. Although tuberculosis and malaria have greater global prevalence and the spread of AIDS is more ominous, the ferocity of staphylococcal infections has remained constant for as long as we can tell. In Shakespeare's Lear (1606) quoted at the beginning of the chapter, the king is not himself infected. He has just chosen two prototype staphylococcal lesions (boil, carbuncle) as the vilest of symbols to characterize his ungrateful daughters and his treatment at their hands. Today, in any hospital in the world *S aureus* still heads the list of pathogens isolated from the bloodstream of seriously ill patients.

EPIDEMIOLOGY

Anterior nares colonization is common

Strains with increased virulence cannot be distinguished

The basic human habitat of *S aureus* is the anterior nares. Ten to thirty percent of the population carry the organism at this site at any given time, and rates among hospital personnel and patients may be much higher. From the nasal site, the bacteria are shed to the exposed skin and clothing of the carrier and others with whom they are in direct contact. Spread is augmented by touching the face and, of course, nose picking. It is blocked by hand washing. Once present on the skin, even transiently, *S aureus* can gain deeper access either through skin appendages or trauma (**Figure 24–3**). Although outbreak investigations show that some strains have enhanced virulence, still no laboratory tests can be used to separate them from the large pool of colonized individuals.

Community infections are endogenous

Most *S aureus* infections acquired in the community are autoinfections with strains that the subject has been carrying in the anterior nares, on the skin, or both. Community outbreaks are usually associated with poor hygiene and fomite transmission from individual to

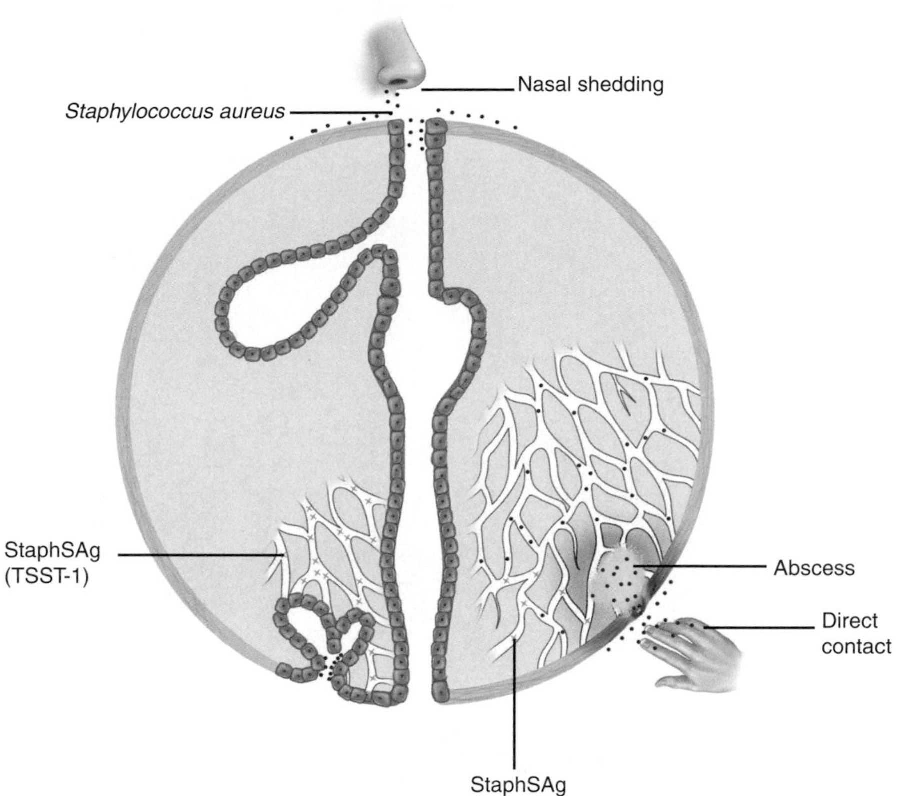

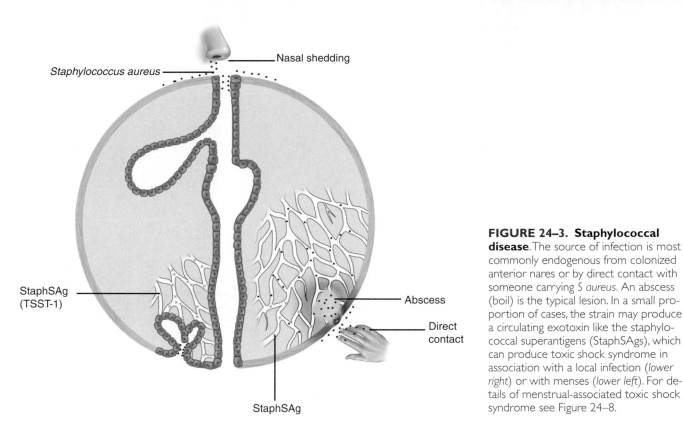

FIGURE 24–3. Staphylococcal disease. The source of infection is most commonly endogenous from colonized anterior nares or by direct contact with someone carrying *S aureus*. An abscess (boil) is the typical lesion. In a small proportion of cases, the strain may produce a circulating exotoxin like the staphylococcal superantigens (StaphSAgs), which can produce toxic shock syndrome in association with a local infection (*lower right*) or with menses (*lower left*). For details of menstrual-associated toxic shock syndrome see Figure 24–8.

individual. Unlike many pathogenic vegetative organisms, *S aureus* can survive periods of drying; for example, recurrent skin infections can result from use of clothing contaminated with pus from a previous infection.

S aureus survives drying

Hospital outbreaks caused by a single strain of *S aureus* most commonly involve patients who have undergone surgical or other invasive procedures. The source of the outbreak may be a patient with an overt or unapparent staphylococcal infection (eg, decubitus ulcer), which is then spread directly to other patients on the hands of hospital personnel. A nasal or perineal carrier among medical, nursing, or other hospital personnel may also be the source of an outbreak, especially when carriage is heavy and numerous organisms are disseminated. The most hazardous source is a medical attendant who works despite having a staphylococcal lesion such as a boil. Hospital outbreaks of *S aureus* infection can be self-perpetuating: infected patients and those who attend them frequently become carriers, and the total environmental load of the causative staphylococcus is increased.

Hospital spread is on the hands of medical personnel

Outbreaks involve nasal carrier or worker with lesion

Staphylococcal food poisoning is one of the most common food-borne illnesses in the world. It has been an unhappy and embarrassing sequel to innumerable group picnics and wedding receptions in which gastronomic delicacies have been exposed to temperatures that allow bacterial multiplication. Characteristically, the food is moist and rich (eg, red meat, poultry, creamy dishes). The food becomes contaminated by a preparer who is a nasal carrier or has a staphylococcal lesion. If the food is not refrigerated for hours between preparation and serving, the staphylococci are able to multiply and produce enterotoxin in the food. Because of the heat resistance of the toxin, toxicity persists even if the food is subsequently cooked before eating.

Enterotoxin is produced in rich foods before they are ingested

PATHOGENESIS

Primary Infection

A boil (or furuncle) is an abscess and a prototype for the purulent lesions produced by many other bacteria. The initial stages of attachment by *S aureus* are mediated by a number of surface proteins, which bind to host cells or elements on their surface. Proteins that

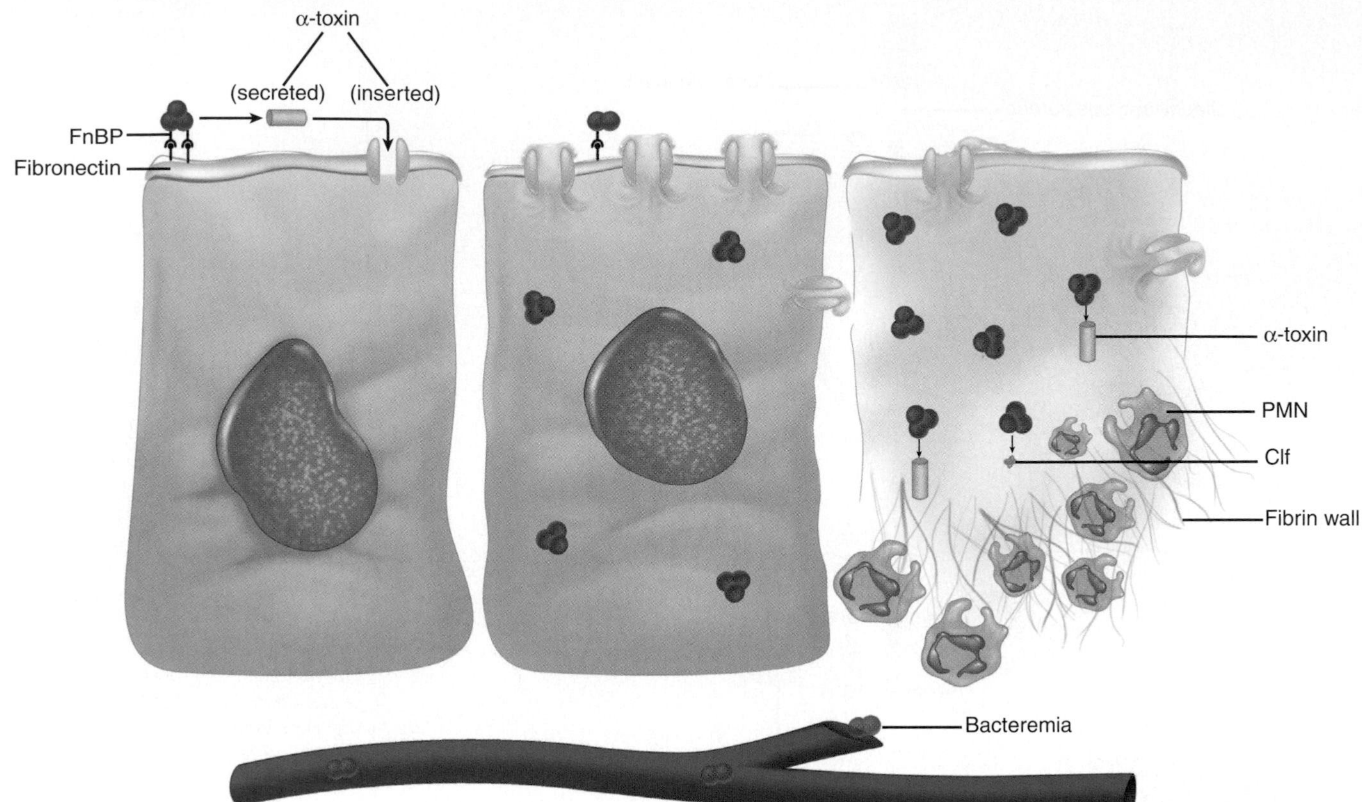

FIGURE 24–4. Staphylococcal disease cellular view. Initial attachment to fibronectin is mediated by fibronectin-binding proteins (FnBP), and the major injury is caused by the pore-forming α-toxin. Cells are destroyed by leaking their cytosol. The α-toxin also inserts in the polymorphonuclear neutrophils. Resistance to phagocytosis and the formation of a wall are aided by fibrinogen-binding Clf.

bind to the glycoprotein fibronectin that is ubiquitous on mucosal surfaces are of particular importance in the early stages of infection. These fibrinectin-binding proteins (FnBPs) mediate adhesion to and perhaps invasion of mammalian cells. This allows *S aureus* to persist and to produce α-toxin and other cytolysins, which injure the cell (**Figure 24–4**). As the lesions become destructive and spread below the surface, other proteins that bind to collagen and other elements of the extracellular matrix may play a role. At this stage, production of fibrinogen-binding Clf, antiphagocytic protein A, and continued production of α-toxin all combine to limit the effectiveness of host phagocytes, allowing the staphylococci to multiply and the lesion to expand. The inflammatory cells, fibrin, and other tissue components form a wall, which becomes the painfully familiar boil (**Figure 24–5**). A carbuncle (**Figure 24–6**) is an extension of this process in which, rather than discharging at the surface, the process forms multiple compartments. There is evidence that *S aureus* can regulate this multifactored process deploying adhesions and extracellular products at the stages they are needed.

FnBPs bind to fibronectin on cell surface

Coagulase, protein A, facilitate α-toxin production

The fate of the lesion depends on the ability of the host to localize the process, which differs depending on the tissue involved. In the skin, spontaneous resolution of the boil by granulation and fibrosis is the rule. In the lung, kidney, bone, and other organs, the process may continue to spread with satellite foci and involvement of broad areas. In all instances, the action of the cytotoxins is highly destructive, creating cavities and massive necrosis with little respect to anatomic boundaries. In the worst cases, the staphylococci are not contained, spreading to the bloodstream and distant organs. Circulating staphylococci may also shed cell wall peptidoglycans, producing massive complement activation, leukopenia, thrombocytopenia, and a clinical syndrome of septic shock.

Destruction and spread are prominent

Peptidoglycan fragments may trigger shock

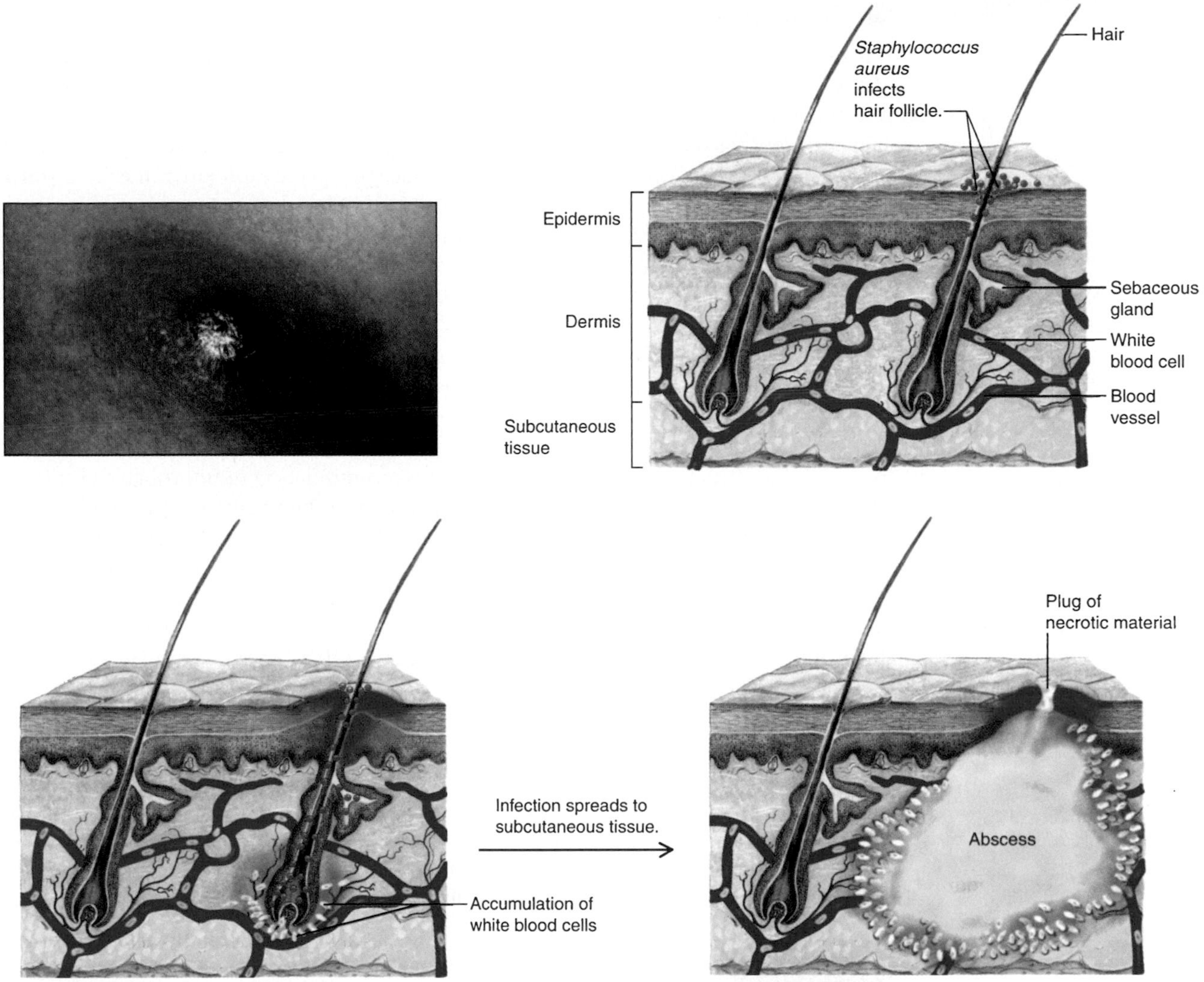

Staphylococcus aureus infects hair follicle.
Hair
Epidermis
Dermis
Subcutaneous tissue
Sebaceous gland
White blood cell
Blood vessel
Plug of necrotic material
Infection spreads to subcutaneous tissue.
Accumulation of white blood cells
Abscess

FIGURE 24–5. Furuncle (boil). Note the focal nature of the lesion. This one appears about to "point" and drain its walled-off pus externally. (Reproduced with permission from Nester EW, Anderson DG, Roberts CE Jr., Nester MT. *Microbiology: A Human Perspective,* 6th ed. New York: McGraw-Hill, 2008.)

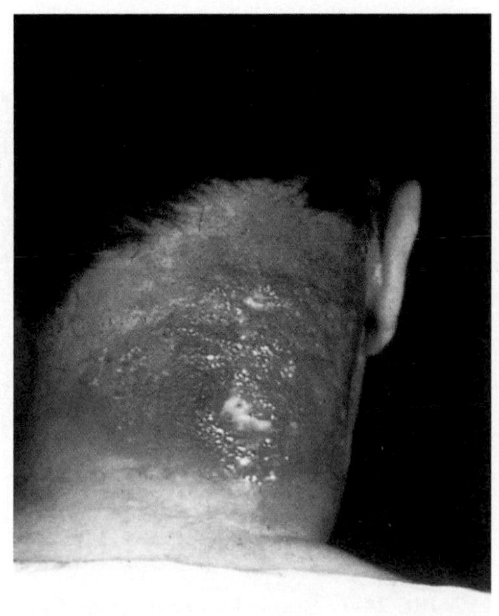

FIGURE 24–6. Staphylococcal carbuncle. Multiple abscesses have coalesced to form this angry cellulitis with draining sinuses. (Reproduced with permission from Connor DH, Chandler FW, Schwartz DQA, Manz HJ, Lack EE (eds). *Pathology of Infectious Diseases,* vol. 1. Stamford, CT: Appleton & Lange; 1997.)

Toxin-mediated Disease

If the strain of *S aureus* causing any of the effects described above also produces one or more of the exotoxins, those actions are added to those of the primary infection. The primary infection serves as a site for absorption of the toxin and need not be extensive or even clinically apparent for the toxic action to occur. In staphylococcal food poisoning, there is no infection at all. The contaminating bacteria produce pyrogenic exotoxin in the food, which can initiate its enterotoxic action on the intestine within hours of its ingestion.

Preformed enterotoxin acts within hours

The in vivo production of exfoliative toxin takes at least a few days and may exert its effect locally or systemically. Toxin absorbed at the infection site reaches its infant stratum granulosum binding site through the circulation causing widespread desquamation by its action on the keratinized epidermis as in the staphylococcal scalded skin syndrome (**Figure 24–7A** and **B**). In older children, exfoliative toxin-producing strains may also cause a localized blister-like lesion called **bullous impetigo**.

Exfoliative toxin causes blisters or scalded skin syndrome

In staphylococcal TSS, the pyrogenic exotoxin TSST-1 is produced during the course of a staphylococcal infection with systemic disease as a result of absorption of toxin from the local site. Menstruation-associated TSS requires a combination of improbable events. At any one time, less than 15% of women carry *S aureus* in their vaginal flora, and less than 10% of these have the potential to produce TSST-1. During menstruation, the relatively high protein level and pH in the vagina favor accelerated growth of these staphylococci. In the presence of such a strain, the combination of menstruation and high-absorbency tampon usage provide conditions that enhance both the growth of the staphylococci and the production of TSST-1. Toxin absorbed from the vagina can then circulate to produce the multiple effects of massive superantigen-mediated cytokine release (**Figure 24–8A** and **B**).

TSST-1–producing strain must colonize vagina

Menstruation and tampons enhance local toxin production

Some cases of full-blown staphylococcal TSS are associated with strains that do not produce TSST-1. This is particularly true of nonmenstrual cases. Other StaphSAgs have been detected in these strains and have been shown to produce experimental toxic shock. TSS may be the result of in vivo production of any of the StaphSAgs, with TSST-1 simply the most common offender. The mechanisms by which the pyrogenic exotoxins produce the multiple renal, cutaneous, intestinal, and cardiovascular manifestations of TSS are not known.

Nonmenstrual TSS cases may have any PTSAg-producing strain

IMMUNITY

Immunity is poorly understood

The natural history of staphylococcal infections indicates that immunity is of short duration and incomplete. Chronic furunculosis, for example, can recur over many years. The

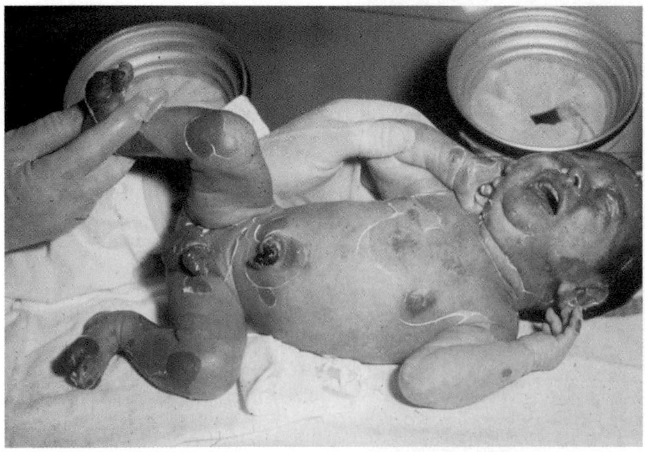

A

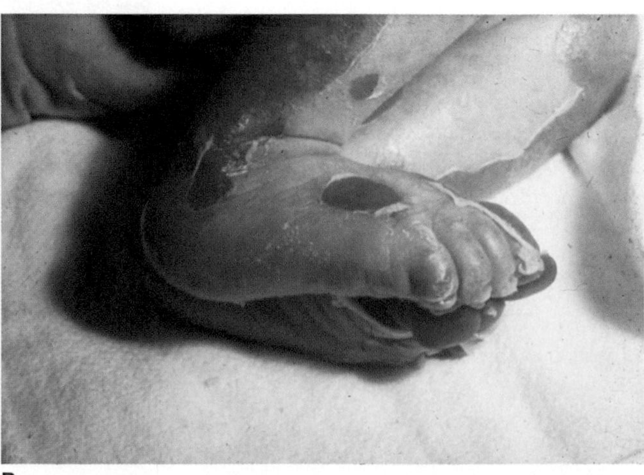

B

FIGURE 24–7. Staphylococcal scalded skin syndrome in a neonate. A. This infant has a small focal staphylococcal breast abscess and looks as if he has been sunburned or dipped in boiling water. **B.** Note the peeling of the superficial layers of the skin as a result of the action of circulating exfolatin.

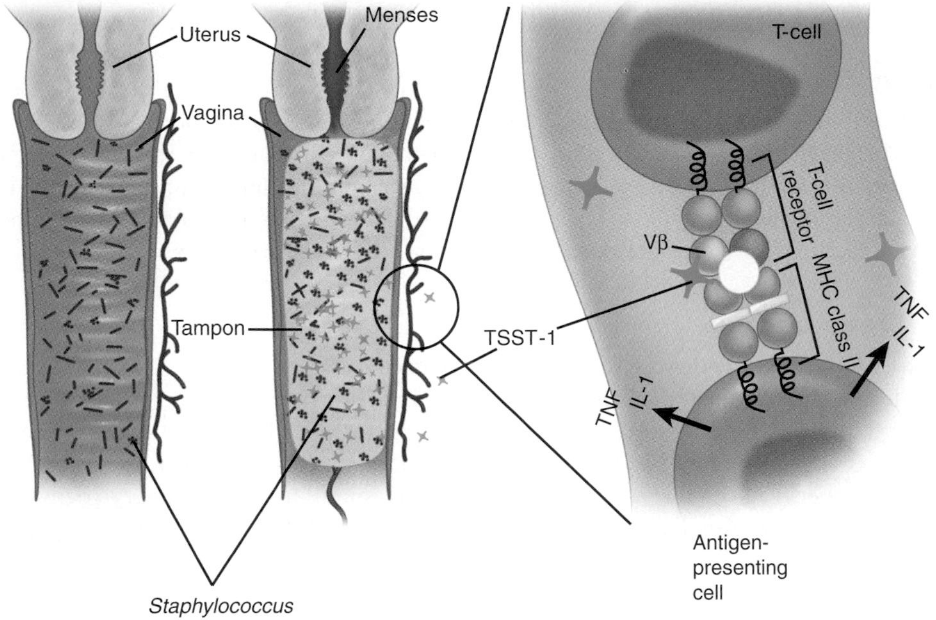

FIGURE 24–8. Pathogenesis of staphylococcal toxic shock syndrome. A. The vagina is colonized with normal flora and a strain of *Staphylococcus aureus* containing the c-1 gene. **B.** The conditions with tampon usage facilitate growth of the *S aureus* and toxic shock syndrome toxin (TSST-1) StaphSAg production. **C.** The toxin is absorbed from the vagina and circulates. The systemic effects may be due to the direct effect of the toxin or via cytokines released by the superantigen mechanism. The toxin is shown binding directly with the Vβ portion of the T-cell receptor and the class II major histocompatibility complex (MHC) receptor. This Vβ stimulation signals the production of cytokines such as interleukin-1 (IL-1) and tumor necrosis factor (TNF).

relative roles of humoral and cellular immune mechanisms are uncertain, and attempts to induce immunity artificially with various staphylococcal products have been disappointing at best. In menstruation-associated TSS, many patients have low or absent antibody levels to TSST-1 and often fail to mount a significant antibody response during the disease. This may be due to SAgs stimulation of T_H1 responses with minimal T_H2 component.

Relapsing infections show little evidence of immunity

STAPHYLOCOCCAL INFECTIONS: CLINICAL ASPECTS

MANIFESTATIONS: PRIMARY INFECTION

Furuncle and Carbuncle

The furuncle or boil (Figure 24–5) is a superficial skin infection that typically develops in a hair follicle, sebaceous gland, or sweat gland. Blockage of the gland duct with inspissation of its contents causes predisposition to infection. Furunculosis is often a complication of acne vulgaris. Infection at the base of the eyelash gives rise to the common stye. The infected patient is often a carrier of the offending *Staphylococcus,* usually in the anterior nares. The course of the infection is usually benign, and the infection resolves upon spontaneous drainage of pus. No surgical or antimicrobic treatment is needed. Infection can spread from a furuncle with the development of one or more abscesses in adjacent subcutaneous tissues. This lesion, known as a carbuncle, occurs most often on the back of the neck (Figure 24–6), but it may involve other skin sites. Carbuncles are serious lesions that may result in bloodstream invasion (bacteremia).

Focal lesions drain spontaneously

Boils develop in hair follicles

Multiple boils become a carbuncle

Chronic Furunculosis

Some individuals are subject to chronic furunculosis, in which repeated attacks of boils are caused by the same strain of *S aureus.* There is little, if any, evidence of acquired immunity to the disease; indeed, delayed-type hypersensitivity to staphylococcal products appears

Links to immune dysfunction are limited

responsible for much of the inflammation and necrosis that develops. Chronic staphylococcal disease may be associated with factors that depress host immunity, especially in patients with diabetes or congenital defects of polymorphonuclear leukocyte function. However, in most instances, predisposing disease other than acne is not present.

Impetigo

Exfoliatin-producing strains cause bullous impetigo

S aureus is most often seen as a secondary invader in group A streptococcal pustular impetigo (see Chapter 25), but it can produce the skin pustules of impetigo on its own. Strains of *S aureus* that produce exfoliatin cause a characteristic form called **bullous impetigo**, characterized by large blisters containing many staphylococci in the superficial layers of the skin. Bullous impetigo is a localized form of staphylococcal scalded skin syndrome. ::: Impetigo, p. 449

Deep Lesions

Acute osteomyelitis is primarily a *S aureus* disease

Pneumonia and deep tissue lesions are highly destructive

Bacteremic spread and endocarditis are most common in drug abusers

S aureus can cause a wide variety of infections of deep tissues by bacteremic spread from a skin lesion that may be unnoticed. These include infections of bones, joints, deep organs, and soft tissues, including surgical wounds. More than 90% of the cases of acute osteomyelitis in children are caused by *S aureus*. Staphylococcal pneumonia is typically secondary to some other insult to the lung, such as influenza, aspiration, or pulmonary edema. At deep sites, the organism has the same tendency to produce localized, destructive abscesses as it does in the skin. All too often the containment is less effective, and spread with multiple metastatic lesions occurs. Bacteremia and endocarditis can develop. All are serious infections that constitute acute medical emergencies. In all these situations, diabetes, leukocyte defects, or general reduction of host defenses by alcoholism, malignancy, old age, or steroid or cytotoxic therapy can be predisposing factors. Severe *S aureus* infections, including endocarditis, are particularly common in drug abusers using injection methods.

MANIFESTATIONS CAUSED BY STAPHYLOCOCCAL TOXINS

Scalded Skin Syndrome

Widespread desquamation in neonates is caused by exfoliatin-producing strains

Staphylococcal scalded skin syndrome results from the production of exfoliatin in a staphylococcal lesion, which can be minor (eg, conjunctivitis). Erythema and intraepidermal desquamation takes place at remote sites from which *S aureus* cannot be isolated (**Figure 24–7**). The disease is most common in neonates and children less than 5 years of age. The face, axilla, and groin tend to be affected first, but the erythema, bullous formation, and subsequent desquamation of epithelial sheets, can spread to all parts of the body. The disease occasionally occurs in adults, particularly those who are immunocompromised. Milder versions of what is probably the same disease are staphylococcal scarlet fever, in which erythema occurs without desquamation, and bullous impetigo, in which local desquamation occurs.

Toxic Shock Syndrome

Fever, vomiting, diarrhea, and muscle pain are early findings

Shock, renal, and hepatic injury may follow

Toxic shock syndrome (TSS) was first described in children, but came to public attention during the early 1980s when hundreds of cases were reported in young women using intravaginal tampons. The disease is characterized by high fever, vomiting, diarrhea, sore throat, and muscle pain. Within 48 hours, it may progress to severe shock with evidence of renal and hepatic damage. A skin rash may develop, followed by desquamation at a deeper level than in scalded skin syndrome. Blood cultures are usually negative. The outbreak receded with the withdrawal of certain brands of highly absorbent tampons.

Staphylococcal Food Poisoning

Vomiting is prominent without fever

Ingestion of staphylococcal enterotoxin–contaminated food results in acute vomiting and diarrhea within 1 to 5 hours. There is prostration, but usually no fever. Recovery is rapid, except sometimes in the elderly and in those with another disease.

DIAGNOSIS

Laboratory procedures to assist in diagnosis of staphylococcal infections are quite simple. Most acute, untreated lesions contain numerous polymorphonuclear leukocytes and large numbers of Gram-positive cocci in clusters. Staphylococci grow overnight on blood agar incubated aerobically. Catalase and coagulase tests performed directly from the colonies are sufficient for identification. Antibiotic susceptibility tests are indicated because of the emerging resistance to multiple antimicrobics, particularly methicillin-resistant *S aureus* (MRSA). Deep staphylococcal infections such as osteomyelitis and deep abscesses present special diagnostic problems when the lesion cannot be directly aspirated or surgically sampled. Blood cultures are usually positive in conditions such as acute staphylococcal arthritis, osteomyelitis, and endocarditis, but less often in localized infection such as deep abscesses.

Gram stain and culture are primary diagnostic methods

Aspirates and blood cultures are necessary for deep infections

TREATMENT

Most boils and superficial staphylococcal abscesses resolve spontaneously without antimicrobial therapy. Those that are more extensive, deeper, or in vital organs require a combination of surgical drainage and antimicrobics for optimal outcome. Penicillins and cephalosporins are active against *S aureus* cell wall peptidoglycan and vary in their susceptibility to inactivation by staphylococcal β-lactamases. Although penicillin G is the treatment of choice for susceptible strains, the penicillinase-resistant penicillins (methicillin, nafcillin, oxacillin) and first-generation cephalosporins are more commonly used because of the high frequency of penicillin resistance (more than 80%). For MRSA strains resistant to these agents or in patients with β-lactam hypersensitivity, the alternatives are vancomycin, clindamycin, or erythromycin. Synergy between cell wall–active antibiotics and the aminoglycosides is present when the staphylococcus is sensitive to both types of agents. Such combinations are often used in severe systemic infections when effective and rapid bactericidal action is needed, particularly in compromised hosts.

Superficial lesions resolve spontaneously

Penicillinase-resistant β-lactams are used pending susceptibility tests

ANTIMICROBIAL RESISTANCE

When penicillin was introduced to the general public after World War II, virtually all strains of *S aureus* were highly susceptible. Since then, the selection of preexisting strains able to produce a penicillinase has shifted these proportions to the point at which 80% to 90% of isolates are now penicillin-resistant. The penicillinase is encoded by plasmid genes and acts by opening the β-lactam ring, making the drug unable to bind with its target. ::: Enzymatic inactivation resistance, p. 422

Most strains of *S aureus* are now penicillin-resistant

Penicillinase production is plasmid mediated

Alterations in the β-lactam target, the peptidoglycan transpeptidases (often called penicillin-binding proteins, or PBPs), are the basis for resistance to methicillin. These MRSA strains are also resistant to the other penicillinase-resistant penicillins such as oxacillin. The most common mechanism is the acquisition of a gene (*mecA*) for a new transpeptidase, which has reduced affinity for β-lactam antibiotics, but is still able to carry out its enzymatic function of cross-linking peptidoglycan. ::: Altered-target resistance, p. 421

Methicillin-resistant strains produce new PBP

The incidence of MRSA has great geographic variation. Most American hospitals report MRSA rates of 5% to 25%, but outbreaks are increasing and resistance rates over 50% have been reported in other countries. Tests are generally performed with methicillin or oxacillin under technical conditions that facilitate detection of what may be a small resistant subpopulation, and the results extrapolated to other relevant agents. For example, oxacillin resistance is considered proof of resistance to methicillin, nafcillin, dicloxacillin, and all cephalosporins. Methods for direct detection of the *mecA* gene have been developed but are not yet practical for widespread use. Vancomycin is often used to treat serious infections with MRSA. The recent emergence of *S aureus* with decreased susceptibility to vancomycin is still uncommon but of great concern. For strains resistant to both methicillin and vancomycin daptomycin is a new alternative.

MRSA rates are variable but increasing

MRSA detection requires special conditions

Vancomycin use for MRSA is threatened

PREVENTION

In patients subject to recurrent infection such as chronic furunculosis, preventive measures are aimed at controlling reinfection and, if possible, eliminating the carrier state. Clothes and bedding that may cause reinfection should be dry-cleaned or washed at a sufficiently high temperature (70°C or higher) to destroy staphylococci. In adults, the use of chlorhexidine or hexachlorophene soaps in showering and washing increases the bactericidal activity of the skin. In such individuals, or persons found to be a source of an outbreak, anterior nasal carriage can be reduced and often eliminated by the combination of nasal creams containing topical antimicrobics (eg, mupirocin, neomycin, and bacitracin) and oral therapy with antimicrobics that are concentrated within phagocytes and nasal secretions (eg, rifampin or ciprofloxacin). Attempts to reduce nasal carriage more generally among medical personnel in an institution are usually fruitless and encourage replacement of susceptible strains with multiresistant ones. ::: Chlorhexidine, hexachlorophene, p. 52

Antistaphylococcal soaps block infection

Elimination of nasal carriage is difficult

Chemoprophylaxis is effective in surgical procedures such as hip and cardiac valve replacements, in which infection with staphylococci can have devastating consequences. Methicillin, a cephalosporin, or vancomycin given during and shortly after surgery may reduce the chance for intraoperative infection while minimizing the risk for superinfection associated with longer periods of antibiotic administration.

Chemoprophylaxis during high-risk surgery is effective

Coagulase-Negative Staphylococci

Other than *S aureus,* there are more than 30 species of staphylococci. In medical practice, the 15 species that have been isolated from human infections are typically lumped together by a negative characteristic—failure to produce coagulase. These coagulase-negative staphylococci (CoNS) also do not produce α-toxin, exfoliatin, or any of the StaphSAg toxins. They have been shown to have surface adhesins and the ability to produce extracellular polysaccharide biofilms. By far the most common CoNS species isolated from human infections is *S epidermidis,* and *S saprophyticus* is a significant cause of urinary tract infections. Clinical laboratories rarely speciate CoNS isolates, although a simple test (novobiocin resistance) is often used to separate *S saprophyticus* from other urinary isolates.

CoNS DISEASE

S epidermidis and many other species of CoNS are normal commensals of the skin, anterior nares, and ear canals of humans. Their large numbers and ubiquitous distribution result in frequent contamination of specimens collected from or through the skin. In the past, they were rarely the cause of significant infections, but with the increasing use of implanted catheters and prosthetic devices, they have emerged as important agents of hospital-acquired infections. Immunosuppressed or neutropenic patients and premature infants have been particularly affected.

Common colonizers of the skin

Commonly colonize implanted medical devices

Organisms may contaminate prosthetic devices during implantation, seed the device during a subsequent bacteremia, or gain access to the lumina of shunts and catheters when they are temporarily disconnected or manipulated. The outcome of the bacterial contamination is determined by the ability of the microbe to attach to the surface of the foreign body and to multiply there. Central to this process is the ability of these species to form a viscous extracellular polysaccharide **biofilm**. Initial adherence is facilitated by the hydrophobic nature of the synthetic polymers used in medical devices, surface proteins, and the polysaccharide itself. As it expands, this biofilm provides additional adhesion, encases the entire bacterial population (**Figure 24–9**), and serves as a barrier to antimicrobial agents and host defense mechanisms.

Polysaccharide biofilm production enhances attachment and survival

The above-mentioned circumstances are found almost exclusively in hospitals and other medical facilities. The most common device colonized is the intravenous catheter, but the same mechanisms apply to any implanted device such as cerebrospinal fluid shunts and artificial heart valves. The ensuing disease is typically low grade with little more than a slowly advancing fever to arouse suspicion. *S aureus* can also produce biofilms, and although a less frequent colonizer of medical devices, it is likely to produce a more aggressive course and metastatic infections. Removal of the contaminated device is the only sure way to avoid these complications.

Catheters, shunts, and artificial valves become colonized

The biology of *S saphrophyticus* infection is entirely different. Its usual habitat is the gastrointestinal tract, and from that location the organism gains access to the urinary tract.

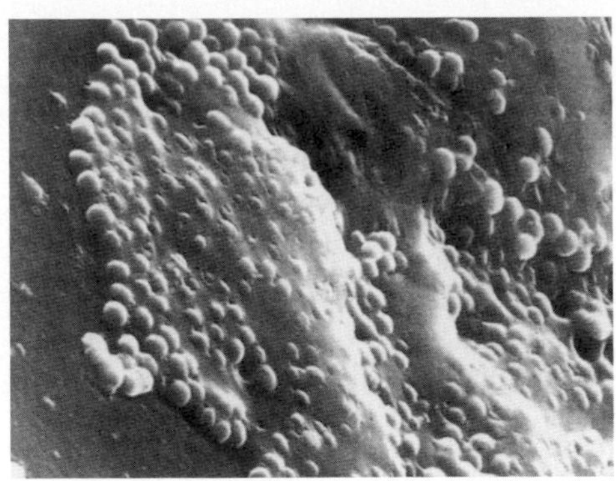
A

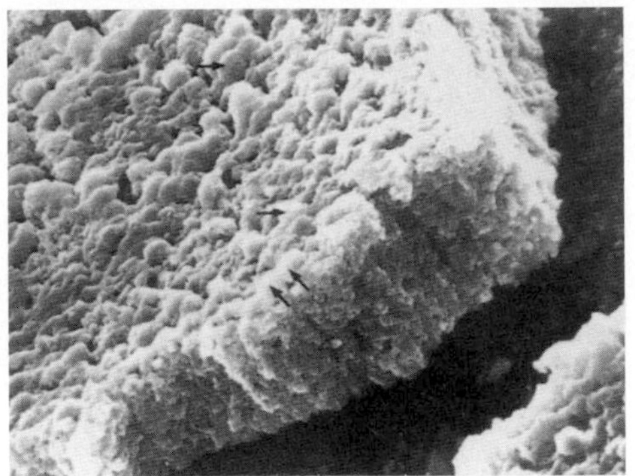
B

FIGURE 24–9. Coagulase negative staphylococcal slime. A. *S epidermidis* cocci are shown attached to the surface of a plastic catheter and are starting to produce extracellular polysaccharide slime. **B.** After 48 hours, the bacteria are fully embedded in the slime glycocalyx. (Reproduced with permission from Connor DH, Chandler FW, Schwartz DQA, Manz HJ, Lack EE (eds). *Pathology of Infectious Diseases,* vol. 1. Stamford, CT: Appleton & Lange; 1997.)

Among sexually active women, *S saphrophyticus* is second only to *Escherichia coli* as a cause of acute urinary tract infection. This process may be aided by surface adhesins to uroepithelial cells and the production of a urease. Thus, although other CoNS are causes of infection among compromised patients in hospitals, *S saphrophyticus* produces community-acquired infection in women who are otherwise healthy.

S saphrophyticus causes urinary infections

The interpretation of cultures that grow CoNS is difficult. In most cases, the finding is attributable to skin contamination during collection of the specimen. The presence of at least moderate numbers of organisms or repeated isolations from the same site argue for infection over skin contamination. Specimens collected directly from catheters and shunts typically show large numbers of staphylococci. Most coagulase-negative staphylococci now encountered are resistant to penicillin, and many are also methicillin-resistant. Resistance to multiple antimicrobics usually active against Gram-positive cocci, including vancomycin, is more common than with *S aureus.* Eradication of coagulase-negative staphylococci from prosthetic devices and associated tissues with chemotherapy alone is very difficult unless the device is also removed.

Repeated positives suggest infection

Multiple antimicrobic resistance is common

CASE STUDY

AFTERMATH OF A BICYCLE FALL

A 14-year-old boy presented with a 3-day history of vomiting, diarrhea, sore throat, headache, weakness, and fever. His temperature was 39.9°C. He had pharyngeal inflammation, and his blood pressure was 60/0 mm Hg while supine and unobtainable when sitting. Initial laboratory findings included white blood cell (WBC) count of 13,600l/mL with a pronounced left shift (ie, many immature forms), blood urea nitrogen (BUN) of 24 mg/dL (normal up to 15 mg/dL), and abnormal urinalysis, with 20 to 30 WBCs and 8 to 10 red blood cells (RBC) per high-power field.

He was treated with large volumes of intravenous fluids and with penicillin; his blood pressure rose, but he had multiple episodes of disorientation, and diffuse erythroderma developed. On admission, a small crusted wound had been noticed on the dorsum of his left foot (the result of a bicycle injury 1 week earlier); 45 hours later the wound became red, warm, and pustular, and a left femoral lymph node became tender and enlarged. A culture of the pustule grew *S aureus* coagulase-positive resistant to penicillin. Several cultures of blood and a throat swab taken before antibiotic therapy was started had been negative. He improved with cephalexin therapy. He had extensive desquamation of the skin of the palms and soles 2 weeks after discharge.

QUESTIONS

1. Which one of the following is most responsible for the nature of the lesion on this boy's foot?

A. Coagulase
B. Catalase
C. Superantigen toxin (StaphSAg)
D. Exfoliatin
E. α-toxin

2. The boy's hypotension and elevated BUN are most probably due to the action of:

A. α-toxin
B. Cytokines
C. Peptidoglycan
D. Catalase
E. Exfoliatin

3. The desquamation of the skin is most probably due to the action of:

A. Exfoliatin
B. Coagulase
C. Superantigen toxin (StaphSAg)
D. Penicillin
E. Fibronectin binding protein (FnBP)

4. The blood culture was negative. What is the best explanation for this?

A. The penicillin may have caused a false-negative
B. There must have been a problem with the blood collection
C. There must have been an error in the laboratory
D. This is typical in staphylococcal TSS. Only the StaphSAg needs to circulate

ANSWERS

1(E), **2**(B), **3**(C), **4**(D)

CHAPTER 25

Streptococci and Enterococci

Scarlet fever awes me, and is above my aim. I leave it to the professional and graduated homicides.

—Sydney Smith, 1833

Bacteria of the genus *Streptococcus* are Gram-positive cocci typically arranged in chains. In addition to relatively harmless members of the oropharyngeal flora, the genus includes three of the most important pathogens of humans. The group A streptococcus (*S pyogenes*) is the cause of "strep throat," which can lead to scarlet fever, rheumatic fever, and rheumatic heart disease; the ability of some strains to cause catastrophic deep tissue infections led British tabloids to apply the gory label "flesh-eating bacteria." The group B streptococcus (*S agalactiae)* is the most common cause of sepsis in newborns and the pneumococcus (*S pneumoniae*) a leading cause of both pneumonia and meningitis in persons of all ages.

STREPTOCOCCI

Group Characteristics

MORPHOLOGY

Streptococci stain readily with common dyes, demonstrating coccal cells that are generally smaller and more ovoid in shape than staphylococci. They are usually arranged in chains with oval cells touching end to end, because they divide in one plane and tend to remain attached (**Figure 25–1**). Length may vary from a single pair to continuous chains of over 30 cells, depending on the species and growth conditions. Medically important streptococci are not acid-fast, do not form spores, and are nonmotile. Some members form capsules composed of polysaccharide complexes or hyaluronic acid.

Oval cells arranged in chains end to end

CULTURAL AND BIOCHEMICAL CHARACTERISTICS

Streptococci grow best in enriched media under aerobic or anaerobic conditions (facultative). Blood agar is preferred because it satisfies the growth requirements and also serves as an indicator for patterns of hemolysis. The colonies are small, ranging from pinpoint size to 2 mm in diameter, and they may be surrounded by a zone where the erythrocytes suspended

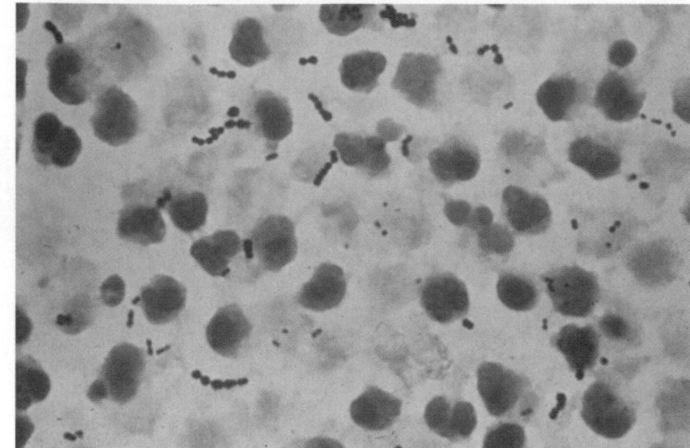

FIGURE 25–1. Group A streptococcus (GAS) Gram stain. Note the oval cocci chaining end to end. (Reprinted with permission of Schering Corporation, Kenilworth, NJ, the copyright owner. All rights reserved.)

β-Hemolysis is clear

α-Hemolysis is greening of blood agar

Catalase negative

in agar have been hemolyzed. When the zone is clear, this state is called **β-hemolysis** (**Figure 25–2**). When the zone is hazy with a green discoloration of the agar, it is called **α-hemolysis.** Streptococci are metabolically active, attacking a variety of carbohydrates, proteins, and amino acids. Glucose fermentation yields mostly lactic acid. In contrast to staphylococci, streptococci are catalase negative.

CLASSIFICATION

Lancefield antigens are cell wall carbohydrates

Presence of Lancefield antigens defines the pyogenic streptococci

At the turn of the 20th century, a classification based on hemolysis and biochemical tests was sufficient to associate some streptococcal species with infections in humans and animals. Rebecca Lancefield, who demonstrated carbohydrate antigens in cell-wall extracts of the β-hemolytic streptococci, put this taxonomy on a sounder basis. Her studies formed a classification by serogroups (eg, A, B, C), each of which is generally correlated with one of the previously established species. Later it was discovered that some nonhemolytic streptococci had the same cell wall antigens. Over the years it has become clear that possession of one of the Lancefield antigens defines a particularly virulent segment of the streptococcal genus regardless of hemolytic patterns. These are called the **pyogenic streptococci,** and in medical circles they are now better known by their Lancefield letter than the older species name. Pediatricians instantly recognize GBS as an acronym for group B streptococcus, but may be confused by use of the proper name, *Streptococcus agalactiae* (**Table 25–1**).

Hemolysis is a practical guide to classification

Only pyogenic streptococci are β-hemolytic

For practical purposes, the type of hemolysis and certain biochemical reactions remain valuable for the initial recognition and presumptive classification of streptococci, and as an indication of what subsequent taxonomic tests to perform. Thus, β-hemolysis indicates that the strain has one of the Lancefield group antigens, but some Lancefield-positive strains or groups may be α-hemolytic or even nonhemolytic. The streptococci are considered here as follows: (1) pyogenic streptococci (Lancefield groups); (2) pneumococci; and (3) viridans and other streptococci (Table 25–1).

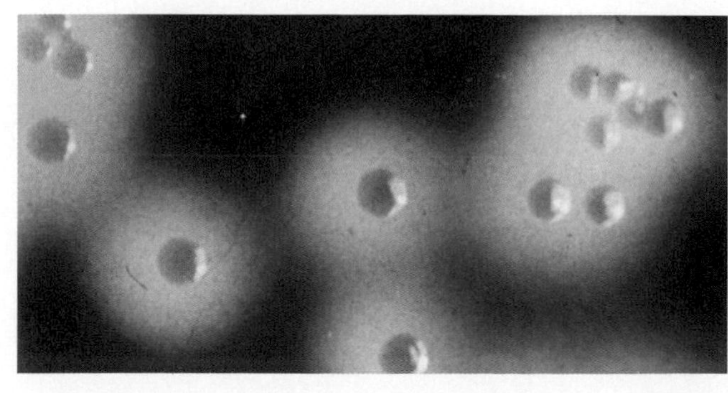

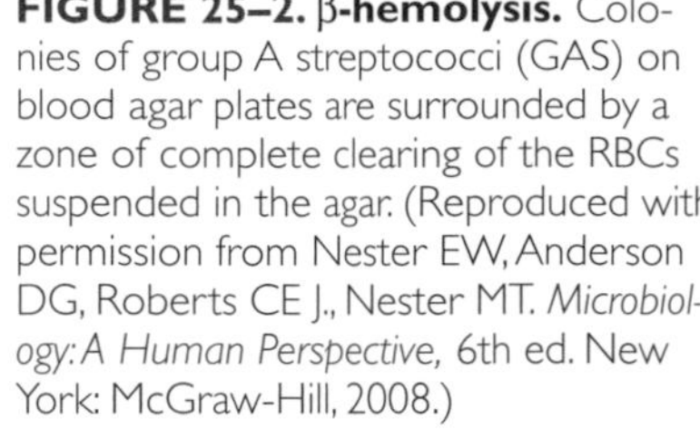

FIGURE 25–2. β-hemolysis. Colonies of group A streptococci (GAS) on blood agar plates are surrounded by a zone of complete clearing of the RBCs suspended in the agar. (Reproduced with permission from Nester EW, Anderson DG, Roberts CE J., Nester MT. *Microbiology: A Human Perspective,* 6th ed. New York: McGraw-Hill, 2008.)

TABLE 25–1 Classification of Streptococci and Enterococci

			MAJOR ANTIGENS/STRUCTURES				
GROUP/SPECIES	**COMMON TERM**	**HEMOLYSIS**	**LANCEFIELD CELL WALL**	**SURFACE PROTEIN**	**CAPSULE**	**VIRULENCE FACTORS**	**DISEASE**
Streptococci							
Pyogenic							
Streptococcus pyogenes	Group A strep (GAS)	β	A	M protein (80+)	Hyaluronic acid	M protein, leipoteichoic acid, streptococcal pyrogenic exotoxins, streptolysin O, streptokinase	Strep throat, impetigo, pyogenic infections, toxic shock, rheumatic fever, glomerulonephritis
S agalactiae	Group B strep (GBS)	β, –	B	–	Sialic acid (9)	Capsule	Neonatal sepsis, meningitis, pyogenic infections
S equi		β	C	–	–	–	Pyogenic infections
S bovis		–, α	D	–	–	–	Pyogenic infections
Other species		β, α, –	E-W	–	–	–	Pyogenic infections
Pneumococcus							
S pneumoniae	Pneumococcus	α	–	Choline-binding protein	Polysaccharide (90+)	Capsule, pneumolysin, neuraminidase	Pneumonia, meningitis, otitis media, pyogenic infections
Viridans and Nonhemolytic							
S sanguis		α	–	–	–	–	Low virulence, endocarditis
S salivarius		α					Low virulence, endocarditis
S mutans		α	–	–			Dental caries
Other species		α, –	–	–	–	–	Low virulence, endocarditis
Enterococci							
Enterococcus faecalis	Enterococcus	–, α	D	–	–	–	Urinary tract, pyogenic infections
E faecium	Enterococcus	–, α	D	–	–	–	Urinary tract, pyogenic infections
Other species		–, α	D, –	–	–	–	Urinary tract, pyogenic infections

Pyogenic Streptococci

Groups A and B streptococci are most common cause of disease

Of the many Lancefield groups, the ones most frequently isolated from humans are A, B, C, F, and G. Of these, groups A *(S pyogenes)* and B *(S agalactiae)* are the most common causes of serious disease. The group D carbohydrate is found in the genus Enterococcus, which used to be classified among the streptococci.

Pneumococci

Pneumococci have an antigenic polysaccharide capsule

This category contains a single species, *S pneumoniae,* commonly called the pneumococcus. Its distinctive feature is the presence of a capsule composed of polysaccharide polymers that vary in antigenic specificity. More than 90 capsular immunotypes have been defined. Although the pneumococcal cell wall shares some common antigens with other streptococci, it does not possess any of the Lancefield group antigens. *S pneumoniae* is α-hemolytic.

Viridans and Other Streptococci

Viridans and nonhemolytic species lack Lancefield antigens or capsules

Viridans streptococci are α-hemolytic and lack both the group carbohydrate antigens of the pyogenic streptococci and the capsular polysaccharides of the pneumococcus. The term encompasses several species, including *S salivarius* and *S mitis.* Viridans streptococci are members of the normal oral flora of humans. They rarely demonstrate invasive qualities. A variety of other streptococci may be encountered, which lack the features of the pyogenic streptococci or pneumococci; these would be classified with the viridans group, except that they are not α-hemolytic. Such strains are usually assigned descriptive terms such as nonhemolytic streptococci or microaerophilic streptococci. They have been less thoroughly studied, but generally have the same biologic behavior as the viridans streptococci.

Group A Streptococci *(Streptococcus pyogenes)*

BACTERIOLOGY

MORPHOLOGY AND GROWTH

Streptolysin O or S cause β-hemolysis

Aerobically, only S is active

Group A streptococci (GAS) typically appear in purulent lesions or broth cultures as spherical or ovoid cells in chains of short to medium length (4 to 10 cells). On blood agar plates, colonies are usually compact, small, and surrounded by a 2- to 3-mm zone of β-hemolysis (Figure 25–2), which is easily seen and sharply demarcated. β-Hemolysis is caused by either of two hemolysins, **streptolysin S** and the oxygen-labile **streptolysin O,** both of which are produced by most group A strains. Strains that lack streptolysin S are β-hemolytic only under anaerobic conditions, because the remaining streptolysin O is not active in the presence of oxygen. This feature is of practical importance, because such strains would be missed if cultures were incubated only aerobically.

STRUCTURE

Wall contains group antigen with multiple surface molecules extending beyond

The structure of GAS is illustrated in **Figure 25–3.** The cell wall is built on a peptidoglycan matrix that provides rigidity, as in other Gram-positive bacteria. Within this matrix lies the group carbohydrate antigen, which by definition is present in all GAS. A number of other molecules such as M protein and lipoteichoic acid (LTA) are attached to the cell wall, but extend beyond often in association with the hair-like pili. GAS are divided into more than 100 serotypes based on antigenic differences in the M protein. Some strains have an overlying nonantigenic hyaluronic acid capsule. ::: Gram-positive cell wall, p. 351

M Protein

The M protein itself is a fibrillar coiled-coil molecule **(Figure 25–4)** with structural homology to myosin. Its carboxy terminus is rooted in the peptidoglycan of the cell wall, and

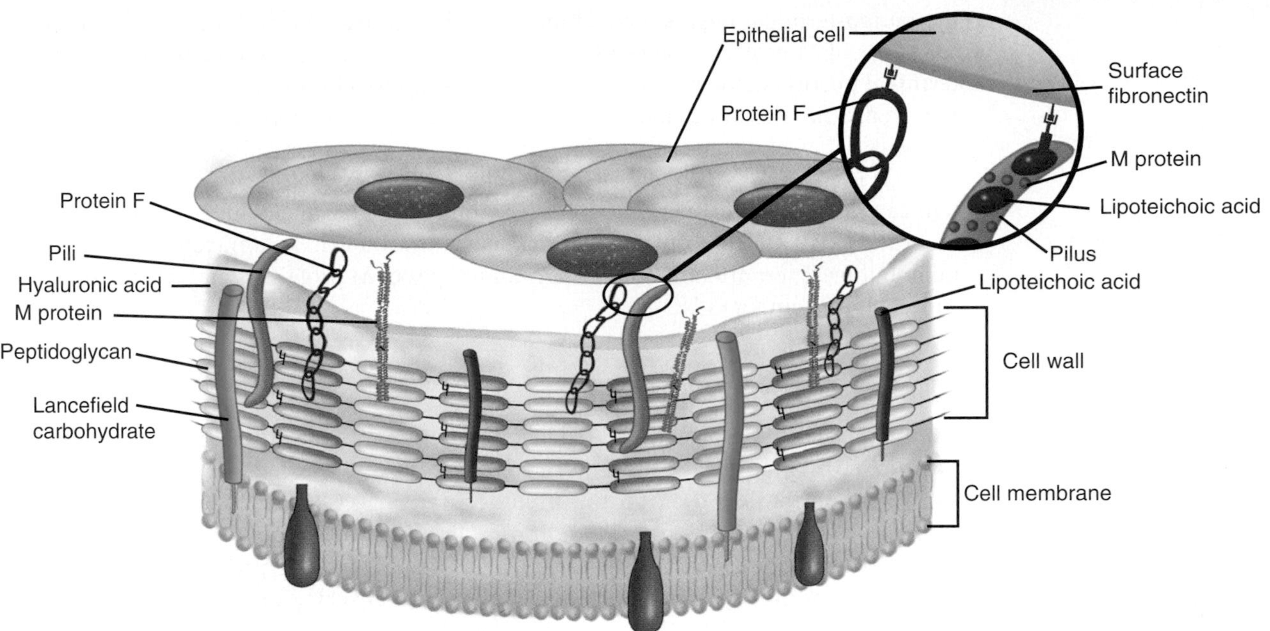

FIGURE 25–3. Antigenic structure of group A streptococci and adhesion to an epithelial cell. The location of peptidoglycan and Lancefield carbohydrate antigens in the cell wall is shown in the diagram. M protein and lipoteichoic acid are associated with the cell surface and the pili. Lipoteichoic acid and protein F mediate binding to fibronectin on the host surface.

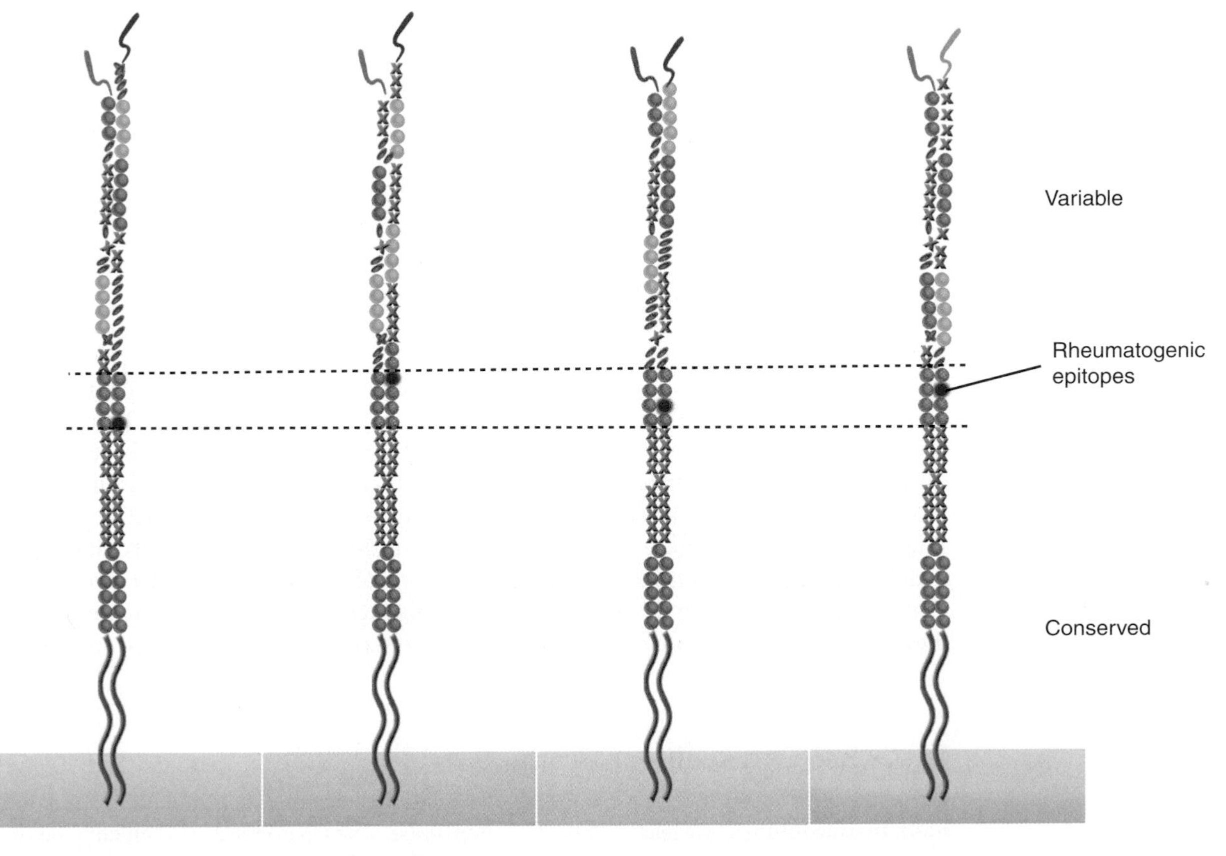

FIGURE 25–4. M protein. The coiled-coil structure of M protein is shown for four hypothetical serotypes. The most variable parts of the molecule are oriented to the outside and provide the antiphagocytic effect and serologic specificity for each type. The conserved potions are rooted in the cell wall and are homologus in structure. All four types contain epitopes that may stimulate the cross-reactive immune reactions seen in rheumatic fever.

the amino-terminal regions extend out from the surface. The specificity of the multiple serotypes of M protein is determined by variations in the amino sequence of the amino-terminal portion of the molecule. Because of its exposed location, this part of the M protein is also the most available to immune surveillance. The middle part of the molecule is less variable, and some carboxy-terminal regions are conserved across many M types. There is increasing evidence that some of the many known biologic functions of M protein can be assigned to specific domains of the molecule. This includes both antigenicity and the capacity to bind other molecules such as fibrinogen, serum factor H, and immunoglobulins. There are more than 80 immunotypes of M protein, which are the basis of a subtyping system for GAS.

Coiled-coil is similar to myosin

Antigenicity and function differ in domains of the molecule

80+ M protein serotypes exist

Other Surface Molecules

A number of surface proteins have been described on the basis of their similarity with M protein or some unique binding capacity. Of these, a fibronectin-binding **protein F** and **LTA** are both exposed on the streptococcal surface (Figure 25–3) and may have a role in pathogenesis. An IgG-binding protein has the capacity to bind the Fc portion of antibodies in much the same way as staphylococcal protein A. In principle, this could interfere with opsonization by creating a covering of antibody molecules on the streptococcal surface that are facing the "wrong way." GAS may have a **hyaluronic acid capsule,** which is a polymer containing repeating units of glucuronic acid and *N*-acetylglucosamine. ::: LTA, p. 353

Protein F and LTA bind fibronectin

Hyaluronic acid capsule may be present

BIOLOGICALLY ACTIVE EXTRACELLULAR PRODUCTS

Streptolysin O

Streptolysin O is a pore-forming cytotoxin, lysing leukocytes, tissue cells, and platelets. The toxin inserts directly into the cell membrane of host cells, forming transmembrane pores in a manner similar to complement and staphylococcal α-toxin. Streptolysin O is antigenic, and the quantitation of antibodies against it is the basis of a standard serologic test called antistreptolysin O (ASO). ::: Pore-forming toxins, p. 396

Streptolysin O is pore-forming and antigenic

Streptococcal Superantigen Toxins

Just as with *Staphylococcus aureus,* approximately 10% of GAS produce one of a family of exotoxins whose major biologic effect is through the superantigen (SAg) mechanism. Over many decades, these toxins have been assigned a number of names linked to their association with **scarlet fever** (erythrogenic toxin) and with streptococcal toxic shock (streptococcal pyrogenic exotoxins [Spe]). As with *S aureus*, there are several antigenically distinct proteins (SpeA, SpeB, and so on). Because the staphylococcal and streptococcal SAgs have similar amino acid structure and biologic activity, in this book they are called **StaphSAgs** and **StrepSAgs.** StrepSAgs have multiple effects, including fever, rash (scarlet fever), T-cell proliferation, B-lymphocyte suppression, and heightened sensitivity to endotoxin. Most of these actions are due to cytokine release through the superantigen mechanism. ::: Superantigen exotoxin, p. 396

StrepSAgs are produced by some strains

StrepSAgs and StaphSAgs are superantigens

Other Extracellular Products

Most strains of GAS produce a number of other extracellular products including **streptokinase, hyaluronidase,** nucleases, and a **C5a peptidase.** The C5a peptidase is an enzyme that degrades complement component C5a, the main factor that attracts phagocytes to sites of complement deposition. The enzymatic actions of the others likely play some role in tissue injury or spread, but no specific roles have been defined. Some are antigenic and have been the basis of serologic tests. Streptokinase causes lysis of fibrin clots through conversion of plasminogen in normal plasma to the protease plasmin.

C5a peptidase degrades complement

Streptokinase converts plasminogen to plasmin

GROUP A STREPTOCOCCAL DISEASE

CLINICAL CAPSULE

Group A streptococci are the cause of "strep throat," an acute inflammation of the pharynx and tonsils that includes fever and painful swallowing. Skin and soft tissue infections range from the tiny skin pustules called impetigo to a severe toxic and invasive disease that can be fatal in a matter of days. In addition to acute infections, GAS are responsible for inflammatory diseases that are not direct infections but represent states in which the immune response to streptococcal antigens causes injury to host tissues. Acute rheumatic fever is a prolonged febrile inflammation of connective tissues, which recurs after each subsequent streptococcal pharyngitis. Repeated episodes cause permanent scarring of the heart valves. Acute glomerulonephritis is an insidious disease with hypertension, hematuria, proteinuria, and edema due to inflammation of the renal glomerulus.

EPIDEMIOLOGY

Pharyngitis

Group A streptococci are the most common bacterial cause of pharyngitis in school-age children 5 to 15 years of age. Transmission is person to person from the large droplets produced by infected persons during coughing, sneezing, or even conversation (**Figure 25–5**). This droplet transmission is most efficient at the short distances (2 to 5 feet) at which social interactions commonly take place in families and schools, particularly in fall and winter months. Asymptomatic carriers (less than 1%) may also be the source of GAS, particularly if colonized in the nose as well as the throat. Although GAS survive for some time in dried secretions, environmental sources and fomites are not important means of spread. Unless the condition is treated, the organisms persist for 1 to 4 weeks after symptoms have disappeared.

Most common bacterial cause of sore throat

Droplets spread over short distances from throat and nasal sites

Impetigo

Impetigo occurs when transient skin colonization with GAS is combined with minor trauma such as insect bites. The tiny skin pustules are spread locally by scratching and to others by direct contact or shared fomites such as towels. Impetigo is most common in summer months when insects are biting and when the general level of hygiene is low. The M protein types of GAS most commonly associated with impetigo are different from those causing respiratory infection.

Skin colonization plus trauma leads to impetigo

Wound and Puerperal Infections

GAS, once a leading cause of postoperative wound and puerperal infections, retain this potential, but the conditions favoring these diseases are now less common in developed countries. As with staphylococci, transmission from patient to patient is by the hands of physicians or other medical attendants who fail to follow recommended handwashing practices. Organisms may be transferred from another patient or may come from the health care workers themselves.

Hospital outbreaks are linked to carriers

Streptococcal Toxic Shock Syndrome

Since the late 1980s, a severe invasive form of GAS soft tissue infection appeared with increased frequency worldwide. Rapid progression to death in only a few days occurred in previously healthy persons, including Muppet creator Jim Henson of Sesame Street fame. The outstanding features of these infections are their multiorgan involvement, suggesting a toxin and rapid invasiveness with spread to the bloodstream and distant organs. The toxic

STSS may be fatal in healthy persons

Strains produce StrepSAgs

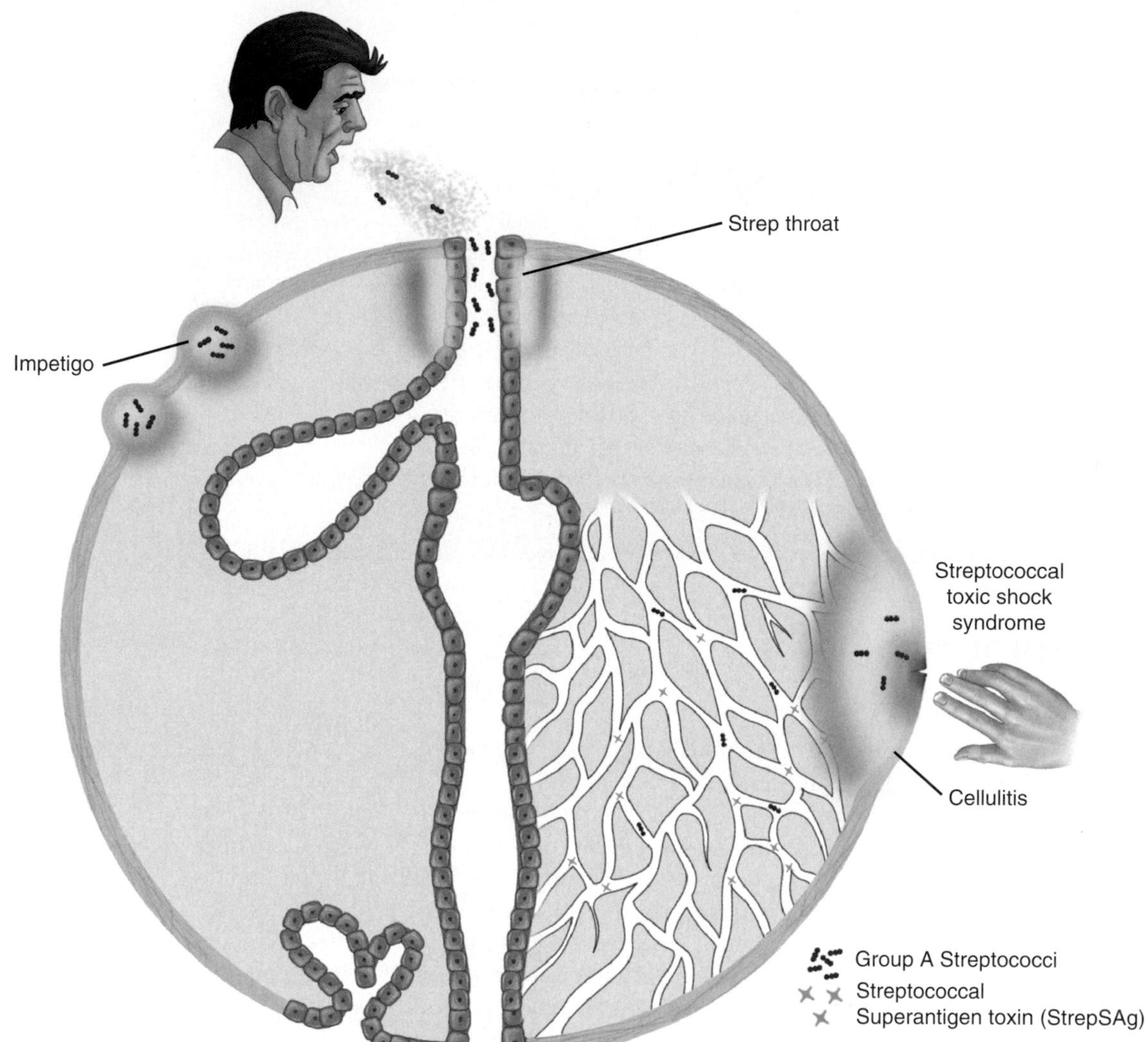

FIGURE 25–5. Group A streptococcal (GAS) disease overview. The primary sources of infection are respiratory droplets or direct contact with the skin. Impetigo results from minor trauma such as insect bites in skin transiently colonized with GAS. In streptococcal toxic shock, StrepSAgs producing GAS in a superficial lesion spread into the bloodstream. Note both toxin and bacteria are circulating.

features together with the discovery that almost all the isolates produce StrepSAgs have caused this syndrome to be labeled **streptococcal toxic shock syndrome (STSS)**.

Poststreptococcal Sequelae

ARF follows respiratory, not skin, infection

Rheumatic heart disease is produced by recurrent ARF

The association between GAS and the inflammatory disease acute rheumatic fever (ARF) is based on epidemiologic studies linking group A streptococcal pharyngitis, the clinical features of rheumatic fever, and heightened immune responses to streptococcal products. ARF does not follow skin or other nonrespiratory infection with GAS. Although some M types are more "rheumatogenic," it is not practical to define risk in advance. The general approach is that recurrences of ARF can be triggered by infection with any GAS. Injury to the heart caused by recurrences of ARF leads to **rheumatic heart disease,** a major cause of heart disease worldwide. Although ARF has declined in developed countries, resurgence in the form of small regional outbreaks began in the late 1980s. These outbreaks involved children of a higher socioeconomic status than that previously associated with ARF and a shift in prevalent M types. The underlying basis of the resurgence is unknown. In contrast, ARF is rampant in many developing countries, particularly in Africa, the Middle East, India, and South America.

Poststreptococcal glomerulonephritis may follow either respiratory or cutaneous GAS infection and involves only certain "nephritogenic" strains. It is more common in temperate climates where insect bites lead to impetigo. The average latent period between infection and glomerulonephritis is 10 days from a respiratory infection, but generally about 3 weeks from a skin infection. Nephritogenic strains are limited to a few M types and seem to have declined in recent years.

Glomerulonephritis follows respiratory or skin infection

Only nephritogenic strains are involved

PATHOGENESIS

Acute Infections

As with other pathogens, adherence to mucosal surfaces is a crucial step in initiating disease (**Figure 25–6**). Along with pili a dozen specific adhesins have been described that facilitate the ability of the group A streptococcus to adhere to epithelial cells of the nasopharynx and/or skin. Of these, the most important are M protein, LTA, and protein F. In the nasopharynx, all three appear to be involved in mediating attachment to the fatty acid binding sites in the glycoprotein fibronectin covering the epithelial cell surface. The role of M protein is not direct, but it appears to provide a scaffold for LTA, which is essential for it to reach its binding site (Figure 25–3).

Surface molecules binding to fibronectin constitutes important first step

M protein supports nasopharyngeal cell adherence

On the other hand, M protein appears to be direct and dominant in binding to the skin through its ability to interact with subcorneal keratinocytes, the most numerous cell type in cutaneous tissue. This adherence takes place at domains of the M protein that bind to receptors on the keratinocyte surface. Protein F is also involved primarily in adherence to antigen-presenting Langerhans cells. Expression of M protein and protein F is regulated in response to environmental conditions (O_2, CO_2), which could play a role in establishing the microbe or in relation to the immune response.

M protein and protein F are involved in epidermis binding

Expression is environmentally regulated

Clinical evidence makes it clear that GAS have the capacity to be highly invasive. The events following attachment that trigger invasion are only starting to be understood. It appears that M protein, protein F, and other fibronectin-binding proteins are required for invasion of non-professional phagocytes. This invasion involves integrin receptors and is accompanied by cytoskeleton rearrangements, but the molecular events do not yet make a coherent story.

Multiple factors are involved in invasion

After the initial events of attachment and invasion, it appears that the concerted activity of the M protein, immunoglobulin-binding proteins, and the C5a peptidase play the key roles in allowing the streptococcal infection to continue (**Figure 25–7**). M protein plays an essential role in GAS resistance to phagocytosis because of the ability of domains of the molecule to bind serum factor H. This leads to a diminished availability of alternative pathway-generated complement component C3b for deposition on the streptococcal surface. In the presence of M type–specific antibody, classical pathway opsonophagocytosis proceeds, and the streptococci are rapidly killed. As a second antiphagocytic mechanism

Antiphagocytic M protein binds factor H

Surface C3b deposition is diminished

C5a peptidase blocks phagocyte chemotaxis

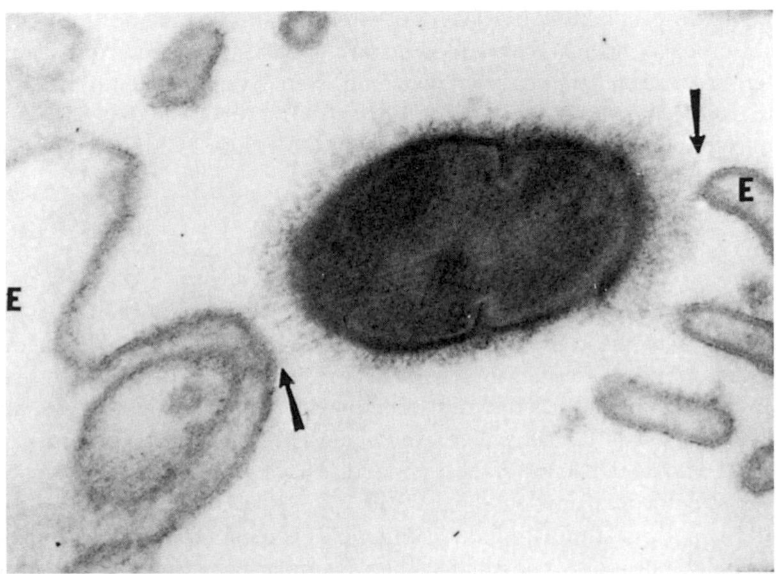

FIGURE 25–6. A group A streptococcus is shown attaching to the cell membrane of a human oral epithelial cell (E). Note the hair-like pili (*arrows*), which mediate the attachment. As in Figure 23–3, both M protein and lipoteichoic acid are associated with the pili. (Reproduced with permission from Beachey EH, Ofek I. J Exp Med 1976;143:764)

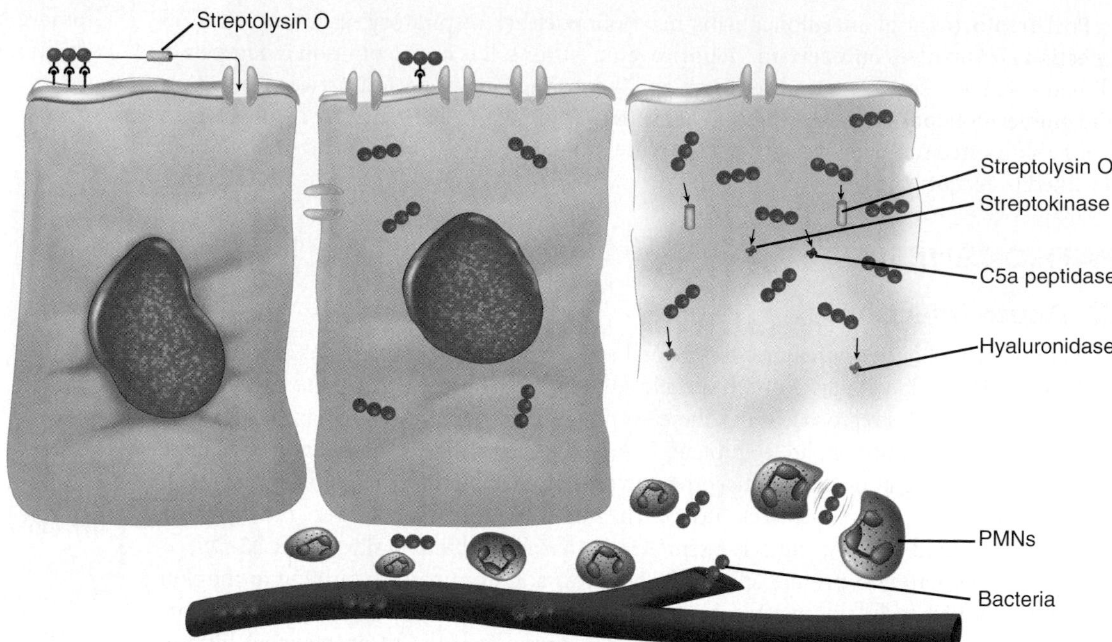

FIGURE 25–7. Group A streptococcal disease, cellular view. The cellular events are similar to *Staphylococcus aureus* (see Figure 24–4). Streptolysin O is a pore-forming toxin (like α-toxin), and there are many extracellular products. A difference is that although *S aureus* tends to be localized, group A streptococci (GAS) tend to spread diffusely, as shown in the cell on the right. This may be due to hyaluronidase (spreading factor) or resistance to phagocytosis. Below the cells, factor H binding is mediating GAS escaping the polymorphonuclear neutrophils (PMNs).

the C5a peptidase inactivates C5a and thus blocks chemotaxis of polymorphonuclear neutrophils (PMNs) and other phagocytes to the site of infection. Although the hyaluronic acid capsule contributes to resistance to phagocytosis, the mechanisms involved are unknown. :::Alternative pathway, p. 26, Antiphagocytic effect, p. 394, Fig. 22–4

Other virulence factors contribute to spread and injury

The precise role of other bacterial factors in the pathogenesis of acute infection is uncertain, but the combined effect of streptokinase, DNAase, and hyaluronidase may prevent effective localization of the infection, whereas the streptolysins produce tissue injury and are toxic to phagocytic cells. Antibodies against these components are formed in the course of streptococcal infection but are not known to be protective.

Superantigenicity of StrepSAgs contributes to STSS

Invasive component is unexplained

In streptococcal toxic shock syndrome (STSS), as with staphylococcal toxic shock syndrome, the findings of shock, renal impairment, and diarrhea seem to be explained by the massive cytokine release stimulated by the superantigenicity of the StrepSAgs. Exotoxin production, however, does not explain the enhanced invasiveness of group A streptococci, which is an added feature of STSS compared to its staphylococcal counterpart. The enzymatic activity of some StrepSAgs have been linked to invasiveness, but the underlying mechanisms are unclear. One theory is that STSS may be due to the horizontal transfer of StrepSAg genes to GAS clones with enhanced invasive potential, a deadly combination.

Poststreptococcal Sequelae

Acute Rheumatic Fever

ARF is an autoimmune state induced by streptococcal infection

Of the many theories advanced to explain the role of GAS in acute rheumatic fever (ARF), an autoimmune mechanism related to antigenic similarities between streptococci antigens and human tissues has the most experimental support. Streptococcal pharyngitis patients who develop ARF have higher levels of antistreptococcal and autoreactive antibodies than those who do not. Some of these antibodies have been shown to react with both heart tissue and streptococcal antigens.

The antigen stimulating these antibodies is most probably M protein, but the group A carbohydrate is also a possibility. There is similarity between the structure of regions of the

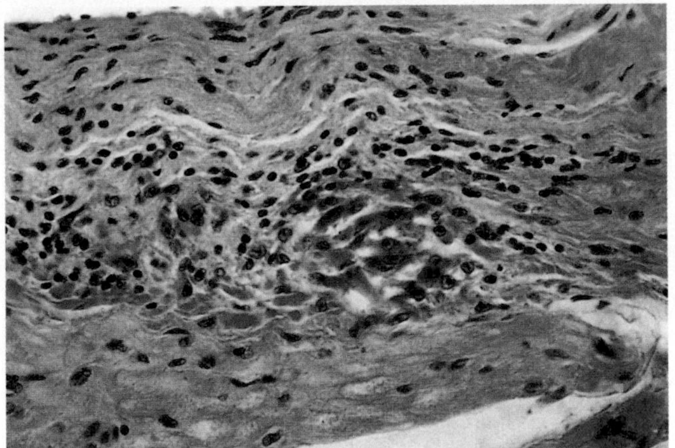

FIGURE 25–8. Aschoff nodule. Reacting lymphocytes and large mononuclear cells in myocardium demonstrate a cellular component to the immune reaction in rheumatic fever. (Reproduced with permission from Connor DH, Chandler FW, Schwartz DQA, Manz HJ, Lack EE (eds). *Pathology of Infectious Diseases,* vol. 1. Stamford, CT: Appleton & Lange; 1997.)

M protein and myosin, and M protein fragments have been shown to stimulate antibodies that bind to human heart sarcolemma membranes, cardiac myosin, synovium, and articular cartilage. ARF is a prime example of the **molecular mimicry** mechanism of Type II autoimmune hypersensitivity. Immunochemical studies of M protein are now directed at locating the epitopes in the large M protein molecule, which stimulate protective antibody (anti-factor H binding sites) and those that stimulate anti-self antibodies. There is evidence these domains are in different locations in the M protein coiled coil (Figure 25–4). If they can be separated, there is hope for an M protein-based vaccine that does not cause the very disease (ARF) it is designed to prevent. A further complication with this approach is establishing the consistency of these relationships among the many M types. ::: Type II hypersensitivity, p. 42

Antibodies react with sarcolemma, myosin, synovium by molecular mimicry

Cross-reactive and protective M protein domains differ

ARF patients also show enhanced T_H1 responses to streptococcal antigens. Cytotoxic T lymphocytes may be stimulated by M protein, and cytotoxic lymphocytes have been observed in the blood of patients with ARF. A cellular reaction pattern consisting of lymphocytes and macrophages aggregated around fibrinoid deposits is found in human hearts. This lesion, called the Aschoff body (**Figure 25–8**), is considered characteristic of rheumatic carditis.

Cell-mediated immunity responses include cytotoxic lymphocytes

Genetic factors are probably also important in ARF because only a small percentage of individuals infected with GAS develop the disease. Attack rates have been highest among those of lower socioeconomic status and vary among those of different racial origins. The gene for an alloantigen found on the surface of B lymphocytes occurs four to five times more frequently in patients with rheumatic fever than in the general population. This further suggests a genetic predisposition to hyperreactivity to streptococcal products.

Alloantigens are associated with hyperreactivity to streptococci

Acute Glomerulonephritis

The renal injury of acute glomerulonephritis is caused by deposition in the glomerulus of antigen-antibody complexes with complement activation and consequent inflammation (type III hypersensitivity). The M proteins of some nephritogenic strains have been shown to share antigenic determinants with glomeruli, which suggests an autoimmune mechanism similar to rheumatic fever. Streptokinase has also been implicated both through molecular mimicry and through its plasminogen activation capacity. ::: Type III hypersensitivity, p. 42

Autoimmune reactions to M protein or streptokinase are implicated

IMMUNITY

It has long been known that antibody directed against M protein is protective for subsequent GAS infections. This protection, however, is only for subsequent infection with strains of the same M type. This is called **type-specific immunity.** This protective IgG is directed against factor H-binding epitopes in the amino-terminal regions of the molecule and reverses the antiphagocytic effect of M protein. Streptococci opsonized with type-specific antibody bind complement C3b by the classical pathway, facilitating phagocyte recognition. There is evidence that mucosal IgA is also important in blocking adherence, whereas the IgG is able to

Type-specific IgG reverses antiphagocytic effect of M protein

Repeated infections and ARF are due to many M types

protect against invasion. Unfortunately, because there are over 80 M types, repeated infections with new M types occur. Eventually, immunity to the common M types is acquired and infections become less common in adults. In ARF patients, it is the hyperreaction seen in each episode that produces the lesions associated with rheumatic heart disease. ::: Classical pathway, p. 28, p. 394 Fig. 22–4

GROUP A STREPTOCOCCAL INFECTIONS: CLINICAL ASPECTS

MANIFESTATIONS

Streptococcal Pharyngitis

Sore throat, fever, malaise

Overlaps with viral pharyngitis

Although it may occur at any age, streptococcal pharyngitis occurs most frequently between the ages of 5 and 15 years. The illness is characterized by acute sore throat, malaise, fever, and headache. Infection typically involves the tonsillar pillars, uvula, and soft palate, which become red, swollen, and covered with a yellow-white exudate. The cervical lymph nodes that drain this area may also become swollen and tender. This clinical syndrome overlaps with viral pharyngitis taking place at the same age.

Spread beyond the pharynx uncommon

GAS pharyngitis is usually self-limiting. Typically, the fever is gone by the third to fifth day, and other manifestations subside within 1 week. Occasionally, the infection spreads locally to produce peritonsillar or retropharyngeal abscesses, otitis media, suppurative cervical adenitis, and acute sinusitis. Rarely, more extensive spread occurs, producing meningitis, pneumonia, or bacteremia with metastatic infection in distant organs. In the preantibiotic era, these suppurative complications were responsible for a mortality rate of 1% to 3% after acute streptococcal pharyngitis. Such complications are much less common now, and fatal infections are rare.

Impetigo

Exposed skin of 2- to 5-year-old children

Tiny pustules may combine to form ulcers

The primary lesion of streptococcal impetigo is a small (up to 1 cm) vesicle surrounded by an area of erythema. The vesicle enlarges over a period of days, becomes pustular, and eventually breaks to form a yellow crust. The lesions usually appear in 2- to 5-year-old children on exposed body surfaces, typically the face and lower extremities. Multiple lesions may coalesce to form deeper ulcerated areas. Although *S aureus* produces a clinically distinct bullous form of impetigo, it can also cause vesicular lesions resembling streptococcal impetigo. Both pathogens are isolated from some cases.

Erysipelas

Spreading erythema of dermal tissues

Erysipelas is a distinct form of streptococcal infection of the skin and subcutaneous tissues, primarily affecting the dermis. It is characterized by a spreading area of erythema and edema with rapidly advancing, well-demarcated edges, pain, and systemic manifestations, including fever and lymphadenopathy. Infection usually occurs on the face (**Figure 25–9**), and a previous history of streptococcal sore throat is common.

Puerperal Infection

GAS causes virulent form of puerperal fever

Infection of the endometrium at or near delivery is a life-threatening form of GAS infection. Fortunately, it is now relatively rare, but in the 19th century, the clinical findings of "childbed fever" were characteristic and common enough to provide the first clues to the transmission of bacterial infections in hospitals. Other organisms can cause puerperal fever, but this form is the most likely to produce a rapidly progressive infection. ::: Childbed fever, p. 53

Disease Associated with Streptococcal Pyrogenic Exotoxins

Scarlet Fever

Infection with strains that elaborate any of the StrepSAgs may superimpose the signs of scarlet fever on a patient with streptococcal pharyngitis. In scarlet fever, the buccal mucosa,

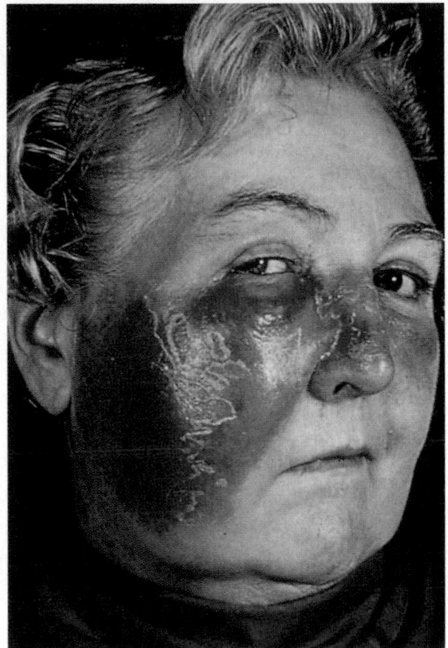

FIGURE 25–9. Streptococcal erysipelas. The diffuse erythema and swelling in the face of this woman are characteristic of group A streptococcal cellulitis at any site. (Reproduced with permission from Connor DH, Chandler FW, Schwartz DQA, Manz HJ, Lack EE (eds). *Pathology of Infectious Diseases*, vol. 1. Stamford, CT: Appleton & Lange; 1997.)

temples, and cheeks are deep red, except for a pale area around the mouth and nose (circumoral pallor). Punctate hemorrhages appear on the hard and soft palates, and the tongue becomes covered with a yellow-white exudate through which the red papillae are prominent (strawberry tongue). A diffuse red "sandpaper" rash appears on the second day of illness, spreading from the upper chest to the trunk and extremities. Circulating antibody to the toxin neutralizes these effects. For unknown reasons, scarlet fever is both less frequent and less severe than early in the 20th century.

Scarlet fever is strep throat with a characteristic rash

Streptococcal Toxic Shock Syndrome

STSS may begin at the site of any GAS infection even at the site of seemingly minor trauma. The systemic illness starts with vague myalgia, chills, and severe pain at the infected site. Most commonly, this is in the skin and soft tissues and leads to necrotizing fasciitis and myonecrosis. The striking nature of this progression when it involves the extremities is the basis of the label "flesh-eating bacteria." STSS continues with nausea, vomiting, and diarrhea followed by hypotension, shock, and organ failure. The outstanding laboratory findings are a lymphocytosis; impaired renal function (azotemia); and, in over half the cases, bacteremia. Some patients are in irreversible shock by the time they reach a medical facility. Many survivors have been left as multiple amputees as the result of metastatic spread of the streptococci.

STSS is a rapidly progressive multisystem disease

Shock, azotemia, and bacteremia are common

Poststreptococcal Sequelae

Acute Rheumatic Fever

ARF is a nonsuppurative inflammatory disease characterized by fever, carditis, subcutaneous nodules, chorea, and migratory polyarthritis. The diagnosis is based on a set of primarily clinical findings (Jones Criteria) recommended by the American Heart Association. Evidence of a previous GAS infection is included in these criteria, but there is no test which is diagnostic of ARF. Cardiac enlargement, valvular murmurs, and effusions are seen clinically and reflect endocardial, myocardial, and epicardial damage, which can lead to heart failure. Attacks typically begin 3 weeks (range 1 to 5 weeks) after an attack of GAS pharyngitis and, in the absence of anti-inflammatory therapy, last 2 to 3 months.

Fever, carditis, nodules, and polyarthritis are clinical criteria

No test is diagnostic

ARF also has a predilection for recurrence with subsequent streptococcal infections as new M types are encountered. The first attack usually occurs between the ages of 5 and 15 years. The risk of recurrent attacks after subsequent GAS infection continues into adult life and then

New M types triggers recurrences

Recurrences lead to rheumatic heart disease

decreases. Repeated attacks lead to progressive damage to the endocardium and heart valves, with scarring and valvular stenosis or incompetence (rheumatic heart disease).

Acute Glomerulonephritis

Children develop a nephritis, which slowly resolves

Poststreptococcal glomerulonephritis is primarily a disease of childhood that begins 1 to 4 weeks after streptococcal pharyngitis and 3 to 6 weeks after skin infection. It is characterized clinically by edema, hypertension, hematuria, proteinuria, and decreased serum complement levels. Pathologically, there are diffuse proliferative lesions of the glomeruli. The clinical course is usually benign, with spontaneous healing over weeks to months. Occasionally, a progressive course leads to renal failure and death.

DIAGNOSIS

Throat culture followed by Lancefield grouping is definitive

Bacitracin susceptibility predicts group A

Although the clinical features of streptococcal pharyngitis are fairly typical, there is enough overlap with viral pharyngitis that a culture of the posterior pharynx and tonsils is required for diagnosis. A direct Gram-stained smear of the throat is not helpful because of the other streptococci in the pharyngeal flora. However, smears from normally sterile sites usually demonstrate streptococci. Blood agar plates incubated anaerobically give the best yield because they favor the demonstration of β-hemolysis (see streptolysins earlier in chapter). β-Hemolytic colonies are identified by Lancefield grouping using immunofluorescence or agglutination methods. In smaller laboratories, an indirect method based on the exquisite susceptibility of GAS to bacitracin and the relative resistance of strains of other groups may be used for presumptive separation of group A strains from the others (**Table 25–2**).

Group A antigen test is rapid and specific but not sensitive

Detection of group A antigen extracted directly from throat swabs is now available in a wide variety of kits marketed for use in physicians' offices. These methods are rapid and specific, but are at best only 90% sensitive compared with culture. Given the importance of the detection of group A streptococci in prevention of ARF (it is the reason physicians culture sore throats), missing 10% of cases is not tolerable. Patients with a positive direct antigen test may be treated without culture, but the American Academy of Pediatrics recommends that negative results must be confirmed by culture before withholding treatment.

TABLE 25–2 Usual Hemolytic, Biochemical, and Cultural Reactions of Common Streptococci and Enterococci[a]

	SUSCEPTIBILITY TO				
	BACITRACIN	**OPTOCHIN**	**BILE SOLUBILITY**	**BILE/ ESCULIN REACTIONB**	**PYR**
Streptococci					
β-Hemolytic					
Lancefield group A	+	–	–	–	+
Lancefield groups B, C, F, G	–	–	–	–	–
α-Hemolytic					
S. pneumoniae	–	+	+	–	–
Viridans group	–	–	–	–	–
Nonhemolytic	–	–	–	–	–
Enterococci	**–**	**–**	**–**	**+**	**+**

PYR, pyrrolidonyl arylamidase test.

[a]All are tests commonly substituted for serologic identification in clinical laboratories.

[b]Tests for the ability to grow in bile and reduce esculin.

Several serologic tests have been developed to aid in the diagnosis of poststreptococcal sequelae by providing evidence of a previous GAS infection. They include the ASO, anti-DNAase B, and some tests that combine multiple antigens. High titers of ASO are usually found in sera of patients with rheumatic fever, so that test is used most widely.

ASO antibodies document previous infection in suspect ARF

TREATMENT

GAS are highly susceptible to penicillin G, the antimicrobal of choice. Concentrations as low as 0.01 μg/mL have a bactericidal effect, and penicillin resistance is so far unknown. Numerous other antimicrobics are also active, including other β-lactams and macrolides, but not aminoglycosides. Patients allergic to penicillin are usually treated with erythromycin or azithromycin, and impetigo is often treated with erythromycin to cover the prospect of *S aureus* involvement. Adequate treatment of streptococcal pharyngitis within 10 days of onset prevents rheumatic fever by removing the antigenic stimulus; its effect on the duration of the pharyngitis is not dramatic because of the short course of the natural infection. Treatment of the acute infection does not prevent the development of acute glomerulonephritis.

GAS remain susceptible to penicillin

Treatment of pharyngitis within 10 days prevents ARF

PREVENTION

Penicillin prophylaxis with long-acting preparations is used to prevent recurrences of ARF during the most susceptible ages (5 to 15 years). Patients with a history of rheumatic fever or known rheumatic heart disease receive antimicrobial prophylaxis while undergoing procedures known to cause transient bacteremia, such as dental extraction. Multivalent vaccines using M protein epitopes that are not cross-reactive to self are in clinical trials with encouraging results.

Prophylactic penicillin prevents ARF recurrences

GROUP B STREPTOCOCCI *(STREPTOCOCCUS AGALACTIAE)*

BACTERIOLOGY

Group B streptococci (GBS) produce short chains and diplococcal pairs of spherical or ovoid Gram-positive cells. Colonies are larger and β-hemolysis is less distinct than with group A streptococci and may even be absent. In addition to the Lancefield B antigen, GBS produce polysaccharide capsules of nine antigenic types (Ia, Ib, II–VIII), all of which contain sialic acid in the form of terminal side chain residues.

Nine capsular types contain sialic acid

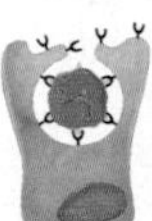

GROUP B STREPTOCOCCAL DISEASE

CLINICAL CAPSULE

The typical GBS case is a newborn in the first few days of life who is not doing well. Fever, lethargy, poor feeding, and respiratory distress are the most common features. Localizing findings are usually lacking, and the diagnosis is revealed only by isolation of GBS from blood or cerebrospinal fluid. The mortality rate is high even when appropriate antibiotics are used.

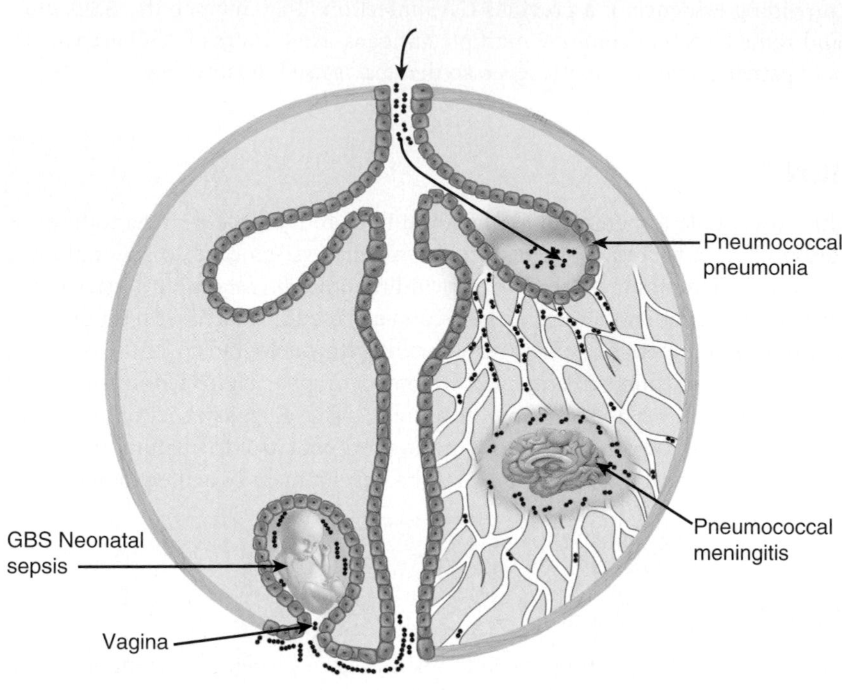

FIGURE 25–10. GBS and pneumococcal disease overview. *S pneumoniae* is aspirated from the normal orophyaryngeal flora to the lung were it produces pneumonia. Bacteremic spread can infect other sites particularly the brain where meningitis is produced. GBS vaginal colonization during pregnancy leads to infection of the fetus either in the uterus or during childbirth.

EPIDEMIOLOGY

Neonatal sepsis is acquired from mother's vaginal flora

Ruptured membranes and prematurity increase risk

GBS are the leading cause of sepsis and meningitis in the first few days of life. The organism is resident in the gastrointestinal tract, with secondary spread to other sites, the most important of which is the vagina. GBS can be found in the lower gastrointestinal and vaginal flora of 10% to 40% of women. During pregnancy and childbirth, these organisms may gain access to the amniotic fluid or colonize the newborn as it passes through the birth canal (**Figure 25–10**). GBS produce disease in approximately 2% of these encounters. The risk is much higher when factors are present that decrease the infant's innate resistance (prematurity) or increase the chances of transmission such as rupture of the amniotic membranes for 18 hours or more before delivery. Some infants are healthy at birth but develop sepsis 1 to 3 months later. It is not known whether the organism in these "late-onset" cases was acquired from the mother, in the nursery, or in the community after leaving the hospital.

PATHOGENESIS

Capsule binds factor H

C3b deposition is disrupted

Transplacental IgG is protective

GBS disease requires the proper combination of organism and host factors. The GBS capsule is the major organism factor. For the initial stages of infection, a number of surface exposed proteins that attach to fibronectin and extracellular matrix proteins have been identified. The sialic acid moiety of the capsule has been shown to bind serum factor H, which in turn accelerates degradation of C3b before it can be effectively deposited on the surface of the organism. This makes alternative pathway-mediated mechanisms of opsonophagocytosis relatively ineffective. Thus, complement-mediated phagocyte recognition requires specific antibody and the classical pathway. Newborns have this antibody only if they receive it from their mother as transplacental IgG. Those who lack the protective "cover" of antibody specific to the type of GBS they encounter must rely on alternative pathway mechanisms, a situation in which the GBS has an advantage over less virulent organisms. GBS have also been shown to produce a peptidase that inactivates C5a, the major chemoattractant of PMNs (polymorphonuclear leukocytes). This may correlate with the observation that serious neonatal infections often show a paucity of infiltrating PMNs. ::: Antiphagocytic capsules, p. 394

IMMUNITY

Antibody is protective against GBS disease, but as with group A streptococcal M protein, the antibody must be specific to the infecting type of GBS. Fortunately, there are only nine types, and type III produces the majority of cases in the first week of life. Antibody is acquired by GBS infection, and specific IgG may be transmitted transplacentally to the fetus, providing protection in the perinatal period. In the presence of type-specific antibody, classical pathway C3b deposition, phagocyte recognition, and killing proceed normally.

Type-specific anticapsular antibody is protective

GROUP B STREPTOCOCCI: CLINICAL ASPECTS

MANIFESTATIONS

The clinical findings of poor feeding, irritability, lethargy, jaundice, respiratory distress, and hypotension are nonspecific and similar to those found in other serious infections in the neonatal period. Fever is sometimes absent, and infants may even be hypothermic. Pneumonia is common, and meningitis is present in 5% to 10% of cases. Most infections have GBS circulating in the bloodstream without localizing findings. The disease onset is typically in the first few days of life, and signs of infection are present at birth in almost 50% of cases. The late-onset (1 to 3 month) cases have similar findings, but are more likely to have meningitis and focal infections in the bones and joints. Even with increased awareness and improved supportive therapy, the mortality rate for early onset GBS infection still approaches 10%.

Nonspecific findings evolve to pneumonia and meningitis

First few days of life or months later

GBS infections in adults are uncommon and fall into two groups. The first group comprises peripartum chorioamnionitis and bacteremia, the mother's side of the neonatal syndrome. Other infections include pneumonia and a variety of skin and soft tissue infections similar to those produced by other pyogenic streptococci. Although adult GBS infections may be serious, they are usually not fatal unless patients are immunocompromised. GBS infections are not associated with rheumatic fever or acute glomerulonephritis.

Maternal and other adult infections can be serious

DIAGNOSIS

The laboratory diagnosis of GBS infection is by culture of blood, cerebrospinal fluid, or other appropriate specimen. Definitive identification involves serologic determination of the Lancefield group by the same methods used for group A streptococci. Maximal detection of vaginal colonization in pregnant women requires selective procedures with selective media and enrichment broths. These must be separately established in the laboratory, since they are used for no other purpose. Methods for direct detection of GBS antigen in vaginal specimens have been evaluated, but their sensitivity is far too low for use in the diagnosis of neonatal infection.

Specialized culture required to detect vaginal colonization

TREATMENT

GBS are susceptible to the same antimicrobics as group A organisms. Penicillin is the treatment of choice, there is no known resistance to β-lactam agents. However, in the initial stage neonatal infections are often initially treated with combinations of penicillin (or ampicillin) and an aminoglycoside because of known synergism and the possibility of other bacterial agents. Once GBS is confirmed, therapy can be completed with penicillin alone.

Penicillin is primary antibiotic

PREVENTION

Intrapartum IV penicillin prophylaxis is protective

Third-trimester culture determines risk

Strategies for prevention of neonatal GBS disease are focused on reducing contact of the newborn with the organism. In colonized women, attempts to eradicate the carrier state have not been successful, but intrapartum (during labor) antimicrobial prophylaxis with intravenous penicillin has been shown to reduce transmission and disease. It is now recommended by expert obstetric and perinatology groups that all newborns at risk receive such prophylaxis. Risk is defined by the presence of vaginal or rectal GBS in a culture taken during the third trimester (35 to 37 weeks). Thus, all expectant mothers must be screened by selective culture (see Diagnosis) and intrapartum prophylaxis administered to all found culture-positive. An alternative risk-based approach (eg, prematurity, prolonged membrane rupture, and fever) is easier for obstetricians to apply, but has now been discarded as much less effective in preventing GBS disease. Implementation of these screening and prophylaxis procedures was followed by a dramatic decrease (more than 70%) in neonatal early-onset GBS disease. For women who present in labor without culture results, a risk-based assessment is all that can be used to decide whether to administer prophylaxis. Prevention by immunization with purified GBS capsular polysaccharide has been shown to be feasible, and considerable effort is now being directed at development of a vaccine.

Other Pyogenic Streptococci

All are virulent but uncommon

None associated with immunologic sequelae

The other pyogenic streptococci occasionally produce various respiratory, skin, wound, soft tissue, and genital infections, which may resemble those caused by group A and B streptococci. Although a few food-borne outbreaks of pharyngitis have been linked to non–group A streptococci, their role as a cause of everyday sore throats is not established. These streptococci are susceptible to penicillin, and infections are managed in a manner similar to that with deep tissue infections caused by group A and B strains. None of the non–group A pyogenic streptococci have been associated with poststreptococcal sequelae.

Streptococcus pneumoniae

BACTERIOLOGY

MORPHOLOGY AND STRUCTURE

Capsule has 90+ serotypes

Choline-binding proteins attach to cells

S pneumoniae (pneumococci) are Gram-positive, oval cocci typically arranged end to end in pairs (diplococcus), giving the cells a bullet shape (**Figure 25–11**). The distinguishing structural feature of the pneumococcus is its capsule (**Figure 25-12**). All virulent strains have surface capsules, composed of high-molecular-weight polysaccharide polymers that are complex mixtures of monosaccharides, oligosaccharides, and sometimes other components.

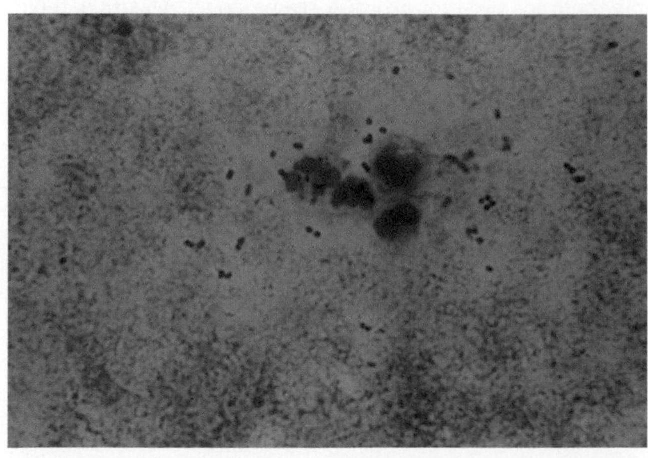

FIGURE 25–11. *Streptococcus pneumoniae* in sputum of patient with pneumonia. Note the marked tendency to form oval diplococci. (Reprinted with permission of Schering Corporation, Kenilworth, NJ, the copyright owner. All rights reserved.)

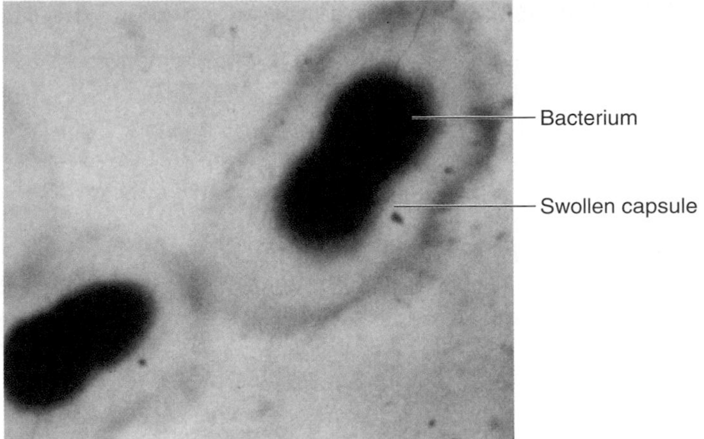

FIGURE 25–12. Pneumococcal capsule. In this test, live *Streptococcus pneumoniae* have been mixed with antibody specific to the capsular polysaccharide. The opsonizing antibody defines the capsule, which appears "swollen" when compared with preparations without antibody. (Reproduced with permission from Willey J, Sherwood L, Woolverton C (eds). *Prescott's Principles of Microbiology.* New York: McGraw-Hill; 2008.)

The exact makeup of the polymer is unique and distinctly antigenic for each of more than 90 serotypes. Pneumococcal cell wall structure is similar to other streptococci, and a variety of surface proteins are rooted in the peptidoglycan extending outward into the capsule. One group of these, the **choline-binding proteins**, is able to bind to both pneumococcal cell wall cholines and carbohydrates that are present on the surface of epithelial cells.

GROWTH

On blood agar, pneumococci produce round, glistening 0.5- to 2.0-mm colonies surrounded by a zone of α-hemolysis. Both colonies and broth cultures have a tendency to undergo autolysis because of their susceptibility to peroxides produced during growth and the action of **autolysins,** a family of pneumococcal enzymes that degrade peptidoglycan. Accelerating the autolytic process with bile salts is the basis of the bile solubility test that separates pneumococci from other α-hemolytic streptococci.

Colonies are α-hemolytic

EXTRACELLULAR PRODUCTS

All pneumococci produce **pneumolysin,** which is a member of the family of transmembrane pore-forming toxins that includes staphylococcal α toxin, *S pyogenes* streptolysin O, and others. The pneumococcus does not secrete pneumolysin, but it is released on lysis of the organisms augmented by autolysins. Pneumolysin has a number of other effects, including its ability to stimulate cytokines and disrupt the cilia of cultured human respiratory epithelial cells. Pneumococci also produce a neuraminidase, which cleaves sialic acid that is present in host mucin, glycolipids and glycoproteins. ::: Pore-forming toxins, p. 396

Pneumolysin forms pores after release by autolysins

PNEUMOCOCCAL DISEASE

CLINICAL CAPSULE

The most common form of infection with *S pneumoniae* is pneumonia, which begins with fever and a shaking chill followed by signs that localize the disease to the lung. These include difficulty breathing and cough with production of purulent sputum, sometimes containing blood. The pneumonia typically fills part or all of a lobe of the lung with inflammatory cells, and the bacteria may spread to the bloodstream and thus other organs. The most important of the latter is the central nervous system, where seeding with pneumococci leads to acute purulent meningitis. Pneumococci are also a leading cause of otitis media the "hot ear" of childhood.

EPIDEMIOLOGY

Pneumonia is common

Young and old are most affected

S pneumoniae is a leading cause of pneumonia, acute purulent meningitis, bacteremia, and other invasive infections. In the United States, it is responsible for an estimated 3000 cases of meningitis, 50,000 cases of bacteremia, and 500,000 cases of pneumonia each year. Worldwide, more than 5 million children die every year from pneumococcal disease. *S pneumoniae* is also the most common cause of otitis media (see Chapter 59), a virtually universal disease of childhood with millions of cases every year. Pneumococcal infections occur throughout life, but are most common in the very young (less than 2 years) and in the elderly (older than 60 years). Alcoholism, diabetes mellitus, chronic renal disease, asplenia, and some malignancies are associated with more frequent and serious pneumococcal infection.

Respiratory colonization is common

Microaerosols transmit person to person

Infections are derived from colonization of the nasopharynx, where pneumococci can be found in 5% to 40% of healthy persons depending on age, season, and other factors. The highest rates are among children in the winter. Respiratory secretions containing pneumococci may be transmitted from person to person by direct contact or from the microaerosols created by coughing and sneezing in close quarters. Such conditions are favored by crowded living conditions, particularly when colonized persons are mixed with susceptible ones, as in child care centers, recruitment barracks, and prisons. As with other bacterial pneumonias, viral respiratory infection and underlying chronic disease are important predisposing factors.

Some serotypes are more common

Surveillance data show that just over 20 of the 90 pneumococcal serotypes produce disease more often than the others. There is also a variation among types in the age and geographic distribution of cases. These differences are presumably due to enhanced virulence factors in these types, but the specific reasons are not known. These features do not influence the medical management of individual cases but are important in devising prevention strategies such as immunization (see following text).

PATHOGENESIS

Aspiration of colonizing bacteria starts the disease process

Impaired clearance mechanisms enhance susceptibility

Pneumococcal adherence to nasopharyngeal cells involves multiple factors. The primary relationship is the bridging effect of the choline binding proteins' attachment to cell wall cholines and carbohydrates covering or exposed on the surface of host epithelial cells. This binding may be aided by the exposure of additional receptors by neuraminidase digestion, viral infection, or pneumolysin-stimulated cytokine activation of host cells. Aspiration of respiratory secretions containing these pneumococci is the initial step leading to pneumonia (**Figure 25–13**). This must be a common event. Normally, aspirated organisms are cleared rapidly by the defense mechanisms of the lower respiratory tract, including the cough and epiglottic reflexes; the mucociliary "blanket;" and phagocytosis by alveolar macrophages. Host factors that impair the combined efficiency of these defenses allow pneumococci to reach the alveoli and multiply there. These include

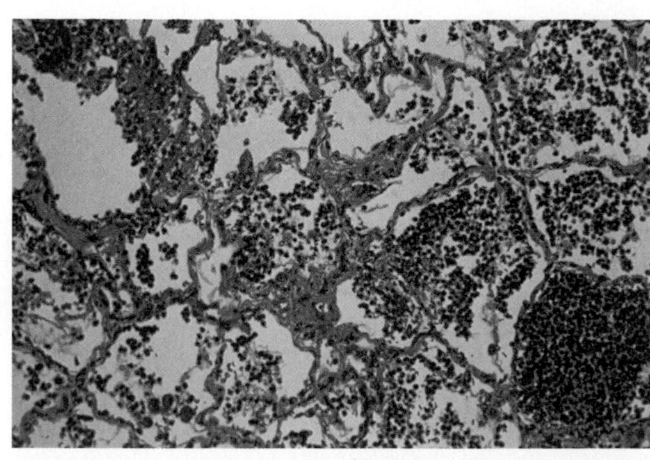

FIGURE 25–13. Pneumococcal pneumonia. In this histologic view of infected lung, note that the alveoli are filled with inflammatory cells. Also note that the alveolar septa are relatively intact despite the high level of cellular infiltrate. The stain used here does not demonstrate the pneumococci, which would be much smaller than the cells at this magnification

chronic pulmonary diseases; damage to bronchial epithelium from smoking or air pollution; and respiratory dysfunction from alcoholic intoxication, narcotics, anesthesia, and trauma.

When organisms reach the alveolus, pneumococcal virulence factors operate in two stages. The first stage is early in infection, when the surface capsule of intact organisms acts to block phagocytosis by complement inhibition. This allows the organisms to multiply and spread despite an acute inflammatory response. The second stage occurs when organisms begin to disintegrate and release a number of factors either synthesized by the pneumococcus or part of its structure, thus causing injury. These include pneumolysin, autolysin, and components of the cell wall.

Capsule interferes with phagocytosis

Pneumolysin causes injury

Capsule

The polysaccharide capsule of *S pneumoniae* is the major determinant of virulence. Unencapsulated mutants do not produce disease in humans or laboratory animals. Like the GBS capsule, pneumococcal polysaccharide interferes with effective deposition of complement on the organism's surface and thus phagocyte recognition and engulfment. This property is particularly important in the absence of specific antibody, when alternative pathway is the primary means for C3b-mediated opsonization. In addition to the capsule, some of the surface choline-binding proteins may participate in this antiphagocytic effect by binding serum factor H. When antibody specific to the capsular polysaccharide appears, classical pathway opsonophagocytosis proceeds efficiently. ::: Antiphagocytic capsules, p. 394

Unencapsulated pneumococci are avirulent

Alternate pathway C3b deposition blocked by capsule

Pneumolysin

Some of the clinical features seen in the course of pneumococcal infections are not explainable by the capsule alone. These include the dramatic abrupt onset, toxicity, fulminant course, and disseminated intravascular coagulation seen in some cases. Pneumolysin's toxicity for pulmonary endothelial cells and direct effect on cilia contributes to the disruption of the endothelial barrier and facilitates the access of pneumococci to the alveoli and eventually their spread beyond into the bloodstream. Pneumolysin also has direct effects on phagocytes and suppresses host inflammatory and immune functions. Because pneumolysin is not actively secreted outside the bacterial cell, the action of the autolysins is required to release it.

Pneumolysin disrupts cells and cilia

Lysis required to release from bacterial cell

The combined effects of pneumococcal and host factors produce a pneumonia, which progresses through a series of stages. Initial alveolar multiplication produces a profuse outpouring of serous edema fluid, which is then followed by an influx PMNs and erythrocytes (Figure 25–13). By the second or third day of illness, the lung segment has increased three- to fourfold in weight through accumulation of this cellular, hemorrhagic fluid typically in a single lobe of the lung. In the consolidated alveoli, neutrophils predominate initially, but once actively growing, pneumococci are no longer present, macrophages replace the granulocytes, and resolution of the lesion ensues. A remarkable feature of pneumococcal pneumonia is the lack of structural damage to the lung, which usually leads to complete resolution on recovery.

PMNs and red blood cells consolidate alveoli

Lesions resolve without structural damage

IMMUNITY

Immunity to *S pneumoniae* infection is provided by antibody directed against the specific pneumococcal capsular type. When antibody binds to the capsular surface, C3b is deposited by classical pathway mechanisms, and phagocytosis can proceed. Because the number of serotypes is large, complete immunity through natural experience is not realistic, which is why pneumococcal infections occur throughout life. Infections are most often seen in the very young, when immunologic experience is minimal, and in the elderly, when immunity begins to wane and risk factors are more common. Antibodies to surface proteins and enzymes, including pneumolysin, are also formed in the course of disease, but their role in immunity is unknown.

Immunity is specific to capsular type

Antibody leads to classical pathway complement deposition

PNEUMOCOCCAL DISEASE: CLINICAL ASPECTS

MANIFESTATIONS

Pneumococcal Pneumonia

Shaking chill is followed by bloody sputum

Lung consolidation is typically lobar

Pneumococcal pneumonia begins abruptly with a shaking chill and high fever. Cough with production of sputum pink to rusty in color (indicating the presence of red blood cells), and pleuritic chest pain are common. Physical findings usually indicate pulmonary consolidation. Children and young adults typically demonstrate a lobular or lobar consolidation on chest radiography, whereas older patients may show a less localized bronchial distribution of the infiltrates. Without therapy, sustained fever, pleuritic pain, and productive cough continue until a "crisis" occurs 5 to 10 days after onset of the disease. The crisis involves a sudden decrease in temperature and improvement in the patient's condition. It is associated with effective levels of opsonizing antibody reaching the lesion. Although infection may occur at any age, the incidence and mortality of pneumococcal pneumonia increase sharply after 50 years.

Pneumococcal Meningitis

Sequelae are slightly higher than other meningeal pathogens

S pneumoniae is one of the three leading causes of bacterial meningitis. The signs and symptoms are similar to those produced by other bacteria (see Chapter 65). Acute purulent meningitis may follow pneumococcal pneumonia or infection at another site or may appear with no apparent antecedent infection. It may also develop after trauma involving the skull. The mortality and frequency of sequelae are slightly higher with pneumococcal meningitis than with other forms of pyogenic meningitis.

Other Infections

Sinusitis and otitis media are common

Pneumococci are common causes of sinusitis and otitis media (see Chapter 59). The latter frequently occurs in children in association with viral infection. Chronic infection of the mastoid or respiratory sinus sometimes extends to the subarachnoid space to cause meningitis. Pneumococci may also cause endocarditis, arthritis, and peritonitis, usually in association with bacteremia. Patients with ascites caused by diseases such as cirrhosis and nephritis may develop spontaneous pneumococcal peritonitis. Pneumococci do not cause pharyngitis or tonsillitis.

DIAGNOSIS

Sputum quality complicates diagnosis

Optochin or bile solubility distinguish from viridans streptococci

Gram smears of material from sputum and other sites of pneumococcal infection typically show Gram-positive, lancet-shaped diplococci (Figure 25–11). Sputum collection may be difficult, however, and specimens contaminated with respiratory flora are useless for diagnosis. Other types of lower respiratory specimens may be needed for diagnosis (see Chapter 61). *S pneumoniae* grows well overnight on blood agar medium and is usually distinguished from viridans streptococci by susceptibility to the synthetic chemical ethylhydrocupreine (optochin) or by a bile solubility (Table 25–2). Bacteremia is common in pneumococcal pneumonia and meningitis, and blood cultures are valuable supplements to cultures of local fluids or exudates. Detection of pneumococcal capsular antigen in body fluids is possible but valuable primarily when cultures are negative.

TREATMENT

Altered transpeptidases decrease penicillin susceptibility

For decades pneumococci were uniformly susceptible to penicillin at concentrations lower than 0.1 μg/mL. In the late 1960s, this began to change, and strains with decreased susceptibility to all β-lactams began to emerge. These strains have penicillin minimal inhibitory concentrations (MICs) of 0.12 to 8.0 μg/mL and are associated with treatment failures in cases of pneumonia and meningitis. The resistance is not absolute and can be overcome with increased dosage, depending on the MIC. The mechanism involves alterations in the β-lactam target, the transpeptidases that cross-link peptidoglycan in cell wall synthesis. Resistant strains have mutations in one or more of these transpeptidases, which cause decreased affinity for penicillin and other β-lactams. Penicillinase is not produced. Resistance rates now exceed 10% in most locales and may be greater than 40% in some areas. Resistance to erythromycin is uncommon but more likely with penicillin-resistant strains. ::: Altered target resistance, p. 421

Antibiotic selection differs with the site of the infection and whether it is to be carried out in an inpatient or outpatient setting. Penicillin is still effective for susceptible strains, but the uncertainty has caused a shift toward third-generation cephalosporins (ceftriaxone, cefotaxime) for primary treatment. Even though penicillin-resistant strains also have decreased susceptibility to cephalosporins, the pharmacologic features of these agents make it easier to achieve blood levels higher than the MIC. Penicillin-resistant strains (MIC greater than 1.0 mcg/mL) are treated with quinolones, erythromycin, or vancomycin. Despite the β-lactam cross resistance, high doses of third-generation cephalosporins are also used in situations such as acute meningitis, for which their added spectrum may be an advantage. The therapeutic response to treatment of pneumococcal pneumonia is often dramatic. Reduction in fever, respiratory rate, and cough can occur in 12 to 24 hours but may occur gradually over several days. Chest radiography may yield normal results only after several weeks.

High doses of third-generation cephalosporins may overcome resistance

PREVENTION

Two pneumococcal vaccines prepared from capsular polysaccharide are now available. The first pneumococcal polysaccharide vaccine (PPV), available since 1977, contains purified polysaccharide extracted from the 23 serotypes of *S pneumoniae* most commonly isolated from invasive disease. It shares the T-cell–independent characteristics of other polysaccharide immunogens and is recommended for use only in those older than 2 years. In 2000, a pneumococcal conjugate vaccine (PCV) was introduced in which polysaccharide was conjugated with protein. This 7-valent vaccine stimulates T-dependent T_H2 responses and is effective beginning at 2 months of age and is thus the standard for childhood immunization. Because of its broader coverage, the 23-valent PPV is recommended after age 2 except for immunocompromised children under 5, who should still receive PCV. ::: T-cell dependent and independent responses, p. 37

23-valent PPV is T-cell independent

7-valent PCV stimulates T_H2 in children

Viridans and Nonhemolytic Streptococci

The viridans group comprises all α-hemolytic streptococci that remain after the criteria for defining pyogenic streptococci and pneumococci have been applied. Characteristically members of the normal flora of the oral and nasopharyngeal cavities, they have the basic bacteriologic features of streptococci but lack the specific antigens, toxins, and virulence of the other groups. Although the viridans group includes many species (Table 25–2), they are usually not completely identified in clinical practice because there is little difference among them in medical significance.

"Left over" α-hemolytic species are in respiratory flora

Although their virulence is very low, viridans strains can cause disease when they are protected from host defenses. The prime example is subacute bacterial endocarditis. In this disease, viridans streptococci reach previously damaged heart valves as a result of transient bacteremia associated with manipulations, such as tooth extraction, which disturb their usual habitat. Protected by fibrin and platelets, they multiply on the valve, causing local and systemic disease that is fatal if untreated. Extracellular production of glucans, complex polysaccharide polymers, may enhance their attachment to cardiac valves in a manner similar to the pathogenesis of dental caries by *S mutans* (see Chapter 60). The clinical course of viridans streptococcal endocarditis is subacute, with slow progression over weeks or months (see Chapter 66). It is effectively treated with penicillin, but uniformly fatal if untreated. The disease is particularly associated with valves damaged by recurrent rheumatic fever. The decline in the occurrence of rheumatic heart disease has reduced the incidence of this particular type of endocarditis.

Low-virulence species may cause bacterial endocarditis

Glucan production enhances attachment

ENTEROCOCCI

BACTERIOLOGY

Former streptococci possess group D antigen

Intestinal inhabitants resist action of bile salts

Until DNA homology studies dictated their separation into the genus *Enterococcus,* the enterococci were classified as streptococci. Indeed, the most common enterococcal species share the bacteriologic characteristics previously described for pyogenic streptococci, including presence of the Lancefield group D antigen. The term "enterococcus" derives from their presence in the intestinal tract and the many biochemical and cultural features that reflect that habitat. These include the ability to grow in the presence of high concentrations of bile salts and sodium chloride. Most enterococci produce nonhemolytic or α-hemolytic colonies that are larger than those of most streptococci. A dozen species are recognized based on biochemical and cultural reactions (Table 25–2) of which *Enterococcus faecalis* and *E faecium* are the most common.

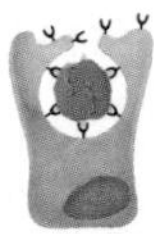

ENTEROCOCCAL DISEASE

CLINICAL CAPSULE

Enterococci cause infection almost exclusively in hospitalized patients with trauma, abdominal surgery, or compromised defenses. The primary sites are the urinary tract and soft tissue sites adjacent to the intestinal flora where enterococcal species are resident. The infections themselves are often low grade and have no unique clinical features.

EPIDEMIOLOGY

Endogenous infection is associated with medical procedures

Enterococci are part of the normal intestinal flora. Although they are capable of producing disease in many settings, the hospital environment is where a substantial increase has occurred in the last two decades. Patients with extensive abdominal surgery, transplantation, or indwelling devices or those who are undergoing procedures such as peritoneal dialysis are at greatest risk. Prolonged hospital stays and prior antimicrobial therapy, particularly with vancomycin, cephalosporins or aminoglycosides, are also risk factors. Most infections are acquired from the endogenous flora but spread between patients has been documented. From 10% to 15% of all nosocomial urinary tract, intra-abdominal, and bloodstream infections are due to enterococci.

PATHOGENESIS

Virulence factors unknown

Enterococci are a significant cause of disease in specialized hospital settings, but they are not highly virulent. On their own, they do not produce fulminant disease and in wound and soft tissue infections are usually mixed with other members of the intestinal flora. Some have even doubted their significance when isolated together with more virulent members of the Enterobacteriaceae or *Bacteroides fragilis. E. faecalis* has been shown to form biofilms and possess surface proteins adherent to urinary epithelium but as a whole virulence factors are lacking among the enterococci.

ENTEROCOCCAL DISEASE: CLINICAL ASPECTS

MANIFESTATIONS

UTIs and soft tissue infections are most common

Enterococci cause opportunistic urinary tract infections (UTIs) and occasionally wound and soft tissue infections, in much the same fashion as members of the Enterobacteriaceae. Infections are often associated with urinary tract manipulations, malignancies, biliary tract disease, and gastrointestinal disorders. Vascular or peritoneal catheters are often points of entry. Respiratory tract infections are rare. There is sometimes an associated bacteremia, which can result in the development of endocarditis on previously damaged cardiac valves.

TREATMENT

The outstanding feature of the enterococci is their high and increasing levels of resistance to antimicrobial agents. Their inherent relative resistance to most β-lactams (especially cephalosporins) and high-level resistance to aminoglycosides can be viewed as a kind of virulence factor in the hospital environment where these agents are widely used. Enterococci also have particularly efficient means of acquiring plasmid and transposon resistance genes from themselves and other species. All enterococci require 4 to 16 μg/mL of penicillin for inhibition owing to decreased affinity of their penicillin-binding proteins for all β-lactams. Higher levels of resistance have been increasing, including the emergence of β-lactamase-producing strains, particularly in *E faecalis*. The β-lactamase genes in these strains are identical with those in *S aureus*. Fortunately, β-lactamase–producing strains have not yet become widely disseminated. Ampicillin remains the most consistently active agent against enterococci.

Inherent resistance is enhanced with β-lactamase emergence

Enterococci share with streptococci a resistance to aminoglycosides based on failure of the antibiotic to be actively transported into the cell. Despite this, many strains of enterococci are inhibited and rapidly killed by low concentrations of penicillin when combined with an aminoglycoside. Under these conditions, the action of penicillin on the cell wall allows the aminoglycoside to enter the cell, where it can then act at its ribosomal site. Some strains show high-level resistance to aminoglycosides based on mutations at the ribosomal binding site or the presence of aminoglycoside-inactivating enzymes. These strains do not demonstrate synergistic effects with penicillin.

Synergy between penicillin and aminoglycosides is based on access to ribosomes

Recently, resistance to vancomycin, the antibiotic most used for ampicillin-resistant strains has emerged. Vancomycin resistance is due to a subtle change in peptidoglycan precursors, which are generated by ligases that modify the terminal amino acids of cross-linking side chains at the point where β-lactams bind. The modifications decrease the binding affinity for penicillins 1000-fold without a detectable loss in peptidoglycan strength. Although hospitals vary, the average rate of resistance in enterococci isolated from intensive care units is around 20%. Enterococci are consistently resistant to sulfonamides and often resistant to tetracyclines and erythromycin. :::Vancomycin resistance, p. 421

Vancomycin resistance is emerging threat

Ligases modify peptidoglycan side chains

Ampicillin remains the agent of choice for most UTIs and minor soft tissue infections. More severe infections, particularly endocarditis, are usually treated with combinations of a penicillin and aminoglycoside. Vancomycin is used for ampicillin-resistant strains in combination with other agents, as guided by susceptibility testing. If the strain is vancomycin resistant, linezolid is an alternative.

Ampicillin or combinations of antimicrobics are used

CASE STUDY

SORE THROAT, MURMUR, AND PAINFUL SWOLLEN JOINTS

An 8-year-old boy presented with a 1-day history of fever (39°C), associated with painful swelling of the right wrist and left knee. The patient had a sore throat 2 weeks before the present illness, which was treated with salicylates. No cultures were obtained. The last medical history was essentially negative, and the boy had no history of drug allergy, weight loss, rash, dyspnea, or illness in siblings.

PHYSICAL EXAMINATION: Temperature (39°C), blood pressure 120/80 mm Hg, pulse 110/min, respirations 28/min. The patient was ill-appearing. He avoided movement of the right wrist and left knee, which were swollen, red, hot, and tender. He had a moderately injected oropharynx without exudate and an enlarged right cervical lymph node estimated to be 1 × 1 cm. The precordium was active and, a systolic thrill could be felt. Auscultation of the heart revealed a heart rate of 120/min, normal heart sounds, and a grade III/VI holosystolic murmur over the apex not transmitting toward the axilla. Lungs were clear. No rush or hepatosplenomegaly was present, and the neurologic exam was normal.

CASE STUDY

LABORATORY DATA:

Hemoglobin 12 g, Hct 37%, WBC 16,500/mm^3

Sedimentation rate 90 mm/h

Urinalysis: Normal

Serology: Antistreptolysin 0 (ASO) titer 666 Todd units (normal <200)

Chest x-ray: Normal (no cardiomegaly)

Throat culture: Negative for group A β-hemolytic streptococci

Blood culture: Negative

Electrocardiogram: Essentially normal except for mild ST depression and nonspecific T-wave changes on V6

Aspirate from left knee: 3 mL of yellow and turbid fluid

WBCs: 3000/mm^3 mainly polymorphonuclear leukocytes

Gram stain: Negative

Culture : No growth

QUESTIONS

■ This patient's condition is most probably a case of:

A. Strep throat
B. Scarlet fever
C. Streptococcal toxic shock
D. Rheumatic fever
E. Poststreptococcal glomerulonephritis

■ This boy's joint and cardiac findings are due to:

A. Circulating streptococcal pyrogenic exotoxin
B. Circulating streptolysin O
C. Antibody directed against M protein
D. Antibody directed against streptolysin O (ASO)
E. Circulating group A streptococci

■ The illness could have been prevented by:

A. Penicillin treatment of the sore throat
B. Penicillin treatment at the onset of joint pain
C. Aspirin at any point
D. Streptococcal vaccine in infancy
E. There is no prevention

■ The etiology of the sore throat would have been best determined by:

A. ASO titer
B. Throat culture
C. Throat antigen detection

D. Exudate on tonsils
E. Presence of cervical lymphadenopathy

ANSWERS

1(D), **2**(C), **3**(A), **4**(B)

CHAPTER 26

Corynebacterium, Listeria, and *Bacillus*

So Asthma Mark would sit on the corner
And he would play his Diphtheria Blues
—Frank Zappa

This chapter includes a variety of highly pathogenic Gram-positive rods that are not currently common causes of human disease. Their medical importance lies in the lessons learned when they were more common, and the continued threat their existence poses. *Corynebacterium diphtheriae,* the cause of diphtheria, is a prototype for toxigenic disease. *Listeria monocytogenes* is a sporadic cause of meningitis and other infections in the fetus, newborn, and immunocompromised host. Occurrences in 2001 have served as a painful reminder that *Bacillus anthracis,* the cause of anthrax, is still the agent with the most potential for use in bioterrorism. The characteristics of these bacilli are presented in **Table 26–1**.

CORYNEBACTERIA

Corynebacteria (from the Greek *koryne*, club) are small and pleomorphic. The genus *Corynebacterium* includes many species of aerobic and facultative Gram-positive rods. The cells tend to have clubbed ends and often remain attached after division, forming "Chinese letter" or palisade arrangements. Spores are not formed. Growth is generally best under aerobic conditions on media enriched with blood or other animal products, but many strains grow anaerobically. Colonies on blood agar are typically small (1 to 2 mm), and most are nonhemolytic. Catalase is produced, and many strains form acid (usually lactic acid) through carbohydrate fermentation. Surface and cell wall structure is similar to other Gram-positive bacteria.

Pleomorphic club-shaped rods

Corynebacterium diphtheriae

C diphtheriae produces a powerful exotoxin that is responsible for diphtheria. Other corynebacteria are nonpathogenic commensal inhabitants of the pharynx, nasopharynx, distal urethra, and skin; they are collectively referred to as "diphtheroids." The species that have disease associations are included in **Table 26–2**.

C diphtheriae produces exotoxin

Other corynebacteria are called diphtheroids

TABLE 26–1 Features of Aerobic Gram-Positive Bacilli

ORGANISM	CAPSULE	ENDOSPORES	MOTILITY	TOXINS	SOURCE	DISEASE
Corynebacterium diphtheriae	–	–	–	DT	Human cases, carriers	Diphtheria
Listeria monocytogenes	–	–	+	LLO	Food, animals	Meningitis, bacteremia
Bacillus						
B anthracis	+	–	–	Exotoxin[a]	Imported animal products	Anthrax
B cereus	–	+	+	Enterotoxin, pyogenic toxin	Ubiquitous	Food poisoning, opportunistic infection
Other species	–	+	+		Ubiquitous	

DT, diphtheria toxin; LLO, listerolysin O.
[a] Exotoxin contains three components: lethal factor, protective antigen, and edema factor.

BACTERIOLOGY

DT gene is in a lysogenic phage

C diphtheriae are differentiated from other corynebacteria by the appearance of colonies on the selective media used for its isolation and a variety of biochemical reactions. Strains of *C diphtheriae* may or may not produce **diphtheria toxin (DT)**. The gene for DT is contained in the genome of a bacteriophage, which is lysogenic in the *C diphtheriae* chromosome. For strains with the gene, DT production is controlled by a repressor protein (DtxR), which responds to iron concentrations and also regulates other toxin-related functions.

DT is an A-B toxin that acts in the cytoplasm to inhibit protein synthesis irreversibly in a wide variety of eukaryotic cells. After binding mediated by the B subunit, both the A and

TABLE 26–2 Other Aerobic and Facultative Gram-Positive Bacilli

ORGANISM	FEATURES	EPIDEMIOLOGY	DISEASE
Corynebacterium ulcerans	Closely related to *C diphtheriae*, including ability to produce small amounts of DT	Similar to diphtheria, also infects animals	Pharyngitis
C jeikeium	Multiresistant, often susceptible only to vancomycin	Acquired from skin colonization	Bacteremia, IV catheter colonization
Erysipelothrix rhusiopathiae	Resembles corynebacteria and *Listeria*	Traumatic inoculation from animal and decaying organic matter	Erysipeloid, painful, slow-spreading, erythematous swelling of skin. Occupational disease of fishermen, butchers, and veterinarians
Lactobacillus spp.	Long, slender rods with squared ends, often chain end to end	Normal oral, gastrointestinal, and vaginal flora	No human infections *L acidophilus* plays role in pathogenesis of dental caries
Propionibacterium	Resemble corynebacteria, anaerobes, or microaerophiles	Normal skin flora	Rare cause of bacterial endocarditis

DT, diphtheria toxin; IV, intravenous.

B subunits enter the cell in an endocytotic vacuole. In the low pH of the vacuole, the toxin unfolds, exposing sites that facilitate translocation of the A subunit from the phagosome to the cytosol. The target is elongation factor 2 (EF-2), which transfers polypeptidyl-transfer RNA from acceptor to donor sites on the ribosome of the host cell. The specific action of the A subunit is to inactivate EF-2, by **ADP-ribosylation** (ADPR), which shuts off protein synthesis. The details of DT action are illustrated in Chapter 1 as a prototype toxin. *C diphtheriae* itself is unaffected because it uses a protein other than EF-2 in protein synthesis. ::: DT, p. 16; ADPR, p. 363

A subunit enters the cytosol from a vacuole

EF-2 is inactivated by ADP-ribosylation

Transfer of tRNA and protein synthesis are stopped

DIPHTHERIA

CLINICAL CAPSULE

Diphtheria is a disease caused by the local and systemic effects of diphtheria toxin, a potent inhibitor of protein synthesis. The local disease is a severe pharyngitis typically accompanied by a plaque-like pseudomembrane in the throat and trachea. The life-threatening aspects of diphtheria are due to the absorption of the toxin across the pharyngeal mucosa and its circulation in the bloodstream. Multiple organs are affected, but the most important is the heart, where the toxin produces an acute myocarditis.

EPIDEMIOLOGY

C diphtheriae is transmitted by droplet spread, by direct contact with cutaneous infections, and, to a lesser extent, by fomites (**Figure 26–1**). Some subjects become convalescent pharyngeal or nasal carriers and continue to harbor the organism for weeks, months, or longer. Diphtheria is rare where immunization is widely used. In the United States, for example, fewer than 10 cases are now reported each year. These usually occur as small outbreaks in populations that have not received adequate immunization, such as migrant workers, transients, and those who refuse immunization on religious grounds. It has been more than 25 years since any outbreak exceeded 50 cases.

Transmitted by respiratory droplets

Most cases are in unimmunized transients

Diphtheria still occurs in developing countries and in places where public health infrastructure has been disrupted. For example, in the former Soviet Union, where the annual number of diphtheria cases had been below 200, over 47,000 cases and 1700 deaths occurred between 1990 and 1995. This outbreak followed the reintroduction of *C diphtheriae* into a population where the public health systems had broken down as a result of the political situation. Reinstitution of effect immunization brought diphtheria rates back to base levels.

Outbreaks occur when immunization rates decrease

PATHOGENESIS

C diphtheriae has little invasive capacity, and diphtheria is due to the local and systemic effects of DT, a protein exotoxin with potent cytotoxic features (**Figure 26–2**). It inhibits protein synthesis in cell-free extracts of virtually all eukaryotic cells, from protozoa and yeasts to higher plants and humans. Its toxicity for intact cells varies among mammals and organs, primarily as a result of differences in toxin binding and uptake. In humans, the B subunit binds to one of a common family of eukaryotic receptors that regulate cell growth and differentiation, thus exploiting a normal cell function.

A subunit inhibits protein synthesis

B-subunit binding determines cell susceptibility

The production of DT has both local and systemic effects. Locally, its action on epithelial cells leads to necrosis and inflammation, forming a pseudomembrane composed of a coagulum of fibrin, leukocytes, and cellular debris. The extent of the pseudomembrane varies from a local plaque to an extensive covering of much of the tracheobronchial tree.

Local effects produce pseudomembrane

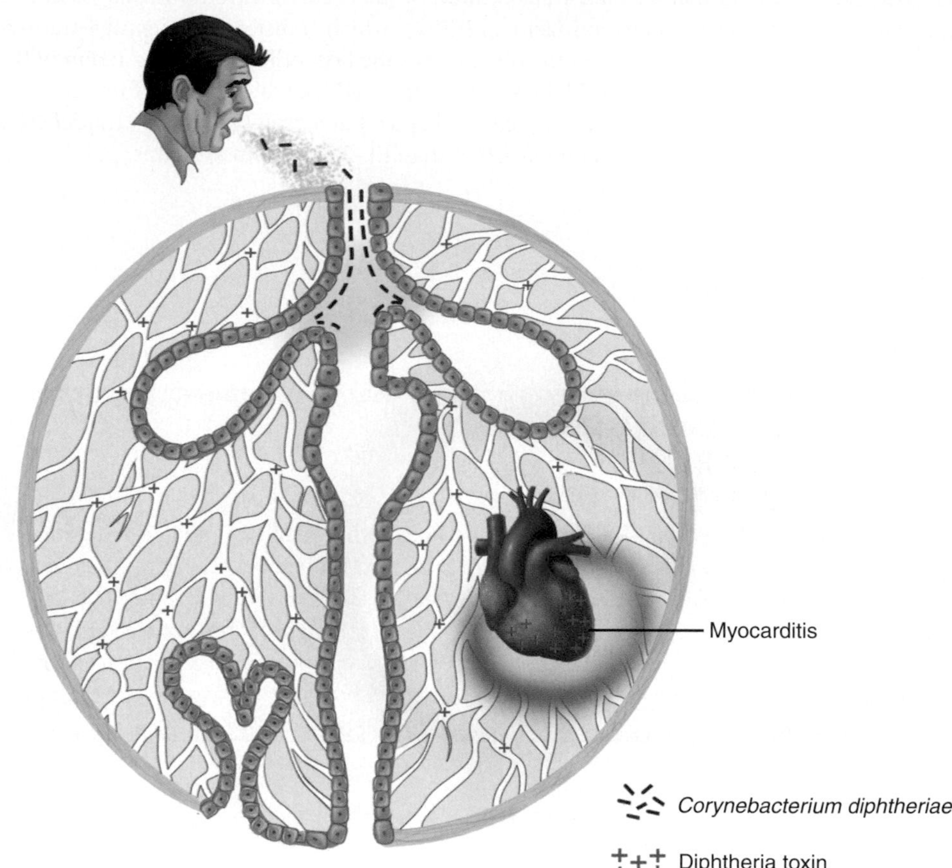

FIGURE 26–1. Diphtheria overview. Infection with *Corynebacterium diphtheriae* is acquired by respiratory droplet spread. The throat and upper airways are infected, but there is no invasion. Diphtheria toxin (DT) produced at the primary side is absorbed into the bloodstream and affects multiple organs, particularly the heart where acute myocarditis is produced.

Absorption of DT leads to myocarditis

Absorption and circulation of DT allow binding throughout the body. Myocardial cells are most affected; eventually, acute myocarditis develops.

IMMUNITY

Antibodies neutralize toxin

Toxoid is formalin inactivated DT

Diphtheria toxin is antigenic, stimulating the production of protective antitoxin antibodies during natural infection. Formalin treatment of toxin produces **toxoid,** which retains the antigenicity but not the toxicity of native toxin and is used in immunization against the disease. It is clear that this process functionally inactivates fragment B. Whether it also inactivates fragment A or prevents its ability to dissociate from fragment B is not known. Molecular studies of the A-subunit structure and action suggest that another approach to immunization may be by genetic engineering of the A subunit so that it fails to bind EF-2 but retains its antigenicity.

DIPHTHERIA: CLINICAL ASPECTS

MANIFESTATIONS

After an incubation period of 2 to 4 days, diphtheria usually manifests as pharyngitis or tonsillitis. Typically, malaise, sore throat, and fever occur, and a patch of exudate or membrane develops on the tonsils, uvula, soft palate, or pharyngeal wall. The gray-white pseudomembrane (**Figure 26–3**) adheres to the mucous membrane and may extend from the oropharyngeal area down to the larynx and into the trachea. Associated cervical

the laboratory of the suspicion of diphtheria in advance. Generally, 2 days are required to exclude *C diphtheriae* (ie, no colonies isolated on Tinsdale agar); however, more time is needed to complete identification and toxigenicity testing of a positive culture.

Laboratory must be notified of suspicion in advance

TREATMENT

Treatment of diphtheria is directed at neutralization of the toxin with concurrent elimination of the organism. The former is most critical and is accomplished by promptly administering a diphtheria antitoxin, an antiserum produced in horses. It must be administered early because it only neutralizes circulating toxin and has no effect on toxin already fixed to or within cells. *C diphtheriae* is susceptible to a variety of antimicrobics, including penicillins, cephalosporins, erythromycin, and tetracycline. Of these, erythromycin has been the most effective. The complications of diphtheria are managed primarily by supportive measures.

Antitoxin therapy aimed at neutralizing free toxin

Erythromycin most effective antimicrobic therapy

PREVENTION

The mainstay of diphtheria prevention is immunization. The vaccine is highly effective. Three to four doses of diphtheria toxoid produce immunity by stimulating antitoxin production. The initial series is begun in the first year of life. Booster immunizations at 10-year intervals maintain immunity. Fully immunized individuals may become infected with *C diphtheriae* because the antibodies are directed only against the toxin, but the disease is mild. Serious infection and death occur only in unimmunized or incompletely immunized individuals. Immunization with DT toxoid prevents serious toxin-medicated disease.

LISTERIA MONOCYTOGENES

BACTERIOLOGY

Listeria monocytogenes is a Gram-positive rod with some bacteriologic features that resemble those of both corynebacteria and streptococci. In stained smears of clinical and laboratory material, the organisms resemble diphtheroids. *Listeria* are not difficult to grow in culture, producing small, β-hemolytic colonies on blood agar. This species is able to grow slowly in the cold even at temperatures as low as 1°C. *Listeria* species are catalase-positive, which distinguishes them from streptococci, and they produce a characteristic tumbling motility in fluid media at temperatures below 30°C, which distinguishes them from corynebacteria.

Rods resemble corynebacteria

Colonies are β-hemolytic

Eleven *L monocytogenes* serotypes are recognized based on flagellar and somatic surface antigens, but most human cases are limited to only three serotypes (1/2a, 1/2b, 4b). The major virulence factors are invasion-associated surface proteins called **internalin** and a pore-forming cytotoxin, **listeriolysin O** (LLO).

Internalin and LLO enhance virulence

LISTERIOSIS

CLINICAL CAPSULE

Listeriosis is often an insidious infection in humans. Infection of the fetus or newborn may result in stillbirth or a fulminant neonatal sepsis. In most adults, there are usually only general manifestations, such as fever and malaise, associated with and eventually traced to bacteremia.

EPIDEMIOLOGY

Widespread in nature and animals

Food-borne transmission is from animal products

Cold growth and biofilms enhance infectivity

Listeria monocytogenes is widespread in nature, in soil, ground water, decaying vegetation, and the intestinal tract of animals including those associated with our food supply (eg, fowl, ungulates). The importance of food-borne transmission of listeriosis (**Figure 26–5**) was not recognized until the early 1980s. A widely publicized 1985 California outbreak involved consumption of Mexican-style soft cheese and included 86 cases and 29 deaths. Most of the cases were among mother–infant pairs. Dairy product outbreaks have been traced to post-pasteurization contamination or deviation from recommended time and temperature guidelines. An important feature of some epidemics has been the ability of *L monocytogenes* to grow at refrigerator temperatures, allowing scant numbers to reach an infectious dose during storage. This persistence is enhanced by its ability to form biofilms, which make surfaces and packages more difficult to decontaminate. Heightened awareness has implicated many other foodstuffs, particularly those prepared from animal products in a ready-to-eat form such as sausages and delicatessen poultry items.

Transplacental and birth canal transmission can occur

L monocytogenes may also be transmitted transplacentally to the fetus, presumably following hematogenous dissemination in the mother. It may also be transmitted to newborns in the birth canal in a manner similar to group B streptococci. Listeriosis is still not a reportable disease in the United States, but active surveillance studies indicate that it may account for more than 1000 cases and 200 deaths each year. Most cases occur at the extremes of life (eg, in infants less than 1 month of age or adults over 60 years of age).

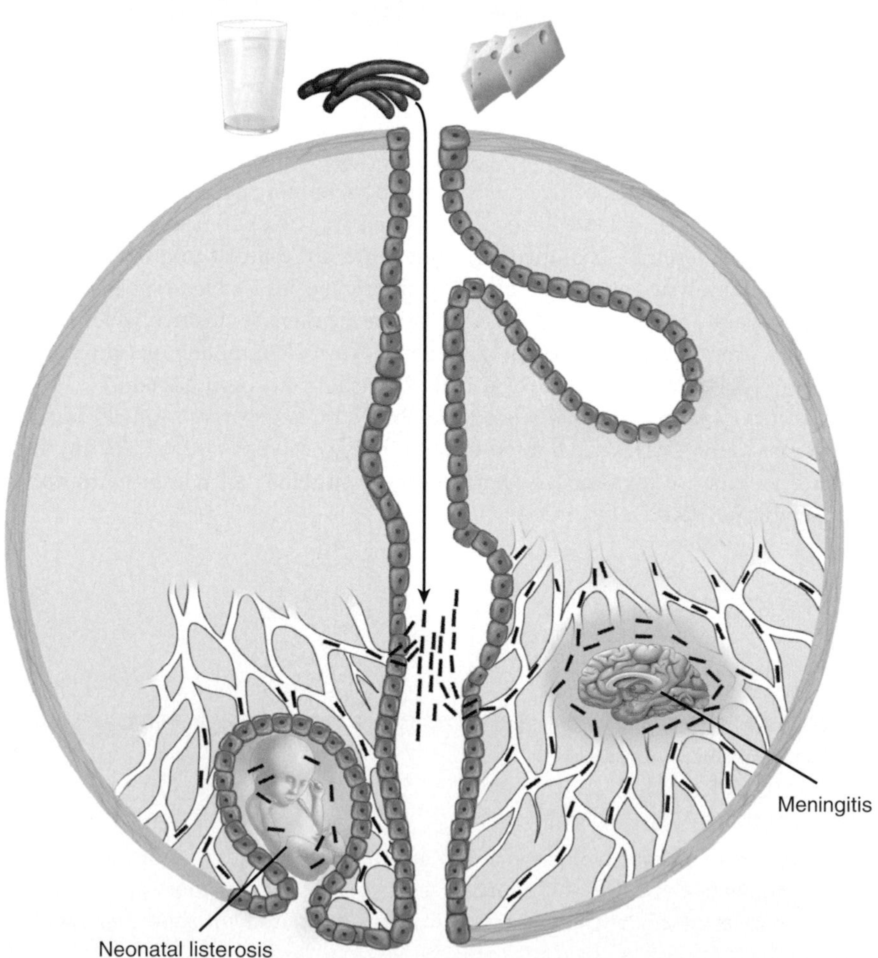

FIGURE 26–5. Listeriosis overview. *Listeria monocytogenes is ingested in dairy and meat products. It invades through the intestinal mucosa producing a bacteremia. The organisms may seed elsewhere particularly the brain (meningitis) or the fetus in pregnancy.*

PATHOGENESIS

L monocytogenes animal models have long been used for the study of cell-mediated immunity because of the ability of the organism to grow in nonimmune macrophages and the requirement for activated macrophages to clear the infection. *L monocytogenes* is able to induce its own uptake by nonprofessional and professional phagocytes including enterocytes, fibroblasts, dendritic cells, hepatocytes, endothelial cells, M cells, and macrophages. The first step in this process takes place when surface **internalin** attaches to a host cell receptor (E-cadherin) and is internalized in an endocytotic vacuole. Once inside the cell, the organism is able to escape from the phagosome to the cytosol in a process mediated by the pore-forming LLO and bacterial phospholipases. ::: Invasion, p. 392

Grows in nonimmune macrophages

Surface internalin starts cell invasion

LLO aids escape from phagosome to cytosol

Once in the cytosol, *L monocytogenes* continues to move through the cell by disrupting the metabolism of the cell's actin and microtubule infrastructure. This process is mediated by LLO and other proteins, particularly one that controls actin polymerization (**Figure 26–6**). In this process actin monomers are sequentially concentrated directly behind the bacterium creating a bacterial "tail" that is connected to the long actin filaments. The addition of new actin units to the tail propels the organisms through the cytosol like a comet through the evening sky (**Figure 26–7**). The motile *Listeria* eventually reach the edge of the cell where, rather than stopping, they protrude into the adjacent cell taking the original cell membrane along with them. When these pinch off, the organisms are surrounded by a double set of host cell membranes that are again dissolved by LLO and phospholipases, releasing the organisms to start the cycle again.

Actin polymerization propels bacteria

Adjacent cells invaded and LLO releases bacteria

This complex strategy allows *L monocytogenes* to survive in macrophages by escaping the phagosome and then to spread from epithelial cell to epithelial cell without exposure to the

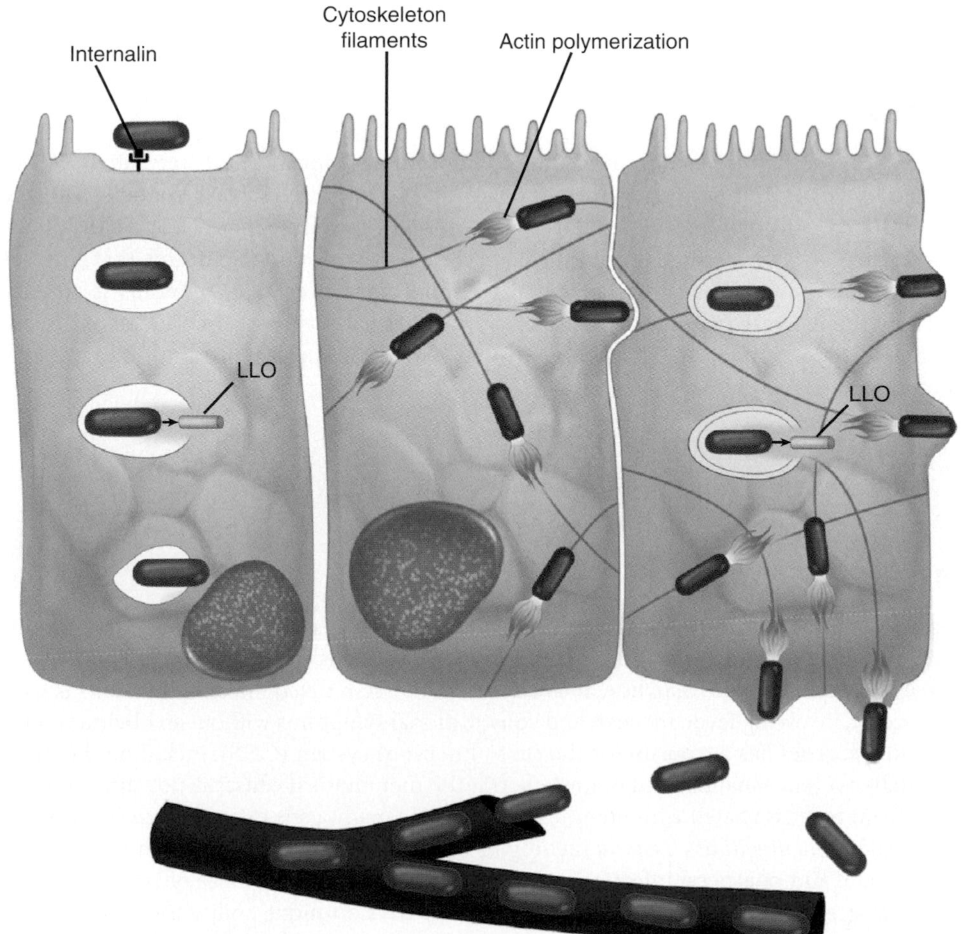

FIGURE 26–6. Listeriosis, cellular view. (*Left*) *Listeria monocytogenes* internalin mediates attachment to an enterocyte and enters in an endocytotic vacuole. Listerolysin O (LLO) lyses the vacuole and the organism escapes to the cytoplasm. (*Middle*) The cytoskeleton is modified and the organisms move along fibers by polymerizing actin (comet tail) invading adjacent cells. (*Right*) *Listeria* has entered another cell now in a double vacuole, which LLO again lyses. The process continues with escape to the submucosa and bloodstream invasion.

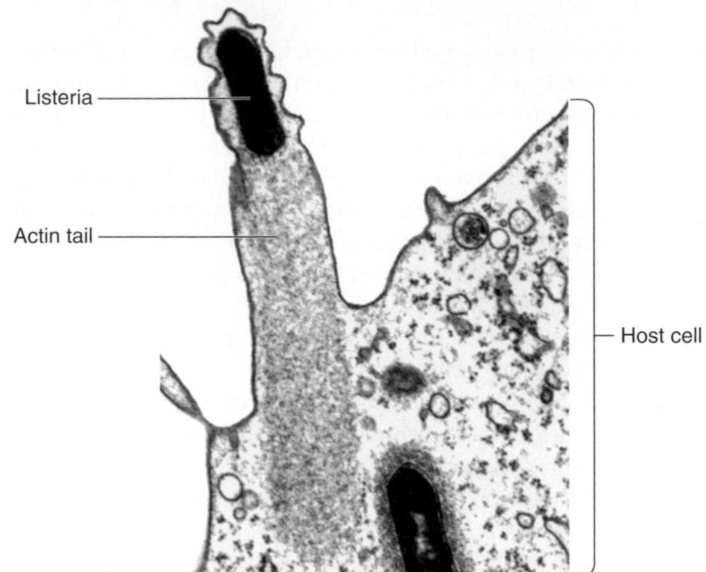

FIGURE 26–7. *Listeria monocytogenes.* This cell is propelled through and soon beyond the cell by the actin tails formed behind it. (Reproduced with permission from Willey J, Sherwood L, Woolverton C (eds). *Prescott's Principles of Microbiology.* New York: McGraw-Hill; 2008.)

Cell-to-cell spread avoids the immune system

Multiple virulence factors are regulated together

immune system. How does *Listeria* keeps its LLO from destroying the host cell membrane from the inside as the pore-forming toxins of other bacteria do from the outside? It appears that *L monocytogenes* may be able to not only regulate the timely production of LLO, but also to trigger its degradation by host cell proteolytic enzymes after it has left the endosome vacuole. LLO and several other genes, including those involved in actin rearrangement, are all part of a virulence regulon contained in a pathogenicity island. The result is a surgically precise deployment of virulence factors.

IMMUNITY

Listeria-specific T-cell activation protects

Immunity to *Listeria* infection owes little to humoral and much to cell-mediated mechanisms. The generation of antigen-specific CD4+ and CD8+ T-cell subsets is required for the resolution of infection and the establishment of long-lived protection. Neutrophils play a role in early stages by lysing *Listeria*-infected cells, but it is cytokine-activation that reverses the intracellular growth in macrophages. The importance of cellular immunity is emphasized by the increased frequency of listeriosis in those with its compromise due to disease such as acquired immunodeficiency syndrome (AIDS), immunosuppressive therapy, age, or pregnancy. ::: Immune activation, p. 35

LISTERIOSIS: CLINICAL ASPECTS

MANIFESTATIONS

Bacteremia is usually occult

Meningitis and encephalitis are produced

Listeriosis usually does not present clinically until there is disseminated infection. In foodborne outbreaks, sometimes gastrointestinal manifestations of primary infection such as nausea, abdominal pain, diarrhea, and fever occur. Disseminated infection in adults is usually occult, involving fever, malaise, and constitutional symptoms without an obvious focus. *L monocytogenes* has a tropism for the central nervous system (CNS), including the brain parenchyma (encephalitis) and brainstem, but the meningitis it causes is not clinically distinct from that associated with other leading bacterial pathogens (*Streptococcus pneumoniae, Neisseria meningitidis*). *Listeria* meningitis does have a particularly high mortality rate.

Neonatal and puerperal infections appear in settings similar to those of infections with group B streptococci. *L monocytogenes* appears to have a unique ability to infect the placenta (**Figure 26–8**), perhaps taking advantage of the mild impairment of cell-mediated

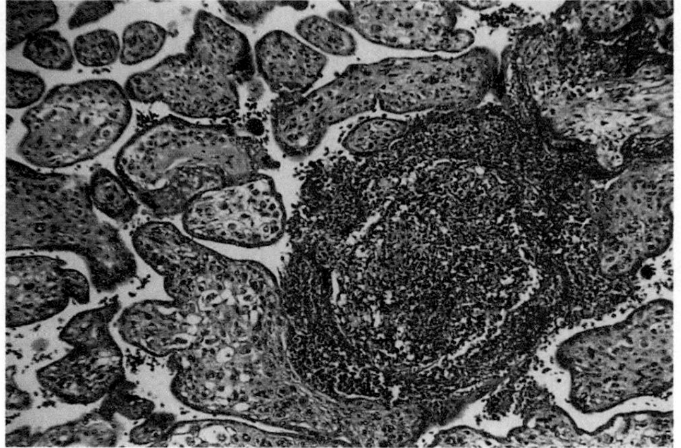

FIGURE 26–8. *Listeria* placentitis. This placental villus has been destroyed by a microabscess due to *L monocytogenes*. The infant was stillborn. (Reproduced with permission from Connor DH, Chandler FW, Schwartz DQA, Manz HJ, Lack EE (eds). *Pathology of Infectious Diseases,* vol. 1. Stamford, CT: Appleton & Lange; 1997.)

immunity during pregnancy. Intrauterine infection leads to stillbirth or a disseminated infection at or near birth. If the pathogen is acquired in the birth canal, the onset of disease is later. The risk of disease is increased in elderly and immunocompromised persons as well as in women in late pregnancy. The number of cases in AIDS patients has been estimated at 300 times that of the general population.

Puerperal infection leads to stillbirth and dissemination

Incidence in AIDS is greatly increased

DIAGNOSIS

Diagnosis of listeriosis is by culture of blood, cerebrospinal fluid (CSF), or focal lesions. In meningitis, CSF Gram stains are usually positive. The first indication that *Listeria* is involved is often the discovery that the β-hemolytic colonies subcultured from a blood culture bottle are Gram-positive rods rather than streptococci.

Blood and CSF culture reveals Gram-positive rods

TREATMENT AND PREVENTION

L monocytogenes is susceptible to penicillin G, ampicillin, and trimethoprim/sulfamethoxazole, (TMP/SMX) all of which have been used effectively. Ampicillin combined with gentamicin is considered the treatment of choice for fulminant cases and in patients with severe compromise of T-cell function. Intense surveillance to prevent the sale of *Listeria*-contaminated ready-to-eat meat products has led to a marked decrease in the incidence of new infections. Avoidance of unpasteurized dairy products and thorough cooking of animal products are wise measures and mandatory for immunocompromised persons. There is no vaccine.

Penicillins and TMP-SMX are effective

BACILLUS

The genus *Bacillus* includes many species of aerobic or facultative, spore-forming, Gram-positive rods. With the exception of one species, *B anthracis,* they are low-virulence saprophytes widespread in air, soil, water, dust, and animal products. *B anthracis* causes the zoonosis anthrax, a disease of animals that is occasionally transmitted to humans.

The genus is made up of rod-shaped organisms that can vary from coccobacillary to rather long-chained filaments. Motile strains have peritrichous flagella. Formation of round or oval spores, which may be central, subterminal, or terminal depending on the species, is characteristic of the genus. *Bacillus* species are Gram positive; however, positivity is often lost, depending on the species and the age of the culture.

Gram-positive spore-forming rods

With *Bacillus,* growth is obtained with ordinary media incubated in air and is reduced or absent under anaerobic conditions. The bacteria are catalase-positive and metabolically active. The spores survive boiling for varying periods and are sufficiently resistant to heat that those of one species are used as a biologic indicator of autoclave efficiency. Spores of *B anthracis* survive in soil for decades.

Aerobic conditions preferred for growth

Heat-resistant spores survive boiling

Bacillus Anthracis

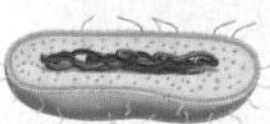

BACTERIOLOGY

Endospores survive in nature

Polypeptide capsule is antiphagocytic

Exotoxin complex has multiple actions

B anthracis has a tendency to form very long chains of rods and in culture is nonmotile and nonhemolytic; colonies are characterized by a rough, uneven surface with multiple curled extensions at the edge resembling a "Medusa head." *B anthracis* has a D-glutamic acid polypeptide capsule of a single antigenic type that has antiphagocytic properties. *B anthracis* endospores are extremely hardy and have been shown to survive in the environment for decades. The organism also produces a potent exotoxin complex, which consists of two enzymes, edema factor (EF) and lethal factor (LF) together with a receptor recognition protein called protective antigen (PA). PA binds to a receptor on the host cell surface forming a pore-like site for EF and LF which allows all three to enter the cell (**Figure 26–9a**). Once in the cytosol multiple toxin actions are expressed including adenylate cyclase activity and host protein inactivation.

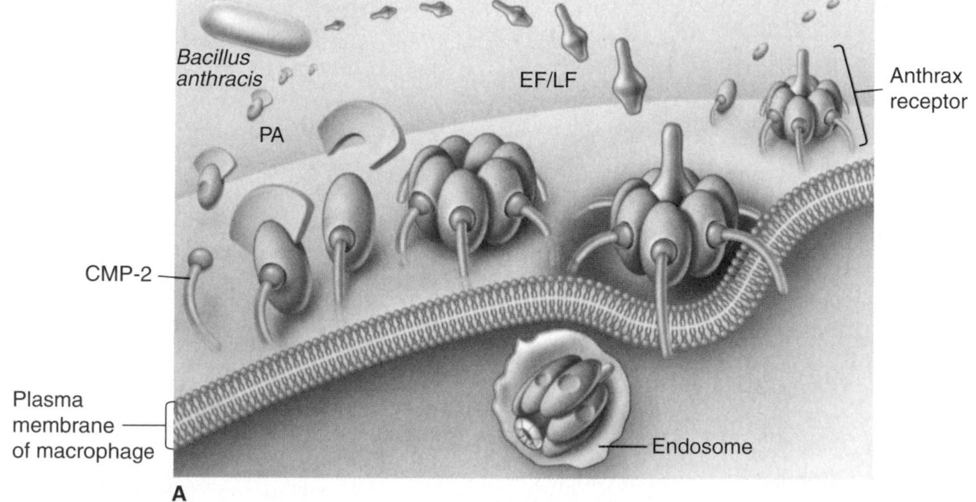

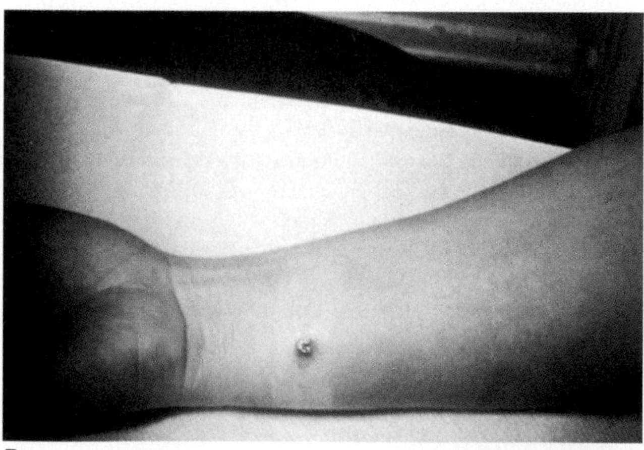

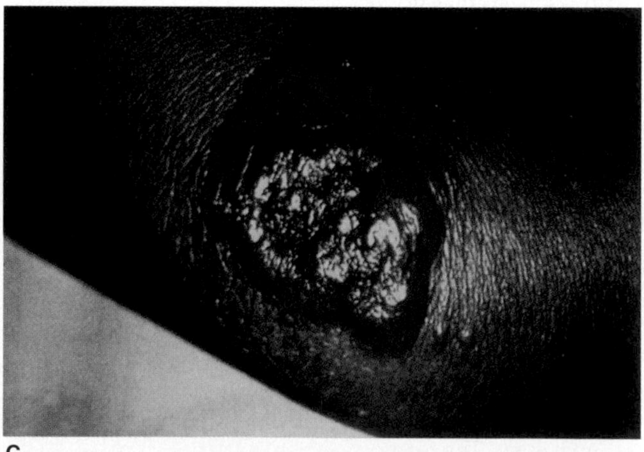

FIGURE 26–9. Anthrax. A. A protein called protective antigen (PA) delivers two other proteins, edema factor (EF) and lethal factor (LF), to the capillary morphogenesis protein-2 (CMP-2) receptor on the cell membrane of a target macrophage, where PA, EF, and LF are transported to an endosome. PA then delivers EF and LF from the endosome into the cytoplasm of the macrophage where they exert their toxic effects. **B.** Early anthrax papule that evolves into **C.** the necrotic eschar called the malignant pustule. (Reproduced with permission from Willey J, Sherwood L, Woolverton C (eds). *Prescott's Principles of Microbiology.* New York: McGraw-Hill; 2008.)

ANTHRAX

CLINICAL CAPSULE

Human anthrax is typically an ulcerative sore on an exposed part of the body. Constitutional symptoms are minimal, and the ulcer usually resolves without complications. If anthrax spores are inhaled, a fulminant pneumonia may lead to respiratory failure and death.

The isolation of *B anthracis,* the proof of its relationship to anthrax infection, and the demonstration of immunity to the disease are among the most important events in the history of science and medicine. Robert Koch rose to fame in 1877 by growing the organism in artificial culture using pure culture techniques. He defined the stringent criteria needed to prove that the organism caused anthrax (Koch's postulates), then met them experimentally. Louis Pasteur made a convincing field demonstration at Pouilly-le-Fort to show that vaccination of sheep, goats, and cows with an attenuated strain of *B anthracis* prevented anthrax. He was cheered and carried on the shoulders of the grateful farmers of the district, an experience now, unhappily, largely restricted to winning football coaches.

Pasteur produced animal vaccine with attenuated anthrax strain

EPIDEMIOLOGY

Anthrax is primarily a disease of herbivores such as horses, sheep, and cattle, who acquire it from spores of *B anthracis* contaminating their pastures. Humans become infected through contact with these animals or their products in a way that allows the spores to be inoculated through the skin, ingested, or inhaled. In the 1920s, more than 100 cases occurred annually in the United States among farmers, veterinarians, and meat handlers, but the control of animal anthrax in developed countries has made human cases rare. A few endemic foci persist in North America and have been the source of naturally acquired disease. Another source is animal products such as wool, hides, or bone meal fertilizer that have been imported from a country where animal anthrax is endemic.

Infection is through injection of spores derived from herbivores into the skin

Contaminated materials are imported from countries with animal anthrax

The real threat associated with anthrax comes from its continuing appeal to those bent on using it as an agent of biologic warfare or terrorism. The long life, stability, and low mass of the dried spores make the prospect of someone producing a "cloud of death" leading to massive pulmonary anthrax a chilling reality. A 1979 episode resulting in more than 60 anthrax deaths in the former Soviet Union is now attributed to an accidental explosion at a biologic warfare research facility that aerosolized more than 20 pounds of anthrax spores. United Nations inspection teams in Iraq uncovered facilities for the production of massive amounts of spores together with plans to create and spread infectious aerosols using missile warheads. The inhalation anthrax among postal workers after the September 11, 2001, terrorist attacks appears to have been due to the mailing of envelopes containing "weapons-grade" anthrax spores stolen from a biologic warfare research facility. Such spores had been treated to enhance their aerosolization and dissemination. The source of this outbreak is still unknown. The forms of anthrax are summarized in **Figure 26–10**.

Use for biologic warfare is a continuing threat

Aerosols could spread pulmonary anthrax widely

Weapons-grade spores are specially treated

PATHOGENESIS

When spores of *B anthracis* reach the rich environment of human tissues, they germinate and multiply in the vegetative state. The antiphagocytic properties of the capsule aid in survival, eventually allowing production of large enough amounts of the exotoxins to cause

Antiphagocytic effect of D-glutamic acid capsule is required for virulence

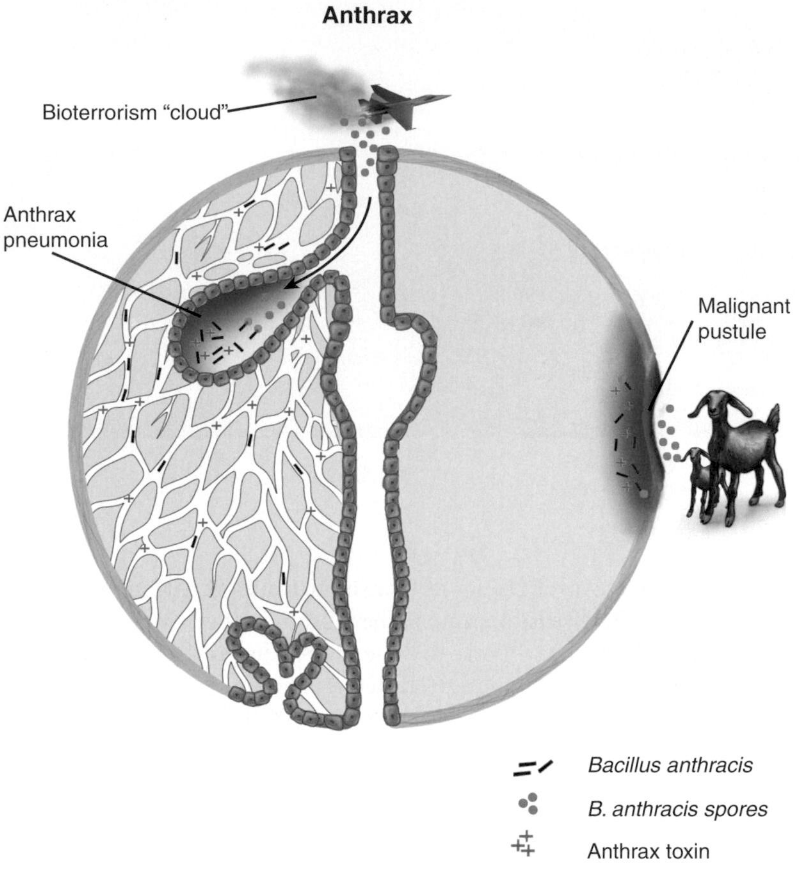

FIGURE 26–10. Anthrax overview. Naturally acquired anthrax (*right*) is from the traumatic inoculation of *Bacillus anthracis* spores derived from animals with anthrax. The lesion is destructive but remains localized. Bioterrorism-acquired anthrax (*left*) would occur by the inhalation of explosive aerosols of *B anthracis* spores. This causes a pneumonia with rapid spread to the bloodstream.

Edema is produced by EF

disease. The tripartite nature of the anthrax exotoxin complex must play an important role, but the timing and relative importance of the components are not known. The EF adenylate cyclase activity is believed to correlate with the striking edema seen at infected sites.

IMMUNITY

Immune mechanisms are unknown

The specific mechanisms of immunity against *B anthracis* are not known. Experimental evidence favors antibody directed against the toxin complex, but the relative role of the components of the toxin is not clear. The capsular glutamic acid is immunogenic, but antibody against it is not protective.

ANTHRAX: CLINICAL ASPECTS

MANIFESTATIONS

Initial papule evolves to malignant pustule

Cutaneous anthrax usually begins 2 to 5 days after inoculation of spores into an exposed part of the body, typically the forearm or hand. The initial lesion is an erythematous papule, which may be mistaken for an insect bite. This papule usually progresses through vesicular and ulcerative stages in 7 to 10 days to form a black eschar (scab) surrounded by edema (Figure 26–9B and C). This lesion is known as the "malignant pustule," although it is neither malignant nor a pustule. Associated systemic symptoms are usually mild, and the lesion typically heals very slowly after the eschar separates. Less commonly, the disease progresses with massive local edema, toxemia, and bacteremia.

Pulmonary anthrax is contracted by inhalation of spores. Historically, this has occurred when contaminated hides, hair, wool, and the like are handled in a confined space (wool-sorter's disease) or after laboratory accidents. Today it is the form we would expect from the dissemination of a spore aerosol in biologic warfare. In the pulmonary syndrome, 1 to 5 days of nonspecific malaise, mild fever, and nonproductive cough lead to progressive respiratory distress and cyanosis. Massive spread to the bloodstream and CNS follow rapidly. Mediastinal edema was a prominent finding in postal workers. If untreated, progression to a fatal outcome is usually very rapid once bacteremia has developed. An intestinal form of anthrax follows ingestion of contaminated food, usually meat. It is characterized by abdominal pain, ascites, and shock.

Pulmonary anthrax is acquired by inhaling spores

Fever and cough progress to cyanosis and death

DIAGNOSIS

Culture of skin lesions, sputum, blood, and CSF are the primary means of anthrax diagnosis. Given some suspicion on epidemiologic grounds, Gram stains of sputum or other biologic fluids showing large numbers of Gram-positive bacilli can suggest the diagnosis. In September of 2001, diagnosis of the first case in Florida was speeded by an infectious disease specialist who knew such rods were extremely rare in the spinal fluid. Large Gram-positive bacilli are also unusual in sputum. *B anthracis* and other *Bacillus* species are not difficult to grow. In fact, clinical laboratories frequently isolate the non-anthrax species as environmental contaminants. The saprophytic species are usually β-hemolytic and motile, features not found in *B anthracis,* but most clinical laboratories are not skilled at separating *Bacillus* species. Blood cultures are positive in most cases of pulmonary anthrax.

Smears with large Gram-positive rods are suggestive

Hemolysis and motility exclude *B anthracis*

Sputum and blood cultures are positive in pneumonia

TREATMENT

Antimicrobial treatment has little effect on the course of cutaneous anthrax but does protect against dissemination. Almost all strains of *B anthracis* are susceptible to penicillin, doxycycline, and ciprofloxacin. Although penicillin has long been the treatment of choice for all forms of anthrax, experience gained during the 2001 outbreak has caused the first-line recommendation to be changed to doxycycline or ciprofloxacin. These antibiotics are also recommended for chemoprophylaxis in the case of known or suspected exposure.

Ciprofloxacin or doxycycline is used for treatment and prophylaxis

PREVENTION

The most important preventive measures are those that eradicate animal anthrax and limit imports from endemic areas. Vaccines are also useful. Pasteur's vaccine used a live strain attenuated by repeated subculture that resulted in the loss of a plasmid encoding toxin production. A similar live vaccine is still effective for animals. The human vaccine licensed in the United States is prepared by extraction from cultures of a nonencapsulated avirulent strain of *B anthracis.* The extract is almost entirely the protective antigen component of the toxin complex. In 2002, the Institute of Medicine issued a detailed analysis of human and animal studies and declared the vaccine both safe and efficacious. Experts also feel that it is very unlikely that the architects of biologic warfare would be able to craft *B anthracis* strains for which this vaccine is not protective. In Russia and China, a live vaccine is used in which spores are inoculated by scarification.

Eradication of animal anthrax is most important

Live and inactivated vaccines are available

Other *Bacillus* Species

Bacillus spores are widespread in the environment, and isolation of one of the more than 20 *Bacillus* species other than *B anthracis* from clinical material usually represents contamination of the specimen. Occasionally *B cereus, B subtilis,* and some other species produce genuine infections, including infections of the eye, soft tissues, and lung. Infection is usually associated with immunosuppression, trauma, an indwelling catheter, or contamination of complex equipment. The relative resistance of *Bacillus* pores to disinfectants aids their survival in medical devices that cannot be heat sterilized.

Spores enhance survival in medical devices

B cereus produces pyogenic toxin and enterotoxin

B cereus deserves special mention. This species is most likely to cause opportunistic infection, which suggests a virulence intermediate between that of *B anthracis* and the other species. A strain isolated from an abscess has been shown to produce a destructive pyogenic toxin. *B cereus* can also cause food poisoning by means of enterotoxins. One enterotoxin acts by stimulating adenyl cyclase production and fluid excretion in the same manner as toxigenic *E coli* and *Vibrio cholerae*.

CLINICAL CASE

SORE THROAT AND CONFUSION AFTER SUMMER CAMP

A 9-year-old girl developed listlessness and a sore throat on 10 days after arriving at a summer camp operated by a religious group that does not accept immunizations. Four days later, the girl returned home on a camp bus along with other unimmunized children and adults who had also attended the camp. A physician evaluated the patient for a sore throat. A throat culture was taken and oral penicillin prescribed. The patient was hospitalized for persistent sore throat, diminished fluid intake, and gingival bleeding. Laboratory tests revealed a white blood cell count of 26,500/mm^3 with 92% polymorphonuclear cells, blood urea nitrogen of 214 mg/dL, creatinine of 12.4 mg/dL, and platelet count of 10,000/mm^3. The throat culture was reported to contain normal flora, group A β-hemolytic streptococci and large numbers of diphtheroids. The patient was transferred to a tertiary care children's hospital.

On admission, she was afebrile and had moderate upper airway obstruction, diffuse ecchymoses, bleeding from the nose and gums, prominent cervical adenopathy and swelling of the jaw and throat. The pharynx revealed severe hemorrhagic and necrotic tonsillitis; no membrane was observed. Treatment with penicillin G, gentamicin, peritoneal dialysis, and platelet transfusions was instituted. The hospital course was complicated by disseminated intravascular coagulation, cardiac conduction abnormalities, and mental confusion. The patient died 2 weeks after the sore throat began. A *Corynebacterium* species isolated from her throat culture was subsequently confirmed to be a toxigenic strain of *C diphtheriae*.

QUESTIONS

■ Attention to what "clue" would have suggested the diagnosis earlier?

A. Hemorrhagic pharyngitis
B. Renal failure
C. Immunization history
D. Group A strep in throat

■ What treatment might have saved this girl's life?

A. Intravenous penicillin
B. Ciprofloxacin
C. Corticosteroids
D. Diphtheria toxoid
E. Diphtheria antitoxin

■ The cardiac conduction abnormalities were probably due to:

A. Infarction
B. Inhibition of protein synthesis
C. Pore-forming toxin
D. Internalin
E. Edema factor

ANSWERS

1(C), **2**(E), **3**(B)

CHAPTER 27

Mycobacteria

A dread disease in which the struggle between soul and body is so gradual, quiet and solemn, and the result so sure that day by day, and grain by grain, the mortal part wastes and withers away. A disease ... which sometimes moves in giant strides and sometimes at a tardy sluggish pace, but, slow or quick, is ever sure and certain.

—Charles Dickens: *Nicholas Nickleby*

Mycobacterium is a genus of Gram-positive bacilli that demonstrate the staining characteristic of acid-fastness. Its most important species, *Mycobacterium tuberculosis,* is the etiologic agent of tuberculosis, the dread disease called consumption in Dickens' time. One of the oldest and most devastating of human afflictions, tuberculosis remains a leading cause of infectious deaths worldwide today. A second mycobacterium, *Mycobacterium leprae,* is the causative agent of leprosy. A large number of less pathogenic species are assuming increasing importance as disease agents in immunocompromised patients, particularly those with acquired immunodeficiency syndrome (AIDS).

MYCOBACTERIUM: GENERAL CHARACTERISTICS

BACTERIOLOGY

MORPHOLOGY AND STRUCTURE

The mycobacteria are slim, poorly staining bacilli, which demonstrate the property of acid-fastness. They are nonmotile, obligate aerobes that do not form spores. The cell wall contains peptidoglycan similar to that of other Gram-positive organisms, to which many branched-chain polysaccharides, proteins, and lipids are attached. Porins and other proteins are found throughout the cell wall. Of particular importance is the presence of long-chain fatty acids called **mycolic acids** (for which the *myco*bacteria are named) and **lipoarabinomannan (LAM),** a lipid polysaccharide complex extending from the plasma membrane to the surface (**Figure 27–1**). LAM is structurally and functionally analogous to the lipopolysaccharide of Gram-negative bacteria. These elements give the

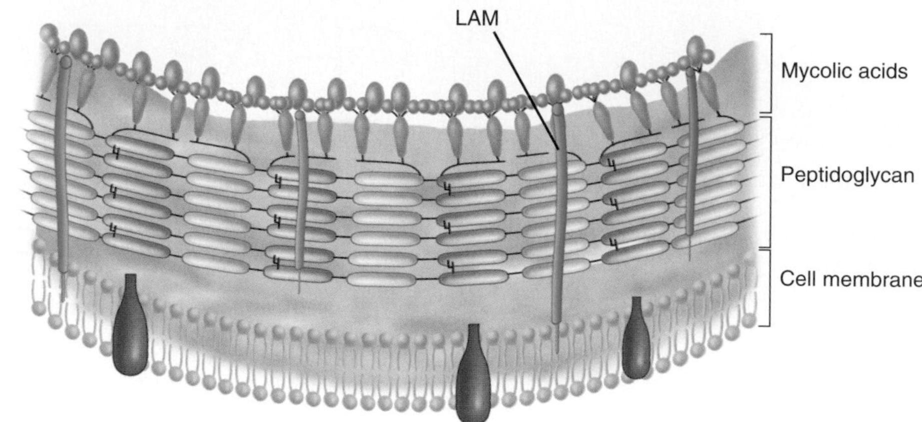

FIGURE 27–1. Mycobacterial cell wall. LAM, lipoarabinomannan. (Reproduced with permission from Willey J, Sherwood L, Woolverton C (eds). *Prescott's Principles of Microbiology.* New York: McGraw-Hill; 2008.)

Cell wall has high lipid content

Mycolic acids and LAM form waxy coat

Acid fastness: Once stained, difficult to decolorize

mycobacteria a cell wall with unusually high lipid content (more than 60% of the total cell wall mass), which accounts for many of their biologic characteristics. It can be thought of as a waxy coat that makes them hardy, impenetrable, and hydrophobic. The staining characteristic of acid-fastness is the most frequently observed of these features. The mycobacterial cell wall can be stained only through the use of extreme measures (heat, penetrating agents) but once in the stain is *fast.* Even the strongest of decolorizing agents (acid and alcohol) do not wash it out (**Figure 27–2**). :::Acid-fast stain, p. 63

GROWTH

Strict aerobes, many species grow slowly

The most important pathogen, *M tuberculosis,* shows enhanced growth in 10% carbon dioxide and at a relatively low pH (6.5 to 6.8). Nutritional requirements vary among species and range from the ability of some nonpathogens to multiply on the washers of water faucets to the strict intracellular parasitism of *M leprae,* which does not grow in artificial media or cell culture. Mycobacteria grow more slowly than most human pathogenic bacteria because of their hydrophobic cell surface, which causes them to clump and limits permeability of nutrients into the cell.

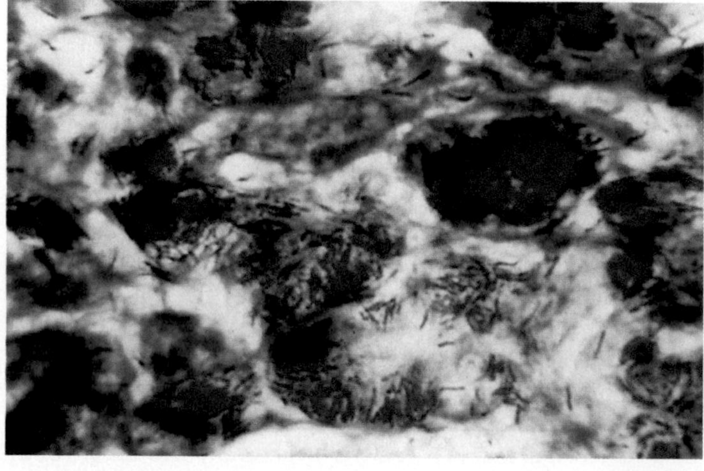

FIGURE 27–2. *Mycobacterium tuberculosis* in sputum stained by the acid-fast technique. The mycobacteria retain the red carbol fuchsin through the decolorization step. The cells, background, and any other organisms stain with the contrasting methylene blue counterstain. (Reproduced with permission from Nester EW, Anderson DG, Roberts CE Jr, Nester MT. *Microbiology: A Human Perspective,* 6th ed. New York: McGraw-Hill, 2008.)

CLASSIFICATION

Classic mycobacterial classification has been based on a constellation of phenotypic characteristics, including nutritional and temperature requirements, growth rates, pigmentation of colonies grown in light or darkness, key biochemical tests, the cellular constellation of free fatty acids, and the range of pathogenicity in experimental animals. The major species and some of their more important characteristics are summarized in **Table 27–1**. Increasingly, this classification system is yielding to molecular-based techniques. The identification of species-specific rRNA and DNA sequences has resulted in the revision and expansion of the older phenotype-based classification system and the provision of an increasing array of species-specific DNA probes to clinical mycobacteriology laboratories.

Distinguished by cultural features, biochemical reactions, and pathogenicity

MYCOBACTERIAL DISEASE

Mycobacteria include a wide range of species pathogenic for humans and animals. Some, such as *M tuberculosis,* occur exclusively in humans under natural conditions. Others, such as *M intracellulare,* can infect various hosts, including humans, but also exist in the free-living state. Most nonpathogenic species are widely distributed in the environment. Diseases caused by mycobacteria usually develop slowly, follow a chronic course, and elicit a granulomatous response. Infectivity of pathogenic species is high, but virulence for healthy humans is low. For example, disease following infection with *M tuberculosis* is the exception rather than the rule. ::: Granuloma, p. 26

Includes human and animal pathogens

Slowly progressive diseases

Mycobacteria do not produce classic exotoxins or endotoxins. Disease processes are thought to be the result of two related host responses. The first, a delayed-type hypersensitivity (DTH) reaction to mycobacterial proteins, results in the destruction of nonactivated macrophages containing multiplying organisms. It is detected by intradermal injections of purified proteins from the mycobacteria. The second, cell-mediated immunity (CMI), activates macrophages, enabling them to destroy mycobacteria contained within their cytoplasm. The balance between these two responses determines the pathology and clinical response to a mycobacterial infection.

Lack exotoxins or endotoxins

MYCOBACTERIUM TUBERCULOSIS

BACTERIOLOGY

M tuberculosis (MTB) is a slim, strongly acid–alcohol–fast rod. It frequently shows irregular beading in its staining, appearing as connected series of acid-fast granules (Figure 27–2). It grows at 37°C, but not at room temperature, and it requires enriched or complex media for primary growth. The classic medium, Löwenstein–Jensen, contains homogenized egg in nutrient base with dyes to inhibit the growth of nonmycobacterial contaminants. Growth is very slow, with a mean generation time of 12 to 24 hours. The dry, rough, buff-colored colonies usually appear after 3 to 6 weeks of incubation. Growth is more rapid in semisynthetic (oleic acid–albumin) and liquid media. The major phenotypic tests for identification are summarized in Table 27–1. Of particular importance is the ability of MTB to produce large quantities of niacin, which is uncommon in other mycobacteria.

Growth takes weeks

Biochemical tests distinguish from other mycobacteria

Because of its hydrophobic lipid surface, MTB is unusually resistant to drying, to most common disinfectants, and to acids and alkalis. Tubercle bacilli are sensitive to heat, including pasteurization, and individual organisms in droplet nuclei are susceptible to inactivation by ultraviolet light. As with other mycobacteria, the MTB cell wall structure is dominated by mycolic acids and LAM. Its antigenic makeup includes many protein and polysaccharide antigens, of which **tuberculin** is the most studied. It consists of heat-stable proteins liberated into liquid culture media. A purified protein derivative (PPD) of tuberculin is used for skin testing for hypersensitivity and is standardized in tuberculin units according to skin test activity.

Unusual resistance to drying and disinfectants but not to heat

PPD is mix of tuberculin proteins

TABLE 27–1 Mycobacteria of Major Clinical Importance[a]

				CHARACTERISTICS					
SPECIES	RESERVOIR	VIRULENCE FOR HUMANS	DISEASE CAUSED	CASE-TO-CASE TRANSMISSION	GROWTH RATE	OPTIMUM GROWTH TEMPERATURE	PIGMENT PRODUCTION[b]	SUBSTANTIAL NIACIN PRODUCTION[c]	VIRULENCE FOR GUINEA PIGS[d]
Mycobacterium tuberculosis	Human	+++	Tuberculosis	Yes	S	37	–	+	+
M bovis	Animals	+++	Tuberculosis	Rare	S	37	–	–	+
Bacillus Calmette-Guérin	Artificial culture	±	Local lesion	Very rare	S	37	–	–	–
M kansasii	Environmental	+	Tuberculosis-like	No	S	37	Photochromogen	–	–
M scrofulaceum	Environmental	+	Usually lymphadenitis	No	S	37	Scotochromogen	–	–
M avium-intracellulare	Environmental; birds	+	Tuberculosis-like	No	S	37	±	–	–
M fortuitum	Environmental	±	Local abscess	No	F	37	±	–	Local abscess
M marinum	Water; fish	±	Skin granuloma	No	S	30	Photochromogen	–	–
M ulcerans	Probably environmental; tropical	+	Severe skin ulceration	No	S	30	–		–
M leprae	Human	+++	Leprosy	Yes	NG	NG	NG	NG	–
M smegmatis	Human, external urethral area	–	None	–	F	37	–	–	–

S, slow (colonies usually develop in 10 days or more); F, fast (colonies develop in 7 days or less); NG, not grown.

[a]Numerous nonpathogenic environmental mycobacteria exist and may contaminate human specimens.

[bc]Yellow-orange pigment. Photochromogen is pigment produced in light; scotochromogen is pigment produced in dark or light.

[c]Many other differential biochemical tests used; eg, nitrate reduction, catalase production, Tween 80 hydrolysis.

[d]Disease following subcutaneous injection of light inoculum (eg, 10^2 cells).

CLINICAL CAPSULE

Tuberculosis is a systemic infection manifested only by evidence of an immune response in most exposed individuals. In some infected persons, the disease either progresses or, more commonly, reactivates after an asymptomatic period (years). The most common reactivation form is a chronic pneumonia with fever, cough, bloody sputum, and weight loss. Spread outside the lung also occurs and is particularly devastating when it reaches the central nervous system. The natural history follows a course of chronic wasting to death aptly called "consumption" in the past.

EPIDEMIOLOGY

A recognized disease of antiquity, tuberculosis first reached epidemic proportions in the Western world during the Industrial Revolution beginning in the 18th and 19th centuries. Associated with urbanization and crowding consumption accounted for 20% to 30% of all deaths in cities winning tuberculosis the appellation of "the captain of all the men of death." Morbidity rates were many times higher. The disease has had major sociologic components, flourishing with ignorance, poverty, and poor hygiene, particularly during the social disruptions of war and economic depression. Under these conditions, the poor are the major victims, but all sectors of society are at risk. Chopin, Paganini, Rousseau, Goethe, Chekhov, Thoreau, Keats, Elizabeth Barrett Browning, and the Brontës, to name but a few, were all lost to tuberculosis in their prime. With knowledge of the cause and transmission of the disease and the development of effective antimicrobial agents, tuberculosis was increasingly brought under control in developed countries. Unfortunately, mortality and morbidity remain at 19th-century levels in many developing countries with over 50,000 deaths every week worldwide (**Figure 27–3**).

Infection of 18th and 19th centuries

Attack rates still high in many developing countries

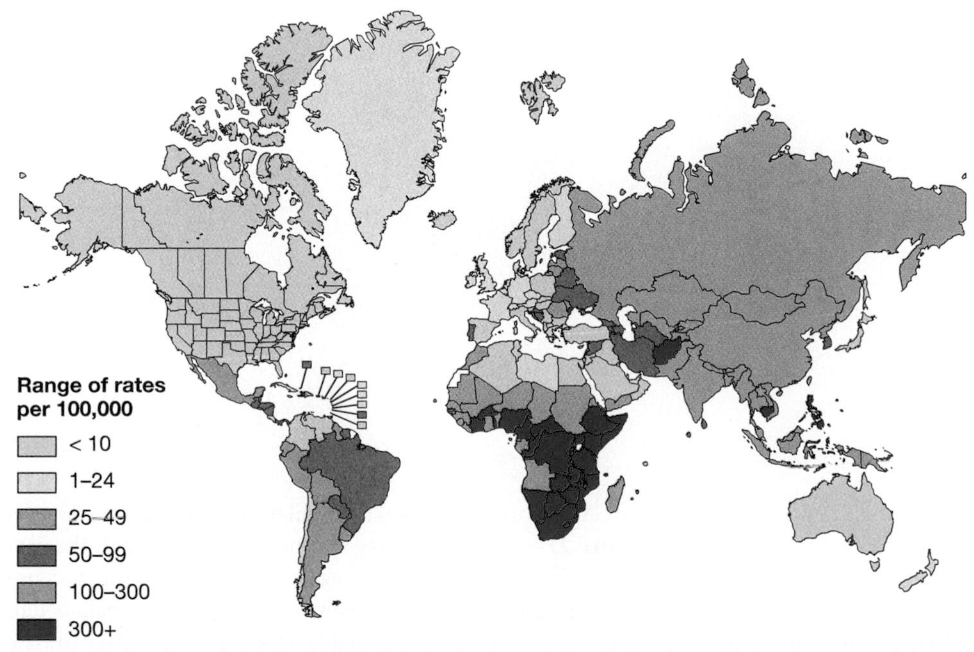

FIGURE 27–3. The worldwide incidence and distribution of tuberculosis. (Reproduced with permission from Willey J, Sherwood L, Woolverton C (eds). *Prescott's Principles of Microbiology.* New York: McGraw-Hill; 2008.)

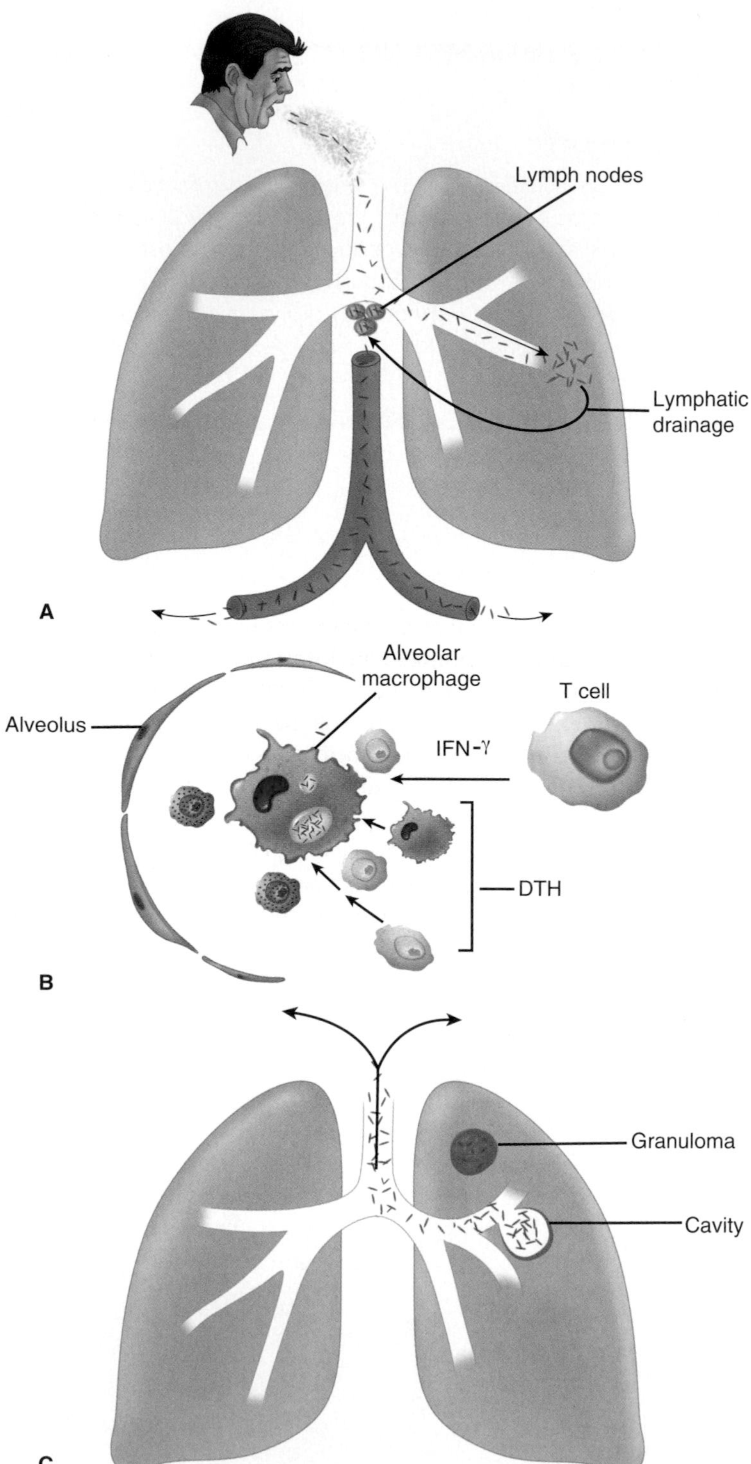

FIGURE 27–4. Tuberculosis. A. Primary tuberculosis. *Mucobacterium tuberculosis* is inhaled in droplet nuclei from an active case of tuberculosis. Initial multiplication is in the alveoli with spread through lymphatic drainage to the hilar lymph nodes. After further lymphatic drainage to the bloodstream, the organisms are spread throughout the body. **B. Alveolar macrophage.** The two-front battle being carried out between A and C is shown. Ingested bacteria multiply in the nonactivated macrophage. (1) T_H1 cellular immune responses attempt to activate the macrophage by secreting cytokines (interferon gamma [IFN-γ]). If successful, the disease is arrested. (2) Inflammatory elements of delayed-type hypersensitivity (DTH) are attracted and cause destruction. If activation is not successful, disease and injury continue. **C. Reactivation tuberculosis.** Reactivation typically starts in the upper lobes of the lung with granuloma formation. DTH-mediated destruction can form a cavity, which allows the organisms to be coughed up to infect another person.

The majority of tuberculous infections are contracted by inhalation of droplet nuclei carrying the causative organism (**Figure 27–4**). Humans may also be infected through the gastrointestinal tract after ingestion of milk from tuberculous cows (now uncommon because of pasteurization) or, rarely, through abraded skin. It has been estimated that a single cough can generate as many as 3000 infected droplet nuclei and that less than 10 bacilli may initiate a pulmonary infection in a susceptible individual. The likelihood of acquiring infection thus relates to the numbers of organisms in the sputum of an open case of the disease, the frequency and efficiency of the coughs, the closeness of contact, and the adequacy of ventilation in the contact area. Epidemiologic data indicate that large doses or prolonged exposure to smaller infecting doses is usually needed to initiate infection in humans. In some closed environments, such as a submarine or a crowded nursing home, a single open case of pulmonary tuberculosis can infect the majority of nonimmune individuals sharing sleeping accommodations. The grandfather who has developed "chronic bronchitis" is a classic source of infection to children.

Most infections are by respiratory route

Repeated coughing generates infectious dose into air

Poor ventilation increases risk

The AIDS pandemic and the spread of MTB strains resistant to multiple drugs have added to the tuberculosis burden. Globally, one third of the world's population is infected, and 30 million people have active disease. It is estimated that patients with latent tuberculosis increase their risk of reactivation disease by a factor 200 to 300 times that with the development of HIV coinfection. With this dark synergy, tuberculosis and AIDS are the leading causes of premature death in the world.

AIDS and drug resistance enhance spread

PATHOGENESIS

Primary Infection

M tuberculosis is an intracellular pathogen whose success depends on avoiding the killing mechanisms of professional phagocytes. Primary tuberculosis is the initial infection in which inhaled droplet nuclei containing tubercle bacilli are deposited in the peripheral respiratory alveoli, most frequently those of the well-ventilated middle and lower lobes. In these lobes they are recognized by alveolar macrophage complement receptors (CR1, CR3, CR4) and phagocytosed. This inaugurates a two-front battle with the macrophage, which may be resolved in days or may last for decades. The first is with the phagosome/lysosome digestive mechanisms of the macrophage. In this process, MTB is able to interfere with the acidification of the phagosome, which renders the lysosomal enzymes (which require acidic pH) less effective. This allows the organisms to multiply freely in the cytoplasm of the nonactivated macrophage (Figure 27–4). The second process is the triggering of T_H1 immune responses, beginning with digestion and surface presentation of mycobacterial components and ending with cytokine activation of the macrophages. The short- and long-term outcomes of the infection depend on the ability of the macrophage activation process to overcome the intracellular edge that MTB has as a result of its ability to block acidification of the phagosome. ::: Phagosome/lysosome, p. 25; T-cell activation, p. 33

MTB multiplies in alveolar macrophages

Acidification of phagosome blocked

T_H1 responses triggered

In the early stages of infection, MTB-laden macrophages are transported through lymphatic channels to the hilar lymph nodes draining the infected site. From there, a low-level bacteremia disseminates the bacteria to a number of tissues, including the liver, spleen, kidney, bone, brain, meninges, and apices or other parts of the lung. Although the primary site of infection and enlarged hilar lymph nodes can often be detected radiologically, the distant sites usually have no findings. In fact, the primary evidence for their existence is reactivation at nonpulmonary sites later in life. Tuberculous meningitis is the most serious of these.

MTB disseminates to lymph nodes and bloodstream

In the primary lesion as MTB cells multiply, macrophages and dendritic cells release cytokines (tumor necrosis factor, interleukin 12, interferon gamma [IFN-γ]), which attract T cells and other inflammatory cells to the site. The recruited CD4 T cells initiate the T_H1-type immune response over the following 3 to 9 weeks in which IFN-γ is the primary activator of macrophages. During these weeks as the bacteria multiply, they may generate quantities of mycobacterial proteins exceeding thresholds required to also trigger a DTH response. The DTH with its phagocytes, fluid, and release of digestive enzymes adds a destructive component to the process and is the sole known source of injury in tuberculosis. If the T_H1 process is effective, the source of DTH stimulation wanes and the disease resolves. The mycobacterial protein-specific DTH sensitization remains, and its elicitation is the basis of the tuberculin skin test (see Diagnosis). ::: T_H1 immunity, p. 35; DTH, p. 42

Cytokines attract T cells and T_H1 response

MTB proteins also trigger DTH and injury

The mixture of the T_H1 immune and DTH responses is manifest in a microscopic structure called a **granuloma,** which is composed of lymphocytes, macrophages, epithelioid cells

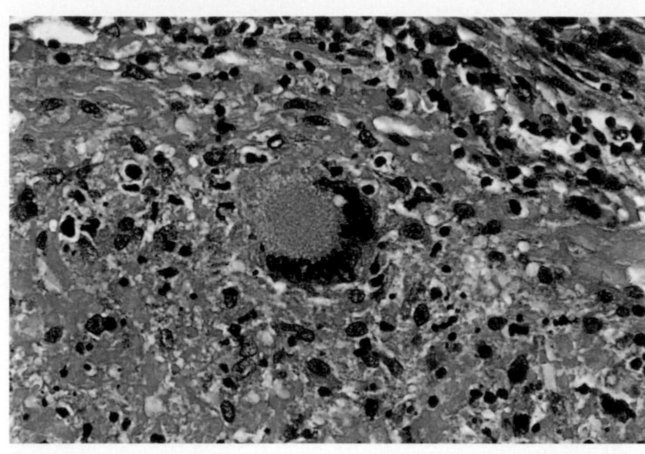

A

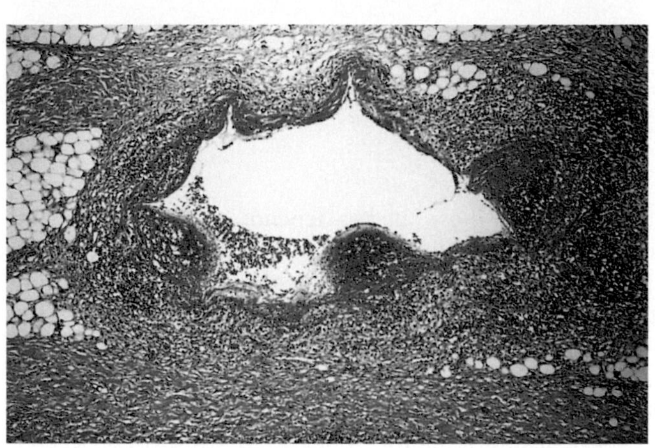

B

FIGURE 27–5. Tuberculous granulomas. A. Early granuloma with lymphocytes, epithelioid cells and fibroblasts organizing around a central focus. The multinucleate giant cell in the center is typical of granulomas but not exclusive to *Mycobacterium tuberculosis*. **B. Multiple** granulomas surround and invade a vein near the lung hilum. Central degeneration is starting to appear and will eventually become caseous necrosis. (Reproduced with permission from Connor DH, Chandler FW, Schwartz DQA, Manz HJ, Lack EE (eds). *Pathology of Infectious Diseases,* vol. 1. Stamford, CT: Appleton & Lange; 1997.)

The granuloma includes macrophages, lymphocytes, **fibroblasts**

Caseous necrosis is due to DTH

(activated macrophages), fibroblasts, and multinucleated giant cells all in an organized pattern (**Figure 27–5**). As the granuloma grows, the destructive nature of the hypersensitivity component leads to necrosis usually in the center of the lesion. This is termed **caseous necrosis** because of the cheesy, semisolid character of material at the center of large gross lesions, but the terms fits the smooth glassy appearance of microscopic granulomas as well. ::: Granuloma, p. 26

Primary lesions heal once immunity develops

Some MTB enter dormant state rather than dying

Primary infections are handled well once the immune response halts the intracellular growth of MTB. Bacterial multiplication ceases, the lesions heal by fibrosis, and the organisms appear to slowly die. This sequence occurs in infections with multiple other infectious agents for which it is the end of the story. In tuberculosis, some of the organisms when faced with oxygen and nutrient deprivation enter a prolonged dormant state rather than dying. Specific factors facilitating this change are not known but the waxy nature of the MTB cell wall must aid survival under these conditions as it does in the environment. These organisms in the lung and elsewhere lie waiting for reactivation months, years, or decades later. For most persons who undergo a primary infection this never happens, either because of the complete killing of the original population or the failure of factors favoring reactivation to materialize

Reactivation (Adult) Tuberculosis

Dormant MTB reactivates at aerobic sites

Destruction forms pulmonary cavities

Although mycobacterial factors have been identified (resuscitation-promoting factor), little is known of the mechanisms of reactivation of these dormant foci. It has generally been attributed to some selective waning of immunity. The new foci are usually located in body areas of relatively high oxygen tension that would favor growth of the aerobe MTB. The apex of the lung is the most common, with spreading, coalescing granulomas, and large areas of caseous necrosis. Necrosis often involves the wall of a small bronchus from which the necrotic material is discharged, resulting in a pulmonary cavity and bronchial spread. Frequently, small blood vessels are also eroded. In all of this the basis for the highly destructive nature of MTB is largely unknown. It produces no exotoxins, and both the intact cell and cellular components are remarkably innocuous to humans and experimental animals not previously sensitized to tuberculin. It appears that with the failure of the host to control growth of MTB the rising load of mycobacterial protein stimulates a progressively autodestructive DTH response.

No toxins

Progressive DTH causes injury

IMMUNITY

Humans generally have a rather high innate immunity to development of disease. This was tragically illustrated in the Lübeck disaster of 1926, in which infants were administered MTB instead of an intended vaccine strain. Despite the large dose, only 76 of 249 died and most of the others developed only minor lesions. Approximately 10% of immunocompetent persons infected with MTB develop active disease at any time in their life. There is epidemiologic and historic evidence for differences in the immunity in certain population groups and between identical and nonidentical twins.

Innate immunity is high and genetically variable

Immunity to tuberculosis is primarily related to the development of reactions mediated through CD4 T lymphocytes via T_H1 pathways (see Chapter 2). Intracellular killing of MTB by macrophages activated by INF-γ and other cytokines is the essential step. The specific components of MTB that are important in initiating these reactions are not known. Cytotoxic CD8 T cells are also generated during infection and may play some role. Although antibodies are formed in the course of disease, there is no evidence they play any role in immunity. ::: Cell-mediated immunity, p.35

T_H1 immunity is most important

Cytotoxic CD8 lymphocytes may participate

TUBERCULOSIS: CLINICAL ASPECTS

MANIFESTATIONS

Primary Tuberculosis

Primary tuberculosis is either asymptomatic or manifest only by fever and malaise. Radiographs may show infiltrates in the mid-zones of the lung and enlarged draining lymph nodes in the area around the hilum. When these lymph nodes fibrose and sometimes calcify, they produce a characteristic picture (Ghon complex) on radiograph. In approximately 5% of patients, the primary disease is not controlled and merges into the reactivation type of tuberculosis, or disseminates to many organs. The latter may result from a necrotic tubercle eroding into a small blood vessel.

Mid-lung infiltrates and adenopathy are produced

Primary infection may progress to reactivation or dissemination

Reactivation Tuberculosis

Approximately 10% of persons recovering from a primary infection develop clinical disease sometime during their lifetime. In Western countries, reactivation of previous quiescent lesions occurs most often after age 50 and is more common in men. Reactivation is associated with a period of immunosuppression precipitated by malnutrition, alcoholism, diabetes, old age, and a dramatic change in the individual's life, such as loss of a spouse. In areas in which the disease is more common, reactivation tuberculosis is more frequently seen in young adults experiencing the immunosuppression that accompanies puberty and pregnancy. Recently, reactivation and progressive primary tuberculosis among younger adults have increased as a complication of AIDS.

Reactivation is most common in older men

Predisposing factors include underlying disease and life events

Cough is the universal symptom of tuberculosis. It is initially dry, but as the disease progresses sputum is produced, which even later is mixed with blood (hemoptysis). Fever, malaise, fatigue, sweating, and weight loss all progress with continuing disease. Radiographically, infiltrates appearing in the apices of the lung coalesce to form cavities with progressive destruction of lung tissue. Less commonly, reactivation tuberculosis can also occur in other organs, such as the kidneys, bones, lymph nodes, brain, meninges, bone marrow, and bowel. Disease at these

Cough is universal

Cavities form in lung apices

Multiple organs are involved

sites ranges from a localized tumor-like granuloma (tuberculoma) to a fatal chronic meningitis. Untreated, the progressive cough, fever, and weight loss of pulmonary tuberculosis creates an internally consuming fire that usually takes 2 to 5 years to cause death. The course in AIDS and other T cell-compromised patients is more rapid.

DIAGNOSIS

Tuberculin Test

PPD test measures DTH to tuberculoprotein

Positive PPD indicates past or current infection

Anergy may develop with immune compromise

The tuberculin skin test (**Figure 27–6**) measures DTH to an international reference tuberculoprotein preparation called purified protein derivative (PPD). The test involves an intradermal injection that is read 48 to 72 hours later. An area of induration of 10 mm or more accompanied by erythema constitutes a positive reaction, and no induration indicates a negative reaction. A positive PPD test indicates that the individual has developed DTH through infection at some time with MTB, but carries no implication as to whether the disease is active. Persons who have been infected with another mycobacterial species or immunized with the bacillus Calmette-Guérin (BCG) vaccine may also be reactive, but the induration is usually less than 10 mm. Patients with severe disseminated disease, those on immunosuppressive drugs, or those with immunosuppressive diseases such as AIDS may fail to react due to anergy.

Predictive value depends on prevalence

BCG immunization compromises public health value

The predictive value of the tuberculin test depends on the prevalence of tuberculosis and other mycobacterial diseases in the population and public health practices, particularly the use of BCG immunization. In the United States, where BCG is not used, and has a low disease prevalence, a positive test is very strong evidence of previous MTB infection. In countries that use BCG, the skin test can only be used selectively. A new test that detects release of MTB-specific IFN-γ from stimulated T cells does not cross-react with BCG or other mycobacteria, but it is expensive and no more sensitive than the tuberculin test in the detection of active disease.

Laboratory Diagnosis

Acid-fast Smears

MTB can be detected microscopically in smears of clinical specimens using one of the acid-fast staining procedures discussed in Chapter 4. Because the number of organisms present is often small, specimens such as sputum and cerebrospinal fluid are concentrated by centrifugation before staining to improve the sensitivity of detection. In one of the acid-fast procedures, the stain is fluorescent, which enhances the chances that a microscopist will

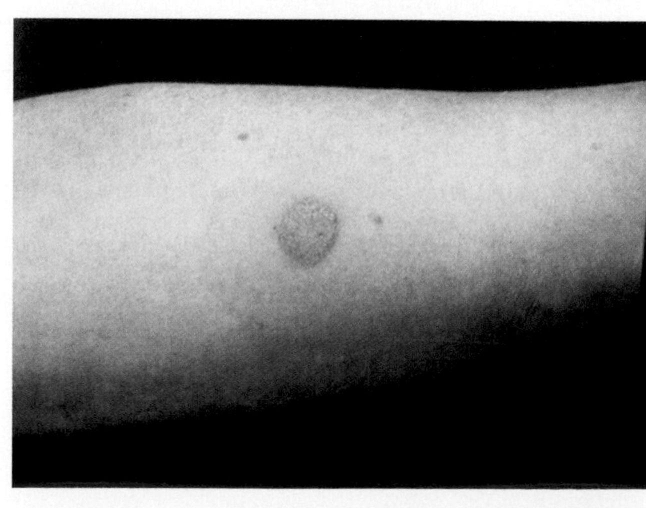

FIGURE 27–6. Tuberculin test. The PPD (purified protein derivative) tuberculoprotein was injected intradermally at this site 48 hours previously. The erythema and induration (>10 mm) that are present indicate the development of delayed-type (type IV) hypersensitivity. (Reproduced with permission from Nester EW, Anderson DG, Roberts CE Jr, Nester MT. *Microbiology: A Human Perspective*, 6th ed. New York: McGraw-Hill, 2008.)

be able to find just a few acid-fast bacilli (AFB) in an entire specimen. Even with the best of concentration and staining methods, little more than half (~ 65%) of culture-positive sputum samples yield positive smears. The yield from other sites is even lower, particularly from cerebrospinal fluid. The presence of AFB is not specific for MTB because other mycobacteria may have a similar morphology.

Mycobacteria are detected in direct smears of clinical material

Additional caution must be exercised in the interpretation of positive smears from urine or from medical devices (bronchoscope, nasotracheal tube) because contamination with environmental AFB is possible. Despite its challenges, the pursuit of a smear diagnosis is well worth the effort because it allows clinical decision making while awaiting the days to weeks required for culture results.

Contaminating mycobacteria may yield "false" positives in some specimens

Culture

Whether the AFB smear is positive or not, culture of the organism is essential for confirmation and for antimicrobial susceptibility testing. Specimens from sites, such as cerebrospinal fluid, bone marrow, and pleural fluid, can be seeded directly to culture media used for MTB isolation. Samples from sites inevitably contaminated with normal flora, such as sputum, gastric aspirations (cultured when sputum is not available), and voided urine, are chemically treated (alkali, acid, detergents) using concentrations, experience has shown to kill the bulk of the normal flora but allow most mycobacteria to survive. Sputum specimens also require the use of agents to dissolve mucus so the specimen can be concentrated by centrifugation or filtration before inoculation onto the culture media just described.

Material contaminated with normal flora is chemically treated

Mucolytic agents used in sputum

Cultures on solid media usually take 3 weeks or longer to show visible colonies. Growth may be detected radiometrically in about half the time by using liquid broth containing ^{14}C-labeled substrates which are metabolized by mycobacteria to liberate $^{14}CO_2$. The labeled CO_2 is detected in the space above the medium using an automated sampling procedure. Identification of the isolated mycobacterium is achieved with a number of cultural and biochemical tests, including those shown in Table 27–1, but the process takes weeks more. Nucleic acid amplification (NAA) procedures targeting both DNA and ribosomal RNA sequences in clinical specimens have been developed, which have high specificity for MTB (>98%), but sensitivities (70% to 80%) that are better than smears but not culture. NAA probes can also be used for identification of mycobacteria isolated in culture. Even with improved sensitivity, direct NAA methods could not substitute for culture because of the need for live bacteria to carry out antimicrobial susceptibility testing, which is essential with newly diagnosed cases. ::: NAA methods, p. 80-81

Cultures take 3+ weeks

Radiometric procedures are faster

NAA lacks sensitivity

Susceptibility testing essential

TREATMENT

Mycobacteria are inherently resistant to many antimicrobial agents based on the unusually impermeable nature of their lipid-rich cell wall. However, several antimicrobics have been shown to be effective in the treatment of MTB infection (**Table 27–2**). The term **first-line** is used to describe the primary drugs of choice (isoniazid, ethambutol, rifampin, pyrazinamide, streptomycin) that have long clinical experience to back up their efficacy and to

TABLE 27–2 Antimicrobics Commonly Used in Treatment of Tuberculosis

FIRST-LINE DRUG	SECOND-LINE DRUG[a]
Isoniazid	*para*-Aminosalicylic acid
Ethambutol	Ethionamide
Rifampin	Cycloserine
Pyrazinamide	Fluoroquinolones
Streptomycin	Kanamycin, amikacin

[a]Second-line drugs added to combinations if resistance or toxicity contraindicates first-line agent.

manage their side effects. **Second-line** agents are less preferred and reserved for use when there is resistance to the first-line agents. The approach with new cases is to start the patient on multiple (usually four) first-line drugs while waiting for the results of susceptibility tests. When these results are available, the regimen is dropped back to two or three agents known to be active against the patient's isolate.

Resistance to first-line causes use of second-line drugs

Isoniazid and rifampin are active against both intra- and extracellular organisms, and pyrazinamide acts at the acidic pH found within cells. Streptomycin does not penetrate into cells and is thus active only against extracellular organisms. MTB is also susceptible to other drugs that may be used to replace those of the primary group if they are inappropriate because of resistance or drug toxicity. The fluoroquinolones, such as ciprofloxacin and ofloxacin, are active against MTB and penetrate well into infected cells. Their role in the treatment of tuberculosis is under evaluation. Isoniazid and ethambutol act on the mycolic acid (isoniazid) and LAM (ethambutol) elements of mycobacterial cell wall synthesis. The molecular targets of the other agents have yet to be defined except for the general antibacterial agents (rifampin, streptomycin, fluoroquinolones) discussed in Chapter 23.

Antimicrobics act intra and extracellularly

Resistance or toxicity may limit some agents

Because of the high bacterial load and long duration of anti-MTB therapy, the emergence of resistance during treatment is of greater concern than with more acute infections. For this reason, the use of multiple drugs each with a different mode of action is the norm. Expression of resistance would then theoretically require a double mutant, a very low probability when the frequency of single mutants is 10^{-7} to 10^{-10}. The percentage of new infections with strains resistant to first-line drugs varies between 5% and 15%, but it appears to be increasing, particularly among those who have been treated previously. Of particular concern is the emergence in the last two decades of multidrug-resistant tuberculosis (MDR-TB) strains, which are resistant to isoniazid and rifampin, the mainstays of primary treatment. MDR-TBs now represent almost 5% of the worldwide cases, and over half of these are concentrated in three countries, China, India, and the Russian Federation. Although still rare, strains that add resistance to one or more second-line drugs (called extensively drug resistant [XDR-TB]) are now being seen.

Multidrug therapy limits expression of resistance

MDR-TB are resistant to isoniazid and rifampin

Effective treatment renders the patient noninfectious within 1 or 2 weeks, which has shifted the care of tuberculous patients from isolation hospitals and sanatoriums to the home or the general hospital. The duration of therapy varies, based on some clinical factors but is usually 6 to 9 months. In patients whose organisms display resistance to one or more of these drugs, and in those with HIV infection, a more prolonged treatment course is used. The effectiveness of chemotherapy on most forms of tuberculosis has been dramatic and has greatly reduced the need for surgical procedures such as pulmonary lobectomy. Failure of chemotherapy is often associated with lack of adherence to the regimen by the patient, the presence of resistant organisms, or both.

Treatment lasts 6 to 9 months

Resistance and HIV require longer

Compliance a major problem

PREVENTION

There are a number of situations in which persons are felt to be at increased risk for tuberculosis even though they have no clinical evidence of disease (healthy, negative chest x-ray, etc.). The most common of these situations are close exposure to an open case (particularly a child) and conversion of the tuberculin skin test from negative to positive. In these instances, prophylactic chemotherapy with isoniazid (alone) is administered for 6 to 9 months. In the exposed person, the goal is to prevent a primary infection. The PPD-positive person has already had a primary infection; therefore, the goal is to reduce the chance of reactivation tuberculosis by eradicating any dormant MTB in the body. This chemoprophylaxis has clear value for the exposed person and recent skin test converters. It is less certain for those whose time of conversion is uncertain and could have been many years ago. Isoniazid may cause a form of hepatitis in adults so its administration carries some risk. ::: Chemoprophylaxis, p. 426

Exposure and PPD conversion warrant isoniazid chemoprophylaxis

BCG is a live vaccine derived originally from a strain of *M bovis* that was attenuated by repeated subculture. It is administered intradermally to tuberculin-negative subjects and leads to self-limiting local multiplication of the organism with development of tuberculin DTH. The latter negates the PPD as a diagnostic and epidemiologic tool. BCG has been used for prevention of tuberculosis in various countries since 1923, with results ranging from ineffective to 80% protection. In most studies, however, BCG has substantially

BCG stimulates tuberculin DTH

decreased disseminated and meningeal forms of tuberculosis among young children. Its use in any country is a matter of public health policy balancing the potential protection against the loss of case tracking through the skin test. BCG is not used in the United States but is in many other counties, particularly those that lack the infrastructure for case tracking. BCG is contraindicated for individuals in whom T-cell–mediated immune mechanisms are compromised, such as those infected with HIV.

Effectiveness is variable

BCG contraindicated in AIDS

MYCOBACTERIUM LEPRAE

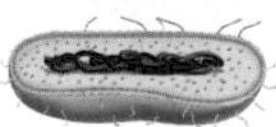

BACTERIOLOGY

Mycobacterium leprae, the cause of leprosy, is an acid-fast bacillus that has not been grown in artificial medium or tissue culture beyond, possibly, a few generations. However, it will grow slowly (doubling time 14 days) in some animals (mice, armadillos). Although lack of in vitro growth severely limits study of the organism, the structure and cell wall components appear to be similar to those of other mycobacteria. One mycoside, (phenolic glycolipid I [PGL-1]), is synthesized in large amounts and found only in *M leprae.*

Fails to grow in culture

Slow growth in animals

LEPROSY

CLINICAL CAPSULE

Leprosy is a chronic granulomatous disease of the peripheral nerves and superficial tissues, particularly the nasal mucosa. Disease ranges from slowly resolving anesthetic skin lesions to the disfiguring facial lesions responsible for the social stigma and ostracism of the individuals with leprosy (lepers).

EPIDEMIOLOGY

The exact mode of transmission is unknown but appears to be by generation of small droplets from the nasal secretions from cases of lepromatous leprosy. Traumatic inoculation through minor skin lesions or tattoos is also possible. The central reservoir is infected humans. The incubation period as estimated from clinical observations is generally 2 to 7 years, but sometimes up to four decades. The infectivity of *M leprae* is low. Most new cases have had prolonged close contact with an infected person. Biting insects may also be involved. Although virtually absent from North America and Europe, still more than 10 million persons are infected in Asia, Africa, and Latin America with more than 40,000 new cases per year.

Nasal droplets transmit infection

Rare in North America

PATHOGENESIS

M leprae is an obligate intracellular parasite that must multiply in host cells to persist. In humans, the preferred cells are macrophages and Schwann cells. PGL-1 and LAM have been implicated in the ability to survive and multiply in these cells. The organism may invade peripheral sensory nerves, resulting in patchy anesthesia. The disease occurs in two major forms with a spectrum of illness in between. In the **tuberculoid** form, few *M leprae* are seen in lesions, which are granulomatous with extensive epithelioid cells, giant cells,

Obligate intracellular parasite of macrophages and Schwann cells and lymphocytic infiltration. In **lepromatous** leprosy, the cellular response is minimal, and growth of *M leprae* is thus relatively unimpeded. Histologically, lesions show dense infiltration with leprosy bacilli, and organisms may reach the bloodstream.

IMMUNITY

T_H1 immunity determines extent of disease

Immunity to *M leprae* is T-cell–mediated. Tuberculoid cases have minimal disease and evidence of T_H1 immune responses including production of typical cytokines (IL-2, IFN-γ). Lepromatous cases have progressive disease and lack T_H1 mediators. In the past, this range of disease also correlated with DTH responsiveness to lepromin, a skin test antigen similar to MTB tuberculin. Lepromin is no longer available, but tuberculoid cases gave a vigorous DTH response whereas lepromatous patients did not respond.

LEPROSY: CLINICAL ASPECTS

MANIFESTATIONS

Tuberculoid Leprosy

Skin and nerve involvement

Anesthetic lesions

Tuberculoid leprosy involves the development of macules or large, flattened plaques on the face, trunk, and limbs, with raised, erythematous edges and dry, pale, hairless centers. When the bacterium has invaded peripheral nerves, the lesions are anesthetic. The disease is indolent, with simultaneous evidence of slow progression and healing. Because of the small number of organisms present, this form of the disease is usually noncontagious.

Lepromatous Leprosy

Lesions are infiltrative and diffuse

In lepromatous leprosy, skin lesions are infiltrative, extensive, symmetric, and diffuse, particularly on the face, with thickening of the looser skin of the lips, forehead, and ears (**Figure 27–7**). Damage may be severe, with loss of nasal bones and septum, sometimes of digits, and testicular atrophy in men. Peripheral neuropathies may produce deformities or nonhealing painless ulcers. The organism spreads systemically, with involvement of the reticuloendothelial system.

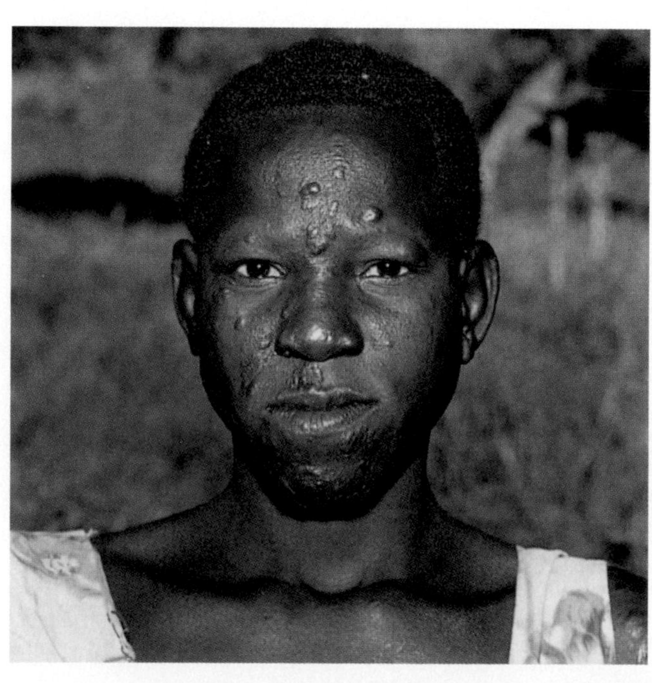

FIGURE 27–7. Lepromatous leprosy. Note the cutaneous plaques, infiltrates, and loss of eyebrows. Scrapings of the ear lobes would reveal numerous acid-fast bacilli. This advanced case will still respond to appropriate chemotherapy. (Reproduced with permission from Connor DH, Chandler FW, Schwartz DQA, Manz HJ, Lack EE (eds). *Pathology of Infectious Diseases,* vol. 1. Stamford, CT: Appleton & Lange; 1997.)

DIAGNOSIS

Leprosy is primarily a clinical diagnosis confirmed by demonstration of AFB in stained scrapings of infected tissue, particularly nasal mucosa or ear lobes. This is readily achieved in lepromatous leprosy because of the typically large numbers of AFB. Tuberculoid leprosy is confirmed by the histologic appearance of full-thickness skin biopsies and hopefully a few AFB.

Acid-fast smears and biopsies are primary diagnostic methods

TREATMENT AND PREVENTION

Treatment has been revolutionized by the development of sulfones, such as dapsone, which blocks *para*-aminobenzoic acid metabolism in *M leprae*. Combined with rifampin, dapsone usually controls or cures tuberculoid leprosy when given for 6 months. In lepromatous leprosy and multibacillary intermediate forms of the disease, a third agent (clofazimine) is added to help prevent the selection of resistant mutants, and treatment is continued at least 2 years. Prevention of leprosy involves recognition and treatment of infectious patients and early diagnosis of the disease in close contacts. Chemoprophylaxis with sulfones has been used for children in close contact with lepromatous cases.

Sulfones combined with rifampin primary treatment

Prevention requires early diagnosis and treatment of cases

A possible diagnosis of leprosy elicits fear and distress in patients and contacts out of all proportion to its risks. Few clinicians in the United States have the experience to make such a diagnosis, and expert help should be sought from public health authorities before reaching this conclusion or indicating its possibility to the patient.

MYCOBACTERIA CAUSING TUBERCULOSIS-LIKE DISEASES

Mycobacteria causing diseases that often resemble tuberculosis are listed in Table 27–1. With the exception of *M bovis,* mycobacteria have become relatively more prominent in developed countries as the incidence of tuberculosis has declined. All have known or suspected environmental reservoirs, and all the infections they cause appear to be acquired from these sources. Immunocompromised individuals or those with chronic pulmonary conditions or malignancies are more likely to develop disease. There is no evidence of case-to-case transmission. Environmental mycobacteria that cause tuberculosis-like infections are usually more resistant than *M tuberculosis* to some of the antimicrobics used in the treatment of mycobacterial diseases, and susceptibility testing is often needed as a guide to therapy.

Acquired from the environment; no case-to-case transmission

Resistance is common

Mycobacterium kansasii

Mycobacterium kansasii is a photochromogenic mycobacterium that usually forms yellow-pigmented colonies after about 2 weeks of incubation in the presence of light. In the United States, infection is most common in Illinois, Oklahoma, and Texas and tends to affect urban residents; it is uncommon in the Southeast. There is no evidence of case-to-case transmission, but the reservoir has yet to be identified. It causes about 3% of non-MTB mycobacterial disease in the United States. *M kansasii* infections resemble tuberculosis and tend to be slowly progressive without treatment. Cavitary pulmonary disease, cervical lymphadenitis, and skin infections are most common, but disseminated infections also occur. They are an important cause of disease in patients with HIV infection and CD4 T lymphocyte counts of less than 200 cells/μL; clinical features closely resemble tuberculosis in patients with AIDS. Hypersensitivity to proteins of *M kansasii* develops and cross-reacts almost completely with that caused by tuberculosis. Positive PPD tests may thus result from clinical or subclinical *M kansasii* infection. Prolonged combined chemotherapy with isoniazid, rifampin, and ethambutol is usually effective.

Resembles tuberculosis

Infection may cause PPD conversion

Mycobacterium avium–intracellulare Complex

Mycobacterium avium–intracellulare (MAC) complex is a group of mycobacteria that grow only slightly faster than *M tuberculosis* and can be divided into a number of serotypes. Among them are organisms that cause tuberculosis in birds (and sometimes swine), but

MAC associated with birds and mammals

rarely lead to disease in humans. Others may produce disease in mammals, including humans, but not in birds. They are found worldwide in soil and water and in infected animals. Human cases are increasingly prominent in developed countries including the United States, Japan, and Switzerland, where they are second only to *M tuberculosis* in significance and frequency of the diseases they cause.

Second MTB cause of disease in developed countries

The most common infection in humans is cavitary pulmonary disease, often superimposed on chronic bronchitis and emphysema. Most individuals infected are white men of 50 years of age or more. Cervical lymphadenitis, chronic osteomyelitis, and renal and skin infections also occur. The organisms in this group are substantially more resistant to antituberculous drugs than most other species, and treatment with the three or four agents found to be most active often requires supplementation with surgery. About 20% of patients suffer relapse within 5 years of treatment.

Wide range of diseases; most common are pulmonary

Relative resistance to antituberculous drugs

Disseminated MAC infections, once considered rare, are now a common systemic bacterial infection in patients with AIDS. They usually develop when the patient's general clinical condition and CD4 T-lymphocyte concentrations are declining. Clinically, the patient experiences progressive weight loss and intermittent fever, chills, night sweats, and diarrhea. Histologically, granuloma formation is muted, and there are aggregates of foamy macrophages containing numerous intracellular AFB. The diagnosis is most readily made by blood culture, using a variety of specialized cultural techniques. Response to chemotherapeutic agents is marginal, and the prognosis is grave.

Common complication of AIDS

Organisms isolated from blood

Mycobacterium scrofulaceum

Mycobacterium scrofulaceum is an acid-fast scotochromogen (Table 27–1), which occurs in the environment under moist conditions. It forms yellow colonies in the dark or light within 2 weeks, and it shares several features with MAC. *M scrofulaceum* is now one of the more common causes of granulomatous cervical lymphadenitis in young children. It derives its name from scrofula, an old descriptive term for tuberculous cervical lymphadenitis. The infection manifests as an indolent enlargement of one or more lymph nodes with little, if any, pain or constitutional signs. It may ulcerate or form a draining sinus to the surface. It does not cause PPD conversion. Treatment usually involves surgical excision.

Granulomatous cervical lymphadenitis in children

MYCOBACTERIAL SOFT TISSUE INFECTIONS

Mycobacterium fortuitum Complex

Mycobacterium fortuitum complex comprises free-living, rapidly growing, AFB, which produce colonies within 3 days. Human infections are rare. Abscesses at injection sites in drug abusers are probably the most common lesions. Occasional secondary pulmonary infections develop. Some cases have been associated with implantation of foreign material (eg, breast prostheses, artificial heart valves). Except in the case of endocarditis, infections usually resolve spontaneously with removal of the prosthetic device.

Rapid growers cause abscesses and infections of prostheses

Mycobacterium marinum

Mycobacterium marinum causes tuberculosis in fish, is widely present in fresh and salt waters, and grows at 30°C but not at 37°C. It occurs in considerable numbers in the slime that forms on rocks or on rough walls of swimming pools and thrives in tropical fish aquariums. It can cause skin lesions in humans. Classically, a swimmer who abrades his or her elbows or forearms climbing out of a pool develops a superficial granulomatous lesion that finally ulcerates. It usually heals spontaneously after a few weeks, but is sometimes chronic. The organism may be sensitive to tetracyclines as well as to some antituberculous drugs.

Cause of fish tuberculosis

Mycobacterium ulcerans

Mycobacterium ulcerans is a much more serious cause of superficial infection. (Like *M marinum, M ulcerans* grows at 30°C but not at 37°C [see Table 27–1].) Cases usually occur in the tropics, most often in parts of Africa, New Guinea, and northern Australia, but have been seen elsewhere sporadically. Children are most often affected. The source of infection and mode of transmission are unknown. Infected individuals develop severe ulceration involving the skin and subcutaneous tissue that is often progressive unless treated effectively. Surgical excision and grafting are usually needed. Antimicrobic treatment is often unsuccessful.

Occurs in tropical areas

Severe, progressive ulcerations require surgical removal

CASE STUDY

JAIL, HIV, AND AFB

A 55-year-old man with a 2-month history of fevers, night sweats, increased cough with bloody sputum production, and a 25-lb weight loss was seen in the emergency room. He reports no intravenous drug use or homosexual activity but has had multiple sexual encounters in the last year. He "sips" a pint of gin a day, was jailed 2 years ago in New York City related to a fight with gunshot and stab wounds. His physical examination revealed bilateral anterior cervical and axillary adenopathy and a temperature of 39.4°C. His chest radiograph showed peritracheal adenopathy and bilateral interstitial infiltrates. His laboratory findings showed a positive HIV serology and a low absolute CD4 lymphocyte count. An acid-fast organism grew from the sputum and bronchoalveolar lavage (BAL) fluid from the right middle lobe.

QUESTIONS

■ The most likely etiologic agent(s) for this patient's infection are:

A. *Mycobacterium tuberculosis* (MTB)
B. *Mycobacterium avium-intracellulare* (MAC)
C. *Mycobacterium leprae*
D. A and B
E. B and C

■ All of the following factors increase this man's risk of developing active tuberculosis *except*:

A. Homosexual relations
B. Jail
C. HIV
D. Alcoholism

■ If the acid-fact bacterium isolated from the man's sputum is identified as *Mycobacterium tuberculosis* and he is placed on a two-drug antituberculous regimen, the resolution of his disease depends primarily on:

A. Antibody to LAM
B. Lifestyle changes
C. T_H1 immune responses
D. T_H2 immune responses
E. Active DTH

ANSWERS

1(D), **2**(A), **3**(C)

CHAPTER 28

Actinomyces and *Nocardia*

Actinomyces and *Nocardia* are Gram-positive rods characterized by filamentous, tree-like branching growth, which has caused them to be confused with fungi in the past. They are opportunists that can sometimes produce indolent, slowly progressive diseases. A related genus, *Streptomyces,* is of medical importance as a producer of many antibiotics, but it rarely causes infections. Important differential features of these groups and of the mycobacteria to which they are related are shown in **Table 28–1.**

ACTINOMYCES

BACTERIOLOGY

Actinomyces are typically elongated Gram-positive rods that branch at acute angles (**Figure 28–1**). They are Gram-positive bacilli that grow slowly (4 to 10 days) under microaerophilic or strictly anaerobic conditions. In pus and tissues, the most characteristic form is the sulfur granule (**Figure 28–2**). This yellow-orange granule, named for its gross resemblance to a grain of sulfur, is a small colony (usually less than 0.3 mm) of intertwined branching *Actinomyces* filaments solidified with elements of tissue exudate.

Slow-growing anaerobic branching Gram-positive rods

Species of *Actinomyces* are distinguished on the basis of biochemical reactions, cultural features, and cell wall composition. Most human actinomycosis is caused by *Actinomyces israelii,* but other species have been isolated from typical actinomycotic lesions. Other species of *Actinomyces* have been associated with dental and periodontal infections (see Chapter 60).

Most infections due to *A israelii*

ACTINOMYCOSIS

CLINICAL CAPSULE

Actinomycosis is a chronic inflammatory condition originating in the tissues adjacent to mucosal surfaces. The lesions follow a slow burrowing course with considerable induration and draining sinuses, eventually opening through the skin. The exact nature depends on the organs and structures involved.

TABLE 28–1 Features of Actinomycetes

GENUS	MORPHOLOGY	ACID-FASTNESS	GROWTH	SOURCE	DISEASE
Actinomyces	Branching bacilli	None	Anaerobic	Oral, intestinal endogenous flora	Chronic cellulitis, draining sinuses
Nocardia	Branching bacilli	Weak[a,b]	Aerobic	Soil	Pneumonia, skin pustules, brain abscess
Rhodococcus	Cocci to bacilli	Variable (weak[a])	Aerobic	Soil, horses[c]	Pneumonia
Streptomyces	Branching bacilli	None	Aerobic	Soil	Extremely rare[d]

[a]Modified stain, fast only to weak decolorizer (1% H_2SO_4).
[b]*N asteroides* and *N brasiliensis;* other species variable.
[c]R equi.
[d]Nonpathogen, but important producer of antibiotics.

Normal flora throughout gastrointestinal tract

Conditions for growth require displacement into tissues

Sinus tracts contain pus and sulfur granules

Actinomyces are normal inhabitants of some areas of the gastrointestinal tract of humans and animals from the oropharynx to the lower bowel. These species are highly adapted to mucosal surfaces and do not produce disease unless they transgress the epithelial barrier under conditions that produce a sufficiently low oxygen tension for their multiplication (**Figure 28–3**). Such conditions usually involve mechanical disruption of the mucosa with necrosis of deeper, normally sterile tissues (eg, following tooth extraction). Once initiated, growth occurs in microcolonies in the tissues and extends without regard to anatomic boundaries. The lesion is composed of inflammatory sinuses, which ultimately discharge to the surface. As the lesion enlarges, it becomes firm and indurated. Sulfur granules are present within the pus, but are not numerous. Free *Actinomyces* or small branching units are rarely seen, although contaminating Gram-negative rods are common. As with other anaerobic infections, most cases are polymicrobial involving other flora from the mucosal site of origin. ::: Mixed anaerobes, p. 519

No evidence of immunity

Human cases of actinomycosis provide little evidence of immunity to *Actinomyces*. Once established, infections typically become chronic and resolve only with the aid of antimicrobial therapy. Antibodies can be detected in the course of infection, but seem to reflect the antigenic stimulation of the ongoing infection rather than immunity. Infections with *Actinomyces* are endogenous, and case-to-case transmission does not appear to occur.

ACTINOMYCOSIS: CLINICAL ASPECTS

MANIFESTATIONS

Actinomycosis exists in several forms that differ according to the original site and circumstances of tissue invasion. Infection of the cervicofacial area, the most common site of

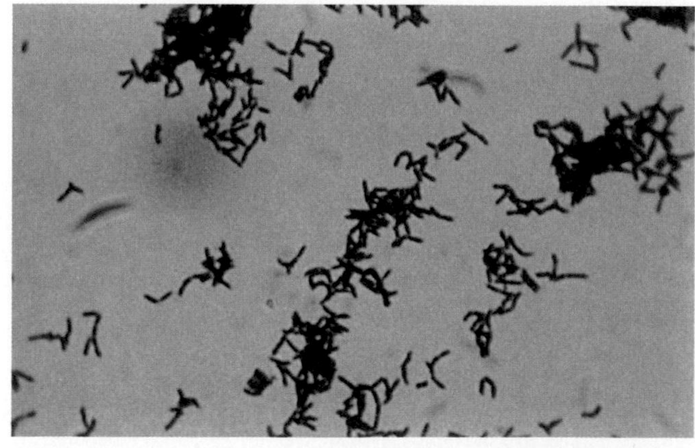

FIGURE 28–1. Actinomyces. Note the angular branching of the Gram-positive bacilli. (Reproduced with permission from Willey J, Sherwood L, Woolverton C (eds). *Prescott's Principles of Microbiology*. New York: McGraw-Hill; 2008.)

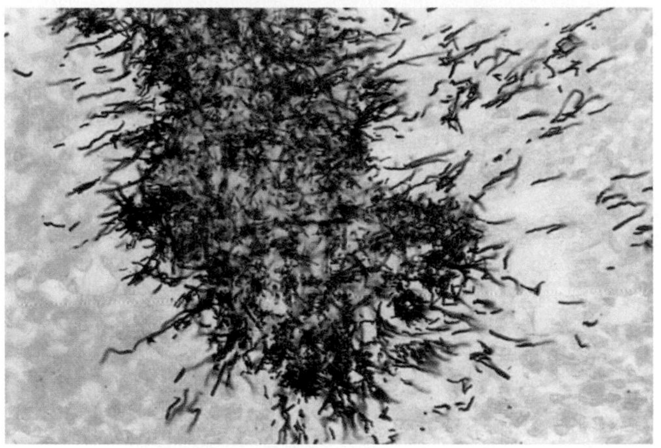

FIGURE 28–2. Sulfur granule. The mass is a microcolony of bacteria Gram-positive bacteria and tissue elements. The branching is clearly seen only at the edge. (Reproduced with permission from Connor DH, Chandler FW, Schwartz DQA, Manz HJ, Lack EE (eds). *Pathology of Infectious Diseases,* vol. 1. Stamford, CT: Appleton & Lange 1997.)

actinomycosis (**Figure 28–4**), is usually related to poor dental hygiene, tooth extraction, or some other trauma to the mouth or jaw. Lesions in the submandibular region and the angle of the jaw give the face a swollen, indurated appearance.

Cervicofacial forms are linked to dental hygiene

Thoracic and abdominal actinomycoses are rare and follow aspiration or traumatic (including surgical) introduction of infected material leading to erosion through the pleura, chest, or abdominal wall. Diagnosis is usually delayed, because only vague or nonspecific symptoms are produced until a vital organ is eroded or obstructed. The firm, fibrous masses are often initially mistaken for a malignancy. Pelvic involvement as an extension from other sites also occurs occasionally. It is particularly difficult to distinguish from other inflammatory

Surgery, trauma, and intrauterine devices provide opportunity

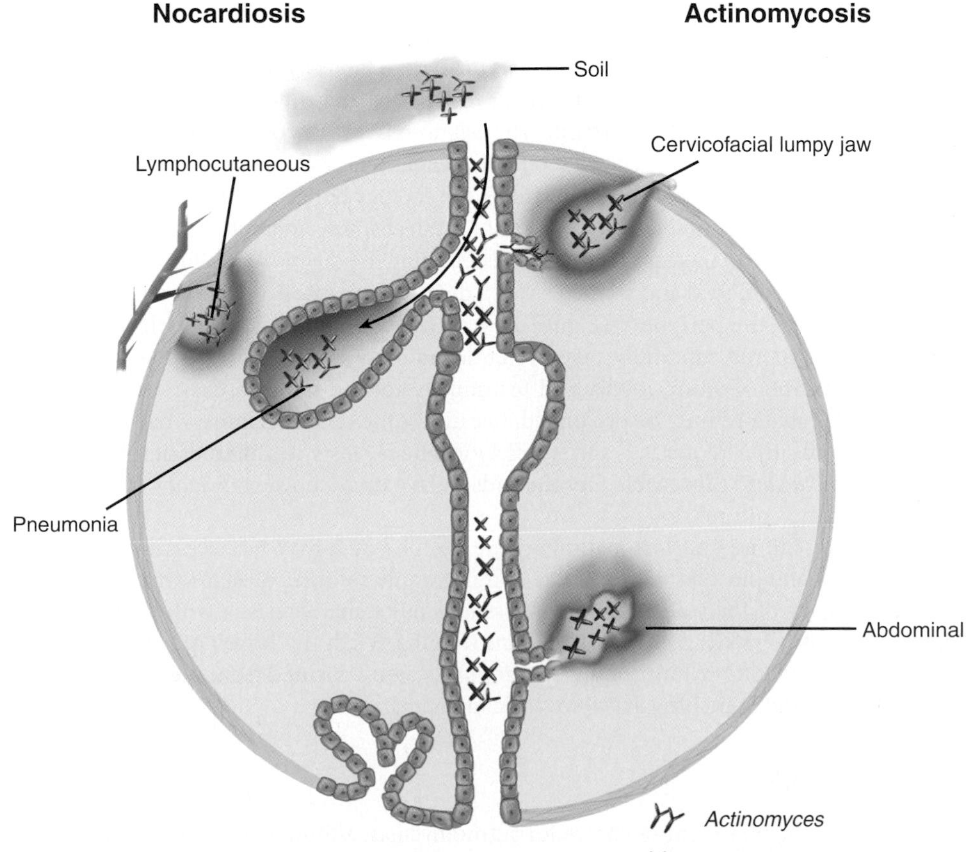

FIGURE 28–3. Actinomycosis and Nocardiosis. (*Right*) *Actinomyces* are members of the normal flora throughout the alimentary tract. Minor trauma allows access to tissues where they create burrowing abscesses that may break trough to the surface. (*Left*) *Nocardia* is present in the soil, where it may be either inhaled to produce a pneumonia or traumatically injected to produce cutaneous pustules and lymphadenitis.

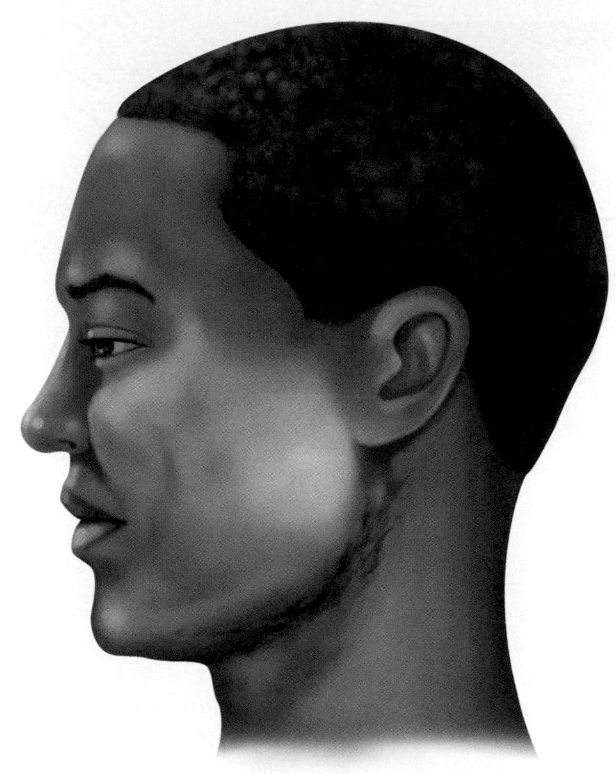

FIGURE 28–4. Cervicofacial actinomycosis. The classic "lumpy jaw" is shown with draining sinuses at the angle of the jaw. The lesion would be very firm on palpation. (Reproduced with permission from Connor DH, Chandler FW, Schwartz DQA, Manz HJ, Lack EE (eds). *Pathology of Infectious Diseases,* vol. 1. Stamford, CT: Appleton & Lange; 1997.)

conditions or malignancies. A more localized chronic endometritis, apparently caused by *Actinomyces*, has been associated with the use of intrauterine contraceptive devices.

DIAGNOSIS

Sinus drainage contains few *Actinomyces*

Drainage is often contaminated with other species

A clinical diagnosis of actinomycosis is based on the nature of the lesion, the slowly progressive course, and a history of trauma or of a condition predisposing to mucosal invasion by *Actinomyces*. The etiologic diagnosis can be difficult to establish with certainty. Although the lesions may be extensive, the organisms in pus may be few and concentrated in sulfur granule microcolonies deep in the indurated tissue. The diagnosis is further complicated by heavy colonization of the moist draining sinuses with other bacteria, usually Gram-negative rods. This contamination not only causes confusion regarding the etiology but interferes with isolation of the slow-growing anaerobic *Actinomyces*. Material for direct smear and culture should include as much pus as possible to increase the chance of collecting the diagnostic sulfur granules.

Gram stains show branching rods

Anaerobic culture is required

Sulfur granules crushed and stained show a dense, Gram-positive center with individual branching rods at the periphery (Figure 28–2). Granules should also be selected for culture, because material randomly taken from a draining sinus usually grows only superficial contaminants. Culture media and techniques are the same as those used for other anaerobes. Incubation must be prolonged, because some strains require 7 days or more to appear. Identification requires a variety of biochemical tests to differentiate *Actinomyces* from *Propionibacteria* (anaerobic diphtheroids), which may show a tendency to form short branches in fluid culture.

Biopsy shows characteristic clubbed lesions

Biopsies for culture and histopathology are useful, but it may be necessary to examine many sections and pieces of tissue before sulfur granule colonies of *Actinomyces* are found. The morphology of the sulfur granule in tissue is quite characteristic with routine hematoxylin and eosin (H&E) or histologic Gram staining. With the histologic H&E stain, the edge of the granule shows amorphous eosinophilic "clubs" formed from the tissue elements and containing the branching actinomycotic filaments.

TREATMENT

Penicillin G is the treatment of choice for actinomycosis, although a number of other antimicrobics (ampicillin, doxycycline, erythromycin, and clindamycin) are active in vitro and have shown clinical effectiveness. High doses of penicillin must be used and therapy prolonged for up to 6 weeks or longer before any response is seen. The initial treatment course is usually followed with an oral penicillin for 6 to 12 months. Although slow, response to therapy is often striking given the degree of fibrosis and deformity caused by the infection. Because detection of the causative organism is difficult, many patients are treated empirically as a therapeutic trial based on clinical findings alone.

Penicillin may have to be used empirically

NOCARDIA

BACTERIOLOGY

Nocardia species are Gram-positive, rod-shaped bacteria that show true branching both in culture and in stains from clinical lesions. The microscopic morphology is similar to that of *Actinomyces,* although *Nocardia* tend to fragment more readily and are found as shorter branched units throughout the lesion rather than concentrated in a few colonies or granules. Many strains of *Nocardia* take the Gram stain poorly, appearing "beaded" with alternating Gram-positive and Gram-negative sections of the same filament (**Figure 28–5A** and **B**). The species most common in human infection (*N asteroides* and *N brasiliensis*) are weakly acid-fast.

Beaded, branching Gram-positive rods are weakly acid-fast

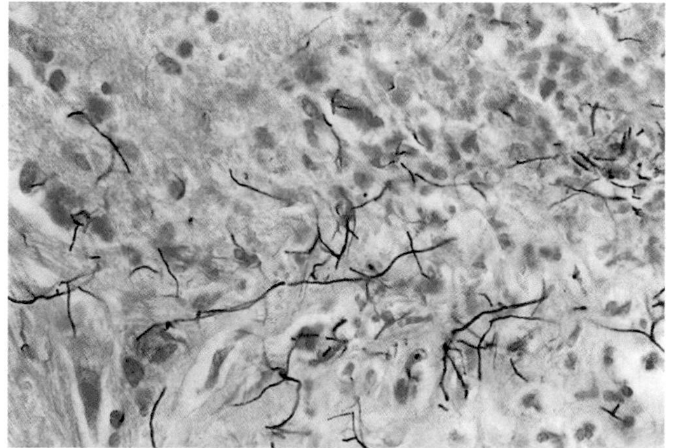

A

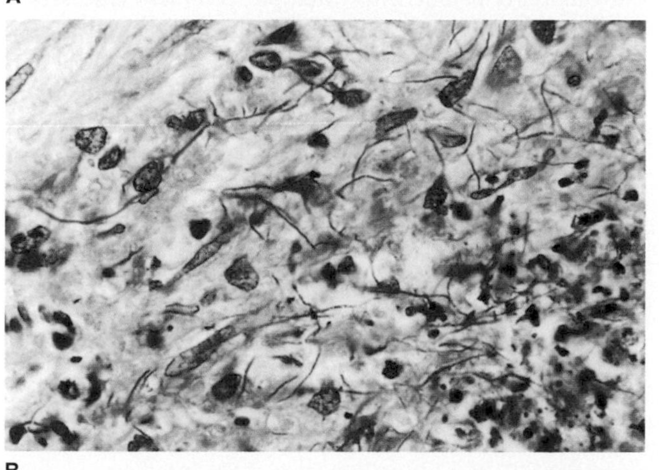

B

FIGURE 28–5. *Nocardia* in sputum. A. Note the filamentous bacteria forming tree-like branches among the neutrophils. The beaded appearance of the rods is typical. **B.** The same sputum stain with the modified (weaker) acid-fast method. Note the red filaments with the same branching pattern as in A. (Reproduced with permission from Connor DH, Chandler FW, Schwartz DQA, Manz HJ, Lack EE (eds). *Pathology of Infectious Diseases,* vol. 1. Stamford, CT: Appleton & Lange; 1997.)

Grow on common media in 2 to 3 days

In contrast to *Actinomyces, Nocardia* species are strict aerobes. Growth typically appears on ordinary laboratory medium (blood agar) after 2 to 3 days incubation in air. Colonies initially have a dry, wrinkled, chalk-like appearance, are adherent to the agar, and eventually develop white to orange pigment. Speciation involves uncommon tests such as the decomposition amino acids and casein.

NOCARDIOSIS

CLINICAL CAPSULE

Nocardiosis occurs in two major forms. The pulmonary form is an acute bronchopneumonia with dyspnea, cough, and sputum production. A cutaneous form produces localized pustules in areas of traumatic inoculation, usually the exposed areas of the skin.

EPIDEMIOLOGY

Primary source is soil

Occurrence in the immunocompromised is increased

Nocardia species are ubiquitous in the environment, particularly in soil. In fact, fully developed colonies of *Nocardia* give off the aroma of wet dirt. The organisms have been isolated in small numbers from the respiratory tract of healthy persons, but are not considered members of the normal flora. The pulmonary form of disease follows inhalation of aerosolized bacteria, and the cutaneous form follows injection by a thorn prick or similar accident (Figure 28–3). Most pulmonary cases occur in patients with compromised immune systems due to underlying disease or the use of immunosuppressive therapy. Transplant patients have been a prominent representative of the latter group. There is no case-to-case transmission.

PATHOGENESIS

Able to survive in phagocytes

Pulmonary infection is usually *N asteroides*

CNS dissemination produces abscesses

Factors leading to disease after inhalation of *Nocardia* are poorly understood. Neutrophils are prominent in nocardial lesions, but appear to be relatively ineffective. The bacteria have the ability to resist the microbicidal actions of phagocytes and may be related to disruption of phagosome acidification or resistance to the oxidative burst. No specific virulence factors are known. The primary lesions in the lung show acute inflammation, with suppuration and destruction of parenchyma. Multiple, confluent abscesses may occur. Unlike *Actinomyces* infections, there is little tendency toward fibrosis and localization. Dissemination to distant organs, particularly the brain, may occur. In the central nervous system (CNS), multifocal abscesses are often produced. The majority of *Nocardia* pulmonary and brain infections are produced by *N asteroides.*

Cutaneous infections follow minor trauma

Skin infections follow direct inoculation of *Nocardia*. This mechanism is usually associated with some kind of outdoor activity and with relatively minor trauma. The species is usually *N brasiliensis*, which produces a superficial pustule at the site of inoculation. If *Nocardia* gain access to the subcutaneous tissues, lesions resembling actinomycosis may be produced, complete with draining sinuses and sulfur granules.

IMMUNITY

There is evidence that effective T-cell–mediated immunity is dominant in host defense against *Nocardia* infection. Increased resistance to experimental *Nocardia* infection in animals has been mediated by cytokine-activated macrophages, and activated macrophages have enhanced capacity to kill *Nocardia* that they have engulfed. Patients with impaired cell-mediated immune responses are at greatest risk for nocardiosis. There is little evidence for effective humoral immune responses.

Cell-mediated immunity mechanisms are dominant

NOCARDIOSIS: CLINICAL ASPECTS

MANIFESTATIONS

Pulmonary infection is usually a confluent bronchopneumonia that may be acute, chronic, or relapsing. Production of cavities and extension to the pleura are common. Symptoms are those of any bronchopneumonia, including cough, dyspnea, and fever. The clinical signs of brain abscess depend on its exact location and size; the neurologic picture can be particularly confusing when multiple lesions are present. The combination of current or recent pneumonia and focal CNS signs is suggestive of *Nocardia* infection. The cutaneous syndrome typically involves a pustule, fever, and tender lymphadenitis in the regional lymph nodes.

Bronchopneumonia and cerebral abscess findings depend on localization

DIAGNOSIS

The diagnosis of *Nocardia* infection is much easier than that of actinomycosis, because the organisms are present in greater numbers and distributed more evenly throughout the lesions. Filaments of Gram-positive rods with primary and secondary branches can usually be found in sputum and are readily demonstrated in direct aspirates from skin or other purulent sites. Demonstration of acid-fastness, when combined with other observations, is diagnostic of *N asteroides* or *N brasiliensis* (Figure 28–5). The acid-fastness of *Nocardia* species is not as strong as that of mycobacteria. The staining method thus uses a decolorizing agent weaker than that used for the classic stain. Culture of *Nocardia* is not difficult because the organisms grow on blood agar. It is still important to alert the laboratory to the possibility of nocardiosis, because the slow growth of *Nocardia* could cause it to be overgrown by the respiratory flora commonly found in sputum specimens. Specific identification can take weeks due the unconventional tests involved.

Gram stain is usually positive

Weak acid fastness is characteristic

Blood agar is sufficient for culture

TREATMENT

Nocardia are usually susceptible to sulfonamides, which have been the mainstay of treatment for decades and which continue to be effective alone or combined with trimethoprim. Technical difficulties in susceptibility testing have hampered the rational selection and study of other antimicrobics. Although most *Nocardia* strains are relatively resistant to penicillin, some of the newer β-lactams (imipenem, cefotaxime) have been effective, as have minocycline and amikacin. Antituberculous agents and antifungal agents such as amphotericin B have no activity against *Nocardia*.

Sulfonamides are active

RHODOCOCCUS

Rhodococcus is a genus of aerobic actinomycetes with characteristics similar to those of *Nocardia*. Morphologically the rods vary from cocci to long, curved, clubbed forms. Some strains are acid-fast. *Rhodococcus* has recently been recognized as an opportunistic pathogen causing an aggressive pneumonia in severely immunocompromised patients, particularly those with acquired immunodeficiency syndrome. The organisms are found in the soil. One species, *Rhodococcus equi*, has an association with horses where it also causes pneumonia in foals. This species is a facultative intracellular pathogen of macrophages with features somewhat similar to those of *Legionella* and *Listeria*. Optimal treatment is unknown, although erythromycin, aminoglycosides, and some β-lactams show in vitro activity.

Morphology varies from cocci to rods

Pneumonia is associated with horses

CLINICAL CASE

LUNG LESIONS AND A BRAIN ABSCESS

The patient was a 34-year-old man with a history of tobacco and alcohol abuse (12 cans of beer per day). Two months before admission, he was seen at an outside hospital, where radiographs revealed a necrotic lesion in his right upper lobe. He was PPD-negative and three sputum cultures analyzed for *Mycobacterium* were negative. He had no risk factors for HIV infection. Four weeks later, he presented with fever, productive cough, night sweats, chills, and a 10-lb (4.5-kg) weight loss.

He was treated with ampicillin for 14 days. Fever, chills, and night sweats decreased. On admission, he presented with a firm right chest wall mass (4 by 4 cm), which was aspirated. The aspirated material was dark green and extremely viscous. Two days later, the nurses found him urinating on the wall of his room. Because of this behavior, it was decided to perform a CT scan of the head; the scan revealed multiple, ring-enhancing lesions. The patient was taken to surgery and the central nervous system (CNS) lesions were drained. A Gram stain of the organism recovered from the brain aspirate showed a branching, beaded Gram-positive rod. The laboratory noted that it was also acid-fast.

QUESTIONS

■ The material in the brain aspirate most likely contains which of the following:

A. *Actinomyces*
B. *Nocardia*
C. *Mycobacterium tuberculosis*
D. Another *Mycobacterium*
E. *Rhodococcus*

■ What risk factor is likely to have contributed the most to this patient's infection?

A. Occupation
B. Alcoholism
C. HIV
D. Smoking

■ The infection was most likely acquired from which of the following:

A. Family member
B. Pet
C. Wild animal
D. Soil
E. Water

ANSWERS

1(B), **2**(B), **3**(D)

CHAPTER 29

Clostridium, Peptostreptococcus, Bacteroides, and Other Anaerobes

Can you watch placidly the horrible struggles of lock-jaw? . . . If you can, you had better leave the profession: cast your diploma into the fire; you are not worthy to hold it.

—Jacob M. Da Costa (1833–1900): *College and Clinical Record*

The bacteria discussed in this chapter are united by a common requirement for anaerobic conditions for growth. Organisms from multiple genera and all Gram stain categories are included. Most of them produce endogenous infections adjacent to the mucosal surfaces, where they are members of the normal flora. The clostridia form spores that allow them to produce diseases such as tetanus and botulism after environmental contamination of tissues or foods. Another anaerobic genus of bacteria, *Actinomyces,* is discussed in Chapter 28.

ANAEROBES AND ANAEROBIC IINFECTION: GROUP CHARACTERISTICS

BACTERIOLOGY

THE NATURE OF ANAEROBIOSIS

Anaerobes not only survive under anaerobic conditions, they require them to initiate and sustain growth. By definition, anaerobes fail to grow in the presence of 10% oxygen, but some are sensitive to oxygen concentrations as low as 0.5% and are killed by even brief exposures to air. However, **oxygen tolerance** is variable, and many organisms can survive briefly in 2% to 8% oxygen, including most of the species pathogenic for humans. The mechanisms involved are incompletely understood, but clearly represent a continuum from species described as **aerotolerant** to those so susceptible to oxidation that growing them in culture requires the use of media prepared and stored under anaerobic conditions.

Anaerobes require low oxygen to initiate growth

Oxygen tolerance is a continuum

Anaerobes lack the cytochromes required to use oxygen as a terminal electron acceptor in energy-yielding reactions and thus to generate energy solely by fermentation (see Chapter 21).

Low redox potential is required

Some anaerobes do not grow unless the oxidation-reduction potential is extremely low (–300 mV); because critical enzymes must be in the reduced state to be active, aerobic conditions create a metabolic block. :::Aerobes and anaerobes, p. 361

Defense against oxygen products is lacking

Pathogens often have catalase and superoxide dismutase

Another element of anaerobiosis is the direct susceptibility of anaerobic bacteria to oxygen. For most aerobic and facultative bacteria, **catalase** and/or **superoxide dismutase** neutralize the toxicity of the oxygen products **hydrogen peroxide** and **superoxide**. Most anaerobes lack these enzymes and are injured when these oxygen products are formed in their microenvironment. As discussed in the following text, many of the virulent anaerobic pathogens are able to produce catalase or superoxide dismutase.

CLASSIFICATION

Biochemical, cultural, and molecular criteria define many species

The anaerobes indigenous to humans include almost every morphotype and hundreds of species. Typical biochemical and cultural tests are used for classification, although this is difficult because the growth requirements of each anaerobic species must be satisfied. Characterization of cellular fatty acids and metabolic products by chromatography has been useful for many anaerobic groups. Nucleic acid base composition and homology have been used extensively to rename older taxonomy. The genera most commonly associated with disease are shown in **Table 29–1** and discussed below.

Anaerobic Cocci

Gram(+) = *Peptostreptococcus*

Gram(–) = *Veillonella*

Virtually all the medically important species of anaerobic Gram-positive cocci are now classified in a single genus, *Peptostreptococcus.* With Gram staining, these bacteria are most often seen as long chains of tiny cocci. *Veillonella,* a Gram-negative genus, deserves mention because of its potential for confusion with *Neisseria,* the only other Gram-negative coccus associated with disease.

Clostridia

The clostridia are large, spore-forming, Gram-positive bacilli. Like their aerobic counterpart, *Bacillus,* clostridia have spores that are resistant to heat, desiccation, and disinfectants. They are able to survive for years in the environment and return to the vegetative form when placed in a favorable milieu. The shape of the cell and location of the spore vary with the species, but the spores themselves are rarely seen in clinical specimens.

TABLE 29–1 Usual Locations of Opportunistic Anaerobes

ORGANISM	GRAM STAIN	MOUTH OR PHARYNX	INTESTINE	UROGENITAL TRACT	SKIN
Peptostreptococcus	Positive cocci	+	+	+	–
Propionibacterium	Positive rods	–	–	–	+
Clostridium	Positive rods (large)	–	+	–	–
Veillonella	Negative cocci	–	+	–	–
Bacteroides fragilis group	Negative rods (coccobacillary)	–	+	–	–
Fusobacterium	Negative rods (elongated)	+	+	–	–
Prevotella	Negative rods	+		+	–
Porphyromonas	Negative rods	+		+	

The medically important clostridia are potent producers of one or more protein exotoxins. The histotoxic group including *Clostridium perfringens* and five other species (**Table 29–2**) produces hemolysins at the site of acute infections that have lytic effects on a wide variety of cells. The neurotoxic group including *C tetani and C botulinum* produces neurotoxins that exert their effect at neural sites remote from the bacteria. *C difficile* produces enterotoxins and disease in the intestinal tract. Many of the more than 80 other nontoxigenic clostridial species are also associated with disease.

Spores vary in shape and location

Hemolysin, neurotoxin, and enterotoxin production cause disease

Nonsporulating Gram-positive Bacilli

Propionibacterium is a genus of small pleomorphic bacilli sometimes called anaerobic diphtheroids because of their morphologic resemblance to corynebacteria. They are among the most common bacteria in the normal flora of the skin. *Eubacterium* is a genus that includes long slender bacilli commonly found in the colonic flora. These organisms are occasionally isolated from infections in combination with other anaerobes, but they rarely produce disease on their own.

Members of the normal flora

Gram-negative Bacilli

Gram-negative, non–spore-forming bacilli are the most common bacteria isolated from anaerobic infections. In the past, most species were lumped into the genus *Bacteroides,*

TABLE 29–2 Features of Pathogenic Anaerobes

ORGANISM	BACTERIOLOGIC FEATURES	EXOTOXINS	SOURCE	DISEASE
Gram-Positive Cocci				
Peptostreptococcus			Mouth, intestine	Oropharyngeal infections, brain abscess
Gram-Negative Cocci				
Veillonella			Intestine	Rare opportunist
Gram-Positive Bacilli				
Clostridium perfringens	Spores	α-toxin, θ-toxin, enterotoxin	Intestine, environment, food	Cellulitis, gas gangrene, enterocolitis
Histotoxic species similar to *C perfringens*[a]	Spores		Intestine, environment	Cellulitis, gas gangrene
C tetani	Spores	Tetanospasmin	Environment	Tetanus
C botulinum	Spores	Botulinum	Environment	Botulism
C difficile	Spores	A enterotoxin, B cytotoxin	Intestine, environment (nosocomial)	Antibiotic-associated diarrhea, enterocolitis
Propionibacterium			Skin	Rare opportunist
Eubacterium			Intestine	Rare opportunist
Gram-Negative Bacilli				
Bacteroides fragilis[b]	Polysaccharide capsule	Enterotoxin	Intestine	Opportunist, abdominal abscess
Bacteroides species			Intestine	Opportunist
Fusobacterium			Mouth, intestine	Opportunist
Prevotella	Black pigment		Mouth, urogenital	Opportunist
Porphyromonas			Mouth, urogenital	Opportunist

[a]*C histolyticum, C noyyi, C septicum,* and *C sordellii.*

[b]The *Bacteroides fragilis* group includes *B fragilis, B distasonis, B ovatus, B vulgatus, B thetaiotaomicron,* and six other species.

which still exists along with five other genera. Of these, *Fusobacterium, Porphyromonas,* and *Prevotella* are medically the most important. The *Bacteroides fragilis* group contains *B fragilis* and 10 similar species noted for their virulence and production of β-lactamases. (Species outside this group generally lack these features and are more similar to the other anaerobic Gram-negative bacilli.) *B fragilis* is a relatively short Gram-negative bacillus with rounded ends sometimes giving a coccobacillary appearance. The lipopolysaccharide (LPS) in its outer membrane has a much lower lipid content and thus lower toxic activity than that of most other Gram-negative bacteria. Almost all *B fragilis* strains have a polysaccharide capsule and are relatively oxygen-tolerant through production of superoxide dismutase. *Prevotella, Porphyromonas,* and *Fusobacterium* are distinguished by biochemical and other taxonomic features. *Prevotella melaninogenica* forms a black pigment in culture, and *Fusobacterium,* as its name suggests, is typically elongated and has tapered ends.

Five genera are medically important

B fragilis group produces β-lactamase and superoxide dismutase

ANAEROBIC INFECTIONS

EPIDEMIOLOGY

Despite our constant immersion in air, anaerobes are able to colonize the many oxygen-deficient or oxygen-free microenvironments of the body. Often these are created by the presence of facultative organisms whose growth reduces oxygen and decreases the local oxidation-reduction potential. Such sites include the sebaceous glands of the skin, the gingival crevices of the gums, the lymphoid tissue of the throat, and the lumina of the intestinal and urogenital tracts. Except for infections with some environmental clostridia, anaerobic infections are almost always endogenous with the infective agent(s) derived from the patient's normal flora. The specific anaerobes involved are linked to their prevalence in the flora of the relevant sites as shown in Table 29–1. In addition to the presence of clostridia in the lower intestinal tract of humans and animals, their spores are widely distributed in the environment, particularly in soil exposed to animal excreta. The spores may contaminate any wound caused by a nonsterile object (eg, splinter, nail) or exposed directly to soil.

Low redox normal flora sites are the origin of most infections

Spore-forming clostridia also come from the environment

PATHOGENESIS

The anaerobic flora normally lives in a harmless commensal relationship with the host. However, when displaced from their niche on the mucosal surface into normally sterile tissues, these organisms may cause life-threatening infections. This can occur as the result of trauma (eg, gunshot, surgery), disease (eg, diverticulosis), or isolated events (eg, aspiration). Host factors such as malignancy or impaired blood supply increase the probability that the dislodged flora will eventually produce an infection. The organisms most often involved are the anaerobes, which are both present at the mucosal site adjacent to the infection and have enhanced virulence. For example, *B fragilis* represents less than 1% of the normal colonic flora but is the organism most frequently isolated from intra-abdominal abscesses.

Anaerobes displaced from normal flora to deeper sites may cause disease

Trauma and host factors create the opportunity for infection

The relation between normal flora and site of infection may be indirect. For example, aspiration pneumonia, lung abscess, and empyema typically involve anaerobes found in the oropharyngeal flora. The brain is not a particularly anaerobic environment, but brain abscess is most often caused by these same oropharyngeal anaerobes. This presumably occurs by extension across the cribriform plate to the temporal lobe, the typical location of brain abscess. In contaminated open wounds, clostridia can come from the intestinal flora or from spores surviving in the environment.

Flora may be aspirated or displaced at a distance

Brain abscess typically involves anaerobic bacteria

Although gaining access to tissue sites provides the opportunity, additional virulence factors are needed for anaerobes to produce infection. Some anaerobic pathogens produce disease even when present as a minor part of the displaced resident flora, and other common members of the normal flora rarely cause disease. Classic virulence factors such as toxins and capsules are known only for the toxigenic clostridia and *B fragilis,* but a feature such as the ability to survive brief exposures to oxygenated environments can also be viewed as a virulence factor. Anaerobes found in human infections are far more likely to produce

Capsules and toxins are known for some anaerobes

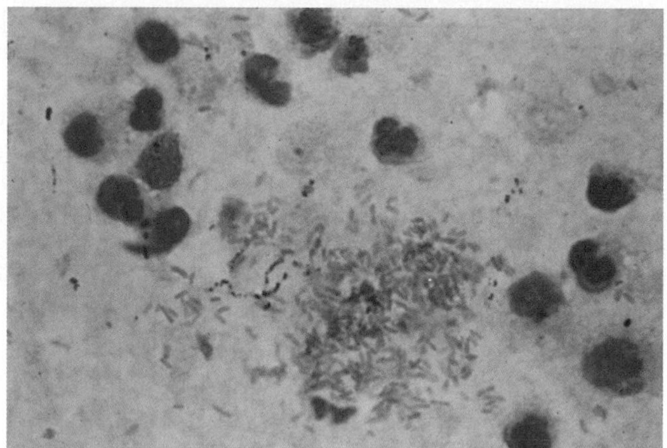

FIGURE 29–1. Gram smear of pus from an abdominal abscess showing polymorphonuclear leukocytes, large numbers of Gram-negative anaerobes, and some peptostreptococci. (Reproduced with permission of Schering Corporation, Kenilworth, NJ, the copyright owner. All rights reserved.)

catalase and superoxide dismutase than their more docile counterparts of the normal flora. Exquisitely oxygen-sensitive anaerobes are seldom involved, probably because they are injured by even the small amounts of oxygen dissolved in tissue fluids.

Survival in oxidized conditions can be a virulence factor

A related feature is the ability of the bacteria to create and control a reduced microenvironment, often with the apparent help of other bacteria. Most anaerobic infections are mixed; that is, two or more anaerobes are present, often in combination with facultative bacteria such as *Escherichia coli* (**Figure 29–1**). In some cases, the components of these mixtures are believed to synergize each other's growth either by providing growth factors or by lowering the oxidation-reduction potential. These conditions may have other advantages such as the inhibition of oxygen-dependent leukocyte bactericidal functions under the anaerobic conditions in the lesion. Anaerobes that produce specific toxins have a pathogenesis all their own, which are discussed in the sections devoted to individual species.

Mixed infections may facilitate an anaerobic microenvironment

ANAEROBIC INFECTIONS: CLINICAL ASPECTS

MANIFESTATIONS

Bacteroides, Fusobacterium, and peptostreptococci, alone or together with other facultative or obligate anaerobes, are responsible for the overwhelming majority of localized abscesses within the cranium, thorax, peritoneum, liver, and female genital tract. As indicated earlier, the species involved relate to the pathogens present in the normal flora of the adjacent mucosal surface. Those derived from the oral flora also include dental infections and infections of human bites.

Abscesses are usually caused by *Bacteroides, Fusobacterium,* or peptostreptococci

In addition, anaerobes play causal roles in chronic sinusitis, chronic otitis media, aspiration pneumonia, bronchiectasis, cholecystitis, septic arthritis, chronic osteomyelitis, decubitus ulcers, and soft tissue infections of patients with diabetes mellitus. Dissection of infection along fascial planes (necrotizing fasciitis) and thrombophlebitis are common complications. Foul-smelling pus and crepitation (gas in tissues) are signs associated with, but by no means exclusive to, anaerobic infections. As with other bacterial infections, they may spread beyond the local site and enter the bloodstream. The mortality rate of anaerobic bacteremias arising from nongenital sources is equivalent to the rates with bacteremias due to staphylococci or Enterobacteriaceae.

Foul-smelling pus suggests anaerobic infection

DIAGNOSIS

The key to detection of anaerobes is a high-quality specimen, preferably pus or fluid taken directly from the infected site. The specimen needs to be taken quickly to the microbiology laboratory and protected from oxygen exposure while on the way. Special anaerobic transport

Specimens must be direct and protected from oxygen

tubes may be used, or any air from the syringe in which the specimen was collected may be expressed. Actually, a generous collection of pus serves as its own best transport medium unless transport is delayed for hours.

A direct Gram-stained smear of clinical material demonstrating Gram-negative and/or Gram-positive bacteria of various morphologies is highly suggestive, often even diagnostic of anaerobic infection (Figure 29–1). Because of the typically slow and complicated nature of anaerobic culture, the Gram stain often provides the most useful information for clinical decision-making. Isolation of the bacteria requires the use of an anaerobic incubation atmosphere and special media protected from oxygen exposure. Although elaborate systems are available for this purpose, the simple anaerobic jar is sufficient for isolation of the clinically significant anaerobes. The use of media that contain reducing agents (cysteine, thioglycollate) and growth factors needed by some species further facilitates isolation of anaerobes. The polymicrobial nature of most anaerobic infections requires the use of selective media to protect the slow-growing anaerobes from being overgrown by hardier facultative bacteria, particularly members of the Enterobacteriaceae. Antibiotics, particularly aminoglycosides to which all anaerobes are resistant, are frequently used. Once the bacteria are isolated, identification procedures include morphology, biochemical characterization, and metabolic end-product detection by gas chromatography.

Gram staining is particularly useful

Anaerobic incubation jar provides atmosphere

Selective media inhibit facultative bacteria

TREATMENT

As with most abscesses, drainage of the purulent material is the primary treatment, in association with appropriate chemotherapy. Antimicrobics alone may be ineffective because of failure to penetrate the site of infection. The selection of antimicrobics is empiric to a degree; such infections typically involve mixed species, and cultural diagnosis is delayed by the slow growth and the time required to distinguish multiple species. In addition, antimicrobial susceptibility testing methods are slow and less standardized than they are for the rapidly growing bacteria. The usual approach involves selection of antimicrobics based on the expected susceptibility of the anaerobes known to produce infection at the site in question. For example, anaerobic organisms derived from the oral flora are often susceptible to penicillin, but infections below the diaphragm are caused by fecal anaerobes such as *B fragilis* which is resistant to many β-lactams. These latter infections are most likely to respond to metronidazole, imipenem, aztreonam, or ceftriaxone, a cephalosporin not inactivated by the β-lactamases produced by anaerobes.

Mixed infections and slow growth dictate empiric therapy

Abdominal infections require β-lactamase–resistant antimicrobics

CLOSTRIDIUM PERFRINGENS

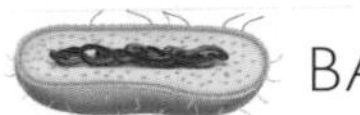

BACTERIOLOGY

C perfringens is a large, Gram-positive, nonmotile rod with square ends. It grows overnight under anaerobic conditions, producing hemolytic colonies on blood agar. In broth containing fermentable carbohydrate, growth of *C perfringens* is accompanied by the production of large amounts of hydrogen and carbon dioxide gas, which can also be produced in necrotic tissues; hence the term gas gangrene.

Hemolysis and gas production are characteristic

C perfringens produces multiple exotoxins that have different pathogenic significance in different animal species and serve as the basis for classification of the five types (A to E). Type A is by far the most important in humans and is found consistently in the colon and often in soil. The most important exotoxin is the **α-toxin**, a phospholipase that hydrolyzes lecithin and sphingomyelin, thus disrupting the cell membranes of various host cells, including erythrocytes, leukocytes, and muscle cells. The **θ-toxin** alters capillary permeability and is toxic to heart muscle. This pore-forming toxin is closely related to streptolysin O. A minority of strains (less than 5%) produce an **enterotoxin,** which, when attached to enterocyte membranes, causes an increase in intracellular calcium and altered membrane permeability. This leads to loss of cellular fluid and macromolecules.

Typing system is based on toxins

Phospholipase α-toxin, pore-forming θ-toxin, and enterotoxin cause disease

C PERFRINGENS DISEASE

CLINICAL CAPSULE

C perfringens produces a wide range of wound and soft tissue infections, many of which are no different from those caused by other opportunistic bacteria. The most dreaded of these, gas gangrene, begins as a wound infection but progresses to shock and death in a matter of hours. Another form of *C perfringens*-caused disease, food poisoning, is characterized by diarrhea without fever or vomiting.

EPIDEMIOLOGY

Gas Gangrene

Gas gangrene develops in traumatic wounds with muscle damage when they are contaminated with dirt, clothing, or other foreign material containing *C perfringens* or another species of histotoxic clostridia. The clostridia can come from the patient's own intestinal flora or spores in the environment. Compound fractures, bullet wounds, or the kind of trauma seen in wartime are prototypes for this infection. A significant delay between the injury and definitive surgical management is an additional requirement. These conditions are more likely to occur in peacetime in a hiking accident in a remote area rather than in an automobile accident on a freeway.

Spores from the host or environment contaminate wounds

Delays allow multiplication

Clostridial Food Poisoning

C perfringens can cause food poisoning if spores of an enterotoxin-producing strain contaminate food. Outbreaks usually involve rich meat dishes such as stews, soups, or gravies that have been kept warm for a number of hours before consumption. This allows time for the infecting dose to be reached by conversion of spores to vegetative bacteria, which then multiply in the food. Clostridial food poisoning is common in developed countries and is third among food-borne illnesses in the United States.

Bacteria multiply in meat dishes

PATHOGENESIS

Gas Gangrene

If the oxidation–reduction potential in a wound is sufficiently low, *C perfringens* spores can germinate and then multiply, elaborating α-toxin. The process passes along the muscle bundles, producing rapidly spreading edema and necrosis as well as conditions that are more favorable for growth of the bacteria. Very few leukocytes are present in the myonecrotic tissue (**Figure 29–2**). As the disease progresses, increased vascular permeability and systemic absorption of the toxin and inflammatory mediators leads to shock. θ-Toxin and oxygen deprivation due to the metabolic activities of *C perfringens* are probable contributors. The basis for the profound systemic effects is not known, but toxin absorption seems probable because fatal cases occur without bacteremia.

Low redox favors multiplication and toxin production

Toxins lead to shock

Clostridial Food Poisoning

The spores of some *C perfringens* strains are often particularly heat-resistant and can withstand temperatures of 100°C for an hour or more. Thus, spores that survive initial cooking can convert to the vegetative form and multiply when food is not refrigerated or is

Spores survive cooking

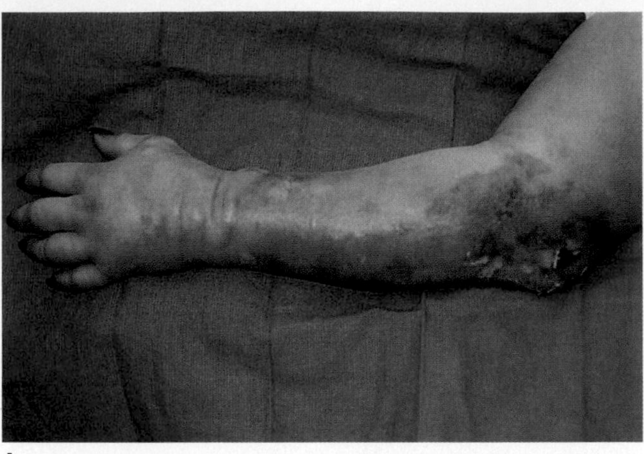

A

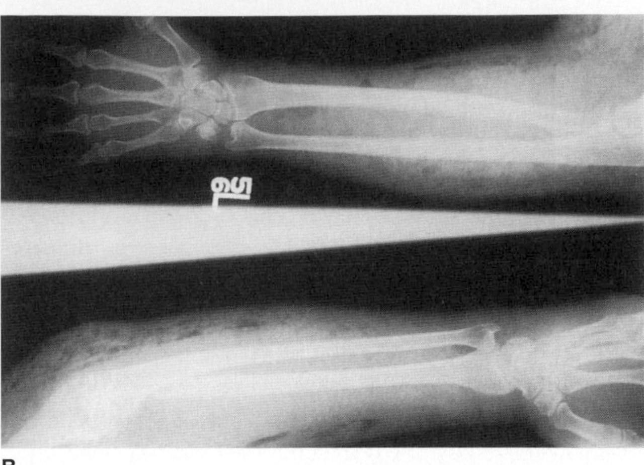

B

FIGURE 29–2. Gas gangrene. A. Arm of a drug abuser with ulcers and swelling traced to needle tracks. **B.** Radiographs from the same patient demonstrating gas (clear spaces) in the tissues. (Reproduced with permission from Connor DH, Chandler FW, Schwartz DQA, Manz HJ, Lack EE (eds). *Pathology of Infectious Diseases,* vol. 1. Stamford, CT: Appleton & Lange; 1997.)

Vegetative cells produce enterotoxin

rewarmed. After ingestion, the enterotoxin is released into the upper gastrointestinal tract, causing a fluid outpouring in which the ileum is most severely involved.:

C PERFRINGENS: CLINICAL ASPECTS

MANIFESTATIONS

Gas Gangrene

Wound pain evolves to edema and shock

Gas gangrene usually begins 1 to 4 days after the injury but may start within 10 hours. The earliest reported finding is severe pain at the site of the wound accompanied by a sense of heaviness or pressure. The disease then progresses rapidly with edema, tenderness, and pallor, followed by discoloration and hemorrhagic bullae. The gas is apparent as crepitance in the tissue, but this is a late sign. Systemic findings are those of shock with intravascular hemolysis, hypotension, and renal failure leading to coma and death. Patients are often remarkably alert until the terminal stages.

Anaerobic Cellulitis

Gas is more likely than in gas gangrene

Anaerobic cellulitis is a clostridial infection of wounds and surrounding subcutaneous tissue in which there is marked gas formation (more than in gas gangrene), but in which the pain, swelling, and toxicity of gas gangrene are absent. This condition is much less serious than gas gangrene and can be controlled with less rigorous methods.

Endometritis

Nonsterile abortion is greatest risk

If *C perfringens* gains access to necrotic products of conception retained in the uterus, it may multiply and infect the endometrium. Necrosis of uterine tissue and bacteremia with massive intravascular hemolysis due to α-toxin may then follow. Clostridial uterine infection is particularly common after an incomplete illegal abortion with inadequately sterilized instruments.

Food Poisoning

Diarrhea without fever or vomiting is most common

The incubation period of 8 to 24 hours is followed by nausea, abdominal pain, and diarrhea. There is no fever, and vomiting is rare. Spontaneous recovery usually occurs within 24 hours.

DIAGNOSIS

Diagnosis is based ultimately on clinical observations. Bacteriologic studies are adjunctive. It is common, for example, to isolate *C perfringens* from contaminated wounds of patients who have no evidence of clostridial disease. The organism can also be isolated from the postpartum uterine cervix of healthy women or from those with only mild fever. Occasionally, *C perfringens* is even isolated from blood cultures of patients who do not develop serious clostridial infection. In clostridial food poisoning, isolation of more than 10° *C perfringens* per gram of ingested food in the absence of any other cause is usually sufficient to confirm an etiology of a characteristic food poisoning outbreak.

Isolation of clostridia alone is not sufficient

TREATMENT AND PREVENTION

Treatment of gas gangrene and endometritis must be initiated immediately because these conditions are almost always fatal if untreated. Excision of all devitalized tissue is of paramount importance because it denies the organism the anaerobic conditions required for further multiplication and toxin production. This often entails wide resection of muscle groups, hysterectomy, and even amputation of limbs. Administration of massive doses of penicillin is an important adjunctive procedure. Because nonclostridial anaerobes and members of Enterobacteriaceae frequently contaminate injury sites, broad-spectrum cephalosporins are often added to the antibiotic regimen. Placement of patients in a hyperbaric oxygen chamber, which increases the tissue level of dissolved oxygen, has been shown to slow the spread of disease, probably by inhibiting bacterial growth and toxin production and by neutralizing the activity of θ-toxin.

Surgical treatment is essential for gas gangrene and endometritis

Antibiotics and hyperbaric oxygen are useful

The most effective method of prevention of gas gangrene is the surgical debridement of traumatic injuries as soon as possible. Thorough cleansing, removal of dead tissue and foreign bodies, and drainage of hematomas limit organism multiplication and toxin production. Antimicrobic prophylaxis is indicated but cannot replace surgical debridement, because the antimicrobics may fail to reach the organism in devascularized tissues.

Debridement of dead tissue is best

Prevention of food poisoning involves good cooking hygiene and adequate refrigeration. There is growing evidence that enterotoxin-producing strains of *C perfringens* may also be responsible for some cases of antimicrobic-induced diarrhea in a setting similar to that from *C difficile.*

CLOSTRIDIUM BOTULINUM

BACTERIOLOGY

C botulinum is a large Gram-positive rod much like the rest of the clostridia. Its spores resist boiling for long periods, and moist heat at 121°C is required for certain destruction. Germination of spores and growth of *C botulinum* can occur in a variety of alkaline or neutral foodstuffs when conditions are sufficiently anaerobic.

The major characteristic of medical importance is that when *C botulinum* grows under these anaerobic conditions, it elaborates a family of neurotoxins of extraordinary toxicity. **Botulinum toxin** is the most potent toxin known in nature, with an estimated lethal dose for humans of less than 1 μg. Botulinum toxin is an enzyme (metalloproteinase) that acts at neuromuscular junctions (**Figure 29–3**). Once bound, it cleaves proteins, which effectively block the release of the neurotransmitter acetylcholine from vesicles at the presynaptic membrane of the synapse. Because acetylcholine mediates activation of motor neurons the blockage of its release causes flaccid paralysis of the motor system.

Cells germinating from spores produce neurotoxin in food

Blockage of synaptic acetylcholine release causes paralysis

C botulinum is classified into multiple types (A to G) based on the antigenic specificity of the neurotoxins. All the toxins are heat-labile and destroyed rapidly at 100°C, but are resistant to the enzymes of the gastrointestinal tract. If unheated toxin is ingested, it is readily absorbed and distributed in the bloodstream.

Toxin is destroyed by boiling

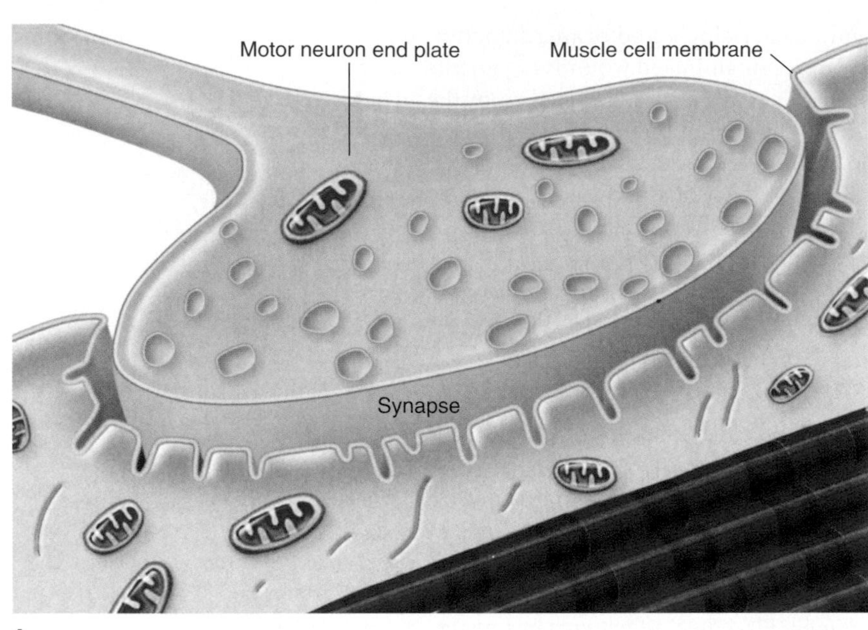

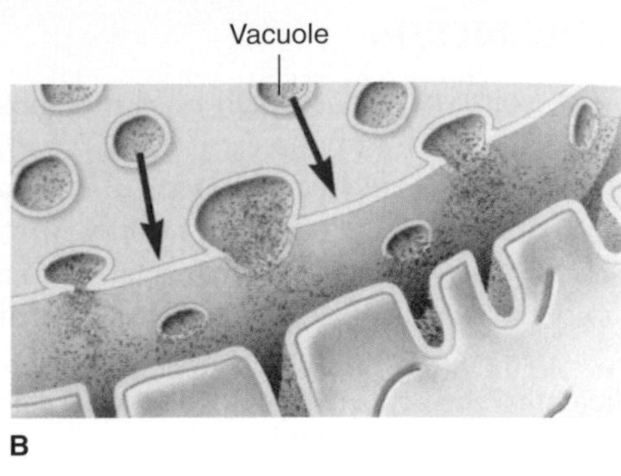

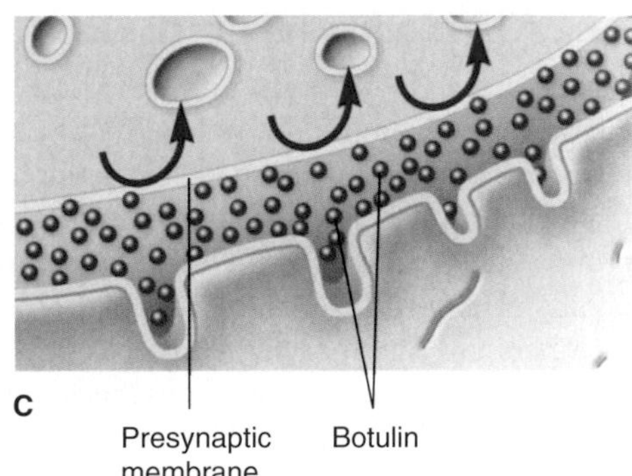

FIGURE 29–3. Clostridial tetanus and botulinum neurotoxins. A. The motor neuron endplate, synapse, and neuromuscular junction are shown. For tetanus toxin, the neurons have an inhibitory function; for botulinum, they are active motor neurons. **B.** Vesicles releasing neurotransmitters across the synapse to the muscle cell membrane are shown. **C.** In the presence of toxin, the release of neurotransmitter vesicles into the synapse is blocked. For botulinum toxin, the neurotransmitter is acetylcholine, and motor neurons are blocked giving flaccid paralysis. For tetanus toxin, release of neurotransmitters activating inhibitory neurons is blocked resulting in spasmodic contractions. (Reproduced with permission from Willey J, Sherwood L, Woolverton C (eds). *Prescott's Principles of Microbiology*. New York: McGraw-Hill; 2008.)

BOTULISM

CLINICAL CAPSULE

Botulism begins with cranial nerve palsies and develops into descending symmetric motor paralysis, which may involve the respiratory muscles. No fever or other signs of infection occur. The time course depends on the amount of toxin present and whether it was ingested preformed in food or produced endogenously in the intestinal tract or a wound.

EPIDEMIOLOGY

Spores are widely distributed

Alkaline foods favor toxin production

Spores of *C botulinum* are found in soil, pond, and lake sediments in many parts of the world, including the United States. If spores contaminate food, they may convert to the vegetative state, multiply, and produce toxin in storage under certain conditions. This may occur with no change in food taste, color, or odor. The alkaline conditions provided by vegetables, such as green beans, and mushrooms and fish particularly support the growth of *C botulinum*. Botulism most often occurs after ingestion of home-canned products that have not been heated at temperatures sufficient to kill *C botulinum* spores, although inadequately sterilized commercial fish products have also been implicated. Because the toxin is heat-labile, food must be ingested uncooked or after insufficient cooking. Botulism often occurs in small family outbreaks in the case of home-prepared foods or less often as isolated cases connected to commercial products. Infant and wound botulism results when the toxin is produced endogenously, beginning with environmental spores that are either ingested or contaminate wounds.

Inadequately heated home-canned foods are most common source

PATHOGENESIS

Food-borne botulism is an intoxication, not an infection. The ingested preformed toxin is absorbed in the intestinal tract and reaches its neuromuscular junction target via the bloodstream. Once bound there, its inhibition of acetylcholine release causes paralysis due to lack of neuromuscular transmission. The specific disease manifestations depend on the specific nerves to which the circulating toxin binds. Cardiac arrhythmias and blood pressure instability are believed to be due to effects of the toxin on the autonomic nervous system. The damage to the synapse once the toxin has bound is permanent, and recovery requires growth of presynaptic axons and formation of new synapses.

Preformed toxin is readily absorbed

Acetylcholine block leads to paralysis and autonomic effects

BOTULISM: CLINICAL ASPECTS

MANIFESTATIONS

Food-borne botulism usually starts 12 to 36 hours after ingestion of the toxin. The first signs are nausea, dry mouth, and, in some cases, diarrhea. Cranial nerve signs, including blurred vision, pupillary dilatation, and nystagmus, occur later. Symmetric paralysis begins with the ocular, laryngeal, and respiratory muscles and spreads to the trunk and extremities. The most serious finding is complete respiratory paralysis. Mortality is 10% to 20%.

Blurred vision progresses to symmetrical paralysis

Infant Botulism

A syndrome associated with *C botulinum* that occurs in infants between the ages of 3 weeks and 8 months is now the most commonly diagnosed form of botulism. The organism is apparently introduced on weaning or with dietary supplements, especially honey, which is virtually impossible to sterilize. Ingested spores yield vegetative bacteria, which multiply and produce small amounts of toxin in the infant's colon. The infant shows constipation, poor muscle tone, lethargy, and feeding problems and may have ophthalmic and other paralyses similar to those in food-borne botulism. Infant botulism may mimic sudden infant death syndrome. The benefits of antitoxin and antimicrobic agents have not been clearly established.

Nonsterile honey introduces spores to intestine

Lethargy, poor feeding occur in addition to adult signs

Wound Botulism

Very rarely, wounds infected with other organisms may allow *C botulinum* to grow. Wound botulism in parenteral users of cocaine and maxillary sinus botulism in intranasal users of cocaine has been reported. Disease similar to that from food poisoning may develop, or it may begin with weakness localized to the injured extremity. Botulism without an obvious food or wound source is occasionally reported in individuals beyond infancy. It is possible that some such cases result from ingestion of spores of *C botulinum* with subsequent in vivo production of toxin in a manner similar to that in infant botulism.

Contaminated wounds of drug users are sites of toxin production

DIAGNOSIS

The toxin can be demonstrated in blood, intestinal contents, or remaining food, but these tests require inoculation of mice and are performed only in reference laboratories. *C botulinum* may also be isolated from stool or from foodstuffs suspected of responsibility for botulism.

Toxin can be detected in some laboratories

TREATMENT AND PREVENTION

The availability of intensive supportive measures, particularly mechanical ventilation, is the single most important determinant of clinical outcome. With proper ventilatory support, mortality rate should be less then 10%. The administration of large doses of horse *C botulinum* antitoxin is thought to be useful in neutralizing free toxin. Frequent hypersensitivity reactions related to the equine origin of this preparation make it unsuitable for use in infants. Antimicrobial agents are given only to patients with wound botulism.

Supportive measures and antitoxin allow survival

Adequate pressure cooking or autoclaving in the canning process kills spores, and heating food at 100°C for 10 minutes before eating destroys the toxin. Food from damaged cans or those that present evidence of positive inside pressure should not even be tasted because of the extreme toxicity of the *C botulinum* toxin.

Cooking food inactivates toxin

In an interesting twist, botulinum toxin as Botox has itself become a therapueutic agent. Originally licensed as a treatment of spasmotic neuromuscular conditions by direct injection into muscle, it has found a far larger use for cosmetic applications. For those that can afford it, a temporary respite from the wrinkles of aging can be gained from Botox injections administered by dermatologists and plastic surgeons.

Botox relieves wrinkles

CLOSTRIDIUM TETANI

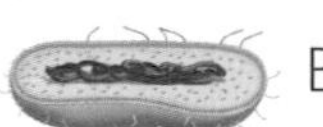

BACTERIOLOGY

C tetani is a slim, Gram-positive rod, which forms spores readily in nature and in culture, yielding a round terminal spore that gives the organism a drumstick appearance. *C tetani* requires strict anaerobic conditions. Its identity is suggested by cultural and biochemical characteristics, but definite identification depends on demonstrating the neurotoxic exotoxin. *C tetani* spores remain viable in soil for many years and are resistant to most disinfectants and to boiling for several minutes.

Gram-positive rods with drumstick-like spore

The most important product of *C tetani* is its neurotoxic exotoxin, **tetanospasmin** or tetanus toxin, a metalloproteinase that has structural and pharmacologic features similar to those of botulinum toxin. Tetanus toxin degrades a protein required for neurotransmitter release from vesicles at the appropriate site on presynaptic membranes (Figure 29–3). The most important difference from botulinum toxin is that the neurotransmitters in this case (glycine and γ-aminobutyric acid) are the ones that affect inhibitory neurons. The result is unopposed firing of the active motor neurons, generating spasms and spastic paralysis, which are the opposite of the botulinum flaccid paralysis. The toxin is heat-labile, antigenic, readily neutralized by antitoxin, and rapidly destroyed by intestinal proteases. Treatment with formaldehyde yields a nontoxic product or **toxoid** that retains the antigenicity of toxin and thus stimulates production of antitoxin.

Toxin blocks release of inhibitory neurotransmitters

Formaldehyde treatment removes toxicity but retains antigenicity

TETANUS

CLINICAL CAPSULE

The striking feature of tetanus is severe muscle spasms (or "lock-jaw" when the jaw muscles are involved). This occurs despite minimal or no inflammation at the primary site of infection, which may be unnoticed even though the outcome is fatal. The disease is caused by in vivo production of a neurotoxin that acts centrally, not locally. Immunization with inactivated toxin prevents tetanus.

EPIDEMIOLOGY

The spores of *C tetani* exist in many soils, especially those that have been treated with manure, and the organism is sometimes found in the lower intestinal tract of humans and animals. The spores are introduced into wounds contaminated with soil or foreign bodies. The wounds are often small, (eg, a puncture wound with a splinter). In many developing countries, the majority of tetanus cases occur in recently delivered infants when the umbilical cord is severed or bandaged in a nonsterile manner. Similarly, tetanus may follow an unskilled abortion, scarification rituals, female circumcision, and even surgery performed with nonsterile instruments or dressings.

Spores from environment germinate in wounds

Nonsterile technique can lead to tetanus

PATHOGENESIS

The usual predisposing factor for tetanus is an area of very low oxidation–reduction potential in which tetanus spores can germinate, such as a large splinter, an area of necrosis from introduction of soil, or necrosis after injection of contaminated illicit drugs. Infection with facultative or other anaerobic organisms can contribute to the development of an appropriate anaerobic nidus for spore germination. Tetanus bacilli multiply locally and neither damage nor invade adjacent tissues. Tetanospasmin is elaborated at the site of infection and enters the presynaptic terminals of lower motor neurons, reaching the central nervous system (CNS) mainly by exploiting the retrograde axonal transport system in the nerves. In the spinal cord, it acts at the level of the anterior horn cells, where its blockage of postsynaptic inhibition of spinal motor reflexes produces spasmodic contractions of both protagonist and antagonist muscles. This process takes place initially in the area of the causative lesion, but may extend up and down the spinal cord. Minor stimuli, such as a sound or a draft, can provoke generalized spasms.

Trauma provides growth conditions

Tetanospasmin produced at the local site ascends through nerves to anterior horn

Blockage of reflex inhibition causes spasmodic contractions

TETANUS: CLINICAL ASPECTS

MANIFESTATIONS

The incubation period of tetanus is from 4 days to several weeks. The shorter incubation period is usually associated with wounds in areas supplied by the cranial motor nerves, probably because of a shorter transmission route for the toxin to the CNS. In general, shorter incubation periods are associated with more severe disease.

The diagnosis is clinical; neither culture nor toxin testing is useful. Although tetanus may be localized to muscles innervated by nerves in the region of the infection, it is usually more generalized. The masseter muscles are often the first to be affected, resulting in inability to open the mouth properly **(trismus);** this effect accounts for the term **lock-jaw**. As other muscles become affected, intermittent spasms can become generalized to include muscles of respiration and swallowing. In extreme cases, massive contractions of the back muscles (opisthotonos) develop (**Figure 29–4**).

Incubation period varies with distance to CNS

Masseter muscle contraction causes lockjaw

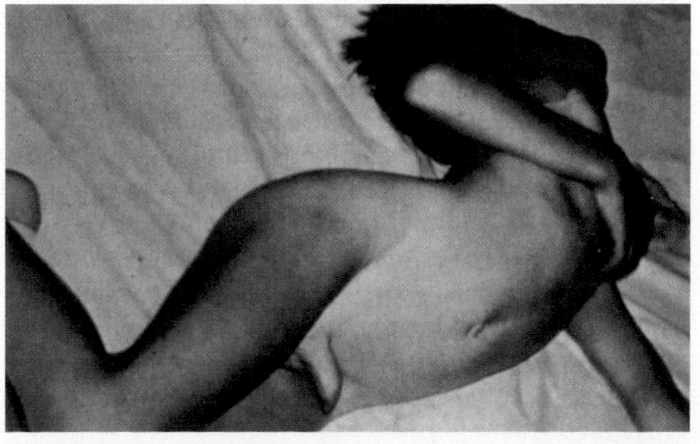

FIGURE 29–4. Tetanus. Opistho tonic posturing caused by involvement of the spinal musculature in a child with generalized tetanus. (Reproduced with permission from Connor DH, Chandler FW, Schwartz DQA, Manz HJ, Lack EE (eds). *Pathology of Infectious Diseases,* vol. 1. Stamford, CT: Appleton & Lange; 1997.)

Untreated patients with tetanus retain consciousness and are aware of their plight, in which small stimuli can trigger massive contractions. In fatal cases, death results from exhaustion and respiratory failure. Untreated, the mortality rate caused by the generalized disease varies from 15% to more than 60%, according to the lesion, incubation period, and age of the patient. Mortality is highest in neonates and in elderly patients.

Respiratory failure leads to death

TREATMENT

Specific treatment of tetanus involves neutralization of any unbound toxin with large doses of human tetanus immune globulin (HTIG), which is derived from the blood of volunteers hyperimmunized with toxoid. Most important in treatment are nonspecific supportive measures, including maintenance of a quiet dark environment, sedation, and provision of an adequate airway. Benzodiazepines are also used to indirectly antagonize the effects of the toxin. The value of antimicrobics is not clear. Because toxin binding is irreversible, recovery requires the generation of new axonal terminals.

Supportive treatment required until axons regenerate

PREVENTION

Routine active immunization with tetanus toxoid, combined with diphtheria toxoid and pertussis vaccine (DTaP) for primary immunization in childhood and DT for adults, can completely prevent tetanus. It has reduced the incidence of tetanus in the United States to less than 50 reported cases per year. Five doses of DT are recommended, to be given at the ages of 2, 4, 6, and 18 months, and once again between the ages of 4 and 6 years. Thereafter, a booster of adult-type tetanus diphtheria toxoid should be given every 10 years. Unfortunately, routine childhood immunization is not administratively and economically feasible in many less well-developed countries, where as many as 1 million cases of tetanus occur annually. In such settings, immunization efforts have been focused on pregnant women, because transplacental transfer of antibodies to the fetus also prevents the highly lethal neonatal tetanus.

Childhood toxoid immunization prevents disease

Boosters required every 10 years

Unimmunized subjects with tetanus-prone wounds should be given passive immunity with a prophylactic dose of HTIG as soon as possible. This immunization provides immediate protection. Those who have had a full primary series of immunizations and appropriate boosters are given toxoid for tetanus-prone wounds if they have not been immunized within the previous 10 years in the case of clean minor wounds or 5 years for more contaminated wounds. If immunization is incomplete or the wound has been neglected and poses a serious risk of disease, HTIG is also appropriate. Penicillin therapy is a prophylactic adjunct in serious or neglected wounds, but in no way alters the need for specific prophylaxis.

Passive immunization used when immunization is neglected

CLOSTRIDIUM DIFFICILE

BACTERIOLOGY

C difficile is a Gram-positive rod that readily forms spores. Like the other clostridia described in this section, *C difficile* has a most important medical feature: its ability to produce toxins. In this species, two distinct large polypeptide toxins, A and B, with similar structure (45% homology) are released during late growth phases of the vegetative organism, perhaps at the time of cell lysis. Both toxins act in the cytoplasm by disrupting signal transduction proteins, particularly those involving the actin cytoskeleton. The A toxin causes cell rounding and the disruption of intercellular tight junctions followed by altered membrane permeability and fluid secretion. The net effect is that of an enterotoxin, although inflammation and cytoxic activity are also present. The B toxin lacks the enterotoxic properties of the A toxin, but has cytotoxic potency at least 10 times higher. The two toxins appear to act synergistically by a mechanism yet to be determined.

A and B toxins disrupt cytoskeleton signal transduction

A is a enterotoxin

B is a cytotoxin

C DIFFICILE DIARRHEA

CLINICAL CAPSULE

C difficile is the most common cause of diarrhea that develops in association with the use of antimicrobial agents. The diarrhea ranges from a few days of intestinal fluid loss to life-threatening pseudomembranous colitis (PMC). This condition is associated with intense inflammation and the formation of a pseudomembrane composed of inflammatory debris on the mucosal surface.

EPIDEMIOLOGY

C difficile is present in the stool of 2% to 5% of the general population, sometimes at higher rates among hospitalized persons and infants. More than two decades of the antibiotic era had elapsed before the medical importance of *C difficile* was recognized through its association with antibiotic-associated diarrhea (AAD). Although infection is endogenous in most cases, hospital outbreaks have clearly established that the environment can be the source as well. In the United States alone, over 300,000 cases are reported each year.

Source is endogenous or environmental

C difficile is not the only cause of AAD, but it is the most common identifiable cause. In simple diarrhea following antimicrobial administration, this organism is responsible for approximately 30% of cases. As the disease progresses to colitis, the association is stronger, rising to 90% if PMC is present. Person-to-person transfer is uncommon except in the instance of hospital-acquired *C difficile* infections, where environmental or hand contamination leads to infection of another patient.

Frequent cause of AAD

Major cause of PMC

PATHOGENESIS

When *C difficile* becomes established in the colon of individuals with normal gut flora, few if any direct consequences result, probably because its numbers are dwarfed by the other flora. Alteration of the colonic flora with antimicrobics (particularly ampicillin, cephalosporins, and clindamycin) favors *C difficile* in two ways. First, strains resistant to the antimicrobial agent can grow in its presence and assume a larger if not dominant position in the flora. Second, in an antimicrobial milieu, the readiness with which *C difficile* forms spores may favor its survival over non–spore-forming bacteria. In either case, the minor niche of the species is improved to the point at which the effect of its toxins on the colonic mucosa becomes significant.

Antimicrobic effect on flora selects for *C difficile*

Increased numbers make toxin more effective

Although most strains produce both toxins, the enterotoxic properties of the A toxin seem to dominate in watery diarrhea cases. In PMC, the colonic mucosa is studded with inflammatory plaques, which may coalesce into an overlying "pseudomembrane" composed of fibrin, leukocytes, and necrotic colonic cells (**Figure 29–5**). This picture fits better with the action of the cytotoxic B toxin.

A enterotoxin stimulates watery diarrhea

IMMUNITY

Antibody against the A toxin is associated with resolution of disease in experimental animals. This feature and the inverse relationship between severity of disease and anti-A antibody both support the importance of humoral immunity in *C difficile* diarrhea. Antibodies directed against the B toxin also appear to offer protection, but the relationship is less clear than with toxin A.

Antitoxin antibodies have protective effect

A

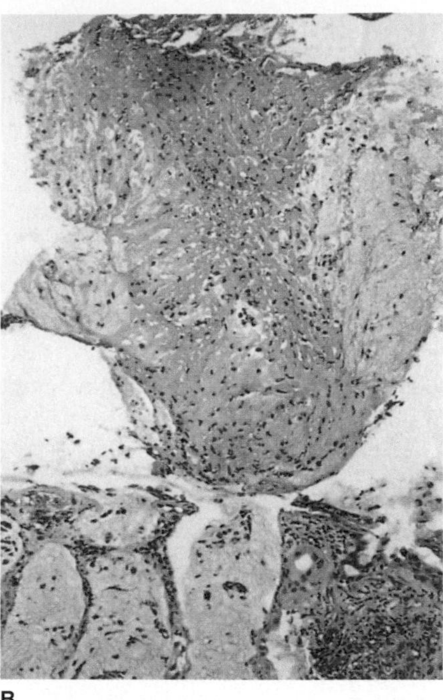

B

FIGURE 29–5. *Clostridium difficile* pseudomembranous colitis. A. Colon with discrete plaques of pseudomembrane. **B.** Histopathology demonstrates the pseudomembrane above the mucosa. It is "pseudo" because it is composed of only fibrin and inflammatory cells. (Reproduced with permission from Connor DH, Chandler FW, Schwartz DQA, Manz HJ, Lack EE (eds). *Pathology of Infectious Diseases,* vol. 1. Stamford, CT: Appleton & Lange; 1997.)

C DIFFICILE DIARRHEA: CLINICAL ASPECTS

MANIFESTATIONS

Diarrhea ranges from mild to PMC

Diarrhea is a common side effect of antimicrobic treatment. In *C difficile*–caused diarrhea, the onset is usually 5 to 10 days into the antibiotic treatment, but the range is from the first day to weeks after cessation. The diarrhea may be mild and watery or bloody and accompanied by abdominal cramping, leukocytosis, and fever. In PMC, it progresses to a severe, occasionally lethal inflammation of the colon that can be demonstrated by endoscopic examination.

DIAGNOSIS

Stool toxin detection is the primary diagnostic tool

Although selective media have been developed for isolation of *C difficile,* direct detection of toxins in the stool has largely replaced culture for diagnostic purposes. *C difficile* is the only pathogen for which detection of its toxin has become routine. The standard toxin assay requires demonstration and neutralization of cytopathic effect in cell culture. Newer enzyme immunoassays, which demonstrate toxin A and/or B in stool, are slightly less sensitive but less expensive and thus more widely available.

TREATMENT

Oral metronidazole or vancomycin reach bacteria in the intestine

Discontinuing the implicated antimicrobic usually results in the resolution of clinical symptoms. If patients are severely ill or fail to respond to drug withdrawal, they should receive metronidazole or vancomycin administered orally. The poor absorption of vancomycin is an advantage in this situation, but its use is now being restricted because of concern about its role in selecting resistant enterococci and other organisms. *C difficile* is susceptible to the penicillins and cephalosporins in vitro, but these drugs are ineffective because of access in the intestinal lumen and the hazard of destruction by β-lactamases produced by other bacteria. Relapses or reinfections requiring retreatment occur in up to 20% of patients. Immunization strategies using a toxin A toxoid are in development.

BACTEROIDES FRAGILIS

BACTERIOLOGY

The *B fragilis* group constitutes the most common opportunistic pathogens of the genus *Bacteroides*. These slim, pale-staining, capsulate, Gram-negative rods form colonies overnight on blood agar medium. The implication of fragility in the name is misleading, because they are actually among the hardier and more easily grown anaerobes. Most strains produce superoxide dismutase and are relatively tolerant to atmospheric oxygen. *B fragilis* has surface pili and a capsule composed of a polymer of two polysaccharides. The LPS endotoxin in the *B fragilis* outer membrane is less toxic than that of most other Gram-negative bacteria, possibly owing to modification or absence of the lipid A portion.

Oxygen-tolerant species that produce superoxide dismutase

Polysaccharide capsule is present

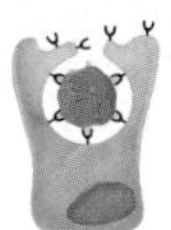

B FRAGILIS DISEASE

CLINICAL CAPSULE

Deep pain and tenderness anywhere below the diaphragm are typical of the onset of *B fragilis* infection. Depending on the extent and spread of the intra-abdominal abscess, fever and widespread findings of an acute abdomen may also be seen.

EPIDEMIOLOGY

Like the other Gram-negative anaerobes, *B fragilis* infections are endogenous, originating in the patient's own intestinal flora. Given the mass and diversity of intestinal anaerobes, the frequent presence of *B fragilis* in clinically significant infections is striking. It is typically mixed with other anaerobes and facultative bacteria. Human-to-human transmission is not known and seems unlikely.

Endogenous infection mixed with other intestinal bacteria

PATHOGENESIS

The relative oxygen tolerance of *B fragilis* probably plays a role in its virulence by aiding its survival in oxygenated tissues in the period between its displacement from the intestinal flora and the establishment of a reduced local microenvironment. Its pili have adhesive properties, and the polysaccharide capsule confers resistance to phagocytosis and inhibits macrophage migration. The most distinguishing pathogenic feature of the organism is its ability to cause abscess formation.

This capsule experimentally stimulates abscess formation, even in the absence of live bacteria. This property is not found in the capsular polysaccharides of organisms such as *Streptococcus pneumoniae*. The active feature is not known, but may be related to the ability of the polysaccharide to trigger T-cell responses that lead to abscess formation. *B fragilis* and other *Bacteroides* species produce a number of extracellular enzymes (collagenase, fibrinolysin, heparinase, hyaluronidase) that may also contribute to the formation of the abscess.

Pili and oxygen-tolerance aid initial stages

Capsule resists phagocytosis and stimulates abscess production

Some strains of *B fragilis* produce an enterotoxin that causes enteric disease in animals, and in some studies they have been associated with a self-limited, watery diarrhea in children. Because these enterotoxin-producing strains are found in up to 10% of healthy individuals, their pathogenic importance is still undetermined.

Diarrheal enterotoxin is possible

IMMUNITY

Cell-mediated immunity may be protective

Although it has been demonstrated that antibody to capsular polysaccharide facilitates classical complement pathway killing, there is no evidence that this confers immunity to reinfection. In contrast, there is some evidence that cell-mediated immunity may be protective.

B FRAGILIS: CLINICAL ASPECTS

MANIFESTATIONS

Some event that displaces *B fragilis* along with other members of the intestinal flora is required to initiate infection; there is no evidence the organism is invasive on its own. This mucosal break may be the result of trauma or other disease states such as diverticulitis.

Abdominal pain and fever may evolve to peritonitis

Abscesses combined with anaerobes and Enterobacteriaceae

The local effects of the developing abscess include abdominal pain and tenderness, often with a low-grade fever. The subsequent course depends on whether the abscess remains localized or ruptures through to other sites such as the peritoneal cavity. This may cause several other abscesses or peritonitis. The course of illness is strongly influenced by the other bacteria in the abscess, particularly members of the Enterobacteriaceae. Spread to the bloodstream is more common with *B fragilis* than any other anaerobe.

TREATMENT

Cephalosporin resistant to β-lactamase is required

Drainage of abscesses and debridement of necrotic tissue are the mainstays of the treatment of *B fragilis* infections, as with anaerobic infections in general. The accompanying antimicrobial therapy is complicated by the fact that abdominal *B fragilis* isolates almost always produce a β-lactamase, which not only inactivates penicillin but other β-lactams, including many cephalosporins. Resistance to tetracycline is also common, but most strains are susceptible to clindamycin, and metronidazole. Among the β-lactams, azthreonam, imipenem, and ceftriaxone have been used effectively, as have combinations of a β-lactamase inhibitor (clavulanate, sulbactam) and a β-lactam (ampicillin, ticarcillin).

CLINICAL CASE

COMPOUND FRACTURE AND A SENSE OF DOOM

A 24-year-old man, an automobile accident victim, was brought to the hospital with a compound fracture of the distal left tibia and fibula. Within 6 hours of the accident, the patient was taken to surgery where the wound was debrided, the leg was immobilized, and therapy was begun (cephalothin sodium IV, 1 g/4 hr). The patient was afebrile. The hematocrit reading was 41%, the WBC count 10,900/mm^3, and blood pressure and pulse rate within normal limits. He did well until the fourth postoperative day when he was noted to have a temperature of 38.3°C orally, a tachycardia rate of 120 bpm, a painful left leg, and a sense of impending doom.

The cast was opened and the entire lower leg was found to be swollen and reddish-brown, and was exuding a serosanguineous foul-smelling discharge. Crepitations were palpable over the anterior tibial and entire gastrocnemius areas. His blood pressure became unstable and then dropped to 70/20 mm Hg. Gram stain of an aspirate from the gastrocnemius demonstrated both Gram-negative and Gram-positive rods, but no spores were seen. At this time, the hematocrit reading had decreased to 35%, and WBC count was 12,000/mm^3, with 85% polymorphonuclear leukocytes.

Therapy was begun with IV penicillin G aqueous, 5 million units every 6 hours. The man was taken to surgery, where an above-knee amputation was performed. While the patient was receiving cephalothin, cultures of the necrotic muscle grew *Escherichia coli* and *Clostridium perfringens*. Within 3 hours after amputation, the patient had a sense of well-being, and complete recovery followed.

QUESTIONS

■ The crepitations in the wound are most likely due to:

A. Production of CO_2 by *Clostridium perfringens*
B. Bowel leakage into the tissue
C. Foreign bodies from the accident
D. Surgical introduction of air
E. Local hematoma

■ The clostridia in the wound most likely came from:

A. Intestinal flora
B. Skin flora
C. Soil
D. Insect bite
E. Water

■ The injury in the tissue is produced by which of the following:

A. ADP-ribosylating toxin
B. Lecithinase α-toxin
C. Pore-forming θ-toxin
D. Enterotoxin
E. Spores

■ The most important treatment for this condition is

A. Antimicrobials
B. Antitoxin
C. Hyperbaric oxygen
D. Surgery
E. Bed rest

ANSWERS

1(A), **2**(C), **3**(B), **4**(D)

CHAPTER 30

Neisseria

Rocky Kilmarry is about as good for you as a dose of clap.

—Adam Diment: *Dolly Dolly Spy*

Neisseria are Gram-negative diplococci. The genus contains two pathogenic and many commensal species, most of which are harmless inhabitants of the upper respiratory and alimentary tracts. The pathogenic species are *Neisseria meningitidis*, a major cause of meningitis and bacteremia, and *Neisseria gonorrhoeae*, the cause of gonorrhea (vulgar, the clap).

NEISSERIA: GENERAL FEATURES

Neisseria are Gram-negative cocci that typically appear in pairs with the opposing sides flattened, imparting a "kidney bean" appearance (**Figure 30–1**). They are nonmotile, non–spore-forming, and non–acid-fast. Their cell walls are typical of Gram-negative bacteria, with a peptidoglycan layer and an outer membrane containing endotoxic glycolipid complexed with protein. The structural elements of *N meningitidis* and *N gonorrhoeae* are the same, except that the meningococcus has a polysaccharide capsule external to the cell wall. ::: Gram-negative envelope, p. 353

Gram-negative diplococci are bean-shaped

Gonococci and meningococci require an aerobic atmosphere with added carbon dioxide and enriched medium for optimal growth. Gonococci grow more slowly and are more fastidious than meningococci, which can grow on routine blood agar. All *Neisseria* are oxidase-positive. Species are defined by growth characteristics and patterns of carbohydrate fermentation. Reagents are also available to distinguish *N gonorrhoeae* and *N meningitidis* from the other *Neisseria* by immunologic methods such as slide agglutination and immunofluorescence.

Gonococci are more fastidious than meningococci

All *Neisseria* are oxidas-positive

Both pathogenic species possess pili and outer membrane proteins (OMPs), which vary in their function and antigenic composition. In the study of these meningococcal and gonococcal proteins, investigators have assigned names to these proteins some of which after further study seem to be for molecules with similar functions in pathogenesis. **Table 30–1** is an attempt to show similarities and differences. It should be understood that the assignment of the same name (eg, PorA) to a protein found in both species does not mean they are identical. It does suggest that they have similar structure and function. This is similar to the situation with the pyrogenic exotoxins of *Staphylococcus aureus* and the group A streptococci.

Similar pili and OMPs are present in both species

The outer membrane of pathogenic *Neisseria* contains a variant of lipopolysaccharide (LPS) in which the side chains are shorter and lack the repeating polysaccharide units found in the LPS of most other Gram-negative bacteria. This short-chain neisserial LPS is called lipooligosaccharide (LOS). The lipid A and core oligosaccharide are structurally and

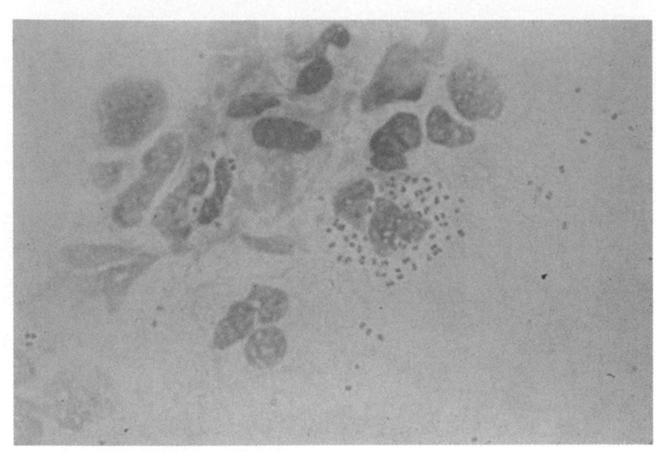

FIGURE 30–1. *Neisseria gonorrhoeae.* Gram stain of urethral exudate. Note the many pairs of Gram-negative bean-shaped diplococcic collected in polymorphonuclear neutrophils (PMNs) and free in the purulent material. The morphology of *N meningitidis* and other *Neisseria* is identical. (Reprinted with permission of Schering Corporation, Kenilworth, NJ, the copyright owner. All rights reserved.)

Outer membrane LOS has short side chains

functionally similar to other Gram-negative LPS. The pili, OMPs, and LOS are antigenic and have been used in typing schemes.

NEISSERIA MENINGITIDIS

BACTERIOLOGY

Meningococci produce medium-sized smooth colonies on blood agar plates after overnight incubation. Carbon dioxide enhances growth, but is not required. Twelve serogroups have been defined on the basis of the antigenic specificity of their polysaccharide capsule. The

TABLE 30–1 Bacteriologic and Pathogenic Features of *Neisseria*

			ANTIGENIC STRUCTURE						
	GROWTH				OUTER MEMBRANE PROTEINS				
ORGANISM	BLOOD AGAR	ML AGAR[a]	CAPSULE	PILI	ADHERENCE-ASSOCIATED	PORINS	BLOCKING AB-ASSOCIATED[b]	TRANSMISSION	DISEASE
N meningitides	+	+	Polysaccharide (12 serogroups)[c]	Class I,[d] II Antigenically diverse	Class 5 (4 variants)	PorA, PorB[e]	Class 4	Inhalation of respiratory droplets	Meningitis, septic shock
N gonorrhoeae	–	+	None[f]	Antigenically diverse[d]	Protein II or Opa (12 variants)	PorA, Por B	Protein III	Sexual contact of mucosal surfaces	Urethritis, cervicitis, PID
Other *Neisseria* species	+	–	None	Present	Unknown	Unknown	Absent	Normal respiratory flora	None

PID, pelvic inflammatory disease.
[a]Martin–Lewis or similar selective medium.
[b]Bind IgG in a way that interferes with bactericidal activity of antibodies directed at other antigens.
[c]A, B, C, H, I, K, L, X, Y, Z, 29E, W-135.
[d]Gonococcal and meningococcal class I are similar to each other and members of a class of bacterial pili with amino-terminal *N*-methylphenylalanine residues (*Bacteroides, Moraxella, Pseudomonas aeruginosa*).
[e]Two antigenic classes.
[f]Lipooligosaccharide sialylation has some of the effects of a capsule (see text).

most important disease-producing serogroups are A, B, C, W-135, and Y. In addition to the group polysaccharides, individual *N meningitidis* strains may contain two distinct classes of pili and multiple classes of OMPs including porins and adherence proteins, some of which have structural and functional similarities to those found in gonococci. The function of other OMPs is unknown.

Serogroups are based on the polysaccharide capsule

Some OMPs are similar to gonococci

MENINGOCOCCAL DISEASE

CLINICAL CAPSULE

Meningococci are usually quiescent members of the nasopharyngeal flora but may produce fulminant infection of the bloodstream and/or central nervous system (CNS). There is little warning; localized infections that precede systemic spread are rarely recognized. The major disease is an acute, purulent meningitis with fever, headache, seizures, and mental signs secondary to inflammation and increased intracranial pressure. Even when the CNS is not involved, *N meningitidis* infections have a marked tendency to be accompanied by rash, purpura, thrombocytopenia, and other manifestations associated with endotoxemia. This bacterium causes one of the few infections in which patients may progress from normal health to death in less than a day. It can also spread quickly in family, school, and even national outbreaks.

EPIDEMIOLOGY

The combination of rapidly progressive disease and obvious person-to-person spread has long made meningococcal disease one of the most feared of all infections. In fact, meningococci are found in the nasopharyngeal flora of approximately 10% of healthy individuals. Transmission occurs by inhalation of aerosolized respiratory droplets. Close, prolonged contact such as occurs in families and closed populations promotes transmission. The estimated attack rate among family members residing with an index case is 1000 times higher than in the general population; this fact is evidence of the contagious nature of meningococcal infection. Other factors that foster transmission are contact with a virulent strain and host susceptibility (lack of protective antibody). Typical settings of larger outbreaks are schools, dormitories, and camps for military recruits. In these close living circumstances, *N meningitidis* spreads readily among newly exposed individuals, but disease develops only in those who lack group-specific antibody.

Nasopharyngeal carrier rate is 10%

Spread in is by respiratory droplets

The incidence of invasive meningococcal infection varies widely depending on age, geographic locale, and serogroup. In the United States, attack rates vary between 0.5 and 1.5 cases per 100,000 population, but in some countries rates as high as 25 per 100,000 have been sustained for some time. Most disease occurs in infants with a second peak at 18 years of age. Most cases are sporadic or in small family or closed-population (school, day-care center) outbreaks. B, C, Y and W-135 are the most common serogroups in developed countries. Serogroup A strains tend to emerge every 10 to 15 years in large epidemics, with attack rates as high as 500 per 100,000. Since the second half of the 20th century, these serogroup A epidemics have been largely confined to tropical countries.

B, C, and Y are the most common serogroups

Group A strains can cause widespread epidemics

PATHOGENESIS

The meningococcus is an exclusively human parasite; it can either exist as an apparently harmless member of the normal flora or produce acute disease. For most individuals, the carrier state is associated with acquisition of protective antibodies, but for some, spread from the nasopharynx to produce bacteremia, endotoxemia, and meningitis takes place too quickly for immunity to develop. Meningococci use pili for initial attachment to a protein (CD46) on the surface of the nonciliated nasopharyngeal epithelium. This is a prelude to invasion. In the invasion process, the microvilli of these cells come in close contact with the bacteria, which then enter the cells in membrane-bound vesicles. Once inside, meningococci quickly pass through the cytoplasm, exiting into the submucosa and eventually the bloodstream (**Figure 30–2**). In the process, they damage the ciliated cells, possibly by direct release of endotoxin.

Meningococci range from carrier state to bacteremia

Pili attach to microvilli as prelude to invasion

Once meningococci gain access to the submucosa, their ability to produce disease is enhanced by several factors that allow them to scavenge essential nutrients and evade the host immune response. One critical nutrient, iron, is supplied by *N meningitidis* proteins, which are able to acquire it from the human iron transport protein transferrin. As with other encapsulated bacteria, the polysaccharide capsule enables meningococci to resist complement-mediated bactericidal activity and subsequent neutrophil phagocytosis. Meningococcal (and gonococcal) LPS/LOS also has features that facilitate evasion of host immune responses. Its chemical structure mimics sphingolipids found in the human brain enough for them to be recognized as self by the immune system. In addition, meningococci are able to incorporate sialic acid from host substrates as terminal substitutions of their LOS side chains. This sialyated LOS is able to down-regulate complement deposition by binding serum factor H in a manner already described in Chapter 22 for bacterial surface molecules. Like the capsule of group B streptococci (Chapter 25), the capsules of group B and C meningococci are polymers of sialic acid. ::: Capsules and factor H, p. 394

Proteins scavenge iron from transferrin

Polysaccharide capsules are antiphagocytic

LOS + sialic acid interferes with complement deposition

The most serious manifestations of meningococcal disease are related to its spread to the bloodstream and, its namesake, the meninges. The exact mechanism of CNS invasion is unclear but is probably related to the level of the bacteremia. It occurs in the choroid plexus with its exceptionally high rate of blood flow. After CNS invasion, an intense subarachnoid space inflammatory response is generated, induced by the release of cell wall peptidoglycan fragments, LPS, and possibly other virulence factors causing the release of inflammatory cytokines. A prominent feature of meningococcal disease with or without CNS invasion

Spread to blood and CNS produce systemic endotoxemia

LPS and peptidoglycan trigger cytokine release

Shedding outer membrane blebs hyperproduces LPS

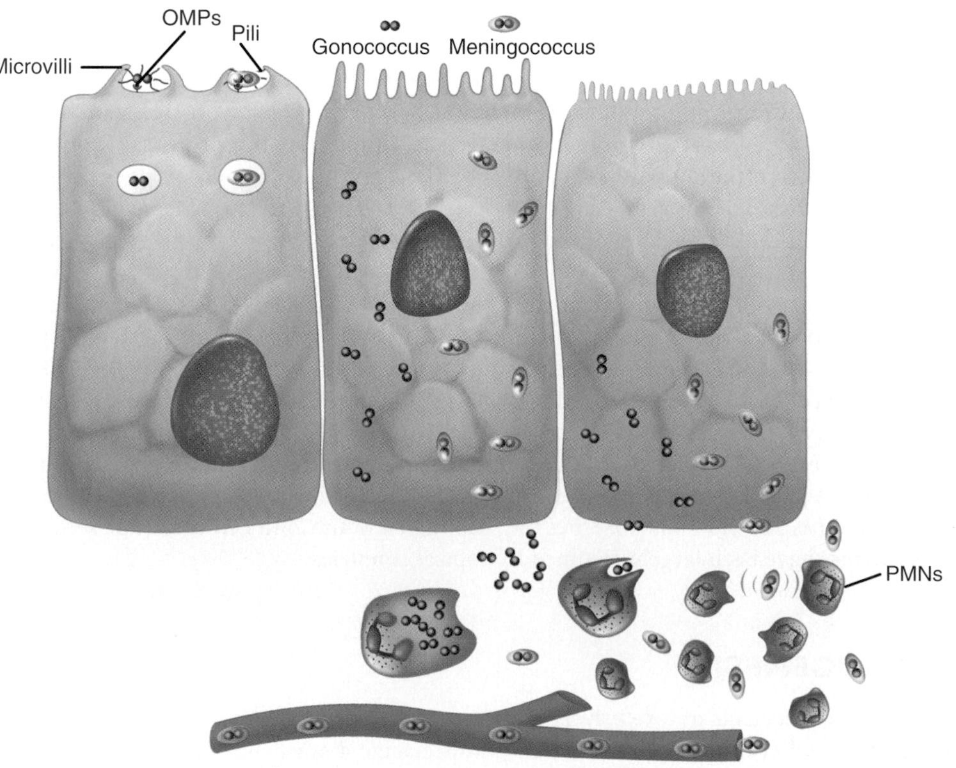

FIGURE 30–2. Gonococcus and meningococcus, cellular view. *Neisseria gonorrhoeae* and *Neisseria meningitidis* differ in that *N meningitidis* has a capsule. *(Left)* Both attach to microvillus cells by outer membrane proteins (OMP) and pili. They are endocytosed in vacuoles. *(Middle)* Both multiply freely in the cytoplasm. *(Right)* Both escape to the submucosa, but the gonococcus is actively phagocytosed and remains localized. The meningococcal capsule allows it to evade phagocytosis and it enters the bloodstream. PMNs, polymorphonuclear neutriphils.

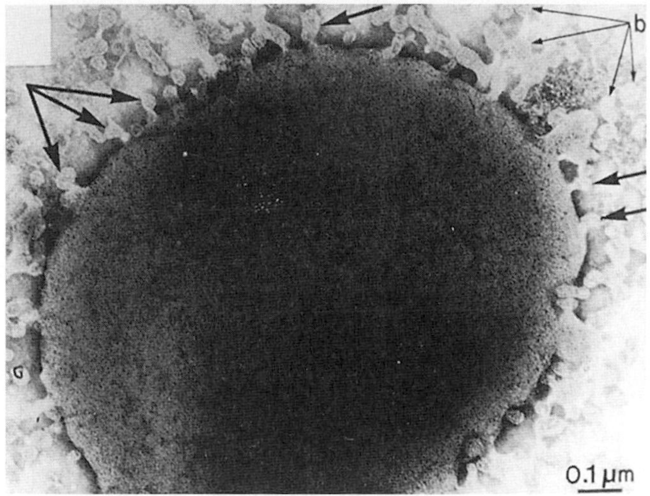

FIGURE 30–3. *Neisseria meningitidis.* Cell wall is shown shedding multiple "blebs" (*arrows*) containing lipopolysaccharide–endotoxin. Note the typical trilamellar Gram-negative cell wall structure in the wall and the blebs. (Reprinted with permission from Devoe IW, Gilcrist JE. J Exp Med 1973;138:1160.)

is disseminated, potent, endotoxic activity (see Manifestations). When grown in culture, *N meningitidis* readily releases endotoxin-containing blebs of its outer membrane from the cell surface as shown in **Figure 30–3**. It is not known whether this occurs in vivo, but the model of the meningococcus as a hyperproducer of LPS endotoxin certainly fits with its most serious disease manifestations. ::: LPS endotoxin, p. 354

IMMUNITY

Group-specific anticapsular antibody is protective

Complement component deficiencies enhance risk

Immunity to meningococcal infections is related to group-specific antipolysaccharide antibody, which is bactericidal and facilitates phagocytosis. The bactericidal activity is due to complement-mediated cell lysis via the classical complement pathway. Individuals with deficiencies in the terminal complement components have an enhanced risk for meningococcal disease but not for other polysaccharide capsule pathogens, such as *Haemophilus influenzae* type b. ::: Classic complement pathway, p. 28

Most common age of infection is 6–24 months

Absence of antibody correlates with susceptibility

The peak incidence of serious infection occurs between 6 months and 2 years of age. This corresponds to the nadir in the prevalence of antibody in the general population, which is the time between loss of transplacental antibody and the appearance of naturally acquired antibody (**Figure 30–4**). By adult life, serum antibody to one or more meningococcal serogroups is usually present, but an immune deficit to the other serogroups remains. Infections appear when populations carrying virulent strains mix (college, summer camp, military barracks) and susceptible individuals acquire a new strain of a serogroup to which they lack group-specific antibody.

Infection, carrier state, or other polysaccharides may stimulate antibody

Protective antibody is stimulated by infection and through the carrier state, which produces immunity within a few weeks. The natural immunization shown in Figure 30–4 may not require colonization with every serogroup or even with *N meningitidis,* because antibody may be produced in response to cross-reactive polysaccharides possessed by other *Neisseria* or even other genera. For example, *Escherichia coli* strains of a particular serotype (K1) have a polysaccharide capsule identical to that of the group B meningococcus. These *E coli* also have enhanced potential to produce meningitis in neonates.

T-cell–independent mechanisms are involved

Group B polysaccharide is not immunogenic

Purified capsular polysaccharides are immunogenic, generating T-cell–independent immune responses in which IgG_2 is the predominant antibody. As with other polysaccharide immunogens, these responses are not strong, particularly in early childhood when there is a relative deficiency of IgG_2. The group B polysaccharide differs from that of the other groups in failing to stimulate bactericidal antibody at all. This is believed to be due to the similarity of its sialic acid polymer to human brain antigens. That is, like the sialated LOS, it may be recognized as self by the immune system. Exposed outer membrane proteins are under study for their potential role in immunity. ::: T-cell independent responses, p. 37

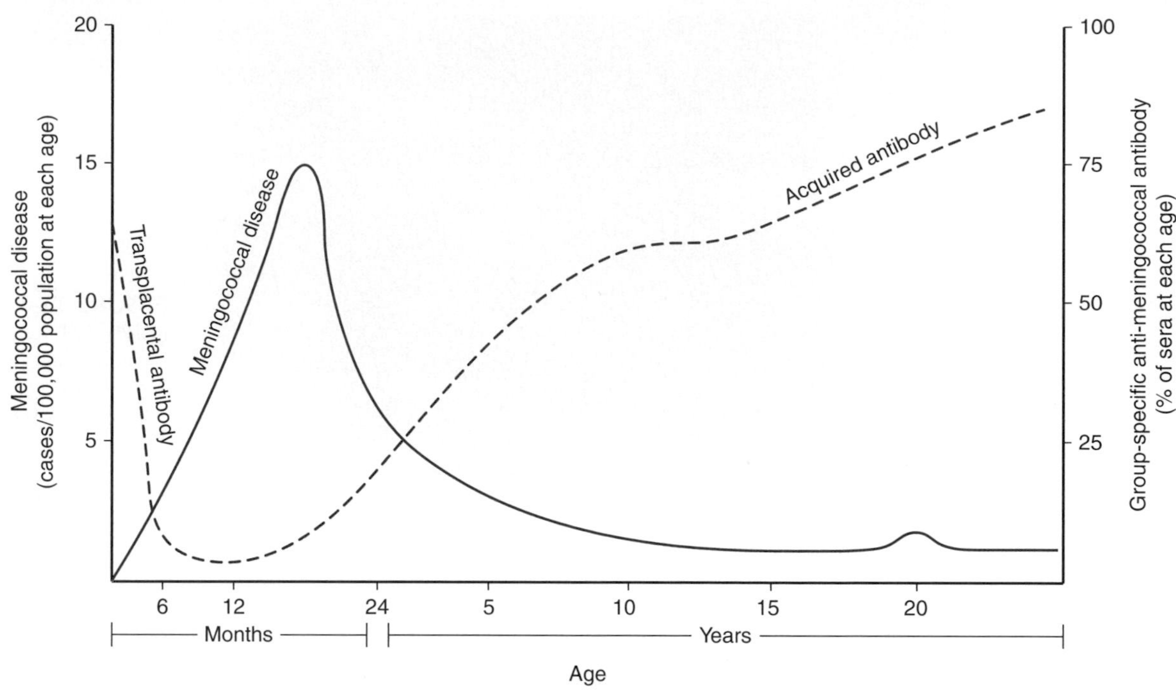

FIGURE 30–4. Immunity to the meningococcus. The inverse relationship between bactericidal meningococcal antibody and meningococcal disease is demonstrated. The "blip" in the disease curve around age 20 is attributable in part to military and other closed-population outbreaks. (Adapted with permission from Goldschneider I, Gotschlich EC, Liu TY, Artenstein MS. Human immunity to the meningococcus I–V. J Exp Med 1969;129:1307–1395.)

MENINGOCOCCAL DISEASE: CLINICAL ASPECTS

MANIFESTATIONS

Meningitis is most common infection

Meningococcemia and rash may progress to DIC

Systemic features resemble endotoxic shock

The most common form of meningococcal infection is acute purulent meningitis, with clinical and laboratory features similar to those of meningitis from other causes (see Chapter 65). A prominent feature of meningococcal meningitis is the appearance of scattered skin petechiae, which may evolve into ecchymoses or a diffuse petechial rash (**Figure 30–5**). These cutaneous manifestations are signs of the disseminated intravascular coagulation (DIC) syndrome, which is part of the endotoxic shock brought on by meningococcal bacteremia (meningococcemia). Meningococcemia sometimes occurs without meningitis and may progress to fulminant DIC and shock with bilateral hemorrhagic destruction of the adrenal glands (Waterhouse–Friderichsen syndrome). However, the disease is not always fulminant, and some patients have only low-grade fever, arthritis, and skin lesions that develop slowly over a period of days to weeks. Meningococci are a rare cause of other infections such as pneumonia, but it is striking that localized infections are almost never recognized in advance of systemic disease.

DIAGNOSIS

Direct CSF Gram smears are diagnostic

Culture requires only blood agar

Direct Gram smears of cerebrospinal fluid (CSF) in meningitis usually demonstrate the typical bean-shaped, Gram-negative diplococci (Figure 30–1). Definitive diagnosis is by culture of CSF, blood, or skin lesions. Although *N meningitidis* is reputed to be somewhat fragile, it requires no special handling for isolation from presumptively sterile sites such as blood and CSF. Growth is good on blood or chocolate agar after 18 hours of incubation. Speciation is based on carbohydrate fermentation patterns or immunologic tests. Serogrouping may be performed by slide agglutination methods but has no immediate clinical importance.

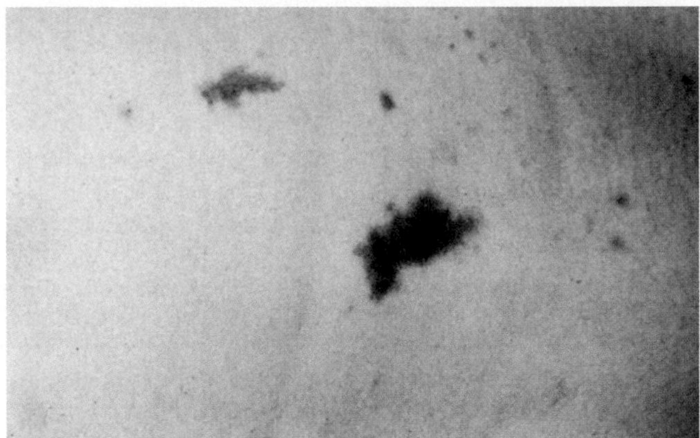

FIGURE 30–5. Meningococcemia. Small and large coalesced petechiae are shown in the skin of a patient with meningococci circulating in the blood. (Reproduced with permission from Nester EW, Anderson DG, Roberts CE Jr, Nester MT. *Microbiology: A Human Perspective,* 6th ed. New York: McGraw-Hill, 2008.)

TREATMENT

Penicillin has long been the treatment of choice for meningococcal infections because of its high antimeningococcal activity and good CSF penetration. Although resistance mediated by both β-lactamase and altered penicillin-binding proteins (PBPs) has been reported, it is still extremely rare. Third-generation cephalosporins such as ceftriaxone and cefotaxime are also effective and are treatments of choice for acute meningitis until the meningococcal etiology is proven. In countries where penicillin resistance is significant, cephalosporins become the first-line treatment.

Penicillin resistance is still rare

PREVENTION

Until the development and spread of sulfonamide resistance in the 1960s, chemoprophylaxis with these agents was the primary means of preventing spread of meningococcal infections. Rifampin is now the primary chemoprophylactic agent, but ciprofloxacin has also been effective. Penicillin is not effective, probably because of inadequate penetration of the uninflamed nasopharyngeal mucosa. Selection of cases to receive prophylaxis is based on epidemiologic assessment. Risk is highest for siblings of the index case and declines with increasing age. The closeness and duration of contact with the index case are also important. For example, an infant sibling sharing a room with a person with meningococcal disease would be at the highest risk. Typically, family members are given prophylaxis, but other adults are not. Common-sense exceptions, such as playmates and healthcare workers with very close contact (eg, mouth-to-mouth resuscitation), are made at the discretion of the physician or epidemiologist. The presence or absence of nasopharyngeal carriage of *N meningitidis* plays no role in this decision because it does not accurately predict risk of disease.

Rifampin is primary antimicrobic for chemoprophylaxis

Close contact with case is indication for prophylaxis

Purified polysaccharide meningococcal vaccines were first developed at Walter Reed Army Institute of Research driven by recruit camp outbreaks caused by sulfonamide-resistant strains. They were shown to stimulate group-specific antibody and to prevent disease in military and adult civilian populations. A vaccine containing A, C, Y, and W-135 polysaccharides was licensed in the United States, but proved poorly immunogenic for infants, the largest group at risk. We now know polysaccharide vaccines only stimulate T-cell–independent responses that develop later in life. As with pneumococcal and *Haemophilus influenzae* polysaccharide vaccines, this problem was overcome by conjugating the polysaccharide to a protein carrier (diphtheria toxoid). This Meningococcal Conjugate Vaccine Quadravalent (MCV4) stimulates T-cell–dependent responses, which are both stronger and present at an earlier age. Initially recommended for 11- to 18-year-olds, the age for receiving vaccine was pushed down to 2 years in late 2007 and will surely go lower with increased experience. :::T-cell dependent and independent responses, p. 37

MCV4 vaccine stimulates T-cell dependent immunity

The protein conjugate approach faces a unique difficulty with the meningococcus—the failure of the group B polysaccharide to be immunogenic at all. If this is due to its similarity to a human neural cell adhesion molecule, as suspected, it may not be overcome simply by

Nonimmunogenic serogroup B polysaccharide remains a problem

Proteins are vaccine candidates

protein conjugation. Group B causes up to one third of all disease, so no vaccine that omits it is likely to be completely successful. For this reason, other approaches such as the use of OMPs (eg, PorA) or serum factor H binding proteins are being pursued. Genetically engineered vaccines based on the sequence of the entire group B meningococcal genome hold the promise of defining proteins that would immunize against all serogroups of *N meningitidis.* ::: Factor H binding, p. 394

NEISSERIA GONORRHOEAE

BACTERIOLOGY

Chocolate agar and CO_2 are required

Fresh isolates have pili

LPS, LOS, and OMPs are in outer membrane

Opa proteins are adherence OMPs

N gonorrhoeae grows well only on chocolate agar and on specialized medium enriched to ensure its growth. It requires carbon dioxide supplementation. Small, smooth, nonpigmented colonies appear after 18 to 24 hours and are well developed (2 to 4 mm) after 48 hours. Gonococci possess numerous pili that extend through and beyond the outer membrane (**Figure 30–6**), which are structurally similar to those of meningococci (Table 30–1).

The gonococcal outer membrane is composed of phospholipids, LPS, LOS, and several distinct OMPs. The OMPs include porins (PorA and PorB) and adherence proteins known as Opa. Opa proteins are a set of at least 12 proteins that get their name from the opaque appearance they give to colonies as a result of adhesion between gonococcal cells. A variable number of the Opa proteins may be expressed at any one time.

ANTIGENIC VARIATION

Pili, OMPs, and LOS vary in gonococci and meningococci

N gonorrhoeae and *N meningitidis* are among several microorganisms whose surface structures are known to change antigenically from generation to generation during growth of a single strain. The mechanisms involved have been more extensively studied in gonococci but appear to be similar in both species. The major structure know to undergo antigenic variation are pili, Opa proteins, and LOS.

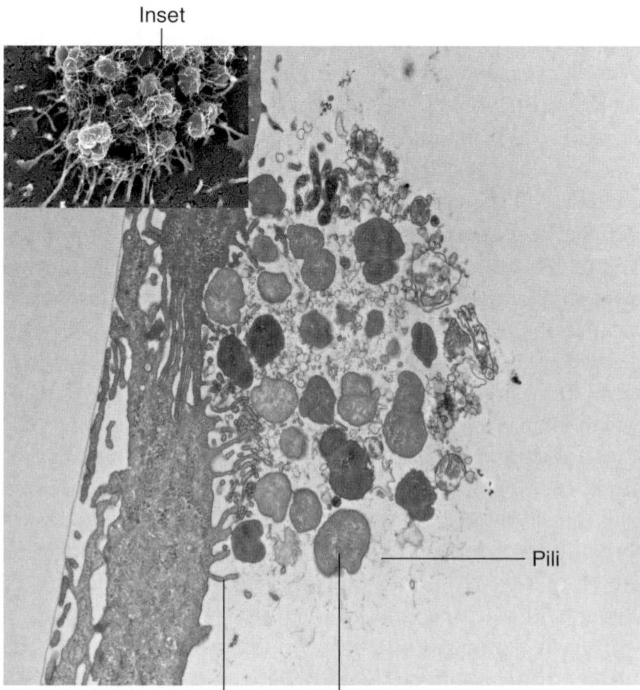

FIGURE 30–6. *Neisseria gonorrhoeae* pili. This view is a cross-section of the microcolony of gonococci on the surface of an epithelial cell orginally shown in Figure 22–2 (inset). Pili are actively attaching to the epithelial cell surface and using a contractile force (twitching motility) to move and modify the surface. (Photomicrographs kindly provided by Dustin L. Higashi and Magdalene So.)

Gonococcal pili are antigenically variable to an extraordinary extent. There are multiple genetic mechanisms, but the most important one appears to be recombinational exchange between the multiple pilin genes present in the chromosome of every strain. Some of these genes are complete and able to express pilin (*pilE*). Others are not due to lack of an effective promoter and are thus silent (*pilS*). When recombination between expression and silent loci results in the donation of new sequences to an expression locus, the result can be expression of a pilin with changes in its amino acid composition and thus its antigenicity. The recombination could also involve exogenous DNA from another cell or strain, because gonococci naturally take up species-specific DNA by transformation. The process is complex, involving other genes that play a role in the assembly of pili and their functional characteristics, such as cellular adhesion. The numerous possible outcomes include no pilin subunits, pilin subunits unable to assemble, mature pili with altered functional characteristics, and fully functional pili with a new antigenic makeup. ::: Pilin genetics, p 395

Genes for pilin subunits may be expressed or silent

Recombination between multiple genes occurs

Outcome may be nonfunctional or antigenically altered pili

The multiple gonococcal Opa proteins are each encoded by separate genes scattered around the genome. Various combinations of these genes may be either "on" or "off" at any one time. The switch is set during the transcription of each Opa gene for the next cell generation. As a result of a process called replicative slippage, the number of repeats of particular gene sequence can vary. When it comes time for translation, the number of repeats determines whether the gene will be in or out of frame to translate its Opa protein. If it is in frame, the gene is "on"; if not, the switch is "off." Variation in gonococcal LOS has been observed in volunteer subjects challenged with intraurethral *N gonorrhoeae,* but the genetic mechanism is unknown. ::: Opa genetics, p. 395

Multiple Opa genes may be "on" or "off"

Translational frame shift controls the switch

LOS also varies antigenically

These changes in the gonococcal surface are random events which may or may not have survival value depending on the circumstances. During the early stages of infection there could be positive selection for the expression of pili and Opas that mediate adherence. If the host has antibodies against one or more of these proteins they would be removed and the infecting population would shift to cells expressing pili or Opas to which there is no immunologic experience. An example of how these antigenic variants could be selected is shown in Figure 30–7. Taken together, these multifactorial, antigenic variations of the gonococcal surface may serve the dual purposes of escape from immune surveillance and timely provision of the ligands required to bind to human cell receptors.

GONORRHEA

CLINICAL CAPSULE

In contrast to meningococcal disease, gonorrhea is primarily localized to mucosal surfaces with relatively infrequent spread to the bloodstream or deep tissues. Infection is sexually acquired by direct genital contact, and the primary manifestation is pain and purulent discharge at the infected site. In men, this is typically the urethra, and in women, the uterine cervix. Direct extension of the infection up the fallopian tubes produces fever and lower abdominal pain, a syndrome called pelvic inflammatory disease (PID). For women, sterility or ectopic pregnancy can be long-term consequences of gonorrhea.

EPIDEMIOLOGY

Gonorrhea is one of our greatest public health problems. The more than 300,000 cases reported in the United States each year are felt to represent less than 50% of the true number and the rates for adolescents are alarmingly high and increasing by 10% a year. The highest rates are in women between the ages of 15 and 19 years and in men between the ages of 20 and 24 years. No truly effective means of control is yet in sight. Our ability to stem the

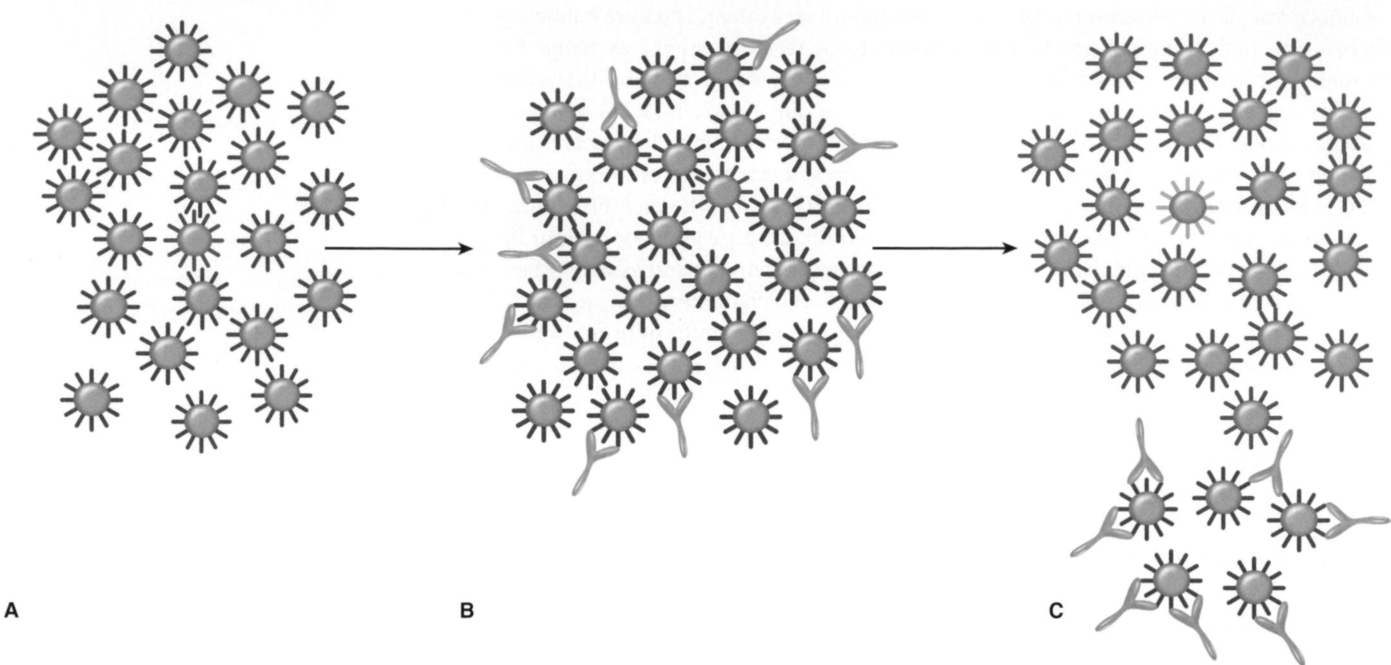

FIGURE 30–7. Gonococcal antigenic variation. A. A population of gonococci is shown with surface pili. There are two antigenic types of pili in the population, one of which is dominant. **B.** IgG against the dominant blue pilin type is introduced and binds all the cells with that pilin type on the surface. **C.** Later, the bound gonococci with their pili are clumped at the bottom. The minor (red) pilin type present in A now predominates, and a new one (green) has appeared but is still a minority member of the population. Antibody directed against the now dominant member would allow the new green one to take over. The same kind of population change occurs based on antigenic variation of outer membrane proteins. The genetic mechanisms involved in generating multiple antigenic types are illustrated in Figure 22–5.

Rates among adolescents are high and increasing

Inability to detect asymptomatic cases hampers control

tide of changed sexual mores continues to be hampered by lack of an effective means to detect asymptomatic cases, resistance of *N gonorrhoeae* to antibiotics (see Treatment), and, to some extent, lack of appreciation of the importance of this disease. The latter is evidenced by failure of patients to seek medical care and reluctance to report cases to public health authorities due to privacy concerns. In the minds of too many, syphilis is dreaded and "unclean," whereas gonorrhea is only "the clap" ("clap" is from the archaic French *clapoir*, "a rabbit warren"; later, "a brothel").

Risk of sexual contact is up to 50%

Asymptomatic cases are highest in women

Nonsexual transmission is rare

Gonorrhea is acquired by genital contact with an infected person. The major reservoir for continued spread is the asymptomatic patient. Screening programs and case contact studies have shown that almost 50% of infected women are asymptomatic or at least do not have symptoms usually associated with venereal infection. Most men (95%) have acute symptoms with infection. Many who are not treated become asymptomatic but remain infectious. Asymptomatic male and female patients can remain infectious for months. The attack rates for those engaging in sexual intercourse with an infected person are estimated to be 20% to 50%. The organism may also be transmitted by oral–genital contact or by rectal intercourse. When all these factors operate in a sexually active population, it is easy to explain the high prevalence of gonorrhea. Although gonococci can survive for brief periods on toilet seats, nonsexual transmission is extremely rare. Virtually all gonococci isolated from children can be traced to sexual abuse by an infected adult.

PATHOGENESIS

Attachment and Invasion

Gonococci are not normal inhabitants of the respiratory or genital flora. When introduced onto a mucosal surface by sexual contact with an infected individual, adherence ligands such as pili and Opa proteins allow initial attachment of the bacteria to receptors (CD46, CD66, integrins) on nonciliated epithelial cells (Figure 30–2). Initial attachment is mediated by the pili, which have been shown to generate an active force with movement in microcolonies across the cell surface (Figure 30–6). This is followed by a tighter attachment owing to Opa proteins. This close binding provides an opportunity for other OMPs (PorA) to trigger signaling cascades activating multiple enzymatic systems within the host cell. These reactions lead to induction of phagocytosis of the gonococci in a process involving microfilaments and microtubules of the invaded cell. The microvilli surround the bacteria and appear to draw them into the host cell in the same manner as meningococci. Thus, after initial attachment the gonococcus appears to induce the host cell to actively take it inside. (Figure 30–2). Once inside, the bacteria transcytose the cell and exit through the basal membrane to enter the submucosa. :::Attachment, p. 391

Pili and Opa proteins mediate attachment to nonciliated epithelium

Gonococci induce their own phagocytosis

Bacteria quickly pass to submucosa

Survival in the Submucosa

Once in the submucosa, the bacteria must survive and resist innate host defenses as well as defenses that may have been acquired from previous infection. As with meningococci, receptors on the gonococcal surface enable the organisms to scavenge iron needed for growth from the human iron transport proteins transferrin and lactoferrin. Although gonococci lack the polysaccharide capsule of the meningococcus, they still have multiple mechanisms that protect them against serum complement and antibody. One of these is LOS sialyation in which the gonococcus is able to incorporate host sialic acid onto its own surface. This provides a mechanism for blocking surface C3b deposition that is the same as that of the encapsulated bacteria. In a sense, the gonococci create their own "capsule" during infection by binding host sialic acid to their LOS. ::: Capsules, p. 394

Receptors scavenge iron

Sialated LOS acts like a capsule

Even when phagocytes do encounter gonococci, surface factors such as pili and Opa proteins interfere with effective phagocytosis. The organisms are also able to defend against oxidative killing inside the phagocyte by up-regulation of catalase production and an efficient antioxidant defense system. Taken together, these factors provide ample evidence that killing by neutrophils is sufficiently retarded to allow prolonged survival of gonococci in mucosal and submucosal locations.

Phagocytosed gonococci resist killing

Spread and Dissemination

In contrast to meningococci, *N gonorrhoeae* bacteria tend to remain localized to genital structures, causing inflammation and local injury, which no doubt facilitate their continued venereal transmission. Purulent exudates containing "sticky" clusters of gonococci held together by Opa proteins could be the primary infectious unit. Infection may spread to deeper structures by progressive extension to adjacent mucosal and glandular epithelial cells. These include the prostate and epididymis in men and the paracervical glands and fallopian tubes in women (Figure 30–8). Spread to the fallopian tubes is facilitated by pilus-mediated attachment to sperm and then to the microvilli of nonciliated fallopian tube cells. Injury to the fallopian epithelium is mediated by the local effect of cell wall LPS/LOS. Gonococci are also known to turn over their peptidoglycan rapidly during growth, releasing peptidoglycan fragments which are also toxic to the ciliated epithelium of the fallopian tube.

Local spread is to epididymis and fallopian tubes

LPS/LOS and peptidoglycan shedding cause local injury

In a small proportion of infections, organisms reach the bloodstream to produce disseminated gonococcal infection (DGI). When this happens, the systemic findings have their own pattern (see Manifestations) and seldom take on the endotoxic shock picture of meningococcemia. Although differences have been noted between *N gonorrhoeae* strains that remain localized and those that produce DGI, their connection to pathogenesis is unknown. Both DGI and salpingitis tend to begin during or shortly after completion of menses. This may relate to changes in the cervical mucus and reflux into the fallopian tubes during menses.

DGI differs from meningococcal endotoxic shock

Reflux during menses may facilitate spread

Genetic Regulation of Virulence

Through all the stages of gonorrhea, gonococci are able to use a particularly rich variety of genetic mechanisms in deployment of the virulence factors previously described at the right time. Some are regulatory responses to environmental cues, such as iron in relation to iron-binding proteins, whereas others involve changes in the genome. Antigenic changes in both

Regulation, recombination, and translational changes deploy virulence factors

pili and Opa proteins have been demonstrated in human infection, including the isolation of antigenic variants from different sites in the same patient. These presumably take place by the recombinational and translational mechanisms described above (see Antigenic Variation) as the organisms replicate in the patient.

IMMUNITY

Antibody response is weak

Gonococcus varies multiple structures to avoid immune surveillance

The apparent lack of immunity to gonococcal infection has long been a mystery. Among sexually active persons with multiple partners, repeated infections are the rule rather than the exception. Both serum and secretory antibodies are generated during natural infection, but the levels are generally low, even after repeated infections. Another aspect is that even when antibodies are formed, antigenic variation defeats their effectiveness and allows the gonococcus to escape immune surveillance. Antigenic variation of pili, Opa proteins, and LOS is particularly likely to be important. Outbreaks have been traced to a single strain that demonstrated multiple pilin variations and Opa types in repeated isolates from the same individual or from sexual partners. In experimental models, passive administration of antibody directed against one pilin type has been followed by emergence of new pilin variants presumably through the sequence illustrated in Figure 30–7. It appears that although some immunity to gonococcal infection is present, its effectiveness is compromised by the ability of the organism to change key structures during the course of infection.

GONORRHEA: CLINICAL ASPECTS

MANIFESTATIONS

Genital Gonorrhea

Urethritis and endocervicitis are primary infections

The clinical spectrum of gonorrhea differs substantially in men and women (**Figure 30–8**). In men, the primary site of infection is the urethra. Symptoms begin 2 to 7 days after infection and consist primarily of purulent urethral discharge and dysuria. Although uncommon, local extension can lead to epididymitis or prostatitis. The endocervix is the primary site in women, in whom symptoms include increased vaginal discharge, urinary frequency, dysuria, abdominal pain, and menstrual abnormalities. As mentioned previously, symptoms may be mild or absent in either sex, particularly women.

Other Local Infections

Rectal and pharyngeal infections relate to sexual practices

Rectal gonorrhea occurs after rectal intercourse or, in women, after contamination with infected vaginal secretions. This condition is generally asymptomatic, but may cause tenesmus, discharge, and rectal bleeding. Pharyngeal gonorrhea is transmitted by oral–genital sex and, again, is usually asymptomatic. Sore throat and cervical adenitis may occur. Infection of other structures near primary infection sites, such as Bartholin's glands in women, may lead to abscess formation.

Transmission at birth causes ophthalmia neonatorum

Inoculation of gonococci into the conjunctiva produces a severe, acute, purulent conjunctivitis. Although this infection may occur at any age, the most serious form is gonococcal ophthalmia neonatorum, a disease acquired by a newborn from an infected mother. The disease was formerly a common cause of blindness, which is now prevented by the use of prophylactic topical eye drops or ointment (silver nitrate, erythromycin, or tetracycline) at birth.

Pelvic Inflammatory Disease (PID)

Salpingitis and pelvic peritonitis cause scaring and infertility

The clinical syndrome of PID develops in 10% to 20% of women with gonorrhea. The findings include fever, lower abdominal pain (usually bilateral), adnexal tenderness, and leukocytosis with or without signs of local infection. These features are caused by spread of organisms along the fallopian tubes to produce salpingitis and into the pelvic cavity to produce pelvic peritonitis and abscesses (**Figure 30–9**). PID is also known to develop when other genital pathogens ascend by the same route. These organisms include anaerobes and *Chlamydia trachomatis,* which may appear alone or mixed with gonococci. The most

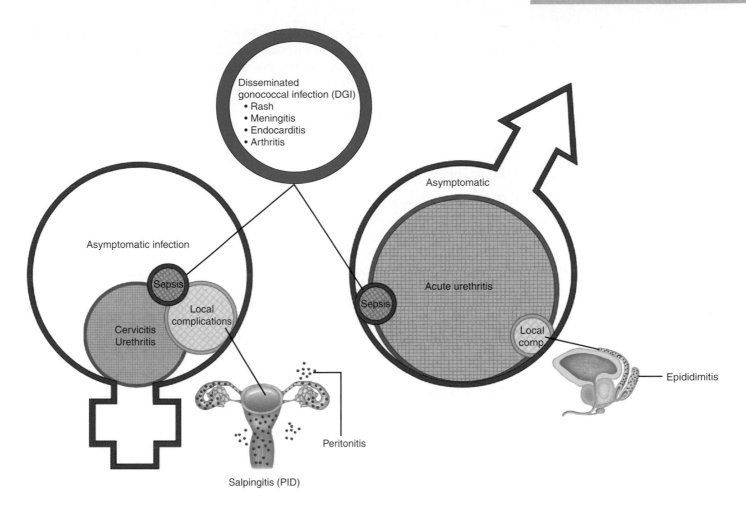

FIGURE 30–8. Gonorrhea in men and women. The majority of cases in women are asymptomatic. Local extension up the fallopian tubes causes salpingitis. The majority of men have acute urethritis, and only a small percentage have local extension the epididimus. A very small part of either spectrum results in bacteremia and disseminated gonococcal infection.

serious complications of PID are infertility and ectopic pregnancy secondary to scarring of the fallopian tubes.

Disseminated Gonococcal Infection (DGI)

Any of the local forms of gonorrhea or their extensions such as PID may lead to bacteremia. In the bacteremic phase, the primary features are fever; migratory polyarthralgia; and a

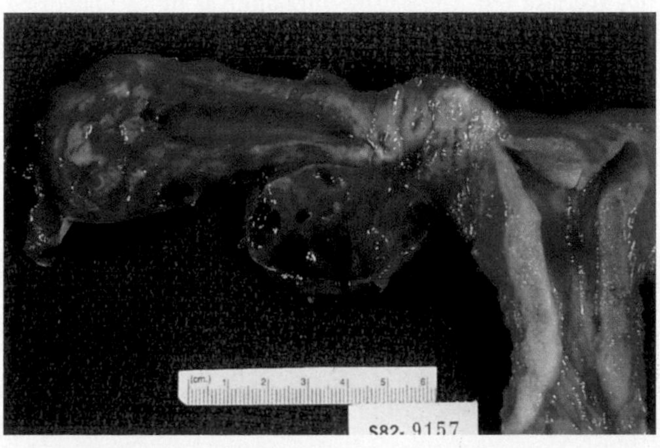

FIGURE 30–9. Tubo-ovarian abscess. This large abscess in the fallopian tube is part of the spectrum of pelvic inflammatory disease (PID) of which *Neisseria gonorrhoeae* is a major cause. (Reproduced with permission from Connor DH, Chandler FW, Schwartz DQA, Manz HJ, Lack EE (eds). *Pathology of Infectious Diseases*, vol. 1. Stamford, CT: Appleton & Lange; 1997.)

Skin rash, arthralgia, and arthritis are associated with bacteremia

Purulent arthritis involves large joints

petechial, maculopapular, or pustular rash. Some of these features may be immunologically mediated; gonococci are infrequently isolated from the skin or joints at this stage despite their presence in the blood. The bacteremia may lead to metastatic infections such as endocarditis and meningitis, but the most common is purulent arthritis. The arthritis typically follows the bacteremia and involves large joints such as elbows and knees. Gonococci are readily cultured from the pus.

DIAGNOSIS

Gram Smear

Direct smear is useful in men

Interfering flora complicates interpretation in women

The presence of multiple pairs of bean-shaped, Gram-negative diplococci within a neutrophil is highly characteristic of gonorrhea when the smear is from a genital site (Figure 30–1). The direct Gram smear is more than 95% sensitive and specific in symptomatic men. Unfortunately, it is only 50% to 70% sensitive in women, and its specificity is complicated by the presence of other bacteria in the female genital flora that may have a similar morphology. Experience is required in reading smears, particularly in women. A positive Gram smear is generally accepted as diagnostic in men. It should not be used as the sole source for diagnosis in women or when the findings have social (divorce) or legal (rape, child abuse) implications.

Culture

Urethra and cervix are preferred culture sites

Attention to detail is necessary for isolation of the gonococcus because it is a fragile organism that is often mixed with hardier members of the normal flora. Success requires proper selection of culture sites, protection of specimens from environmental exposure, culture on appropriate media, and definitive laboratory identification. In men, the best specimen is urethral exudate or urethral scrapings (obtained with a loop or special swab). In women, cervical swabs are preferred over urethral or vaginal specimens. The highest diagnostic yield in women is with the combination of a cervical and an anal canal culture; this is because some patients with rectal gonorrhea have negative cervical cultures. Throat or rectal cultures in men are needed only when indicated by sexual practices.

Transport media required unless plating is immediate

Selective medium inhibits competing flora

Swabs may be streaked directly onto culture medium or promptly transmitted (in less than 4 hours) to the laboratory in a suitable transport medium. Laboratory requests must specify the suspicion of gonorrhea so that media that satisfy the nutritional requirements of the gonococcus and inhibit competing normal flora can be seeded. The selective medium (eg, Martin–Lewis agar) is an enriched selective chocolate agar with antibiotics. The exact formulation has changed over the years, but includes agents active against Gram-positive bacteria (vancomycin), Gram-negative bacteria (colistin, trimethoprim), and fungi (nystatin, anisomycin) at concentrations that do not inhibit *N gonorrhoeae*.

Isolates are identified by fermentation or immunoassay

Colonies appear after 1 to 2 days of incubation in carbon dioxide at 35°C. They may be identified as *Neisseria* by demonstration of typical Gram stain morphology and a positive oxidase test. Classically, speciation is by carbohydrate degradation pattern, but this approach has been replaced by immunologic procedures (immunofluorescence, coagglutination, enzyme immunoassay) using monoclonal antibodies to unique antigens. *Neisseria* species other than *N gonorrhoeae* are unusual in genital specimens, but speciation is the only way to be certain of the diagnosis.

Direct Detection

DNA amplification methods are combined with *Chlamydia* detection

Much effort has been directed at developing immunoassay and nucleic acid hybridization methods that detect gonococci in genital and urine specimens without culture. Such methods have particular importance for screening populations in which culture is impractical. Of these, only the nucleic acid amplification methods have the sensitivity to substitute for culture and are now widely used in public health laboratories. The cost/benefit ratio of these tests has been improved by combining them with *Chlamydia* detection (Chapter 39), which targets the same clinical population. ::: Nucleic acid amplification, p. 80-81

Serology

Attempts to develop a serologic test for gonorrhea have not yet achieved the needed sensitivity and specificity. A test that would detect the disease in asymptomatic patients would be very useful in control of this disease.

No serologic test

TREATMENT

The treatment of gonorrhea, as with other sexually transmitted diseases, includes individual patient issues as well as public health concerns. Patients who do not complete a course of treatment once they begin to feel better present a risk of continued transmission and selection of resistant strains. For this reason, definitive treatment at the time of the initial visit has been the favored approach. For decades, this was easily accomplished with a single intramuscular injection of penicillin G.

Compliance dictates treatment on first encounter

Penicillin, which once was active against all know gonococci at extremely low concentrations (less than 0.1 μg/mL), is no longer used. This is due to the development of multiple mechanisms of resistance. Mutations that altered the affinity of penicillin for its transpeptidase (or penicillin-binding protein [PBP]) target were the first to be recognized. Other mutations in porins—either restricting penicillin transport into the cell or efflux systems pumping it out—have also been discovered either alone or in combination with the transpeptidase alterations. Over decades, a subpopulation of gonococci slowly emerged that required ever higher minimum inhibitory concentrations (up to 8.0 μg/mL). For a time this was managed by increasing the penicillin dose, which for the single injection treatment approached the maximum volume that could be humanely administered (even injecting both buttocks). Finally, the most powerful resistance mechanism, penicillinase production, appeared during the Vietnam War and spread throughout the world. These strains produce a plasmid-encoded β-lactamase identical with that of members of Enterobacteriaceae and are resistant at a level that far exceeds achievable therapeutic levels. This was the end for penicillin and gonorrhea. ::: PBP alteration, p. 421; β-lactamase, p. 420

PBP alterations cause incremental resistance

β-Lactamase–producing strains are highly resistant

This situation has caused a shift in treatment of genital gonorrhea to third-generation cephalosporins because of their resistance to the β-lactamases prevalent in gonococci. The recommended agents have high enough activity to still be used as single-dose treatment either intramuscularly (ceftriaxone) or orally (cefixime). Other agents recommended for primary treatment include fluoroquinolones (ciprofloxacin or ofloxacin) and azithromycin. Doxycycline is also effective but must be given orally for 7 days. Doxycycline and azithromycin have the additional advantage of also being effective against *Chlamydia*, which may also be present in up to one third of gonorrhea cases. Resistance to quinolones is frequent enough to limit their use in some parts of the world. Azithromycin resistance is just beginning to be reported.

Ceftriaxone, quinolones, and azithromycin are recommended therapy

Quinolone and azithromycin resistance is still uncommon

PREVENTION

Condoms provide a high degree of protection against both infection with *N gonorrhoeae* and transmission to a sexual partner. Spermicides and other vaginal foams and douches are not reliable protection. The classic public health methods of case–contact tracing and treatment are important, but difficult because of the size of the infected population. The availability of a good serologic test would greatly aid control, as it has for syphilis. The development of a vaccine is a high but distant goal. Achieving it awaits further understanding of immunity and its relationship to the shifting target provided by the gonococcus.

Condoms block transmission

Vaccine strategies await better understanding of immunity

CLIINICAL CASE

RECRUIT WITH FEVER, BACKACHE, AND RASH

A 20-year-old man presented to the emergency room because of fever and backache. A basic trainee on leave from a naval training station, he was perfectly well until the day of admission when he awakened with fever, malaise, and lumbar backache, all of which gradually worsened over the ensuing 6 hours.

Examination revealed an acutely ill man with blood pressure of 105/65 mm Hg, pulse rate 120/min, and temperature 104°F. A few small petechiae were on the volar surfaces of each forearm. The muscles of the back, arms, and legs were tender to palpation. The remainder of the examination was normal. A lumbar puncture showed 1500 white blood cells/mL, 95% of which were PMNs. CSF cultures were obtained.

QUESTIONS

■ Which factor would most influence the likely etiologic agents?

A. Height of fever
B. Number of PMNs in CSF
C. Immunization status
D. Extent of petechiae
E. Prior antibiotics

■ What is the primary cause of the patient's petechiae?

A. Superantigen production
B. Pore-forming toxin
C. LPS endotoxin
D. Pili
E. OMPs

■ In addition to culture of the CSF, culture of what other site would be most valuable?

A. Throat
B. Sputum
C. Petechiae
D. Blood

■ If the CSF cultures are positive for *N meningitidis*, is any preventive action appropriate for the man's contacts?

A. Conjugate vaccine for family
B. Conjugate vaccine for healthcare workers
C. Chemoprophylaxis for family
D. Chemoprophylaxis for healthcare workers
E. No action required

ANSWERS

1(C), **2**(C), **3**(D), **4**(C)

CHAPTER 31

Haemophilus and *Bordetella*

Whooping cough, — why, he nearly whooped himself to death.

—R. N. Carey: *Uncle Max*

Haemophilus and *Bordetella* are small, Gram-negative rods that tend to assume a coccobacillary shape. Members of both genera contain species exclusively found in humans that cause respiratory tract infections. The major species are *Haemophilus influenzae,* a major cause of purulent meningitis and *Bordetella pertussis,* the cause of whooping cough.

HAEMOPHILUS

Haemophilus are among the smallest of bacteria. The curved ends of the short (1.0 to 1.5 μm) bacilli make many appear nearly round; hence the term coccobacilli (**Figure 31–1**). The cell wall has a structure similar to that of other Gram-negative bacteria. The most virulent strains of *H influenzae* have a polysaccharide capsule, but other species of *Haemophilus* are not encapsulated.

Tiny Gram-negative coccobacilli

The cultivation of *Haemophilus* species requires the use of culture media enriched with blood or blood products (Greek *haema,* blood, and *philos,* loving) for optimal growth. This requirement is attributable to the need for exogenous hematin and/or nicotinamide adenine dinucleotide (NAD). These growth factors, also termed X factor (hematin) and V factor (NAD), are both present in erythrocytes. In culture media, optimal concentrations are not available unless the red blood cells are lysed by gentle heat (chocolate agar) or added separately as a supplement. Although erythrocytes are the only convenient source of hematin, sufficient amounts of NAD may be provided by some other bacteria and yeasts. This is responsible for the "satellite phenomenon," in which colonies of *Haemophilus* grow on blood agar only in the vicinity of a colony of *Staphylococcus aureus.* The several species of Haemophilus are defined by their requirement for hematin and/or NAD, CO2 dependence, and other cultural characteristics (**Table 31–1**). Species of Haemophilus other than *H influenzae* have the same biology described below for the nonencapsulated strains of *H influenzae.*

Require hematin and/or NAD

Staphylococcus aureus may provide NAD

Species other than *H influenzae* are similar

Haemophilus Influenzae

BACTERIOLOGY

Haemophilus that meet the species requirements for *H influenzae* may or may not have a capsule. Those that do are divided into six serotypes (a to f) based on the capsular polysaccharide antigen. Type b capsule is made up of a polymer of ribose, ribitol, and phosphate,

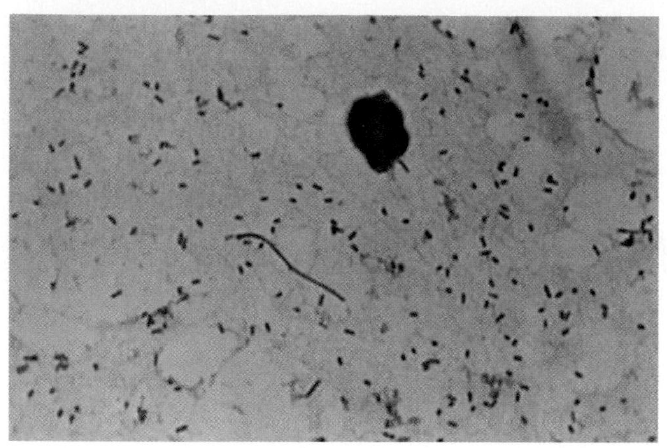

FIGURE 31–1. *Haemophilus influenzae* Gram stain. The Gram-negative bacilli are small and so short that some appear almost round. This is the basis of the term coccobacilli. The morphology of *Bordetella pertussis* is the same. (Reproduced with permission from Connor DH, Chandler FW, Schwartz DQA, Manz HJ, Lack EE (eds). *Pathology of Infectious Diseases*, vol. 1. Stamford, CT: Appleton & Lange; 1997.)

Six serotypes are based on capsular polysaccharide

Hib capsule is PRP

called **polyribitol phosphate (PRP).** These surface polysaccharides are strongly associated with virulence, particularly *H influenzae* type b (**Hib**). The surface of *H influenzae* includes pili and an outer membrane similar to the structure of other Gram-negative bacteria. The outer membrane includes proteins (HMW1, HMW2) and lipooligosaccharides (LOS). The nonencapsulated, and thus nontypable, *H influenzae* can be classified by a number of typing schemes based on outer membrane proteins and other factors, but have no known association with virulence. *H influenzae* produces no known exotoxins.

TABLE 31–1 Features of *Haemophilus* and *Bordetella*

SPECIES	TYPE	GROWTH REQUIREMENT	CAPSULE	ADHERENCE FACTORS	TOXINS	EPIDEMIOLOGY	DISEASE
Haemophilus							
H influenzae	a–f	Hematin and NAD	Polysaccharide	Pili, HMW	—	Normal flora, respiratory droplet spread	Meningitis, epiglottitis, arthritis, sepsis, otitis media
H influenzae	—	Hematin and NAD	—	Pili, HMW	—	Normal flora, respiratory droplet spread	Otitis media, bronchitis, sinusitis
H ducreyi	—	Hematin	—	Pili	Cytolethal distending toxin	Sexual contact	Chancroid
Other species[a]	—	Hematin or NAD	—	—	—	Normal flora	Bronchitis
Bordetella							
B pertussis	—	Nicotinamide[b]	—	Pili, FHA, PT, pertactin	PT, AC, TCT	Strict pathogen, respiratory droplet spread	Whooping cough
B bronchiseptica	—	Nicotinamide	—	Pili, FHA	PT[c], AC, TCT	Dogs, rabbits	Rhinitis, cough

HMW, high-molecular-weight proteins (HMW1, HMW2); FHA, filamentous hemagglutinin; PT, pertussis toxin; AC, adenylate cyclase; TCT, tracheal cytotoxin

[a]*H parainfluenzae, H aphrophilus, H hemolyticus.*

[b]Also requires additives such as charcoal to neutralize toxicity in standard media.

[c]The PT gene is present but expression of the protein is variable.

H INFLUENZAE DISEASE

CLINICAL CAPSULE

Hib produces acute, life-threatening infections of the central nervous system, epiglottis, and soft tissues, primarily in children. Disease begins with fever and lethargy, and in the case of acute meningitis, can progress to coma and death in less than 1 day. In affluent countries, Hib disease has been controlled by immunization. *H influenzae* also produces common, but less fulminant infections of the bronchi, respiratory sinuses, and middle ear. The latter are usually associated with nonencapsulated strains.

EPIDEMIOLOGY

H influenzae is a strictly human pathogen and has no known animal or environmental sources. It can be found in the normal nasopharyngeal flora of 20% to 80% of healthy persons, depending on age, season, and other factors. Most of these are nonencapsulated, but capsulated strains, including Hib, are not rare. Spread is by respiratory droplets, as with streptococci. Before the introduction of effective vaccines, approximately 1 in every 200 children developed invasive disease by the age of 5 years. Meningitis is the most common form and most often attacks those under 2 years of age. Cases of epiglottitis and pneumonia tend to peak in the 2- to 5-year age group. Over 90% of these cases are due to a single serotype, Hib.

Nasopharyngeal colonization is common

Meningitis develops in children under 2 years of age

The introduction of universal immunization with the Hib protein conjugate vaccine (see Prevention) has reduced invasive disease rates by 99%. Most of the cases in immunized populations are caused by serotypes other than b and appear in people at all ages. Unfortunately, Hib disease continues as before in countries and populations unable to afford the vaccine.

Immunization (where implemented) has dramatically reduced disease

At one time *H influenzae*–caused meningitis was believed to be an isolated endogenous infection, but reports of outbreaks in closed populations and careful epidemiologic studies of secondary spread in families have changed this view. The risk of serious infection for unimmunized children under 4 years of age living with an index case is more than 500-fold that for nonexposed children. This risk indicates a need for protection of susceptible contacts (see Prevention).

Person-to person spread requires prophylaxis

PATHOGENESIS

Invasive Disease

For unknown reasons, *H influenzae* strains commonly found in the normal flora of the nasopharynx occasionally invade deeper tissues. Bacteremia then leads to spread to the central nervous system and metastatic infections at distant sites such as bones and joints (**Figure 31–2**). These events seem to take place within a short period (less than 3 days) after an encounter with a new virulent strain. Systemic spread is typical only for capsulated *H influenzae* strains, and over 90% of invasive strains are type b. Even among Hib strains there are distinct clones, which account for about 80% of all invasive disease worldwide.

Only capsulated strains are invasive

Certain clones account for most disease

Attachment to respiratory epithelial cells is mediated by pili and outer membrane proteins. Some evidence suggests that this is a complex regulatory cascade, coordinating capsular biosynthesis and adherence factors that act cooperatively in establishing the microbe within susceptible hosts. *H influenzae* can be seen to invade between the cells of the respiratory epithelium (**Figure 31–3**), and for a time resides between and below them. Once past the mucosal barrier, the antiphagocytic capsule confers resistance to C3b deposition in the same manner as it does with other encapsulated bacteria. As with the pathogenic *Neisseria*, there is evidence that *H influenzae* LOS may provide an antiphagocytic effect by binding host components like sialic

Pili and other adhesins bind to epithelial cells

Invasion goes between cells

Capsule prevents phagocytosis

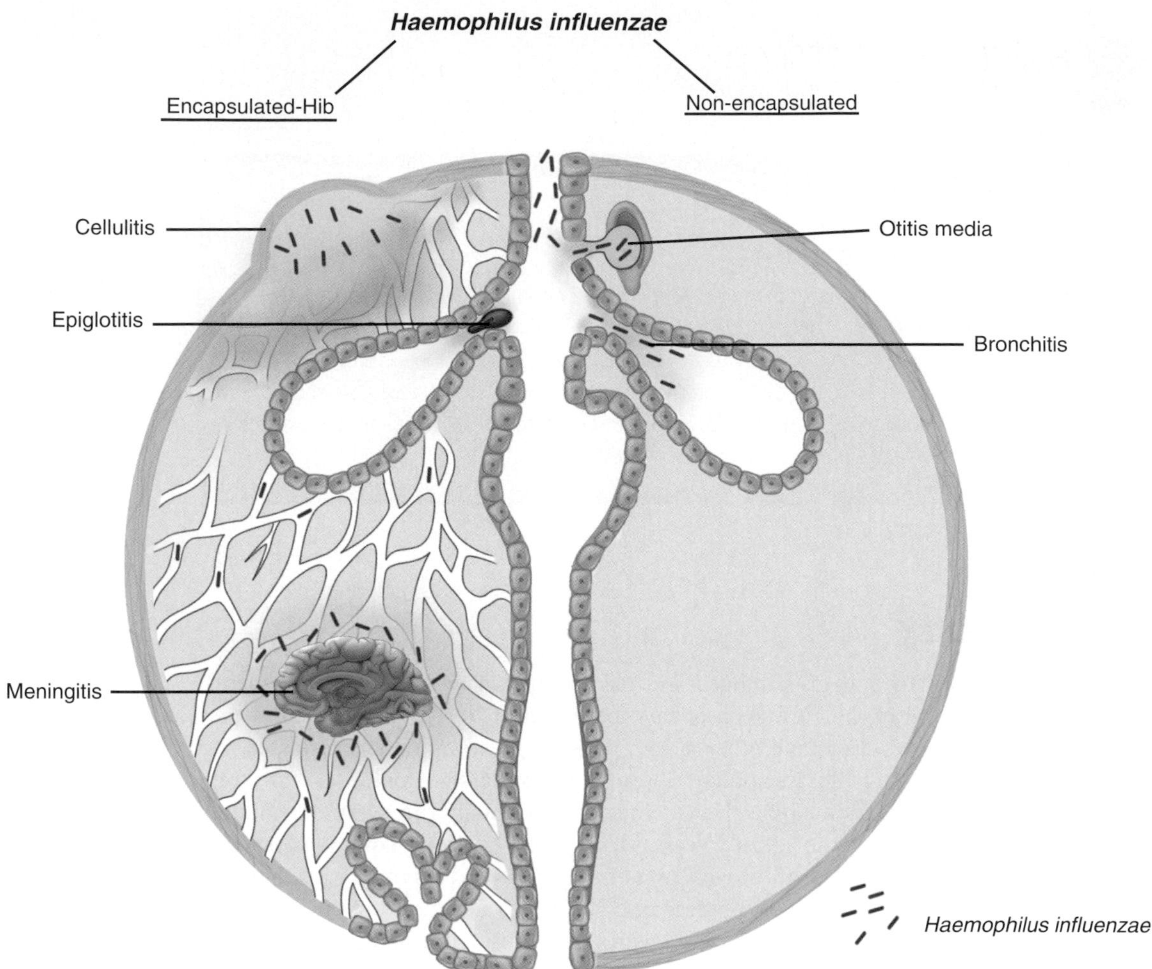

FIGURE 31–2. Haemophilus disease overview. *(Left)* Invasive disease is caused by encapsulated strains, mostly type b (Hib). From a naspharyngeal colonization site, the organisms invade locally to produce cellulitis or epiglottitis. Invasion of the blood occurs in all Hib forms and most frequently leads to meningitis. *(Right)* Localized disease is produced when nonencapsulated strains from the nasopharynx are trapped in the middle ear paranasal sinuses or compromised bronchi.

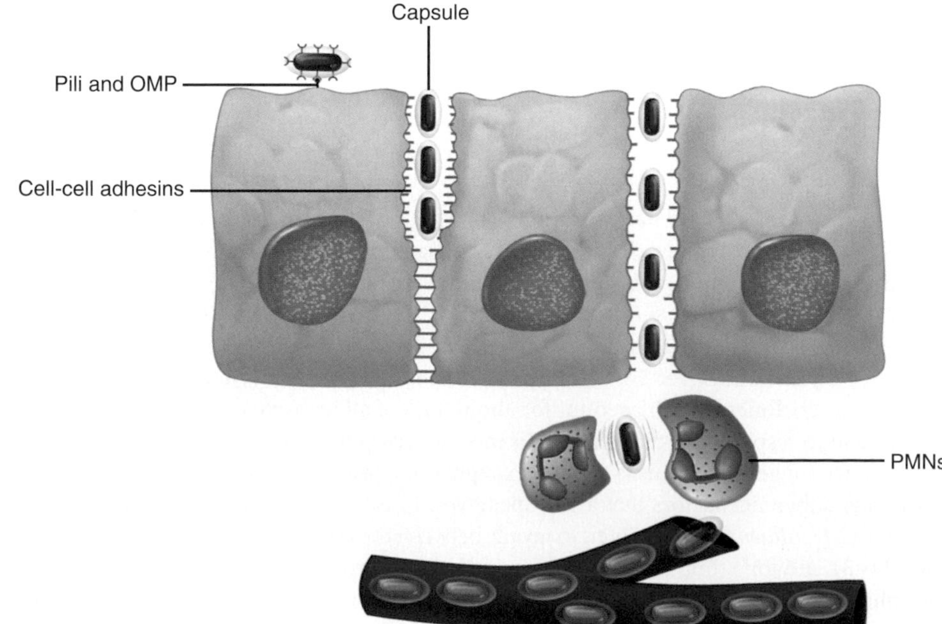

FIGURE 31–3. *Haemophilus influenzae* disease, cellular view. Organisms attach to epithelial cells using pili and outer membrane proteins (OMP). Invasion takes place between cells by disruption of cell-cell adhesion molecules. In the submucosa, the capsule allows the bacteria to evade phagocytosis and enter the bloodstream. PMNs, polymorphonuclear neutrophils.

acid. Lipopolysaccharide endotoxin in the cell wall is toxic to ciliated respiratory cells, and when circulating in the bloodstream produces all the features of endotoxemia. ::: Capsules and C3b, p. 394

Localized Disease

Nonencapsulated *H influenzae* produce disease under circumstances in which they are entrapped at a luminal site adjacent to the normal respiratory flora such as the middle ear, sinuses, or bronchi (Figure 31–2). This is usually associated with some compromise of normal clearing mechanisms, which is caused by a viral infection or structural damage. Consistent with their relative prevalence in the respiratory tract, nontypable organisms account for more than 90% of localized *H influenzae* disease, particularly otitis media, sinusitis, and exacerbations of chronic bronchitis. Nonencapsulated *H influenzae* attach to host epithelial cell using pili and the outer membrane proteins.

Bacterial trapped in middle ear, sinuses, and bronchi produce localized infections

Most are nonencapsulated strains

Adherence is by pili and surface proteins

IMMUNITY

Immunity to Hib infections has long been associated with the presence of anticapsular (PRP) antibodies, which are bactericidal in the presence of complement. The infant is usually protected by passively acquired maternal antibody for the first few months of life. Thereafter, actively acquired antibody increases with age; it is present in the serum of most children by 10 years of age. The peak incidence of Hib infections in unimmunized populations occurs at 6 to 18 months of age, when serum antibody is least likely to be present. This inverse relationship between infection and serum antibody is similar to that for *N meningitidis* (see Figure 30–4). The major difference is that substantial immunity is provided by antibody directed against a single type (Hib) rather than the multiple immunotypes of other bacteria. Thus, systemic *H influenzae* infections (meningitis, epiglottitis, cellulitis) are rare in adults. When such infections develop, the immunologic deficit is the same as that with meningococci—lack of circulating antibody. ::: Meningococcal immunity, p. 541

Anticapsular antibody is bactericidal and protective

Hib infections occur at ages when antibody is absent

Like other polysaccharides, Hib PRP behaves as a T-cell–independent antigen, and antibody responses from immunization are poor in a child younger than 18 months of age. Significant secondary responses from boosters are not elicited. Conjugation of PRP to protein dramatically improves the immunogenicity by eliciting the T-cell-dependent responses typical while preserving the specificity for PRP. :::T-cell dependent and independent responses, p. 37

T-cell–independent response to PRP is poor at less than 18 months of age

Protein conjugate vaccine elicits T-cell response in infants

H INFLUENZAE DISEASE: CLINICAL ASPECTS

MANIFESTATIONS

Of the major acute Hib infections, meningitis accounts for just over 50% of cases. The remaining cases are distributed among pneumonia, epiglottitis, septicemia, cellulitis, and septic arthritis. Localized infections can be caused by capsulated strains including Hib, but most are noncapsulated *H influenzae.*

Meningitis

Hib meningitis follows the same pattern as other causes of acute purulent bacterial meningitis (see Chapter 65). Meningitis is often preceded by signs and symptoms of an upper respiratory infection, such as pharyngitis, sinusitis, or otitis media. Whether these represent a predisposing viral infection or early invasion by the organism is not known. Just as often, meningitis is preceded by vague malaise, lethargy, irritability, and fever. Mortality is 3% to 6% despite appropriate therapy, and roughly one third of all survivors have significant neurologic sequelae.

Acute purulent meningitis may follow sinusitis or otitis media

Mortality and neurologic sequelae are significant

Acute Epiglottitis

Acute epiglottitis is a dramatic infection in which the inflamed epiglottis and surrounding tissues obstruct the airway. Hib is one of a number of causes. The onset is sudden, with fever,

Cherry-red, swollen epiglottis, and stridor are hallmarks

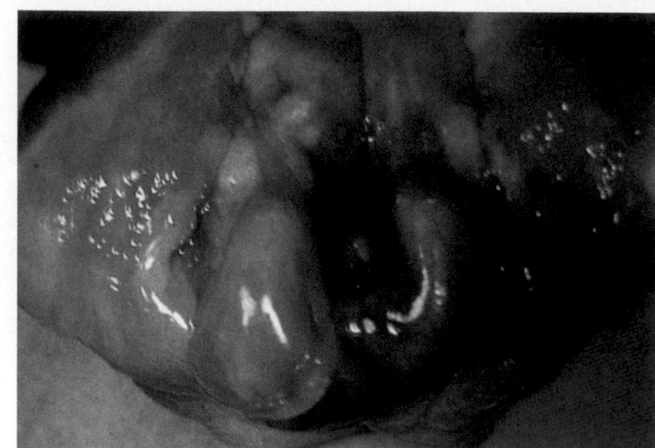

FIGURE 31–4. The swollen epiglottis characteristic of *Haemophilus influenzae* acute epiglottitis. FS:(Reproduced with permission from Connor DH, Chandler FW, Schwartz DQA, Manz HJ, Lack EE (eds). *Pathology of Infectious Diseases,* vol. 1. Stamford, CT: Appleton & Lange; 1997.)

Airway maintenance is needed

sore throat, hoarseness, an often muffled cough, and rapid progression to severe prostration within 24 hours. Affected children have air hunger, inspiratory stridor, and retraction of the soft parts of the chest with each inspiration. The hallmark of the disease is an inflamed, swollen, cherry-red epiglottis that protrudes into the airway (**Figure 31–4**) and can be visualized on lateral x-rays. As with meningitis, this infection is treated as a medical emergency, with prime emphasis on antimicrobics and maintenance of an airway (tracheostomy or endotracheal intubation). Manipulations, including routine examination or attempting to take a throat swab, can trigger a fatal laryngospasm and acute obstruction.

Cellulitis and Arthritis

Cellulitis is usually facial

Large joints are involved

A tender, reddish-blue swelling in the cheek or periorbital areas is the usual presentation of Hib cellulitis. Fever and a moderately toxic state are usually present, and the infection may follow an upper respiratory infection or otitis media. Joint infection begins with fever, irritability, and local signs of inflammation, often in a single large joint. *Haemophilus* arthritis is occasionally the cause of a more subtle set of findings, in which fever occurs without clear clinical evidence of joint involvement. Bacteremia is often present in both cellulitis and arthritis.

Other Infections

Nonencapsulated strains are common in otitis media, sinusitis, and bronchitis

Pneumonia is linked to underlying damage

H influenzae is an important cause of conjunctivitis, otitis media, and acute and chronic sinusitis. It is also one of several common respiratory organisms that can cause and exacerbate chronic bronchitis. Most of these infections are caused by nonencapsulated strains and usually remain localized without bacteremia. Disease may be acute or chronic, depending on the anatomic site and underlying pathology. For example, otitis media is acute and painful because of the small, closed space involved, but after antimicrobial therapy and reopening of the eustachian tube, the condition usually clears without sequelae. The association of *H influenzae* with chronic bronchitis is more complex. There is evidence that *H influenzae* and other bacteria play a role in inflammatory exacerbations, but a unique cause-and-effect relationship has been difficult to prove. The underlying cause of the bronchitis is usually related to chronic damage resulting from factors such as smoking. *Haemophilus* pneumonia may be caused by either encapsulated or nonencapsulated organisms. Encapsulated strains have been observed to produce a disease much like pneumococcal pneumonia; however, unencapsulated strains may also produce pneumonia, particularly in patients with chronic bronchitis.

DIAGNOSIS

Blood cultures are useful in systemic infections

The combination of clinical findings and a typical Gram smear is usually sufficient to make a presumptive diagnosis of *Haemophilus* infection. The tiny cells are usually of uniform shape except in cerebrospinal fluid, where some may be elongated to several times their usual length (Figure 31–1). The diagnosis must be confirmed by isolation of the organism from the site of infection or from the blood. Blood cultures are particularly useful in systemic *H influenzae* infections because it is often difficult to obtain an adequate specimen directly from the site of infection. Bacteriologically, small coccobacillary Gram-negative rods that grow on chocolate agar but not blood agar strongly suggest *Haemophilus*. Confirmation and speciation depend on demonstration of the requirement for hematin (X factor) and/or NAD (V factor) and/or biochemical tests. Serotyping is unnecessary for clinical purposes, but important in epidemiologic and vaccine studies.

Demonstrating X and V requirement defines species

TREATMENT

H influenzae is often susceptible in vitro to ampicillin and amoxicillin and usually susceptible to the newer cephalosporins, tetracycline, aminoglycosides, and sulfonamides. It is less susceptible to other penicillins and to erythromycin. Since the 1970s, the therapy of systemic infections has been complicated by the emergence of resistance in a pattern similar to that of *Neisseria gonorrhoeae*. The major mechanism is production of a β-lactamase identical with that found in *Escherichia coli*. The frequency of β-lactamase–producing strains varies between 5% and 50% in different geographic areas. Ampicillin-resistant strains due to alterations in the transpeptidase binding site also occur, but are less common. Current practice is to start empiric therapy with a third-generation cephalosporin (eg, ceftriaxone, cefotaxime), which can be changed to ampicillin if susceptibility tests indicate that the infecting strain is susceptible.

Ampicillin-resistant strains produce β-lactamase

Third-generation cephalosporin is initial treatment

PREVENTION

Purified PRP vaccines became available in 1985; however, owing to the typically poor immune response of infants to polysaccharide antigens, their use was limited to children 24 months of age and older. Because immunization at this age misses the group most susceptible to Hib invasive disease, a new vaccine strategy was needed to include improved stimulation of T-cell–dependent immune responses in infants. To achieve this, the first protein conjugate vaccines were developed by linking PRP to proteins derived from bacteria (diphtheria toxoid, *N meningitidis* outer membrane protein). The first PRP–protein conjugate vaccines were licensed in 1989; by late 1990, they were recommended for universal immunization in children beginning at 2 months of age. As illustrated in **Figure 31–5**, the impact has been dramatic. This 99% reduction in what was once one of the most feared diseases of childhood is one of the greatest achievements in medical history. Fortunately, the decline in Hib has not been accompanied by compensatory rise in the numbers of non-b

PRP vaccine missed peak age of disease

PRP conjugated to bacterial proteins stimulates T cells

Dramatic reduction in Hib disease has been sustained

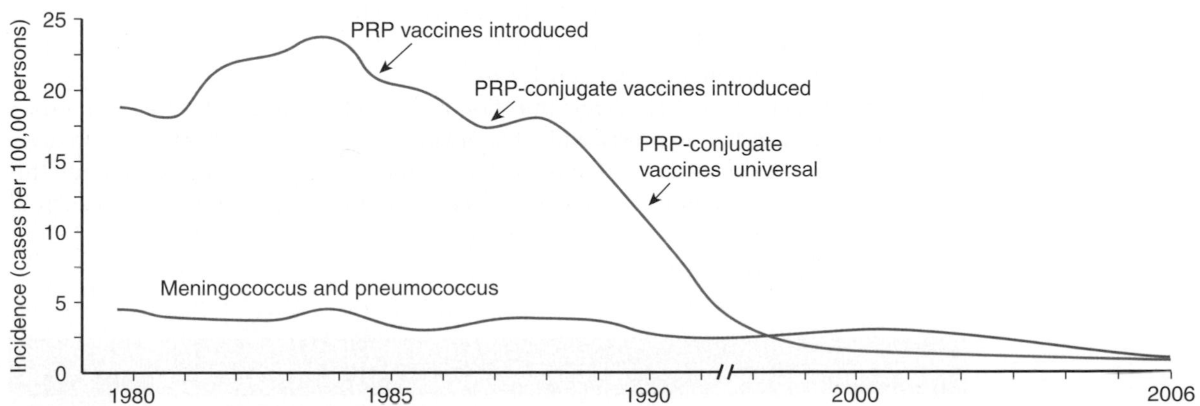

FIGURE 31–5. The decline in *Haemophilus influenzae* type b (Hib) meningitis in association with the introduction of new vaccines is shown. Note also the steady state of the other major causes of childhood meningitis. They did not increase to "fill in the gap" nor did *H influenzae* invasive disease caused by other serotypes.

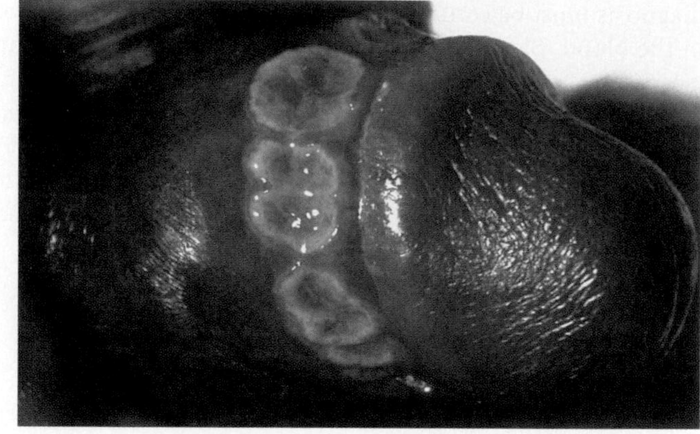

FIGURE 31–6. Chancroid. These penile ulcers are caused by *Haemophilus ducreyi*. In contrast to the ulcers of syphilis they are soft and painful. (Reproduced with permission from Nester EW, Anderson DG, Roberts CE Jr, Nester MT. *Microbiology: A Human Perspective*, 6th ed. New York: McGraw-Hill, 2008.)

cases or in the other causes of acute purulent meningitis. An unexpected concomitant finding has been a dramatic drop in *H influenzae* colonization rates in immunized populations. Under the direction of the World Health Organization, government and philanthropic efforts like those of the Bill and Melinda Gates Foundation are underway to implement Hib immunization of children throughout the world.

Rifampin prophylaxis indicated

As with *Neisseria meningitidis,* rifampin chemoprophylaxis is indicated for unimmunized close contacts. This includes children and adults when there is a child in the family who has not had a full course of the Hib conjugate vaccine.

Haemophilus ducreyi

Soft chancre is a genital ulcer with satellite lesions

May contribute to spread of AIDS in Africa

H ducreyi causes chancroid, a common cause of genital ulcer in Africa, Southeast Asia, India, and Latin America. Occasional outbreaks in North America have most often been associated with the exchange of sex for drugs or money. The typical lesion is a tender papule on the genitalia that develops into a painful ulcer with sharp margins (**Figure 31–6**). Satellite lesions may develop by autoinfection, and regional lymphadenitis is common. The incubation period is usually short (2 to 5 days). The lack of induration around the ulcer has caused the primary lesion to be called "soft chancre" to distinguish it from the primary syphilitic chancre, which is typically indurated and painless. The presence of open genital sores due to *H ducreyi* greatly enhances the risk of transmission of HIV either by providing a portal of entry or by the recruitment of CD4+ cells to the site. This may contribute to the heterosexual spread of acquired immunodeficiency syndrome (AIDS) on the African continent, where chancroid is common. Candidate *H ducreyi* virulence factors include pili and an outer membrane protein (DsrA), which mediate attachment to epithelial cells resistance to complement-mediated killing. A seeming lack of immunity may be due to the action of a toxin (cytolethal distending toxin) on T cells.

Culture requires selective medium

The specific diagnosis of *H ducreyi* infection is difficult. Although the organism grows on chocolate agar, it does so slowly, and other organisms in the genital flora are apt to overgrow the plates. Incorporating antibiotics (usually vancomycin) in the agar overcomes this problem, but few laboratories in the United States have this medium on hand. Chancroid is effectively treated with azithromycin, ceftriaxone, ciprofloxacin, or erythromycin. Condoms are effective in blocking transmission.

BORDETELLA

Species similar to *B pertussis* do not cause classic whooping cough

The genus *Bordetella* contains seven species. *B pertussis* is by far the most important because it is the cause of classic pertussis (whooping cough). Nucleic acid homology and other analyses indicate that *B parapertussis* and *B bronchiseptica* are close enough to *B pertussis* to be considered variants of the same species. *B parapertussis* occasionally causes a disease similar to, but milder than, pertussis. This is probably because it does not produce pertussis toxin even though it has a silent copy of the toxin gene. The remainder of this section focuses solely on *B pertussis.*

Bordetella Pertussis

BACTERIOLOGY

GROWTH AND STRUCTURE

B pertussis is a tiny (0.5 to 1.0 μm), Gram-negative coccobacillus morphologically much like *Haemophilus*. Growth requires a special medium with nutritional supplements (nicotinamide), additives (charcoal) to neutralize the inhibitory effect of the compounds in standard bacteriologic media, and antibiotics to inhibit other respiratory flora. Under the best conditions, growth is still slow, requiring 3 to 7 days for isolation. The organism is also very susceptible to environmental changes and survives only briefly outside the human respiratory tract.

Morphologically similar to *Haemophilus*

Nicotinamide required for slow growth

The cell wall of *B pertussis* has the structure typical of Gram-negative bacteria, although the outer membrane lipopolysaccharide differs significantly in structure and biologic activity from that of the Enterobacteriaceae. The surface exhibits a rod-like protein called the **filamentous hemagglutinin (FHA)** because of its ability to bind to and agglutinate erythrocytes. FHA has strong adherence qualities, based on domains in its structure that interact with an amino acid sequence (arginine, glycine, aspartic acid) present in host integrins, epithelial cells, and macrophages. In addition to its adherence functions, FHA also stimulates cytokine release and interferes with T_H1 immune responses. The organism surface also contains other adhesive structures including **pili** and an outer membrane protein called **pertactin.**

FHA binds amino acid sequences found in host cells

Pili and pertactin are adhesins

EXTRACELLULAR PRODUCTS

Pertussis Toxin

Pertussis toxin (PT) is the major virulence factor of *B pertussis*. It is an A-B toxin produced from a single operon as an enzymatic subunit and five binding subunits that are assembled into the complete toxin on the bacterial surface. The binding subunits mediate attachment of the toxin to carbohydrate moieties on the host cell surface. The enzymatic subunit is then internalized and ADP-ribosylates a G protein that affects adenylate cyclase activity. Unlike cholera toxin, which in essence keeps cyclase activity "turned on," pertussis toxin freezes the opposite side of the regulatory circuit and cripples the capacity of the host cell to inactivate cyclase activity. Multiple intracellular signaling pathways are disrupted by this G protein modification. Among the results of this action are lymphocytosis, insulinemia, and histamine sensitization. :::A-B toxins, p. 396

A-B toxin ADP-ribosylates G protein

Adenylate cyclase and cell regulation are disrupted

Other Toxins

Another potent toxin, a pore-forming **adenylate cyclase (AC),** enters host cells and catalyzes the conversion of host cell ATP to cyclic AMP at levels far above what can be achieved by normal mechanisms. This activity interferes with cellular signaling, chemotaxis, superoxide generation, and microbicidal function of immune effector cells, including polymorphonuclear leukocytes and monocytes. It can also induce programmed cell death (apoptosis). **Tracheal cytotoxin (TCT)** is a monomer of *B pertussis* peptidoglycan generated during cell wall synthesis. The fragments are released into the environment by multiplying bacterial cells because *B pertussis* lacks mechanisms for recycling these monomers present in other bacteria. Tracheal cytotoxin is directly toxic to ciliated tracheal epithelial cells causing their extrusion from the mucosa and eventual death. There is little or no effect on the nonciliated cells.

Bacterial adenylate cyclase disrupts immune cell function

Peptidoglycan fragments injure ciliated tracheal cells

PERTUSSIS (WHOOPING COUGH)

CLINICAL CAPSULE

Pertussis is a prolonged illness caused by toxins produced by *B pertussis* bacteria attached to the cilia of respiratory epithelial cells. It progresses in stages over many weeks beginning with a rhinorrhea (runny nose), which evolves into a persistent paroxysmal cough lasting weeks more. The name "whooping cough" comes from children who exhibit an inspiratory "whoop" following an exhausting series of retching coughs.

EPIDEMIOLOGY

Highly contagious and spread by airborne droplet nuclei

Immunization reduces disease but outbreaks continue

B pertussis is spread by airborne droplet nuclei and remains localized to the trachobronchial tree. It is highly contagious, infecting more than 90% of exposed susceptible persons. Secondary spread in families, schools, and hospitals is rapid. Sporadic epidemics occur, but there is no strong seasonal pattern. *B pertussis* is a strictly human pathogen. It is not found in animals and survives poorly in the environment. Asymptomatic carriers are rare except in outbreak situations. The introduction of immunization in the 1940s produced a dramatic reduction in disease, but outbreaks persist in 3- to 5-year cycles. Large outbreaks occurred in populations where the immunization rates fell as a result of concerns about febrile reactions to the original pertussis vaccine.

Atypical adult disease facilitates spread

Infants have high mortality

Waning immunity needs boosting

Immunization also produced a change in the age distribution of the residual cases. Previously a disease of toddlers and young children, pertussis began to appear in infants and adults beginning in late adolescence. This is felt to be due to the relatively short duration (10 to 12 years) of immunity provided by the vaccine. These adults are susceptible if exposed but usually have a milder form of the disease, which is often not recognized as pertussis. These unwitting adults are the major source for outbreaks in highly susceptible populations, such as infants. In preimmunization days, newborns were usually infused with maternal transplacental IgG stimulated by exposure to the constant circulation of *B pertussis* in children. In an immunized population with waning immunity, this antibody has dropped below protective levels by the childbearing years. In a cruel twist, infants turn out to have the most severe disease. Over 70% of fatal cases occur in children under 1 year of age. The current strategy is to reverse this trend by giving vaccine booster doses later in life (see Prevention).

PATHOGENESIS

Attachment to cilia provides site for toxin production

Mucosa becomes devoid of ciliated cells

When introduced into the respiratory tract, *B pertussis* has a remarkable tropism for ciliated bronchial epithelium attaching to the cilia themselves. This adherence is mediated by FHA, pili, pertactin, and the binding subunits of PT. Once attached, the bacteria immobilize the cilia and begin a sequence in which the ciliated cells are progressively destroyed and extruded from the epithelial border (**Figures 31–7** and **31–8**). This local injury is caused primarily by the action of tracheal cytotoxin. It eventually produces an epithelium devoid of the ciliary blanket, needed to move foreign matter away from the lower airways. Persistent coughing is the clinical correlate of this deficit. Although considerable local inflammation and exudate are produced in the bronchi, *B pertussis* does not directly invade the cells of the respiratory tract or spread to deeper tissue sites.

Virulence Factors

PT and adenylate cyclase attack immune cells

In addition to the local effects on the bronchial epithelium, other virulence factors of *B pertussis* contribute to the disease in diverse ways. The combined action of PT and AC on neutrophils, macrophages, and lymphocytes creates paralysis and even death of these crucial

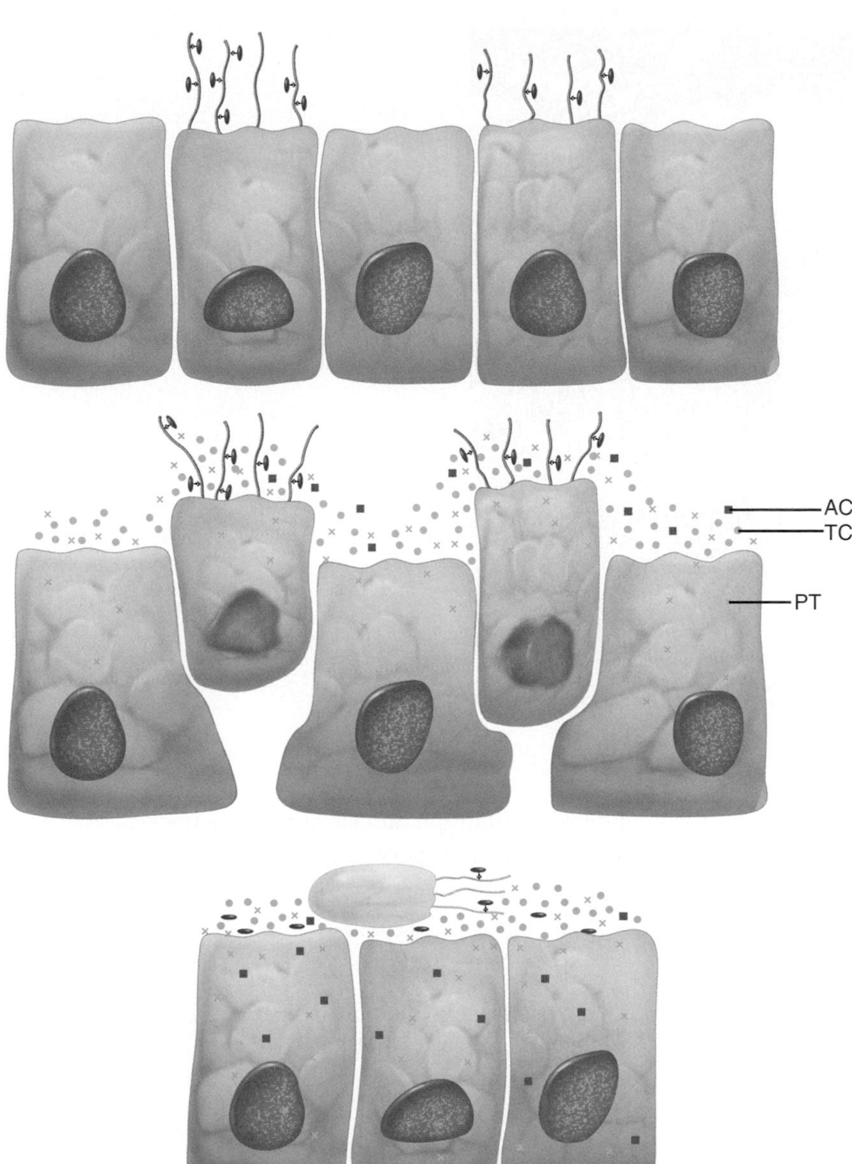

FIGURE 31–7. Whooping cough, cellular view. (*Top*) *Bordetella pertussis* attaches to the cilia of cells in the respiratory epithelium. Attachment is mediated by pili, filamentous hemagglutinin, and pertactin. (*Middle*) Regulatory systems initiate production of pertussis toxin (PT), and adenylate cyclase (AC), which injure the cells and they begin to be extruded. Additional injury is from the peptidoglycan fragments of tracheal cytotoxin (TCT). (*Bottom*) The ciliated cells are destroyed, leaving a denuded mucosa without protective cilia. PT is absorbed into the bloodstream to act throughout the body.

effector cells of the immune system. Many of the systemic manifestations of the disease such as lymphocytosis, histamine sensitization, and insulin secretion are due to the action of circulating PT absorbed at the primary infection site. The specific biologic effect depends on how disruption of G-protein regulation by PT is manifested by the host cell type that the toxin reaches. Pertussis is the result of a well-orchestrated delivery by *B pertussis* of toxic and adhesive factors to host cells at local and distant sites to produce a disease that persists for many weeks.

Absorbed PT acts on multiple cell types

Genetic Regulation of Pathogenicity

How *B pertussis* deploys its repertoire of virulence genes is a model for the control of bacterial pathogenicity. *B pertussis* regulates the synthesis of PT, AC, FHA, pili, and many other genes through genetic loci that control the expression of at least 20 unlinked chromosomal genes at the transcriptional level. Expression is modulated in a two-compartment system by changes in specific environmental parameters, including temperature. The induction of virulence factors in *B pertussis* is sequential, with adhesin expression (FHA and pili) preceding expression of factors involved in tissue injury (PT, AC). The finely honed responses of *B pertussis* virulence factors to changes in temperature and ionic conditions presumably play a role in the pathogenesis of infection and help the organism adapt in a stepwise fashion

Multiple virulence genes respond to temperature and ionic changes

Virulence genes are regulated in two-compartment model

Adherence factors precede injury products

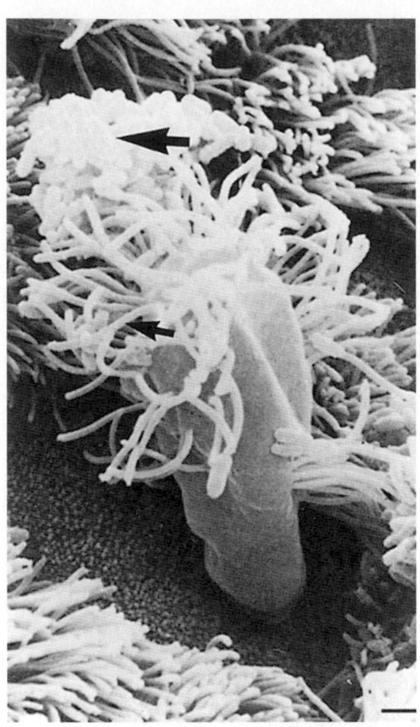

FIGURE 31–8. A tracheal organ culture 72 hours after infection with *Bordetella pertussis*. The organisms have attached to the cilia of some cells and killed them. These balloon-like cells with attached bacteria are extruded from the epithelium. The large arrow shows the *Bordetella*, and the small arrow shows cilia. Note the background of uninfected ciliated cells and denuded epithelium where nonciliated cells remain. (Reproduced with permission from Muse KE, Collier AM, Baseman JB. J Infect Dis 136:768–777. Figure 3, copyright 1977 by University of Chicago, publisher.)

to the diverse local conditions within the human respiratory tract. Details of the genetic mechanisms involved are discussed in Chapter 22 and Figure 22–8. ::: Pertussis virulence regulation, p. 400

IMMUNITY

Immunity is not long-term

Although IgG antibodies are produced to PT, pili, and pertactin during the course of natural infection and by immunization, they are not longlasting, and their role in immunity is not well understood. Naturally acquired immunity is not lifelong, although second attacks, when recognized, tend to be mild.

PERTUSSIS: CLINICAL ASPECTS

MANIFESTATIONS

Catarrhal phase is most communicable

After an incubation period of 7 to 10 days, pertussis follows a prolonged course consisting of three overlapping stages: (1) catarrhal, (2) paroxysmal, and (3) convalescent. In the catarrhal stage, the primary feature is a profuse and mucoid rhinorrhea, which persists for 1 to 2 weeks. Nonspecific findings such as malaise, fever, sneezing, and anorexia may also be present. The disease is most communicable at this stage, because large numbers of organisms are present in the nasopharynx and the mucoid secretions.

Paroxysmal coughing phase lasts for weeks

Inspiratory whoop and coughing may lead to apnea

Lymphocytosis is marked

The appearance of a persistent cough marks the transition from the catarrhal to the paroxysmal coughing stage. At this time, episodes of paroxysmal coughing occur up to 50 times a day for 2 to 4 weeks. The characteristic inspiratory whoop follows a series of coughs as air is rapidly drawn through the narrowed glottis. Vomiting frequently follows the whoop. The combination of mucoid secretions, whooping cough, and vomiting produces a miserable, exhausted child barely able to breathe. Apnea may follow such episodes, particularly in infants. Marked lymphocytosis reaches its peak at this time, with absolute lymphocyte counts of up to 40,000/mm^3.

Convalescent phase is a gradual fading

During the 3- to 4-week convalescent stage, the frequency and severity of paroxysmal coughing and other features of the disease gradually fade. Partially immune persons and infants younger than 6 months of age may not show all the typical features of pertussis. Some evolution through the three stages is usually seen, but paroxysmal coughing and lymphocytosis may be absent.

The most common complication of pertussis is pneumonia caused by a superinfecting organism such as *Streptococcus pneumoniae*. Atelectasis is also common but may be recognized only by radiologic examination. Other complications, including convulsions and subconjunctival or cerebral bleeding, are related to the venous pressure effects of the paroxysmal coughing and the anoxia produced by inadequate ventilation and apneic spells.

Atelectasis and superinfection are major complications

DIAGNOSIS

A clinical diagnosis of pertussis is best confirmed by isolation of *B pertussis* from nasopharyngeal secretions or swabs. Throat swabs are not suitable, because the cilia to which the organism attaches are not found there. Specimens collected early in the course of disease (during the catarrhal or early paroxysmal stage) provide the greatest chance of successful isolation. Unfortunately, the diagnosis is frequently not considered until paroxysmal coughing has been present for some time, and the number of organisms has decreased significantly. The nasopharyngeal specimens are plated onto a special charcoal blood agar medium made selective by the addition of a cephalosporin. This allows the slow-growing *B pertussis* to be isolated in the presence of more rapidly growing members of the normal upper respiratory flora. The characteristic colonies appear after 3 to 7 days of incubation and look like tiny drops of mercury. Immunologic methods (agglutination, immuofluorescence) are required for specific identification.

Nasopharyngeal swab is plated on charcoal blood agar

Organisms are often gone by later paroxysmal phase

A direct immunofluorescent antibody (DFA) technique has been successfully applied to nasopharyngeal smears for rapid diagnosis of pertussis. DFA is particularly helpful in pertussis because of the many days required for culture results. Because the sensitivity and specificity of DFA can vary with the quality of the reagents, these results should always be confirmed by culture, if possible. Nucleic acid amplification tests have been developed, which show potential for being highly sensitive but are not yet practical for most laboratories. Serologic tests are widely used for epidemiologic studies but not for diagnosis of individual clinical cases. ::: DFA, p. 66

DFA allows rapid diagnosis

TREATMENT

Once the paroxysmal coughing stage has been reached, the treatment of pertussis is primarily supportive. Antimicrobial therapy is useful at earlier stages and for limiting spread to other susceptible individuals. Of a number of antimicrobics active in vitro against *B pertussis,* the macrolides are preferred for treatment. Of these, erythromycin has the greatest experience demonstrating its clinical effectiveness and relative lack of toxicity.

Erythromycin is most effective in catarrhal phase

PREVENTION

Active immunization is the primary method of preventing pertussis. The original vaccine, which produced a dramatic reduction in disease, was prepared from inactivated whole cell suspensions and given together with diphtheria and tetanus toxoids as DTP. The undoubted efficacy of this vaccine was colored by a high rate of side effects due to the crude nature of the whole cell preparation. These included local inflammation, fever and, rarely, febrile seizures. Although permanent neurologic sequelae were never convincingly linked to pertussis immunization, there were those who argued that the vaccine was worse than the disease. This led to the development of acellular vaccines, guided by knowledge of the pathogenesis of pertussis. These vaccines contain virulence factors purified from whole cell preparations inactivated where appropriate. Another vaccine strategy is the production of recombinant components, genetically engineered to be immunogenic but nontoxic.

Whole cell vaccine was effective but had side effects

Acellular vaccines are purified preparations

The multiple acellular vaccines have different combinations of virulence factors. All contain PT and FHA, and some add pertactin or pili (vaccine manufacturers use the term fimbriae). The efficacy of these vaccines has now been established, and all have dramatically less frequent side effects compared with the whole cell preparations. In combination with diphtheria and tetanus toxoids, they have now replaced the whole cell DTP as DTaP ("a" for acellular). This vaccine is now recommended for the full primary immunization (at 2, 4,

Vaccines include PT, FHA, and other virulence factors

DTaP has replaced DTP

and 6 months) and boosters (at 15 to 18 months, 4 to 6 years). For booster immunizations, vaccines combining tetanus toxoid with lower doses of the diphtheria and pertussis components (Tdap) are recommended every 10 years beginning at age 11.

CLINICAL CASE

A CHOKING, COUGHING INFANT

A male infant born prematurely was still in the pediatric intensive care unit at 12 days old. On the eighth day, he began to exhibit repetitive coughing, which progressed to his turning red, choking, and gasping for breath. The episodes were sometimes followed by vomiting. On the tenth day, he suffered apnea and now requires ventilatory assistance. His physical examination was significant for a pulse of 160 bpm and respiratory rate of 72/min (both highly elevated). The child's chest radiograph was clear. There was no evidence of tracheal abnormalities. The infant's white cell count was 15,500/mm^3 with 70% lymphocytes.

QUESTIONS

■ Which of this patient's findings are most unique for pertussis?

A. Cough
B. Choking
C. Vomiting
D. Leukocytosis
E. Lymphocytosis

■ Which of the following would yield the most rapid confirmation of a whooping cough diagnosis?

A. Throat culture
B. Nasopharyngeal culture
C. Nasopharyngeal direct fluorescent antibody smear
D. Throat direct fluorescent antibody smear
E. *B pertussis* serology

■ What is the most likely source of this child's infection?

A. Sibling
B. Parent
C. Delivery room environment
D. Healthcare worker carrier
E. Healthcare worker with disease

ANSWERS

1(D), **2**(C), **3**(E)

CHAPTER 32

Vibrio, Campylobacter, and *Helicobacter*

I am poured out like water, and all my bones are out of joint: my heart is like wax; it is melted in the midst of my bowels.

—The Bible: *Psalms* 22:14

This group of curved Gram-negative rods includes *Vibrio cholerae,* the cause of cholera, one of the first proven infectious diseases, and two newcomers incriminated as pathogens late in the 20th century (**Table 32–1**). The peptic ulcer disease now known to be caused by *Helicobacter pylori* had been long accepted to be due to stress and disturbed gastric acid secretion. *Campylobacter jejuni* is one of the most common causes of diarrhea in virtually every country of the world. Cholera has undergone a resurgence in recent decades, spreading from its historic Asiatic locale to the Americas, including the coastline of the United States.

VIBRIO

Vibrios are curved, Gram-negative rods (**Figure 32–1**) commonly found in saltwater. Cells may be linked end to end, forming S shapes and spirals. They are highly motile with a single polar flagellum, non-spore-forming, and oxidase-positive, and they can grow under aerobic or anaerobic conditions. The cell envelope structure is similar to that of other Gram-negative bacteria. *Vibrio cholerae* is the prototype cause of a water-loss diarrhea called **cholera.** Other species causing diarrhea, wound infections, and, rarely, systemic infection are listed in **Table 32–2.**

Rapidly motile curved rods are found in seawater

VIBRIO CHOLERAE

BACTERIOLOGY

GROWTH AND STRUCTURE

V cholerae has a low tolerance for acid, but grows under alkaline (pH 8.0 to 9.5) conditions that inhibit many other Gram-negative bacteria. It is distinguished from other vibrios by

TABLE 32–1 Features of *Vibrio, Campylobacter,* and *Helicobacter*[a]

	BACTERIOLOGY			PATHOGENESIS		
ORGANISM	**GROWTH**	**UREASE**	**EPIDEMIOLOGY**	**ADHERENCE**	**TOXINS**	**DISEASE**
Vibrio cholerae	Facultative	–	Fecal–oral, water-borne, pandemics	Surface protein[b], pili	CT[c]	Watery diarrhea (cholera)
Campylobacter jejuni	Microaerophilic	–	Animals, unpasteurized milk	Unknown[d]	Unknown[e]	Dysentery, watery diarrhea
Helicobacter pylori	Microaerophilic	+	Human, gastric secretions[f]	OMPs[g]	VacA[h], urease, Cag[i]	Chronic gastritis, ulcers, adenocarcinoma, lymphoma

[a]All are curved Gram-negative rods with similar morphology.
[b]Surface protein able to bind to chitin and human intestine.
[c]Cholera toxin.
[d]Lipooligosaccharide, flagellin, a major outer membrane protein (MOMP), and a fibronectin-binding protein CadF are candidate adhesions.
[e]Cytolethal-distending toxin is a candidate toxin.
[g]OMPs, outer membrane proteins (BabA, SabA, AlpA, AlpB, HopZ).
[h]Vacuolating cytotoxin.
[i]Not technically a toxin. but Cag is strongly associated with virulence.

Growth prefers alkaline over acid conditions

Cholera is limited to O1 and O139 serotypes

Serotype O139 is encapsulated

biochemical reactions, lipopolysaccharide (LPS) O antigenic structure, and production of cholera toxin (CT). There are over 150 O antigen serotypes, only two of which (O1 and O139) cause cholera. *V cholerae* biogroup El Tor, an O1 variant, is a biotype of the classic strain. The O139 strains phenotypically resemble O1 El Tor strains. *V cholerae* possess long filamentous pili that form bundles on the bacterial surface and belong to a family of pili whose chemical structure is similar to those of the gonococcus and a number of other bacterial pathogens. All strains capable of causing cholera produce a colonizing factor known as the toxin-coregulated pilus (TCP) because its expression is regulated together with CT.

CHOLERA TOXIN

The structure and mechanism of action of CT have been studied extensively (**Figure 32–2**). CT is an A-B type ADP-ribosylating toxin. Its molecule is an aggregate of multiple polypeptide chains organized into two toxic subunits (A1, A2) and five binding (B) units. The B

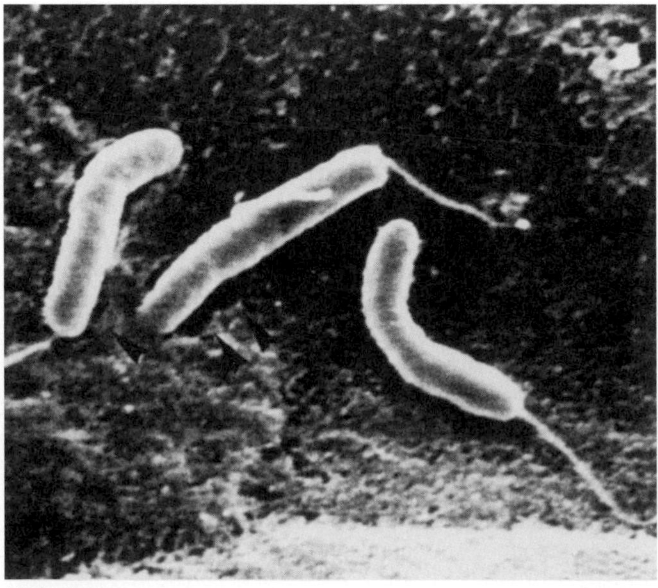

FIGURE 32–1. Vibrio cholerae (scanning electron micrograph). Note the curved rods and polar flagella. (Reproduced with permission from Willey J, Sherwood L, Woolverton C (eds). *Prescott's Principles of Microbiology.* New York: McGraw-Hill; 2008.)

TABLE 32–2 Features of Less Common *Vibrio* and *Campylobacter* Species

ORGANISM	FEATURES	EPIDEMIOLOGY	DISEASE
Vibrio			
V mimicus	Closely related to *V cholerae*, cholera-like enterotoxin	Ingestion of raw seafood	Watery diarrhea
V parahaemolyticus	Produces bowel inflammation, enterotoxin unclear	Coastal seawater; ingesting raw seafood; outbreaks on cruise ships; common in Japan	Watery diarrhea, occasionally dysentery
V vulnificus	Produces powerful siderophores which scavenge iron from host transferrin and lactoferrin	Coastal seawater, particularly when water temperatures rise; ingesting raw seafood or contamination of wound with seawater	Fulminant bacteremia following ingestion, cellulitis from wound contamination, high fatality rate in those with Fe^{+} storage disease
V alginolyticus		Wounds contaminated by seawater	Cellulitis
Campylobacter			
C fetus	Fails to grow on selective medium used for *C jejuni*	Cause of abortion in cattle and sheep	Bacteremia, thrombophlebitis
C upsaliensis	Fails to grow on selective medium used for *C jejuni*	Associated with dogs and cats	Diarrhea similar to *C jejuni*
C hyointestinalis		Enteritis in swine	Diarrhea in immunocompromised and homosexual men
C lari		Associated with birds	Diarrhea, bacteremia in immunocompromised

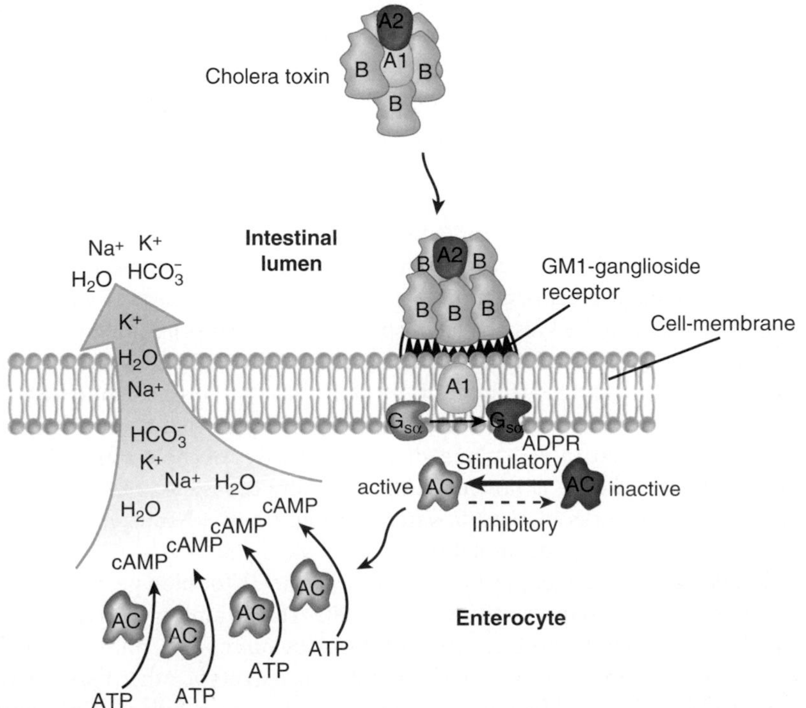

FIGURE 32–2. The action of cholera toxin. The complete toxin is shown binding to the GM1-ganglioside receptor on the cell membrane via the binding (B) subunits. The active portion (A1) of the A subunit catalyzes the ADP-ribosylation (ADPR) of the G_s (stimulatory) regulatory protein, "locking" it in the active state. Because the G_s protein acts to return adenylate cyclase from its inactive to active form, the net effect is persistent activation of adenylate cyclase. The increased adenylate cyclase (AC) activity results in accumulation of cyclic adenosine 3′,5′-monophosphate (cAMP) along the cell membrane. The cAMP causes the active secretion of sodium (Na^+), chloride (Cl^-), potassium (K^+), bicarbonate (HCO_3^-), and water out of the cell into the intestinal lumen.

B subunit receptor is a ganglioside on cell surface

A1 enters cytoplasm and ADP ribosylates regulatory G protein

Adenylate cyclase becomes locked in active state

Hyperproduction of cAMP causes hypersecretion of water and electrolytes

units bind to a GM1-ganglioside receptor found on the surface of many types of cells. Once bound, the A1 subunit is released from the toxin molecule by reduction of the disulfide bond that binds it to the A2 subunit, and it enters the cell by translocation. In the cell, it exerts its effect on the membrane-associated adenylate cyclase system at the basolateral membrane surface. The target of the toxic A1 subunit is a guanine nucleotide (G) protein, Gsα, which regulates activation of the adenylate cyclase system. CT catalyzes the ADP ribosylation of the G protein, rendering it unable to dissociate from the active adenylate cyclase complex. This causes persistent activation of intracellular adenylate cyclase, which in turn stimulates the conversion of adenosine triphosphate to cyclic adenosine 3′, 5′-monophosphate (cAMP). The net effect is excessive accumulation of cAMP at the cell membrane, which causes hypersecretion of chloride, potassium, bicarbonate, and associated water molecules out of the cell. Strains of *V cholerae* other than the two epidemic serotypes may or may not produce CT. :::A-B toxin, p. 396; ADPR, p. 363

CHOLERA

CLINICAL CAPSULE

Cholera produces the most dramatic watery diarrhea known. Intestinal fluids pour out in voluminous bowel movements; this eventually leads to dehydration and electrolyte imbalance. These effects come from the action of cholera toxin secreted by *V. cholerae* in the bowel lumen. Despite the profound physiologic effects, there is no fever, inflammation, or direct injury to the bowel mucosa.

EPIDEMIOLOGY

Transmission is through untreated water supply

Incubation period is 2 days

Epidemic cholera is spread primarily by contaminated water under conditions of poor sanitation, particularly where sewage treatment is absent or defective. Even though convalescent human carriage is brief, if the numerous vibrios purged from the intestines of those infected with cholera are able to reach the primary water supply, the conditions for spread are established. The short incubation period (2 days) ensures that organisms ingested by others quickly enter the epidemic cycle. Even so, modern travel makes imported cases of cholera possible. One man developed diarrhea in Florida after eating ceviche (marinated uncooked fish) just before departure from an airport in Ecuador.

Cholera is endemic in India and Africa

Pandemics span decades

Gulf Coast cases result from undercooked shellfish

Latin American epidemic is widespread

Cholera is endemic in the Indian subcontinent and Africa. Over the last two centuries, its spread beyond this historic locale to other parts of Asia, Indonesia, and Europe has been described in eight great pandemics, each lasting 5 to 25 years. The current pandemic has brought cholera to the Western Hemisphere for the first time since 1911. Sporadic cases of cholera in the United States first appeared in the early 1970s and were traced to inadequately cooked crabs and shrimp caught off the Gulf Coast of Louisiana and Texas. In 1991, Latin America was hit with epidemic cholera with cases reported from 21 countries from Peru to northern Mexico. In Peru alone, over 500,000 cases of cholera and 4500 deaths occurred in 2 years. The disease is now endemic, claiming thousands of lives every year. Virulent *V. cholerae* now lurks in coastal waters throughout the hemisphere and in the drinking water of locales with poor sanitation.

El Tor biotype dominated 20th century

New O139 serotype is spreading

The dominant strain of the 20th century was the El Tor biotype, first isolated from Mecca pilgrims at the El Tor quarantine camp in 1905. This strain survives slightly longer in nature and is more likely to produce subclinical cases of cholera, both of which facilitated its spread. In 1992, the first cases of cholera due to a serotype other than O1 were detected in India and Bangladesh. The new serotype (O139 Bengal) is fully virulent with the additional threat of enhanced ability to produce disease in persons whose immunity is due to exposure to the old serotype. This development is important for the global spread of cholera and for the vaccine strategies designed to prevent it.

The epidemic potential of *V cholerae* depends on its ability to survive in both aquatic environments and human hosts. In the environment, it persists in a dormant state in association with shellfish and plankton by attaching to their chitinous exoskeleton and forming biofilms. This dual life is facilitated by a surface protein able to bind a constituent of chitin as well as glycoproteins and lipids on the intestinal epithelium. Satellite tracking has linked periodic climate changes (warming seawater), plankton blooms, and cholera epidemics along the coast of South America. Otherwise, the organism is fragile, surviving only a few days in the environment outside its human or crustacean hosts. ::: Dormant cells, p. 373

Survival in shellfish and plankton facilitates epidemics

PATHOGENESIS

To produce cholera, *V cholerae* must reach the small intestine, swim to the intestinal crypts, multiply, and produce virulence factors. In healthy people, ingestion of large numbers of bacteria is required to offset the acid barrier of the stomach. Colonization of the entire intestinal tract from the jejunum to the colon by *V cholerae* requires adherence to the epithelial surface by the above-mentioned protein and surface pili. The outstanding feature of *V cholerae* pathogenicity is the ability of virulent strains to secrete CT, which is responsible for the disease cholera. The water and electrolyte shift from the cell to the intestinal lumen is the fundamental cause of the watery diarrhea of cholera.

Large doses required to pass stomach acid barrier

Pili and proteins mediate epithelial adherence

CT-stimulated intestinal hypersecretion causes diarrhea

Fluid Loss

The fluid loss that results from the adenylate cyclase stimulation of cells depends on the balance between the amount of bacterial growth, toxin production, fluid secretion, and fluid absorption in the entire gastrointestinal tract. The outpouring of fluid and electrolytes is greatest in the small intestine, where the secretory capacity is high and absorptive capacity low. The diarrheal fluid can amount to many liters per day, with approximately the same sodium content as plasma, but two to five times the potassium and bicarbonate concentrations. The result is dehydration (isotonic fluid loss), hypokalemia (potassium loss), and metabolic acidosis (bicarbonate loss). The intestinal mucosa remains unaltered except for some hyperemia, because *V cholerae* does not invade or otherwise injure the enterocyte.

Small intestine loses liters of fluid

K^+ and bicarbonate losses cause hypokalemia and acidosis

Intestinal mucosa is structurally unaffected; no invasion

Genetic Regulation of Virulence

The expression of the multiple virulence factors of *V cholerae* is controlled in a coordinated system involving environmental sensors and as many as 20 chromosomal genes divided between a pathogenicity island (PAI) containing CT and one containing TCP. The chief regulator is a transmembrane protein (ToxR) that "senses" environmental changes in pH, osmolarity, and temperature, which convert it to an active form. In the active state, ToxR can directly turn on CT genes as well as activate transcription of a second regulatory protein, ToxT. ToxT, whose natural effector may be bile, then activates transcriptional of virulence genes in both PAIs, including TCP and CT. Quorum-sensing systems appear to deploy this virulence gene expression at a point when a critical mass of *V cholerae* is present to sustain it. ::: Environmental sensing, p. 373; PAI, p. 399

Regulatory system turns on CT and TCP in response to environmental changes

IMMUNITY

Nonspecific defenses such as gastric acidity, gut motility, and intestinal mucus are important in preventing colonization with *V cholerae*. For example, in persons who lack gastric acidity (gastrectomy or achlorhydria from malnutrition), the attack rate of clinical cholera is higher. Natural infection provides longlasting immunity. The immune state has been associated with IgG directed against the cell wall LPS and with the production of secretory IgA by lymphocytes in the subepithelial areas of the gastrointestinal tract. The precise protective mechanisms remain to be established.

Attack rate is higher with achlorhydria

Immunity is associated with sIgA

CHOLERA: CLINICAL ASPECTS

MANIFESTATIONS

Extreme watery diarrhea causes large fluid loss

Dehydration and electrolyte imbalance are the major problems

Typical cholera has a rapid onset, beginning with abdominal fullness and discomfort, rushes of peristalsis, and loose stools. Vomiting may also occur. The stools quickly become watery, voluminous, almost odorless, and contain mucus flecks, giving it an appearance called **rice-water stools.** Neither white blood cells or blood are in the stools, and the patient is afebrile. Clinical features of cholera result from the extensive fluid loss and electrolyte imbalance, which can lead to extreme dehydration, hypotension, and death within hours if untreated. No other disease produces dehydration as rapidly as cholera.

DIAGNOSIS

Stool culture using selective media is required

The initial suspicion of cholera depends on recognition of the typical clinical features in an appropriate epidemiologic setting. A bacteriologic diagnosis is accomplished by isolation of *V cholerae* from the stool. The organism grows on common clinical laboratory media such as blood agar and MacConkey agar, but its isolation is enhanced by a selective medium (thiosulfate-citrate-bile salt-sucrose agar). Once isolated, the organism is readily identified by biochemical reactions. Outside cholera endemic areas, the selective medium is not routinely used for stool cultures, so clinical laboratories must be alerted to the suspicion of cholera.

TREATMENT

Oral or intravenous fluid and electrolyte replacement is crucial

Antimicrobial therapy can reduce duration and severity

The outcome of cholera depends on balancing the diarrheal fluid and ionic losses with adequate fluid and electrolyte replacement. This is accomplished by oral and/or intravenous administration of solutions of glucose with near physiologic concentrations of sodium and chloride and higher than physiologic concentrations of potassium and bicarbonate. Exact formulas are available as dried packets to which a given volume of water is added. Oral replacement, particularly if begun early, is sufficient for all but the most severe cases and has substantially reduced the mortality from cholera. Antimicrobial therapy plays a secondary role to fluid replacement. Doxycycline shortens the duration of diarrhea and magnitude of fluid loss. Trimethoprim-sulfamethoxazole, erythromycin, and ciprofloxacin are alternatives for use in children and pregnant women.

PREVENTION

Water sanitation and cooking shellfish prevent infection

Vaccines are disappointing

Epidemic cholera, a disease of poor sanitation, does not persist where treatment and disposal of human waste are adequate. Because good sanitary conditions do not exist in much of the world, secondary local measures such as boiling and chlorination of water during epidemics are required. Cholera associated with ingestion of crabs and shrimp can be prevented by adequate cooking (10 minutes) and avoidance of recontamination from containers and surfaces. Vaccines prepared from whole cells, lipopolysaccharide, and CT B subunit have been disappointing, providing protection that is not longlasting. Current interest includes live attenuated vaccine strains because of their potential to stimulate the local sIgA immune response.

Other *Vibrios*

V parahaemolyticus in undercooked or raw seafood causes diarrhea

V vulnificus sepsis and wound infections linked to raw oysters and iron overload

Species of *Vibrio* other than *V cholerae* may still produce disease, but are uncommon and typically restricted to seacoast locales. Most, such as *V parahaemolyticus,* produce a diarrheal illness after ingestion of raw or inadequately cooked seafood. They do not produce cholera toxin, but some have been shown to produce their own enterotoxins. Of these, *V vulnificus* stands out because it can produce a rapidly progressive cellulitis in wounds sustained in seawater. It also produces a bacteremic infection after ingestion of raw seafood, which has been common enough in Florida to threaten the local oyster trade. Cases were also seen in the area devastated by hurricane Katrina. *V vulnificus* is also a spectacular scavenger of host iron stores and produces particularly fulminant disease in persons with iron-overload states (eg, thalassemia, hemochromatosis). Features of the less common vibrios are included in Table 32–2.

CAMPYLOBACTER

Campylobacters are motile, curved, oxidase-positive, Gram-negative rods similar in morphology to vibrios. The cells have polar flagella and are often attached at their ends giving pairs "S" shapes or a "seagull" appearance. More than a dozen *Campylobacter* species have been associated with human disease. Of these, *C jejuni* is by far the most common and is discussed here as the prototype for intestinal disease. The features of other species are summarized in Table 32–2.

BACTERIOLOGY: *CAMPYLOBACTER JEJUNI*

Before 1973, *C jejuni* was not recognized as a cause of human disease. Not until selective methods for its isolation were developed was it recognized as one of the most common causes of infectious diarrhea. Like other campylobacters, *C jejuni* grows well only on enriched media under microaerophilic conditions. That is, it requires oxygen at reduced tension (5–10%), presumably because of the vulnerability of some of its enzyme systems to superoxides. Growth usually requires 2 to 4 days, sometimes as much as 1 week. *C jejuni* has the structural components found in other Gram-negative bacteria (eg, outer membrane, LPS). In contrast to the vibrios, it does not break down carbohydrates, but uses amino acids and metabolic intermediates for energy.

Microaerophilic atmosphere is required for growth

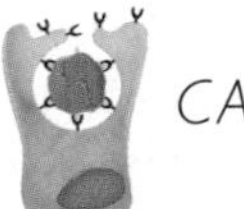

CAMPYLOBACTER ENTERITIS

CLINICAL CAPSULE

C jejuni infection typically begins with lower abdominal pain, which evolves into diarrhea over a matter of hours. The diarrhea may be watery or dysenteric, with blood and pus in the stool. Most patients are febrile. The illness resolves spontaneously after a few days to 1 week.

EPIDEMIOLOGY

It is humbling to consider how a pathogen as common as *C jejuni* could have been missed for decades. Rates of campylobacteriosis vary widely around the world but at 4% to 30% of diarrheal stools, it is the leading cause of gastrointestinal infection in developed countries. Over 2 million cases occur each year in the United States, at a rate roughly double that of the second most common bacterial enteric pathogen, *Salmonella.* This high rate of disease is facilitated by the low infecting dose of *C jejuni*—only a few hundred cells.

Causes diarrhea worldwide

Infecting dose is low

The primary reservoir is in animals, and the bacteria are transmitted to humans by ingestion of contaminated food or by direct contact with pets. Campylobacters are commonly found in the normal gastrointestinal and genitourinary flora of warm-blooded animals, including sheep, cattle, chickens, wild birds, and many others. Domestic animals such as

Reservoir is animals

Undercooked poultry and unpasteurized milk are major sources

dogs may also carry the organisms and probably play a significant role in transmission to humans. The most common source of human infection is undercooked poultry, but outbreaks have been caused by contaminated rural water supplies and unpasteurized milk often consumed as a "natural" food. Sometimes a direct association can be made as with a household pet, particularly a new puppy just brought home from a kennel.

PATHOGENESIS

Intracellular movement is associated with microtubules

Infection is established by oral ingestion, followed by colonization of the intestinal mucosa. The bacteria have been shown to adhere to endothelial cells and then enter cells in endocytotic vacuoles. Once inside, they move in association with the cell's microtubule structure, rather than the actin microfilaments associated with many other invasive bacteria. The search for enterotoxins associated with *C jejuni* has yielded some candidates but none yet seem to explain significant aspects of the disease. All in all, the virulence determinants of this major pathogen remain uncertain.

Guillain-Barré syndrome may follow infection

Antiganglioside antibodies cross-react with neural tissue

There is an association between *C jejuni* infection and **Guillain-Barré syndrome,** an acute demyelinating neuropathy that is frequently preceded by an infection. Although *C jejuni* is not the only antecedent to this syndrome, it is the most common of identifiable causes. Up to 40% of patients have culture or serologic evidence of *Campylobacter* infection at the time the neurologic symptoms occur. The mechanism is believed to involve antibody elicited by lipooligosaccharides in the *C jejuni* cell envelope that cross-react with similar molecules in the host peripheral nerve myelin. These antiganglioside antibodies are found in the serum of patients with Guillain-Barré syndrome motor neuropathies. This molecular mimicry is similar to the mechanism of rheumatic fever stimulated by the group A streptococcus. ::: Molecular mimicry, p. 42

IMMUNITY

Immune mechanisms are unclear

Acquired immunity after natural infection with *C jejuni* has been demonstrated in volunteer studies, but the mechanisms involved are unknown. Secretory and serum IgA are formed in the weeks after infection but decline thereafter. The high rate of *Campylobacter* infection in patients with acquired immunodeficiency syndrome suggests the importance of cellular immune mechanisms.

MANIFESTATIONS AND DIAGNOSIS

Abdominal pain and dysentery are present

Selective medium is incubated in microaerophilic atmosphere

The illness typically begins 1 to 7 days after ingestion, with fever and lower abdominal pain that may be severe enough to mimic acute appendicitis. These are followed within hours by dysenteric stools that usually contain blood and pus. The illness is typically self-limiting after 3 to 5 days but may last 1 to 2 weeks. The diagnosis is confirmed by isolation of the organism from the stool. This requires a special medium made selective for *Campylobacter* by inclusion of antimicrobials that inhibit the normal facultative flora of the bowel. Plates must be incubated in a microaerophilic atmosphere, which can now be conveniently generated in a sealed jar by hydration of commercial packs similar to those used for anaerobes.

TREATMENT

Erythromycin may shorten course

Since less than 50% of patients clearly benefit from antimicrobial therapy, cases of *Campylobacter* infection are usually not treated unless the disease is severe or prolonged (lasting longer than 1 week). *C jejuni* is typically susceptible to macrolides and fluoroquinolones but resistant to β-lactams. Erythromycin is considered the treatment of choice but must be given early for maximal effect. Fluoroquinolones are also effective, but resistance is becoming more common.

HELICOBACTER

In 1983, a pair of Australian microbiologists (Warren and Marshall) suggested that gastritis and peptic ulcers were infectious diseases, contradicting long-held beliefs concerning their epidemiology, pathogenesis, and treatment. In the same year, the 10th edition of *Harrison's Principles of Internal Medicine* described peptic ulcers as due to an unfavorable balance between gastric acid–pepsin secretion and gastric or duodenal mucosal resistance. Underlying causes cited included genetic and lifestyle (smoking) as well as psychological factors (anxiety, stress). Treatment with bismuth salts, antacids, and inhibitors of acid secretion gave relief but not cure. Relapsing patients (50% to 80%) were subjected to surgical treatments (vagotomy, partial gastrectomy), which had their own set of complications (reflux, afferent loop syndrome, dumping syndrome). All of this was logical and supported by clinical observations and research studies, but was simply incorrect. The bacteria now called *Helicobacter* had been observed but dismissed because they were so common and its urease was once considered a secretory product of the stomach itself. The Nobel Prize-winning studies that stimulated the reversal of this dogma have led to cures using antimicrobics and new ideas linking *Helicobacter* infection to cancer. This experience has also left us with a sense that we can never be smug about what we "know" in medicine.

Everything we once knew about ulcers was wrong

BACTERIOLOGY: *HELICOBACTER PYLORI*

H pylori has morphologic and growth similarities to the campylobacters, with which they were originally classified. The cells are slender, curved rods with polar flagella. The cell wall structure is typical of other Gram-negative bacteria. Growth requires a microaerophilic atmosphere and is slow (3 to 5 days).

Features are similar to *Campylobacter*

A number of unique bacteriologic features have been found in *H pylori*. The most distinctive is a **urease** whose action allows the organism to persist in low pH environments by the generation of ammonia. The urease is produced in amounts so great (6% of bacterial protein) that its action can be demonstrated within minutes of placing *H pylori* in the presence of urea. Another secreted protein called the **vacuolating cytotoxin** (VacA) causes apoptosis in eukaryotic cells it enters generating multiple large cytoplasmic vacuoles (**Figure 32–3**). The vacuoles are felt to be generated by the toxin's formation of channels in lysosomal and endosomal membranes.

Urease raises pH rapidly

VacA injures lysosomal and endosomal membranes

Most *H pylori* strains also contain a PAI with 30+ genes, most of which code for elements of an **injection secretion system.** The secretion system injects VacA and a protein Cag, also coded in the PAI, into epithelial cells. Once in the cell, Cag induces changes in multiple cellular proteins and has a strong association with virulence (see Pathogenesis).

PAI contains genes for secretion system

HELICOBACTER GASTRITIS

CLINICAL CAPSULE

Helicobacter infections are limited to the mucosa of the stomach, and most are asymptomatic even after many years. Burning pain in the upper abdomen, accompanied by nausea and sometimes vomiting, is a symptom of gastritis. Ulcers may cause additional symptoms, depending on their anatomic location. It is common for gastric and duodenal ulcers to be unrecognized by the patient until they cause frank bleeding or rupture.

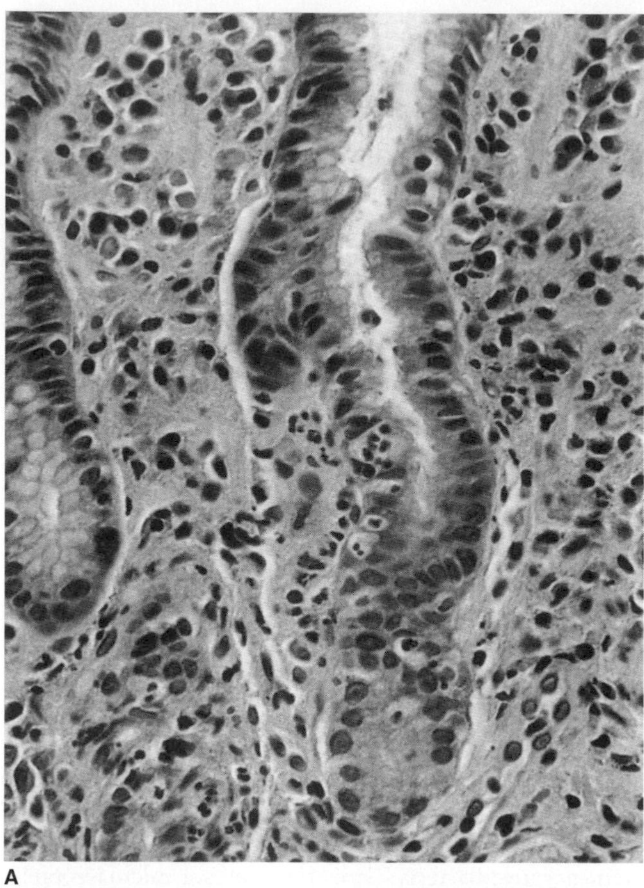

A

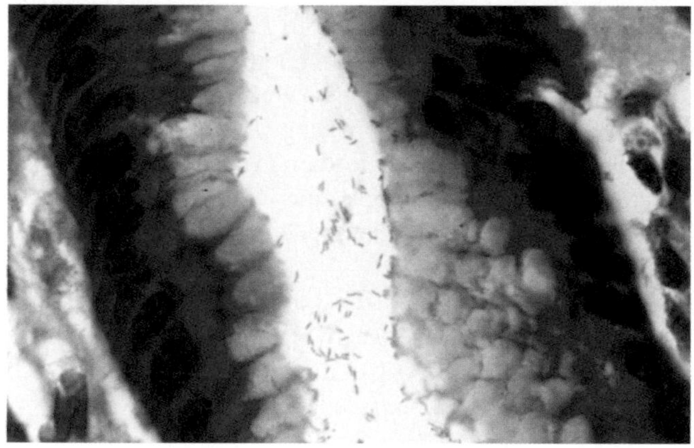

B

FIGURE 32–3. *Helicobacter* gastritis. A. Gastric mucosa shows infiltration of neutrophils and destruction of epithelial cells. **B.** High magnification shows curved bacilli and vacuolization of some cells. (Reproduced with permission from Connor DH, Chandler FW, Schwartz DQA, Manz HJ, Lack EE (eds). *Pathology of Infectious Diseases*, vol. 1. Stamford, CT: Appleton & Lange; 1997, A, Figure 60-2; B, Figure 60-3.)

EPIDEMIOLOGY

Infection is transmitted by human fecal or gastric secretions

Gastric colonization is prevalent worldwide

Infection with *H pylori* causes what is perhaps the most prevalent disease in the world. The organism is found in the stomachs of 30% to 50% of adults in developed countries, and it is almost universal in developing countries. The exact mode of transmission is not known, but is presumed to be person to person by the fecal–oral route or by contact with gastric secretions in some way. Colonization increases progressively with age, and children are believed to be the major amplifiers of *H pylori* in human populations. A declining prevalence in developed countries may be due to decreased transmission because of less crowding and frequent exposure to antimicrobial agents.

Once established, the same strain persists for years, decades, even for life. Molecular epidemiologic analysis indicates the strains themselves have strong linkages to ethnic origins that can be traced back to the earliest known patterns of human migration. *H pylori* has been called an "accidental tourist," which was established in the stomachs of humans thousands of years ago and remained bound to the original population as it dispersed from continent to continent.

Colonization persists indefinitely

Ethnic links are strong

H pylori is the most common precursor of gastritis, gastric ulcer, and duodenal ulcer cases which are not due to drugs. In addition *Helicobacter* gastritis caused by Cag^{+} strains is acknowledged to be an antecedent of gastric adenocarcinoma, one of the most common causes of cancer death in the world. It is also linked to a gastric mucosa-associated lymphoid tissue (MALT) lymphoma, which is less common but shows the striking property of regressing with antimicrobial therapy. *H pylori* gained the dubious distinction of being the first bacterium declared a class I carcinogen by the World Health Organization.

H. pylori is the sole non-drug cause of gastritis and ulcers

Adenocarcinoma and lymphoma are preceded by infection

H pylori is exclusive to humans, but other species have been found in the stomachs of a wide range of animals, where they are also associated with gastritis. It is difficult to imagine the old "stress ulcer" theories surviving the discovery of a cheetah with *Helicobacter* gastritis. Speculation that domestic animals may serve as a reservoir for human infection has not been confirmed.

Other *Helicobacter* species occur in animals

PATHOGENESIS

To persist in the hostile environs of the stomach, *H pylori* uses many mechanisms to adhere to the gastric mucosa and survive the acid milieu of the stomach (**Figure 32–4**). Motility provided by the flagella allows the organisms to swim to the less acid pH locale beneath the gastric mucus, where the urease further creates a more neutral microenvironment by ammonia production. At the mucosa, adherence is mediated by multiple outer membrane proteins which bind to the surface of gastric epithelial cells.

Urease ammonia production neutralizes acid

Motility facilitates surface microenvironment

H pylori colonization is almost always accompanied by a cellular infiltrate ranging from minimal mononuclear infiltration of the lamina propria to extensive inflammation with

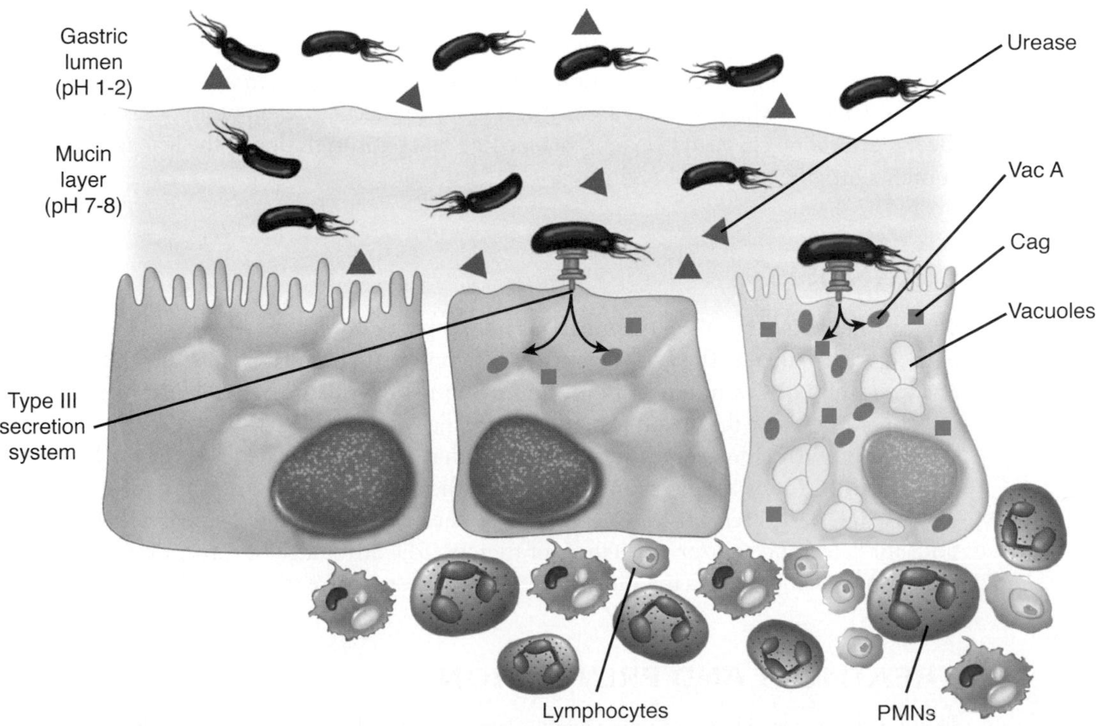

FIGURE 32–4. Helicobacter gastritis, cellular view. From the low pH gastric lumen *H pylori* swims beneath the mucus layer, produces urease, and persists in a more physiologic environment. A type III secretion system injects the vacuolating cytotoxin (VacA), and Cag, into the gastric cells. Acute and chronic inflammatory cells gather in the submucosa. PMNs, polymorphonuclear neutrophils.

Multiple factors stimulate inflammation

VacA directly induces cellular changes and death

Cag alters cytoskeleton

neutrophils, lymphocytes, and microabscess formation. This inflammation may be due to toxic effects of the urease or the VacA transported into the gastric epithelial cells by the type III secretion system. Inside the cell, VacA causes vacuolization of endosomal compartment and has other effects including altered T-cell function. The Cag protein is injected into the gastric epithelial cell by the secretion system, where it triggers multiple enzymatic reactions including those that cause reorganization of the actin cytoskeleton. Cag may also contribute by stimulation of cytokines and by a protein that recruits neutrophils to the gastric mucosa. Added together urease, Cag, and VacA provide ample explanation for the gastritis that is universal in *H pylori* infection. This prolonged and aggressive inflammatory response could lead to epithelial cell death and ulcers, but the progression from gastritis to ulcer has not been adequately explained. ::: Secretion systems, p. 368

Carcinogenic mechanisms are unknown

That decades of inflammation and assault by the virulence factors just described could cause metaplasia, and eventually cancer seems logical, but the specific mechanisms of carcinogenesis are unknown. Cag is a leading candidate. A curious paradox is that although Cag^+ strains are associated with ulcers and adenocarcinoma of the lower stomach, they are associated with a decreased incidence of adenocarcinoma of the upper stomach (cardia) and esophagus. The gastric lymphomas may represent neoplastic transformation of B-lymphocyte clones proliferating in response to chronic antigenic stimulation.

IMMUNITY

There is obviously little evidence of natural immunity in an infection that typically lasts for many years. The immunosuppressive effect of virulence factors such as VacA may be responsible in combination with yet to be discovered mechanisms.

HELICOBACTER DISEASE: CLINICAL ASPECTS

MANIFESTATIONS

Epigastric pain and nausea are signs of gastritis

Primary infection with *H pylori* is either silent or causes an illness with nausea and upper abdominal pain lasting up to 2 weeks. Years later, the findings of gastritis and peptic ulcer disease include nausea, anorexia, vomiting, epigastric pain, and even less specific symptoms such as belching. Many patients are asymptomatic for decades, even up to perforation of an ulcer. Perforation can lead to extensive bleeding and peritonitis due to the leakage of gastric contents into the peritoneal cavity.

DIAGNOSIS

Culture or urease detection is diagnostic

Serologic tests demonstrate chronic infection

The most sensitive means of diagnosis is endoscopic examination, with biopsy and culture of the gastric mucosa. The *H pylori* urease is so potent that its activity can be directly demonstrated in biopsies in less than an hour. Noninvasive methods include serology and a urea breath test. For the breath test, the patient ingests ^{13}C- or ^{14}C-labeled urea, from which the urease in the stomach produces products that appear as labeled CO_2 in the breath. A number of methods for detection of antibody directed against *H pylori* are now available. Because IgG or IgA remains elevated as long as the infection persists, these tests are valuable both for screening and for evaluation of therapy. The advantage of direct detection of the organism is that culture is the most sensitive indicator of cure following therapy.

TREATMENT AND PREVENTION

Antimicrobics and bismuth salts achieve lasting cures

H pylori is susceptible to a wide variety of antimicrobial agents. Bismuth salts (eg, Pepto-Bismol), which in the past were believed to act by coating the stomach, also have antimicrobial activity. Cure rates approaching 90% have been achieved with various combinations of bismuth salts and two antibiotics. Metronidazole, tetracycline, clarithromycin, and amoxicillin have been effective. Relapse rates are low, particularly when acid secretion is also controlled with the use of a proton pump inhibitor. These combination regimens must be continued for at least 2 weeks and may be difficult for some patients to tolerate. Prevention of *H pylori* disease awaits further understanding of transmission and immune mechanisms. Prophylactic treatment of asymptomatic persons colonized with *H pylori* is not yet recommended.

Regimen may be difficult to tolerate

CLINICAL CASE

RAW OYSTERS IN RIFLE

On August 17, 1988, a 42-year-old man was treated for profuse, watery diarrhea, vomiting, and dehydration at an emergency room in Rifle, Colorado. On August 15, he had eaten approximately 12 raw oysters from a new oyster-processing plant in Rifle. Approximately 36 hours after eating the oysters, he had sudden onset of symptoms and passed 20 stools during the day before seeking medical attention. Stool culture subsequently yielded toxigenic *Vibrio cholerae* 01, biotype eltor. The patient had no underlying illness, was not taking medications, and had not traveled outside the region during the month before onset.

The oysters had been harvested on August 8, 1988, in a bay off the coast of Louisiana. Approximately 1000 bushels (200,000 oysters) arrived by refrigerator truck at the plant in Rifle on August 11. The patient purchased three dozen of these oysters on August 15. During a 6-day period, eight other persons shared the oysters purchased by the patient. None became ill. Although one of seven tested had a vibriocidal antibody titer of 1:640, none had elevated antitoxic antibody titers, and none had *V cholerae* 01 isolated from stool. Physicians and local health departments were asked to notify the Colorado Department of Health about similar cases, but no cases were reported.

QUESTIONS

■ What is the probable source of this patient's *V cholerae* infection?

A. Oyster bar employee
B. An imported case from Asia
C. Gulf of Mexico
D. Rifle groundwater
E. South America

■ What would you expect a biopsy of this patient's small intestine to show?

A. Hyperemia
B. Pseudomembrane
C. Flask-shaped ulcers
D. Enterocyte necrosis
E. Focal hemorrhage

■ Which of the following measures would be the *least effective* in preventing a recurrence of this outbreak?

A. Disinfecting the plant
B. A new source for oysters
C. Prophylactic rifampin
D. Cooking the oysters

ANSWERS

1(C), **2**(A), **3**(C)

CHAPTER 33

Enterobacteriaceae

She died of a fever
And no one could save her
And that was the end of sweet Molly Malone
But her ghost wheels her barrow
Through streets broad and narrow
Crying cockles and mussels alive, alive o!

—James Yorkston: Irish Ballad

The Enterobacteriaceae are a large and diverse family of Gram-negative rods, members of which are both free-living and part of the indigenous flora of humans and animals. A few are adapted strictly to humans. The Enterobacteriaceae grow rapidly under aerobic or anaerobic conditions and are metabolically active. They are by far the most common cause of urinary tract infections (UTIs), and a limited number of species are also important etiologic agents of diarrhea**.** Spread to the bloodstream causes Gram-negative endotoxic shock, a dreaded and often fatal complication. In 19th century literature and song, dying "of a fever" usually meant typhoid (*Salmonella* ser. Typhi), which, because of its prolonged course and lack of localizing signs, unfortunates like Molly Malone seemed to be dying of fever alone.

GENERAL CHARACTERISTICS

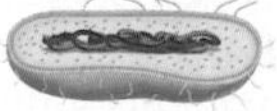

BACTERIOLOGY

The Enterobacteriaceae are among the largest bacteria, measuring 2 to 4 μm in length with parallel sides and rounded ends. Forms range from large coccobacilli to elongated, filamentous rods. The organisms do not form spores or demonstrate acid-fastness.

Rods are large

The cell wall, cell membrane, and internal structures are morphologically similar for all Enterobacteriaceae, and follow the cell plan described in Chapter 21 for Gram-negative bacteria. Components of the cell wall and surface, which are antigenic, have been extensively studied in some genera and form the basis of systems dividing species into serotypes. The outer membrane lipopolysaccharide (LPS) is called the **O antigen.** Its antigenic specificity is determined by the composition of the sugars that form the long terminal polysaccharide side chains linked to the core polysaccharide and lipid A. Cell surface polysaccharides may form a well-defined capsule or an amorphous slime layer and are termed the **K antigen** (from the

O = LPS

K = polysaccharide capsule

H = flagellar protein

Danish Kapsel, capsule). Motile strains have protein peritrichous flagella, which extend well beyond the cell wall and are called the **H antigen.** Many Enterobacteriaceae have surface pili (fimbriae), which are antigenic proteins, but not yet part of any formal typing scheme.

Facultative growth is rapid

Enterobacteriaceae grow readily on simple media, often with only a single carbon energy source. Growth is rapid under both aerobic and anaerobic conditions, producing 2- to 5-mm colonies on agar media and diffuse turbidity in broth after 12 to 18 hours of incubation. All Enterobacteriaceae ferment glucose, reduce nitrates to nitrites, and are oxidase negative.

CLASSIFICATION

Biochemical characteristics establish species

Antigenic characters define serotypes within species

Genus and species designations are based on phenotypic characteristics, such as patterns of carbohydrate fermentation, and amino acid breakdown. The O, K, and H antigens are used to further divide some species into multiple **serotypes.** These types are expressed with letter and number of the specific antigen, such as *Escherichia coli* **O**157:**H**7, the cause of numerous food-borne outbreaks. These antigenic designations have been established only for the most important species and are limited to known antigenic structures. For example, many species lack capsules and/or flagella. In recent years, DNA and rRNA homology comparisons have been used to validate these relationships and establish new ones. The genera containing the species most virulent for humans are *Escherichia, Shigella, Salmonella, Klebsiella,* and *Yersinia.* Other less common but medically important genera are *Enterobacter, Serratia, Proteus, Morganella,* and *Providencia.*

TOXINS

All have LPS

Cytotoxins kill cells

Enterotoxins cause secretion and diarrhea

In addition to the **LPS endotoxin** common to all Gram-negative bacteria, some Enterobacteriaceae also produce **protein exotoxins,** which act on host cells by damaging membranes, inhibiting protein synthesis, or altering metabolic pathways. The end result of these actions may be cell death (cytotoxin) or a physiologic alteration, the net effect of which depends on the function of the affected cell. For example, enterotoxins act on intestinal enterocytes, causing the net secretion of water and electrolytes into the gut to produce diarrhea. Although these toxins are most strongly associated with *E coli, Shigella,* and *Yersinia,* others with the same or very similar actions have now been discovered in other species. Toxins found in another species may differ slightly in protein structure and genetic regulation but still have the same biologic action on host cells. Details of these toxins are discussed later in this chapter in relation to their prototype species.

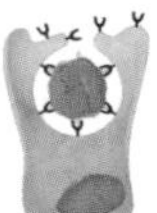

DISEASES CAUSED BY ENTEROBACTERIACEAE

EPIDEMIOLOGY

Present in nature and the intestinal tract

Shigella and *S Typhi* are found only in humans

Most Enterobacteriaceae are primarily colonizers of the lower gastrointestinal tract of humans and animals. Many species survive readily in nature and live freely anywhere that water and minimal energy sources are available. In humans, they are the major facultative components of the colonic bacterial flora and are also found in the female genital tract and as transient colonizers of the skin. Enterobacteriaceae are scant in the respiratory tract of healthy individuals; however, their numbers may increase in hospitalized patients with chronic debilitating diseases. *E coli* is the most common species of Enterobacteriaceae found among the indigenous flora, followed by *Klebsiella, Proteus,* and *Enterobacter* species. *Salmonella* and *Shigella* species are not considered members of the normal flora, although carrier states can exist. *Shigella* and *Salmonella* serovar Typhi are strict human pathogens with no animal reservoir. An overview of these infections is illustrated in **Figure 33–1**.

PATHOGENESIS

Opportunistic Infections

Enterobacteriaceae are often poised to take advantage of their common presence in the environment and normal flora to produce disease when they gain access to normally sterile body sites.

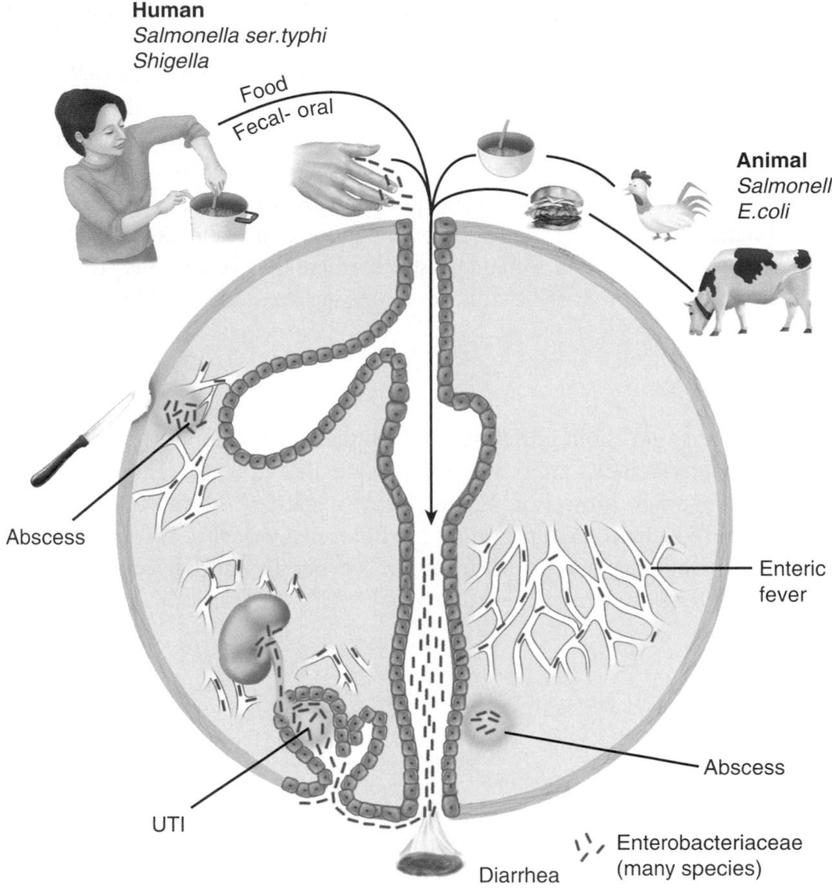

FIGURE 33–1. Enterobacteriaceae disease overview. The external sources of infection are frequently from animal sources, but some pathogens are strictly human (*Shigella*, *Salmonella* ser. Typhi). Endogenous flora is the source of opportunistic infection particularly urinary tract infection (UTI). Bacteria from any source entering the blood may cause endotoxic shock.

Surface structures such as pili are known to aid this process for some species and surely do for many others. Once in deeper tissues, their ability to persist and cause injury is little understood except for the action of LPS endotoxin and the species known to produce exotoxins or capsules. The prototype opportunistic infection is the UTI, in which Enterobacteriaceae gain access to the urinary bladder due to minor trauma or instrumentation. Strains able to adhere to uroepithelial cell can persist and multiply in the nutrient-rich urine, sometimes spreading through the ureters to the renal pelvis and kidney (pyelonephritis). Likewise, mucosal or skin trauma can allow access to soft tissues and aspiration to the lung when the relevant site is colonized with Enterobacteriaceae.

Colonization presents opportunity when defense barriers open

UTI follows access and adherence to bladder mucosa

Intestinal Infections

Salmonella, Shigella, Yersinia enterocolitica, and certain strains of *E coli* are able to produce disease in the intestinal tract. These intestinal pathogens have invasive properties or virulence factors such as cytotoxins and enterotoxins, which correlate with the type of diarrhea they produce. In general, the invasive and cytotoxic strains produce an inflammatory diarrhea called **dysentery** with white blood cells (WBCs) and/or blood in the stool. The enterotoxin-producing strains cause a **watery diarrhea** in which fluid loss is the primary pathophysiologic feature. For a few species, the intestinal tract is the portal of entry, but the disease is systemic as a result of spread of bacteria to multiple organs. **Enteric (typhoid) fever** caused by *Salmonella* ser. Typhi is the prototype of this form of infection.

Cell destruction causes dysentery

Enterotoxins cause watery diarrhea

Enteric fever is a systemic illness

Regulation of Virulence

In addition to adherence pili, LPS, and exotoxins, the Enterobacteriaceae produce a myriad of other virulence factors to cause disease. Many of them are deployed in a complex and sequential

Virulence factors respond to environment

fashion in response to environmental cues (temperature, iron, calcium) or as yet unknown factors. Some members have **injection (type III) secretion systems** that target human cells by delivering a syringe-like injection of virulence factors into the cytoplasm of host cells. ::: Secretion systems, p. 368

Contact secretion systems inject factors

Virulence genes are organized into gene clusters

Expression may be stimulated by environmental cues

PAIs contain foreign DNA

The genes for these factors, located in the chromosome, plasmids, or both, are controlled by interactive regulators that seem to produce each virulence factor exactly when it is needed. The genes themselves are often organized into clusters, which include the genes for the effector molecules as well as their regulatory proteins. This is particularly true for complex characteristics such as invasiveness, which involve multiple sequential steps. Some of these gene clusters are **pathogenicity islands** acquired from another bacterium in the genetically distant past. In particular, PAIs are associated with injection secretion systems, where they contain the structural genes for the injection apparatus, as well as the virulence factors injected. ::: PAI, p. 399

IMMUNITY

Immunity is short-lived

Little is understood about immunity to the broad range of opportunistic infections caused by Enterobacteriaceae. Antibody directed against an LPS core antigen has been shown to provide a degree of protection against Gram-negative endotoxemia, but the diversity of antigens and virulence factors among the Enterobacteriaceae is too great to expect broad immunity. Immunity to intestinal infection is generally short-lived and is discussed where it is relevant to specific intestinal pathogens.

ENTEROBACTERIACEAE: CLINICAL ASPECTS

MANIFESTATIONS

UTI and acute diarrhea are most common

The Enterobacteriaceae produce the widest variety of infections of any group of microbial agents, including two of the most common infectious states, UTI and acute diarrhea. UTIs are manifested by dysuria and urinary frequency when infection is limited to the bladder, with the addition of fever and flank pain when the infection spreads to the kidney. Enterobacteriaceae are by far the most common cause of UTIs, and the most common species involved is *E coli.* The features of UTIs are discussed in more detail in Chapter 63.

DIAGNOSIS

MacConkey agar demonstrates lactose fermentation

Selective media required for *Salmonella* and *Shigella* in stools

Gene probes allow direct detection

Culture is the primary method of diagnosis; all Enterobacteriaceae are readily isolated on routine media under almost any incubation conditions. Special indicator media such as MacConkey agar are commonly used in primary isolation to speed separation of the many species. For example, the common pathogens *E coli* and *Klebsiella* typically ferment lactose rapidly, producing acid (pink) colonies on MacConkey agar, whereas the intestinal pathogens *Salmonella* and *Shigella* do not. Separation of the intestinal pathogens from all the other Enterobacteriaceae in stool requires highly selective media designed solely for this purpose. These are discussed as they relate to individual pathogens. Improved understanding of the genetic and molecular basis for virulence has led to the development of direct nucleic acid and immunodiagnostic techniques for direct detection of toxin, adhesin, or invasin genes in clinical material (eg, stool). These methods are still too expensive for use in clinical laboratories but are of extraordinary value in epidemiologic work and clinical research.

TREATMENT

Susceptibility to antimicrobials is highly variable

Antimicrobial therapy is crucial to the outcome of infections with members of the Enterobacteriaceae. Unfortunately, combinations of chromosomal and plasmid-determined resistance render them the most variable of all bacteria in susceptibility to antimicrobial agents. They are usually resistant to high concentrations of penicillin G, erythromycin, and clindamycin, but may be susceptible to the broader-spectrum β-lactams, aminoglycosides, tetracycline, chloramphenicol, sulfonamides, quinolones, nitrofurantoin, and the polypeptide antibiotics. Because the probability of resistance varies among genera and in different epidemiologic settings, the susceptibility of any individual strain must be determined by in vitro tests. Typical patterns of resistance for some of the more common Enterobacteriaceae appear in Appendix 23–1.

ESCHERICHIA COLI

BACTERIOLOGY

Most strains of *E coli* ferment lactose rapidly and produce indole. These and other biochemical reactions are sufficient to separate it from the other species. There are over 150 distinct O antigens and a large number of K and H antigens, all of which are designated by number. The antigenic formula for serotypes is described by linking the letter (O, K, or H) and the assigned number of the antigen(s) present (eg, **O**111:**K**76:**H**7).

Serotypes use O, H, K antigens

PILI

Pili play a role in virulence as mediators of attachment to human epithelial surfaces. They show marked tropism for different epithelial cell types, which is determined by the availability of their specific receptor on the host cell surface. Most *E coli* express **type 1** or common pili. Type 1 pili bind to the D-mannose residues commonly present on epithelial cell surfaces and thus mediate binding to a wide variety of cell types. More specialized pili are found in subpopulations of *E coli*. **P pili** bind to digalactoside (Gal–Gal) moieties on certain mammalian cells, including uroepithelial cells and erythrocytes of the P blood group. Pili that mediate binding to enterocytes are found among the diarrhea-causing *E coli* and are specific to the pathogenic type as shown in **Figure 33–2** and listed in **Table 33–1**. *E coli* also causes diarrhea in animals, and different sets of pili exist with host-specific tropism for their enterocytes. The receptor(s) for the enteric pili are not known in detail but include glycolipids and glycoproteins on the enterocyte surface.

Type I pili bind mannose

P pili bind uroepithelial cells

Pili of diarrhea strains bind enterocytes

The genetics of pilin expression is complex. The genes are organized into multicistronic clusters that encode structural pilin subunits and regulatory functions. Pili of different types may coexist on the same bacterium, and their expression may vary under different

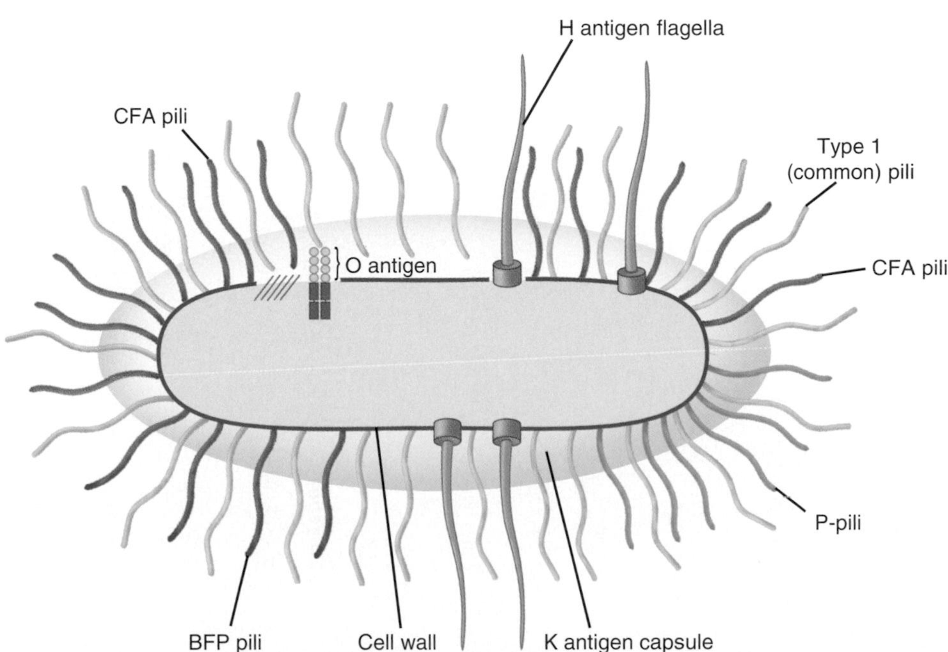

FIGURE 33–2. Antigenic structure of *Escherichia coli*. The O antigen is contained in the repeating polysaccharide units of the lipopolysaccharide (LPS) in the outer membrane of the cell wall. The H antigen is the flagellar protein. The K antigen is the polysaccharide capsule present in some strains. Most *E coli* have type I (common) hair-like pili extending from the surface. Some *E coli* have specialized P, colonization factor antigens (CFAs), or bundle-forming pili (Bfp), as well as type I pili.

TABLE 33–1 Characteristics of Pathogenic Enterobacteriaceae

		VIRULENCE FACTORS							
		SURFACE							
	DIAGNOSTIC ANTIGENS	PILI	ADHESIN OR CAPSULE	EXOTOXIN (S)	PATHOGENIC LESIONS	SECRETED PROTEINSA	GENETICS	TRANSMISSION	DISEASE
Escherichia coli	**O, H, K**								
Common	>150 types	Type I[b]	K1 polysaccharide	α-Hemolysin	Inflammation			Adjacent flora	Opportunistic
Uropathic (UPEC)		Type I[b], P (Gal-Gal)		α-Hemolysin	Inflammation			Fecal flora, ascending	Urinary tract
Enterotoxigenic (ETEC)		CFs		LT, ST	Hypersecretion		Plasmid (CF, LT, ST)	Fecal-oral	Watery diarrhea (travelers)
Enteropathogenic (EPEC)		Bfp	Intimin		A/E, small intestine	Esps	PAI	Fecal-oral	Watery diarrhea
Enteroinvasive (EIEC)			Ipas		Invasion, inflammation, ulcers	Ipas	Large plasmid, PAI	Fecal-oral	Dysentery
Enterohemorrhagic (EHEC)	0157;H7	Lpf, Sfp	Intimin	Stx	A/E, colon, hemorrhage	Esps	PAI	Fecal-oral direct, low dose, cattle	Bloody diarrhea, HUS
Enteraggregative (EAEC)		AAFs			Adherent biofilm				watery diarrhea
Shigella	O serogroups								
S dysenteriae	A (10 types)		Ipas	Stx (AI potent)	Invasion, inflammation, colonic ulcers	Ipas	Large plasmid, PAI	Fecal-oral, direct, low dose	Dysentery (severe), HUS
S flexneri	B (6 types)		Ipas	Stx (variable)	Invasion, inflammation, colonic ulcers	Ipas	Large plasmid, PAI	Fecal-oral, direct, low dose	Dysentery, HUS
S boydii	C (15 types)		Ipas	Stx (variable)	Invasion, inflammation, colonic ulcers	Ipas	Large plasmid, PAI	Fecal-oral, direct, low dose	Dysentery, HUS
S sonnei	D		Ipas	Stx (variable)	Invasion, inflammation, colonic ulcers	Ipas	Large plasmid, PAI	Fecal-oral, direct, low dose	Dysentery, HUS

Salmonella enterica	**O, H_1, H_2, K**								
Serotypes	>2000 serovars	Type I[b]			Ruffles, invasion, inflammation	Inv, Spa, others	PAI	Fecal-oral, animals and humans	Gastroenteritis, sepsis
Typhi	O group D	Type I[b]	Vi polysaccharide		Macrophage survival, RES growth	As in serotypes[c]	PAI	Fecal-oral, moderate dose, humans only	Enteric (typhoid) fever
Yersinia	***O, H***								
Y pestis			Invasin	Protease, fibrinolysin	RES growth, bacteremia, pneumonia	Yops	PAI	Rats, flea bite, aerosol (human)	Plague
Y pseudotuberculosis	10 types		Invasin		RES growth, microabscesses	Yops	PAI	Fecal-oral, animal	Mesenteric adenitis
Y enterocolitica	>50 types		Invasin		RES growth, microabscesses	Yops	PAI	Fecal-oral, animals	Mesenteric adenitis, enteric fever
Klebsiella	70 capsular types	Pili	Polysaccharide					Adjacent flora	Opportunistic, pneumonia
Enterobacter								Adjacent flora	Opportunistic
Serratia								Adjacent flora	Opportunistic
Citrobacter								Adjacent flora	Opportunistic
Proteus								Adjacent flora	Opportunistic

A/E, attaching and effacing lesion; Bfp, bundle-forming pili; CFs, colonizing factor antigens; Esps, *E coli*–secreted proteins; HUS, hemolytic uremic syndrome; Ipas, invasion protein antigens; LT, labile toxin; PAI, pathogenicity island; RES, reticuloendothelial system; ST, stable toxin; Yops, *Yersinia* outer membrane proteins.

[a]Delivered by injection (type III) secretion system.

[b]Bind to mannose.

[c]No animal model, presumed to be similar to *S enterica* serotypes.

Type I has on–off switch

environmental conditions. Type 1 pilin expression can be turned "on" or "off" by inversion of a chromosomal DNA sequence containing the promoter responsible for initiating transcription of the pilin gene. Other genes control the orientation of this switch.

TOXINS

α-Hemolysin is pore-forming cytotoxin

CNF disrupts intracellular signaling

E coli can produce every kind of protein exotoxin found among the Enterobacteriaceae. These include a pore-forming cytotoxin, inhibitors of protein synthesis, and a number of toxins that alter messenger pathways in host cells. The **α-hemolysin** is a pore-forming cytotoxin that inserts into the plasma membrane of a wide range of host cells in a manner similar to streptolysin O (Chapter 25) and *Staphylococcus aureus* α-toxin (Chapter 24). The toxin causes leakage of cytoplasmic contents and eventually cell death. The more recently discovered **cytotoxic necrotizing factor (CNF)** is often produced in concert with α-hemolysin. CNF is an AB toxin that disrupts G proteins regulating the switching of signaling pathways in the cell cytoplasm between active and inactive states. ::: Pore-forming toxins, p. 397

Shiga toxin is produced by *Shigella* and *E coli*

Inhibits protein synthesis by ribosomal modification

Shiga toxin (Stx) is named for the microbiologist who discovered *Shigella dysenteriae,* and this toxin was once believed to be limited to that species. It is now recognized to exist in at least two molecular forms released by multiple *E coli* and *Shigella* strains on lysis of the bacteria. In the years after the discovery of this toxin, the term Shiga toxin was reserved for the original toxin, and others were called Shiga-like. In this book, the term Stx is used for all molecular variants that have the same mode of action regardless of the species under consideration. Stx is an A-B type toxin. The B unit directs binding to a specific glycolipid receptor (Gb_3) present on eukaryotic cells and is internalized in an endocytotic vacuole. Inside the cell, the A subunit crosses the vacuolar membrane in the trans-Golgi network, exits to the cytoplasm and enzymatically modifies the ribosome site (28S-RNA of 60S subunit) where amino acyl tRNA binds. This alteration blocks protein synthesis, leading to cell death (**Figure 33–3**). This action is very similar to the plant toxin Ricin. :::A-B toxins, p. 396

LT ADP-ribosylates G protein

Adenylate cyclase stimulation similar to cholera

Labile toxin (LT) is also an A-B toxin. Its name relates to the physical property of heat lability, which was important in its discovery, and contrasts with the heat-stable toxin described in the text that follows. The B subunit binds to the cell membrane, and the A subunit catalyzes the ADP-ribosylation of a regulatory G protein located in the membrane of the intestinal epithelial cell. This inactivation of part of the G protein complex causes permanent activation of the membrane-associated adenylate cyclase system and a cascade of events, the net effect of which depends on the biologic function of the stimulated cell. If the cell is an enterocyte, the result is the stimulation of chloride secretion out of the cell and the blockage of NaCl absorption. The net effect is the secretion of water and electrolytes into the bowel lumen. The structure and action of LT are nearly identical with that already described for cholera toxin, but LT is less potent than CT. ::: Cholera toxin, p. 566

ST stimulates guanylate cyclase

Stable toxin (ST) toxin is a small peptide that binds to a glycoprotein receptor, resulting in the activation of a membrane-bound guanylate cyclase. The subsequent increase in cyclic GMP concentration causes an LT-like net secretion of fluid and electrolytes into the bowel lumen.

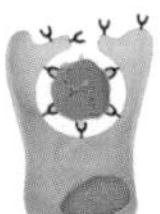

E COLI OPPORTUNISTIC INFECTIONS

CLINICAL CAPSULE

The term UTI encompasses a range of infections from simple cystitis involving the bladder to full-blown infection of the entire urinary tract, including the renal pelvis and kidney (pyelonephritis). The primary feature of cystitis is frequent urination, which often has a painful burning quality. In pyelonephritis, symptoms include fever, general malaise, and flank pain in addition to frequent urination. Cystitis is usually self-limiting, but infection of the upper urinary tract carries a risk of spread to the bloodstream. It is the leading cause of Gram-negative sepsis and septic shock.

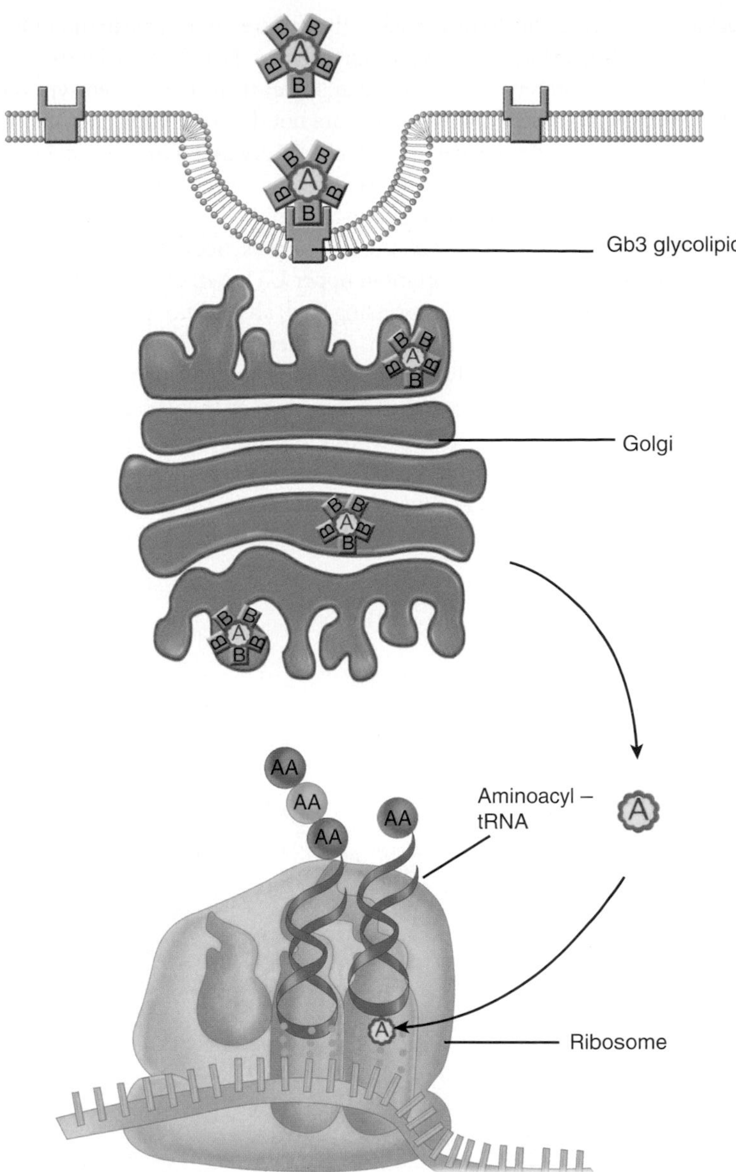

FIGURE 33–3. Stx (Shiga) toxin. The A-B toxin binds to the cytoplasmic membrane, enters in an endocytotic vacuole, and enters the Golgi network. Exiting to the cytoplasm, it combines at ribosome sites involved with tRNA binding. The result is interference with protein synthesis.

URINARY TRACT INFECTION (UTI)

Epidemiology

E coli accounts for more than 90% of the more than 7 million cases of cystitis and 250,000 of pyelonephritis estimated to occur in otherwise healthy individuals every year in the United States. UTIs are much more common in women, 40% of whom have an episode in their lifetime, usually when they are sexually active. The reservoir for these infections is the patient's own intestinal *E coli* flora, which contaminate the perineal and urethral area. In individuals with urinary tract obstruction or instrumentation, environment sources assume some importance.

Perineal flora is reservoir of common cystitis

Pathogenesis

Relatively minor trauma or mechanical disruptions can allow bacteria colonizing the periurethral area brief access to the urinary bladder. These bacteria originally derived from the fecal flora are frequently present in the bladder of women after sexual intercourse. In most instances, they are purged by the flushing action of voiding, but may persist to cause a UTI, depending on host and bacterial factors. Host situations that violate bladder integrity (urinary catheters) or that obstruct

Minor trauma admits *E coli* to the bladder

UPEC cause most UTIs

urine outflow (enlarged prostate) allow the bacteria more time to multiply and cause injury. However, most UTIs are in otherwise healthy women. Here bacterial virulence factors are important, and *E coli* is the prototype UTI pathogen. Fewer than 10 *E coli* serotypes account for the majority of UTI cases, and these UTI serotypes are not the most common ones in the fecal flora. These *E coli* with enhanced potential to produce UTI are called **uropathic *E coli* (UPEC).**

Type I pili adhere to periurethral and bladder cells

P pili prominent in pyelonephritis

The ability of UPEC to produce UTI begins with type 1 pili, which are important for periurethral colonization and for attachment to epithelial cells in the bladder. The presence of P pili adds an additional component because of the presence of their Gal-Gal receptor on uroepithelial cells. This is particularly important in upper UTIs, which extend to the renal pelvis and kidney to cause pyelonephritis. *E coli* possessing P pili are a minor percentage of the fecal flora (less than 20%), but the proportion of P^+ strains is higher in UTI and progressively increases with severity of infection. It is particularly high (70%) in pyelonephritis isolates. Motility driven by flagellar motors may also play a role, and there is evidence that the on/off switching of type 1 pili may allow UPEC to alternate between swimming and adherent phases. Once established, LPS and the production of other virulence factors such as α-hemolysin and CNF cause injury. Spread to the bloodstream leads to LPS-induced septic shock. The pathogenic features that allow UPEC to play such a prominent role in this disease are illustrated in **Figure 33–4.**

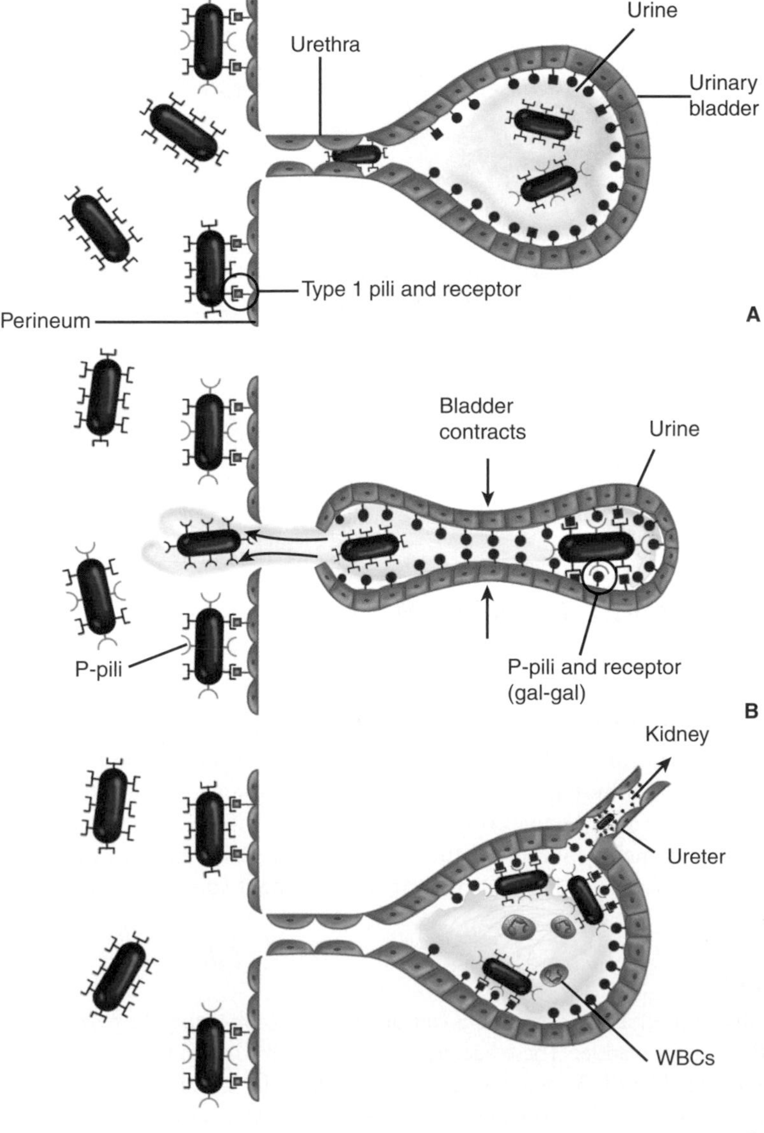

FIGURE 33–4. Urinary tract infection due to *Escherichia coli*. The urinary bladder, perineal mucosa, and short female urethra are shown. *E coli* from the nearby rectal flora have colonized the perineum, utilizing binding by type I (common) pili. *E coli* with P pili are also present but are of no use at this site. **A.** A few *E coli* have gained access to the bladder owing to mechanical disruptions such as sexual intercourse or instrumentation (catheters). Note that receptors for the P pili not present on the perineal mucosa are found on the surface of bladder mucosal cells. **B.** During voiding, the bladder has expelled the *E coli*, which have only type I pili. The P pili–containing bacteria remain behind due to the strong binding to the P (Gal-Gal) receptor. **C.** The remaining *E coli* have multiplied and are causing a urinary tract infection (cystitis) with inflammation and hemorrhage. In some cases, the bacteria ascend the ureter to cause pyelonephritis. WBCs, white blood cells.

OTHER OPPORTUNISTIC INFECTIONS

Meningitis

E coli is one of the most common causes of neonatal meningitis; many features of which are similar to group B streptococcal disease. The pathogenesis involves vaginal *E coli* colonization of the infant via ruptured amniotic membranes or during childbirth. Failure of protective maternal IgM antibodies to cross the placenta and the special susceptibility of newborns surely play a role. Fully 75% of cases are caused by strains possessing the K1 capsular polysaccharide that contains sialic acid and is structurally identical to the group B polysaccharide of *Neisseria meningitidis,* another cause of meningitis. ::: GBS disease, p. 458

Infection from vaginal flora such as group B streptococcus

K1 capsule identical to meningococcus

With the exception of UTIs, extraintestinal *E coli* infections are uncommon unless there is a significant breach in host defenses. Opportunistic infection may follow mechanical damage such as a ruptured intestinal diverticulum, trauma, or involve a generalized impairment of immune function. The virulence factors involved are likely the same as with UTI (eg, pili, α-hemolysin), but have been less specifically studied. Failure of local control of infection can lead to spread and eventually Gram-negative septic shock. A significant proportion of blood isolates have the K1 surface polysaccharide. The particular diseases that result depend on the sites involved and include many of the syndromes covered in Chapters 57 through 66.

Non-UTI infections require some breach of defenses

E COLI INTESTINAL INFECTIONS

CLINICAL CAPSULE

Diarrhea is the universal finding with *E coli* strains that are able to cause intestinal disease. The nature of the diarrhea varies depending on the pathogenic mechanism. Enterotoxigenic and enteropathogenic strains produce a watery diarrhea, the enterohemorrhagic strains produce a bloody diarrhea, and the enteroinvasive strains may cause dysentery with blood and pus in the stool. The diarrhea is usually self-limiting after only 1 to 3 days. The enterohemorrhagic *E coli* are an exception, with life-threatening manifestations outside the gastrointestinal tract due to Shiga toxin production.

Diarrhea-causing *E coli* are conveniently classified according to their virulence properties as **enterotoxigenic (ETEC), enteropathogenic (EPEC), enteroinvasive (EIEC), enterohemorrhagic (EHEC),** or **enteroaggregative (EAEC).** Each group causes disease by a different mechanism, and the resulting syndromes usually differ clinically and epidemiologically. For example, ETEC and EIEC strains infect only humans. Food and water contaminated with human waste and person-to-person contact are the principal means of infection. A summary of the pathogenesis of infection, clinical syndromes, and epidemiology of infection for each enteropathogen is shown in Table 33–1.

Multiple pathogenic mechanisms have their own epidemiologic and clinical features

ENTEROTOXIGENIC *E COLI* (ETEC)

Epidemiology

ETEC are the most important cause of traveler's diarrhea in visitors to developing countries. ETEC also produce diarrhea in infants native to these countries, where they are a leading cause of morbidity and mortality during the first 2 years of life. Repeated bouts of diarrhea caused by ETEC and other infectious agents are an important cause of growth retardation,

Traveler's diarrhea affects children of developing countries

malnutrition, and developmental delay in third-world countries where ETEC are endemic. ETEC disease is rare in industrialized nations, although recent outbreaks suggest that it may be underestimated.

High dose in uncooked foods required

Transmission is by consumption of food and water contaminated by infected human or convalescent carriers. Uncooked foods such as salads or marinated meats and vegetables are associated with the greatest risk. Direct person-to-person transmission is unusual, because the infecting dose is high. Animals are not involved in ETEC disease.

Pathogenesis

LT and/or ST cause fluid outpouring in small intestine

CF pili are required

ETEC diarrhea is caused by strains of *E coli* that produce LT and/or ST enterotoxins in the proximal small intestine. Strains that elaborate both LT and ST cause more severe illness. Adherence to surface microvilli mediated by the colonizing factor (CF) pili is essential for the efficient delivery of toxin to the target enterocytes. The genes encoding the ST, LT, and the CF pili are borne in plasmids. A single plasmid can carry all three sets of genes. The bacteria remain on the surface, where the adenylate cyclase–stimulating action of the toxin(s) creates the flow of water and electrolytes from the enterocyte into the intestinal lumen. The mucosa becomes hyperemic but is not injured in the process. There is no invasion or inflammation.

Immunity

sIgA to LT and CFs may provide some protection

Although there can be more than one episode of diarrhea, infections with ETEC can stimulate immunity. Travelers from industrialized nations have a much higher attack rate than adults living in the endemic area. This natural immunity is presumably mediated by sIgA specific for LT and CFs. The small ST peptides are nonimmunogenic. The disease is of very low incidence in breast-fed infants, underscoring the protective effect of maternal antibody and the importance of transmission by contaminated food and water.

ENTEROPATHOGENIC *E COLI* (EPEC)

Epidemiology

Nursery outbreaks and endemic diarrheas occur in developing world

EPEC strains were first identified as the cause of explosive outbreaks of diarrhea in hospital nurseries in the United States and Great Britain during the 1950s. The link to *E coli* was established on epidemiologic grounds alone using serotyping of stool isolates, no small task. The World Health Organization still recognizes a group of 12 EPEC serotypes. The disease seems to have disappeared in industrialized nations, although it may be underestimated because of the difficulty of diagnosis. In developing countries throughout the world, EPEC account for up to 20% of diarrhea in bottle-fed infants younger than 1 year of age. The reservoir is infant cases and adult carriers with transmission by the fecal–oral route. Nursery outbreaks demonstrate the importance of spread by fomites, which suggests that the infecting dose for infants is low. Adult cases are felt to require a very high infecting dose (10^8 to 10^{10} bacteria).

Pathogenesis

Intimin receptor and Esps are injected

Cytoskeleton modification produces A/E lesion

EPEC initially attach to small intestine enterocytes using bundle-forming (Bfp) pili to form clustered microcolonies on the enterocyte cell surface. The lesion then progresses with localized degeneration brush border, loss of the microvilli and changes in the cell morphology including the production of dramatic "pedestals" with the EPEC bacterium at their apex. The combination of these actions is called the **attachment and effacing (A/E)** lesion (**Figures 33–5**). The many steps involved in the formation of the A/E lesion are genetically controlled in a PAI, which includes the genes for the major EPEC attachment protein, **intimin,** and an injection (type III) secretion system. The secretion system injects over 30 ***E coli* secretion proteins (Esps)** into the host cell cytoplasm including—remarkably—the surface receptor (Tir) for intimin. The other *E coli* secretion proteins perturb intracellular signal transduction pathways, one effect of which is the induction of modifications in enterocyte cytoskeleton proteins (actin, talin). The cytoskeleton accumulates beneath the attached bacteria to form the pedestals and complete the A/E lesion (**Figure 33–6**). The

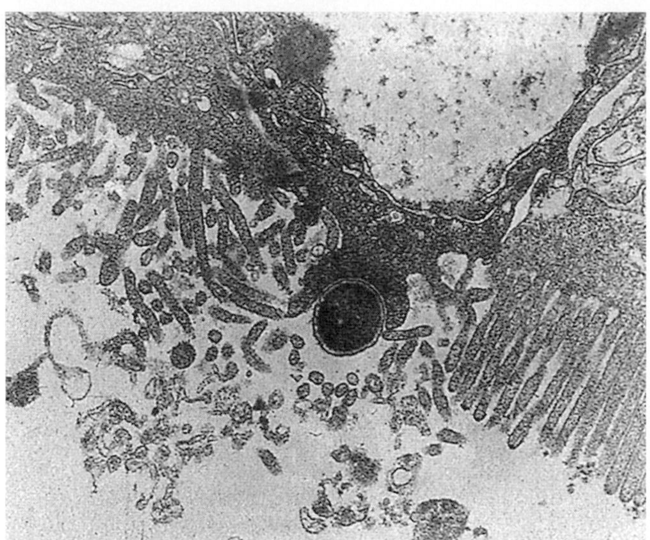

FIGURE 33–5. Enteropathogenic *Escherichia coli* (EPEC) attachment to epithelial cells. The EPEC are attaching to and effacing the microvilli on the epithelial cell surface. The cell's filamentous actin is rearranged at the attachment point. Note the pedestal below the EPEC cell.

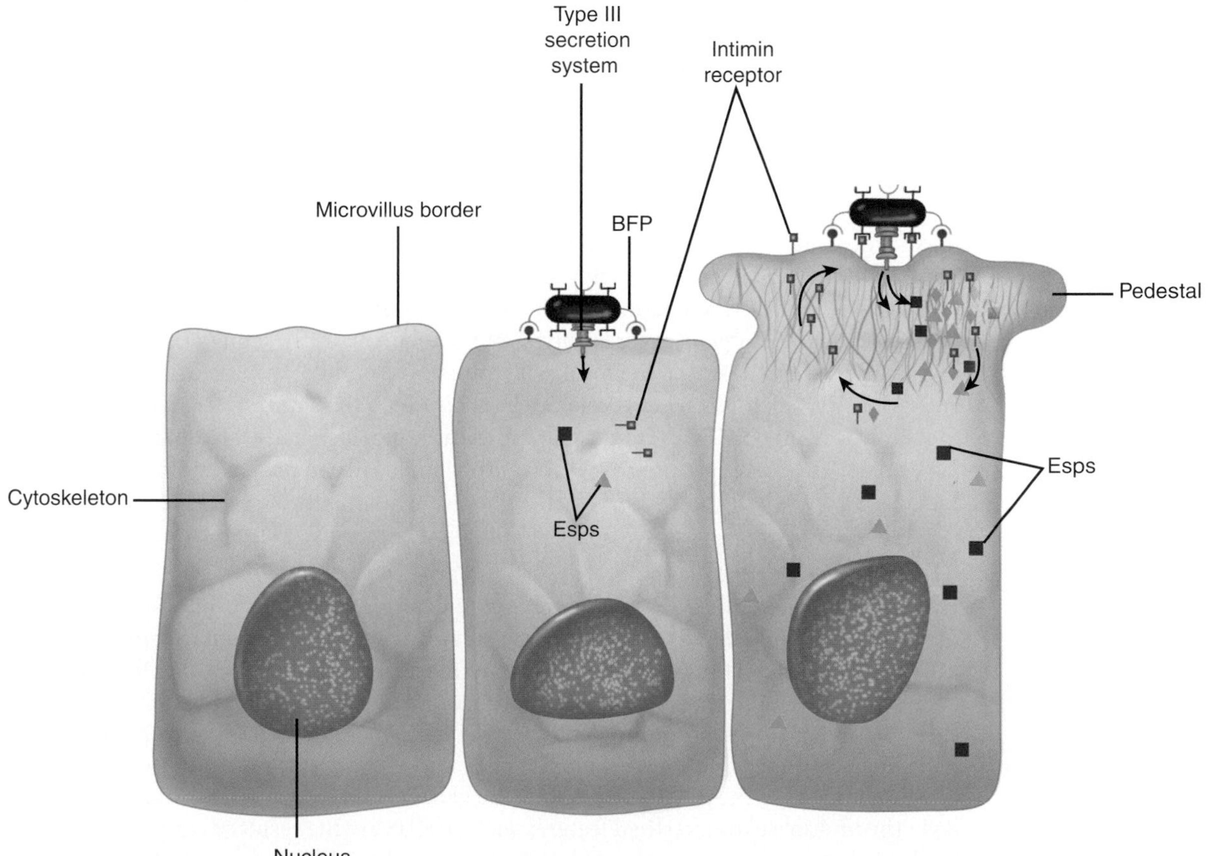

FIGURE 33–6. Enteropathogenic *Escherichia coli* (EPEC) contact secretion system. *(Left)* An enterocyte is shown with a microvillus border and a delicate supporting cytoskeleton. *(Middle)* An EPEC has attached to the cell surface by binding of the bundle-forming pili to receptors on the host cell surface. A type III secretion system apparatus has been inserted into the cell and is exporting secretion proteins (Esps) into the cytoplasm. One of these is the receptor for intimin. *(Right)* The intimin receptor has been inserted below the host cell membrane and is now mediating tight binding to the surface. The other Esps have disrupted multiple cellular functions, including the structure of the cytoskeleton. Cytoskeleton elements have been concentrated to form a pedestal cradling the EPEC (Figure 33–5). Bfp, bundle-forming pili. ::: Secretion systems, p. 368

Esps cause a host of other intracellular disruptions, including mitochondrial injury and induction of apoptosis. The link between these morphologic changes of the A/E and diarrhea is not known, but the injected Esps have also been shown to change electrolyte transport across the luminal membrane. ::: PAI, p. 399

Immunity

Little evidence for immunity

In endemic areas, EPEC can be isolated often from the stool of asymptomatic adults, but unlike ETEC, these strains do not seem to cause traveler's diarrhea in individuals new to the area. This casts doubt on whether adults have acquired immunity or resistance based on physiologic factors.

ENTEROHEMORRHAGIC *E COLI* (EHEC)

Epidemiology

Consumption of contaminated animal products is the main source

EHEC disease and the accompanying **hemolytic uremic syndrome (HUS)** are the result of consumption of products from animals colonized with EHEC strains. It is also clear from secondary cases in families during outbreaks that person-to-person transmission also occurs. This disease occurs more in developed than developing countries.

Bloody diarrhea and HUS linked to O157:H7

EHEC was first recognized in the early 1980s when outbreaks of HUS (hemolytic anemia, renal failure, and thrombocytopenia) were linked to a single *E coli* serotype, O157:H7. Since then, EHEC disease has emerged as an important cause of **bloody diarrhea** in industrialized nations and retained a remarkable but not exclusive relationship with the O157:H7 serotype. Regional and national outbreaks associated with hamburger, unpasteurized juices, and fresh vegetables have caught the attention of the public, the press, and the government.

Low infecting dose facilitates transmission

Modern meat processing facilitates widespread outbreaks

Unpasteurized beverages are another risk

The emergence of EHEC is related to its virulence (see following text), low infecting dose, common reservoir (cattle), and changes in the modern food processing industry that provide us with fresher meat (and bacteria). The infecting dose, estimated to be as low as 100 organisms, is particularly important. This is a level at which food need not come directly from the infected animal, but only be contaminated by it. For example, large modern meat-processing plants can mix EHEC from colonized cattle at one ranch into beef from hundreds of other farms and quickly ship it all over the country. Therefore, the worst outbreaks have been seen in countries with the most advanced food production and distribution systems. If the organisms are ground into hamburger, an infecting dose of EHEC may remain even after cooking if the meat is left rare in the middle. Unpasteurized milk carries an obvious risk, but fruits and vegetables have also been the source for EHEC infection. In these instances, the EHEC from the manure of cattle grazing nearby has contaminated fruit in the field. The bacterial dose from a few "drop" apples (those picked up from the ground) included in a batch of cider has been enough to cause disease.

Pathogenesis

Produce both Stx and A/E lesions

Lesions are in colon

EHEC strains cause the A/E lesions previously described for EPEC, but also produce the Stx toxin. Another difference is that EHEC primarily attacks the colon rather than the small intestine. The interaction of EHEC with enterocytes is much the same as that of EPEC, except that EHEC strains do not form localized microcolonies on the mucosa and specific adhesive pili have not been identified. The outer membrane protein intimin mediates adherence, and the injection secretion system injects the *E coli* secretion proteins, which cause alterations in the host cytoskeleton. The genes for these properties are also found in a PAI. The multiple extraintestinal features such as HUS are the result of circulating Stx.

Stx causes capillary thrombosis and inflammation

Circulating Stx leads to HUS

The A/E features alone are sufficient to cause nonbloody diarrhea. On top of this, Stx production causes capillary thrombosis and inflammation of the colonic mucosa, leading to a hemorrhagic colitis. Although it has not been detected in the blood of human cases, Stx is presumed to be absorbed across the denuded intestinal mucosa. Circulating Stx binds to renal tissue, where its glycoprotein receptor is particularly abundant, causing glomerular swelling and the deposition of fibrin and platelets in the microvasculature. How Stx causes hemolysis is less clear; perhaps the erythrocytes are simply damaged as they attempt to traverse the occluded capillaries. Cases and outbreaks caused by Stx–producing *E coli* of other serotypes are common in many countries.

ENTEROINVASIVE *E COLI* (EIEC)

The biochemistry, genetics, and pathogenesis of EIEC strains are so close to those of *Shigella* (see following text) that our understanding of disease is generally extrapolated from that genus. EIEC disease is essentially a mild version of shigellosis. Epidemiologically, EIEC infections are primarily seen in children younger than 5 years living in developing nations. The occasional documented outbreaks in industrialized nations are usually linked to contaminated food or water. There is a lower incidence of person-to-person transmission of EIEC, which correlates with the observation that the infecting dose is higher than it is for *Shigella*. Humans are the only known reservoir.

EIEC closely resemble *Shigella*

ENTEROAGGREGATIVE *E COLI* (EAEC)

EAEC is associated with a protracted (more than 14 days) watery diarrhea occasionally with blood and mucus. It is seen in infants and children in developing countries. The EAEC strains are defined on the basis of the pattern the bacteria make (stacked-brick) when adhering to cultured mammalian cells. Even though EAEC adheres tightly to the intestinal mucosa, the A/E lesions of the EPEC and EHEC are not present. The pathogenesis of diarrhea involves formation of a thick mucus–bacteria biofilm on the intestinal surface. Inflammatory cells are not seen. *E coli* with a diffuse aggregating pattern (DAEC) have also been observed but their pathogenic significance is not yet established.

Adherence and biofilm cause diarrhea

E COLI INFECTIONS: CLINICAL ASPECTS

MANIFESTATIONS

Opportunistic Infections

The most common symptoms of *E coli* UTI are dysuria and urinary frequency and do not differ significantly in character from those produced by the other less common Gram-negative urinary pathogens discussed in Chapter 63. If the infection ascends the ureters to produce pyelonephritis, fever and flank pain are common and bacteremia may develop. Although *E coli* may have enhanced virulence in the production of pneumonia as well as soft tissue and other infections, no clinical features distinguish these cases from those caused by other members of the Enterobacteriaceae.

Dysuria and frequency are features of UTIs

Intestinal Infections

Infections caused by all of the *E coli* virulence types usually begin with a mild watery diarrhea starting 2 to 4 days after ingestion of an infectious dose. In most instances, the duration of diarrhea is limited to a few days, with the exception of EAEC diarrhea, which can last for weeks. With ETEC and EPEC, the diarrhea remains watery, but with EIEC and EHEC, a dysenteric illness follows. Some EPEC cases may also become chronic. EHEC disease begins like the others but often also includes vomiting. In 90% of cases, this is followed in 1 to 2 days by intense abdominal pain and bloody diarrhea, but fever is not prominent. Some EHEC cases develop into a dysentery that is less severe than that seen in shigellosis. Colonoscopy reveals edema, hemorrhage, and pseudomembrane formation. Resolution usually takes place over a 3- to 10-day period, with few residual effects on the bowel mucosa.

ETEC and EPEC diarrhea is watery

EHEC diarrhea is bloody

HUS develops as a complication in about 10% of cases of EHEC hemorrhagic colitis, primarily in children under 10 years of age. The disease begins with oliguria, edema, and pallor, progressing to the triad of microangiopathic hemolytic anemia, thrombocytopenia, and renal failure. The systemic effects are often life-threatening, requiring transfusion and hemodialysis for survival. The mortality rate is 5%, and up to 30% of those who survive suffer sequelae such as renal impairment or hypertension.

HUS begins as oliguria and may progress to renal failure

DIAGNOSIS

Like the rest of the Enterobacteriaceae, *E coli* is readily isolated in culture. In UTIs, the bacteria typically reach high numbers (higher than 10^5/mL), which makes them readily detectable

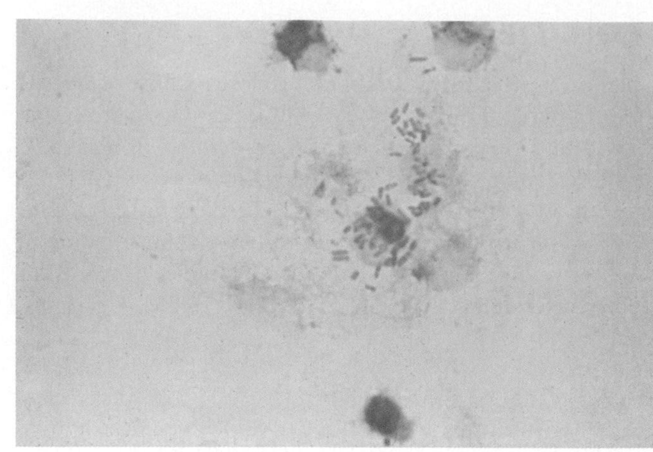

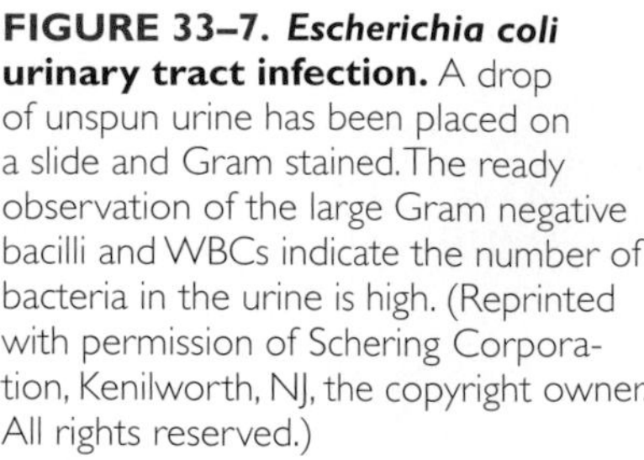

FIGURE 33–7. *Escherichia coli* urinary tract infection. A drop of unspun urine has been placed on a slide and Gram stained. The ready observation of the large Gram negative bacilli and WBCs indicate the number of bacteria in the urine is high. (Reprinted with permission of Schering Corporation, Kenilworth, NJ, the copyright owner. All rights reserved.)

Numbers in urine are high

Diarrhea requires immunoassay or gene probe

Sorbitol agar screens for O157:H7

by Gram stain even in an unspun urine specimen (**Figure 33–7**). For the diagnosis of intestinal disease, separating the virulent types discussed previously from the numerous other *E coli* strains universally found in stool presents a special problem. A myriad of immunoassay and nucleic acid amplification methods have been described that are able to detect the toxins (LT, ST, Stx) or genes associated with virulence. These methods work but are still too expensive to be practical, especially in developing countries where ETEC, EIEC, EPEC, and EAEC are prevalent. A screening test for EHEC takes advantage of the observation that the O157:H7 serotype typically fails to ferment sorbitol. Incorporating sorbitol in place of lactose in MacConkey agar provides an indicator medium from which suspect (colorless) colonies can be selected and then confirmed with O157 antisera. This procedure has become routine in areas where EHEC is endemic but does not detect the non-O157 EHEC strains.

TREATMENT

Resistance influences antimicrobial selection

Acute uncomplicated UTIs are often treated empirically. Because of widespread resistance to earlier agents like ampicillin, trimethoprim/sulfamethoxazole (TMP-SMX) or fluoroquinolones are used for this purpose. Selection of other antimicrobials must be guided by antimicrobial susceptibility testing of the patient's isolate. ::: Antimicrobial susceptibility testing, p. 416

Antimicrobics may help all but EHEC

Because most *E coli* diarrheas are mild and self-limiting, treatment is usually not an issue. When it is, rehydration and supportive measures are the mainstays of therapy, regardless of the causative agent. In the case of EHEC with hemorrhagic colitis and HUS, heroic supportive measures such as hemodialysis or hemapheresis may be required. Treatment with TMP-SMX or fluoroquinolones reduces the duration of diarrhea in ETEC, EIEC, and EPEC infection. Because the risk of HUS may be increased by the use of antimicrobial agents, their use is contraindicated when EHEC is even suspected. Antimotility agents are not helpful and are contraindicated when EIEC or EHEC could be the etiologic agent.

PREVENTION

Avoid uncooked foods

Chemoprophylaxis works for defined periods

Traveler's diarrhea is usually little more than an inconvenience. Because the infecting dose is high, the incidence of the disease can be greatly reduced by eating only cooked foods and peeled fruits and drinking hot or carbonated beverages. Avoiding uncertain water, ice, salads, and raw vegetables is a wise precaution when traveling in developing countries. High-priced hotel accommodations have no protective effect. Chemoprophylaxis against traveler's diarrhea is not routinely recommended. TMP-SMX or ciprofloxacin have been recommended for a short-term (less than 2 weeks) for those at high risk for disease resulting from such chronic conditions as achlorhydria, gastric resection, prolonged use of H_2 blockers or antacids, and underlying immunosuppressive diseases.

These public health measures apply equally to EHEC, but here prevention is more difficult because the infecting dose is so low. Cooking hamburgers all the way through is sensible, but no one is recommending abstinence from salads when at home. Recent U.S. recommendations for the irradiation of meats and the extension of pasteurization requirements to fruit juices are largely designed to stem the spread of EHEC. ::: Pasteurization, p. 50

Rare hamburgers carry risk for EHEC

SHIGELLA

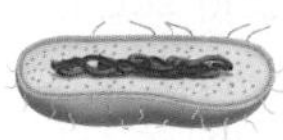

BACTERIOLOGY

Shigella species are closely related to *E coli*. Most fail to produce gas when fermenting glucose and do not ferment lactose. Their antigenic makeup has been characterized in a manner similar to that of *E coli* with the exception that they lack flagella and thus H antigens. All *Shigella* species are nonmotile. The genus is divided into four species, which are defined by biochemical reactions and specific O antigens organized into serogroups. The species are *Shigella dysenteriae* (serogroup A), *Shigella flexneri* (serogroup B), *Shigella boydii* (serogroup C), and *Shigella sonnei* (serogroup D). All but *S sonnei* are further subdivided into a total of 38 individual O antigen serotypes specified by numbers. *Shigella* is the prototype invasive bacterial pathogen. All species are able to invade and multiply inside a wide variety of epithelial cells, including their natural target, the enterocyte. *S dysenteriae* type A1, the Shiga bacillus, is the most potent producer of Stx. Other *Shigella* species produce various molecular forms and quantities of Stx.

O antigens and biochemicals define four species

Invasiveness and Stx production are virulence factors

SHIGELLOSIS

CLINICAL CAPSULE

Shigella is the classic cause of dysentery, which is typically spread person to person under poor sanitary conditions. The illness begins as a watery diarrhea but evolves into an intense colitis with fever and frequent small-volume stools that contain blood and pus. Despite the invasive properties of the causal organism, the infection usually does not spread outside the intestinal tract.

EPIDEMIOLOGY

Shigellosis is a strictly human disease with no animal reservoirs. Worldwide, it is consistently one of the most common causes of infectious diarrhea with over 150 million cases and 600,000 deaths per year. As with almost all infectious diarrheas, the incidence is related to general levels of sanitation, but *Shigella* disease remains important in both developed and developing countries. This high prevalence despite lack of a non-human reservoir is primarily due to the highly efficient transmission by the fecal–oral route. This spread by person-to-person contact is so effective because the infecting dose is extremely low, as few as 10 organisms in some studies. The secondary attack rates among family members are as high as 40%. *Shigella* is also spread by food or water contaminated by humans.

Strictly human disease

Low infecting dose facilitates fecal–oral spread

The incidence and spread of shigellosis is directly related to personal and community sanitary practices. In developed countries, it is largely a pediatric disease. In countries where the sanitary infrastructure is inadequate and in institutions plagued by crowding

Personal and community sanitary practices determine incidence

Wars and disasters create outbreaks

and poor hygienic conditions, the disease may be more widespread. Wartime and natural disasters create similar circumstances. The most common species are *S sonnei* and *S flexneri,* with *S dysenteriae* largely limited to underdeveloped tropical areas. *S dysenteriae,* type 1 produces the most severe disease, historically known as "bacillary dysentery." This condition has slowed the march of many an army; it was the leading cause of death in the notorious Andersonville prison camp during the American Civil War.

PATHOGENESIS

Bacteria pass stomach acid and invade colon

Shigella, unlike *Vibrio cholerae* and most *Salmonella* species, is acid-resistant and survives passage through the stomach to reach the intestine. Once there, the fundamental pathogenic event is invasion and destruction of the human colonic mucosa. This triggers an intense acute inflammatory response with mucosal ulceration and abscess formation. The steps involved in this process describe one of the richest tales in bacterial pathogenesis (**Figure 33–8**). Most of the research has been done with *Shigella flexneri* but there is no reason to believe it does not apply equally to the three other species and to EIEC.

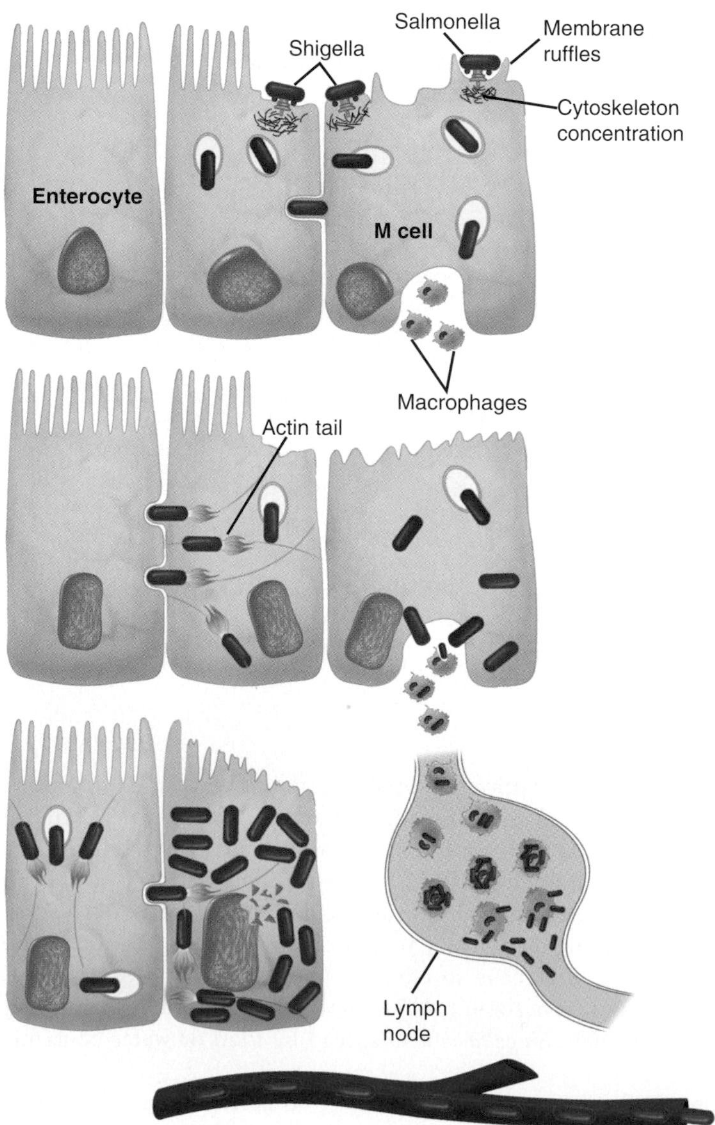

FIGURE 33–8. Invasion by *Shigella flexneri* and *Salmonella* serotype Typhi. The *Shigella* and *Salmonella* are shown invading the intestinal M cells, but taking different paths after escaping the endocytotic vacuole. The *Shigella* multiplies in the cell and propels itself through the cytoplasm to invade adjacent cells, and the *Salmonella* passes through the cell to the submucosa, where it is taken up by macrophages. Serotype Typhi is able to multiply in the macrophages in the lymph node and other reticuloendothelial sites. Both organisms induce apoptosis in their host cells. In the case of *Shigella,* this produces a mucosal ulcer; in the case of Typhi, it leads to seeding of the bloodstream and typhoid fever.

Shigella initially cross the mucosal membrane by entering the follicle-associated M cells of the intestine, which lack the highly organized brush borders of absorptive enterocytes. The *Shigella* adhere selectively to M cells, enter and then transcytose through them into the underlying collection of macrophages. Inside macrophages, the organisms escape from the phagosome to the cytoplasm and activate programmed cell death (apoptosis) in the macrophage. Bacteria released from the dead macrophage contact the basolateral side of enterocytes and initiate a multistep invasion process mediated by a set of **invasion plasmid antigens** (IpaA, IpaB, IpaC). On contact with the enterocyte, these proteins are injected by an injection secretion system and induce cytoskeleton reorganization, actin polymerization, and other changes particularly at the cell surface. Rather than create the A/E lesions of the EPEC and EHEC, this cytoskeleton modification process induces engulfment and internalization of *Shigella* into the host cell by endocytosis. ::: M cell, p. 22

Transcytose M cells to macrophages

Invade enterocytes from dead macrophages

Injected Ipa proteins induce endocytosis

Shigella brought into cells are highly adapted to the intracellular environment and make unique use of it to continue the infection. Although initially the bacteria are surrounded by a phagocytic vacuole, they quickly escape and enter the cytoplasmic compartment of the host cell. Almost immediately, they orient parallel with the filaments of the actin cytoskeleton of the cell and initiate a process in which they control polymerization of the monomers that make up the actin fibrils. This process creates an actin "tail" at one end of the microbe, which appears to propel it through the cytoplasm like a comet. This exploitation of the cytoskeletal apparatus allows nonmotile *Shigella* to not only replicate in the cell but to move efficiently through it. Apparently, the cell's microtubule network is an obstruction so the bacteria produce an enzyme that digests them. One microbiologist called this strategy "bushwhacking through a microtubule jungle."

Escape phagosome to cytoplasm

Polymerization of cytoskeletal actin propels bacteria

Microtubules are digested

Eventually, the bacteria encounter the host cell membrane, much of which is adjacent to the neighboring enterocytes. At this point some *Shigella* rebound, but others push the membrane as much as 20 μm into the adjacent cell. This invasion of the neighboring enterocyte forms finger-like projections, which eventually pinch off, placing the bacterium within a new cell but surrounded by double membrane. The organisms then lyse both membranes and are released into the cytoplasm, free to begin their relentless invasion anew.

Adjacent enterocytes are invaded directly

Double-membrane lysis restarts process

The cell-by-cell extension of this process radially destroys enterocytes and creates focal ulcers in the mucosa, particularly in the colon. The ulcers add a hemorrhagic component and allow *Shigella* to reach the lamina propria, where they evoke an intense acute inflammatory response. Extension of the infection beyond the lamina is unusual in healthy individuals. The diarrhea created by this process is almost purely inflammatory, consisting of small-volume stools containing WBCs, red blood cells (RBCs), bacteria, and little else. This is classic dysentery. The disease remains localized to the colonic mucosa. Spread to the bloodstream is uncommon.

Enterocyte invasion creates ulcers

Diarrhea + WBCs + RBCs = dysentery

Some *Shigella* also produce Stx, which is not essential for disease, but does contribute to the severity of the illness. The original and most potent producer of Stx, *S dysenteriae* type 1, is the only *Shigella* with a significant mortality rate in previously healthy individuals. This is probably due to systemic effects of the toxin, which can be the same as previously described for the EHEC, including HUS. The role, if any, of Stx in enterocyte injury and diarrhea is uncertain.

Stx increases severity of disease

All virulent *Shigella* and EIEC carry a very large plasmid that has several genes essential for the attachment and entry process, including the Ipa genes. The characteristics of *Shigella* entry and interaction with cellular elements are very similar to those observed with *Listeria monocytogenes*, which is Gram-positive and motile and prefers livestock to humans. Finding that such dissimilar bacteria use such similar tactics to infect their preferred host suggests that this represents a common thread in the selective pressures for a microbe to become a "successful" enteric pathogen. ::: Listeria invasion, p. 479

Large plasmid-containing Ipa genes is required for virulence

IMMUNITY

Shigella infection produces relatively short-lived immunity to reinfection with homologous serogroups. There is no consensus on the mechanisms involved.

Immunity is brief

SHIGELLOSIS: CLINICAL ASPECTS

MANIFESTATIONS

Watery diarrhea is followed by fever, bloody mucoid stools, and cramping

Mortality significant with *S dysenteriae* type 1

Most infections are self-limiting

Shigella organisms cause an acute inflammatory colitis and bloody diarrhea, which in the most characteristic state presents as a dysentery syndrome—a clinical triad consisting of cramps, painful straining to pass stools (tenesmus), and a frequent, small-volume, bloody, mucoid fecal discharge. However, most clinical shigellosis due to *S sonnei* is a watery diarrhea that is often indistinguishable from that of other bacterial or viral diarrheal illness. The disease usually begins with fever and systemic manifestations of malaise, anorexia, and sometimes myalgia. These nondescript symptoms are followed by the onset of watery diarrhea containing the large numbers of leukocytes detectable by light microscopy. The diarrhea may turn bloody with or without the other classic signs of dysentery. The manifestations may be more severe when *S flexneri,* the species that predominates in the developing world, is involved and most severe with *S dysenteriae* type 1 (*Shiga bacillus*). Although most cases of shigellosis resolve spontaneously after 2 to 5 days, the mortality rate in Shiga epidemics in Asia, Latin America, and Africa has been as high as 20%.

DIAGNOSIS

Selective media are routinely used

O antigens confirm species

All *Shigella* species are readily isolated using selective media (eg, Hektoen enteric agar), which are part of the routine stool culture in all clinical laboratories. These media contain chemical additives empirically shown to inhibit facultative flora (eg, *E coli, Klebsiella*), with relatively little effect on *Shigella* (or *Salmonella*). They also contain indicator systems that use typical biochemical reactions to mark suspect *Shigella* colonies among the other flora. Isolates are identified with further biochemical tests. Slide agglutination tests using O group-specific antisera (A, B, C, D) confirm both the species and the *Shigella* genus.

TREATMENT

Treatment shortens illness and excretion

Ampicillin resistance is common

Several antimicrobial agents have proved effective in the treatment of shigellosis. Because the disease is usually self-limiting, the beneficial effect of treatment is in shortening the duration of the illness and the period of excretion of organisms. Ampicillin was once the treatment of choice, but resistance rates of 5% to 50% have caused a shift to other agents. In recent years, TMP-SMX, fluoroquinolones, ceftriaxone, azithromycin, and nalidixic acid have been used depending on susceptibility testing. Ampicillin is still effective if the patient's isolate is susceptible. Antispasmodic agents may aggravate the condition and are contraindicated in shigellosis and other invasive diarrheas.

PREVENTION

Sanitation, insect control, handwashing, and cooking block transmission

Live attenuated vaccines are under investigation

Standard sanitation practices such as sewage disposal and water chlorination are important in preventing the spread of shigellosis. In certain circumstances, insect control may also be important, because flies can serve as passive vectors when open sewage is present. Good individual sanitary practices, such as handwashing and proper cooking of food, are highly protective. Parenteral vaccines have proved disappointing, and current efforts are directed toward finding orally administered live vaccines that can stimulate mucosal IgA. Many strains, including attenuated *Shigella* mutants, *E coli–Shigella* genetic hybrids and *E coli* with genes for some (but not all) the invasive (Ipa) proteins, are vaccine candidates. The general idea is to find a strain that will go through enough of the multistage process (see Pathogenesis) to stimulate an immune response but stop short of full penetration and spread.

SALMONELLA

BACTERIOLOGY

More than any other genus, *Salmonella* has been a favorite of those who love to subdivide and apply names to biologic systems. At one time, there were over 2000 names for various members of this genus, often reflecting colorful aspects of place or circumstances of the original isolation (eg, *S budapest, S seminole, S tamale, S oysterbeds*). This has now been reduced to a single species, *Salmonella enterica,* with the previous species names relegated to the status of serotypes. All of this is made particularly robust by the fact that, in addition to a large number of the LPS O and some capsular K antigens, the flagellar H antigens of most *Salmonella* undergo phase variation. This adds the prospect of two sets of H antigens to the already complex system. As in *Shigella,* the specific O antigens are organized into serogroups (eg, A, B, . . . K, and so on) to which the two H and K (if present) antigen designations are appended to achieve the full antigenic formula. It is not difficult to understand why microbiologists, when confronted with a salmonella with the antigenic formula O:group B [1,4,12] H:I;1,2, still prefer to call it *Salmonella typhimurium.* The proper name for this organism is *Salmonella enterica* serovar Typhimurium, but indulging in the convenience of elevating the serotype to species status is still common.

Complexity of O, K, and H antigens leads to many serotypes

Historic names persist as serotypes of *S enterica*

Another feature distinguishing *Salmonella* serotypes is their host range. Some are highly adapted to particular mammals or amphibians, and others infect a broad range of hosts. Of interest for medical microbiology are those that infect humans and other animals and those strictly adapted to humans and higher primates. *S enterica* serovar Typhimurium is the prototype for the former and *S enterica* serovar Typhi for the latter. In the following discussions, Typhi is used for the strictly human species that produce enteric (typhoid) fever. Unless otherwise specified, *S enterica* is used for serotypes such as Typhimurium, which are able to infect animals or humans and typically cause gastroenteritis in the latter.

Salmonella species vary in preferred host

S Typhi infects only humans

Salmonellae possess multiple types of pili, one of which is morphologically and functionally similar to the *E coli* type 1 pili, and bind D-mannose receptors on various eukaryotic cell types. Most strains are motile through the action of their flagella. *S Typhi* has a surface polysaccharide called the Vi antigen, but capsules have not been important in the other *Salmonella.*

Type 1 pili and flagella present

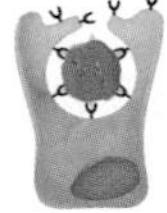

SALMONELLA GASTROENTERITIS (*S ENTERICA*)

CLINICAL CAPSULE

The typical example of *Salmonella* "food poisoning" is the community picnic or bazaar, in which volunteers prepare poultry, salads, and other potential culture media to be eaten later in the day. Because the refrigerators are filled with beer and soda, the food is left out in covered pans. A near physiologic incubation temperature is provided by the still-warm contents and the afternoon sun. This allows the organisms to enter logarithmic growth during the softball game. The bacteria usually produce no noticeable change in the food. One to two days after the feast, a significant portion of the revelers develop abdominal pain, nausea, vomiting, and diarrhea lasting for 3 or 4 days. An investigation points to a particular food such as potato salad or turkey dressing, which is found to have a correlation with both attack rate and severity of illness.

EPIDEMIOLOGY

S enterica gastroenteritis is predominantly a disease of industrialized societies and improper food handling, which allows the transmission from the animal reservoir to humans. The infecting dose or *S enterica* infection varies widely with the serotype (200 to 10^6 bacteria), but is generally considerably higher than *Shigella*. This makes human-to-human transmission by direct contact unlikely, so these infections are transmitted by circumstances in which the bacteria increase their numbers by growth in contaminated foods before ingestion. Achlorhydric individuals or those taking antacids can be infected with smaller inocula. Consistently, salmonellae are a leading cause of food-borne intestinal infection under conditions similar to those described in the preceding capsule.

Infecting dose is higher than *Shigella*

Poultry products, including eggs infected transovarially, are most often implicated as the vehicle of infection of *Salmonella* gastroenteritis. Food preparation practices that allow achievement of an infecting dose by growth of the bacteria in the food before ingestion are most commonly involved. The incidence in the United States is approximately double that of *Shigella,* with 40,000 to 50,000 reported cases per year. This is believed to reflect only about 1% to 5% of the actual infections. The number of cases varies seasonally, with peak incidence in summer and fall.

Poultry products are common source

The highest rates of infection are in children less than 5 years old, persons aged 20 to 30, and those older than 70. If one household member becomes infected, the probability that another will become infected approaches 60%. Nearly one third of all *Salmonella* epidemics occur in nursing homes, hospitals, mental health facilities, and other institutions. Increases in the popularity of raw milk have been associated with outbreaks of *Salmonella* (and *Campylobacter*) infection. Exotic pets such as turtles have also been the source of infection. Humans can also be the source of disease. Fully 5% of patients recovering from gastroenteritis still shed the organisms 20 weeks later. Chronic carriers who are food handlers are an important reservoir in the epidemiology of food-borne disease.

Outbreaks in institutions are common

Human carriers can be a source

In recent years, the number of multistate outbreaks has increased, often through the contamination of foodstuffs during large-scale production at a single plant. A recent outbreak involving peanut butter and paste products is an example. Efficient interstate and international distribution systems that deliver large amounts of the contaminated food over a wide area facilitate spread. Under these conditions, an attack rate as low as 0.5% can still produce many infections because of the large number of persons at risk. It is of concern that relatively small numbers of cases sprinkled over a massive area will be missed by local surveillance systems crippled by budgetary cutbacks.

Modern delivery systems can spread disease efficiently

PATHOGENESIS

Ingested *S enterica* cells that pass the stomach acid and swim through the intestinal mucous layer eventually reach the small bowel. It is not clear whether the initial contact there is with M cells, enterocytes, or both, but initial adherence is probably mediated by pili. On engagement of one of *S enterica's* injection (type III) secretion systems, the creation of membrane "ruffles" dramatically alters the normal host cell architecture within minutes (**Figure 33–9**). These "ruffles" are specialized plasma membrane sites of filamentous actin cytoskeletal rearrangement normally induced by physiologic molecules such as growth factors. The secretion systems inject multiple other effectors coded by genes located in PAIs inserted into the *Salmonella* genome. The virulence factors coded by the genes in the PAIs are either components of the apparatus of the injection secretion system or the effector proteins it injects.

Adherence triggers surface ruffles

Secretion system genes are in PAIs

The ruffles seem to engulf the organism in an endocytotic vacuole and allow it to transcytose from the apical surface to the basolateral membrane. Once in the cell, the organisms continue on through and enter the lamina propria, where they induce a profound inflammatory response and are phagocytosed by neutrophils and macrophages. Persistence in the lamina propria is aided by their ability to kill macrophages intracellularly by multiple mechanisms including induction of apoptosis. This process contrasts with *Shigella,* which escapes the endocytotic vacuole (and double vacuole) to the cytoplasm and prefers to invade adjacent enterocytes rather than move through to the submucosa.

Ruffles induce endocytosis

Progress to lamina propria

Macrophage apoptosis aids survival

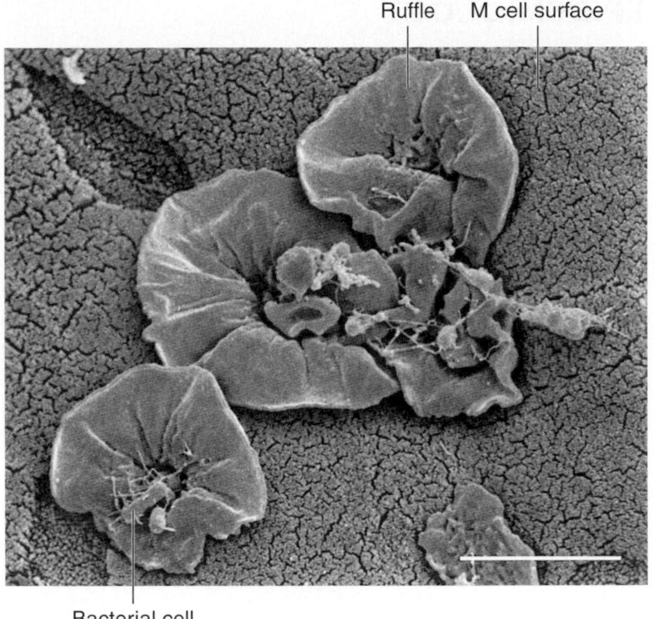

FIGURE 33–9. *Salmonella* ruffles. S serotype Typhimurium is shown inducing wave-like ruffles on an intestinal M cell. This leads to induction of uptake of the bacteria by the M cell. (Reproduced with permission from Nester EW, Anderson DG, Roberts CE Jr, Nester MT. *Microbiology: A Human Perspective*, 6th ed. New York: McGraw-Hill, 2008, Figure 19.5.)

Although some enterotoxins have been described in *Salmonella,* their role in diarrhea is unclear. The best estimate is that the invasion and transcytosis of enterocytes together with the associated increased vascular permeability and inflammatory response are enough to account for the diarrhea. The release of prostaglandins and chemotactic factors may trigger inflammation and biochemical changes in enterocytes. Although the process remains localized to the mucosa and submucosa with most *S enterica* strains, some invade more deeply, reaching the bloodstream and distant organs. Some serotypes (eg, choleraesuis) even invade so rapidly that they produce minimal diarrhea and are isolated more frequently from the blood than stool.

Enterotoxin role is unclear

Invasion and inflammation cause diarrhea

IMMUNITY

Evidence that both humoral and cell-mediated immune responses are stimulated by infection with *S enterica* is ample. How these processes relate to immunity and control of the bacterial infection is largely unknown.

Immune mechanisms unclear

ENTERIC (TYPHOID) FEVER (*SALMONELLA* SEROVAR TYPHI)

CLINICAL CAPSULE

Typhoid fever has a slow, insidious onset, and, if untreated, lasts for weeks. The primary symptom is a slowly rising fever often accompanied by abdominal pain but little else. It ends either by a gradual resolution or in death due to complications (eg, rupture of the intestine or spleen). Family members may note only the extended fever, although physicians may observe a subtle rash or feel an enlarged spleen. Diarrhea may occur once or twice during the course but is not a consistent feature.

EPIDEMIOLOGY

Cases are traceable to a human source

Fecal–oral transmission requires moderate dose

Typhoid fever is a strictly human disease. Chronic carriers of serotype Typhi are the primary reservoir. Some patients become chronic carriers for years (hence the infamous "Typhoid Mary" Mallon), usually because of chronic infection of the biliary tract when gallstones are present. All cases can and should be traced back to their human source. If a patient with typhoid has not traveled to an endemic area, the source must be a visitor or someone else who prepared food. The pathogen can be transmitted in the water supply in developing endemic areas or where defects in any system allow sewage from carriers to contaminate drinking water. Transmission is by the fecal–oral route. The infecting dose of 10^5 to 10^6 bacteria is intermediate between *Shigella* and most *S enterica* and decreases in the presence of the capsular Vi antigen. Three serotypes called Paratyphi (A, B, C) have features similar to *S* Typhi, including the production of an enteric fever syndrome similar to typhoid.

Prevalence is linked to sanitary infrastructure

Typhoid fever is still an important cause of morbidity and mortality worldwide. In developed countries, it is mostly seen in travelers returning from endemic areas such as Latin America, Asia, and India. Visitors from these areas who are carriers are often the source of isolated cases. The decline in disease in industrialized nations largely reflects the availability of clean water supplies and improved disposal of fecal waste.

PATHOGENESIS

Typhi invades M cells and macrophages

Vi polysaccharide limits PMN (neutrophil) phagocytosis

There is no animal model for the strictly human Typhi. The details of the cellular events are inferred from studies of Typhimurium, which in mice produces a disease similar to typhoid (thus the name). The invasion and killing of intestinal M cells and macrophages are presumed to follow the same pattern as that of *S enterica*. Two differences are the Vi surface polysaccharide and the extended multiplication of Typhi in macrophages. In the submucosa, Vi (for virulence) retards neutrophil phagocytosis by interfering with complement deposition in a manner similar to that of other bacterial surface polysaccharides. This may favor uptake by macrophages where at least some Typhi cells establish a privileged niche and the Vi^+ phenotype favors intracellular multiplication. Like other serotypes of *Salmonella*, Typhi remains within a membrane-bound vacuole, but unlike them, rather than killing the macrophage, it enters a stage of extended replication.

Macrophage oxidative burst inhibited

Infection spreads through RES

The primary difference between Typhi and the other serotypes is this prolonged intracellular survival in macrophages. This is due to the organism's ability to inhibit the oxidative metabolic burst and continue to multiply. As the bacteria proliferate in macrophages, they are carried through the lymphatic circulation to the mesenteric nodes, spleen, liver, and bone marrow, all elements of the reticuloendothelial system (RES). At the RES sites, Typhi continues to multiply, infecting new host macrophages. Rather than the acute inflammatory response seen with *S enterica*, *S* Typhi generates a mononuclear response and often not enough irritation to cause diarrhea. This may be due to down-regulation of innate toll-like receptor responses in the intestinal mucosa by the Vi antigen. :::Toll-like receptors, p. 24; Oxidative burst, p. 25

RES sites seed the bloodstream and other organs

Endotoxin produces the fever

Eventually, the increasing bacterial population begins to overflow into the bloodstream (Figure 33–8). The entry of Gram-negative bacteria and their LPS endotoxin into the blood starts the fever, which slowly increases and persists with the continued seeding of *S* Typhi. This sometimes results in metastatic infection of other organs including the urinary tract and the biliary tree. The latter causes reinfection of the bowel. This cycle beginning and ending in the small intestine takes approximately 2 weeks to complete.

IMMUNITY

Immunity follows natural infection

Natural infection with *S* Typhi confers immunity, and reinfection is rare unless the course was shortened by early administration of antimicrobics. The immune response is both T_H1- and T_H2-mediated. In nonfatal cases, antibody and activated macrophages eventually subdue the untreated infection over a period of about 3 weeks. Which antigens stimulate this immunity is not clearly understood. The Vi antigen is usually credited, but various surface proteins are also candidates.

SALMONELLOSIS: CLINICAL ASPECTS

MANIFESTATIONS

The clinical patterns of salmonellosis can be divided into gastroenteritis, bacteremia with and without focal extraintestinal infection, enteric fever, and the asymptomatic carrier state. Any *Salmonella* serotype can probably cause any of these clinical manifestations under appropriate conditions, but in practice the *S enterica* serotypes are associated primarily with gastroenteritis. Typhi and related serotypes (Paratyphi) cause enteric fever.

S enterica = gastroenteritis

Typhi = enteric fever

Gastroenteritis

Typically, the episode begins 24 to 48 hours after ingestion, with nausea and vomiting followed by, or concomitant with, abdominal cramps and diarrhea. Diarrhea persists as the predominant symptom for 3 to 4 days and usually resolves spontaneously within 7 days. Fever (39°C) is present in about 50% of the patients. The spectrum of disease ranges from a few loose stools to a severe dysentery-like syndrome.

Diarrhea, vomiting, and cramps are common

Bacteremia and Metastatic Infection

The acute gastroenteritis caused by *S enterica* can be associated with transient or persistent bacteremia. Frank sepsis is uncommon, except in those with a compromised cell-mediated immune system. *Salmonella* infection in patients with acquired immunodeficiency syndrome (AIDS) is common and often severe. Bacteremia occurs in 70% of these patients and can cause septic shock and death. Despite adequate antimicrobial coverage, relapses are common. Patients with lymphoproliferative disease, perhaps owing to T-cell defects similar to those in patients with AIDS, are also highly susceptible to disseminated salmonellosis. Metastatic spread by salmonellae is a significant risk when bacteremia occurs. These organisms have a unique ability to colonize sites of preexisting structural abnormality including atherosclerotic plaques, sites of malignancy, and the meninges (especially in infants). *Salmonella* infection of the bone typically involves the long bones; in particular, sites of trauma, sickle cell injury, and skeletal prosthesis are at risk.

Bacteremia is most common and severe in the immunocompromised

Metastatic sites linked to previous injury particularly sickle-cell

Enteric Fever

Enteric fever is a multiorgan system *Salmonella* infection characterized by prolonged fever, sustained bacteremia, and profound involvement of the mesenteric lymph nodes, liver, and spleen. The manifestations of typhoid (**Figure 33–10**) have been well documented in

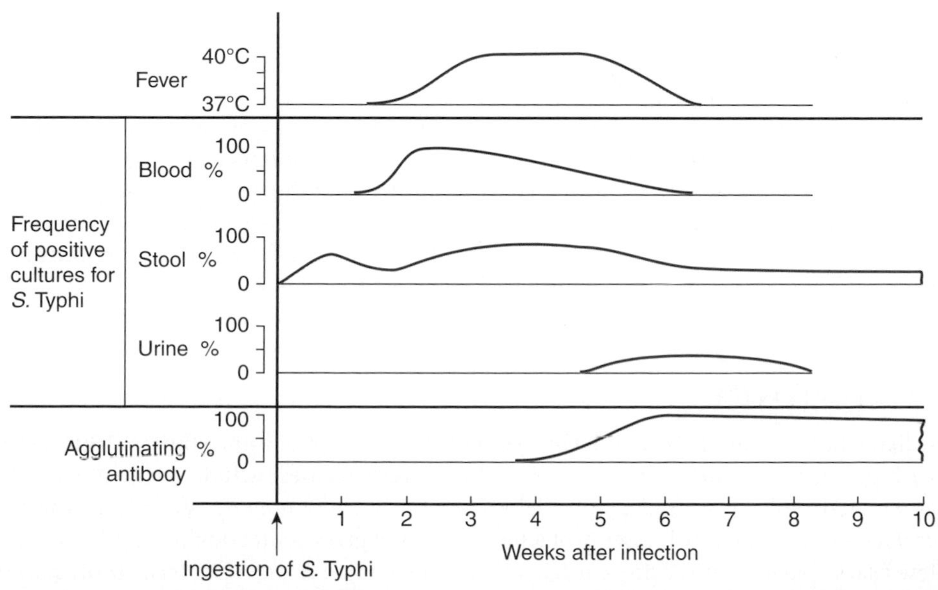

FIGURE 33–10. Natural history of enteric (typhoid) fever. The course of disease without antimicrobial therapy. Fever chart shows time course for typical patient. Culture and agglutinating antibody show timing and probability of positive results in a group of typhoid fever patients.

human volunteer studies conducted during vaccine trials. The mean incubation period is 13 days, and the first sign of disease is fever associated with a headache. The fever rises in a stepwise fashion over the next 72 hours. A relatively slow pulse is characteristic and out of character with the elevated temperature. In untreated patients, the elevated temperature persists for weeks. A faint rash (rose spots) appears during the first few days on the abdomen and chest. Few in number, these spots are readily overlooked, especially in dark-skinned persons. Many patients are constipated, although perhaps one third of patients have a mild diarrhea. As the untreated disease progresses, an increasing number of patients complain of diarrhea.

Slowly increasing fever lasts for weeks

Diarrhea is intermittent or absent

Obviously, chronic infection of the bloodstream is serious, and the effects of endotoxin can lead to myocarditis, encephalopathy, or intravascular coagulation. Moreover, the persistent bacteremia can lead to infection at other sites. Of particular importance is the biliary tree, with reinfection of the intestinal tract and diarrhea late in the disease. UTI and metastatic lesions in bone, joint, liver, and meninges may also occur. However, the most important complication of typhoid fever is hemorrhage from perforations through the wall of the terminal ileum or proximal colon at the site of necrotic Peyer's patches. These occur in patients whose disease has been progressing for 2 weeks or more.

Biliary tree infection reseeds intestine

Urinary tract, bone, and joints are metastatic sites

DIAGNOSIS

Culture of *Salmonella* from the blood or feces is the primary diagnostic method. Early in the course of enteric fever, blood is far more likely to give a positive culture result than culture from any other site. The media used for stool culture are the same as those used for *Shigella*. Failure to ferment lactose and the production of hydrogen sulfides from sulfur-containing amino acids are characteristic features used to identify suspect colonies on the selective isolation media. Characteristics of biochemical tests are used to identify the genus, and O serogroup antisera are available in larger laboratories for confirmation. Typhi has a pattern of biochemical reactions that are sufficient to characterize it without reference to its serogroup (D). All isolates should be referred to public health laboratories for confirmation and epidemiologic tracing. Serologic tests are no longer used for diagnosis.

Stool and blood culture are routine

Typhi has characteristic features

TREATMENT

The primary therapeutic approach to *Salmonella* gastroenteritis consists of fluid and electrolyte replacement and the control of nausea and vomiting. Antibiotic therapy is usually not appropriate because it has a tendency to increase the duration and frequency of the carrier state. When used to eradicate the carrier state, it meets with erratic success and usually fails in the presence of coexisting biliary tract disease. Therefore, the use of antimicrobial agents in *S enterica* gastroenteritis is restricted to those with severe infections or underlying risk factors, particularly children. In these instances, antimicrobials are viewed as a measure to prevent systemic spread.

Antimicrobics are of limited use in gastroenteritis

Antimicrobial therapy is clearly indicated in typhoid fever. Chloramphenicol and then ampicillin were the first antibiotics used and reduced the mortality rate from 20% to less than 2%. These drugs are now restricted by resistance, and the newer cephalosporins (ceftriaxone, cefixime) and ciprofloxacin have taken their place as first-line agents. With proper antimicrobial therapy, patients feel better in 24 to 48 hours, their temperature returns to normal in 3 to 5 days, and they are generally well in 10 to 14 days.

Antibiotics effective but resistance is common

PREVENTION

Killed whole bacterial vaccines have been available for typhoid since the late 19th century with protection in the range of 50% to 70%. Newer vaccines include one that uses a live attenuated Typhi strain and a polysaccharide vaccine containing the Vi antigen. The newer vaccines give slightly higher protection, but none gives protection lasting more than a few years. The newest vaccine contains Vi antigen conjugated to a bacterial protein in the manner of Hib, meningococcal, and pneumococcal vaccines. It shows promise for both a higher efficacy and use in children less than 5 years of age. No human vaccine is available for the other *Salmonella* serotypes. When all is said and done, the provision of clean water supplies and the treatment of carriers will lead to the disappearance of typhoid. The importance of carriers and sanitation was emphasized by a 1973 typhoid outbreak among migrant workers in Florida. The source was traced to leakage of sewage into the water supply, failure of chlorination, and a chronic carrier. All three are required to sustain an outbreak when adequate sanitary infrastructure is in place.

Typhoid vaccines are only moderately effective

Sanitation and public health measures can eliminate Typhi

YERSINIA

BACTERIOLOGY

Morphologically, *Yersinia* tend to be coccobacillary and to retain staining at the ends of the cells (bipolar staining). Growth and metabolic characteristics are the same as those of other Enterobacteriaceae, although some strains grow more slowly or have optimal growth temperatures lower than 37°C. The genus includes 11 species, of which *Yersinia pestis, Yersinia pseudotuberculosis,* and *Yersinia enterocolitica* are the pathogens for humans. *Y pestis* is antigenically homogenous, but *Y pseudotuberculosis* and *Y enterocolitica* have multiple O and H antigen serotypes. *Yersinia* are primarily animal pathogens, with occasional transmission to humans through direct or indirect contact. *Y pestis,* the cause of plague, is discussed in Chapter 36, although features of its pathogenesis common to other *Yersinia* are included below.

Coccobacillary and grow at variable temperatures

Human pathogens linked to animals

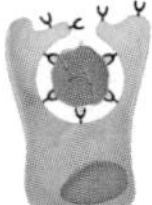

YERSINIA DISEASES (*Y PSEUDOTUBERCULOSIS* AND *Y ENTEROCOLITICA*)

EPIDEMIOLOGY

In animals, *Y pseudotuberculosis* causes pseudotuberculosis, a disease characterized by local necrosis and granulomatous inflammation in the lymph nodes, spleen, and liver. The portal of entry for humans is the gastrointestinal tract, presumably by consumption of contaminated food or water. In most cases, animals, including wild animals, are the most likely source of infection, but the exact mode of transmission is unknown. Geographic variation in the frequency of *Y enterocolitica* infections is marked. The highest rates are reported from Scandinavian and other European countries, with much lower rates in the United Kingdom and the United States. Low isolation rates may be partially attributable to the difficulty of isolating *Y enterocolitica* from stool specimens.

Transmitted by ingestion from animal source

Geographic variation is great

PATHOGENESIS

Enteropathogenic *Yersinia* enter the human host in contaminated food and invade the M cells of the Peyer's patch. The invasive process and its effect on the host cell are driven by a large array of virulence factors that are deployed under complex genetic and environmental regulation. These proteins include **invasin,** which binds to integrins on the surface of host cells, and the major effector proteins called **Yops**. The Yops are delivered by yet another injection (type III) secretion system. When injected into the host cell, they trigger cytotoxic events, including disruption of biochemical pathways (dephosphorylation, serine kinase), sensor functions, and the actin cytoskeleton.

Intestinal M cells are invaded

Secreted Yops disrupt cellular function

Some of the virulence factors produced by *Yersinia* are regulated in a system in which expression responds to either temperature or free calcium (Ca^{2+}) concentration. The physiologic temperature in a mammalian host is different from that in an insect or the environment, and the intracellular calcium concentration is markedly different from that of extracellular fluids. By sensing the environment, *Yersinia* is able to express or suppress virulence factors at different stages of the pathogenic process. The results seem timed to support the pathogenic strategy of *Yersinia,* which is to paralyze the phagocytic activity of defending macrophages and neutrophils to nullify the host cellular immune response. The virulence determinants are encoded both on the bacterial chromosome and on a plasmid that contains genes for the secretion apparatus and the Yops. Another genetic component is a PAI, which is found only in the three pathogenic species and not the other *Yersinia*.

Ca^{2+} and temperature regulate virulence factor expression

Plasmid and PAI contain virulence genes

The biological outcome of this extraordinary multifactorial process is the enhanced capacity of the pathogenic *Yersinia* to enter and replicate within the RES and to delay the cellular immune response. This leads to the formation of microabscesses and destruction of the cytoarchitecture of Peyer's patches and the mesenteric lymph nodes. The systemic symptoms seen with dissemination can largely be attributed to the effects of endotoxin.

Spread leads to microabscesses in lymph nodes

Y pestis is a specialized variant closely related to *Y pseudotuberculosis*. Instead of entering the intestinal tract, *Y pestis* reaches the dermal lymphatics by the bite of an infected flea. It has its own adhesin similar to invasin and two plasmids not found in the enteropathogenic *Yersinia*. Unique virulence factors for *Y pestis* include a capsular protein antigen with antiphagocytic properties, a plasminogen activator protease that promotes adherence to basement membranes, and a fibrinolysin that may play a survival role in the flea.

Y pestis has capsule, plasminogen activator, and fibrinolysin

YERSINIA INFECTIONS: CLINICAL ASPECTS

Both *Y enterocolitica* and *Y pseudotuberculosis* cause acute mesenteric lymphadenitis, a syndrome involving fever and abdominal pain that often mimics acute appendicitis. *Y enterocolitica* also produces a wider variety of manifestations. The most common of these is an enterocolitis, which usually occurs in children. It is characterized by fever, diarrhea, and abdominal pain. It also causes enteric fever, terminal ileitis, and a polyarthritic syndrome associated with its diarrheal manifestations. Few laboratories in the United States routinely screen stools for *Yersinia* because yield has been low and good selective media are not available.

Mesenteric lymphadenitis creates abdominal pain

Yersinia are not routinely sought in stools

The role of antimicrobial therapy in the enteric *Yersinia* infections is uncertain, because they are usually self-limiting. *Y pseudotuberculosis* is susceptible to ampicillin, cephalosporins, aminoglycosides, and tetracyclines, but *Y enterocolitica* is usually resistant to penicillins and cephalosporins through the production of β-lactamases.

Antimicrobics have variable effect

OTHER ENTEROBACTERIACEAE

All the Enterobacteriaceae are capable of producing opportunistic infections of the type discussed under *E coli.* None is considered a proven cause of enteric disease, although no doubt some will be in the future. The genera isolated in at least moderate frequency are discussed briefly below. There are many other less common species.

KLEBSIELLA

The most distinctive bacteriologic features of the genus *Klebsiella* are the absence of motility and the presence of a polysaccharide capsule. This gives colonies a glistening, mucoid character and forms the basis of a serotyping system. Over 70 capsular types have been defined, including some that cross-react with those of other encapsulated pathogens, such as *Streptococcus pneumoniae* and *Haemophilus influenzae.* Limited studies suggest that the capsule interferes with complement activation in a way similar to the other encapsulated pathogens. Several types of pili are also present on the surface and probably aid in adherence to respiratory and urinary epithelium.

Polysaccharide capsule blocks complement deposition

K pneumoniae, the most common species, is able to cause classic lobar pneumonia, a characteristic of other encapsulated bacteria. Most *Klebsiella* pneumonias are indistinguishable from those produced by other members of the Enterobacteriaceae. Of all the Enterobacteriaceae, *Klebsiella* species are now among the most resistant to antimicrobics.

Often are multiresistant

ENTEROBACTER

Enterobacter species generally ferment lactose promptly and produce colonies similar to those of *Klebsiella,* though not as mucoid. A differential feature is motility by peritrichous flagella, which are generally present in *Enterobacter* species but uniformly absent in *Klebsiella. Enterobacter* species, which are generally less virulent than *Klebsiella,* are usually found in mixed infections, in which their significance must be decided on clinical and epidemiologic grounds. Several hospital outbreaks traced to contaminated parenteral fluid solutions have implicated *Enterobacter* species. In addition to ampicillin, most isolates are resistant to first-generation cephalosporins, but may be susceptible to later generation cephalosporins; however, mutants derepressed for β-lactamase production occur relatively frequently and confer resistance to many cephalosporins.

Modest virulence but are linked to hospital contamination

SERRATIA

Serratia strains ferment lactose slowly (3 to 4 days), if at all. Some produce distinctive brick-red colonies. Although less common, this genus produces the same range of opportunistic infections seen with the remainder of the Enterobacteriaceae. *Serratia* strains show consistent resistance to ampicillin and cephalothin, with the frequent addition of resistance to many other antimicrobics, including the aminoglycosides. Sporadic infections and nosocomial outbreaks with multiresistant strains have often been difficult to control.

Red pigment and multiresistance are characteristic

CITROBACTER

The genus *Citrobacter,* though biochemically and serologically similar to *Salmonella,* is an uncommon cause of opportunistic infection. Like many other Enterobacteriaceae, *Citrobacter* strains may be present in the normal intestinal flora and cause opportunistic infections. Despite reports of association with diarrheal disease, present evidence does not indicate that *Citrobacter* should be considered an enteric pathogen. *C freundii* has been associated with neonatal meningitis and brain abscess.

Opportunistic infection and brain abscess are uncommon

PROTEUS, PROVIDENCIA, AND MORGANELLA

Proteus, Morganella, and *Providencia* are also opportunistic pathogens found with varying frequencies in the normal intestinal flora. *Proteus mirabilis,* the most commonly isolated member of the group, is one of the most susceptible of the Enterobacteriaceae to the penicillins; this characteristic includes moderate susceptibility to penicillin G. Other Proteeae are regularly resistant to ampicillin and the cephalosporins. *P mirabilis* and *Proteus vulgaris* share the ability to swarm over the surface of media, rather than remaining confined to discrete colonies. This characteristic makes them readily recognizable in the laboratory—often with dismay because the spreading growth covers other organisms in the culture and thus delays their isolation. Swarming along with motility could facilitate the production of UTIs by movement of *Proteus* up urinary catheters. *Proteus* and *Morganella* differ from other Enterobacteriaceae in the production of a very potent urease, which aids their rapid identification. It also contributes to the formation of urinary stones and produces alkalinity and an ammoniac odor to the urine. *Providencia* species do not produce urease, are the least frequently isolated, and are generally the most resistant of the group to antimicrobics.

Swarming is a feature of some species

Urease production is linked to urinary stones

CLINICAL CASE

HAMBURGERS AND HEMORRHAGE

A 24-year-old woman was seen in a hospital emergency department with a history of nausea, vomiting, and nonbloody diarrhea, which progressed to bloody diarrhea. Four days earlier she had eaten a hamburger at a fast-food restaurant. To replace fluid lost from diarrhea, she was given 2 liters of IV fluid. Her condition improved and she was sent home with anti-nausea medication.

Two days later, the symptoms had not resolved; the vomiting, nausea, and bloody diarrhea persisted with abdominal cramps and orthostatic dizziness. She returned to the emergency department, was admitted, again given IV fluids, and discharged after 2 days of hospitalization. A stool sample was taken for culture.

Three days later the patient awoke with vomiting and contacted her private physician. Laboratory tests were done with the following results: blood urea nitrogen 67.0 mg/dL (ref. 7-19); white blood cells 13,100/mL; hemoglobin 7.0 g/dL (ref. 11.5-15.5); platelet count 75,000/μL (ref. >150,000). The stool culture taken earlier was positive for *E coli* 0157:H7.

The patient was transferred to the ICU the same day and was described as severely ill. She was fatigued, very dehydrated, with abdominal tenderness and back pain but no neurologic problems. Steroids were the only additional medication given in addition to plasmapheresis, which was done five times during her hospitalization. She gradually recovered and was discharged.

QUESTIONS

■ Which of the following is probably the source of this patient's infection?

A. Colonized cow
B. Colonized restaurant worker
C. Contaminated restaurant water
D. Family member
E. Restaurant air

■ What bacterial product was primarily responsible for the hemorrhage and renal injury?

A. Endotoxin
B. α-toxin
C. Labile toxin (LT)
D. Stable toxin (ST)
E. Shiga toxin (Stx)

■ If hamburger is the source, this infection could have been prevented by which of the following?

A. Screening the restaurant workers
B. Hand-washing
C. Disinfectants
D. Complete cooking
E. Antibiotic prophylaxis

ANSWERS

1(A), **2**(E), **3**(D)

CHAPTER 34

Legionella

The death toll in the outbreak of the mysterious respiratory disease in Philadelphia rose by two to 25 as medical detectives accelerated efforts today to seek a chemical or poison as the possible cause.

—*The New York Times,* August 7, 1976

Legionella is a genus of Gram-negative bacilli that take its name from the outbreak at the American Legion convention where they were first discovered. The species designation of the prime human pathogen, *Legionella pneumophila,* reflects its propensity to cause pneumonia. *Legionella* species are now known to be widespread in the environment.

BACTERIOLOGY

MORPHOLOGY AND STRUCTURE

Legionella pneumophila is a thin, pleomorphic, Gram-negative rod that may show elongated, filamentous forms up to 20 μm long. In clinical specimens, the organism stains poorly or not at all by Gram stain or the usual histologic stains; however, it can be demonstrated by certain silver impregnation methods (Dieterle stain) and by some simple stains without decolorization steps. Polar, subpolar, and lateral flagella may be present. Most species of *Legionella* are motile. Spores are not found.

Gram-negative rod that stains with difficulty

Structurally, *L pneumophila* has features similar to those of Gram-negative bacteria with a typical outer membrane, thin peptidoglycan layer, and cytoplasmic membrane. The toxicity of *L pneumophila* lipopolysaccharide (LPS) is significantly less than that of other Gram-negative bacteria such as *Neisseria* and the Enterobacteriaceae. This has been attributed to chemical makeup of the LPS side chains, which are a homopolymer of an unusual sugar (legionaminic acid), which renders the cell surface highly hydrophobic. It has been postulated that this hydrophobicity may promote adherence of bacterial cells to membranes or their concentration in aerosols.

LPS is less toxic than that of most Gram-negative species

Side chains are hydrophobic

GROWTH AND CLASSIFICATION

Legionella species fail to grow on common enriched bacteriologic media such as blood agar. This is due to requirements for certain amino acids (L-cysteine), ferric ions, and slightly acidic conditions (optimal pH 6.9). Even when these requirements are met, growth under

Growth requires iron and low pH

aerobic conditions is slow, requiring 2 to 5 days to produce colonies that have a distinctive surface resembling ground glass.

Few phenotypic properties are demonstrable

Classification is based on antigenic structure and DNA homology

Because of the difficulty in growing *Legionella.* there are few phenotypic properties to use in its classification. It is possible to directly demonstrate some enzymatic actions (catalase, oxidase, β-lactamase), but the other cultural and metabolic taxonomic tests used to classify other bacteria cannot be applied to *Legionella.* Thus, the classification depends largely on antigenic features, chemical analysis, and nucleic acid homology comparisons.

Multiple *L pneumophila* serogroups and *Legionella* species exist

L pneumophila has multiple serogroups (16) and there are almost 50 other *Legionella* species (eg, *L longbeachae, L bozemanii, L dumoffii, L micdadei*). The original Philadelphia strain (serogroup 1) is still the most common, and a limited number of *L pneumophila* serogroups (1 to 4) account for 80% to 90% of cases. Less than half of the non–*L pneumophila* species have been isolated from human infections.

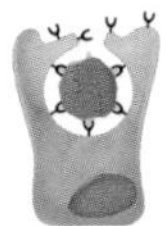

LEGIONNAIRES DISEASE

CLINICAL CAPSULE

Legionella are inhaled into the lung from an aquatic source in the environment. Once there, they produce a destructive pneumonia marked by headache, fever, chills, dry cough, and chest pain. Although there may be multiple foci in both lungs and extension to the pleura, spread outside the respiratory tree is very rare.

EPIDEMIOLOGY

1976 outbreak lead to discovery of new bacterium

Earlier outbreaks have been solved

The widely publicized outbreak of pneumonia among attendees of the 1976 American Legion convention in Philadelphia led to the isolation of a previously unrecognized infectious agent, *L pneumophila.* The event was unique in medical history. For months the American public entertained theories of its cause that ranged from chemical sabotage to viroids and fears that something like Michael Crichton's 1969 novel *The Andromeda Strain* was ahead. It was almost a letdown to find that a Gram-negative rod that could not be stained or grown by the common methods was responsible. The Centers for Disease Control investigation was an outstanding example of the benefits of pursuing sound epidemiologic evidence until it is explained by equally sound microbiologic findings. We now know the disease had occurred for many years. Specific antibodies and organisms have been detected in material preserved from the 1950s, and a mysterious hospital outbreak in 1965 has been solved retrospectively by examination of preserved specimens. Today, most cases of Legionnaires disease in the United States are caused by just a few *L pneumophila* serotypes, including the original Philadelphia strain, but there is considerable variation worldwide. In Western Australia, *L longbeachae* is the predominant species.

Amoebas in fresh water habitat act as reservoir

Infections are associated with aerosols distributed by humidifying and cooling systems

In nature, *Legionella* species are ubiquitous in freshwater lakes, streams, and subterrestrial groundwater sediments. They are also found in moist potting soil, mud, and riverbanks. In these sites, they also exist as parasites of protozoa including numerous species of amoebas, which appear to be the environmental reservoir. Transmission to humans occurs when aerosols are created in manmade water supplies that harbor *Legionella.* Most outbreaks have occurred in or around large buildings such as hotels, factories, and hospitals with cooling towers or some other part of an air-conditioning system as the spreading vehicle. Some hospital outbreaks have implicated respiratory devices and potable water coming from parts of the hot water system such as faucets and shower heads. Even the mists used in supermarkets to make the vegetables look shiny and fresh have been the source of outbreaks.

Legionella can persist in a water supply despite standard disinfection procedures, particularly when the water is warm and the pipes contain scale or low-flow areas that compromise chlorine access.

It is difficult to ascertain the overall incidence of *Legionella* infections because most information has been from outbreaks that constitute only a small part of the total cases. Estimates based on seroconversions suggest approximately 25,000 cases in the United States each year. The attack rate among those exposed is estimated at less than 5% and serious cases are generally limited to immunocompromised persons. Person-to-person transmission has not been documented, and the organisms have not been isolated from healthy individuals. Growth in free-living amoebas produces *Legionella* cells that are more resistant to environmental stress (acid, heat, osmotic) and have enhanced infectivity.

Person-to-person transmission or carriers are unknown

Disease rate among exposed is low

PATHOGENESIS

L pneumophila is striking in its propensity to attack the lung, producing a necrotizing multifocal pneumonia. Microscopically, the process involves the alveoli and terminal bronchioles, with relative sparing of the larger bronchioles and bronchi (**Figure 34–1**). The inflammatory exudate contains fibrin, polymorphonuclear neutrophils (PMNs), macrophages, and erythrocytes. A striking feature is the preponderance of bacteria within phagocytes and the lytic destruction of inflammatory cells.

Strong tropism for the lung

Necrotizing multifocal pneumonia with intracellular bacteria

L pneumophila is a facultative intracellular pathogen. Its pathogenicity depends on its ability to survive and multiply within cells of the monocyte–macrophage series. Inhaled *Legionella* bacteria reach the alveoli, where they attach to their pathogenic target the alveolar macrophage. In this process, they are aided by pili and outer membrane proteins (OMP), which bind complement components. Another OMP called **macrophage invasion potentiator (Mip)** facilitates the early stages of cell entry into an endocytotic vacuole.

Facultative intracellular pathogen multiplies in alveolar macrophages

OMPs facilitate phagocyte entry to specialized vacuole

Inside the vacuole, the bacteria continue to replicate by preventing phagosome-lysosome fusion and avoiding the destructive acidification and enzymatic digestion found there. Instead, the *Legionella*-containing endosome recruits secretory vesicles from the endoplasmic reticulum (ER) remodeling it into rough ER. The morphology of the replicative vacuole created is shown in **Figure 34–2**. *L. pneumophila* appears to accomplish this control of the phagocyte by use of a system that secretes proteins able to modulate host cell vesicle traffic and prevent transport into the lysosome. Other elements of the organism's intracellular success include its ability to extract iron from transferrin and a peptide toxin that inhibits activation of the oxidative killing mechanisms of phagocytes. Thus, instead of being killed by the bactericidal mechanisms of phagocytes, *L pneumophila* multiplies freely (**Figure 34–3**). Induction of apoptosis and formation of a pore-forming toxin eventually lead to death of the macrophage. Lysis of the dying macrophages releases a new population of *Legionella* cells, which then repeat this infective cycle. The multiple degradative enzymes released in this process lead to destructive lesions in the lung and a systemic toxicity that may be related to cytokine release.

Secreted proteins block phagosomal fusion with lysosomes

Control of vesicular traffic creates replicative vacuole

Intracellular events are similar in amoeba

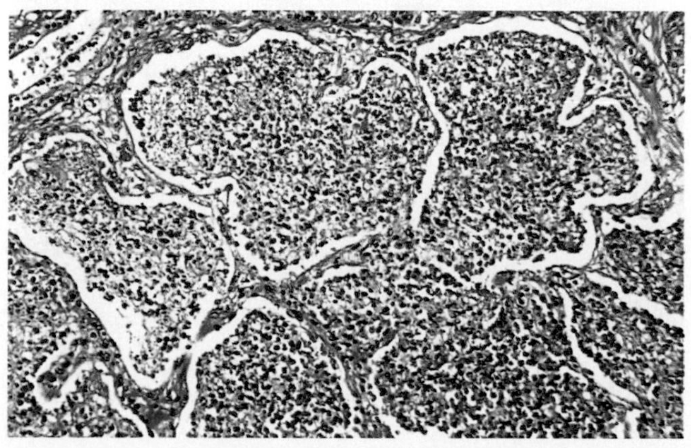

FIGURE 34–1. *Legionella* pneumonia. Note the filling of alveoli with exudate. Some of the alveolar septa are starting to degenerate. (Reproduced with permission from Connor DH, Chandler FW, Schwartz DQA, Manz HJ, Lack EE (eds). *Pathology of Infectious Diseases,* vol. 1. Stamford, CT: Appleton & Lange; 1997.)

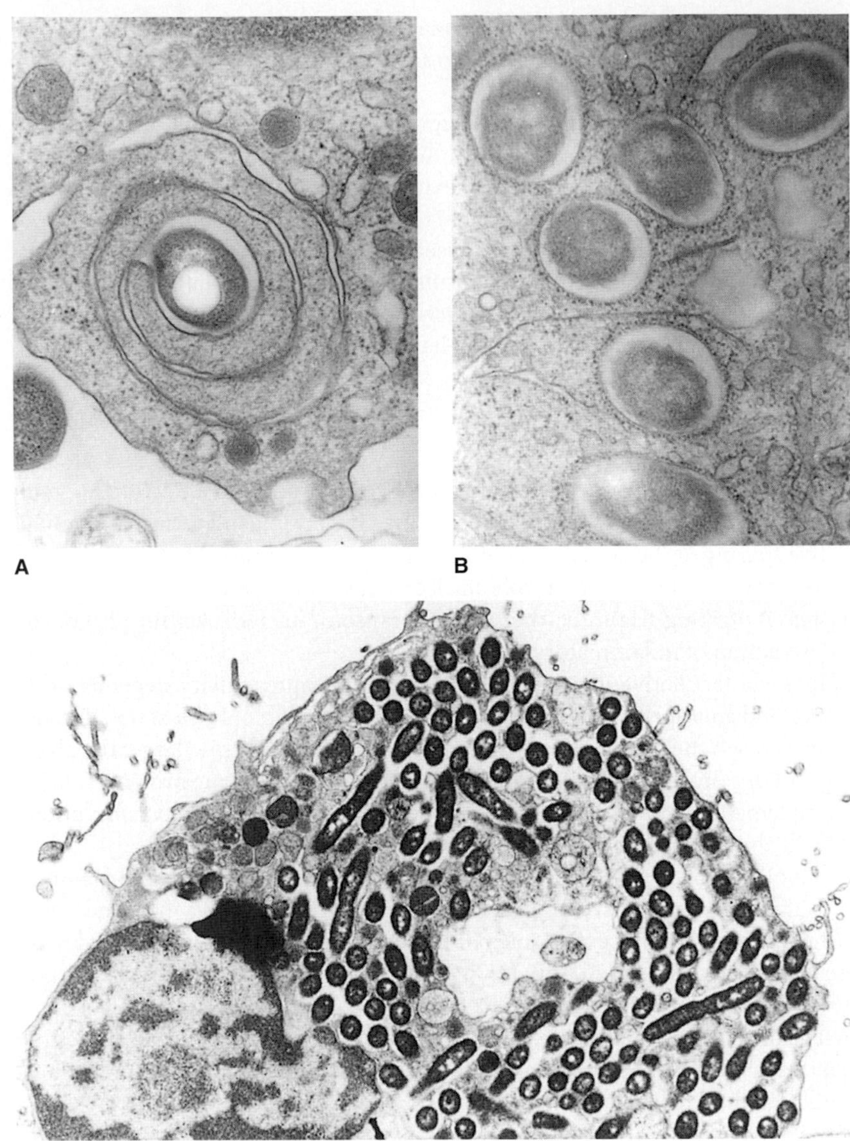

FIGURE 34–2. Multiplication of *Legionella pneumophila* in human macrophages. *L pneumophila* enters the cell by coiling phagocytosis **A.** and the phagosome created is lined by ribosomes and mitochondria **B.** The bacteria multiply within the macrophages to reach very high numbers **C.** (Courtesy of Dr. Marcus Horwitz.)

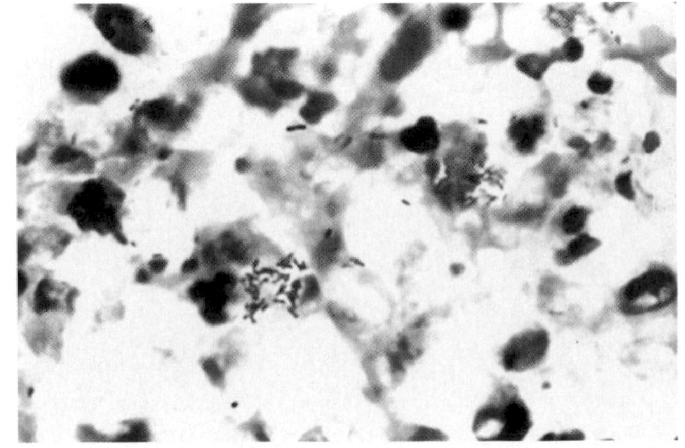

FIGURE 34–3. Legionnaires disease. Imprint smear of lung shows *L pneumophila* (*stained red*) mostly inside alveolar macrophages. (Reproduced with permission from Connor DH, Chandler FW, Schwartz DQA, Manz HJ, Lack EE (eds). *Pathology of Infectious Diseases*, vol. 1. Stamford, CT: Appleton & Lange; 1997.)

IMMUNITY

Just as intracellular multiplication is the key to *L pneumophila* virulence, its arrest by innate and adaptive mechanisms is the most important aspect of immunity. The high level of innate immunity to *Legionella* infection in most persons is related to brisk pattern recognition responses triggered by toll-like receptors (TLRs) in macrophages and dendritic cells (see Chapter 2). These include TLRs that recognize foreign molecular patterns associated with bacteria in general (lipopolysaccharide, peptidoglycan, and others) and may also include a TLR located in endocytotic vesicles that recognizes patterns found in intracellular pathogens including viruses and some fungi. The activation of the T_H1 adaptive immune response and its associated cytokines (interferon, IL-12, IL-18) completes the process of macrophage activation and intracellular killing of the invading *Legionella*. Failure of this aspect of the immune response is the primary reason for most cases of progressive Legionnaires disease in the immunocompromised. Antibodies formed in the course of *Legionella* infection are useful for diagnosis, but do not appear to be important in immunity. It is unknown whether humans who have had Legionnaires disease are immune to reinfection and disease. ::: Toll-like receptors, p. 24; T-cell response, p. 33

Innate defenses triggered by TLRs

Cytokine activated macrophages limit intracellular growth

Antibody is less important

LEGIONNAIRES DISEASE: CLINICAL ASPECTS

MANIFESTATIONS

Legionnaires disease is a severe toxic pneumonia that begins with myalgia and headache, followed by a rapidly rising fever. A dry cough may develop and later become productive, but sputum production is not a prominent feature. Chills, pleuritic chest pain, vomiting, diarrhea, confusion, and delirium all may be seen. Radiologically, patchy or interstitial infiltrates with a tendency to progress toward nodular consolidation are present unilaterally or bilaterally. Liver function tests often indicate some hepatic dysfunction. In the more serious cases, the patient becomes progressively ill and toxic over the first 3 to 6 days, and the disease terminates in shock, respiratory failure, or both. The overall mortality rate is about 15%, but it has been higher than 50% in some hospital outbreaks. Mortality is particularly high in patients with serious underlying disease or suppression of cell-mediated immunity.

Severe toxic pneumonia occurs in 5% of those exposed

Mortality is high among the immunocompromised

A less common form of disease called **Pontiac fever** (named for a 1968 Michigan outbreak), is a non-pneumonic illness with fever, myalgia, dry cough, and a short incubation period (6 to 48 hours). Pontiac fever is a self-limiting illness and may represent a reaction to endotoxin or hypersensitivity to components of the *Legionella* or their protozoan hosts.

Pontiac fever may be hypersensitivity response

DIAGNOSIS

The established approach to diagnosis combines direct fluorescent antibody (DFA) with culture of infected tissues. For this purpose, a high-quality specimen such as lung aspirates, bronchoalveolar lavage, or biopsies are preferred, because the organism may not be found in sputum. Typically, the Gram smear fails to show bacteria owing to poor staining, but organisms are revealed by DFA using *L pneumophila*–specific conjugates. These conjugates use monoclonal antibodies, which bind to all serotypes of *L pneumophila,* but not the non–*L pneumophila* species. DFA is rapid, but it yields a positive result in only 25% to 50% of culture-proved cases. ::: DFA, p. 66

High-quality specimens are needed

DFA is rapid but only 50% sensitive

Cultures must be made on buffered charcoal yeast extract (BCYE) agar medium that includes supplements (amino acids, vitamins, L-cysteine, ferric pyrophosphate), which meets the growth requirements of *Legionella*. It is buffered to meet the acidic conditions—optimal for *Legionella* growth (pH 6.9). The isolation of large Gram-negative rods on BCYE after 2 to 5 days that have failed to grow on routine media (blood agar, chocolate agar) is presumptive evidence for *Legionella.* Diagnosis is confirmed by DFA staining of bacterial smears prepared from the colonies. BCYE also allows isolation of species of *Legionella* species other than *L pneumophila.*

Culture on BCYE is required for isolation

Cultures will isolate other species

The difficulty and slow speed of culture together with the low sensitivity of DFA have spurned searches for other methods. This has led to the development of nucleic acid amplification procedures for use in respiratory specimens and immunoassay methods for detection of antigen in urine. Amplification methods like such as polymerase chain reaction (PCR) have proved to be rapid and much more sensitive than DFA. A simple card-based antigenuria detection test has also proved to be sensitive for the most common *L pneumophila* serogroup 1, but less so for other serogroups and *Legionella* species. The primary barrier to making these methods more widely used is that Legionnaires disease is uncommon except in immunocompromised populations. This tends to limit their availability to reference laboratories and hospitals serving immunocompromised patients. Demonstrating a significant rise in serum antibody is used primarily for retrospective diagnosis and in epidemiologic studies.

PCR is rapid and sensitive

Antigenuria is detects serogroup 1

TREATMENT

The best information on antimicrobial therapy is still provided by the original Philadelphia outbreak. Because the cause of Legionnaires disease was completely obscure at the time, the cases were treated with many different regimens. Patients treated with erythromycin clearly did better than those given the penicillins, cephalosporins, or aminoglycosides. Subsequently, it was shown that most *Legionella* produce β-lactamases. In vitro susceptibility tests and animal studies have confirmed the activity of erythromycin and have shown that tetracycline, rifampin, and the newer quinolones are also active. Although the other antimicrobics are sometimes used in combination, erythromycin and the newer macrolides (azithromycin, clarithromycin) remain the agents of choice.

Erythromycin is treatment of choice

Tetracycline, rifampin, and quinolones are alternatives

PREVENTION

The prevention of legionellosis involves minimizing production of aerosols in public places from water that may be contaminated with *Legionella.* Prevention is complicated by the fact that, compared with other environmental bacteria, *Legionella* bacteria are relatively resistant to chlorine and heat. The bacteria have been isolated from hot water tanks held at over 50°C. Methods for decontaminating water systems are still under evaluation. Some outbreaks have been aborted by hyperchlorination, by correcting malfunctions in water systems, or by temporarily elevating the system temperature above 70°C. The installation of silver and copper ionization systems similar to those used in large swimming pools has been effective as a last resort in hospitals plagued with recurrent nosocomial legionellosis.

Preventing *Legionella* aerosols is primary goal

Heat, hyperchlorination, and metal ions may be needed in institutions

CLINICAL CASE

FATAL PNEUMONIA WITH MYSTERY GRAM-NEGATIVE BACILLUS

A 54-year-old man with multiple myeloma was admitted with a 2-day history of fever, nausea, and diarrhea. His lungs were initially clear, but during the first 3 days of his hospitalization he developed a progressive right lower lobe pneumonia and pleural effusion. Initial antibiotic therapy included cephalothin, tobramycin, and ticarcillin. On day 3, intravenous erythromycin, was added.

Initial cultures of blood, sputum, urine, cerebrospinal fluid, and stool failed to reveal an etiologic agent. A transtracheal aspirate was obtained also with negative results, including a *Legionella* DFA. There was no resolution of the pneumonia, and spiking fevers continued. On day 13, his respiratory difficulties increased, with frank bleeding from the upper respiratory tract, and he died.

At autopsy, the most prominent findings were bronchopneumonia with focal organization and hemorrhage in the right lung. Stains of the lung tissue were negative by Gram, methenamine silver, and acid-fast methods, but Dieterle silver stains revealed short bacilli. Lung cultures yielded Gram-negative bacilli, which grew aerobically on buffered charcoal-yeast extract, but not on blood or chocolate agar. The organisms resembled *Legionella,* but failed to stain with immunofluorescence conjugates for *Legionella pneumophila* and multiple other species (*L micdadei, L longbeachae, L gormanii, L dumoffi, L bozemanii*). The organism was sent to the Centers for Disease Control and Prevention, where it was eventually identified as a new species of *Legionella*.

QUESTIONS

■ What is the most probable source of this man's infection?

A. Family member
B. Water
C. Food
D. Insect
E. Bioterrorism

■ What cell type did the organism initially infect in this patient?

A. Ciliated epithelial cell
B. Squamous epithelial
C. Microvillous cell
D. M cell
E. Alveolar macrophage

■ Which of the following contributes most to the ability of *Legionella* to multiply in host phagocytes?

A. Pore-forming toxin
B. Superantigen action
C. Cytokine stimulation
D. Inhibition of lysosome fusion
E. Inhibition of protein synthesis

ANSWERS

1(B), **2**(E), **3**(D)

CHAPTER 35

Pseudomonas and Other Opportunistic Gram-negative Bacilli

A number of opportunistic Gram-negative rods of several genera not considered in other chapters are included here. With the exception of *Pseudomonas aeruginosa,* they rarely cause disease, and all are frequently encountered as contaminants and superficial colonizers. The significance of their isolation from clinical material thus depends on the circumstance and site of culture and on the clinical situation of the patient.

PSEUDOMONAS

There are a large number of *Pseudomonas* species, the most important of which is *P aeruginosa.* The number of human infections produced by the other species together is far lower than that produced by *P aeruginosa* alone. *Pseudomonas* species are most frequently seen as colonizers and contaminants, but are able to cause opportunistic infections. The assignment of species names has little clinical importance beyond differentiation from *P aeruginosa.* Reports vary regarding the frequency of their isolation from cases of bacteremia, arthritis, abscesses, wounds, conjunctivitis, and urinary tract infections. In general, unless isolated in pure culture from a high-quality (direct) specimen, it is difficult to attach pathogenic significance to any of the miscellaneous *Pseudomonas* species. ::: Direct specimen, p. 60

P aeruginosa most important

Other *Pseudomonas* species cause opportunistic infection

Pseudomonas Aeruginosa

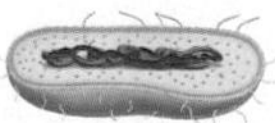

BACTERIOLOGY

Pseudomonas aeruginosa is an aerobic, motile, Gram-negative rod that is slimmer and more pale staining than members of the Enterobacteriaceae. Its most striking bacteriologic feature is the production of colorful water-soluble pigments. *P aeruginosa* also demonstrates the most consistent resistance to antimicrobics of all the medically important bacteria.

Pigment-producing rod is resistant to many antimicrobics

P aeruginosa is sufficiently versatile in its growth and energy requirements to use simple molecules such as ammonia and carbon dioxide as sole nitrogen and carbon sources. Thus, it does not require enriched media for growth and can survive and multiply over a wide temperature range (20° to 42°C) in almost any environment, including one with a high salt content. The organism uses oxidative energy-producing mechanisms and has high levels of cytochrome oxidase (oxidase-positive). Although an aerobic atmosphere is necessary for optimal growth and metabolism, most strains multiply slowly in an anaerobic environment if nitrate is present as an electron acceptor.

Grows aerobically with minimal requirements

Colonies are oxidase-positive

Growth on all common isolation media is luxurious, and colonies have a delicate, fringed edge. Confluent growth often has a characteristic metallic sheen and emits an intense "fruity" odor. Hemolysis is usually produced on blood agar. The positive oxidase reaction of

P aeruginosa differentiates it from the Enterobacteriaceae, and its production of blue, yellow, or rust-colored pigments differentiates it from most other Gram-negative bacteria. The blue pigment, **pyocyanin,** is produced only by *P aeruginosa*. **Fluorescin,** a yellow pigment that fluoresces under ultraviolet light is produced by *P aeruginosa* and other free-living less pathogenic *Pseudomonas* species. Pyocyanin and fluorescin combined produce a bright green color that diffuses throughout the medium.

Blue pyocyanin produced only by *P aeruginosa*

Yellow fluorescin and pyocyanin combine for green color

Lipopolysaccharide (LPS) is present in the outer membrane, as are porin proteins, which differ from those of the Enterobacteriaceae family in offering much less permeability to a wide range of molecules, including antibiotics. Pili composed of repeating monomers of the pilin structural subunit extend from the cell surface. A single polar flagellum rapidly propels the organism and assists in binding to host tissues.

Outer membrane protein porins are relatively impermeable

A mucoid exopolysaccharide slime layer is present outside the cell wall in some strains. This layer is created by secretion of **alginate,** a copolymer of mannuronic and glucuronic acids. It is created by the action of several enzymes that effectively channel carbohydrate intermediates into the alginate polymer. All *P aeruginosa* produce moderate amounts of alginate, but those with mutations in regulatory genes overproduce the polymer. These mutants appear as striking mucoid colonies in cultures from the respiratory tract of patients with cystic fibrosis.

Secreted alginate forms a slime layer

Overproduction is due to regulatory mutations

Most strains of *P aeruginosa* produce multiple extracellular products, including **exotoxin A (ExoA)** and other enzymes with phospholipase, collagenase, adenylate cyclase or elastase activity. ExoA is a secreted protein that inactivates eukaryotic elongation factor 2 (EF-2) by ADP-ribosylation (ADPR). This arrests translation leading to shutdown of protein synthesis and cell death. Although this action is the same as diphtheria toxin, the two toxins are otherwise unrelated. The **elastase** acts on a variety of biologically important substrates, including elastin, human IgA and IgG, complement components, and some collagens. **Exoenzyme S (ExoS)** and a number of other proteins (ExoT, ExoY, ExoU) are transported directly into host cells by an injection (type III) secretion system. Inside the cell, ExoS acts on regulatory G proteins affecting the cytoskeleton, signaling pathways, and inducing apoptosis. :::ADPR, p. 363; Secretion systems, p. 368

Multiple extracellular enzymes are produced

ExoA action same as diphtheria toxin

ExoS injected by secretion system

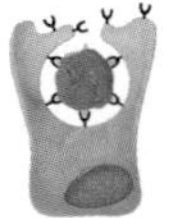

P AERUGINOSA DISEASE

CLINICAL CAPSULE

P aeruginosa produces infection at a wide range of pulmonary, urinary, and soft tissue sites, much like the opportunistic Enterobacteriaceae. The clinical manifestations of these infections reflect the organ system involved and are not unique for Pseudomonas. However, once established, infections are particularly virulent and difficult to treat. Affected patients almost always have some form of debilitation or compromise of immune defenses.

EPIDEMIOLOGY

The primary habitat of *P aeruginosa* is the environment. It is found in water, soil, and various types of vegetation throughout the world. *P aeruginosa* has been isolated from the throat and stool of 2% to 10% of healthy persons. Colonization rates may be higher in hospitalized patients. Infection with *P aeruginosa,* rare in previously healthy persons, is one of the most important causes of invasive infection in hospitalized patients with serious underlying disease, such as leukemia, cystic fibrosis (CF), and extensive burns (**Figure 35–1**).

Primary habitat is environmental

Colonizes humans

The ability of *P aeruginosa* to survive and proliferate in water with minimal nutrients can lead to heavy contamination of any nonsterile fluid, such as that in the humidifiers of respirators. Inhalation of aerosols from such sources can bypass the normal respiratory defense mechanisms and initiate pulmonary infection. Infections have resulted from the growth of *Pseudomonas* in medications, contact lens solutions, and even some disinfectants. Sinks and faucet aerators may be heavily contaminated and serve as the environmental source for

Multiplies in humidifiers, solutions, and medications

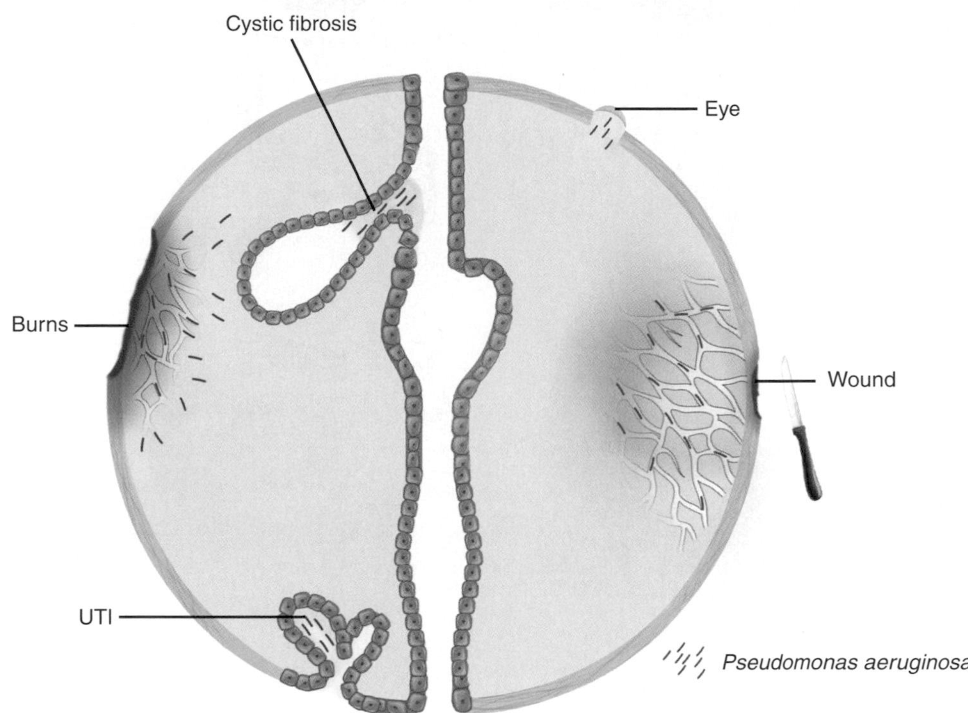

FIGURE 35–1. *Pseudomonas* disease overview. *P aeruginosa* is a leading cause of opportunistic infection in the eye (contact lenses), wounds, urinary tract, and burns. In a special situation it colonizes the respiratory tract of persons with cystic fibrosis by formation of a biofilm (see Figure 35–4). UTI, urinary tract infection.

contamination of other items. The presence of *P aeruginosa* in drinking water or food is not a cause for alarm. The risk lies in the proximity between items susceptible to contamination and persons uniquely predisposed to infection.

Risk for immunocompromised persons is high

P aeruginosa is now the most common bacterial pathogen to complicate the management of patients with cystic fibrosis (CF), an inherited defect in chloride ion transport that leads to a buildup of thick mucus in ducts and the tracheobronchial tree. In a high percentage of cases, the respiratory tract becomes colonized with *P aeruginosa,* which, once established, becomes almost impossible to eradicate. This infection is a leading cause of morbidity and eventual death of these patients.

Respiratory colonization of CF patients becomes chronic

PATHOGENESIS

Although *P aeruginosa* is an opportunistic pathogen, it is one of particular virulence. The organism usually requires a significant break in first-line defenses (such as a wound) or a route past them (such as a contaminated solution or intratracheal tube) to initiate infection. Attachment to epithelial cells is the first step in infection and is likely mediated by pili, flagella, and the extracellular polysaccharide slime. The receptors include sialic acid and *N*-acetylglucosamine borne by cell surface glycolipids. Attachment is favored by loss of surface fibronectin, which explains in part the propensity for debilitated persons.

Needs break in first-line defenses

Pili, flagella, and slime mediate adherence

Once established, the virulence of *P aeruginosa* seems obvious, given its myriad enzymes and other factors (**Figure 35–2**). The importance of ExoA is supported by studies in humans and animals, which correlate its presence with a fatal outcome and an antibody against it with survival. The effect of ExoA is not immediate, since it is one of a number of factors activated through a **quorum-sensing** system in which chemicals (lactones, quinolones) signal cell-to-cell. When the *P aeruginosa* cell population reaches a certain threshold, the signals direct the cytotoxin gene to be transcribed, and the toxin is then secreted by the entire population at once. No diphtheria-like systemic effect of ExoA has been demonstrated, but its action correlates with the primarily invasive and locally destructive lesions seen in *P aeruginosa* infections.

ExoA secreted through quorum-sensing

ExoA correlates with invasion, destruction

The elastase and phospholipase degrade proteins and lipids, respectively, allowing the organism to acquire nutrients from the host and disseminate from the local site. The many biologically important substrates of **elastase** argue for its importance, particularly its namesake, elastin. Elastin is found at some sites that *P aeruginosa* preferentially attacks, such as the lung and blood vessels. Hemorrhagic destruction, including the walls of blood vessels (**Figure 35–3**), is the histologic hallmark of *Pseudomonas* infection. The intracellular

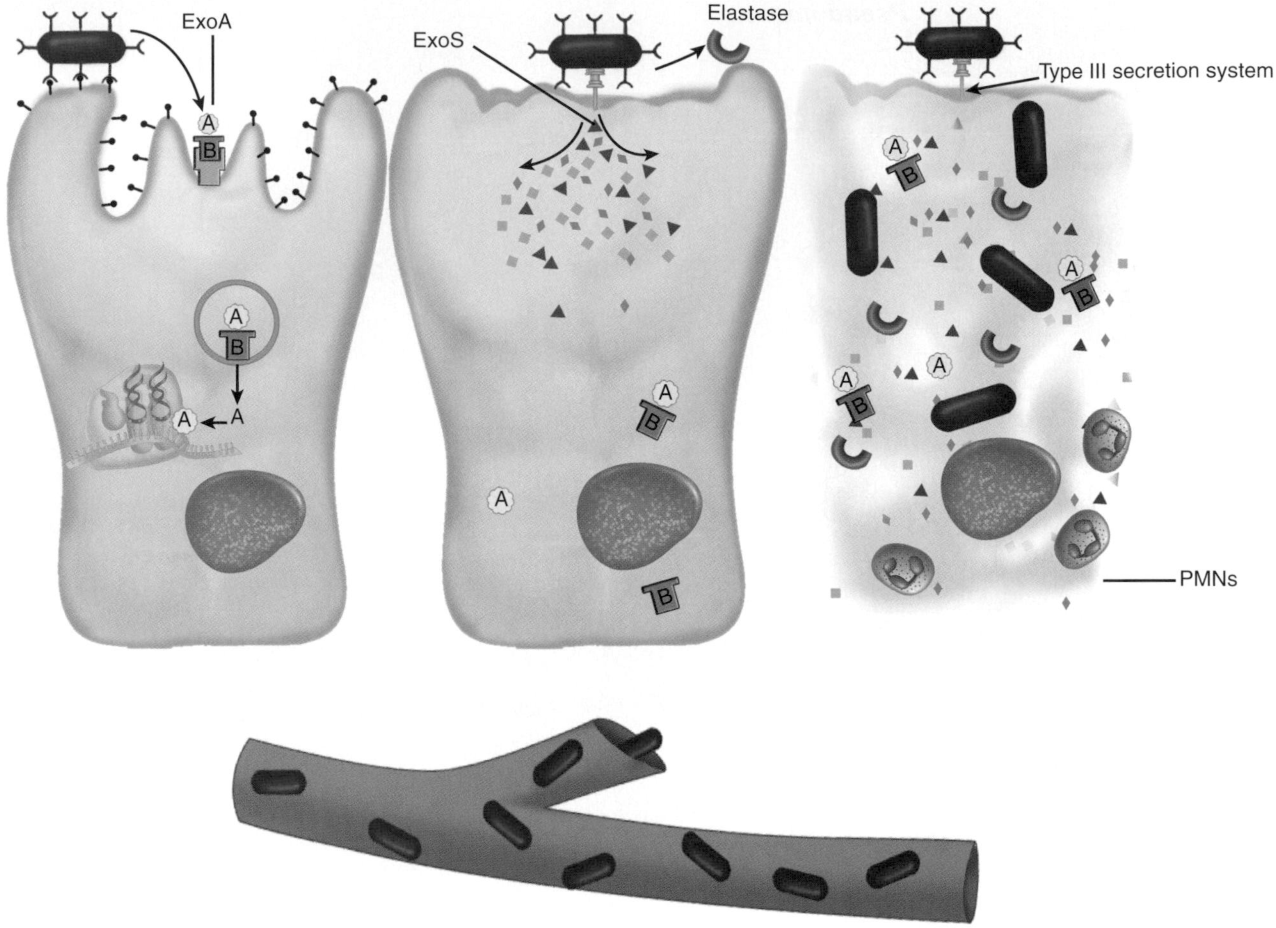

FIGURE 35–2. *Pseudomonas* disease, cellular view. (*Left*) *P aeruginosa* binds and secretes the A-B exotoxin A (ExoA), which acts on protein synthesis by the same mechanism as diphtheria toxin. (*Middle*) A type III injection secretion system delivers exoenzyme S (ExoS) to the cell cytoplasm. Elastase is secreted extracellularly. (Right) All toxins act to destroy the cell and the bacteria may enter the blood.

Elastin is attacked in lung and blood vessels

Injected ExoS disrupts cells

dysfunction caused by ExoS and other factors injected by the secretion system begin immediately upon contact with the host cell. ExoS is associated with dissemination from burn wounds and with actions destructive to cells, including its action on the cytoskeleton. The blue pigment pyocyanin has been detected in human lesions and shown to have a toxic effect on respiratory ciliary function.

P aeruginosa and cystic fibrosis

P aeruginosa is the most persistent of the infectious agents that complicate the course of CF. Initial colonization may be aided by the fact that cells from CF patients are less

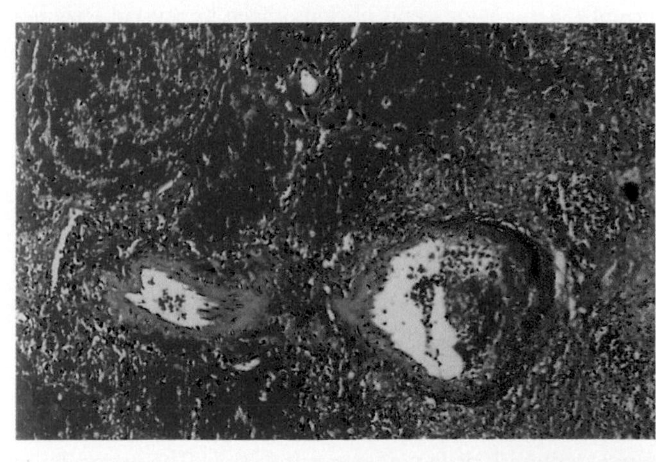

FIGURE 35–3. *Pseudomonas aeruginosa* pneumonia. This blood vessel in the lung of a fatal case is infected with *P aeruginosa* and is undergoing destruction. A thrombus is forming in the lumen as well. (Reproduced with permission from Connor DH, Chandler FW, Schwartz DQA, Manz HJ, Lack EE (eds). *Pathology of Infectious Diseases,* vol. 1. Stamford, CT: Appleton & Lange; 1997.)

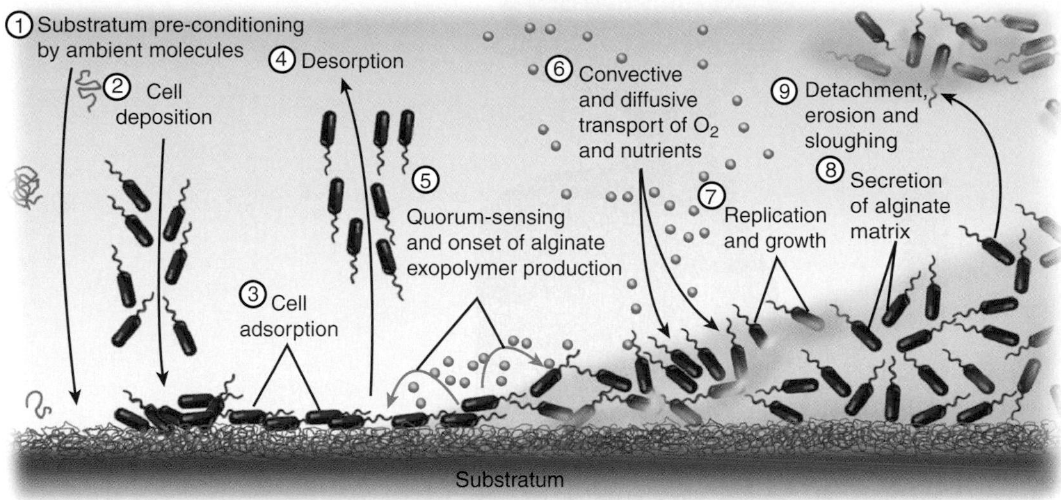

FIGURE 35–4. *Pseudomonas aeruginosa* alginate biofilm in cystic fibrosis. (Modified with permission from Willey J, Sherwood L, Woolverton C (eds). *Prescott's Principles of Microbiology.* New York: McGraw-Hill; 2008.)

highly sialylated than normal epithelial cells, providing increased receptors for *P aeruginosa* attachment. Defects in the epithelia of CF patients may also retard their clearing by desquamation. Once the bronchi are colonized, the organisms remain, forming a biofilm containing microcolonies of bacteria, which together form a **biofilm** (**Figure 35–4**). The most striking feature of this association is the unique presence of strains with multiple mutations in regulatory genes whose end result is overproduction of the alginate polymer. The high osmolarity of the thick CF secretions facilitates expression of these alginate hyperproducing mutants, and quorum-sensing is involved in biofilm formation. The selective advantages of this biofilm include inaccessibility of the immune system (complement, antibody, phagocytes) and antimicrobial agents.

Mutants overproduce alginate polymer

Biofilm protects bacteria

IMMUNITY

Human immunity to *Pseudomonas* infection is not well understood. Inferences from animal studies and clinical observations suggest that both humoral and cell-mediated immunity are important. The strong propensity of *P aeruginosa* to infect those with defective cell-mediated immunity indicates that these responses are particularly important.

Humoral and cellular immune responses both important

P AERUGINOSA DISEASE: CLINICAL ASPECTS

MANIFESTATIONS

P aeruginosa can produce any of the opportunistic extraintestinal infections caused by members of the Enterobacteriaceae. Burn, wound, urinary tract, skin, eye, ear, and respiratory infections all occur and may progress to bacteremia. *P aeruginosa* is also one of the most common causes of infection in environmentally contaminated wounds (eg, osteomyelitis after compound fractures or nail puncture wounds of the foot).

Infects burns and environmentally contaminated wounds

P aeruginosa pneumonia is a rapid, destructive infection particularly in patients with granulocytopenia. It is associated with alveolar necrosis, vascular invasion, infarcts, and bacteremia. Pulmonary infection in CF patients is different; it is a chronic infection that alternates between a state of colonization and more overt bronchitis or pneumonia (**Figure 35–5**). Although the more aggressive features of *Pseudomonas* infection in the immunocompromised are not common, the infection is still serious enough to be a leading cause of death in CF patients.

Pneumonia is aggressive in the immunocompromised and chronic in CF patients

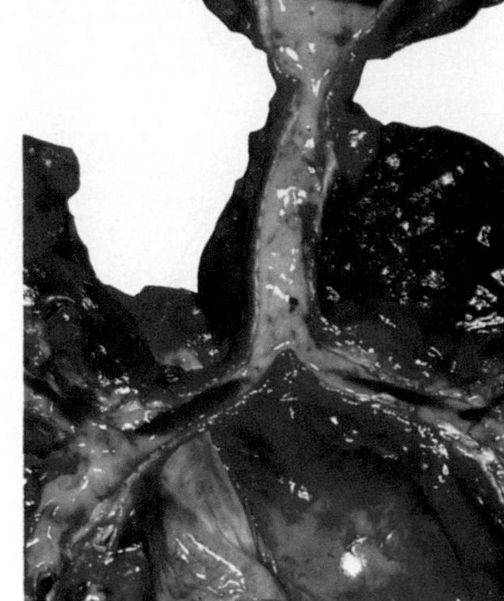

FIGURE 35–5. *Pseudomonas aeruginosa* and cystic fibrosis. The lungs of a young adult are shown at autopsy. There is both extensive inflammation and thick biofilm throughout. (Reproduced with permission from Connor DH, Chandler FW, Schwartz DQA, Manz HJ, Lack EE (eds). *Pathology of Infectious Diseases*, vol. 1. Stamford, CT: Appleton & Lange; 1997.)

Common cause of otitis externa

Contamination of contact lenses leads to keratitis

Bacteremia may cause ecthyma gangrenosum

P aeruginosa is also a common cause of otitis externa, including "swimmer's ear" and a rare but life-threatening "malignant" otitis externa seen in patients with diabetes. Folliculitis of the skin may follow soaking in inadequately decontaminated hot tubs that can become heavily contaminated with the organism. The organism can cause conjunctivitis, keratitis, or endophthalmitis when introduced into the eye by trauma or contaminated medication or contact lens solution. Keratitis can progress rapidly and destroy the cornea within 24 to 48 hours. In some cases of *P aeruginosa* bacteremia, cutaneous papules develop, which progress to black, necrotic ulcers. This is called **ecthyma gangrenosum** and is the result of direct invasion and destruction of blood vessel walls by the organism.

DIAGNOSIS

Pigments typically produced in culture

P aeruginosa is readily grown in culture. The combination of characteristic oxidase positive colonies, pyocyanin production (**Figure 35–6**), and the ability to grow at 42°C is sufficient to distinguish *P aeruginosa* from other *Pseudomonas* species. Although biochemical test can identify other species, such tests are usually not done unless the clinical evidence for infection is very strong. A clinically important organism initially resembling *P aeruginosa* usually turns out to be one of the species discussed below.

TREATMENT

Multiresistance is by restricting permeability

Resistance to penicillins and aminoglycosides is common

Third-generation cephalosporins are often active

Of the pathogenic bacteria, *P aeruginosa* is the organism most consistently resistant to many antimicrobials. This is primarily due to the porins that restrict their entry to the periplasmic space. *P aeruginosa* strains are regularly resistant to penicillin, ampicillin, cephalothin, tetracycline, chloramphenicol, sulfonamides, and the earlier aminoglycosides (streptomycin, kanamycin). Much effort has been directed toward the development of antimicrobials with anti-*Pseudomonas* activity. The newer aminoglycosides—gentamicin, tobramycin, and amikacin—all are active against most strains despite the presence of mutational and plasmid-mediated resistance. Carbenicillin and ticarcillin are active and can be given in high doses, but permeability mutations occur more frequently than with the aminoglycosides. The most prized feature of some of the third-generation cephalosporins (ceftazidime, cefepime, cefoperazone), carbapenems (imipenem, meropenem), and monobactams (aztreonam) is their activity against *Pseudomonas*. In general, urinary infections may be treated with a single drug, but more serious systemic *P aeruginosa* infections are usually treated with a combination of an anti-*Pseudomonas* β-lactam and an aminoglycoside, particularly in neutropenic patients. Ciprofloxacin is also used in treatment of such cases. In all instances, susceptibility must be confirmed by in vitro tests. ::: Susceptibility tests, p. 416

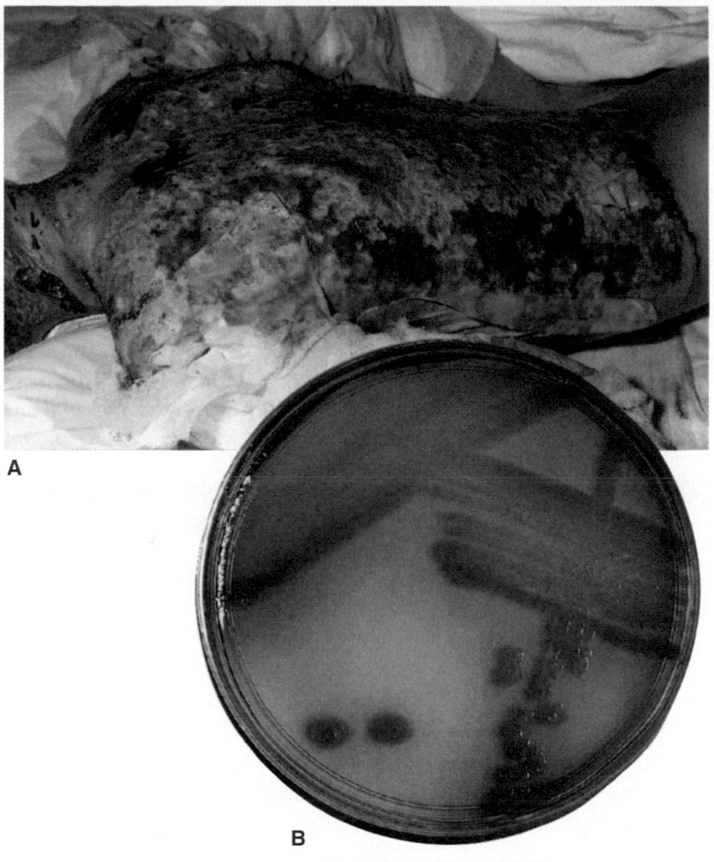

FIGURE 35–6. ***P aeruginosa* pigment production.** The blue color of pyocyanin when mixed with yellow tissue or media components typically produces a green discoloration. This is sometimes seen in clinical cases **A.** and regularly seen on culture plates **B.** (Reproduced with permission from Nester EW, Anderson DG, Roberts CE Jr, Nester MT. *Microbiology: A Human Perspective,* 6th ed. New York: McGraw-Hill, 2008.)

The treatment for *P aeruginosa* infection in CF presents special problems because most of the effective antimicrobics are only given intravenously. There is a reluctance to hospitalize in many patients, and oral agents are used instead. There is less experience with their efficacy under these conditions, and the chronic nature of CF is a setup for development of resistance during therapy. This has already been seen with ciprofloxacin and aztreonam. Aerosolized tobramycin has also been used in some CF patients, with some evidence of clinical improvement.

Effective oral agents are scarce

PREVENTION

Vaccines incorporating somatic antigens from multiple *P aeruginosa* serotypes have been developed and proved immunogenic in humans. The primary candidates for such preparations are patients with burn injuries, CF, or immunosuppression. Although some protection has been demonstrated, these preparations are still experimental.

Vaccines are experimental

BURKHOLDERIA

Burkholderia pseudomallei is a saprophyte in soil, ponds, rice paddies, and vegetables located in Southeast Asia, the Philippines, Indonesia, and other tropical areas. Infection is acquired by direct inoculation or by inhalation of aerosols or dust containing the bacteria. The disease, **melioidosis,** is usually an acute pneumonia; however, it is sufficiently variable that subacute, chronic, and even relapsing infections may follow systemic spread. Some soldiers relapsed years after their return from Vietnam. The clinical and radiologic features may resemble tuberculosis. In fulminant cases of melioidosis, rapid respiratory failure may ensue and metastatic abscesses develop in the skin or other sites. Tetracycline, chloramphenicol, sulfonamides, and trimethoprim-sulfamethoxazole have been effective in therapy. *B cepacia* is an opportunistic organism that has been found to contaminate reagents, disinfectants, and medical devices in much the same manner as does *P aeruginosa.* It has also complicated the course of CF but does not produce the mucoid colony type seen with *P aeruginosa.*

Melioidosis is a tropical pneumonia that relapses

B cepacia is a nosocomial, CF pathogen

ACINETOBACTER

Respiratory and urinary infection come from soil and water

The genus *Acinetobacter* comprises Gram-negative coccobacilli that occasionally appear sufficiently round on Gram smears to be confused with *Neisseria*. On primary isolation, they closely resemble Enterobacteriaceae in growth pattern and colonial morphology, but are distinguished by their failure to ferment carbohydrates or reduce nitrates. As with most of the organisms discussed in this chapter, the isolation of *Acinetobacter* from clinical material does not define infection because they appear most frequently as skin and respiratory colonizers. They are most frequently found as contaminants of almost anything wet, including soaps and some disinfectant solutions. Pneumonia is the most common infection, followed by urinary tract and soft tissue infections. Nosocomial respiratory infections have been traced to contaminated inhalation therapy equipment, and bacteremia to infected intravenous catheters. Treatment is complicated by frequent resistance to penicillins, cephalosporins, and occasionally aminoglycosides.

MORAXELLA

Bronchitis and otitis come from respiratory flora

Moraxella is another genus of coccobacillary, Gram-negative rods that are usually paired end-to-end. Some species require enriched media, such as blood or chocolate agar. Their morphology, fastidious growth, and positive oxidase reaction can result in confusion with *Neisseria*. This is particularly true for *M catarrhalis,* which for many years was classified with *Neisseria.* More recently, it was called *Branhamella catarrhalis*, and it is an occasional cause of otitis media and lower respiratory tract infection. Both infections relate to the presence of *M catarrhalis* in the normal oropharyngeal flora. With the exception of *M catarrhalis*, which frequently produces β-lactamase, *Moraxella* species are generally susceptible to penicillin.

AEROMONAS AND *PLESIOMONAS*

Resemble other enteric bacteria

The genera *Aeromonas* and *Plesiomonas* have bacteriologic features similar to those of the Enterobacteriaceae, *Vibrio,* and *Pseudomonas.* They are aerobic and facultatively anaerobic, attack carbohydrates fermentatively, and demonstrate various other biochemical reactions. *Aeromonas* colonies are typically β-hemolytic. The major taxonomic resemblance to *Pseudomonas* is that both *Aeromonas* and *Plesiomonas* are oxidase-positive with polar flagella. Their habitat is basically environmental (water and soil), but they can occasionally be found in the human intestinal tract.

Rapid cellulitis follows injury in water

Diarrheas relate to enterotoxin production

Aeromonas is an uncommon, but highly virulent cause of wound infections acquired in fresh or saltwater. The onset can be as rapid as 8 hours after the injury, and the cellulitis progresses rapidly to fasciitis, myonecrosis, and bacteremia in less than a day. *Aeromonas* is also the leading cause of infections associated with the use of leeches, owing to its regular presence in the leech foregut. In addition to opportunistic infection, some evidence suggests an occasional role for *Aeromonas* in gastroenteritis through production of toxins with enterotoxic and cytotoxic properties. *Plesiomonas* is also associated with an enterotoxic diarrhea. These associations are not yet strong enough to justify attempts to routinely isolate *Aeromonas* and *Plesiomonas* from diarrheal stools. Resistance to penicillins and cephalosporins is common. Most strains show susceptibility to tetracycline, with variable susceptibility to aminoglycosides, including gentamicin.

OTHER GRAM-NEGATIVE RODS

Rare species are interpreted on the basis of their clinical setting

There are many other Gram-negative rods that rarely cause disease in humans. Some are members of the normal flora, and others come from the environment. Because many of these do not ferment carbohydrates or react in many of the tests routinely used to characterize bacteria, their identification is frequently delayed while additional tests are tried or the organism is sent to a reference laboratory. The clinical significance of all these organisms is essentially the same; the clinician usually receives a report of a "nonfermenter" or another descriptive term and a susceptibility test result. The significance of the isolate is then determined on clinical grounds. The major characteristics of some of these organisms are shown in **Table 35–1**. The types of

TABLE 35–1 *Pseudomonas* and Other Opportunistic Gram-negative Rods

	BACTERIOLOGIC FEATURES						
SPECIES	**MACCONKEY GROWTH**	**CO_2 REQUIRED**	**PIGMENTS**	**ADHERENCE**	**VIRULENCE FACTORS**	**EPIDEMIOLOGY**	**DISEASE**
Pseudomonas							
P aeruginosa	+	–	Pyocyanin, fluorescin	Pili, flagella, alginate slime	Exotoxin A, exoenxyme S, elastase, alginate slime	Environmental, normal flora, mucosal breaks, nosocomial	Wounds, pneumonia, burns, otitis externa, cystic fibrosis
P fluoresces	+	–	Fluorescin			Environmental	Opportunistic
Other species	+	–	Fluorescin			Environmental	Opportunistic
Stenotrophomonas maltophilia	+	–	–		Protease	Environmental, mucosal breaks, water, nosocomial	Pneumonia, bacteremia
Acinetobacter	+	–	–		Capsule	Environmental, skin colonization, water, nosocomial	Respiratory, urinary catheter bacteremia
Burkholderia							
B mallei	+	–	–			Contact with horses	Glanders
B pseudomallei	+	–	–		Facultative intracellular growth	Environmental in Southeast Asia and tropical regions	Melioidosis
B cepacia	+	–	–	Pili	Invasion, elastase, biofilm	Environmental, mucosal breaks, water, nosocomial	Wounds, pneumonia, cystic fibrosis
B mallei	+	–	–			Contact with horses	Glanders

(*Continued*)

TABLE 35–1 *Pseudomonas* and Other Opportunistic Gram-negative Rods (*Continued*)

Aeromonas	+	–	–		Enterotoxin, cytotoxin	Environmental, fresh and salt water, leeches, intestinal flora	Wounds, diarrhea
Plesiomonas	+	–	–		Enterotoxin	Water, seafood, soil	Diarrhea
Alkaligenes	+	–	–			Respiratory, intestinal flora	Blood, urine, wounds
Cardiobacterium	+	+	–			Nasopharyngeal, intestinal flora	Endocarditis
Chromobacterium	+	+	Violet			Water, soil, (tropical)	Cellulitis, bacteremia
Flavobacterium	+	+	Yellow			Environmental, nosocomial	Meningitis
Eikenella	+	+	–			Respiratory flora	Oropharyngeal abscess, draining sinuses
Actinobacillus	+	+	–			Respiratory flora, animals	Endocarditis, periodontal disease
Moraxella	+	+	–	Pili		Respiratory flora,	Bronchitis, pneumonia

infection listed represent the most common among scattered case reports and should not be interpreted as typical for each organism.

Some Gram-negative bacilli fail to conform to any of the species currently recognized. If clinically important, such strains are sent to reference centers, such as the Centers for Disease Control and Prevention (CDC) in Atlanta, Georgia. Eventually, some are given designations such as "CDC group IIF," which may appear in clinical reports. Much later, a new genus and/or species name may be issued if agreement among taxonomists is sufficient.

Some bacteria remain unnamed for years

CLINICAL CASE

LEUKEMIA AND BLACK SKIN ULCERS

An 8-year-old boy with recently diagnosed acute leukemia was treated with potent cytotoxic drugs in an effort to induce a remission. Within 5 days of initiation of chemotherapy, his total white blood cell count had fallen from 60,000/mm^3 pretreatment to 300/mm^3, with no granulocytes present. On the sixth day, the boy developed a high fever (40.1°C) with no focal findings except for the appearance of several faintly erythematous nodules on the thighs.

Over the next 2 days, his skin lesions became purple, then black and necrotic, eventually forming multiple deep ulcers. Chest radiographs taken at the onset of fever were clear, but the following day showed diffuse infiltrates in both lungs. All three blood cultures taken on day 6 were positive for an oxidase-positive Gram-negative rod that produced blue-green discoloration of the culture plates.

QUESTIONS

■ This infection is most likely due to which of the following:

A. *Pseudomonas aeruginosa*

B. *Burkholderia pseudomallei*

C. *Burkholderia cepacia*

D. *Aeromonas*

E. *Acinetobacter*

■ Which is the most important predisposing feature for this infection?

A. Hospital environment

B. Antibiotic treatment

C. Neutropenia

D. Age

■ The skin lesions are most likely due to the action of:

A. Alginate

B. Pyocyanin

C. Oxidase

D. Elastase

E. Flagella

ANSWERS

1(A), **2**(C), **3**(D)

CHAPTER 36

Plague and Other Bacterial Zoonotic Diseases

Dr. Rieux resolved to compile this chronicle…
to state quite simply what we learn
in a time of pestilence: that there are more things
to admire in men than to despise.

—Albert Camus: *The Plague*

Many bacterial, rickettsial, and viral diseases are classified as zoonoses because they are acquired by humans either directly or indirectly from animals. This chapter considers bacteria causing four zoonotic infections that are not discussed in other chapters. All four species, *Brucella, Yersinia pestis, Francisella tularensis,* and *Pasteurella multocida,* are Gram-negative bacilli that are primarily animal pathogens. The diseases they cause, brucellosis, plague, tularemia, and pasteurellosis, are now rare in humans and develop only after unique animal contact. The full range of zoonoses considered in this and other chapters is shown in **Table 36–1**.

BRUCELLA

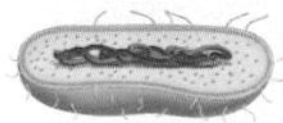

BACTERIOLOGY

Brucella species are small, coccobacillary, Gram-negative rods that morphologically resemble *Haemophilus* and *Bordetella.* They are nonmotile, non–acid-fast, and non–spore-forming. The cells have a typical Gram-negative structure, and the outer membrane contains proteins. Although long assigned to multiple species, the three most common in humans (*B abortus, B melitensis,* and *B suis*) are now considered variants of *Brucella abortus.* Their growth is slow, requiring at least 2 to 3 days of aerobic incubation in enriched broth or on blood agar. They produce catalase, oxidase, and urease, but do not ferment carbohydrates.

Coccobacilli resemble *Haemophilus*

TABLE 36–1 Some Important Bacterial and Rickettsial Zoonotic Infections

DISEASE	ETIOLOGIC AGENT	USUAL RESERVOIR	USUAL MODE OF TRANSMISSION TO HUMANS	TRANSMISSION BETWEEN HUMANS	MODE OF TRANSMISSION BETWEEN HUMANS	SPECIAL CHARACTERISTICS
Anthrax	*Bacillus anthracis*	Cattle, sheep, goats	Infected animals or products	No[a]		Resistant spores
Bovine tuberculosis	*Mycobacterium bovis*	Cattle	Milk	No[a]		
Brucellosis	*Brucella abortus*	Cattle, swine, goats	Milk, infected carcasses	No[a]		
Campylobacter infection	*C jejuni*	Wild mammals, cattle, sheep, pets	Contaminated food and water	Yes	Fecal-oral	
Leptospirosis	*Leptospira* spp.	Cattle, rodents	Water contaminated with urine	No[a]		
Lyme disease	*Borrelia burgdorferi*	Deer, rodents	Ticks; transplacentally	No[a]		Relapsing disease
Pasteurellosis	*Pasteurella multocida*	Animal oral cavities	Bites, scratches	No[a]		
Plague	*Yersinia pestis*	Rodents	Fleas	Yes	Droplet (pneumonic) spread	Great epidemic potential
Other *Yersinia* infections	*Y enterocolitica, Y pseudotuberculosis*	Wild mammals, pigs, cattle, pets	Fecal-oral	Yes	Fecal-oral	
Relapsing fever	*Borrelia* spp.	Rodents, ticks	Ticks	Yes	Body louse[b]	Epidemic potential
Salmonellosis	*Salmonella* serotypes	Poultry, livestock	Contaminated food	Yes	Fecal contamination of food	
Rickettsial spotted fevers	*R rickettsii*[c]	Rodents, ticks, mites	Ticks, mites	No[a]		
Murine typhus	*Rickettsia typhi*	Rodents	Fleas	No[a]		
Q fever	*Coxiella burnetii*	Cattle, sheep, goats	Contaminated dust and aerosols	No[a]		

[a]"What never?" "No never." "What *never*?" "Well, hardly ever!" (W. S. Gilbert, from *H.M.S. Pinafore*).
[b]The relationship between tick-borne relapsing fever and epidemic relapsing fever by the body louse remains uncertain.
[c]One of several etiologic agents.

BRUCELLOSIS

CLINICAL CAPSULE

Brucellosis is a genitourinary infection of sheep, cattle, pigs, and other animals. Humans such as farmers, slaughterhouse workers, and veterinarians become infected directly by occupational contact or indirectly by consumption of contaminated animal products such as milk. In humans, brucellosis is a chronic illness characterized by fever, night sweats, and weight loss lasting weeks to months. Because the infection is localized in reticuloendothelial organs, there are few physical findings unless the liver or spleen becomes enlarged. When patients develop a cycling pattern of nocturnal fevers, the disease has been called undulant fever.

EPIDEMIOLOGY

Brucellosis, a chronic infection that persists for life in animals, is an important cause of abortion, sterility, and decreased milk production in cattle, goats, and hogs. It is spread among animals by direct contact with infected tissues and ingestion of contaminated feed; it causes chronic infection of the mammary glands, uterus, placenta, seminal vesicles, and epididymis.

Causes abortion in cattle, goats, and pigs

Humans acquire brucellosis by occupational exposure or consumption of unpasteurized dairy products. The bacteria may gain access through cuts in the skin, contact with mucous membranes, inhalation, or ingestion. In the United States, the number of cases has dropped steadily from a maximum of more than 6000 per year in the 1940s to the current level of less than 100 per year. Of these cases, 50% to 60% are in abattoir employees, government meat inspectors, veterinarians, and others who handle livestock or meat products. Consumption of unpasteurized dairy products, which accounts for 8% to 10% of infections, is the leading source in persons who have no connection with the meat-processing or livestock industries. Some recent cases of this type have been associated with "health" foods. In the United States, the distribution of human cases of brucellosis includes virtually every state, but is concentrated in those with large livestock industries or in those in proximity to Mexico (California, Texas). An outbreak in Texas was traced to unpasteurized goat cheese brought in from Mexico.

Occupational disease for veterinarians

Unpasteurized dairy products and "health" foods are a risk

PATHOGENESIS

All *Brucella* are facultative intracellular parasites of epithelial cells and professional phagocytes. After they penetrate the skin or mucous membranes, they enter and multiply in macrophages in the liver sinusoids, spleen, bone marrow, and other components of the reticuloendothelial system eventually forming granulomas (**Figure 36–1**). Intracellular survival is facilitated by inhibition of both the myeloperoxidase system and of phagosome–lysosome fusion. This is followed by formation of a replicative phagosome and continued multiplication in vesicles associated with the endoplasmic reticulum. *Brucella* is also able to *inhibit* apoptosis, thus prolonging the life of the host cell where it is replicating. In cows, sheep, pigs, and goats, erythritol, a four-carbon alcohol present in chorionic tissue, markedly stimulates growth of *Brucella*. This stimulation probably accounts for the tendency of the organism to locate in these sites. The human placenta does not contain erythritol. ::: Phagosome-lysosome. p. 25

Inhibits myeloperoxidase, lysosome fusion, and apoptosis

Multiplies in endoplasmic reticulum vesicles

Animal placental erythritol stimulates growth

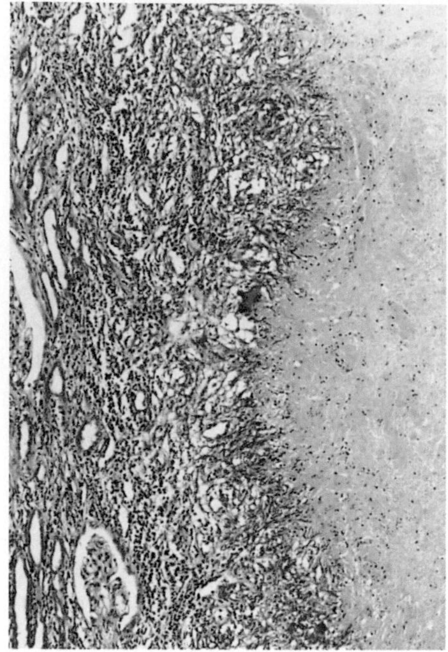

FIGURE 36–1. Brucellosis. Caseating granuloma in the kidney of a midwestern cattle farmer. The giant and epithelioid cells are pallisaded around the caseating area on the right. Glomeruli are compressed on the left. (Reproduced with permission from Connor DH, Chandler FW, Schwartz DQA, Manz HJ, Lack EE (eds). *Pathology of Infectious Diseases*, vol. 1. Stamford, CT: Appleton & Lange; 1997.)

If not controlled locally, infection progresses with the formation of small granulomas in the reticuloendothelial sites of bacterial multiplication and with release of bacteria back into the systemic circulation. These bacteremic episodes are largely responsible for the recurrent chills and fever of the clinical illness. These events resemble the pathogenesis of typhoid fever (see Chapter 33). ::: Granuloma, p. 26

IMMUNITY

Macrophage killing requires TH1-type responses

Although antibodies are formed in the course of brucellosis, there is little evidence that they are protective. Control of disease is due to T-cell–mediated cellular immune responses. Development of TH1-type responses with the production of cytokines (tumor necrosis factors [TNF-α, TNF-γ, IL-1] and interleukin [IL-12]) are associated with the elimination of *Brucella* from macrophages. ::: Cell-mediated immunity, p. 35

BRUCELLOSIS: CLINICAL ASPECTS

MANIFESTATIONS

Recurrent bacteremia comes from reticuloendothelial sites

Night sweats and periodic fever continue without obvious organ focus

Brucellosis starts with malaise, chills, and fever 7 to 21 days after infection. Drenching sweats in the late afternoon or evening are common, as are temperatures in the range of 39.4° to 40°C. The pattern of periodic nocturnal fever (undulant fever) typically continues for weeks, months, or even 1 to 2 years. Patients become chronically ill with associated body aches, headache, and anorexia. Weight loss of up to 20 kg may occur during prolonged illness. Despite these dramatic effects, physical findings and localizing signs are few. Less than 25% of patients show detectable enlargement of the reticuloendothelial organs, the primary site of infection. Of such findings, splenomegaly is most common, followed by lymphadenopathy and hepatomegaly. Occasionally, localized infection develops in the lung, bone, brain, heart, or genitourinary system. These cases usually lack the pronounced systemic symptoms of the typical illness.

DIAGNOSIS

Definitive diagnosis of brucellosis requires isolation of *Brucella* from the blood or from biopsy specimens of the liver, bone marrow, or lymph nodes. The slow growth of some strains requires prolonged incubation of culture media to achieve isolation. Blood cultures may require 2 to 4 weeks for growth, although most are positive in 2 to 5 days. The diagnosis is often made serologically, but is subject to the same interpretive constraints as are all serologic tests. Antibodies that agglutinate suspensions of heat-killed organisms typically reach titers of 1:640 or more in acute disease. Lower titers may reflect previous disease or cross-reacting antibodies. Titers return to the normal range within 1 year of successful therapy.

Blood culture is primary method

Serologic tests may be useful

TREATMENT AND PREVENTION

Doxycycline in combination with an aminoglycoside (streptomycin or gentamicin) is the primary treatment for brucellosis. Rifampin, ciprofloxacin, and timethoprim-sulfamethoxazole are also used in combinations. Although β-lactams are active in vitro, clinical response is poor, probably as a result of failure to penetrate the intracellular location of the bacteria. The therapeutic response is not rapid; 2 to 7 days may pass before patients become afebrile. Up to 10% of patients have relapses in the first 3 months after therapy. Prevention is primarily by measures that minimize occupational exposure and by the pasteurization of dairy products. Control of brucellosis in animals involves a combination of immunization with an attenuated strain of *B abortus* and eradication of infected stock. No human vaccine is in use.

Doxycycline plus aminoglycoside is effective

Pasteurization is primary prevention

YERSINIA PESTIS

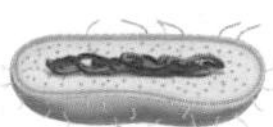

BACTERIOLOGY

Y pestis is a nonmotile, non–spore-forming, Gram-negative bacillus with a tendency toward pleomorphism and bipolar staining. It is a member of the Enterobacteriaceae family and shares features of the other *Yersinia* pathogenic for humans (*Y pseudotuberculosis, Y enterocolitica*), such as virulence plasmids and multiple *Yersinia* outer membrane proteins (Yops). In addition, *Y pestis* has two additional virulence plasmids, which code for a glycoprotein capsular antigen called F1 and enzymes with phospholipase, protease, and fibrinolytic and plasminogen-activating activity. *Y pestis* also has its own adhesin similar to the invasins of the other *Yersinia*. ::: General features of Yersinia, p. 605

Member of Enterobacteriaceae

Yops, protein capsule, and enzymes are present

PLAGUE

CLINICAL CAPSULE

Plague, an infection of rodents transmitted to humans by the bite of infected fleas, is the most explosively virulent disease known. Most cases begin with a painful swollen lymph node (bubo) from which the bacteria rapidly spread to the bloodstream. Plague pneumonia (Black Death) is produced by pulmonary seeding from the bloodstream or directly from another patient with pneumonia. All forms cause a toxic picture with shock and death within a few days. No other disease regularly kills previously healthy persons so rapidly.

EPIDEMIOLOGY

The term **plague** is often used generically to describe any explosive pandemic disease with high mortality. Medically, it refers only to infection caused by *Y pestis,* and this appellation was justly earned because *Y pestis* was the cause of the most virulent epidemic plague of recorded human history, the Black Death of the Middle Ages. In the 14th century, the estimated population of Europe was 105 million; between 1346 and 1350, 25 million died of plague. Pandemics continued through the end of the 19th century and the early 20th century despite elaborate quarantine measures developed in response to the obvious communicability of the disease. Yersin isolated the etiologic agent in China in 1894 and named it after his mentor, Pasteur (*Pasteurella pestis*). The name was later changed to honor Yersin (*Yersinia pestis*).

Black Death continued into 20th century

Plague is a disease of rodents transmitted by the bite of rat fleas (*Xenopsylla cheopis*) that colonize them. It exists in two interrelated epidemiologic cycles, the **sylvatic** and the **urban** (**Figure 36–2**). Endemic transmission among wild rodents in the sylvatic (*L sylvaticus,*

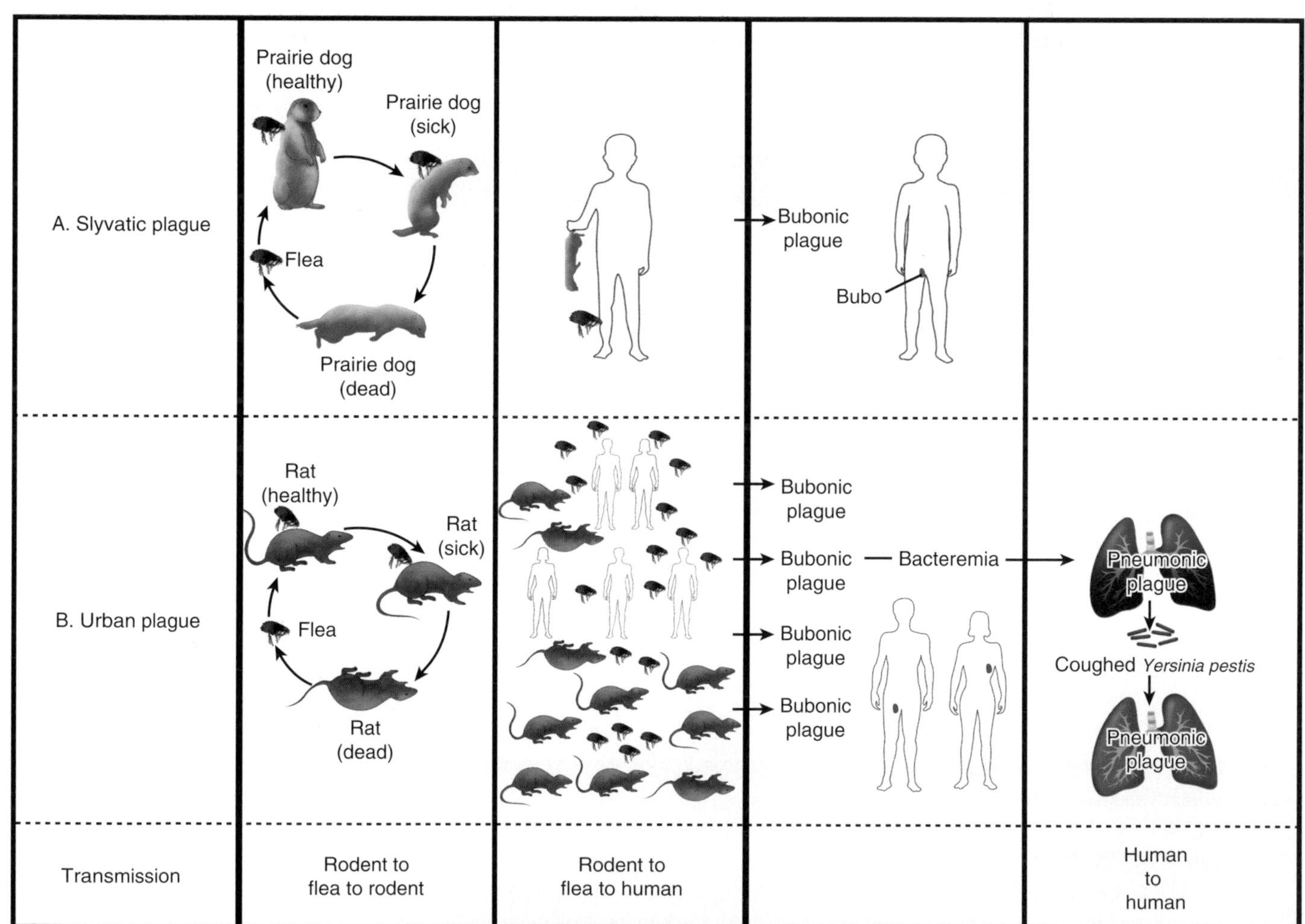

FIGURE 36–2. The epidemiology of plague. A. In the sylvatic cycle, fleas leaving infected rodents, such as mice and prairie dogs, pass the infection to others in the population. Humans rarely contact these rodents but when they do, the flea bite transmits plague. **B.** In the urban cycle, masses of rats are in closer contact with humans, and bites from infected fleas transmit the infection to many. In both cycles initial transmissions result in bubonic plague. Bacteremia with *Yersinia pestis* may infect the lungs to cause pneumonic plague. Pneumonic plague is transmitted human to human by the respiratory route without the involvement of fleas.

belonging to or found in the woods) is the primary reservoir of plague. When infected rodents enter a city, circumstances for the urban cycle are created. Humans can enter the cycle from the bite of the flea in either environment. However, chances are greater in the urban setting, particularly with crowding and poor sanitation.

Sylvatic transmission among rodents is primary reservoir

The plagues of the Middle Ages are examples of the urban cycle involving rats and humans. When food is scarce in the countryside, rats migrate to cities. This facilitates rat-to-rat transmission and brings the primary reservoir into closer contact with humans. When the number of nonimmune rats is sufficient, epizootic plague develops among them, with bacteremia and high mortality. Fleas feeding on the rats become infected, and the bacteria multiply in their intestinal tract eventually blocking the proventriculus a valve-like organ connecting the esophagus to the midgut. When the rat dies, the fleas seek a new host, which is usually another rat but may be a nearby human. Because of the intestinal blockage, the infected flea regurgitates *Y pestis* into the new bite wound. Therefore, the probability of transmission to humans is greatest when both rat population and rat mortality are high.

Rat migration to cities increases human risk

Fleas regurgitate into bite wounds

The bite of the flea is the first event in the development of a case of **bubonic plague,** which, even if serious enough to kill the patient, is not contagious to other humans. However, some patients with bubonic plague develop a secondary pneumonia by bacteremic spread to the lungs. This **pneumonic plague** is highly contagious person to person by the respiratory droplet route. It is not difficult to understand how rapid spread proceeds in conjunction with crowded unsanitary conditions and continued flea-to-human transmission. An urban plague epidemic is vividly described through the eyes of a physician in Albert Camus' novel, *The Plague.*

Bubo is initial lesion

Pneumonia is contagious

Although urban plague epidemics have been essentially eliminated by rat control and other public health measures, sylvatic transmission cycles persist in many parts of the world, including North America. These cycles involve nonurban mammals such as prairie dogs, deer mice, rabbits, and wood rats. Transmission between them involves fleas. Coyotes or wolves may be infected by the same fleas or by ingestion of infected rodents. By their nature, the reservoir animals rarely come in contact with humans; when they do, however, the infected fleas they carry can transmit *Y pestis.* The most common circumstance is a child who is exploring the outdoors, comes across a dead or dying prairie dog, and pokes, carries, or touches it long enough to be bitten by the fleas leaving the animal. The result is a sporadic case of bubonic plague, which occasionally becomes pneumonic.

Nonepidemic disease is linked to animal contact

Sylvatic plague, which exists in most continents, is common in Southeast Asia, but is not found in Western Europe or Australia. In the United States, the primary enzootic areas are the semiarid plains of the western states. Infected animals and fleas have been detected from the Mexican border to the eastern half of Washington State. The geographic focus of human plague in the United States is in the "four corners" area, where Arizona, New Mexico, Colorado, and Utah meet, but cases have occurred in California, west Texas, Idaho, and Montana. Most years, as many as 15 cases of plague are reported, although this number rose to 30 to 40 in the mid-1980s. These variations are strongly related to changes in the size of the sylvatic reservoir.

Most US cases are in arid western states

PATHOGENESIS

It should not be surprising that the molecular pathogenesis of plague is quite complex, given its extremely high virulence in both insect and mammalian environments, and is incompletely understood. A general pattern is that of the more than 20 known virulence factors; some are deployed primarily in the flea, whereas others are produced only in the rodent or human victim. *Y pestis* has regulatory systems that sense temperature, calcium, and surely other environmental triggers to turn the production of appropriate virulence factors on or off. At ambient temperature (20° to 28°C) in the flea, factors that facilitate multiplication of the organism (fibrinolysin, phospholipase) and blockage of the proventriculus (coagulase, polysaccharide biofilm) are produced. The flea, sensing starvation, feeds voraciously and eventually regurgitates blood and

Multiplication in flea foregut is aided by low temperature virulence factors

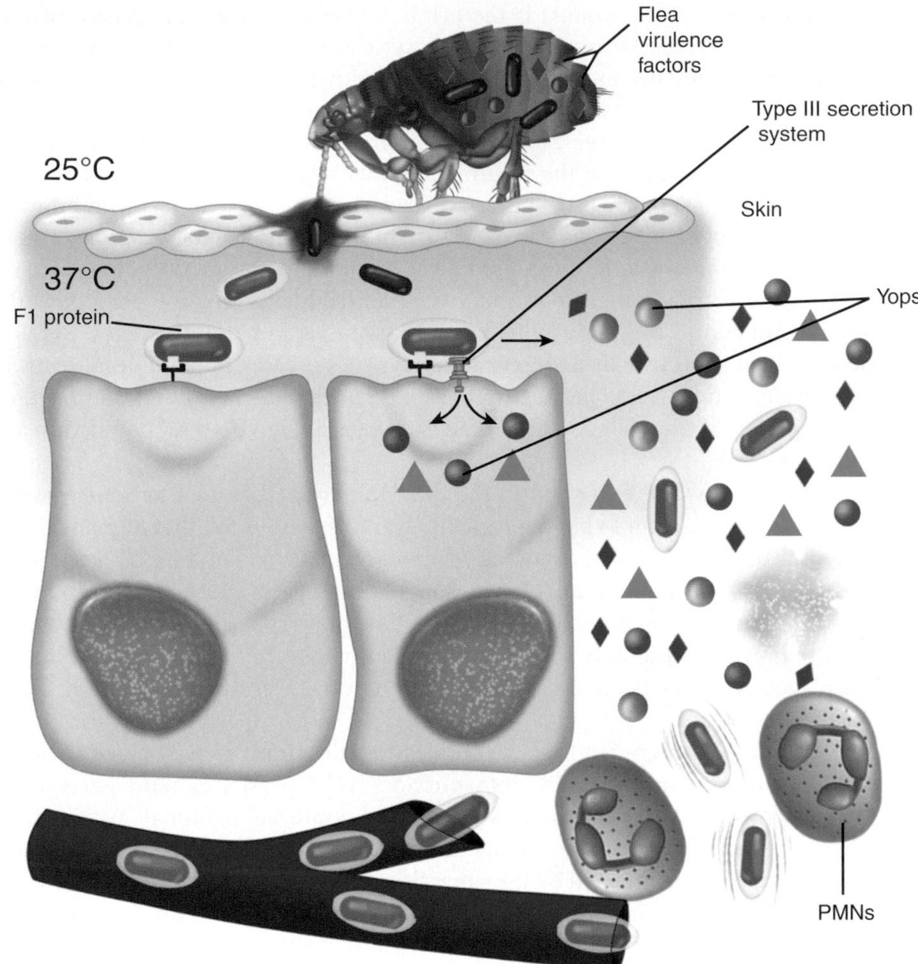

FIGURE 36–3. Plague, cellular view. *(Top) Yersinia pestis* is growing in the flea and producing virulence factors unique to that environment. Bacteria are regurgitated as part of the flea's feeding on human skin and reach the subepithelial tissues. Here, triggered by environmental cues such as a new warmer temperature (37°C), they start to produce a new set of virulence factors unique to mammalian victims such as the F1 protein capsule. *(Left) Y pestis* attaches to and epithelial cell. *(Middle) Yersina* outer membrane proteins (Yops) begin to be produced. Some are injected by a type III secretion system, others are secreted on the surface. *(Right)* The cell is destroyed and the organisms evade phagocytosis to enter the bloodstream. PMNs. polymorphonuclear neutrophils.

bacteria into the bite wound. In this wound (rat or human), *Y pestis* is suddenly moved into a new environment.

F1 protein, Pla, Yops produced at 37°C

F1 is antiphagocytic

Yops destroy and disrupt

In a new warm-blooded (35° to 37°C) host, *Y pestis* produces a second set of virulence factors including the F1 protein, a plasminogen activator (Pla), and the Yops (**Figure 36–3**). The F1 protein forms a gel-like capsule with antiphagocytic properties that allow the bacteria to persist and multiply. The plasminogen activator facilitates metastatic spread through enzymatic activity and adhesion to extracellular matrix proteins. The Yops, though named as a protein family (YopA, YopB, and so on), have diverse biologic activities, which fall into two categories. The first category is direct destructive enzymatic activity directed at host cells. The other set of actions disrupt intracellular function and are mediated through injection (type III) secretion systems. Once inside host cells, including professional phagocytes, these secreted proteins disrupt signaling pathways, destroy cytoskeleton structure, inhibit cytokine production, and trigger apoptosis. ::: Secretion systems, p. 368

Bubo progresses to bacteremia

LPS and other products produce shock

The organisms eventually reach the regional lymph nodes through the lymphatics, where they multiply rapidly and produce a hemorrhagic suppurative lymphadenitis known clinically as the **bubo.** Spread to the bloodstream quickly follows. The extreme systemic toxicity that develops with bacteremia appears to be due to lipopolysaccharide (LPS) endotoxin combined with the many actions of Yops, proteases, and other extracellular products. The bacteremia causes seeding of other organs, most notably the lungs, producing a necrotizing hemorrhagic pneumonia known as pneumonic plague.

IMMUNITY

Recovery from bubonic plague appears to confer lasting immunity, but for obvious reasons the mechanisms in humans have not been extensively studied by modern immunologic methods. Animal studies suggest that antibody against the F1 capsular protein is protective by enhancing phagocytosis, but cell-mediated mechanisms are required for intracellular killing.

Anticapsular antibody may be protective

PLAGUE: CLINICAL ASPECTS

MANIFESTATIONS

The incubation period for bubonic plague is 2 to 7 days after the flea bite. Onset is marked by fever and the painful bubo, usually in the groin (bubo is from the Greek *boubon* for "groin") or, less often, in the axilla (**Figure 36–4**). Without treatment, 50% to 75% of patients progress to bacteremia and die in Gram-negative septic shock within hours or days of development of the bubo. About 5% of victims develop pneumonic plague with mucoid, then bloody sputum. Primary pneumonic plague has a shorter incubation period (2 to 3 days) and begins only with fever, malaise, and a feeling of tightness in the chest. Cough, production of sputum, dyspnea, and cyanosis develop later in the course. Death on the second or third day of illness is common, and there are no survivors without antibiotic therapy. A terminal cyanosis seen with pneumonic plague is responsible for the term Black Death. Even today, plague pneumonia is almost always fatal if appropriate treatment is delayed more than a day from the onset.

Bubonic plague mortality is 50% to 75% in untreated cases

Pneumonic plague is fatal if untreated

Terminal cyanosis = Black Death

DIAGNOSIS

Gram smears of aspirates from the bubo typically reveal bipolar-staining Gram-negative bacilli. An immunofluorescence technique is available in public health laboratories for immediate identification of smears or cultures. *Y pestis* is readily isolated on the media used for other members of the Enterobacteriaceae (blood agar, MacConkey agar), although growth may require more than 24 hours of incubation. The appropriate specimens are bubo aspirate, blood, and sputum. Laboratories must be notified of the suspicion of plague to avoid delay in the bacteriologic diagnosis and to guard against laboratory infection.

Immunofluorescent staining is rapid

Cultures grow on routine media

TREATMENT

Streptomycin or gentamicin is the treatment of choice for both bubonic and pneumonic plague. Doxycycline, ciprofloxacin, and chloramphenicol are alternatives. Timely treatment

Streptomycin or gentamicin primary treatment

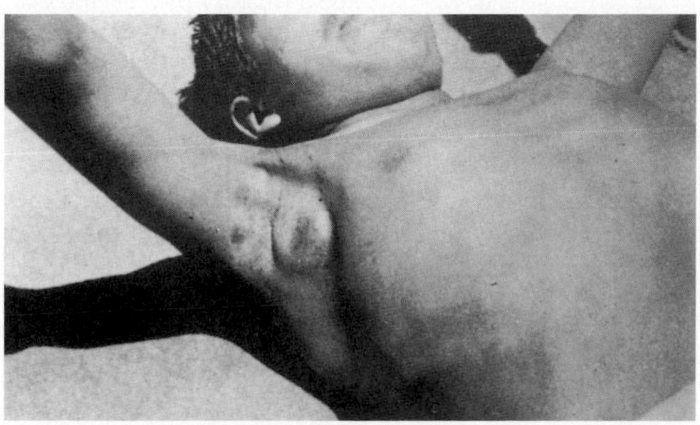

FIGURE 36–4. Bubonic plague. A swollen bubo is seen in the axilla of this child. (Reproduced with permission from Connor DH, Chandler FW, Schwartz DQA, Manz HJ, Lack EE (eds). *Pathology of Infectious Diseases*, vol. 1. Stamford, CT: Appleton & Lange; 1997.)

reduces the mortality of bubonic plague to less than 10%, but the mortality rate of human cases of plague reported in developed countries is still around 20% because of delays in initiation of appropriate therapy.

PREVENTION

Avoid sick or dead wild rodents

Chemoprophylaxis for respiratory exposure

Urban plague has been prevented by rat control and general public health measures such as use of insecticides. Sylvatic plague is virtually impossible to eliminate because of the size and dispersion of the multiple rodent reservoirs. Disease can be prevented by avoidance of sick or dead rodents and rabbits. Eradication of fleas on domestic pets, which have been known to transport infected fleas from wild rodents to humans, is recommended in endemic areas. The continued presence of fully virulent plague in its sylvatic cycle poses a risk of extension to the urban cycle and epidemic disease in the event of major disaster or social breakdown. Chemoprophylaxis with doxycycline or ciprofloxacin is recommended for those who have had close contact with a case of pneumonic plague. It is also used for the household contacts of a person with bubonic plague, because they may have had the same flea contact. A formalin-killed plague vaccine once used for those in high-risk occupations is no longer available.

FRANCISELLA

BACTERIOLOGY

Gram-negative coccobacilli have requirement for –SH compounds

Francisella tularensis is a small, facultative, coccobacillary, Gram-negative rod with much the same morphology as *Brucella.* Virulent strains possess a lipid-rich capsule. *F tularensis* is one of the few bacterial species of medical importance that does not grow on the usual enriched media. This characteristic is due to a special requirement for sulfhydryl compounds, and growth occurs best on a cysteine–glucose blood agar medium after 2 to 10 days of incubation.

TULAREMIA

CLINICAL CAPSULE

Tularemia is a disease of wild mammals caused by *F tularensis*. Humans become infected by direct contact with infected animals or through the bite of a vector (tick or deer fly). The illness is characterized by a local ulcer with high fever and severe constitutional symptoms. The epidemiology of tularemia and many features of the clinical infection are similar to those of plague.

EPIDEMIOLOGY

Humans most often acquire *F tularensis* by contact with an infected mammal or a blood-feeding arthropod. Because the infecting dose is very low (less than 100 organisms), many routes of infection are possible. A tick bite or direct contact with minor skin abrasion are the most common mechanisms of infection. Many wild mammals can be infected, including squirrels, muskrats, beavers, and deer. A common history is that of skinning wild rabbits on a hunting trip. Inhalation may also lead to disease. In a recent outbreak of pulmonary tularemia on Cape Cod, experts believed that lawn mowing and brush cutting facilitated inhalation. Occasionally, the bite or scratch of a domestic dog or cat has been implicated when the animal has ingested or mouthed an infected wild mammal. Infected animals may not show signs of infection, because the organism is well adapted to its natural host. The usual vectors in animals are ticks and deer flies. Ticks may also serve as a reservoir of the organism by transovarial transmission to their offspring.

Infecting dose is low

Acquired by tick bites or directly from wild mammal

Tularemia is distributed throughout the Northern Hemisphere, although there are wide variations in specific regions. The highly virulent tick/rabbit-associated strains are common only in North America and cases have declined steadily since World War II. In the United States, 100 to 200 cases are reported each year, half of which are in the lower midwestern states (Arkansas, Missouri, Oklahoma). Tularemia is not found in the British Isles, Africa, South America, or Australia.

Distribution throughout Northern Hemisphere

PATHOGENESIS

Relatively little is known of the events that occur during the 2- to 5-day incubation period. A lesion often develops at the site of infection, which becomes ulcerated. The organism then infects the reticuloendothelial organs, often forming granulomas, and the disease may sometimes follow a chronic relapsing course. These properties suggest a facultative intracellular pathogen and indeed the virulence of *F tularensis* has been linked to its ability to multiply within many cell types, including hepatocytes and macrophages. Although the organisms enter cells in a phagocytic vacuole, they are able to prevent acidification of the phagosome/lysosome and eventually escape into the cytosol. Early bacteremic spread probably occurs, although it is rarely detected. Other areas of multiplication are characterized by necrosis or granuloma production, and a mixture of abscesses and caseating granulomas may be seen in the same organ.

Intracellular multiplication is key to virulence

Escapes phagosome to cytosol

IMMUNITY

Naturally acquired infection appears to confer longlasting immunity. Antibody titers remain elevated for many years, but cellular immunity plays the major role in resistance to reinfection. T-cell–dependent reactions involving either CD4+ or CD8+ cell are detectable even before antibody responses. ::: CD4+ and CD8+ pathways, p. 33

Cell-mediated immunity is dominant

TULAREMIA: CLINICAL ASPECTS

MANIFESTATIONS

After an incubation period of 2 to 5 days, tularemia may follow a number of courses, depending on the site of inoculation and extent of spread. All begin with the acute onset of fever, chills, and malaise. In the ulceroglandular form, a local papule at the inoculation site becomes necrotic and ulcerative (**Figure 36–5**). Regional lymph nodes become swollen and painful. The oculoglandular form, which follows conjunctival inoculation, is similar except that the local lesion is a painful purulent conjunctivitis. Ingestion of large numbers of *F tularensis* (more than 10^8) leads to typhoidal tularemia, with abdominal manifestations and a prolonged febrile course that is similar to that of typhoid fever. Inhalation of the organisms can result in pneumonic tularemia or a more generalized infection similar to the typhoidal form. Like plague pneumonia, tularemic pneumonia may also develop through seeding of the lungs by bacteremic spread of one of the other forms. Any form of tularemia may progress to a systemic infection with lesions in multiple organs.

Ulceroglandular, oculoglandular, typhoidal, and pneumonic forms exist

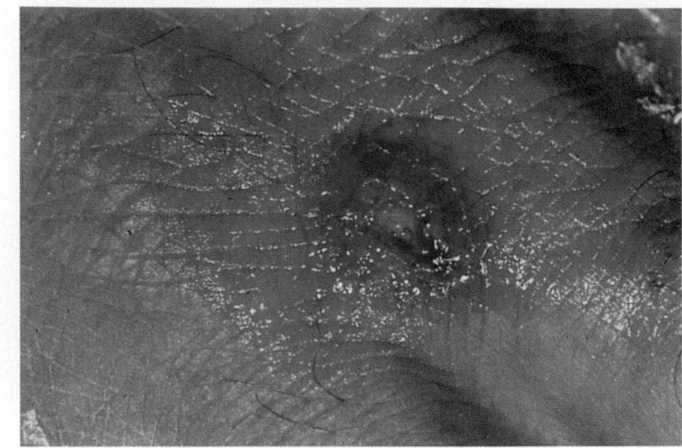

FIGURE 36–5. Tularemia. Ulcer on the hand of a trapper is infected with *F. tularensis*. (Reproduced with permission from Connor DH, Chandler FW, Schwartz DQA, Manz HJ, Lack EE (eds). *Pathology of Infectious Diseases*, vol. 1. Stamford, CT: Appleton & Lange; 1997.)

Ulceroglandular has lowest mortality

Without treatment, mortality rate ranges from 5% to 30%, depending on the type of infection. Ulceroglandular tularemia, the most common form, generally carries the lowest risk of a fatal outcome estimated at 2%.

DIAGNOSIS

Special media are needed for culture

Serodiagnosis is common

Because tularemia is uncommon and *F tularensis* has unique growth requirements, the diagnosis is easily overlooked. Although some strains grow on chocolate agar, laboratories must be alerted to the suspicion of tularemia so that specialized media supplemented with cysteine can be prepared and precautions taken against the considerable risk of laboratory infection. An immunofluorescent reagent is available in reference laboratories for use directly on smears from clinical material. Because of the difficulty and risk of cultural techniques, many cases of tularemia are diagnosed by serologic tests. Agglutinating antibodies are usually present in titers of 1:40 by the second week of illness, increasing to 1:320 or greater after 3 to 4 weeks. Unless previous exposure is known, single high antibody titers are considered diagnostic.

TREATMENT AND PREVENTION

Aminoglycosides are effective

Streptomycin or gentamicin is the drug of choice in all forms of tularemia. Doxycycline, ciprofloxacin and chloramphenicol have also been effective, but relapses are more common than with the aminoglycosides. Prevention mainly involves the use of rubber gloves and eye protection when handling potentially infected wild mammals. Prompt removal of ticks is also important. A live attenuated vaccine exists, but it is used only in laboratory workers and those who cannot avoid contact with infected animals.

PASTEURELLA MULTOCIDA

P multocida, one of many species of *Pasteurella* in the respiratory flora of animals, is a cause of respiratory infection in some. This small, coccobacillary, Gram-negative organism grows readily on blood agar but not on MacConkey agar. It is oxidase-positive and ferments a variety of carbohydrates. Unlike most Gram-negative rods, *P multocida* is susceptible to penicillin. Humans are usually infected by the bite or scratch of a domestic dog or cat.

Infection develops at the site of the lesion, often within 24 hours. The typical infection is a diffuse cellulitis with a well-defined erythematous border. The diagnosis is made by culture of an aspirate of pus expressed from the lesion. Frequently, too few organisms are present to be seen on a direct Gram smear. *P multocida* is by far the most common cause of an infected dog or cat bite. For unknown reasons, *P multocida* is occasionally isolated from the sputum of patients with bronchiectasis. Infections are treated with penicillin.

Penicillin-sensitive, Gram-negative rods

Most common cause of infected animal bites or scratches

CLINICAL CASE

DOWNHILL TO DEATH FOLLOWING CAT EXPOSURE

A 31-year-old man had just returned from visiting a friend in Chaffee County, Colorado. While there, he helped remove an obviously ill domestic cat from the crawl space under a friend's cabin. They also noticed a number of dead chipmunks in a nearby arroyo. Two days after returning to his home in Tucson, the man began to have abdominal cramps. The next day, he had the onset of fever, nausea, vomiting, severe diarrhea, and cough. On the third day, he consulted a physician because of diarrhea and vomiting. On examination, he was febrile (104°F) and dehydrated; no abnormal chest sounds were heard, and he had no lymphadenopathy. The man was treated for gastroenteritis with clindamycin and given oral ciprofloxacin to be taken the following day. The next day, he was hospitalized with cyanosis and septic shock. Chest radiographs revealed a right upper lobar pneumonia. A Gram stain of a sputum sample obtained at hospital admission showed numerous Gram-negative rods. Antibiotic therapy with ceftazidime, erythromycin, and one dose each of penicillin and gentamicin was initiated for treatment of overwhelming sepsis and pneumonia. He died 24 hours after admission.

Investigation by Chaffee County public health officials indicated that the cat, reported to have submandibular abscesses and oral lesions consistent with feline plague, died on August 19 before being evaluated by a veterinarian. The cat was cremated without diagnostic studies. A dead chipmunk found in the area where the cat lived was culture-positive for *Y pestis*.

QUESTIONS

■ This man most probably had which disease?

A. Brucellosis
B. Bubonic plague
C. Pneumonic plague
D. Typhoidal tularemia
E. Pneumonic tularemia

■ What is the most probable source of his infection?

A. Flea
B. Cat
C. Chipmunk
D. Rat
E. Human

■ Which of the following contributed to his death?

A. Yops
B. Biofilm
C. Erythritol
D. Adenylate cyclase
E. ADP-ribosylation

ANSWERS

1(C), **2**(B), **3**(A)

CHAPTER 37

Spirochetes

The French disease, for it was that, remained in me more than four months dormant before it showed itself, and then it broke out over my whole body at one instant ... with certain blisters, of the size of six-pence, and rose colored.

—Benvenuto Cellini (1500-1571): *The Life of Benvenuto Cellini*

Spirochetes are bacteria with a spiral morphology ranging from loose coils to a rigid corkscrew shape. The three medically important genera include the cause of syphilis, the ancient scourge of sexual indiscretion, and Lyme disease, a newly discovered consequence of an innocent walk in the woods.

BACTERIOLOGY

MORPHOLOGY AND STRUCTURE

Spiral structure is wound around endoflagella

Motility includes rotation and flexion

The spiral morphology of spirochetes (**Figure 37–1**) is produced by a flexible, peptidoglycan cell wall around which several axial fibrils are wound. The cell wall and axial fibrils are completely covered by an outer bilayered membrane similar to the outer membrane of other Gram-negative bacteria. In some species, a hyaluronic acid slime layer forms around the exterior of the organism and may contribute to its virulence. Spirochetes are motile, exhibiting rotation and flexion; this motility is believed to result from movement of the axial filaments, although the mechanism is not clear.

Many are thin and take stains poorly

Darkfield demonstrates spirochetes

Many spirochetes are difficult to see by routine microscopy. Although they are Gram-negative, many either take stains poorly or are too thin (0.15 μm or less) to fall within the resolving power of the light microscope. Only darkfield microscopy (**Figure 37–2**), immunofluorescence, or special staining techniques can demonstrate these spirochetes. Other spirochetes such as *Borrelia* are wider and readily visible in stained preparations, even routine blood smears. ::: Darkfield illumination, p. 62

GROWTH AND CLASSIFICATION

Parasitic spirochetes grow more slowly in vitro than most other disease-causing bacteria. Some species, including the causative agent of syphilis, have not been grown beyond a few generations in cell culture. Some are strict anaerobes, others require low concentrations of oxygen, and still others are aerobic. Compared with other bacterial groups, the taxonomy

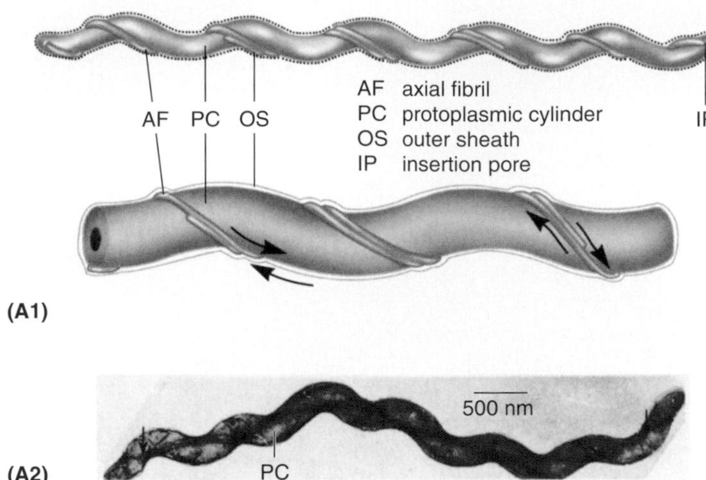

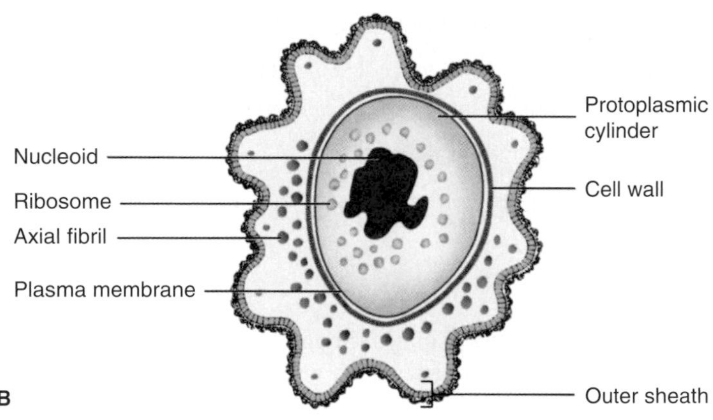

FIGURE 37–1. Spirochete morphology. A1. Longitudinal surface view of typical spirochete. **A2.** Electron Micrograph of *Treponema* with axial filaments extending most of cell length. **B.** Cross-section of typical spirochete. (Reproduced with permssion from Willey J, Sherwood L, Woolverton C (eds). *Prescott's Principles of Microbiology*. New York: McGraw-Hill; 2008.)

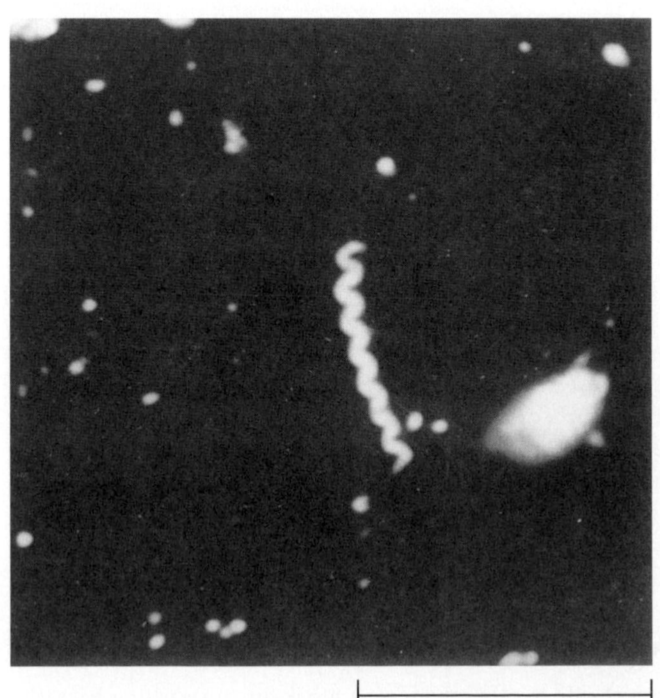

FIGURE 37–2. *Treponema pallidum* seen by darkfield microscopy. The darkfield method creates a bright halo around the corkscrew-shaped spirochetes. (Reproduced with permission from Nester EW, Anderson DG, Roberts CE Jr, Nester MT. *Microbiology: A Human Perspective*, 6th ed. New York: McGraw-Hill, 2008.)

of the spirochetes is underdeveloped. Because spirochetes are difficult to grow, they are difficult to study; thus, there are relatively few phenotypic properties on which to base a classification. The medically important genera *Treponema, Leptospira,* and *Borrelia* have been distinguished primarily by morphologic characters such as the nature of their spiral shape and the arrangement of flagella. Modern DNA homology and ribosomal RNA analyses have supported these groupings. ::: Molecular classification, p. 82

Some have not been isolated in culture

May be aerobic or anaerobic

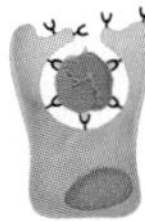

SPIROCHETAL DISEASES

Some spirochetes are free living; some are members of the normal flora of humans and animals. The oral cavity, particularly the dental crevice, harbors a number of nonpathogenic species of *Treponema* and *Borrelia* as part of its normal flora. Under unusual conditions, these spirochetes, together with anaerobes in the normal flora, can cause necrotizing, ulcerative infection of the gums, oral cavity, or pharynx (Vincent's infection, trench mouth). The pathogenesis of these opportunistic infections is not understood, but they are correlated with immunocompromise, severe malnutrition, and neglect of basic hygiene (**Table 37–1**). The term "trench mouth" refers to the occurrence of these infections in troops under the appalling conditions that existed in the trenches during World War I.

Many are part of oropharyngeal flora

Overgrowth causes trench mouth

The major spirochetal diseases are caused by selected species of three genera that are not found in the normal flora, *Treponema (T pallidum), Leptospira (L interrogans),* and *Borrelia (B recurrentis, B hermsii,* and *B burgdorferi).* Most *Borrelia* and *Leptospira* infections are zoonoses transmitted from wild and domestic animals. *T pallidum* is a strict human pathogen transmitted by sexual contact. Some rare nonvenereal treponemal diseases are summarized in **Appendix 37–1**.

Diseases are zoonoses or venereal

TABLE 37–1 Features of Spirochetal Diseases

				DIAGNOSIS			
ORGANISM	**MORPHOLOGY**	**TRANSMISSION**	**RESERVOIR**	**MICROSCOPY**	**CULTURE**	**SEROLOGY**	**DISEASE**
Treponema pallidum	Corkscrew spirals	Sexual, transplacental, transfusion	Humans	Darkfield of chancre or secondary lesions	None	VDRL, RPR, FTA-ABS, MHA-TP	Syphilis
Leptospira interrogans	Close spirals, hooked ends	Ingestion of contaminated water	Rodents, cattle, dogs	Not recommended[a]	Rarely performed[b]	MAT	Fever, meningitis, hepatitis
Borrelia recurrentis	Loose spirals	Lice	Humans	Giemsa or Wright stain of blood smear	Rarely performed[c]	None	Relapsing fever
Borrelia hermsii	Loose spirals	Ticks[d]	Rodents	Giemsa or Wright stain of blood smear	Rarely performed[c]	None	Relapsing fever
Borrelia burgdorferi	Loose spirals	Ticks[e]	White-footed mice, other rodents, (deer)[f]	Not recommended[a]	Rarely performed[c]	EIA + Immuno blot	Lyme disease

EIA, enzyme immunoassay; FTA-ABS, fluorescent treponemal antibody; MAT, microagglutination test; MHA-TP, microhemagglutination test for *T pallidum;* RPR, rapid plasma reagin; VDRL, Venereal Disease Research Laboratory.

[a]Organisms are small in number and rarely seen in clinical lesions.

[b]Culture of blood or urine in semisolid Fletcher's medium takes 1 to many weeks and is generally not available.

[c]Culture of blood in liquid Barbour-Stoener-Kelly medium takes 1 to many weeks and is generally not available.

[d]*Ornithodoros hermsi.*

[e]*Ixodes scapularis* in the eastern and central United States, *I pacificus* in the western United States.

[f]Transmitting ticks mature on deer that are not actually a reservoir.

TREPONEMA PALLIDUM

Syphilis represents an extended balance of parasitism and disease

T pallidum is the causative agent of syphilis, a venereal disease first recognized in the 16th century as the "great pox," which rapidly spread through Europe in association with urbanization and military campaigns. Some argue that it was brought back from the New World by the sailors with Christopher Columbus. Its extended course and the protean, often dramatic nature of its findings (genital ulcer, ataxia, dementia, ruptured aorta) are due to a state of balanced parasitism that spans decades. The cause of syphilis is actually a subspecies (*T pallidum* subsp. *pallidum*) closely related to other agents that cause rare nonvenereal treponematoses. *T pallidum* is used here to indicate the *pallidum* subspecies.

BACTERIOLOGY

Corkscrew spirals spin and bow

Heat, drying, and disinfectants kill quickly

T pallidum is a slim spirochete 5 to 15 μm long with regular spirals whose wavelength and amplitude resemble a corkscrew (Figure 37–2). The organism is readily seen only by immunofluorescence, darkfield microscopy, or silver impregnation histologic techniques. Live cells show characteristic slow, rotating motility with sudden 90-degree angle flexion, which suggests a gentleman quickly bowing at the waist. *T pallidum* is extremely susceptible to any deviation from physiologic conditions. It dies rapidly on drying and is readily killed by a wide range of detergents and disinfectants. The lethal effect of even modest elevations of temperature (41° to 42°C) was the basis for the technique of fever therapy for syphilis introduced in Vienna a century ago (patients were infected with malaria parasites!).

Prolonged growth only in animals

Genes compared to other pathogens

Beyond these observations, the study of the biology and pathogenesis of *T pallidum* is severely impeded by our inability to grow the organism in culture. It multiplies for only a few generations in cell cultures and is difficult to subculture. Sustained growth is achieved only in animals (rabbit testes), which are the sole source of bacteria for diagnostic reagents and scientific study. The *T pallidum* genome is amenable to study, and much of what follows is based on extrapolations comparing genes found there with those in other pathogenic bacteria. This genome, however, is among the smallest known and several times less than other bacterial pathogens discussed in this book. The unfolding picture of the syphilis spirochete is that it is a minimalist pathogen, growing very slowly and producing few definitive structures or products.

Lacks common enzymes

No LPS and few proteins in outer membrane

The sluggish growth (mean generation time more than 30 hours) of *T pallidum* is felt to be due to lack of enzymes that detoxify reactive oxygen species (catalase, oxidase) and the absence of efficient energy (ATP)-producing pathways such as the tricarboxylic acid cycle and electron transport chain. *T pallidum* shares the Gram-negative structural style of other spirochetes, but its outer membrane lacks lipopolysaccaride (LPS) and contains few proteins.

SYPHILIS

CLINICAL CAPSULE

Syphilis is typically acquired by the direct contact of mucous membranes during sexual intercourse. The disease begins with a lesion at the point of entry, usually a genital ulcer. After healing of the ulcer, the organisms spread systemically, and the disease returns weeks later as a generalized maculopapular rash called secondary syphilis. The disease then enters a second eclipse phase called latency. The latent infection may be cleared by the immune system or reappear as tertiary syphilis years to decades later. Tertiary syphilis is characterized by focal lesions whose locale determines the injury. Isolated foci in bone or liver may be unnoticed, but infection of the cardiovascular or nervous systems can be devastating. Progressive dementia or a ruptured aortic aneurysm are two of many fatal outcomes of untreated syphilis.

EPIDEMIOLOGY

T pallidum is an exclusively human pathogen under natural conditions. In most cases, infection is acquired from direct sexual contact with a person who has an active primary or secondary syphilitic lesion (**Figure 37–3**). Partner notification studies suggest transmission occurs in over 50% of sexual contacts in which a lesion is present. Less commonly, the disease may be spread by nongenital contact with a lesion (eg, of the lip), sharing of needles by intravenous drug users, or transplacental transmission to the fetus within approximately the first 3 years of the maternal infection. Late disease is not infectious. Modern screening procedures have essentially eliminated blood transfusion as a source of the disease. The incidence of new cases of primary and secondary syphilis in developed countries declined to an all time low at the end of the 20th century, but since then has risen more than 10%. Worldwide, syphilis remains a major public health problem, with an estimated 12 million new cases annually. There is evidence that syphilitic lesions are a portal for HIV transmission.

Transmission is by contact with mucosal surfaces or blood

Congenital infection is transplacental

Tertiary syphilis is not infectious

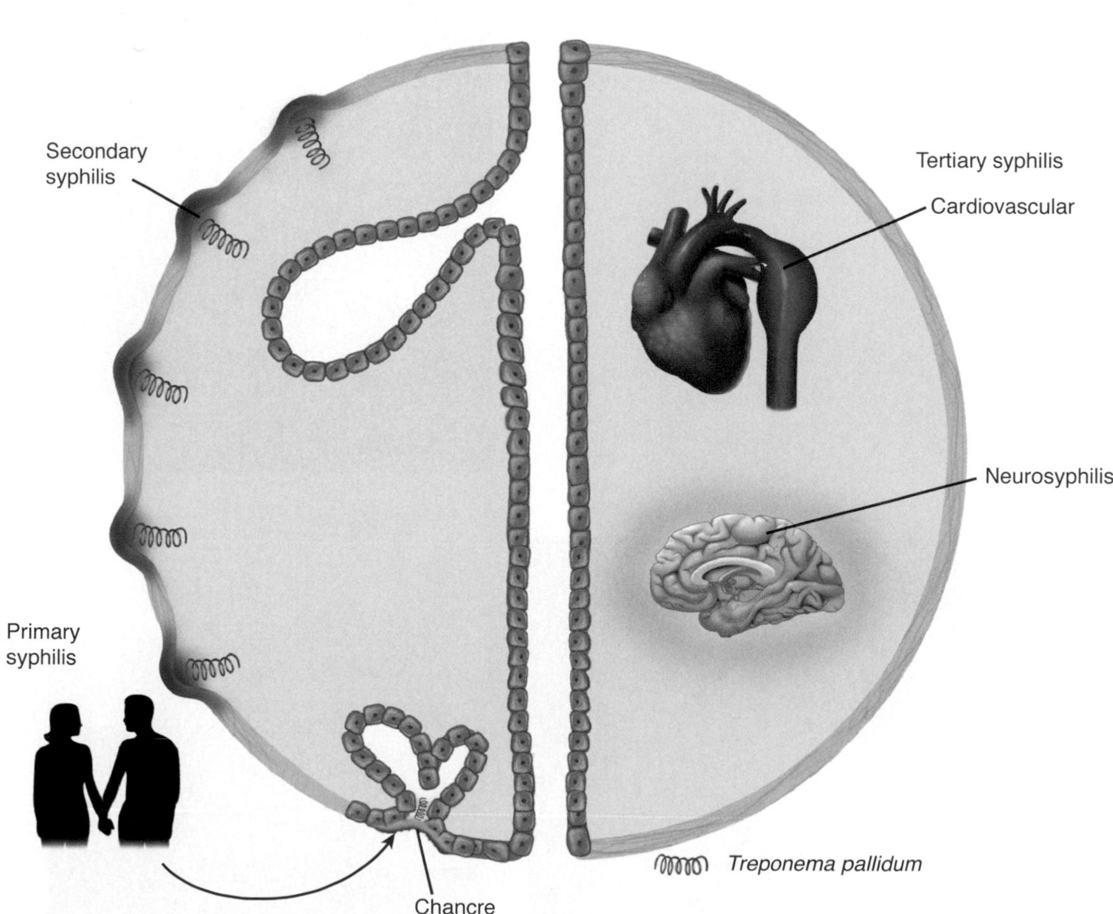

FIGURE 37–3. Syphilis overview. Infection is acquired by sexual contact, and the primary lesion is an ulcer on the genitalia called the chancre. The major feature of secondary syphilis is a maculopapular rash that is teeming with spirochetes. Tertiary syphilis (*right*) involves multiple organ systems. Shown are an aortic aneurysm as part of cardiovascular syphilis and inflammation of the brain in neurosyphilis.

PATHOGENESIS

Spread from mucosal breaks to blood is rapid

Slow multiplication produces endarteritis, granulomas

Ulcer heals but spirochetes disseminate

The spirochete reaches the subepithelial tissues through unapparent breaks in the skin or possibly by passage between the epithelial cells of mucous membranes aided by binding to fibronectin and elements of the extracellular matrix. In the submucosa, it multiplies slowly stimulating little initial tissue reaction. This is probably due to the relative paucity of antigens in the *T pallidum* outer membrane that would be exposed to the immune system. In experimental infections, the organisms spread from the primary site to the bloodstream within minutes and are established in distant tissues within hours. As lesions develop, the basic pathologic finding is an endarteritis. The small arterioles show swelling and proliferation of their endothelial cells. This reduces or obstructs local blood supply, probably accounting for the necrotic ulceration of the primary lesion and subsequent destruction at other sites (**Figure 37–4A–C**). Dense, granulomatous cuffs of lymphocytes, monocytes, and plasma cells surround the vessels. There is no evidence that this injury is due to any toxins or other classic virulence factors produced by *T pallidum.* Although the primary lesion heals spontaneously, the bacteria have already disseminated to other organs by way of local lymph nodes and the bloodstream.

The disease is clinically silent until the disseminated secondary stage develops and then is silent again with entry into latency. Although evasion of host defenses is clearly taking place,

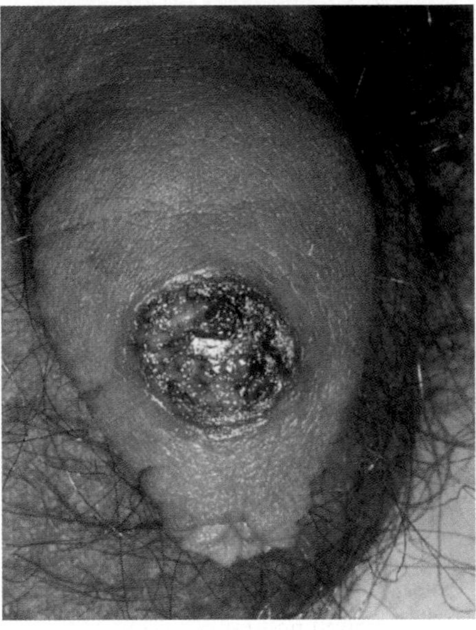

A

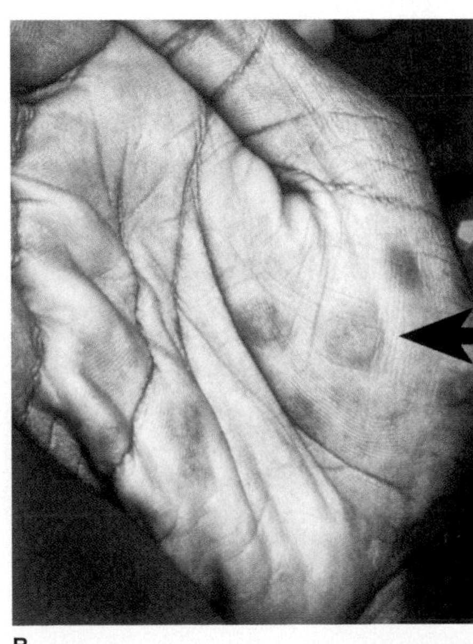

B

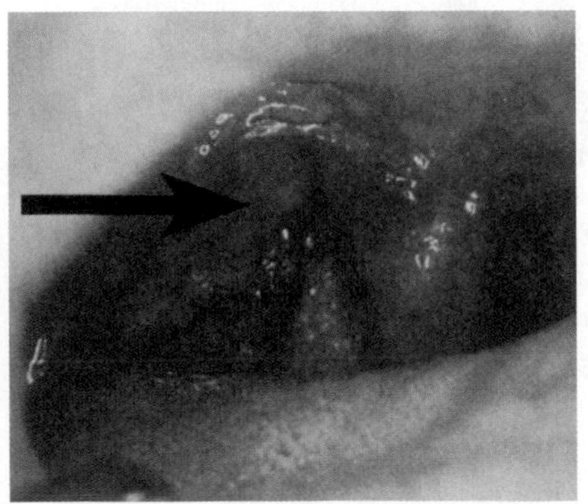

C

FIGURE 37–4. Syphilitic lesions. **A.** Primary syphilis. A syphilitic chancre is shown on the foreskin of the penis. Note the sharp edge and raw base of the ulcer. **B.** Secondary syphilis. The maculopapular rash appears on the palm. **C.** Tertiary syphilis. A ruptured gumma appears as a lump and ulcer in the hard palate of the mouth. (A, Reproduced with permission from Nester EW, Anderson DG, Roberts CE Jr, Nester MT. *Microbiology: A Human Perspective,* 6th ed. New York: McGraw-Hill, 2008, Figure 26.12. B and C, Reproduced with permission from Willey J, Sherwood L, Woolverton C (eds). *Prescott's Principles of Microbiology.* New York: McGraw-Hill; 2008.)

the mechanisms involved are unknown. *T pallidum* strains found in secondary lesions have not been demonstrated to differ antigenically from those in the primary chancre. It may be that the combination of the low antigen content of its outer membrane combined with the extremely slow multiplication rate allows the organism to stay below whatever critical antigenic mass is required to trigger an effective immune response. Without virulence factors to explain the tissue destruction, we are left with injury due to a prolonged delayed-type hypersensitivity (DTH) response to the persistent bacteria. ::: DTH, p. 42

Minimal triggers for immune response

Injury is due to prolonged hypersensitivity responses

IMMUNITY

Clinical observations suggest an immune response in syphilis that is slow and imperfect. Immunity to reinfection does not appear until early latency, and for at least one third of those infected the subsequent host response is successful in clearing most but not all of the treponemes.

Immunity develops slowly and incompletely

The immune mechanisms involved are far from clear, but appear to involve both humoral and cell-mediated responses. Resistance to reinfection is correlated with appearance of antitreponemal antibody, which is able to immobilize and kill the organism. Exposed treponemal outer membrane proteins (OMPs) are the most probable target of these antibodies. Cell-mediated responses appear to be dominant in syphilitic lesions with T lymphocytes (CD4+ and CD8+) and macrophages, the primary cell types, present. Activated macrophages play a major role in the clearance of *T pallidum* from early syphilitic lesions. The relapsing course of primary and secondary syphilis may reflect shifts in the balance between developing cellular immunity and suppression of T lymphocytes. Syphilis in immunocompromised patients such as those with acquired immunodeficiency syndrome may present with unusually aggressive or atypical manifestations. T-cell mechanisms, p. 33

Antibodies to OMP are associated with reinfection resistance

Development of cell-mediated immunity clears lesions

Variable T-lymphocyte suppression may link to stages

SYPHILIS: CLINICAL ASPECTS

MANIFESTATIONS

Primary Syphilis

The primary syphilitic lesion is a papule that evolves to an ulcer at the site of infection. This is usually the external genitalia or cervix, but could be in the anal or oral area depending on the nature of sexual contact. The lesion becomes indurated and ulcerates but remains painless, though slightly sensitive to touch. The fully developed ulcer with a firm base and raised margins is called the chancre (Figure 37–4A). Firm, nonsuppurative, painless enlargement of the regional lymph nodes usually develops within 1 week of the primary lesion and may persist for months. The median incubation period from contact until appearance of the primary lesion is about 3 weeks (range 3 to 90 days). It heals spontaneously after 4 to 6 weeks.

Painless, indurated ulcer starts the disease

Heals spontaneously after weeks

Secondary Syphilis

Secondary or disseminated syphilis develops 2 to 8 weeks after the appearance of the chancre. The primary lesion has usually healed but may still be present. This most florid form of syphilis is characterized by a symmetric mucocutaneous maculopapular rash and generalized nontender lymph node enlargement with fever, malaise, and other manifestations of systemic infection. Skin lesions are distributed on the trunk and extremities, often including the palms (Figure 37–4B), soles, and face, and can mimic a variety of infectious and noninfectious skin eruptions. About one third of patients develop painless mucosal warty erosions called **condylomata lata.** These erosions usually develop in warm, moist sites such as the genitals and perineum. All the lesions of secondary syphilis are teeming with spirochetes and are highly infectious. They resolve spontaneously after a few days to many weeks, but the infection itself has resolved in only one third of patients. In the remaining two thirds, the illness enters the latent state.

Lymphadenopathy and maculopapular rash are generalized

Spirochetes are abundant

Lesions resolve, but disease continues in one third of patients

Latent Syphilis

Latent syphilis is by definition a stage in which no clinical manifestations are present, but continuing infection is evidenced by serologic tests. In the first few years, latency may be

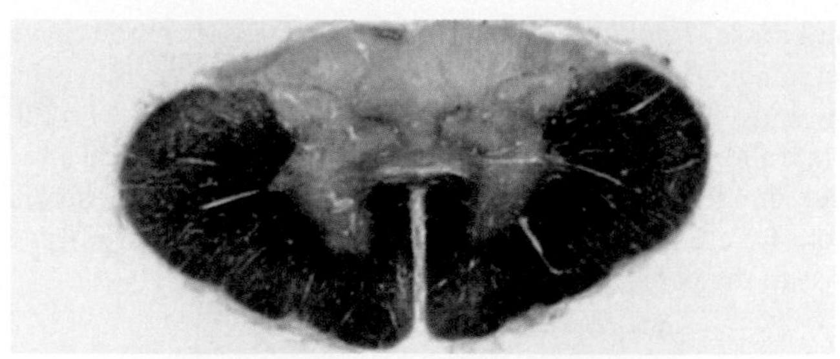

FIGURE 37–5. Tabes dorsalis. Loss of axons and myelin is evident in the posterior columns of the spinal cord (Woelke stain). (Reproduced with permission from Connor DH, Chandler FW, Schwartz DQA, Manz HJ, Lack EE (eds). *Pathology of Infectious Diseases,* vol. 1. Stamford, CT: Appleton & Lange; 1997.)

Secondary relapses interrupt latency

Blood-borne transmission risk continues

interrupted by progressively less severe relapses of secondary syphilis. In late latent syphilis (more than 4 years), relapses cease, and patients become resistant to reinfection. Transmission to others is possible from relapsing secondary lesions and by transfusion or other contact with blood products. Mothers may transmit *T pallidum* to their fetus throughout latency. About one third of untreated cases do not progress beyond this stage.

Tertiary Syphilis

Another one third of patients with untreated syphilis develop tertiary syphilis. The manifestations may appear as early as 5 years after infection but characteristically occur after 15 to 20 years. The manifestations depend on the body sites involved, the most important of which are the nervous and cardiovascular systems.

Chronic meningitis leads to degenerative changes and psychosis

Demylination causes peripheral neuropathies

Syphilitic paresis has many signs

Neurosyphilis is due to the damage produced by a mixture of meningovasculitis and degenerative parenchymal changes in virtually any part of the nervous system. The most common entity is a chronic meningitis with fever, headache, focal neurologic findings, and increased cells and protein in the cerebrospinal fluid (CSF). Cortical degeneration of the brain causes mental changes ranging from decreased memory to hallucinations or frank psychosis. In the spinal cord, demyelination of the posterior columns, dorsal roots, and dorsal root ganglia produces a syndrome called **tabes dorsalis** (**Figure 37–5**), which includes ataxia, wide-based gait, foot slap, and loss of the sensation. The most advanced central nervous system (CNS) findings include a combination of neurologic deficits and behavioral disturbances called **paresis,** which is also a mnemonic (**p**ersonality, **a**ffect, **r**eflexes, **e**yes, **s**ensorium, **i**ntellect, **s**peech) for the myriad of changes seen.

Aortitis leads to aneurysm

Gummas are destructive, localized granulomas

Cardiovascular syphilis is due to arteritis involving the vasa vasorum of the aorta and causing a medial necrosis and loss of elastic fibers. The usual result is dilatation of the aorta and aortic valve ring. This in turn leads to aneurysms of the ascending and transverse segments of the aorta and/or aortic valve incompetence. The expanding aneurysm can produce pressure necrosis of adjacent structures or even rupture. A localized, granulomatous reaction to *T pallidum* infection called a **gumma** (Figure 37–4C) may be found in skin, bones, joints, or other organ. Any clinical manifestations are related to the local destruction as with other mass-producing lesions, such as tumors.

Congenital Syphilis

Rhinitis, rash, and bone changes are common

Serologic screening and treatment is preventive

Fetuses are susceptible to syphilis only after the fourth month of gestation, and adequate treatment of infected mothers before that time prevents fetal damage. Because active syphilitic infection is devastating to infants, routine serologic testing is performed in early pregnancy and should be repeated in the last trimester in women at high risk for acquiring syphilis. Untreated maternal infection may result in fetal loss or congenital syphilis, which is analogous to secondary syphilis in the adult. Although there may be no physical finding at all, the most common are rhinitis and a maculopapular rash. Bone involvement produces characteristic changes in the architecture of the entire skeletal system (saddle nose, saber shins). Anemia, thrombocytopenia, and liver failure are terminal events.

DIAGNOSIS

Microscopy

T pallidum can be seen by darkfield microscopy in primary and secondary lesions, but the execution of this procedure requires experience and attention to detail. The suspect lesion must be cleaned and abraded to produce a serous transudate from below the surface of the ulcer base. This material can be captured in a capillary tube or placed directly on a microscope slide if a darkfield setup is close at hand. The microscopist must observe the corkscrew morphology and characteristic motility to make a diagnosis (Figure 37–2). A negative result from examination does not exclude syphilis; to be readily seen, the fluid must contain thousands of treponemes per milliliter. Darkfield microscopy of oral and anal lesions is not recommended because of the risk of misinterpretation of other spirochetes present in the normal flora. Direct fluorescent antibody methods have been developed but are available only in certain centers.

Darkfield requires experience and fluid from deep in lesion

May be negative owing to small numbers

Serologic Tests

Most cases of syphilis are diagnosed serologically using serologic tests that detect antibodies directed at either lipid or specific treponemal antigens. The former are called nontreponemal tests, and the latter are referred to as treponemal tests. Their use in screening, diagnosis, and therapeutic evaluation of syphilis has been refined over many decades (**Figure 37–6**).

Tests may or may not use treponemes

Nontreponemal Tests

Nontreponemal tests measure antibody directed against **cardiolipin,** a lipid complex so called because one component was originally extracted from beef heart. Anticardiolipin antibody is called **reagin,** and the tests that detect it depend on immune flocculation of cardiolipin in the presence of other lipids. The most common nontreponemal tests are the rapid plasma reagin (RPR) and the Venereal Disease Research Laboratory (VDRL). The results become positive in the early stages of the primary lesion and, with the possible exception of some patients with advanced human immunodeficiency virus (HIV) infection, are uniformly positive during the secondary stage. They slowly wane in the later stages of the disease. In neurosyphilis, VDRL test results on CSF may be positive when the serum VDRL has reverted to negative. Nontreponemal tests are nonspecific; they may become positive in a variety of autoimmune diseases or in diseases involving substantial tissue or liver destruction, such as lupus erythematosus, viral hepatitis, infectious mononucleosis, and malaria. False-positive results can also occur occasionally in pregnancy and in patients with HIV infection.

Reagin antibody reacts with cardiolipin, a lipid complex

Antibody level peaks in secondary syphilis

Nonspecfic reactions linked to autoimmune diseases

Sensitivity and low cost make nontreponemal tests preferred for screening, but positive results must be confirmed by one of the more specific treponemal tests described in the following text. The tests are also valuable for monitoring treatment because the height of the antibody titer is directly related to activity of disease. With successful antibiotic therapy, nontreponemal serologies slowly revert to negative.

Titer is used to follow therapy

Treponemal Tests

Treponemal tests detect antibody specific to *T pallidum,* such as an indirect immunofluorescent procedure called the fluorescent treponemal antibody (FTA-ABS), which uses

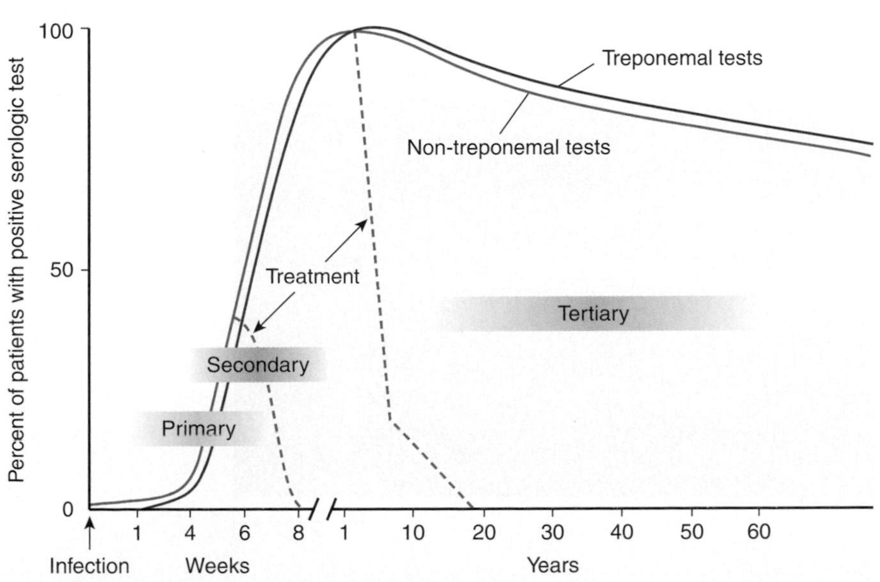

FIGURE 37–6. Syphilis serology. The time course of treponemal and nontreponemal tests in treated and untreated syphilis is shown. The nontreponemal test results (VDRL, RPR) rise during primary syphilis and reach their peak in secondary syphilis. They slowly decline with advancing age. With treatment, they revert to normal over a few weeks. The treponemal tests (FTA-ABS, MHA TP) follow the same course but remain elevated even after successful treatment. FTA-ABS, fluorescent treponemal antibody; MHA-TP, microhemagglutination test for *T pallidum*; RPR, rapid plasma reagin; VDRL, Venereal Disease Research Laboratory.

T pallidum is used as the antigen

spirochetes fixed to slides. ABS refers to an absorption step that removes nonspecific antispirochetal antibodies often found in normal serum. Another method, the microhemagglutination test for *T pallidum* (MHA-TP), uses antigens attached to the surface of erythrocytes, which then agglutinate in the presence of specific antibody. ::: Hemagglutination, p. 74

Positive result confirms RPR or VDRL

Remain positive for life

Treponemal tests are considerably more specific than the cardiolipin-based nontreponemal tests. Their primary role in diagnosis is to confirm positive RPR and VDRL results obtained in the evaluation of a patient suspected of having syphilis or in screening programs. These tests are not useful for screening or after therapy because, once positive, they usually remain so for life except for the immunocompromised. Thus, if nontreponemal tests can be thought of as the measure of active syphilis, treponemal tests are the indelible print of sin. The time course of serologic tests in the various stages of syphilis is illustrated in Figure 37–6.

IgM is used to diagnose congenital syphilis

The use of serologic tests in the diagnosis of congenital syphilis is complicated by the presence of IgG antibodies in infants, who acquire it transplacentally from their mothers. If available, treponemal IgM tests are useful in establishing the presence of an acute infection in infants.

TREATMENT AND PREVENTION

Penicillin is preferred

Safe sex blocks transmission

T pallidum remains exquisitely sensitive to penicillin, which is the preferred treatment in all stages. In primary, secondary, or latent syphilis persons hypersensitive to penicillin may be treated with doxycycline. The efficacy of agents other than penicillin has not been established in tertiary or congenital syphilis. It is recommended that penicillin-hypersensitive patients with neurosyphilis or congenital syphilis be desensitized rather than use an alternate antimicrobial. Safe sex practices are as effective for prevention of syphilis as they are for other sexually transmitted diseases. The development of a vaccine awaits greater understanding of pathogenesis and immunity.

LEPTOSPIRA INTERROGANS

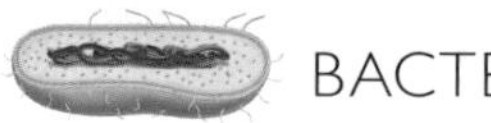

BACTERIOLOGY

L interrogans is the member of the genus *Leptospira* that is pathogenic to humans and animals. There are other free-living species of *Leptospira*. This species is a slim spirochete 5 to 15 μm long, with a single axial filament; fine, closely wound spirals; and hooked ends (**Figure 37–7**).

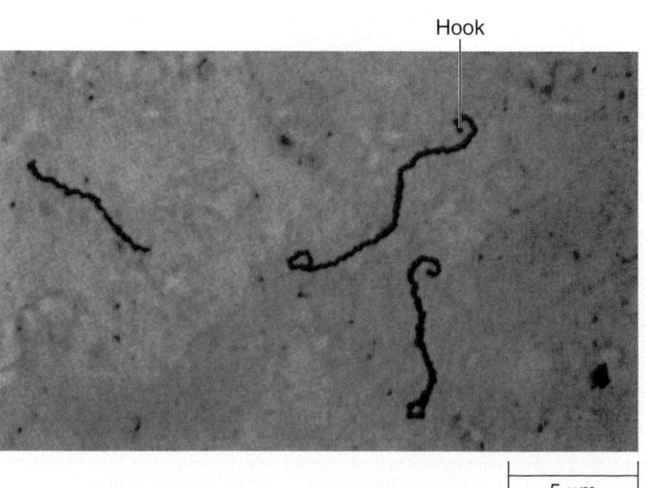

FIGURE 37–7. *Leptospira interrogans*. Note the tight primary coiling, loose loops, and hooked ends of the spirochete. (Reproduced with permission from Nester EW, Anderson DG, Roberts CE Jr, Nester MT. *Microbiology: A Human Perspective*, 6th ed. New York: McGraw-Hill, 2008.)

It is not visualized with the usual staining procedures, and detection is best accomplished using darkfield microscopy. It can be grown in aerobic culture using certain special enriched semisolid media. The outer membrane contains LPS and OMPs with adhesive or factor H-binding properties. ::: Factor H binding, p. 394

Loose spirals seen in darkfield

L interrogans has multiple serogroups and over 200 serotypes, many of which were previously accorded species status (eg, *L icterohaemorrhagiae, L canicola, L pomona*) based on geographic occurrence, differences in host species, and associated clinical syndromes. The distinction among serogroups and serotypes is of epidemiologic and epizoologic importance but has no clinical significance. *L interrogans* can survive days or weeks in some waters in the environment at a pH above 7.0. Acidic conditions, such as those that may be found in urine, rapidly kill the organism. It is highly sensitive to drying and to a wide range of disinfectants.

Multiple serogroups have geographic associations

Survives in water

LEPTOSPIROSIS

CLINICAL CAPSULE

Leptospirosis is a systemic flu-like illness associated with water contaminated by animal urine. It begins with fever, nausea, vomiting, headache, abdominal pain, and severe myalgia. In severe cases, a second phase is characterized by impaired hepatic and renal function with jaundice, prostration, and circulatory collapse. The CNS is often involved, with stiff neck and inflammatory changes in the CSF.

EPIDEMIOLOGY

Leptospirosis is a worldwide disease of a variety of wild and domestic animals, particularly rodents, cattle, and dogs. It is usually transmitted to humans through water contaminated with animal urine. Secondary human-to-human transmission occurs rarely. Individuals who are exposed to animals (eg, farmers, veterinarians, slaughterhouse employees) are at increased risk, although most clinical cases are now associated with recreational exposure to contaminated water (eg, irrigation ditches or other bodies of water receiving farmland drainage). In tropical areas, leptospirosis may account for up to 10% of hospital admissions, particularly after rains or floods.

Animals are reservoirs

Water is transmission route

PATHOGENESIS AND IMMUNITY

The organism gains entrance to the tissues through small skin breaks, the conjunctiva, or, most commonly, ingestion and the upper alimentary tract mucosa. The active motility of the hooked ends driven by periplasmic flagella may allow the organism to burrow into tissues. A few OMPs mediate adherence in much the way seen with invasive members of the Enterobacteriaceae and one with factor H-binding properties interferes with complement mediated killing. The organisms spread widely through the bloodstream to all parts of the body including the CSF. In animals, they colonize the proximal renal tubule, from which they are shed into the urine, facilitating transmission to new hosts. The kidney is also a target organ in human disease, causing tubular infection and interstitial nephritis.

Enters through small mucosal breaks

Blood and CNS spread is common

Clearing of the bacteremia is associated with the appearance of circulating antibody but little else is known of immune mechanisms. Antibody is also rising during the second phase

Antibody may be part of disease

of the disease, which suggests an immunologic component to its pathogenesis. This is supported by the absence of response to antimicrobics when given at this stage and the typical failure to recover the organism from the CSF in cases of leptospiral meningitis.

LEPTOSPIROSIS: CLINICAL ASPECTS

MANIFESTATIONS

Initial disease is flu-like

Meningitis and muscle aches last for weeks

Hemorrhagic rash is linked to fatal outcome

Most infections are subclinical and detectable only serologically. After an incubation period of 7 to 13 days, an influenza-like febrile illness with fever, chills, headache, conjunctival suffusion, and muscle pain develops in persons who become ill. This disease is associated with bacteremia. Leptospires are also found in the CSF at this stage, but without clinical or cytologic evidence of meningitis. The fever often subsides after about a week coincident with the disappearance of the organisms from the blood, but it may recur with a variety of clinical manifestations depending partly on the serogroup involved. This second phase of the disease usually lasts 3 or more weeks and may manifest as an aseptic meningitis resembling viral meningitis (see Chapter 65) or as a more generalized illness with muscle aches, headache, rash, pretibial erythematous lesions, biochemical evidence of hepatic and renal involvement, or all of these. In its most severe form (Weil's disease), there is extensive vasculitis, jaundice, renal damage, and sometimes a hemorrhagic rash. The mortality rate in such cases may be as high as 10%.

DIAGNOSIS

Serologic tests are limited to reference laboratories

The diagnosis of leptospirosis is primarily serologic. Although the spirochetes can theoretically be detected, darkfield examination of body fluids is not recommended. The yield is very low and the chance for confusion with fibrin and debris is significant. Likewise, leptospires can be isolated from the blood, CSF, or urine, but culture is rarely attempted because the organisms take weeks to grow in a special medium that few laboratories bother to stock. The standard serologic test (microscopic agglutination) is also limited to reference laboratories. A simpler slide agglutination is less specific, but may be suggestive of infection in the presence of a compatible clinical picture.

TREATMENT AND PREVENTION

Penicillin is primary treatment

Rodent and water control are important

Penicillin is the primary treatment for all forms of leptospirosis. Doxycycline and ceftriaxone are alternatives. Doxycycline is recommended as chemoprophylaxis for individuals engaging in high-risk activities, such as swimming in jungle rivers or kayaking in developing countries. Other measures include rodent control, drainage of waters known to be contaminated, and care on the part of those subject to occupational exposure to avoid ingestion or contamination with *L interrogans.* Vaccines are used in cattle and household pets to prevent the disease, and this has reduced its occurrence in humans.

BORRELIA

Relapsing fever and Lyme disease caused by different species

More than 15 species of *Borrelia* have been associated with human disease, and other species are responsible for similar diseases in animals. *B burgdorferi* is the cause of Lyme disease. Other members of the genus cause relapsing fever, an illness with intermittent fevers and little else. The relapsing fevers differ in their specific vector and geographic distribution. The human body louse is the vector for *B recurrentis,* but the remainder of the relapsing fevers are linked to several ticks and species of *Borrelia;* these are discussed together here as *B hermsii,* the most common cause of relapsing fever in North America.

Borrelia are long (10 to 30 μm), slender spirochetes containing multiple (7 to 20) axial flagella. In contrast to *Treponema* and *Leptospira*, its spirals form loose, irregular waves. The basic organizational structure of the cell and its motility conform to that of the other Gram-negative spirochetes, but unlike the others, *Borrelia* are readily demonstrated by common staining methods such as the Giemsa or Wright stains. *Borrelia* are microaerophilic and have been successfully grown in specially supplemented (*N*-acetylglucosamine, fatty acids) liquid, or semisolid media. The organisms are generally deficient in genes for synthesis of many essential nutrients (amino acids, fatty acids, nucleic acids) and thus must obtain them from external sources. A distinct feature of *Borrelia* is the partitioning of the genome between the chromosome and multiple circular and linear plasmids. In some species, a large proportion (more than 40%) of the genome is in the plasmids, including genes important in animal and human disease.

Loose, irregular spirals take common stains

Nutrients taken from external sources

Many genes in plasmids

Relapsing Fever *Borrelia* (*B recurrentis, B hermsii*)

Borrelia hermsii and *Borrelia recurrentis*

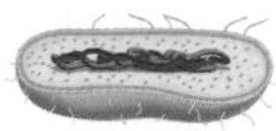

BACTERIOLOGY

The outer membrane of all *Borrelia* species contains abundant OMPs and lipoproteins. In some species, these surface proteins have been observed to vary antigenically too abundantly to be explained by simple mutation. Experiments with *B hermsii* have demonstrated up to 40 antigenically distinct variants of the same protein arising from a single cell. The genetic mechanism for this antigenic variation involves recombination between genes located in the distinctive linear plasmids. Multiple copies of the genes for these proteins are present. Some genes express the protein, whereas others are "silent" because they lack crucial promoter sequences. When structural sequences from a silent gene are transferred by recombination to an expressing gene on another plasmid, the protein expressed is altered, which may make it antigenically different. This recombination mechanism resembles that described for antigenic variation of gonococcal pili. :::Antigenic variation, p. 395

Surface proteins undergo antigenic variation

Recombination between linear plasmids leads to altered protein

RELAPSING FEVER

CLINICAL CAPSULE

Relapsing fever is an illness with fever, headache, muscle pain, and weakness but no signs pointing to any organ system. It lasts about 1 week and returns a few days later. The relapses may continue for as many as four cycles. During each relapse, spirochetes are present in the bloodstream. The causative *Borrelia* species are transmitted to humans from ticks or body lice.

EPIDEMIOLOGY

Relapsing fever occurs in two forms linked to the mode of transmission and the *Borrelia* species involved. The louse-borne form usually appears in epidemics, because of circumstances connected with body lice, whereas the tick-borne form does not. For this reason, the two forms are sometimes called epidemic (louse-borne) and endemic (tick-borne) relapsing fever. Here they are identified simply by the insect involved.

Body lice or ticks transmit the spirochete

The occurrence and distribution of tick-borne relapsing fever are determined by the biology of multiple species of a single tick genus (*Ornithodoros*) and their relation to the primary *Borrelia* reservoir in rodents and other small animals (rabbits, birds, lizards). *B hermsii* is one of at least 15 *Borrelia* species associated with this cycle. Ticks may remain infectious

Tick reservoir feeds on rodents and small animals

Night-time painless tick bite transmits bacteria

for several years even without feeding, and transovarial passage to their progeny extends the infectious chain even further. Humans are infected when they accidentally enter this cycle and are bitten by an infected tick. The bite is painless and the feeding period is brief (less than 20 minutes). Because the ticks usually feed at night, cases of relapsing fever are most often associated with overnight recreational forays into wild, wooded areas. A large outbreak in the United States involved National Park employees and tourists who slept in tick- and rodent-infested cabins on the Northern Rim of the Grand Canyon.

Body lice infected from human blood

Lice must be transferred from human to human

The epidemiologic conditions associated with louse-borne relapsing fever are much more exacting. The human body louse has no other host, infected lice live no more than 2 months, and there is no transovarial passage to progeny. *B recurrentis* is the only species involved. Lice are infected from human blood, but the spirochetes multiply in their hemolymph, not any of the feeding parts or excrement. This means they can infect another human only if the louse is crushed by scratching and the *Borrelia* reach a superficial wound or mucosal surface. Infected lice must be passed human to human for the disease to persist. These conditions are met by circumstances that combine overcrowding with extremely low levels of general hygiene. War, other kinds of social breakdown, and dire poverty are the prime associates. Currently, this variety of relapsing fever appears to be limited to East and Central Africa and the Peruvian Andes.

PATHOGENESIS

Spirochetes appear in blood

Altered OMPs occur with relapse

The disease manifestations develop at times when thousands of spirochetes are circulating per milliliter of blood. The febrile illness has endotoxin-like features, but the exact mechanisms of disease are unknown. Between episodes, the organisms disappear from the blood and are sequestered in internal organs only to reappear during relapses. The OMPs are antigenically different with each relapse. The relapsing cycles correlate with antibody production to the new protein followed by clearing followed by emergence of a new antigenic type.

IMMUNITY

Antibody eventually controls disease

Immunity to relapsing fever is largely humoral and appears to involve lysis of the organism in the presence of complement. The disease is controlled when variants from the antigenic repertoire are no longer able to escape the immune response.

RELAPSING FEVER: CLINICAL ASPECTS

MANIFESTATIONS

Fever, headache, and muscle pain last 2 to 4 days

After a mean incubation period of 7 days, massive spirochetemia develops, with high fever, rigors, severe headache, muscle pains, and weakness. The febrile period lasts about 1 week and terminates abruptly with the development of an adequate immune response. The disease relapses 2 to 4 days later, usually with less severity, but following the same general course. Tick-borne relapsing fever is usually limited to one or two relapses, but with louse-borne disease three or four may occur.

Louse-borne is more severe

Louse-borne relapsing fever is more severe than tick-borne disease, possibly because of predisposing social conditions. Fatalities are rare in tick-borne disease but may be as high as 40% in untreated louse-borne fever. Fatal outcomes are due to myocarditis, cerebral hemorrhage, and hepatic failure.

DIAGNOSIS

Blood smears demonstrate *Borrelia*

Diagnosis of relapsing fever is readily made during the febrile period by Giemsa or Wright staining of blood smears. The appearance of the spirochete among the red cells is characteristic. Cultural and animal inoculation procedures are also used for recovery of the infecting organism. Serodiagnostic tests are unhelpful.

TREATMENT

Patients with relapsing fever respond well to doxycycline therapy, with erythromycin and ceftriaxone as alternatives. If the level of spirochetes is high at the time treatment is initiated, a systemic febrile reaction (Jarisch-Herxheimer) resembling Gram-negative sepsis may ensue. This is felt to be due to rapid lysis of the organisms with release of outer membrane LPS. It is more common in louse-borne than tick-borne relapsing fever.

Doxycycline is primary treatment

PREVENTION

Prevention of tick-borne relapsing fever involves attention to deticking, insecticide treatment, and rodent control around habitations such as mountain cabins, which are shown to be associated with infection. Control of louse-borne relapsing fever involves delousing, particularly dusting of clothing with appropriate insecticides. Ultimately, improved hygiene stops outbreaks and prevents further occurrences.

Attention to ticks and general hygiene are important

Borrelia burgdorferi

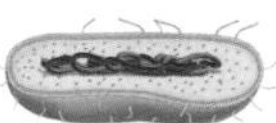

BACTERIOLOGY

B burgdorferi consists of at least 10 subspecies (eg, *B burgdorferi* sensu stricto, *B afzelii, B garini,* and others), which differ in geographic distribution and some clinical manifestations. All are here referred to as *B burgdorferi.* As with other species of *Borrelia,* there are multiple classes of OMPs, many of which undergo antigenic variation. Recent studies have focused on a class called outer surface proteins (Osps), which have been linked to aspects of pathogenesis and immunity. In response to environmental signals (temperature, pH) two of these proteins, OspA and OspC, are differentially expressed, depending on the stage of tick or mammalian infection. Other Osps have been shown to bind to fibronectin and serum factor H.

Grows in microaerophilic atmosphere

Osps differ at stages of infection

LYME DISEASE

CLINICAL CAPSULE

Acute Lyme disease is characterized by fever, a migratory "bull's eye" skin rash, muscle and joint pains, often with evidence of meningeal irritation. In a chronic form evolving over several years, meningoencephalitis, myocarditis, and a disabling recurrent arthritis may develop. *B burgdorferi* is transmitted to humans by *Ixodes* ticks.

EPIDEMIOLOGY

B burgdorferi exists in a complex cycle involving ticks, mice, and deer (**Figure 37–8**). Lyme disease occurs when the ticks feed on humans who enter their wooded habitat. The disease is endemic in several regions of the United States, Canada, and temperate Europe and Asia. Approximately 90% of the 10,000 to 15,000 cases reported each year in the United States

Spirochetes are transmitted in tick–mouse–deer cycle

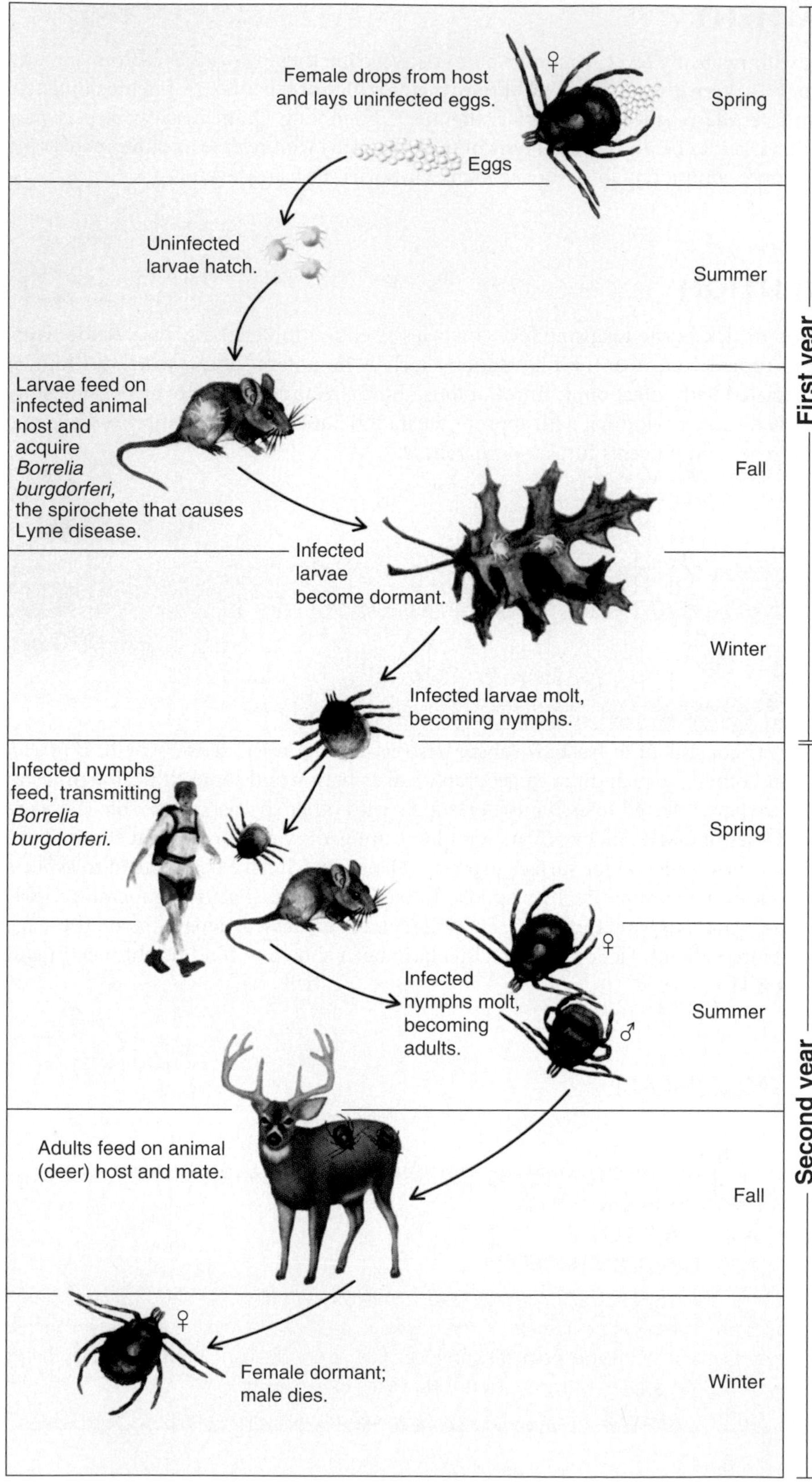

FIGURE 37–8. Lyme disease life cycle. The life cycle covers 2 years during which the tick obtains three blood meals. The males die soon after mating; the females die after depositing their eggs in the following spring. Variations depend on climate and food availability for the natural hosts. (Reproduced with permission from Nester EW, Anderson DG, Roberts CE Jr, Nester MT. *Microbiology: A Human Perspective,* 6th ed. New York: McGraw-Hill, 2008, Figure 22-16.)

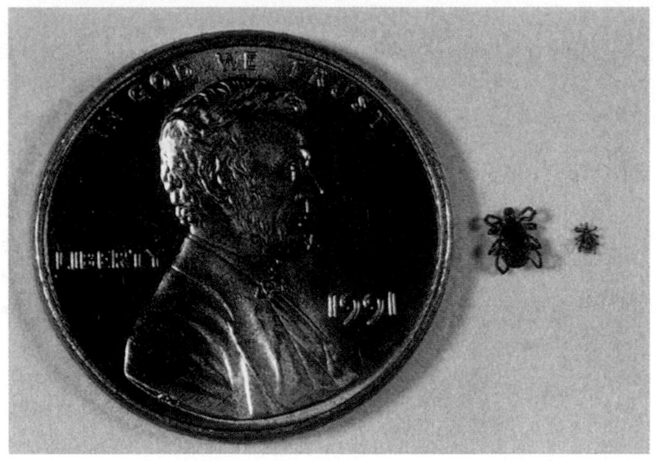

2 mm

FIGURE 37–9. The deer tick (*Ixodes scapularis*) adult and nymph. (Reproduced with permission from Nester EW, Anderson DG, Roberts CE Jr, Nester MT. *Microbiology: A Human Perspective*, 6th ed. New York: McGraw-Hill, 2008.)

occur in areas along the northeastern and mid-Atlantic seaboard, including Old Lyme, Connecticut, where the disease was first recognized. Most cases probably go unreported, particularly outside the primary endemic regions.

Ticks must feed on humans in the woods

The primary reservoir of *B burgdorferi* is rodents, particularly white-footed mice. Infection is transmitted by *Ixodes* ticks (**Figure 37–9**), whose complete life cycle involves rodents for the early stages and deer for adult maturation. In the spring, fertile female ticks, engorged from their blood meals, fall from their deer hosts to the ground and deposit their eggs. During the summer, the tick larvae seek out and obtain a blood meal from mice and the *B burgdorferi* ingested by the larvae are maintained through the subsequent development stages of the tick. The following spring or summer, the small (1 to 2 mm) nymphs feed again on vertebrate hosts to obtain the blood required for maturation to adulthood. The engorged, satiated nymphs fall off their hosts and mature into adults by parasitizing available deer, thus completing a life cycle that has occupied a full 2 years. Vertebrates other than deer can be infected by both the adult and nymph stages of the tick, but human Lyme disease is acquired primarily from nymphs, because they are active at the time of year when humans are most likely to invade their ecosystem. The infecting dose is very low (less than 20 organisms), making even a single tick bite a risk for disease. Deer are essential to the mating and survival of the tick, and thus the disease does not occur in areas in which deer are not abundant.

Ticks feed on mice and then deer

Adult and nymph stages can infect humans

No deer, no disease

PATHOGENESIS

Because Lyme disease is a recently discovered disease with a complex biology, it is not surprising that the pathogenic mechanisms in humans remain to be established clearly. Studies in ticks have shown changes in the antigenic makeup of *B burgdorferi* as it migrates from the midgut and salivary glands and again after it reaches mammalian tissue. OspA is the major outer surface protein expressed when *B burgdorferi* resides in ticks, where it mediates binding to midgut cells. OspA expression diminishes during tick feeding and engorgement, whereas OspC increases, so that by the time of transmission to animal hosts, OspC predominates. Although OspC has been shown to stimulate protective antibody in animals, its role in disease is unknown.

OspA predominates in ticks

Shift to OspC is completed at vertebrate transmission

After infection, the *B burgdorferi* surface proteins that mediate adhesion to fibronectin or elements of the extracellular matrix could be important in the early stages of disease. By analogy with other bacterial proteins that bind serum factor H, similar Osps of *B burgdorferi* are likely to facilitate persistence by interference with effective complement deposition. The spirochete is not known to produce digestive enzymes, but tissue spread and dissemination may

Surface proteins bind to fibronectin, factor H

Peptidoglycan stimulates inflammation

be facilitated by the utilization of host proteases. As the organism spreads, inflammation is stimulated by the cell wall peptidoglycan and possibly by elements of the outer membrane, although *B burgdorferi* lacks classic LPS. When deposited in joint tissues, these elements may contribute to the arthritis of Lyme disease.

Down-regulation of immune function contributes to chronicity

Antibody may have autoimmune action

Clinical investigations in patients with Lyme disease have noted modulation of immune responses, including inhibition of mononuclear and natural killer cell function, lymphocyte proliferation, and cytokine production. The ability of *B burgdorferi* to down-regulate deleterious immune responses could serve as a survival strategy or play a role in chronic disease. Chronic disease, particularly Lyme arthritis, has aspects of autoimmunity. The sera of individuals with Lyme arthritis, but not other forms of arthritis react with multiple elements of the organism.

IMMUNITY

Target of protective antibody is unclear

The immune response to *B burgdorferi* infection develops slowly, with IgM followed by IgG antibody over weeks to months. Although immune-mediated killing by the classical complement pathway has been demonstrated, the molecular target is unknown. Host neutrophils and macrophages can phagocytose opsonized spirochetes and induce a metabolic burst leading to spirochetal death. OspC elicits protective immunity in rodents, but this protection is short-lived and ineffective against challenge with heterologous *B burgdorferi* isolates. Antigens capable of eliciting broadly protective immune responses have not been identified. ::: Complement pathways, p. 27

LYME DISEASE: CLINICAL ASPECTS

MANIFESTATIONS

Spreading lesion from bite site is most characteristic finding

Lyme borreliosis is a highly variable disease involving many body systems. It occurs in overlapping patterns that come and go at different times. The skin lesion spreading from the site of the tick bite is its most distinctive feature. Relapsing arthritis is the most persistent finding and the one most likely to become chronic. Lyme disease is rarely fatal, but if untreated, it is often a source of chronic ill health.

Erythema migrans and febrile aches mark acute disease

The primary lesion begins sometime in the first month after a tick bite, which is often unnoticed. A macule or papule appears at the site of the bite and expands to become an annular lesion with a raised, red border and central clearing forming a bull's-eye pattern. As the bull's-eye ring expands, the lesion known as **erythema migrans** forms (**Figure 37–10**). Along with the skin lesions, fever, fatigue, myalgia, headache, joint pains, and mild neck stiffness are often present. Approximately 50% of untreated patients develop secondary skin lesions that closely resemble the primary one, but are not at the site of the tick bite. In untreated patients, the skin lesions usually disappear over a period of weeks, but constitutional symptoms may persist for months.

Nerve palsies and cardiac findings appear later

Days, weeks, even months after the onset of the primary lesion, a second stage may develop in which involvement of the nervous or cardiovascular system is superimposed. Neurologic abnormalities include a fluctuating meningitis, cranial nerve palsies, and peripheral neuropathy. Cardiac disease is usually limited to conduction abnormalities (atrioventricular block), but in some cases acute myocarditis can lead to cardiac enlargement. Both neurologic and cardiac abnormalities fluctuate in intensity, but generally resolve completely in a matter of weeks.

Fluctuating arthritis may become chronic

Weeks to years after the onset of infection, arthritis marks the continuing state of the disease. It develops in almost two thirds of untreated patients. Typically, it too follows a fluctuating or intermittent course, generally involving the large joints, particularly the knees. The arthritis may become chronic with erosion of the bone and cartilage, although the spirochetes are rarely demonstrable in the lesions. Less common chronic neurologic dysfunctions include subtle encephalitis affecting memory, mood, or sleep, and peripheral neuropathies.

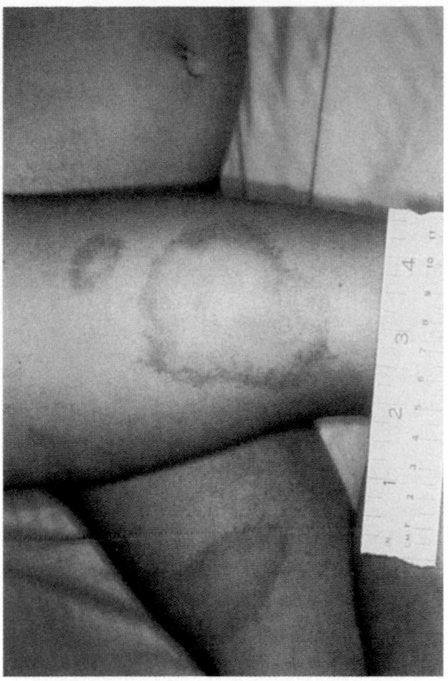

FIGURE 37-10. Erythema migrans. The typical rash of Lyme disease is shown evolving in concentric rings around the site if the tick bite. (Reproduced with permssion from Willey J, Sherwood L, Woolverton C (eds). *Prescott's Principles of Microbiology.* New York: McGraw-Hill; 2008.)

DIAGNOSIS

Presently, the diagnosis of early Lyme disease is based on exposure and typical clinical findings. Although *B burgdorferi* can be cultured from erythema migrans skin lesions, blood, joint fluid, and CSF, few laboratories have the skill to accomplish this or even stock the special medium required. The spirochetes are seldom detected on any kind of direct microscopic examination. Polymerase chain reaction (PCR) procedures able to detect *B burgdorferi*–specific DNA sequences in body fluids (joint, CSF) have been developed, but are not yet specific enough to be used alone for diagnosis. ::: PCR, p. 84

Culture is not practical

With culture generally unavailable, the diagnosis in later stages of disease usually rests on the demonstration of circulating antibodies to *B burgdorferi*. Despite considerable progress, these tests still lack the sensitivity and specificity to be considered more than supportive of a clinical diagnosis. The current recommendation is to first perform a sensitive screening test (enzyme immunoassay) followed by an immunoblot, which detects specific antigens of the organism. Unfortunately, there are many procedures in use that are not particularly well standardized. Even with the two-step approach, patients in the early stages of Lyme disease may be seronegative, and cross-reactive antigens may cause false-positive results. For persons who lack a typical clinical or epidemiologic history, great caution should be exercised before making a diagnosis of Lyme disease based on positive serologic or even PCR results.

Serologic tests are not definitive

TREATMENT

Doxycycline and amoxicillin are the first-line antimicrobics for the treatment of early Lyme disease and arthritis. For persons who cannot tolerate either of these agents, cefuroxime is a much more expensive alternative for oral therapy. Intravenous therapy with ceftriaxone or penicillin G is recommended for patients with neurologic involvement or cardiovascular findings such as atrioventricular heart block. The response to treatment is typically slow, requiring the continuation of antimicrobics for 30 to 60 days. So called "chronic Lyme disease" is most probably an autoimmune state, and thus antimicrobial agents are not appropriate treatment.

Doxycycline and β-lactams are recommended

PREVENTION

Preventing bites and removing ticks are important

The most useful preventive measures in endemic areas are the use of clothes that reduce the likelihood of the infected nymph reaching the legs or arms, careful search for nymphs after potential exposure, and removal of the tick by its head with tweezers. Duration of tick attachment to humans is also a factor in transmission; the risk is greatest when the tick has been feeding for at least 48 to 72 hours. Some insect repellents may provide added protection. The risk of Lyme disease after a random tick bite is too low to justify administration of antimicrobics prophylactically.

Vaccine no longer available

A vaccine for Lyme disease was developed, but is no longer available. The manufacturer discontinued production in 2002, citing insufficient consumer demand. The vaccine was unique in that it was composed of recombinant OspA and thus designed to act in the feeding tick, not the human.

CLINICAL CASE

A RASH AND FACIAL PARALYSIS

This 39-year-old man was in his usual state of good health and had just returned from a summer trip to Rhode Island. One week after returning home, he developed a fever and muscle aches, which resolved and were followed 2 weeks later with a rash on his right forearm, right hip, and left knee. At each site, the rash was initially localized but then over a few days moved outward forming large erythematous rings. Two weeks after the rash started, he felt a numbness on the left side of his face followed by a sagging and inability to move the facial muscles below his eye.

On physical examination, the patient was afebrile and had normal vital signs. A skin examination demonstrated the three skin lesions noted above, which had, according to the patient, faded significantly. A neurologic examination demonstrated left facial nerve weakness. The remainder of the examination was normal.

Laboratory studies included a normal complete blood count. A lumbar puncture was performed. CSF contained 78 nucleated cells/mm^3 with 88% lymphocytes and 12% monocytes. CSF glucose level was 60 mg/dL, and protein level was 55 mg/dL.

QUESTIONS

■ To consider a diagnosis of Lyme disease, what additional history would be most helpful from this patient?

A. Food consumption
B. Swimming in lakes or streams
C. Sexual contact
D. Hiking locales
E. Illness of friends

■ What laboratory test would be most likely to confirm this diagnosis?

A. *Borrelia burgdorferi* immunoassay
B. *Borrelia burgdorferi* immunoblot
C. *Borrelia burgdorferi* immunoassay plus immunoblot
D. Darkfield examination of rash
E. PCR of CSF

■ What molecular structure of *B burgdorferi* facilitates its life cycle in ticks?

A. OspA
B. OspB
C. OspC
D. LPS
E. Peptidoglycan

ANSWERS

1(D), **2**(C), **3**(A)

APPENDIX 37–1 Nonvenereal Treponemes

DISEASE	CAUSE	MAJOR GEOGRAPHIC LOCATION	PRIMARY LESION	SECONDARY LESIONS	TERTIARY LESIONS
Bejel	*T pallidum*, subspecies *endemicum* [a]	Middle East; arid, hot areas	Oral cavity[b]	Oral mucosa	Rare; gummatous lessions of skin, periosteum, bone, and joint
Yaws	*T pallidum*, subspecies *pertenue*	Humid, tropical belt	Skin, papillomatous	Systemic; resemble syphilis	Rare; gummatous lesions of skin, periosteum, bone, and joint[c]
Pinta	*T carateum*	Central and South America	Skin, erythematous papule	Skin; merge into primary lesion; altered pigmentation	Areas of altered skin pigmentation and hyperkeratoses

[a]Probably a variant of that causing venereal syphilis.
[b]Often inapparent.
[c]Neurologic manifestations usually absent.

CHAPTER 38

Mycoplasma and *Ureaplasma*

Mycoplasma and *Ureaplasma* are two genera of unique microbes that lack a cell wall and are therefore distinct from other bacteria. They differ from viruses by having both DNA and RNA and by the ability to grow in cell-free media. They are ubiquitous in nature as the smallest of free-living microorganisms. Numerous *Mycoplasma* species have been isolated from animals and humans, but only two species have been significantly associated with human disease (**Table 38–1**). *Mycoplasma pneumoniae* is a lower respiratory tract pathogen. *Mycoplasma genitalium* causes genitourinary tract infections.

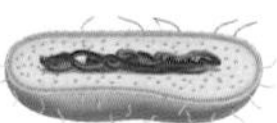

MICROBIOLOGY

The organisms have diameters of about 0.2 to 0.3 mm, but they are highly plastic and pleomorphic and may appear as coccoid bodies, filaments, and large multinucleoid forms. They do not have a cell wall and are bounded only by a single triple-layered membrane (**Figure 38–1**), which, unlike bacteria, contains sterols. The sterols are not synthesized by the organism, but are acquired as essential components from the medium or tissue in which the organism is growing. Lacking a cell wall, *Mycoplasma* and *Ureaplasma* stain poorly or not at all with the usual bacterial stains. Their double-stranded DNA genome is small, probably because of lack of genes encoding a complex cell wall. *M pneumoniae* is an aerobe, but most other species are facultatively anaerobic. All grow slowly in enriched liquid culture medium and on special *Mycoplasma* agar to produce minute colonies only after several days of incubation. The center of the *M pneumoniae* colony grows into the agar and appears denser, giving the appearance of an inverted "fried egg." Growth in culture is inhibited by specific antisera directed at the particular species. Colonies of *M pneumoniae* bind red blood cells (RBCs) onto the surface of agar plate cultures (hemadsorption). This is due to binding by the mycoplasma to sialic acid–containing oligosaccharides present on the RBC surface.

No cell walls

Cell membrane contains sterols

Not stained well by common methods

Slow growth in specialized media

Hemadsorption is a feature of *M pneumoniae*

MYCOPLASMA PNEUMONIAE

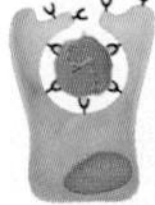

MYCOPLASMA PNEUMONIAE

CLINICAL CAPSULE

M pneumoniae produces a common form of pneumonia, which tends to occur in any season and has a predilection for younger persons. The illness is characterized by a nonproductive cough, fever, and headache, with radiologic and clinical evidence of scattered areas of pneumonia. The course is almost always benign, but improvement is accelerated by treatment with non–cell wall–active antimicrobials.

TABLE 38–1	Pathogenic *Mycoplasma* and *Ureaplasma* Species of Humans		
ORGANISM	**SITE**	**PREVALENCE**	**DISEASE**
M pneumoniae	Upper and lower respiratory tract	Common	Primary atypical pneumonia
M genitalium	Genitourinary tract	Common	Urethritis, postpartum fever; pelvic inflammatory disease

EPIDEMIOLOGY

Infecting dose is very low

Found worldwide most often in teenagers

Outbreaks occur in families and closed communities

M pneumoniae accounts for approximately 10% of all cases of pneumonia. Infection is acquired by droplet spread. Experimental challenges indicate that the human infectious dose is very low, possibly less than 100 colony-forming units. Infections with *M pneumoniae* occur worldwide, but they are especially prominent in temperate climates. Epidemics at 4- to 6-year intervals have been noted in both civilian and military populations. The most common age range for symptomatic *M pneumoniae* infection is between 5 and 15 years, and the disease accounts for more than one third of all cases of pneumonia in teenagers (but is also seen in older persons). Infections in children younger than 6 months are uncommon. The disease often appears as a sporadic, endemic illness in families or closed communities because its incubation period is relatively long (2 to 3 weeks) and because prolonged shedding in nasopharyngeal secretions may cause infections to be spread over time. In families, attack rates in susceptible persons approach 60%. Asymptomatic infections occur, but most studies have suggested that more than two thirds of infected cases develop some evidence of respiratory tract illness.

PATHOGENESIS

M pneumoniae infection involves the trachea, bronchi, bronchioles, and peribronchial tissues and may extend to the alveoli and alveolar walls. Initially, the organism attaches to the cilia and microvilli of the cells lining the bronchial epithelium. This attachment is mediated by a

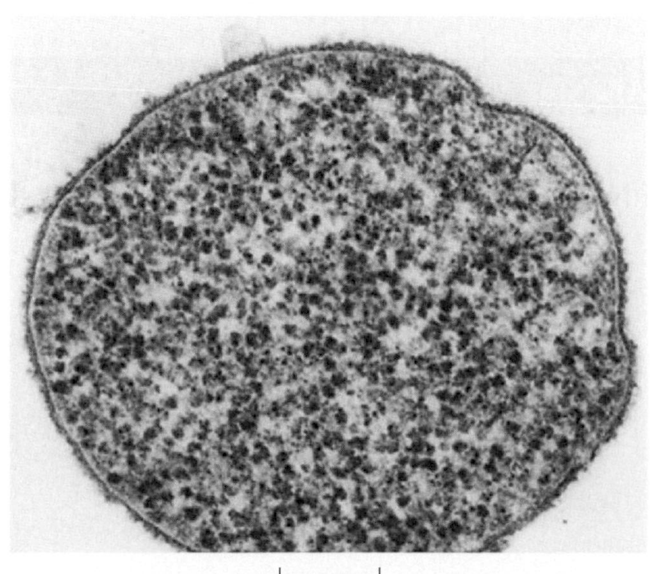

FIGURE 38–1. Electron micrograph of *Mycoplasma*. Note cytoplasmic membrane ribosomes and surface amorphous material with absence of cell wall. (Courtesy of the late Dr. E. S. Boatman.)

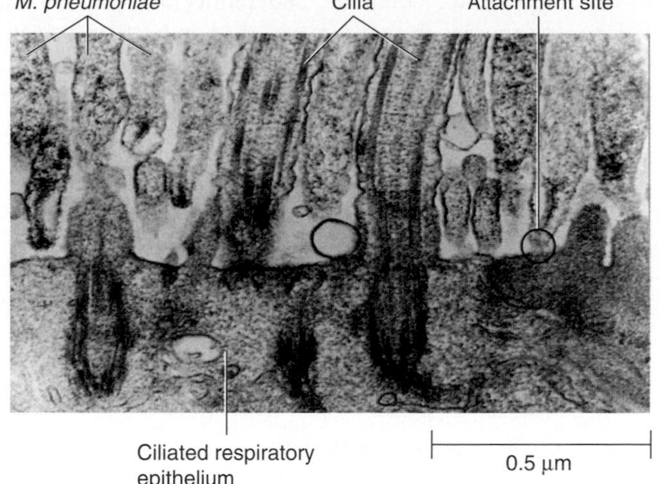

FIGURE 38–2. *Mycoplasma pneumoniae* infecting respiratory epithelium. Transmission electron micrograph. Note the distinctive appearance of the tips of the mycoplasmas adjacent to the host epithelium. The tips probably represent a site on the microorganism that is specialized for attachment. (Reproduced with permission from Nester EW, Anderson DG, Roberts CE Jr, Nester MT. *Microbiology: A Human Perspective,* 6th ed. New York: McGraw-Hill, 2008.)

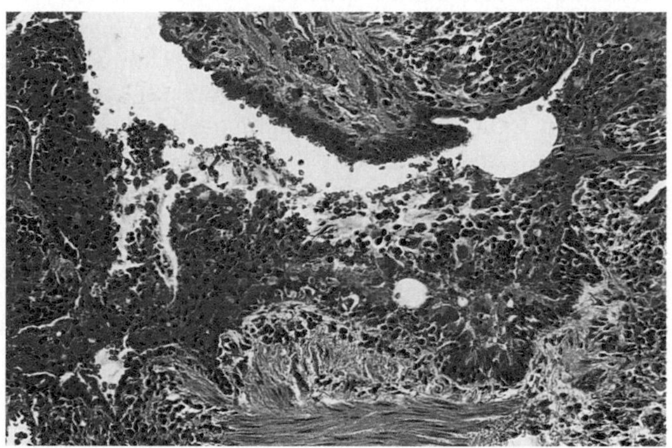

FIGURE 38–3. *Mycoplasma pneumoniae* bronchiolitis. This lung section shows destruction of the bronchiolar wall and mucosal ulceration. (Reproduced with permission from Connor DH, Chandler FW, Schwartz DQA, Manz HJ, Lack EE (eds). *Pathology of Infectious Diseases,* vol. 1. Stamford, CT: Appleton & Lange; 1997.)

surface mycoplasmal cytadhesin (P1) protein that binds to complex oligosaccharides containing sialic acid found in the apical regions of bronchial epithelial cells (**Figure 38–2**). The oligosaccharide receptors are chemically similar to the I antigen on the surface of erythrocytes and are not found on the nonciliated goblet cells or mucus, to which *M pneumoniae* does not bind. The organisms interfere with ciliary action and initiate a process that leads to desquamation of the involved mucosa and a subsequent inflammatory reaction and exudates (**Figure 38–3**). The inflammatory response is at first most pronounced in the bronchial and peribronchial tissue and is composed of lymphocytes, plasma cells, and macrophages, which may infiltrate and thicken the walls of the bronchioles and alveoli. Organisms are shed in upper respiratory secretions for 2 to 8 days before the onset of symptoms, and shedding continues for as long as 14 weeks after infection.

Adherence to bronchial epithelial cells is mediated by P1 protein

Interferes with ciliary action and leads to desquamation

IMMUNITY

Both local and systemic specific immune responses occur, which generally appear to be effective in preventing reinfection. Complement-fixing serum antibody titers reach a peak 2 to 4 weeks after infection and gradually disappear over 6 to 12 months. Also, nonspecific immune responses to the glycolipids of the outer membrane of the organism often develop, which can be detrimental to the host. For example, cold hemagglutinins are IgM antibodies that react with an altered I antigen on human RBCs and are seen in about two thirds of symptomatic patients infected with *M pneumoniae.*

Complement-fixing antibody titers peak at 2 to 4 weeks

Cold agglutinins are IgM

Immunity is incomplete, and reinfection may occur

Immunity is not complete, and reinfection with *M pneumoniae* may occur. Clinical disease appears to be more severe in older than in younger children, which has led to the suggestion that many of the clinical manifestations of disease are the result of immune responses rather than invasion by the organism. High titers of cold agglutinins may be associated with hemolysis and Raynaud's phenomena. Antibodies may develop in response to an alteration of the I antigen by the organism or may represent cross-reacting antibodies.

MYCOPLASMAL PNEUMONIA: CLINICAL ASPECTS

MANIFESTATIONS

"Walking" pneumonia has insidious onset

Cough is usual

A mild tracheobronchitis with fever, cough, headache, and malaise is the most common syndrome associated with acute *M pneumoniae* infection. The pneumonia is typically less severe than other bacterial pneumonias. It has been described as "walking" pneumonia because most cases do not require hospitalization. The disease is of insidious onset, with fever, headache, and malaise for 2 to 4 days before the onset of respiratory symptoms. Pulmonary symptoms are generally limited to a non- or minimally productive cough. X-rays reveal a unilateral or patchy pneumonia, usually in a lower lobe, although multiple lobes are sometimes involved. Small pleural effusions are seen in up to 25% of cases. The average duration of untreated illness is 3 weeks. The severity of pulmonary involvement is particularly great in patients with immune deficiencies, sickle cell disease, or Down's syndrome. The reason for the latter is not understood.

Pharyngitis and otitis are also common

Pharyngitis with fever and sore throat may also occur. Nonpurulent otitis media or myringitis may occur concomitantly in up to 15% of patients with *M pneumoniae* pneumonitis, but bullous myringitis is rare.

Other extrapulmonary complications sometimes occur

A variety of other extrapulmonary complications have been described, involving skin (erythema multiforme, sometimes severe), peripheral vasospasm (Raynaud's phenomenon), central nervous system (encephalitis, myelitis), joints (arthralgias), and other sites.

DIAGNOSIS

Diagnosis is usually serologic

Single high complement fixation or IgM-specific antibody titer supports diagnosis

Clinical diagnosis of *M pneumoniae* infection may be difficult because the manifestations overlap with those of other "atypical" respiratory infections. Gram-stained sputum usually shows some mononuclear cells, but because it lacks a cell wall, *M pneumoniae* is not seen. The absence of organisms, however, may help to suggest an atypical etiology. The organism can be isolated from throat swabs or sputum of infected patients using special culture media and methods, but because of its slow growth, isolation usually requires incubation for a week or longer. Thus, serologic tests rather than cultures are more commonly used for specific diagnosis. A fourfold rise of serum antibody titer or seroconversion in acute and convalescent sera indicates *M pneumoniae* infection. The most widely used serologic method is complement fixation. With the relatively long incubation period and insidious onset of the disease, many patients already have high antibody titers at the time they are first seen. In these situations, a single high titer, such as a complement fixation titer greater than 1:128 or IgM-specific antibody (measured by enzyme immunoassay or immunofluorescence), indicates recent or current infection because these antibodies are generally of short duration.

Cold agglutinins are nonspecific, but helpful if present

Because more than two thirds of patients with symptomatic lower respiratory *M pneumoniae* infection develop high titers of cold hemagglutinins, their demonstration can be useful in some clinical situations. It must be remembered that cold hemagglutinins are nonspecific and have been observed in adenovirus infections, infectious mononucleosis, and some other illnesses. The test is simple, however, and can be performed rapidly in any clinical laboratory or even at the bedside. Direct detection of the organism in respiratory secretions can be accomplished using immunoassay methods, DNA hybridization, and the polymerase chain reaction. These methods are not commonly used for routine diagnosis.

TREATMENT

Macrolides or tetracycline are the usual agents used for treatment of *M pneumoniae* pneumonia. Almost all patients with *M pneumoniae* pneumonia recover, but treatment markedly shortens the course of illness. Azithromycin and clarithromycin are comparable to erythromycin, but clindamycin is not effective. Most quinolones are also active.

Erythromycin, tetracycline, clarithromycin, or azithromycin used in treatment

MYCOPLASMA GENITALIUM AND MYCOPLASMA HOMINIS

M genitalium and related organisms are common inhabitants of the genitourinary tract, which has made it difficult to define their contribution to genitourinary disease. Eight species of *Mycoplasma* and *Ureaplasma* have been identified in the genital tract, with *M genitalium* as one of the most prevalent and the one most often implicated in causing urethritis. This Mycoplasma species does not grow on conventional *Mycoplasma* media, so its contribution to urethritis has been defined by molecular methods, such as PCR.

Genitourinary inhabitant

Grows rapidly on specialized agar

Clinical studies indicate that pelvic inflammatory disease syndromes in women may be associated with *M genitalium* infection. The organism is sensitive to tetracycline and erythromycin. *M hominis* may be associated with postpartum fever, or localized abscesses in immune compromised patients. It is resistant to erythromycin.

Association with pelvic inflammatory disease

M hominis resistant to erythromycin

UREAPLASMA

The genus *Ureaplasma* contains two species of human importance, *U urealyticum* and *U parvum*. *Ureaplasma* are distinguished from *Mycoplasma* by production of urease and growth only in media with an acid pH. On special *Ureaplasma* agar media, colonies are small and circular and grow downward into the agar. In liquid media containing urea and phenol red, growth of *Ureaplasma* results in production of ammonia from the urea, with a resultant increase in pH and a change in color of the indicator.

Urease production marks the species

EPIDEMIOLOGY

The main reservoir of human species of *Ureaplasma* is the genital tract of sexually active men and women; it is rarely found before puberty. Colonization, which probably results primarily from sexual contact, occurs in more than 80% of individuals who have had three or more sexual partners.

Most commonly acquired by sexual contact

MANIFESTATIONS

Because of the high colonization rate, it has been difficult to associate specific illness with *Ureaplasma*. As with *M hominis*, *Ureaplasma* are found as often in controls as in male patients with urethritis. In women, *Ureaplasma* have been shown to cause chorioamnionitis and postpartum fever. The organism has been isolated from 10% of women with the latter syndrome.

Association with urethritis in men

DIAGNOSIS AND TREATMENT

Men with nongonococcal urethritis should be treated because *Ureaplasma* infection may be involved. A tetracycline is the treatment of choice because it is also active against *Chlamydia*, but tetracycline-resistant strains of *Ureaplasma* have been reported, and have been associated with recurrences of nongonococcal urethritis in men. In such cases, treatment with azithromycin or a quinolone should be effective.

Tetracyclines, azithromycin or quinolones are effective

CLINICAL CASE

A TEENAGER WITH RESPIRATORY COMPLAINTS

In July, a 14-year-old girl presents with cough and fever to 102°F. She does not appear seriously ill. Chest examination is abnormal and chest x-ray reveals bilateral, patchy infiltrates. Her brother, aged 12, had a similar illness three weeks earlier.

QUESTIONS

■ Whch is the most likely cause of this girl's illness?

A. *Legionella pneumophila*
B. *Chlamydia pneumoniae*
C. *Mycoplasma pneumoniae*
D. Influenza A virus
E. *Metapneumovirus*

■ Which is the most appropriate diagnostic test?

A. Culture
B. Imnmunofluorescent assay on sputum
C. Serology
D. PCR

■ Which is the treatment of choice for this patient?

A. Penicillin
B. Ribavirin
C. Oseltamivir
D. Erythromycin
E. A cephalosporin

ANSWERS

1(C), **2**(C), **3**(D)

CHAPTER 39

Chlamydia

Members of the genus *Chlamydia* are classified as bacteria but differ by replicating only within cells and by lacking peptidoglycan in their cell wall. Of the three species causing disease in humans, *Chlamydia trachomatis* is the most common as a major cause of genital infection and conjunctivitis. A chronic form of *C trachomatis* conjunctivitis, called trachoma, is the leading preventable cause of blindness in the world. *Chlamydia pneumoniae* and *Chlamydia psittaci* are respiratory pathogens. Our knowledge of biology and pathogenesis of these bacteria is based primarily on the study of *C trachomatis.*

CHLAMYDIA TRACHOMATIS

BACTERIOLOGY

MORPHOLOGY

C trachomatis are round cells between 0.3 and 1 μm in diameter depending on the replicative stage (see following text). The envelope surrounding the cells includes a trilaminar outer membrane that contains lipopolysaccharide and proteins similar to those of Gram-negative bacteria. A major difference is that chlamydiae lack the thin peptidoglycan layer between the two membranes. They are obligate intracellular parasites and have not been grown outside eukaryotic cells. *C trachomatis* has ribosomes and is able to carry out the common energy-producing pathways of other bacteria.

Envelope has no peptidoglycan layer between membranes

Obligate intracellular bacteria, which fail to grow in artificial media

DNA homology between *C trachomatis, C psittaci,* and *C pneumoniae* is less than 30%, although rRNA sequence analysis suggests they share a common origin. The three species share a common group antigen. Their major differential features are shown in **Table 39–1**. *C trachomatis* has two biovars; multiple outer membrane proteins further divide the two biovars into multiple serovars, or strains (**Table 39–2**). Three serovar groups of *C trachomatis* affect humans: serovars A–C cause trachoma; D–K cause nongonococcal urethritis, mucopurulent cervicitis, and inclusion conjunctivitis; and L1–3 cause lymphogranuloma venereum (LGV).

REPLICATIVE CYCLE

The replicative cycle of chlamydiae is illustrated in **Figure 39–1**. It involves two forms of the organism: a small, hardy infectious form termed the elementary body (EB), and a larger fragile intracellular replicative form termed the reticulate body (RB). The major

TABLE 39–1 Major Differential Features of *Chlamydia* Species That Cause Human Disease

FEATURE	*C TRACHOMATIS*	*C PSITTACI*	*C PNEUMONIAE*
Natural host	Humans	Birds, including turkeys, ducks, geese, pheasants	Humans
Disease	Conjunctivitis, genital tract infections, lymphogranuloma venereum, pneumonia (infants)	Pneumonia, endocarditis	Bronchitis, pneumonia, atherosclerosis
Glycogen-containing inclusion bodies	Yes	No	No
Sulfonamide susceptibility	Yes	No	No

TABLE 39–2 Epidemiologic Associations Between Chlamydial Species, Serovars (Strains), and Diseases

SPECIES	SEROVARS (STRAINS)	MODES OF TRANSMISSION	DISEASES
Chlamydia trachomatis	A,B,Ba,C	Hand to eye, fomites, flies	Trachoma
	D–K	Sexual, intrapartum, hand to eye	Inclusion conjunctivitis; genital infection
	L_1, L_2, L_3	Sexual	Lymphogranuloma venereum
C psittaci	Many	Aerosol	Psittacosis
C pneumoniae	TWAR[a]	Human to human	Respiratory infection

[a]TW and AR were the laboratory designations for the first conjunctival and respiratory isolates, respectively.

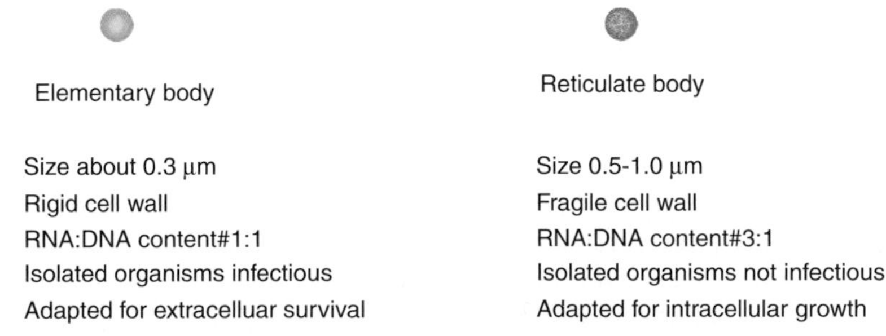

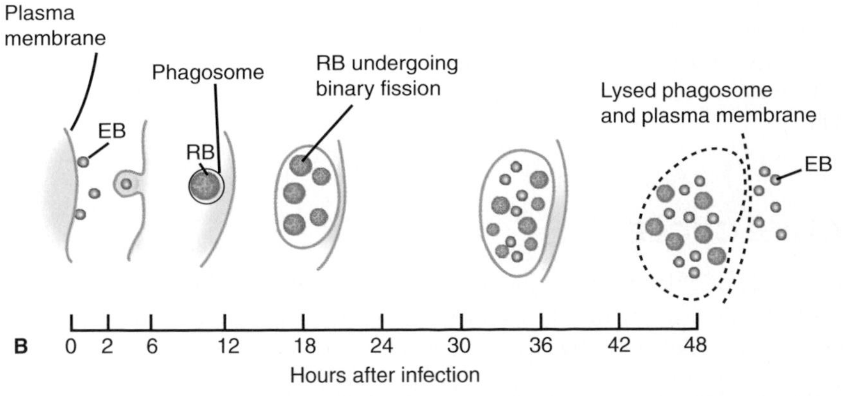

FIGURE 39–1. Chlamydia life cycle. A. An electron micrograph of an inclusion body containing a mixture of small, blue elementary bodies (EB), larger reticulate bodies (RB) and an intermediate body (IB), a chlamydial cell intermediate in morphology in morphology between EB and RB. The RBs appear tan and granulated because of their high concentration of ribosomes. **B.** A schematic representation of the infectious cycle of chlamydiae. (Reproduced with permission from Willey J, Sherwood L, Woolverton C (eds). *Prescott's Principles of Microbiology.* New York: McGraw-Hill; 2008.)

biochemical difference between the EB and the RB is the extent of cross-linking of the major outer membrane protein (MOMP); EB proteins are highly linked by disulfide bonds, and RB proteins are less so. The EB is a metabolically inert form that neither expends energy nor synthesizes protein. The cycle begins when the EB attaches to unknown receptors on the plasma membrane of susceptible target cells (usually columnar or transitional epithelial cells). It then enters the cell in an endocytotic vacuole and begins the process of converting to the replicative RB. There is evidence that pinocytosis may also occur. Endosomes containing *C trachomatis* EBs maintain a near neutral pH and fuse with each other but not with lysosomes. The ability to inhibit lysosomal fusion is unique to chlamydia and enables the EB to survive in a vesicle referred to as an inclusion. As the RBs increase in number, the endosomal membrane expands by fusing with lipids of the Golgi apparatus, eventually forming a large inclusion body, called the reticulate body. (RB) After 24 to 72 hours, the process reverses and the RBs reorganize and condense to yield multiple EBs. The endosomal membrane then either disintegrates or fuses with the host cell membrane, releasing the EBs to infect adjacent cells. The metabolic changes that lead the EB to reorganize into the larger reticulate body are incompletely understood, but involve protein synthesis and modification of MOMPs between the monomeric and cross-linked state. *C trachomatis* also inhibits apoptosis of epithelial cells, thus enabling completion of its replicative cycle.

Elementary body enters epithelial cells by endocytosis

Host cell metabolism used for growth and replication

Transiently inhibits apoptosis of infected cells

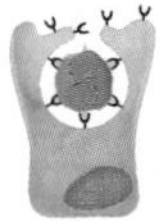

CHLAMYDIA TRACHOMATIS DISEASE

CLINICAL CAPSULE

Ocular trachoma, with progressive inflammation and scarring leading to blindness, has been recognized since antiquity, but the role of chlamydiae in conjunctivitis and pneumonia in young infants, and in a variety of genital infections was only clarified during the past 45 years. Like trachoma, the genital infections can persist or recur, with chronic sequelae.

EPIDEMIOLOGY

C trachomatis causes disease in several sites, including the conjunctiva and genital tract. In the United States over 4,000,000 new infections occur each year, Humans are the sole reservoir (Table 39–1). Inclusion conjunctivitis is seen among population groups in which the strains causing *C trachomatis* genital infections are common. *C trachomatis* also causes the most common form of neonatal conjunctivitis in the United States. The infection results from direct contact with infective cervical secretions of the mother at delivery.

Neonatal conjunctivitis contracted from maternal genital infection

High attack rate

Trachoma, a chronic follicular conjunctivitis, afflicts an estimated 500 million persons worldwide and blinds 7 to 9 million, particularly in Africa. The disease is usually contracted in infancy or early childhood from the mother or other close contacts. Spread is by contact with infective human secretions, directly via hands to the eye or via fomites transmitted on the feet of flies.

Fomites, fingers, and flies involved in transmission of trachoma

The prevalence of chlamydial urethral infection in US men and women ranges from 5% in the general population to 20% in those attending sexually transmitted disease clinics. Approximately one third of male sexual contacts of women with *C trachomatis* cervicitis develop urethritis after an incubation period of 2 to 6 weeks. The proportion of men with mild to absent symptoms is higher than in gonorrhea. Nongonococcal urethritis is most commonly caused by *C trachomatis* and less frequently by *Ureaplasma urealyticum.*

High rate of sexual transmission

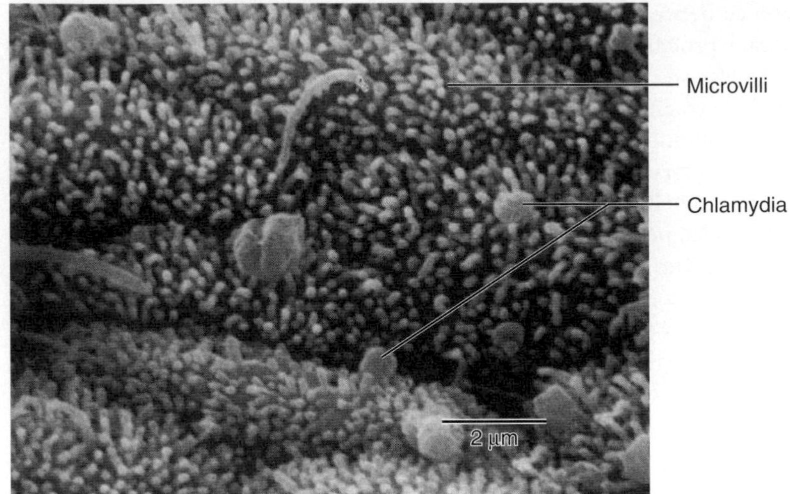

FIGURE 39–2. Scanning electron micrograph of *Chlamydia trachomatis* attached to fallopian tube mucosa. (Reproduced with permission from Nester EW, Anderson DG, Roberts CE Jr, Nester MT. *Microbiology: A Human Perspective,* 6th ed. New York: McGraw-Hill, 2008.)

PATHOGENESIS

Early release of proinflammatory cytokines

Later development of fibrosis and scarring

Chlamydiae have a tropism for epithelial cells of the endocervix and upper genital tract of women (**Figure 39–2**), and the urethra, rectum and conjunctiva of both sexes. The LGV serovars can also enter through breaks in the skin or mucosa. Once infection is established, there is a release of proinflammatory cytokines such as interleukin-8 by infected epithelial cells. Chlamydial lipopolysaccharides probably also play an important role in initiation of the inflammatory process. This results in early tissue infiltration by polymorphonuclear leukocytes, later followed by lymphocytes, macrophages, plasma cells, and eosinophils. If the infection progresses further (because of lack of treatment and/or failure of immune control), aggregates of lymphocytes and macrophages (lymphoid follicles) may form in the submucosa; these can progress to necrosis, followed by fibrosis and scarring.

Persistent or recurrent infections cause chronic eye or genital sequelae

Autoimmunity may play an important role

The chronic sequelae of progressive inflammation with scarring that are seen in trachoma and some female genital tract infections are commonly due to persistent or recurrent infections, which may, in turn, be controlled by host cell immune responses. One theory is that this may result from molecular mimicry, involving epitopes found on the chlamydial 60-kd heat shock protein and also on human cells.

IMMUNITY

Immunity is short-lived

Secretory IgA and CD4+ lymphocytes may influence severity

C trachomatis infections do not reliably result in protection against reinfection, although there is evidence that secretory immunoglobulin A may confer at least some partial immunity against genital tract reinfection. Any strain-specific protection that may result is short-lived. Local production of antibody, along with CD4+ lymphocytes of the T_H1 type that traffic to the genital mucosa may together play a role in mitigating most acute infections. This would at least partially explain why most untreated chlamydial genital tract infections are persistent, but often subclinical in character.

CHLAMYDIA TRACHOMATIS: CLINICAL ASPECTS

MANIFESTATIONS

Eye Infections

Trachoma and inclusion conjunctivitis are distinct diseases of the eye that have some overlap in their clinical manifestations. Trachoma, a chronic conjunctivitis caused by *C trachomatis*

FIGURE 39–3. Trachoma. An active infection showing follicular hypertrophy. The inflammatory nodules cover the thickened conjunctiva.

serovars A, B, Ba, and C, is usually seen in less developed countries and often leads to blindness. Inclusion conjunctivitis, an acute infection commonly caused by serovars D to K, is usually not associated with chronicity or permanent eye damage. It occurs in newborns and adults worldwide.

Trachoma and inclusion conjunctivitis due to different serotypes

Trachoma

Chronic inflammation of the eyelids and increased vascularization of the corneal conjunctiva are followed by severe corneal scarring and conjunctival deformities (**Figure 39–3**). Visual loss often occurs 15 to 20 years after the initial infection as a result of repeated scarring of the cornea.

Leading cause of blindness in some developing countries

Inclusion Conjunctivitis

Neonatal inclusion conjunctivitis usually presents as an acute, watery then mucopurulent eye discharge 5 to 12 days after birth. Infection occurs in roughly one third of infants born vaginally to infected mothers. The infection is not prevented by prophylaxis with topical erythromycin or tetracycline. Untreated, it may persist for 3 to 12 months. Inclusion conjunctivitis is clinically similar in adults and is usually associated with concomitant genital tract disease. Diagnosis can be made by demonstrating characteristic cytoplasmic inclusions in smears of conjunctival scrapings (**Figure 39–4**), by demonstration of antigen using direct immunofluorescence, by culture or specific nucleic acid testing of conjunctival swabs. In both neonates and adults, systemic therapy is preferred because the nasopharynx, rectum,

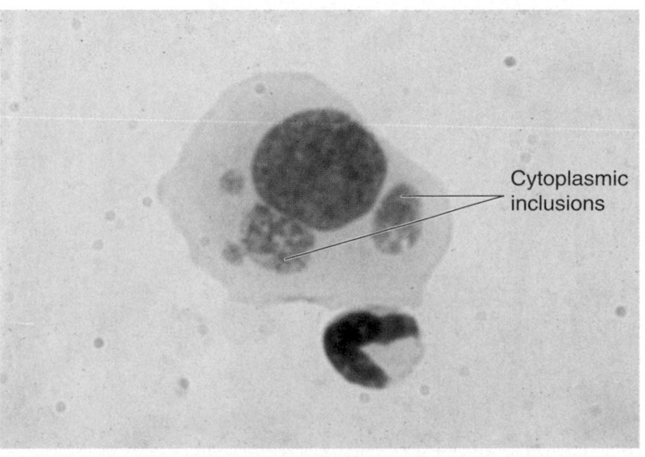

FIGURE 39–4. *Chlamydia trachomatis* cytoplasmic inclusion bodies in a conjunctival epithelial cell. (Reproduced with permission from Willey J, Sherwood L, Woolverton C (eds). *Prescott's Principles of Microbiology.* New York: McGraw-Hill; 2008.)

and vagina may also be colonized and other forms of disease may develop, such as an infant pneumonia syndrome. More than 50% of all infants born to mothers excreting *C trachomatis* during labor show evidence of infection during the first year of life. Most develop inclusion conjunctivitis (see earlier discussion), but 5% to 10% develop the infant pneumonia syndrome. *C trachomatis* accounts for about one third to one half of all cases of interstitial pneumonia in infants. The illness usually develops in a child between 6 weeks and 6 months of age and has a gradual onset. The child is usually afebrile, but develops difficulty in feeding, a characteristic staccato (pertussis-like) cough, and shortness of breath. The disease is rarely fatal, but may be associated with decreased pulmonary function later in life.

Infant pneumonia syndrome has delayed, gradual onset

Genital Infections

The clinical spectrum of sexually transmitted infections with *C trachomatis* is similar to that of *Neisseria gonorrhoeae*. *C trachomatis* can cause urethritis and epididymitis in men and cervicitis, salpingitis, and a urethral syndrome in women. In addition, three strains of *C trachomatis* cause LGV, a distinctly different sexually transmitted disease (Table 39–2).

Clinical spectrum is similar to *N gonorrhoeae*

Urethritis in men, but many asymptomatic

C trachomatis urethritis is manifested by dysuria and a thin urethral discharge. Infections of the uterine cervix may produce vaginal discharge but are usually asymptomatic. Ascending infection in the form of salpingitis and pelvic inflammatory disease (PID) occurs in an estimated 5% to 30% of infected women. The scarring produced by chronic or repeated infection is an important cause of sterility and ectopic pregnancy. ::: PID in gonorrhea, p. 546

Salpingitis and pelvic inflammatory disease can cause permanent sequelae

LGV is a sexually transmitted infection caused by *C trachomatis* strains L_1, L_2, or L_3. It occurs principally in South America Africa, Southeast Asia ,India, and Caribbean countries. The clinical course is characterized by a transient genital lesion followed by multilocular suppurative involvement of the inguinal lymph nodes (**Figure 39–5**). The primary genital lesion is usually a small painless ulcer or papule, which heals in a few days and may go unnoticed. The most common presenting complaint is inguinal adenopathy. Nodes are initially discrete, but as the disease progresses, they become matted and suppurative (bubos). The skin over the node may be thinned, and multiple draining fistulas develop. Systemic symptoms such as fever, chills, headaches, arthralgia, and myalgia are common. Late complications include urethral or rectal strictures and perirectal abscesses and fistulas. In homosexual men, LGV strains can cause a hemorrhagic ulcerative proctitis. Lymph nodes may need to be aspirated to prevent rupture. A further consideration of genital tract infections is given in Chapter 64.

Papule and inguinal adenopathy

Abscesses, strictures, and fistulas with chronic infection

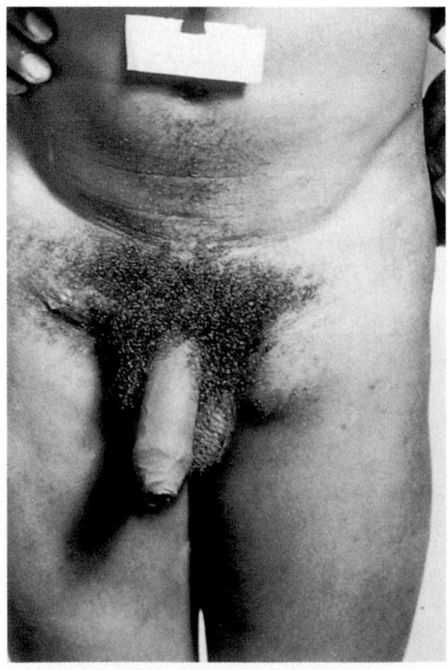

FIGURE 39–5. Lymphogranuloma venerum. Ulcerated inguinal lymph node. (Reproduced with permission from Connor DH, Chandler FW, Schwartz DQA, Manz HJ, Lack EE (eds). *Pathology of Infectious Diseases,* vol. 1. Stamford, CT: Appleton & Lange; 1997,)

DIAGNOSIS

Most *C trachomatis* diagnostic tests require the collection of epithelial cells from the site of infection, but urine is suitable when nucleic acid assays are used. Inflammatory cells are not useful and should be cleaned away as much as possible. For genital infections, cervical specimens are preferred in females and urethral scrapings in males. Eye infections require conjunctival scrapings.

Epithelial cells are required for detection

Isolation of *C trachomatis* has been the "gold standard" for diagnosis, but it is clear that it is less sensitive than specific nucleic acid amplification (NAA) assays. Culture is performed using idoxuridine- or cycloheximide-treated McCoy cells. Treatment of the cells with antimetabolites inhibits host cell replication, but allows chlamydiae to use available cell nutrients for growth. After inoculation with samples and incubation for 3 to 7 days, the cells are stained with fluorescein-labeled monoclonal antibodies to detect intracytoplasmic chlamydiae. *C trachomatis* reticulate bodies synthesize large amounts of glycogen, and the inclusion bodies in the cell thus stain reddish-brown with iodine. *C psittaci* and *C pneumoniae* inclusions do not contain glycogen.

Isolation requires special treatment of cell lines

C trachomatis inclusions contain glycogen and are stainable with iodine

A large number of procedures are now available for noncultural direct detection of *C trachomatis* in clinical specimens. These include direct fluorescent antibody (DFA) methods using monoclonal antibodies directed against outer membrane proteins of elementary bodies in epithelial cells, enzyme immunoassays that detect chlamydial lipopolysaccharide, and *C trachomatis* NAA methods. These tests are faster and, in the case of NAA, also more sensitive than culture. Ligase chain reaction (LCR) or polymerase chain reaction (PCR) are the most sensitive NAA methods. The latter procedures can be used on urine samples, which are easier to obtain than cervical or urethral samples. This greatly aids the detection of chlamydial genital infections, especially in adolescents and ultimately facilitates control of the spread of this sexually transmitted disease. ::: NAA tests, p. 80-81

LCR or PCR are most sensitive methods of noncultural diagnosis

Serodiagnostic methods have little use in diagnosis of chlamydial genital infection because of the difficulty of distinguishing current from previous infection. Detection of IgM antibodies against *C trachomatis* is helpful in cases of infant pneumonitis. Chlamydial serology is also useful in the diagnosis of LGV, where a single high complement fixation antibody titer (higher than 1:32) or a fourfold rise supports a presumptive diagnosis. The most satisfactory method for diagnosis of LGV is isolation of an LGV strain of *C trachomatis* from aspirated bubos or tissue biopsies. In 80% to 90% of patients, the LGV complement fixation test is positive (titer higher than 1:64) shortly after the appearance of the bubo.

Serodiagnosis not helpful for most genital infections

TREATMENT

Strains of *C trachomatis* are sensitive to tetracyclines, macrolides and related compounds, and some fluoroquinolones. Azithromycin is the first-line therapy because it is given as a single oral dose for non-LGV *C trachomatis* infection. Erythromycin is used for pregnant women and infants because of the tooth staining that may result from tetracycline therapy and less experience with the newer agents. Doxycycline is an alternative for *C trachomatis* and is the drug of choice for treating LGV.

Effective antimicrobics include erythromycin, azithromycin, and doxycycline

For trachoma, a single dose of azithromycin is now the treatment of choice, although a tetracycline for 14 days is an alternative. Corrective surgery may prevent blindness and is required for severe corneal and conjunctival scarring. Control of trachoma is directed toward prevention of continued reinfection during early childhood. Improvement in general hygienic practices is the most important factor in decreasing transmission of infection within families and, of course, one of the most difficult to implement on a broad scale.

Prevention of reinfection most important control measure

PREVENTION

Prophylaxis for infants using topical erythromycin or silver nitrate on the conjunctiva has limited effectiveness for *Chlamydia* because 15% to 25% of exposed infants still develop inclusion conjunctivitis. The primary approach to prevention of all forms of genital and infant *C trachomatis* infection comprises detection of this infection in sexually active individuals and appropriate treatment, including infected women with erythromycin late in pregnancy. No vaccine is available or under development.

Primary approach is detection and treatment of infection in high-risk individuals

No vaccine available

CHLAMYDIA PSITTACI

EPIDEMIOLOGY

Pneumonia contracted from birds

Human psittacosis (ornithosis) is a zoonotic pneumonia contracted through inhalation of respiratory secretions or dust from droppings of infected birds. It was initially described in psittacines, such as parrots and parakeets, but was subsequently shown to occur in a wide range of avian species, including turkeys. The disease is usually latent in its natural host, but may become active, particularly with the stress of recent captivity or transport; *C psittaci* is then excreted in large amounts.

Associated with poultry processing and captive psittacine birds

Psittacosis in humans is seen mainly as an occupational hazard of poultry workers and bird fanciers, particularly owners of psittacine birds. Reported cases of human psittacosis in the United States decreased during the 1950s, in association with the use of antimicrobials in poultry feeds and quarantine regulations for imported psittacine birds. Currently, 100 to 200 cases of psittacosis are reported each year. Some strains of *C psittaci* are highly contagious and pose a hazard for laboratory workers processing specimens for *C psittaci* isolation. Human-to-human transmission is rare.

CLINICAL DISEASE AND TREATMENT

Interstitial pneumonia is bilateral

Diagnosis is primarily serologic

Treatment with tetracycline or doxycycline

The incubation period psittacosis is 5 to 15 days. Psittacosis in humans is an acute infection of the lower respiratory tract, usually presenting with acute onset of fever, headache, malaise, muscle aches, dry hacking cough, and bilateral interstitial pneumonia. Occasionally, systemic complications such as myocarditis, encephalitis, endocarditis, and hepatitis may develop. The liver and spleen are often enlarged. The diagnosis of psittacosis should be suspected in any patient with acute onset of febrile lower respiratory illness who gives a history of close exposure to birds. Indeed, a history of bird exposure should be especially sought in patients who appear to have a bilateral pneumonia not proven to be caused by other agents. It must be remembered that spread can occur from both symptomatic and asymptomatic infections of birds. The specific diagnosis is usually made by demonstrating seroconversion or a fourfold rise in the titer of complement-fixing or indirect fluorescent antibody to chlamydial group antigen. Although *C psittaci* can be isolated from blood or sputum early in the disease, these methods are attempted only in specialized laboratories because of the risk of laboratory infection. Treatment with tetracycline or doxycycline is effective if given early in the course of illness.

CHLAMYDIA PNEUMONIAE

Clinical manifestations are similar to those of *M pneumoniae*

Possible role in atherosclerosis is proposed

Treatment is tetracycline erythromycin or newer macrolides

C pneumoniae has been shown to be a cause of "walking pneumonia" in adults worldwide. It is estimated that 10% of pneumonia and 5% of bronchitis cases are due to this agent. Epidemiologic evidence indicates that infection occurs throughout the year and is spread between humans by person-to-person contact. Unlike psittacosis, birds are not the reservoir. Outbreaks of community-acquired pneumonia caused by *C pneumoniae* have been reported, as well as apparent nosocomial spread. Reinfections occur, and clinically evident *C pneumoniae* infection may be more evident in the elderly than in younger individuals. Most infections manifest as pharyngitis, lower respiratory tract disease, or both, and the clinical spectrum is similar to that of *Mycoplasma pneumoniae* infection. Pharyngitis or laryngitis may occur 1 to 3 weeks before bronchitis or pneumonia, and cough may persist for weeks. The diagnosis is established by serologic testing or culture, but these tests are not routinely available. Treatment with tetracycline, erythromycin, or newer macrolides is effective in ameliorating the signs and symptoms of *C pneumoniae* infection. Currently, there is ongoing scientific interest in the potential role of persistent infection by *C pneumoniae* in the pathogenesis of human vascular endothelial and intimal diseases, such as atherosclerosis.

CLINICAL CASE

AN UNANTICIPATED RESULT

A 29-year-old man presents with a 2-day history of burning on urination and a thin, watery urethral discharge. He had unprotected sex with a new female partner 4 weeks ago. A Gram stain reveals 50% polymorphonuclear (PMN) and 50% mononuclear leukocytes. No microorganisms are visible.

QUESTIONS

■ Which is the most likely cause of this man's urethritis?

A. *Neisseria gonorrhoeae*
B. *Ureaplasma urealoyticum*
C. *Chlamydia trachomatis*
D. *Trichomonas vaginalis*
E. *Mycoplasma hominis*

■ Which is the most sensitive test to detect the pathogen?

A. Culture
B. Serology
C. Immunofluorescent assay
D. Nucleic acid amplification assay

■ To which is the causative microbe susceptible?

A. Not susceptible to antibiotics
B. Most susceptible to β-lactam antibiotics
C. Resistant to quinolones
D. Susceptible to macrolides

ANSWERS

1(C), **2**(D), **3**(D)

CHAPTER 40

Rickettsia, Ehrlichia, Coxiella, and *Bartonella*

Rickettsia are Gram-negative, intracellular bacteria that have been recently classified into three phyla: *Rickettsia, Ehrlichia,* and *Coxiella*. They are the causes of spotted fevers, typhus, and similar illnesses; a related organism, the cause of scrub typhus, has been renamed *Orientia tsutsugamushi*. *Ehrlichia* are distinct from true rickettsiae and have four genera, of which *Ehrlichia* and *Anaplasma* are the most important in humans. *Coxiella burnetii* is the cause of Q fever. *Bartonella* are similar bacteria but are not members of the same taxonomic family. All are Gram-negative bacilli, and all but *Bartonella* are strict intracellular pathogens. The reservoir is animals, and, with the exception of *R prowazekii,* the cause of epidemic typhus, they are animal pathogens that infect humans only incidentally. Almost all are transmitted by arthropod vectors. The diseases are typically fevers, often with vasculitis. The most common infections are the various spotted fevers found throughout the world.

Obligate intracellular parasites

RICKETTSIA

BACTERIOLOGY

MORPHOLOGY AND STRUCTURE

Rickettsiae are small, intracellular coccobacilli that often have a transverse septum between two bacilli, reflecting division by binary fission. They commonly measure no more than 0.3 to 0.5 μm. Although the Gram reaction is negative, they take the usual bacterial stains poorly and are better demonstrated by specific immunofluorescence. The ultrastructural morphology, which is similar to that of other Gram-negative bacteria, includes a Gram-negative type of cell envelope, ribosomes, and a nuclear body. Chemically, the cell wall contains lipopolysaccharide and at least two large proteins in the outer membrane, as well as peptidoglycan. The outer membrane proteins extend to the cell surface, where they are the most abundant protein present.

Small, Gram-negative coccobacilli stained best by immunofluorescence

Abundant outer membrane proteins at surface

GROWTH AND METABOLISM

Rickettsia grow freely in the cytoplasm of eukaryotic cells to which they are highly adapted, in contrast to *Ehrlichia* and *Coxiella,* which replicate in cytoplasmic vacuoles. Rickettsiae can be grown only in living host cells, such as cell cultures and embryonated eggs. Infection of the host cell begins by induction of an endocytic process, which is analogous

Grow in cytoplasm following induced endocytosis

Growth slow compared with most bacteria

to phagocytosis, but requires expenditure of energy by the rickettsiae. Penetration of infected cells appears to be facilitated by production of a rickettsial phospholipase. The organisms then escape the phagosome or endocytic vacuole to enter the cytoplasm, possibly aided by elaboration of the phospholipase. Intracytoplasmic growth eventually produces lysis of the cell. The estimated generation time of rickettsiae is much longer than that of bacteria such as *Escherichia coli,* but more rapid than that of *Mycobacterium tuberculosis.*

Exogenous cofactors and ATP required for survival

Rapidly loses infectivity outside of host cell

The obligate intracellular parasitism of rickettsiae has several interesting features. Failure to survive outside the cell is apparently related to requirements for nucleotide cofactors (coenzyme A, nicotinamide adenine dinucleotide) and adenosine triphosphate (ATP). Outside the host cell, rickettsiae not only cease metabolic activity, but leak protein, nucleic acids, and essential small molecules. This instability leads to rapid loss of infectivity because the penetration of another cell requires energy. In summary, rickettsiae have the metabolic capabilities of other bacteria, but must borrow some essential elements from host cells for adequate growth and thus do not survive well in the environment.

RICKETTSIAL DISEASE

CLINICAL CAPSULE

The classic example of rickettsial disease is epidemic typhus, but the most important rickettsiosis year to year is Rocky Mountain spotted fever (RMSF). Both types of rickettsial disease are characterized by fever, rash, and myalgias/myositis. In RMSF, the rash appears first on the palms and soles, wrists, and ankles, and it migrates centripetally; in epidemic typhus, the rash begins on the trunk and spreads to the extremities, traveling in the opposite direction. Both diseases may be fatal as the result of severe vascular collapse. The vectors also differ; for RMSF, the vector is a tick, and for epidemic typhus, a louse.

EPIDEMIOLOGY AND PATHOGENESIS

Infect vascular endothelium with resultant vasculitis and thrombosis

Multiple vascular lesions, including adrenal glands

Most rickettsiae have animal reservoirs and are spread by insect vectors, which are prominent components of their life cycles (**Table 40–1**). Rickettsial infections of humans usually result in clinical illness. Rickettsiae have a tropism for vascular endothelium, and the primary pathologic lesion is a vasculitis in which rickettsiae multiply in the endothelial cells lining the small blood vessels (**Figure 40–1**). Focal areas of endothelial proliferation and perivascular infiltration leading to thrombosis and leakage of red blood cells into the surrounding tissues account for the rash and petechial lesions. Vascular lesions occur throughout the body, producing the systemic manifestations of the disease. They are obviously most apparent in skin but most serious in the adrenal glands. An endotoxin-like shock has been demonstrated in animals on injection of whole rickettsial cells, but the nature and role of any toxin in human disease are unknown.

DIAGNOSIS

Culture of rickettsiae is both difficult and hazardous. Their isolation in fertile eggs or cell cultures is generally attempted only in reference centers with special facilities and personnel experienced in handling the organisms. For this reason, serologic tests are the primary

TABLE 40–1 Examples of Pathogenic Rickettsiae

			ZOONOTIC CYCLE	
DISEASE	**ORGANISM**	**MOST COMMON GEOGRAPHIC DISTRIBUTION**	**VECTOR**	**RESERVOIR**
Spotted fever group: Rocky Mountain spotted fever	*Rickettsia rickettsii*	North, Central, and South America	Tick	Rodents, dogs
Rickettsialpox	*Rickettsia akari*	Worldwide	Mite	Mouse
Other spotted fevers	*Rickettsia conorii, R africae, R australis*, etc	Worldwide	Tick	Rodents, dogs
Typhus group epidemic	*Rickettsia prowazekii*	Worldwide	Body louse	Humans
Murine typhus	*Rickettsia typhi*	Worldwide	Flea	Rodents, esp. rats
Scrub typhus	*Orientia tsutsugamushi*	Far East, China, India	Mite larvae (chiggers)	
Q fever	*Coxiella burnetii*	Worldwide	None[a]	Sheep, goats, other hoofed animals, cats, dogs
Trench fever	*Bartonella quintana*	Worldwide	Body louse	Humans
Cat scratch fever	*Bartonella henselae*	Worldwide	Cat to cat by fleas	Cats
Oroya fever, verruga peruana	*Bartonella bacilliformis*	South America	Sandfly	
Human ehrlichiosis	*Ehrlichia chaffeensis*	United States	Tick	Deer,
	Anaplasma phagocytophila	United States, Europe, Asia	Tick	Deer
	E ewingii, Neorickettsia	Japan		Dogs
	Wohlbachia			Filarial worms

[a]Transmission by inhalation of infected aerosols

means of specific diagnosis. A number of test systems using specific rickettsial antigens have been developed, of which the indirect fluorescent antibody (IFA) method is generally the most sensitive and specific. This test is usually available only in reference laboratories. For rapid diagnosis, examination of biopsies such as skin lesions by immunofluorescence or immunoenzyme methods to detect antigens can be used. ::: IFA, p. 66

In vitro cultivation is hazardous

IFA method usually employed for serologic diagnosis

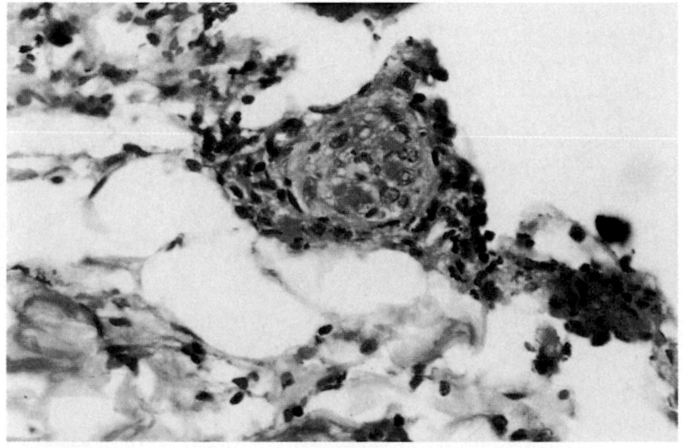

FIGURE 40–1. ***Rickettsia* vasculitis.** The endothelial cells of this small vessel in the skin are swollen and injured by infection with *Rickettsia typhi*. Note also the perivascular lymphocytic infiltrate. (Reproduced with permission from Connor DH, Chandler FW, Schwartz DQA, Manz HJ, Lack EE (eds). *Pathology of Infectious Diseases*, vol. 1. Stamford, CT: Appleton & Lange; 1997.)

SPOTTED FEVER GROUP

Many tick-borne rickettsioses occur throughout the world

The most important rickettsial disease in North America is RMSF, which is caused by *Rickettsia rickettsii*. A number of other spotted fever rickettsioses are found in other parts of the world (Table 40–1); the name often reveals the locale (eg, Mediterranean spotted fever, Marseilles fever). They are caused by *Rickettsia species,* serologically related to, but distinct from, *R rickettsi,* (eg, *R conorii, R africae, etc).* Another less severe spotted fever, rickettsialpox, also occurs in North America.

Rocky Mountain Spotted Fever

RMSF is an acute febrile illness that occurs in association with residential and recreational exposure to wooded areas where infected ticks exist. The disease has a significant mortality rate (25%) if untreated.

Epidemiology

Ticks naturally infected

Transovarial spread perpetuates tick infection

R rickettsii is primarily a parasite of ticks. In the western United States; the wood tick (*Dermacentor andersoni*) is the primary vector. In the East, the dog tick (*Dermacentor variabilis*) is the natural carrier and vector of the disease, and in the Southwest and Midwest, the vector is the Lone Star tick (*Amblyomma americanum*). Recently, another dog tick, *Rhipicephalus sanguineus,* has been implicated in cases occurring in rural eastern Arizona. *R rickettsii* does not kill its arthropod host, so the parasite is passed through unending generations of ticks by transovarial spread. Adult females require a blood meal to lay eggs and thus may transmit the disease. Infected adult ticks have been shown to survive as long as 4 years without feeding.

R rickettsii is found in North, Central, and South America. The United States has over 500 cases per year, and the highest attack rate is in the mid-Atlantic states (**Figure 40–2**).

Rocky mountain spotted fever. Number of reported cases, by county — United States, 2006

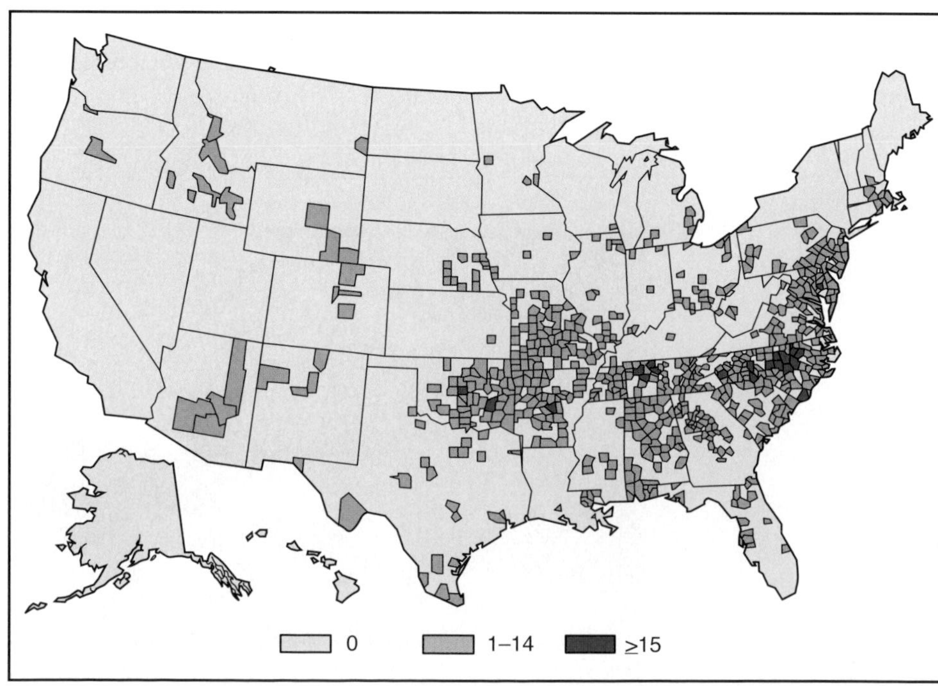

FIGURE 40–2. Distribution of Rocky Mountain spotted fever. Number and distribution of cases in the United States in 2006. (Reprinted with permission from Centers for Disease Control and Prevention. Summary of Notifiable Diseases, *MMWR* 2008;55(53):63.)

More than two thirds of cases are in children younger than 15 years. The illness is generally seen between April and September because of increased exposure to ticks. A history of tick bite can be elicited in approximately 70% of cases.

Most cases in children

Manifestations

The incubation period between the tick bite and the onset of illness is usually 6–7 days, but it may be from 2 days to 2 weeks. Fever, headache, rash, toxicity, mental confusion, and myalgia are the major clinical features. The rash is the most characteristic feature of the illness, but may not occur in up to one third of cases. Rash usually develops on the second or third day of illness as small erythematous macules that rapidly become petechial (**Figure 40–3**). The lesions appear initially on the wrists and ankles and then spread up the extremities to the trunk in a few hours. A diagnostic feature of RMSF is the frequent appearance of the rash on the palms and soles, a finding not usually seen in maculopapular eruptions associated with other infections, including typhus. Muscle tenderness, especially in the gastrocnemius, is characteristic and may be extreme. If untreated, or occasionally in patients despite therapy, complications such as disseminated intravascular coagulation, thrombocytopenia, encephalitis, vascular collapse, and renal and heart failure may ensue.

Incubation period 2 to 14 days after tick bite

Rash spreads from extremities to trunk and often involves palms and soles

Diagnosis

Because serologic testing is the primary diagnostic approach, it is often difficult to establish the diagnosis of RMSF early in the course of illness. However, antibodies may appear by the sixth or seventh day of illness, and a fourfold rise in antibody titer between acute serum and convalescent serum establishes the diagnosis. A skin lesion biopsy may be stained with specific immunofluorescent antibody to provide rapid confirmation of the diagnosis. Most often specific therapy must be started solely on the basis of clinical signs, symptoms, and epidemiologic considerations.

Rising antibody titers or DFA of skin biopsy confirm diagnosis

Prompt initiation of therapy based on clinical and epidemiologic considerations

Treatment

Appropriate antibiotic therapy is highly effective if given during the first week of illness. If delayed into the second week or when pathologic processes such as disseminated intravascular coagulation are present, therapy is much less effective. The antibiotic of choice is doxycycline. Sulfonamides may worsen the disease process and are thus contraindicated. Before specific therapy became available, the mortality rate associated with RMSF was approximately 25%. Treatment has reduced this figure to between 5% and 7%. Death results primarily in patients in whom diagnosis and therapy are delayed into the second week of illness.

Treatment during first week most effective

Doxycycline is the treatment of choice

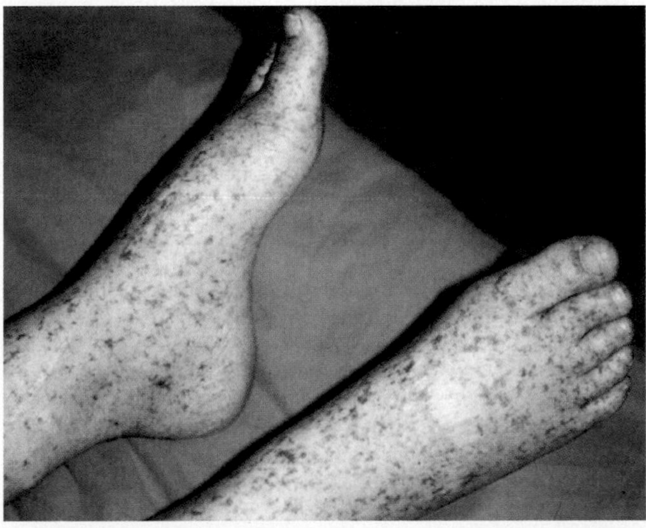

FIGURE 40–3. Rocky Mountain spotted fever. The rash begins on the arms and legs and spreads centrally. (Reproduced with permission from Nester EW, Anderson DG, Roberts CE Jr, Nester MT. *Microbiology: A Human Perspective,* 6th ed. New York: McGraw-Hill, 2008.)

Prevention

Frequent deticking, avoidance, and protective clothing important in prevention

The major means of preventing RMSF is avoidance or reduction of tick contact. Frequent deticking in tick-infested areas is important, because ticks generally must feed for 6 hours or longer before they can transmit the disease. Tick surveys in the Carolinas have shown infection in about 5% of samples. Killed vaccines prepared from infected ticks, or rickettsiae grown in embryonated eggs and cell cultures have been developed. None is licensed for clinical use at present.

Rickettsialpox

Benign disease transmitted by rodent mites

Rickettsialpox was first recognized in New York City in 1946 where an average of 5 cases per year continue to occur. It has been reported in other US cities and in eastern Europe, Korea, and South Africa. It is a benign rickettsial illness caused by *Rickettsia akari* and transmitted by a rodent mite. Distinguishing features of the disease include an eschar at the site of the bite and a vesicular rash. The house mouse and other semidomestic rodents are the primary reservoirs. Humans acquire infection when the mite seeks an alternative host.

Local eschar followed by fever and vesicular rash

Doxycycline therapy

Rickettsialpox is a biphasic illness. The first phase is the local lesion at the bite, which starts as a papulovesicle and develops into a black eschar in 3 to 5 days. Fever and constitutional symptoms appear as the organism disseminates. The second phase of the disease is a diffuse rash distributed randomly in the body, which, like the local lesion, becomes vesicular and develops into eschars. However, the rash does not occur in palms or soles. Rickettsialpox is self-limiting after 1 week, and no deaths have been reported. Doxycycline therapy shortens the course to 1 to 2 days.

TYPHUS GROUP

Epidemic Louse-Borne Typhus Fever

Severe louse-borne disease due to *R prowazekii*

Endemic foci in the homeless population

Primary louse-borne typhus fever is caused by *Rickettsia prowazekii,* which is transmitted to humans by the body louse. Historically, it has appeared during times of misery (war, famine) that create conditions favorable to human body lice (crowding, infrequent bathing). This is the only rickettsial disease that can occur as an epidemic. Foci of typhus persist in parts of Africa and Latin America, Russia, United States, and France. In the latter countries, the homeless population is a focus. Epidemic typhus has not been seen in the United States for more than half a century. *R prowazekii* has been recovered from flying squirrels and their ectoparasites in the southeastern United States, and a few human cases of sylvatic typhus have occurred in these areas.

Infection involves feeding and defecation by louse

The chain of epidemic typhus infection starts with *R prowazekii* circulating in a patient's blood during an acute febrile infection. The human body louse becomes infected during one of its frequent blood meals, and after 5 to 10 days of incubation, large numbers of rickettsiae appear in its feces. Since the louse defecates while it feeds, the organisms can be rubbed into the louse bite wounds when the host scratches the site. Dried louse feces are also infectious through the mucous membranes of the eye or respiratory tract. The louse dies of its infection in 1 to 3 weeks, and the rickettsiae are not transmitted transovarially.

Fever, headache, and rash with high mortality rate

Louse control is primary prevention

Fever, headache, and rash begin 1 to 2 weeks after the bite. A maculopapular rash occurs in 20% to 80% of patients and appears first on the trunk and then spreads centrifugally to the extremities, a pattern opposite to that of RMSF. Headache, malaise, and myalgia are prominent components of the illness. Complications include myocarditis and central nervous system dysfunction. In untreated disease, the fatality rate increases with age from 10% to as high as 60%. The diagnostic test of choice is serology, but therapy must be initiated immediately on clinical suspicion. Treatment with doxycycline is effective. Louse control is the best means of prevention and is particularly important in controlling epidemics. No effective vaccine is available.

Endemic (Murine) Typhus

Endemic or murine typhus is caused by *Rickettsia typhi* and transmitted to humans by the rat flea (*Xenopsylla cheopis*). Human illness is incidental to the natural transmission of the disease among urban rodents, which serve as the reservoir. The disease occurs worldwide but only 50 to 100 cases of murine typhus are reported in the United States each year. These typically occur along the Gulf Coast of Texas and in Southern California.

Transmitted by rat fleas

The pathogenesis is similar to that of louse-borne typhus, but the history includes exposure to rats, rat fleas, or both. The flea defecates when it takes a blood meal, and the infected feces gain access through the bite wound. After an incubation period of 1 to 2 weeks, illness begins with headache, myalgia, and fever. The rash is maculopapular, not petechial; it starts on the trunk and then spreads to the extremities in a manner similar to typhus. Because of antigens shared by *R typhi* and *R prowazekii,* serologic tests may not separate the two diseases. In the untreated patient, fever may last 12 to 14 days. With doxycycline therapy, the course is reduced to 2 to 3 days. Mortality and complications are rare, even if the disease is untreated.

Resembles typhus but less severe

R typhi shares antigens with *R prowazekii*

Scrub Typhus

Scrub typhus is found predominantly in the Far East, China, and India. The causative organism is *Orientia tsutsugamushi,* a rickettsial organism. Mites that infest rodents are the reservoir and vectors, transmitting the rickettsiae to their own progeny via infected ova. Humans pick up the mites as they pass by low trees or brush. The mite larvae (chiggers) deposit rickettsiae as they feed.

Scrub typhus transmitted by rodent mite larvae (chiggers)

The typical initial lesion, a necrotic eschar at the site of the bite on the extremities, develops in only 50% to 80% of cases. Fever increases slowly over the first week, sometimes reaching 40.5°C. Headache, rash, and generalized lymphadenopathy follow later.

Local eschar followed by fever, headache, rash, and lymphadenopathy

The maculopapular rash, which appears after about 5 days, is more evanescent than that seen with louse-borne or murine typhus. Hepatosplenomegaly and conjunctivitis may also appear. Specific diagnosis requires demonstration of a serologic response using the IFA test or polymerase chain reaction (PCR) on blood or biopsy. The prognosis is good with doxycycline therapy, but the mortality rate of untreated patients is as high as 30%.

Serologic diagnosis by IFA

COXIELLA

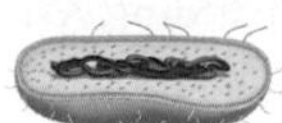

BACTERIOLOGY

Coxiella burnetii, a Gram negative bacterium and the cause of **Q fever,** has morphologic features similar to those of rickettsiae, but differs in DNA composition and a number of other features. Phase variation of surface polysaccharide in response to environmental conditions has been observed and linked to virulence. The organism is taken into host cells by a phagocytic process that in contrast to rickettsiae does not involve expenditure of energy by the parasite. It multiplies in the phagolysosome primarily because it is adapted to growth at low pH and resists lysosomal enzymes. *C burnetii* is much more resistant to drying and other environmental conditions than rickettsiae, which substantially accounts for its ability to produce infection by the respiratory route. ::: Phagolysosome, p. 25

Multiplies in phagolysosome

Resistant to drying

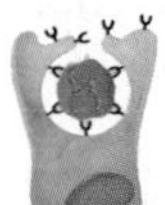

COXIELLA INFECTION: Q FEVER

Q fever is primarily a zoonosis transmitted from animals to humans by inhalation rather than by arthropod bite. Its distribution is worldwide among a wide range of mammals, of which cattle, sheep, and goats are most associated with transmission to humans. *C burnetii* grows particularly well in placental tissue, attaining huge numbers (more than 10^{10} per gram), which at the time of parturition contaminate the soil and fomites, where it may survive for years. Q fever occurs in those who are exposed to infected animals or their products, particularly farmers, veterinarians, and workers involved in slaughtering. Another high-risk environment is animal research facilities that have not provided adequate protection for personnel.

Transmission usually by inhalation; occasionally by ingestion

Occupational exposure in abattoirs and research facilities

Infection in all of these circumstances is believed to result from inhalation, which may be at some distance from the site of generation of the infectious aerosols. Infection can also occur from ingestion of animal products such as unpasteurized milk.

Q FEVER: CLINICAL ASPECTS

Systemic infection without rash

Pneumonia and endocarditis may occur

C burnetii has an affinity for the reticuloendothelial system, but little is known of the pathology, because fatal cases are rare. As in livestock, most human infections are inapparent. When clinically evident, Q fever usually begins an average of 20 days after inhalation, with abrupt onset of fever, chills, and headache. A mild, dry, hacking cough and patchy interstitial pneumonia may or may not be present. There is no rash. Hepatosplenomegaly and abnormal liver function tests are common. Complications such as myocarditis, pericarditis, and encephalitis are rare. Chronic infection is also rare, but particularly important when it takes the form of endocarditis. There is evidence that the strains associated with endocarditis constitute an antigenic subgroup of *C burnetii.*

Diagnosis is serologic

Diagnosis of Q fever is usually made by demonstrating high or rising titers of antibody to Q fever antigen by complement fixation, IFA, or enzyme immunoassay procedures. Although most infections resolve spontaneously, doxycycline therapy is believed to shorten the duration of fever and reduce the risk of chronic infection. Vaccines have been shown to stimulate antibodies, and some studies have suggested a protective effect for heavily exposed workers.

EHRLICHIA

Tick-borne and WBC-associated

The *Ehrlichia* genus includes several species of white blood cell (WBC)–associated bacteria that cause human disease. In the United States, two genera are the principal causes of two distinct diseases: (1) human monocytic ehrlichiosis (HME), which is due to *Ehrlichia chaffeensis;* and human granulocytic anaplasmosis (HGA), which is due to *Anaplasma phagocytophilum. E chaffeensis* infections tend to occur in the southeastern and lower midwestern United States, whereas HGA tends to cluster in the northern states, with a distribution similar to Lyme disease. It has also reported from other areas of the world, including Asia and Europe. HGA is the predominant form of ehrlichiosis and is second only to Lyme disease as a tick-borne infection in the United States. HME is transmitted by deer ticks, and the white-tailed deer is the animal reservoir. HGA is transmitted by *Ixodes* ticks, as is Lyme disease, and the animal reservoir is small mammals (eg mice, rats, voles). The findings are clinically similar to RMSF, but rashes are less commonly seen. Other ehrlichieae are shown in Table 40–1. ::: Lyme disease, p. 660

Intracytoplasmic inclusions (morulae) in monocytes or granulocytes

Treatment is doxycycline

On occasion, the diagnosis of ehrlichiosis may be suggested by observation of characteristic ehrlichial intracytoplasmic inclusions (morulae) in granulocytes (HGA) or mononuclear cells (HME) (**Figure 40–4**). The diagnosis is usually made serologically by a fourfold or greater rise in IFA antibody or a titer greater than or equal to 1:64 to the specific antigen. These tests require the assistance of specialized laboratories. Another diagnostic test for detection of ehrlichia DNA is PCR. Laboratory clues to human ehrlichiosis include a falling leukocyte count, thrombocytopenia, anemia, and impaired liver and renal function. Doxycycline is the drug of choice for ehrlichiosis. The risk of infection can be reduced by avoiding wooded areas and tick bites.

BARTONELLA

Bartonella species differ from rickettsiae in that they can be cultured on artificial media. By 16s ribosomal comparison, they are actually more closely related to *Brucella* than to rickettsiae. *Bartonella quintana,* the best-known species of this genus, causes **trench fever,** which

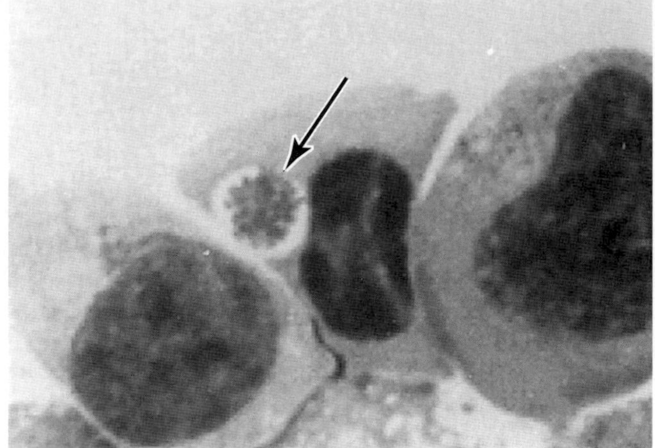

FIGURE 40–4. *Erlichia* inclusions. Mononuclear cell in the cerebrospinal fluid containing *Ehrlichia* intracytoplasmic inclusions (*arrow*). (Reprinted with permission from Dunn BE, Monson TP, Dumler JS, et al. Identification of *Ehrlichia chaffeensis morulae* in cerebrospinal fluid mononuclear cells. *J Clin Microbiol* 1992;30:2207–2210.)

has a worldwide distribution. The name derives from its prominence in the trenches of World War I. This disease has a reservoir in humans, and its vector is the body louse. Most cases are mild or subclinical. When symptomatic, the patient has sudden onset of chills, headache, relapsing fever, and a maculopapular rash on the trunk and abdomen. Illness can last for 4 to 5 days, can recur in repeated 4- to 5-day bouts, or can persist uninterruptedly for up to 6 weeks. The disease is suggested by a history of louse contact. More recently, *B quintana* bacteremia and endocarditis have been described in homeless alcoholic men in both France and the United States. The diagnosis can be made by culturing the organism on special agar medium or by demonstrating seroconversion.

B quintana causes trench fever; it is also associated with alcoholism

B bacilliformis, a related organism, is the cause of acute Oroya fever and, in its chronic phase, verruga peruana. Infections with this agent are seen only in South America at intermediate altitudes, in keeping with the distribution of its sandfly vector.

Another species, *B henselae,* has been associated with a number of diseases, the most common of which is **catscratch disease.** Catscratch disease is a febrile lymphadenitis with systemic symptomatology that sometimes persists for weeks to months. Approximately 25,000 cases occur in the United States each year. The disease is thought to be transmitted by cat scratches or bites and perhaps by the bites of cat fleas. Manifestations may include skin rashes, conjunctivitis, encephalitis, and prolonged fever. Occasionally, retinitis, endocarditis, and granulomatous or suppurative hepatosplenic and osseous lesions have also been seen. *B henselae* has been isolated directly from the blood of cats, although the latter do not appear ill. It can also be isolated from human blood, lymph nodes, and other materials using special media. Organisms can sometimes be directly demonstrated in infected tissues by using the Warthin–Starry silver impregnation stain. A serologic response to *B henselae* antigens is the primary method of diagnosis. Azithromycin or erythromycin may reduce the duration of lymph node enlargement and symptoms.

Catscratch disease is common in children

Persistent lymphadenitis is the usual finding

Bacillary angiomatosis, a proliferative disease of small blood vessels of the skin and viscera, seen in acquired immunodeficiency syndrome (AIDS) patients and other immunocompromised hosts, has been associated with *Bartonella* by molecular methods. PCR was used to amplify ribosomal RNA gene fragments directly from tissue samples. Subsequently, both *B henselae* and *B quintana* have been cultured from AIDS patients with bacillary angiomatosis. Other conditions seen primarily in AIDS patients, such as peliosis hepatis and bacteremia with fever, have also been associated with *B henselae*. *Bartonella* infections in AIDS and other immunosuppressed patients, as well as the bacteremia observed in alcoholic and homeless men, generally respond to prolonged courses of erythromycin or doxycycline. *Bartonella* endocarditis usually requires valve replacement as well. ::: PCR, p. 83

AIDS and other immunocompromised states are associated with more severe, protracted infections

CLINICAL CASE

FEVER AND RASH FOLLOWING TICK BITE

A 6-year-old girl from North Carolina was in her usual state of good health until 10 days before admission, when she had a tick removed from her scalp. She developed a sore throat, malaise, and a low-grade fever 8 days after tick removal. She was seen by her pediatrician when she began developing a pink, macular rash, which started on her palms and lower extremities and spread to cover her entire body. The pediatrician's diagnosis was viral exanthem. One day before admission, she developed purpura, emesis, diarrhea, myalgias, and increased fever. On the day of admission, she was taken to her local hospital emergency room because of mental status changes. Her physical examination was significant for diffuse purpura; periorbital, hand and foot edema, cool extremities with weak pulses, and hepatosplenomegaly. Her laboratory studies revealed: Na^+ level of 125 mmol/L, platelet count 26,000/mm^3 WBC count 14,900/mm^3 hemoglobin level of 8.8 g/L, and greatly increased coagulation times. Ampicillin therapy was begun, and she was intubated but died soon after transfer to another institution.

QUESTIONS

■ What feature in this patient's history is most helpful?

A. Sore throat
B. Rash
C. Tick bite
D. Diarrhea
E. Leukocytosis

■ To confirm a diagnosis of Rocky Mountain spotted fever, what would be the most useful laboratory test?

A. Culture
B. Gram stain
C. Serology
D. Darkfield examination

■ The primary cause of the fatal outcome in this patient is the tropism of *Rickettsia* for:

A. Skin
B. WBCs
C. Enterocytes
D. Muscle
E. Blood vessels

ANSWERS

1(C), **2**(C), **3**(E)

PART IV

Pathogenic Fungi

Kenneth J. Ryan

The Nature of Fungi CHAPTER 41

Pathogenesis of Fungal Infection CHAPTER 42

Antifungal Agents and Resistance CHAPTER 43

Dermatophytes, *Sporothrix,* and Other Superficial and Subcutaneous Fungi CHAPTER 44

Candida, Aspergillus, Pneumocystis, and Other Opportunistic Fungi CHAPTER 45

Cryptococcus, Histoplasma, Coccidioides, and Other Systemic Fungal Pathogens CHAPTER 46

CHAPTER 41

The Nature of Fungi

Fungi are a distinct class of microorganisms, most of which are free-living in nature where they function as decomposers in the energy cycle. Of the more than 200,000 known species, fewer than 200 have been reported to produce disease in humans. These diseases, the mycoses, have unique clinical and microbiologic features and are increasing in immunocompromised patients.

MYCOLOGY

Fungi are eukaryotes with a higher level of biologic complexity than bacteria. They may be unicellular or may differentiate and become multicellular by the development of branching filaments. They reproduce sexually or asexually. The mycoses vary greatly in their manifestations but tend to be subacute to chronic with indolent, relapsing features. Acute disease, such as that produced by many viruses and bacteria, is uncommon with fungal infections.

Cell organization is eukaryotic

STRUCTURE

The fungal cell has typical eukaryotic features, including a nucleus with a nucleolus, nuclear membrane, and linear chromosomes (**Figure 41–1**). The cytoplasm contains a cytoskeleton with actin microfilaments and tubulin-containing microtubules. Ribosomes and organelles, such as mitochondria, endoplasmic reticulum, and the Golgi apparatus, are also present. Fungal cells have a rigid cell wall external to the cytoplasmic membrane, which differs in its chemical makeup from that of bacteria and plants. An important difference from mammalian cells is the sterol makeup of the cytoplasmic membrane. In fungi, the dominant sterol is ergosterol; in mammalian cells, it is cholesterol. Fungi are usually in the haploid state, although diploid nuclei are formed through nuclear fusion in the process of sexual reproduction.

Presence of a nucleus, mitochondria, and endoplasmic reticulum

Ergosterol, not cholesterol, makes up cell membrane

The chemical structure of the cell wall in fungi is markedly different from that of bacterial cells in that it does not contain peptidoglycan, glycerol or ribitol teichoic acids, or lipopolysaccharide. In their place are the polysaccharides **mannan, glucan,** and **chitin** in close association with each other and with structural proteins (**Figure 41–2**). Mannoproteins are mannose-based polymers (mannan) found on the surface and in the structural matrix of the cell wall, where they are linked to protein. They are major determinants of serologic specificity because of variations in the composition and linkages of the polymer side chains. Glucans are glucosyl polymers, some of which form fibrils that increase the strength of the fungal cell wall, often in close association with chitin. Chitin is composed of long, unbranched chains of poly-*N*-acetylglucosamine. It is inert, insoluble, and rigid and provides structural support in a manner analogous to the chitin in crab shells or cellulose in plants. It is a major component of the cell wall of filamentous fungi. In yeasts, chitin appears

Cell wall mannan linked to surface proteins

Chitin and glucans give rigidity to cell wall

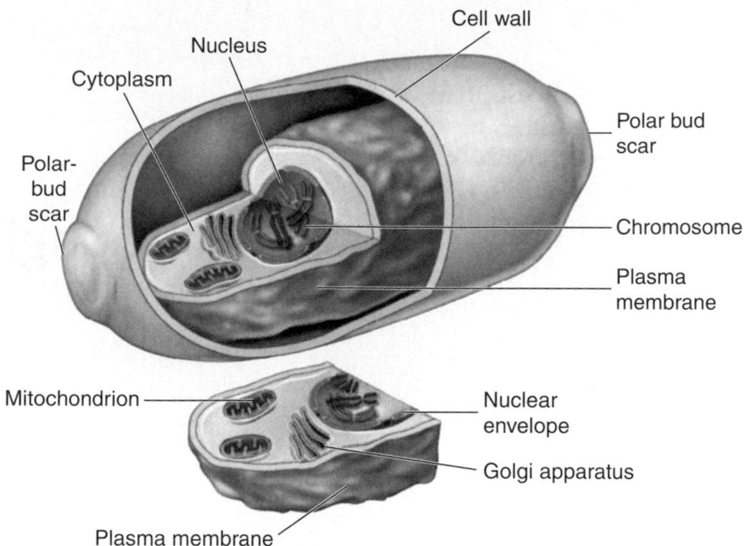

FIGURE 41–1. A yeast cell showing the cell wall and internal structures of the fungal eukaryotic cell plan. (Reproduced with permission from Willey J, Sherwood L, Woolverton C (eds). *Prescott's Principles of Microbiology.* New York: McGraw-Hill; 2008.)

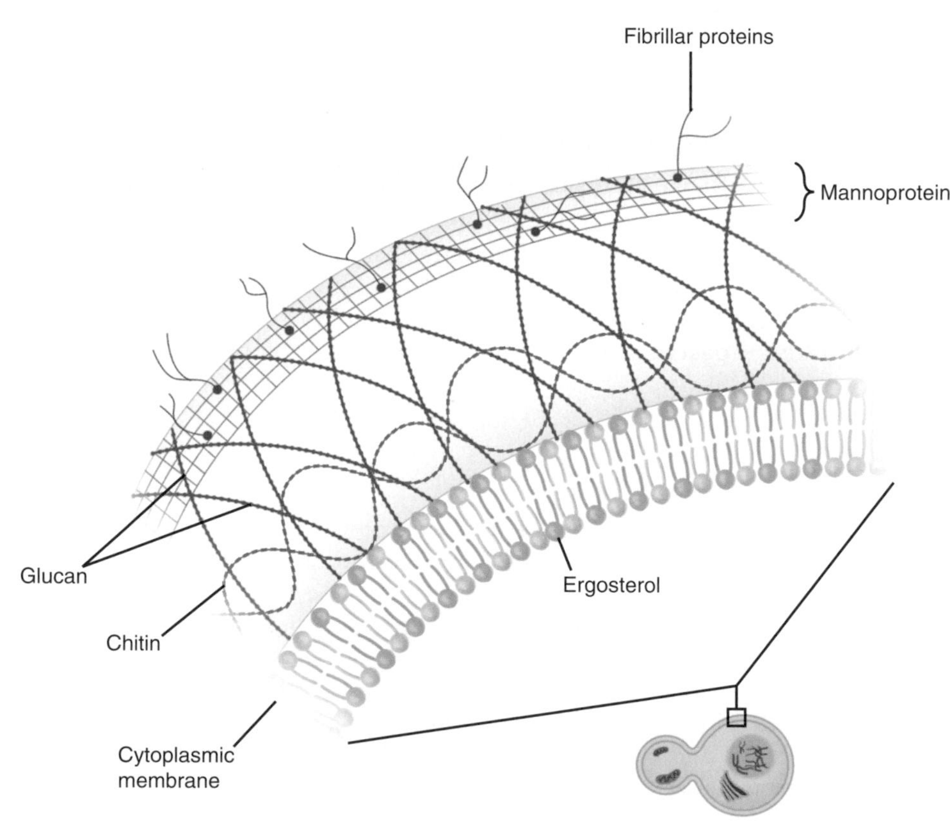

FIGURE 41–2. The fungal cell wall. The overlapping mannan, glucan, chitin, and protein elements are shown. Proteins complexed with the mannan (mannoproteins) extend beyond the cell wall.

to be of most importance in forming cross-septa and the channels through which nuclei pass from mother to daughter cells during cell division.

METABOLISM

Heterotrophic metabolism uses available organic matter

Fungal metabolism is heterotrophic, degrading organic substrates as an exogenous source of carbon. Metabolic diversity is great, but most fungi grow with only an organic carbon source and ammonium or nitrate ions as a nitrogen source. In nature, nutrients for free-living fungi are derived from decaying organic matter. A major difference between fungi and plants is that fungi lack chloroplasts and photosynthetic energy-producing mechanisms. Most are strict aerobes, although some can grow under anaerobic conditions. Only a few are anaerobes, none of which are human pathogens.

Photosynthetic mechanisms are lacking

REPRODUCTION

Fungi may reproduce by either asexual or sexual process. The asexual form is called the anamorph, and its reproductive elements are termed **conidia.** The sexual form is called the teleomorph, and its reproductive structures are called **spores** (eg, ascospores, zygospores, basidiospores). Asexual reproduction involves mitotic division of the haploid nucleus and is associated with production by budding spore-like conidia or separation of hyphal elements. In sexual reproduction, the haploid nuclei of donor and recipient cells fuse to form a diploid nucleus, which then divides by classic meiosis. Some of the four resulting haploid nuclei may be genetic recombinants and may undergo further division by mitosis. Highly complex specialized structures may be involved. Detailed study of this process in fungal species such as *Neurospora crassa* has been important in gaining an understanding of basic cellular genetic mechanisms.

Asexual reproduction forms conidia by mitosis

Meiosis forms sexual spores in specialized structures

FUNGAL MORPHOLOGY AND GROWTH

The size of fungi varies immensely. A single cell without transverse septa may range from bacterial size (2 to 4 μm) to a macroscopically visible structure. The morphologic forms of growth vary from colonies superficially resembling those of bacteria to some of the most complex, multicellular, colorful, and beautiful structures seen in nature. Mushrooms are an example and can be regarded as a complex organization of cells showing structural differentiation.

Vary from bacterial size to multicellular mushrooms

Mycology, the science devoted to the study of fungi, has many terms to describe the morphologic components that make up these structures. The terms and concepts that must be mastered can be limited by considering only the fungi of medical importance and accepting some simplification.

YEASTS AND MOLDS

Initial growth from a single cell may follow either of two courses, yeast or mold (**Figure 41–3A** and **B**). The first and simplest is the formation of a bud, which extends out from a round or oblong parent, constricts, and forms a new cell, which separates from the parent. These buds are called **blastoconidia**, and fungi that reproduce in this manner are called **yeasts.** On plates, yeasts form colonies that resemble those of bacteria. In fluids they are much more portable than molds due to the retention of a single cell nature.

Yeasts produce blastoconidia by budding

Fungi may also grow through the development of **hyphae** (singular, hypha), which are tube-like extensions of the cell with thick, parallel walls. As the hyphae extend, they form an intertwined mass called a **mycelium.** Most fungi form hyphal **septa** (singular, septum), which are cross-walls perpendicular to the cell walls that divide the hypha into subunits (**Figure 41–4**). These septa vary among species and may contain pores and incomplete walls that allow movement of nutrients, organelles, and nuclei. Some species are nonseptate; they form hyphae and mycelia as a single, continuous cell. In both septate and nonseptate hyphae, multiple nuclei are present, with free flow of cytoplasm along the hyphae or through pores in any septum. A portion of the mycelium (vegetative mycelium) usually grows into the medium or organic substrate (eg, soil) and functions like the roots of plants as a collector of nutrients and moisture. The more visible surface growth may assume a fluffy character as the mycelium becomes aerial. The hyphal walls are rigid enough to support this extensive, intertwining network, commonly called a **mold.** The aerial hyphae bear the reproductive structures of this class of fungi. Some fungi form structures called **pseudohyphae**, which differ from true hyphae in having recurring bud-like constrictions and less rigid cell walls.

Molds produce septate or nonseptate hyphae

Vegetative mycelium acts as a root

Aerial mycelium bears reproductive conidia or spores

Pseudohyphae are less rigid

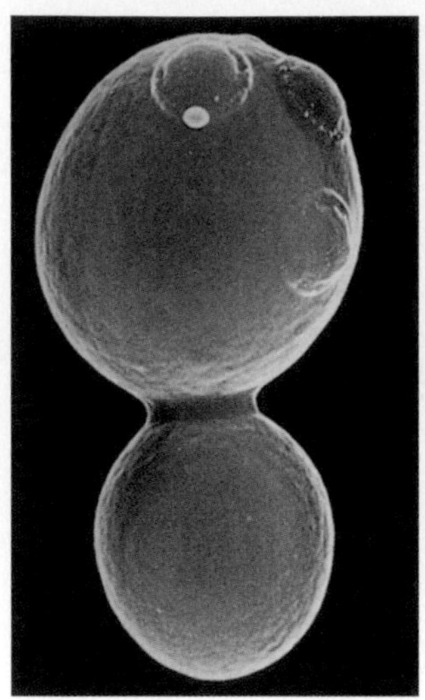

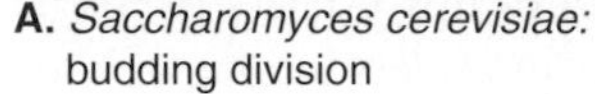

FIGURE 41–3. Yeast and mold forms of fungal growth. A. This oval yeast cell is budding to form a blastoconidium. Scars from the separation of other blastoconidia can be seen on other parts of the cell. **B.** The mold form is highly variable. Here tubular stalks called condiophores arising from hyphae (not seen) bear a "Medusa head" crop of reproductive conidia. (Reproduced with permission from Willey J, Sherwood L, Woolverton C (eds). *Prescott's Principles of Microbiology*. New York: McGraw-Hill; 2008.)

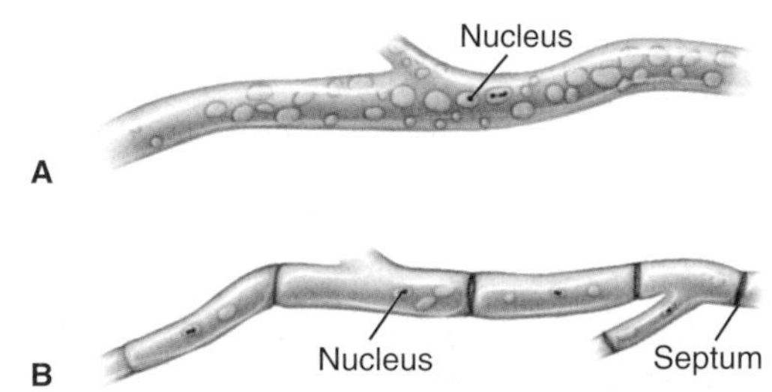

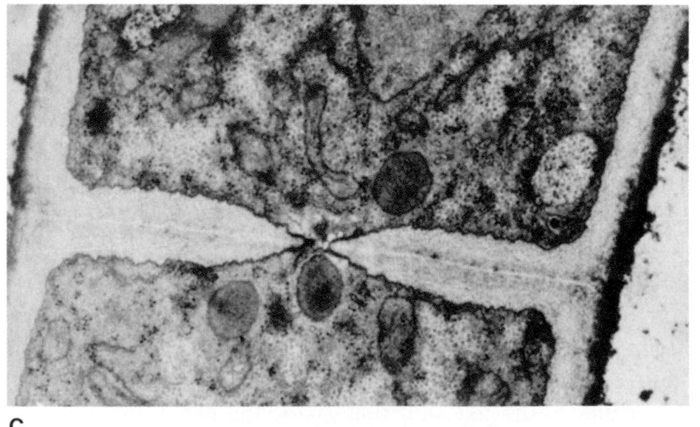

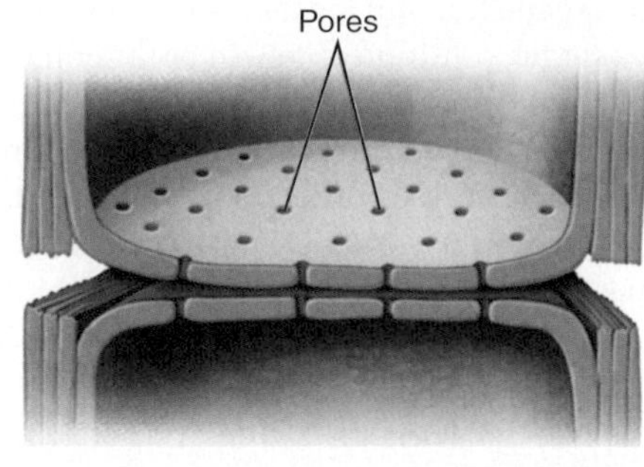

FIGURE 41–4. Hyphae. A. Nonseptate hyphae with multiple nuclei. **B.** Septate hyphae divide nuclei into separate cells. **C.** Electron micrograph of septum with a single pore. **D.** Multipore septum structure. (Reproduced with permission from Willey J, Sherwood L, Woolverton C (eds). *Prescott's Principles of Microbiology*. New York: McGraw-Hill; 2008.)

The reproductive conidia and spores of the molds and the structures that bear them assume a great variety of sizes, shapes, and relationships to the parent hyphae, and the morphology and development of these structures are the primary basis of identification of medically important molds. The mycelial structure plays some role in identification, depending on whether the hyphae are septate or nonseptate, but differences are not sufficiently distinctive to identify or even suggest a fungal genus or species.

Morphology of reproductive conidia and spores used for identification

Exogenously formed conidia may arise directly from the hyphae or on a special stalk-like structure, the **conidiophore** (Figure 41–3B). Occasionally, terms such as **macroconidia** and **microconidia** are used to indicate the size and complexity of these conidia. Conidia that develop within the hyphae are called either **chlamydoconidia** or **arthroconidia.** Chlamydoconidia become larger than the hypha itself; they are round, thick-walled structures that may be borne on the terminal end of the hypha or along its course. Arthroconidia conform more to the shape and size of the hyphal units but are thickened or otherwise differentiated. Arthroconidia may form a series of delicately attached conidia that break off and disseminate when disturbed. Some of the asexual reproductive forms are illustrated in **Figure 41–5A–D**.

Conidia and conidiophore arrangements determine names

Ascospores are borne in ascus sac

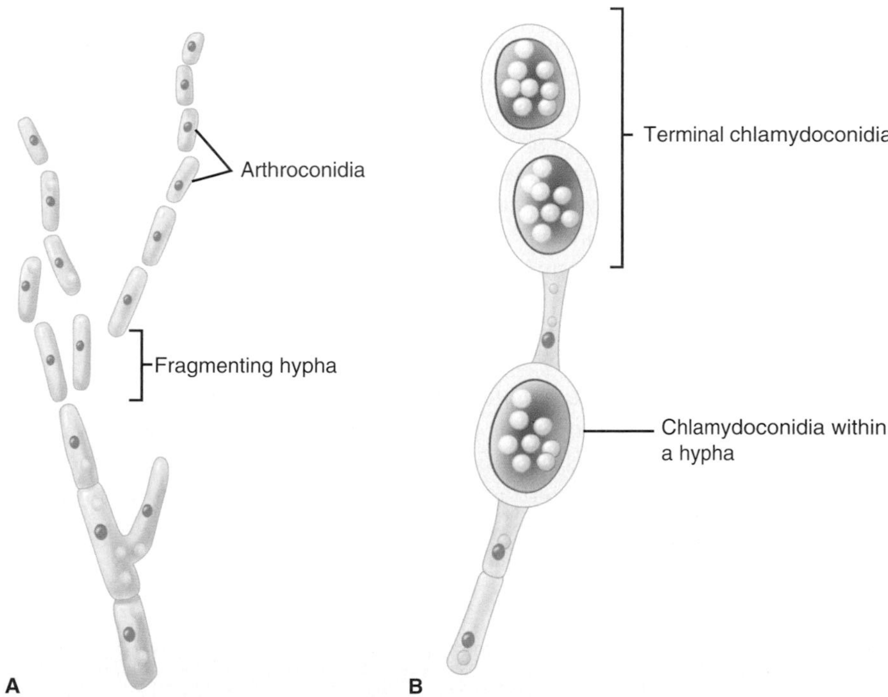

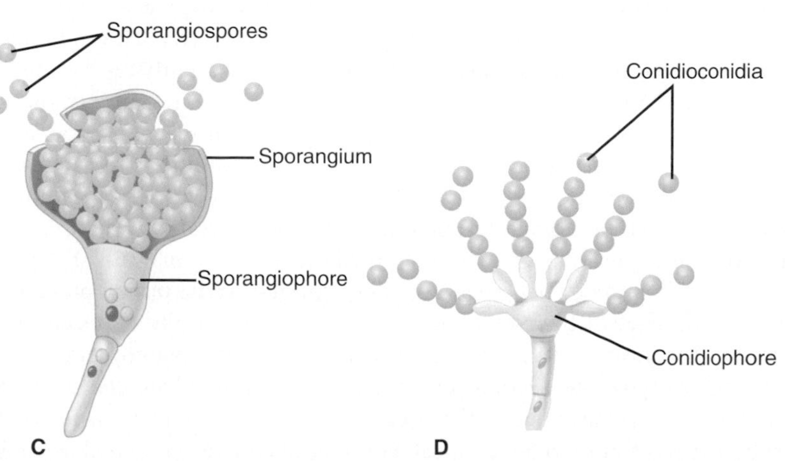

FIGURE 41–5. Asexual mold forms. A. Arthroconidia develop within the hyphae and eventually break off. **B.** Chlamydoconidia are larger than the hyphae and develop with the cell or terminally. **C.** Sporangiocondia are borne terminally in a sporangium sac. **D.** Simple conidia arise directly from a conidiophore (Reproduced with permission from Willey J, Sherwood L, Woolverton C (eds). *Prescott's Principles of Microbiology*. New York: McGraw-Hill; 2008.)

The most common sexual spore is termed an ascospore. Four or eight ascospores may be found in a sac-like structure, the ascus.

DIMORPHISM

Grow in yeast or mold form

Temperature triggers shift between phases

In general, fungi grow either as yeasts or as molds; mold forms show the greatest diversity. Some species can grow in either a yeast or a mold phase, depending on environmental conditions. These species are known as **dimorphic fungi.** Several human pathogens demonstrate dimorphism; they grow in the mold form in their environmental reservoir and in culture at ambient temperatures but convert to the yeast or some other form in infected tissue. For most, it is possible to manipulate the cultural conditions to demonstrate both yeast and mold phases in vitro. Yeast phase growth requires conditions similar to those of the physiologic in vivo environment, such as 35° to 37°C incubation and enriched medium. Mold growth requires minimal nutrients and ambient temperatures. The conidia produced in the mold phase may be infectious and serve to disseminate the fungus.

Shift from mold to yeast begins with heat shock response

Metabolic shift is toward sulfhydryl compounds in yeast form

Global regulator controls process

The morphologic and physiologic events associated with conversion from the mold to the yeast phase have been most extensively studied in the human pathogen *Histoplasma capsulatum*. They are understandably complex, given the dramatic change of milieu encountered by the fungus when its mold conidia float from their soil habitat to the pulmonary alveoli. Conversion to the yeast phase is then triggered by the host temperature (37°C) and possibly by other aspects of the new environment. In vitro studies show that the earliest events in this shift from the mold to yeast form involve induction of the **heat shock response** and uncoupling of oxidative phosphorylation. These are followed by a shutdown of RNA synthesis, protein synthesis, and respiratory metabolism. The cells then pass through a metabolically inactive state, emerging with enhanced enzymatic capacities involving sulfhydryl compounds (eg, cysteine, cystine) that are exclusive to the yeast stage. In the yeast stage, there is recovery of mitochondrial activity and synthetic capacity, but a new constellation of oxidases, polymerases, proteins, cell wall glucans, and other compounds are present. In all, more than 500 genes are differentially expressed in the mold and yeast phases. A global regulating gene controls mold to yeast process as well as the expression of some virulence genes. ::: Heat shock response, p. 371

Dimorphism is reversible and linked to virulence

Dimorphism in fungi is reversible, a feature that distinguishes it from developmental processes such as embryogenesis seen in higher eukaryotes. The importance of the conversion to virulence of *Histoplasma* is shown by animal studies using strains biochemically blocked from converting to the yeast phase. They neither produce disease nor persist in the host. To the extent known, these features are similar in the other dimorphic fungi. ::: Histoplasma dimorphism, p. 743

CLASSIFICATION

Taxonomy is based on sexual spores and septation of hyphae

Asexual form is unknown for most pathogens

rRNA genes are used for classification

Although conidia are more readily observed, the major classification of fungi primarily depends on the nature of the teleomorph spores and septation of hyphae as its differential characteristics. On this basis, fungi have been organized into four phyla: Chlytridiomycota, Zygomycota, Ascomycota, and Basidiomycota. A major issue in classifying the medically important fungi using these groups is that for most species, no teleomorph form has been demonstrated. It may be that because of an advantage for pathogenicity, persistently anamorph clones were selected during evolution, or that the conditions for natural production of spores are just yet to be met. One approach has been to park these fungi in their own artificial class (Deuteromycetes, or fungi imperfecti) awaiting the discovery of their teleomorph. The application of molecular methods such as analysis of ribosomal RNA genes has made this somewhat irrelevant, since species can be placed on genotypic grounds without knowledge of their reproductive forms. The medically important genera fall mostly into the Ascomycota, with a few in Basidiomycota, and Zygomycota, as shown in **Table 41–1**. Discovery of the teleomorph may not bring immediate clarity from the student's standpoint; for instance, when the sexual stage of *Trichophyton mentagrophytes* was demonstrated, it was found to be identical with that of an already named ascomycete (*Arthroderma benhamiae*).

TABLE 41–1 Classification of Medically Important Fungi

GENUS	TYPICAL GROWTH	SEPTATION[a]	SEXUAL FORM	PHYLUM	MEDICAL CLASSIFICATION
Aspergillus	Mold	+	?	Ascomycota	Opportunistic
Blastomyces	Dimorphic	+	?	Ascomycota	Systemic
Candida	Dimorphic	+	?	Ascomycota	Opportunistic
Coccidioides	Dimorphic	+	?	Ascomycota	Systemic
Cryptococcus	Yeast		+	Basidiomycota	Systemic
Epidermophyton	Mold	+	+	Ascomycota	Superficial
Histoplasma	Dimorphic	+	+	Ascomycota	Systemic
Microsporum	Mold	+	+	Ascomycota	Superficial
Mucor	Mold	–	+	Zygomycota	Opportunistic
Pneumocystis	Cysts[b]		?	Ascomycota	Opportunistic
Rhizopus	Mold	–	+	Zygomycota	Opportunistic
Sporothrix	Dimorphic	+	?	Ascomycota	Subcutaneous
Trichophyton	Mold	+	+	Ascomycota	Superficial

[a]For those that form hyphae.
[b]Tissue forms but does not grow in culture.

Medical grouping organized by biologic behavior in humans

Systemic fungi infect previously healthy persons

The grouping of medically important fungi used in the following chapters is based on the types of tissues they parasitize and the diseases they produce, rather than on the principles of basic mycologic taxonomy. The **superficial** fungi, such as the dermatophytes, cause indolent lesions of the skin and its appendages, commonly known as ringworm and athlete's foot. The **subcutaneous** pathogens characteristically cause infection through the skin, followed by subcutaneous spread, lymphatic spread, or both. The **opportunistic** fungi are those found in the environment or in the normal flora that produce disease under certain circumstances and in the compromised host. The **systemic** pathogens are the most virulent fungi and may cause serious progressive systemic disease in previously healthy persons. They are not found in the normal human flora. Although their major potential is to produce deep-seated visceral infections and systemic spread (systemic mycoses), they may also produce superficial infections as part of their disease spectrum or as the initiating event. The superficial mycoses do not spread to deeper tissues. As with all clinical classifications, overlaps and exceptions occur. In the end, the organism defines the disease, and it must be isolated or otherwise demonstrated.

LABORATORY DIAGNOSIS

Direct Examination

KOH digests tissue, but not fungal wall

Some yeasts are Gram-positive

Calcifluor white enhances detection

Because of their large size, fungi often demonstrate distinctive morphologic features on direct microscopic examination of infected pus, fluids, or tissues. The simplest method is to mix the specimen with a 10% solution of potassium hydroxide (KOH) and place it under a coverslip. The strong alkali digests the tissue elements (epithelial cells, leukocytes, debris), but not the rigid cell walls of either yeasts or molds. After digestion of the material, the fungi can be observed under the light microscope with or without staining (**Figure 41–6**). Direct examinations can be aided by the use of calcifluor white, a dye that binds to polysaccharides in cellulose and chitin. Under ultraviolet light, calcifluor white fluoresces, enhancing detection of fungi in fluids or tissue sections. A few yeasts take the Gram stain, including *Candida albicans* (Gram-positive).

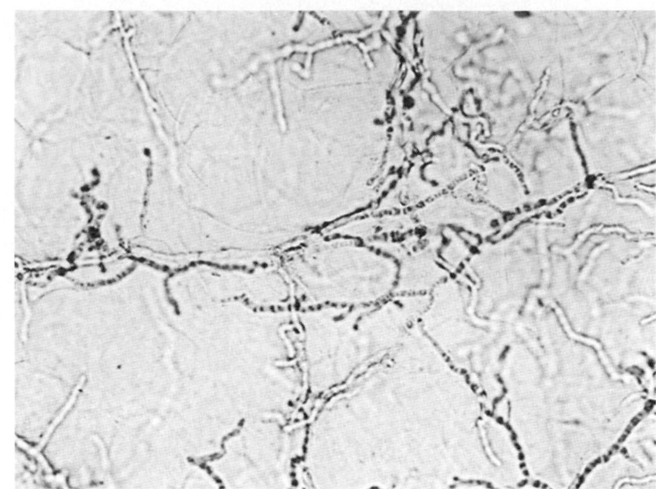

FIGURE 41–6. KOH (potassium hydroxide) preparation. Scalp scrapings from a suspected ringworm lesion have been mixed with 10% KOH and viewed under low power. The skin has been dissolved, revealing tubular branching hyphae.

Often visible in H&E preparations

Silver stains enhance detection

Histopathologic examination of tissue biopsy specimens is widely used and shows the relation of the organism to tissue elements and responses (blood vessels, phagocytes, granulomatous reactions). Most fungi can be seen in sections stained with the basic hematoxylin and eosin (H&E) method used in histology laboratories (**Figures 41–7**). Specialized staining procedures such as the silver impregnation methods are frequently used because they stain almost all fungi strongly, but only a few tissue components (**Figure 41–8**). The

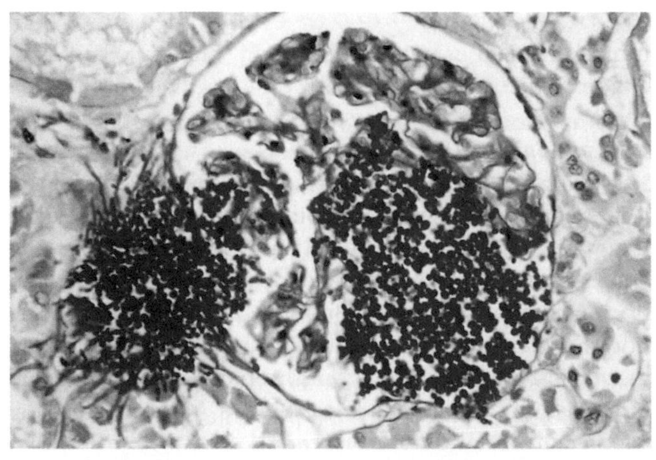

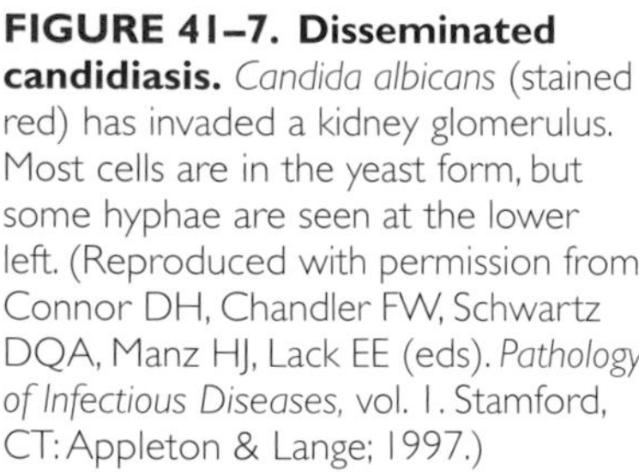

FIGURE 41–7. Disseminated candidiasis. *Candida albicans* (stained red) has invaded a kidney glomerulus. Most cells are in the yeast form, but some hyphae are seen at the lower left. (Reproduced with permission from Connor DH, Chandler FW, Schwartz DQA, Manz HJ, Lack EE (eds). *Pathology of Infectious Diseases,* vol. 1. Stamford, CT: Appleton & Lange; 1997.)

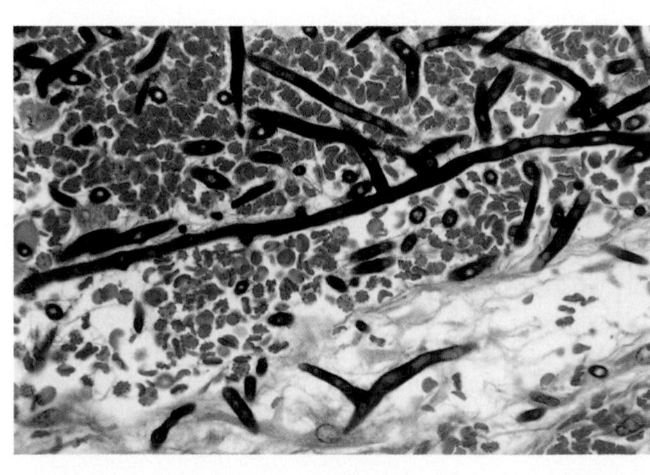

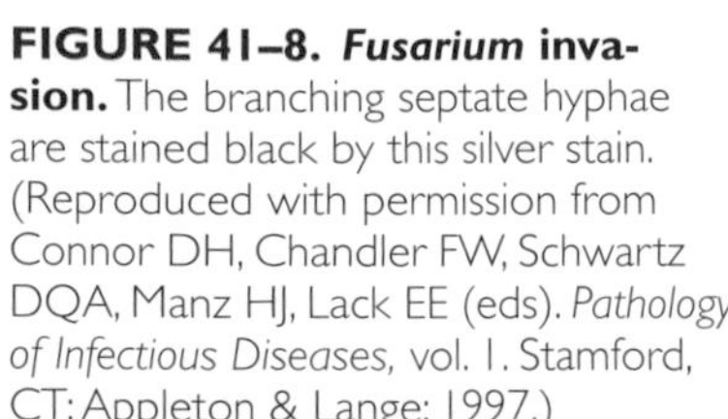

FIGURE 41–8. *Fusarium* invasion. The branching septate hyphae are stained black by this silver stain. (Reproduced with permission from Connor DH, Chandler FW, Schwartz DQA, Manz HJ, Lack EE (eds). *Pathology of Infectious Diseases,* vol. 1. Stamford, CT: Appleton & Lange; 1997.)

pathologist should be alerted to the possibility of fungal infection when tissues are submitted, because special stains and searches for fungi are not made routinely.

Culture

Fungi can be grown by methods similar to those used to isolate bacteria. Growth occurs readily on enriched bacteriologic media commonly used in clinical laboratories (eg, blood agar and chocolate agar). Many fungal cultures, however, require days to weeks of incubation for initial growth; bacteria present in the specimen grow more rapidly and may interfere with isolation of a slow-growing fungus. Therefore, the culture procedures of diagnostic mycology are designed to favor the growth of fungi over bacteria and to allow incubation to continue for a sufficient time to isolate slow-growing strains.

Growth in culture is simple but slow

Selective media allow isolation in the presence of bacteria

The most commonly used medium for cultivating fungi is Sabouraud's agar, which contains only glucose and peptones as nutrients. Its pH is 5.6, which is optimal for growth of dermatophytes and satisfactory for growth of other fungi. Most bacteria fail to grow or grow poorly on Sabouraud's agar. A wide variety of other media are in use, many of which use either Sabouraud's or brain-heart infusion as their base.

Sabouraud's agar optimal for fungi but poor for bacteria

Blood agar or another enriched bacteriologic agar medium is used when pure cultures would be expected. It is made selective for fungi by the addition of antibacterial antibiotics such as chloramphenicol and gentamicin. Cycloheximide, an antimicrobic that inhibits some saprophytic fungi, is sometimes added to Sabouraud's agar to prevent overgrowth of contaminating molds from the environment, particularly for skin cultures. Media containing these selective agents cannot be relied on exclusively because they can interfere with growth of some pathogenic fungi or because the "contaminant" may be producing an opportunistic infection. For example, cycloheximide inhibits *Cryptococcus neoformans,* and chloramphenicol may inhibit the yeast forms of some dimorphic fungi. Selective media are not needed for growing fungi from sterile sites such as cerebrospinal fluid or tissue biopsy specimens. In contrast to most pathogenic bacteria, many fungi grow best at 25° to 30°C, and temperatures in this range are used for primary isolation. Paired cultures incubated at 30° and 35°C may be used to demonstrate dimorphism.

Selective media make use of antimicrobics

Cultures incubated at 30°C for primary isolation

Once a fungus is isolated, identification procedures depend on whether it is a yeast or a mold. Yeasts are identified by biochemical tests analogous to those used for bacteria, including some that are identical (eg, urease production). The ability to form pseudohyphae is also taxonomically useful among the yeasts.

Yeast identified biochemically

Molds are most often identified by the morphology of their conidia and conidiophores. Other features such as the size, texture, and color of the colonies help characterize molds, but without demonstrating conidiation they are not sufficient for identification. The ease and speed with which various fungi produce conidia vary greatly. Minimal nutrition, moisture, good aeration, and ambient temperature favor development of conidia.

Molds identified by morphology and culture features

Microscopic fungal morphology is usually demonstrated by methods that allow in situ microscopic observation of the fragile asexual conidia and their shape and arrangement. Morphology may also be examined in fragments of growth teased free of a mold and examined moist in preparations containing a dye called lactophenol cotton blue. The dye stains the hyphae, conidia, and spores. Conidium production may not occur for days or weeks after the initial growth of the mold. It is somewhat like waiting for flowers to bloom, and it can be frustrating when the result has immediate clinical application.

Lactophenol cotton blue stains mycelia, conidia, and spores

It is desirable, but not always possible, to demonstrate both the yeast and mold phases with dimorphic fungi. In some cases, this result can be achieved with parallel cultures at 30° and 35°C. The tissue form of *Coccidioides immitis* is not readily produced in vitro. Demonstration of dimorphism has become less important with the development of specific DNA probes for the major systemic pathogens. These probes are rapid and can be applied directly to the mycelial growth of the readily grown mold phases of these fungi. ::: DNA probes, p. 81

Temperature variation demonstrates dimorphism

DNA probes are more rapid

Antigen and Antibody Detection

Serum antibodies directed against a variety of fungal antigens can be detected in patients infected with those agents. Except for some of the systemic pathogens, the sensitivity, specificity, or both, of these tests have not been sufficient to recommend them for use in

Serologic tests are useful for systemic fungi

Antigen detection shows promise

diagnosis or therapeutic monitoring of fungal infections. Immunoassays and oligonucleotide probes to detect fungal antigens have been pursued for some time. The major targets are mannans, mannoproteins, glucan, chitin, or some other structure unique to the fungal pathogen(s). Recent work on a panfungal probe that detects antigens of multiple fungi has been encouraging. The only established test of this type is one that detects the polysaccharide capsule of *Cryptococcus neoformans*. The serologic and antigen detection tests of value are discussed in sections on specific agents.

CHAPTER 42

Pathogenesis of Fungal Infection

We all have regular contact with fungi. They are so widely distributed in our environment that thousands of fungal spores are inhaled or ingested every day. Other species are so well adapted to humans that they are common members of the normal flora. Despite this ubiquity, clinically apparent systemic fungal infections are uncommon, even among persons living within the geographic habitat of the more pathogenic species. However, progressive systemic fungal infections pose some of the most difficult diagnostic and therapeutic problems in infectious disease, particularly among immunocompromised patients to whom they are a major threat. The purpose of this chapter is to give an overview of the pathogenesis and immunology of fungal infections. Details relating to specific fungi are given in Chapters 44 to 46.

GENERAL ASPECTS OF FUNGAL DISEASE

EPIDEMIOLOGY

Fungal infections are acquired from the environment or may be endogenous in the few instances where they are members of the normal flora (**Figure 42–1**). Inhalation of infectious conidia generated from molds growing in the environment is a common mechanism. Some of these molds are ubiquitous, whereas others are restricted to geographic areas whose climate favors their growth. In the latter case, disease can be acquired only in the endemic area. Some environmental fungi produce disease after they are accidentally injected past the skin barrier. The pathogenic fungi represent only a tiny percentage of those found in the environment. Endogenous infections are restricted to a few yeasts, primarily *Candida albicans.* These yeasts have the ability to colonize by adhering to host cells and, given the opportunity, invade deeper structures.

Environmental conidia are inhaled or injected

Endogenous yeasts may invade

PATHOGENESIS

Compared with bacterial, viral, and parasitic disease, less is known about the pathogenic mechanisms and virulence factors involved in fungal infections. Analogies to bacterial diseases come the closest because of the apparent importance of adherence to mucosal surfaces, invasiveness, extracellular products, and interaction with phagocytes (Figure 42–2). In general, the principles discussed in Chapter 22 apply to fungal infections. Most fungi are opportunists, producing serious disease only in individuals with impaired host defense systems. Only a few fungi are able to cause disease in previously healthy persons.

Fungal pathogenesis is similar to bacteria

Most fungi are opportunists

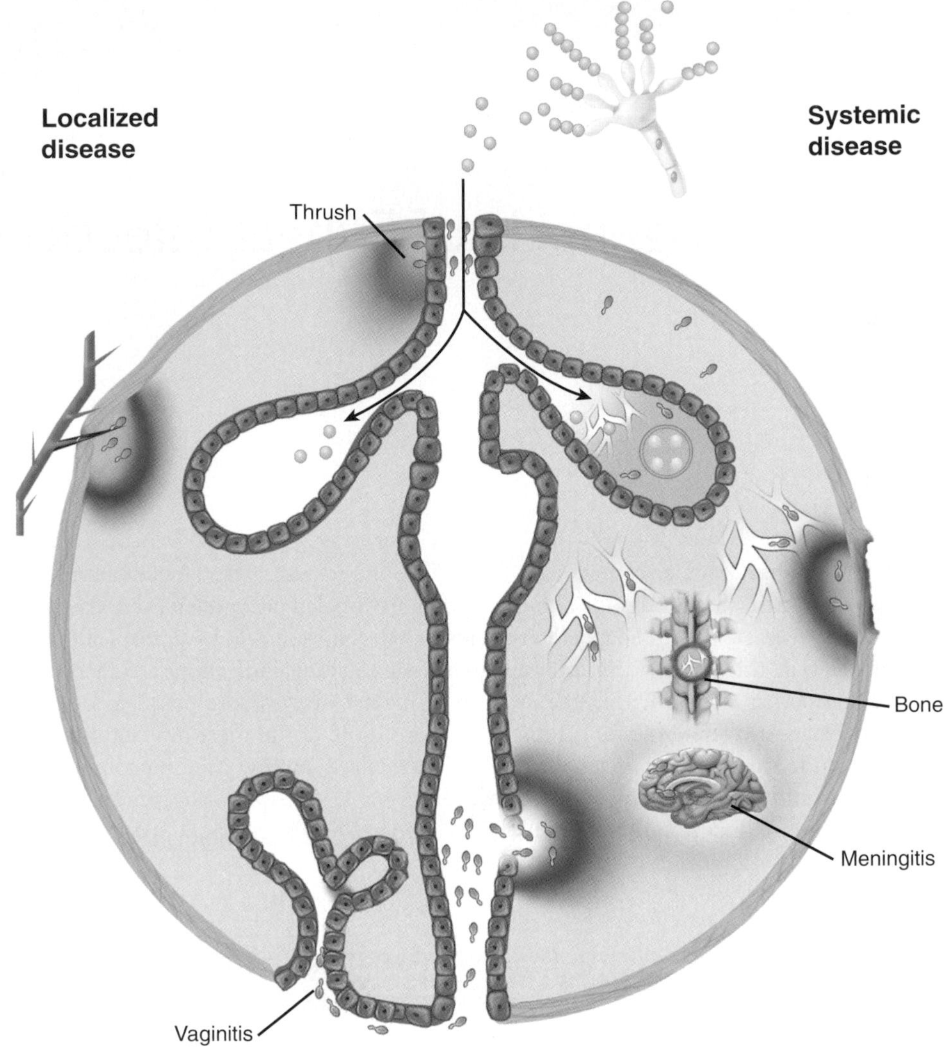

FIGURE 42–1. Fungi system view. Localized disease (left) is produced by local trauma or the superficial invasion of flora resident on the oropharyngeal (thrush), gastrointestinal, or vaginal mucosa. Systemic disease (right) begins with inhalation of conidia followed by dissemination to other sites.

Adherence

Adherence is mediated by fungal adhesins and host cell receptors

Mannoprotein is an adhesin, and fibronectin a receptor

A number of fungal species, particularly the yeasts, are able to colonize the mucosal surfaces of the gastrointestinal and female genital tracts. It has been shown experimentally that the ability to adhere to buccal or vaginal epithelial cells is associated with colonization and virulence. Within the genus *Candida*, the species that adhere best to epithelial cells are those most frequently isolated from clinical infections. Adherence usually requires a surface adhesin on the microbe and a receptor on the epithelial cell. In the case of *C albicans,* mannoprotein components extending from the cell wall have been implicated as the adhesin and fibronectin, and components of the extracellular matrix as the receptor(s). A few binding mediators have been identified for other fungi, usually a surface mannoprotein. ::: Candida adherence, p. 726

Invasion

Traumatic injection is linked to trauma

Small conidia may pass airway defenses

Passing an initial surface barrier—skin, mucous membrane, or respiratory epithelium—is an important step for most successful pathogens. Some fungi are introduced through mechanical breaks. For example, *Sporothrix schenckii* infection typically follows a thorn prick or some other obvious trauma. Fungi that initially infect the lung must produce conidia small enough to be inhaled past the upper airway defenses. For example, arthroconidia of *Coccidioides immitis* (2 to 6 μm) can remain suspended in air for a considerable time and can reach the terminal bronchioles to initiate pulmonary coccidioidomycosis. ::: Sporothrix trauma, p. 719

Triggered by temperature and possibly other cues, dimorphic fungi from the environment undergo a metabolic shift similar to the heat shock response and completely change their morphology and growth to a more invasive form. Invasion directly across mucosal barriers by the endogenous yeast *C albicans* is similarly associated with a morphologic change, the formation of hyphae. The triggering mechanisms of this change are unknown, but the new form is able to penetrate and spread. Extracellular enzymes (eg, proteases, elastases) are associated with the advancing edge of the hyphal form of *Candida* and with the invasive forms of many of the dimorphic and other pathogenic fungi. Although these enzymes must contribute to some aspect of invasion or spread, their precise role is unknown for any fungus.

Invasion across mucosal barriers may involve enzymes

Injury

None of the extracellular products of opportunistic fungi or dimorphic pathogens has been shown to injure the host directly during infection in a manner analogous to bacterial toxins. Although the presence of necrosis and infarction in the tissues of patients with invasion by fungi such as *Aspergillus* suggests a toxic effect, direct evidence is lacking. A number of fungi do produce exotoxins, called **mycotoxins,** in the environment but not in vivo. The structural components of the cell do not cause effects similar to those of the endotoxin of Gram-negative bacteria, although mannan is known to circulate widely in the body. The circulating products of *Cryptococcus neoformans* have been shown to down-regulate immune functions. The injury caused by fungal infections seems to be due primarily to the destructive aspects of delayed-type hypersensitivity (DTH) responses as a result of the inability of the immune system to clear the fungus. In this respect, fungal infections resemble tuberculosis more than any other disease. ::: DTH, p. 42; DTH in tuberculosis, p. 495

No classic exotoxins are produced in vivo

Injury is due to inflammatory and immunologic responses

IMMUNITY

Innate Immunity

There is considerable evidence that healthy persons have a high level of innate immunity to most fungal infections. This is particularly true of opportunistic molds. This resistance is mediated by the professional phagocytes (neutrophils, macrophages, and dendritic cells), the complement system, and pattern recognition receptors. For fungi, the most important receptors include a lectin-like structure on phagocytes (dectin-1) that binds glucan, and Toll-like receptors (TLR2, TLR4). In most instances, neutrophils and alveolar macrophages are able to kill the conidia of fungi if they reach the tissues. A small number of species, all of which are dimorphic, are able to produce mild to severe disease in otherwise healthy persons. In vitro studies have shown these fungi to be more resistant to killing by phagocytes than the opportunists possibly because of a change in surface structures subject to pattern recognition. *C albicans* is able to bind complement components in a way that interferes with phagocytosis.

Most fungi are readily killed by neutrophils

C immitis, one of the best-studied species, has been shown to contain a component in the wall of its conidial (infective) phase that is antiphagocytic. As the hyphae convert to the spherule (tissue) phase, they also become resistant to phagocytic killing because of their size and surface characteristics. Some fungi produce substances such as melanin, which interfere with oxidative killing by phagocytes. The tissue yeast form of *Histoplasma capsulatum* multiplies within macrophages by interfering with lysosomal killing mechanisms in a manner similar to that of some bacteria. These mechanisms of avoiding phagocytic killing appear to allow many dimorphic fungi to multiply sufficiently to produce an infection that can be controlled only by the immune response.

Tissue phases of dimorphic fungi resist phagocytic killing

Adaptive Immune Response

A recurrent theme with fungal infections is the importance of an intact immune response in preventing infection and progression of disease. Most fungi are incapable of producing even a mild infection in immunocompetent individuals. A small number of species are able to cause clinically apparent infection that usually resolves once there is time for activation of normal immune responses. In most instances in which it has been investigated, the actions of neutrophils and T_H1-mediated immune responses have been found to be of primary importance in this resolution. Progressive, debilitating, or life-threatening disease with these agents is commonly

T-cell–mediated responses of primary importance

Progressive fungal diseases occur in the immunocompromised

associated with depressed or absent cellular immune responses, and the course of any fungal disease is worse in immunocompromised than previously healthy persons.

Humoral Immunity

Opsonizing antibody is effective in some yeast infections

Antibodies can be detected at some time during the course of almost all fungal infections, but for most there is little evidence that they contribute to immunity. The only encapsulated fungus, *Cryptococcus neoformans,* is an example of a fungus against which antibody plays a role in controlling infection. Although the polysaccharide capsule of *C neoformans* has antiphagocytic properties similar to those of encapsulated bacterial pathogens, it is less antigenic. Anticapsular antibody plays a role in resolving cryptococcal infection, but T_H1 responses are still dominant. Antibody also plays a role in control of *C albicans* infections by enhancing fungus–phagocyte interactions, and this is probably true for other yeasts. In some other fungal infections, the lack of protective effect of antibody is striking. In coccidioidomycosis, for example, high titers of *C immitis*–specific antibodies are associated with dissemination and a worsening clinical course.

Cellular Immunity

Systemic disease associated with deficiencies in neutrophils and T-cell–mediated immunity

Considerable clinical and experimental evidence points toward the importance of cellular immunity in fungal infections. Most patients with severe systemic disease have neutropenia, defects in neutrophil function, or depressed T_H1 immune reactions. These can result from factors such as steroid treatment, leukemia, Hodgkin's disease, and acquired immunodeficiency syndrome. In other cases, an immunologic deficit can usually be demonstrated by absence of delayed-type hypersensitivity responses or T_H1-stimulated cytokines specific to the fungus in question. In the latter case, it is possible that hyporesponsiveness is due at least in part to activation of suppressor cells or continued circulation of fungal antigen. ::: Cellular immunity, p. 35

Fungi that escape neutrophils grow slowly in macrophages

Although not all fungi have been studied to the same degree, a unified picture emerging from clinical and experimental animal studies is illustrated in **Figure 42–2**. When hyphae or yeast cells of the fungus reach deep tissue sites, they are either killed by neutrophils or resist destruction by one of the antiphagocytic mechanisms described earlier. Surviving cells continue to grow slowly or, if they are dimorphic, convert to their yeast, hyphal, or spherule tissue phases. The growth of these invasive forms may be slowed but not killed by macrophages, which ingest them. A feature of the fungal pathogens is to resist the killing mechanisms of the macrophage and continue to multiply. In healthy persons, the extent of infection is small, and any symptoms are caused by the inflammatory response. ::: Phagocytic killing, p. 25

Growth is restricted when macrophages activated by cytokines

Immune defects lead to progressive disease

Everything awaits the specific adaptive immune response to the invader. In fungal infections, it is the interaction between dendritic cells and macrophages that favors production of interleukin 12 (IL-12) and interferon XX (INF-γ) leading the CD4 cells to differentiate to T_H2 cells that has the dominant effect. The turning point comes when local macrophages containing multiplying fungi are activated by cytokine mediators produced by T lymphocytes that have interacted with the fungal antigen. The activated macrophages are then able to restrict the growth of the fungus, and the infection is controlled. Defects that disturb this cycle lead to progressive disease. To the extent that they are known, the specifics of these reactions are discussed in the following chapters. ::: T-cell responses, p. 33

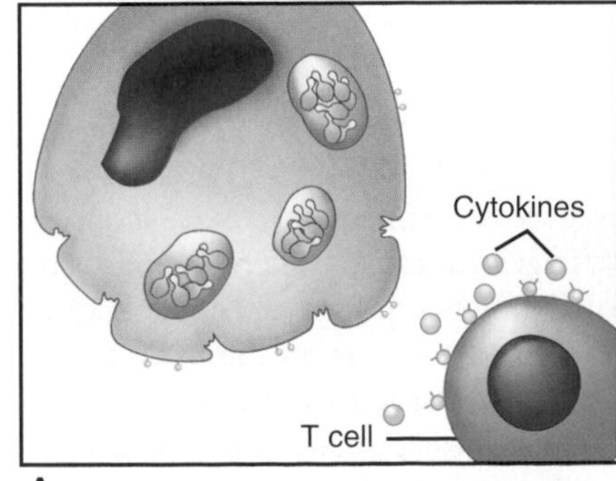

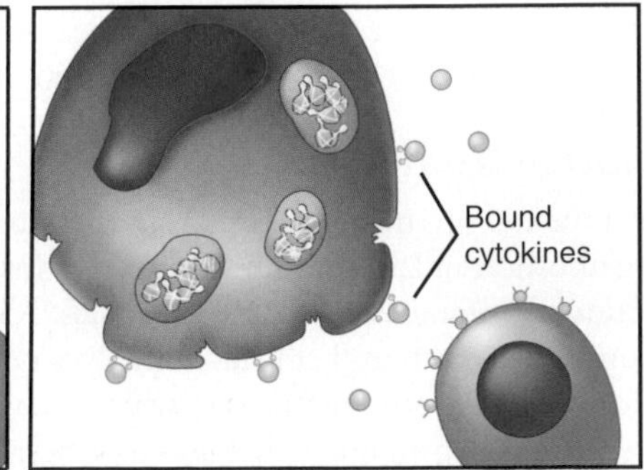

FIGURE 42–2. Immunity to fungal infections. **A.** Pathogenic fungi are able to survive and multiply slowly in non-activated macrophages. **B.** When macrophages are activated by cytokines from T-cells the growth is restricted and the fungi digested.

CHAPTER 43

Antifungal Agents and Resistance

Compared with antibacterial agents, relatively few antimicrobials are available for treatment of fungal infections. Many substances with antifungal activity have proved to be unstable, to be toxic to humans, or to have undesirable pharmacologic characteristics, such as poor diffusion into tissues. Of the agents in current clinical use, the newer azole compounds have the broadest spectrum with significantly lower toxicity than earlier antifungal agents. An even newer class of cell wall active agents offers hope for the selective toxicity that β-lactams provide for antibacterial therapy.

Many antifungals are too toxic for use

Fortunately, most fungal infections are self-limiting and require no chemotherapy. Superficial mycoses are often treated, but topical therapy can be used, thus limiting toxicity to the host. The remaining small group of deep mycoses that are uncontrolled by the host's immune system require the prolonged use of antifungals. This, combined with the fact that most of the patients have underlying immunosuppression, makes them among the most difficult of all infectious diseases to treat successfully. The characteristics of currently used antifungal agents are discussed next and summarized in **Table 43–1**. They are discussed in the text that follows in relation to their target of action, as illustrated in **Figure 43–1**.

Treatment is most needed for dissemination in immunocompromised persons

ANTIFUNGAL AGENTS

CYTOPLASMIC MEMBRANE

Polyenes

The polyenes **nystatin** and **amphotericin B** are lipophilic and bind to ergosterol, the dominant sterol in the cytoplasmic membrane of fungal cells. After binding, they form annular channels, which penetrate the membrane and lead to leakage of essential small molecules from the cytoplasm and cell death. Their binding affinity for the ergosterol of fungal membranes is not absolute and includes sterols such as cholesterol, which are present in human cells. This is the basis of the considerable toxicity that limits their use. Almost all fungi are susceptible to amphotericin B, and the development of resistance is too rare to be a consideration in its use.

Ergosterol binding forms membrane channels

Active against most fungi

At physiologic pH, amphotericin B is insoluble in water and must be administered intravenously as a colloidal suspension. It is not absorbed from the gastrointestinal tract. The major limitation to amphotericin B therapy is the toxicity created by its affinity for mammalian as well as fungal membranes. Infusion is commonly followed by chills, fever, headache, and dyspnea. The most serious toxic effect is renal dysfunction and is seen in virtually every patient receiving a therapeutic course. Experienced clinicians learn to titrate the dosage for each patient to minimize the nephrotoxic effects. For obvious reasons, amphotericin B is limited to progressive, life-threatening fungal infections. In such cases, despite its toxicity it

Insoluble compound must be infused in suspension

TABLE 43–1 Features of Antifungal Agents

AGENT	MECHANISM OF ACTION	MECHANISM OF RESISTANCE	ROUTE	CLINICAL USE
Polyenes				
Nystatin	Cytoplasmic membrane pores	Sterol modification	Topical	Most fungi
Amphotericin B	Cytoplasmic membrane pores	Sterol modification	Intravenous	Most fungi
Azoles				
Ketoconazole	Ergosterol synthesis (demethylase)	Efflux, demethylase alteration, bypass, overproduction[a]	Oral	*Candida*, dermatophytes, dimorphic fungi[b]
Fluconazole	Ergosterol synthesis (demethylase)	Efflux, demethylase alteration, bypass, overproduction[a]	Oral, intravenous	*Candida*, *Cryptococcus*, dimorphic fungi[e]
Itraconazole	Ergosterol synthesis (demethylase)	Efflux, demethylase alteration, bypass, overproduction[a]	Oral, intravenous	*Aspergillus*, *Sporothrix*, *Candida*, dimorphic fungi
Voriconazole	Ergosterol synthesis (demethylase)		Oral, intravenous	*Candida*, *Aspergillus*, some other yeasts and molds
Posaconazole	Ergosterol synthesis (demethylase)		Oral	*Candida*, *Aspergillus* prophylaxis
Clotrimazole	Ergosterol synthesis (demethylase)	Unknown[c]	Topical	*Candida*, dermatophytes
Miconazole	Ergosterol synthesis (demethylase)	Unknown[c]	Topical	*Candida*, dermatophytes
Allylamines				
Terbinafine	Ergosterol synthesis (squalene epoxidase)	?Efflux	Oral	Dermatophytes, combined with azoles for *Candida*, *Aspergillus*
Naftifine	Ergosterol synthesis (squalene epoxidase)	Unknown	Topical	dermatophytes
Flucytosine				
	DNA synthesis, RNA transcription	Permease or modifying enzymes[d] mutation	Oral	*Candida* and *Cryptococcus*[f]
Echinocandins				
Caspofungin	Glucan synthesis (glucan synthetase)	Altered synthetase	Intravenous	*Candida*, *Aspergillus* (other fungi[g])
Micafungin	Glucan synthesis (glucan synthetase)	Altered synthetase	Intravenous	*Candida* (*Aspergillus*)
Anidulafungin	Glucan synthesis (glucan synthetase)	Altered synthetase	Intravenous	*Candida* (*Aspergillus*)
Nikkomycins	Chitin synthesis (chitin synthetase)			Developmental
Griseofulvin	Microtubule disruption	Unknown	Oral	Dermatophytes
Potassium iodide	Unknown	Unknown	Oral	*Sporothrix schenckii*
Tolnaftate	Unknown	Unknown	Oral	Dermatophytes

5FC, 5-flucytosine.

[a]Most work is with fluconazole and *Candida*; other azoles are to be assumed similar.

[b]Generally less absorbed and less active than fluconazole or itraconazole.

[c]Probably similar to other azoles, but resistance to the concentrations in topical preparations may differ.

[d]Cytosine deaminase and uracil phosphoribosyltransferase (the enzymes that modify 5FC to active forms).

[e]Itraconazole generally preferred.

[f]Only in combination with amphotericin B owing to resistance mutation.

[g]Empiric therapy in neutropenic patients.

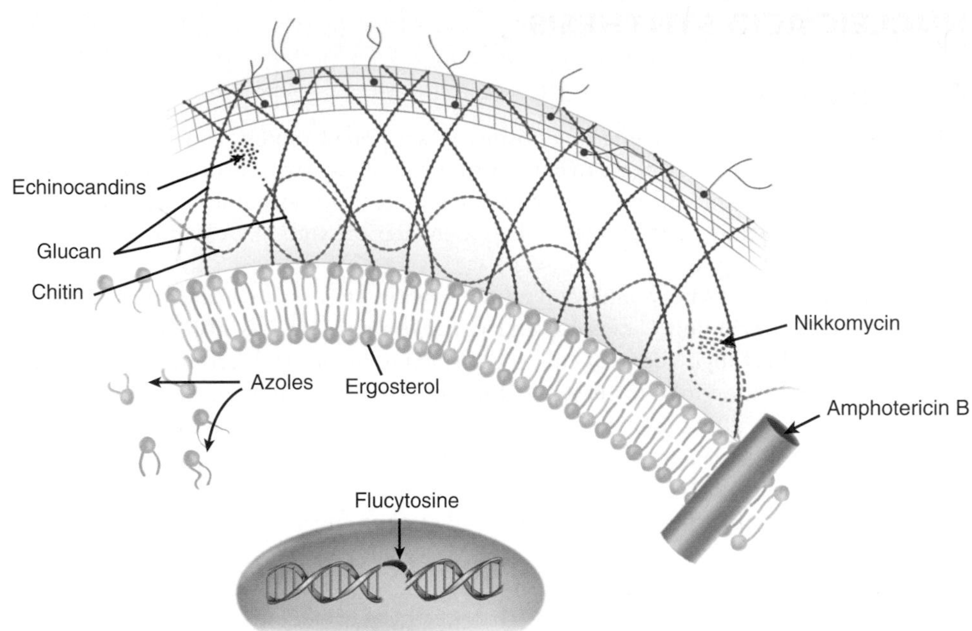

FIGURE 43–1. Action of antifungal agents. The sites where the major antifungal agents act in the cell wall (echinocandins, nikkomycin, cytoplasmic membrane (azoles, amphotericin B) and genome (flucytosine) are shown.

retains a prime position in treatment often by administration of an initial course of amphotericin followed by a less toxic agent. Preparations that complex amphotericin B with lipids have been used as a way to limit toxicity. The even greater toxicity of nystatin limits its use to topical preparations.

Therapy must be titrated against toxicity

Azoles

The azoles are a large family of synthetic organic compounds, which includes members with antibacterial, antifungal, and antiparasitic properties. The important antifungal azoles for systemic administration are ketoconazole, fluconazole, itraconazole, and voriconazole. Clotrimazole and miconazole are limited to topical use. Others azoles are under development or evaluation. Their activity is based on inhibition of the enzyme (14 α-demethylase) responsible for conversion of lanosterol to ergosterol, the major component of the fungal cytoplasmic membrane. This leads to lanosterol accumulation and the formation of a defective membrane. Effects on the precursors of some hormones may cause endocrine side effects and restricts use in pregnancy. All antifungal azoles have the same mechanism of action. The differences among them are in avidity of enzyme binding, pharmacology, and side effects.

Inhibit enzyme crucial for synthesis of membrane ergosterol

Ketoconazole, the first azole, has now been supplanted by the later azoles for most systemic mycoses. Although nausea, vomiting, and elevation of hepatic enzymes complicate the treatment of some patients, the azoles are much less toxic than amphotericin B. **Fluconazole** was the first azole with good central nervous system penetration, but **itraconazole** is now generally preferred for fungal meningitis. Azoles are also effective for superficial and subcutaneous mycoses in which the initial therapy either fails or is not tolerated by the patient. In general, itraconazole and, more recently, **voriconazole** are the primary azoles used together with, or instead of, amphotericin B for serious fungal infections. **Clotrimazole** and **miconazole** are available in over-the-counter topical preparations.

Less toxic than amphotericin B

Itraconazole and voriconazole prime systemic agents

Allylamines

The allylamines are a group of synthetic compounds that act by inhibition of an enzyme (squalene epoxidase) in the early stages of ergosterol synthesis. The allylamines include an oral agent, **terbinafine,** and a topical agent, **naftifine.** Both are used in the treatment of dermatophyte (ringworm) infections.

Inhibit ergosterol synthesis

NUCLEIC ACID SYNTHESIS

Flucytosine

Enzymatically modified form makes defective RNA

Inhibits DNA synthesis

5-Flucytosine (5FC) is an analog of cytosine. It is a potent inhibitor of RNA and DNA synthesis. 5FC requires a permease to enter the fungal cell, where its action is not direct but through its enzymatic modification to other compounds (5-fluorouracil, 5-fluorodeoxyuridyic acid, 5-fluoruridine). These metabolites then interfere with DNA synthesis and cause aberrant RNA transcription.

Active against yeasts but not molds

Resistance develops during therapy if used alone

Flucytosine is well absorbed after oral administration. It is active against most clinically important yeasts, including *C albicans* and *C neoformans,* but has little activity against molds or dimorphic fungi. The frequent development of mutational resistance during therapy limits its application to mild yeast infections or it use in combination with amphotericin B for cryptococcal meningitis. The combination reduces the probability of expression of resistance and allows a lower dose of amphotericin B to be used. The primary toxic effect of flucytosine is a reversible bone marrow suppression that can lead to neutropenia and thrombocytopenia. This effect is dose related and can be controlled by drug monitoring.

CELL WALL SYNTHESIS

The unique chemical nature of the fungal cell wall, with its interwoven layers of mannan, glucan, and chitin (Figure 43–1), make it an ideal target for chemotherapeutic attack. Although such antifungal agents have only recently (2002) entered the armamentarium, they are most welcome. The echinocandins, which block glucan synthesis, are now in clinical use and the nikkomycins, which block chitin synthesis, are in development.

Echinocandins

Inhibit synthease crucial for glucan synthesis

Current use is *Candida, Aspergillus*

Echinocandins act by inhibition of a glucan synthetase (1,3-β-D-glucan synthetase) required for synthesis of the principal cell wall glucan of fungi. Its action causes morphologic distortions and osmotic instability in yeast and molds that are similar to the effect of β-lactams on bacteria. The first such agent to be licensed is **caspofungin,** which has good activity against *Candida* and *Aspergillus* and a wide range of other fungi. *Cryptococcus neoformans* whose cell wall glucans have a slightly different structure is resistant. Since there are no similar human structures, toxicity is minimal. The newest echinocandins, **micafungin** and **andiulafungin,** have the same mode of action and similar spectrum.

Nikkomycins

Inhibit chitin synthesis

Nikkomycins have a mechanism of action analogous to echinocandins. They inhibit chitin synthetase, which polymerizes the *N*-acetylglucosamine subunits that make up chitin. The result is inhibition of chitin synthesis. The agent in development, nikkomycin Z, has activity against dimorphic fungi such as *Coccidioides immitis* and *Bastomyces dermatitidis* but not against yeast or *Aspergillus.*

Other Antifungal Agents

Microtubule disruption interferes with cell division

Active against dermatophytes

Griseofulvin is a product of a species of the mold *Penicillium.* It is active only against the agents of superficial mycoses. Griseofulvin is actively taken up by susceptible fungi and acts on the microtubules and associated proteins that make up the mitotic spindle. It interferes with cell division and possibly other cell functions associated with microtubules. Griseofulvin is absorbed from the gastrointestinal tract after oral administration and concentrates in the keratinized layers of the skin. Clinical effectiveness has been demonstrated for all causes of dermatophyte infection, but the response is slow. Difficult cases may require 6 months of therapy to effect a cure.

Iodide inhibits *Sporothrix*

Potassium iodide is the oldest known oral chemotherapeutic agent for a fungal infection. It is effective only for cutaneous sporotrichosis. Its activity is somewhat paradoxical, because the mold form of the etiologic agent, *Sporothrix schenckii,* can grow on medium containing 10% potassium iodide. The pathogenic yeast form of this dimorphic fungus appears to be susceptible to molecular iodine. **Tolnaftate** is a derivative of naphthiomate. It has activity against dermatophytes, but not against yeasts. It has been effective in topical treatment of dermatophytoses and is available in over-the-counter preparations. ::: Dermatophytes, p. 719

RESISTANCE TO ANTIFUNGAL AGENTS

DEFINITION OF RESISTANCE

The concepts, definitions, and laboratory methods described in Chapter 23 for bacterial resistance are generally applicable to fungi. Quantitative susceptibility is measured by the minimal inhibitory concentration (MIC) under conditions that favor the growth of fungi. The wide range of growth rates and diversity of growth forms (yeast, hyphae, conidia) in the various fungi have added technical variables to testing, but standardized methods are now available. Comparison of MICs with drug pharmacology allows classification of fungi as susceptible or resistant, but these results do not yet predict clinical outcome with the same certainty they do with bacteria. Because of its specialized nature, the availability of antifungal susceptibility testing is restricted to major centers and reference laboratories. ::: Bacterial resistance testing, p. 415

Concepts are similar to bacterial resistance

Laboratory methods are variable

MECHANISMS OF RESISTANCE

The same resistance mechanisms seen in bacteria are also found in fungi. A major addition is the much greater use of metabolic means such as efflux pumps and changes in synthetic pathways by fungi. The most glaring difference is the complete absence of enzymatic inactivation of antifungals as resistance mechanism. Perhaps related to this is the absence in fungi of powerful means for gene transfer such as conjugation and transposition.

Polyene Resistance

Since amphotericin B binds directly to the cytoplasmic membrane, the only means to resist this action is to change the membrane composition. The uncommon strains that have been studied show a decrease in the ergosterol content of the membrane. This limits the primary binding sites.

Membrane ergosterol decreased

Flucytosine Resistance

Flucytosine requires a permease for entry into the cell and then multiple enzymes to modify it to the active metabolites. Mutation in any one of these enzymes renders the drug ineffective. This happens readily under the selective pressure of 5FC use. It is one of the few antimicrobials in which emergence of resistance *during* therapy of an acute infection is predictable. It is the reason its use is limited to combinations with other antifungals.

Multiple enzymes can mutate

Azole Resistance

There are four major mechanisms of resistance that cross all the azole agents. The two most prominent are efflux pumps and altered target. The efflux pumps transport drug that has entered the cell back outside. Some pumps act for all azoles and others act on only one. Alteration of subunits of the demethylase enzyme by mutation decreases the affinity of the azole for its enzyme target. Multiple mutations can have an additive effect. ::: Efflux pumps, p. 419

Azole pumped out

Enzyme target altered

Two metabolic mechanisms compensate for the drug's presence without altering its target or directly inactivating it. Up-regulation of the target demethylase allows its action to continue despite binding of some of the enzyme by the azole. Some resistant strains have been shown to accomplish ergosterol synthesis by an alternate pathway, thus bypassing the azole affected mechanism.

Demethylase up-regulated or bypassed

Echinocandin Resistance

Although the echinocandins are relatively new, resistance has already been observed with their use. The mechanism is altered target. Mutations in subunits of the glucan synthetase target have been correlated with increases in MIC of up to a thousandfold. ::: Altered target, p. 419-420

Mutant synthetase

SELECTION OF ANTIFUNGALS

As with all chemotherapy, the selection of antifungal agents for treatment of superficial, subcutaneous, and systemic mycoses involves balancing probable efficacy against toxicity. The factors to be considered are (1) the threat of morbidity or mortality posed by the specific infection, (2) the immune status of the patient, (3) the toxicity of the antifungal, and (4) the probable activity of the antifungal agent against the fungus. In the case of superficial mycoses, the risks of appropriate therapy are small, and a number of topical agents may be tried. At the other extreme, an immunocompromised patient will most likely be treated aggressively with systemic agents for proven or even suspected systemic fungal infection. Since susceptibility tests are usually not available, the decisions regarding which agents to use are usually made and sustained on an empiric basis. Even when guided by in vitro testing, treatment failures are common particularly in the immunocompromised. It is hoped that the addition of the new cell wall active agents to the regimen will have a favorable effect on these outcomes.

CHAPTER 44

Dermatophytes, *Sporothrix*, and Other Superficial and Subcutaneous Fungi

The least invasive of pathogenic fungi are the dermatophytes and other superficial fungi that are adapted to the keratinized outer layers of the skin. The subcutaneous fungi go a step farther by extending to the tissue beneath the skin but rarely invade deeper. Both are discussed here and summarized in **Table 44–1**.

SUPERFICIAL FUNGI

Dermatophytes

Dermatophytoses are superficial infections of the skin and its appendages, commonly known as ringworm (**Figure 44–1**), athlete's foot, and jock itch. They are caused by species of three genera collectively known as dermatophytes. These fungi are highly adapted to the nonliving, keratinized tissues of nails, hair, and the stratum corneum of the skin. The source of infection may be humans, animals, or the soil.

MYCOLOGY

The three genera of medical important Dermatophytes (literally, skin-plants) are *Epidermophyton, Microsporum*, and *Trichophyton*. They are separated primarily by the morphology of their macroconidia and the presence of microconidia. Many species cause dermatophyte infections; the most common of these are shown in Table 44–1. They require a few days to a week or more to initiate growth. Most grow best at 25°C on Sabouraud's agar, which is usually used for culture. The hyphae are septate, and their conidia may be borne directly on the hyphae or on conidiophores. Small microconidia may or may not be formed; however, the larger and more distinctive macroconidia are usually the basis for identification.

Form septate hyphae, macroconidia, and microconidia

Epidermophyton, Microsporum, and *Trichophyton* are major genera

Grow best at 25°C

TABLE 44–1 Agents of Superficial and Subcutaneous Mycoses

FUNGUS	FUNGAL GROWTH IN LESION	IN CULTURE (25°C)	INFECTION SITE	DISEASE
Dermatophytes				
Microsporum canis	Septate hyphae	Mold	Hair,[a] skin	Ringworm
Microsporum audouini	Septate hyphae	Mold	Hair[a]	Ringworm
Microsporum gypseum	Septate hyphae	Mold	Hair, skin	Ringworm
Trichophyton tonsurans	Septate hyphae	Mold	Hair, skin, nails	Ringworm
Trichophyton rubrum	Septate hyphae	Mold	Hair, skin, nails	Ringworm
Trichophyton mentagrophytes	Septate hyphae	Mold	Hair, skin	Ringworm
Trichophyton violaceum	Septate hyphae	Mold	Hair, skin, nails	Ringworm
Epidermophyton floccosum	Septate hyphae	Mold	Skin	Ringworm
Other superficial fungi				
Malassezia furfur[b]	Yeast (mycelia)[c]	Yeast	Skin (pink to brown)[d]	Pityriasis (tinea) versicolor
Hortaea werneckii[e]	Septate hyphae, ellipsoidal cells	Yeast (mold)	Skin (brown–black)[d]	Tinea nigra
Trichosporon cutaneum	Septate hyphae	Mold	Hair (white)[b]	White piedra
Piedraia hortae	Septate hyphae	Mold, ascospores	Hair (black)[b]	Black piedra
Subcutaneous fungi				
Sporothrix schenckii	Cigar-shaped yeast (rare)	Mold	Subcutaneous, lymphatic spread	Sporotrichosis
Fonsecaea pedrosoi	Muriform body[f]	Mold	Wart-like foot lesions	Chromoblastomycosis
Phialophora verrucosa	Muriform body[f]	Mold	Wart-like foot lesions	Chromoblastomycosis
Cladophialophora (Cladosporium) carrionii	Muriform body[f]	Mold	Wart-like foot lesions	Chromoblastomycosis

[a]Specimens fluoresce under ultraviolet light.
[b]Previously known as *Pityrosporum orbiculare*.
[c]Denotes less frequent findings.
[d]Color of clinical lesions.
[e]Previously known as *Cladosporium werneckii*.
[f]Multicompartment yeast-like structure.

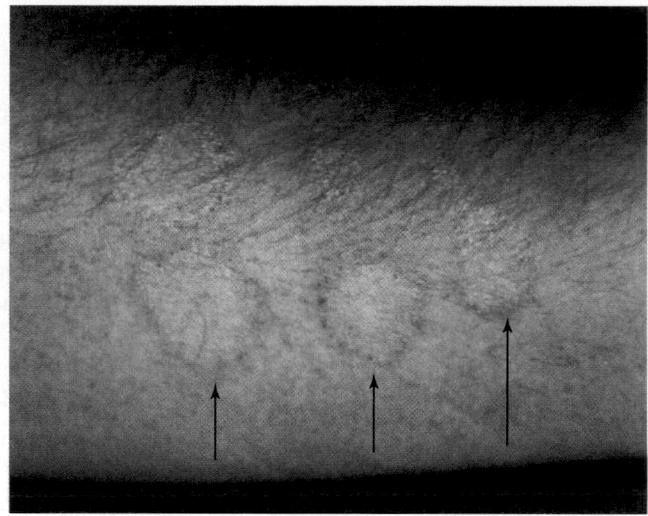

FIGURE 44–1. Ringworm. The ring-like lesions on this forearm are due to advancing growth of *Trichophyton mentagrophytes*. (Reproduced with permission from Willey J, Sherwood L, Woolverton C (eds). *Prescott's Principles of Microbiology*. New York: McGraw-Hill; 2008.)

DERMATOPHYTE DISEASE

CLINICAL CAPSULE

Dermatophytoses are slowly progressive eruptions of the skin and its appendages that may be unsightly, but are not painful or life threatening. The manifestations (and names) vary, depending on the nature of the inflammatory response in the skin, but they typically involve erythema, induration, itching, and scaling. The most familiar is "ringworm," which gets its name from the annular shape of creeping margin at the advancing edge of dermatophyte growth.

EPIDEMIOLOGY

There are both ecologic and geographic differences in the occurrence of the various dermatophyte species. Some are primarily adapted to the skin of humans, others to animals, and others to the environment. All may serve as the source for human infection. Many wild and domestic animals, including dogs and cats, are infected with certain dermatophyte species and represent a large reservoir for infection of humans. There are differences between temperate and tropical climates in the number of cases and isolations from nonhuman sources of the different species. Many of these differences are changing with shifts in population.

Reservoir may be human, animal, or soil

Human-to-human transmission usually requires close contact with an infected subject or infected person or animal, because dermatophytes are of low infectivity and virulence. Transmission usually takes place within families or in situations involving contact with detached skin or hair, such as barber shops and locker rooms. No special precautions beyond handwashing need be taken by the medical attendant after contact with an infected patient.

Transmission requires contact with intact or detached skin or hair

PATHOGENESIS

Dermatophytoses begin when minor traumatic skin lesions come in contact with dermatophyte hyphae shed from another infection. Susceptibility may be enhanced by local factors such as the composition surface fatty acids. Once the stratum corneum is penetrated, the organism can proliferate in the keratinized layers of the skin aided by a variety of proteinases. The course of the infection depends on the anatomic location, moisture, the dynamics of skin growth and desquamation, the speed and extent of the inflammatory response, and the infecting species. For example, if the organisms grow very slowly in the stratum corneum and if turnover by desquamation of this layer is not retarded, the infection will probably be short-lived and cause minimal signs and symptoms. Inflammation tends to increase skin growth and desquamation rates and helps limit infection, whereas immunosuppressive agents such as corticosteroids decrease shedding of the keratinized layers and tend to prolong infection. Invasion of any deeper structures is extremely rare.

Initial infection is through minor skin breaks

Balance between fungal growth and skin desquamation determines outcome

Most infections are self-limiting, but those in which fungal growth rates and desquamation are balanced and in which the inflammatory response is poor tend to become chronic. The lateral spread of infection and its associated inflammation produce the characteristic sharp advancing margins that were once believed to be the burrows of worms. This characteristic is the origin of the common name **ringworm** and the Latin term *tinea* (worm), which is often applied to the clinical forms of the disease (Figure 44–1).

Poor inflammatory response leads to chronic infection

Infection may spread from skin to other keratinized structures, such as hair and nails, or may invade them primarily. The hair shaft is penetrated by hyphae (**Figure 44–2**), which extend as arthroconidia either exclusively within the shaft (endothrix) or both within and outside the shaft (ectothrix). The end result is damage to the hair shaft structure, which often breaks off. Loss of hair at the root and plugging of the hair follicle with fungal elements may result. Invasion of the nail bed causes a hyperkeratotic reaction, which dislodges or distorts the nail.

Hair shaft is penetrated and broken by hyphae

IMMUNITY

Most dermatophyte infections pass through an inflammatory stage to spontaneous healing. Phagocytes are able to use oxidative pathways to kill the fungi both intracellularly and extracellularly. Little is known about the factors that mediate the host response in these self-limiting infections or whether they confer immunity to subsequent exposures. Antibodies may be formed during infection but play no known role in immunity. Most clinical and experimental evidence points to the importance of T-cell–mediated T_H1 responses, as with other fungal infections. The timing of the inflammatory response to infection correlates with appearance of delayed-type hypersensitivity, and resolution of infection is associated with the blastogenic T-lymphocyte responses. Enhanced desquamation with the inflammatory response helps remove infected skin.

Delayed hypersensitivity responses occur

Cell-mediated immune responses are the most important

Occasionally, dermatophyte infections become chronic and widespread. This progression has been related to both host and organism factors. Approximately half of these patients have underlying diseases affecting their immune responses or are receiving treatments that compromise T-lymphocyte function. These chronic infections are

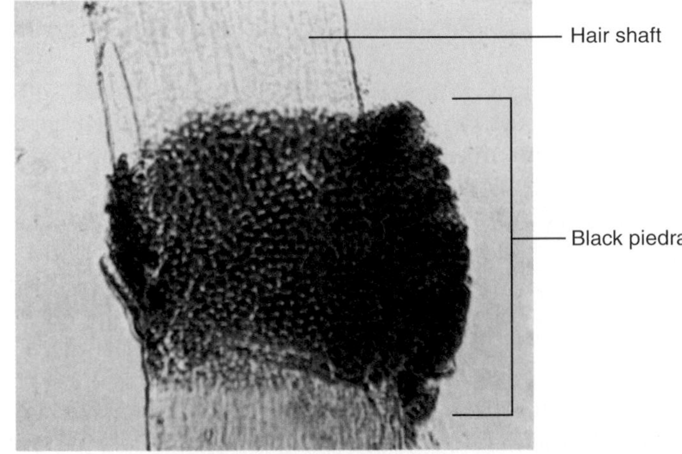

FIGURE 44–2. Black piedra. Note invasion by *Piedraia hortae* both within (endothrix) and outside (exothrix) the hair shaft. Dermatophyte invasion would be similar. (Reproduced with permission from Willey J, Sherwood L, Woolverton C (eds). *Prescott's Principles of Microbiology.* New York: McGraw-Hill; 2008.)

particularly associated with *Trichophyton rubrum*, to which both normal and immunocompromised persons appear to be hyporesponsive. Although a number of mechanisms have been proposed, how this organism is able to grow without stimulating much inflammation is unexplained.

Widespread infection is associated with T-lymphocyte defects and *T rubrum*

DERMATOPHYTOSES: CLINICAL ASPECTS

MANIFESTATIONS

Dermatophyte infections range from inapparent colonization to chronic progressive eruptions that last months or years, causing considerable discomfort and disfiguration. Dermatologists often give each infection its own "disease" name, for example, tinea capitis (scalp; **Figure 44–3A**), tinea pedis (feet, athlete's foot), tinea manuum (hands), tinea cruris (groin), tinea barbae (beard, hair), and tinea unguium (nail beds). Skin infections not included in this anatomic list are called tinea corporis (body). There are some general clinical, etiologic, and epidemiologic differences among these syndromes, but they are all the same disease in different locations. The primary differences among etiologic agents that infect different sites are shown in Table 44–1.

Various skin sites are labeled as tinea "diseases"

Infection of hair begins with an erythematous papule around the hair shaft, which progresses to scaling of the scalp, discoloration, and eventually fracture of the shaft. Spread to adjacent hair follicles progresses in a ring-like fashion, leaving behind broken, discolored hairs, and sometimes black dots where the hair is absent but the infection has gone into the follicle. The degree of inflammatory response markedly affects the clinical appearance and, in some cases, can cause constitutional symptoms. In most cases, symptoms beyond itching are minimal.

Hair infection leads to itching and hair loss

Skin lesions begin in a similar pattern and enlarge to form sharply delineated erythematous borders with skin of nearly normal appearance in the center. Multiple lesions can fuse to form unusual geometric patterns on the skin. Lesions may appear in any location, but are particularly common in moist, sweaty skin folds. Obesity and the wearing of tight apparel increase susceptibility to infection in the groin and beneath the breasts. Another form of infection, which involves scaling and splitting of the skin between the toes, is commonly known as **athlete's foot**. Moisture and maceration of the skin provide the mode of entry.

Skin infection favors moist areas and skin folds

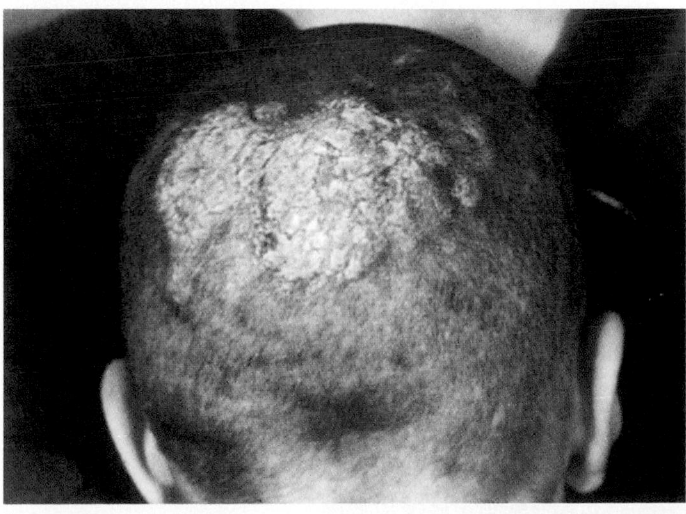

A

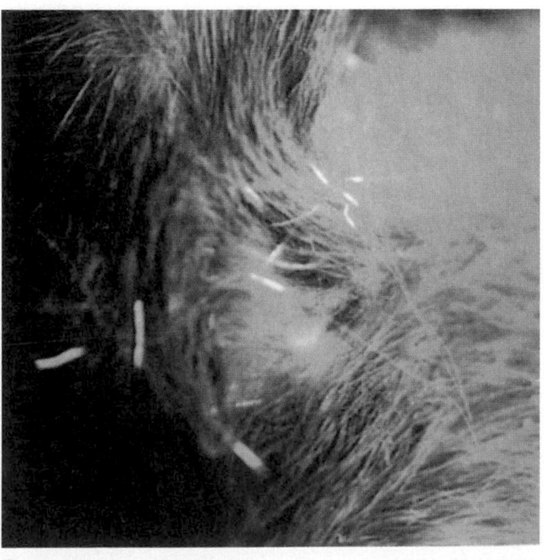

B

FIGURE 44–3. Tinea capitis. A. Ringworm of the scalp with superficial lesions and loss of hair. **B.** Close-up using an ultraviolet lamp (Wood's light) reveals fluorescing hair fragments. The culture grew *Microsporum audouinii*. (Reproduced with permission from Willey J, Sherwood L, Woolverton C (eds). *Prescott's Principles of Microbiology.* New York: McGraw-Hill; 2008.)

Hyperkeratosis can dislodge the nail bed

Nail bed infections first cause discoloration of the subungual tissue, then hyperkeratosis and apparent discoloration of the nail plate by the underlying infection follow. Direct infection of the nail plate is uncommon. Progression of hyperkeratosis and associated inflammation cause disfigurement of the nail but few symptoms until the nail plate is so dislodged or distorted that it exposes or compresses adjacent soft tissue.

DIAGNOSIS

KOH mounts of skin scrapings and infected hairs demonstrate hyphae

Some species fluoresce

The goal of diagnostic procedures is to distinguish dermatophytoses from other causes of skin inflammation. Infections caused by bacteria, other fungi, and noninfectious disorders (eg, psoriasis and contact dermatitis) may have similar features. The most important step is microscopic examination of material taken from lesions to detect the fungus. Potassium hydroxide (KOH) or calcifluor white preparations of scales scraped from the advancing edge of a dermatophyte lesion demonstrate septate hyphae. Examination of infected hairs reveals hyphae and arthroconidia penetrating the hair shaft. Broken hairs give the best yield. Some species of dermatophyte fluoresce, and selection of hairs for examination can be aided by the use of an ultraviolet lamp (Wood's) lamp (Figure 44–3B). ::: KOH preparation, p. 700

Culture is used when KOH preparations negative

The same material used for direct examination can be cultured for isolation of the offending dermatophyte and demonstration of typical conidia (**Figure 44–4**) which are not produced in clinical lesions. Mild infections with typical clinical findings and positive KOH preparations are often not cultured because clinical management is not influenced significantly by the identity of the etiologic species. Clinically typical infections with negative KOH preparations require culture. The major reason for false-negative KOH results, however, is failure to collect the scrapings or hairs properly. Nucleic acid amplification procedures have been successfully applied to skin and nail scrapings, but their use is limited.

TREATMENT AND PREVENTION

Topical tolnaftate, allylamines, or azoles usually sufficient

Systemic griseofulvin or azoles used in refractory cases

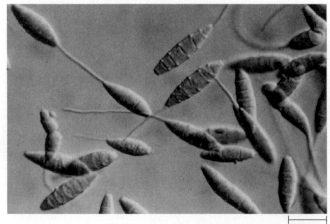

Many local skin infections resolve spontaneously without chemotherapy. Those that do not may be treated with topical tolnaftate, allylamines, or azoles (miconazole, clotrimazole, ketoconazole, oxiconazole). Nail bed and more extensive skin infections require systemic therapy with griseofulvin or itraconazole and terbinafine, often combined with topical therapy. Therapy must be continued over weeks to months, and relapses may occur. Keratolytic agents (Whitfield's ointment) may be useful for reducing the size of hyperkeratotic lesions. Dermatophyte infections can usually be prevented simply by observing general hygienic measures. No specific preventive measures such as vaccines exist.

FIGURE 44–4. Large boat-shaped macroconidia of *Microsporum gypseum*. (Reproduced with permission from Nester EW, Anderson DG, Roberts CE Jr, Nester MT. *Microbiology: A Human Perspective,* 6th ed. New York: McGraw-Hill, 2008.)

Other Superficial Mycoses

Pityriasis (tinea) versicolor occurs in tropical and temperate climates; it is characterized by discrete areas of hypopigmentation or hyperpigmentation associated with induration and scaling. Lesions are found on the trunk and arms; some assume pigments ranging from pink to yellow-brown—hence the term *versicolor*. Members of the genus *Malassezia*, of which *M furfur* is the most common, are the cause of pityriasis versicolor; these organisms can be seen in skin scrapings as clusters of budding yeast cells mixed with hyphae. They grow in the yeast form in culture media enriched with lipids.

M furfur requires lipids for growth

Tinea nigra, another tropical infection, is characterized by brown to black macular lesions, usually on the palms or soles. There is little inflammation or scaling, and the infection is confined to the stratum corneum. The cause, *Hortaea werneckii*, is a black-pigmented fungus found in soil and other environmental sites. Scrapings of the lesion show brown-black–pigmented septate hyphae. In culture, initial growth is in the yeast form, with slow development of hyphal elements.

H werneckii causes black lesions

Piedra is an infection of the hair characterized by black or white nodules attached to the hair shaft. White piedra (caused by *Trichosporon cutaneum*) infects the shaft in hyphal forms, which fragment with occasional buds. Black piedra (caused by *Piedraia hortae*) shows branched hyphae in sections of the hair (Figure 44–2).

Black or white piedra are infections of hair shaft

SUBCUTANEOUS FUNGI

Assignment of fungal organisms to the category of subcutaneous fungi is somewhat arbitrary, because fungal pathogens can produce many subcutaneous manifestations as part of their disease spectrum. Those considered here are introduced traumatically through the skin and are typically limited to subcutaneous tissues, lymphatic vessels, and contiguous tissues. They rarely spread to distant organs. The diseases they cause include sporotrichosis, chromoblastomycosis, and mycetoma. Only sporotrichosis has a single specific etiologic agent, *Sporothrix schenckii*. Chromoblastomycosis and mycetoma are clinical syndromes with multiple fungal etiologies.

Sporothrix

SPOROTHRIX SCHENCKII

S schenckii is a dimorphic fungus that grows as a cigar-shaped, 3- to 5-mm yeast in tissues and in culture at 37°C. The mold, which grows in culture at 25°C, is presumably the infectious form in nature. The hyphae are thin and septate, producing clusters of conidia at the end of delicate conidiophores. *S schenckii* is able to synthesize melanin which is present in the conidia.

Mold conidiophores convert to cigar-shaped yeast

SPOROTRICHOSIS

CLINICAL CAPSULE

Sporothrix schenckii is widely present in soil and other organic matter in the environment. Sporotrichosis begins with injection of one of the organism's conidia into the subcutaneous tissue. A thorn prick or sliver in the hand is a typical event. *S schenckii* then begins a slow inflammatory process that follows the lymphatic drainage from the original site. Superficial ulcer are produced, but the organism rarely invades deeper.

EPIDEMIOLOGY

Soil saprophyte is introduced by trauma

Occupational disease of gardeners and farmers

Outbreaks involve wood and moss

S schenckii is a ubiquitous saprophyte particularly found in hay, moss, soil (including potting soil), and decaying vegetation, and on the surfaces of various plants. Infection is acquired by traumatic inoculation through the skin of material containing the organism. Exposure is largely occupational or related to hobbies. The skin of gardeners, farmers, and rural laborers is frequently traumatized by thorns or other material that may be contaminated with conidia of *S schenckii*. An unusual outbreak of sporotrichosis involving nearly 3000 miners was traced to *S schenckii* in the timbers used to support mine shafts. A 1988 outbreak covered 15 states and was traced to sphagnum moss. Infection is occasionally acquired by direct contact with infected pus or through the respiratory tract; these modes of infection, however, are much less common than the cutaneous route.

PATHOGENESIS

Surface binds to extracellular matrix

Melanin resists oxidative killing

Both the conidia and yeast cells of *S schenckii* are able to bind to extracellular matrix proteins such as fibronectin, laminin, and collagen. Local multiplication of the organism stimulates both acute pyogenic and granulomatous inflammatory reactions. The presence of melanin in the infectious conidia may facilitate survival in the early stages of infection, since it is known to protect against oxidative killing in tissues and macrophages. Proteinases similar to those seen in other fungal pathogens are present, but no connection to virulence has been established. The infection spreads along lymphatic drainage routes and reproduces the original inflammatory lesions at intervals. The organisms are scanty in human lesions.

IMMUNITY

CMI is primary immune mechanism

Some studies indicate that exposure to *S schenckii* is fairly common and there is a high level of innate immunity. The cellular response to infection is mixed. The increased frequency and greater severity of disseminated disease in patients with T-cell defects points to T_H1 responses as the primary immune mechanism. Antibody plays no known role in immunity.

SPOROTRICHOSIS: CLINICAL ASPECTS

MANIFESTATIONS

Skin papule eventually ulcerates

Lymphatic involvement creates multiple lesions

A skin lesion begins as a painless papule that develops a few weeks to a few months after inoculation. Its location can usually be explained by occupational exposure; the hand is most often involved. The papule enlarges slowly and eventually ulcerates, leaving an open sore. Draining lymph channels are usually thickened, and pustular or firm nodular lesions may appear around the primary site of infection or at other sites along the lymphatic drainage route (**Figure 44–5**). Once ulcerated, lesions usually become chronic. Multiple ulcers often develop if the disease is untreated. Symptoms are those directly related to the local areas of infection. Constitutional signs and symptoms are unusual.

Deep infection is rare

Occasionally, spread occurs by other routes. The bones, eyes, lungs, and central nervous system are susceptible to progressive infection if the organisms reach these organs; such spread, however, occurs in less than 1% of all cases. Primary pulmonary sporotrichosis occurs but is also rare.

DIAGNOSIS

Direct microscopic examination for *S schenckii* is usually unrewarding because there are too few organisms to detect readily with KOH preparations. Even specially stained biopsy samples and serial sections are usually negative, although the presence of a histopathologic structure, the asteroid body, is suggestive. This structure is composed of *S schenckii* yeast cells surrounded by amorphous eosinophilic "rays." Definitive diagnosis depends on culture of infected pus or tissue. The organism grows within 2 to 5 days on all media commonly used in medical mycology. Identification requires demonstration of the typical conidia and of dimorphism.

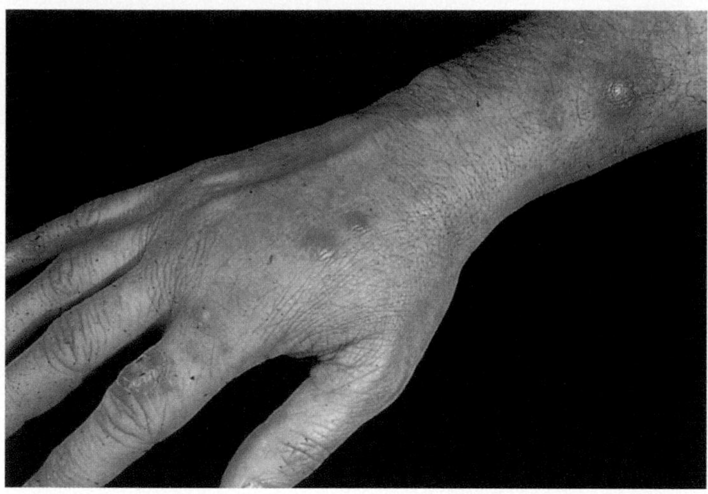

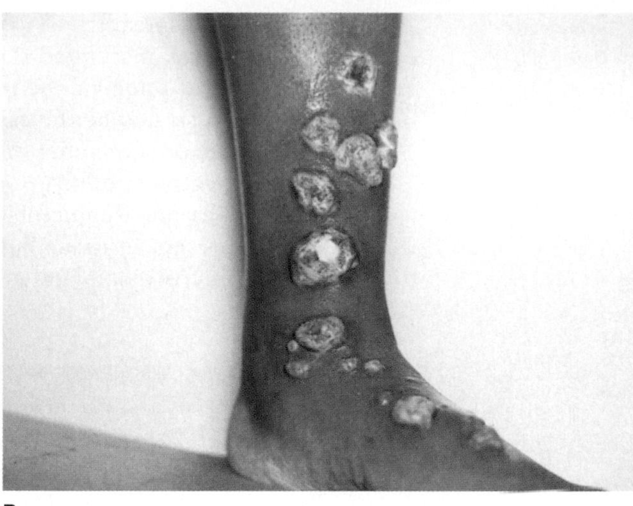

A B

FIGURE 44–5. Sporotrichosis. A. This infection began on the finger and has started to spread up the arm, leaving satellite lesions behind. If untreated, these lesions will evolve into ulcers. **B.** A more advanced case beginning with inoculation in the foot. (Reproduced with permission from Connor DH, Chandler FW, Schwartz DQA, Manz HJ, Lack EE (eds). *Pathology of Infectious Diseases,* vol. 1. Stamford, CT: Appleton & Lange; 1997.)

TREATMENT AND PREVENTION

Cutaneous sporotrichosis was long treated with a saturated solution potassium iodide (SSKI) administered orally. This treatment has now been supplanted by itraconazole for all forms of disease. Pulmonary and systemic infections may require the additional use of amphotericin B. Eradication of the environmental reservoir of *S schenckii* is not usually practical, although the mine outbreak mentioned previously was stopped by applying antifungal agents to the mine shaft timbers.

Potassium iodide works for cutaneous fungi

Amphotericin B or itraconazole required for progressive disease

CHROMOBLASTOMYCOSIS

Chromoblastomycosis is primarily a tropical disease caused by multiple species of *Fonsecaea, Phialophora*, and *Cladophialophora* (*Cladosporium*). The disease occurs typically on the foot or leg. It appears as papules that develop into scaly, wart-like structures, usually under the feet. Fully developed lesions have been likened to the tips of a cauliflower. Extension is by satellite lesions; it is slow and painless and does not involve the lymphatic vessels. The organisms are found in the soil of endemic areas, and most infections occur in individuals who work barefoot.

Multiple species produce wart-like pigmented lesions in tropics

The outstanding mycologic feature is the presence of brown-pigmented, thick-walled, multiseptate, 5- to 12-mm globose structures called muriform bodies on histologic section. Branching septate hyphae may also be demonstrated in KOH preparations of scrapings. Cultures grow as dark molds, but may take weeks to appear and longer for demonstration of characteristic conidia. Surgery and antifungal therapy have been used in chromoblastomycosis, but results in advanced disease are disappointing. Flucytosine or itraconazole have been the antifungal agents most frequently used.

Brown pigmented bodies are seen in tissues

MYCETOMA

Mycetoma is a clinical term for an infection associated with trauma to the foot that causes inoculation of any of a dozen fungal species. Actinomycetes such as *Nocardia* may produce a similar disease. The typical clinical appearance is of massive induration with draining sinuses. Some of the fungi that cause mycetoma are geographically widespread; most cases, however, occur in the tropics, probably because the chronically damp, macerated skin of the feet that causes predisposition toward mycetoma occurs most often among those who go barefoot in the tropical environment. This finding is illustrated by the case of a college rower in Seattle who developed mycetoma; he was the only member of his shell who insisted on rowing barefoot. Once established, the treatment of mycetoma is difficult. No antimicrobic stands out as particularly helpful. The precise microbiologic features depend on the agent involved. Hyphae are usually present in tissue but, may be difficult to demonstrate because of a tendency to form microcolonial granules. ::: Cutaneous Nocardia, p. 512

Multiple species are involved

Trauma to bare feet injects the fungi

CLINICAL CASE

HEAD BUMP

A 4-year-old boy was taken by his mother to the family doctor for evaluation of a 2-month history of a slowly growing "bump" on the back of his head. The boy had no other siblings nor any pets at home. He attended a day care center weekdays while his mother worked. Examination revealed a happy, alert child in no distress. A raised, scaling lesion 3.5 cm in diameter with a few pinpoint pustules was present on the posterior scalp. A KOH preparation of material from the lesion was negative. A fungal culture of material from the lesion was later positive for a fungus with numerous microconidia and macroconidia typical of *Microsporum* species.

QUESTIONS

■ What is the most likely source of this child's infection?

A. Parents
B. Child at day care center
C. Animal
D. Insect
E. Food

■ What is the human niche where this organism proliferates best?

A. Fibronectin
B. Macrophages
C. M cells
D. Keratin

■ What additional examination might have revealed this infection while the child was in the doctor's office?

A. X-ray
B. Serologic test
C. Ultraviolet light
D. Biopsy
E. DNA probe

ANSWERS

1(B), **2**(D), **3**(C)

CHAPTER 45

Candida, Aspergillus, Pneumocystis, and Other Opportunistic Fungi

The fungi considered in this chapter are usually found as members of the normal flora or as saprophytes in the environment. With breakdown of host defenses, they can produce disease ranging from superficial skin or mucous membrane infections to systemic involvement of multiple organs. The most common opportunistic infections are caused by the yeast *Candida albicans*, a normal inhabitant of the gastrointestinal and genital floras, and the mold *Aspergillus*, commonly found in the environment. *Pneumocystis*, a prominent cause of pneumonia in AIDS patients, used to be considered a parasite on morphologic grounds. The diseases caused by these opportunistic fungi are summarized in **Table 45–1**.

CANDIDA: GENERAL CHARACTERISTICS

Candida species grow as typical 4- to 6-µm, budding, round, or oval yeast cells (**Figure 45–1**) under most conditions and at most temperatures. Under certain conditions, including those found in infection, they can form hyphae. Some species form chlamydoconidia in culture. *Candida* species identification is based on a combination of biochemical, enzymatic, and morphologic characteristics, such as carbohydrate assimilation; fermentation; and the ability to produce hyphae, germ tubes, and chlamydoconidia. Of the over 150 *Candida* species, fewer than 10 appear in human disease. Particular attention is given to the differentiation of *C albicans* from other species, because it is by far the most common cause of disease.

Formation of hyphae and chlamydoconidia are distinguishing features

Carbohydrate assimilation and fermentation determine species

Most *Candida* species grow rapidly on Sabouraud's agar and on enriched bacteriologic media such as blood agar. Smooth, white, 2- to 4-mm colonies resembling those of staphylococci are produced on blood agar after overnight incubation. Aeration of cultures favors their isolation. The primary identification procedure involves presumptive differentiation of *C albicans* from the other *Candida* species with the germ tube test (see text that follows). Germ tube–negative strains may be further identified biochemically or reported as "yeast not *C albicans*," depending on their apparent clinical significance. :::Yeasts and molds, p. 695

Rapidly produce colonies resembling staphylococci

C albicans produces germ tubes

Candida albicans

MYCOLOGY

C albicans grows in multiple morphologic forms, most often as a yeast with budding by formation of blastoconidia. *C albicans* is also able to form hyphae triggered by changes in conditions such as temperature, pH, and available nutrients. When observed in their initial stages when still attached to the yeast cell, these hyphae look like sprouts and are called germ tubes (**Figure 45–2A**).

TABLE 45–1 Agents of Opportunistic Mycoses

		GROWTH			
ORGANISM	**TISSUE**	**CULTURE AT 25°C**	**CULTURE AT 37°C**	**SOURCE**	**INFECTION**
Candida	Yeast (hyphae)[a]	Yeast (hyphae)[a]	Yeast	Endogenous	Skin, mucous membranes, urinary, disseminated
Aspergillus	Hyphae (septate)	Mold	Mold	Environment	Lung, disseminated
Zygomycetes[b]	Hyphae (nonseptate)	Mold	Mold	Environment	Rhinocerebral, lung, disseminated
Pneumocystis	Elliptical spores	None[c]	None[c]	Unknown	Pneumonia

[a]Less common feature; pseudohyphae are produced as well.
[b]Such genera as *Absidia, Mucor,* and *Rhizopus.*
[c]Has not been grown in culture.

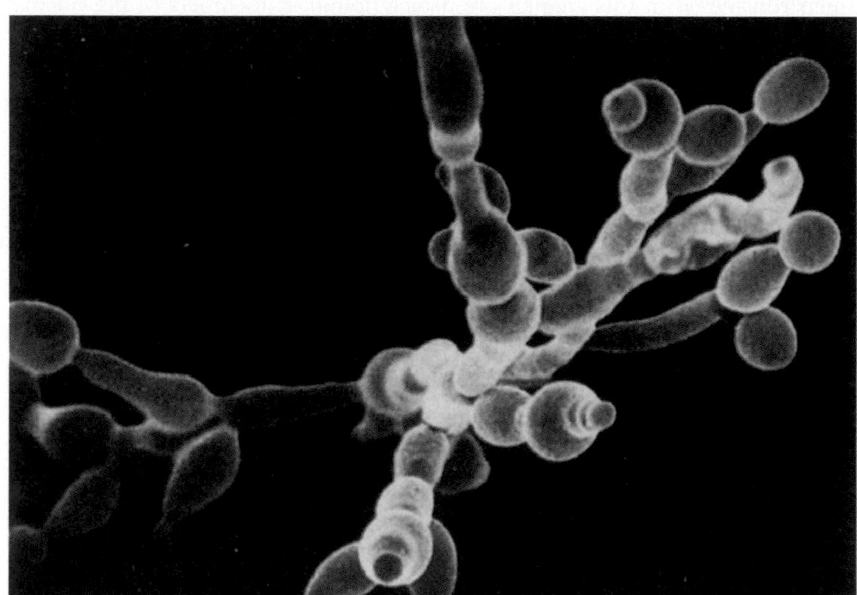

FIGURE 45–1. *Candida albicans.* This scanning electron micrograph demonstrates dimorphism with both blastoconidia and hyphae. (Reproduced with permission from Willey J, Sherwood L, Woolverton C (eds). *Prescott's Principles of Microbiology.* New York: McGraw-Hill; 2008.)

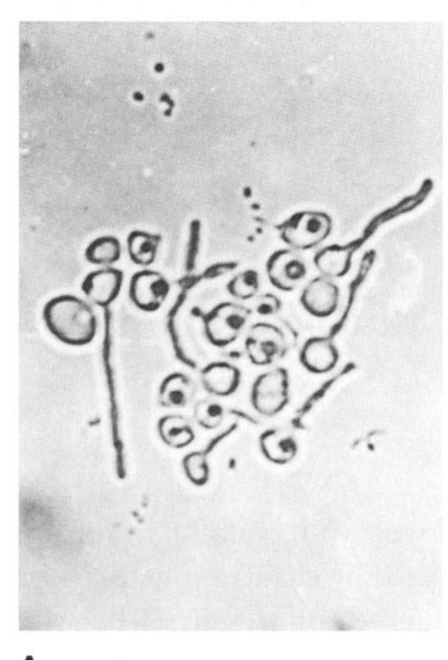

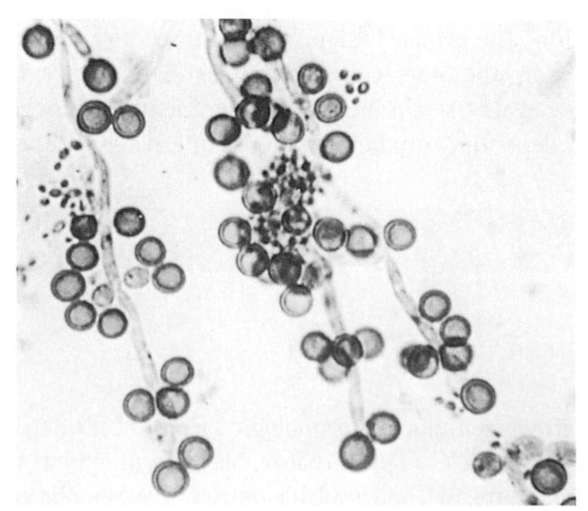

FIGURE 45–2. *Candida albicans.* **A.** When incubated at 37°C, *C albicans* rapidly forms elongated hyphae called germ tubes. **B.** On specialized media, *C albicans* forms thick-walled chlamydoconidia, which differentiate it from other *Candida* species. (Reprinted with permission from Dr. E. S. Beneke and the Upjohn Company: Scope Publications, *Human Mycoses.*)

Other elongated forms with restrictions at intervals are called **pseudohyphae** because they lack the parallel walls and septation of the true hyphae. There is evidence that these three forms have distinct stimuli and genetic regulation, making *C albicans* a polymorphic fungus. Unless otherwise specified, the term **hyphae** is used here to encompass both the true and pseudohyphal forms. The hyphal form also develops characteristic terminal thick-walled **chlamydoconidia** under certain cultural conditions (Figure 45–2B).

Yeast, hyphae, and pseudohyphae are formed

Chlamydoconidia develop from hyphae in culture

The *C albicans* cell wall is made up of a mixture of the polysaccharides mannan, glucan, and chitin alone or in complexes with protein. A fibrillar outer layer extending to the surface contains a number of glycoproteins and complexes of mannan with protein called mannoproteins. The exact composition of the cell wall and surface components varies under different growth conditions. ::: Fungal cell wall, p. 694

Cell wall includes surface mannoproteins

CANDIDIASIS

CLINICAL CAPSULE

Candidiasis occurs in localized and disseminated forms. Localized disease is seen as erythema and white plaques in moist skinfolds (diaper rash) or on mucosal surfaces (oral thrush). It may also cause the itching and thick white discharge of vulvovaginitis. Deep tissue and disseminated disease are limited almost exclusively to the immunocompromised. Diffuse pneumonia and urinary tract involvement are especially common.

EPIDEMIOLOGY

C albicans is a common member of the oropharyngeal, gastrointestinal, and female genital flora. Infections are endogenous except in cases of direct mucosal contact with lesions in others (eg, through sexual intercourse). Although *C albicans* is a common cause of nosocomial infections, the fungi are also derived more frequently from the patient's own flora than from cross-infection. Invasive procedures and indwelling devices may provide the portal of entry, and the number of available *Candida* may be enhanced by the used of antibacterial agents.

Infections are from endogenous flora

PATHOGENESIS

Because *C albicans* is regularly present on mucosal surfaces, disease implies a change in the organism, the host, or both. The change from the yeast to the hyphal form is strongly associated with enhanced pathogenic potential of *C albicans*. In histologic preparations, hyphae are seen only when *Candida* starts to invade, either superficially or in deep tissues. This switch can be controlled in vitro by the manipulation of a wide variety of environmental conditions (serum, pH, temperature, amino acids). Although a number of sensors and signaling pathways have been discovered, it is still not known what triggers these changes in human disease. What is known is that the morphologic change is also associated with the appearance of a number factors associated with tissue adherence and digestion.

Shift from yeast to hyphae is associated with invasion

Switch is triggered by environmental conditions

C albicans hyphae have the capacity to form strong attachments to human epithelial cells. One mediator of this binding is a surface **hyphal wall protein (Hwp1)**, which is found only on the surface of germ tubes and hyphae. Other mannoproteins that have similarities to vertebrate integrins may also mediate binding to components of the **extracellular matrix (ECM)**, such as fibronectin, collagen, and laminin. Hyphae also secrete

Hwp1 binds to epithelial cells

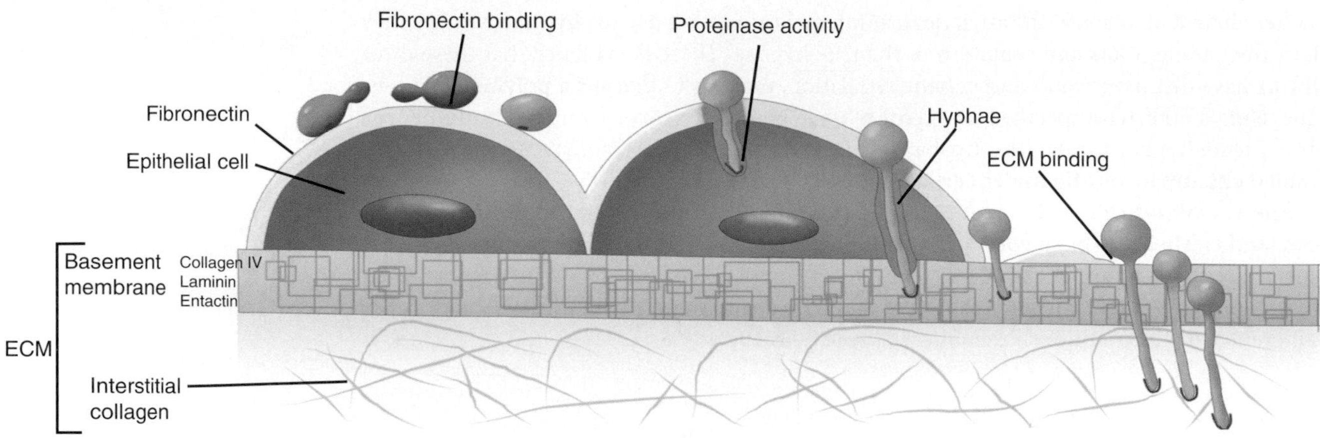

FIGURE 45–3. Pathogenesis of *Candida albicans* infections. Proposed mechanisms of *C albicans* attachment and invasion are shown. Surface glucomannan receptor(s) on the yeast may bind to fibronectin covering the epithelial cell or to elements of the extracellular matrix (ECM) when the epithelial surface is lost or when the *Candida* have invaded beyond it. Invasion is associated with formation of hyphae and production of proteinases, which may digest tissue elements.

Mannoproteins bind to ECM

Hyphae produce Saps, other enzymes

proteinases and phospholipases that are able to digest epithelial cells and probably facilitate invasion (**Figures 45–3** and **45–4A** and **B**). One family of hyphal enzymes, the secreted aspartic proteinases (Saps), are able to digest keratin and collagen, which would facilitate deep tissue invasion. The pattern of Sap production may be tissue-specific. For example, *C albicans* invading vaginal epithelium produce a particular set of Saps. Taken together, these factors represent a rich armamentarium of virulence factors all linked to the change from yeast to hyphal growth.

C albicans hyphae also have surface proteins that resemble the complement receptors (CR2, CR3) on phagocytes. This seems likely to confuse their ability to recognize C3b bound to the candidal surface. Enhanced production of these receptors under various conditions,

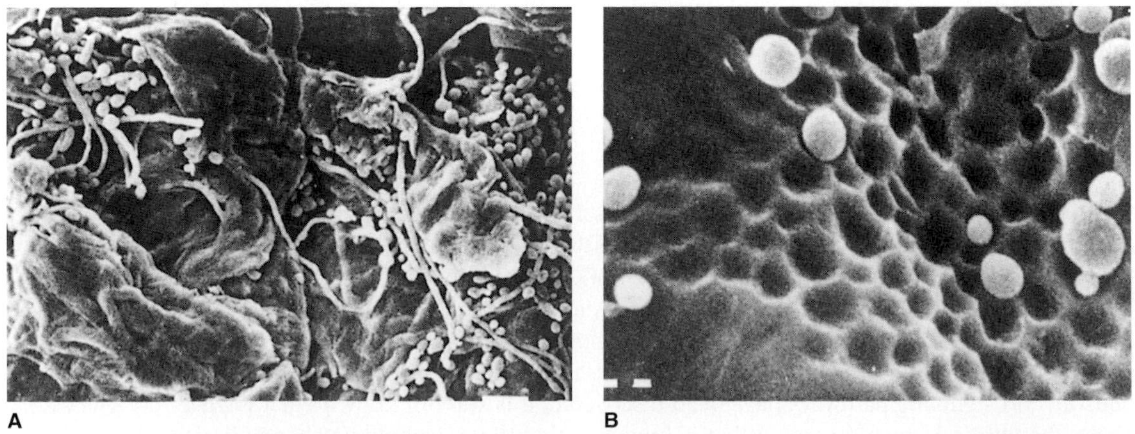

FIGURE 45–4. Invasiveness of *Candida albicans*. Two features of invasiveness are seen in these scanning electron micrographs taken from experiments with murine corneocytes. **A.** Both blastoconidia and mycelial elements are present. The mycelial elements spread over the surface and invade the cell cuticle. **B.** A *C albicans* strain that produces a protease is seen producing cavity-like depressions in the cell surface. This action could play a role in invasion of the cell. (Reprinted with permission of Thomas L. Ray and Candia D. Payne. *Infect Immunol* 1988;56:1945–1947, Figures 4,6B. Copyright American Society for Microbiology.)

such as elevated glucose concentration, is associated with resistance to phagocytosis by neutrophils.

Surface proteins resemble complement receptors

Factors that allow *C albicans* to increase its relative proportion of the flora (antibacterial therapy), that compromise the general immune capacity of the host (leukopenia or corticosteroid therapy), or that interfere with T-lymphocyte function (acquired immunodeficiency syndrome; AIDS) are often associated with local and invasive infection. The disruptions of the mucosa associated with chronic disease and their treatments (in-dwelling devices, cancer chemotherapy) may enhance the invasion process by exposing *Candida* binding sites in the ECM. *C albicans* has also demonstrated a capacity to form biofilms on the plastics used in medical devices. Diabetes mellitus also predisposes to *C albicans* infection, possibly because of the known greater production of the surface mannoproteins in the presence of high glucose concentrations.

Antimicrobics and immunosuppression increase risk

Mechanical disruptions may provide access to ECM

IMMUNITY

Both humoral immunity and cell-mediated immunity (CMI) are involved in defense against *Candida* infections. Neutrophils are the primary first-line defense. Yeast forms of *C albicans* are readily phagocytosed and killed when opsonized by antibody and complement. In the absence of specific antibody, the process is less efficient, but a naturally occurring antimannan IgG is able to activate the classical complement pathway and facilitate the alternate pathway. Hyphal forms may be too large to be ingested by polymorphonuclear neutrophils (PMNs), but they can still kill the fungi by attaching to the hyphae and discharging metabolites generated by the oxidative metabolic burst. A deficit in neutrophils or neutrophilic function is the most common correlate of serious *C albicans* infection. ::: Complement pathways, p. 26

Opsonized yeast forms are killed by PMNs

Antimannan IgG activates complement

The association of chronic mucocutaneous candidiasis (see text that follows) with a number of T-lymphocyte immunodeficiencies emphasizes the importance of this arm of the immune system in defense against *Candida* infections. The increased frequency of oral and vaginal candidiasis in AIDS patients suggests that even superficial infections involve T-lymphocyte–mediated T_H1 immune responses. In animal studies, *Candida* cell wall mannan has been shown to play an immunoregulatory function by down-regulating CMI responses. A possible explanation for the association between AIDS and *Candida* infection is the up-regulation of CD4 receptors on monocytes by *Candida* products. As with other fungi, cytokine activation of macrophages enhances their ability to kill *C albicans*. A favorable outcome appears to require the proper balance between T_H1- and T_H2-mediated cytokine responses. The cytokines associated with T_H1 (interleukin-2 [IL-2], IL-12, interferon-γ, tumor necrosis factor-α) are correlated with enhanced resistance against infection in which T_H2 responses (IL-4, IL-6, and IL-10) are associated with chronic disease. ::: T-cell responses, p. 33

Compromised CMI is associated with progressive infection

Candida mannan may down-regulate CMI responses

Balance between T_H1 and T_H2 cytokines is necessary

CANDIDIASIS: CLINICAL ASPECTS

MANIFESTATIONS

Superficial invasion of the mucous membranes by *C albicans* produces a white, cheesy plaque that is loosely adherent to the mucosal surface. The lesion is usually painless, unless the plaque is torn away and the raw, weeping, invaded surface is exposed. Oral lesions, called **thrush,** occur on the tongue, palate, and other mucosal surfaces as single or multiple, ragged white patches (**Figure 45–5**). A similar infection in the vagina, vaginal candidiasis, produces a thick, curd-like discharge and itching of the vulva. Although most women have at least one episode of **vaginal candidiasis** in a lifetime, a small proportion suffer chronic, recurrent infections. No general or specific immune defect has yet been linked to this syndrome.

White mucosal plaque is called thrush

Vaginitis may be recurrent

C albicans skin infections occur in crural folds and other areas in which wet, macerated skin surfaces are opposed. For example, one type of diaper rash is caused by *C albicans* (**Figure 45–6A**). Other infections of the skinfolds and appendages occur in association

Macerated skin is a common site

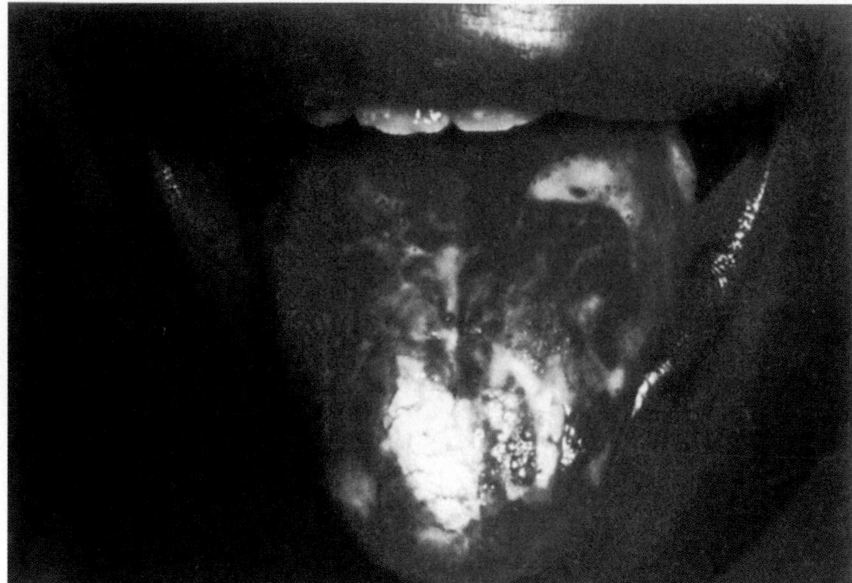

FIGURE 45–5. Trush. The white plaques on this AIDS patient's tongue are caused by *Candida albicans*. (Reproduced with permission from Willey J, Sherwood L, Woolverton C (eds). *Prescott's Principles of Microbiology*. New York: McGraw-Hill; 2008.)

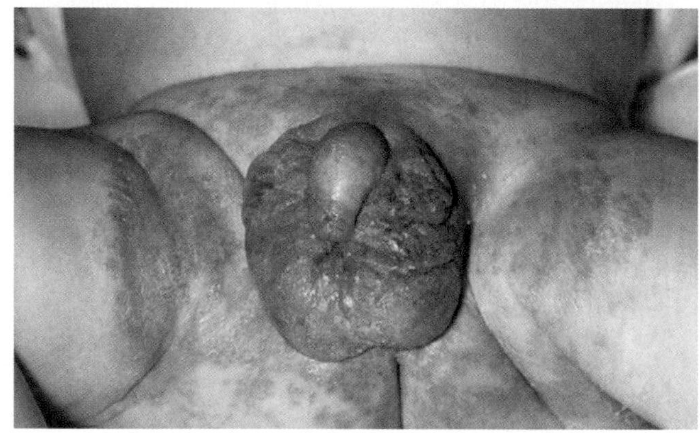

A

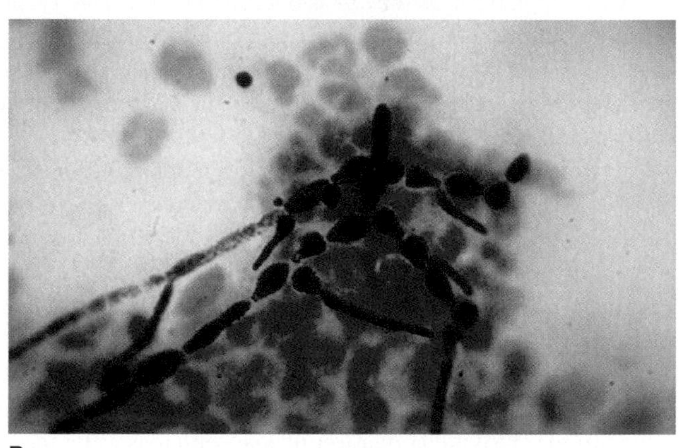

B

FIGURE 45–6. *Candida albicans* skin infection. A. This rash is preceded by chronically damp skin in the diaper area. **B.** This Gram stain demonstrates yeast cells and pseudohyphae. (Reproduced with permission from Nester EW, Anderson DG, Roberts CE Jr, Nester MT. *Microbiology: A Human Perspective*, 6th ed. New York: McGraw-Hill, 2008.)

with recurrent immersion in water (eg, dishwashers). The initial lesions are erythematous papules or confluent areas associated with tenderness, erythema, and fissures of the skin. Infection usually remains confined to the chronically irritated area, but may spread beyond it, particularly in infants.

Chronic mucocutaneous candidiasis is associated with specific T-cell defects

In rare persons with specific defects in T_H1 immune defenses against *Candida*, a chronic, relapsing form of candidiasis known as **chronic mucocutaneous candidiasis** develops. Infections of the skin, hair, and mucocutaneous junctions fail to resolve with adequate therapy and management. There is considerable disfigurement and discomfort, particularly when the disease is accompanied by a granulomatous inflammatory response. Although lesions may become extensive, they usually do not disseminate. To some degree, candidiasis may represent a clinical example of immunologic tolerance. Cutaneous anergy to *C albicans* antigens is commonly seen in these patients and is often reversed during antifungal chemotherapy, suggesting that it is due to chronic antigen excess.

Inflammatory patches similar to those in thrush may develop in the esophagus with or without associated oral candidiasis. Painful swallowing and substernal chest pain are the most common symptoms. Extensive ulcerations, deformity, and occasionally perforation of the esophagus may ensue. In immunocompromised patients, similar lesions may also develop in the stomach, together with deep ulcerative lesions of the small and large intestine.

Esophagitis and intestinal candidiasis are similar to thrush

Infection of the urinary tract via the hematogenous or ascending routes may produce cystitis, pyelonephritis, abscesses, or expanding fungus ball lesions in the renal pelvis. The clinical findings in disseminated infections of the kidneys, brain, and heart are generally not sufficiently characteristic to suggest *C albicans* over the bacterial pathogens, which more commonly produce infection of deep organs. ***Candida* endophthalmitis** has the characteristic funduscopic appearance of a white cotton ball expanding on the retina or floating free in the vitreous humor. Endophthalmitis and infections of other eye structures can lead to blindness.

Urinary tract infections are ascending or hematogenous

Endophthalmitis appears as white cotton on retina

DIAGNOSIS

Superficial *C albicans* infections provide ready access to diagnostic material. Exudate or epithelial scrapings examined by KOH preparations or Gram smear (Figure 45–6B) demonstrate abundant budding yeast cells; if associated hyphae are present, the infection is almost certainly caused by *C albicans*. *C albicans* is readily isolated from clinical specimens including blood if aerobic conditions are provided. Cultures from specimens such as sputum run the risk of contamination from the normal flora or a superficial mucous membrane lesion. A direct aspirate, biopsy, or bronchoalveolar lavage is often required to establish the diagnosis.

KOH and Gram smears of superficial lesions show yeast and hyphae

Lung involvement requires bronchoalveolar lavage

Deep organ involvement is difficult to prove without a direct aspirate or biopsy. Even positive blood cultures must be interpreted with caution if they could represent colonization of intravenous catheters. *Candida* endocarditis represents a special diagnostic problem because the yeasts seeding the blood from the valve may be filtered out in the capillary beds as a result of their large size. Arterial blood cultures may be required in this situation.

Endocarditis may require arterial cultures

TREATMENT

C albicans is usually susceptible to amphotericin B, nystatin, flucytosine, caspofungin, and the azoles. Superficial infections are generally treated with topical nystatin or azole preparations. Measures to decrease moisture and chronic trauma are important adjuncts in treating *Candida* skin infections. Deeper *C albicans* infections may resolve spontaneously with elimination or control of predisposing conditions. Removal of an infected catheter, control of diabetes, or an increase in peripheral leukocyte counts is often associated with recovery without antifungal therapy. Persistent relapsing or disseminated candidiasis is treated with various combinations of fluconazole, amphotericin B, and caspofungin. Fluconazole has been the most effective treatment for chronic mucocutaneous candidiasis.

Topical nystatin or azoles for superficial lesions

Amphotericin B, fluconazole, and caspofungin reserved for invasive disease

Other Candida Species

Species of *Candida* other than *C albicans* produce infections in circumstances similar to those described previously, but do so less frequently. When contamination of an indwelling device is the portal of entry, the probability of infection by these other species increases. Little is known of the pathogenesis of these species with the exception of *Candida tropicalis*. Both experimental and clinical evidence indicate that *C tropicalis* has virulence at least equal to that of *C albicans*. *C tropicalis* produces extracellular proteinases similar to those of *C albicans*, which may enhance its invasiveness.

C tropicalis is highly virulent

C glabrata is another common species. This species is very small for a yeast (2 to 4 μm) and does not produce hyphae. It is a member of the normal gastrointestinal and genital flora. The most common infections are in the urinary tract, but deep tissue involvement and fungemia occur. The organisms are small enough to be confused with *Histoplasma capsulatum* in histologic preparations. Therapy is similar to that for *C albicans* infections, although *C glabrata* is more resistant to fluconazole. Other species of *Candida*, which lack

C glabrata is a very small yeast

any distinguishing morphologic or clinical characteristics, may produce disease. Some of these fungi are inherently resistant to the antifungal azoles.

ASPERGILLUS

MYCOLOGY

Species are based on arrangement of conidia on the conidiophore

Aspergillus species are rapidly growing molds with branching **septate hyphae** and characteristic arrangement of conidia on the conidiophore (**Figure 45–7A–C**). Fluffy colonies appear in 1 to 2 days; by 5 days, they may cover an entire plate with pigmented growth. Species are defined on the basis of differences in the structure of the **conidiophore** and the arrangement of the **conidia**. The most common in human infections are *A fumigatus* and *A flavus*, but others, such as *A niger*, may be involved. ::: Conidia and conidiophores, p. 697

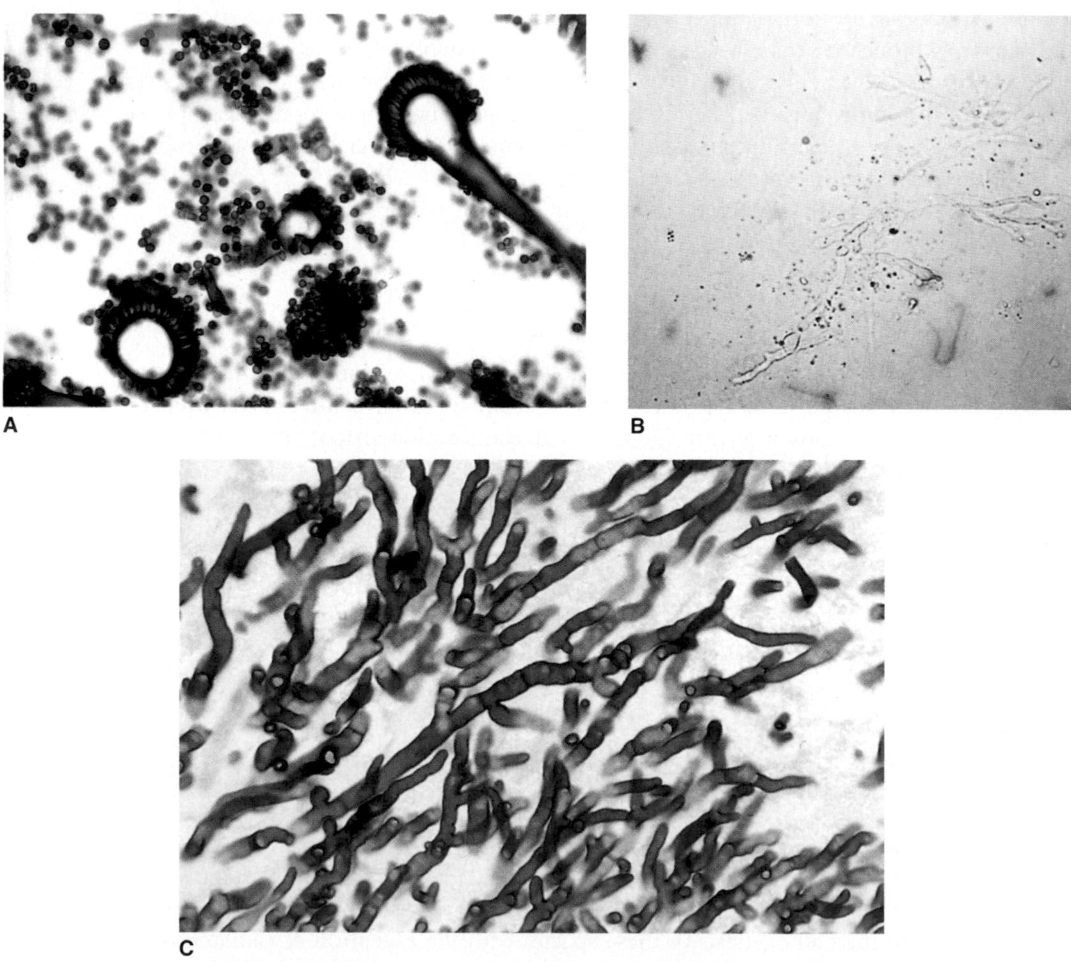

FIGURE 45–7. *Aspergillus*. A. This asexual conidium-forming structure is characteristic of *Aspergillus* species. The conidia are borne at the end of the finger-like extensions at the end of the conidiophore. These structures are rarely produced in vivo. **B.** This tissue aspirate mixed with KOH shows branching, septate hyphae. **C.** Histologic sections also show branching, septate hyphae, but because the conidia shown in A are not seen the findings are not diagnostic of *Aspergillus*. (A and C, Reproduced with permission from Connor DH, Chandler FW, Schwartz DQA, Manz HJ, Lack EE (eds). *Pathology of Infectious Diseases*, vol. 1. Stamford, CT: Appleton & Lange; 1997.)

ASPERGILLOSIS

CLINICAL CAPSULE

Invasive aspergillosis is distinguished by its setting in immunocompromised persons and its rapid progression to death. The typical patient is one who has leukemia or is under immunosuppression for a bone marrow transplantation. Fever and a dry cough may be the only signs until pulmonary infiltrates are demonstrated radiologically. Until *Aspergillus* hyphae are demonstrated, almost any of the causes of pneumonia could be responsible.

EPIDEMIOLOGY

Aspergillus species are widely distributed in nature and found throughout the world. They seem to adapt to a wide range of environmental conditions, and the heat-resistant conidia provide a good mechanism for dispersal. Like bacterial spores, the conida survive well in the environment, and their inhalation is the mode of infection. Hospital air and air ducts have received attention as sources of nosocomial *Aspergillus* isolates. Occasionally, construction, remodeling, or other kinds of major environmental disruption have been associated with increased frequency of *Aspergillus* contamination, colonization, or infection.

Conidia may be spread by construction projects

PATHOGENESIS

Aspergillus conidia are small enough to readily reach the alveoli when inhaled, but disease is rare in those without compromised defenses. Factors that aid the fungus in the initial stages are not known, but the ability of proteins on the surface of the conidia to bind fibrinogen and laminin probably contribute to adherence. Gliotoxin, a molecule that inhibits steps in the oxidative killing mechanisms of phagocytes, may also assist early progression. The more virulent species produce extracellular elastase, proteinases, and phospholipases. The appearance of antibodies to these enzymes during and following invasive aspergillosis argues for their importance, but their specific pathogenic role remains to be demonstrated. Most species produce aflatoxins and other toxic secondary metabolites, but their role in infection is also unknown.

Adherence, gliotoxin aid early survival

Extracellular proteases may cause injury

IMMUNITY

Macrophages, particularly pulmonary macrophages, are the first line of defense against inhaled *Aspergillus* conidia, phagocytosing and killing them by nonoxidative mechanisms. For the conidia that survive and germinate, PMNs become the primary defense. They are able to attach to the growing hyphae, generate an oxidative burst, and secrete reactive oxygen intermediates. Little is known of adaptive immunity in humans. Antibodies are formed, but their protective value is unknown. Although AIDS patients do develop *Aspergillus* infections, the association with T-cell deficiencies is not strong enough to draw conclusions about their importance.

Alveolar macrophages kill conidia, and PMNs attack hyphae

ASPERGILLOSIS: CLINICAL ASPECTS

MANIFESTATIONS

Aspergillus can cause clinical allergies or occasional invasive infection. In both cases, the lung is the organ primarily involved. Allergic aspergillosis, which can be a mechanism of exacerbation in patients with asthma, is characterized by transient pulmonary infiltrates, eosinophilia, and a rise in *Aspergillus*-specific antibodies. These conditions follow direct inhalation of fungal elements or colonization of the respiratory tract. Areas of the bronchopulmonary tree with poor drainage because of underlying disease or anatomic abnormalities may serve as a site for growth of organisms and continuous seeding with antigen.

Allergic disease marked by eosinophilia and specific IgG

Invasive aspergillosis occurs in the settings of preexisting pulmonary disease (bronchiectasis, chronic bronchitis, asthma, tuberculosis) or immunosuppression. Colonization with *Aspergillus* can lead to invasion into the tissue by branching septate hyphae (Figure 45–7C). In patients who already have a chronic pulmonary disease, mycelial masses can form a radiologically visible fungus ball (aspergilloma) within a preexisting cavity. Lung tissue invasion may penetrate blood vessels, causing hemoptysis or erosion into other structures with development of fistulas. Invasive disease outside the lung is rare unless patients are immunocompromised.

Highly invasive, including blood vessels

Fungus ball in cavities

An acute pneumonia may occur in severely immunocompromised patients, particularly those with phagocyte defects or depressed neutrophil counts due to immunosuppressive drugs. Multifocal pulmonary infiltrates expanding to consolidation are present with high fever. The prognosis is grave and dissemination to other organs common, which is not the case in immunocompetent hosts.

PQ: Pneumonia in immunocompromised host has grave prognosis

DIAGNOSIS

Aspergillus is relatively easy to isolate and identify. Its rapidly spreading mold growth and all too frequent contamination of cultures cause it to be regarded by microbiologists as a kind of weed. The diagnostic problem is distinguishing contamination and colonization with *Aspergillus* from invasive disease. The diagnosis cannot be made for certain without the use of lung aspiration, biopsy, or bronchoalveolar lavage. With material directly from the lesion, the presence of large, branching, septate hyphae (Figure 45–7B and C) and a positive culture are diagnostic. Occasionally, the complete fruiting bodies are produced in vivo, creating a striking and diagnostic histologic picture (Figure 45–7A). Serologic methods have been developed to demonstrate *Aspergillus* antibodies. Although these tests may be helpful in suggesting allergic aspergillosis, they have little value in invasive disease because anti-*Aspergillus* antibody is common in healthy persons. Detection of circulating *Aspergillus* galactomannan or glucan by various immunoassays shows promise for earlier diagnosis in immunocompromised patients.

Direct aspirate or biopsy is required to distinguish colonization from invasion

Serodiagnosis is useful only for allergic disease

TREATMENT AND PREVENTION

The newer azoles (voriconazole, itraconazole), caspofungin, and amphotericin B in various combinations are the recommended antifungals for invasive aspergillosis. No regimen is considered highly effective, because the mortality rate of invasive disease approaches 100%. In patients with pulmonary structural abnormalities and fungus balls, chemotherapy has little effect. Surgical removal of localized lesion is sometimes helpful, even in the brain. Construction of rooms with filtered air has been attempted to reduce exposure to environmental conidia.

Amphotericin B, azoles, caspofungi, and surgery are used

ZYGOMYCETES AND ZYGOMYCOSIS

Absidia, Rhizopus, and *Mucor* are soil saprophytes

Zygomycosis (mucormycosis) is the term applied to infection with any of a group of zygomycetes, the most common of which are *Absidia, Rhizopus,* and *Mucor*. These fungi are

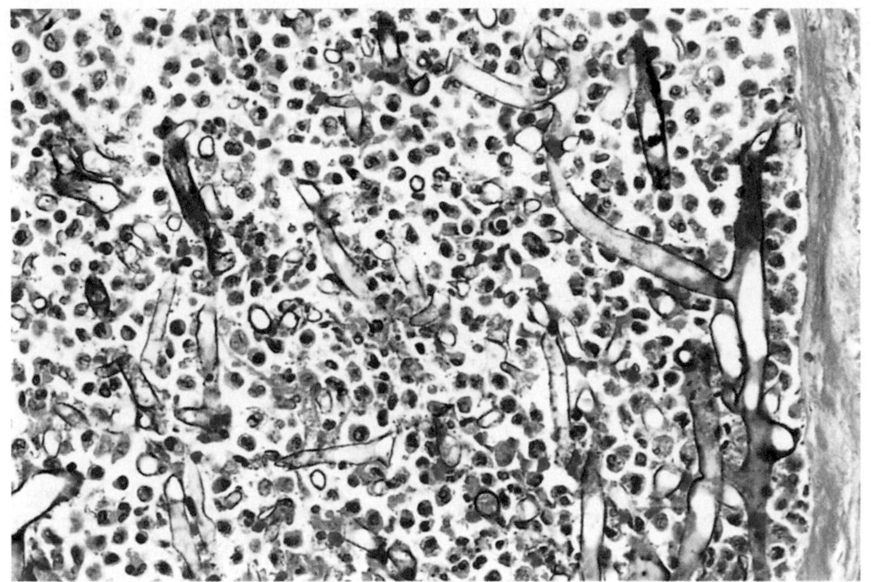

FIGURE 45–8. Zygomycosis. This zygomycete has invaded a blood vessel. Note the ribbon-like hyphae without septation. (Reproduced with permission from Connor DH, Chandler FW, Schwartz DQA, Manz HJ, Lack EE (eds). *Pathology of Infectious Diseases*, vol. 1. Stamford, CT: Appleton & Lange; 1997.)

ubiquitous saprophytes in soil and are commonly found on bread and many other foodstuffs. They occasionally cause disease in persons with diabetes mellitus and in immunosuppressed patients receiving corticosteroid therapy. Diabetic acidosis has a particularly strong association with zygomycosis.

Immunocompromised hosts with diabetes are infected

Pulmonary or rhinocerebral disease is acquired by inhalation of conidia. The pulmonary form has clinical findings similar to those of other fungal pneumonias; the rhinocerebral form, however, produces a dramatic clinical syndrome in which agents of zygomycosis show striking invasive capacity. They penetrate the mucosa of the nose, paranasal sinuses, or palate, often resulting in ulcerative lesions. Once beyond the mucosa, they progress through tissue, nerves, blood vessels, fascial planes, and often the vital structures at the base of the brain. The clinical syndrome begins with headache and may progress through orbital cellulitis and hemorrhage to cranial nerve palsy, vascular thrombosis, coma, and death in less than 2 weeks.

Pulmonary disease is similar to that from other fungi

Sinus infections erode straight to the brain

The pathologic cerebral and pulmonary findings are distinctive: the zygomycetes involved all show ribbon-like **nonseptate hyphae** in tissue which are so large their branch points can be difficult to visualize (**Figure 45–8**). Conidia are not seen. As with *Aspergillus*, tissue biopsies are necessary to demonstrate the invasive hyphae, unless they can be seen on scrapings from palatal or nasal ulcers. For reasons that are obscure, cultures are sometimes negative, even those from tissue containing characteristic hyphae. Therapy involves control of underlying disease, amphotericin B, and occasionally surgery.

Large ribbons of nonseptate hyphae are seen in tissues

PNEUMOCYSTIS

Pneumocystis is the cause of a lethal pneumonia of immunocompromised persons, particularly those with AIDS. The organism has not been grown in culture and was long considered a parasite based on the morphology of forms seen in infected tissue.

MYCOLOGY

Because it has not been possible to cultivate *Pneumocystis*, our knowledge is limited. Observations on its nature rest on morphology and the study of organisms purified from infected lungs of humans and animals. Even the appropriate species name is debated. Recent recommendations say the term long used (*P carinii*) is a rat pathogen and should be replaced by the proper human species (*P jiroveci*); others disagree. The *Pneumocystis* "life cycle" is deduced from static images seen in infected tissues. The observed stages include a delicate 5- to 8-μm cystic structures (**Figure 45–9**) within which elliptical subunits grow and repeat the cycle on rupture. These have been placed in three stages called trophic, precyst, and cysts. No filamentous form has been observed.

Life cycle is deduced from static images

Elliptical spores in sporocyte form spore case

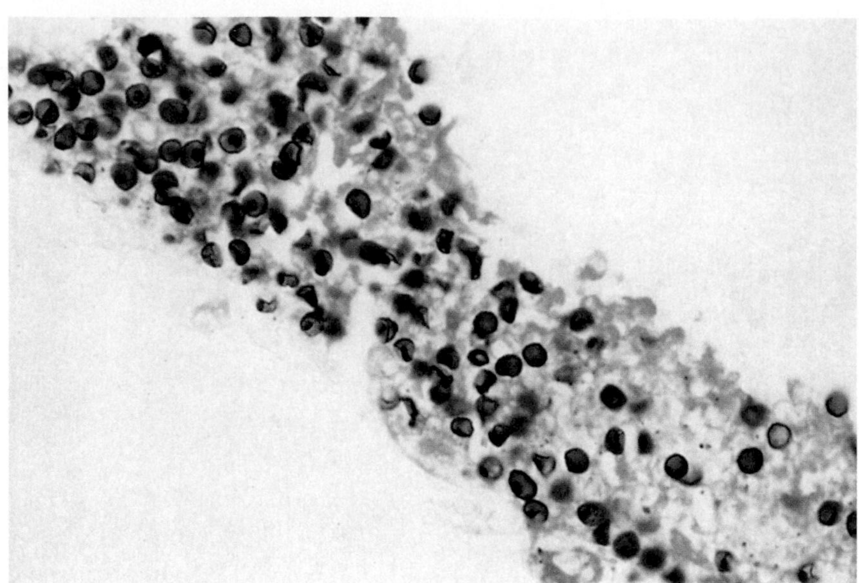

FIGURE 45–9. *Pneumocystis* pneumonia. A silver stain of this material from the lung reveals folded cysts some of which contain comma-shaped spores. (Reproduced with permission from Connor DH, Chandler FW, Schwartz DQA, Manz HJ, Lack EE (eds). *Pathology of Infectious Diseases*, vol. 1. Stamford, CT: Appleton & Lange; 1997.)

Eight sporocytes each have nucleus and mitochondria

The trophic form is bounded by a cell wall and cytoplasmic membrane that enclose a nucleus and several mitochondria. As the precyst matures, the nuclei divide to form eight “spores” within the original structure to form the cyst. The spores have an eccentric nucleus, a nucleolus, and a single mitochondrion in the cytoplasm. If this is sexual reproduction, the surrounding structure would be called a spore case or ascus and the subunits would be called sporocytes.

Cell wall is thin, but glucan and chitin elements are present

rRNA and mitochondrial gene sequences homologous with fungi

The cell wall lacks the rigidity typical of other fungi; however, biochemical elements of the fungal cell wall appear to be present. These include glucan and *N*-acetylglucosamine, the major subunit of chitin. The dominant sterol of the cytoplasmic membrane is cholesterol, rather than the ergosterol characteristic of fungi. Other biochemical analyses, however, support the fungal nature such as the presence of elements of protein synthesis (elongation factor 3), which is unique to fungi. The fungal classification of *Pneumocystis* is most strongly supported by sequence analysis of the genes coding for ribosomal RNA, mitochondrial proteins, and major enzymes. These sequences show the closest homology with fungi and molecular phylogenic analysis, which places *Pneumocystis* in the ascomycetes.

PNEUMOCYSTOSIS

CLINICAL CAPSULE

Pneumocystis pneumonia is insidious, beginning with mild fever or malaise in persons whose immune system is compromised. Signs referable to the lung come later with nonproductive cough and shortness of breath. Radiographs reveal symmetric alveolar pulmonary infiltrates, which spread from the hili. Progressive cyanosis, hypoxia, and asphyxia can lead to death in a 3- to 4-week period.

EPIDEMIOLOGY

Worldwide distribution in humans and animals

Pulmonary infection with *Pneumocystis* occurs worldwide in humans and in a broad spectrum of animal life. Exposure must be common; specific antibodies are present in nearly all children by the age of 4. The reservoir and mode of transmission remain unknown, but the view that most *Pneumocystis* pneumonia (PcP) cases represent reactivation of latent infection is no longer held. *Pneumocystis* is not found in the respiratory tract of asymptomatic persons, even among HIV-infected individuals, and the strains involved in second and third episodes are frequently antigenically different. Animal studies have shown that airborne transmission is possible, and the circumstances of hospital outbreaks point to active cases as a probable source.

Antibodies are common

Airborne transmission is probable

Before the AIDS pandemic, PcP occurred sporadically among infants with congenital immunodeficiencies and in older children and adults as a complication of immunosuppressive therapy. Now AIDS has become the most common predisposing condition, and PcP is often the presenting manifestation of AIDS. In fact, before the development of effective chemoprophylactic regimens (see Treatment and Prevention), it was present in approximately 50% of all AIDS patients at the time of initial diagnosis. Depending on the effectiveness of their HIV treatment regimen, AIDS patients may develop one or more bouts of PcP, often in conjunction with another opportunistic infection.

PcP is a complication of immunodeficient states

AIDS patients are at high risk

PATHOGENESIS

Pneumocystis is an organism of low virulence, which seldom produces disease in a host with normal T-lymphocyte function. In experimental animals, progressive infection can be initiated with starvation or corticosteroid administration, and in AIDS patients the risk of developing pneumocystosis increases dramatically once the CD4+ T-lymphocyte count has fallen to 200 cells/mm^3 or lower. Concurrent viral, bacterial, fungal, and protozoan infections are found frequently in humans with PcP, suggesting that *Pneumocystis* may require the presence of another microbial agent for its multiplication.

Low CD4 counts increase the risk in AIDS

Little is known about the early stages of disease. A **major surface glycoprotein (MSG)** abundant on the surface of *P carinii* may act as an attachment ligand to several host proteins, including fibronectin, vitronectin, and surfactant proteins. MSG undergoes antigenic variation, which could aid in its persistence in human hosts. Histologically, PcP is characterized by alveoli filled with desquamated alveolar cells, monocytes, organisms, and fluid, producing a distinctive foamy, honeycombed appearance (**Figure 45–10**); hyaline membranes may be present, and round cell infiltrates may be visible in the septa.

MSG attaches to pneumocytes

Alveoli are filled with foamy exudate

IMMUNITY

The nature of the immunodeficiencies in patients with pneumocystosis points to the primacy of T_H1 immune responses in resolution of infection with *Pneumocystis.* Alveolar macrophages are the first line of defense, with activated macrophages and CD4+ lymphocytes

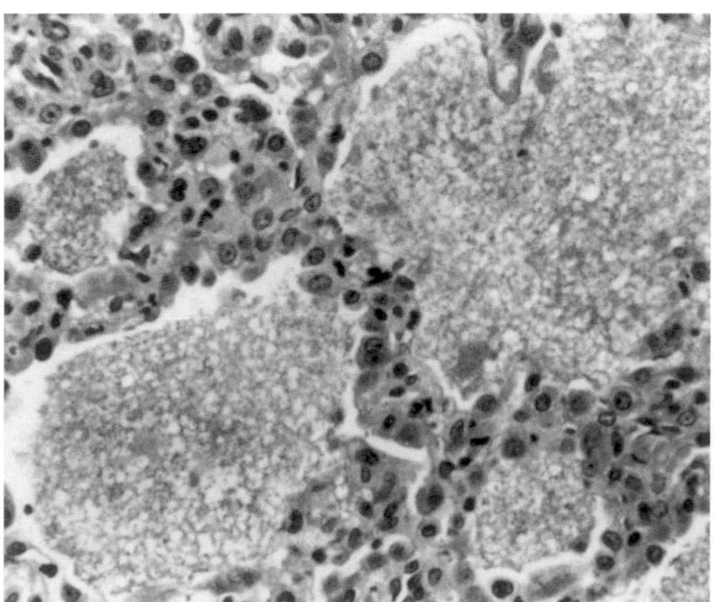

FIGURE 45–10. Lung biopsy specimen from *Pneumocystis* pneumonia, showing "foamy" contents of alveoli. (Reproduced with permission from Connor DH, Chandler FW, Schwartz DQA, Manz HJ, Lack EE (eds). *Pathology of Infectious Diseases*, vol. 1. Stamford, CT: Appleton & Lange; 1997.)

Activated macrophages and cytokines mediate CMI

playing essential roles in the resolution of the infection. Activated macrophages release several cytotoxic factors, including O_2-derived radicals, reactive nitrogen intermediates, and cytokines (tumor necrosis factor-α, IL-2).

Antibody plays a role in protection

Specific antibody responses to the MSG and other antigens appear in the course of pneumocystosis. A significant role for humoral immunity is suggested by the ability of MSG antibody to protect against experimental PcP in animals.

PNEUMOCYSTOSIS: CLINICAL ASPECTS

MANIFESTATIONS

Diffuse pneumonitis with insidious onset

Nonproductive cough, dyspnea, and cyanosis develop later

Alveolar infiltrates spread out from the hili

In the immunocompromised host, the disease presents as a progressive, diffuse pneumonitis. Illness may begin after discontinuation or a decrease in the dose of corticosteroids or, in the case of acute lymphatic leukemia, during a period of remission. In infants and AIDS patients, onset is typically insidious, and the clinical course is 3 to 4 weeks in duration. Fever is mild or absent. In older persons and patients who have previously been on high doses of corticosteroids, the onset is more abrupt, and the course is both febrile (38° to 40°C) and abbreviated. In both populations, the cardinal manifestations are progressive dyspnea and tachypnea; cyanosis and hypoxia eventually supervene. A nonproductive cough is present in 50% of all patients. Clinical signs of pneumonia are usually absent, despite the presence of infiltrates on x-ray. These infiltrates are alveolar in character and spread out symmetrically from the hili, eventually affecting most of the lung. Occasionally, unilateral infiltrates, coin lesions, lobar infiltrates, cavitary lesions, or spontaneous pneumothoraces are observed. Pleural effusions are uncommon. Clinical and radiographic abnormalities are generally accompanied by a decrease in arterial oxygen saturation, diffusion capacity of the lung, and vital capacity. Death occurs by progressive asphyxia.

Extrapulmonary lesions are seen in AIDS

Lesions outside the lung were rarely seen before the AIDS epidemic, but are now seen with some regularity. The sites most often involved are lymph nodes, bone marrow, spleen, liver, eyes, thyroid, adrenal glands, gastrointestinal tract, and kidneys. The extrapulmonary clinical manifestations range from incidental autopsy findings to progressive multisystem disease.

DIAGNOSIS

Diagnostic yield from sputum is low

BAL is the best of the invasive procedures

Definite diagnosis of pneumocystosis depends on finding organisms of typical morphology in appropriate specimens. Because the pathologic process is alveolar rather than bronchial, the organisms are not readily seen in expectorated specimens such as sputum. The diagnostic yield is much better from specimens obtained by more invasive procedures. Of these, bronchoalveolar lavage (BAL) gives the best results with the least morbidity. Percutaneous needle aspiration of the lung, transbronchial biopsy, and open lung biopsy, though somewhat more sensitive techniques, are accompanied by more complications, including pneumothorax and hemothorax.

Silver and other stains readily demonstrate *P carinii*

DFA is sensitive

Pneumocystis can be demonstrated by a wide variety of staining procedures. The standard stain is methenamine silver (Figure 45–9), but direct fluorescent antibody (DFA) method, if available, is slightly more sensitive. Laboratories often perform a rapid stain (Wright, Giemsa, Papanicolaou) first and confirm by methenamine silver or DFA later. Methods developed for detection of *Pneumocystis* DNA in BAL and other specimens by polymerase chain reaction may soon be practical for clinical laboratories.

TREATMENT AND PREVENTION

TMP-SMX is treatment of choice

The fixed combination of trimethoprim and sulfamethoxazole (TMP-SMX) is the treatment of choice for all forms of pneumocystosis. It is administered orally or intravenously for 14 to 21 days. Patients with AIDS receive the longer course because they start with a higher organism burden, respond more slowly, and suffer relapse more often. Unfortunately, AIDS patients have a high incidence of adverse effects to TMP-SMX, particularly the sulfonamide component. This requires the use of other antimicrobics (eg, clindamycin, primaquine, dapsone) alone or in combination with TMP.

Low-dose administration of TMP-SMX has been shown to significantly decrease the incidence of PcP in high-risk patients and prevents relapse in AIDS patients. This chemoprophylaxis is indicated for patients who have CD4+ lymphocyte counts lower than 200/mm^3, unexplained fever, or a previous episode of PcP. Chemoprophylaxis is continued as long as the immunosuppressive conditions persist.

Treatment is extended in AIDS

Chemoprophylaxis prevents PcP in high-risk groups

CLINICAL CASE

A BUDDING BLOOD CULTURE

This 71-year-old woman was admitted with a recurrence of poorly differentiated squamous cell carcinoma of the cervix. She underwent extensive gynecologic surgery (excision of the organs of the anterior pelvis) and was maintained postoperatively on broad-spectrum intravenous antibiotics. The woman had a central venous catheter placed on the day of the surgery.

Beginning 3 days postoperatively, the woman had temperatures of 38.0° to 38.5°C, which persisted without a clear source. On day 8 postoperatively, she had a temperature of 39.2°C. Cultures of blood and of the tip of the central line both grew an agent with large ovoid cells, some of which had constricted buds at their ends. When incubated in serum these cell sprouted long tubes with parallel sides.

QUESTIONS

■ Which organism is most likely to be identified in this patient's blood culture?

A. *Candida albicans*
B. *Candida glabrata*
C. *Aspergillus*
D. *Mucor*
E. *Pneumocystis*

■ What feature of the organism might have facilitated its infection in these circumstances?

A. Mannoprotein
B. Glucan
C. Germ tube formation
D. Biofilm formation
E. Sporocytes

■ Which is the probably origin of the infecting agent?

A. Animals
B. Hospital air
C. Medical devices
D. Patient's flora
E. Health care workers

ANSWERS

1(A), **2**(D), **3**(D)

CHAPTER 46

Cryptococcus, Histoplasma, Coccidioides, and Other Systemic Fungal Pathogens

The fungi discussed in this group cause a variety of infections, each ranging in severity from subclinical to progressive, debilitating disease. Most species are dimorphic, growing in the infectious mold form in the environment but switching to a yeast form in tissues to produce infection. They differ from the opportunistic fungi in their ability to cause disease in previously healthy persons, but the most serious disease still occurs in immunocompromised persons. With the exception of *Cryptococcus neoformans*, each of these species is restricted to a geographic niche corresponding to the environmental habitat of the mold form of the species. None is transmitted from human to human. The major features of the systemic pathogens are summarized in **Table 46–1**.

CRYPTOCOCCUS

CRYPTOCOCCUS NEOFORMANS

Cryptococcus neoformans (cryptococcus) is a yeast 4 to 6 μm in diameter that produces a characteristic **capsule** (**Figure 46–1**), extending the overall diameter to 25 μm or more. It is a basidiomycete, and what was once considered a uniform species has now been divided into four serotypes (A-D) and three varieties (*neoformans, grubii, gattii*). Although there are some epidemiologic differences among the varieties, their pathogenic biology is essentially the same. Here, all will be called *Cryptococcus neoformans,* or simply the cryptococcus.

Serotypes and varieties have same biology

The capsule is unique among pathogenic fungi and is a complex polysaccharide polymer, the major component of which is **glucuronoxylomannan (GXM).** Capsule production is repressed under environmental conditions and stimulated in the physiologic conditions found in tissues and in culture on the common clinical laboratory media, such as blood agar, chocolate agar, and Sabouraud's agar. At either 25° or 37°C, mucoid yeast colonies are produced in 2 to 3 days. The teleomorph (sexual) forms with hyphae and basidiospores have been produced only in the laboratory under specialized conditions. It is suspected but not observed that this is the environmental growth form. In addition to the capsule, extracellular products include urease and laccase enzymes. A melanin pigment is the product of laccase activity.

GXM capsule in tissues

Urease, laccase, and melanin produced

TABLE 46–1 Features of Systemic Fungal Pathogens

	GROWTH					
ORGANISM	**CULTURE AT 25°C**	**CULTURE AT 37°C**	**TISSUE**	**SOURCE**	**PRIMARY DISEASE**	**DISSEMINATED DISEASE**
Cryptococcus neoformans	Encapsulated yeast	Encapsulated yeast	Encapsulated yeast	Environment, worldwide	Pneumonia	Chronic meningitis
Histoplasma capsulatum	Mold, tuberculate macroconidia[a]	Small yeast	Small intracellular yeast[b]	Environment, US Midwest[d]	Pneumonia, hilar adenopathy	RES enlargement
Blastomyces dermatitidis	Mold[a]	Yeast		Environment, US Midwest[c]	Pneumonia	Skin and bone lesions
Coccidioides immitis	Mold, arthroconidia	(Spherules)[e]	Spherules	Environment, Sonoran desert[c,f]	Valley fever	Pneumonia, meningitis, skin, bone
Paracoccidioides brasiliensis	Mold	Yeast, multiple blastoconidia		Environment, Latin America	Pneumonia	Mucocutaneous, RES

RES, reticuloendothelial system (lymph nodes, liver, spleen, bone marrow).
[a]Micoconidia are formed but are not distinctive.
[b]Typically multiple yeast within macrophages.
[c]Ecologic "islands" are found throughout the Americas.
[d]Ecologic islands are found worldwide.
[e]It is difficult to grow the spherule phase in culture.
[f]In the United States and includes parts of Arizona, California, Nevada, and western Texas.

CRYPTOCOCCOSIS

CLINICAL CAPSULE

The primary disease caused by cryptococci is a chronic meningitis. The onset is slow, even insidious, with low-grade fever and headache progressing to altered mental state and seizures. In the cerebrospinal fluid (CSF) and in tissues, the inflammatory response is often remarkably muted. Most patients have some obvious form of immune compromise, although some show no demonstrable immune defect.

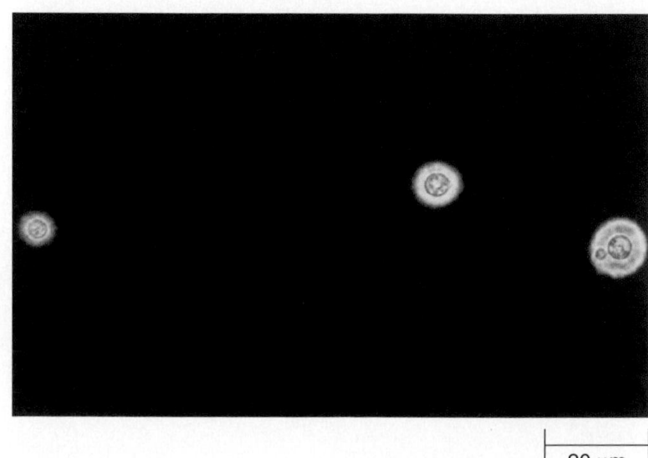

FIGURE 46–1. *Cryptococcus neoformans.* This India ink preparation was made by mixing cerebrospinal fluid containing cryptococci with India ink. The yeast cells can be seen within the clear space caused by the large polysaccharide capsule excluding the ink particles. Note that the one on the right is budding. (Reproduced with permission from Nester EW, Anderson DG, Roberts CE Jr, Nester MT. *Microbiology: A Human Perspective*, 6th ed. New York: McGraw-Hill, 2008.)

EPIDEMIOLOGY

C neoformans is ubiquitous throughout the world, particularly in soil contaminated with avian droppings and decaying vegetable matter. One environmental niche is the hollowed-out areas of trees, where laccase is involved in the degradation of wood. The infectious form is felt to be either desiccated yeast cells or basidiospores stirred up from these sites and inhaled. Cases appear sporadically, with no particular occupational predisposition, including bird fanciers and those who work with the cryptococcus in the laboratory. Cryptococcosis in immunocompromised patients occurs primarily in those with defects in T-lymphocyte function, particularly in those with AIDS, in whom it is the most common fungal infection seen. In counties with well-developed antiretroviral therapy programs, cryptococcal disease has declined in AIDS patients but remains persistent in other immunocompromised persons. Disease can occur in persons with no known immune defect and is said to be more likely with certain variants. Case-to-case transmission has not been documented.

Associated with soil and bird droppings

Yeasts or basidiospores inhaled

PATHOGENESIS

After being inhaled, cryptococci reach the alveoli, where production of the polysaccharide capsule is the prime determinant of virulence. Intracellular survival in phagocytes is aided by melanin production, which interferes with oxidative killing mechanisms. The GXM capsule is antiphagocytic and has a number of other immunomodulating effects such as interference with antigen presentation, leukocyte migration, specific antibody responses, and the development of T_H1 immune responses. These immune suppressing effects may act at both a local and systemic level because cryptococci produce enough capsule that the GXM is readily detected in the blood and other body fluids. This muting of the first lines of defense may be what allows the organisms to spread outside the lung. The affinity of *C neoformans* for the central nervous system (CNS) is striking. Proposed explanations include crossing the blood-brain barrier inside macrophages (Trojan horse) and the ability of laccase to convert the abundant catecholamines in the CNS to melanin.

Antiphagocytic capsule is prime factor

Circulating GXM interferes with immune function

Melanin provides oxidative protection in CNS

Tissue reaction to *C neoformans* varies from little or none to purulent or granulomatous. Many cases of pulmonary, cutaneous, and even meningeal cryptococcal infection show a remarkable paucity of inflammatory cells. This certainly fits for a fungus that not only blocks its own phagocytosis but is able to down-regulate multiple aspects of the immune response.

Tissue reaction is often minimal

IMMUNITY

In immunocompetent persons, alternate pathway binding of complement by the capsule is probably sufficient for opsonophagocytosis. The capsule is not particularly antigenic, and anticryptococcal antibodies are not usually detected in the course of infection. Some antibodies can protect the organism, but their role in immunity is unknown.

Antibody plays some role

Animal studies and the strong clinical association of cryptococcosis with T-cell defects indicate that T_H1 type immune responses are most important in the outcome of infection. Cryptococci phagocytosed by macrophages may not be killed, and cytokine activation is needed to complete the clearing of the organisms. Immunodominant mannoproteins have been identified which use dendritic cells as the primary presenter to CD4+ T cells. In patients with cryptococcosis who have no known immune defects, it is often possible to detect subnormal T_H1 immune functions in laboratory testing. Clinical recovery in such cases is associated with return of these immune functions.

T_H1 responses are dominant

Dendritic cells present mannoproteins

CRYPTOCOCCOSIS: CLINICAL ASPECTS

MANIFESTATIONS

Meningitis is the most commonly recognized form of cryptococcal disease; it usually has a slow, insidious onset with relatively nonspecific findings until late in its course. Intermittent headache, irritability, dizziness, and difficulty with complex cerebral functions appear

Meningitis is insidious and chronic

Course is more rapid with AIDS

over weeks or months with no consistent pattern. Behavioral changes have been mistaken for psychoses. Fever is usually, but not invariably, present. Seizures, cranial nerve signs, and papilledema may appear later in the clinical course, as may dementia and decreased levels of consciousness. A more rapid course may be seen in AIDS patients, 5% to 15% of whom become infected with *C neoformans*.

Cryptococcal pneumonia often asymptomatic

Cryptococcal pneumonia is often asymptomatic or mild. Sputum production is minimal, and no findings are sufficiently specific to suggest the etiology. Skin and bone are the sites most frequently involved in disseminated disease; skin lesions are sometimes the presenting sign and are often remarkable for their lack of inflammation. The diagnosis is sometimes made when lesions are biopsied as suspected neoplasms.

DIAGNOSIS

Cells and glucose depression in CSF may be minimal

India ink prep is positive in 50% of cases

Typical CSF findings in cryptococcal meningitis are increased pressure, pleocytosis (usually 100 cells or more) with predominance of lymphocytes, and depression of glucose levels. In some cases, one or all of these findings may be absent, yet cryptococci are isolated on culture. Cryptococcal capsules are demonstrable in CSF in roughly 50% of cases by mixing centrifuged sediment with **India ink** and examining the mixture under the microscope (Figure 46–1). Some experience is necessary to avoid confusion of lymphocytes with cryptococci. *C neoformans* stains poorly or not at all with routine histologic stains; thus, it is easily missed unless special fungal stains are used (**Figure 46–2**).

Few cryptococci may be present in CSF

GXM is detectable in CSF and serum

In the isolation of *C neoformans*, the volume of CSF sampled is important. The number of organisms present may be small enough to require a substantial volume of fluid (more than 30 mL) to yield a positive culture. If cryptococcosis is suspected and cultures are negative, detection of the GXM polysaccharide antigen in the CSF or serum by latex agglutination or enzyme immunoassay methods is recommended. These tests are very sensitive and specific, and their quantitation has prognostic significance. A rising antigen level indicates progression and a declining titer is a favorable sign.

TREATMENT

Amphotericin, flucytosine, and fluconazole used in combination

Amphotericin B plus flucytosine followed by an extended course of fluconazole is the primary treatment for systemic cryptococcal disease. Although 75% of persons with meningitis respond to treatment, a significant percentage suffer relapses after antifungal therapy is stopped; many become chronic and require repeated courses of therapy. One half of those cured have some kind of residual neurologic damage.

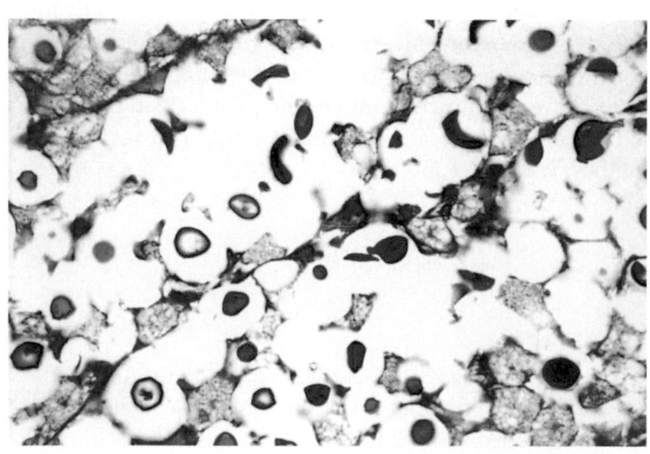

FIGURE 46–2. Cryptococcal meningitis. The *C neoformans* cells are stained red by this PAS (periodic acid-Schiff) stain. The capsule is not stained but is creating the halo around the organisms. Note the lack of inflammatory cells. (Reproduced with permission from Connor DH, Chandler FW, Schwartz DQA, Manz HJ, Lack EE (eds). *Pathology of Infectious Diseases*, vol. 1. Stamford, CT: Appleton & Lange; 1997.)

HISTOPLASMA

HISTOPLASMA CAPSULATUM

Histoplasma capsulatum is a dimorphic fungus **(Figure 46–3B)** that grows in the yeast phase in tissue and in cultures incubated at 37°C. The mold phase grows in cultures incubated at 22° to 25°C and as a saprophyte in soil. There are three varieties of *Histoplasma* (*capsulatum, duboisii, farciminosum*), which vary in their geographic distribution. The yeast forms are small for fungi (2 to 4 μm) and reproduce by budding (blastoconidia). The mycelia are septate and produce **microconidia** and macroconidia. The diagnostic structure is termed the **tuberculate macroconidium** because of its thick wall and radial, finger-like projections (Figure 46–3A). Growth is obtained on blood agar, chocolate agar, and Sabouraud's agar, but may take many weeks. The designation *H capsulatum* is actually a misnomer, because no capsules are formed. It comes from the halos seen around the yeasts in tissue sections, which are caused by a shrinkage artifact of routine histologic methods.

Small dimorphic fungus producing tuberculate macroconidia

Growth may take weeks

HISTOPLASMOSIS

CLINICAL CAPSULE

Histoplasmosis is limited to the endemic area, where most patients are asymptomatic or show only a fever and cough. If affected persons are seen by a physician, a pulmonary infiltrate and hilar adenopathy may or may not be evident on a radiograph. Progressive cases show extension in the lung or enlargement of lymph nodes, liver, and spleen.

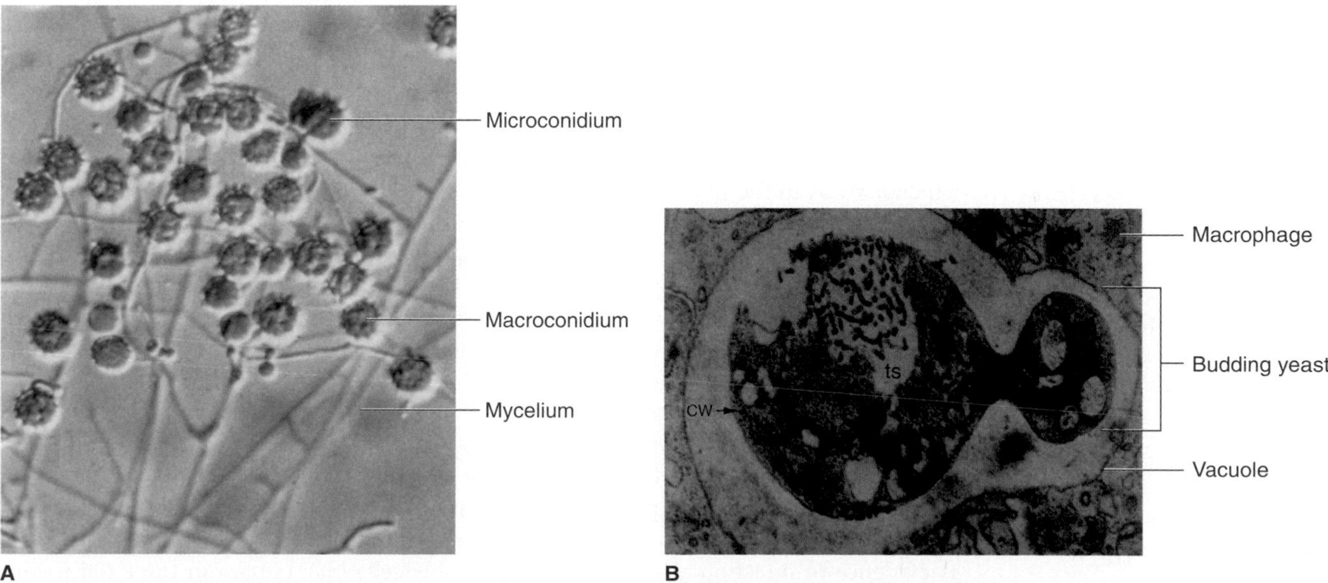

FIGURE 46–3. Histoplasma capsulatum. A. Mold phase with hyphae, microconidia, and tuberculate microconidia. **B.** A yeast cell is multiplying (note budding) within a macrophage phagocytic vacuole. (Reproduced with permission from Willey J, Sherwood L, Woolverton C (eds). *Prescott's Principles of Microbiology*. New York: McGraw-Hill; 2008.)

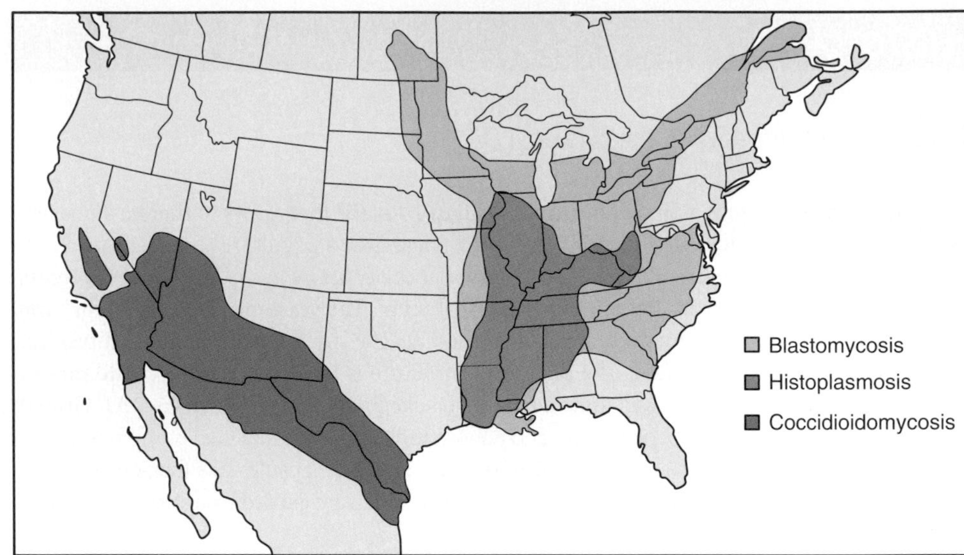

FIGURE 46–4. Geographic distribution of systemic fungal infections in the United States.

EPIDEMIOLOGY

Microconidia are infectious

Mold grows in humid soil with bird droppings

High prevalence in central United States

H capsulatum grows in soil under humid climatic conditions, particularly soil containing bird or bat droppings. Inhalation of the mold microconidia, which are small enough (2 to 5 μm) to reach the terminal bronchioles and alveoli, is believed to be the mode of infection. The organism is particularly prevalent in certain temperate, subtropical, and tropical zones, and endemic areas are present in all continents of the world except Antarctica. The largest and best defined is the United States region drained by the Ohio and Mississippi Rivers (**Figure 46–4**). Over 50% of the residents of states in this area show radiologic evidence of previous infection, and in some locales, up to 90% demonstrate delayed-type hypersensitivity to *Histoplasma* antigens. Disturbances of bird roosts, bat caves, and soil have been associated with point source outbreaks. Persons in endemic areas whose employment (agriculture, construction) or avocation (spelunkers) brings them in contact with these sites are at increased risk. The infection is not transmitted from person to person. Disease is more common in men, but there are no racial or ethnic differences in susceptibility.

PATHOGENESIS

Reticuloendothelial system is focus of infection

Grows in macrophages by controlling lysosomal pH

The hallmark of histoplasmosis is infection of the lymph nodes, spleen, bone marrow, and other elements of the reticuloendothelial system with intracellular growth in phagocytic macrophages. The initial infection is pulmonary, through inhalation of infectious conidia, which convert to the yeast form in the host. They attach to integrin and fibronectin receptors and are readily taken up by professional phagocytes. Dendritic cells kill the invading yeast cells, but inside neutrophils and macrophages they survive the effects of the oxidative burst and inhibit phagosome-lysosome fusion. Key features in this survival and multiplication are the ability of *H capsulatum* to capture iron and calcium from the macrophage and to modulate phagolysosomal pH. The acidic pH required for optimal killing effect in the lysosome is elevated by *H capsulatum* toward the less effective neutral range (pH 6.0 to 6.5). ::: Phagosome-lysosome, p. 25

Lymphatic spread and reactivation are similar to tuberculosis

With continued growth, there is lymphatic spread and development of a primary lesion similar to that seen in tuberculosis. The extent of spread to the reticuloendothelial system within macrophages during primary infection is unknown, but such spread is presumed to occur. Most cases never advance beyond the primary stage, leaving only a calcified node as evidence of infection. As in tuberculosis, viable cells may remain in these old lesions and reactivate later, particularly if the person becomes immunocompromised. ::: Primary tuberculosis, p. 495

Granulomatous response seen in liver, spleen, and bone marrow

Pathologically, granulomatous inflammation with necrosis is prominent in pulmonary lesions, but *H capsulatum* may be difficult to detect, even with special fungal stains. Extrapulmonary spread involves the reticuloendothelial system, with enlargement of the liver

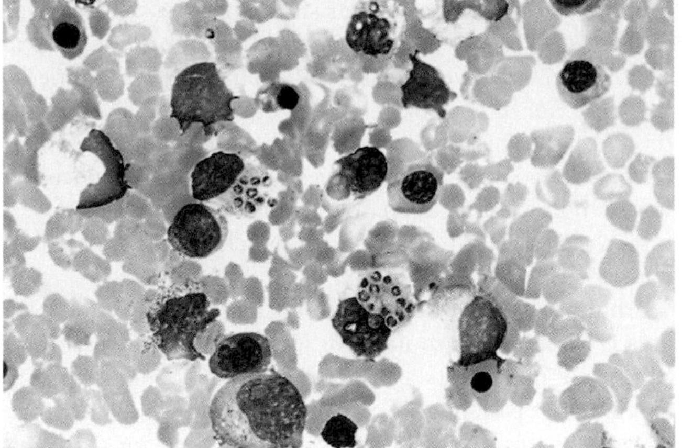

FIGURE 46–5. *Histoplasma capsulatum*. This peripheral blood smear shows two monocytes with multiple organisms are stuffed within their cytoplasm. Note the size of the yeast cells, which is very small for fungi. (Reproduced with permission from Connor DH, Chandler FW, Schwartz DQA, Manz HJ, Lack EE (eds). *Pathology of Infectious Diseases*, vol. 1. Stamford, CT: Appleton & Lange; 1997.)

and spleen. Numerous organisms within macrophages may be found in these organs, in lymph nodes, bone marrow, or even peripheral blood (**Figure 46–5**).

IMMUNITY

Infection with *H capsulatum* is associated with the development of cell-mediated immunity, as demonstrated by a positive result of a delayed hypersensitivity skin test to *H capsulatum* mycelial antigens. Infection is believed to confer long-lasting immunity, the most important component of which is T_H1-mediated. In experimental infections, macrophages activated by T-lymphocyte–derived cytokines are able to inhibit intracellular growth of *H capsulatum* and thus control the disease. Neither B cells nor antibody have a significant influence on resistance to reinfection. Immunocompromised persons, particularly those with T-lymphocyte–related defects, are unable to stop growth of the organism and tend to develop progressive, disseminated disease. :::T-cell responses, p. 33

Skin test demonstrates delayed-type hypersensitivity

Immunity is T_H1-mediated

HISTOPLASMOSIS: CLINICAL ASPECTS

MANIFESTATIONS

Most cases of *H capsulatum* infection are asymptomatic or show only fever and cough for a few days or weeks. Mediastinal lymphadenopathy and slight pulmonary infiltrates may be seen on x-rays. More severe cases may have chills, malaise, chest pain, and more extensive infiltrates, which usually resolve nonetheless. A residual nodule may continue to enlarge over a period of years, causing a differential diagnostic problem with pulmonary neoplasms. Progressive pulmonary disease occurs in a form similar to that of pulmonary tuberculosis, including the development of cavities, with sputum production, night sweats, and weight loss. The course is chronic and relapsing, lasting many months to years.

Most cases are asymptomatic or with fever and cough

Progressive pulmonary disease shows cavities and weight loss

Disseminated histoplasmosis generally appears as a febrile illness with enlargement of reticuloendothelial organs. The CNS, skin, gastrointestinal tract, and adrenal glands may also be involved. Painless ulcers on mucous membranes are a common finding. The course of histoplasmosis is typically chronic, with manifestations that depend on the organs involved. For example, chronic bilateral adrenal failure (Addison's disease) may develop when the adrenal glands are involved.

Dissemination involves reticuloendothelial organs, mucous membranes, and adrenal glands

DIAGNOSIS

In most forms of pulmonary histoplasmosis, the diagnostic yield of direct examinations or culture of sputum is low. In disseminated disease, blood culture or biopsy samples of a reticuloendothelial organ are the most likely to contain *Histoplasma*. Bone marrow culture has the highest yield. Because of their small size, the yeast cells are difficult to see in potassium hydroxide (KOH) preparations, and their morphology is not sufficiently distinctive to be diagnostic. Selective fungal stains such as methenamine silver demonstrate the organism but may not differentiate it from other yeasts. Hematoxylin and eosin (H&E)–stained tissue or Wright-stained bone marrow often demonstrates the organisms in their intracellular location in macrophages (Figure 46–5). Specimens must be examined carefully under high magnification. Identification of culture isolates requires demonstration of the typical conidia and dimorphism. Nucleic acid probes have been developed for culture confirmation.

Blood and bone marrow examination require special stains

Immunodiffusion and probes used with cultures

Antibodies can be detected during and after infection, but their usefulness in the endemic area is limited by false-negative results and cross-reactions in patients with blastomycosis. Rising antibody titers are suggestive of dissemination or relapse. The skin test has been useful in the past, but the reagents are no longer commercially available. Cultural isolation or clear histologic demonstration is necessary for a firm diagnosis. A circulating polysaccharide antigen has been demonstrated in serum and urine by enzyme immunoassay (EIA) in more than 90% of patients with disseminated disease.

Culture is required for firm diagnosis

EIA detects circulating antigen

TREATMENT

Primary infections and localized lung lesions usually resolve without treatment. For mild disease localized to the lung, itraconazole is used. For more severe or disseminated disease, a course of amphotericin B is followed by the itraconazole regimen. Itraconazole can be effective for prophylaxis of persons with a high risk of disease. These include AIDS patients with low CD4 counts and other immunocompromised patients in a endemic area.

Amphotericin B and itraconazole

BLASTOMYCES

BLASTOMYCES DERMATITIDIS

Blastomyces dermatitidis is a dimorphic fungus with some characteristics similar to those of *Histoplasma*. Growth develops in the yeast phase in tissues and in cultures incubated at 37°C. The yeast cells are typically larger (8 to 15 mm) than those of *H capsulatum*, with broad-based buds and a thick wall (**Figure 46–6**). The mold phase appears in culture at 25°C. Hyphae are septate and produce round to oval conidia sufficiently similar to the microconidia produced by *H capsulatum* to cause confusion between the two in young

Large yeast cells have broad-based buds

Mold has small oval conidia-like *Histoplasma*

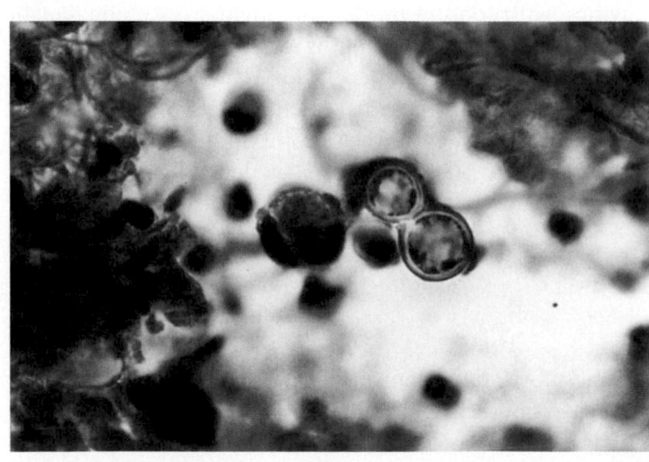

FIGURE 46–6. *Blastomyces dermatitidis.* Large thick-walled yeast cells are shown in this sputum. Note how the blastoconidia retain a broad attachment to the mother cell before separating. (Reproduced with permission from Connor DH, Chandler FW, Schwartz DQA, Manz HJ, Lack EE (eds). *Pathology of Infectious Diseases*, vol. 1. Stamford, CT: Appleton & Lange; 1997.)

cultures. Although older cultures may produce chlamydoconidia, *B dermatitidis* produces no structure as distinctive as the tuberculate macroconidium of *Histoplasma*.

BLASTOMYCOSIS

CLINICAL CAPSULE

Most clinical features of blastomycosis are similar to histoplasmosis. Patients are asymptomatic or have only mild fever and cough unless the disease progresses outside the lung. Skin lesions are the most common manifestation of disseminated disease. The reticuloendothelial system is not involved.

EPIDEMIOLOGY

Cases of blastomycosis have a geographic distribution and conditions for maturation of conidia in the soil, which are similar to that of histoplasmosis but no associations with birds or mammals have been established. Most infections occur in the middle and eastern portions of North America (Figure 46–4), but they have been reported in Africa, the Middle East, and Europe. A specific skin test for blastomycosis has never been available; this limits mapping of the endemic area. It is assumed that inhalation of environmental microconidia is the means of infection.

Geographic distribution similar to *Histoplasma*

PATHOGENESIS

Much less is known about blastomycosis than the more common systemic mycoses, such as histoplasmosis and coccidioidomycosis. The lower frequency of disseminated infections and the nonspecificity of skin and serologic tests are partly responsible for this lack of information. Much of what is believed to be true of blastomycosis is based on analogy with histoplasmosis.

The primary infection is pulmonary after inhalation of conidia, which develop in soil. Surface glucans and a glycoprotein adhesin (BAD1) have been identified, which bind the fungi to receptors on host cells, macrophages, and the extracellular matrix. A mixed inflammatory response results, which ranges from neutrophil infiltration to well-organized granulomas with giant cells. The organisms grow in tissue as large yeasts with thick double walls with blastospores attached. A significant difference from *Histoplasma* is that the yeast cells are primarily extracellular rather than within macrophages. This may be due to their relatively large size, but there is little to suggest that *B dermatitidis* shares the propensity for intracellular parasitism that is characteristic of *H capsulatum*.

Surface adhesin binds to host cells

Large yeast are primarily outside cells

IMMUNITY

The principal host defense mechanisms against *B dermatitidis* have not been clearly defined. The fungal cells activate the complement system by both the classical and alternate pathways, and antibodies directed against a glucan component of the cell wall have been identified. These antibodies decline as the infection resolves. As with other fungi, T_H1-mediated responses appear to be the most important determinants of immunity. Macrophages activated with cytokines have enhanced capacity to kill *B dermatitidis*.

Complement, antibody, and cell-mediated immunity are involved

BLASTOMYCOSIS: CLINICAL ASPECTS

MANIFESTATIONS

Pulmonary blastomycosis is similar to other mycoses

Skin lesions are on exposed surfaces

Because mild cases of blastomycosis are difficult to diagnose, most infections are recognized at advanced or disseminated stages of the disease. This problem was also posed by the other systemic mycoses before the development of sensitive and specific diagnostic procedures. Pulmonary infection is evidenced by cough, sputum production, chest pain, and fever. Hilar lymphadenopathy may be present, as may nodular pulmonary infiltrates with alveolar consolidation. The total picture may mimic a pulmonary tumor, tuberculosis, or some other mycosis. Skin lesions are common and were once considered a primary form of the disease. In contrast to histoplasmosis, lesions develop on exposed skin; mucous membrane infection is uncommon. Extensive necrosis and fibrosis may produce considerable disfigurement. Bone infection has features similar to those of other causes of chronic osteomyelitis. The urinary and genital tracts are the most commonly affected visceral sites; the prostate is especially prone to infection.

DIAGNOSIS

KOH and biopsy show budding yeast

Culture takes weeks and conidia not distinctive

Direct demonstration of typical large yeasts with broad-based buds (blastoconidia) in KOH preparations is the most rapid means of diagnosis (Figure 46–6). Biopsy specimens also have a high yield, and the organisms are visible with either H&E or special fungal stains. *B dermatitidis* grows on routine mycologic media, but culture may take as long as 4 weeks. Conidia are not particularly distinctive, and demonstration of dimorphism and typical yeast morphology is essential to avoid confusion with other fungi. A DNA probe is particularly useful in differentiating cultures from *Histoplasma*. Serologic tests are available but may be negative in up to 50% of cases.

TREATMENT

Amphotericin B and azoles are effective

As with histoplasmosis, itraconazole is used for mild to moderate disease and preceded by amphotericin B for more serious or disseminated disease. Fluconazole or voriconazole may be used in meningitis. As with other systemic mycoses, response to treatment is slow, and relapse is common.

COCCIDIOIDES

COCCIDIOIDES IMMITIS

Dimorphism involves unique spherule

Spherules differentiate to form and release endospores

Coccidioides immitis is also a dimorphic fungus, but instead of a yeast phase, a large (12- to 100-μm), distinctive, round-walled **spherule** (**Figure 46–7A**) is produced in the invasive tissue form. This structure is unique among the pathogenic fungi. Its formation takes place in a process illustrated in **Figure 46–8.** Spherule development requires simultaneous invagination of the fungal membrane (plasmalemma) and production of new cell wall to form the large multicompartmental structure. The compartments differentiate into uninucleate structures called **endospores**, each with a thin wall layer. Multiple endospores develop within each spherule and the entire structure is surrounded by an extracellular matrix. The spherule eventually ruptures, releasing 200 to 300 endospores (**Figure 46–9**), each of which can differentiate into another spherule.

In the environment, *C immitis* grows under harsh conditions in sandy alkaline soil with high salinity. Both in the environment and in the laboratory it grows as a mold regardless

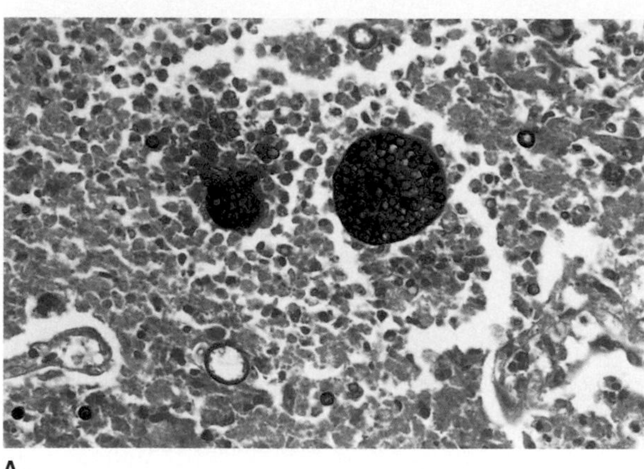

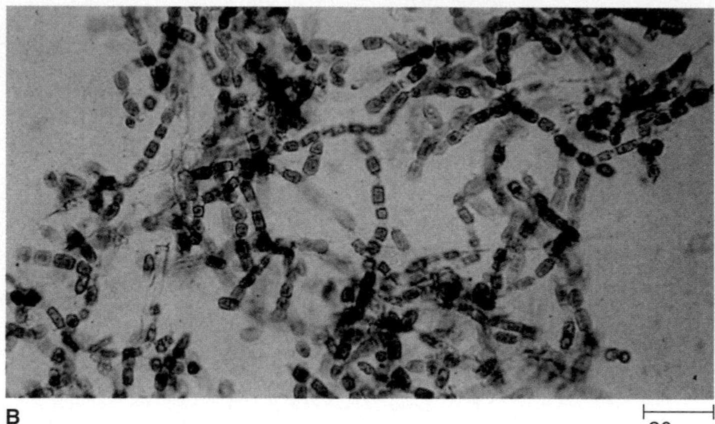

FIGURE 46–7. ***Coccidioides immitis*. A.** Lung tissue with a large thick-walled spherule containing multiple endospores. The smaller spherule to its left has ruptured releasing endospores. **B.** Mold phase in which alternate cells have differentiated to form barrel-shaped arthroconidia. (A, Reproduced with permission from Connor DH, Chandler FW, Schwartz DQA, Manz HJ, Lack EE (eds). *Pathology of Infectious Diseases*, vol. 1. Stamford, CT: Appleton & Lange; 1997. B, Reproduced with permission from Nester EW, Anderson DG, Roberts CE Jr, Nester MT. *Microbiology: A Human Perspective*, 6th ed. New York: McGraw-Hill, 2008.)

of temperature. Growth becomes visible in 2 to 5 days. The hyphae are septate and produce thick-walled, barrel-shaped **arthroconidia** (Figure 46–7B) in about 1 week. Mature arthroconidia readily separate from the hyphae and survive for long periods in the environment. When airborne, they are the infectious units in nature. Arthroconidia can be converted to spherules in the laboratory, but only under very specialized conditions. As with other fungi, the application of modern genotyping methods has led to some splitting within *Coccidioides*. The original *C immitis* isolated in California's San Joaquin Valley appears to be a distinct clone, and most strains from elsewhere in the Americas belong to another species (*C posadasii*). Because there are no differences in disease and clinical laboratories cannot distinguish the two, the more familiar *C immitis* is used for both species here.

Barrel-shaped arthroconidia form in hyphae

Conidia are readily airborne

COCCIDIOIDOMYCOSIS

CLINICAL CAPSULE

Acute primary infection with *C immitis* either is asymptomatic or manifests as a complex called valley fever by residents of the endemic areas. Valley fever includes fever, malaise, dry cough, joint pains, and sometimes a rash. There are few physical or radiologic findings, but the illness persists for weeks. Disseminated disease involves lesions in the bones, joints, skin, and a progressive chronic meningitis.

EPIDEMIOLOGY

Coccidioidomycosis is the most geographically restricted of the systemic mycoses because *C immitis* grows only in the alkaline soil of semiarid climates known as the Lower Sonoran life zone (Figure 46–4). These areas are characterized by hot, dry summers,

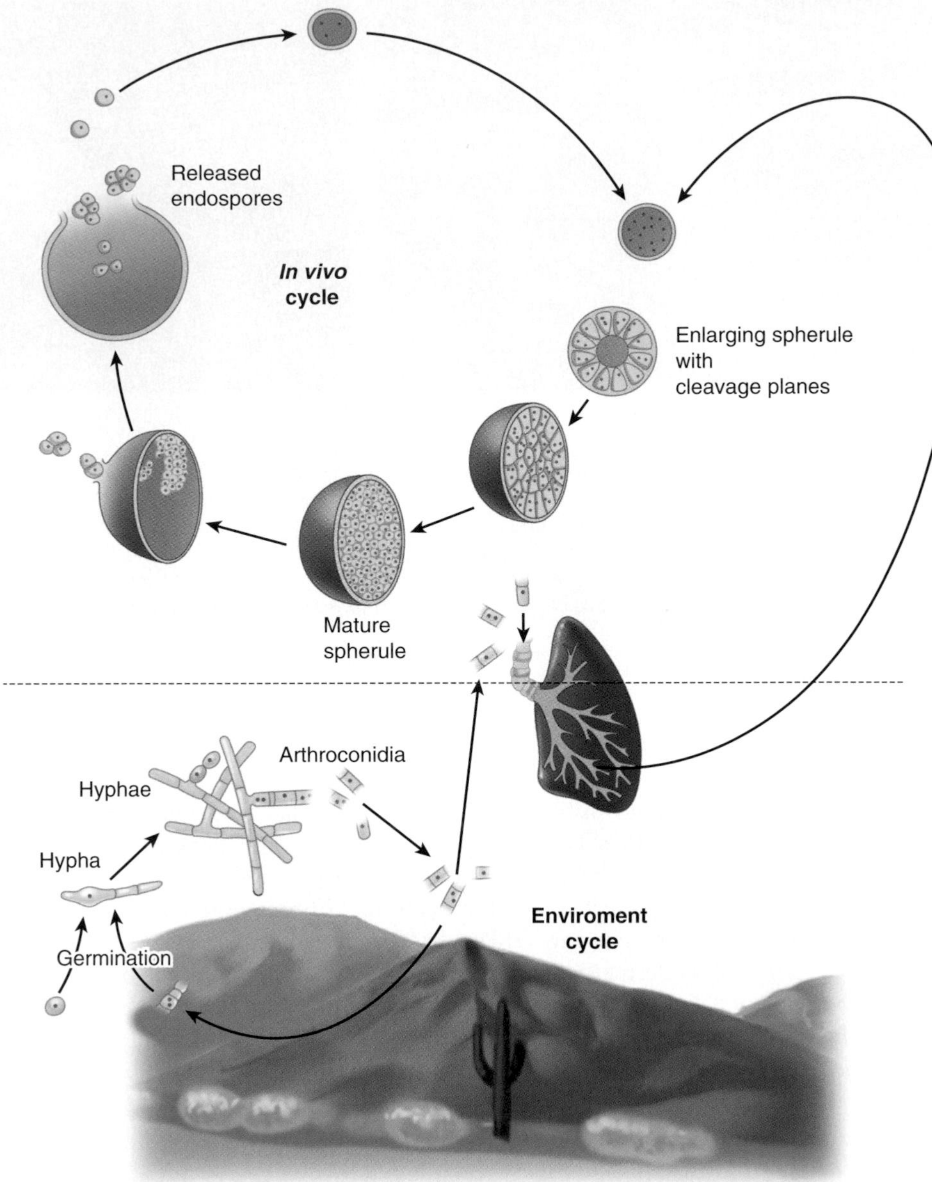

FIGURE 46–8. Life cycle of *Coccidioides immitis*. The nature cycle takes place in desert climates with modest rainfall. Hyphae differentiate into arthroconidia, which break loose and may be suspended in the air. Soil disruptions and wind facilitate spread and the probability of inhalation into human lungs. In the human host environment, in vivo differentiation produces cleavage planes and eventually huge spherules. The spherules rupture releasing endospores, which can then repeat the in vivo cycle.

Geographically restricted to Sonoran desert

High proportion of locals have been infected

mild winters with few freezes, and annual rainfall of about 10 inches during brief rainy seasons. Ecologic "islands" with these conditions are found scattered throughout Central and South America. The primary endemic zones in the United States are in Arizona, Nevada, New Mexico, western Texas, and the arid parts of central and southern California. Persons living in the endemic areas are at high risk of infection, although disease is much less common. Positive skin test rates of 50% to 90% occur in long-time residents of highly endemic areas. Coccidioidomycosis is not transmissible from person to person.

Arthroconidia can spread long distances

Infection cannot be acquired without at least visiting an endemic area, although some interesting examples of the endemic zone itself paying a visit have been recorded. In 1978, a storm originating in Bakersfield, California (endemic zone), carried a thick coat of dust all the way to San Francisco. This was followed by cases of coccidioidomycosis in persons who had never left the Bay Area. In 1992, a tenfold increase in disease in California followed an unusually wet winter in which the storms created a drought–rain–drought pattern just right for growth of the mold (and wildflowers). When the Sonoran desert blooms, an arthroconidium "crop" is not far behind. Coccidioidomycosis is increasing in

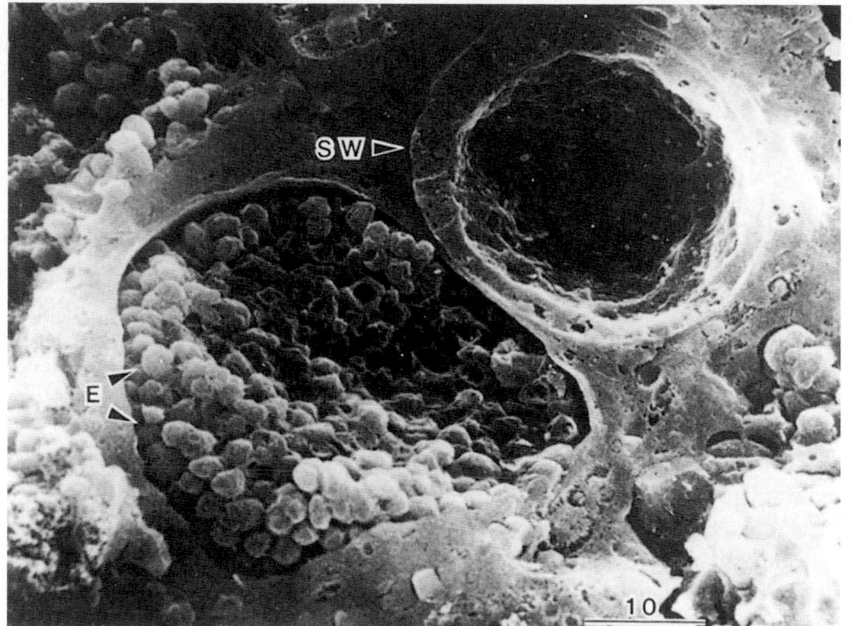

FIGURE 46–9. *Coccidioides immitis.* This electron micrograph of infected mouse lung shows a spherule filled with endospores (E) and one that has discharged its endospores into the surrounding tissue. Note the thickness of the spherule wall (SW). (Reprinted with permission from Drutz DJ, Huppert M. J Infect Dis 1983;147:379, Figure 7. Copyright University of Chicago Publisher.)

the United States primarily because of the influx of population into the sunbelt states where *C immitis* is endemic. Ninety percent of new cases coccidioidomycosis are in California or Arizona.

Rainfall pattern influences attack rate

C immitis is also a notorious cause of infection in laboratory workers. The high infectivity of cultured arthroconidia has caused it to be classified as a bioweapon.

Considered a potential bioweapon

PATHOGENESIS

Inhaled arthroconidia are small enough (2 to 6 μm) to bypass the defenses of the upper tracheobronchial tree and lodge in the terminal bronchioles. The incubation period is 1 to 3 weeks. Human monocytes can ingest and kill some arthroconidia on initial exposure, although the outer portion of the wall of the arthroconidium has antiphagocytic properties, which persist into the early stages of spherule development. Surviving arthroconidia convert to the spherule stage, which begins its slow growth to a size that makes effective phagocytosis difficult. Although polymorphonuclear neutrophils are able to digest the spherule wall, their access appears to be restricted by the extracellular matrix surrounding it. The young endospores are released in packets that include the extracellular matrix derived from the parent spherule, which may protect them until they develop into new spherules.

Arthroconidial wall resists phagocytosis

Spherules produce endospores with extracellular matrix

A number of proteases found in the conidial cell wall or in spherules have been proposed as *C immitis* virulence factors. In addition to their role in the fungal life cycle, some of these enzymes attack host substrates such as collagen, elastin, and immunoglobulins, but no direct specific contribution to disease has been defined. Components of the spherule outer wall and a metalloproteinase found there have been linked to virulence in animals and to survival of developing endospores.

Proteases and spherule outer wall may be linked to virulence

IMMUNITY

Lifelong immunity to coccidioidomycosis clearly develops in most of those who become infected. This immunity is associated with strong polymorphonuclear leukocyte and T_H1-mediated responses to coccidioidal antigens. In most cases, a mixed inflammatory response

Cell-mediated immunity is of prime importance

Progressive disease develops in patients with AIDS or defects in cell-mediated immunity

is associated with early resolution of the infection and development of a positive delayed hypersensitivity skin test. Progressive disease is associated with weak or absent cellular immunity and loss of delayed-type hypersensitivity to coccidioidal antigens. In most infected persons, the infection is controlled after mild or unapparent illness. The disease progresses when cell-mediated immunity and consequent macrophage activation do not develop. Such immune deficits may be a result of disease (AIDS) or immunosuppressive therapy, but may occur in persons with no other known immune compromise.

Endospores must be destroyed by cytokine-activated macrophages

The central event appears to be the reaction to arthroconidia or to endospores released from ruptured spherules. Arthroconidia can be phagocytosed and killed by polymorphonuclear leukocytes even before an adaptive immune response is mounted. The handling of endospores requires the additional participation of macrophages that do not become maximally effective until activated by cytokines produced by the T_H1 subsets. Prior to this, *C immitis* endospores may be able to impair phagosome–lysosome fusion in the phagocyte.

Antibody production is inversely related to disease progress

Humoral mechanisms are not known to play any role in immunity. In fact, *C. immitis* is resistant to complement-mediated killing, and levels of complement-fixing antibody are inversely related to the process of disease resolution. Persons with minimal objective indications of tissue involvement (eg, lesions, radiographs) have strong T-lymphocyte responses to *C immitis* antigens and little if any detectable antibody. Those with disseminated disease and absent cellular immunity have high titers of antibody. Thus, the levels of antibody seem to indicate the degree of antigenic stimulation rather than any known contribution to resolution of the infection.

COCCIDIOIDOMYCOSIS: CLINICAL ASPECTS

MANIFESTATIONS

Valley fever usually self-limiting

Erythema nodosum common in women

Chronic and disseminated disease less than 1%

More than 50% of those infected with *C immitis* suffer no symptoms, or the disease is so mild that it cannot be recalled when evidence of infection (serology, skin test) is discovered. Others develop malaise, cough, chest pain, fever, and arthralgia 1 to 3 weeks after infection. This illness dubbed **valley fever** by the local San Joaquin Valley residents lasts 2 to 6 weeks with few objective findings. The chest x-ray is usually clear or shows only hilar adenopathy. Erythema nodosum may develop midway through the course, particularly in women. In most cases, resolution is spontaneous but only after considerable discomfort and loss of productivity. In more than 90% of cases, there are no pulmonary residua. A small number of cases progress to a chronic pulmonary form characterized by cavity formation and a slow relapsing course that extends over years. Less than 1% of all primary infections and 5% of symptomatic cases disseminate to foci outside the lung.

Racial orientation and immune status are risk factors for dissemination

Meningitis is chronic sign

Disseminated disease is more common in men, dark-skinned races, particularly Filipinos, and in AIDS patients and other immunosuppressed persons. Evidence of extrapulmonary infection almost always appears in the first year after infection. The most common sites are bones, joints, skin, and meninges. Coccidioidal meningitis develops slowly with gradually increasing headache, fever, neck stiffness, and other signs of meningeal irritation. The CSF findings are similar to those in tuberculosis and other fungal causes of meningitis, such as *C neoformans*. Mononuclear cells predominate in the cell count, but substantial numbers of neutrophils are often present. If untreated, the disease is slowly progressive and fatal.

DIAGNOSIS

Direct examination for spherules is diagnostic

With enough persistence, direct examinations are usually rewarding. The thick-walled spherules are so large and characteristic (Figure 46–7A) that they are difficult to miss in wet mounts (KOH, calcifluor) or biopsy section. Skin and visceral lesions are most likely to demonstrate spherules; CSF is least likely. Spherules released into expectorated sputum are often small (10 to 15 mm) and immature without well-developed endospores. Spherules stain well in histologic sections with either H&E or special fungal stains. ::: Calciflour white, p. 699

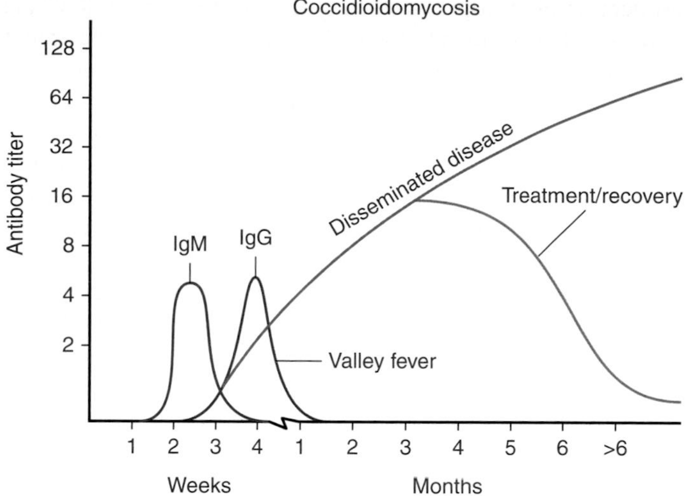

FIGURE 46–10. Serologic tests in coccidioidomycosis.

Culture of *C immitis* from sputum, visceral lesions, or skin lesions is not difficult, but must be undertaken only by those with experience and proper biohazard protection. Cultures of CSF are positive in less than half the cases of meningitis. Laboratories must be warned of the possibility of coccidioidomycosis to ensure diagnosis and prevent inadvertent laboratory infection. The latter is particularly significant outside the endemic areas, where routine precautions may not be in place. Identification requires observation of typical arthroconidia and confirmation using a DNA probe. Nucleic acid amplification procedures for direct detection are in development. ::: DNA probes, p. 81

Culture from CSF may be difficult

Substantial risk of laboratory infection with arthroconidia

Serologic tests are particularly useful in diagnosis and management of coccidioidomycosis (**Figure 46–10**). One half to three quarters of patients with primary infection develop serum IgM antibody in the first 3 weeks of illness. IgG antibodies (measured by complement fixation) appear in the third week or later, and their amount and duration depend on the extent of disease. IgG disappears with resolution and persists with continuing infection. In an appropriate clinical setting, the detection of specific antibodies can confirm the diagnosis of coccidioidomycosis, but their absence does not exclude it. In managing disseminated disease, the height of the IgG titer is a measure of the extent of disease and the direction of any change an indication of prognosis. For example, a high (greater than 1:32) and rising titer indicates a poor prognosis. The presence of IgG in the CSF is also important in the diagnosis of coccidioidal meningitis because cultures are frequently negative. The coccidioidin skin test was also a useful tool but is no longer commercially available.

Coccidioidin skin test remains positive for life

Precipitating IgM indicates acute infection

IgG detected by complement fixation quantitates disease

TREATMENT

Primary coccidioidomycosis is self-limiting, and no antifungal therapy is indicated except to reduce the risk of dissemination in patients with risk factors, such as immunocompromise and pregnancy. Progressive pulmonary disease and disseminated infection require antifungal agents, usually some combination of amphotericin B and one of the azoles. As with other systemic fungi, a course of amphotericin is often followed with itraconazole. Fluconazole is sometimes favored in meningitis because of its CSF penetration and clinical experience. In refractory meningitis, amphotericin B may be infused directly into the CSF.

Primary disease treated only with risk factors

Amphotericin B and azoles in progressive disease

PARACOCCIDIOIDES BRASILIENSIS

Paracoccidioides brasiliensis is the cause of paracoccidioidomycosis (South American blastomycosis), a disease limited to tropical and subtropical areas of Central and South America. The organism is a dimorphic fungus, the most noteworthy feature of which is the

Yeast with multiple blastoconidia are seen in ulcerative lesions

production of multiple blastoconidia from the same cell. Characteristic 5- to 40-µm cells covered with budding blastoconidia may be seen in tissue or in yeast-phase growth at 37°C. The disease manifests primarily as chronic mucocutaneous or cutaneous ulcers. The ulcers spread slowly and develop a granulomatous mulberry-like base. Regional lymph nodes, reticuloendothelial organs, and the lungs may also be involved.

Disease has a strong predilection for men

Little is known of the pathogenesis of paracoccidioidomycosis, although the route of infection is believed to be inhalation. Progression in experimental animals is associated with depressed T-lymphocyte–mediated immune responses. Paracoccidioidomycosis has a striking predilection for men, despite skin test evidence that subclinical cases occur at the same rate in both sexes. This may be related to the experimental observation that estrogens but not androgens inhibit conversion of mold-phase conidia to the yeast phase. Treatment is with sulfonamides, amphotericin B, and, more recently, the azole compounds.

CLINICAL CASE

A FORGETFUL FARMER

A 64-year-old white male farmer in Montana was hospitalized because of dementia. He had been in excellent health and working full time until 8 months before admission, when he became sloppy, careless, forgetful, and at times confused. These symptoms remitted somewhat, and he was able to perform his work on the farm. His family felt he was normal except for mild impairment of recent memory. One month before admission the symptoms recurred, and he complained of headache. After this, he became progressively worse and was finally brought to the hospital.

Physical examination revealed a well-developed man who did not appear ill. His blood pressure, pulse and respiratory rate were normal, and his temperature was 99.2°F. The rest of the examination was normal except for mild nuchal rigidity, disorientation to time and place, and marked confusion.

Lumbar puncture revealed clear fluid under an opening pressure of 250 mm; 100 white blood cells, all of which were mononuclear; protein of 85 mg% and sugar of 45 mg% (concomitant blood sugar was 90 mg%). Gram stain and India ink preparations of the CSF were negative.

QUESTIONS

- If this is a case of fungal meningitis, the most likely etiologic agent is:

A. *Candida albicans*
B. *Cryptococcus neoformans*
C. *Histoplasma capsulatum*
D. *Coccidioides immitis*
E. Any of the above

- If blood and CSF cultures for bacteria, mycobacteria, and fungi are negative, what test might reveal the diagnosis?

A. GMX antigen detection
B. GMX antibody detection
C. Germ tube test
D. Silver stain
E. *C immitis* IgG

■ Which route of infection is most likely in this case?

A. Inoculation
B. Ingestion
C. Insect vector
D. Inhalation
E. Animal bite

ANSWERS

1(B), **2**(A), **3**(D)

PART V

Pathogenic Parasites

C. George Ray
and James J. Plorde

The Nature of Parasites CHAPTER 47

General Principles of Pathogenesis, Immunology, and Diagnosis of Parasitic Infection CHAPTER 48

Antiparasitic Antimicrobics and Resistance CHAPTER 49

Sporozoa CHAPTER 50

Rhizopods CHAPTER 51

Flagellates CHAPTER 52

Intestinal Nematodes CHAPTER 53

Tissue Nematodes CHAPTER 54

Cestodes CHAPTER 55

Trematodes CHAPTER 56

CHAPTER 47

The Nature of Parasites

The discipline of Medical Parasitology encompasses a broad, diverse spectrum of agents that, while often quite dissimilar, do share some important traits. These include extremely high prevalence rates for many, with significant morbidity and mortality. The purpose of this chapter, and Chapters 48 and 49 that follow, is to lay a foundation of basic definitions and principles that hopefully will aid the student in better understanding the specific diseases that will be described in succeeding chapters.

DEFINITION

Within the context of this section of the book, the term **parasite** refers to organisms belonging to one or two major taxonomic groups: protozoa and helminths. Protozoa are microscopic, single-celled eukaryotes superficially resembling yeasts in both size and simplicity. Helminths, in contrast, are macroscopic, multicellular worms possessing differentiated tissues and complex organ systems; they vary in length from a meter to less than 1 millimeter. The majority of both protozoa and helminths are free-living, play a significant role in the ecology of the planet, and seldom inconvenience the human race. The less common disease-producing species are typically obligate parasites, dependent on vertebrate hosts, arthropod hosts, or both for their survival. When their level of adaptation to a host is high, their presence typically produces little or no injury. Less complete adaptation leads to a more serious disturbance of the host and, occasionally, to death of both host and parasite.

Eukaryotic single-celled protozoa and multicellular macroscopic helminths

Most are free living

Disease-producing species usually obligate parasites

SIGNIFICANCE OF HUMAN PARASITIC INFECTIONS

The relative infrequency of parasitic infections in the temperate societies of the industrialized world with strict sanitation has sometimes led to the parochial view that knowledge of parasitology has little relevance for physicians practicing in these areas. The continuing presence of parasitic disease among the impoverished, immunocompromised, sexually active, and peripatetic segments of industrialized populations, however, means that most physicians throughout the world regularly encounter those pathogens. Parasitic diseases remain among the major causes of human misery and death in the world today and, as such, are important obstacles to the development of economically less favored nations (**Table 47–1**). Moreover, a number of recent medical, socioeconomic, and political phenomena have combined to produce a dramatic recrudescence of several parasitic diseases with important consequences to both the United States and the developing world.

Major cause of disease and death worldwide

Currently, 2.5 billion people live in malarious areas; of these, approximately 500 million are infected at any given time. Between 1 and 3 million people, predominantly children, die of malaria each year. *Plasmodium falciparum,* the most deadly of the malarial organisms, has developed resistance to several categories of antimalarial agents, and resistant strains are

Resistance of malarial parasites to chemotherapeutics

TABLE 47–1 Prevalence of Parasitic Infections

DISEASE	ESTIMATED POPULATION AFFECTED
Amebiasis	10% of world population
Annual deaths	40–110 thousand
Giardiasis	200 million
Malaria	500 million
Population at risk	3 billion
Annual deaths	2–3 million
Leishmaniasis	12 million
African trypanosomiasis	
Population at risk	50 million
New cases per year	100,000
Annual deaths	5000
American trypanosomiasis	24 million
Population at risk	65 million
New cases per year	60,000
Schistosomiasis	207 million
Population at risk	600 million
Annual deaths	0.5–1.0 million
Clonorchiasis and opisthorchiasis	13.5 million
Paragonimiasis	2.1 million
Fasciolopsiasis	10 million
Filiariasis	128 million
Onchocerciasis	18 million
Dracunculiasis	<100,000
Ascariasis	1.3 billion
Hookworm	1.3 billion
Trichuriasis	0.9 billion
Strongyloidiasis	35 million
Enterobius vermicularis	400 million
Cestodiasis	65 million

Resistance of insect vectors to insecticides

now found throughout Southeast Asia, parts of the Indian subcontinent, southeast China, large areas of tropical America, and tropical Africa. Growing resistance of the mosquito vector of malaria to the less toxic and less expensive insecticides has resulted in a cutback of many malaria control programs. In countries such as India, Pakistan, and Sri Lanka, where eradication efforts had previously interrupted parasite transmission, the disease incidence has increased 100-fold in recent years. In tropical Africa, the intensity of transmission defies current control measures. Of direct interest to American physicians is the spillover of this phenomenon to the United States. Presently, approximately 1000 cases of imported malaria are reported annually.

Recent increases in imported malaria

Entamoeba spp. are intestinal protozoa that infect 10% of the world's population, including 2% to 3% in the United States. Most individuals are infected with the noninvasive *E dispar*. The invasive *E histolytica* produces amebiasis, a disease characterized by intestinal ulcers and liver abscesses. It is more commonly seen in areas of the world with poor sanitation, but occurs in the United States as well, particularly in institutions for the mentally retarded and among migrant workers and some male homosexuals.

Amebic infections in 10% of world population

In the poor, rural areas of Latin America, *Trypanosoma cruzi* infects an estimated 16 million individuals annually, leaving many with the characteristic heart and gastrointestinal lesions of Chagas' disease. In Africa, from the Sahara Desert in the north to the Kalahari in the south, a related organism, *Trypanosoma brucei*, causes one of the most lethal of human infections, sleeping sickness. Animal strains of this same organism limit food supplies by making the raising of cattle economically unfeasible.

Trypanosomiasis produces disease and limits food supplies

Leishmaniasis, a disease produced by another intracellular protozoan, is found in parts of Europe, Asia, Africa, and Latin America. Clinical manifestations range from a self-limiting skin ulcer, known as oriental sore, through the mutilating mucocutaneous infection of espundia, to a highly lethal infection of the reticuloendothelial system (kala azar).

Leishmaniasis can cause cutaneous or disseminated disease

In 1947, in an article entitled "This Wormy World," Stoll estimated that between the Tropic of Cancer and the Tropic of Capricorn, there were many more intestinal worm infections than people. The prevalence was judged to be far lower in temperate climates. The most serious of the helminthic diseases, schistosomiasis, affects an estimated 200 million individuals in Africa, Asia, and the Americas. Persons with heavy worm levels develop bladder, intestinal, and liver disease, which may ultimately result in death. Unfortunately, the disease is frequently spread as a consequence of rural development schemes. Irrigation projects in Egypt, the Sudan, Ghana, and Nigeria have significantly increased the incidence of the disease in these areas, often mitigating the economic gains of the development program itself.

Parasitic worm infections prevalent, may be spread by irrigation projects

Two closely related filarial worms, *Wuchereria bancrofti* and *Brugia malayi*, which are endemic in Asia and Africa, interfere with the flow of lymph and can produce grotesque swellings of the legs, arms, and genitals. Another filaria produces onchocerciasis (river blindness) in millions of Africans and Americans, leaving thousands blind.

Filariasis produces swellings

Toxoplasmosis, giardiasis, trichomoniasis, and pinworm (enterobiasis) infections are four cosmopolitan parasitic infections well known to American physicians. Toxoplasmosis, a protozoan infection of cats, infects possibly one third of the world's human population. Although it is usually asymptomatic, infection acquired in utero may result in abortion, stillbirth, prematurity, or severe neurologic defects in the newborn. Asymptomatic infection acquired either before or after birth may subsequently produce visual impairment. Immunosuppressive therapy may reactivate latent infections, producing severe encephalitis.

Multiple parasitic diseases common in the United States

BIOLOGY, MORPHOLOGY, AND CLASSIFICATION

Protozoa

Morphology

Protozoa range in size from 2 to more than 100 μm. Their protoplasm consists of a true membrane-bound nucleus and cytoplasm. The former contains clumped or dispersed chromatin and a central nucleolus or **karyosome.** The shape, size, and distribution of these structures are useful in distinguishing protozoan species from one another.

The cytoplasm is frequently divided into an inner endoplasm and a thin outer ectoplasm. The granular **endoplasm** is concerned with nutrition and often contains food reserves, contractile vacuoles, and undigested particulate matter. The **ectoplasm** is organized into specialized organelles of locomotion. In some species, these organelles appear as blunt, dynamic extrusions known as pseudopods. In others, highly structured thread-like cilia or flagella arise from intracytoplasmic basal granules. Flagella are longer and less numerous

Endoplasm contains nutrients

Ectoplasm has organelles of locomotion

TABLE 47–2 Classes of Protozoa

CLASS	ORGANELLES OF LOCOMOTION	METHOD OF REPRODUCTION
Rhizopods (amebas)	Pseudopods	Binary fission
Ciliates	Cilia	Binary fission
Flagellates	Flagella	Binary fission
Sporozoa	None	Schizogony/sporogony

than cilia and possess a structure and a mode of action distinct from those seen in prokaryotic organisms.

Classification

Most amebas are either free-living or commensal

Modes of reproduction and type of locomotive organelles are used to divide the protozoa into four major classes (**Table 47–2**). Although most **rhizopods** (amebas) are free-living, several are found as commensal inhabitants of the intestinal tract in humans. One of these organisms, *E histolytica*, may invade tissue and produce disease. Occasionally, free-living amebas may gain access to the body and initiate illness. The majority of **ciliates** are free-living and seldom parasitize humans. **Flagellates** of the genera *Trypanosoma* and *Leishmania* are capable of invading the blood and tissues of humans, where they produce severe chronic illness. Others, such as *Trichomonas vaginalis* and *Giardia lamblia*, inhabit the urogenital and gastrointestinal tracts and initiate disease characterized by mild to moderate morbidity but no mortality. **Sporozoan** organisms, in contrast, produce two of the most potentially lethal diseases of humans—malaria and toxoplasmosis.

Physiology

Protozoa are facultative anaerobes

Nutrients engulfed by phagocytosis or pinocytosis

Most parasitic protozoa are facultative anaerobes. They are heterotrophic and must assimilate organic nutrients. This assimilation is accomplished by engulfing soluble or particulate matter in digestive vacuoles, processes termed **pinocytosis** and **phagocytosis,** respectively. In some species, food is ingested at a definite site, the peristome or cytostome. Food may be retained in special intracellular reserves, or vacuoles. Undigested particles and wastes are extruded at the cell surface by mechanisms that are the reverse of those used in ingestion.

Reproduction usually by binary fission

Many protozoa form resistant cysts as survival form

Survival is ensured by highly developed protective and reproductive techniques. Many protozoa, when exposed to an unfavorable milieu, become less active metabolically and secrete a cyst wall capable of protecting the organism from physical and chemical conditions that would otherwise be lethal. In this form, the parasite is better equipped to survive passage from host to host in the external environment. Immunoevasive mechanisms described later contribute to survival within the host. Reproduction is accomplished primarily by simple binary fission. In one class of protozoa, the Sporozoa, a cycle of multiple fission (schizogony) alternates with a period of sexual reproduction (sporogony).

Helminths

Morphology and Classification

Worms are elongated, bilaterally symmetric animals that vary in length from less than a millimeter to a meter or more. The body wall is covered with a tough acellular **cuticle,** which may be smooth or possess ridges, spines, and tubercles. At the anterior end, there are often suckers, hooks, teeth, or plates used for the purpose of attachment. All helminths have differentiated organs. Primitive nervous and excretory systems and a highly developed reproductive system are characteristic of the entire group. Some have alimentary tracts; none possesses a circulatory system. The common helminthic parasites of humans can be

TABLE 47–3 Classification of Helminthic Parasites of Humans

CHARACTERISTIC	ROUNDWORM (NEMATODE)	TAPEWORM (CESTODE)	FLUKE (TREMATODE)
Morphology	Spindle-shaped	Head with segmented body (proglottids)	Leaf-shaped with oral and ventral suckers
Sex	Separate sexes	Hermaphroditic	Hermaphroditic[a]
Alimentary tract	Tubular	None	Blind
Intermediate host	Variable[b]	One[c]	Two[d]

[a]*Schistosoma* group has separate sexes.
[b]Tissue nematodes have intermediate hosts; intestinal nematodes do not.
[c]*Diphyllobothrium* group has two hosts.
[d]*Schistosoma* group has one host.

placed in one of three classes on the basis of body and alimentary tract configuration, nature of the reproductive system, and need for more than a single host species for the completion of the life cycle (**Table 47–3**).

Roundworms, or nematodes, have a cylindrical fusiform body and a tubular alimentary tract that extends from the mouth at the anterior end to the anus at the posterior end. The sexes are separate, and the male worm is typically smaller than the female. These worms can be divided into those that dwell within the gastrointestinal tract and those that parasitize the blood and tissues of humans. Unlike the latter, those in the gastrointestinal tract generally do not require intermediate hosts.

Differentiated organs: no circulatory system

Tapeworms, or cestodes, have flattened, ribbon-shaped bodies. The anterior end, or **scolex,** is armed with suckers and frequently with **hooklets,** which are used for attachment. Immediately behind the head is a neck that generates a chain of reproductive segments, or **proglottids.** Each segment contains both male and female gonads. The worm lacks a digestive tract and presumably absorbs nutrients across its cuticle. One or sometimes two intermediate hosts are required for completion of the life cycle.

Proglottids contain male and female gonads

Flukes, or trematodes, are leaf-shaped organisms with blind, branched alimentary tracts. Particulate waste is regurgitated through the mouth. Two suckers, one surrounding the mouth and the second located more distally on the ventral aspect of the body, serve as organs of attachment and locomotion. Most are hermaphroditic and require two intermediate hosts. The blood-dwelling schistosomes, however, are unisexual and require but a single intermediate.

Most require two intermediate hosts

Physiology

Helminthic parasites are nourished by ingestion or absorption of the body fluids, lysed tissue, or intestinal contents of their hosts. Carbohydrates are rapidly metabolized, and the glycogen concentration of the worms is high. Respiration is primarily anaerobic, although larval offspring frequently require oxygen. A large part of the energy requirement is devoted to reproductive needs. The daily output of offspring can be as high as 200,000 for some worms. Typically, helminths are **oviparous** (excrete eggs), but a few species are **viviparous** (give birth to living young). The egg shells of many parasites with an aquatic intermediate host possess a lidded opening, or **operculum,** through which the embryo escapes once the egg reaches water. Whether hatched or freeborn, the resulting larva is morphologically distinct from the mature worm and undergoes a series of changes or molts before it achieves adulthood.

Anaerobic respiration of adult worm

High fertility producing thousands of eggs

Protection from the host's digestive and body fluids is afforded by the tough cuticle and the secretion of enzymes. Some worms, such as the schistosomes, can protect themselves from immunologic attack by the incorporation of host antigens into their cuticles. The life span of the adult helminth is often measured in weeks or months, but some, such as the hookworms, filaria, and flukes, can survive within their hosts for decades.

Mechanical protection from environment and immunologic attack

LIFE CYCLES, TRANSMISSION, AND DISTRIBUTION

Single-Host Parasites

As is evident from the previous discussion, many parasites require but a single host species for the completion of their life cycles. The method by which the parasite is transmitted from individual to individual within that species is determined in large part by its viability in the external environment and, in the case of helminths, by the conditions required for the maturation of offspring. The mode of transmission, in turn, determines the social, economic, and geographic distribution of the parasite. A few examples are described in **Table 47–4**.

Transmission by direct sexual contact

The protozoan *T vaginalis* does not produce protective cyst forms. Although its active, or trophozoite, form is relatively hardy, it can survive only a few hours outside of its normal habitat, the human genital tract. Thus, for all practical purposes, transmission requires the direct genital contact of sexual intercourse. As a result, trichomoniasis is cosmopolitan, occurring wherever human hosts engage in sexual activity with multiple partners.

Fecal–oral transmission common for less fragile intestinal parasites

Another protozoan, *E histolytica*, inhabits the human gut and produces hardy **cysts** that are passed in the stool. Transmission occurs when another individual ingests these cysts. Like *T vaginalis*, the organism can be passed by direct physical contact, in this case by oral–anal sexual activity. This mode of transmission, in fact, accounts for the high incidence of amebic infections in male homosexuals. Unlike *T vaginalis*, however, the cysts can survive prolonged periods in the external environment, where they may eventually contaminate food or drinking water. Thus, in environments such as mental institutions, where the level of personal hygiene is low, or in populations in which methods for the sanitary disposal of human wastes are not available, amebiasis is common.

Some require infectivity development in soil

The intestinal helminth *Ascaris lumbricoides* illustrates still another transmission pattern. In this infection, highly resistant eggs are passed in the human stool. Unlike the situation with *E histolytica* described previously, the eggs are not immediately infective but must incubate in soil under certain conditions of temperature and humidity before they are fully embryonated and infectious. As a result, this parasite cannot be transmitted directly from host to host. The organism spreads only when indiscriminate human defecation results in deposition of eggs on soil and subsequent exposure of that soil to the climatic conditions required for embryonation of the eggs. For this reason, *Ascaris* infections are most prevalent in areas of the tropics and subtropics with poor sanitation.

Multiple-Host Parasites

Multiple hosts may be involved in life cycle

A few protozoa and many helminths require two or more host species in their life cycle. To avoid confusion, it is customary to refer to the species in which the parasite reproduces sexually as the **definitive host** and that in which asexual reproduction or larval development takes place as the **intermediate host.** When there is more than one intermediate, they are known simply as the first and second intermediate hosts. In some cases, such as that of *Taenia saginata*, the beef tapeworm, both host species are vertebrates; humans serve as the definitive host and cattle as the intermediate. Among parasites that inhabit the blood and tissues of humans, it is more common for a blood-feeding arthropod to serve as a second

TABLE 47–4 Transmission and Distribution of Four Representative Parasites

ORGANISM	INFECTIVE FORM	MECHANISM OF SPREAD	DISTRIBUTION
Trichomonas vaginalis	Trophozoite	Direct (venereal)	Worldwide
Entamoeba histolytica	Cyst/trophozoite	Direct (venereal)	Worldwide
	Cyst	Indirect (fecal–oral)	Areas of poor sanitation
Ascaris lumbricoides	Egg	Indirect (fecal–oral)	Areas of poor sanitation
Plasmodium falciparum	Sporozoite	*Anopheles* mosquito	Tropical and subtropical areas

host and as the transmitting vector. An example is malaria, in which the causative plasmodium is transmitted from person to person by the bite of an infected female mosquito of the genus *Anopheles*. In this instance, sexual reproduction occurs in the mosquito, making it the definitive host and relegating the human host to the role of a mere intermediate.

Definitive host and one or more intermediate hosts

The distribution of parasites requiring a nonhuman host is limited to the ecologic niche occupied by this second host. Thus, the areas in which malaria is endemic are restricted by the distribution of *Anopheles* mosquitoes. The area of disease distribution is, in fact, generally smaller than that of the nonhuman host, because conditions favoring parasite transmission may also differ. For example, both the abundance of *Anopheles* and the speed with which the malarial parasite completes its development within them are directly related to the ambient temperature and humidity. Among temperate zone *Anopheles*, the number of infected mosquitoes may be insufficient to sustain parasite transmission. In tropical areas, transmission is more likely to be constant and intense. In another more obvious example, infections with *T saginata* are found only in areas where cattle are raised for human consumption and, within those areas, only where indiscriminate human defecation and the ingestion of raw or undercooked beef are common.

Distribution of nonhuman hosts influences disease occurrence

CHAPTER 48

General Principles of Pathogenesis, Immunology, and Diagnosis of Parasitic Infection

In helminthic infections, humans may serve as the definitive host to the sexually mature adult worms (eg, *Taenia saginata*) or as the intermediate host to the larval stages (eg, *Echinococcus granulosus*). Occasionally, they serve as both the definitive and the intermediate host to the same worm (eg, *Trichinella spiralis, Taenia solium*). Unlike protozoan parasites, most adult helminths are incapable of increasing their numbers within their definitive host. As a result, the severity of clinical illness is related to the total number of worms acquired by the host over time. Most small worm loads are in fact asymptomatic and may not require therapy. Many worms are long-lived, however, and repeated infections can result in very high worm loads with subsequent disability.

Most adult helminths do not multiply within host

Disease severity related to worm load

The pathogenesis of both protozoan and helminthic disease is highly variable. The fish tapeworm *Diphyllobothrium latum* competes with the host for nutrients. The protozoan *Giardia lamblia* and the helminth *Strongyloides stercoralis* interfere with the absorption of food across the intestinal mucosa. Hookworm infections cause loss of iron, an essential mineral.

Other helminths, such as *Clonorchis sinensis* and *Schistosoma haematobium*, compromise the function of important organs by obstruction, secondary bacterial infection, and induction of carcinomatous changes. Occasionally, as in the case of echinococcosis, disease results from pressure and displacement of normal tissue by the slow growth of the parasitic cyst.

In malaria, the primary pathogenic mechanism appears to be the invasion and subsequent alteration or destruction of human erythrocytes. Similarly, many helminthic larvae are capable of tissue invasion and destruction. *Entamoeba histolytica* can destroy host cells without actual cellular invasion.

Wide range of direct pathogenic mechanisms

Finally, immunologic mechanisms are responsible for tissue damage and clinical manifestations in many diseases. Allergic or anaphylactic reactions play a major role in the cutaneous reactions to invading hookworm, strongyloides, and schistosome larvae (ground itch, swimmers' itch) and in the fever, rash, and lymphadenopathy that accompany the therapeutic destruction of onchocercal microfilariae (Mazzotti reaction). Transient pneumonias induced by the pulmonary migration of *Ascaris* and other nematode larvae (Loeffler's syndrome), nocturnal paroxysms of asthma in some patients with filariasis (tropical pulmonary eosinophilia), and the shock, asthma, and urticaria that follow rupture of a hydatid cyst all are immunologically mediated. Hemolysis in malaria and cardiac damage in Chagas' disease are thought, at least in part, to reflect antibody-mediated cytotoxicity. Immune complex diseases are seen in schistosomiasis (Katayama syndrome) and malaria (nephrosis). The granulomatous reaction to schistosomal eggs, the muscle damage in trichinosis, and the entire clinicopathologic spectrum of the leishmanial infections appear to be caused by cell-mediated immune responses.

Immunopathologic mechanisms contribute to parasitic diseases

IMMUNITY

Immune response to parasites vigorous but often relatively ineffective

The large size, complex structure, varied metabolic activity, and synthetic prowess of most parasites provide their human host with an intense antigenic challenge. Generally, the resulting immunologic response is vigorous, but its role in modulating the parasitic invasion differs significantly from that in viral and bacterial infections. It is apparent from the chronic course and frequent recurrences typical of many parasitic diseases that acquired resistance is often absent. When present, it is generally incomplete, serving to moderate the intensity of the infection and its associated clinical manifestations rather than to destroy or expel the causative pathogen. In fact, clinical recovery and resistance to reinfection in some instances require the persistence of viable organisms at low concentration within the body of the host **(premunition).** Complete sterilizing immunity with prolonged resistance to reinfection is exceptional.

All elements of immune response mobilized

IgE response to worm infections attracts eosinophils

Eosinophils bind to IgG-coated parasite and release toxic protein

This pusillanimous response does not result from any dearth of immunologic mechanisms available to the host. All those generally exercised against the more primitive microorganisms, including antibodies, cytotoxic T lymphocytes, activated macrophages, natural killer cells, antibody-dependent cell-mediated toxicity, lymphokines, and complement activation, have been shown to play a part in moderating parasitic infection. In worm infections, some of these mechanisms find unique implementation. On invasion of tissue, helminths stimulate the production of IgE, the Fc portion of which binds to mast cells and basophils. Interaction of the antibody with parasitic antigen triggers the release of histamine and other mediators from the attached cells. These may injure the worm directly or, by increasing vascular permeability and stimulating the release of chemotactic factors, may lead to the accumulation of other cells and IgG antibodies capable of initiating antibody-dependent, cell-mediated destruction of the parasite. The specific killer cell involved is often the eosinophil. These cells attach by their Fc receptor site to IgG antibody-coated parasites and degranulate, releasing a major basic protein that is directly toxic to the worm. ::: adverse immunologic reactions, p. 41-42

Some intracellular protozoa avoid phagolysosome destruction

The techniques by which parasites have been shown to evade the consequences of the host's specific acquired immunity are numerous. Included among them are seclusion within immunologically protected areas of the body, continual alteration of surface antigens, and active suppression of the host's effector mechanisms. A number of protozoa are shielded from humoral defenses by virtue of their intracellular location. Some have even found ways to avoid or survive the normally lethal environment of the phagolysosome of the macrophage. *T cruzi*, for example, lyses the phagosomal membrane, providing escape into the cytoplasm, whereas *Toxoplasma gondii* inhibits fusion of the phagosome with lysosomes. *Leishmania* species, capable of neither of these feats, are resistant to the action of lysosomal enzymes and survive in the phagolysosomes.

Encysted and intestinal parasites relatively inaccessible to host defenses

Toxoplasma, cestode larvae, and *Trichinella spiralis* armor themselves against immunologic attack by encysting within the tissue of the host. The gut lumen is perhaps the largest immunologic sanctuary within the body, because, unless the integrity of the intestinal mucosa is breached by injury or inflammation, this barrier protects lumen-dwelling parasites from most of the effective humoral and cellular immune mechanisms of the host, allowing almost unfettered growth and multiplication.

Antigenic shifts occur with developmental changes in parasite

Trypanosomal antigenic variation outpaces immunologic response

Antigenic glycoprotein variants of trypanosomes selected from preexisting genetic repertoire

Most immune effector mechanisms are directed against the surface antigens of the parasite, and alteration of these antigens may blunt the immunologic attack. Many parasites undergo developmental changes within their hosts that are generally accompanied by alterations in surface antigens. Immune responses directed at an early developmental stage may be totally ineffective against a later stage of the same parasite. Such stage-specific immunity has been demonstrated in malaria, schistosomiasis, and trichinosis, accounting for the seeming paradox of parasite survival in a host resistant to reinfection with the same strain of organism. Even more intriguing is the ability of some parasites to vary the antigenic characteristics of a single developmental stage. The trypanosomes that cause African sleeping sickness circulate in the bloodstream coated with a thick layer of glycoprotein. The development of humoral antibody to this coating results in the elimination of the parasite from the blood. This is followed by successive waves of parasitemia, each associated with a new glycoprotein antigen on the parasite against which the previously produced antibody is ineffective. The parasite is capable of producing more than 100 glycoprotein variants, each encoded by a different structural gene. The expression of individual genes from this large genetic repertoire is controlled by the sequential transfer of a duplicate copy of each gene to an area of the parasite responsible for gene expression.

Several protozoan and helminthic pathogens are thought to be capable of neutralizing antibody-mediated attack by shedding and, later, regenerating specific surface antigens. Adult schistosomes, in addition, may immunologically hide from the host by masking themselves with host blood group antigens and immunoglobulins.

Antigenic shedding and masking with host antigens

A number of parasites can destroy or inactivate immunologic mediators. Tapeworm larvae produce anticomplementary chemicals, and *T cruzi* splits the Fc component of attached antibodies, rendering it incapable of activating complement. Several protozoa, most notably *T brucei*, the etiologic agent of African sleeping sickness, induce polyclonal B-cell activation leading to the production of nonspecific immunoglobulins and eventual exhaustion of the antibody-producing capacity of the host. This and other protozoa can produce nonspecific suppression of both cellular and humoral effector mechanisms, enhancing the host's susceptibility also to a variety of unrelated secondary infections. Patients with disseminated leishmaniasis display a specific inability to mount a cellular immune response to parasitic antigens in the absence of evidence of generalized immunosuppression.

Parasites may destroy immunologic mediators

Some parasites cause immune suppression

Finally, the thick, tough cuticle of many adult helminths renders them impervious to immune effector mechanisms designed to deal with the less robust microbes.

Cuticle helps resist immune effectors

DIAGNOSIS

Although parasitic diseases are not as common in the United States as elsewhere, they do occur and may at times be life-threatening. In addition, the continuous arrival of travelers and immigrants from endemic areas necessitates consideration of these diseases in differential diagnoses. Unfortunately, the clinical manifestations of parasitic infections are seldom sufficiently characteristic to raise this possibility in the clinician's mind. Moreover, routine laboratory tests are seldom helpful. Although eosinophilia has been recognized as an important clue to the diagnosis of parasitic disease, this phenomenon is characteristic only of helminthic infection, and even in these cases it is frequently absent. Eosinophilia, which presumably reflects an immunologic response to the complex foreign proteins possessed by worms, is most marked during tissue migration. Once migration ceases, the eosinophilia may decrease or disappear entirely. Thus, the clinician must usually rely on a detailed travel, food intake, transfusion, and socioeconomic history to raise the possibility of parasitic disease.

Need to consider indigenous and imported infections

Eosinophilia seen in helminthic infections

Once considered, diagnosis is usually straightforward. Typically, it rests on the demonstration and morphologic identification of the parasite or its progeny in the stool, urine, sputum, blood, or tissues of the human host.

Morphologic demonstration of parasites primary diagnostic means

In intestinal infections, a simple wet mount or stained smear, or both, of the stool is often adequate. Some parasites, however, are passed in the feces intermittently or in fluctuating numbers, and repeated specimens are needed. Ova of worms and cysts of protozoa may be concentrated by sedimentation or flotation techniques to increase their numbers for diagnosis. Occasionally, specimens other than stool must be examined. In the case of small bowel infections such as giardiasis and strongyloidiasis, aspirates of the duodenum or a small bowel biopsy may be required to establish the diagnosis. Similarly, the recovery of large bowel parasites such as *E histolytica* and *Schistosoma mansoni* may require proctoscopy or sigmoidoscopy, with aspiration or biopsy of suspect lesions. Eggs of pinworms (*Enterobius*) and tapeworms (*Taenia*) may be found on the perineal skin when they are absent from the stool.

Stool concentration techniques used for intestinal parasites

Parasites dwelling within the tissue and blood of the host are more difficult to identify. Direct examination of the blood is useful for the detection of malarial parasites, leishmania, trypanosomes, and filarial progeny (microfilariae). The concentration of organisms in the bloodstream often fluctuates, however, requiring the collection of multiple specimens over several days. Both wet mount and stained preparations of thin and thick blood smears (see Chapter 50) are used. Lung flukes and occasionally other helminths discharge their offspring in the sputum and may be found there with appropriate concentration techniques. In others, larvae can be recovered with skin (onchocerciasis) or muscle (trichinosis) biopsy.

Demonstration of blood and tissue parasites requires proper timing

In some infections, parasite recovery is uncommon. Immunodiagnostic and nucleic acid hybridization techniques provide diagnostic alternatives for these situations. Although tests for circulating antibodies have long been available for a number of parasitic diseases, they have often lacked sensitivity and specificity. The replacement of crude, antigenically complex parasitic extracts with purified homologous antigens, together with the adaptation of highly reactive test systems, has significantly increased the sensitivity and specificity of such tests. Currently, reliable serologic procedures are available for amebiasis, cysticercosis, echinococciasis, paragonimiasis, schistosomiasis, strongyloidiasis, toxocariasis, toxoplasmosis, and trichinosis. More will undoubtedly follow in the near future.

Serologic tests available for some parasites

Techniques for the detection of parasitic antigens in blood, body fluids, tissues, and excreta also have been developed. Commercial immunofluorescent and immunosorbent kits for *Pneumocystis carinii* (pulmonary secretions), *T vaginalis* (genitourinary fluids), and *E histolytica, Giardia*, and *Cryptosporidium* (feces) are now commonly found in clinical laboratories. Less generally available are systems for the detection of malaria antigens in blood and *T gondii* in tissue.

Antigen detection becoming available

DNA probes are available for the detection of *P falciparum, T cruzi, T brucei, Onchocerca* species, and the etiologic agents of lymphatic filariasis. The probes for *P falciparum* and lymphatic filariae have demonstrated sensitivities that match or exceed those of traditional techniques. The major limitations of DNA probes as diagnostic tools relate to the technical aspects of the hybridization procedure, which should soon be overcome.

Molecular methods for DNA detection being used with increasing frequency

CHAPTER 49

Antiparasitic Antimicrobics and Resistance

The study and management of parasitic disease were seminal to the initiation of the chemotherapeutic era. Amazonian Indians first used quinine-containing extracts of cinchona tree bark to treat malarious patients more than 300 years ago. It was in the attempt to synthesize this same antimalarial compound that 19th-century German chemists discovered aniline dyes. The circle closed in the early years of this century when Ehrlich, while investigating the suitability of these dyes as protozoan stains, developed the concept that chemicals might be found that have the capacity to destroy microbial pathogens selectively without damage to the tissues of the human host. Although the most dramatic confirmation of that concept came with the introduction of arsenical compounds for the treatment of syphilis, his first successful chemotherapeutics were directed against protozoan agents. By 1930, chemically synthesized drugs had been marketed for the treatment of malaria, trypanosomiasis, and schistosomiasis.

Antiparasitic agents among first antimicrobics

The introduction and explosive increase in the number and variety of antimicrobic agents introduced in the latter three fourths of the 20th century forever changed the face of medicine. Unfortunately, however, few were effective against parasites because they share the eukaryotic characteristics of their hosts. With the resources of the pharmaceutical companies directed toward the development and introduction of antibacterial agents, work on antiparasitic agents lagged. Because of the lack of safer alternatives, chemotherapeutics synthesized in the preantibiotic era remained critical elements of the parasitologist's therapeutic armamentarium until very recently. Most required prolonged or parenteral administration, the effectiveness of many was restricted to particular disease stages, and the toxicity of a few mandated that use be limited to very severe or life-threatening conditions. With time, and at a pace much slower than that seen for the antibacterial agents, newer antiparasitic agents were developed that overcame many of these problems. Their numbers are still limited, and only recently has their safety and efficacy begun to match those of their antibacterial equivalents.

Newer antiparasitics have broader spectrum and are less toxic

THERAPEUTIC GOALS

The process of antiparasitic drug development and use has been shaped to a significant degree by the concentration of parasitic diseases in the impoverished areas of the world. Community-based public health measures aimed at interrupting pathogen transmission, such as provision of sanitary facilities and clean water supplies, are still often beyond the capacity of tightly constrained budgets, and the major burden of mitigating the impact of parasitic illnesses in endemic areas often falls on medical auxiliaries or village health workers who, operating in remote and relatively primitive conditions, must examine, diagnose, and treat sick patients with whom they have only fleeting contact. Given these limitations

Treatment programs difficult in underdeveloped economies

Ideal agents would be inexpensive, of low toxicity, and effective in single doses; few of these exist

and the large numbers of the afflicted, optimal therapy requires drugs that are effective in a single dose, easily administered, safe enough to be dispensed with limited medical supervision, and sufficiently inexpensive to be widely used. Few such agents exist. Pharmaceutical companies, faced with the enormous costs of drug development and approval, have been reluctant to expend resources they are unlikely to recover. Until the international community provides the resources needed for the development of more suitable agents, the full potential of antiparasitic chemotherapy will not be realized.

For worms, treatment efforts should be concentrated on the most heavily parasitized individuals

The practical aspects of antiparasitic therapy are illustrated in the principles governing the treatment of worm infections, which differ significantly from those applied to prokaryotic or protozoan infections. Helminths, with few exceptions, do not multiply within the human host, and severe infections thus require the repeated acquisition of infectious parasites. It is interesting that the intensity of infection or worm burden does not follow a normal distribution in human populations. Most infected persons harbor fewer than a dozen adult worms; a small minority harbor very large worm numbers. Because there is a direct correlation between worm burden and clinical disease, only this minority suffers significant morbidity. Concentrating treatment on those few clinically ill patients moderates the medical impact of a helminthic disease on a community at a cost dramatically lower than that required for mass treatment. Moreover, it is usually unnecessary to eradicate all worms from treated patients; a significant decrease in the worm burden is adequate to alleviate clinical symptoms. This can often be accomplished with short, subcurative doses that further reduce cost and minimize the likelihood of drug toxicity. Because this approach can dramatically decrease the total community worm burden, the number of worm progeny shed into the environment is similarly reduced and the transmission of the disease slowed or, at times, eliminated.

STRUCTURE AND ACTION

Most antiparasitics are synthetic

With few exceptions, antiparasitic agents have been synthesized de novo rather than developed from naturally occurring substances. Most are relatively simple and often contain benzene or other ring structures.

Differential toxicity based on uptake, metabolic factors

It is believed that most antiprotozoan drugs interfere with nucleic acid synthesis or, less commonly, with carbohydrate metabolism. Anthelmintics, on the other hand, apparently act by compromising the worm's glycolytic pathways or neuromuscular function. In most cases, the parasite and host cells have functionally equivalent target sites. Differential toxicity is achieved by preferential uptake, metabolic alteration of the drug by the parasite, or differences in the susceptibility of functionally equivalent sites in parasite and host.

Acquired mutational resistance usually involves reduced uptake of drug

As has been the case for antibacterial agents, the impact of many antiparasitic agents has been compromised by the development of resistance in the parasite. This seems to have resulted from mutation and selection in the face of intensive, often prophylactic, drug use. The mechanisms responsible have been studied for only a few parasites, but appear to be related to reduced uptake of drug.

DRUGS

Heavy Metals

Arsenic and antimonial compounds inactivate –SH groups

Differential toxicity based on enhanced uptake by parasite

Melarsoprol active against all stages of trypanosomiasis

Arsenic and antimonial compounds have been used since ancient times. They form stable complexes with sulfur compounds and probably exert their biologic effects by binding to sulfhydryl (–SH) groups. They are toxic to the host as well as to the parasite and have their greatest impact on cells that are most metabolically active such as neuronal, renal tubular, intestinal epithelial, and bone marrow stem cells. Their differential toxicity and therapeutic value are due to enhanced uptake by the parasite and its intense metabolic activity. Only one trivalent arsenical, melarsoprol (Mel B), is now widely used. It is capable of penetrating the blood–brain barrier and is effective in all stages of trypanosomiasis. Because of its toxicity, it is used only when less toxic agents have failed or when the central nervous system is involved. The recently introduced less toxic trypanocides that penetrate the blood–brain barrier may soon replace this drug.

Antimonial agents are now restricted to the management of leishmanial infections. Two pentavalent compounds, sodium stibogluconate (Pentostam) and meglumine antimoniate* (Glucantime), are used for all forms of leishmaniasis. In disseminated disease, prolonged therapy is usually required, and relapses often occur. In localized cutaneous leishmaniasis, cure is usually achieved with a relatively brief course. Toxic side effects are similar to those of the arsenicals.

Antimonials used only for leishmanial infections

Antimalarial Quinolines

Cinchona bark was used in Europe for the treatment of fever as early as 1640. Only after Pelletier and Caventou isolated quinine from cinchona in 1820 did this alkaloid gain widespread acceptance as an antimalarial. Synthesis of new quinolines was stimulated by the interruption of quinine supplies during World War I and World War II and, after 1961, by the growing impact of drug-resistant falciparum malaria in several areas of the world. Among the most effective agents are those that share the double-ring structure of quinine.

Quinine and quinoline analogs active against malaria

Current analogs fall into three major groups: 4-aminoquinolines, 8-aminoquinolines, and 4-quinolinemethanols. Selective destruction of intracellular parasites results from accumulation of the quinolines by parasitized host cells. Most of these agents appear to block nucleic acid synthesis by intercalation into double-stranded DNA. However, the failure of the 4-quinolinemethanols to intercalate indicates that other mechanisms, perhaps inhibition of heme polymerase, with the build up of toxic hemoglobin metabolites within the malarial parasite, are involved.

Accumulate in parasitized cells and block DNA synthesis

Quinine, 4-aminoquinolines, and 4-quinolinemethanols are preferentially concentrated in parasitized erythrocytes and rapidly destroy the erythrocytic stage of the parasite that is responsible for the clinical manifestations of malaria. Thus, these agents can be used either prophylactically to suppress clinical illness if infection occurs or therapeutically to terminate an acute attack. They do not concentrate in tissue cells, and thus organisms sequestered in exoerythrocytic sites, particularly the liver, survive and may later reestablish erythrocytic infection and produce a clinical relapse. The 8-aminoquinolines accumulate in tissue cells, destroy hepatic parasites, and effect a radical cure.

Quinine, 4-aminoquinolines (eg, chloroquine), and 4-quinolinemethanols suppress malarial infection

8-Aminoquinolines (eg, primaquine) effect radical cure

Chloroquine phosphate, a 4-aminoquinoline, is the most widely used of the blood schizonticidal drugs. In the doses used for long-term malarial prophylaxis, it has proved to be remarkably free of untoward effects. Primaquine phosphate, the 8-aminoquinoline used to eradicate persistent hepatic parasites, has toxic effects related to its oxidant activity. Methemoglobinemia and hemolytic anemia are particularly frequent in patients with glucose-6-phosphate dehydrogenase deficiency, because they are unable to generate sufficient quantities of the reduced form of nicotinamide adenine dinucleotide to respond to this oxidant stress. Typically, the anemia is severe in patients of Mediterranean and Far Eastern ancestry and mild in black patients.

Primaquine has hematologic toxicity

Quinine is the most toxic of the quinolines and is currently used primarily to treat the strains of *Plasmodium falciparum* resistant to several blood schizonticidal agents that are spreading rapidly through Asia, Latin America, and Africa. Chloroquine resistance is the most common and worrisome because suitable alternatives to this safe and highly effective agent are few. Quinidine, a less cardiotoxic optical isomer of quinine, is more readily available in the United States and is preferred to quinine when parenteral administration is required. Mefloquine, a more recently developed oral 4-quinolinemethanol, originally displayed a high level of activity against most chloroquine-resistant parasites; however, mefloquine-resistant strains of *P falciparum* are now widespread in Southeast Asia and are present to a lesser degree in South America. Resistant strains have recently been identified in Africa.

Quinine is active against many chloroquine-resistant malarial strains

Phenanthrene methanols are not in the strict sense quinine analogs. Nevertheless, they are structurally similar to this group of agents and, together with them, were discovered to have antimalarial activity during the World War II. Halofantrine†, the most effective of the

*Not available in the United States.
†Not available in the United States.

group, has only recently become available. In vitro and in vivo studies demonstrated that it is an effective blood schizonticide against both sensitive and multidrug-resistant strains of *P falciparum*. Its mechanism of action was originally thought to differ from that of quinine and mefloquine. Recently, mefloquine-resistant strains of *P falciparum* have demonstrated decreased sensitivity to halofantrine, raising the possibility of cross-resistance between these two agents. Rarely, halofantrine has produced fatal heart arrhythmias, and it should not be given to patients with cardiac conduction abnormalities. It is otherwise well tolerated and appears to be free of teratogenicity. Oral absorption is both slow and erratic, reaching maximum concentrations in 5 to 7 hours; its half-life is relatively short (1 to 3 days). Clinical studies have demonstrated high failure rates when the drug is given in a single dose; cure rates with multiple-dose regimens, however, have been high.

Phenanthrene methanols active against multidrug-resistant malaria

Quinones

Atovaquone is a novel hydroxynaphthoquinone that shows promise in the treatment of malaria and toxoplasmosis. In the search for effective antimalarial agents during World War II, a number of hydroxynaphthoquinones were found to have antimalarial activity in experimental animals; however, all were rapidly metabolized in humans and proved ineffective in the treatment of malarious patients. In the 1980s, a single hydroxynaphthoquinone, atovaquone, was found to be both highly effective in vitro against *P falciparum* and metabolically stable in humans when administered orally. Its antiparasitic activity appears to result from the specific blockade of pyrimidine biosynthesis secondary to the inhibition of the parasite's mitochondrial electron transport chain at the ubiquinol–cytochrome c reductase region (complex III). Its long half-life (70 hours) and lack of serious adverse reactions suggested that it would be of great value in the treatment of malaria.

Efficacy trials established its capacity to effect rapid clearance of parasitemia in patients with chloroquine-resistant falciparum malaria. Frequent parasitic recrudescences were eliminated when atovaquone was administered in combination with proguanil or tetracycline. Subsequently, this agent has shown to be effective for the treatment of toxoplasmosis in patients with acquired immunodeficiency syndrome (AIDS). Unlike other antitoxoplasma agents, atovaquone has been found to be active against *Toxoplasma gondii* cysts as well as tachyzoites, suggesting that this agent may produce radical cure. Supporting this is the infrequency with which cessation of atovaquone treatment of toxoplasmic cerebritis in AIDS patients has resulted in relapse. Relapse after atovaquone treatment of pneumocystosis in this same patient population appears similarly uncommon.

Atovaquone stable and active against malaria and toxoplasmosis

Folate Antagonists

Folic acid serves as a critical coenzyme for the synthesis of purines and ultimately DNA. In protozoa, as in bacteria, the active form of folic acid is produced in vivo by a simple two-step process. The first, the conversion of *para*-aminobenzoic acid to dihydrofolic acid, is blocked by sulfonamides. The second, the transformation of dihydro- to tetrahydrofolic acid, is inhibited by folic acid analogs (folate antagonists), which competitively inhibit dihydrofolate reductase. Used together with sulfonamides, folate antagonists are very effective inhibitors of protozoan growth.

Sulfonamide and folate antagonists inhibit protozoa

Trimethoprim, an inhibitor of dihydrofolate reductase, is used in combination with sulfamethoxazole to treat toxoplasmosis. Another folate antagonist, pyrimethamine, has a high affinity for sporozoan dihydrofolate reductase and has been particularly effective, when used with a sulfonamide, in the management of clinical malaria and toxoplasmosis. In East Africa, a third folate antagonist, proguanil, is commonly taken in combination with chloroquine for malaria prophylaxis. Acquired protozoal resistance to folate antagonists is mutational and generally has been limited to particular species of malarial parasites.

Trimethoprim effective in *Toxoplasma* infections

Folate antagonists may result in folate deficiency in individuals with limited folate reserves, such as newborns, pregnant women, and the malnourished. This is of great concern when large doses are used for prolonged periods, as in the treatment of acute toxoplasmosis. When folate antagonists are used with sulfonamides, the entire range of sulfonamide toxic effects may be seen. Patients with AIDS appear to suffer an unusually high incidence of toxic side effects to trimethoprim–sulfamethoxazole.

Folate deficiency and sulfonamide toxicities occur

Qinghaosu (Artemisinin)

This natural extract of the plant *Artemisia annua* (qing hao, sweet wormwood) is a sesquiterpenelactone peroxide that is structurally distinct from all other known antiparasitic compounds. Extracts of qing hao were recommended for the treatment of fevers in China as early as AD 341; their specific antimalarial activity was defined in 1971. Although qinghaosu has also been shown to be active against the free-living ameba *Naegleria fowleria* and several trematodes, including *Schistosoma japonicum, Schistosoma mansoni*, and *Clonorchis sinensis*, its greatest impact to date has been in the treatment of malaria. Extensive investigations showed it to be schizonticidal for both chloroquine-sensitive and chloroquine-resistant strains of *P falciparum*. Several derivatives, among them artemether and artesunate, are significantly more active than the parent compound. All are concentrated in parasitized erythrocytes, where they decompose, releasing free radicals, which are thought to be damaging to parasitic membranes. Artemisinin compounds act more rapidly than other antimalarial agents, stopping parasite development and preventing cytoadherence in falciparum malaria. Although depression of reticulocyte counts has been noted, these agents appear significantly less toxic than quinoline antimalarials. Since there is some evidence that they may possess teratogenic properties, they should not be used in pregnancy. Note that they may be given orally, rectally (by suppository), or parenterally. Relapses can occur unless they are given for several days or combined with a second agent such as mefloquine or tetracycline.

Plant derivative active against malaria, amebas, and *Schistosoma*

Concentrated in parasitized erythrocytes

Nitroimidazoles

Metronidazole, a nitroimidazole, was introduced in 1959 for the treatment of trichomoniasis. Subsequently, it was found to be effective in the management of giardiasis, amebiasis, and a variety of infections produced by obligate anaerobic bacteria. Energy metabolism in all of them depends on the presence of low-redox-potential compounds, such as ferredoxin, to serve as electron carriers. These compounds reduce the 5-nitro group of the imidazoles to produce intermediate products responsible for the death of the protozoal and bacterial cells, possibly by alkylation of DNA. Resistance, though uncommon, has been noted in strains of *Trichomonas vaginalis* lacking nitroreductase activity. Of greater concern is in vitro evidence of mutagenicity.

Metronidazole is the drug of choice for trichomoniasis and invasive amebiasis. It is effective in giardiasis, although it is not yet approved by the Food and Drug Administration for use in this infection. Tinidazole, a newer nitroimidazole not yet available in the United States, appears to be both a more effective and less mutagenic antiprotozoal agent. Its greater lipid solubility improves cerebrospinal fluid levels and in vitro activity.

Active against protozoa at low redox potential

Benzimidazoles

As the name benzimidazole implies, the basic structure of these antiparasitic agents consists of linked imidazole and benzene rings. Unlike their antiprotozoal cousins discussed previously, the benzimidazoles are broad-spectrum anthelmintic agents. The prototype drug, thiabendazole, acts against both adult and larval nematodes and was shown to be useful in the management of cutaneous larva migrans, trichinosis, and most intestinal nematode infections soon after its introduction in the early 1960s. The mechanism by which it exerts its anthelmintic action is uncertain. It is known to inhibit fumarate reductase, an important mitochondrial enzyme of helminths. The primary mode of action, however, may derive from the known capacity of all benzimidazoles to inhibit the polymerization of tubulin, the eukaryotic cytoskeletal protein, as described for mebendazole in the following text. Side effects are mild, are related to the gastrointestinal tract or liver, and rapidly disappear with the discontinuation of the drug. Hypersensitivity reactions, induced either by the drug or by antigens released from the damaged parasite, may occur.

Broad-spectrum anthelmintics

Inhibit helminth fumarate reductase

Mebendazole, a carbamate benzimidazole introduced in 1972, has a spectrum similar to that of thiabendazole, but also has been found to be effective against a number of cestodes, including *Taenia, Hymenolepsis*, and *Echinococcus*. It irreversibly blocks glucose uptake of both adult and larval worms, resulting in glycogen depletion, cessation of ATP formation, and paralysis or death. It does not appear to affect glucose metabolism in humans and is thought to exert its effect in worms by binding to tubulin, thus interfering with the assembly

Mebendazole blocks glucose uptake by adult and larval worms

of cytoplasmic microtubules, structures essential to glucose uptake. Unlike thiabendazole, the drug is not well absorbed from the gastrointestinal tract and may owe part of its effectiveness against intestine-dwelling adult worms to its high concentrations in the gut. Toxicity is uncommon. Teratogenic effects have been observed in experimental animals; its use in infants and pregnant women is contraindicated.

Interferes with tubulin and cytoplasmic microtubules

Albendazole is a benzimidazole carbamate that has a somewhat broader spectrum than that of its close relative, mebendazole, being more active against *Strongyloides stercoralis* and several tissue nematodes. In addition to the vermicidal and larvicidal properties that it shares with other benzimidazoles, it is ovicidal, enhancing its effectiveness in tissue cestode infections such as echinococciasis and cysticercosis. Its activity against *Giardia*, one of the most common intestinal protozoa, makes it an appealing candidate for the treatment of polyparasitism. Although it shares the teratogenic potential of other benzimidazoles, it is otherwise extremely well tolerated. Single-dose therapy is effective in the management of many intestinal nematode infections.

Albendazole has broader spectrum

Avermectins

Avermectins are macrocyclic lactones produced as fermentation products of *Streptomyces avermitilis*. Structurally similar to the macrolide antibiotics, they are effective at extremely low concentration against a wide variety of nematodes and arthropods. The avermectins appear to induce neuromuscular paralysis by acting on a receptor of the parasites-peripheral neurotransmitter, γ-aminobutyric acid (GABA). In mammals, GABA is confined to the central nervous system, and because the avermectins do not cross the blood–brain barrier in significant concentration, they do not appear to produce significant untoward effects in the mammalian host. Ivermectin, a derivative of avermectin B1, is currently the drug of choice for the treatment of onchocerciasis and is undergoing evaluation for the treatment of other human filarial infections. Its usefulness in other parasitic infections of humans remains to be established.

Antibiotics that influence nematode neurotransmitters

Activity against filariae

Praziquantel

Praziquantel, a heterocyclic pyrazinoisoquinoline, is an important new anthelmintic effective against a broad range of cestodes and trematodes, many of which had been poorly responsive to previously available agents. It is given in one to three doses. The drug is rapidly taken up by susceptible helminths, in which it appears to induce the loss of intracellular calcium, tetanic muscular contraction, and destruction of the tegument. The differential toxicity of this agent may be related to the inability of susceptible worms to metabolize the drug. Aside from transient, mild gastrointestinal symptoms, praziquantel appears remarkably free of side effects in humans. It is currently the drug of choice for the treatment of schistosomiasis, clonorchiasis, opisthorchiasis, and neurocysticercosis. Good activity has been demonstrated against other common trematode and cestode infections. Its high level of safety suggests that it may well play a significant role in worldwide mass therapy campaigns.

Causes loss of intracellular calcium in cestodes and trematodes

Safety of praziquantel allows use in mass therapy campaigns

Eflornithine (Difluoromethylornithine)

Eflornithine is a specific, enzyme-activated, irreversible inhibitor of ornithine decarboxylase (ODC). In mammalian cells, decarboxylation of ornithine by ornithine decarboxylase is a mandatory step in the synthesis of polyamines, compounds thought to play critical roles in cell division and differentiation. Originally developed as an antineoplastic agent, eflornithine proved ineffective in cancer chemotherapy trials. With the discovery that polyamines of *Trypanosoma* species were also synthesized from ornithine, eflornithine was successfully tested in the treatment of animal trypanosomiasis. Host survival was high and associated with decreases in parasitic polyamines and inhibition of nucleic acid synthesis. In the dosage required to treat trypanosomiasis, mammals tolerated the agent well, presumably because *T brucei* is 100 times more sensitive to the effects of eflornithine than are mammalian cells. Eflornithine appears to be cytostatic and requires an intact host immune system for maximum effect.

Originally an anticancer drug

Active against trypanosomes

Other Antiparasitic Agents

A number of antiparasitic agents used in therapy, their properties, and their clinical uses are listed in **Table 49–1**.

TABLE 49–1 Miscellaneous Antiparasitic Agents

COMPOUND	DRUG CLASS	ROUTE	MECHANISM OF ACTION	CLINICAL USE	COMMENT
Benznidazole	Nitroimidazole	Oral	DNA binder	Acute Chagas' disease	Bone marrow depression peripheral neuropathy, rash, itching
Bithionol	Phenol	Oral	Uncouples phosphorylation	Paragonimiasis	Not commercially available in United States
Diethylcarbamazine	Piperazine	Oral	Neuromuscular paralysis	Filarial infections	Allergic reactions to filarial antigens
Diloxanide furoate	Acetanilide	Oral	Unknown	Intestinal amebiasis	Used only for asymptomatic carriers
Lodoquinol (diiodohydroxyquin)	Halogenated quinoline	Oral	Unknown	Intestinal amebiasis *Dientamoeba* infections	Related drug has caused optic atrophy
Nifurtimox	Nitrofuran	Oral	Oxidative stress by production of free radicals	Acute Chagas' disease	Toxicity Prolonged therapy Marginal effectiveness
Nitazoxanide	Nitrothiazolyl-salicylamide	Oral	Inhibits anaerobic metabolism	*Cryptosporidium, Giardia*	Occasional vomiting, abdominal pain, diarrhea
Paromomycin	Aminoglycoside	Oral	Similar to other aminoglycosides	Intestinal cryptosporidiosis	Not absorbed Marginal effectiveness
Pentamidine	Diamidine	IV	Binds DNA	Leishmaniasis Trypanosomiasis	Toxic
Pyrantel pamoate	Tetrahydropyrimidine	Oral	Neuromuscular blockade; inhibits fumarate reductase	Pinworm infection, hookworm infection, ascariasis	Single-dose therapy
Spiramycin	Macrolide	Oral	Blocks protein synthesis	toxoplasmosis	Used to treat pregnant women
Suramin	Sulfated naphthylamine	IV	Inhibits glycerophosphate oxidase and dehydrogenase	African trypanosomiasis Onchocerciasis	Not effective in central nervous system disease Renal toxicity

IV, intravenous.

ANTIPARASITIC RESISTANCE

The major problems with resistance are related to *Plasmodium* species, specifically *P falciparum*. This problem is widespread throughout sub-Saharan Africa, Asia, and Latin America, but resistance has also appeared in lesser numbers in other areas as well. The most common is chloroquine resistance, wherein the parasite reduces the amount of drug that accumulates in its digestive vacuoles. This involves mutations in a transport molecule of digestive membrane called PfCRT (*P falciparum* chloroquine resistance transporter). Other parasite point mutations can similarly result in resistance to sulfadoxine-pyrimethamine and atovaquone-proguanil (the latter by mutations in the cytochrome b gene) and reduced susceptibility to mefloquine, quinine, and quinidine.

P falciparum resistance is a major problem

PfCRT mutation frequently associated

CHAPTER 50

Sporozoa

He is so shak'd of a burning quotidian tertian that it is most lamentable to behold.

—Shakespeare, *Henry V*

Sporozoa are a unique class of intracellular protozoa distinguished by their alternating cycles of sexual and asexual reproduction. Asexual multiplication occurs by a process of multiple fission termed schizogony. The nucleus of a trophozoite divides into several parts, forming a multinucleated schizont. Cytoplasm then condenses around each nuclear portion to form new daughter cells, or merozoites, which burst from their intracellular location to invade new host cells. After the completion of one or more of these asexual cycles, some merozoites differentiate into male and female gametocytes, initiating the cycle of sexual reproduction known as sporogony. The gametocytes mature and effect fertilization, forming a zygote. On encysting, the zygote is known as an **oocyst**. Sporozoites formed within the oocyst are released, penetrate host tissue cells, and begin another asexual cycle as trophozoites.

Intracellular protozoa with alternating sexual and asexual cycles

Two sporozoan infections, malaria and toxoplasmosis, are common diseases of humans; together, they affect more than one third of the world's population and kill or deform perhaps a million neonates and children each year. A third infection, cryptosporidiosis, has only recently been found to be an important cause of diarrhea, particularly in immunocompromised hosts.

Cause malaria, toxoplasmosis, and cryptosporidiosis

PLASMODIUM

Of all infectious diseases there is no doubt that malaria has caused the greatest harm to the greatest number.

—Laderman, 1975

PARASITOLOGY

DEFINITION

The plasmodia are sporozoa in which the sexual and asexual cycles of reproduction are completed in different host species. The sexual phase occurs within the gut of mosquitoes. These arthropods subsequently transmit the parasite while feeding on a vertebrate host. Within the red blood cells (RBCs) of the vertebrate, the plasmodia reproduce asexually; they eventually

Sexual phase in mosquito and asexual phase in humans

Four species infect humans

burst from the erythrocyte and invade other uninvolved RBCs. This event produces periodic fever and anemia in the host, a disease process known as malaria. Of the many species of plasmodia, four are known to infect humans and are considered here: *Plasmodium vivax, P ovale, P malariae,* and *P falciparum*.

MORPHOLOGY

Morphology of the parasite and the infected RBC vary by stage and species

The morphology of the stained intraerythrocytic plasmodia parasites is shown in **Figure 50–1**. In stained smears, three characteristic features aid in the identification of plasmodia: red nuclear chromatin; blue cytoplasm; and brownish-black malarial pigment, or hemozoin, consisting largely of a hemoglobin degradation product, ferriprotoporphyrin IX. The change in the shape of the cytoplasm and the division of the chromatin at different stages of parasite development are obvious. Gametocytes can be differentiated from

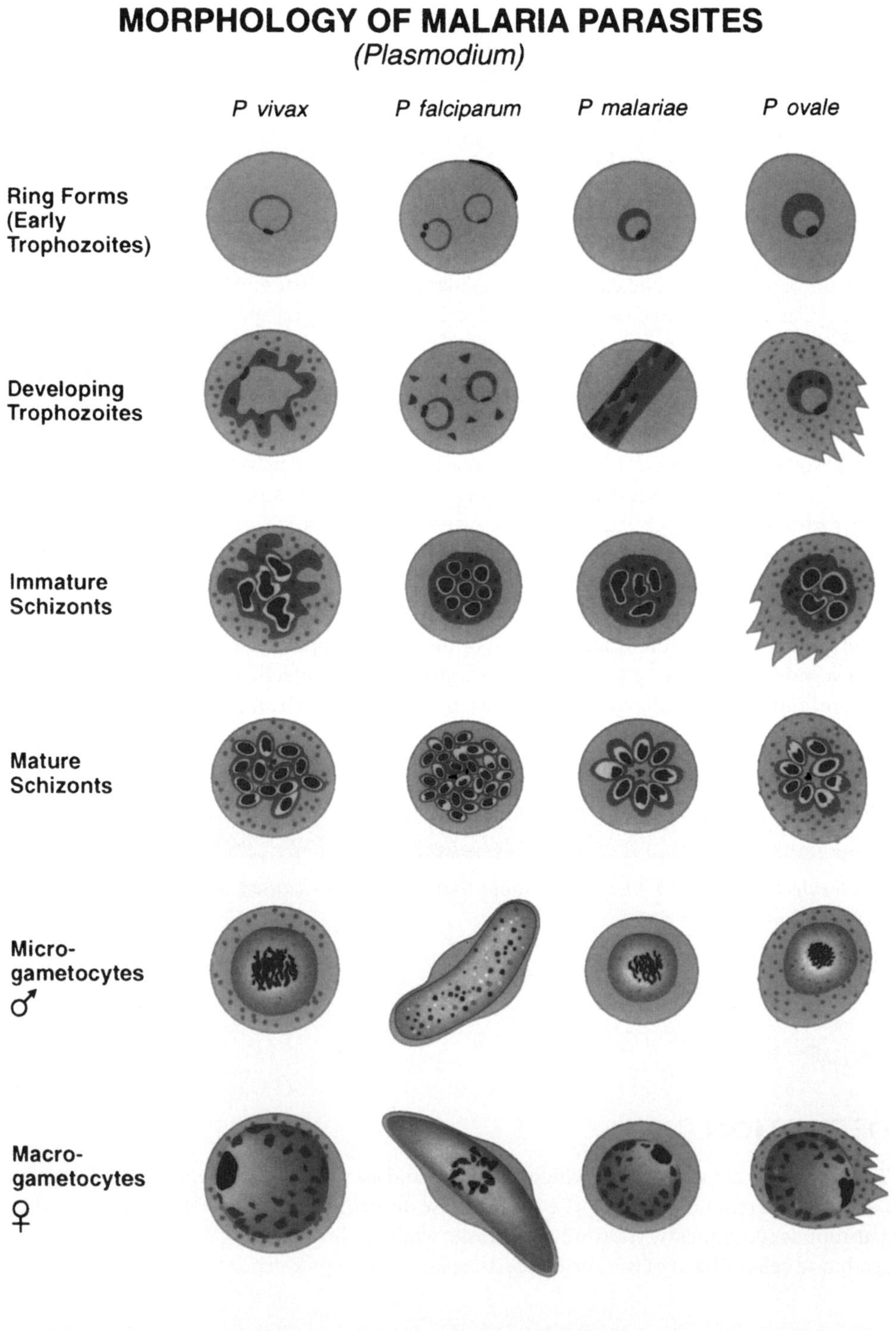

MORPHOLOGY OF MALARIA PARASITES
(Plasmodium)

P vivax | *P falciparum* | *P malariae* | *P ovale*

Ring Forms (Early Trophozoites)

Developing Trophozoites

Immature Schizonts

Mature Schizonts

Micro-gametocytes ♂

Macro-gametocytes ♀

FIGURE 50–1. Drawings of erythrocytic stages of malarial parasites. Note that trophozoite and schizont forms of *Plasmodium falciparum* occur in visceral capillaries rather than in blood. Female gametophytes have morphologic differences from the male forms shown. (Reproduced with permission from Connor DH, Chandler FW, Schwartz DQA, Manz HJ, Lack EE (eds). *Pathology of Infectious Diseases*, vol. 1. Stamford, CT: Appleton & Lange; 1997.)

TABLE 50–1 Differential Characteristics of *Plasmodium* Species

CHARACTERISTICS	*P VIVAX*	*P OVALE*	*P MALARIAE*	*P FALCIPARUM*
Erythrocyte				
Enlarged, pale	+	+	–	–
Oval, fimbriated	–	+	–	–
Schüffner's dots	+	+	–	–
Maurer's dots	–	–	–	+
Parasite				
All asexual stages seen	+	+	+	–
Band forms	–	–	+	–
Double infections	–	–	–	+
Double chromatin dots	–	–	–	+
Banana-shaped gametocytes	–	–	–	+

the asexual forms by their large size and lack of nuclear division. Some of the infected erythrocytes develop membrane invaginations or caveolae–vesicle complexes, which are thought to be responsible for the appearance of the pink Schüffner's dots or granules (see following text).

The appearance of each of the four species of plasmodia that infect humans is sufficiently different to allow their differentiation in stained smears. The parasitized erythrocyte in *P vivax* and *P ovale* infections is pale and enlarged and contains numerous Schüffner's dots. All asexual stages (trophozoite, schizont, merozoite) may be seen simultaneously. Cells infected by *P ovale* are elongated and frequently irregular or fimbriated in appearance. In *P malariae* infections, the RBCs are not enlarged and contain no granules. The trophozoites often present as "band" forms, and the merozoites are arranged in rosettes around a clump of central pigment. In *P falciparum* infections, the rings are very small and may contain two chromatin dots rather than one. There is often more than one parasite per cell, and parasites are frequently seen lying against the margin of the cell. Intracytoplasmic granules known as Maurer's dots may be present but are often cleft-shaped and fewer in number than Schüffner's dots. Schizonts and merozoites are not present in the peripheral blood. Gametocytes are large and banana shaped. These characteristics are summarized in **Table 50–1**.

Morphologic differences are the primary means of diagnosis

LIFE CYCLE OF MALARIAL PARASITES

Sporogony, or the sexual cycle, begins when a female mosquito of the genus *Anopheles* ingests circulating male and female gametocytes while feeding on a malarious human. In the gut of the mosquito, the gametocytes mature and effect fertilization. The resulting zygote penetrates the mosquito's gut wall, lodges beneath the basement membrane, and vacuolates to form an oocyst. Within this structure, thousands of sporozoites are formed. The enlarging cyst eventually ruptures, releasing the sporozoites into the body cavity of the mosquito. Some penetrate the salivary glands, rendering the mosquito infectious for humans. The time required for the completion of the cycle in mosquitoes varies from 1 to 3 weeks, depending on the species of insect and parasite as well as on the ambient temperature and humidity.

Mosquito ingests gametocytes from blood of infected human

Sporozoites from oocyst reach mosquito salivary glands

Schizogony, the asexual cycle, occurs in the human and begins when the infected *Anopheles* takes a blood meal from another individual. Sporozoites from the mosquito's salivary glands are injected into the human's subcutaneous capillaries and circulate in the peripheral blood. Within 1 hour they attach to and invade liver cells (hepatocytes), a process thought to be mediated by a ligand present in the sporozoites' outer protein coat (circumsporozoite

Humans infected by mosquito bite

Rapid infection of hepatocytes starts asexual cycle in humans protein). In *P vivax* and *P ovale* infections, some of the sporozoites enter a dormant state immediately after cell invasion. The remaining sporozoites initiate exoerythrocytic schizogony, each producing about 2000 to 40,000 daughter cells, or merozoites. One to two weeks later, the infected hepatocytes rupture, releasing merozoites into the general circulation.

Erythrocytic cycle begins with merozoite attachment to RBC receptor

Trophozoites multiply in RBC to form new merozoites

In 48 to 72 hours, RBCs rupture, releasing merozoites to infect new RBCs

Intrahepatic dormancy causes relapses with *P vivax* and *P ovale*

The erythrocytic phase of malaria starts with the attachment of a released hepatic merozoite to a specific receptor on the RBC surface. After attachment, the merozoite invaginates the cell membrane and is slowly endocytosed. The intracellular parasite initially appears as a ring-shaped trophozoite, which enlarges and becomes more active and irregular in outline. Within a few hours, nuclear division occurs, producing the multinucleated schizont. Cytoplasm eventually condenses around each nucleus of the schizont to form an intraerythrocytic cluster of 6 to 24 merozoite daughter cells. About 48 (*P vivax, P ovale,* and *P falciparum*) to 72 (*P malariae*) hours after initial invasion, infected erythrocytes rupture, releasing the merozoites and producing the first clinical manifestations of disease. The newly released daughter cells invade other RBCs, where most repeat the asexual cycle. Other daughter cells are transformed into sexual forms or gametocytes. These latter forms do not produce RBC lysis and continue to circulate in the peripheral vasculature until ingested by an appropriate mosquito. The recurring asexual cycles continue, involving an ever-increasing number of erythrocytes until finally the development of host immunity brings the erythrocytic cycle to a close. The dormant hepatic sporozoites of *P vivax* and *P ovale* survive the host's immunologic attack and may, after a latent period of months to years, resume intrahepatic multiplication. This leads to a second release of hepatic merozoites and the initiation of another erythrocytic cycle, a phenomenon known as relapse. The life cycle of malarial parasites is summarized in **Figure 50–2**.

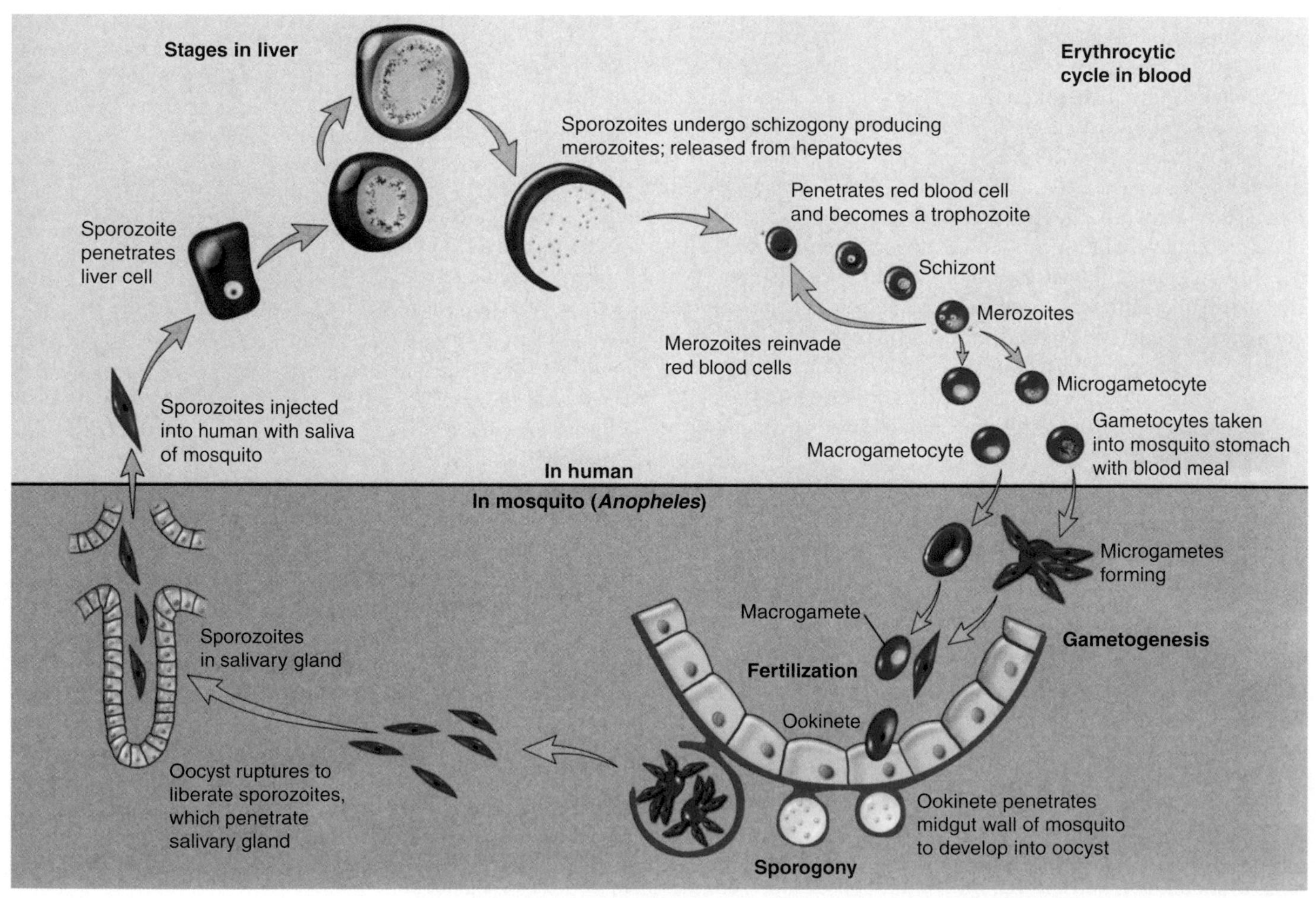

FIGURE 50–2. Malaria. Life cycle of *Plasmodium vivax*. (Reproduced with permission from Willey J, Sherwood L, Woolverton C (eds). *Prescott's Principles of Microbiology.* New York: McGraw-Hill; 2008.)

PHYSIOLOGY

Species of plasmodia differ significantly in their ability to invade subpopulations of erythrocytes; *P vivax* and *P ovale* attack only immature cells (reticulocytes), whereas *P malariae* attacks only senescent cells. During infection with these species, therefore, no more than 1% to 2% of the cell population is involved. *P falciparum,* in contrast, invades RBCs, regardless of age and may produce very high levels of parasitemia and particularly serious disease. In part, these differences may be related to the known differences in the RBC receptor sites available to the individual *Plasmodium* species. In the case of *P vivax,* the site is closely related to the Duffy blood group antigens (Fy^a and Fy^b). Duffy-negative individuals, who constitute the majority of people of West African ancestry, are therefore resistant to vivax malaria. RBC sialoglycoprotein, particularly glycoprotein A, has been implicated as the *P. falciparum* receptor site.

Parasites vary in ability to attack subpopulations of erythrocytes

RBC Duffy antigen and glycoprotein A are RBC receptors

Certain RBC abnormalities may also effect parasitism. The altered hemoglobin (hemoglobin S) associated with the sickle cell trait limits the intensity of the parasitemia caused by *P falciparum* and thereby provides a selective advantage to individuals who are heterozygous for the sickle cell gene. As a result, the sickle cell gene, which would otherwise be disadvantageous, is very common in populations living in malarious areas. Parasite growth appears to be retarded in RBCs heterozygous for hemoglobin S (SA) when they are exposed to conditions of reduced oxygen tension such as might be present in the visceral capillaries. Sickling may also render the erythrocyte more susceptible to phagocytosis or directly damage the parasite. A similar protective effect may be exerted by hemoglobins C, D, and E; thalassemias; and glucose-6-phosphate dehydrogenase (G6PD) or pyridoxal kinase deficiencies, because these abnormalities have also been found more frequently in malarious areas. The protection in these conditions may be related to the increased susceptibility of such RBCs to oxidant stress. In thalassemia, the protection may also be related in part to the production of fetal hemoglobin, which retards maturation of *P falciparum,* as well as an increased binding of antibodies to modified parasitic antigens (neoantigens) presenting on the surface of the erythrocytes.

Sickle cell trait limits intensity of *P falciparum* infection

Other hemoglobinopathies can also exert protection

Once invasion has occurred, malaria parasites may induce a number of changes in the erythrocytic membrane. These include alteration of its lipid concentration, modification of its osmotic properties, and incorporation of parasitic neoantigens, rendering the RBC susceptible to immunologic attack. *P vivax* and *P ovale* stimulate the production of caveolae–vesicle complexes, which are visualized as Schüffner's dots in stained smears. In *P falciparum* infections, electron-dense elevated knobs or excrescences form on the RBC surface. These produce a strain specific, high-molecular-weight adhesive protein (PfEMP1), which mediates binding to receptors on the endothelium of capillaries and postcapillary venules of the brain, placenta, and other organs, where they can produce obstruction and microinfarcts.

Changes induced in erythrocyte membrane

Binding to endothelium may cause microinfarcts

Malarial parasites generate energy by the anaerobic metabolism of glucose. They appear to satisfy their protein requirements by the degradation of hemoglobin within their acidic food vacuoles, resulting in the formation of the malarial pigment (hemozoin) mentioned previously. It has been estimated that the average plasmodium destroys between 25% and 75% of the hemoglobin of its host erythrocyte. Unlike their vertebrate hosts, malarial parasites synthesize folates de novo. As a result, antifolate antimicrobics such as pyrimethamine are effective antimalarial agents.

Malarial parasites metabolize anaerobically, synthesize their own folate

GROWTH IN THE LABORATORY

Continuous in vitro cultivation of plasmodia in human erythrocytes was first achieved in 1976. More recently, the successful in vitro completion of the entire sporogonic cycle, from ookinete to sporozoite, has been achieved. These twin developments provide new opportunities for studying the biology, immunology, and chemotherapy of human malaria. The most immediate impact of these advances has been on the introduction of methods for testing the sensitivity of *P falciparum* to chemotherapeutic agents. Ultimately, these agents will play critical roles in the development of effective antimalarial vaccines.

MALARIA

CLINICAL CAPSULE

Malaria is a febrile illness caused by a parasitic infection of human erythrocytes transmitted by the bite of a mosquito. The fevers are accompanied by headache, sweats, and malaise and typically appear in paroxysmal episodes lasting hours and recurring for weeks. Complications due to capillary blockade can be fatal, particularly in the brain.

EPIDEMIOLOGY

Distribution in tropical areas worldwide

Malaria has a worldwide distribution between 45°N and 40°S latitude, generally at altitudes below 1800 m. *P vivax* is the most widely distributed of the four species, and together with the uncommon *P malariae,* is found primarily in temperate and subtropical areas. *P falciparum* is the dominant organism of the tropics. *P ovale* is rare and found principally in Africa.

Clinical manifestations muted with hyperendemicity

The intensity of malarial transmission in an endemic area depends on the density and feeding habits of suitable mosquito vectors and the prevalence of infected humans, who serve as parasite reservoirs. In hyperendemic areas (areas where more than half of the population is parasitemic), transmission is usually constant, and disease manifestations are moderated by the development of immunity. Mortality is largely restricted to infants and to nonimmune adults who migrate into the region. When the prevalence of disease is lower, transmission is typically intermittent. In this situation, solid immunity does not develop and the population suffers repeated, often seasonal, epidemics, the impact of which is shared by people of all ages.

Presently, an estimated 2 billion people live in malaria endemic areas in 103 of the poorest countries of Africa, Asia, Latin America, and Oceania (**Figure 50–3**). Between 25% and 50% of these persons are thought to be carrying the malaria parasite at any given time. From 1 to 3 million individuals, primarily African children, die of malaria annually. A recent study concluded that the development of resistance to chloroquine, the single most widely used antimalarial agent, has increased mortality four- to eightfold. Although endemic malaria disappeared from the United States three decades ago, imported cases continue to be reported, and the recent worldwide resurgence of malaria combined with an increase in international travel has resulted in an increase in the number of US cases to approximately 1000 annually. Forty-five percent of patients with imported malaria have acquired the disease in Africa, 30% in Asia, and 10% in the Caribbean or Latin America. Fifty percent of recent infections have involved American travelers: nearly 60% of these acquired their infection in Africa.

Malaria kills 1 to 3 million annually; mostly children

Imported malaria may develop months after travel

Clinical manifestations of malaria typically develop within 6 months of arrival of cases in the United States; however, 25% of cases caused by *P vivax* are delayed beyond that time. Approximately 40% of imported cases and almost all associated fatalities have been caused by the virulent *P falciparum*. Tragically, most of these cases could have been prevented or successfully treated. Congenital malaria in infants born in the United States of mothers from malarious areas is occasionally observed. Infections transmitted by transfusions of whole blood, leukocytes, or platelets, or by organ transplantation are, fortunately, now unusual in this country due to the improved screening procedures of blood banks.

Anopheline mosquitoes capable of transmitting malaria are present in the United States, and rarely malaria is transmitted from an imported case to individuals who have never traveled outside of the country.

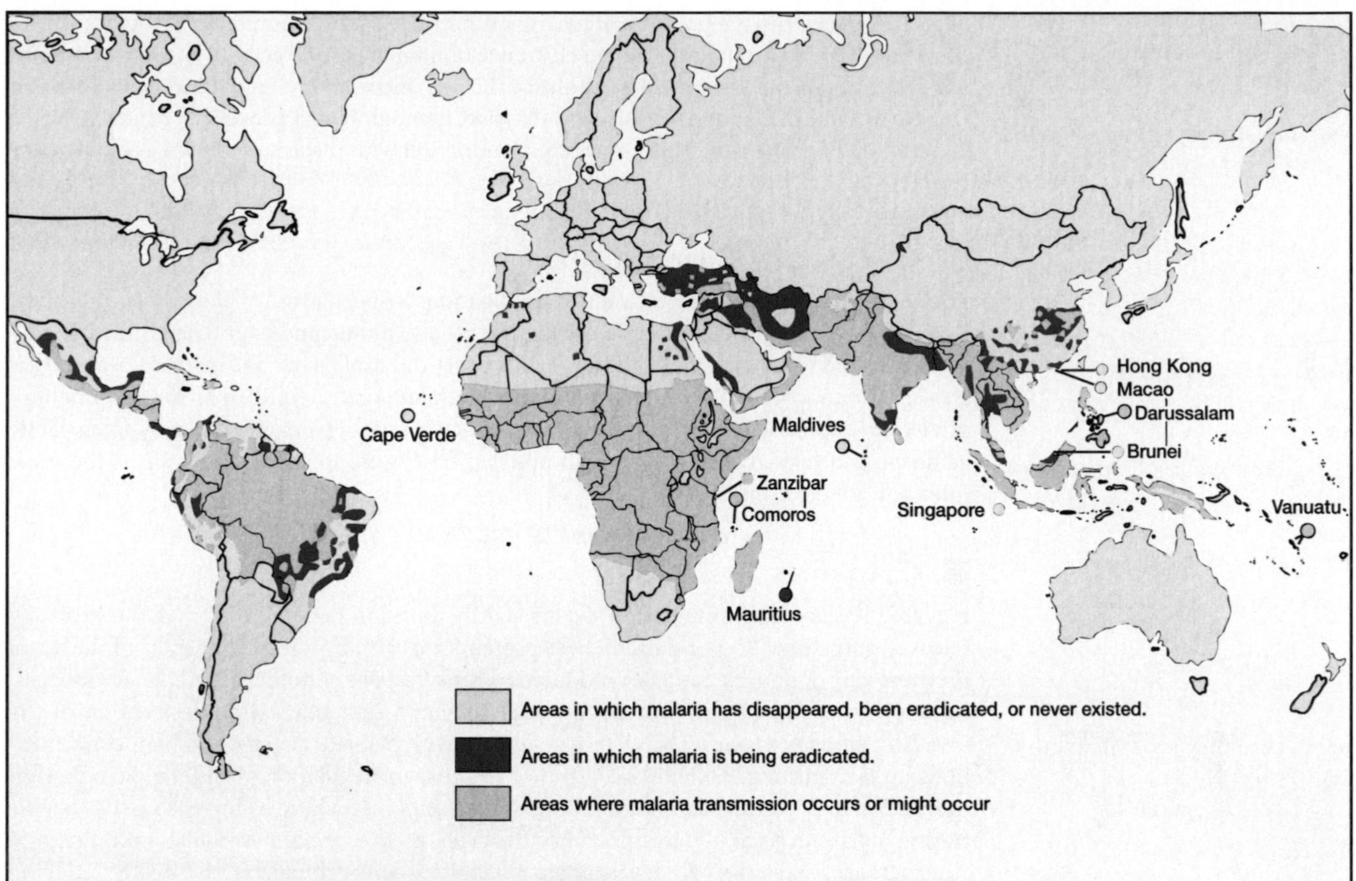

FIGURE 50–3. Geographic distribution of malaria. (Reproduced with permission from Willey J, Sherwood L, Woolverton C (eds). *Prescott's Principles of Microbiology.* New York: McGraw-Hill; 2008; Data from *World Health Statistics Quarterly,* 41:69 1988.)

PATHOGENESIS

The fever, anemia, circulatory changes, and immunopathologic phenomena characteristic of malaria are all the result of erythrocytic invasion by the plasmodia.

Fever

Fever, the hallmark of malaria, appears to be initiated by the process of RBC rupture that leads to the liberation of a new generation of merozoites (sporulation). To date, all attempts to detect the factor(s) mediating the fever have been unsuccessful. It is possible that parasite-derived pyrogens are released at the time of sporulation; alternatively, the fever might result from the release of interleukin-1 (IL-1) and/or tumor necrosis factor (TNF) from macrophages involved in the ingestion of parasitic or erythrocytic debris. Early in malaria, RBCs appear to be infected with malarial parasites at several different stages of development, each inducing sporulation at a different time. The resulting fever is irregular and hectic. Because temperatures higher than 40°C destroy mature parasites, a single population eventually emerges, sporulation is synchronized, and fever occurs in distinct paroxysms at 48-hour or, in the case of *P malariae,* 72-hour intervals. Periodicity is seldom seen in patients who are rapidly diagnosed and treated.

Fever associated with RBC rupture

Synchronization of sporulation causes cyclic fever

Anemia

Parasitized erythrocytes are phagocytosed by a stimulated reticuloendothelial system or are destroyed at the time of sporulation. At times, the anemia is disproportionate to the

Destruction of normal and parasitized RBCs causes anemia

Massive intravascular hemolysis can occur

degree of parasitism. Depression of marrow function, sequestration of erythrocytes within the enlarging spleen, and accelerated clearance of nonparasitized cells all appear to contribute to the anemia. Intravascular hemolysis, though uncommon, may occur, particularly in falciparum malaria. When hemolysis is massive, hemoglobinuria develops, resulting in the production of dark urine. This process in conjunction with malaria is known as **blackwater fever.**

Circulatory Changes

Blood flow decreased to vital organs

The high fever results in significant vasodilatation. In falciparum malaria, vasodilatation leads to a decrease in the effective circulating blood volume and hypotension, which may be aggravated by other changes in the small vessels and capillaries. The intense parasitemias *P falciparum* is capable of producing and the adhesion of infected RBCs to the endothelium of visceral capillaries can impair the microcirculation and precipitate tissue hypoxia, lactic acidosis, and hypoglycemia. Although all deep tissues are involved, the brain is the most intensely affected (**Figure 50–4**).

Cytokines

Elevated cytokine levels contribute to injury

Elevated levels of IL-1 and TNF are consistently found in patients with malaria. Probably released at the time of sporulation, these proteins are certainly an essential part of the host's immune response to malaria. By modulating the effects of endothelial cells, macrophages, monocytes, and neutrophils, they may play an important role in the destruction of the invading parasite. However, TNF levels increase with parasite density, and high concentrations appear harmful. TNF has been shown to cause up-regulation of endothelial adhesion molecules; high concentrations might precipitate cerebral malaria by increasing the sequestration of *P falciparum*–parasitized erythrocytes in the cerebral vascular endothelium. Alternatively, excessive TNF levels might precipitate cerebral malaria by directly inducing hypoglycemia and lactic acidosis. ::: cytokines, p.28

Other Pathogenic Phenomena

Thrombocytopenia and nephritis common

Thrombocytopenia is common in malaria and appears to be related to both splenic pooling and a shortened platelet lifespan. Both direct parasitic invasion and immune mechanisms may be responsible. There may be an acute transient glomerulonephritis in falciparum malaria and progressive renal disease in chronic *P malariae* malaria. These phenomena probably result from the host immune response, with deposition of immune complexes in the glomeruli.

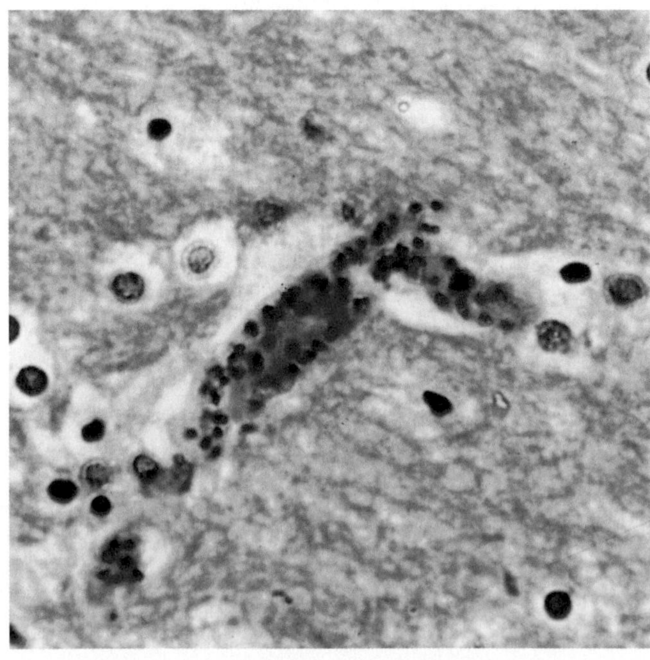

FIGURE 50–4. Central nervous system malaria. This small cerebral blood vessel is blocked with many parasitized erythrocytes adherent to the endothelium. (Reproduced with permission from Connor DH, Chandler FW, Schwartz DQA, Manz HJ, Lack EE (eds). *Pathology of Infectious Diseases*, vol. 1. Stamford, CT: Appleton & Lange; 1997.)

IMMUNITY

Once infected, the host quickly mounts a species- and strain-specific immunologic response that typically limits parasite multiplication and moderates the clinical manifestations of disease, without eliminating the infection—a phenomenon referred to as **premunition.** A prolonged recovery period marked by recurrent exacerbations in both symptoms and number of erythrocytic parasites follows. With time, these recrudescences become less severe and less frequent, eventually stopping altogether.

Initial immune response limits parasite multiplication, but does not eliminate infection (premunition)

The exact mechanisms involved in this recovery are uncertain. In simian and probably in human malaria, recovery is known to require the presence of both T and B lymphocytes. It is probable that the T lymphocytes act partially through their helper effect on antibody production. Some authorities have suggested that they also play a direct role through lymphokine production by stimulating effector cells to release nonspecific factors capable of inhibiting intraerythrocytic multiplication. The B lymphocytes begin production of stage- and strain-specific antiplasmodial antibodies within the first 2 weeks of parasitemia. With the achievement of high levels of antibodies, the number of circulating parasites decreases. The infrequency with which malaria occurs in young infants has been attributed to the transplacental passage of such antibodies. It is uncertain whether they are directly lethal, act as opsonizing agents, or block merozoite invasion of RBCs.

Antibody-mediated immunity important

In simian malaria, the parasite can undergo antigenic variation and thereby escape the suppressive effect of the antibodies. This antigenic variation leads to cycles of recrudescent parasitemia but ultimately to production of specific antibodies to the variants, and cure. It seems probable that similar changes occur in humans, leading to the eventual disappearance of erythrocytic parasites. With *P falciparum* and *P malariae,* which have no persistent hepatic forms, this results in cure. With *P falciparum,* the disease typically does not exceed 1 year, but with *P malariae* the erythrocytic infection can be extremely persistent, lasting in one case up to 53 years. How erythrocytic parasites circulating in numbers too small to be detected on routine blood films escape immunologic destruction remains a puzzle. In a closely related simian malaria, splenectomy results in rapid cure, suggesting that suppressor T lymphocytes in the spleen may play a protective role. In infection with *P vivax* and *P ovale,* latent hepatic infection may result in the discharge of fresh merozoites into the bloodstream after the disappearance of erythrocytic forms. This phenomenon, known as **relapse**, is capable of maintaining infection for 3 to 5 years.

Antigenic variation could play a role in persistence

Suppressor *T lymphocytes* in spleen may protect parasites

MALARIA: CLINICAL ASPECTS

MANIFESTATIONS

The incubation period between the bite of the mosquito and the onset of disease is approximately 2 weeks. With *P malariae* and with strains of *P vivax* in temperate climates, however, this period is often more prolonged. Individuals who contract malaria while taking antimalarial suppressants may not experience illness for many months. In the United States, the interval between entry into the country and onset of disease exceeds 1 month in 25% of *P falciparum* infections and 6 months in a similar proportion of *P vivax* cases.

Incubation period prolonged by suppressant use

The clinical manifestations of malaria vary with the species of plasmodia but typically include chills, fever, splenomegaly, and anemia. The hallmark of disease is the malarial paroxysm. This manifestation begins with a cold stage, which persists for 20 to 60 minutes. During this time, the patient experiences continuous rigors and feels cold. With the consequent increase in body temperature, the rigors cease and vasodilatation commences, ushering in a hot stage. The temperature continues to rise for 3 to 8 hours, reaching a maximum of 40° to 41.7°C before it begins to fall. The wet stage consists of a decrease in fever and profuse sweating. It leaves the patient exhausted but otherwise well until the onset of the next paroxysm.

Malarial paroxysm: cold, hot, wet stages

Typical paroxysms first appear in the second or third week of fever, when parasite sporulation becomes synchronized. In falciparum malaria, synchronization may never take place, and the fever may remain hectic and unpredictable. The first attack is often severe

Typical paroxysms after 2 to 3 weeks when sporulation is synchronized

and may persist for weeks in the untreated patient. Eventually the paroxysms become less regular, less frequent, and less severe. Symptoms finally cease with the disappearance of the parasites from the blood.

Cerebral falciparum malaria often lethal

In falciparum malaria, capillary blockage can lead to several serious complications. When the central nervous system is involved (cerebral malaria), the patient may develop delirium, convulsions, paralysis, coma, and rapid death. Acute pulmonary insufficiency frequently accompanies cerebral malaria, killing about 80% of those involved. When splanchnic capillaries are involved, the patient may experience vomiting, abdominal pain, and diarrhea with or without bloody stools. Jaundice and acute renal failure are also common in severe illness. These pernicious syndromes generally appear when the intensity of parasitemia exceeds 100,000 organisms per cubic millimeter of blood. Most deaths occur within 3 days.

DIAGNOSIS

Thick and thin blood smears detect parasites

Malarial parasites can be demonstrated in stained smears of the peripheral blood in virtually all symptomatic patients. Typically, capillary or venous blood is used to prepare both thin and thick smears, which are stained with Wright or Giemsa stain and examined for the presence of erythrocytic parasites. Thick smears, in which erythrocytes are lysed with water before staining, concentrate the parasites and allow detection of very mild parasitemia. Nonetheless, it may be necessary to obtain several specimens before parasites are seen. Artifacts are numerous in thick smears, and correct interpretation requires experience. The morphologic differences among the four species of plasmodia allow their speciation on the stained smear by the skilled observer.

A number of attempts have been made to improve the standard thin and thick smear method. One such procedure involves acridine orange staining of centrifuged parasites in quantitative buffy coat (QBC) tubes. Although it is expensive, this requires a fluorescence microscope and permits less reliable parasite speciation; its rapidity and ease of use make it attractive to laboratories that are only occasionally called on to identify patients with malaria. Simple, specific card antigen detection procedures are now available. The most widely used test, ParaSight F, detects a protein (HRP2) excreted by *P falciparum* within minutes. The test can be performed under field conditions and has a sensitivity of more than 95%. A second rapid test, OptiMAL, detects parasite lactate dehydrogenase, and, unlike ParaSight F, can distinguish between *P falciparum* and *P vivax*.

Acridine orange stains and other rapid detection methods available

Serologic tests for malaria are offered at a few large reference laboratories but are used primarily for epidemiologic purposes. They are occasionally helpful in speciation and detection of otherwise occult infections. The recently completed sequencing of the malaria genome will lead to newer diagnostic methods.

TREATMENT

The indications for treatment rest on two factors. The first is the infecting species of *Plasmodium,* and the second is the immune status of the afflicted patient. Falciparum malaria is potentially lethal in nonimmune individuals such as new immigrants or travelers to a malarious area and immunosuppressed indigenous individuals such as pregnant women. These individuals must be treated emergently.

Need to destroy all forms of the parasite

The complete treatment of malaria requires the destruction of three parasitic forms: the erythrocytic schizont, the hepatic schizont, and the erythrocytic gametocyte. The first terminates the clinical attack, the second prevents relapse, and the third renders the patient noninfectious to *Anopheles* and thus breaks the cycle of transmission. Unfortunately, no single drug accomplishes all three goals. The present strategy of chemotherapy is shown in **Table 50–2**.

Termination of Acute Attack

Several agents can destroy asexual erythrocytic parasites. Chloroquine, a 4-aminoquinoline, has been the most commonly used. It acts by inhibiting the degradation of hemoglobin, thereby

TABLE 50–2 Chemotherapy of Malaria

STAGE OF PARASITE	CLINICAL GOAL	DRUG
Erythrocytic schizont	Treat clinical attack:	
	All species	Chloroquine
	CRFM	Quinine, antifolates, sulfonamides, artemisinin (regionally dependent)
	Suppress clinical attack:	
	All species	Chloroquine
	CRFM	Antifolates, sulfonamides (regionally dependent)
Erythrocytic gametocyte	Prevent transmission:	
	Relapsing malaria	Chloroquine
	Falciparum malaria	Primaquine, artemisinin
Hepatic schizont	Radical cure:	
	Relapsing malaria	Primaquine
	Falciparum malaria	None required

CRFM, chloroquine-resistant falciparum malaria.

limiting the availability of amino acids necessary for growth. It has been suggested that the weak basic nature of chloroquine also acts to raise the pH of the food vacuoles of the parasite, inhibiting their acid proteases and effectiveness. When originally introduced, it was rapidly effective against all four species of plasmodia and, in the dosage used, free of serious side effects. However, chloroquine-resistant strains of *P falciparum* are now widespread in Africa and Southeast Asia; they are also found, though less frequently, in other areas of Asia and in Central America and South America.

Other schizonticidal agents include quinine/quinidine, antifolate–sulfonamide combinations, mefloquine, halofantrine, and the artemisinins. Unfortunately, resistance to all of these agents is increasing. The artemisinins are also unique in their capacity to reduce transmission by preventing gametocyte development.

Chloroquine inhibits hemoglobin degradation by parasite

Artemisinins prevent gametocyte development

Strains of *P malariae, P ovale,* and *P vivax* (except for some acquired in the South Pacific and South America) remain sensitive to chloroquine and may be treated with this agent. *P vivax* infections acquired in New Guinea and Sumatra, however, should be assumed to be chloroquine-resistant and managed with mefloquine alone or in combination with other agents. *P falciparum* has now become variably resistant to all drug groups, including the artemisinin compounds.

Resistance of chloroquine and other drugs now common with *P falciparum*

There is a growing consensus that the most effective way to slow the further development of drug-resistant strains of *P falciparum* is to use one of the artemisinins in combination with quinine/quinidine, antifolate–sulfonamide compounds, mefloquine, or halofantrine.

Combination therapy may be necessary

Radical Cure

In *P vivax* and *P ovale* infections, hepatic schizonts persist and must be destroyed to prevent reseeding of circulating erythrocytes with consequent relapse. Primaquine, an 8-aminoquinaline, is used for this purpose. Some *P vivax* infections acquired in Southeast Asia and New Guinea fail initial therapy owing to relative resistance to this 8-aminoquinaline. Retreatment with a larger dose of primaquine is usually successful. Unfortunately, primaquine may induce hemolysis in patients with glucose-6-phosphate dehydrogenase deficiency. Persons of Asian, African, and Mediterranean ancestry should thus be screened for this abnormality before treatment. Chloroquine destroys the gametocytes of *P vivax, P ovale,* and *P malariae* but not those of *P falciparum.* Primaquine and artemisinins, however, are effective for this latter species.

Primaquine used to destroy hepatic schizonts of *P vivax* and *P ovale*

PREVENTION

Personal Protection

Mosquito protection with screens and repellents

Chemoprophylaxis choice must consider resistance in area

In endemic areas, mosquito contact can be minimized with the use of house screens, insecticide bombs within rooms, and/or insecticide-impregnated mosquito netting around beds. Those who must be outside from dusk to dawn, the period of mosquito feeding, should apply insect repellent and wear clothing with long sleeves and pants. In addition, it is possible to suppress clinical manifestations of infection, if they occur, with a weekly dose of chloroquine. In areas where chloroquine-resistant strains are common, an alternative schizonticidal agent should be used. Mefloquine or doxycycline are usually preferred. The antifolate pyrimethamine plus a sulfonamide can be taken as well. However, use of this combination is occasionally accompanied by serious side effects, so it is recommended only when mefloquine- and doxycycline-resistant strains are present in the area, and then only for individuals residing in areas of intense transmission for prolonged periods of time. On leaving an endemic area, it is necessary to eradicate residual hepatic parasites with primaquine before discontinuing suppressive therapy.

General

Reduce human reservoir and eradicate mosquitoes

Attempts at eradication have failed

Malaria control measures have been directed toward reducing the infected human and mosquito populations to below the critical level necessary for sustained transmission of disease. The techniques used include those mentioned previously, treatment of febrile patients with effective antimalarial agents, chemical or physical disruption of mosquito breeding areas, and residual insecticide sprays. An active international cooperative program aimed at the eradication of malaria resulted in a dramatic decline in the incidence of the disease between 1956 and 1968. Eradication was not achieved, however, because mosquitoes became resistant to some of the chemical agents used, and today malaria still infects 200 to 300 million inhabitants of Africa, Latin America, and Asia. Tropical Africa alone accounts for 100 million of the afflicted and for most of the 1 to 3 million deaths that occur annually as a result of this disease. The long-term hope for progress in these areas now depends on the development of new technologies.

Vaccines

Subunit vaccines fused with a hepatitis B protein have shown promise

Three advances in the last decade have produced the hope for the first time that an effective malaria vaccine might be within reach of medical science. The establishment of a continuous in vitro culture system provided the large quantities of parasite needed for antigenic analysis. Development of the hybridoma technique allowed the preparation of monoclonal antibodies with which antigens responsible for the induction of protective immunity could be identified. Finally, recombinant DNA procedures enabled scientists to clone and sequence the genes encoding such antigens, permitting the amino acid structure to be determined and peptide sequences suitable for vaccine development to be identified. In 2008, several extensive field studies from areas in Africa that had endemic *P falciparum* malaria have been reported to show encouraging efficacy (50% or greater reduction in clinical disease) with the use of a new vaccine in infants and young children. This vaccine consists of a protein fragment from the outer surface of *P falciparum*, fused with a hepatitis B virus protein, and combined with an immune adjuvant. Studies are continuing, with development of new adjuvants that may be even more potent. Hopefully, this may lead to vaccine strategies that are sorely needed throughout the developing world.

TOXOPLASMA GONDII

PARASITOLOGY

Asexual and sexual cycles in felines

Like the plasmodia, *Toxoplasma gondii,* the cause of toxoplasmosis, is an obligate intracellular sporozoan. It differs from *Plasmodium* in that both sexual and asexual reproductive cycles occur within the gastrointestinal tract of felines, the definitive host. The disease is transmitted to other host species by the ingestion of oocysts passed in the feces of infected felines.

Three forms of human disease

MORPHOLOGY

T gondii was first demonstrated in 1908 in the gondi, an African rodent, by Nicolle and Marceaux. Its name, derived from the Greek *toxo* (arc), is based on the characteristic shape of the organism. All strains of this parasite appear to be closely related antigenically. The major morphologic forms of the parasite are the oocyst, trophozoite, and tissue cyst (**Figure 50–5**).

Spread to humans from felines via fecal–oral route

Tissue cysts killed by cooking as well as freezing and thawing

Oocyst

The oocyst is ovoid, measures 10 to 12 μm in diameter, and possesses a thick wall that makes it resistant to most environmental challenges. It may be destroyed by heat higher than 66°C and by chemicals such as iodine and formalin. In its immature form, the center of the cyst lacks internal structure. With maturation, two sporocysts appear, and later four sporozoites may be discerned within each sporocyst. Sporulation does not occur at temperatures lower than 4°C or higher than 37°C. This form is responsible for the spread of the parasites from felines to other warm-blooded animals via the fecal–oral route.

Tachyzoite (Trophozoite)

The term "trophozoite" is used in its broadest sense to refer to the asexual proliferative forms responsible for cell invasion and clinical disease. In different stages of the asexual cycle, it is referred to by several other terms, including merozoite and tachyzoite. It is crescent or

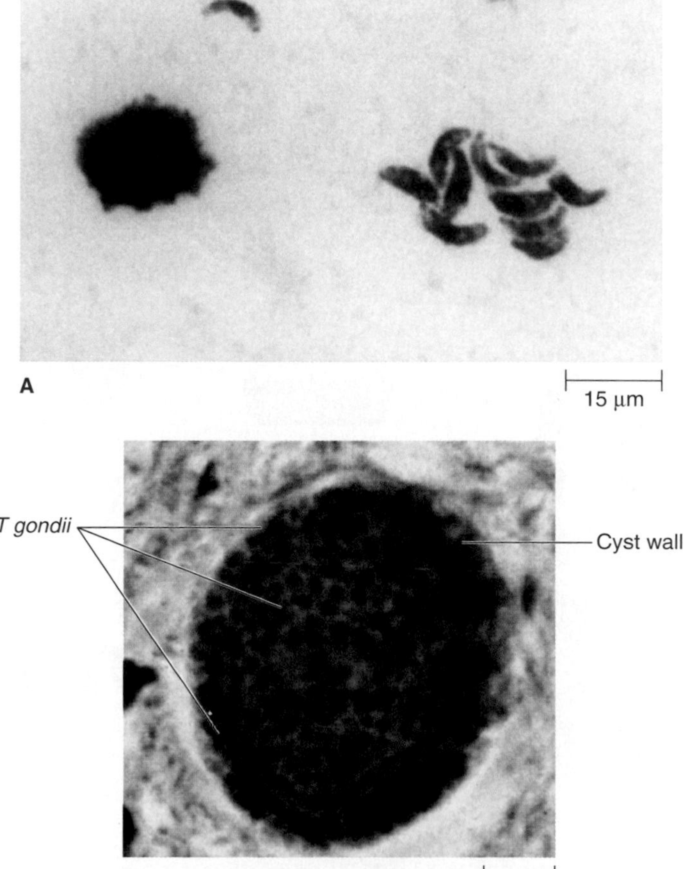

FIGURE 50–5. *Toxoplasma gondii*. A. Invasive trophozoite forms. **B.** Cyst in tissue. (Reproduced with permission from Nester EW, Anderson DG, Roberts CE Jr, Nester MT. *Microbiology: A Human Perspective*, 6th ed. New York: McGraw-Hill, 2008.)

arc-shaped, measures 3 by 7 µm, and can invade all nucleated cell types. Although tachyzoites are obligate intracellular organisms, they may survive extracellularly in a variety of body fluids for periods of hours to days. They cannot, however, survive the digestive activity of the stomach and therefore are not infective on ingestion.

Tissue Cysts

Cysts measure 10 to 200 µm in diameter. The contained organisms, referred to as bradyzoites, are similar to tachyzoites, but are smaller and divide more slowly. Tissue cysts are resistant to digestive enzymes and, like oocysts, are infectious to the animal that ingests them. They survive normal refrigerator temperatures but are killed by freezing and thawing and by normal cooking temperatures.

LIFE CYCLE (Figure 50–6)

Definitive Host

Sexual reproduction of *T gondii* occurs only in the intestinal tract of felines, most commonly in the domestic cat. Ingested parasites enter the epithelial cells of the ileum by mechanisms that remain poorly defined. Intracellularly, the trophozoites reside within a membrane-bound vacuole and undergo schizogony. With cell rupture, merozoites are released. The merozoites infect adjacent epithelial cells; they then repeat another asexual

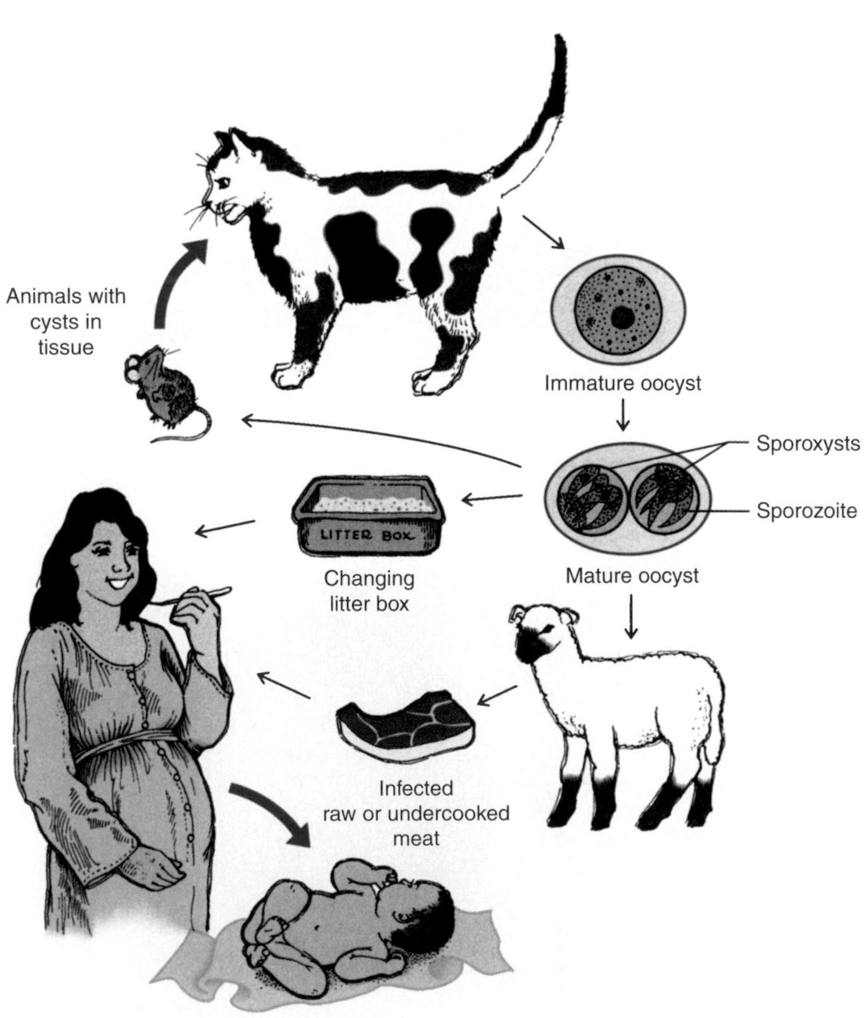

FIGURE 50–6. Toxoplasmosis. *Toxoplasma gondii* life cycle shows oocysts from cat feces or cysts from inadequately cooked meat as infectious to humans and other animals. (Reproduced with permission from Nester EW, Anderson DG, Roberts CE Jr, Nester MT. *Microbiology: A Human Perspective*, 6th ed. New York: McGraw-Hill, 2008.)

cycle or eventually differentiate into gametocytes, initiating sexual reproduction. Fusion of the mature male and female gametes leads to the formation of an oval, thick-walled oocyst that is then shed in the feces. In the typical infection, millions of these structures are released daily for 1 to 3 weeks. The oocysts are immature at the time of shedding and must complete sporulation in the external environment. In this process, two sporocysts, each containing four sporozoites, develop within each oocyst. The time required for sporulation varies from 1 day to 3 weeks, depending on the ambient temperature and moisture. Once mature, the resistant oocysts may remain viable and infectious for many years in soil.

Infection in cat ileal cells

Fusion of gametes leads to oocyst formation; shed in feces

Sporulate in external environment

Intermediate Hosts

After ingestion by a susceptible warm-blooded animal, sporozoites are released from the disrupted oocyst and enter macrophages. Within these cells, they are transported through the lymphohematogenous system to all organ systems. Continued intracellular schizogony results in macrophage rupture and release of new parasites, which may invade any adjacent nucleated host cell and continue the asexual cycle. With the development of host immunity, many of the parasites are destroyed. Within the cells of certain organs, particularly the brain, heart, and skeletal muscle, the trophozoites produce a membrane that surrounds and protects them: within this tissue cyst, multiplication continues at a more leisurely pace. Eventually, cysts that measure up to 200 μm in diameter and contain more than 1000 organisms are produced. These cysts persist intact for the life of the host or rupture, producing parasitologic relapse. If they are ingested by a carnivore, they survive the digestive enzymes and initiate infection in the new host.

Mature oocysts infect hosts orally

Released sporozoites invade macrophages

Cysts develop and can persist for life of host

TOXOPLASMOSIS

CLINICAL CAPSULE

Toxoplasma can infect most warm-blooded animals, both domestic and wild; it is thus the most cosmopolitan of parasites. Approximately 50% of the human population of the United States has been infected. In the overwhelming majority of persons, infection is chronic, asymptomatic, and self-limiting. Clinical disease manifests in three major forms: (1) self-limiting febrile lymphadenopathy, (2) highly lethal infection of immunocompromised patients, and (3) congenital infection of infants.

EPIDEMIOLOGY

Prevalence and Distribution

Toxoplasmosis occurs in almost all mammals and many birds. Human infections are found in every region of the globe; in general, the incidence is higher in the tropics and lower in cold and/or arid regions. In the United States, the prevalence of positive serologic evidence for the disease increases with age. By adulthood, approximately 50% of Americans can be shown to have circulating antibodies against *T gondii.*

Worldwide distribution among mammals and birds

Transmission

Although it is known that humans may acquire toxoplasmosis in a variety of ways, data on their relative frequency are both meager and conflicting. It is likely that the route of transmission varies from population to population, and perhaps from age to age, within any given area. The most important transmission mechanisms of toxoplasmosis are discussed below.

Ingestion of Oocysts

Increased hazard to children by close contact with contaminated areas

Felinophobes are inclined to the view that the deposition of oocysts in the feces of cats and their subsequent ingestion by the unsuspecting owner is the most common way in which humans acquire this important infection. Disease epidemics of toxoplasmosis associated with exposure to infected cats have been reported. Unfortunately, data from studies relating the frequency of feline exposure to the prevalence of positive serologic tests are conflicting. Acutely infected cats shed oocysts for only a few weeks. It has been shown, however, that chronically infected felines can occasionally re-shed oocysts, and prevalence studies have demonstrated that 1% of domestic cats excrete oocysts at any given time. The large number of these structures passed during active shedding and their prolonged survival in the external environment greatly enhance their chance of transmission. Particularly at risk are children at play, who may come in close contact with areas likely to be contaminated with cat feces, and adults responsible for changing a cat's litter box. It is also possible that insects can mechanically transfer oocysts to human food.

Ingestion of Tissue Cysts

Cysts present in meat

Tissue cysts have been frequently demonstrated in meat produced for human consumption. They are most common in pork (25%) and mutton (10%) and less so in beef and chicken (less than 1%). Although such cysts are killed at normal (well-done) cooking temperatures, an impressive array of epidemiologic information links the handling and/or ingestion of raw or undercooked meat with serologic and, occasionally, clinical evidence of disease. Confounding these data is an Indian study that demonstrated no difference between meat eaters and vegetarians in the incidence of positive serologic tests.

Congenital

Transplacental transmission highest in third trimester

Approximately 1 of every 500 pregnant women acquires acute toxoplasmosis, and approximately 10% to 20% of the involved women become symptomatic. Regardless of the clinical status of the infected mother, the parasite involves the fetus in 33% to 50% of all acute maternal infections. The risk of transplacental transmission is independent of the clinical severity of the disease in the mother, but does correlate with the stage of gestation at which she is exposed. Fetal involvement occurs in 17% of first-trimester and 65% of third-trimester infections. Conversely, the earlier a fetal infection is acquired, the more severe it is likely to be. Overall, 20% of fetuses experienced severe consequences; a similar proportion develop mild disease. The remainder are asymptomatic.

Miscellaneous

Transmitted by transfusions and organ transplants

In addition to causing congenital infection, trophozoites have been responsible for disease transmission in a number of other situations, including laboratory accidents, transfusions of whole blood and leukocytes, and organ transplantation. Because trophozoites may survive for several hours in body fluids or exudates of acutely infected humans, it is possible for infection to occur after contact with such materials.

PATHOGENESIS AND IMMUNITY

Dissemination in immunosuppressed subjects

In primary infection, the proliferation of trophozoites results in the death of involved host cells, stimulation of a mononuclear inflammatory reaction, and a parasite-specific secretory IgA response. In immunodeficient hosts, rapid organism proliferation continues, producing numerous widespread foci of tissue necrosis. The consequences are most serious in organs such as the brain, where the potential for cell regeneration is limited.

In normal hosts, however, acute infection is rapidly controlled with the development of humoral and cellular immunity. Extracellular parasites are destroyed, intracellular multiplication is hindered, and tissue cysts are formed. With the exception of lysis of extracellular parasites by antibody and complement, cell-mediated immunity appears to play the principal role in this process, mediated in part by IL-2, interferon-α, and cytotoxic T cells. Immunity appears to be lifelong, possibly because of survival of the parasite in the tissue cysts.

The cysts, which are found most frequently in the brain, retina, heart, and skeletal muscle, normally produce little or no tissue reaction. The suppression of cell-mediated immunity that accompanies serious illness, or the administration of immunosuppressive agents, may lead to the rupture of a cyst and the release of trophozoites. Their subsequent proliferation and the intense antibody reaction to their presence result in an acute exacerbation of the disease.

Immunity is primarily cell-mediated

TOXOPLASMOSIS: CLINICAL ASPECTS

MANIFESTATIONS

In most patients, infection with *T gondii* is completely asymptomatic. Clinical manifestations, when they do appear, vary with the type of host involved. In general, they may be grouped into one of the three syndromes listed below.

Congenital Toxoplasmosis

Immune mechanisms are poorly developed in utero. As a result, a large proportion of fetal infections results in clinical illness. If the infection spreads to the central nervous system, the outcome is often catastrophic. Abortion and stillbirth are the most serious consequences. Liveborn children may demonstrate microcephaly, hydrocephaly, cerebral calcifications, convulsions, and psychomotor retardation. Disease of this severity is usually accompanied by evidence of visceral involvement, including fever, hepatitis, pneumonia, and skin rash. Infants infected with toxoplasmosis later in prenatal development demonstrate milder disease. Many appear healthy at birth but develop epilepsy, retardation, or strabismus months or years later. Probably the most common delayed manifestation of congenital toxoplasmosis is chorioretinitis. This condition, which is thought to result from the reactivation of latent tissue cysts, typically presents during the second or third decade of life as recurrent bouts of eye pain and loss of visual acuity. The lesions are usually bilateral but focal. If the retinal macula is not involved, vision improves as the inflammation subsides. *T gondii* accounts for 25% of all cases of granulomatous uveitis seen in the United States.

Infection in utero can produce malformations, chorioretinitis, and stillbirth

Normal Host

The most common clinical manifestation of toxoplasmosis acquired after birth is asymptomatic localized lymphadenopathy. The cervical nodes are most frequently involved, but nontender enlargement of other regional groups, including the retroperitoneal nodes, also occurs. At times, adenopathy is accompanied by fever, sore throat, rash, hepatosplenomegaly, and atypical lymphocytosis, thus mimicking the clinical and laboratory manifestations of infectious mononucleosis. Occasionally, the normal host develops severe visceral involvement, which may be manifested as meningoencephalitis, pneumonitis, myocarditis, or hepatitis. Chorioretinitis after postnatally acquired infection, though documented, is uncommon. Unlike congenitally acquired ocular disease, it occurs during midlife and is generally unilateral.

Fever and lymphadenopathy can mimic infectious mononucleosis

Immunocompromised Host

In the immunocompromised host, toxoplasmosis is a serious, often fatal disease. If primary infection is acquired while a patient is undergoing immunosuppressive therapy for malignancy or organ transplantation, widespread dissemination of the infection with necrotizing pneumonitis, myocarditis, and encephalitis may occur. More commonly, acute disease in this population results from the activation of chronic, latent infection by immunosuppressive therapy, or from the acquisition of a concurrent immunosuppressive infection, particularly acquired immunodeficiency syndrome (AIDS). Encephalitis occurs in 50% of such cases and in more than 90% of fatal cases. Toxoplasmic encephalitis is particularly common in AIDS patients; it is seen in approximately 10% of those with circulating toxoplasma

Primary infection or reactivation of latent infections can produce severe, widespread disease

AIDS patients develop encephalitis

antibodies. As such, it is a major cause of morbidity and mortality in this patient population. Clinically, encephalitis may present as a meningoencephalitis, diffuse encephalopathy, or mass lesion.

DIAGNOSIS

Demonstration of parasite in histopathologic specimens

The diagnosis of toxoplasmosis may be established by a variety of methods. In acute toxoplasmic lymphadenitis, the histologic appearance of the involved nodes is often pathognomonic. The trophozoite may be demonstrated in tissue with Wright or Giemsa stain. Electron microscopy and indirect fluorescent antibody techniques have also been used successfully on heart transplant or brain tissue obtained by biopsy. Although tissue cysts are selectively stained by periodic acid-Schiff, their presence is not indicative of acute disease. Isolation of the organism can be accomplished by inoculating blood or other body fluids into mice or tissue cultures. Inoculation of other tissues is not usually helpful because a positive result may only reflect the presence of latent tissue cysts.

Serodiagnosis is the primary approach

Serologic procedures are the primary method of diagnosis. To establish the presence of acute infection, it is usual to demonstrate a fourfold rise in the IgG antibody titer between acute and convalescent serum specimens. Peak titers are often reached within 4 to 8 weeks, so the acute serum must be collected early in the course of illness. Of the many tests developed for the detection of IgG antibodies, indirect hemagglutination, indirect fluorescent antibody, or enzyme immunoassay (EIA) tests are the tests most frequently used. Titers may remain high for many years.

Rising titers of IgG or detection of IgM suggest acute infection or reactivation

The detection of IgM antibodies provides a more rapid confirmation of acute infection. These antibodies appear within the first week of infection, peak in 2 to 4 weeks, and may slowly revert to negative. It also appears that IgM antibodies are produced after reactivation of latent disease. EIA for IgM antibody is now commonly used. Examination of tissues, urine, and other body fluids for the presence of toxoplasma antigen, or DNA by the polymerase chain reaction, have been shown to be useful adjunctive tests in immunocompromised individuals and in the diagnosis of congenital infections.

TREATMENT AND PREVENTION

Usually, patients infected with toxoplasmosis do not require therapy unless symptoms are particularly severe and persistent or unless vital organs, such as the eye, are involved. Immunocompromised and pregnant women, however, should be treated if acute infection (or reactivation) is documented (**Table 50–3**). Routine serial serologic testing of such individuals would allow early detection of infected persons and enhance the prospects of a successful outcome. It is now clear that early treatment of acutely infected pregnant women

TABLE 50–3 Indications for Treatment of Toxoplasmosis[a]

SEROLOGIC CRITERIA	CLINICAL CRITERIA
Elevated IgM titers	Potential laboratory-acquired infection
Fourfold rise in IgG titers	Pregnant woman
Very high IgG titers (>1:1000)	Neonate
	Immunocompromised patient (including AIDS)
	Severe constitutional symptoms
	Vital organ involvement (including active chorioretinitis)

Ig, immunoglobulin.
[a]Must satisfy one serologic plus one clinical criterion.

significantly reduces the incidence of severe congenital infections and reduces the ratio of benign to subclinical forms in infants. At present, the most commonly used therapeutic regimen in the United States for toxoplasmosis is the combination of pyrimethamine and sulfonamides. Unfortunately, the former drug is teratogenic and should not be used in the first trimester of pregnancy; spiramycin, a cytostatic macrolide, is often substituted in this setting.

Spiramycin used to prevent congenital infection

Although the pyrimethamine–sulfonamide combination is very effective against trophozoites, it is inactive against the cyst forms. Since both parasitic forms are present in patients with toxoplasmic encephalitis, recrudescence of illness generally follows completion of standard therapy in AIDS patients. This may be prevented by initiating chronic, low-dose suppressive therapy after completion of the standard regimen. Atovaquone, a recently introduced hydroxynaphthoquinone, possesses activity against both trophozoites and cysts. Its use, therefore, may result in radical cure of toxoplasma encephalitis, eliminating the need for chronic suppression.

Atovaquone is active against tachyzoites and cysts

Prevention of toxoplasmosis should be directed primarily at pregnant women and immunologically compromised hosts. Hands should be carefully washed after handling uncooked meat. Cysts in meat can be destroyed by proper cooking (56°C for 15 minutes) or by freezing to −20°C. Cat feces should be avoided, particularly the changing of litter boxes.

CRYPTOSPORIDIUM

Cryptosporidia ("hidden-spore") are small parasites that can infect the intestinal tract of a wide range of mammals, including humans. Like other sporozoan parasites, they are obligate intracellular organisms that exhibit alternating cycles of sexual and asexual reproduction. As with *Toxoplasma,* both cycles are completed within the gastrointestinal tract of a single host. Long recognized as an important cause of diarrhea in animals, cryptosporidia were not identified as causes of human enteritis until 1976.

PARASITOLOGY

MORPHOLOGY

Regardless of animal host, all strains of this tiny (2 to 6 μm) parasite appear morphologically identical. Although all strains can reasonably be regarded as a single species, the one that infects humans and cattle is often referred to as *C parvum.* The organisms appear as small spherical structures arranged in rows along the microvilli of the epithelial cells. They are readily stained with Giemsa and hematoxylin-eosin. Although they remain external to the cytoplasm of the intestinal epithelial cell, the organisms are covered by a double membrane derived from the reflection, fusion, and attenuation of the microvilli and are thus, by definition, intracellular organisms. Oocysts shed into the intestinal lumen mature to contain four sporozoites; their cell wall provides the unusual property of acid-fastness, allowing them to be visualized with stains generally employed for mycobacteria.

Small spherical particles associated with microvilli

Oocysts are acid-fast

LIFE CYCLE

Infective oocysts are excreted in the stool of the parasitized animal. Unlike those of *Toxoplasma,* cryptosporidia oocysts are fully mature and immediately infective on passage in the feces. After ingestion by another animal, sporozoites are released from the oocyst and attach to the microvilli of the small bowel epithelial cells, where they are transformed into trophozoites. These divide asexually by multiple fission (schizogony) to form schizonts containing eight daughter cells known as type 1 merozoites. On release from the schizont, each daughter cell attaches itself to another epithelial cell, where it repeats the schizogony cycle, producing another generation of type 1 merozoites.

Mature, infective oocysts excreted in stools

Protective cell wall ensures survival of oocysts

Eventually, schizonts containing four type 2 merozoites are seen. Incapable of continued asexual reproduction, these develop into male (microgamete) and female (macrogamete) sexual forms. After fertilization, the resulting zygote develops into an oocyst that is shed into the lumen of the bowel. The majority possess a thick protective cell wall that ensures their intact passage in the feces and survival in the external environment.

Some thin-walled oocysts can autoinfect

Approximately 20% fail to develop the thick protective wall. The cell membrane ruptures, releasing infective sporozoites directly into the intestinal lumen and initiating a new "autoinfective" cycle within the original host. In the normal host, the presence of innate or acquired immunity dampens both the cyclic production of type 1 merozoites and the formation of thin-walled oocysts, halting further parasite multiplication and terminating the acute infection. In the immunocompromised, both presumably continue, explaining why such individuals develop severe, persistent infections in the absence of external reinfection.

CRYPTOSPORIDIOSIS

CLINICAL CAPSULE

Cryptosporidiosis is an intestinal illness acquired from domestic animals. The course includes profuse watery diarrhea, vomiting, and weight loss. Spontaneous complete recovery is the usual outcome, except in immunocompromised persons, in whom debilitating illnesses can occur.

EPIDEMIOLOGY

Animal reservoirs and person-to-person transmission both important

Cryptosporidiosis appears to involve most vertebrate groups. In all species, infection rates are highest among the young and immature. Experimental and epidemiologic data suggest that domestic animals constitute an important reservoir of disease in humans. However, outbreaks of human disease in day care centers, hospitals, and urban family groups indicate that most human infections result from person-to-person transmission. In Western countries, between 1% and 4% of small children presenting to medical centers with gastroenteritis have been shown to harbor cryptosporidia oocysts. In Third World countries, the rates have varied from 4% to 11%. In some outbreaks of diarrhea in day care centers, the majority of attendees were found to have oocysts in their stool.

Infection rates highest in young children

Infection rates cryptosporidiosis in adults suffering from gastroenteritis is approximately one third of that reported in children; it has been highest in family members of infected children, medical personnel caring for patients with cryptosporidiosis, male homosexuals, and travelers to foreign countries. In the United States, the parasite has been identified in 15% of patients with AIDS and diarrhea; in Haiti and Africa, 50% of such individuals may be involved. Asymptomatic carriage is uncommon. Other enteric pathogens, particularly *Giardia lamblia,* are recovered from a significant minority of infected patients.

Can be transmitted via contaminated water

Because oocysts are found almost exclusively in stool, the principal transmission route of cryptosporidiosis is undoubtedly by direct fecal–oral spread. Transmission via contaminated water has been documented, and the hardy nature of the oocysts makes it likely that there is also indirect transmission via contaminated food and fomites.

PATHOGENESIS AND IMMUNITY

Although the jejunum is most heavily involved, cryptosporidia have been found throughout the gastrointestinal tract, particularly in immunocompromised subjects. Cryptosporidial cholecystitis is seen with some frequency in AIDS patients with enteritis. By light microscopy, bowel changes appear minimal, consisting of mild to moderate villous atrophy, crypt enlargement, and a mononuclear infiltrate of the lamina propria. The pathophysiology of the diarrhea is unknown, but its nature and intensity suggest that a cholera-like enterotoxin may be involved. The vital role played by the host's immune status in the pathogenesis of the disease is indicated by both the enhanced susceptibility of the young to infection and the prolonged severe clinical disease seen in immunocompromised patients. Indirect evidence suggests antibodies in the intestinal lumen exert a protective effect against initial *C parvum* infection. Experimental animal studies indicate that CD4+ T lymphocytes and interferon play independent roles in the immunologic clearance of the parasite.

Minimal intestinal pathology

Prolonged disease in AIDS patients

CRYPTOSPORIDIOSIS: CLINICAL ASPECTS

MANIFESTATIONS

Immunocompetent patients usually note the onset of explosive, profuse, watery diarrhea 1 to 2 weeks after exposure. Typically, cryptosporidiosis persists for 5 to 11 days and then rapidly abates. Occasionally, purging, accompanied by a mild malabsorption and weight loss, continues for up to 1 month. A few patients complain of nausea, anorexia, vomiting, and low-grade fever. Except for its shorter duration, more prominent abdominal pain, and relative lack of flatulence, the clinical manifestations of cryptosporidiosis closely resemble those produced by *G lamblia*. Radiographic and endoscopic examinations of the gut are either normal or demonstrate mild, nonspecific abnormalities. Recovery is complete, and neither relapse nor reinfection has been reported.

Self-limiting diarrhea in normal hosts

Cryptosporidiosis has been described in patients with a broad range of immunodeficiencies, including childhood malnutrition in third world countries, AIDS, and congenital hypogammaglobulinemia, and in those resulting from cancer chemotherapy and immunosuppressive management of organ transplantations. In such patients, cryptosporidiosis is usually indolent in onset, and manifestations are similar to those seen in normal hosts, but the diarrhea is more severe. Fluid losses of up to 25 L/day have been described. Patients with biliary cryptosporidiosis present with typical manifestations of cholecystitis and cholangitis. Unless the immunologic defect is reversed, the disease usually persists for the duration of the patient's life. Weight loss is often prominent. The prognosis depends on the nature of the underlying immunologic abnormality; 50% of patients with AIDS die within 6 months. Although other intercurrent infections are usually the direct cause of death, malnutrition and complications of parenteral nutrition contribute.

DIAGNOSIS

The diagnosis of cryptosporidiosis is established by the recovery and identification of *Cryptosporidium* oocysts in a recently passed or preserved diarrheal stool. Oocyst excretion is most intense during the first week of illness, tapers during the second week, and generally stops with the cessation of diarrhea. Because cryptosporidia oocysts are one of the few acid-fast particles found in feces, a definitive identification can be established with any one of the acid-fast staining procedures developed for mycobacteria (**Figure 50–7**). A direct immunofluorescence antibody stain using a monoclonal antibody to oocyst wall has been recently introduced, which appears to be superior to acid-fast stains. When direct examinations are negative, concentration procedures are used and the concentrate restained. Immunofluorescence and EIAs for the detection of anticryptosporidial antibodies are now available.

Detection of oocysts by acid-fast or immunofluorescent stains

TREATMENT AND PREVENTION

In the immunocompetent patient, the disease is self-limited and attempts at specific antiparasitic therapy are not warranted; rehydration may be required in small children. In the immunocompromised host, the severity and chronicity of the diarrhea warrant therapeutic intervention. Unfortunately, there is no uniformly effective anticryptosporidial agent

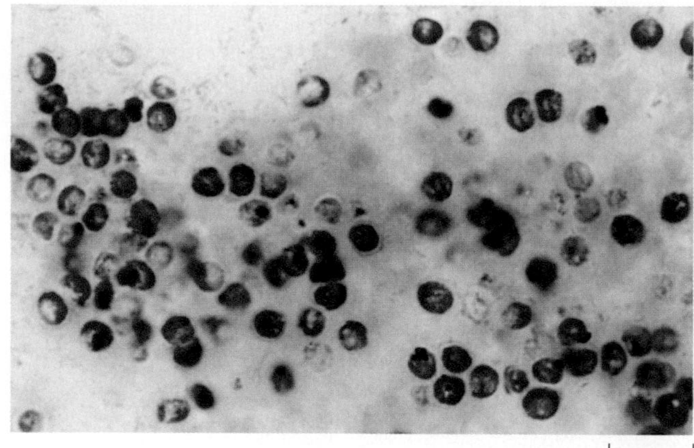

FIGURE 50–7. *Cryptosporidium parvum*. This acid-fast stain demonstrates oocysts in the feces of a diarrheal patient. (Reproduced with permission from Nester EW, Anderson DG, Roberts CE Jr, Nester MT. *Microbiology: A Human Perspective*, 6th ed. New York: McGraw-Hill, 2008.)

Specific treatment remains problematic

available at this time. Paromomycin, a luminal antimicrobic, has been shown to reduce the intensity of diarrhea in some patients, and parenteral octreotide acetate, a somatostatin analog, has been useful in decreasing stool volumes. Macrolide antimicrobials have also been suggested in difficult cases. The only uniformly successful approach has been the reversal of underlying immunologic abnormalities. When appropriate, withdrawal of cancer chemotherapy agents or immunosuppressive drugs may result in a cure.

Strict stool precautions should be used for symptomatic patients

The stools of patients with cryptosporidiosis are infectious. Stool precautions should be instituted at the time the diagnosis is first suspected; for the immunosuppressed patient, this should be whenever diarrhea, regardless of the presumed cause, is first noted. This is particularly important in cancer chemotherapy and transplantation units, where spread of the disease from a symptomatic patient to other immunosuppressed patients can have life-threatening consequences.

OTHER INTESTINAL PROTOZOA

Other protozoa that can infect and cause diarrheal illnesses in humans include *Cyclospora, Isospora,* and other somewhat related species. They are usually found in fecally contaminated groundwater and food. Clinical illnesses are similar to cryptosporidiosis, but are usually self-limited. Diagnoses are established by microscopic examination of diarrheal stool samples.

CASE STUDY

FEVER AFTER AN EXCURSION

A 30-year-old man returned to the United States 3 weeks ago from a guided tour of Thailand. On advice of his physician, he took oral chloroquine prophylaxis beginning 1 week before departure and ending 1 week after his return. Over the last 4 days, he has developed repeated episodes of fever to 40°C, preceded by chills and associated with a severe headache. The duration of these symptoms has been about 8 hours, ending with profuse sweating, only to recur again within 48 hours.

Physical examination is unremarkable, except for fever.

Laboratory studies reveal only a mild anemia, with a platelet count of 100,000/mm^3 (normal 200,000 to 400,000).

QUESTIONS

■ Which is the most likely diagnosis for this patient?

A. Vivax malaria
B. Falciparum malaria
C. Toxoplasmosis
D. Ovale malaria
E. Malariae malaria

■ The diagnostic test of choice is:

A. Peripheral blood smears
B. PCR of red blood cells
C. IgM ELISA serology
D. Paired sera for IgG antibody quantitation

■ In some malaria infections, treatment to prevent relapse (by destroying persistent hepatic schizonts) is necessary in which two of the following?

A. *P malariae*
B. *P ovale*
C. *P falciparum*
D. *P vivax*

■ After primary infection, *Toxoplasma gondii* may persist as cyst forms in all of the following tissues **except** which of the following:

A. Brain
B. Heart
C. Skin
D. Skeletal muscle
E. Retina

ANSWERS

1(B), **2**(A), **3**(B) and (D), **4**(C)

CHAPTER 51

Rhizopods

Amoebas at the start
Were not complex;
They tore themselves apart
And started sex.

—Arthur Guiterman, *The Light Guitar*

Rhizopods, or amebas, are the most primitive of the protozoa. They multiply by simple binary fission and move by means of cytoplasmic organelles called **pseudopodia**. These projections of the relatively solid ectoplasm are formed by streaming of the inner, more liquid endoplasm. They move the ameba forward and, incidentally, engulf and internalize food sources found in its path. Most amebas, when faced with a hostile environment, can produce a chitinous, external wall that surrounds and protects them. These forms are referred to as cysts and may survive for prolonged periods under conditions that would rapidly destroy the motile trophozoite. The majority of amebas belong to free-living genera. They are widely distributed in nature, being found in literally all bodies of standing fresh water. Few free-living amebas produce human disease, although two genera, *Naegleria* and *Acanthamoeba,* have been implicated occasionally as causes of meningoencephalitis and keratitis.

Several genera of amebas, including *Entamoeba, Endolimax,* and *Iodamoeba,* are obligate parasites of the human alimentary tract and are passed as cysts from host to host by the fecal–oral route. Several are devoid of mitochondria, presumably because of the anaerobic conditions under which they exist in the colon. Only one, *Entamoeba histolytica,* regularly produces disease; it has been recently subdivided into two morphologically identical but genetically distinct species, an invasive pathogen that retains the species appellation "histolytica" and a commensal organism, now designated *E dispar*. The two species can be differentiated by isoenzyme analysis, antibodies to surface antigens, and DNA markers.

ENTAMOEBA HISTOLYTICA

PARASITOLOGY

MORPHOLOGY AND PHYSIOLOGY

E histolytica possesses both trophozoite and cyst forms (**Figure 51–1**). The trophozoites are microaerophilic, dwell in the lumen or wall of the colon, feed on bacteria and tissue cells, and multiply rapidly in the anaerobic environment of the gut. When diarrhea occurs,

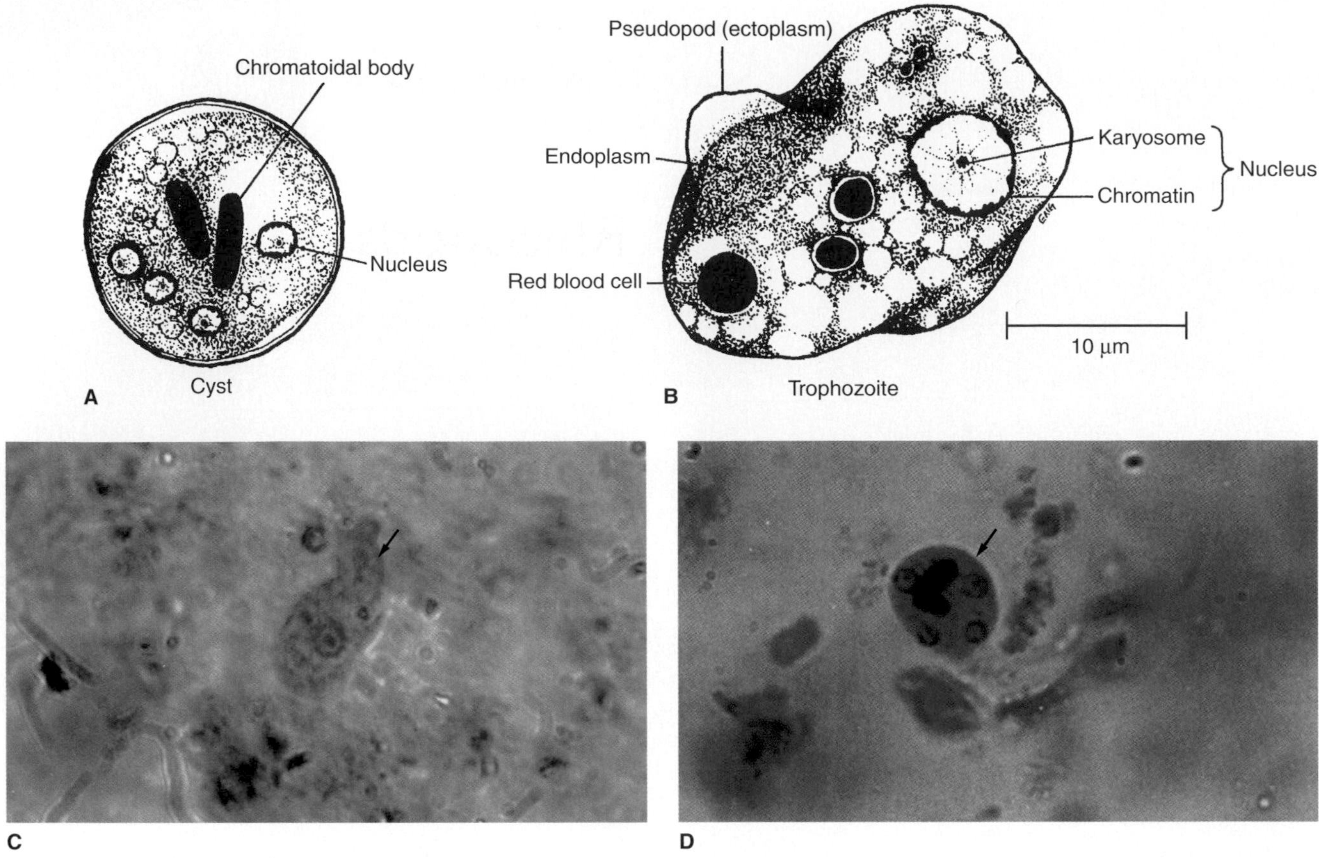

FIGURE 51–1. *Entamoeba histolytica*. A. Cyst structures. **B.** Trophozoite structures. **C.** Trophozoite in stool (arrow). **D.** Cyst (arrow) in stool iodine preparation and cysts in stool iodine preparation. **B.** Trophozoite structures and trophozoite in stool.

the trophozoites are passed unchanged in the liquid stool. Here they can be recognized by their size (12 to 20 μm in diameter); directional motility; granular, vacuolated endoplasm; and sharply demarcated, clear ectoplasm with finger-like pseudopods. Invasive strains tend to be larger and may contain ingested erythrocytes within their cytoplasm (**Figure 51–2**). Appropriate stains reveal a 3- to 5-μm nucleus with a small central karyosome or nucleolus and fine regular granules evenly distributed around the nuclear membrane (peripheral chromatin). Electron microscopic studies demonstrate microfilaments, an external glycocalyx, and cytoplasmic projections thought to be important for attachment.

Trophozoites multiply rapidly in the gut

With normal stool transit time, trophozoites usually encyst before leaving the gut. Initially, a cyst contains a single nucleus, a glycogen vacuole, and one or more large, cigar-shaped ribosomal clusters known as chromatoid bodies. With maturation, the cyst becomes quadrinucleate, and the cytoplasmic inclusions are absorbed. In contrast to the fragile trophozoite, mature cysts can survive environmental temperatures up to 55°C, chlorine concentrations normally found in municipal water supplies, and normal levels of gastric acid. *E histolytica* can be differentiated from the other amebas of the gut by its size, nuclear detail, and cytoplasmic inclusions (**Table 51–1**).

Cysts are hardy; can survive in chlorinated water supply

LIFE CYCLE

Humans are the principal hosts and reservoirs of *E histolytica*. Transmission from person to person occurs when a parasite passed in the stool of one host is ingested by another. Because the trophozoites die rapidly in the external environment, successful passage is achieved only by the cyst. Human hosts may pass up to 45 million cysts daily. Although the average infective dose exceeds 1000 organisms, ingestion of a single cyst has been known

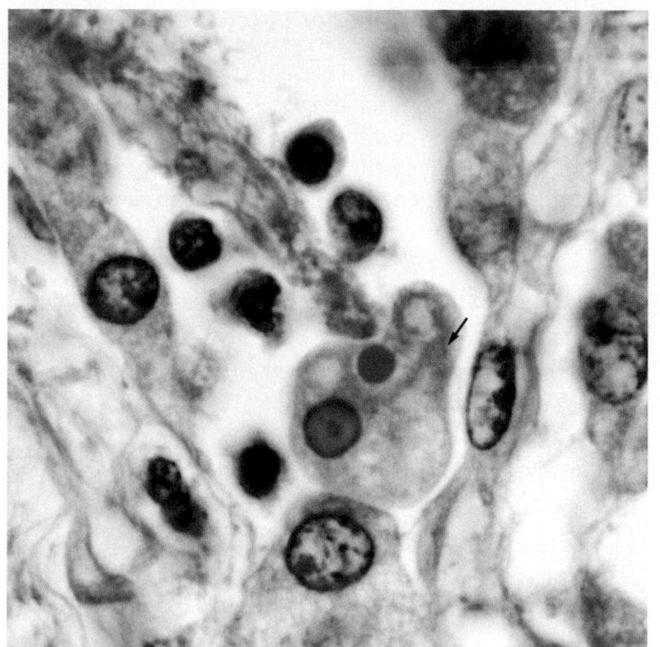

FIGURE 51–2. Amebiasis. An *Entamoeba histolytica* trophozoite (arrow) is invading tissue. Note the extending pseudopod and engulfed erythrocyte. (Reproduced with permission from Connor DH, Chandler FW, Schwartz DQA, Manz HJ, Lack EE (eds). *Pathology of Infectious Diseases,* vol. 1. Stamford, CT: Appleton & Lange; 1997.)

to produce infection. After passage through the stomach, the cyst eventually reaches the distal small bowel. Here, the cyst wall disintegrates, releasing the quadrinucleate parasite, which divides to form eight small trophozoites that are carried to the colon. Colonization is most intense in areas of fecal stasis such as the cecum and rectosigmoid, but may be found throughout the large bowel.

Humans are the hosts and reservoir; fecal–oral transmission

TABLE 51–1 Some Differential Characteristics of *Entamoeba* Species

CHARACTERISTICS	*E HISTOLYTICA*	*E HARTMANNI*	*E COLI*
Trophozoites			
Cytoplasm	Differentiated[a]	Differentiated	Undifferentiated
Nucleus			
Peripheral chromatin	Fine	Fine	Coarse, irregular
Karyosome	Small, central	Small, central	Large, eccentric
Ingested particles			
Bacteria	No	—	Yes
Red blood cells	Yes	No	No
Size	>12 μm	<12 μm	>12 μm
Cysts			
Nuclei[b]	1–4	1–4	1–8
Chromatoid bodies	Rods	Rods	Splinters
Size	>10 μm	<10 μm	>10 μm

[a]Sharp differentiation between ectoplasm and endoplasm.

[b]Fine structure similar to that of trophozoites.

LABORATORY GROWTH

Facultative anaerobes

Trophozoites are facultative anaerobes that require complex media for growth. Most require the addition of live bacteria for successful isolation. Sterile culture techniques (axenic) have been developed, however, and are essential for the preparation of the purified antigens required for serologic testing, zymodeme typing, and characterization of virulence factors. Such techniques are generally available only in research laboratories.

Amebiasis

CLINICAL CAPSULE

Amebiasis may be asymptomatic or produce intermittent diarrhea with abdominal pain. Occasionally, severe dysentery can occur with abdominal cramping and a high fever. Invasion of the colonic mucosa is typical and may spread to the liver, where an abscess is produced.

EPIDEMIOLOGY

Worldwide infection; highest rates in warmer climates

E histolytica infection rates are higher in warm climates, particularly in areas where the level of sanitation is low. Worldwide, this organism is thought to produce more deaths than any other parasite, except those that cause malaria and schistosomiasis. Reports of amebic liver abscess, for instance, emanate primarily from Mexico, western South America, South Asia, and West and South Africa. For reasons apparently unrelated to exposure, symptomatic illness is much less common in women and children than in men.

Invasive disease rare in United States

Although stool surveys in the United States indicate that 1% to 5% of the population harbors *Entamoeba,* the majority of these are now known to be colonized with the nonpathogenic *E dispar*. The incidence of invasive amebiasis in the United States decreased sharply over several decades, reaching a nadir in 1974. Since then, the numbers have increased steadily. It is now seen particularly in institutionalized individuals, Indian reservations, migrant labor camps, victims of acquired immunodeficiency syndrome (AIDS), and travelers to endemic areas.

Fecal–oral spread linked to poor hygiene

Food and water are other modes of transmission

Symptomatic amebiasis is usually sporadic, the result of direct person-to-person fecal–oral spread under conditions of poor personal hygiene. Venereal transmission is seen in male homosexuals, presumably the result of oral–anal sexual contact. Food- and waterborne spread occur, occasionally in epidemic form. Such outbreaks, however, are seldom as explosive as those produced by pathogenic intestinal bacteria. One outbreak of intestinal amebiasis was due to colonic irrigation at a chiropractic clinic.

PATHOGENESIS

Virulence determinants include lectin-mediated adherence to mucosa and capacity to lyse host cells

A number of virulence factors have been identified in *E histolytica*. In an experimental setting, invasiveness correlates well with endocytic capacity, the production of extracellular proteinases capable of activating complement and degrading collagen, the presence of a galactose-specific lectin apparently capable of mediating attachment of the organism to colonic mucosa, and—perhaps most important—the capacity to lyse host cells on contact. The latter phenomenon is initiated by the galactose-specific lectin-mediated adherence of the trophozoite to a target cell. After adherence, the ameba releases a pore-forming protein that polymerizes in the target cell membrane, forming large tubular lesions. Cytolysis rapidly follows.

In most cases of *E histolytica* infections, however, tissue damage is minimal, and the host remains symptom-free, suggesting that host factors may modulate the invasiveness of virulent strains. These factors are still poorly understood, but changes in host resistance, the colonic milieu, or the parasite itself may amplify tissue damage and clinical manifestations. Protein malnutrition, high-carbohydrate diets, corticosteroid administration, childhood, and pregnancy all appear to render the host more susceptible to invasion. Certain colonic bacteria appear to enhance invasiveness, possibly by providing a more favorable redox potential for survival and multiplication or by facilitating the adherence of the parasite to colonic mucosa. Finally, it is known that the pathogenic strains in the tropics are more invasive than those isolated in temperate areas, possibly because poor sanitation results in more frequent passage through humans.

Most infected individuals are symptom-free

Colonic microflora may influence invasiveness

Virulence is increased with passage through humans

PATHOLOGY

Amebas contact and lyse colonic epithelial cells, producing small mucosal ulcerations. There is little inflammatory response other than edema and hyperemia, and the mucosa between ulcers appears normal. Trophozoites are present in large numbers at the junction between necrotic and viable tissue. Once the lesion penetrates below the superficial epithelium, it meets the resistance of the colonic musculature and spreads laterally in the submucosa, producing a flask-like lesion with a narrow mucosal neck and a large submucosal body. It eventually compromises the blood supply of the overlying mucosa, resulting in sloughing and a large necrotic ulcer. Extensive ulceration leads to secondary bacterial infection, formation of granulation tissue, and fibrotic thickening of the colon. In approximately 1% of patients, the granulation tissue is organized into large, tumor-like masses known as **amebomas**. The major sites of involvement, in order of frequency, are the cecum, ascending colon, rectum, sigmoid, appendix, and terminal ileum. Amebas may also enter the portal circulation and be carried to the liver or, more rarely, to the lung, brain, or spleen. In these organs, liquefaction necrosis leads to the formation of abscess cavities.

Mucosal ulceration with little inflammatory response

Flask-like ulcers extend to submucosa

Amebomas and metastatic amebic abscesses in a few cases

IMMUNITY

Although *E histolytica* elicits both humoral and cellular immune responses in humans, it is still not clear which, and to what degree, these responses are capable of modulating initial infection or thwarting reinfection. In endemic areas, the prevalence of gastrointestinal colonization increases with age, suggesting that the host is incapable of clearing *E histolytica* from the gut. However, the relative infrequency with which populations living in these areas suffer repeated bouts of severe amebic colitis or liver abscess indicates that those who experience such infections have protection against recurrent disease.

Patients with invasive disease are known to produce high levels of circulating antibodies. Nevertheless, no correlation exists between the presence or concentration of such antibodies and protective immunity, possibly because pathogenic *E histolytica* trophozoites have the capacity to aggregate and shed attached antibodies and are resistant to the lytic action of complement. The susceptibility to invasive amebiasis of malnourished populations, pregnant women, steroid-treated individuals, and AIDS patients indicates that cell-mediated immune mechanisms may be directly involved in the control of tissue invasion.

Immunity is incomplete and does not correlate with antibody response

Trophozoites shed antibody and resist complement lysis

Pathogenic *E histolytica* strains produce a lectin-like substance that is mitogenic for lymphocytes. It has been suggested that this substance could stimulate viral replication of human immunodeficiency virus-infected lymphocytes as does another mitogen, phytohemagglutinin.

AMEBIASIS: CLINICAL ASPECTS

MANIFESTATIONS

Individuals who harbor *E histolytica* are usually clinically well. In most cases, particularly in the temperate zones, the organism is avirulent, living in the bowel as a normal commensal inhabitant. Spontaneous disappearance of amebas, over a period of weeks to months, among such patients is common and perhaps universal. Serologic data, however, suggest that some asymptomatic carriers possess virulent strains and incur minimal tissue invasion. In this population, the infection may eventually progress to produce overt disease.

Relationship usually commensal

Diarrhea, flatulence, and abdominal pain most common

Ulcerations with mucus and blood in stool occur in fulminant disease

Diarrhea, flatulence, and cramping abdominal pain are the most common complaints of symptomatic patients. The diarrhea is intermittent, alternating with episodes of normality or constipation over a period of months to years. Typically, the stool consists of one to four loose to watery, foul-smelling passages that contain mucus and blood. Physical findings are limited to abdominal tenderness localized to the hepatic, ascending colonic, and cecal areas. Sigmoidoscopy reveals the typical ulcerations with normal intertwining mucosa.

Hepatic abscess may have acute or insidious onset

Fulminating amebic dysentery is less common. It may occur spontaneously in debilitated or pregnant individuals or maybe precipitated by corticosteroid therapy. Its onset is often abrupt, with high fever, severe abdominal cramps, and profuse diarrhea. Most commonly, abscesses occur singly and are localized to the upper outer quadrant of the right lobe of the liver. This localization results in the development of point tenderness overlying the cavity and elevation of the right diaphragm. Liver function is usually well preserved. Isotopic or ultrasound scanning confirms the presence of the lesion. Needle aspiration results in the withdrawal of reddish-brown, odorless fluid free of bacteria and polymorphonuclear leukocytes; trophozoites may be demonstrated in the terminal portion of the aspirate.

Hepatic abscess may extend to other tissues

Approximately 5% of all patients with symptomatic amebiasis present with a liver abscess. Ironically, fewer than one half can recall significant diarrheal illness. Although *E histolytica* can be demonstrated in the stools of 72% of patients with amebic liver abscess when a combination of serial microscopic examinations and culture is used, routine microscopic examination of the stool detects less than half of these. Complications relate to the extension of the abscess into surrounding tissue, producing pneumonia, empyema, or peritonitis. Extension of an abscess from the left lobe of the liver to the pericardium is the single most dangerous complication. It may produce rapid cardiac compression (tamponade) and death or, more commonly, a chronic pericardial disease that may be confused with congestive cardiomyopathy or tuberculous pericarditis.

DIAGNOSIS

Stools examined for trophozoites and cysts in stained or wet preparations

E histolytica trophozoites ingest erythrocytes; *E dispar* trophozoites do not

The microscopic diagnosis of intestinal amebiasis depends on the identification of the organism in stool or sigmoidoscopic aspirates. Because trophozoites appear predominantly in liquid stools or aspirates, a portion of such specimens should be fixed immediately to ensure preservation of these fragile organisms for stained preparations. The specimen may then be examined in wet mount for typical motility, concentrated to detect cysts, and stained for definitive identification. If trophozoites or cysts are seen, they must be carefully differentiated from those of the commensal parasites, particularly *E hartmanni* and *E coli* (Table 51–1). *E histolytica* trophozoites can be differentiated from those of *E dispar* only by the presence of ingested erythrocytes in the former; the cysts appear identical.

Enzyme immunoassay and other methods can detect antigen in stool

Recently, sensitive and specific stool antigen tests for *E histolytica* have become commercially available; their value in the clinical diagnosis of amebiasis, when compared with microscopic examination, is now clear. Although cultural and polymerase chain reaction techniques are somewhat more sensitive, they are not widely available in most clinical laboratories.

Extraintestinal amebiasis usually demonstrates high antibody levels

The diagnosis of extraintestinal amebiasis is more difficult, because the parasite usually cannot be recovered from stool or tissue. Serologic tests are therefore of paramount importance. Typically, results are negative in asymptomatic patients, suggesting that tissue invasion is required for antibody production. Most patients with symptomatic intestinal disease and more than 90% with hepatic abscess have high levels of antiamebic antibodies. Unfortunately, these titers may persist for months to years after an acute infection, making the interpretation of a positive test difficult in endemic areas. At present, the indirect hemagglutination test and enzyme immunoassays using antigens derived from axenically grown organisms appear to be the most sensitive. Several rapid tests, including latex agglutination, agar diffusion, and counterimmunoelectrophoresis, are available to smaller laboratories.

TREATMENT

Treatment is directed toward relief of symptoms, blood and fluid replacement, and eradication of the organism. The need to eliminate the parasite in asymptomatic carriers remains uncertain. The drug of choice for eradication is metronidazole. It is effective against all forms of amebiasis, but should be combined with a second agent, such as diloxanide, to improve cure rates in intestinal disease and diminish the chance of recrudescent disease in hepatic amebiasis. In severe extraintestinal infections, parenteral dehydroemetine treatment may be considered.

Metronidazole combined with other agents

PREVENTION

Because the disease is transmitted by the fecal–oral route, efforts should be directed toward sanitary disposal of human feces and improvement in personal hygienic practices. In the United States, this applies particularly to institutionalized patients and to camps for migrant farm workers. Male homosexuals should be made aware that certain sexual practices substantially increase their risk of amebiasis and other infections.

NAEGLERIA AND *ACANTHAMOEBA* INFECTIONS

AMEBIC MENINGOENCEPHALITIS

Primary amebic meningoencephalitis is caused by free-living amebas belonging predominantly to the *Naegleria* and *Acanthamoeba* genera. The disease produced by the former has been better defined; it affects children and young adults, appears to be acquired by swimming in fresh water, and is almost always fatal. *Acanthamoeba* meningoencephalitis is a subacute or chronic illness that also is usually fatal. *Naegleria* species are found in large numbers in shallow fresh water, particularly during warm weather. *Acanthamoeba* species are found in soil and in fresh and brackish water, and they have been recovered from the oropharynx of asymptomatic humans.

Meningoencephalitis due to free-living amebas

Warm weather and brackish water favor amebas

Approximately 140 cases of *Naegleria* meningoencephalitis have been reported, primarily in Great Britain, Belgium, Czechoslovakia, Australia, New Zealand, India, Nigeria, and the United States. Serologic studies suggest that inapparent infections are much more common. Most cases in the United States have occurred in the southeastern states. Characteristically, patients have fallen ill during the summer after swimming or water-skiing in small, shallow, freshwater lakes. The Czechoslovakian cases followed swimming in a chlorinated indoor pool, and several have occurred after bathing in hot mineral water. A recent report from Africa suggests that the disease may have been acquired by inhaling airborne cysts during the dry, windy season in the sub-Sahara.

Naegleria infections associated with freshwater swimming

Histologic evidence suggests that *Naegleria* traverses the nasal mucosa and the cribriform plate to the central nervous system. Here, the organism produces a severe purulent, hemorrhagic inflammatory reaction, which extends perivascularly from the olfactory bulbs to other regions of the brain. The infection is characterized by the rapid onset of severe bifrontal headache, seizures, and at times abnormalities in taste or smell. The disease runs an inexorably downhill course to coma, ending fatally within a few days.

Passage to central nervous system across cribriform plate

A careful examination of the cerebrospinal fluid often provides a presumptive diagnosis of *Naegleria* infection. The fluid is usually bloody and demonstrates an intense neutrophilic response. The protein level is elevated and the glucose level decreased. No bacteria can be demonstrated on stain or culture. Early examination of a wet mount preparation of unspun spinal fluid reveals typical trophozoites. Staining with specific fluorescent antibody confirms the identification. The organism can usually be isolated on agar plates

Purulent bloody cerebrospinal fluid containing *Naegleria* trophozoites

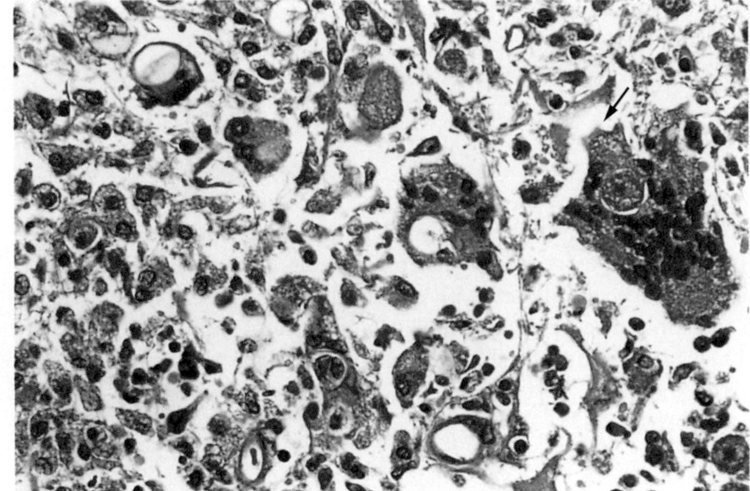

FIGURE 51–3. Acanthamoebic granulomatous encephalitis. A trophozoite (arrow) entering an epithelioid cell is seen at the right. The empty ovals in other cells are collapsed cysts. (Reproduced with permission from Connor DH, Chandler FW, Schwartz DQA, Manz HJ, Lack EE (eds). *Pathology of Infectious Diseases*, vol. 1. Stamford, CT: Appleton & Lange; 1997.)

seeded with a Gram-negative bacillus (to feed the amebas) or grown axenically in tissue culture. To date, there are reports of only four patients who have survived a *Naegleria* infection. All were diagnosed early and treated with high-dose amphotericin B along with rifampin.

Acanthamoeba affects older immunocompromised persons

Granulomatous encephalitis with cysts and trophozoites

The epidemiology of *Acanthamoeba* encephalitis has not been clearly defined. Infections usually involve older, immunocompromised persons, and a history of freshwater swimming is generally absent. The ameba probably reaches the brain by hematogenous dissemination from an unknown primary site, possibly the respiratory tract, skin, or eye. Metastatic lesions have been reported. Histologically, *Acanthamoeba* infections produce a diffuse, necrotizing, granulomatous encephalitis (**Figure 51–3**), with frequent involvement of the mid-brain. Both cysts and trophozoites can be found in the lesions. Cutaneous ulcers and hard nodules containing amebas have been detected in AIDS patients.

More prolonged disease with occasional spontaneous recovery

The clinical course of *Acanthamoeba* disease is more prolonged than that of *Naegleria* infection and occasionally ends in spontaneous recovery; the disease in immunocompromised hosts is invariably fatal. The spinal fluid usually demonstrates a mononuclear response. Amebas can occasionally be visualized in or cultured from the cerebrospinal fluid or biopsy specimens. Fluorescein-labeled antiserum is available from the Centers for Disease Control and Prevention. Definitive diagnosis is usually made histologically after death. *Acanthamoeba* species are sensitive to a variety of agents, but studies of clinical efficacy have not been performed.

OTHER *ACANTHAMOEBA* INFECTIONS

Corneal ulcerations associated with contact lens use

Skin lesions, uveitis, and corneal ulcerations have also been reported with *Acanthamoeba* disease. The latter are serious, producing a chronic progressive ulcerative lesion that may result in blindness. Infection commonly follows mild corneal trauma; most recently reported cases have been in users of soft contact lenses. Clinically, severe ocular pain, a paracentral ring infiltrate of the cornea, and recurrent epithelial breakdown are helpful in distinguishing this entity from the more common herpes simplex keratitis. The diagnosis can be confirmed by demonstrating typical wrinkled, double-walled cysts in corneal biopsies or scrapings using wet mounts, stained smears, and/or fluorescent antibody techniques. Culture of corneal tissue and contact lenses is frequently successful when the laboratory is given time to prepare satisfactory media. Chemotherapy has generally been ineffective unless given very early in the course of infection. Although a combination of corneal transplantation and chemotherapy may be successful later in the course of the disease, enucleation of the eye may be necessary to cure advanced infections. The drugs of choice are propamidine and neomycin eyedrops administered alternately for a period of several months. Successful use of clotrimazole has been recently reported.

CASE STUDY

WEIGHT LOSS, ABDOMINAL DISCOMFORT, AND A TENDER LIVER

A 21-year-old college student volunteered for a 2-year assignment as a missionary in a rural area of Central Mexico. Within 4 months of arrival, he developed a mild diarrheal illness with flatulence and abdominal discomfort that subsided spontaneously within a few weeks. Six months later, he noted progressive weight loss over several weeks, a low-grade fever, and right upper abdominal tenderness.

He returned to the United States for medical consultation. The primary physical finding was an enlarged right lobe of the liver, which was tender on palpation. An ultrasound study confirmed the presence of an abscess at that site.

The diagnosis of an amebic hepatic abscess was seriously considered.

QUESTIONS

■ Which of the following laboratory findings would be most likely to be helpful in supporting this patient's diagnosis?

A. Demonstration of cyst forms in the stool
B. Demonstration of trophozoites containing erythrocytes in the stool
C. Isolation of the organism from the abscess
D. Demonstration of high serum antibody titers to *E histolytica*

■ Your choice of treatment would usually be:

A. Tetracycline
B. Amphotericin B
C. Clotrimazole
D. Metronidazole

■ A diagnosis of amebic meningoencephalitis is suggested by a recent history of the following, **except:**

A. Exposure to a household contact with a similar illness
B. Swimming in a fresh water lake
C. Bathing in hot springs
D. Swimming in a chlorinated pool

ANSWERS

1(D), **2**(D), **3**(A)

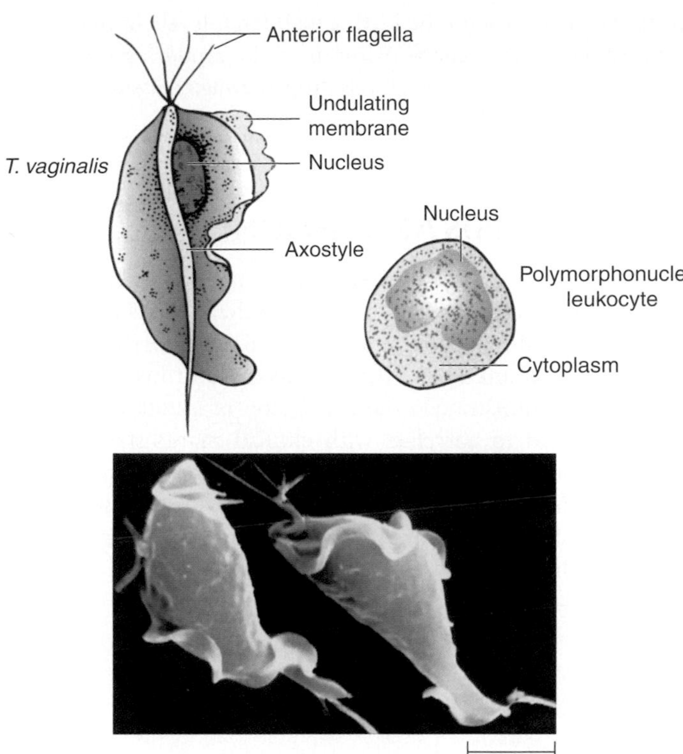

FIGURE 52–1. ***Trichomonas vaginalis.*** The parasite and its structures are shown in relation to the size of a polymorphonuclear leukocyte (top). The micrograph below illustrates their use for motility. (Reproduced with permission from Nester EW, Anderson DG, Roberts CE Jr., Nester MT. *Microbiology: A Human Perspective*, 6th ed. New York: McGraw-Hill, 2008.)

TRICHOMONIASIS

CLINICAL CAPSULE

Trichomoniasis is a sexually transmitted disease, which produces a vaginitis with pain, discharge, and dysuria. The infection fluctuates over weeks to months. Men are usually asymptomatic but may have urethritis or prostatitis.

EPIDEMIOLOGY

Trichomoniasis is a cosmopolitan disease usually transmitted by sexual intercourse. It is estimated that 3 million women in the United States and 180 million worldwide acquire this disease annually, and 25% of sexually active women become infected at some time during their lives; 30% to 70% of their male sexual partners are also parasitized, at least transiently. As would be expected, the likelihood of acquiring the disease correlates directly with the number of sexual contacts. Infection is rare in adult virgins, whereas rates as high as 70% are seen among prostitutes, sexual partners of infected patients, and individuals with other venereal diseases. In women, the peak incidence of trichomoniasis is between 16 and 35 years of age, but there is still a relatively high prevalence in the 30- to 50-year age group.

Transmission usually sexual

Prevalence linked to sexual activity

Nonvenereal transmission is uncommon

Nonvenereal transmission is uncommon. Transfer of organisms on shared washcloths may explain, in part, the high frequency of infection seen among institutionalized women. Female neonates are occasionally noted to harbor *T vaginalis,* presumably acquiring it during passage through the birth canal. High levels of maternal estrogen produce a transient decrease in the vaginal pH of the child, rendering it more susceptible to colonization. Within a few weeks, estrogen levels drop, the vagina assumes its premenarcheal state, and the parasite is eliminated.

PATHOGENESIS AND IMMUNITY

Parasite damages epithelial cells on contact

Direct contact of *T vaginalis* with the squamous epithelium of the genitourinary tract results in destruction of the involved epithelial cells and the development of a neutrophilic inflammatory reaction and petechial hemorrhages. The precise pathogenesis of these changes is unknown. The organism is not invasive, and extracellular toxins have never been demonstrated. The expression of a 200-kd parasitic glycoprotein, however, has been found to correlate with clinical manifestations. Changes in the microbial, hormonal, and pH environment of the vagina as well as factors inherent to the infecting parasite are thought to modulate the severity of the pathologic changes. Although humoral, secretory, and cellular immune reactions can be demonstrated in most infected women, they are of little diagnostic help and do not appear to produce clinically significant immunity.

TRICHOMONIASIS: CLINICAL ASPECTS

MANIFESTATIONS

Chronic vaginitis lasting weeks to months

In women, *T vaginalis* produces a persistent vaginitis. Although up to 50% are asymptomatic at the time of diagnosis, most develop clinical manifestations within 6 months. Approximately 75% develop a discharge, which is typically accompanied by vulvar itching or burning (50%), dyspareunia (50%), dysuria (50%), and a disagreeable odor (10%). Although fluctuating in intensity, symptoms usually persist for weeks or months. Commonly, manifestations worsen during menses and pregnancy. Eventually, the discharge subsides, even though the patient may continue to harbor the parasite. In symptomatic patients, physical examination reveals reddened vaginal and endocervical mucosa. In severe cases, petechial hemorrhages and extensive erosions are present. A red, granular, friable endocervix (strawberry cervix) is a characteristic but uncommon finding. An abundant discharge is generally seen pooled in the posterior vaginal fornix. Although classically described as thin, yellow, and frothy in character, the discharge more frequently lacks these characteristics. Trichomoniasis may increase the risk of preterm birth and enhance susceptibility to human immunodeficiency virus (HIV) infections.

Urethral and prostatic infection in men usually asymptomatic

The urethra and prostate are the usual sites of trichomoniasis in men; the seminal vesicles and epididymis may be involved on occasion. Infections are usually asymptomatic, possibly because of the efficiency with which the organisms are removed from the urogenital tract by voided urine. Symptomatic men complain of recurrent dysuria and scant, nonpurulent discharge. Acute purulent urethritis has been reported rarely. Trichomoniasis should be suspected in men presenting with nongonococcal urethritis, or a history of either prior trichomonal infection or recent exposure to trichomoniasis.

DIAGNOSIS

The diagnosis of trichomoniasis rests on the detection and morphologic identification of the organism in the genital tract. Identification is accomplished most easily by examining a wet mount preparation for the presence of motile organisms. In women, a drop of vaginal discharge is the most appropriate specimen; in men, urethral exudate or urine sediment after prostate massage may be used. Although highly specific when positive, wet mounts have a sensitivity of only 50% to 60%. They are most likely to be negative in asymptomatic or mildly symptomatic patients and in women who have douched in the previous 24 hours. Giemsa- and Papanicolaou-stained smears provide little additional help. The recent introduction of a commercial system that allows direct, rapid microscopic examination without the need for daily sampling may ameliorate this situation. Direct immunofluorescent antibody staining has a sensitivity of 70% to 90%. Parasitic culture, though more sensitive, requires several days to complete and is frequently unavailable.

Wet mount examination for motile trophozoites sufficient in most symptomatic cases

TREATMENT

Oral metronidazole is extremely effective in recommended dosage, curing more than 95% of all *Trichomonas* infections. It may be given as a single dose or over 7 days. Simultaneous treatment of sexual partners may minimize recurrent infections, particularly when single-dose therapy is used for the index case. Because of the disulfiram-like activity of metronidazole, alcohol consumption should be suspended during treatment. The drug should never be used during the first trimester of pregnancy because of its potential teratogenic activity. Use in the last two trimesters is unlikely to be hazardous but should be reserved for patients whose symptoms cannot be adequately controlled with local therapies. High-dose, long-term metronidazole treatment has been shown to be carcinogenic in rodents. No association with human malignancy has been described to date, and in the absence of a suitable alternative drug, metronidazole continues to be used.

Metronidazole cures 95% of cases

Giardia lamblia

PARASITOLOGY

Giardia lamblia was first described by Anton von Leeuwenhoek 300 years ago when he examined his own diarrheal stool with one of the first primitive microscopes. It was not until the last several decades, however, that this cosmopolitan flagellate became widely regarded in the United States as a pathogen. Of the six other flagellated protozoans known to parasitize the alimentary tract of humans, only one, *Dientamoeba fragilis,* has been credibly associated with disease. Definitive confirmation or refutation of its pathogenicity will, it is hoped, not require the passage of another three centuries.

Unlike *T vaginalis, Giardia* possesses both a trophozoite and a cyst form (**Figure 52–2**). It is a sting-ray–shaped trophozoite 9 to 21 μm in length, 5 to 15 μm in width, and 2 to 4 μm in thickness. When viewed from the top, the organism's two nuclei and central parabasal bodies give it the appearance of a face with two bespectacled eyes and a crooked mouth. Four pairs of flagella—anterior, lateral, ventral, and posterior—reinforce this image by suggesting the presence of hair and chin whiskers. These distinctive parasites reside in the duodenum and jejunum, where they thrive in the alkaline environment and absorb nutrients from the intestinal tract. They move about the unstirred mucous layer at the base of the microvilli (**Figure 52–3**) with a peculiar tumbling or "falling leaf" motility or, with the aid of a large ventral sucker, attach themselves to the brush border of the intestinal epithelium. Unattached organisms may be carried by the fecal stream to the large intestine.

Trophozoite and cyst stages

Move about duodenum and jejunum with tumbling motility

In the descending colon, if transit time allows, the flagella are retracted into cytoplasmic sheaths and a smooth, clear cyst wall is secreted. These forms are oval and somewhat smaller than the trophozoites. With maturation, the internal structures divide, producing a quadrinucleate organism harboring two sucking discs, four parabasal bodies, and eight axonemes. When fixed and stained, the cytoplasm pulls away from the cyst wall in a characteristic fashion. The mature cysts, which are the infective form of the parasite, may survive in cold water for more than 2 months and are resistant to concentrations of chlorine generally used in municipal water systems. They are transmitted from host to host by the fecal–oral route. In the duodenum of a new host, the cytoplasm divides to produce two binucleate trophozoites.

Cystic forms develop in colon

Resistant cysts transmitted from host to host

Organisms of the genus *Giardia* are among the most widely distributed of intestinal protozoa; they are found in fish, amphibians, reptiles, birds, and mammals. At first, it was assumed that *Giardia* strains found in different animals were host-specific; on this basis,

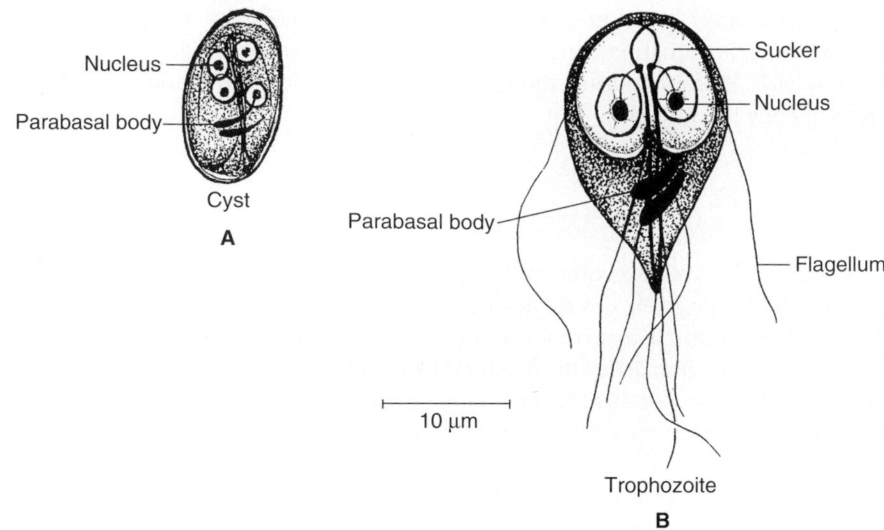

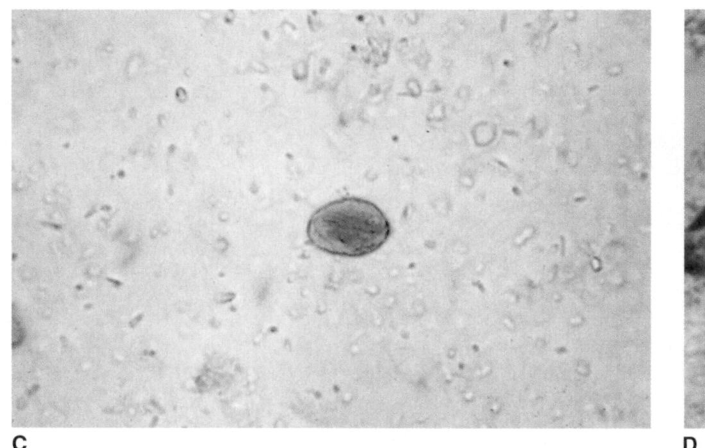

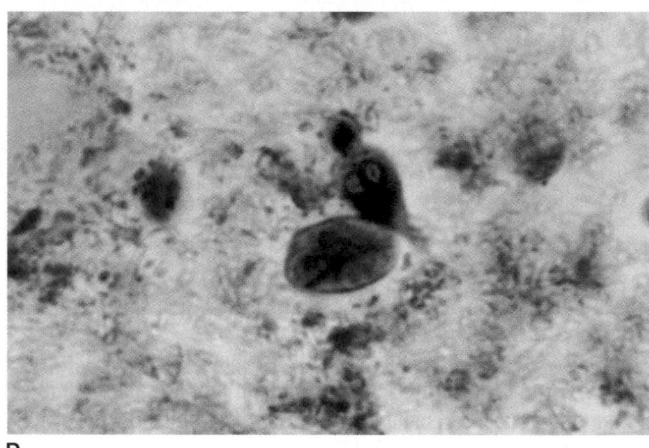

FIGURE 52–2. ***Giardia lamblia.*** **A.** Cyst structures. **B.** Trophozoite structures. **C.** Cyst in stool iodine preparation. **D.** Trophozoite in stool. (C and D, Reproduced with permission from Connor DH, Chandler FW, Schwartz DQA, Manz HJ, Lack EE (eds). *Pathology of Infectious Diseases*, vol. 1. Stamford, CT: Appleton & Lange; 1997.)

Wide distribution in animal kingdom

some 40 different species were described. Since it is now recognized that some strains can infect multiple animal hosts, the practice of assigning species status by the host from which the parasite was recovered is considered invalid. Unfortunately, there is still no general agreement on an alternate method of speciation. Three morphologically distinct groups of *Giardia* have been described on the basis of their central parabasal body morphology.

GIARDIASIS

CLINICAL CAPSULE

Giardiasis, an intestinal infection acquired from untreated water sources, is most often symptomatic. When disease occurs, it is in the form of a diarrhea lasting up to 4 weeks with foul-smelling, greasy stools. Abdominal pain, nausea, and vomiting are also present.

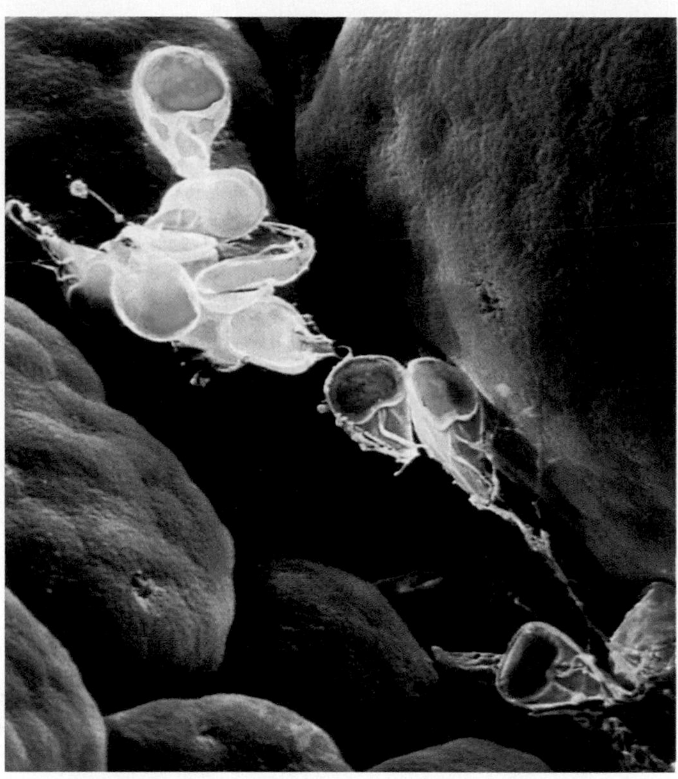

FIGURE 52–3. Giardiasis. Scanning electron micrograph of *G lamblia* trophozoites in human intestine. (Reproduced with permission from Nester EW, Anderson DG, Roberts CE Jr, Nester MT. *Microbiology: A Human Perspective*, 6th ed. New York: McGraw-Hill, 2008.)

EPIDEMIOLOGY

Giardiasis has a cosmopolitan distribution; its prevalence is highest in areas with poor sanitation and among populations unable to maintain adequate personal hygiene. In developing countries, infection rates may reach 25% to 30%; in the United States, *G lamblia* is found in 4% of stools submitted for parasitologic examination, making it this country's most frequently identified intestinal parasite. All ages and economic groups are represented, but young children and young adults are preferentially involved. Children with immunoglobulin deficiencies are more likely to acquire the flagellate, possibly because of a deficiency in intestinal immunoglobulin A. Giardiasis is also common among attendees of day care centers. Attack rates of over 90% have been seen in the ambulatory non–toilet-trained population (age 1 to 2 years) of these institutions, suggesting direct person-to-person transmission of the parasite. The frequency with which secondary cases are seen among family contacts reinforces this probability. Undoubtedly, direct fecal spread is also responsible for the high infection rate among male homosexuals. In several recent studies, the prevalence of giardiasis and/or amebiasis in that population has ranged from 11% to 40% and is correlated closely with the number of oral–anal sexual contacts.

Transmission facilitated by poor hygiene and IgA deficiency

High attack rates in day-care centers

Giardiasis common among male homosexuals

Water-borne and, less frequently, food-borne transmission of *G lamblia* has also been documented, and probably accounts for the frequency with which American travelers to Third World nations acquire infection. Unlike the typical bacterial diarrhea syndrome seen in travelers, the diarrhea begins late in the course of travel and may persist for several weeks. More than 20 water-borne outbreaks of giardiasis have also been reported in the United States. The sources have included untreated pond or stream water, sewage-contaminated municipal water supplies, and chlorinated but inadequately filtered water. In a few of these outbreaks, epidemiologic data have suggested that wild mammals, particularly beavers, served as the reservoir hosts. Domestic cats and dogs, which have recently been shown to have a high prevalence of *G lamblia*, may also act as reservoirs for human infections.

Water- or food-borne traveler's diarrhea lasts for weeks

Beavers and other mammals possible sources

PATHOGENESIS

Disease manifestations appear related to intestinal malabsorption, particularly of fat and carbohydrates. Disaccharidase deficiency with lactose intolerance, altered levels of intestinal peptidases, and decreased vitamin B_{12} absorption have been demonstrated. The precise pathogenetic mechanisms responsible for these changes remain poorly understood. Mechanical blockade of the intestinal mucosa by large numbers of *Giardia,* damage to the brush border of the microvilli by the parasite's sucking disc, organism-induced deconjugation of bile salts, altered intestinal motility, accelerated turnover of mucosal epithelium, and mucosal invasion have all been suggested. None of these correlates well with clinical manifestations. Patients with severe malabsorption have jejunal colonization with enteric bacteria or yeasts, suggesting that these organisms may act synergistically with *Giardia.* Eradication of the associated microorganism, however, has not uniformly resulted in clinical improvement. Jejunal biopsies sometimes reveal a flattening of the microvilli and an inflammatory infiltrate, the severity of which correlates roughly with that of the clinical disease. Generally, both malabsorption and the jejunal lesions have been reversed with specific treatment. The demonstration of occasional trophozoites in the submucosa raises the possibility that these changes reflect T-lymphocyte–mediated damage.

Basis for malabsorption and jejunal pathology remains uncertain

IMMUNITY

Susceptibility to giardiasis has been related to several factors, including strain virulence, inoculum size, achlorhydria or hypochlorhydria, and immunologic abnormalities. In one experimental study, humans were challenged with varying doses from as few as 10 cysts. They were uniformly parasitized when 100 or more were ingested. Several workers have noted the frequency with which giardiasis occurs in achlorhydric and hypochlorhydric individuals. Although reinfection is common, the frequent occurrence of giardiasis in patients with immunologic diseases, plus the rarity with which it is seen in older adults, suggests that protective immunity, albeit incomplete, does develop in humans. Animal studies have demonstrated that *Giardia*-specific, secretory IgA (sIgA) antibodies inhibit attachment of trophozoites to intestinal epithelium, perhaps by blocking parasite surface lectins. Moreover, antitrophozoite IgM or IgG antibodies, plus complement, are known to be capable of killing *Giardia* trophozoites.

Predisposing factors include hypochlorhydria and immunocompromise

MANIFESTATIONS

In endemic situations, over two thirds of persons infected with giardiasis are asymptomatic. In acute outbreaks, this ratio of asymptomatic to symptomatic patients is usually reversed. When they do occur, symptoms begin 1 to 3 weeks after exposure and typically include diarrhea, which is sudden in onset and explosive in character. The stool is foul-smelling, greasy in appearance, and floats. It is devoid of blood or mucus. Upper abdominal cramping is common. Large quantities of intestinal gas produce abdominal distention, sulfuric eructations, and abundant flatus. Nausea, vomiting, and low-grade fever may be present. The acute illness generally resolves in 1 to 4 weeks; in children, however, it may persist for months, leading to significant malabsorption, weight loss, and malnutrition.

Subclinical infections common in endemic areas

Diarrhea, cramping, flatus, and greasy stools

In many adults, the acute phase of giardiasis is often followed by a subacute or chronic phase characterized by intermittent bouts of mushy stools, flatulence, "heartburn" and weight loss that persist for weeks or months. At times, patients presenting in this fashion deny having experienced the acute syndrome described previously. In the majority, symptoms and organisms eventually disappear spontaneously. It is not uncommon for lactose intolerance to persist after eradication of the organisms. This condition may be confused with an ongoing infection, and the patient may be subjected to unnecessary treatment.

Subacute and chronic infections with weight loss in adults

Lactose intolerance may persist

DIAGNOSIS

The diagnosis of giardiasis is made by finding the cyst in formed stool or the trophozoite in diarrheal stools, duodenal secretions, or jejunal biopsy specimens. In acutely symptomatic patients, the parasite can usually be demonstrated by examining one to three stool specimens after appropriate concentration and staining. In chronic cases, excretion of the organism is often intermittent, making parasitologic confirmation more difficult. Many of these patients can be diagnosed by examining specimens taken at weekly intervals over 4 to 5 weeks. Alternatively, duodenal secretions can be collected and examined for trophozoites in trichrome or Giemsa-stained preparations. There are now a number of reliable, commercially available, enzyme immunoassays (EIAs) for the direct detection of parasite antigen in stool. They appear to be as sensitive and specific as microscopic examinations. The organism can be grown in culture, but the methods are not currently adaptable to routine diagnostic work.

Demonstration of trophozoites and cysts in stool or duodenal aspirates diagnostic

EIAs detect *Giardia* antigen in stool

TREATMENT

Four drugs are currently available for the treatment of giardiasis in the United States: quinacrine hydrochloride, metronidazole, furazolidone, and paromomycin. Quinacrine and metronidazole are somewhat more effective (70% to 95%) and are preferred for patients capable of ingesting tablets. Furazolidone is used by pediatricians because of its availability as a liquid suspension, but it has the lowest cure rate. These three agents require 5 to 7 days of therapy. Tinidazole, an oral agent, is safe and effective in single-dose treatment. Because of the potential for person-to-person spread, it is important to examine and, if necessary, treat close physical contacts of the infected patient, including playmates at nursery school, household members, and sexual contacts. None of the aforementioned agents should be used in pregnant women because of their potential teratogenicity. Paromomycin, a nonabsorbed but somewhat less effective agent, may be used in this circumstance.

Several drugs available

Close contacts should be examined

PREVENTION

Hikers should avoid ingestion of untreated surface water, even in remote areas, because of the possibility of contamination by feces of infected animals. Adequate disinfection can be accomplished with halogen tablets yielding concentrations higher than that generally achieved in municipal water systems. The safety of the latter results from additional flocculation and filtration procedures.

Avoid drinking untreated surface water

BLOOD AND TISSUE FLAGELLATES

Two of the many genera of hemoflagellates are pathogenic to humans, *Leishmania* and *Trypanosoma*. They reside and reproduce within the gut of specific insect hosts. When these vectors feed on a susceptible mammal, the parasite penetrates the feeding site, invades the blood and/or tissue of the new host, and multiplies to produce disease. The life cycle is completed when a second insect ingests the infected mammalian blood or tissue fluid. During the course of their passage through insect and vertebrate hosts, flagellates undergo developmental change. Within the gut of the insect (and in culture media), the organism assumes the promastigote (*Leishmania*) or epimastigote (*Trypanosoma*) form (**Figure 52–4**). These protozoa are motile and fusiform and have a blunt posterior end and a pointed anterior from which a single flagellum projects. They measure 15 to 30 μm long and 1.5 to 4.0 μm wide. In the promastigote, the kinetoplast is located in the anterior extremity, and the flagellum exits from the cell immediately. The kinetoplast of the epimastigote, in contrast, is located centrally, just in front of the vesicular nucleus. The flagellum runs anteriorly in the free edge of an undulating membrane before passing out of the cell. In the mammalian host, hemoflagellates appear as trypomastigotes (*Trypanosoma*)

Life cycle includes insect host stage

Promastigote and epimastigote forms in insects

Trypomastigote and amastigote forms in humans

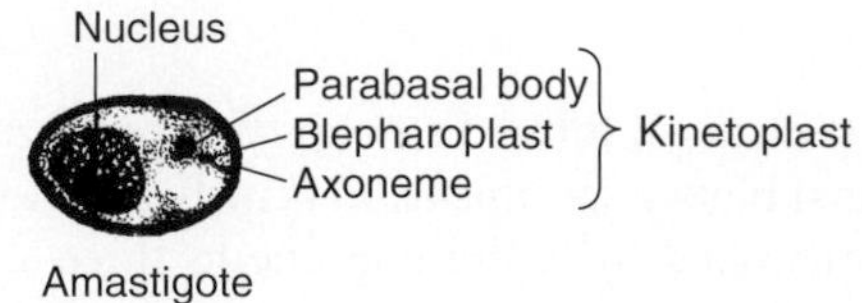

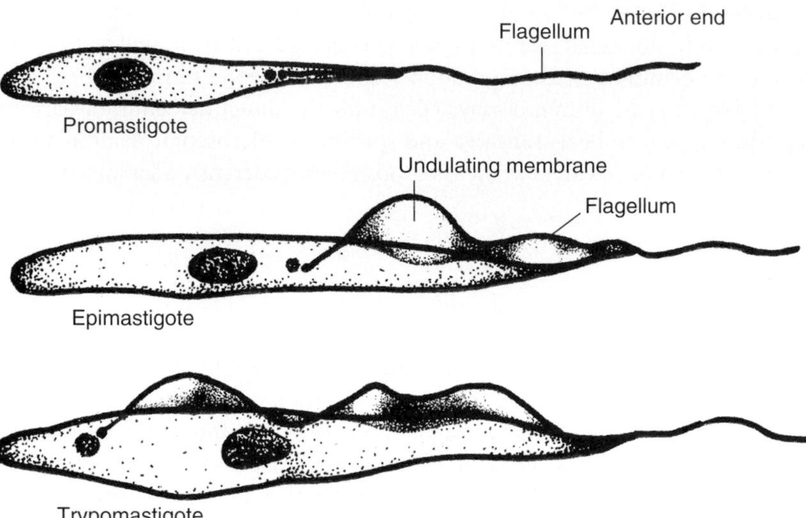

FIGURE 52–4. Stages in the life cycle of the hemoflagellates (Trypanosomidae).

or amastigotes (*Leishmania, T cruzi*). The former circulate in the bloodstream and closely resemble the epimastigote form, except that the kinetoplast is in the posterior end of the parasite. The amastigote stage is found intracellularly. It is round or oval, measures 1.5 to 5.0 μm in diameter, and contains a clear nucleus with a central karyosome. Although it has a kinetoplast and an axoneme, there is no free flagellum.

The flagellated forms move in a spiral fashion, and all reproduce by longitudinal binary fission. The flagellum itself does not divide; rather, a second one is generated by one of the two daughter cells. The organisms use carbohydrate obtained from the body fluids of the host in aerobic respiration.

Leishmania

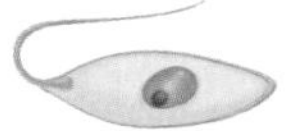

PARASITOLOGY

Species morphologically similar; differ in molecular features

Leishmania species are obligate intracellular parasites of mammals. Several strains can infect humans; they are all morphologically similar, resulting in some confusion over their proper speciation. Definitive identification of these strains requires isoenzyme analysis, monoclonal antibodies, kinetoplast DNA buoyant densities, DNA hybridization, and DNA restriction endonuclease fragment analysis or chromosomal karyotyping using pulse-field electrophoresis. The many strains can be more simply placed in four major groups based on their serologic, biochemical, cultural, nosologic, and behavioral characteristics. For the sake of clarity, these groups are discussed as individual species. Each, however, contains a variety of strains that have been accorded separate species or subspecies status by some authorities. The organisms can be propagated in hamsters and in a variety of commercially available liquid media.

DISEASE TRANSMISSION

Cutaneous ulcer or visceral infection (kala azar) the primary diseases

It is estimated that over 20 million people worldwide suffer from leishmaniasis, and 1 to 2 million additional individuals acquire the infection annually. *Leishmania tropica* in the Old World and *L mexicana* in the New World produce a localized cutaneous lesion or ulcer, known popularly as oriental sore or chiclero ulcer; *L braziliensis* is the cause of American mucocutaneous leishmaniasis (espundia); and *L donovani* is the etiologic agent of kala azar, a disseminated visceral disease.

All four are transmitted by phlebotomine sandflies. These small, delicate, short-lived insects are found in animal burrows and crevices throughout the tropics and subtropics. At night, they feed on a wide range of mammalian hosts. Amastigotes ingested in the course of a meal assume the flagellated promastigote form, multiply within the gut, and eventually migrate to the buccal cavity. When the fly next feeds on a human or animal host, the buccal promastigotes are injected into the skin of the new host together with salivary peptides capable of inactivating host macrophages. Here, they activate complement by the classic (*L donovani*) or alternative pathway and are opsonized with C3, which mediates attachment to the CR1 and CR3 complement receptors of macrophages. After phagocytosis, the promastigotes lose their flagella and multiply as the rounded amastigote form within the phagolysosome. In stained smears, the parasites take on a distinctive appearance and have been termed Leishman–Donovan bodies. Intracellular survival is mediated by a surface lipophosphoglycan and an abundance of membrane-bound acid phosphatase, which inhibit the macrophage's oxidative burst and/or inactivate lysosomal enzymes. Continued multiplication leads to the rupture of the phagocyte and release of the daughter cells. Some may be taken up by a feeding sandfly; most invade neighboring mononuclear cells.

All four groups transmitted by nocturnally feeding sandflies

Complement activation mediates attachment to macrophages

Intracellular survival by inhibiting macrophage killing mechanisms

Amastigotes released from macrophages can infect feeding sandfly

Continuation of this cycle results in extensive histiocytic proliferation. The course of the disease at this point is determined by the species of parasite and the response of the host's T cells. CD4+ T cells of the T_H1 type secrete interferon-γ in response to leishmanial antigens. This, in turn, activates macrophages to kill intracellular amastigotes by the production of toxic nitric oxide. In the localized cutaneous forms of leishmaniasis, this immune response results in the development of a positive delayed skin (leishmanin) reaction, lymphocytic infiltration, reduction in the number of parasites, and, eventually, spontaneous disappearance of the primary skin lesion. In infections with *L braziliensis,* this sequence may be followed weeks to months later by mucocutaneous metastases. These secondary lesions are highly destructive, presumably as a result of the host's hypersensitivity to parasitic antigens.

In localized cutaneous disease, cellular immune responses produce spontaneous cure

Mucocutaneous metastases in *L braziliensis* infections

Some strains of *L tropica* and *L mexicana* fail to elicit an effective intracellular immune response in certain hosts. Such patients appear to have a selective suppressor T-lymphocyte–mediated anergy to leishmanial antigens. Consequently, there is no infiltration of lymphocytes or decrease in the number of parasites. The skin test remains negative, and the skin lesions disseminate and become chronic (diffuse cutaneous leishmaniasis). In infections with *L donovani,* there is a more dramatic inhibition of the T_H1 response. The leishmanial organisms are able to disseminate through the bloodstream to the visceral organs, possibly because of a relative resistance of *L donovani* to the natural microbicidal properties of normal serum, and/or their ability to better survive at 37°C than strains of *Leishmania,* causing cutaneous lesions. Although dissemination is associated with the development of circulating antibodies, they do not appear to serve a protective function and may, via the production of immune complexes, be responsible for the development of glomerulonephritis. A simplified outline of the immune responses in different forms of leishmaniasis is presented in **Table 52–2**.

Lack of cellular immune response in disseminated and chronic infections

LOCALIZED CUTANEOUS LEISHMANIASIS

EPIDEMIOLOGY

Cutaneous leishmaniasis is a zoonotic infection of tropical and subtropical rodents. It is particularly common in areas of Central Asia, the Indian subcontinent, Middle East, Africa, the Mediterranean littoral, and Central and South America. In the latter area, *L mexicana* infects several species of arboreal rodents. Humans become involved when they enter forested areas to harvest chicle for chewing gum and are bitten by infected sandflies. In the Eastern Hemisphere, the desert gerbil and other burrowing rodents serve as the reservoir hosts of *L tropica.* Human infection occurs when rural inhabitants come in close contact

Geographic distribution related to human and rodent reservoirs

Canine reservoir in urban disease

TABLE 52–2 Immune Response to Leishmaniasis

HUMAN DISEASE	PARASITE	LEISHMANIN SKIN TEST	NUMBER OF LYMPHOCYTES	NUMBER OF PARASITES	PROGNOSIS	HUMORAL ANTIBODY TITER
Localized skin ulcer (oriental sore, chiclero ulcer, uta)	*L tropica* *L mexicana*	Positive	Many	Few	Good	Low
Mucocutaneous lesions (espundia)	*L braziliensis*	Positive	Many	Few	Poor	Low
Disseminated cutaneous						
Ethiopian	*L tropica*[a]	Negative	Few	Many	Poor	High
American	*L mexicana*[a]					
Disseminated visceral (kala azar)	*L donovani*	Negative	Few	Many	Poor	High

[a]Different subspecies from those causing localized skin ulcers.

with the burrows of these animals. In the Mediterranean area, southern Russia, and India, human disease involves urban dwellers, primarily children. In this setting, the domestic dog serves as the reservoir, although sandflies may also transmit *L tropica* directly from human to human.

LOCALIZED CUTANEOUS LEISHMANIASIS

MANIFESTATIONS

Chronic, self-limiting skin ulceration

Strain-specific immunity

Lesions usually appear on the extremities or face (the ear in cases of chiclero ulcer) weeks to months after the bite of the sandfly (**Figure 52–5**). They first appear as pruritic papules, often accompanied by regional lymphadenopathy. In a few months, the papules ulcerate, producing painless craters with raised erythematous edges, sharp walls, and a granulating base. Satellite lesions may form around the edge of the primary sore and fuse with it. Multiple primary lesions are seen in some patients. Spontaneous healing occurs in 3 to 12 months, leaving a flat, depigmented scar. Occasionally, the lesions fail to heal, particularly on the ears, leading to progressive destruction of the pinna. A permanent strain-specific immunity usually follows healing. Multiple, disseminated nonhealing lesions may be seen in patients with acquired immunodeficiency syndrome (AIDS).

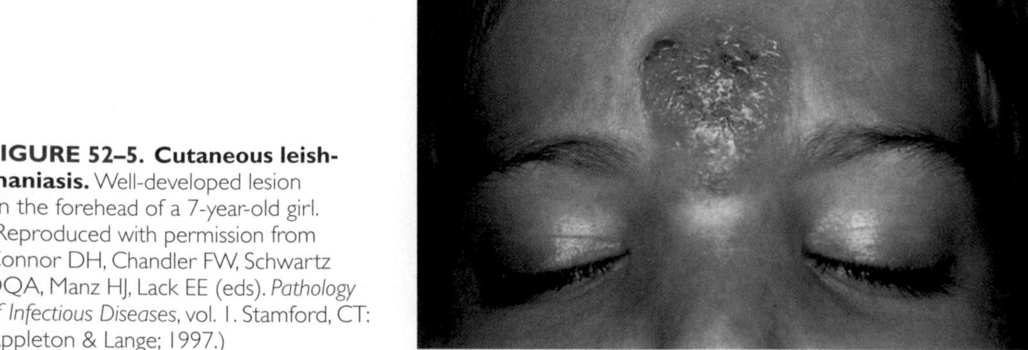

FIGURE 52–5. Cutaneous leishmaniasis. Well-developed lesion on the forehead of a 7-year-old girl. (Reproduced with permission from Connor DH, Chandler FW, Schwartz DQA, Manz HJ, Lack EE (eds). *Pathology of Infectious Diseases*, vol. 1. Stamford, CT: Appleton & Lange; 1997.)

TREATMENT

In endemic areas, the diagnosis of localized cutaneous leishmaniasis is made on clinical grounds and confirmed by the demonstration of the organism in the advancing edge of the ulcer. Material collected by biopsy, curettage, or aspiration is smeared and/or sectioned, stained, and examined microscopically for the pathognomonic Leishman–Donovan bodies. Material should also be cultured in liquid media. The leishmanin skin test becomes positive early in the course of the disease and remains so for life. Recently, it has been demonstrated that small numbers of *Leishmania* may be detected in tissue by the polymerase chain reaction (PCR), and strains distinguished with probes to kinetoplast DNA. These techniques, though not widely available, permit direct, rapid, and specific diagnosis of all leishmanial infections.

Demonstration of Leishman–Donovan bodies or culture from tissue biopsy

Patients with small, cosmetically minor lesions that do not involve the mucous membrane may be carefully followed without treatment. Pentavalent antimonial agents and liposomal amphotericin B have proved to be effective chemotherapeutic agents for individuals with more consequential lesions. Recently, ketoconazole and itraconazole, alone or in combination with the previously mentioned agents, have been found to be effective in some forms of cutaneous leishmaniasis. Bacterial superinfections are treated with appropriate antibiotics. Prophylactic measures include the control of the sandfly vector by use of insect repellents and fine mesh screening on dwellings.

MUCOCUTANEOUS LEISHMANIASIS

EPIDEMIOLOGY

L braziliensis causes a natural infection in the large forest rodents of tropical Latin America. Sandflies transmit the infection to humans engaged in military activities, road builders, opening jungle areas for new settlements, and others.

Rodent reservoir of *L braziliensis*

MUCOCUTANEOUS LEISHMANIASIS

MANIFESTATIONS

A primary skin lesion similar to oriental sore develops 1 to 4 weeks after sandfly exposure. Occasionally, it undergoes spontaneous healing. More commonly, it progressively enlarges, often producing large vegetating lesions. After a period of weeks to years, painful, destructive, metastatic mucosal lesions of the mouth, nose, and occasionally the perineum, appear in 2% to 50% of patients. Sometimes, decades pass and the primary lesion totally resolves before the metastases manifest themselves. Destruction of the nasal septum produces the characteristic tapir nose. Erosion of the hard palate and larynx may render the patient aphonic. In blacks, the lesions are often large, hypertrophic, polypoid masses that deform the lips and cheeks. Fever, anemia, weight loss, and secondary bacterial infections are common. Mucosal lesions caused by other *Leishmania* species may be seen after visceral dissemination in AIDS patients.

Primary lesion metastasizes to oral and nasal areas

TREATMENT

The diagnosis of mucocutaneous leishmaniasis is made by finding the organisms in the lesions as described for localized cutaneous leishmaniasis. Because the propensity to metastasize to mucocutaneous sites is specific to certain species and subspecies, precise identification of the responsible organism as described in the introduction is of clinical importance. The leishmanin skin test yields positive results, and most patients have detectable antibodies. As described for cutaneous leishmaniasis, it is now possible to provide a rapid, direct, species-specific diagnosis through the use of the PCR and probes to kinetoplast DNA.

Treatment is accomplished with the agents described later in the chapter for kala azar. Advanced lesions are often refractory, and relapse is common. Cured patients are immune to reinfection. Control measures, other than insect repellents and screening of dwellings, are impractical because of the sylvatic nature of the disease.

Detection of organisms as with cutaneous leishmaniasis

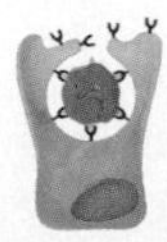

DISSEMINATED VISCERAL LEISHMANIASIS (KALA AZAR)

EPIDEMIOLOGY

Marked geographic differences in reservoirs and disease severity

Kala azar, which is caused by *L donovani,* occurs in the tropical and subtropical areas of every continent except Australia. Its epidemiologic and clinical patterns vary from area to area. In Africa, rodents serve as the primary reservoir. Human cases occur sporadically, and the disease is often acute and highly lethal. In Eurasia and Latin America, the domestic dog is the most common reservoir. Human disease is endemic, primarily involves children, and runs a subacute to chronic course. In India, the human is the only known reservoir, and transmission is carried out by anthropophilic species of sandflies. The disease recurs in epidemic form at 20-year intervals, when a new cadre of nonimmune children and young adults appears in the community. There appears to be a high incidence of visceral leishmaniasis in patients with HIV infection. Presumably, HIV-induced immunosuppression either facilitates acquisition of the disease and/or allows reactivation of latent infection.

PATHOGENESIS

Parasites invade macrophages of reticuloendothelial system

After the host is bitten by an infected sandfly, the parasites disseminate in the bloodstream and are taken up by the macrophages of the spleen, liver, bone marrow, lymph nodes, skin, and small intestine. Histiocytic proliferation in these organs produces enlargement with atrophy or replacement of the normal tissue.

KALA AZAR: CLINICAL ASPECTS

MANIFESTATIONS

Delayed onset, recurrent fever, chronic disease, diarrhea

Severe systemic manifestations

Immune complex glomerulonephritis

The majority of kala azar infections are asymptomatic; these become symptomatic years later during periods of host immunocompromise. Symptomatic disease most commonly manifests itself 3 to 12 months after acquisition of the parasite. It is often mild and self-limited. A minority of infected individuals develop the classic manifestations of kala azar. Fever, which is usually present, may be abrupt or gradual in onset. It persists for 2 to 8 weeks and then disappears, only to reappear at irregular intervals during the course of the disease. A double-quotidian pattern (two fever spikes in a single day) is a characteristic but uncommon finding. Diarrhea and malabsorption are common in Indian cases, resulting in progressive weight loss and weakness. Physical findings include enlarged lymph nodes and liver, massively enlarged spleen, and edema. In light-skinned persons, a grayish pigmentation of the face and hands is commonly seen, which gives the disease its name (kala azar, black disease). Anemia with resulting pallor and tachycardia are typical in advanced cases. Thrombocytopenia induces petechial formation and mucosal bleeding. The peripheral leukocyte count is usually less than 4000/mm^3; agranulocytosis with secondary bacterial infections contributes to lethality. Serum immunoglobulin G levels are enormously elevated, but play no protective role. Circulating antigen–antibody complexes are present and are probably responsible for the glomerulonephritis seen so often in this disease.

DIAGNOSIS AND TREATMENT

Demonstration of Leishman–Donovan bodies or culture

The diagnosis of kala azar is made by demonstrating the presence of the organism in aspirates taken from the bone marrow, liver, spleen, or lymph nodes. In the Indian form of kala azar, *L donovani* is also found in circulating monocytes. The specimens may be smeared, stained, and examined for the typical Leishman–Donovan bodies (amastigotes in mononuclear phagocytes) or cultured in artificial media and/or experimental animals. As described for cutaneous leishmaniasis, a limited number of reference laboratories can provide a rapid, direct, species-specific diagnosis through the use of the PCR and probes to kinetoplast DNA. Results of the leishmanin skin test are negative during active disease but become positive after successful therapy.

The mortality rate in untreated cases of kala azar is 75% to 90%. Treatment with pentavalent antimonial drugs lowers this rate dramatically. Initial therapy, however, fails in up to 30% of African cases, and 15% of those that do respond eventually relapse. Resistant cases are treated with the more toxic pentamidine, amphotericin B, or liposomal amphotericin B. Allopurinol and interferon-γ have proved to be useful adjunctive therapies in resistant cases. Control measures are directed at the *Phlebotomus* vector, with the use of residual insecticides, and at the elimination of mammalian reservoirs by treating human cases and destroying infective dogs.

Up to 90% mortality rate without treatment

African *Trypanosoma*

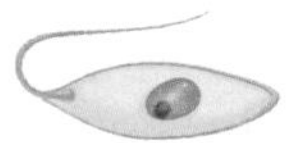

PARASITOLOGY

The trypanosomes that produce these diseases are morphologically and serologically identical. Accordingly, they are considered varieties of a single species, *Trypanosoma brucei.* The three subspecies, known as *T brucei gambiense, T brucei rhodesiense,* and *T brucei brucei,* can be distinguished by their biologic characteristics, zymodeme types, mitochondrial morphology, and DNA hybridization patterns. All undergo similar developmental changes in the course of their passage between their insect and mammalian host. On ingestion by the tsetse fly (*Glossina* spp.) and after a period of multiplication in the midgut, the parasites migrate to the insect's salivary glands and assume the epimastigote form. After a period of weeks, they are transformed into metacyclic trypomastigotes, rendering them infectious to mammals. When the fly again takes a meal, the parasites are inoculated with the fly's saliva. In the mammalian host, they acquire a highly variable surface glycoprotein (VSG), multiply extracellularly, and eventually invade the bloodstream. During the initial stages of parasitemia, some trypomastigotes elongate to become graceful, slender organisms 30 μm or more in length and divide every 5 to 10 hours. For reasons apparently independent of the host's immune response, multiplication eventually slows. Some forms lose their flagella and assume a short, stumpy appearance. The latter forms have more developed mitochondria and are thought to be particularly infective to the insect host. Near the end of the episode of parasitemia, both morphologic types may be seen in a single blood specimen. Individual strains of *T brucei* can change the antigenic character of their glycoprotein coat in a sequential and at times predictable fashion. A single strain is capable of producing dozens, perhaps hundreds, of these variable antigen types, each of which is encoded in its own structural gene. The genetic repertoire seems to be strain-specific. Expression of individual genes appears to be controlled by the sequential duplication and subsequent transfer of each gene (expression-linked copy) to one or more areas of the genome responsible for gene expression.

Three recognized subspecies of *T brucei*

Epimastigote and trypomastigote forms develop in tsetse fly

Infectious trypomastigote form injected into the bloodstream of mammalian host from fly's saliva

Antigenic variation of glycoprotein coat of trypomastigotes is due to shifting expression of preexisting genes

AFRICAN TRYPANOSOMIASIS (SLEEPING SICKNESS)

CLINICAL CAPSULE

African trypanosomiasis is a highly lethal meningoencephalitis transmitted to humans by bloodsucking flies of the genus *Glossina*. It occurs in two distinct clinical and epidemiologic forms: West African or Gambian sleeping sickness and East African or Rhodesian sleeping sickness. Nagana, a disease of cattle caused by a closely related trypanosome, renders over 10 million square kilometers of Central Africa unsuitable for animal husbandry.

EPIDEMIOLOGY

Tsetse fly confined to Central Africa

The tsetse fly, and consequently sleeping sickness, is confined to the central area of Africa between that continent's two great deserts, the Sahara in the north and the Kalahari in the south. Approximately 50 million people live in this area, and 10,000 to 20,000 acquire sleeping sickness annually. Major outbreaks have been reported in several locations within the endemic area over the past two decades, partly as a result of the internecine wars in this area that have interrupted control programs. Although an estimated 20,000 Americans travel to endemic areas each year, less than two dozen cases of African trypanosomiasis have been diagnosed in Americans since 1967.

Humans major reservoir of West African sleeping sickness; chronicity ensures maintenance

Riverine tsetse flies found in the forest galleries that border the streams of West and Central Africa serve as the vectors of the Gambian disease. Although these flies are not exclusively anthropophilic, humans are thought to be the major reservoir of the parasite. The infection rate in humans is affected by proximity to water but seldom exceeds 2% to 3% in nonepidemic situations. Nevertheless, the extreme chronicity of the human disease ensures its continued transmission.

Savanna antelopes are reservoirs of East African trypanosomiasis; humans infected incidentally

Rhodesian sleeping sickness, in contrast, is transmitted by flies indigenous to the great savannas of East Africa that feed on the blood of the small antelope inhabiting these areas. The antelope serves as the major parasite reservoir, although human-to-human and cattle-to-human spread has been documented. Humans typically become infected only when they enter the savanna to hunt or to graze their domestic animals. Currently, Sudan is the only country where both the Gambian and Rhodesian forms of sleeping sickness are still found. At present, there is little evidence of coinfections with African trypanosomes and HIV, possibly because the former is primarily rural in distribution and the latter is concentrated in cities.

PATHOGENESIS

Local chancre at site of inoculation and lymphadenitis

Intermittent parasitemia with antigenic shifts

Parasites localize in blood vessels of heart and CNS with local vasculitis

Multiplication of the trypomastigotes at the inoculation site produces a localized inflammatory lesion. After the development of this chancre, organisms spread through lymphatic channels to the bloodstream, inducing a proliferative enlargement of the lymph nodes. The subsequent parasitemia is typically low grade and recurrent. As host antibodies (predominantly IgM) are produced to the surface antigen characteristic of a particular parasitemic wave, they bind to the organism, leading to its destruction by lysis and opsonization. The trypomastigotes disappear from the blood, reappearing 3 to 8 days later as new antigenic variants arise. The recurrences gradually become less regular and frequent, but may persist for weeks to years before finally disappearing. During the course of the parasitemia, trypanosomes localize in the small blood vessels of the heart and central nervous system (CNS). This localization results in endothelial proliferation and a perivascular infiltration of plasma cells and lymphocytes. In the brain, hemorrhage and a demyelinating panencephalitis may follow.

High levels of IgM include specific and nonspecific antibodies

Immune complexes may cause anemia and vasculitis

The mechanism by which the trypanosomes elicit vasculitis is uncertain. The infection stimulates a massive, nonspecific polyclonal activation of B cells, the production of large quantities of immunoglobulin M (typically 8 to 16 times the normal limit), and the suppression of other immune responses. Most of this reaction represents specific protective antibodies that are ultimately responsible for the control of the parasitemia. Some, however, consist of nonspecific heterophile antibodies, antibodies to DNA, and rheumatoid factor. Antibody-induced destruction of trypanosomes releases invariant nuclear and cytoplasmic antigens with the production of circulating immune complexes. Many authorities believe that these complexes are largely responsible for the anemia and vasculitis seen in this disease.

MANIFESTATIONS

Raised red papule on exposed surface

The trypanosomal chancre appears 2 to 3 days after the bite of the tsetse fly as a raised, reddened nodule on one of the exposed surfaces of the body. With the onset of parasitemia 2 to 3 weeks later, the patient develops recurrent bouts of fever, tender lymphadenopathy, skin rash, headache, and impaired mentation. In the Rhodesian form of disease, myocarditis and CNS involvement begin within 3 to 6 weeks. Heart failure, convulsions, coma, and death follow in 6 to 9 months. Gambian sleeping sickness progresses more slowly. Bouts of fever often persist for years before CNS manifestations gradually appear. Spontaneous activity progressively diminishes, attention wavers, and the patient must be prodded to eat or talk. Speech grows indistinct, tremors develop, sphincter control is lost, and seizures with transient bouts of paralysis occur. In the terminal stage, the patient develops a lethal intercurrent infection or lapses into a final coma.

Parasitemic manifestations 2 to 3 weeks later

Late CNS involvement

DIAGNOSIS

A definitive diagnosis is made by microscopically examining lymph node aspirates, blood, or cerebrospinal fluid for the presence of trypomastigotes (**Figure 52–6**). Early in the disease, actively motile organisms can often be seen in a simple wet mount preparation smear; identification requires examination of an appropriately stained smear. If these tests prove negative, the blood can be centrifuged and the stained buffy coat examined. Inoculation of rats or mice can also prove helpful in diagnosing the Rhodesian disease. The patient may also be screened for elevated levels of IgM in the blood and spinal fluid or specific trypanosomal antibodies by a variety of techniques. A simple card agglutination test, which can be performed on fingerstick blood, can provide serologic confirmation within minutes. Subspecies-specific DNA probes may eventually prove useful for the identification of organisms in clinical specimens.

Trypomastigotes sought in lymph node aspirates, blood, and cerebrospinal fluid

Animal inoculation may be required in Rhodesian disease

TREATMENT

Lumbar puncture must always be performed before initiation of therapy for sleeping sickness. If the specimen reveals evidence of CNS involvement, agents that penetrate the blood–brain barrier must be included. Unfortunately, the most effective agent of this type is a highly toxic arsenical, melarsoprol (Mel B). Although this agent occasionally produces a lethal hemorrhagic encephalopathy, the invariably fatal outcome of untreated CNS disease warrants its use. The ornithine decarboxylase inhibitor, eflornithine appears capable, when used alone, or together with suramin, of curing CNS disease caused by *T brucei gambiense* without the serious side effects associated with melarsoprol. Unfortunately, it is very expensive and is only variably effective in *T brucei rhodesiense* infections. If the CNS is not yet involved, less toxic agents, such as suramin, pentamidine, or eflornithine, can be used. In such cases, the cure rate is high and recovery complete.

Selection of drugs dependent on whether CNS is involved

Without CNS involvement, recovery often complete

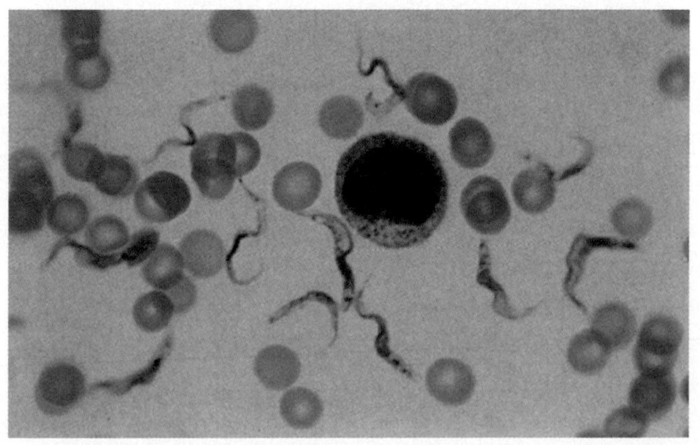

FIGURE 52–6. African sleeping sickness. *Trypanosoma brucei* in a routine blood smear. (Reproduced with permission from Nester EW, Anderson DG, Roberts CE Jr, Nester MT. *Microbiology: A Human Perspective*, 6th ed. New York: McGraw-Hill, 2008.)

PREVENTION

Neither vector or reservoir control has been successful

Although a variety of tsetse fly control measures, including the use of insecticides, deforestation, and the introduction of sterile males into the fly population, have been attempted, none has proved totally practicable. Similarly, eradication of disease reservoirs by the early detection and treatment of human cases and the destruction of wild game has had limited success. Attempts to develop effective vaccines are currently underway but are complicated by the antigenic variability of most trypomastigotes. A degree of personal protection can be achieved with insect repellents and protective clothing. Although prophylactic use of pentamidine was once advocated, enthusiasm for this treatment has waned.

American *Trypanosoma*

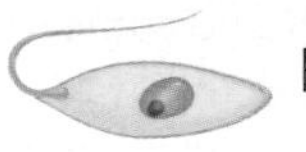

PARASITOLOGY

Mammalian cycle with nondividing extracellular trypomastigotes and dividing intracellular amastigotes

Invertebrate cycle produces trypomastigotes in bug

Reduviid bug may remain infectious for up to 2 years

The trypomastigotes of *Trypanosoma cruzi* closely resemble those of *T brucei,* and like them, disseminate from the site of inoculation to circulate in the peripheral blood of their mammalian hosts. Their developmental cycle, however, differs in several respects. Most significant, *T cruzi* does not multiply extracellularly. The circulating trypomastigotes must invade tissue cells, lose their flagella, and assume the amastigote form before binary fission can occur. Continued multiplication leads to distention and eventual rupture of the tissue cell. Released parasites revert to trypomastigotes and regain the bloodstream. This new generation of trypomastigotes may invade other host cells, thus continuing the mammalian cycle. Alternatively, they may be ingested by a feeding reduviid and develop into epimastigotes within its midgut. On completion of the invertebrate cycle, the parasites migrate to the hindgut and are discharged as infectious trypomastigotes when the reduviid defecates in the process of taking another blood meal. This process can recur at each feeding for as long as 2 years. Infection in the new host is initiated when the trypomastigotes contaminate either the feeding site or the mucous membranes.

T cruzi comprises a number of strains, each with its own distinct geographic distribution, tissue preference, and virulence. These strains may be distinguished from one another with specific antisera and by differences in their isoenzyme and DNA restriction patterns. All are morphologically identical. In blood specimens, the trypomastigotes can be distinguished from those of *T brucei* by their characteristic C or U shape, narrow undulating membrane, and large kinetoplast.

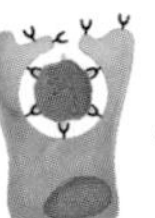

AMERICAN TRYPANOSOMIASIS (CHAGAS DISEASE)

CLINICAL CAPSULE

American trypanosomiasis is a disease produced by *T cruzi* and transmitted by true bugs of the family Reduviidae. Clinically, the infection presents as an acute febrile illness in children and a chronic heart or gastrointestinal malady in adults.

EPIDEMIOLOGY

Chagas disease affects 16 to 18 million people in a geographic area extending from Central America to southern Argentina, producing death in 50,000 annually. Within these areas, it is the leading cause of heart disease, accounting for 25% of all deaths in the 25- to 44-year age group. Transmission occurs primarily in rural settings, where the reduviid can find harborage in animal burrows and in the cracked walls and thatch of poorly constructed buildings. This large (3-cm) winged insect leaves its hiding place at night to feed on its sleeping hosts. Its predilection to bite near the eyes or lips have earned this pest the nicknames of "kissing bug" and "assassin bug." Most new infections in these areas occur in children. Infection can also be acquired in utero and, less frequently, through breastfeeding.

Chagas' disease in South and Central America

"Kissing bug" feeds at night in rural areas

In addition to humans, a number of wild and domestic animals, including rats, cats, dogs, opossums, and armadillos, serve as reservoirs for Chagas disease. The close association of many of these hosts with human dwellings tends to amplify the incidence of disease in humans and the difficulty involved in its control.

Other wild and domestic animal reservoirs amplify transmission

Organ transplantation and transfusion-related infections are rapidly increasing problems in urban settings within endemic areas. Recrudescence of the latent infection is increasingly seen in immunosuppressed individuals, including patients with HIV infections. More effective blood bank screening provides hope that transmission of this disease will be substantially curtailed in the near future.

An estimated 50,000 infected Latin American immigrants are currently living in the United States. Because *T cruzi* has been found in both vertebrate and invertebrate hosts in the southwestern United States, there is a possibility of sustained transmission of this organism within this country. Although serologic evidence suggests that the acquisition of human infection in this area is not uncommon, clinically apparent autochthonous cases have been rare. The majority of these acquired the infection through blood–blood transfusions.

PATHOGENESIS

Multiplication of the parasite at the portal of entry stimulates the accumulation of neutrophils, lymphocytes, and tissue fluid, resulting in the formation of a local chancre or chagoma. The subsequent dissemination of the organism with invasion of tissue cells produces a febrile illness that may persist for 1 to 3 months and result in widespread organ damage. Any nucleated host cell may be involved, but those of mesenchymal origin, especially the heart, skeletal muscle, smooth muscle, and ganglion neural cells, are particularly susceptible. Cell entry is facilitated by binding to host cell fibronectin; a 60-kd *T cruzi* surface protein (penetrin) appears to promote adhesion. After penetration, the trypomastigote escapes the phagolysosome via the production of a pore-forming protein, transforms to the amastigote form, and multiplies freely within the cytoplasm to produce a pseudocyst, a greatly enlarged and distorted host cell containing masses of organisms (**Figure 52–7**).

Local chancre at site of inoculation

Entry to mesenchymal cells facilitated by fibronectin-binding surface protein

Pore-forming protein aids escape from phagolysosome

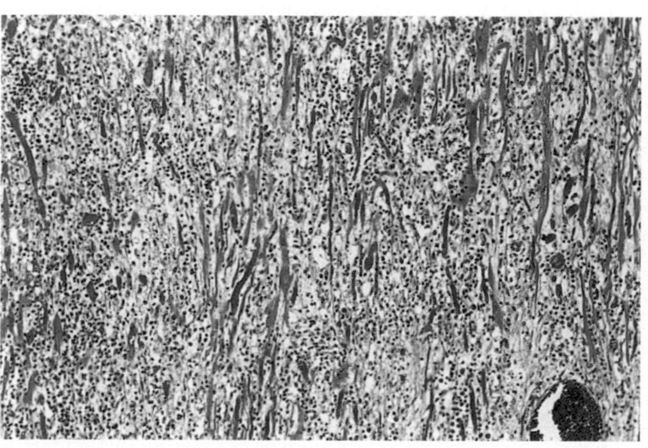
A

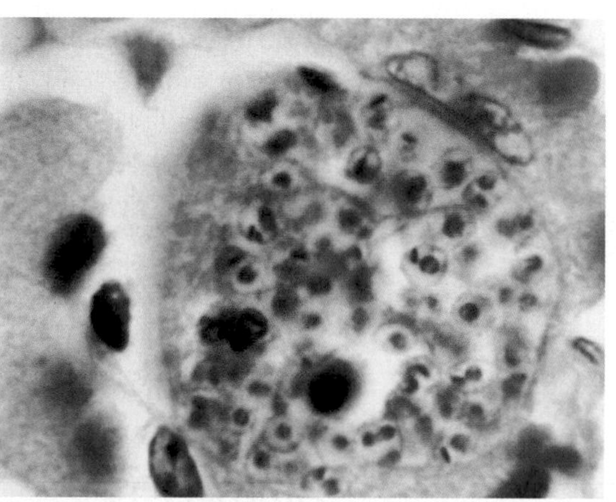
B

FIGURE 52–7. Chagas disease. A. Acute myocarditis with atrophic myofibers separated by inflammatory cells. **B.** *Trypanosoma cruzi* amastigotes clustered in myofiber from the same case. (Reproduced with permission from Connor DH, Chandler FW, Schwartz DQA, Manz HJ, Lack EE (eds). *Pathology of Infectious Diseases*, vol. 1. Stamford, CT: Appleton & Lange; 1997.)

Pseudocysts formed from cytoplasmic multiplication in host cells

With the rupture of the pseudocyst, many of the released parasites disintegrate, eliciting an intense inflammatory reaction with destruction of surrounding tissue. The development of an antibody-dependent, cell-mediated immune response leads to the eventual destruction of the *T cruzi* parasites and the termination of the acute phase of illness.

Damage to heart may have immune mechanism

Ganglionic and smooth muscle cells lost in digestive tract

Parasitic antigens released during this acute phase may bind to the surface of tissue cells, rendering them susceptible to destruction by the host's immune response. It has been suggested by some that this results in the production of antibodies that cross-react with host tissue, initiating a sustained autoimmune inflammatory reaction in the absence of systemic manifestation of illness. In the heart, this reaction leads to changes in coronary microvasculature, loss of muscle tissue, interstitial fibrosis, degenerative changes in the myocardial conduction system, and loss of intracardiac ganglia. In the digestive tract, loss of both ganglionic nerve cells and smooth muscle results in dilatation and loss of peristaltic movement, particularly of the esophagus and colon.

AMERICAN TRYPANOSOMIASIS (CHAGAS DISEASE): CLINICAL ASPECTS

MANIFESTATIONS

Most infections asymptomatic; acute disease usually in children

Myocardial injury indicated by tachycardia and electrocardiographic changes

Serologic studies suggest that only one third of persons newly infected with Chagas' disease develop clinical illness. Acute manifestations, when they occur, are seen primarily in children. They begin with the appearance of the nodular, erythematous chagoma 1 to 3 weeks after the bite of the reduviid. If the eye served as a portal of entry, the patient presents with Romaña's sign: reddened eye, swollen lid, and enlarged preauricular lymph node. The onset of parasitemia is signaled by the development of a sustained fever; enlargement of the liver, spleen, and lymph nodes; signs of meningeal irritation; and the appearance of peripheral edema or a transient skin rash. In a small percentage of symptomatic patients, heart involvement results in tachycardia, electrocardiographic changes, and occasionally arrhythmia, enlargement, and congestive heart failure. Newborns may experience acute meningoencephalitis. Clinical manifestations persist for weeks to months. In 5% to 10% of untreated patients, severe myocardial involvement or meningoencephalitis leads to death.

Chronic cardiomyopathy in adults leads to heart block and/or congestive heart failure

Dilatation of esophagus and colon seen in southern latitudes

Chronic disease, the result of end-stage organ damage, is usually seen only in adulthood. Ironically, most patients with late manifestations have no history of acute illness. The most serious of the late manifestations is heart disease. Studies of asymptomatic, seropositive patients in endemic areas have shown that a significant proportion have cardiac abnormalities demonstrated by electrocardiographic, echocardiographic, or cineangiographic techniques, suggesting that Chagas' cardiomyopathy is a progressive, focal disease of the myocardium and conduction system, leading eventually to clinical disease. This may present as arrhythmia, thromboembolic events, heart block, enlargement with congestive heart failure, and cardiac arrest. In some areas of rural Latin America, up to 10% of the adult population may show cardiac manifestations. In the United States, chagasic heart disease in immigrants is usually initially misdiagnosed as coronary artery disease or idiopathic dilated cardiomyopathy. Megaesophagus and megacolon, which are less devastating than the heart disease, are typically seen in more southern latitudes. This geographic variation in clinical manifestations is thought to be attributable to a difference in tissue tropism between individual strains of *T cruzi*. Megaesophagus leads to difficulty in swallowing and regurgitation, particularly at night. Megacolon produces severe constipation with irregular passage of voluminous stools. *T cruzi* brain abscess has been described in a small number of AIDS patients.

DIAGNOSIS

The diagnosis of acute Chagas' disease rests on finding the trypomastigotes in the peripheral blood or buffy coat, and their morphologic identification as *T cruzi.* The methods are similar to those described for diagnosis of African trypanosomiasis. If the results are negative, a laboratory-raised reduviid can be fed on the patient, then dissected and examined for the presence of parasites, a procedure known as **xenodiagnosis**. Alternatively, the blood may be cultured in a variety of artificial media or experimental animals. In the diagnosis of chronic disease, recovery of the organisms is the exception rather than the rule, and diagnosis depends on the clinical, epidemiologic, and immunodiagnostic findings. A variety of serologic tests are available; small numbers of false-positive results limit their usefulness, particularly when used as screening procedures in nonendemic areas. The recent production of specific recombinant proteins and synthetic peptides for use as antibody targets may improve the reliability of these procedures. PCR techniques for the amplification of trypomastigote DNA are available in a small number of research laboratories.

Demonstration of trypomastigotes in peripheral blood

Xenodiagnosis involves allowing bugs to feed

Organisms difficult to recover in chronic disease

TREATMENT

The role of treatment in Chagas' disease remains unsettled. Two agents, nifurtimox and benznidazole, effectively reduce the severity of acute disease but appear to be ineffective in chronic infections. Both drugs must be taken for prolonged periods of time, may cause serious side effects, and do not always result in parasitologic cure. Allopurinol, a hypoxanthine oxidase inhibitor devoid of serious side effects, has recently been shown to be capable of suppressing parasitemia and reversing the serostatus of patients with acute disease. Additional studies to confirm these encouraging results are necessary.

Treatment may reduce acute disease

PREVENTION

The reduviid vector can be controlled by applying residual insecticides to rural buildings at 2- or 3-month intervals. The addition of latex to the insecticide creates a colorless paint that prolongs activity. Fumigants can be used to prevent reinfection. Patching wall cracks, cementing floors, and moving debris and woodpiles away from human dwellings reduces the number of reduviids within the home. Transfusion-induced disease, a major problem in endemic areas, has been partially controlled by the addition of gentian violet to all blood packs before use or by screening potential donors serologically for Chagas' disease. The large number of infected immigrants now entering nonendemic countries presents an increasing risk of transfusion-mediated parasite transmission in these areas as well. Cases of acute Chagas' disease have been reported in the United States in immunosuppressed patients who received blood from donors unaware of their infection status; the resulting diseases were particularly fulminant. Immunodiagnostic tests for Chagas' disease are neither readily available nor sufficiently specific for use in nonendemic areas; prevention will probably require deferral of blood donations from persons who have recently emigrated from endemic areas. Immunoprophylaxis is not available at present.

Control of reduviid bugs in rural homes most important measure

CASE STUDY

A CHILD WITH RECURRENT FEVER AND DIARRHEA

This 3-year-old girl who resides in Central Africa has had recurrent fevers for the last 6 weeks, accompanied by persistent diarrhea and weight loss. A physical examination reveals her to be alert, but with significant generalized weakness, widespread lymphadenopathy, hepatomegaly and massive splenomegaly.

Laboratory findings include anemia, leukopenia, thrombocytopenia, and hematuria.

QUESTIONS

■ Which is the most likely cause of this child's illness?

A. *Leishmania donovani*

B. *Leishmania tropica*

C. *Trypanosoma cruzi*

D. *Trypanosoma brucei*

■ Which is the insect vector involved?

A. Mosquito

B. Tsetse fly

C. Sandfly

D. Reduviid bug

■ *T cruzi* can significantly affect all of the following tissues, except:

A. Heart

B. Smooth muscle

C. Skin

D. Skeletal muscle

E. Neural tissue

ANSWERS

1(A), **2**(C), **3**(C)

CHAPTER 53

Intestinal Nematodes

The intestinal nematodes have cylindric, fusiform bodies covered with a tough, acellular cuticle. Sandwiched between this integument and the body cavity are layers of muscle, longitudinal nerve trunks, and an excretory system. A tubular alimentary tract consisting of a mouth, esophagus, midgut, and anus runs from the anterior to the posterior extremity. Highly developed reproductive organs fill the remainder of the body cavity. The sexes are separate; the male worm is generally smaller than its mate. The female, which is extremely prolific, can produce thousands of offspring, generally in the form of eggs. Typically, the eggs must incubate or embryonate outside of the human host before they become infectious to another person; during this time, the embryo repeatedly segments, eventually developing into an adolescent form known as a **larva**. In some species of nematodes, offspring develop to the larval stage in the uterus of the worm. The duration and site of embryonation differ with each worm species and determine how it will be transmitted to the new host. In many cases, eggs of nematodes that dwell within the human gastrointestinal tract are carried to the environment in the feces and embryonate on the soil for a period of weeks before becoming infectious. The egg may then be ingested with contaminated food. In some species, the egg hatches outside of the host, releasing a larva capable of penetrating the skin of a person who comes in direct physical contact with it. Obviously, intestinal nematodes are principally found in areas where human feces are deposited indiscriminately or used for fertilizer.

Six intestinal nematodes commonly infect humans: *Enterobius vermicularis* (pinworm), *Trichuris trichiura* (whipworm), *Ascaris lumbricoides* (large roundworm), *Necator americanus* and *Ancylostoma duodenale* (hookworms), and *Strongyloides stercoralis*. Together they infect more than 25% of the human race, producing embarrassment, discomfort, malnutrition, anemia, and occasionally death. Other closely related nematodes of animals that may occasionally infect humans are also listed in **Table 53–1**, but are discussed here.

The adults of each of the six nematodes listed previously can survive for months or years within the lumen of the gut. The severity of illness produced by each depends on the level of adaptation to the host it has achieved. Some species have a simple life cycle that can be completed without serious consequences to the host. Less well-adapted parasites, on the other hand, have more complex cycles, often requiring tissue invasion and/or production of enormous numbers of offspring to ensure their continued survival and dissemination. Within a given species, disease severity is related directly to the number of adult worms harbored by the host. The greater the worm load or worm burden, the more serious the consequences. Because nematodes do not multiply within the human, small worm loads may remain asymptomatic and undetected throughout the lifespan of the parasite. Repeated infections, however, progressively increase the worm burden and at some point induce symptomatic disease. Although humans can mount an immune response that eventually leads to the expulsion of worms, it is slow to develop and incomplete. It is therefore the frequency and intensity of reinfection more than the host's immune response that determine the worm burden. This burden is seldom uniform within affected populations, but rather "aggregated" within subgroups related to their hygienic practices.

Long survival in gut lumen

Worm load and repeated infection important to disease severity

TABLE 53–1 Intestinal Nematodes

HUMAN PARASITE	ANIMAL PARASITE	HUMAN DISEASE
Enterobius vermicularis (pinworm)		Enterobiasis
Trichuris trichiura (whipworm)		Trichuriasis
	Capillaria philippinensis	Intestinal capillariasis
Ascaris lumbricoides (large roundworm)		Ascariasis
	Ascaris suum	Ascariasis
	Anisakis spp.	Anisakiasis
	Toxocara canis	Toxocariasis (visceral larva migrans)
	Toxocara cati	
Necator americanus (hookworm)		Hookworm disease
Ancylostoma duodenale (hookworm)	*Ancylostoma braziliense*	Cutaneous larva migrans
Strongyloides stercoralis		Strongyloidiasis

LIFE CYCLES

Enterobius vermicularis is the best adapted intestinal nematode

The life cycles of the intestinal nematodes are summarized in **Table 53–2**. *E vermicularis* (pinworm), the best adapted of the intestinal nematodes, has the simplest life cycle. It feeds, grows, and copulates within the gut of its host before transiting the anus to deposit its eggs on the perineal skin. The eggs embryonate within hours and are subsequently transported to the same, or a new, host via fingers or dust. After their inhalation or ingestion, the eggs are swallowed and hatch in the bowel lumen, completing the cycle. The only significant difference between this and the life cycle of *T trichiura* (whipworm) is that the eggs of the latter are passed in the stool and must incubate on soil before becoming infectious. This relatively minor difference has profound epidemiologic ramifications because *Trichuris* can be passed only in populations that practice indiscriminate defecation and live in climates suitable for the maturation of eggs in the soil.

TABLE 53–2 Life Cycles of Intestinal Nematodes

PARASITE	ROUTE OF INFECTION	MIGRATION IN BODY	DIAGNOSTIC FORM	SITE OF EMBRYONATION	INFECTIVE FORM	FREE-LIVING CYCLE
Enterobius vermicularis	Mouth	Intestinal	Egg	Perineum	Egg	No
Trichuris trichiura	Mouth	Intestinal	Egg	Soil	Egg	No
Ascaris lumbricoides	Mouth	Pulmonary	Egg	Soil	Egg	No
Necator americanus[a]	Skin	Pulmonary	Egg	Soil	Filariform larvae	No
Stronglyoides stercoralis	Skin	Pulmonary	Rhabditiform larvae	Soil; intestine[b]	Filariform larvae	Yes

Reproduced with permission from Plorde JJ. In Isselbacher KJ, et al. *Harrison's Principles of Internal Medicine*, 9th ed. New York, McGraw-Hill, 1980, Table 206–3, p. 891.

[a]Also *Ancylostoma duodenale*.

[b]Intestine only in cases of autoinfection.

A lumbricoides is transmitted in a manner similar to *T trichiura.* However, after hatching from the egg in the gut lumen, ascarid larvae penetrate the bowel wall and migrate through the host's liver and lung before returning older and more sedentary to the protective environment of the gut lumen. This maladaptive sojourn of juvenile worms through the host tissue is also seen in the life cycles of the hookworms and *S stercoralis.* In contrast to *Ascaris,* however, the eggs of the latter two nematodes hatch shortly before or after they are passed in the stool of the original host, resulting in the seeding of the external environment with larval forms capable of penetrating human skin. Transmission is effected when a new host comes into physical contact with the contaminated soil. The adaptation of *S stercoralis* is the least satisfactory of the intestinal nematodes and, in an evolutionary sense, appears to have occurred quite recently. In addition to the hookworm-like cycle described above, *S stercoralis* has the twin capacities to complete its life cycle entirely within the body of the host or to survive in the external environment as a free-living soil organism.

Other nematodes have increasingly complex life cycles

S stercoralis is least well adapted

PARASITES AND DISEASES

Enterobius

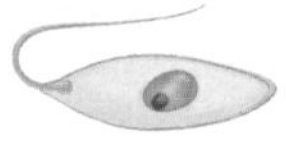

ENTEROBIUS VERMICULARIS (PINWORM): PARASITOLOGY

The adult female pinworm is 10-mm long, cream-colored, with a sharply pointed tail; such characteristics have given rise to the common name pinworm. Running longitudinally down both sides of the body are small ridges that widen anteriorly to fin-like alae. The seldom-seen male is smaller (3 mm) and possesses a ventrally curved tail and copulatory spicule. The clear, thin-shelled, ovoid eggs are flattened on one side and measure 25 by 50 μm (**Figure 53–1**).

Common name is pinworm

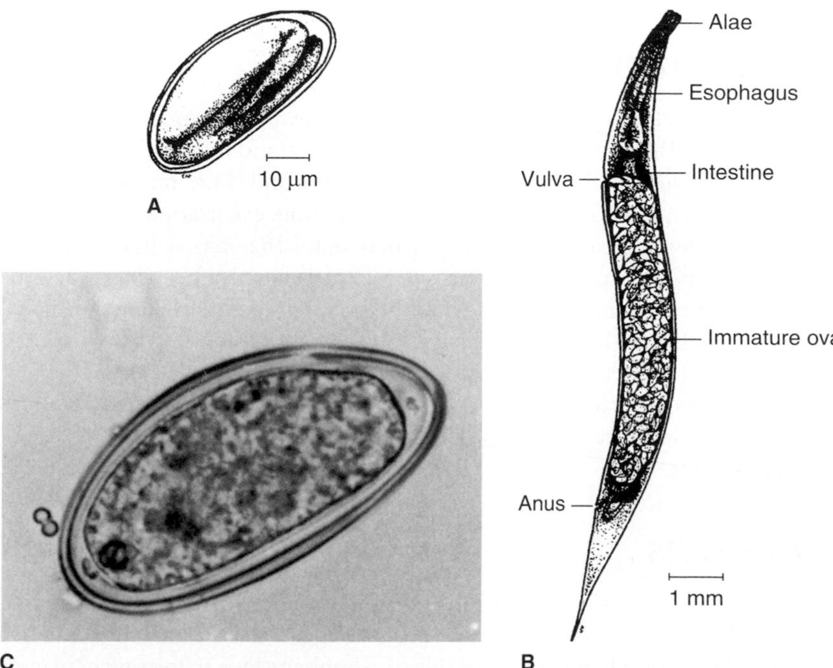

FIGURE 53–1. ***Enterobius vermicularis.*** **A.** Egg structure. **B.** Structure of adult female pinworm. **C.** Embryonated egg recovered from stool. (C, Reproduced with permission from Connor DH, Chandler FW, Schwartz DQA, Manz HJ, Lack EE (eds). *Pathology of Infectious Diseases,* vol. 1. Stamford, CT: Appleton & Lange; 1997.)

LIFE CYCLE

Adults inhabit cecum

Female transits anus at night to deposit eggs on perineum

Eggs infectious to host and others shortly after deposition

Ingested eggs hatch and larvae mature to adults in intestine

The adult worms lie attached to the mucosa of the cecum. As its period of gravidity draws to a close, the female migrates down the colon, slips unobserved through the anal canal in the dark of the night, and deposits as many as 20,000 sticky eggs on the host's perianal skin, bedclothes, and linens. The eggs are near maturity at the time of deposition and become infectious shortly thereafter. Handling of bedclothes or scratching of the perianal area to relieve the associated itching results in adhesion of the eggs to the fingers and subsequent transfer to the oral cavity during eating or other finger–mouth maneuvers. Alternatively, the eggs may be shaken into the air (eg, during making of the bed), inhaled, and swallowed. The eggs subsequently hatch in the upper intestine, and the larvae migrate to the cecum, maturing to adults and mating in the process. The entire adult-to-adult cycle is completed in 2 weeks.

EPIDEMIOLOGY

Infects 30 to 40 million in United States

The pinworm is the oldest and most widespread of the helminths. Eggs have been found in a 10,000-year-old coprolith, making this nematode the oldest demonstrated infectious agent of humans. It has been estimated to infect at least 200 million people, particularly children, worldwide, and 40 million in the United States alone. Despite evidence that its prevalence is now decreasing in the United States, in both the United States and in western Europe it remains the single most common cause of human helminthiasis. Infection is more common among the young and poor, but may be found in any age or economic class.

Resistant infective eggs

The eggs are relatively resistant to desiccation and may remain viable in linens, bedclothes, or house dust for several days. Once infection is introduced into a household, other family members are rapidly infected.

PATHOGENESIS AND IMMUNITY

The adult worms produce no significant intestinal pathology and do not appear to induce protective immunity.

MANIFESTATIONS

Nocturnal pruritus ani

Occasional infection of female genitourinary tract

E vermicularis seldom produces serious disease. The most common symptom is pruritus ani (anal itching). This symptom is most severe at night and has been attributed to the migration of the gravid female. It may lead to irritability and other minor complaints. In severe infections, the intense itching may lead to scratching, excoriation, and secondary bacterial infection. In female patients, the worm may enter the genital tract, producing vaginitis, granulomatous endometritis, or even salpingitis. It has also been suggested that migrating worms might carry enteric bacteria into the urinary bladder in young women, inducing an acute bacterial infection of the urinary tract. Although this worm is frequently found in the lumen of the resected appendix, it is doubtful that it plays a causal role in appendicitis. Perhaps the most serious effect of this common infection is the psychic trauma suffered by the economically advantaged when they discover that they, too, are subject to intestinal worm infection.

DIAGNOSIS

Eosinophilia is usually absent. The diagnosis is suggested by the clinical manifestations and confirmed by the recovery of the characteristic eggs from the anal mucosa. Identification is accomplished by applying the sticky side of cellophane tape to the mucocutaneous junction, then transferring the tape to a glass slide and examining the slide under the low-power lens of a microscope. Occasionally, the adult female is seen by a parent of an infected child or recovered with the cellophane tape procedure.

Anal cellophane tape test detects ova

TREATMENT AND PREVENTION

Several highly satisfactory agents, including pyrantel pamoate and mebendazole, are available for treatment for enterobiasis. Many authorities believe that all members of a family or other cohabiting group should be treated simultaneously. In severe infections, retreatment after 2 weeks is recommended. Although cure rates are high, reinfection is extremely common. It need not be treated in the absence of symptoms.

All family members may need treatment

Reinfection common

Trichuris

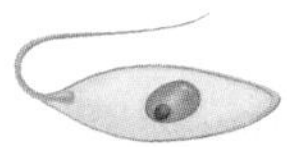

TRICHURIS TRICHIURA (WHIPWORM): PARASITOLOGY

The adult whipworm is 30 to 50 mm in length. The anterior two thirds is thin and threadlike, whereas the posterior end is bulbous, giving the worm the appearance of a tiny whip. The tail of the male is coiled; that of the female is straight. The female produces 3000 to 10,000 oval eggs each day. They are of the same size as pinworm eggs, but have a distinctive thick brown shell with translucent knobs on both ends (**Figure 53–2**).

Whipworm produces up to 10,000 eggs a day

LIFE CYCLE

Trichuris trichiura has a life cycle that differs from that of the pinworm only in its external phase. The adults live attached to the colonic mucosa by their thin anterior end. While retaining its position in the cecum, the gravid female releases its eggs into the lumen of the gut. These

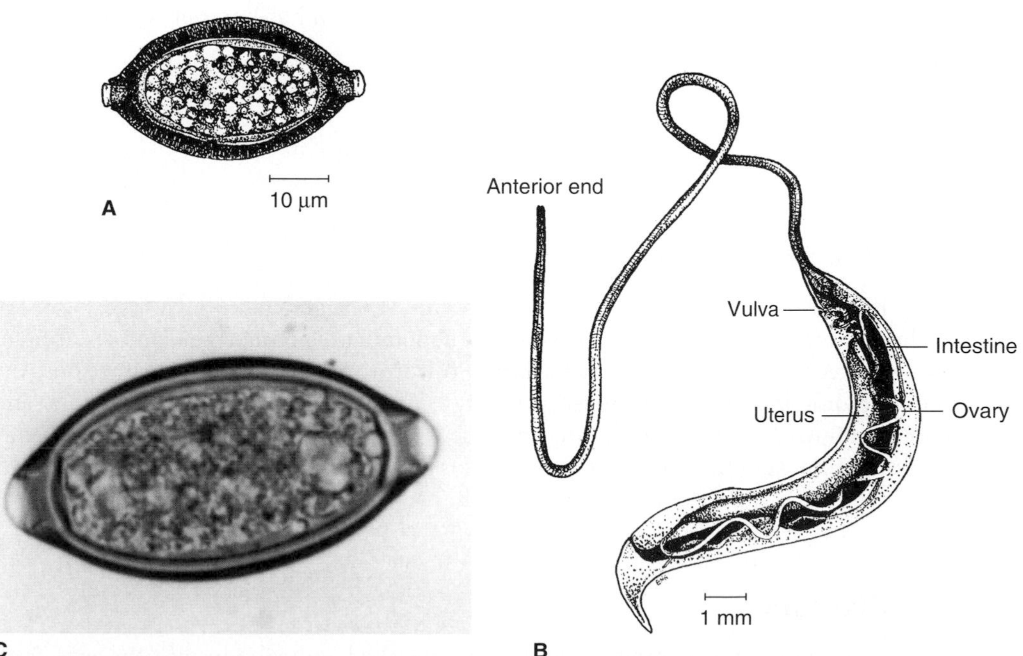

FIGURE 53–2. *Trichuris trichiura*. A. Egg structure. **B.** Structure of female adult whipworm. **C.** Embryonated egg with bipolar plugs from stool. (C, Reproduced with permission from Connor DH, Chandler FW, Schwartz DQA, Manz HJ, Lack EE (eds). *Pathology of Infectious Diseases*, vol. 1. Stamford, CT: Appleton & Lange; 1997.)

Adults inhabit cecum and release eggs to lumen

Eggs must mature in soil for 10 days

pass out of the body with the feces and, in poorly sanitated areas of the world, are deposited on soil. The eggs are immature at the time of passage and must incubate for at least 10 days (longer if soil conditions, temperature, and moisture are suboptimal) before they become fully embryonated and infectious. Once mature, they are picked up on the hands of children at play or of agricultural workers and passed to the mouth. In areas where human feces are used as fertilizer, raw fruits and vegetables may be contaminated and later ingested. After ingestion, the eggs hatch in the duodenum, and the released larvae mature for approximately 1 month in the small bowel before migrating to their adult habitat in the cecum.

TRICHURIASIS

EPIDEMIOLOGY

Associated with defecation on soil and warm, humid climate

Adult worms live for years

Although it is less widespread than the pinworm, the whipworm is a cosmopolitan parasite, infecting approximately 1 billion people throughout the world. It is concentrated in areas where indiscriminate defecation and a warm, humid environment produce extensive seeding of soil with infectious eggs. In tropical climates, infection rates may be as high as 80%. Although the incidence is much lower in temperate climates, trichuriasis affects 2 million individuals throughout the rural areas of the southeastern United States. Here, it occurs primarily in family and institutional clusters, presumably maintained by the poor sanitary habits of toddlers and the mentally retarded. Although the intensity of infection is generally low, adult worms may live 4 to 8 years.

PATHOGENESIS AND IMMUNITY

Local colonic ulceration provides entry point to bloodstream for bacteria

Attachment of adult worms to the colonic mucosa and their subsequent feeding activities produce localized ulceration and hemorrhage (0.005 mL blood per worm per day). The ulcers provide enteric bacteria with a portal of entry to the bloodstream, and occasionally a sustained bacteremia results. A decrease in the prevalence of trichuriasis in the post-adolescent period and the demonstration of acquired immunity in experimental animal infections suggest that immunity may develop in naturally acquired human infections. An IgE-mediated immune mucosal response is demonstrable in humans, but is insufficient to cause appreciable parasite expulsion.

TRICHURIASIS: CLINICAL ASPECTS

MANIFESTATIONS

Colonic damage with abdominal pain and diarrhea

Colonic or rectal prolapse with heavy worm load

Light infections of trichuriasis are asymptomatic. With moderate worm loads, damage to the intestinal mucosa may induce nausea, abdominal pain, diarrhea, and stunting of growth. Occasionally, a child may harbor 800 worms or more. In these situations, the entire colonic mucosa is parasitized, with significant mucosal damage, blood loss, and anemia (**Figure 53–3**). The shear force of the fecal stream on the bodies of the worms may produce prolapse of the colonic or rectal mucosa through the anus, particularly when the host is straining at defecation or during childbirth.

DIAGNOSIS

Stools examined for characteristic eggs

In light infections, stool concentration methods may be required to recover the eggs. Such procedures are almost never necessary in symptomatic infections, because they inevitably produce more than 10,000 eggs per gram of feces, a density readily detected by examining 1 to 2 mg of emulsified stool with the low-power lens of a microscope. A moderate eosinophilia is common in such infections.

FIGURE 53–3. Whipworm infestation. Terminal ileum covered with adult *Trichuris trichiura.* (Reproduced with permission from Connor DH, Chandler FW, Schwartz DQA, Manz HJ, Lack EE (eds). *Pathology of Infectious Diseases,* vol. 1. Stamford, CT: Appleton & Lange; 1997.)

TREATMENT AND PREVENTION

Infections should not be treated unless they are symptomatic. Mebendazole is the drug of choice; albendazole is thought to be equally effective. Although the cure rate is only 60% to 70%, more than 90% of adult worms are usually expelled, rendering the patient asymptomatic. Prevention requires the improvement of sanitary facilities.

Ascaris

ASCARIS LUMBRICOIDES: PARASITOLOGY

A lumbricoides, a short-lived worm (6 to 18 months), is the largest and most common of the intestinal helminths. Measuring 15 to 40 cm in length, it dwarfs its fellow gut roundworms and brings an unexpected richness to our mental image of a parasite. Its firm, creamy cuticle and more pointed extremities differentiate it from the common earthworm, which it otherwise resembles in both size and external morphology. The male is slightly smaller than the female and possesses a curved tail with copulatory spicules. The female passes 200,000 eggs daily, whether or not she is fertilized. Eggs are elliptical; measure 35 by 55 μm; and have a rough, mamillated, albuminous coat over their chitinous shells (**Figure 53–4**). These roundworms are highly resistant to environmental conditions and may remain viable for up to 6 years in mild climates.

Earthworm-sized roundworm produces elliptical eggs

Eggs viable up to 6 years

LIFE CYCLE

The adult ascarids live high in the small intestine, where they actively maintain themselves by dint of muscular activity. The eggs are deposited into the intestinal lumen and passed in the feces. Like those of *Trichuris,* the eggs must embryonate in soil, usually for a minimum of 3 weeks, before becoming infectious. The similarity to *Trichuris* ends, however, with the ingestion of the eggs by the host. After hatching, the larvae penetrate the intestinal mucosa and invade the portal venules. They are carried to the liver, where they are still small enough to squeeze through that organ's capillaries and exit in the hepatic vein. They are then carried to the right side of the heart and subsequently pumped out to the lung. In the course of this migration, the larvae increase in size. By the time they reach the pulmonary capillaries, they are too large to pass through to the left side of the heart. Finding their route blocked, they rupture into the alveolar spaces, are coughed up, and subsequently swallowed. After regaining access to the upper intestine, they complete their maturation and mate.

Adults inhabit small intestine

Eggs must mature for 3 weeks in soil

Larvae from ingested eggs enter bloodstream and pass through alveoli and via respiratory tract and esophagus to intestines

Fertile ovum
Infertile ovum
10 μm
A
B
Esophagus
Anus
10 mm
C

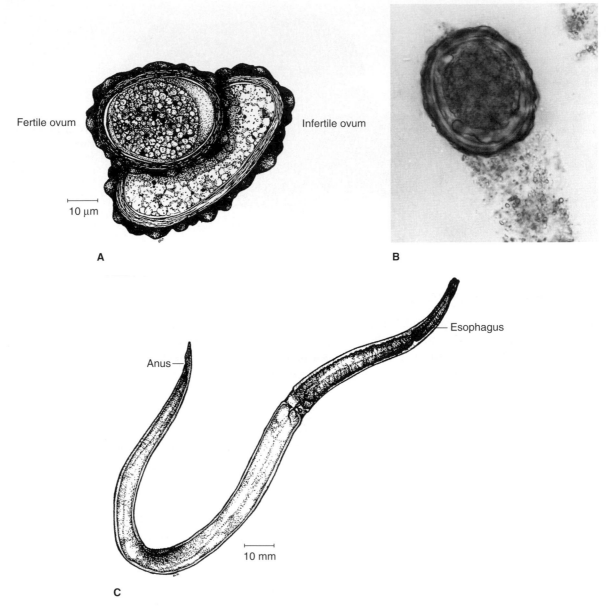

FIGURE 53–4. *Ascaris lumbricoides*. A. Structure of fertile and infertile egg. **B.** Fertilized egg in stool. **C.** Adult female worm. (B, Reproduced with permission from Connor DH, Chandler FW, Schwartz DQA, Manz HJ, Lack EE (eds). *Pathology of Infectious Diseases*, vol. 1. Stamford, CT: Appleton & Lange; 1997.)

ASCARIASIS

EPIDEMIOLOGY

More than 1 billion of the world's population, including 4 million Americans, are infected with *A lumbricoides*. Together they have been estimated to pass more than 25,000 tons of *Ascaris* eggs into the environment annually. Like trichuriasis, with which it is coextensive, ascariasis is a disease of warm climates and poor sanitation. It is maintained by small children who defecate indiscriminately in the immediate vicinity of the home and pick up infectious eggs on their hands during play. Geophagia may result in massive worm loads. The parasite may also be acquired through ingestion of egg-contaminated food by the host; in dry, windy climates, eggs may become airborne and be inhaled and swallowed. In tropical areas, the entire population may be involved; most worms, however, appear to be aggregated in a minority of the population, suggesting that some individuals are predisposed to heavy infections. Isolated infected family clusters are more common in temperature climates.

Epidemiology similar to that of *Trichuris*

PATHOGENESIS AND IMMUNITY

There is convincing evidence that ascariasis induces a protective immune response in the host. Moreover, the severity of pulmonary damage induced by the migration of larvae through the lung appears to be related in part to an immediate hypersensitivity reaction to larval antigens.

Hypersensitive pulmonary reactions to larval migration

ASCARIASIS: CLINICAL ASPECTS

MANIFESTATIONS

Clinical manifestations of ascariasis may result from either the migration of the larvae through the lung or the presence of the adults in the intestinal lumen. Pulmonary involvement is usually seen in communities where transmission is seasonal; the severity of symptoms is related to the degree of hypersensitivity induced by previous infections and the intensity of the current exposure. Fever, cough, wheezing, and shortness of breath are common. Laboratory studies reveal eosinophilia, oxygen desaturation, and migratory pulmonary infiltrates. Death from respiratory failure has been noted occasionally.

If the worm load is small, infections with adult worms may be completely asymptomatic. They come to clinical attention when the parasite is vomited up or passed in the stool. This situation is most likely during episodes of fever, which appear to stimulate the worms to increase motility. Most physicians who have worked in developing countries have had the disconcerting experience of observing an ascarid crawl out of a patient's mouth, nose, or ear during an otherwise uneventful evaluation of fever. Occasionally, an adult worm migrates to the appendix, bile duct, or pancreatic duct, causing obstruction and inflammation of the organ. Heavier worm loads may produce abdominal pain and malabsorption of fat, protein, carbohydrate, and vitamins. In marginally nourished children, growth may be retarded. Occasionally, a bolus of worms may form and produce intestinal obstruction, particularly in children (**Figure 53–5**). Worm loads of 50 are not uncommon, and as many as 2000 worms have been recovered from a single child. In the United States, where worm loads tend to be modest, obstruction occurs in 2 per 1000 infected children per year. The mortality rate in cases of ascariasis is 3%. Estimates of deaths from ascariasis range from 8000 to 100,000 annually worldwide.

Infections asymptomatic with small worm loads

Malabsorption and occasional obstruction produced with heavy worm loads

FIGURE 53–5. Ascariasis intestinal obstruction. Mass of adult worms recovered from infant at autopsy. (Reproduced with permission from Connor DH, Chandler FW, Schwartz DQA, Manz HJ, Lack EE (eds). *Pathology of Infectious Diseases*, vol. 1. Stamford, CT: Appleton & Lange; 1997.)

DIAGNOSIS

Stool examination readily reveals characteristic eggs

The diagnosis of ascariasis is generally made by finding the characteristic eggs in the feces. The extreme productivity of the female ascarid generally makes this task an easy one, except when the atypical-appearing unfertilized eggs predominate. The pulmonary phase of ascariasis is diagnosed by the finding of larvae and eosinophils in the sputum.

TREATMENT AND PREVENTION

Albendazole, mebendazole, and pyrantel pamoate are highly effective; the first two are preferred when *T trichiura* is also present. Community-wide control of ascariasis can be achieved with mass therapy administered at 6-month intervals. Ultimately, control requires adequate sanitation facilities.

Hookworms

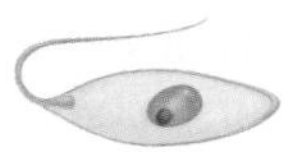

ANCYLOSTOMA AND *NECATOR*: PARASITOLOGY

N americanus and *A duodenale* infect humans

Species differentiated by morphology of oral cavity

Two species, *Necator americanus* and *Ancylostoma duodenale*, infect humans. Adults of both species are pinkish-white and measure about 10 mm in length (**Figure 53–6**). The head is often curved in a direction opposite that of the body, giving these worms the hooked appearance from which their common name is derived. The males have a unique fan-shaped copulatory bursa, rather than the curved, pointed tail common to the other intestinal nematodes. The two species can be readily differentiated by the morphology of their oral cavity. *A duodenale*, the Old World hookworm, possesses four sharp tooth-like structures, whereas *N americanus*, the New World hookworm, has dorsal and ventral cutting plates. With the aid of these structures, the hookworms attach to the mucosa of the small bowel and suck blood. The fertilized female releases 10,000 to 20,000 eggs daily. They measure 40 by 60 μm, possess a thin shell, and are usually in the two- to four-cell stage when passed in the feces (Figure 53–6).

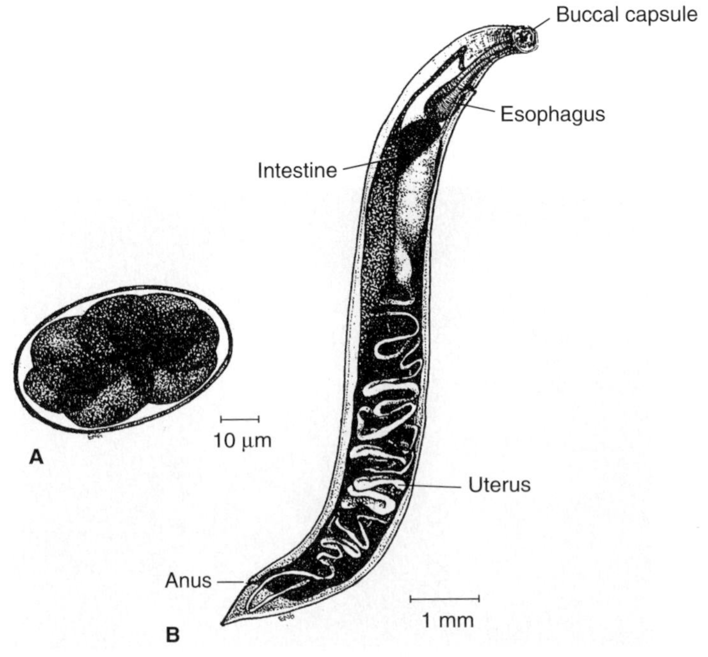

FIGURE 53–6. *Necator americanus*. A. Structure of hookworm egg and egg in stool. **B.** Structure of female adult hookworm.

LIFE CYCLE

For all practical purposes, the life cycles of the two hookworms, *N americanus* and *A duodenale*, are identical. The eggs are passed in the feces at the 4- to 8-cell stage of development and, on reaching soil, hatch within 48 hours, releasing rhabditiform larvae. These move actively through the surface layers of soil, feeding on bacteria and debris. After doubling in size, they molt to become infective filariform larvae, which may survive in moist conditions without feeding, for up to 6 weeks. On contact with human skin, these hookworms penetrate the epidermis, reach the lymphohematogenous system, and are passively transported to the right side of the heart and onward to the lungs. Here, they rupture into alveolar spaces and, like juvenile ascarids, are coughed up, swallowed, and pass into the small intestine, where they mature to adulthood. Larvae of *A duodenale*, if swallowed, can survive passage through the stomach and develop into adult worms in the small intestine.

In soil, eggs mature and release rhabditiform larvae that molt to produce infective filariform larvae

Filariform larvae penetrate skin and then follow same path as *Ascaris* larvae to gut

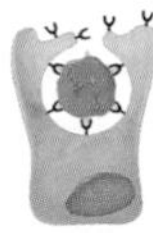

HOOKWORM DISEASE

EPIDEMIOLOGY

Hookworm infection is found worldwide between the latitudes of 45°N and 30°S. Transmission requires deposition of egg-containing feces on shady, well-drained soil; development of larvae under conditions of abundant rainfall and high temperatures (23° to 33°C); and direct contact of unprotected human skin with resulting filariform larvae. Infections become particularly intense in closed, densely populated communities, such as tea and coffee plantations. *N americanus* is found in the tropical areas of South Asia, Africa, and America, as well as the southern United States, where it was introduced with the African slave trade. *A duodenale* is seen in the Mediterranean basin, the Middle East, northern India, China, and Japan. It has been estimated that together these two worms extract over 7 million L of blood each day from 700 million people scattered around the globe, including 700,000 in the United States, leading to 50,000 to 60,000 deaths annually.

Larvae require hot, moist conditions

Limited to tropical areas and southern United States

PATHOGENESIS AND IMMUNITY

Each adult *A duodenale* extracts 0.2 mL of blood daily and *N americanus* 0.03 mL of blood. Additional blood loss may be related to the tendency of the worms to migrate within the intestine, leaving bleeding points at old sites of attachment. Because the adults may survive 2 to 14 years, the accumulated blood loss may be enormous. The infection elicits both a humoral antibody response and immediate hypersensitivity reaction in the host, but evidence that these moderate the infection is lacking. The peripheral and gut eosinophilia characteristic of this disease may play a role in the destruction of worms and/or modulation of the immediate hypersensitivity reaction.

Adult worms live in gut for years

Blood loss significant

Produce peripheral and gut eosinophilia

HOOKWORM DISEASE: CLINICAL ASPECTS

MANIFESTATIONS

In the overwhelming majority of persons infected with hookworm disease, the worm burden is small and the infection asymptomatic. Clinical manifestations, when they do occur, may be related to the original penetration of the skin by the filariform larva, the migration of the larva through the lung, and/or the presence of the adult worm in the gut. Skin penetration may produce a pruritic erythematous rash and swelling, popularly known as ground itch. This manifestation is more common in infection with *N americanus*, generally occurs between the toes, and may persist for several days. It is probably the result of prior sensitization to larval antigens.

Most infections asymptomatic depending on worm load

Pruritus at site of skin penetration

Pulmonary manifestations of hookworm disease may mimic those seen in ascariasis, but are generally less frequent and less severe. In the gut, the adult worm may produce epigastric pain and abnormal peristalsis. The major manifestations, however—anemia and hypoalbuminemia—are the result of chronic blood loss. The severity of the anemia depends on the worm burden and intake of dietary iron. If iron intake exceeds iron loss resulting from hookworm infection, a normal hematocrit will be maintained. Commonly, however, dietary iron is ingested in a form that is poorly absorbed. As a result, severe anemia may develop over a period of months or years. In children, this condition may often precipitate heart failure or kwashiorkor. Mental, sexual, and physical development may be retarded.

Iron deficiency anemia caused by blood loss from intestinal worms

DIAGNOSIS

The diagnosis of hookworm disease is made by examining direct or concentrated stool for the distinctive eggs. As these eggs are nearly identical in the two species, precise identification of the causative worm is generally not attempted. Quantitative egg counts can permit accurate estimation of worm load. If the stool is allowed to stand too long before it is examined, the eggs may hatch, releasing rhabditiform larvae. These larvae closely resemble those of *S stercoralis* and must be differentiated from them.

Eggs of both species look the same

TREATMENT AND PREVENTION

The anemia must be corrected. When it is mild or moderate, iron replacement is adequate. More severe anemia may require blood transfusions. The three most widely used anthelmintic agents, pyrantel pamoate, mebendazole and albendazole, are all highly effective. Prevention requires improved sanitation.

STRONGYLOIDES

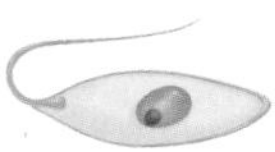

STRONGYLOIDES STERCORALIS: PARASITOLOGY

Strongyloides stercoralis adults measure only 2 mm in length, making them the smallest of the intestinal nematodes. The male is seldom seen within the human host, leading some authorities to believe that the female can conceive parthenogenetically in this environment. Be that as it may, the gravid female penetrates the mucosa of the duodenum, where she deposits her eggs. In severe infections, the biliary and pancreatic ducts, the entire small bowel, and the colon may be involved. The eggs hatch quickly, releasing rhabditiform larvae that reenter the bowel lumen and are subsequently passed into the stool. These larvae, which measure about 16 by 200 μm, can be distinguished from the similar larval stage of the hookworms by their short buccal cavity and large genital primordium (**Figure 53–7**).

Larvae differ slightly from hookworm

LIFE CYCLE

Three different life cycles have been described for the *Strongyloides* nematode. The first, or direct cycle, is similar to that observed with the hookworms. After rhabditiform larvae are passed in the stool, they molt on soil to become filariform larvae. Filariform larvae can penetrate human skin. After transport to the lung in the vascular system, they are coughed up and swallowed and then mature to adults in the small bowel. In the second, or autoinfective cycle, the rhabditiform larva's passage through the colon to the outside world is delayed by constipation or other factors, allowing it to transform into an infective filariform larva while still within the body of its host. This larva may then invade the internal mucosa (internal autoinfection) or perianal skin (external autoinfection) without an intervening soil phase. Thus, *S stercoralis,* unlike any of the other intestinal nematodes, has the capacity to multiply within the body of the host. The worm burden may increase dramatically, and the infection may persist indefinitely, without the need for reinfection from the environment and often

Primary cycle resembles hookworm except rhabditiform larvae develop in gut

Development of filariform stage in gut produces autoinfection

Adults can develop in soil, producing sustained life cycle

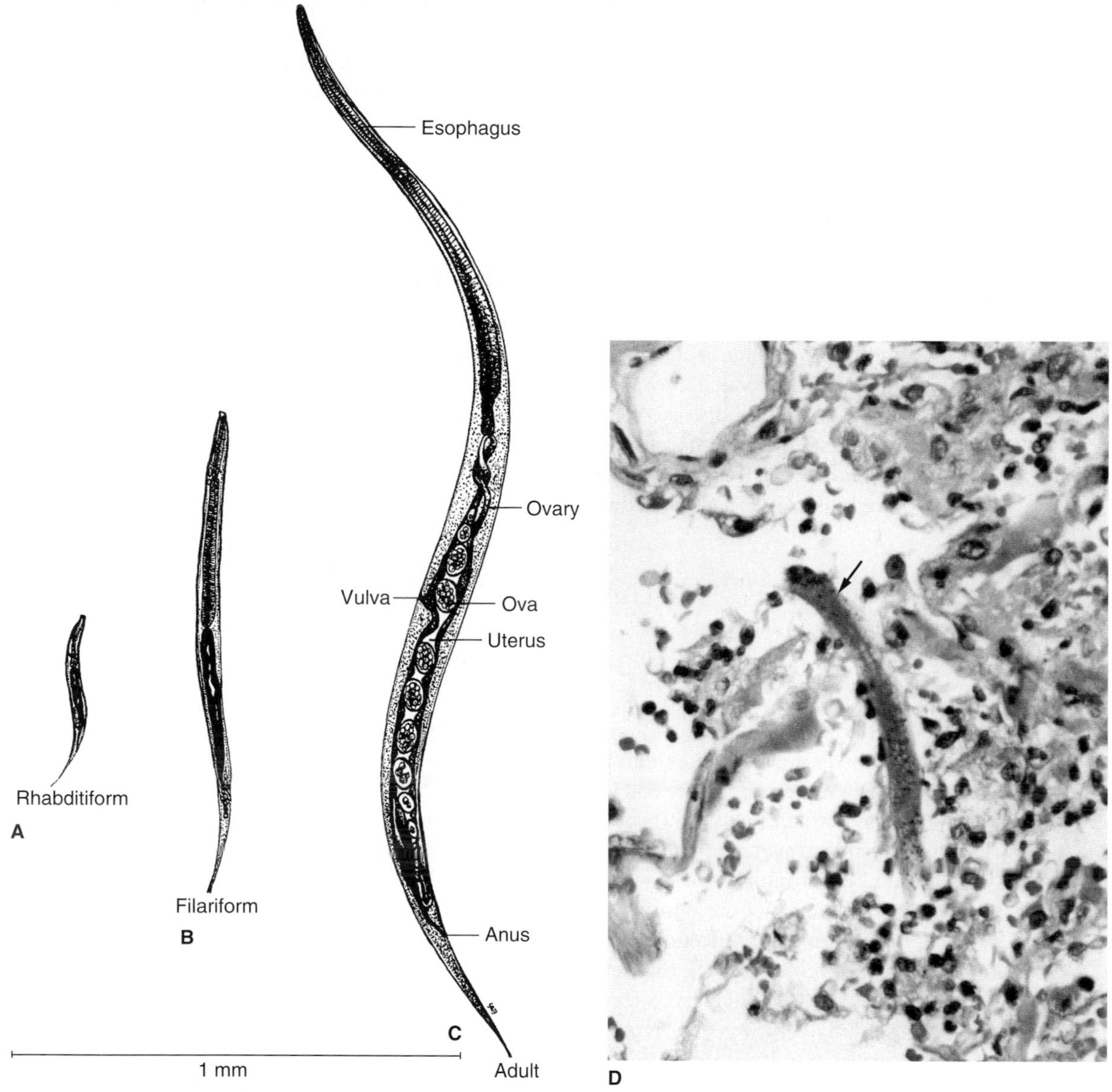

FIGURE 53–7. *Strongyloides stercoralis*. A-C. Structure of rhabditiform larvae, filariform larvae, and adult worm. **D.** Filariform larvae in lung surrounded by fibrin and inflammatory cells. (D, Reproduced with permission from Connor DH, Chandler FW, Schwartz DQA, Manz HJ, Lack EE (eds). *Pathology of Infectious Diseases*, vol. 1. Stamford, CT: Appleton & Lange; 1997.)

with dire consequence to the host. In the third, or free-living cycle, the rhabditiform larvae, after passage in the stool and deposition on the soil, develop into free-living adult males and females. These adults may propagate through several generations of free-living worms before infective filariform larvae are again produced. This cycle creates a soil reservoir that may persist even without continued deposition of feces.

STRONGYLOIDIASIS

EPIDEMIOLOGY

The distribution of *S stercoralis* parallels that of the hookworms, although it is less prevalent in all but tropical areas. It infects 90 million individuals worldwide, including 400,000 throughout the rural areas of Puerto Rico and the southeastern sections of the continental

Distribution similar to hookworm but less common

Infection by ingestion of filariform larvae also occurs

United States. Although, like hookworm infection, *S stercoralis* is generally acquired by direct contact of skin with soil-dwelling larvae, infection may also follow ingestion of filariform-contaminated food. Transformation of the rhabditiform larvae to the filariform stage within the gut can result in seeding of the perianal area with infectious organisms. These larvae may be passed to another person through direct physical contact or autoinfect the original host. In debilitated and immunosuppressed patients, transformation to the filariform stage occurs within the gut itself, producing marked autoinfection or hyperinfection.

PATHOGENESIS AND IMMUNITY

Damage to intestinal mucosa may cause malabsorptive syndrome

Immunosuppression enhances risk of autoinfection by accelerating larval development

Invasion of the intestinal epithelium may accelerate epithelial cell turnover, alter intestinal motility, and induce acute and chronic inflammatory lesions, ulcerations, and abscess formation, all of which may play a role in the malabsorptive syndrome that frequently characterizes clinical disease. Steroid- or malnutrition-related immunosuppression appears to accelerate the metamorphosis of rhabditiform to filariform larvae within the bowel lumen, enhancing the frequency and intensity of autoinfection. There is little evidence that protective immunity develops in the infected host.

STRONGYLOIDIASIS: CLINICAL ASPECTS

MANIFESTATIONS

Pulmonary and intestinal manifestations can be similar to hookworm, *Ascaris* infections

Patients with strongyloidiasis do not generally give a history of "ground itch." They do, however, manifest the pulmonary disease seen in both ascariasis and, less often, in hookworm infection (Figure 53–7). The intestinal infection itself is usually asymptomatic. With heavy worm loads, however, the patient may complain of epigastric pain and tenderness, often aggravated by intake of food. In fact, peptic ulcer-like pain associated with peripheral eosinophilia strongly suggests strongyloidiasis. With widespread involvement of the intestinal mucosa, vomiting, diarrhea, paralytic ileus, and malabsorption may be seen.

External autoinfection causes lesions over buttocks and back

External autoinfection produces transient, raised, red, serpiginous lesions over the buttocks and lower back that reflect larval invasion of the perianal area. If the patient is not treated, these lesions may recur at irregular intervals over a period of decades; they are particularly common after recovery from a febrile illness. Over 25% of British and American servicemen imprisoned in Southeast Asia during World War II continued to demonstrate such lesions before diagnosis and treatment some 40 years after exposure.

Massive hyperinfection occurs in immunosuppressed but uncommon in AIDS

Massive hyperinfection with strongyloidiasis may occur in immunosuppressed patients, especially in those receiving glucocorticoid therapy, producing severe enterocolitis and widespread dissemination of the larvae to extraintestinal organs, including the heart, lungs, and central nervous system. Inexplicably, this phenomenon has been unusual in acquired immunodeficiency syndrome (AIDS) patients, even in areas where strongyloidiasis is highly endemic. The larvae may carry enteric bacteria with them, producing Gram-negative bacteremia and occasionally Gram-negative meningitis that may result in death.

DIAGNOSIS

Rhabditiform larvae detected in stool or duodenal aspirates

The diagnosis of strongyloidiasis is usually made by finding the rhabditiform larvae in the stool. Preferably, only fresh specimens should be examined to avoid the confusion induced by the hatching of hookworm eggs with the release of their look-alike larvae. The number of larvae passed in the stool varies from day to day, often requiring the examination of several specimens before the diagnosis of strongyloidiasis can be made. When absent from the stool, larvae may sometimes be found in duodenal aspirates or jejunal biopsy specimens. If the pulmonary system is involved, the sputum should be examined for the presence of larvae. Agar plate culture methods may recover organisms that go undetected by microscopic examination. Enzyme-linked immunosorbent assays for antibodies to excretory–secretory or somatic antigens are now available in reference laboratories.

TREATMENT AND PREVENTION

All infected patients should be treated to prevent the buildup of the worm burden by autoinfection and the serious consequences of hyperinfection. The drugs of choice for strongyloidiasis are ivermectin and thiabendazole. In hyperinfection syndromes, therapy must be extended for 1 week. The cure rate is significantly less than 100%, and stools should be checked after therapy to see whether retreatment is indicated. Patients who have resided in an endemic area at some time in their lives should be examined for *S stercoralis* both before and during steroid treatment or immunosuppressive therapy. Medical personnel caring for patients with hyperinfection syndromes should wear gowns and gloves because stool, saliva, vomitus, and body fluids may contain infectious filariform larvae.

Treatment essential to prevent autoinfection cycle

Medical personnel can be infected with filariform larvae

CASE STUDY

A WORM IN THE THROAT!

This 4-year-old boy, who resides in the rural Southeastern United States, likes to play barefooted in the summer. He is brought to the physician's office with a 3-day history of fever, cough, and mild wheezing. On initial examination, the physician is startled to observe two small worm-like objects in the posterior oropharynx.

QUESTIONS

■ Which of the following is the least likely cause?

A. *Ascaris*
B. *Trichuris*
C. *Necator*
D. *Ancylostoma*

■ Stool examination is the usual initial diagnostic approach in all of the following, except:

A. *Enterobius*
B. *Trichuris*
C. *Ascaris*
D. *Necator*

■ Which of the following can multiply within the human host (autoinfection)?

A. *Ascaris*
B. *Ancylostoma*
C. *Trichuris*
D. *Strongyloides*

ANSWERS

1(B), **2**(A), **3**(D)

CHAPTER 54

Tissue Nematodes

There is the elephant disease which is generated beside the streams of the Nile in the midst of Egypt and nowhere else. In Attica the feet are attacked and the eyes in Achean lands. And so different places are hurtful to different parts and members.

—Lucretius (99-55 B.C.)

The nematodes discussed in this chapter induce disease through their presence in the tissues and lymphohematogenous system of the human body. They are a heterogeneous group. Three of them, *Toxocara canis, Trichinella spiralis,* and *Ancylostoma braziliense,* are natural parasites of domestic and wild carnivores. Although they are capable of infecting humans, they cannot complete their life cycle in the human host. Humans therefore serve only as injured bystanders rather than major participants in the life cycle of these parasites (**Table 54–1**).

The remaining four major nematodes, *Wuchereria bancrofti, Brugia malayi, Onchocerca volvulus,* and *Loa loa,* are members of a single superfamily (Filarioidea), and all use humans as their natural definitive host (Table 54–1). The thin, thread-like adults live for years in the subcutaneous tissues and lymphatic vessels, where they discharge their live-born offspring or microfilariae. These progeny circulate in the blood or migrate in the subcutaneous tissues until they are ingested by a specific bloodsucking insect. Within this vector, they transform into filariform larvae capable of infecting another human when the invertebrate host again takes a blood meal.

The nematodes considered, diseases caused, and usual routes of infection in humans are listed in Table 54–1.

TOXOCARA

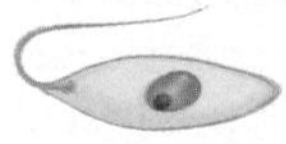

TOXOCARA CANIS: PARASITOLOGY

Toxicara canis is a large, intestinal ascarid of canines, including dogs, foxes, and wolves. Occasionally, a related organism found in cats (*T cati*) can behave in a similar fashion. Each female worm discharges approximately 200,000 thick-shelled eggs daily into the fecal stream. After reaching the soil, these eggs embryonate for a minimum of 2 to 3 weeks. Thereafter, the eggs are infectious to both canines and humans, and in moist soil they may remain so for months to years. When ingested by a young dog, the larvae exit from the

Cycle in canines resembles ascariasis in humans

TABLE 54–1 General Characteristics of Tissue Nematodes

PARASITE	DISEASE	USUAL SOURCE OF HUMAN INFECTION
Toxocara canis	Toxocariasis (visceral larva migrans)	Ingestion of ova from canine stools
Baylisascaris procyonis	Raccoon roundworm	Ingestion of ova from raccoon stools
Trichinella spiralis	Trichinosis	Ingestion of improperly cooked pork
Ancylostoma braziliense	Cutaneous larva migrans	Soil contaminated with dog or cat feces
Major filarial worms		
Wuchereria bancrofti, Brugia malayi	Lymphatic filariasis (elephantiasis)	Mosquito
Onchocerca volvulus	Onchocerciasis (river blindness)	*Simulium* flies
Loa loa (eye worm)	Loiasis (Calabar swellings)	Deer flies

eggshell, penetrate the intestinal mucosa, and migrate through the liver and the right side of the heart to the lung. Here, like the offspring of *Ascaris lumbricoides,* they burst into the alveolar airspaces and are coughed up and swallowed; thereafter, they mature in the small bowel. In fully grown dogs, most of the migrating larvae pass through the pulmonary capillaries and reach the systemic circulation. These larvae eventually are filtered out and encyst in the tissues. Hormonal changes and/or diminished immunity in the pregnant bitch stimulate the larvae to resume development, migrate across the placenta, and infect the unborn pups. Larvae may also pass to the newborn puppies in their mother's milk. Approximately 4 weeks after parturition, both the puppies and the lactating mother begin to pass large numbers of eggs in their stools. The mother may be superinfected by ingesting the newly passed eggs and can redevelop clinical symptoms.

Eggs embryonate 2 to 3 weeks in soil

Transplacentally infected puppies and infected lactating bitches excrete numerous ova

When humans ingest infectious eggs, the liberated larvae are small enough to pass through the pulmonary capillaries and reach the systemic circulation. Rarely does the organism break into the alveoli and reach the intestine to complete its maturation to adulthood. Larvae in the systemic circulation continue to grow. When their size exceeds the diameter of the vessel through which they are passing, they penetrate its wall and enter the tissue. The larvae induce a T_H2-type CD4+ response characterized by eosinophilia and IgE production.

Transmission to humans by ingestion of ova, and larvae invade tissues

TOXOCARIASIS

EPIDEMIOLOGY

T canis is a cosmopolitan parasite. The infection rate in the 50 million dogs inhabiting the United States is very high; over 80% of puppies and 20% of older animals are involved. "Man's best friend" deposits more than 3500 tons of feces daily in the streets, yards, and parks of America, and there is a real health risk. In some areas, between 10% and 30% of soil samples taken from public parks have contained viable *Toxocara* eggs. Moreover, serologic surveys of humans indicate that approximately 4% to 20% of the population has ingested these eggs at some time. The incidence of infection appears to be higher in the southeastern sections of the United States; presumably the warm, humid climate prolongs survival of the eggs, thereby increasing exposure. Indeed, seroprevalence rates of more than 50% have been noted in some developing nations. Puppies in the home increase the risk of infection. Clinical manifestations occur predominantly among children 1 to 6 years of age; many have a history of geophagia, suggesting that disease transmission results from direct ingestion of eggs in the soil. Most infections are subclinical, but the incidence of overt disease,

Soil extensively contaminated with ova deposited by domestic animals

Children are most often infected

Infection much more common than disease, but disease underreported

though difficult to assess, is certainly underreported. Serious ocular infection by larvae is frequently seen by ophthalmologists.

TOXOCARIASIS: CLINICAL ASPECTS

MANIFESTATIONS

The larvae of *Toxocara* that reach the systemic circulation may invade any tissue of the body, where they can induce necrosis, bleeding, and eosinophilic granulomas and, subsequently, fibrosis. The liver, lungs, heart, skeletal muscle, brain, and eye are involved most frequently. The severity of clinical manifestations is related to the number and location of these lesions and the degree to which the host has become sensitized to larval antigens. Children with more intense infection may have fever and an enlarged, tender liver. Those who are seriously ill may develop a skin rash, an enlarged spleen, asthma, recurrent pulmonary infiltrates and abdominal pain, sleep and behavioral changes, focal neurologic defects, and convulsions. Illness often persists for weeks to months, a condition frequently referred to as visceral larva migrans. Death may result from respiratory failure, cardiac arrhythmia, or brain damage. In older children and adults, systemic manifestations are uncommon. Eye invasion by larvae (ocular larva migrans) is more common. Typically, unilateral strabismus (squint) or decreased visual acuity causes the patient to consult an ophthalmologist. Examination reveals granulomatous endophthalmitis, which is usually a reaction to a larva that is already dead; it is sometimes mistaken for malignant retinoblastoma, and an unnecessary enucleation is performed.

Any tissue invaded by larvae

Disease results from organ invasion and hypersensitivity

Ocular invasion produces granulomatous endophthalmitis

DIAGNOSIS

Stool examination is not helpful because the parasite seldom reaches adulthood in humans. Definitive diagnosis requires demonstration of the larva in a liver biopsy specimen or at autopsy. A presumptive diagnosis may be made based on the clinical picture; eosinophilic leukocytosis; elevated levels of IgE; and on elevated antibody titers to blood group antigens, particularly the group A antigen. An enzyme immunoassay (EIA) using larval antigens has been developed, providing clinicians with a reasonably sensitive (75%) and specific (90%) serologic test. A Western blot procedure is somewhat more sensitive but is not widely available. Unfortunately, many patients with related ocular infections remain seronegative; some demonstrate elevated aqueous humor titers.

Tissue biopsy required for detection

Serodiagnosis using EIA reliable

TREATMENT AND PREVENTION

Corticosteroid treatment may be lifesaving if the patient has serious pulmonary, myocardial, or central nervous system involvement. Anthelmintic therapy with albendazole or mebendazole is generally administered, although the efficacy of these drugs remains uncertain. Prevention requires control of indiscriminate defecation by dogs and repeated worming of household pets. Worming must begin when the animal is 3 weeks of age and must be repeated every 3 months during the first year of life and twice a year thereafter.

Corticosteroids helpful in serious disease

Worming of household pets important

BAYLISASCARIS

Recently, another nematode that shares clinical and epidemiologic similarities with *Toxocara* has been increasingly recognized. *Balysascaris procyonis* (raccoon roundworm) has predominantly affected children playing in wooded areas that are frequented by raccoons. Raccoon "latrines" may be teeming with infective eggs, which, when accidentally ingested, may cause an evolution of disease that mimics toxocariasis. Unfortunately, this organism has a particular predilection for neural and eye tissue, and can lead to devastating eosinophilic meningoencephalitis and retinitis. The diagnostic and therapeutic approaches are similar to those for *Toxocara*, but the clinical outcome is usually fatal , especially when therapy is delayed.

Raccoon roundworm mimics *Toxocara*, but can especially lethal

TRICHINELLA

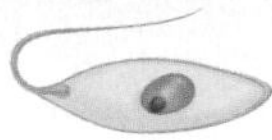

TRICHINELLA SPIRALIS: PARASITOLOGY

Intestinal parasite of many flesh-eating mammals

Adult *Trichinella* live in the duodenal and jejunal mucosa of flesh-eating animals throughout the world, particularly swine, rodents, bears, canines, felines, and marine mammals. Originally thought to be members of a single species, arctic, temperate, and tropical strains of *Trichinella* demonstrate significant epidemiologic and biologic differences and have recently been reclassified into seven distinct species. Only two species, *T spiralis* and the arctic species *T nativa,* display a high level of pathogenicity for humans. This discussion focuses on the former, while highlighting the unique epidemiologic and clinical characteristics of the latter.

The tiny (1.5-mm) male copulates with his outsized (3.5-mm) mate and, apparently spent by the effort, dies. Within 1 week, the inseminated female begins to discharge offspring. Unlike those of most nematodes, these progeny undergo intrauterine embryonation and are released as second-stage larvae. The birthing continues for the next 4 to 16 weeks, resulting in the generation of some 1500 larvae, each measuring 6 by 100 μm.

Larvae reach striated muscle and encapsulate but are still viable

Eating infected flesh spreads the disease

From their submucosal position, the larvae find their way into the vascular system and pass from the right side of the heart through the pulmonary capillary bed to the systemic circulation, where they are distributed throughout the body. Larvae penetrating tissue other than skeletal muscle disintegrate and die. Those finding their way to striated muscle continue to grow, molt, and gradually encapsulate over a period of several weeks. Calcification of the cyst wall begins 6 to 18 months later, but the contained larvae may remain viable for 5 to 10 years (**Figure 54–1**). The muscles invaded most frequently are the extraocular muscles of the eye, the tongue, the deltoid, pectoral, and intercostal muscles, the diaphragm, and the gastrocnemius. If a second animal feeds on the infected flesh of the original host, the encysted larvae are freed by gastric digestion, penetrate the columnar epithelium of the intestine, and mature just above the lamina propria.

TRICHINOSIS

EPIDEMIOLOGY

Swine infected by eating rats or meat in garbage

Trichinosis is widespread in carnivores. Among domestic animals, swine are most frequently involved (**Figure 54–2**). They acquire the infection by eating rats or garbage containing cyst-laden scraps of uncooked meat. Human infection, in turn, results largely from the consumption of improperly prepared pork products. In the United States, most outbreaks have been traced to ready-to-eat pork sausage prepared in the home or in small, unlicensed butcheries. Disease incidence is highest in Americans of Polish, German, and

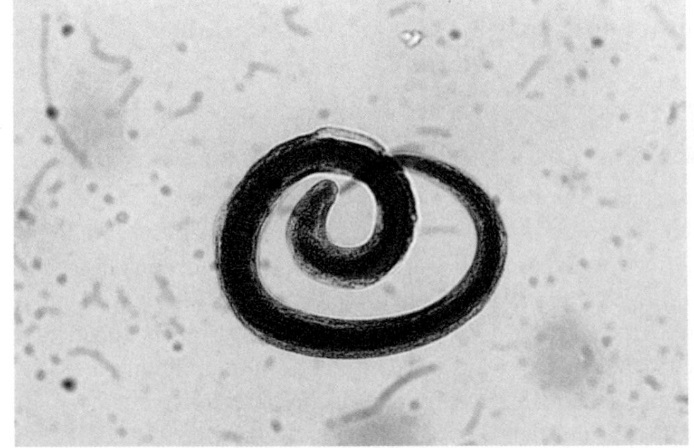

FIGURE 54–1. A coiled larva of *Trichinella spiralis* from a muscle digest. (Reproduced with permission from Connor DH, Chandler FW, Schwartz DQA, Manz HJ, Lack EE (eds). *Pathology of Infectious Diseases*, vol. 1. Stamford, CT: Appleton & Lange 1997.)

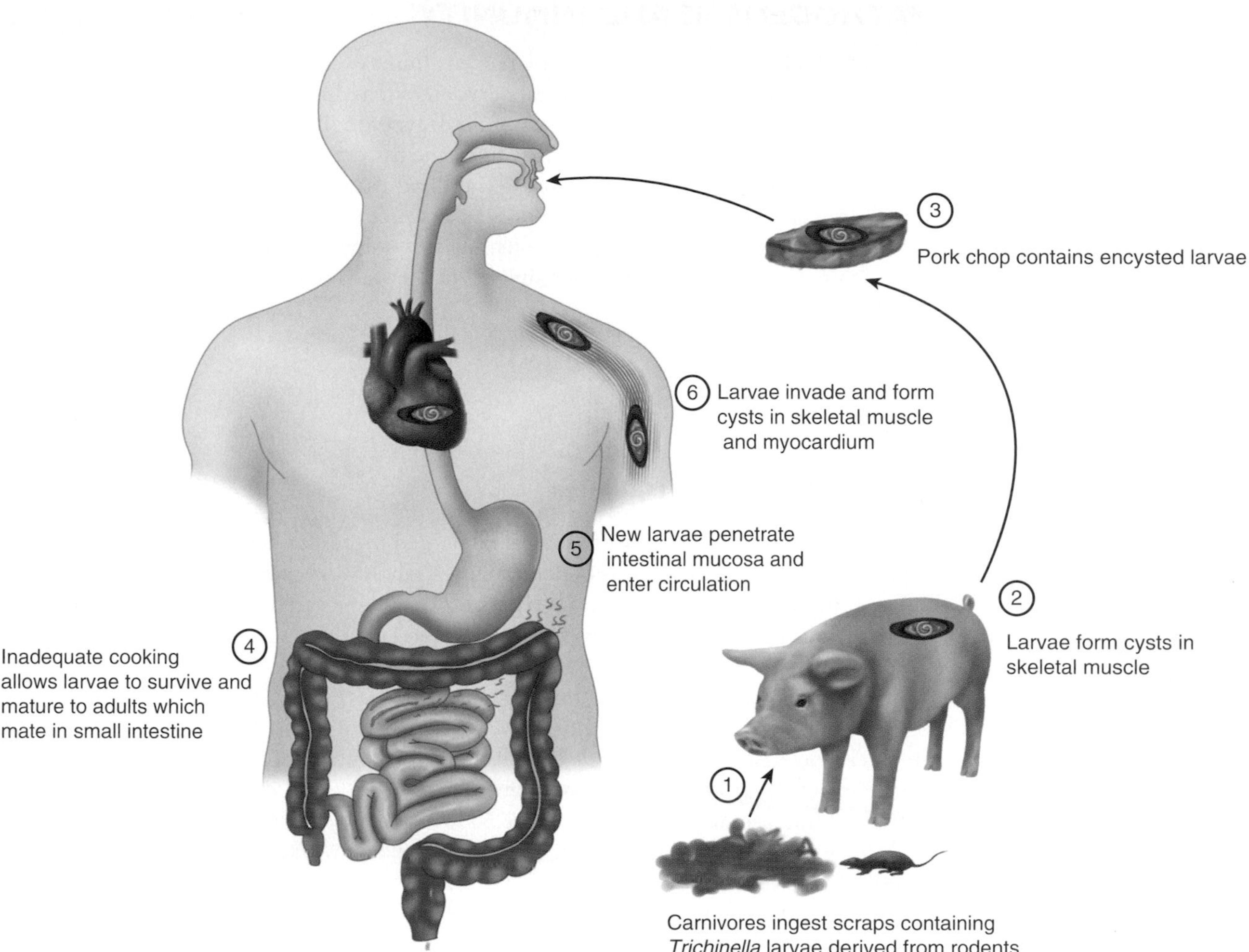
3
Pork chop contains encysted larvae
6
Larvae invade and form cysts in skeletal muscle and myocardium
5
New larvae penetrate intestinal mucosa and enter circulation
2
Larvae form cysts in skeletal muscle
4
Inadequate cooking allows larvae to survive and mature to adults which mate in small intestine
1
Carnivores ingest scraps containing *Trichinella* larvae derived from rodents

FIGURE 54–2. Trichinosis. T *spiralis* larvae ingested by pig (1) eventually end up as human cysts (6).

Italian descent, presumably because of their custom of producing and eating such sausage during holidays. Recent outbreaks have been reported among Indochinese refugees, apparently related to undercooking of fresh pork. Clusters have also followed feasts of wild pig in California and Hawaii. At present, nearly one third of human cases in the United States, particularly those in Alaska and other western states, have been attributed to consumption of the meat of wild animals, particularly bears. Outbreaks among Alaskan and Canadian Inuit populations have followed the ingestion of raw *T nativa*–infected walrus meat. Several recent outbreaks in Europe have involved horse meat or wild boar. Each year, a few cases of trichinosis are acquired from ground beef intentionally but illegally adulterated with pork.

Human infection most often from undercooked pork

Wild animals (eg, bear, walrus) also a risk

Human infections occur worldwide. In the United States, the prevalence of cysts found in the diaphragms of patients at autopsy has declined from 16.1% to 4.2% over a period of 30 years. This decline has been attributed to decreased consumption of pork and pork products; federal guidelines for the commercial preparation of such foodstuffs; the widespread practice of freezing pork, which kills all but arctic strains of *Trichinella*; and legislation requiring the thorough cooking of any meat scraps to be used as hog feed. Nevertheless, it is estimated that more than 1.5 million Americans carry live *Trichinella* in their musculature and that 150,000 to 300,000 acquire new infection annually. Fortunately, the overwhelming majority are asymptomatic, and only about 100 clinically recognized cases are reported annually to federal officials. In other areas of the world, infection is more commonly acquired from sylvatic sources, including wild boar, bush pigs, and warthogs.

Prevalence declining as a result of cooking and freezing of pork

Human infections usually are subclinical

PATHOGENESIS AND IMMUNITY

Larvae in striated muscle, heart, and central nervous system

Acute inflammatory reaction with eosinophil-mediated destruction of larvae

The pathologic lesions of trichinosis are related almost exclusively to the presence of larvae in the striated muscle, heart, and central nervous system. Invaded muscle cells enlarge, lose their cross-striations, and undergo a basophilic degeneration. Surrounding the involved area is an intense inflammatory reaction consisting of neutrophils, lymphocytes, and eosinophils. With the development of specific IgG and IgM antibodies, eosinophil-mediated destruction of circulating larvae begins, production of new larvae is slowed, and the expulsion of adult worms is hastened. A vasculitis demonstrated in some patients has been attributed to deposition of circulating immune complexes in the walls of the vessels. ::: Immune complexes, p. 42

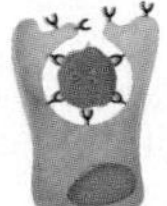

TRICHINOSIS: CLINICAL ASPECTS

MANIFESTATIONS

Initial abdominal pain and diarrhea as adults penetrate

Symptoms depend on number and extent of larval muscle invasion

Severe complications include hemoptysis and heart failure

One or two days after the host has ingested tainted meat, the newly matured adults penetrate the intestinal mucosa, producing nausea, abdominal pain, and diarrhea. In mild infections, these symptoms may be overlooked, except in a careful retrospective analysis; in more serious infections, they may persist for several days and render the patient prostrate. Diarrhea persisting for a period of weeks has been characteristic of *T nativa* outbreaks after ingestion of walrus meat by the Inuit population of northern Canada. Larval invasion of striated muscle begins approximately 1 week later and initiates the longer (6 weeks) and more characteristic phase of the disease. Patients in whom 10 or fewer larvae are deposited per gram of tissue are usually asymptomatic; those with 100 or more generally develop significant disease; and those with 1000 to 5000 have a very stormy course that occasionally ends in death. Fever, muscle pain, muscle tenderness, and weakness are the most prominent manifestations of trichinosis. Patients may also display eyelid swelling, a maculopapular skin rash, and small hemorrhages beneath the conjunctiva of the eye and the nails of the digits. Hemoptysis and pulmonary consolidation are common in severe infections. If there is myocardial involvement, electrocardiographic abnormalities, tachycardia, or congestive heart failure may be seen. Central nervous system invasion is marked by encephalitis, meningitis, and polyneuritis. Delirium, psychosis, paresis, and coma can follow.

DIAGNOSIS

Eosinophilia up to 50% from second week on

The most consistent abnormality is an eosinophilic leukocytosis during the second week of illness, which persists for the remainder of the clinical course. Eosinophils typically range from 15% to 50% of the white cell count, and in some patients, this may induce extensive damage to the cardiac endothelium. In severe or terminal cases, the eosinophilia may disappear altogether. Serum levels of IgE and muscle enzymes are elevated in most clinically ill patients.

Antibody usually appears after 2 weeks and then persists

There are a number of valuable serologic tests, including indirect fluorescent antibody, bentonite flocculation, and enzyme-linked immunosorbent assay. Significant antibody titers are generally absent before the third week of illness, but may then persist for years.

Muscle biopsy reveals larvae

Biopsy of the deltoid or gastrocnemius muscles during the third week of illness often reveals encysted larvae.

TREATMENT

Corticosteroids used in severe cases

Patients with severe edema, pulmonary manifestations, myocardial involvement, or central nervous system disease are treated with corticosteroids. The value of specific anthelmintic therapy remains controversial. The mortality rate of symptomatic patients is 1%, rising to 10% if the central nervous system is involved. Mebendazole and albendazole halt the production of new larvae, but in severe infection, the destruction of tissue larvae may provoke a hazardous hypersensitivity response in the host. This may be moderated with corticosteroids.

PREVENTION

Control of trichinosis requires adherence to federal feeding regulations for pigs, and limiting contact between domestic pigs and wild animals, particularly rodents, who might be carrying *Trichinella* larvae in their tissues. Domestically, care should be taken to cook pork to an internal temperature of at least 76.6°C, freeze it at −15°C for 3 weeks, or thoroughly smoke it before it is ingested. *T nativa* in the flesh of arctic animals may survive freezing for a year or more. All strains may survive apparently adequate cooking in microwave ovens due to the variability in the internal temperatures achieved.

Primary prevention involves thorough cooking

CUTANEOUS LARVA MIGRANS

Cutaneous larva migrans, or creeping eruption, is an infection of the skin caused by the larvae of a number of animal and human parasites, most commonly the dog and cat hookworm *Ancylostoma braziliense.* Eggs discharged in the feces of infected animals and deposited on warm, moist, sandy soil develop filariform larvae capable of penetrating mammalian skin on contact. In the United States, parasite transmission is particularly common in the beach areas of the southern Atlantic and Gulf states.

Caused usually by larvae of dog and cat hookworms

Filariform larvae penetrate and migrate in human skin

Although larvae do not develop further within humans, they may migrate within the skin for a period of weeks to months. Clinically, the patient notes a pruritic, raised, red, irregularly linear lesion 10 to 20 cm long. Skin excoriation from scratching enhances the likelihood of secondary bacterial infection. Half of infected patients develop Löffler's syndrome of transient, migratory pulmonary infiltrations associated with peripheral eosinophilia. The syndrome most probably reflects pulmonary migration of larvae. Larvae are rarely found in either sputum or skin biopsies, and the diagnosis must be established on clinical grounds.

Adult forms do not develop in humans

Cutaneous larva migrans responds well to albendazole, ivermectin, or topical thiabendazole. Antihistamines and antibiotics may be helpful in controlling pruritus and secondary bacterial infection, respectively.

LYMPHATIC FILARIA

Lymphatic filariasis encompasses a group of diseases produced by certain members of the superfamily Filarioidea that inhabit the lymphatic system of humans. Their presence induces an acute inflammatory reaction, chronic lymphatic blockade, and, in some cases, grotesque swellings of the extremities and genitalia known as **elephantiasis**.

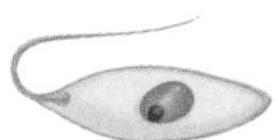

WUCHERERIA AND *BRUGIA*: PARASITOLOGY

The two agents most commonly responsible for lymphatic filariasis are *Wuchereria bancrofti* and *Brugia malayi.* Both are thread-like worms that lie coiled in the lymphatic vessels, male and female together, for the duration of their decade-long lifespan. The female *W bancrofti* measures 100 mm in length, and the male 40 mm. *B malayi* adults are approximately half these sizes. The gravid females produce large numbers of embryonated eggs. At oviposition, the embryos uncoil to their full length (200 to 300 μm) to become microfilariae. The shell of the egg elongates to accommodate the embryo and is retained as a thin, flexible sheath. Although the offspring of the two species resemble each other, they may be differentiated on the basis of length, staining characteristics, and internal structure (**Table 54–2**). The microfilariae eventually reach the blood (**Figure 54–3**). In most *W bancrofti* and *B malayi* infections, they accumulate in the pulmonary vessels during the day. At night, in response to changes in oxygen tension, they spill out into the peripheral circulation, where they are found in greatest numbers between 9 PM and 2 AM. A Polynesian strain of *W bancrofti* displays a different periodicity, with the peak concentration of organisms occurring in the early evening. Periodicity has an important epidemiologic consequence,because it determines the species of mosquito to serve as vector and intermediate host. Within the thoracic muscles of the mosquito, microfilariae are transformed first into rhabditiform and then into filariform larvae. The latter actively penetrate the feeding site when the mosquito takes its next meal. Within the new host, the parasite migrates to the lymphatic vessels, undergoes a series of molts, and reaches adulthood in 6 to 12 months (**Figure 54–4**).

Adult worms live in lymphatic vessels for a decade

Microfilariae develop from ova

Microfilariae circulate in peripheral blood once each day

Mosquito is essential vector and intermediate host

TABLE 54–2 Differentiation of Microfilariae

PARASITE	LOCATION	SHEATH	SIZE (μM)	NUCLEI OF TAIL	PERIODICITY
Wuchereria bancrofti	Blood	Yes	360	None	Usually nocturnal
Brugia malayi	Blood	Yes	220	Two	Nocturnal
Loa loa	Blood	Yes	275	Continuous	Diurnal
Onchocerca volvulus	Skin	No	300	None	None

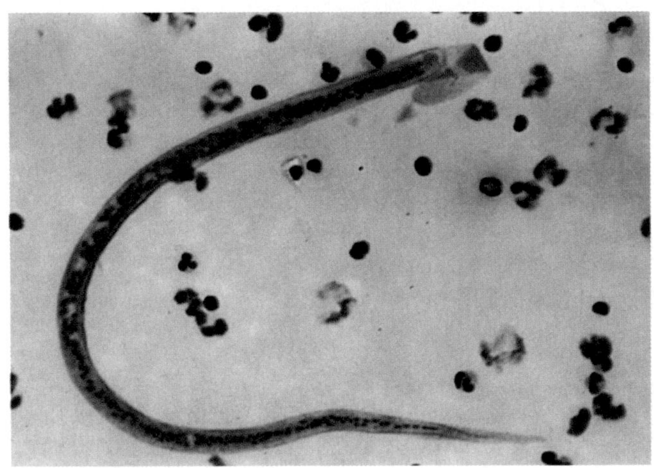

FIGURE 54–3. Microfilaria of *Wuchereria bancrofti* in blood film. (Reproduced with permission from Connor DH, Chandler FW, Schwartz DQA, Manz HJ, Lack EE (eds). *Pathology of Infectious Diseases*, vol. 1. Stamford, CT: Appleton & Lange; 1997, Figure 147-6.)

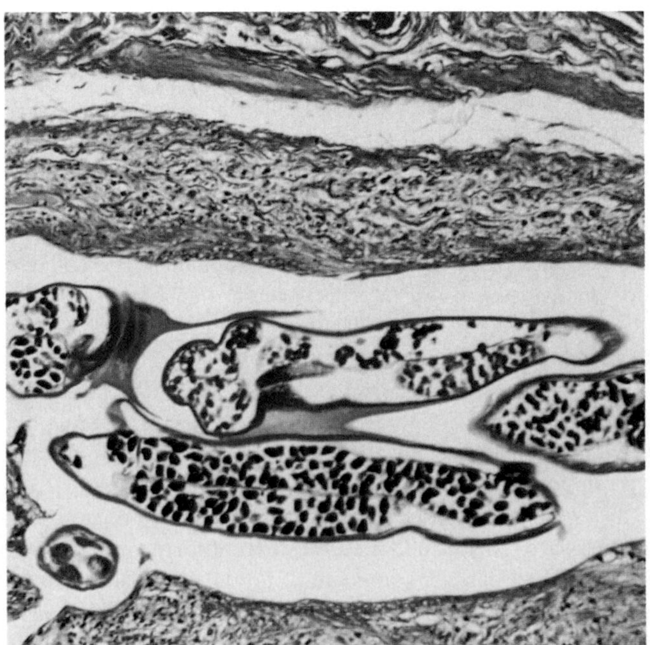

FIGURE 54–4. Lymphatic filariasis. These dilated lymphatics are filled with a gravid adult *W bancrofti* female. Eggs and developing microfilaria are within the paired uterine tubes. Note the surrounding thickened fibrous tissue. (Reproduced with permission from Connor DH, Chandler FW, Schwartz DQA, Manz HJ, Lack EE (eds). *Pathology of Infectious Diseases,* vol. 1. Stamford, CT: Appleton & Lange; 1997, Figure 147-4.)

LYMPHATIC FILARIASIS

EPIDEMIOLOGY

Lymphatic filariasis currently infects about 120 million people in Africa, Latin America, the Pacific Islands, and Asia; more than 75% of these cases are concentrated in Asia. *W bancrofti,* transmitted primarily by mosquitoes of the genera *Anopheles* or *Culex,* is the more cosmopolitan of the two species; it is found in patchy distribution throughout the poorly sanitated, densely crowded urban areas of all three continents. A small endemic focus once existed near Charleston, South Carolina, but died out in the 1920s. Moreover, some 15,000 *W bancrofti* infections were acquired by American servicemen during World War II. The same infection has recently been found in approximately 7% of Haitian refugees to the United States.

Primarily in Asia and other tropical areas

B malayi, transmitted by mosquitoes of the genus *Mansonia,* is confined to the rural coastal areas of Asia and the South Pacific. Strains with an unusual periodicity have been found in animals. Humans are the only known vertebrate hosts for most strains of *B malayi* and for *W bancrofti.* In the eastern Indonesian archipelago, a closely related species, *B timori,* is transmitted by night-feeding anopheline mosquitoes.

Humans are the only vertebrate hosts for *Wuchereria*

PATHOLOGY AND PATHOGENESIS

Pathologic changes, which are confined primarily to the lymphatic system, can be divided into acute and chronic lesions. In acute disease, the presence of molting adolescent worms and dead or dying adults stimulates dilatation of the lymphatics, hyperplastic changes in the vessel endothelium, infiltration by lymphocytes, plasma cells, and eosinophils, and thrombus formation (ie, acute lymphangitis). These developments are followed by granuloma formation, fibrosis, and permanent lymphatic obstruction. Repeated infections eventually result in massive lymphatic blockade. The skin and subcutaneous tissues become edematous, thickened, and fibrotic. Dilated vessels may rupture, spilling lymph into the tissues or body cavities. Bacterial and fungal superinfections of the skin often supervene and contribute to tissue damage.

Lymphatic blockade with repeated infections

LYMPHATIC FILARIASIS: CLINICAL ASPECTS

MANIFESTATIONS

Individuals who enter endemic areas as adults and reside therein for months to years often present with acute lymphadenitis, urticaria, eosinophilia, and elevated serum IgE levels; they seldom go on to develop lymphatic obstruction. A significant proportion of indigenous populations present with asymptomatic microfilaremia. Some of these spontaneously clear their infection, and others go on to experience "filarial fevers" and lymphadenitis 8 to 12 months after exposure. The fever is typically low grade; in more serious cases, however, temperatures as high as 40°C, chills, muscle pains, and other systemic manifestations may be seen. Classically, the lymphadenitis is first noted in the femoral area as an enlarged, red, tender lump. The inflammation spreads centrifugally down the lymphatic channels of the leg. The vessels become enlarged and tender, the overlying skin red and edematous. In bancroftian filariasis, the lymphatic vessels of the testicle, epididymis, and spermatic cord are frequently involved, producing a painful orchitis, epididymitis, and funiculitis; inflamed retroperitoneal vessels may simulate acute abdomen. Epitrochlear, axillary, and other lymphatic vessels are involved less frequently. The acute manifestations last a few days and resolve spontaneously, only to recur periodically over a period of weeks to months.

With repeated infection, permanent lymphatic obstruction develops in the involved areas. Edema, ascites, pleural effusion, hydrocele, and joint effusion result. The lymphadenopathy persists and the palpably swollen lymphatic channels may rupture, producing an abscess or draining sinus. Rupture of intra-abdominal vessels may give rise to chylous ascites or

Lymphadenitis, urticaria, and eosinophilia are early findings

Acute manifestations can recur

Elephantiasis may be end result

urine. In patients heavily and repeatedly infected over a period of decades, elephantiasis may develop. Such patients may continue to experience acute inflammatory episodes.

In southern India, Pakistan, Sri Lanka, Indonesia, Southeast Asia, and East Africa, an aberrant form of filariasis is seen. This form, termed tropical pulmonary eosinophilia, is characterized by an intense eosinophilia, elevated levels of IgE, high titers of filarial antibodies, the absence of microfilariae from the circulating blood, and a chronic clinical course marked by massive enlargement of the lymph nodes and spleen (children) or chronic cough, nocturnal bronchospasm, and pulmonary infiltrates (adults). Untreated, lymphatic filariasis may progress to pulmonary interstitial fibrosis. Microfilariae have been found in the tissues of such patients, and the clinical manifestations may be terminated with specific antifilarial treatment. It is believed that this syndrome is precipitated by the removal of circulating microfilariae by an IgG-dependent, cell-mediated immune reaction. Microfilariae are trapped in various tissue sites, where they incite an eosinophilic inflammatory response, granuloma formation, and fibrosis.

For tropical eosinophilia syndrome, microfilaria not found in blood

DIAGNOSIS

Eosinophilia is usually present during the acute inflammatory episodes, but definitive diagnosis requires the presence of microfilaria in the blood or lymphatic, ascitic, or pleural fluid. They are sought in Giemsa- or Wright-stained thick and thin smears. The major distinguishing features of these and other microfilariae are listed in Table 54–2. Because the appearance of the microfilariae is usually periodic, specimen collection must be properly timed. If this procedure proves difficult, the patient may be challenged with the antifilarial agent diethylcarbamazine (DEC). This drug stimulates the migration of the microfilariae from the pulmonary to the systemic circulation and enhances the possibility of their recovery. If the parasitemia is scant, the specimen may be concentrated before it is examined. Once found, the microfilariae must be differentiated from those produced by other species of filariae. A number of serologic tests have been used for the diagnosis of microfilaremic disease, but until recently they have lacked adequate sensitivity and specificity; even the more recent tests are of little diagnostic significance in individuals indigenous to the endemic area, because many people have experienced a prior filarial infection. Circulating filarial antigens can be found in most microfilaremic patients and also in some seropositive amicrofilaremic individuals. Antigen detection may thus prove to be a specific indicator of active disease. Tropical eosinophilia is diagnosed as described previously.

Eosinophilia during acute episodes

Search for microfilariae in the blood requires careful timing

TREATMENT

Diethylcarbamazine eliminates the microfilariae from the blood and kills or injures the adult worms, resulting in long-term suppression of the infection or parasitologic cure. Frequently, the dying microfilariae stimulate an allergic reaction in the host. This response is occasionally severe, requiring antihistamines and corticosteroids. The role of ivermectin in the treatment of lymphatic filariasis has not yet been established. Early studies have demonstrated a high level of effectiveness in clearing microfilaremia after the administration of a single dose. The tissue changes of elephantiasis are often irreversible, but the enlargement of the extremities may be ameliorated with pressure bandages or plastic surgery. Control programs combine mosquito control with mass treatment of the entire population.

Killing of microfilariae may stimulate allergic response

ONCHOCERCA

Onchocerciasis or river blindness, produced by the skin filaria *Onchocerca volvulus,* is characterized by subcutaneous nodules, thickened pruritic skin, and blindness.

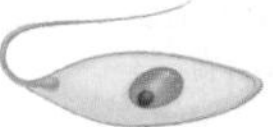

ONCHOCERCA VOLVULUS: PARASITOLOGY

The 40- to 60-cm thread-like female adults lie, together with their diminutive male partners, in coiled masses within fibrous subcutaneous and deep tissue nodules. The female

CASE STUDY

A TODDLER WHO LOVES DOGS

This 2-year-old boy loves to go to the public park and play with other people's pets. He also has a history of pica (eating dirt). Over the last week, he has developed fever and wheezing, along with some vague complaints of abdominal discomfort. Physical findings include wheezes and a moderately enlarged, tender liver.

Laboratory findings: scattered interstitial pulmonary infiltrates; white blood cell count of 29,000/mm^3 with 40% eosinophils, and a mild anemia.

QUESTIONS

■ What is the most likely cause of this child's illness?

A. *Trichinella*
B. *Toxocara*
C. *Baylisascaris*
D. *Ancylostoma*

■ *Ancylostoma* infections are acquired by:

A. Mosquito transmission
B. Black fly bites
C. Deer fly bites
D. Direct larval penetration of skin

■ Thick skin snips are used to diagnose:

A. *Loa*
B. *Wuchereria*
C. *Brugia*
D. *Onchocerca*

ANSWERS

1(B), **2**(D), **3**(D)

CHAPTER 55

Cestodes

Cestodes are long, ribbon-like helminths that have gained the common appellation of tapeworm from their superficial resemblance to sewing tape. Their appearance, number, and exaggerated reputation for inducing weight loss have made them the best known of the intestinal worms. Although improvements in sanitation have dramatically reduced their prevalence in the United States, they continue to inhabit the bowels of many of its citizens. In some parts of the world, indigenous populations take purgatives monthly to rid themselves of this, the largest and most repulsive of the intestinal parasites.

MORPHOLOGY

Like all helminths, tapeworms lack vascular and respiratory systems. In addition, they are devoid of both gut and body cavity. Food is absorbed across a complex cuticle, and the internal organs are embedded in a solid parenchyma. The adult is divided into three distinct parts: the "head" or scolex; a generative neck; and a long, segmented body, the strobila. The scolex typically measures less than 2 mm in diameter and is equipped with four muscular sucking discs used to attach the worm to the intestinal mucosa of its host. (In one genus, *Diphyllobothrium,* the discs are replaced by two grooves, or bothria.) As a further aid in attachment, the scolex of some species possesses a retractable protuberance, or rostellum, armed with a crown of chitinous hooks. Immediately posterior to the scolex is the neck from which individual segments, or proglottids, are generated one at a time to form the chain-like body. Each proglottid is a self-contained hermaphroditic reproductive unit joined to the remainder of the colony by a common cuticle, nerve trunks, and excretory canals. Its male and female gonads mature and effect fertilization as the segment is pushed farther and farther from the neck by the formation of new proglottids. When the segment reaches gravidity, it releases its eggs by rupturing, disintegrating, or passing them through its uterine pore. The eggs of the genus *Taenia* possess a solid shell and contain a fully developed, six-hooked (hexacanth) embryo. The eggs of *Diphyllobothrium latum,* in contrast, are immature at the time of deposition and possess a covered aperture, or operculum, through which the embryo exits once fully developed.

Without gut; food absorbed from host

Divided into scolex, neck, and segmented body parts

Each proglottid a hermaphroditic unit releasing eggs via rupture or through uterine pore

LIFE CYCLE

With the exception of *Hymenolepis nana,* further development of all cestodes requires the passage of the larvae through one or more intermediate hosts. Eggs of the genus *Taenia* pass in the stool of their definitive host, reach the soil, and are ingested by the specific intermediate. They hatch within its gut, and the released embryos penetrate the intestinal mucosa,

Eggs of *Taenia* must be ingested by intermediate host

Infectious cysts of *Taenia* form in tissues of intermediate

Definitive host ingests cysts in flesh of intermediate hosts to yield adult intestinal worms

find their way through the lymphohematogenous system to the tissues, and encyst therein. From the germinal lining of this cyst, immature scolices or protoscolices are formed. A cyst with a single such structure is known as a cysticercus (or, in the case of *H nana,* a cysticercoid); a cyst with multiple protoscolices is known as a coenurus. In some species of tapeworm, daughter cysts, each containing many protoscolices, are formed within the mother or hydatid cyst. The cycle for all is completed when the definitive host ingests the cyst-ridden flesh of the intermediate host. After digestion of the surrounding meat in the stomach, the cyst is freed, and the protoscolex everts to become a scolex. Following attachment to the mucosa, a new strobila is generated.

D latum requires two intermediates—a copepod and a freshwater fish—to complete cycle

D latum, whose eggs are immature on release, requires two intermediates to complete its larval development. The egg must reach fresh water before the operculum opens and a ciliated, free-swimming larva, or coracidium, is released. The coracidium is then ingested by the first intermediate host, a copepod, in which it is transformed into a larva (procercoid). When the copepod is, in turn, ingested by a freshwater fish, the larva penetrates the musculature of the fish to form an elongated and infectious larva, the plerocercoid. Life cycles and characteristics of important intestinal and tissue tapeworms infecting humans are summarized in **Table 55–1**.

CLINICAL DISEASE

Clinical effects depend on whether humans are definitive hosts or intermediate hosts

The clinical consequences of tapeworm infection in humans depend on whether the patient serves as the primary or the intermediate host. In the former case, the adult worm is confined to the lumen of the gut, and the consequences of the infection are typically minor. Taeniasis saginata and diphyllobothriasis are prime examples. In contrast, when the patient

TABLE 55–1 Intestinal and Tissue Tapeworms

STAGE	*DIPHYLLOBOTHRIUM LATUM*	*TAENIA SAGINATA*	*TAENIA SOLIUM*	*HYMENOLEPIS NANA*	*ECHINOCOCCUS GRANULOSUS*	*ECHINOCOCCUS MULTILOCULARIS*
Adult						
Definitive host	Humans, cats, dogs	Humans	Humans	Humans, rodents	Dogs, wolves	Foxes
Location	Gut lumen[a]	Gut lumen[a]	Gut lumen[a]	Gut lumen[a]	Gut lumen	Gut lumen
Length (m)	3-10	4-6	2-4	0.02-0.04	0.005	0.005
Attachment device	Grooves	Discs	Discs, hooklets	Discs, hooklets	Discs, hooklets	Discs, hooklets
Mature segment	Broad	Elongated	Elongated	Broad	Elongated	Elongated
Egg						
Maturation status	Nonembryonated	Embryonated	Embryonated	Embryonated	Embryonated	Embryonated
Distinguishing characteristics	Operculated	Radial striations	Radial striations	Polar filaments	Radial striations	Radial striations
Larval development in humans	No	No	Yes	Yes	Yes	Yes
Larva						
Intermediate host	Copepods, fishes	Cattle	Swine, humans	Humans, rodents	Herbivores, humans	Field mice, humans
Location	Tissue	Tissue	Tissue[a]	Gut mucosa[a]	Tissue[a]	Tissue[a]
Form	Procercoid (copepod)	Cysticercus	Cysticercus	Cysticercoid	Hydatid cyst	Hydatid cyst
	Plerocercoid (fish)					

[a]Site of human infection.

serves as the intermediate host (eg, for *Echinococcus granulosus*), larval development produces tissue invasion and frequently serious disease. The capacity of *H nana* and *T solium* to use humans as both primary and intermediate hosts is unique.

BEEF TAPEWORM

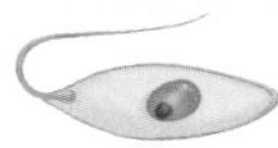

TAENIA SAGINATA: PARASITOLOGY

T saginata inhabits the human jejunum, where it may live for up to 25 years and grow to a maximum length of 10 m. Its 1-mm scolex lacks hooklets but possesses the four sucking discs typical of most cestodes (**Figure 55–1A**). The creamy white strobila consists of 1000 to 2000 individual proglottids. The terminal segments are longer (20 mm) than they are wide (5 mm) and contain a large uterus with 15 to 20 lateral branches; these characteristics are useful in differentiating them from those of the closely related pork tapeworm, *T solium.* When fully gravid, strings of six to nine terminal proglottids, each containing approximately 100,000 eggs, break free from the remainder of the strobila. These muscular segments may crawl unassisted through the anal canal or be passed intact with the stool. Proglottids reaching the soil eventually disintegrate, releasing their distinctive eggs. These eggs are 30 to 40 μm in diameter, spherical, and possess a thick, radially striated shell (Figure 55–1B). In appropriate environments, the hexacanth embryo may survive for months. If ingested by cattle or certain other herbivores, the embryo is released, penetrates the intestinal wall, and is carried by the vascular system to the striated muscles of the tongue, diaphragm, and hindquarters. Here it is transformed into a white, ovoid (5 by 10 mm) cysticercus (*Cysticercus bovis*). When present in large numbers, cysticerci impart a spotted or "measly" appearance to the flesh. Humans are infected when they ingest inadequately cooked meat containing these larval forms.

T saginata inhabits human jejunum

Gravid proglottids passed in stool

Eggs ingested by herbivore intermediates

Cysticerci in bovine striated muscle

Humans infected by eating inadequately cooked infected meat

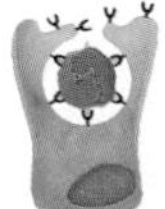

BEEF TAPEWORM DISEASE

EPIDEMIOLOGY

In the United States, sanitary disposal of human feces and federal inspection of meat have nearly interrupted transmission of *T saginata.* At present, less than 1% of examined carcasses are infected. Nevertheless, bovine cysticercosis is still a significant problem in the southwestern area of the country, where cattle become infected in feedlots or while pastured on land irrigated with sewage or worked by infected laborers without access to sanitary facilities. Shipment of infected carcasses can result in human infection in other areas of the United States. In countries where sanitary facilities are less comprehensive and undercooked or raw beef is eaten, *T saginata* is highly prevalent. Examples include Kenya, Ethiopia, the Middle East, the former Yugoslavia, and parts of the former Soviet Union and South America.

Indigenously acquired disease rare in United States

BEEF TAPEWORM DISEASE: CLINICAL ASPECTS

MANIFESTATIONS

Most persons infected with beef tapeworm are asymptomatic and become aware of the infection only through the spontaneous passage of proglottids. The proglottids may be observed on the surface of the stool or appear in the underclothing or bed sheets of the alarmed host. Passage may occur very irregularly and can be precipitated by excessive alcohol consumption. Some patients report epigastric discomfort, nausea, irritability (particularly after passage of segments), diarrhea, and weight loss. Occasionally, the proglottids may obstruct the appendix, biliary duct, or pancreatic duct.

Clinical symptoms usually mild

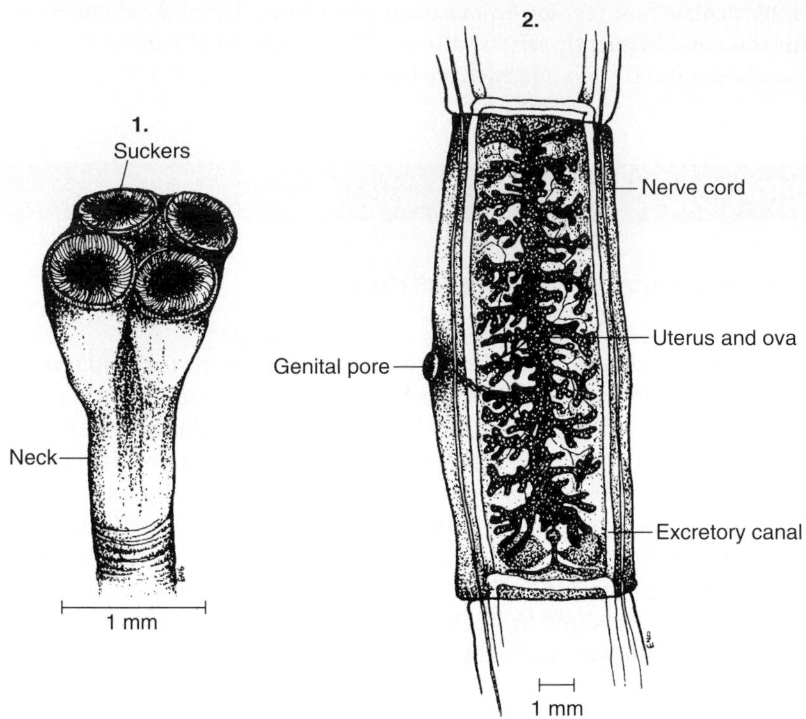

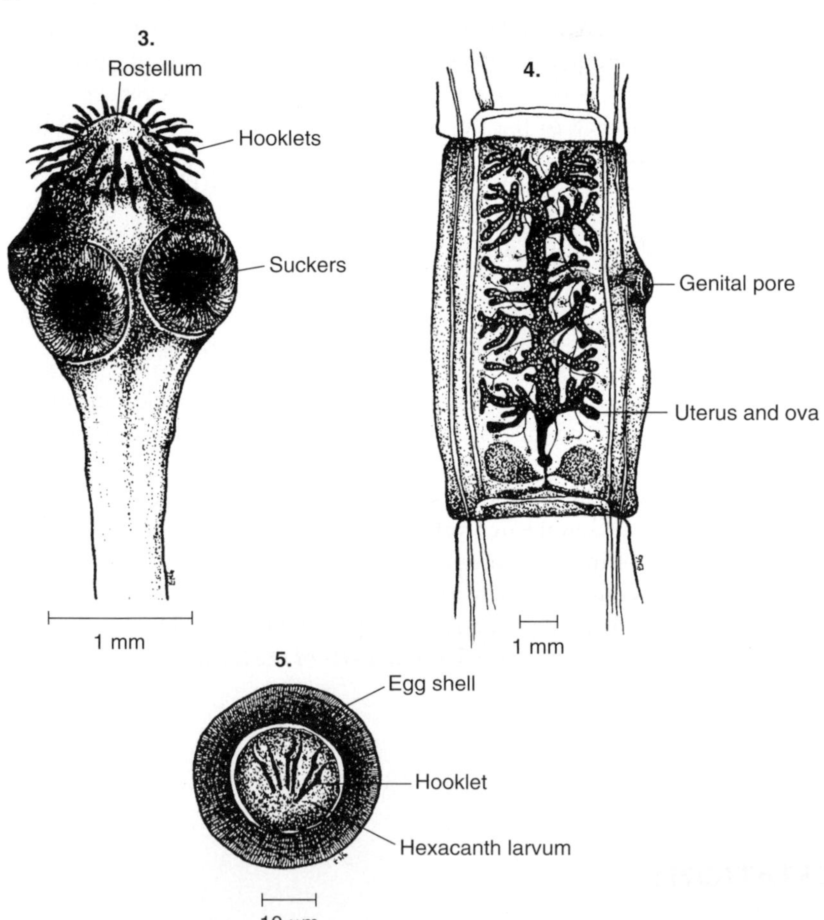

FIGURE 55–1. Tapeworm structures. A. *Taenia saginata.* **B.** *Taenia solium.* (1, 3) scolices; (2, 4) gravid proglottids; (5) ova (indistinguishable between species).

DIAGNOSIS

The diagnosis of beef tapeworm disease is made by finding eggs or proglottids in the stool. Eggs may also be distributed on the perianal area secondary to rupture of proglottids during anal passage. The adhesive cellophane tape technique described for pinworm can be used to recover the worms from this area. With this procedure, 85% to 95% of infections are detected, in contrast to only 50% to 75% by stool examination. Because the eggs of *T solium* and *T saginata* are morphologically identical, it is necessary to examine a proglottid to identify the species correctly.

Adhesive cellophane tape technique and stool examination detect eggs and proglottids

TREATMENT AND PREVENTION

The drugs of choice are praziquantel or niclosamide, which act directly on the worm. Both are highly effective in single-dose oral preparations. Ultimately, control is best effected through the sanitary disposal of human feces. Meat inspection is helpful; the cysticerci are readily visible. In areas where the infection is common, thorough cooking is the most practical method of control. Internal temperatures of 56°C or more for 5 minutes or longer destroy the cysticerci. Salting or freezing for 1 week at –15°C or less is also effective.

Sewage disposal, meat inspection, and adequate cooking

PORK TAPEWORM

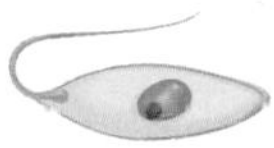

TAENIA SOLIUM: PARASITOLOGY

Like the beef tapeworm, which it closely resembles, *T solium* inhabits the human jejunum, where it may survive for decades. It can be distinguished from its close relative only by careful scrutiny of the scolex and proglottids; *T solium* possesses a rostellum armed with a double row of hooklets (Figure 55–1B3). The strobila is generally smaller than that of *T saginata,* seldom exceeding 5 mm in length or containing more than 1000 proglottids. Gravid segments measure 6 by 12 mm and thus appear less elongated than those of the bovine parasite (Figure 55–1B4). Typically, the uterus has only 8 to 12 lateral branches. Although the eggs appear morphologically identical to those of *T saginata,* they are infective only to swine and, perhaps reflecting a genetic proximity we would prefer to overlook, humans. Both pigs and people become intermediate hosts when they ingest food contaminated with viable eggs (**Figure 55–2**). Some authorities have suggested that humans may be autoinfected when gravid proglottids are carried backward into the stomach during the act of vomiting, initiating the release of the contained eggs. It seems more likely that autoinfection results from the transport of the eggs from the perianal area to the mouth on contaminated fingers.

T solium strobila shorter than in *T saginata*

Eggs infective to swine and to humans

Regardless of the route, an egg reaching the stomach of an appropriate intermediate host hatches, releasing the hexacanth embryo. The embryo penetrates the intestinal wall and may be carried by the lymphohematogenous system to any of the tissues of the body. Here it develops into a 1-cm, white, opalescent cysticercus over 3 to 4 months (**Figure 55–3**). The cysticercus may remain viable for up to 5 years, eventually infecting humans when they ingest undercooked and "measly" flesh. The scolex everts, attaches itself to the mucosa, and develops into a new adult worm, thereby completing the cycle.

Tissue cysticerci develop in humans and swine

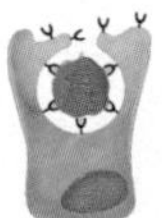

PORK TAPEWORM DISEASE

EPIDEMIOLOGY

Although infected swine are still occasionally found in the United States, most human disease is found in immigrants from endemic areas. Although pork tapeworm disease is widely distributed throughout the world, it is particularly common in south and southeast Asia, Africa, Latin America, and Eastern Europe.

T solium rarely found in United States

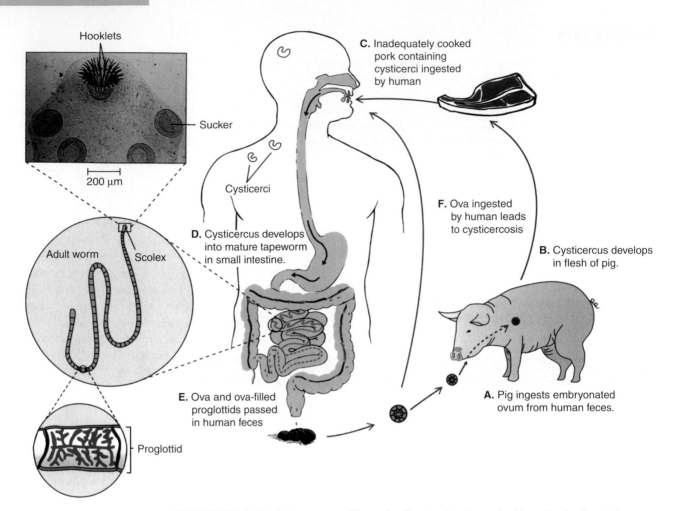

Hooklets
Sucker
200 μm
Adult worm
Scolex
Proglottid
Cysticerci
C. Inadequately cooked pork containing cysticerci ingested by human
D. Cysticercus develops into mature tapeworm in small intestine.
E. Ova and ova-filled proglottids passed in human feces
F. Ova ingested by human leads to cysticercosis
B. Cysticercus develops in flesh of pig.
A. Pig ingests embryonated ovum from human feces.

FIGURE 55–2. Pork tapeworm life cycle. *Taenia solium* is acquired by eating inadequately cooked pork. If worm ova hatch in tissues, they form cysticerci. (Reproduced with permission from Nester EW, Anderson DG, Roberts CE Jr, Nester MT. *Microbiology: A Human Perspective*, 6th ed. New York: McGraw-Hill, 2008.)

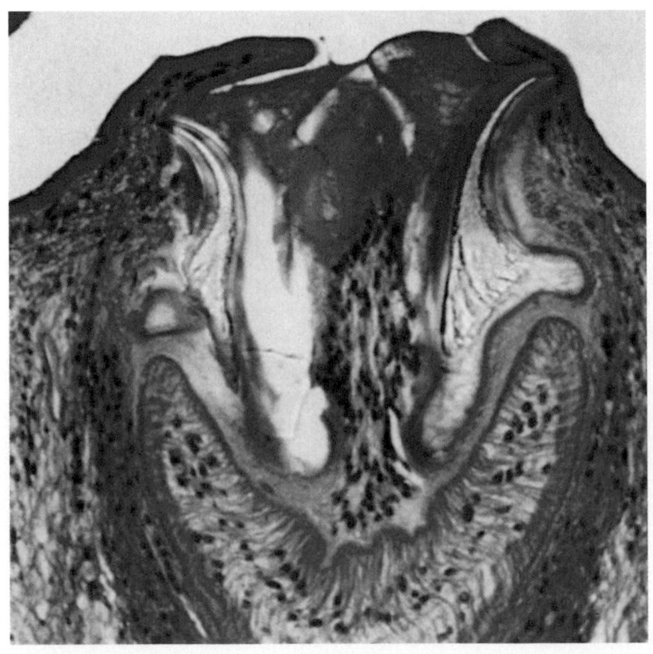

FIGURE 55–3. Cysticercosis of muscle. This section shows a cysticercus with the hooklets of a worm scolex. (Reproduced with permission from Connor DH, Chandler FW, Schwartz DQA, Manz HJ, Lack EE (eds). *Pathology of Infectious Diseases*, vol. 1. Stamford, CT: Appleton & Lange; 1997.)

PORK TAPEWORM DISEASE: CLINICAL ASPECTS

MANIFESTATIONS

The signs and symptoms of infection with the adult worm are similar to those of taeniasis saginata. Clinical manifestations are totally different when humans serve as intermediate hosts. Cysticerci develop in the subcutaneous tissues, muscles, heart, lungs, liver, brain (**Figure 55–4**), and eye. As long as the number is small and the cysticerci remain viable, tissue reaction is moderate and the patient is asymptomatic. The death of the larva, however, stimulates a marked inflammatory reaction, fever, muscle pains, and eosinophilia.

Major clinical manifestations caused by reaction to cysticerci

The most important and dramatic clinical presentation of cysticercosis results from lesions in the central nervous system (CNS). During the acute invasive stage, patients experience fever, headache, and eosinophilia. In heavy infections, a meningoencephalitic syndrome with cerebrospinal fluid (CSF) eosinophilic pleocytosis may be present. Established cysts can be found in the cerebrum, ventricles, subarachnoid space, spinal cord, and eye. Cerebral cysts are usually small, often measuring 2 cm or less in diameter; racemose (clustered) lesions may be threefold larger. These parenchymal infections can induce focal neurologic abnormalities, personality changes, intellectual impairment and/or seizures; in many endemic areas, cysticercosis is the leading cause of epilepsy. Subarachnoid lesions and cysticerci located within the fourth ventricle may obstruct the flow of CSF, producing increased intracranial pressure with its associated headache, vomiting, visual disturbances, or psychiatric abnormalities. Multiple racemose lesions have a predilection for the basal cisterns, particularly in young women, from whence they rapidly spread around the base of the brain and cerebrum with catastrophic result. Spinal involvement produces cord compression or meningeal inflammation. Eye lesions incite pain and visual disturbances.

Meningoencephalitic syndrome with eosinophilia produced by CNS invasion

Multiple small cysts formed

Focal neurologic signs and epilepsy related to cysts

DIAGNOSIS

Infection with the adult worm is diagnosed as described for *T saginata*. Cysticercosis is suspected when an individual who has been in an endemic area presents with neurologic manifestations or subcutaneous nodules. Roentgenograms of the soft tissues often reveal dead, calcified cysticerci. Viable lesions may be detected as low-density masses by computed tomography (CT) or magnetic resonance imaging (MRI). Brain cysticerci typically are 5 to 10 mm in diameter (Figure 55–4). Subarachnoid lesions are often larger, may be lobulated, and are often "isodense," making them difficult to identify radiographically. The diagnosis is confirmed by demonstrating the larva in a biopsy sample of a subcutaneous

Presence of adult worm diagnosed from proglottids

Biopsy required for cysticerci

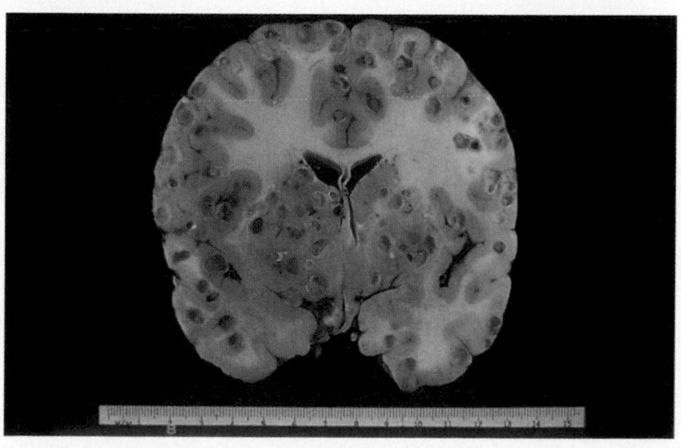

FIGURE 55–4. Cysticercosis of brain. This brain from a 16-year-old girl shows multiple cysticercal cysts primarily at the junction of white and gray matter. (Reproduced with permission from Connor DH, Chandler FW, Schwartz DQA, Manz HJ, Lack EE (eds). *Pathology of Infectious Diseases*, vol. 1. Stamford, CT: Appleton & Lange; 1997.)

nodule or specific antibodies in the circulating blood. Serum and CSF enzyme immunoassays and Western blot testing for specific anticysticercal antibodies have a sensitivity of 80% to 95%. The presence of IgG antibodies alone may reflect the presence of past or inactive disease.

TREATMENT AND PREVENTION

Surgery occasionally needed for cysticercosis

Infection with the adult worm is approached in the manner described for *T saginata*. Because the mortality rate in patients with symptomatic neurocysticercosis approaches 50%, aggressive management is warranted. Patients with parenchymal lesions usually respond to prolonged treatment with praziquantel or albendazole. Concomitant corticosteroid administration helps minimize the inflammatory response to dying cysticerci. Intraventricular subarachnoid and eye lesions appear relatively refractory to chemotherapy; surgery, CSF shunts, and corticosteroids may help ameliorate symptoms.

FISH TAPEWORM

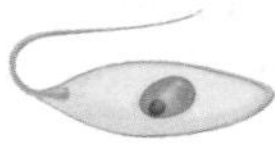

DIPHYLLOBOTHRIUM LATUM: PARASITOLOGY

The adult *D latum* attaches to the ileal mucosa with the aid of two sucking grooves (bothria) located in an elongated fusiform scolex (**Figure 55–5**). In lifespan and overall length, it resembles the *Taenia* species (discussed previously). The 3000 to 4000 proglottids,

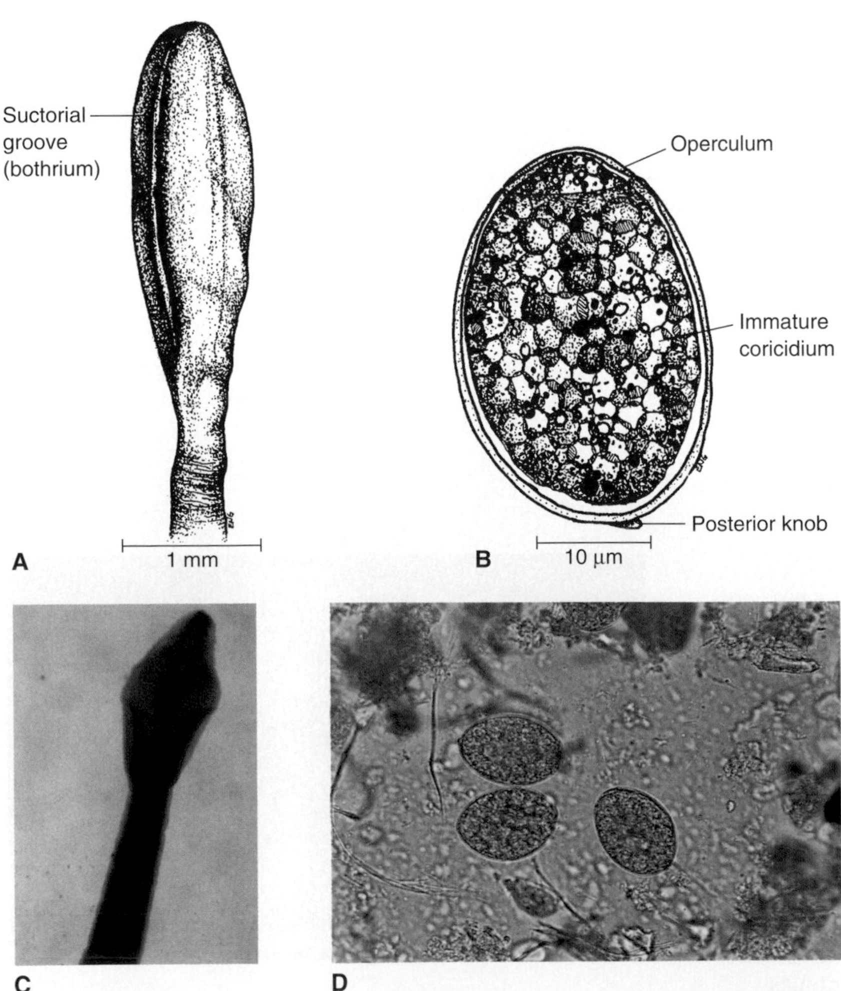

FIGURE 55–5. *Diphyllobothrium latum*. A. Structure of scolex. **B.** Structure of egg. **C.** Scolex from a human case. **D.** Ova in stool stained with iodine. (C and D, Reproduced with permission from Connor DH, Chandler FW, Schwartz DQA, Manz HJ, Lack EE (eds). *Pathology of Infectious Diseases*, vol. 1. Stamford, CT: Appleton & Lange; 1997.)

however, are uniformly wider than they are long, accounting for this cestode's species designation as well as one of its common names, the broad tapeworm. The gravid segments contain a centrally positioned, rosette-shaped uterus unique among the tapeworms of humans. Unlike those of the *Taenia* species, ova are released through the uterine pore. Over 1 million oval (55 by 75 mm) operculated eggs are released daily into the stool (Figure 55–5).

D latum has broad proglottids

On reaching fresh water the eggs hatch, releasing ciliated, free-swimming larvae or coracidia. If ingested within a few days by small freshwater crustaceans of the genera *Cyclops* or *Diaptomus,* they develop into procercoid larvae. When the crustacean is ingested by a freshwater or anadromous marine fish, the larvae migrate into the musculature of the fish and develop into infectious plerocercoid larvae. Humans are infected when they eat improperly prepared freshwater fish containing such forms.

Eggs release motile coracidia in water

Crustacean and fish intermediates; humans infected by ingesting inadequately cooked fish

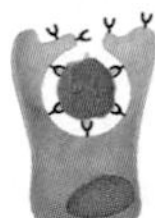

FISH TAPEWORM DISEASE

EPIDEMIOLOGY

Fish tapeworms are found wherever raw, pickled, or undercooked freshwater fish from fecally contaminated lakes and streams is eaten by humans. Other fish-eating mammals may also serve as reservoir hosts. Human infections have been described in the Baltic and Scandinavian countries, Russia, Switzerland, Italy, Japan, China, the South Pacific, Chile, and Argentina. The worm, brought to North America by Scandinavian immigrants, is now found in Alaska, Canada, the midwestern states, California, and Florida. It was shown recently that infectious plerocercoid larvae may develop in anadromous salmon, and human cases have been traced to the ingestion of fish freshly taken from Alaskan waters. The increasing popularity of raw fish dishes such as Japanese sushi and sashimi may lead to increased prevalence of this disease in the United States. Among Ontario Indians, infection is acquired by eating fresh salted fish. Even when fish is appropriately cooked, individuals may become infected by sampling the flesh during the process of preparation.

Worldwide distribution

Worm found in Alaska, Canada, midwestern states, California, and Florida

Eating raw fish increases risk

FISH TAPEWORM DISEASE: CLINICAL ASPECTS

MANIFESTATIONS

Most infected patients are asymptomatic. On occasion, however, they have complained of epigastric pain, abdominal cramping, vomiting, and weight loss. Moreover, the presence of several adult worms within the gut has been known to precipitate intestinal or biliary obstruction. Forty percent of fish tapeworm carriers demonstrate low serum levels of vitamin B_{12}, apparently as a result of the competition between the host and the worm for ingested vitamin. Studies have shown that a worm located high in the jejunum may take up 80% to 100% of vitamin B_{12} given by mouth. Approximately 0.1% to 2% of patients develop macrocytic anemia. They tend to be elderly, to have impaired production of intrinsic factor, and to have worms located high in the jejunum. In many, folate absorption is also diminished. Lysolecithin, a tapeworm product, may also contribute to the anemia. Neurologic manifestations of vitamin B_{12} deficiency occur, sometimes in the absence of anemia. They include numbness, paresthesia, loss of vibration sense, and, rarely, optic atrophy with central scotoma.

Occasional intestinal obstruction

Vitamin B_{12} deficiency related to consumption by worm

DIAGNOSIS

The diagnosis is established by finding the typical eggs in the stool. As *D latum* produces large numbers of ova, identification is usually accomplished without the need for concentration techniques.

Eggs demonstrated in stool

TREATMENT AND PREVENTION

Fish rendered noninfectious at −10°C for 48 hours

Treatment is carried out as described for *T saginata* infections. When anemia or neurologic manifestations are present, parenteral administration of vitamin B_{12} is also indicated. Personal protection can be accomplished by thorough cooking of all salmon and freshwater fish. Devotees of raw fish may choose to freeze their favorite dish at −10°C for 48 hours before serving. Ultimately, control of diphyllobothriasis is accomplished only by prohibiting the discharge of untreated sewage into lakes and streams.

HYMENOLEPIS

Can be transmitted directly human to human

The so-called "dwarf tapeworm" is the only tapeworm that can be transmitted directly from human to human. Endemic areas include parts of Asia, Europe, Central and South America and Africa. Occasionally, it is found in institutionalized persons in North America. Transmission is fecal-oral, either directly or via contaminated food. Most persons are asymptomatic, but heavy worm burdens may be associated with diarrhea, abdominal cramping, and anorexia. Treatment is similar to that for other tapeworms, but may need to be prolonged to fully eradicate the cysts.

ECHINOCOCCUS

Echinococcosis is a tissue infection of humans caused by larvae of *Echinococcus granulosus* and *E multilocularis*. The former is a more common cause of human disease.

Echinococcus Granulosus

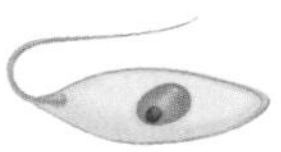

PARASITOLOGY

Adult in small intestine of canines

Herbivores and humans serve as intermediates

Larvae penetrate to portal or systemic circulation

Cysts and daughter cysts develop in tissues

Cycle completed with ingestion of cysts by canine

The adult *E granulosus* inhabits the small bowel of dogs, wolves, and other canines, where it survives for a scant 12 months. The scolex, like that of the genus *Taenia,* possesses four sucking discs and a double row of hooklets. The entire strobila, however, measures only 5 mm in length and contains but three proglottids; one immature, one mature, and one gravid. The latter segment splits either before or after passage in the stool, releasing eggs that appear identical to those of *T saginata* and *T solium.* A number of mammals may serve as intermediates, including sheep, goats, camels, deer, caribou, moose, and, most important, humans. When one of these hosts ingests eggs, they hatch. The embryos penetrate the intestinal mucosa and are carried by the portal blood to the liver. Here, many are filtered out in the hepatic sinusoids. The rest traverse the liver and are carried to the lung, where most lodge. A few pass through the pulmonary capillaries, enter the systemic circulation, and are carried to the brain, heart, bones, kidneys, and other tissues. Many of the larvae are phagocytosed and destroyed. The survivors form a cyst wall composed of an external laminated cuticle and an internal germinal membrane. The cyst fills with fluid and slowly expands, reaching a diameter of 1 cm over 5 to 6 months (**Figure 55–6**). Secondary or daughter cysts form within the original hydatid. Within each of these daughter cysts, new protoscolices are produced from the germinal lining. Some break free, dropping to the bottom of the cyst to form hydatid sand. When hydatid-containing tissues of the intermediate host are ingested by a canine, thousands of scolices are released in the intestine to develop into adult worms.

ECHINOCOCCOSIS

EPIDEMIOLOGY

There are two major epidemiologic forms of *E granulosus*–induced echinococcosis: pastoral and sylvatic. The more common pastoral form has its highest incidence in Australia,

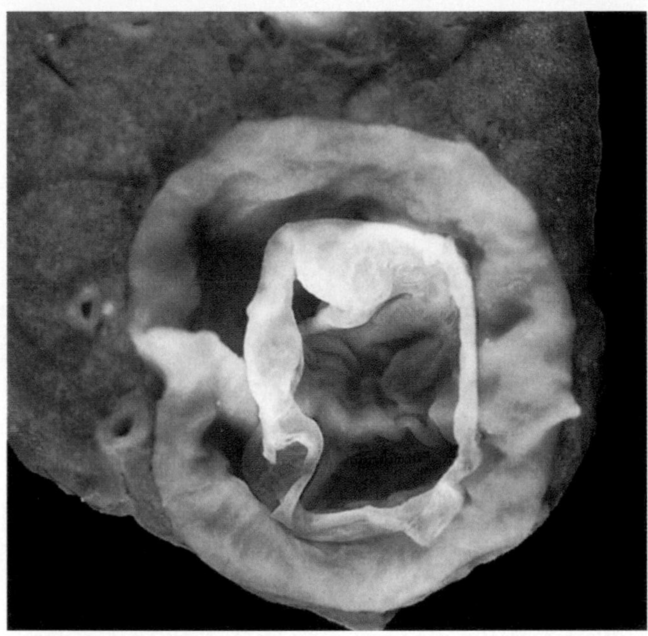
A

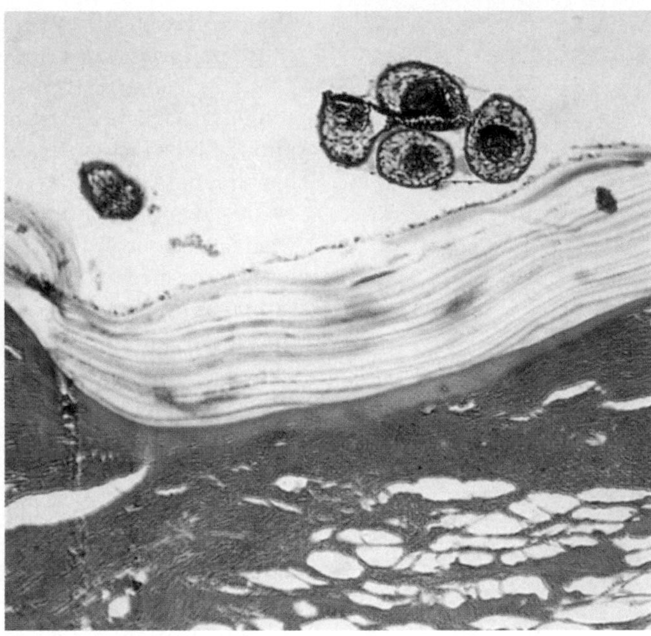
B

FIGURE 55–6. Echinococcosis. A. Echinococcal cyst in lung with white lining membrane. **B.** Echinococcal cyst wall with lung parynchema below and five scolices above. (Reproduced with permission from Connor DH, Chandler FW, Schwartz DQA, Manz HJ, Lack EE (eds). *Pathology of Infectious Diseases*, vol. 1. Stamford, CT: Appleton & Lange; 1997.)

New Zealand, South and East Africa, the Middle East, Central Europe, and South America, where domestic herbivores such as sheep, cattle, and camels are raised in close contact with dogs. Although approximately 200 human cases are reported each year in the United States, most were acquired elsewhere. Indigenous cases do occur, however, particularly among Basque sheep farmers in California, southwestern Native Americans, and some Utah shepherds. Animal husbandry practices that permit dogs to feed on the raw viscera of slaughtered sheep allow the cycle of transmission to continue. Shepherds become infected while handling or fondling their dogs. Eggs retained in the fur of these animals are picked up on the hands and later ingested. Sylvatic echinococcosis is found principally in Alaska and western Canada, where wolves act as the definitive host and moose or caribou as the intermediate. In two counties in California, a second cycle involving deer and coyotes has been described. When hunters kill these wild deer and feed their offal to accompanying dogs, a pastoral cycle may be established.

Pastoral infections maintained by allowing dogs to feed on sheep viscera

Hand-to-mouth infection of humans by dog contact

Sylvatic cycle in Alaska and western Canada

ECHINOCOCCOSIS: CLINICAL ASPECTS

MANIFESTATIONS

The enlarging *E granulosus* cysts produce tissue damage by mechanical means. The clinical presentation depends on their number, site, and rate of growth. Typically, a latent period of 5 to 20 years occurs between acquisition of infection and subsequent diagnosis. Intervals as long as 75 years have been reported occasionally.

Disease caused by mechanical effects of cysts after many years

In sylvatic infections, two thirds of the cysts are found in the lung, the remainder in the liver. Most patients are asymptomatic when the lesion is discovered on routine chest x-ray or physical examination. Occasionally, the patient may present with hemoptysis, pain in the right upper quadrant of the abdomen, or a tender hepatic mass. Significant morbidity is uncommon, and death extremely rare. In the pastoral form of disease, 60% of the cysts are found in the liver, 25% in the lung. One fifth of all patients show involvement of multiple sites. The hydatid cysts, which grow more rapidly (0.25 to 1 cm/year) than the sylvatic lesions, may reach enormous size. Twenty percent eventually rupture, inducing fever, pruritus, urticaria, and, at times, anaphylactic shock and death. Release of thousands of scolices may lead to dissemination of the infection. Rupture of pulmonary lesions also induces cough, chest pain, and hemoptysis. Liver cysts may break through the diaphragm or rupture into the bile duct or peritoneal cavity. The majority, however, present as a tender, palpable hepatic mass. Intrabiliary extrusion of calcified cysts may mimic the signs of acute cholecystitis; complete obstruction results in jaundice. Bone cysts produce pathologic fractures, whereas lesions in the CNS often manifest as blindness or epilepsy. Cardiac lesions have been associated with conduction disturbances, ventricular rupture, and embolic metastases. It has been suggested that circulating antigen–antibody complexes may be deposited in the kidney, initiating membranous glomerulonephritis.

Pulmonary cysts predominate in sylvatic disease, hepatic in pastoral

Cysts may attain large size

Rupture leads to hypersensitivity manifestations and dissemination

DIAGNOSIS

In *E granulosus*–infected patients, chest x-rays reveal pulmonary lesions as slightly irregular, round masses of uniform density devoid of calcification. In contrast, more than one half of hepatic lesions display a smooth, calcific rim. CT, ultrasonography, and MRI may reveal either a simple fluid-filled cyst or daughter cysts with hydatid sand. Endoscopic retrograde cholangiography has been valuable for determining cyst location and possible communication with the biliary tree. Because of the potential for an anaphylactoid reaction and dissemination of infection, diagnostic aspiration has been considered contraindicated. Nevertheless, in the hands of some investigators, ultrasonically guided percutaneous drainage, followed by the introduction of ethanol to kill protoscoleces and germinal layer, has proved safe and useful both diagnostically and therapeutically (see following text). In patients with ruptured pulmonary cysts, scolices may be demonstrated in the sputum.

Radiologic and scanning appearance characteristic

In most cases, confirmation of the diagnosis requires serologic testing. Unfortunately, current procedures are not totally satisfactory. Indirect hemagglutination and latex agglutination tests are positive in 90% of patients with hepatic lesions and 60% of those with pulmonary hydatid cysts. When using hydatid cyst fluid or soluble scolex antigen, the presence of a precipitin line in the immunoelectrophoresis test appears to be more specific. An adaption of this test to an enzyme-linked immunoelectrodiffusion technique appears to provide a rapid, sensitive diagnostic test. Other serologic tests are in the process of evaluation. Polymerase chain reaction assay has been shown capable of detecting picogram quantities of *Echinococcus* genomic DNA in fine-needle biopsy material from patients with suspected echinococcosis.

Serologic diagnosis important, but needs improved sensitivity

TREATMENT AND PREVENTION

For years, the only definitive therapy available was surgical extirpation. Patients with pulmonary hydatid cysts of the sylvatic type and small calcified hepatic lesions underwent surgery only when they became symptomatic or the cysts increased dramatically in size over time. For other lesions, **P**ercutaneous **A**spiration, **I**nfusion of scolicidal and **R**easpiration (PAIR) can be used in lieu of surgery. Presently, it is recommended that high-dose albendazole be administered before and for several weeks (or years in the case of *E multilocularis* infection) after surgery and/or aspiration. Infected dogs should be wormed, and infected carcasses and offal burned or buried. Hands should be carefully washed after contact with potentially infected dogs.

Treatment may include careful aspiration with concomitant albendazole

Echinococcus Multilocularis

E multilocularis is found primarily in subarctic and arctic regions in North America, Europe, and Asia. The adult worms are found in the gut of foxes and, to a lesser extent, coyotes. Their larval forms find harborage in the tissues of mice and voles, the canines' rodent prey. Domestic dogs may acquire adult tapeworms by killing and ingesting these larval-infected sylvatic rodents. Humans are infected with larval forms through the ingestion of eggs passed in the feces of their domestic dogs or ingestion of egg-contaminated

Larvae bud externally; produce multilocular cysts

vegetation. Unlike the larval forms of *E granulosis*, those of *E multilocularis* bud externally, producing proliferative, multilocular cysts that slowly but progressively invade and destroy the affected organs and adjacent tissues.

The clinical course in humans is characterized by epigastric pain; obstructive jaundice; and, less frequently, metastasis to the lung and brain, thus closely mimicking a hepatoma.

CASE STUDY

SEIZURES ON THE TENNIS COURT

A 26-year-old professional tennis player from Mexico suddenly developed a left-sided epileptic seizure, lasting 5 minutes, while competing in an international tournament. He had no history of such occurrences and had been well before this episode.

Physical examination was normal, but brain MRI imaging revealed a round, calcified 3-cm lesion in the right parietal lobe.

QUESTIONS

■ Which of the following is most likely responsible for this patient's condition?

A. *Taenia saginata*
B. *Taenia solium*
C. *Echinococcus granulosus*
D. *Diphyllobothrium latum*

■ Vitamin B_{12} deficiency, with macrocytic anemia is associated with which of the following parasites:

A. *Echinococcus multilocularis*
B. *Diphyllobothrium latum*
C. *Taenia saginata*
D. *Taenia solium*

■ Which is the most common intermediate host for transmission of echinococcosis?

A. Pig
B. Cow
C. Fish
D. Dog

ANSWERS

1(B), **2**(B), **3**(D)

CHAPTER 56

Trematodes

Of the myriad relationships that have developed between helminths and humans over the millennia of our mutual existence, none has proved more destructive to our health and productivity than that forged by the indomitable flukes. Typically, the adults live for decades within human tissues and vascular systems, where they resist immunologic attack and produce progressive damage to vital organs. Morphologically, trematodes are bilaterally symmetric, vary in length from a few millimeters to several centimeters, and possess two deep suckers from which they derive their name ("body with holes"). One surrounds the oral cavity, and the other is located on the ventral surface of the worm. These organs are used for both attachment and locomotion; movement is effected in a characteristic inchworm fashion.

The digestive tract begins at the oral sucker and continues as a muscular pharynx and esophagus before bifurcating to form bilateral ceca that end blindly near the posterior extremity of the worm. Undigested food is vomited out through the oral cavity. The excretory system consists of a number of hollow, ciliated flame cells that excrete waste products into interconnecting ducts terminating in a posterior excretory pore.

Persistent flukes move through tissue and vasculature with inchworm locomotion

The reproductive systems vary and serve as a means of dividing the trematodes into two major categories: the hermaphrodites and the schistosomes. The adult hermaphrodite contains both male and female gonads and produces operculated (defined as a lid or covering flap) eggs. The schistosomes have separate sexes, and the fertilized female deposits only nonoperculated offspring. The two groups have similar life cycles. The major differential features are summarized in **Table 56–1.** Eggs are excreted from the human host and, if they reach fresh water, hatch to release ciliated larvae called **miracidia**. These larvae find and penetrate a snail host specific for the trematode species. In this intermediate host, they are transformed by a process of asexual reproduction into thousands of tail-bearing larvae or cercariae, which are released from the snail over a period of weeks and swim about vigorously in search of their next host. In the case of schistosomal cercariae, this host is the human. When they come in contact with the skin surface, they attach, discard their tails, and invade, thereby completing their life cycle. The cercariae of the hermaphroditic flukes encyst in or on an aquatic plant or animal, where they undergo a second transformation to become infective metacercariae. Their cycle is completed when the second intermediate host is ingested by a human.

Two types of reproductive systems

Snails release motile cercariae in water

Of the many trematodes that infect humans, only five are discussed: the blood flukes, all of which are members of the genus *Schistosoma* (*S mansoni, S haematobium,* and *S japonicum*); and the lung (*Paragonimus* spp.) and liver (*Clonorchis sinensis*) flukes, which are hermaphroditic. Basic details of other hermaphroditic tissue and intestinal flukes are listed in **Table 56–2**.

Schistosoma cercariae infect humans through skin

Paragonimus and *Clonorchis* have second intermediate host

TABLE 56–1 General Characteristics of Trematodes

	TREMATODE TYPE	
CHARACTERISTIC	**BLOOD**	**TISSUE/INTESTINAL**
Genus	*Schistosoma*	*Paragonimus, Clonorchis, Opisthorchis, Fasciola*
Morphology		
Adult	Oral and ventral suckers	Oral and ventral suckers
	Blind gastrointestinal tract	Blind gastrointestinal tract
	Slender, worm-like	Flat, leaf-like
Egg	Nonoperculated	Operculated
Biology		
Sexes	Separate	Hermaphroditic
Intermediates	One	Two
Life span	Long	Long

TABLE 56–2 Intestinal and Tissue Trematodes

	PARAGONIMUS	***CLONORCHIS***	***OPISTHORCHIS***	***FASCIOLA***	***FASCIOLOPSIS***	***HETEROPHYES/ METAGONIMUS***
Distribution						
Geographic	Asia, Africa, Central America	Japan, China, Taiwan, Vietnam	Asia, Eastern Europe	Worldwide	East and Southeast Asia	Asia, former USSR, Mediterranean
Infected population (in millions)	3	20	4	—	10	—
Adult worms						
Reservoir hosts	Domestic and wild animals	Cats, dogs	Domestic and wild animals	Sheep and other herbivores	Pigs	Fish-eating mammals
Location in body	Lungs, CNS	Biliary tract	Biliary tract	Biliary tract	Small intestine	Small intestine
Length (mm)	7-12	10-25	10	20-30	20-75	1-2
Life span (years)	4-6	20-30	20-30	10-15	0.5	1
Eggs						
Characteristics	Operculated	Operculated	Operculated	Operculated	Operculated	Operculated
Size (μm)	80-100	26-30	26-30	130-150	130-150	26-30
Location[a]	Sputum, stool	Bile, stool	Bile, stool	Bile, stool	Stool	Stool
Larvae						
First intermediate	Snail	Snail	Snail	Snail	Snail	Snail
Second intermediate	Freshwater crab and crayfish	Freshwater fish	Freshwater fish	Watercress and other aquatic plants	Water chestnut and other aquatic plants	Freshwater fish

CNS, central nervous system.
[a]Diagnostic specimens.

PARAGONIMUS

PARAGONIMUS SPECIES: PARASITOLOGY

Several *Paragonimus* species may infect humans. *P westermani,* which is widely distributed in East Asia, is the species most frequently involved. The short, plump (10 by 5 mm), reddish-brown adults are characteristically found encapsulated in the pulmonary parenchyma of their definitive host. Here they deposit operculated, golden-brown eggs, which are distinguished from similar structures by their size (50 by 90 μm) and prominent periopercular shoulder. When the capsule erodes into a bronchiole, the eggs are coughed up and spat out or swallowed and passed in the stool. If they reach fresh water, they embryonate several weeks before the ciliated miracidia emerge through the open opercula. After invasion of an appropriate snail host, 3 to 5 months pass before cercariae are released. These larval forms invade the gills, musculature, and viscera of certain crayfish or freshwater crabs; over 6 to 8 weeks, the larval forms transform into metacercariae. When the raw or undercooked flesh of the second intermediate host is ingested by humans, the metacercariae encyst in the duodenum and burrow through the gut wall into the peritoneal cavity. The majority continue their migration through the diaphragm and reach maturity in the lungs 5 to 6 weeks later. Some organisms, however, are retained in the intestinal wall and mesentery or wander to other foci such as the liver, pancreas, kidney, skeletal muscle, or subcutaneous tissue. Young worms migrating through the neck and jugular foramen may encyst in the brain, the most common ectopic site. In addition to humans, other carnivores, including the rat, cat, dog, and pig, may serve as definitive hosts. Immature ectopic adults in the striated muscles of the pig may infect humans after ingestion of undercooked pork.

Adults encapsulate in lung

Capsule erodes into bronchiole and eggs are coughed up; cycle continues if eggs reach water with susceptible snail

Crayfish and freshwater crabs are second intermediate hosts

Other carnivores are also definitive hosts

PARAGONIMIASIS (LUNG FLUKE INFECTION)

EPIDEMIOLOGY

Although most of the 5 million human infections are concentrated in the Far East (eg, Korea, Japan, China, Taiwan, the Philippines, and Indonesia), paragonimiasis has recently been described in India, Africa (*P africanus*), and Latin America (*P mexicanus*). *P kellicotti,* a parasite of mink, is widely distributed in eastern Canada and the United States but rarely produces human infection. Approximately 1% of recent Indochinese immigrants to the United States are found to be infected with *P westermani.* Infection of the snail host, which is typically found in small mountain streams located away from human habitation, is probably maintained by animal hosts other than humans. Human disease occurs when food shortages or local customs expose individuals to infected crabs. When these crustaceans are prepared for cooking, juice containing metacercariae may be left behind on the working surface and contaminate other foods subsequently prepared in the same area. Fresh crab juice, which is used for the treatment of infertility in Cameroon and of measles in Korea, may also transmit the disease. In the Far East, crabs are frequently eaten after they have been lightly salted, pickled, or immersed briefly in wine (drunken crab), practices that are seldom lethal to the metacercariae. Children living in endemic areas may be infected while handling or ingesting crabs during the course of play.

Infected snails often found in mountain streams

Humans infected by ingesting infected crustaceans

PARAGONIMIASIS (LUNG FLUKE INFECTION): CLINICAL ASPECTS

MANIFESTATIONS

Adult worms in the lung elicit an eosinophilic inflammatory reaction and, eventually, the formation of a 1- to 2-cm fibrous capsule that surrounds and encloses one or more parasites.

Multiple lung cysts are formed

Secondary infection of ruptured cysts produces bronchitis

Chronic pulmonary abscess may resemble tuberculosis

The infected patient may harbor as many as 25 such lesions. With the onset of oviposition, the capsule swells and erodes into a bronchiole, resulting in expectoration of the brownish eggs, blood, and an inflammatory exudate. Secondary bacterial infection of the evacuated cysts is common, producing a clinical picture of chronic bronchitis or bronchiectasis. When cysts rupture into the pleural cavity, chest pain and effusion can result.

Early in infection, chest x-rays demonstrate small segmental infiltrates; these are gradually replaced by round nodules that may cavitate. Eventually, cystic rings, fibrosis, and calcification occur, producing a picture closely resembling that of pulmonary tuberculosis. The confusion is compounded by the frequent coexistence of the two diseases.

Adult flukes in the intestine and mesentery produce pain, bloody diarrhea, and on occasion, palpable abdominal or cutaneous masses; the latter is characteristic of a second Chinese fluke, *P skrjiabini.* In approximately 1% of cases of paragonimiasis in the Far East, more commonly in children, parasites lodge in the brain and produce a variety of neurologic manifestations, including epilepsy, paralysis, homonymous hemianopsia, optic atrophy, and papilledema.

DIAGNOSIS

Eggs difficult to find in sputum, pleural fluid, and feces

Serodiagnosis, antigen detection procedures available

Eggs are usually absent from the sputum during the first 3 months of overt infection; however, repeated examinations eventually demonstrate them in more than 75% of infected patients. When a pleural effusion is present, it should be checked for eggs. Stool examination is frequently helpful, particularly in children who swallow their expectorated sputum. Approximately 50% of patients with brain lesions demonstrate calcification on x-ray films of the skull. The cerebrospinal fluid in such cases shows elevated protein levels and eosinophilic leukocytosis. A diagnosis in these cases, however, often depends on the detection of circulating antibodies. Their presence usually correlates well with acute disease and disappears with successful therapy. Recently developed antigen detection techniques have been proved to be both highly sensitive and specific and may soon displace antibody detection procedures.

TREATMENT AND PREVENTION

Lung fluke infection responds well to praziquantel or bithionol therapy. Control requires adequate cooking of shellfish before ingestion.

CLONORCHIS

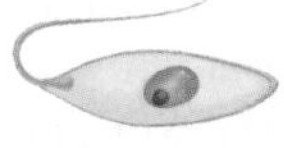

CLONORCHIS SINENSIS: PARASITOLOGY

Adults survive decades in biliary tract

Eggs discharged in bile ducts appear in feces

Snails are first intermediate host; fish second

Metacercariae from ingested fish migrate to biliary system

Flukes of the genera *Fasciola, Opisthorchis,* and *Clonorchis* all may infect the human biliary tract and at times produce manifestations of ductal obstruction. *C sinensis,* the Chinese liver fluke, is the most important and is discussed here (Table 56–2). The small, slender (5 by 15 mm) adult survives up to 50 years in the biliary tract of its host by feasting on the rich mucosal secretions. A cone-shaped anterior pole, a large oral sucker, and a pair of deeply lobular testes arranged one behind the other in the posterior third of the worm distinguish it from other hepatic parasites. Approximately 2000 tiny (15 by 30 μm) ovoid eggs are discharged daily and find their way down the bile duct and into the fecal stream. The exquisite urn-shaped shells have a discernible shoulder at their opercular rim and a tiny knob on the broader posterior pole (**Figure 56–1**). On reaching fresh water, they are ingested by their intermediate snail host, transformed into cercariae (**Figure 56–2A**), and released to penetrate the tissues of freshwater fish, in which they encyst to form metacercariae. If the latter host is ingested by a fish-eating mammal, the larvae are released in the duodenum, ascend the common bile duct, migrate to the second-order bile ducts, and mature to adulthood over 30 days.

In addition to humans, rats, cats, dogs, and pigs may serve as definitive hosts.

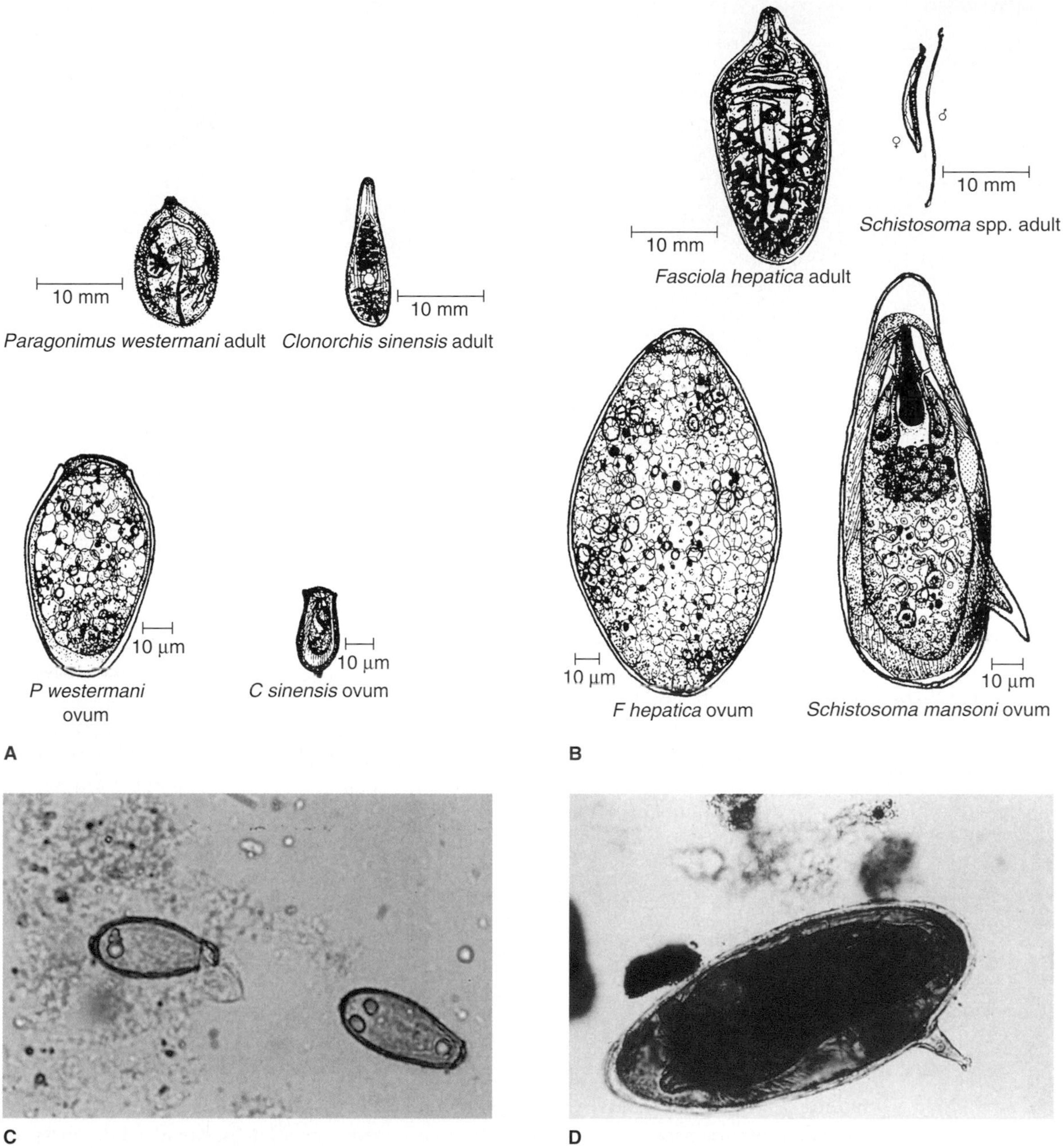

Fasciola hepatica adult
10 mm
Schistosoma spp. adult
10 mm
Paragonimus westermani adult
10 mm
Clonorchis sinensis adult
10 mm
P westermani ovum
10 μm
C sinensis ovum
10 μm
F hepatica ovum
10 μm
Schistosoma mansoni ovum
10 μm
A
B
C
D

FIGURE 56–1. Trematode eggs. A. Structure of *Paragonimus* and *Clonorchis* adults and ova. **B.** Structure of *Fasciola* and *Schistosoma* adults and ova. **C.** Two *Clonorchis* sinensis eggs in stool. The left egg has an open operculum to hatch a transparent miracidium. **D.** Mature *Schistosoma mansoni* egg in stool. (C and D, Reproduced with permission from Connor DH, Chandler FW, Schwartz DQA, Manz HJ, Lack EE (eds). *Pathology of Infectious Diseases*, vol. 1. Stamford, CT: Appleton & Lange; 1997.)

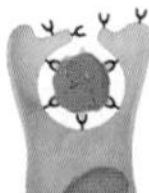

CLONORCHIASIS (LIVER FLUKE INFECTION)

EPIDEMIOLOGY

Clonorchiasis is endemic in the Far East, particularly in Korea, Japan, Taiwan, the Red River Valley of Vietnam, the Southern Chinese province of Kwantung, and Hong Kong. In previous years, parasite transmission was perpetuated by the practice of fertilizing commercial fish ponds with human feces. Recent improvements in the disposal of human

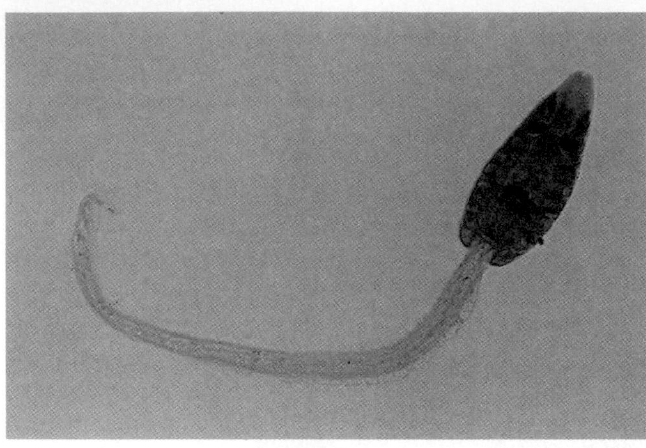

A

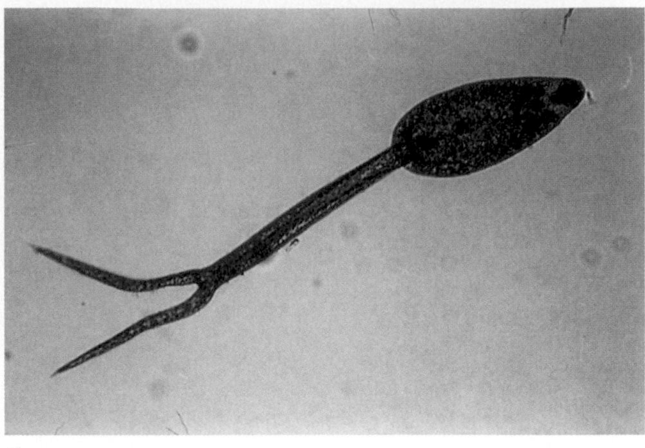

B

FIGURE 56–2. Trematode cercarial larvae. A. *Clonorchis sinensis.* **B.** *Schistosoma mansoni.* (Reproduced with permission from Connor DH, Chandler FW, Schwartz DQA, Manz HJ, Lack EE (eds). *Pathology of Infectious Diseases*, vol. 1. Stamford, CT: Appleton & Lange; 1997.)

Endemic in Far East

Transmission to humans related to waste disposal

Ingestion of uncooked fish infects humans

waste have diminished acquisition of the disease in most countries. However, the extremely long lifespan of these worms is reflected in a much slower decrease in the overall infection rate. In some villages in southern China, the entire adult population is infected. A recent survey of stool specimens from immigrants from Hong Kong to Canada showed an infection rate of more than 15% overall and 23% in adults between 30 and 50 years of age. Clonorchiasis is acquired by eating raw, frozen, dried, salted, smoked, or pickled fish. Commercial shipment of such products outside of the endemic area may result in the acquisition of worms far from their original source.

CLONORCHIASIS (LIVER FLUKE INFECTION): CLINICAL ASPECTS

MANIFESTATIONS

Light infection usually asymptomatic

Severe hepatic and biliary manifestations from heavy worm loads

Migration of the larvae from the duodenum to the bile duct may produce fever, chills, mild jaundice, eosinophilia, and liver enlargement. The adult worm induces epithelial hyperplasia, adenoma formation, and inflammation and fibrosis around the smaller bile ducts. In light infection, clinical disease seldom results. However, numerous reinfections may produce worm loads of 500 to 1000, resulting in the formation of bile stones and sometimes bile duct carcinoma in patients with severe, long-standing infections. Calculus formation is often accompanied by asymptomatic biliary carriage of *Salmonella typhi.* Dead worms may obstruct the common bile duct and induce secondary bacterial cholangitis, which may be accompanied by bacteremia, endotoxin shock, and hypoglycemia. Occasionally, adult worms are found in the pancreatic ducts, where they can produce ductal obstruction and acute pancreatitis.

DIAGNOSIS

Distinctive eggs present in feces and duodenal aspirates

Definitive diagnosis of clonorchiasis requires the recovery and identification of the distinctive egg from the stool or duodenal aspirates. In mild infections, repeated examinations may be required. Because most patients are asymptomatic, any individual with clinical manifestations of disease in whom *Clonorchis* eggs are found must be evaluated for the presence of other causes of illness. In acute symptomatic clonorchiasis, there is usually leukocytosis, eosinophilia, elevation of alkaline phosphatase levels, and abnormal computed tomography and ultrasonographic liver scans. Cholangiograms may reveal dilatation of the intrahepatic ducts, small filling defects compatible with the presence of adult worms, and occasionally cholangiocarcinoma.

Eosinophilia common in acute disease

TREATMENT AND PREVENTION

Praziquantel and albendazole have proved to be effective therapeutic agents. Prevention requires thorough cooking of freshwater fish and sanitary disposal of human feces.

SCHISTOSOMA

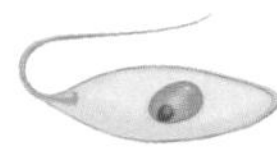

SCHISTOSOMA SPECIES: PARASITOLOGY

The schistosomes are a group of closely related flukes that inhabit the portal vascular system of a number of animals. Of the five species known to infect humans, *S mansoni, S haematobium,* and *S japonicum* are of primary importance. They infect 200 to 300 million persons in Africa, the Middle East, Southeast Asia, the Caribbean, and South America and kill 1 million annually. The remaining two species are found in limited areas of West Africa (*S intercalatum*) and Southeast Asia (*S mekongi*), and are not discussed here in detail.

Inhabit portal vascular system

The adult worms can be distinguished from the hermaphroditic trematodes by the anterior location of their ventral sucker, by their cylindric bodies, and by their reproductive systems (ie, separate sexes). They are differentiated from one another only with difficulty. The 1- to 2-cm male possesses a deep ventral groove, or gynecophoral canal, in which it carries the longer, more slender female in lifelong copulatory embrace. The schistosome life cycle (**Figure 56–3**) begins after mating in the portal vein when the conjoined couple uses their suckers to ascend the mesenteric vessels against the flow of blood. Guided by unknown stimuli, *S japonicum* enters the superior mesenteric vein, eventually reaching the venous radicals of the small intestine and ascending colon; *S mansoni* and *S haematobium* are directed to the inferior mesenteric system. The destination of the former is the descending colon and rectum; the latter, however, passes through the hemorrhoidal plexus to the systemic venous system, ultimately coming to rest in the venous plexus of the bladder and other pelvic organs.

Different morphology and separate sexes

S mansoni reaches colon and rectum, and *S haematobium* reaches veins of bladder and pelvic organs

On reaching the submucosal venules, the worms initiate oviposition. Each pair deposits 300 (*S mansoni, S haematobium*) to 3000 (*S japonicum*) eggs daily for the remainder of its 4- to 35-year life span. Enzymes secreted by the enclosed miracidium diffuse through the shell and digest the surrounding tissue. Ova lying immediately adjacent to the mucosal surface rupture into the lumen of the bowel (*S mansoni, S japonicum*) or bladder (*S haematobium*) and are passed to the outside in the excreta. Here, with appropriate techniques, they may be readily observed and differentiated. The eggs of *S mansoni* are oval, possess a sharp lateral spine, and measure 60 by 140 μm (Figure 56–1). Those of *S haematobium* differ primarily in the terminal location of their spine. The eggs of *S japonicum,* in contrast, are more nearly circular, measuring 70 by 90 μm. A minute lateral spine can be visualized only with care.

Eggs deposited submucosally, rupture to lumina, and pass outside

When the eggs are deposited in fresh water, the miracidia hatch quickly. On finding a snail host appropriate for their species, they invade and are transformed over 1 to 2 months into thousands of forked-tailed cercariae (Figure 56–2B). When released from the snail, these infectious larvae swim about vigorously for a few days. Cercariae coming in contact with human skin during this time attach, discard their tails, and penetrate. During a 1- to 3-day sojourn in the skin, the outer cercarial membrane is transformed from a trilaminar to a heptalaminar structure, an adaption that is thought to be critical to the survival of the parasite within the human body. The resulting schistosomula enter small venules and find their way through the right side of the heart to the lung. After a delay of several days, the parasites enter the systemic circulation and are distributed to the gut. Those surviving passage through the pulmonary and intestinal capillary beds return to the portal vein, where they mature to sexually active adults over 1 to 3 months.

In water, eggs hatch to form miracidia, which invade snail

Cercariae from snail traverse human skin and vascular system

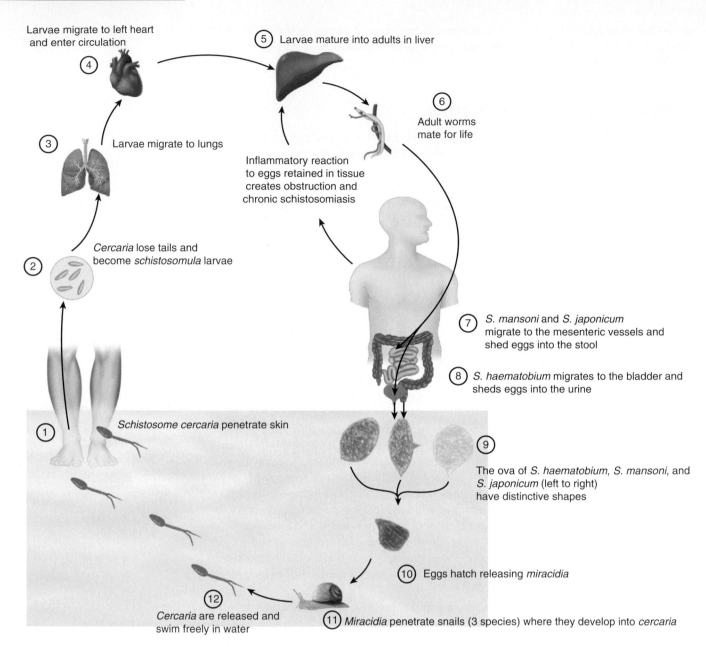

(5) Larvae mature into adults in liver
(6) Adult worms mate for life
(3) Larvae migrate to lungs

(2) Cercaria lose tails and become schistosomula larvae

(8) S. haematobium migrates to the bladder and sheds eggs into the urine

(12) Cercaria are released and swim freely in water
(11) Miracidia penetrate snails (3 species) where they develop into cercaria

FIGURE 56–3. Life cycle of schistosomes.

SCHISTOSOMIASIS (BLOOD FLUKE INFECTION)

EPIDEMIOLOGY

Most important of helminthic infections would be stopped by modern waste disposal

Larvae migrate to left heart and enter circulation

The widespread distribution and extensive morbidity of schistosomiasis make it the single most important helminthic infection in the world today. Currently, more than 200 million people in 74 countries are infected. The continued presence of the parasite depends on the disposal of infected human excrement into fresh water, the availability of appropriate snail hosts, and the exposure of humans to water infected with cercariae. The construction of modern sanitation and water purification facilities would break this cycle of transmission but exceeds the economic resources of most endemic nations. Paradoxically, several massive land irrigation projects launched over the last two decades for the express purpose of speeding economic development have resulted in the dispersion of infected humans and snails to previously uninvolved areas. *S mansoni,* the most widespread of the blood flukes, is the only one present in the Western Hemisphere. Originally thought to have been introduced by African slaves, *S mansoni* is now found in Venezuela, Brazil, Surinam, Puerto Rico, the Dominican Republic, St. Lucia, and several other Caribbean islands.

Spread to areas caused by new irrigation projects

Because a suitable snail host is lacking, transmission of *S mansoni* does not occur within the continental United States; however, nearly a half-million individuals residing in the United States have acquired schistosomiasis elsewhere. Puerto Rican, Yemenite, and Southeast Asian populations are predominantly involved. In the Eastern Hemisphere, the prevalence of *S mansoni* infection is highest in the Nile Delta and the tropical section of Africa. Isolated foci are also found in East and South Africa, Yemen, Saudi Arabia, and Israel.

Geographic distribution varies with species and depends on presence of snail host

S haematobium is largely confined to Africa and the Middle East, where its distribution overlaps that of *S mansoni. Schistosoma japonicum* affects the agricultural populations of several Far Eastern countries, including Japan, China, the Philippines, and the Celebes. The closely related *S mekongi* is found in the Mekong and Mun River valleys of Vietnam, Thailand, Cambodia, and Laos.

Within endemic areas of schistosomiasis, wide variations in both age-specific infection rates and worm loads exist. In general, both peak in the second decade of life and then decrease with advancing age. This finding has been explained in part by changes in the intensity of water exposure and in part by the slow development of IgE-mediated immunity. Most infected patients carry fewer than 10 pairs of worms in the vascular system and, accordingly, lack clinical manifestations of disease. Individuals who develop much heavier loads as a result of repeated infections may experience serious morbidity or mortality. Patients with concomitant *S mansoni* and human immunodeficiency virus infections excrete substantially fewer eggs in their stool.

Age-related susceptibility with peak in second decade

PATHOGENESIS

There are three major clinicopathologic stages in schistosomiasis. The first stage is initiated by the penetration and migration of the schistosomula. The second or intermediate stage begins with oviposition and is associated with a complex of clinical manifestations. The third or chronic stage is characterized by granuloma formation and scarring around retained eggs.

IMMUNITY

The major clinicopathologic manifestations of schistosomiasis result from the host's cell-mediated immune response to the presence of retained eggs. With time, the intensity of this reaction is muted; granulomas formed in the later stages of infection are smaller and less damaging than those formed early. The mechanisms responsible for this modulation are not fully understood. Present evidence suggests that both suppressor T-lymphocyte activity and antibody blockade are involved. The correlation in humans between human leukocyte antigen (HLA) types A1 and B5 and the development of hepatosplenomegaly suggests that the extent of the immunoregulation is influenced, at least in part, by the genetic background of the host. ::: cell-mediated immune response, p. 35

Major manifestations from cell-mediated immune response to eggs

As evidenced by their prolonged survival, the adult worms are remarkably well tolerated by their hosts. In part, this tolerance may be attributable to the formation of IgG4 blocking antibodies early in the course of infection. Tolerance may also reflect the ability of the developing parasites to disguise themselves by adsorbing host molecules, including immunoglobulins, blood group glycolipids, and histocompatibility complex antigens. Nevertheless, as mentioned earlier, the prevalence and intensity of human infection begin to abate during adolescence, despite continuing exposure to infective cercariae. It has been suggested that schistosomula penetrating the skin after the primary infection are coated with specific antibody, bound to eosinophils, and destroyed before they can reach the portal system. Although protection is not complete, the 60% to 80% kill rate is highly effective in controlling the intensity of parasitism. This condition, in which adult worms from a primary infection can survive in a host resistant to reinfection, has been termed **concomitant**

Blocking antibodies and adsorption of host molecules provide antigenic disguise

Concomitant immunity prevents new infections

immunity. Eventually, production of blocking antibodies wanes, and that of protective IgE antibodies active against adult worms increases, leading to a decrease in the host's total worm population.

SCHISTOSOMIASIS (BLOOD FLUKE INFECTION): CLINICAL ASPECTS

EARLY STAGE

Within 24 hours of penetrating the skin, a large proportion of the schistosomula die. In *S mansoni* and *S haematobium* infections, immediate and delayed hypersensitivity to parasitic antigens results in an intensely pruritic papular skin rash, which increases in severity with repeated exposures to cercariae. As the viable schistosomula begin their migration to the liver, the rash disappears and the patient experiences fever, headache, and abdominal pain for 1 to 2 weeks.

Local and systemic hypersensitivity reactions produce rash

Note: In the United States, cercariae of avian schistosomes can penetrate human skin and die, producing an intensely pruritic, transient rash known as "swimmers' itch." No further disease occurs.

INTERMEDIATE STAGE

One to two months after primary exposure, patients with severe *S mansoni* or *S japonicum* infections may experience an acute febrile illness that bears a striking resemblance to serum sickness. The onset of oviposition leads to a state of relative antigen excess, the formation of soluble immune complexes, and the deposition of these in the tissues of the host. Indeed, high levels of such complexes have been demonstrated in the peripheral blood and correlate well with the severity of illness. In addition to fever and chills, patients experience cough, urticaria, arthralgia, lymphadenopathy, splenomegaly, abdominal pain, and diarrhea. Sigmoidoscopic examination reveals an inflamed colonic mucosa and petechial hemorrhages; occasionally, patients with *S japonicum* infection develop clinical manifestations of encephalitis. Typically, leukocytosis, marked peripheral eosinophilia, and elevated levels of IgM, IgG, and IgE immunoglobulins are present. This symptom complex is commonly termed **Katayama syndrome.** It is more common and more severe in visitors to endemic areas, in whom it may persist for 3 months or more and occasionally result in death.

Prolonged febrile period with circulating immune complexes

Intestinal inflammation and encephalitis occur acutely

CHRONIC STAGE

Approximately one half of all deposited eggs reach the lumen of the bowel or bladder and are shed from the body. Eggs that are retained induce inflammation and scarring, initiating the final and most morbid phase of schistosomiasis. Soluble antigens excreted by the eggs stimulate the formation of T-lymphocyte–mediated eosinophilic granulomas. Early in the infection, the inflammatory response is vigorous, producing lesions more than 100-fold larger than the inciting egg itself. Obstruction of blood flow is common. With time, the host's inflammatory response moderates, leading to a significant decrease in granuloma size. Fibroblasts stimulated by factors released by both retained eggs and the granulomas lay down scar tissue, rendering the earlier, granuloma-induced vascular obstruction permanent. As would be expected, the severity of tissue damage is directly related to the total number of eggs retained.

Inflammatory and fibrotic reactions to retained eggs cause chronic disease

In *S haematobium* infection, the bladder mucosa becomes thickened, papillated, and ulcerated. Hematuria and dysuria result; repeated hemorrhages produce anemia. In severe infections, the muscular layers of the bladder are involved, with loss of bladder capacity and contractibility. Vesicoureteral reflux, ureteral obstruction, and hydronephrosis may follow. Progressive obstruction leads to renal failure and uremia. Calcification of the bladder wall is occasionally seen, and approximately 10% of patients harbor urinary tract calculi.

S haematobium produces bladder lesions with hemorrhage and obstruction

Secondary bacterial infections are common. Chronic *Salmonella* bacteriuria with recurrent bouts of bacteremia have been reported from Egypt, where bladder carcinoma is frequently seen as a late complication of disease.

Chronic urinary carriage of *Salmonella* may cause bacteremia

In *S mansoni* and *S japonicum* infections, the bowel mucosa is congested, thickened, and ulcerated. Polyposis has been reported from Egypt but nowhere else. Patients experience abdominal pain, diarrhea, and blood in the stool. Eggs deposited in the larger intestinal veins may be carried by the portal blood flow back to the liver, where they lodge in the presinusoidal capillaries. The resulting inflammatory reaction leads to the development of periportal fibrosis and hepatic enlargement. The frequency and severity with which the liver is involved are genetically determined and associated with the patient's HLA type. In most cases, liver function is well preserved. Infected persons who subsequently acquire hepatitis B or C viruses develop chronic active hepatitis more frequently than those free of schistosomes. The presinusoidal obstruction to blood flow can result in the serious manifestations of portal obstruction. Eggs carried around the liver in the portosystemic collateral vessels may lodge in the small pulmonary arterioles, where they produce interstitial scarring, pulmonary hypertension, and right ventricular failure. Immune complexes shunted to the systemic circulation may induce glomerulonephritis. Occasionally, eggs may be deposited in the central nervous system, where they may cause epilepsy or paraplegia.

Severity of liver involvement linked to HLA type

Hepatitis B or C superinfection may progress to chronic active hepatitis

Some differences between the clinical presentation of schistosomiasis mansoni and that of schistosomiasis japonicum have been noted. Manifestations of the latter disease typically occur earlier in the course of the infection and tend to be more severe. When involvement of the central nervous system develops, it is more likely to occur in the brain than in the spinal cord. On the other hand, immune complex nephropathy and recurrent *Salmonella* bacteremia are more likely to be seen in hepatosplenic *S mansoni* infections. The latter phenomenon is apparently related to the ability of *Salmonella* to parasitize the gut and integument of the adult fluke, providing a persistent bacterial focus within the portal system of the infected patient. This focus cannot be eradicated without treatment of the schistosomal infection.

Elimination of *Salmonella* focus requires eradication of parasite

DIAGNOSIS

Definitive diagnosis of schistosomiasis requires the recovery of the characteristic eggs in urine, stool, or biopsy specimens. In *S haematobium* infections, eggs are most numerous in urine samples obtained at midday. When examination of the sediment yields negative results, eggs may sometimes be recovered by filtering the urine through a membrane filter. Cystoscopy with biopsy of the bladder mucosa may be required for the diagnosis of mild infection. Eggs of *S mansoni* and *S japonicum* are passed in the stool. Concentration techniques such as formalin–ether or gravity sedimentation are necessary when the ova are scanty. Results of rectal biopsy may be positive when those of repeated stool examinations are negative.

S haematobium eggs found in urine

S mansoni and *S japonicum* eggs in stool; rectal biopsy

Because dead eggs may persist in tissue for a long time after the death of the adult worms, active infection is confirmed only when the eggs are shown to be viable. This confirmation may be obtained by observing the eggs microscopically for movement of flame cell cilia or by hatching them in water. Quantitation of egg output is useful in estimating the severity of infection and in following response to treatment.

Determination of egg viability and output useful

Conventional serologic tests detect circulating antibodies with sensitivities exceeding 90% but cannot distinguish active from inactive infection. Recently introduced enzyme immunoassay (EIA)–based reagent strip (dipstick) tests capable of detecting circulating, genus-specific, adult-worm antigens in blood and urine are rapid, simple, and sensitive. They are particularly helpful in the diagnosis of Katayama syndrome in those returning from endemic areas. Moreover, because antigen levels drop rapidly after successful therapy, these tests may prove helpful in distinguishing active from inactive disease.

EIA detection of antigens in blood and urine

TREATMENT

No specific therapy is available for the treatment of schistosomal dermatitis or Katayama syndrome. Antihistamines and corticosteroids may be helpful in ameliorating their more severe manifestations. In the late stage of schistosomiasis, therapy is directed at interrupting egg deposition by killing or sterilizing the adult worms. Because the severity of clinical and pathologic manifestations is related to the intensity of infection, therapy of long-term residents of endemic areas is often reserved for patients with moderate or severe active infections.

Multiple anthelmintic drugs are used

Several anthelmintic agents may be used. Praziquantel, which is active against all three species of schistosomes, is the agent of choice. Unfortunately, several recent reports have suggested increased resistance to this single-dose oral agent in areas where it has been used in mass therapy programs. *S mansoni* infections acquired in such areas may be treated with oxamniquine. Use of this agent is contraindicated in pregnancy.

PREVENTION

Sanitary disposal of feces often limited by economic status

Molluscicides effective but large-scale application difficult

It has proved both difficult and expensive to control this deadly disease. Programs aimed at interrupting transmission of the parasite by the provision of pure water supplies and the sanitary disposal of human feces are often beyond the economic reach of the nations most seriously affected. Similarly, measures to deny snails access to newly irrigated lands are expensive. Chemical molluscicides have been shown to be effective in limited trials, but have been less successful when used over large areas for prolonged periods. Mass therapy of the infected human population has until recently been severely limited by the toxicity of effective agents. Newer agents, particularly praziquantel, has proved to be more suitable for this purpose. Nevertheless, discontinuation of mass therapy, without other control measures, can result in a rapid rebound of active disease.

Multi-pronged approach is necessary

In 2009, a report of an extensive controlled study in an area of Southeastern China that was hyperendemic for *S japonicum* yielded remarkable results that are most instructive. These included removal of cattle from snail-infested grasslands, providing mechanized equipment to farmers, improving sanitation of drinking water, building lavatories and latrines, providing boats with fecal-matter containers, and implementing intensive health education programs. The infection rates fell dramatically in the intervention villages as compared to non-intervention areas. Thus, a multi-pronged approach such as this offers the best hope for lasting control.

Vaccines under development

Currently, there is intense interest in developing a vaccine suitable for human use. A vaccine made from irradiated *S bovis* cercariae, which was developed for cattle, appears to confer a significant degree of protection against infection. Although a similar live vaccine would not be suitable for human populations, the success of the animal vaccine has provided clues to potential immunoprotective mechanisms in human schistosomiasis. Monoclonal antibodies have been used to identify a number of schistosomula and adult antigens thought to be capable of inducing protective immunity; the World Health Organization has selected six of these for further evaluation.

CASE STUDY

THE RISKS OF ADVENTUROUS TOURISM

A 35-year-old American adventurer returned from a 3-week tour of rural areas in Southeast Asia, which involved hiking forays and sharing meals with local residents. One month after his return to the United States, he developed fever and chills, accompanied by cough, urticaria, arthralgia, abdominal pain, and diarrhea.

Laboratory studies demonstrated leukocytosis and marked eosinophilia, with elevated immunoglobulin levels.

Sigmoidoscopic examination revealed mucosal inflammation and petechial hemorrhages.

QUESTIONS

■ Which of the following is the most likely cause of this man's symptoms?

A. Paragonimiasis
B. Schistosomiasis
C. Clonorchiasis
D. Fascioliasis

■ Which of the following is *not true* regarding paragonimiasis?

A. Ingestion of crayfish and freshwater crabs is risky
B. Chest x-ray can mimic tuberculosis
C. Praziquantel is effective treatment.
D. Biliary tract involvement is prominent

■ Perpetuation of transmission in *Clonorchis* infections is primarily due to which of the following:

A. Ingestion of salted fish
B. Refusal to treat with albendazole
C. Lack of careful handwashing
D. Use of human waste as fertilizer

ANSWERS

1(B), **2**(D), **3**(D)

PART VI

Clinical Aspects of Infection

C. George Ray, Kenneth J. Ryan, and W. Lawrence Drew

Skin and Wound Infections CHAPTER 57
Bone and Joint Infections CHAPTER 58
Eye, Ear, and Sinus Infections CHAPTER 59
Dental and Periodontal Infections CHAPTER 60
Respiratory Tract Infections CHAPTER 61
Enteric Infections and Food Poisoning CHAPTER 62
Urinary Tract Infections CHAPTER 63
Genital Infections CHAPTER 64
Central Nervous System Infections CHAPTER 65
Intravascular Infections, Bacteremia, and Endotoxemia CHAPTER 66

CHAPTER 57

Skin and Wound Infections

SKIN INFECTIONS

Infections of the skin can result from microbial invasion from an external source or from organisms reaching the skin through the bloodstream as part of a systemic disease. Blood-borne involvement is evidenced by rashes in many viral and bacterial infections, such as measles and secondary syphilis, or may yield more chronic granulomatous skin lesions in blastomycosis, tuberculosis, and syphilis. Skin lesions remote from sites of infection can be produced by some bacterial toxins, such as the pyrogenic exotoxins of group A streptococci (GAS) and *Staphylococcus aureus*. They can also result from immunologic responses to microbial antigens that have reached the skin. Thus, there are manifold skin manifestations of infections; however, this chapter is restricted to the discussion of direct infections that may occur in the Western Hemisphere.

Primary, blood-borne, or due to toxins

The skin is an organ system with multiple functions, including protection of the tissues from external microbial invasion. Its keratinized stratified epithelium prevents direct microbial invasion under normal conditions of surface temperature and humidity, and its normal flora, pH, and chemical defenses tend to inhibit colonization by many pathogens. However, the skin is subject to repeated minor traumas that are often unnoticed, but that destroy its integrity and allow organisms to gain access to its deeper layers from the external environment. The surface is also penetrated by ducts of pilosebaceous units and sweat glands, and microbial invasion can occur along these routes, particularly when the ducts are obstructed.

Trauma and the appendages of the skin provide access

INFECTIONS IN HAIR FOLLICLES, SEBACEOUS GLANDS, AND SWEAT GLANDS

Folliculitis

Folliculitis is a minor infection of the hair follicles and is usually caused by *S aureus*. It is often associated with areas of friction and of sweat gland activity and is thus seen most frequently on the neck, face, axillae, and buttocks. Blockage of ducts with inspissated sebum, as in acne vulgaris, predisposes to the condition. Folliculitis can also be caused by *Pseudomonas aeruginosa*, and this form of the disease has become more common with the popularity of hot tubs and whirlpool baths. Unless these facilities are thoroughly cleansed and adequately chlorinated, they can grow large numbers of pseudomonads at their normal operating temperatures, causing extensive folliculitis on areas of the body that have been immersed. The lesions subside rapidly when the insult is discontinued. Occasionally, folliculitis may be caused by infection with *Candida albicans*. Such cases are particularly common in immunocompromised hosts.

Staphylococci and *Pseudomonas* infect hair follicles

Acne vulgaris also involves inflammation of hair follicles and associated sebaceous glands. The comedo of acne results from multiplication of *Propionibacterium acnes*, the predominant anaerobe of the normal skin, behind and within inspissated sebum. Organic acids produced by the organism are believed to stimulate an inflammatory response and

Propionibacterium acnes contributes to inflammation of acne

thus contribute to the disease process. However, the primary cause of the disease is hormonal influences on sebum secretion that occur at puberty, and the disease usually resolves in early adult life.

Furuncles

Staphylococcal furuncles are skin abscesses that can spread

The furuncle is a small staphylococcal abscess that develops in the region of a hair follicle. Furuncles may be solitary or multiple and may constitute a troublesome recurrent disease. Spread of infection to the dermis and subcutaneous tissues can result in a more extensive multiloculated abscess, the **carbuncle.** These lesions and their treatment are considered in Chapter 24.

Treatment

Skin care and tetracycline may be used

Folliculitis and individual furuncles are normally treated locally by measures designed to establish drainage without the use of antibiotics. Chronic furunculosis may require attempts to eliminate nasal carriage of *S aureus,* which is sometimes the source of the infection. Antimicrobics are not usually required unless surrounding cellulitis or carbuncles develops. Severe acne can often be treated effectively with topical drying agents. Prolonged administration of low oral doses of a tetracycline or macrolide is often effective, although the reason for the therapeutic response is uncertain.

INFECTIONS OF OTHER SKIN LAYERS

Minor or unapparent skin lesions serve as the route of infection in many localized skin infections and in some systemic diseases, such as syphilis and leptospirosis.

Infection of Keratinized Layers

Inflammatory response is important with dermatophytes

Cell-mediated immunity defects in chronic candidiasis

The only organisms that can use the keratin on cells, hairs, and nails are the dermatophyte fungi. The dermatophytes are particularly well adapted to these sites, cannot grow at 37°C, and fail to invade deeper layers. The clinical manifestations of these infections result from the inflammatory and delayed hypersensitivity responses of the host, and the desquamation induced by these processes is a major factor in the ultimate control of the infection by removing infected skin. In candidiasis, control involves cell-mediated immune mechanisms, and chronic *Candida* skin and nail infections are often associated with defects in cellular immunity.

Impetigo

Group A streptococci are primary cause

S aureus may colonize or act as primary pathogen

Pyoderma, also termed impetigo, is a common, sometimes epidemic skin lesion. This disease is caused primarily by group A streptococci. The initial lesion is often a small vesicle that develops at the site of invasion and ruptures with superficial spread characterized by skin erosion and a serous exudate, which dries to produce a honey-colored crust. The exudate and crust contain numerous infecting streptococci. *S aureus* may occasionally produce pustular impetigo or contaminate the lesions caused by streptococci. Epidemic impetigo is most common in childhood and under conditions of heat, humidity, poor hygiene, and overcrowding. The infection may be spread by fomites such as shared clothing and towels. It is sometimes caused by nephritogenic strains of GAS, particularly in the tropics, and acute glomerulonephritis may result. Rheumatic fever is not associated with streptococcal lesions of the skin. Treatment is usually with penicillin or erythromycin and topical antimicrobics or skin antiseptics to limit spread.

Bullous impetigo is caused by exfoliation-producing *S aureus*

Bullous impetigo is a distinct disease caused by strains of *S aureus* that produce exfoliation. It is most common in small children, but may occur at any age. The infection is characterized by large serum-filled bullae (blisters) within the skin layers at the site of infection. Minor infections are treated topically; however, bullous impetigo in infants is a serious disease that usually requires systemic antimicrobic treatment. Epidemic spread may occur under conditions similar to those described for streptococcal impetigo.

Erysipelas

Erysipelas is a rapidly spreading infection of the deeper layers of the dermis that is almost always caused by group A streptococci. It is associated with edema of the skin, marked erythema, pain, and systemic manifestations of infection including fever and lymphadenopathy.

Because the infection is intradermal, the streptococci cannot usually be isolated from the skin surfaces. The disease can progress to septicemia or local necrosis of skin. It is serious and requires immediate treatment with penicillin or erythromycin.

Group A streptococcal erysipelas is a spreading cellulitis with risk of bacteremia

Cellulitis

Cellulitis is not a skin infection as such, but it can develop by extension from skin or wound infections. It usually manifests as an acute inflammation of subcutaneous connective tissue with swelling and pain and often with marked constitutional signs and symptoms. It can be caused by many pathogenic bacteria, but *S aureus* and group A streptococci are most common. *Haemophilus influenzae* type b is a cause in infants and children. Enteric Gram-negative rods, clostridia, and other anaerobes may also cause cellulitis as a complication of wound infections, particularly in immunocompromised hosts and individuals with uncontrolled diabetes.

Most often caused by pyogenic cocci **or** *H influenzae* in children

SKIN ULCERS AND GRANULOMATOUS LESIONS

Many acute and subacute skin infections are characterized by ulceration or a granulomatous response. Some are sexually transmitted and are discussed in Chapter 64. Others derive from systemic infection and are not direct infections of skin. A few examples of direct infections, which pose special diagnostic problems, are considered in the text that follows. Herpes simplex virus can invade through the skin to produce a local vesicular lesion followed by ulceration. The lesion may then recur in the infected area. Primary herpetic lesions of the finger can mimic staphylococcal paronychia very closely, as well as produce lymphangitis and local lymph node enlargement with pain and fever.

Herpetic paronychia can mimic staphylococcal infections

Skin diphtheria, which remains common in some tropical areas, also occurred endemically among the transient population of the West Coast of the United States during the 1970s and early 1980s. The organism gains access through a wound or insect bite and causes chronic erosion and ulceration of the skin, sometimes with evidence of the systemic effects of diphtheria toxin.

Skin diphtheria is seen in transients

Mycobacterium marinum produces a self-limiting granuloma, usually of the forearms and knees. The organism generally enters through superficial abrasions from rocks or swimming pool walls. Infections with *M ulcerans* are more serious and produce progressive ulceration, but are limited to tropical areas and do not occur in the United States or Europe. Several rare forms of necrotic spreading skin ulceration tend to develop in immunosuppressed hosts, in diabetics, and as complications of abdominal surgery. These lesions include bacterial synergistic gangrene, caused by mixtures of peptostreptococcus, *S aureus,* and group A streptococci. Variants of these conditions produce extensive and spreading necrotic cellulitis. The major form of treatment is to excise the infected tissues widely and supplement such surgery with massive chemotherapy.

Mycobacterial species cause granulomas

Synergistic gangrene may require surgery

Several primary fungal diseases are associated with cutaneous ulceration or cellulitis, including mycetoma and chromoblastomycosis, which involve the feet, and sporotrichosis, in which ulceration often develops from infected subcutaneous lymph nodes and vessels. Likewise, some parasites directly infect and ulcerate the skin, as in cutaneous leishmaniasis and cutaneous amebiasis. The latter two diseases are not contracted in the United States.

Fungal and parasitic ulcerations are usually related to trauma

WOUND INFECTIONS

Wounds subject to infection can be surgical, traumatic, or physiologic. The latter include the endometrial surface, after separation of the placenta, and the umbilical stump in the neonate. Traumatic wounds comprise such diverse damage as deep cuts, compound fractures, frostbite necrosis, and thermal burns. Sources of infection include (1) the patient's own normal flora; (2) material from infected individuals or carriers that may reach the wound on fomites, hands, or through the air; and (3) pathogens from the environment that can contaminate the wound through soil, clothing, and other foreign material. Examples of such infections include contamination of a penetrating stab wound to the abdomen by colonic flora, contamination of a clean surgical wound in the operating room with *S aureus* spread from the flora of a perineal carrier, and introduction of spores of *Clostridium tetani* into the tissues on a splinter.

Related to trauma, surgery, childbirth

Sources of infection include patient, environment, and infected persons

CLASSIFICATION OF WOUNDS

Wounds vary in risk of bacterial exposure

Surgical and traumatic wounds are classified according to the extent of potential contamination and thus the risk of infection. These criteria carry important implications regarding surgical treatment and chemoprophylaxis. **Clean wounds** are surgical wounds made under aseptic conditions that do not traverse infected tissues or extend into sites with a normal flora. **Clean contaminated wounds** are operative wounds that extend into sites with a normal flora (except the colon) without known contamination. **Contaminated wounds** include fresh surgical and traumatic wounds with a major risk of contamination, such as incisions entering nonpurulent infected tissues. **Dirty and infected wounds** include old, infected traumatic wounds; wounds substantially contaminated with foreign material; and wounds contaminated with spillage from perforated viscera.

Infection rates in clean surgical wounds should be less than 1%, whereas untreated dirty wounds have a higher probability of infection. Similar considerations apply to the chance of infection developing in a placental site or on the umbilicus. A normal delivery without retained products is rarely followed by endometrial infection. A prolonged delivery after rupture of the membranes with retained placental fragments poses an increased risk. In some rural cultures in Africa, soil is applied to the umbilical stump, and neonatal tetanus is common, whereas it is almost unknown in other cultures.

FACTORS CONTRIBUTING TO WOUND INFECTION

Infectious risk increases with contaminating dose of organisms

Vascular integrity is important for defense

Various factors, in addition to those indicated previously, contribute to the probability of a wound becoming infected. The contaminating dose of microorganisms and their virulence can be critical and, other things being equal, the chance of infection increases progressively with the contaminating dose. The physical and physiologic condition of the wound also influences the probability of infection. Areas of necrosis, vascular strangulation from excessively tight sutures, hematomas, excessive edema, poor blood supply, and poor oxygenation all compromise normal defense mechanisms and substantially reduce the dose of organisms needed to initiate infection. Thus, removal of necrotic tissue and the surgeon's skill, gentleness, and attention to detail are major factors in preventing the development of infection.

Nutritional and immunologic status and inflammatory response of the host

The general health, nutritional status, and ability of patients to mount an inflammatory response are also major determinants of whether a wound infection develops. Infection rates are higher in the elderly, the obese, those with uncontrolled diabetes, and those on immunosuppressive or corticosteroid therapy. Nutritional deficiencies enhance the risk of infection, and new approaches to avoid protein–calorie malnutrition in patients with severe burns, for example, have led to substantial reductions in serious clinical infections.

First 3 hours is critical period for surgical wounds

There is strong evidence that the critical period determining whether contamination of surgical wounds proceeds to infection lies within the first 3 hours after contamination. For this reason, prophylactic chemotherapy of some surgical wounds and procedures can be restricted to the operative and immediate perioperative period. There is general agreement that extending such prophylaxis beyond 24 hours increases the chance of complications without reducing the risk of infection.

TREATMENT AND PREVENTION

Chemoprophylaxis is mainstay

Severe wound infections are almost always treated with a combination of surgical and chemotherapeutic approaches. Necrotic tissue and contaminated foreign bodies, such as sutures, must be removed, pockets of pus opened, and drainage established. This approach permits access of the appropriate antibiotics to viable tissues in which they can act. Under certain conditions of high risk (contaminated wounds) or high vulnerability (heart valve surgery) chemoprophylaxis is appropriate.

ETIOLOGIC AGENTS

S aureus and Gram-negative bacteria are most common

Some major causes of skin and wound infections are shown in **Table 57–1**. *S aureus* remains the single most common cause of infection of clean surgical wounds; however, the number

TABLE 57–1 Major Causes of Skin and Wound Infections

SYNDROME	BACTERIA	FUNGI	OTHER
Impetigo	Group A streptococci		
	Staphylococcus aureus		
Folliculitis	*Pseudomonas aeruginosa*	*Candida albicans*	
	Staphylococcus aureus		
Acne	*Propionibacterium acne*		
Furuncle	*Staphylococcus aureus*		
Cellulitis	Group A streptococci[a]		
	Staphylococcus aureus		
	Haemophilus influenzae		
Intertrigo	*Staphylococcus aureus*	*Candida albicans*	
	Enterobacteriaceae		
Chronic ulcers[b]	*Treponema pallidum*	*Sporothrix*	Herpesvirus
	Haemophilus ducreyi		
	Corynebacterium diphtheriae		
	Bacillus anthracis		
	Nocardia		
	Mycobacterium		
Wounds			
Trauma	*Clostridium*		
	Enterobacteriaceae		
	Pseudomonas aeruginosa		
Surgical (clean)	*Staphylococcus aureus*		
	Enterobacteriaceae		
	Group A streptococci		
Surgical (dirty)[c]	*Staphylococcus aureus*		
	Enterobacteriaceae		
	Anaerobes		
Burns	*Pseudomonas aeruginosa*	*Candida albicans*	
	Staphylococcus aureus		
	Enterobacteriaceae		
Animal bites	*Pasteurella multocida*		

[a]Including "erysipelas," an infection primarily involving the deeper layers of the dermis.
[b]Usually begin as nodules or pustules.
[c]Etiology determined by the origin of the contaminating flora (eg, abdominal vs. gynecologic surgery).

Streptococcal toxic shock begins with wound

of infections caused by opportunistic Gram-negative organisms is increasing. This finding reflects the extension of surgical intervention to more patients whose defenses are compromised or who would have been unacceptable surgical risks before the introduction of new technical and therapeutic procedures. Severe invasive group A streptococcal infections with the toxic shock syndrome often begin with a simple skin or wound infection.

Bacteroides and anaerobic Gram-positive coccal infections derived from patient's flora

Anaerobic Gram-negative wound infections have been reported increasingly as a result of the higher incidence of such infections in immunocompromised patients and better laboratory recognition. Most infecting organisms derive from normal floral sites, and the majority are *Bacteroides,* often in combination with anaerobic Gram-positive cocci and facultative bacteria. They tend to be associated with necrosis, which may spread subcutaneously, and with thrombophlebitis, which may lead to bacteremia. Most postpartum uterine infections are caused by Gram-negative anaerobes or anaerobic Gram-positive cocci; they can range from self-limiting infections to severe infections of the uterus with pelvic thrombophlebitis. Human bite wounds are particularly subject to anaerobic infections. In contrast, infected bites of domestic animals (dogs, cats) are almost always due to *Pasteurella multocida*.

P aeruginosa is a virulent cause of burn infections

Burns and areas of necrosis resulting from vascular stasis or insufficiency are subject to infection with the same organisms that predominate in postsurgical wound infections. However, *P aeruginosa* causes particularly serious infections in burns, with loss of skin grafts and a high risk of septicemia and death. If the fluid electrolyte and nutritional deficiencies of a burned patient can be controlled, the greatest hazard to life is infection.

C perfringens Tetanus is derived from the environment

Gas gangrene requires surgical intervention

Tetanus remains a threat to the unimmunized or inadequately immunized individual, particularly from heavy contamination of puncture wounds or introduction of foreign bodies such as splinters, soil, or clothing into the subcutaneous tissues. *C tetani* never spreads beyond the site of the local lesion, and adequate circulating antibody from tetanus toxoid immunization prevents the development of the disease. Gas gangrene (clostridial myositis) can develop within a few hours of traumatic injury and lead to rapid death. *C perfringens* is the most common cause, and its α-toxin produces the spreading tissue damage and muscle death. The disease is always associated with muscle trauma and necrosis, which provide the conditions for anaerobic multiplication. Compound fractures, gunshot wounds, and similar extensive injuries that allow entry of clostridial spores set the stage for the disease. Prevention involves surgically debriding all necrotic or potentially necrotic tissue as soon as possible and administering high-dose penicillin.

CHAPTER 58

Bone and Joint Infections

Infections of bones and joints may exist separately or together. Both are most common in infancy and childhood. They are usually caused by blood-borne (hematogenous) spread to the infected site but can also result from local trauma with secondary infection. Sometimes there may be local spread from a contiguous soft tissue infection, such as chronic ulcers in diabetic patients. The presence of a foreign body at the site of the primary wound, such as joint prostheses, also predisposes to infection.

The local effect of such infections can be devastating if they are inadequately treated, because inflammation and resultant tissue necrosis may produce irreparable damage. The presence of pus under pressure can compromise normal blood flow and even cause destruction of blood vessels with avascular necrosis of tissue. When this condition develops, a **sequestrum** can result, in which a part of the cartilage or bone becomes totally separated from its blood supply and cannot be incorporated into the healing process. In some patients, sequestrum formation can lead to a smoldering chronic infection with draining sinuses and loss of functional integrity. Normal growth of the affected site can be severely impaired in the infant or child, particularly when the epiphysis is involved. In the acute phase of infection, bacteremia may also cause sepsis and metastatic infections in sites such as the lungs and heart. The result may be fatal.

Sequestrum formation can lead to chronic infection with draining sinuses

Infection can cause growth impairment in children

Bacteremia and metastatic spread from bone and joint infections is common

OSTEOMYELITIS

The onset of acute hematogenous osteomyelitis is usually abrupt, but can sometimes be insidious. It is classically characterized by localized pain, fever, and tenderness to palpation over the affected site. More than one bone or joint may be involved as a result of hematogenous spread to multiple sites. With progression, the classic signs of heat, redness, and swelling may develop. Laboratory findings often include leukocytosis and elevated acute-phase reactants, such as C-reactive protein and sedimentation rate. Osteomyelitis caused by a contiguous focus of infection is usually associated with the presence of local findings of soft tissue infection, such as skin abscesses and infected wounds.

Local pain and signs of inflammation

When osteomyelitis occurs in close proximity to a joint, septic arthritis may develop by direct spread through the epiphysis (usually in infants) or by lateral extension through the periosteum into the joint capsule. Such extension is particularly common in hip and elbow infections.

May come from contiguous focus

Extend to joints through epiphysis and adjacent periosteum

COMMON ETIOLOGIC AGENTS

The most common causes of acute osteomyelitis and those associated with special circumstances are shown in **Table 58–1**. It is clear that age plays a significant role in influencing the relative frequency of the various infective agents, particularly in early infancy; however, most infections are caused by *Staphylococcus aureus*.

Age-related etiologies, but staphylococcal osteomyelitis most common

TABLE 58–1 Common Causes of Acute Osteomyelitis

SITUATION	USUAL CAUSATIVE ORGANISM
Age group	
Neonates (<1 mo)	*Staphylococcus aureus*, group B streptococci, Gram-negative rods (eg, *Escherichia coli*, *Klebsiella*, *Proteus*, *Pseudomonas*)
Older infants, children, adults	*S aureus*, *S pneumoniae*, *Kingella kingae*
Special problems	
Chronic hemolytic disorders (eg, sickle cell disease)	*S aureus*, *S pneumoniae*, *Salmonella* species
Infection after trauma or surgery	*S aureus*, group A streptococci, Gram-negative aerobic or anaerobic bacteria
Infection after puncture wound of foot	*Pseudomonas aeruginosa*, *S aureus*

Chronic granulomatous osteomyelitis suggests mycobacteria or fungi

Low-grade smoldering infections may also occur with the organisms listed in Table 58–1; however, chronic granulomatous processes must also be considered, including tuberculosis, coccidioidomycosis, histoplasmosis, and blastomycosis. The latter infections usually result from systemic dissemination, and the lesions develop slowly over a period of months. Occasionally, bone tumors or cysts and leukemia must also be considered in the differential diagnosis.

DIAGNOSTIC APPROACHES

The primary goals of diagnosis are to establish the existence of infection and to determine its cause. The following procedures are generally used:

Blood cultures, direct aspirates, and bone scans

X-rays may be normal in early stages of infection

1. Blood cultures, because many infections are associated with bacteremia.
2. Radionuclide scanning or magnetic resonance imaging to demonstrate evidence of localized infection.
3. Direct staining, culture, and histology of needle aspirates or biopsies of periosteum or bone.
4. X-rays of affected sites, which often appear normal in the early stages of infection.

The first changes seen are swelling of surrounding soft tissues, followed by periosteal elevation. Demineralization of bone may not become apparent for 2 weeks or more after the onset of symptoms; calcification of the periosteum and surrounding soft tissues is usually delayed even longer.

MANAGEMENT PRINCIPLES

Bactericidal antimicrobics, sometimes continued for weeks

Surgery and prolonged therapy required for chronic osteomyelitis

In acute infections, early intervention is important. Management includes vigorous use of bactericidal antimicrobics, which may need to be continued for several weeks to ensure a bacteriologic cure and prevent progression to chronic osteomyelitis. Surgical drainage is also essential if there is significant pressure from the localized, purulent process. In chronic osteomyelitis, sequestrum formation is frequent, and sinuses may develop that drain the bone abscess to the skin surface. The infection is persistent, and treatment becomes extremely difficult. Such patients often require long-term antibiotic treatment (months to years) combined with surgical procedures to drain the abscesses and remove necrotic, infected tissues in an attempt to control infection while preserving the integrity of the affected bone.

SEPTIC ARTHRITIS

The usual clinical features of septic arthritis include onset of pain, which is often abrupt and accompanied by fever. Single or multiple joints may be involved. Tenderness and swelling of the affected joints and frequently other signs of local inflammation are present. Attempts to move the joints, either actively or passively, result in severe pain. In infants, the symptoms may be somewhat nonspecific; local swelling or excessive irritability with unwillingness to move the affected extremity (pseudoparalysis) may be the only clues to the diagnosis.

Pain on movement with swelling and fever

COMMON ETIOLOGIC AGENTS

The major causes of septic arthritis are listed in **Table 58–2**. Although *S aureus* infection can occur at any age, there are some significant age-specific relationships to other bacterial causes. There is a high incidence of group B streptococcal infections in neonates, whereas in children between 1 month and 4 years of age, pneumococci are more likely to be involved. *Haemophilus influenzae* type b disease, which was once common in this age group, has been markedly diminished in recent years; this is believed to be due to widespread use of an effective vaccine. *Neisseria gonorrhoeae* is implicated in most cases of septic arthritis in young adults. Subacute or chronic infective arthritis should prompt consideration of tuberculosis, Lyme disease, syphilis, and fungal infections such as coccidioidomycosis or *Candida.* Arthritis attributable to *Candida* is particularly likely in immunocompromised patients.

S aureus appears at any age

Other pyogenic cocci are related to age and behavior

Tuberculous, spirochetal, and fungal arthritis have subacute or chronic course

Viruses and *Mycoplasma* can also cause acute arthritis in single or multiple joints. Such illnesses have been associated with rubella, hepatitis B, mumps, parvovirus B19, varicella, Epstein–Barr virus, coxsackievirus, and adenovirus infections, as well as with *M pneumoniae* and *M hominis.* These arthritides are usually self-limiting and rarely require specific therapy. Some bacterial infections of sites other than joints may be associated with noninfectious (reactive) arthritis, possibly resulting from deposition of circulating immune complexes and complement in synovial tissues, leading to inflammation. This has occurred with intestinal infections caused by *Yersinia enterocolitica, Campylobacter jejuni,* and some *Salmonella* species and also as a delayed sequela after successful treatment of sepsis due to *N meningitidis* or *H influenzae.*

Viral or *Mycoplasma* arthritis is usually self-limiting

Immune complexes from other sites may cause reactive arthritis

Noninfectious causes of arthritis must also be considered in the differential diagnosis. They can closely mimic septic arthritis. Examples include inflammatory collagen vascular disease such as rheumatoid arthritis, gout, traumatic arthritis, and degenerative arthritis.

DIAGNOSTIC APPROACHES

In acute cases, blood cultures are often useful because bacteremia may be present. The definitive diagnosis is established by examination of synovial fluid removed from the joint by needle aspiration (arthrocentesis). Because other noninfectious causes must be considered, it is important to analyze the chemical and cellular characteristics of the fluid

Blood culture is particularly useful

Needle aspiration of synovial fluid is used for analysis and culture

TABLE 58–2 Common Causes of Septic Arthritis

AGE GROUP	USUAL CAUSATIVE ORGANISM
Neonate (<1 mo)	*Staphylococcus aureus*, group B streptococci, Gram-negative rods (eg, *Escherichia coli, Klebsiella, Proteus, Pseudomonas*)
1 mo-4 y	*S aureus*, group A streptococci, *Streptococcus pneumoniae, Neisseria meningitidis, Haemophilus influenzae*
4-16 y	*S aureus, Streptococcus pyogenes*
16-40 y	*Neisseria gonorrhoeae, S aureus*
>40 y	*S aureus*

TABLE 58–3 Findings in Synovial Fluid in Various Forms of Arthritis

LABORATORY TEST	NORMAL	SEPTIC ARTHRITIS	TRAUMA, DEGENERATIVE JOINT DISEASE	RHEUMATOID ARTHRITIS, GOUT
Clarity and color	Clear	Opaque, yellow to green	Clear, yellow	Translucent, yellow; or opalescent
Viscosity	High	Variable	High	Low
White blood cells/mm^3	<200	25,000-100,000	200-2000	2000-20,000
Polymorphonuclear cells (%)	<25	>75	25-50	≥50
Glucose level (relative to simultaneous blood glucose level)	Nearly equal	<25%	Nearly equal	50-80%

in addition to performing a Gram stain and culture. **Table 58–3** summarizes the major findings in synovial fluid in normal and various disease states. Septic bacterial arthritis is usually associated with grossly purulent fluid containing more than 25,000 white blood cells per cubic millimeter, predominantly polymorphonuclear cells. The glucose level in the synovial fluid is usually less than 25% of that in the blood.

Biopsy is especially useful in chronic cases

In viral, tuberculous, and fungal arthritis, as well as in partially treated bacterial arthritis, cell counts are usually lower, and mononuclear cells may constitute a greater proportion of the inflammatory cells. Occasionally, biopsy of the synovial membrane may be required to resolve the diagnosis. Histologic examination and culture of the tissue are particularly helpful in distinguishing granulomatous from rheumatoid disease.

Gonococci may be difficult to isolate from joint fluid

In most cases of acute septic arthritis, the blood culture and/or synovial fluid culture yields the specific etiologic agent. One major exception is *N gonorrhoeae,* which can be difficult to isolate from these sources. When this organism is suspected, it is wise to include cultures of other sites of potential infection, such as the urethra, cervix, rectum, and pharynx, as well as skin lesions.

MANAGEMENT PRINCIPLES

Drainage of hip infections often necessary

Prompt, vigorous, systemic antimicrobial therapy is required as soon as diagnostic tests suggest a bacterial cause. This treatment may need to be continued for 3 to 6 weeks, depending on the etiologic agent and the clinical response to therapy. Drainage of pus under pressure is also an important aspect of management. In cases of hip joint involvement, open surgical drainage is often necessary because collateral blood supply to the hip joint is relatively limited, and pus under pressure can lead to irreversible avascular necrosis of the tissues with permanent crippling. It is also difficult to evaluate the amount of pus that may be present because of the overlying muscles. Other joints can usually be managed by simple aspiration of pus whenever it reaccumulates significantly during the acute phase of infection. In infections associated with prosthetic devices, it often becomes necessary to remove the device in order to successfully treat the infection.

CHAPTER 59

Eye, Ear, and Sinus Infections

EYE INFECTIONS

Ocular infections can be divided into those that primarily involve the external structures—eyelids, conjunctiva, sclera, and cornea—and those that involve internal sites. The major defense mechanisms of the eye are the tears and the conjunctiva, as well as the mechanical cleansing that occurs with blinking of the eyelids. The tears contain secretory IgA and lysozyme, and the conjunctiva possesses numerous lymphocytes, plasma cells, neutrophils, and mast cells, which can respond quickly to infection by inflammation and production of antibody and cytokines.

Defenses of the eye include tears, conjunctiva, and blinking

Tears have sIgA and lysozymes

The internal eye is protected from external invasion primarily by the physical barrier imposed by the sclera and cornea. If these are breached (eg, by a penetrating injury or ulceration), infection becomes a possibility. In addition, infection may reach the internal eye via the blood-borne route to the retinal arteries and produce chorioretinitis and/or uveitis. Such infections are a particularly common problem in immunocompromised patients.

Other causes of inflammation of the external or internal eye can involve autoimmune or allergic mechanisms, which may be provoked by infectious agents or diseases such as rheumatoid arthritis.

Autoimmune and allergic causes of inflammation

COMMON CLINICAL CONDITIONS

Blepharitis is an acute or chronic inflammatory disease of the eyelid margin. It can take the form of a localized inflammation in the external margin (hordeolum or stye) or a granulomatous reaction to infection and plugging of a sebaceous gland of the eyelid (chalazion).

Dacryocystitis is an inflammation of the lacrimal sac. It usually results from partial or complete obstruction within the sac or nasolacrimal duct, where bacteria may be trapped and initiate either an acute or a chronic infection.

Conjunctivitis is a term used to describe inflammation of the conjunctiva; it may extend to involve the eyelids, cornea (keratitis), or sclera (episcleritis). Extensive disease involving the conjunctiva and cornea is often called keratoconjunctivitis. Progressive keratitis can lead to ulceration, scarring, and blindness. **Ophthalmia neonatorum** is an acute, sometimes severe, conjunctivitis or keratoconjunctivitis of newborn infants.

Endophthalmitis is rare, but often leads to blindness even when treated aggressively. The term refers to infection of the aqueous or vitreous humor, usually by bacteria or fungi.

Uveitis consists of inflammation of the uveal tract—iris, ciliary body, and choroid. Although most inflammations of the iris and ciliary body (iridocyclitis) are not of infectious origin, some agents have been implicated. The acute disease may be associated with severe eye pain, redness, and photophobia; other cases may progress silently, with decreased visual acuity as the only symptom in the late stages. The most common infective involvement of the uveal tract is **chorioretinitis,** in which inflammatory infiltrates are seen in the retina; this infection can lead to destruction of the choroid and inflammation of the optic nerve (optic neuritis) and may extend into the vitreous humor to cause endophthalmitis. If the disease is not treated adequately, the end result can be blindness.

TABLE 59–1 Major Infectious Causes of Eye Disease

DISEASE	BACTERIA	VIRUSES	FUNGI	PARASITES
Blepharitis	*Staphylococcus aureus*			
Dacryocystitis	*Streptococcus pneumoniae, S aureus*			
Conjunctivitis, keratitis, keratoconjunctivitis	*S pneumoniae, Haemophilus influenzae, H aegyptius, Streptococcus pyogenes, S aureus, Chlamydia trachomatis, Neisseria gonorrhoeae, Neisseria meningitidis*	Adenoviruses, herpes simplex; measles, varicella–zoster	*Fusarium* species, *Aspergillus* species	*Acanthamoeba* (keratitis)
Ophthalmia neonatorum	*N gonorrhoeae, Chlamydia trachomatis*	Herpes simplex		
Endophthalmitis	*S aureus, Pseudomonas aeruginosa*, other Gram-negative organisms		*Candida* species, *Aspergillus* species	
Iridocyclitis	*Treponema pallidum*	Herpes simplex, varicella–zoster		
Chorioretinitis	*Mycobacterium tuberculosis*	Cytomegalovirus, herpes simplex, varicella–zoster	*Histoplasma capsulatum, Coccidioides immitis, Candida* species	*Toxoplasma gondii, Toxocara canis*

COMMON ETIOLOGIC AGENTS

Blepharitis often staphylococcal

Acute conjunctivitis: age-related etiologies

Chronic conjunctivitis: *C trachomatis* and herpes simplex

Epidemic adenovirus conjunctivitis related to swimming pools and eye drops

The major infectious causes of various inflammatory diseases of the eye are listed in **Table 59–1**. *Staphylococcus aureus* is the principal offender in bacterial infections of the eyelid and cornea. *Haemophilus* species and *Streptococcus pneumoniae* are common causes of acute bacterial conjunctivitis. In young infants, *Neisseria gonorrhoeae* and *Chlamydia trachomatis* are significant causes of external eye disease, contracted from the mother's birth canal, which must be diagnosed and treated promptly. Chronic conjunctivitis or keratoconjunctivitis at any age must also prompt consideration of *C trachomatis* infection. Herpes simplex is also a major cause of chronic or recurrent conjunctivitis, especially in infections of the external structures, and specific therapy is available. Epidemic acute conjunctivitis or keratoconjunctivitis is most commonly associated with a variety of adenovirus serotypes. Outbreaks have been associated with inadequately chlorinated swimming pools, contaminated equipment or eye drops in physicians' offices, and communal sharing of towels, which facilitates direct transmission.

Chorioretinitis usually linked to systemic disease, congenital infections, or immunocompromise

Endophthalmitis is from blood-borne or contiguous spread

Chorioretinitis is frequently a manifestation of systemic disease (eg, histoplasmosis, tuberculosis) and congenital infections. It is particularly common in immunocompromised patients, who are liable to develop disseminated *Candida, cytomegalovirus,* or *Toxoplasma gondii* infections. Endophthalmitis may also result from blood-borne dissemination or by contiguous spread as a result of injury (eg, corneal ulcerations). In the latter situation, iatrogenic infection by agents such as Pseudomonas species can be induced by contaminated eye drops and ophthalmologic examination equipment.

Infection of the soft tissues surrounding the eye (periorbital or orbital cellulitis) is potentially severe and can spread to involve the functions of the eye itself. Major causes are *S aureus, H influenzae, Streptococcus pyogenes,* and *S pneumoniae.*

DIAGNOSTIC APPROACHES

Gram stain and cultures of surface scrapings

Careful ophthalmologic examination may suggest cause

In external bacterial infections of the eye, etiologic diagnoses can usually be established by Gram stain and culture of surface material or, in the case of viral infections, by tissue culture. Conjunctival scrapings for *C trachomatis* can be prepared for immunofluorescent or cytologic examination and for appropriate culture. Infections of internal sites pose a more difficult problem. Some, such as acute endophthalmitis, may require removal of infected aqueous humor for microbiologic studies. Infections involving the uveal tract may require indirect methods of diagnosis, such as serologic tests for toxoplasmosis and deep mycoses, blood cultures to demonstrate evidence of disseminated disease (eg, *Candida* sepsis), and efforts to demonstrate infection in other sites (eg, chest radiography and sputum culture to diagnose tuberculosis). Careful ophthalmologic examination using slit lamps and retinoscopy often helps suggest specific etiologic agents based on the morphology of the lesions observed.

MANAGEMENT PRINCIPLES

Various topical antimicrobial agents have been used effectively in external eye infections of presumed or proven bacterial origin. In addition, topical antiviral treatment is available for herpes simplex infections but has not proved efficacious for other viral diseases of the eye. Severe infections, whether external or internal, require specialized treatment that nearly always includes ophthalmologic consultation because they may threaten vision. Systemic infection associated with eye disease (eg, fungemia, tuberculosis) must be treated vigorously with appropriate antimicrobial agents.

Topical agents used for superficial bacterial and herpes simplex infections

Ophthalmologic consultation needed with severe or deep infection

EAR INFECTIONS

Most infections of the ear involve the external otic canal (otitis externa) or the middle ear cavity (otitis media), which contains the ossicles and is enclosed by bony structures and the tympanic membrane. Factors of importance in the pathogenesis of otitis externa include local trauma, furunculosis, foreign bodies, and excessive moisture, which can lead to maceration of the external ear epithelium (swimmer's ear). Occasionally, external otitis occurs as an extension of infection from the middle ear, with purulent drainage through a perforated tympanic membrane.

Otitis externa linked to ear canal trauma and excessive moisture

The **eustachian tube,** which vents the middle ear to the nasopharynx, appears to play a major role in predisposing patients to otitis media. The tube performs three functions: ventilation, protection, and clearance via mucociliary transport. Viral upper respiratory infections or allergic conditions can cause inflammation and edema in the eustachian tube or at its orifice. These developments disturb its functions, of which ventilation may be the most important. As ventilation is lost, oxygen is absorbed from the air in the middle ear cavity, producing negative pressure. This pressure in turn allows entry of potentially pathogenic bacteria from the nasopharynx into the middle ear, and failure to clear these normally can result in colonization and infection. Other factors that can lead to compromise of eustachian tube function include anatomic abnormalities, such as tissue hypertrophy or scarring around the orifice, muscular dysfunction associated with cleft palate, and lack of stiffness of the tube wall. The latter is common in infancy and early childhood and improves with age. It may explain in part why otitis media occurs most often in infants 6 to 18 months of age and then decreases in frequency as patency of the eustachian tube becomes established.

Viral infections and allergy predispose to otitis media

Microbes enter middle ear by the eustachian tube

Failure to clear leads to otitis media

MANIFESTATIONS

Otitis externa is characterized by inflammation of the ear canal, with purulent ear drainage. It can be quite painful, and cellulitis can extend into adjacent soft tissues. A common form is associated with swimming in water that may be contaminated with aerobic Gram-negative organisms such as *Pseudomonas* species. "Malignant" otitis externa is a considerably more severe form of external ear canal infection that can progress to invasion of cartilage and adjacent bone, sometimes leading to cranial nerve palsy and death. It is seen most frequently in elderly patients with diabetes mellitus and in immunocompromised hosts of any age. *Pseudomonas aeruginosa* is the most common causative pathogen.

P aeruginosa causes swimming pool and malignant otitis externa

Otitis media is arbitrarily classified as acute, chronic, or serous (secretory). Acute otitis media, nearly always caused by bacteria, is often a complication of acute viral upper respiratory illness. Fever, irritability, and acute pain are common, and otoscopic examination reveals bulging of the tympanic membrane, poor mobility, and obscuration of normal

Acute otitis media usually bacterial

Extension to deeper structures leads to mastoiditis and sometimes CNS involvement

anatomic landmarks by fluid and inflammatory cells under pressure. In some cases, the tympanic membrane is also acutely inflamed, with blisters (bullae) on its external surface (myringitis). If treated inadequately, the infection can progress to involve adjacent structures such as the mastoid air cells (mastoiditis) or can lead to perforation with spontaneous drainage through the tympanic membrane. Potential acute, suppurative sequelae include extension into the central nervous system (CNS) and sepsis.

Chronic otitis media follows unresolved acute infections

Chronic otitis media is usually a result of acute infection that has not resolved adequately, either because of inadequate treatment in the acute phase or because of host factors that perpetuate the inflammatory process (eg, continued eustachian tube dysfunction, caused by allergic or anatomic factors or immunodeficiency). Sequelae include progressive destruction of middle ear structures and a significant risk of permanent hearing loss. Serous otitis media may represent either a form of chronic otitis media or allergy-related inflammation. It tends to be chronic, causing hearing deficits, and is associated with thick, usually nonpurulent secretions in the middle ear.

COMMON ETIOLOGIC AGENTS

S pneumoniae most common cause

H influenzae strains usually nontypeable

The common causes of ear infections are listed in **Table 59–2.** *S pneumoniae* is the single most common cause of acute otitis media after the first 3 months of life, accounting for 35% to 40% of all cases. *H influenzae* is also common, particularly in patients less than 5 years of age. The majority of *H influenzae* isolates from the middle ear are nontypeable; thus, the current vaccine against type b strains would not be expected to markedly reduce the incidence of acute otitis media. Viruses and *Mycoplasma* are rare primary causes of acute or chronic otitis media; however, they predispose patients to superinfection by the bacterial agents.

DIAGNOSTIC APPROACHES

External ear canal cultures often confusing

Middle ear aspirate cultures reliable but reserved for difficult cases

The diagnosis is established on the basis of clinical examination. Tympanometry can be performed in suspected cases of otitis media to detect the presence of fluid in the middle ear and to assess tympanic membrane function. The specific cause of otitis externa can be determined by culture of the affected ear canal; however, one must keep in mind that surface contamination and normal skin flora may lead to mixed cultures, which can be confusing. In otitis media, the most precise diagnostic method is careful aspiration with a sterile needle through the tympanic membrane after decontamination of the external canal. Gram stain and culture of such aspirates are highly reliable; however, this procedure is generally reserved for cases in which etiologic possibilities are extremely varied, as in young infants, or when

TABLE 59–2 Common Causes of Ear Infection

DISEASE	CAUSE
Otitis externa	*Pseudomonas aeruginosa* is common; occasionally *Proteus* species, *Escherichia coli*, and *Staphylococcus aureus;* bacteria found in otitis media may also be recovered if the process is secondary to middle ear infection with perforation and drainage through the tympanic membrane; fungi, such as *Aspergillus* species, are occasionally implicated
Acute otitis media	
<3 months of age	*Streptococcus pneumoniae*, group B streptococci, *S aureus*, *P aeruginosa*, and Gram-negative enteric bacteria
>3 months of age	*S pneumoniae* and *Haemophilus influenzae* are most common; others include *Streptococcus pyogenes*, *Moraxella catarrhalis*, and *S aureus*
Chronic otitis media	Mixed flora in 40% of cases cultured. Common organisms include *P aeruginosa*, *H influenzae*, *S aureus*, *Proteus* species, *Klebsiella pneumoniae*, *Moraxella catarrhalis*, and Gram-positive as well as Gram-negative anaerobic bacteria
Serous otitis media	Same as chronic otitis media; however, many more of these effusions are sterile, with relatively few acute inflammatory cells

clinical response to the usual antimicrobial therapy has been inadequate. Respiratory tract cultures, such as those from the nasopharynx, cannot be relied on to provide an etiologic diagnosis.

Respiratory tract cultures unhelpful

MANAGEMENT PRINCIPLES

Except in severe cases, otitis externa can usually be managed by gentle cleansing with topical solutions. The Gram-negative bacteria most commonly involved are often susceptible to an acidic environment, and otic solutions buffered to a low pH (3.0 or less), as with 0.25% acetic acid, are often effective. Various preparations are available, many of which also contain antimicrobics.

Otitis externa treated with topical agents

Acute otitis media often improves spontaneously without specific treatment; however, if symptoms persist for 2 days or more, antimicrobial therapy and careful follow-up are recommended to ensure that the disease has resolved. The choice of antimicrobic agent is usually empirical, designed specifically to cover the most likely bacterial pathogens, because direct aspiration for diagnostic purposes is usually unnecessary. In the usual case, these pathogens would be *S pneumoniae* and *H influenzae*. If there is extreme pressure with severe pain, drainage of middle ear exudates by careful incision of the tympanic membrane may be necessary. In patients with chronic or serous otitis media, management can be more complex, and it is often advisable to seek otolaryngologic consultation to determine further diagnostic procedures as well as to plan medical and possible surgical measures.

Antimicrobic therapy for otitis media directed at common agents for age group

Drainage may be required

SINUS INFECTIONS

The paranasal sinuses (ethmoid, frontal, sphenoid, and maxillary) all communicate with the nasal cavity. In healthy persons, these sinuses are air-filled cavities lined with ciliated epithelium and are normally sterile. They are poorly developed in early life and, in contrast to otitis media, sinus infections are a rare problem in infancy. The pathogenesis of sinus infection can involve several factors, most of which act by producing obstruction or edema of the sinus opening, impeding normal drainage. Consequently, bacterial infection and inflammation of the mucosal lining tissues develop. Predisposing factors may be (1) local, such as upper respiratory infections producing edema of antral tissues, mucosal polyps, deviation of the nasal septum, enlarged adenoids, or a tumor or foreign body in the nasal cavity; or (2) systemic, such as allergy, cystic fibrosis, or immunodeficiency. Occasionally, maxillary sinusitis can result from extension of a maxillary dental infection.

Factors predisposing to sinusitis involve obstruction of drainage or extension from other sites

MANIFESTATIONS

Signs and symptoms of sinus infection vary according to which sinuses are affected and whether the illness is acute or chronic. Fever is sometimes present. In addition, nasal or postnasal discharge, daytime cough that may become worse at night, fetid breath, pain over the affected sinus, headache, and tenderness to percussion over the frontal or maxillary sinuses are all features that may appear in different combinations and suggest the diagnosis. Complications of sinusitis can include extension of infection to nearby soft tissues, such as the orbit, and occasionally spread, either directly or via vascular pathways, into the CNS.

Fever and tenderness in local area are common

COMMON ETIOLOGIC AGENTS

Table 59–3 summarizes the typical causes of sinus infections. Respiratory viruses are also occasional direct causes, but are most important as predisposing factors to bacterial superinfection of inflamed sinuses and their antral openings. Together, *S pneumoniae* and unencapsulated *H influenzae* account for more than 60% of cases of acute sinusitis. Opportunistic, saprophytic fungi, such as *Mucor, Aspergillus,* and *Rhizopus* species, are

Opportunistic fungi are increasingly found in immunocompromised patients

TABLE 59–3 Common Causes of Sinus Infection

DISEASE	CAUSE
Acute sinusitis	*Streptococcus pneumoniae* and *Haemophilus influenzae* are most common; also *Streptococcus pyogenes, Staphylococcus aureus,* and *Moraxella catarrhalis*
Chronic sinusitis	Same as for acute sinusitis; also Gram-negative enteric bacteria and anaerobic Gram-negative and Gram-positive bacteria; mixed aerobic and anaerobic infections are relatively common; opportunistic fungi may be found in compromised patients (eg, those with diabetes mellitus)

being increasingly seen in compromised hosts, such as those with severe diabetes mellitus or immunodeficiency. These have a particular tendency to spread progressively to adjacent tissues and to the CNS and are very difficult to treat.

DIAGNOSTIC APPROACHES

Gram stain and cultures of direct sinus aspirates most accurate

Cultures of sinus drainage unreliable

Radiographic studies of the sinuses confirm the diagnosis. If it becomes necessary to determine the specific infectious agent, fluid should be obtained directly from the affected sinus by needle puncture of the sinus wall or by catheterization of the sinus antrum after careful decontamination of the entry site. Gram smears and cultures are then made. Cultures of drainage from the antral orifices or nasal secretions are unreliable because of contaminating aerobic and anaerobic normal flora.

MANAGEMENT PRINCIPLES

Antimicrobic choice is usually empirical in uncomplicated cases

Direct cultures may be required in severe, chronic cases

In uncomplicated acute sinusitis, the choice of antimicrobics is usually empirical, based on the most likely bacterial causes and their usual susceptibility. Severe, complicated acute infections and chronic sinusitis often require otolaryngologic consultation. In such cases, it is often necessary to obtain cultures directly from the sinuses to select specific antimicrobial therapy, to consider the need for surgical procedures to adequately remove the pus and inflammatory tissues, and to correct any anatomic obstruction that may exist.

CHAPTER 60

Dental and Periodontal Infections

Dental caries, periodontitis, and the tooth loss and other sequelae that follow are secondary to the microbial build up on teeth called plaque. The prevention and/or halting of the progression of these diseases relies on the elimination of dental plaque from the tooth surfaces. In addition to causing caries and chronic periodontitis, the bacteria of dental plaque play a role in more aggressive forms of periodontitis and necrotizing periodontal diseases.

DENTAL PLAQUE

Dental plaque is an adherent dental deposit that forms on the tooth surface composed almost entirely of bacteria derived from the normal flora of the mouth. From a microbial pathogenesis standpoint, dental plaque is the most prevalent and densest of human biofilms (**Figure 60–1**). The biofilm first forms in relation to the dental pellicle, which is a physiologic thin organic film covering the mineralized tooth surface composed of proteins and glycoproteins derived from saliva and other oral secretions. As the plaque biofilm evolves, it does so in relation to the pellicle, not the mineralized tooth itself. The formation of plaque takes place in stages and layers at two levels. The first is the anatomic location of the plaque in relation to the gingival line. The earliest plaque is supragingival, which may then extend to subgingival plaque. The second level is the layering within the plaque, the bacterial species involved, and the bacteria/pellicle and bacteria/bacteria binding mechanisms required.

Dental plaque is a bacterial biofilm

Plaque forms in stages

The initial supragingival plaque primarily involves Gram-positive bacteria using specific ionic and hydrophobic interactions as well as lectin-like (carbohydrate binding) surface structures to adhere to the pellicle and to each other. The prototype early colonizer is *Streptococcus sanguis,* but other streptococci (*S mutans, S mitis, S salivarius, S oralis, S gordonii*), lactobacilli, and *Actinomyces* species are usually present. If the early colonizers are undisturbed, the late colonizers appear in the biofilm in as little as 2 to 4 days. These are primarily Gram-negative anaerobes including anaerobic spirochetes. These include *Fusobacterium, Porphyromonas, Prevotella, Veillonella, Treponema denticola,* and more *Actinomyces* species. These bacteria use similar mechanisms to bind to the early colonizers and to each other. This sets up a highly complex biofilm in which co-aggregation involves structures that the bacteria brought with them (lectins), quorum sensing, and new metabolic activity. An example of the latter is the formation of extracellular glucan polymers, which act like a cement binding the plaque biofilm together. The biofilm also fastens nutrient and growth regulatory relationships between its members and provides a shield from the outside. In all, there are thought to be 300 to 400 bacterial species present in mature dental plaque. The structure of the involved bacteria is shown in Figure 60–1 and its gross and microscopic appearance in **Figure 60–2.** ::: Quorum sensing, p. 619

Attachment of bacteria to dental pellicle begins colonization

Early and late colonizers differ

Adhesion mechanisms create biofilm

Dental plaque would coat the tooth surfaces uniformly but for its physical removal during chewing and other oral activities. Characteristically, plaque remains in the non–self-cleansing

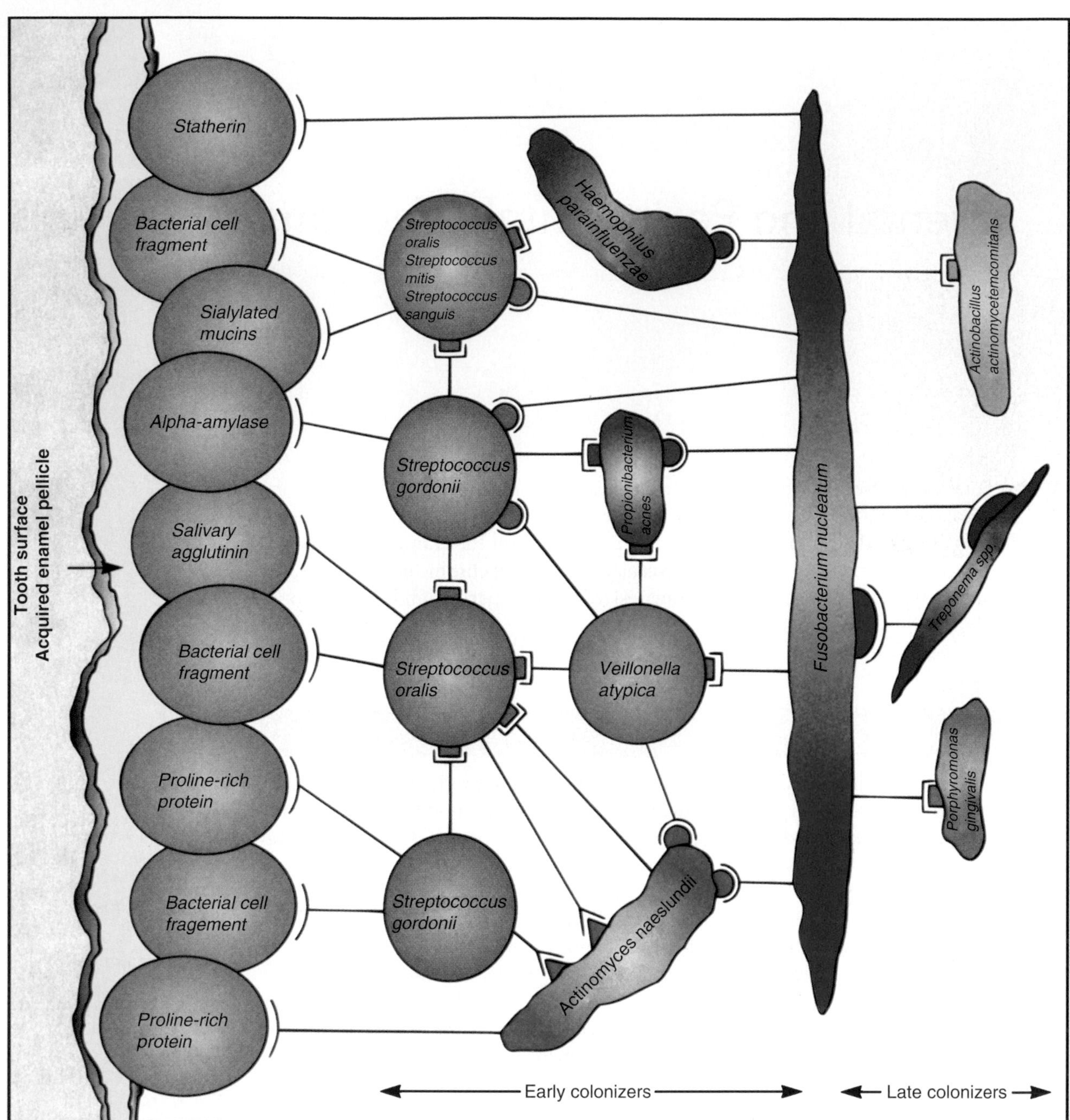

FIGURE 60–1. Dental plaque biofilm. The stages of formation the bacterial biofilm called dental plaque are shown. Early colonizers bind to the enamel pellicle and late colonizers bind to the other bacteria. (Reproduced with permission from Willey J, Sherwood L, Woolverton C (eds). *Prescott's Principles of Microbiology*. New York: McGraw-Hill; 2008.)

Plaque accumulates in non–self-cleansing areas

Subgingival plaque differs in bacterial composition

areas of the teeth such as pits and fissures, along the margins of the gingiva, and between the teeth. For this reason, the plaque-related diseases—caries, gingivitis, and periodontitis—occur most frequently and most severely at these locations. Subgingival plaque extends below the gum line to the sulcus around the tooth and periodontal pockets, which are pathologic extensions of the sulcus. This plaque has a thin adherent layer attached to the tooth surface and a nonadherent bacterial zone between that and the epithelial cells lining the sulcus. Supragingival plaque lacks such a distinct nonadherent zone. The bacterial composition of subgingival plaque is shifted toward the Gram-negative anaerobic bacteria and spirochetes. In addition to the late colonizers cited above, it may also include members of the *Campylobacter, Capnocytophagia,* and *Eikenella* genera.

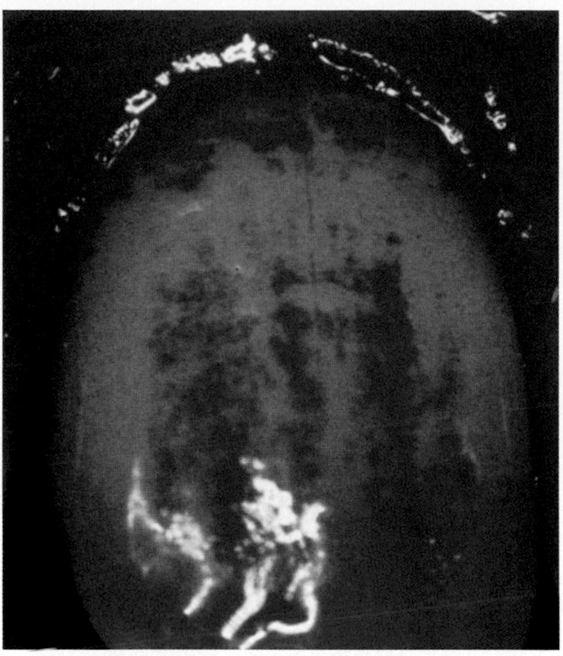

A B

FIGURE 60–2. Dental plaque. A. Disclosing tablets containing vegetable dye stain heavy plaque accumulation at the junction of the tooth and gingival. **B.** Scanning electron micrograph of supragingival plaque. (A, Reproduced with permission from Willey J, Sherwood L, Woolverton C (eds). *Prescott's Principles of Microbiology*. New York: McGraw-Hill; 2008, Figure 38.30a. B, Courtesy of Dr. W. Fischlsweiger and Dr. Dale Birdsell.)

Because the causative organisms of both dental caries and chronic periodontitis are believed to be in the dental plaque, a prime method for maintaining oral health is regular home care practices for plaque removal. Dental plaque cannot be effectively removed from the teeth solely by chemical or enzymatic means, and the use of antibiotics for prophylactic inhibition of plaque formation cannot be clinically justified, although patients undergoing long-term antibiotic treatment for other medical reasons demonstrate a lower incidence of caries and periodontal disease. Antiseptic substances that bind to tooth surfaces and inhibit plaque formation, such as the bis-biguanides, chlorhexidine, and alexidine, have been shown to be effective in reducing plaque, caries, and gingival inflammation. A commercial preparation containing 0.12% chlorhexidine can be used in controlling dental plaque and associated disease. Toothpaste and mouth rinse additives such as phenolic compounds, essential oils, triclosan, fluorides, herbal extracts, and quaternary ammonium compounds have been shown to have some plaque-reducing ability as well. The use of these substances must be accompanied by proper tooth brushing, flossing and periodic professional cleaning to ensure effective disease prevention.

Removal of plaque prime element of oral hygiene

Chemicals may be used along with brushing and flossing

DENTAL CARIES

Dental caries is the progressive destruction of the mineralized tissues of the tooth, primarily caused by the acid products of glycolytic metabolic activity when the plaque bacteria are fed the right substrate. The basic characteristic of the carious lesion is that it progresses inward from the tooth surface, either the enamel-coated crown or the cementum of the exposed root surface, involving the dentin and finally the pulp of the tooth (**Figures 60–3** and **60–4**). From there, infection can extend into the periodontal tissues at the root apex or apices.

Caries produced by plaque bacteria

The microbial basis of dental caries has been long established based on work first with *Lactobacillus acidophilus* and then *Streptococcus mutans.* Although *S mutans* is now

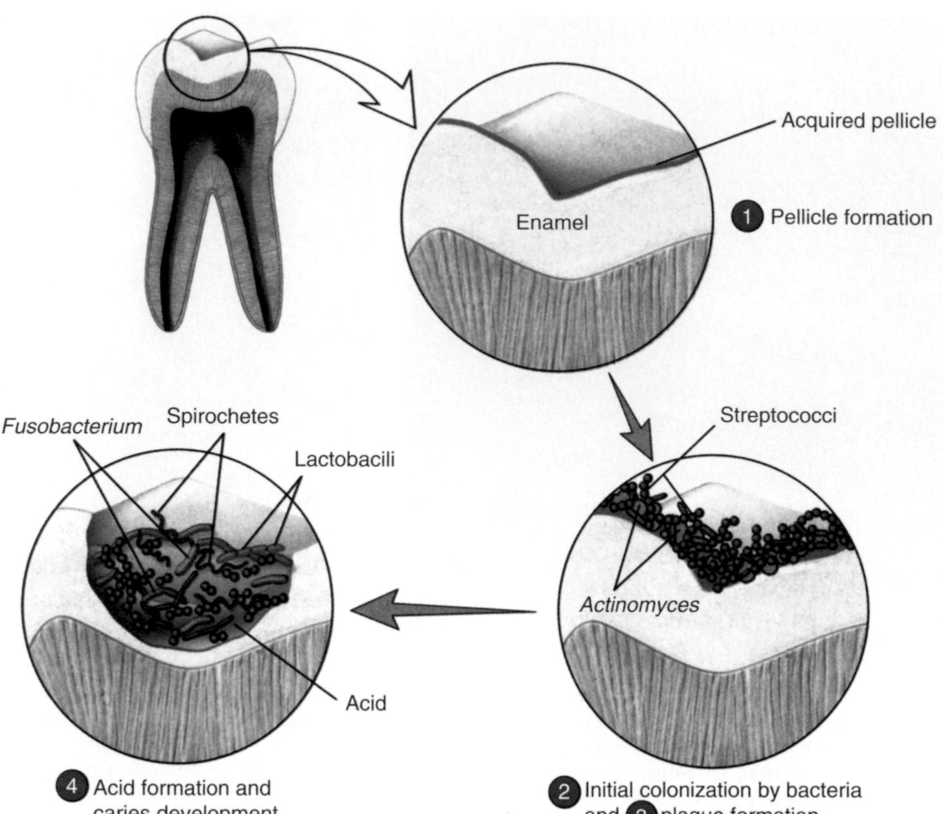

FIGURE 60–3. Cariogenesis. A microscopic view of pellicle and plaque formation, acidification, and destruction of tooth enamel. (Reproduced with permission from Willey J, Sherwood L, Woolverton C (eds). *Prescott's Principles of Microbiology*. New York: McGraw-Hill; 2008.)

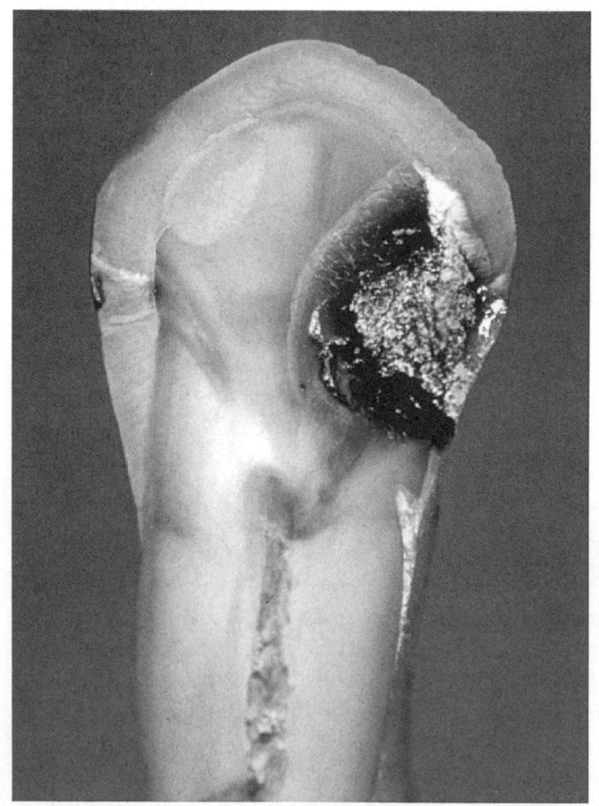

FIGURE 60–4. Hemisected human tooth showing an advanced carious lesion on the right side of the crown and a much smaller lesion on the left side. Note the progression of the lesion through the enamel and dentin, pointing toward the pulp chamber in the center of the tooth.

regarded as the dominant organism for the initiation of caries, multiple members of the plaque biofilm participate in the evolution of the lesions. These include other streptococci (*S salivarius, S sanguis, S sobrinus*), lactobacilli (*L acidophilus, L casei*), and actinomycetes (*A viscosus,* and *A naeslundii*). The acid products produced by the interaction of *S mutans* with multiple species in the biofilm are the underlying cause of dental caries.

Members of biofilm produce acid

S mutans is most cariogenic

Dietary monosaccharides and disaccharides such as glucose, fructose, sucrose, lactose, and maltose provide an appropriate substrate for bacterial glycolysis and acid production to cause tooth demineralization. A possible edge for *S mutans* is its ability to metabolize sucrose more efficiently than other oral bacteria. Ingested carbohydrates permeate the dental plaque, are absorbed by the bacteria, and are metabolized so rapidly that organic acid products accumulate and cause the pH of the plaque to drop to levels sufficient to react with the hydroxyapatite of the enamel, demineralizing it to soluble calcium and phosphate ions. Production of acid and the decreased pH are maintained until the substrate supply is exhausted. Obviously, foods with high sugar content, particularly sucrose, which adhere to the teeth and have long oral clearance times are more cariogenic than less retentive foodstuffs such as sugar-containing liquids. Once the substrate is exhausted, the plaque pH returns slowly to its more neutral pH resting level and some recovery can take place. This sets up a demineralization-remineralization cycle, which depends on carbohydrate refueling from the diet. With repeated snacking between meals, the plaque pH may never return to normal and demineralization dominates.

Demineralization is by acid production from dietary carbohydrate

Acid production facilitated by sticky carbohydrates

Demineralization-remineralization related to snacking

An additional factor with sucrose is that it is also used in the synthesis of extracellular polyglycans such as dextrans and levans by transferase enzymes on the bacterial cell surfaces. This polyglycan production by *S mutans* contributes to aggregation and accumulation of the organism on the tooth surface. Extracellular polyglycan may also increase cariogenicity by serving as an extracellular storage form of substrate. Certain microorganisms synthesize extracellular polyglycan when sucrose is available but then break it down into monosaccharide units to be used for glycolysis when dietary carbohydrate is exhausted. Some oral bacteria also use dietary monosaccharides and disaccharides internally to form glycogen, which is stored intracellularly and used for glycolysis after the dietary substrate has been exhausted; thus, the period of acidogenesis is again prolonged and the cariogenicity of the microorganism increased. These microorganisms can prolong acidogenesis beyond the oral clearance time of the substrate.

Extracellular polyglycans from sucrose important in adherence and carbohydrate storage

Acidogenesis prolonged by intracellular glycogen stores

The most common complications of dental caries are extension of the infection into the pulp chamber of the tooth (pulpitis), necrosis of the pulp, and extension of the infection through the root canals into the periapical area of the periodontal ligament. Periapical involvement may take the form of an acute inflammation (periapical abscess), a chronic nonsuppurating inflammation (periapical granuloma), or a chronic suppurating lesion that may drain into the mouth or onto the face via a sinus tract. A cyst may form within the chronic nonsuppurating lesion as a result of inflammatory stimulation of the epithelial rests normally found in the periodontal ligament. If the infectious agent is sufficiently virulent or host resistance is low, the infection may spread into the alveolar bone (osteomyelitis) or the fascial planes of the head and neck (cellulitis). Alternatively, it may ascend along the venous channels to cause septic thrombophlebitis. Because most carious lesions represent a mixed infection by the time cavities have developed, it is not surprising that most oral infections resulting from the extension of carious lesions are mixed and frequently include anaerobic organisms.

Extension to pulp and periapical locations complicate infections

Severe complications spread to bone or local fascia

Dental caries is the single greatest cause of tooth loss in the child and young adult. Its onset can occur very soon after the eruption of the teeth. The first carious lesions usually develop in pits or fissures on the chewing surfaces of the deciduous molars and result from the metabolic activity of the dental plaque that forms in these sites. Later in childhood, the incidence of carious lesions on smooth surfaces increases; these lesions are usually found between the teeth. The factors involved in the formation of a carious lesion are (1) a susceptible host or tooth, (2) the proper microflora on the tooth, and (3) a substrate from which the plaque bacteria can produce the organic acids that result in tooth demineralization.

Greatest cause of tooth loss in children and young adults

Require microflora and suitable substrates for organic acid production

The newly erupted tooth is most susceptible to the carious process. It gains protection against this disease during the first year or so by a process of posteruptive maturation believed to be attributable to improvement in the quality of surface mineral on the tooth. Saliva provides protection against caries, and patients with dry mouth (xerostomia) suffer from high caries attack rates unless suitable measures are taken. In addition to the mechanical flushing and diluting action of saliva and its buffering capacity, the salivary glands also secrete several antibacterial products. Thus, saliva is known to contain lysozyme, a thiocyanate-dependent sialoperoxidase, and immunoglobulins, principally those of the secretory IgA class. The individual importance of these antibacterial factors is unknown, but they clearly play some role in determining the ecology of the oral microflora.

Saliva protects by mechanical flushing and multiple chemical actions

Proper levels of fluoride, either systemically or topically administered, result in dramatic decreases in the incidence of caries (50% to 60% reduction by water fluoridation, 35% to 40% reduction by topical application). In the case of systemic fluoridation, the protective effect is thought to result from the incorporation of fluoride ions in place of hydroxyl ions of the hydroxyapatite during tooth formation, producing a more perfect and acid-resistant mineral phase of tooth structure. Topical application of fluoride is believed to achieve the same result on the surface of the tooth by initial dissolution of some of the hydroxyapatite, followed by recrystallization of apatite, which incorporates fluoride ions into its lattice structure. Another important mode of action, namely, the inhibition of demineralization, and the promotion of remineralization of incipient carious lesions by fluoride ions in the oral fluid, has more recently been proposed as an important anticaries mechanism of fluoride, perhaps more important than the other proposed mechanisms. In any event, fluoridation represents the most effective means known for rendering the tooth more resistant to the carious process.

present Fluoride produces more acid-resistant mineral phase of tooth

CHRONIC PERIODONTITIS

Plaque-induced periodontal disease encompasses two separate disease entities: gingivitis and chronic periodontitis. These diseases are believed to be related, in that gingivitis, although a reversible condition, is thought to be an early stage leading ultimately to chronic periodontitis in the susceptible subject. The term **gingivitis** is used when the inflammatory condition is limited to the marginal gingiva and bone resorption around the necks of teeth has not yet begun. Gingivitis develops within 2 weeks in individuals who fail to practice effective tooth cleansing. **Chronic periodontitis** is used to connote the stage of chronic periodontal disease in which there is progressive loss of tooth support owing to resorption of the alveolar bone and periodontal ligament. Periodontitis can also lead to periodontal abscess when the chronic inflammatory state around the necks of the teeth becomes acute at a specific location.

Causes destruction of supporting tissues

Both gingivitis and chronic periodontitis are caused by bacteria in the dental plaque that lie in close proximity to the necks of the teeth and marginal gingival tissues. Thus, subgingival plaque found within the gingival crevice or the sulcus around the necks of the teeth is thought to house the etiologic agent(s). The characteristic histopathologic picture of gingivitis is of a marked inflammatory infiltrate of polymorphonuclear leukocytes, lymphocytes, and plasma cells in the connective tissue that lies immediately adjacent to the epithelium lining the gingival crevice and attached to the tooth. Collagen is lost from the inflamed connective tissue. There does not seem to be any direct invasion of the gingival tissues by large numbers of intact bacteria, at least in the early stages of the disease.

Subgingival plaque causes collagen loss

All forms of periodontitis are polymicrobial infections primarily involving anaerobic bacteria in much the same way described for other anaerobes in Chapter 29. The agents involved are derived from the predominantly Gram-negative anaerobic flora of the subgingival plaque (see previous text) lead by *Porphyromonas gingivalis* and *Treponema denticola.* Just as bacteria-bacteria interactions determine the plaque, cross-feeding and growth stimulation have been observed between these two organisms when grown together. This kind of synergism between *P gingivalis, T denticola,* and other plaque members is felt to foster progression of gingivitis to chronic periodontitis. Some of these organisms have also been shown to produce virulence factors similar to those associated with other invasive bacterial pathogens. *T denticola* is able to bind serum factors that interfere with complement deposition, and *P gingivalis* is a potent producer of extracellular proteases. The former facilitates survival in tissues and the latter injury to those tissues.

Polymicrobial anaerobic infection from subgingival plaque

Synergistic interaction facilitate growth

Virulence factors cause disease

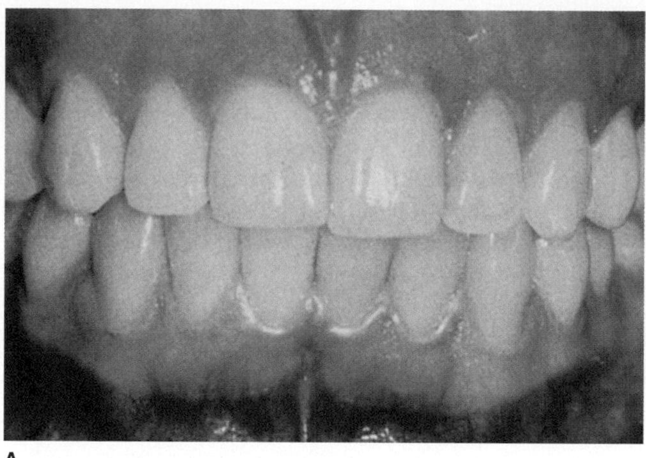

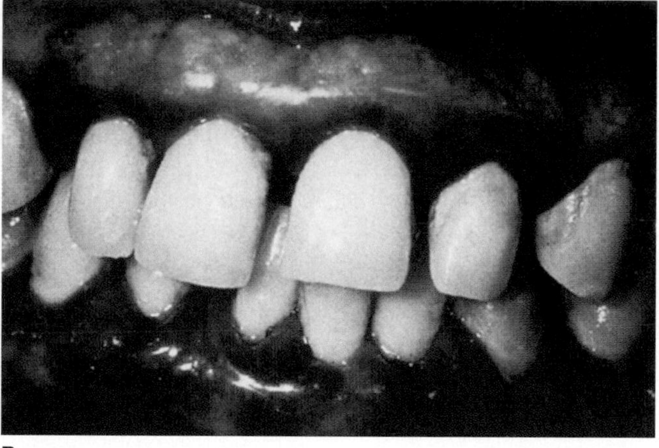

A B

FIGURE 60–5. Periodontitis. A. Normal gingival. **B.** Periodontal disease, with plaque, inflammatory changes, bleeding, and shortening of the gingival between the teeth. (Reproduced with permission from Nester EW, Anderson DG, Roberts CE Jr, Nester MT. *Microbiology: A Human Perspective*, 6th ed. New York: McGraw-Hill, 2008.)

Chronic periodontitis is responsible for most tooth loss in people older than 35 to 40 years. The disease progresses slowly and results in the progressive destruction of the supporting tissues of the tooth (periodontal ligament and alveolar bone) from the margins of the gingiva toward the apices of the roots of the teeth. Progression may occur as a series of acute episodes separated by quiescent periods of indeterminate duration. More aggressive forms of periodontitis result in more rapid loss of tooth support. Aggressive types of disease called localized aggressive periodontitis occur in adolescents, and generalized aggressive periodontitis occurs in young adults. There is some evidence that the causative agents may differ in this form of periodontitis. A small capnophilic (carbon dioxide–requiring) Gram-negative rod (*Actinobacillus actinomycetemcomitans*) has been indicted based on studies of the flora of disease sites. A virulence factor found in those strains of *A actinomycetemcomitans* that are associated with this disease is the production of a leukotoxin by the bacteria.

Chronic periodontitis causes tooth loss

Acute juvenile periodontitis associated with *Actinobacillus*

As the disease progresses, a point may be reached at which the alveolar bone around the necks of the teeth is resorbed; the condition is then no longer termed gingivitis, but periodontitis. With resorption of the bone, the attachment of the periodontal ligament is lost and the gingival sulcus deepens into a periodontal pocket. Periodontitis is not considered to be a reversible disease in that the lost alveolar bone and periodontal ligament do not regenerate with cessation of the inflammation, even though further progression may be halted. If unchecked, bone resorption progresses to loosening of the tooth, which may ultimately be exfoliated. **Figure 60–5** shows a case of advanced chronic periodontitis. Occasionally, the neck of a periodontal pocket becomes constricted, the bacteria proliferate causing an acute inflammatory response in the occluded pocket, and a periodontal abscess results. This acute exacerbation requires drainage in the same way as abscesses elsewhere for the patient to obtain symptomatic relief.

With continued progress, periodontitis and bone resorption develop

Periodontal abscess may result

NECROTIZING PERIODONTAL DISEASES

Necrotizing ulcerative gingivitis (also called acute necrotizing ulcerative gingivitis, Vincent's infection, or trench mouth) and necrotizing ulcerative periodontitis represent a spectrum of acute inflammatory disease starting with destruction limited to the soft tissues (gingivitis) and extending to destruction of the alveolar bone and periodontal ligament (periodontitis). This disease spectrum is distinctly different from gingivitis–chronic periodontitis. It has an acute onset, frequently associated with periods of stress and poor oral hygiene. Rapid ulceration of the interdental areas of the gingiva results in destruction of the interdental papillae. The inflammatory condition initially confined to the gingival tissues can quickly extend into pathologic bone resorption. Unlike gingivitis and chronic periodontitis, acute necrotizing periodontal disease is painful. As the oral epithelium is destroyed, the causative bacteria come into direct contact with the underlying tissues and may invade them. Spirochetes and fusiform bacteria have been implicated; thus, the term **fusospirochetal disease** has been used to describe this infection, which can also be manifested as ulceration in other areas of the pharynx or oral cavity. *Prevotella intermedia* has also been found in high numbers in the lesions. Morphologic studies have shown that the spirochetes actually appear to invade the tissues. The disease may be treated with systemic antibiotics and topical antimicrobials for immediate relief of symptoms, but resolution depends on thorough professional cleaning of the teeth and institution of good home care.

Acute onset with painful ulcerative lesions

Fusospirochetal etiology together with other anaerobes

CHAPTER 61

Respiratory Tract Infections

Worldwide, an estimated 3 to 5 million children die annually as a result of acute respiratory disease. Morbidity from respiratory infections constitutes the most common issue in humans. In this chapter, the illness types are discussed in the contexts of the sites where major clinical manifestations of involvement are expressed: upper, middle, and lower respiratory disease. Although there is considerable overlap wherein multiple contiguous sites may be simultaneously affected, such a classification is useful to highlight etiologic differences.

UPPER RESPIRATORY TRACT INFECTION

Upper respiratory infections usually involve the nasal cavity and pharynx, and most (more than 80%) are caused by viruses. Like middle and lower respiratory illnesses, the diseases of the upper respiratory tract are named according to the anatomic sites primarily involved. **Rhinitis** (or coryza) implies inflammation of the nasal mucosa, **pharyngitis** denotes pharyngeal infection, and **tonsillitis** indicates an inflammatory involvement of the tonsils. Because of the close proximity of these structures to one another, infections may simultaneously involve two or more sites (eg, rhinopharyngitis or tonsillopharyngitis).

Most upper respiratory infections are caused by viruses

Other infections considered are peritonsillar abscess (quinsy), retrotonsillar abscess, and retropharyngeal abscess. These infections are the result of direct invasion from mucosal sites and localization in deeper tissues to produce inflammation and abscess formation.

CLINICAL FEATURES

Rhinitis is the most common manifestation of the common cold. It is characterized by variable fever, inflammatory edema of the nasal mucosa, and an increase in mucous secretions. The net result is varying degrees of nasal obstruction; the nasal discharge may be clear and watery at the onset of illness, becoming thick and sometimes purulent as the infection progresses over 5 to 10 days.

The common cold is characterized by rhinitis

Pharyngitis and tonsillitis are associated with pharyngeal pain (sore throat) and the clinical appearance of erythema and swelling of the affected tissues. There may be exudates, consisting of inflammatory cells overlying the mucous membrane, and petechial hemorrhages; the latter may be seen in viral infections but tend to be more prominent in bacterial infections. Viral infections, particularly herpes simplex, may also lead to the formation of vesicles in the mucosa, which quickly rupture to leave ulcers. Pharyngeal candidiasis can also erode the mucosa under the plaques of "thrush." On rare occasions, the local inflammation may be sufficiently severe to produce **pseudomembranes,** which consist of necrotic tissue, inflammatory cells, and bacteria. This finding is particularly common in pharyngeal diphtheria, but may be mimicked by fusospirochetal infection (Vincent's angina) and sometimes by

Inflammatory exudates and hemorrhages more common in bacterial infections

Vesicles and ulcerated lesions more common in viral disease

Pharyngeal pseudomembranes in diphtheria

infectious mononucleosis. In acute tonsillitis or pharyngitis of any etiology, regional spread of the infecting agents with inflammation and tender swelling of the anterior cervical lymph nodes is also common.

Tonsillar asymmetry a sign of peritonsillar abscess

Peritonsillar or retrotonsillar abscesses are usually a complication of tonsillitis. They are manifested by local pain, and examination of the pharynx reveals tonsillar asymmetry with one tonsil usually displaced medially by the abscess. This infection is most common in children older than 5 years and in young adults. If not properly treated, the abscess may spread to adjacent structures. It can involve the jugular venous system, erode into branches of the carotid artery to cause acute hemorrhage, or rupture into the pharynx to produce severe aspiration pneumonia.

Retropharyngal abscess causes anterior bulging of pharyngeal wall

Retropharyngeal or lateral pharyngeal abscesses occur most frequently in infants and children less than 5 years of age. They can result from pharyngitis or from accidental perforation of the pharyngeal wall by a foreign body. The infection is characterized by pain, inability or unwillingness to swallow, and, if the pharyngeal wall is displaced anteriorly near the palate, a change in phonation (nasal speech). The neck may be held in an extended position to relieve pain and maintain an open upper airway. Examination of the pharynx usually reveals anterior bulging of the pharyngeal wall; if this finding is not apparent, lateral x-rays of the neck may demonstrate a widening of the space between the cervical spine and the posterior pharyngeal wall. The complications of such abscesses are basically the same as those described for peritonsillar abscesses. In addition, the suppurative process can extend posteriorly to the cervical spine to produce osteomyelitis or inferiorly to cause acute mediastinitis.

Oral and pharyngeal lesions accentuated in immunocompromised hosts

May be portal of entry for systemic infection

In the immunocompromised patient, all of the various forms of stomatitis and pharyngitis described previously can be accentuated. Leukemia, agranulocytosis, chronic ulcerative colitis, congenital or acquired immunodeficiency (eg, AIDS), and treatment with cytotoxic or immunosuppressive drugs are commonly associated with such lesions. The marked damage to mucosal tissues that sometimes occurs can provide a portal of entry into deeper structures and then to the systemic circulation, creating a risk of bacterial or fungal sepsis. Conversely, oral lesions may also result from dissemination of infection from other remote sites. Examples include disseminated histoplasmosis and sepsis caused by *Pseudomonas* species.

COMMON ETIOLOGIC AGENTS

Table 61–1 lists the more common causes of upper respiratory infections and stomatitis. Viral infections predominate. The most common bacterial cause to be considered is the group A streptococcus (GAS). *Corynebacterium diphtheriae,* though rare in the United States, is a major pathogen that continues to cause infection in many other countries and must not be

TABLE 61–1 Major Infectious Causes of Upper Respiratory Disease

DISEASE	VIRUSES	BACTERIA AND FUNGI
Rhinitis	Rhinoviruses, adenoviruses, coronaviruses, parainfluenza viruses, influenza viruses, respiratory syncytial virus, some coxsackie A viruses	Rare
Pharyngitis or tonsillitis	Adenoviruses, parainfluenza viruses, influenza viruses, rhinoviruses, coxsackie A or B virus, herpes simplex virus, Epstein–Barr virus	Group A streptococcus, *Corynebacterium diphtheriae, Neisseria gonorrhoeae*
Peritonsillar or retropharyngeal abscess	None	Group A streptococcus (most common), oral anaerobes such as *Fusobacterium* species, *Staphylococcus aureus, Haemophilus influenzae* (usually in infants)

overlooked, particularly if clinical and epidemiologic findings (immunization status) suggest this possibility. *Neisseria gonorrhoeae,* isolated from adults with symptomatic pharyngitis in whom no other etiologic agent can be demonstrated, is now considered a pharyngeal pathogen that is usually transmitted by oral–genital contact. Occasionally, other bacteria have been implicated as causes of acute pharyngitis (eg, *Corynebacterium ulcerans, Francisella tularensis,* and streptococci of groups B, C, and G). These are listed here for the sake of completeness but are not routinely sought except in unusual circumstances.

Viral infections predominate

Streptococcus pyogenes and C diphtheriae are bacterial pathogens

Gonococcal pharyngitis occurs with oral–genital contact

In patients with purulent rhinitis, sinusitis should also be considered in the differential diagnosis. Unilateral and foul-smelling purulent discharge suggests the presence of a foreign body in the nose. ::: Sinusitis, p. 909

GENERAL DIAGNOSTIC APPROACHES

Although viruses cause most upper respiratory infections, they are generally not amenable to specific therapy, and laboratory tests for viral infections are usually reserved for investigating outbreaks or when the illness seems unusually severe or atypical.

The primary diagnostic approach in pharyngitis and tonsillitis is to determine whether there is a bacterial cause requiring specific treatment. The only reliable method is to collect a throat swab for culture, taking care to thoroughly swab the tonsillar fauces as well as the posterior pharynx, and to include any purulent material from inflamed areas. Cultures are usually made only to detect the presence or absence of GAS. Direct antigen tests for rapidly detecting the group A antigen in throat swabs have gained popularity in recent years. These tests are rapid and very specific when positive but only about 90% sensitive. This means that a positive result may be accepted without culture but negative results must be confirmed by culture before withholding treatment.

Approach is to determine whether there is a bacterial etiology by culture

Direct detection methods have false-negative results

For the laboratory diagnosis of diphtheria or pharyngeal gonorrhea, the clinical suspicion should be indicated to the laboratory so that specific cultures for *C diphtheriae* or *N gonorrhoeae* may be made. *Candida* species, fusospirochetal bacteria, *Pseudomonas* species, and other Gram-negative organisms are often found in pharyngeal or oral specimens from healthy individuals as well as in certain infections. Their probable pathogenic significance in association with disease in these sites, largely based on the appearance of the lesions and the presence of the organisms in large numbers, can be supported by histologic demonstration of tissue invasion by the organisms. It is important to remember that other bacterial pathogens such as *Streptococcus pneumoniae, Staphylococcus aureus, Haemophilus influenzae,* and even *Neisseria meningitidis* may be present in the pharynx. These organisms are not primary etiologic agents in rhinitis, pharyngitis, and tonsillitis, and their presence in the throat does not implicate them as causes of the illnesses; they should instead be regarded as colonizers.

Evidence for pathogenic role of opportunists assessed by multiple means

Pathogens may be present in normal flora but not cause pharyngitis

The laboratory diagnosis of causes of peritonsillar and retropharyngeal abscesses is based on Gram staining and culture of purulent material obtained directly from the lesion, including anaerobic cultures.

GENERAL PRINCIPLES OF MANAGEMENT

Viral infections of the upper respiratory tract can only be treated symptomatically. If GAS is the cause, penicillin therapy is required; if the patient is allergic to penicillin, an alternative is chosen (eg, erythromycin or a cephalosporin). Such treatment prevents suppurative or toxigenic complications (eg, pharyngeal abscess, cervical adenitis, and scarlet fever) and the development of acute rheumatic fever. The latter, a serious complication, may occur in 1% to 3% of patients in certain population groups if they are not adequately treated. In addition, treatment of acute streptococcal infections can aid in reducing spread of the organisms to other persons.

Penicillins or cephalosporins necessary to treat *S pyogenes* infections

Macrolides used in penicillin allergic patients

C diphtheriae infections involve more complex management, which includes antitoxin as well as antimicrobic treatment. Infections caused by *N gonorrhoeae* are treated with appropriate antimicrobics. The management of stomatitis includes maintenance of adequate oral hygiene. If invasive *Candida* infection is present, topical and/or systemic antifungal therapy is sometimes necessary. Vincent's angina and other fusospirochetal infections are usually treated with systemic penicillin therapy as well as with appropriate dental and periodontal care.

Peritonsillar and retropharyngeal abscesses often require surgical drainage

There is no specific, widely accepted treatment for aphthous stomatitis. Peritonsillar and retropharyngeal abscesses are treated aggressively with antimicrobics and often require surgical drainage, taking care to prevent accidental aspiration of the abscess contents into the lower respiratory tract. ::: Diphtheria treatment, p. 477

MIDDLE RESPIRATORY TRACT INFECTION

For the purpose of this discussion, the middle respiratory tract is considered to comprise the epiglottis, surrounding aryepiglottic tissues, larynx, trachea, and bronchi. Inflammatory disease involving these sites may be localized (eg, **laryngitis**) or more widespread (eg, laryngotracheobronchitis). The majority of severe infections occur in infancy and childhood. Disease expression varies somewhat with age, partly because the diameters of the airways enlarge with maturation and because immunity to common infectious agents increases with age. For example, an adult with a viral infection of the larynx (laryngitis) who was exposed to the same virus in childhood has a relatively better immune response; in addition, the larger diameter of the larynx in the adult permits greater air flow in the presence of inflammation. An infant or child with the same infection in the same site can develop a much more severe illness, known as **croup,** which can lead to significant obstruction of air flow.

Most severe middle tract infections occur in infancy and childhood

CLINICAL FEATURES

Epiglottitis is often characterized by the abrupt onset of throat and neck pain, fever, and inspiratory stridor (difficulty in moving adequate amounts of air through the larynx). Because of the inflammation and edema in the epiglottis and other soft tissues above the vocal cords (supraglottic area), phonation becomes difficult (muffled phonation or aphonia), and the associated pain leads to difficulty in swallowing. If epiglottitis is not treated promptly, death may result from acute airway obstruction.

Epiglottitis carries risk of acute airway obstruction

Laryngitis or its more severe form, croup, may have an abrupt onset (spasmodic croup) or may develop more slowly over hours or a few days as a result of spread of infection from the upper respiratory tract. The illness is characterized by variable fever, inspiratory stridor, hoarse phonation, and a harsh, barking cough. In contrast to epiglottitis, the inflammation is localized to the subglottic laryngeal structures, including the vocal cords. It sometimes extends to the trachea (laryngotracheitis) and bronchi (laryngotracheobronchitis), where it is associated with a deeper, more severe cough that may provoke chest pain and variable degrees of sputum production. When vocal cord inflammation is severe, transient aphonia may result.

Laryngitis and croup involve subglottic laryngeal structures

Bronchitis or **tracheobronchitis** may be a primary manifestation of infection or a result of spread from upper respiratory tissues. It is characterized by cough, variable fever, and sputum production, which is often clear at the onset but may become purulent as the illness persists. Auscultation of the chest with the stethoscope often reveals coarse bubbling rhonchi, which are a result of inflammation and increased fluid production in the larger airways.

Bronchitis involves larger airways

Chronic bronchitis is a result of longstanding damage to the bronchial epithelium. A common cause is cigarette smoking, but a variety of environmental pollutants, chronic infections (eg, tuberculosis), and defects that hinder normal clearance of tracheobronchial secretions and bacteria (eg, cystic fibrosis) can be responsible. Because of the lack of functional integrity of their large airways, such patients are susceptible to chronic infection with members of the oropharyngeal flora and to recurrent, acute flare-ups of symptoms when they become colonized and infected by viruses and bacteria, particularly *Streptococcus pneumoniae* and nontypeable *Haemophilus influenzae.* A vicious cycle of recurrent infection may evolve, leading to further damage and increasing susceptibility to pneumonia.

Chronic bronchitis associated with smoking, air pollution, and other diseases

Nontypeable *H influenzae* and *S pneumoniae* found in exacerbations of chronic bronchitis

COMMON ETIOLOGIC AGENTS

With the exception of epiglottitis, acute diseases of the middle airway are usually caused by viral agents (**Table 61–2**). When acute airway obstruction is present, noninfectious

TABLE 61–2 Major Causes of Acute Middle Respiratory Tract Disease

SYNDROME	VIRUSES	BACTERIA	PERCENTAGE CAUSED BY VIRUSES
Epiglottitis	Rare	*Haemophilus influenzae, Streptococcus pneumoniae, Corynebacterium diphtheriae, Neisseria meningitidis*	10
Laryngitis and croup	Parainfluenza viruses, influenza viruses, adenoviruses; occasionally respiratory syncytial virus, metapneumovirus, rhinoviruses, coronaviruses, echoviruses	Rare	90
Tracheitis[a]	Same as for laryngitis and croup	*H influenzae, Staphylococcus aureus*	90
Bronchitis and bronchiolitis	Parainfluenza viruses, influenza viruses, respiratory syncytial virus, adenoviruses, measles	*Bordetella pertussis, H influenzae, Mycoplasma pneumoniae, Chlamydia pneumoniae*	80

[a]Often in combination with laryngitis and/or bronchitis.

possibilities, such as aspirated foreign bodies and acute laryngospasm or bronchospasm caused by anaphylaxis, must also be considered.

Most subglottic middle airway infections are viral

GENERAL DIAGNOSTIC APPROACHES

When a viral etiology is sought, the usual method of obtaining a specific diagnosis is by inoculation of cell cultures with material from the nasopharynx and throat, or by PCR. Acute and convalescent sera can also be collected to determine antibody responses to the common respiratory viruses and *Mycoplasma pneumoniae*. In bacterial infections, the approaches noted below are valuable.

Epiglottitis

H influenzae type b, once the most common cause of epiglottitis, produces an associated bacteremia in 85% of cases or more. Attempts to obtain cultures from the epiglottis or throat may provoke acute reflex airway obstruction in patients who have not undergone intubation to ensure proper ventilation; furthermore, the yield is lower than that of blood culture. In addition, other bacterial agents that cause epiglottitis can often be isolated from the blood. The exception is *Corynebacterium diphtheriae* infection, in which cultures of the nasopharynx or pharynx are required.

Incidence of bacteremia is high in epiglottitis

Laryngotracheitis and Laryngotracheobronchitis

Although most cases of laryngotracheitis and laryngotracheobronchitis have a viral etiology, a severe purulent process is seen occasionally. The latter is often referred to as acute **bacterial tracheitis,** and it can be rapidly fatal if not managed aggressively. Gram staining and culture of sputum, or better yet, of purulent secretions obtained by direct laryngoscopy, help to establish the causative agent. Blood cultures are again useful in such cases when a bacterial etiology is suspected.

Bacterial tracheitis is best diagnosed by direct laryngoscopy specimens

Acute Bronchitis

A major bacteriologic consideration in acute bronchitis, especially in infants and preschool children, is *Bordetella pertussis*. Deep nasopharyngeal cultures plated on the appropriate media constitute the best specimens. Examination of nasopharyngeal smears or aspirates

Nasopharyngeal specimens are appropriate for diagnosis of pertussis

by direct fluorescent antibody or PCR methods are also useful adjuncts to establishing the diagnosis. When purulent sputum is produced, Gram staining and culture may be useful in suggesting other bacterial causes (Table 61–2). Exceptions include *M pneumoniae* and *Chlamydia pneumoniae* infections, which are usually diagnosed by serologic testing of acute and convalescent sera.

Serodiagnosis commonly used for *M pneumoniae* and *C pneumoniae* infections

GENERAL PRINCIPLES OF MANAGEMENT

The primary initial concern is ensuring an adequate airway. It is particularly crucial in epiglottitis but can also become a major issue in laryngitis or laryngotracheobronchitis. Thus, some patients require placement of a rigid tube that provides communication between the tracheobronchial tree and the outside air (a nasotracheal tube or a surgically placed tracheostomy). Other adjunctive measures, such as highly humidified air and oxygen, may also provide relief in acute diseases involving the structures in and around the larynx. In proved or suspected bacterial infections, specific antimicrobic therapy is required; other treatment, such as antitoxin administration in diphtheria, may also be necessary.

Maintenance of airway patency required

Antimicrobic therapy for bacterial infections

LOWER RESPIRATORY TRACT INFECTION

Lower respiratory tract infection develops with invasion and disease of the lung, including the alveolar spaces and their supporting structure, the interstitium, and the terminal bronchioles. Bronchiolitis, an inflammatory process primarily affecting the small terminal airways in infants, is discussed extensively in Chapter 9. Infection may occur by extension of a middle respiratory tract infection, aspiration of pathogens past the upper airway defenses, or less commonly by hematogenous spread from a distant site such as an abscess or an infected heart valve. When infection develops through the respiratory tract, some compromise of the upper airway mechanisms for filtering or clearing inhaled infectious agents usually occurs. The most common compromises are those that impair the epiglottic and cough reflexes, such as drugs, anesthesia, stroke, and alcohol abuse. Toxic inhalations and cigarette smoking may also interfere with the normal mucociliary action of the tracheobronchial tree. In healthy persons, the most common antecedent to lower respiratory infection is infection of the middle respiratory structures (usually viral), allowing an otherwise innocuous aspiration of oropharyngeal flora to reach the lower tract and progress to disease rather than undergo rapid clearance. Some small infectious particles can accomplish airborne passage through the middle airway and bypass mucociliary defenses; if they can survive or multiply in alveolar macrophages, they may produce a primary infection. Examples include arthroconidia of *Coccidioides immitis* and cells of *Mycobacterium tuberculosis* and spores of *Bacillus anthracis.*

Infection can be by inhalation, aspiration, extension from middle tract, or blood-borne

Infection through air passages is associated with compromised local clearance defenses

CLINICAL FEATURES

Acute Pneumonia

Acute pneumonia is an infection of the lung parenchyma that develops over hours to days and, if untreated, runs a natural course lasting days to weeks. The onset may be gradual, with malaise and slowly increasing fever, or sudden, as with the bed-shaking chill associated with the onset of pneumococcal pneumonia. The only early symptom referable to the lung may be cough, which is caused by bronchial irritation. In adults, the cough becomes productive of **sputum,** which is purulent material generated in the alveoli and small air passages. In some cases, the sputum may be blood-streaked, rusty in color, or foul-smelling. Labored or difficult breathing (dyspnea), rapid respiratory rate, and sometimes cyanosis are signs of increasing loss of alveolar air-exchange surface through spread of exudate. Chest pain from inflammatory involvement of the pleura is common. Physical signs on auscultation reflect the filling and eventual consolidation of alveoli by fluid and inflammatory cells.

Sputum is purulent material generated in the bronchi and alveoli

Fever, respiratory distress, and sputum production are signs of acute pneumonia

The radiologic pattern of inflammatory changes in the lung is very useful in the diagnosis of pneumonia and for clinical differentiation into likely etiologic categories. The most common pattern is patchy infiltrates related to multiple foci centering on small bronchi (bronchopneumonia), which may progress to a more uniform consolidation of one or more lobes (lobar pneumonia). A more delicate, diffuse, or "interstitial" pattern, which is also common, is particularly associated with viral pneumonia.

Radiologic changes confirm and refine diagnosis

Chronic Pneumonia

Chronic pneumonia has a slow insidious onset that develops over weeks to months and may last for weeks or even years. The initial symptoms are the same as those of acute pneumonia (fever, chills, and malaise), but they develop more slowly. Cough can develop early or late in the illness. As the disease progresses, appetite and weight loss, insomnia, and night sweats are common. Cough and sputum production may be the first indication of a vague constitutional illness referable to the lung. Bloody sputum (hemoptysis), dyspnea, and chest pain appear as the disease progresses. The physical findings and radiologic features can be similar to those of acute pneumonia, except that the diffuse interstitial infiltrates of viral pneumonia are uncommon. There may be parenchymal destruction and the formation of abscesses or cavities communicating with the bronchial tree. The clinical features of chronic pneumonia may be due to a number of infectious agents or noninfectious causes such as neoplasms, vasculitis, allergic conditions, infarction, radiation or toxic injury, and diseases of unknown etiology (eg, sarcoidosis).

Chronic pneumonia develops over weeks to months

Abscesses and cavities may develop

Chronic pneumonia may have noninfectious causes

Pleural effusion is the transudation of fluid into the pleural space in response to an inflammatory process in adjacent lung parenchyma. It may result from a wide variety of causes, both infectious and noninfectious. **Empyema** is a purulent infection of the pleural space that develops when the infectious agent gains access by contiguous spread from an infected lung through a bronchopleural fistula or, less often, by extension of an abdominal infection through the diaphragm. Symptoms are usually insidious and related to the primary infection until enough exudate is formed to produce symptoms referable to the chest wall or to compromise the function of the lung. The physical and radiologic findings are characteristic, with dullness to percussion and localized opacities on x-ray. In contrast to noninfectious effusions, empyema is frequently loculated.

Pleural effusions may be infectious or noninfectious

Empyema is a purulent infection of pleural space usually by extension of bacterial infection

Lung Abscess

Lung abscess is usually a complication of acute or chronic pneumonia caused by organisms that can cause localized destruction of lung parenchyma. It may occur as part of a chronic process or as an extension of an acute, destructive pneumonia, often after aspiration of oral or gastric contents. The symptoms of lung abscess, which are usually not specific, resemble those of chronic pneumonia or an acute pneumonia that has failed to resolve. Persistent fever, cough, and the production of foul-smelling sputum are typical. Lung abscess can be diagnosed and localized with certainty only radiologically; it appears as a localized area of inflammation with single or multiple excavations or as a cavity with an air–fluid level. Multiple abscesses may develop as a result of blood-borne infection.

Lung abscess frequently follows aspiration pneumonia

Blood-borne infection may cause multiple abscesses

COMMON ETIOLOGIC AGENTS

The infectious agents that most frequently cause lower respiratory infection are listed in **Table 61–3**. The cause of acute pneumonia is strongly dependent on age. More than 80% of pneumonias in infants and children are caused by viruses, whereas less than 10% to 20% of pneumonias in adults are viral. The reasons are probably the same as those indicated previously for middle respiratory tract infections. Influenza and other viruses, however, may provide the initial predisposition toward bacterial infection. Viruses are extremely rare as a cause of chronic as opposed to acute lower respiratory tract infections, although some symptoms of the acute infection, such as cough, may persist for weeks until the bronchial damage has healed. Influenza virus is noteworthy as a cause of acute life-threatening pneumonia, even in previously healthy young adults. Pneumonia caused by bacteria such as enteric Gram-negative rods, *Pseudomonas,* and *Legionella* is primarily limited to patients with serious debilitating underlying disease or as a complication of hospitalization and its procedures (nosocomial infection). At any age, the pneumococcus is the most common bacterial cause of acute pneumonia, and Gram-negative infections other

Most pneumonias are viral in infants and children

Viral infections predispose to acute bacterial pneumonia

Gram-negative pneumonias occur in debilitated hosts

TABLE 61–3 Major Causes of Lower Respiratory Tract Infection

SYNDROME	VIRUSES	COMMON BACTERIA	FUNGI	OTHER AGENTS
Acute pneumonia	Influenza,[a] parainfluenza, adenovirus, respiratory syncytial virus (infants and elderly)[a], metapneumovirus	*Streptococcus pneumoniae*, *Staphylococcus aureus*, *Haemophilus influenzae*, Enterobacteriaceae, *Legionella*, mixed anaerobes (aspiration), *Pseudomonas aeruginosa*[b]	*Candida albicans*,[b] *Aspergillus* species *Pneumocystis*[b]	*Mycoplasma pneumoniae*, *Chlamydia trachomatis* (infants), *Chlamydia pneumoniae*
Chronic pneumonia	Rare	*Mycobacterium tuberculosis*, other mycobacteria, *Nocardia*	*Coccidioides immitis*,[c] *Blastomyces dermatitidis*,[c] *Histoplasma capsulatum*,[c] *Cryptococcus neoformans*	*Paragonimus westermani*[c]
Lung abscess	None	Mixed anaerobes, *Actinomyces*, *Nocardia*, *S aureus*,[d] *Enterobacteriaceae*,[d] *P aeruginosa*[b,d]	*Aspergillus* species	*Entamoeba histolytica*
Empyema	None	Mixed anaerobes, *S aureus*,[d] *S pneumoniae*,[d] Enterobacteriaceae, *P aeruginosa*[d]	Rare	

[a]Occurrence limited to seasonal epidemics.
[b]Primarily infects the immunologically compromised host.
[c]Geographically limited.
[d]Infection develops during or after acute pneumonia.

Pneumoccoccus is most common cause of acute bacterial pneumonia

than *Haemophilus* are rare in children unless they have cystic fibrosis or immunodeficiency. Acute and subacute pneumonia may be due to *Chlamydia. C trachomatis* is almost exclusively limited to infants less than 7 months of age, whereas *C pneumoniae* commonly affects school children and young adults, producing both bronchitis and pneumonia.

Lung abscess has different patterns

Lung abscess and empyema follow infections with the more destructive organisms or aspiration of mixed anaerobic flora from the oropharynx. Several clinical clues can suggest some of the etiologic agents, given a typical clinical syndrome. For example, *Nocardia* and mycobacteria, which are strict aerobes, tend to produce upper lobe infiltrates, whereas aspiration pneumonia caused by anaerobes tends to develop in the most dependent parts of the lung. Textbooks on infectious disease should be consulted for further details regarding these features.

GENERAL DIAGNOSTIC APPROACHES

Interpretation depends on whether agent is found in normal flora

The degree of difficulty in establishing an etiologic diagnosis for a lower respiratory tract infection depends on the number of organisms produced in respiratory secretions, whether the causative species is normally found in the oropharyngeal flora, and how easily it is grown. In the presence of typical clinical findings, the isolation of influenza virus from the throat or of *M tuberculosis* from sputum is sufficient for diagnosis of influenza or tuberculosis, because these organisms are not normally found in such sites. The same cannot be said for *S pneumoniae* and most bacterial pathogens, because they may be found in the throat in a significant number of healthy persons.

Sputum collection has problems of quality and specificity

The examination of expectorated sputum has been the primary means of diagnosing the causes of bacterial pneumonia, but this approach has several advantages and disadvantages. The advantages are ease of collection and absence of risk to the patient. The primary disadvantage is the confusion that results from contamination of the sputum with oropharyngeal flora in the process of expectoration and excessive contamination with saliva. Efforts have been unsuccessful to remove saliva from sputum by washing or to accomplish interpretive differentiation of infective from normal flora by quantitative culture as with urine specimens. The quality of a sputum sample can be enhanced by collection early in the morning (just after the patient arises), careful instruction of the patient, and occasionally by the use of saline aerosols (induced sputum) under the supervision of an inhalation therapy specialist. The worst results can be expected when the physician's only involvement is writing an order, which is then passed down the ward chain of command to an orderly, who directs the patient to put his "sputum" in a cup placed at the bedside. ::: Quantitative urine culture, p. 942

Contamination with oropharyngeal secretions is primary problem

Microscopic examination before culture of direct Gram smears of specimens alleged to be sputum has proved useful. Polymorphonuclear leukocytes and large numbers of a single morphologic type of organism are typical findings in sputum from patients with bacterial pneumonia. Squamous epithelial cells from the oropharynx and a mixed bacterial population are characteristic of saliva (**Figure 61–1A** and **B**). Unfortunately, most specimens are a mixture of both, which makes interpretation more difficult. Studies have shown that more than 10 to 25 squamous epithelial cells per low-power microscopic field are evidence of excessive salivary contamination, and such specimens should not be cultured because the results may be misleading. Thus, the direct Gram smear is crucial to the use of expectorated sputum for diagnosis of acute bacterial pneumonia. The smear may be useful in the absence of cultural results, but cultures are useless without a Gram smear to assess specimen quality.

Microscopic characteristics of sputum can differentiate from saliva

Salivary specimens should not be cultured

Another approach is to attempt a more direct collection from the lung using methods that bypass the oropharyngeal flora. This approach may be used in patients who are not producing sputum or in cases where analysis of expectorated sputum has been inconclusive. The major techniques include transtracheal aspiration, bronchoalveolar lavage (BAL), direct aspiration, and open biopsy. In transtracheal aspiration, an incision is made in the cricothyroid membrane and a catheter advanced deep into the tracheobronchial tree to aspirate sputum directly. This method is useful in diagnosis of both pneumonia and lung abscess. BAL is a modification of bronchoscopy in which the bronchi and alveoli are infused with saline, which is aspirated back through the bronchoscope.

Transtracheal and direct lung aspiration bypass oral flora

Specimens obtained by BAL have been increasingly useful for demonstration of organisms such as *Pneumocystis carinii,* which were previously seen only in open lung biopsies. Because BAL involves initial passage of the instrument through the upper airway, interpretation must take into account the possibility of some contamination with oropharyngeal secretions. Aspirates taken through tracheostomies or endotracheal tubes are of almost no value, because these sites become colonized with Gram-negative bacteria within hours of their implantation. Direct aspiration through the chest wall can be used for diagnosis of pneumonia or empyema if the involved area can be well localized and is at the lung periphery. In some cases, an open lung biopsy is the only way to obtain diagnostic material. Bacteremia may occur in acute pneumonia, particularly in its early stages. A blood culture should be part of the evaluation of every acute pneumonia. If positive, it can confirm or overrule a diagnosis based on expectorated sputum culture.

BAL washes material from deep in the lung

Blood culture is valuable in acute pneumonia

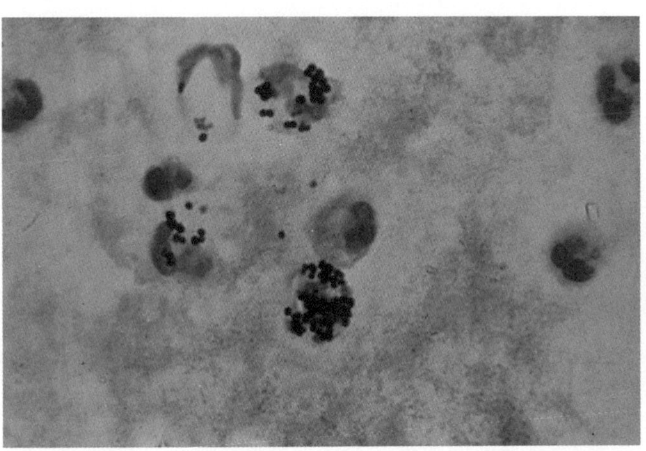

A

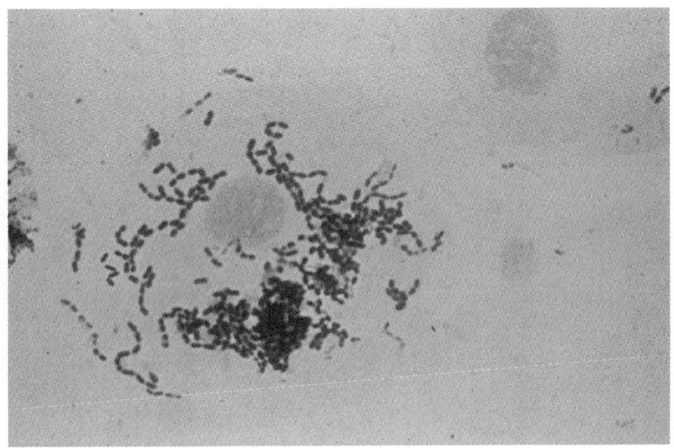

B

FIGURE 61–1. Comparison of findings in sputum and saliva. A. True sputum shows an abundance of inflammatory cells and no squamous epithelial cells. In acute bacterial pneumonia, large numbers of a single organism are usually present. This Gram smear shows polymorphonuclear leukocytes and *Staphylococcus aureus.* **B.** Saliva typically contains squamous epithelial cells and a mixed bacterial population, some of which can look like pathogens. (Reprinted with permission of Schering Corporation, Kenilworth, NJ, the copyright owner. All rights reserved.)

Anaerobic infections cannot be diagnosed from expectorated sputum

Once an appropriate specimen is obtained, diagnosis is usually readily made by culture using the methods described in Chapter 4 and in the sections on the individual etiologic agents. Only specimens collected by one of the invasive techniques should be used for anaerobic culture, because expectorated sputum is invariably contaminated with oropharyngeal anaerobes and results are meaningless.

GENERAL PRINCIPLES OF MANAGEMENT

The general principles of management of lower respiratory tract infections are similar to those of middle tract infections. Drainage or surgical measures are needed more often as adjuncts to antimicrobial therapy in cases of chronic pneumonia, lung abscess, and empyema. When bacterial infection is considered, empiric therapy is usually given until the results of cultures and antimicrobial susceptibility tests are available. Treatment may vary from penicillin alone for a previously healthy person, in whom the most reasonable nonviral possibility is *S pneumoniae,* to multiple drugs for a debilitated or immunocompromised patient, in whom the possibilities are much broader.

CHAPTER 62

Enteric Infections and Food Poisoning

Acute infections of the gastrointestinal tract are among the most common of all illnesses, exceeded only by respiratory tract infections such as the common cold. Diarrhea is the most common manifestation of these infections. However, because it is usually self-limiting within hours or days, most of those afflicted with gastrointestinal infections do not seek medical care. Nonetheless, gastrointestinal infection remains one of the three most common syndromes seen by physicians who practice general medicine. Worldwide, diarrheal disease remains one of the most important causes of morbidity and mortality among infants and children. It has been estimated that in Asia, Africa, and Latin America, depending on socioeconomic and nutritional factors, a child's chance of dying of a diarrheal illness before the age of 7 years can be as high as 50%. In developed countries, mortality rate is very much lower, but it is still significant. This chapter summarizes the known etiologies and epidemiologic circumstances of these infections, as well as diagnostic methods and some aspects of management. Chapters on the individual etiologic agents should be consulted for details.

Diarrheal diseases in developing countries cause death of children

CLINICAL FEATURES

The most prominent clinical features of gastrointestinal infections are fever, vomiting, abdominal pain, and diarrhea. Their presence varies with different diseases and different stages of infection. The occurrence of diarrhea is a central feature, and its presence and nature form the basis for classification of gastrointestinal infections into three major syndromes: watery diarrhea, dysentery, and enteric fever.

Watery Diarrhea

The most common form of gastrointestinal infection is the rapid development of frequent intestinal evacuations of a more or less fluid character known as diarrhea (derived from the Greek *dia* for through, and *rhein*, meaning to flow like a stream). Nausea, vomiting, fever, and abdominal pain may also be present, but the dominant feature is intestinal fluid loss. Diarrhea is produced by pathogenic mechanisms that attack the proximal small intestine, the portion of the bowel in which more than 90% of physiologic net fluid absorption occurs. The purest form of watery diarrhea is that produced by enterotoxin-secreting bacteria such as *Vibrio cholerae*, and enterotoxigenic *Escherichia coli* (ETEC), which cause fluid loss without cellular injury. Other common pathogens that damage the epithelium, such as rotaviruses and calcivirus also cause fluid loss, but are more likely to cause fever and vomiting as well. Most cases of watery diarrhea run an acute but brief (1 to 3 days) self-limiting course. Exceptions are those caused by *V cholerae,* which usually produces a more severe illness, and those caused by *Giardia lamblia,* which produces a watery diarrhea that may last for weeks. ::: Cholera, p. 570

Fluid loss from proximal small intestine is the primary mechanism

Dysentery

Inflammation, cytotoxins, or invasion produce pus and blood

Colon is primary location

Dysentery begins with the rapid onset of frequent intestinal evacuations, but the stools are of smaller volume than in watery diarrhea and contain blood and pus. If watery diarrhea is the "runs," dysentery is the "squirts." Fever, abdominal pain, cramps, and tenesmus are common complaints. Vomiting occurs less often. The focus of pathology is the colon. Organisms causing dysentery can produce inflammatory and/or destructive changes in the colonic mucosa either by direct invasion or by production of cytotoxins. This damage produces the pus and blood seen in the stools, but does not result in substantial fluid loss because the absorptive and secretory capacity of the colon is much less than that of the small bowel. Dysenteric infections generally last longer than the common watery diarrheas, but most cases still resolve spontaneously in 2 to 7 days.

Enteric Fever

Systemic disease begins in the intestine

Focus often becomes lymphoid and reticuloendothelial invasion

Enteric fever is a systemic infection, the origin and focus of which are the gastrointestinal tract. The most prominent features are fever and abdominal pain, which develop gradually over a few days in contrast to the abrupt onset of the other syndromes. Diarrhea is usually present but may be mild and not appear until later in the course of the illness. The pathogenesis of enteric fever is more complex than that of watery diarrhea or dysentery. It generally involves penetration by the organism of the cells of the distal small bowel with subsequent spread outside the bowel to the biliary tract, liver, mesentery, or reticuloendothelial organs. Bacteremia is common, occasionally causing metastatic infection in other organs. Typhoid fever caused by *Salmonella enterica* serovar Typhi is the only infection for which these events have been well studied. Although it is usually self-limiting, enteric fever carries a significant risk of serious disease and significant mortality. ::: Typhoid fever, p. 603

COMMON ETIOLOGIC AGENTS

Clinical syndromes overlap for specific etiologic agents

Great advances have been made in our understanding of gastrointestinal infections. Before the late 1960s, less than 20% of the infectious syndromes just described could be linked to a specific etiologic agent by any known diagnostic method. The organisms listed in **Table 62–1** now account for 80% to 90% of cases, although diagnostic methods for all of them are not yet practical for clinical laboratories. The primary clinical syndrome listed for each agent in Table 62–1 should not be regarded as absolute because of individual variations and overlap; some pathogens cause more than one syndrome. For example, *Shigella* infections frequently go through a brief watery diarrhea stage before localizing in the colon, and *Campylobacter* enteritis usually begins with fever, malaise, and abdominal pain, followed by dysentery. In any single case, the clinical findings may suggest a range of etiologic agents, but none is sufficiently specific to be diagnostic of any single organism.

EPIDEMIOLOGIC SETTING

Epidemiologic setting narrows the diagnostic possibilities

The epidemiologic setting of the infection is of great importance in assessing the relative probability of the infectious agents. When combined with clinical findings, the differential diagnosis can often be limited to two or three organisms. The major epidemiologic settings are (1) endemic infection, (2) epidemic infection, (3) traveler's diarrhea, (4) food poisoning, and (5) hospital-associated diarrhea.

Endemic Infections

High incidence in children is related to fecal—oral spread and lack of immunity

By definition, endemic diarrheas are those that occur sporadically in the usual living circumstances of the patient (from the Greek *endemos*, dwelling in a place). Some organisms are endemic worldwide, whereas others are geographically limited. There are also seasonal variations and age-related attack rates within the endemic foci. In developed countries, the most common causes of endemic gastrointestinal infections are rotaviruses, caliciviruses, *Campylobacter, Salmonella,* and *Shigella.* All are more common in infants and children because they are more prone to fecal–oral spread and because development of immunity is related to age. Rotaviruses account for 40% to 60% of diarrheal infections occurring during the cooler months in infants and children less than 2 years of age but are uncommon in older persons. Calciviruses produce gastroenteritis in an older population and have been responsible for multiple outbreaks in closed populations such as cruise ships.

Geographic distributions change

The geographically limited agents are common only in the areas listed (Table 62–1). These distributions are not fixed, making it necessary to keep abreast of geographic changes in the distribution of established agents as well as the recognition of new ones. For example, cholera has long been limited to warm-climate river deltas in Asia, Africa, and the Middle East, but recently it has spread to South and Central America and the Gulf Coast of Louisiana and Texas.

Epidemic Infections

Typhoid, cholera, and shigellosis spread where hygiene is poor or after major disasters

EHEC more common in developed countries

Under certain epidemiologic conditions, some of the organisms responsible for endemic infections can spread beyond the family unit to cause epidemics involving regional, national, and even international populations. The diarrheal diseases most frequently associated with epidemics are typhoid fever, cholera, and shigellosis. For all three, epidemics are related to the failure of basic public health sanitary measures. For example, *Salmonella* serovar Typhi and *Vibrio cholerae* may be spread for some distance through the water supply, a route blocked by modern sewage and water treatment practices. When these procedures are not used or are interrupted by equipment failure or natural disasters (floods, earthquakes), these diseases can and do recur in epidemic form. Epidemics of shigellosis may be water-borne under the same conditions, but *Shigella* dysentery is more typically a disease "of wars and armies, and of crowds and movement."* The very low infecting dose of *Shigella* can make spreading through direct contact reach epidemic proportions when crowding and poor sanitary facilities are combined. *Giardia* and *Cryptosporidium* have been frequently identified causes of recent water-borne epidemics. *E coli* O157:H7 and other enterohemorrhagic *E coli* (EHEC) has been the cause of outbreaks hemorrhagic colitis related to meats, fruit juices, and fresh vegetables distributed widely by modern packing and transport systems. It is the only diarrheal agent more common in developed than developing counties.

Most recent epidemic is cholera in South America

Although large epidemics are usually associated with the 19th century, it is clear that the potential remains. In the late 1970s, large epidemics of both typhoid fever and shigellosis spread through Central and South America. In 1973, more than 200 cases of typhoid fever in Florida were associated with a defective chlorinator in the local water system. The current cholera pandemic claimed thousands of lives in South America in the last decade of the 20th century.

Traveler's Diarrhea

Visits to developing countries are frequently marred

From 20% to 50% of travelers from developed countries who go to less developed countries experience a diarrheal illness in the first week that is usually brief but can be serious. The common names applied to this syndrome, such as "Delhi belly" and "Montezuma's revenge," reflect geographic associations and the cumulative frustration of those forced to spend part of their vacation next to the toilet rather than the swimming pool.

ETEC is the predominant cause of traveler's diarrhea

Travelers should avoid salads and other uncooked foods

The most extensive studies of traveler's diarrhea have involved travelers from the United States to Latin American countries, particularly Mexico. In nearly 50% of these cases, the diarrhea is caused by enterotoxigenic strains of *E coli* (ETEC) acquired during travel. *Shigella* infections account for another 10% to 20%, and the remaining cases are attributable to various pathogens or unknown causes. Ingestion of uncooked or incompletely cooked foods is the most likely source of infection, but most epidemiologic studies have not shown specific food associations. An exception is the strong relation between ETEC diarrhea and the consumption of salads containing raw vegetables. "Don't drink the water" still seems like sound advice for travelers to countries where hygiene remains poor, but the adage is not well supported by studies relating infection to water or ice consumption.

---

*Christie AB. *Infectious Disease, Epidemiology and Clinical Practice,* 2nd ed. New York: Churchill Livingstone; 1974, p. 137.

TABLE 62–1 Features of Infectious Gastrointestinal Syndromes

					LABORATORY DIAGNOSIS[a]				
					CULTURE			SEROLOGY	
ORGANISM	COMMON DISTRIBUTION	CLINICAL SYNDROME	PATHOGENIC MECHANISM	STOOL MICROSCOPY	STOOL[b]	BLOOD	TOXIN IN STOOLS	ANTIBODY DETECTION	ANTIGEN DETECTION
Salmonella serotypes	Worldwide	Dysentery	Mucosal invasion	PMNs	+	–	–	–	–
Salmonella serovar Typhi	Tropical, developing countries	Enteric fever	Penetration, spread	Monocytes	+	+	–	+	–
Shigella spp.	Worldwide	Dysentery	Mucosal invasion, cytotoxin	PMNs, RBCs	+	–	–	–	–
Shigella dysenteriae (Shiga)	Tropical, developing countries	Dysentery	Mucosal invasion, cytotoxin	PMNs, RBCs	+	+	–	–	–
Campylobacter jejuni	Worldwide	Dysentery	Unknown	PMNs, RBCs	+	–	–	–	–
Escherichia coli (EIEC)	Worldwide	Dysentery	Mucosal invasion	PMNs, RBCs	+[c]	–	–	–	–
E coli (ETEC)	Worldwide[d]	Dysentery	Enterotoxin(s)	–	+[c]	–	–	–	–
E coli (EHEC)	Worldwide	Watery diarrhea	Cytotoxin	RBCs	+[c]	–	–	–	–
E coli (EPEC)	Worldwide[d]	Watery diarrhea	Adherence	–	+[c]	–	–	–	–
Vibrio cholerae	Asia, Africa, Middle East, Central and South America, Louisiana, Texas	Watery diarrhea	Enterotoxin	–	+	–	–	–	–
Vibrio parahaemolyticus	Seacoast	Watery diarrhea	Unknown	–	+	–	–	–	–

Yersinia enterocolitica	Worldwide	Enteric fever[c]	Penetration, spread	–	+	+	–	–	–
Clostridium difficile	Worldwide	Dysentery	Cytotoxin, enterotoxin	–	+	–	+	–	–
Clostridium perfringens	Worldwide	Watery diarrhea	Enterotoxin	–	+	–	–	–	–
Bacillus cereus	Worldwide	Watery diarrhea	Enterotoxin	–	+	–	–	–	–
Rotavirus	Worldwide	Watery diarrhea	Mucosal destruction	Electron microscopy[f]	–	–	–	–	+
Caliciviruses	Worldwide	Watery diarrhea	Mucosal destruction	Electron microscopy[f]	–	–	–	–	–
Giardia lamblia	Worldwide	Watery diarrhea	Mucosal irritation	Flagellates, cysts	–	–	–	–	–
Entamoeba histolytica	Worldwide[d]	Dysentery	Mucosal invasion	Amebas, PMNs	–	–	–	+	–
Cryptosporidium	Worldwide	Watery diarrhea	?toxin	Acid-fast oocysts	–	–	–	–	–

RBCs, red blood cells; EIEC, enteroinvasive *E coli*; EHEC, enterohemorrhagic *E coli;* EPEC, enteropathogenic *E coli;* ETEC, enterotoxigenic *E coli;* PMNs, polymorphonuclear leukocytes.
[a]Positive sign indicates procedure is useful and usually available in clinical laboratories.
[b]Which cultures are done routinely depends on the laboratory and/or physician's request.
[c]Organism may be isolated in culture, but demonstration of pathogenic potential (toxin production, etc.) is limited to specialized laboratories.
[d]Organism is more common in developing countries.
[e]Infection may also manifest watery diarrhea or dysentery.
[f]Appropriate methods may be available in only a limited number of laboratories.

Food Poisoning

Single-source outbreaks are becoming more widespread with modern food processing and distribution

Many gastrointestinal infections involve food as a vehicle of transmission. The term "food poisoning," however, is usually reserved for instances in which a single meal can be incriminated as the source. This situation typically arises when multiple cases of the same gastrointestinal syndrome develop at the same time among persons whose only common experience is a meal shared at a social event or restaurant. The probable etiologic agent can usually be assessed from knowledge of the incubation period, the food vehicle, and the clinical findings. Changes in the importation, processing, and distribution of foods have increased the complexity and potential for food-borne transmission of enteric pathogens. Outbreaks that in the past might have been limited may now be widely distributed by fast-food chains or airline catering services.

Diseases from ingestion of preformed toxin have short incubation periods

The most common causes of food poisoning are shown in **Table 62–2**. Some are not infections but intoxications, caused by ingestion of a toxin produced by bacteria in the food before it was eaten. Intoxications have shorter incubation periods than infections and may involve extraintestinal symptoms (eg, the neurologic damage in botulism). Infectious food poisoning does not differ from endemic diarrheal infections caused by the same species. The length of the incubation period and the severity of the symptoms are generally related to the number of organisms ingested. ::: Botulism, p. 525

The epidemiologic circumstances of food poisoning vary with the etiologic agent, but almost always involve a breach in the recommended procedures for handling food. The organisms may be present as contaminants in raw food before cooking or introduced by a carrier or contaminated utensil involved in preparation. Causes of bacterial food

TABLE 62–2 Clinical and Epidemiologic Features of Food Poisoning

ETIOLOGY	PERCENTAGE OF CASES[a]	TYPICAL INCUBATION PERIOD	PRIMARY CLINICAL FINDINGS	CHARACTERISTIC FOODS
Intoxication[b]				
Bacillus cereus (vomiting toxin)	1-2	1-6 h	Vomiting, diarrhea	Rice, meat, vegetables
Clostridium botulinum	5-15	12-72 h	Neuromuscular paralysis	Improperly preserved vegetables, meat, fish
Staphylococcus aureus	5-25	2-4 h	Vomiting	Meats, custards, salads
Chemical[c]	20-25	0.1-48 h	Variable	Variable
Infection[d]				
Clostridium perfringens	5-15	9-15 h	Watery diarrhea	Meat, poultry
Salmonella	10-30	6-48 h	Dysentery	Poultry, eggs, meat
Shigella	2-5	12-48 h	Dysentery	Variable
Vibrio parahaemolyticus	1-2	10-24 h	Watery diarrhea	Shellfish
Trichinella spiralis	5-10	3-30 days	Fever, myalgia	Meat, especially pork
Hepatitis A	1-3	10-45 days	Hepatitis	Shellfish

[a]Based on documented outbreaks reported to the Centers for Disease Control and Prevention, Atlanta, GA (variable from year to year).
[b]Disease caused by toxin in food at time of ingestion.
[c]Includes heavy metals, monosodium glutamate, mushrooms, and various toxins of nonmicrobial origin.
[d]Disease caused by infection after ingestion.

poisoning include failure to kill the organisms by adequate cooking, almost always followed by a period of warming (incubation) long enough for the organisms to multiply to infectious numbers or, in the case of toxigenic disease, to produce sufficient toxin to cause disease. In 80% to 90% of investigated outbreaks of bacterial food poisoning, the most important contributing factor is the use of improper storage temperatures for the food. This factor may obtain in home-cooked meals as well as those prepared in restaurants, in schools, or at large social events such as community picnics.

Infection is associated with improper cooking and/or storage

The relative frequency of each etiologic agent and the foods most often involved are also shown in Table 62–2. This information is based on outbreaks investigated by public health agencies, but it is generally accepted that these represent the "tip of the iceberg" owing to underreporting. Large outbreaks, restaurant-associated outbreaks, and outbreaks involving serious illness with hospitalization or death are more likely to be reported to health authorities than are mild diarrheas after a dinner party or airline meal. Of the 400 to 500 outbreaks (10,000 to 15,000 cases) reported each year in the United States, fewer than 200 are "solved." Food poisoning characterized by a short incubation period (eg, *Staphylococcus aureus*) is more likely to be recognized because it can be easily associated with a specific meal and because the food itself may still be available for examination.

There are also large geographic differences in reporting. For example, in one year, New York City, in which 50% of the state population resides, reported 98% of New York State's food-borne outbreaks, and Connecticut reported more outbreaks than all of the southeastern states combined.

Reporting of outbreaks varies greatly

Sampling problems aside, the food poisoning syndromes listed in Table 62–2 are well recognized, with *Salmonella*, *Clostridium perfringens,* and *S aureus* accounting for more than 70% of those for which a microbial cause can be found. For bacterial infections such as *Salmonella* and *Shigella,* which are not normal members of the stool flora, establishing the diagnosis by isolating the causative organism is relatively easy. If the circumstances indicate *C perfringens* or *S aureus* food poisoning, investigation involves cultures of vomitus, stool from several cases, and the suspect food. In some cases, toxin detection is required to establish the etiology and source. Such investigations are best coordinated by public health authorities, who can also address the legal and community implications of the outbreak. For example, one investigation of *Salmonella* food poisoning led to the discovery that the owner of a restaurant was keeping and slaughtering chickens at the restaurant. Although this practice may have provided very fresh chicken, it guaranteed *Salmonella* contamination of the entire kitchen.

Determining the cause of microbial food poisoning is best done by public health authorities

Hospital-Associated Diarrhea

The hospital environment should not allow spread of the usual causes of endemic intestinal infection. When such infection occurs, it can usually be traced to an employee who continues working while ill or to contaminated food prepared outside the hospital that is "smuggled" in by the patient's friends. Two special causes of hospital-associated diarrhea are caused by enteropathogenic *E coli* (EPEC) in infants and *Clostridium difficile* in patients treated with antimicrobial agents. Fortunately, EPEC outbreaks have become rare. *C difficile* accounts for more than 90% of cases of a syndrome that ranges from mild diarrhea to fulminant pseudomembranous colitis during or after treatment with antibiotics. The responsible toxigenic *C difficile* may be resident in the patient's intestinal flora before administration of antimicrobics or be acquired by spread from other patients in the hospital. Rotaviruses can also cause hospital outbreaks in infants.

EPEC, *C difficile*, and rotaviruses can cause hospital outbreaks

GENERAL DIAGNOSTIC APPROACHES

Laboratory diagnostic procedures (summarized in Table 62–1) include microscopic examination, culture, toxin detection, and serologic procedures. The relative value of each is different for the various etiologic agents. The diagnostic approach therefore requires that the physician assess the clinical and epidemiologic features of the case, decide which organisms are potential causes, and provide this assessment to the laboratory so that appropriate procedures will be used.

Microscopic Examination

Stool microscopy demonstrates white blood cells and parasites

EM and antigen detection for Rotavirus

Microscopic examination is of value in the assessment of bacterial infections when results are positive. The presence of polymorphonuclear leukocytes or blood in the stool correlates with organisms that produce disease by invasion, but false-negative results are common. The leukocytes may be seen in unstained or methylene blue–stained wet mount preparations; the absence of fecal leukocytes, however, does not exclude invasive diarrhea. The observation and morphologic characterization of amebas and flagellates on wet or stained preparations are the primary means by which amebic (*Entamoeba histolytica*) and flagellate (*Giardia lamblia*) infections are diagnosed. The viruses of diarrhea cannot be grown in cell culture but can be detected by electron microscopy or for Rotavirus by antigen detection.

Culture

Blood cultures are positive in early stages of enteric fever

Stool culture requires selective media for common agents

Isolation of the etiologic agent is the primary means by which bacterial enteric infection is diagnosed. In enteric fever, the organism is typically present in the blood in the early stages of disease. Blood cultures are, however, usually negative in watery diarrhea and dysenteric infections, and stool culture must be relied on for diagnosis. Fortunately, several good selective media have been developed for both direct plating and enrichment culture, which allow isolation of the infecting organism in the presence of a predominant normal flora. Selective media are then used for the various enteric pathogens (see Chapter 4). Media routinely used may vary among clinical laboratories but should include those appropriate for *Salmonella, Shigella,* and *Campylobacter jejuni.* Diarrhea caused by *E coli* is a special problem, because the methods that define the enterotoxigenic, invasive, or other pathogenic mechanisms are not yet practical for clinical laboratories.

Toxin Assay

Immunoassays detect *C difficile* toxin

The B cytotoxins of *C difficile* can be detected by its cytopathic effect in a cell culture system. In most clinical cases, enough toxin is present for direct detection in a stool specimen. This assay is currently available only in reference laboratories. Methods that detect the *C difficile* A and B toxins by latex agglutination and immunoassays are now in common use.

Antigen and Antibody Detection

Serology is generally ancillary

Antigen detection available for a rotaviruses

At present, antibody detection is useful in the diagnosis of amebic dysentery caused by *E histolytica* and of typhoid fever. Both are considered ancillary to the primary diagnostic tests, which involve specific detection of the organism by microscopic and cultural methods. Reagents are commercially available for the detection of rotavirus antigen in stool by latex agglutination or enzyme immunoassay. These methods have a sensitivity roughly comparable to that of electron microscopy. Serologic methods have been described for many other causes of gastrointestinal infection, but are not generally used because of lack of sensitivity, specificity, or availability of reagents.

OTHER CAUSES OF INTESTINAL INFECTION

Candidate agents await proof

Despite recent advances in defining the causes of enteric infections, there are surely more to be discovered. Organisms not listed in Table 62–1, such as *Aeromonas, Citrobacter,* and *Plesiomonas,* have occasionally been associated with intestinal infections, but the evidence for their enteropathogenicity is not yet strong enough to interpret their isolation from individual cases. At our present state of knowledge, it is not useful to attempt isolation of these organisms unless strong epidemiologic evidence, such as a food-borne outbreak, supports interpretation of the results.

General Principles of Management

In most gastrointestinal infections, the primary goal of treatment is relief of symptoms, with particular attention to maintaining fluid and electrolyte balance. The effects of common antidiarrheal medications such as subsalicylate-containing compounds (Pepto-Bismol) or antispasmodics (loperamide) are variable, depending on the cause. In general, these medications may be helpful for the watery diarrhea caused by enterotoxins, but not for dysentery caused by mucosal invasion, and antispasmodics may be harmful in the latter instance. Antimicrobial agents are usually not indicated for self-limited watery diarrhea, but are required for more severe dysenteric infections. Some enteric infections, such as typhoid fever, are always treated with antimicrobics. Prophylactic regimens for traveler's diarrhea have been effective if it is recognized they do not cover all potential causes. More information on therapy is given in the individual chapters, but texts on infectious diseases should be consulted for specific recommendations.

Maintenance of fluid and electrolyte balance always important

Antimicrobic therapy is primarily for invasive disease

CHAPTER 63

Urinary Tract Infections

Bacterial colonization of the urine within this tract (**bacteriuria**) is common and can at times result in microbial invasion of the tissues responsible for the manufacture, transport, and storage of urine. Infection of the upper urinary tract, consisting of the kidney and its pelvis, is known as **pyelonephritis.** Infection of the lower tract may involve the bladder (**cystitis**), urethra (**urethritis**), or prostate (**prostatitis**), the genital organ that surrounds and communicates with the first segment of the male urethra. Because all portions of the urinary tract are joined by a fluid medium, infection at any site may spread to involve other areas of the system.

EPIDEMIOLOGY

Urinary tract infection (UTI) is among the most common of diseases, particularly among women. Prevalence is age- and sex-dependent. Approximately 1% of children, many of whom demonstrate functional or anatomic abnormalities of the urinary tract, develop infection during the neonatal period. It is estimated that 20% or more of the female population suffers some form of UTI in their lifetime. Infection in the male population remains uncommon through the fifth decade of life, when enlargement of the prostate begins to interfere with emptying of the bladder. In the elderly of both sexes, gynecologic or prostatic surgery, incontinence, instrumentation, and chronic urethral catheterization push UTI rates to 30% to 40%. A single bladder catheterization carries an infectious risk of 1%, and at least 10% of individuals with indwelling catheters become infected.

Young women are commonly infected

Prostate hypertrophy is linked to male disease

PATHOGENESIS

The urine produced in the kidney and delivered through the renal pelvis and ureters to the urinary bladder is sterile in health. Infection results when bacteria gain access to this environment and are able to persist. Access primarily follows an ascending route for bacteria that are resident or transient members of the perineal flora. These organisms are derived from the large intestinal flora, which is uncomfortably nearby. Conditions that create access are varied, but the most important is sexual intercourse, which has been shown to transiently displace bacteria into the bladder. This puts the female partner at risk because of the short urethral distance. Steps in pathogenesis for *E coli*, the most common and best understood pathogen are generally felt to represent the bacterial UTI pathogens. Other manipulations of the urethra carry risk as well, particularly medical ones such as catheterization. Bacteria may also reach the urinary tract from the bloodstream. This is obviously much less common, because it requires an uncontrolled infection at another site. ::: *E coli* UTI, p. 588

Bacteria ascend from perineal flora

Intercourse is common association

Catheters increase risk

For bacteria that reach the urinary tract, the major competing forces are the rich nutrient content of the urine itself and the flushing action of bladder voiding. Persistence is favored by host factors that interrupt or retard the urinary flow such as instrumentation,

Obstruction of urine flow increases risk

Bacterial adherence favors persistence

E coli is virulence model

obstruction, or structural abnormalities. In youth, factors are congenital malformations, and with age these include changes that alter the mechanics of outflow, such as prostatic hypertrophy. Bacterial factors include the ability to adhere to the perineal and uroepithelial mucosa and to produce other classic virulence factors, such as exotoxins. *Escherichia coli* is by far the most common and potent UTI pathogen. Urease-producing members of the genus *Proteus* are associated with urinary stones, which themselves are predisposing factors for infection.

ETIOLOGIC AGENTS

Vast majority due to *E coli*

Enterobacteriaceae and Gram-positive bacteria appear with complications

Over 95% of UTIs are caused by Gram-negative rods, and 90% of these are *E coli.* Other Enterobacteriaceae, *Pseudomonas,* and Gram-positive bacteria become increasingly common with chronic, complicated, and hospitalized patients. Of the Gram-positive bacteria enterococci are the most important. *Staphylococcus saprophyticus,* a coagulase-negative staphylococcus, is now recognized as the cause in a significant minority of symptomatic infections in young, sexually active women. Yeasts, particularly species of *Candida,* may be isolated from catheterized patients receiving antibacterial therapy and from diabetic individuals, but they seldom produce symptomatic disease. ::: *S saprophyticus* UTI, p. 440

MANIFESTATIONS

Some cases are asymptomatic

The clinical manifestations of UTI are variable. Approximately 50% of infections do not produce recognizable illness and are discovered incidentally during a general medical examination. Infections in infants produce symptoms of a nonspecific nature, including fever, vomiting, and failure to thrive. Manifestations in older children and adults, when present, often suggest the diagnosis and sometimes the localization of the infection within the urinary tract.

Cystitis

Urethral irritation differs from genital infections

Fever is usually absent

The symptoms of cystitis are **dysuria** (painful urination), **frequency** (frequent voiding), and **urgency** (an imperative "call to toilet"). These findings are similar to those of urethritis caused by sexually transmitted agents. The cystitis complex is, in fact, produced by irritation of the mucosal surface of the urethra as well as the bladder. It is clinically distinguished from pure urethritis by a more acute onset, more severe symptoms, the presence of bacteriuria, and in approximately 50% of cases—hematuria. The urine is often cloudy and malodorous and occasionally frankly bloody. Cystitis patients also experience pain and tenderness in the suprapubic area. Fever and systemic manifestations of illness are usually absent unless the infection spreads to involve the kidney.

Pyelonephritis

Fever and flank pain mark upper tract disease

Prematurity is risk in pregnancy

Chronic pyelonephritis is not linked to UTI

The typical presentation of upper urinary infection consists of **flank pain** and **fever** that exceeds 38.3°C. These findings may be preceded or accompanied by manifestations of cystitis. Rigors, vomiting, diarrhea, and tachycardia are present in more severely ill patients. Physical examination reveals tenderness over the costovertebral areas of the back and, occasionally, evidence of septic shock. In the absence of obstruction, the clinical manifestations usually abate within a few days, leaving the kidneys functionally intact. It has been estimated, however, that 20% to 50% of pregnant women with acute pyelonephritis give birth to premature infants, one of the most serious consequences of UTI. In the presence of obstruction, a neurogenic bladder, or vesicoureteral reflux, clinical manifestations are more persistent, occasionally leading to necrosis of the renal papillae and progressive impairment of kidney function with chronic bacteriuria. If a renal calculus or necrotic renal papilla impacts in the ureter, severe flank pain with radiation to the groin occurs. The term chronic pyelonephritis is used to describe inflamed, scarred, contracted kidneys often in association with compromised renal function. There is no known connection between UTI and chronic pyelonephritis.

Prostatitis

Infection of the prostate is typically manifested as pain in the lower back, perirectal area, and testicles. The same bacteria that cause cystitis and pyelonephritis are involved. In acute infection, the pain may be severe and accompanied by high fever, chills, and the signs and symptoms of cystitis. Inflammatory swelling can lead to obstruction of the neighboring urethra and urinary retention. On rectal palpation, the prostate is boggy and exquisitely tender. Response to antibiotic therapy is good, but occasionally abscess formation, epididymitis, and seminal vesiculitis or chronic infection develop. Typically, acute prostatitis develops in young adults; however, it can also follow placement of an indwelling catheter in an older man. Patients with chronic prostatitis seldom give a history of an acute episode. Many are totally without symptoms; others experience low-grade pain and dysuria. Periodic spread of prostatic organisms to the urine in the bladder produces recurrent bouts of cystitis. In fact, chronic prostatitis is probably the major cause of recurrent bacteriuria in men. The etiologic agents are the same as in cystitis and pyelonephritis.

Back and perirectal pain are signs

Chronic disease a source for cystitis

DIAGNOSIS

Specimen Collection

The diagnosis of UTI is based on examination of the normally sterile urine for evidence of bacteria or an accompanying inflammatory reaction. Critical to this examination is the use of appropriate techniques for specimen collection. Urine is most easily obtained by spontaneous micturition. Unfortunately, voided urine is invariably contaminated with urethral flora and, in female patients, perineal and vaginal flora, which can confound the results of laboratory testing. Although the contaminants can never be completely eliminated, their quantity may be diminished by carefully cleansing the periurethrum before voiding and allowing the initial part of the stream to flush the urethra before collecting a specimen for examination. This **clean-voided midstream urine** collection procedure is preferred to catheterization for routine purposes because it prevents the introduction of organisms into the bladder. When the laboratory examination of such a specimen produces equivocal results or the patient cannot comply with the requirements of the clean-voided technique, catheterization or suprapubic aspiration from the distended bladder may be necessary.

Midstream collection intended to bypass contamination

Direct collections are confirmatory

For the diagnosis of prostatitis, urine is collected in three segments by interrupting a single bladder excavation. The first voiding is considered a urethral washout. The midstream specimen that follows is used to assess cystitis. The prostate is then massaged, and the final urine is a prostatic secretions washout. The quantitative culture results are then compared. In prostatitis, it is expected that the third specimen contains the largest numbers of the pathogen.

Prostatitis requires three-component voiding

Microscopic Examination

Approximately 90% of patients with acute symptomatic UTI have pyuria (that is, more than 10 white cells/mm^3 of urine). This finding is also common, however, in a number of noninfectious diseases. More specific is the presence of white cell casts, which occur primarily in patients with acute pyelonephritis. A more sensitive and specific microscopic procedure is a Gram-stained smear of uncentrifuged urine. The presence of at least one organism per oil-immersion field is almost always indicative of bacterial infection. The absence of white cells and bacteria in several fields makes the diagnosis of UTI unlikely. However, this finding does not rule it out, especially in young women with acute, symptomatic infection who may be infected with smaller numbers of organisms. ::: Urine Gram stain, p. 585

Pyuria suggests UTI but is not specific

Bacteria on unspun smear correlates with bacteriuria

Chemical Screening Tests

A number of nonmicroscopic urinary screening tests have been commercially marketed. The most successful detects leukocyte esterase from inflammatory cells and nitrite produced from urinary nitrates by bacterial metabolism. Although technically simpler, the sensitivity and specificity of these products are similar to that of microscopic examination. Like microscopic examination, they do not reliably detect bacteriuria below the level of 10^5 organisms/mL.

Leukocyte esterase detects pyuria

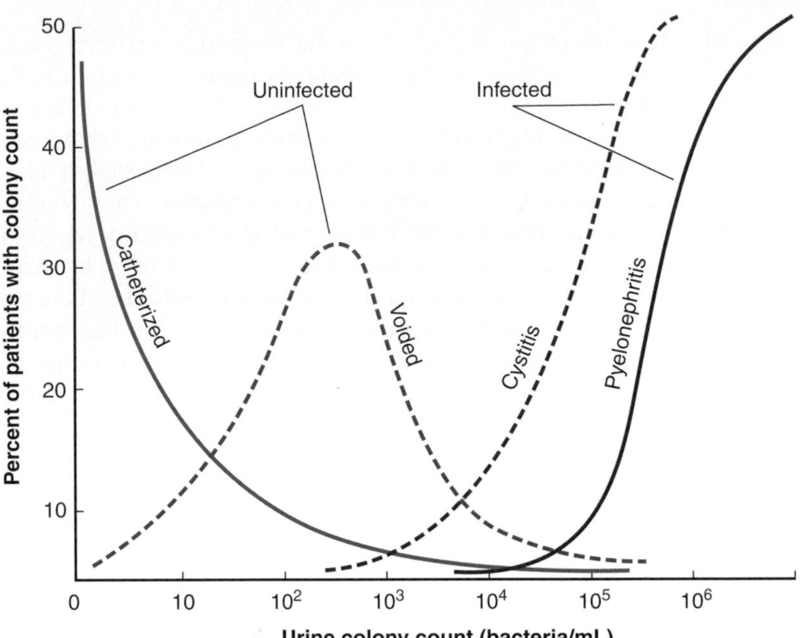

FIGURE 63–1. Quantitative urine culture. Bacteria are routinely quantitated in the range of 10 to more than 10^5. Uninfected persons may show bacteria in the urine due to contamination from the perineal flora. The numbers are small if the specimen is collected by catheterization, but voided (midstream method) specimens contain larger numbers. Patients with pyelonephritis have very high numbers of bacteria but those with only cystitis often have numbers less than 10^5.

Urine Culture

Higher than 10^5 bacteria/mL is typical for UTI

Contaminants can be higher than 10^5 bacteria/mL

Bacterial counts may be lower than 10^5 bacteria/mL in UTIs

Based on studies done half a century ago demonstrating that the number of bacteria in infected urine is large, quantitative bacteriology has been the gold diagnostic standard for UTI. Perhaps no number in medicine is better known or more slavishly adhered to than 10^5 bacteria/mL of urine. Higher than it is UTI, lower than it is contamination. We now know that it is possible to void more than 10^5 of contaminants and to have a genuine UTI with less than 10^5 bacteria as illustrated in **Figure 63–1**. Virtually no woman with sterile bladder urine, as determined by suprapubic aspiration, can void a sterile specimen even with periurethral cleansing. Voided contaminants are most often mixtures of vaginal flora not associated with UTI such as lactobacilli, diphtheroids, and streptococci, but can include urinary pathogens. Conversely, we now know that bacterial counts in UTI represent a spectrum from 10^2 to more than 10^6 bacteria/mL. The lower counts are typical for simple cystitis and the high counts for pyelonephritis. Fully one third of women with UTI limited to the bladder demonstrate counts less than 10^5 bacteria/mL.

Presence of both pathogens and symptoms is diagnostic

Asymptomatic positives should be repeated

Given the overlap, application of these findings to clinical practice requires linking the epidemiologic probability to the clinical findings. If a woman has symptoms of cystitis and a culture positive for a urinary pathogen, the probability that she has a UTI is 90%, even if the count is as low as 10^3 bacteria/mL. If the woman is asymptomatic, the probability drops to 80% even if the count is more than 10^5/mL. In the latter case, the culture must be repeated before concluding that a UTI is present. Voiding more than 10^5 of the same contaminant twice in a row is unlikely. There is no reason to repeat positive cultures from symptomatic patients. Catheterized and suprapubic specimens may be accepted at face value, because they come directly from the bladder.

TREATMENT

Empiric treatment is common

Resistance limits ampicillin use

The treatment of UTI is best guided by the results of cultures and antimicrobial susceptibility tests. In simple isolated instances of cystitis in a young woman, the etiology is often assumed to be *E coli* and the antimicrobic selected empirically based on knowledge of the susceptibility of local strains. Sulfonamides and trimethoprim alone or in combination with sulfamethoxazole, a fluoroquinolone, and nitrofurantoin are the agents most commonly used. In most areas, ampicillin is precluded by resistance rates exceeding 25%. For children and patients with risk factors or recurrent infections, empiric therapy should always be confirmed by culture and susceptibility testing. Likewise, the duration of therapy depends on the severity of the infection and the risk status of the patient. Success of treatment may be tested by a follow-up urine culture 1 to 2 weeks after therapy is completed.

PREVENTION

Those with several symptomatic episodes annually may be helped with long-term, low-dose chemoprophylaxis. In women whose recurrences are related to sexual activity, administration of the chemoprophylactic agent may be limited to immediately after intercourse. Infected children, men, and those who experience UTI relapse should be investigated with intravenous pyelography to allow detection and correction of any factor causing predisposition to infection.

Chemoprophylaxis may be effective

CHAPTER 64

Genital Infections

Genital infections are most often sexually transmitted infections (STDs). Examples of genital infections that are not STDs include vaginitis secondary to antibiotic treatment and epididymitis in older men. The most common agents of STDs are *Chlamydia trachomatis,* papillomavirus, herpes simplex virus, *Neisseria gonorrhoeae,* and the most worrisome, human immunodeficiency virus (HIV). Additional agents spread by sexual contact include hepatitis B, cytomegalovirus, syphilis, chancroid, and lymphogranuloma venereum. **Table 64–1** lists the major sexually transmitted pathogens and the disease syndromes associated with them. These infections are discussed in detail in chapters related to the etiologic agents.

Depending on the pathogen, the disease produced may be local or systemic. For the localized STDs, due to *Chlamydia,* for example, the most common manifestations are inflammation (eg, urethritis, cervicitis), which may or may not be noticed by the patient. In some cases, deeper structures become involved when the infection spreads beyond the local site by direct extension (eg, epididymitis, salpingitis). As with other infectious diseases, some of these can gain access to the bloodstream and produce systemic symptoms and spread to other organs. The systemic STDs produce infection beyond the genital site as part of their basic pathogenesis (eg, HIV, hepatitis B, and syphilis); syphilis produces and HIV and hepatitis B do not produce a local genital lesion. The most common clinical syndromes are discussed next.

Some STDs begin as localized infections; others are primarily systemic

GENITAL ULCERS

Single or multiple ulcerative lesions on the genitalia constitute one of the most common manifestations of STDs. Infection may begin as a papule or pustule and evolve into an ulcer. **Table 64–2** lists the causes and major features of genital ulcerations. The nature of the ulcer and whether it is painful are significant differential features. The ulcer (chancre) of syphilis is typically single, firm, and indurated but painless, whereas genital herpes ulcers are often multiple and painful. The evaluation of genital ulcers usually focuses on the separation of genital herpes, the most common cause in industrialized nations, and syphilis from lesions due to other causes. In the laboratory workup, it should be emphasized that direct microscopy and serologic tests may be negative at the time of presentation of the syphilitic chancre and that cultures for herpes simplex virus are usually positive from vesicular, pustular, or ulcerative lesions but may be negative from crusted areas. Chancroid, caused by *Haemophilus ducreyi,* is relatively rare in the developed world; it may be suggested by direct microscopy but requires a special selective medium for culture. Granuloma inguinale, a disease also seen primarily in developing countries, is characterized by chronic, persistent genital papules or ulcers. It is caused by *Calymmatobacterium granulomatis,* an encapsulated Gram-negative bacillus, which has not been grown in artificial medium. The diagnosis is usually made by examination of Wright- or Giemsa-stained impression smears from biopsy specimens that demonstrate clusters of encapsulated coccobacilli in the cytoplasm of mononuclear cells.

Pain and induration are major differential features

C granulomatis shows encapsulated Gram-negative bacilli on smear

TABLE 64–1 Sexually Transmitted Agents and Diseases Caused

AGENT	DISEASE OR SYNDROME
Bacteria	
Neisseria gonorrhoeae	Urethritis, cervicitis, proctitis, pharyngitis, conjunctivitis, endometritis, pelvic inflammatory disease, perihepatitis, bartholinitis, disseminated gonococcal infection
Chlamydia trachomatis	Nongonococcal urethritis, epididymitis, cervicitis, salpingitis, inclusion conjunctivitis, infant pneumonia, trachoma, lymphogranuloma venereum
Mycoplasma genitalium	Nongonococcal urethritis
Treponema pallidum	Syphilis, condylomata lata
Haemophilus ducreyi	Chancroid
Calymmatobacterium granulomatis	Granuloma inguinale
Viruses	
HIV	AIDS, AIDS-related complex, perinatal and congenital AIDS, aseptic meningitis, subacute neurologic syndromes, persistent generalized adenopathy, asymptomatic infection
Herpes simplex virus	Primary and recurrent genital herpes, aseptic meningitis, neonatal herpes
Papillomavirus	Condylomata accuminata, laryngeal papilloma of newborn, cervical carcinoma
Cytomegalovirus	Heterophil-negative infectious mononucleosis, congenital birth defects
Hepatitis B virus	Hepatitis B, acute and chronic infection
Molluscum contagiosum virus	Genital molluscum contagiosum
Protozoa	
Trichomonas vaginalis	Trichomonal vaginitis
Fungi	
Candida albicans	Vulvovaginitis, penile candidiasis
Ectoparasites	
Phthirus pubis	Pubic louse infestation
Sarcoptes scabiei	Scabies

AIDS, acquired immunodeficiency syndrome; HIV, human immunodeficiency virus.

GENITAL WARTS

Many genotypes of papillomaviruses

Genital warts may be caused by human papillomavirus (condyloma acuminatum) or *Treponema pallidum* (condyloma latum). Of the more than 100 genotypes of human papillomavirus (HPV), types 6, and 11 are the predominant causes of genital warts. Types 16 and 18, are highly associated with cervical cancer, and are less common causes of warts. Condylomata lata are painless mucosal warty erosions that develop in warm, moist sites such as the genitals and perineum in about one third of cases of secondary syphilis. Darkfield examinations are invariably positive as are both nontreponemal and treponemal serologic tests.

TABLE 64–2 Causes of Genital Ulcerations

DISEASE	TYPE OF LESION	TYPE OF INGUINAL ADENOPATHY[a]	DIAGNOSIS
Genital herpes	Multiple grouped vesicles to coalesced ulcers, painful	Tender, discrete, nonsuppurative	Viral culture, enzyme immunoassay, PCR
Chancroid	Tender, shallow, painful ulcer, not indurated ulcer	Suppurative	Special culture
Syphilis	Nontender, indurated ulcer	Rubbery consistency	Darkfield or FA exam, serology
Lymphogranuloma venereum	Painless, small ulcer or papule, usually healed at time of presentation	Discrete progressing to suppurative, draining fistulas	Special culture, serology
Granuloma inguinale	Papular to nodular to ulcerative lesion(s), painless	"Pseudobubo" caused by induration of subcutaneous tissue in inguinal area	Giemsa stain of biopsy

[a]Involvement of inguinal lymph nodes

URETHRITIS

Urethritis usually manifests as dysuria, urethral discharge, or both. The discharge may be prominent enough to be the chief complaint or may have to be milked from the urethra. The major causes of urethritis are *N gonorrhoeae* and *C trachomatis*, followed by *Mycoplasma genitalium* and herpes simplex virus. Infection with more than one organism is common, particularly dual gonococcal and chlamydial infection. Up to 20% of cases have no established etiology but are probably infectious.

C trachomatis and *N gonorrhoeae* often coinfect

The diagnosis of gonorrhea is established primarily by culture, although direct examinations (Gram stain, DNA assays) may suffice in symptomatic patients. DNA-based assays are comparable to culture for screening. Newly developed nonculture techniques (eg, Nucleic acid amplification) are superior to culture for *C trachomatis*, whereas culture is the most appropriate test for herpes simplex virus. Treatment depends on the etiologic agent and whether the disease has progressed beyond the local site. Empiric regimens are directed at the two most common causes, *N gonorrhoeae* and *C trachomatis.* In cases of gonorrhea, concurrent treatment for chlamydia is recommended, unless the latter has been specifically excluded. In general, the same approach is followed for epididymitis and cervicitis.

Culture and DNA-based assays available

Combined treatment often recommended

EPIDIDYMITIS

Unilateral swelling of the epididymis is a common clinical illness seen in sexually active men. It is usually painful, with fever and acute unilateral swelling of the testicle that is sometimes confused with testicular torsion. In the preantibiotic era, approximately 10% to 15% of untreated gonococcal infections resulted in epididymitis. In developed countries, the two most common causes of epididymitis are *N gonorrhoeae* and *C trachomatis,* especially in younger men. In men older than 35 and in homosexual men, Enterobacteriaceae and coagulase-negative staphylococci may also cause the disease, probably from reflux of infected urine into the epididymis. Treatment depends on demonstration of the etiologic agent in urethral specimens or epididymal aspirates (see treatment of urethritis for additional considerations).

Gonococcal and chlamydial infections more common in men 35 years and younger

Enterobacteriaceae and *S epidermidis* more common in older men

CERVICITIS

The microbial etiology of cervical infections is varied; *N. gonorrhoeae* and *C. trachomatis* cause endocervicitis, and herpes simplex virus can infect the stratified squamous epithelium

Gonococcal, chlamydial, and herpes simplex virus infections most common

of the ectocervix. The major clinical manifestation of cervicitis is a mucopurulent vaginal discharge. The cervix is friable and inflamed, and polymorphonuclear leukocytes are present in the exudate. Chlamydial, gonococcal, and viral cultures are needed to demonstrate the etiologic agent. Therapy depends on the etiologic agent involved (see treatment of urethritis for additional considerations).

VAGINITIS AND VAGINAL DISCHARGE

Pelvic examination helps define important infection sites

Symptomatic vaginal discharge may occur alone or accompany salpingitis, endometritis, or cervicitis. Evaluation includes pelvic examination, cervical cultures for *N gonorrhoeae* and *C trachomatis,* and microscopic examination of the discharge. Measurement of the pH of the discharge may also be helpful. Pelvic examination is valuable in determining whether uterine, adnexal, or cervical tenderness is present and whether the source of the discharge is the cervix or the vagina.

Candida vaginitis causes itching, thick discharge

Trichomonas infection produces foamy discharge

The clinical and laboratory findings vary with the etiologic agent. *Candida albicans* generally produces a vulvovaginitis associated with pruritus and erythema of the vulvar area and a discharge with the consistency of cottage cheese. Microscopic demonstration of yeast and pseudomycelia in a potassium hydroxide or Gram stain preparation of the exudate confirms the diagnosis. *Trichomonas vaginalis* typically produces a foamy, purulent vaginal discharge. The pH is variable (usually higher than 5.0), and numerous polymorphonuclear cells and motile trichomonads are seen on wet mount examination.

Bacterial vaginosis is a shift in flora with anaerobic overgrowth

KOH added to discharge produces a fishy smell (amines)

Clue cells are present and lactobacilli are absent

Bacterial vaginosis (BV), previously termed "nonspecific vaginitis," is the most common form of vaginitis in women. BV is associated with overgrowth of multiple members of the vaginal anaerobic flora, genital mycoplasmas, and a small Gram-negative rod (*Gardnerella vaginalis*), once believed to be the sole cause of the disease. The vaginal discharge of BV is yellowish, homogeneous, and adherent to the vaginal wall. The pH is greater than 5.0. Addition of potassium hydroxide (KOH) to the vaginal secretions produces a fishy smell as a result of volatilization of amines. The Gram stain shows a shift from the usual lactobacillary flora to one of many Gram-negative coccobacilli. Clue cells, which are vaginal epithelial cells heavily coated with *G vaginalis,* may also be seen. Therapy depends on the etiologic agent.

PELVIC INFLAMMATORY DISEASE

Multiple etiologic agents; gonococcus predominant

Incidence higher with use of intrauterine devices

Clinical manifestations of pelvic inflammatory disease (PID) vary but generally include lower abdominal pain elicited by movement of the cervix or palpation of the adnexal or endometrial areas. About 50% of cases are caused by *N gonorrhoeae.* Nongonococcal PID has a complex and sometimes polymicrobial etiology, including *C trachomatis, Bacteroides,* anaerobic streptococci, and *Mycoplasma hominis* alone or in various combinations. In general, nongonococcal PID is milder than that associated with *N gonorrhoeae* infection. The incidence of PID is five to ten times higher in women with intrauterine devices than in those not using this form of contraception. The diagnosis is established most reliably by culture of peritoneal aspirates from the vaginal cul-de-sac. Treatment of PID is complex because of the multiple etiologies and relative inaccessibility of the definitive diagnostic specimen.

LYMPHADENITIS

Generalized adenopathy in secondary syphilis

Inguinal lymphadenitis may be seen with several STDs, especially primary herpes simplex infection and lymphogranuloma venereum. The latter is caused by specific strains of *C trachomatis.* It may begin as a small genital ulcer, which is frequently unnoticed. More often, the first evidence of lymphogranuloma venereum is a tender swollen inguinal lymphadenitis, which may suppurate and drain spontaneously if not treated. Primary syphilis may be associated with unilateral or bilateral inguinal lymph node enlargement, but these nodes are not usually tender. Secondary syphilis may be associated with generalized lymphadenopathy. Primary herpes simplex infection of the genitalia may be associated with tender inguinal lymphadenitis, but recurrent genital herpes infection is not.

SYSTEMIC SYNDROMES

As indicated earlier, some STDs may manifest important pathology outside the genital tract, including diseases such as syphilis, hepatitis B, and AIDS, whose most devastating consequences are at nongenital sites. These diseases can be highly complex, involving multiple organs and lifelong illness. These organisms and diseases are best reviewed by referring back to the specific chapters that deal with each agent.

Most serious effects of syphilis, hepatitis B, and AIDS are outside the genital tract

CHAPTER 65

Central Nervous System Infections

The cerebrum, cerebellum, brainstem, spinal cord, and their covering membranes (meninges) constitute the central nervous system (CNS). Because of the unique anatomic and physiologic features of the CNS, infections of this site can represent special challenges to the microbiologist and clinician. The CNS is encased in a rigid, bony vault, and it is highly vulnerable to the effects of inflammation and edema: its critical life regulatory functions and the metabolic requirements to sustain these functions can also be easily disrupted by infection, with resultant local acidosis, hypoxia, and destruction of nerve cells. Thus, the effects of increased pressure, biochemical abnormalities, and tissue necrosis can be profound and sometimes irreversible. One specialized defense mechanism of the CNS is the blood–brain barrier, which serves to minimize passage of infectious agents and potentially toxic metabolites into the cerebrospinal fluid (CSF) and tissues, as well as to regulate the rate of transport of plasma proteins, glucose, and electrolytes. When CNS infection develops, however, this barrier also poses difficulties in control; some antimicrobial agents and host immune factors, such as immunoglobulins and complement, do not pass as readily from the blood to the site of infection as they do to other tissues.

Blood–brain barrier affects access of microbes, immune factors, and antimicrobics

Within the brain are the ventricles, which are cavities in which CSF is actively produced, primarily by specialized structures called the **choroid plexuses**. The CSF fills the lateral ventricles in each half of the brain, circulates into a central third ventricle, and then passes through the cerebral aqueduct to emerge through foramina at the brainstem. From cisterns at the base of the brain, the CSF circulates in the subarachnoid space over the entire CNS, including the spinal cord, to supply nutrients and serve as a hydraulic cushion for these tissues. It is reabsorbed primarily by the major venous system in the meninges. Obstruction of the normal flow of CSF in either the internal (ventricular) or external (subarachnoid) systems can result in increased intracranial pressure because production of CSF by the choroid plexuses continue within the ventricles. Such impairment of flow or normal reabsorption can occur as a result of inflammation or subsequent fibrosis, leading to dilatation of the ventricles, compression of brain tissue, and a condition known as **hydrocephalus.**

CSF continuously produced by choroid plexus

Obstruction of CSF flow or reabsorption causes hydrocephalus

ROUTES OF INFECTION

Most CNS infections appear to result from blood-borne spread; for example, bacteremia or viremia resulting from infection of tissue at a site remote from the CNS may result in penetration of the blood–brain barrier. Examples of infectious agents that commonly infect the CNS by this route are *Haemophilus influenzae, Neisseria meningitidis, Streptococcus pneumoniae, Mycobacterium tuberculosis,* and viruses such as enteroviruses and mumps (**Tables 65–1** and **65–2**). The initial source of infection leading to bloodstream invasion may be occult (eg, infection of reticuloendothelial tissues) or overt (eg, pneumonia, pharyngitis, skin abscess or cellulitis, or bacterial endocarditis). Occasionally, the route

Blood-borne spread most common access to CNS

Direct spread occurs from adjacent infected focus such as middle ear

TABLE 65–1 Common Causes of Purulent Central Nervous System Infections

AGE GROUP	AGENT
Newborns (<1 mo old)	Group B streptococci (most common), *Escherichia coli, Listeria monocytogenes, Klebsiella* species, other enteric Gram-negative bacteria
Infants and children	*Streptococcus pneumoniae, Neisseria meningitidis, Haemophilus influenzae*
Adults	*S pneumoniae, N meningitidis*
Special circumstances	
Meningitis or intracranial abscesses associated with trauma, neurosurgery, or intracranial foreign bodies	*Staphylococcus aureus, Staphylococcus epidermidis, S pneumoniae;* anaerobic Gram-negative and Gram-positive bacteria; *Pseudomonas* species
Intracranial abscesses not associated with trauma or surgery	Microaerophilic or anaerobic streptococci, anaerobic Gram-negative bacteria (often mixed aerobic and anaerobic flora of upper respiratory tract origin)

TABLE 65–2 Primary Acute Viral Infections of the Central Nervous System

AGENT	MAJOR AGE GROUP AFFECTED	SEASONAL PREDOMINANCE
Enteroviruses	Infants, children	Summer–fall
Mumps	Children	Winter–spring
Herpes simplex		
Type 1	Adults	None
Type 2	Neonates, young adults	None
Arboviruses		
Western equine encephalitis	Infants, children	Summer–fall
St. Louis encephalitis	Adults >40 years	Summer–fall
California encephalitis	School-aged children	Summer–fall
Eastern equine encephalitis	Infants, children	Summer–fall
West Nile encephalitis	Adults	Summer–fall
Rabies	All ages	Summer–fall
Measles	Infants, children	Spring
Varicella–zoster	Infants, children	Spring
Lymphocytic choriomeningitis	Adults, children	None
Epstein–Barr virus	Children, young adults	None
Other (eg, myxoviruses, human immunodeficiency virus, cytomegaloviruses)	All ages	Variable

of infection is from a focus close to or contiguous with the CNS. These possible sources include middle ear infection (otitis media), mastoiditis, sinusitis, or pyogenic infections of the skin or bone. Infection may extend directly into the CNS, indirectly via venous pathways, or in the sheaths of cranial and spinal nerves.

Traumatic, surgical, or congenital lesions may give direct access

Implanted foreign bodies such as shunts increase risk

In some cases, a contiguous or distant infectious focus may not be necessary to produce CNS infection. If an anatomic defect exists in the structures encasing the CNS, infectious agents may readily gain access to the vulnerable site and establish themselves. Such defects may be traumatically or surgically induced or result from congenital malformations. For example, fractures of the base of the skull may produce an opening between the CNS and the sinuses, nasal passages (defects in the cribriform plate), mastoid, or middle ear. All of these sites are contiguous with the upper respiratory tract, which enables a potentially pathogenic member of the respiratory flora to gain ready access to the CNS. Neurosurgical procedures also create transient communications between the external environment and the CNS, which can be readily contaminated. This risk can be compounded when foreign bodies, such as shunts or external drainage tubes, must be left in place for the treatment of hydrocephalus. These foreign bodies, when colonized, can serve as chronic foci of infection.

Congenital defects, such as meningomyeloceles or sinus tracts through the cranium or spine, may also be sources of CNS infection. The latter may be overlooked; the orifice of the sinus may be a small cleft on the skin surface, or occasionally it may open internally into the intestinal tract. Recurrent purulent meningitis or unusual pathogens in an otherwise healthy host should prompt a careful search for such defects.

Intraneural pathways operate with a few viruses

Perhaps the least common route of CNS infection is by intraneural pathways. Agents capable of intraneural spread to the CNS include rabies virus (presumably along peripheral sensory nerves), herpes simplex virus (often, but not exclusively, via the trigeminal nerve root or sacral nerves), polioviruses, and perhaps some togaviruses.

Abscesses of the CNS deserve special mention. Although relatively uncommon compared with other CNS infections, they represent a special microbiologic and clinical problem. Such abscesses may be within the tissues of the CNS (eg, brain abscess; **Figure 65–1**)

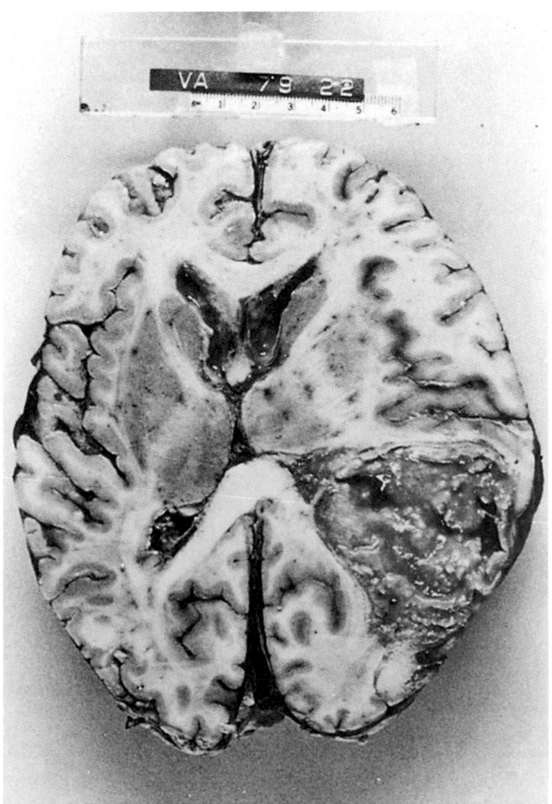

FIGURE 65–1. Coronal section of a brain demonstrating a poorly encapsulated abscess.

Abscesses present diagnostic problems and may localize in brain, at subdural or epidural sites

or localized in the subdural or epidural spaces. They sometimes develop as a complication of pyogenic meningitis. More commonly, abscesses of the CNS result from embolization of bacteria or fungi from a distant focus, such as endocarditis or pyogenic lung abscess; from extension from a contiguous focus of infection (eg, sinusitis or mastoiditis); or from a complication of surgery or nonsurgical trauma.

CLINICAL FEATURES

Acute onset and progression with stiff neck and neurologic dysfunction

Usually fatal if untreated

Several terms commonly applied to CNS infections need to be understood. **Purulent meningitis** refers to infections of the meninges associated with a marked, acute inflammatory exudate and is usually caused by a bacterial infection. Such infections frequently involve the underlying CNS tissue to a variable degree, and often the ventricular system is also involved (ventriculitis). Most cases of purulent meningitis are acute in onset and progression and are characterized by fever, stiff neck, irritability, and varying degrees of neurologic dysfunction, which, if untreated, usually progress to a fatal outcome. Large numbers of polymorphonuclear leukocytes are present in the CSF of established cases.

Granulomatous infections are chronic

Chronic meningitis has a more insidious onset, with progression of signs and symptoms over a period of weeks. This is usually caused by mycobacteria or fungi that produce granulomatous inflammatory changes, but occasionally protozoal agents are responsible (**Table 65–3**). The cellular response in the CSF reflects the chronic inflammatory nature of the disease.

Aseptic meningitis is a term used to describe a syndrome of meningeal inflammation associated mostly with an increase of cells (pleocytosis), primarily lymphocytes and other

TABLE 65–3 Other Causes of Central Nervous System Infections

DISEASE	AGENT
Chronic granulomatous infection	*Mycobacterium tuberculosis*[a]
	Coccidioides immitis
	Cryptococcus neoformans
	Histoplasma capsulatum
Parasitic infection	
Protozoa	*Toxoplasma gondii*[b]
	Trypanosoma
	Acanthamoeba species
Nematodes	*Toxocara* species
	Trichinella spiralis
	Angiostrongylus cantonensis
Cestodes	*Taenia solium* (cysticercosis)
Other	*Leptospira* species
	Treponema pallidum
	Borrelia burgdorferi

[a]Tuberculous meningitis can appear as acute or chronically progressive disease.
[b]Toxoplasmosis of the central nervous system is usually seen in congenital infections or immunocompromised hosts.

mononuclear cells in the CSF, and absence of readily cultivable bacteria or fungi. It is associated most commonly with viral infections and is often self-limiting. The syndrome can also occur in syphilis and some other spirochetal diseases, as a response to the presence of drugs or radiopaque substances in the CSF, or from tumors or bleeding involving the meninges or subarachnoid space. The primary site of inflammation is in the meninges without clinical evidence of involvement of the neural tissue. Such patients may have fever, headache, a stiff neck or back, nausea, and vomiting.

Aseptic meningitis is commonly of viral etiology

Other causes include syphilis

Encephalitis also implies a primary viral etiology; however, acute or chronic demyelinating diseases with or without inflammation must also be considered. The latter group includes the postinfectious or allergic encephalomyelitis syndromes, in which the cause and pathogenesis are not always clearly defined. Clinically, the diagnosis of encephalitis is applied to patients who may or may not show signs and CSF findings compatible with aseptic meningitis, but also show objective evidence of CNS dysfunction (eg, seizures, paralysis, and disordered mentation). Many clinicians use the term **meningoencephalitis** to describe conditions of patients with both meningeal and encephalitic manifestations.

Viral and postinfectious etiology most common

Poliomyelitis refers to the selective destruction of anterior motor horn cells in the spinal cord and/or brainstem, which leads to weakness or paralysis of muscle groups and occasionally respiratory insufficiency. It is usually associated with aseptic meningitis, sometimes with encephalitis. The polioviruses are the major causes of this syndrome, although coxsackieviruses (primarily type A7) and other enteroviruses, such as enterovirus 71, have been implicated. The hallmark of poliomyelitis is asymmetric flaccid paralysis.

Viral destruction of anterior horn cells causes paralysis

Two other nervous system syndromes presumably associated with infection deserve brief mention. **Acute polyneuritis,** an inflammatory disease of the peripheral nervous system, is characterized by symmetric flaccid paralysis of muscles. In most cases, no specific etiology is found; some, however, have been associated with *Corynebacterium diphtheriae* toxin and infections by bacterial enteric pathogens, cytomegalovirus or Epstein–Barr virus. **Reye's syndrome** (encephalopathy with fatty infiltration of the viscera) is an acute, noninflammatory process, usually observed in childhood, in which cerebral edema, hepatic dysfunction, and hyperammonemia develop within 2 to 12 days after onset of a systemic viral infection. Although the influenza A and B and varicella–zoster viruses have been most frequently implicated in this syndrome, the precise pathogenesis is not yet known. Concomitant salicylate therapy is known to be an important contributory factor.

Acute polyneuritis involves peripheral nervous system

Reye's syndrome precipitated by salicylate treatment of systemic viral infection

COMMON ETIOLOGIC AGENTS

The causes of CNS infections are numerous, as illustrated in Tables 65–1 through 65–3. Acute purulent meningitis is usually caused by one of three organisms: *H influenzae* type b, *N meningitidis,* or *S pneumoniae.* The incidence of *H influenzae* meningitis has now fallen sharply in the United States as a result of routine immunization. In neonatal infections, group B streptococci or *Escherichia coli* are most frequently implicated. However, many other bacteria can occasionally cause the disease if they gain access to the meninges.

Acute purulent meningitis caused by encapsulated pathogens

Of the viral causes of acute CNS disease, the categories most commonly encountered are the enteroviruses, human immunodeficiency virus, herpes simplex, Epstein–Barr virus, and arthropod-borne viruses. In the United States, enteroviruses account for the greatest proportion of infections. Viral CNS infections can be manifested clinically as aseptic meningitis, encephalitis, poliomyelitis, or any combination of these. The age of the patient and the season of occurrence help somewhat in predicting some of the agents that may be involved, (Table 65–2); other epidemiologic, ecologic, and clinical factors associated with these infections are discussed in the individual chapters on specific virus groups.

Acute viral disease has a variety of manifestations

Seasonality and patient age important clues

Slow viral infections of the CNS, such as subacute sclerosing panencephalitis (due to measles or sometimes congenitally acquired rubella virus), acquired immunodeficiency syndrome encephalopathy, progressive multifocal leukoencephalopathy (due to JC polyomavirus), and Creutzfeldt–Jacob disease ("unconventional" viruses), are discussed in Chapters 10, 18, 19, and 20, respectively. Other important causes of CNS infections (Table 65–3) that must not be overlooked include *Mycobacterium tuberculosis* and the deep mycoses (especially *Cryptococcus neoformans* and *Coccidioides immitis*). These chronic infections can be insidious in onset and mimic other processes, thus delaying consideration of the proper diagnosis.

Chronic meningitis is caused by slow-growing agents

Noninfectious diseases may mimic infections

Finally, noninfectious causes of CNS disease need to be considered in the differential diagnosis. These include (1) metabolic disturbances, such as hypoglycemia, diabetic coma, and hepatic failure; (2) toxic conditions, such as those caused by bacterial toxins (diphtheria, tetanus, botulism), insect toxins (tick paralysis), poisons (lead), and drug abuse; (3) mass lesions, such as acute trauma, hematoma, and tumor; (4) vascular lesions, such as intracranial embolus, aneurysm, and subarachnoid hemorrhage; and (5) acute psychiatric episodes.

GENERAL DIAGNOSTIC APPROACHES

Lumbar puncture for pressure, cells, protein, and glucose in CSF

CSF glucose should be compared with simultaneous blood level

Except in unusual circumstances, in which severe increases in intracranial pressure make the procedure dangerous, a lumbar puncture is the first step in the workup of a patient with suspected CNS infection. The CSF pressure is determined at the time of the procedure, and CSF is removed for analysis of cells, protein, and glucose. Ideally, the glucose content of the peripheral blood is determined simultaneously for comparison with that in the CSF. **Table 65–4** presents guidelines for interpretation of results of CSF analysis; these guidelines represent generalizations, however, and must not be considered as absolute findings in all cases. For example, although a patient with bacterial, mycobacterial, or fungal meningitis usually has a glucose level in the CSF of less than 40 mg/dL, or less than half the blood glucose level (hypoglycorrhachia), this finding may not be present in the early stages of infection. Viral infections of the CNS can occasionally produce low glucose values in the CSF; in addition, the early stages of viral infection may be associated with a preponderance of polymorphonuclear leukocytes. It is clearly important to recognize that viral CNS infections can exist with a negligible CSF cell count. This sometimes also occurs in the early stages of bacterial meningitis.

Viral and granulomatous infections often cause predominance of lymphocytes in CSF

Realizing the limitations, it is possible to make some general interpretations that are helpful in the diagnosis. Viral CNS infections are usually associated with a preponderance of lymphocytes, a normal glucose value, and a normal or moderately elevated protein level in the CSF. In contrast, acute bacterial meningitis usually causes a CSF pleocytosis consisting primarily of polymorphonuclear cells, a low glucose value, and a high protein level. Mycobacterial and fungal infections are more commonly associated with lymphocytosis (and sometimes moderate eosinophilia) in the CSF; like the acute bacterial infections, however, they tend to lower glucose and increase protein levels markedly.

TABLE 65–4 Findings of Cerebrospinal Fluid Analysis: Normal versus Infection

CLINICAL SITUATION	LEUKOCYTES/ MM^3	% POLYMORPHONUCLEARS	GLUCOSE (% OF BLOOD)	PROTEIN (MG/DL)
Children and adults				
Normal	0-5	0	≥60	≤30
Viral infection	2-2000 (80)[a]	≤50	≥60	30-80
Pyogenic bacterial infection	5-5000 (800)	≥60	≤45[b]	>60
Tuberculosis and mycoses	5-2000 (100)	≤50	≤45	>60
Neonates				
Normal (term)	0-32 (8)	≤60	≥60	20-170 (90)
Normal (preterm)	0-29 (9)	≤60	≥60	65-150 (115)

[a]Numbers in parentheses represent mean values.
[b]Usually very low.

Normal values for CSF are also shown in Table 65–4. Polymorphonuclear cells are not usually seen in normal CSF, but as many as five lymphocytes/mm^3 may be found in healthy individuals. Neonatal CSF is considerably more difficult to interpret because cell counts are often elevated when there is no infection. Glucose values, however, should be within the normal range.

Polymorphonuclear cells not often found in normal CSF

The other major procedures that must be performed on all CSF samples in which any infection is suspected include bacterial cultures and Gram staining. If the CSF is grossly purulent and the patient is untreated, a Gram stain of the uncentrifuged CSF or of its centrifuged sediment frequently shows the infecting organism and indicates the probable diagnosis. According to the clinical indications and results of CSF cytology and chemistry, other microbiologic tests may be used, including viral cultures, special stains and cultures for fungi and mycobacteria, immunologic methods to detect fungal or bacterial antigens (eg, latex agglutination for *Cryptococcus*), and polymerase chain reactions to detect viral or bacterial nucleic acids.

Direct staining and culture are definitive diagnostic methods

Tests for free antigens useful in some circumstances

Tests on specimens other than CSF are selected on the basis of the clinical diagnostic possibilities. If acute bacterial meningitis is suspected, blood cultures should also be used to ensure the diagnosis. Viral cultures of the pharynx, stool, or rectal swabs may provide indirect evidence of CNS infection. In encephalitis, CSF is often used to detect viral nucleic acid by polymerase chain reaction and/or to perform serologic testing for IgM specific antiviral antibodies. Other studies may include acute and convalescent sera for viral serology and serologic tests to detect antibodies to certain fungi, such as *C immitis.*

Culture of blood and other sites depend on suspected etiology

Biopsy and serology useful for some agents

Intracranial abscesses can often be detected with radiologic techniques, such as computed tomography or magnetic resonance imaging. A definitive etiologic diagnosis is established by careful aerobic and anaerobic culture of the contents of the abscess.

Imaging methods useful to detect abscesses

GENERAL PRINCIPLES OF MANAGEMENT

In bacterial, mycobacterial, and fungal infections of the CNS, prompt and aggressive antimicrobial therapy is required. The duration of treatment varies from as little as 10 days for uncomplicated bacterial meningitis to 12 months or longer for tuberculous meningitis and to several years for some cases of fungal meningitis.

Antimicrobial therapy is administered immediately

In addition to antimicrobial therapy, correction of associated metabolic defects (acidosis, hypoxia, saline depletion, inappropriate antidiuretic hormone secretion) is necessary. Increased intracranial pressure as a result of vasogenic edema or hydrocephalus must be monitored and controlled accordingly; osmotic agents such as intravenous mannitol are often used to control acute cerebral edema, and neurosurgical shunting procedures may be needed to treat progressive hydrocephalus. Abscesses often require drainage. Except for patients with herpes simplex encephalitis, who often respond to early treatment with antiviral agents, most viral infections of the CNS can only be managed supportively. This includes specific attention to the metabolic and respiratory problems that may develop in severe cases.

Correction of metabolic defects and raised intracranial pressure important

Viral infections are managed supportively

CHAPTER 66

Intravascular Infections, Bacteremia, and Endotoxemia

Frequently, the presence of circulating microorganisms in the blood is either a part of the natural history of the infectious disease or a reflection of serious, uncontrolled infection. Depending on the class of agent involved, this process is described as viremia, bacteremia, fungemia, or parasitemia. The terms **sepsis** and **septicemia** refer to the major clinical symptom complexes generally associated with bacteremia. The clinical findings may develop acutely, as in septic shock, or slowly, as in most forms of infective endocarditis. Viremia is usually a very early, even prodromal, event accompanied by fever, malaise, and other constitutional symptoms such as muscle aches. With the exception of a few specific infections (eg, cytomegalovirus), the detection of viremia does not play a role in the diagnosis or management of viral infections. The presence of bacteremia defines some of the most serious and life-threatening situations in medical practice, and it has a marked impact on the management and outcome of bacterial infections. This chapter focuses on the causes and implications of bacteremia and, to a lesser extent, fungemia. Diseases in which parasitemia is a feature are covered in Chapters 50 to 52.

Sepsis refers to clinical findings

Onset can be insidious or dramatic

Bacteremia or fungemia may also result from microbial growth on the inner or outer surfaces of intravenous devices. Clinical manifestations may be minor initially, but may later become severe. Because the bloodstream is sterile in healthy individuals, bacteremia is considered potentially serious regardless of the symptoms present; however, transient bacteremia may occur when there is manipulation or trauma to a body site that has a normal flora. After such events, species indigenous to the site may appear briefly in the blood, but they are soon cleared. Such transient bacteremias usually have no immediate clinical significance, but they are important in the pathogenesis of infective endocarditis.

Bloodstream is sterile in health

Transient, benign bacteremia is common

INTRAVASCULAR INFECTION

Intracardiac infections (endocarditis) and those primarily involving veins (thrombophlebitis) or arteries (endarteritis) are usually caused by bacteria, although other agents including fungi and viruses occasionally have been implicated. Bacteria are by far the most common causes of these three syndromes. Infections of the cardiovascular system are usually extremely serious, and if not promptly and adequately treated, they can be fatal. They commonly produce a constant shedding of organisms into the bloodstream that is often characterized by continuous, low-grade bacteremia (1 to 20 organisms/mL of blood) in untreated patients.

Primarily caused by bacteria

Infective Endocarditis

The term infective endocarditis is preferable to the commonly used term bacterial endocarditis, simply because not all infections of the endocardial surface of the heart are caused by bacteria. Most infections occur on natural or prosthetic cardiac valves, but can also develop on septal defects, shunts (eg, patent ductus arteriosus), or the mural endocardium.

Sites of endocardial infection include prosthetic valves

Infections involving coarctation of the aorta are also classified as infective endocarditis because the clinical manifestations and complications are similar.

Pathogenesis

The pathogenesis of infective endocarditis involves several factors that, if concurrent, result in infection:

Cardiac abnormalities create sites for attachment

1. Endothelium altered by previous disease (rheumatic fever) or congenital malformations is more susceptible to deposition of bacteria, platelets, and fibrin. Most infections involve the mitral or aortic valves, which are particularly vulnerable when abnormalities such as valvular insufficiency, stenosis, intracardiac shunts (eg, ventricular septal defect), or direct trauma (eg, catheters) exist. The turbulence of intracardiac blood flow that results from such abnormalities can lead to further irregularities of the endothelial surfaces that facilitate platelet and fibrin deposition. These factors produce a potential nidus for colonization.

Normal flora usual organism source for transient bacteremia

2. Transient bacteremia is common, but it is usually of no clinical importance. Often seen for a few minutes after a variety of dental procedures, it has also been shown to develop after normal childbirth and manipulations such as bronchoscopy, sigmoidoscopy, cystoscopy, and some surgical procedures. Even simple activities such as tooth brushing or chewing candy can cause such bacteremia. The organisms responsible for transient bacteremia are the common surface flora of the manipulated site such as viridans streptococci (oropharynx) and are usually of low virulence. Other more virulent strains may also be involved, however; for example, intravenous drug abuse may lead to transient bacteremia with *Staphylococcus aureus* or a variety of Gram-negative aerobic and anaerobic bacteria. Whether or not the organisms causing bacteremia (or fungemia) are of high virulence, they can colonize and multiply in the heart if local endothelial changes are suitable.

Bacteria adhere and start development of vegetation

Embolization created by dislodged parts of vegetation

3. Circulating organisms adhere to the damaged surface, followed by complement activation, inflammation, fibrin, and platelet deposition and further endothelial damage at the site of colonization. The resulting entrapment of organisms in the thrombotic "mesh" of platelets, fibrin, and inflammatory cells leads to a mature vegetation, which protects the organisms from host humoral and phagocytic immune defenses and to some extent from antimicrobial agents (**Figure 66–1A** and **B**). As a result, the infection can be exceedingly difficult to treat. The vegetation can also create greater hemodynamic alterations in terms of obstruction to flow and increased turbulence. Parts of vegetations may break off and

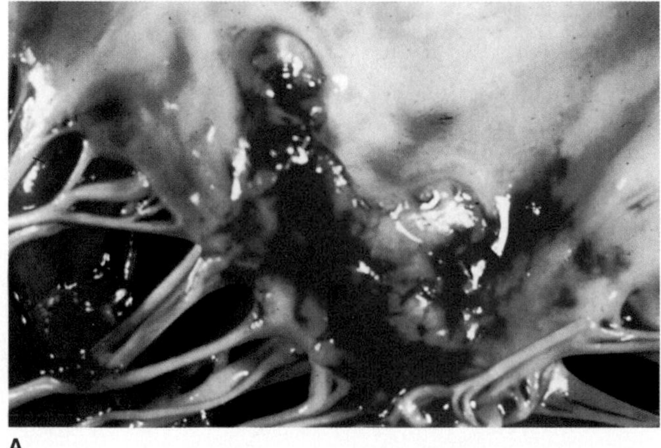

A

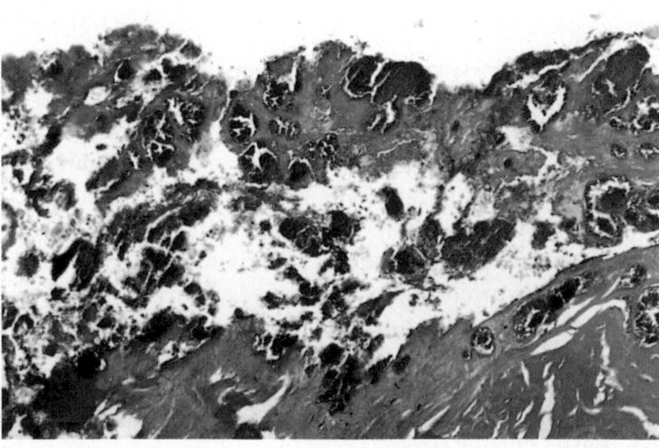

B

FIGURE 66–1. Bacterial endocarditis. A. Mitral valve with vegetations on the closing surface. **B.** Microscopic view of vegetations. The dark purple areas are composed almost entirely of bacterial microcolonies. (Reproduced with permission from Connor DH, Chandler FW, Schwartz DQA, Manz HJ, Lack EE (eds). *Pathology of Infectious Diseases*, vol. 1. Stamford, CT: Appleton & Lange; 1997.)

be deposited in smaller blood vessels (embolization) with resultant obstruction and secondary sites of infection. Emboli may be transported to the brain or coronary arteries, for example, with disastrous results.

Circulating immune complexes cause peripheral manifestations

Another phenomenon shown to contribute to the infective endocarditis syndrome is the development of circulating immune complexes of microbial antigen and antibody. These complexes can activate complement and contribute to many of the peripheral manifestations of the disease, including nephritis, arthritis, and cutaneous vascular lesions. ::: Immune complexes, p. 42

Rheumatoid factor and antinuclear antibodies contribute to pathogenesis

Frequently, there is a widespread stimulus to host cellular and humoral immunity, particularly if the infection continues for more than about 2 weeks. This condition is characterized by hyperglobulinemia, splenomegaly, and the occasional appearance of macrophages in the peripheral blood. Some patients develop circulating rheumatoid factor (IgM anti-IgG antibody), which may play a deleterious role by blocking IgG opsonic activity and causing microvascular damage. Antinuclear antibodies, which also appear occasionally, may contribute to the pathogenesis of the fever, arthralgia, and myalgia that is often seen.

In summary, infective endocarditis involves an initial complex of endothelial damage or abnormality, which facilitates colonization by organisms that may be circulating through the heart. This colonization, in turn, leads to the propagation of a vegetation, with its attendant local and systemic inflammatory, embolic, and immunologic complications.

Clinical Features

Acute, subacute, and chronic infective endocarditis determined by virulence of organism

Infective endocarditis has often been classified by the progression of the untreated disease. **Acute endocarditis** is generally fulminant with high fever and toxicity, and death may occur in a few days or weeks. **Subacute endocarditis** progresses to death over weeks to months with low-grade fever, night sweats, weight loss, and vague constitutional complaints. The clinical course is substantially related to the virulence of the infecting organism; *S aureus*, for example, usually produces acute disease, whereas infections by the less virulent viridans streptococci are more likely to be subacute. Before the advent of antimicrobial therapy, death was considered inevitable in all cases of infective endocarditis. Physical findings often include a new or changing heart murmur, splenomegaly, various skin lesions (petechiae, splinter hemorrhages, Osler's nodes, Janeway's lesions), and retinal lesions.

Cardiac, embolic, and immunologically mediated complications lead to death without treatment

Complications of infective endocarditis include the risk of congestive heart failure as a result of hemodynamic alterations, rupture of the chordae tendinea of the valves, or perforation of a valve. Abscesses of the myocardium or valve ring can also develop. Other complications relate to the immunologic and embolic phenomena that can occur. The kidney is commonly affected, and hematuria is a typical finding. Renal failure, presumably from immune complex glomerulonephritis, is possible. Left-sided endocarditis can readily lead to coronary artery embolization and "mycotic" aneurysms; the latter is discussed later in this chapter. In addition, more distant emboli to the central nervous system can lead to cerebral infarction and infection. Right-sided endocarditis often causes embolization and infarction or infection in the lung.

Etiologic Agents

Streptococci are most common cause

Many explanations for culture-negative endocarditis

Table 66–1 summarizes the most common causes of infective endocarditis. Viridans streptococci and enterococci are involved in just over 50% of cases. In the so-called culture-negative group, infective endocarditis is diagnosed on clinical grounds, but cultures do not confirm the etiologic agent. This group of patients is difficult to treat, and the overall prognosis is considered poorer than when a specific etiology has been determined. Negative cultures may result from (1) prior antibiotic treatment; (2) fungal endocarditis with entrapment of these relatively large organisms in capillary beds; (3) fastidious, nutritionally deficient, or cell wall–deficient organisms that are difficult to isolate; (4) infection caused by obligate intracellular parasites, such as chlamydiae (*Chlamydia psittaci*), rickettsiae (*Coxiella burnetii*), *Rochalimaea* species, or viruses; (5) immunologic factors (eg, antibody acting on circulating organisms); or (6) subacute endocarditis involving the right side of the heart, in which the organisms are filtered out in the pulmonary capillaries.

TABLE 66–1 Common Etiologic Agents in Infective Endocarditis

AGENT	APPROXIMATE PERCENTAGE OF CASES
Viridans streptococci (several species)	30-40
Enterococci	5-18
Other streptococci	15-25
Staphylococcus aureus	15-40
Coagulase-negative staphylococci	4-30
Gram-negative bacilli	2-13
Fungi (eg, *Candida, Aspergillus*)	2-4

Some special circumstances alter the relative etiologic possibilities, such as intravenous drug addiction, prosthetic valves, and immunocompromise. The major associations in these cases are summarized in **Table 66–2**.

General Diagnostic Approaches

Blood culture the most important diagnostic test

The diagnosis of infective endocarditis is usually suspected on clinical grounds; however, the most important diagnostic test for confirmation is the blood culture. In untreated cases, the organisms are generally present continuously in low numbers (1 to 20/mL) in the blood. If an adequate volume of blood is obtained, the first culture is positive in over 95% of culturally confirmed cases. Most authorities recommend three cultures over 24 hours to ensure detection, and an additional 3 if the first set is negative. Multiple cultures yielding the same organism support the probability of an intravascular or intracardiac infection. In acute endocarditis, the urgency of early treatment may require collection of only two or three cultures within a few minutes so that antimicrobial therapy can begin.

Echocardiography defines vegetations

Cardiologic procedures such as transthoracic or transesophageal echocardiography can delineate the nature and size of the vegetations and progression of disease. They are also helpful in the prediction of some complications such as embolization.

General Principles of Management

Bactericidal antimicrobics required because of protective effect of the vegetation

Because of the nature of the lesions and their pathogenesis, response to therapy may be slow and cure is sometimes difficult. Therefore, specific antimicrobial therapy must be aggressive, using agents that are bactericidal (rather than bacteriostatic) and can be given in amounts that achieve high continuous blood levels without causing toxicity to the patient. Treatment may involve a single antimicrobial if the organism is highly susceptible in vitro, or antimicrobial

TABLE 66–2 Endocarditis Agents Observed in Special Circumstances

SITUATION	AGENT
Intravenous drug abuse	*Staphylococcus aureus;* enterococci; Enterobacteriaceae and *Pseudomonas;* fungi
Prosthetic valve infection	Coagulase-negative streptococci; *S aureus;* Enterobacteriaceae and *Pseudomonas;* diphtheroids; *Candida* and *Aspergillus* spp.
Immunocompromise, chronic illness	Any of the above organisms

combinations if synergistic effects are possible (eg, a penicillin and an aminoglycoside for enterococcal endocarditis). Parenteral therapy is begun to produce adequate blood levels, and the patient may need to be monitored frequently to ensure antimicrobial activity in the serum sufficient to kill the organisms without causing unnecessary toxicity. Therapy is usually prolonged, lasting longer than 4 weeks in most cases. In some cases, surgery may be required to excise the diseased valve and replace it with a valvular prosthesis. The decision for surgery is sometimes difficult, requiring consultation with both a cardiologist and a surgeon. ::: Bactericidal testing, p. 418

Antimicrobic combinations often used for synergistic effect

Mycotic Aneurysm

The term mycotic aneurysm is somewhat misleading, because it suggests infection by fungi. Originally used by Sir William Osler to describe the mushroom-shaped arterial aneurysm that can develop in patients with infective endocarditis, the term now applies to infection with any organism that causes inflammatory damage and weakening of an arterial wall with subsequent aneurysmal dilatation. This sequence can progress to rupture, with a fatal outcome.

Intra-arterial infection occurs at sites of vascular injury

Arterial infection can result from direct extension of an intracardiac infection or from septic microemboli from a cardiac focus, with seeding of vasa vasorum within the arterial wall. In addition to infective endocarditis, other predisposing factors include damaged arterial intima by atherosclerotic plaques, vascular thrombi, congenital malformations, trauma, or spread from a contiguous focus of infection directly into the artery. The clinical features vary according to the site of involvement. Common findings may include pain at the site of primary arterial supply (eg, back or abdominal pain in abdominal aortic infections) and fever. In many cases, the initial presentation is the result of a catastrophic hemorrhage, particularly intracerebral aneurysms. The etiologic agents, diagnosis, and management are similar to infective endocarditis.

Etiologic agents similar to those of infective endocarditis

Suppurative Thrombophlebitis

Suppurative (or septic) thrombophlebitis is an inflammation of a vein wall frequently associated with thrombosis and bacteremia. There are four basic forms: superficial, pelvic, intracranial venous sinus, and portal vein infection (pylephlebitis). With the steadily increasing use of intravenous catheters, the incidence of superficial thrombophlebitis has risen and represents a major complication in hospitalized patients.

Thrombotic site may become seeded with organisms from blood

The pathogenesis involves thrombus formation, which may result from trauma to the vein, extrinsic inflammation, hypercoagulable states, stasis of blood flow, or combinations of these factors. The thrombosed site is then seeded with organisms, and a focus of infection is established. In superficial thrombophlebitis, an intravenous cannula or catheter may cause local venous wall trauma, as well as serve as a foreign body nidus for thrombus formation. Infection develops if bacteria are introduced by intravenous fluid, local wound contamination, or bacteremic seeding from a remote infected site.

Intravenous catheter often associated with thrombophlebitis

Thrombophlebitis of pelvic, portal, or intracranial venous systems most often occurs as a result of direct extension of an infectious process from adjacent structures or from venous and lymphatic pathways near sites of infection. For example, infections of intracranial venous sinuses usually result from orbital or sinus infections (causing cavernous sinus thrombophlebitis) or from infections of the mastoid and middle ear (causing lateral and sagittal sinus thrombophlebitis). Pelvic thrombophlebitis is a potential result of intrauterine infection (endometritis), particularly after pelvic surgery or 2 to 3 weeks after childbirth. Pelvic or intra-abdominal infections may also spread to the portal venous system to produce pylephlebitis.

Local infection may extend to veins

Clinical Features

Common features of suppurative thrombophlebitis often include fever and inflammation over the infected vein. Pelvic or portal vein thrombophlebitis is usually associated with high fever, chills, nausea, vomiting, and abdominal pain. Jaundice may develop in portal vein infections. Intracranial thrombophlebitis varies in its presentation. Headache, facial or orbital edema, and neurologic deficits are variably present; for example, cavernous sinus thrombophlebitis often causes palsies of the third through sixth cranial nerves. Complications include extension of suppurative infection into adjacent structures, further propagation of thrombi, bacteremia,

Signs and symptoms depend on anatomic site involved

TABLE 66–3 Common Etiologic Agents in Suppurative Thrombophlebitis

SITE	AGENT
Superficial veins (eg, saphenous, femoral, antecubital)	*Staphylococcus aureus;* Gram-negative bacilli
Pelvic veins, portal veins	*Bacteroides* spp; *Peptostreptococcus; Escherichia coli;* group A or B streptococci
Intracranial venous sinuses (cavernous, sagittal, lateral)	*Haemophilus influenzae, Streptococcus pneumoniae;* group A streptococcus; *Peptostreptococcus; S aureus*

and septic embolization. Embolization from pelvic or leg veins is to the lungs, and pulmonary embolism with infarction may be the presenting manifestation of the remote infection.

Etiologic Agents

The major infectious causes of suppurative thrombophlebitis are outlined in **Table 66–3**. In superficial thrombophlebitis, which often follows intravenous therapy, organisms that are common nosocomial offenders predominate (*S aureus,* Gram-negative aerobes). Deeper infections are more frequently caused by organisms that reside on adjacent mucous membranes (eg, *Bacteroides* species in intestinal and vaginal sites) or commonly infect adjacent sites (eg, *Haemophilus influenzae* and *S pneumoniae* in acute otitis media and sinusitis).

General Diagnostic Approaches

Direct culture or blood culture is usually positive

The diagnosis is often suspected on clinical grounds and from associated events known to create predisposition to such infections (eg, surgery, presence of indwelling venous cannulas). Direct cultures of the infected site or blood cultures usually yield the infecting organism because bacteremia is often present. Radiologic procedures, including scanning methods, may be necessary to localize the process and support the diagnosis. In some cases, surgical exploration is required, both for definitive treatment and to obtain specimens for cultures.

General Principles of Management

Antimicrobic therapy and removal of catheters

The choice of antimicrobial agents is based on culture and susceptibility test results, or without microbiologic data, on the most likely possibilities listed in Table 66–3. Other important aspects of management include prompt removal of possible offending sources, such as intravenous catheters, vigorous treatment of adjacent infections, and sometimes surgical excision and drainage. Severe cases may also benefit from systemic anticoagulant therapy to prevent further propagation of thrombi and embolization. ::: Susceptibility testing, p. 415-416

Many cases are preventable. Unnecessary, long-term intravenous cannulation should be avoided. Whenever possible, it is better to use short needles such as "scalp vein" cannulas than venous catheters or plastic cannulas. Careful asepsis is essential with all intravenous procedures to prevent contamination of intravenous fluids, tubing, and the site of venous entry. ::: Asepsis, p. 56

Intravenous Catheter Bacteremia

Significant endocarditis and metastatic infection risk

A variant of intravascular infection develops when a medical device such as an intravenous catheter or any of several types of monitoring devices placed in the bloodstream becomes colonized with microorganisms. The event itself does not have immediate clinical significance but, unlike transient bacteremia from manipulation of normal floral sites, the bacteremia continues. This persistence greatly increases the chances of secondary complications such as infective endocarditis and metastatic infection, depending on any underlying disease and the virulence of the organism involved.

Skin flora most commonly involved

The organisms involved are usually those found in the skin flora, such as *Staphylococcus epidermidis,* or *S aureus.* In debilitated patients already on antimicrobial therapy, *Candida*

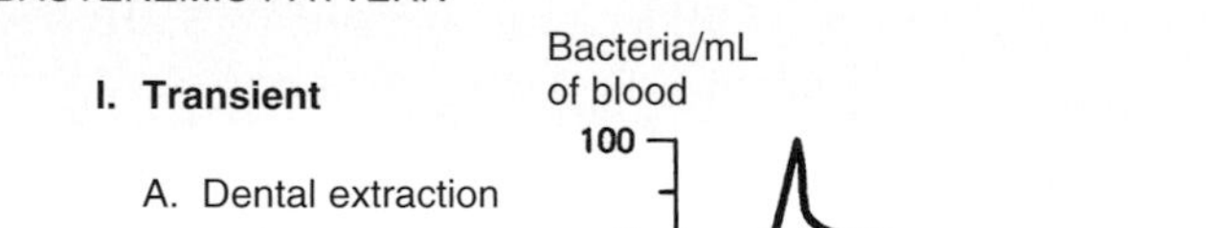
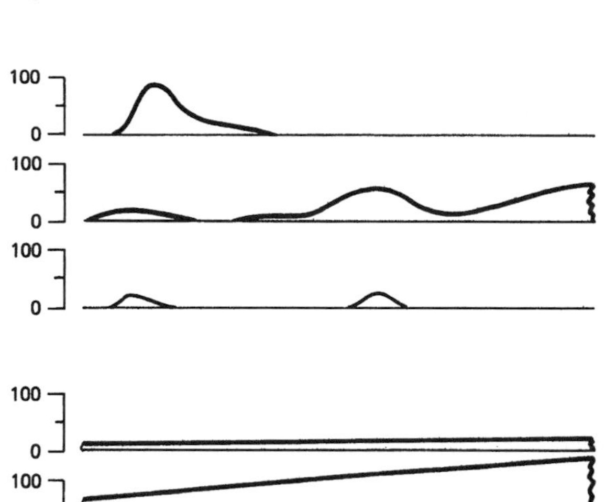

FIGURE 66–2. Patterns of bacteremia. The magnitude and timing of bacteremia for six typical patients (A–F) are depicted. These findings have implications for blood culture sampling plans. Cases such as A and B are detected only by cultures taken early in their course. Cases such as C and particularly D are more variable and more likely to be detected by cultures spaced over the time period shown. Continuous bacteremia (E and F) should be detected by any sampling plan. It could be confused with transient bacteremia on single blood cultures because both are caused by organisms of low virulence (viridans streptococci, *Staphylococcus epidermidis*); in cases such as E and F, however, bacteremia is sustained, whereas cases of transient bacteremia yield multiple positive results only if they are collected at or near the same time.

species may be involved. Occasionally, the sources of contamination are the intravenous solutions themselves rather than the skin. In these cases, members of the Enterobacteriaceae, *Pseudomonas,* or other Gram-negative rods are more likely.

Removal of contaminated catheter usually necessary

The clinical findings in catheter bacteremia are usually mild despite large numbers of organisms in the bloodstream (**Figure 66–2**). In addition to low-grade fever, signs of inflammation may or may not be present. Management includes removal of the contaminated catheter. Antimicrobial therapy alone often does not eradicate the organisms in the presence of a foreign body (the catheter).

BACTEREMIA FROM EXTRAVASCULAR INFECTION

Bacteremia may be high despite mild manifestations

Bacteremia is more variable than with intravascular infection

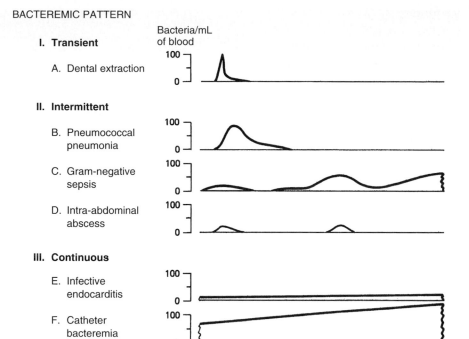

Although bacteremia is an integral feature of intravascular infection, most cases of clinically significant bacteremia are the result of overflow from an extravascular infection. In these cases, the organisms drained by the lymphatics or otherwise escaping from the infected focus reach the capillary and venous circulation through the lymphatic vessels. Depending on the magnitude of the infection and the degree of local control, these organisms may be filtered in the reticuloendothelial system or circulate more widely, producing bacteremia or fungemia. The process is dependent on the timing and interaction of multiple events and is thus much less predictable than intravascular infection. If the infection is extensive and uncontrolled, as with an overwhelming staphylococcal pneumonia, there may be hundreds or even thousands of organisms per milliliter of blood—a poor prognostic sign. An intra-abdominal abscess may seed only a few organisms intermittently until it is discovered and drained. Most infections that produce bacteremia fall between these extremes, with bloodstream invasion more common in the acute phases and intermittent at other times.

Frequently associated with severe infections such as meningitis

The causative organisms and the frequencies with which they usually produce bacteremia (or fungemia) are listed in **Table 66–4**. There is considerable overlap, and the probability of bacteremia is dependent on the site as well as the organism. Any organism producing meningitis is likely to produce bacteremia at the same time. Infections with *H influenzae* type b are usually bacteremic whether the site is the meninges, epiglottis, or periorbital tissues. Meningitis caused by *S pneumoniae* can be expected to be bacteremic, but only 20% to 30% of patients with pneumococcal pneumonia have positive blood cultures.

TABLE 66–4 Frequency of Detection of Bloodstream Invasion by Bacteria and Some Fungi During Significant Infections at Extravascular Sites

Large (>90%) proportion of cases	
Haemophilus influenzae type b	*Brucella*[a]
Neisseria meningitidis	*Salmonella* serovar Typhi
Streptococcus pneumoniae (meningitis)	*Listeria*
Variable (10-90%) depending on stage and severity of infection	
Pyogenic streptococci	Enterobacteriaceae
S pneumoniae	*Pseudomonas*
Staphylococcus aureus	*Bacteroides*
Neisseria gonorrhoeae	*Clostridium* (myositis and endometritis)
Leptospira[a]	*Peptostreptococcus*
Borrelia[a]	*Candida*
Acinetobacter	*Cryptococcus neoformans*[a]
Shigella dysenteriae	
Small (<10%) proportion of cases	
Shigella (except *S dysenteriae*)	*Pasteurella multocida*
Salmonella enterica	*Haemophilus*, nonencapsulated
Campylobacter jejuni[a]	
Isolation too rare to justify attempt	
Vibrio (intestinal infections)	*Clostridium tetani*
Corynebacterium diphtheriae	*Clostridium botulinum*
Bordetella pertussis	*Clostridium difficile*
Mycobacterium[b]	*Legionella*[c]

[a]Isolation and/or demonstration requires special methods or prolonged incubation.
[b]*Mycobacterium avium-intracellulare* infections in AIDS patients often yield positive results.
[c]Infrequent isolation may be due to inadequate cultural methods.

Bacteremia is overflow from respiratory, urinary, wound, and other primary sites of infection

The most common sources of bacteremia are urinary tract infections, respiratory tract infections, and infections of skin or soft tissues, such as wound infections or cellulitis. The frequency with which any organism causes bacteremia is related to both its propensity to invade the bloodstream (Table 66–4) and how often it produces infections. For example, cases of *Escherichia coli* bacteremia are common, attributable in part to the fact that *E coli* is the most common cause of urinary tract infection.

SEPSIS AND SEPTIC SHOCK

Associated with bacteremic Gram-negative and Gram-positive infections

Bacteremia is the presence of viable bacteria circulating in the blood. When signs and symptoms result, further terms are used to delineate the progression of potential consequences that may occur. Both Gram-negative and Gram-positive organisms can produce the same findings, as well as fungi, protozoa, and even some viruses.

Sepsis is the suspicion (or proof) of infection and evidence of a systemic response to it (eg, tachycardia, tachypnea, hyperthermia, or hypothermia). The **sepsis syndrome** includes findings of sepsis plus evidence of altered organ perfusion. These can include reduction in urine output, mental status changes, systemic acidosis, and hypoxemia. If the process remains uncontrolled, there is subsequent progression to **septic shock** (development of hypotension); **refractory septic shock** (hypotension not responsive to standard fluid and pharmacologic treatment); and **multiorgan failure,** including major target organs such as the kidneys, lungs, and liver, and disseminated intravascular coagulation. Mortality is exceedingly high when patients develop refractory septic shock or multiorgan failure.

Sepsis syndrome progresses through shock to organ failure

The initial events in the sepsis syndrome appear to be vasodilatation with resultant decreased peripheral resistance and increased cardiac output. The patient is flushed and febrile. Capillary leakage and reduced blood volume follow, leading to a whole series of events identical to those seen in shock resulting from blood loss. These manifestations include vasoconstriction, reflex capillary dilatation, and local anoxic damage. Once this stage is reached, the patient may develop hypotension and hypothermia; acidosis, hypoglycemia, and coagulation defects ensue with failure of highly perfused organs such as the lungs, kidneys, heart, brain, and liver.

Vasodilatation is followed by complex response

The mechanisms involved in the development of septic shock have been studied extensively in experimental animals. Most of the features seen in humans can be produced with the lipopolysaccharide endotoxin of the Gram-negative cell wall, although there is some variation among animal species and with different preparations. The various events that occur are complex. They include (1) release of vasoactive substances such as histamine, serotonin, noradrenaline, and plasma kinins, which may cause arterial hypotension directly and facilitate coagulation abnormalities; (2) disturbances in temperature regulation, which may be due to direct central nervous system effects or, in the case of the early febrile response, mediated by interleukin 1 (IL 1) and tumor necrosis factor (TNF) released from macrophages; (3) complement activation and release of other inflammatory cytokines by macrophages (eg, IL-2, IL-6, IL-8, and interferon γ); (4) direct effects on vascular endothelial cell function and integrity; (5) depression of cardiac muscle contractility by TNF, myocardial depressant factor, and other less well-defined serum factors; and (6) impairment of the protein C anticoagulation pathway, resulting in disseminated intravascular coagulation. The resultant alterations in blood flow and capillary permeability lead to progressive organ dysfunction. ::: LPS endotoxin, p. 354

Endotoxin causes release of vasoactive substances

Cytokines, complement, and other mediators have physiologic effects

Early recognition of the problem is critical, and management obviously requires considerably more than antimicrobial therapy. Other primary therapeutic measures include maintenance of adequate tissue perfusion through careful fluid and electrolyte management and the use of vasoactive amines. There is also evidence that protein C replacement may ameliorate the coagulopathy.

Antimicrobic, fluid, and coagulation management are crucial

BLOOD CULTURE

The primary means of establishing a diagnosis of sepsis is by blood culture. The microbiologic principles involved are the same as with any culture. A sample of the patient's blood is obtained by aseptic venipuncture and cultured in an enriched broth or, after special processing, on plates. Growth is detected, and the organisms are isolated, identified, and tested for antimicrobial susceptibility. Because of the importance of blood cultures in the diagnosis and therapy of most bacterial and fungal infections, considerable attention must be paid to details of sampling if the prospects of obtaining a positive culture are to be maximized. The approach to blood culture must be tailored to the individual patient; no single procedure is best for everyone. The important features are described below.

Importance of blood culture demands attention to details

Blood Culture Sampling

Venipuncture

Before venipuncture, the skin over the vein must be carefully disinfected to reduce the probability of contamination of the blood sample with skin bacteria. Although it is not possible to "sterilize" the skin, quantitative counts can be markedly reduced with a combination of 70% alcohol and an iodine-based antiseptic. Mechanical cleansing is as important as use of

Skin decontamination removes bulk of skin flora

Some anticoagulants have antimicrobial properties

the antiseptic. Poor phlebotomy technique such as repalpating the vein after the preparation is related to introduction of contaminants. Blood is ideally drawn directly into a blood culture bottle or a sterile blood collection vacuum tube containing an anticoagulant free of antimicrobial properties such as sodium polyanethol sulfonate. Other anticoagulants commonly used in laboratories such as citrate and ethylenediaminetetraacetic, acid have antibacterial activity. Blood should not be drawn through indwelling venous or arterial catheters unless it cannot be obtained by venipuncture.

Volume

Number of organisms in blood often less than 1 organism/mL

The number of organisms present in blood is often low (less than 1 organism/mL) and cannot be predicted in advance. Thus, small samples yield fewer positive cultures than larger ones. For example, as the volume sampled increases from 2 to 20 mL, the diagnostic yield increases by 30% to 50%. Samples of at least 10 mL should be collected from adult patients. The same principles apply with infants and young children, but the sample size must be reduced to take account of the smaller total blood volume of a child. Although it should be possible to obtain at least 1 mL, smaller volumes should still be cultured because bacteremia at levels of more than 1000 bacteria/mL is found in some infants.

Number

Two or three blood cultures usually adequate

If the volume is adequate, it is rarely necessary to collect more than two or three blood cultures to achieve a positive result. In intravascular infections (eg, infective endocarditis), a single blood culture is positive in more than 95% of cases. Studies of sequential blood cultures from bacteremic patients without endocarditis have yielded 80% to 90% positive results on the first culture, more than 90% to 95% with two cultures, and 99% in at least one of a series of three cultures.

Timing

Timing of intermittent bacteremia not predictable

Antimicrobic therapy may interfere with blood culture results

The best timing schedule for a series of two or three blood cultures is dependent on the bacteremic pattern of the underlying infection and the clinical urgency of initiating antimicrobial therapy. Figure 66–2 illustrates some typical bacteremic patterns that can be related to the probability of obtaining positive blood cultures. Transient bacteremia is usually not detected because organisms are cleared before the appearance of any clinical findings suggesting sepsis. The continuous bacteremia of infective endocarditis is usually readily detected, and timing is not critical. Intermittent bacteremia presents the greatest challenge because fever spikes generally occur after, rather than during, the bacteremia. Little is known about the periodicity of bloodstream invasion, except that the bacteremia is more likely to be present and sustained in the early acute stages of infection. Closely spaced samples are less likely to detect the organism than those spaced an hour or more apart. In urgent situations, when antimicrobial therapy must be initiated, two or three samples should be collected at brief intervals and therapy begun as soon as possible. It is generally not useful to collect blood cultures while the patient is receiving antimicrobics unless none were collected before therapy or there is a change in the clinical course, suggesting superinfection. The laboratory should be advised when such cultures are submitted, because it is sometimes possible to inactivate an antimicrobic, for example, with β-lactamases.

Laboratory Processing

Blood added to enriched broth

Automated and direct plating procedures now available

The basic blood culture procedure of incubating blood in an enriched broth is simple, but considerable effort must be expended to ensure detection of the broadest range of organisms in the least possible time. Daily examination of cultures for 1 week or more and a routine schedule of stains and/or subcultures of apparently negative cultures are required to detect organisms such as *H influenzae* or *Neisseria meningitidis,* which usually do not produce visual changes in the broth. Direct plating of blood onto blood or chocolate agar is accomplished in a system that concentrates the blood by centrifugation following lysis of the erythrocytes. This is particularly useful for bacterial quantification and rapid identification. Automated blood culture systems detect metabolic activity in broth culture by CO_2 generation or metabolism of labeled substrates in place of the conventional visual and staining examinations. These systems detect growth sooner than conventional methods, but still require subculture for confirmation, identification, and susceptibility testing.

Isolation of fungi is favored by ensuring maximum aerobic conditions in direct plating systems and broth bottles. Conversely, anaerobes are recovered best when a highly reduced environment is provided for plates and broths. Some bacteria, such as *Leptospira,* are not isolated by routine blood culture procedures. The laboratory must be notified in advance so special media can be used.

Special cultural conditions required for yeasts and anaerobes

Because the blood is normally sterile, the interpretation of blood cultures growing a pathogenic organism is seldom a problem. The major problem is the differentiation of skin contamination and agents causing transient bacteremia from opportunists associated with an intravascular or extravascular infection. Transient bacteremia is of short duration (Figure 66–2), is associated with manipulation of or trauma to a site possessing a normal flora, and involves species indigenous to that site. Despite skin disinfection, 2% to 4% of venipunctures result in contamination of the culture with small numbers of cutaneous flora such as *S epidermidis,* corynebacteria (diphtheroids), and propionibacteria. The presence of these organisms in blood cultures can be considered a result of skin contamination unless quantitative procedures indicate large numbers (more than 5 organisms/mL) or repeated cultures are positive for the same organism. These findings should suggest diseases such as infective endocarditis or catheter bacteremia.

Interpretation involves distinguishing infection from normal skin flora contamination

GLOSSARY

The glossary is intended as an adjunct to the index for rapid reference. The emphasis is on medical and biologic words and phrases that have not been defined in the text or are used frequently. The specific names of microorganisms, antimicrobial agents, and infectious diseases are in the index and not repeated here. Where a word has multiple uses, the one relevant to this text is emphasized.

The prefixes and suffixes in each alphabetical section include word elements used in combined form. The meaning of many words can be derived from the prefixes and suffixes and therefore have not been included in the glossary.

A-, An- Without.

AB toxin A bacterial toxin with separate binding and active units.

Acanthosis Hyperplasia and thickening of prickle cell layer of skin.

Accessory sinuses Blind-ended cavities in bone draining into nasal cavity.

Acetyl-coenzyme A An energy-rich combination of acetic acid and coenzyme A.

Achlorhydria Absence of hydrochloric acid in stomach.

Acid-fast Describes an organism that resists acid decolorization after straining.

Acidosis Increased acidity of body fluids.

Aciduric Resistant to effects of acid.

Acquired immunity Immunity developed following exposure to infectious agents or by infusion of antibodies.

Acquired immunodeficiency syndrome (AIDS) A disease caused by viral infection of key components of the immune system.

Actin Major structural protein of the eukaryotic cell cytoskeleton.

Acute viral gastroenteritis Condition characterized by vomiting and diarrhea.

Addison's disease Result of primary deficiency of production of adrenal hormones.

Adenocarcinoma Malignant tumor derived from glandular epithelium.

Adhesin Surface component of a microbe that binds to a cell receptor.

Adnexa (uterine) Fallopian tubes and ovaries.

ADP-ribosylation An enzymatic reaction that attaches the ADP-ribose moiety from NAD to the target protein.

Adrenal glands Important endocrine glands situated above the kidneys.

Aerobactin A hydroxame siderophore produced by many bacteria.

Aneorobe Microorganism that is inhibited or killed by oxygen and utilizes fermentation exclusively.

Agammaglobulinemia Absence of immunoglobulins in the blood.

Agar A polysaccharide derived from algae used as a solidifying agent in culture media.

Agarose gel Highly purified agar.

Agglutinate Clumping.

Agranulocytosis Failure of white blood cell production in bone marrow.

-algia Pain.

Allele Alternate forms of a gene at the same chromosomal locus.

Alloantigen An antigen that exists in alternate allelic forms.

Allosteric Property of a protein that leads to a change in conformation and function associated with attachment of a smaller effector molecule.

Alternative pathway An antibody-independent mechanism of complement activation.

Alveoli (lung) Microscopic air sacs in lung.

Ameboma A local inflammatory mass caused by an amebic infection.

Aminoglycosides A group of antibiotics which inhibit protein synthesis by ribosomal binding.

Amniotic fluid Fluid in amniotic sac surrounding the fetus.

Anaerobe Microorganism that grows and survives only in the absence of oxygen.

Analog Structurally or functionally similar substance or property.

Anamnestic Enhanced immunologic memory response on reexposure to antigen.

Anaphylaxis Immediate and severe antibody-mediated hypersensitivity reaction.

Anergic Absence of ability to respond to antigen.

Anergy A state of unresponsiveness to antigens.

Aneurysm Localized abnormal dilatation of blood vessel.

Anicteric Absence of clinical jaundice.

Anneal Subject to controlled heating and cooling to achieve a particular property.

Anorexia Loss of appetite.

Anoxia Lack of adequate oxygenation of blood or tissues.

Anterior horn cell Motor neuron in the anterior gray matter of the spinal cord.

Anthropo- Related to humans.

Antibiogram Pattern of in vitro susceptibilities to different antimicrobial agents.

Antibody A glycoprotein molecule produced by plasma cells in response to introduction of an antigen; can bind to the antigen with exact specificity.
Antigen A substance that elicits a specific immunologic response or reacts with antibody.
Antigenic drift Random mutation of a virus leading to new variants not recognized by the immune system.
Antiseptic Chemical agent that inhibits or kills pathogenic microorganisms.
Antiserum Serum containing specific antibodies.
Antitoxin An antibody that neutralizes an exotoxin.
Antitussive Substance that helps control coughing.
Aphonia Loss of speech.
Aplastic anemia Failure of red cell production in bone marrow.
Apnea Temporary absence of breathing.
Apoptosis Programmed cell death.
Aqueduct of Sylvius Canal connecting the third and fourth ventricles of the brain.
Arachidonic acid Precursor of prostaglandins.
Arachnoid The middle of three membranes that cover the brain and spinal cord (meninges).
Arrythmia Irregularity of heartbeat.
Arteriole Smallest artery leading to capillary.
Arthralgia Pain in a joint.
Arthro- Pertaining to joints.
Arthroconidia Conidia that develop within the hyphae and eventually break off.
Aryepiglottis Related to the epiglottis and the arytenoid cartilage.
Ascites Fluid in a peritoneal cavity.
Ascus A sac. In mycology, a specialized structure containing spores termed ascospores.
Asepsis Exclusion of pathogenic organisms.
Aseptic meningitis Meningeal inflammation associated mostly with an increase of cells (pleocytosis).
Asphyxia Suffocation.
Astrocyte Connective tissue cell of the central nervous system.
Ataxia Disturbance of muscular coordination.
Ataxia telangiectasia Hereditary disorder causing ataxia and permanent dilatation of some blood vessels.
Atelectasis Collapse of part of lung.
Atherosclerosis Hardening of the arteries.
Athlete's foot *See* **Tinea pedis**.
Atrophy Wasting.
Attenuated Reduced in virulence, (eg, organisms in a live vaccine).
Auto- Self, or arising from within.
Autochthonous flora Organism with intimate and permanent association with an epithelial surface.
Autoclave Sophisticated pressure cooker used to kill microorganisms.
Autoimmunity An immune response against the body's own tissues.
Autolysis Lysis of a cell by its own enzymes.
Autonomic Relates to involuntary nervous system controlling cardiac, vascular, intestinal, and other functions.
Auxo- Pertaining to growth.
Auxotroph Bacterial mutant that has lost the ability to synthesize an essential nutrient or metabolite.
Avascular Absence of blood vessels or blood supply.
Axenic Refers to pure cultures of a microorganism without presence of a contaminating or symbiotic organism.
Axon The extension of a neuron that conducts nerve impulses.

B cell Bone marrow–derived lymphocyte that can differentiate a plasma cell and produce antibody.
β-lactam Antibiotic class that inhibits synthesis of peptidoglycan for the bacterial cell wall.
Bacillus Rod-shaped bacterial cell.
Bacteremia Bacteria in the blood.
Bacteriocins Proteins produced by one bacterium that kill another of the same or other species.
Bacteriophage Bacterial virus.
Bacteriostatic Inhibition of bacterial growth without killing.
Bacteriuria Bacteria in the urine.
Bartholin's glands Lubricating glands on either side of the vaginal opening.
Basophil Polymorphonuclear leucocyte with basophilic granules.
Basophilic Stains with a basic dye.
Biliary Pertaining to the bile and bile ducts.
Bilirubin A bile pigment.
Bio- Pertaining to life.
Biofilm Extracellular film produced by an organized community of microbial cells.
Bioterrorism Use of infectious agents to deliberately produce disease.
Biotype Subtype within a species characterized by physiologic properties.
Blackwater fever Condition in which hemoglobinuria develops, resulting in the production of dark urine, along with malaria.
-blast Precursor cell.
Blastoconidia Buds that form from a single cell.
Bleb *See* **Bulla**.
Blepharal Pertaining to the eyelids.
Blepharo- Pertaining to the eyelid.
Blepharoplast Basal body of a cilium or flagellum.
Blood–brain barrier Functional barrier preventing passage of large molecules to the brain parenchyma.
Blood fluke infection Schistosomiasis.
Bolus Rounded mass that may obstruct (eg, fecal bolus) or a concentrated mass (eg, an antibiotic) given rapidly and intravenously.
Bothria Paired sucking grooves in the head of the fish tapeworm (*Diphyllobothrium*).
Brady- Slowing.
Bradycardia Unusually slow heartbeat.
Broad-spectrum agent Inhibiting a wide range of Gram-negative and Gram-positive bacteria.
Bronchial tree Bronchi and bronchioles that conduct gases to and from the lung alveoli.
Bronchiectasis Pathologic dilatation of terminal bronchi.
Bronchiole Smallest subdivision of bronchial tree.
Broncho- Pertaining to the bronchial tree.

Bubo Swollen, inflamed, infected lymph node.
Bubonic plague Infection produced by *Yersinia pestis* from rodents and transmitted to humans by the bite of infected fleas, is the most explosively virulent disease known, which begins with a bubo and spreads to the bloodsteam. Also called Black Death.
Buccal Pertaining to the inside of the cheek.
Bulla Blister or large vesicle containing fluid.
Bursa Sac filled with fluid (eg, protecting a joint or tendon).

Calabar swelling Localized area of allergic inflammation.
Calculus Pathologic stone (eg, renal or gallbladder calculus).
Calmodulin A protein present in eukaryotic cells that activates some essential enzymes when it has bound calcium.
Capillary The smallest blood vessel connecting the arterial and venous systems.
Capsid The outer protein coat of a virus that protects its nucleic acid.
Capsomeres Protein subunits of viral capsids.
Carbuncle A necrotic staphylococcal infection of skin and subcutaneous tissue that is formed of coalesced furuncles (boils).
Carcinoma Malignant growth of epithelial cells.
Cardio- Pertaining to the heart.
Cardiolipin A phospholipid occurring naturally in mitochondrial membranes against which antibodies are formed in syphilitic infection.
Cardiomyopathy Disease of heart muscle.
Caries Progressive destruction of the mineralized tissues of the tooth.
Caseous Cheesy in consistency.
Catalase Enzyme that catalyzes the reduction of toxic hydrogen peroxide to oxygen and water.
Cathelicidins Antimicrobial peptides produced by a variety of epithelial and inflammatory cells.
Cat scratch disease Lymphadenitis that causes fever and systemic symptoms that may persist for weeks to months.
Cell-mediated immunity Immune reactions in which T lymphocytes secrete cytokines to modify or destroy foreign or infected cells.
Cell strain Culture that consists of diploid cells, commonly fibroblastic, which can be redispersed and regrown a finite number of times.
Cellulitis Inflammation of subcutaneous tissue.
Cementum Layer of modified bone on tooth root.
Cerebrospinal fluid Fluid that fills spaces within and surrounding the central nervous system.
Cervical Pertaining to the neck or uterine cervix.
Cervix The constricted portion of an organ. Usually refers to the lower part of the uterus.
Cestode Tapeworm.
Chancre Sore or ulcer that develops at the site of an infection. Most often used to describe the primary syphilitic lesion.
Chelator Compound that binds metallic ions.
Chemokines Proteins or glycoproteins that are involved in cell-to-cell communication
Chemoprophylaxis Use of antimicrobics to prevent infection.
Chemotaxis Attraction of a motile cell to a chemical.
Childbed fever Puerperal endometritis caused primarily by group A streptococci.
Chitin Polysaccharide forming exoskeletons of some insects or cell walls of fungi.
Chlamydoconidia Conidia that develop within hyphae.
Chlorhexidine Routine hand and skin disinfectant and used for other topical applications.
Cholangitis Inflammation of the bile ducts.
Chole- Pertaining to bile.
Cholecystitis Inflammation of the gallbladder.
Cholestasis Interruption of the flow of bile.
Cholinergic nerves Nerve fibers that release acetylcholine as a mediator at their effector terminals.
Chordae tendinae Small tendons that connect papillary muscles of the heart to the cusps of the atrioventricular valves.
Chorea Rapid purposeless involuntary movements.
Chorioallantoic membrane The outer membrane surrounding an avian embryo within the egg shell.
Chorionic membrane The outer extraembryonic membrane from which the placenta originates.
Chorioretinitis Inflammation of choroid and retina of the eye.
Choroid plexus Vascular invagination into the cerebral ventricles. Produces the cerebrospinal fluid
Chromatin Complex of DNA and histones making up the chromosomes of eukaryotic cells.
Chronic granulomatous disease Genetic disorder causing absence of H_2O_2 production and myeloperoxidase activity of phagocytes. Results in repeated infections with catalase positive bacteria.
Chronic mucocutaneous candidiasis Chronic, relapsing candidiasis.
-cidal Killing.
Cilia Surface structures of some eukaryotic cells that beat rhythmically to move mucus over surfaces or confer motility on some single-celled organisms.
Cirrhosis Fibrosis and nodular regeneration of the liver with loss of function.
Cistron A segment of DNA that encodes a polypeptide.
Clade Subtype of HIV-1 class.
Clone Identical progeny of a single cell, gene, or genes.
Co-agglutination Agglutination involving two organisms, one of which acts as an inert particle coated with specific antibody to the other.
Coarctation Stricture or narrowing (eg, of the aorta).
Cocci Spherical or oval bacteria typically arranged in clusters or chains.
Coccus Spherical bacterial cell.
Co-cultivation Process that can be used for unmasking latent virus by growing susceptible cells with those from affected tissue.
Codon The three nucleotides encoding an amino acid or a chain termination signal.
Cold sore Lesion on a specific area of the lip and immediate adjacent skin. also called fever blister.
Coliform An imprecise term referring to Gram-negative facultative bacteria generally resident in the intestine.
Collagen Fibrous component of connective tissue.

Coloboma A defect of the iris of the eye.
Colonial morphology Features of isolated colonies of bacteria that vary greatly, such as shape, texture, color, and other features
Colostrum Initial secretion of the breast after delivery (contains antibodies and lymphocytes).
Comedo Blocked sebaceous duct with retention of sebum (blackhead).
Commensal Organism of the normal flora that has a symbiotic relationship with the host.
Communicability Ability of an the organism to shed in secretions.
Competent A bacterial cell able to take up free DNA fragments.
Complement A system of serum proteins that act in sequence to mediate inflammatory and some immune responses.
Concatemers Long linear DNA molecules that are the products of replication.
Concomitant immunity The ability of adult worms from a primary infection to survive in a host resistant to reinfection.
Condyloma acuminatum A wart-like infectious benign growth that occurs on the genitalia and in the anal canal.
Conidia Asexual fungal reproductive spore-like bodies.
Conidiophore Stalk-like fungal structure that bears conidia.
Conjugation A process that transfers DNA from one bacterial cell to another.
Conjunctivitis Inflammation of the conjunctiva, which may involve cornea, sclera, or sclera.
Copepod Minute fresh water fleas that serve as intermediate hosts for some parasites.
Coprolith Stony, hard stool.
Coracidium The ciliated free swimming embryo of certain tapeworms.
Core polysaccharide Component of lipopolysaccharide that contains some unusual carbohydrate residues and fairly constant in structure among related species of bacteria.
Cornea Clear, anterior portion of the eyeball.
Cortex The outer layer of an organ.
Corticosteroid Steroid hormone from adrenal gland; some are anti-inflammatory.
Coryza Catarrhal rhinitis (eg, from the common cold).
Councilman bodies Cytoplasmic eosinophilic masses produced from hyaline necrosis of hepatocytes caused by yellow fever or another arbovirus infection.
Counterimmunoelectrophoresis A technique for increasing the sensitivity and speed of the immunodiffusion procedure by the application of an electrophoretic field.
Crepitation A crackling or rattling sound elicited by palpation of tissues.
Cribriform plate Area of bone above nasal cavity through which pass the olfactory nerves.
Croup Manifestations of laryngeal obstruction from inflammation or other causes.
Crustacean Hard shelled invertebrates such as crabs, shrimp, and lobsters.
CSF *See* **Cerebrospinal fluid**.
Curare A plant extract that produces generalized paralysis by acting at neuromuscular junctions.
Cuticle Skin or surface layer.
Cyanosis Blue color of skin caused by lack of oxygen.
Cystic fibrosis Congenital disease of secreting glands affecting pancreas, respiratory tract, and sweat glands. Associated with viscid respiratory mucus and chronic respiratory infections.
Cysticercus Larval form of tapeworm enclosed in a cyst.
Cysto- Pertaining to the bladder.
Cystoscope Instrument for examining inside the urinary bladder.
Cyto- Pertaining to the cell.
Cytokine Messenger proteins released by cells (lymphocytes, monocytes, etc) that mediate activities in other cells.
Cytokine storm Caused by the secretion of cytokines from viral infection, producing cell damage rather than direct viral replication.
Cytology The study of cells rather than of tissues and organs.
Cytopathic effect Common effect in which lytic or cytopathic viruses, as they replicate in cells, produce alterations in cellular morphology (or cell death).
Cytoplasm Cellular contents excluding the nucleus.
Cytoskeleton Network of microfilaments in the cytoplasm of eukaryotic cells that gives shape and structural support.
Cytosol Cytoplasm of prokaryotic cells.
Cytosome The body of a cell apart from its nucleus.
Cytostome The mouth opening of certain ciliated protozoa.

Dacrocystitis Inflammation of the lacrimal sac.
Dalton Atomic mass unit that gives the same number as atomic weight.
Dane particle The complete and intact hepatitis B virus particle.
Darkfield microscopy Method in which a condenser focuses light diagonally on the specimen in such a way that only light is reflected from particulate matter.
Debridement Removing foreign matter and dead tissue.
Decelerating phase The period of time in the culture growth cycle in which nutrients are depleted, waste products are accumulated, and growth becomes progressively limited.
Decubitus ulcer Pressure sore (bed sore).
Defective interfering particles Noninfectious genomes that interfere with the replication of the infectious virus.
Defensins A family of microbial, cationic, cystine rich polypeptides abundant in the azurophilic granules of polymorphonuclear leukocytes.
Definitive host Species in which the parasite reproduces sexually.
Demyelination Loss of nerve sheaths.
Dendritic Branched.
Dendritic cell An antigen-presenting cell found in lymph nodes, spleen and thymus.
Dermatophyte Fungus that causes skin infections.
Dermis Skin connective tissue immediately below the epidermis.
Dermo- Pertaining to the skin.
Desquamation Loss of skin epithelial cells.
Dextran A polymer of D-glucose.
Dimorphism Occurring in two morphologic forms under different conditions.

Diphtheria toxin An A-B toxin that acts in the cytoplasm to inhibit protein synthesis irreversibly in a wide variety of eukaryotic cells.

Diploid Possessing two sets of chromosomes.

Direct fusion Method by which certain viruses enters a cell.

Direct transposition Excision of the transposon from its original location and insertion in a simple cut-and-paste manner into its new site without replication.

Disease index Number of persons who develop a disease divided by total number infected.

Disseminated infection Infection that spreads throughout the body.

Disseminated intravascular coagulation (DIC) A clinical syndrome with multiple causes. Thrombocytopenia and complex coagulation abnormalities are prominent.

Diverticulum Blind-ended extrusion from a hollow organ.

DNA polymerase An enzyme that synthesizes new DNA using the parental strand as a template.

Ductus arteriosus Fetal blood vessel connecting the pulmonary artery to the descending aorta.

Duplication The production of a redundant segment of DNA, usually adjacent to the original segment.

Dys- Difficult or painful.

Dysentery Pain and frequent defecation resulting from inflammation of the colon or other intestines, with blood and pus in the stool.

Dyspareunia Difficult or painful intercourse.

Dysphagia Difficulty in swallowing.

Dysplasia Histologic evidence of possible premalignant changes in cells.

Dyspnea Shortness of breath.

Dysuria Difficult or painful urination.

Ecchymosis A large area of hemorrhage into the skin, often a coalescence of petechiae.

Eclipse phase Period of infection in which no infectious viruses are found inside the cell.

Ecthyma Eroded, scabbed lesion of the skin.

Ecto- Outside or outer.

-ectomy Surgical removal of.

Ectopic pregnancy Fetal development outside the uterus (usually in the fallopian tubes).

Ectoplasm Clear layer of cytoplasm near the cell membrane of amebas.

Edema Excessive fluid in tissues.

EIA. *See* **Enzyme immunoassay**.

Elastosis Disorder of fibroelastic proteins.

Electrophoresis Procedure for separating charged particles by differences in their migration in an electric field.

Elephantiasis Grotesque swelling of the extremities and genitalia.

ELISA Enzyme-linked immunosorbent assay (*See* **Enzyme immunoassay**).

Embolism Sudden blockage of an artery.

-emia Of the blood.

Emphysema (pulmonary) Irreversible enlargement of alveolar sacs of lung.

Empyema Pus in a body cavity (eg, pleural cavity).

Encapsidation Process of enclosing the viral genome in a protein capsid.

Encephalitis Inflammation of brain tissue.

End problem In DNA replication, constraint on the completion of DNA chains on a linear template.

Endarteritis Inflammation of the inner coat of an artery or arteriole.

Endemic A disease that is continuously present at subepidemic levels in a particular region, locality, or group.

Endo- Within.

Endogenote DNA from the same chromosome.

Endogenous Originating within an organism.

Endometrium Interior epithelial lining of the uterus.

Endonuclease Enzyme of a class that hydrolyzes internal bonds of DNA or RNA. Involved in synthesis and breakdown of nucleic acids.

Endophthalmitis Inflammation of interior tissues of the eye.

Endoplasm Central portion of cytoplasm of some cells, beneath the ectoplasm.

Endoplasmic reticulum A system of membranes and tubules within the cytoplasm of eukaryotic cells.

Endosome A vesicle formed by endocytosis.

Endospore A heat- and chemical-resistant spore within some Gram-positive bacteria.

Endotoxin A toxic lipopolysaccharide moiety of the Gram-negative bacterial cell wall outer membrane.

Entactin Protein component of the extracellular matrix.

Enteric Pertaining to the intestinal tract.

Enteric fever Prolonged febrile illness originating in the gastrointestinal tract. Typhoid fever is the prototype.

Entero- Pertaining to intestines.

Enterobactin A phenolate siderophore produced by *E coli* and some other enteric species of bacteria.

Enterochelin Synonym for Enterobactin.

Enterotoxin Bacterial exotoxin that affects the intestinal mucosa causing vomiting and/or diarrhea.

Enucleation (ocular) Removal of an eye.

Enzootic Disease present at low levels at all times in an animal community.

Enzyme immunoassay A method for detecting antigen–antibody reactions by labeling one of the reagents with detectable enzyme marker.

Eosinophil Polymorphonuclear leukocyte with eosinophilic granules.

Epi- Upon or additional to.

Epicardium Outer lining of the heart.

Epidemic A disease that rapidly affects many people in a circumscribed period of time.

Epididymis Tubular structure attached to the testes in which spermatozoa mature.

Epigastrium Upper central region of the abdomen overlying the stomach.

Epiglottis Movable structure overlying and protecting the larynx.

Epiphysis Growing end of bone.

Episome Plasmid or viral DNA that can replicate extrachromosomally or can integrate into chromosome.

Epitope Structural part of a protein that determines antigenic specificity (also called antigenic determinant).

Epitrochlear node Lymph node above inner side of elbow.
Erysipelas Rapid-spreading infection of the deep layers of the dermis caused by group A streptococci with risk of bacteremia.
Erythema Red color in tissues and skin caused by dilatation of blood vessels.
Erythema migrans The expanded bulls-eye lesion associated with Lyme borreliosis.
Erythema nodosum Red raised skin nodules usually on the legs, which is typically a manifestation of a hypersensitivity reaction.
Erythro- Red.
Erythrocyte Red blood cell.
Eschar Necrotic scab-like area of injured skin.
Ethylene oxide An inflammable and potentially explosive gas, which is an alkylating agent that inactivates microorganisms.
Etiology Cause of a disease.
Eukaryote Organism comprising one or more cells containing true nuclei and cytoplasmic organelles.
Eustachian tube Tube connecting the middle ear and the nasopharynx.
Exanthem Disease in which skin rashes are the major manifestations.
Exocrine glands Glands excreting their products to skin, intestinal, respiratory, or genitourinary tracts.
Exogenote An external molecule of DNA introduced into a recipient.
Exotoxin Toxic protein secreted by a bacterial cell.
Exponential (or logarithmic) growth period The period of time in the culture growth cycle in which the growth rate is maximal and constant.
Extrinsic incubation period The period of time required for virus multiplication to enhance the capacity to transmit infection to vertebrates by bite.

Facultative Bacteria able to grow aerobically or anaerobically.
Fallopian tubes Tubes extending from ovaries to uterus.
Fascia Sheets of specialized connective tissue.
Fauces Area between the mouth and the pharynx. Bounded by the tonsils, soft palate, and base of tongue.
Fc fragment The stalk of the Y-shaped antibody structure.
Febrile Having a raised temperature.
Fecal–oral spread Direct or finger-to-mouth spread of infection, the use of human feces as a fertilizer, or fecal contamination of food or water.
Feedback inhibition Process in which the end product of the pathway controls the activity of the first enzyme in the pathway.
Fermentation Energy-producing metabolic process using an endogenous electron acceptor, usually pyruvate.
Fever blister *See* **Cold sore**.
Fibrin Insoluble protein of blood clots.
Fibrinogen Precursor of fibrin.
Fibroblast Specialized cell producing collagen and elastic connective tissue.
Fibronectin A glycoprotein widely distributed in connective tissue and coating cells at mucosal surfaces.
Fibrosis Formation of collagenous connective tissue.
Filament Structure that consists of polymerized molecules of a single protein species called flagellin.
Filamentous hemagglutinin (FHA) A rod-like protein with the ability to bind to and agglutinate erythrocytes in *Bordetella.*
Filtration Method by which both live and dead microorganisms can be removed from liquids, by positive- or negative-pressure filtration.
Fimbriae Very fine fibrils on the surface of a bacterium often referred to as pili.
Fistula An abnormal passage from a hollow organ (eg, intestine).
Flaccid Loose; absence of muscle tone.
Flagellum Whip-like appendage of motion used by bacteria and some parasites.
Fluke Flat parasitic worm (trematode).
Fluorescein isothiocyanate Fluorescent dye.
Fluorescence Light emitted by a substance when irradiated with light of a shorter wavelength.
Fluorescin Yellow dye produced by *P aeruginosa* and other free-living less pathogenic *Pseudomonas* specie
Fluorochrome A fluorescent dye.
Follicle A small sac or cavity.
Folliculitis Usually describes localized inflammation of hair follicles without the purulence of furuncles.
Fomites Inanimate objects transmitting infectious agents.
Foramina Outlets to cavities.
Formaldehyde An alkylating agent whose vapor can be used without pressure to decontaminate larger areas such as rooms.
Frameshift mutation Change in the reading frame by which the ribosomes translate the mRNA from the mutated gene.
Fulminant Rapid and severe development (eg, of an infection).
Fungemia Fungi in the bloodstream.
Funiculitis Inflammation a cord-like structure, usually the spermatic cord.
Furuncle Purulent infection of a hair follicle; a boil.
Fusiform Tapering at both ends.

Gametocyte Male or female sexual cell of the malarial parasite found in the blood of humans and transmissible to mosquitoes.
Ganglion Group of nerve cells outside the spinal cord.
Gangrene Death of tissue.
Gastro- Pertaining to the stomach.
Gene A DNA sequence that codes for a polypeptide or RNA molecule.
General secretory pathway (GSP) The simplest and most common mechanism for protein secretion used by both Gram-positive and Gram-negative bacteria.
-genic Arising from, origin.
Genome The total gene complement of a cell or microbe.
Genotype Classification based on genetic constitution.
Genus A well-defined group of species that is clearly separate from other micoorganisms.
Geophagia Eating soil.
Germination The production of progeny.

Giemsa stain A combination of basic and acidic dyes used to stain blood smears and to demonstrate some protozoa.

Gingival crevice Area between the tooth and the gums.

Gingivo- Pertaining to the gums.

Glaucoma Excessive pressure in eyeball that can lead to blindness.

Glia Supporting cells of the central nervous system (neuroglia).

Glomerulonephritis Inflammatory disease of the kidney glomeruli.

Glomerulus Microscopic organ of specialized capillaries in the kidney that filters waste products from the blood.

Glottis The sound-producing area of the larynx.

Glucans Polymers of glucose.

Glutaraldehyde Alkylating agent highly lethal to essentially all microorganisms.

Gnotobiotic Animals reared under aseptic conditions, which may either be sterile ("germ free") or in which defined microflora are introduced.

Golgi Eukaryotic cellular organelle composed stacks of folded sacs, which prepare materials for secretion and other cellular processes.

Gonads Ovaries or testes.

Gram-negative shock or **endotoxic shock** Fever and shock syndrome brought on by an endotoxin.

Granulocyte Polymorphonuclear leukocyte of the neutrophil, basophil, or eosinophil series.

Granuloma Chronic inflammatory lesion infiltrated with macrophages and lymphocytes and accompanied by fibroblast activity.

Gravid Pregnant.

Group translocation Process that involves the chemical conversion of the solute into another molecule as it is transported.

Guillain-Barré syndrome Febrile polyneuritis with muscle weakness; may lead to paralysis.

Gumma Soft, gummy granulomatous lesion which is one of the features of tertiary syphilis.

H antigen Antigenic term for the flagella of bacteria of the Enterobacteriaceae family.

Halophilic Preferring or requiring a high salt content (eg, for growth).

Haploid Half the number of chromosomes of eukaryotic tissue cells (see **Meiosis**) or number of chromosomes in asexual organisms.

Hapten A molecule not immunogenic by itself but with the ability to elicit antibody production when attached to a larger molecule.

Heat-shock response A phenomenon in which up to 20 genes may be transcriptionally activated on an upward shift in temperature or on imposition of several kinds of chemical stress.

Helminth A parasitic worm.

Helper T (T_H) cell T cell needed for effective presentation to B cells.

Hemadsorption Adherence of red blood cells to a surface.

Hemagglutination Agglutination of erythrocytes by binding of antibody or microorganisms.

Hemagglutinin The virion attachment protein on the influenza virion.

Hematocrit Volume of erythrocytes in blood as a percentage of the total volume of blood.

Hematogenous Derived from blood. Spread by the bloodstream.

Hematoma Extravasation of blood into the tissues causing a swelling.

Hematopoietic system Precursor cells that produce blood cells.

Hematoxylin–eosin stain Commonly used histologic stain. Hematoxylin stains nuclei blue. Eosin is a red counterstain.

Hematuria Blood in the urine.

Hemianopsia Loss of vision in half the visual field.

Hemo-, Hema- Pertaining to blood.

Hemoglobulinemia Free hemoglobin in the blood.

Hemolysin A substance or enzyme causing lysis of erythrocytes.

Hemolysis Disruption of red blood cells with liberation of hemoglobin.

Hemolytic–uremic syndrome A syndrome that includes hemolytic anemia, thrombocytopenia, and renal dysfunction.

Hemoptysis Coughing up of blood.

Hemothorax Blood in the pleural cavity of the chest.

Hepato- Pertaining to the liver.

Hepatocellular Pertaining to liver cells (hepatocytes).

Hepatocytes Liver cells.

Hepatoma Malignant tumor of liver cells.

Hetero- Of different origin.

Heterologous Derived from a different clone, strain, species, or tissue.

Heterophil antibody Antibody reacting with an antigen other than that which elicited its production.

Heteroploid Eukaryotic cell with abnormal number of chromosomes.

Heterotroph An organism that requires organic carbon for nutrition.

Heterozygous Possessing different alleles at a particular genetic locus in a diploid cell.

Hexacanth A tapeworm embryo containing six pairs of hooklets.

Hexamer In virology, a capsomer comprising six subunits.

Hilar lymph nodes Nodes at the root of the lung.

Histiocyte Tissue macrophage.

Histocompatibility Antigens on tissue cells that are recognized by the host as self or foreign.

Hodgkin disease A malignant lymphoma initially affecting groups of lymph nodes.

Homeostasis Tendency to stability of conditions within a complex biologic system.

Homonymous hemianopsia Blindness affecting the same half of the visual field in each eye.

Homozygous Possessing the same alleles at a particular genetic locus in a diploid cell.

Horizontal transmission Spread of infection through an animate insect vector.

Host range The limited spectrum of cell types that a virus is capable of infecting.

Humoral Mediated by fluids. In immunology relates to antibody-mediated immunity as opposed to cellular immunity.
Hyaline Clear and transparent.
Hyaluronic acid Acid mucopolysaccharide comprising the ground substance of connective tissue. Also found on bacterial surfaces.
Hyaluronic acid capsule A polymer containing repeating units of glucuronic acid and *N*-acetylglucosamine.
Hybridization Process in which denatured, single-stranded nucleic acids from different sources are annealed.
Hybridoma A clone derived from fused cells of different origin (eg, from an antibody producing lymphocyte and a tumor cell).
Hydrocele Fluid accumulation within the scrotum.
Hydrocephalus Pathologic accumulation of cerebrospinal fluid in the ventricles of brain.
Hydronephrosis Accumulation of urine in the renal pelvis due to obstruction of urinary flow. Associated with atrophy of the renal parenchyma.
Hyper- Greater than, above normal.
Hyperalimentation Intravenous administration of nutrients for treatment of actual or potential malnutrition.
Hyperammonemia Excessive amounts of ammonia in the blood.
Hyperbaric oxygen Oxygen under increased pressure relative to the atmosphere.
Hyperemia Increased blood flow to a tissue.
Hypernatremia Increased serum sodium.
Hyperplasia Increase in the number of cells in a tissue.
Hypersensitivity Exaggerated and harmful immune response to a normally innocuous antigenic stimulus.
Hypertension Elevated blood pressure.
Hypertonic Of higher osmotic pressure than fluid on the other side of a semipermeable membrane (eg, cell membrane).
Hypertrophy Enlargement of an organ as a result of an increase in size of its cells. Note distinction from hyperplasia.
Hypha A fungal filament.
Hypo- Less than, below normal.
Hypochlorhydria Reduced hydrochloric acid in the stomach.
Hypoglycemia Blood sugar below normal levels.
Hypotension Low blood pressure.
Hypothalamus Portion of the brain that forms the floor and part of the lateral wall of the third ventricle.
Hypothermia Serious reduction in body temperature.
Hypoxia Decreased oxygen supply to the tissues.

Icosahedron A solid geometric shape having 12 vertices. Serves as the structural basis for many viruses.
Icteric Pertaining to jaundice.
Idiopathic Of unknown origin.
Idiotype Variation in the hypervariable region of the Fab-combining site due to mutations
Ig Abbreviation for immunoglobulin antibodies. Classes include IgG, IgM, IgA, IgD, IgE, and sIgA.
Ileitis Inflammation of the lower ileum.
Ileum Portion of the small intestine between the jejunum and the cecum.
Immunocompromise Deficiency in some components of the body's immune mechanisms.
Immunocyte Cell of the lymphoid series that responds to an antigenic stimulus by producing antibodies or initiating cell mediated immune processes.
Immunodiffusion A procedure involving diffusion of antigen and antibody toward each other in a gel. A visible precipitate develops where optimal concentrations interact.
Immunofluorescence A microscopic procedure using antibody labeled with a fluorescent dye that allows visible detection of sites of reaction with antigen.
Immunogen An antigen that induces an immune response.
Immunoglobulins *See* **Antibody**.
Impetigo Superficial pustular skin infection.
Intermediate host Species in which the parasite reproduces asexually.
In vitro Occurring in the test tube.
In vivo Occurring in the living animal.
Incidence The number of new cases of a disease within a specified period.
Inclusion body A morphologically distinct intracellular mass of viruses or virus components.
Incubation period Time between exposure to an organism and appearance of the first symptoms.
Indirect samples Specimens of inflammatory exudates that have passed through sites known to be colonized with normal flora.
Inducer Regulatory molecule that turns on transcription.
Infarct Interference with the blood supply producing local death of tissue.
Infectivity Rate of attack of disease.
Insertion sequence *See* **Transposon**.
Insertional mutagenesis Mechanism in which the viral promoter or enhancer is sufficient to cause the inappropriate expression of a cellular gene residing in the immediate vicinity of the integrated provirus.
Integument Enveloping layer, eg, skin, membrane, or cuticle.
Integrins Family of transmembrane proteins of eukaryotic cells that interact with extracellular matrix and cytoskeleton proteins
Inter- Between.
Interference Method of viral detection in cell culture in which the infecting virus can be detected by challenging the cell culture with a different virus.
Interferon Class of cytokines that have nonspecific antiviral activity.
Interleukin Class of cytokines produced by macrophages or T cells that mediate growth and differentiation of cells, particularly lymphocytes.
Interstitial Spaces between the cells of a tissue.
Intertriginous Pertaining to area between folds of the skin.
Intima Inner lining of a blood vessel.
Intimin Major enteropathogenic *E coli* attachment protein.
Intra- Within.
Intrapartum Occurring during the process of childbirth.
Intrathecal Within the membranes of the spinal cord.
Introitus An opening.
Invasin A class of molecules that either directs bacterial entry into cells or provides an intimate direct contact between the bacterial surface and the host cell plasma membrane.

Inversion Change in the direction of a segment of DNA by splicing each strand of the segment into the complementary strand.
Iodine An effective disinfectant that acts by iodinating or oxidizing essential components of the microbial cell.
Iodophors Agents that are combined with carriers (povidone) or nonionic detergents that gradually release small amounts of iodine
Ionizing radiation Light that carries greater energy than ultraviolet light, causes direct damage to DNA, and produces toxic free radicals and hydrogen peroxide from water within the microbial cells.
Isoantigen Normal substance present in one individual that may elicit an antibody response in another.
Isotonic Of the same osmotic pressure as a solution on the other side of a semipermeable membrane.
-itis inflammation.

Janeway's lesions Painless macular lesions of palms and soles seen in acute bacterial endocarditis.
Jejunum Portion of small intestine between duodenum and ileum.

K antigen Antigenic term for surface polysaccharides of the Enterobacteriaceae bacteria.
Kaposi's sarcoma Multiple malignant vascular tumor. Occurs most commonly as a complication of AIDS.
Karotype Size, structure, and organization of chromosomes within a cell.
Karyosome Area of chromatin concentration in a cell nucleus.
Katayama syndrome A condition of persons with schistosomiasis in which leukocytosis, marked peripheral eosinophilia, and elevated levels of IgM, IgG, and IgE immunoglobulins are present.
Keratin Major protein of the skin, hair, and nails.
Keratitis Inflammation of the cornea of the eye.
Kinetoplast Structure at the base of a protozoal flagellum.
Kupffer cells Fixed phagocytic cells of the liver sinusoids. Part of the reticulo-endothelial system.
Kwashiorkor Condition caused by severe protein malnutrition in children.

Labia Structures of the external female genitalia.
Labile toxin (LT) An AB toxin that has the physical property of heat lability.
Lactoferrin Iron-binding protein present in milk, other secretions, and granules of neutrophils.
Lag period The period of time during the culture growth cycle in which growth is not detectable.
Lamina propria Connective tissue supporting the epithelial cells of a mucous membrane.
Laminin Major protein component of basal lamina.
Latent period The length of time from the beginning of infection until progeny virions are found outside the cells.
Latex beads Used to adsorb soluble antigens. The treated beads agglutinate with specific antibody.
Lectin Mechanism that binds carbohydrate moieties and protein–protein interactions based on a specific peptide sequence.
Lentiviruses HIV-1 and HIV-2, which cause AIDS, are lentiviruses.
Leukemia Malignant tumor of white blood cells.
Leuko- White; relating to a leukocyte.
Leukocyte White blood cells including granulocytes, lymphocytes, and monocytes.
Leukocytosis Increased blood leukocyte count.
Leukopenia Abnormally low leukocyte count.
Leukotrienes Products of arachidonic acid that mediate inflammatory and allergic reactions.
Ligand One component of a complex involving the binding of molecules or structures.
Lipid A A phospholipid containing glucosamine.
Lipo- Relating to fats or lipids.
Lipoarabinomannan (LAM) Lipid polysaccharide complex.
Lipopolysaccaride Special molecular in the outer leaflet of the Gram-negative cell membrane, which is toxic to humans.
Lipoteichoic acid A type of teichoic acid linked to a glycolipid in the underlying Gram-positive cell membrane.
Lobar Related to a lobe of the lung.
Lophotrichous Describing several flagella at one or both ends of a bacillus.
Lumen Cavity within a tubular organ.
Lupus erythematosus (systemic) Autoimmune inflammatory disease of skin, joints, and other tissues.
Lymph Tissue fluid derived from the bloodstream and passing to the lymphatics.
Lymphadenitis Enlarged, inflamed lymph nodes.
Lymphangitis Inflammation of lymphatic vessels.
Lympho- Pertaining to the lymphatic system.
Lymphocytosis Increased blood lymphocyte count.
Lymphokine Cytokine produced by lymphocytes.
Lymphoma Tumor of lymphatic tissues.
Lymphoreticular Relating to the reticuloendothelial system.
Lysate Supernatant resulting from lysis.
Lysis Dissolution of cells.
Lysogeny State in which a viral genome remains in and replicates with a bacterial genome.
Lysosome Intracellular organelle containing hydrolytic digestive enzymes.
Lysozyme Enzyme that breaks down bacterial peptidoglycan.
-lytic Pertaining to lysis.
Lytic infection Results in cell death.

M cell Specialized antigen delivery cell of the intestinal mucosa.
Macro- Large.
Macrocytic anemia Anemia characterized by large erythrocytes.
Macrophage Tissue and blood phagocyte derived from mononuclear cells.
Macule A flat lesion of skin rash.
Major histocompatibility complex Collection of genes that control integrity and homeostasis.
Masseter Major muscle controlling movement of the lower jaw.

Mast cell Connective tissue cell analogous to the blood basophil. Granules contain heparin, histamine, and other vasoactive mediators.
Mastitis Inflammation of the breast.
Mastoid Process of temporal bone behind the ear that contains air cells.
Meatus Orifice.
Meckel's diverticulum Congenital diverticulum of the lower part of the ileum.
Mediastinum Mid-portion of the chest including heart, bronchial bifurcation, and esophagus.
Medulla The inner portion of an organ within the cortex.
Medulla oblongata Portion of central nervous system between the brain and spinal cord.
Mega- Large.
Megacolon Dilatation of the colon.
-megaly Enlargement, usually of an organ.
Meiosis Cellular division process yielding haploid gametes.
Melioidosis A tropical, often relapsing, pneumonia.
Membrane attack complex Complement proteins inserting in membrane.
Memory cell Immunce T-cell which recalls past experience.
Meninges The membranes covering the brain and the spinal cord.
Meningomyelocele Malformation of vertebral column with protrusion of meninges.
Mentation Mental activity; thinking.
Merozoite A stage in the life cycle of a sporozoan parasite resulting from asexual division; a daughter cell.
Mesenchymal Derived from the embryonic mesoderm layer.
Mesentery Fold of peritoneum surrounding the intestinal tract and attaching it to the posterior abdominal wall.
Mesophile A microbe that grows best at temperatures of approximately those of the body.
Mesosome A complex invagination of the bacterial cell membrane.
Metastases Satellite tumors or infections spread through lymphatics or the bloodstream from a primary site.
-metry measure.
Micro- Small.
Microaerophilic Growth is best at oxygen concentrations between atmospheric and anaerobiasis.
Microcephaly Small head with failure of development of the brain.
Microdeletion The removal of a single nucleotide and its complement in the opposite strand.
Microfilaments Protein filaments forming internal structure of eukaryotic cells
Microinsertion The addition of a single nucleotide and its complement in the opposite strand.
Microphthalmia Failure to develop normal sized eyes.
Microtubule Cylindrical cytoskeletal elements made of tubulin proteins.
Minimal bactericidal concentration (MBC) The least amount required to kill a predetermined portion of the inoculum.
Miracidia Ciliated larvae.
Missense mutation Replacement mutation in a codon that changes the mRNA transcript to a different amino acid.
Mitochondria Complex cytoplasmic organelles of eukaryotic cells involved in oxidative phosphorylation.
Mitogen Substance that increases the normal frequency of mutations.
Mitral valve Valve between the left atrium and ventricle of the heart.
Molecular mimicry Epitopes of infectious agents that stimulate immune reactions to host tissues as well as the homologous antigen.
Monoclonal Derived from a single cell.
Monocyte Large mononuclear phagocyte of blood and tissues that eventually becomes a macrophage.
Monolayer A single layer of cultured eukaryotic cells on a glass or plastic surface.
Monotrichous Possessing a single flagellum.
Mordant Substance that enhances the effect of a stain.
Morphology The shape, size, and form of an organism or cell.
Mucolytic Substance that dissolves mucus.
Multiple sclerosis Demyelinating disease of brain and spinal cord that can progress to neurologic impairment.
Mutagen Substance that increases the mutation rate of cells or organisms.
Mutant allele Mutated, usually inactive, form of a gene.
Mutation Permanent, heritable change in the genome.
Myalgia Pain in the muscles.
Mycelium A mass of fungal hyphae.
Mycetoma A localized granuloma or lesion caused by a fungus.
Mycolic acid Long-chain fatty acid.
Mycology Science devoted to the study of fungi.
Mycosis A fungal infection.
Mycotoxin Toxin formed by fungi in the environment.
Myelin Component of the sheath around the axon of a neuron that increases the conduction velocity of the nerve impulse.
Myelitis Inflammation of the spinal cord.
Myeloma Malignant tumor derived from bone marrow cells.
Myeloperoxidase Intracellular enzyme of professional phagocytes.
Myo- Pertaining to muscle.
Myocardium Heart muscle.
Myringitis Inflammation of the tympanic membrane of the ear.

Nares Interior of the nostrils.
Narrow-spectrum agent Highly active against many Gram-positive and Gram-negative cocci, but little activity against others, such as enteric Gram-negative bacilli.
Nasal turbinates Three scroll-like bony projections from the lateral wall of the nasal cavity (nasal conchae).
Nasolacrimal duct Duct draining the conjunctiva into the nasal cavity.
Necrosis Death of tissue.
Nematode Roundworm.
Neo- New.
Neoplasm Tumor.
Nephrito- Pertaining to the kidney.
Nephritogenic Producing inflammation of the kidneys.

Neucleoid Nuclear body.
Neuraminidase An antigenic hydrolytic enzyme that acts on the hemagglutinin receptors by splitting off their terminal neuraminic acid.
Neuro- Pertaining to the central nervous system or nerves.
Neuromotor synapses Connections between nerve endings and muscle.
Neurone Nerve and its nerve cell. Also spelled **neuron**.
Neutropenia Reduced number of circulating neutrophils.
Neutrophils Major class of polymorphonuclear phagocytic leukocytes.
Neurosyphilis A condition in tertiary syphilis in which chronic meningitis is the most common manifestation.
Nidus Focus of infection, a cluster.
Noma A gangrenous condition spreading from the oral cavity to the skin; seen in undernourished children.
Non A, non B hepatitis Term used to identify hepatitis not due to hepatitis A or B, but now rarely used because of the discovery of other specific hepatitis viruses.
Noncommunicable infection Not transmitted from human to human.
Nonpermissive cell Cell that does not allow virus to replicate, but may be able to transform the cell.
Nonsense mutation Replacement mutation that changes a codon specifying an amino acid to one specifying none.
Normal flora Microorganisms frequently found in body sites in normal healthy persons.
Norwalk agent The original virus causing enteritis; later called calicivirus or norovirus.
Nosocomial Acquired within a hospital.
Nucleocapsid The nucleic acid and surrounding protein coat (capsids) that form the basic structure of viruses.
Nucleoid The double-stranded circular DNA genome of a bacterium.
Nucleolus Round body within a eukaryotic nucleus that is the site of synthesis of ribosomal RNA.

O antigen Antigenic term for outer membrane lipopolysaccharide of the Enterobacteriaceae family of bacteria.
O antigen polysaccharide side chain The major surface antigen of Gram-negative cells.
Occult Hidden, inapparent.
Olfactory Pertaining to the sense of smell.
Olfactory bulb Terminal enlarged portion of the olfactory tract from which the olfactory nerves emerge.
Oligo- Small, few.
Oligodendroglia Specialized connective tissue of the central nervous system.
Onco- Pertaining to tumors.
Oncogene Gene whose activation is associated with malignant change and progression.
Oncoretrovirus One of two major groups of retroviruses that infect humans. They transform cells and produce new virus indefinitely.
Ontogeny Origin and course of development of an individual organism.
Oocyst Structure formed when a zygote encysts.
Operator One of the components of the structure of a typical operon.
Operculum A lid or cover.
Operon An operator and the adjacent structural gene(s) that it controls.
Ophthalmia Severe inflammation of the eye.
Opisthotonos Severe spasm of back muscles leading to hyperextension of the spine.
Opportunist A microorganism that causes disease only when the body's defenses are compromised or bypassed.
Opsonin, opsonization Antibody or complement coating of microbes, which facilitates their phagocytosis.
Orbit Skull cavity that contains the eyeball.
Orchitis Inflammation of a testis.
Organelles Membrane-bound cytoplasmic structures of eukaryotic cells, which perform specific functions.
Organogenesis Formation of the organs of the body.
Oro- Pertaining to the mouth.
-oscopy Use of an instrument to see within a viscus or vessel.
Osler's nodes Skin papules, usually of hands and feet, seen in bacterial endocarditis.
Ossicles Small bones (eg, of hearing).
Osteo- Pertaining to bone.
Osteomyelitis Inflammation of bone marrow and adjacent bone.
Otitis externa Unflammation of the ear canal with purulent ear drainage.
Oto- Pertaining to the ear.
Outer membrane A second membrane outside the peptidoglycan found only in Gram-negative bacteria.
Overwintering The phenomenon of a virus surviving between transmission periods.
Oviparous Producing eggs from which the embryo is released outside the body.
Oxidase Oxidation-reduction enzyme that catalyzes transfer of electrons to molecular oxygen with formation of water.

Packaging site Assembly often initiates at a particular locus on the genome
Pan- All, throughout.
Pandemic Worldwide severe epidemic.
Panencephalitis Inflammation of all tissues of the brain.
Papilla Small nipple-like swelling.
Papilledema Edema of the optic nerve and adjacent retina.
Papilloma Warty tumor of the epithelium.
Papule Small, firm, elevated nodule on the skin.
Para- Beside, abnormal.
Parasite An organism that lives on and at the expense of another organism.
Parasitism Describes the relationship between parasite and host.
Parenchymal Substance of body organs in contrast to their covering.
Parenteral Administration by injection rather than by mouth.
Paresis Paralysis.
Paresthesias Disorders of sensation; tingling.
Paronychia Infection of nail fold.
Parotid glands Salivary glands beneath the cheek.
Parturition The process of giving birth.
Passive immunity The transfer of antibodies from one person to another.

Pasteurization Process of heating milk or other liquids to destroy microorganisms.
Pathogenic Capable of causing disease.
Pathogenicity island Large block of genes found on the bacterial chromosome, which have fundamental characteristics that are different from the rest of the genome of the current host organism.
Pathognomonic Diagnostic, distinctive.
-pathy Denoting disease.
-penia Decreased numbers.
Penicillin *See* **β-lactam**.
Penicillin-binding proteins (PBPs) Peptidoglycan cross-linking enzymes so named for their property of binding penicillin.
Pentamer A polymer of viral capsid having five structural units.
Peptidoglycan High-molecular-weight cross-linked polymer forming the rigid structure of the bacterial cell wall.
Peptone Protein hydrolysed product used as a source of amino acids in bacterial culture media.
Peri- Around, covering.
Periapical Beside the root of a tooth.
Pericardium Membranous lining around the heart.
Perineum Area between vulva or scrotum and the anus.
Periodontal Area around the tooth including supporting tissues.
Periosteum Membrane around the bone.
Periplasm Area between the outer and cell membranes of a Gram-negative bacterium. Contains the peptidoglycan layer.
Peristalsis Normal contractile waves of a hollow organ.
Peristome The mouth and surrounding areas of certain ciliated protozoa.
Peritrichous Presence of multiple flagella around a bacterial cell.
Permease A protein of the bacterial cell membrane transport system.
Permissive cell Cell that permits production of progeny virus particles or viral transformation.
Petechiae Small (<3 mm) hemorrhages in the skin containing red blood cells or hemoglobin.
Peyer's patches Lymphoid follicles in the ileum.
Phage Common abbreviation for bacteriophage.
Phagocyte A cell that ingests foreign material.
Phagolysosome The digestive vacuole formed by fusion of the cell lysosomes with the phagocytic vacuole.
Pharyngitis Infection of the pharynx.
Phenol A potent protein denaturant and bactericidal agent.
Phenotype The properties expressed by the complete genome under particular conditions.
Pheromone Hormone-like substance that elicits a favorable or attraction response in an individual of the same species.
-phobia ear of, repulsion.
Phonation Speech.
Photophobia Intolerance of light.
-phylia affection for.
Phylogeny Pertaining to the evolution of a species.
PID Pelvic inflammatory disease.
Piedra Infection of the hair characterized by black or white nodules attached to the hair shaft.
Pilosebaceous Unit of hair follicle and sebaceous gland.
Pilot proteins Accompany the phage genome into the bacterial cell and serve the function of "piloting" the nucleic acid to a particular target.
Pilus Fibrillar structure on the surface of a bacterial cell.
Pinocytosis Uptake of fluids into a cell by a mechanism analogous to phagocytosis.
Pityriasis (tinea) versicolor Discrete areas of skin hypopigmentation or hyperpigmentation associated with induration and scaling, occurring in tropical and temperate climates.
Plankton Minute free-floating organisms, vegetable and animal, which live in natural waters.
Plaque A patch or flat area. An area of lysis in fixed host cells by an infecting virus.
Plasma Noncellular component of whole blood.
Plasmid Extrachromosomal circular double stranded DNA molecule.
Plasmin Derived from plasminogen; dissolves fibrin.
Platelet Small anucleate cell involved in filling small holes in blood vessels and in clotting mechanisms.
Pleo- More.
Pleocytosis Increased number of cells in a particular area.
Pleomorphism Variation in shape and size.
Pleura Membrane covering the lungs and thoracic cavity enclosing the pleural space.
Pleurisy Inflammation of the pleura.
Pleuro- Relating to the pleura.
Pleurodynia Pain caused by inflammation or irritation of the pleura.
Pneumonic plague Highly contagious pneumonia secondary to bubonic plague that is transmitted person to person by the respiratory droplet route.
Pneumonitis Inflammation of the lung.
Pneumothorax Air in the pleural cavity.
Polar mutation Prevention of the expression of all genes away from the promoter of the mutated gene.
Poliomyelitis Selective destruction of anterior motor horn cells in the spinal cord and/or brainstem.
Poly- Many, repeated.
Polyarthralgia Pain in several joints.
Polycistronic Encoding two or more proteins (eg, polycistronic mRNA).
Polyclonal activation Simultaneous activation of different antibody producing clones of lymphocytes.
Polymerase chain reaction Continuous enzyme-mediated amplification of a nucleotide sequence that allows its detection and analysis.
Polymorphonuclear Two or more lobes to the nucleus.
Polymyositis Inflammation of many muscles.
Polyneuritis Inflammation of many nerves.
Polyp A sessile benign or malignant tumor of a mucous membrane (usually of colon).
Polyposis Presence of many polyps.
Polyprotein Long polypeptide chain.
Pontiac fever A form of Legionnaire's disease.
Porin Protein of outer membrane pores of Gram-negative bacteria.
Portal venous system Veins carying blood from the intestinal tract to the liver.

Premenarcheal Prepubertal years in the female (before onset of menses).

Premunition Host-mounted immunologic response that limits parasite multiplication and moderates the clinical manifestations without eliminating the infection.

Prepuce Foreskin.

Prevalence Indicates the total number of cases existing in a population.

Primary response The result of an initial contact with a new antigen.

Prion Infectious agent composed only of proteins and appears to be responsible for some transmissible and inherited spongiform encephalopathies in animals and humans.

Pro- Before, a precursor.

Probe A short, labeled nucleic acid segment capable of detecting the same sequence in an unknown target.

Proctoscopy Use of an instrument to examine interior of the rectum.

Prodromal Initial symptoms before the characteristic manifestations of disease develop.

Proglottid One of the segments of the body of a tapeworm.

Prokaryote Organism lacking a true nucleus. Possesses a single chromosome.

Promoter DNA region at start of gene, which binds RNA polymerase and initiates transcription.

Prophage Complete bacterial virus genome integrated in the chromosome.

Prophylaxis Measures or treatments designed to prevent disease.

Prostaglandins Derivatives of arachidonic acid that mediate a variety of biologic reactions including inflammation.

Prostate gland Gland surrounding the male urethra that produces part of the seminal fluid.

Prosthesis Artificial replacement of a missing part of the body.

Proteasome Large protein complex that digests proteins to peptides.

Proteinuria Protein in the urine indicating a renal abnormality.

Prothrombin Precursor of thrombin; thrombin activates the terminal blood clotting mechanism.

Protomer Protein subunit of a viral capsomere.

Proto-oncogene Normal cell that possesses homologs of oncogene

Protoplasm The viscid colloidal solution that makes up living matter.

Protoplast A Gram-positive bacterium that has lost its cell wall.

Prototroph Bacterial strains with complete synthetic pathways from which auxotrophs may be derived.

Protozoan A unicellular member of the animal kingdom.

Proventriculus An enlargement of the alimentary tract of an invertebrate that precedes the stomach.

Provirus Complete viral genome integrated into a eukaryotic genome.

Pruritus Itching.

Pseudo- False.

Pseudohypha A structure that has recurring bud-like constrictions and less rigid cell walls than a hypha.

Pseudomembrane Membrane that consists of necrotic tissue, inflammatory cells, and bacteria.

Pseudopod A pseudopodium. Moving extrusion of the cytoplasm of an amoeboid cell that brings about movement or ingestion of food particles.

Psychrophile A microorganism that grows best or exclusively at low temperatures.

Puerperal Following childbirth.

Purpura Multiple hemorrhages in the skin, mucous membrane, or other organs.

Purulent meningitis Infections of the meninges associated with a marked, acute inflammatory exudate usually caused by a bacterial infection.

Pustule Pus in a vesicle, infected hair follicle, or sweat gland producing a visible inflammatory swelling.

Pyelonephritis Infection of the pelvis and tissues of the kidney.

Pylephlebitis Inflammation in the portal venous system.

Pyo- Producing pus.

Pyocyanin Blue pigment produced by *P aeruginosa*.

Pyoderma Impetigo.

Pyogenic Producing pus and pustular lesions.

Pyuria Pus in the urine.

Quaternary ammonium compounds (quats) Agents such as benzalkonium chloride that are highly bactericidal in the absence of contaminating organic matter.

Quinolones Class of antimicrobial agents that bind to bacteria DNA gyrase, inhibiting DNA replication.

Quorum sensing Process by which bacteria use signal molecules to monitor their population density.

Radioimmunoassay A method for detecting antigenantibody reactions that uses a radioisotope as a readily detectable label.

Rales Crackling respiratory sounds heard with the stethoscope.

Reactive oxygen intermediates Superoxide, hydrogen peroxide, singlet oxygen, produced by phagocytes.

Reading frame The way in which nucleotides in DNA and mRNA are grouped into codons of three for reading the message.

Reassortment Whole gene "swapping."

Receptor Component of the cell surface to which another substance or organism attaches specifically.

Receptor-mediated endocytosis A type of endocytosis triggered by the binding of a ligand on the pathogen to a receptor on a cell surface.

Recombination The combining of genetic material from different souces.

Reducing agents Chemicals placed in culture media to lower the oxidation-reduction potential.

Reduviid A large, winged "cone-nosed" insect.

Regulon A collection of genes or operons that is controlled by a common regulatory protein.

Replication The process by which an exact copy of parental or viral DNA is made using the parental molecule as a template.

Renal Pertaining to the kidney.

Replicative transposion Moving of a transposon to a new site while leaving a copy behind.

Repressor A regulatory protein that binds to an operator sequence and inhibits transcription of the adjacent gene(s).
Reservoir Natural habitat or source of an infecting organism.
Respiration Fueling metabolism whch utilizes oxygen.
Restriction enzymes Enzymes that cleave DNA at specific points.
Reticuloendothelial system System of phagocytic monocytes, particularly those in the spleen, bone marrow, and lymph nodes.
Retinoblastoma Malignant tumor of the retina.
Retrovirus RNA virus, the genome of which is transcribed into DNA by the enzyme reverse transcriptase.
Reverse transcriptase RNA-dependent DNA polymerase that uses a viral RNA genome as a template to synthesize a DNA copy.
Reverse transcription Use of a viral RNA genome to synthesize a DNA copy.
Reye's syndrome Encephalopathy with fatty infiltration of the viscera.
Rhino- Pertaining to the nose.
Rhinorrhea Continuous discharge of watery mucus from the nose.
Rhizopod Ameba.
Rhonchi Coarse snoring or rattling respiratory sounds heard with a stethoscope.
RIA *See* **Radioimmunoassay**.
Ribonucleic acid (RNA) A polynucleotide composed of ribonucleotides joined by phosphodiester bridges.
Ribosomal RNA (rRNA) RNA present in ribosomes including transfer RNA (tRNA) involved in protein synthesis.
Ribotyping The use of rRNA to probe chromosomes for typing.
RNA polymerase Enzyme that catalyzes the synthesis of mRNA under the direction of a DNA template.
Romana's sign Unilateral ophthalmia, edema of the eyelids, and enlarged draining lymph nodes.
Rostellum Portion of tapeworm head that contains hooklets or other attachment organs.
R plasmid (R factor) Plasmid containing antimicrobial drug resistance gene.

Salpingitis Inflammation of the fallopian tubes.
Saprophyte Organism living on dead organic material in the environment.
Sarcoidosis Disease of unknown etiology characterized by granulomatous lesions of many tissues and organs.
Sarcolemma Membrane surrounding muscle fibers.
Scaffolding proteins Constituents intermediate in the assembly of a bacteriophage head.
Schistosomiasis Infection with blood flukes.
Schizogony Asexual reproduction in sporozoa producing merozoites by multiple nuclear fusion followed by cytoplasmic segregation.
Schizont The multinucleated stage of a sporozoan undergoing schizogony.
Sclera White part of the eyeball.
Scolex The attachment organ or head of a tapeworm.
-scopy Denotes use of an instrument for visual examination of a hollow viscus (eg., bronchoscopy).
Scotoma A blind spot in the visual field.
Sebaceous Relating to sebum and sebum production.
Sebum Waxy secretion of sebaceous glands.
Secretory IgA A complex of two IgA molecules and the secretory piece.
Secretory piece Protein on the surface of local epithelial cells.
Selective media Culture media designed to suppress the growth of common organisms to allow isolation of a targeted pathogen.
Seminal vesicles Sacs in which semen is stored before ejaculation.
Septic shock Sepsis with progression to hypotension.
Sepsis A clinical term used synonymously with septicemia.
Sepsis syndrome Findings of sepsis in addition to altered perfusion.
Septicemia A clinical term indicating evidence of systemic disease associated with presence of organisms in the blood (*see* **Bacteremia**).
Sequelae Results occurring subsequent to an infection or other disease.
Sequestrum Necrotic bony fragment.
Seroconversion Development of antibodies in response to an infection.
Serodiagnosis Diagnosis of an infection by serologic procedures.
Serotonin Vasoconstricting amine usually derived from platelets
Serotype Subtype of species detectable with specific antisera.
Serpiginous Moving irregularly from one place to another, snake-like.
Serum Liquid part of blood separable after clotting.
Sex pilus Specialized structure involved in exchange of genetic material between some Gram-negative bacteria.
Shunt Deviation of blood or other body fluids (eg, from artery to vein) or a device designed to do so.
Sickle cell anemia Hereditary anemia associated with crescent-shaped erythrocytes resulting from an abnormal hemoglobin.
Siderophore A small molecule that binds iron and aids in its transport across membranes.
Sigmoid colon Lower portion of the colon between descending colon and rectum.
Sinus A tract leading from an infected area or hollow viscus to the surface; a wide venous blood channel; accessory nasal sinuses that are blind sacs draining to the nasopharynx.
Sinusoid A wide thin-walled venous passage. Smaller than a sinus.
Slim disease The severe intractable wasting and diarrhea of AIDS.
Slime layer Term sometimes used for polysaccharide surface components of bacteria that do not constitute a morphologic capsule.
Southern hybridization Method in which the DNA is separated by agarose gel electrophoresis before binding to the membrane.
Spasticity Excessive tone of muscles leading to awkward movement.

Spheroplast A circular, osmotically unstable, Gram-negative rod that has lost its peptidoglycan layer.
Sphincter Circular muscle controlling a natural orifice.
Splanchnic Pertaining to the viscera.
Spleno- Relating to the spleen.
Splicing Removal of internal sequences in the synthesis of eukaryotic mRNA.
Spore A specialized microbial form that facilitates survival and dissemination.
Sporogony Sexual reproduction process in sporozoan parasites leading to formation of oocysts and sporozoites.
Sporozoite Motile, elongated, infective stage of sporogony.
Sporulation One bacterial cell forms one spore under adverse conditions.
Sprue A chronic form of intestinal malabsorption.
Sputum Purulent material generated in the alveoli and small air passages.
Squamous epithelium Composed of layers of flattened cells.
Stable toxin (ST) A small peptide that binds to a glycoprotein receptor
Standard precautions Measures recommended by the Centers for Disease Control and Prevention, including use of gowns and gloves when in contact with patient blood or secretions.
Stasis Stagnation or cessation of flow of body fluids.
Stationary phase The period of time in the culture growth cycle in which growth stops.
Stenosis Reduction in diameter of a blood vessel or tubular organ.
Sterilization Complete killing, or removal, of all living organisms from a particular location or material.
Steroids Derivatives of cholesterol including hormones, some of which have anti-inflammatory effects.
Sterol Lipid-soluble steroid with long aliphatic side chains. Present in eukaryotic cell membranes as cholesterol or ergosterol.
Stevens-Johnson syndrome A serious allergic reaction, characterized by multiple blister-like lesions of skin and mucous membrane.
Stomatitis Inflammation of the mouth.
Strabismus Squint.
Stratum corneum Outer keratinized part of the skin.
Stridor Harsh respiratory sound due to partial respiratory obstruction.
Strobila Chain of segments making up the body of a tapeworm.
Sub- Below.
Subarachnoid Cerebrospinal fluid containing area between the middle (arachnoid) and inner (pia mater) layers of the meninges.
Subdural Between the outer (dura mater) and middle (arachnoid) layers of the meninges.
Submandibular Below the jaw.
Subphrenic Below the diaphragm.
Sulcus Groove.
Superantigen An antigen able to stimulate massive cytokine release by simultaneous interaction (without processing) with class II MHC and T-cell receptors.
Superoxide dismutase An enzyme found in all organisms that survive the presence of oxygen.
Suppurative Producing pus.
Supra- Above.
Surfactant A substance that acts on a surface to reduce surface tension (eg, a detergent).
Sylvatic Pertaining to the woods. Commonly applied to nonurban plague or arboviruses whether occurring in wooded or prairie land.
Symbiont An organism living on or in close association with another.
Synapse A connection between neurons for nerve impulse transmission.
Syncytium A multinucleate mass of fused cells.
Syndrome Group of clinical manifestations characterizing a particular disease or condition.
Synergistic Enhanced rather than additive effect of two agents or processes acting together.
Synovium Lining membrane of a joint, tendon, or bursa.

T cells Bone marrow–derived, thymus-matured lymphocytes; involved in a variety of cell-mediated immune reactions, eg, helper, suppressor, and cytotoxic.
T-dependent antigen Antigen that incorporates T cells in the process of activating B cells.
T_H1 cell CD4+ lymphocyte that initiates cell-mediated immune responses.
T_H2 cell CD4+ lymphocyte that initiates antibody-mediated immune responses.
T-independent antigen Antigen that directly stimulates B cells without involvement of T cells.
Tabes dorsalis A syndrome in tertiary syphilis that produces ataxia, wide-based gait, foot slap, and loss of the sensation.
Tachy- Increased rate, swift.
Tachypnea Abnormally rapid rate of breathing.
Talin One of the proteins that connects integrins to the actin cytoskeleton of eukaryotic cells.
Tamponade (cardiac) Increased fluid or constriction around the heart leading to interference in cardiac function.
Tegument The protein-filled region that fills the space between capsid and envelope.
Teichoic acid A major component of the Gram-positive cell wall.
Tenesmus Ineffective and painful straining at stool or urination.
Tenosynovitis Inflammation of a tendon sheath.
Teratogenic Causing abnormalities of fetal development.
Terminator One of the components of the structure of a typical operon.
Tetanospasmin A neurotoxic exotoxin, a product of *Clostridium tetani.* Also called tetanus toxin.
Thalassemia Hereditary hemolytic anemia resulting from abnormal hemoglobin synthesis.
Thermo- Pertaining to heat.
Thermophile Bacteria with an optimal growth temperature of over 50°C.
Thrombo- Pertaining to thrombosis.
Thrombocyte *See* **Platelet**.
Thrombocytopenia Abnormally low platelet count.
Thrombophlebitis Inflammation of a vein with thrombosis; may release infected emboli.

Thrombus A blood clot developing in vivo.

Thrush Oral white fungal patches on the tongue, palate, and other mucosal surfaces.

Thymus A lymphoid organ located in the anterior upper mediastinum, which is required for early development of immune functions and the maturation of T cells.

Tinea nigra Skin infection characterized by brown to black macular lesions, usually on the palms or soles and occurring in tropical climates.

Tinea pedis Infection involving scaling and splitting of the skin between the toes. Also called athlete's foot.

Titer Highest dilution of an active substance (eg, antibody in serum) that still causes a discernible reaction (eg, an agglutination reaction).

Toll-like receptor A pattern recognition receptor on the surface of phagocytes that triggers responses to pathogens.

Tonsillitis Inflammation of the tonsils.

Toxic shock syndrome Potentially fatal illness caused by staphylococcal or streptococcal toxins.

Tracheo- Pertaining to the trachea.

Tracheostomy Surgically produced artificial air passage to the trachea.

Trans- Across.

Transcriptase DNA-directed RNA polymerase.

Transcription The process by which single-stranded RNA with a base sequence complementary to the template strand of DNA or RNA is synthesized.

Transduction The transfer of genes to bacteria by bacteriophages.

Transfer RNA (tRNA) Small RNA that binds an amino acid and delivers it to the ribosome for incorporation into a protein.

Transferrin Serum protein that binds and transports iron.

Transformation A process whereby bacteria can acquire genes by direct uptake of free DNA.

Transforming retrovirus Transforming virus that carries cellular genes.

Translation The process by which the genetic message carried by mRNA directs protein synthesis.

Transovarial Passage of infectious agents to progeny by way of the egg. Usually occurs in ticks and mites.

Transposition The movement of genes by transposons.

Transposon A DNA segment containing insertion sequences able to mediate movement between plasmids and chromosome; may also contain one or more recognizable genes.

Trematode Fluke.

Trimester A 3-month period of pregnancy in humans.

Trismus Spasm of the masseter muscle; lockjaw.

Trophozoite The motile feeding stage of a protozoan parasite.

Tropism Having an affinity for a particular organ, or moving toward or away from a particular stimulus.

Tubulin Protein subunit of microtubules.

Tumor necrosis factors Cytokines that play an important role in inflammation and other aspects of immunity.

Tumorigenesis The property of causing tumors.

Turgor pressure Osmotic pressure of the cellular contents.

Tympanic membrane Eardrum.

Type-specific immunity Protection from subsequent infection with strains of the same M type of streptococci, for example.

Ultrasonogram Picture of deep organs of the body derived from reflection of ultrasonic waves.

Ultraviolet (UV) light Has the wavelength range of 240 to 280 nm, is absorbed by nucleic acids, and causes genetic damage.

Uremia Toxic accumulation of nitrogenous metabolites due to renal insufficiency.

Ureter Tube carrying urine from the kidney to bladder.

Urethra Tube carrying urine from the bladder to the exterior.

-uria Pertaining to urine.

Uropathic Causing disease of the urinary tract.

Urticaria Local edema and itching of the skin.

Uvea Inner vascular coat of the eyeball, including the iris.

Uvula Small extension hanging from the back of the soft palate.

Vacuolate Forming small holes or vacuoles.

Vacuole Microscopic hole or cavity.

Vagotomy Surgical cutting of the vagus nerve.

Valley fever Usually self-limiting fever, malaise, dry cough, joint pains, and sometimes a rash; caused by *Coccidioides immitis* infection.

Varicella Chickenpox.

Vasa vasorum Small blood vessels in walls of veins and arteries.

Vasculitis Inflammation of blood vessels.

Vaso- Pertaining to blood vessels.

Vector An animate transmitter of disease (eg, an insect). In molecular biology a genetically engineered molecule able to transport foreign DNA.

Venipuncture Insertion of a hypodermic needle into a vein—usually to draw blood.

Ventricle Fluid cavity (eg, chamber of the heart).

Vertical transmission Spread of infection from mother to fetus.

Vesicle Small fluid-filled cavity (eg, a blister-like lesion of the skin).

Vesicoureteral junction Junction of ureter with the urinary bladder.

Vestibular function Function of the vestibular branch of the eighth cranial nerve concerned with the body's equilibrium.

Vinculin One of the proteins that connects integrins to the actin cytoskeleton of eukaryotic cells.

Viremia Presence of a virus in the bloodstream.

Virion A complete virus particle.

Viroid Infectious circular RNA molecule that lacks protein shell.

Virokine Protein secreted from infected cells that acts as a cytokine, helping cells to proliferate and increase virus production.

Viropexis Viral entry into the cell by endocytosis.

Virulence A term expressing degrees of pathogenicity.

Viruria Viruses in the urine.
Viscera Interior organs of the body (eg, the intestinal tract).
Vitreous humor The clear viscous fluid in the posterior chamber of the eye.
Vitronectin Protein component of extracellular matrix.
Viviparous Developing young within the body as opposed to outside the body (oviparous).
von Magnus phenomenon A phenomenon in which noninfectious RNA or DNA viral genomes interfere with replication of the infectious virus. *See also* **Defective interfering particles**.
VPg Viral protein genome.

Western blot Test for antibodies to specific proteins separated by gel electrophoresis.
Whitlow Abscess of the terminal pulp of the finger. Also paronychia.
Wild-type allele Normal, usually active, form of a gene.
Wright's stain Stain for blood cells that has properties similar to those of Giemsa stain.

Xenodiagnosis Recovery of a parasite by allowing an arthropod to feed on the patient and seeking the parasite in the arthropod.
Xerostomia Dry mouth from dysfunction of the salivary glands.

Yeast Simple fungal cell which reproduces by budding.

Zoonotic infection A disease transmissible to humans from an animal host or reservoir.
Zygote The cell that results from fusion of male and female gamete.
Zymodeme An isoenzyme typing pattern.

INDEX

Page numbers followed by italic *f* or *t* denote figures or tables, respectively. Page numbers in **boldface** refer to major discussions.

A
A subunits, 473
AAV. *See* adeno-associated virus
A-B exotoxins, **396**, 620*f*
abdominal abscess, 519*f*
abdominal actinomycosis, 509-510
abortion, 522
abortive infection, 110, 143
abortive poliomyelitis, 219
abscesses, causes of, 519
Absidia, 732-733
AC. *See* adenylate cyclase
acanthamebic granulomatous encephalitis, 810*f*
Acanthamoeba, 803, **809-810**, 906*t*
  epidemiology of, 810
accessory proteins, HIV, 311-312
acetylcholine, 523, 525
acid-fast bacilli (AFB), 499
  culture, 499
acid-fast smears
  culture, 499
  tuberculosis, 498-499
acid-fast stain, **63-64**, 64*f*
  Cryptosporidium parvum, 800
  for *Mycobacterium*, 490
acidogenesis, 915
Acinetobacter, 56, **624**, 625*t*
  bacteremia from extravascular infection and, 966*t*
acne vulgaris, **895**, 899*t*
acquired immunity, development, 31*f*
acquired immunodeficiency syndrome (AIDS), 3, 96, 116, 305. *See also* human immunodeficiency virus
  Bartonella and, 689
  Candida albicans, 727
  CD4+ helper T lymphocytes, 317
  CDC on, 316
  clinical capsule, 312
  CMV and, 263
  coccidioidomycosis and, 752
  Cryptococcus neoformans and, 742
  dementia complex, **338**
  diagnosis, 319-320
  EBV in, **266**
  epidemiology of, 14, 313-315
  Haemophilus ducreyi and, 558
  hepatitis G in, 244
  herpes simplex virus in, 256
acquired immunodeficiency syndrome (AIDS) (*Cont.*):
  immune deficiency in, 316
  listeriosis in, 480, 481
  molluscum contagiosum, 210*f*
  mortality rates, 314
  Mycobacterium avium-intracellulare in, 504
  Mycobacterium in, 489
  number of cases, 317*f*
  occurrence, 314
  opportunistic infections in, 318*t*
  parvovirus B19 and, 202
  Pneumocystis and, 733, 735
  pneumocystosis in, 736
  prevention, 321
  reactivation tuberculosis and, 497-498
  survival of, 319
  toxoplasmosis and, 774, 795-796
  treatment, 160
  tuberculosis, 495
  varicella-zoster virus in, 258
acquired resistance, **423**, 424*f*
acridine orange stains, 788
actin filaments, 392
Actinobacillus, 626*t*
Actinobacillus actinomycetemcomitans, 917
Actinomyces, **507-511**, 911, 915
  bacteriology, 507
  chronic, 508
  features of, 508*t*
  lower respiratory tract infections from, 926*t*
Actinomyces israelii, 507
actinomycosis, 507-508, 509*f*
  abdominal, 509-510
  ampicillin for, 510-511
  cervicofacial, 510*f*
  cervicofacial forms, 509
  clinical aspects, 508-511
  clinical capsule, 507-508
  diagnosis, 510
  doxycycline for, 510-511
  erythromycin for, 510-511
  manifestations, 508-510
  penicillin for, 510-511
  surgery and, 509
  thoracic, 509-510
  treatment, 510-511
activator proteins, 371
active immunity, hepatitis A, **229**
active transport, 359, 360*f*
acute bacterial conjunctivitis, 906
acute bronchitis, 923-924
acute endocarditis, 961
acute epiglottitis, *Haemophilus influenzae*, 555
acute glomerulonephritis, 453
acute hematogenous osteomyelitis, 901-902
acute herpes simplex viruses, 251
acute inflammation, 26
acute juvenile periodontitis, 917
acute necrotizing periodontal disease, 918
acute otitis media, 907, 909
acute pneumonia, **924-925**, 926*t*
acute polyneuritis, 955
acute pyelonephritis, 941
acute respiratory disease (ARD), 167
acute retroviral syndrome, 145
acute rheumatic fever (ARF), 450, 452-453, 455-456
acute sinusitis, 910, 910*t*
acute transforming viruses, 148
acute viral gastroenteritis, 272
acute viral hepatitis, 227*f*
acycloguanosine, 123
acyclovir, 158*t*, **159**
  pharmacology, **160**
  prophylaxis with, **160**
  toxicity, **160**
  treatment with, **160**
  for varicella-zoster virus, 259
adaptive immunity, 19, 30-40, 150
  antibodies in, 36-40
  antigens in, 32-33
  B cells in, 36-40
  epitopes in, 32-33
  fungi and, 705-706
  hepatitis C, 241
  memory of, 30
  superantigens in, 35
  T-cell response, 33-36
  viral infections, **152**
ADCC. *See* antibody directed cell-mediated cytotoxicity
Addison's disease, 745
adenine arabinoside, 122
adeno-associated virus (AAV), 142
adenocarcinoma, 575

adenovirus, 106, 129*f*, **182-185**, **277**
classification of, 111*t*
clinical syndromes associated with, 184*t*
death protein, 184
diagnosis, **184**
enteric, 272
epidemiology, **182-183**
eye infection from, 906, 906*t*
immunity, **184**
lower respiratory tract infections from, 926*t*
manifestations, **184**
pathogenesis, **183-184**
prevention, **185**
receptors, 114*t*
treatment, **185**
upper respiratory infection from, 920*t*
virology, **182**
adenylate cyclase (AC), 559, 567*f*, 568
adherence
bacteria, **389-392**
Candida albicans, 704
fungi, 704
Salmonella enterica, 600
adherence factors, 9
adhesin, 389, 559, 561
surface, 747
ADPR. *See* ADP-ribosylation
ADP-ribosylation (ADPR), 16*f*, 363*f*, 396, 473, 567*f*, 618
adrenal glands, 682, 745
adsorption, in virus replication, **113-115**
adult T-cell leukemia, 149, 305, **322-324**
pathogenesis, 323
adventurous tourism, 890
A/E lesion. *See* attachment and effacing lesion
Aedes aegypti, 290
Aedes triseriatus, 290
aerial mycelium, 695
aerobes, **361-362**, 362*t*
aerobic atmospheric conditions, 69
Aerobic Gram-positive bacteria, features, 471*t*
Aeromonas, **624**, 626*t*
tetracycline for, 624
aerosol spread, of rabies, 299
aerotolerance, 515
AFB. *See* acid-fast bacilli
African Burkitt lymphoma, 265, 267
African trypanosomiasis, **827-830**
anemia and, 828
antigenic shifts, 828
clinical aspects, 828-830
clinical capsule, 827
diagnosis, 829
epidemiology, 828
IgM and, 828
melarsoprol for, 829
parasitology, 827
pathogenesis, 828
prevalence of, 760*t*
prevention, 830
treatment, 829
vasculitis and, 828
agar, 67, 86
agarose, 80
agarose gel electrophoresis, **80**
agglutination, 74*f*, 563
agglutinins, cold, 668
agranulocytes, 20*f*
AIDS. *See* acquired immunodeficiency syndrome
airborne transmission precautions, 57
airway obstruction, 922-923
alanine aminotransferase (ALT), 228*f*
alanyl alanine, 421
alastrim, 207
albendazole, 776
for echinococcosis, 876
for fluke infections, 885
for pork tapeworm, 872
for trichinosis, 856
alcohol
disinfection with, 51
for sterilization, 47*t*
alexidine, 913
alginate biofilm, 621*f*
alkaline foods, 524
alloantigens, 453
allopurinol, for American trypanosomiasis, 833
allosteric enzymes, 370
allosteric regulation, 369*f*
allylamines, **709-710**, 718
features of, 708*t*
Alphavirus, 279
virion structure of, 281*f*
ALT. *See* alanine aminotransferase
altered targets, **419-421**, 439
alveolar macrophage, 494*f*
amantadine, 157, 158*t*, 176, 177
pharmacology, 159
toxicity, 159
amastigote, 821
Trypanosoma cruzi, 831*f*
Amblyomma americanum, 684
amebiasis, 805*f*, **806-807**
clinical aspects, 807-809
cutaneous, 897
diagnosis, 808-809
diarrhea in, 808
epidemiology, 806
extraintestinal, 808
hepatic abscess in, 808
immunity, 807
manifestations, 807-808
pathogenesis, 806-807
pathology, 807
prevalence of, 760*t*
prevention, 809
treatment, 809
virulence, 806
American Academy of Pediatrics, 456
American trypanosomiasis, 767, **830-833**
allopurinol for, 833
benznidazole for, 833
chronic, 832
clinical aspects, 832-833
clinical capsule, 830
diagnosis, 832-833
epidemiology, 830-831
manifestations, 832
myocarditis in, 831*f*
nifurtimox for, 833
pathogenesis, 831-832
prevalence of, 760*t*
prevention, 833
treatment, 833
amikacin, 410
for *Nocardia*, 513
for *Pseudomonas aeruginosa*, 622
susceptibility patterns to, 427*t*
amines, 176
amino nitrogen, 362
4-aminoquinolines, 773
aminoglycosides, 405*t*, **410-411**
for brucellosis, 633
for enterococcal disease, 466
for group B streptococci, 459
resistance, 419*t*, 467, 622
for tularemia, 640
ammonium, 4
amoebas, 8
amoxicillin, 405*t*
for *Helicobacter pylori*, 576-577
for Lyme disease, 661
amphotericin B, 707, 709, 721
action of, 709*f*
for aspergillosis, 732
for blastomycosis, 748
for *Candida albicans*, 729
for coccidioidomycosis, 753
for *Cryptococcus neoformans*, 742
features of, 708*t*
for histoplasmosis, 746
for *Paracoccidioides brasiliensis*, 753
ampicillin, 405*t*, 408, 422, 425, 529
for actinomycosis, 510-511
for *Bacteroides fragilis*, 532
for enterococcal disease, 467
for group B streptococci, 459
for listeriosis, 481
resistance, 557, 942-943
for salmonellosis, 604
for shigellosis, 598
susceptibility patterns to, 427*t*
anaerobes, **361-362**, 362*t*
bacteriology, 515-518
capsules, 518
classification of, 516-518
facultative, 806
flora, 518
group characteristics, **515-520**
opportunistic, 516*t*
pathogenic, 517*t*
virulence of, 518-519
wound infections from, 899*t*
anaerobic atmospheric conditions, 69
anaerobic cellulitis, 522
anaerobic cocci, 516
anaerobic diphtheroids, 510
anaerobic endometritis, 522
anaerobic Gram-negative wound infections, 900
anaerobic incubation jar, 520
anaerobic infections
aztreonam for, 520
ceftriaxone for, 520
cephalosporins for, 520
clinical aspects, 519-520
diagnosis, 519-520
epidemiology, 518
group characteristics, **515-520**
manifestations, 519
metronidazole for, 520
mixed, 520
pathogenesis, 518-519
treatment, 520
anaerobic media, 87
anaerobiosis, 515-516
catalase and, 516
hydrogen peroxide in, 516
superoxide in, 516
anal neoplasia, 329
anamnestic response, 40

anaphylaxis, 97
Anaplasma, 681
Anaplasma phagocytophila, 683*t*, 688
Ancylostoma braziliense, 851
  general characteristics of, 852*t*
Ancylostoma duodenale, 835
  life cycles of, 845
  parasitology, 844
Andrews, Christopher, 185
anemia
  African trypanosomiasis and, 828
  fetal, 202
  iron deficiency, 846
  in malaria, 785-786
aneurysm, mycotic, **963**
Angiostrongylus cantonensis, 954*t*
anicteric hepatitis A, 228
anidulafungin, 710
  features of, 708*t*
animal anthrax, 485
animal bites, 899*t*
animal rotaviruses, 274
animal viruses, 101. *See also specific viruses*
anisomycin, for gonorrhea, 548
Anopheles, 765, 781, 859
anopheline mosquitoes, 784
anorexia, 228
anthrax, 426, 471, 482*f*, **483-485**
  animal, 485
  bioterrorism and, 483
  clinical aspects, 484-485
  clinical cases, 486
  contaminated materials, 483
  diagnosis, 485
  doxycycline for, 485
  edema factor in, 484
  epidemiology, 14, 483
  immunity, 484
  malignant pustule, 484
  manifestations, 484-485
  overview, 484*f*
  pathogenesis, 483-484
  penicillin for, 485
  prevention, 485
  pulmonary, 484-485, 485
  treatment, 485
  vaccines, 483, 485
  weapons-grade, 483
Anthroderma benhamiae, 698
antibacterial agents, **403-427**. *See also specific agents*
  cell wall synthesis and, **406-410**
  empiric therapy, **426**
  prophylaxis, **426**
  specific therapy, **426**
  spectrum of, **404**
antibiotics, 12. *See also specific drugs*
  to antibiotics, 17
  definition, 403
  for plaque inhibition, 913
  for *Streptococcus pneumoniae*, 465
  Streptomyces and, 404
antibodies, 16, 152
  in adaptive immunity, 36-40
  anamnestic response, 40
  anticapsular, 459, 538
  antihemagglutinin, 169, 174
  arboviruses, 288*f*
  complement-fixing, 667
  detection, **78-79**
antibodies (*Cont.*):
  diphtheria toxin and, 474
  direct fluorescent, 736
  EBV, 268*t*
  fungal, **701-702**
  HBsAg, 234
  hepatitis B, 235*t*
  hepatitis C, 241
  Histoplasma capsulatum, 746
  kinetics, 41*f*
  measles, 194
  monoclonal, 74, 255
  primary response, 40
  production, 40, 41*f*
  response, 78*f*
  in rubella, 198*f*
  secondary response, 40
  structure, 38-39
  to surface proteins, 214
  in varicella-zoster virus, 258
antibody detection, 936
antibody directed cell-mediated cytotoxicity (ADCC), 144, 252-253, 316
antibody-mediated hypersensitivity, 42
anticapsular antibodies, 459, 538
antifungals, 17, **707-712**
  action of, 709*f*
  cell wall synthesis, 710
  cytoplasmic membrane, 707-709
  features of, 708*t*
  nucleic acid synthesis, 710
  resistance, **711**
  selection of, 712
antigen binding sites (Fab), 38
antigen detection, 936
antigen-antibody reaction
  complement fixation, **76**
  detecting, **74-77**
  labeling methods, **76-77**
  neutralization, **75-76**
antigenic drift, 130, 169
  in influenza, 172*f*
  in influenza A, 171
  minor, 172
  of retroviruses, 131
antigenic masking, 769
antigenic shedding, 769
antigenic shift, 169
  African trypanosomiasis, 828
  in influenza, 172*f*
  in influenza A, 171-172
  major, 173*t*
  in parasites, 768
antigenic shifts, 132
antigenic structure
  bacteria, 71
  Escherichia coli, 583*f*
  fungi, 71
  group A streptococci, 447*f*
antigenic variation, **377**, 395*f*
  of bacteria, **394**
  in gonorrhea, 544
  malaria, 787
  Neisseria gonorrhoeae, 542
  in parasites, 768
antigens, 19, 28
  in adaptive immunity, 32
  detection, 79
  Duffy blood group, 783
  early, 264
antigens (*Cont.*):
  EBV nuclear, 264
  fungal, **701-702**
  group A, 456
  H, 580
  hepatitis B, 235*t*
  Histoplasma capsulatum, 744
  invasion plasmid, 597
  K, 579
  Lancefield, 444
  O, 579, 595
  presentation, 34*f*
  processing, 34*f*, 38
  protective, 482
  T-dependent reactions, 37
  T-independent reactions, 37-38
  Vi, 602
antigenuria, 614
antihemagglutinin antibodies, 169, 174
antihistamines, 889
antimalarial quinolines, **773-774**
antimannan IgG, 727
antimicrobial. *See also specific agents*
  automated tests, **417**
  bactericidal testing, **418**
  definition, 403
  diffusion tests, **416-417**
  dilution tests, **416**
  glycopeptide, 409-410
  laboratory control of, **416-418**
  molecular testing, **417-418**
  outer membrane acting, 415
  peptidoglycan synthesis and, 407*f*
  resistance, **415-415**
    features of, 419*t*
    mechanisms, **418-423**, 420*f*
    Staphylococcus aureus, 439
  sources of, **404**
  susceptibility patterns to, 427*t*
antiparasitic antimicrobics, **771-777**
  action, 772
  for helminths, 772
  resistance, 777
  structure, 772
  therapeutic goals of, 771-772
  toxicity of, 771, 772
antiphagocytic activity, **394**
antiphagocytic capsules, 458
antiprotozoals, 17
antisepsis, 56
antiseptics, 45-46
antistreptolysin O (ASO), 448
antivirals, 17, 150, **157-165**
  for influenza, 176*t*
  interferon as, 29*f*
  resistance, **164-165**
  selected, **157-161**
  summary of, 158*t*
aphthous stomatitis, 922
aplastic crisis, 202
APOBEC3G, 151, 312
Apodemus, 293
apoptosis, 127, 174, 393
  induction of, **394-395**
  macrophage, 600
apparent infection, 144
apparent responses, 144
appendages, 347
  bacteria, **348-356**, 350*t*
arachidonic acid, 26

arboviruses, **285**
antibody response, 288*f*
arthropod-sustained cycle, **287**
clinical aspects, **291-292**
CNS infections and, 952*t*
diagnosis, **291**
epidemiology, **285-287**
immunity, **288**
pathogenesis, 287-288
prevention, **291-292**
selected, 280*t*
sylvatic cycle, **286-287**
transmission of, 285, 287*f*
treatment, **291-292**
urban cycle, **286**
vertical transmission, 285
Archaebacteria, 7
ARD. *See* acute respiratory disease
arenaviruses, 120, 153*t*, **282-283**, **292-293**
classification of, 110*t*
with hemorrhagic fevers, **292-293**
virion structure, 283*f*
ARF. *See* acute rheumatic fever
Aristotle, 297
arsenical compounds, 403
Artemisia annua, 775
artemisinin, 775
for malaria, 789*t*
arterial culture, for endocarditis, 729
arthralgia, 200, 668
arthritis, 198, 200
fluctuating, 660
Haemophilus influenzae, 556
septic, 519, 901, **903-904**
diagnosis of, **903-904**, 904*t*
etiologic agents, 903, 903*t*
management of, **904**
osteomyelitis and, 901
synovial fluid findings and, 904*t*
arthrocentesis, 903
arthroconidia, 697, 704
Coccidioides immitis, 749
spread, 750
arthropod-borne viruses, **279-295**
arthropod-sustained cycle, **287**
artificial transformation, 380
ascariasis, **842-844**
clinical aspects, 843-844
diagnosis, 844
epidemiology, 842-843
immunity, 843
intestinal obstruction, 843*f*
iron deficiency anemia from, 846
malabsorption and, 843
manifestations, 843
mortality rate, 843
pathogenesis, 843
prevalence of, 760*t*
prevention, 844
treatment, 844
Ascaris lumbricoides, 764, 835, 836*t*, **841-844**
disease caused by. *See* ascariasis
distribution of, 764*t*
eggs of, 841
fertile, 842*f*
infertile, 842*f*
life cycles of, 836*t*, 841
parasitology, 841
transmission of, 764*t*
Aschoff nodule, 453
Ascomycota, 698
ascospores, 697
asepsis
definition of, 46
in hospital ward, 57
in infection control, **56-57**
in operating room, 56-57
in outpatient clinic, 57
aseptic meningitis, 219, 221, 946*t*
Asian flu, 173
ASO. *See* antistreptolysin O
aspergillosis, **731-732**
amphotericin B for, 732
caspofungin for, 732
clinical capsule, 731
diagnosis, 732
epidemiology, 731
immunity, 731
itraconazole for, 732
manifestations, 732
pathogenesis, 731
treatment, 732
voriconazole for, 732
Aspergillus, 705, 723, **730-732**
classification of, 699*t*
conidia, 730, 731
conidiophore, 730
ear infections from, 908*t*
eye infection from, 906*t*
gliotoxin, 731
lower respiratory tract infections from, 926*t*
sinus infections from, 909
Aspergillus flavus, 730
Aspergillus fumigatus, 730
Aspergillus niger, 730
aspiration pneumonia, 519
aspiration treatment, for echinococcosis, 876
assassin bug, 831
asthma, 12, 41
respiratory syncytial virus and, 181
astrovirus, **277**
ataxia, 340
atelectasis, 563
athlete's foot, 717
atmospheric conditions, **69**
aerobic, 69
anaerobic, 69
atovaquone, 774
resistance, 777
attachment
enteropathogenic *Escherichia coli*, 591*f*
gonorrhea, 544-545
inhibitors, **157**
in virus replication, **113-115**
attachment and effacing (A/E) lesion, 590
attL, 134, 135
attR, 134, 135
atypical lymphocytosis, 267
Auden, W.H., 8
AUG initiation, 120
autoclave, 48-49, 526
downward displacement, 48-49, 48*f*
flash, 49
for sterilization, 47*t*
autolysins, in group B streptococci, 461
automated tests, **417**
avermectins, **776**
avian influenza virus (H5N1), 143, 171
azidothymidine, **162**
azithromycin, 405*t*, 412
for *Haemophilus influenzae*, 558
resistance, 549
for shigellosis, 598
for *Ureaplasma*, 669
azoles, **709**, 718
action of, 709*f*
for blastomycosis, 748
for *Candida albicans*, 729
for coccidioidomycosis, 753
features of, 708*t*
for *Paracoccidioides brasiliensis*, 753
resistance, 711
azotemia, 455
aztreonam, 405*t*, 409
for anaerobic infections, 520
for *Bacteroides fragilis*, 532
susceptibility patterns to, 427*t*

B

B cells, 20*f*, 24*f*, 30*t*
in adaptive immunity, 36-40
T-dependent reactions, 37
T-independent reactions, 37-38
B lymphocytes, 453
B subunits, 473
bacilli, 347
Bacillus, **481-486**
Bacillus anthracis, 471, **482-485**, 630*t*, 924
bacteriology, 482-484
ciprofloxacin, 485
ciprofloxacin for, 426
doxycycline for, 485
growth of, 485
penicillin for, 485
ulcers from, 899*t*
Bacillus Calmette-Guérin, 492*t*
vaccine, 498, 500
Bacillus cereus, 933*t*-934*t*
bacitracin, 405*t*, 440, 456, 456*t*
bacteremia, 636, 959
from extravascular infection, **965-966**, 966*t*
intravenous catheter, **964-965**, 965*f*
patterns, 965*f*
recurrent, 642
Salmonella and, 889
salmonellosis, 603
sepsis and, 959, **966-967**
transient, 9
bacteria, 6*f*, 7
adaptation, **369-371**
adherence, **389-392**
aerobes, **361-362**, 362*t*
aerobic Gram-positive, 471*t*
anaerobes, **361-362**, 362*t*
antigenic structure, 71
antigenic variation, **394**
antiphagocytic activity, **394**
appendages, 347, **348-356**, 350*t*
capsule, **348-349**, 350*f*
cell division, **366-368**
cell membrane, **355**
cell stress regulons, 371
cell survival, **371-374**
cell wall, **349**
chemotaxis, **373-374**
chromosomes, 82
classifying, 70-71, 362*t*, **384-385**
taxonomic methods, **385**
colonial morphology, 68*f*

bacteria (*Cont.*):
components of, 350*t*
conjugation, **382-385**, 382*f*
core, 350*t*, **356-357**
cultural characteristics, 70
culture, **66-71**
cytosol, 347
DNA replication in, 364*f*
endocarditis, 960, 960*f*
endospore, 350*t*
entry of, **389**
envelope, 347, **348-356**, 350*t*
facultative, 362
flagella, **355-356**
fueling reactions, 359-361
gene expression, **371-372**
genetic exchange, **378-384**
genetics, 374-384
genomes, 81-82
genomic structure, 71
Gram-negative, 61-62, 349, 365, 366, **517-518**
conjugation in, **383-384**
ear infections from, 908*t*, 909
eye infection from, 906*t*
osteomyelitis from, 902*t*
plaque colonized by, 911-912, 916
protein secretion in, 365, 366
septic arthritis from, 903
sinus infections from, 910*t*
Gram-positive, 61-62, 349, 365, 485
aerobic, features, 471*t*
conjugation in, **384**
ear infections from, 908*t*
nonsporulating, **517**
as plaque colonizer, 911
protein secretion in, 365
sinus infections from, 910*t*
in sputum, 485
growth, **358-374**
curve, 369*f*
decelerating phase, 368
exponential phase, 368
lag period, 368
logarithmic phase, 368
stationary phase, 368
humans and, **388**
immune system and, 393-396
injury, **395-398**
invasion, **392-393**, 393*f*
media for isolation of, 86
metabolism, **358-374**
simplicity, 358
speed, 358
uniqueness, 358
versatility, 358
microaerophilic, 362
motility, **373-374**
mutations, **374-375**
nature of, **347-385**
nonspecific defenses against, 390*t*
nucleoid, 347, **357**
pathogenicity, 70-71
persistence of, 393-398
plasma membrane, 355
plasmids, **357**
polymerization reactions, 363-365
prokaryotic cells, 347, 349*f*
protein secretion, 365-366
recombination, **376-377**
regulation, **369-371**

bacteria (*Cont.*):
regulatory proteins, 372*f*
replication of, 7
shapes, 348*f*
spores, **357**
formation, 358*f*
stationary phase cells, **373**
strategies for survival, 392
structure, **347-357**
toxin production, 70-71
transcription in, 365*f*
translation in, 365*f*
transposition in, **377-378**
bacterial infections, **387-401**
attributes of, **388-398**
dose required, 391*t*
endotoxins in, **398**
exotoxins in, **396-398**
inflammation and, **398**
injury, **395-398**
misdirected immune responses and, **398**
pathogenicity, 398-401
clonality, 399-401
islands, 399
plasmids, 399
virulence gene regulation, 399
bacterial tracheitis, 923
bacterial vaginosis (BV), 948
bactericidal, definition, 404
bacteriologic plate streaking, 67*f*
bacteriophages, 5, 101, 129*f*, 380
assembly of, 126*f*
classification of, 111*t*
entry, 116*f*
release, **126**
strategy of, **115-116**
bacteriostatic, 404
bacteriuria, 939
Bacteroides fragilis, 466, 517*t*, 519, **531-532**
ampicillin for, 532
aztreonam for, 532
ceftriaxone for, 532
clavulanate for, 532
clindamycin for, 532
clinical aspects, 532
clinical capsule, 531
disease, 531-532
epidemiology, 531-532
immunity, 531
pathogenesis, 531-532
manifestations, 532
metronidazole for, 532
sulbactam for, 532
ticarcillin for, 532
treatment, 532
Bacteroides spp., 11, 531, 948
suppurative thrombophlebitis and, 964*t*
bactoprenol, 365, 366*f*
BAD1, 747
BAL. *See* bronchoalveolar lavage
Baltimore, David, 305
B-aminoquinolines, 773
Barré-Sinoussi, Françoise, 305-306
bartholinitis, 946*t*
Bartonella, **688-689**
AIDS and, 689
doxycycline for, 689
erythromycin for, 689
lymphadenitis, 689
Bartonella bacilliformis, 683*t*, 689

Bartonella henselae, 683*t*, 689
Bartonella quintana, 683*t*, 688-689
Basidiomycota, 698
basophils, 23
Baylisascaris procyonis, **853**
general characteristics of, 852*t*
B-cell lymphomas, 265
BCYE. *See* buffered charcoal yeast extract
bDNA. *See* branched-chain DNA
bear, 855
beef tapeworm, **867-869**
clinical aspects, 867
diagnosis, 869
disease, 867-869
epidemiology, 867
manifestations, 867
prevention, 869
treatment, 869
Bejel, 663
benzimidazoles, **775-776**
benznidazole, 777*t*
for American trypanosomiasis, 833
benzodiazepines, for tetanus, 528
benzyl penicillin, 404
susceptibility patterns to, 427*t*
biased random walks, 373
bicarbonate, 569
biochemical characteristics
bacteria, 70
fungi, 70
biofilm, 440, 620
alginate, 621*f*
Listeria monocytogenes, 478
biofilm, plaque, 911, 912*f*, 915
biopsy, of synovial membrane, 904
biosynthesis, 362-363
folate, 406*t*
bioterrorism, anthrax and, 483
bis-biguanides, 913
bismuth salts, for *Helicobacter pylori*, 576-577
bithionol, 777*t*
therapy, 882
BK virus (BKV), 333, 335
BKV. *See* BK virus
Black Death. *See* plague
black eschar, 484
black piedra, 716*f*, 719
bladder, 11
catheterization, 939
blastoconidia, 695
Histoplasma capsulatum, 743
Blastomyces dermatitidis, 710, **746-747**
disease caused by. *See* blastomycosis
features of, 740*t*
lower respiratory tract infections from, 926*t*
Blastomyces spp., **746-747**
classification of, 699*t*
blastomycosis, **747-748**
amphotericin B for, 748
azoles for, 748
clinical aspects, 748
clinical capsule, 747
diagnosis, 748
epidemiology, 747
fluconazole for, 748
immunity, 747
manifestations, 748
pathogenesis, 747
pulmonary, 748
treatment, 748

blepharitis, 905, 906*t*
blindness, 254
  Chlamydia trachomatis and, 675
blinking, 905
blood
  in blood-brain barrier, 951
  culture, 902-903, 936, 962-963, **967-969**
  microbial flora in, 9
  nosocomial infections from, 56
blood agar, 701
blood and tissue flagellates, **821-834**
blood cells, 20*f*
blood flukes. *See Schistosoma haematobium*; *Schistosoma japonicum*; *Schistosoma mansoni*
bloodborne transmission, **94**, 139*t*
bloody diarrhea, 592
Bocavirus, **186**
body fluids, microbial flora in, 9
boiling, 48
  for sterilization, 47*t*
boils, 433
bone
  demineralization, 902
  infections, **901-902**
  resorption, 917
bone marrow, 20*f*, 31*t*
Bordetella bronchiseptica, 552*t*
Bordetella pertussis, 371, 388, 399, 412, 551, 552*t*, **559-564**
  bacteremia from extravascular infection and, 966*t*
  extracellular products, 559
  growth, 559
  middle respiratory infections and, 923
  structure, 559
  virulence factors, 400*f*
Bordetella spp., **558-564**
  features of, 552*f*
Borrelia burgdorferi, 630*t*, 657-662, 954*t*
  features of, 645*t*
  macrolides and, 412
Borrelia hermsii, **655-657**
  bacteriology, 655
  features of, 645*t*
  immunity, 656
  manifestations, 656
  pathogenesis, 656
Borrelia recurrentis, **655-657**
  bacteriology, 655
  features of, 645*t*
  immunity, 656
  manifestations, 656
  pathogenesis, 656
Borrelia spp., 630*t*, **654-655**
  bacteremia from extravascular infection and, 966*t*
Botox, 526
botulinum neurotoxins, 524*f*
botulinum toxin, 523
botulism, 49, **524-525**
  Botox and, 526
  clinical aspects, 525-526
  clinical capsule, 524
  diagnosis, 525-526
  epidemiology, 524-525
  infant, **525**
  manifestations, 525-526
  pathogenesis, 525
  prevention, 526
botulism (*Cont.*):
  treatment, 526
  wound, **525**
bovine papular stomatitis, 206*t*
bovine spongiform encephalopathy, 101, 145, **342-343**
bovine tuberculosis, 630*t*
bradycardia, 290
bradykinin, 26, 185
brain abscess, 518
brain-blood barrier, 951
branched-chain DNA (bDNA), 320
broad-spectrum agents, 404
bronchiectasis, 519
bronchiolitis, 181
  Mycoplasma pneumoniae, 667*f*
bronchitis, 922-924, 923*t*
bronchoalveolar lavage (BAL), 729, 736, 927
broth dilution tests, 417*f*
Brucella abortus, 630*t*
Brucella spp., **629-633**
  bacteremia from extravascular infection and, 966*t*
  bacteriology, 629
  disease caused by. *See* brucellosis
brucellosis, 630*t*, **631-633**
  aminoglycosides for, 633
  ciprofloxacin for, 633
  clinical aspects, 642-643
  clinical capsule, 631
  diagnosis, 633
  doxycycline for, 633
  epidemiology, 631
  immunity, 642
  manifestations, 642
  pasteurization and, 633
  pathogenesis, 631-632
  prevention, 633
  rifampin for, 633
  treatment, 633
  for trimethoprim-sulfamethoxazole, 633
Brugia malayi, 761, 851, 858*t*, 859
  disease caused by. *See* filariasis
  general characteristics of, 852*t*
  parasitology, 857
bubo, 636, 676
bubonic plague, 635
budding, in human viruses, **127-128**
buffalopox, 208
buffered charcoal yeast extract (BCYE), 613
bulbar polio, 219
bullous impetigo, 436
bunyaviruses, 120, 280*t*, **282**, 285
  classification of, 110*t*
  virion structure, 282*f*
Burkholderia cepacia, 625*t*
Burkholderia mallei, 625*t*
Burkholderia pseudomallei, 625*t*
Burkholderia spp., **623**
Burkitt lymphoma, **266**
burns, infection of, 899*t*, 900
BV. *See* bacterial vaginosis

C

C3b, 27, 28, 394, 451, 453, 553
C5a peptidase, 448, 451
cadavers, 53
calcium dipicolinate, 357
calcofluor, 752
caliciviruses, **276-277**
  classification of, 110*t*
caliciviruses (*Cont.*):
  clinical aspects, **277**
  epidemiology, **276-277**
  fecal-oral spread, 277
  gastrointestinal infections from, 933*t*
  immunity, **277**
  pathogenesis, **277**
  reinfection, 277
California encephalitis, 952*t*
California virus, **290**
Calymmatobacterium granulomatis, 945, 946*t*
cAMP. *See* cyclic adenosine 3',5'-monophosphate
Campylobacter fetus, 567*t*
Campylobacter hyointestinalis, 567*t*
Campylobacter jejuni, **571-572**, 630*t*, 903, 932*t*, 936
  bacteremia from extravascular infection and, 966*t*
  clinical aspects, 572
  clinical capsule, 571
  diagnosis, 572
  diarrhea, 571
  enteritis, 571-572
  epidemiology, 571-572
  erythromycin for, 572
  fluoroquinolones for, 572
  immunity, 572
  pathogenesis, 572
  treatment, 572
Campylobacter lari, 567*t*
Campylobacter spp., 60, 388, 412, **571-572**
  enteritis, 930, 932*t*
  features of, 566*f*
  plaque colonized by, 912
Campylobacter upsaliensis, 567*t*
Camus, Albert, 629, 635
Candida albicans, 318, 411, 699, 706, **723-729**, 724*f*
  adherence, 704
  in AIDS, 727
  amphotericin B for, 729
  azoles for, 729
  caspofungin for, 729
  diagnosis, 729
  endophthalmitis, 729
  folliculitis from, 895, 899*t*
  genital infection from, 946*t*, 948
  intertrigo from, 899*t*
  invasiveness of, 726*f*
  mycology, 723-725
  nystatin for, 729
  pathogenesis of, 726*f*
  skin infection, 728*f*
  treatment, 729
  wound infections from, 899*t*
Candida glabrata, 729
Candida mannan, 727
Candida spp., **723-730**
  bacteremia from extravascular infection and, 966*t*
  classification of, 699*t*
  eye infection from, 906, 906*t*
  opportunistic, 724*f*
  septic arthritis from, 903
  upper respiratory infection and, 921
  UTI from, 940
candidate viruses, **277**
candidiasis, 318*t*, 411, 725-729
  clinical aspects, 727-729
  clinical capsule, 725

candidiasis (*Cont.*):
  disseminated, 700*f*
  epidemiology, 725
  pathogenesis, 725-727
  vaginal, 727
cannibalism, 341
capillary morphogenesis protein (CMP-2), 482*f*
Capnocytophaga, plaque colonized by, 912
capsids, structure, **102-108**
  cylindrical architecture, **106**
  special, **106**
  spherical architecture, **106**
capsomeres, 106
capsular glutamic acid, 484
capsule
  anaerobes, 518
  antiphagocytic, 458
  bacteria, **348-349**, 350*f*
  Cryptococcus neoformans, 739
  Hib, 552
  hyaluronic acid, 448
  hydrophilic, 349
  phagocytosis and, 553
  pneumococcal, 461*f*
  in pneumococcal disease, 463
  Streptococcus pneumoniae, 460, 463
  synthesis, 349
carbapenems, 405*t*, 407, **409**
  for *Pseudomonas aeruginosa*, 622
carbenicillin, for *Pseudomonas aeruginosa*, 622
carbohydrate breakdown, 88
carbuncle
  cause of, 896
  Staphylococcus aureus, 435*f*, 437
Cardiobacterium, 626*t*
cardiomyopathy, chronic, 832
cardiovascular syphilis, 650
caries. *See* dental caries
cariogenesis, 914*f*
carriage, 92
carrier state, 8
carriers, 92, 144
caseous necrosis, 496
caspofungin, 710
  for aspergillosis, 732
  for *Candida albicans*, 729
  features of, 708*t*
cat scratch fever, 683*t*, 689
catalase, 471
  anaerobiosis and, 516
  production, 88
catheters, 939, 963, **964-965**, 965*f*
CC4, 312
CCR5, 149, 306, 310
  coreceptor, 308
CD4+ helper T lymphocytes, 33-34, 154, 268, 312
  in AIDS, 317
CD8 T cells, 152
CD46, 193, 538
CD81, 240
CDC. *See* Centers for Disease Control and Prevention
cDNA. *See* complementary DNA
cefaclor, 408
cefazolin, 408
cefepime, 405*t*, 409
cefixime
  for gonorrhea, 549
  for salmonellosis, 604
cefotaxime, 405*t*, 408, 541
  for *Nocardia*, 513
cefotetan, susceptibility patterns to, 427*t*
cefoxitin, 408
ceftazidime, 408
  susceptibility patterns to, 427*t*
ceftriaxone, 408
  for anaerobic infections, 520
  for *Bacteroides fragilis*, 532
  for gonorrhea, 549
  for *Haemophilus influenzae*, 558
  for leptospirosis, 654
  for salmonellosis, 604
  for shigellosis, 598
cell culture, **71**
  primary, 71
  secondary, 71
cell death, in human viruses, **127**
cell membrane, bacteria, **355**
cell strain, 71
cell survival, **128**
  human viruses, **128**
cell wall
  bacteria, **349**
  fungi, 694*f*
  Gram negative, 351*f*, **353-355**
  Gram positive, **351-352**, 351*f*
  Mycobacterium, 490*f*
  Neisseria gonorrhoeae, 538*f*
  Pneumocystis, 734
  synthesis
    antibacterial agents, **406-410**
    antifungals, 710
  yeast, 694*f*
Cellini, Benvenuto, 643
cell-mediated immunity, 15, 30, 35-36, 497
  in mycobacterial disease, 491
  tuberculosis, 497
cellular immune response, 198
cellular immunity, fungal infections, 706
cellulitis, 624, **897**, 899*t*, 915
  anaerobic, 522
  Haemophilus influenzae, 556
Centers for Disease Control and Prevention (CDC), 208, 627
  on AIDS, 316
central nervous system (CNS), 215
  infection, **951-957**
    abscesses, 953-954, 953*f*
    clinical features, **954-955**, 954*t*
    CSF and, 951, 956-957, 956*t*
    diagnosis, **956-957**, 956*t*
    etiologic agents, **955-956**
    management, **957**
    routes of, **951-954**, 952*t*, 953*f*
  malaria in, 786*f*
  noninfectious disease in, 956
  viral infections, **337-344**
    viruses causing, 338*t*
cephalexin, 405*t*, 408
cephalosporins, 405*t*, 407, 408, **408-409**, 529
  for anaerobic infections, 520
  Enterobacteriaceae and, 409
  for enterococcal disease, 466
  for *Haemophilus influenzae*, 557
  for pertussis, 563
  for *Pseudomonas aeruginosa*, 622
  resistance, 421, 607
  third-generation, 465
cephalothin, susceptibility patterns to, 427*t*
cephradine, 405*t*
cercariae, 879, 886*f*
cerebellar ataxia, 221*t*
cerebral falciparum malaria, 788
cerebrospinal fluid (CSF), 128, 192, 335, 540
cervicitis, 946*t*, **947-948**
cervicofacial actinomycosis, 510*f*
cervix carcinoma, 329
Cesarian section delivery, 321
cestodes, **865-877**
  case study, 877
  classification of, 763*t*
  clinical disease, 866-867
  life cycle, 865-866
  morphology, 865
  parasitology, 865-866
cestodiasis, prevalence of, 760*t*
CF. *See* colonizing factor; cystic fibrosis
Chagas' disease. *See* American trypanosomiasis
chancroid, 558*f*, 945, 946*t*-947*t*
chemical mediators, in innate immunity, 26-30
chemical screening tests, **941**
chemokines, 26, 28, 150
  receptors, 114
chemoprophylaxis, 898, 943
chemotaxis, **373-374**
chemotherapeutic
  definition, 404
  for malaria, 789*t*
chickenpox. *See* varicella-zoster virus
chiggers, 687
Chikungunya fever, **290-291**
childbed fever, 53-54, 54*f*, 454
childbirth, wound infections from, 897-898
childhood exanthems, **189-204**
  incubation period, 141*t*
Chilomastix mesnili, 814*t*
chitin, 693
Chlamydia pneumoniae, 671, **678**, 924
  clinical disease, 678
  epidemiologic associations, 672*t*
  epidemiology, 678
  erythromycin for, 678
  lower respiratory tract infections from, 926, 926*t*
  tetracycline for, 678
  treatment, 678
Chlamydia psittaci, 671, **678**
  doxycycline for, 678
  epidemiologic associations, 672*t*
  tetracycline for, 678
Chlamydia spp., 66, 79-80, 94, 349, 410, 549, **671-678**
  chloramphenicol and, 412
  clinical case, 679
  elementary body, 672*f*
  epidemiologic associations, 672*t*
  features of, 672*t*
  life cycle, 672*f*
  ofloxacin for, 413
  reticulate body, 672*f*
  sexual transmission of, 673
  tetracycline and, 411
Chlamydia trachomatis, 94, 546, **671-677**
  bacteriology, 671-673
  blindness and, 675
  clinical aspects, 674-677

Chlamydia trachomatis (*Cont.*):
diagnosis, 677
disease, 673-677
clinical capsule, 673
epidemiology, 673
epidemiologic associations, 672*t*
epididymis, 676
eye infections, 674-675, 906, 906*t*
genital infections, 676, 945, 946*t*, 947, 948
immunity, 674
inclusion bodies, 675*f*
inclusion conjunctivitis and, 675-676
LCR in, 677
lower respiratory tract infections from, 926, 926*t*
manifestations, 674-676
morphology, 671
pathogenesis, 674
PCR in, 677
prevention, 677
replicative cycle, 671-673
scanning electron micrograph, 674*f*
treatment, 677
urethritis, 676
chlamydoconidia, 697, 723, 725, 747
chloramphenicol, 404, 405*t*, **411-412**, 423
Chlamydia and, 412
resistance, 419*t*
Rickettsia and, 412
for salmonellosis, 604
susceptibility patterns to, 427*t*
chlorhexidine, 52, 440, 913
chlorine, 51
for sterilization, 47*t*
chloroquine, 773
for malaria, 789*t*
resistance, 773, 784
chloroquine phosphate, 773
Chlytridiomycota, 698
chocolate agar, 701
cholecystitis, 519
cholera, **568-571**
ciprofloxacin, 570
clinical aspects, 570
diagnosis, 570
diarrhea in, 569
doxycycline for, 570
endemic, 568
epidemiology of, 14, 568-569
erythromycin for, 570
fluid loss from, 569
immunity, 569
manifestations, 570
pandemic, 568
pathogenesis, 569
prevention, 570
toxin, 566-568, 567*f*, 586
treatment, 570
trimethoprim-sulfamethoxazole for, 570
virulence, 569
cholestasis, 234
choline-binding proteins, 460, 461
chorioamnionitis, 458
choriomeningitis, 283
chorioretinitis, 795, 905-906, 906*t*
choroid plexuses, 951
Chrichton, Michael, 610
chromatin, 143
Chromobacterium, 626*t*
chromoblastomycosis, 719, **721**, 897
chronic bronchitis, 922
chronic cardiomyopathy, 832
chronic conjunctivitis, 906
chronic endocarditis, 961
chronic furunculosis, *Staphylococcus aureus*, 437-438
chronic granulomatous infection, 954, 954*t*
chronic granulomatous osteomyelitis, 901-902
chronic infection, 101
chronic inflammation, 26
chronic measles, 196
chronic meningitis, 954-955
chronic osteomyelitis, 504, 519
chronic otitis media, 519, 908, 908*t*
chronic periodontitis, **916-917**, 917*f*
chronic pneumonia, **925**, 926*t*
chronic pyelonephritis, 940
chronic sinusitis, 519, 910, 910*t*
chronic ulcers, 899*t*
Chrysops, 862
cidofovir, 158*t*, **163**, 335
CIE. *See* counterimmunoelectrophoresis
cilastatin, 409
cilia, 22
pneumolysin and, 463
ciliates, classes of, 762*t*
cinchona bark, 773
ciprofloxacin, 406*t*, 413, 414, 440
for *Bacillus anthracis*, 426, 485
for brucellosis, 633
for cholera, 570
for gonorrhea, 549
for *Haemophilus influenzae*, 558
for salmonellosis, 604
susceptibility patterns to, 427*t*
for tuberculosis, 500
for *Yersinia pestis*, 426
circumcision, 321
cirrhosis, 234
in hepatitis B, 234*f*
cistrons, 370
citrate utilization, 88
Citrobacter, 585*t*, **607**
Citrobacter freundii, 607
clades, HIV, 314-315
Cladophialophora carrionii, 714*t*, 721
clarithromycin, 405*t*, 412
for *Helicobacter pylori*, 576-577
classic pathway, 28
clavulanate, for *Bacteroides fragilis*, 532
clean contaminated wounds, **898**
clean wounds, **898**
clean-voided midstream urine, 941
Clf. *See* clumping factor
clindamycin, 405*t*, **412**, 529
for actinomycosis, 510-511
for *Bacteroides fragilis*, 532
resistance, 419*t*, 421
susceptibility patterns to, 427*t*
clinical microbiology systems, **69**
clofazimine, for leprosy, 503
clonal activation, 266
clonal structure, 401
clonality, 399-401
clonorchiasis, prevalence of, 760*t*
Clonorchis sinensis, 767, 879, **882-883**
cercarial larvae, 884*f*
characteristics of, 880*t*
eggs, 883*f*
parasitology, **882**, **883*f*-884*f***
clostridia, **516-517**
enterotoxins in, 517
hemolysin in, 517
neurotoxin in, 517
clostridial food poisoning, 521, 522
clostridial myositis, 900
clostridial tetanus, 524
Clostridium, 11, 396
bacteremia from extravascular infection and, 966*t*
endometritis, 522
Clostridium botulinum, 517, 517*t*, **523-526**
bacteremia from extravascular infection and, 966*t*
bacteriology, 523
food poisoning from, 934*t*
spores of, 49
Clostridium difficile, 12, 517, 517*t*, **528-531**
bacteremia from extravascular infection and, 966*t*
bacteriology, 528
cytotoxin, 528
diarrhea, **529-531**
clinical aspects, 530-531
clinical capsule, 529
diagnosis, 530
epidemiology, 529
immunity, 529
manifestations, 530
pathogenesis, 529
treatment, 530-531
enterotoxin, 528
gastrointestinal infections from, 933*t*
hospital-associated diarrhea from, 935
metronidazole for, 530-531
pseudomembranous colitis, 530*f*
treatment, 410
vancomycin for, 530-531
Clostridium perfringens, 517, 517*t*, **520-523**, 900
α-toxin, 520
bacteriology, 520
clinical aspects, 522-523
clinical capsule, 521
diagnosis, 523
disease, 521-522
enterotoxin, 520
epidemiology, 521
gastrointestinal infections from, 932*t*-933*t*, 935
manifestations, 522-523
pathogenesis, 521
penicillin for, 523
prevention, 523
spores, 521-522
β-toxin, 520
treatment, 523
Clostridium tetani, 517, 517*t*, **526-528**, 897, 899*t*, 900, 966*t*
bacteriology, 526
epidemiology, 527
pathogenesis, 527
clotrimazole, 709, 718
features of, 708*t*
clumping factor (Clf), 430
CMP-2. *See* capillary morphogenesis protein
CMV. *See* cytomegalovirus
CNF. *See* cytotoxic necrotizing factor
CNS. *See* central nervous system
coagulase, 88, 430, 434
coagulase-negative staphylococci, 440-441
slime A, 441*f*

cocci, 347
Coccidioides, 710, **748-753**
classification of, 699*t*
Coccidioides immitis, 701, 704, **748-753**, 906*t*, 924
arthroconidia, 749
chronic granulomatous infection from, 954*t*
conidia, 749
culture of, 753
dimorphism in, 748
disease caused by. *See* coccidioidomycosis
endospore in, 748
features of, 740*t*
geographic restriction of, 750-751
life cycle of, 750*f*
lower respiratory tract infections from, 926*t*
spherules, 749
Coccidioides posadasii, 749
coccidioidomycosis, 318*t*, **749-753**
AIDS and, 752
amphotericin B for, 753
clinical aspects, 752-753
clinical capsule, 749
diagnosis, 752-753
endospores, 752
epidemiology, 749-750
erythema nodosum in, 752
geographic restriction of, 750
immunity, 751-752
manifestations, 752
pathogenesis, 751
proteases, 751
serologic tests in, 753*f*
skin test, 750
treatment, 753
virulence of, 751
cockroaches, 216
coenzyme A, 682
cohesive ends, 124
cold agglutinins, 668
cold hemagglutinins, 667
cold sores, 253
colistin, 415
for gonorrhea, 548
collagen, lost, 916
colon, flora in, 11
colonial morphology, 67
bacterial, 68*f*
colonizing factor (CF), 590
Colorado tick fever, **291**
Coltivirus, 282
Columbus, Christopher, 646
commensal, 388
common cold, 919
communicability, **92**
communicable infections, **91**
endemic, 91
epidemic, 91
pandemic, 91
competence, 380
complement component deficiencies, 538
complement fixation, **76**, 668
complement system, 26-28
classic pathway, 28
components of, 27*f*
lectin pathway, 26-27
complementary DNA (cDNA), 308
complement-fixing antibody, 667
complications, measles, **195**
computed tomography (CT), 871
concatemers, 126
concomitant immunity, 887-888
condoms, 321, 333, 549
condyloma latum, 946
condylomata, 331*f*
condylomata acuminata, 946*t*
condylomata lata, 649
congenital malaria, 784
congenital syphilis, **650**
congenital toxoplasmosis, 795
conidia, 695
in *Aspergillus*, 730, 731
Coccidioides immitis, 749
conidiophore, 697
in *Aspergillus*, 730
conjugation, 378, 424
bacterial, **382-385**
in Gram-negative bacteria, **383-384**
in Gram-positive bacteria, **384**
resistance and, **423-424**
conjugative plasmids, 382
conjunctiva, 10, 390*t*
conjunctivitis, 183, 184, 221*t*, 438, 687, 905-906, 906*t*
inclusion, 675-676
contact precautions, 57
contaminated wounds, **898**
contract secretion system, enteropathogenic *Escherichia coli*, 591*f*
conventional agents, **338**
copy choice mechanisms, 132
cor pulmonale, 181
core, **356-357**
bacteria, 350*t*
core polysaccharides, 354
coreceptors, 114
co-repressors, 371
corneal ulcerations, 810
coronavirus, 89, 120, **185-186**
classification of, 110*t*
receptors, 114*t*
upper respiratory infection from, 920*t*
virion structure of, 186*f*
corticosteroids, 889
for toxocariasis, 853
for trichinosis, 856
corynebacteria, 10, **471-477**
Corynebacterium diphtheriae, 69, 108, 381, **471-477**
bacteriology, 472-473
bacteriophages, 111*f*
clinical aspects, 474-477
clinical capsule, 473
disease caused by. *See* diphtheria
manifestations, 474-476
middle respiratory infections and, 923
nonrespiratory infections, 476
respiratory obstructions, 475
ulcers from, 899*t*
upper respiratory infection from, 920-921, 920*t*
Corynebacterium ulcerans, upper respiratory infection from, 921
coryza. *See* rhinitis
Councilman bodies, 287
counterimmunoelectrophoresis (CIE), 73
cowpox, 206*t*, **211**
Coxiella burnetii, 630*t*, 681, 683*t*, **687-688**. *See also* Q fever
bacteriology, 687
Coxsackie virus, 154
coxsackieviruses, 213, **220-222**, 277
epidemiology, **220-221**
group A, 215
group B, 215, 217, 221
manifestations, **221-222**
CPE. *See* cytopathic effect
CR1, 495
CR3, 495
CR4, 495
Creutzfeldt-Jakob disease, 101, 145, 339*t*, **341-342**
pathology, 342
prevention, 342
therapy, 342
variant, 339*t*, **342-343**
cribriform plate, 809
cross-infection, 54
croup, 178, 922
cryptococcal meningitis, 320
cryptococcosis, **740-742**
clinical aspects, 741-742
clinical capsule, 740
epidemiology, 741
immunity, 741
manifestations, 741-742
pathogenesis, 741
Cryptococcus neoformans, 701, 702, 705, 706, **739-742**, 740*f*
AIDS and, 742
amphotericin B for, 742
bacteremia from extravascular infection and, 966*t*
capsule, 739
chronic granulomatous infection from, 954*t*
dendritic cells, 741
disease caused by. *See* cryptococcosis
features of, 740*t*
GXM in, 739, 741
lower respiratory tract infections from, 926*t*
meningitis, 742*f*
treatment, 742
Cryptococcus spp., 318, **739-742**
classification of, 699*t*
cryptosporidiosis, 89, 318*t*, **798-800**
clinical aspects, 799-800
clinical capsule, 798
diagnosis, 799
diarrhea in, 799, 800
epidemiology, 798
immunity, 798-799
manifestations, 799
paromomycin for, 800
pathogenesis, 798-799
prevention, 799-800
stool precautions in, 800
treatment, 799-800
Cryptosporidium, 770, **797-800**, 931, 933*t*
disease caused by. *See* cryptosporidiosis
life cycle, 797-798
morphology, 797
oocysts, 797
parasitology, 797-798
Cryptosporidium parvum, 797, 799
acid-fast stain, 800
crystal violet stains, 64*f*
CSF. *See* cerebrospinal fluid
CT. *See* computed tomography
CTL. *See* cytotoxic T lymphocytes
cubic symmetry, viruses with, **126**
Culex tarsalis, 289, 859
Culiseta melanura, 289

cultural characteristics
bacteria, 70
fungi, 70
culture, **66-73**
acid-fast bacilli, 499
arterial, 729
atmospheric conditions, **69**
aerobic, 69
anaerobic, 69
bacteria, **66-71**
blood, 902-903, 936, 962-963, **967-969**
clinical microbiology systems, **69**
Coccidioides immitis, 753
fungi, **66-71**, **701-702**
identification, **69-71**
media, **68-69**
indicator, **69**
nutrient, **68**
selective, **69**
Trichomonas vaginalis, 814
urine, **942**, 942*f*
viruses, **71-73**
primary, 112
tissue, 112
cutaneous amebiasis, 897
cutaneous larva migrans, 775, **857**
cutaneous leishmaniasis, 824*f*, 897
cuticle, 769
CXCR4, 151, 306, 308, 310, 315
cyclic adenosine 3',5'-monophosphate (cAMP), 568
cycloheximide, 701
Cyclops, 873
Cyclospora, 800
cylindrical architecture, **106**
cysteine, 520
cystic fibrosis (CF), *Pseudomonas aeruginosa*, 619, 620-621, 622*f*
cysticercosis
of brain, 871*f*
of muscle, 870*f*
surgery for, 872
Cysticercus bovis, 867
cystitis, 939-942, **940**
cytochrome oxidase, 617
cytokines, 26, 28-30
in infection, 29*t*
innate immunity, 28-30
in malaria, 786
storm, 153*f*, 241
cytology, 73
cytomegalovirus (CMV), 94, 142, 159, 247, 248*t*, 318
AIDS and, 263
clinical capsule, 261
congenital infection, 263
diagnosis, **263**
epidemiology, **261**
eye infection from, 906, 906*t*
ganciclovir and, 263
HIV and, 263
immunity, **262**
in immunocompromised patients, 263
latent infection, 261
manifestations, **262-263**
maternal infection, 262
pathogenesis, **262**
perinatal infection, 262, 263
prevention, **264**
receptors, 114*t*
cytomegalovirus (CMV) (*Cont.*):
STD from, 945, 946*t*
treatment, 161, 163, **263-264**
cytomegaly, 260
cytopathic effect (CPE), 71, 72*f*, 143, 145
of viruses, 112
cytopathogenicity, of viral infections, **143-144**
cytoplasmic membrane, antifungals, 707-709
cytosol, 347, **356-357**, 365
Listeria monocytogenes in, 479
cytotoxic necrotizing factor (CNF), *Escherichia coli*, 586
cytotoxic T cell, 30*t*
destruction, 37*f*
cytotoxic T lymphocytes (CTL), 34, 152, 241, 302
cytotoxin
Clostridium difficile, 528
tracheal, 559
vacuolating, 573

D

Da Costa, Jacob M., 515
dacryocystitis, 905, 906*t*
DAEC. *See* diffuse aggregating pattern
dairy, unpasteurized, 631
dalfopristin, 405*t*
Dane particle, 230
daptomycin, 406*t*
dark-field microscopy, **65**
daughter viruses, 101
Davulanic acid, 409
deaminases, 88
death, 45
decarboxylases, 88
decelerating phase, 368
decubitus ulcer, 433, 519
deep lesions, *Staphylococcus aureus*, 438
deer tick, 659*f*
defective interfering (DI) particles, 131-132
defensins, 25
definitive host, 764
Toxoplasma gondii, 792-793
dehydration, 569
delavirdine, 158*t*, 162
delayed-type hypersensitivity (DTH), 42, 398, 649
dermatophytes and, 716
fungi and, 705
mycobacteria and, 491
tuberculosis and, 495
deletions, 374, 375*t*
dementia, 343
AIDS and, **338**
demethylase, 711
demineralization, 902
bone, 902
tooth, 915-916
dendritic cells, 20*f*, 23
Cryptococcus neoformans, 741
dengue, 203, 288, **290**
dengue shock syndrome (DSS), 153
dental caries, **913-916**, 914*f*
causes of, 913, 914*f*, 915
complications from, 915
dental infections
chronic periodontitis, **916-917**, 917*f*
dental caries, **913-916**, 914*f*
dental plaque and, **911-913**, 912*f*-913*f*
necrotizing periodontal diseases, **917-918**
dental plaque, **911-913**
as biofilm, 911, 912*f*, 915
dental plaque (*Cont.*):
colonizers, 911-912, 915
dental infections caused by, 911-913, 912*f*-913*f*
inhibition of, 913
pH, 915
subgingival, 911-912, 916
supragingival, 911-912, 913*f*
Dermacentor andersoni, 291, 684
Dermacentor variabilis, 684
dermatomes, 258
dermatophyte fungi, 896
dermatophytes, 709, 713-718, 714*t*
clinical aspects, 717-718
diagnosis, 718
disease, 715-717
clinical capsule, 715
epidemiology, 715
pathogenesis, 715-716
DTH and, 716
immunity, 716-717
manifestations, 717
Deuteromycetes, 698
DFA. *See* direct fluorescent antibody
DGI. *See* disseminated gonococcal infection
DHHS. *See* United States Department of Health and Human Services
DI particles. *See* defective interfering
diagnosis
culture, **66-73**
immunologic systems, **73-80**
of infectious diseases, 17
laboratory, 59-88
of fungi, **699-702**
nucleic acid analysis, **80-85**
Diaptomus, 873
diarrhea, **271-288**, 274*f*
in amebiasis, 808
bloody, 592
Campylobacter jejuni, 571
cholera, 569
Clostridium difficile, **529-531**
clinical aspects, 530-531
clinical capsule, 529
diagnosis, 530
epidemiology, 529
immunity, 529
manifestations, 530
pathogenesis, 529
treatment, 530-531
in cryptopsoridiosis, 799, 800
from *E coli*, 936
giardiasis, 820
hospital-associated, **935**
incubation period, 141*t*
traveler's, 594, **931**
trichinosis and, 856
watery, 581, **929**, 932*t*-933*t*
diarrheal diseases. *See* enteric infections
DIC. *See* disseminated intravascular coagulation
Dickens, Charles, 489
didanosine, **162**
dideoxycytidine, 158*t*
dideoxyinosine, 123, 158*t*
Dientamoeba fragilis, 814*t*, 817
diet, flora and, 11
diethylcarbamazine, 777*t*, 860
diffuse aggregating pattern (DAEC), 593
diffuse pneumonitis, 736

diffusion
  facilitated, 359, 360*f*
  simple, 359
  tests, **416-417**
dihydrofolic acid, 774
dihydropteroate synthetase, 414
diloxanide furoate, 777*t*, 809
dilution tests, **416**
  broth, 417*f*
Diment, Adam, 535
dimorphism, 698, 701
  in *Coccidioides immitis*, 748
diphtheria, 42
  cellular view, 475*f*
  clinical capsule, 473
  complications, 475
  cutaneous, 476
  diagnosis, 476-477, 921
  epidemiology, 473
  immunity, 474
  manifestations, 474-476
  molecular view of, 16*f*
  myocarditis, 476
  overview, 474*f*
  pathogenesis, 15, 473-474
  pharyngeal, 919, 921
  prevention, 477
  pseudomembrane, 476*f*
  respiratory obstruction in, 475
  skin, 897
  treatment, 477
  vaccines, 18
diphtheria toxin (DT), 472, 474*f*
  antibodies and, 474
  myocarditis and, 474, 475
diphtheria toxoid, 541, 557
diphtheria toxoid and pertussis vaccine (DTaP), 528
diphyllobothriasis, 874
Diphyllobothrium, 767
Diphyllobothrium latum, 865, 866*t*, **872-874**
  eggs of, 865-866
    structure, 872*f*
  parasitology, 872
diplococcus, 460
direct aspirates, 902
direct droplet spread, of influenza A, **173-174**
direct examination, **61-65**
  light microscopy, **61-65**
direct fluorescent antibody (DFA), 563, 736
direct fusion, 116
  entry, 117*f*
direct immunofluorescence, 66*f*
direct rapid progression, 175
direct tissue, 59-60
direct transposition, 378
dirty and infected wounds, **898**
disaccharides, 915
disease, **91-92**
disease index, 95
  viral infections, 137
disinfection, **50-52**
  with alcohol, 51
  chemical methods, 50-52
  definition of, 45-46
  filtration, 50
  with formaldehyde, 52
  with glutaraldehyde, 52
  with halogens, 51
  with hydrogen peroxide, 51
disinfection (*Cont.*):
  microwaves, 50
  pasteurization, 50
  with phenolics, 52
  physical methods, 50
  with surfactants, 51
disseminated candidiasis, 700*f*
disseminated gonococcal infection (DGI), 545, 547-548, 946*t*
disseminated histoplasmosis, 745
disseminated infection, viral, 140
disseminated intravascular coagulation (DIC), 398, 540
disseminated visceral leishmaniasis, **826-827**
DNA hybridization, **80**, 81*f*
DNA polymerase, 119, 123
  inhibition of, 158*t*
DNA probes, **82-83**
DNA replication, 123*f*
  in bacteria, 364*f*
DNA viruses, **121-124**
  oncogenicity of, 147*t*
  transformation by, **147-148**
Domagk, 4, 403
donor cells, 378
double-stranded break model, 376*f*
double-stranded RNA viruses, 273
Down's syndrome, 668
downward displacement autoclave, 48-49, 48*f*
doxycycline, 405*t*, 411, 790
  for actinomycosis, 510-511
  for anthrax, 485
  for *Bacillus anthracis*, 485
  for *Bartonella*, 689
  for brucellosis, 633
  for *Chlamydia psittaci*, 678
  for cholera, 570
  for leptospirosis, 654
  for Lyme disease, 661
  for rickettsialpox, 686
  for Rocky Mountain spotted fever, 685
dracunculiasis, prevalence of, 760*t*
droplet nuclei, 93
droplet precautions, 57
DSS. *See* dengue shock syndrome
DT. *See* diphtheria toxin
DTaP. *See* diphtheria toxoid and pertussis vaccine
DTH. *See* delayed-type hypersensitivity
DtxR, 472
Duffy blood group antigens, 783
duplications, 374
dura mater grafts, 342
dwarf tapeworm, 874
dysentery, 581, **930**, 932*t*-933*t*
dysuria, 184

E

E1A. *See* early proteins
E6, 327, 330
E7, 327, 330
EAEC. *See* enteroaggregative *Escherichia coli*
ear infections, **907-909**
  diagnosis, **908-909**
  etiologic agents, **908**, 908*t*
  management, **909**
  manifestations, **907-908**
  otitis externa, 907-909, 908*t*
  otitis media, 907-909, 908*t*
early antigens (EAs), 264
early proteins (E1A), 183
EAs. *See* early antigens
eastern equine encephalitis, **289**, 952*t*
EBNAs. *See* EBV nuclear antigens
Ebola virus, 283-284, 293
EBV. *See* Epstein-Barr virus
EBV nuclear antigens (EBNAs), 264
E-cadherin, 479
echinocandins, 710
  action of, 709*f*
  features of, 708*t*
  resistance, 711
echinococcosis, **874-876**
  albendazole for, 876
  aspiration treatment for, 876
  clinical aspects, 875-876
  cysts, 875*f*
  diagnosis, 876
  in lung, 875*f*
  manifestations, 875-876
  prevention, 876
  sylvatic cycle, 875
  transmission of, 875
  treatment, 876
Echinococcus granulosus, 767, 866*t*, **874-876**
  disease caused by. *See* echinococcosis
  parasitology, **874**
Echinococcus multilocularis, 866*t*, **876-877**
echocardiography, 962
echoviruses, 213, 215, 216, **220-222**
  epidemiology, **220-221**
  manifestations, **221-222**
eclipse phase, 113
ectoparasites, 946*t*
ectoplasm, of protozoa, 761
ectothrix, 716
edema factor (EF), 482, 482*f*
  in anthrax, 484
EF. *See* edema factor
EF-2. *See* elongation factor 2
efavirenz, 158*t*, 162
efflux, 419
eflornithine, **776**
EHEC. *See* enterohemorrhagic *Escherichia coli*
Ehrlich, Paul, 403, 771
Ehrlichia, 681, **688**
  inclusions, 689*f*
Ehrlichia chaffeensis, 683*t*, 688
Ehrlichia ewingii, 683*t*
ehrlichiosis, 683*t*
EIA. *See* enzyme immunoassay
EIEC. *See* enteroinvasive *Escherichia coli*
Eikenella, plaque colonized by, 912
El Tor, 568
elastase, 618
elastin, 620
electron microscopy, **65**, 73
elephantiasis, 857
elongation factor 2 (EF-2), 16*f*, 473, 474
Embden-Meyerhof glycolytic pathway, 360
emerging diseases, 90*f*
empiric therapy
  antibacterial agents, **426**
  UTI, 942-943
empyema, 925, 926*t*
enamel pellicle, 911, 912*f*
encephalitis, 191, 200, 668
  acanthamebic granulomatous, 810*f*
  California, 952*t*
  eastern equine, **289**, 952*t*
  granulomatous, 810

encephalitis (*Cont.*):
herpes simplex virus 1, 254
Japanese B, **290**
in rabies, 301
St. Louis, **289**, 952*t*
West Nile, 952*t*
western equine, 288-289, 952*t*
encephalopathies, subacute spongiform, 339-343
end problem, 123*f*, 124
solutions, 124*f*
endarteritis granuloma, 648
endemic infections, 91
cholera, 568
enteric, **930-931**
malaria, 784
endemic typhus, **686-687**
endocarditis, 438
acute, 961
arterial cultures for, 729
bacterial, 960, 960*f*
chronic, 961
infective, **959-963**
clinical features, **961**
diagnosis, **962**
etiologic agents, **961-962**, 962*t*
management, **962**
pathogenesis, **960**, 960*f*
subacute, 961
subacute bacterial, 465
endocervicitis, 546
endocytosis, receptor-mediated, 116
endogenote, 376, 378
Endolimax, 803
endometritis, 510
anaerobic, 522
Clostridium, 522
endonuclease activity, 170
endophthalmitis, 905-906, 906*t*
Candida albicans, 729
endoplasm, of protozoa, 761
endoplasmic reticulum (ER), 611
endosomal vesicle, 116
endospore, **371-372**
bacteria, 350*t*
in *Coccidioides immitis*, 748
endospore stain, 64*f*
endospores, coccidioidomycosis, 752
endothrix, 716
endotoxic shock, 354, 540
endotoxins, 354, **398**
lipopolysaccharide, 398, 580, 602, 636
enfuvirtide, 158*t*, 320
Entamoeba coli, differential characteristics, 805*t*
Entamoeba hartmanni, differential characteristics, 805*t*
Entamoeba histolytica, 764, 769, **803-809**, 926*t*, 933*t*, 936
cysts, 804
differential characteristics, 805*t*
distribution of, 764*t*
fecal-oral transmission, 805
immunity, 807
laboratory growth, 806
life cycle, 804-805
metronidazole for, 809
morphology, 803-804
parasitology, 803-804
pathology, 807
physiology, 803-804
transmission of, 764*t*
trophozoites, 804, 807, 808
Entamoeba spp., 761
differential characteristics, 805*t*
enteric adenovirus, 272
enteric fever, 581, 601-602, **930**, 932*t*-933*t*
salmonellosis, 603-604
enteric infections, **929-931**
candidate agents of, 936
clinical features, **929-930**, 932*t*-933*t*, 933, 935
diagnosis, **935-936**
endemic, **930-931**
epidemic, **931**
epidemiologic setting, **930-934**
etiologic agents, **930**, 932*t*
features of, 932*t*
food poisoning, **933-935**, 934*t*
management, **936-937**
enteroaggregative *Escherichia coli* (EAEC), 584*t*, 593
Enterobacter, 580, 585*t*, **607**
Enterobacteriaceae, 408, 417, 520
bacteriology, 579-580
cephalosporins and, 409
characteristics of, 584*t*-585*t*
classification, 580
clinical aspects, 582-583
clinical case, 608
diagnosis, 582
diseases
epidemiology, 580
intestinal infections, 581
opportunistic infections, 580-581
overview, 581*f*
pathogenesis, 580-581
general characteristics, **579-583**
immunity, 582
intertrigo from, 899*t*
LPS endotoxins, 580
manifestations, 582
protein exotoxins, 580
toxins, 580
treatment, 582-583
UTI from, 940
virulence, 581-582
wound infections from, 899*t*
Enterobacteriaceae, lower respiratory tract infections from, 926*t*
enterobiasis, 761, **838-839**
clinical aspects, 838-839
epidemiology, 838
immunity, 838
manifestations, 838-839
pathogenesis, 838
prevention, 839
treatment, 839
Enterobius vermicularis, 769, 835, 836*t*, **837-839**
disease caused by. *See* enterobiasis
egg structure, 837*f*
life cycles of, 836*t*, 838
parasitology, 837
prevalence of, 760*t*
enterococci, **466-467**
biochemical reactions, 456*t*
classification, 445*t*
cultural reactions, 456*t*
disease
aminoglycosides for, 466
ampicillin for, 467
case study, 467
cephalosporins for, 466
clinical aspects, 466-467
enterococci, disease (*Cont.*):
clinical capsule, 466
epidemiology, 466
manifestations, 466-467
pathogenesis, 466
treatment, 467
vancomycin for, 466
hemolytic reactions, 456*t*
infective endocarditis and, 961, 962*t*
vancomycin-resistant, 421
Enterococcus faecalis, 384, 445*t*, 466, 467
Enterococcus faecium, 445*t*, 466, 467
enterocytes, 597
enterohemorrhagic *Escherichia coli* (EHEC), 584*t*, 592, 931, 932*t*
epidemiology, 592
pathogenesis, 592
Shiga toxin, 592
enteroinvasive *Escherichia coli* (EIEC), 584*t*, 593
Enteromonas hominis, 814*t*
enteropathogenic *Escherichia coli* (EPEC), 584*t*, 590-592, 932*t*, 935
attachment, 591*f*
contract secretion system, 591*f*
epidemiology, 590
immunity, 592
pathogenesis, 590-592
enterotoxigenic *Escherichia coli* (ETEC), 584*t*, 589-590, 929, 931, 932*t*
epidemiology, 589-590
immunity, 590
pathogenesis, 590
enterotoxin
Clostridium difficile, 528
Clostridium perfringens, 520
Salmonella enterica, 601
enterotoxins, 275
in clostridia, 517
Staphylococcus aureus, 432
enteroviruses, 71, **213-224**
biological features, **213-214**
clinical aspects, **217-218**
clinical capsule, 215
clinical syndromes associated with, 221*t*
CNS infections and, 951, 952*t*
diagnosis, **217-218**
distribution of, 215
epidemiology, **215-216**
group characteristics, **213-214**
growth, **214-216**
human, 215*t*
immunity, **217**
incubation period, 141*t*, 216
morphological features, **213-214**
pathogenesis, **216-217**
persistent infection, **338**
prevention, **218**
specific groups, **218-223**
treatment, **218**
Entner-Doudoroff pathway, 360
entry, **115-117**
bacteria, **389**
bacteriophages, 116*f*
direct fusion, 117*f*
inhibition of, 158*t*
retroviruses, 306-308
viral infection, **137-139**
env, 306, 307*t*, 310

envelope, 102
  bacteria, 347, **348-356**, 350*t*
  Gram negative, 353*f*
  Gram positive, 352*f*
  structure, **108**
enveloped human viruses, **116**
enveloped paramyxoviruses, 177
enveloped togaviruses, 197
enveloped viruses, 109*f*
enzymatic inactivation, **421-423**
  modifying enzymes and, 423
  resistance, 439
enzyme activity, **369-370**
  allosteric regulation, 369*f*
enzyme immunoassay, 76, **77**, 178-179, 192, 255, 271, 853
  for toxoplasmosis, 796
enzyme immunoassay (EIA), 746
enzyme linked immunosorbent assay (ELISA), 128, 192, 319
  for HIV detection, 319
  in human T-cell lymphotropic virus detection, 324
eosinophils, 23, 856, 859
EPEC. *See* enteropathogenic *Escherichia coli*
epidemic adenovirus conjunctivitis, 906
epidemic infections, 91, **95-96**
  control of, **96**
  disease index, 95
  enteric, **931**
  impetigo, 896
  incidence, 95
  infectivity, 95
  prevalence, 95
epidemic louse-borne typhus fever, **686-687**
epidemic myalgia, 222
epidemiology, of infectious diseases, 13-14
Epidermophyton, 713
  classification of, 699*t*
epididymis, 545
  Chlamydia trachomatis, 676
epididymitis, 945, 946*t*, **947**
epiglottitis, 922-924, 923*t*
  acute, 555-556
epimastigote, 821
epiphysis, 901
epitopes, in adaptive immunity, 32
Epstein-Barr virus (EBV), 72, 150, 247, 248*t*, 264-265, **264-268**
  AIDS and, **266**
  clinical aspects, **266-268**
  CNS infections and, 952*t*
  diagnosis, **266-267**
  epidemiology, **265**
  immunity, **265-266**
  in immunocompromised patients, 265
  manifestations, **266**
  pathogenesis, **265**
  prevention, **268**
  receptors, 114*t*
  treatment, **267**
  upper respiratory infection from, 920*t*
  virology, **264**
  virus-specific antibodies, 268*t*
ergosterol, 693, 707
  inhibition of, 709
erysipelas, **896-897**
  group A streptococci, 454
erythema infectiosum, **202-203**
erythema migrans, 660, 661*f*
erythema nodosum, in coccidioidomycosis, 752
erythrogenic toxin, 448
erythromycin, 404, 405*t*, 412, 423
  for actinomycosis, 510-511
  for *Bartonella*, 689
  for *Campylobacter jejuni*, 572
  for *Chlamydia pneumoniae*, 678
  for cholera, 570
  for *Haemophilus influenzae*, 558
  for Legionnaires disease, 614
  for pertussis, 563
  resistance, 467
  susceptibility patterns to, 427*t*
ESBL. *See* extended spectrum β-lactamase
Escherichia coli, 11, 122, 134, 275, 358, 366, 367, 380, 383, 387, 441, 486, 519, 538, 557, 580, **583-595**
  antigenic structure of, 583*f*
  bacteremia from extravascular infection and, 966, 966*t*
  bacteriophages, 111*f*
  clinical aspects, 593-595
  CNS infections and, 952*t*
  common, 584*t*
  cytotoxic necrotizing factor, 586
  diagnosis, 593-594
  diarrhea from, 936
  ear infections from, 908*t*
  enteroaggregative, 584*t*, 593
  enterohemorrhagic, 584*t*, 931, 932*t*
    epidemiology, 592
    pathogenesis, 592
  enteroinvasive, 584*t*, 593
  enteropathogenic, 584*t*, 590-592, 932*t*, 935
    attachment, 591*f*
    contract secretion system, 591*f*
    epidemiology, 590
    immunity, 592
    pathogenesis, 590-592
  enterotoxigenic, 584*t*, 929, 932*t*
    epidemiology, 589-590
    immunity, 590
    pathogenesis, 590
  epidemiology, 14
  etiologic agents, **940**
  intestinal infections, 589-592, 593
    clinical capsule, 589
  labile toxin, 586
  manifestations, 593-595
  opportunistic infection, 593
  opportunistic infections, 586-589, 593
    clinical capsule, 586
    meningitis, 589
    urinary tract infection, 587-588
  pili, 583-586
  prevention, 594-595
  Shiga toxin, 586
  stable toxin, 586
  suppurative thrombophlebitis and, 964*t*
  toxins, 586
  treatment, 594
  trimethoprim-sulfamethoxazole for, 594
  uropathic, 584*t*, 588
  UTI from, 939-940, 942
Escherichia coli secretion proteins (Esps), 590
Esps. *See Escherichia coli* secretion proteins
ETEC. *See* enterotoxigenic *Escherichia coli*
ethambutol
  for *Mycobacterium kansasii*, 503
  for tuberculosis, 499, 500
ethylene oxide gas, 49
Eubacterium, 11, 517, 517*t*
eukaryotic cells, features of, 7*t*
eustachian tube, 907
exanthem subitum, **203**
excess mortality, 174
exclusion, **418-419**
  barrier resistance, 420*f*
exclusionary effect, flora in, 12
exfoliatin, 431
exfoliative toxin, 436
exoenzyme S (ExoS), 618, 620*f*
exoerythrocytic schizogony, 782
exogenote, 378
ExoS. *See* exoenzyme S
exosporium, 357
exotoxin(s), **396-398**
  A-B, **396**, 620*f*
  membrane-active, **396**
  pore forming, 397*f*, 586
  protein, 580
  pyrogenic, 454-455
  superantigen, **396**, 397*f*
exotoxin A, 618
exponential kinetics, 46*f*
exponential phase, 368
extended spectrum β-lactamase (ESBL), 422
extensively drug-resistant tuberculosis (XDR-TB), 500
extracellular matrix, [illegible]
extracellular polyglycans, 915
extravascular infection, bacteremia from, **965-966**, 966*t*
extrinsic incubation period, 285
eye infections, **905-907**
  Chlamydia trachomatis and, 674-675
  common clinical conditions, **905**, 906*t*
  diagnosis, **906-907**
  etiologic agents, **906**, **906*t***
  inflammation, 905
  management, **907**
eye-to-eye transmission, **94**

F

F factor, 383
Fab. *See* antigen binding sites
Fab fragment, 389
facilitated diffusion, 359, 360*f*
factor H, 27, 538
facultative anaerobes, 806
facultative bacteria, 362
facultative Gram-positive bacilli, 472*t*
fallopian tubes, 545
famciclovir, 158*t*, **160**
families, 384
Fasciola spp., 880*t*, 882, 883*f*
fasciolopsiasis, prevalence of, 760*t*
fatal familial insomnia, 339*t*, **343**
Fc fragment, 38
Fc receptors, 39
fecal-oral spread, **94**
  caliciviruses, 277
  Entamoeba histolytica, 805
  hepatitis A, 227
  of rotavirus, 273
feedback inhibition, 370
fermentation, 360
  pathways, 361*f*
fetal anemia, 202
fetus, 9

fever, pyelonephritis and, 940
fever blisters, 253
FHA. *See* filamentous hemagglutinin
fibrin, 529
fibrin clots, 430
fibrinogen, 430
fibroblasts, 201
  granuloma, 496
fibronectin, 430, 434, 458, 659
fibronectin-binding proteins (FnBP), 430, 434
fibrosis, 674
filament, 356
filamentous hemagglutinin (FHA), 559
filariasis, prevalence of, 760*t*
Filarioidea, 851
filoviruses, 120, **283-284**, **293**
  classification of, 110*t*
  virion, 284
filtration, 50
fish tapeworm, **872-874**
  disease, **873-874**
    clinical aspects, 873-874
    epidemiology, 873
    manifestations, 873-874
  prevention, 874
  treatment, 874
fish tuberculosis, 504
FITC. *See* fluorescein isothiocyanate
flagella, 356*f*
  bacteria, **355-356**
  lophotrichous, 355
  monotrichous, 355
  polar, 355
  rotation, 373
flagellar stain, 64*f*
flagellates, **813-834**
  blood and tissue, **821-834**
  case study, 833
  classes of, 762*t*
  noninvasive luminal, **813-821**
flagellin, 356
flank pain, 940
flash autoclaves, 49
flavivirus, 153*t*, 239, **280-281**, 280*t*, 285, 291
Flavobacterium, 626*t*
Fleming, 4, 403
flies, 216
flora, 8-12, 508
  anaerobic, 518
  in blood, 9
  in body fluids, 9
  carrier state, 8
  in colon, 11
  diet and, 11
  at different sites, 9-12
  in exclusionary effect, 12
  in genitourinary tract, 11-12
  good, 12
  in immune system, 12
  interfering, 548
  in intestinal tract, 10-11
  in mouth, 10-11
  nature of, 9
  in opportunistic infection, 12
  origin of, 9
  in pharynx, 10-11
  potentially pathogenic, 10*f*
  residents, 8
  in respiratory tract, 11
  role of, 12
flora (*Cont.*):
  samples from, 60-61
  skin, 10
  in skin, 9-10
  stool, 11*f*
  in tissues, 9
  transients, 8
  in vagina, 11
fluconazole, 709
  for blastomycosis, 748
  for *Candida albicans*, 729
  features of, 708*t*
fluctuating arthritis, 660
flucytosine
  action of, 709*f*
  features of, 708*t*
  resistance, 711
fluid samples, 59-60
flukes. *See* trematode(s)
FluMist, 176
fluorescein, 618
fluorescein isothiocyanate (FITC), 76
fluorescence microscopy, 62*f*, **65**
fluorides, 913, 916
5-fluorouracil, 333
fluorochrome stain, 64
fluoroquinolones, 406*t*, 414, 942
  for *Campylobacter jejuni*, 572
  resistance, 419*t*
  for shigellosis, 598
  for tuberculosis, 500
FnBP. *See* fibronectin-binding proteins
folate antagonists, **774**
  for malaria, 789*t*
folate biosynthesis, 406*t*
folate deficiency, 774
folate inhibitors, **414**
  resistance, 419*t*
follicular hypertrophy, 675*f*
folliculitis, **895-896**, 899*t*
fomivirsen, 158*t*, **164**
Fonsecaea pedrosoi, 714*t*, 721
food poisoning, 388, **933-935**, 934*t*
  clinical and epidemiologic features of, 933, 934*t*, 935
  clostridial, 521, 522
  Staphylococcus aureus, 433, 438
food-borne transmission, of listeriosis, 478
Fore culture, 341
foreignness, 32-33
formaldehyde, 49, 214
  disinfection with, 52
  treatment, 526
foscarnet, 158*t*, **163**
frameshift mutation, 374, 375*f*
Francisella, **638-640**
  bacteriology, 638
  clinical capsule, 638
Francisella tularensis, upper respiratory infection from, 921
frequency, cystitis and, 940
fungal culture, **66-71**
fungal infections, **703-706**. *See also* antifungals
  cellular immunity, 706
  epidemiology, 703
  general aspects of, 703-706
  humoral immunity, 706
  pathogenesis, 703-704
fungal stains, 65
fungemia, 959
fungi. *See also* antifungals
  adherence, 704
  antibodies, **701-702**
  antigens, **701-702**
  arthroconidia in, 697
  cell wall, 694*f*
  chlamydoconidia in, 697
  classification, **698-702**
  conidia in, 695
  conidiophore in, 697
  culture, **701-702**
  dimorphism, 698, 701
  DTH and, 705
  growth, **695-698**
  immunity, **705-706**
    adaptive, 705-706
    innate, 705
  injury, 705
  invasion, 704-705
  laboratory diagnosis, **699-702**
    direct examination, 699-701
  macroconidia in, 697
  medically important, 699*t*
  metabolism, 694-695
  microconidia in, 697
  morphology, **695-698**
  mycotoxins, 705
  nature of, 7-8, 693-702
    antigenic structure, 71
    classifying, 70-71
    cultural characteristics, 70
    genomic structure, 71
    pathogenicity, 70-71
    replication of, 8
    toxin production, 70-71
  opportunistic, 699
  reproduction, 695
  spores, 695
  structure, 693-694
  subcutaneous, 699, **719-722**
  superficial, 699, **713-719**
  system view, 704*f*
  systemic, 699
    clinical case, 754
    features of, 740*t*
    geographic distribution of, 744*f*
furazolidone, for giardiasis, 821
furuncles, 433, **896**
  Staphylococcus aureus, 435*f*, 437
Fusarium
  eye infection from, 906*t*
  invasion, 700*f*
fusion protein, 177
Fusobacterium, 11, 518, 519
  plaque colonized by, 911
fusospirochetal disease, 918, 919, 921

G

G glycoproteins, 297
G6PD. *See* glucose-6-phosphate dehydrogenase
GABA, 776
GAD. *See* glutamic acid decarboxylase
gag, 306, 307*t*, 310
gamma globulin, 97
ganciclovir, 158*t*, **160-161**
  clinical use, **161**
  CMV and, 263
  oral, 161
  resistance, **161**
gangliosides, 568

gangrene, 897
gas, 900
Gardnerella vaginalis, 948
GAS. *See* group A streptococci
gas gangrene, 521, 522, 522*f*, 900
gastritis, *Helicobacter pylori*, **573-577**, 574f
epidemiology, 574-*575*
immunity, 576
gastroenteritis
acute viral, 272
salmonellosis, 603
winter, 272
gastrointestinal infections, **929-937**
candidate agents of, 936
clinical features, **929-930**, 932*t*-933*t*, 933, 935
diagnosis, **935-936**
endemic, **930-931**
epidemic, **931**
epidemiologic setting, **930-934**
etiologic agents, **930**, 932*t*-933*t*
features of, 932*t*-933*t*
food poisoning, **933-935**, 934*t*
management, **936-937**
GBS. *See* group B streptococci
gene expression, 372*f*
bacteria, **370-371**
general secretory pathway (GSP), 365, 366, 367*f*
generalized transduction, 381
genetic exchange, in bacteria, **378-384**
genetics
bacteria, 374-384
of resistance, **423-425**
transposition, **424-425**
transposons, **424-425**
of viruses, **130-133**
defective interfering particles, 131-132
mutation, **130-131**
recombination, **132-133**
von Magnus phenomenon, **131-132**
genital gonorrhea, 546
genital herpes, 946*t*-947*t*
genital infections, 676, **945-949**
cervicitis, 946*t*, **947-948**
diagnosis, 947, 947*t*
epididymitis, 945, 946*t*, **947**
genital ulcers, **945**, 947*t*
genital warts, **946**
lymphadenitis, **948**
PID, 946*t*, **948**
systemic, 945, **949**
urethritis, 939, 946*t*, **947**
vaginitis, 945, **948**
genital transmission, **94**, 139*t*
herpes simplex 2, 254-255
genital ulcers, **945**, 947*t*
genital warts, 331, **946**
genitourinary tract, microbial flora in, 11-12
genomes
bacterial, 81-82
viral, 81-82, **118-120**
replication, **121-124**
structure, **102**
genomic structure
bacteria, 71
fungi, 71
genotypic resistance, **164**
gentamicin, 405*t*, 410
for plague, 637-638
for *Pseudomonas aeruginosa*, 622
gentamicin (*Cont.*):
susceptibility patterns to, 427*t*
for tularemia, 640
genus, 384
German measles, 200
germination, 357
Gerstmann-Straussler-Scheinker syndrome, 101, 339*t*, **343**
Ghon complex, 497
Giardia, 770
Giardia lamblia, 813, 814*t*, **817-821**, 929, 931, 933*t*
cyst structures, 818*f*
disease caused by. *See* giardiasis
motility, 817
parasitology, 817-818
scanning electron micrograph of, 819*f*
trophozoite, 817, 818*f*
giardiasis, 761, **818-821**
clinical aspects, 820-821
clinical capsule, 818
diagnosis, 821
diarrhea, 820
epidemiology, 819
furazolidone for, 821
in homosexual men, 819
immunity, 820
lactose intolerance and, 820
manifestations, 820
metronidazole for, 821
paromomycin for, 821
pathogenesis, 820
prevalence of, 760*t*
prevention, 821
quinacrine hydrochloride for, 821
tinidazole for, 821
transmission of, 819
treatment, 821
Giemsa stain, 655, 788
gingivitis, **916**
gingivostomatitis, 253
gliotoxin, *Aspergillus*, 731
globoside, 201
glomerulonephritis, 786
immune complex, 826
Glossina, 827
glucan, 693, 734
glucose-6-phosphate dehydrogenase (G6PD), 783
glucuronoxylomannan (GXM), *Cryptococcus neoformans*, 739, 741
glutamate, 374, 414
glutamic acid decarboxylase (GAD), 154
glutaraldehyde
action of, 53*f*
disinfection with, 52
for sterilization, 47*t*
glycine, 526
glycolipids, 501, 586, 667
glycopeptide, 409-410
resistance, 419*t*
glycoproteins, 26, 179, 240
surface, 306
goats, 211
goblet cells, 667
Golgi apparatus, 127
gonococcus, cellular view, 537*f*
gonorrhea, 15, 394, **543-549**
anisomycin for, 548
antigenic variation in, 544
attachment, 544-545
cefixime for, 549
gonorrhea (*Cont.*):
ceftriaxone for, 549
ciprofloxacin for, 549
clinical aspects, 546-549
clinical capsule, 543
colistin for, 548
culture, 548
diagnosis, 548-549
direct detection, 548-549
dissemination, 545
epidemiology, 543-544
genital, 546
Gram smear, 548
immunity, 546
invasion, 544-545
manifestations, 546-547
in men, 547*f*
nystatin for, 548
ofloxacin for, 549
pathogenesis, 544-546
penicillin for, 549
pharyngeal, 921
prevention, 549
serology, 549
spread, 545
in submucosa, 545
treatment, 549
trimethoprim for, 548
vancomycin for, 548
virulence of, 545-546
in women, 547*f*
gp41 protein, 308, 310, 320
gp120, 310
Gram, Hans Christian, 61
Gram negative cell wall, 351*f*, **353-355**
outer membrane, 353
Gram negative envelope, 353*f*
Gram negative shock, 354
Gram positive cell wall, **351-352**, 351*f*
Gram positive envelope, 352*f*
Gram smear, 548
Gram stains, **61-62**, 64*f*
Haemophilus influenzae, 552*f*
Gram-negative bacteria, 61-62, 349, **517-518**
conjugation in, **383-384**
ear infections from, 908*t*, 909
eye infection from, 906*t*
osteomyelitis from, 902*t*
plaque colonized by, 911-912, 916
protein secretion in, 365, 366
septic arthritis from, 903
sinus infections from, 910*t*
Gram-negative secretion systems, 368*f*
Gram-positive bacteria, 61-62, 349
aerobic, features, 471*t*
conjugation in, **384**
ear infections from, 908*t*
nonsporulating, **517**
as plaque colonizer, 911
protein secretion in, 365
sinus infections from, 910*t*
in sputum, 485
granulocytes, 20*f*
innate immunity in, 23
granulocytopenia, 621
granuloma, 26
endarteritis, 648
fibroblasts, 496
inguinale, 946*t*-947*t*
lymphocytes, 496

granuloma (*Cont.*):
  macrophages, 496
  mycobacterial disease and, 491
  tuberculosis, 495, 496*f*
granulomatous encephalitis, 810
granulomatous lesions, **897**
granzymes, 34, 151
Gregg, Norman, 197
griseofulvin, 710, 718
  features of, 708*t*
group A antigen, 456
group A coxsackievirus, 215
  upper respiratory infection from, 920*t*
group A streptococci (GAS), 15, **446-457**
  acute, 451-452
  antigenic structure of, 447*f*
  bacteriology, 446-454
  biologically active extracellular products, 448
  cellular view of, 452
  cellulitis, 899*t*
  diagnosis, 456-457
  disease, 449-454, 450*f*
  epidemiology, 449-451
  erysipelas, 454
  Gram stain, 444*f*
  immunity, 453-454
  impetigo, 449, 454
  M protein and, 446-448
  manifestations, 454-456
  nephritogenic strains, 451
  osteomyelitis from, 902*t*
  pathogenesis, 451-453
  penicillin for, 425
  pharyngitis, 449, 454
  prevention, 457
  puerperal infections, 449, 454
  pyrogenic exotoxins, 454-455
  septic arthritis from, 903
  structure, 446-448
  suppurative thrombophlebitis and, 964*t*
  surface molecules, 448
  toxic shock syndrome, 35
  treatment, 457
  upper respiratory infection from, 920-921, 920*t*
  wounds, 449
group B coxsackievirus, 215, 217, 221
  upper respiratory infection from, 920*t*
group B streptococci (GBS), 426, **457-465**
  aminoglycosides for, 459
  ampicillin for, 459
  autolysins in, 461
  bacteriology, 457-459
  clinical capsule, 457
  CNS infections and, 952*t*
  diagnosis, 459
  ear infections from, 908*t*
  epidemiology, 458
  neonatal sepsis, 458
  osteomyelitis from, 902*t*
  pathogenesis, 458-459
  penicillin for, 459
  pneumococcal disease, 458*f*, **461-465**
  pore forming toxins in, 461
  prevention, 460
  septic arthritis from, 903
  shape, 348*f*
  suppurative thrombophlebitis and, 964*t*
  treatment, 459
group translocation, 359
growth impairment, 901
GSP. *See* general secretory pathway
Guarnieri bodies, 208
Guillain-Barré syndrome, 301, 572
gumma, cardiovascular, 650
GXM. *See* glucuronoxylomannan

H
H antigen, 580
H5N1. *See* avian influenza virus
HAART. *See* highly active antiretroviral therapy
Haemophilus aegyptius, 906, 906*t*
Haemophilus ducreyi, 552*t*, **558**, 899*t*, 946*t*
  AIDS and, 558
Haemophilus influenzae, 409, 425, 538, 551-558
  acute epiglottitis, 555
  arthritis, 556
  azithromycin for, 558
  bacteremia from extravascular infection and, 966*t*
  ceftriaxone for, 558
  cellulitis, 556, 899*t*
  cephalosporins for, 557
  ciprofloxacin for, 558
  clinical aspects, 555-558
  clinical capsule, 553
  CNS infections and, 951, 952*t*, 955
  diagnosis, 556-557
  disease, 553-555
    cellular view, 554
    epidemiology, 553
    invasive, 553-555
    pathogenesis, 553-555
  ear infection from, 908-909, 908*t*
  erythromycin for, 558
  eye infection from, 906*t*
  Gram stain, 552*f*
  influenza A and, 175
  localized, 555
  lower respiratory tract infections from, 926*t*
  manifestations, 555-556
  meningitis, 555
  middle respiratory infections and, 922-923
  pneumonia, 556
  prevention, 557
  rifampin for, 415, 558
  septic arthritis from, 903
  sinus infections from, 909, 910*t*
  suppurative thrombophlebitis and, 964*t*
  treatment, 557
  vancomycin for, 558
Haemophilus spp., 11, 79, **551-558**
  bacteriology, 551-552
  disease overview, 554*f*
  eye infection from, 906, 906*t*
  features of, 552*f*
hairy leukoplakia, 267
hakuri, 275
halofantrine, 773-774, 774
halogens, disinfection with, 51
HAM. *See* HTLV-associated myelopathy
hamsters, 293
hand-foot-and-mouth disease, 222*f*
handwashing, 54
hantavirus hemorrhagic fever, **293-294**
hantaviruses, **293-295**
  radiographs in, 294*f*
  in United States, 294*f*
HBcAg. *See* hepatitis B core antigen
HBIG. *See* hepatitis B immune globulin
HBsAg. *See* hepatitis B surface antigen
HCC. *See* hepatocellular carcinoma
HCV. *See* hepatitis C virus
H&E. *See* hematoxylin and eosin
heat-shock response, 371, 698
heavy lines, 123
heavy metals, **772-773**
helical symmetry, viruses with, **125**
Helicobacter, **573-577**
  features of, 566*f*
Helicobacter pylori, 3
  amoxicillin for, 576-577
  bacteriology, 573
  clarithromycin for, 576-577
  disease
    clinical aspects, 576-577
    diagnosis, 576
    manifestations, 576
    prevention, 576-577
    treatment, 576-577
  gastritis, **573-577**, 574*f*
    clinical capsule, 573
    epidemiology, 574-*575*
    immunity, 576
  injection secretion system, 573
  pathogenesis, 575-576
  tetracycline for, 576-577
  urease, 573
  vacuolating cytotoxin in, 573
helminths, 759, **762-763**
  antiparasitics for, 772
  classification, 762-763, 763*f*
  cuticle of, 769
  morphology, 762-763
  oviparous, 763
  physiology, 763
  viviparous, 763
helper T cells, 30*t*
hemadsorption, 72, 128
  inhibition, 169
hemagglutination inhibition (HI), 175, 178-179
hemagglutinins, 71-72, 143, 168, 189, 193
  assay, **128**
  cold, 667
  in influenza, 171
  inhibition, 169
  viral, 76*f*
hematin, 551, 557
hematopoietic stem cells, 20*f*
hematoxylin and eosin (H&E), 510, 700, 746
hematuria, 184
hemoglobin S, 783
α-hemolysis, 444
hemolysin, in clostridia, 517
hemolysis, 69
β-hemolysis, 444
β-hemolytic streptococci, 456*t*
α-hemolytic streptococci, 456*t*
hemolytic uremic syndrome (HUS), 592
hemoptysis, 497
hemozoin, 783
Hendra, 284
Hendra virus, 295
henipaviruses, **284**, **295**
Hepacivirus, 239
hepatic abscess, amebiasis, 808
hepatitis A, **225-229**
  active immunization, **229**
  acute, 227*f*
  anicteric, 228
  clinical aspects, **228-229**

hepatitis A (*Cont.*):
- comparison of, 226*t*
- diagnosis, **229**
- epidemiology, **227**
- fecal-oral transmission of, 227
- food poisoning from, 934*t*
- hepatitis E and, 244
- manifestations, **228-229**
- passive immunization, **229**
- pathogenesis, **227**
- prevention, **229**
- receptors, 114*t*
- replication of, 225-226
- structure, 226*f*
- treatment, **229**
- vaccine, 229
- virology, **225-226**

hepatitis B, 56, 153*t*, **229-237**
- antibodies, 235*t*
- antigens, 235*t*
- chronic, 235
- chronic carriers, 232
- cirrhosis in, 234*f*
- classification of, 111*t*
- clinical aspects, **234-237**
- clinical capsule, 232
- comparison of, 226*t*
- diagnosis, **235-236**
- epidemiology, **232-233**
- hepatitis D and, 238
- HIV and, 232
- manifestations, **234-235**
- needlestick transmission, 233
- nomenclature, 235*t*
- pathogenesis, **233-234**
- prevention, **236-237**
- replication, **230-232**, 231*f*
- schematic diagram, 230*f*
- self-limited cases, 236*f*
- serotypes of, 231
- STD from, 945, 946*t*
- structure, **229-230**
- treatment, **236**
- vertical transmission, 233
- virology, **229-230**
- worldwide distribution, 233*f*, 238*f*

hepatitis B core antigen (HBcAg), 230, 233

hepatitis B immune globulin (HBIG), 236
- postexposure prophylaxis with, 237

hepatitis B surface antigen (HBsAg), 152, 209, 230, 233, 236, 237, 238
- antibody to, 234

hepatitis C, 56, **239-244**
- adaptive immunity, 241
- antibodies, 241
- clinical aspects, **242-243**
- clinical capsule, 240
- comparison of, 226*t*
- diagnosis, **242-243**
- epidemiology, **240-241**
- manifestations, **242**
- mutations, 240
- pathogenesis, 241-242
- prevention, 243
- transmission, 240
- treatment, 243
- virology, **239**

hepatitis C virus (HCV), 150

hepatitis D, **237-239**
- clinical aspects, **238**
- comparison of, 226*t*
- diagnosis, **239**
- hepatitis B and, 238
- manifestations, **238**
- prevention, **239**
- risk, 238
- treatment, **239**
- virology, **237-238**
- worldwide distribution, 238*f*

hepatitis E, **243-244**
- comparison of, 226*t*
- distribution of, 244*f*
- hepatitis A and, 244

hepatitis G, **244-245**
- in AIDS, 244

hepatitis viruses, **225-245**
- comparison of, 226*t*
- incubation period, 141*t*

hepatocellular carcinoma (HCC), 149, 232, 233, 234

hepatocytes, 232, 237

hepatosplenomegaly, 687, 688

hermaphrodite trematodes, 879, 885

herpangina, 222, 222*f*

herpes, genital, 946*t*-947*t*

herpes simplex virus
- AIDS in, 256
- CNS infections and, 952*t*
- eye infection from, 906-907, 906*t*
- genital infection from, 945, 946*t*, 948
- herpetic paronychia and, 897
- upper respiratory infection from, 920*t*

herpes simplex virus 1, 108, 109*f*, 122, 248*t*
- acute infections, 251
- clinical aspects, **253**
- conjunctival infection, 254
- corneal infection, 254
- diagnosis, **255**
- encephalitis, 254
- epidemiology, 251
- herpes simplex virus 2 and 250
  - cross protection, 252
- immunity, 252-253
- latent infection, 252
- lesion, 253*f*
- manifestations, **253-254**
- multinucleated giant cells from, 252*f*
- neonatal, **255**
- prevention, **256**
- primary infection, 253
- receptors, 114*t*
- replication, **249**, 250*f*
- treatment, 159, **256**
- virion structure of, 248*f*

herpes simplex virus 2, 248*t*
- acute infections, 251
- diagnosis, **255**
- genital transmission, 254-255
- herpes simplex virus 1 and, 250
  - cross protection, 252
- immunity, 252-253
- latent infection, 252
- manifestations, **254-255**
- neonatal, **255**
- prevention, **256**
- primary infection, **254-255**
- recurrent, **255**
- treatment, **256**

herpes zoster, 257
- of thorax, 259*f*

herpesviruses, 108, 203, **247-270**
- classification of, 111*t*
- clinical capsule, 251
- human, 248*t*
- incubation period, 141*t*
- morphology, 247
- mutations, 164
- replication, **249**
- strains, 250
- ulcers from, 899*t*
- virology, **247-249**

herpetic paronychia, 897

herpetic whitlow, 253

Heterophyes spp., 880*t*

heterotrophic metabolism, 694-695

hexachlorophene, 52

HFR. *See* high-frequency recombination

HGA. *See* human granulocytic anaplasmosis

HI. *See* hemagglutination inhibition

Hib capsule, 552, 555

high-frequency recombination (HFR), 133

highly active antiretroviral therapy (HAART), 317

highly selective media, 87

hip infections, 901, 904

histamine, 26

histology, 73

Histoplasma capsulatum, 698, **743-746**, 906*t*
- antibodies, 746
- antigens, 744
- blastoconidia, 743
- chronic granulomatous infection from, 954*t*
- disease caused by. *See* histoplasmosis
- features of, 740*t*
- granulomatous response in, 744-745
- growth of, 744
- lower respiratory tract infections from, 926*t*
- microconidia, 743
- tuberculate macroconidia, 743

Histoplasma duboisii, 743-746

Histoplasma farciminosum, 743-746

Histoplasma spp., **743-746**
- classification of, 699*t*

histoplasmosis, 318*t*, **743-746**
- amphotericin B for, 746
- clinical aspects, 745-746
- disseminated, 745
- epidemiology, 744
- immunity, 745
- itraconazole for, 746
- manifestations, 745-746
- pathogenesis, 744-745

HIV. *See* human immunodeficiency virus

HIV-2. *See* human immunodeficiency virus 2

HLA. *See* human leukocyte antigen

HME. *See* human monocytic ehrlichiosis

hMPV. *See* human metapneumovirus

HMW1, 552

HMW2, 552

homologous recombination, 376*f*, **377**

homosexual men, 313, 329
- giardiasis in, 819

honey, 525

Hong Kong flu, 171

hooklets, 763

hookworm, 836*t*, **844-846**
clinical aspects, 845-846
disease, 845-846
diagnosis, 845-846, 846
epidemiology, 845
immunity, 845
pathogenesis, 845
prevention, 846
treatment, 846
manifestations, 845-846
prevalence of, 760*t*
horizontal transmission, 92, 139
Hortaea werneckii, 714*t*
hospital personnel, nosocomial infections from, 54-55
hospital ward, asepsis in, 57
hospital-associated diarrhea, **935**
hosts
defenses in viral infection, **150-151**
definitive, 764, 792-793
factors in viral infection, **149-150**
intermediate, 764
in *Toxoplasma gondii*, 793
range, 115
HPV. *See* human papilloma virus
HRP2, 788
HTIG. *See* human tetanus immunoglobulin
HTLV-associated myelopathy (HAM), 323
human granulocytic anaplasmosis (HGA), 688
human herpes 6, 247, 248*t*, **268-269**
diagnosis, **269**
epidemiology, **268**
in immunosuppression, 269
manifestations, **268-269**
treatment, **269**
human herpes 7, 247, 248*t*, **269**
receptors, 114*t*
human herpes 8, 247, 248*t*, **269-270**
human immunodeficiency virus (HIV), 72, 109*f*, 305. *See also* acquired immunodeficiency syndrome
accessory proteins, 311-312
clades, 314-315
clinical aspects, 316-322
clinical latency, 315-316
CMV and, 263
ELISA, 319
geographic distribution, 314-315
hepatitis B and, 232
immune control in, 315
immune deficiency in, 316
immunosuppression and, 154-155
infection with, 315
inhibitors of, **162-163**
manifestations, 316-319
mortality rates and, 317*f*
mutant forms, 315
occurrence, 314
pathogenesis, 315-316
plasma levels, 315
prevention, 321
receptors, 114*t*
regulatory proteins, 311-312
resistance, 321
reverse transcriptase, 310
sexual transmission of, 313
structure of, 307*f*
surface glycoprotein, 306
syphilis and, 647, 651
temporal changes in viral load, 316*f*
human immunodeficiency virus (HIV) (*Cont.*):
transmission of, 56, 313-314
treatment, 320-321
Trichomonas vaginalis, 816
tuberculosis and, 500
in United States, 314
Western blot detection of, 319*f*
human immunodeficiency virus 2 (HIV-2), 314
human leukocyte antigen (HLA), 150
human metapneumovirus (hMPV), 167, **182**
human monocytic ehrlichiosis (HME), 688
human papilloma virus (HPV), 145, 327, 329, 946
diagnosis of, 332
electron micrograph of, 328*f*
external genital, 331
manifestations, 330-332
prevention, 333
replication cycle of, 327-328
treatment, 333
vaccines, 333
human poxviruses, 206*t*
human T-cell lymphotropic virus, 145, 305, **322-324**
diagnosis, 324
ELISA and, 324
epidemiology, 323
incubation period, 142*t*
latency period, 323
manifestations, 323
pathogenesis, 323
prevention, 324
transmission, 323
treatment, 324
virology, 322-323
human tetanus immunoglobulin (HTIG), 528
human viruses, 101, **127-135**
budding, **127-128**
cell death, **127**
cell survival, **128**
classification of, 110*t*
enveloped, **116**
latent state, **133**
lysogeny, **133-135**
naked capsid, **116-117**
unclassified, 110*t*
humoral immunity, 15, 30
fungal infections, 706
HUS. *See* hemolytic uremic syndrome
hwp1. *See* hyphal wall protein
hyaluronic acid capsule, 448
hyaluronidase, 448
hydrocephalus, 951
hydrochloric acid, 389
hydrogen peroxide, 25, 361, 516
in anaerobiosis, 516
disinfection with, 51
for sterilization, 47*t*
hydrogen sulfide, 88
hydrolytic enzymes, **396**
hydrophobia, 301
hydrops fetalis, 202
hydroxyapatite, 916
hydroxynaphthoquinones, 774
Hymenolepis nana, 865, 866*t*, **874**
hyperbaric oxygen, 523
hypercapnia, 181
hyperexpansion, 181
hyperplasia, 210
hyperreflexia, 340
hypersensitivity, 41
antibody-mediated, 42
delayed-type, 42, 398, 491, 495, 649
immune-complex, 42
to microfilariae, 862
hypertrophy, follicular, 675*f*
hypervariable regions, 240
hyphae, 695, 725, 726
nonseptate, 696*f*, 733
septate, 696*f*
hyphal wall protein (hwp1), 725
hypochlorhydria, 820
hypochlorite, 51, 214
hypokalemia, 569
hypoxemia, 181

I

ICAM-1. *See* intercellular adhesion molecule 1
icosahedral symmetry, **126**
viruses with, **126**
icosahedron, 106, 107*f*
ICTV. *See* International Committee for Taxonomy of Viruses
idiotypes, 39
idoxuridine, **159**
IFA. *See* indirect fluorescent antibody
IFN-γ. *See* interferon-gamma
IgA. *See* immunoglobulin A
IgE. *See* immunoglobulin E
IgG. *See* immunoglobulin G
IgG4 blocking antibodies, 887
IgM. *See* immunoglobulin M
IgM/IgG switch, 40
IL-12. *See* interleukin 12
imipenem, 405*t*
for anaerobic infections, 520
for *Bacteroides fragilis*, 532
for *Nocardia*, 513
susceptibility patterns to, 427*t*
immune complex glomerulonephritis, 826
immune deficiency
in HIV/AIDS, 316
upper respiratory infections and, 920
immune response, 19-43
cellular, 198
favorable use of, 43
misdirected, **398**
virulence and, 15
immune serum globulin (ISG), 97, 229
immune suppression, 769
immune system, flora in, 12
immune-complex hypersensitivity, 42
immunity, 19
acquired, 31*f*
active, **229**
adaptive, 19, 150, **152**, 241
fungi and, 705-706
cell-mediated, 15, 30, 35-36, 491, 497
cellular, 706
humoral, 15, 30
fungal infections, 706
infectious diseases, 15-16
innate, 19, 21-30, 150
fungi and, 705
natural, 43
passive, 43, **229**
transient, 178
immunization, 17-18
general principles of, **96-97**

immunization (*Cont.*):
  pertussis, 18, 560
  strategies, 18
immunoassays
  enzyme, 76, **77**
  radio, 76, **77**
  Western blot, **79**
immunodiffusion, 746
immunodominant mannoproteins, 741
immunofluorescence, 65, 76-77, 79, 563
  direct, 66*f*
  indirect, 66*f*
immunoglobulin A (IgA), 39, 199
  secretory, 40, 275, 389, 674
immunoglobulin E (IgE), 768
  schistosomiasis and, 888
immunoglobulin G (IgG), 39-40, 79, 97, 199, 216, 228*f*, 239, 267, 394, 430, 538, 560, 652, 753
  antimannan, 727
  schistosomiasis and, 888
  in toxoplasmosis, 796
  transplacental, 43
  type-specific, 453
immunoglobulin M (IgM), 39, 40, 79, 201, 216, 228*f*, 239, 667
  African trypanosomiasis and, 828
  schistosomiasis and, 888
  structure, 39
  in toxoplasmosis, 796
immunoglobulins, 38
  functional properties of, 39-40
immunologic assay, **129**
immunologic reactions, adverse effects of, 41-42
immunologic systems, **73-80**
  antibody detection, **78-79**
  antigen-antibody reaction, **74-77**
  serologic classification, **77-78**
immunopathology, virus-induced, **152-154**
immunoreconstitution inflammatory syndrome (IRIS), 320
immunoresponsive cells and organs, innate immunity, 22-25
immunosuppression
  HIV and, 154-155
  rubella and, 155
  virus-induced, **154-155**
impetigo, **896**
  epidemic, 896
  group A streptococci, 449, 454
  Staphylococcus aureus, 438
IN. *See* integrase
in vivo isolation, for viruses, 72-73
inactivated vaccines, 17-18
  polio, 220
incidence, 95
  of viral infections, 137
inclusion conjunctivitis, 946*t*
  Chlamydia trachomatis and, 675-676
incubation period, **92**
  childhood exanthems, 141*t*
  diarrhea viruses, 141*t*
  enteroviruses, 141*t*
  hepatitis viruses, 141*t*
  herpesviruses, 141*t*
  human T-cell lymphotropic virus, 142*t*
  papovaviruses, 142*t*
  poxviruses, 141*t*
  respiratory viruses, 141*t*
  retroviruses, 142*t*
incubation period (*Cont.*):
  viral infection, 140
  zoonotic viruses, 142*t*
India ink, 742
India ink capsule stain, 64*f*
indicator media, **69**
indinavir, 158*t*, 163
indirect fluorescent antibody (IFA), 683
indirect immunofluorescence, 66*f*
indirect samples, 60
indole, 88
inducers, 371
inducible genes, 372*f*
Industrial Revolution, 493
INF. *See* interferons
infant botulism, **525**
infant pneumonia syndrome, 676
infantile laryngeal papillomas, 329
infections, 3-18, **91-92**. *See also specific types*
  eye, 674-675
  low-grade smoldering, 902
  rickettsial zoonotic, 630*t*
infectious agents, 4-8, 6*f*. *See also specific agents*
  features of, 5*t*
infectious diseases, 12-18
  clinical aspects of, 16-18
  communicability, **91**
  diagnosis, 17
  emerging, 90*f*
  epidemiology, 13-14
  immunity, 15-16
  manifestations, 16-17
  mortality rates for, 1*f*
  nucleic acid methods, **81-83**
  pathogenesis, 14-15
  prevention, 17-18
  sources, **91**
  treatment, 17
infectious mononucleosis, **266**, 267, 920
infectious subviral particle (ISVP), 273
infective endocarditis, **959-963**
  clinical features, **961**
  diagnosis, **962**
  etiologic agents, **961-962**, 962*t*
  management, **962**
  pathogenesis, **960**, 960*f*
infectivity, 95
  of viral infections, 137
inflammation
  acute, 26
  eye, 905
  injury from, 15
  in innate immunity, 26
  persistent, **398**
  in respiratory syncytial virus, 180*f*
inflammatory exudates, 919
influenza, 96, 109*f*, **168-177**
  antigenic drift, 172*f*
  antigenic shift, 172*f*
  antivirals for, 176*t*
  differences among, 168*t*
  excess mortality, 174
  group characteristics, **168-169**
  hemagglutinins in, 171
  life cycle, 170*f*
  lower respiratory tract infections from, 926*t*
  neuraminidases in, 171
  pandemic, 173
  reassortment of, 133*f*
  upper respiratory infection and, 920*t*
influenza A, 168, **169-177**
  antigenic drift in, 171
  antigenic shift in, 171-172
  diagnosis, **175**
  diagrammatic view of, 169*f*
  direct droplet spread, 173-174
  epidemiology, **173-174**
  Haemophilus influenzae and, 175
  immunity, **174**
  major antigenic shifts, 173*t*
  manifestations, **175**
  pathogenesis, **174**
  prevention, **176-177**
  receptors, 114*t*
  Staphylococcus aureus and, 175
  Streptococcus pneumoniae and, 175
  superinfection in, 175
  treatment, **175-176**
  vaccines, 176-177
  virus-coded proteins, 170*t*
influenza B, 168, 173
influenza C, 168, 173
inguinal lymphadenitis, **948**
inhibition zone, 416
inhibitors of attachment, **157**
inhibitors of entry, 158*t*
inhibitors of penetration, **157-159**
inhibitors of uncoating, **157-159**
injection secretion system, 573, 582
innate immunity, 19, 21-30, 150
  cell response in, 23-25
  chemical mediators in, 26-30
  complement system in, 26-28
  cytokines in, 28-30
  features of, 21*t*
  fungi and, 705
  granulocytes in, 23
  immunoresponsive cells and organs, 22-25
  inflammation in, 26
  monocytes in, 23
  mucosa in, 21
  physical barriers in, 22
  skin in, 21
insecticides, 760, 790
insects, 216
insertion sequences, **377**
insertional mutagenesis, 148, 322
insertions, 374, 375*t*
integrase (IN), 306
integrins, 216
intercellular adhesion molecule 1 (ICAM-1), 185
interference, 72, 143
interferon α, 158*t*, 162, 164, 236
interferon gamma, 491, 498, 502
interferon XX, 706
interferon-gamma (IFN-γ), 152
interferons (INF), 28, **150-151**, **163-164**
  antiviral action of, 29*f*
  pathway, 151*f*
interleukin 2, 502, 727
interleukin 12 (IL-12), 491, 706
interleukins, 28, 152, 180, 241, 397
intermediate hosts, 764
  Toxoplasma gondii, 793
internalin, 477
  Listeria monocytogenes, 479*f*, 480
International Committee for Taxonomy of Viruses (ICTV), 108
intertrigo, 899*t*

intestinal infections
  Enterobacteriaceae, 581
  Escherichia coli, 589-592, 593
    clinical capsule, 589
intestinal nematodes, **835-849**
  case study, 849
  life cycles of, **836-837**
intestinal tapeworms, 866*t*
intestinal tract, 390*t*
  microbial flora in, 10-11
intimin receptor, 590
intra-arterial infection, 963
intracardiac infection. *See* endocarditis
intracranial abscess, 952*t*
intracranial pressure, 951, 956-957
intranuclear incursions, 261*f*
intrauterine devices, 948
intravascular infection, **959-965**
  bacteremia from extravascular infection, **965-966**, 966*t*
  blood culture for, **967-969**
  infective endocarditis, **959-963**
    clinical features, **961**
    diagnosis, **962**
    etiologic agents, **961-962**, 962*t*
    management, **962**
    pathogenesis, **960**, 960*f*
  intravenous catheter bacteremia, **964-965**, 965*f*
  mycotic aneurysm, **963**
  sepsis, 959, **966-967**
  suppurative thrombophlebitis, **963-964**, 964*t*
intravenous catheter bacteremia, **964-965**, 965*f*
intravenous drug use, 313
intravenous immune globulin (IVIG), 240
intrinsic resistance, **423**
Inuits, 855
invasins, 392, 605
invasion
  of bacteria, **392-393**, 393*f*
  Candida albicans, 726*f*
  fungi, 704-705
  Fusarium, 700*f*
  gonorrhea, 544-545
  Salmonella, 596*f*
  Shigella flexneri, 596*f*
invasion plasmid antigens, 597
inversions, 375*t*
invertible element, 377
Iodamoeba, 803
iodine, 51, 677, 872*f*
  stain, 65*f*
5'-iododeoxyuridine, 122
iodophors, 51
  for sterilization, 47*t*
iodoquinol, 777*t*
ionizing radiation, for sterilization, 47*t*, 49-50
Iraq, 483
iridocyclitis, 906*t*
IRIS. *See* immunoreconstitution inflammatory syndrome
iron, 393, 537
  deficiency, 846
ISG. *See* immune serum globulin
isolation procedures, 57
isoniazid
  for *Mycobacterium kansasii*, 503
  for tuberculosis, 499, 500
Isospora, 800
ISVP. *See* infectious subviral particle
itraconazole, 709, 721, 825
  for aspergillosis, 732
  features of, 708*t*
  for histoplasmosis, 746
ivermectin
  for lymphatic filaria, 860
  onchocerciasis for, 862
IVIG. *See* intravenous immune globulin
Ixodes, 659, 688

J
Japanese B encephalitis, **290**
jaundice, 228, 234
JC virus (JCV), 333, 335
JCV. *See* JC virus
Jefferson, Thomas, 205
Jenner, Edward, 208
joint infections
  hip, 901, 904
  osteomyelitis, **901-902**
  septic arthritis, **903-904**, 903*t*-904*t*
Jones criteria, 455

K
K antigen, 579
kala azar, 761, 822, **826-827**
  diagnosis, 826-827
  epidemiology, 826
  manifestations, 826
  mortality rate, 827
  pathogenesis, 826
  treatment, 826-827
kallikrein, 26
Kaposi sarcoma-associated herpes virus (KSHV), 248*t*, 269, 318, 318*t*
karyosome, 761
Katayama syndrome, 767, 888-889
keratinized skin layers, infection of, **896**
keratinocytes, 328
keratitis, 906*t*
keratoconjunctivitis, 184, 906, 906*t*
ketoconazole, 709, 718, 825
  features of, 708*t*
KHF. *See* Korean hemorrhagic fever
killed vaccines, 43
  viral, 176
killing, 45
  microbial, **46-47**
Kingella kingae, osteomyelitis from, 902*t*
kissing bug, 831
Klebsiella, 580, 585*t*, **606-607**
  CNS infections and, 952*t*
Klebsiella pneumoniae, 908*t*
Koch, Robert, 3-4, 483
KOH. *See* potassium hydroxide
Koplik's spots, 194, 195
  oral, 195*f*
Korean hemorrhagic fever (KHF), 293
Krebs cycle, 360
KSHV. *See* Kaposi sarcoma-associated herpes virus
Kupffer cells, 227
kuru, 101, 339*t*, **340-341**

L
L protein, 282
L1, 327
L2, 327
labeling methods, **76-77**
labile toxin (LT), *Escherichia coli*, 586
laboratory diagnosis, 59-88
  of fungi, **699-702**
    direct examination, 699-701
  specimen, **59-61**
  tuberculin, 498-499
laboratory processing, **968-969**
β-lactamase inhibitors, **409**, **422-423**
  resistance, 419*t*
β-lactams, 405*t*, 406-409. *See also specific drugs*
  clinical use, **409**
  resistance, 467, 572
  structure of, 406, 407*f*
  toxicity of, 409
lactate dehydrogenase, 788
lactobacilli, 12
  plaque colonized by, 911, 915
Lactobacillus, 11
Lactobacillus acidophilus, 913
Lactobacillus rhamnosus, 12
lactoferrin, 359, 389
lactophenol, 701
lactose intolerance, giardiasis and, 820
lag period, 368
LAIV. *See* live attenuated influenza vaccine
LAM. *See* lipoarabinomannan
lamina propria, 854
lamivudine (3TC), 158*t*, **162**, 236
Lancefield, Rebecca, 444
Lancefield antigens, 444
laryngeal papilloma, 946*t*
laryngitis, 922, 923*t*, 924
laryngotracheitis, 923
laryngotracheobronchitis, 923-924
Lassa fever, 161, 292
latent infection, 143
  CMV, 261
  herpes simplex virus 1, 252
  herpes simplex virus 2, 252
  HIV, 315-316
latent period, 113
latent state, **133**
  human viruses, **133**
latent syphilis, **649-650**
LCR. *See* ligase chain reaction
LDL. *See* low-density lipoprotein
lectins, 24
  pathway, 27-28
Lederberg, J., 383
Legionella, **609-615**, 926*t*, 966*t*
  bacteriology, 609
  classification, 609-610
  growth, 609-610
  morphology, 609
  pneumonia, 611*f*
  structure, 609
Legionella bozemanii, 610
Legionella dumoffii, 610
Legionella longbeachae, 610
Legionella micdadei, 610
Legionella pneumophila, 51, 78, 79, 388, 412, 513
  multiplication of, 612*f*
  pneumonia from, 925, 926*t*
Legionnaire's disease, 14, 78, 388, **610-613**, 612*f*
  clinical aspects, 613-614
  clinical capsule, 610
  diagnosis, 613
  epidemiology, 610-611
  erythromycin for, 614
  immunity, 613
  manifestations, 613

Legionnaire's disease (*Cont.*):
- pathogenesis, 611
- prevention, 614
- quinolones for, 614
- rifampin for, 614
- tetracycline for, 614
- treatment, 614

Leishman-Donovan bodies, 825, 826
Leishmania, 768, 821, **822-827**
- chronic infection with, 823
- parasitology, 822
- transmission, 822-823

Leishmania braziliensis, 823
Leishmania donovani, 823
Leishmania mexicana, 823
leishmaniasis, 761
- cutaneous, 824*f*, 897
- disseminated visceral, **826-827**
- immune response to, 824*t*
- localized cutaneous, **823-825**
  - epidemiology, 823-824
  - manifestations, 824-825
- mucocutaneous, **825-826**
- prevalence of, 760*t*

lentiviruses, 305
lepromatous leprosy, 502, 502*f*
leprosy, **501-503**
- clinical aspects, 502-503
- clinical capsule, 501
- clofazimine for, 503
- diagnosis, 503
- epidemiology, 501
- immunity, 502
- lepromatous, 502, 502*f*
- manifestations, 502-503
- pathogenesis, 501-503
- prevention, 503
- rifampin for, 503
- sulfones for, 503
- treatment, 503
- tuberculoid, 502

Leptospira, 630*t*, 954*t*
- bacteremia from extravascular infection and, 966*t*

Leptospira interrogans, **652-654**
- bacteriology, 652-653
- features of, 645*t*
- serogroups, 653

leptospirosis, 630*t*, **653-655**
- ceftriaxone for, 654
- clinical aspects, 654
- clinical capsule, 653
- diagnosis, 654
- doxycycline for, 654
- epidemiology, 653
- immunity, 653-654
- manifestations, 654
- pathogenesis, 653-654
- penicillin for, 654
- prevention, 654
- treatment, 654

lethal factor (LF), 482, 482*f*
leukocyte esterase, 941
leukocytes, 194, 529
- polymorphonuclear, 30*t*, 458

leukocytosis, 211
LF. *See* lethal factor
LGV. *See* lymphogranuloma venereum
ligase chain reaction (LCR), in *Chlamydia trachomatis*, 677
light microscopy, **61-65**
- acid-fast stain, **63-64**
- Gram stain, **61-62**

linezolid, 405*t*, 412
lipid A, 354
lipid bilayers, 177
lipoarabinomannan (LAM), 489
lipooligosaccharide (LOS), 545
- meningococcal disease, 537
- *Neisseria*, 535
- *Neisseria gonorrhoeae*, 542, 543

lipopolysaccharide (LPS), 23, 354, 518, 545
- endotoxin, 398, 580, 602, 636
- meningococcal disease, 537
- *Neisseria*, 535
- *Neisseria gonorrhoeae*, 542
- *Pseudomonas aeruginosa*, 618
- structure, 354*f*
- *Treponema pallidum*, 646

lipoteichoic acid (LTA), 446, 451
liquid nitrogen, 333
Listeria, 392, 513
- bacteremia from extravascular infection and, 966*t*

Listeria monocytogenes, 471, **477-481**, 480*f*, 597, 952*t*
- bacteriology, 477
- biofilm, 478
- in cytosol, 479
- diagnosis, 481
- epidemiology, 478
- immunity, 480
- internalin, 478*f*, 480
- manifestations, 480-481
- pathogenesis, 479-480
- prevention, 481
- treatment, 481
- virulence, 477, 480

listeriolysin O (LLO), 477, 479
listeriosis, 477-480
- AIDS in, 480, 481
- ampicillin for, 481
- cellular view, 479*f*
- clinical aspects, 480-481
- diagnosis, 481
- food-borne transmission of, 478
- manifestations, 480-481
- overview, 478*f*
- pathogenesis, 479-480
- penicillin for, 481
- prevention, 481
- transplacental transmission, 478
- treatment, 481
- trimethoprim-sulfamethoxazole for, 481

live attenuated influenza vaccine (LAIV), 176
live vaccines, 17-18
liver flukes. *See Clonorchis sinensis*
LLO. *See* listeriolysin O
Loa loa, 851, 858*t*, **862**
- general characteristics of, 852*t*

localized cutaneous leishmaniasis, **823-825**
- epidemiology, 823-824
- manifestations, 824-825

localized infection, viral, 140
localized STDs, 945
lock-jaw, 527
logarithmic phase, 368
loiasis, 862
long terminal repeats (LTRs), 309
loperamide, 937
lophotrichous flagella, 355
lopinavir, 158*t*
LOS. *See* lipooligosaccharide
louse-borne relapsing fever, 656
louse-borne typhus fever, 686-687
low virulence, 387
low-density lipoprotein (LDL), 240
lower respiratory tract, 390*t*
lower respiratory tract infections, **924-928**
- clinical features, **924-925**
- diagnosis, **926-928**, 927*f*
- etiologic agents, **925-926**, 926*t*
- management, **928**
- sputum and, 924-928, 927*f*

low-grade smoldering infections, 902
LPS. *See* lipopolysaccharide
LT. *See* labile toxin
LTA. *See* lipoteichoic acid
LTRs. *See* long terminal repeats
Lubeck disaster, 497
Lucretius, 851
lumbar puncture, 956
lung abscess, **925**, 926*t*
lung biopsy, 735*f*
lung flukes. *See Paragonimus* spp.
Lyme disease, 412, 630*t*
- amoxicillin for, 661
- clinical aspects, 660-662
- diagnosis, 661
- doxycycline for, 661
- immunity, 660
- life cycle, 658*f*
- manifestations, 660
- pathogenesis, 659-660
- prevention, 662
- treatment, 661
- vaccine, 662

lymphadenitis, 504, 828, 859, **948**
- in *Bartonella*, 689

lymphadenopathy, 195, 200, 649
lymphatic filaria, **857-860**
- diagnosis, 860
- epidemiology, 859
- ivermectin for, 860
- manifestations, 859-860
- mosquitoes and, 857
- pathogenesis, 859
- pathology, 859
- treatment, 860

lymphoblasts, 20*f*
lymphocytes, 20*f*, 299
- atypical, 267*f*
- CD4+ helper T, 33-34, 154
- CD8+ cytotoxic T, 34
- granuloma, 496

lymphocytic choriomeningitis virus, **292-293**
- CNS infections and, 952*t*

lymphocytosis, 267
- atypical, 267
- in pertussis, 562

lymphogranuloma venereum (LGV), 676*f*, 945, 946*t*-947*t*, 948
lymphoid hyperplasia, 288
lymphoid stem cells, 20*f*
lymphoma, 265
- African Burkitt, 265, 267
- B-cell, 265
- Burkitt, **266**
- MALT, 575
- non-Hodgkins, 319

lymphonodules, 222
lymphoproliferative syndrome, **266**
lysates, 112
lysine, 374
lysis, 352, 463
lysogenic conversion, 135
lysogenic cycles, of temperate phages, 381*f*
lysogeny, 101, 108, 381
  human viruses, **133-135**
lysosomes, 495, 641, 744
lysozyme, 22, 126, 389
lytic cycle, of temperate phages, 381*f*
lytic infection, 143, 144
lytic phages, 380
lytic response, 108

M
M cells, 22, 597
  intestinal, 605
M protein, 15, 127, 297, 451
  group A streptococci and, 446-448
M strand, 282
MAC. *See* membrane attack complex
MacConkey agar, 582
macroconidia, 697
  tuberculate, 743
macrolides, 405*t*, **412**
  Borrelia burgdorferi and, 412
  for *Mycoplasma pneumoniae*, 669
  resistance, 419*t*
macrophage invasion potentiator (Mip), 611
macrophages, 20*f*, 30*t*, 747
  alveolar, 494*f*
  apoptosis, 600
  granuloma, 496
macular rash, 200
magnetic resonance imaging, 871, 902
major histocompatibility complex (MHC), 32, 174, 241, 312
  class I, 32, 33*f*
  class II, 32, 33*f*, 397, 431
major outer membrane protein (MOMP), 673
major surface glycoprotein (MSG), 735
malabsorption, 820
  ascariasis and, 843
  strongyloidiasis and, 848
malaria, 767, 784-790
  anemia in, 785-786
  antigenic variation, 787
  artemisinin for, 789*t*
  central nervous system, 786*f*
  cerebral falciparum, 788
  chemotherapy of, 789*t*
  chloroquine for, 789*t*
  circulatory changes in, 786
  clinical aspects, 787-790
  clinical capsule, 784
  clinical manifestations of, 784
  congenital, 784
  cytokines in, 786
  diagnosis, 788
  endemic areas, 784
  epidemiology, 784
  erythrocytic stages of, 780*f*
  fever in, 785
  folate antagonists for, 789*t*
  geographic distribution of, 785*f*
  immunity, 787
  imported, 784
  manifestations, 787-788
malaria (*Cont.*):
  mefloquine for, 789
  morphology of, 780*f*
  mortality from, 784
  nephritis in, 786
  paroxysm, 787
  pathogenesis, 785-786
  personal protection from, 790
  premunition in, 787
  prevention, 790
  primaquine for, 789*t*
  quinine for, 789*t*
  radical cures for, 789
  relapse, 787
  resistance of, 759-760
  serologic tests for, 788
  simian, 787
  sulfonamides for, 789*t*
  thrombocytopenia in, 786
  treatment, 788-789
  vaccines, 790
Malassezia furfur, 714*t*
malignant otitis externa, 907
malignant pustule, 484
malnutrition, protein-calorie, 898
MALT lymphoma. *See* mucosa-associated lymphoid tissue lymphoma
mannan, 693
mannitol, 957
mannoprotein, 704, 726
mannose, 27
Mansonia, 859
maraviroc, 158*t*
Marburg virus, 283-284, 293
Martin-Lewis medium, 87, 548
masseter muscle, 527
mast cells, 20*f*, 23
mastoiditis, 908
matrix proteins, 102
Mauer's dots, 781
Mazzotti reaction, 767
MBC. *See* minimum bactericidal concentration
MBP. *See* myelin basic protein
McCoy cells, 677
MCV4. *See* Meningococcal Conjugate Vaccine Quadravalent
MDR-TB. *See* multidrug-resistant tuberculosis
measles, **193-197**
  antibodies, 194
  chronic, 196
  clinical aspects, **195**
  CNS infections and, 952*t*
  comparison, 190*t*
  complications, **195**
  diagnosis, **196-197**
  epidemiology, **193-194**
  German, 200
  immunity, **194**
  manifestations, **195-196**
  pathogenesis, **194**
  prevention, **197**
  rash, 196*f*
  receptors, 114*t*
  treatment, **197**
  vaccines, 196
  virology, **193**
measles, mumps, rubella, and varicella vaccine (MMRV), 192, 201, 260
mebendazole, 775
  for trichinosis, 856
Mecca, 568
Medawar, Peter, 101
media
  indicator, **69**
  nutrient, **68**
  selective, **69**
medical devices, nosocomial infections from, 55-56
mefloquine, 774, 775, 790
  for malaria, 789
megacolon, 832
megaesophagus, 832
megakaryocytes, 201
Meister, Joseph, 302
melanin, 720, 739
melarsoprol, 772
  for African trypanosomiasis, 829
melioidosis, 623
membrane attack complex (MAC), 28*f*
membrane-active exotoxins, **396**
memory cells, 34, 37
men, gonorrhea in, 548*f*
meninges, 191
meningitis, 293, 553
  aseptic, 219, 221, 954-955
  chronic, 954-955
  cryptococcal, 320
  Cryptococcus neoformans, 742*f*
  Escherichia coli, 589
  Haemophilus influenzae, 555
  pneumococcal, **464**
  purulent, 953-955
  treatment, 957
  vaccines, 18
Meningococcal Conjugate Vaccine Quadravalent (MCV4), 541
meningococcal disease, **537-542**
  cellular view, 537*f*
  clinical aspects, 540
  clinical capsule, 537-542
  diagnosis, 540
  epidemiology, 537
  immunity, 538, 540*f*
  lipooligosaccharide, 537
  lipopolysaccharide, 537
  manifestations, 540
  pathogenesis, 537-538
  prevention, 541-542
  rifampin for, 541
  treatment, 541
meningococcemia, 540, 541*f*
meningococci, 11
meningoencephalitis, 809, 871, 955
menstruation, 436
meropenem, 405*t*, 409
merozoites, 781
  in red blood cells, 782
messenger RNA (mRNA), 16*f*, **118-120**, 309
  monocistronic, **120-121**, 121*f*
  pathways, 119*f*
metabolic acidosis, 569
Metagonimus spp., 880*t*
metalloprotease, 186
metapneumovirus
  human, 167, **182**
  lower respiratory tract infections from, 926*t*
metastatic infection, salmonellosis, 603
Metchnikoff, Elie, 19
methicillin, 405*t*, 408, 440

methicillin-resistant *Staphylococcus aureus* (MRSA), 421, 439
methylene blue stain, 64*f*
metronidazole, **414**, 775
  for anaerobic infections, 520
  for *Bacteroides fragilis*, 532
  for *Clostridium difficile*, 530-531
  for *Entamoeba histolytica*, 809
  for giardiasis, 821
  for *Helicobacter pylori*, 576-577
  for trichomoniasis, 817
Mexico, 568
MHC. *See* major histocompatibility complex
MIC. *See* minimal inhibitory concentration
micafungin, 710
  features of, 708*t*
mice, 293, 294
  nude, 330
miconazole, 709, 718
  features of, 708*t*
microaerophilic bacteria, 362
microaerosols, 462
microbes
  in environment, 4
  features of, 5*t*
  relative size of, 5*f*
microbial flora, 8-12, 508
  in blood, 9
  in body fluids, 9
  carrier state, 8
  in colon, 11
  diet and, 11
  at different sites, 9-12
  in exclusionary effect, 12
  in genitourinary tract, 11-12
  good, 12
  in immune system, 12
  in intestinal tract, 10-11
  in mouth, 10-11
  nature of, 9
  in opportunistic infection, 12
  in pharynx, 10-11
  potentially pathogenic, 10*f*
  residents, 8
  in respiratory tract, 11
  role of, 12
  samples from, 60-61
  skin, 10
  in skin, 9-10
  stool, 11*f*
  in tissues, 9
  transients, 8
  in vagina, 11
microbial killing, **46-47**
  kinetics of, 46*f*
microbiology, 4-8
microconidia, 697
  Histoplasma capsulatum, 743
microfilaments, 804
microfilariae, 857
  differentiation of, 858*t*
  hypersensitivity reaction to, 862
microscopy
  dark-field, **65**
  electron, **65**, 73
  fluorescence, 62*f*, **65**
  light, 61-65
  stool, 936
  for UTI diagnosis, **941**
Microsporum, 713
  classification of, 699*t*
Microsporum audouini, 714*t*
Microsporum canis, 714*t*
Microsporum floccosum, 714*t*
Microsporum gypseum, 714*t*
Microsporum mentagrophytes, 714*t*
Microsporum rubrum, 714*t*
Microsporum tonsurans, 714*t*
Microsporum violaceum, 714*t*
microwaves, 50
middle ear, 390*t*
middle respiratory tract infection, **922-924**
  clinical features, **922**
  etiologic agents, **922-923**, 923*t*
  management, **924**
milker's nodules, **211**
Milne, A.A., 189
minimal inhibitory concentration (MIC), 417*f*, 464
  definition, 404
minimum bactericidal concentration (MBC), 418
minocydine, for *Nocardia*, 513
Mip. *See* macrophage invasion potentiator
miracidia, 879, 886*f*
misdirected immune response, **398**
missense mutation, 374
mite larvae, 687
MMRV. *See* measles, mumps, rubella, and varicella vaccine
moderate virulence, 387
MOI. *See* multiplicity of infection
molds, 7-8, **695-698**
  asexual, 697*f*
  forms, 696*f*
  shifts from, 698
molecular assay, **129**
molecular diagnostic methods, 83*f*
molecular mimicry, 42, 154, 217, 453
Molluscipoxvirus, 206*t*
molluscum contagiosum, 206*t*, **210**
  in AIDS, 210*f*
  of skin, 210*f*
Molluscum contagiosum virus, 946*t*
MOMP. *See* major outer membrane protein
monkeypox, 206*t*, 208, **209-210**
  cases of, 210
monobactams, 405*t*, 407, **409**
  for *Pseudomonas aeruginosa*, 622
monoblast, 20*f*
monocistronic mRNA rule, **120-121**, 121*f*
monoclonal antibodies, 74, 255
monocytes, 20*f*
  in innate immunity, 23
mononucleosis, 920, 946*t*
monosaccharides, 915
monotrichous flagella, 355
Montagnier, Luc, 306
Moraxella, 10, **624**, 626*t*
  penicillin for, 624
Moraxella catarrhalis, 908*t*, 910*t*
morbidity
  rotavirus, 273*f*
  tuberculosis, 493
Morganella, 580, **607**
mortality rates, 90*f*. *See also* morbidity
  AIDS, 314
  ascariasis, 843
  excess, 174
  HIV and, 317*f*
mortality rates (*Cont.*):
  infectious diseases, 4*f*
  kala azar, 827
  of malaria, 784
  of neonatal herpes, 255
  respiratory syncytial virus, 181
  tuberculosis, 493
mosquitoes
  anopheline, 784
  eradication of, 790
  lymphatic filaria and, 857
motility, **373-374**
motor neuron cells, 219
motor neuron endplate, 524*f*
mouse retrovirus (MPMV), 310*f*
mouth, microbial flora in, 10-11
MPMV. *See* mouse retrovirus
mRNA. *See* messenger RNA
MRSA. *See* methicillin-resistant *Staphylococcus aureus*
MSG. *See* major surface glycoprotein
mucociliary action, 11
mucocutaneous leishmaniasis, **825-826**
  manifestations, 825
  treatment, 825
Mucor, 732-733, 909
  classification of, 699*t*
mucosa, in innate immunity, 21
mucosa-associated lymphoid tissue (MALT) lymphoma, 576
multicistronic operons, 370
multidrug-resistant tuberculosis (MDR-TB), 500
multinucleated giant cells, 252*f*
multiorgan failure, 967
multiple-host parasites, 764-765
multiplicity of infection (MOI), 112
mumps, **189-192**
  clinical aspects, **191-192**
  CNS infections and, 951, 952*t*
  comparison, 190*t*
  complications, 191
  diagnosis, **192**
  epidemiology, **190-191**
  immunity, **191**
  infection, 190
  manifestations, **191-192**
  pathogenesis, **191**
  prevention, **192**
  virology, **189-190**
mupirocin, 440
Murine typhus, 630*t*
murine typhus, 683*t*
mutation, **130-131**, 214
  frameshift, 374, 375*f*
  herpesviruses, 164
  HIV, 315
  missense, 374
  nonsense, 374
  point, 131*f*
  types of, 374-375
mutational resistance, **423**
mutations
  bacteria, **374-375**
  hepatitis C, 240
  polar, 375
myc gene, 148
mycetoma, **721-722**, 897
mycobacteria, 367
  case study, 505
  of clinical importance, 492*t*

mycobacteria (*Cont.*):
disease, 491
cell-mediated immunity in, 491
granuloma and, 491
DTH and, 491
soft tissue infections, 504-505
tuberculosis-like diseases caused by, **503**
Mycobacterium, **489-506**
acid-fast stain for, 490
AIDS and, 489
bacteremia from extravascular infection and, 966*t*
bacteriology, 489-491
cell wall, 490*f*
classification, 491
growth, 490
morphology, 489-490
structure, 489-490
ulcers from, 899*t*
Mycobacterium avium-intracellulare, 318, 318*t*, 492*t*, **503-504**
in AIDS, 504
Mycobacterium bovis, 492*t*, 630*t*
Mycobacterium fortuitum, **504-505**
complex, 504
Mycobacterium kansasii, 492*t*, **503**
ethambutol for, 503
isoniazid for, 503
rifampin for, 503
Mycobacterium leprae, 492*t*, **501-503**
bacteriology, 501
tuberculoid form, 501
Mycobacterium marinum, 492*t*, **504**, 897
Mycobacterium scrofulaceum, 492*t*, 504
Mycobacterium smegmatis, 492*t*
Mycobacterium tuberculosis, 3, 42, 63, 89, 318, 318*t*, 351, 392, **491-501**, 492*t*, 682
chronic granulomatous infection from, 954*t*
CNS infections and, 951, 955
dormant, 496
eye infection from, 906*t*
growth of, 398
lower respiratory tract infections from, 924, 926, 926*t*
primary infection, 495-496
rifampin for, 415
in sputum, 490*f*
Mycobacterium ulcerans, 492*t*, **504-505**, 897
mycolic acids, 489, 490, 491
mycology, **693-695**
Mycoplasma, 408, **665-670**
clinical cases, 670
ear infections caused by, 908
electron micrograph, 666*f*
microbiology, 665
septic arthritis from, 903
tetracycline and, 411
Mycoplasma genitalium, 666*t*, **669**, 946*t*, 947
Mycoplasma hominis, **669**, 903, 948
Mycoplasma pneumoniae, **665-669**, 666*t*, 903
bronchiolitis, 667*f*
clinical aspects, 668-669
clinical capsule, 665
diagnosis, 668
epidemiology, 666
immunity, 667-668
infecting dose, 666
infecting respiratory epithelium, 667*f*
lower respiratory tract infections from, 926*t*
macrolides for, 669
Mycoplasma pneumoniae (*Cont.*):
manifestations, 668
middle respiratory infections and, 923-924
pathogenesis, 666-667
tetracycline for, 669
treatment, 669
mycoplasmal pneumonia, 668-669
mycotic aneurysm, **963**
mycotoxins, 705
myelin basic protein (MBP), 154
myelitis, 668
myeloid stem cell, 20*f*
myocarditis, 221*t*
in American trypanosomiasis, 831*f*
diphtheria and, 476
diphtheria toxin and, 474, 475
myoclonic jerks, 196
myosin, 453
myringitis, 668, 908

N

NAA. *See* nucleic acid amplification
N-acetylglucosamine (NAG), 351, 365, 366*f*, 655, 734
NAD. *See* nicotinamide adenine dinucleotide
NADPH. *See* nicotinamide adenine dinucleotide phosphate
Naegleria fowleri, 775, **809-810**
trophozoites, 809
Naegleria spp., 803
nafcillin, 408
naftifine, 709
features of, 708*t*
NAG. *See N*-acetylglucosamine
Nairovirus, 282
naked capsid viruses, 102
assembly of, **125-126**
human, **116-117**
nalidixic acid, 413, 420
for shigellosis, 598
Napoleon, 861
narrow-spectrum agents, 404
nasopharyngeal carcinoma (NC), **266**, 267
natural immunity, 43
natural killer (NK) cells, 20*f*, 23, 30*t*, 151, 241
NC. *See* nasopharyngeal carcinoma; nucleocapsid protein
NCAM. *See* neural cell adhesion molecule
Necator americanus, 835, 836*t*
egg structure, 844*f*
life cycles of, 836*t*, 845
parasitology, 844
structure of, 844*f*
necrosis, caseous, 496
necrotic colonic cells, 529
necrotic spreading skin ulceration, 897
necrotizing fasciitis, 519
necrotizing periodontal diseases, **917-918**
needle sharing, 240
needlestick transmission, of hepatitis B, 233
Nef, 311, 312
negative-sense RNA viruses, 297
Negri body, 299, 300*f*
Neisseria, **535-554**
bacteriologic features of, 536*t*
general features, 535
lipopolysaccharide, 535
pathogenic features of, 536*t*
Neisseria gonorrhoeae, 61, 94, 377, 392, 394, 409, 412, 425, 535, 536*f*, **542-549**
Neisseria gonorrhoeae (*Cont.*):
antigenic variation, 542
bacteremia from extravascular infection and, 966*t*
bacteriology, 542
cell wall, 538*f*
eye infection from, 906, 906*t*
genital infection from, 945, 946*t*, 947-948
lipooligosaccharide, 542
lipopolysaccharide, 542
Opa proteins, 395*f*
pili, 542*f*
plasmids, 424
porins, 542
septic arthritis from, 903-904, 903*t*
upper respiratory infection from, 920*t*, 921
Neisseria meningitidis, 8, 535, **536-542**
bacteremia from extravascular infection and, 966*t*
bacteriology, 536-537
CNS infections and, 951, 952*t*, 955
eye infection from, 906*t*
genotypes, 401
in oropharynx, 10
plasmids, 424
rifampin for, 415
septic arthritis from, 903
nelfinavir, 158*t*, 163
nematodes
classification of, 763*t*
intestinal, **835-849**
case study, 849
life cycles of, **836-837**
tissue, **851-864**
case study, 863
general characteristics of, 852*t*
neomycin, 411, 440
neonatal herpes, **255**, 946*t*
nephritis, 456
in malaria, 786
nephrotoxicity, 163
neural cell adhesion molecule (NCAM), 297
neuraminidase, 168, 189, 193, 461
in influenza, 171
inhibition, 158*t*, **159**, 176, 177
neuromuscular junction, 524*f*
Neurospora crassa, 695
neurosyphilis, 650
neurotoxin
botulinum, 524*f*
in clostridia, 517
neutralization, 73, **75-76**, 152
nevirapine, 158*t*, 162
nicotinamide adenine dinucleotide (NAD), 363, 551, 557, 682, 773
nicotinamide adenine dinucleotide phosphate (NADPH), 362
nifurtimox, 777*t*
for American trypanosomiasis, 833
nikkomycins, 710
action of, 709*f*
features of, 708*t*
Nipah, 284
nitazoxanide, 777*t*
nitrate, 25, 361
nitrate reduction, 88
nitric oxide, 25
nitrite, 25
nitrofurantoin, 942
nitroimidazoles, **775**

NK cells. *See* natural killer cells
NNRTIs. *See* non-nucleoside reverse transcriptase inhibitors
Nobel Prize, 305, 306
Nocardia, 414, **511-514**
  amikacin for, 513
  bacteriology, 511
  cefotaxime for, 513
  cutaneous, 722
  features, 508*t*
  imipenem for, 513
  lower respiratory tract infections from, 926, 926*t*
  minocydine for, 513
  in sputum, 512*f*
  sulfonamides for, 513
  ulcers from, 899*t*
Nocardia asteroides, 511, 512, 513
Nocardia brasiliensis, 511, 512, 513
nocardiosis, 509*f*, **512-514**, 513
  clinical aspects, 513
  clinical capsule, 512
  diagnosis, 513
  epidemiology, 512
  immunity, 512-513
  manifestations, 513
  pathogenesis, 512
  treatment, 513
non-β-lactams, 405*t*
noncommunicable infections, 91
nonconjugative plasmids, 382
noncytocidal viruses, 322
nongonococcal urethritis, 946*t*
nonhemolytic streptococci, 456*t*, **465**
non-Hodgkin's lymphoma, 319
noninvasive luminal flagellates, 813-821
non-nucleoside reverse transcriptase inhibitors (NNRTIs), **162**, 320
nonpermissive cells, 110, 143
nonproductive response, 108
nonpurulent otitis media, 668
nonsense mutation, 374
nonseptate hyphae, 696*f*, 733
nonsporulating Gram-positive bacteria, **517**
nonsterile honey, 525
nonvenereal treponemes, 663
norfloxacin, 413, 414
Norovirus, 276
Norwalk agent, 276
nosocomial infections, **53-54**
  from blood, 56
  Creutzfeldt-Jakob disease, 342
  environment, 55
  from hospital personnel, 54-55
  from medical devices, 55-56
  precautions for, 58*t*
  prevention, 58
  from respirators, 56
  sources, **54-56**
  from urinary catheters, 55
  from vascular catheters, 55-56
novobiocin, resistance, 440
nuclear inclusions, 260
nucleic acid amplification (NAA), **80-81**, 499, 677
nucleic acid analysis, **80-85**
  methods, **80-83**
    for infectious diseases, **81-83**
nucleic acid synthesis, 406*t*
  antifungals, 710
  inhibitors, **159-161**, **413-415**
nucleocapsid, 102, 169, 281*f*, 284
  assembly of, **125-126**
nucleocapsid protein (NC), 306
nucleoid, 347
  bacteria, **357**
nucleoside reverse transcriptase inhibitors, **162**, 320
nucleotide analogs, **163**
nude mice, 330
nutrient broths, 87
nutrient media, **68**
nutritional deficiencies, 898
nystatin, 707
  for *Candida albicans*, 729
  features of, 708*t*
  for gonorrhea, 548

O

O antigen, 579, 595
O antigen polysaccharide side chains, 354
oculoglandular tularemia, 639
oculomotor muscles, 475
ODC. *See* ornithine decarboxylase
ofloxacin, 406*t*, 413
  for *Chlamydia*, 413
  for gonorrhea, 549
  for *Pseudomonas aeruginosa*, 413
  for tuberculosis, 500
oleic acid albumin, 401
2',5'-oligoadenylate synthetase, 151
oligosaccharides, 353
OMP. *See* outer membrane protein
Onchocerca spp., 770, **860-862**
Onchocerca volvulus, 851, 858*t*, **860-862**
  general characteristics of, 852*t*
onchocerciasis, 769, **861-862**
  clinical aspects, 861-862
  diagnosis, 862
  epidemiology, 861
  ivermectin for, 862
  manifestations, 861-862
  prevalence of, 760*t*
  prevention, 862
  treatment, 862
oncogenes, 322
oncogenic transformation, 108
oncogenic viruses, 146
oncogenicity
  of DNA viruses, 147*t*
  of RNA viruses, 147*t*
oncoretroviruses, 305, 306, 321-322
O-nitrophenyl-β-D-galactoside, 88
oocyst
  Cryptosporidium, 797
  ingestion, 794
  Toxoplasma gondii, 791
    ingestion of, 794
oophoritis, 191
Opa proteins, 395*f*, 542, 543, 545, 546
operating room, asepsis in, 56-57
operator region, 370
operculum, 763
operon, 370, 371*f*
ophthalmia neonatorum, 546, 905, 906*t*
Opisthorchis spp.
  characteristics, 880*t*
  infections from, 882
opportunistic anaerobes, 516*t*
opportunistic fungi, 699
opportunistic infections
  in AIDS, 318*t*
  Enterobacteriaceae, 580-581
  Escherichia coli, 586-589, 593
    clinical capsule, 586
    meningitis, 589
    urinary tract infection, 587-588
  flora in, 12
opportunistic pathogens, 388
opsonization, 24, 26
opsonophagocytosis, 394*f*, 741
optic neuritis, 905
OptiMAL, 788
Optochin, 456*t*
OPV. *See* oral polio vaccine
oral hygiene, 913, 917
oral polio vaccine (OPV), 144, 220
Orbivirus, 280*t*
orchitis, 191
orf, 206*t*, **211**
organ transplantation, 831
organogenesis, 199
Orientia tsutsugamushi, 681, 683*t*, 687
ornithine, 776
ornithine decarboxylase (ODC), 776
ornithosis, 678
oropharynx, 390*t*
  Neisseria meningitidis in, 10
  streptococci in, 10
Oroya fever, 686*t*, 689
orthomyxoviruses, 168, **295**
  classification of, 110*t*
Orthopoxvirus, 206*t*
oseltamivir, 158*t*, 159, 176, 177
Osler, William, 3, 225
Osps. *See* outer surface proteins
osteomyelitis, 438, **901-902**
  chronic, 519
  from dental caries, 915
  diagnosis of, **902**
  etiologic agents, **901**, **902*t***
  management of, **902**
  septic arthritis and, 901
otitis externa, 907-909, 908*t*
otitis media, 907-909, 908*t*
outer membrane, 353
outer membrane protein (OMP), 611
  Treponema pallidum, 649
outer membrane protein porins, 418
outer surface proteins (Osps), 657, 659
outpatient clinic, asepsis in, 57
overwintering, 285
oviparous, 763
owl eye cells, 260, 261*f*
oxacillin, 408
oxazolidinones, **412**
oxiconazole, 718
oxygen tolerance, 515

P

P pili, 583
p53, 148, 327
PA. *See* protective antigen
PABA. *See* para-aminobenzoic acid
packaging site, 124
PAIR. *See* Percutaneous, Aspiration, Infusion of scolicidal and Reaspiration
PAMPs. *See* pathogen-associated molecular patterns
pancreas, 191

pandemic infections, 91
antigenic shifts associated with, 173*t*
cholera, 568
influenza, 173
Papanicolaou smear, 332*f*, 333
papillomaviruses, **327-333**
characteristics of, 328*t*
clinical aspects, 330-333
clinical capsule, 329
epidemiology, 329
genomes, 327
manifestations, 330-332
pathogenesis, 329
STD from, 945-946, 946*t*
virology, 327-333
papovaviruses
classification of, 111*t*
incubation period, 142*t*
para-aminobenzoic acid (PABA), 414
Paracoccidioides brasiliensis, **753-754**
amphotericin B for, 753
azoles for, 753
disease, 753
features of, 740*t*
sulfonamides for, 753
paragonimiasis, **881-882**
diagnosis, **882**
epidemiology, **881**
manifestations, **881-882**
treatment and prevention, **882**
Paragonimus africanus, 881
Paragonimus kellicotti, 881
Paragonimus mexicanus, 881
Paragonimus skrjabini, 881
Paragonimus spp., 879, **881-882**
characteristics of, 880*t*
disease caused by. *See* paragonimiasis
parasitology, 881
Paragonimus westermani, 881, **881-882**
lower respiratory tract infections from, 926*t*
parainfluenza, **177-179**
clinical aspects, **178-179**
diagnosis, **178-179**
diagram, 177*f*
manifestations, **178-179**
prevention, **178-179**
treatment, **178-179**
upper respiratory infection and, 920*t*
parainfluenza 1, **178**
parainfluenza 2, **178**
parainfluenza 3, **178**
parainfluenza 4, **178**
paralytic poliomyelitis, 219
paramyxoviruses, 116, 120, 153*t*, 177, 189, 284
classification of, 110*t*
diagram, 177*f*
enveloped, 177
paranasal sinuses, 390*t*
Parapoxvirus, 206*t*, 211
ParaSight F, 788
parasites, 6*f*, 8, **759-765**
antigenic shifts in, 768
antigenic variation in, 768
definition, 759
definitive hosts of, 764
diagnosis, 769
distribution of, 764*t*
immune suppression by, 769
immunity, **768-769**
intermediate hosts of, 764
parasites (*Cont.*):
single-host, 764
transmission of, 764*t*
parasitic infections, **759-761**
prevalence of, 760*t*
parasitic stains, 65
parechoviruses, 213
paresis, 650
paromomycin, 777*t*
for cryptosporidiosis, 800
for giardiasis, 821
paronychia, 253
paroxysm, malarial, 787
parvovirus B19, **201-203**
AIDS and, 202
erythema infectiosum, **202-203**
parvoviruses
classification of, 111*t*
comparison, 190*t*
receptors, 114*t*
passive immunity, 43
hepatitis A, **229**
Pasteur, Louis, 3-4, 295, 302, 483
Pasteurella multocida, 630*t*, **640-641**
bacteremia from extravascular infection and, 966*t*
wound infections from, 899*t*
Pasteurella pestis, 634
pasteurellosis, 630*t*
pasteurization, 50. *See also* unpasteurized dairy
brucellosis and, 633
definition of, 45-46
for sterilization, 47*t*
pathogen-associated molecular patterns (PAMPs), 23
pathogenesis, of infectious diseases, 14-15
pathogenicity, 387
bacterial, **398-401**
islands, 399-401, 401*f*, 569, 582
pathogens
opportunistic, 388
primary, 388
pathology, rubella, **199**
PBPs. *See* penicillin-binding proteins
PCR. *See* polymerase chain reaction
pelvic examination, 948
pelvic inflammatory disease (PID), 543, **546-547**, 548*f*, 676, 946*t*, **948**
penciclovir, 158*t*, **160**
penetration, **115-117**
inhibitors, **157-159**
penetrin, 831
penicillin, 4, 17, 352, 403, 404, 405*t*, 407, **408**, 426. *See also specific drugs*
for actinomycosis, 510-511
for anthrax, 485
for *Bacillus anthracis*, 485
benzyl, 404
susceptibility patterns to, 427*t*
for *Clostridium perfringens*, 523
for gonorrhea, 549
for group A streptococcus, 425
for group B streptococci, 459
for leptospirosis, 654
for listeriosis, 481
for *Moraxella*, 624
for pneumococcal disease, 464
Pseudomonas aeruginosa and, 408
resistance, 421, 622
for *Streptococcus pneumoniae*, 464
penicillin (*Cont.*):
susceptibility patterns to, 427*t*
for tetanus, 528
for *Treponema pallidum*, 425, 652
for wound infections, 900
penicillin-binding proteins (PBPs), 365, 406, 420, 421, 541, 549
Penicillium, 404, 406, 710
penile candidiasis, 946*t*
pentamer, 106
Penton projections, 183
peplomers, 185, 193
peptidoglycan, 24, 351, 353, 406
fragments, 434
structure, 352*f*
synthesis, 364-365, 366*f*
antimicrobials acting on, 407*f*
peptidyl transferase, 411
Peptostreptococcus, 517*t*, 519, 964*t*, 966*t*
Percutaneous, Aspiration, Infusion of scolicidal and Reaspiration (PAIR), 876
perforins, 151
periapical abscess, 915
periapical granuloma, 915
pericarditis, 221*t*
perihepatitis, 946*t*
perinuclear inclusions, 260
periodontal abscess, 917
periodontal diseases, necrotizing, **917-918**
periodontitis, chronic, **916-917**, 917*f*
periosteum, 901
peripheral nerves, 191
periplasm, 353, 365
periplasmic gel, 353
perirectal pain, **941**
peritonsillar abscess, 919-922, 920*t*
permissive cells, 110, 143
peroxidase, 362
persistent infection, 110, 143, 145
enterovirus, **338**
viral, 145
CNS, **337-344**
persistent inflammation, **398**
persistent mucocutaneous herpes simplex, 318*t*
persistent viruses, 101
pertactin, 559
pertussis, 558, **560-564**
cellular view, 561*f*
cephalosporins for, 563
clinical aspects, 562-564
clinical case, 564
convalescent phase, 562
diagnosis, 563
epidemiology, 560
erythromycin for, 563
immunity, 562
immunization, 18, 560
in lymphocytosis, 562
manifestations, 562
pathogenesis, 560-562
genetic regulation, 561-562
prevention, 563-564
treatment, 563
virulence, 560-561, 561-562
pertussis toxin (PT), 559
Peru, 568
pets, toxocariasis and, 853
phage genes, 135
phagocytes, 24-25

phagocytosis, 24-25, 150
  capsule and, 553
  in protozoa, 762
phagolysosome, 25, 687
phagosome, 392, 495, 641, 744
pharmacology
  acyclovir, **160**
  amantadine, 159
  rimantadine, 159
pharyngeal diphtheria, 919, 921
pharyngeal gonorrhea, 921
pharyngeal pseudomembranes, 919-920
pharyngitis, 184, 426, 475
  group A streptococci, 449, 454
pharyngoconjunctival fever, 184
pharynx, microbial flora in, 10-11
phenanthrene, 774
phenol, 52
phenolics, 501
  disinfection with, 52
  for sterilization, 47*t*
phenotypic resistance, **164**
pheromones, 384
Phialophora verrucosa, 714*t*, 721
Phlebotomus, 279, 827
Phlebovirus, 282
phospholipase, 619
photoreactivation, 45
Phthirus pubis, 946*t*
phylogenetic relationships, 385
physical barriers, in innate immunity, 22
physiologic wound infections, 897, 899*t*
picornaviruses, 115, 213
  classification of, 110*t*
  replication cycle of, 214*f*
PID. *See* pelvic inflammatory disease
piedra, 719
  black, 716*f*, 719
  white, 719
Piedraia hortae, 714*t*, 719
pilE, 543
pili, **356**, 356*f*, 392*f*
  Escherichia coli, 583-586
  Neisseria gonorrhoeae, 542*f*
  P, 583
  sex, 383
  type I, 583, 588
pilot proteins, 117
pilS, 543
pinocytosis, in protozoa, 762
Pinta, 663
pinworm, 761, 836*t*
piperacillin, 405*t*, 408
  susceptibility patterns to, 427*t*
pityriasis versicolor, 719
plague, **629-641**, 630*t*. *See also Yersinia pestis*
plaque, 112, 129. *See also* dental plaque
  assay, 128-129
plasma cells, 37
plasma membrane, 355
plasmids, 382
  bacteria, **357**
  conjugative, 382
  fingerprinting, 82*f*
  mobilization, 382
  Neisseria gonorrhoeae, 424
  Neisseria meningitidis, 424
  nonconjugative, 382
  R, **384**
  in resistance, **423-424**
Plasmodium, **779-770**
  asexual phase of, 779
  characteristics of, 781*t*
  definition, 779
  growth in laboratory, 783
  intrahepatic dormancy, 782
  life cycle of, 781-782
  morphology, 780-781
  physiology, 783
  sexual phase of, 779
Plasmodium falciparum, 759, 773, 774, 775
  characteristics of, 781*t*
  distribution of, 764*t*
  growth in laboratory, 783
  resistance of, 777, 789
  transmission of, 764*t*
Plasmodium malariae, 779
  characteristics of, 781*t*
Plasmodium ovale, 779
  characteristics of, 781*t*
  radical cure, 789
Plasmodium vivax, 779
  characteristics of, 781*t*
  life cycle of, 782*f*
  radical cure, 789
Plesiomonas, **624**, 626*t*
  tetracycline for, 624
pleural effusion, 925
pleuritic pain, 222
PMC. *See* pseudomembranous colitis
PMN. *See* polymorphonuclear neutrophil
pneumococcal capsule, 461*f*
pneumococcal disease
  capsule in, 463
  clinical capsule, 461
  diagnosis, 464
  epidemiology, 462
  GBS and, 458*f*
  group B streptococci, **461-465**
  immunity, 463
  manifestations, 464
  pathogenesis, 462-463
  pneumolysin, 463
  in polymorphonuclear leukocytes, 463
  prevention, 465
  treatment, 464-465
pneumococcal meningitis, **464**
pneumococcal pneumonia, **464**
pneumococcal polysaccharide vaccine (PPV), 465
pneumococci, 11, 446
Pneumocystis carinii, 414, 733, 770
Pneumocystis jirovecii, 318*t*, 733
Pneumocystis spp., **733-737**
  AIDS and, 733, 735
  classification of, 699*t*
  disease caused by. *See* pneumocystosis
  immunity, 735-736
  pathogenesis, 735
  pneumonia, 734*f*, 735
  sporocytes, 734
pneumocystosis, **734-737**
  in AIDS, 736
  clinical aspects, 736-737
  clinical capsule, 734
  diagnosis, 736
  epidemiology, 734-735
  immunity, 735-736
  manifestations, 736
  opportunistic, 724*f*
  pathogenesis, 735
pneumocystosis (*Cont.*):
  prevention, 736-737
  treatment, 736-737
  trimethoprim-sulfamethoxazole for, 736-737
pneumolysin, 461
  cilia and, 463
  in pneumococcal disease, 463
  Streptococcus pneumoniae, 463
pneumonia, 438
  acute, **924-925**, 926*t*
  chronic, **925**, 926*t*
  epidemiology of, 462
  etiologic agents, 925-926, 926*t*
  Haemophilus influenzae, 556
  infant, 676
  Legionella, 611*f*
  mycoplasmal, 668-669
  pneumococcal, **464**
  Pneumocystis, 734*f*, 735
  Pseudomonas aeruginosa, 620*f*
  walking, 668, 678
pneumonic plague, 635, 637
pneumonic tularemia, 639
pneumonitis, 736
  diffuse, 736
pneumovirus, 179
podophyllin, 333
podophyllotoxin, 333
poikilocytosis, 332*f*
point mutation, 131*f*
pol, 306, 307*t*, 310
polar flagella, 355
polar mutations, 375
poliomyelitis, 115, 213, 219*f*
  abortive, 219
  bulbar, 219
  inactivated vaccine, 220
  paralytic, 219
  pathogenesis, 138*f*
  receptors, 114*t*
  recombinants, 132
  subclinical, 219
  vaccine, 17-18, 144
  vaccine-associated, 220
polioviruses, 215
  clinical aspects, **219-220**
  epidemiology, **218**
  manifestations, **219**
  pathogenesis, **218-219**
  prevention, **219-220**
polyenes, **707-709**
  features of, 708*t*
  resistance, 711
polyglycan production, 915
polymerase chain reaction (PCR), 80, 129, 175, 186, 202, 241, 254, 271, 330, 661, 825
  applications of, **83-85**
  in *Chlamydia trachomatis*, 677
  diagnostic applications of, 84*f*
polymerization reactions, 363-365
  peptidoglycan synthesis, 364-365
  transcription, 363
polymorphonuclear leukocytes, 30*t*, 452
  infiltrating, 458
  in pneumococcal disease, 463
polymorphonuclear neutrophil (PMN), 23, 251, 398, 536, 727
polymyxin B, 406*t*, 415

polyomaviruses, 150, **333-335**
characteristics of, 328*t*
clinical aspects, 334-335
clinical capsule, 334
diagnosis, 335
epidemiology, 334
manifestations, 334-335
pathogenesis, 334
virology, 333-334
polyproteins, 120-121
polyribitol phosphate (PRP), 552
polysaccharides
core, 354
O antigen, 354
Pontiac fever, 613
pore forming exotoxins, 397*f*, 586
pore forming proteins, 151, 831
pore forming toxins, 396
in group B streptococci, 461
porins, 354, 373*f*
Neisseria gonorrhoeae, 542
outer membrane protein, 418
pork
freezing, 855
undercooked, 881
tapeworms and, 870*f*
trichinosis and, 854
pork tapeworm, **869-872**
albendazole for, 872
disease, 869-872
diagnosis, 871-872
epidemiology, 869
manifestations, 871
life cycle, 870*f*
praziquantel for, 872
prevention, 872
treatment, 872
Porphyromonas, 517*t*, 518, 911, 916
Porphyromonas gingivalis, 916
posaconazole, features of, 708*t*
poststreptococcal acute glomerulonephritis, 398
poststreptococcal sequelae, 450-451, 452-453, 455-456
potassium hydroxide (KOH), 699, 700*f*, 718, 720, 729, 746, 752, 948
potassium iodide, features of, 708*t*
potassium tellurite, 476
Powassan virus, **291**
poxviruses, **205-212**
electron microscopic appearance, 206*f*
group characteristics, **205**
human, 206*t*
incubation period, 141*t*
replication, 205, 207*f*
virion structure, 206*f*
PPD. *See* purified protein derivative
PPV. *See* pneumococcal polysaccharide vaccine
praziquantel, **776**
for pork tapeworm, 872
for schistosomiasis treatment, 882, 885, 890
precautions
airborne transmission, 57
contact, 57
droplet, 57
for nosocomial infections, 58*t*
standard, 57
transmission-based, 57
precipitation, **74**
pregnancy, pyelonephritis and, 940
prematurity, pyelonephritis and, 940
premunition, 768
in malaria, 787
prevalence, 95
of viral infections, 137
prevention. *See also* immunization; vaccines
of infectious diseases, 17-18
Prevotella intermedia, 918
Prevotella melaninogenica, 518
Prevotella spp., 517*t*, 518, 911, 918
primaquine, 773
for malaria, 789*t*
primary cell culture, 71
primary culture, of viruses, 112
primary pathogens, 388
primary response, 40
primary syphilis, 648*f*, **649**, 948
primary tuberculosis, 493*f*, 497, 744
primary viremia, 142
prion(s), 101, 145, 337
protein conversion, 341*f*
prion diseases, 339*t*
proctitis, 946*t*
progeny virions, 101
proglottids, 763, 867, 871
progressive fibrosis, 234
progressive multifocal leukoencephalopathy, 318, 334-335, **338**
progressive postrubella panencephalitis, **338**
proguanil, 774
resistance, 777
proinflammatory cytokines, 674
prokaryotic cells, 347, 349*f*
features of, 7*t*
promastigote, 821
promoter region, 370
propagation, of viral infections, 137
prophage, 133, 380
prophylactic chemotherapy, 898
prophylaxis, 97
with acyclovir, **160**
antibacterial agents, **426**
hepatitis B immune globulin, 237
rabies, 302
rifampin, 558
propiolactone, 302
Propionibacterium acnes, 895, 899*t*
Propionibacterium spp., 10, 510, 517*t*
prostaglandin, 26
prostate hypertrophy, 939
prostatitis, 939-941, **941**
protease inhibitors, **163**, 320
proteases, 306, 309
coccidioidomycosis, 751
proteasome, 32
protective antigen (PA), 482
protein carriers, 38
protein exotoxins, 580
protein F, 448, 451
protein secretion, 365-366
transport systems, 366
protein synthesis inhibitors, **410-413**
proteinase, 88
protein-calorie malnutrition, 898
Proteus mirabilis, 607
Proteus spp., 580, 585*t*, **607**, 908*t*, 939
protomer, 106
proton pump inhibitors, 577
proto-oncogenes, 148, 322
protoplast, 352
protozoa, 759, **761-762**
classes of, 762, 762*t*
ectoplasm, 761
endoplasm, 761
morphology, 761-762
phagocytosis in, 762
physiology, 762
pinocytosis in, 762
Providencia, 580, **607**
provirus, 133, 309
PRP. *See* polyribitol phosphate
PrPc, 340
PrPsc, 342
pseudocowpox, 206*t*, **211**
pseudocysts, 831
pseudohyphae, 695, 725
pseudomembranes, 473, 475, 529, 919-920
diphtheria, 476*f*
pseudomembranous colitis, *Clostridium difficile*, 530*f*
pseudomembranous colitis (PMC), 529
Pseudomonas, 55, 56, 417, **617-627**
clinical case, 627
pneumonia from, 925, 926*t*
treatment, 411
upper respiratory infection and, 920
UTI from, 940
Pseudomonas aeruginosa, **617-623**, 625*t*
alginate biofilm, 621*f*
amikacin for, 622
bacteriology, 617-618
carbapenems for, 622
carbenicillin for, 622
cephalosporins for, 622
cystic fibrosis and, 619, 620-621, 622*f*
diagnosis, 622
disease, 618-623
cellular view, 620*f*
clinical aspects, 621-623
epidemiology, 618-619
manifestations, 621-622
overview, 619*f*
ear infections, 907, 907*t*, 908*t*
eye infections, 906*t*
folliculitis caused by, 895, 899*t*
gentamicin for, 622
immunity, 621
lipopolysaccharide, 618
lower respiratory tract infections from, 926*t*
monobactams for, 622
ofloxacin for, 413
osteomyelitis from, 902*t*
penicillins and, 408
pigment production, 623*f*
pneumonia, 620*f*
prevention, 623
ticarcillin for, 622
tobramycin for, 622
treatment, 622-623
wound infections from, 899*t*, 900
Pseudomonas fluoresces, 625*t*
pseudopodia, 803
psychrophiles, 367
PT. *See* pertussis toxin
pteridine, 41
Pteropus, 295
pubic lice, 946*t*
puerperal infections, group A streptococci, 449, 454
pulmonary anthrax, 484-485
pulmonary blastomycosis, 748

pulpitis, 915
purified protein derivative (PPD), 498, 503
purulent meningitis, 953-955
purulent rhinitis, 921
pyelonephritis, 939-942, **940-941**
pyocyanin, 618
pyoderma. *See* impetigo
pyogenic streptococci, 444, 446, **460**
pyrantel pamoate, 777*t*
pyrazinamide, for tuberculosis, 499
pyrimethamine, 774, 790
  resistance, 777
pyrogenic exotoxins, group A streptococci, 454-455
pyuria, 941

Q
Q fever, 630*t*, 683*t*, 687-688
  clinical aspects, 688
QBC. *See* quantitative buffy coat
Qinhaosu, **775**
quantitative buffy coat (QBC), 788
quaternary ammonium compounds, 52
  for plaque inhibition, 913
  for sterilization, 47*t*
quinacrine hydrochloride, for giardiasis, 821
quinidine, 773
quinine, 773
  for malaria, 789*t*
quinoline methanols, 773
quinolones, 413-414, 420
  antimalarial, **773-774**
  for Legionnaires disease, 614
  resistance, 549
  for *Ureaplasma*, 669
quinones, **774**
quinsy. *See* peritonsillar abscess
quinupristin dalfopristin, 413
quorum-sensing system, 619

R
R factors, 384
R plasmids, **384**
R5, 308, 315
RA 27/3, 201
rabies, 109*f*, **297-304**
  acute neurologic stage, 301*t*
  aerosol spread of, 299
  clinical capsule, 299
  clinical stages of, 301*t*
  CNS infections and, 952*t*, 953
  diagnosis, 302
  electron micrograph of, 298*f*
  encephalitis in, 301
  epidemiology of, 298-299
  in humans, 299-300
  incubation period, 299, 301*t*
  manifestations, 301
  pathogenesis, 299
  postexposure prophylaxis, 302
  prevention, 302
  prodrome stage, 301*t*
  receptors, 114*t*
  replication of, 299
  sequential steps in, 300*f*
  transmission of, 297-298
  treatment, 302
  in United States, 299*f*
  vaccine, 302
  virology, 297-298
raccoons, 299, 853
radioimmunoassay, 76, **77**
radionuclide scanning, 902
raltegravir, 158*t*
rapid plasma reagin (RPR), 651
Raynaud's phenomenon, 668
reactivation tuberculosis, 494*f*, 496-498
  AIDS and, 497-498
  predisposing factors, 497
reactive nitrogen intermediates, 25
reactive oxygen intermediates, 25
reassortment, 169
receptor-mediated endocytosis, 116
receptors, 113
recipient cell, 378
recombination, **132-133**
  antigenic variation and, **377**
  in bacteria, **376-377**
  high-frequency, 133
  homologous, 376*f*, **377**
  in RNA viruses, 133
  site-specific, **377**
recurrent genital herpes, 255
red blood cells, 597
  binding, 128
  merozoites in, 782
  trophozoites in, 782
reducing agents, 69
reduviid bug, 830-831
refractory septic shock, 967
regulator protein, 370
regulatory proteins, HIV, 311-312
regulon, 371, 373
  cell stress, 371
relapsing fever, 630*t*, **655-657**
  clinical aspects, 656-657
  louse-borne, 656
  prevention, 657
  tick-borne, 656
  treatment, 657
relatedness, 385
release, **126**
  bacteriophages, **126**
reoviruses, 117, **187**, 280*t*, **282**
  case study, 187
  classification of, 110*t*
  receptors, 114*t*
replacements, 374, 375*t*
replicative transposition, 378
repressible genes, 372*f*
repressor, 134, 370
reproductive systems, of trematodes, 879
RES. *See* reticuloendothelial system
residents, 8
resistance
  acquired, **423**, 424*f*
  aminoglycosides, 419*t*, 467, 622
  ampicillin, 557
  to antibiotics, 17
  antifungals, **711**
  antimicrobial, **415-415**
    features of, 419*t*
    mechanisms, **418-423**, 420*f*
    Staphylococcus aureus, 439
  antiparasitic antimicrobics, 777
  antiviral, **164-165**
  atovaquone, 777
  azithromycin, 549
  azole, 711
  binding sites and, 420
resistance (*Cont.*):
  cephalosporins, 421, 607
  chloramphenicol, 419*t*
  chloroquine, 773, 784
  clindamycin, 419*t*, 421
  conjugation and, **423-424**
  definition, 404, 711
  echinocandins, 711
  enzymatic inactivation, **421-423**, 439
  epidemiology of, 425
  erythromycin, 467
  flucytosine, 711
  fluoroquinolones, 419*t*
  folate inhibitors, 419*t*
  ganciclovir, **161**
  genetics of, **423-425**
    transposition, **424-425**
    transposons, **424-425**
  genotypic, **164**
  glycopeptides, 419*t*
  HIV, 321
  intrinsic, **423**
  β-lactamase inhibitors, 419*t*
  β-lactams, 467, 572
  macrolides, 419*t*
  of malaria, 759-760
  mechanisms, 711
  moderate, 415
  mutational, **423**
  novobiocin, 440
  penicillin, 421, 622
  phenotypic, **164**
  plasmids in, **423-424**
  of *Plasmodium falciparum*, 777, 789
  polyenes, 711
  proguanil, 777
  pyrimethamine, 777
  quinolones, 549
  rifampin, 419*t*
  sulfadoxine, 777
  sulfonamides, 421, 467
  susceptibility and, **415-416**
  tetracycline, 419*t*, 467, 532
  trimethoprim, 421
  vancomycin, 421
  viral, **157-165**
  virulence v., 418
respiration, 360, 361
respirators, nosocomial infections from, 56
respiratory spread, **93**, 139*t*
respiratory syncytial virus (RSV), 116, 167, **179-182**
  antigenic subgroups, 179
  asthma and, 181
  diagnosis, **181**
  epidemiology, **179-180**
  immunity, **180**
  inflammation in, 180*f*
  lower respiratory tract infections from, 926*t*
  manifestations, **181**
  mortality, 181
  pathogenesis, **180**
  prevention, **182**
  treatment, **182**
  virology, **179**
respiratory tract, microbial flora in, 11
respiratory tract infections
  lower, **924-928**
    clinical features, **924-925**
    diagnosis, **926-928**, 927*f*

respiratory tract infections, lower (*Cont.*):
- etiologic agents, **925-926**, 926*t*
- management, **928**
- sputum and, 924-928, 927*f*

middle, **922-924**
- clinical features, **922**
- etiologic agents, **922-923**, 923*t*
- management, **924**

upper, **919-922**
- clinical features, **919-920**
- diagnosis, **921**
- etiologic agents, **920-921**, 920*t*
- immune deficiency and, 920
- management, **921-922**

reticulate body, 673
reticulocytes, 783
reticuloendothelial system (RES), 602
retinoblastoma, 327
retinoblastomaprotein, 148
Retortamonas intestinalis, 814*t*
retropharyngeal abscess, 919-922, 920*t*
retrotonsillar abscess, 919-920
retroviruses, 128, **305-324**
- antigenic drift of, 131
- classification of, 110*t*
- diploid nature of, 132
- DNA replication, 309*f*
- entry, 306-308
- genes, 310-311
- incubation period, 142*t*
- life cycle, 308*f*
- major genes, 307*t*
- post-entry events, 308-310
- replication cycle, 306-310
- structure, 306, 310*f*
- transducing, 148
- transformation by, 148-149, 321-322
- virology, **306-316**

Rev, 311
reverse transcriptase, 120, 305, 306, 309
- HIV, 310

reverse transcriptase PCR (RT-PCR), 217
reverse transcription, 309
Rev-responsive element (RRE), 311
Reyes syndrome, 175
RGD receptors, 24
Rhabdoviridae, 120, 297
- classification of, 110*t*
- disease caused by. *See* rabies

rheumatic fever, 42
rheumatic heart disease, 450, 456
rhinitis, 919-921, 920*t*
rhinoviruses, 115, **185**, 214
- prevention, **185**
- receptors, 114*t*
- treatment, **185**
- upper respiratory infection from, 920*t*

Rhipicephalus sanguineus, 684
rhizopods, 803-811
- case study, 811
- classes of, 762*t*

Rhizopus, 732-733, 909
- classification of, 699*t*

Rhodococcus, **513**
- features of, 508*t*

ribavirin, 158*t*, **161**
- aerosol administration, 161

ribotyping, **85**

Rickettsia, 66, 410, 630*t*, **681-687**
- bacteriology, 681-682
- chloramphenicol and, 412
- disease, 682-687
  - clinical capsule, 682
  - diagnosis, 682-683
  - epidemiology, 682
  - pathogenesis, 682
  - spotted fever group, 684-686
  - typhus group, **686-687**
- examples of, 683*t*
- growth, 681-682
- metabolism, 681-682
- morphology, 681
- structure, 681
- tetracycline and, 411
- vasculitis, 683*f*

Rickettsia africae, 683t
Rickettsia akari, 683*t*, 686
Rickettsia australis, 683*t*
Rickettsia conorii, 683*t*
Rickettsia prowazekii, 683*t*, 687
Rickettsia rickettsii, 683*t*, **684-686**
Rickettsia typhi, 630*t*, 683*t*
rickettsial spotted fevers, 630*t*
rickettsial zoonotic infections, 630*t*
rickettsialpox, 683*t*, 686
- doxycycline for, 686

rifampin, 363, 406*t*, **414-415**, 440
- for brucellosis, 633
- for *Haemophilus influenzae*, 415, 558
- for Legionnaires disease, 614
- for leprosy, 503
- for meningococcal disease, 541
- for *Mycobacterium kansasii*, 503
- for *Mycobacterium tuberculosis*, 415
- for *Neisseria meningitidis*, 415
- prophylaxis, 558
- resistance, 419*t*
- for tuberculosis, 499

rifamycins, 406*t*
rimantadine, 157, 158*t*, 176, 177
- pharmacology, 159
- toxicity, 159

ringworm, 715*f*, 716
ritonavir, 158*t*, 163
RNA polymerase, 363
RNA synthesis inhibition, **161**
RNA viruses, **124-125**
- double-stranded, 273
- error rates, 130
- negative-sense, 297
- oncogenicity of, 147*t*
- recombination in, 133
- transformation by, 149
- zoonotic, **292-293**

RNase H, 309
RNase L, 151
Rocky Mountain spotted fever, 683*t*, **684-686**, 684*f*
- diagnosis, 685
- doxycycline for, 685
- epidemiology, 684-685
- manifestations, 685
- prevention, 686
- treatment, 685

Romaña's sign, 832
roseola, 203, 268, 269
- comparison, 190*t*

roseola infantum, **203**

rotavirus, 12, **272-276**, **273**
- animal, 274
- clinical aspects, **275**
- diagnosis, **276**
- epidemiology, **275**
- fecal-oral spread, 273
- gastrointestinal infections from, 933*t*, 936
- immunity, **275**
- manifestations, **275**
- morbidity, 273*f*
- pathogenesis, **275**
- prevention, **276**
- receptors, 114*t*
- structure, 273*f*
- treatment, **276**
- vaccines, 274
- virology, **272**

roundworms, 836*t*. *See also Ascaris lumbricoides*; nematodes
RPR. *See* rapid plasma reagin
RRE. *See* Rev-responsive element
RSV. *See* respiratory syncytial virus
RT-PCR. *See* reverse transcriptase PCR
rubella, **197-201**
- antibody response in, 198*f*
- in childbearing women, 198
- clinical aspects, **200-201**
- clinical capsule, 198
- comparison, 190*t*
- congenital infection, 199, 200
- diagnosis, 201, 203
- epidemiology, **198**
- fetal infection, 199, 200
- immunity, **199**
- immunosuppression and, 155
- isolation, 199
- manifestations, **200-201**
- pathogenesis, **198**
- pathology, **199**
- persistence of, 199*f*
- prevention, 201
- rash, 200*f*, **203**
- treatment, 201
- vaccine, 201
- virology, **197-199**

rubeola, 195
ruffles, *Salmonella enterica*, 600, 601*f*
Russian flu, 171-172

S

S phase, 123
Sabouraud's agar, 701
Saccharomyces cerevisiae, 696*f*
saliva, 916
salivary spread, **94**, 139*t*
Salmonella, 60, 388, 392, 571, 580, **599-605**, 630*t*
- bacteriology, 599
- enteric infections from, 930-931, 932*t*, 936
- food poisoning from, 934*t*, 935
- gastroenteritis, 599-601
  - clinical capsule, 599
- invasion, 596*f*
- osteomyelitis from, 902*t*
- septic arthritis from, 903

Salmonella enterica, 585*t*, **599-601**, 930, 966*t*
- adherence, 600
- enterotoxin, 601
- epidemiology, 600
- immunity, 601

Salmonella enterica (*Cont.*):
pathogenesis, 600-601
ruffles, 600, 601*f*
Salmonella typhi, 416, 599, 601-602
clinical capsule, 601
salmonellosis, **603-605**, 630*t*
ampicillin for, 604
bacteremia, 603
cefixime for, 604
ceftriaxone for, 604
chloramphenicol for, 604
ciprofloxacin for, 604
clinical aspects, 603-605
diagnosis, 604
enteric fever, 603-604
gastroenteritis, 603
manifestations, 603-604
metastatic infection, 603
prevention, 604-605
treatment, 604
salpingitis, 546, 676, 946*t*
San Joaquin Valley, 750-752
sanitization, definition of, 46
Sapovirus, 276
saquinavir, 158*t*, 163
sarcolemma, 453
Sarcoptes scabiei, 946*t*
SARS. *See* severe acute respiratory syndrome
saturated solution potassium iodide (SSKI), 721
scabies, 946*t*
scaffolding proteins, 126
scalded skin syndrome, 436*f*, 438
scanning electron micrograph
Chlamydia trachomatis, 674*f*
Vibrio cholerae, 566*f*
scarlet fever, 203, 448, 454-455
Schistosoma bovis, 890
Schistosoma haematobium, 767, 879
epidemiology, **887**
infections, 888-889
life cycle, 886*f*
parasitology, **885**
Schistosoma intercalatum, 885
Schistosoma japonicum, 775, 879
epidemiology, **887**
infections, 888-889
life cycle, 886*f*
parasitology, **885**
prevention, **890**
Schistosoma mansoni, 769, 775, 879
eggs, 883*f*
epidemiology, **886-887**
infections, 888-889
larvae, 884*f*
life cycle, 886*f*
parasitology, **885**
Schistosoma mekongi, 885, 887
Schistosoma spp., 879
disease caused by. *See* schistosomiasis
eggs, 883*f*, 885
life cycle of, 883*f*, 885
parasitology, **885-890**
schistosomes, 879, 885, 886*f*
schistosomiasis, 761
clinical aspects, **888-890**
diagnosis, **889**
epidemiology, **886-887**
immunity, **887-888**
Katayama syndrome and, 888-889
pathogenesis, **887**
schistosomiasis (*Cont.*):
prevalence of, 760*t*
prevention, **890**
stages of, **888-889**
treatment, 882, 885, **889-890**
vaccine development for, 890
schizogony, 781
exoerythrocytic, 782
schizonts, 781
Schuffner's dots, 781
Schwann cells, 501
scrapie, 101, 342
scrub typhus, 683*t*, **687**
secondary cell culture, 71
secondary response, 40
secondary syphilis, **649**, 948
secondary viremia, 142
secretory component, 39
secretory IgA, 40, 275, 389, 674
secretory piece, 40
selectins, 26
selective media, **69**
Semmelweis, Ignaz, 13, **53-54**
sensitivity
definition, 404
moderate, 415
sepsis, 959, **966-967**
sepsis syndrome, 967
septate hyphae, 696*f*, 730
September 11, 2001, 403
septic arthritis, 519, **903-904**
diagnosis of, **903-904**, 904*t*
etiologic agents, 903, 903*t*
management of, **904**
osteomyelitis and, 901
synovial fluid findings and, 904*t*
septic shock, **966-967**
septicemia, 959
sequelae, 191
sequestrum, **901**
serine proteases, 151
serologic classification, **77-78**
serologic detection, 73
serotypes, 177, 580
serous otitis media, 908, 908*t*
Serratia, 580, 585*t*, **607**
serum sickness, 42
severe acute respiratory syndrome (SARS), 89, 186
sex pilus, 383
sexual transmission
of *Chlamydia*, 673
of HIV, 313
syphilis, 647
sexually transmitted infections (STDs), **945-949**, 946*t*. *See also* genital infections
Shakespeare, 429, 432
shaking chill, 464
sheep, 211
shellfish, undercooked, 881-882
Shiga toxin, 587*f*, 592
enterohemorrhagic *Escherichia coli*, 592
Escherichia coli, 586
Shigella, 597
Shigella, 22, 60, 392, 580, 593, **595-598**, 602
food poisoning from, 934*t*, 935
immunity, 597-598
Shiga toxin, 597
Shigella boydii, 584*t*, 595
Shigella dysenteriae, 584*t*, 595, 596, 932*t*
bacteremia from extravascular infection and, 966*t*
Shigella flexneri, 584*t*, 595, 596
invasion, 596*f*
Shigella sonnei, 584*t*, 596
shigellosis, **595-598**
ampicillin for, 598
azithromycin for, 598
ceftriaxone for, 598
clinical aspects, 598
clinical capsule, 595
diagnosis, 598
epidemiology, 595-596
fluoroquinolones for, 598
manifestations, 598
nalidixic acid for, 598
pathogenesis, 596-597
prevention, 598
treatment, 598
shingles, 257
shunts, 953
sialic acid, 461, 537
sickle cell disease, 668
Sickle cell trait, 783
siderophores, 359, 393
simian malaria, 787
simple diffusion, 359
simple transposition, 379*f*
Simulium, 861, 862
single-host parasites, 764
single-stranded RNA, 168
singlet oxygen, 25
sinus infections, **909-910**
diagnosis, **910**
etiologic agents, **909-910**, 910*t*
management, **910**
manifestations, **909**
site-specific recombination, **377**
skin, 390*t*
Candida albicans infection, 728*f*
diphtheria, 897
flora, 10
infections
acne, **895**, 899*t*
cellulitis, **897**, 899*t*
erysipelas, **896-897**
etiologic agents of, **898-900**, **899*t***
folliculitis, **895-896**, 899*t*
furuncle, **896**, 899*t*
granulomatous lesions, **897**
impetigo, **896**, 899*t*
intertrigo, 899*t*
of keratinized layers, **896**
treatment of, **896**
ulcers, **897**, 899*t*
in innate immunity, 21
integrity, 895
microbial flora in, 9-10
skin-to-skin transfer, **94**
sleeping sickness. *See* African trypanosomiasis
slim disease, 319
slime layer, 349
slow viral diseases, 337
smallpox, **207-208**. *See also* variola
clinical aspects, **208**
diagnosis, **208**
epidemiology, 14
facial lesions of, 209*f*
immunity, 209

- smallpox (*Cont.*):
  - manifestations, **208**
  - pathogenesis, **208**
  - prevention, **208**
  - vaccine, 43
  - virology, **207**
- Smith, Sydney, 443
- soft palate, 475
- soft tissue infection, 901, 906
- sources, **91**
- Southern hybridization, 80
- Southern transfer, 83
- Soviet Union, 483
- Spanish flu, 173
- spasmodic croup, 922
- spasticity, 340
- specialized transduction, 135, 381
- species, 384
- specific therapy, antibacterial agents, **426**
- specimen, **59-61**
  - collection, 61, **941**
  - direct examination, **61-65**
  - direct tissue, 59-60
  - fluid samples, 59-60
  - indirect samples, 60
  - transport, 61
- spectrum
  - of antibacterial agents, **404**
  - broad, 404
  - definition, 404
  - narrow, 404
- spelunkers, 744
- spherical architecture, **106**
- spherules, 748
  - *Coccidioides immitis*, 748
- spikes, 102, 113, 142, 145
- spinal cord, 191
- spiramycin, 777*t*, 797
- spirochetes, **643-648**, 918
  - bacteriology, 643-645
  - classification, 643-645
  - clinical case, 662
  - disease, 645
    - features of, 645*t*
  - growth, 643-645
  - morphology, 644*f*
- spongiform changes, 340*f*
- spongiform encephalopathies, subacute, **339-343**
- spores
  - bacteria, **357**
  - in fungal reproduction, 695
  - membrane, 357
- sporocytes, *Pneumocystis*, 734
- *Sporothrix*, **719-721**, 899*t*
  - classification of, 699*t*
- *Sporothrix schenckii*, 704, 714*t*, **719-721**
- sporotrichosis, **719-720**, 897
  - clinical aspects, 720-721
  - clinical capsule, 719
  - diagnosis, 720
  - epidemiology, 720
  - immunity, 720
  - manifestations, 720
  - pathogenesis, 720
  - prevention, 721
  - treatment, 721
- sporozoa, **779-802**
  - classes of, 762*t*
  - clinical case, 800-801
- sporulation, 357, 371
- spotted fever group, **684-686**
- sputum, 924-928, 927*f*
  - Gram-positive bacteria in, 485
  - *Mycobacterium tuberculosis* in, 490*f*
  - *Nocardia* in, 512*f*
- SSKI. *See* saturated solution potassium iodide
- SSPE. *See* subacute sclerosing panencephalitis
- ST. *See* stable toxin
- St. Louis encephalitis, **289**, 952*t*
- stable toxin (ST), *Escherichia coli*, 586
- stains
  - acid-fast, **63-64**, 64*f*
  - acridine orange, 788
  - crystal violet, 64*f*
  - endospore, 64*f*
  - flagellar, 64*f*
  - fluorochrome, 64
  - fungal, 65
  - Gram, **61-62**, 64*f*
  - India ink capsule, 64*f*
  - iodine, 65*f*
  - methylene blue, 64*f*
  - parasitic, 65
- standard precautions, 57
  - for nosocomial infections, 58*t*
- staphylococcal scalded skin syndrome, 436*f*, 438
- staphylococcal toxic shock syndrome, 437*f*, 438
- staphylococci, 10, **429-442**
  - coagulase-negative, 440-441
  - superantigen toxins, 431
- *Staphylococcus aureus*, 11, 351, 384, 388, **429-440**, 467, 551
  - antimicrobial resistance, 439
  - bacteremia from extravascular infection and, 966*t*
  - bacteriology, 429-432
  - carbuncle, 435*f*, 437
  - carriers, 54
  - cellular view, 434*f*
  - cellulitis, 899*t*
  - chemoprophylaxis, 440
  - chronic furunculosis, 437-438
  - clinical aspects, 437-440
  - CNS infections and, 952*t*
  - deep lesions, 438
  - diagnosis, 439
  - disease, 433*f*
  - drying, 433
  - ear infections from, 908*t*
  - enterotoxins, 432
  - epidemiology, 432-433
  - exfoliatin and, 431
  - extracellular enzymes, 431-432
  - eye infection from, 906, 906*t*
  - features, 430*t*
  - folliculitis caused by, 895
  - food poisoning, 433, 438
  - furuncles, 435*f*, 437
  - hospital outbreaks, 433
  - identifying, 430-431
  - immunity, 436-437
  - impetigo, 438
  - infective endocarditis and, 960, 962*t*
  - influenza A and, 175
  - intertrigo, 899*t*
  - lower respiratory tract infections from, 926*t*, 927*f*
  - manifestations, 437-438
  - methicillin-resistant, 421
  - morphology, 429-430
- *Staphylococcus aureus* (*Cont.*):
  - osteomyelitis from, 901, 902*t*
  - pathogenesis, 433-436
  - prevention, 439-440
  - primary infection, 433-435, 437-438
  - relapsing, 437
  - septic arthritis from, 903, 903*t*
  - shape, 348*f*
  - sinus infections from, 910*t*
  - structure, 429-430
  - subtyping, 430-431
  - superantigen toxins, 431
  - suppurative thrombophlebitis and, 964*t*
  - toxic shock syndrome, 35
  - α-toxin, 431
  - toxin-mediated disease, 436
  - toxins, 431-432, 438
  - treatment, 439
  - wound infections from, 897-898, 899*t*
- *Staphylococcus epidermidis*, 440
  - CNS infections and, 952*t*
- *Staphylococcus saprophyticus*, 440, 441
  - UTI from, 940
- stationary phase, 368
  - cells, **373**
- stavudine, 158*t*, **162**
- STDs. *See* sexually transmitted infections
- steam, 48-49
- stem cells, 20*f*
  - hematopoietic, 20*f*
  - lymphoid, 20*f*
  - myeloid, 20*f*
- *Stenotrophomonas maltophilia*, 625*t*
- sterile swab, 61
- sterilization, **47-50**
  - achieving, 46
  - chemicals, 47*t*
  - definition of, 45-46
  - ethylene oxide gas, 47*t*, 49
  - heat, 47*t*, **48-49**
  - ionizing radiation for, 47*t*, 49-50
  - methods, 47*t*
  - radiation, 47*t*
  - ultraviolet radiation for, 47*t*, 49-50
- stillbirth, 795
- stomach, 389, 390*t*
- stomatitis, 920, 922
- stool, flora, 11*f*
- stool microscopy, 936
- strep throat, 15
- StrepSAgs, 448
- streptococcal superantigen toxins, 448
- streptococcal toxic shock syndrome, 449-450, 452, 455
- streptococci, **443-465**
  - biochemical characteristics, 443-444, 456*t*
  - classification, 444-446, 445*t*
  - cultural characteristics, 443-444, 456*t*
  - group A, 15, **446-457**
    - acute, 451-452
    - antigenic structure of, 447*f*
    - bacteriology, 446-454
    - cellular view of, 452
    - diagnosis, 456-457
    - disease, 450*f*
    - epidemiology, 449-451
    - erysipelas, 454
    - Gram stain, 444*f*
    - immunity, 453-454

Streptococcus pneumoniae (*Cont.*):
impetigo, 449, 454
M protein and, 446-448
manifestations, 454-456
nephritogenic strains, 451
pathogenesis, 451-453
penicillin for, 425
pharyngitis, 449, 454
prevention, 457
puerperal infections, 449, 454
pyrogenic exotoxins, 454-455
structure, 446-448
surface molecules, 448
toxic shock syndrome, 35
treatment, 457
wounds, 449
group B, 426, **457-465**
aminoglycosides for, 459
ampicillin for, 459
autolysins in, 461
bacteriology, 457-459
clinical capsule, 457
diagnosis, 459
epidemiology, 458
neonatal sepsis, 458
pathogenesis, 458-459
penicillin for, 459
pneumococcal disease, 458*f*, **461-465**
pore forming toxins in, 461
prevention, 460
shape, 348*f*
treatment, 459
group characteristics, 443-446
hemolytic reactions, 456*t*
α-hemolytic, 456*t*
β-hemolytic, 456*t*
infective endocarditis and, 961-962, 962*t*
morphology, 443
nonhemolytic, 456*t*, **465**
in oropharynx, 10
plaque colonized by, 911
pyogenic, 444, 446, **460**
upper respiratory infection from, 921
viridans, 446, 456*t*, **465**
Streptococcus agalactiae, 444, 445*t*
Streptococcus bovis, 445*t*
Streptococcus equi, 445*t*
Streptococcus mitis, 446
Streptococcus mutans, 349, 445*t*, 465, 913, 915
Streptococcus pneumoniae, 8, 458*f*, **460-465**, 531, 563
antibiotic selection, 465
bacteremia from extravascular infection and, 966*t*
capsule, 460, 463
CNS infections and, 951, 952*t*
diagnosis, 464
ear infection from, 908*t*
epidemiology, 462
extracellular products, 461
eye infection from, 906, 906*t*
growth, 461
immunity, 463
influenza A and, 175
lower respiratory tract infections from, 926, 926*t*
manifestations, 464

Streptococcus pneumoniae (*Cont.*):
middle respiratory infections and, 922
morphology, 460-461
pathogenesis, 14, 462-463
penicillin for, 464
pneumolysin, 463
prevention, 465
septic arthritis from, 903*t*
sinus infections from, 909, 910*t*
structure, 460-461
suppurative thrombophlebitis and, 964*t*
treatment, 464-465
Streptococcus pyogenes, 445*t*
ear infections from, 908*t*
eye infection from, 906, 906*t*
penicillin for, 921
septic arthritis from, 903*t*
sinus infections from, 910*t*
Streptococcus salivarius, 387, 445*t*, 446
Streptococcus sanguis, 445*t*
plaque colonized by, 911
streptogramins, 405*t*, **413**
streptokinase, 448, 453
streptolysin O, 446, 461
streptolysin S, 446
Streptomyces, 406, 507
antibiotics and, 404
features of, 508*t*
streptomycin, 410, 420
for plague, 637-638
for tuberculosis, 499, 500
for tularemia, 640
Strongyloides stercoralis, 835, 836*t*, 837, **846-849**
disease caused by. *See* strongyloidiasis
life cycles of, 836*t*, 846
structure of, 847*f*
strongyloidiasis, **847-849**
clinical aspects, 848-849
diagnosis, 848
epidemiology, 847-848
immunity, 848
malabsorption and, 848
manifestations, 848
pathogenesis, 848
prevalence of, 760*t*
prevention, 849
treatment, 849
structural subunits, 106
subacute bacterial endocarditis, 465
subacute endocarditis, 961
subacute sclerosing panencephalitis (SSPE), **195-196, 338**
subacute spongiform encephalopathies, **339-343**
subclinical infection, 144
subclinical poliomyelitis, 219
subcutaneous fungi, 699, **719-722**
subgingival plaque, 911-912, 916
submucosa, 545
subsalicylate-containing compounds, 937
succinate, 361
sucrose, 915
sulbactam, 409
for *Bacteroides fragilis*, 532
sulfadoxine, resistance, 777
sulfamethoxazole, 942
sulfhydryl compounds, 698, 772
sulfonamides, 4, 403, 406*t*, **414**, 774, 942
for malaria, 789*t*
for *Nocardia*, 513
for *Paracoccidioides brasiliensis*, 753

sulfonamides (*Cont.*):
resistance, 421, 467
for toxoplasmosis, 797
sulfones, for leprosy, 503
sulfur, 4
granule, 508, 509*f*
superantigen exotoxins, **396**, 397*f*
superantigens
in adaptive immunity, 35
staphylococcal, 431
superficial fungi, 699, **713-719**
superinfection, 174
in influenza A, 175
superoxide, 25
in anaerobiosis, 516
anion, 361
dismutase, 361, 362, 516
suppurative thrombophlebitis, **963-964**, 964*t*
supragingival plaque, 911-912, 913*f*
suramin, 777*t*
surface adhesin, 747
surface proteins, 214
surfactants, disinfection with, 51
surgical wound infections, 897, 899*t*
susceptibility
definition, 404
resistance and, **415-416**
SV40, 334
receptors, 114*t*
sylvatic cycle, **286-287**
echinococcosis, 875
sylvatic plague, 635
symmetry
cubic, **126**
helical, **125**
icosahedral, **126**
synapse, 524*f*
syncytia, 145
syncytium formation, 179
synovial membrane, biopsy of, 904
synovium, 453
synthetic orvirion component production, **117-118**
syphilis, 403, 544, **646-652**, 945, 946*t*-947*t*, 948
cardiovascular, 650
clinical aspects, 649-652
clinical capsule, 646
congenital, **650**
diagnosis, 650-652
microscopy, 650-651
nontreponemal tests, 651
serologic tests, 651
treponemal, 651-652
epidemiology, 647
HIV and, 647, 651
immunity, 649
latent, **649-650**
lesions, 648
manifestations, 649-650
overview, 647*f*
pathogenesis, 648-649
primary, 648*f*, **649**
secondary, **649**
tertiary, 647
transmission, 647
systemic fungi, 699
clinical case, 754
features of, 740*t*
geographic distribution of, 744*f*
systemic STDs, 945, **949**

T
T antigen, 333
T cell receptors (TCR), 32
T cells, 20*f*, 24*f*
cytotoxic, 30*t*
helper, 30*t*
response, 33, 36*f*
in adaptive immunity, 33-36
T1 pathway, 34
T2 pathway, 34
tabes dorsalis, 650, 650*f*
tachycardia, 832
tachyzoite, *Toxoplasma gondii*, 791-792
Taenia saginata, 764, 765, 767, 769, 865, 866*t*
eggs of, 865-866
parasitology, 867
structures, 868*f*
Taenia solium, 767, 866*t*
acquisition of, 870*f*
parasitology, 869
structures, 868*f*
tampons, 436
tanapox, 206*t*, 209-210
tapeworms, 8. *See also* cestodes
beef, **867-869**
prevention, 869
treatment, 869
clinical aspects, 867
diagnosis, 869
dwarf, 874
epidemiology, 867
fish, **872-874**
disease, **873-874**
prevention, 874
treatment, 874
intestinal, 866*t*
manifestations, 867
pork, **869-872**
albendazole for, 872
life cycle, 870*f*
praziquantel for, 872
prevention, 872
treatment, 872
undercooked meat and, 870*f*
structures, 868*f*
tissue, 866*t*
TAR, 311
target proteins, 396
Tat, 311, 322
Tatum, E., 383
Tax, 322
taxonomic methods, 385
tazobactam, 409
TCP. *See* toxin-coregulated pilus
TCR. *See* T cell receptors
TCT. *See* tracheal cytotoxin
Tdap, 564
T-dependent reactions, 37
tears, 905
tegument, 247
teichoic acid, 352
teicoplanin, 406, 409
Temin, Howard, 305
temperate phages, 380
lysogenic cycles of, 381*f*
lytic cycle of, 381*f*
temperate viruses, 112
terbinafine, 709
features of, 708*t*
terminators, 370
tertiary syphilis, 647, **650**
tetanospasmin, 526, 527
tetanus, 42, **526-528**
benzodiazepines for, 528
clinical aspects, 527-528
clinical capsule, 526
clostridial, 524*f*
manifestations, 527
pathogenesis, 527
penicillin for, 528
treatment, 528
vaccine, 18
wound infections and, 900
tetracycline, 404, 405*t*, **411**, 775
for *Aeromonas*, 624
Chlamydia and, 411
for *Chlamydia pneumoniae*, 678
for *Chlamydia psittaci*, 678
for *Helicobacter pylori*, 576-577
for Legionnaires disease, 614
Mycoplasma and, 411
for *Mycoplasma pneumoniae*, 669
for *Plesiomonas*, 624
resistance, 419*t*, 467, 532
Rickettsia and, 411
susceptibility patterns to, 427*t*
for *Ureaplasma*, 669
tetrahydrofolic acid, 774
thiabendazole, 775
thiogycollate, 520
Thomas, Lewis, 387
thoracic actinomycosis, 509-510
3TC. *See* lamivudine
thrombocytopenia, in malaria, 786
thrombocytopenic purpura, 195, 200
thrombophlebitis, 519, 959, **963-964**, 964*t*
thrombosis, 592
thrush, 727, 728*f*, 919
thymidine kinase, 159
thymus, 31*t*
ticarcillin, 405*t*, 408
for *Bacteroides fragilis,* 532
for *Pseudomonas aeruginosa*, 622
tick-borne relapsing fever, 656
ticks. *See specific types*
tigecycline, 411
T-independent reactions, 37
tinea barbae, 717
tinea capitis, 717
tinea corporis, 717
tinea cruris, 717
tinea manuum, 717
tinea nigra, 719
tinea pedis, 717
tinea unguium, 717
tinidazole, 775
for giardiasis, 821
Tinsdale medium, 476
tissue culture, of viruses, 112
tissue cysts, ingestion, 794
tissue nematodes, **851-864**
case study, 863
general characteristics of, 852*t*
tissue tapeworms, 866*t*
tissues, microbial flora in, 9
TLR. *See* toll-like receptors
TM. *See* transmembrane
TMP-SMX. *See* trimethoprim-sulfamethoxazole
TMV. *See* tobacco mosaic virus
TNF. *See* tumor necrosis factor
tobacco mosaic virus (TMV), 106
assembly, 125*f*
tobramycin, 405*t*, 410, 623
for *Pseudomonas aeruginosa*, 622
togaviruses, 120, **279-280**, 280*t*, 285
classification of, 110*t*
enveloped, 197
toll-like receptors (TLR), 24, 613
tolnaftate, 718
features of, 708*t*
tonsillitis, 919-921, 920*t*
tooth demineralization, 915-916
tooth loss, 915
topoisomerase, 413
tourism, 890
toxic shock syndrome (TSS), 35
group A streptococci, 35
staphylococcal, 437*f*, 438
Staphylococcus aureus, 35
streptococcal, 449-450, 452, 455
toxic shock syndrome toxin (TSST), 436
toxicity
acyclovir, **160**
amantadine, 159
of antiparasitic agents, 771, 772
rimantadine, 159
toxin assay, 936
α-toxin, 431, 434
Clostridium perfringens, 520
β-toxin, *Clostridium perfringens*, 520
toxin-coregulated pilus (TCP), 566
Toxocara canis, 851-853, 906*t*
disease caused by. *See* toxocariasis
eggs of, 852
general characteristics of, 852*t*
parasitology, 851-852
transmission, 852
toxocariasis, **852-853**
clinical aspects, 853
corticosteroids for, 853
diagnosis, 853
epidemiology, 852-853
manifestations, 853
pets and, 853
prevention, 853
treatment, 853
toxoid, 474, 526
Toxoplasma gondii, 412, 768, 774, 790-797, 906*t*
asexual cycle, 790-791
definitive host, 792-793
intermediate hosts, 793
life cycle, 792-793
morphology, 791-792
oocyst, 791
parasitology, 790-791
relapse, 793
sexual cycle, 790-791
tachyzoite, 791-792
tissue cysts, 791
trophozoites, 791-792
toxoplasmosis, 318*t*, 412, 761, 774, **793-797**
AIDS and, 774, 795-796
atovaquone for, 797
clinical aspects, 795-797
clinical capsule, 793
congenital, 795
diagnosis, 796
distribution, 793
enzyme immunoassay for, 796
epidemiology, 793-794

toxoplasmosis (*Cont.*):
IgG in, 796
IgM in, 796
immunity, 794-795
in immunocompromised host, 795-796
manifestations, 795-796
pathogenesis, 794-795
prevalence, 793
sulfonamides for, 797
transmission, 793-794
congenital, 794
oocyst ingestion, 794
tissue cyst ingestion, 794
treatment, 796
trimethoprim for, 797
tra, 383
tracheal cytotoxin (TCT), 559
tracheal organ culture, 562*f*
tracheitis, 923, 923*t*
bacterial, 923
tracheobronchitis, 178, 922
trachoma, 673
Chlamydia trachomatis and, 675
transactivating, 322
retroviruses, 149
transcription
in bacteria, 365*f*
polymerization reactions, 363
viruses, **118-121**
transducing retroviruses, 148
transduction, 378, **380-381**, 381*f*
generalized, 381
specialized, 381
transferrin, 359, 537
transformation, 378, **380**
artificial, 380
transient bacteremia, 9
transient immunity, 178
transient viremia, 285
transients, 8
translation, 363-364
in bacteria, 365*f*
transmembrane (TM), 306
transmembrane proteins, 569
transmission routes, **92-95**
aerosol, 299
bloodborne, **94**, 139*t*
common, 93*t*, 139*t*
eye-to-eye, **94**
fecal-oral spread, **94**, 227, 273, 277
Entamoeba histolytica, 805
food-borne, 478
genital transmission, **94**, 139*t*
horizontal, 92, 139
human T-cell lymphotropic virus, 323
Leishmania, 822-823
needlestick, 233
respiratory, **93**, 139*t*
salivary spread, **94**, 139*t*
sexual, 313, 647, 673
skin-to-skin transfer, **94**, 139*t*
Toxocara canis, 852
toxoplasmosis, 793-794
congenital, 794
tissue cyst ingestion, 794
transplacental, 478
vertical, 92, **95**, 139, 140, 233, 285
viruses, 139*t*
water-borne, giardiasis, 819
zoonotic, **95**, 139*t*
transmission-based precautions, 57
airborne, 57
for nosocomial infections, 58*t*
transovarial transmission, 287
transpeptidases, 365
transpeptidation, 367*f*
transplacental IgG, 43
transplacental transmission, listeriosis, 478
transport media, 61
transposable elements, 378*f*
transposases, 377
transposition
bacteria, **377-378**
direct, 378
replicative, 378
in resistance, **424-425**
simple, 379*f*
transposons, 374, 377, **378**
in resistance, **424-425**
traumatic wound infections, 897, 899*t*
traveler's diarrhea, 594, **931**
trematode(s), **879-890**
characteristics of, 880*t*
classification of, 763*t*
eggs, 883*f*, 885
hermaphrodite, 879, 885
intestinal, 880*t*
locomotion of, 879
reproductive systems of, 879
schistosome, 879, 885, 886*f*
tissue, 880*t*
trench fever, 683*t*, 688-689
trench mouth, 645
Treponema carateum, 663
Treponema denticola, 911, 916
Treponema pallidum, 65, 94, 351, 644*f*, **646-652**, 946, 946*t*
bacteriology, 646
epidemiology, 647
eye infection from, 906*t*
features of, 645*t*
growth, 647
lipopolysaccharide, 646
outer membrane proteins, 649
penicillin for, 425, 652
prevention, 652
treatment, 652
ulcers from, 899*t*
Treponema pallidum subspecies *endemicum*, 663
Treponema pallidum subspecies *pertenue*, 663
Trichinella nativa, 854
Trichinella spiralis, 767, 768, 851, **854-857**, 934*t*, 954*t*
disease caused by. *See* trichinosis
general characteristics of, 852*t*
larva, 854
parasitology, 854
trichinosis, 769, 775, **854-857**
albendazole for, 856
clinical aspects, 856-857
corticosteroids for, 856
cysts, 855*f*
diagnosis, 856
diarrhea and, 856
epidemiology, 854-855
immunity, 856
manifestations, 856
mebendazole for, 856
pathogenesis, 856
prevention, 857
trichinosis (*Cont.*):
treatment, 856
undercooked pork and, 854
trichloroacetic acid, 333
Trichomonas hominis, 814*t*
Trichomonas tenax, 814*t*
Trichomonas vaginalis, 813, **814-817**, 814*t*, 946*t*, 948
culture, 814
diagnosis, 816-817
disease caused by. *See* trichomoniasis
distribution of, 764*t*
HIV and, 816
immunity, 816
manifestations, 816
parasitology, 814
pathogenesis, 816
structure, 815*f*
transmission of, 764*t*
trichomoniasis, 761
clinical aspects, 816-817
clinical capsule, 815
epidemiology, 815-816
manifestations, 816
prevalence, 815
treatment, 817
Trichophyton, 713
classification of, 699*t*
Trichophyton mentagrophytes, 698
Trichophyton rubrum, 717
Trichosporon cutaneum, 714*t*, 719
trichuriasis, **840-841**
clinical aspects, 840-841
diagnosis, 840
epidemiology, 840
immunity, 840
manifestations, 840
pathogenesis, 840
prevalence of, 760*t*
prevention, 841
treatment, 841
Trichuris trichiura, 835, 836*t*, **839-841**
disease caused by. *See* trichuriasis
egg structure, 839*f*
embryonated egg, 839*f*
life cycles of, 836*t*, 839-840
triclosan, 913
trifluorothymidine, **159**
trifluridine, 158*t*
trimethoprim, 406*t*, 774, 942
for gonorrhea, 548
resistance, 421
for toxoplasmosis, 797
trimethoprim-sulfamethoxazole (TMP-SMX), **414**
brucellosis for, 633
for cholera, 570
for *Escherichia coli*, 594
for listeriosis, 481
for pneumocystosis, 736-737
susceptibility patterns to, 427*t*
trismus, 527
tRNA, 364, 399
Trojan horse, 741
trophozoites
Entamoeba histolytica, 804, 807, 808
Giardia lamblia, 817, 818*f*
multiplication of, 804
Naegleria fowleri, 809
in red blood cells, 782
Toxoplasma gondii, 791-792

tropical spastic paraparesis (TSP), 323
tropism, **142-143**
Trypanosoma, 821
  life cycle, 822*f*
Trypanosoma brucei, 761, 770
Trypanosoma brucei gambiense, 829
Trypanosoma brucei rhodesiense, 829
Trypanosoma cruzi, 761, 770
  amastigotes, 831*f*
Trypanosoma spp., 776
trypanosomiasis
  African, **827-830**
    anemia and, 828
    antigenic shifts, 828
    clinical aspects, 828-830
    clinical capsule, 827
    diagnosis, 829
    epidemiology, 828
    IgM and, 828
    melarsoprol for, 829
    parasitology, 827
    pathogenesis, 828
    prevalence of, 760*t*
    prevention, 830
    treatment, 829
    vasculitis and, 828
  American, 767, **830-833**
    allopurinol for, 833
    benznidazole for, 833
    chronic, 832
    clinical aspects, 832-833
    clinical capsule, 830
    diagnosis, 832-833
    epidemiology, 830-831
    manifestations, 832
    myocarditis in, 831*f*
    nifurtimox for, 833
    pathogenesis, 831-832
    prevalence of, 760*t*
    prevention, 833
    treatment, 833
trypomastigote, 821, 830
tsetse fly, 828
TSP. *See* tropical spastic paraparesis
TSS. *See* toxic shock syndrome
TSST. *See* toxic shock syndrome toxin
tuberculate macroconidia, 743
  Histoplasma capsulatum, 743
tuberculin, 491
  skin test, 498, 498*f*
tuberculoid leprosy, 502
tuberculoma, 498
tuberculosis, 319, 373, **493-501**
  acid-fast smears, 498-499
  AIDS and, 495
  bovine, 630*t*
  cell-mediated immunity, 497
  ciprofloxacin for, 500
  clinical aspects, 497-501
  clinical capsule, 493
  in developing countries, 493
  diagnosis, 498-499
  DTH and, 495
  in eighteenth century, 493
  epidemiology, 493-495
  ethambutol for, 499, 500
  extensively drug-resistant, 500
  fish, 504
  fluoroquinolones for, 500
  granuloma, 495, 496*f*
tuberculosis (*Cont.*):
  HIV and, 500
  immunity, 497
  isoniazid for, 499, 500
  manifestations, 497-498
  morbidity, 493
  mortality rates, 493
  multidrug-resistant, 500
  in nineteenth century, 494
  ofloxacin for, 500
  pathogenesis, 495-497
  prevention, 500-501
  primary, 493*f*, 497, 744
  primary infection, 495-496
  pyrazinamide for, 499
  reactivation, 494*f*, 496-498
  rifampin for, 499
  second-line agents, 500
  streptomycin for, 499, 500
  treatment, 499-500, 499*t*
  worldwide distribution of, 493*f*
tubo-ovarian abscess, 547*f*
tularemia, **638-640**
  aminoglycosides for, 640
  clinical aspects, 639-640
  diagnosis, 640
  gentamicin for, 640
  immunity, 639
  manifestations, 639-640
  oculoglandular, 639
  pathogenesis, 639
  pneumonic, 639
  prevention, 640
  streptomycin for, 640
  treatment, 640
  typhoidal, 639
  ulceroglandular form, 639
tumor necrosis factor (TNF), 28, 152, 241, 397, 632, 727
tympanic membrane, 906-907
tympanometry, 908
type II hypersensitivity. *See* antibody-mediated hypersensitivity
type III hypersensitivity. *See* immune-complex hypersensitivity
type IV hypersensitivity. *See* delayed-type hypersensitivity
type-specific immunity, 453
type-specific immunoglobulin G, 453
Typhimurium, 599
typhoid fever, 581, 601-602, 930-931
  epidemiology, 602
  immunity, 602
  pathogenesis, 602
typhoidal tularemia, 639
typhus group, **686-687**
tyrosine, 374
Tzanck test, 255

U

UL97, 160
ulceroglandular tularemia, 639
ulcers, **897**, 899*t*
  chronic, 899*t*
  decubitus, 433, 519
  genital, **945**, 947*t*
ultraviolet (UV) irradiation, 45
  for sterilization, 47*t*, 49-50
unapparent infections, 144
uncoating, **115-117**
  inhibitors, **157-159**
United Nations, 483
United States Department of Health and Human Services (DHHS), 320
unpasteurized dairy, 631
UPEC. *See* uropathic *Escherichia coli*
upper respiratory infection (URI), 178, 185, **919-922**
  clinical features, **919-920**
  diagnosis, **921**
  etiologic agents, **920-921**, 920*t*
  immune deficiency and, 920
  management, **921-922**
upper respiratory tract, 390*t*
upper urinary tract, 11
urban cycle, **286**
Ureaplasma, **665-670**
  azithromycin for, 669
  diagnosis, 669
  epidemiology, 669
  microbiology, 665
  quinolones for, 669
  tetracycline for, 669
  treatment, 669
Ureaplasma urealyticum, 673
urease, 88
  Helicobacter pylori, 573
urease jaccase, 739
urethritis, 546, 939, 946*t*, **947**
  Chlamydia trachomatis, 676
urgency, cystitis and, 940
URI. *See* upper respiratory infection
urinary catheters, nosocomial infections from, 55
urinary stones, 940
urinary tract, 390*t*
urinary tract infection (UTI), 335, 466-467, **939-943**
  diagnosis, **941-942**, 942*f*
  epidemiology, 587, **939**
  Escherichia coli, 587-588
  manifestations, **940-941**
  pathogenesis, 587-588, **939-940**
  prevention, **943**
  treatment, **942-943**
urine
  clean-voided midstream, 941
  culture, **942**, 942*f*
uropathic *Escherichia coli* (UPEC), 584*t*, 588
urticaria, 859
UTI. *See* urinary tract infection
UV irradiation. *See* ultraviolet irradiation
uveitis, 905

V

V factor, 557
V3 loop. *See* variable region 3
VacA. *See* vacuolating cytotoxin
vaccines, 43
  anthrax, 483, 485
  bacillus Calmette-Guerin, 498, 500
  diphtheria, 18
  diphtheria toxoid and pertussis, 528
  genetically engineered, 18
  hepatitis A, 229
  human papilloma virus, 333
  inactivated, 17-18
  influenza A, 176-177
  killed, 43
    viral, 176

vaccines (*Cont.*):
live, 17-18
live attenuated influenza, 176
Lyme disease, 662
malaria, 790
measles, 196
meningitis, 18
MMRV, 192, 201, 260
polio, 17-18, 144
poliomyelitis associated with, 220
rabies, 302
rotavirus, 274
rubella, 201
schistosomiasis, 890
smallpox, 43
tetanus, 18
varicella-zoster virus, 260*t*
vaccinia, 206*t*, **209-210**
immunity, 209
receptors, 114*t*
scientific interest in, 209
vaccinia complement control protein (VCP), 144
vacuolating cytotoxin (VacA), in *Helicobacter pylori*, 573
vagina, 390*t*
candidiasis of, 727
flora in, 11
vaginal discharge, **948**
vaginitis, 727, 945, **948**
vaginosis, bacterial, 948
valacyclovir, 158*t*, **160**, 256
valganciclovir, 161
valley fever, 752
vancomycin, 405*t*, 406, 409, 441
for *Clostridium difficile*, 530-531
for enterococcal disease, 466, 467
for gonorrhea, 548
for *Haemophilus influenzae*, 558
resistance, 421
susceptibility patterns to, 427*t*
variable domains, 38
variable region 3 (V3 loop), 310
variable surface glycoprotein (VSG), 827
varicella-zoster virus (VZV), 150, 247, 248*t*, **256-260**, 318*t*
acyclovir in, 259
in AIDS, 258
antibodies, 258
clinical aspects, **258-260**
CNS infections and, 952*t*
diagnosis, **259**
epidemiology, **257**
eye infection from, 906*t*
immunity, **257-258**
manifestations, **258-259**
maternal, 259
pathogenesis, **257**
prevention, **259-260**
primary, 258*f*
reactivation, 258
treatment, 160, **259**
vaccine, 260*t*
variola, 206*t*, **207-208**. *See also* smallpox
clinical aspects, **208**
diagnosis, **208**
major, 207
manifestations, **208**
minor, 207
pathogenesis, **208**
variola (*Cont.*):
prevention, **208**
virology, **207**
vascular catheters, nosocomial infections from, 55-56
vasculitis
African trypanosomiasis and, 828
Rickettsia, 683*f*
VCA. *See* viral capsidantigen
VCP. *See* vaccinia complement control protein
VDRL. *See* Venereal Disease Research Laboratory
vegetative DNA replication, 328
vegetative mycelium, 695
Veillonella, 516, 517*t*
plaque colonized by, 911
Venereal Disease Research Laboratory (VDRL), 651
venipuncture, **967-968**
verruga peruana, 683*t*
vertical transmission, 92, **95**, 139
arboviruses, 285
of hepatitis B, 233
of viruses, 139*t*
vesicular stomatitis virus, **295**
Vi antigen, 602
Vibrio, **565-571**
features of, 566*f*
Vibrio alginolyticus, 567*t*
Vibrio cholerae, 486, **565-571**, 929, 931, 932*t*
0139 strain, 566
disease caused by. *See* cholera
growth, 565-566
scanning electron micrograph, 566*f*
structure, 565-566
Vibrio mimicus, 567*t*
Vibrio parahaemolyticus, 567*t*, 570, 932*t*, 934*t*
Vibrio vulnificus, 567*t*, 570
Vietnam War, 549
Vif, 311, 312
Vincent's angina. *See* fusospirochetal disease
viral capsidantigen (VCA), 264
viral infections
adaptive immunity, **152**
CNS, **337-344**
cytopathogenicity of, **143-144**
disseminated, 140
endemic, 137
entry, **137-139**
epidemic, 137
host defenses, **150-151**
host factors, **149-150**
immune-mediated, 153*t*
incidence, 137
incubation period, 140
interferons and, **150-151**
localized, 140
pandemic, 137
pathogenesis, **137-155**
patterns, **144-145**, 146*f*
spread in host, **140-142**
transmission, **137-139**
tropism, **142-143**
viral transformation, **146-147**
virulence, **143-144**
viral set point, 145
viral transformation, **146-147**
by DNA viruses, **147-148**
by retroviruses, 148-149
by RNA viruses, 149
viremia, 142, 194, 285, 959
primary, 142
secondary, 142
transient, 285
viridans streptococci, 446, 456*t*, **465**, 962*t*
virulence, 465
virions
Alphavirus, 281*f*
arenaviruses, 283*f*
attachment proteins, 113
bunyaviruses, 282*f*
coronavirus, 186*f*
herpes simplex 1, 248*f*
poxvirus, 206*f*
schematic drawing of, 104*f*
structure, 186*f*
viroids, 101
virokines, 144
viropexis, 116, 118*f*, 213
viroreceptors, 144
virulence, 13, 387
amebiasis, 806
of anaerobes, 518-519
Bordetella pertussis, 400*f*
cholera, 569
of coccidioidomycosis, 751
Enterobacteriaceae, 581-582
genetic regulation of, 569
of gonorrhea, 545-546
immune response and, 15
Listeria monocytogenes, 477, 480
low, 387
moderate, 387
pertussis, 561-562
plasmid, 399
regulation of, 399, 581-582
resistance v., 418
of viral infections, 137, **143-144**
viridans streptococci, 465
virulent phages, 380
virulent viruses, 112
viruses, 5-7, 6*f*. *See also specific viruses*
virus-induced immunopathology, **152-154**
virus-induced immunosuppression, **154-155**
vitamin B_{12}, 873
viviparous, 763
Voges-Proskauer test, 88
vomiting, 432
von Magnus phenomenon, **131-132**
voriconazole, 709
for aspergillosis, 732
features of, 708*t*
Vpr, 311, 312
Vpu, 311, 312
Vpx, 312
VSG. *See* variable surface glycoprotein
vulvovaginitis, 946*t*
VZV. *See* varicella-zoster virus

W

walking pneumonia, 668, 678
walrus, 855
Warthin-Finkeldey cells, 194
warts, 330*f*
genital, 331, **946**
waste disposal, 884, 886, 890
water fluoridation, 916
water-borne transmission, giardiasis, 819
watery diarrhea, 581, **929**, 932*t*-933*t*
Weil's disease, 654

West Nile encephalitis, 952*t*
West Nile virus, 150, 280*t*, **289-290**
  in United States, 289*f*
Western blot immunoassay, **79**
  for HIV detection, 319*f*
western equine encephalitis, 288-289, 952*t*
whipworm, 836*t*
white blood cells, 20*f*
white piedra, 719
Whitfield's ointment, 718
whooping cough. *See* pertussis
wild-type allele, 374
winter gastroenteritis, 272
Wohlbachia, 683*t*
women, gonorrhea in, 548*f*
Woolf, Virginia, 167
wool-sorter's disease, 485
World Health Organization, 207-208, 558
World War I, 689
wound botulism, **525**
wound infections, **897-898**
  anaerobic Gram-negative, 900
  animal bites, 899*t*
  burns, 899*t*, 900
  causes of, 897-898, 899*t*
  etiologic agents of, **898-900**, **899*t***
  factors contributing to, **898**
  prevention of, **898**
  tetanus and, 900
  treatment of, **898**, 900
wounds
  animal bites, 899*t*
  burns, 899*t*
  classification of, **898**
  group A streptococci and, 449
wounds (*Cont.*):
  physiologic, 897, 899*t*
  surgical, 897, 899*t*
  traumatic, 897
Wright stains, 655, 788
Wuchereria bancrofti, 761, 851, 858*t*, 859
  general characteristics of, 852*t*
  parasitology, 857

X

X factor, 557
XDR-TB. *See* extensively drug-resistant tuberculosis
xenodiagnosis, 833
Xenopsylla cheopis, 635
xerostomia, 916
Xis proteins, 134

Y

yabapox, 206*t*, 209-210
Yatapoxvirus, 206*t*, 209-210
Yaws, 663
yeasts, 7-8, **695-698**
  cell wall, 694*f*
  conversion to, 698
  forms, 696*f*
  opportunistic, 724*f*
  opsonized, 727
yellow fever, 280*t*, **290**
Yersinia, **605-606**
  bacteriology, 605
  diseases
    epidemiology, 605
    pathogenesis, 605-606
  infections, 606
Yersinia enterocolitica, 585*t*, **605-606**, 630*t*, 903, 933*t*
  clinical aspects, 605
  epidemiology, 605
  pathogenesis, 605-606
Yersinia pestis, 13, 388, 580, 585*t*, 630*t*, 633-638. *See also* plague
  bacteriology, 633
  ciprofloxacin for, 426
Yersinia pseudotuberculosis, 585*t*, **605-606**, 630*t*
  clinical aspects, 605
  epidemiology, 605
  pathogenesis, 605-606
yogurts, 12
Yops, 605, 636
Yorkston, James, 579

Z

Z protein, 283
zalcitabine, **162**
zanamivir, 158*t*, 159, 176, 177
Zappa, Frank, 471
zidovudine, 123, 158*t*
zoonotic diseases, **629-641**
zoonotic transmission, **95**, 139*t*, 140
zoonotic viruses
  incubation period, 142*t*
  RNA, **292-293**
zur Hausen, Harold, 330
zygomycetes, **732-733**
  opportunistic, 724*f*
zygomycosis, **732-733**, 733*f*
Zygomycota, 698